75th World Foundry Congress (WFC 2024)

Deyang, China
25-30 October 2024

Volume 1 of 2

ISBN: 979-8-3313-1254-1

Printed from e-media with permission by:

Curran Associates, Inc.
57 Morehouse Lane
Red Hook, NY 12571

Some format issues inherent in the e-media version may also appear in this print version.

World Foundry Organization Ltd
Winton House, Lyonshall
Kington, Herefordshire
HR5 3JP, United Kingdom

Phone: +44 (0) 1544 340332
Fax: +44 (0) 1544 340332

www.thewfo.com

Additional copies of this publication are available from:

Curran Associates, Inc.
57 Morehouse Lane
Red Hook, NY 12571 USA
Phone: 845-758-0400
Fax: 845-758-2633
Email: curran@proceedings.com
Web: www.proceedings.com

TABLE OF CONTENTS

VOLUME 1

PART 1 : CAST IRON AND CAST STEEL TECHNOLOGY

PART 2 : NON-FERROUS ALLOY

PART 3 : HIGH ENTROPY ALLOY

PART 4 : HIGH-TEMPERATURE ALLOY

PART 5 : METAL MATRIX COMPOSITE MATERIALS

VOLUME 2

PART 6 : RARE EARTH MATERIALS

PART 7 : FUNCTIONAL MATERIALS

PART 8 : MATERIALS GENOME ENGINEERING

PART 9 : KEY CASTINGS MANUFACTURING FOR RAIL TRANSPORTATION

PART 10 : DIE-CASTING TECHNOLOGY

PART 11 : LOST FOAM CASTING TECHNOLOGY

PART 12 : PRECISION CASTING TECHNOLOGY

PART 13 : ADDITIVE MANUFACTURING

PART 14 : DIGITAL INTELLIGENT CASTING TECHNOLOGY

PART 15 : MOULDING MATERIAL AND TECHNOLOGY

WELCOME MESSAGE

Yanchun Lou

Chairman, The World Foundry Organization; Chairman, The Scientific Committee of the 75th World Foundry Congress

Chief Scientist, China Academy of the Machinery Science and Technology Group; Director and Researcher, State Key Laboratory of Advanced Casting Technology; Chairman, Foundry Machinery Technical Committee of the International Standard Organization (ISO/TC306)

Distinguished guests, ladies and gentlemen,

First, on behalf of the Executive Committee of the World Foundry Organization (WFO), I would like to extend a warm welcome to all of you to the 75th World Foundry Congress (WFC).

Today, over a thousand of colleagues from global foundry industry gathered here in Deyang, a city which boasts a remarkable bronze metallurgy legacy that spans over 3,000 years and now is known as a major equipment manufacturing base in China. We are here to explore and exchange ideas on innovations in foundry technology, share the latest advancements in the industry from our respective countries, foster friendship and collaboration, and work together towards healthier and more sustainable development for the global foundry industry.

This WFC 2024 has received an impressive total of 473 paper submissions from 25 countries and regions around the globe. After thorough review by the Scientific Committee, we are proud to present 9 reports for the congress and 229 reports for the sub-forums. Additionally, we have set up a dedicated poster paper presentation area to encourage lively discussions among all attendees. So, I firmly believe that this WFC 2024 will be a vibrant and high-level academic feast. On behalf of the 75th WFC Scientific Committee, I would like to extend my heartfelt respect and sincere gratitude to all the authors of the papers, all the members of the Scientific Committee, and all the organizers from the Secretariat of the Foundry Institution of Chinese Mechanical Engineering Society (FICMES) for your tremendous efforts and contributions!

In today's era, the global foundry industry is making rapid strides towards integrated development with the trend to pursue digital, smart and green transition. The high-quality development of the foundry industry is a kind of integrated development led by high-quality and high-level talents and based on the advancement in casting technology, which offers a win-win scenario for various stakeholders. By promoting talent cultivation, particularly nurturing young professionals worldwide, new energy and vitality has been injected into industry development. Former WFO President Dr. Carsten Kuhlgatz has taken the initiative to organize the "WFO Foundry Young Researchers and Early Career Professionals Conference," which has significantly contributed to the development of young talents in the global foundry industry and has garnered high recognition from the industry. This year's 2nd WFO Foundry Young Researchers and Early Career Professionals Conference has recognized three outstanding winners. The FICMES has generously supported and invited them to attend the Global Youth Forum during this WFC 2024, where they will each share their insights.

In 1926, the WFO was founded. Now, nearly a century has passed. Currently, we boast representatives from over thirty member countries, encompassing major foundry-producing nations and regions in the world, thereby establishing an international network that serves the global foundry industry. Remarkably, the annual production of castings worldwide has reached an impressive 100 million tons. Through diverse events like the biennial WFC, technology forums, and foundry summits, we provide valuable support to the industry. As an advocate of and a leader in sharing foundry knowledge and driving technological progress worldwide, the WFO shoulders the significant mission of promoting and boosting the development of the foundry industry. Besides, the WFO stands ready and willing to make its due contribution to tackling global climate change, facilitating a low-carbon transition in the industry and promoting the advancement of human civilization.

As the President of the WFO Executive Committee for 2024-2025, I will work closely with the WFO Executive Committee, Secretariat, working committees, and foundry organizations from over thirty member countries to build a high-quality and high-level platform for global exchange and cooperation, promote talent development and contribute to the advancement of the global foundry industry.

Thanks to Deyang Municipal People's Government and relevant departments in Sichuan Province, enterprises, universities, research institutions, and related service organizations engaged in the foundry industry from around the globe for your generous support of this congress!

I believe that WFC 2024, hosted in China, will leave everyone with wonderful memories.

Wenjiang Ding

Chairman, The Scientific Committee of the 75th World Foundry Congress

Academician of Chinese Academy of Engineering; Professor, Shanghai Jiao Tong University; Director, Light Alloy Net Forming National Engineering Research Center

Distinguished guests, ladies and gentlemen,

First, please allow me to extend a warm welcome and heartfelt thanks on behalf of the Organizing Committee and the Scientific Committee of the 75th World Foundry Congress (WFC) to over a thousand colleagues around the globe.

As we all know, China is an ancient country in the East with a history of over five thousand years and a brilliant foundry civilization that spans more than three thousand years. A prime example of this is the renowned Sanxingdui cultural relics, located right here in Deyang. As a part of this year's congress, a special visit to this cultural site will be arranged for you to experience the awe and inspiration that our ancestors have imparted to modern civilization and modern foundry.

Today, China has become the world's largest producer of castings, with an annual output of approximately 55 million tons, accounting for nearly 50% of the global total. As Chinese foundry enterprises continue to improve and upgrade technology and equipment, the quality of castings made in China has steadily increased. Particularly in industrial application fields such as aerospace, automotive, rail transportation, machine tools, power and energy equipment, and internal combustion engines, a number of foundry enterprises have emerged that are notable for their high quality, substantial scale, and specialization. These enterprises span the entire spectrum of ferrous and non-ferrous metal castings, driving the overall development and continuous technological advancement of China's foundry industry.

At present, the foundry industry is grappling with multiple pressures from rising labor costs and climbing raw material prices, compounded by the global economic downturn, all the vicissitudes in the international landscape, and a need for energy conservation and environmental protection. Against the background, China's foundry industry is still steadily transitioning toward green and smart development while maintaining a strong focus on high-quality, innovation-driven development. This progress is guided by the Chinese government's commitment to further deepen reform and opening-up and pursue high-quality development driven by new quality productive forces.

At the same time, it is important to recognize that China is not only the largest casting producer in the world but also a massive consumer market for castings. An increasing number of overseas enterprises, research institutions, universities, and other related service organizations are choosing to invest in and seek cooperation with China. China's foundry industry is integrating into overseas markets at a faster pace. I believe that as the global foundry industry continues to develop in an integrated manner, China's foundry industry will undoubtedly contribute even more to global foundry development.

Taking WFC 2024 as an opportunity, I sincerely invite foundry professionals worldwide to visit, exchange and cooperate with China's foundry enterprises, universities, research institutions, suppliers across the foundry industrial chain, and relevant service organizations to develop the foundry industry and foundry technology in China and globally.

Since its establishment in 1962, the Foundry Institution of Chinese Mechanical Engineering Society (FICMES) has been dedicated to advancing the development of China's foundry industry, driving continuous transformation and high-quality growth. Meanwhile, the FICMES has consistently focused on promoting exchanges in casting technology between China and foreign countries, fostering talent development, and supporting industry cooperation and related services. Back in 1978, the FICMES represented China in joining the WFO, marking that China officially became a member country of the WFO. Currently, the secretariats of the four WFO working committees on ferrous metals, non-ferrous materials, moulding materials, and die casting are all located in the FICMES which represents them in planning, organizing, and coordinating global activities with stakeholders from member countries. On the occasion, I would like to extend our gratitude to the WFO for your long-standing trust and support. In the future, the FICMES will continue to support the work of the WFO and strive to contribute more to the development of the foundry industry in China and around the world!

Thank you for attending WFC 2024. I sincerely wish you all good health and success in your work!

ORGANIZATION

Host

The World Foundry Organization (WFO)

Organizer

Foundry Institution of Chinese Mechanical Engineering Society (FICMES)

China Foundry Association

National Key Laboratory of Advanced Casting Technologies

Joint Organizer

Shenyang Research Institute of Foundry Co., Ltd. CAM

Co-organizer

National Technical Committee 54 on Foundry of Standardization Administration of China

Tsinghua University

Shanghai Jiao Tong University

Huazhong University of Science and Technology

Sichuan University

Harbin Institute of Technology

Southeast University

Shandong University

Northwestern Polytechnical University

South China University of Technology

Jilin University

Dalian University of Technology

Chongqing University

University of Science and Technology Beijing

Nanjing University of Science and Technology

Northeastern University

Beijing Jiaotong University

Zhengzhou University

Shanghai University

Jinan University

Wuhan University of Technology

China University of Mining and Technology

Hohai University

China University of Geosciences

Donghua University

Hefei University of Technology

University of Shanghai for Science and Technology

Hebei University of Technology

Qilu University of Technology

Nanchang Hangkong University

Henan University of Science and Technology

North University of China

Shenyang University of Technology

Lanzhou University of Technology

College of Nuclear Equipment and Nuclear Engineering of Yantai University

Harbin University of Science and Technology

Dongguan University of Technology

Taiyuan University of Science and Technology

Minjiang University

Hubei University of Automotive Technology

Taiyuan University of Technology

Xi'an Jiaotong University

Xi'an University of Technology

Sichuan Polytechnic University

Institute of Metal Research of Chinese Academy of Sciences

No.59 Institute of China Ordnance Industry

Beijing Institute of Aeronautical Materials

Ningbo Institute of Materials Technology and Engineering of Chinese Academy of Sciences

Baotou Research Institute of Rare Earths

Deyang Guangda Dongqi New Material Co., Ltd.

Erzhong (Deyang) Heavy Equipment Co., Ltd.

FAW Foundry Co., Ltd.

Sichuan Dawn Precision Technology Co., Ltd.

National Key Laboratory of Clean and Effective Turbine

Foundry Professional Committee of Jiangsu Mechanical Engineering Society

Jiangsu Foundry Association

Foundry Institution of Zhejiang Mechanical Engineering Society

Zhejiang Foundry Industry Association

Foundry Professional Committee of Shanxi Mechanical Engineering Society

Shanxi Foundry Association

Sichuan Province Foundry Association

Foundry Institution of Sichuan Mechanical Engineering Society

Foundry Institution of Henan Mechanical Engineering Society

Henan Casting & Forging Industrial Association

Shandong Foundry Association

Foundry Institution of Liaoning Mechanical Engineering Society

Liaoning Casting & Forging Industrial Association

Foundry Professional Committee of Fujian Mechanical Engineering Society

Fujian Foundry Industry Association

Hong Kong Foundry Association

Foundry Professional Committee of Hubei Mechanical Engineering Society

Hubei Foundry Society

Foundry Institution of Guangxi Mechanical Engineering Society

Guangxi Foundry Association

Foundry Institution of Guangdong Mechanical Engineering Society

Foundry Industry Association, Guangdong Province

Foundry Professional Committee of Inner Mongolia Mechanical & Electrical Engineering Society

Foundry Professional Committee of Anhui Mechanical Engineering Society

Foundry Institution of Guizhou Mechanical Engineering Society

Foundry Institution of Yunnan Mechanical Engineering Society

Foundry Professional Committee of Gansu Mechanical Engineering Society

Foundry Professional Committee of Xinjiang Mechanical Engineering Society

Jilin Foundry Association

Foundry Institution of Chongqing Mechanical Engineering Society

Chongqing Foundry Association

Xi'an Foundry Society

CRRC Qishuyan Institute Co., Ltd.

SIPPR Engineering Group Co., Ltd.

Jinan Shengquan Group Share-Holding Co., Ltd.

Vesuvius Foundry Technology (Jiangsu) Co., Ltd.

voxeljet China Co., Ltd.

MAGMA Engineering (Suzhou) Co., Ltd.

Guangxi Lanco Resources Recycling Co., Ltd.

International Co-organizer

Polish Foundrymen's Association (STOP)

Romanian Foundry Technical Association (ATTR)

Casting Exporters' Association of Spain (FUNDIGEX)

Turkey Foundry Association (TUDOKSAD)

Institute of Cast Metals Engineers (U.K.)

Indonesian Association of Foundry Industries (APLINDO)

Slovenian Foundrymen Society (DLS)

Croatia Foundry Association

Vietnam Foundry & Metalurgy Science and Techology Asoociation (VFMSTA)

The Institute of Indian Foundrymen

COMMITTEE

Domestic Scientific Committee

Honorary Chairman

Hengzhi Fu	China
Baicheng Liu	China
Yuanyuan Li	China
Fusheng Pan	China
Guang Chen	China
Zhongde Shan	China
Xinya Li	China
Rongde Li	China
Weidong Huang	China

Chairman

Wenjiang Ding	China
Yanchun Lou	China

Vice Chairman

Huasheng Xie	China
Xiaoguang Yuan	China
Shigen Zhu	China
Shoumei Xiong	China
Zitian Fan	China
Tongmin Wang	China
Baode Sun	China
Guohua Wu	China
Yanxiang Li	China
Qingyan Xu	China
Zhigang Lv	China
Ruirun Chen	China
Jianxin Zhou	China
Yingdong Qu	China
Hua Hou	China
Qian Li	China

Member

The names are listed in alphabetical order.

Anping Dong	China
Bin Jiang	China
Bin Yang	China
Bo Yu	China
Chen Lu	China
Da Shu	China
Dawei Yin	China
Daxin Zeng	China
Dayong Li	China
Dezhi Zhu	China
Gaosheng Fu	China
Guoxiang Wang	China
Hai Hao	China
Hai Nan	China
Haidong Zhao	China
Hongchao Kou	China
Hongyuan Fan	China
Huizong Song	China
Huan Yu	China
Jianfei Sun	China
Jianjun Wang	China
Jianxun Zhu	China
Jiguang Liu	China
Jincheng Liu	China
Jingpei Xie	China
Jinguo Li	China
Jinshan Li	China
Jiuzhou Zhao	China
Jufu Jiang	China
Jun Zhang	China
Jun Zhao	China
Junpin Lin	China
Liming Peng	China
Ning Li	China
Qichi Le	China
Qijie Zhai	China
Qingyou Han	China
Qiwei Wang	China
Rui Yang	China
Runxia Li	China
Shaokui Yin	China
Shifang Su	China
Shusen Wu	China
Tongzhou Liu	China
Wei Li	China
Weiwen Zhang	China
Wentao Yan	China
Xiang Chen	China
Xiangfa Liu	China
Xiangshang Liu	China
Xiaobing Yang	China
Xin Lin	China
Yanqing Su	China
Ye Pan	China
Yejian Gao	China
Ying Xiong	China
Yiqing Chen	China
Yongfeng Liang	China
Yuan Liu	China
Yuan Wu	China

Yuandong Li	China
Yufu Sun	China
Yuhong Zhao	China
Yusheng Shi	China
Yuyong Chen	China
Yuzeng Chen	China
Zehua Wang	China
Zhengfeng Zhu	China
Zhijun Feng	China
Zhiqiang Han	China
Zhongliang Lu	China
Zhongli Liu	China

Domestic Youth Committee

The names are listed in alphabetical order.

Binguo Fu	China
Bobo Li	China
Dequan Shi	China
Donghong Wang	China
Fuchu Liu	China
Gong Zheng	China
Guoliang Zhu	China
Guoqiang You	China
Hailong Jia	China
Hongze Fang	China
Hucheng Pan	China
Huijun Kang	China
Jiangfeng Song	China
Jide Liu	China
Jieren Yang	China
Jingchang Cheng	China
Jun Wang	China
Kehui Hu	China
Lei Jin	China
Liang Wang	China
Likun Jiang	China
Mengwu Wu	China
Pei Liu	China
Qi Wang	China
Qiaolei Li	China
Qiongyuan Zhang	China
Ruilong Yu	China
Shaohua Xia	China
Shibing Liu	China
Shulin Lv	China
Tong Gao	China
Wei Guo	China
Wenbo Yu	China
Wenming Jiang	China
Wenxin Hu	China
Xiaoyuan Ji	China
Xin Tong	China
Xinwang Liu	China
Xiufang Gong	China
Xixi Dong	China
Xuecheng Cai	China
Xuewei Yan	China
Yajun Yin	China
Yanpeng Wei	China
Yanwei Sui	China
Yiping Lu	China
Yufei Li	China
Zengrui Wang	China
Zhiqiang Fu	China
Zhixiang Qi	China

International Scientific Committee

The names are listed in alphabetical order.

Ayax Rangel	USA
Abdul Razaq	Pakistan
Adel Nofal	Egypt
Alan A. Luo	USA
Ali Emre Zobaroğlu	Turkey
Andrea Montagnoli	Italy
Angelika Kmita	Poland
Anil Kumar Birru	India
Anna Brudny	Poland
Anwar Hamasaiid	Spain
Arron Rimmer	U.K.
Barbara Juszczyk	Poland
Cathrine Hartung	Norway
Christian Kneissl	Austria
Christian Lustig	Germany
Christoph Schendera	Germany
Christa Zengerer	Austria
Chuan Zhang	USA
D. Benny Karunakar	India
Daniel Liang	Australia
David Abedul Moreno	Spain
Dawid Myszka	Poland
Dorota Wilk-Kołodziejczyk	Poland
Doru Michael Stefanescu	Romania
Eitaro Koya	Japan
Ekaterina Potaturina	Germany

Ewa Olejnik	Poland	Mathias Manser	Switzerland
G. Alonso	Spain	Ming Chen	Denmark
Giuliano Angella	Italy	Mitja Petrič	Slovenia
Gorka Páramo Argote	Spain	Moritz Weidt	Germany
Hayati Sahin	The Netherlands	Morten Bechlund	Denmark
Hiroto Kato	Japan	Mustafa Ata	Germany
Hongbiao Dong	UK	Natalia Sobczak	Poland
Hou Yabo	Belarus	Olaia Gordo-Burgoa	Spain
Ian Andrews	UK	Pablo Garcia Michelena	Spain
Ibán Vicario Gómez	Spain	Patricia Caballero	Spain
Ingo Rudolf Ederer	Germany	Pawel Malinowski	Poland
Iulian Riposan	Romania	Pawel Popielarski	Poland
Jack Ho	Australia	Per Larsen	UK
Jadwiga Kamińska	Poland	Philippe Kientzler	France
James Xi	Singapore	Primož Mrvar	Slovenia
Jan Jezierski	Poland	Qigui Wang	USA
Javier Nieves	Spain	Rafal Cyga	Poland
Jean Amadeo Brambila	Brazil	Rafał Dańko	Poland
Jerzy J. Sobczak	Poland	Ramon Suarez	Spain
Jianyue Zhang	USA	Rodolfo González-Martínez	Spain
Jianzheng Guo	USA	Saif Haider Kayani	Korea
Jiehua Li	Austria	Saria Akhtar	Pakistan
Joanna Borowiecka-Jamrozek	Poland	Shi Li	Australia
Joanna Kulasa	Poland	Shouxun Ji	UK
John Samson Khalkho	India	Shrivatsav Shankar	USA
Justyna Kasińska	Poland	Tohru Mizuki	Japan
Karol Janus	Poland	Weidong Li	USA
Katarzyna Liszka	Poland	Xiaobo Chen	Australia
Kento Miwa	Japan	Yanfei Gao	USA
Kun Wang	USA	Yasuhiko Okimura	Japan
Łukasz Rakoczy	Poland	Yasuhiro Maeda	Japan
Lukasz Szymanski	Poland	Yooseob Song	USA
Marcin Górny	Poland	Yuefeng Gu	Japan
Marek Bruna	Slovakia	Yury Nikalaichyk	Belarus
Marek Matejka	Slovakia	Zhanhu Guo	UK
Maria Pammer	Austria	Zhipeng Liu	The Netherlands
Menghuai Wu	Austria	Zdenka Zovko Brodarac	Croatia

Organizing Committee

Chairman

Huasheng Xie	China

Domestic Members

Zhiyong Zhang	China
Shuang Xu	China
Zihua Yang	China
Yusong Liu	China
Hongyuan Fan	China
Siqing Gong	China
Yanmei Hu	China
Zhiyong You	China
Jiang Zhang	China
Dongjie Pan	China
Shiya Liang	China
Anyuan Wan	China
Lucai Wang	China
Dawei Yin	China
Yue Liu	China
Yu Fang	China
Guoqu Shen	China
Diyuan Tang	China
Jinxing Wang	China
Yunqiao Xu	China
Wei Zhao	China
Wucheng Liu	China
Zhipeng Guo	China
Bing Li	China
Qinghai Yang	China
Shengli Han	China
Shuzhong Xie	China
Chunlei Liu	China
Deli Zhao	China
Zekun Yang	China
Jianru Fang	China
Chaoyang An	China
Jinde Yang	China
Jiang Fan	China

International Members

Katarzyna Liszka	Poland
Tunçağ Cihangir Şen	Turkey
Marina Giacopinelli	Spain
Mirjam Jan-Blazic	Slovenia
Ion Alexandru Băcanu	Romania
Franz Friedrich Butz	Germany
A. Rangel	USA
Ingo Rudolf Ederer	Germany

WFO Executive Committee

President

Yanchun Lou	China

Immediate Past President

Carsten Kuhlgatz	Germany

Vice President

Rafal Danko	Poland

Treasurer

Xabier Gonzalez Azpiri	Spain

General Secretary

José Javier Gonzalez	Spain

Executive Secretary

Andrew Turner	UK

Executive Members

Lionel Alves	France
Mark Fenyes	UK
Lee Jenkins	UK
Sang Mok Lee	Korea
Chris Norch	USA
Mehmet Ozalp	Turkey
Kazumichi Shimizu	Japan

SPECIAL THANKS TO

Platinum Sponsors

圣泉集团 SQ GROUP

埃克斯邦 xBang Technology Co.,LTD

Golden Sponsors

FOSECO

鑫刚新材 XINGANG NEW MATERIAL

HA GROUP

Silver Sponsors

voxeljet 维捷

龍鳳山

SINYE 蘇州興業材料科技 MATERIALS TECHNOLOGY

适创科技 SUPREIUM

卓越高新材料 JoYo CARBON

FA

Bronze Sponsors

DEC 东方电气 四川东树新材料有限公司 SICHUAN DONGSHU NEW MATERIALS CO., LTD.

興龙射线 Xiong Radiative

MAGMA

草原狼 CAO YUAN LANG

铸力

lanco 兰科

中国机械总院轻量化研究院

Special Sponsors

ALBERTUS-STIFTUNG

ANY SOFTWARE

This page intentionally left blank.

Development of an Austempered Ductile Iron Unimog 3 Point Linkage Power Lift Arm

Arron Rimmer[1], Edward Dale[2]
1. ADI Treatments Ltd, West Bromwich, West Midlands U. K.
2. Atkinson Vos Ltd, Bentham, Lancashire, U. K

Abstract: This paper summarizes the investigation of the conversion of an 11-piece steel fabrication to an ADI casting. It was a collaborative effort involving the component manufacturer, foundry, machinist, local university, and an ADI heat treater. A number of design concepts were developed, with the final, significantly stronger, ADI design being compared to the fabrication using FEA analysis. The resulting conversion, at similar weight and cost, eliminated the failures that were experienced in the fabricated components in addition to reducing the production lead-time.

Keywords: 3-point linkage, power lift arm, austempered ductile iron

1 Introduction

The component manufacturer, Atkinson Vos produces aftermarket 3-point linkages for UGN and UGE Unimogs; none of which are standard. The power lift arm, the key component in this system, transfers the forces from the hydraulic cylinders to the link arms and the pick-up hitch.
Previously, this was a fabricated component made from 9 laser cut pieces of 20 and 30mm S355 steel (Figure 1). This version of the component experienced a torsional failure under eccentric loading which was temporarily mitigated by the addition of 2 x 20mm plates that were welded on to increase the torsional cross section. This resulted in a total fabricated weight of 55kg. However, this left an ugly design and was an unsatisfactory solution. In addition, the geometry of the fabricated design meant that the telescopic rods used to lift the pick-up hitch were set at an angle which occasionally caused them to experience problems with binding internally, which would eventually cause them to bend.

Through a component manufacturer partnership with Lancaster University, two student projects were completed to look at the feasibility of converting the 3 Point Linkage Arm into a cast part. This paper will discuss the results of both student lead programs.

Figure 1 Fabricated power lift arm with additional strengthening plates (11 pieces)

2 Experimental procedure

DESIGN DEVELOPMENT

University Project 1

The first project with a group of final year students focused on producing a design that would be cast. A concept was developed that would initially be cast using steel (Figure 2 - Top) with a calculated weight of 51kg. On further investigation, the students found that by using ADI, they could optimize the geometry further and reduce the section size on the arms (Figure 2 - Bottom). This design had a calculated weight of 34kg, much less than both the fabricated and cast steel designs.

University project 2

The second student project focused on the binding issue. It was determined that the flexibility of casting enabled the use of a more complex geometry than was possible with a fabrication. This allowed for the telescopic lift rods to be

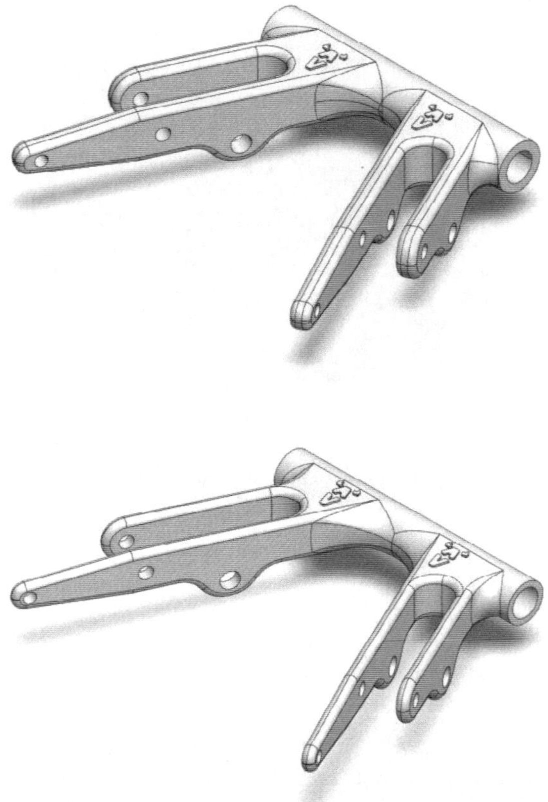

Figure 2 Initial cast steel concept (Top) – Initial concept optimized for ADI (Bottom)

brought into line, which mitigates any binding (Figure 3). The downside of this design was that the more complex geometry resulted in a weight increase to 64kg for a cast steel component which was more than the original fabricated weight. This design was then optimized to use ADI; allowing for the section sizes to be reduced. This, coupled with ADI's lower density (10% less than steel), meant that the design in ADI had a calculated weight of 54kg which was equivalent to the competitive fabricated part (55kg).

The use of ADI was selected as it solved both the binding issue and reduced the weight. The general casting dimensions are 660 mm Wide, 520 mm Long and 120 mm High, with a maximum section thickness of 80 mm for heat treatment (to through harden i.e. austemper).

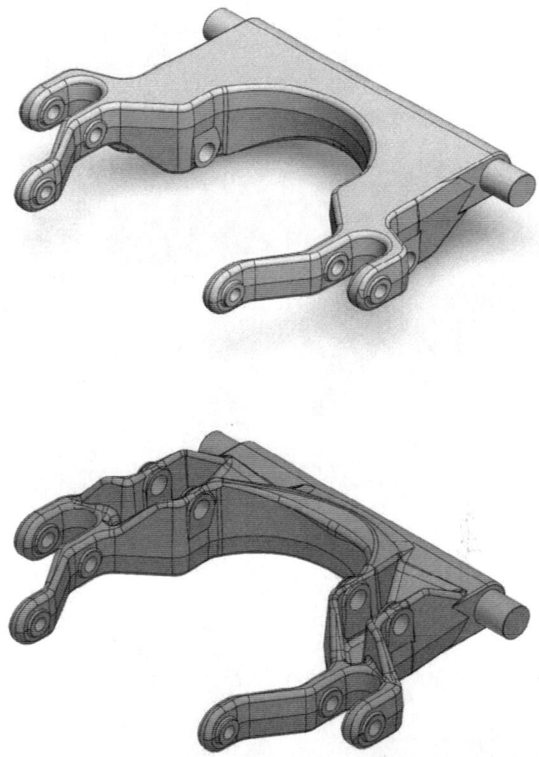

Figure 3 Final cast design in ADI Left – Top View, Right – Bottom View).

3 Result and discussion

Analysis

FEA analysis was conducted on the final design in ADI to compare it to the reinforced fabricated design. Both designs were simulated with an extreme eccentric load case that caused the original fabricated design to fail (Ref. 1). Figures 4 and 5 show a factor of safety plot with this load application on both designs. The fabricated design has some stress concentrations that are not present after the component has been welded: however, the general picture is still clear. The factor of safety is much greater in the ADI cast design. This is replicated in the Von Mises stress plots Figures 6 and 7 where the stress observed across the entire cast design is significantly less than that in the fabrication.

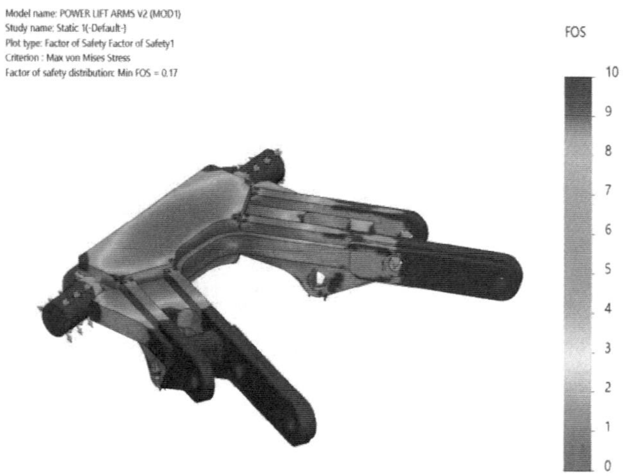

Figure 4 Factor of Safety, Original Fabricated Design

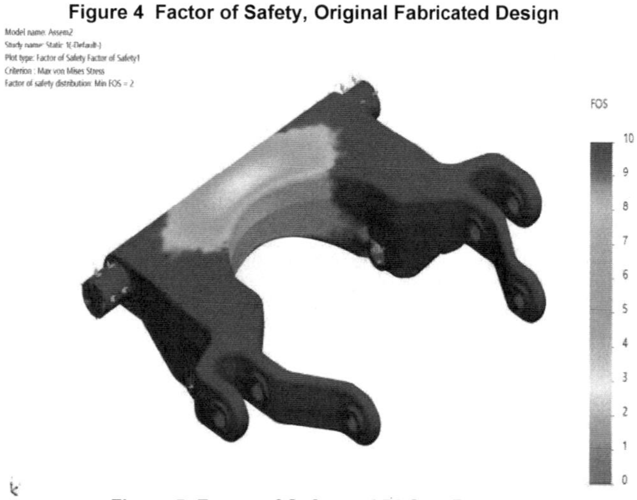

Figure 5 Factor of Safety – ADI Cast Design

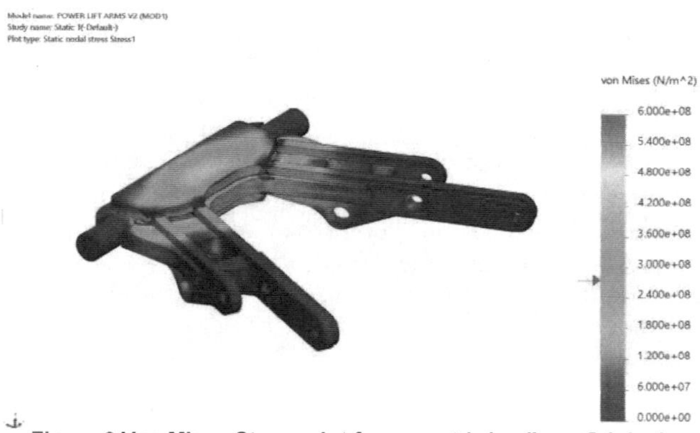

Figure 6 **Von Mises Stress plot for eccentric loading -** Original Fabricated Design

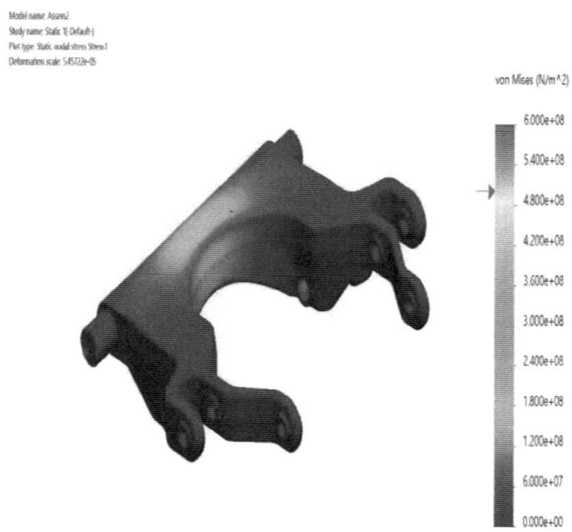

Figure 7 Von Mises Stress plot for eccentric loading ADI cast design.

The ADI design, even with the inclusion of the patterns, tooling and jigs came in at the same cost as the fabricated design.

The initial batch lead time was extended due to pattern making and machining jigs; however, the typical lead time will be approximately 6 weeks which is a reduction from the fabricated design.

PRODUCTION AND RESULTS.

A loose pattern with an odd side was manufactured using 3D printing combined with traditional wood pattern making.

The castings were floor molded at a commercial foundry, using an alkaline phenolic sand with mold painting to improve the surface finish. The weight of iron poured was 88kg with a 70% yield. The casting was poured at 1380°C. The ductile iron was alloyed with copper and nickel to ensure the 80mm section would through austemper (Table 1).

Table 1: Chemical Composition

Melt	C	Si	Mn	Cu	Ni
1	3.58	2.31	0.36	0.76	1.89
2	3.58	2.21	0.31	0.75	1.99

A tie bar was incorporated into the casting design and pattern to prevent movement of the arms during the austemper heat treatment. The tie bar was then removed post-austempering.

After shot blasting and fettling, castings were austempered to EN1564 900-8 (Ref. 2). Austempering was carried out in an UBQA (Universal Batch Quench Austemper) by a commercial ADI heat treater. The castings were hung, using the tie bar for support, ensuring optimum quenching characteristics. The UBQA working envelope for heat treatment is 920 mm Wide, 1840 mm Long and 920 mm

High. With the castings being hung, the optimum loading pattern with the casting dimensions was also achieved.

The austemper cycle was designed based on 80mm section and required EN1564 900-8. Hardness and tensile tests (Table 2) were carried out to approve the material prior to shipping for machining. The tensile tests were carried out bars machined from separately cast Y blocks, representing the maximum component cast section of 80 mm. Hardness was carried out on the Y blocks and the cast component.

The typical microstructure is shown in Figure 8 at x200 and x500 magnification (4 % Nital).

Table 2: Mechanical Properties

Melt	Tensile Strength MPa	Yield Strength MPa	Elongation %	Brinell Hardness
1	1000	732	10	333
2	970	674	8	333
EN1564	900	600	8	280 - 340

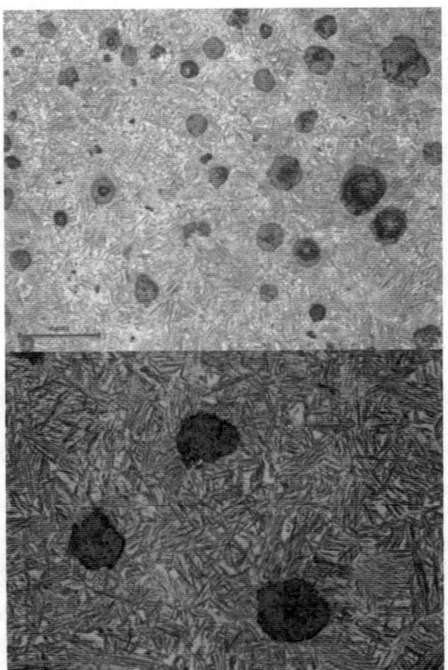

Figure 8 – Microstructure at x200 (left) and x500 (right) (Etchant – 4% Nital)

Machining was completed in the austempered state. The reasoning for machining post austemper was not only to remove the tie bar but also to address the two bores (50 mm diameter) where the steel inserts are fitted as they have an H7 tolerance. All other through holes (10 off) have relatively open tolerances. The key to successful machining was the clamping of the Linkage Arm, to ensure minimal vibration, especially in the arms when machining the through holes. Increasing the feed rate while reducing the cutting speed, as is commonly advised when machining ADI was also employed. The machining source also machined the 2 EN24T (AISI 4340) steel inserts and press fitted so the component manufacturer received a fully assembled component. Final assembly was carried out at the component manufacturer and the first working unit is shown in Figure 9.

Figure 9 – Assembled Power Lift Arm on Working Vehicle

4 Conclusion

The component manufacturer, working with engineering students from Lancaster University successfully re-designed an 11 piece 3-Point Linkage Power Lift Arm into a 1-piece ADI casting. The ADI casting was 1 kg lighter (a weight saving of just under 2%) than the steel fabrication. Even after the production costs for the patterns and machine fixtures are accounted for, the ADI casting, while having a similar production cost to the steel fabrication, is a superior product. The ADI design reduces the potential of torsional failure that the fabricated steel design could encounter under eccentric loading and, the design allowed for the telescopic lift rods to be brough into line, mitigating binding issues. Moving from a fabrication to one piece cast component has also reduced the lead-time for a typical batch and will allow for future changes, including casting on the customer logo.

Acknowledgments

The authors would like to thank the component manufacturer, Atkinson Vos Ltd, Lancaster University, who conducted the component design and FEA analysis, also the commercial foundry, The Boro Foundry Ltd, the machinist, Partline Ltd, the pattern maker, Cerve Ltd, and finally ADI Treatments Ltd who conducted the ADI heat treatment.

References

1. Ihor Dzioba, Sebastian Lipiec; "Fracture Mechanisms of S355 Steel – Experimental Research, FEM Simulation and SEM Observation"; Materials; pp 1- 17, 12 (23), 2019.
2. BS EN1564 – 2011: Founding – Ausferritic spheroidal graphite cast irons – Classification.

A Generic Calculation Method of Choked Section of Gating Systems Based on the Golden Ratio and K-Means Clustering Algorithm

Jianchen Cong[1,2,*], Haixiao Wang[1], Jinling Wang[3], Hongri Cong[2], Meizhen Feng[4]

1. School of Mechanical Engineering, Shandong University of Technology, Zibo, China
2. Tianrun Industry Technology Co., Ltd, Weihai, China
3. School of Intelligent Manufacturing, Zibo Vocational Institute, Zibo, China
4. Caterpillar Research & Development Center (China) Co., Ltd, Wuxi, China
*Corresponding address: e-mail: jchcong@tianrun.com

Abstract: The calculation results of the choked section have been recognized as one of the most significant factors in the design of a gating system. However, various difficulties in existing methods, e.g., the complicated calculation procedure, unknowable physical parameters, and poor calculation accuracy, limit their widespread use in engineering practices. To bridge the gap, a generic formula of the choked section is derived according to the Karsay chart and the experience data. A judgment standard (α) using the golden ratio and two newly defined distances are proposed to distinguish whether a gating system belongs to a high-pressure head casting system or low-pressure one. In the two kinds of gating systems, the correlation coefficients are determined through the K-means clustering of the experience casting data. The generic formula of the choked section is thus modified. To further validate the novel method, various castings in different shapes, weights, and sizes are cast. The experimental results show that all of the castings exhibit good quality and no macroscopical defects. The casting efficiency and finished product ratio has been significantly improved. Besides, it shows that the proposed method could be generalized in practice workshops.

Keywords: choked section; K-Means clustering algorithm; golden ratio; gating system

1 Introduction

Two types of calculation methods for the choked section of a gating system are referred to in existing literature [1]. The first is a theoretical style calculation method based on the principle of hydraulics. A series of problems, e.g., lots of physical variables, a complicated calculation process, and a significant influence of variables on the calculation results, have occurred in engineering applications. The range of the flow coefficient varies greatly (the difference is about 6 times). Especially, for inexperienced designers, it is impossible to accurately determine the values of various physical parameters in practice production.

The second method is called Karsay chart, which was proposed by Dr. Karsay [1-2]. The Karsay chart based on practical experience is more straightforward and user-friendly. However, no definition or foundation for dividing the low and high sprue height curves is included in the Karsay chart. Inexperienced designers haven't any idea on how to appropriately select the low or high sprue height curve. Besides, the Karsay chart is only some suitability for ductile iron castings in the sand casting process.

Thus, a straightforward, accurate, and user-friendly method for calculating the choked section is developed. It is suitable for all kinds of casting procedures for ductile iron and gray iron produced under sand casting circumstances.

The calculation method of the choked section

The generic calculation formula

The nonlinear fitting curve is established based on the Karsay chart, and the formula for fitting curve is Eq. (1). The fitting curve lies between the low sprue height curve and high sprue height curve, as shown in Figure 1a. Figure 1b shows the choked section curves of the gray iron castings at different pouring speeds [3]. The nonlinear fitting method is used to fit the curve under the condition of medium speed pouring, and the curve formula is shown in Eq. (2).

$$\sum F_Z = 0.5\left(W_{JA} + W_M + W_{JO}\right)^{0.43} + 1.31 \quad (1)$$

$$\sum F_Z = 1.06\left(W_{JA}\right)^{0.43} \quad (2)$$

where $\sum F_Z$ is the choked section of the gating system; W_{JA} is the weight of the casting; W_M is the weight of the riser; W_{JO} is the gate weight.

The formulas for calculating the choked section obtained utilizing the Karsay chart and experience date are shown in Eq. (1) and (2), respectively. The generic formula for calculating the choked section is created by combining Eq. (1) and (2), as shown in Eq. (3).

$$\sum F_Z = k\left(W_{JA} + W_M + W_{JO}\right)^{0.43} \quad (3)$$

where k is the correction coefficient.

The judgment criteria for high and low-pressure head gating systems

The distance between the casting's parting face and the highest point of the metal liquid filling is defined as h. Δh is the distance between the highest point of the metal liquid filling and that of the pouring cup. Figure 2 shows the schematic diagrams of h and Δh.

As shown in Eq. (4), the ratio of h to Δh is defined as the golden ratio (0.618).

$$\frac{h}{\Delta h} = 0.618 \quad (4)$$

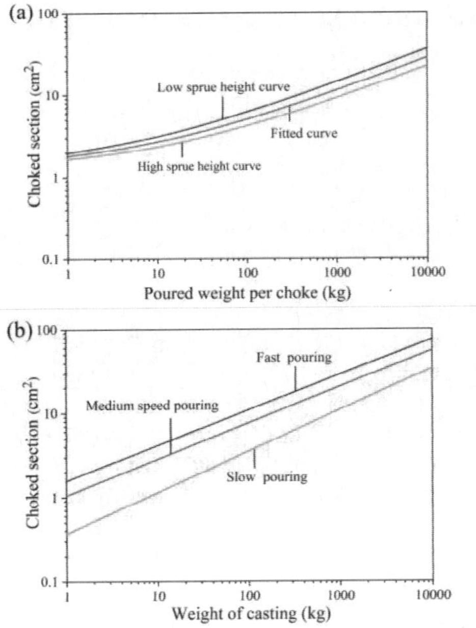

Fig. 1 Choked section curves of gating systems: (a) The choked section curves of the Karsay chart [2]; (b) The choked section curves of gray iron the casting from experience data[3]

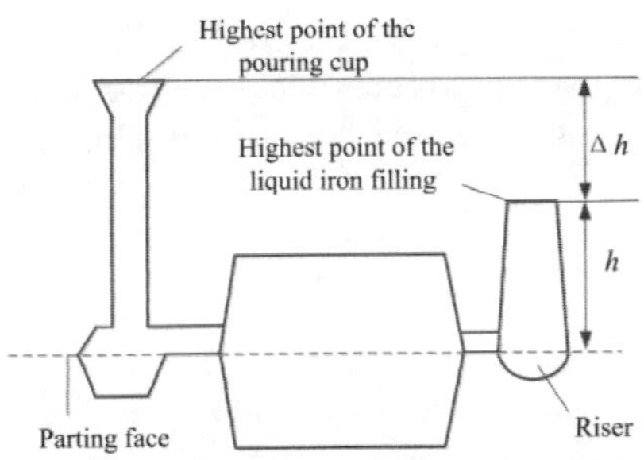

Fig. 2 Schematic diagram of different gating systems

The cube root is calculated concurrently for both sides of the equal sign equation in Eq. (4), as shown in Eq. (5).

$$\sqrt[3]{h/\Delta h} = \sqrt[3]{0.618} = 0.85 \quad (5)$$

The judgment standard (α) is thus got at 0.85. This gating system is identified as a high-pressure head gating system. The generic formula is modified as Eq. (6).

$$\sum F_Z = k_1 \left(W_{JA} + W_M + W_{JO}\right)^{0.43} \quad (6)$$

where k_1 is the correction coefficient for the high-pressure head gating system.

On the one hand, the generic formula is updated to Eq. (7) for the case of low-pressure head gating system.

$$\sum F_Z = k_2 \left(W_{JA} + W_M + W_{JO}\right)^{0.43} \quad (7)$$

where k_2 is the correction coefficient for the low-pressure head gating system.

Parametric study of correction coefficients based on the K-means clustering algorithm

The K-means clustering method is an unsupervised learning algorithm [4]. The K-means clustering algorithm was utilized to search the centroid in the experience data. The clustering results and centroid coordinates are shown in Figure 3.

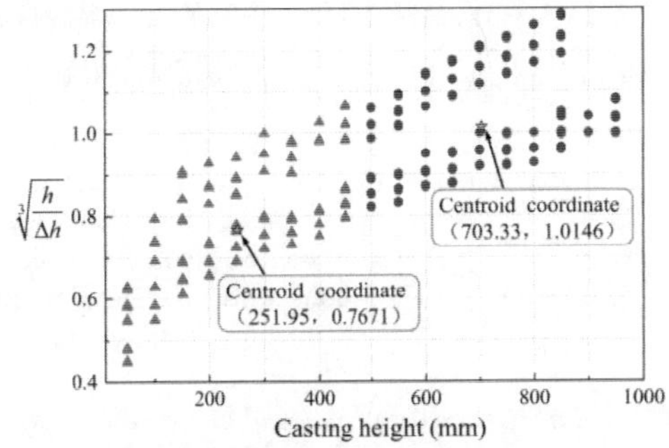

Fig. 3 Results of K-means clustering

According to the clustering results, k_1 is defined as 0.767, and k_2 is determines to be 1.016. The calculation formulas of the choked section for the two types of gating system are shown in Eq. (8) and (9), respectively.

$$\sum F_Z = 0.767 \left(W_{JA} + W_M + W_{JO}\right)^{0.43} \quad (8)$$

$$\sum F_Z = 1.014 \left(W_{JA} + W_M + W_{JO}\right)^{0.43} \quad (9)$$

2 Results and discussions of pouring experiments

Various products in different shapes and sizes are cast to further explore the validity of the proposed calculation method. The results show that the surface quality of the castings is good, and no macroscopical defects were found. The calculation method proposed in this paper is reasonable and effective, with good applicability to gray cast iron and ductile iron.

3 Conclusion

A novel judgment standard to classify the casting system belonging to the low or high-pressure one is proposed, where the golden ratio judgment method is employed. The correction coefficients in the generic formulas are corrected through the K-means clustering of numerous experience data. Experimental results have demonstrated that the casting efficiency is improved and the proposed method could be generalized in practice workshops.

Acknowledgments

The work is financially supported by the Major Science and Technology Innovation Engineering project of Shandong Province, China (Grant No. 2019JZZY010445).

References

[1] Peng X. The exploration of the minimum cross-sectional area which is the important parameters of nodular iron casting foundry process design. Applied Mechanics and Materials, 2012, 157: 673-679.
[2] Karsay S I. Ductile Iron III: Gating and Risering, QIT-Fer et Titane Inc., Montreal, 1981.
[3] Chinese Machinery Project Academy Casting Professional Academy, Casting handbook, Vol.5, Machinery Industry Publishing Company, Beijing, 1996.
[4] Liu Y, Li B F. Bayesian hierarchical K-means clustering. Intelligent Data Analysis, 2020, 24(5): 977-992.

Study on Optimization of Casting Process and Microstructure of Large CNC Machine Tool Ductile Iron Bed

Jianhua Huang[1,] *, Chen Yang[2], Chentao Li[1], Wei Li[1], Shilei Li[3], Hongyuan Fan[3]
1. Sichuan Jianyang Haite Co., Ltd, 641400 Chengdu, China
2. Center of Medical Product Technical Inspection， 610015 Chengdu， China
3. School of Mechanical Engineering, Sichuan University, 610065 Chengdu, China
*Corresponding address: E-mail: 20296317@qq.com

Abstract: The bed structure of high-grade large-scale CNC machine tools is crucial for maintaining their long-term service accuracy. This study focused on the key casting process of large nodular cast iron beds. AnyCasting software was employed to simulate and analysis of the casting scheme, and then the mold filling and solidification processes were proposed based on the simulation results. The proposed casting scheme indicate that the designed gating system exhibited excellent filling capability.

Keywords: High-Grade CNC Machine; Nodular Cast Iron; Casting Process

1 Introduction

Machine tools are often referred to as the "mother machines of industry." High-grade CNC machine tools require a high level of dimensional precision control. Compared to the use of gray iron in the casting of bed bodies for large machine tools in China, foreign counterparts frequently utilize nodular cast iron (NCI) such as QT600-3 for producing high-precision machine tool beds. This choice is based on NCI's ability to control the elastic modulus of the castings, thereby imparting higher rigidity and resistance to deformation. Consequently, NCI beds outperform gray cast iron counterparts in terms of precision retention and noise reduction.

2 Experimental procedure

In this paper, the gravity sand mold casting process is used to complete the casting of a large integrated lathe bed for car grinding. The material used for the molding is resin sand, which has good flowability and is easy to compact. The sand core is made of alkaline phenolic resin self-hardening sand, which shapes the internal control cavity and parts of the casting that cannot be directly cast into shape. The core bone is embedded in the sand core to improve its strength and stiffness. The core bone material is round steel and cast iron. The mold used in this process is a wooden mold. The casting material is ductile iron QT600-3, and the molten iron is poured after smelting, spheroidizing, and inoculation processes. A commercial software (AnyCasting) is used to simulate mold filling and solidification in casting and assist in optimizing casting processes.

3 Result and discussion

a) Design of Bed Casting Process Plan

The three-dimensional models of the machine tool bed at various angles are shown in Figure 1, and the overall size of the bed is 4580 mm × 1659 mm × 744mm, with a guide rail length of 4515mm and a single guide rail width of 111.5mm. The grinding wheel bed is perpendicular to the machine tool guide rail and is set to the left. The interior is composed of rib plates, crossbeams, and cavities. The material used for casting is QT600-3, and resin sand with good rigidity and suitable yielding properties is selected for the sand mold, which can resist the impact of molten iron on the sand mold and can be combined with graphitization expansion.

b) Casting diagram and obtained bed

During the casting process, in order to ensure timely conduction of accumulated heat on the guide rail surface and reduce internal stress caused by temperature differences, conformal cold iron is set at intervals of 15mm near the inner gate along the guide rail direction to weaken stress and increase the strength and hardness of the guide rail. The on-site design is shown in Figure 2. Figure 1 shows the casting diagram obtained from on-site production of the pouring system and cold iron arrangement.

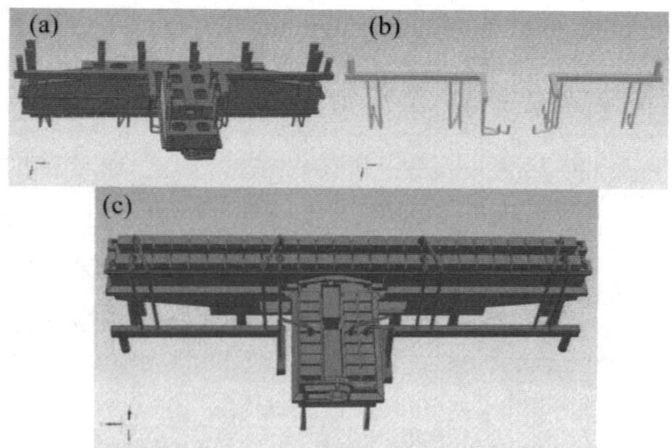

Fig. 1 Design models of the gating system and chills.
(a) gating system and castings, (b) pouring system (at both ends); (c) castings and chills.

The specific casting parameters and simulation condition settings are shown in Table 1.

Table 1 Pouring process parameters

Material	Discharging temp. of molten iron /°C	Pouring temp. /°C	Pouring time /s	Cooling time /h
QT600-3	1420	1360	48	72

Figure 2 presents the finished product image of the castings after shakeout. In Figure 2(a), it can be observed that the pouring system and risers were connected to the castings after pouring, and irregularly solidified thin sheet-like structures remained at the parting line. The physical appearance of the bed after removing the gating system and other residual structures is shown in Figure 2(b).

Fig. 2 Site view of the castings (a) before and (b) after the removal of the gating system.

The microstructures are shown in Figure 3. Scanning electron microscope (SEM) images in Figures 3(a)-(c) reveal well-dispersed spherical graphite within the microstructure, with good sphericity and a spheroidization grade of 2 to 3. Statistical analysis indicates a spheroidization rate exceeding 85%. Fine graphite nodules were uniformly distributed on the lamellar pearlite matrix, and no porosity defects were found in the matrix. The matrix structure was dense, and there was a close bond between the graphite nodules and the matrix, with no signs of detachment or gaps between them. Optical microscope images in Figures 3(d)-(f) show that the matrix structure mainly consisted of lamellar pearlite, and the content of phosphorus eutectic + cementite was ≤0.5%.

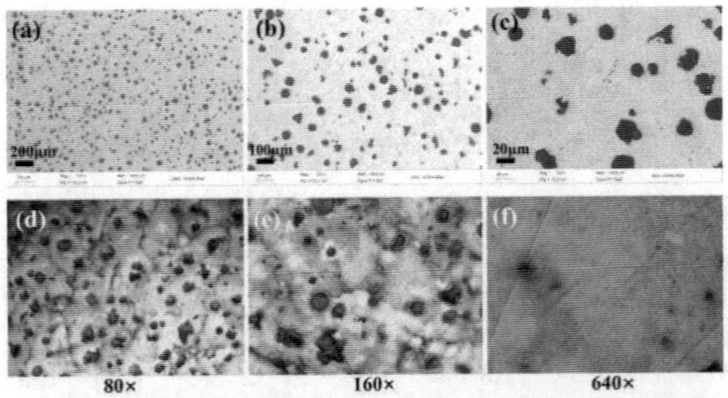

Fig. 3 Microstructure diagrams: (a)-(c) showing SEM images at different magnifications, and (d)-(f) showing metallographic microscope images at different magnifications. The spheroidization rate of graphite exceeded 85%, with a spheroidization grade of 2 to 3, and the matrix structure consisted primarily of lamellar pearlite.

4 Conclusion

The simulation analysis is conducted on the casting process by using AnyCasting for large-scale machine tool bed casting. So it is possible to prepare good casting process parameters to the molten iron filling pattern of the gating system, it provides guidance for the actual production process. The finding indicates that this process exhibits uniform and stable filling capabilities. The results of the filling and solidification simulation demonstrate that the inclusion of chills in the machine tool lead rails and grinding machine bed facilitates timely heat transfer, reducing the heat accumulation from the bottom-pouring casting process on both components. This structure may effectively resist deformation and make this material is suitable for high-strength low-stress bed castings.

Acknowledgments

This work was partially supported by the Sichuan Science and Technology Program (2022YFG0066).

References

[1] Fragassa C. Material selection in machine design: The change of cast iron for improving the high-quality Iin woodworking, Proceedings of the Institution of Mechanical Engineers, Part C: Journal of Mechanical Engineering Science 231 (2016) 18-30.

[2] Jeyaprakash N, Yang C, Duraiselvam M, et al. Investigation of high temperature wear performance on laser processed nodular iron using optimization technique, Results in Physics 15 (2019) 102585.

Effect of Composition on the Corrosion Resistance of Cast Iron to Concentrated Sulfuric Acid

Tohru Mizuki[1, *], Yuichi Hiramoto[1], Takeaki Takahashi[2]
1. Research and Development, Kimura Foundry Co., ltd, Japan
2. Materials Science and Bioengineering, Nagaoka University of Technology, Japan
*Corresponding address: e-mail: t-mizuki@kimuragrp.co.jp

Abstract: Cast iron is used in a variety of fields, including automotive parts, industrial machinery, and the energy industry. One of them is a product used in an environment that comes into contact with concentrated sulfuric acid (hereinafter referred to as CSA). While there are several reports about the corrosion resistance of cast iron in the literature or handbooks, there are few reports where the CSA. On the other hand, there is an experimental report in which the concentration of sulfuric acid is changed in stainless steel, but according to this, the corrosion mechanism is different between dilute sulfuric acid (hereinafter referred to as DSA) and CSA, so the experimental results of DSA may not be applicable in a CSA environment. Therefore, in this study, we investigated about a corrosion resistance of cast iron in a CSA. As a result, it is found that the corrosion resistance is changed by compositions in a cast iron. For example, it is getting worse with the increasing carbon, and it is improved with the increasing silicon.

Keywords: Cast iron; Sulfuric acid; Flake graphite cast iron; Carbon, Silicon

1 Introduction

Cast iron is used in various fields, including automobile parts, industrial machinery and the energy industry. Among its applications, are there parts that come into contacts with CSA. While there are several reports about the corrosion resistance of cast iron in the literature or handbooks, there are few reports where the CSA [1,2]. According to these reports of stainless steel (hereinafter referred to as SST), the corrosion mechanism of the DSA is different from that of the CSA [3]. Therefore, from these experimental results of SST, the experimental results obtained with CSA could not be applied to the results obtained with DSA. In this study, corrosion resistance tests with using CSA and investigated the resistance of cast iron to CSA

2 Experimental procedure

Table 1 shows the composition of the tested materials. As a basic test, we did comparative tests on flake graphite cast iron (equivalent to ISO185 JL300, hereinafter referred to as FC), general structural steel (equivalent to ISO630-2 SG250A, hereinafter referred to as SS) and stainless steel (equivalent to ISO15510 X5CrNi18-10, hereinafter referred to as SST). FC samples were cast into a separately cast specimens with a diameter of 30 mm, and were then machined to a diameter of 20 mm and a thickness of 10 mm for testing. SS and SST were obtained as commercially available products with a diameter of 20 mm, and were cut to a thickness of 10 mm for testing.

Table1 Composition of the Test Materials, mass%

	C	Si	Mn	P	S	Cr	Ni
Equivalent to ISO185 JL300	3.1	1.7	0.75	0.07	0.05	0.02	0.01
Equivalent to ISO630 SG250A	0.15	0.25	0.65	0.02	0.01	0.02	0.01
Equivalent to ISO15510 X5CrNi18-10	0.06	0.42	1.06	0.04	0.02	18.1	8.1

The test materials were polished up to #600 grit using wet grinding and then followed by corrosion tests. For the corrosion tests, the specimens were immersed into 60 ml of 97wt% CSA for 1 hour, then were immediately washed under running water. Next, they were immersed for 10 minutes into a corrosion remover solution prepared by adding 3.5 g/L of hexamethylenetetramine, a corrosion inhibitor, to 50 vol% hydrochloric acid. After immersion, the specimens were taken out from the cleaning solution and were washed again under running water. After the test, the weight before immersion in CSA and that after the removal of corrosion products were compared, and the difference was evaluated as the corrosion loss. The basic tests were conducted under two conditions: with changing the concentration of sulfuric acid from 0 to 97wt% and with an immersion time of 1 hour, and changing the immersion time from 1 to 144 hours at a concentration of 97wt%. Subsequently, the carbon composition of FC was changed from 2.5 to 3.5 mass% and the silicon composition of FC from 1.6 to 2.5 mass%, as shown in Table1. The specimen conditions and corrosion conditions were the same as the basic tests.

3 Result and discussion

Fig.1 shows the relationship between sulfuric acid concentration and corrosion loss. The order of corrosion loss with sulfuric acid is as follows.

With DSA; SS > FC > SST (1)

With CSA; FC > SS > SST (2)

About FC, corrosion loss increases with the increasing concentration of sulfuric acid, shows peak at about 30%, then decreases, and stabilizes at approximately 1 mg/cm^2 for concentrations above 60%. SS shows a similar trend to

FC, but the corrosion amount is higher than FC near 30% and slightly less than FC at concentrations above 60%. Different from FC and SS, SST shows a maximum value at around 60% concentration. Although the overall corrosion amount for SST is smaller compared to FC and SS, it is much more than others in the range of 60%-90%.

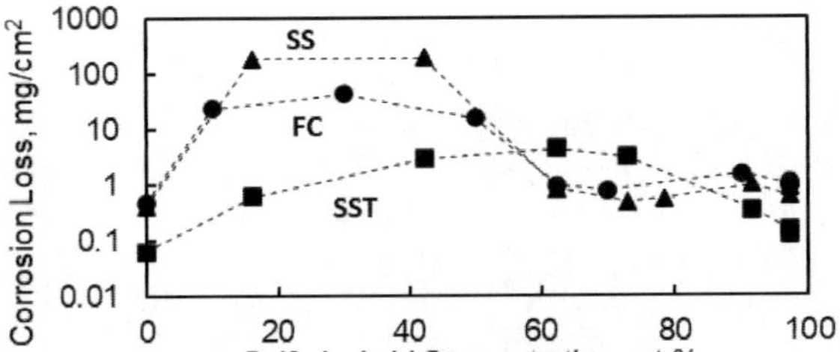

Fig.1 Relationship between Sulfuric Acid Concentration and Corrosion Loss (immersion time 1 hour)

Fig.2 shows the relationship between immersion time and corrosion loss. For FC and SS, the corrosion loss increases with time being, but its increasing rate becomes low with the lapse of time. This suggests that a protective layer should form and enhance its corrosion resistance. Additionally, since the corrosion loss shows continually increasing trend even after 144 hours, it can be said that the protective effect of the protective layer is not complete and the corrosion continues to progress beyond this time. For SST, the corrosion loss does not increase over time. It is thought that the formation of sufficient passive film at the initial stages effectively hinders further corrosion.

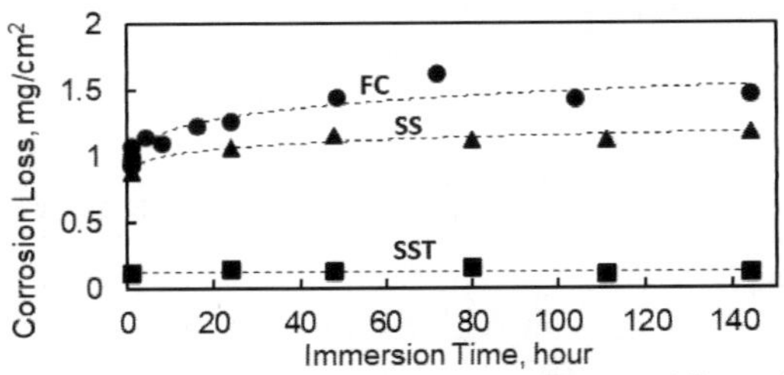

Fig.2 Relationship between Immersion Time and Corrosion Loss (97wt.% sulfuric acid)

Fig.3 shows the results of corrosion tests for FC materials with changing carbon content. As the carbon content increases, the corrosion loss increases slightly. The reason is thought as follows; Unlike steel, cast iron contains graphite in its matrix. As the iron in the matrix has lower

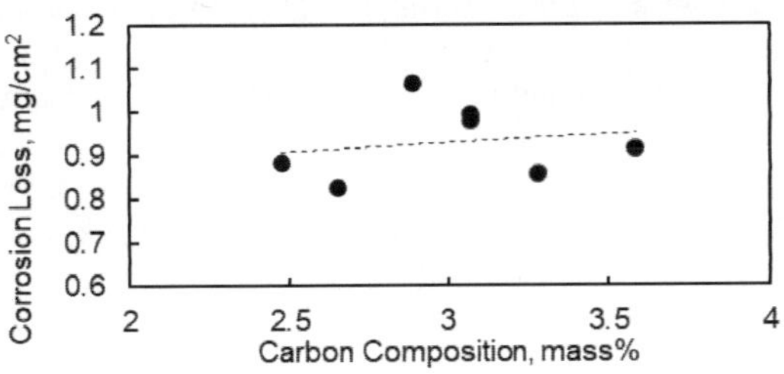

Fig.3 Effect of Carbon composition on Corrosion Loss(FC)

potential than graphite, the iron in the matrix corrodes preferentially. As the amount of graphite increases, the interface area between the graphite and the iron in the matrix also increases, therefore the amount of corrosion loss increases.

Fig.4 shows the results of corrosion tests in FC materials with changing silicon (Si) content. As the silicon content increases, there is a tendency for the corrosion resistance to be improved. That is, the corrosion loss decreases. It is known that cast iron containing 15% silicon shows very strong acid resistance. While the mechanism for the improved acid resistance remains unclear, there are reports suggesting that silicon oxides should enhance oxidation resistance [4]. According to reports, silicon acts to both enhance the corrosion resistance or to promote corrosion [5]. It is thought that the stability of silicon oxides influences to the corrosion resistance of silicon.

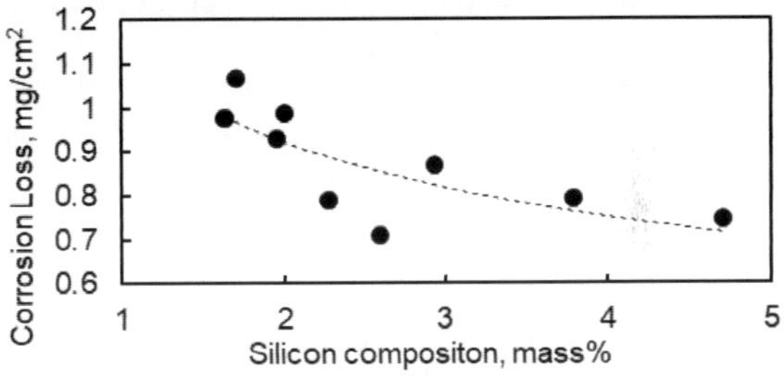

Fig.4 Effect of Silicon composition on Corrosion Loss (FC)

4 Conclusion

The corrosion behavior of flake graphite cast iron in CSA was investigated through immersion tests. The corrosion loss of cast iron increases with the increasing concentration of sulfuric acid, shows peak at about 30%, then decreases, and stabilizes at approximately 1 mg/cm^2 for concentrations above 60%. SS shows a similar trend to FC, but the corrosion amount is higher than FC near 30% and slightly less than FC at concentrations above 60%. Different from FC and SS, SST shows a maximum value at around 60% concentration. Although the overall corrosion amount for SST is smaller compared to FC and SS, it is much more than others in the range of 60%-90%. Compared to stainless steel, cast iron shows higher corrosion resistance in the 60%-90% concentration range.

References

[1] Kuwahara, Miyata, Asakura, et al: Journal of Japan Foundry Engineering Society 85(2013)03, 161-165.
[2] American Foundry Society: Iron Castings Engineering Handbook, 2003, 257.
[3] O' M Bockris, J Drasic, D. and Despic, A.R.: Electrochim Acta, 1961, Vol.4, 325.
[4] The Corrosion Engineering Association: Corrosion and corrosion resistance Handbook, 2000, 295.
[5] Journal of The Japan Institute of Metals 34(1970)10, 1017-1022.

Analysis of the Effect of Scrap on the Refractory Lining and the Application of New DVR Material

Bing Fan[1,*], Chenglong Peng[1], Wei Cheng[2], Chaoliang Wang[2], Ping Ke[2]
1. Xiangyang Juli High Technology Material Co., Ltd. Xiangyang, Hubei, China
2. Xiangyang Public Inspection and Testing Center, Xiangyang, Hubei, China
* Corresponding address: fanbing_940@163.com

Abstract: The erosion mechanism of the most common scrap steel on the silica-based dry vibration refractory (DVR) of the induction furnace (IF) in the casting and melting process was analyzed by using FactSage phase diagram software.

The results show that the eutectic point of the binary compound formed by FeO from the corroded scrap steel and SiO_2 in the siliceous furnace lining is 1205 ℃. The eutectic point of the binary compounds formed by MnO from manganese-containing scrap and SiO_2 is 1345 ℃, and these two types of scrap are easy to react with the furnace lining to form low-melting point compounds, thereby reducing the refractory of the furnace lining. The eutectic point of the binary compound formed by ZnO from galvanized scrap and SiO_2 is 1510 ℃, so it will prevent the sintering of the lining material. In addition, because the melting point of zinc metal is about 420 ℃ and the vaporization temperature is about 906 ℃, the addition of galvanized scrap may also penetrate the entire furnace lining in the form of zinc vapor at low temperatures, resulting in thermal erosion damage of the coil.

According to the analysis of the causes of damage to the DVR lining caused by different types of scrap steel, a new type of DVR suitable for high scrap steel melting conditions was developed, which can effectively extend the service life of the furnace lining and improve the safety of the melting equipment.

Keywords: Scrap steel; Induction Furnace (IF); Phase diagram; Dry vibration refractory (DVR)

1 Introduction

In recent decades, the rapid expansion of steel production has provided the foundry industry with abundant reserves of scrap steel[1]. Scrap steel has emerged as a primary raw material for casting production, and its share in casting feedstock has been steadily increasing. Consequently, there is a growing interest in investigating the influence of scrap steel on casting performance[2]. Induction furnaces are widely utilized for casting production, and extensive research has been conducted on the design and enhancement of induction furnace linings[3-6]. However, there is a lack of systematic study regarding the impact of different types of scrap steel on the lining material used in induction furnaces.

2 Experimental procedure

In this work, FactSage phase diagram software was used to analyze the erosion mechanism of typical scrap steel on silica lining that widely using in induction furnace. According to the investigation, a new type of silica-based lining material suitable for high scrap ratio melting environment is developed and applied in practice. In the end, the sample of used silica lining was analyzed by SEM-EDS.

3 Result and discussion

1. Influence of different types of scrap steel on silica lining materials

1.1 Rusted scrap steel

As shown in Figure 1, the eutectic point of the binary compound formed by FeO in the rusted scrap steel and SiO_2 lining is 1205 ℃, that is much lower than the melting temperature. Therefore, the silica lining will be eroded obviously by the rusted scrap due to the formation of a low melting point compound.

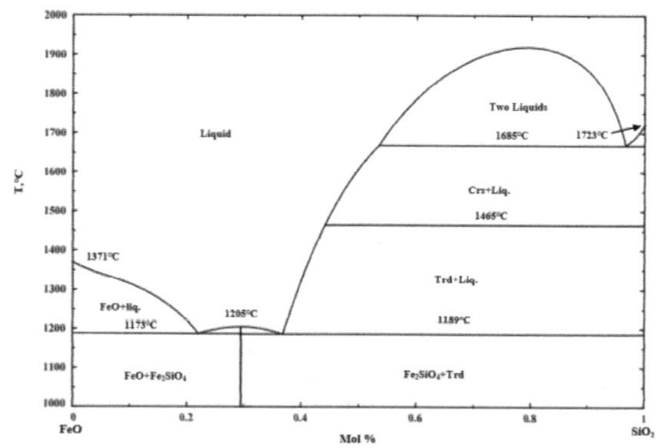

Fig. 1 SiO_2-FeO diagram

1.2 Manganese scrap steel

As shown in Figure 2, the eutectic point of the binary compound formed by MnO and SiO_2 in manganese-containing scrap steel is 1345 ℃. Therefore, manganese-containing scrap steel is also prone to react with the furnace lining to form low-melting compounds, thereby reducing the service life of the furnace lining. Additionally, high-manganese steel exhibits extremely high permeability and often infiltrates into the furnace lining along cracks during its use, leading to shutdown alarms.

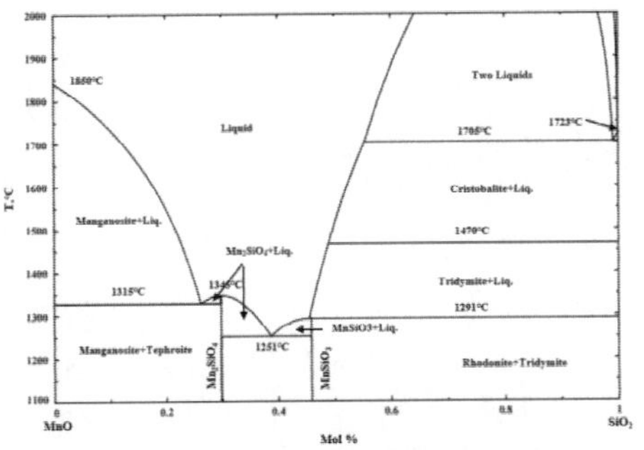

Fig. 2 SiO_2-MnO diagram

1.3 Galvanized scrap steel

The melting point of zinc is approximately 420 ℃, the vaporization temperature is approximately 906 ℃. The zinc in galvanized scrap steel may penetrate the entire furnace lining at low temperatures in the form of zinc vapor, causing thermal erosion damage to the coil. Additionally, as shown in Figure 3, the melting point of ZnO formed by the oxidation of zinc is up to 1976 ℃, the eutectic point of the binary compound formed by ZnO and SiO_2 is 1510 ℃. Therefore, after penetrating the furnace lining, the zinc will adhere to the surface of the silica grains, blocking the normal sintering of the furnace lining material. Therefore, zinc-containing scrap steel should be avoided as much as possible during in the foundry melting.

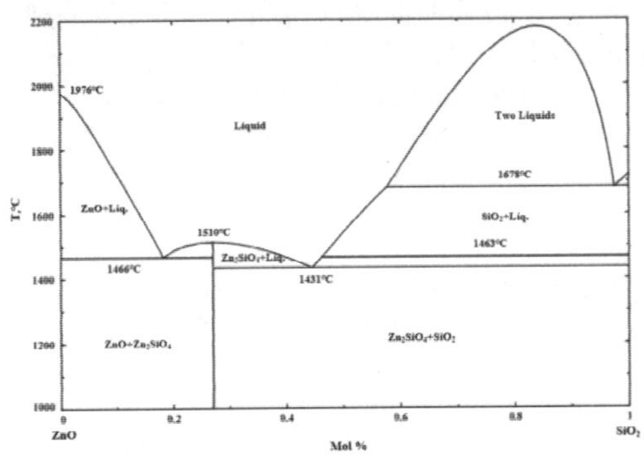

Fig. 3 SiO_2-ZnO diagram

2. Development and application of new silica lining materials

According to the analysis of the damage mechanism of different types of scrap steel to the lining, it is not difficult to see that in the melting environment of high scrap ratio, higher requirements are put forward for the sintering performance and corrosion resistance of silicon-based lining materials.

A new type of silica lining material suitable for high scrap ratio was developed, and applied in a 10 t induction furnace of a local foundry. The customer mainly produces small cast pipes, and the melting temperature is about 1560 ℃. The types of scrap steel are mainly rusted scrap steel and high-manganese scrap steel, with the total consumption of scrap steel is 60%. Finally, the service life of the new furnace lining reaches more than 25 days, while the service life of the conventional material is only about 15 days.

3. Investigation of used silica lining material

Further microscopic analysis was carried out on the sample of the new furnace lining after use. Fig. 4 shows the SEM microstructure of the hot surface in direct contact with hot metal.

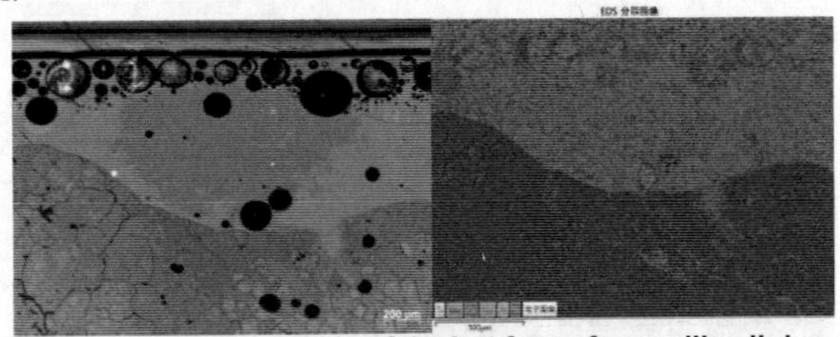

Fig. 4 SEM-EDS analysis of the hot face of new silica lining after used

According to EDS component analysis, the main chemical components of surface attachments are shown in Table 1, in which the contents of MnO and FeO are 2.30% and 1.33%, respectively.

Table 1. Compositions of slag coating /%

SiO_2	Al_2O_3	MgO	CaO	MnO	TiO_2	FeO
47.79	28.35	11.45	6.05	2.30	2.12	1.33

In general, the erosion layer of the used lining is very thin, and no obvious penetration can be found, indicating that the new lining has a good resistance to the corrosion of scrap steel and its low melt on the lining, which is also the fundamental reason for its higher service life.

4 Conclusions

(1) Based on the analysis of phase diagram software, it can be seen that different types of scrap have different effects on silica lining materials.

(2) A new DVR lining is developed for the high scrap ratio melting environment and its performanc is significantly improved compared with the conventional lining material.

References

[1] Çiftçi B B. The Future of Global Scrap Availability, World Steel Association, 2018, https: // worldsteel. org/media-centre/blog/2018/future-of-global-scrap-availability.

[2] ZOU Rong-jian, Production of Cast Iron only by Scrap Steel. Foundry Technology, Vol.29, No.12.

[3] Dötsch E. Refractory Demands on Inductive Melting of Cast Iron. Refractories WORLDFORUM 3 (2011).

[4] Soudier J. Study of dry vibratable mixes design for optimization of the product behaviour during installation. Aachen 49th Coolloquium on Refractories.

[5] ZHAO Zhongkui, SUN Qingzhou, ZHANG Puqing, et al. Study on phase transition and expansion of quartz sand when heated. Journal of Materials Engineering, 2006(10): 25-27.

[6] Yan Fenglin. Distribution characteristics, occurrence status and purification method of harmful impurities in quartz. Chinese Journal of Geology, 2009, 33(3): 277-279.

Production of Steel Castings in Green Sand

Per Larsen
DISA Industries A/S, Højager 8, 2630 Taastrup, Denmark
per.larsen@noricangroup.com

Abstract: Traditionally, steel castings are produced in shell moulding and different no-bake processes. A limited amount of steel castings is produced in green sand today.
However, the shell moulding and different no-bake processes are much less efficient than the green sand moulding process. That means there is considerable potential for cost savings if steel castings can be moved to the green sand process.

There are many different reasons why steel castings are being produced in shell and other no-bake processes, some of which are technically valid while others are more about simply following tradition – "we've always done it this way".

This paper will discuss the opportunities and challenges involved in producing steel castings in green sand. Opportunities include significant cost savings driven by productivity, flexibility, low manpower requirements and green sand's inherent recyclability; simply by adding a little fresh bentonite, sand and water, and mixing it again, green sand can be easily reused.

Some examples of steel castings being poured in green sand will also be given in the paper, such as container corners.

Keywords: green sand; steel; shell moulding; digger teeth; automotive

1 Introduction

The green sand process and especially vertically parted green sand moulding is well known for producing iron parts in high quality, with outstanding efficiency and at minimum cost.

This paper describes some of the opportunities and challenges in producing steel castings in green sand from moulding and materials through to pouring and feeding. Many different castings have been produced successfully in green sand in different steel alloys ranging from plain cast steel to highly wear-resistant alloys.

If the benefits of the green sand process enjoyed by iron foundries can be transferred to the casting of steel parts, it offers the opportunity for lower costs compared to alternative production options.

2 Moulding options

While the most commonly known vertical moulding lines achieve high moulding speeds of up to 555 moulds per hour, slower vertical moulding lines with a maximum speed of 250 moulds per hour are now available at approximately a third of the price of conventional vertical lines.

It is feasible to operate these machines at moulding rates far below 100 moulds per hour; several foundries around the world do this regularly. This allows cost-effective production at lower volumes while retaining the flexibility to easily increase production output if required.

Horizontal moulding employing the well-known matchplate principle is an attractive alternative, especially when production volumes are lower and greater flexibility is needed. Existing pattern plates for horizontal equipment can often be transferred to matchplate moulding machines with little effort.

Green sand moulding material

The moulding material is similar to cast iron production with some modifications. The base sand can be quartz, ceramic or chromite sand with bentonite and water added. Quartz sand with a high silica content (98% or more) is recommended for steel casting. There is no need to add coal dust. With no chemical binders required, the sand costs less.

Due to steel's higher pouring temperatures compared to cast iron, more new sand must be added during production than is usual for cast iron. Typically, around 4% new sand and 1.2% new bentonite must be added, depending on the steel to sand ratio in each mould. This is slightly higher than the typical 2%-3% for cast iron but is still much lower than the chemical binder systems typically used for casting steel.

Green sand process advantages

The DISA green sand process offers many advantages over alternative steel casting processes using chemical binder systems and also over other green sand processes like floor moulding or jolt/squeeze

These include very high productivity and low manpower requirements, automated moulding and core setting, and high production capacity from a single, affordable line.

Core setting is a simple, fast and proven process that can also be easily automated. The green sand process can currently produce up to 470 cored moulds per hour – much faster than shell/no bake processes.

Tool changes take as little as 1-2 minutes. Pattern changes can be manual or are easily automated. Depending on the required pattern plate lifetime, they can be made from a wide variety of materials from wood to tool steel.

Castings with less complex geometries and shorter pattern lifetime requirements can use low-cost wood and plastic patterns. Pattern lifetime can be more than 500,000 moulds if tool steel is used. Due to this long lifetime, normally only one set of pattern plates and possibly one back-up set is needed for each part. Pattern plate maintenance costs are low.

Moulds produced on DISAMATIC or MATCH are harder than other methods like jolt/squeeze, hand or floor moulding, and so are well suited for steel casting.

Green sand's safety and environmental benefits

Compared to shell and no-bake processes, no chemical binders are used for moulding and only water and bentonite are used as a bonding material. There is less sand not reused, fewer dust creation points and it is easy to use hoods to collect and exhaust dust and fine particulates.

Pouring

Slag-induced gas porosity is a risk during pouring. A "teapot"-shaped pouring ladle or similar can be used to keep slag in the ladle and so reduce slag inclusions. Stopper rod-based solutions can also be used.

Semi- or fully-automated pouring can also be applied. Pouring will be faster and more precise, with less metal spilled and lost. Fewer operators are required and there will be fewer mistakes due to human error.

Feeder design

Due to the metallurgy of steel, it tends to shrink more than cast iron during solidification and so requires more feeding than cast iron.

Isolated casting sections can be difficult to feed on vertical machines. Both insulated feeder sleeves set with the core setter and spot feeding with ram-up sleeves can be used as on horizontal lines.

Through trials and finetuning, feeder diameter can be optimized to increase yield while still avoiding shrinkage. If a solution with 2 castings in the mould is chosen, the yield will also improve due to sharing the gating system and feeders between 2 castings.

Core setting

A core binder system with low gas development is important when casting steel. Automated core setting is seldom seen on shell/no bake lines. They must use manual core placement and so core handling is slow, inaccurate and prone to mistakes. DISA green sand lines offer excellent core handling and multiple automation options.

Any type of cores suitable for steel and compatible with green sand can be used on the DISAMATIC and DISA MATCH lines. Cores can be placed manually or automatically using a core mask.

Casting example – container corners

Many different steel casting types can be made in green sand. Steel shipping container corners were selected for testing and made in accordance with ISO 1161. Specifically, a bottom corner casting was used for the test.

A chromite core was used with an ester-cured binder in testing. But any type of core binder with low gas development that is suitable for steel and compatible with green sand can be used.

Low-cost tooling was made in wood and plastic for the steel casting trials. All tooling is machined based on 3D CAD files. This makes tooling production fast and accurate while keeping the price low. This way of making patterns is suitable for prototype tooling and also for tooling where low production volumes are needed.

3 Conclusion

Many steel castings can be feasibly produced in green sand. With equipment currently available, both high-volume production of hundreds of moulds per hour and low-volume production down to 80 or even fewer moulds per hour can be delivered efficiently.

Using green sand can add capacity and flexibility to steel production while reducing per-casting cost for many applications. For parts that are naturally suited to it, green sand offers significant cost advantages that benefit foundries and manufacturers in sectors like mining, railways, transportation and construction.

Casting Process and Pre Deformation Prevention of Large Diameter Retaining Rings for Steam Turbines in Production Practice

Dihua Zhou[1, *], Yi Li[2], Qiao Ma[3], Zezhong Wang[4]
1. Department of Materials Engineering,Sichuan Polytechnic University [1, 4], Deyang, Sichuan, China;
2. Sichuan GuanghanLeshi Machinery Casting Co., Ltd. [2], Deyang , Sichuan, China ;
3. Dongfang Steam Turbine Co., Ltd. [3], Deyang, Sichuan, China.
*Corresponding address:801, Section 2, Mount Taishan South Road, Jingyang District, Deyang City, Sichuan Province, China; e-mail: zdh_zh@126.com

Abstract: The retaining ring is one of the important components of the same flow section on the reaction turbine, and is the load-bearing body of the multi-stage stationary blades. The casting shape is simple, but the contour size is large, which is prone to deformation and the casting difficulty is high. By optimizing the casting process of the holding ring and setting process ribs at the corresponding positions of the inner runner, sand inclusion defects and deformation can be prevented during the casting process; Strengthen the quality inspection process control, use a laser line projector to level the ring before heat treatment, ensure that the deformation generated during the heat treatment process is within a controllable range, and ensure that the product quality meets the requirements.

Keywords: Holding ring; Casting process; Pre deformation treatment; Laser line projector inspection

1 Introduction

The stationary blade holding ring is an annular component that serves as a fixed stationary blade, and its quality directly affects the steam pressure ratio of the flow section of the steam turbine, thereby affecting the working efficiency of the steam turbine. Therefore, ensuring the quality of the holding ring casting is crucial.

The material of the holding ring is ZG230-450. The cleanliness of the inner and outer surfaces of the casting blank should meet the requirements of Cleanliness -2 in JB/T4058—2017. The machining surface should reach UT level II or above, MT level I or above, and the outer diameter of the casting should meet the requirements of Cleanliness -2 in JB/T4058—2017 φ Greater than 4800mm, the middle spoke plate has a thickness of 50mm, a length of 1100mm, and an angle of 83° with the vertical plane, forming a fan-shaped angle. Therefore, size control is very difficult.[4]

2 Development of process plan

2.1 Technical difficulties of holding rings

The holding ring is a semi-circular structure with an outer diameter of Φ4824mm, with an inner diameter of Approximately Φ2390mm, with a main wall thickness of 50mm and curvature. The wall thickness of the amplitude board is 50mm, with an error of within +5mm, and the size requirements are relatively strict. The shape of the retaining ring component is shown in Figure 1.

Fig.1 Shape of Holding Ring Parts

2.2 Determination of Casting Process Plan for Holding Rings

(1) Sand mold: Organic ester hardened water glass sand is used as the surface sand, and the filling sand is mixed with old sand.

(2) Design of gating and riser system scheme: Determine the hot nodes based on the characteristics of the holding ring casting, that is, at the inner and outer ring flanges, evenly place 9 waist shaped risers at the outer ring flange, and place 5 waist shaped risers at the inner ring position as shown in the diagram. As shown in Figure 2.

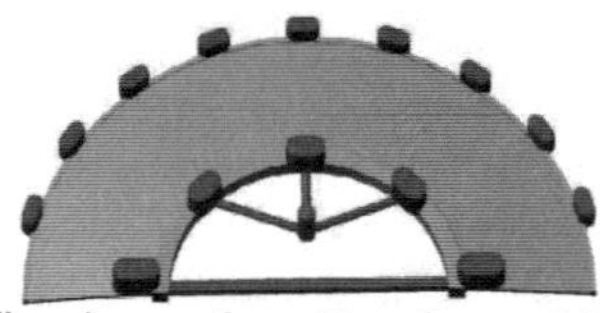

Fig.2 Holding ring casting with gating and riser system

(1) Pouring system setting: adopts a central pouring system, with 3 pouring channels in the middle and lower parts of the inner flange waist type riser, using a pouring channel Φ60 mm fire-resistant ceramic pipe, used for vertical pouring Φ100 mm fire-resistant ceramic pipes.

(2) Method of molding and core making: Using the method of assembling cores, two core boxes are made and then assembled into cores.

(3) Pouring temperature: controlled between 1540 and 1560 ℃.

(4) Placement method for heat treatedworkpieces: Four pieces are stacked horizontally, with 8 tie bars welded on the outer ring and 6 tie bars welded on the inner ring, as shown in Figure 5.

3 Quality issues arising from the original process

The quality problem that arises is: firstly, the height difference between the outer circle and the inner hole (see Figure 3) reaches more than 35mm and undergoes irregular wave like deformation; Secondly, there are surface defects of large areas of sand sticking and sand inclusion on the surface of the spoke plate.

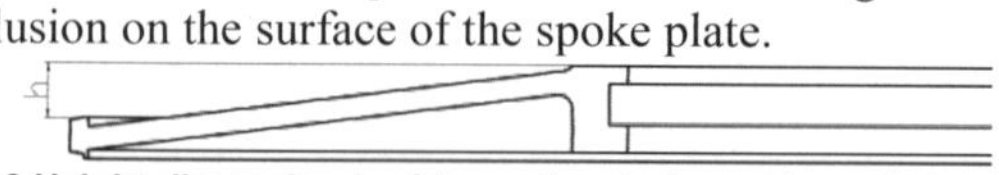

Fig.3 Height dimension h of the outer circle and inner hole of the retaining ring

Based on the defects generated by the casting and the analysis of the entire process flow, the following problems were identified: firstly, the position of the inner runner severely eroded the mold, causing surface defects such as sand sticking and sand inclusion; Secondly, during heat treatment, due to the size of the heat treatment furnace, the castings can only be placed flat. However, if the time distance between the height blocks is too large, the average spacing is between 400mm and 500mm, which ultimately leads to wavy deformation of the product; Finally, there was a lack of inspection of the dimensions of the castings before loading, and the opportunity for heat treatment was not utilized properly to adjust the dimensions of the castings.

4 Improvement measures for casting process and heat treatment process

The improvement measures for casting process include adding tie bars at the corresponding inner runner position, totaling three bars (as shown in Figure 4), to prevent significant deformation of the casting during solidification shrinkage. At the same time, it serves as a runner to avoid defects such as sand inclusion and sticking;

Fig.4 Improvement process for holding ring

The optimization measures for heat treatment process include adding a dimension inspection during the furnace loading of heat treatment. The inspection method is to use two 3D laser line projectors to find two horizontal lines, align the upper line of the inner hole, and the lower line is the horizontal height of the equal height cushion block. The spacing between the equal height blocks should be less than 200mm.

Through the above optimization process measures, a high-quality, low deformation, and compliant holding ring casting was ultimately obtained.

5 Conclusion

Adding tie bars at the corresponding inner runner position can effectively prevent significant deformation of the casting during solidification, control the deformation within a small range, and lay the groundwork for size adjustment in the next heat treatment step; At the same time, it can also serve as a pouring channel, avoiding the defects of sand inclusion and sticking caused by the erosion of the metal liquid on the sand mold.

When using a 3D laser projection instrument for furnace loading during heat treatment, it can quickly and accurately lay out contour pads, ensuring the dimensional accuracy of the castings after heat treatment.

References

[1] Yuan Cheng.300MW Steam Turbine Flow Transformation and Performance Test Research. Suzhou: Suzhou University, 2014.

[2] Wangkou Liu. Experimental and computational analysis of the impact of suction and leakage between the static and dynamic blades of a positively curved stationary blade turbine stage on the aerodynamic performance of the stage. Shanghai: Shanghai Power Generation Equipment Design and Research Institute, 2009.

[3] Bao Yang, Suining Yue, Domestic Processing Technology for 2 # Static Blade Carrier Ring of Steam Turbine, Modern Manufacturing Technology and Equipment, 2017,11:113-116.

[4] Haibiao Huang, Research and Development of Steel Casting Process Technology for Steam Turbine Diaphragm Bushing. Collection of Essays on the 13th National Foundry Annual Conference and 2016 China Foundry Activity Week, 2016.

Effect of Ni, Ti and W Alloying Elements on the Microstructure, Mechanical and Tribological Properties

Abdul Razaq, Muhammad Arslan Hafeez, Xiao-yuan Ji, Ya-jun Yin, Jian-xin Zhou*
School of Materials Science and Engineering, Huazhong University of Science and Technology, Wuhan 430074, China
*Corresponding address: School of Materials Science and Engineering, Huazhong University of Science and Technology, Wuhan 430074, China. e-mail: zhoujianxin@hust.edu.cn

Abstract: Gray cast iron is one of the most widely used materials in a variety of applications due to their low cost and tremendous properties, i.e., mechanical and tribological properties, cast ability, machinability, thermal conductivity, fluidity, and damping ability. Herein, the effect of alloying elements like Ni, W, and Ti on the microstructure and mechanical, thermal, and tribological properties of gray cast iron was investigated. The microstructure was evaluated by light optical microscopy, scanning electron microscopy, transmission electron microscopy, energy dispersive spectroscopy, X-ray diffraction analysis, electron back scattered diffraction analysis, and image analysis. Mechanical, tribological, and thermal properties were explored with the help of Brinell hardness and tensile tests, fractography, ball-on-disc tribometer test, and thermal conductivity test, respectively. The results show that Ni, Ti, and W addition produced graphite flaky A + E + D-type microstructure with coarsened grain size, improved pearlite fractions, enhanced graphite contents, and increased average graphite length. The alloying of Ni, Ti, and W improved the Brinell hardness as well as tensile properties of gray cast iron and facilitated smooth brittle fracture. Their addition in small quantities significantly improved the wear resistance by reducing the wear rate and surface roughness of gray cast iron.

Keywords: Cast Iron; Microstructure; Tribological properties; Flakes, Surface Finish

In-situ Surface Modification of High Silicon Ductile Cast Iron Produced by Sand Mold Casting

R. González Martinez[1]*, Garikoitz Artola[1], Gurutze Arruebarrena[2], Juan José Trujillo[2], Enara Mardaras[1]
1. AZTERLAN, Basque Research and Technology Alliance (BRTA), Durango, Spain
2. Faculty of Engineering, Mondragon Unibertsitatea, 20500 Arrasate/Mondragon, Spain
*Corresponding address: rgonzalez@azterlan.es

Abstract: Corrosion is a critical and common problem in cast iron castings used in industry. To improve the corrosion resistance different strategies could be used, however, most of these methods lead to expensive casting.

In this work, a surface modification layer was produced by in situ casting on high-silicon ductile cast iron using "Duplex powder". The microstructure of the generated surface layer was characterized by using optical microscopy (MO), scanning electron microscopy (FE-SEM), electron backscatter diffraction (EBSD) and the phases were identified by X-ray diffraction.

The corrosion behavior of the samples was analyzed by potentiodynamic polarization test in 3.5% NaCl solution. The Vickers hardness measurements were performed to determine the properties of the modified surface layer. The results showed that the layer formed consists of carbides, ferrite, and γ-Fe. The modified surface layer created has higher hardness and better corrosion resistance compared with the high-silicon ductile cast iron. EBSD characterization shows that the morphology of grains.

Keywords: In-situ surface modification, Electrochemical tets, Vickers hardness, high strength ductile iron.

1 Introduction

High silicon ductile cast irons with silicon contents in the range 3.2wt.%,-4.3wt.%, is an attractive material for automotive and agriculture industry, and most recently, an increased production of wind turbine parts such as hubs and rotor housing castings has been observed [1].

However, the corrosion resistance of high silicon cast iron is limited due to the formation of micro-galvanic cells between the graphite and the ferrite matrix [2]. Therefore, the application of corrosion resistance coatings on the surface of cast iron components has become an important means to reduce maintenance cost and to prolong their service life.

Currently, to extend the demand and working life of ductile cast iron components, new surface in situ reaction are being developed to increase the wear resistance [3], hardness [4], corrosion resistance [5] and high temperature oxidation [6] in cast iron. However, this method requires additional processes, such as heat treatment, that significantly increases the cost of the final product.

Therefore, the aim of the present work was to create an in-situ modified surface layer by using the heat content of a molten ductile cast iron alloy and investigate mechanical and corrosion properties of the created surface layer.

2 Experimental procedure

The high silicon ductile cast iron alloy was prepared in an induction furnace. Then, a special mold with two areas was prepared for the in-situ reaction, the preparation method is described in detail elsewhere [5]. The as-cast specimens were cut in cross section to examine the created surface layer by MO, FE-SEM, EBSD and electrochemical test were realized on the surface.

The chemical compositions of the as-cast ductile iron part and its stainless-steel surface layer are shown in Table 1.

Table 1 Chemical composition of the reference high silicon cast iron (HSDCI) and the duplex stainless steel 316L (wt%)

Material	C	Si	Cr	Ni	Mo	Mg
HSDCI	3.16	4.11	0.04	0.06	—	0.039
316L	0.03	0.50	25.50	6.90	3.99	—

3 Result and discussion

Microstructure characterization

The figure 1 shows a scanning electron micrograph and EDS elemental mapping of the cross-sectional microstructure produced by in-situ surface modification.

As can see in figure 1a, three distinct zones can be observed: (1) zone 1 corresponds to a ferrite-perlite microstructure with spheroidal graphite, (2) zone 2 consists of a pearlite structure containing carbide blocks and some lamellar graphite, (3) zone 3 contains a γ-austenite matrix and chromium carbides. Figure 1b shows the elemental mapping analysis of the Cr, Ni, C and Fe. The distribution of Ni and Cr was relatively uniform in the zone 3 and it gradually decreased from zone 3 to zone 2 and zone 1.

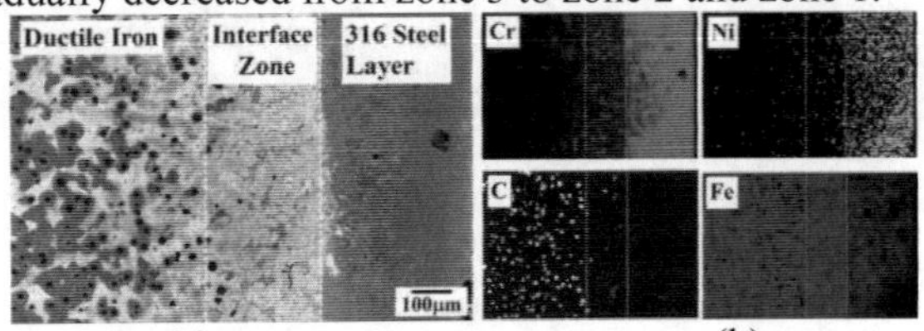

(a) (b)

Fig. 1 SEM microstructure (a) and EDS elemental mapping analysis (b)

Figure 2 shows the microstructure obtained. The inverse pole figure (IPF) map in Fig. 2a shows a bimodal distribution of the grain size and crystallographic phase distribution (Fig. 2b). The red and blue colors refer to ferrite and austenite phase, the green and yellow colors correspond to chromium carbide and cementite.

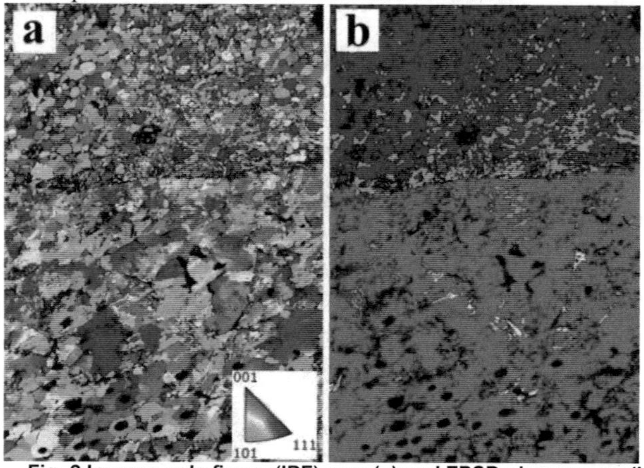

Fig. 2 Inverse pole figure (IPF) map (a) and EBSD phase maps (b)

Electrochemical results

The figure 3 shows the polarization curve of the reference HSDCI and the HSDCI with γ-austenite matrix and carbide layer. The HSDCI with layer presents a more noble potential (E_{corr}) and higher corrosion current density (I_{corr}) than those of the reference HSDCI. The Tafel analysis of the polarization curve is shown in Table 2.

Table 2. Electrochemical parameters obtained from the polarization curves presented in Fig. 2

Material	E_{corr}(V)	I_{corr}(A • cm^{-2})
HSDCI	-0.79	3.0×10^{-6}
HSDCI+316L	-0.68	1.5×10^{-6}

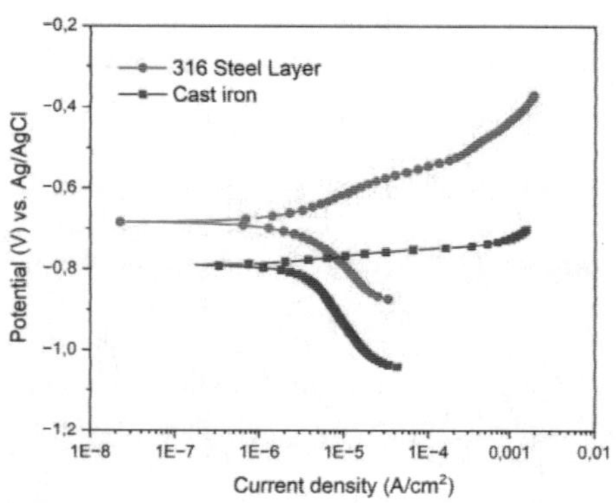

Fig. 3 Polarization curves of cast iron and cast iron with 316L in 3.5% NaCl

Conclusion

The suggested in-situ casting technique allows the formation of a coated cast iron that consists of three zones.

The layer consists of γ-austenite matrix and chromium carbides.

The corrosion current density decreases with the modified layer when compared to the HSDCI.

References

[1] Werner H, Lappat I, and Aurich B. Mischkristallverfestigte EN-GJS-Werkstoffe für Gross- und Schwergussteile, Giesserei, 2016, 103(07): 38-42.

[2] Ochoa N, Mardaras E, González-Martiez R and Artola G. Pseudo-passive films on cast Irons: A strategy to mitigate corrosion by acting directly on microstructure. Corros. Sci. 2022, 206: 110480.

[3] Ye F, Hojamberdiev M, Xu Y, Zhong L, Zhao N, Li Y and Huang X. Microstructure, microhardness and wear resistance of VC_P/Fe surface composite fabricated in situ, Appl. Surf. Sci. 2013, 280: 297-303.

[4] Fischer, S F, Muschna S, Bührig-Polaczek A, and Bünck M. In-situ surface hardening of cast iron by surface layer metallurgy. Mater. Sci. Eng. A, 2014, 615: 61-69.

[5] Mardaras E, González-Martinez R, Bayon R, Nastac L and Méndez S. Surface modification of ductile iron produced by an innovative in-situ casting technique. Int. J. Cast. Metal. Res., 2020, 33(2-3): 103-111.

[6] Lin, M-B and Wang C-J. Microstructure and high temperature oxidation behavior of hot-dip aluminized coating on high silicon ductile iron. Surf. Coat. Technol. 2010, 205: 1220-1224.

Evaluation of Optimum Niobium Addition in Heat Resistant Cast Steel (1.4837) and Development of a Low Nickel Grade with Optimized Niobium

Jean Amadeo Brambila[1], Francisco Manoel Perez do Nascimento[2], Nico Scheidhauer[3], Gotthard Wolf[3], Erico França[4]

1. Nb Consulting, Av. Itú 124, Barueri - São Paulo, Brazil; 2. Metaprol, Rua 20 de Novembro 278, Centro Timóteo, Brazil; 3. Technical University Freiberg, Foundry Department, Bernhard-von-Cotta-Str. 4, 09599 Freiberg, Germany; 4. Companhia Brasileira de Metalurgia e Mineração, Av. Brigadeiro Faria Lima, 4285, São Paulo- SP-Brazil

Abstract: The present study analyses four different Niobium (Nb) additions in Heat Resistant Cast Steel (Din 1.4837) for optimization purposes in this alloy grade. In addition to this evaluation, this study also suggests a new alloy design to reduce material costs, due to high pressure from the automotive industry claiming high nickel (Ni) price to produce turbine housings.

Therefore, the baseline alloy (Din 1.4837) was thermomechanical evaluated with 0%Nb, 0.6%Nb, 1.2%Nb, and 1.8%Nb additions to define the optimum %Nb addition to promote the best performance considering a complete list material property.

Keywords: Heat Resistant Cast Steel; JMatPro; Niobium Impact; Low Nickel; HH-alloy; Nb

1 Introduction

The influence of alloying elements on the thermomechanical properties of Heat Resistant Cast Steel is related to the effect of such elements on the ferrite and austenite content in the matrix [1] and the carbide precipitation behavior.

Several factors cause the failure of heat-resistant steel castings such as oxidation, sulfidation, carburization, creep, thermal fatigue, and sigma phase embrittlement. Intermetallic, such as the sigma phase are important sources of failure in high-temperature materials [2].

This research was separated in two phases, phase 1 the evaluation of the best thermomechanical performance of HH alloy considering only different Nb additions of 0%Nb, 0.6%Nb, 1.2%Nb and 1.8%Nb, subsequently an alloy optimization was proposed in order to increase the alloy performance and also reduce the alloy price keeping the same temperature range application.

2 Experimental Procedure

Phase 1 covered the manufacturing procedure as well as the properties achievable by adding four different niobium content according to the Table 1 all other elements were kept the same.

All the HH alloys (1.4837), see Table 1, were cast at a temperature of 1550 ℃ ± 5 ℃ after inductive melting of low alloyed steel scrap, Nickel, Ferrosilicon, Chromium, Manganese, and Ferroniobium Grade 111 in a 10kg lifting inductive furnace.

Table 1 HH-alloys after ASTM A 297 (1.4837)

Alloy	C	Si	Mn	Ni	Cr	Nb	Mo
Nb00	0.42	1.51	1.25	12.55	24.55	0.00	0.02
Nb06	0.39	1.43	1.20	12.38	24.69	0.60	0.02
Nb12	0.40	1.51	1.22	12.53	24.67	1.20	0.02
Nb18	0.41	1.51	1.18	12.54	24.56	1.80	0.02

Considering the test list mentioned above one decision matrix was created to quantify the thermomechanical and thermal properties to define the best performance of %Nb addition in the HH alloy.

Phase 2 consists of use the best performance of %Nb defined by Phase 1 and optimize it considering price reduction and improving thermomechanical performance creating a new alloy called WP3.

3 Results and Discussions

The best performance of the Nb addition in the standard HH (1.4837) alloy was 1.2%Nb due to a quantification table including all the listed parameters mentioned above, from now on, this investigation will consider the new alloy called WP3 with the following chemical composition:

Table 2 Chemical composition of the casted WP3 alloy

Alloy	C	Si	Mn	Ni	Cr	Nb	Mo	N
WP3	0.55	1.25	4.00	9.60	24.2	1.20	0.02	0.20

For the Machinability test three specimens each were machined at a cutting speed of 50m/min and at a cutting speed of 60m/min. Machinability and tool wear is impacted by the Nb addition. The biggest tool wear is visible at 1.2%Nb. After exceeding 1.2% the tool wear decreases slightly, which is addressed to higher ferrite contents in the alloy. WP3 demonstrated better machinability performance compared with HH 1.2%Nb.

For the evaluation of the Castability respectively Mold filling the cast spirals, were casted. The evaluation of the casting spirals is a comparative procedure. There is no clear tendency for improvement or decrease of flow length (component 1 of mold filling capability). Niobium has no significant impact on flow length of 1.4837Nb.

During the Creep resistant tests, a total of three different temperature/load combinations were tested. Once at 275 MPa and 600 ℃, at 150MPa and 700 ℃ as well as at 90MPa and 800 ℃. At 700 ℃, only the 0.6 % and 1.2 % Nb alloys were tested. For the WP3-alloy a higher load (300 MPa) was defined for the test at 600 ℃. The WP3-alloy shows a huge increase in creep resistance.

Table 3 Rupture time during creep testing

Temp.[℃] / Stress [MPa]	0%Nb [h]	0.6%Nb [h]	1.2%Nb [h]	1.8Nb% [h]	WP3 [h]
600 / 300	-	-	-	-	2274
600 / 275	387	971	398	266	-
700 / 150	-	697	296	-	4720
800 / 90	82	130	238	92	1121

During the LCF (Low Cycle Fatigue) tests, the strain amplitude was constant at 0.3%. The alloy with 1.2% niobium shows a high lifetime at all temperatures, whenever the differences are not significant. A factor 2 in lifetime at the LCF tests anyway has to be judged as a normal scattering. The 1.8%Nb was not done due to low values of Creep resistant test.

Temp / Cycles	0%Nb	0.6%Nb	1.2%Nb	1.8Nb%	WP3
600 °C	1474	1153	2205	-	1109
800 °C	2515	2289	2079	-	729
900 °C	1047	926	1598	-	740

The Hot Cracking behavior is strongly influenced by the Nb content. Additions from above 1.2%Nb demonstrates that the hot cracking tendency significantly decreased. This specific topic was presented in the 74th WFC in Busan South Korea 2022 "New Evaluation Method for Hot Cracking Tendency of High Alloyed Steels containing Niobium".

Fig. 1 Results of 1.4837 with 0% Nb (left) to 1.80% Niobium (right).

The cracks are represented by red circles, the samples are always positioned in the same direction, the cracks appear when the dimension of the geometry changes drastically, the lesser the Niobium the higher the cracks, additions of Niobium around 1.2%Nb is efficient for preventing hot cracks. For more complex geometries higher levels of Niobium might be necessary.

4 Conclusion

The new developed material is out of any found existing patents or specifications. The nickel was successfully replaced by carbon, manganese, and nitrogen, without having a negative impact on the ferrite content. This change makes the new material 10% cheaper in comparison to the baseline material 1.4837 with 1.2 % niobium.

The impact of Nb addition is huge when the Hot cracking is taken into consideration the lesser the Niobium the higher the cracks, additions of Niobium around 1.2%Nb suggest successful preventing of hot cracks.

This was leading to a slight decrease in lifetime of the samples under cyclic load. This decrease in cyclic lifetime can be addressed to the, especially at 800℃ and 900℃, significantly higher tensile amplitude achieved in the tests. The creep properties were significantly improved after the alloy modification.

Acknowledgments

The authors would like to acknowledge the project partner CBMM Brazil and special thanks to TU Freiberg Germany.

References

[1] Lamb S. Practical Handbook of Stainless Steels and Nickel Alloys, ASM International, Materials Park, Ohio, 1999.

[2] Prager M, Svoboda J. Cast High Alloy Metallurgy, Steel Casting Metallurgy, Steel Founder's Society of America, Rocky River, OH, 1984.

Design and Production of Diesel Engine Body Blank Parts

Bobo Lu*, Shiyu Yang, Zibin Tang, Gang Yang
School of Mechanical Engineering, Sichuan University, Chengdu 610065, China
*Corresponding address: No. 24 South Section 1, Yihuan Road. e-mail: luboboscu@163.com

Abstract: The internal structure of diesel engine body is extremely complex, with uneven wall thickness distribution, and defects such as shrinkage and porosity are prone to occur during casting production, affecting the compactness of the body. Studies have shown that producing diesel engine bodies using a combination of welding and casting can avoid the aforementioned issues. However, welding repair of oil and water passages can lead to deformation, leakage, and stress concentration, which in turn affect the normal operation and lifespan of diesel engines. To address these issues, we adopt integral casting to design and produce four different types of diesel engine body blanks. The design of core making, core assembly box, metal melting, and casting processes, along with associated process parameters, were developed, and subsequent treatments were carried out based on application requirements, including the design of heat treatment processes and shot blasting cleaning. Furthermore, 3D simulations were used for core making and core assembly box to validate the rationality of the processes, leading to actual core making and shortening the new product development cycle effectively, thus saving on research and development costs. The results indicate that rational parameter design and process flow can not only avoid shrinkage and porosity issues, as well as stress concentration, but also provide the basic conditions for the scale production of different body types.

Keywords: diesel engine body; combination of welding and casting; 3D simulation

1 Introduction

The wide application of diesel engines has promoted the development of all walks of life [1-3]. In this paper, two kinds of V-type and two kinds of L-type diesel engine blocks are selected as research objects, and a series of studies are carried out as described later. The wall thickness of diesel engine block has a direct influence on the cooling rate of metal liquid during solidification. The heat dissipation effect of the body is poor in the area with larger wall thickness, so the solidification speed of the metal liquid in is slower and is the more conducive to graphite precipitation [4-6]. However, the body wall thickness is too large, easy to produce isolated heat node and C element is easy to exist in the form of graphite floating. The smaller the wall thickness of the body, the better the heat dissipation effect, the faster the solidification rate of liquid metal, the more conducive to the precipitation of cementite. However, the wall thickness is too small, which is easy to cause defects such as cold insulation and insufficient pouring. Therefore, the body wall thickness should be as thin as possible on the basis of meeting the minimum wall thickness.

2 Experimental procedure

The 3D casting process model of J320 body was established in UG software, the boundary conditions were set by entering the Visual-Cast module. The filling method was set as gravity filling, the sand was set as furan resin sand, the casting material was set as HT250, the cold iron material was set as HT300, the pouring temperature was set as 1360 ℃, and the pouring time was set as 38 s. The initial temperature of mold and sand was set at room temperature 20 ℃, the interface heat transfer coefficients of casting and sand, casting and cold iron, sand and cold iron were selected as: 1000 Wm^2/K, 2000 Wm^2/K. 1000 Wm^2/K and the cooling method of the outer surface of the sand box was selected as air natural cooling.

3 Result and discussion

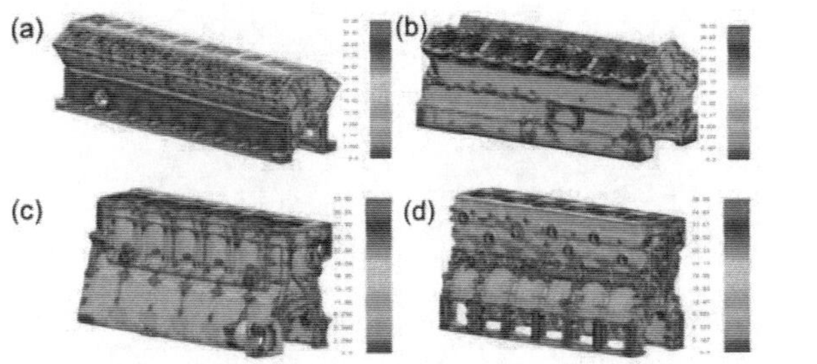

Fig. 1 Overall wall thickness distribution of four body models, (a) J320 Body model, (b) V16 Body model, (c) K19 Body model, (d) ZTD100 Body model

The casting production of diesel engine block is mainly divided into two parts, one is the casting process design, and the other is the control of the two physical state changes of metal. The casting process design of diesel engine block mainly includes pouring system, riser, exhaust system and cold iron design. For diesel engine block with complex structure and large wall thickness difference, the distribution of flow field and temperature field should be considered in the casting process design to avoid internal defects. The casting process design is shown in Figure 2.

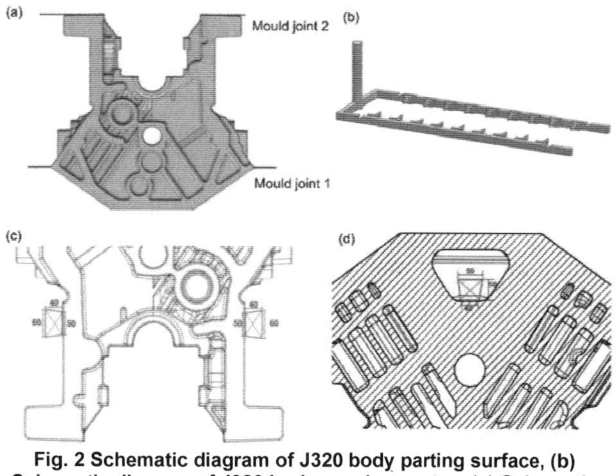

Fig. 2 Schematic diagram of J320 body parting surface, (b) Schematic diagram of J320 body pouring system (c) Schematic diagram of 1 cold iron (d) schematic diagram of 2 cold iron

Based on the problems of the original scheme, the original scheme is optimized. In the original pouring system, a straight runner was set on the end face of the body to introduce the metal liquid, which dispersed to both sides at the bottom of the straight runner into the cross runner, and then flowed to each inner runner in turn. In order to balance the time of metal liquid entering the inner runner, the optimization scheme chooses to set one straight runner and one cross runner on each side of the body, and the straight runner is located in the middle of the cross runner. 3D simulation [7] was used to predict defects, as shown in Figure 3.

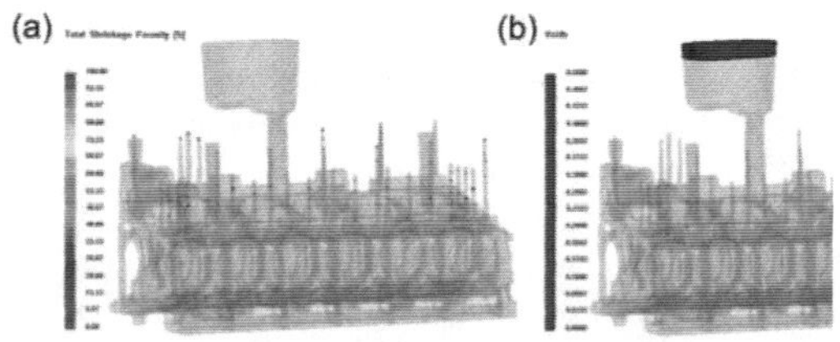

Figure 3. Prediction of (a) porosity and porosity and (b) trapped gas distribution

In order to eliminate the residual stress inside the body, the body billet is annealed for stress relief. The heat treatment process curve is determined according to the characteristics of the body. After the heat treatment, the follow-up treatment such as shot blasting and hardness testing is carried out, and the body is accepted into the warehouse after passing the quality inspection.

4 Conclusion

The design of core making, core assembly box, metal melting, and casting processes, along with associated process parameters, were developed, and subsequent treatments were carried out based on application requirements, including the design of heat treatment processes and shot blasting cleaning. Furthermore, 3D simulations were used for core making and core assembly box to validate the rationality of the processes, leading to actual core making and shortening the new product development cycle effectively, thus saving on research and development costs. The results indicate that rational parameter design and process flow can not only avoid shrinkage and porosity issues, as well as stress concentration, but also provide the basic conditions for the scale production of different body types.

Acknowledgments

Thanks to Mr. Yang Gang for the financial support and the scientific compass testing organization

References

[1] Klaus M. Heim, Marco Coppo. Common Rail Technology for Low-speed Engines[J]. MTZ industrial, 2017, 7(2):26-33.

[2] Awadhesh Kumar Tiwari, R K Mandloi. Impact of Diesel Engine Technology Advancement on Particulate Exhaust Emission - A Review [J]. International Journal of Vehicle Structures & Systems, 2019,11(5):608-613.

[3] Du Yu, Qian Cheng, Zhao Yue, et al. Numerical Simulation of Pressure Fluctuation in High-pressure Common-rail Fuel Injection System [J]. American Journal of Mechanical and Industrial Engineering, 2019, 4(3):45-51.

[4] Jiang J, Ning G E, Huang M, et al. Numerical Simulation of Squeeze Casting of Aluminum Alloy Flywheel Housing with Large Wall Thickness Difference and Complex Shape [J]. Transactions of Nonferrous Metals Society of China, 2023, 33(5): 1345-1360.

[5] Guzik, Edward. Evaluation of the number of primary grains in hypoeutectic chromium cast iron with different wall thickness using the ProCAST program [J]. Materials, 2023, 16(8): 3217.

[6] Kaymak, Kübra. Effects of Ti addition and wall thickness on microstructure, hardness, and wear properties of spheroidal graphite cast irons [J]. Tribology International 2024,195: 109612.

[7] Li, Guodong, et al. Dynamic evolution of residual stress upon manufacturing Al-based diesel engine diaphragm. High Temperature Materials and Processes, 2024, 43 (1): 20240032.

Geometry Conversion of Cast Elements of Heating Devices

Kamińska Jadwiga[1]*, Żuczek Robert, Krzysztof Sanocki[1], Michał Angrecki[2], Marcin Małysza[2]
1. Odlewnia KAW-MET Marek Kawiński Sp. z o.o., Zadąbrowie 311, 37-716 Orły, Poland
2. Łukasiewicz Research Network – Krakow Institute of Technology, Zakopiańska 73, 30-418 Krakow, Poland
*Corresponding address: e-mail: j.kaminska@kawmet.pl

Abstract: The reliability of exploitation of heating devices are determine by the durability of their components which are exposed to elevated temperatures conditions [1]. Among the commonly used components are grate elements and deflectors. Due to cyclic temperature changes, these castings are exposed to the effects of fatigue processes that contribute to unfavorable changes in the material structure, which can ultimately lead to the formation of cracks and deformations in the shape of castings, ultimately rendering the casting unusable [2-4].

The aim of the conducted research was to determine the effect of applied shape conversion on the elimination of areas of concentration of thermal stresses causing the occurrence of fatigue cracking mechanisms both during casting process and exploitation. The model of the element operating over a wide range of temperatures in the heating device was subjected to structural conversion. The main assumption was to reduce the weight of the element while simultaneously achieving strength parameters similar to those currently present in the construction.

Keywords: geometry conversion, heating device components; numerical calculations; weight reduction

1 Introduction

Computer simulations are currently a tool not only at the design stage but also for improving production efficiency and exploitation properties of parts and devices in use. In the analyzed issue, a structural conversion was used, which directly affects the weight of the part, and thus affects the liquid metal consumption and the amount of energy consumed during the melting process of given alloy. The conversion is carried out based on the exploitation conditions of the casting. A model of thermal exploitation cycle was assumed based on information from the manufacturer. It was assumed that the upper part of the casting could be heated to a maximum temperature of $T_{max} = 600°C$. For the numerical model, a material database with properties of the cast iron currently used for the castings of heating device grates was assumed. The mass of the analyzed grate structure was approximately 4.38 kg.

2 Experimental procedure

The preliminary analysis of the grate structure indicates that for the assumed thermal exploitation scheme, the maximum stresses in the structure do not exceed σ_{max}=2.15e^8 Pa, with a global deformation of the grate u=0.0019 m. The deformation pattern of the grate is similar to the characteristic of the plates, with vertical deformation at its corners. The maximum stress values are located in the lower part of the casting and accumulated in the area of reinforcing ribs.

Several modifications to the grate structure were made in order to reduce weight and potentially lowering the maximum stress values. The most favorable results of these changes were presented for two variants, V02 and V03. Changes in the V02 grate structure involve altering the thickness of the casting wall and modeling pockets in the lower part of the casting. The weight of the casting after these changes is approximately 4.11 kg. The distribution of reduced stress field is presented in Figure 1.

Fig. 1 The distribution of reduced stresses in the grate casting for version V02

The reduction of weight for version V02 by 6% resulted in a decrease in the maximum deformation of the casting corners to approximately u=0.0017 m, however, the maximum stress value increased to σ= 2.58e^8 Pa. The stress distribution is more uniform in the plate casting. The maximum stress values accumulates in the reinforcing ribs.

The next iteration of changes in the construction includes removing of problematic reinforcing ribs and introducing relieving holes in the nodal areas of the grate plate (Figure 2a and b). The changes allowed for a reduction in the casting mass to approximately 3.89 kg (a reduction in weight 11%).

a)

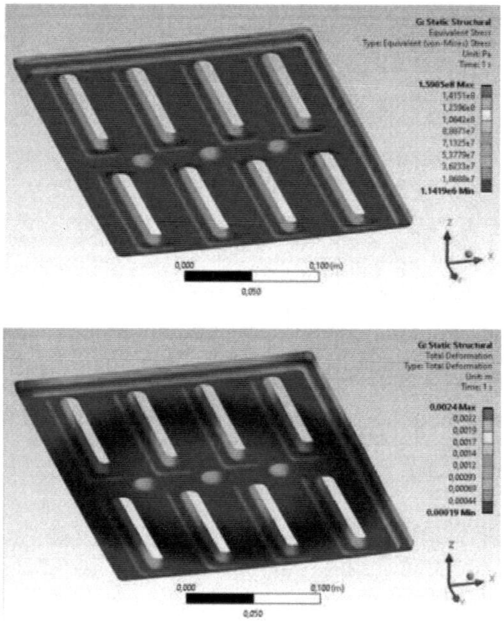

b)

Fig. 2 The distribution of reduced stresses (a) and displacements (b) in the grate casting for version V03

3 Result and discussion

The introduced structural changes allowed for the reduction of maximum stress values to σ=1.59e^8 Pa. Additionally the changes results in an increase in maximum displacements in the casting to approximately u = 0.0024 m. Additionally, the introduced changes allowed for increased stiffness in the lower part of the structure.

4 Conclusion

The implementation of relieving holes in the central part of the casting and the elimination of flexible reinforcing ribs allowed for a 10% reduction in the casting weight and relief of the structural stress of the grate, which should positively impact the thermal-resilience characteristics of the grate structure.

Acknowledgments

The work was carried out as part of project No. FENG.01.01-IP.02-1570/23 entitled "Material and structural conversion of components operating at elevated temperatures along with the implementation of an innovative technology for refining liquid cast iron in casting form".

References

[1] Słowik J A, Drotlew A, Piekarski B. Mechanical Stresses Induced by Compression in Castings of the Load-carrying Grate. Archives of Foundry Engineering, 2016, 16(2): 51-54.

[2] Słowik J, Drotlew A and Piekarski B. The Flexibility of Pusher Furnace Grate. Archives of Foundry Engineering, 2016, 16(4): 137-140.

[3] Piekarski B, Drotlew A. Cast Grates Used in Heat Treatment Furnaces. Archives of Foundry Engineering, 2019, 19(3): 49-54.

[4] Kuder M, Jaśkowiec K. Thermal shock resistance of solution-hardened spheroidal cast iron with selected alloy elements. Transactions of Foundry Research Institute, 2017, LVII(3): 169-179.

Research and Analysis of the International Standards System for Cast Steels

Jiahui Zhu [1, 2], Guanghua Qin [1], Yin Zhang [1], Shifang Su [1]

1. National Technical Committee 54 on Foundry of Standardization Administration of China, Shenyang 110022, Liaoning, China;
2. Shenyang Research Institute of Foundry Co., Ltd., CAM, National Key Laboratory of Advanced Casting Technologies, Shenyang 110022, Liaoning, China
*Corresponding address: e-mail: zjh@chinasrif.com

Abstract: This paper introduces management work of the organization, ISO/TC17, Steel, Subcommitee SC11, Steel castings, and its comprising participating members and observing members. ISO/TC17/SC11 has developed 21 International Standards and all are included in the Standards catalogue. This paper introduces the overrall of the published the International Standards and Standards under development, expounds the characteristics and typical applications of cast steel materials, research and analyzes the present status, and makes a future outlook of International Standard of cast steels.

Keywords: cast steel; International Standard system; analysis study

1 Introduction

As the reference to the products design, manufacture, purchase, inspection, use and maintenance in foundry, standards play a key role in determining the quality and competitiveness of worldwide foundry products[1], due to the advancement, coordination, and systematic nature. Steel castings are basic and key components of modern industry and major engineering equipment. Steel castings is promoted to large-scale, complicated, precise, high strength and high-toughness, and high corrosion resistance with the development of material technology and manufacturing processes.

International Standard of cast steel, as a technical document is used to guide the production and manufacture, with additions made, including the achievements of advanced materials and manufacturing technology, to form a complete technical standard system, which can meet the needs of many industries and different service environments for steel castings. International standard of cast steels is not only used to asist production, quality control and inspection of steel castings, but also as a judgment for the international trade and acceptance of steel castings.

2 Organization structure of international standards for cast steel

ISO/TC17/SC11 comprises 10 Participating members and 22 Observing member. ISO/TC17/SC11 Secretariat is direct responsible for developing and revising international standards in the field of cast steel, established and dissolved by technical management board(TMB).

3 The current status and analysis of the international standard system for cast steel

There are 21 International Standards published and 4 standards under development. Published International Standards, including 9 cast steel material standards, 8 detection methods standards for the chemical composition, surface quality, and internal quality of cast steel, as well as 4 basic standards for welding processes and general technical delivery requirements of cast steel castings, specify the chemical composition and mechanical properties of cast steel materials, respectively, and the Information Annex given in standards is refered to aid product designers and for specifiers in the successful execution of their designs as steel castings.

The ISO cast steel standard system is mainly classified based on material chemical composition, material characteristics, applications, and testing methods. The ISO cast steel standard system is similar to Chinese standard system. China has done pioneering work on the development of basic standards, such as digital radiography, metallographic examination, and repair welding.

China pays more attention to the development of basic and general standards for digital radiography, compensating for the lack of standards in the cast steel industry, popularize and enhance digital radiography technology, and promote the application in the cast steel industry.

(1) Cast steel material standards

There are 9 published International Standards for cast steel materials. ISO/TC17/SC11 updates cast steel material standards in time, with new grades additions. These standards specify the chemical composition range of materials, with strictly limitations on harmful elements, achieving performance improvements of cast steel materials through new grade and advanced technology, stimulating growth and improvement in the cast steel industry.

(2) Detection method

(1) The following cast steels standards is classified by non-destructive testing techniques. Two standards is classified by ultrasonic testing, including ISO 4992-1:2020,

Steel castings — Ultrasonic testing — Part 1: Steel castings for general purposes, and ISO 4992-2:2020, Steel castings — Ultrasonic testing — Part 2: Steel castings for highly stressed components. ISO 4993:2024, Steel and iron castings — Radiographic testing is classified by radiographic inspection. ISO 4986:2020, Steel and iron castings — Magnetic particle testing is classified by magnetic method. ISO 4987:2020, Steel and iron castings — Liquid penetrant testing is classified by penetrant testing. With the development of eddy current testing technology, Chinese experts should actively put forward new proposal of International Standards for eddy current testing, promoting the application of eddy current testing technology in the cast steel industry.

(2) ISO 11971:2020, Steel and iron castings — Visual testing of surface quality and ISO 19959:2020, Steels, nickel alloys and cobalt alloys investment castings — Visual testing of surface quality visual testing are classified by visual testing. With the rapid growth of foreign trade, higher demands for excellent products, it not only requires superior performance and inner quality, but also puts forward strict requirements for the surface quality of steel castings.

(3) ISO 13520, Determination of ferrite content in austenitic stainless steel castings is classified by metallographic, magnetic and chemical methods.

With the development of detection technology and equipment, digital radiography technologies, CT and DR have developed rapidly. Intelligent metallographic detection, such as automated defect recognition technologies, widely used in large-scale foundries. New technologies such as eddy current testing, acoustic emission testing, infrared testing, and time-of-flight diffraction technique have been popularized and applied.

(3) General standards

ISO/TC17/SC11 pays great attention to update basic standards, strictly manage the cycle of the standards, elapsed time before systematical review is usually 3 years, then a systematic review will be initiated. To better facilitate exchanges and cooperation among countries, Within 5 years of International Standards since publication will initiated revised.

4 Future development analysis of cast steel international standards

(1) Facilitating the exchanges and cooperation among relevant industrial standardization organizations at home and abroad.

(2) Paying attention to the application of new technologies in the production of steel castings, and further efforts will be put into making related standards proposal.

(3) Actively proposing new international standards proposals for intelligent inspection, with ISO/TC17/SC11 paying close attention to the technical development and industrial application of intelligent inspection. (4) Efforts should be made to strengthen international exchanges, keep up with advanced foreign technical standards, actively promote the transformation of Chinese national standards in the field of cast steels.

5 Conclusion

(1) The application of new technologies, new processes, and new methods, promotes the rapid development of cast steel standardization work.

(2) ISO standards provide a large amount of basic data on the mechanical and physical properties of cast steel materials, which can be referenced by design engineers. By understanding the technical development of cast steel international standards, technical trade barriers caused by misunderstandings of standards can be eliminate.

(3) Focus should be placed on the development of key standards related to cast steel materials, purification of molten metal, high-precision defect prediction technology, numerical simulation of casting process, and digital management, in order to improve the quality and competitiveness of cast steel products.

References

[1] ZHU Jiahui, ZHANG Yin, LU Jun, etc. Current Situation and Existing Problems of National and Industrial Foundry Standards in China [J].Foundry, 2020, 69(12): 1370-1377.

Graphite Growth and Shrinkage Formation During Solidification of CGI

JinhaiLiu[1,2], BinguoFu[1], HaitaoXue[1], XueboZhao[1], Guixian Zhang[2], Hongchen Liu[2]
1. School of Materials Science and Engineering, Hebei University of Technology, Tianjin 300401, China
2. Key Laboratory of Research and Application of Glass , Rubber and Plastic Mold Materials in Hebei Province，Huanghua 061100, Hebei, China
*Corresponding address: Xiping Road 5340, Beichen District Tianjin City, 300401, China. E-mail: jhliu57@126.com

Abstract: The morphology and evolution ofthe graphite have an important impact on the mechanical properties and shrinkage tendency ofthecompacted graphite cast iron (CGI). Compared to gray cast iron and ductile iron, there is relatively little understanding of the graphite precipitation behavior during the solidification process of CGI. Therefore, the precipitation and evolution behavior of graphite during the solidification process of hypoeutectic, eutectic, and hypereutectic CGI was systematically investigated based on thermal analysis, liquid quenching experiments, and metallographic analysis. Results demonstrate that the thermal analysis curve characteristics of three types of CGI are significantly different from the ideal curves. The closer the composition of hypoeutectic CGI is to the eutectic point, the higher the vermicularity and the smaller the shrinkage tendency. The primary graphite precipitated during solidification of hypereutectic CGI is spherical or exploded shaped, and the further away from the eutectic point, the larger the volume fraction of graphite balls, the lower the vermicularity, and there is theaustenite dendrite precipitation before eutectic reaction, which increases the probability of shrinkage porosity in hypereutectic CGI. The solidification of eutectic CGI is prone to the formation of a large amount of vermicular graphite, with a higher vermicularity and the smallest shrinkage tendency. The tendency of eutectic CGI to produce leakage is relatively small.

Keywords:compacted graphite cast iron;carbon equivalent; graphite; solidification; thermal analysis; liquid quenching; shrinkage tendency

Introduction

During the production of compacted graphite cast iron (CGI), the stability of compacted graphite rate and the shrinkage porosity of castings are two key technical problems to be solved. As for present control techniques of the CGI castings, the stability of CG rate is easy relatively. But the complete elimination of the shrinkage porosity is difficult. The porosity of castings is related to its solidification to a great extent. Therefore, the investigation of the solidification process of CGI is an important to the reduction of shrinkage defects in CGI castings.

Experimental procedure

Based on the thermal analysis curve of CGI and the quenching technology, the compacted graphite growth and shrinkage rules of CGI was studied at different stages of solidification, as shown in the Fig 1. The curve of thermal analysis charaterized the centre temperature of samples during solidifying. Therefore, the CG growth and the shrinkage of sample centre part are observed mainly.

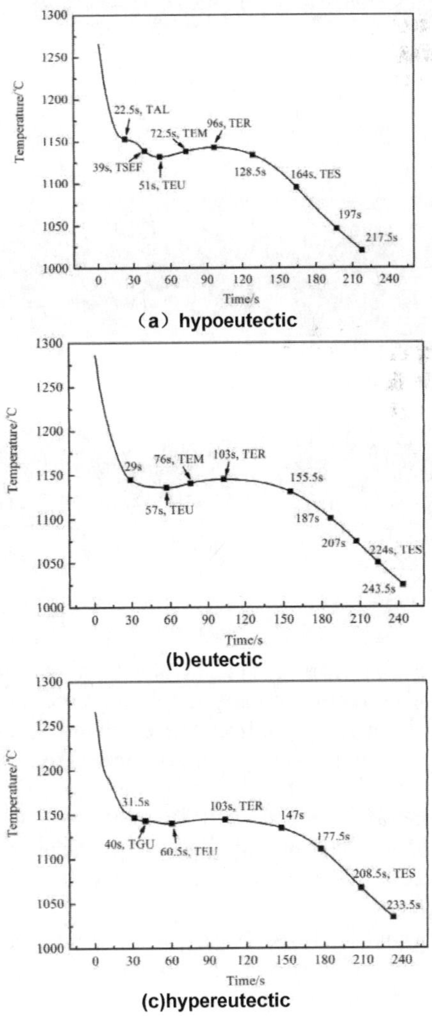

Fig. 1 thermal analysis curves of CGI and quenching point

Results and Discussion

Table 1 shows the graphite growth rules of CGI at different stage of solidification and the shrinkage at end of solidification. As for hypoeutectic CGI, the primary austenite grows before eutectic and in the range of solidifying the shrinkage is great. At the point of TEU, a few of eutectic cells formed. Along with solidifying, the latent heat of crystallization gives out and the centre temperature of sample increased. When the temperature reaches the highest TER point, the size of cell grows up and the cell number increased. The large part of eutectic micro-structure crystallized between the TEU and the TES. When the sample solidified completely, there are a few of porous hole in the centre of sample and the side surface sink.

To the hypereutectic CGI, the thick primary graphite crystallized in the shape of nordular before TEU and the proportion of primary graphite is bigger, and at the same time the austenite could form around the primary graphite. At the TEU, the primary graphite become large size and some of eutectic cell happened. As the same to hypoeutectic CG iron, the temperature increased after TEU. The large proportion of eutectic melt crystallized in range of TER and TES. But in this stage the proportion of eutectic graphite becomes less. Therefore, the central porosity and the upper surface sink of sample was produced at the end of solidification.

Table 1. Results of Testing

	Before TEU	TEU	TER	After TER	TES	shrinkage
hypoeutectic	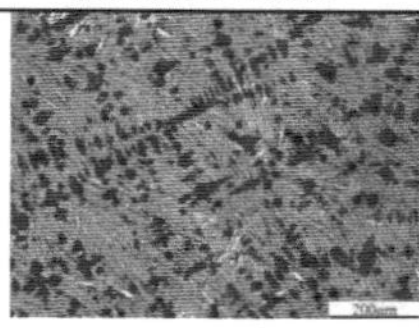 TAL-22.5s	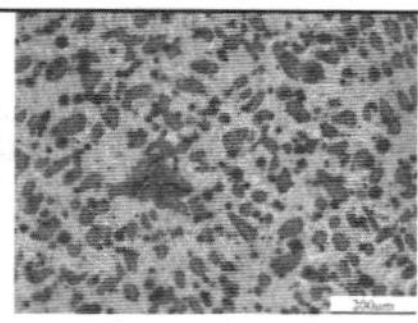 51s	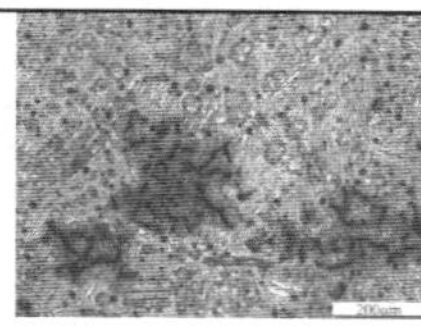 96s	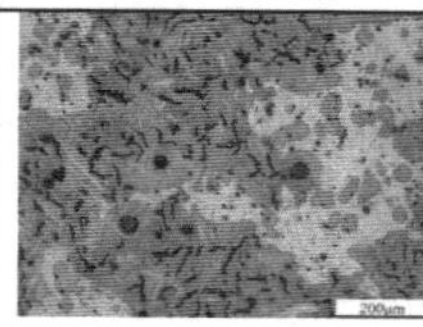 128s	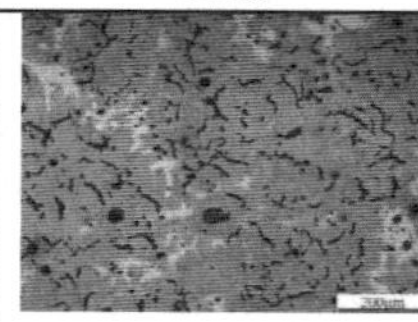 164s	
eutectic	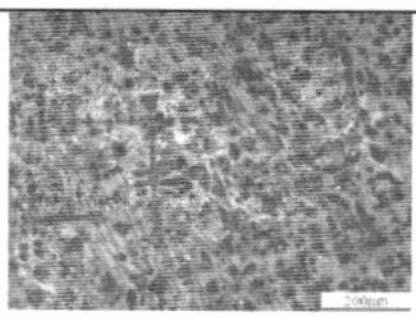 29s	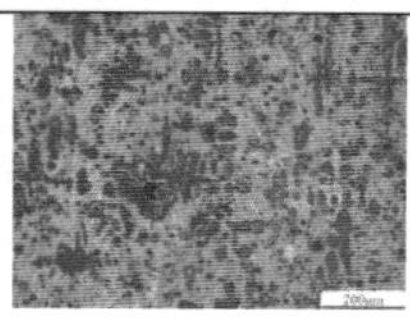 57s	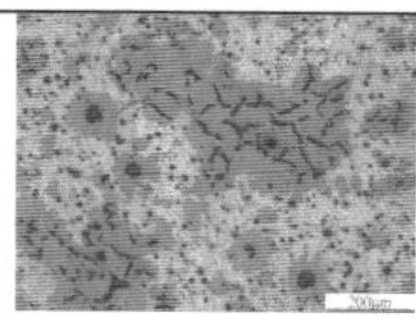 103s	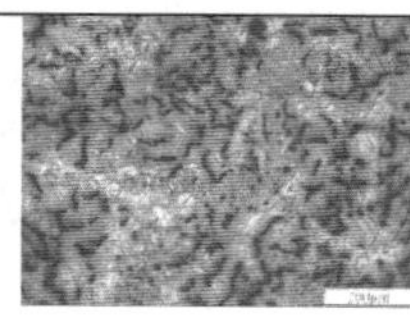 155.5s	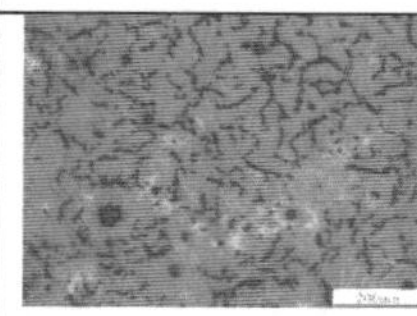 224s	
hypereutectic	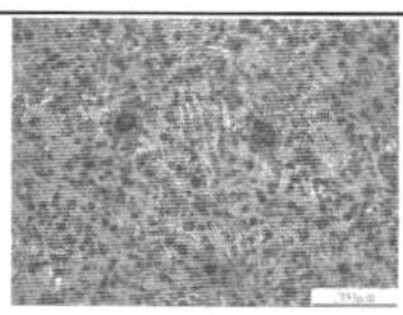 TGU-40s	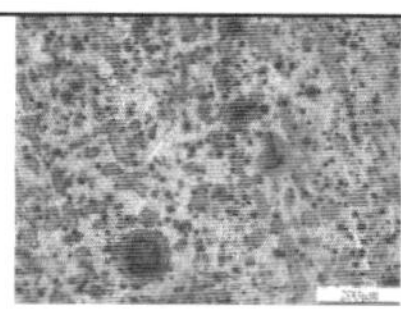 60.5s	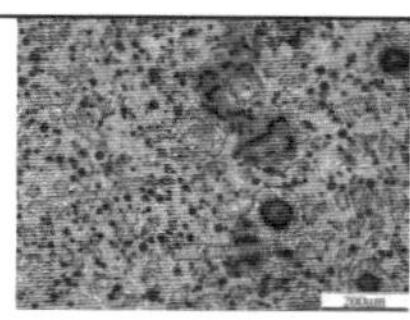 103s	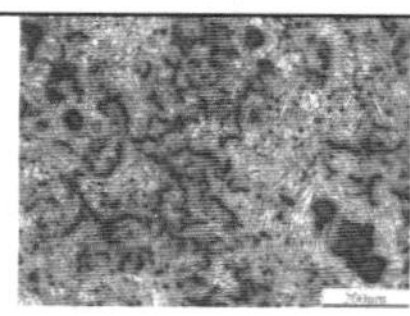 147s	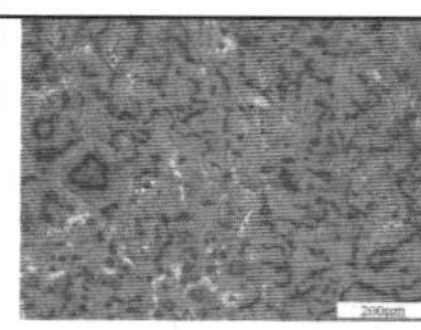 208.5s	

To the eutectic CG iron, It can be observed that there is no primary austenite and primary graphite before TEU point. Basically, the melt crystallized the eutectic microstructure of compact graphite and austenite completely. The large proportion of compacted graphite formed and the more graphitizing expansion can compensate the solidifying shrinkage in the late period of eutectic process. So when the solidification is finished there is no porosity in sample.

conclusion

(1) The thermal analysis curve of CGI could characterize its solidification process. The characteristic point can predict the situation of compacted graphite, austenite and eutectic microstructures.

(2) The large proportion of eutectic graphite in sample centre crystallized after the highest eutectic temperature TER.

(3) The eutectic CGI has the most eutectic graphite expansion and the lowest tendence of shrinkage in solidification. This leads to the biggest ability of self-compensation and the favor of elimination of porosity in CGI castings.

Refferences

[1] Zhang Zhong-chou, Li Ke-rui, Zeng Yi-cheng, Status and Prospect of Compacted / Vermicular Graphite Cast Iron in China. Foundy, 20126, 1(11): 1303-1307.

[2]DIC Heisser, -Ing Jrg, C Sturm. Casting Processing Simulation of Compacted Graphite Iron. American Foundry Society, 2003.

[3] Battezzati L, Baricco M, Maronglu F. Melting and Solidification Studies By Advanced Thermal Aanalysis of Cast Iron. Fonderie sous Pression International, 2002(3): 23-24.

Regression Analysis of Graphite Characteristics and Properties of Gray Cast Iron

Qi X Y[1], Fan J H[1*], Zhou J[2], Diao X R[3], Wang W[4]
1. Mechanical Engineering College,Donghua University, Shanghai, China
2. Jiangsu Liyuan Jinhe Casting Co., Ltd, Nantong, China
3. Tianchang Cylinder Head Co., Ltd, Chuzhou, China
4. Zhejiang Asia Pacific Electromechanical Co., Ltd, Xuancheng, China
* Fan J H:Male, Associate Professor.His research direction is casting quality control.E-mail: fanjinhui@dhu.edu.cn

Abstract: The microstructure and properties of gray cast iron are closely related. Properties of gray cast iron can be improved by changing the graphite characteristics when the pearlite content is constant. It is of great significance to analyze the relationship between graphite characteristics and properties of gray cast iron for the development of gray cast iron with high tensile strength and high thermal conductivity. Regression models of graphite characteristics and properties of gray cast iron was established, including tensile strength and thermal conductivity. By calculating the coefficients of the two models, the different effects of graphite characteristics on the properties of gray cast iron were obtained. The analysis results are applied to the actual production, and the performance of the castings is improved.

Keywords: gray cast iron, regression analysis, tensile strength, thermal conductivity, graphite

1 Introduction

Gray cast iron is the most widely used material in casting production. For example, a large number of parts such as brake discs and cylinder heads on vehicles are produced by casting. These parts are in a high temperature environment in addition to being subjected to large loads during operation, and have high requirements for the mechanical properties and thermal conductivity of the parts. A large number of studies have shown that it is very difficult to improve the mechanical properties and thermal conductivity of castings at the same time. To improve the performance of gray cast iron, it is necessary to change the structure of gray cast iron. High-strength gray cast iron is usually a full pearlite matrix, so it is particularly important to study the effect of graphite on the performance of gray cast iron.

2 Method

The grade of pearlite and the type of graphite in the sample can be evaluated according to GB/T 7216-2023. The pearlite content of the sample is 98%, and the graphite type is A-type. The metallographic image is imported into Image J for measurement. In order to ensure that the measured graphite can accurately reflect the graphite morphology in the sample, three longest graphite sheets in the image are selected for measurement.

The length l, width w of the graphite sheet and the length a of the line between the two ends, the length b from the farthest point to the vertical line segment of the connection bare measured. The ratio of b to a is the angularity K of the graphite sheet. Graphite content is also an important aspect of graphite characteristics. The proportion of graphite area in the total area of the picture is used to represent the graphite content G in the sample.

3 Result and discussion

Results of regression analysis

The measured data were input into SPSS for multiple linear regression analysis.

The regression equations of tensile strength and thermal conductivity of gray cast iron with graphite characteristics can be obtained, such as equation (1) and (2):

$$R_m = 501.033 - 0.111l - 7.648w + 0.124K - 13.819G \quad (1)$$

$$\lambda = 55.858 + 0.024l - 5.927w - 0.468K + 0.553G \quad (2)$$

For tensile strength, only the angularity is positively correlated with the tensile strength, and the rest is negatively correlated. This is because the strength of graphite sheet is much lower than that of gray cast iron matrix, and the bonding force between graphite and gray cast iron matrix is small, so graphite can be regarded as a small crack that splits the matrix [1-3]. During the tensile process of the casting, the graphite tip is prone to stress concentration and micro cracks initiation. The continuous propagation of cracks will lead to the fracture of the casting. The longer and wider the graphite sheet, the easier the crack initiation and propagation at the graphite tip, and the easier it is to connect different cracks [4]. The crack consumes energy to grow along the graphite sheet. When the graphite is bent, the crack growth direction changes, and more energy needs to be consumed, resulting in higher difficulty in crack growth and propagation. So the tensile strength is improved.

For thermal conductivity, only the length of graphite sheet and graphite content are positively correlated with the thermal conductivity, and the rest is negatively correlated. It's due to the different thermal conductivity of graphite and basic microstructure.The thermal conductivity in the direction of graphite base surface is much larger than that of other microstructures. However, the thermal conductivity of graphite in the direction perpendicular to the base surface is much smaller than other microstructures of cast iron. Therefore, increasing the graphite content of gray cast iron,increasing the graphite length and reducing the graphite width can allow as much heat to be transmitted along the graphite base surface as possible, thereby increasing the speed of heat conduction in cast iron. The fastest way to transfer heat on cast iron is to transfer heat from one point to another through graphite.When the graphite is bent, the heat needs to be transferred along the curve, and the thermal conductivity of the casting decreases [5,6]. It can also improve the thermal conductivity of castings.

The guidance of results to production

Taking a product as an example, it is prepared to improve the tensile strength of the product. According to the results of the regression analysis, the tensile strength of the casting can be improved by reducing the graphite content, refining the graphite, and increasing the bending degree. The addition of Mo affects the precipitation and growth process of graphite, thereby changing the morphology of graphite. The graphite characteristics are measured, and the results are shown in Fig. 2. The results show that molybdenum plays a role in refining graphite and shortening the length of graphite.

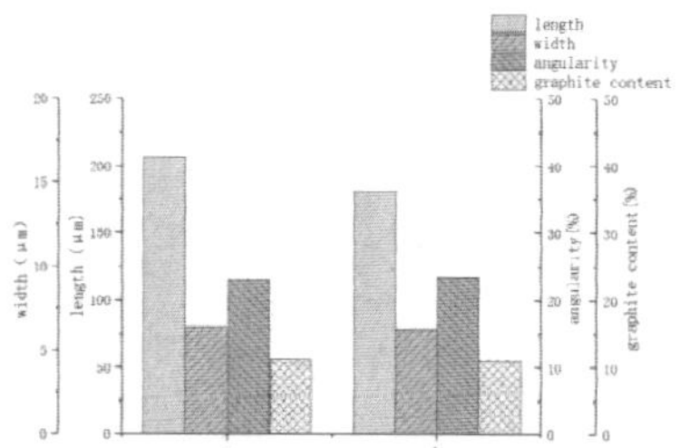

Fig.1 Comparison of graphite characteristics before and after Mo content adjustment

According to the conclusions obtained from the previous regression analysis, it can be inferred that the tensile strength of the castings with increased Mo content is higher. The actual measurement results are 351 MPa and 380 MPa, which are in line with the analysis results.

Similarly, according to the analysis results, the improvement of thermal conductivity can be achieved by increasing the graphite content, reducing the graphite width and increasing the graphite length. A product adjusts the thermal conductivity of the casting by changing the type of inoculant used. Fig.2is the comparison of graphite characteristics of castings using two inoculants.

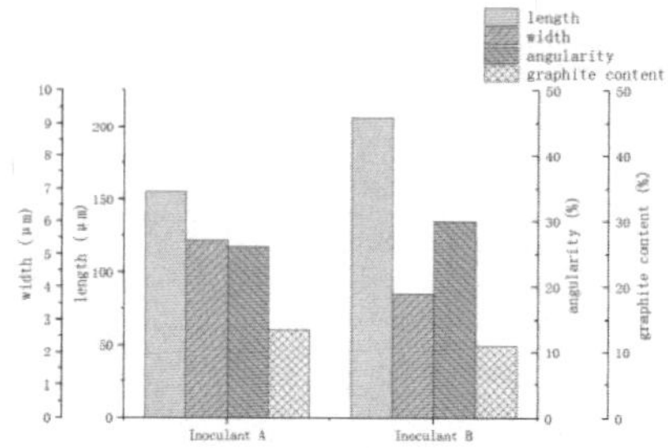

Fig.2 Comparison of graphite morphology of castings using two inoculants

From the measurement results, the graphite structure obtained by inoculation with different inoculants has different morphologies. This is because the (Mn, X)S compound formed by the inoculant element after the addition of the inoculant is the main nucleation position of the graphite sheet, and the number of compounds formed by different elements is also different, and the number of graphite sheets is also different. Under the premise of equal carbon content, the more the number of graphite flakes precipitated, the shorter the graphite flakes. Combined with the analysis results, the thermal conductivity of the castings using the first inoculant is slightly worse. The actual measured values are 35W / (m · K) and 39W / (m · K), respectively, which are in line with the predicted results.

References

[1] Nga P T H, Thien T N, Pritadewi P J, et al. Research on Factors Influencing the Formation Graphite and Effect of Graphite on Mechanical Properties of Grey Cast Iron. International Conference on System Science and Engineering. IEEE, 2019: 619-629.

[2] Wei Wang, Jing Tianfu, Gao Yuwei,et al. Properties of a gray cast iron with oriented graphite flakes. Journal of Materials Processing Technology, 2008, 182: 593-597.

[3] Agus Suprihanto, Darmawan Harsokoesoemo, Rochim Suratman. The Effect of Cr & Cu on the Fatigue Strength of Grey Cast Irons. Key Engineering Materials, 2006, 306: 953-958.

[4] Lu Haiqiu. Effects of graphite types and carbon content on thermal fatigue and wear properties of bionic laser-processed gray cast iron. Changchun: Jilin University, 2019.

[5] Wang W, Jing T, Gao Y, et al. Properties of a gray cast iron with oriented graphite flakes. Journal of Materials Processing Technology, 2007, 182(1-3): 593-597.

[6] Holmgren D. Review of thermal conductivity of cast iron. International Journal of Cast Metals Research, 2005, 18(6): 331-345.

Improving the Anti-wear Properties of Cast Steel for Working Parts of Mining Industry Equipment

Justyna Kasińska
Department of Metal Science and Manufacturing Processes, Kielce University of Technology, Al. Tysiąclecia Państwa Polskiego 7, 25 314 Kielce, Poland; kasinska@tu.kielce.pl

Abstract: Tribology has become a crucial field in both science and industry in recent decades. Abrasive wear is also increasingly being considered in terms of the functionality of materials used in many industries. Tribological wear of machinery and equipment components for mining depends on surface roughness and hardness. At low roughness, adhesion has the greatest influence on wear performance, and as the roughness of the metal component increases, the proportion of micro-cutting increases and abrasive wear dominates. To improve the service life of castings for high-wear operation, padded layers are applied using welding techniques. The selection of appropriate parameters for the process itself and the materials used is crucial. Padded layers are currently the subject of innovation, for example through modification processes or the creation of composite layers.

Keywords: mining industry; cast; wear resistant; pad welds

1 Introduction

Steel castings are widely used in the mining industry. Elements of crushers and mills work under high operational loads therefore, they wear out quickly and require frequent replacement or regeneration. Repairs and modifications are thus made to improve their wear resistance [1, 2]. One of the methods to extend the working cycles of castings is the application of wear-resistant coatings using different fabrication techniques [3, 4].

Rare earth metals (and their compounds) have been used as an additive to improve tribological properties, corrosion resistance, mechanical properties and resistance to oxidation [5 – 11].

In this article, the authors present the results of research on abrasive wear of L20HGSNM and L120G13 (Table 1) steel castings and pad welded layers of FeCrVW with the addition of rare earth metal oxides. A decrease in average wear [μm] and an increase in hardness were demonstrated.

2 Experimental procedure

The coatings were deposited using metal active gas welding (MAG) combined with plasma arc welding (PAW). Each weld consisted of a base layer (buffer) and a wear-resistant working layer. The working layer was made with binders doped with CeO_2, La_2O_3, Y_2O_3. Tribological tests were performed for the prepared pad weld variants under dry friction. Tribological characteristics were determined.

Table 1 Chemical Composition of casts

	C	Si	Mn	Cr	Mo	Ni
L20HGSNM	0.18-0.25	0.7-1	0.8-1.1	0.6-0.9	0.1-0.2	0.9-1.2
L120G13	1.0 – 1.4	0.3-1.0	12-14	≤ 1,00	-	≤ 1,00

3 Result and discussion

After modification with the rare earth metal oxide, an increase in hardness and a decrease in the friction coefficient were observed for all the pad welds (samples 2, 3, 4). Linear wear decreased significantly. Similar results were obtained for dry friction (Table 2). Observations of the obtained wear track profiles (Figure 1) confirmed the differences in the wear mechanism after the modification.

Table 2 Tribological properties and hardness of the pad welds

Sample	RE oxygen	Tribological Tests		Hardness *
		Dry Friction		Vickers hardness [HV]
		Coefficient of Friction	Linear Wear [μm]	
L20HGSNM quenching and tempering	-	0.18	52.73	510.37
L120G13 supersaturated state	-	0.53	63.35	603.17
1	-	0.46	38.06	683.86
2	CeO_2	0.21	64.34	826.12
3	La_2O_3	0.39	26.68	744.19
4	Y_2O_3	0.54	28.57	655.87

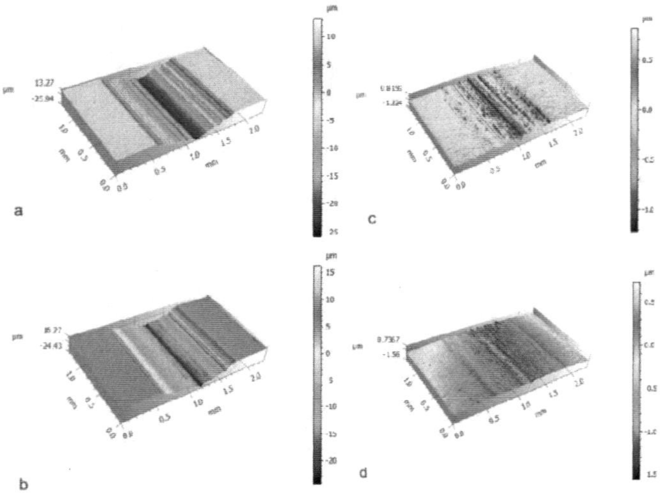

Fig.1 The figure shows isometric images of wear tracks on the cross-section a) L20HGSNM, b) L120G13, c) sample 1, d) sample 2

4 Conclusion

The addition of rare earth metal oxides into the pad welds significantly improved hardness and the tribological properties of the welds. The most favourable result under dry friction was obtained after adding CeO_2. The rare earth oxides lowered the wear rates, and in the case of CeO_2, the wear track area reduction was multifold to comparison with casts steels without welded pads.

Acknowledgments

Acknowledgments to PW, TSA" M. Górski S. Rutkowski Sp.j (Stalowa Wola, Poland) for the opportunity to cooperate within the POIR.01.01.01-00-0098/18 project.

References

[1] Yakushev AA, Dukhanin SA, BSTU. Improving the wear resistance of the working bodies of a rotary-vortex mill. Energy-resource-saving technologies and equipment in the road and construction industries: proceedings of the international scientific and practical conference, Belgorod, 2020, 444–447.

[2] Franc Zupanič. Wear resistance of coke-crushing hammers. Canadian Metallurgical Quarterly, 2013, 52:4, 458-464.

[3] Teixeira M F, Pacheco J T, da Silva L J, *et al.,* Wear resistance of a Metco 1030A hard coating deposited on Hadfield steel by laser cladding for ore comminution application, Int J Adv Manuf Technol, 2021, 112, 1873–1884.

[4] Ganapathi M, Eliseeva S V, Brooks N R, et al. Electrodeposition of luminescent composite metal coatings containing rare-earth phosphor particles. J. Mater. Chem., 2012, 22(12), 5514.

[5] Zhang L M, Sun D B, Yu H Y. Characteristics of plasma cladding Fe-based alloy coatings with rare earth metal elements. , Mater. Sci. Eng. A, 2007, 619, 452-453.

[6] Yi W, Zheng C Q, Fan P, et al, Effect of rare earth on oxidation resistance of iron base flux-ing alloy spray-welding coating. J. Alloys Compd., 2000, 311(1): 65.

[7] Kasińska. Wide–ranging influence of mischmetal on properties of G17CrMo5-5 cast steel. J. Metalurgija, 2015, 54 (1), 135-138.

[8] Satet R L, Hoffmann M J. Experimental evidence of the impact of rare-earth elements on particle growth and mechanical behaviour of silicon nitride. J. Am. Ceram. Soc., 2000, 88 (9), 2485–2490.

[9] Wu C F, Ma M X, Liu W J, et al. Laser cladding in-situ carbide particle reinforced Fe-based composite coatings with rare earth oxide addition. J. Rare Earths, 2009, 27(6): 997.

[10] Xing X G, Han Z J, Wang H F, et al. Electrochemical corrosion resistance of CeO_2-Cr/Ti coatings on 304 stainless steel via pack cementation. J. Rare Earths, 2015, 33(10): 1122.

Oxidation Behaviour of Inoculated Si/SiMo Spheroidal Graphite (Ductile) Cast Iron

Iuliana Stan, Mihai Chisamera, Robert Lascu, Codrut Cariga, Eduard Stefan, Stelian Stan*, Denisa Anca, Iulian Riposan
National University of Science and Technology POLITEHNICA Bucharest, Materials Science and Engineering Faculty, 313 Splaiul Independentei, Sector 6, 060042 Bucharest, ROMANIA
*Corresponding address: constantin.stan@upb.ro

Abstract: The present work aims to investigate the effect of heating temperature (400, 600 and 800℃, 8 h) and inoculating elements (Ca/Ca,Ba/Ca,RE) on 5.25%Si and 4.8%Si/2.3%Mo ductile irons' oxidation behaviour in dry air and combustion atmospheres containing water vapour. The oxidation is influenced by the considered influencing factors, with different contributions depending on the temperature (higher temperature, higher specific effects of these factors). Depending on its level, the water vapour in the combustion atmosphere is an important factor at 600 ℃ and especially at 800 ℃.

Keywords: Ductile iron; Inoculation; Si/SiMo; Graphite; Oxidation; Gas atmospheres; Water vapour; Structure.

1 Introduction

Si-alloying of ferrite in ductile irons leads to higher hardness and tensile properties, but with a decreasing level of ductility and toughness, especially at more than 4.3%Si. Higher content of Si and Si-Mo association promotes resistance to oxidation and corrosion at higher temperature applications, such as exhaust manifolds, furnace parts, turbocharger housing, etc., by the facilitation of a protective oxides layer. Typical for high Si-content (especially > 4%Si) is the presence of the nodules at lower compactness degree (Form V, ISO 945, 0.6-0.8 Roundness Shape Factor), detrimental for the cast irons properties, possibly improved by inoculation [1].

2 Experimental procedure

Si and SiMo alloyed cast irons were melted using a 10 kg induction furnace (8000 Hz). FeSiCaMgRE treated iron melts were subjected to an in-the-mould inoculation. A pattern for four work positions includes a central downsprue that supplies Mg-treated iron simultaneously to four separate reaction chambers, including Ca/Ca,Ba/Ca, RE-FeSi alloys, in a furan resin silica sand mould. Static oxidation tests were performed in dry air (atmosphere A) and in combustion gasses (natural gas burning), B (0.55 excess air coefficient) and C (1.05 excess air coefficient) at 400, 600 and 800 ℃ (8h); electric furnace, with SiC heating bars; 10 mm×10 mm×10 mm specimens, placed into an alumina crucible. The oxidation is expressed by the specimens' weight gain (wt.%) as a result of the oxidation process.

3 Result and discussion

Two cast irons (0.045%Mg residual) are tested, specifically for high Si content (I) (5.05%CE) and high Mo content associated with Si (II) (4.75%CE), with minor elements typically for commercial ductile irons (Table 1). Structure (Fig. 1) is characterised by spheroidal graphite mainly as Form V - ISO 945, with Roundness Shape Factor *RSF*=0.62-0.71, matrix predominantly ferritic (I) or pearlitic (II), with free carbides restricted by inoculation.

Table 1 Chemistry and Structure characteristics*

Iron	C %	Si %	Mo %	NG %	F %	K %	RSF
I	3.46	5.25	0.04	70-75	85-90	-	0.68-0.71
II	3.30	4.80	2.31	62-67	< 15	< 5	0.62-0.66

*F-NG-nodularity (ISO 945-4-2019) F-ferrite; K-carbides; RSF-Roundness Shape Factor (RSF = 4AG/πFmax)

The resulted CO from the graphite particles oxidation leaves the oxidised layer, emphasising its discontinuity, and, consequently, favouring the oxygen penetration inside the iron casting to produce metal oxides. These oxides lead to the weight gain of the iron casting, while the oxidised carbon (graphite) (leaving the casting body) contributes to the weight loss. Generally, the weight change is a result of these two opposite processes [2].

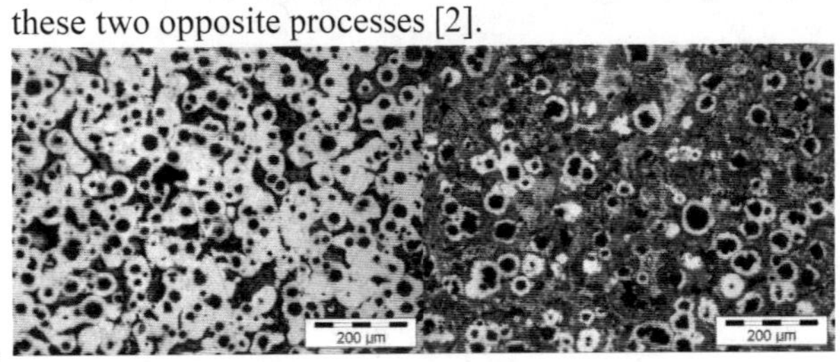

Fig. 1 Structure of I-Si (a) and II-SiMo (b) irons (Nital)

Figure 2 summarised the data obtained, as a result of the two mentioned processes, and referring to the four influencing factors considered, with different contributions depending on the heating temperature. The weight gain (wt.%) increases from 0.001-0.1% (400 ℃) to 0.05-0.70% (600 ℃) and up to 0.10-2.15% (800 ℃).

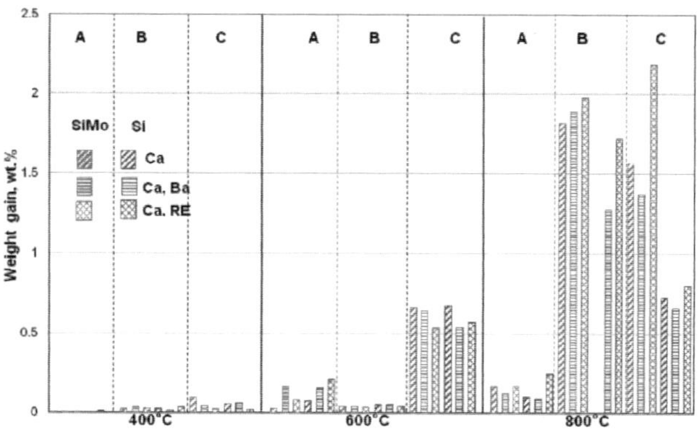

Fig. 2. Weight gain of the Si/SiMo inoculated (Ca/Ca,Ba/Ca,RE) DI, in dry air (A) and combustion atmospheres (B, C), at 400, 600 and 800°C.

The water vapour is present only in atmospheres B (lower level) and C (higher level), with effects depending on the heating temperature. No specific effects are found at 400°C heating. At 600°C heating, up to 0.05% weight gain in an atmosphere obtained by natural gas burning with lower excess air coefficient (0.55, atmosphere B), up to 0.21% in dry air (atmosphere A) and 0.5-0.7%% in combustion gas resulted from a burning with high excess air coefficient (1.05, atmosphere C). At the highest heating temperature (800°C), a limited increase of the weight gain is found in dry air (up to 0.25%), but it drastically increases in combustion atmospheres (0.65 - 2.15%), without a clear differentiation of the two atmospheres (B and C), as was found at 600 ℃ heating. The water vapour presence in a combustion atmosphere decreased the protection efficiency of the silica film formed on the metal/scale boundary, and the resistance to oxidation, too [3]. In atmosphere B, at up to 600 ℃ heating, a lower oxidation rate is observed as compared to dry air (atmosphere A, even at 800 ℃). No difference in the oxidation behaviour is observed at 400 ℃, depending on the water vapour content between the atmospheres B and C, but it is strongly visible at 600 ℃, with water vapour presence at higher level in atmosphere C, as an important oxidising factor. The highest oxidation rate characterizes irons at 800 ℃, in the atmospheres with water vapour presence. The present experiments confirmed that the water vapour is an important favouring factor of HighSi/SiMo oxidation, but depending on its level and temperature: mainly at high water vapour level at 600 ℃, and unrelated to its level at 800 ℃. The cast iron alloying system appears to influence the oxidation behaviour mainly at 800 ℃ heating in the combustion atmosphere (5.25%Si cast irons at lower weight gain). The increasing silicon content in the cast iron chemical composition confers a higher oxidation resistance, especially in the combustion gas atmospheres. The difference between the inoculating elements' effects increases with the heating temperature, generally with Ca, Ba-FeSi inoculation in better position and with CaRE-FeSi inoculated irons at higher oxidation degree. As previously reported data, the Ca-Ba appear to be better than simple Ca for improving graphite parameters, while the RE-bearing alloy negatively affects the compactness degree of graphite particles.

Acknowledgments
This work was supported by a grant from the National Program for research of the National Association of Technical Universities - GNAC ARUT 2023.

References

[1] Riposan I, Stan S, Anca D, et al. Structure Characteristics of High-Si Ductile Cast Irons. Int. J. Metalcasting, 2023, 17(4): 2389-2412.

[2] Lekakh S N, Bofan A, Osei R, et al. High Temperature Oxidation and Decarburization of SiMo Cast Iron in Air and Combustion Atmospheres. Oxidation of Metals, 2021, 95(3-4): 251-268.

[3] Ebel A, Brou SY, Mallard B, et al. High Temperature Oxidation of a High Si SiMo Spheroidal Cast Iron in Air with In Situ Change in H_2O Content. Materials Science Forum 2018, 925:353.

Effect of Inoculation, Section Size and Cooling Rate on Microstructure and Inclusion Characteristics in Spheroidal Graphite Iron

Cathrine Hartung[1*], Leander Michels[1], Eivind G. Hoel[1], Bente Kroka[2]
1. Foundry Innovation – Elkem Silicon Products (ESP) – Fiskaaveien 100, NO-4621 Kristiansand Norway
2. Elkem Technology (ET) – Fiskaaveien 100, NO-4621 Kristiansand Norway
*Corresponding address: cathrine.hartung@elkem.com

Abstract: The cooling rate substantially affects the graphite nucleation in spheroidal graphite irons. Faster cooling, in general, requires more potent inoculants to prevent eutectic cementite. However, the effect of cooling rate on the population of micro-inclusions and graphite nucleation frequency is still not completely understood. In this context, two different casting sections with thicknesses of 75 mm (slow cooling) and 35 mm (fast cooling) were studied regarding final microstructure and micro inclusions populations with and without inoculation. Higher nodule and micro-inclusion density are seen with both inoculation and slower cooling. In contrast, the difference in type and size of micro-inclusions can be observed between inoculated and uninoculated conditions.

Keywords: SGI, Inoculation, Modulus, Holding time

1 Introduction

Inoculation is regarded as a critical process step when producing SGI to ensure the nucleation of graphite and reduce the risk for carbides [1]. To have the best effect of inoculation, it is best to inoculate as close to pouring [2] as possible to reduce the risk of fading [3]. All inoculants fade, and the effect of inoculation is sometimes unclear for slow cooling conditions.

2 Experimental procedure

A furnace with 280 kg base iron was tapped at 1500 ℃ into a tundish ladle where 1.30 wt% MgFeSi (6.0% Mg,1.8% Ca, 0.5% Al) was added to the bottom of the pocket and covered by 0.73 wt% steel. After treatment and removal of slag the treated iron was split into two 32 kg capacity alumina crucibles with inoculation added to the bottom of one and pure Si-metal to the other to have the same final Si for the inoculated and uninoculated sample. The composition of the treated iron but uninoculated (UN) was (3.60% C, 2.18% Si, 0.20% Mn, 0.033% P, 0.010% S, 0.048% Mg).

A Ce-inoculant (73.1% Si, 0.97% Ca, 1.67% Ce, 0.88% Al) was used, and the addition rate was 0.20 wt%. The Ce-inoculated (Ce) composition was (3.63% C, 2.24% Si, 0.19% Mn, 0.034% P, 0.010% S, 0.048% Mg). Both inoculated (Ce) and uninoculated (UN) crucible was held for 60 s before casting of Y-block type IV equipped with a thermocouple, collection sample for chemical analysis with immersion dip sampler, and pouring of thermal analysis sample in a sand mould with 36×36×40 mm dimensions containing a type K thermocouple in the geometrical center.

3 Result and discussion

Microstructure results from thermal analysis cup (fast) and Y-block (slow) show a higher nodule density, nodularity, and amount of graphite and ferrite for Ce-fast and Ce-slow. The slow cooled samples show a larger average nodule size.

Table 1 Microstructure results.

	Nodule density /Nmm^{-2}	Nodularity /%	Graphite /%	Ferrite /%	Pearlite /%	Nodule size Average /μm
UN-fast	58±6	52±6.7	3.6±0.4	-	-	19.8
Ce-fast	174±8	78±4.2	9.6±0.5	43.9 ±1.3	46.5±1.3	24.3
UN-slow	15±2	69±6.2	8.9±1.0	43.2 ±2.4	47.9±2.4	64.9
Ce-slow	89±7	88±2.4	11.1±0.5	81.7 ±1.2	7.2±1.3	38.7

The etched microstructures in figure 1 show that the UN-fast cooled is mostly carbidic, while UN-slow is not. The Ce-inoculated samples do not present carbides, and the difference between Ce-fast and Ce-slow is the size and distribution of nodules. The cooling curves of these respective samples are shown in figure 2. The effect of the cooling rate is directly observed in the curves. The same melt composition can yield substantially different undercoolings as indicated in figure 2. All Ce-inoculated samples have a higher LET than the UN counterparts.

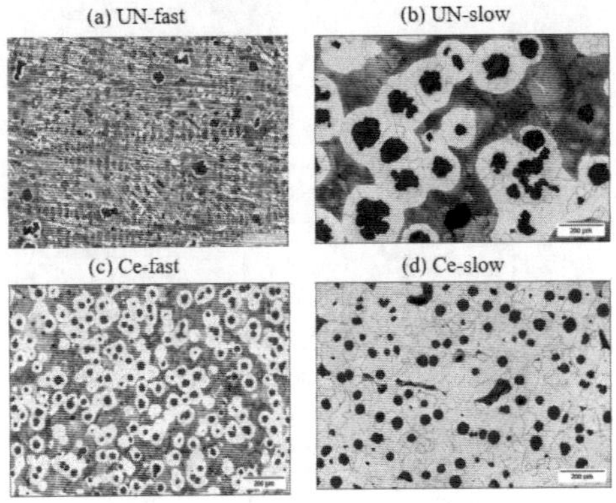

Fig. 1 Microstructure in etched condition for the two conditions and cooling rates compared.

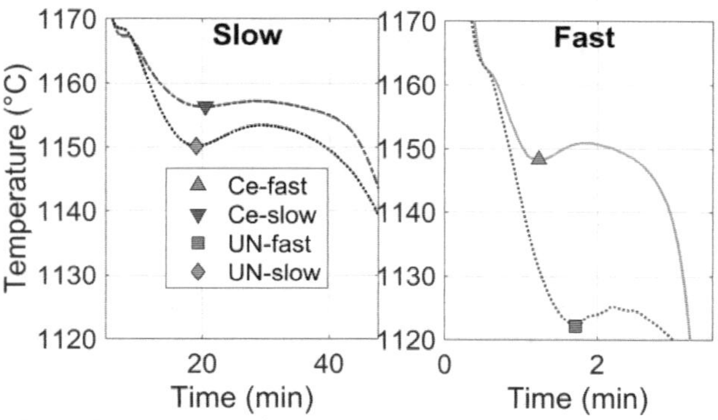

Fig. 2 Cooling curves for the UN and Ce. The symsbols are the low eutectic temperature (LET).

In figure 3 and figure 4, the details of a automated SEM/EDX analysis can be seen for both graphite and micro-inclusion phases. This method is described in ref. [3].

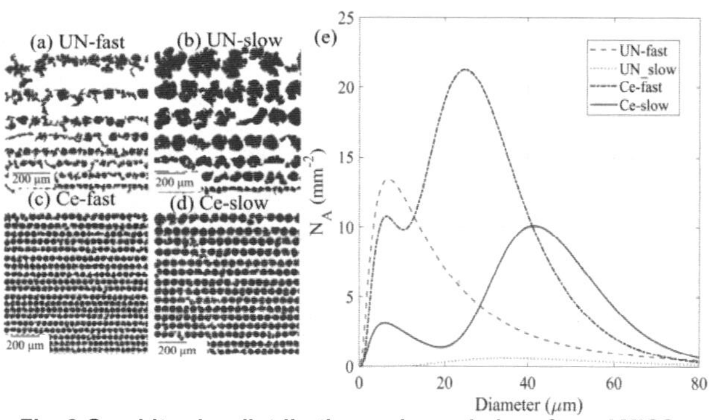

Fig. 3 Graphite size distribution and morphology from AMICS.

For the Ce-fast and Ce-slow, a dual peak size distribution of graphite particles can be seen, while only a single peak distribution is seen in the samples without inoculation. The nodularity and the number of the nodules is also higher with inoculation.

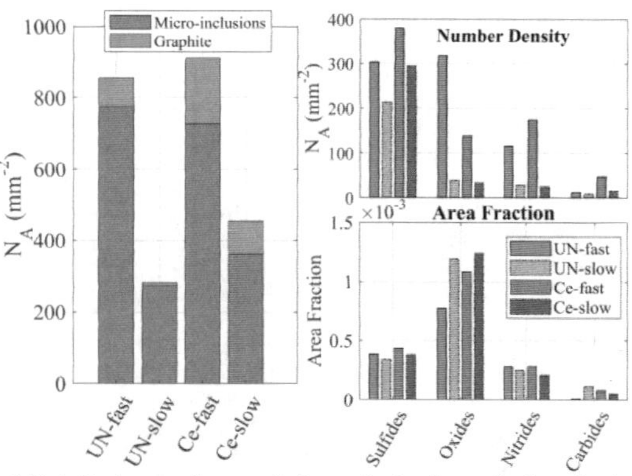

Fig. 4 Detailed overview and characterisation of micro inclusions from SEM/EDX.

Figure 4 shows the effect of the cooling rate in the micro-inclusions formation. All fast-cooled samples have a significantly higher number density of micro inclusions, i.e., higher sulfides, oxides, and nitrides. Further, there is also a substantially lower number density of graphite in both UN-slow and Ce-slow.

4 Discussions and Conclusion

The present study shows the effect of inoculation in two cooling conditions. The Ce-samples have a bimodal size distribution where each peak can be related to an undercooling for nucleation [4] i.e., Ce-slow has a lower undercooling for graphite nucleation. However, fast cooling shows that for UN-fast, the number of good nucleation sites is too low to avoid the formation of Fe_3C. In contrast, the UN-slow does not show Fe_3C, but the number of nucleation sites is also low in UN-slow, resulting in a very low nodule density with poor nodularity.

Fast cooling increases the number density of sulfides, oxides, and nitrides. However, Ce-slow has more sulfides than UN-slow, which shows that inoculation modifies these micro-inclusions.

Acknowledgments

We would like to thank Gro Eide, Emmanuelle Ott and Ilpo Sarikka

References

[1] Morgan H L. Inoculation of cast iron. BCIRA Journal 1984, (32): 339-350.
[2] Pearce J. Inoculation of cast irons: practices and developments. Foundry Trade Journal, 2008 : 28-32.
[3] Michels L, Pires AJF, Ribeiro CAS, et al. Effect of Holding Time on Populations of Microparticles in Spheroidal Graphite Irons. Metallurgical and Materials Transactions B, 2022: 1-12
[4] Michels L, Cygan B, Miroslawa P, et al. Graphite nucleation on (Al, Si, Mg)-nitrides: Elucidating the chemical interactions and turbostratic structures in spheroidal graphite cast irons. Metallurgical and Materials Transactions B, 2024(220): 118848.

Research on Strain Partitioning and Mechanical Properties of a Cast Multiphase Stainless Steel

Zixiang Wu[1], Xiang Chen[1,2,*]
1. School of Materials Science and Engineering, Tsinghua University, Beijing 100084, China;
2. Key Laboratory for Advanced Materials Processing Technology, Ministry of Education, Beijing 100084, China;
*Correspondence: Xiang Chen; E-mail: xchen@tsinghua.edu.cn

Abstract: Multiphase stainless steels can offer very attractive combinations of high strength, high toughness and good corrosion resistance due to the characteristics of multiphase, metastable, and multi-scale. In this study, the strain partitioning in ferrite of a cast multiphase stainless steel was investigated by EBSD. The deformation occurs first in the softer ferrite and later transfers to the martensite/austenite regions. As the strain increases, subgrains form within the ferrite, effectively accommodating the deformation of different phases.

Keywords: multiphase stainless steel; strain partitioning; geometrically necessary dislocations; subgrains

1 Introduction

Multiphase stainless steel containing ferrite, martensite, and a moderate amount of austenite can combine high strength and ductility with excellent formability, making it a widely choice for turbine blades, valve bodies, and ship propellers. The enhanced strength and plasticity are attributed mainly to the transformation-induced plasticity (TRIP) resulting from the metastable austenite and strain partitioning into different constituent phases[1-2].

During the deformation, the softer phase yields first, and after appropriate work hardening, sufficient strain is transferred to the harder phases to cause them to yield later[3]. For dual-phase steel, local plasticity generated in ferrite adjacent to martensite can lead to the generation of geometrically necessary dislocations (GNDs) to maintain lattice continuity and thus improving work hardening[4]. As the deformation continues, the dislocation subgrain structures are more likely to be generated within grains due to the increased GNDs and divide the grain into different subgrain regions[5-6]. Therefore, strain partitioning in ferrite is important to mechanical properties.

In this study, a novel cast stainless steel was prepared and processed by intercritical tempering (IT) and aging treatment to obtain a multi-phase microstructure. The role of ferrite in strain partitioning during deformation was carefully investigated.

2 Experimental procedure

The tested steel with a composition of Fe-0.03C-13.88Cr-6.88Ni-1.54Mo-1.08Mn-2.60Si-1.01Cu-1.07Al-0.30Nb (wt.%) was prepared in a 10 kg vacuum induction furnace. Samples were heated at 1050 ℃ for 30 minutes and quenched to 25 ℃ in water (labeled as S1050), followed by IT at 620 ℃ for 1h (IT620) and aging at 500 ℃ for 4 h (AG500). The relevant testing methods can be found in Ref[7].

3 Result and discussion

Mechanical Properties

Fig. 1a shows the SEM micrograph of the S1050 sample characterized by the presence of δ-ferrite and martensite, while there is barely no retained austenite (RA) (as shown in Fig. 1c). Besides, NbC particles are found to distribute at the phase boundaries. Fig. 1b shows the SEM micrograph of the AG500 sample which combines δ-ferrite, lath martensite and RA (11.8%, as shown in Fig. 1c). Fig. 1d shows the tensile properties of the investigated samples. The AG500 sample has the best combination of mechanical properties (UTS:1260 MPa, YS: 1072 MPa, TE: 17.7%).

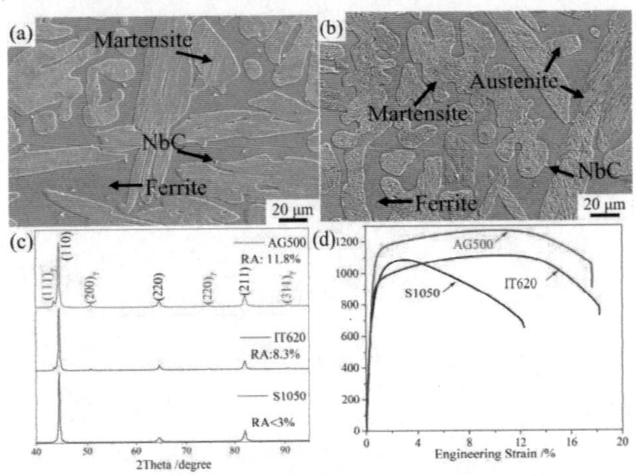

Fig. 1 (a-b) Sem micrograph of the investigated samples; (c) XRD patterns of the investigated samples; (e) tensile curves of the investigated samples.

Strain partitioning in ferrite

Fig. 2a shows the lateral side of fracture in the AG500 sample which exhibits the characteristics of ductile fracture and region B experiences more severe deformation to region A. Fig. 2b-c show the image quality (IQ) map and the kernel average misorientation (KAM) map taken from region A. The tensile direction is vertical and slip bands occur in the deformed ferrite region (as shown by red arrows in Fig. 2b). Fig. 2c shows that relatively higher KAM along the slip bands, which suggests that the slip bands are the regions with relatively higher dislocation density[8]. This is related to the formation of GNDs accommodating the strain gradient[7]. The density of GNDs on the phase boundary is smaller compared to the ferrite interior which indicates that the deformation occurs first within ferrite. When deformation encounters the hard martensite, it turns and deforms along the martensite plate (as shown by yellow arrows in Fig. 2c). This leads to the accumulation of GNDs on the phase boundaries.

Fig. 2d-f show the IQ map, the KAM map and the inverse pole figure (IPF) color map taken from region B. The tensile direction is also vertical and significant necking occurs in this region, corresponding to large plastic deformation. Fig. 2d indicates that the higher dislocations created within the ferrite region rearranged itself as subgrains due to more drastic deformation (as shown by green arrows). Fig. 2e shows that the substructures have a higher density of GNDs. In contrast to Fig. 2c, the martensite regions in Fig. 2e have a greater degree of deformation due to the very high misorientations especially on the phase boundary. This indicates that as the strain increases, the deformation within the ferrite is transmitted to the martensite through the accumulation of GNDs at the phase boundaries, which in turn leads to the TRIP effect of RA in martensite regions. Fig. 2f shows the IPF map in Y direction, subgrains form within ferrite and distribute in bands parallel to each other, thus refining the ferrite grains. Strain partitioning and subgrains formation in ferrite are beneficial for accommodating the strain of different phases and retarding strain to fracture[8].

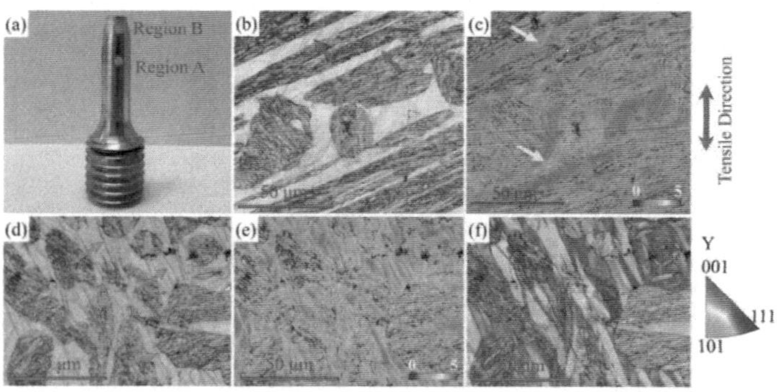

Fig. 2 (a) Lateral side of the fracture of the AG500 sample; (b) IQ map of the deformed AG500 sample in region A; (c) KAM map of the deformed AG500 sample in region A; (d) IQ map of the deformed AG500 sample in region B; (e) KAM map of the deformed AG500 sample in region B; (f) IPF map of the deformed AG500 sample in region B.

4 Conclusion

In this study, a novel cast stainless steel was prepared and processed by IT and aging treatment to obtain a multi-phase microstructure. The deformation first occurs in the softer ferrite and later transfers to the martensite/austenite regions. Strain partitioning and subgrains formation in ferrite can accommodate the deformation of different phases. The AG500 sample has the best combination of mechanical properties (UTS:1260 MPa, YS: 1072 MPa, TE: 17.7%).

Acknowledgments

This work was supported by the Inner Mongolia Autonomous Region Science and Technology Major Special Project (Grant No.: 2021SZD0082).

References

[1] Han G, Shang C J, Xie Z J, et al. On the thermal and mechanical stability of reverted austenite by intercritical tempering. Mater. Lett., 2021, 291: 129457.

[2] Yang Z N, Liu C B, Zhang C Y, et al. Microplasticity behavior of multiphase high-strength nanobainitic steel based on a modified law of mixtures. Mater. Sci. Eng. A, 2021, 825:141848.

[3] Song C, Yu H, Lu J, et al. Stress partitioning among ferrite, martensite and retained austenite of a TRIP-assisted multiphase steel: An in-situ high-energy X-ray diffraction study. Mater. Sci. Eng. A, 2018,726:1-9.

[4] Cong Z H, Jia N, Sun X, et al. Stress and Strain Partitioning of Ferrite and Martensite during Deformation. Metall. Mater. Trans. A, 2009,40(6) :1383-1387.

[5] Badkoobeh F, Mostaan H, Rafiei M, et al. Microstructural Characteristics and Strengthening Mechanisms of Ferritic–Martensitic Dual-Phase Steels: A Review. Metals, 2022, 12(1):101-130.

[6] Kratochvíl J, Sedláček R. Energetic approach to subgrain formation. Mater. Sci. Eng. A, 2004, 387-389: 67-71.

[7] Wu Z, Liu Y, Zhang H, et al. Microstructure and mechanical properties of a novel cast multi-phase stainless steel. J. Mater. Res. Technol. 2022 (19): 4177-4182.

[8] Wang N, Chen Y, Wu G, et al. Investigation on micromechanism involved in ferrite hardening after prestraining of dual-phase steel. Mater. Sci. Eng. A, 2021,800:140387.

Effect of Mn Content on Damping Capacity and Mechanical Properties of Cast FeMn Binary Alloys

Wei He, Xin Zeng, Haozhe Qiu, Yuhua Wen *
School of Mechanical Engineering, Sichuan University, Chengdu, 610065, China
*Corresponding address: e-mail: wenyh@scu.edu.cn

Abstract: There were few works on cast FeMn damping alloys although many studies on deformation processed FeMn counterparts have been conducted. Cast Fe(15-21)Mn alloys were prepared to investigate the effects of Mn content and solution treatment on the damping capacity and mechanical properties of cast FeMn alloys. Significant compositional segregation existed in the cast FeMn alloys, with α '-martensite readily forming in Mn-depleted regions. With the increase in Mn and after the ST, the content of the α '-martensite decreased. The optimal Mn contents for damping capacity in the cast FeMn binary alloys was around 19%, 2% higher than in the deformation-processed counterparts. The damping capacity strongly inversely depended on the amount of BCC phase in the as-cast Fe(15-19)Mn binary alloys but weakly on that in the solution-treated ones. During tensile deformation, apparent deformation-induced α '-martensite and TRIP occurred. As Mn content increased, the transformation kinetics of deformation-induced α '-martensite and the TRIP effect diminished, reducing the plasticity of the as-cast alloys. The ST enhanced the TRIP effect, thereby improving the plasticity.

Keywords: cast FeMn alloys; damping capacity; mechanical properties; α '-martensite; TRIP effect

1 Introduction

FeMn damping alloys attracted considerable attention due to their high strength and cost-effectiveness. The Mn content notably influenced the microstructures and properties of FeMn damping alloys. For the deformation processed (DP) binary FeMn alloys, the Fe17Mn (wt%, here and after) alloy exhibited excellent damping capacity among the DP Fe(5-30) Mn alloys [1].

However, there were few works on cast FeMn damping alloys. Compared to the DP FeMn binary alloys made from homogenized cast ingots, serious composition segregations and their resultant non-equilibrium phases must occur in the cast counterparts [2-3]. Therefore, the dependence of damping capacity on the Mn content in the cast FeMn binary alloys must be different from that in the DP ones. In addition, the mechanical properties of FeMn alloys also strongly depend on the microstructure [21-23]. Therefore, it is of significance to investigate the relationship between the mechanical properties, damping capacity, and the Mn content in the cast FeMn alloys.

In this paper, the effects of Mn contents and solution treatment (ST) on the microstructure, mechanical properties and damping capacity were investigated in the as-cast Fe(15-21)Mn alloys.

2 Experimental procedure

The cast FeMn alloys were melted by a vacuum induction melting furnace in an Ar atmosphere. The molten alloys were cast into wedge-shaped silica-sand moulds. The chemical compositions of the as-prepared alloys are shown in Table 1. Some of the specimens were solution treated at 1050 ℃ for 0.5 h. The damping capacity was tested by the DMA Q850. Mechanical properties were examined by a WDW-20S tensile testing machine. The microstructures were analyzed by OM, SEM equipped with EDS and EBSD systems, and an XRD with a Co target and a FeritScope FMP30C.

Table 1 Chemical composition (wt%) of the alloys

Alloys	Mn	C	P	S	N	Fe
Fe15Mn	15.61	0.003	0.001	0.005	0.001	Bal.
Fe17Mn	17.43	0.002	0.001	0.002	0.005	Bal.
Fe19Mn	18.66	0.002	0.001	0.001	0.002	Bal.
Fe21Mn	20.77	0.007	0.001	0.001	0.008	Bal.

3 Result and discussion

Phases in as-cast FeMn binary alloys and evolution after solution treatment

The as-cast Fe(15-21)Mn alloys were composed of ε-martensite, γ-austenite, and α'-martensite. Significant compositional segregation existed in all alloys, with α'-martensite primarily distributed in Mn-depleted regions. As the Mn content increased and after the ST, the amount of α'-martensite significantly decreased (Fig. 1).

Damping capacity and mechanical properties in cast FeMn binary alloys

The optimal Mn contents for damping capacity in the cast FeMn binary alloys was around 19%, 2% higher than that in the DP counterparts (Fig. 2A and B) [1]. The rapid decrease in the amount of α'-martensite with the rise in Mn content can account for the strong inverse dependence of damping in the as-cast Fe(15-19)Mn alloys. The further decrease in the amount of α'-martensite after the solution treatment also explained the further improvements of damping capacity in the Fe15Mn alloy. However, it seemed that the moderate amount of α'-martensite contributed to improving damping capacity.

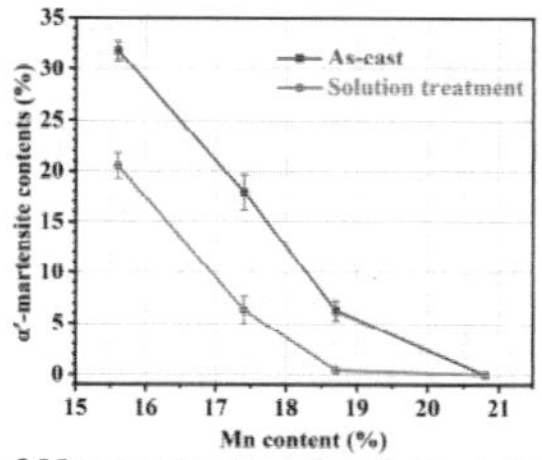

Fig. 1 Effects of Mn content and solution treatment on the amount of α'-martensite in as-cast Fe(15-21)Mn alloys.

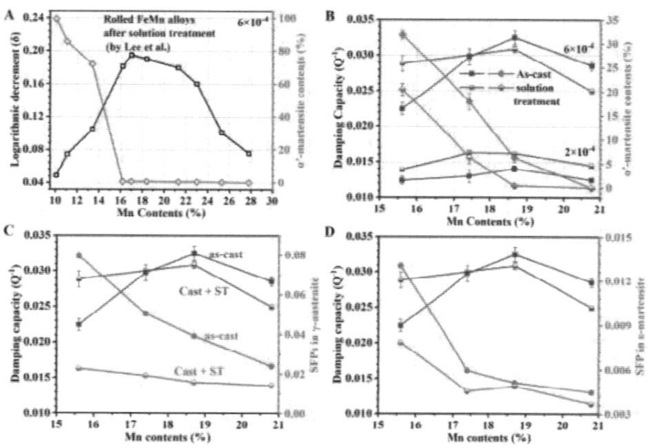

Fig. 2 Damping capacity in (A) DP and (B) cast FeMn alloys as a function of Mn content. (C) and (D) stacking faults probabilities in γ-austenite and ε-martensite.

The yield strength ($\sigma_{0.2}$) in the as-cast FeMn binary alloys increased almost linearly with the rise in the Mn content. The ST hardly changed their $\sigma_{0.2}$ (Fig. 3A). In contrast, the ultimate tensile strength (σ_b) in the as-cast FeMn binary alloys showed a negative dependence on the Mn content. The ST slightly enhanced the σ_b. The total elongation (ε_T) in the as-cast FeMn binary alloys also showed a negative dependence on the Mn content. The ST improved the ε_T, especially in the Fe19Mn alloy. Consequently, the ε_T was the highest in the solution-treated Fe19Mn alloy (Fig. 3B). The improvement in the ε_T corresponded to those in the uniform elongation (ε_U). Significant deformation-induced α'-martensite was produced (Fig. 3C). The plasticity of the as-cast FeMn alloys showed a positive correlation with the increment of α'-martensite. After the ST, the increment of α'-martensite increased. Thus, the plasticity was further enhanced (Fig. 3D).

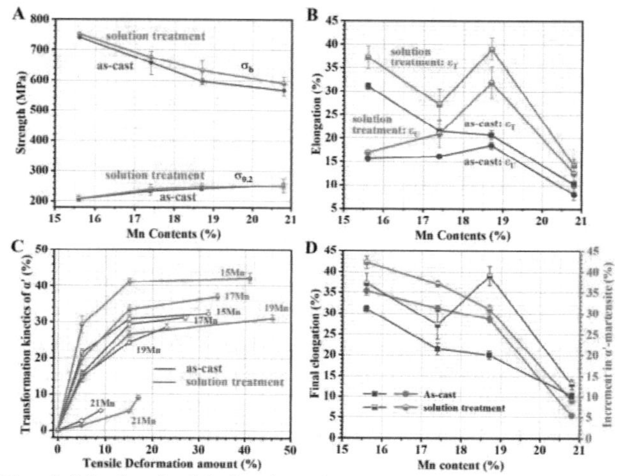

Fig. 3 Mechanical behaviour in cast FeMn binary alloys before and after the ST. (A) ultimate tensile strength (σ_b) and yield strength ($\sigma_{0.2}$); (B) total elongation (ε_T) and uniform elongation (ε_U); (C) transformation kinetics of α'-martensite; (D) relationship between plasticity and increment of α'-martensite.

4 Conclusion

1) The optimal Mn contents for damping capacity in the cast Fe(15-21)Mn binary alloys was around 19%, 2% higher than that in the deformation-processed counterparts.

2) The damping capacity inversely depended on the amount of α '-martensite in the as-cast Fe(15-19)Mn alloys but weakly on that in the solution-treated ones.

3) Both the ultimate tensile strength and total elongation in the as-cast Fe(15-21)Mn binary alloys also inversely depended on the Mn content. The ST could remarkably improve the total elongation in the as-cast Fe(15-21)Mn binary alloys. The enhanced uniform elongation could account for the improvement in the total elongation.

Acknowledgments

This work was supported financially by the Sichuan Science and Technology Program (No. 2022YFG0097).

References

[1] Lee Y K, Jun J H, Choi C S. Damping capacity in Fe-Mn binary alloys. ISIJ Int. 1997,37(10): 1023-1030.

[2] Alves C, Rezende J, Senk D. Comparison Between Segregation of High - Manganese Steels from 2 - D Phase - Field Simulations, 1 - D Analytical Theories, and Solidification Experiments. Steel Res. Int. 2016, 87(9): 1179-1189.

[3] He W, Wang Y, Yuan Y, et al. Comparison of microstructure, damping capacity and mechanical properties of cast and deformation processed Fe17Mn alloys. Mater Today Commun. 2024, 39:108972.

Insights into Carbides in Cast Iron

M. Górny [1*], M. Kawalec [1], J. Marosz [1], M. Bork [1], G. Angella [2]
1. AGH University of Krakow, Faculty of Foundry Engineering, 23 Reymonta St., 30–059 Krakow, Poland
2. Institute of Condensed Matter Chemistry and Technologies for Energy (ICMATE), National Research Council of Italy (CNR), Via R. Cozzi 53, 20125 Milan, MI, Italy
*Corresponding address: e-mail: mgorny@agh.edu.pl

Abstract: This work focuses on carbides in high-alloyed, gray, vermicular, and spheroidal cast iron. The presence of complex carbides containing elements such as V, Ti, Mo, and Cr is crucial for strengthening cast iron and improving specific functional properties, including wear resistance, hardness, and increased thermal stability. In addition to the classic method of introducing carbide-forming elements into cast iron, the high-temperature synthesis reaction in a metal bath of the SHSB (Self-propagating High-temperature Synthesis in Bath) reaction was used to produce ceramic phases (carbide phases). The paper explores crystallization, structure formation, and the properties of cast iron with particular emphasis on the influence of carbides with different morphological characteristics and their distribution. The study also analyzes the impact of carbides on the formation of the primary structure, i.e., austenite dendrites, and their role in enhancing the thermal stability of cast iron. Metallographic examinations, including electron backscattering diffraction (EBSD), were conducted to reveal macro- and micro-structural features during the primary and eutectic solidification of cast iron. Furthermore, this work demonstrates that melt quality is closely linked to the resultant morphology and number of carbides, graphite, and matrix structure.

Keywords: Cast Iron; Carbides; Solidification; SHSB; Austenite Dendrites; Eutectic; Wear Resistance

1 Introduction

Cast iron belongs to the group of the most popular casting alloys, which has an attractive combination of mechanical and usable properties that can compete with many alloys, including the so-called light alloys [1–3]. The main factors that influence the microstructure and properties of cast iron are its chemical composition, cooling rate, liquid metal treatment, and heat treatment [1,4]. Considering that cast iron, like many casting alloys, is sensitive to the cooling rate, the key issue in attaining homogeneous structures and properties is the control of graphite, carbides and metallic matrix, especially when entering the high cooling rates that take place in the production of thin-walled castings [1]. This paper focuses on the issue of carbides in cast iron. It has been analysed (1) vanadium cast iron - in which carbides with different morphology can be obtained, (2) Cast iron (CGI, FGI) with small additions of titanium, and (3) Cast Iron (SGI, CGI, D-5S) with carbides produced by SHSB method. The SHSB method, makes it possible to obtain a wide group of compounds, i.e.: carbides, borides and nitrides [5]. This paper presents the synthesis process for titanium carbides, selected for their very favourable enthalpy of formation and attractive functional properties such as high hardness and melting point, and high thermodynamic stability. The SHSB reaction, due to its versatility, offers the possibility of transforming any material (metal) into a cast composite material.

2 Experimental procedure

The first series of melts consist of high vanadium cast iron (15% of V) with different carbon equivalent. These iron was made in a vacuum-induction furnace (Balzers VSG 02 type) under argon protective atmosphere as described in [4].

In a second series of experiments the cast iron was produced in an electrical induction furnace of intermediate frequency, with a 15 kg capacity crucible. Fe-Ti was added to the liquid metal up to 0.15% Ti to achieve satisfactory percentage of compacted graphite in the microstructure of thin-walled castings and to enhance the strength of flake graphite cast iron.

Third series of cast iron was produced in a similar way as in second series. In this case, however, an SHSB method was used to obtain titanium carbides in ductile, vermicular and grey cast iron. The SHSB reaction procedure is provided in the work [5].

3 Result and discussion

Vanadium cast iron

Depending on the carbon and vanadium content, it is possible to achieve hypoeutectic, eutectic, and hypereutectic compositions in white cast iron. The C/V ratio as well as Si content are the main important factor which determines the type of metal matrix.

Cast iron (CGI, FGI) with small additions of titanium

In the second series, Fe-Ti additives cause the transformation in CGI from columnar to equiaxed austenite grains (exogenous to endogenous crystallization). The addition of another carbides promoting elements eg. zirconium and niobium did not change the morphology of austenite dendrites. Additions of Ti to FGI cast iron creates carbides, whose degree of dispersion depends, among other things, on the time elapsed since the introduction of titanium into the liquid metal.

Cast Iron with carbides produced by SHSB method

Fig. 1 shows an example of the microstructure in vermicular and ductile iron's in a thin-walled casting (g=3mm) strengthened with TiC by SHSB reaction.

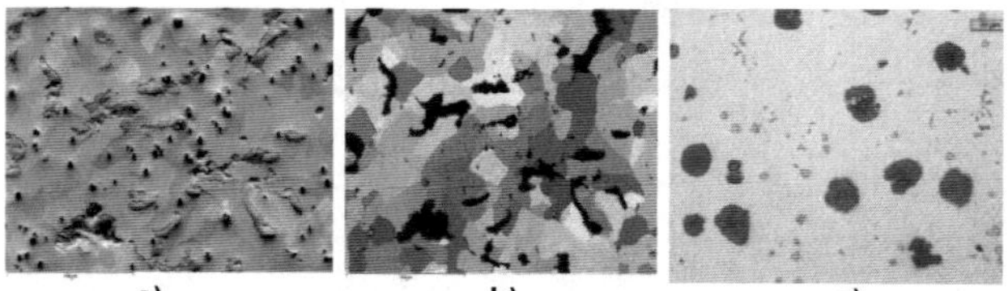

a) b) c)

Fig. 1 Microstructure of CGI + TiC EBSD (a, b) and SGI + TiC in thin walled casting (3mm) (c)

Studies of cast iron have shown that the SHSB reaction produces thermodynamically stable carbides that do not dissolve even after prolonged exposure at temperatures above 1000°C. The microstructures in Fig. 1 also demonstrate that the SHSB reaction does not limit cast iron in terms of graphite formation. It is possible to achieve both vermicular and spheroidal graphite, even in the presence of titanium carbides.

4 Conclusion

a) It is possible to obtain high-vanadium cast iron with different carbides morphology with attractive mechanical and utility properties (eg. High abrasive wear resistance).

b) Titanium increases the fraction of D-type graphite, which may result in the formation of ferrite due to the reduction of the distance between the graphite particles. This may result in a reduction of the mechanical properties. There is a certain range of Ti content below which tensile strength and hardness slightly decrease.

c) Titanium can cause "mesh graphite" in cast iron (similar to Widmanstatten graphite).

d) Titanium influences the crystallization of austenite. No increase in the number of eutectic grains was observed after adding titanium.

e) It is possible to produce SGI, CGI (also FGI) cast iron with stable carbides using the SHSB method.

f) SHSB does not interfere with other metallurgical procedures, e.g. Graphite spheroidization or modification.

g) The arrangement of carbides can be even, which has a positive effect on the mechanical and utility properties (eg. machinability).

h) Carbides produced through SHSB synthesis, characterized by covalent bonds, exhibit thermal stability and are effective for reinforcing cast iron, eg. ADI or D-5S.

Acknowledgments

This research was conducted within Horizon Europe Project No. 101159771 — NetCastPL4.0.

References

[1] Górny M, Stefanescu D M. Thin-wall ductile iron castings. In ASM Handbook; Stefanescu, D.M., Ed. ASM International:Materials Park, OH, USA, 2017, 1A: 617–628.

[2] Stefanescu D M, Dix L P, Ruxanda R E, et al. Tensile Properties of ThinWall Ductile Iron. AFS Trans. 2002, 110: 1149–1161.

[3] Druschitz A P, Fitzgerald D C. Lightweight Iron and Steel Castings for Automotive Applications; SAE Tech. Pap. 2000-01-0679; SAE International: Warrendale, PA, USA, 2000.

[4] Kawalec M, Fraś E. Structure, Mechanical Properties and Wear Resistance of Highvanadium Cast Iron. ISIJ International. 2008, 48 (4), 518-524.

[5] Fraś E, Wierzbiński S, Janas, Andrzej, et al. SHSB processing and properties of Al/TiC "in-situ" composites. Archives of Metallurgy. 2001, 46: 407-423.

Influence of Active Oxygen on Vermicularizing Effect and Method of Evaluating the Vermicularity

Dequan Shi[1]*, Guili Gao[2], Xuhui Wang[1]
1. School of Materials and Chemistry, University of Shanghai for Science and Technology, Shanghai, China
2. School of Mechanical and Energy Engineering, Shanghai Technical Institute of Electronics and Information, Shanghai, China
*Corresponding address: e-mail: shidequan2008@163.com

Abstract: The influences of oxygen activity and oxygen potential on vermicularizing effect were analyzed. The smaller the oxygen activity and oxygen potential are, the lower the vermicularity gets. When the oxygen activity is 0.1ppm and the oxygen potential is -237mV, the deoxidation effect is the best, and the vermicularity is only about 13%. When the oxygen activity is more than 0.8ppm and the oxygen potential is higher than -166 mV, the deoxidation effect is suitable, and the vermicularity can reach more than 85%. A method to evaluate the vermicularity by the oxygen activity α_O, oxygen potential E_O and temperature t was proposed, and the corresponding criterion was established. When the metallurgical state (α_O, E_O, t) of molten iron was got, the vermicularity can be evaluated roughly according to this criterion.

Keywords: oxygen activity; oxygen potential; Compacted cast iron; vermicularity

1 Introduction

The compacted cast iron has been widely used such as large diesel engine [1]. However, the production is relatively complex due to 0.008% narrow range of residual magnesium. So, it is necessary to accurately control the content of vermicular agent in order to obtain qualified products with 80% or higher vermicularity. Oxygen plays an important role in molten iron mainly in the formation of graphite crystallization [2]. According to thermodynamics, there should be certain relationship between the spheroidizing elements in molten iron and the oxygen activity [3]. Early research found that the oxygen activity in molten iron was lower than that of molten steel, and there was a corresponding relationship between the oxygen activity and dissolved Mg. Many researches have been done on the behavior of oxygen in molten cast iron [4-5]. However, there are few reports on using active oxygen content to evaluate the vermiculating effect of compacted cast iron.

2 Experimental procedure

FeSiMg5RE1 and ferrosilicon were used as the vermicular agent and the inoculant, respectively. Five schemes were designed where the vermicular agent was 0.4wt.% (marked 1#, 2#), 0.6wt.% (marked 3#), 0.8wt.% (marked 4#) and 1.0wt.% (marked 5#). An induction furnace was used for smelting iron, and after the molten iron was overheated to 1550±10 ℃, the furnace was turned off. The oxygen activity α_O, oxygen potential E_O and temperature t were measured once every 2 min by NSP3603 with disposable probe. After five measurements, the molten iron was spheroidized and inoculated by Sandwich method. Then, three parameters were measured four times at 10min, 12min, 14min and 16min. In addition, some data between vermicularity and active oxygen were collected from the production site.

3 Result and discussion

Influence of active oxygen on vermicularizing effect

Figure 1 showed the change of the oxygen activity and oxygen potential of molten iron with the vermicular agent. Both oxygen activity and oxygen potential decrease with the increasing vermicular agent. That is, the more the vermicular agent is, the more the spheroidizing elements are, the greater the deoxidation degree of molten iron is, and the less the active oxygen of molten iron is.

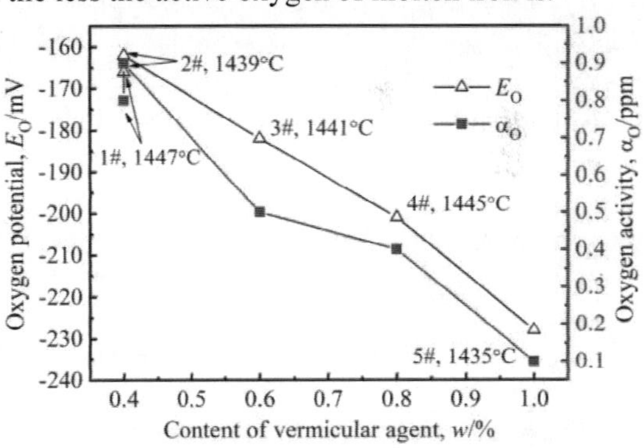

Fig. 1 Change of oxygen activity and oxygen potential with the vermicular agent.

Figure 2 showed the change of oxygen activity and oxygen potential with the vermicularity. The oxygen activity and oxygen potential will increase gradually with the vermicularity. When the oxygen activity and oxygen potential are α_O=0.1ppm and E_O=-237mV, the level of active oxygen is very low, and the deoxidation effect is the best. The vermicularity is about 13%. When the oxygen activity is over 0.8ppm and the oxygen potential is higher -166mV, and the degree of deoxidation is suitable for crystallization and solidification of iron, and the vermicularity reaches more than 85%.

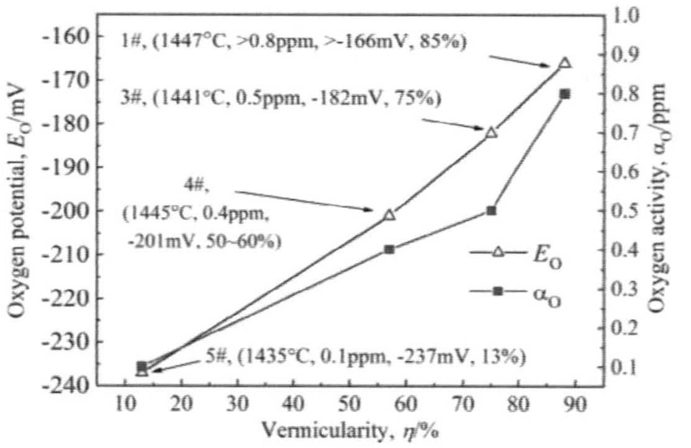

Fig. 2 Change of oxygen activity and oxygen potential near same temperature with the vermicularity.

Method of evaluating vermicularity

Figure 3 showed all data and model. It is difficult to distinguish the verimicularity only by oxygen activity or oxygen potential. So, the oxygen activity, oxygen potential and temperature were proposed to evaluate the vermicularity η. The areas were divided by five "oxygen activity lines" with different threshold. The criteria of evaluating the vermicularity are as follows. If $\alpha_O \leqslant$ 0.1ppm, η<20%; If 0.1ppm<$\alpha_O \leqslant$0.3ppm, 20%<η<55%; If 0.3ppm<$\alpha_O \leqslant$0.6ppm, 55%<η<70%, most higher than 60%; If 0.6ppm<$\alpha_O \leqslant$ 0.8ppm, 70%<η<80%; If 0.8ppm<α_O<1.3ppm, η>80%, most higher than 85%; If $\alpha_O \geqslant$1.3ppm, the vermicularity may be suddenly reduced, and the tendency to precipitate flake increases.

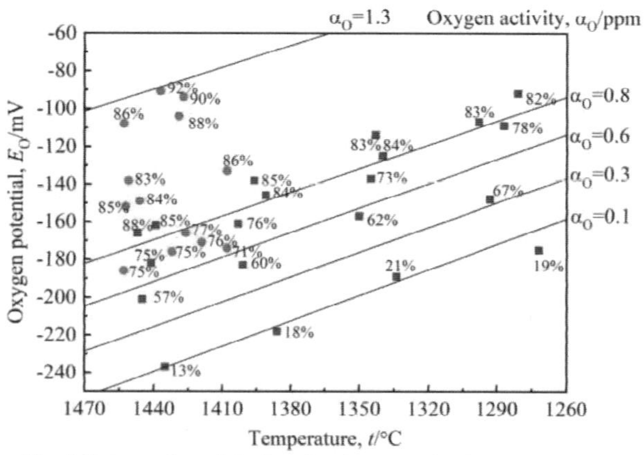

Fig. 3 Data and model of evaluating vermicularity by oxygen activity, oxygen potential and temperature.

4 Conclusion

(1) The lower the oxygen activity and oxygen potential are, the lower the vermicularity is. When oxygen potential is more than 0.8ppm or oxygen activity is more than -166mV, the deoxidation degree is suitable and the vermicularity reaches more than 85%. When oxygen potential is 0.1ppm and oxygen activity is -237mV, the deoxidation degree is the best and the vermicularity is only about 13%, which is ductile graphite cast iron.

(2) Five lines with oxygen activity α_O thresholds of 0.1ppm, 0.3ppm, 0.6ppm, 0.8ppm and 1.3ppm were selected to divide different areas of vermicularity, and the evaluation model by multi parameters is obtained. The vermicularity was evaluated by using oxygen potential, oxygen activity and temperature.

Acknowledgments

It was supported by the Chinese Government Guidance on Local Scientific and Technological Development of Heilongjiang (SBZY2021E071).

References

[1] Dawson S. Compacted graphite iron-a material solution for modern diesel engine cylinder blocks and heads. China Foundry, 2009, 6(3): 241-246.
[2] Ten E B. Oxygen state forms in cast iron and their effect on graphite crystallization. Key Eng. Mater., 2011, 457: 43-47.
[3] Fray D J. The use of solid electrolytes in the determination of activities and the development of sensors. Metall. Mater. Trans B, 2003, 34(5): 589-594.
[4] Hampl J and Elbel T. On modeling of effect of oxygen on graphite morphology and properties of modified cast irons. Arch. Foundry Eng., 2010, 10(4): 55-60.
[5] Ekengard J, Dioszegi A and Jonsson P G. A study of oxygen activities before the start of solidification of cast irons. Int. J. Metalcast., 2016, 10(4): 500-515.

High-Temperature Deformation Behavior and Processing Maps of As-cast Borated Stainless Steel

Jian Shi*, Tianze Di, Bo Yu
National Key Laboratory of Advanced Casting Technologies, China Academy of Machinery Shenyang Research Institute of Foundry Co., Ltd., Shenyang, China
*Corresponding address: e-mail: shijian@chinasrif.com

Abstract: In response to the cracking problem during hot working of borated stainless steel, physical simulation technology was used to study the high-temperature deformation behavior of borated stainless steel billets. Using high-temperature compression thermal simulation experiments to simulate thermal deformation parameters, based on the Zhou Jihua-Guan Kezhi model and DMM model, establish a borated stainless steel deformation resistance model and processing maps.

Keywords: as-cast Borated stainless steel; Deformation behavior; DMM mode; Processing maps

1 Introduction

Borated stainless steel is a special type of stainless steel made by adding boron element with a larger neutron absorption cross-section to it. Currently, it is mainly used for the transportation and storage of spent fuel in nuclear power plants and has broad application prospects in spent fuel treatment[1]. Boron is difficult to dissolve in the steel matrix, and during the solidification process, it forms low melting point brittle compounds with Fe and Cr, which are continuously distributed at grain boundaries and cut through the matrix structure. Due to the fact that borated stainless steel plates are mainly formed by hot working, the problem of cracking during thermal deformation has become a difficult point in the manufacturing of borated stainless steel. Lian[2] studied the high-temperature deformation behavior of borated stainless steel, fitted the thermal deformation constitutive equation of borated stainless steel based on the Arrhenius equation, and analyzed its dynamic recrystallization conditions.

2 Experimental procedure

Borated stainless steel alloy ingots were melted in a 25 kg vacuum induction furnace, the crucible was made of magnesium oxide material, and the casting mold was ductile iron metal mold. The composition range of the melted alloy is shown in Table 1.

Table 1 Content of borated stainless steel ingot

C	Si	Mn	Cr	Ni	B
0.03	<1.0	<2.0	18-20	12-15	1.8

Conduct isothermal constant strain rate single pass compression experiments using the Gleeble-3800 thermal simulation testing machine. The experimental temperature is 900~1150 ℃; The strain rate is 0.1~10 s^{-1}; The compression rate in the height direction is 70%. Heat the sample at a speed of 10 ℃/s to 1150 ℃, keep it warm for 3 minutes, then lower it to the test temperature at a speed of 10 ℃/s, keep it warm for 30 seconds, and compress it at the specified rate to the specified deformation amount.

3 Result and discussion

True stress-strain curve

Figures 1a, b, and c show the true stress-strain curves of as cast borated stainless steel under high temperature single pass compression at different strain rates at deformation temperatures of 900 ℃, 1000 ℃, 1100 ℃, and 1150 ℃, respectively. Based on the true stress-strain curve as cast, the deformation resistance of borated stainless steel decreases with increasing temperature and decreasing strain rate.

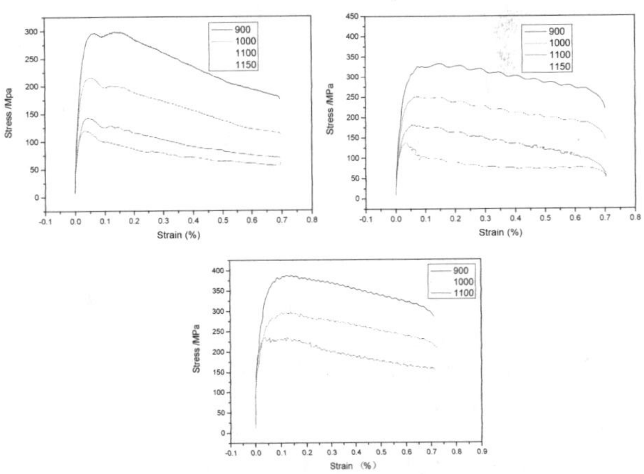

Fig.1 True stress-strain curves of borated stainless steel at different deformation temperatures as cast
(a) $\dot{\varepsilon}=0.1s^{-1}$; (b) $\dot{\varepsilon}=1s^{-1}$; (c) $\dot{\varepsilon}=10s^{-1}$

Establishment of processing maps

In the dynamic material model, the hot working material system is regarded as a nonlinear energy dissipator, and the energy P input per unit volume of the system per unit time is mainly divided into two parts during transfer: one is the energy G that causes deformation and dissipation of the workpiece, and the other is the energy J that causes changes in the microstructure and dissipation.

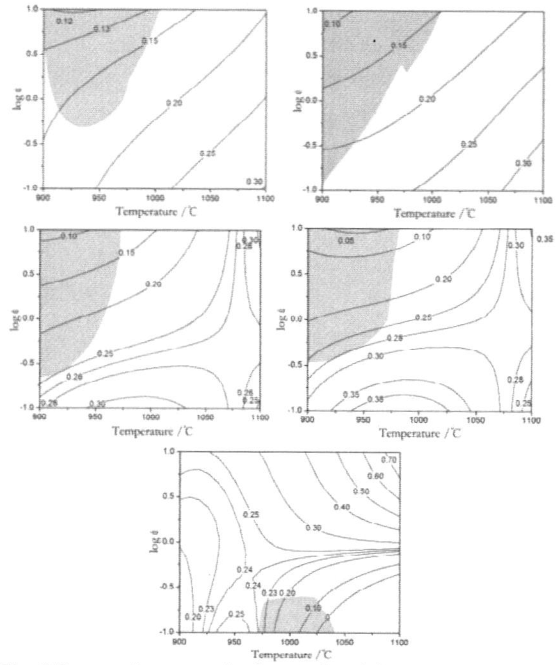

Fig. 2 Processing maps for borated stainless steel as cast at different hot deformation conditions
(a) ε=0.3, (b) ε=0.4, (c) ε=0.5, (d) ε=0.6, (e) ε=0.7

When the true strain ε of as cast borated stainless steel is not greater than 0.6, the instability zone during hot working is located in the low-temperature and high-speed region of the processing maps, indicating that as cast borated stainless steel is prone to instability at lower deformation temperatures and higher speed conditions. When the true strain ε reaches 0.7 (Figure 2e), the unstable zone shifts from the low-temperature and high-speed zone to the temperature and low-speed deformation zone in the middle of the processing maps. It indicates that as cast borated stainless steel is prone to cracking under low deformation rate conditions in the temperature range of approximately 970-1040 ℃ when the degree of deformation is large.

The contour lines of cast borated stainless steel transfer to the low-temperature and high-speed region with the increase of true strain ε, but the position of the contour lines varies greatly. The true strain ranges from ε=0.4 (Figure 4b) to 0.5 (Figure 4c), and in addition to the original shape, a new shape of the contour line appears. As cast borated stainless steel, the contour line is sensitive to the degree of deformation. Therefore, when the degree of deformation of as cast borated stainless steel changes, the deformation temperature and deformation rate should be adjusted accordingly. The deformation rate of as cast borated stainless steel should be below $0.2s^{-1}$ (log ε̇<-0.7).

Establishment of deformation resistance model

In each true stress-strain curve, take the true strain of 0.1 as the step value to obtain the stress value corresponding to the strain. The obtained stress-strain data was input into the Zhou- Guan model, and regression analysis was performed on the single pass compression experimental data of borated stainless steel. The deformation resistance model for borated stainless steel as cast was established as follows:

$$\sigma = 263.13\exp(-2.967T + 3.777)(\frac{\dot{\varepsilon}}{10})^{0.331T+0.308}\left[1.378(\frac{\varepsilon}{0.4})^{0.079} - 0.378\frac{\varepsilon}{0.4}\right]$$

4 Conclusion

Based on the Zhou - Guan model, regression analysis was conducted on the true stress true strain data, and a deformation resistance model for borated stainless steel as cast was established ;Through high-temperature compression experiments, a processing diagram of as cast borated stainless steel was established based on a dynamic material model, providing an experimental basis for formulating the hot deformation process of the alloy; Borated stainless steel has good deformation ability at a temperature of 1100 ℃ and a deformation rate below $0.2s^{-1}$.

Acknowledgments

This project was financially supported by National Natural Science Foundation of China (No. U1908219), Natural Science Foundation of Liaoning Province of China (No. 20170540890).

References

[1] Orive R, Diego J L D, Lasema J A, et al. Used fuel management strategy in Spain: increase of used fuel storage capacity in cofrentes NPP.C. Barcelona, Spain: 2010.

[2] Lian X J. Hot deformation behavior of stainless steel containing Borin. J. Materials for Mechanical Engineering, 2021,45(04):81-87.

Optimizing the Impact Performance of High Manganese Steel Casting by Controlling the Pouring Temperature and Soaking Time

Biju Karakkunnummal[1]*, Kranthi Peddeti[2], Ł. Szymański[3]
1. FLSmidth Private Limited, 34 Egatoor, Kalembakkam, Old Mahabalipuram Road, 603103 Chennai, Tamil Nadu, India
2. FLSmidth Inc., 1100 American Parkway, Allentown, PA, 18109, USA
3. FLSmidth Mining Technology GmBH, 12 Schleebergstrasse, 59320 Ennigerloh, Germany
*Corresponding address: e-mail: bkl-in@flsmidth.com

Abstract: The wear resistance of high manganese steel is significantly affected by its ability to undergo work hardening under impact loads. This property is influenced by factors such as composition, pouring temperature, and heat treatment. Achieving optimal ductility and tensile properties requires a fine grain size. As carbon content increases from 0.7% to 1.3%, higher soaking and quenching temperatures are necessary. However, excessively high soaking temperatures can lead to decarburization and grain growth. In this study, 16 austenitic manganese steel test blocks with a Mn:C ratio greater than 10.5 were produced using sand molds. Two pouring temperatures (1425 ℃ and 1520 ℃) and two heat treatment temperatures (1050 ℃ and 1100 ℃) were investigated, with soaking durations of 2 and 12 hours. The castings were evaluated for microstructure, hardness, and impact resistance. The most significant increase in impact energy was observed at 1100 ℃ with a 12-hour soak, highlighting the critical relationship between heat treatment parameters and the mechanical properties of the castings.

Keywords: work hardening; pouring temperature; soaking time; grain size

1 Introduction

Austenitic manganese steel, renowned for its toughness and wear resistance, is crucial in the crushing industry. High manganese content enhances durability against impacts and wear, reducing maintenance costs and downtime. Components like hammer crusher, jaw plates, and cone crusher liners are essential for efficient crushing operations. With a Mn: C ratio exceeding 11, this steel exhibits exceptional wear resistance through surface work hardening while maintaining core toughness, ideal for challenging applications in cement and mining industries. However, its wear performance and work hardening ability depend on chemical composition, casting parameters, and heat treatment process [1]. Effective heat treatment is essential for achieving desired properties, such as a carbide-free, fully austenitic structure[3]. The selection of quenching temperature and soaking time in heat treatment significantly influences the final properties [4]. Prolonged soaking can lead to decarburization and grain growth, while slow cooling accelerates austenite decomposition. Water quenching from solution temperatures typically results in hardness below 230 BHN. Under high impact, surface hardness can reach around 550 BHN, varying with work hardening and impact intensity. The initial impact properties of Austenitic manganese steel casting after heat treatment are influenced by pouring temperature, heat treatment soaking temperature, and soaking time[2].

2 Experimental procedure

This research involved casting fourteen samples of dimensions 50×50×100 mm into sand molds, as shown in Fig 1. The chemical composition of these castings is detailed in Table 1, highlighting crucial elements for understanding their material properties. The experiment focused on exploring the influence of different pouring temperatures: 1506 ℃, 1480 ℃, 1432 ℃, and 1425 ℃, on the resulting castings. Following solidification and cooling, the castings underwent solution heat treatment stages. This process was meticulously carried out in a muffle furnace, shown in Fig 1b. Two key parameters were varied during heat treatment :soaking temperature and soaking time. Specifically, two soaking temperatures, 1050 ℃ and 1100 ℃, were applied along with soaking times of 2 hours and 12 hours. These variations were chosen to investigate their influence on the final mechanical and microstructural properties of the castings.

Table 1 Composition of Austenitic manganese steel

%C	%Si	%Mn	%P	%S	%Cr	%Ni	%V
1.33	0.273	13.95	0.034	0.008	2.29	0.179	0.013

Fig. 1 a) Casting Mould, b) Heat treatment furnace

Following heat treatment, the castings underwent tests for Charpy V-notch impact toughness, surface hardness, and microstructural analysis, correlating with pouring and heat treatment parameters.

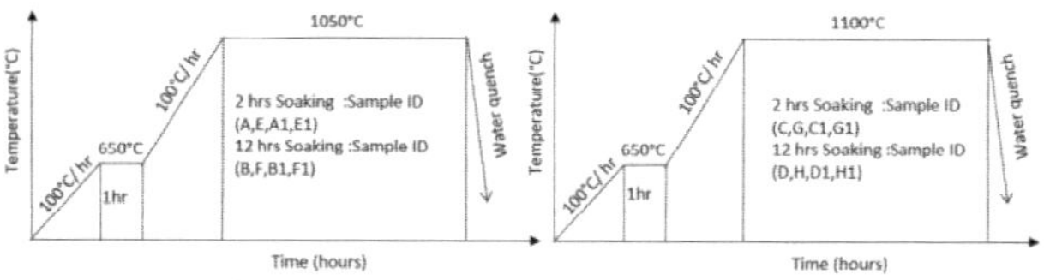

Fig. 2 Heat treatment cycle.

3 Result and discussion

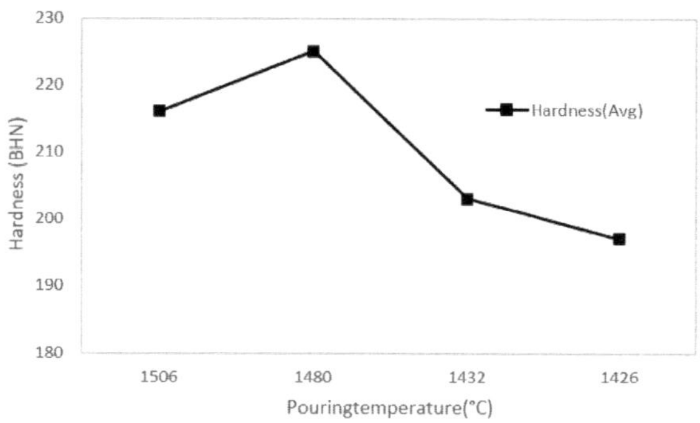

Fig. 3 Hardness variation with pouring temperature in as cast Austenitic manages steel.

The experiment used a high-Manganese composition (Mn:C > 10.5), enhancing toughness and wear resistance. With a chromium content of 2.29%, the alloy is predisposed to carbide precipitation, contributing to improving wear performance and structural stability. Increased pouring temperatures lead to grain coarsening, which may potentially decrease hardness. As-cast data, Fig 3, shows decreasing hardness from 216 HB at 1506 ℃ to 197 HB at 1426 ℃. Higher temperatures may segregate chromium, forming carbides. Figure 4.a, b; microstructure reveals significant grain boundary precipitation whereas lower temperatures reduce segregation. Lower hardness enhances ductility and work-hardening in the as-cast stage, supporting preferred conditions.

The data presents, Fig.5, a clear relationship between soaking temperature, soaking time, and impact values at different pouring temperature. At 1050 ℃, shorter soak times of 2 hours generally result in increased impact values compared to higher 12-hour soaks, with noticeable variations across different pouring temperatures. However, lower soaking temperatures with longer soaking times did not yield any variation, regardless of the pouring temperature variations. For instance, lower pouring temperatures like 1426 ℃ show a more pronounced increase in impact values with shorter soaking durations. At 1100 ℃, shorter soak times show varying impact values, while longer 12-hour soaks consistently elevate impacts across all pouring temperatures, indicating a significant temperature-dependent effect on steel properties during casting processes.

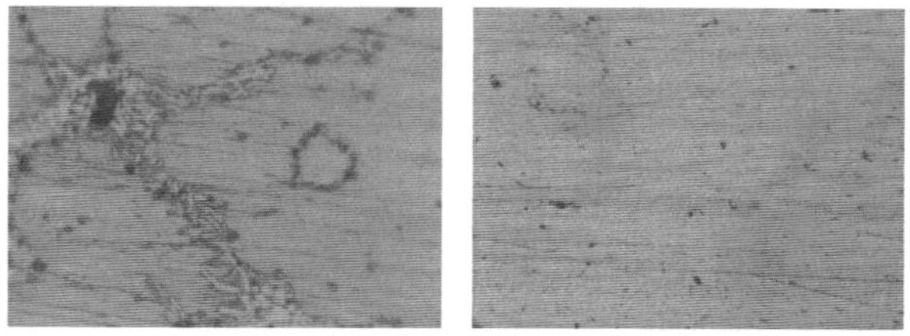

Fig. 4 a) Microstructure in as cast state,500X b) microstructure after solution treatment at 500X

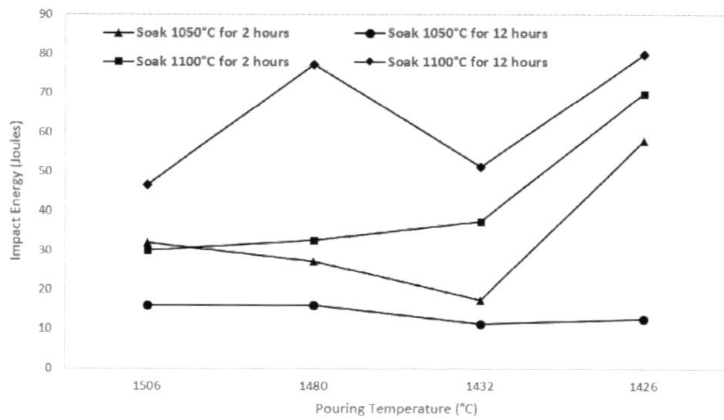

Fig. 5 The relation between the pouring temperature, solution treatment soaking temperature and time.

4 Conclusion

In conclusion, the study underscores the pivotal role of soaking temperature and time in the solution heat treatment of Austenitic manganese steel, notably affecting its impact values. Higher soaking temperatures and prolonged times generally correlate with enhanced impact resistance, indicating superior toughness and fracture resilience. To optimize impact strength and mitigate carbide precipitation in chromium-alloyed Austenitic manganese steel, the recommended approach involves employing higher soaking temperatures alongside lower casting pouring temperatures. This strategy aims to achieve both superior mechanical properties and structural stability in the final castings.

References

[1] Mahalami C S, Pan X. An overview of high manganese steelcasting, Advanced sustainable foundry, 19-21 May 2014, Foundry congress, Bilbao 2014.
[2] David Havel, P. E. Austenitic manganese steel -a complete overview Columbia steel casting Co Inc.
[3] Selçuk Kuyucak, Renata Zavadil, et al. Heat-treatment Processing of Austenitic Manganese Steels, CANMET – Materials Technology Laboratory, Ottawa, Ontario, Canada.
[4] Tęcza G, Sobula S. Microstructure of Cast High-manganese Hadfield Steel, Archive of foundry engineering, 2014, 14 (3): 67 – 70.

Effect of Fe Content on the Corrosion Behavior of FCC Structured NiCrFeCo-Based MPEAs

Shiyu Liu, Zhiqiang Fu
Guangdong Key Laboratory for Advanced Metallic Materials Processing, South China University of Technology, Guangzhou, Guangdong 510641, China
*Corresponding address: e-mail: zhiqiangfu2019@scut.edu.cn

Abstract: In this work, $NiCrCoAlMoFe_x$ multi-principal element alloys (MPEAs) were fabricated by arc melting method. The influence of Fe content on the corrosion behavior in 0.6 M NaCl solution was investigated in solid solution FCC-structured samples without any precipitation phase. The experimental results showed the corrosion resistance are firstly enhanced and then deteriorated with increased Fe content. The sample with 8 at.% Fe exhibited the best corrosion resistance due to the better passive film with less defect density. With the Fe increased to 16 at.%, more defects were formed within the corresponding passive film resulting in worse corrosion resistance.

Keywords: Multi- principal element alloys; microstructure; corrosion behavior; microstructure.

1 Introduction

Almost metallic materials are subjected to corrosion from exterior environment where they are in service, which usually leads to the loss efficiency for these materials. In deal with this predicament, alloys with well corrosion resistance were designed to be applied in aggressive environment. In recent years, Multiple-principal elements alloys (MPEAs)[1-3], consisting of three or more principal elements, also exhibited well corrosion resistance. No matter what kind of alloys, their well corrosion resistance is closely related to a thin film formed on the alloy matrix during corrosion, namely passive film as well known. The passive film isolates alloys matrix from aggressive medium, thereby impeding further corrosion. For the most part of corrosion resistance alloys, the concentration of Cr is the key for the excellent anticorrosion properties. Because high Cr content can ensure there are enough Cr_2O_3, a compact production generated by the oxidation of Cr, within passive film resulting in better protective ability[4-5]. However, Fe also plays an important role in corrosion defending in alloys. It can fastly be oxidized to FeO, FeOOH and Fe_2O_3 to form a protective layer on the alloy surface[6-7]. To pity, the corresponding study that how Fe content effect the corrosion behavior in NiCrFeCo-based MPEAs is still unclear. Hence, in this work, we investigated the effect of Fe on corrosion behavior in 0.6M NaCl solution of solid solution FCC-structured MPEAs.

2 Experimental procedure (Bold, 10 pt., Arial)

Material preparation

Ingots of $Ni_{36.85}Cr_{29.95}Al_{7.4}Mo_{1.8}Co_{24-x}Fe_x$ (x=0, 8, 16) MPEAs were fabricated by vacuum arc melting (VAM) in Ar protecting atmosphere. The mixtures were melted and then solidified in a button-shaped water-cooled copper mold to obtain bulk ingots. To ensure uniform distribution of each element, all ingots were flipped and remelted at least five times. Finally, these ingots were obtained by molding in copper mold with baguette shape. In order to obtain the alloys without any precipitation, homogenization treatment was used at 1200 ℃ for 12 h and subsequently water quenched.

3 Microstructure characterization

Microstructures of the samples under different states were characterized by a Zeiss Merlin scan electron microscopy (SEM).

4 Electrochemical testing

Electrochemical measurements in this work were performed using a Zahner Pro workstation. The electrolyte used was a 0.6M NaCl solution. To achieve a stable state in the electrochemical system, a 3000 s open circuit potential (OCP) was applied to the samples. Then electrochemical impedance spectroscopy (EIS) measurement was conducted using an applied AC amplitude of 10 mv, with a frequency ranging from 100 kHz to 10 mHz. Potentiodynamic polarization test was carried out by sweeping the potential from -0.6 v to 0.6 v (vs. SCE) at a scan rate of 1mv/s. Measurements were repeated three times for each sample. In addition, Mott-Schottky method was used to make clear the semi-conductor character of these alloys with a test frequency of 10 Hz, and additionally potential scan rate of 10 mV/step from 1.0 V to -1.0 V.

5 Result and discussion

Microstructure

There is no precipitation phase for three samples from SEM images just a little residual cast defects, namely black holes.

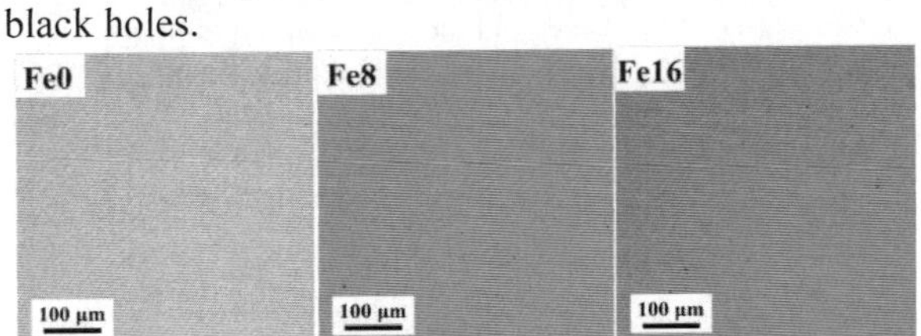

Fig. 1 Backscatter electron SEM images of Fe0, Fe8 and Fe16 MPEAs at homogenized condition.

Corrosion behavior

From the potentiodynamic polarization curves in Fig.2, all samples exhibited passivation behavior in anodic region. The Fe8 sample possesses the lowest corrosion current density deduced from the curves among three samples and inversely the Fe16 sample has the highest one.

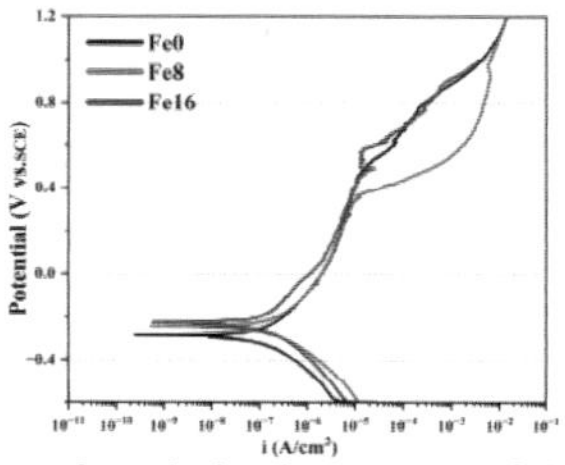

Fig. 2 Potentiodynamic polarization curves of three homogenized MPEAs

In Fig. 3(a), nyquist plot demostrates the passive film formed the Fe8 alloy presents the best quality due to the large radius[8]. However, the Fe16 alloy exhibits the smallest radius, namly meaning the worst passive film for it. The bode plots showed the same trend in passive film.

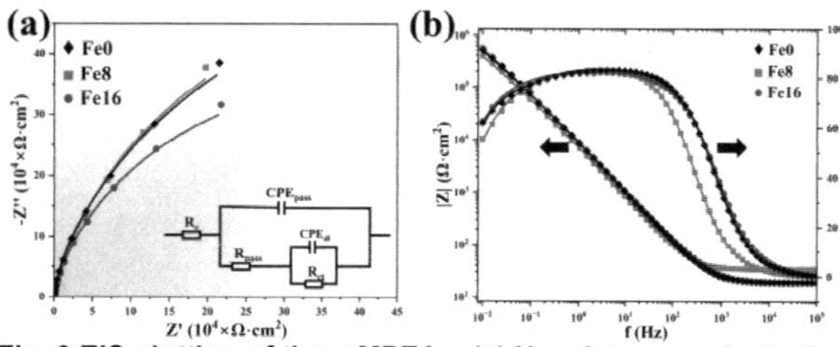

Fig. 3 EIS plotting of three MPEAs: (a) Nyquist curves including in equivalent circuit; (b) Bode curves.

In order to make clear the difference of the passive film formed on these samples, motty-schottky curves are obtained shown in Fig. 4. The defects within the passive fim, such as vacancies, can be calculated from the linear region to reflect the quality of the passive film[9-10]. From the Table 2, the most defects are formed within the passive film of Fe16 alloy, and inversely the lowest defects for Fe8 alloy.

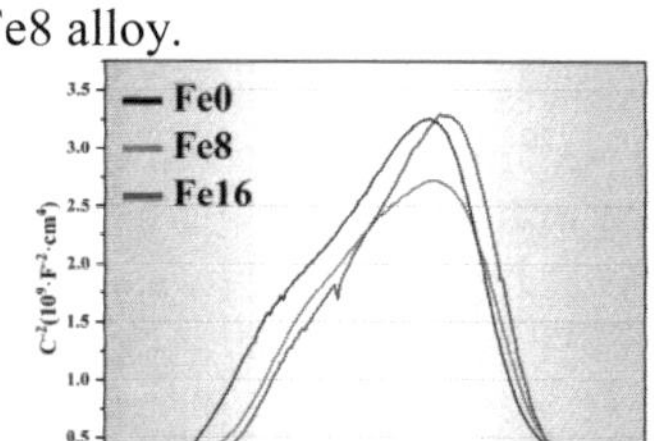

Fig. 4 Mott-Schottky curves of four samples at homogenized condition divided into two regions, I and II from -1.0 V to 1.0 V.

Table 1 defects density of three samples

samples	N_d (10^{19}/cm^3)
Fe0	1.35
Fe8	1.14
Fe16	1.56

6 Conclusion

Appropriate addition of Fe can enhance the corrosion resistance in NiCrFeCoAlMo MPEAs due to the formation of better passive film with less defects

Acknowledgments

This project was financially supported by the Natural Science Foundation of China (Grant No. 52103360), the Basic and Applied Basic Research Foundation of Guangdong Province (Grant No. 2023A1515012363).

References

[1] Gludovatz B, Hohenwarter A, Catoor D. A fracture-resistant high-entropy alloy for cryogenic applications. Science, 2014, 345: 1153–1158.

[2] Miracle D B, Senkov O N, A critical review of high entropy alloys and related concepts. Acta Mater. 2017, 122: 448–511.

[3] Wetzel A, von der Au M, Dietrich P M, The comparison of the corrosion behavior of the CrCoNi medium entropy alloy and CrMnFeCoNi high entropy alloy. Appl. Surf. Sci. 2022, 601: 154171.

[4]Torbati-Sarraf H, Shabani M, Jablonski P D, The influence of incorporation of Mn on the pitting corrosion performance of CrFeCoNi High Entropy Alloy at different temperatures. Mater. Des. 2019 (4):108170.

[5] Fu A, Liu B, Lu W. A novel supersaturated medium entropy alloy with superior tensile properties and corrosion resistance. Scr. Mater. 2020 (6) 381–386.

[6] Z H Han, W N Ren, J Yang. The corrosion behavior of ultra-fine grained CoNiFeCrMn high-entropy alloys. J. Alloys Compd. 2020, 816: 152583.

[7] L Xue, Y Ding, K G Pradeep. The grain size effect on corrosion property of $Al_2Cr_5Cu_5Fe_{53}Ni_{35}$ high-entropy alloy in marine environment. Corros. Sci, Volume 2022, 208: 110625.

[8] H Luo, S Gao, C Dong, X Li. Characterization of electrochemical and passive behaviour of Alloy 59 in acid solution. Electrochim. Acta, 2014, 135: 412–419.

[9] Z Wang, G H Zhang, X H Fan, Corrosion behavior and surface characterization of an equiatomic CoCrFeMoNi high-entropy alloy under various pH conditions. J. Alloy. Compd. 2022, 900: 163432.

[10] C B Nascimento, U Donatus, C T Ríos. Electronic properties of the passive films formed on CoCrFeNi and CoCrFeNiAl high entropy alloys in sodium chloride solution. J. Mater. Res Techonol. 2020 (9): 13879–13892.

TheCurrent Status, Problems and Prospects of ADI in China: A Review

QidongYan[1,3], WenbangGong[2,3], Jincheng Liu[4,*]
1. YuzhouHenglilai New Materials Co., Ltd., Xuchang Henan, China 452570
2. School of Mechanical Engineering and Automation, Wuhan Textile University, Wuhan China 430200
3. ADI Committee of China Foundry Association, Beijing China
4. WFO Ferrous Metals Commission, Shenyang Liaoning, China
*Corresponding address: e-mail: liujincheng64@tsinghua.org.cn

Abstract: In 2023 the total production of ADI in China was approximately 140,000 tons, supplied to a range of industries in various grades of ADI and CADI. The growth of ADI in China is promising, compared to the total ductile iron production and the need for high-strength, lightweight components in China. The development capability of new lightweight and innovative products is insufficient, and there is an urgent need for cooperation between universities, designers, foundries and heat treaters to develop new ADI components.

Keywords: ADI; development; foundry; heat treater; lightweight innovation; application; prospects

1 Introduction

The research and application of ADI began in the 1970s, synchronized with America, Finland etc. However, in the early stages, ductile iron production technology and austempering processes were relatively backward. Significant progress was made, but the relatively poor stability and consistency of ADI quality, resulted in difficulty to realize industrial scale application. (1-2)

Since 2000 in China, significant improvements in the raw materials and melting technology have been achieved. During the same period, spheroidization and inoculation techniques have been improved, also new treatment alloys and inoculants have been rapidly developed, resulting in increased stability and consistency of SG iron quality. All these have laid down the foundation for the production and promotion of ADI parts. [1-2]

In 2003, the Shanghai Powermax Furnace Corp. was formed, and the advanced American AFC-Holcraft Universal Batch Ausquench furnace was introduced into China. In 2006, the first commercial austempering service plant, Austempered Component (Suzhou) Manufacture Co., was established and successfully put into operation in China. Later, more commercial ADI treatment service plants were established and more foundries started to install their own austempering furnaces. In 2009, China published the first ADI national standard, which was consistent with the ISO standard[3]. In 2023, China published a new national standard with performance higher than that of the ISO standard [4].At present, many more manufacturers and foundries can produce high quality, safety-critical ADI components.

2 Current status of ADI in China

At present in China, there are 5 commercial ADI heat-treatment plants operating 26 professional austempering furnaces, and approximately 17 foundries having their own austempering facilities. Among these, some are commercial austempering furnaces and others are self-made. SEW (Tianjin) is sourcing SG iron castings, heat treating these in their own Powermax UBQA furnaces, and then machining and assembling them into the final products. There are 26 Powermax UBQA furnaces among all the above furnaces, each having a capacity from 1.5t – 4.5t.

Currently, there are hundreds of manufacturers and foundries, big and small, developing and producing ADI parts. All the ADI parts are austempered in commercial ADI treatment plants or in their own ADI treatment furnaces.

According to incomplete statistics from the ADI Committee of CFA, a total of 135,699 tons of ADI were produced in China in 2023. This included 56,916 tons (42%) of structural ADI castings and 78,783 tons (58%) of CADI wear-resistance parts.

In 2023, the ADI structural castings by application fields: road transportation comprising 35.05%, of the production, followed by PV power generation 18.45%, engineering machinery 13.78%, and agricultural machinery 9.15%. Defenseaccounted for 7.03%, and passenger cars, 6.35%. Mining industry is at 4.57% and large passenger cars 3.26%. Wind power turbine is at 1.06%, railway transportation, 0.74%, petroleum machinery 0.43% and robotics 0.08%, ADI product exports are very small, only approximately 0.04%, see figure 1. In 2023, the majority of ADI castings were made in grade QTD1050-7, a total of 3,3415 tons, representing 59% of all production, followed by QTD900-9 9,385 tons at 17% and QTD1200-4, 6,607 tons at 12%, then is QTD1400-2, 3,550 tons, accounting for 6%, and QTD1600-1, 1,365 tons, accounting for 2%; QTD850-10 has been minimal, 518 tons, only 1%, as shown in figure 2.

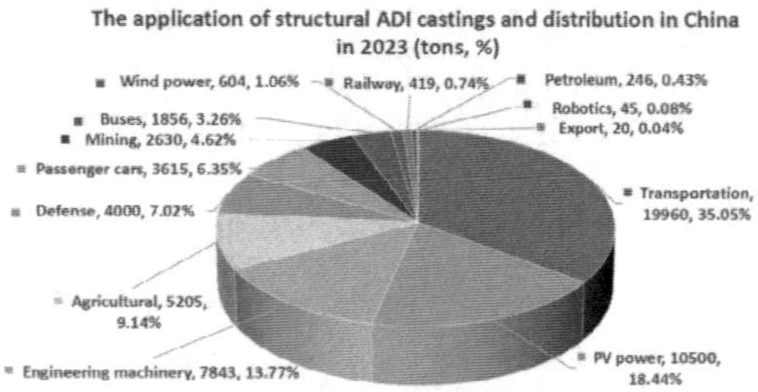

Fig. 1 Applicationsof ADI structural castings in 2023.

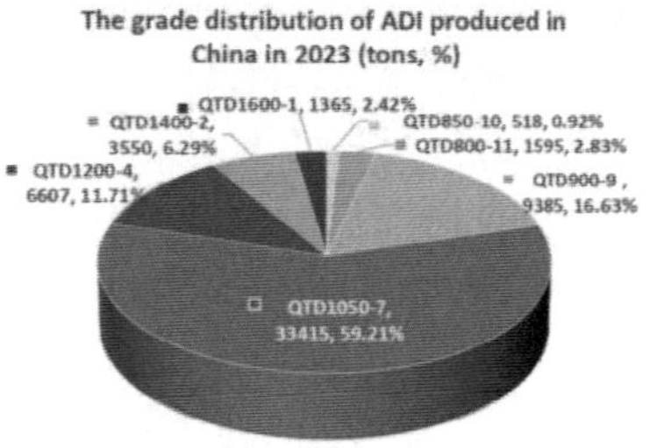

Fig. 2 ADI grades and distribution in 2023.

The application of ADI has progressed from simply replacing ductile iron castings, steel castings and steel forgings to designing lightweight, innovative components. Numerous new ADI components, weighing from a few hundred grams to 4.5 tons, have been used in traditional industries, new energy and new technology industries, reducing weight and cost, enhancing performance, extending service life and improving the environment by saving energy. It is not possible to list them all here.

2 Problems and Prospects of ADI in China

China produced 15.95 million tons of ductile iron in 2021, and 14.95 million tons in 2022, which marks the two highest production years in the history of China. In 2023, ADI production in China was only about 140,000 tons, representing less than 1% of total ductile iron production. This figure is quite low, andshould be increased to 3%, or even more. In addition, structural ADI is less than half the total ADI inChina. Therefore, in China, there is great opportunity for the development and application of ADI, especially for ADI structural parts.

Now, China has excellent conditions to develop high-quality ADI. Some foreign users are willing to cooperate with China to develop new lightweight ADI components, so ADI has a bright future in China.

In China, ADI development capability is still insufficient. It requires close cooperation between universities, designers, casting producers, and heat treaters to continuously develop new lightweight innovativeADI products to meet the demand for high-performance, safety-critical key components.

3 Conclusions

In 2023 in China, there are five commercial ADI treatment service plants and seventeen foundries with world-class ADI treatment furnaces or other austempering furnaces. More than a hundred companies produce ADI parts which are either teat-treated in commercial plants or in their own ADI furnaces.

The current total production of ADI in China is only about 140,000 tons, used in various industries with various grades of ADI and CADI. The ADI casting ratio to total ductile iron production is very low, and there is still considerable growth opportunity, especially for ADI structural castings.

The rate of innovation, and optimization of new, lightweight ADI products is insufficient, and there is an urgent need for cooperation between colleges, design departments, casting manufacturers, and heat treaters units to increase the development of ADI new products. The ADI in China has bright prospects.

Acknowledgments

The authors wish to acknowledge the professors and researchers from universities, research institutions, and enterprises for their contributions to the development of ADI in China. The authors also thank Mr. John Keough and Andy Chen for their introduction of the world's advanced AFC UBQA treatment furnace and for establishing the first commercial heat treatment plant, promoting high quality ADI industrial-scale production in China. Finally, the author must acknowledge all the companies, foundries, and their employees, who produce ADI components. Without their hard work and contribution, this paper would be impossible.

Reference

[1]Gong Wenbang, Liu Jincheng, Xiang Gangyu.Austempered Ductile Iron (ADI) -Theory, Production Technology and Application(in Chinese). China Machine Press, September 2020 ISBN 978-7-111-66450-5.

[2] Yan Qidong etc., Progress of ADI Production Technology Viewed from the Revision of the Standard Foundry (in Chinese), 2024, 73(1): 53-59.

[3] Austempered ductile iron (ADI) castings, GB/T 24733-2009, Standards Press of China 2009.

[4] Austempered ductile iron (ADI) castings, GB/T 24733-2023, Standards Press of China 2023.

Influence of Graphite Morphology on the Frequency Response Function of Brake Discs

G. Alonso[1], B. Bravo[1,*], I. Quintana[2], D. M. Stefanescu[3], P. Rodriguez[2], R. Suárez[1,4]
1. AZTERLAN, Basque Research and Technology Alliance (BRTA), Durango, Bizkaia, Spain
2. EDERTEK Technology Centre, 20500 Arrasate-Mondragon, Gipuzkoa, Spain
3. University of Alabama and Ohio State University, Columbus, OH, USA
4. Veigalan Studio 2010 SLU, Askatasun Etorbidea 16, 48200 Durango, Bizkaia, Spain
* Corresponding address: e-mail: galonso@azterlan.es

Abstract: The automotive industry has evolved hugely in recent years. Components such as brake discs not only ensure the customer safety, but also provide comfort and a sense of well-being. This new concept of quality is defined by an absence of noise which is related to the natural frequencies of the brake discs. This issue is of significant concern in future years. As for the mechanical properties, these natural frequencies are directly influenced by the amount, shape and type of the graphite. To further understand the influence of graphite morphology on the frequency response function (FRF) of brake discs, six gray cast irons with carbon equivalents (CE) between 4.30 and 4.47 were produced. A total of 25 brake discs were cast from each melt. An inoculant rich in Ba was also used. The FRF was measured by an impact test. Graphite flake shape parameters were determined by a commercial image analysis software. Hardness (HB) and tensile strength were also evaluated.

As expected, the type of lamellar graphite changes from A to C when CE increases, decreasing significantly the FRF. This type C is characterized by a lamellae thickening, but also by the precipitation of graphite with starfish shape (kish graphite) and by the formation of big clusters of graphite. A clear inverse relation between the number of primary kish graphite particles and natural frequencies was found. The drop in FRF is also related to a loss of mechanical properties.

Keywords: brake discs; frequency response function; lamellar graphite; kish graphite

1 Introduction

The high level of competition in the automotive world demands that components such as calipers, pads, forks, and brake discs satisfy a series of requirements that guarantee the quality of the vehicles, from the point of view of security and of good feelings of customers. Nowadays, it is not enough to verify certain mechanical properties as requirements related to noise, vibrations and the frequency response of the castings are mandatory.

Gray cast iron is the material most widely used for automotive brake discs because of its excellent thermal and damping properties [1], which are affected by the matrix, chemical composition and graphite flake morphology. This type and distribution of graphite are function of the initial content of carbon in the melt and they are directly related to the natural frequencies of brake discs [2]. The present investigation was designed to clarify the effect of the graphite morphology on the FRF and on the final mechanical properties.

Table 1 FRF, type of graphite and chemical composition (%mass) of experimental cast irons

Melt	C	Si	CE	Mn	S	Cu	Sn	Ti	Type Gr.	FRF (Hz)
1	3.62	2.14	4.33	0.67	0.127	0.29	0.093	0.01	A	987
2	3.73	2.21	4.47	0.68	0.125	0.32	0.112	0.01	A+C	932
3	3.60	2.14	4.31	0.67	0.133	0.17	0.049	0.01	A	987
4	3.60	2.09	4.30	0.56	0.081	0.27	0.100	0.02	A	975
5	3.72	2.18	4.45	0.68	0.082	0.29	0.103	0.02	A+C	927
6	3.73	2.16	4.45	0.68	0.116	0.25	0.062	0.02	A+C	914

2 Experimental Procedure

Six heats were produced in a 12 t, 7000 kw industrial induction furnace. Additions were done directly in a 2 t ladle to obtain the desired compositions. The chemical analysis of the experimental irons is reported in Table 1. CE was calculated as CE= %C+1/3 %Si. From each melt, a series of 25 brake discs were poured. Ladle inoculations was performed with a commercial Ba-inoculant (9% Ba, 1.25% Al, 1.05% Ca). Natural frequencies were measured in all castings by an impact test. Hardness was also evaluated taking four different values per disc. A final average value of FRF was obtained for each heat as listed Table 1. One disc from each melt was used to obtain the tensile strength and another one for metallographic characterization in order to identify the type, area, length and size of the graphite.

3 Result and Discussion

Significant changes are obtained in the microstructure when the CE increases from 4.3 to 4.4-4.5. Type -A graphite, where the graphite lamellae are randomly oriented and uniformly distributed (Fig. 1a), is replaced by type-C graphite typical for hypereutectic compositions showing coarse and straight grey sheets. This thickening of the lamellae is accompanied by the formation of big clusters of graphite (Fig. 1b), as well as by the precipitation of primary particles of graphite with spider/starfish shape, known as kish graphite (Fig. 1c).

Fig. 1 Formation of graphite for different CE: a) type-A (CE=4.30); b) big agglomerates of graphite (CE=4.45); c) kish graphite (CE=4.47)

The proeutectic graphite precipitates (kish) that appear to be formed on the free surface of molten irons, are characterized by a bigger size and a foliated dendritic structure with a substantial tilting and branching of the platelets [3,4]. Their apparition seems to be related to a clear decrease of the frequency response function (FRF) of the castings (Fig. 2). Similar conclusions were obtained when the % area of graphite was analyzed, as the % area increased the natural frequencies decreased.

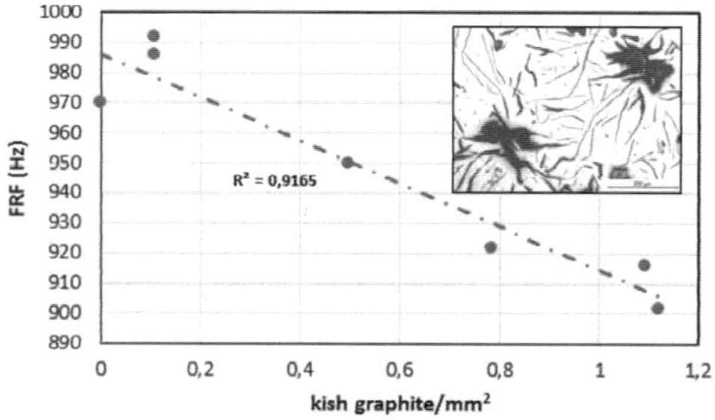

Fig. 2 Evolution of FRF in function of the number of kish particles /mm2

This variation on the natural frequencies caused by a change on the morphology of graphite has also an important effect on the final mechanical properties of the brake discs producing a significant drop of both hardness and yield strength (Fig. 3).

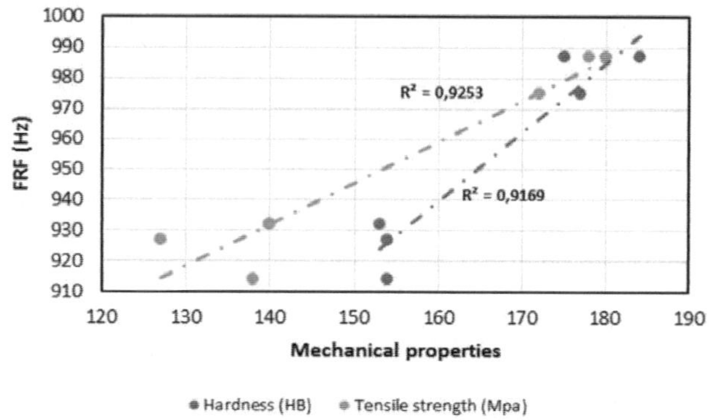

Fig. 3 Evolution of hardness and tensile strength (MPa) in function of FRF

4 Conclusion

The role played by the flake graphite morphology on the frequency response function (FRF) on the brake discs was investigated. Six heats with different CE, three eutectic and three hypereutectic, were analyzed.

The formation of graphite type-C is linked to an increase in CE and to a significant drop of the natural frequencies. The apparition of this graphite includes the formation of big graphite clusters accompanied by the precipitation of spider-shape graphite, known as kish graphite, which is related to a relevant loss of the mechanical properties.

5 Acknowledgments

This work was supported by Diputación Foral de Bizkaia, Spain.

References

[1] Hecht R L, Dinwiddie R B and Wang H. The Effect of Graphite Flake Morphology on the Thermal Diffusivity of Gray Cast Irons used for Automotive Brake Discs. Journal of Materials Science, 1999, 34: 4775-4781.

[2] Lee J Y, Park H D, Kim S J, et al. Effects of Chemical Components and Manufacturing Process of Cast Iron Brake Disc on its Resonant Frequency Variation. SAE International Journal of materials and Manufacturing, 2010 (2): 19-24.

[3] Liu S and Loper CR Jr, "The Formation of Kish Graphite", Carbon, 29 (1991) 547-555

[4] López M, Massone JM and Boeri RE, "Evolution of the Macrostructure of Gray Cast Iron from Eutectic to Hypereutectic Composition", Materials Science Forum, 2018, 925: 110-117.

Novel Processing Techniques and Applications of ADI at CMRDI

Adel Nofal

Central Metallurgical R&D institute, Cairo, Egypt
e-mail: adelnofal@hotmail.com

Keywords: Austempered Ductile Iron (ADI), Structural Refinement, Dynamic Solidification, Cold Working, Thermomechanical Treatment, High Strain Rate Deformation, Ballistic Properties

1 Introduction

Austempered Ductile Iron ADI is a relatively new engineering material with exceptional combination of mechanical properties and marked potential for numerous applications. The attracting properties of ADI return to its distinct and unique microstructure, which consists of spheroidal graphite in a fine acicular ferrite within C-enriched stabilized austenite (ausferrite).

This work is a review of recent research programs aiming at producing ultrafine or nanoscale level microstructures of ADI, which often leads to superior properties. This goal has been achieved applying different processing techniques; dynamic solidification under ultrasonic oscillation, cold rolling as well as thermo-mechanical processing during different stages of austempering cycle.

2 Experimental procedure

Treatment of solidifying iron with ultrasonic oscillations:
A schematic drawing of the experimental setup is shown in fig. (1). The ultrasonic system with potential power of 1Kw with 20 KHz frequency was used. Dilatometry experiments were conducted to produce different grades of ADI and study the influence of graphite nodule count on the ausferritic transformation. Dilatometry tests were performed using Quenching LINEIS L87/RITA dilatometer.

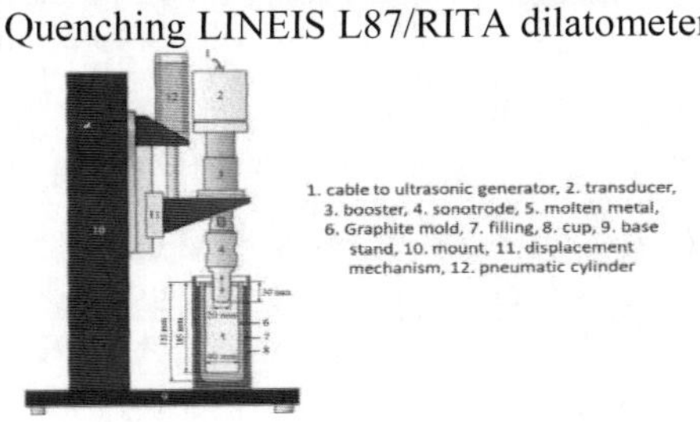

Figure (1): A schematic drawing of the experimental setup for dynamic solidification.

I.

hermomechanical treatment and dilatometric study:
A BaehrDil 80SD thermomechanical simulator was used (fig. 2). The specimens were subjected to two thermo-mechanical schedules, where they were heated up to 960°C and subjected to two deformation steps at 960°C and 940°C. Two types of matrices were produced in the ductile iron, ausferritic and ferrite-ausferritic matrices. The ductile was deformed applying a total true strain of 0.3 in the austenite region one time and 0.2 in the austenite region and 0.1 during austempering (ausforming) other time (fig. 3).

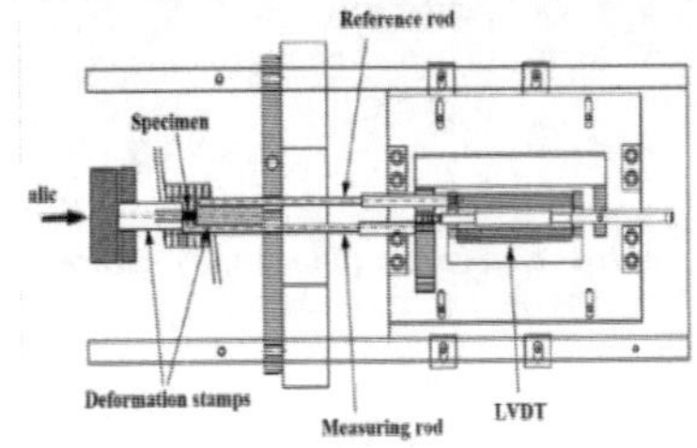

Figure (2): Baehr Dill 805D thermomechanical simulator.

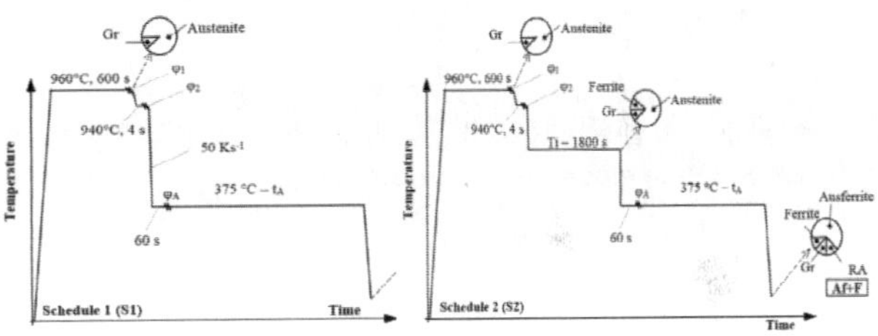

Figure (3): The thermomechanical schedules.

II.

igh strain rate dynamic deformation (crash loading):Tensile tests were performed in a servo hydraulic testing machine at crash-line strain rates ofε^0 = 50 S^{-1}. The elongation of the gauge length of the specimens was measured by a high-speed video camera with a time resolution of 90.000 frames Sec^{-1} (fig. 4).

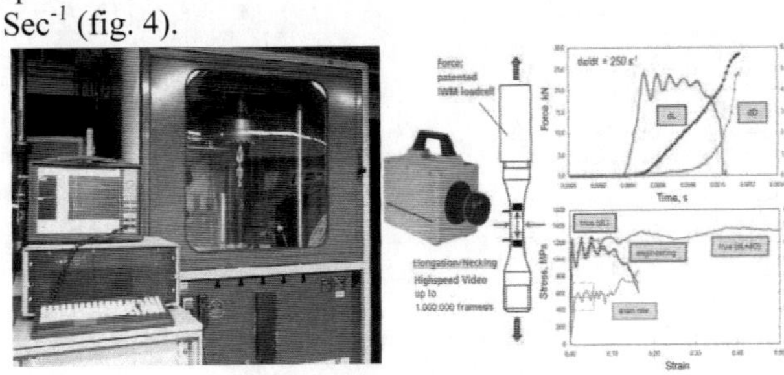

T
Fig. (4): Servo hydraulic testing machine.

III.

3Results and discussion:

Subjecting the solidifying parent ductile iron to ultrasonic oscillations resulted in considerable refinement of the microstructure with nodule count reaching 2000 N/mm^2, fig.

(5), as well as ultrafine ausferritic structure with ferritic needles less than 100 nm, fig. (6).
Dilatometric studies showed that the time required for the completion of ausferritic transformation was four times shorter than that required for statistically solidified ductile iron,fig.(7). Such shortening of austempering cycle would reflect in significant economic production cost [1].

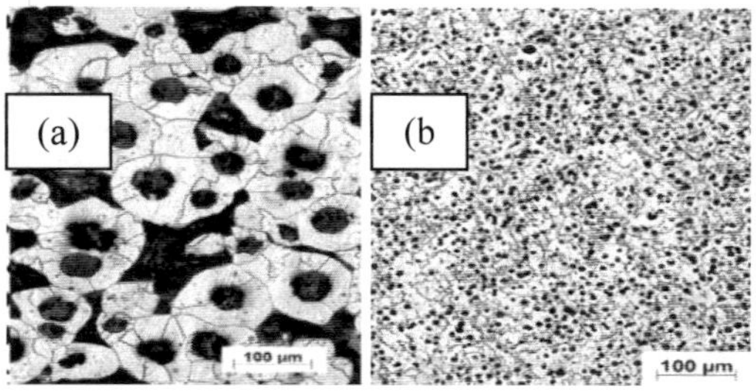

Figure (5): Microstructure of ductile Iron (a) static condition, (b) dynamic condition.

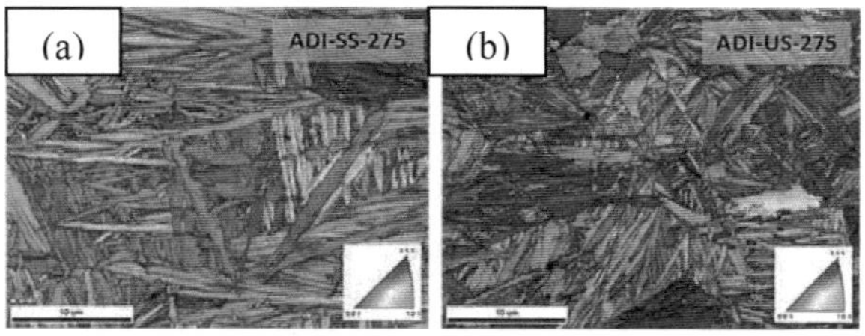

Figure (6): Refinement of ausferritic matrix after ultrasonic treatment, (a) static condition, (b) after ultrasonictreatment.

Cold Rolling (CR) to produce as thin as 3mm sheets, with increasing CR reduction ultimate strength and hardness increased whereas elongation and impact toughness decreased,which was related to the hardening of the investigated ADI with cold deformation bands and twins as

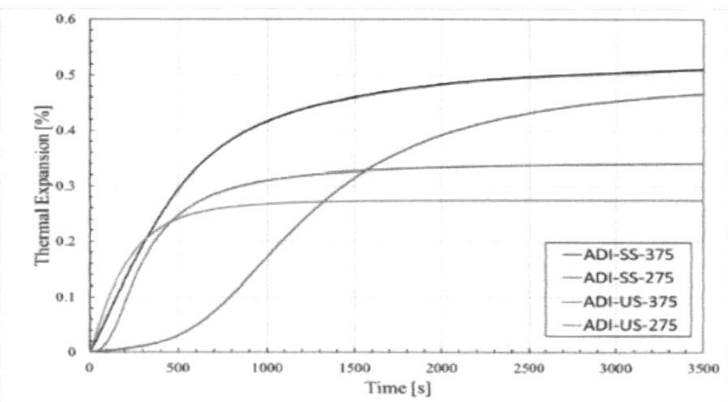

Figure (7): Dilatometry curves of samples subjected to different austempering temperatures.

well as the formation of the deformation-induced martensite[2].
Thermo-mechanical treatment (ausforming) of ADI was introduced during different stages of the austempering. The finer scale and more homogeneous ausferrite resulted dramatic increase in strength, hardness, fig. (8), and wear resistance [3,4].
The report explains how the new developments inmetallurgy have opened new horizons for ADI applications such as high strain rate loads dynamic deformation (crash loading) and ballistic applications[5,6]. Finally, examples of novel applications of ADI castings produced at the experimental foundry of CMRDI such as (b high-performance gears, components of agricultural equipment and static parts of the rolling lines of steel plants are discussed.

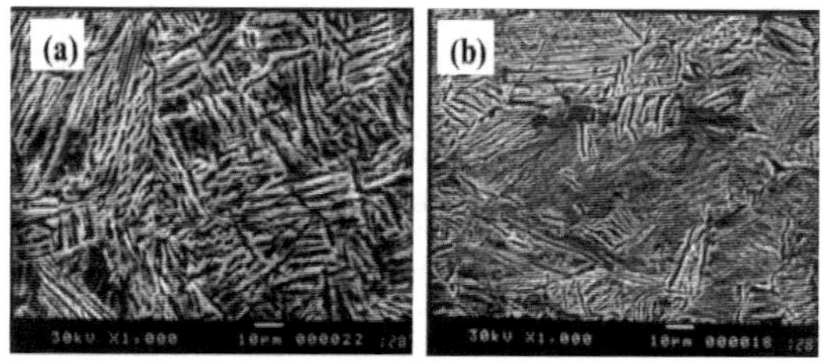

Figure (8):Effect of ausforming on structure refinement of ADI. (a) without ausforming, (b) after ausforming.

References

[1]A. Nofal,M. Ahmed, R. Bähr, V. Rubanik, M. Kovalko, A. Volochko.,Effect of Ultrasonic Vibrations on the Structure Refinement of the Ductile and Austempered Ductile Iron Alloys, International symposium "Perspective materials and technologies", Belarus, 2021.
[2]A.A. Nasr El-din H., Nofal, A., Ramadan, A. and Ibrahim, M., Structure Characteristics and Mechanical Properties of Cold Rolled Austempered Ductile Iron Alloyed with Nickel, Canadian Metallurgical Quarterly, 2004, 43(2), 203-218.
[3] Nofal, A.A., Nasr El-din, H., and Ibrahim, M.M., Thermomechanical Treatment of Austempered Ductile Iron, Proceedings of the 8th Symposium on Science and Processing of Cast Iron SPCI-8, Beijing, China, Oct. 16-19, 2006, 397-402.
[4] Soliman, M., Nofal, A., &Palkowski, H., Effect of Thermo-Mechanical Processing on Structure and Properties of Dual-Phase Matrix ADI with Different Si-Contents, International Journal of Metalcasting, 2020,14, 853-860.
[5] Reissig, L., Nofal, A., Boehme, W., Reichert, T., Al Masry, I.H.,Crash Behavior of ADI Steering Knuckles, International Symposium on the Science and Processing of Cast Iron (SPCI 10), Argentina: Mar del Plata 10 to 13th of November, 2014.
[6] D. Myszka, A. Nofal, M. Ahmed and E. Skolek, High Strain Rate Dynamic Deformation of ADI, Materials Science Forum, 2018, 925,210-217), Trans Tech Publications.

Research on Lost Foam Casting Technique and Wear Resistance of Ductile Iron Reinforced by TiC/CeO_2 Particles

R. D. Li, Y. D. Qu*, G. L. Li, W. Zhang, S. R. Zhang, S. L. Dong, Q. W. Zhou, Y. Zhao
School of Materials Science and Engineering, Shenyang University of Technology, No.111, Shenliao West Road, Shenyang, China
*Corresponding address: quydong@sut.edu.cn

Abstract: To improve the wear resistance of ductile iron, different contents of TiC microparticles and CeO_2 nanoparticles were added through lost foam casting. Their wear resistance was studied, and the results showed that when the addition of TiC particles was 1 wt.% and CeO_2 particles was 0.5 wt.%, the wear resistance of ductile iron significantly improved, with the TiC microparticles addition showing better results. Finally, dual-scale particles were added to the ductile iron using the same method. By comparing the wear volumes of four types of samples, it was found that the wear resistance of ductile iron with dual-scale particles was the best, with a wear volume reduction of 89%. This improvement is related to the decrease in friction coefficient.

Keywords: ductile iron, particles, wear resistance

Introduction

Ductile iron, due to its excellent mechanical and thermal conductivity properties, is widely used in structural components such as crankshafts and pumps[1-2]. However, wear is often the primary cause of failure during service. Heat treatment methods such as isothermal quenching can often enhance the strength, hardness, and wear resistance of ductile iron[3], but they also complicate the production process and increase production costs. Therefore, improving the wear resistance of ductile iron under as-cast conditions is crucial. Adding ceramic particles to ductile iron can enhance its wear resistance[4]. This is because microparticles can enhance the overall hardness of the material, while nanoparticles typically serve as solid lubricants during sliding friction[5]. In addition, nanoparticles can often refine grains and improve the strength and hardness[6]. In this study, TiC was selected as the reinforcing microparticle due to its high hardness and excellent chemical stability. It exhibits good wettability with iron and does not react with it[7]. CeO_2 was chosen as the reinforcing nanoparticle because of its low disregistry with the primary phase (γ -Fe) of ductile iron, which is conducive to promoting heterogeneous nucleation[8]. The lost foam casting technique is an efficacious approach for incorporating particles into ductile iron[9]. TiC and CeO_2 particles are introduced by pre-dispersing them within the foam model. During the casting process, the foam model is vaporized and subsequently removed by a negative pressure system, thereby ensuring that the particles remain in the molten metal[10]. Firstly, the effects of TiC microparticles and CeO_2 nanoparticles on the wear resistance of ductile iron were studied separately to determine their optimal contents. Subsequently, both types of particles were added simultaneously. Finally, the wear resistance of ductile iron with no particle addition, with only microparticles or nanoparticles addition, and with dual-scale particles addition was compared to identify the most effective particles addition method.

Experimental procedure

The addition content of TiC particles was 0 wt.%, 0.1 wt.%, 0.2 wt.%, 0.5 wt.%, and 1 wt.% of the casting, which were named as 0TiC, 0.1TiC, 0.2TiC, 0.5TiC and 1TiC. The addition content of CeO_2 particles was 0 wt.%, 0.25 wt.%, 0.5 wt.% and 0.75 wt.% of the casting, which were named as sample 0Ce, 0.25Ce, 0.50Ce and 0.75Ce. And the casting with dual-scale particles was named as sample Dual. The particles to be added ware pre-dispersed with Fe particles through mechanical milling process. The mixed particles were mixed with pre-expanded polystyrene beads to prepare a casting pattern. The ductile iron was melted in a 50 kg capacity medium frequency induction electric furnace using pig iron, scrap steel and ferrosilicon. The TRUSCO CSA-2000 carbon-silicon analyzer was utilized to determine the initial molten iron composition. A nodulizer (FeSiMgRE, 1.6 wt.%) was added to the ladle, which was covered by an inoculant (FeSi, 0.5 wt.%). The melting temperature was controlled at approximately 1530 ℃ . Following spheroidization, the molten iron was poured into the foam model cluster at 1450 ℃. Finally, ductile iron castings containing various types of particle obtained. Tribological tests were conducted on an MZF-1 tribometer using a ball-on-disk setup. The OLS4100 laser scanning confocal microscope was utilized to reconstruct the three-dimensional wear groove, and its volume was measured.

Result and discussion

Wear resistance of ductile iron with TiC particles

Figure 1 shows the local and typical three-dimensional morphologies of the worn surface. As the addition amount of TiC particles increased from 0 wt.% to 1 wt.%, both of the width and depth of the grooves on ductile iron gradually decreased, with the wear volume decreasing from $14.16\times10^{-2}mm^3$ to $2.56\times10^{-2}mm^3$, a reduction of 82%. This indicates that the wear resistance of ductile iron was significantly improved after the addition of TiC particles.

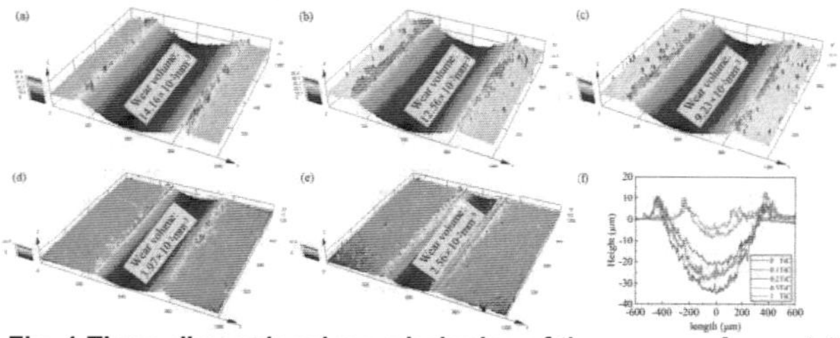

Fig. 1 Three-dimensional morphologies of the worn surfaces: (a) 0TiC; (b) 0.1TiC; (c) 0.2TiC; (d) 0.5TiC; (e) 1TiC; (f) Two-dimensional profile of wear groove.

Wear resistance of ductile iron with CeO2 particles

The three-dimensional reconstruction of the worn surface is depicted in Figure 2. After the addition of CeO_2, both the depth and width of the wear grooves decreased. When the addition amount is 0.5 wt.%, the wear volume was the smallest, only $4.67\times10^{-2}mm^3$, which is about 70% less than that of sample 0Ce, indicating the wear resistance of ductile iron was also significantly improved after the addition of CeO_2 particles, but not to the same extent as with the addition of TiC particles.

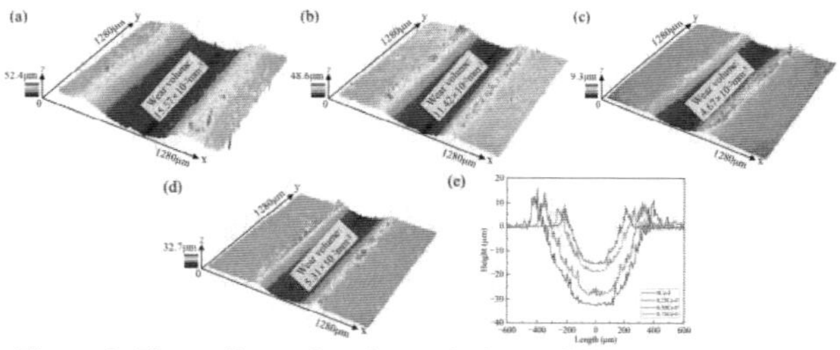

Figure 2. Three-dimensional morphology of the worn surfaces: (a) 0Ce; (b) 0.25Ce; (c) 0.50Ce; (d) 0.75Ce; (e) Two-dimensional profile of wear groove.

Wear resistance of ductile iron with dual-scale particles

When 1 wt.% TiC microparticles and 0.5 wt.% CeO_2 nanoparticles were simultaneously added to ductile iron, and the wear resistance was tested under the same conditions. The three-dimensional reconstruction of the worn surface is shown in Figure 3a. The wear volumes of ductile iron without particle addition, with microparticles, nanoparticles, and dual-scale particles are presented in Figure 3b. It was found that sample Dual the wear volume of sample Dual was the smallest, only $1.59\times10^{-2}mm^3$, which was 89% less than that of sample 0TiC. This phenomenon is closely related to the friction coefficient of friction. After adding different particles, the distance traveled to reach the stable friction coefficient during sliding friction varied, as done the final friction coefficient. The friction coefficient of sample Dual was the smallest, only 0.244.

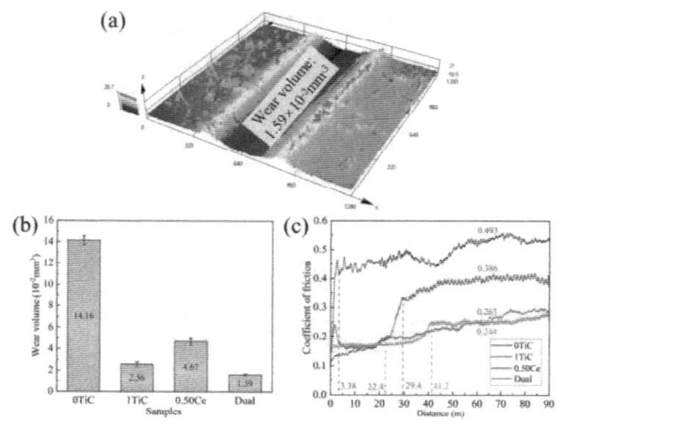

Fig. 3 The wear resistance of ductile iron with different scales particles: (a) Two-dimensional profile of sample Dual; (b) Wear volume; (c) Coefficient of friction.

Conclusion

In this study, different contents of TiC microparticles and CeO_2 nanoparticles were added to ductile iron using lost foam casting. When the content of TiC was 1 wt.% and that of CeO_2 was 0.5 wt.%, the wear resistance of ductile iron was significantly improved, with the microparticle addition showing better results. Subsequently, dual-scale particles were simultaneously added to the ductile iron. By comparing the wear volumes of four types of samples, it was found that the wear resistance of ductile iron with dual-scale particles was the best, with a wear volume reduction of 89%. This improvement is related to the decrease in friction coefficient.

References

[1] Xu C L , et al. China Foundry (2019), 16 (2): 118–125.
[2] Bayata F, et al. Frontiers of Mechanical Engineering, 2023, 18 (3): 251–263.
[3] Gundlach R B. International Journal of Metalcasting, 2020, 14: 1065–1077.
[4] Wang B X, et al. Wear, 2022, 490-491, 204206.
[5] Dong B X, et al. Tribology International, 2023, 177: 107943.
[6] Zhang Y, et al. Journal of Materials Research and Technology, 2023, 24: 653–662.
[7] Qiu F, et al. China Foundry, 2020, 17 (2), 111–126.
[8] Bramfitt B L. Metallurgical Transactions, 1970(1): 1987–1995.
[9] Xiao B T, et al. Journal of Iron and Steel Research International, 2014, 21, 1049–1054.
[10] Zhao Y, et al. Tribology Letters, 2023, 71: 27.

Effect of N Addition on Microstructure and Mechanical Properties of Austenitic Heat Resistant Steel

Guanyu Jiang1, Mengwu Wu1,*, Xiaoguang Yang2, Hui Wang2, Yuyuan Zhu2
1. School of Automotive Engineering, Wuhan University of Technology, Wuhan 430070, China
2. Wescast Industries (China) Co., Ltd, Wuhan 430100, China
*Corresponding address: e-mail: wumw@whut.edu.cn

Abstract: Four types of steels were designed on the basis of GX40CrNiSi25-12 austenitic heat-resistant steel by adding N and reducing Ni contents. The effect of N replacing Ni on the microstructure and mechanical properties of austenitic heat-resistant steels was investigated. Results show that all four test steels have austenite and $M_{23}C_6$ carbides in their microstructure. After the addition of N, the carbide NbC in the test steels is transformed into Nb(C,N), and the amount of $M_{23}C_6$ carbides is reduced. Due to the solid solution strengthening and precipitation strengthening of the N, the mechanical properties of the test steels are improved, and the tensile strength, yield strength and elongation are increased by 19%, 16% and 72%, respectively.

Keywords: austenitic heat resistant steel; nitrogen; microstructure; mechanical properties

1 Introduction

Austenitic heat-resistant steels have been widely used in automotive, electric power, chemical and other industrial fields due to its advantages of excellent high-temperature performance, oxidation resistance and corrosion resistance[1]. It is necessary to add a large amount of Ni element to stabilize the austenite structure in austenitic heat resistant steel. However, the high cost of Ni limits its application. N element is also an austenitising element, which can stabilise the austenite structure in austenitic heat-resistant steels[2]. Therefore, it is of great significance to study the effect of using N instead of Ni on the microstructure and mechanical properties of austenitic heat-resistant steels.

2 Experimental procedure

Scrap returns with similar composition to GX40CrNiSi25-12 were melted in a medium frequency induction furnace. The N content of the liquid metal was adjusted by adding Ferrochrome Nitride. Other intermediate alloys were added to adjust the composition of the experimental steel. When the composition of the metal liquid was uniform, the metal liquid was poured into the Y-block at 1620℃~1660℃. The final composition of the test steels was shown in Table 1. The mechanical properties of the steels were tested by using an electronic universal testing machine and a HV-10004 Vickers hardness tester. Microstructure characterization was conducted by using a ZEISS Scope A1 optical microscope (OM). Phase analysis was carried out on a SmartLab SE X-ray diffractometer (XRD).

Table 1. Composition of the test steels (wt.%)

Steels	C	Si	Mn	Ni	Cr	Nb	N
N0	0.38	1.41	2.00	11.65	24.65	0.82	0
N0.26	0.41	1.36	1.53	9.24	23.16	1.00	0.26
N0.30	0.38	1.27	1.25	8.68	23.26	0.87	0.30
N0.32	0.36	1.27	1.54	8.82	23.67	0.86	0.32

3 Result and discussion

Microstructure characterization

From the XRD results in Fig. 1, it is clear that austenite, $M_{23}C_6$ carbides are present in all four test steels. MX is NbC phase in the N0 test steel, and MX is Nb(C,N) phase in the three N-substituted Ni test steels. When N element is used to replace Ni element in the austenitic heat-resistant steels, the NbC phase in the test steels is converted to Nb(C,N) and the peak intensity of the $M_{23}C_6$ phase is reduced.

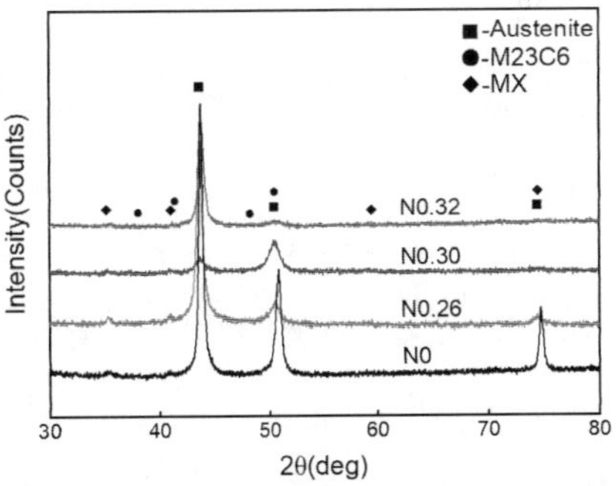

Fig.1 XRD patterns of the four types of steel

Figure 2 shows the microstructure of the four test steels. It can be seen that all test steels have a dendritic structure, consisting of an austenitic matrix and precipitated phases between the dendrites. Under the optical microscope, the austenite phase is in light gray, while the precipitations are in dark black. The degree of aggregation of the precipitated phases increases slightly after the substituting Ni with N, and the number of precipitated phases remains basically unchanged.

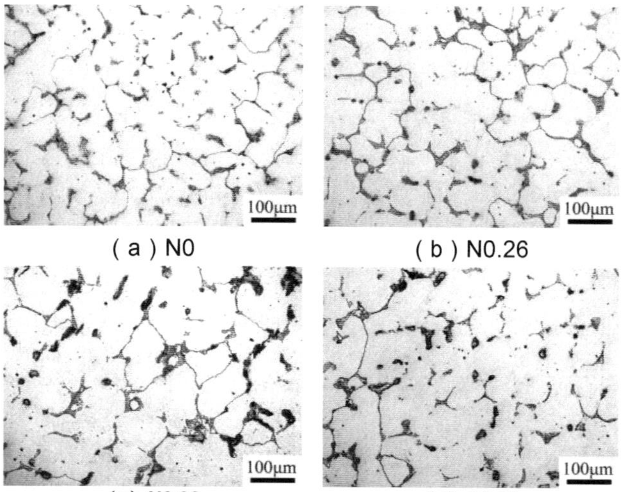

(a) N0 (b) N0.26
(c) N0.30 (d) N0.32

Fig.2 OM images showing microstructure of the steels

Mechanical properties

Figure 3 shows the mechanical properties of the test steels with different N and Ni contents. When the N content of the test steel increases from 0% to 0.32%, the tensile strength increased by 19%, the yield strength increased by 16%, the elongation increased by 72%, and the hardness increased by 9.2%.

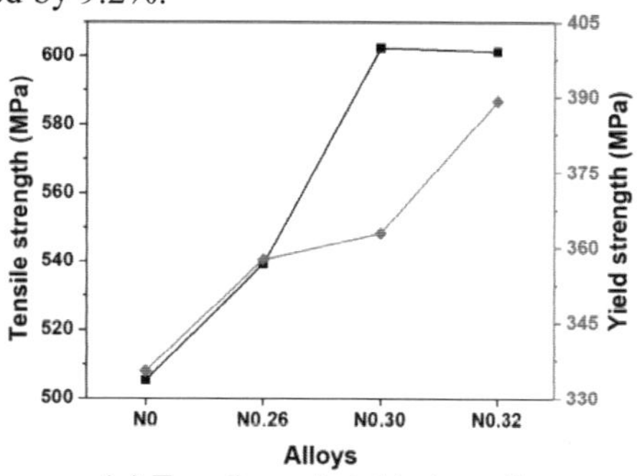

(a) Tensile and yield strength

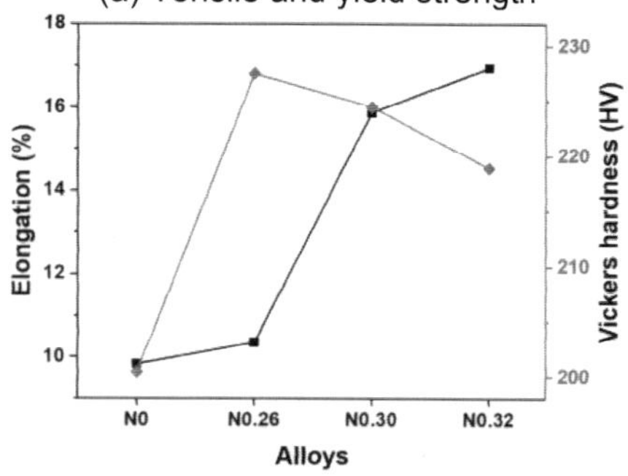

(b) Elongation and hardness

Fig.3 Mechanical properties of the steels

After the substituting Ni with N, the matrix of the test steels is still a fully austenitic structure. N is an interstitial solid solution element, which increases the strength of the austenitic heat-resistant steel due to the strong solid solution strengthening effect. After replacing Ni with N, the NbC phase in the test steels is transformed into the precipitation of the Nb(C,N). The fine lamellar Nb(C,N) between the dendrites can effectively pin the grain boundaries and prevent the grain boundaries from slipping during deformation, which is conducive to improving the strength of the test steel between the dendrites[3]. At the same time, the Nb element preferentially combines with the N element to precipitate as Nb(C,N), which inhibits the precipitation of $M_{23}C_6$ carbides and weakens the cutting effect of large-size $M_{23}C_6$ carbides on the matrix.

3 Conclusion

The effect of substituting Ni with N on the microstructure and mechanical properties of austenitic heat-resistant steel was investigated. It is found that the tensile strength, yield strength, and elongation of the test steels increase gradually. After the addition of N element, NbC in the test steels is transformed into Nb(C,N) phase, and the amount of $M_{23}C_6$ carbides between dendrites is reduced, which reduces the cutting effect of large-size $M_{23}C_6$ between dendrites relative to the matrix. Meanwhile, the mechanical properties of the test steels are also improved due to the solid solution strengthening effect of N.

4 Acknowledgments

This work was financially supported by the National Natural Science Foundation of China (Grant No. 52275370), and the Key R&D Program of Hubei Province, China (Grant No. 2022BAD100).

References

[1] Huang J, Wang J, Yang L, et al. Comparative study on microstructure and mechanical properties of a novel nano-composite strengthening heat-resistant steel and two typical heat-resistant steels[J]. Materials Today Communications, 2023, 36: 106679.

[2] Lee Y, Lee J H, Moon J, et al. Correlation Between Constituent Phase and Weld Metal Properties in Ni-Reduced Duplex Stainless Steel[J]. Metals and Materials International, 2022, 28(8): 1971-1980.

[3] Zhang Y, Li M, Godlewski L, et al. Creep Behavior at 1273 K (1000 A degrees C) in Nb-Bearing Austenitic Heat-Resistant Cast Steels Developed for Exhaust Component Applications[J]. Metallurgical and Materials Transactions A, 2016, 47(7): 3289-3294.

Simulation of Electro-slag Overflow Forming and the Appearance of the Isolated Liquid Phase

Heng Cao*, Yu Wang, Yanchun Lou
National Key Laboratory of Advanced Casting Technologies, Shenyang, China
*address: No.17 Yunfeng South Street, Tiexi District, Shenyang, China. E-mail: caohengheng123@gmail.com

Abstract: The electroslag remelting process integrates 3 crucial stages, namely, refining and purification, solidification and crystallization, and final shaping, resulting in products that possess not only superior solidification quality but also exhibit shape dimensions closely approaching or replicating those of the final product. While electroslag remelting has found extensive commercial utilization for simple non-uniform castings, it remains a challenge to employ this technique for manufacturing three-dimensional variable curved surface components lacking a common-type mold. One proposed solution involves the strategic insertion of multiple electrodes at locations with a common-type mold, facilitating the overflow of molten steel and its horizontal flow around the electrodes, thus termed as "electroslag overflow forming." To ascertain the rational range of process parameters for effective electroslag overflow forming, a comprehensive investigation is conducted through a combination of experimental study and numerical simulation. The outcomes of this study demonstrate that, when the input current is 8000A and the input voltage is 75V, the overflow molten steel effectively fills the designated 250mm*50mm overflow region and exhibits commendable filling quality upon solidification.

Keywords: ESRC; no common cavity problem;

1 Introduction

Electro-slag casting[1] is a near-net shape manufacturing technology developed on the basis of electric slag remelting. The products using this technology are single-piece, requiring minimal post-processing, with mechanical properties similar to forged components and high surface quality.

Electro-slag overflow forming is a shaping technique that, for lacking a common cavity, relies on the overheating of the molten steel itself to extend its flow into the overflow area, as the electrode is difficult to insert into the crystallizer's internal cavity.

The intense cooling conditions within the water-cooled crystallizer cavity continuously dissipate heat during the forward flow of molten steel. The extensibility of molten steel in the overflow area is limited by the input power[2].

This study employs numerical simulation methods to investigate the extendability of molten steel in the overflow channel and determine the most suitable casting process parameters under experimental conditions. The parameters derived from the simulation are subsequently validated through experiments to ensure their rationality.

2 Experimental procedure

In order to investigate the spreading capacity of molten steel within overflow channels, a dual-electrode electric slag overflow model, as depicted in Figure 1, was designed. At both ends of the model are electrode channels used to introduce the electrodes into the bottom of the slag bath. The deep blue segments represent the self-consuming electrodes. In the middle of the two electrode channels, there is a slender overflow channel measuring 50mm by 250mm. The pale blue portions signify the already solidified steel ingot, while the deep red areas denote the slag bath and the molten pool.

To investigate the extension capacity of molten steel in overflow channels under different input power levels, a controlled variable method was employed for grouping during the simulation. Initial melt speed and voltage were kept constant, while the overflowing of electric slag was simulated at various current levels. Table 1 presents the electrical parameters utilized in the simulated experiments, with a current frequency of 50Hz.

Table 1 The simulation employs electrical parameters and input power.

Current (DC)/ A	Voltage / V	Starting Melting Rate/ $kg{\cdot}s^{-1}$	Power/ kW
8000	75	0.015	600
6000	75	0.015	450
3000	75	0.015	225

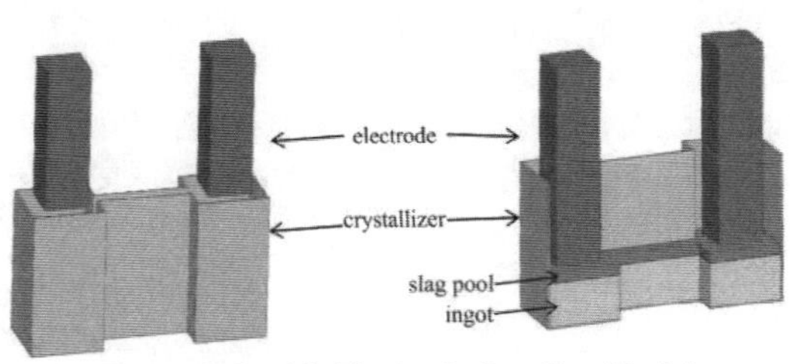

Figure 1 Double Electrode Overflow Model.

3 Result and discussion

In the overflow channel, there is an isolated liquid phase region at a certain distance from the leading edge of molten steel movement in the direction of flow. Typically, as molten steel moves away from the heat source within the overflow channel, its temperature gradually decreases, and its mobility decreases as well, ultimately leading to sequential solidification under intense cooling conditions, without the occurrence of isolated liquid phase regions similar to sand casting processes. However, simulation results indicate that within the overflow channel, while molten steel farthest from the electrode (heat source) has not yet solidified, molten steel closer to the electrode (heat source) solidifies first.

Extracting a section of the overflow channel in the model and analyzing the formation process of slag overflow within the overflow area, as depicted in Figure 2

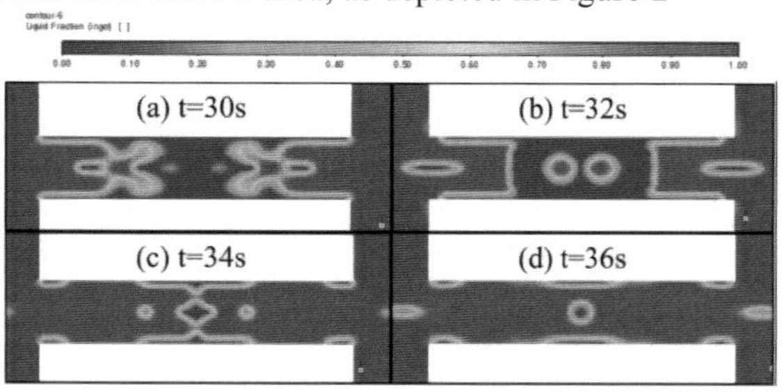

Figure 2 Liquid Phase Distribution of Overflowing Molten Steel.

Figure 3 illustrates four stages of slag overflow at the center of the overflow channel, divided into phases a to d.As shown in Figure 3a, molten steel from the overflow flows toward the center from the two entrances of the overflow channel simultaneously, with decreasing superheat as it gets closer to the center of the channel.When the input power is insufficient, both streams of molten steel start solidifying before they meet.As shown in Figure 3b, since the molten steel solidifies and does not advance further, the liquid slag phase between the two solidified steel streams begins to solidify, forming a pasty slag zone.As shown in Figure 3c, as the solid slag phase begins to develop, new liquid metal flows along the solid steel phase. The separation between the new molten steel and the slag shell attached to the interior of the crystallizer, created by the solid metal phase and solid slag phase, weakens the intense cooling conditions provided by the crystallizer. This allows the new molten steel to flow to the center of the channel and solidify once more.

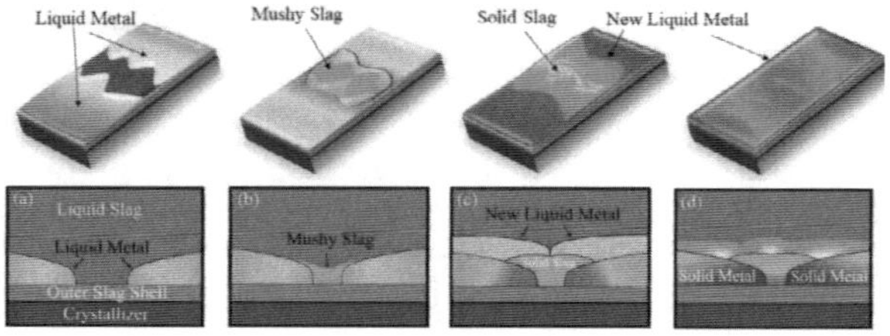

Figure 3 The Formation Principle of Isolated Liquid Phase Regions.

4 Conclusion

In this paper, an ESR overflow forming method for complex components without common cavity was proposed, and the extension of molten steel in the overflow channel was studied by numerical simulation. The numerical simulation results show that the flow of molten steel with low superheat stops earlier, and the front molten slag cools earlier, which results in non-uniform cooling of the subsequent molten steel solidification. Based on the above, the formation mechanism of isolated liquid phase region in the process of low power electroslag overflow was proposed. The experiment of overflow channel with input power of 450-520kW and 50*250mm verifies the applicable electrical parameters and realizes the completed forming.

Acknowledgments

The authors appreciate the financial support by the National Nature Science Foundation of China(No. 52175352).

References (Bold, 10 pt., Arial)

[1] Li Z B. Electroslag Metallurgy Theory and Practice. Beijing: Metallurgical Industry Press, 1,1996,15

[2] Ginzburg V.L., Landau L.D. Towards the theory of superconductivity [J]. Original in Journal of Experimental and Theoretical Physics, 20,1950, 1064-1082.

Study on Electroslag Casting and Welding Of Special Ingot

Yu Wang, Heng Cao, Yanchun Lou
National Key Laboratory of Advanced Casting Technologies, ShenYang, China
*Corresponding address: 17 Yunfeng South Street, Tiexi District, Shenyang, China
e-mail: 593339375@qq.com

Abstract: The Electro-Slag Remelting (ESR) process has been used widely to produce high quality ingots based on the controlled solidification and chemical refining. With the increase of ingot size, especially for super large high alloy ingot, the traditional ESR method still has limitations. The key problems were: large equipment size, poor ingot composition uniformity, and difficult to control the solidification quality. In this study, based on the additive manufacturing concept, a new method of Electro-Slag Remelting Casting-Welding of large high alloy ingot was proposed. How to control the shape of the fusion zone and the solidification quality of the fusion zone is the key problem to be solved in the production of large ingots by ESR Casting-Welding, to control the overall quality of ingots. First, based on the distribution of temperature field in fusion zone, a model of energy equidensity for molten bath was developed to achieve an uniform structure density in fusion zone. In addition, the conductive electroslag remelting technology of crystallizer is proposed to improve the forming quality and solidification structure of ingot. Furthermore, according to the designed experimental model and the simulated process parameters, the electroslag casting welding experiment is carried out, and the simulation results are compared.

Keywords: electroslag casting-welding, fusion zone

1 Introduction

The Electro-Slag Remelting (ESR) process has been used widely to produce high quality ingots based on the controlled solidification and chemical refining [1]. ESR can replace electric furnace melting and sand molding of traditional casting [2]. With the increase of ingot size, especially for super large high alloy ingot, the traditional ESR method still has limitations. The key problems were: large equipment size, poor ingot composition uniformity, and difficult to control the solidification quality.

In this study, based on the additive manufacturing concept, a new method of Electro-Slag Remelting Casting-Welding of large high alloy ingot was proposed. How to control the shape of the fusion zone and the solidification quality of the fusion zone is the key problem to be solved in the production of large ingots by ESR Casting-Welding, to control the overall quality of ingots.

2 Experimental procedure

ZG0Cr13Ni4Mo was the material of mould conductive electroslag remelting, electroslag casting-welding and physical simulation experiments. Two conduction modes, traditional and double-circuit, are designed, which can be realized by the switch on and off of S1-S3 in Figure 1. The physical model of electroslag casting welding as shown in Figure 2. The governing equations and boundary conditions used in the calculation refer to previous studies [3].

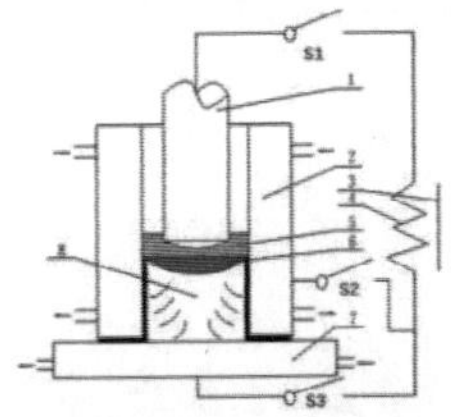

Figure 1. Schematic of traditional and double-circuit

Figure 2. Physical model of electroslag casting welding

3 Result and discussion

Mould conductive electroslag remelting

The first step of electroslag casting welding was to prepare basic ingot by electroslag remelting. The control of molten pool morphology and columnar crystal production direction is particularly important to improve the solidification quality of the base ingot, which can be achieved by forming a conductive circuit in the mold. The distribution law of its temperature field is shown in Figure 2(a-traditional, b-double-circuit), and the melting morphology of the electrode tip at the stable stage is shown in Figure 3(c-traditional, d-double-circuit). In the traditional conduction mode, the melting morphology of the electrode tip is a shallow cone, and it is conical in the isothermal range of 1500-1800℃ in the temperature field distribution. The high temperature region covers the actual morphology of the melting tip and expands outward. In the double-circuit conduction mode, the melting shape of the electrode tip is shallow and flat, and the temperature field is narrow longitudinally.

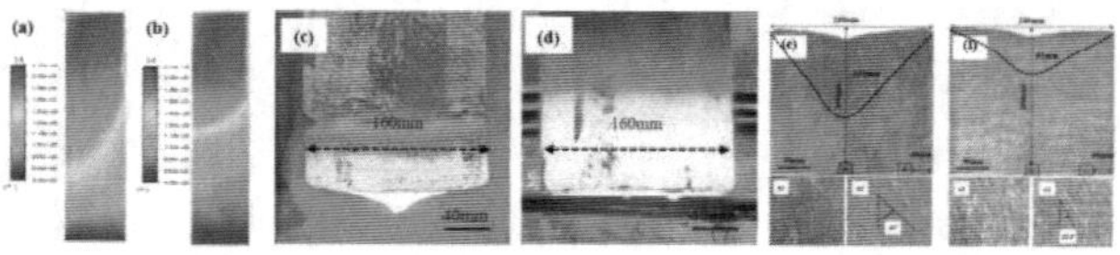

Figure 3. Results of mould conductive ESR

The metal melt pool morphology of the two kinds of conductive ingots is shown in Figure 2(e-traditional, f-double-circuit), and the depth of the metal melt pool is 133mm and 91mm, respectively. The growth direction of the columnar crystal was observed by sampling at the 2/3R position of the ingot. The angles between the columnar crystal growth direction and the axial direction were 45° and 23.5°, respectively. The smaller the angle between the growth direction of columnar crystals and the ingot axis, the shallower the molten pool. The results show that the temperature of the slag pool and the melting amount of the slag skin are increased by the mold branch during the remelting process, which makes the temperature field distribution of the slag pool more uniform and improves the temperature field distribution of the whole system.

Electroslag casting welding simulation

The distribution results of temperature field at different times in the process as shown in Figure 4. The temperature is unevenly distributed and reaches the highest at the tip of electrode in casting and welding area below the slag pool, then decreases outward. The surface area of the electrode inserted into the slag pool in this region is the largest. The difference in joule heat results in a great temperature difference between the slag region and the slab ingot region. In the molten metal pool and the solidification ingot, the temperature decreases along the axial direction with the lowest temperature appearing at the periphery of the solidified bottom. The reason is that the crystallizer cooling plays a paramount role in creating a large radial temperature gradient and makes the temperature of slag near to the mold lower. The maximum temperature in the model can reach 2,087 K, while the minimum temperature is 370 K. With the progress, the high temperature region of the fusion zone between the basic ingot and the cast-welded ingot gradually expands, and the high temperature region is like a parabola. The reason is that with the process of casting and welding, the heat accumulates at the bottom of the electrode inserted into the slag pool.

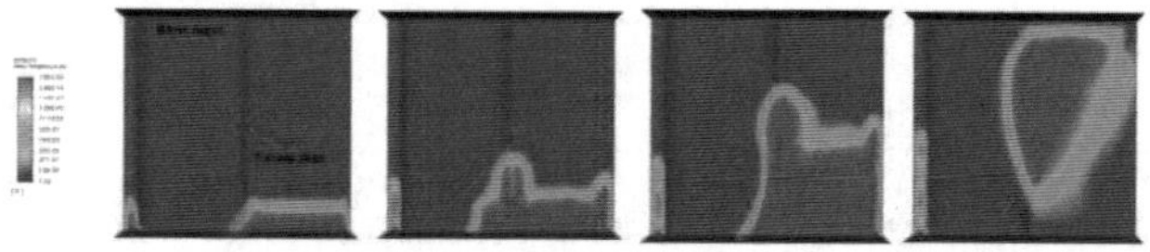

Figure 4. Results of temperature field in the process

4 Conclusion

In this paper, an electroslag casting-welding method of high alloy ingot was developed through combining electroslag casting and electroslag welding. This method can reduce the composition segregation caused by the large size of the metal melt pool, which is better than the microstructure of the mold casting ingot. The uniform microstructure and properties of the fusion zone can be achieved through the conduction technology of the mold and the control of electrical parameters, which effectively ensures the integrity of the ingot made by electroslag casting-welding. At the same time, the batch forming manufacturing mode of the method can reduce the equipment limitation required for the manufacture of medium and large ingot.

Acknowledgments

Appreciate the financial support by the National Nature Science Foundation of China(No. 52175352).

References

[1] Li Zhengbang . China Foundry, "Development of electroslag metallurgy and casting in China", 1,2004, 10.

[2] Li Baokuan, Wang Fang, Chen Mingqiu, et al. Journal of Iron and Steel Research (International), "The Cellular Automata Model of Electroslag Remelting Ingot Structure" 18, 2011, 159-165.

[3] Wang Fang, Wang Qiang, Lou Yanchun, et al. JOM: The Journal of The Minerals, Metals & Materials Society, "Investigation of Heat Transfer and Magnetohydrodynamic Flow in Electroslag Remelting Furnace Using Vibrating Electrode", 1, 2016, 410-420.

Fracture Analysis of QY26N Heat-Resistant Steel Exhaust Pipe Volute Casting

Xianzhi Li1, Han Deng1, Jianfeng Cao1 , Yufeng Qi
1. CRRC QiShuYan Institute Co.,Ltd., No.258 Wuyi Road,Changzhou City, Jiangsu Province, China
*Corresponding address: e-mail: lixianzhi.qs@crrcgc.cc

Abstract: The exhaust pipe volute often has problems such as cracks and deformation during the casting production process. This article analyzes the crack defects of exhaust pipe volutes using methods such as X-ray inspection, penetration inspection, scanning electron microscopy, and casting stress simulation. The results show that the maximum tensile stress at the crack site is significantly higher than the high-temperature yield limit of the material, and there is segregation of niobium elements and precipitation of dendritic niobium carbide at the defect site, resulting in the formation of hot tearing.

Keywords: heat-resistant steel, fracture analysis, exhaust pipe volute casting, segregation of niobium

1 Introduction

QY26N heat-resistant stainless steel has become the excellent material for hot end components such as integrated exhaust pipe volutes in turbochargers due to its excellent high-temperature mechanical properties, corrosion resistance, and low cost. The use of film coated sand shell casting technology for the production of integrated exhaust pipe snail shells can significantly improve the production efficiency and market competitiveness of enterprises. The shell casting process of QY26N heat-resistant steel snail shell is a complex dynamic solidification process, which is prone to casting problems such as structural segregation, compositional segregation, shrinkage porosity, cracks and hot tears during the casting process. This article analyzes the crack defects of exhaust pipe volutes .

2 Experimental procedure

Phenomenon description

The exhaust casing produced has linear crack particle marks on the machining surface during crack particle inspection. As shown in the part blank quality inspection drawing in Figure 1, the material is changed from 1.4837 heat-resistant stainless steel to QY26N. The red area is the location where the linear magnetic trace exceeds the standard. Based on Figure 1, it can be inferred that the defect occurs in the area of sudden wall thickness change on the casting.

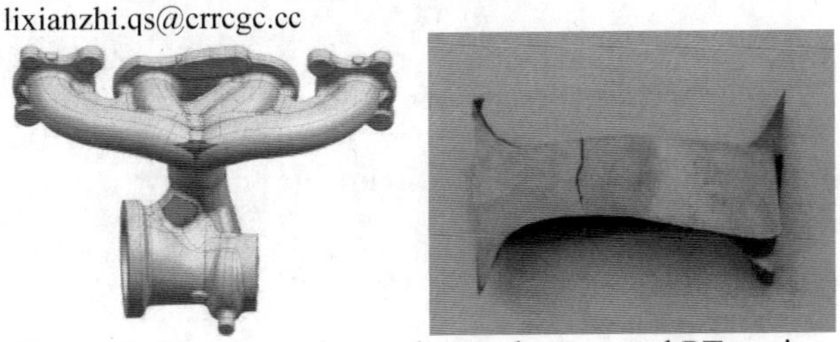

Figure 1: Linear crack area in rough parts and PT testing

3 Result and discussion (Bold, 10 pt., Arial)

Chemical composition detection

The chemical composition was extracted from the matrix structure near the defect of the exhaust casing, and the measured values were compared with the QY26N composition range. It was found that the Nb element content near the defect was extremely low, and there was segregation of Nb elements in the defect area (up to 12.23%); The remaining chemical elements are within the composition range of QY26N.

Table 1. Composition of QY26N exhaust casing

QY26N	C	Si	Cr	Ni	Nb
Composition range	0.3 ~ 0.5	1.0 ~ 1.5	21 ~ 23	9 ~ 11	0.5 ~ 0.9
matrix structure	0.42	1.41	22.35	9.24	0.65
segregation of Nb	/	2.16	23.02	8.81	12.23

Simulation analysis of solidification temperature field

Simulate the casting process of exhaust casing made of QY26N material to investigate the temperature changes and residual isolated melt distribution at the defect location of the partition during the solidification process. The temperature field from the end of pouring to the end of solidification is shown in Figure 2, and the red circle in the figure represents the local high-temperature area. As the solidification process progresses, the elliptical red circle first forms a residual isolated liquid phase, followed by the rectangular red circle area. Due to the inability to obtain riser shrinkage in the isolated liquid phase region, there is a relatively high tendency for looseness in the area shown in the coil.

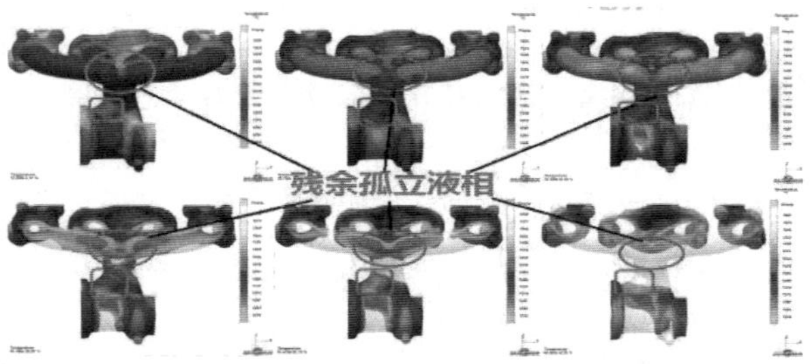

Figure 2: QY26N Solidification Temperature Distribution

Simulation of hot spot and shrinkage porosity

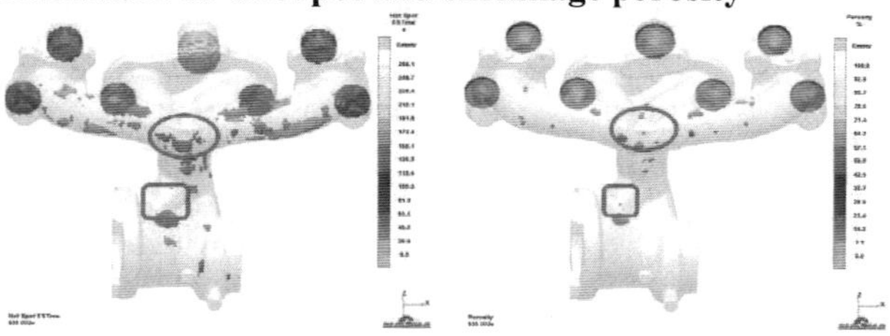

Figure 3: exhaust casing hot spot and shrinkage porosity

Casting thermal stress and cracking prediction

It is necessary to analyze the maximum tensile stress during the solidification process of castings. Figure 4 shows the distribution of maximum principal stress during the solidification process of the casting. When the casting solidifies for 60 seconds (30% solidification), the two parts of the circle bear tensile stress, with a maximum tensile stress of 36 MPa, which is higher than the stress limit that the casting can withstand during the solidification process. At this point, cracks will appear in the casting.When the casting solidifies for 110 seconds (60% solidification), the maximum tensile stress on the two parts of the circle is 55MPa, and the cracks in the casting will expand accordingly.

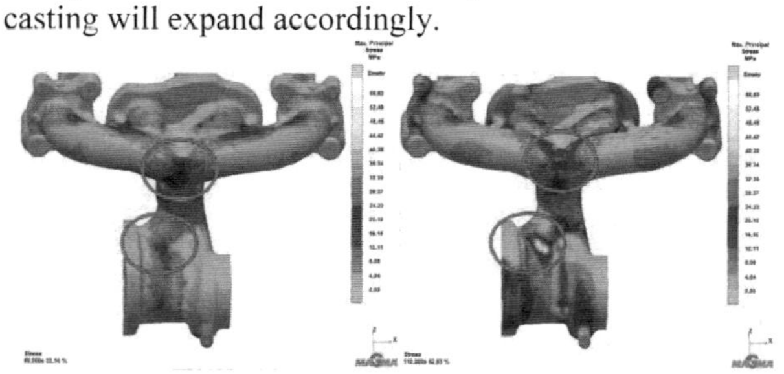

Figure 4: Maximum principal stress distribution

Observation by scanning electron microscopy

ETD-SE secondary scanning mode to observe the defects in Figure 1 at high magnification using scanning electron microscopy. The result was identified as a casting hot crack. There are dendrites present at the micro cracks of the casting, and the cracks propagate along the grain boundaries in a tortuous manner. Dendritic carbides can be observed on both sides of the cracks.

元素	重量百分比	原子百分比
N K	0.03	0.04
O K	14.23	39.52
Al K	0.15	0.25
Si K	3.39	5.37
Cl K	0.52	0.65
Ca K	0.04	0.05
Cr K	23.57	20.15
Mn K	3.44	2.78
Fe K	45.69	36.37
Ni K	6.78	5.13
Cu K	4.50	3.15
Nb L	2.27	1.08
总量	100.00	

Figure 6: Morphology of cracked oxide

4 Conclusion (Bold, 10 pt., Arial)

1. Crack display area forms a little isolated melt.
2. Cooling and crystallization rate of the cracked area is slow, and this area is subjected to casting tensile stress.
3. defects in the areas where the magnetic traces exceed the standard are casting hot cracks. The hot cracks exhibit a high-temperature oxidation morphology of the matrix, and there is Nb element segregation near the hot cracks. Dendritic structures and niobium carbide aggregation appear in the segregated areas.

References (Bold, 10 pt., Arial)

[1] Hui Ruituo, Xu Jinfeng, Research on the Thermal Fatigue Properties of Cr26Ni14Nb2 Heat Resistant Stainless Steel [C]. 2013 China Foundry Activity Week Collection. Jinan.

[2] Li Jie, Wang Jingcheng, Zhao Fangxin, et al. Causes and Countermeasures of Cracks in Duplex Stainless Steel Castings [J]. Casting, 2015 (6): 590-592

Microstructure and Properties of Compacted Cast Iron Inlaid with Carbon Steel

Fenglong Hai, Han Deng, Rui Wang, Bin Zhang
CRRC Qishuyan Institute Co. Ltd, No.258 Wuyi Road Qishuyan, Changzhou, Jiangsu, China
*Corresponding address: e-mail: haifenglong.qs@crrcgc.cc

Abstract: The microstructure and mechanical properties of compacted cast iron inlaid with carbon steel were studied by metallographic microscope, scanning electron microscope, micro-Vickers hardness tester and electronic universal testing machine. The results show that the metallurgical bond between compacted cast iron and carbon steel is formed by atomic diffusion. The thickness of the transition region of the bond is about 304μ m, which is full pearlite structure. The transition zone is divided into two sub-layers. The pearlite lamellae on the carbon steel side have a larger spacing and have a hardness between 231HV and 344HV; the pearlite lamellae on the compacted cast iron side have a smaller spacing and have a hardness between 344HV and 362HV. The shear strength of the compacted cast iron and carbon steel composite structure is 317MPa.

Keywords: compacted cast iron; carbon steel; insert casting; mechanical properties

1 Introduction

Cast iron and carbon steel have their own advantages and disadvantages. Because of the similar microstructure, the two can be combined into one by welding, casting, etc., so as to obtain a better overall performance of the composite parts. However, the welding performance of cast iron is poor, and it is easy to become white cast iron when welding. Therefore, many researchers use composite casting to connect cast iron and carbon steel, and can achieve solid-liquid composite, liquid-liquid composite and other composite casting process. Compacted cast iron is widely used in automotive, rail transport and other fields because of its excellent thermal fatigue properties, but its toughness and plasticity still limit its application. Therefore, this study tries to use the inlay casting process to achieve the metallurgical combination of the two, and at the same time to study the bonding properties of compacted cast iron and carbon steel for the production of compacted cast iron-carbon steel composite castings to provide a basis for the production and application.

2 Experimental procedure

The grade of the peristaltic cast iron is RuT350 and the grade of the carbon steel is 45 steel. The size of the 45 steel is $30\times10\times20mm^3$ and the volume ratio of the compacted iron to the carbon steel is 70:1. Two pieces of the pre-treated 45 steel are fixed in a standard 25mm thick Y-shaped test block. Then the pregnant and creep-treated iron was poured into the Y-shaped test block. The pouring temperature was 1400°C. After cooling the samples were prepared and analysed.

3 Result and discussion

Metallographic microstructure

The matrix organisation of carbon steel is ferrite + 25% pearlite. The composite interface region is a good metallurgical bond with a fully pearlitic organisation. The matrix organisation of compacted cast iron is ferrite + 45%pearlite. In the pearlitic organisation of the composite interface, it is obvious that there is a relatively straight line separating the two sublayer structures, in which the thickness of the pearlitic layer on the carbon steel side is about 251μm, and the thickness of the pearlitic layer on the other side is about 53μm. In addition, the graphite morphology of the compacted cast iron close to the composite interface side shows obvious spheroidisation phenomenon.

Due to the difference in cooling rate around the composite interface, the creep rate of peristaltic cast iron will appear to change the situation. Within 1 mm from the interface, the nodularity is about 60%; as the distance increases, the nodularity increases rapidly to about 70% at about 2 mm from the interface; subsequently, the tendency of nodularity to decrease slows down, and the nodularity basically stabilizes at 15% at a distance of more than 8 mm from the interface.

SEM Analysis

The SEM morphology and energy spectrum line scan results of the composite interface are shown in Fig.1. From the figure, it can be seen that the dotted line 2 divides the composite interface into two regions, A and B. The region A is near the 45 steel side, and the region B is near the compacted iron side. Among them, region A is close to the 45 steel side, which has a larger pearlitic lamellae spacing; region B is close to the compacted cast iron side, which has a smaller pearlitic lamellae spacing. It can be seen from the elemental distribution that the Si element content changes abruptly at the dotted line 2, the Cu element has obvious diffusion phenomenon in region B, and there is no obvious change in the C and Mn elements.

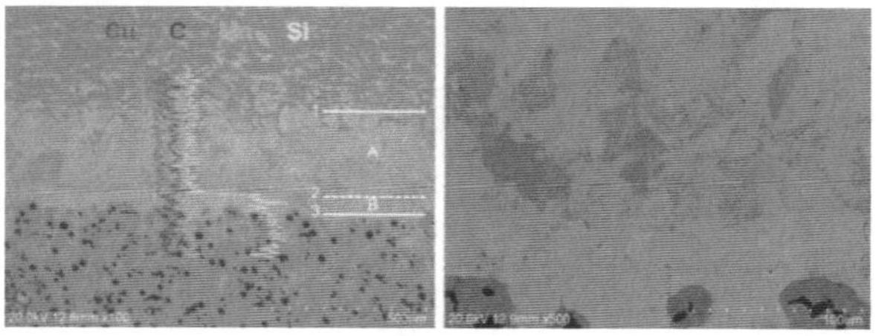

Fig.1 SEM morphology and energy spectrum analysis results of composite interface

Mechanical property

The matrix hardness of 45 steel is about 230 HV and that of compacted cast iron is about 210 HV. And in the full pearlite zone of the composite interface, the hardness varies from 231HV to 362HV.The smaller the concentration of C atoms in 45 steel, the larger the spacing of the pearlite lamellae, and therefore the higher the hardness at this position, which is consistent with the previous analysis.

The composite specimen was subjected to compression shear test, and its average shear strength was 317 MPa. After shear, the specimen was observed in side sections, and the composite interface remained intact after the shear test, and the carbon steel portion cracked after severe plastic deformation during the shear process, which indicated that the composite interface of the compacted cast iron - 45 steel had a rather excellent bonding strength.

4 Conclusion

(1) The metallurgical bond was formed by atomic diffusion between the creep cast iron and 45 steel, and the thickness of the bonded transition zone was about 304 μm, which was a fully pearlitic organisation.

(2) The transition zone is divided into two sub-layers, near the 45 steel side of the pearlite lamellae spacing is larger, the hardness in 231HV to 344HV between; near the peristaltic cast iron side of the pearlite lamellae spacing is smaller, the hardness in 344HV to 362HV between.

(3) The shear strength of the composite structure is 317MPa after the casting of compacted cast iron and 45 steel.

References

[1] Fang Jian, Yu yun, Wang Hongjin. Interface Investigation in Gray Cast lron/ Carbon Steel Bimetal Composite Casting. Foundry, 2015, 64(2):3. 122-124.

[2] Taleff E M, Lewandowski J J, Pourladian B. Microstructure-property relationships in pearlitic eutectoid and hypereutectoid carbon steels. JOM, 2002, 54(7):25-30.

Casting Process Design And Simulation of Ductile Iron Gear Box

Shuai Yang 1, Yanqing Mu 1, Peng Dong 1 , Yulong Fu 1
1. Changzhou Longrui Casting Co., LTD., Changzhou 213011, China
*6 Longpan Road, Wujin District, Changzhou City, Jiangsu Province: 364949990@qq.com

Abstract: The casting process design of ductile iron gear box was carried out according to the customer's technical requirements. The flow field, solidification field and shrinkage criterion during the casting solidification process were simulated by the MAGMA software, and the rationality of the riser chills design was analyzed. Finally, the process verification was carried out.
Keywords: Gear housing; Casting process; MAGMA; Gating system

1 Introduction

Chongqing is located in the southwest of inland China, is a unique "mountain city, river city", the Yangtze River runs through the whole, the climate is humid, more rain. The special geographical environment and climate have special requirements for subway and other public transportation, and the ball iron gearbox plays an important role in the subway transmission system, so the requirements for the subway gearbox are more stringent. In this paper, we designed the casting process for the subway gear box independently developed by our company in Chongqing, and simulated the flow field and solidification field of its castings in the melting process by applying MAGMA, and obtained the castings that meet the requirements of customers, hoping to help the casting development of domestic ductile iron gear boxes, and provide a reference for similar products.

1 Product technical requirements and box structure

Because the structure of the ductile iron gearbox is more complex, the overall size of the box is 872×520, the wall thickness is larger, the maximum wall thickness is 56, the minimum wall thickness is 8, and there are many reinforcement bars on the surface of the box, most of the cooling speed is slow during the solidification process of the box, the filling is slow, it is easy to produce shrinkage, shrinkage and other defects, and the casting process requirements are higher at the reinforcement bar. The overall process design of the box is more complicated, affecting the quality and performance of the casting blank. Its technical requirements mainly include the following aspects: the material is QT450-10, the casting size tolerance and machining allowance meet the CT11 level specified in GB/T6414-1999;Magnetic particle inspection is required for key parts, and the defect grade must meet the requirements of level 3 of EN 1369-1996. In addition, the key parts must be inspected by X-ray inspection, and defect grade requirements are not greater than Class A, Class B Level 2, Class C level 3 as defined in ASTM E446 and ASTM E186. No D, E, F and G 4 defects are allowed. The surface of the box is smooth and beautiful, there shall be no cracks and slag inclusion, and there shall be no oil leakage.

2. Casting process design and simulation results analysis

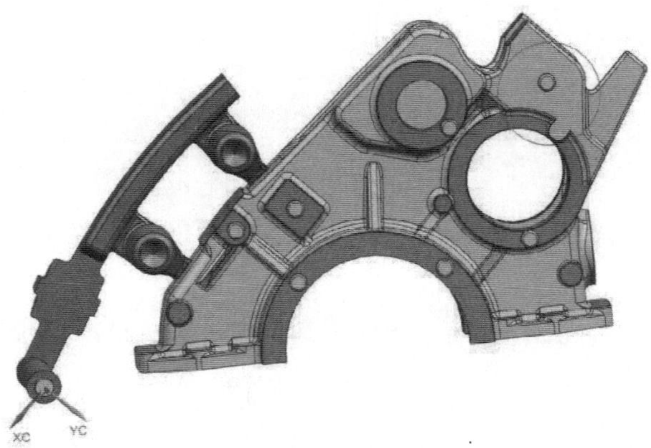

Figure 1 Box casting process diagram

3 Casting Process Results Verification

According to the above casting process, two sets of housings were designed and manufactured for trial production. Magnetic particle testing and X-ray testing were conducted on the castings. The magnetic particle testing showed that there were no cracks, inclusions, etc. in the castings, which met the technical requirements. The X-ray testing did not reveal any excessive defects, and the internal structure of the castings met the requirements of the customer. The microstructure of the casting samples was analyzed, and the mechanical properties of the castings, as shown in Table 1, met the corresponding national standard requirements for the material. Subsequent rough and fine machining of the castings did not reveal any problems with dimensions or internal defects, and the castings met the customer's requirements and could be produced and delivered.

Table 1 Mechanical properties of samples

separately cast test bar	*R*m/MPa	*R*p0.2/MPa	*A*/%	(HBW10/3000)
QT450-10	487	350	17.5	167

2 Conclusion (Bold, 10 pt., Arial)

Box casting process design is reasonable, casting process verification qualified, in line with customer requirements. The gating system design is reasonable, the casting solidification process is stable, no turbulence, insufficient pouring and other problems occur; Riser and cold iron design is reasonable, in the casting solidification process, the formation of the cold iron to the riser temperature gradient field, conducive to the casting sequence solidification, the use of riser reduces the production of shrinkage and other defects, the use of cold iron speeds up the cooling of the thick wall of the casting, providing the quality of the casting.

The correctness of casting process design was verified through casting trial production, X-ray testing and magnetic particle testing, and the casting products met customer requirements, which provided a reference for similar products.

3 References

[1] Han Wen. Study on self-hardening furan resin with middle nitrogen and low free formaldehyde for casting [J]. Foundry,2012,61(12):1452-1454.

[2] Li Xinya. Foundry Manual: Foundry Process [M]. 3rd edition. Beijing: China Machine Press,2011.

[3] Mu Yanqing. Casting Process Design of Ductile Iron Gear Box based on MAGMA [J]. Casting Technology,2014,35(12):2.

The Application of Wire-Feeding Inoculation Technology in the Production of Vermicular Graphite Cast Iron

Zhongteng Zhang
Changzhou Langrui Foundry Co.Ltd
*Corresponding address: e-mail: 1473711989@qq.com

Abstract: Our company is dedicated to the mass production of RuT400 brand cast iron parts. Confronted with issues such as insufficient stability during the in-mold process, substantial fume emissions during graphitization, and the high costs associated with alloy usage, we have adopted an economical and environmentally friendly wire feeding nodularization process to enhance the production stability of ductile iron. By precisely controlling the composition of molten iron and nodularization temperatures, and designing specialized wire feeding packages, wire feeding process design, and inoculation process design, we have established a PLC-based automated wire feeding station. This development has enabled our company to achieve environmentally friendly and one-time stable nodularization of molten iron. Dissection and sampling analysis of the castings confirm that both the nodularization rate and mechanical properties meet the technical specifications.

Keywords: wire feeding nodulization ; Component Control

1 Introduction

In modern foundry technology, Compacted Graphite Iron (CGI), commonly referred to as vermicular iron, has become a crucial material widely applied in the automotive, construction machinery, and other industrial sectors due to its superior mechanical properties and casting characteristics. Vermicular iron combines the excellent casting performance of gray cast iron with the high strength of ductile iron, thereby offering unique advantages in various applications. To achieve the optimal performance of CGI, the vermicularization treatment process is essential.

The wire feeding vermicularization process, as an advanced treatment method, has garnered extensive attention and application in CGI production in recent years. This process involves the precise addition of vermicularizing agents into the molten iron, either continuously or intermittently. By doing so, it ensures efficient control over the vermicularization process, thereby enhancing the quality and stability of the CGI.

2 Experimental procedure

1.1 Control of Sulfur Content
1.2 Control of Mg Content
1.3 Feeding Wire Station Design
1.4 Design of Wire Feeding Ladle
1.5 Feeding Wire Temperature

1.6 Calculation of Wire Feeding Quantity
1.7 Wire Feeding Speed
1.8Inoculation Treatment

3 Result and discussion

Using the wire feeding spheroidization process, a trial production was conducted on the vermicular graphite cast iron brake discs for railway passenger cars, as shown in Figure 1. These discs feature two friction surfaces connected by uniformly arranged cylindrical heat dissipation fins, with connecting lugs positioned at the inner bore. The casting material is RuT400, requiring a predominant vermicular graphite morphology, with the amount of vermicular graphite being ≥65%. The matrix structure should consist of pearlite and ferrite, with pearlite content ≥30%. After finishing, the disc should undergo magnetic particle inspection, and defects must meet the SM2, LM2, and AM2 grade requirements as specified in EN1369. Additionally, the disc should undergo ultrasonic testing. Within an 8mm distance from the friction surface, no shrinkage defects are allowed. Beyond the 8mm range, the defect area must be ≤ $2000mm^2$.

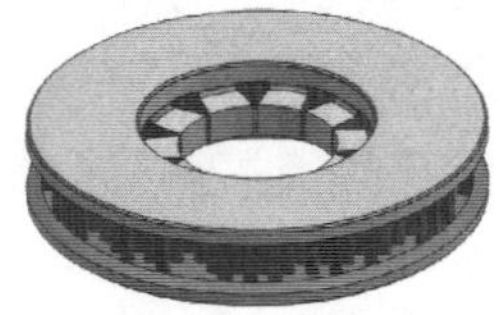

After the physical dissection of the casting, the parameters of each part, as shown in Table, meet the technical requirements. Additionally, the friction surfaces of the casting satisfy the ultrasonic inspection criteria, and all machined surfaces meet the magnetic inspection requirements.

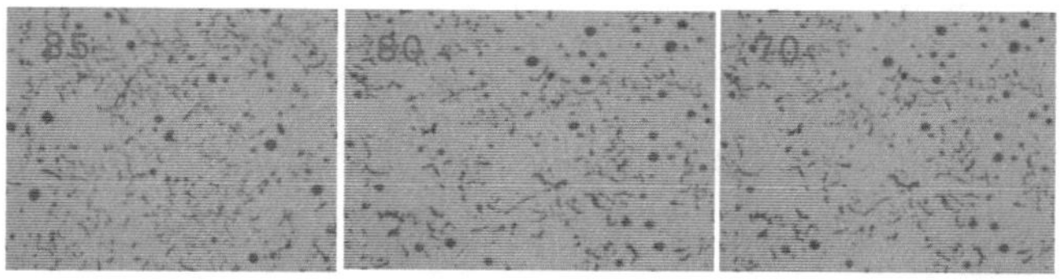

Position Parameter	Thickness (mm)	Spheroidization Rate (%)	Tensile Strength Rm (MPa)	Elongation A (%)	Hardness (HBW)
Upper plate	30	85	430	2	210
Lower plate	30	85	418	3	197
Heat dissipation fins	15	70	420	1.5	220

Table : Main Parts Inspection Results of the Casting

4 Conclusion

By employing the wire-feeding vermicularization process and ensuring control over raw materials and a reasonable vermicularization procedure, the technical specifications for the vermicular graphite cast iron brake discs can be satisfactorily met.

Analysis of experimental data indicates that the optimal vermicularization process for our company includes the following key parameters:

i) Managing the furnace charge to control sulfur content between 0.015% and 0.02%;

ii)Controlling the vermicularization treatment temperature within the range of 1390-1410℃;

iii) Maintaining a molten iron weight of 1500 kg;

iv) Setting the wire feeding speed to 20 m/min;

v) Ensuring a molten iron surface-to-neck ratio of 1.5;

vi) Utilizing a 0.7% silicon-barium-calcium inoculant for ladle inoculation.

Compared to the plunge method of vermicularization, the wire-feeding vermicularization process demonstrates higher alloy utilization efficiency, reduced overall costs, and improved working conditions.

References

[1] Gong Hongtao. Study on the Inoculation Wire Feeding Process of Vermicular Graphite Cast Iron[D]. Harbin University of Science and Technology, 2011.

[2] Li Rongde, Yu Haipeng. Quality and Control Technology of Cast Iron[M]. Beijing: China Machine Press, 1998: 83-86.

[3] STEFANECU D, ROXAANA R. Lightweight Iron Castings-can they Replace a Luminumc Astings[J]. Foundryman, 2003, 1(9): 221-224.

Multi-Factor Process Parameter Optimization of S30432 Continuous Casting Secondary Cooling Process

Zhiqiang Li1, Yingxuan Shan1, Hua Hou1,3, Yuhong Zhao1,2,4*
[1] School of Materials Science and Engineering, Collaborative Innovation Center of Ministry of Education and Shanxi Province for High-performance Al/Mg Alloy Materials, North University of China, Taiyuan, 030051, China
[2] Beijing Advanced Innovation Center for Materials Genome Engineering, University of Science and Technology Beijing, Beijing, 100083, China
[3] School of Materials Science and Engineering, Taiyuan University of Science and Technology, Taiyuan, 030024, China
[4] Institute of Materials Intelligent Technology, Liaoning Academy of Materials, Shenyang, 110004, China
*Corresponding address: e-mail: zhaoyuhong@nuc.edu.cn

Abstract: Niobium-containing austenitic stainless steel is widely used in ultra-supercritical thermal power units due to its excellent high temperature stability and corrosion resistance. Due to the high Nb content in S30432, the coarse primary NbC phase is produced during the solidification process of continuous casting, which can easily lead to the central crack of continuous casting billet and the perforation deformation crack of subsequent seamless pipe. In this study, high-temperature laser confocal scanning electron microscopy and macroscopic continuous casting simulation methods were used to study the solidification structure at different cooling rates and optimize the secondary cooling continuous casting process for the secondary cooling solidification process of continuous casting. The results show that the high cooling rate can greatly reduce the size of the primary NbC phase. At the same time, considering the characteristics of the solidification structure of continuous casting, the cooling intensity of the secondary cooling zone of continuous casting is designed by segmented gradient reduction, and the optimum process parameters are obtained by orthogonal process optimization of casting speed, superheat and amount of cooling water.

Keywords: S30432, Continuous Casting, NbC, Cooling Rate, Process Optimization

1 Introduction

Austenitic heat-resistant stainless steel has become an important choice material for seamless steel tubes for supercritical and ultra-supercritical utility boilers due to its excellent high temperature stability[1]. However, due to the solidification conditions of the continuous casting process and the low solute equilibrium distribution coefficient of Nb element, serious grain boundary segregation occurs in the solidification structure of S30432 with high Nb content, and large blocks of unevenly distributed niobium compounds are formed, which greatly reduces its thermal processing performance and causes pore initiation and cracks[2]. Since the bulk NbC primary phase is difficult to be removed by subsequent heat treatment methods, it can be considered to change the process parameters in the continuous casting process to realize the regulation of the solidification structure of the alloy. The main factors are superheat, casting speed and cooling water. Researchers have widely used macroscopic simulation methods to study the continuous casting process, but did not actually consider the close relationship between specific process parameters and solidification structure. Therefore, in this study, firstly, based on the influence of cooling rate on the primary phase, considering the characteristics of the actual continuous casting solidification structure, the cooling water distribution in the secondary cooling zone was designed in a targeted manner, and finally the optimized process parameters were obtained.

2 Experimental procedure

Firstly, a cylindrical sample with a diameter of 7mm and a height of 3mm was processed from the S30432 round billet. It was placed in a ceramic crucible with a diameter of 7.8mm and a height of 3.5mm for HT-CSLM experiment. The sample was heated from room temperature to 1500°C for 5 minutes to ensure its complete melting. Then they were cooled to room temperature at cooling rates of 1, 10, 30, 50, 70 and 90 °C/s, respectively. Observe and record the evolution of microstructure during solidification under high temperature conditions, and use Shimadzu EPMA-8050G to further accurately analyze and determine the element distribution in dendrites and interdendritic regions. On this basis, the Procast casting simulation software was used to simulate the continuous casting process. The casting speed was set at 0.8, 0.9, 1.0, 1.1, 1.2 m/min, the superheat was set at 10, 20, 30, 40, 50°C, and the specific water amount was set at 0.1, 0.2, 0.3, 0.4, 0.5 L/Kg. The orthogonal method was used to design 25 groups of process schemes to evaluate the final slab quality.

3 Result and discussion

Solidification microstructure characteristics

Fig.1 shows the solidification microstructure at different cooling rates. With the increase of cooling rate, the grain structure is obviously refined, and exists in the form of fine dendrites, and the size of NbC particles is significantly reduced. However, during the solidification process of continuous casting billet, the cross-section structure of the billet presents a typical three-crystal zone morphology. According to the simulation results of continuous casting structure, the increase of secondary cooling water can directly increase the cooling rate, but at the same time, it will lead to the expansion of the columnar crystal area, which is obviously inconsistent with the ultimate goal.

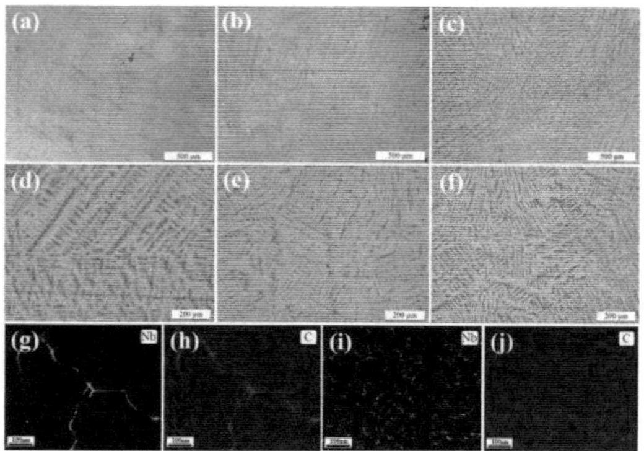

Fig.1 Solidification structure and element distribution at different cooling rates. (a)1°C/s, (b)10°C/s, (c)30°C/s, (d) 50°C/s, (e)70°C/s, (f)90°C/s, (g)~(h)1°C/s, (i)~(j)90°C/s.

Secondary cooling process design

According to the characteristics of the solidification structure of continuous casting, the cooling rate can be reduced by multi-stage gradient in the secondary cooling zone, which is beneficial to the increase of the equiaxed crystal region in the core. The secondary cooling segmentation scheme adopted is shown in Table 1 :

Table 1. Cooling area of continuous casting secondary cooling zone

Cooling Zone	T_0	T_1	T_2	T_3
The actual area length L_i (m)	0.62	1.7	2	2.3
Crystallizer liquid level to the regional center length S_i (m)	0.97	2.13	3.98	6.13

The water allocation Q_i of each region is :

$$Q_0 : Q_1 : Q_2 : Q_3 = \frac{1}{\sqrt{S_0}} : \frac{1}{\sqrt{S_1}} : \frac{1}{\sqrt{S_2}} : \frac{1}{\sqrt{S_3}} \quad (1)$$

In the formula, Q_i is the amount of water in the foot roller zone, the first zone of secondary cooling, the second zone of secondary cooling and the third zone of secondary cooling, and S_i is the length from the liquid level of the crystallizer to the cooling center point of each zone..

Fig.2 shows the evaluation results of 25 orthogonal process schemes. It can be seen that under the influence of multiple factor parameters, the optimal continuous casting process parameters are obtained based on the straightening point temperature above 900 ℃, lower shape factor, and lower sectional crack rate of billet. The casting speed is 0.9 m/min, the superheat is 20~30 ℃, and the amount of secondary cooling water is 0.3~0.4 L/Kg.

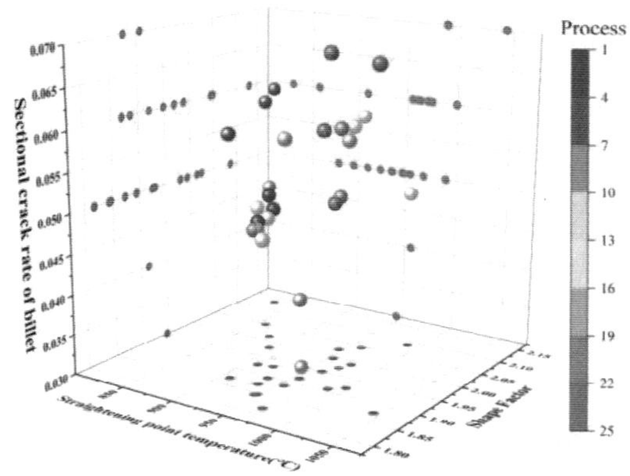

Fig.2 Multi-factor evaluation results of 25 groups of process schemes

4 Conclusion

(1) High cooling rate can greatly reduce the size of primary NbC phase.

(2) Considering the characteristics of solidification structure in continuous casting, the equiaxed grain ratio of the square billet can be improved by reducing the cooling water gradient in the secondary cooling zone of continuous casting.

(3) The optimum continuous casting process parameters were obtained by orthogonal process optimization of casting speed, superheat and cooling water. The casting speed was 0.9m/min, the superheat was20 ~ 30 °C, and the amount of secondary cooling water was0.3~0.4 L/Kg.

Acknowledgments

This work was supported by National Natural Science Foundation of China (Nos. 52375394, 52074246, 52275390, 52201146).

References

[1] Oliveira M P, Zhang W, Yu H Y, et al. Recent developments in niobium containing austenitic stainless steels for thermal power plants[J], Energy Materials, 2014.

[2] Deardo A J. Niobium in modern steels[J]. International Materials Reviews, 2013, 48: 371-402.

The Role of Microstructure Regulation on Magnetic Properties of Non-Oriented Silicon Steels

Lingzi An 1,2, Haitao Liu 2, Guodong Wang 2
1. School of Materials Science and Engineering, Shenyang University of Technology, No.111, Shenliao West Road, Economic & Technological Development Zone, 110870, PR China
2. State Key Laboratory of Rolling and Automation, Northeastern University, No. 11, Lane 3, Wenhua Road, Heping District, Shenyang, 110819, PR China
*Corresponding address: e-mail: anlingzi@sut.edu.cn

Abstract: In this work, a new method was proposed to obtain large grain size before cold rolling and it was applied to 1.0%Si non-oriented silicon steel, i.e., appropriate low-finishing-temperature hot rolling combined with batch annealing. The grain growth mechanism during batch annealing was investigated using a quasi in-situ electron backscattered diffraction (EBSD) technique. It was demonstrated that significant grain growth took place through strain-induced boundary migration mechanism. Then a large number of shear bands formed in the cold rolled microstructure and provided nucleation sites for Goss ({110}<001>) and λ-fiber (<001>//ND) recrystallized grains, thus resulting in a coarse and homogeneous recrystallization microstructure with weakened γ-fiber (<111>//ND) texture, strengthened Goss and λ-fiber texture. Consequently, superior magnetic properties (B_{50}=1.763 T, $P_{1.5/50}$=4.30 W/kg) were obtained. This work provided a new way to improve magnetic properties by low temperature batch annealing instead of the conventional high temperature normalization.

Keywords: Non-oriented silicon steel, Batch annealing, Grain size, Texture, Magnetic properties.

1 Introduction

Non-oriented silicon steels (NOSs) are mainly used for manufacturing iron core of electric motors and generators. Improving the efficiency of electric motors can save a lot of energy. However, in the conventional producing routes, the sheets ordinarily exhibit small grains before cold rolling. This leads to the formation of small recrystallized grains with strong γ -fiber texture in the finally annealed sheets, which deteriorates the magnetic properties. It has been widely accepted that increasing the initial grain size before cold rolling can improve magnetic properties of non-oriented silicon steels. Higher annealing temperature and longer holding time after hot rolling can improve the grain size [1, 2]. However, these methods may deteriorate the magnetic properties by affecting the precipitation of fine particles [3]. In this paper, a new method to obtain large grains before cold rolling was proposed, that was, low-finishing-temperature hot rolling combined with batch annealing was used to promote grain growth. Microstructure, texture evolution along the whole process and magnetic properties were carefully studied.

2 Experimental procedure

Fe-1.0Si non-oriented silicon steel ingot was produced by vacuum induction smelting furnace. The 70 mm-thick ingot was reheated to 1150 ℃ and hot rolled to 2.41 mm with the finishing temperature of 855 ℃. One hot-rolled sheet was directly cold rolled to 0.5 mm. Three hot-rolled sheets were annealed at 750, 800 and 850 ℃ for 3 h in a chamber furnace to simulate batch annealing, and then cold rolled to 0.5 mm. The cold-rolled sheets were finally annealed at 930 ℃ for 3 min in Ar atmosphere.

3 Result and discussion

1. The mechanism of grain growth during batch annealing after hot rolling

The hot rolled sheet was characterized by a quite heterogeneous microstructure with an average grain size of 40 μm. Due to the low-finishing-temperature, some darkly etched grains with high stored energy formed in the hot rolled sheet. After batch annealing, small equiaxed grains formed in the surface zone and lots of grains significantly grew up to 140-200 μm in the center zone. A quasi in situ EBSD technique was utilized to track the grain boundary migration process in the same observation area. The hot rolled sheet was annealed at 750 °C for different times, and the quasi in-situ EBSD results of hot-rolled sheet were shown in Fig.1. 10 grains that grew up obviously during annealing were marked. It demonstrated that most of the original grains could grow up larger by swallowing surrounding grains and maintained original orientation during annealing. The grain boundaries of these initial grains with lower KAM (lower dislocation density) gradually migrated to grains with higher KAM (higher dislocation density). It indicated that pronounced grain growth occurred during batch annealing through the strain induced boundary migration (SIBM) mechanism and resulted in large grains. The driving force for SIBM was caused by the difference in dislocation content on opposite sides of the grain boundary.

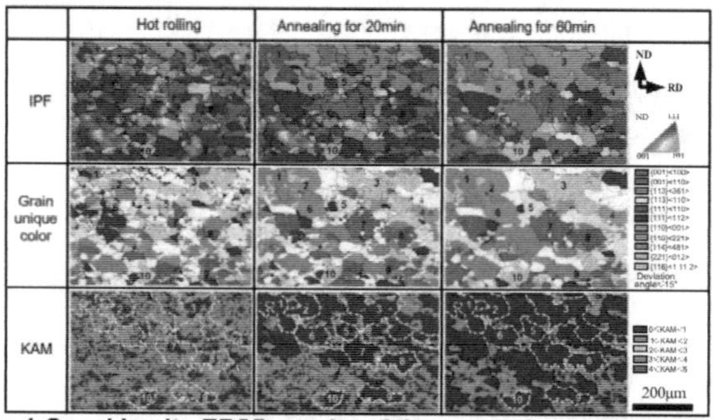

Fig.1 Quasi in-situ EBSD results of the same observation area in the center region of hot-rolled sheet with different annealed time.

2. Effects of grain size prior to cold rolling on subsequent microstructure and texture evolution

The grain size of hot-rolled sheet was relatively small, resulting in a higher grain boundary density after direct cold rolling. It promoted the development of γ-grains and formation of small grains in the final annealed sheet. By comparison, large grains formed by introducing batch annealing, which were conductive to form in-grain shear bands during cold rolling. These shear bands could provide preferential nucleation sites for Goss and λ-grains during final annealing, leading to the nucleation and growth of γ-grains were inhibited. Meanwhile, due to inheritance of microstructure, coarse grains were obtained after final annealing in the case of batch annealing.

3. Effects of final recrystallization microstructure and texture on magnetic properties

It is well known that grain boundaries imped the domain movement during magnetization. Coarse grains meant less boundaries, which is helpful for reducing hysteresis losses. Fig. 2 showed magnetic properties of the final annealed sheets under different batch annealing temperatures. Apparently, iron losses were significantly reduced and the magnetic inductions were remarkably increased by introducing batch annealing, which was attributed to the coarse grain sizes and favorable textures in the final annealing sheets. However, as the batch annealing temperature reached 800 and 850 ℃, magnetic inductions were deteriorated. This was because higher temperature promoted the formation of MnS particles, and γ-grains were easily formed near the small particles. As the batch annealing temperature was 750 ℃, the strong λ-fiber and Goss textures as well as fewer MnS particle led to the highest magnetic induction (1.763 T). At the same time, the combination of larger grain size and favorable textures resulted in lower iron loss (4.30 W/kg).

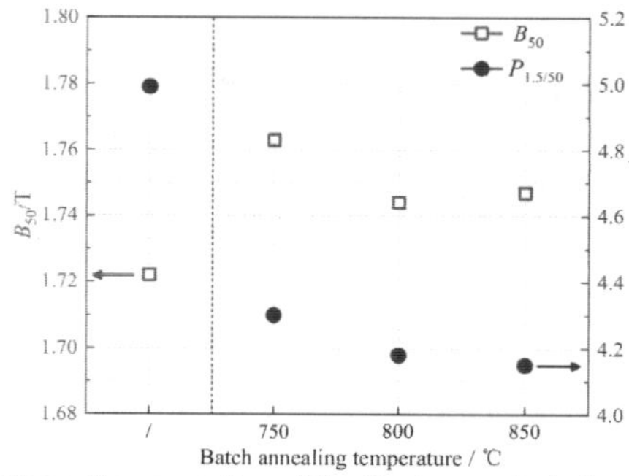

Fig.2 Magnetic properties of the final annealed sheets under different batch annealing temperatures.

4 Conclusion

Appropriate low-finishing-temperature hot rolling could produce inhomogeneous microstructure across the sheet thickness, resulting in different stored energies among adjacent grains. The quasi in-situ EBSD results indicated that during batch annealing, the difference in stored energy triggered SIBM, leading to significant grain growth. However, high-temperature batch annealing led to more MnS particles and strong γ-fiber texture in the final sheets. Thus, appropriate low-finishing-temperature hot rolling combined with low-temperature batch annealing was an effective method to manufacture non-oriented silicon steels with superior magnetic properties.

References

[1] Park J T, Szpunar J A. Effect of initial grain size on texture evolution and magnetic properties in nonoriented electrical steels. J. Magn. Magn. Mater, 2009, 321(13): 1928-1932.

[2] Lee K M, Huh M Y, Lee H J, et al. Effect of hot band grain size on development of textures and magnetic properties in 2.0% Si non-oriented electrical steel sheet. J. Magn. Magn. Mater, 2015, 396: 53-64.

[3] Huneus H, Günther K, Kochmann T, Plutniok V, Schoppa A. Nonoriented magnetic steel with improved texture and permeability. J. Mater. Eng. Perform. 1993, 2(2), 199-203.

Low-Alloy Steel Refining Process

PengchengZhang[1], XinfengLiu[1], YongliangHe[1], WanqingLi [1], BoDou [1], ShaoweiWen[1], HuiYang[2], RonghengCui[2], MinShen[3], TiechenWang[4], XingguoHu[5], LeiChen[5],

1.Shenyang Research Institute of Foundry/State Key Laboratory of Advanced Casting Technologies,,No.17,YunFeng South Street, Tiexi District, Shenyang City, Liaoning Province, China
2.Jiangsu Province Foundry Association, Room 207, No.40, General Avenue, Jiangning District, Nanjing city ,China
3.Taicang TEMC Metal & Chemical Co,. Ltd, TaicangShaxi Town Yuewang Taiwan funded Science and Technology Innovation Industrial Park, China
4.ARES Industrial Furnace (Tianjin) Co., Ltd., No.408 Building 10, Highland Industrial Park, No.35 Intelligent road, Xiqing District, Tianjin, P.R.China
5.Anhui Yingliu Group Huoshan Casting Co., LTD, No.96, PI Hexi Road, Hengshan Town, Huoshan County, Lu'an City, Anhui Province, China
*Corresponding address: e-mail: poncinz@163.com

Abstract: The machinery industry requires high-quality, low-alloy steels to enhance the performance and durability of mechanical components. However, conventional steelmaking methods often lead to impurities and inclusions that can compromise the integrity and functionality of these steels. The AOD (Argon Oxygen Decarburization) refining process has emerged as a promising solution to this problem, offering a more efficient and effective way to produce low-alloy steels with superior properties.This paper introduces the comparison of LF, LF + VD and AOD of low alloy steel refining, and the advantage of AOD during low alloy steel refining process.

Keywords:LF, VD, AOD, refining, degassing, desulfurization, inclusions,

1 Introduction

Low-alloy cast steel generally has good comprehensive mechanical properties. It has higher strength, atmospheric corrosion resistance and wear resistance, good toughness and quenching properties. Now the application of low alloy wear-resistant steel materials, because of its high hardness and wear resistance,good toughness, good technology, and good economy. Widely used in metallurgy, coal, building materials, non-metallic mines, power system and other industries. Such as: crusher hammer head, ball mill lining board, loader shovel teeth and agricultural machinery anti-grinding accessories.The conventional smelting method leads to the insufficient purity of molten steel (the inclusion affects the product performance and the final use effect), the high gas content, the product toughness index can not reach the ideal value and other problems. This has become the development bottleneck of improving the quality and performance of low-alloy steel wear-resistant materials. This makes it necessary to introduce the refining process, remove impurities and improve the purity of molten steel.LF, LF + VD and AOD are three typical cast steel refining processes, we will compare the experiments respectively.

2 Experimental procedure

Steel refining process has four functions: 1)Precise control of carbon and alloy elements;2) Desulfurization, The inclusions in liquid steel are mostly various oxides or sulfide. As a harmful element in low alloy wear-resistant steel, it is hoped that the content of S in steel is lower, so as to reduce the generation of inclusions. 3) Degassing ([O], [N], [H]). As mentioned above, oxide is the main component of the inclusions, and the oxygen content in the steel is required to be as low as possible. In low alloy steel, [N] and [H] are also recognized harmful elements and hope to be removed as much as possible. 4) Remove the inclusions. Through effective refining methods, the impurities produced during reduction and desulfurization (various oxides and sulfide inclusion) are separated from the liquid steel.

LF(ladle furnace) practice, The commonly used low-alloy steel refining process in China is LF ladle refining furnace, see Figure 1. This is the process invented by the Japanese special steel in 1971, the original intention is to move the reduction operation in the electric arc furnace to the outside. By adding synthetic white slag, the arc insulation can maintain the reduction period for a long enough time (> 120min). The steel slag is well mixed and maintain a strong reduction atmosphere through argon mixing, and the liquid steel solution can be deeply desulfurization to 10ppm and the oxygen is 25ppm. However, the conventional LF process has almost no ability to remove [H] and [N] (some material introduces that the hydrogen content in LF can be removed to 1.5ppm, but this is due to the good boiling of the electric arc furnace in the early stage).

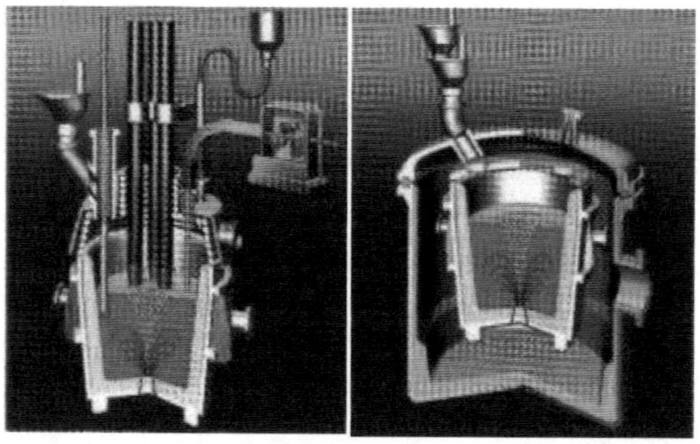

Fig.1LF（left）andVD（right）

LF+VD(Vacuum Degassing) practice, To improve the degassing effect, a vacuum degassing station is added, also known as VD. Deep desulfurization by LF + VD to below 20ppm[1] in experimental production. When the vacuum is extracted, the three breathable elements at the bottom of the ladle are blown into argon gas for stirring, and the sulfur can be removed to less than 0.001% within 30min. The desulfurization rate is as high as 95% [2]. For degassing within the VD, the overall period was> 30 min. The high vacuum retention period is 10-12 minutes. Three products were tested using the EAF(electric Arc Furnace) + LF + VD refining process. The results are shown in Table 1.

Table 1.EAF + LF + VD process, oxygen, nitrogen, hydrogen, sulfur content in steel10^{-6}

Product	[O]	[N]	[H]	S
Electroblank	6.0～8.5	60～70	0.70～1.0	<100
Steel product	9.8～15	68～87	0.49～0.57	<100

AOD(Argon Oxygen Decarburization) practice, Low alloy steel AOD refining cycle, the initial US refining cycle is about 40 minutes, see Figure 2.

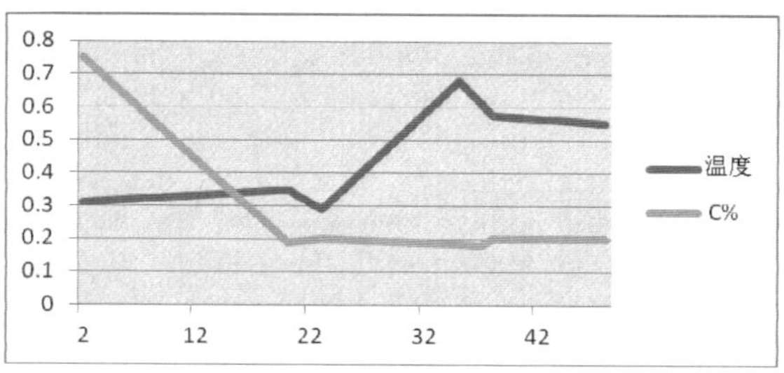

Figure. 2 The AOD refining process of low-alloy steel

A company in Ningbo adopts the complete set of equipment and process of 5-ton AOD refining furnace provided by SRIF to produce low alloy steel steel castings for mining machinery. The minimum AOD refining cycle is 29min, with excellent mechanical performance.Theresult of the final product are shown in Table 2.

Table. 2 IF+AOD process, oxygen, nitrogen, hydrogen, sulfur content(10-6) and properties in steel

Product	[O]	[N]	[H]	S	HRC	Ak/j -40℃
Casting	30	21	1	4-6	50 52	35 25

3 Result and discussion

LF process and LF+VD process

LF process can reach a low inclusion level in the steel, but it can not remove the gas effectively. LF+VD process can remove gas in the steel, but need a long period.

AOD process

AOD refining process of low alloy steel, generally adopts EAF + AOD double process, can meet all the functions needed by steel refining：within 20-40 minutes under the continuous hot furnace production conditions. See table 3.

Table 3 Comparison of refining process of low alloy steel（AOD，LF，LF+VD）

Refining process		AOD	LF	LF+VD
Conventional capacity		2-180t	>20t	>20t
Refining cycle(min)		20-40	35-55	50-80
Function	decarburization	●	○	○
	desulfurization	●	●	●
	Deoxidizationde	●	●	●
	hydrogenation	●	○	●
	denitrification	●	○	●

Conclusion

AOD is the best choice for low-carbon steel casting making, It can finish all the 4 functions which refining needs, just within 20-40 minutes, especially for Small and medium-sized foundry enterprises

Acknowledgments

This article will thank Anhui Yingliu Group for sharing valuable refining experience, and Jiangsu Casting Association, TEMC, Ares for providing relevant consulting support.

References

[1] Hao Xin, Wang Xinhua, Huang Fuxiang, Yang Guangwei, Zhao Heming. The LF-VD highly efficient desulfurization process [J]. Steel, 2014,49 (7): 72-75.
[2] Liu Xiaofeng, An Chang xia, Du Yawei, Chen Dengguo. Development status of VD vacuum refining technology and equipment [J]. China Metallurgy, 2013,23 (5): 8-9
[3] TianXiaoxia, Pei Fengjuan, Huang Zhong, Wang Xinzhi. Research on water steel desulfurization process of LF-VD in Anyang [J]. Henan Metallurgy, 2012,20 (4), 17

Effect of Nb Addition on Microstructure, Mechanical Properties and Wear Resistance of Medium Carbon Low Alloy

Yanliang Yi 1*, Chaozhong Wang 1, Wei Li 1*
1 Institute of Advance Wear & Corrosion Resistant and Functional Materials, Jinan University, Guangzhou 510632, China
*Corresponding address: y_yanliang@163.com; Lwxasn @ sohu.com

Abstract: TMicrostructure, mechanical properties and impact abrasive wear resistance of medium carbon low alloys with different Nb contents were investigated. The alloys are composed of tempered martensite and residual austenite. Adding Nb facilitates NbC forming and dispersing in the martensite, and refines the martensite structure. This improves the mechanical properties and impact abrasive wear resistance of the alloy, as a result the hardness, impact toughness, tensile strength and impact abrasive wear resistance of the alloy increase with the amplifications of 1.1%, 18.3%, 28.0% and 7.9%, respectively.

Keywords: Medium carbon low alloy, Niobium, Impact abrasive wear resistance.

Introduction

Medium carbon low alloy is considered as one of the ideal materials for hammer head because of its high hardness, high wear resistance and good toughness [1]. In order to further improve the service efficiency of the alloy, scholars often use the method of adding alloying elements. For example, Vahid Javaheri et al. [2] studied the effect of Nb addition on themicrostructure of medium carbon low alloy steel with 0.4% carbon content, and found that the microstructure of medium carbon low alloy steel with 0.013wt.%Nb was refined and the microhardness of steel was increased. It can be seen that the addition of alloying elements can improve the microstructure optimization, mechanical properties and wear resistance of steel. Niobium (Nb) is a strong carbide forming element, which can form stable carbides with C in steel, and adding Nb to steel can effectively improve the mechanical properties of steel [3].

In this paper, the influence of Nb addition on the microstructure, mechanical properties and impact abrasive wear resistance of medium carbon low alloy used for hammer head was studied.

Experimental procedure

Table 1. Composition of Alloys A and B

Samples	C	Si	Mn	Cr	Mo	Ni	Nb
A	0.51	1.22	0.63	0.67	0.36	1.53	-
B	0.52	1.13	0.65	0.72	0.41	1.56	0.18

A medium frequency induction furnace is used to melt the two components of molten steel, and then the alloy liquid steel is respectively poured into the same cast, followed by removing and polishing to obtain the Y-shaped block. The main chemical components of the alloys were examined and showed in Table 1.

According to experiences of relative enterprises, heat treatment of the alloys were designed as follows: heated at 1000 ℃ for 6 h, followed by quenching to room temperature by using PAG water-soluble liquid, and then tempering at 330 ℃ for 8 h.

According to the T/CFA 010604-3-2016, using MLD-10 dynamic load impact abrasive wear testing machine to examine the wear resistance of the alloy. The upper wear sample is alloy, and the lower sample is 45# steel with a hardness of ~49 HRC.

To ensure the accuracy of the data, pre-grinding for 30 min and then formal test for 150 min were conducted. After each test, ultrasonic cleaning was performed with alcohol in a high-power CNC ultrasonic cleaner (KQ-800KDE), and weighed with an analytical balance (GL124i-1SCN) with an accuracy of 0.0001g.

Result and discussion (Bold, 10 pt., Arial)

Microstructure

As seen in Fig.2, samples A and B are mainly composed of tempered martensite (Fig.2a) and residual austenite (Fig.2b). From Figs.2c and 2d, the precipitated phase in the sample B is NbC with lattice parameters of $a=b=c=4.4713$ Å. This shows that adding Nb facilitates NbC forming and dispersing in the martensite. Moreover, the lath width of tempered martensite in the sample A is in the range of 40~60 nm, with an average value of 51.3 nm. However, the lath width of tempered martensite in the sample B distributes between 15~20 nm and 30~35 nm, with an average value of 28.1 nm. The added Nb refines the martensite structure.

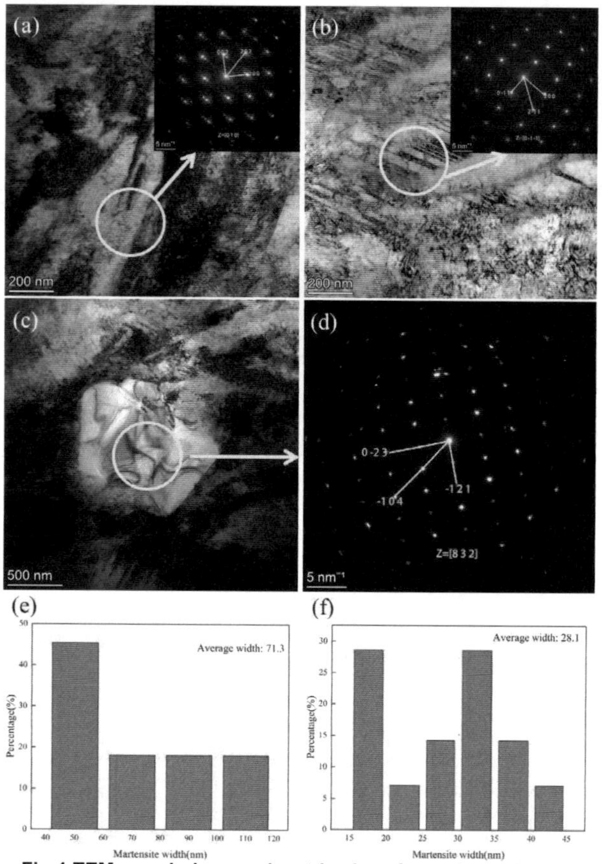

Fig.1 TEM morphology and matrix size of samples: (a) and (b) show samples A and B, respectively; (c) and (d) NbC; (e) and (f) show martensite lath width of the samples A and B, respectively.

Mechanical property

Table 2. Mechanical properties of the alloys of A and B

Samples	Rockwell hardness/HRC	Impact energy/J	Tensile Strength/ MPa
A	55.9	6.0	1532.2
B	56.5	7.1	1960.9

From Table 2, the hardness, impact energy, and tensile strength of the alloy A are 55.9 HRC, 6.0 J, and 1532.2 MPa respectively, while that of the alloy B are 56.5 HRC, 7.1 J, and 1960.9 MPa respectively. The Nb addition causing fine grain strengthening and second phase strengthening improves the mechanical property of the alloy.

Impact abrasive wear resistance

From Fig.2a, the weight losses of samples A, B, U (ZG50Ni2SiCrMo alloy), and G (ZG50Ni2Si2CrMo alloy) are 0.1566, 0.1451, 0.1568 and 0.1422g, respectively. From Figs.2b~2e, the wear mechanisms are mainly micro-cutting, and fatigue spalling is auxiliary. Moreover, the deepest distance of spalling pit in the samples A, B, U, and G reaches 20, 18, 22, and 15 μm, respectively. It can be concluded that sample B has better impact abrasive wear resistance compared to the samples A and U, and approaches the sample G.

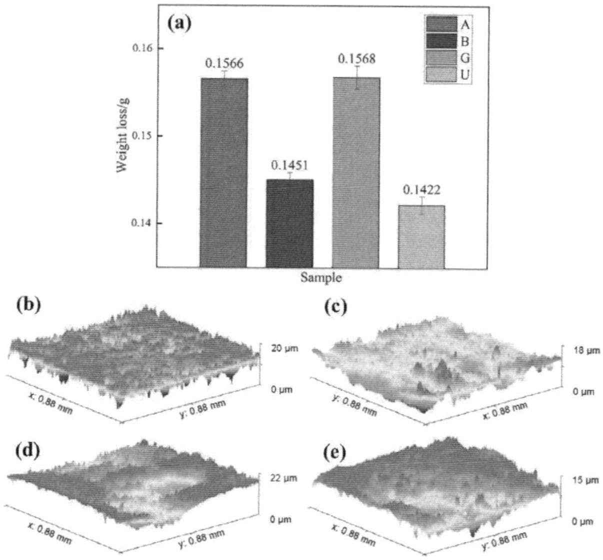

Fig.2 Weight loss and 3D microscopies: (a) Weight loss; (b), (c), (d), and (e) show samples A, B, U and G, respectively.

Conclusion

1. Adding Nb facilitates NbC forming and dispersing in the martensite, and refines the martensite structure.
2. The hardness, impact energy and tensile strength of the alloy containing Nb are 56.5HRC, 7.1J and 1960.9MPa, respectively, which are 1.1%, 8.3% and 7.59% higher than the alloy without Nb addition, respectively.
3. The Nb addition improves the wear resistance of alloy, and impact abrasive wear resistance of the alloy containing Nb is 7.9% higher than that of the alloy without Nb.

Acknowledgments

This work was financially supported by Guangdong Basic and Applied Basic Research Fundation (2024A1515012353 and 2022A1515010091) and Science and Technology Projects in Guangzhou (2024A04J4063).

References

[1] Taher E B, Mohammed G, Ibrahim M, Fouad H. Development of carbon-Low alloy steel grades for low temperature applications. Mater. Sci. Eng. A, 2011,528(18): 6039-6044.
[2] Vahid J, Nasseh K, Antti K, David P. Effect of niobium and phase transformation temperature on the microstructure and texture of a novel 0.40% C thermomechanically processed steel. Mater. Charact., 2018, 142: 295-308.
[3] Lin L, Li B S, Zhu G M, Kang Y L, Liu R D. Effects of Nb on the microstructure and mechanical properties of 38MnB5 steel. Int. J. Min. Met. Mater., 2018, 25(10):1181-1190.

Corrosion Behavior of Austenitic Cast Iron in 3.5wt.%NaCl Solution and 10wt% Quartz Sand

Lan Sun 1, Qin Wu 1
School of Manufacturing Science and Engineering, Sichuan University, Chengdu 610065, Sichuan, China
*Corresponding address: e-mail: sunlan@scu.edu.cn

Abstract: This research focused on adding low-cost manganese element to the cast iron instead of part of the expensive nickel and adjusting the contents of other elements to cast a series of the nickel iron of the different alloy composition. A series of cast irons with different compositions were prepared. The nickel and manganese contents in cast iron were reduced and increased, respectively. The other alloying elements were adjusted appropriately. The electrochemical measurements and the erosion-corrosion resistances of cast iron in 3.5wt.%NaCl solution and in 10wt% quartz sand were investigated comparatively. The results showed that: the electrochemical corrosion type of nickel cast iron in NaCl solution was oxygen absorption and the pitting corrosion happened in the matrix region near the phases of the graphite and metal cementite. The nickle cast iron at the process of erosion-corrosion in the sand solution mainly was shown as the plastic deformation and micro cutting forms. The low Ni cast iron in which Ni content was 8.36 wt.%, and Mn content was 6.57 wt.% exhibited better corrosion resistance than other low Ni cast irons and demonstrated comparable performance with high Ni austenite cast iron.

Keywords: nickel austenitic cast iron, erosion-corrosion behavior, saline slurry medium, manganese

1 Introduction

High nickel austenite cast iron has good wear resistance, good mechanical and physical properties and excellent manufacturability [1,2]. This cast iron generally contains 13.5%-36% nickel. The price of nickel is higher, resulting in higher cost of raw materials. Therefore, low-cost substitute elements should be used instead of nickel to ensure wear resistance and corrosion resistance of cast iron. Decreasing the nickel content and increasing the manganese content in high-nickel cast iron are desirable because of the lower cost of manganese. Both the Ni and the Mn elements can promote the formation of austenite. Nickel is dissolved in the austenite matrix, which can not only enhance the solid solution strengthening, but also increase the electrode potential of the matrix, and improve the wear resistance and corrosion resistance. Mn element can form the solid solution in the matrix and carbides to ensure higher hardness to improve the wear resistance of cast iron[3,4]. In this paper, a series of cast irons with different compositions were prepared. The nickel and manganese contents in cast iron were reduced and increased, respectively. The other alloying elements were adjusted appropriately. The electrochemical measurements and the erosion-corrosion resistances of cast iron in 3.5wt.%NaCl solution and in 10wt% quartz sand were investigated comparatively.

2 Experimental procedure (Bold, 10 pt., Arial)

Four groups of austenitic cast iron containing different nickel contents have been prepared, and their composition has been given in Table 1. The content of Ni increased gradually from A1-A3, and A4 as the high nickel austenitic cast iron was used as the compared group. Use "body" text style (10 pt., Times New Roman).

Table 1 Main alloy compositions /wt.%

Specimen	Mn	Ni	Cu	Cr
A1	7.88	5.57	2.87	0.28
A2	7.90	6.42	3.08	0.22
A3	6.57	8.36	4.33	0.41
A4	1.09	13.27	6.86	2.35

This electrochemical experiment was carried out in 3.5 wt.% NaCl solution at room temperature (25 ±3°C). The erosion-corrosion tests were carried out using a mud wear device, mainly composed of one sample holder, one drum and a stirring blade. The radius of rotation was 100 millimeters. The impact angle was 90 degrees. The liquid solid corrosion test was carried out under 25 + 3°C. In this study, the eroded corrosive fluid consists of 3.5 wt.% NaCl, 10 wt.% silica sand and distilled water. The erosion-corrosion tests were conducted for 30min in a drum, in which the coupons were placed on blades spinning at 360 r/min. The samples were weighed before and after scouring experiments to evaluate their erosion resistance. Scanning electron microscopy (JSM-5900LV) was used to observe the corroded samples.

3 Result and discussion

Electrochemical studies

From the potentiodynamic polarisation curves of the samples, the passivation region of the sample A2 is small and that of the sample A3, A4 is obvious which shows the characteristics of the typical activation - passivation metal polarization curve in the NaCl solution.

SEM was used to observe the morphology of the surface of nickel cast iron A3 and A4 by electrochemical etching, as shown in Figure 2. In the vicinity of the second phase, there are many different size and honeycomb round holes, which was resulted in the pitting corrosion.

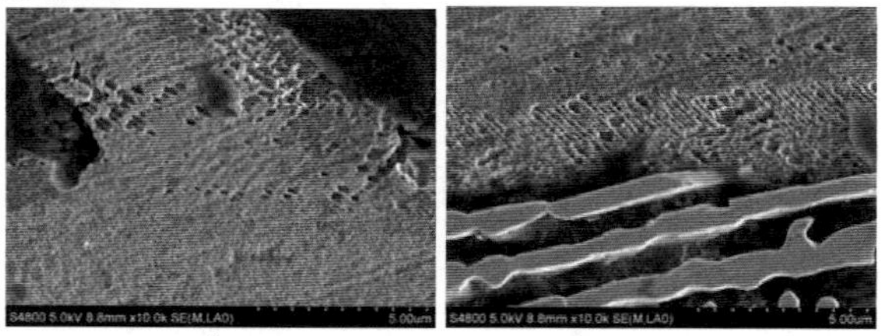

(a)A3 (b) A4

Fig.2 Images (SEM) of the samples A3 and A4 by electrochemical etching

Erosion–corrosion studies

The corrosion coefficient is determined and shown in three medium conditions in Fig. 3. The low-nickel and high-nickel cast irons show a good corrosion resistance in the 3.5 wt.%NaCl solution. The samples have a certain degree of weight loss in the sand solution of 10 wt.%, while the weight loss of those are the most serious in the mixed solution of 3.5 NaCl wt.% and 10 wt.% sand. By Fig.3, the corrosion coefficient of A3 sample in low nickel cast iron is the most close to A4 of high nickel cast iron, and lower than that of low nickel cast iron A1 and A2. The erosion–corrosion resistance of low nickel cast iron A3 and high nickel cast iron A4 is almost close.

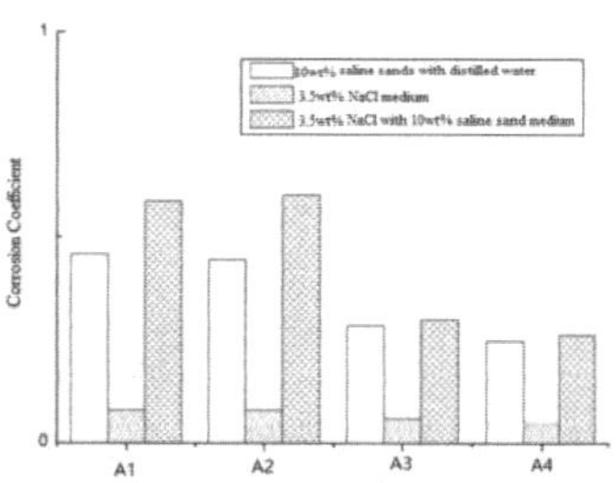

Fig.3 Images (SEM) of the samples after 30 min erosion–corrosion in 10 wt-% sand condition

The corrosion coefficient (Fig. 3) of the samples in saline slurry medium was not significantly increased compared with that in the pure NaCl solution and sand solution. Therefore, the interaction of the corrosion and wear is not obvious.

Because of the low erosion-corrosion coefficient, the erosion-corrosion resistance of the sample A3 in mixed solution was the best. The surface morphologies of A3 and A4 observed by SEM in the mixed medium were shown in Fig.4. The experimental results showed that the erosion-corrosion process of cast iron in saline slurry medium mainly made the impacted pits, corrosion pits, the cutting surface, and so on.

In Figure 4, the corrosion morphology of the samples is mainly showed the form of wear, and the corrosion of the small hole caused by electrochemical corrosion is not observed. This may be because the nickel cast iron in the NaCl environment will be a passivation phenomenon, form a layer of dense and hard oxide film on the substrate surface, and improve the corrosion resistance of the material.

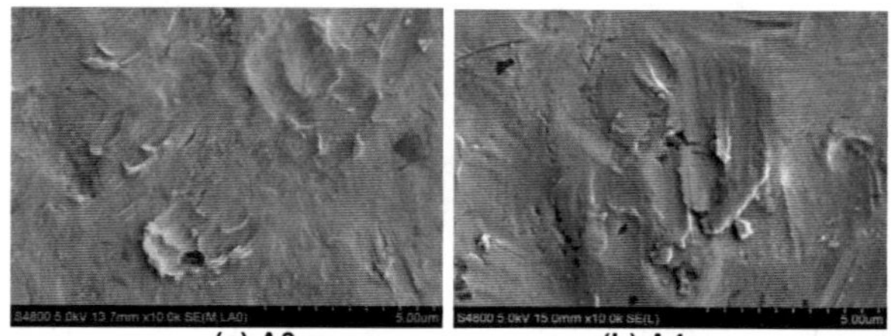

(a) A3 (b) A4

Fig.4 Images (SEM) of the samples after 30 min erosion-corrosion in 3.5wt.%NaCl solution and in 10wt% quartz sand condition

4 Conclusion

In 3.5wt.% NaCl solution, when Ni content in the low nickel cast iron are 6.42 wt.% and 8.36 wt.%, the shapes of the polarization curve are similar to the high nickel cast iron. The pitting corrosion has happened in the low nickel cast iron with 8.36 wt.%Ni and the high nickle cast iron during the process of electrochemical corrosion. The pitting corrosion preferentially occurs in the matrix region near the phases of the graphite and metal cementite. The nickle cast iron at the process of erosion-corrosion in the sand solution mainly was shown as the plastic deformation and micro cutting forms. The low Ni cast iron in which Ni content was 8.36 wt.%, and Mn content was 6.57 wt.% exhibited better corrosion resistance than other low Ni cast irons and demonstrated comparable performance with high Ni austenite cast iron.

References

[1] Alzafin Y A, Mourad A H, Zour M A, and Abuzeid O A. A study on the failure of pump casings made of ductile Ni-resist cast irons used in desalination plants, Eng. Fail. Anal., 2007, 14 (7):1294-1300.

[2] Yoganandh J, Natarajan S, and Babu S K. Erosive wear behavior of nickel-based high alloy white cast iron under mining conditions using orthogonal array, J. Mater. Eng. Perform., 2013, 22 (9): 2534–2541.

[3] Fatahalla N, AbuElEzz A and Semeida M. C, Si and Ni as alloying elements to vary carbon equivalent of austenitic ductile cast iron: microstructure and mechanical properties, Mater. Sci. Eng. A, 2009, A504 (1): 81–89.

[4] Milititsky M, Matlock D K, Regully A, and Dewispelaere N, Penning J, and Hanninen H. Impact toughness properties of nickel-free austenitic stainless steels, Mater. Sci. Eng. A, 2008, A496 (1):189–199.

Key Forming Technologies and Applications of Large Steel Castings for Heavy Equipment

Enxian Xu 1, Lei Tian 1, Jingpei Xie 2, Jiazan Wei 1, Zhonghua Hu 1
1. Luoyang CITIC Casting and Forging Co., Ltd, Luoyang, China
2. School of Materials Science and Engineering, Henan University of Science and Technology, Luoyang, China

*Corresponding address: e-mail: tianlei0522311@163.com

Abstract: This paper focuses on the key forming technology and application characteristics of large steel castings. By using casting steel melting and purification technology, numerical simulation analysis, combined with the filling and solidification laws of steel castings, the design principles of casting process for large steel castings are formulated. A riser design and verification method has been proposed, including a gating system that can control the timing of steel feeding, Multi ladle collaborative pouring and supplementary pouring technology, to ensure the internal quality of steel castings.

Keywords: Large scale steel castings; Smelting technology; Casting process; Numerical analysis

1 Introduction

Currently, China's manufacturing capacity for large steel castings is in a leading position in the world, but there are few manufacturers that can produce steel castings with a capacity of over 100 tons, and many large steel castings do not have direct production experience to learn from. This article takes 100-600t large steel castings as a research case, focusing on the key forming technology and application characteristics of large steel castings. By using casting steel melting and purification technology, numerical simulation analysis technology, combined with the filling and solidification laws of steel castings, the design principles of casting process for large steel castings are formulated. A riser design and verification method was proposed, using a pouring system that can control the timing of steel feeding, multi package collaborative pouring and supplementary pouring technology, and other measures to ensure the internal quality of large steel castings, providing experience and theoretical support for the production of 100 ton steel castings in the heavy equipment manufacturing industry.

2 Experimental procedure

The large-scale steel castings described in this article adopt the modeling method of sodium silicate-bonded sand pattern+core assembly. The smelting process adopts electric arc furnace initial smelting +LF ladle refining, and MAGMA numerical simulation analysis software is used to assist in casting process design. The rationality of the process plan and parameters is verified through actual production.

3 Result and discussion

1 Casting Steel Melting and Purification Technology

The combination of external refining and vacuum treatment with electric arc furnace and induction furnace forms a joint smelting and refining technology. Aluminum is added for final deoxidation before steelmaking, supplemented by bottom blowing argon and cavity blowing argon technology to control S $\leqslant$ 0.01% and P $\leqslant$ 0.015%, avoid steel oxidation, greatly improve the purity of steel, and reduce micro porosity defects. Adding 0.025% Nb, 0.1% Ti, or a small amount of rare earths during the smelting process can effectively refine the grain size of thick and thick wall castings, reduce the risk of coarse grains, and improve the comprehensive mechanical properties of castings.

2 Pad and Riser Design

The riser has the functions of filling, venting, and slag discharge. The riser is generally set at the position where the casting is thicker, and the pad and riser are matched to ensure that the casting meets the sequential solidification. The pad size for simple sections can be directly determined by a slope of 1:10, while for complex sections, the pad size is determined by a rolling hot spot circle, with a rolling coefficient of 1.03-1.05 for the hot spot circle. The size of the riser is generally 1.5-2.2 times that of the hot spot circle, and the height to diameter ratio of the riser must be greater than 1. After the riser setting is completed, the feasibility of the process plan can be verified through the casting modulus and yield rate. The modulus of the riser is usually 1-1.35 times that of the casting, and the yield of large steel castings is generally controlled at 50% to 65%. The riser is separated by a cold iron partition to achieve partition and contraction effect, and the cold iron is set at the bottom for the best cooling effect. For thick and large castings with inspection requirements, the effective feeding distance of the riser is controlled to be $L \leq 4d$ for hot spot, and the number of risers can be determined by the effective feeding distance. Try to use large risers instead of dispersed small risers for thick walled components.

3 Gating system design

The design of the gating system should strive to minimize the flow of molten steel into the mold and the process inside the mold. During the filling process, all gating systems should be unchoked, and the cross-sectional area of the sprue should be greater than 1.6 times the cross-sectional area of the ladle nozzle. The steel flow rate at the inner gate should not be too high to ensure that the steel is in a slow flow state, and the flow rate and state of the steel in the mold cavity are consistent. The use of a blocking gating system can effectively control the timing of molten steel entering the upper and lower layers, and has the advantages of both bottom gating system and step gating systems. It can effectively improve the temperature gradient of molten steel and ensure the sequential solidification of castings.

4 Collaborative pouring of multiple steel ladles

The total weight of molten steel in steel castings with a weight of over 100 tons is large, and the rising speed of the molten steel level is slow. Therefore, it is necessary to use multiple ladle combined casting and supplementary casting. When selecting the distribution of molten steel and the size of the water outlet, the rising speed of the molten steel level in the casting body cavity should not be less than 5mm/s, while ensuring that the time difference after pouring each ladle is as small as possible. Due to equipment limitations, it is not possible to ensure that the pouring amount reaches the required height of the riser process. Therefore, it is necessary to perform additional pouring, but it is necessary to ensure that the pouring height reaches 1/2 to 2/3 of the riser height, and the interval between each additional pouring is controlled within 0.5 to 1.5 hours. To prevent cracking during gas cutting caused by segregation under the riser, the content of carbon, phosphorus, and sulfur in the steel used for teeming should be lower than that of these elements in the poured steel. If there are no special requirements, ZG200-400 should be used for teeming.

5 Numerical simulation analysis

Casting is a complex discipline with many influencing factors in process design. With the help of computer solidification simulation technology, coupled with flow field, temperature field, and stress field, combined with the formation and deformation laws of defects in large steel castings, it can assist in the formulation of casting processes. After long-term numerical simulation and actual production tracking, the Niyama criterion analyzed through MAGMA software numerical simulation can effectively predict casting porosity and shrinkage defects, and the MAGMA simulation time is shorter, making it more suitable for production. When the Niyama criterion value meets the general UT testing requirements for large steel castings within the range of 0.5-0.7, it can be used to assist in guiding the process.

4Conclusion

(1) By using joint smelting and refining technology, supplemented by bottom blowing argon and cavity blowing argon technology, and adopting final deoxidation technology, the purity of molten steel is improved; Controlling the content of harmful elements such as S and P, and adding Nb, Ti, or a small amount of rare earths during the smelting process can effectively refine the grain size of castings, reduce the risk of coarsening, and improve the comprehensive mechanical properties of castings.

(2) Verify the feasibility of the process plan through casting modulus and yield. For thick castings with certain inspection requirements, control the effective feeding distance of the riser. Try to use large risers instead of scattered small ones. The use of a blocking gating system can effectively improve the temperature gradient of molten steel and ensure the sequential solidification of castings. The use of multiple ladle combined casting and supplementary casting is an effective method to ensure stable horizontal filling, rapid casting, and sequential solidification of steel.

(3) The Niyama criterion analyzed through numerical simulation using MAGMA software can effectively predict casting porosity and shrinkage. When the Niyama criterion value meets the general inspection requirements of large steel castings in the range of 0.5-0.7, it can be used to guide production processes.

References

[1] Xie J P. The Development and Current Situation of Steel Casting Technology in China [J]. Foundry, 2022, 71 (4): 395-420.

[2] Wei L Y, Dong Y H, Li X Y, etc. Process research and manufacturing of the largest domestic steel castings [J] Casting Equipment and Process, 2018 (3): 7-13.

[3] Xie J P, Wang A Q, Wang P F, etc. Casting technology for 100 ton steel castings [M] Beijing: Science Press, 2019.

[4] Foundry Branch of China Mechanical Engineering Society. Casting Handbook Volume 5, Casting Technology [M]. Beijing: Machinery Industry Press, 2011.

[5] Fan K X, Zhu S G and Yao X M. The Application of Blocking Pouring System in Large Steel Castings [J] Metal Processing (Hot Working), 2018 (12): 25-26.

Research on the Excitation Control Method of Induction Heating Power Supply

Xiaoming Kang 1, Fuli Zhang 1, Jingyan Liu 1, Dou BO1
1.Shenyang Research Institute of Foundry/State Key Laboratory of Advanced Casting Technologies,,No.17,YunFeng South Street, Tiexi District, Shenyang City, Liaoning Province, China
*Corresponding address: e-mail: 35873576@qq.com

Abstract (Bold, 10 pt., Arial): The study of using series inverter excitation control method for medium and high frequency induction heating, especially in special load parameter applications, can effectively avoid the impact of various interferences in the feedback loop caused by significant changes in load parameters on the control system, and ensure the stable operation of the equipment. It is particularly suitable for special induction heating methods such as vacuum suspension, heat transfer, quenching, and sintering.

Keywords: series inverter; separate excitation; load parameters; suspension.

Introduction (Bold, 10 pt., Arial)

Intermediate frequency induction heating equipment is widely used in the casting field, with control methods divided into series inverter and parallel inverter. Currently, the mainstream in the market is mainly series inverter and closed-loop control.

However, in practical applications, the production process is complex and variable, and the closed-loop control method requires a high level of load feedback signal. In application scenarios with significant changes in load parameters, due to the strong interference of the feedback signal, a large number of irregular clutter signals can directly affect the accuracy and effectiveness of the inverter power device, leading to the collapse of the entire inverter system and the inability to operate normally. As a result, the stability of the closed-loop control method cannot fully meet the stable operation requirements of the equipment.

Therefore, the original phase tracking feedback loop is cancelled, and the control board automatically generates a commutation signal that triggers the inverter element based on the configuration of the main circuit system. The continuously adjustable oscillation frequency output is used to control the normal operation of the equipment and achieve a pulse output signal of separate excitation.

During device operation, the true intermediate frequency voltage feedback signal of the load part does not directly affect the resonant sampling signal of the main control board. The control circuit is in an open-loop feedback form and operates independently. To avoid the feedback of noise interference generated by the load during the high-voltage operation of intermediate frequency power supply equipment, which may cause abnormal changes in control signal parameters, and ensure stable operation of the equipment.

Experimental procedure(Bold, 10 pt., Arial)

The excitation control method does not change the original structure of the main circuit, but only focuses on the original closed-loop feedback circuit on the main control board. After improvement and corresponding parameter adjustment, the goal of open-loop feedback control is achieved. The entire control system includes a power supply unit, start delay and power regulation unit, rectification trigger unit, inverter trigger unit, digital phase shifting unit, turn off time control unit, and line protection unit (see Figure 1).

Among them, the power supply unit, startup delay and power regulation unit, rectification trigger unit, and series inverter conventional circuit work in the same way.

The inverter triggering unit, digital phase shifting unit, turn off time control unit, and line protection need to be adjusted and modified according to the type of inverter component (thyristor or IGBT) used.

When the device is in its initial state, the power regulation voltage in the power regulation unit is set to zero. When the control board receives a startup signal at the startup port, a set of high-level signals are output to the startup delay unit, and the rectifier unit starts first. After being processed by the rectification unit, the pulse signal forms a pulse waveform with a frequency of 5kHz, which is output to the rectification port. By delaying, the charging of the rectifier main circuit is completed, and the rectifier thyristor remains fully open. Subsequently, the inverter unit starts and outputs the processed analog inverter pulse signal in the circuit to the inverter output port, driving the inverter main circuit to work and putting the equipment into normal working state. As the power regulation potentiometer gradually increases the control voltage, the absolute value of the power regulation bus voltage gradually increases. At this time, the width of the inverter waveform narrows, the operating frequency increases, and the output power also increases. When the output frequency of the control board exceeds the designed operating frequency of the load, the equipment maintains the highest resonant state of the designed operating frequency of the load, and the power automatically stops increasing. It enters the limited power protection state through line control. During operation, the feedback signals of the load do not affect the overall operation of the circuit and are only used for protection purposes. The line protection unit collects feedback signals from various channels and compares them with the set limit values to output limiting equipment operation parameter signals or alarm signals. Through the reset end, the power adjustment line is quickly reset to zero, ensuring a smooth shutdown of the equipment. Protection signals include intermediate frequency voltage limiting, intermediate frequency voltage overvoltage, DC current limiting, DC current overcurrent, intermediate frequency current limiting, intermediate frequency current overcurrent, water temperature, water pressure, etc.

For the selection of different inverter devices, the control circuit composition of the corresponding inverter trigger unit is different.

If the IGBT module is selected as the inverter device, the CD4046 lock loop integrated circuit is used as the inverter unit to adjust the operating range of the inverter frequency composed of R181, R182, and C42, and output a set of square wave pulses (currently the experimental frequency can reach 200kHz). Through an inverter, it is divided into two pulse signals. A group of square waves output by the inverter unit is divided into two pulses through the circuit. Through the inverter in the unit, it maintains a phase difference of 180 degrees with the other channel. At the same time, it passes through a differential capacitor to form two groups of square wave pulses with a phase difference of 180 degrees. After being driven and amplified by a transistor, it is output to the inverter pulse output port (see Figure 1).

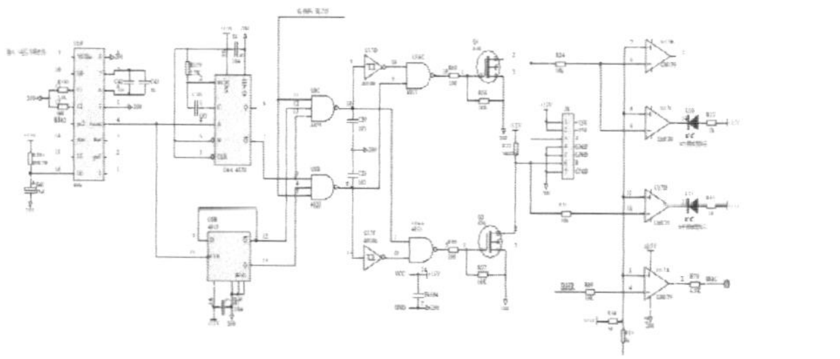

Figure 1. Schematic diagram of IGBT inverter trigger signal circuit

When using controllable silicon as an inverter device, a set of pulse waveforms in the circuit are passed through a serial counter in the digital phase shifting unit, selecting different waveform front and rear edges, and then cleared by the counter and inverter to form the required frequency band waveform, achieving digital phase shifting with a phase shifting accuracy of half a clock pulse. The time input terminal uses a Schmitt trigger, which has no limit on the rise and fall time of the pulse. This waveform is selected by a jumper switch in the circuit to output a feedback waveform close to the intermediate frequency voltage during the current main circuit operation. Together with the load intermediate frequency current waveform collected by external sensors, the resonance angle is compared through the turn off time control unit to obtain a phase difference signal. Then, the signal is compared with the limit voltage value set in the control board to control the thyristor turn off time and achieve stable operation of the equipment (see Figure 2).

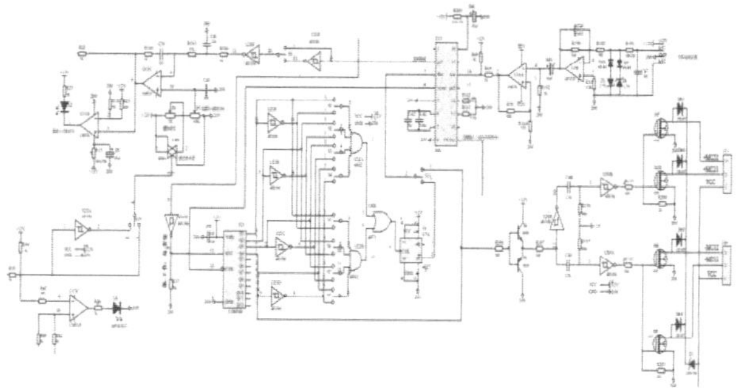

Figure 2. Circuit schematic diagram of thyristor inverter trigger, digital phase shift, and turn off time

Conclusion (Bold, 10 pt., Arial)

The above situations have been verified by actual equipment operation, and the use of separate excitation signal triggering signals. During the operation of the equipment, the feedback signal of clutter interference generated on the load side does not have a direct impact on the control circuit, and the entire control circuit is in an open-loop operation state. It basically avoids the feedback of noise interference generated by the load during the high-voltage operation of intermediate frequency power supply equipment, which causes abnormal changes in control signal parameters and ensures stable operation of the equipment.

Acknowledgments (Bold, 10 pt., Arial)

In terms of application methods, the design of excitation signals can be based on the selection and combination of inverter devices, and multiple comparison methods of pulse waveform signals can be set up to meet the control requirements of various main circuits for trigger signals, with a wide range of application prospects.

References(Bold, 10 pt., Arial)

[1] Fu Zhengbo. Induction Heating and Energy Conservation - Design and Application of Induction Heaters (Furnaces). Beijing: Machinery Industry Press, September 2008.

Digital Three-Phase Constant Melting Speed Protected Atmosphere Electroslag Melting and Casting Furnace Technology Research

Xuezhuo Hao[1], Rui Chen[2], Ruixin Li[3], Jingyan Liu[4], Wanqing Li[5], Xu Zheng[6], Yanping Yao[6]
1. China National Machinery Academy Group Shenyang Foundry Research Institute Co.
*Corresponding address: e-mail: 13904050858@139.com

Abstract: Electroslag melting and casting is a key melting process for the preparation of high-end pure steel materials, with obvious advantages in high-end product quality and material yield. Promote the rapid development of China's equipment manufacturing industry. Electroslag melting and casting equipment is divided into single-phase electroslag melting and casting furnace, double-pole tandem electroslag melting and casting furnace and three-phase electroslag melting and casting furnace. Among them, three-phase electroslag melting and casting furnace in more than 20 tons of large-scale equipment occupies a dominant position. However, the traditional three-phase electroslag furnace is almost always smelting in the atmosphere, the use of traditional power control, equipment operation is not stable enough, the product of [H], [O] and other gases cannot be effectively controlled. For this reason, in-depth study of the three-phase electroslag melting and casting furnace of the whole closed atmosphere protection function, large current direct supply type short network, position control accuracy, digital constant melting speed control and other technologies is imperative.

Keywords: Digitalization; Electroslag melting and casting; Fully enclosed protective atmosphere; Three-phase direct-supply short grid; Constant melting speed control

1 Introduction

Electroslag melting and casting is a key smelting process for the preparation of high-end special steel, which has obvious advantages in product quality and material yield and promotes the rapid development of China's equipment manufacturing industry. At present, China has about 1500 industrial electroslag furnaces, and the annual output of electroslag steel is more than 2 million tons [1]. The production of electroslag steel ranks in the forefront of the world. At present, the process form of large electroslag melting and casting equipment above 20 tons in China is mainly three-phase electroslag furnace. However, the existing three-phase electroslag melting and casting furnace due to the specificity of the structure are low control accuracy, poor reliability, short network arrangement is unreasonable and other problems. There are three electrodes and cannot realize the closed atmosphere protection, smelting in the atmosphere caused by gas content exceeds the fatal weaknesses, resulting in the production of 316L nuclear power pipeline and Cr12 rotor and other large-scale products with high requirements for the gas content of the qualification rate is low. Therefore, the three-phase electroslag melting and casting furnace is very important for the in-depth study of the three-phase electroslag melting and casting furnace of the fully enclosed atmosphere protection, high-current direct-supply type short network and digital constant melting speed control and other key technologies.

2 Three-phase protected atmosphere electroslag melting and casting furnace structure

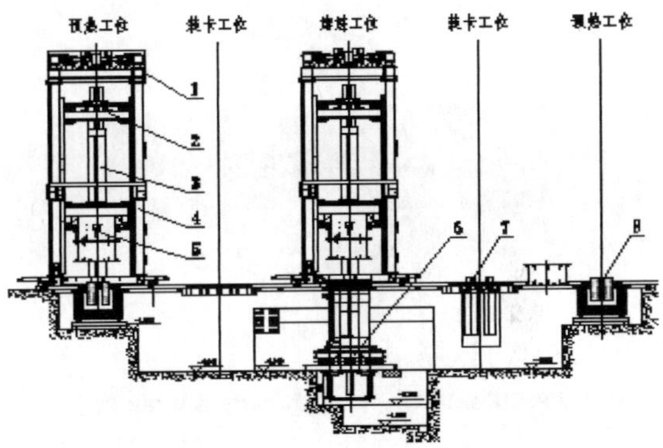

Fig.1 Equipment Process Layout

To ensure the precision of constant melting speed control and realize the full closed atmosphere protection, the new three-phase atmosphere protection electroslag furnace completely abandoned the original cantilever structure. The structural forms adopted are: double-gantry frame-type furnace head, furnace head carrying electrode linear moving exchange electrode, space-symmetric compact wiring direct supply type three-phase high-current feeder short network [2], single melting position, crystallizer fixed melting, fully enclosed inert atmosphere protection, 10kV AC50HZ three-phase power supply, unique and accurate three-point weighing platform, solid slag automatic arc-inducing, digitized constant melting speed control of the new type electroslag melting and casting furnace.

Inert atmosphere protection system

To ensure that the slag pool melt pool and air isolation to prevent the increase of [H], [O]. Equipment inert protective atmosphere system using a fully enclosed structure. Slag pool above the water-cooled stainless steel protective cover on the top of the crystallizer. The use of bellows to achieve soft sealing while ensuring the accuracy of the weighing data, the casting process in the protective cover to achieve a micro-positive pressure, higher than the standard atmospheric pressure of 200PPm. casting process to avoid the formation of self-consumption electrode surface iron oxide caused by ingot increase [O], and at the same time to avoid contact with the atmosphere to absorb the [H], [O] in the air.

The inert gas protection system includes: 2 sets of water-cooled stainless steel protective covers, inert gas flow and pressure monitoring control panels, oxygen concentration monitoring panels (inert gas parameters can be reflected and controlled on the computer interface), various types of flow and pressure instruments and piping.

Digital constant melt speed control system

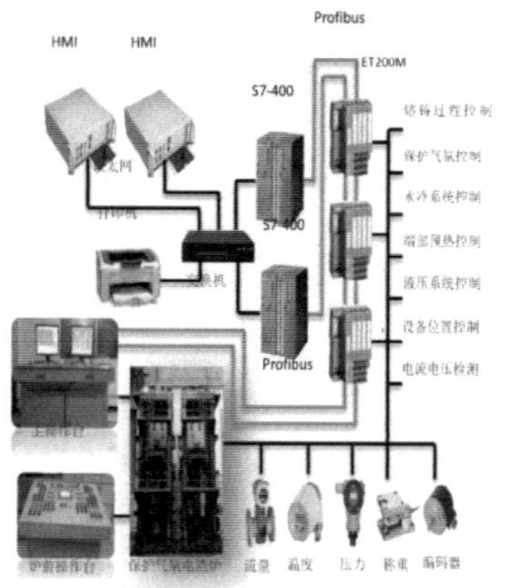

Fig.2 Control system structure diagram

The redundant CNC electroslag furnace PLC control system mainly consists of two industrial computers (HMI) equipped with WinCC monitoring software connected to Siemens high-end product S7-400 redundant controller CPU through industrial Ethernet and connected to three ET200M remote workstations through redundant Profibus-DP network, and the motion control is performed through the FM357-2 motion positioning drive module on each ET200M substation. FM357-2 motion positioning drive module on each ET200M substation connects servo drives and servo motors to perform motion control, timing and positioning functions.

Practical Applications

Equipment in a nuclear power pipeline company put into production, the production of varieties are mainly made of 0Cr18Ni9 (304) stainless steel, 1Cr9MoVNb (F91), 00Cr17Ni14Mo2 (316L), MC5 10 tons, 20 tons, 30 tons of electroslag ingot products are shown in Figure 7, the product's qualification rate of 100%, the product is mainly used in the military industry, nuclear power, and other fields.

Fig.3 Electroslag melting ingot

MC5 electroslag fusion casting ingot after forging after sampling inspection, self-consumption electrode and electroslag fusion casting ingot chemical composition comparison is basically unchanged, there is no compositional segregation problem, the crystalline organization is dense. Comparative data are shown in Table 1.

Table 1. Comparison of inclusions and gas content

	Gas(PPm)		Inclusions							
	H	O	A		B		C		D	
			Thin	Thick	Thin	Thick	Thin	Thick	Thin	Thick
D	2.5	19	-	-	-	-	-	-	-	-
C	3.0	23	1.0	0.5	1.0	0	0	0	1.5	0.5
M	2.0	18	0.5	0	1.0	0	0	0	0.5	0

References

[1] Study on primary carbides precipitation in H13 tool steel regarding cooling rate during solidification[J]. Guang-di Zhao; Xi-min Zang; Wan-ming Li; Zhuo Zhao; De-jun Li. China Foundry,2020(03).

[2] Effect of Ce on the cleanliness, microstructure and mechanical properties of high strength low alloy steel Q690E in industrial production process[J]. Hao Wang; Yan-ping Bao; Ming Zhao; Min Wang; Xiao-ming Yuan; Shuai Gao. International Journal of Minerals Metallurgy and Materials,2019(11).

[3] Chemical composition and structural identification of primary carbides in as-cast H13 steel[J]. Ming-tao Mao; Han-jie Guo; Fei Wang; Xiao-lin Sun. International Journal of Minerals Metallurgy and Materials,2019(07).

[4] Effect of slag on oxide inclusions in carburized bearing steel during industrial electroslag remelting[J]. Shi-jian Li; Guo-guang Cheng; Zhi-qi Miao; Lie Chen; Xin-yan Jiang. International Journal of Minerals Metallurgy and Materials,2019(03).

Research on Low Residual-Stress & High Dimension-Accuracy of Iron Castings for High-grade CNC Machine Tools

Kerui Li 1, Zengli Li 1, Donghai Wei 1, Yu Cui 1, Zhao Chen 1, Bingsheng Yin 1, Dunming Liao 2
（1. Zhengzhou Research Institute of Mechanical Engineering Co., Ltd., 450001 Zhengzhou, Henan; 2. Huazhong University of Science and Technology, 430074, Wuhan，Hubei）
*Corresponding address: e-mail: zrimpc@126.com

Abstract: Study on casting structure analysis, stress and deformation numerical simulation, alloyed cast iron with high carbon equivalent, residual stress test etc. has been conducted, its aging techniques has been developed. Machine tool castings with improved technology were produced, their residual stress and dimension accuracy were measured. The results show that residual stress of gray iron castings with new technology can be less than 20Mpa, its dimensional stability has been improved by 33% compared to traditional process.

Keywords: Machine Tool Castings, alloyed cast iron, residual stress, Dimensional Accuracy, Aging techniques

1 Introduction:

Material stiffness, residual stress and deformation of machine tool castings are important factors that affect the accuracy and accuracy retention of machine tools. Deformation and even cracking caused by residual stress of castings are main technical problems in foundry industry. Through systematic research, a set of stress and accuracy control technology of iron castings for high-grade CNC machine tool was proposed in the paper.

2 Experimental procedure:

Stress-boxes and T-type samples were poured, residual stress was measured with blind hole method, and dimension was measured with accurate laser interferometer.

3 Result and discussion:

1. Structure Analysis of Machine tool Castings

Among the factors affecting the casting accuracy, such as gravity, residual stress, assembly stress and anchor bolts, the residual stress is the key factor that affects the accuracy and accuracy maintenance of the casting,

The elastic modulus has a significant effect on the deformation of the casting, increasing the elastic modulus of the material is conducive to reducing the deformation of the casting. When the elastic modulus is increased from 120 GPa(gray cast iron) to 160 GPa(ductile cast iron), the deformation of a beam casting is reduced from 0.03576 mm to 0.02384 mm, with a reduction scale of 33.3%.

2. Stress and Deformation Numerical Simulation

A numerical simulation software has been developed, which was used to analyze and predict the stress distribution of the castings, and the stress results were consistent with the actual measurement.

The multi-point-dispersion-bottom filling casting method, anti-deformation design, heat preservation riser, slag gate basin, balance type, core strength and yield are important measures to avoid casting defects and reduce the residual stress and deformation of castings.

3. Development of a new alloyed Cast Iron

A new alloyed cast iron with high carbon equivalent has been developed. The test shows that residual stress decreases with the increase of carbon equivalent. When the carbon equivalent of ductile iron stress frame samples increases from 3.94% to 4.59%, the maximum residual tensile stress decreases from 119.5 MPa to 53.7 MPa, which decreases by 55%. High carbon equivalent combined composite alloying is the basis for obtaining high stiffness and low stress cast iron. At high carbon equivalent (3.7-3.9%), high stiffness and low stress HT300 gray cast iron with tensile strength ≥300MPa and elastic modulus ≥120GPa can be obtained by 0.5~0.7%Cu and 0.25~0.3%Cr composite alloying. High stiffness and low stress QT600-3 ductile iron with tensile strength ≥600MPa and elastic modulus ≥160GPa can be obtained by composite alloying of 0.5~0.7%Cu, 0.04~0.06%Sn and 0.4~0.6%Mn at high carbon equivalent (4.4~4.6%).

4. Residual stress testing and aging treatment technology for bed castings

The blind hole method is more suitable for residual stress measurement of cast iron than the other methods. The main frequency vibration aging treatment system and equipment have been developed, the precise temperature-controlled hot aging furnace has been developed, and the composite aging treatment process of "temperature-controlled aging+ precise heat treatment aging + rough machining + main frequency vibration aging +(natural aging)" has been formulated to to remove the overall internal stress, vibration aging to eliminate the peak stress and increase dimensional stability. The residual stress of gray cast iron and ductile cast iron can be reduced to 20MPa and 50MPa respectively.

5. Trial manufacture and dimensional precision test of cast iron for high-grade CNC machine tools

The machine tool samples were cast by traditional process (low carbon equivalent without alloying) and new process (high carbon equivalent + alloying + heat-treatment aging). The alloyed gray cast iron or ductile iron samples with high CE have higher tensile strength and elastic modulus at the same time: gray cast iron reaches tensile strength > 300MPa and elastic modulus > 120GPa; Ductile iron reached tensile strength>700MPa, elastic modulus > 170GPa. The maximum residual stress of low stress gray iron castings and ductile iron beam castings is 18.1 MPa and 49.4Mpa respectively, the deformation of guide rail is less than 4μm/4000mm, and the accuracy retention of the casting is more than 33% higher than that of the traditional process.

4 Conclusion:

This research has achieved remarkable results in the development and application of numerical simulation software in casting process. High quality and low stress alloyed cast iron has been developed, a complete set of manufacturing technology for high-precision and stable machine tool castings has been achieved. Finally, the guide rail deformation of the casting is less than 4μm/4000mm, and its geometric accuracy retention has been improved by 30%.

The deformation caused by the residual stress release is a long process, and in the future, it is necessary to strengthen the research on the change theory of the residual stress on the deformation of castings and the accuracy retention of machine tools, to establish a database, and to improve the accuracy and accuracy retention for China's high-grade CNC machine tools.

Acknowledgments:

Many Thanks to Professor Zeng Yi-cheng, and Cast Iron Expert Ma Jing-zhong for their guide during the research work.

References:

[1] Ma Jing-zhong, Li Ke-rui. Development Direction of High Quality Numerical Control Precision Machine Tool Casting [J]. FOUNDRY, 2014, 63(12) : 1217-1225..

[2] Ma Jing-zhong, Li Ke-rui. High Carbon Equivalent, High Strength, High Rigidity and Low Stress is the Development Direction of High Quality Numerical Control Precision Machine Tool Casting. National Science & Technology Key Projects of Numerical Control, 2014, 3:42-46.

[3] Ma Jing-zhong. Mastering Key Technologies to Improve Quality of Machine Tool Castings[J]. FOUNDRY, 2011, 60(7):666-672.

[4] Wei Dong-hai, Li Ke-rui, Wu Xian-long, et al. Study on High Strength, High Elastic Modulus and Low Residual Stress of As-cast Ductile Iron [J]. FOUNDRY, 2014, 63(2) : 120-127.

[5] Wei Dong-hai, Li Ke-rui, Wu Xian-long, et al. Numerical Simulation and Experiment on Stress and Strain in Iron Castings[J]. FOUNDRY, 2013, 62(2):134-137.

[6] Wei Dong-hai, Li Ke-rui, Wu Xian-long, et al. Study on Reducing the Residual Stress of Iron Casting and the Application in Machine Tool Lathe Bed [J]. MODERN CAST IRON, 2015, (5): 62-69.

[7] China national standardization management committee . GB/T 22315-2008 Metallic materials-Detennination of modulus of elasticity and Poisson's ratio[S]. Beijing: China Standards Press, 2008.

[8] Ministry of Water Resources of the People's Republic of China. SL 499-2010 Standard test method for determining residual stresses by the hole-drilling strain-gage method[S]. Beijing: China WaterPower Press, 2011.

[9] Wei Dong-hai, Li Ke-rui, Wu Xian-long, et al. Numerical Simulation of Machine Tool Iron Castings and the Reseach of Residual Stress [J]. FOUNDRY TECHNOLOGY, 2013, 35(259) : 221-223.

[10] Refaey A, Fatahalla N. Effect of microstructure on properties of ADI and low alloyed ductile iron[J]. Journal of Materials Science, 2003, 38(2) : 351-362.

[11] Hu Yong, Rao Li, Chen Guo-xiang. Effects of Mn, Cu on the microstructures and mechanical properties of nodular cast iron[J]. Ordnance Material Science and Engineering, 2012, 35(4):5-7.

Application of 3D Design and Simulation Technology in the Manufacturing of Steel Castings for Extra Large Bridges

PingLin, Xiaobing Yang,Youqing Lv,Zhangyu Xiao
Erzhong (Deyang) Heavy Equipment Co., Ltd., Deyang, Sichuan,China
*Corresponding address: e-mail: lpj908@163.com

Abstract: This article studies the application of 3D design and simulation technology in the manufacturing of steel castings for 300 ton ultra large complex spatial structure bridges. Through structural optimization, CNC machining and modular design methods, as well as 3D photography technology, we have successfully solved technical challenges such as dimensional accuracy control, internal quality assurance, uniformity of organizational performance, and machining alignment. This study not only improves the machinability of the product, but also ensures that its internal quality and dimensional accuracy meet high-tech requirements, filling the manufacturing gap in this field in China.

Keywords:3D design, simulation, CNC machining, 3D photography, bridge steel castings

1 Introduction

The manufacturing of ultra large bridge steel castings faces many technical challenges, such as complex spatial structures, strict dimensional accuracy requirements, and high standards of internal quality assurance. This study uses 3D design and simulation technology, combined with CNC machining and 3D photography technology, to successfully manufacture the largest, heaviest, and most complex load-bearing steel castings in the domestic cable-stayed bridge industry.

2 Main technical difficulties

(1)Large spatial dimensions, complex shapes, and large wall thickness differences make it difficult to control dimensions and ensure internal quality;

(2)High overall quality requirements, requiring 100% UT and MT testing, with a high acceptance level;

(3)The size and weight of a single sand core are large, making it difficult to ensure the strength and fixation of the sand core;

(4)The performance requirements for ultra large carbon manganese steel castings are high, and heat treatment is difficult to ensure;

(5)Large volume, overweight, and irregular polygonal structures make flipping, installation, and marking extremely difficult, and precision machining tolerances are difficult to ensure.

3 Main research contents

1.Optimized design structure

The original part structure design mainly considered the functionality and appearance of the product (as shown in Figure 1), without considering the manufacturing process of the product. The maximum wall thickness reached 1800mm, and the design weight was about 390t, reaching the limit of production and manufacturing. Not only is production organization difficult, but internal quality is also difficult to ensure. Through 3D design and simulation technology, the design structure was optimized from the perspective of manufacturing technology, with the maximum wall thickness adjusted to 700mm (as shown in Figure 2), improving the processability and non-destructive testing detectability of the product, and reducing the difficulty of production organization and later transportation and lifting.

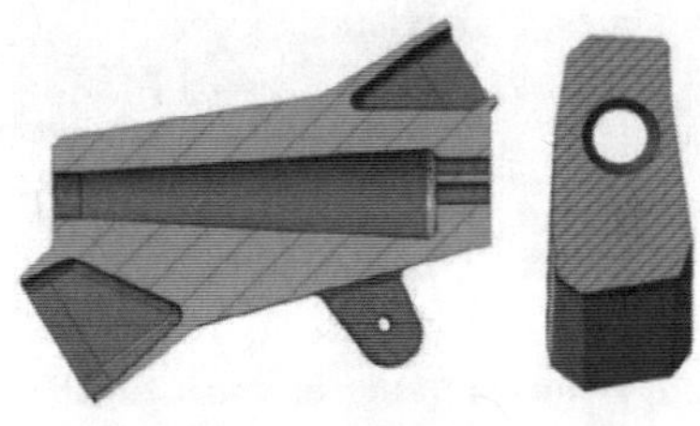

Figure 1 Original Design Structure

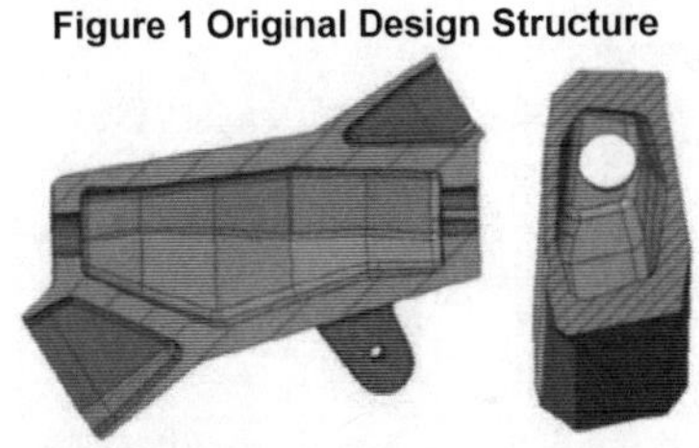

Figure 2 Optimized Design Structure

2. Design of casting process

The external dimensions of the steel castings are relatively large, and the hotspots are relatively scattered.Advanced MAGMA numerical simulation software was used for solidification simulation analysis to determine the final casting process plan(as shown in Figure 3).

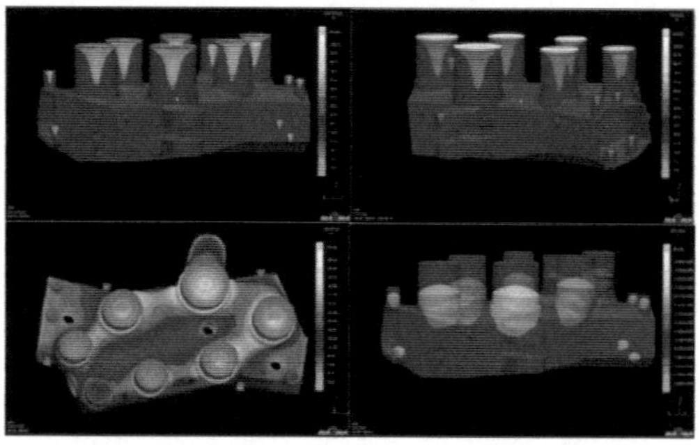

Figure 3 Solidification simulation analysis

3. Model design and production

Due to the complex spatial and multi-faceted structure of the product, traditional model making and inspection methods are difficult to ensure the correctness of the model shape and size accuracy. The modular design is carried out using the advantages of virtual manufacturing using 3D technology (as shown in Figure 4), combined with CNC machining and 3D photography detection methods, effectively ensuring the shape, structure, and dimensional accuracy of the model.

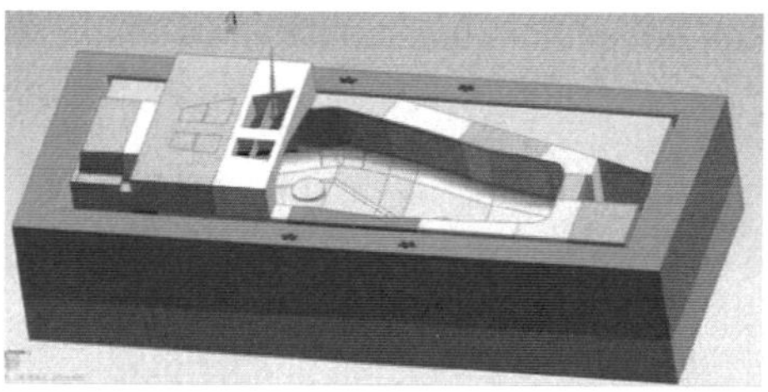

Figure 4 Modular Design of Model Structure

4.Molding process

Due to the large size and complex shape of the product contour, as well as the lack of easily measurable benchmarks during the molding process, traditional operating methods are difficult to control the cavity size. Using 3D virtual manufacturing technology and 3D photography technology, control the size of the sand mold cavity to ensure that the product size meets the process requirements.

5.Heat treatment process

Due to the large contour size of the product, various parts are subjected to uneven heating, making it difficult to control performance and uniformity. optimize the chemical composition of materials, use time-sharing and subsection control technology for spray cooling, effectively ensure that the product performance indicators meet the technical requirements.

6.Processing techniques

Using a 3D photography system to determine processing benchmarks, developing CNC programs, optimizing processing parameters, and ensuring product size and accuracy.

4 Conclusion

Through the application of 3D design and simulation technology, this study successfully manufactured 300 ton ultra large complex spatial structure bridge steel castings that meet high-precision requirements. The optimized design structure and advanced manufacturing technology not only improve the machinability of the product, but also ensure that its internal quality and dimensional accuracy meet high technical requirements. By solving multiple technical difficulties, this study provides an effective technical path for the manufacturing of ultra large bridge steel castings, which is of great significance for the development of this field in China.

References

[1] Principles of Metal Solidification. Beijing: Machinery Industry Press, 2000
[2] Simulation and Quality Control of Casting Engineering. Beijing: Machinery Industry Press, 2001
[3] Foundry Branch of China Mechanical Engineering Society Foundry Handbook - Volume 5, Foundry Processes

Effect of Silicon on Microstructure and Properties of Cr12 High Chromium Cast Iron

Shitao Zhu [1], Dahai Zeng [1], Jinsong Gui [2], Yongmao Yao [2], Yongzhe Wang [1*], Wei Li [1]
1. Institute of Advance Wear & Corrosion Resistant and Functional Materials, Jinan University, Guangzhou, Guangdong Province,China
2. Ningguo Huafeng Wear Resistant Materials Co.,Ltd, Ningguo City, Anhui Province, China
*Corresponding address: e-mail: wangyongzhe@jnu.edu.cn

Abstract: This study aims to investigate the influence of varying silicon content on the microstructure and mechanical properties of both as-cast and heat-treated high chromium cast iron with 12 wt.% Cr. The findings demonstrate that an increase in silicon content leads to a higher proportion of eutectic carbide in the as-cast structure, accompanied by a gradual refinement in carbide size. Moreover, an elevated silicon content enhances the hardness of as-cast samples, with 2.7Si exhibiting a hardness of 46.8 HRC, which is 2.2 HRC higher than the lowest hardness sample- 1.3Si. Following salt bath heat treatment, the microstructure of high chromium cast iron primarily consists of bainite, pearlite, retained austenite, secondary carbide, and eutectic carbide. Furthermore, an increased silicon content improves both surface hardness and wear resistance. Specifically, at 2 mm from salt bath treatment for the 0.8Si sample exhibits a hardness value of 59.9 HRC; whereas when the silicon content increases to 2.7 wt%, the sample's hardness reaches up to 60.8 HRC after salt bath heat treatment.

Keywords: high chromium cast iron, Si, hardness, wear resistance

1 Introduction

High chromium cast iron exhibits exceptional hardness and superior wear resistance, making it a crucial component in various applications such as ball mills, crushers, and metal rolling mills[1,2]. However, the elevated cost of chromium often leads to higher expenses associated with high chromium cast iron. To address this issue, incorporating silicon into the alloy can be considered as a viable strategy to enhance both the utilization rate of chromium and the wear resistance of chromium white cast iron. Silicon is a common element in high chromium cast iron, which is introduced into castings along with casting raw materials. Not only does silicon possess excellent deoxidation properties but it also effectively prevents the loss of chromium during oxidation processes. Moreover, silicon influences the cooling process by shifting the eutectic point towards lower temperatures, narrowing down the temperature range for eutectic reactions while refining eutectic carbides and austenite dendrites. Numerous studies have demonstrated that an increase in silicon content results in higher carbide content within high chromium cast iron along with finer carbide structures, thereby contributing to improved wear resistance [3-5]

This study focuses on investigating the effects of varying silicon content on the microstructure and mechanical properties of both as-cast and heat treated Cr12 high chromium cast iron. Furthermore, the impact of salt bath isothermal treatment on the microstructure and properties of high chromium cast iron is examined. The goal is to obtain high chromium cast iron with improved wear resistance by adding appropriate silicon elements.

2 Experimental procedure

Five high chromium cast irons with 12 wt.% Cr and varying silicon contents were cast into Y-shaped test blocks. Their chemical compositions are indicated in Table 1. Subsequently, these test blocks underwent salt bath isothermal heat treatment.

The microstructure and properties of the cast iron are characterized and analyzed using X-ray diffractometry, optical microscopy, scanning electron microscopy, and laser confocal microscopy.

The friction and wear behavior of salt bath treated specimens was measured by MFT-5000 Rtec multifunctional friction machine. Finally, MLD-10 dynamic load impact abrasive wear testing machine was used to characterize the impact abrasive wear resistance of salt bath treated samples. The impact abrasive wear sample was sampled from Cr12 high chromium cast iron samples 2 mm from the ionized salt bath interface.

Table 1. Chemical Composition of the studied high chromium cast iron

No.	C	Cr	Si	Ni	Mn	P	S
S1	2.75	12.43	0.78	0.103	0.549	0.019	0.023
S2	2.71	12.27	1.27	0.102	0.547	0.019	0.023
S3	2.67	12.21	1.62	0.076	0.525	0.020	0.016
S4	2.63	12.20	2.14	0.075	0.528	0.020	0.016
S5	2.58	12.05	2.70	0.074	0.525	0.020	0.016

3 Result and discussion

The cast high chromium cast iron exhibits a typical hypoeutectic structure, consisting of dendritic pearlite and interdendritic eutectic clusters. The eutectic structure is composed of M_7C_3 carbide and pearlite. With an increase in silicon content, the carbide content increases, and the size of the carbides gradually refines, thereby enhancing the hardness of the cast iron. Among the as-cast samples, 1.3Si exhibited the lowest hardness at 44.6 HRC, while 2.7Si demonstrated the highest hardness at 46.8 HRC

Ater undergoing salt bath treatment, the matrix structure of high chromium cast iron was achieved through isothermal quenching involving bainite and residual austenite. The granular secondary carbides present in the matrix are identified as $M_{23}C_6$ carbides. On the quenched surface, the matrix consists of bainite and residual austenite. As the depth of quenching increases, pearlite emerges within the matrix. In low silicon and high chromium cast iron, there is a relatively low content of pearlite observed. However, in high silicon and high chromium cast iron, an increase in quenching depth leads to a corresponding increase in pearlite formation. Notably, at 25m from the quenching interface for the 2.7Si sample, bainite disappears entirely while pearlite becomes predominant within the matrix.

The silicon content increase leads to a gradual enhancement in the hardness of the cast iron surface. The 0.8Si sample shows a measured hardness of 59.9 HRC at 2 mm from the quenching interface, while increasing the silicon content to 2.7 wt. % results in a hardness of 60.8 HRC at the same distance. Additionally, there is a continuous decline in hardness as samples move away from the quenching interface for similar silicon content levels. When the silicon content exceeds 2.1 wt. %, there is an abrupt decline in hardness indicating poor hardenability for cast iron with high silicon content ≥2.1 wt. %. The difference in hardness between samples with 0.8Si at distances of 2 mm and 25 mm from the quenching interface is only measured to be 0.5 HRC; however, this difference significantly increases to around 1.1 HRC when silicon content rises to approximately 1.6 wt. %. When the silicon content increases again, the hardness of the sample decreases significantly with the increase of depth, and the hardness of the 2.7Si sample at 25mm from the quenching interface is only 49.5HRC.

The results of the impact abrasive wear test demonstrate a gradual decrease in sample wear weight with increasing silicon content, followed by a slight increase at 2.7 wt.% silicon. Notably, when the silicon content reaches 2.1 wt.%, the sample exhibits the lowest wear weight loss of 102 mg, which accounts for approximately 77.3% of the wear weight loss observed in the 0.8Si sample. Subsequently, as the silicon content continues to increase, there is a small rise in sample wear weight loss while maintaining overall stability.

4 Conclusion

(1) The microstructure of cast iron is mainly composed of pearlite and M_7C_3 type eutectic carbide. With the increase of silicon content and eutectic carbide content, the hardness of cast iron is increased.

(2) After salt bath isothermal heat treatment, the microstructure of high chromium cast iron is mainly composed of bainite, residual austenite, secondary carbide and eutectic carbide. With the increase of silicon content, the hardness of cast iron sample surface gradually increased, but the hardenability gradually decreased. When the silicon content ≥2.1 wt.%, the hardenability of cast iron was poor.

(3) After undergoing salt bath heat treatment, the impact abrasive wear resistance of high chromium cast iron gradually improves with an increase in Si content, reaching a stable state when the Si content exceeds 2.1 wt.%.

Acknowledgments

The authors would like to thank the support of the National Nature Science Foundation of China (51204073). The National Nature Science Foundation of Guangdong province (2021A1515011728).

References

[1] Jones L C, Llewellyn R J. Sliding abrasion resistance assessment of metallic materials for elevated temperature mineral processing conditions[J]. Wear, 2009, 267(11): 2010.

[2] Aldrich C. Consumption of steel grinding media in mills–A review[J]. Minerals Engineering, 2013, 49: 77.

[3] Bedolla-Jacuinde A, Rainforth M W, Mejia I. The role of silicon in the solidification of high-Cr cast irons[J]. Metallurgical and Materials Transactions A, 2013, 44(2): 856.

[4] Lai J P, Pan Q L, Sun Y W, et al. Effect of Si content on the microstructure and wear resistance of high chromium cast iron[J]. Isij International, 2018: ISIJINT-2018-099.

[5] Lai J P, Pan Q L, Peng H J, et al. Effects of Si on the microstructures and mechanical properties of high-chromium cast iron[J]. Journal of Materials Engineering and Performance, 2016, 25(11): 4617.

Early Stages of Ba- Containing Inoculantion in Ductile Cast Iron

Mitja Petrič[1], Marica Tonkovič Prijanovič[2], Primož Mrvar[1], Črtomir Donik[3], Jaka Burja[3]
1. Faculty of Natural Sciences and Technology, University of Ljubljana, Aškerčeva 12, Ljubljana, Slovenia
2. Short-cycle Higher Vocational College - Novo mesto, Šegova ulica 112, Novo mesto, Slovenia
3. Institute of Metals and Technology, Lepi Pot 11, Ljubljana, Slovenia
*Corresponding address: e-mail: mitja.petric@ntf.uni-lj.si

Abstract: The basis for the production of high-quality cast iron and ductile cast iron is the use of complex inoculants. In addition to iron and silicon, inoculants usually also contain alloying elements such as zirconium, strontium, barium, calcium and rare earth metals. The addition of these elements reduces carbide solidification and increases the number of eutectic cells, which improves the quality of the cast iron produced.

The present study investigates the process of melting a grain of complex barium inoculant and its effects on graphite nucleation. In the study, the sample was prepared by inserting the inoculant grain into the ductile cast iron melt. The area between the inoculant grain and the less inoculated matrix was metallographically analysed using light and scanning electron microscopy (SEM). Energy dispersion spectroscopy (EDS) was used to determine the phases in the microstructure formed. It was found that graphite particles can already nucleate and grow in solid silicon particles when the inoculant is still dissolving and also from the melt, where the Ba and Ca concentrations are high, and BaO·CaO phases form, which serve as nuclei for graphite growth.

Keywords: ductile cast iron, graphite nucleation, barium

1 Introduction

Development of spheroidal cast iron origins from 1940's and the inoculation process is the matter of research ever since. Inoculants for ductile iron are ferro-silicon based alloys with alloying elements such as: zirconium, strontium, barium, calcium and rare earth metals. These types of inoculants are called complex inoculants. The addition of the inoculant has a large effect on the number of eutectic cells [1] resulting in improving the mechanical properties and machinability of the solidified casting, as the inoculant reduces the possibility of precipitation of free cementite in the casting and microporosity [2]. Among these complex inoculants, commonly used in the foundry industry is also the barium-containing inoculant (FeSiBa). Skaland [3] reported about nucleation mechanisms and the heterogeneous nucleation of graphite nuclei which occurs in the Si rich regions formed in the dissolved inoculant. This promotes the precipitation of graphite in the region where the inoculant begins to melt, precipitating tiny graphite beads. The presence of Ba has a positive effect on the precipitation graphite. Since a lot of barium-based inoculants are used in the foundry industry, in our research we were interested in the role of Ba in the dissolution of the inoculant in spheroidal cast iron and which Ba-containing phases can act as nuclei.

2 Experimental procedure

To observe how the barium inoculant dissolves and what role barium plays in inoculation, we used a base cast iron from a cupola furnace. After desulfurization by CaC_2 and nodularization by introduction of a wire filled with FeSiMg alloy, a ductile cast iron was obtained with the composition given in Table 1. Cast iron was inoculated with a complex inoculant (1.3 wt.%) containing, in addition to iron, 72 - 77 wt% Si, 1 - 2 wt% Ca, 0.8 - 1.6 wt% Al, 2 - 3 wt% Ba. Inoculant particle was put into a thermal analysis cup and overpoured by the melt. After solidification the sample was analyzed by optic and scanning electron microscopy (SEM) including EDS analysis. Area of interest was the surrounding of inoculant particle.

Table 1. Composition of alloy after nodularization

wt.%	C	Si	Mn	S	Cr	Mg
	3.90	2.20	0.18	0.009	0.04	0.06

3 Result and discussion

SEM and EDS analyses of inoculant particle showed that the microstructure consist of light grey and dark grey phases and eutectic regions. A dark phase represents silicon. The grey phase represents $FeSi_2$. Eutectic regions are rich in Ba, Al and Ca, next to Si. Some of these phases can reach above 50 wt.% Ba.

Figure 1 is presenting optic micrograph of area around dissolving inoculant particle. One can see the inoculant particle, the dissolving zone and the reaction zone. Furthermore, these zones are analyzed by SEM and EDS.

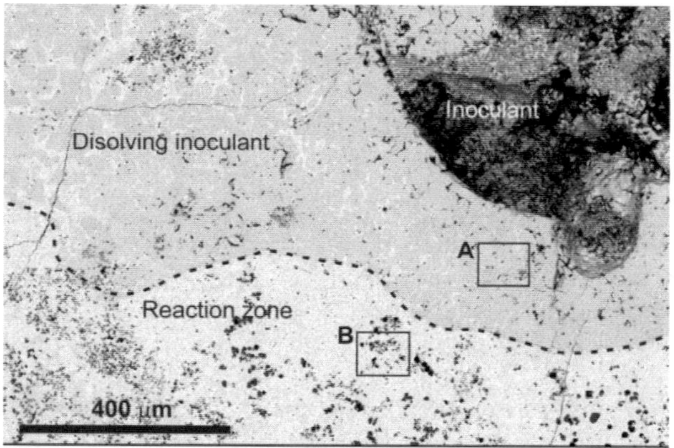

Figure 1: Optic micrograph of surrounding of inoculant particle with marked analysed areas

Figure 2 is presenting the complex particle from which graphite is growing. The centre of the particle is of white colour and according to line EDS analysis also presented in the figure it contains Ba, Ca and O. it is supposed to be BaO·CaO [4, 5].

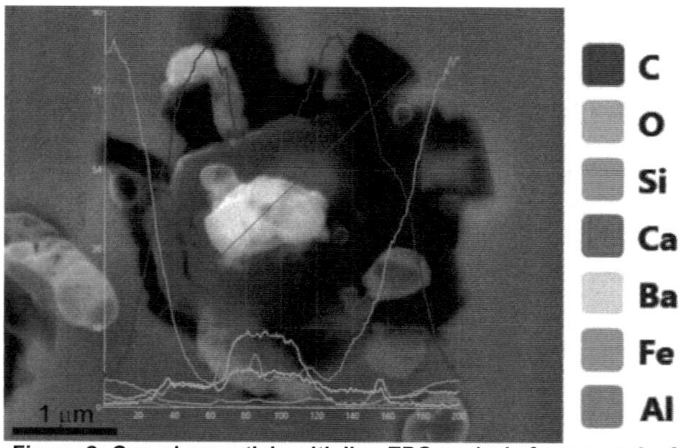

Figure 2: Complex particle with line EDS analysis from area A of Figure 1

Figure 3 is presenting a complex particle from area B (Figure 1). The rounded particle is nodular graphite surrounded by silicon. Inside the silicon also iron particle is present. From the Figure 3 it can be concluded that at just after the inoculation process when inoculant is still dissolving the graphite can nucleate inside the silicon particles present in the microstructure of the inoculant itself. During this stage it seems that silicon phase was still solid and by the diffusion the C entered the silicon particle and formed graphite.

b

wt.%	Fe	C	Si	Al
1	80.6	-	19.0	0.4
2	1.9	-	96.6	1.5
3	80.6	3.2	15.9	0.3

Figure 3: Complex particle with EDS analyses

4 Conclusion

The paper describes the early processes after inoculation when the Ba containing inoculant is still dissolving. It was confirmed that BaO·CaO particles can act as nuclei for graphite growth. Secondly it was found that when inoculant is dissolving and disintegrating the solid silicon particles are present in the melt which is rich in carbon. Carbon diffuses into the solid silicon and forms graphite. Presumably the authors suggest that after time, the silicon is dissolved in the melt and graphite, if the temperatures are low enough stays solid and grows further during solidification.

References

[1] Harding R A, Campbell J, Saunders N J. An assessment of our current understanding of the inoculation of cast iron, AFS Conference proceedings, 1998

[2] Lekakh S N. Effect of Nonmetallic Inclusions on Solidification of Inoculated Spheroidal Graphite Iron. International J. Metalcasting, 2019, 13: 47-57

[3] Skaland_T. Nucleation Mechanisms in Ductile Iron, in Proceedings of the AFS Cast Iron Inoculation Conference, 2005, 12-21

[4] NagendraNaik K, Yogendra K, Mahadevan K M. Solution combustion synthesis of $BaCaO_2$ nanoscale metal oxide and photocatalysis of methyl violet 10b dye. J. Critical Review, 2020, 4(7): 2259-2265

[5] Zhang R. Taskinen P. A thermodynamic assessment of the BaO-MgO, BaO-CaO, BaO-Al_2O_3 and BaO-SiO_2 systems. Aalto University publication series, SCIENCE+ TECHNOLOGY4, 2014

Casting Process of Large Guideway Grinding Machine Bed

Xiaoliang Pu, Qiang Zhang, Liuhua Shi, Siqing Gong
1. Changzhou Kaiyuan Casting CO., Ltd., Changzhou Jiangsu 213169
2. Jiangsu Foundry Association, Changzhou Jiangsu 213169
*Corresponding address:kyzz@k-yuan.com.cn

Abstract: The structural characteristics, technical requirements and casting difficulties of the large guideway grinding machine bed casting are introduced, and the key control technology in the casting production process is elaborated in detail from both the casting process and melting process, Including segmented molding, combining matching moulds, setting the anti-deformation, taking measures to prevent porosity, designing pouring system, controlling the melting l, etc. The production verification results show that the dimensional accuracy meets the requirements of CT10, the straightness of the guideway is within 2mm, and the hardness of the bed guideway is HB190-204.

Keywords: large machine tool bed, segmental molding combination molding, guideway straightness, guideway hardness

1 Introduction

Large guideway grinding machine bed castings must have the characteristics of high strength, high stiffness, high dimensional stability, good vibration damping and good wear resistance, etc., but it also requires high dimensional accuracy.

2 Experimental procedure

Casting structure & technical requirements

The external dimension of the bed is 15650*2204*800 mm, the weight is 28540 Kg, the thickness of the guideway surface is 60, and the size of the sand outlet hole on the bottom surface is 80*130. The guide surface hardness is required as HB180-220 and the dimensional tolerance class is CT10.No defects are allowed on the guide surfaces.

Confirm of casting process

(1) The bed guide surface is facing down, and the foot surface (bottom surface) is facing up, to ensure the quality of the guide surface and the smooth exhaust of the casting inner cavity of the cement core;

(2) According to our lifting equipment for the maximum lifting capacity of 50 tons and as the castings are narrow and long, the length of the mold will be divided into two segments, and then combinate them together, pouring it as a whole;

(3) Due to the small size of the air outlet on the lid is small, the head of the cement core is extended to set into the cover box surface.

(4) Setting of anti-deformation amount

According to the experience of our company, 2 ends of the low box will padded up when matching the box, so that the two top heads are higher than the middle point of 1 ‰ - 2 ‰.

(5) Setting of pouring channel

a). inner pouring channel

In order to ensure that the iron can fill smoothly, the anti-rain shower pouring system was adopted, Φ30 inner channel is distributed on the guideway surface, and the space shall not be more than 900mm

b). pouring time

This long bed decided that two sets of pouring systems along the length of the casting were arranged with the pouring weight and calculated the pouring time to be 65 seconds.

c). The proportions of the cross-section & each channel

$$Sobstruction\ of\ cross-section) = (22.6 * W)/ptfy * (Hp)^{\frac{1}{2}} = 184cm^2$$

The proportions of the cross-section of each sprue are as follows:

$$Sstraight\ channel: Scross\ channel: Stransition\ inner = 1:1.74:1.28:(>1.28)$$

The obstruction of the cross-section is a straight pouring channel, considering the specifications of the specialized ceramic tube, Φ150 ceramic tube was chosen.

Melting process

A set of the 6-ton melting furnace (1 to 2), 3.5 t &10 t medium frequency induction holding furnace.

(1) Target chemical composition

C 3.0-3.10; Si (former) ;1.3-1.5;Si (Final)1.47-1.8;Mn 0.8-1.1;P < 0.12 ; S 0.06-0.08 ; Cu 0.4-0.6 ; Cr 0.15-0.25

(2) Ratio of basic raw material

Scrap70%、pig iron5%、Recovery Iron 25%、Silicon Carbide 1%

(3) Inoculation Treatment for one time

The amount of the inoculant at the bottom of the package is in the range is 0.3-0.4%. Take a sample with the check of the spectral specimen and triangular specimen until it conforms to the target, and the triangular specimen with a white mouth width is in the range of 4 to 7 mm, then it is lifted to the casting pouring point.

(4) Pouring & Metal-stream Inoculation

Start pouring after detecting the temperature of molten iron in the package in the range of 1380 ~ 1390 ℃, at the same time with the metal-stream inoculation, silicon barium inoculants, with the inoculant of 0.1-0.15%.

3Result and discussion

Chemical Composition

Elements: C 3.09; Si1.72; Mn1.07; P0.041; S0.06; Cu 0.52; Cr0.22

Single-cast test bar

Tensile strength: 334 MPa, Test bar hardness HB244.

Metallographic organization: A-type graphite (grade 4), 100% pearlite.

Hardness of guideway surface:

Testing: On the two guide rails, 11 points are taken which are distributed uniformly, with two points at the head of the top of the rail each 25mm from the end, space between points is 1500mm.

TEST POINT	1	2	3	4	5	6
GUIDE1	190	190	191	191	200	200
GUIDE2	190	200	191	191	191	200
TEST POINT	7	8	9	10	11	
GUIDE1	191	191	191	204	204	
GUIDE2	200	204	204	204	191	

The difference between the highest and the lowest hardness on the guideway surface is HB14, the uniformity is quite good.

Metallurgical Quality Indicators

Maturity RG=1.245,Hardening Degree HG =1.12， Quality Factor Quality Factor Qi=1.11

The quality factor is over 1, which means the process performance is good.

2Dimensional Accuracy of Castings

Straightness of guide rail 2mm and the casting dimensional accuracy reaches the class of CT10.

4 Conclusion

(1) For large and long beds, the following difficulties are solved by modeling in two sections and combining with boxes ,such as solve the problem of insufficient lifting capacity of traveling crane and reduce the difficulty and cost of transportation and stacking for patterns and sand boxes, etc..

(2) For the large long bed, two sets of pouring system are adopted, and the pouring time and obstruction of cross-section area are calculated according to the table by half of the casting weight, which is in line with the requirements of casting technology.

(3) In the melting, the use of high scrap steel addition, plus 1% of silicon carbide, with low alloying, and a inoculation and metal-stream inoculation can obtain the castings with high strength, excellent processing performance.

References

[1]. Peng Fan, Yuan Xiaolei, Ruili Xue, Modern Cast Iron Technology, Mechanical Industry Press.

Analysis and Mitigation of Microporosity in Spheroidal Graphite Cast Iron Castings

Andrea Montagnoli[1], Adriana Mura[1], Linda Mazzocco[1]
1. Proservice srl, Via Marco Polo 3 Borgoricco (PD), Italy
*Corresponding address: e-mail: a.montagnoli@proservicetech.it

Extended abstract: Microporosity in spheroidal graphite cast iron castings presents a critical defect that adversely affects mechanical properties and overall functionality. The unpredictable emergence of this defect complicates its analysis and mitigation. Despite rigorous process controls aimed at maintaining stability and consistency, foundries frequently encounter microporosity. These defects can appear in established castings or tooling even when the chemical composition and pouring temperatures are within specified limits.

1 Introduction

This study investigates the causes and solutions for microporosity defects that unexpectedly appeared in isolated areas of spheroidal graphite cast iron castings. The investigation involves a comprehensive examination of the underlying causes of these defects, followed by the implementation of corrective actions to mitigate the issue.

2 Methodology

In-depth analysis and mitigation of microporosity defects were facilitated by the integration of advanced process control systems. Thermal analysis was pivotal, providing real-time data on the cooling curves and solidification patterns of the molten iron. Additionally, the monitoring of late stream inoculation was crucial, using systems designed for accurate and consistent inoculant delivery during the pouring process. These advanced monitoring systems, combined with the foundry's existing control measures, enabled a detailed examination and a deeper understanding of the factors contributing to microporosity.

Findings

The investigation identified two primary factors contributing to the occurrence of microporosity: excessive inoculant addition and a high percentage of remelted cast iron returns.

In the case of excessive inoculant addition, inoculants were introduced to prevent carbide-related defects in thin-walled castings, since produced with vertical pouring lines. These castings were produced using pearlitic spheroidal graphite with a high carbon equivalent within the eutectic range. However, the increased use of inoculants resulted in over-inoculation, shifting the cast iron into the hyper-eutectic range. This shift disrupted the proper functioning of the feeding system, leading to the emergence of microporosity in isolated areas of the casting. Over time, this phenomenon resulted in a gradual increase in defect incidence, thereby increasing the scrap rate and, consequently, the amount of cast iron returns.

The high percentage of remelted cast iron returns further exacerbated the issue. As the scrap rate increased, a larger proportion of cast iron returns were remelted, becoming the predominant part of the metallic charge mix. This created a feedback loop, where the introduction of inoculants, both through external addition and the remelting of cast iron returns, perpetuated the defect.

3 Corrective Actions

To address the issue of microporosity, a series of corrective actions were implemented. The first corrective action involved reducing the introduction of inoculants through the metallic charge. A balanced charge mix of cast iron returns, steel scrap, and pig iron was developed to minimize the introduction of hereditary inoculants. This "clean-up" of the metal aimed to eliminate any residual effects of the inoculants.

The second corrective action focused on optimizing the amount and flow of inoculants introduced during the casting process. The quantity of inoculants was reduced, and a more precise and centered flow of inoculants with the cast iron was ensured. This adjustment was critical in preventing over-inoculation and maintaining the iron within the eutectic range.

Additionally, an operating procedure was established to guarantee precise and consistent eutectic solidification during the pouring process. This procedure involved the appropriate management of the melting furnaces, supported by thermal analysis systems.

4 Results

The introduction of this new operating procedure was instrumental in reducing the incidence of microporosity defects from percentages between 8% and 12% to below 2%. The advanced process control systems, combined with meticulous monitoring and adjustments to the charge mix and inoculant flow, proved effective in mitigating the issue. This case study underscores the importance of stringent process control in foundries, demonstrating that metallurgical defects such as microporosity can be resolved without modifying tooling or the feeding system, which could potentially worsen the situation.

5 Conclusion

The study highlights the critical role of advanced process control and meticulous, continuous monitoring in addressing microporosity defects in spheroidal graphite cast iron castings. By reducing inoculant usage and optimizing the charge mix, foundries can significantly improve casting quality and reduce defect rates. The introduction of a balanced charge mix and precise control of the inoculant flow were decisive in mitigating the microporosity issue. Furthermore, the establishment of an operating procedure to ensure consistent eutectic solidification was crucial in achieving the desired outcome.

In conclusion, this study demonstrates that rigorous process control and advanced monitoring systems are essential in addressing metallurgical defects in foundries. The successful mitigation of microporosity in spheroidal graphite cast iron castings without the need for tooling modifications highlights the effectiveness of the implemented corrective actions. By maintaining stringent process controls and continuously monitoring key parameters, foundries can ensure the production of high-quality castings with minimal defects.

Ranking of Selected Production Parameters in Aspect Of Gas Defects Prediction in Cast Iron Castings

Pawel Popielarski[1], Robert Sika[1], Michal Rogalewicz[1]
1. Faculty of Mechanical Engineering, Poznan University of Technology, Poznan, Poland
*Corresponding address: e-mail: pawel.popielarski@put.poznan.pl

Abstract: The paper presents a method of analysis of datasets containing a very high number of parameters coming from data acquisition of partial processes of production of iron alloys casting in industrial conditions. It is synonymous to not considering classic methods of experiment planning. In such cases, it is advisable to reasonably limit data quantity by taking actions described in the paper. It may be achieved, for example, by ordering (ranking) relevance of influence of parameters on a selected feature of casting quality. In the initial stage, a knowledge of a group of experts was utilized for that purpose. Basing on a ranking, assumed on this basis, an influence of parameters selected in such a way on formation of a selected casting defect (gas porosity, GP) was studied, with use of statistical tools. An assessment of relation of production parameters with a selected feature characterizing quality of a cast product was performed. Certain dangers of selecting typical statistical tools for identification of selected relations were shown. Moreover, guidelines worth considering while defining topography of information sources about casting processes were presented, as well as conditions of acquisition and gathering of such data.

Keywords: castings defects, gas porosity, foundry data acquisition, cast iron

1 Introduction

Casting processes subjected to monitoring can generate a great amount of data, related to course of the realized processes. It may therefore lead to gathering of data in quantities not corresponding to needs of process control, neither qualitatively, nor quantitatively. In this situation, one of the most fundamental tasks of data acquisition is selection of relevant sources of information on these processes.

In a situation of an excess amount of parameters and in order to avoid false interpretations of influence of the insignificant ones, it is important to select the most important ones and order them by a criterion of predicted influence on a given type of defects.

2 Experimental procedure

The data analyzed was obtained in one of European foundries. The production is realized on an automated moulding line with a horizontal division of mould made of green sand, with presence of cores in a mould. The database initially contained more than 250 partial parameters referred to production processes (INput data), as well as several dozen related to the final casting quality (OUTput data).

It was proposed to divide the INput data into three categories (sets): MOULD, METALLURGY and POURING, while the OUTput data was assigned the name of QUALITY. In these sets, sub-categories were distinguished along with parameters representing them. A part of a proposed, ordered structure of the full database for the METALLURGY category and the "Final chemical composition" sub-category is shown in Figure 1.

A basic assumption was obtaining a single complete data record (meaning a record containing a complete set of recorded data), grouped according to the categories/sub-categories, representing particular partial processes. Such a record (a single row in a data table) includes values of all parameters: first the INput, then the OUTput, assigned to a single series of castings for a specific assortment. The series may contain several, more than a dozen or up to several dozen castings.

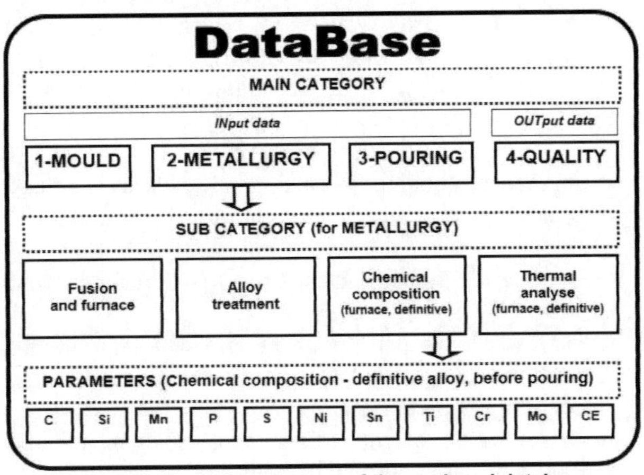

Figure 1. Part of a structure of the analyzed database (METALLURGY) with division into main categories, sub-categories and parameters.

Must be noted specific for the small-sized castings – defects of a single, cast piece can rarely be assigned to order of its pouring, meaning its „number" in a set of castings of a single series. Hence, some measured parameters have to be averagely assigned to a certain number of castings of a given series, sometimes to castings of 2 or 3 short series

In scope of the studies, the above mentioned ranking of significance of influences on the gas porosity defect was created. As a result, 12 parameters characterizing properties of a green mould sand (classic bentonite sand) were selected and one parameter (pouring temperature) was added, measured by a pyrometer for each poured form, as a temperature of a stream from a ladle.

Moreover, certain assumptions were taken, related to necessity of assigning a recorded daily percentage of defect fraction of a total number of castings to particular series of castings of the same assortment. The authors applied two distinct approaches:

- Procedure P1 – distribution of the total percentage of defective castings on a given day for each record containing a series of pouring moulds (one series consisted of several to several dozen castings). In this case, weighting factors related to the size of the series were used) and associating this share individually with the parameters of each record,
- Procedure P2 – assigning the averaged daily defect share (all series of a given assortment) to the averaged parameter values (categories/sub-categories) from individual records, without their individual association with this share.

3 Result and discussion

It was attempted to expose the known high influence of high pouring temperature on possibility of occurrence of a defect of the gas porosity type. For this purpose, for each particular series the Taver+1sigma parameter was determined (an average value of pouring temperature increased by one standard deviation), expressed as a percentage. Description of a procedure of calculation of this parameter is presented below:

- for all pours of particular days, a number of temperatures equal or greater than the day average increased by one standard deviation (+1 sigma) was counted,
- the above-mentioned, in practice, is translated into values of 1405°C +27°C = 1432°C,
- in such a way, a percentage of pours of temperature exceeding the 1432°C was determined (the Taver+1sigma parameter).

Change of value of the r-Pearson coefficient for the selected pairs of parameters (Diff between 0,52 to 1,01 – in some cases even the sign was changed) made the authors perform a more detailed analysis of simultaneous influence of two parameters, chosen in such a way, on the analyzed defect of gas porosity.

After conducting a thorough analysis, it was found that in a general case, approx. 35 pours (molds) are made per one series on average. There are few cases, where a number of pours is much smaller (approx. 20 molds), while a percentage of castings with the gas porosity defect is higher than 50%. And vice versa – for a greater amount of pours (approx. 45 molds), very small percentage of defects was noted (approx. 5%). These cases are relatively rare, not corresponding to a tendency of presence of the gas porosity defect in case of a greater amount of pours. Eventually, they were discarded from the further analysis. Usual percentage of the defects varied between 16 and 30% (Figure 2). After conducting the described procedure, it turned out that it is still possible to indicate the expected and knowledge-compatible influence of a high pouring temperature on increase of intensity of forming of the gas porosity in castings (Figure 3).

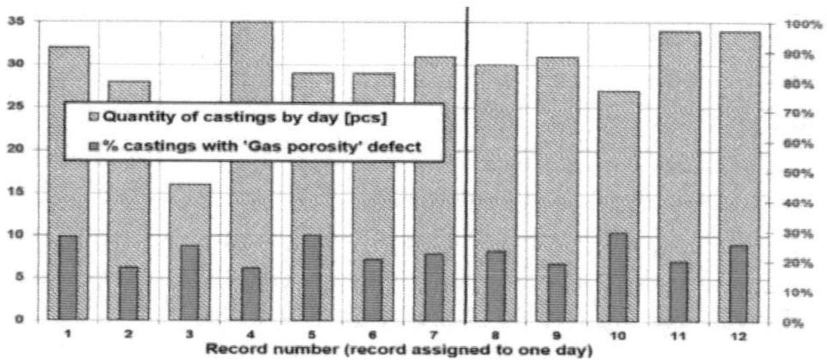

Figure 2. Cases for the most frequently cast series (the left axis) with reference to a number of defects.

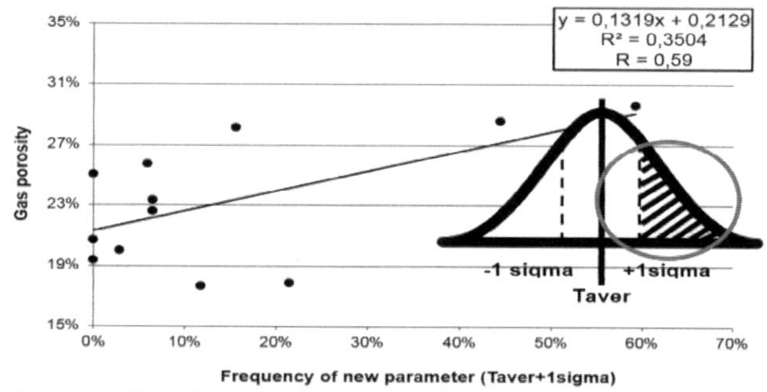

Figure 3. Final diagram of influence of the Taver+1sigma parameter on a percentage of defect of the gas porosity type.

4 Conclusion

Formal relations between the gas porosity defect (and other defects as well) and some parameters from data acquisition, without limitation of their quantity was indicated. The methods presented in the paper were preceded with ordering of the datasets. After applying an authors' method of determination of a parameter considering the average pouring temperature and its deviation (Taver+1sigma), as well as assuming frequency of occurrence of a percentage range of defects, it was found that the pouring temperature, particularly its high value outside the average, has an expected influence of occurrence of the gas porosity in the analyzed type of assortment.

Precision Spray Forming (PSF) of High Performance Alloys

Y. Yang [1]
1. Foshan Fenghe PSF Technology Limited, 70 Leaguer Science Park, CN-528225 Foshan, China
*Corresponding address: e-mail: yunfeng.yang@fenghe-psf.com

Abstract: Features of Precision Spray Forming (PSF) in production of high-performance alloys are described with example of industrial applications. Its supper rapid solidification and simple process makes uniform and fine structure in the metals, results in better mechanical properties and cost performance.
The contents cover ferrous as well as nonferrous metals. Industrial application results reveal that PSF makes good techno-economic benefits.

Keywords: Precision Spray Forming, PSF, High Performance Alloys

1 Introduction

Spray forming (SF) is invented in GB in early 1970's. Advantages of SF are that its solidification cooling rate is around 10k °C/s, about 100k time as high as that in casting. As a result, so solidified metals have fine and uniform microstructure with hardly any macro-segregation. SFed metals have normally 50-100% higher impact toughness and about 20% high tensile strength, better cold and hot work-ability, and better machinability. Such microstructure and property advantages fascinated massive R&D for SF of near-net-shape for nearly 30 years. However, such efforts interrupted at the end of last century due to a few technical obstacles, such as metallurgical processes, atomization method, substrate materials, etc. Through an R&D project 2000-2003, the author dissolved those obstacles and established Precision Spray Forming (PSF) process [1, 2]. An example of PSF forging die is shown in Fig. 1, with spontaneous improvement in life-time, costs and lead-time.

Besides near-net-shape, PSF covers stable production of many high-performance alloys, bimetallic component and particle reinforced MMCs.

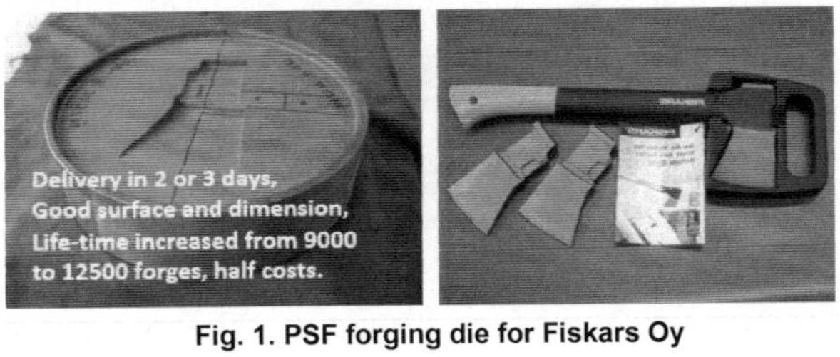

Fig. 1. PSF forging die for Fiskars Oy

2 Production of special steels

Because of systematic optimization, PSF is more stable in production special steels, especially high alloyed steels, such as cold work steel and high-speed steel. A PSF ingot can be 6 tons.

Fig. 2. shows microstructure difference between ESR- and PSF-made M2 high speed steel.

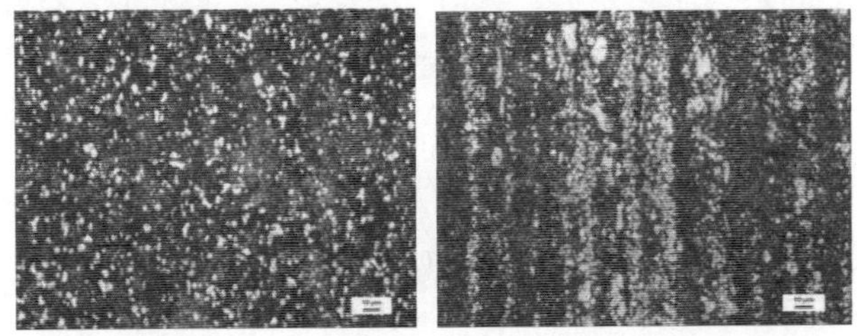

Fig. 2. Microstructure of PSF-M2 (L) against ESR-M2 (R), 100x, 4% nital etched.

As a result of the fine structures, PSF-M2 increases impact strength by 100-150% at the same hardness HRC66, as can be seen in Fig. 3. Cutting tool lifetime of PSF-M2 increased by about 4 times.

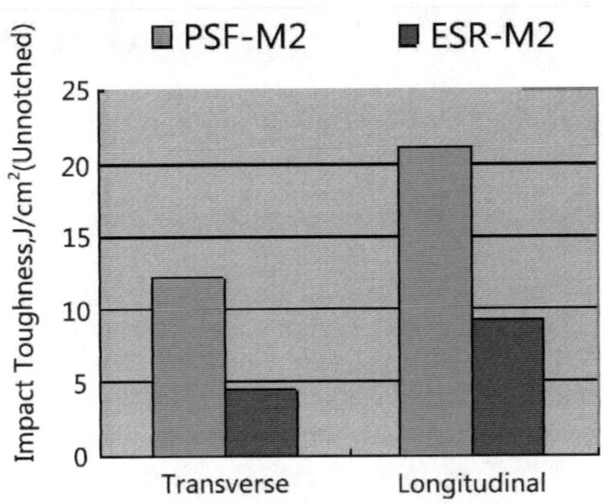

Fig. 3. Impact toughness of PSF-M2 & ESR-M2, sampled from Dia. 200 forged bar, HRC66.

Fig. 4. shows microstructure difference between ESR-D2 and PSF-D2 cold work steel. Impact toughness of PSF-D2 is about doubled over ESR-D2 at HRC60, and such made wear parts' lifetime is more than doubled.

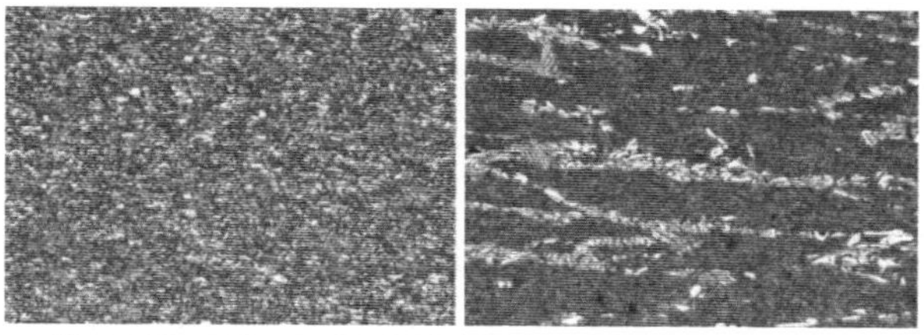

Fig. 4. Microstructure of PSF-D2 (L) against ESR-D2 (R), 100x, 4% nital etched.

Not only outstanding properties, PSF steels have also excellent cost performance. Good industrial application results also received with corrosion and wear resistant steel, high-grade mirror steel, etc.

3 Bimetallic composite structured materials

As discussed above, PSF steels have much better mechanical properties and tooling lifetime. PSF also can reduce material cost by spay cladding to make bimetallic materials. Fig. 5 shows a bimetallic gear hob. Its outer part is made of PSF-T15M, and inner part is of H13, with metallurgical bonding between the two materials. That way, better life-time is kept and raw material costs are reduced by half.

Fig. 5. Microstructure of the bimetallic boundary, 500x (R), PSF-T15M on the left PSF-H13 on the right; and the gear hob (L).

4 Particle reinforced MMCs

PSF is an efficient way to make particle reinforced MMCs. Fig. 6 shows 10% SiC particles of ≤10 μm uniformly distributed in Al-alloy. It is of the same easiness to make particle reinforced steel or other alloys.

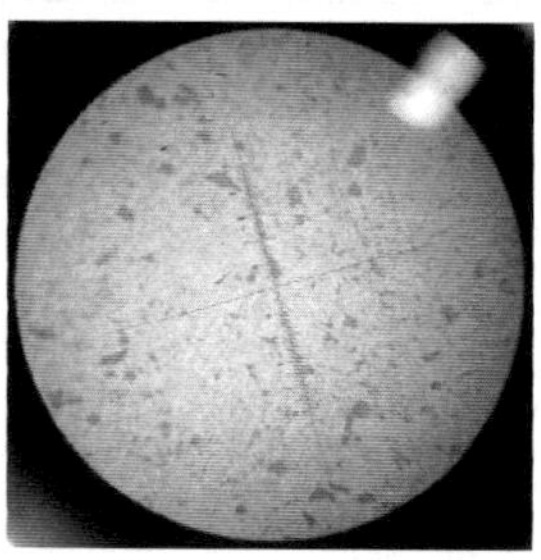

Fig. 6. Microstructure of 10%wt SiC particle reinformced in 6066 Al-alloy, 500x

5 Production of special Al-alloys

Tens of special Al-alloys are in our production scope. An example is high strength Al alloy 7055. Composition of 7055 is listed in Table 1, and microstructure of 7055 of different processes is shown in Fig. 7, mechanical properties are listed in Table 2. Advantages of PSF should be clear.

Table 1. Composition of 7055 Al-alloy (wt%)

Cu	Mg	Zn	Fe	Zr	Al
2.3	2.0	8.0	5.01	0.15	Bal.

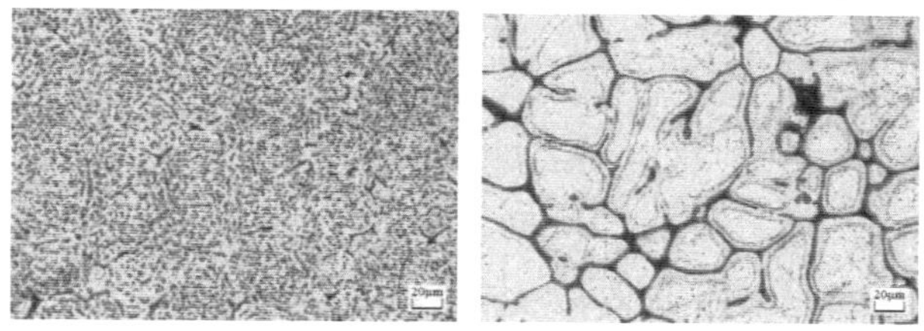

Fig. 7. Microstructure of as-sprayformed (L) and as-cast 7055 (R) Al-alloy, 500x

Table 2. Mechanical properties of different 7055 Al-alloy

Grades	σ_b / MPa	$\sigma_{0.2}$ / MPa	δ / %
Cast-7055, T6	580-630	550-600	≤10
PSF-7055, T6	720-740	690-705	14-16

6 Production of other alloys

Production of other high-performance alloys, such as supper alloys and copper alloys are to be presented as well.

7 Conclusion

High speed solidification and capabilities to make near-net-shape, bimetallic composite structures and MMCs make PSF a unique technology that can simultaneously improve properties or lifetime, costs, and lead-time of high-performance alloy products.

References

[1] Yang Y. and Hannula S-P. Precision Spray Forming – A New Approach to Rapid Tooling. TCT 2003, 11th – 13th November 2003, NEC, Birmingham, UK, 6 p.

[2] Yang Y. and Hannula S-P. Development of Precision Spray Forming for Rapid Tooling. Materials Science and Engineering A, 2008, 477 63-68.

Influence of Intercritical Quenching Treatment on Microstructure And Mechanical Properties in a High-Strength Low-Alloy Steel

Changfu Li[1, 2], Xu Wang[1, 2], Yunbao Gao[1, 2], *Zengrui Wang[1, 2], Ling Zhao[1, 2], Ruyi Guo[2]
1.National Key Laboratory of Advanced Casting Technologies, Shenyang 110022, China
2.Shenyang Research Institute of Foundry Co., Ltd. CAM, Shenyang 110022, China
*Corresponding address: e-mail: wangzr_srif@foxmail.com

Abstract: The present work aims to investigate the effect of intercritical quenching treatment between traditional quenching and tempering on microstructure and mechanical properties in a high-strength low-alloy (HSLA) steel. The microstructure of investigated steel subjected to quenching and tempering (QT) treatment was composed of tempered bainite, while the investigated steel subjected to quenching, intercritical quenching, and tempering (QIQT) treatment consisted of intercritical ferrite and tempered bainite. Results indicated that QIQT treatment not only led to significant grain-size refinement, but also resulted in a lower yield ratio (YR) and higher impact toughness.

Keywords: HSLA steel; intercritical quenching treatment; microstructure; yield ratio; impact toughness

1 Introduction

High-strength low-alloy (HSLA) steels with excellent yield strength (YS> 500 MPa) are in great demand in bridges, construction, and shipbuilding due to exceptional strength-toughness balance, low yield ratio (YR) (ratio of YS to tensile strength), good weather resistance, and remarkable weldability [1]. To obtain superior comprehensive mechanical properties, alloy design plays a significant role. For example, Ultra-low carbon design (<0.10 wt.%) is carried out for ensuring remarkable weldability, Cu is added for improving corrosion resistance and precipitation strengthening, Ni, Cr, and Mo are added for improving hardenability, and Nb, V, and Ti are added for thermomechanical processing [2]. Generally, quenching and tempering (Q&T) treatment is widely used in the production of HSLA steels. Quenching process is used to obtain the bainitic or martensitic microstructure to increase the strength while tempering process can significantly improve toughness and ductility. However, the YR of HSLA steels subjected to Q&T process is usually higher than 0.90 [3]. Hence, how to ensure that HSLA steels possess high strength and toughness while sustaining a low YR have received increasing attention.

2 Experimental procedure

The chemical composition of the investigated steel is shown in Table 1. The investigated steel was smelted by induction furnace and argon-oxygen decarburization furnace (AOD). To improve the hardenability of the investigated steel, an appropriate amount of element Cr and Mo was added to the steel. The 150 mm limit wall thickness equivalent test block which can represent the whole properties of HSLA steel is selected for casting. Moreover, the mechanical properties were tested by sampling the core. The chemical composition of investigated steel is shown in Table 1. Moreover, the treatment process is shown in Fig. 1. After austenitizing at 910 °C for 60 min, the samples were quenched (Q) to room temperature to obtain bainite structure, and then tempered (T) at 670 °C for 60 min. For convenience, the process route was referred to as QT. To further improve mechanical properties and prepare the heterogeneous microstructure in this steel, a new process was designed. Intercritical quenching (IQ) treatment was added between traditional QT process, i.e., the investigated steel was first quenched at 910 °C for 60 min, and then quenched at 780 °C for 60 min followed by tempering at 450 °C for 30 min, which was referred to as QIQT.

Table 1: Chemical composition of investigated steel (wt.%)

C	Si	Mn	P	Ni+Cr+ Mo+V
0.13~ 0.18	0.17~ 0.4	0.3~ 0.6	≤0.02	3.7~ 4.4

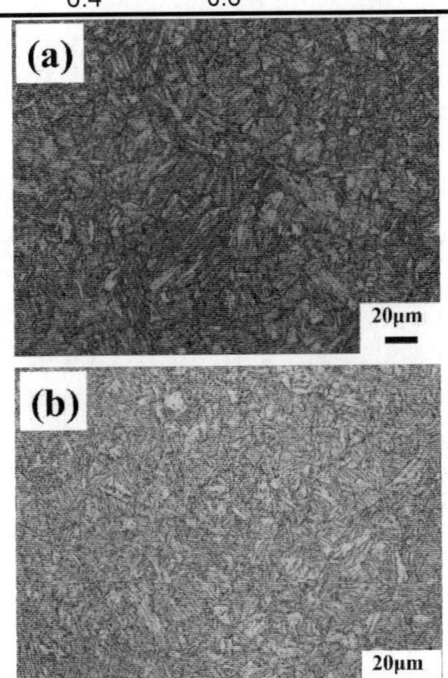

Fig. 1: Metallographic micrographs of (a) QT sample and (b) QIQT sample

3 Result and discussion

The microstructure and mechanical properties of QT and QIQT samples were compared, as shown in Fig. 1 and Fig. 2. A typical tempered bainite microstructure can be observed in the QT sample, as shown in Fig. 1(a). After QIQT treatment, a heterogeneous microstructure of intercritical ferrite and tempered bainite was obtained in the QIQT sample, which was much finer than full tempered bainite microstructure (Fig. 1(b)). In other words, there is a higher density of high angle grain boundaries in QIQT sample. The mechanical properties of the QT and QIQT samples are presented in Fig. 2. The QT sample exhibited tensile properties with a YS of 711 MPa, tensile strength of 780.5 MPa, YR of 0.91, total elongation of 18.8%, and -80 °C, -100 °C, -120 °C impact energy of 157.9 J, 60.9 J and 36.8 J. In comparison to the QT sample, there was a significant change in strength, ductility, and toughness. The yield strength and tensile strength of the QIQT sample were decreased to 556.5 and 679 MPa respectively, thus lowering the YR. Meanwhile, total elongation was 23.5%, an increase of 25%, while the impact energy was significantly increased to 216 J, 207.8 J and 183.2 J. The excellent ductility and toughness of QIQT samples was mainly attributed to the reasonably ratio of the soft intercritical ferrite and hard tempered bainite and the high density of high angle grain boundaries. These, in turn, contributed to decreasing the probability of microcrack initiation and increasing the resistance of microcrack propagation, thus enhancing the ductility and impact toughness.

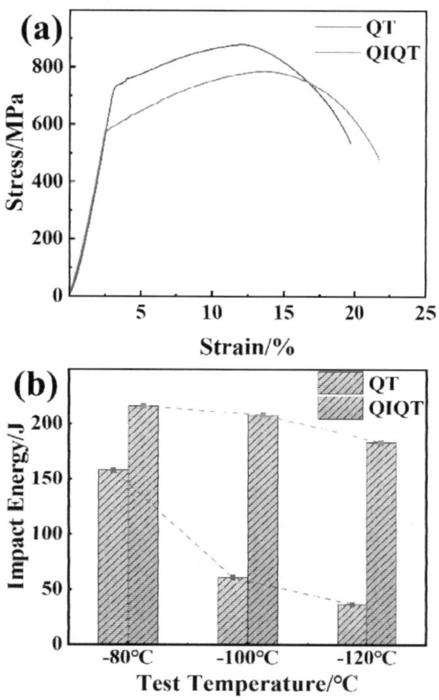

Fig. 2: Mechanical properties of QT sample and QIQT sample: (a) engineering stress–strain curves; (b) impact energy at −80 °C, −100 °C and −120 °C.

4 Conclusion

A better balance of strength and toughness was achieved in
sample subjected to quenching, intercritical quenching, and tempering (QIQT) treatment, which is appropriate for structural applications. The QIQT sample exhibited tensile properties with a YS of 556.5 MPa, tensile strength of 679 MPa, YR of 0.80, total elongation of 23.5%, and -80 °C, -100 °C, -120 °C impact energy of 216 J, 207.8 J and 183.2J. In comparison to QT treatment, QIQT treatment not only led to significant grain-size refinement, but also resulted in a lower yield ratio (YR) and higher impact toughness.

References

[1] Liu D, Cheng B, Chen Y. Strengthening and Toughening of a Heavy Plate Steel for Shipbuilding with Yield Strength of Approximately 690 MPa[J]. Metallurgical and materials transactions. A, Physical metallurgy and materials science, 2013,44(1): 440-455.

[2] Yan W, Shan Y Y, Yang K. Effect of Ti N Inclusions on the Impact Toughness of Low 2 Carbon Microalloyed Steel[J]. Metallurgical and Materials Transactions, 2006, 37(7): 2147-2158.

[3] Siyuan Z, Kaixuan C, Wuqikun Y, et al. Effect of Heat Treatment on Microstructure and Mechanical properties of high strength low alloy (HSLA) steel[J]. Research and Application of Materials Science, 2019,1(2).

Research, Development, and Application of Manufacturing Technology for Oversized High-Strength and Tough Alloy Steel Trackpad

Xiang Zhou [1], Zhonggui Guo [1], Shujun Huang [1], Yu Dai [1], Gang Yang [1]
1.Zhongshi Huajin Casting Co. LTD. Yuncheng City

*Corresponding address:e-mail:huangsj@machine.citic.com

Abstract: The oversized, high-strength and tough alloy steel track plate serves as the load-bearing component of the electric shovel's travelling mechanism in large open-pit mines. These track shoes are tasked with moving around 700-2000 tons of vehicle body for mobile work, operating in a minimum working temperature of -50℃. Given the poor road conditions, the force distribution on the track shoes during operation is uneven, resulting in extremely high local stress, which can easily lead to failures such as breakage, deformation, and wear. Extending the service life of electric shovel track shoes in mining operations is currently the primary goal of developing track shoes. This article focuses on tracking and developing various aspects of the track shoe production process, including component selection, casting process design, heat treatment process design, induction hardening process design, and production process operation control. The aim is to ensure that the internal structure of the track shoe is dense, strong, tough, and wear-resistant, with the ability to operate in low temperature environments, handle high loads, and sustain long periods of operation, thus meeting the current service requirements of track shoes.

Key words: alloy steel trackpad; thermal centre; induction hardening

1 Introduction

Due to its large profile size, high load-bearing capacity, and significant stress during service, the oversized alloy steel track pad features a complex structure and experiences high local wear and significant thermal stress points [1]. Compared to high manganese steel track plates and smaller alloy steel track plates, oversized alloy track plates pose greater challenges in manufacturing due to their high hardenability, elevated carbon equivalent, poor casting properties, elevated mechanical requirements, and specific wear resistance demands. Through component design, casting process research, heat treatment process research, and induction hardening process research tailored to the working conditions of the track pad, a comprehensive manufacturing process plan for this type of track pad is finally obtained. This plan meets the requirements for dense, crack-free, high-strength, high-toughness, and high wear resistance of the track pad, enabling continuous supply of batch track pads.

2 Experimental procedure

For the XYZ-1 track shoe model, we have identified and analyzed the risk points of the track shoe based on production experience, and determined the key tasks for the early development and design stage. Based on our company's experience in producing track shoes, we have preliminarily determined the material composition control range, metallographic structure requirements, and heat treatment plan. We poured test blocks for composition and performance verification tests, and completed composition determination and heat treatment process plan determination through metallographic analysis and performance testing. Based on the design drawings, modeling is carried out using MAGMA software for CAE simulation to preliminarily determine the process plan. Subsequently, the track pad castings were cast for casting process, heat treatment process, and induction process verification to determine the chemical composition, casting process plan, heat treatment process plan, and induction treatment plan. Ultimately, we obtain a reasonable and comprehensive production process plan, achieving mass supply.

3 Results and Discussion

(1) We adopt the process scheme of riser partition shrinkage, local chills, and open pouring system for castings. In conjunction with strict process operation control in the production workshop, we achieve stable filling and sequential solidification of castings, thereby ensuring the compactness, absence of shrinkage cavities, absence of sand inclusions, and absence of cracks in the crawler shoe castings, ultimately guaranteeing the quality of these castings.

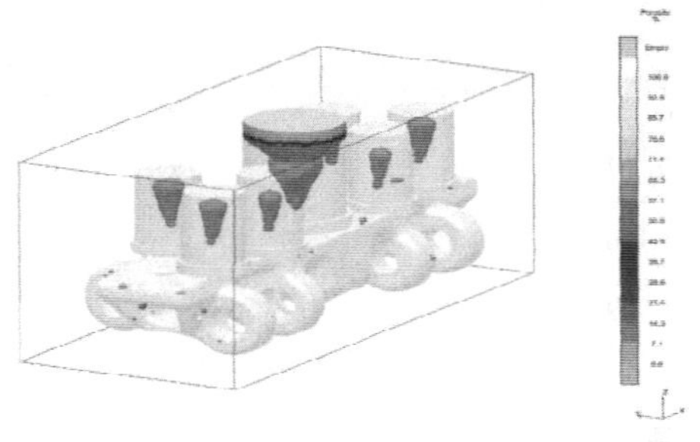

Fig.1 MAGMA Porsity simulation results

(2) We adopted a heat treatment plan consisting of normalizing and tempering, coupled with strict process control in the production workshop. This ensures that the casting achieves a full-section tempered sorbite structure, thereby guaranteeing the qualified performance of the casting matrix.

Fig. 2 Tempered sorbite structure

(3) Thanks to our adoption of a specific power intermediate frequency induction preheating + induction + low-temperature tempering process, we ultimately achieved a induction hardening area with a qualified depth of the hardened layer, qualified hardness, no cracks, and no coarse grains, fulfilling the special requirement of high wear resistance in the local area of the trackpad.

4 Conclusion

Through research on the components, casting process, heat treatment process, and induction process of the XYZ-1 oversized high-strength and tough alloy steel trackpad, we have obtained a comprehensive production scheme for this model of trackpad. This ensures that the dimensions,

Fig. 3 Tempered martensite structure

performance, internal and external quality, and hardness of the induction area, among other parameters, meet the required standards. As a result, we are able to supply the trackpad in large quantities. Through continuous use over many years, our customers have expressed great satisfaction, and our factory has been recognized as their gold supplier for several consecutive

Acknowledgments

Thanks to all the people involved in this study, whose contributions have had a crucial impact on the results of this study.

References

[1] Chen Guozhen. Casting Solidification Theory. University of Mechanical Engineers.1989.01.
[2] Cao Wenlong. Foundry Technology. Mechanical Technology Press, 1997.10.
[3] Cui Zhongqi. Metal Science and Heat Treatment. Harbin Institute of Technology.1995.10.

Casting Process Design and Production of Ultra Heavy Ductile Iron Crossbeam

Hongcheng Liang [1], Chun Liu [1], Yajun Yin [2], Guyu Deng [1], Mi Pan [1], Chunsheng Chen [1], Yuke Ma [1], Wenhua Ye [1], HuaYang[1], Bin Wang [1], Yan Zhou [1]

1. Wuhan Wuzhong Casting and Forging Co., Ltd. Wuhan, 430205, China 2. State Key Laboratory of Material Processing and Die & Mould Technology, Huazhong University of Science & Technology, Wuhan, 430074, China
*Corresponding address: e-mail: fdlsjy@163.com

Abstract: This article focuses on the structural characteristics, difficulties in manufacturing, and technical requirements of the 18.5-meter ductile iron crossbeam. The key control technologies for producing beams are introduced from both the casting process and the melting process perspectives, which includes pouring system design, core design and chemical composition control, etc. Through casting simulation, the filling and solidification process of the crossbeam is simulated and analyzed, and the pouring system of the crossbeam is optimized. A QT600-3 crossbeam with a length of 18.5 meters was successfully produced, featuring a clear outline and no defects such as cold shut or shrinkage porosity. The overall deformation of the long guide rail of the crossbeam is controlled at 16mm, the hardness of the guide rail is about HB192, the tensile strength of the body test block is 632Mpa, and the elongation rate is 3%.

Keywords: Crossbeam; Simulation; Deformation

1 Introduction

With the development of industries such as nuclear power equipment, large ships, aerospace, and military products, higher requirements have been placed on heavy-duty and ultra heavy machine tools, requiring them to develop towards high-speed, high-precision, and multifunctional composite directions. This has brought about new demands to machine tool castings: the castings have become larger and heavier, with a single piece reaching a length of up to 22 meters and possibly weighing over 140 tons. Additionally, the structure of the castings has become increasingly complex.In the past, these castings were assembled from several sections, but now these products require integral casting; The strength requirements for castings have also significantly increased. Also, HT250 and HT300 grades of gray cast iron could meet the requirements for castings previously, but now QT600-3 grade is required. [1]

The XKC56145 CNC machine tool manufactured by the company is the largest turning and milling composite machining machine tool in China, with a gantry span of 14.5m. The crossbeam is a key foundational component on the machine tool, requiring high strength and stiffness. The gross weight of the crossbeam is about 96 tons, with dimensions of 18948 * 2707 * 1815, made of QT600-3 material. The elastic modulus of the body is required to be no less than 160GPa, and the hardness of the guide rail body is required to be no less than HB190. Defects such as cracks, shrinkage cavities, and slag inclusions are not allowed. The external structure of the casting is shown in Figure 1.

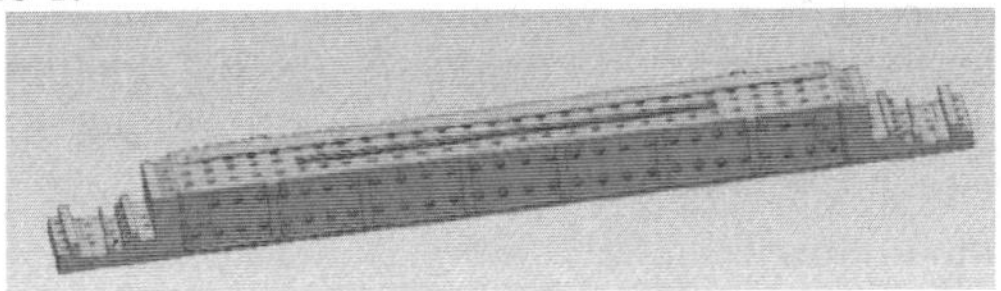

Fig1 Outline of the crossbeam

The crossbeam has many production difficulties due to its large size and heavy: (1) the total length of the crossbeam is 18.5m, the guide rail is relatively thick, while the inner cavity walls are thinner, making it prone to bending deformation; (2) Due to the presence of diagonal bars in the inner cavity of the crossbeam and multiple layers of clay cores on both sides of the cantilever, choking may occur; (3) The crossbeam guide rail has a wall thickness of approximately 150mm, while the wall thickness at the joint with the column exceeds 200mm. These thick parts are prone to shrinkage and loosening defects; (4) The amount of molten iron poured into the crossbeam is large as much as 130 tons of molten iron. On site, multiple ladle spheroidization is required; otherwise, this could lead to uneven temperature and composition of the molten iron, as well as the decline of spheroidization level.

2 Experimental Procedure

In order to obtain qualified crossbeams, research was conducted on both the molding process and melting process, and casting simulation analysis was carried out.

1.Design of Molding Process

(1) Process analysis of crossbeam. In response to the characteristic of the crossbeam being prone to bending deformation, anti-deformation has been designed in the process, and process correction has been added locally to the guide rail. In addition, there are many internal chambers of the crossbeam and multiple layers of core

material, which can easily lead to poor air flow and cause choking. On one hand, the moisture content of the core material in the internal chamber needs to be reduced; on the other hand, it is necessary to increase the number and diameter of vent holes to facilitate the core material exhaust and reduce the probability of choking. (2) Modeling materials and modeling methods. The crossbeam is shaped and cored with furan resin sand, and based on the structural characteristics, size, and actual production conditions of the crossbeam, a pit mold was used for casting.

(3) Selection of parting surface. To ensure the quality of the crossbeam, place the crossbeam guide rail at the bottom and create a parting surface on the back of the crossbeam.

(4) Core design. The inner cavity of the crossbeam is designed as a core for each level, with two layers at both ends.

(5) Gating system design. The size of the crossbeam is relatively large, and a large amount of iron water is required for pouring. Therefore, a stepped pouring system is used. Meanwhile the internal pouring channel is arranged along one side of the length direction of the crossbeam and enters the mold cavity from the side of the guide rail.

(6) Riser and cold iron setting. The wall thickness of the crossbeam is uneven, and the wall thickness of the guide rail surface and the joint surface is large, which may lead to the appearance of shrinkage and looseness. It is necessary to use risers to eliminate shrinkage defects. At the same time, in order to improve the density of the casting structure, external cooling irons are installed in the thick parts of the guide rail surface and the joint surface for quenching.

2.Melting and pouring process design

(1) The chemical composition is determined based on the material and structure of the casting, and the designed chemical composition is shown in Table 1.

Table 1. Composition of Crossbeam(wt%)

C	Si	Mn	P	S	Mg
3.7-3.8	1.9-2.1	0.3-0.6	≤0.03	≤0.020	0.035-0.055

(2) Due to the gross weight of the crossbeam being 96000kg and the pouring weight reaching 110000kg, molten iron was poured from three positions using three ladles. The temperature of the three packs of molten iron should be as consistent as possible, with a temperature difference not exceeding 30 ℃.

(3) The furnace temperature is greater than 1460 ℃, and the pouring temperature is controlled at 1350-1370 ℃. To avoid fertility decline, efficient composite inoculants are used and multiple pregnancies are carried out.

3.Crossbeam simulation and process optimization

Using casting simulation software, the filling process of the crossbeam was simulated and analyzed. Through observation of the filling process, it was found that under the existing pouring system layout, some gates of the molten iron could not be filled at all, as shown in the green rectangular box in Figure 2; In response to this issue, the layout of the inner runners was optimized, reducing the number of inner runners. Then, simulation software was used again to simulate, as shown in Figure 3. From the simulation results, after seconds, all inner runners were filled with molten iron, which can evenly enter the mold cavity and distribute the temperature of the molten iron evenly, without problems such as cold shut and insufficient pouring.

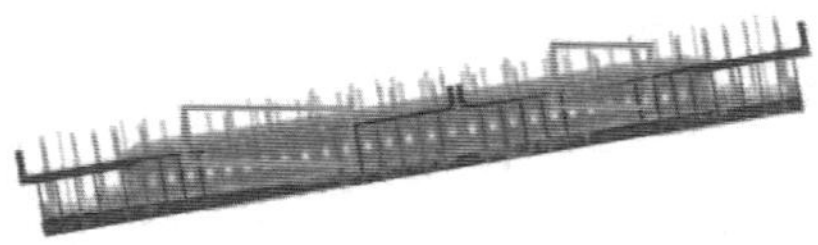

Fig2 Filling process of crossbeam before process optimization

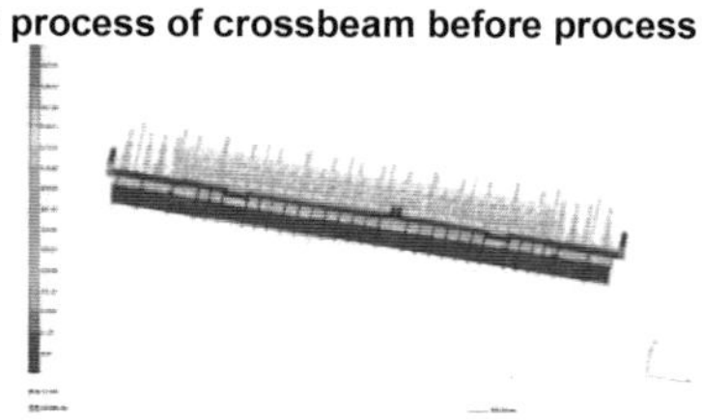

Fig3 Filling process of crossbeam after process optimization

3 Result and discussion

The 18.5-meter crossbeam was produced using optimized technology. The appearance of the crossbeam is complete, the contour is clear, the size and shape are consistent with the design structure, and the organization and mechanical properties also meet the material requirements of the crossbeam:

(1) The crossbeam has no defects such as cold insulation, shrinkage, or looseness.The overall deformation of the crossbeam is about 16mm, which is within the machining allowance of the crossbeam and will not affect the shape and size of the parts.

(2) The average hardness of the crossbeam guide rail is HB192.

(3) By observing the microstructure of the crossbeam, graphite appears spherical, as shown in the figure 4, with a spheroidization level of 3 or above; Tensile strength 632Mpa, elongation rate 3%.

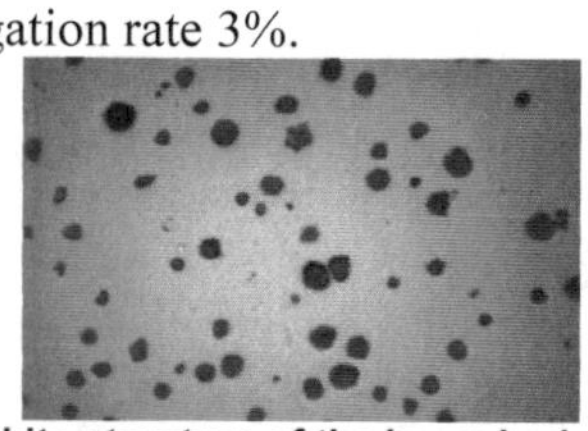

Fig4 Graphite structure of the beam body test block

References

[1].Ren H C, Zhao J K and Li S Y. Production of Ductile Iron Crossbeam Components for Ultra Heavy Machine Tools. Modern Cast Iron, 2010(4):24-29.

Discussion on the Current Research Status of Manufacturing Technology for Large High-Temperature Alloy Steel Castings in 650℃ Ultra Supercritical Units

PingLin, Xiaobing Yang, Zhangyu Xiao,Youqing Lv
Erzhong (Deyang) Heavy Equipment Co., Ltd., Deyang, Sichuan,China
*Corresponding address: e-mail: lpj908@163.com

Abstract: In the field of energy, ultra supercritical units are considered the future development trend due to their high efficiency and low emissions. At present, the most advanced commercial ultra supercritical unit is the 620 ℃ unit, but there is no substantial breakthrough in the manufacturing technology of 10 ton large nickel based high-temperature alloy cast steel required for higher parameter 650 ℃ units at home and abroad. By studying the smelting, casting, heat treatment, welding, processing, and non-destructive testing techniques of these steel castings, we have revealed the process characteristics of the materials, identified existing problems, and laid the foundation for future research directions and potential breakthroughs. However, improving machining efficiency and developing more advanced non-destructive testing methods are key challenges in achieving engineering applications for these castings. Clean and efficient energy technology is an inevitable direction for future development, and continuous innovation and breakthroughs are the responsibilities of practitioners in the foundry industry.

Keywords: steel casting of Nickel based alloy, 650 ℃ ultra supercritical unit, manufacturing technology, Clean and efficient energy technology

1 Introduction

The data from the 2023 World Energy Statistical Yearbook shows that although coal still plays an important role in the global energy structure as the main source of electricity for many countries or regions, its development is facing challenges from multiple aspects. Especially with the rapid advancement of renewable energy technologies and attention to climate change, the regulatory pressure on coal-fired power plants continues to increase.

In the European Union and the United States, coal-fired power generation is gradually declining, but in developing countries such as China and India, the demand for electricity due to economic growth continues to increase, and coal-fired power generation is still growing. However, these countries are also actively seeking technologies to improve energy efficiency and reduce emissions to address environmental and climate challenges. High parameter ultra supercritical units are regarded as the future development direction of the coal-fired power industry due to their high energy conversion efficiency and low pollutant emission levels.

2 Main technical difficulties

(1)Poor material fluidity can easily lead to defects such as slag inclusion, shrinkage porosity, and porosity;;

(2)The material is difficult to process and requires precise control of the size of the casting blank;

(3)The content of impurity elements Si and Mn is strictly limited;

(4)Fe is an impurity element with high requirements for the furnace body and smelting ladle;

(5)Gas elements O and N have high requirements and are difficult to remove.

(6)There are many alloying elements with large sizes, and there is a certain degree of compositional segregation;

(7)The heat treatment parameters (insulation temperature, insulation time, cooling rate, etc.) have a significant impact on the properties of related precipitates and nickel based alloy castings, and there is sensitivity to process parameters;

(8)Ni Cr Mo alloy has a unidirectional structure, and macroscopic and microscopic cracks are prone to occur in the weld seam during welding;

(9)Ni Cr Mo alloy has a thick melt pool, poor fluidity, and is prone to defects such as porosity and slag inclusion.

3 Main research contents

1.Foundry Technology

1)Research on Material Process Characteristics and Thermal Properties Parameters.

Through material testing and simulation analysis, the solidification shrinkage characteristics and thermal parameters of nickel based materials have been preliminarily mastered.

2)Conduct (solidification, filling, stress) simulation to optimize casting process design.

To ensure the rationality and efficiency of the casting

process for the test piece, we developed a casting process plan based on the solidification characteristics of the material and the structural characteristics of the test piece. We used MAGMASOFT simulation software for detailed calculation and analysis to evaluate the risk of potential defects during the casting process, such as shrinkage, cracks, and segregation, and verify the effectiveness of the casting process plan.

2.Smelting technology

1) Research on Precise Control of Material Chemical Composition

Based on the analysis of C, O, and N in the chemical composition standards, the EF → LF (VOD) process flow is adopted to achieve precise control of chemical composition by studying the order and timing of alloy addition.

2) Research on Material Slag System

According to metallurgical principles, the melting point of slag is usually 50-200 ℃ lower than that of molten steel, while the melting temperature range of nickel based high-temperature alloys is 1290-1350 ℃. The existing stainless steel iron-based alloy slag system needs to be readjusted to adapt to the production of high-temperature alloys.

3. Heat treatment technology

1) Conduct process experiments to determine process parameters

By conducting heat treatment process experiments with different parameters, analyzing the influence of heat treatment process parameters (temperature, cooling medium, etc.) on mechanical properties, and selecting the optimal heat treatment process parameters

2) Develop heat treatment process route

By analyzing the microstructure and properties under different heat treatment conditions, a reasonable process route is formulated.

4.Welding technology

1) Selection of welding materials.

According to the principle of equal composition and equal strength matching, choose ENiCrMo-3 welding rod and ERNiCrMo-3 welding wire.

2) Welding Procedure Qualification Test.

Conduct welding process qualification tests in accordance with ASME BPVC. IX-2021 "Welding, Brazing, and Adhesive Qualification" standard.

5 Conclusion

Through research, although we have successfully manufactured large high-temperature alloy cylinder castings of 10 tons, we still need to continue to address two key challenges in order to achieve engineering applications: first, improving the machining efficiency of the castings; Secondly, develop more advanced non-destructive testing technologies.

Clean and efficient energy technologies are leading the future development direction of the coal-fired power generation industry, and innovation and breakthroughs are the mission of every foundry practitioner.

References

[1] Principles of Metal Solidification. Beijing: Machinery Industry Press, 2000

[2] Simulation and Quality Control of Casting Engineering. Beijing: Machinery Industry Press, 2001

[3]Foundry Branch of China Mechanical Engineering Society Foundry Handbook - Volume 5, Foundry Processes

[4]Peng Jianqiang. Selection of Key Components for 650 ℃ Ultra supercritical Steam Turbine [J]. Thermal Turbine, 2016 (01): 51-55

[5]Jiang Xinliang, Men Zhengxing, Mou Chenghai, et al. Current Status and Feasibility Analysis of Manufacturing Technology for Large High Temperature Alloy Castings of 700 ℃ Ultra supercritical Units [J]. Mechanical, 2014, 41 (03): 4-7

[6]Tian Yu, Peng Jianqiang, Wei Shuangsheng, et al. Development trend of high-temperature casting materials for ultra supercritical steam turbines in China [J]. Large scale castings and forgings, 2016 (01): 1-5

This page intentionally left blank.

Effect of Convection Solidification on the Halo Formation in Zn-Al Alloys

Wenchao Yang[1,2], Shouxun Ji[2*]
1. State Key Laboratory of Solidification Processing, Northwestern Polytechnical University, Xi'an, 710072, China
2. Brunel Centre for Advanced Solidification Technology (BCAST), Brunel University London, Uxbridge, UB8 3PH, UK
*Corresponding address: e-mail: Shouxun.Ji@brunel.ac.uk

Abstract: The formation of halos was investigated for hypoeutectic, eutectic and hypereutectic Zn-Al alloys under conventional and under convection solidification. It was found that the Zn-rich halos occurred in the surroundings of the Al-rich primary phase for the hypereutectic Zn-Al alloys at Al>5 wt.% and no halos occurred for the hypoeutectic and the eutectic Zn-Al alloys at Al<5 wt.% under conventional solidification. However, the Zn-rich halos were completely vanished from the Al-rich phase under intensive convection solidification. The interaction among the solid phases and the liquid phase was found to be responsible for the halo formation.

Keywords: Solidification; Microstructure; Crystal growth; Alloys; Halo formation

1 Introduction

Halo formation is generally characterized by an envelope of non-faceted phase around a faceted primary phase, not vice versa. By controlling the morphology of halo, the alloys with required mechanical properties may be developed, such as the deformation behaviour of ductile iron. This unique feature is important to affect other performance such as corrosion resistant and the heat treatment process of casting alloys. The study for elucidating the origin of halos are generally categorised to nucleation-dominant [1] and/or growth-dominant [2] according to solidification mechanism. The present study aims to find the features of halos in Zn-Al alloys formed in conventional solidification and convection solidification through isolating the growth process. The halos can be then identified as the halo is formed by the simultaneous interaction of the primary solid phase, the halo phase and the melt encompassing the solid phases. The halos were studied on hypoeutectic, eutectic and hypereutectic alloys. The discussion focuses on the effect of convection on the halo formation.

2 Experimental procedure

Pure elemental Zn (99.95 wt.%) and Al (99.995wt.%) were used to make Zn-Al alloy ingots. The chemical composition was measured by optical mass spectroscopy. In the experiments of conventional solidification, the melt were poured into a wedge-shape copper mould at a fixed superheat. The casting was cut at the same thickness for microstructural examination. For the solidification under convection, the melt was sheared using a twin-screw extruder as described in [3] at different temperatures and different times, followed by extruding a φ8mm bar with immediate water quench.

3 Results and discussion

1. Halos under conventional solidification

Fig. 1 shows the Zn-rich halos formed around the primary Al-rich phase. It is seen that three alloys exhibited different characteristics. In the hypereutectic Zn-6.3 wt.% Al alloy, the primary Al-rich phase was surrounded by a halo of Zn-rich phase, and the eutectic microstructure was clearly a lamellar morphology (Fig. 1a). In the eutectic Zn-5 wt.% Al alloy, the eutectic microstructure showed lamellar eutectic cell morphology all over the matrix (Fig. 1b). There was no halo in the microstructure. In the hypoeutectic Zn-4.1 wt.% Al alloy, the primary Zn-rich phase showed no halo in the surroundings of Zn-rich dendrites (Fig. 1c). The results confirmed that halos only occurred on the primary Al-rich phase with Zn-rich halo at the hypereutectic Zn-Al alloys with Al>5 wt.%, and no halos were formed for the hypo eutectic and eutectic Zn-Al alloys at 5 wt.% Al.

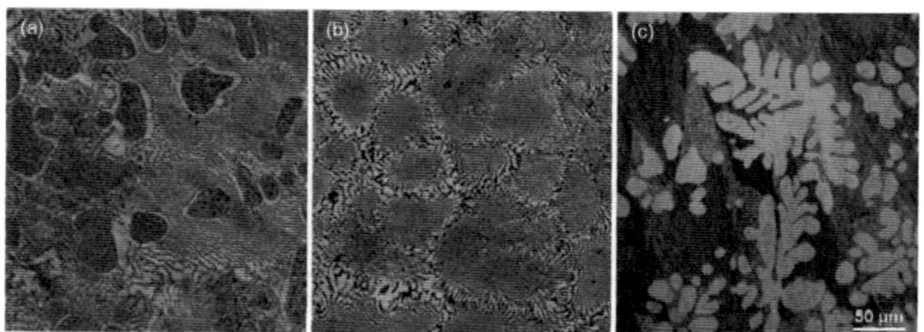

Fig. 1. The optical micrographs showing the typical microstructures and halos of (a) hypereutectic Zn-6.3 wt.% Al, (b) eutectic Zn-5 wt.% Al, (c) hypoeutectic Zn 4.1 wt.% Al alloys (white: Zn-rich phase, black: Al-rich phase).

2. Halos under convection solidification

The halo characteristics of hypereutectic Zn-6.3 wt. %Al alloy under convection solidification are shown in Fig. 2, which were quenched into water from different temperatures. It was seen that only primary Al-rich phase was formed at the initial stage of solidification (Fig. 2a). No halos existed around the primary Al-rich phase before commencing the eutectic solidification, during which two solid phases were created simultaneously in the form of round particles for both the Zn-rich phase and the Al-rich phase (Fig. 2b). The micrograph showed the complete absence of halos around the Zn-rich particles. The primary Al-rich phase had no apparent halos adjacent to the liquid and no halos adjacent to the solid Zn-rich phase.

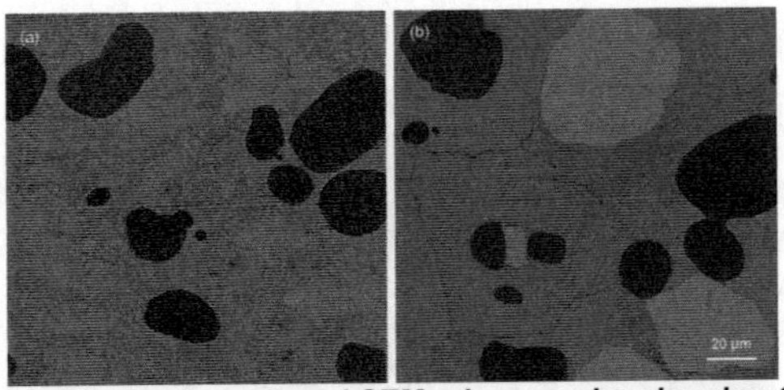

Fig. 2. The back-scattered SEM micrographs showing the microstructures and halo features of water-quenched hypereutectic Zn-6.3 wt.% Al alloy under convection solidification, (a) 389.4 oC, and (b) 380.6 oC.

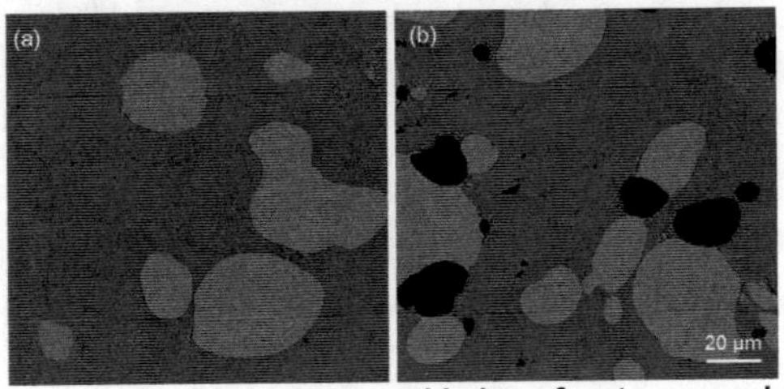

Fig. 3. The microstructures and halos of water-quenched hypereutectic Zn-4.1 wt.% Al alloy under convection solidification, (a) 387.3 oC, and (b) 380.6 oC.

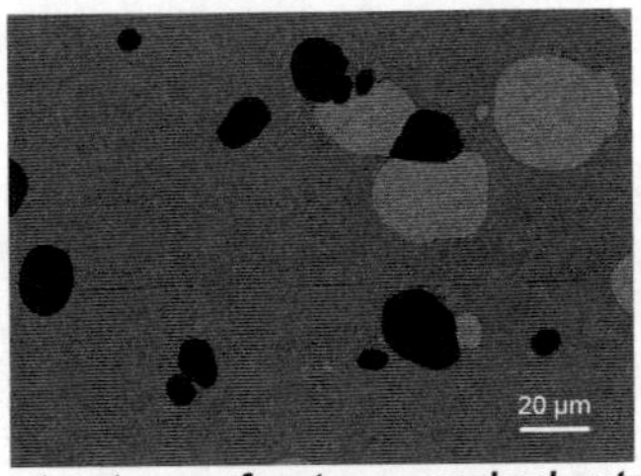

Fig. 4. The microstructures of water-quenched eutectic Zn-5wt.%Al alloy at 380.2 oC under convection solidification.

Fig. 3 shows the halos of the hypoeutectic Zn-4.1 wt.% Al alloy under convection solidification. No halos were found around the Zn-rich particles formed during the primary solidification. For the eutectic solidification of the hypo-eutectic Zn-4.1 wt.% Al alloy, Al-rich particles and Zn-rich particles were formed simultaneously under intensive convection. The eutectic Zn-rich particles exhibited complete absence of halos around the solid phase. The eutectic Al-rich phase showed either no apparent halos adjacent to liquid or completely absent adjacent to the solid Zn-rich phase. Fig. 4 shows the halos of the eutectic Zn-5wt.%Al alloy under convection solidification. The eutectic Zn-rich and Al-rich particles also showed no halos around the another solid phase and no apparent halos adjacent to liquid. Overall, the experimental results confirmed that no any halos were formed under convection around the Al-rich phase, which was contradictory to the conventional solidification.

There are three phases essentially existing simultaneously under conventional solidification when halos are formed. At the initial stage of solidification to form Al-rich primary phase, the Zn-rich phase starts to nucleate when the melt temperature is lowered to the eutectic point, and attempts to grow on the surface of the Al-rich particles. When applying strong convection on a eutectic alloy, particles of two eutectic phases are precipitated at the same time. Both temperature and composition fields in the melt are uniform with small fluctuations. Both phases nucleate simultaneously at the eutectic temperature and the suppression from one phase to the other phase is much less. Once the nuclei survive in the melt, the subsequent growth is immediately commenced. The growth of solid particles in the eutectic melt is enhanced both the thermal and mass transport, resulting from a very small undercooling, thermal and compositional gradient at the solid/liquid interface. The diffusion boundary layer roughly equals to the lamellar distance for a normal eutectic reaction and it is at the same order during the growth of halos. And, the boundary layer is expected to be thinner than the diffusion boundary layer for a strong convection. Therefore, the halo is not formed because of the significant decrease or even disappear of the diffusion layer, by which the reaction is faster by precipitating the two phases separately. As a result, the solidification process under intensive convection is pretty close for all systems, regardless of their physical properties under conventional condition, which causes the disappearance of halos on the prior Al-rich phase during solidification.

4 Conclusions

(1) In conventional solidification, the Zn-rich halos occurs around the Al-rich primary phase for the hypereutectic Zn-Al alloys and no halos for the hypoeutectic and eutectic Zn-Al alloys.

(2) In convection solidification, the Zn-rich halos are absent from the Al-rich phase formed under the primary solidification and the eutectic solidification of the Zn-Al alloys because of the uniform temperature distribution and the enhanced mass transport by intensive convection.

(3) The halo formation is a growth dominant phenomenon. The interaction among the solid phases and the liquid phase is responsible for the halo formation.

References

[1] Bluni ST, Notis MR, Marder AR. Nucleation characteristics and microstructure in off-eutectic Al-Zn alloys. Acta Metall. Mater. 1995, 43 (5): 1775-1782.

[2] Nave MD, Dahle AK, StJohn DH. Halo formation in directional solidification. Acta Mater. 2002, 50 (11):2837-2849.

[3] Ji S, Fan Z. Solidification behaviours and microstructural evolution of near-eutectic Zn-Al alloys under intensive shear. Metall. Mater. Trans. A, 2009, 40 (11):185-195

Effect of As-Cast Microstructure on Precipitation Behavior and Thermal Conductivity of Al-Si-Mg Alloys

Saif Haider Kayani, Kwangjun Euh, Jung-Moo Lee, Young-Hee Cho
Department of Aluminum, Advanced Metallic Division, Korea Institute of Materials Science, Changwon 51508, Republic of Korea
*Corresponding address: e-mail: saifkayani@kims.re.kr

Abstract: This study examines the precipitation evolution during T5 heat treatment of an Al–7Si–0.35Mg alloy component with varying thicknesses. We employ a step mold casting technique to develop an Al–7Si–0.35Mg alloy component with cavity thickness ranging from 4-40 mm. Our experimental investigations unveil varying solidification times (from 5.0 to 152.2 sec) and cooling rates (ranging from 2 to 50 K/s) due to the step mold casting. The thermal conductivity of component sections increase from 174.1 to 178.9 W/mK as the cooling rate surged to 50 K/s. Microstructure analysis of 50 K/s alloy reveals a large number density of coarse Mg-Si-rich pre-precipitates near dendritic cell boundaries, alongside fine GPII and β″ pre-precipitates within the cell interior. The pre-precipitation behavior in 50 K/s alloy is attributed to the retention in the mold until thick sections fully solidify, leading to low solute supersaturation and higher conductivity. Contrarily, 2 K/s alloy exhibits the formation of fine dot-like β″ and a few coarse Mg-Si-rich pre-precipitates along the cell boundaries. This alloy demonstrates superior age-hardening behavior and its low thermal conductivity increases during aging. This increment is associated with a small lattice misfit between precipitates and matrix in the as-cast state, inducing significant lattice distortion which lowers the thermal conductivity. Upon peak aging, these distortions are replaced by dislocations at the precipitate-matrix interface due to large lattice misfits. Finally, we offer an insight into how the solidification microstructure affects T5 treatment considering hardness and conductivity synergy.

Keywords: Al–Si–Mg alloy; Cooling rate; Pre-precipitation; Age-hardening; Thermal conductivity

1 Introduction

The automotive industry is driven by the need for lighter, stronger, and more environmentally friendly vehicles due to increased CO_2 emissions and fuel costs. To meet these demands, manufacturers utilize structural components made of lightweight alloys like Al–Si–Mg, known for their high-strength-to-weight ratio and excellent castability. However, these components face challenges during operation due to large temperature variations, often leading to non-uniform temperature distribution caused by uneven thicknesses [1].

Artificial aging, particularly the T5 treatment, presents a promising solution to enhance the thermal conductivity of Al–Si–Mg alloys and achieve uniform temperature distribution. Unlike the T6 treatment, which incurs high costs and distortion issues, T5 treatment after casting is preferred. Understanding the casting parameters' influence on microstructure is crucial for successful T5 treatment implementation, as it directly impacts age-hardening, primarily achieved through precipitation nucleation and growth [2].

Pre-precipitation, occurring during casting under specific solidification conditions, can deplete mobile solute atoms and vacancies, hindering subsequent strengthening precipitate formation during aging. This phenomenon is more pronounced in thick sections experiencing slower cooling rates. Recent studies have shown that water quenching during casting enhances solute supersaturation, while slow cooling promotes pre-precipitation, affecting the alloy's age-hardening capability [3].

Although strengthening mechanisms typically lower thermal conductivity by impeding heat carrier movement, recent findings present conflicting results, suggesting precipitation-induced increases in thermal conductivity [4]. The coherency of precipitates formed during aging, influenced by pre-precipitation, remains unexplored. Therefore, analyzing the effect of as-cast microstructure on thermal conductivity and age-hardening behavior is imperative. This study investigates the impact of as-cast microstructure on precipitation evolution during T5 treatment in an Al–7Si–0.35Mg alloy, evaluating age-hardenability, thermal and electrical conductivity. Ultimately, a trade-off relationship between hardness and conductivity is established, elucidating the influence of as-cast microstructure on alloy aging behavior.

2 Experimental procedure

An Al-7Si-0.35Mg alloy was prepared by melting the raw materials in an electric resistance furnace. The melt underwent degassing with Ar gas for 10 minutes using a rotary impeller, followed by stabilization at 710 °C for 15 minutes. The melt was then poured into a pre-heated permanent steel step mold, featuring four steps with varying cavity thicknesses. After solidification, the component was extracted and water-quenched to room temperature. Cooling rates were determined using a tapered cooling method with K-type thermocouples. The microstructure was analyzed using optical, SEM, and TEM. Age-hardenability was assessed through matrix hardness and conductivity measurements, while thermal conductivity was determined using a light flash apparatus.

3 Result and discussion

The cooling curves indicated varying durations to reach the extraction temperature of 350 °C, notably 152.2 and 5.0 sec for steps 1 and 4, respectively, suggesting retention at high temperatures before water quenching, potentially affecting microstructure. Optical micrographs revealed primary α-Al dendritic cells with eutectic Si and intermetallic phases, with finer Si phases and smaller secondary dendrite arm spacing at higher cooling rates. Backscattered electron images identified β-AlFeSi intermetallic, with high-resolution scans revealing small pre-precipitates in slower-cooled alloy and needle-like pre-precipitates in faster-cooled alloy. TEM analysis confirmed fine pre-precipitates β″ and solute clusters in slower-cooled alloy and GP zones and β phases in faster-cooled alloy. As a result, faster-cooled sections showed higher conductivity in as-cast state compared to slower-cooled sections. Matrix hardness and electrical conductivity increased with aging time, with slower-cooled sections showing higher response. After aging, thermal conductivity increased, except for a slight decrease in faster-cooled specimens. TEM analysis revealed evolution of precipitates during aging, with slower-cooled alloy exhibiting finer precipitates and faster-cooled alloy experiencing rapid coarsening due to pre-precipitates.

Contrary to literature, our study shows that a fast cooling rate of 50 K/s results in higher thermal and electrical conductivities in as-cast state. This is attributed to coarse pre-precipitation, particularly near dendritic cell arms, due to retention in mold at high temperature. In contrast, slower cooling at 2 K/s promotes fine dot-like β″ phase and needle-like pre-precipitation near cell boundaries.

These differences in pre-precipitate distribution affect conductivity. For instance, coherent interfaces between β″ pre-precipitates and the matrix, observed at slower-cooled specimen in as-cast state, cause lattice distortions, impeding electron and phonon movement and resulting in low conductivity.

Aging causes a loss of coherency between precipitates and the matrix, leading to misfit strain and dislocation loop formation. The interspacing among solute clusters and precipitates exceed the mean free path of electrons (~100 nm) for beneficial conductivity effects [5]. However, the high density of pre-precipitation in the 50 K/s alloy limits precipitation coarsening and impedes conductivity enhancement during aging. Our results show that achieving synergy between hardness and conductivity is easier at slower cooling rates (2-16 K/s) due to finer pre-precipitates. In contrast, the presence of coarse pre-precipitation at 50 K/s hampers this balance. Overall, pre-precipitation not only reduces solute supersaturation but also limits conductivity and age-hardenability in Al–Si–Mg alloys.

4 Conclusion

This study highlights the significant impact of as-cast microstructure on the age-hardenability and thermal conductivity of Al–Si–Mg alloys. The faster cooling rate of 50 K/s resulted in higher conductivity in the as-cast state due to the formation of pre-precipitates, albeit at the expense of inferior age-hardening response compared to the slower cooling rate of 2 K/s. The superior age-hardening response and thermal conductivity improvement in the 2 K/s alloy after peak aging were attributed to the formation of cuboidal β″, stable β, and cubic Si precipitates, which replaced lattice distortions with misfit dislocations. This suggests that alloys solidified at slower cooling rates achieve a better balance between age-hardenability and conductivity, offering valuable insights for optimizing the properties of Al–Si–Mg alloy.

References

[1] S.A. Sajjadi, H.R. Ezatpour, M. Torabi Parizi, Comparison of microstructure and mechanical properties of A356 aluminum alloy/Al_2O_3 composites fabricated by stir and compo-casting processes. Mater. Des., 2012, 34: 106–111.

[2] K. Yamamoto, M. Takahashi, Y. Kamikubo, Y. Sugiura, S. Iwasawa, T. Nakata, S. Kamado, Influence of process conditions on microstructures and mechanical properties of T5-treated 357 aluminum alloys, J. Alloy. Compd., 2020, 834: 155133.

[3] S.-B. Kim, J.-M. Lee, T.-M. Koo, S.-U. Lee, J.-Y. Lee, K.-S. Son, Y.-H. Cho, Influence of cooling condition after solidification on T5 heat treatment response of hypoeutectic Al-7Si-0.4Mg casting alloy, J. Alloy. Compd, 2022, 906: 164339.

[4] B. Li, L. Hou, R. Wu, J. Zhang, X. Li, M. Zhang, A. Dong, B. Sun, Microstructure and thermal conductivity of Mg-2Zn-Zr alloy, J. Alloy. Compd, 2017, 722: 772–777.

[5] A.R. Eivani, H. Ahmed, J. Zhou, J. Duszczyk, Correlation between Electrical Resistivity, Particle Dissolution, Precipitation of Dispersoids, and Recrystallization Behavior of AA7020 Aluminum Alloy, Metall. Mater. Trans. A., 2009, 40: 2435–2446.

New Viewpoints on the Oxide Film Defects in Advanced Alloy Casting

Lunyong Zhang*, Fuyang Cao, Jianfei Sun
School of materials science and enginnering, Harbin Institute of Technology, 150001, Harbin, Heilongjiang, China
*Corresponding address: e-mail: zhangly@hit.edu.cn

Abstract: Multiple scale oxide bifilm defects have been a fatal problem recent years for high-qualitively casting large size and complex aerospace components of varied advanced materials including the aluminum, magnesium, copper alloys. This talk will introduce the most recent development of oxide bifilm studies and new viewpoints we achieved in the past years on different materials system, on the basis of reviewing the existing understandings of the oxide bifilm defects in a full-screen, so prompting the investigation of such an old but emergent important casting defects.

Keywords: complex structural castings, aluminum and magnesium alloys castings, oxide bifilms,

Introduction

As a minor casting defects constructed by double layers oxide films, the named oxide bifilm defect in casting, has attracted increased attentions because which was found more and more in the aerospace casting products, Such bifilm defects in fact has been proposed for around several decades in the Campel's theory, it was believed to cause all the metalogical defects in castings. However, traditionally they were not emphasized before because they are deemed as a kind defects similar to the traditional oxide inclusions which can be easily removed during the casting process and the requirement on the casting quality was also not so high standard which means the oxide bifilm defects with small size could not seriously degrade the casting quality under to the requirement thus they were ignored. However, the casting quality is emphasized higher and higher due to the increased performance demanding for the modern aerospace equipment and also others, small size bifilms has been an emergent casting defect necessary to be controlled.

With a survey on the oxide bifilm defects in casting, Campel and the successive researchers have proposed the surface oxide film involving mechanism due to flow turbulence during the melt filling, and a large number of experiments have been carried out to prove there are oxide biflms in casting by creating negative pressure or filling gas in to the melt. These experiments in generally demonstrated defects formed by two layers of oxide film could form in castings. Some further studies also shew that oxide bifilm could reduce the strength of castings because which would be a source of crack.

Recent years, such kind oxide bifilms defects were observed by us in the casting products of varied alloys including aluminum alloys, magnesium alloys, copper alloys etc. giving us opportunities to investigate such kind of emergent defects from different sides. Our work have revealed the complex components of the bifillm defects, the varied morphology and so might distinct mechanisms for the defect dependent on the materials and also casting processes. It was also shown that the relationship between the bifilm defect formation and the sodification process defined by the temperature field and also the filling process outlined by the flow field.

References

[1] Song H, Zhang L, Shen H, Cao F, Zhao X, Gu X, et al. Morphology and fracture of oxide bifilm defects in nickel-aluminium bronze. Journal of Materials Research and Technology-Jmr&T. 2021;10:926-34.
[2] Song H, Zhang L, Cao F, Gu X, Sun J. Oxide bifilm defects in aluminum alloy castings. Materials Letters. 2021;285.
[3] Song H, Zhang L, Sun J, Zhang D, Zhang G, Xu H, et al. Research on the formation mechanism and fracture behavior of bifilm defects. Journal of Materials Research and Technology. 2023;26:6919-27.
[4] Song H, Zhang L, Ning Z, Wang Y, Yi J, Cao F, et al. Shrinked bifilms in Mg-Gd-Y-Zr alloy. Materials Letters. 2022;306.
[5] Song H, Zhang L, Cao F, Shen H, Ning Z, Huang Y, et al. Structure of oxide bifilms in nickel-aluminium bronze alloys. Applied Surface Science. 2021;541.
[6] Song H, Zhang L, Cao F, Shen H, Ning Z, Huang Y, et al. Three-dimensional reconstruction of bifilm defects. Scripta Materialia. 2021;191:179-84.

Research on the Application of A356.0 Aluminum Alloy Semi Solid Slurry Technology in Gravity Casting

Lijun Hu *, Yuxiao Zhu, Haichao Zhang
Wuxi Beier Machiney Co., Ltd. , 227Changjiang road , Xinwu District, Wuxi, China
*Corresponding address: E-mail: 51394018@qq.com

Abstract: This article mainly studies the application of A356.0 aluminum alloy semi-solid slurry in gravity casting process. The process selects thin-walled aluminum alloy Valve body castings with complex structures as the research object. Trial production was carried out using two processes, the process of using aluminum liquid with a certain solid phase ratio prepared by semi-solid slurry to tilting gravity casting and the other process of conventional liquid tilting gravity casting. Conducting multiple rounds of experiments and collecting data from multiple sets of experiments. Comparative analysis of multiple data such as semi-solid and liquid interfacial fluidity, metallographic structure, and product properties, which laid a theoretical and practical foundation for the application of semi-solid slurry in aluminum alloy gravity casting.

Keywords:
semi-solid slurry; fluidity; flow trace; misrun; grain

1 Introduction

Aluminum alloy gravity casting is a common aluminum alloy casting process, which happens to be used for products with worse work condition and high mechanical properties request. Because it works by gravity, this process has some limits in product Structures. Such as same products with thickness less than 5mm, complex structures, high requirement for pressure tolerance, when using gravity casting process, porosity, shrinkage and poor molding may occur. These defects lead to a low product qualification rate and an increase in manufacturing costs. Semi-solid slurry preparation technology involves the use of specialized methods to obtain a mixture of liquid and solid slurry, where the metal acquires a certain amount of non-dendritic primary solid phase near the liquids curve. Semi solid slurry due to its unique preparation method, results in finer grain size [1], increased fluidity, and reduced deformation resistance due to the presence of a certain amount of liquid phase. This enhances the fluidity of complex parts in gravity casting.

In this study, a common aluminum alloy casting material, A356.0, was proposed to use semi-solid slurry technology and carry out tilting gravity casting, in order to improve product qualification rate, reduce manufacturing costs, shorten process cycles, and explore mass production applications.

2 Experimental procedure

The supercharged valve body was selected as the research object in the experiment. The structure of the valve body was complex, and the inner cavity with an airway and inlays, it need high mechanical properties and tightness. The supercharged valve body was verified by the same tilting gravity casting machine, the same set of molds, the same A356.0 aluminum alloy. Experiments were conducted under two production process conditions: tilting gravity casting of semi-solid slurry obtained by semi-solid slurry technology and tilting gravity casting of conventional liquid slurry. Each process designed 3 groups process parameters with 20pcs products. Products are inspected for appearance, X-ray inspection, and tightness separately by group, and the comprehensive pass rate were separately calculated. According to the results of the comprehensive pass rate, randomly sampling a sample from each process of the two highest qualification rate groups, and making comparisons in appearance, internal quality (CT scan detection), metallographic structure and mechanical properties.

3 Results

Comprehensive pass rate

Through comprehensive pass rate comparison, it was found that the comprehensive qualification rate of semi-solid slurry gravity casting product was higher.

Appearance comparison

Through the appearance comparison, it was found that the surface flow trace and misrun of semi-solid slurry gravity casting product was significantly reduced

Internal quality

Through the CT scanning comparison, it was found that the internal micro shrinkage of semi-solid slurry gravity casting product was smaller.

Metallographic structure

Through the metallographic analysis and comparison, it was found that the grain size of semi-solid slurry gravity casting product was smaller.

Mechanical properties

Through the mechanical properties testing comparison, it was found that semi-solid slurry gravity casting product had better mechanical properties.

Research Progress on Semi-Continuous Casting of Magnesium Alloys under External Field

Qichi Le[1*], Qiyu Liao[2], Dazhi Zhao [2], Lei Bao [1], Tong Wang [1], Yonghui Jia [3]
1. Key Laborator of Electromagnetic Processing of Materials, Ministry of Education, Northeastern University, Shenyang 110819, China
2. School of Materials Science and Engineering, Northeastern University, Shenyang 110819, China
3. Key Laboratory of Light-weight Materials, Nanjing Tech University, Nanjing 210009, China
*Corresponding address: e-mail: qichil@mail.neu.edu.cn

Abstract: High-performance magnesium alloy shows a trend of large-scale and integrate and the key foundation for its preparation is high-quality casting ingots. Magnesium alloy ingots prepared by traditional semi-continuous casting process inevitably have casting defects, making it difficult to produce high-quality ingots. The external field assisted controlled solidification technology developed by combining physical fields such as electromagnetic and ultrasonic fields with traditional semi-continuous casting processes can prepare high-quality magnesium alloy ingots with uniform microstructure and no cracks. This article mainly summarizes the technical principles of external field assisted casting process, including electromagnetic field, ultrasonic field, and combined physical field. The focus is on elaborating the refinement mechanism of different types of external field in magnesium alloy solidification.

Keywords: Magnesium alloy; Semi-continuous casting; Electromagnetic fields; Ultrasonic fields;

Introduction

Semi-continuous casting (DC casting) is an effective method for stable production of magnesium alloy ingots. Due to the characteristics of low heat capacity, low melting heat, and low thermal conductivity of magnesium alloy, it is difficult to dissipate heat during the solidification process. The large temperature difference between the center and edge of the casting cavity results in problems such as coarse and uneven microstructure, severe segregation, poor surface quality, large cutting amount, and serious tendency for solidification cracking. With the improvement of product specifications and quality requirements, traditional DC casting technology as the main method for magnesium alloy ingots can no longer meet the needs of fine-grained and homogeneous microstructure.

Research has shown [1-3] that applying physical field during the solidification process can significantly improve the solidification conditions of the melt, thereby achieving the goal of refining the microstructure and improving casting defects. For example, electromagnetic stirring, electromagnetic oscillation, ultrasonic treatment, electric pulse technology, etc. This method mainly achieves the goals of degassing, refining grain size, improving ingot density and mechanical properties by introducing appropriate stirring or fluctuation during the solidification process of the melt. The team led by Qichi Le from Northeastern University has conducted in-depth research on the solidification behavior of magnesium alloy ingots under external physical fields for over twenty years. This article summarizes the principles and latest research progress of magnesium alloy DC casting technology under external field, elucidates the mechanism of external field in the casting process.

The application of electromagnetic field in DC casting of magnesium alloy

The researchers found that the as-cast microstructure of magnesium alloys can be significantly refined by applying an electromagnetic field. Le [4] found that the electromagnetic DC process of magnesium alloy can refine the microstructure, improve the grain size uniformity and inhibit the generation of macroscopic segregation and cracks.

In recent years, the Le's team has further improved the magnetic field type of magnesium alloy electromagnetic DC casting and proposed differential pulse magnetic field (DPMF). Jia [5] studied the effects of harmonic magnetic field (HMF), pulsed magnetic field (PMF), differential harmonic magnetic field (DHMF) and DPMF on metal crystalline liquid region and paste region by numerical simulation, and found that pulsed current generates stronger magnetic field than harmonic current under the same conditions. It was found that differential magnetic field can effectively reduce the temperature of the melt in the liquid pool and DPMF has both larger velocity vibration amplitude and lower temperature, making the temperature distribution more uniform [6]. Fig. 1 displays the microstructure of AZ31 magnesium alloy ingot under different electromagnetic fields. Therefore, the DMF shows great potential in refining the microstructure and suppressing columnar crystals in large-sized ingot casting.

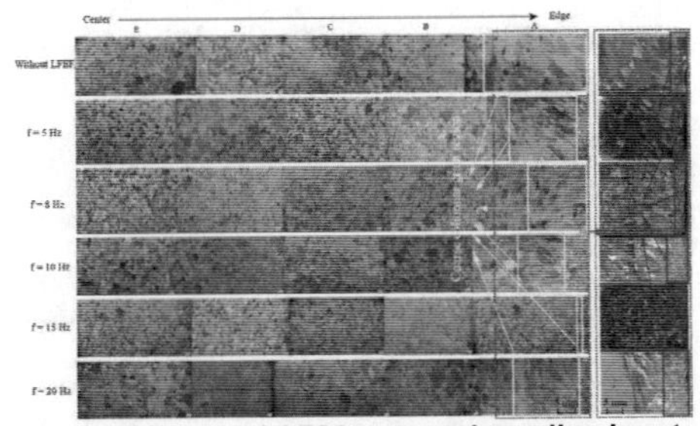

Fig. 1 Microstructure of AZ31 magnesium alloy ingot under different electromagnetic fields [6]

The application of ultrasonic field in DC casting of magnesium alloy

Ultrasonic fields can refine the solidification structure of metals because the cavitation effect and acoustic flow effect generated by the introduction of ultrasonic waves into the melt can change the solidification behavior of the melt. Shao [7] used ultrasonic treatment in AZ80 alloy melt with different intensities and found that the average grain size of the alloy decreased from 303 μm to 148 μm. Chen found that traditional single-frequency ultrasound has problems such as difficult resonance and serious attenuation in metal melt. On this basis, they developed a way to apply dual-frequency ultrasound combinations. Chen [8] found that the as-cast microstructure was significantly more refined after dual frequency ultrasonic field (DUF) treatment, as shown in Fig. 2. Moreover, the yield strength, ultimate tensile strength and elongation after 1400w DUF treatment reached 153 MPa, 239 MPa and 13.9%, respectively, which were 20.5%, 20.7% and 30.0% higher than those after single frequency ultrasonic field (SUF) treatment.

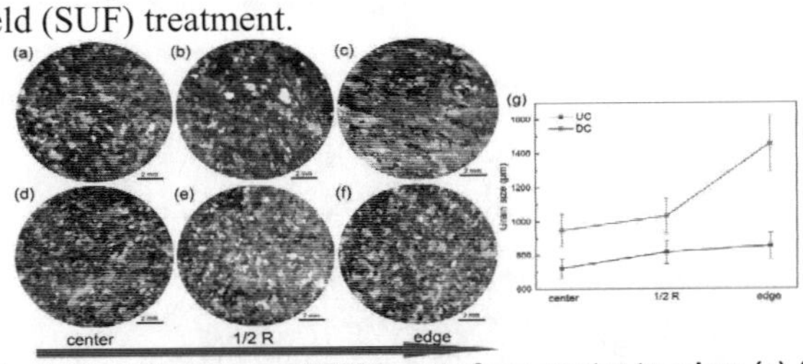

Fig. 2 Microstructure of AZ80 ingots from center to edge: (a)-(c) for DC casting, (d)-(f) for DUF casting and (g) grain size. [8]

The application of combined external field in DC casting of magnesium alloy

The researchers applied the two external fields in the magnesium alloy melt at the same time to achieve better refinement effect. Shao [9] found that simultaneous application of ultrasonic vibration to the melt resulted in significant refinement and uniform distribution of the as-cast microstructures, and ultimately improved mechanical properties. Chen [10] studied the effects of traditional fixed frequency ultrasonic field (FUF), LFEF, variable frequency ultrasonic field (VUF), and ultrasonic+electromagnetic combination field (VUF+LEF) and found that the combination of ultrasonic and electromagnetic fields had the best refining effect and the grain size refined from 679~1454 μm to 116~141 μm.

Conclusions

(1) Applying electromagnetic and ultrasonic fields during the semi-continuous casting process of magnesium alloys can significantly refine the microstructure, as well as achieve degassing and reduce segregation.

(2) The refinement effect of LFEF is remarkable, but the effect is mostly limited by the size of the ingot. For large-sized ingots, the refinement effect of DPM on the entire domain is better.

(3) Applying DUF can significantly improve the refining ability of the as-cast microstructure and expand the refining range.

(4) Compared to applying electromagnetic or ultrasonic fields separately, the combination of electromagnetic and ultrasonic fields has the best refining effect on the as-cast microstructure of magnesium alloy.

Acknowledgments

This research was financially supported by the National Natural Science Foundation of China (No.52274377 and No.52304391).

References

[1] Zuo Y B, Cui J Z, Zhao Z H, et al. Effect of low frequency electromagnetic field on casting crack during DC casting superhigh strength aluminum alloy ingots. Microstructure and Processing, 2005, A406: 286-292

[2] Vives C. Electromagnetic refining of aluminum alloys by the CREM process: Part II. Specific practical problems and their solutions. Metall. Trans. B., 1989, 20(5): 631-643

[3] Sarou-Kanian V, Millot F, Rifflet J C. Surface tension and density of oxygen-free liquid aluminum at high temperature. Int. J. Thermophys., 2003, 24(1): 277-286.

[4] Le Q C, Guo, S J, Zhao Z H, et al. Numerical simulation of electromagnetic DC casting of magnesium alloys, J. Mater. Process. Tech., 2007, 183: 194-201.

[5] Jia Y H, Wang H, Le Q C, et al. Transient coupling simulation of multi-physical field during pulse electromagnetic direct-chill casting of AZ80 magnesium alloy, Int. J. Heat and Mass Tran., 2019, 143: 118524.

[6] Jia Y H, Hu C L, Le Q C, et al. Effect of electromagnetic parameters on melt flow and heat transfer of AZ80 Mg alloy during differential phase electromagnetic DC casting based on numerical simulation, China Foundary, 2021, 19(5): 191-200.

[7] Shao Z W, Le Q C, Zhang Z Q, et al. Numerical simulation of acoustic pressure field for ultrasonic grain refinement of AZ80 magnesium alloy, T. Nonferr. Metal. Soc., 2011, 21(11): 2476-2483.

[8] Chen X R, Ning F K, Hou J, et al. Dual-frequency ultrasonic treatment on microstructure and mechanical properties of ZK60 magnesium alloy, Ultrason. Sonochem., 2018, 40: 433-441.

[9] Shao Z W, Le Q C, Zhang Z Q, et al. A new method of semi-continuous casting of AZ80 Mg alloy billets by a combination of electromagnetic and ultrasonic fields. Mater. Des., 2011, 32: 4216-4224.

[10] Chen X R, Jia Y H, Jia W T, et. al. The simultaneous application of variable frequency ultrasonic and low frequency electromagnetic fields in semi continuous casting of AZ80 magnesium alloy. J. Alloy. Compd., 2019, 774: 710-720.

The Application of X-ray Flaw Detection Machine in Casting Quality Control

Yibin Li, Dianpeng Jia
Dandong Aolong Radiative Instrument Group Co., Ltd.
*Corresponding address: No.46 Fumin Street, Lingang Industrial Zone, Dandong City, China
e-mail: sale@alndt.com

Abstract: Industrial X-ray is a very important non-destructive testing technology that utilizes the characteristics of X-ray that can penetrate materials, and obtains internal information of objects by detecting the absorption and scattering of X-rays.

Industrial X-ray can penetrate the tested object without causing damage to its internal structure, hence it is called Non-Destructive Testing (NDT). When X-ray penetrate through an object, two phenomena occur: absorption and scattering. Absorption refers to the process of X-ray being absorbed by an object, and the degree of absorption is related to the composition and density of the object. Substances with different densities and components have varying degrees of absorption of X-ray, so the internal structure of an object can be determined by observing the absorption of X-ray. Scattering refers to the process of X-ray changing direction in an object, and the degree of scattering is related to the density and thickness of the object. By observing the scattering of X-ray, information such as the density and thickness of the object can be inferred.

Industrial X-ray is widely used in non-destructive testing and quality control in the industrial fields. High resolution imaging and quantitative analysis can be performed on the internal defects and structures of various industrial products such as welds, castings, pipelines, and aviation engine blades, providing important detection methods and quality assurance for industrial production.

This speech will mainly show the widespread application of X-ray in the field of casting through practical testing cases.

1 Introduction

At present, there are two types of X-ray non-destructive testing, one is two-dimensional imaging and the other is three-dimensional imaging, commonly known as ICT, to perform non-destructive testing on internal defects of objects.

2 Experimental procedure

Ⅰ Examples of X-ray testing of castings

1. X-ray detection examples for castings

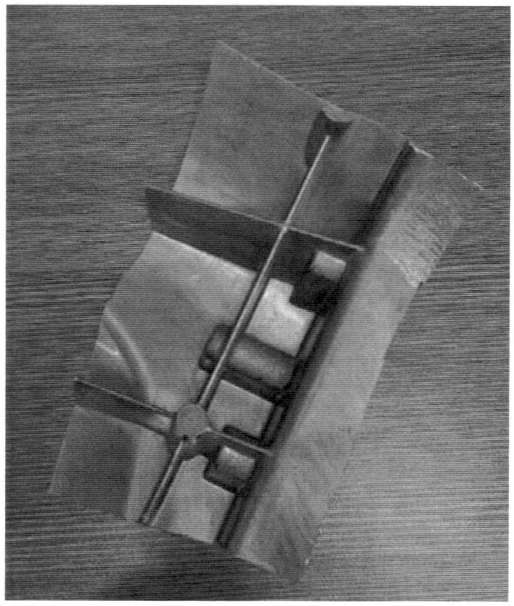

1. 3D X-ray detection examples for castings

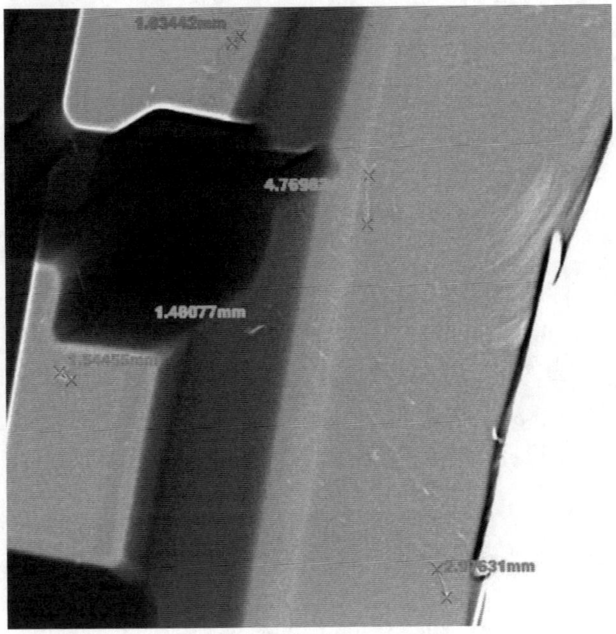

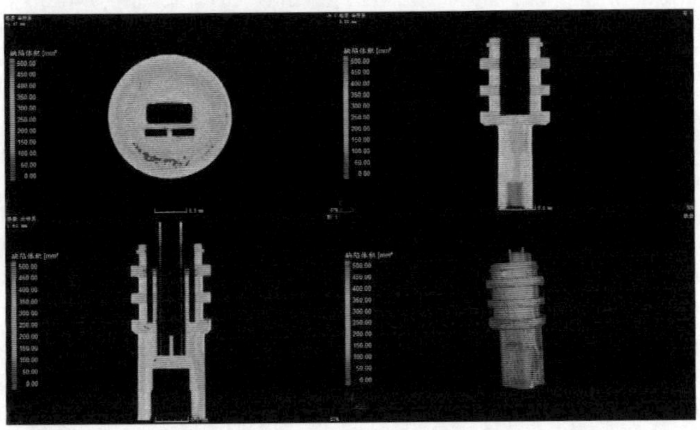

3 Result and discussion

From the above testing examples, we can see that X-ray can clearly find the pores, porosity, and slag in the castings. And the sizes of defects can be measured, with measurement accuracy reaching the micrometer level. Fully meeting the testing requirements of castings, providing protection for enterprises to improve product quality and processing technology.

The other example is the result of using ICT to detect internal defects in castings. ICT is a non-destructive technology that uses X-ray scanning objects to obtain three-dimensional images of their interior, thereby detecting internal defects and structural information.

4 Conclusion

X-ray is a very advanced non-destructive testing technology, which can effectively improve the quality of castings through X-ray machines. It plays a very important role in controlling the quality of casting products and improving the manufacturing process of casting products. For workpieces with large detection volume and simple structure, two-dimensional detection equipment can be used. For castings with complex internal structures that require high detection accuracy, a three-dimensional X-ray machine is required for inspection.

Research and Optimization of the Conditions of the Melting and CastingProcess of Bearing Alloys Intended for Sliding Layers of Bearing Shells

A. Brudny*, M. Maleta, B. Cwolek, J. Kulasa, W. Malec
Łukasiewicz Research Network – Institute of Non-Ferrous Metals, Sowińskiego 5, Gliwice, Poland
*Corresponding address: anna.brudny@imn.lukasiewicz.gov.pl

ABSTRACT: The main aim of the work was to study the influence of variable melting and casting process conditions on the microstructure and hardness of a commercial tin-based alloy with the addition of copper and antimony. The research was aimed at checking the possibility of controlling the share and morphology of brittle Sb-Sn-based phases in the microstructure by using a variable heat removal rate during the solidification process. The research included chemical and metallographic analyzes of the produced castings, along with the analysis of the present phases and hardness measurements.

Keywords: Melting and casting; Bearing alloy; Babbit; Microstructure

1 Introduction

The motivation to undertake this research topic was the increased interest in the production of bearing shells by domestic enterprises due to the difficult geopolitical situation, which influenced the availability and disruption of supply chains of these products from the Eastern market, which was previously leading in this segment [1]. The future introduction of an innovative technology for the production and regeneration of bearings based on Wire Arc Additive Manufacturing (WAAM) 3D printing will significantly improve the current situation and will also improve the quality of the layer applied to the bearing, reduce the consumption of expensive bearing material, reduce removal machining and improve the process yield. The introduction of 3D printing technology, in turn, first requires the development of a technology for producing wires from SnSbCu bearing alloys. In turn, to produce wires, it is necessary to prepare a feedstock in the form of cast material.

In addition to the melting and casting process parameters, the chemical composition has a key influence on the formation of hard reinforcing phases and the final properties of the SnSb11Cu6 bearing alloy. The main alloy additions in the form of Sb and Cu added to the Sn matrix increase the strength and bearing properties while reducing ductility. The ductility of bearing alloys depends strictly on the amount of antimony and copper added to the chemical composition. Increasing the copper content at a constant antimony content causes a drastic decrease in elongation and ductility, which is an undesirable effect from the point of view of plastic working [2,3]. The standard chemical composition of the SnSb11Cu6 alloy (Ł83) developed so far is aimed at ensuring appropriate casting properties enabling the production of sliding layers with the required level of mechanical and sliding properties in the casting process of bearings.

As part of the work, conducted research aimed at optimizing the melting and casting process of the commercial SnSb11Cu6 alloy in terms of developing conditions enabling the elimination of the negative impact of brittle phases on the susceptibility to plastic working (by changing the morphology and reducing the size of brittle phases), while maintaining the appropriate bearing properties of the alloy, defining the possibility of its application on the sliding layers of bearing shells. Changing the crystallization conditions of the alloy by using appropriate casting conditions resulted in a reduction in the size of the brittle phases.

2 Experimental procedure

Melting and casting tests were conducted using a crucible furnace. The temperature of the castingmold and liquid metal was recorded using a multi-channel recorder connected to a computer with LabMini software. Chemical composition analyzes were performed using anoptical spark emission spectrometer SPECTROMAXx LMX09. Hardness measurements using the Brinell method were performed using a universal Wolpert hardness tester (HBW2.2/15.625/180). Metallographic observations of the specimens were carried out using a Keyence VHX-7000 digital microscope. The specimens were observed in both polished and etched states. Additionally, to identify phases and inclusions, examination of sections of the produced castings was carried out using a Zeiss EVO MA10 scanning electron microscope, and energy dispersive analysis (EDS) was used to determine the chemical composition in micro-areas using a Bruker Xflash® 5010 EDS spectrometer.

3 Result and discussion

The paper presents the results of tests on casting the alloy into a casting mold made of B555 bronze. Incase of variant 1, the material was cast into a cold mold, while in the case of variant 2, before casting the ingot mold was heated to a temperature of 400℃. The casting temperature was 490℃.

The results of the analysis of the chemical composition (average value) of the produced castings are presented in Table 1. The chemical composition was within the range specified in the PN 82/H 87111 standard [4] for the Ł83 alloy (SnSb11Cu6).

Table 1.Chemical composition of the produced castings, %

Sn	Sb	Cu	As	Cd	Ni
82,7	10,74	6,47	<0,0025	0,0014	0,0019
Co	**Pb**	**Zn**	**Al**	**Fe**	**Bi**
0,00042	0,0504	0,00022	<0,00010	0,0481	0,0114

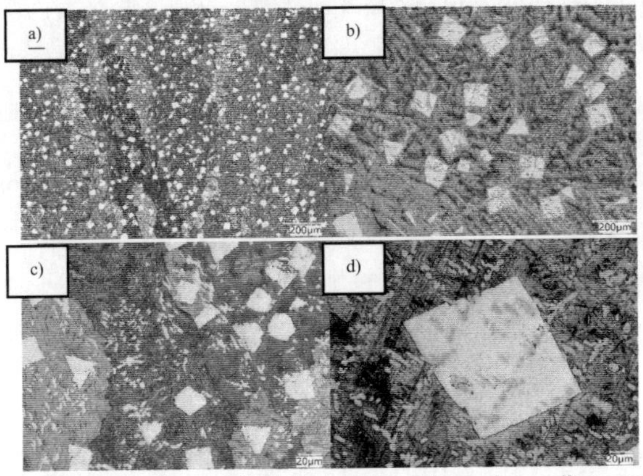

Fig.1 Microstructure of the produced castings depending on the process conditions: a) c) variant 1, b) d) variant 2

No internal casting defects were observed for the tested samples. In the case of faster heat removal, which took place in the case of variant 1, a significant reduction in the size of the SbSn phase precipitates is visible (Fig. 1. a,c). The average size of these precipitates is approximately 468 μm^2, while for variant 2 these precipitates are approximately 18926 μm^2.

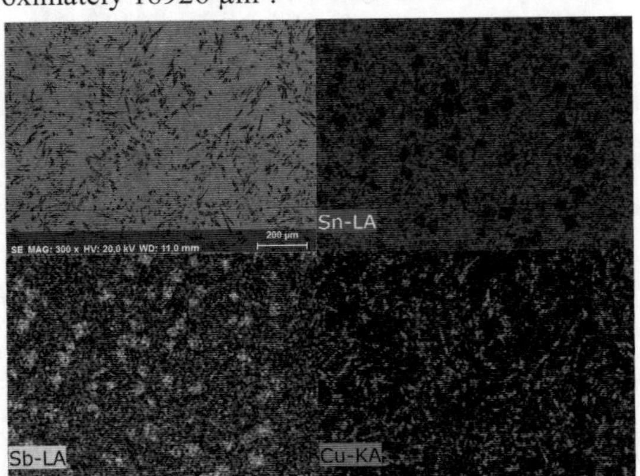

Fig.2 Elements distribution map – variant 1

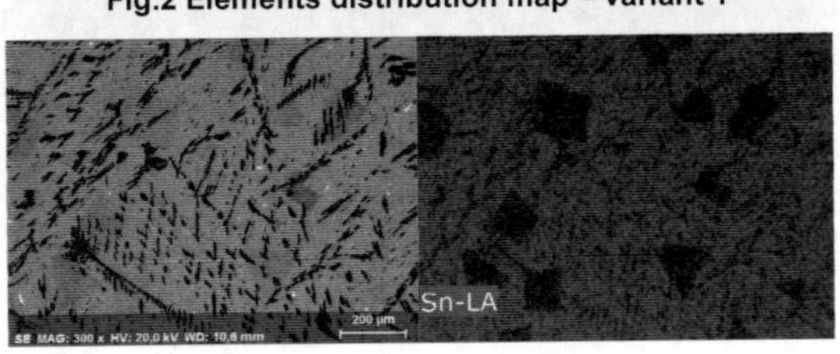

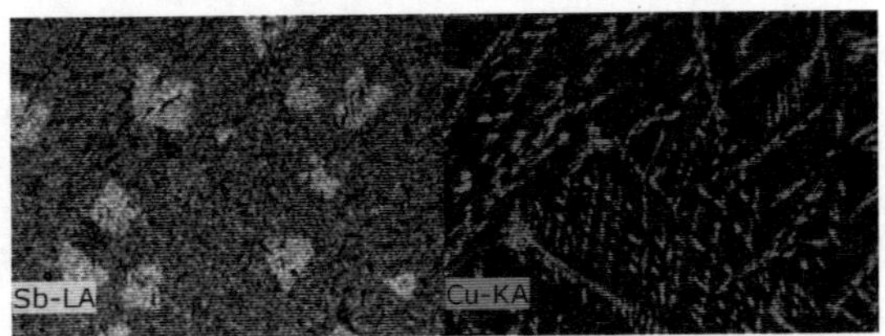

Fig.3 Elements distribution map – variant 2

The microstructure of the produced castings consists of rhomboid precipitates with a stoichiometry corresponding to the SnSb phase and numerous needle-shaped precipitates in the case of variant 2 and much finer precipitates in the case of variant 1 with a stoichiometry corresponding to the Cu_6Sn_5 phase distributed against the background of a tin-rich matrix.

The structure of tin-based bearing alloys therefore includes hard antimony-rich crystals and a network of copper-rich acicular crystals. Hard crystals are dispersed in the soft matrix, thus increasing the hardness of the alloy. The average hardness of the cast material does not differ between the cooling variants - in the case of variant 1 it is - 25.2 HBW, while for variant 2 - 24.7 HBW, so the difference in the obtained hardness values is within the measurement error. The results of the distribution of hardness measurements show relatively good homogeneity across the cross-section of the tested samples.

4 Conclusion

The use of a variable heat removal rate during the solidification process of the SnSb11Cu6 alloy changed the share and morphology of brittle Sb-Sn-based phases in the microstructure. The fragmentation of the precipitates and their even distribution in the matrix did not result in an increase in Brinell hardness.

It is possible to control the morphology and size of the SnSb and Cu_6Sn_5 phase precipitatesby appropriately selecting the parameters of the melting and casting process.

Acknowledgments

The research is carried out as part of the LIDER project (LIDER14/0127/2023) financed by the National Centre for Research and Development.

References

[1] https://www.nakolei.pl/polski-rynek-producentow- taboru-w-tarapatach-lozyska-sprowadzano-z-rosji-stal-z-ukrainy/ (20 June, 2024).

[2] J. Hucińska, Materiały do ćwiczeń laboratoryjnych, Metaloznawstwo, Gdańsk, Wyd. PolitechnikiGdańskiej.

[3] Potekhin B.A., Il'yushin V.V., Khristolyubov A.S.,Effect of casting methods on the structureand properties of tin babbit, Metal Science and Heat Treatment, 2009, Vol. 51, No. 7 – 8.

[4] PN-82/H-87111 - Stopy łożyskowe cyny i ołowiu.

Study on the Microstructure and Hardness of Mg-4Al-4RE-0.3Mn Alloy in High Vacuum Die Casting under Aging

Yongfeng Li, Ang Zhang*, Chuangming Li, Hecong Xie, Bin Jiang* *, Zhihua Dong, Fusheng Pan
National Engineering Research Center for Magnesium Alloys, National Key Laboratory of Advanced Casting Technologies, College of Materials Science and Engineering, Chongqing University, Chongqing 400044, China
* Corresponding author: Ang Zhang, E-mail: angzhang@cqu.edu.cn (Ang Zhang)
** Corresponding author: Bin Jiang, E-mail: jiangbinrong@cqu.edu.cn (Bin Jiang)

Abstract: This article studies the microstructures and properties of Mg-4Al-4RE-0.3Mn (wt.%) (AE44-2) magnesium alloy before and after aging. The alloy is prepared by high vacuum die casting (HVDC). The aging treatment on HVDC-AE44-2 is performed at 100 ° C (0h, 4h, 8h, 12h, 16h, 20h, 24h, and 28h) and 150 ° C (0h, 4h, 8h, 12h, 16h, 20h, 24h, and 28h), respectively. The results show that there is precipitation of the Al-Mn phase in the HVDC-AE44-2 alloy after aging. The Al-Mn phase can effectively hinder the slip of dislocations and make the alloy achieve the hardness peak after aging at 150 ° C for 16 hours. The hardness peak is 66.2HV.

Keywords: Magnesium alloy; Die casting; Microstructure; Hardness

1 Introduction

Magnesium alloy is the lightest structural metal material, widely used in the automotive industry, aerospace, and electronic communication due to its high specific strength and stiffness, good electrical and thermal conductivity, electromagnetic shielding, damping, and shock absorption [1, 2]. Compared with other processing technologies, high-pressure die casting injects molten metal into the mold rapidly, resulting in castings with smooth surfaces and near-net shape, which effectively saves time and reduces costs.

However, liquid melt splashes backward during the die-casting process, creating a highly turbulent flow that entraps a large amount of air. The trapped gas shrinks and deforms, remaining on the surface or inside the castings, resulting in more micropores in the die castings. Currently, many scholars have conducted research on the aging treatment of magnesium alloys [3]. However, research on die-cast magnesium alloys mainly focuses on Mg-Al-RE alloys without high vacuum. Mg-Al-RE alloys without high vacuum cannot be aged due to the presence of pores trapped inside or on the surface of the casting, making it difficult to further improve the alloy properties. There is little research on the aging treatment of Mg-Al-RE alloy by high vacuum die casting (HVDC).

This paper investigates the effects of aging temperature and time on the microstructures and properties of the HVDC-AE44-2 alloy. The Al-Mn precipitates are characterized to study the change in the hardness of HVDC-AE44-2 alloy.

2 Experimental procedure

The alloy used in this work is Mg-4Al-4RE-0.3Mn (mixed rare earth (RE) element of La and Ce) provided by Haimagnete Co., Ltd. (Shaanxi Province, China). The die-casting parts are prepared using the precision horizontal cold chamber die-casting machine DM300 under high vacuum and high pressure.

3 Result and discussion

Microstructure

Fig. 1 shows the metallographic microstructures under different heat treatment conditions. Under aging treatment (T5), the grain size of the HVDC-AE44-2 alloy remains basically unchanged before and after aging. The alloy is mainly composed of α-Mg matrix and second phases, where α-Mg is surrounded by the second phases. A large number of the second phases are distributed at the grain boundaries and a small number of fine particles are distributed inside the grains. Compared to the die-cast state, HVDC-AE44-2 showed a significant increase in Al-Mn precipitates within the crystal and at grain boundaries after 16 hours of aging at 100 ° C and 150 ° C.

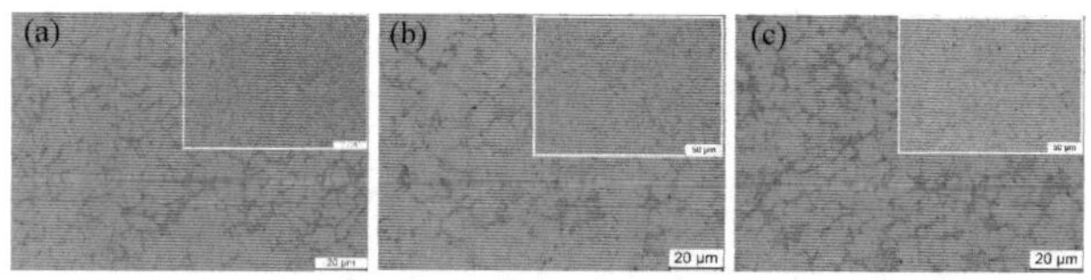

Fig. 1. Metallographic image of HVDC-AE44-2 alloy. (a) as cast, (b) at 100 ° C × 16h, (c) at 150 ° C × 16h.

From the phase diagram in Fig. 2, the precipitated phase Al-Mn is a high-temperature stable phase. When the aged HVDC-AE44-2 alloy undergoes deformation at 150 ° C, the Al-Mn phase still exists. This phase can effectively hinder the slip of dislocations and improve the strength of the HVDC-AE44-2 alloy.

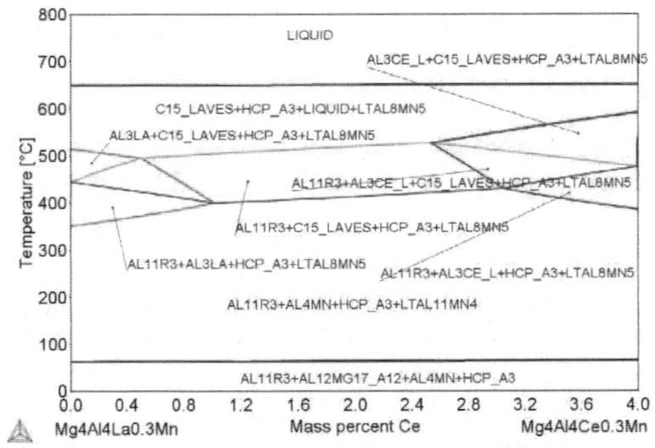

Fig. 2. Phase diagram of HVDC-AE44-2 alloy.

Hardness

Fig. 3 shows the age-hardening curve of HVDC-AE44-2 alloy. The specimens are subjected to medium temperature aging at 100 °C and 150 °C. With the increase of aging time, the hardness of the specimens shows a trend of first increasing and then decreasing at both 100 °C and 150 °C and the hardness reached the peak at 16 hours of aging. The hardness after 16 hours of aging at 150 °C is 66.2 HV, and the hardness after 16 hours of aging at 100 °C is 65.4 HV. The hardness after 16 hours of aging at 150 °C is higher than that after 16 hours of aging at 100 °C. When the hardness peak is reached, the hardness value shows a rapid decrease, manifested as over-aging.

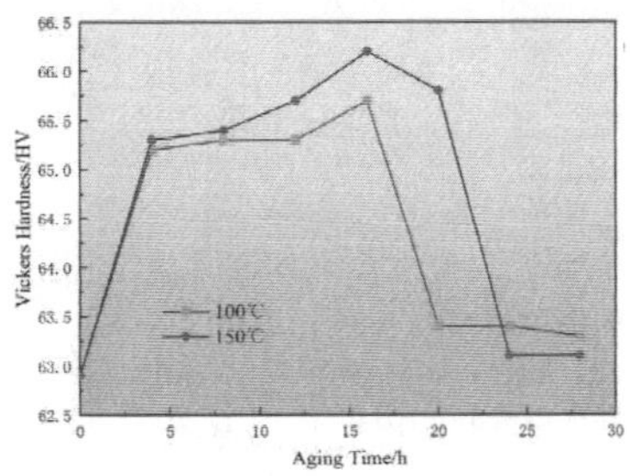

Fig. 3. Age hardening curve of HVDC-AE44-2 alloy.

4 Conclusion

In the aged HVDC-AE44-2 alloy, the precipitation of high-temperature stable Al-Mn phase can effectively prevent dislocation slip, resulting in a trend of first increasing and then decreasing hardness of the aged alloy. The hardness after 16 hours of aging at 150 °C is 66.2 HV, which is the hardness peak.

Acknowledgments

This work is financially supported by the National Key Research and Development Program of China (grant number 2021YFB3701000), and National Natural Science Foundation of China (grant numbers 52101125, U2037601, and U21A2048).

References

[1] D. Zhang, J. Zhang, Y. Zhang, D. Zhang, T. Xu, B. Li, Y. Zhao, J. Meng, Superior high-temperature strength in a low RE containing Mg extrusion alloy with nano-spaced stacking faults, Materials Science and Engineering: A 854 (2022) 143791.

[2] J. Xie, J. Zhang, Z. You, S. Liu, K. Guan, R. Wu, J. Wang, J. Feng, Towards developing Mg alloys with simultaneously improved strength and corrosion resistance via RE alloying, Journal of Magnesium and Alloys 9(1) (2021) 41-56.

[3] Q. Yang, K. Guan, B. Li, S. Lv, F. Meng, W. Sun, Y. Zhang, X. Liu, J. Meng, Microstructural characterizations on Mn-containing intermetallic phases in a high-pressure die-casting Mg–4Al–4RE–0.3Mn alloy, Materials Characterization 132 (2017) 381-387.

Microstructural Characterization and Optimization of CoCrCuFeMnNi High Entropy Alloy to be in Contact with Hydrogen

O. Gordo-Burgoa[1*], P. Garcia-Michelena[1], L. Armendariz[2], I. Vicario[3], E. Mardaras[4], G. Arruebarrena[1], I. Hurtado[1]
1. Mechanical and Manufacturing Department, Mondragon University, Arrasate-Mondragon, Spain
2. Department of Mining & Metallurgical Engineering and Materials Science, Faculty of Engineering of Bilbao (UPV/EHU), Bilbao, Spain
3. Metal processing platform, TECNALIA, Basque Research and Technology Alliance (BRTA), Derio, Spain
4. Azterlan, Basque Research and Technology Alliance (BRTA), Durango, Spain

*Corresponding address: ogordo@mondragon.edu

Abstract: This study aims to develop a High Entropy Alloy (HEA) with reduced susceptibility to Hydrogen embrittlement (HE) and improved resistance to oxidation, potentially surpassing existing materials in hydrogen-rich environments.

Microstructural analysis revealed a dendritic structure where two FCC phases have been formed: a dendritic zone with the designed composition and a Cu-rich interdendritic structure with reduced Mn content. The second phase has been eliminated with the application of heat treatment but Cr-rich segregation in grain boundary appeared during a prolongated exposure at 1100℃, which could correspond to σ phase. Segregation of σ phase might be eliminated with an optimization of the composition or heat treatment conditions, achieving a single-phase solid solution.

Keywords: High Entropy Alloys; Hydrogen Embrittlement; Single-phase solid solution; Interdendritic structure

1 Introduction

The hydrogen industry is expanding rapidly to meet increasing energy demands, utilizing existing gas pipelines to transport hydrogen. However, current steels suffer irreversible damage from hydrogen exposure, known as Hydrogen Embrittlement (HE).

High Entropy Alloys (HEAs) are promising due to their excellent mechanical properties, even at cryogenic temperatures, making them suitable for hydrogen applications. The Cantor alloy, developed by B. Cantor [1], is a well-known HEA composed of five metallic elements in varying atomic percentages (5-35%) and a single-phase solid solution structure. This alloy exhibits improved ductility and strength resistance in hydrogen-rich environments [2]. Despite extensive research on HEA mechanical properties, there is limited information about their behavior concerning hydrogen damage.

This study focuses on modifying the Cantor alloy to enhance its performance when exposed to hydrogen. Mn affects oxidation behavior, leading to rapid oxide formation at high temperatures [3]. Some researchers have replaced all the Mn with Cu to create a more oxide-resistant material, resulting in a two-phase microstructure with a primary FCC structure enriched in Cu [3]. Given copper's low permeation [4] with respect to hydrogen, this work explores the possibility to develop an HEA variant that retains the beneficial properties of the original Cantor alloy while exhibiting improved resistance to hydrogen-induced damage and oxidation.

2 Experimental procedure

The CAPHAD method is employed to determine the target composition of the alloy optimizing the balance between Mn and Cu to achieve a composition window that results in a single-phase FCC solid solution.

Following multiple calculations using FactSage software and a steel database, a specific chemical composition has been determined that exhibits a single FCC phase window between 1000 and 1200 ℃. Quenching from a temperature within this range allows the retention of the FCC phase at room temperature.

The casting process to produce the HEA was conducted using an induction furnace under an argon atmosphere. Raw metals with a purity exceeding 99.5% were utilized to achieve a precise composition and clean alloy. The melt was poured into a stair-shape H13 metallic mold.

The microstructural characterization was done using a Field Emission Scanning Electron Microscope (SEM), while thermophysical properties were assessed using a Differential Scanning Calorimeter (DSC).

3 Result and discussion

All as-cast samples exhibit a dendritic structure. Two types of dendrites can be distinguished: primary solidified dendrites, which nucleate and grow first, and secondary solidified dendrites, which are finer and form between the primary ones. Samples from the thickest stair exhibit higher Secondary Dendrite Arm Spacing (SDAS) values compared to those from the thinnest stair.

In this material, two principal phases can be identified: FCC#1 and FCC#2. FCC#1 is a solid solution with the design, while FCC#2 is a Cu-rich phase.

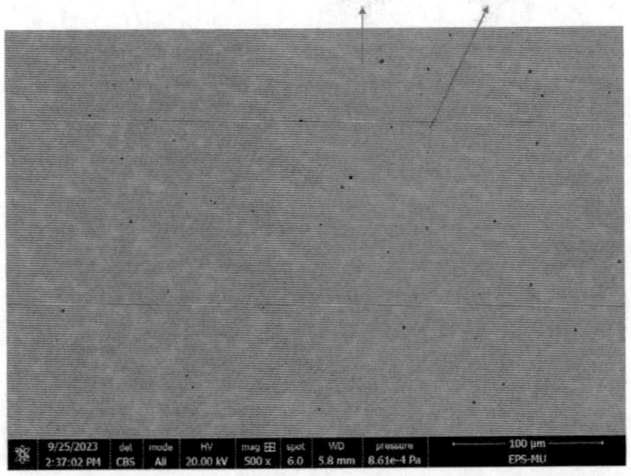

Figure 1 Disposition of FCC#1 and FCC#2 in the alloy

The results show a Cu-rich interdendritic structure. Moreover, the presence of Mn in interdendritic areas is evident but in a lower amount compared to Cu. From the equilibrium solidification diagram, it is known that the FCC#2 phase is the Cu-rich phase, and, taking into account the EDS results, the FCC#2 phase could be the second solidifying phase.

For the elimination of the FCC#2 and to obtain a single-phase FCC#1 solid solution, a heat treatment has been done: 900℃ for 24h + 1000℃ for 24h + 1100℃ for 72h and final quenching. The dendritic structure of the alloy disappeared where FCC#2 phase has been eliminated and FCC#1 solid solution was homogenized. However, due to the prolonged exposure at high temperature, Cr-rich grain boundary segregation appeared. This segregation probably corresponds to the σ phase as it is gathered by A.J. Maldonado [5] in Cantor alloy at a lower temperature. Additionally, it is proven that Cr segregation destabilize the FCC solid solution which could be stabilize or eliminate increasing Ni and Co [6] or optimizing the heat treatment procedure.

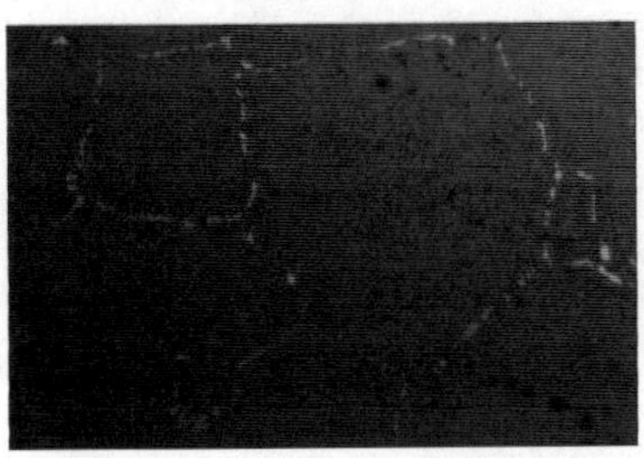

Figure 2 Cr-rich grain boundary segregation

4 Conclusion

The development of a CoCrCuFeMnNi high entropy alloy to be in contact with hydrogen has been done. The main results are the following:

- It is shown FCC#1 solid solution with a similar composition to the defined one, while FCC#2 is a Cu-rich phase.
- FCC#2 phase has been eliminated and FCC#1 solid solution has homogenized. However, a Cr-rich segregation appeared.
- Cr-rich segregation is associated to σ phase that could destabilize the FCC solid solution.
- The stabilization of FCC solid solution could be controlled increasing Ni and Co in the developed HEA or optimizing the heat treatment procedure.

References

[1] [1] B. Cantor, I. T. H. Chang, P. Knight, and A. J. B. Vincent, "Microstructural development in equiatomic multicomponent alloys," Materials Science and Engineering A, vol. 375–377, no. 1-2 SPEC. ISS., pp. 213–218, Jul. 2004.

[2] T. Zhu et al., "Influence of hydrogen behaviors on tensile properties of equiatomic FeCrNiMnCo high-entropy alloy," J Alloys Compd, vol. 892, Feb. 2022.

[3] M. L. Bürckner, L. Mengis, E. M. H. White, and M. C. Galetz, "Influence of copper and aluminum substitution on high-temperature oxidation of the FeCoCrNiMn 'Cantor' alloy," Materials and Corrosion, vol. 74, no. 1, pp. 79–90, Jan. 2023.

[4] J.-M. Chen and J.-K. Wu, "Hydrogen diffusion through copper-plated AISI 4140 Steel," Corros Sci, vol. 33, no. 5, pp. 657–666, 1992.

[5] A. J. Maldonado, K. P. Misra, and R. D. K. Misra, "Grain boundary segregation in a high entropy alloy," Materials Technology, vol. 38, no. 1, 2023.

[6] G. Bracq, M. Laurent-Brocq, L. Perrière, R. Pirès, J. M. Joubert, and I. Guillot, "The fcc solid solution stability in the Co-Cr-Fe-Mn-Ni multi-component system," Acta Mater, vol. 128, pp. 327–336, Apr. 2017.

Research of Hybrid Aluminium Castings with the Use of Porous Cores

Marek Bruna[1]*, Martin Medňanský[1], Miloš Mičian[1], Peter Oslanec[2]
1. Faculty of mechanical engineering, University of Žilina, Univerzitná 1, 01026, Žilina, Slovakia
2. Institute of Materials and Machine Mechanics, Dúbravská cesta 9/6319, 845 13 Bratislava, Slovakia
*Corresponding address: e-mail: rapture.marko@gmail.com

Abstract: The paper focuses on the research of hybrid aluminum castings produced by overcasting technology of the foamed cores. The aim is to optimize the pressure during the foaming process on the resulting porosity evaluated by X-ray tomograph. Alternative using an initial pressure of 0.3 MPa appear to be the most satisfactory. The challenge of this technology is to ensure adequate layer between the porous core and the solidified metal without penetrating the coating layer. For this reason, the surface treatment of foamed cores with various etchants has been proposed.

Keywords: Hybrid Castings, Aluminium Foams, Lighweighting, porous Materials

1 Introduction

In an effort to achieve carbon neutrality, the automotive industry is using a variety of technologies to reduce the weight of its products. The proposed alternative approach in this paper is the overcasting technology.

Overcasting is defined as a manufacturing technology in which two metals, one in the solid state and the other in the liquid state, are brought into contact with each other so that a diffusion reaction zone is formed between the two materials, resulting in a smooth transition from one metal to the other.

The main challenge of this process is the bonding of materials during the casting of metal foam with molten metal of similar chemical composition, because on the surface of the solid core, as well as on the surface of the melt encapsulating the porous core, oxide layers of Al_2O_3 occurs. In this paper a principle of chemical stabilisation of the oxide layer is used, by etching the surface [1].

2 Experimental procedure

The basis of the powder mixture used was Al99.5 powder (with an average fraction < 400 μm) with the addition of 10 wt% Si powder (< 20 μm) and 0.8 wt% foaming agent TiH_2 powder (< 5 μm).

In the first step, samples in the shape of a plate with dimensions of 15 x 55 x 55 mm were foamed in order to determine the influence of the applied pressure when foaming cores with simple geometry. The foaming temperature was 680 °C. The pressure of the atmosphere (N_2) in the foaming chamber changed rapidly from the initial pressure to the stabilization pressure, which acts as a counterpressure against the bubbles of released hydrogen. The variants of foaming pressures during the realization of individual samples are shown in Table 1.

Table 1.Foaming pressure variants

Variant	Starting pressure [MPa]	Stabilising pressure [MPa]
A0	Atmospheric	Atm.
A1	0.1	Atm.
A2a/b/c	0.2	Atm. / 0.05 / 0.1
A3a/b/c	0.3	0.1 / 0.15 / 0.2

3 Results and Discussion

Eight samples were foamed, while three different stabilization pressures were used for sub-variants A2 and A3.

The porosity was analyzed using a Nikon XT H225 ST computer tomograph - the average Feret's pore diameter and its standard deviation were determined.

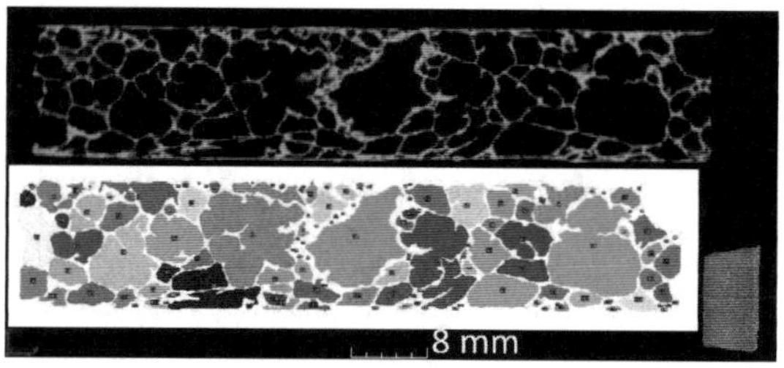

Figure. 1. Pore amount in the cross-section of A3a sample

Based on these results (figure 2), three variants of foamed cores were selected for overcasting. B1 as a reference sample foamed at atmospheric pressure, then B7 as a sample foamed under conditions similar to plate sample A3a (0.1 - 0.3 / 0.15 MPa), and sample B5, which will serve as a control sample when using foaming pressures approximately half (0.1 - 0.2 / 0.05 MPa). Foamed cores intended for overcasting are more complex, cylindrical shapes, with geometry of which is shown schematically in Figure 3.

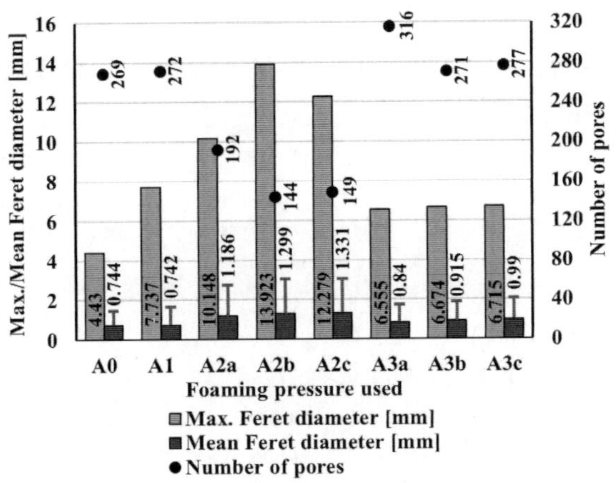

Figure 2. Dimensions of the foamed core

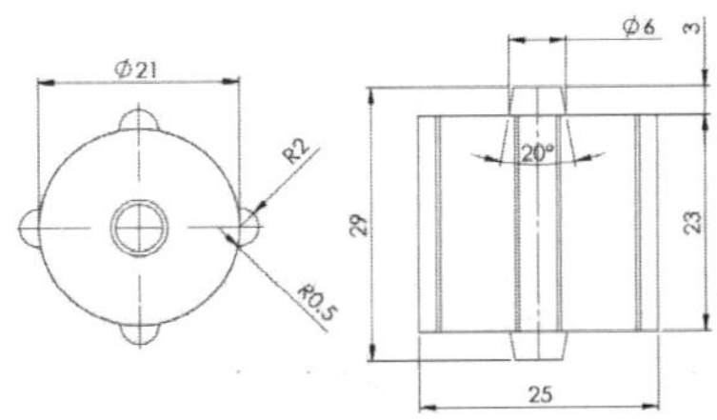

Figure. 3. Dimensions of the foamed core

The molds used for casting were made from sand blocks using a patternless process. AlSi7Mg0.3 alloy was selected for overcasting foamed cores by using standard gravity casting technology. The mold was designed with a vertical parting plane, a naturally pressurized gating system, three cavities and atmospheric sprues (Figure 4). The material was melted in an electric resistance furnace at a casting temperature of 760 ± 10 °C. Macroscopic observation of the selected hybrid castings was performed using an Olympus SZX16 stereomicroscope.

The goal of the overcasting process was to obtain a hybrid casting with a foam core bonded to the surface with a 5 mm thick solid wall of the casting. 4 molds were made, one for each core surface modification. B1, B5 and B7 foamed cores were inserted into each of these molds. Four variants were determined for macroscopic observation: hybrid casting with a non-etched B5 core, B1 core etched with HF, B7 core etched with H_3PO_4 for 1:30 min and B1 core etched with H_3PO_4 for 4 minutes.

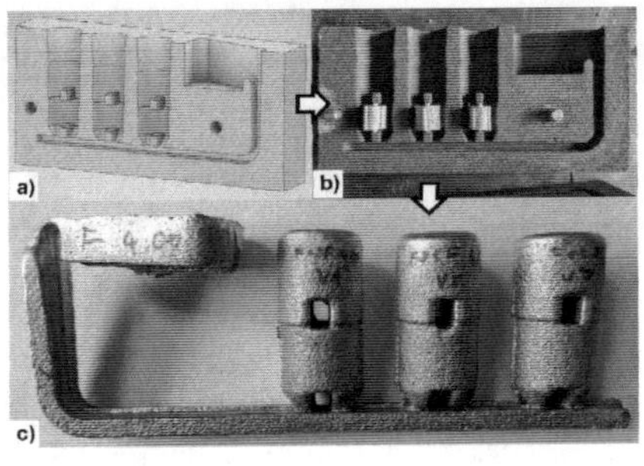

Figure 4. a) model, b) mold with fixed cores, c) casting

Macroscopic observation of these four samples revealed an undesirable penetration of the molten Al alloy into the pore volume of the foamed core almost to the full extent of the casting with an unetched core (Figure 5a). In the image of the HF-etched core variant (Figure 5b), the penetration of the melt can be seen from the bottom. The distribution of the penetrated metal indicates that the melt flow disturbed the surface of the core as it entered the mold cavity. Predicted core/casting interfaces are indicated by orange lines. The best potting result seems to be the variant etched with phosphoric acid for 1:30 min (Figure 5c), where there was no unwanted penetration of the core by the melt, longer etching time did not change the penetration character (figure 5d).

Figure 5. Macroscopic evaluation of the samples: a) nonetched B5 core, b) B1 core etched with HF, c) B7 core etched with H3PO4 for 1:30 min, d) B1 core etched with H3PO4 for 4 min

4 Conclusions

Based on the results obtained, it can be concluded that:

- The least suitable foaming pressure was the starting pressure of 0.2 MPa, which produced the largest pores in the board samples and the widest range of their sizes.
- The smallest variability of the results of the relative density of the molded specimens was obtained with the core variants B6 and B7, i.e. when using a starting/stabilizing pressure of 0.1 - 0.2 MPa / atm. pressure and 0.1 - 0.3 / 0.15 MPa, respectively.
- From the results of the experimental casting, it seems necessary to use a surface treatment to prevent the liquid metal from penetrating into the pores of the core. The best result was obtained by surface treatment with H_3PO_4.

Acknowledgments

The research was created within the projects: VEGA 1/0241/23 and KEGA 029ŽU-4/2023. The authors thank for the support.

References

[1] Papis, K., Hallstedt, B., Löffler, J., & Uggowitzer, P. 2008. Interface formation in aluminium–aluminium compound casting. Acta Materialia, 56(13), 3036–3043. doi: 10.1016/j.actamat.2008.02.042

Effect of Electro-Pulse on Microstructure of Al-Cu-Mn-Zr-V Alloy During Aging Treatment

Dequan Shi[1], Wenbo Yu[1], Guili Gao[2]
1. School of Materials and Chemistry, University of Shanghai for Science and Technology, No. 516 Jungong Road, Yangpu District, Shanghai, China
2. School of Mechanical and Energy Engineering, Shanghai Technical Institute of Electronics and Information, No. 3098 Wahong Road, Fengxian District, Shanghai, China
*Corresponding address: e-mail: shidequan2008@163.com

Abstract: The effects of electro-pulse on microstructure and mechanical properties of Al-Cu-Mn-Zr-V alloy were investigated. As the current density increases, the size and quantity of precipitates gradually transit from continuous aggregation to dispersion at grain boundaries, and the mechanical properties are improved. When the current density is 15 A·mm^{-2}, the precipitates are smallest and the mechanical properties are best. The tensile strength is 443.5 MPa and the elongation is 8.1%, which are 51.7% and 42.1% higher than those of conventional ageing treatment, respectively. An electro-pulse with a current density of 15 A·mm^{-2} can rapidly nucleate the second phase at 463K while the precipitates are relatively small after growth.

Keywords: Electro-pulse; Current density; Ageing treatment; Al-Cu-Mn-Zr-V alloy

1 Introduction

Al-Cu alloy is one of the representative high-strength cast aluminum alloys, and was widely used in the fields of aerospace, defense industry and civil tools [1]. Their mechanical properties can be greatly improved by the heat treatment [2]. The electro-pulse can not only improve microstructure and mechanical properties of alloys, but also reduce the heat treatment time. Some researches have been conducted to improve mechanical properties by adopting electro-pulse [3]. Bian *et al.* [4] synchronously applied an electro-pulse to the steady-state creep aging of Al-Cu alloy, and the hardness and corrosion resistance of alloy are improved. Kang *et al.* [5] concluded when the electro-pulse is used for the ageing of Al-Zn-Mg-Cu alloy, the tensile strength and elongation are 14.1 MPa and 4.45% higher than those of conventional ageing.

In this study, the conventional ageing (CA) and the ageing with electro-pulse (AEP) was performed on Al-Cu-Mn-Zr-V cast alloy, respectively. The mechanical properties and microstructures were analyzed.

2 Experimental procedure

The alloy for experiments is Al-5.02Cu-0.33Mn-0.15Zr-0.11V, and the sample size was shown in Figure 1. By DSC analysis, the solid solution process is set as 495±5°C/5±0.5 h and 530±5°C/24±0.5 h. The CA process is 190±2°C/26±0.5 h, and AEP process is CA + 10 s electro-pulse every 5 min. The density of electro-pulse is set as 0, 5, 10, 15, 20, 25 A·mm^{-2} with fixed duty ratio 1% and frequency 100 Hz.

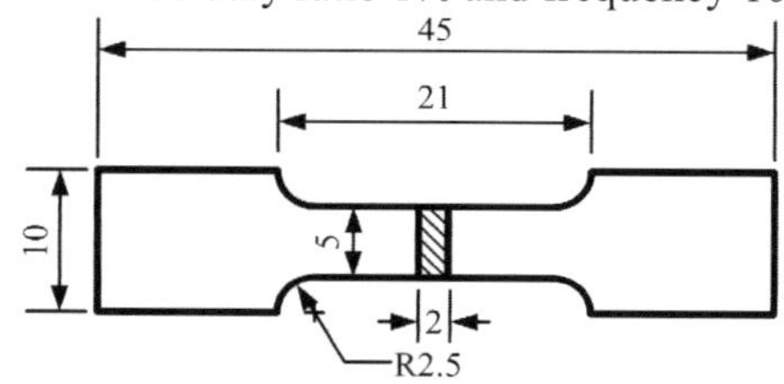

Fig. 1 Sample size of Al-5.02Cu-0.33Mn-0.15Zr-0.11V alloy

3 Result and discussion

Microstructure

Figure 2 indicates there is a Cu-gathering at the grain boundary while there are small amounts of Mn, Zr, V, Ti-gathering at the grain boundary. All elements are uniformly and diffusely distributed within the grains. The EDS expressed that point 1 is taken as θ (Al_2Cu) phase, point 2 may be a mixed phase of Al_3(Ti, Zr, V), and point 3 is a dispersed phase distributed within the grain and it should be $Al_{12}CuMn_2$.

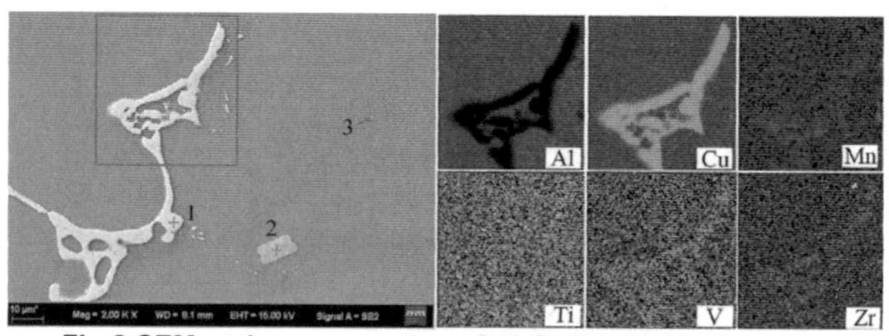

Fig.2 SEM and map scanning of Al-Cu-Mn-Zr-V alloy after conventional ageing treatment

The microstructures indicate the electro-pulse has a significant influence on the size and number of θ (Al_2Cu) phase during ageing process. For the CA, the size of the precipitates is larger, and they are continuously aggregated at the grain boundary. For AEP, the size of the precipitates decreases with the increase of the current density when the current density is below 15 A·mm^{-2}, and the precipitates gradually changes from continuous aggregation distribution to the dispersed distribution at the grain boundary.

Figure 3a and 3b express that the θ' (Al_2Cu) phase is uniformly distributed in the α(Al) phase after ageing treatment, and it is needle-like and vertically precipitated along the two directions of $[010]_{Al}$ and $[001]_{Al}$, which is semi-coherent with α(Al) phase. The precipitates in Figure 3b are very concentrated and even some overlap partially. If the current density further increases, the precipitates will be layered, causing the stacking distribution and weakening the strengthening effect of precipitates on the matrix. Figure $3b_1$ indicates that the θ' phase after CA has better coherent relationship with the α(Al) phase, which results in greater degree of lattice distortion and better strengthening on the alloys.

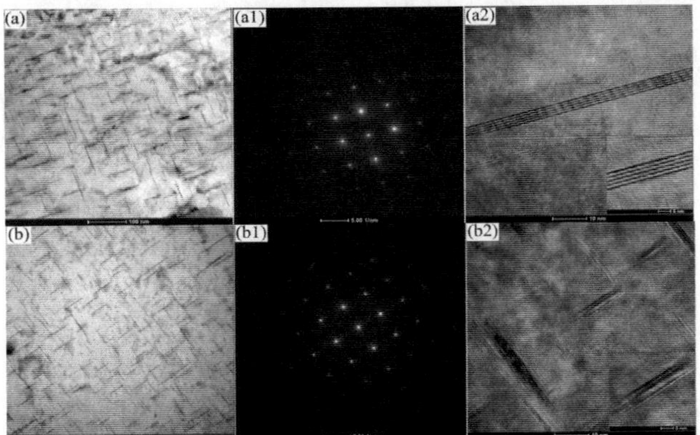

Fig.3 TEM for CA and AEP treatment: (a)CA treatment, (b) AEP treatment with current density of 15 A·mm^{-2}

Mechanical properties

The mechanical properties were shown in the Figure 4. With the increase of current density, the tensile strength and elongation of samples generally increase to the maximum and then decrease. When the current density is 15 A·mm^{-2}, the tensile strength and elongation of samples reach the peak. The tensile strength and elongation for CA treatment are 292.4 MPa and 5.7%. However, for AEP treatment with 15 A·mm^{-2}, they are 443.5 MPa and 8.1%, respectively, which are 51.7% and 42.1% higher than those of CA treatment.

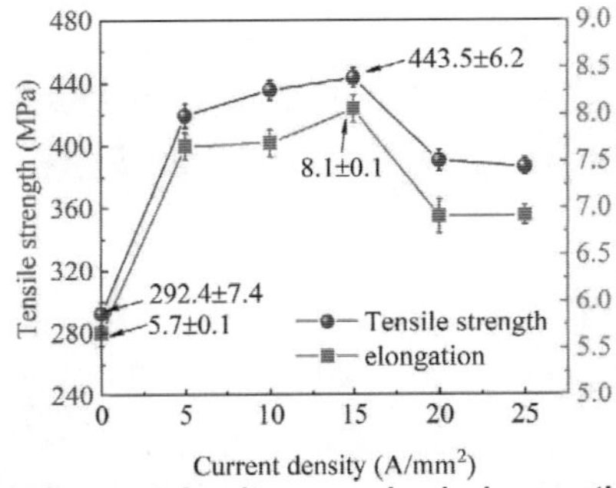

Fig.4 Effect of current density on mechanical properties of Al-Cu-Mn-Zr-V alloy

4 Conclusion

As the current density increases, the precipitates gradually transform from continuous aggregated distribution at grain boundary to dispersed distribution, and mechanical properties increase. When the current density is 15 A·mm^{-2}, the mechanical properties is the best with tensile strength of 443.5 MPa and elongation of 8.1%, which are 51.7% and 42.1% higher than CA, respectively. When the current density exceeds 15 A·mm^{-2}, the size and quantity of precipitates begin to increase again and aggregate at grain boundaries. The electro-pulse of 15 A·mm^{-2} makes the second phases to more rapidly nucleate while growing into smaller particles.

References

[1] Su R M, Jia Y X, Xiao J and Li G L. Effect of secondary aging on microstructure and properties of cast Al-Cu-Mg alloy. China Foundry, 2023, 20(1): 71-77.

[2] Cai Z, Liu H, Wang R, Peng C, Feng Y and Wang X. Microstructure and mechanical properties of the extruded Al-Cu-Mn-Sc-Zr alloy during single-stage and two-stage aging. J. Mater. Eng. Perform., 2023, 32(1): 185-198.

[3] Liu X Y, Li M J, Gao F, Liang S X, Zhang X L and Cui H X. Effects of aging treatment on the intergranular corrosion behavior of Al-Cu-Mg-Ag alloy. J. Alloy. Compd., 2015, 639: 263-267.

[4] Bian T J, Li H, Lei C, Wu C H and Zhang L W. Microstructures and properties evolution of Al-Zn-Mg-Cu alloy under electrical pulse assisted creep aging. Adv. Manuf., 2022, 10: 596-609.

[5] Kang K J, Li D Y, Shi D Q and Gao G L. Evolution of microstructure and mechanical properties in Al-Zn-Mg-Cu alloy by electric pulse aging treatment. Trans. Indian Inst. Met., 2021, 74(11): 2835-2842.

Effects of Interface on Mechanical Propertiesof High-Entropy Alloy Particle-ReinforcedStainless Steel Through Molecular Dynamic Simulation

Dezhu Zhang, Haijun Huang
School of Materials and Chemistry, University of Shanghai for Science and Technology, No. 516 Jungong Road, Yangpu District, Shanghai, China
*Corresponding address: e-mail: haijun.huang@usst.edu.cn

Abstract: This study utilized the Atomsk software to establish molecular dynamics models and conducted simulations of uniaxial tensile deformation processes at interfaces of 316L SS/HEA composite materials using the large-scale atomic/molecular massively parallel simulator Lammps. The interface structures' impact on the tensile performance of composite materials was analyzed using the Ovito visualization software. Compared to the SS(001)/HEA(111) interface combination, SS(110)/HEA(111) and SS(111)/HEA(111) interfaces are more conducive to achieving high-performance composite materials. This is attributed to their more stable interfaces, capable of triggering the interface semi-blocking effect. The occurrence of martensitic phase transformation and twinning can promote synergistic enhancement of strength and plasticity. Additionally, higher dislocation density can increase material strength.

Keywords: Highentropyalloyreinforcedstainlesssteelcomposite; Interfacestructure; Interfacestability; Moleculardynamicssimulation

1 Introduction

To gain deeper insights into the deformation behavior of composite material interfaces, many researchers have resorted to Molecular Dynamics (MD) simulations to replicate the dynamic processes of interface micro-deformation and analyze the internal mechanisms of deformation at the atomic level. Zhang et al. incorporated FeCoNiCr high-entropy alloy nanoparticles as reinforcing phases into 316L austenitic stainless steel to fabricate composites using laser powder bed fusion. MD simulations were employed to study the evolution of layer faults and dislocations with strain in MMCs, illustrating that interfaces hinder dislocation motion to some extent and allow dislocations to propagate smoothly with increasing deformation. However, the establishment of the stainless steel model and the selection of modeling directions have not been thoroughly investigated. To gain a clearer understanding of the influence of different interface structures on the tensile properties and mechanisms of composite materials, it is necessary to investigate the micro-mechanical behavior of high-entropy alloy particle-reinforced stainless steel composites during uniaxial tensile deformation.

2 Experimental procedure

ATOMSK was utilized to model the SS and HEA systems separately. MD calculations were performed using the Large-scale Atomic/Molecular Massively Parallel Simulator. Initially, the models were energy-minimized using the Conjugate Gradient (CG) method, and then relaxed for 50 ps at room temperature and 0 atm in the NPT ensemble. They were combined into a composite material with an interfacial spacing greater than one atomic distance. Subsequently, the composite material models were melted by increasing the temperature from 300 K to 3000 K within 300 ps in the NVT ensemble, and the melting point in the simulation was determined to be around 2600 K. The models were then annealed by gradually cooling from 2600 K to 300 K at a rate of 0.46 K/ps in the NPT ensemble after relaxation for 20 ps. Finally, the models were re-equilibrated for 100 ps at 300 K and 0 atm pressure in the NPT ensemble to obtain stable relaxed models. Tensile forces were uniformly applied along the z-axis at a rate of 0.001 Å/ps until fracture occurred. Ovito software was used for visual analysis of crystal structures, and the evolution of dislocations during deformation was deeply studied using the Dislocation Extractor Algorithm.

3 Result and discussion

Interface Stability

Calculations reveal that the Gibbs free energy of the three crystal orientations of the composite materials differs insignificantly (the Gibbs free energy ΔG_1 for 316L SS, ΔG_2 for HEA, and ΔG_3 for the SS/HEA composite, $\Delta G=\Delta G_3-(\Delta G_1+\Delta G_2)$. The Gibbs free energy values for the SS(111)/HEA(111) (marked S111) composite (-2.395 kJ/mol) and SS(110)/HEA(111) (marked S110) composite (-2.280 kJ/mol) are close, compared to the SS(001)/HEA(111) (marked S001) composite (-2.076 kJ/mol), indicating that these two crystal orientations are more stable.

Tensile Property MD Simulations

Figure 1 illustrates the atomic configuration, atomic types, and distribution of dislocations with strain rate in composite materials. Prior to tensile loading, partial Shockley dislocations and a small number of Stair-rod dislocations were already formed at the interface of the S001 composite material (Figure 1a_1). At a strain of 0.091,

when the stress reaches its peak, the dislocations become tangled and complex, and fracture occurs after an extremely brief plastic deformation, leading to a sudden decrease in stress (Figure 2). When the SFE of FeCoNiCr is extremely low, the lattice structure transforms from FCC to HCP, resulting in Strain-Induced Martensitic Transformation (SIMT), forming finer martensitic grains. Fine grains can increase the interfacial density of the material, impeding dislocation movement, and thereby enhancing material strength. HEAs may also produce twinning during plastic deformation, with lower SFE favoring twin growth. Twinning initially occurs in the high-entropy alloy, and as the strain increases, the Stacking Fault Tetrahedron dissociates into Shockley dislocations (Figure 1b_3). The occurrence of twinning and SIMT leads to Twinning Induced Plasticity and Transformation Induced Plasticity effects, allowing for strong plasticity enhancement of such composite structures. However, at the onset of plastic deformation, the formation of Hirth dislocations causes stress reduction (seen in Figure 1b_5). Therefore, the S110 composite material exhibits higher plasticity, but its strength is somewhat inferior to that of the S111 composite material. In the S111 composite material, two Shockley dislocation loops were formed in the high-entropy alloy section at a strain of 0.063 (Figure 1c_2).
In this study, it is evident that the S111 composite material exhibits a higher elastic modulus compared to the other two composite materials (Figure 2).

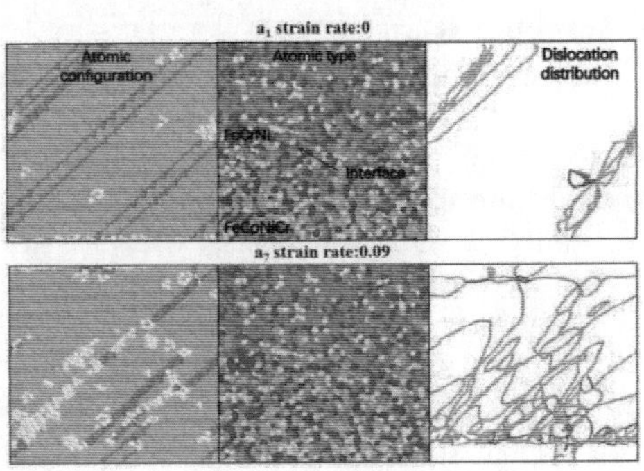

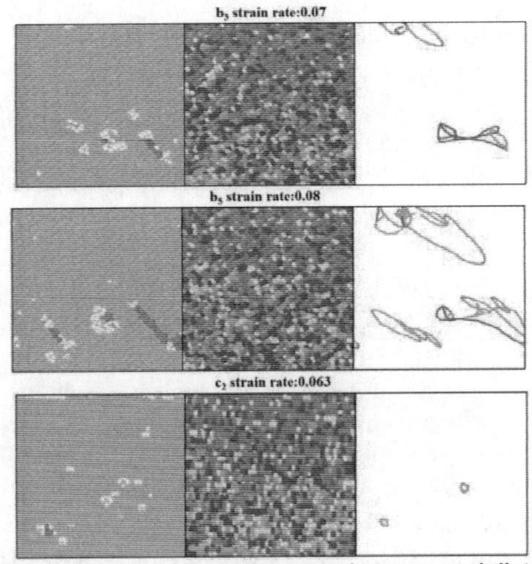

Figure 1.The atomic configuration, atomic types, and distribution ofdislocationswith strain rate: SS(001)/HEA(111); (b) SS(110)/HEA(111); (c) SS(111)/HEA(111).

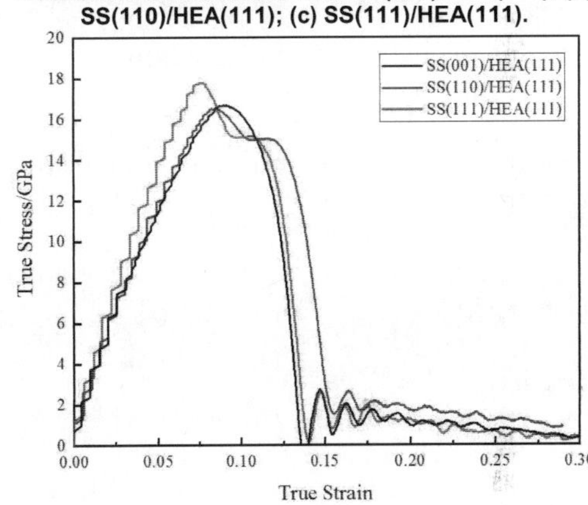

Figure 2.Depicts the true stress-strain curves of the composite materials.

4 Conclusion

The S110and S111 composite materials are more prone to forming stable interfaces, and the atomic arrangement in the S111 is more regular, exhibiting better crystal structure and higher elastic modulus. Under uniaxial tension, the interfaces of two composite materials impede dislocation motion, triggering an interface semi-blocking effect. The S111composite material exhibits the highest strength, while the S110 composite material demonstrates the highest toughness.

Unveiling the Origin of Severe Hot Cracking Susceptibility in Al-Li Alloys

Liang Zhang [1,*], Youjie Guo [1], Guohua Wu [1]
1. School of Materials Science and Engineering, Shanghai Jiao Tong University, Shanghai 200240, China
*Corresponding address: e-mail: liangzhang08@sjtu.edu.cn

Abstract: This work reports a new mechanism for the severe hot cracking susceptibility (HCS) of cast Al-Li alloys. In contrast to the narrow vulnerable solidification temperature range increasing cracking resistance, alloys with high Li content or solute content, which are supposed to show lower HCS, instead exhibit anomalously high HCS. Results reveal that this phenomenon arises from a variety of Li-rich inclusions distributed in the interdendritic paths, including Li_2O, $LiAlO_2$, Li_2CO_3 and LiH. These particles can form spontaneously with O_2, CO_2 and H_2O during solidification, deteriorating the interfacial bonding capacity. Meanwhile, the occupation of refilling channels for residual liquid increases the potential risk of microcrack initiation. These unfavorable factors promote the rapid propagation of cracks from the casting surface inward, which eventually leads to catastrophic fractures.

Keywords: Metals and alloys; Solidification; Microstructure; Cast

1 Introduction

Al-Li alloys provide unrivalled advantages when applied to aerospace and military industries due to their low density and high stiffness [1]. At the same time, cast Al-Li alloys are more suitable for manufacturing large-sized complex-shaped components and thin-walled parts than wrought counterparts [2]. However, these alloys suffer from severe hot cracking susceptibility (HCS) during casting, which limits their broader application. Although several methods to reduce the HCS of Al alloys were shown to be feasible in recent years [3], the high HCS of cast Al-Li alloys remains a long-standing issue.

Researchers have long believed it is the wide vulnerable temperature range that is responsible for the high HCS in Al-Li castings. A narrow vulnerable temperature range indicates the alloy has a shorter residence time in the weak mushy zone and is therefore less susceptible to hot cracking. However, subsequent tests showed that although the variation of HCS for alloys with low Li content is consistent with the predictions, it fails to explain the anomalously high HCS for alloys with high Li content or solute content, as they possess much narrower temperature ranges. Therefore, it is rational to speculate the existence of other underlying mechanism(s) affecting the hot cracking behavior of Al-Li alloys. To the best of our knowledge, there have been no reports available on this interesting phenomenon. Hence, the present work attempts to explore the possible reason for the anomalously severe HCS in Al-Li alloys.

2 Experimental procedure

Four alloys were designed to investigate the HCS variations, which were labelled as Al-1Li-3Cu-0.5Mg (Alloy 1), Al-2.5Li-1.5Cu-0.5Mg (Alloy 2), Al-2.5Li-3Cu-0.5Mg (Alloy 3), Al-4Li-1.5Cu-0.5Mg (Alloy 4). These alloys were prepared from pure Al, Li, Mg and Al-50Cu master alloys under the protection of Ar.

Hot cracks were non-destructively detected and reconstructed using an X-ray micro-tomography (XRM, ZEISS). The fracture morphology was characterized using scanning electron microscopy (SEM, TESCAN) at a working voltage of 25 kV, equipped with time-of-flight secondary ion mass spectrometry (TOF-SIMS). X-ray photoelectron spectroscopy (XPS, Nexsa) was used to detect the compositions of the fracture surface. Numerical simulations were performed using the computational fluid dynamics (CFD) software ANSYS Fluent 18.0.

3 Result and discussion

Fig. 1a illustrates the f_S *vs.* T curves for Alloys 1-4. According to the classic theory, the alloy is most susceptible to cracking when f_S is in the range of 0.85-0.98, as it has not yet formed a rigid solid skeleton while liquid feeding is impeded. On this basis, Alloys 1 & 2 are supposed to exhibit a higher HCS while Alloys 3 & 4 are associated with a lower HCS. However, experimental results were contrary to the predictions. Macro-photographs of castings indicate that Alloys 3 & 4 possess larger open cracks compared to Alloys 1 & 2. The crack volumes of Alloys 3 & 4 are significantly larger than those of Alloys 1 & 2. Meanwhile, the top view shows the cracks of Alloys 3 & 4 have propagated inward from the surface to form planar, penetration-type cracks. The cracks of Alloys 1 & 2, on the other hand, appear only in the surface area. This indicates that the conventional hot cracking theory based on vulnerable solidification temperature range fails to explain the variation pattern of HCS in some Al-Li alloys.

Hence, we hypothesized that the high HCS may be associated with the intrinsic properties of Al-Li alloys, i.e., extremely high reactivity at elevated temperatures. Low magnification fracture morphology images of Alloys 1-4 are presented in Fig. 2a, which exhibit identical characteristics. However, higher magnification images reveal clean interdendritic regions for Alloys 1 & 2, whereas a large number of particles smaller than 1 μm in

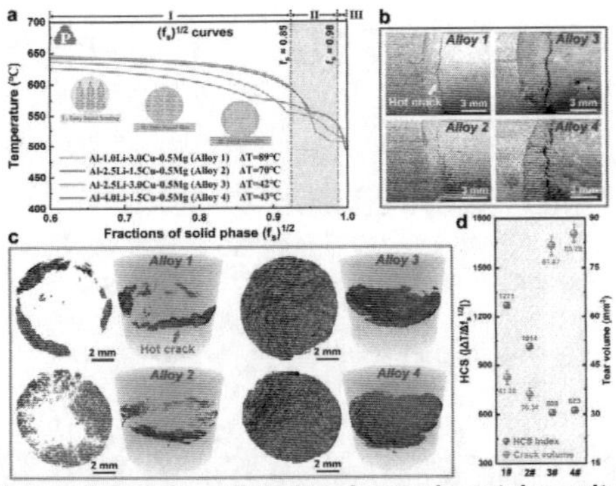

Fig. 1. Comparison of predicted and experimental results of HCS for Alloys 1-4: (a) $(f_S)^{1/2}$ - T curves; (b) Macroscopic cracks at the hot-spot locations; (c) XRM analysis of hot cracks in 3D visualization; (d) Comparison between the predicted HCS with the crack volume.

size are distributed within the interdendritic channels of Alloys 3 & 4. TOF-SIMS results indicate these particles are enriched in Li. Corresponding XPS spectra show that they are composed of Li_2O, $LiAlO_2$, Li_2CO_3 and LiH (Fig. 2d). Compared to Alloys 1 & 2, Alloy 4 has a higher Li addition, and the higher content of other solutes in Alloy 3 reduces the solid solution limit of Li in the Al matrix. This accelerates the tendency of the melt to react with O_2, CO_2 and H_2O during casting. These Li-rich inclusions all possess negative reaction free energy and high melting points, allowing them to form spontaneously during solidification. Moreover, their thermal expansion coefficient and elastic modulus differ substantially from those of the Al matrix, resulting in strong strain concentration at the interface, which increases the potential risk of microcrack initiation.

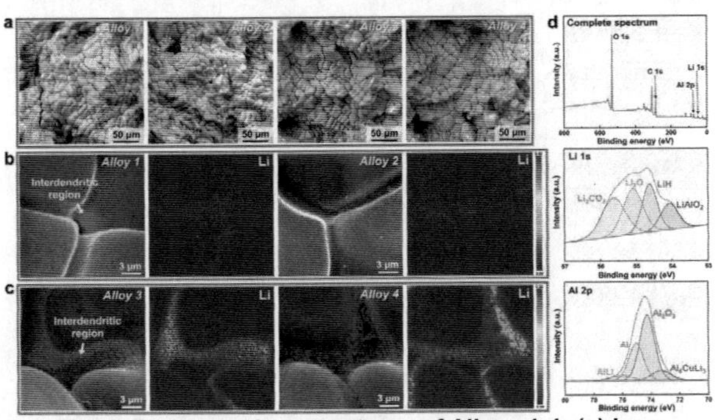

Fig. 2. Fracture microstructures of Alloys 1-4: (a) Low magnification SEM images of fracture surface; (b, c) TOF-SIMS images of the interdendritic regions; (d) Corresponding XPS spectra.

In addition to reducing the interfacial bonding capacity, such inclusions also occupy the narrow interdendritic channels and thereby reduce the feeding efficiency of residual liquid. Fig. 3a and b compare the CFD results of the pressure drop variations in the interdendritic regions with and without Li-rich inclusions, accompanied by meshing details. When Li-rich inclusions are present, the pressure drop within the interdendritic channels is an order of magnitude greater than in the absence of inclusions. The insufficient pressure prevents the residual liquid from feeding the empty space, exacerbating the initiation and propagation of microcracks and ultimately leading to irreversible cracking.

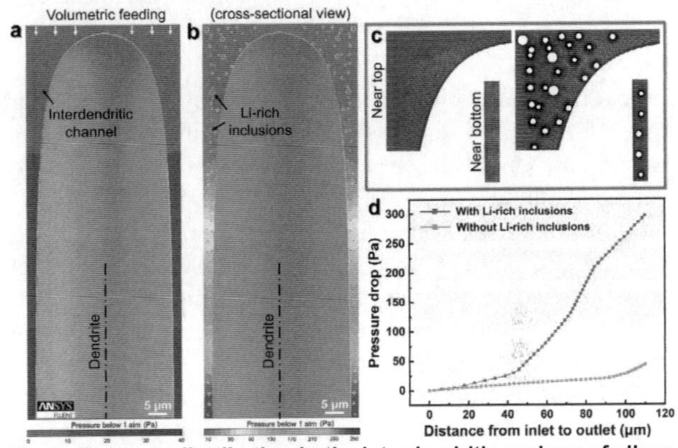

Fig. 3. Pressure distribution in the interdendritic regions of alloys with and without Li-rich inclusions: (a, b) CFD results; (c) Mesh system; (d) Comparison of pressure distribution.

4 Conclusion

When the Li content is close to the solid solution limit of the Al matrix, a large number of Li-rich inclusions can form in the interdendritic channels in the mushy zone owing to the high reactivity of Al-Li alloys at high temperatures. These particles not only induce strong strain concentration at the interface but hinder the refilling of residual liquid, which makes the alloy exhibit severe HCS even with a narrow vulnerable temperature range. Cracks that originally existed only in the hot-spot regions on the surface may thus rapidly propagate internally, leading to catastrophic penetrating large-sized cracks.

Acknowledgments

This work was financially sponsored by the National Natural Science Foundation of China (Nos. 51821001 and 51871148).

References

[1] Y.X. Wang, G.H. Wu, L. Zhang, et al., Mater. Lett. 349 (2023) 134854.
[2] Y. Li, H.X. Li, L. Katgerman, et al., Prog. Mater. Sci. 117 (2021) 100741.
[3] D.G. Eskin, Suyitno, L. Katgerman, Prog. Mater. Sci. 49 (2004) 629-711.

The Effect of Fe Content on the Microstructure and Properties of T6-Treated A356 Aluminum Alloy

Dongyang Yang[1], Dongfu Song[2], Chuanyao Tang[1], Weiwen Zhang[1,*]
1. School of Mechanical and Automotive Engineering, South China University of Technology, Guangzhou, 510640, China
2.Guangdong Provincial Key Laboratory of Metal Toughening Technology and Application,Institute of New Materials Guangdong Academy of Sciences, Guangzhou 510650, China
*Corresponding address: e-mail: mewzhang@scut.edu.cn

Abstract: The efficient utilization of recycled aluminum promotes the transition of the aluminum industry towards low-carbon and green manufacturing. This study investigates the evolution of the morphology of the Fe-rich phase in recycled T6 A356 aluminum alloy. With increasing Fe content, the volume fraction of the Fe-rich phase increases, and its morphology transforms from granular and short rod-like to acicular and Chinese character-like. While the ultimate tensile strength remains relatively stable, the yield strength gradually increases, whereas the elongation significantly decreases.

Keywords: A356 alloy; Fe-rich phase; Critical Fe content;

1 Introduction

A356 alloy is widely used in automotive, electronics, aerospace, and other fields with improved performance through T6 heat treatment [1]. Aluminum alloys maintain good mechanical properties for repeated use, but introduce harmful elements. Fe content significantly influences the microstructure and mechanical properties of A356 alloy, resulting in varied structures of the Fe-rich phase at different iron levels. When Fe content reaches a critical value, beneficial Al-Si-Fe ternary eutectic formation reduces the harmful effects of the Fe-rich phase [2]. This study investigates the influence of different Fe contents on the microstructure and properties of T6 A356 alloy, optimizing the permissible Fe content range to enhance the efficiency and value of recycled resources.

2 Experimental procedure

The alloy used in the experiment is Al-7Si-0.3Mg-xFe, with the addition of an Al-8Sr-8Ce intermediate alloy during the melting process. T6 heat treatment includes solution treatment and aging treatment. The solution treatment process involves a 540 ° C soak for 8 hours followed by water quenching, while the aging treatment process involves a 175°C soak for 10 hours. Microstructure and composition of the samples were characterized and analyzed using OM, SEM, and EDS.

Table 1. Actual chemical compositions of the A356 alloys(wt.%)

Alloy	Si	Fe	Sr	Mg	Al
0.15Fe	7.45	0.15	0.012	0.27	bal.
0.25Fe	6.84	0.26	0.012	0.33	bal.
0.40Fe	7.05	0.39	0.017	0.32	bal.
0.60Fe	7.20	0.58	0.019	0.35	bal.

3 Result and discussion

Microstructure

Fig. 1 illustrates the metallographic structure of T6 A356 alloy at different Fe contents, including α-Al (white), eutectic silicon (black), and Fe-rich phase (light gray). With increasing Fe content, the distribution of eutectic silicon remains largely unchanged, but larger-sized eutectic silicon slightly increases. The quantity and size of the Fe-rich phase increase, exhibiting coarse acicular morphology. It is noteworthy that eutectic silicon near the α-Al dendrites is larger in size.

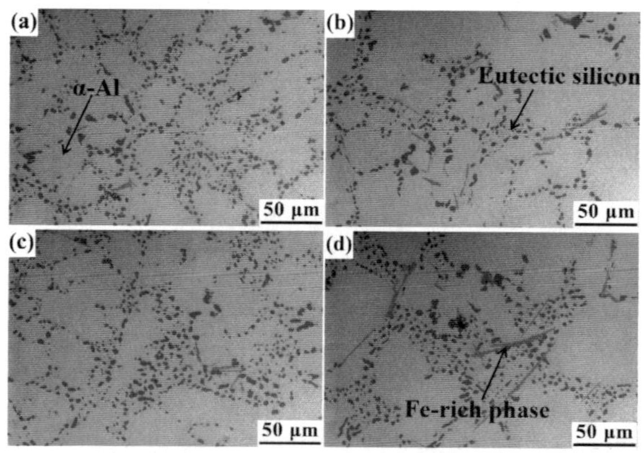

Figure 1 The Optical microstructure of T6 A356 alloy with different Fe content:(a) 0.15%; (b) 0.25%; (c) 0.4 %; (d) 0.6%

Figs. 2 depicts the SEM-BSE microstructure of different samples. EDS point scans indicate that the microstructure of the alloy consists of α-Al8Fe2Si and β-Al5FeSi phases. With increasing iron content, the Fe-rich phase transitions from fine granular and short rod-like morphology to a Chinese character-like and needle-like morphology. At an iron content of 0.4%, Chinese character phases appear, accompanied by a small amount of short rod-like phases. Subsequently, when the iron content reaches 0.6%, the quantity of Chinese character phases decreases, while the quantity of needle-like phases with larger aspect ratios increases.

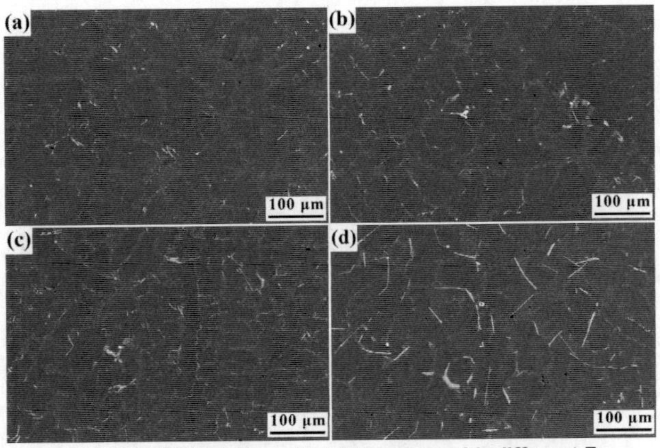

Figure 2 SEM-BSE photos of T6 A356 alloy with different Fe content:(a) 0.15%; (b) 0.25%; (c) 0.4 %; (d) 0.6%

Mechanical properties

Figure 3 illustrates the influence of Fe content on the mechanical properties of the alloy. Analysis of the graph reveals that the tensile strength of the 0.15% Fe alloy is 289 MPa with an elongation of 11.9%. As the Fe content increases, the yield strength of the alloy gradually rises, while the ultimate tensile strength stabilizes at 290±3 MPa and the elongation significantly decreases from 11.9% to 6.2%. This indicates that the formation of the Fe-rich phase has minimal effect on the tensile strength of the alloy, but the transition in Fe-rich phase types results in a substantial decrease in the alloy's ductility.

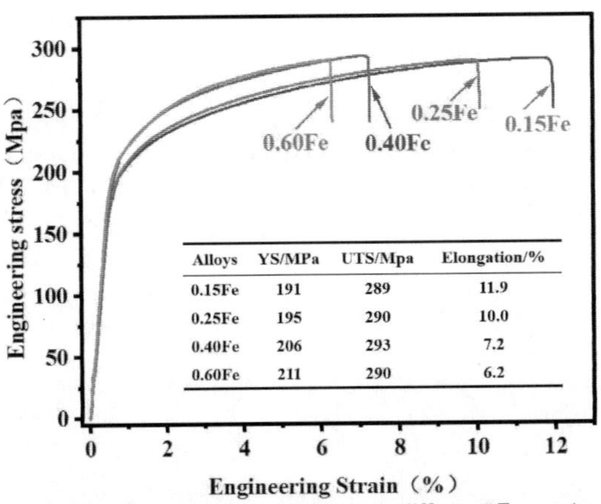

Figure 4 Tensile curves of alloys with different Fe content

Fig. 4 presents the fracture morphologies of T6-treated A356 alloys with varying Fe contents. The hard phases consist of Fe-rich phases and eutectic silicon phases. Al-Sr-Ce modification leads to a refinement of the eutectic silicon. At low Fe contents, eutectic silicon serves as the primary crack initiation site, resulting in the formation of ductile dimples due to plastic deformation. With increasing Fe content, Fe-rich phases become the crack initiation sites, leading to brittle fracture characteristics. At Fe contents of 0.4% and 0.6%, the area of brittle fracture surfaces increases, while the number of dimples decreases, indicating a decrease in ductility.

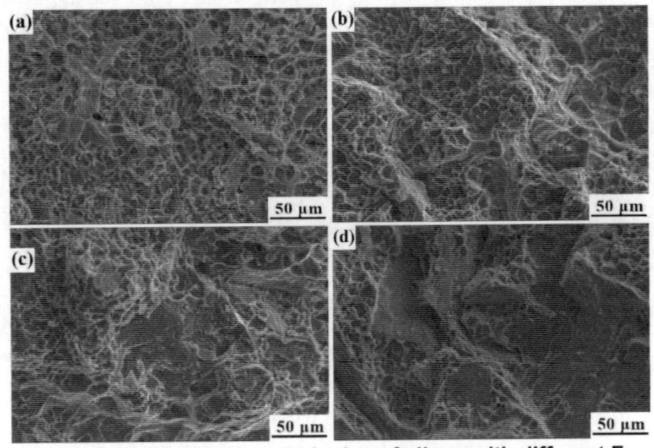

Fig. 4 Surface fracture morphologies of alloys with different Fe content:(a) 0.15%; (b) 0.25%; (c) 0.4 %; (d) 0.6%

Conclusion

With the increase in iron content, both the size and volume fraction of Fe-rich phases increase. The granular and short rod-like Fe-rich phases transform into coarse and acicular Chinese script-like phases. The alloy's tensile strength shows little variation, while the yield strength gradually increases. However, the elongation rate significantly decreases as a result.

Acknowledgments

We would like to acknowledge the financial support from the National Natural Science Foundation of China (No. 52074131)

References

[1] Yi JZ, Gao YX, Lee PD, et al. Materials Science & Engineering A, "Effect of Fe-content on fatigue crack initiation and propagation in a cast aluminum–silicon alloy (A356–T6).", 386.1-2(2004):396-407.

[2] Taylor JA, Schaffer GB, StJohn DH, The role of iron in the formation of porosity in Al-Si-Cu-based casting alloys: Part III.A microstructural model. 30(1999):1657–1662.

Latest Advancements and Prospects in TIG Repair Welding of Sand-Cast Magnesium Rare-Earth Alloys

Xin Tong, Guohua Wu*, Wenjiang Ding
National Engineering Research Center of Light Alloy Net Forming and State Key Laboratory of Metal Matrix Composites, Shanghai Jiao Tong University, 800 Dongchuan Road, Min Hang, Shanghai, China
*Corresponding address: e-mail: ghwu@sjtu.edu.com

Abstract: The large-scale sand castings of Mg-RE alloys have broad application prospects in the aerospace field. However, some casting defects are often generated in the component, which seriously increases the scrap rate. As a welding technology with high utility and economy, TIG welding can be performed manually, exhibiting unique advantages for repairing some inner structures and cambered surfaces. This report presents a comprehensive summary of the repair welding of sand-cast Mg-RE alloys using TIG welding. The weldability of Mg-RE alloys and the microstructure evolution during repair welding will be discussed. In addition, the effect of the pre- and post-weld heat treatment on the microstructure and mechanical properties of the repair welded joints are also summarized. In particular, the abnormal grain coarsening that occurred in post-weld solid-solution treatment of Mg-RE alloy joints and its regulating method are systematically analyzed. Drawing upon these studies, the existing problems and development trends of repair welding of cast Mg-RE alloys are summarized based on the actual applications.

Keywords: Mg-RE alloys; Cast; Repair welding; heat treatment; Microstructure evolution

1 Introduction

Magnesium rare-earth (Mg-RE) alloys have attracted considerable attention because of their low density, high specific strength, and good creep resistance. The application of Mg-RE alloys in aerospace industries can achieve an obvious weight reduction of the component. Until now, the large-scaled components containing complex structures are mainly fabricated by low- or differential-pressure sand casting process. However, due to the high oxygen affinity of the RE alloying elements, RE is easily oxidized and forms inclusions in the castings. In addition, the mushy zone solidification induced by the low-temperature gradient within the sand mold cavity would also result in the formation of shrinkage porosities and microcracks. Therefore, the development of defect-repair technology for sand-cast Mg-RE alloy is of great importance.

As compared to other welding processes, TIG welding can be performed without a complex welding tool setup, exhibiting higher operability and convenience for repairing large-scale components with complex structures.

Nevertheless, TIG welding shows larger heat input with lower energy density, so the grain size of the fusion zone (FZ) is larger than some other welding methods (e.g., laser welding). Moreover, it is quite difficult to simultaneously achieve the desired strengthening state by conventional heat treatment in the whole joint, because there are various zones with completely different microstructure characteristics.

Given these facts, this report mainly focuses on the TIG repair welding of the sand-cast Mg-RE alloys. The microstructure evolution during the repaired welding as well as the pre- and post-weld heat treatment would be summarized in detail.

Microstructure of TIG-repair welded joint

TIG welding has been used for repairing sand-cast Mg-Gd-Y-Zr [1], Mg-Y-Nd-Gd-Zr [2,3], and Mg-Gd-Nd-Zr [4] alloys. In general, TIG repair welded joints of sand-cast Mg-RE alloy can be divided into four different zones due to their different thermal cycles: FZ, partially melted zone (PMZ), "true heat-affected zone"(T-HAZ), and base material (BM). As compared to the BM, an effective grain refinement can be always achieved in FZ (Fig. 1), because of the combination of Zr dissolution at high temperatures and the rapid cooling rate in the molten pool. Due to the pinning effect of eutectics on grain boundaries, grain coarsening can be negligible in the HAZ of sand-cast Mg-RE alloys, which is quite different from that of the wrought alloys. As reported by our previous study, the grain size of the FZ of the TIG repair welded WE43 alloy is ~13 μm, which is only about 1/4 of that of the sand-cast BM.

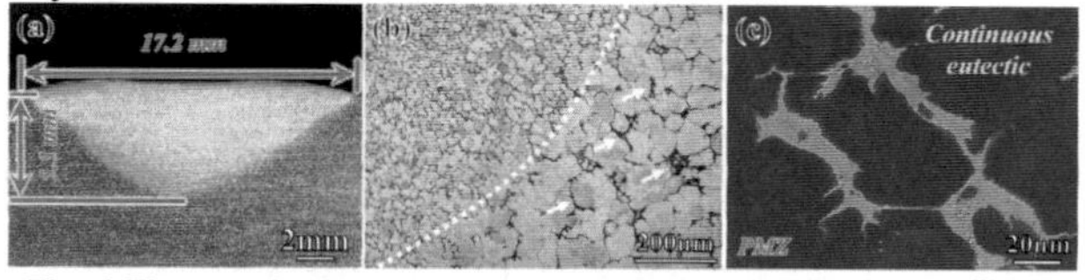

Fig. 1 The repaired welding joint of sand-cast WE43 alloy [2]: (a) macrostructure; (b) fusion line; (c) PMZ

Microstructure stability of repaired joint

The repaired welding process is usually associated with an extremely fast cooling rate, resulting in high residual stress. Therefore, the density of dislocation and vacancy is relatively high within the FZ, and this could decrease the microstructure stability. For instance, during the post-weld solution treatment at high temperatures, the nucleation and growth of recrystallized grains occur, and some abnormally coarsened grains even with an average size of up to 1 mm are formed. The abnormal grain coarsening has been observed in the FZ of the repaired joint of Mg-Y-RE-Zr, Mg-Gd-Y-Zr, Mg-Gd-Nd-Zr series alloys, etc. Tong et al. [2] reported that the abnormal grain coarsening occurred in repair welded Mg-4Y-2Nd-1Gd-0.5Zr alloy even after solution treated at 480℃ for 8 h (Fig. 2a). Thus, it leaves a significant contradiction between the grain refinement and the solid-solution effect.

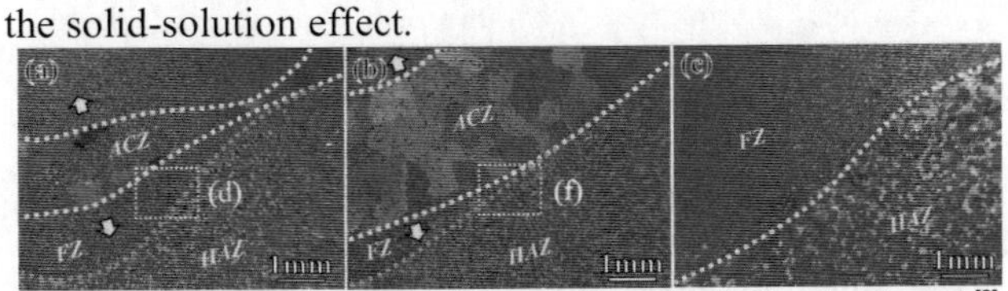

Fig. 2 Grain coarsening during the post-weld solution treatment [2]: (a) 480°C × 8 h; (b) 520℃ × 8 h; (c) combination of pre- and post-weld solution treatment

Pre- and post-weld heat treatment

To avoid abnormal grain coarsening, several strategies, including the application of the pre-weld solution treatment [2] and the multiple-stage post-weld solution treatment [3,4], have been proposed. The pre-weld solution treatment could eliminate the eutectic structures in sand-cast BM in advance, so the heat-treated time required for the post-weld solution treatment can be shortened (Fig. 2c). But this could weaken the pinning effect of RE-containing phase on grain boundaries, resulting in an increased grain growth in HAZ. On the other hand, as compared to the conventional single-stage treatment (520°C × 8 h), it was found that the eutectics in sand-cast WE43 alloy can be dissolved without abnormal grain coarsening under the two-stage solution treatment (480°C × 6 h + 520°C × 2 h) [3]. However, there were still a few residual eutectics that can be found in HAZ. Therefore, more efforts are still required for the balance between the grain refinement and eutectic dissolution.

Mechanical properties of repair-welded joint

As compared to the sand-cast BM, the repaired welded joint prepared under optimized welding parameters and without heat treatment usually exhibits lower strength but a higher elongation. The improvement of the EL could be attributed to that the fine grains in FZ restrained the crack initiation by affording more deformation before failure [4]. However, due to the weak bonding between the continuous eutectics and the matrix within PMZ, the strength of the joint is reduced significantly, which can be confirmed by the broken eutectic phases exposed on the fracture surface. It was found that the elongation of the repaired joint of sand-cast WE43 alloy was 24.4% higher than the BM, although the tensile strength of the repaired joint was only 87.8% of the BM. However, the elongation of the joint after T6 treatment was significantly reduced (Fig. 3c), and the tensile fracture was located in FZ, because the abnormal grain growth occurred in FZ during the solid solution treatment.

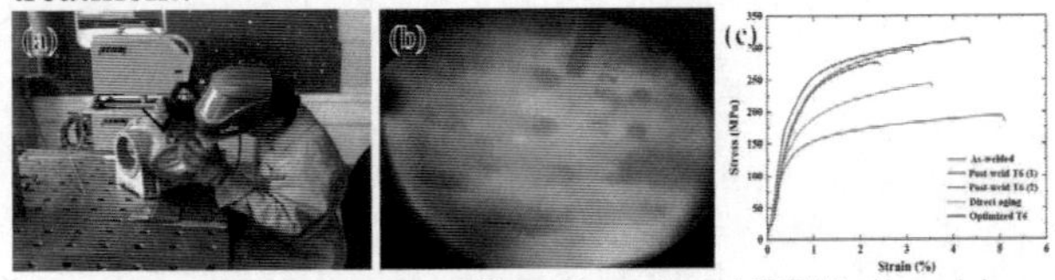

Fig. 3 The repaired welding of sand-cast WE43 alloy: (a) operational process; (b) X-ray detection; (c) tensile curves

2 Conclusion and prospect

The casting defects can be effectively repaired using TIG welding, which is of great value for promoting the application of Mg-RE alloy castings in aerospace. However, the microstructure characteristics of FZ, PMZ and T-HAZ are quite different from that of BM. It is still a big challenge to realize the microstructure coordination between these different regions and improve the stability of the repaired regions in the future.

References

[1] Hou G J, Xu Y L, Wang S W, et al., Effects of welding wire composition on the repair welds of sand-cast Mg–Gd–Y alloy: Microstructure and mechanical properties[J]. Vacuum, 2022, 199: 110919.

[2] Tong X, Zhang G Q, Wu G H, et al., Addressing the abnormal grain coarsening during post-weld heat treatment of TIG repair welded joint of sand-cast Mg-Y-RE-Zr alloy[J]. Mater. Charact., 2021, 176: 111125.

[3] Zhang G Q, Tong X, Wu G H, et al., Research on the post-weld heat treatment of TIG repair welded joint of sand-cast Mg-Y-RE-Zr alloy[J]. Mater. Sci. Eng. A, 2021, 821: 141577.

[4] Wang Q M, Tong X, Wu G H, et al., Microstructure and strengthening mechanism of TIG welded joints of a Mg-Nd-Gd alloy: Effects of heat input and pulse current[J]. Mater. Sci. Eng. A, 2023, 869: 144816.

Microstructure Evolution of Immiscible Alloy Solidified Under the Effect of Composite Electric and Magnetic Fields

Hongxiang Jiang[1, 2], Yanqiang Li[1, 2], Jiuzhou Zhao[1, 2], Lili Zhang[1, 2], Jie He[1, 2]
1. Shi-changxu Innovation Center for Advanced Materials, Institute of Metal Research, Chinese Academy of Sciences, 72 Wenhua Road, Shenyang, Liaoning, 110016, China
2. School of Materials Science and Engineering, University of Science and Technology of China, 72 Wenhua Road, Shenyang, Liaoning, 110016, China
*Corresponding address: e-mail: hxjiang@imr.ac.cn

Abstract: Solidification experiments were performed with Lead–Aluminum immiscible alloy under the effect of composite electric and magnetic fields (CEMFs). The results demonstrate that CEMFs not only decrease the size of minority phase particles (MPPs), but also promote a more uniform distribution of the MPPs. A theoretical model was built to describe the microstructure evolution during cooling the immiscible alloy. The solidification process of Pb-Al alloy under the effect of the CEMFs was simulated. The numerical results are well consistent with the experimental data. These results demonstrate that CEMFs affect the solidification process through changing melt convection and the nucleation behavior of minority phase droplets (MPDs). On one hand, the CEMFs can inhibit the convection and lead to the homogeneous distribution of MPPs along the radial direction of the sample. On the other hand, the CEMFs can increase the nucleation driving force for the MPDs, which decreases the average radius of MPDs and boosts the formation of dispersed solidification structures. This research indicates that the application of CEMFs is a promising strategy for controlling the microstructure of immiscible alloys.

Keywords: Phase separation, Solidification, Simulation, Composite electric and magnetic fields

1 Introduction

Liquid-liquid phase separation phenomenon during solidifying an immiscible alloy usually leads to the formation of two liquid phases with distinctive, even opposing characteristics. It provides a unique opportunity for the manufacture of in-situ metal matrix composites with a dispersed microstructure or core/shell microstructure, which may possess exceptional properties [1-5]. For instance, Pb-Al immiscible alloy is a potential grid material for high-performance lead-acid battery [6]. However, the liquid-liquid phase separation phenomenon usually causes a serious phase segregated microstructure [6].

The application of external fields can effectively control the microstructure evolution during cooling immiscible alloys [6-10 31-41]. To date, plenty of studies have been carried out to investigate the effect of a single physical field on the solidification behaviors of immiscible alloys, but the studies on the effect of composite fields are limited. In this study, solidification experiments are carried out with Pb-Al alloy under the effect of composite electric and magnetic fields (CEMFs). A numerical model is built to describe the microstructure evolution during cooling the immiscible alloy. The solidification process of immiscible alloys and the affecting mechanisms of CEMFs are analyzed in detail.

2 Experimental procedure

Solidification experiments were performed with Pb-0.4 wt.%Al immiscible alloy by using a continuous solidification setup equipped with CEMFs producing device. The detailed experimental procedures are as follows: the weighted lead and aluminum were first put into a crucible, then they were melted and heated to 1223 K by using an electric resistance furnace. After that, the melt was stirred for obtaining a homogeneous melt. Subsequently, the pulse current, the static magnetic field, or the CEMFs were applied to the alloy melt. Afterwards, the crucible was pulled out of the furnace at the rate of 10 mm/s, and liquid Ga-In-Sn metal which has good thermal conductivity and high heat capacity was used to cool down the specimen. Finally, cylindrical samples with a diameter of 4 mm and a length of 100 mm were obtained. The microstructures were analyzed by using a scanning electron microscope. SISC IAS V8.0 software was employed to determine the average radius and size distribution of MPPs.

3 Result and discussion

Experimental results

Fig. 1 shows the microstructures of different radial positions in the Pb-0.4 wt.%Al alloy samples solidified under the comprehensive effect of static magnetic field and pulse current. The results indicate that the CEMFs not only decrease the average radius of MPPs but also promote a more homogenous distribution of MPPs along the radial direction.

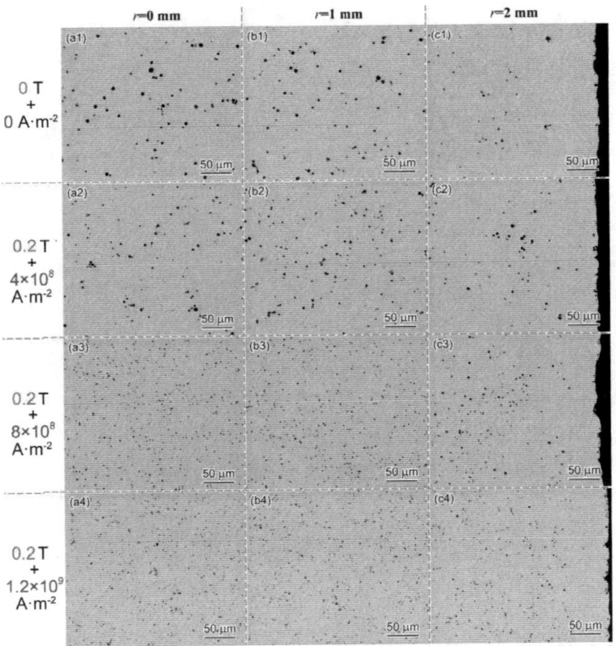

Fig. 1. Microstructures of different radial positions in the Pb-0.4 wt.%Al alloy samples solidified under the effect of CEMFs.

Discussion

The simulation results indicate that the application of CEMFs has an ignorable effect on the temperature profile, while it substantially inhibits melt convection. Meanwhile, the CEMFs not only lead to an increase in the nucleation rate of MPDs but also lead to a uniform distribution of MPDs along the radial direction, as shown in Fig.2.

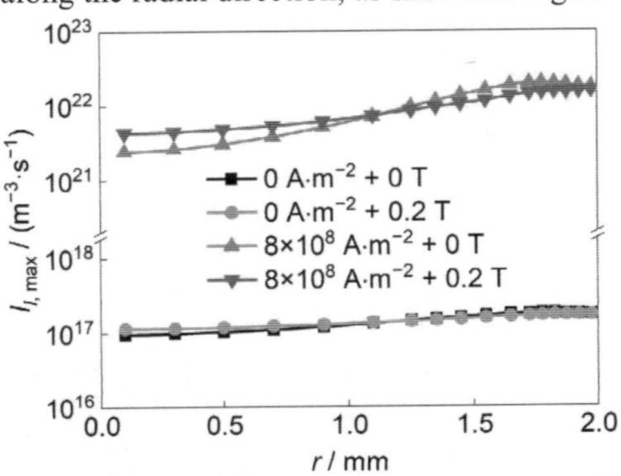

Fig. 2. Effect of CEMFs on the maximum nucleation rate of MPDs in front of solidification interface along radial direction for the Pb-0.4 wt.%Al alloy.

4 Conclusion

.

Continuous solidification experiments were performed with Pb-Al alloy, and the effects of pulse current, the static magnetic field, and the CEMFs were investigated. A model has been built to describe the microstructure evolution. The nucleation, growth, and migration of MPDs/MPPs, as well as the effect of CEMFs, have been taken into account. The solidification process and microstructure evolution were simulated. These results present that the pulse current promotes the nucleation of MPDs, and the static magnetic field inhibits melt convection. The application of CEMFs takes both advantages of pulse current and static magnetic field. It is conducive to the formation of a microstructure with refined minority phase particles uniformly distributed along the sample radial direction. The application of CEMFs is a promising strategy for controlling the microstructure of immiscible alloys.

Acknowledgments

The authors acknowledge the financial support from the National Natural Science Foundation of China (grant Nos. 51971227, 51974288, and 52174380), the National Key Research and Development Program of China (grant No. 2021YFA0716303).

References

[1] Kaban I, Köhler M, Ratke L, Hoyer W, Mattern N, Eckert J, Greer A L. Interfacial tension, wetting and nucleation in Al-Bi and Al-Pb monotectic alloys[J]. Acta Materialia, 2011 59: 6880-6889.

[2] Wang W L, Li Z Q, Wei B. Macrosegregation pattern and microstructure feature of ternary Fe-Sn-Si immiscible alloy solidified under free fall condition[J]. Acta Materialia, 2011, 59: 5482-5493.

[3] Cao C Z, Liu W Q, Liu Z W, Xu J Q, Hwang I, Rosa I D, Li X C. Scalable manufacturing of immiscible Al-Bi alloy by self-assembled nanoparticles[J]. Materials & Design, 2018, 146: 163-171.

[4] Liu Y G, Zhang J Q, Sun Q, Li M, Yan M, Cheng X, Li M Q, Zhang M X. Laser powder bed fusion of copper matrix iron particle reinforced nanocomposite with high strength and high conductivity[J]. Journal of Materials Science & Technology, 2023, 134: 50-59.

[5] Zhong C X, Liang D S, Wan T, He S, Yang L, Fang J, Zhang G, Ren F Z. Ultrafine-grained Nb-Cu immiscible alloy implants for hard tissue repair: Fabrication, characterization, and in vitro and in vivo evaluation[J]. Journal of Materials Science & Technology, 2022, 127: 214-224.

[6] Li Y Q, Jiang H X, Sun H, Zhang L L, He J, Zhao J Z. Microstructure evolution of immiscible alloy solidified under the effect of composite electric and magnetic fields[J]. Journal of Materials Science & Technology, 2023, 162: 247-259

Al-Si Divorced Eutectic Solidification in Hypereutectic Al-Si Alloy by Counter-Gravity Casting

Chengcheng Han[1], Hao Huang[1], Yuna Wu[1*], Hengcheng Liao[2]
College of Materials Science and Engineering, Hohai university, Changzhou, China
2. School of Materials Science and Engineering, Southeast university, Nanjing, China
*Corresponding address: e-mail: wuyuna@hhu.edu.cn

Abstract: Counter-gravity casting, which involves sucking A390 melt into various mold tubes, enables the eutectic growth to divorce and the lamellar eutectic Si to disappear. The A390 alloy melt experiences chilling and convection effects, resulting in greatly refined and uniformly distributed primary silicon particles (PSPs). The average equivalent diameter of PSPs in A390 alloy through brass tube suction casting (BTSC) is the smallest, measuring 5.3 μm. The divorced eutectic and fine PSPs enhance the wear resistance of A390 alloy. These findings provide new insights into the development of a simple and effective manufacturing process to refine the primary and eutectic Si phases in hypereutectic Al-Si alloys.

Keywords: Hypereutectic Al-Si alloy; Counter-gravity casting; Divorced eutectic; Primary silicon particles.

1 Introduction

Hypereutectic Al-Si alloys are widely used in aerospace, automobile manufacturing and electronic packaging [1,2], due to their excellent wear resistance [3] and thermal conductivity [4]. However, the coarse primary Si and lamellar eutectic Si lead to stress concentrations that weaken the material's performance. Traditional methods [5-9] to modify the Si phases have limitations in refinement effect, complexity of processing technologies, cost, and environmental concerns. A previous study [10] discovered that glass tube suction casting (GTSC) can induce divorced eutectic solidification in A390 alloy, eliminating eutectic Si and forming fine and uniform primary silicon particles (PSPs). GTSC is a simple and effective counter-gravity casting method for refining PSPs without the need of any modifier. To promote the process of divorced eutectic solidification, this study utilized molds made of different materials (glass, brass and 304-SUS) to suck A390 melt, and investigated the mechanism of divorced eutectic and PSPs refinement.

2 Experimental procedure

The counter-gravity casting is depicted in Fig. 1. Glass, brass and 304 stainless steel tubes are used as molds, suction-casted samples of which are marked as GTSC, BTSC and STSC, respectively. The A390 melt is suction-cast, with the solidification cooling curve recorded using a temperature measuring device. The melt was also casted by permanent mold casting for reference, which is marked as PMC. The average equivalent diameter (AED) of PSPs was determined through image processing software. The suction-casted samples were subjected to a rotational friction and wear test at 15N load and 200 r/min speed.

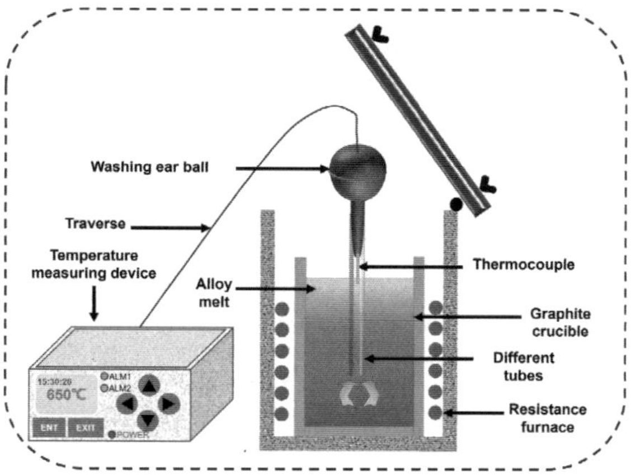

Fig. 1 Counter-gravity casting schematic diagram

Table 1. Actual chemical composition of A390 alloy (wt.%)

Al	Si	Cu	Fe	Zn	Mg	Mn
Bal.	16.1	4.5	0.8	0.7	0.4	0.2

3 Result and discussion

Microstructure and Solidification analysis

Fig. 2 displays the optical microstructures of A390 alloy from different casting methods. In permanent mold casting, PSPs appear as coarse blocks with numerous eutectic Si particles. Counter-gravity casting results in a refined microstructure with minimal eutectic Si and significantly smaller PSPs (AED reduced to 5.3 μm by BTSC). For Al-Si divorced growth during eutectic solidification in A390 alloy, prerequisites include abundant PSPs precipitation during solidification and inadequate undercooling within the eutectic temperature range. Counter-gravity casting's chilling effect induces supercooling for PSPs nucleation, shown in Fig. 3 solidification curves, with vigorous convection promoting rapid PSPs proliferation. This depletes Si atoms, lowering Si content in the remaining melt below eutectic composition and hindering the eutectic reaction, causing Al-Si eutectic solidification divorced growth.

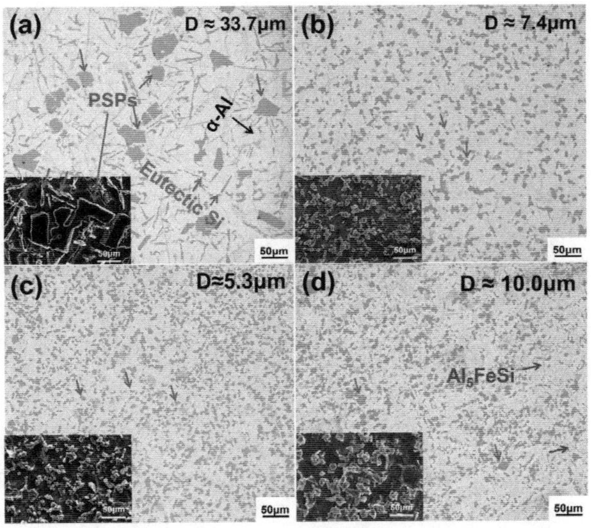

Fig. 2 Different casting methods: (a) permanent mold casting, and suction casting by (b) glass tube, (c) brass tube, (d) 304-SUS tube.

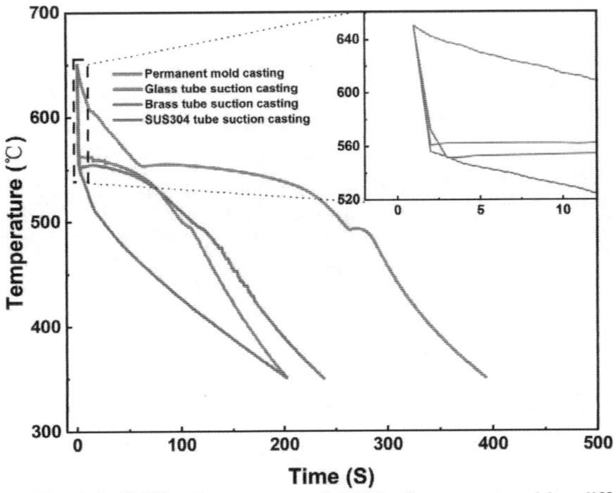

Fig. 3 Solidification curves of A390 alloy prepared by different casting methods.

Wear resistance analysis

The friction and wear test results show that small PSPs increase the abrasion resistance of A390 alloy, with the average friction coefficient and wear rate of BTSC being the lowest. However, in the wear stage of A390 alloy cast by metal mold, the coarse PSPs and lamellar eutectic Si were seriously worn and peeled off, causing a decrease in wear resistance.

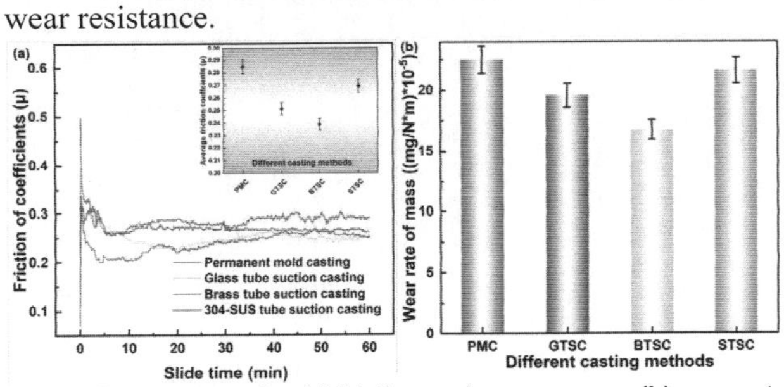

Fig. 4 Wear test results: (a) friction and wear curves; (b) wear rate.

4 Conclusion

1. Counter-gravity casting is achieved by sucking the A390 melt into the mold tube. In this process, Al-Si undergoes divorced eutectic growth.
2. The average equivalent diameter of the PSPs in A390 alloy by brass tube suction casting is the smallest, measuring 5.3 μm.
3. The divorced eutectic and fine PSPs enhance the wear resistance of A390 alloy.

Acknowledgments

The authors would like to acknowledge the financial supports of the Fundamental Research Funds for the National Natural Science Foundation of China (52303390) and Jiangsu Provincial Key Research and Development Program (BE2021027).

References

[1] K. Godbole, et. al, Al-Si controlled expansion alloys for electronic packaging applications[J]. Progress in Materials Science. 144 (2024) 1001268.
[2] Y. Li, et. al, Understanding grain refining and anti Si-poisoning effect in Al-10Si/Al-5Nb-B system[J]. Journal of Materials Science & Technology, 65 (2021) 190-201.
[3] Wu Y, et al. Effect of Si content on dynamic recrystallization of Al-Si-Mg alloys during hot extrusion[J]. Journal of Materials Science & Technology, 2014, 30(12): 1271-1277.
[4] Ding C, et al. Fabrication of hypereutectic Al-Si alloy with improved mechanical and thermal properties by hot extrusion[J]. Materials Characterization, 2023, 202: 113026.
[5] Balasubramani N, et al. A review of the origin of equiaxed grains during solidification under mechanical stirring, vibration, electromagnetic, electric-current, and ultrasonic treatments[J]. Journal of Materials Science & Technology, 2023, 144: 243-265.
[6] Lien H H, et al. Ultrahigh strength and plasticity in laser rapid solidified Al-Si nanoscale eutectics[J]. Materials Research Letters, 2020, 8(8): 291-298.
[7] Zhang J, et al. Al-Si nano clusters and short-range orders induced fibrous modified eutectic Si[J]. Materials Characterization, 2022, 194: 112348.
[8] Zhang L, et al. Formation mechanism and conditions of fine primary silicon being uniformly distributed on single α-Al matrix in Al-Si alloys[J]. Materials & Design, 2020, 193: 108853.
[9] Liang C, et al. Optimizing microstructures and mechanical properties of hypereutectic Al-18% Si alloy via manipulating its parent liquid state[J]. Materials Science and Engineering: A, 2017, 690: 387-392.
[10] Liao H, et al. Divorced Eutectic Solidification in Hypereutectic Al-Si Alloys[J]. Metallurgical and Materials Transactions A, 2022, 53(7): 2346-2350.

Effect of Trace Mounts of Zr on Microstructure and Properties of K4169 Alloy

Jingyu Zhang*, Qian Du, Qinrui Zhang, Keyan Xia, Yamin Li, Yi Chen
College of Materials Science and Engineering, Lanzhou University of Science and Technology, Lanzhou, 730000, China
*Corresponding address: e-mail: zhangjy@lut.com

The casting K4169 alloy, exhibited high mechanical properties, good welding performance, and casting process performance, in a wide range of low and medium temperature range.This paper examines the impact of Zr addition on the casting state organization, standard heat treatment state and mechanical properties of the casting K4169 alloy.

The composition of alloy element for K4169 (metal powder, Shanghai Sinopharm Group) used in this experiment is shown in Table 1. In Table 1, alloy No. 0 was the original composition without the addition of Zr. The Zr contents of alloys No.1, 2 and 3 were 0.03%, 0.05% and 0.09%, respectively.

The alloy solid solution and ageing treatment are based on the heat treatment regime of the standard K4169 alloy (1150°C &3.5h, air-cooled; 955°C & 1h, air-cooled; 720°C & 8h, followed by a cooling rate of 50°C/h (FC) down to 620°C & 8h, air-cooled).

Table 1 Chemical composition of experimental alloys (wt %)

Sample	C	Al	Zr	Ti	Nb	Mo	Cr	Ni
No.0	0.05	0.81	0	1.05	5.4	3.05	18.4	52.0
No.1	0.05	0.81	0.03	1.05	5.4	3.05	18.4	52.0
No.2	0.05	0.81	0.05	1.05	5.4	3.05	18.4	52.0
No.3	0.05	0.81	0.09	1.05	5.4	3.05	18.4	52.0

1 Effect of Zr element on the microstructure of cast K4169 alloy

Based on XRD analysis, it is shown that an increase in the Zr element caused a significant increase in the precipitation phase and serious alloy segregation, which result in the formation of new phases in the alloy. Figure 2 displays the secondary electronic micrograph of the microstruture for the as-cast alloy, the width of the white γ matrix region diminished,the phase content of the corresponding grain boundary precipitation increased, and the density of the dendritic and secondary dendritic arms increased from alloy No. 0 to alloy No. 3 as the Zr content increased. It suggested that the Zr element had the potential to facilitate the precipitation of the grain boundary, which in turn led to a mobility of the alloy' s liquid phase. This is due to the pinning of the precipitated granular phase and grain boundary precipitates during the solidification process of the alloy, which imped the growth of grain boundaries.

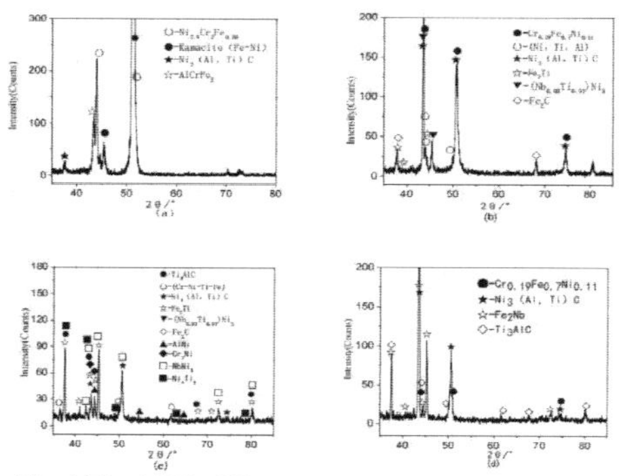

Fig.1 The XRD diffraction spectrum of as-cast Alloy (a) No. 0 (b) No.1 (c) No.2 (d) No.3

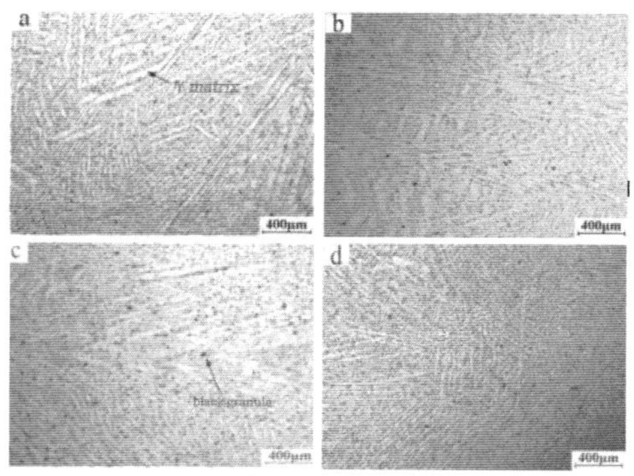

Fig. 2 The microstructure of as-cast K4169 alloy with different Zr contents(50×) (a) No. 0 (b) No. 1 (c) No. 2 (d) No. 3

2 Effect of Zr on the microstructure of alloys in the standard heat-treated state

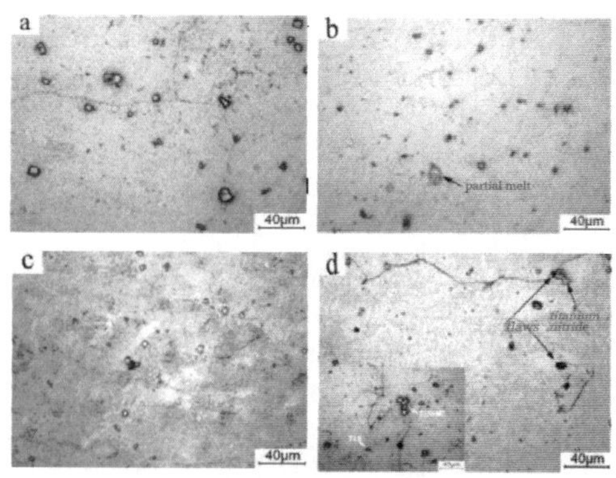

Fig.3 The microstructure of alloys in standard heat-treated state (50×) (a) No. 0 (b) No. 1 (c) No. 2 (d) No. 3

The $Ni_3(Al, Ti)$ C structure remained unaltered both prior to and following the application of heat treatment, due to the highly stable structure formed by the solid dissolution of carbon elements within the γ′ phase. As shown in Figure 3, as the Zr content increased, the granular precipitated phases of alloys No. 0 to 2 exhibited a more uniform distribution and a robust intergranular bonding. This resulted in a notable enhancement in the performance of the alloys. In contrast to the as-cast alloys, the laves phase was no longer present in the alloys after the standard heat treatment.

3 Effect of Zr on mechanical properties and fracture morphology of K4169 alloy

3.1 Hardness and tensile properties

Figure 4 displays the hardness values of the tested alloys after standard heat treatment.The figure showed that the alloy's hardness was observed to decrease when the Zr element content was reduced, indicating that the Zr content of the alloy has a detrimental effect on performance.The properties of alloys following heat treatment are shown in Table 2. An increase in Zr content resulted in a trend in opposite directions for the alloy's tensile strength and hardness value in response to changes in elongation.

Table 2 Strength and elongation of alloys after standard heat treatment

Number\Performance	σ_b(Mpa)	$\sigma_{0.2}$(Mpa)	δ(%)
No.0	1203.91	818.93	8.51
No.1	1178.89	930.60	8.07
No.2	1206.47	1035.74	6.97
No.3	1156.53	743.06	7.24

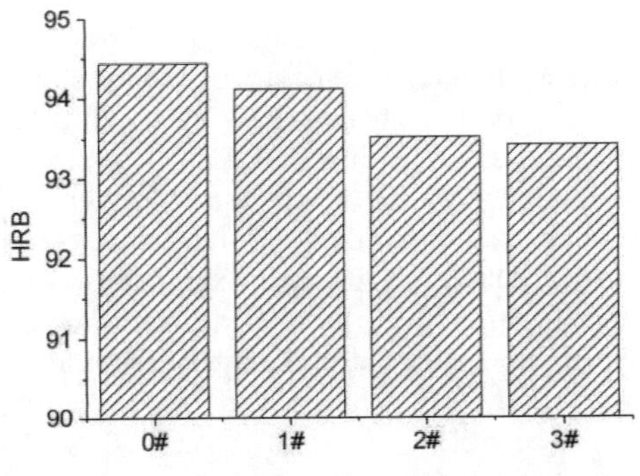

Fig.4 Histogram of hardness for the standard heat-treated state alloy

3.2 Fracture analysis

As shown in Figure 5, The toughness and fracture characteristics of the alloy decreased as the Zr content increased, and accompanied by a reduction in grain boundaries. The black line in the image indicated the location of the tearing bump, which was responsible for the specimen's overall breakage. The crack then penetrated the interior of the grain along the tearing bump, establishing a tough indention region for crack extension. The tensile stress induced significant grain deformation and specimen edge contraction, which resulted in necking of the material.

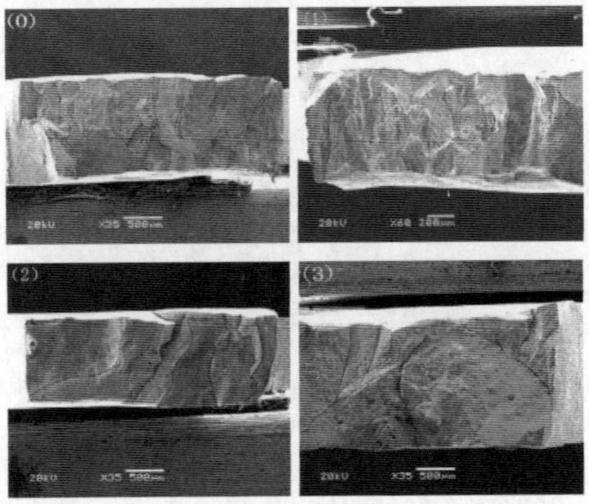

Fig.5 Macroscopic section and crack diagram of tensile specimen

The experimental analysis revealed that the second phase particles in the alloy increased with the increase of Zr content in the specimens, from No. 0 to 2. The second phase strengthening effect was pronounced, resulting in an elevated tensile strength of the alloy. Furthermore, when the Zr content of alloy No. 3 added to 0.09%, the alloy's comprehensive properties, such as strength, were found to be inferior.

4 Conclusion

In this study, the effect of the Zr elements on the organization and mechanical properties of K4169 alloys was investigated by adjusting the content of tensile fracture. The main conclusions are as follows:

(1) After the addition of Zr elements to K4169 alloy, the precipitation phase of the casting alloy is essentially the same as the original composition of the alloy organization.

(2) In comparison to as-cast alloys, alloys after heat treatment exhibited a more uniform composition, reduced segregation, superior crystallization, and a narrower range of phases.

(3) As the concentration of Zr in the alloy increases, the hardness and yield strength of the alloy in its typical heat-treated form will initially increase and then decrease, while the elongation will initially decrease and then increase.

Microstructure and Mechanical Properties of Magnetic Levitation Fabricated Cu-Ni-Co Multi-principal Element Alloys

Chengyuan Zhu, Zhiqiang Fu*
Guangdong Key Laboratory for Advanced Metallic Materials Processing, South China University of Technology, Guangzhou 510640, China
*Corresponding address: e-mail: zhiqiangfu2019@scut.edu.cn

Abstract: The $Cu_{35}Ni_{35}Co_{30}$ multi-principal element alloy (MPEA) with dual-FCC phase structure was prepared by magnetic levitation melting. Single-phase face-centered cubic (FCC) structure was obtained by heat treatment at 1030 °C for 15 h. Cold rolling and post-deformation annealing (PDA) heat treatments were used to produce heterogeneous grain structures (HGS). X-ray diffraction (XRD), scanning electron microscope (SEM) equipped with an energy-dispersive spectrometry (EDS) detector were used to characterize the microstructure and phase composition of the alloys and to determine their room temperature tensile properties in different states. The room temperature tensile test was used to determine their mechanical properties in different states. HGS could be achieved in PDA-treated sample at 1030 °C for 60 s. This sample exhibits excellent combination of strength and ductility, showing a tensile yield strength of 562 MPa, an ultimate strength of 758 MPa and a total elongation of 29.9 %.

Keywords: Multi-principal element alloy, magnetic levitation melting, Mechanical properties

1 Introduction

The concept of multi-principal element alloys (MPEA) was introduced by Yeh et al [1] in 2004, which consists of solid solution phases. Until now, the FCC MPEA have been widely studied for their properties such as good plasticity and excellent corrosion resistance [2]. However, FCC-structured MPEA exists the strength-ductility trade-off [3]. Fortunately, heterogeneous structural architectural strategies, such as bimodal structures, heterogeneous layered structures, gradient structures， have been reported to be effective strategies for realizing strength-toughness synergies in metals and alloys. Heterogeneous structures introduced by cold rolling and post-deformation annealing (PDA) heat treatment is an effective and simple way to realize the perfect combination of strength and ductility in FCC structured MPEA.

Therefore, $Cu_{35}Ni_{35}Co_{30}$ MPEA was chosen as an ideal prototype in this study. The material was cast by magnetic levitation melting, followed by heat treatment and cold rolling, and finally PDA heat treatment to obtain a heterogeneous grain structure (HGS) that produces a synergistic strengthening effect. The phase compositions, microstructures and mechanical properties of the as-cast and PDA states were investigated.

2 Experimental procedure

The $Cu_{35}Ni_{35}Co_{30}$ MPEA was prepared by magnetic levitation melting. After solidification, the ingot was homogenized in an electric-resistant furnace at 1030 ℃ for 15 h followed by water quenching. Then, cold rolling process was conducted on the homogenized ingot with a thickness reduction of 75 %. Finally, PDA heat treatments were performed on the as-rolled samples at 1030 °C for 60 s, followed by water quenching. The crystalline structure was identified by X-ray diffraction (XRD). The microstructure was characterized by a scanning electron microscope (SEM) equipped with an energy-dispersive spectrometry (EDS) detector and a backscattered electron (BSE) detector. Tensile tests were carried out on a universal testing machine at a strain rate of 1×10^{-3} s_{-1} at room temperature.

3 Result and discussion

Fig. 1 (a) shows the XRD patterns of the as-cast $Cu_{35}Ni_{35}Co_{30}$ MPEA, and the high-magnification image of (111) plane is shown in Fig. 1 (b). As shown by the bimodal (111) plane and (220) plane, the as-cast MPEA exhibit two sets of FCC diffraction peaks. Fig. 1 (c) shows BSE images of the as-cast sample, consisting of two distinct regions with light and dark contrast, showing a typical equiaxed dendritic structure. The inset of Fig. 1 (c) presents the chemical composition of the dendritic regions (DR) and inter-dendritic regions (IR) of the as-cast sample as measured by EDS/BSE. A visual examination of the inset reveals that Cu is rich in the inter-dendritic regions, while Ni and Co are concentrated in the inter-dendritic regions. In order to eliminate segregation and obtain a single FCC phase, the as-cast alloy was held at 1030 °C for 15 h. And to optimize the mechanical properties of $Cu_{35}Ni_{35}Co_{30}$ MPEA, the homogenized samples were cold-rolled with a deformation amount of 75 % and annealed at 1030 ℃ for 60 s. XRD patterns of the cold-rolled and PDA-treated $Cu_{35}Ni_{35}Co_{30}$ sample was shown in Fig. 1 (a). As expected, it is composed of a single FCC phase, suggesting that no phase transformation or precipitation occurred at 1030 ℃. Therefore, XRD results reveal that the $Cu_{35}Ni_{35}Co_{30}$ MPEA possesses a good structural stability against plastic deformation and PDA heat treatments. BSE images of the PDA-treated samples are showed in fig.1 (d). The PDA-1030-60 s sample exhibit notable heterogeneous grain structures. It can be seen that PDA-1030-60 s sample is composed by micro-sized recrystallized grains (MRGs) with an average grain size of ~4 μm and many residual deformed grains (RDGs) in the non-recrystallized regions. The volume fraction of MRGs and RDGs of the sample is estimated to be ~ 77.9 % and ~ 22.1 %, respectively.

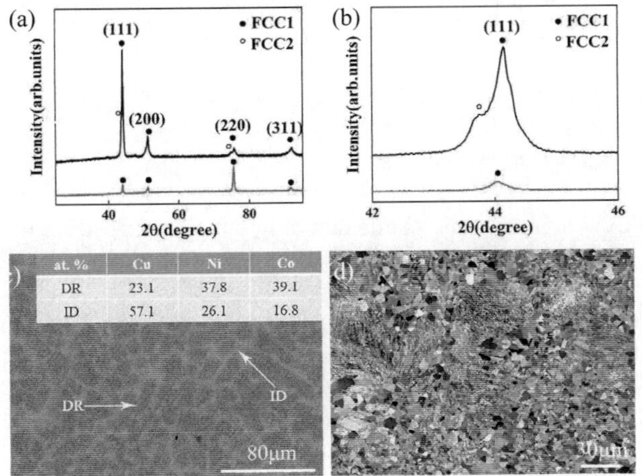

Fig. 1 (a) XRD patterns of as-cast and PDA-treated Cu35Ni35Co30 MPEA, (b) high-magnification of (111) plane (c) BSE images of as-cast Cu35Ni35Co30 MPEA, (d) BSE images of PDA-treated samples (PDA-1030-60 s)

Table 1. mechanical properties at ambient temperature of Cu35Ni35Co30 samples

Process	$\sigma_{0.2}$ MPa	σ_{UTS} MPa	El %	$\sigma_{t0.2}$ /MPa	σ_{tUTS} MPa	El_t %
As-cast	504	686	11.0	507	761	10.34
PDA	562	758	29.9	568	964	24.22

Representative engineering tensile stress–strain curves of the as-cast and PDA heat-treated $Cu_{35}Ni_{35}Co_{30}$ samples at room temperature are shown in Fig. 2. The engineering tensile yield strength ($\sigma_{0.2}$), ultimate strength (σ_{UTS}) and total elongation (El) of the samples are summarized in Table 1. As shown in Fig. 2 (a) and Table 1, the as-cast sample possesses a good ductility strength but an insufficient ductility, in detail: it shows a $\sigma_{0.2}$ of ~ 504 MPa, a σ_{UTS} of ~ 686 MPa and a total elongation of ~ 11.0%. Compared with the cold-rolled sample, PDA-treated sample at 1030 ℃ for 60 s shows distinct recovery of ductility. Specifically, the total elongation of the sample increases dramatically from ~ 11.0 to ~ 29.9 %, while the $\sigma_{0.2}$ was also slightly improved, from ~ 504 to ~ 562 MPa. The excellent combinations of strength and ductility have been achieved in PDA-1030-60 s. The true stress–strain curves of the two samples at room temperature are shown in Fig. 2 (b), while the results of the true tensile yield strength ($\sigma_{t0.2}$), true ultimate strength (σ_{tUTS}) and total elongation (El_t) were listed in table 1. The true stress–strain curves of the two samples at room temperature are shown in Fig. 2 (b), while the results of the true tensile yield strength ($\sigma_{t0.2}$), true ultimate strength (σ_{tUTS}) and total elongation (El_t) were listed in table 1. It can be seen that compared with as-cast sample, PDA heat-treated sample with HGS displays a significant improvement in ductility and $\sigma_{t0.2}$. Figure 2 (c) depicts the strain-hardening rate curves of as-cast and PDA heat-treated samples. It can be seen that the strain hardening rate of as-cast sample first monotonically decreases and then maintains at a relatively low-level work hardening rate as strain increases. However, the strain hardening rates of the PDA-1030-60 s samples first decreases sharply, then tends to increase, and finally decreases linearly with increasing true strain. The sudden increase is due to the presence of twin deformation in the alloy, which increases the work hardening rate of the alloy.

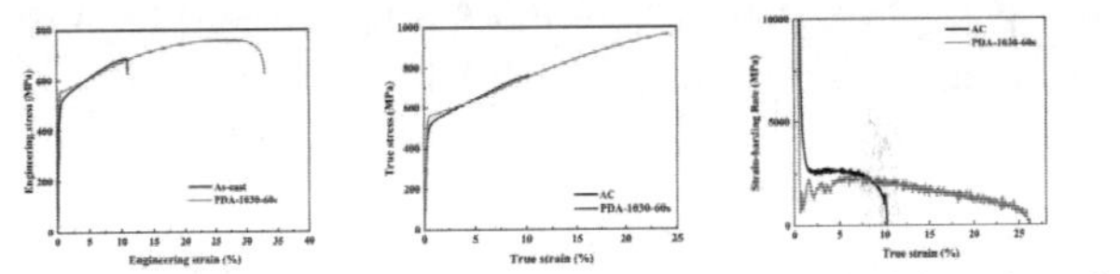

Fig. 2 Mechanical behavior of Cu35Ni35Co30 MPEA (a) engineering tensile stress–strain curves, (b) representative true stress–strain curves, (c) strain-hardening rate as a function of true strain

4 Conclusion

$Cu_{35}Ni_{35}Co_{30}$ MPEA with dual-FCC phase structure was prepared by magnetic levitation melting. Single-phase FCC sample with heterogeneous microstructure was successfully fabricated via the combination of cold rolling and short-time PDA heat treatments. The PDA-1030-60 s sample exhibits a well combination of strength and ductility, showing a tensile yield strength of ~ 562 MPa, an ultimate strength of ~ 758 MPa and a total elongation of ~ 29.9%. Furthermore, the PDA-1030-60 s sample displays a notable enhancement in ductility when compared with its as-cast counterpart, while simultaneously maintaining a high level of strength.

References

[1] Yeh J W, Chen S K, Lin S J, et al. Nanostructured high-entropy alloys with multiple principal elements: novel alloy design concepts and outcomes[J]. J. Advanced Engineering Materials, 2004, 6(5): 299-303.

[2] Wu Z, Bei H, Pharr G M, et al. Temperature dependence of the mechanical properties of equiatomic solid solution alloys with face-centered cubic crystal structures[J]. J. Acta Materialia, 2014, 81: 428-441.

[3] Wu X, Yang M, Yuan F, et al. Heterogeneous lamella structure unites ultrafine-grain strength with coarse-grain ductility[J]. J. Proc Natl Acad Sci USA. 2015, 112(47):14501.

Enhancing Mechanical Performance and Cavitation Erosion-Corrosion Resistance in fcc Multi-Principal Element Alloy Through L12 Precipitates

Jiacheng Niu[1,2], Zhiqiang Fu[1,2,*]
1. Guangdong Key Laboratory for Advanced Metallic Materials Processing, South China University of Technology, Guangzhou 510640, China
2. National Engineering Research Center of Near-net-shape Forming for Metallic Materials, South China University of Technology, Guangzhou 510640, China
*Corresponding address: e-mail: zhiqiangfu2019@scut.edu.cn

Abstract: The effects of $L1_2$ nano-precipitates on the mechanical properties and resistance to cavitation erosion-corrosion (CE-C) of multi-principal element alloy (MPEA) with a fcc structure were investigated. Heat treatment involving homogenization and aging introduces abundant spherical $L1_2$ precipitates into the fcc MPEA matrix. This not only significantly enhances the alloy's strength but also reduces volume loss in deionized water and NaCl solution, thereby improving its CE-C resistance.

Keywords: Multi-principal element alloy; Cavitation erosion-corrosion; Mechanical properties.

1 Introduction

In the field of marine vessels and water-based equipment, critical flow components such as propellers and pumps experience significant changes in fluid pressure during operation. These variations lead to erosion. The resulting high-speed, high-pressure microjets or shock waves repeatedly impact the surface of these components, causing plastic deformation, fatigue cracking, and eventual failure. When the surrounding fluid is corrosive (such as seawater), corrosion also occurs. Consequently, marine propellers rotating in oceanic environments are susceptible to severe cavitation erosion-corrosion (CE-C) [1-4].

Currently, the primary materials for marine propellers are nickel-aluminum bronze (NAB) alloys and stainless steel. Although these alloys exhibit some resistance to CE-C in seawater (due to Cl ions), their inherent strength remains relatively low. As a result, propellers face fracture risks after encountering impacts, leading to both operational delays and substantial losses [5].

In recent years, researchers have explored multi-principal element alloys (MPEAs) due to their ability to combine various properties into a single material. MPEAs can simultaneously offer excellent mechanical strength, corrosion resistance, oxidation resistance, and wear resistance. Consequently, they may hold promise in the field of CE-C-resistant materials. However, studies specifically investigating the CE-C performance of MPEAs are lacking [6].

Directly cast multi-principal element alloys often suffer from coarse grain structures and compositional segregation, limiting their overall performance. To address this, we investigated the controlled mechanisms of ordered coherent nanoscale precipitates ($L1_2$) in face-centered cubic (fcc) MPEA produced through casting and heat treatment. Our study focused on understanding the microstructure, mechanical properties, and CE-C behavior of MPEA.

2 Experimental procedure

The $Cr_{20}Fe_{44}Ni_{30}Al_3Ti_3$ alloy was prepared using a vacuum arc melting furnace. Subsequently, it underwent homogenization treatment at 1200°C for 10 hours, followed by water quenching. Aging treatment was then performed at 680°C for 4 hours. The as-cast sample was named "As-cast," while the aged sample was named "Aged." Microstructural observations were conducted using X-ray diffraction (XRD), scanning electron microscopy (SEM), and transmission electron microscopy (TEM). Tensile tests were performed using a universal testing machine, and erosion-corrosion behavior was evaluated using an ultrasonic cavitation apparatus.

3 Result and discussion

Fig. 1(a) shows the XRD patterns of the as-cast and aged samples. It is evident that both exhibit a fcc structure. The similarity in phase structure between the aged and as-cast samples is primarily due to the $L1_2$ phase, which also belongs to the fcc structure and cannot be distinguished in XRD. Fig. 1(b) displays SEM images of the aged sample, revealing significantly coarse grain sizes and straight twin boundaries. Additionally, no other precipitate phases are observed at the grain boundaries. Subsequently, the aged sample was examined using TEM, as shown in Fig. 1(c) and (d). Fig. 1(c) presents the bright-field (BF) image of the aged sample, while the inset shows the selected area electron diffraction (SAED) pattern along the [001] crystallographic axis. The presence of weak superlattice spots (highlighted by yellow circles) indicates that the aged sample possesses an ordered fcc structure ($L1_2$), confirming the precipitation of the $L1_2$ phase. Further observations of the precipitate morphology were conducted using dark-field (DF) imaging in Fig. 1(d), revealing numerous spherical precipitates with an average size of approximately 7 nm.

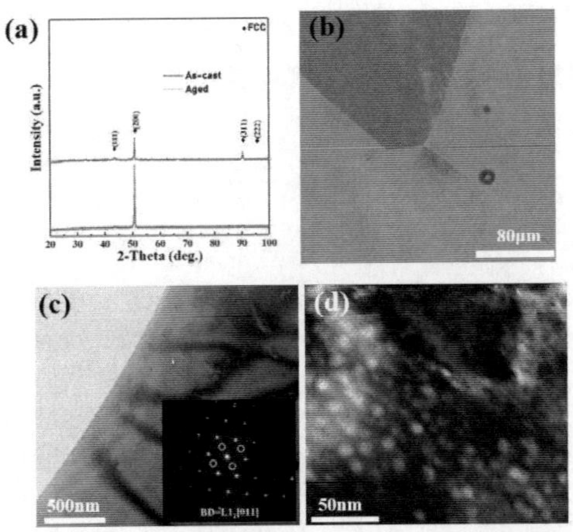

Fig. 1 (a) represents the XRD patterns; (b) shows SEM images of the aged sample; (c) displays the bright-field image of the aged sample along with the corresponding SAED (selected-area electron diffraction); and (d) exhibits the dark-field image of the aged sample.

Subsequently, mechanical properties and resistance to CE-C were tested for both samples, and the results are shown in Fig. 2. Fig. 2(a) displays the tensile curves for the two samples. Notably, the aged sample exhibits twice the strength of the as-cast sample, with only a 3% reduction in ductility. The grain sizes of both alloys are nearly identical, and the consistent solid solution strengthening and lattice friction stress further emphasize the pronounced strengthening effect of the $L1_2$ phase.

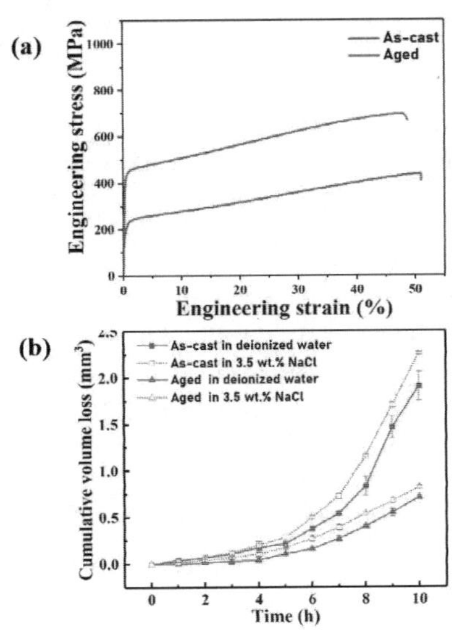

Fig. 2 (a) represents the tensile curves of the as-cast and aged samples; (b) shows the cumulative volume loss curves for both sample types in deionized water and NaCl solution during CE-C.

In Fig. 2(b), we observe the cumulative volume loss curves for the two samples in deionized water and 3.5 wt.% NaCl solution. Comparing the data in deionized water, it's evident that the aged sample experiences significantly reduced volume loss compared to the as-cast sample, likely attributed to the improved mechanical properties. When comparing the data in the NaCl solution, we find that the volume loss for the aged alloy remains lower than that of the as-cast sample. Overall, this suggests that corrosion plays a role in volume loss during the CE-C process.

References

[1] J. Ma, G. Hou, H. Cao, Y. An, H. Zhou, J. Chen, W. Duan, Why does seawater corrosion significantly inhibit the cavitation erosion damage of nickel-aluminum bronze?, Corros. Sci. 209 (2022) 110700.

[2] H. Cao, G. Hou, T. Xu, J. Ma, H. Wan, Y. An, H. Zhou, J. Chen, Effect of seawater temperature on the corrosion and cavitation erosion-corrosion resistance of $Al_{10}Cr_{28}Co_{28}Ni_{34}$ high-entropy alloy coating, Corros. Sci. 228 (2024) 111822.

[3] Z. Wang, B. Zhang, Cavitation erosion mechanism and microstructural design of pure titanium, Wear 490–491 (2022) 204189.

[4] Z.X. Li, L.M. Zhang, I.I. Udoh, A.L. Ma, Y.G. Zheng, Deformation-induced martensite in 304 stainless steel during cavitation erosion: Effect on passive film stability and the interaction between cavitation erosion and corrosion, Tribol. Int. 167 (2022) 107422.

[5] J. Niu, Z. Fu, G. Hou, Q. Chen, N. Yao, W. Chen, T. Lu, C. Chu, H. Wen, Exceptional combination of mechanical properties and cavitation erosion-corrosion resistance in a $Fe_{23.7}Co_{23.8}Ni_{23.8}Cr_{23.7}Mo_5$ multi-principal element alloy, Tribol. Int. 196 (2024) 109691.

[6] R.B. Nair, H.S. Arora, H.S. Grewal, Microwave synthesized complex concentrated alloy coatings: Plausible solution to cavitation induced erosion-corrosion, Ultrason. Sonochem. 50 (2019) 114–125.

Effect of Al Content on Microstructure and Mechanical Behavior of As-CastNi35(CoCr)62-xAlxTi3High-Entropy Alloys

Liran Huang, Zhiqiang Fu*
Guangdong Key Laboratory for Advanced Metallic Materials Processing, South China University of Technology, Guangzhou, 510641, China
*Corresponding address: e-mail: zhiqiangfu2019@scut.edu.cn (Z.Q. Fu)

Abstract: The effects of different Al content on the microstructure and mechanical properties of$Ni_{35}(CoCr)_{62-x}Al_xTi_3$ high entropy alloys (HEAs) were studied and analyzed. The results show that the microstructure of the alloys is the FCC main phase and a small amount of the B2 phase. And the mechanical properties of the alloys also change, yield strength and tensile strength at room temperature increase, while the plasticity becomes worse.

Keywords: High entropy alloy, Microstructure,B2 phase,Mechanical properties

1 Introduction (Bold, 10 pt., Arial)

EquiatomicCoCrNi is one of the typical HEAs characterized by a single FCC solid-solution phase, exhibiting a disappointing yield strength, which typically falls below 450 MPa at room temperature[1-3].Lu et al [4]investigated $NiCoCrAl_x$ (x = 0-30, at.%) alloy by adding Al elements, the results showed that the solid solution strengthening of Al element and the formation of hard BCC phase promotes the strengthening of this alloy system, and the FCC+BCC dual-phase can significantly enhance the CoCrNiHEA.Previous studies proved that by adding Al and Ti elements and adjusting their proportion, the strength of alloys can be greatly improved, while ensuring that the reduction of plasticity is within the acceptable range, resulting in HEAs with excellent mechanical properties [5]. In this paper, Al and Ti elements were selected as alloying elements, and the effects of Al content on phase composition, microstructure, and mechanical properties of $Ni_{35}(CoCr)_{62-x}Al_xTi_3$HEAs were studied systematically.

3 Experimental procedure

Three designed HEAs ingots with the nominal chemical composition $Ni_{35}(CoCr)_{62-x}Al_xTi_3$ (x=7, 10, 12, denoted by Al_7, Al_{10}, and Al_{12} HEAs, respectively) were synthesized by vacuum arc-melting with high-purity metals (purities above 99.9 wt. %) in a water-cooled copper mold.

Phase constitutions of three studied HEA samples were identified, using a X'Pert Pro (PANalytical) X-ray diffractometer (XRD) with Cu Kα radiation (λ = 0.154 nm). The morphology and microstructure were characterized by a Zeiss Sigma 500 scanning electron microscopy (SEM) operating at 15 kV equipped with a backscattering electron detector (BSE).

A Zwick/Roell Z020 universal testing machine was employed to conduct the tensile test at ambient temperature to investigate mechanical properties of studied alloys.

2 Result and discussion

Phase constitution and microstructure

Fig. 1 shows the XRD patterns of Al_7, Al_{10} and Al_{12} HEAs under as-cast. As can be seen, the main diffraction peaks of Al_7 HEA are identified as a single FCC phase (based on the typical lattice of (111), (200) and (311) for the FCC structure). While the BSE images (Figs. 2(a) and (b)) clearly indicate the presence of a few second phase in Al_7-HEA, the diffraction peaks of the secondary phase in Al_7-HEA cannot be discovered may due to its too small volume fraction. When the addition of Al increases to 10, the diffraction peaks corresponding to (100) and (110) can be distinguished demonstrating the formation of the $L1_2$ precipitates and B2 secondary phase, therefore the Al10-HEA primarily consists by the FCC matrix with B2 secondary phase and $L1_2$ precipitates occupied in it. The existence of diffraction peaks corresponding to the ordered BCC structure suggests that the content of Al element is related to the phase transformation from the FCC structure to B2 structure. The Al_{12}-HEA, as with Al_{10}-HEA, consists of the FCC major phase, B2 secondary phase, and ordered $L1_2$ precipitates within the FCC matrix. It worth noting that the number of diffraction peaks for the B2 structure is significantly increased compared to the Al_{10}-HEA, which indicates the increment of the B2 secondary phase volume fraction.

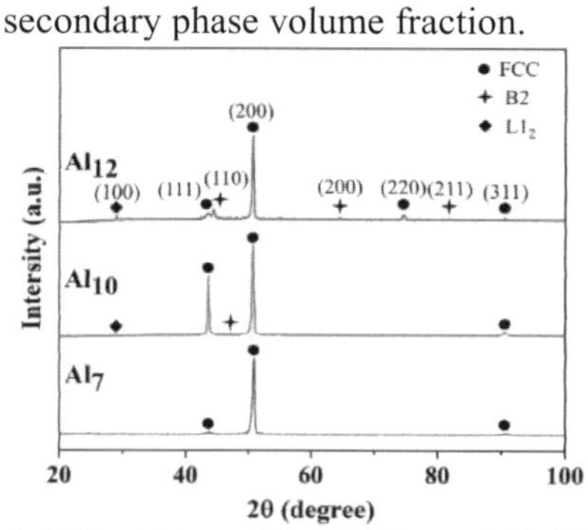

Fig.1. XRD patterns of the Al_7, Al_{10} and Al_{12} HEAs.

To further investigate the microstructure evolution and phase transformation, macro and micro BSE (as shown in Fig.2) were employed. Three studied samples all contain the FCC matrix (bright region) and the B2 secondary phase (dark region) in grain interiors. However, the proportion of the B2 secondary phase in Al_7, Al_{10} and Al_{12} HEAs is different, it can be distinctly seen from the BSE images in Fig. 2 that the volume fraction of the B2 secondary phase increases with the increment of the Al addition in these studied HEA system, which is consistent with the XRD results in Fig. 1. The volume fractions of the B2 secondary phase in the Al_7, Al_{10} and Al_{12} HEAs are measure to be ~0.7%, ~5.6% and ~14.0% by ImageJ software, respectively. In addition, some precipitates are visible in the B2 secondary phase of three studied HEAs. The precipitates in the B2 secondary phases of Al_7-HEA exhibit a needle-like structure and are interspersed with each other, nevertheless the precipitates of Al_{10} and Al_{12}HEAs are cubic or spherical particles discretely distributed in the B2 secondary phases. Moreover, some uncertain particles discontinuously located in the boundaries of the B2 secondary phases. The specific structure of precipitates in the B2 secondary phases and uncertain particles located in the boundaries need to be combined with high-end characterization methods for in-depth analysis.

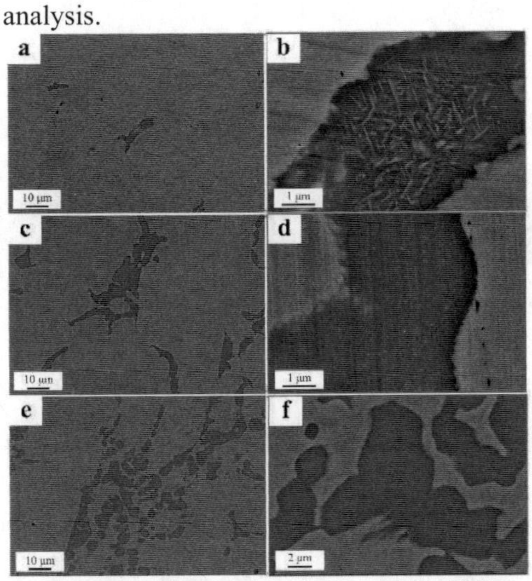

Fig.2. SEM images of the Al_7, Al_{10} and Al_{12} HEAs.

Mechanical properties

The mechanical properties are demonstrated by representative tensile engineering stress-strain curves of the three studied alloys at ambient temperature in Fig. 3.

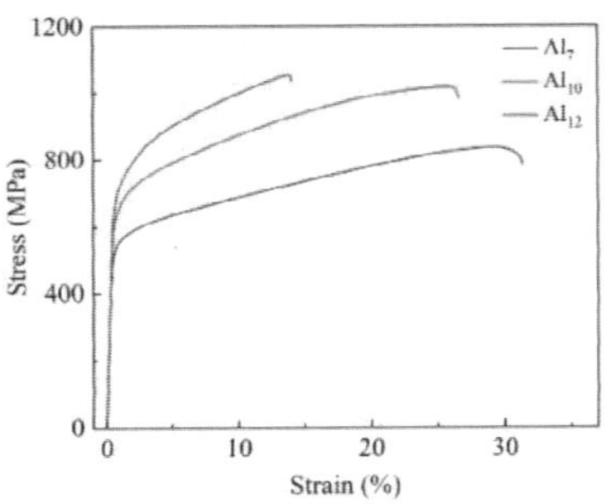

Fig.3. Engineering stress-strain curves of the Al_7, Al_{10} and Al_{12} HEAs at ambient temperature.

The Al_7-HEA represents a moderate yield strength ($\sigma_{0.2}$) of ~531 MPa, and a good ultimate tensile strength (σ_{UTS}) of ~836 MPa with an excellent total elongation of ~32.1%.The strength of the Al_{10}-HEA has a significant improvement, the values of the yield strength and ultimate tensile strength increase to ~639 MPa (~20.4%) and ~1016 MPa (~21.5%), respectively, while the ductility of the Al_{10}-HEA decreases to ~25.4%. The Al_{10}-HEA shows a further increase in strength and decrease in ductility, with the $\sigma_{0.2}$ of the Al_{10}-HEA reaching ~683 MPa (~8.6%), the σ_{UTS} of the Al_{10}-HEA reaching ~1053 MPa (~3.6%). Although the enhancement in strength of the Al_{10}-HEA is slight, the ductility critically deteriorates, the total elongation dramatically decreases to ~13.6% (46.5%). It can draw a conclusion that the strength of the studied alloy system improves with the increment of the Al content, while the ductility becomes worse.

4 Conclusion

With the increase of Al content, the microstructures of the studied alloys change. Both the size and volume fraction of the B2 phase show a significant increasing trend. Meanwhile, the mechanical properties of the studied alloys also change, the strength increases with the Al content, while the plasticity decrease. Therefore, it can be concluded that the Al element promotes the formation of the B2 phase, which improves the strength of alloys, but results in a loss of plasticity simultaneously.

Acknowledgments

This project was financially supported by the Natural Science Foundation of China (Grant No. 52103360), the Basic and Applied Basic Research Foundation of Guangdong Province (Grant No. 2023A1515012363).

References

[1] Wu Z, Bei H, Otto F, et al. Recovery, recrystallization, grain growth and phase stability of a family of FCC-structured multi-component equiatomic solid solution alloys[J].Intermetallics, 2014,46:131-140.
[2] Otto F, Yang Y, Bei H, et al. Relative effects of enthalpy and entropy on the phase stability of equiatomic high-entropy alloys[J]. ActaMaterialia. 2013, 61(7): 2628-2638.
[3] Miao J, Slone C E, Smith T M, et al. The evolution of the deformation substructure in a NiCoCr solid solution alloy[J]. ActaMaterialia. 2017, 132: 35-48.
[4] Lu W, Luo X, Yang Y, et al. Effects of Al addition on structural evolution and mechanical properties of the CrCoNiMEA[J]. Materials Chemistry and Physics. 2019, 238: 121841.
[5] Zhou Y J, Zhang Y, Wang F J, et al. Effect of Cu addition on the microstructure and mechanical properties of $AlCoCrFeNiTi_{0.5}$ solid-solution alloy[J]. Journal of Alloys and Compounds, 2008,466(1-2):201-204.

The Influence Of Chemical Composition on TiC Particle Distribution in In-Situ Al-Based Cast Composites

Ewa Olejnik[1,3,*], Paweł Kurtyka[3], Agnieszka Czajka[1], Robert Chulist[2], Karol Janus[1], Łukasz Szymański[1], Wojciech Maziarz[2]
1.Faculty of Foundry Engineering, AGH University of Krakow, Reymonta 23, 30-059 Krakow, Poland
2. Institute of Metallurgy and Materials Science, Polish Academy of Science, Reymonta 25, 30-059 Krakow, Poland
3. Innerco Sp. z o.o., Władysława Siwka 17, 31-588 Krakow, Poland
*Corresponding address: eolejnik@agh.edu.pl

Abstract: To improve fuel efficiency while maintaining the safety and performance of modern vehicles, lightweight Aluminum-based Metal Matrix Composites (AMMCs) are essential. These materials offer significant potential for enhancing vehicle performance due to their high strength-to-weight ratio, which is further improved by the addition of TiC, enhancing their mechanical properties. However, challenges such as the manufacturing process, the pushing-engulfment phenomenon, particle interaction, and particle-matrix reactivity need to be addressed to fully realize their potential. Main goal of this work was to determine the influence of the Al-base alloy's chemical composition on the pushing-engulfment phenomenon of the TiC particles and their interaction. The Al 1000/TiC composite exhibits an inter-dendritic distribution of TiC particles, resulting from the pushing phenomenon at the solid-liquid interface. In contrast, the Al-7Si/TiC composite displays an inter-eutectic arrangement, with TiC particles being engulfed within the eutectic regions and the formation of the Al_3Ti phase, indicating additional reactions within the Al-Si-TiC system. It was found that the TiC particle size increases were observed for Al-7Si compared to Al 1000 alloy.

Keywords:AMMC, casting, in situ, TiC, pushing-engulfment phenomenon, tensile strength

1 Introduction

The requirement to reduce the weight of automobiles has become increasingly important due to growing environmental concerns. To improve fuel efficiency while maintaining the safety and performance of modern vehicles, lightweight AMMC are essential and offer significant potential for increasing vehicle efficiency. Although the application of this group of materials is often discussed, commercial solutions are still rare. One of the main issues is the manufacturing process, pushing-engulfment phenomenon, particle interaction and particle-matrix reactivity. Aluminum matrix composites reinforced with titanium carbide (TiC) particles represent a significant area of research in materials science and foundry technology, driven by the properties of TiC such as high hardness, low density, and good wettability by molten aluminum [1]. The incorporation of TiC strengthens the composite, enabling it to withstand higher loads and stresses, thereby contributing to both fuel efficiency and durability enhancements in automotive applications. It has been demonstrated that TiC particles in aluminum not only function as a strengthening phase but also contribute to grain refinement by being pushed to the grain boundaries, thereby inhibiting grain growth [2]. The main goal of this work was to determine the influence of the Al-base alloy's chemical composition on the pushing-engulfment phenomenon of the TiC particles and their interaction.

2 Experimental procedure

This study focused on two types of MMC matrices: Al 1000 and Al with 7 wt% Si. The raw material used for TiC synthesis consisted of Ti and C powders. These powders were subsequently compacted to obtain a reactive green compact, which was then introduced at 10 wt.% into a furnace during the casting process. The schematic of the materials manufacturing process is shown in Fig. 1.

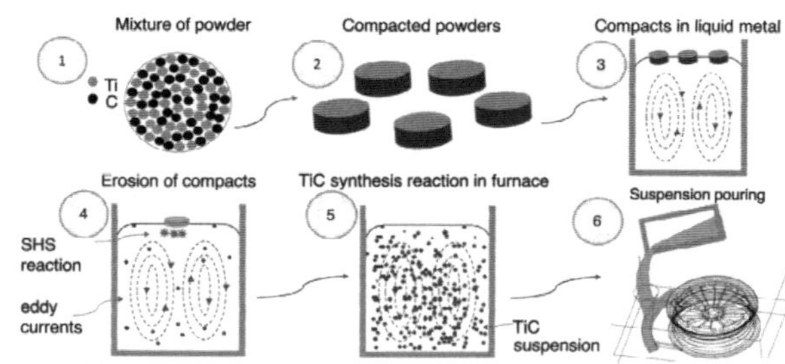

Figure 1. Schema of the in-situ TiC/Al-base composites manufacturing process

Finally, Al 1000/TiC and Al-7%Si/TiC composites were successfully fabricated. Microstructure, structure, chemical composition in the micro area and tensile strength of composites were examined.

3 Result and discussion

Differences in the Al 1000/TiC microstructure and Al-7Si/TiC microstructure is visible in Fig. 2. The Al 1000/TiC composite is characterized by the inter-dendritic distribution of TiC particles forming a coating around each grain. In contrast, the Al-7Si/TiC material exhibits an inter-eutectic arrangement of TiC particles. The inter-dendritic distribution is a consequence of the pushing phenomenon, which results in the accumulation of TiC particles at the solid-liquid interface (SLI) during dendritic grain growth. Only a few TiC particles are incorporated directly within the grains, confirming the criterion that the critical velocity, above which particles are engulfed by the SLI, is inversely proportional to the particle radius. This implies that the SLI will engage larger particles and push away smaller ones for specific particle-matrix system.

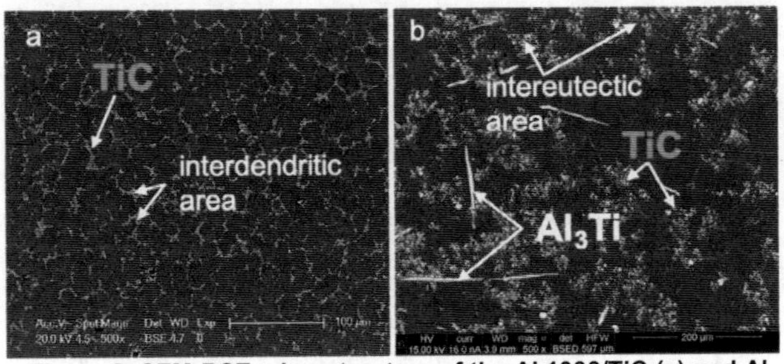

Figure 2. SEM-BSE microstructure of the Al 1000/TiC (a) and Al-7Si/TiC (b)

Another particle distribution observed in the Al-7Si/TiC composite involves the engulfment phenomenon within the eutectic area and the pushing around the (Al) dendrites (Fig. 3). This indicates that the Al-Si-TiC system represents a different mechanism. Additionally, the microstructure of this composite shows the formation of the oblong Al_3Ti phase.

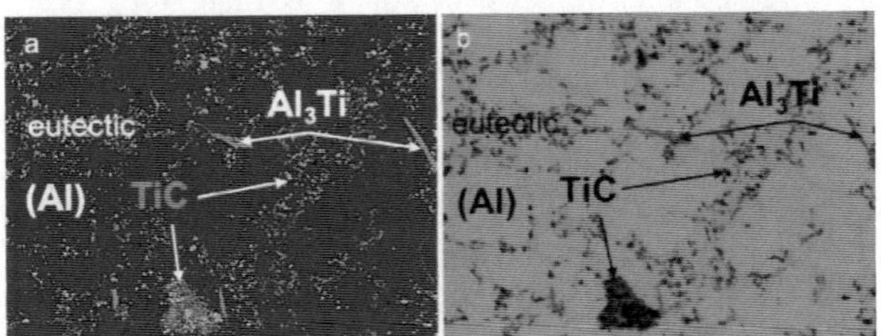

Figure 3. SEM-BSE microstructure of the Al-7Si/TiC composite (a) and corresponding SEM-EDS elements distribution map (b)

This suggests that additional reactions take place in the discussed system, which may form the basis for the Al-Si-Ti-C system's reactivity. This reactivity was reported by Guo and al. [4] during the wettability examination. In addition to different particle distribution in the matrix, the size of TiC particles also changes in both analyzed composites. Therefore, in Al 100/TiC the particle size is much smaller than in Al7Si/TiC. Bearing in mind that the chemical composition is the only variable factor, we can confidently assume a change in the physicochemical properties. Then we can consider a higher probability of coalescence in the Al7Si/TiC composite compared to Al 1000/TiC. The TiC agglomerates are visible on the SEM/EDS Al distribution map (Fig. 2b), highlighting a general tendency observed in the Al-7Si alloy. The formation of these TiC agglomerates indicates the initial stage of the coalescence process.

4 Conclusion

The Al 1000/TiC composite shows an inter-dendritic distribution of TiC particles, attributed to the pushing phenomenon at the solid-liquid interface. In contrast, the Al-7Si/TiC composite exhibits an inter-eutectic arrangement, with TiC particles engulfed within the eutectic area and the formation of the Al_3Ti phase, indicating additional reactions in the Al-Si-TiC system. Furthermore, TiC particles are smaller in Al 1000/TiC than in Al-7Si/TiC, suggesting a higher probability of coalescence in the latter due to differences in physicochemical properties.

Acknowledgments

Financial support of the National Science Centre, Poland (project no. 2021/43/B/ST8/03271) is greatly acknowledged.

References

[1] Amosov A P, Luts A R, Ermoshkin A A. Nanostructured aluminum matrix composites of Al-10% TiC obtained in situ by the SHS method in the melt, Key Engineering Materials, 2016, 684: 281-286.

[2] Maziarz W, Wójcik A, Chulist R, Bigos A, Kurtyka P, Szymański Ł, Jimenez Zabaleta A, García de Cortázar M, Olejnik E. Microstructure and mechanical properties of Al/TiC and Al/(Ti,W)C nanocomposites fabricated via in situ casting method. Journal of Materials Research and Technology, 2024, 28: 1852–1863.

[3] Shangguan D, Ahuja S, Stefanescu D M. An analytical model for the interaction between an insoluble particle and an advancing solid/liquid interface. Metallurgical Transactions A, 1992, 23A: 669-680.

[4] Guo R F, Wang Y, Ma Y H, Shen P. Role of Si in the wetting of TiC by Al. J Mater Sci, 2021, 56: 7791–7798.

Formation Mechanism of Cellular Structure of NiCoCrFeAlEutectic High-Entropy Alloys by Selective Laser Melting

An Xie[1], Huan Yu[1], Zhijun Wang[1], Junjie Li[1], Jincheng Wang[1], Feng He[1,2,3*]
1. State Key Laboratory of Solidification Processing, Northwestern Polytechnical University, Xi'an 710072, China
2. Research & Development Institute of Northwestern Polytechnical University in Shenzhen, Shenzhen 518063, China
3. Collaborative Innovation Center of Northwestern Polytechnical University, Shanghai 201100, China
*Corresponding address: e-mail: fenghe1991@nwpu.edu.cn

Abstract: In this study, a NiCoCrFeAl eutectic high entropy alloy (EHEA) was prepared by selective laser melting (SLM). The sample was a cellular eutectic structure composed of BCC/B2 cells and FCC cell boundaries. The results show that under the extremely high-temperature gradient and cooling rate of selective laser melting, the microstructure characteristics of cellular eutectic produced by strong non-equilibrium solidification will be greatly affected by scanning speed and laser power. The reasons for the formation of the cellular structure are as follows: When the alloy composition slightly deviates from the eutectic point and becomes hypereutectic, it is beneficial to the formation of B2 cells/dendrites rather than eutectic under the condition of rapid growth caused by SLM rapid cooling. In addition, the size of the cellular structure decreases with the increase in scanning speed. However, at a lower scanning speed, the cyclic thermal loading time of the deposited materials is prolonged, which is beneficial to the diffusion and enrichment of elements and promotes the amplitude modulation decomposition process of the BCC phase in the cell. These results provide new insight into the formation and evolution behavior of the rapidly solidified structure of SLMed EHEAs.

Keywords: additive manufacturing, high entropy alloy, cellular structure, rapid solidification

1 Introduction

Eutectic high-entropy alloys show excellent adaptability to solidification-controlled processing technologies, such as casting and additive manufacturing, due to its narrow crystallization temperature range and good fluidity [1]. In particular, additive manufacturing not only efficiently processes EHEAs with complex geometry but also improves their mechanical properties significantly.

Ren et al.[2] achieved a tensile yield strength of 1300MPa and uniform elongation of 14% in the $AlCoCrFeNi_{2.1}$ EHEA. The high strength and good ductility have been attributed to the rapid solidification caused by ultrafine BCC and FCC nanolamellae microstructures. Whereas, in some other articles, the SLMed $AlCoCrFeNi_{2.1}$ EHEA showed cellular structures with different mechanical performances[3, 4]. These pioneering works indicated that the microstructure of SLMed EHEA was of great importance for tuning mechanical properties, but the formation mechanisms of the unique microstructure during SLM were unfortunately not clear. Therefore, in this work, the formation mechanism and evolutionary behavior of SLMed EHEAs cellular structures were systematically investigated by adjusting the SLM process parameters and analyzing the alloy composition.

2 Experimental procedure

$Ni_{35}Fe_{36}Cr_{10}Al_{17}Mo_{2}$ and $Ni_{30}Co_{30}Cr_{10}Fe_{10}Al_{18}Mo_{1}W_{1}$ pre-alloyed powders with particle size distribution ranging from 15 to 53 μm prepared by gas atomization were mixed for 12 hours according to the mass ratio of 1: 1, to achieve a nominal $Ni_{32.5}Co_{15}Cr_{10}Fe_{23}Al_{17.5}Mo_{1.5}W_{0.5}$ composition. All parts are prepared by BLT-S200 SLM machine and protected by argon. The microstructures were characterized by scanning electron microscopy (SEM, TESCAN MIRA3) and transmission electron microscopy (TEM, Talos F200X) equipped with an energy dispersive spectrometer (EDS).Cell size was measured by image processing software (Image-Pro).

3 Result and discussion

Fig. 1(a1-a3) show the cellular structure morphology perpendicular to the building direction at different laser scanning speeds. With the increase in scanning speed, the cellular structure size becomes smaller. The statistical results in Fig. 1(b) show that with the increase of the volume energy density (VED) from 49.5J/mm^3 to 145.8 J/mm^3, the wall thickness of the FCC phase increases from ~100nm to over 180nm, while the volume fraction of BCC phase decreases from ~63% to ~48%. Fig. 1(c) shows the average diameter of cellular eutectic decreases from 615.3±53.7nm at 600mm/s to 487.4±19.0nm at 1200mm/s, which is because the cooling rate of the molten pool increases with the increase of scanning speed, and the growth rate of dendrites accelerates, thus reducing the primary arm spacing of dendrites. The results show that under the extremely high-temperature gradient and cooling rate of selective laser melting, the microstructure characteristics of cellular eutectic produced by strong non-equilibrium solidification will be greatly affected by the scanning speed and laser power.

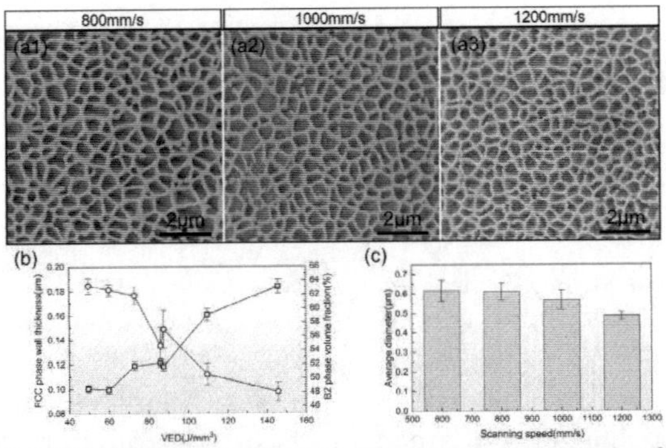

Fig. 1 (a1-a3) SEM images of samples at scanning speeds of 800 mm/s, 1000 mm/sand 1200 mm/s; (b) Changes of FCC phase wall thickness and BCC phase volume fraction with VED; (c) The cell diameters with the scanning speed.

To understand the formation mechanisms of the cellular structures, we in the following theoreticallyanalyze its solidification behaviors by referring to the binary Ni-Al FCC-B2 eutectic system. As reported by Kurz et al.[5] in the directional solidification structure of Ni-Al alloy,under certain growth conditions, the structure with the highest interface growth temperature among FCC dendrite, B2 dendrite and FCC-B2 eutectic will become the desired solidification structure of the alloy.

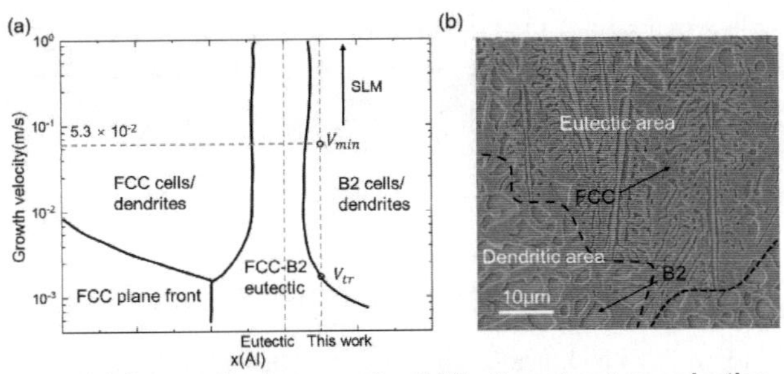

Fig. 2(a) Schematic diagram of solidification structure selection of Ni50-xCo15Cr10Fe23AlxMo1.5W0.5 alloy.[5]; (b) The as-cast microstructure shows alloy is hypereutectic.

Fig.2(a)sketches the selection of the solidification structure of $Ni_{50-x}Co_{15}Cr_{10}Fe_{23}Al_xMo_{1.5}W_{0.5}$alloy under a certain temperature gradient by following the results of Kurz et al.[5]. When the alloy is in eutectic composition, no matter how the growth rate changes, the eutectic solidification structure of FCC-B2 will be obtained. When the alloy composition slightly deviates from the eutectic point and becomes hypereutectic, the final solidification structure begins to be related to the growth rate. The eutectic can be maintained at a low growth rate, but it becomes B2 cells/dendrites at a high growth rate. Thus, two possible reasons, i.e., growth rate and compositions, should be considered when we understand the formation mechanism of cellular structure in SLM EHEAs.

Here, the as-cast microstructure in Fig.2(b) shows that the $Ni_{32.5}Co_{15}Cr_{10}Fe_{23}Al_{17.5}Mo_{1.5}W_{0.5}$EHEA in this study is actually a hypereutectic with B2 primary phase.Therefore, the growth rate during SLM plays an important role in its solidificationstructures. In this work, the minimum growth rate calculated exceeds the maximum growth rate that can maintain eutectic, so the alloy tends to form B2 cells/dendrites.

4 Conclusion

In this paper, the evolution behavior of the cellular structure of SLM EHEAs is studied by adjusting SLM process parameters. When the laser scanning speed is high, the cooling rate of the molten pool increases and the size of the cellular structure is refined. This study also explains the formation mechanism of the cellular structure by sketching the solidification structure selection of the alloy: Under the fast growth rate caused by the rapid cooling of SLM,The composition slightly deviating from the eutectic point is easy to form B2cells/dendrites.

Acknowledgments

The authors express their gratitude for the financial support by the the Shanghai "Phosphor" Science Foundation, China (23YF1450900), the Basic and Applied Basic Research Foundation (Grant No. 2023A1515012703) of Guangdong Province, Young Elite Scientists Sponsorship Program by CAST, Grant No. 2023QNRC001, National Natural Science Foundation of China (Grant No. 52001266), the Fundamental Research Funds for the Central Universities (Grant No. G2022KY05109), and the Research Fund of the State Key Laboratory of Solidification Processing (NPU), China (Grant No. 2023-QZ-02).

References

[1] Lu, Y., Dong, Y., Guo, S., Jiang, L., Kang, H., Wang, T., Wen, B., Wang, Z., Jie, J., Cao, Z., Ruan, H. and Li, T. A Promising New Class of High-Temperature Alloys: Eutectic High-Entropy Alloys. Scientific Reports, 4, 1 (2014).

[2] Ren, J., Zhang, Y., Zhao, D., Chen, Y., Guan, S., Liu, Y., Liu, L., Peng, S., Kong, F., Poplawsky, J. D., Gao, G., Voisin, T., An, K., Wang, Y. M., Xie, K. Y., Zhu, T. and Chen, W. Strong yet ductile nanolamellar high-entropy alloys by additive manufacturing. Nature, 608, 7921 (Aug 2022), 62-68.

[3] Geng, Z., Chen, C., Li, R., Luo, J. and Zhou, K. Composition inhomogeneity reduces cracking susceptibility in additively manufactured AlCoCrFeNi2.1 eutectic high-entropy alloy produced by laser powder bed fusion. Additive Manufacturing, 56 (2022).

[4] Guo, Y., Su, H., Yang, P., Shen, Z., Zhao, D., Zhao, Y., Liu, Y. and Zhou, H. New insight into tailorable eutectic high entropy alloys with remarkable strength–ductility synergy and ample shaping freedom fabricated using laser powder bed fusion. Additive Manufacturing, 60 (2022).

[5] Hunziker, O. and Kurz, W. Solidification microstructure maps in Ni-Al alloys. Acta Materialia, 45, 12 (1997), 4981-4992.

Cooling Rate Dependence of Iron-Rich Phase in A356 Aluminum Alloy and Its Effect on Mechanical Properties

Yunxia Niu, Lizhi Zhang, Jie Wan*, Yuzeng Chen*
State Key Laboratory of Solidification Processing, Northwestern Polytechnical University, Xi'an, Shaanxi 710072, People's Republic of China
*Correspondence: wan@nwpu.edu.cn (Jie Wan); yzchen@nwpu.edu.cn (Yuzeng Chen)

Abstract: In the process of recycling waste aluminum, the formation of a large number of iron-rich phases seriously affects the properties of A356. The formation of iron-rich phase in A356 is closely related to the melt cooling rate in the casting process. In this work, two groups of A356 with iron content of 0.18 wt.% and 0.36 wt.% were used as research objects to prepare alloy samples with different melt cooling rates, and then the correlation between iron-rich phase characteristics and melt cooling rates in the as-cast alloy was studied. On this basis, the casting defects of the alloy were reduced by hot isostatic pressing. Then the microstructure of A356 was regulated by solution aging heat treatment, and the influence of iron-rich on the mechanical properties of A356 at different cooling rates was investigated. It is found that the cooling rate affects not only the microstructure of the alloy, but also the nucleation and growth of iron-rich phase. The iron-rich phase content increases with the increase of iron content in the alloy. In addition, when iron content is less than 0.36 wt.%, iron content has little effect on the yield strength, tensile strength, fracture elongation and impact toughness at room temperature of A356.

Keywords: Cast aluminum alloy; Cooling rate; Iron-rich phase; Microstructure; Mechanical properties

1 Introduction

A356 is a commonly used cast aluminum alloy, which is widely used in aerospace and electronic information equipment and other industries in the industrial production process. In this alloy, iron (Fe) is the main impurity element that is difficult to be completely removed, and it is often combined with aluminum (Al), silicon (Si) and other elements to form an iron-rich phase. These iron-rich phases usually exist in the alloy in the form of needles or sheets, which can split the matrix of the alloy and cause more serious stress concentration, resulting in a decrease in the mechanical properties of the material. In particular, the hard and brittle sheet β-Fe phase is extremely destructive to the mechanical properties of aluminum alloys [1-3]. In order to improve the morphology of iron-rich phase and reduce its harm to alloy, two kinds of A356 ingot containing iron were obtained by adding Al-Fe intermediate alloy, and the alloys were solidified by adjusting different melt cooling rates during the casting process. After that, the microstructure of the product was observed and the properties of the alloy were evaluated. The correlation of melt cooling rate of iron-rich phase in A356 aluminum alloy and its influence on the mechanical properties of the alloy were revealed, which provides a reference for the recycling and practical production of recycled aluminum.

2 Experimental procedure

In this study, samples of A356.2 aluminum alloy (Table 1) and Al-Fe intermediate alloy with different thickness were prepared by square copper casting method. The combination of UG modeling and ProCAST simulation software overcame the measurement problem caused by the rapid temperature drop during the experiment, and a more reliable cooling rate curve was obtained. T6 heat treatment and HIP+T6 heat treatment were performed on A356 alloy tensile specimens to verify the effect of HIP on eliminating internal hole defects. The sample was then extracted from the cast alloy and prepared into a sample block, the metallographic structure was observed and the tissue photographs were taken. In addition, the tensile properties at room temperature, V-notch impact toughness at room temperature and other mechanical properties were tested and the fracture morphology was analyzed.

Table 1 The chemical component of A356.2 alloy（wt.%）

Si	Mg	Fe	Cu	Zn	Mn	Ti	Al
7.36	0.329	0.100	0.002	<0.005	0.007	0.141	

3 Result and discussion

Two groups of A356 with 0.18 wt.% and 0.36 wt.% iron(A356-0.18Fe and A356-0.36Fe) were studied. Alloy samples with different melt cooling rates were prepared by adjusting the cavity size of the copper mold. The relationship between iron-rich phase and melt cooling rate was studied, and the mechanism of iron-rich alloy with different iron-content on microstructure and mechanical properties was expounded.

Microstructure

The bright white iron-rich phase in Fig.1 (a) is mainly π-Fe phase containing magnesium elements and a small amount of β-Fe phase. (b) The bright white iron-rich phase in the figure is mainly β-Fe phase without magnesium and a small amount of π-Fe phase. In addition, it is found that the cooling rate of melt significantly affects the type and size of iron-rich phase: in the range of 15~108 K s^{-1}, the increase of cooling rate leads to the decrease of the size of iron-rich phase, and the increase of iron content is conducive to the growth of iron-rich phase.

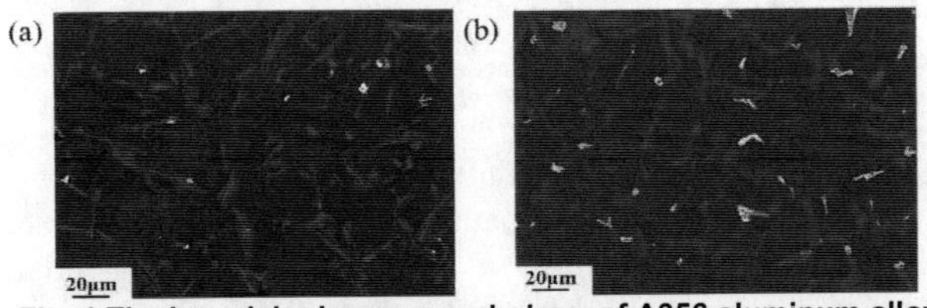

Fig. 1 The iron-rich phases morphology of A356 aluminum alloy: (a) is A356-0.18Fe; (b) is A356-0.36Fe

Tensile behavior

Fig 2 (a) shows the tensile yield strength test values of each group, (b) the tensile tensile strength test values of each group, and (c) the tensile elongation at break test values of each group. It can be seen from the figure that under the same cooling rate, the yield strength, tensile strength and elongation at break of A356-0.18Fe and A356-036Fe alloys are almost the same. The yield strength fluctuates at about 160 MPa, the tensile strength fluctuates at about 260 MPa, and the elongation at break increases with the increase of cooling rate. At the same cooling rate, when iron content is 0.18 wt.% and 0.36 wt.%, the yield strength, tensile strength and elongation at break of the alloy are not affected by iron content.

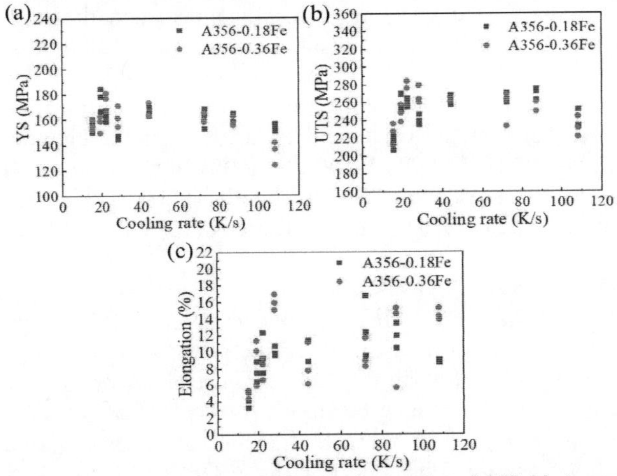

Fig. 2 Tensile properties of A356 aluminum alloy: (a) Yield strength; (b)Ultimate tensile strength; (c) Elongation

At the same cooling rate, with the increase of iron content, the average size of iron-rich phase increases, the number of small-size iron-rich phase decreases, and the number of large-size iron-rich phase increases. At different iron content, melt cooling rate has little effect on the yield strength and tensile strength of the alloy, but can significantly increase the elongation at break and plasticity of the alloy. The increase of cooling rate refines the microstructure and improves the distribution of eutectic silicon phase and other particles, thus increasing the plasticity and elongation at break of the alloy.

4 Conclusion

In this study, the properties of two A356 with different iron content were studied at different cooling rates. The results show that the cooling rate has a significant effect on the plasticity and elongation at break of the alloy, and the higher the iron content in the alloy, the more prominent this effect is. In addition, hot isostatic pressing (HIP) can repair the defects in the alloy, making the eutectic silicon phase smaller and smoother, which helps to reduce the yield strength of the alloy, and significantly improve its toughness and elongation at break, which also has an important impact on the performance of A356 .

Acknowledgments

This work was supported by the National Natural Science Foundation of China (Nos. 52071262; 52301197), the Natural Science Basic Research Program of Shaanxi Province, China (No. 2023-JC-QN0421) and the Advanced Development of Aluminum Alloy Filler Materials Used in Welded High-end Lightweight Constructions - Research & Industrial Implementation (No. 2022Z109).

References

[1] L. Kucharikovà, D. Medvecká, E. Tillová, et al. The Effect of the β-Al5FeSi Phases on Microstructure, Mechanical and Fatigue Properties in A356.0 Cast Alloys with Higher Fe Content without Additional Alloying of Mn. Materials. 2021, 14(08): 1943.

[2] L. Ceschini, A. Morri, G. Sambogna,The effect of hot isostatic pressing on the fatigue behaviour of sand-cast A356-T6 and A204-T6 aluminum alloys,Journal of Materials Processing Technology,2008, 204(1–3), 231-238.

[3] S.Hafenstein, E. Werner, Direct aging of a hot isostatically pressed A356 aluminum cast alloy,Materials Science and Engineering: A,2019, 768.

Evolution of Divorced Eutectic Structure inNiCoCrFeAl Eutectic High Entropy Alloy by Selective Laser Melting

An Xie, Zhijun Wang, Jincheng Wang, Feng He*
State Key Laboratory of Solidification Processing, Northwestern Polytechnical University, Xi'an 710072, PR China
*Corresponding address: e-mail: fenghe1991@nwpu.edu.cn

Abstract: In this study, a NiCoCrFeAl eutectic high entropy alloy (EHEA) was prepared by selective laser melting (SLM). The sample was a cellular eutectic structure composed of BCC/B2 cells and FCC cell boundaries. The results show that under the extremely high-temperature gradient and cooling rate of selective laser melting, the microstructure characteristics of cellular eutectic produced by strong non-equilibrium solidification will be greatly affected by scanning speed and laser power. The reasons for the formation of the cellular structure are as follows: The growth rate of B2 dendrites is much faster than that of eutectic at high undercooling, resulting in hypoeutectic structure, which evolves into divorced eutectic under certain conditions. When the laser scanning speed is high, the cooling rate of the molten pool is increased, and the primary arm spacing of dendrites is shortened; thus, the cell structure size is refined. However, at a lower scanning speed, the cyclic thermal loading time of the deposited materials is prolonged, which is beneficial to the diffusion and enrichment of elements and promotes the amplitude modulation decomposition process of the BCC phase in the cell. These results provide new insight into the formation and evolution behavior of the rapidly solidified structure of SLMed EHEAs.

Keywords:additive manufacturing, high entropy alloy, cellular structure, rapid solidification

1 Introduction

Eutectic high entropy alloy (EHEA) is a multi-element alloy proposed by Lu et al.[1]in 2014, which combines the advantages of eutectic alloy and high entropy alloy. It usually has a microstructure of hard BCC phase and soft FCC phase, which can achieve a good combination of strength and plasticity. As a eutectic alloy, EHEAs have a narrow crystallization temperature range and good fluidity and are suitable for various preparation processes such as casting, powder metallurgy, and additive manufacturing.

During the rapid solidification of SLMedNiCoCrFeAl EHEAs, in addition to the formation of the conventional lamellar eutectic structure, a cellular structure resulting from non-equilibrium solidification also occurs[2-4]. At present, there is no unified explanation for why the cellular eutectic structure different from the traditional lamellar structure is formed in SLMedNiCoCrFeAl EHEAs, and this cellular structure is very similar to the divorced eutectic that the traditional binary eutectic alloy evolved from hypoeutectic during rapid solidification[5].Therefore, in this study, $Ni_{32.5}Co_{15}Cr_{10}Fe_{23}Al_{17.5}Mo_{1.5}W_{0.5}$ eutectic high entropy alloy was designed based on Thermo-Calc thermodynamic software and prepared by SLM. The formation mechanism and evolution behavior of the cellular structure of SLMed EHEAs were systematically studied by adjusting SLM process parameters.

2 Experimental procedure

$Ni_{35}Fe_{36}Cr_{10}Al_{17}Mo_2$ and $Ni_{30}Co_{30}Cr_{10}Fe_{10}Al_{18}Mo_1W_1$ pre-alloyed powders with particle size distribution ranging from 15 to 53μm prepared by gas atomization were mixed for 12 hours according to the mass ratio of 1: 1, to achieve a nominal $Ni_{32.5}Co_{15}Cr_{10}Fe_{23}Al_{17.5}Mo_{1.5}W_{0.5}$ composition. All parts are manufactured by BLT-S200 SLM machine and protected by argon. The experimental parameters used in this work are as follows: the laser beam diameter is 80 μm, the layer thickness (40μm) and the hatch spacing (50μm) are unchanged, and the laser power (190-330 W) and scanning speed (600-1200mm/s) are changed at the same time. Adjacent layers are rotated by 67°.The microstructure and chemical composition were characterized by a field emission scanning electron microscope (SEM, TESCAN MIRA3) and transmission electron microscope (TEM, FEI Talos F200X) equipped with an energy dispersive spectrometer (EDS).

3 Result and discussion

Fig. 1(a1-a3) show the cellular structure morphology perpendicular to the building direction at different laser scanning speeds. With the increase in scanning speed, the cellular structure size becomes smaller. The statistical results in Fig. 1(b) show that with the increase of the volume energy density (VED) from 49.5J/mm^3 to 145.8 J/mm^3, the wall thickness of the FCC phase increases from ~100nm to over 180nm, while the volume fraction of BCC phase decreases from ~63% to ~48%. Fig. 1(c) shows the average diameter of cellular eutectic decreases from 615.3±53.7nm at 600mm/s to 487.4±19.0nm at 1200mm/s, which is because the cooling rate of the molten pool increases with the increase of scanning speed, and the growth rate of dendrites accelerates, thus reducing the primary arm spacing of dendrites. The results show that under the extremely high-temperature gradient and cooling rate of selective laser melting, the microstructure characteristics of cellular eutectic produced by strong non-equilibrium solidification will be greatly affected by the scanning speed and laser power.

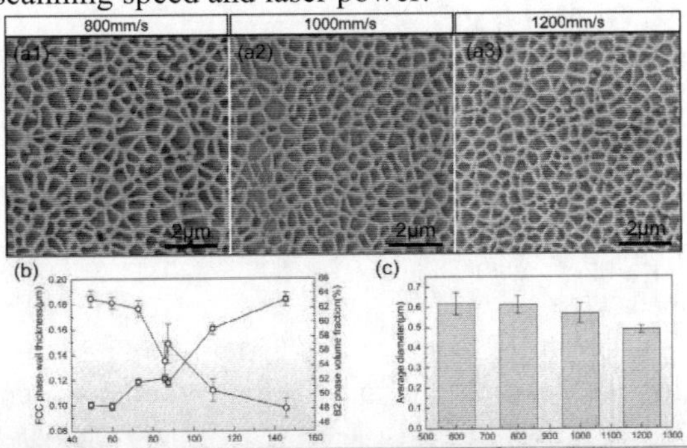

Fig. 1(a1-a3) showthe morphology of cellular eutectic at scanning speeds of 800 mm/s, 1000 mm/sand 1200 mm/s; (b) The variation of FCC phase thickness and BCC phase volume fraction with VED; (c) The change of the average diameter of the cellular eutectic with the scanning speed.

According to the HADDF images at different scanning speeds shown in Fig. 2(a1-c1), the BCC precipitates all show spherical and noodle-like morphology. Compared with the enlarged dark-field illustrations, there is no obvious difference in the size of BCC precipitates, with a diameter of 8-20 nm. In the high-magnification HADDF images and its TEM-EDS maps shown in Fig. 2(a2-c2), Cr gradually enriched in the BCC phase with the decrease of scanning speed, reflecting that the precipitation of BCC phase is a process of spinodal decomposition.With the decrease of scanning speed, the time for circulating heat input by laser to the deposited material will be longer, which promotes the diffusion and enrichment of elements and eventually leads to the formation of BCC precipitation phase.

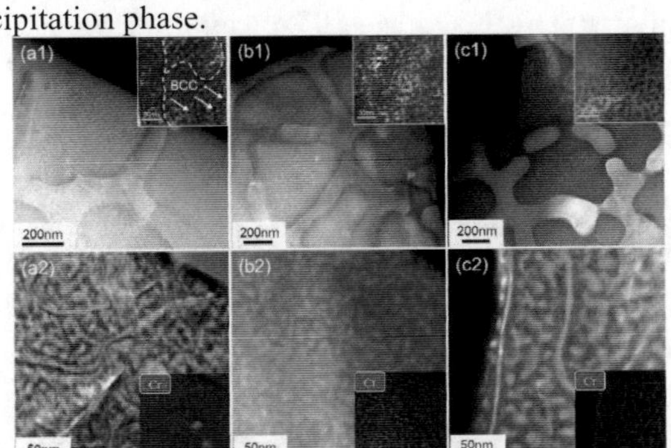

Fig. 2(a1-c1) show the HADDF images and the enlarged DF imagesof spinodal decompositionat the scanning speed of 1200mm/s, 1000mm/s and 800mm/s respectively; (a2-c2) HAADF images at high magnification and its EDS maps.

Study shows that dendrites and eutectic in eutectic alloys will compete for growth under different undercooling degrees[6]. At lower undercooling, the growth rate of eutectic is the fastest, so a complete eutectic structure can be formed. When the undercooling reaches a certain degree, the growth rate of dendrites exceeds the growth rate of eutectic, and the trend is more obvious with the increase of supercooling.In the rapid solidification process of laser selective melting, with large undercooling, B2 dendrites grow rapidly and become the main structure in the alloy. When the eutectic transformation volume of the remaining liquid phase is small, and its thickness is equivalent to the eutectic spacing, divorced eutectic may occur[5]. The B2 phase in eutectic is attached to the primary B2 dendrite and continues to grow, while the FCC phase is left alone between dendrites.

4 Conclusion

In this paper, the evolution behavior of the cellular structure of SLM EHEAs is studied by adjusting SLM process parameters. When the laser scanning speed is high, the cooling rate of the molten pool increases, and the size of the cellular structure is refined. With the decrease in scanning rate, the time of cyclic thermal loading of deposited materials by laser is prolonged, which promotes the amplitude modulation decomposition process of the BCC phase in cells.

References

[1] Lu, Y., Dong, Y., Guo, S., Jiang, L., Kang, H., Wang, T., Wen, B., Wang, Z., Jie, J., Cao, Z., Ruan, H. and Li, T. A Promising New Class of High-Temperature Alloys: Eutectic High-Entropy Alloys. Scientific Reports, 4, 1 (2014).

[2] Fu, Z., Zhang, Y., Li, Z., Lu, Y., Wu, X. and Wang, H. Hierarchically heterogeneous microstructure enables ultrahigh-strength and good ductility in selective laser melted eutectic high-entropy alloys. Journal of Materials Research and Technology, 24 (2023), 3300-3308.

[3] Geng, Z., Chen, C., Li, R., Luo, J. and Zhou, K. Composition inhomogeneity reduces cracking susceptibility in additively manufactured AlCoCrFeNi2.1 eutectic high-entropy alloy produced by laser powder bed fusion. Additive Manufacturing, 56 (2022).

[4] Guo, Y., Su, H., Yang, P., Shen, Z., Zhao, D., Zhao, Y., Liu, Y. and Zhou, H. New insight into tailorable eutectic high entropy alloys with remarkable strength–ductility synergy and ample shaping freedom fabricated using laser powder bed fusion. Additive Manufacturing, 60 (2022).

[5] Hu, L., Li, Y., Luo, G., Shi, Q., Wang, Y., Ma, S., Cui, Y., Wang, H. and Chen, Z. Understanding cellular structure formation in rapidly solidified Al–Ni alloys by constructing solidification microstructure selection maps. Additive Manufacturing, 73 (2023).

[6] Wang, H., Liao, H., Hu, L., Zheng, C., Chang, J., Liu, D., Li, M., Zhao, J., Xie, W. and We, B. Freezing Shrinkage Dynamics and Surface Dendritic Growth of Floating Refractory Alloy Droplets in Outer Space. Advanced Materials (2024).

Effect of Al Content on Microstructure and Mechanical Properties of As-Cast Mg-Gd-Y-Yb-Zn Alloys

Yamin Wang[1], Yicheng Feng[1,2*], Yuanke Fu[1],Lei Wang[1], Erjun Guo[1]
1. School of Materials Science and Chemical Engineering, Harbin University of Science and Technology, Harbin 150040, China
2. Key Laboratory of Advanced Manufacturing and Intelligent Technology, Ministry of Education, Harbin University of Science and Technology, Harbin 150080, China
*Corresponding address: e-mail: fyc7806067@163.com

Abstract: Grain refinement is an effective method to enhance the mechanical properties of cast Mg-RE alloys. Zr is the most commonly used magnesium alloy refiner, but the high cost and sedimentation of Zr refiner limits its application, and the Al refiner solves these determinations, and the grain refinement effect is almost the same as that of Zr. This paper study the effect of Al contents on the microstructure and mechanical properties of Mg-6Gd-2Y-2Yb-1Zn alloy, which were investigated by optical microscopy (OM), scanning electron microscopy (SEM), X-ray diffraction (XRD), and a universal tensile testing machine. The results of microstructure analyses showed that the addition of Al changed the morphology of second phase from single blocky phase to a combination of blocky, granular, needle-like, and plate-like phases. The addition of Al also refined the grain size of alloys, and with the increase of Al content from 0% wt. % to 2.0 wt. %, the average grain size was reduced from 300 μm to 41 μm. The results of mechanical properties tests showed that with the increase of Al content, the strength of alloys increased and then decreased, the optimum mechanical properties occurred at the Al addition of 2.0 wt. %. The yield strength, tensile strength, and elongation of alloy with 2.0 wt. % Al are 169.0 MPa, 265.0 MPa, and 23.75%, respectively, which improved 8.3%, 61.3%, and 125.9%, respectively compared to alloy without Al.

Keywords: Mg-6Gd-2Y-2Yb-1Zn-xAl alloy; Al refiner; Microstructure; Mechanical properties

1 Introduction

Are-earth (RE) magnesium alloys have recently trig-gered extensive research because of their high strength,corrosion resistance, heat resistance, and excellent cast-ing and mechanical properties [1-3]. Grain refinement often leads to a refinement of sec-ondary phases and a reduction in microsegregation and improves the ductility, strength, machinability, and cast-ability of alloys[4,5]. Mg alloys have a high Hall-Petch coefficient[6], thus grain refinement increases their strength considerably. Zr is a very powerful grain refiner for Al-free alloys[4], and it is often introduced through adding commercial binary Mg-Zr master alloys containing Zr particles ranging from submicrometer to about 50 lm in size[7]. The use of Mg-Zr master alloys has some disad-vantages. First, Mg–Zr master alloys are expensive and increase the cost of production[8]. Second, the utilization rate of Zr is low since the undissolved Zr particles spontaneously settle due to the much higher density of Zr (6.52 g/cm^3) compared to pure Mg melt (~1.58 g/cm3)[9].Therefore, cheaper and efficient grain refiners for these alloys are still under research. Recent studies revealed that additions of Al can lead to substantial grain refinement of Mg-Gd, Mg-Y and Mg-Gd-Y alloys by the in-situ formation of Al2RE (Al2Y, Al2Gd and Al2(Gd0.5Y0.5), respectively) particles acting as nucleation sites for a-Mg[10–12]. In the present study the effect of the Al content on the mechanical properties and refinement mechanisms of Mg-6Gd-2Y-2Yb-1Zn alloy.

2 Experimental procedure

Mg-6Gd-2Y-2Yb-1Zn-xAl (x = 0, 0.5, 1, 1.5, 2.0, 3.0%) alloy was prepared by pure Mg (>99.95%), pure Zn (>99.90%), Mg-30Y master alloy (~30%), Mg-30Gd mster alloy (~30%) and Mg-20Yb master alloy (~20%), pure Al (>99.90%). All alloys were melted in 20# steel crucible using an electric resistant furnace. Pure Al and pure Zn was added into the melt at 750°C, and the melt was stirred manually for 2 min, held for 30 min and then poured into a permanent mold (Φ60 × 120 mm) which was preheated to 200°C.

3 Result and discussion

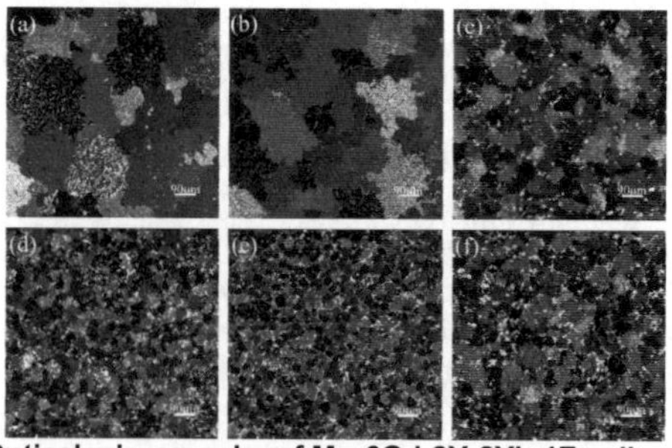

Fig.1 Optical micrographs of Mg-6Gd-2Y-2Yb-1Zn alloys with different Al additions: (a) 0%, (b) 0.5%,(c) 1.0%, (d) 1.5%, (e) 2.0% (f)3.0%

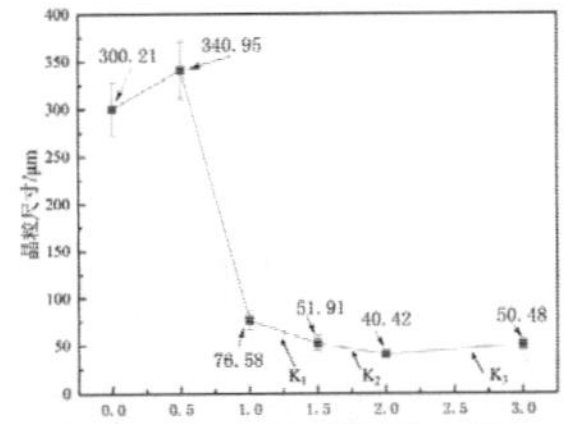

Fig.2 Grain size of Mg-6Gd-2Y-2Yb-1Zn alloys with different Al additions

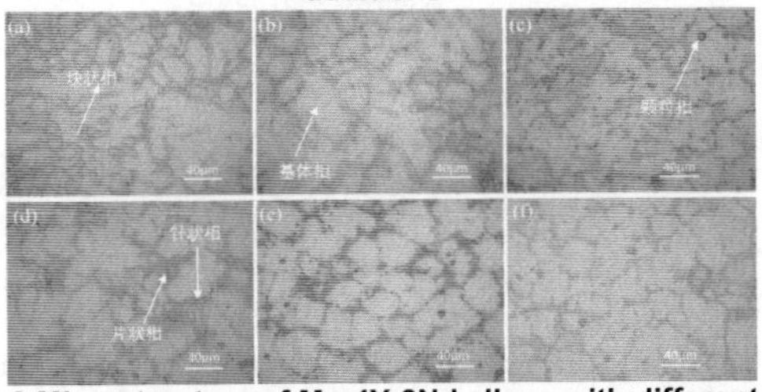

Fig.3 Microstructure of Mg-4Y-3Nd alloys with different Al additions: (a) 0%, (b) 0.5%, (c)1.0%, (d) 1.5%, (e) 2.0%, (f) 3.0%

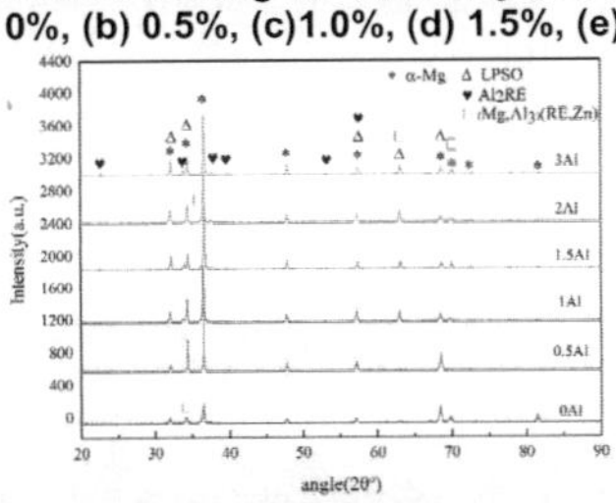

Fig.4 XRD patterns of Mg-6Gd-2Y-2Yb-1Zn alloys with different Al additions

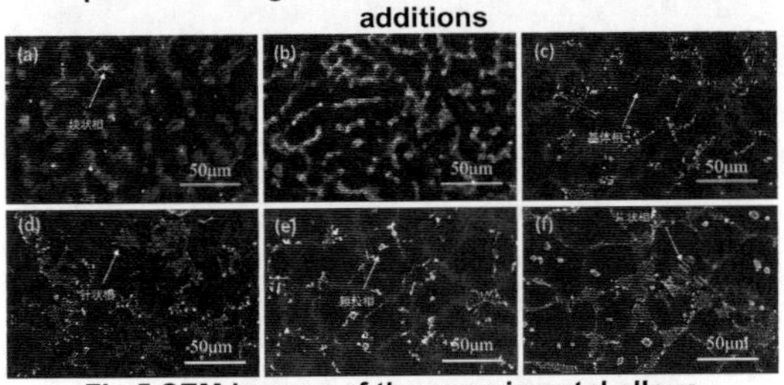

Fig.5 SEM images of the experimental alloys

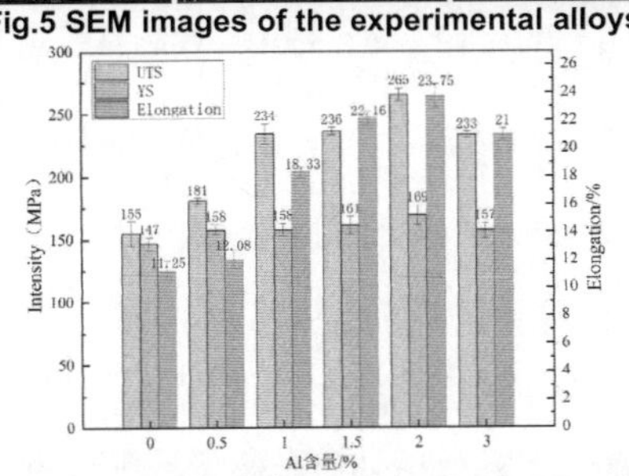

Fig.6 Mechanical properties of Mg-6Gd-2Y-2Yb-1Zn alloys refined different Al contents

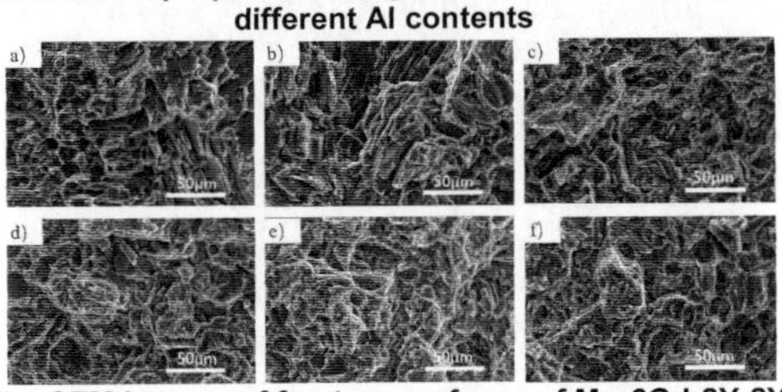

Fig.7 The SEM images of fracture surfaces of Mg-6Gd-2Y-2Yb-1Zn alloys refined by different Al contens

4 Conclusion

(1) Al addition can significantly refine the grain size of as-cast Mg-6Gd-2Y-2Yb-1Zn alloy, and the refining effect increases with the increasing Al addition.The in-situ formed Al2RE can act as a heterogeneous nucleation site for α-Mg. With the increase of Al content from 0% wt. % to 2.0 wt. %, the average grain size was reduced from 300 μm to 41 μm.

(2) The yield strength, tensile strength, and elongation of alloy with 2.0 wt. % Al are 169.0 MPa, 265.0 MPa, and 23.75%, respectively, which improved 8.3%, 61.3%, and 125.9%, respectively compared to alloy without Al.

(3) The tensile fracture surface of alloy without Al is consist of large cleavage facets and numerous tear ridges with cleavage steps. With increasing Al content, the size of the cleavage facets decreased until they disappeared, and the number of dimples increased.

References

[1] Pan H, Pan F, Wang X, et al. High conductivity and high strength Mg-Zn-Cu alloy. Mater Sci Technol. 2014, 30: 759-764.
[2] Wang C L, Dai J C, Liu W C, et al. Effect of Al additions on grainrefinement and mechanicalproperties of Mg-Sm alloys. J Alloy Compd. 2015 ,620: 172-179.
[3] Arrabal R, Mingo B, Pardo A, et al. Role of alloyed Nd in the microstructure and atmospheric corrosion of as-cast magnesium alloy AZ91. Corros Sci. 2015, 97: 38-48.
[4] Mert F, Özdemir A, Kainer KU, et al. Influence of Ce addition on microstructure and mechanical properties of high pressure die cast AM50 magnesium alloy. Trans Nonferrous Metal Soc China. 2013, 23: 66-72.
[5] Wu J, Chiu Y L, Jones I P. Effect ofGd on the microstructure of as-cast Mg-4.2Zn-0.8Y (at.%) alloys. J Alloy Compd. 2016, 661: 455-460.
[6] D.H. StJohn, M. Qian, M.A. Easton, P. Cao, and Z. Hildebrand:Grain refinement of magnesium alloys. Metall. Mater. Trans. A 36, 1669 (2005).
[7] K. Kubota, M. Mabuchi, and K. Higashi: Review processing and mechanical properties of fine-grained magnesium alloys. J. Mater.Sci. 34, 2255 (1999).
[8] C.J. Bettles, M.A. Gibson, and S.M. Zhu: Microstructure and mechanical behavior of an elevated temperature Mg-rare earth based alloy. Mater. Sci. Eng., A 505, 6 (2009).
[9] M. Qian, D.H. StJohn, and M.T. Frost: Heterogeneous nuclei size in magnesium-zirconium alloys. Scr. Mater. 50,1115 (2004).
[10] M. Qian, Z.C.G. Hildebrand, and D.H. StJohn: The loss of dissolved zirconium in zirconium-refined magnesium alloys after remelting. Metall. Mater. Trans. A 40, 2470 (2009).
[11] M. Qian, L. Zheng, D. Graham, M.T. Frost, and D.H. StJohn:Settling ofundissolved zirconium particles in puremagnesiummelts.J. Light Met. 1, 157 (2001).
[12] D. Qiu, M.X. Zhang, J.A. Taylor, P.M. Kelly, Acta Mater. 57 (2009) 3052-3059.

Influence of Ca on Thermal Expansion and Mechanical Properties of Mg

Cuihong Wang[1,2], Zhihua Dong[1,2*], Bin Jiang[1,2*]
1. National Engineering Research Center for Magnesium Alloys, Chongqing University, Chongqing, 400044, China
2. National Key Laboratory of Advanced Casting Technologies, Chongqing University, Chongqing, 400044, China
*Corresponding address: dzhihua@cqu.edu.cn (Dong Z). jiangbinrong@cqu.edu.cn (Jiang B)

Abstract: The influence of Ca on the thermal expansion and mechanical properties of Mg are investigated in the present work. Alloying Ca is demonstrated to noticeably decrease the coefficient of thermal expansion (CTE) and increase the strength of Mg. A low CTE of ~18.7×10^{-6} K^{-1} in combination with the high ultimate compression strength of 291 MPa are achieved within the studied concentration interval. The reduced CTE and enhanced strength of Mg-Ca alloys is primarily emerging from the increased volume fraction of Mg_2Ca phase showing the relatively high melting point, elastic/shear moduli and low CTE. The low thermal expansion and high strength achieved in the Mg-Ca alloys provide a feasible route for the development advanced Mg alloys.

Keywords: Mg alloys; thermal expansion; mechanical properties, alloying

1 Introduction

Magnesium (Mg) alloys are very promising light metal materials applied in aerospace, transportation, and electronic industries [1-3]. However, Mg alloys generally have relatively high coefficient of thermal expansion (CTE) and low mechanical properties. The large thermal stress and volume change emerging from the high CTE of Mg alloys may cause hot-cracking during solidification and cooling process when preparing large and complex Mg alloy structural components. It is thus important to develop low-CTE Mg alloys with the improved mechanical properties. While the mechanical properties of Mg alloys have attracted extensive attention during last two decades [2, 3], the study on the thermal expansion behavior remains scarce. Among the available reports, the precipitation of low-expansion second phase is effective to depress the CTE of the alloys [4-6]. For instance, Si was reported to efficiently reduce the CTE of Mg owing to the precipitation of low CTE Mg_2Si phase [6]. Nevertheless, the brittle dendritic Mg_2Si phase was deemed to deteriorate mechanical properties of the alloys.

As a cheap non-rare earth element, low CTE and good strengthening effect of Ca [7-9], the influence of Ca on the thermal expansion and mechanical properties of Mg are investigated. Addition of Ca is demonstrated to noticeably decrease the CTE of Mg in a wide temperature interval and increase the strength simultaneously. The present advance is intended to provide a feasible route for the development of low-CTE and high-strength Mg alloys.

2 Experimental procedure

The Mg-*x*Ca (*x* = 0, 4.1, 6.7, 10.1 and 12.4 wt.%) binary alloys were prepared in an electric resistance furnace using the commercial pure Mg (99.95 wt.%) and Mg-25 wt.% Ca master alloy. Second phases were characterized by a scanning electron microscope (SEM, JEOL JSM-7800F). Linear thermal expansion (dL/L_0) were measured using the NETZSCH DIL 402C dilatometer. The compressive mechanical properties were measured using the CMT6305-300 KN universal tensile testing machine.

3 Result and discussion

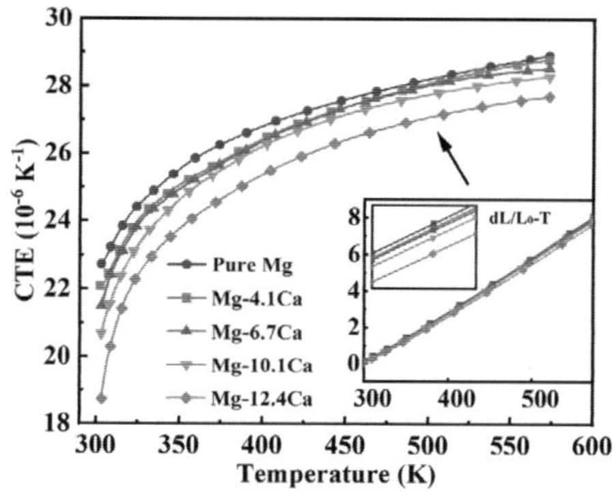

Fig.1 The CTE as a function of temperature

It is clear that linear thermal expansion (dL/L_0) increases with temperature while decreasing with Ca addition from the inserts of Figure 1. The CTE of Mg-Ca alloys gradually increase with temperature at a decreasing rate, which consists well with the Debye model [10]. As Ca content increased from 0 to 12.4 wt.%, the CTE noticeably decreases from ~22.7 to 18.7×10^{-6} K^{-1} at room temperature. At 573 K, the CTE slightly decreases from ~29 to 27×10^{-6} K^{-1} with Ca addition, indicating a rather weakened alloying influence at high temperature.

Figure 2 shows that with the addition of Ca, the compression yield strength (CYS) and ultimate compression strength (UCS) of the Mg-Ca alloy increases apparently. The CYS and UCS of Mg-12.4Ca alloy reach 252 and 291 MPa, respectively Nevertheless, the compression strain of Mg-Ca alloy gradually decreases with increasing Ca content.

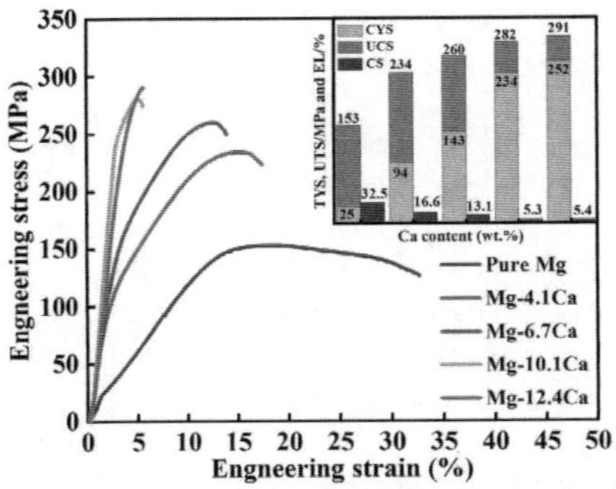

Fig.2 The compression curve of the Mg-Ca alloys

As mentioned above, Ca noticeably decreases the CTE of Mg and increase the strength simultaneously. To clarify the rationale behind, the physical parameters of pure Mg and Mg_2Ca phase are compared. It is found that Mg_2Ca phase exhibits the relatively high melting point, elastic/shear moduli, which are the physical quantities strongly correlating with the atomic bonding energy [11-13]. This results in a lower CTE for Mg_2Ca. Therefore, the increased volume fraction of Mg_2Ca phase induced by Ca addition (see Figure 3) will noticeably decrease the CTE of Mg-Ca alloys according to the rule-of-mixtures. In addition, Mg_2Ca phase is an efficient strengthening phase, the fraction of which noticeably increases with Ca addition, ultimately leading to the improved strength.

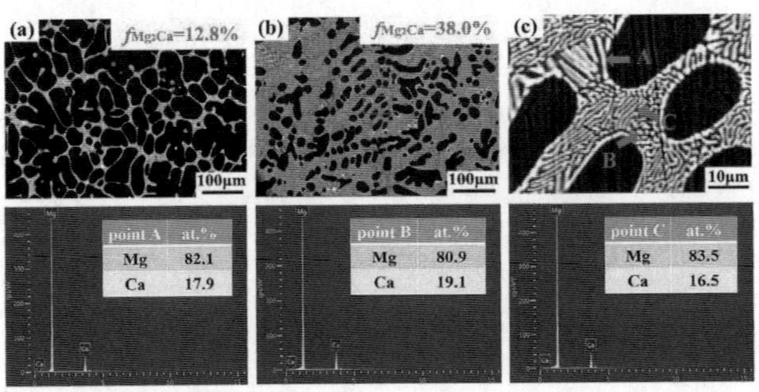

Fig.3 SEM-BSE images of the (a) Mg-4.1Ca and (b) Mg-12.4Ca alloys; (c) is the high magnification view for Mg-12.4Ca alloy

4 Conclusion

In summary, alloying Ca is demonstrated to noticeably decrease the coefficient of thermal expansion whilst increasing the strength of Mg. The lowest CTE of ~18.7×10^{-6} K^{-1} and highest UTS of 291 MPa are achieved at Ca content of 12.4 wt.%, respectively. The reduced CTE and enhanced strength of Mg-Ca alloys primarily emerge from the increased volume fraction of Mg_2Ca phase with Ca addition, which exhibits the relatively high elastic/shear moduli and low CTE.

Acknowledgments

The authors are grateful for the National Key Research and Development Program of China (No. 2021YFB3701000), the National Natural Science Foundation of China (52101126, U2037601, and U21A2048), and the China Postdoctoral Science Foundation (2021M700566).

References

[1] Wang X, Xu D, Wu R, Chen X, Peng Q, Jin L, et al. What is going on in magnesium alloys? J Mater Sci Technol 2018;34(2):245-247.

[2] Jayasathyakawin S, Ravichandran M, Baskar N, Chairman CA, Balasundaram R. Mechanical properties and applications of Magnesium alloy – Review. Mater Today Proc 2020;27:909-913.

[3] Bamberger M, Dehm G. Trends in the Development of New Mg Alloys. Annu Rev Mater Res 2008;38(1):505-533.

[4] Kumar A, Meenashisundaram GK, Manakari V, Parande G, Gupta M. Lanthanum effect on improving CTE, damping, hardness and tensile response of Mg-3Al alloy. J Alloys Compd 2017;695:3612-3620.

[5] Ma H, Li Z, Wang J. Effects of Y content on microstructures and thermal expansion behavior of Mg-Al-Y alloys. Mater Res Express 2021;8(7):076518.

[6] Guo T, Wu S, Zhou X, Lü S, Xia L. Effects of Si content and Ca modification on microstructure and thermal expansion property of Mg-Si alloys. Mater Chem Phys 2020;253:123260.

[7] Jiang Z, Jiang B, Yang H, Yang Q, Dai J, Pan F, Influence of the Al_2Ca phase on microstructure and mechanical properties of Mg-Al-Ca alloys. J Alloys Compd 2015;647:357-363.

[8] Zhang J, Tong L, Xu C, Jiang Z, Cheng L, Kamado S, et al. Influence of Ca-Ce/La synergistic alloying on the microstructure and mechanical properties of extruded Mg-Zn alloy. Mater Sci Eng, A 2017;708:11-20.

[9] Zhou P, Gong H. Phase stability, mechanical property, and electronic structure of an Mg-Ca system. J Mech Behav Biomed Mater 2012;8:154-64.

[10] Roder C, Einfeldt S, Figge S, Hommel D. Temperature dependence of the thermal expansion of GaN. Phys Rev B 2005;72(8):085218.

[11] Greeff CW, Moriarty JA. Ab initio thermoelasticity of magnesium. Phys Rev B 1999;59(5):3427-3433.

[12] Mao P, Yu B, Liu Z, Wang F, Ju Y. Mechanical, electronic and thermodynamic properties of Mg_2Ca Laves phase under high pressure: A first-principles calculation. Comput Mater Sci 2014;88:61-70.

[13] Zhang H, Wang S. A first-principles study on hot crack mechanism in Mg-Al-Ca alloys. J Mater Res 2012;27(12):1631-1637.

Quantitative Analysis of Graphite in Cast Iron Based on Deep Learning and Digital Image Processing

Pingping LI
CRRC Qishuyan Locomotive and Rolling Stock Technology Research Institute Co., Ltd., Changzhou 213011, Jiangsu, China

Abstract: The inspection of cast iron graphite currently faces several challenges, including multiple formulas, multiple fields of view, multiple types, multiple grades, and difficulties with manual assessment.The cast iron quantitative analysis module developed by the authors' team, as a multi-threaded, low-latency, non-blocking, and highly portable intelligent analysis system, has been successfully applied in several domestic enterprises. Through a simple operational process, accurate quantitative analysis and batch processing of images can be achieved, with results output in seconds.The successful implementation of the new standards and the development of the intelligent analysis system have greatly improved the detection efficiency and accuracy of results in the cast iron field, effectively promoting the high-quality development of the cast iron industry.

Keywords: Analysis of graphite; Deep Learning; Image Processing

1 Introduction

There are various classification methods for cast iron, generally categorized into five types based on the state of carbon in the iron, the morphological characteristics of graphite, and the performance characteristics of the cast iron: gray cast iron, ductile iron, vermicular iron, white cast iron, and malleable cast iron. Among these, the first three are the most commonly used in engineering applications and are widely used in industries such as mechanical manufacturing, metallurgy, mining, petrochemicals, and transportation.

In recent years, with the increasing requirements for cast iron technical indicators and the development of modern testing technologies, corresponding standards have been revised and improved to better guide and serve product production and inspection. A series of standards revised under the leadership of CRRC Qishuyan Institute Co., Ltd. (ISO 945-4-2019, GB/T 9441-2021, GB/T 7216-2023, GB/T 26656-2023) have not only clarified the definition of spherical graphite and standardized the boundaries and evaluation methods for spheroidization rate and vermicularization rate but also made field selection more reasonable. The application of these new standards has shifted from traditional visual judgment to image processing, changing the output from qualitative and semi-quantitative to quantitative, rendering manual visual methods inadequate for quantitative evaluation of cast iron graphite. In response to this situation, the authors' team developed a quantitative analysis module for cast iron graphite using deep learning and digital image processing technologies.

2 Experimental procedure

To perform quantitative analysis of graphite in cast iron, this study adopted the technical route shown in Figure 1, which mainly includes three parts: dataset construction, model training, and image post-processing. The following is a detailed introduction to each part.

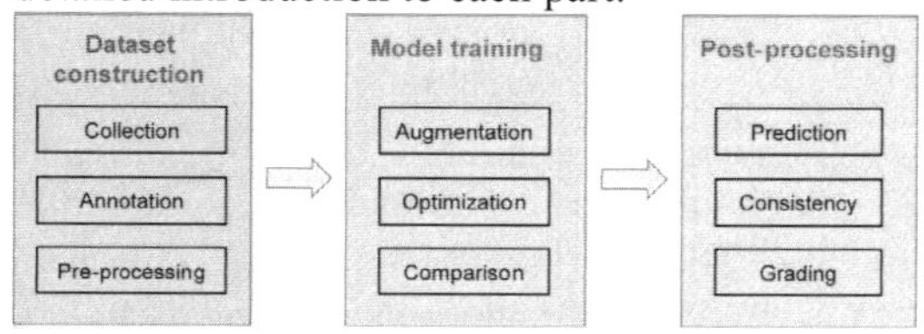

Figure 1. Main Technical Route of This Study

Data Construction

This study selected spherical, nodular, clumpy, vermicular, and flake (types A~F) graphite as the research objects. During the collection of metallographic images, special attention was paid to capturing various conditions such as different magnifications, lighting conditions, filter types, contrast, and the presence of impurities, while ensuring the inclusion of special samples containing multiple types of graphite. Ultimately, we screened out 3000 metallographic images with resolutions of 2580×1944 and 2752×2208. Figure 2 shows examples of several typical graphite morphologies.

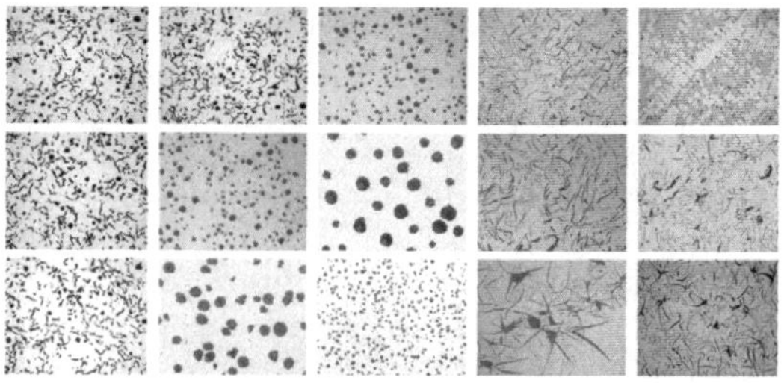

Figure 2. Examples of Original Metallographic Images

For the supervised training required by deep learning, we meticulously annotated the collected metallographic images, clearly marking the types of graphite. The annotation work was conducted under the guidance of experts with rich experience in metallographic image analysis, ensuring the accuracy and consistency of the annotations. To meet the needs of model training, we further binarized, skeletonized, and dilated the annotated metallographic images to generate binary label maps containing only graphite and background. Figure 3 shows examples of original metallographic images and their corresponding annotations.

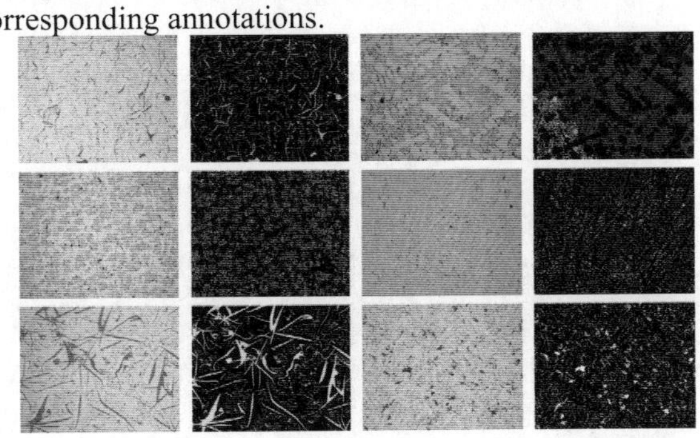

Figure 3. Examples of Metallographic Image Annotations Model training

Currently, many deep learning network models can be used for cast iron graphite analysis tasks, including Fully Convolutional Networks (FCN), DeepLab, U-net, and its variants. U-net, designed specifically for image segmentation, consists of an encoder (for feature extraction) and a decoder (for mapping features back to the original size), forming a U-shaped structure. The network structure of U-net adopted in this study is shown in Figure 4, with the encoder on the left and the decoder on the right.

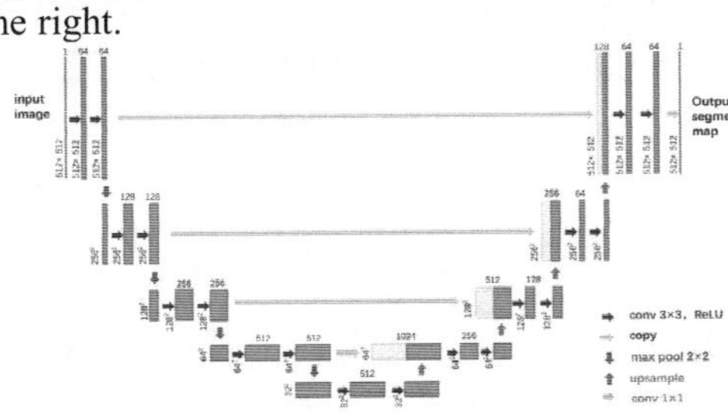

Figure 4. Schematic Diagram of the U-net Network Structure Used in This Study

When training the segmentation model using the metallographic image dataset, we divided the data into training, validation, and test sets in a 6:2:2 ratio, using stratified sampling to ensure data distribution uniformity.

1. Image Post-Processing

To ensure prediction accuracy, when using the above-mentioned optimal model to predict new cast iron metallographic images, it is necessary to preprocess them in the same way as the training images, and the predicted images must undergo graphite consistency processing before being used for subsequent rating analysis.

During rating, the contours and sizes of all graphite are calculated first, and graphite that is too small (e.g., less than 10μm) or located at the edge of the image is excluded according to the standard. Then, the corresponding graphite contours are selected based on the type of graphite and subjected to size statistics and grade calculation. Finally, the analysis results of all graphite are integrated to achieve quantitative rating of the entire image.

3 Result and discussion

The inspection of cast iron graphite currently faces several challenges, including multiple formulas, multiple fields of view, multiple types, multiple grades, and difficulties with manual assessment. In arbitration situations, computer-aided quantitative rating can effectively assist inspection engineers in grade determination.

The cast iron quantitative analysis module developed by the authors' team, as a multi-threaded, low-latency, non-blocking, and highly portable intelligent analysis system, has been successfully applied in several domestic enterprises. Through a simple operational process, accurate quantitative analysis and batch processing of images can be achieved, with results output in seconds (less than 1 second). Figure 5 shows examples of quantitative analysis of graphite in ductile iron, compacted graphite iron, and gray iron.

4 Conclusion

The inspection of cast iron graphite currently faces several challenges, including multiple formulas, multiple fields of view, multiple types, multiple grades, and difficulties with manual assessment. In arbitration situations, computer-aided quantitative rating can effectively assist inspection engineers in grade determination.

The cast iron quantitative analysis module developed by the authors' team, as a multi-threaded, low-latency, non-blocking, and highly portable intelligent analysis system, has been successfully applied in several domestic enterprises. Through a simple operational process, accurate quantitative analysis and batch processing of images can be achieved, with results output in seconds (less than 1 second). Figure 5 shows examples of quantitative analysis of graphite in ductile iron, compacted graphite iron, and gray iron.

In conclusion, the successful implementation of the new standards and the development of the intelligent analysis system have greatly improved the detection efficiency and accuracy of results in the cast iron field, effectively promoting the high-quality development of the cast iron industry.

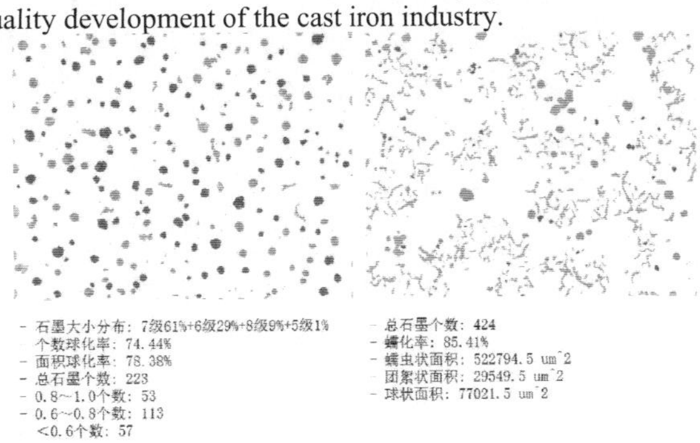

Figure 5. Examples of Graphite Quantitative Analysis

Research on Double Stage Solid Solution/Aging Heat Treatment Process of High Strength AlCu5Mn Alloy

Xiangping LI, Juncheng Ma, Shaohua XIA, Jialin GONG,Yongxiang Xu
(CRRC Qishuyan Locomotive and Rolling Stock Technology Research Institute Co., Ltd., Changzhou City, Jiangsu Province, 213011)

Abstract: Single stage solid solution/aging cannot meet the needs of material properties in industrial applications. In or der to further improve the mechanical properties of AlCu5 Mn alloy, a dual stage solid solution/aging heat treatment p rocess was adopted to study the influence of heat treatment process on the properties of AlCu5Mn alloy. Through ortho gonal experiments and range analysis, the optimal dual stag e solid solution process parameters were determined to be 5 28 ℃/2h,538 ℃/13h, 155 ℃/9h, 60 ℃/1h+175 ℃/8h. Und er this heat treatment process, the mechanical properties of AlCu5Mn alloy were optimal, with tensile strength reachin g 480MPa, yield strength reaching 385MPa, and elongation rate reaching 385MPa. Reached 14.5%, Brinell hardness re ached 142, tensile strength and yield strength increased by 3.7% and 7.5% respectively compared to singlestage solid s olution/aging heat treatment process. Compared with single -stage solid solution, two-
stage solid solution increases the supersaturation of the soli d solution without increasing the total content of alloy elem ents, while reducing the coarse undissolved crystalline phas e. This has a positive significance for improving the degree of aging precipitation and improving the fracture resistance, and is an effective way to improve the comprehensive perf ormance of alloys.

Keywords: High strength ; AlCu5Mn Alloy ; Double Stage Solid Solution; Double Stage Aging

1 Introduction

With the increasing demand for lightweight in the aerospac e, rail transit, and automotive industries. Lightweight and hi ghstrength aluminum alloys have become an inevitable tren d in the development of highstrength aluminum alloys in te rms of optimizing casting performance and heat treatment p rocesses. A1 Cu series alloys have excellent comprehensive properties. In order to further improve the performance and quality of aluminum copper castings, deepen further resear ch on highstrength aluminum alloy casting and heat treatme nt processes, and continuously study new technical theories, it will help improve the casting and heat treatment process es of highstrength aluminum alloys, and enhance the qualit y and performance of highstrength aluminum alloy compon ents [1-3].

2 Experimental procedure

The experiment used gravity metal casting method to pour AlCu5Mn single cast test bars, and the composition range o f the alloy was shown in Table 1. According to literature, w hen the solid solution temperature was 538 ℃ and the solid solution time was 15 hours, the lattice constant of Δ -Al is small. In order to further improve the mechanical prop erties of the alloy, it was only necessary to explore the twos tage solid solution process on the basis of the optimal singl e-stage solid solution process: 528 ℃/2h-538. ℃/13h. In order to obtain the optimal dual stage aging heat treatme nt process parameters, a dual stage factor level table as sho wn in Table 2 was designed, which includes three factors: p rimary aging temperature, secondary aging time, and secon dary aging temperature. Due to the fixed total aging time of 9 hours, the secondary aging time only needed to be subtra cted from the primary aging time. Four horizontal variables were taken for each experimental factor, and 16 sets of exp eriments are required.

Table. 1 Chemical composition of aluminum alloy (wt.%)

element	Cu	Mn	Ti	Cd	V	Zr	B
content	4.6 ~5. 3	0.3~ 0.5	0. 15 ~0 .3 5	0.1 5~ 0.2 5	0. 05 ~0 .3	0.1 5~0 .25	0.00 5~0. 6

Table.2 Double level timeliness factor level table

lev el	Experimental factors First level aging temperature T_1/℃	First level timeliness time t_1/h	Secondary aging temperature T_2/℃	Secondary level timeliness time t_2/h
1	60	1	145	
2	80	2	155	
3	100	3	165	t_2=9-t_1
4	120	4	175	

3 Result and discussion

Subheads 1

After comparing the singlestage solid solution and two-stage solid solution treatments, the test bar showed slight o verburning and severe pinhole defects in the tissue. After th e twostage solid solution treatment, the overburning pheno menon improved and the pinhole degree level was reduced from level 3 to level 1, as shown in Figure 1.

Single stage solid solution Secondary solid solution

Figure.1The tissue state and pinhole degree level after different solid solution processes

The mechanical properties of the test rod after single-stage and twostage solid solution treatment are shown in Table 3. As shown in the table, the secondary solid solution process can enhance the tensile strength and specified plastic elongation strength by 2.1% and 6.8%, respectively; The elongation will decrease to a certain extent, by 12.8%; The hardness value has little effect.

Table. 3 Mechanical properties after treatment with different solid solution processes

Solid solution process	Tensile strength σ_b MPa	Yield strength $\sigma_p 0.2$ MPa	Elongation A %	Brinell hardness HBW
Single stage solid solution	470.25	361.5	16.63	142.5
Secondary solid solution	480.33	386.33	14.5	142.67

Subheads 2

The range analysis results of tensile strength are shown in Figure 2. From the graph, it can be seen that the tensile strength of AlCu5Mn aluminum alloy slowly decreases with the increase of the first stage aging temperature and time, and sharply increases with the increase of the second stage aging temperature. The degree of influence of various factors on tensile strength: secondary aging temperature>primary aging temperature>primary aging time. The target process for obtaining maximum tensile strength is: 60 ℃/1h+175 ℃/8h; The range analysis results of yield strength are shown in Figure 3. From the graph, it can be seen that the yield strength of AlCu5Mn aluminum alloy slowly decreases with the increase of first order aging temperature and first order aging time, and sharply increases with the increase of second order aging temperature. The degree of influence of various factors on yield strength is as follows: secondary aging temperature>primary aging temperature>primary aging time. The target process for obtaining maximum yield strength is: 60 ℃/1h+175 ℃/8h.

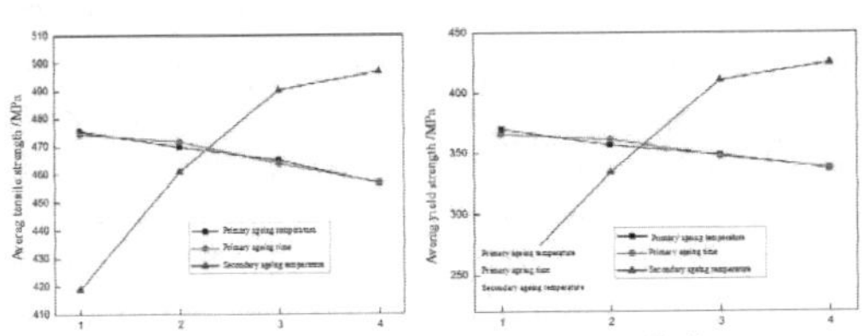

Figure.2/3Effect curve of tensile strength/yield strength factor indicators

The range analysis results of elongation are shown in Figure 4. From the graph, it can be seen that the elongation of AlCu5Mn aluminum alloy slowly increases with the increase of first stage aging temperature and first stage aging time. However, with the increase of second stage aging temperature, there is no significant change at first, and it sharply decreases between 80℃ to 100 ℃, and slowly decreases between 100℃ to 120 ℃. The degree of influence of various factors on elongation: secondary aging temperature>primary aging time>primary aging temperature. The target process for obtaining maximum elongation is: 120 ℃/4h+145 ℃/5h; The range analysis results of hardness are shown in Figure 5. From the graph, it can be seen that the hardness of AlCu5Mn aluminum alloy does not show a significant change with the increase of the first level aging temperature; Slowly decrease with the increase of the first level aging time; As the secondary aging temperature increases, it first increases sharply and then slowly. The degree of influence of various factors on hardness: secondary aging temperature>primary aging temperature>primary aging time. The target process for obtaining maximum hardness is: 100 ℃/1h+155 ℃/8h. Overall, our optimal process for two-stage aging is: 60 ℃/1h+175 ℃/8h.

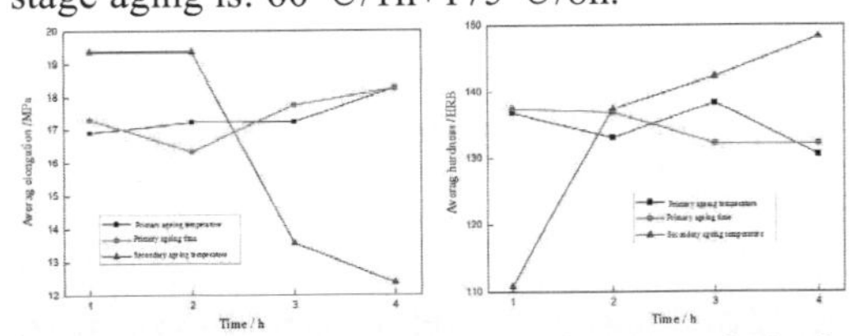

Figure.4/5Effect curve of elongation/hardness factor indicators

4 Conclusion

1. Double stage solid solution can improve the properties of AlCu5Mn alloy. The hardness value has little effect,and the optimal two-stage solid solution process is 528℃/2h,538℃/13h; the optimal two-stage aging process for AlCu5Mn alloy was determined to be 60 ℃/1h+175 ℃/8h;
2. The tensile strength of the alloy under the optimal two-stage solid solution/aging process is 501 MPa, the yield strength is 431 MPa, the elongation is 10%, and the Brinell hardness is 151.

References

[1] Qiu Shuwen Research and analysis on heat treatment technology of high-strength aluminum alloys [J] Association for Science and Technology Forum (second half of the month), 2019, (8): 34-36
[2] Zhou Zhenghai, Zheng Pengfei, Zhao Lei High strength aluminum alloy casting and heat treatment process [J] China High tech Zone, 2019 (15): 151
[3] Shen Guozhu The research status and development trend of high-strength aluminum alloys [J] Technology and Economic Market, 2018 (3): 30

Divorced Growth of Eutectic Solidification of Hypereutectic Al-Si Alloy in Sand Casting and Its Influence on Mechanical Properties

Hengcheng Liao[1], Tianrui Yan[1], Yuna Wu[2]
1. School of Materials Science and Engineering, Southeast University, Nanjing 211189, China
2. College of Materials Science and Engineering, Hohai University, Nanjing 210098, China
*Corresponding address: e-mail: hengchengliao@seu.edu.cn

Abstract: In this study, divorced growth of eutectic solidification in Al-18Si-5Cu alloy was successively achieved in sand casting. The pouring temperature plays an important role in degree of divorce. At 720°C, the eutectic solidification isn't divorced, and at 700°C it is basically divorced, but completely at 680 and 650°C, companying with significant refinement of Si particles and considerable improvement in its shape factor. The divorced growth of eutectic significantly improves the mechanical property both in ultimate tensile strength (UTS) and elongation (EL). In as-cast state, compared with the un-divorced alloy, the values of UTS and EL of the completely divorced alloy are increased by 60.0% and 92.3%, and, in T6 state, by 113.9% and 64.7%.

Keywords: Hypereutectic Al-Si Alloy, Divorced Growth of Eutectic Solidification, Microstructure, Mechanical Property

1 Introduction

Hypereutectic Al-Si alloys, such as s A390 and ZL108, have been used in Engine cylinders, pistons and other parts due to excellent castability and corrosion resistance, good strength and wear resistance, and low thermal expansivity and excellent dimensional stability. The smaller the size, the higher the shape factor of primary silicon particles, the better the mechanical properties of the alloy. Refining the primary Si particles has been a necessary practice, such as by modification, semi-solid casting, rapid solidification and melt superheating treatment. however, the size of primary Si particles is not enough small and the keen-edged eutectic Si flakes still exist.

In this study, divorced growth of eutectic solidification in Al-18Si-5Cu alloy was investigated in sand casting and the mechanical property was examined.

2 Experimental procedure

The nominal Al-18Si-5Cu-0.1P alloy was prepared in a crucible resistance furnace using Al-24Si, Al-50Cu and Cu-7P master alloys and commercial purity Al ingot. After holding at 850°C for 2h, the melt was cooled to 720°C for de-gassing and de-slagging, and then to the designed temperatures (720, 700, 680, and 650 °C) and holding for 10-15min. Finally, the melt was gravity-poured into a sand mold with a cavity of $220\times100\times15mm^3$.

Metallographic samples ($20\times10\times10mm^3$) and tensile specimens (with a gauge size of $18\times3\times3mm^3$) were cut from the cast plates. Some were experienced a T6 heat treatment (520℃/8h + 180°C/5h). After eroded in Killer reagent, microstructure was examined by optical microscope and SEM. Tensile test was carried out at room temperature with a rate of 1mm s^{-1}. Three specimens were tested to obtain the average of mechanical property.

3 Result and discussion

Microstructure

Fig. 1 shows the microstructure of Al-18Si-5Cu-0.1P alloy cast in sand mold at different temperatures. At 720°C of pouring temperature, the primary Si particles in Fig. 1 (a)is refined, due to the addition of P. But a great amount of eutectic Si flakes or needles exist in microstructure. In this case, the eutectic growth is coupled, not divorced. However, when the pouring temperature was at 700°C, the amount of Si particles is significantly increased and the size is greatly refined, as shown in Figure 1(b). Very few Si flakes are observed locally. Thus, the eutectic solidification is basically divorced. At 680°C and 650°C, no Si flakes are observed and Si particles become more and smaller, as seen in Figures 1(c-d). The black-grey filiform structure in Figures 1(b-d) was examined by SEM. It is β-Fe phase (demonstrated by Figure 2) along with eutectic θ-Al_2Cu phase. In these two cases, the eutectic solidification is completely divorced.

When pouring at 720°C, the superheat degree of melt is large. As the melt entered into the temperature range of primary Si phase, the chilling effect of sand mold is greatly decreased that results in a low nucleation rate of primary Si crystal. Divorced growth of eutectic solidification requires that: at the stage of eutectic reaction ($L\rightarrow\alpha$-Al+Si), Si atom should epitaxially grow on the surface of Si crystals which have formed at the filling and subsequent cooling stages. In the case of pouring at 720°C, before the melt enters the temperature range of the eutectic reaction, the formed Si crystals aren't abundant that makes the melt oblige to be undercooled to lower temperature where normal and coupled eutectic nucleation and growth of α-Al+Si phases occur, thus, flake-like or needle-like eutectic Si phase is formed.

With the decrease of pouring temperature, when the melt went thorough the runner and mold walls, the chill-cooling nucleation effect of primary Si phase at the filling stage was greatly enhanced. A great amount of fine Si crystals have formed in the melt. At the subsequent cooling stage, growth of these Si crystals in the temperature range of primary Si phase rejectes Cu solute in front of liquid at the liquid/solid interface, which causes consitutional supercooling that leads to the continous formation of new Si crystals. It is also important for the divorced eutectic growth. When the liquid enters the temperature range of the eutectic reaction, there are enough abundant and fine Si crystals in liquid. At the same time, the mold walls havre been heated and the cooling ability has been weakened that can't provide the required undercooling for the coupled nucleation of eutectic. Thus, in this case, the Si atoms in liquid prefer to grow on the pre-existed Si crystals epitaxially, leaving Al atoms along. In this way, divorced growth of eutectic occurs. At 680°C and 650°C, the chill-cooling nucleation effect becomes more great, thus the amount of Si particles becomes more and the size becomes smaller.

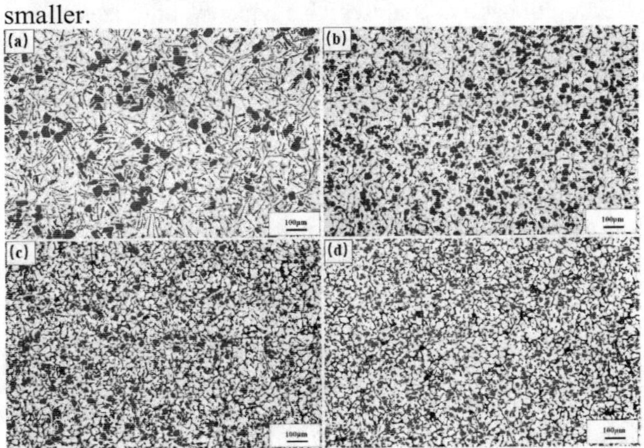

Fig. 1Microstructure of Al-18Si-5Cu-0.1P alloy cast in sand mold at different pouring temperatures: (a) 720oC; (b) 700oC; (c) 680 oC; (d) 650 oC

4 Mechanical Property

Table 1 lists the obtained data of tensile mechanical property. In both as-cast and T6 states, divorcing of eutectic growth of hypereutectic Al-18Si-5Cu-0.1P alloy results in a significant simultaneous improvement in strength and plasticity. In as-cast state, UTS of the un-divorced sample (at 720°C) is 110MPa and EL is 1.3%, however, UTS and EL of the completely divorced sample (at 650°C) are 176MPa and 2.5%, respectively, significantly increased by 60.0% and 92.3%. As seen in Figure 1, the divorced eutectic growth greatly refines Si particles, from 39.17μm (in sample at 720°C) to 16.82μm (in sample at 650°C) and the shape factor of Si particles is greatly improved from 0.64 to 0.78. At the same time, it makes needle-like eutectic Si disappear completely. During tensile deformation, spiculate edges of primary Si particles and eutectic Si needles severely cut the matrix and cause severe stress concentration that leads to a quick failure, thus the un-divorced sample has lower strength and elongation. After the divorcing of eutectic growth, the eutectic Si needles completely disappear and Si particles are significantly refined and rounded. The cutting effect is greatly reduced. Thus, the matrix can bear more deformation. Therefore, both the strength and elongation are greatly increased simultaneously.

Table 1 Data of Tensile Mechanical Property of Specimens of Al-18Si-5Cu Alloy in Diffirent States

Specimens		R_m /MPa	Elongation%
720 °C	As-cast	110±8	1.3±0.2
	T6	137±8	1.7±0.2
700 °C	As-cast	124±3	1.8±0.2
	T6	179±3	2.2±0.2
680 °C	As-cast	146±7	1.9±0.3
	T6	234±7	2.7±0.3
650 °C	As-cast	176±6	2.5±0.1
	T6	293±6	2.8±0.1

T6 treatment results in a great increase in both UTS and EL. The increase in strength is due to precipitation hardening effect of nano-size sub-stable particles of θ phase. The improvement in elongation might be related to the divorced eutectic growth. In T6 state, UTS and EL of the un-divorced sample (at 720°C) are 137MPa and 1.7%, respectively, and of the completely sample (at 650°C) are 293MPa and 2.8%, respectively. The divorcing effect of eutectic growth greatly improves the strength and elongation of hypereutectic Al-18Si-5Cu-0.1P alloy, by 113.9% and 64.7%, respectively.

Conclusion

(1) Divorced eutectic growth in Al-18Si-5Cu alloy was successively achieved in sand casting. The pouring temperature plays an important role in it. At 680 °C and 650°C, it is completely divorced, companying with significant refinement of Si particles and considerable improvement in its shape factor.

(2) The divorced eutectic growth significantly improves the mechanical property both in UTS and EL. In as-cast state, compared with the un-divorced alloy, the values of UTS and EL of the completely divorced alloy are increased by 60.0% and 92.3%, and, in T6 state, by 113.9%和 64.7%.

Recent Developments and Prospects on the Research and Application of Cast Magnesium Rare-Earth Alloys

Guohua Wu, Xin Tong, Wenjiang Ding
National Engineering Research Center of Light Alloy Net Forming and State Key Laboratory of Metal Matrix Composites,Shanghai Jiao Tong University, 800 Dongchuan Road, Min Hang, Shanghai, China
*Corresponding address: e-mail: ghwu@sjtu.edu.com

Abstract: Magnesium (Mg) alloy exhibits the characteristics of low density, high specific strength and stiffness, good damping capacity, etc. The addition of rare earth (RE) elements can effectively improve the creep resistance and corrosion resistance of Mg alloys, showing broad application prospects in aerospace, national defense, and military industries. This report systematically reviews the latest research progress of high-performance casting Mg-RE alloy materials. Based on the alloy characteristic, a comprehensive summary of the progress of the melt preparation of high-quality Mg-RE alloy is provided from the perspective of melt purification and grain refinement, and the impact of melt treatment on the preparation quality of Mg-RE alloys is discussed. Various kinds of casting processes of Mg-RE alloys are summarized, and the key technical difficulties and countermeasures in the casting and repair welding processes of Mg-RE alloys are concluded. Finally, the practical engineering applications of Mg-RE alloys in aerospace and other fields are introduced, and the key problems and development trends of Mg-RE alloy casting are prospected.

Keywords: Mg-RE alloys, high-performance, melt preparation,casting processes, application

Introduction

In recent years, magnesium rare earth (Mg-RE) alloys have received increasing attention because of their low densities, high specific strengths and creep resistances, good ignition-proof performance, and corrosion resistance, which exhibits wide application prospects in aerospace, military industry, railway, and 3C products. In comparison with other traditional magnesium alloys (Mg-Al, Mg-Zn alloys, etc.), Mg-RE alloys show different characteristics, which bring completely different alloying strategies, melt preparation, casting processes, and application scenarios [1]. For instance, precipitation strengthening contributes the most to the strength of cast Mg-RE alloys, so its alloying strategy should be developed based on a promoted aging precipitation behavior. The melt treatment (including purification and grain refinement) of Mg-RE alloys is more complicated since both the loss of RE elements with high chemical activities and the settling of the heterogeneous nucleation site of Zr should be given extra consideration. The die-casting process commonly used for Mg-Al alloy is not suitable for Mg-RE alloy, because it is difficult for the die-cast Mg-RE alloy to reach its full potential on precipitation strengtheningthrough heat treatment. Moreover, in the actual production, the repair welding of cast Mg-RE alloys is urgently required because casting defects are inevitable even in an upgraded casting process. Based on the above analyses, this paper summarizes some noteworthy aspects related to the cast Mg-RE alloys including alloy development, melt preparation, precision liquid processing, repair welding, and some practical applications.

Development of Mg-RE casting alloys

High-strength Mg casting alloys are generally considered to be alloys with tensile strength above 350 MPa, and they are mainly Mg-Gd series alloys with a high total RE content. The high-strength Mg-Gd alloys can be further divided into Mg-Gd-RE-Zr,Mg-Gd-Zn-Zr, Mg-Gd-Ag-Zr, Mg-Gd-RE-Zn-Zr, Mg-Gd-RE-Ag-Zr series. In the peak-aged Mg-Gd-RE-Zr alloys, the nanoscale β' precipitates with dense distribution are considered as the key strengthening phase, while some researchers also suggest that the coexistence of $\beta''+\beta'$ or $\beta'+\beta_1$ leads to the peak aging response. The β' phase precipitated in a triangular arrangement on $\{11\text{-}20\}_{\alpha\text{-}Mg}$ prismatic planes can serve as effective barriers to the basal dislocation slip, resulting in higher strength. The addition of Zn or Ag to Mg-RE series alloys can further promote the precipitation strengthening effect due to the introduction of the basal precipitates γ''. It has been reported that the cast Mg-15.6Gd-1.8Ag-0.4Zr exhibits an extremely high tensile strength of 423 MPa with an elongation of 2.6% [2].

Meltpreparation of Mg-RE alloys

The melt treatment is one of the key processes to determine the service performance of cast Mg-RE alloys, including grain refinement and melt purification.

Zr is the most effective grain refiner for casting Mg-RE alloy, which is always introduced into the melt by Mg-Zr master alloy. The existence type (undissolved particle or dissolved solute) and distribution of Zr within the melt have significant impacts on grain refinement efficiency [3]. Large plastic deformation and high-frequency re-melting were used as pretreatments to break up Zr clusters and increase the number of effective Zr nuclei in Mg-Zr master alloys. Moreover, it was found that the external electrical or ultrasonic field can also promote the dispersion and dissolution of Zr in the melt, resulting in a higher grain refinement efficiency.

The purification of Mg-RE alloy melt can be divided into flux purification, non-flux purification (e.g., filtration, gas bubbling), and compound purification. Up to now, flux purification is still the most effective and widely used method for Mg-RE alloys in industry. However, the RE elements usually react with $MgCl_2$ in the flux and cause RE loss. A new type of flux containing $RECl_3$ has been developed to reduce the RE loss. Moreover, it has been reported that both purification and grain refinement of Mg-RE alloy can be realized in only one step by flux refining using a new compound flux composed of $ZrCl_4$ and JDMJ purifying flux, and the melt preparation process is significantly simplified.

Casting process for Mg-RE alloys

The casting methods of Mg-RE alloy components include high-pressure die casting, low-pressure casting, squeeze casting, semi-solid forming, etc. Different casting methods contain different temperature distributions, solute fields, and melt flows during solidification, which plays an important role in the alloy microstructure. For example, die-cast Mg-RE alloy involves an extremely fast cooling rate, containing a grain size of only 3~10 μm. However, the properties of die-cast Mg-RE alloys cannot be further improved by heat treatments. The internal quality of low-pressure casting and squeeze casting of Mg-RE alloys is higher, so the advantages of aging hardening of Mg-RE alloys can be exerted by heat treatments. In practical production, the large-scale Mg-RE alloy components with complex structures used in aerospace are usually fabricated by a low-pressure sand-casting (LPSC) process. For instance, the thin-walled Mg-RE castings with a length of 1700 mm and thickness of 2 ~ 5 mm were successfully fabricated by LPSC [1]. The antigravity melt flow and applied pressure during solidification are beneficial to the surface quality, dimension accuracy and reduced shrinkage porosity.

Repair welding of Mg-RE alloy castings

In actual production of Mg-RE alloy castings, repair welding technologies is usually used to remove the casting defects, exhibiting significant value in reducing the rejection rate and production cost. Various welding methods including tungsten inert gas (TIG) welding, laser beam welding, and friction stir welding have been studied for repairing Mg-RE alloy castings. The casting defects are often located in complex structures such as reinforcing rib and lug boss, thus the TIG repair welding with good operability is the main method used in actual applications. However, it was found that abnormal grain coarsening is usually occurred in the TIG-repaired Mg-RE alloys during the conventional post-weld solution treatment, which can be alleviated by using pre-weld solution treatment or progressive post-weld solution treatment. After heat treatment, the joint efficiency of TIG-repaired Mg-RE alloy can reach up to 80%, and the elongation is even higher than that of the base material.

Applications of Mg-RE alloy castings

Mg-RE alloys have the characteristics of light weight, high creep resistance, good damping capability as well as good castability. Therefore, Mg-RE alloy castings are widely used in aerospace and aviation industries, weapon and defense industries, transportation and other civil fields. For instance, commercial Elektron® 43 (WE43C) and Electron® 21 (EV31A), are firstly been accepted for applications in jet engines and military aircrafts. Now, the Mg-RE alloy castings are also used as aircraft seats, shell of the reduction gearboxes, light missilescabin, radar components and car engine block.

Conclusion

This paper reviewed the recent developments of high-performance Mg-RE casting alloys, especially on alloy design, melt preparation, precision liquid processing and repair welding techniques, and engineering applications.

References

[1] Wu G H, Wang C L, Sun M and Ding W J.Recent developments and applications on high-performance cast magnesium rare-earth alloys. J. Magnes. Alloy., 2021, 9(1):1-20.

[2] Zhang Y, Wu Y J, Peng L M, Fu P H, Huang F and Ding W J.Microstructure evolution and mechanical properties of an ultra-high strength casting Mg–15.6Gd–1.8Ag–0.4Zr alloy.J. Alloy. Compd., 2014, 615:703-711.

[3] Wu G H, Tong X, Wang C L, Jiang R and Ding W J.Recent advances on grain refinement of magnesium rare-earth alloys during the whole casting processes: A review. J. Magnes. Alloy., 2023, 11(10):3463-3483.

Effect of Al Solution and Precipitation on Strength and Plasticity in Mg-Li-Al

Guoning Bai[1], Hua Hou[1,3], Yuhong Zhao[1,2,4*]
1 School of Materials Science and Engineering, Collaborative Innovation Center of Ministry of Education and Shanxi Province for High-performance Al/Mg Alloy Materials, North University of China, Taiyuan, 030051, China
2 Beijing Advanced Innovation Center for Materials Genome Engineering, University of Science and Technology Beijing, Beijing, 100083, China
3 School of Materials Science and Engineering, Taiyuan University of Science and Technology, Taiyuan, 030024, China
4 Institute of Materials Intelligent Technology, Liaoning Academy of Materials, Shenyang, 110004, China

*Corresponding address: e-mail: zhaoyuhong@nuc.edu.cn

Abstract: In the fields of aerospace and military defense, ultra-light Mg-Li alloy with sufficient strength are urgently needed to improve energy utilization. In this work, we investigate solid solution strengthening and precipitation strengthening of Al element in Mg-11Li alloy. Firstly, with Al content increase, the modulus and tensile strength of MgLiAl was significantly improved, but its effect on the elongation is not obvious. Secondly, the appropriate concentration (about 3~4wt.%) of Al element reduces the stacking fault energy of solid solution, which improve the plastic deformation ability of alloy. Finally, the precipitates with lower Li content or higher Al content resulted in higher strength. This work provided theoretical basis for the design of MgLiAl alloys.

Keywords: MgLiAl alloys; solid solution strengthening; precipitation strengthening; DFT

1 Introduction

Al is one of the most important alloying elements in Mg-Li alloys, which can improve the strength without impairment Sof density of the Mg-Li alloy obviously ,but sacrificed elongation, which due to its high solubility in Mg and low density [1]. Solid solution strengthening (SSS) and precipitation strengthening are commonly accepted as an efficient approach to improve the strength performance of Al in Mg-Li system [2]. After long-term aging at room temperature, Al element not only dissolves in the alloy matrix, but also gradually precipitates from the matrix in the form of spinodal decomposition, forming regularly distributed metastable clusters in the Al-rich region and coherent metastable phase $MgAlLi_2$, which can significantly improve the strength of the alloy [3]. In this paper, we employ first principles, cluster expansion, and molecular dynamics techniques to accurately calculate the separate influences of solid solution and precipitated phases on strength and plasticity The result provided a theoretical basis for the development of new ultra-light Mg alloys.

2 Experimental procedure

The Vienna ab initio simulation package (VASP) [4] based on density functional theory (DFT) has been used in first principles calculations, in which the local interactions between ions and electrons were modeled using the Projector Augmented-Wave (PAW) method. The plane wave energy cut-off of all elements was set to 500 eV. A Γ-centered k-point density equivalent to 0.02/Å per cell were determined by the Monkhorst-Pack scheme. The electronic self-consistent loops were considered converged when the difference of total energies between two steps fell below 1×10^{-6} eV per cell.

The maps program in the (Alloy Theoretic Automated Toolkit, ATAT) [5] was used to predict the phase structures in Mg-Li and Mg-Li-Al alloy systems over the entire composition interval.

3 Result and discussion

Table 1. shows the elastic modulus of solid solution formed after Al element solution into Mg-Li alloy. Bulk modulus, Young's modulus and Shear modulus all show an obvious increasing trend with the increase of Al element. Moreover, the tensile stress-strain results in Fig. 1(a) also show that the increase of Al content improves the tensile strength of the solid solution structure, but the effect on the elongation is not obvious.

Table 1. Lattice Constant a(Å), Formation energy ΔH(eV) and Bulk modulus B(GPa), Young's modulus E(GPa), Shear modulus G(GPa) of Mg-11Li-(1~5)Al solid solution

	a(Å)	ΔH(eV)	B(GPa)	E(GPa)	G(GPa)
Mg11LiAl	3.488	-4.93	43.08	11.11	3.81
$Mg11Li_2Al$	3.478	-4.91	43.56	14.07	4.86
$Mg11Li_3Al$	3.477	-4.88	43.31	15.25	5.29
$Mg11Li_4Al$	3.475	-4.85	43.18	16.37	7.79
$Mg11Li_5Al$	3.472	-4.83	44.08	18.05	7.99

Based on the most common {110}<111> slip system in BCC structure, a slip layer fault model was constructed, and the unstable slip and intrinsic slip layer fault energy of

BCC solid solution close-packed planes were calculated, so as to reflect the influence of Al element on alloy plasticity as shown in Fig. 1 (b).

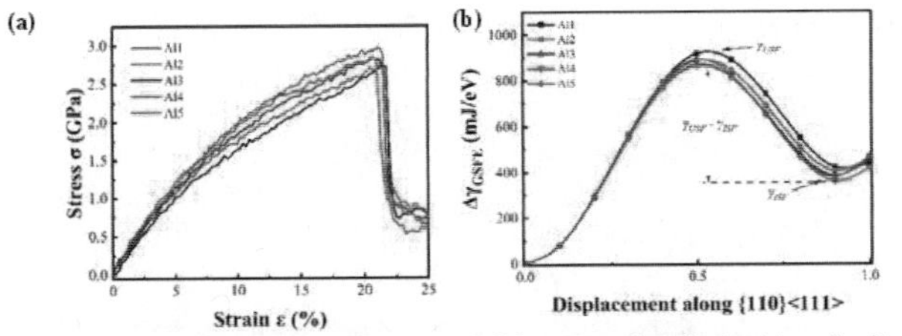

Fig. 1. (a) Stress–strain curve and (b) generalised stacking fault energy of BCC Mg-11Li-(1~5)Al solid solution

The results show that the stacking fault energy of γ_{ISF} and γ_{USF} does not change linearly with the Al content, and there is a minimum point in the Al content (3~4) wt.%, that is, the effect of Al element on the plasticity of solid solution structure is not linear. There is a certain deviation between this content and the optimal Al content of 5 wt.% in the experiment, which is mostly caused by the precipitation phase in the experiment.

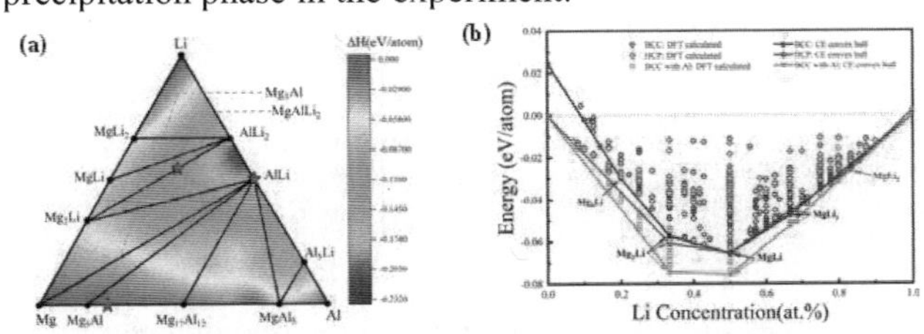

Fig. 2. Mg-Li-Al ternary convex hull at 0 K without zero-point energy(a). Stable competing phases (black-filled circles) are connected by tie lines, and the the metastable phase observed in the experiment are marked by colorful symbols

The cluster expansion method was used to predict the possible presence of precipitated phases in Mg-Li and Mg-Li-Al alloy, as shown in Fig. 2. And the elastic modulus of the precipitated phase is calculated based on the predicted structure as shown in Table 2. The precipitates with lower Li content or higher Al content showed higher strength, and the $Mg_{17}Al_{12}$, $MgAl_5$ and Mg_3Al phases showed obvious strengthening effect. However, the lattice structure of $Mg_{17}Al_{12}$ in I-43m structure and $MgAl_5$ in Cmmm structure does not have a good coherence with BCC Mg-Li. And it tends to split the lattice, affecting its integrity and harmful to the plasticity of Mg-Li alloy.

Table 2. Symmetry group, Lattice constant(Å), Formation energy(eV/atom), Bulk modulus (GPa), Young's modulus (GPa) and Shear modulus (GPa) of precipitated phase in Mg-Li-Al system

	Symmetry group	ΔH (eV/atom)	B (GPa)	E (GPa)	G (GPa)
$MgLi_2$	I4/mmm	-0.431	21.77	17.74	6.5
MgLi	Pm-3m	-0.062	26.23	52.20	22.34
Mg_2Li	I4/mmm	-0.056	30.15	53.39	22.15
Mg_3Al	Fm-3m	0.071	44.11	87.15	37.22
Mg_5Al	Cmmm	-0.298	41.55	60.05	23.84
$Mg_{17}Al_{12}$	I-43m	-0.009	50.01	47.25	17.59
$MgAl_5$	Cmmm	-0.0003	68.45	109.09	44.18
Al_5Li	P2/m	-0.047	66.51	76.96	29.43
AlLi	Fd-3m	-0.186	48.78	69.97	27.74
$AlLi_2$	Cmcm	-0.169	33.53	71.58	31.28
$MgAlLi_2$	Pmma	-0.143	34.21	58.48	24.06

4 Conclusion

(1) The Al addition can improve the strength of Mg-Li solid solution, but the effect on its elongation is not obvious. The effect of Al element on the plasticity of solid solution structure is not linear, The best results were achieved when the Al content was about (3~4) wt.%

(2) The modulus of the precipitated phase with high Al content is significantly higher.

(3) The effect of Al element on the alloy plasticity is partly due to the reduction of the stacking fault energy of the solid solution, and partly due to the strengthening of the coherency precipitation phase.

References

[1] Kalimullin R, Valuev V, Berdnikov A. The effect of surface laser treatment on the creep of the magnesium-lithium alloy MA21 [J]. Metal Science and Heat Treatment, 1986, 28: 668-70.

[2] Su H, Zhang C, Wang S, et al. Local atomic ordering strategy for high strength Mg alloy design by first-principle calculations [J]. Journal of Alloys and Compounds, 2022, 907: 164491.

[3] Mcdonald J. Age hardening of magnesium alloy LA 141 A(Mg-Li-Al alloy precipitation hardening mechanism as function of temperature from hardness and metallographic tests) [J]. AMS Transactions Quarterly, 1968, 61: 505-18.

[4] Kresse G, Furthmüller J. Efficient iterative schemes for ab initio total-energy calculations using a plane-wave basis set [J]. Physical review B, 1996, 54: 11169-86.

[5] Walle A, Asta M, Ceder G. The alloy theoretic automated toolkit: A user guide [J]. Calphad, 2002, 26:539-553

Phase Field Method to Study The Mechanism of Cu-Rich Phase Precipitation in AlxCuMnNiFe High-Entropy Alloy

Kaile Wang[1], Wenkui Yang[1], Xincheng Shi[1], Hua Hou[1,3], Yuhong Zhao[1,2,4*]
1 School of Materials Science and Engineering, Collaborative Innovation Center of Ministry of Education and Shanxi Province for High-performance Al/Mg Alloy Materials, North University of China, Taiyuan, 030051, China
2 Beijing Advanced Innovation Center for Materials Genome Engineering, University of Science and Technology Beijing, Beijing, 100083, China
3 School of Materials Science and Engineering, Taiyuan University of Science and Technology, Taiyuan, 030024, China
4 Institute of Materials Intelligent Technology, Liaoning Academy of Materials, Shenyang, 110004, China

*Corresponding address: e-mail: zhaoyuhong@nuc.edu.cn

Abstract: High-entropy alloys with coexisting BCC and FCC structures usually have excellent comprehensive mechanical properties, and Al and Mn elements can promote the transformation of Cu-containing high-entropy alloys from FCC to BCC structures. In this paper, based on Chan-Hilliard and Allen-Cahn equations, a three-dimensional phase field model of AlxCuMnNiFe high-entropy alloy was established to simulate the microevolution of nano-Cu-rich phases of AlxCuMnNiFe high-entropy alloys (x=0.4, 0.5, 0.6, and 0.7) during isothermal aging at 823K. The results show that the aging of AlxCuMnNiFe high-entropy alloys produces two kinds of complex core-shell structures: Cu-rich nuclei/B2s shells and B2c nuclei/FeMn shells. Combined with empirical formulas to predict the properties of the high-entropy alloys, the alloy compositions and aging times for optimal properties are obtained, and this work can provide new references for the process design of high-entropy alloys.

Keywords: Phase field method, high entropy alloy, Cu-rich phase, elastic energy.

1 Introduction

High-entropy alloys have high strength, high plasticity, high wear resistance, and high fatigue resistance[1].Phase field modelling can significantly reduce the experimental burden and provide a sustainable framework for alloy design [2].The element Cu has a tendency to promote liquid phase separation before solidification and precipitate to form a nano Cu-rich phase at the nanoscale. The element Al has a large atomic radius which leads to the transition from FCC to BCC structure, while the elements Al and Ni also have the potential to form ordered phases to improve the strength of the alloy. Therefore, it is proposed to use the phase field method to design a high entropy alloy with co-existence of ordered and disordered phases in a cost-effective manner.

In this paper, the phase field method is used to simulate the microstructure evolution of the Cu-rich phase of AlxCuMnNiFe high-entropy alloy. The formation process of nano-rich Cu phase by Al element is analysed, and the evolution curves of yield strength and aging time of the Cu-rich phase are calculated by combining with empirical formulae to predict the mechanical properties under different aging times, which provide reference for the process design of the alloy.

2 Phase field method

The phase field method is expressed by equations (1) and (2) [3]:

$$\frac{\partial c_i(\boldsymbol{r},t)}{\partial t} = \nabla \cdot M_{ij} \cdot \nabla \cdot \frac{\partial F}{\partial c_i(\boldsymbol{r},t)} + \xi_{c_i}(\boldsymbol{r},t) \#(1)$$

$$\frac{\partial \eta(\boldsymbol{r},t)}{\partial t} = - L_\eta \frac{\partial F}{\partial \eta(\boldsymbol{r},t)} + \xi_\eta(\boldsymbol{r},t) \#(2)$$

M_{ij} denotes the chemical mobility of the alloying elements, F is the total free energy, and L_η is the kinetic coefficient characterising the phase transition.

3 Result and discussion

Double core-shell structure

As shown in Fig. 1(a), we take the 3D simulation results of t*=20000 as the basis, and intercept the spherical and annular B2 phases, as well as the spherical short rod-shaped Cu-rich phase, respectively.

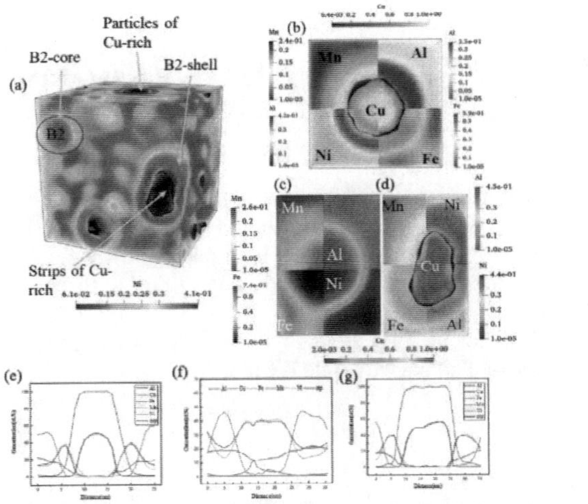

Fig. 1. Morphology of $Al_{0.7}Cu_{1.5}Mn_1Ni_1Fe_{1.5}$ high-entropy alloy at t*=20,000, (a) 3D simulation results; (b) Distribution of core-shell structure of Cu-rich nanophase; (c) Structural distribution of the B2 phase core and shell; (d) Short-rod Cu-rich phase and the elemental distributions; (e) Compositional curves of spherical Cu-rich phase; (f) Spherical B2 phase composition curve; (g) Short rod-shaped Cu-rich phase composition curve

Fig. 1(b) shows the 3D elemental distribution of a single spherical Cu-rich phase, and Fig. 1(e) shows the compositional profile of the line sweep along the central axis, which shows that the particle diameter of a single spherical Cu-rich phase is about 15 nm, and in the Cu-rich region, the Cu content is 100%, and surrounded by the Cu-rich NiAl shells with about 10 nm, and the NiAl content is up to 40%, with an elemental ratio of 1:1 for a B2 shell,. While Fe-Mn is distributed in the outermost layer, forming a spherical Cu-rich nano-core-shell structure with FeMn in the outermost layer, B2-NiAl shells in the middle and Cu in the core. The results of our simulations have similarities with those reported in the literature, as the Cu atoms tend to separate in clusters during the cooling process due to the small binding energy of Cu atoms with atoms of Al, Fe, Ni, etc., which leads to the formation of these FCC Cu-rich nano-precipitated phases. Fig. 1(d)(g) represents the morphology map and compositional profile of the Cu-rich nano phases in the form of short rods, with a formation mechanism similar to that of the spherical ones. Effect of Cu-rich relative mechanical properties of yield strength over time areshown in Fig. 2, and all four systems have two peak yield strengths, the peak yield strength of the nano-rich Cu phases of Al0.4 can reach 1,166 MPa at t*=4,500, and the comparison of the peak yield strengths of Al0.7 can be seen, with the increase of Al content, the nano-Cu-rich phase reaches the peak for a longer period of time, the peak yield strength decreases, and reaches 775 MPa at t* = 7500, and reaches the peak after the decline is obvious, may be the emergence of the B2-NiAl phase leads to the dissolution of Cu particles to form the Cu-depleted zone.

.

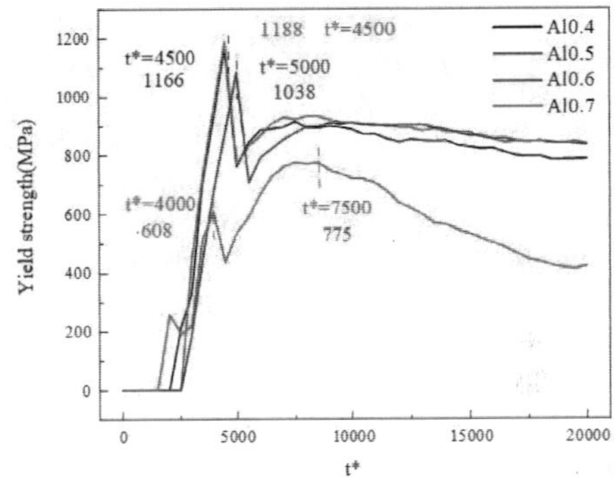

Fig. 2. Variation of yield strength of nano-Cu-rich phase with aging time during aging.

4 Conclusion

A double core-shell structure of nano-rich Cu and B2NiAl phases is generated during the formation of nano-rich Cu phases, and the statistical simulation results and the combination of empirical formulas can be used to obtain the yield strength over time.

Acknowledgments

This work was supported by the National Natural Science Foundation of China (Grant No. 52074246). Shanxi Graduate Innovation Project, China (Grant No. 2021Y592).

References

[1] Tsai C W, Chen Y L, Tsai M H, et al. Deformation and annealing behaviors of high-entropy alloy Al0.5CoCrCuFeNi [J]. J. Alloys Compd, 2009, 486(1-2): 427.

[2]Chen L Q, Zhao Y H. From classical thermodynamics to phase-field method[J]. Prog. Mater. Sc, 2021, 124: 100868.

[3]Zhao Y H, Sun Y Y, Hou H. Core-shell structure nano precipitates in Fe-xCu-3.0Mn-1.5Ni-1.5Al alloys: A phase field study[J]. Prog. Nat. Sci.: Mater. Int, 32(2022):358-368

BCC Precipitates Induced Strength-Ductility Balance in $Al_5Ti_{2.5}Fe_{25}Cr_{25}Ni_{42.5}$ High Entropy Alloy

Xiaotao Xu[1], Hua Hou[1,3], Yuhong Zhao[1,2,4*]
1 School of Materials Science and Engineering, Collaborative Innovation Center of Ministry of Education and Shanxi Province for High-performance Al/Mg Alloy Materials, North University of China, Taiyuan, 030051, China
2 Beijing Advanced Innovation Center for Materials Genome Engineering, University of Science and Technology Beijing, Beijing, 100083, China
3 School of Materials Science and Engineering, Taiyuan University of Science and Technology, Taiyuan, 030024, China
4 Institute of Materials Intelligent Technology, Liaoning Academy of Materials, Shenyang, 110004, China

*Corresponding address: e-mail: zhaoyuhong@nuc.edu.cn

Abstract: FCC-structured $Al_5Ti_{2.5}Fe_{25}Cr_{25}Ni_{42.5}$ high entropy alloy (HEA) was melted, cryo-rolled and annealing. The influences of cryo-rolling and annealing on microstructural evolution and mechanical properties were investigated. The cryo-rolled $Al_5Ti_{2.5}Fe_{25}Cr_{25}Ni_{42.5}$ HEA is comprised of HCP- and BCC-martensite induced by deformation, along with a residual FCC matrix. After annealing at 1000℃, the cryo-rolled $Al_5Ti_{2.5}Fe_{25}Cr_{25}Ni_{42.5}$ HEA completely recrystallized and precipitated 0.2 % BCC phase. The tensile strength of annealed $Al_5Ti_{2.5}Fe_{25}Cr_{25}Ni_{42.5}$ HEAs are higher than 700 MPa, and the elongation are more than 50 %. The better combination of strength and ductility of annealed $Al_5Ti_{2.5}Fe_{25}Cr_{25}Ni_{42.5}$ HEAs is due to grain boundary hardening and precipitate hardening.

Keywords: high entropy alloy, cryo-rolled, microstructural, precipitated, grain boundary hardening.

1 Introduction

High-entropy alloys (HEAs) have a nearly infinite multi-phase space due to their unique design concept, which makes it possible to design alloys with the required strength-ductility combination [1]. Single-phase FCC HEAs exhibit many excellent functional properties and are expected to become a new generation of engineering materials. However, single-phase FCC HEAs have obvious strength-ductility tradeoff, which rejects their practical application. An extremely effective method is to introduce the BCC phase into the FCC matrix to improve the mechanical properties of HEAs. There are two ways to introduce the BCC precipitates into the FCC structure HEA: (1) adjusting the additional Al and Ti elements to introduce the BCC phase into the FCC matrix, (2) precipitating the BCC phase during the heat treatment process [2] Compared with cold-rolled materials, the introduction of cryo-rolling and subsequent annealing in the thermomechanical processing route can significantly increase strength while helping to enhance or maintain ductility. Cryo-rolling has a better ability than cold rolling to refine grains and induce additional deformation nanostructures such as twins and martensite lamellae, which has been applied to achieve high performance in HEAs [3].

2 Experimental procedure

The $Al_5Fe_{25}Cr_{25}Ni_{42.5}Ti_{2.5}$ HEA used in the present study was prepared by arc-melting high-purity metal pellets (⩾ 99.9% purity) in an inert argon atmosphere. The button-shaped ingot (~ 40 g) was subjected to five cycles of re-melting and flipping to achieve enhanced chemical homogeneity, followed by suction casting into sheets with dimensions of 60 × 10 × 6 mm^3 and homogenized at 1250℃ for 2h, water-quenched. In the cryo-rolling process, the samples were immersed in liquid nitrogen bath before and after each pass. The initial alloy plate was subsequently cryo-rolled along the longitudinal direction with a thickness reduction from 6.0 to 1.5 mm (rolling reduction ~ 75%). The cryo-rolled samples were subsequently short-annealed at 1000 °C with varying times, followed by water quenching, hereafter denoted as 1000-5, 1000-10, 1000-60 samples, respectively.

3 Result and discussion

The EBSD results of $Al_5Fe_{25}Cr_{25}Ni_{42.5}Ti_{2.5}$ HEA samples after recrystallization annealing are shown in Fig.1 The IPF diagram in Fig.1(a_1-a_3) confirms that all annealed $Al_5Fe_{25}Cr_{25}Ni_{42.5}Ti_{2.5}$ HEA samples have fully recrystallized microstructures, including equiaxed grain structures rich in annealing twins (ATs), which are transformed from deformed grains in cryo-rolled $Al_5Fe_{25}Cr_{25}Ni_{42.5}Ti_{2.5}$ samples. The phase map in Fig.1(b_1-b_3) shows that after recrystallization annealing, there are still about 0.2 % BCC phase in the three samples, indicating that the BCC phase produced by cryo-rolling does not disappear due to recrystallization annealing. The corresponding EBSD statistics show that the average grain sizes of 1000-5,1000-10 and 1000-60 are 7.89 μm, 9.47 μm and 17.03 μm, respectively, and the AT are also regarded as HAGBs. In other words, the increase of annealing time has a significant effect on the grain size of the cryo-rolled and subsequently annealed $Al_5Fe_{25}Cr_{25}Ni_{42.5}Ti_{2.5}$ samples.

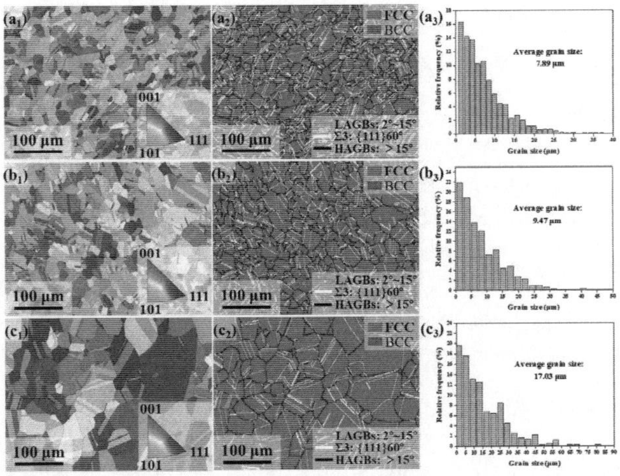

Fig.1 Microstructural evolution with increasing annealing time for (a) 1000-5, (b) 1000-10, (c) 1000-60. For phase maps, the gray line means the low-angle grain boundary, the black line means the high-angle grain boundary, and the white line means the ATs (Σ3, 60 ° < 111 >).

Fig.2(a) and (b) show the tensile engineering stress-strain curves and true stress-strain curves of annealed $Al_5Fe_{25}Cr_{25}Ni_{42.5}Ti_{2.5}$ HEA samples, respectively, revealing the strong dependence of mechanical properties on annealing time. Ultimate tensile strength (UTS) is the maximum stress that the sample is subjected to during tension or before fracture. The 1000-5 sample exhibits a combination of high strength and low ductility, low uniform elongation of ~ 53.9 % and high tensile strength of 847.1 MPa. With the increase of annealing time, the strength of the alloy decreases and the ductility increases, which is related to the grain coarsening of the alloy. Fig.2 (c) is the relationship curve between strain hardening rate and true strain. The three samples show three-stage characteristics in the tensile deformation process: in the first stage, the work hardening rate decreases sharply with the increase of true strain; in the second stage, with the increase of true strain, the work hardening rate decreases slowly, indicating that the material has undergone continuous work hardening. In the third stage, the work hardening rate decreases rapidly, indicating that the material enters an unstable state.

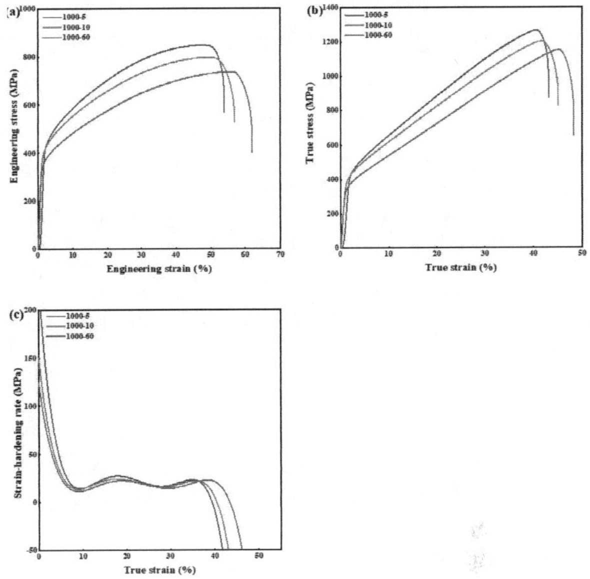

Fig.2(a) Engineering stress-strain curves; (b) True stress-strain curves and (c) Work hardening rate-True strain of 1000-5, 1000-10, 1000-60 samples, respectively.

4 Conclusion

1. After annealing at 1000 °C, completely recrystallized microstructures with different grain sizes were obtained, and all of them contained 0.2 % BCC phase.
2. The strength of $Al_5Fe_{25}Cr_{25}Ni_{42.5}Ti_{2.5}$ annealed HEAs decreases with the increase of annealing time, while UE decreases.

Acknowledgments

This work was supported by the National Natural Science Foundation of China (Nos. 52275390, 52375394), Key Research and Development Program of Shanxi Province (202102050201011, 2022ZDYF035)

References

[1] Su J, Raabe D, Li Z M. Hierarchical microstructure design to tune the mechanical behavior of an interstitial TRIP-TWIP high-entropy alloy[J]. Acta Materialia, 2019, 163:40-54.

[2] Mehranpour M S, Shahmir H, Kim H S. Microstructure tailoring by manipulating chemical composition in novel CoNiMnCrAl high-entropy alloys[J]. Journal of Alloys and Compounds, 2023, 944:169207.

[3] Won J W, Lee S W, Park S H, et al. Ultrafine-grained CoCrFeMnNi high entropy alloy produced by cryogenic multi-pass caliber rolling[J]. Journal of Alloys and Compounds, 2018, 742:290–295

Multiple Strengthening Mechanisms Lead to an Excellent Strength-Ductility Combination of Ni-Rich High-Entropy Alloy for Multiple Temperature Applications

Xiaoyang Zhang[1], Jinchen Yang[1], Huanqing Li[1], Ruifeng Dong[1], Hua Hou[1], [3], Yuhong Zhao[1,2,4*]
1 School of Materials Science and Engineering, Collaborative Innovation Center of Ministry of Education and Shanxi Province for High-performance Al/Mg Alloy Materials, North University of China, Taiyuan, 030051, China
2 Beijing Advanced Innovation Center for Materials Genome Engineering, University of Science and Technology Beijing, Beijing, 100083, China
3 School of Materials Science and Engineering, Taiyuan University of Science and Technology, Taiyuan, 030024, China
4 Institute of Materials Intelligent Technology, Liaoning Academy of Materials, Shenyang, 110004, China

*Corresponding address: e-mail: zhaoyuhong@nuc.edu.cn

Abstract: In this paper, a new high-entropy alloy with component of $Ni_{44}Fe_{19}Co_{16}Cr_9V_9Al_3$shows excellent strength and ductility across a wide temperature range (77 K to 1073 K). It achieves this through various strengthening mechanisms depending on temperature, including back stress, dislocation, coherent precipitates, stacking faults, and TWIP effects.Phase field simulation indicated that, besides the TWIP effect caused by the lower stacking fault energy, the subsequent deformation mechanisms, including the breakage of the wider twins into nanotwins, and the non-turning transfer of intracrystalline dislocations by the nanotwin boundaries, are important reasons for the high strength and high ductility of the alloy at 77 K.This study offers valuable insights into designing HEAs for diverse temperature applications.

Keywords: High-entropy alloy; Heterogeneous structure; Multiple strengthening; Elevated temperatures; Phase field simulation.

1 Introduction

HEAs are defined by unique properties such as the high-entropy effect, sluggish diffusion, lattice distortion, and cocktail effects, which contribute to their performance [1]. Contemporary research mainly focuses on enhancing strength or toughness within narrow temperature ranges, overlooking the potential of HEAs to maintain superior properties across a wide temperature spectrum, which is crucial for applications like outer space.

Reinforcement methods for HEAs include grain refinement, interstitial solid solution, phase precipitation, and twinning or transformation-induced plasticity (TWIP/TRIP) [2]. However, TWIP/TRIP effects are difficult to achieve at high temperatures due to dynamic recovery processes. Alternative strengthening mechanisms like heterogeneous structure and short-range ordering are also important.

Nickel-based alloys are favored for high-temperature applications due to their stable microstructure. This study presents a nickel-rich HEA ($Ni_{44}Fe_{19}Co_{16}Cr_9V_9Al_3$) designed for excellent high-temperature performance, with low stacking fault energy and a heterogeneous structure achieved through controlled heat treatment. The alloy shows superior strength and ductility across a wide temperature range (77 K to 1073 K) and demonstrates the feasibility of multiple strengthening mechanisms for HEAs.

Experimental procedure

High-purity metal powders were melted in a vacuum furnace under high-purity argon atmospheres, then the alloys were homogenized at 1000°C for 4 hours. 70% thickness reduction cold rolling was performed, followed by static recrystallization at 800°C for 30 minutes with water quenching to room temperature. Tensile mechanical properties from low temperature to high temperature were tested on an electronic universal material testing machine with a tensile rate of 1.0×10^{-3}/s. Cyclic loading-unloading tensile tests were conducted at 77 K and 293 K. The microstructure and crystal orientation were analyzed using a Gemini 300 field emission scanning electron microscope. The dislocation/twin substructures were characterized using a FEI Talos F200X transmission electron microscope. The crystal phase field method is used to construct a polycrystalline system containing twins and discuss the response of the polycrystalline system to external forces.

Result and discussion

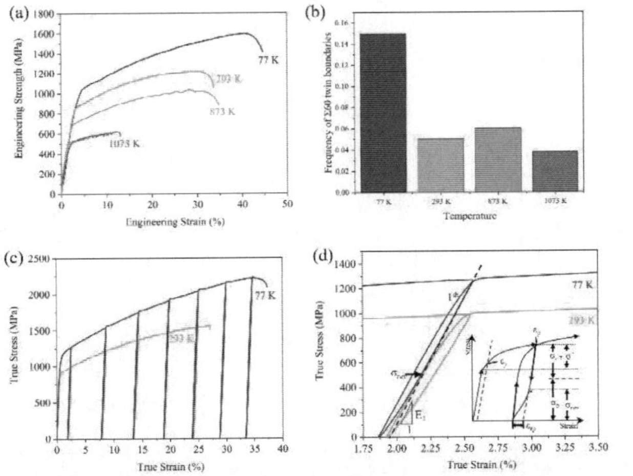

Fig. 1(a) The tensile curves showing engineering stress versus strain; (b) Frequency of Σ60 twin boundaries after tensile failure at different temperatures; (c) True stress-strain curves from loading-unloading-reloading (LUR) tests; (d) An enlarged view of two LUR cycles.

Heterostructure and twin induced strength

The engineering stress-strain curves is depicted in Fig. 1(a), indicating that the HEA maintains considerable tensile strength and ductility from 77 K to 1073 K. In Fig. 1(b), the proportion of twin boundaries at the three temperatures is the same while the proportion of twin boundaries at 77 K is significant large, indicating that many deformation twins are generated during the deformation process, and TWIP occupies the dominant position of deformation, which strongly suggest that the formation of planar faults plays a vital role in achieving enhanced strength at 77 K.Additionally, loading-unloading-reloading (LUR) tests were conducted across different strain-hardening stages to explore the Bauschinger effect.

Deformation mechanisms at 77 K

Fig. 2(a) portrays a bright-field image of multiple coexistence of recrystallized and deformed grains of alloy after deforming to failure at 77K. Dislocations glide from the deformed grains towards the interior of the recrystallized grains, as indicated by the white open arrows in the Fig. 2(a). The glide of dislocations is impeded when encountering a grain boundary or a twin boundary, resulting in their accumulation at these interfaces and forming denser dislocation walls, as indicated by the closed white arrows. On the other hand, twins can also serve as pathways for dislocation glide, as depicted by the red circles in the image. When a dislocation crosses the twin boundary, there may be movable dislocation, immovable dislocation and stacking faults. Suppose the slip dislocation meets the twin boundary and is decomposed into a partial dislocation entering the twin and a partial dislocation remaining at the twin boundary. In that case, the twin boundary will absorb the partial dislocation at the twin boundary and then slip, resulting in the migration of the twin boundary. According to the reports by Fu et al.,[3] dislocation accumulation at the twin boundary increases the twin thickness, namely, twin growth. In Fig. 2(a), two twins that appear to have thickened due to dislocation accumulation are marked by the green-yellow gradient arrows. Dislocation movement within the twin can be seen in Fig. 2(b).

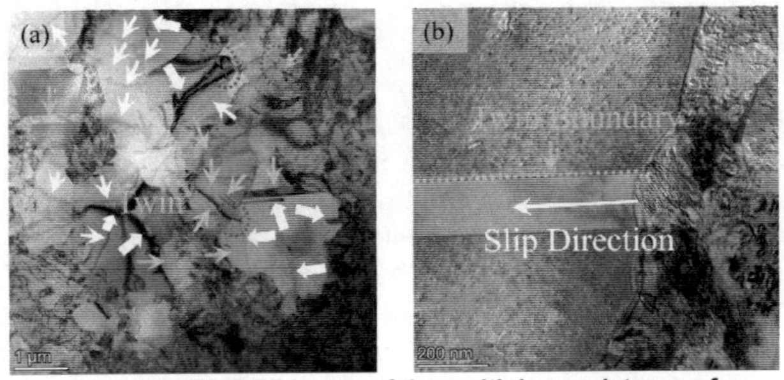

Fig. 2(a) TEM-BF image of the multiple coexistence of recrystallized and deformed grains of alloy after deforming to failure at 77K; (b) TEM-BF image of the enlarged region in (a).

Conclusion

By controlling stacking fault energy and adding nickel element, an FCC high entropy alloy with excellent mechanical properties from high to cryogenic temperatures was developed. At 77 K, deformation mechanisms include dislocation accumulation, stacking faults, coherent precipitation, twins, and back stress. At 293 K to 1073 K, hetero-deformation hardening and GNDs dominate. High stacking fault energy prevents the TWIP effect. This intentional design enables exceptional mechanical performance across various temperatures.

Acknowledgments

This work was supported by the National Natural Science Foundation of China (Nos.52074246, 52275390, 52375394, U23A20628).

References

[1] Sonar T, Ivanov M, Trofimov E, et al. An overview of microstructure, mechanical properties and processing of high entropy alloys and its future perspectives in aeroengine applications[J]. Materials Science for Energy Technologies, 2024;7:35–60.

[2] Li Q, Li X, Jin C, et al. Trace B doping Fe50Mn30Co10Cr10 high entropy alloy: Mechanical response and multi-microstructure evolution under TWIP and TRIP effects[J]. Materials Science and Engineering: A, 2023;873:145037.

[3] Fu X, Wu X, Yu Q. Dislocation plasticity reigns in a traditional twinning-induced plasticity steel by in situ observation[J]. Materials Today Nano, 2018;3:48–53.

Segregation Defect Analysis for Magnesium Alloy Structural Parts in Squeeze Casting Technology

Yuan Zhao [1], Liwen Chen[1], Hua Hou [1,3], Yuhong Zhao [1,2,4]*, Jianhui Jing[1]
1.School of Materials Science and Engineering, Collaborative Innovation Center of Ministry of Education and Shanxi Province for High-performance Al/Mg Alloy Materials, North University of China, Taiyuan, 030051, China
2.Beijing Advanced Innovation Center for Materials Genome Engineering, University of Science and Technology Beijing, Beijing, 100083, China
3.School of Materials Science and Engineering, Taiyuan University of Science and Technology, Taiyuan, 030024, China
4.Institute of Materials Intelligent Technology, Liaoning Academy of Materials, Shenyang, 110004, China
*Corresponding address: e-mail: zhaoyuhong@nuc.edu.cn

Abstract: Magnesium alloy squeeze casting technology has been widely used in aerospace, automotive, and military industries, with promising application prospects. However, issues such as magnesium alloy's susceptibility to oxidation, combustion, cold shut defects, and solute segregation hinder the progress of this process. This study focuses on the macroscopic segregation defects that occur during magnesium alloy squeeze casting. Taking ZM5 magnesium alloy partition components as an example, we investigate the distribution of segregation in castings at different pouring temperatures. The results indicate that the higher the pouring temperature, the greater the macrosegregation area in the casting.

Keywords: Squeeze casting; magnesium alloy; Macroscopic segregation

1 Introduction

Magnesium alloy possesses numerous excellent properties such as low density, good seismic performance, high specific strength, high specific stiffness, and good recyclability [1]. Squeeze casting is a near-net-shape forming technology that can reduce macroscopic defects and achieve excellent mechanical properties when used to prepare magnesium alloy products [2-3].

This study focuses on investigating the design of pouring temperature in the Squeeze casting process. Indirect Squeeze casting experiments were conducted according to the designed experimental scheme, and the microstructure of ZM5 magnesium alloy baffle components produced were analyzed [4]. The microstructure is mainly composed of α-Mg matrix and the second phase $Mg_{17}Al_{12}$. The Squeeze casting process has a certain crushing effect on dendritic crystals [5].

2 Experimental procedure

In this experiment, commercial ZM5 magnesium alloy was used as the experimental material. During the experiment, 4 sets of process parameters varying with pouring temperature were designed for comparison, and the key process parameters used are shown in Table 1. The same squeeze pressure and casting rate were used in all processes, and the same amount of protective gas was introduced.

Table 1. Casting process parameters of ZM5 partition components

Group number	Pouring rate ((L/m)	Mold temperature ((℃	Pouring temperature ((℃
A1	9.0	240	700
A2	9.0	240	720
A3	9.0	240	740
A4	9.0	240	760

3 Result and discussion

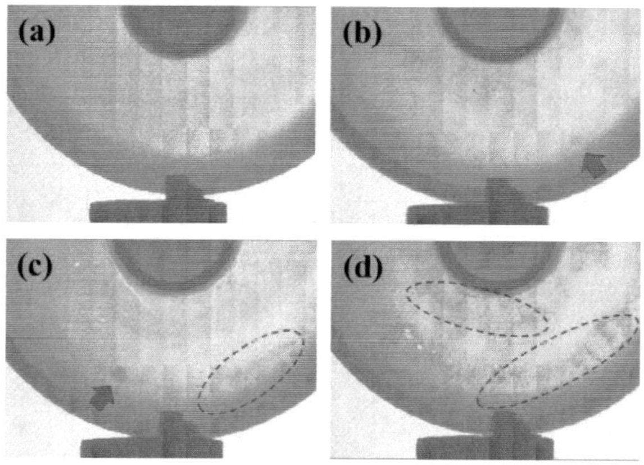

Fig.1 X-ray flaw detection results of partition plates 700 °C, (b) 720 °C, (c) 740 °C, (d) 760 °C

Fig. 1 is an X-ray flaw detection diagram of four groups of separators. The X-ray inspection image reveals that the normal part of the casting appears as a bright area, while the segregation bands appear black, exhibiting a band-like and dot-like distribution. it can be inferred that segregation phenomena often occur at the edges of castings, and the microstructure is observed by optical microscope at this location. As can be seen in Fig. 1, as the casting temperature increases, the area proportion of the segregation zone increases significantly.

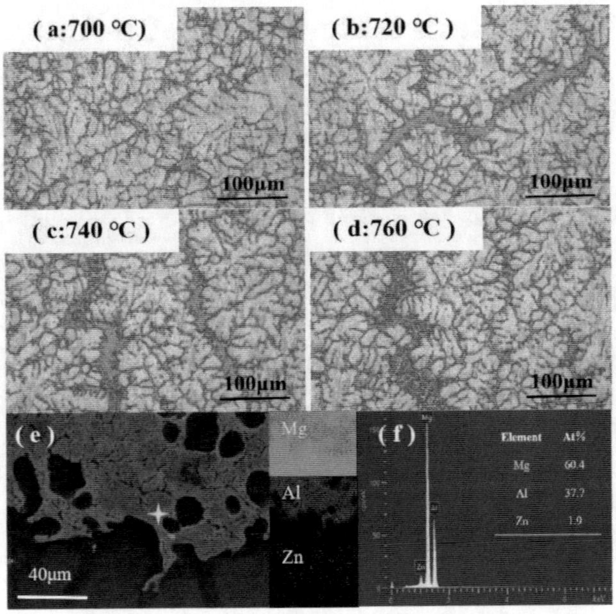

Fig.2 (a)(b)(c)(d) Morphology of segregated regions under light microscopy (OM), (c) Surface scanning, (f) Point scanning

Fig. 2 shows the microstructure of the four sets of partition plates observed using an optical microscope. As can be seen from the figure, the grain growth near the segregation zone is significantly affected, the grain size near the segregation zone is smaller than in the normal area. At the same time, it was found that with the increase of temperature, the grain size increased, the width of the segregation zone increased, and the color became darker. Fig. 2 is a scanning electron microscope (SEM) scan, and the defect band can be seen as a solute separation band enriched by the Al element. As shown in Fig. 2(e) and (f), the energy spectrum analysis of the segregation site showed that the components of the two tissues were mainly Mg, Al and Zn. The results indicate that the ratio of Mg to Al in the segregation area is approximately 17:12, confirming that the segregation phase primarily exists in the form of $Mg_{17}Al_{12}$.

Combining the distribution characteristics of segregation bands, it can be inferred that in the early stages of solidification, the surface layer first contacts the low-temperature mold, resulting in a higher degree of undercooling, which increases the solidification rate and leads to rapid surface cooling. Due to the shear stress induced by the flow of melt beneath the semisolid region, the grain network formed in the semisolid region gradually deforms. Once the shear stress exceeds the shear strength limit of the grain network, it collapses as the grains slide and rotate. Consequently, the semisolid region expands, allowing grains and dendrites to enter this region, ultimately forming segregation bands.

4 Conclusion

(1) For ZM5 magnesium alloy, the segregation band is enriched with Mg and Al elements, mainly in the form of $Mg_{17}Al_{12}$ phase, and the grain size near the segregation is smaller than in normal regions.

(2) Selecting the appropriate casting temperature plays a significant role in suppressing segregation. At 700 °C, the cooling rate is faster, resulting in fewer segregation phenomena and higher performance of the castings. As the temperature increases, the cooling rate slows down, leading to coarser grains in the casting. Meanwhile, the duration of the melt in the semisolid region prolongs, making segregation more likely to occur, thus affecting the mechanical properties of the casting.

Acknowledgments

This work was supported by National Natural Science Foundation of China (Nos. 52375394, 52275390, 52205429, 52201146, 52305429), National Defense Basic Scientific Research Program of China (JCKY2020408B002), Key Research and Development Program of Shanxi Province (Nos. 202102050201011).

References

[1] Song J F, She J, Chen D L and Pan F S. Latest research advances on magnesium and magnesium alloys worldwide. J. Magnes. Alloy, 2020, 01: 1-41.

[2] Zhao Y H, Zhang B, Hou H, Chen W P and Wang M. Phase-field simulation for the evolution of solid/liquid interface front in directional solidification process. J. Mater. Sci. Technol, 2019, 35: 1052.

[3] Yahia Ali, Qiu D, Jiang B, Pan F S and Zhang M X. Current research progress in grain refinement of cast magnesium alloys: a review article. J. Alloys. Compd, 2015, 619: 639–51.

[4] Hou Y ,Wu M ,Huang F and Ding W. Defect band formation in high pressure die casting AE44 magnesium alloy. China. Foundry, 2022, 19(03): 201-210.

[1] [5] Zeng X, Zhu Q, Li Y X and Ding W. Second Phase Particle Strengthening in Magnesium Alloys. Materials. China, 2019.

Microstructure and Mechanical Properties of Extruded Mg-Y-Zn-Ni-Co Alloy

Binxin An[1,2], Jing Jiang[1,2], Haobin Xu[1,2], Yuandong Li[1,2], Tijun Chen[1,2], Guangli Bi[1,2*]
1. Key Laboratory of Advanced Processing and Recycling of Nonferrous Metals, Lanzhou University of Technology, Lanzhou 730050, China
2. School of Material Science and Engineering, Lanzhou University of Technology, Lanzhou 730050, China
Corresponding authors: Email addresses: glbi@163.com (G. Bi), Tel/Fax: +86-931-2976378

Abstract: Microstructure and mechanical properties of Mg-2Y-0.5Zn-0.5Ni-0.5Co (MYZNC, at.%) alloy were investigated in the as-cast, homogeneous and extruded states. The as-cast MYZNC alloy mainly consisted of α-Mg matrix, gray lamellar 18R LPSO phase, and white irregular Mg_2Y, $MgY(Co,Zn,Ni)_4$, and granular $Mg_3(Co,Zn,Ni,Y)$ phases. After homogenization treatment, a large number of strip-like 14H LPSO phases precipitated in the grain interior. After hot extrusion, the alloy occured obvious dynamic recrystallization with a significant grain refinement from 46.5 μm in the homogenized state to 2.98 μm in the extruded state, while the second phases were fragmented and distributed along the extrusion direction. The tensile testing results indicated that the extruded MYZNC alloy exhibited the good tensile strengths as compared with the as-cast and homogenized alloys, whose ultimate tensile strength (σ_b), yield strength ($\sigma_{0.2}$) and elongation to failure (ε_f) of were 280 MPa, 225 MPa and 11.7%, respectively. The high tensile properties of the extruded alloy were mainly attributed to the grain refinement and the second-phase strengthening by the high volume fraction of LPSO phase. In addition, the homogenized appeared a more pronounced strain hardening behavior, attributed to its large grain size and higher dislocation density arised from the presence of numerous fine particles and LPSO phases.

1 Introduction

Mg alloys are favored by researchers for their low density and high specific strength, however, low room-temperature strength and poor plasticity limit their development and application in aerospace, automotive and electronic 3C and other fields. How to improve the tensile strengths, room-temperature plasticity and processing performance of Mg alloys has become an urgent problem.

Recently, rare-earth Mg alloys containing a long-period stack-ordered (LPSO) structure phase have exhibited exceptional mechanical properties both at room and high-temperatures [1]. Considering the formation conditions, including structural and chemical order, of the LPSO phase, it is inevitable that the properties of the transition elements therein (atomic radius, melting point, and solid solubility, etc.) will exert a significant influence on the aforementioned properties of the LPSO phase. Consequently, these factors collectively contribute to enhancing the overall mechanical properties of the alloy. Previous studies [2,3] have demonstrated that the solid solubility of Ni or Co in α-Mg matrix is lower than as compared to Zn, indicating a higher propensity for the LPSO phase formation. Therefore, in this paper, the composite addition of the elements of Zn, Ni, and Co in Mg-2Y alloys, the extruded Mg-2Y-0.5Zn-0.5Ni-0.5Co (at.%) alloy (MYZNC) was prepared and the corresponding relation between microstructure and mechanical properties was discussed. The strengthening mechanisms of extruded alloy was also clarified. The experimental results would provide theoretical support for the design and development of high-strength and high-toughness extruded Mg alloys.

2 Experimental procedure

The experimental Mg-2Y-0.5Zn-0.5Ni-0.5Co (at.% , MYZNC) alloy was prepared by pure Mg (99.90 wt.%), Zn (99.95 wt.%) and Mg-15Ni (wt.%), Mg-15Co (wt.%), and Mg-20Y (wt.%) master alloys by a gravity casting under the protection of the molten salt covering agent (NaCl and KCl). The melting in iron crucible located in a resistance furnace was heated up to 750 °C and held for 60 min. After the pure metal and alloy ingots were was completely melted, the refining agent was added into the melts and stirred slowly for about 5 min until the surface of the melts appeared to a metallic luster, and held for about 10 min. Subsequently, the melts were poured into a water-cooled mold and formed the cylindrical ingot with a diameter of 90 mm and a length of 200 mm. The ingot was homogenized at 400 °C for 12 h before extrusion. The homogenized ingot was extruded in to extrusion sheet with a thickness of 10 mm and a width of 40 mm. The extrusion temperature and extrusion ratio were 366 °C and 16:1, respectively.

The microstructures of the alloys were characterized using optical microscopy (OM, LSM800Laser confocal microscope), X-ray diffractometer (XRD, Rigaku D/max 2500 PC), scanning electron microscope (SEM, JEM-2100F), the electron backscatter diffraction (EBSD, HRFESEM, TESCANMAIA3) and transmission electron microscope (TEM, FEI Talos F200X).

3 Result and discussion

Fig. 1 shows the hardness and tensile stress-strain curves of the as-cast, homogenized and extruded MYZNC alloys. it can be concluded that the extruded alloy exhibits a higher hardness value (64.7 HV) compared to both the as-cast alloy (53.4 HV) and the homogenized alloy (56.4 HV). Furthermore. the tensile test results also reveal that the extruded alloy exhibits the higher tensile strengths than as-cast and homogenized alloys, whose ultimate tensile strength (σ_b), yield strength ($\sigma_{0.2}$), and elongation (ε_f) of the extruded alloy are 280 MPa, 225 MPa, and 11.7%, respectively. The σ_b, $\sigma_{0.2}$ and ε_f of extruded alloy increase by 131%, 257% and 154%, respectively as compared with that of as-cast alloy. The σ_b, $\sigma_{0.2}$ and ε_f of extruded alloy increase by 131%, 257% and 154%, respectively as compared with that of as-cast alloy. Combining Figure 2 and Figure 3, It can be seen that the hot extrusion is indeed an effective method to improve the mechanical properties of the alloy through grain size reduction and refinement of the LPSO phase.

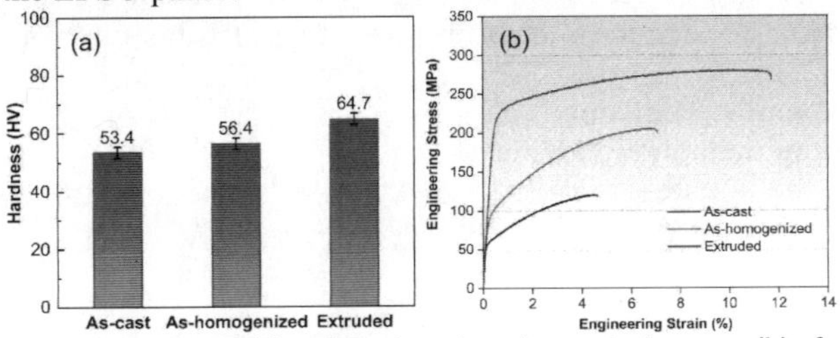

Fig. 1 Hardness (a) and Engineering stress-strain curve (b) of MYZNC alloy as cast, homogenized, and extruded

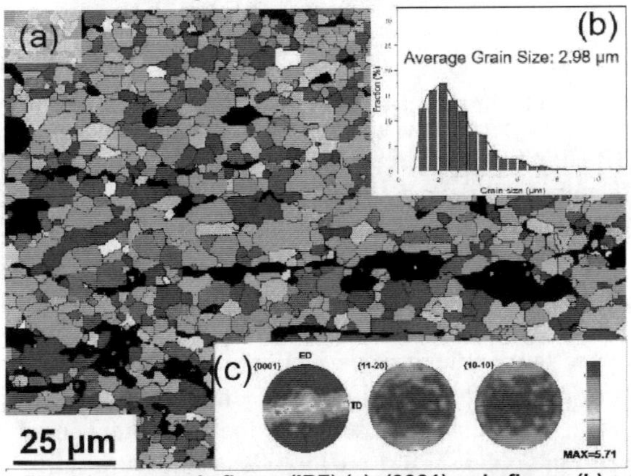

Fig. 2 Inverse pole figure (IPF) (a), (0001) pole figure (b) and average grain size distribution of extruded alloy (c)

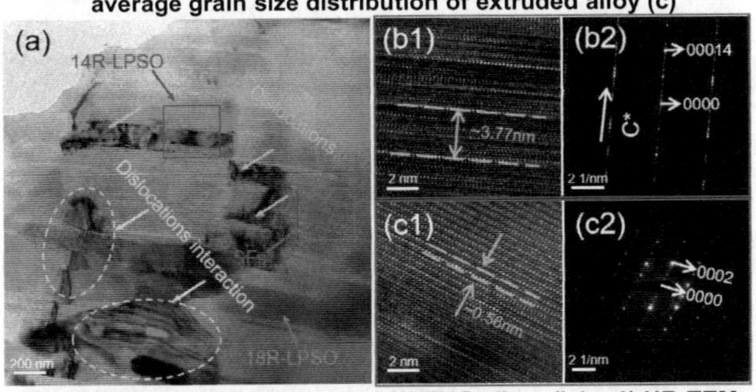

Fig. 3 TEM images of extruded MYZNC alloy; (b1, c1) HR-TEM images; (b2, c2) select the Electron diffraction (SAED) images; Where (b1, b2) corresponds to the blue box area of images(a), and (c1, c2) corresponds to the red box area of images (a)

4 Conclusion

(1) The microstructure of the as-cast MYZNC alloy is composed of $MgY(Co,Zn,Ni)_4$, $Mg_3(Co,Zn,Ni,Y)$, and Mg_2Y granular phases, as well as lamellar 18R LPSO phase. After homogenization, some of the 18R LPSO phases transforms into finely striated 14H LPSO phases parallel to each other and with a slight increase in grain size. After hot extrusion, the grain size of the MYZNC alloy was refined to 2.98 μm, the second phase is fragmented to form extruded bands along the extrusion direction, and the grain occurs to rotate and generate a basal texture with a maximum texture strength of 5.71.

(2) The extruded alloy exhibits a high tensile properties as compared with as-cast and homogenized alloys. The ultimate tensile strength, yield strength, elongation to failure and hardness of the extruded MYZNC alloy are 280 MPa, 225 MPa,11.7% and 64.7 HV at room temperature. The high tensile strengths and hardness of the extruded alloy is mainly related to the grain refinement and second phase strengthening from LPSO phase.

Acknowledgments

This work was financially supported by National Natural Science Foundations of China (Nos. 52261027, 51961021and 52001152), Open Project of State Key Laboratory for Mechanical Behavior of Materials (20192102), Undergraduate Innovation and Entrepreneurship Training Program (Nos. DC20231188, DC20231482, DC20231558, DC20231469 and DC20231441) and Sinoma Institute of Materials Research (Guang Zhou) Co., Ltd (SIMR).

References

[1] X.H. Shao, Z.Q. Yang, X.L. Ma, Strengthening and toughening mechanisms in Mg-Zn-Y alloy with a long period stacking ordered structure, Acta Materialia 58(14) (2010) 4760-4771.
[2] G. Bi, H. Man, J. Jiang, Y. Li, T. Chen, X. Zhang, D. Fang, X. Ding, Effects of Zn addition on microstructure and tensile properties of Mg-Y-Co alloy, Journal of Materials Research and Technology 20 (2022) 590-605.
[3] Q.-Q. Jin, S.-B. Mi, Intermetallic phases in Mg–Co–Y alloys, Journal of Alloys and Compounds 582 (2014) 130-134.

Morphology, Structure and Composition of Precipitated Phase in As-Cast Fe35Ni35Cr20Mn10Ti6 High Entropy Alloy

Mengchuan Shi, Hengcheng Liao
School of Materials Science and Engineering, Southeast University, Nanjing 211189, China

Abstract: In this study, Fe35Ni35Cr20Mn10Ti6 high entropy alloy was prepared by arc-melting. The microstructure and chemical composition of the alloy were characterized by FIB double beam scanning electron microscope. The result show that there is a typical dendritic structure in the as-cast Fe35Ni35Cr20Mn10Ti6 HEA, which is composed of grey dendritic tissue, white and black island-like inter-dendritic structure. The morphology, crystal structure and composition of the precipitated phases in the as-cast Fe35Ni35Cr20Mn10Ti6 HEA were further determined by transmission electron microscopy. The formation sequence and formation mechanism of various precipitated phases during the cooling process were discussed by microstructure observation and phase calculation.

Keywords: Fe35Ni35Cr20Mn10Ti6 high entropy alloy, Microstructure, Solidification process

1 Introduction

The novel alloy design concept of high entropy alloys has been widely concerned since it was proposed in 2004, and related research work has been vigorously developed [1, 2]. Efforts have been made to improving the comprehensive mechanical properties of FCC HEA, such as fine-grain strengthening, second-phase strengthening, solid-solution strengthening and deformation strengthening, etc., but in the most cases, a significant degree of plasticity is lost [3, 4]. Dual-phase HEA can be strengthened and toughened by using softer matrix phase and harder second phase with sufficient fraction, so it becomes an important point in the research field of HEA.

Liao et al. [5] pointed out that Ti alloying can effectively solve the strength-toughness balance problem of Fe35Ni35Cr20Mn10 HEA. However, it is unclear how the second phase is formed during solidification of dual-phase HEA. Therefore, it is a key scientific problem to elucidate the solidification and crystallization law of dual-phase HEA, especially the formation process of the second phase.

In this study, the as-cast Fe35Ni35Cr20Mn10Ti6 HEA was prepared by arc-melting. The microstructure and chemical composition of HEA was studied, and the formation sequence and formation mechanism of various precipitates during solidification were discussed. It has positive significance for the development of HEA solidification theory.

2 Experimental procedure

The as-cast Fe35Ni35Cr20Mn10Ti6 HEA was prepared by arc-melting. The microstructure and chemical composition of the alloy were characterized by FIB double beam scanning electron microscope and transmission electron microscopy.

3 Result and discussion

Pandat phase diagram

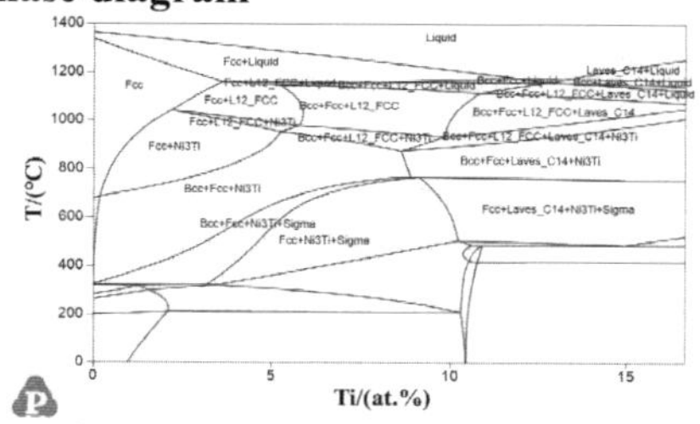

Fig.1 The pseudo-binary phase diagram of (Fe35Ni35Cr20Mn10)-Ti system, calculated by Pandat software

Fig. 1 shows the pseudo-binary phase diagram of (Fe35Ni35Cr20Mn10)-Ti system, calculated by Pandat software. Because of hysteretic diffusion effect of HEA, the phase transformation reactions do not occur at low temperature, and the phase composition above 600°C is shown in Fig.1. It has been shown in literature that Ni and Ti have the largest negative mixing enthalpy and are more likely to aggregate with each other, causing element segregation. This also indicates that the Ti content in each region of Fe35Ni35Cr20Mn10Ti6 alloy is not at a uniform level during the solidification process, so the phase transition in the process is complex. However, it is worth noting that according to the phase diagram, in the range of 3.5%-15% Ti content (Ti content is about 4.8% in the Fe35Ni35Cr20Mn10Ti6), the final solid phases were FCC phase, Ni_3Ti phase, σ phase and Laves phase.

Microstructures

Fig. 2 shows the images of Fe35Ni35Cr20Mn10Ti6 HEA in as cast state (a: low magnification; b: high magnification). The prepared HEA shows a typical dendritic structure, which is composed of grey dendritic trunks, white and black island-like inter-dendritic structure. Spot-scanning results of areas A, B, and C in Fig. 2(b) reveals that the dendritic trunk lacks of Ni and Ti elements compared with the design composition; the inter-dendritic tissue is enriched in Ni and Ti elements; and the content of Ni and Ti in the black island-like structure is further high.

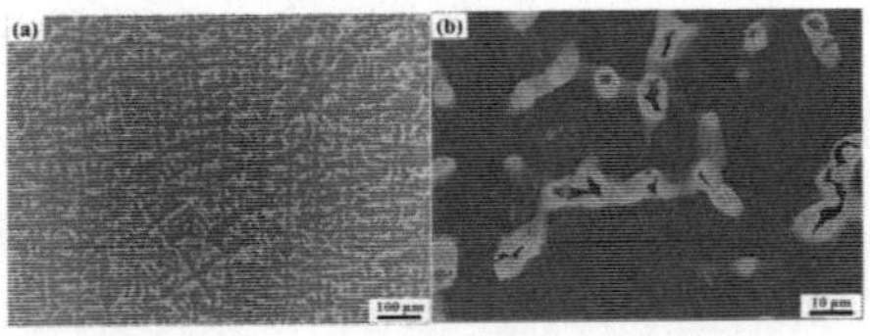

Fig.2 SEM images of as-cast Fe35Ni35Cr20Mn10Ti6 HEA

In order to further analyze the microstructure of the as-cast Fe35Ni35Cr20Mn10Ti6 HEA, the transmission samples of the point C (dendritic trunk) and point A (inter-dendritic zone) in Fig. 2 were prepared by GAI FIB. Fig. 3 shows the TEM bright field image of HEA and the SAED pattern. The dendritic trunk of HEA is a single FCC phase structure as shown in Fig.3(a). As seen in Fig. 3(b), the black island-like structure in Fig. 2 is composed of multiple intermetallic compounds. EDS results of these compounds reveals that the A block structure is rich in Fe, Ni and Ti elements; B block structure is rich in Cr and Fe elements; C block structure is rich in Ni and Ti elements. Combined with the corresponding SADE patterns in Fig. 3, it can be determined that the block structures A, B and C are Laves, σ, and η phases, respectively.

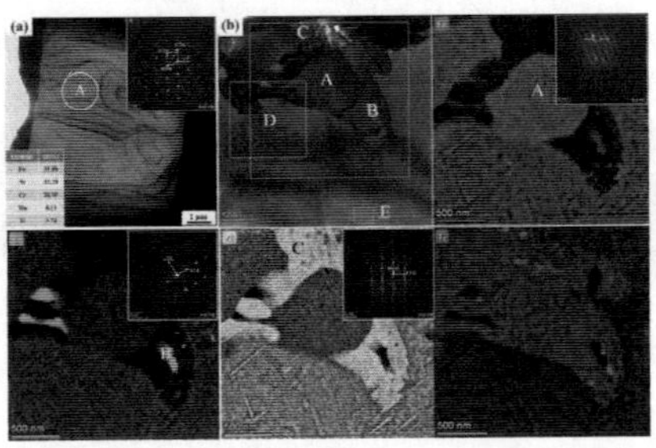

Fig.3 TEM bright-field image and SADE pattern of as-cast Fe35Ni35Cr20Mn10Ti6 HEA: (a) dendritic trunk; (b) inter-dendritic zone

Fig. 4 shows the EDS face-scanning results of areas D and E in the Fig.3 and the SADE patterns of related areas. By the EDS face-scanning results, spherical nanoparticles γ' phase and needle-like η phase can be observed in region D in Fig. 3. Whereas, only spherical nanoparticle γ' phase was seen in region E in Fig. 3. In addition, the Ti content in region C is higher than that in region A in the Fig. 4. This indicates that the level of Ti content has an important effect on the formation of η phase.

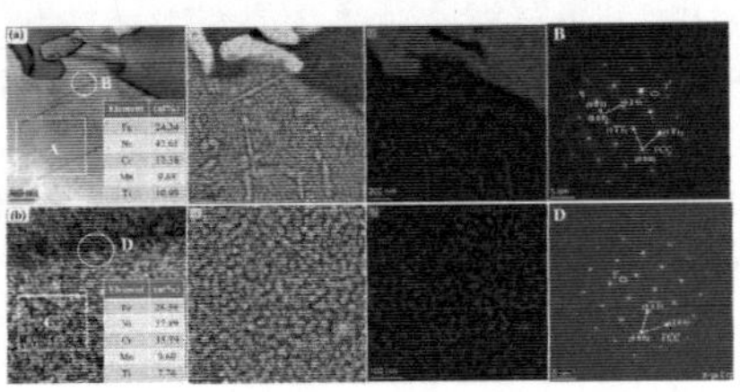

Fig. 4 EDS results and SADE pattern of area D (a) and area E (b) in the Fig. 3

In summary, the as-cast Fe35Ni35Cr20Mn10Ti6 high-entropy alloy has a typical dendritic structure, and the specific structure can be summarized as follows: the phase distribution centered on the Ti rich island-like up to the dendrite region is in the order of Laves + η + σ, FCC + η + γ', FCC + γ' and FCC, as shown in Fig.5.

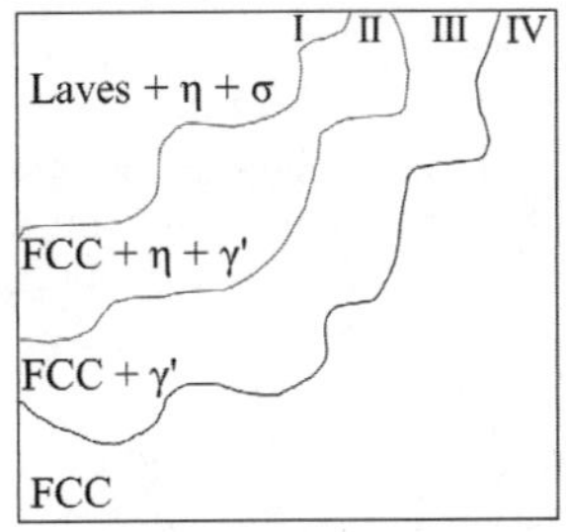

Fig. 5 The microstructure distribution diagram of as-cast Fe35Ni35Cr20Mn10Ti6 HEA.

4 Conclusion

There is a typical dendritic structure in the as-cast Fe35Ni35Cr20Mn10Ti6 HEA, which is composed of grey dendritic trunk, white and black island-like inter-dendritic zone.

The specific structure can be briefly summarized as the phase distribution centered on the Ti rich island-like up to the dendritic region is in the order of Laves + η + σ, FCC + η + γ', FCC + γ' and FCC.

The gradient distribution of Ti elements from dendrites to inter-dendritic zone is the main reason for the gradient distribution of phases

References

[1] Yeh J W et al. Nanostructured high-entropy alloys with multiple principal elements: Novel alloy design concepts and outcomes. Advanced Engineering Materials, 2004, 6(5): 299-303.

[2] Cantor B, et al. Microstructural development in equiatomic multicomponent alloys. Materials Science and Engineering A, 2004, 375: 213-218.

[3] H Zhang, et al. Synthesis and Characterization of NiCoFeCrAl3 High Entropy Alloy Coating by Laser Cladding. Advanced Materials Research, 2010, 97: 1408-1411.

[4] Z Wang, I Baker. Interstitial strengthening of an FCC FeNiMnAlCr high entropy alloy. Materials Letters, 2016, 180(10): 153-156.

[5] Jun Zhou, Hengcheng Liao, et al. Excellent strength-ductility in Ti-alloyed Fe35Ni35Cr20Mn10 high-entropy alloy. Intermetallics, 2022, 148: 107637.

Exploring Microstructure Evolution and Machine-learning Method During Hot Deformation of 56Ni-32Ti-12Hf Alloy

Chunzhe Fu, Hongjun Huang*, Ge Zhou, Xiaojiao Zuo, Bowen Zheng, Xiaoguang Yuan**
School of Materials Science and Engineering, Shenyang University of Technology
Shenyang 110870, No.111, Shenliao West Road, Economic & Technological Development Zone, China
*Corresponding address: e-mail: huanghong1977@163.com

Abstract: Hot compression simulation tests were carried out on a 56Ni-32Ti-12Hf alloy at temperatures ranging from 800 to 1000 °C and strain rates ranging from 0.001 to 1 s^{-1}. The study aimed to explore the hot deformation behavior and microstructure evolution mechanism of the alloy. A flow stress prediction models was developed: a chaotic mapping adaptive inertia weight whale optimization algorithm-optimized back propagation neural network (CIWOA-BPNN) model based on machine learning. The hot processing map of the 56Ni-32Ti-12Hf alloy was developed using dynamic material modeling (DMM) theory and flow stress data predicted by the CIWOA-BPNN model. The optimal hot processing range was found to be between temperatures of 875–975 ℃ and strain rates of 0.001–1 s^{-1}. The study found that within the optimal processing interval, the softening mechanisms observed were discontinuous dynamic recrystallization (DDRX) and continuous dynamic recrystallization (CDRX).

Keywords: Machine learning; NiTiHf alloys; Constitutive model; CIWOA-BP algorithm

1 Introduction

NiTi shape memory alloys are vital functional materials used in various industries such as aerospace (e.g., turbojet nozzles) [1], space communications (e.g., antennas) [2], biomedicine (e.g., brackets) [3], and fluid transmission (e.g., actuators) [4]. Specifically, NiTiHf alloys are known for their high hardness and yield strength. However, due to the limited plasticity, its processing is a challenge [5, 6]. Additionally, the variation law of flow stress with deformation parameters (e.g., strain, strain rate, and temperature) during hot deformation of alloys has been of great interest [7]. The processing of NiTiHf alloys presents challenges, and its microstructure and properties are highly sensitive to deformation parameters [5, 6]. Furthermore, the hot deformation mechanism and constitutive relationship of NiTiHf ternary alloy remain unclear. This paper aims to address these research gaps by utilizing 56Ni-32Ti-12Hf (wt.%) alloy as the experimental material. Through hot simulation experiments, true stress-strain data was obtained and used to develop a CIWOA-BPNN model. The CIWOA-BPNN model was to verify its accuracy. Subsequently, the hot processing of the alloy was mapped based on the data predicted by the CIWOA-BPNN model. Finally, the microstructure after hot deformation was analyzed to validate the reliability of the hot processing map. A model based on an improved machine learning algorithm for predicting the flow stress is proposed in this work. It also provides a theoretical guide for the selection of process parameters for the hot processing of NiTiHf alloys.

2 Experimental procedure

The nominal composition of 56Ni-32Ti-12Hf (wt.%) alloy was prepared by vacuum-induction melting. The hot compression test was implemented through the Gleeble 3800D hot simulation testing machine. After the thermal compression test, the samples were cut from the middle along the axis by wire cutting. The obtained samples were inlaid, ground, polished, and corroded. Then, Zeiss optical microscope (OM) and a scanning electron microscope (SEM) equipped with electron backscatter diffraction (EBSD) and transmission electron microscopy (TEM) were used for microstructure observation and analysis.

3 Result and discussion

1 Modeling of flow behavior

To ensure the flow stress prediction accuracy remains unaffected by the model structure, Liu et al. [8] adopted a backpropagation neural network model. This method is more apt for handling intricate and fluctuating data, accurately forecasting the material's flow stress during hot deformation. This study adopts a three-layer neural network with a single hidden layer, as depicted in Fig. 1. The structure of Fig. 1 shows the weights and thresholds between layers, which directly determine the learning effect of the prediction model. However, the traditional BP neural network model recognizes the samples to be tested mainly through the back-propagation algorithm and gradient descent method to optimize the network's weights and thresholds. Therefore, in this paper, the optimization method of the BP neural network weights and thresholds search using the optimized WOA algorithm (CIWOA) is to regard each neural network as an individual. Fig. 2 illustrates the correlation between the training values in the CIWOA-BPNN model's network output. It demonstrates the accurate capture of the connection between input and output, as evidenced by a high correlation coefficient of 0.99812. Consequently, the CIWOA-BPNN model exhibits effective predictive capability for the flow stress of the 56Ni-32Ti-12Hf alloy.

2 Establishment of the processing map

By substituting the predicted value by the CIWOA-BPNN model into Eq. (1), the instability coefficient contours maps of the 56Ni-32Ti-12Hf alloy at true strains of 0.1 and 0.7 are produced, as illustrated in Fig.3a and 3b. The instability region mainly appears under high ε value conditions, with the area of the instability region increasing with the increase of strain. A hot processing map based on the CIWOA-BPNN model of 56Ni-32Ti-12Hf with a strain of 0.7 is obtained by superimposing Fig. 3a and 3b, as shown in Fig. 4. In this representation, the gray part represents the instability regions (T = 825–875 ℃, ε = 0.5–1 s^{-1}; T = 925–1000 ℃ and ε= 0.5–1 s^{-1}).

3 Deformation mechanism

To explain the softening mechanism of, the deformed microstructure was observed by EBSD. Local misorientations (point-to-point) and cumulative misorientations (point-to-origin) of the two deformed grains extracted from Fig. 5a and 5b are analyzed, respectively. Measurement lines are drawn along the compression direction and vertical directions (Fig. 5). It can be observed that the misorientation accumulated along GBs and inside GBs exceeds 10° in both compression and vertical directions. This indicates that the migration and accumulation of dislocations leads to the gradual rotation of sub-grain boundaries into high-angle grain boundaries (HAGB). The sub-grain boundary orientation angle of B_2 is 12.1° (Fig. 5b), which falls within the medium angle grain boundary (MAGB) range of 10°–15°. Therefore, it is easier to transform into a HAGB [9]. Both phenomena are typical of CDRX [10, 11].

4 Conclusion

(1) The true stress-strain curve of 56Ni-32Ti-12Hf alloy exhibits obvious DRX softening characteristics. The peak stress rises with increasing strain rate or decreasing deformation temperature.

(2) The CIWOA-BPNN model has higher accuracy in predicting the high-temperature flow behavior of 56Ni-32Ti-12Hf alloy.

(3) The optimal processing parameters for the 56Ni-32Ti-12Hf alloy are also determined as a hot deformation temperature range of 875–975 °C and a strain rate range of 0.001–1 s^{-1}.

(4) The softening mechanisms of the alloy in the optimal processing interval are DDRX and CDRX within the optimal machining interval determined.

Acknowledgments

This work was supported the Liaoning Provincial Natural Science Foundation Joint Fund-Doctoral Research Initiation Program(2023-BSBA-258), Project of Liaoning Provincial Department of Education(JYTMS20231200).

References

[1] D.J. Hartl, D.C. Lagoudas, Aerospace applications of shape memory alloys, SAGE Publications (4) (2007)

[2] J. Mohd Jani, M. Leary, A. Subic, M.A. Gibson, A review of shape memory alloy research, applications and opportunities, Mater. Des. 56 (2014) 1078-1113.

[3] T. Duerig, The metallurgy of Nitinol as it pertains to medical devices, Titanium in Medical and Dental Applications (2018) 555-570.

[4] D. Stoeckel, Shape Memory Actuators for Automotive Applications - ScienceDirect, Engineering Aspects of Shape Memory Alloys (1990) 283-294.

[5] M. Belbasi, M.T. Salehi, S.A.A.A. Mousavi, S.M. Ebrahimi, A study on the mechanical behavior and microstructure of NiTiHf shape memory alloy under hot deformation, Mater. Sci. Eng.A 560 (2013) 96-102.

[6] X. Huo, M. Jin, K.M. Reddy, Y. Liu, S. Lahkar, Q. Jin, X. Wang, Strain glass transition in Nb nanowire toughened NiTiHf shape memory alloy composite wires, J. Alloys Compd. 958 (2023) 170393.

[7] Y.V.R.K. Prasad, S. Sasidhara, V.K. Sikka, Characterization of mechanisms of hot deformation of as-cast nickel aluminide alloy, Intermetallics 8(9) (2000) 987-995.

[8] J. Liu, H. Chang, T.Y. Hsu, X. Ruan, Prediction of the flow stress of high-speed steel during hot deformation using a BP artificial neural network, J. Mater. Process. Technol. 103(2) (2000) 200-205.

[9] Y. Wang, Y. Tan, Y. Zhang, R. Bai, X. Li, P. Li, Hot deformation behavior and dynamic recrystallization mechanisms of GH4975 alloy prepared by electron beam smelting layered solidification, Mater. Charact. 202 (2023) 113034.

[10] H. Cui, Y. Tan, R. Bai, L. Ning, X. You, C. Cui, P. Li, Deformation and recrystallization behavior of the new Ni-Co based superalloy ingot material prepared by electron beam smelting layered solidification, J. Alloys Compd. 934 (2023) 167880.

[11] Z. Zhou, R. Zhang, C. Cui, Y. Zhou, X. Sun, J. Qu, Y. Gu, J. Du, Y. Tan, Deformation micro-twinning arising at high temperatures in a Ni-Co-based superalloy, J. Mater. Sci. Technol. 115 (2022) 10-18.

Study on High Temperature Deformation Behavior and Deformation Mechanism of a Ti-47.5Al-2.5V-1.0Cr-0.2Zr Alloy

Xuejian Lin, Hongjun Huang*, Xiaoguang Yuan, Bowen Zheng, XiaojiaoZuo, Ge Zhou
School of Materials Science and Engineering, Shenyang University of Technology
No.111, Shenliao West Road, Economic & Technological Development Zone, Shenyang 110870, P.R.China
*Corresponding address: e-mail: huanghong1977@163.com

Abstract: The hot compression test of a Ti-47.5Al-2.5V-1.0Cr-0.2Zr alloy was carried out. The constitutive equation and hot processing maps were established, and combined with the deformation mechanical model and microstructure characterization, the deformation behavior and deformation mechanism were studied. The results indicated that the correlation coefficient between the calculated stress and the experimental data was 0.991; the deformation temperature and strain rate of the optimal hot working parameter region are 1132-1185 ℃ and 0.001-0.002 s^{-1}. The deformation mechanism is dominated by the DRX. Meantime, the increment of dislocations will induce to the generation of sub-grain boundaries. In addition to the mechanism of dislocation-induced recrystallized grain formation, twinning deformation is also a typical deformation mechanism.

Keywords: TiAl alloy, hot deformation, constitutive equation, processing map, deformation mechanism

1 Introduction

TiAl alloys are promising structural material and are widely used in aviation and automotive engines[1, 2]. However, due to the limited plasticity, its processing is a challenge[3-5]. Studies have shown that hot working deformation can effectively improve the microstructure of the alloy, reduce casting defects, and thus significantly improve the plasticity and processing properties of the alloy[3-5]. The high temperature deformation behavior is significantly affected by the deformation process parameters, and the microstructure and properties are hard to regulate after deformation. To guarantee the stability of the quality of plastic forming parts, many scientists have studied the thermal temperature deformation behavior and microstructural change characteristics of TiAl alloys[6]. Additionally, the processing map creation and microstructure changes under thermal deformation have only been studied separately, and there is a lack of systematic research considering both[7-9]. Therefore, it is essential to systematically study the constitutive relationship, processing window, and the corresponding deformation mechanism. In this study, the constitutive equation and hot processing maps were established, and combined with the deformation mechanical model and microstructure characterization, the deformation behavior and deformation mechanism were studied.

2 Experimental procedure

The nominal composition of the alloy used in the experiment was Ti-47.5Al-2.5V-1.0Cr-0.2Zr(at.%), hereinafter referred to as TiAl alloy, the preparation method is through induction suspension furnace. The hot compression test was implemented through the Gleeble 3800 thermal simulation testing machine. After the thermal compression test, the samples were cut from the middle along the axis by wire cutting. The obtained samples were inlaid, ground, polished, and corroded. Then, transmission electron microscopy (TEM) was used for microstructure observation and analysis.

3 Result and discussion

1Establishment of the constitutive equation

In the thermal deformation process of metallic materials, the Zener–Hollomon parameter is usually used to describe the influence of the deformation parameters [10-12]:

The constitutive equation was solved using measured stress–strain data. The relationship diagrams of $\sigma - ln\dot{\varepsilon}$ and $ln\sigma - ln\dot{\varepsilon}$ are plotted. A good linear relationship was observed under various deformation conditions.The mean values of the reciprocal slopes of the straight line at each temperature are the n_1 and β values of the material, respectively. Consequently, the valuesof n_1 and β were 4.228 and 0.0215, respectively, which were obtained by linear fitting. The constant α was 0.00509 MPa^{-1}.Presuming that Q is unrelated to T in a certain temperature range, a partial differentiation on both sides of Eq. (1) results in the following equation:

$$Q = R\left[\frac{\partial \ln\dot{\varepsilon}}{\partial \ln[\sinh(\alpha\sigma)]}\right]\Bigg|_T \left[\frac{\partial \ln[\sinh(\alpha\sigma)]}{\partial T^{-1}}\right]\Bigg|_{\dot{\varepsilon}} \quad (1)$$

Diagrams of $ln[sinh(\alpha\sigma)]$-$ln\ \dot{\varepsilon}$ and $ln[sinh(\alpha\sigma)]$-T^{-1} are plotted Taking the average slope of the curves, the value of Q is determined to be 456.6 kJ/mol.

Based on the study of C. Zener and H. Hollomon [26], the Z can be expressed as:

$$\ln Z = n ln[sinh(\alpha\sigma)] + lnA \quad (2)$$

The linear regression of $\ln Z - \ln[\sinh(\alpha\sigma)]$ from Eq. (10) is shown; n was 3.0646 and $\ln A$ was 35.1.

Therefore, the constitutive equation can be written as:

$$\dot{\varepsilon} = 1.752816[sinh\,(0.00509\sigma)]^{3.0646} \exp\left[-\frac{456600}{RT}\right] (3)$$

2Establishment of the processing map

According to the true stress–strain data and the DMM model, the hot processing map of TiAl alloy with true strain of 0.5 is demonstrated. The grey region corresponding the instability region, the corresponding thermal deformation parameters were as follows: the temperature ranges were 1050-1090 °C and 1120-1200 °C, and the strain rate range was 0.3-1 s^{-1}. These correspond to regions of medium-high and high strain rates. The deformation conditions corresponding to the peak η were a temperature of 1140-1180 °C and a strain rate of 0.001-0.006 s^{-1}. The peak energy dissipation rate (η) was 0.51; at this thermal deformation condition, the thermal processing performance was the best.

3Deformation mechanism

To explain the softening mechanism of, the deformed microstructure was observed by TEM.In the early stage of deformation, the lamellar was bent, and dislocation stacking and pile-up appeared. With further deformation, the lamellar transforms into DRX grains. Massive dislocations are plugged. It shows the existence of curved subgrain boundaries. In addition to the mechanism of dislocation induced DRX grain formation, twinning deformation is also a typical deformation mechanism. It shows a twin lamellar microstructure with extremely fine spacing and a pattern of selected area electron diffraction (SAED). In addition, another kind of twin morphology. These twin boundaries have a certain angle with the lamellar interface, and most of them are entangled with dislocations. There are γ grains in the lamellar, which are formed by the $\alpha \rightarrow \gamma$ phase transformation or the recrystallization grains. The transition from (α/γ) lamellar to γ+α+β phase also leads to the decomposition of the lamellar and the formation of fully DRX microstructure. The dislocation density inside the alloy decreases significantly due to the DRX. The hot compression deformation eliminates the α_2/γ lamellar microstructure, refines the grain size, and greatly improves the thermal deformation ability of the material. Higher deformation temperature and a lower strain rate will correspond to more DRX time and driving force, so DRX will be more complete.

4 Conclusion

1. The constitutive equation and its material constants were established, and the constitutive relationship could accurately predict the thermal deformation behavior.
2.The processing map was constructed, and the deformation instability and optimal deformation regions were determined.
3.The deformation mechanism largely depends on the DRX. The stacking faults in the γ lamellar can provide the driving force for subsequent DRX. And twinning deformation is also a typical deformation mechanism.

Acknowledgments
This work was supported the Liaoning Provincial Natural Science Foundation Joint Fund-Doctoral Research Initiation Program(2023-BSBA-258), Project of Liaoning Provincial Department of Education(JYTMS20231200).

References

[1] Wu X. Review of alloy and process development of TiAl alloys[J]. Intermetallics, 2006, 4(11): 1114-1122.
[2] Clemens H, Mayer S. Design, Processing, Microstructure, Properties, and Applications of Advanced Intermetallic TiAl Alloys[J]. Advanced Engineering Materials, 2013, 15(4): 191-215.
[3] Hwang B C, Lee C G, Lee S H. Microstructure and Mechanical Properties of Ultra-High Strength Steel Plates with High Deformability[J]. Materials Science Forum, 2010, 638-642: 3266-3271.
[4] Xu X J, Lin J P, Wang Y L, et al. Deformability and microstructure transformation of pilot ingot of Ti–45Al–(8–9)Nb–(W,B,Y) alloy[J]. Materials Science & Engineering A, 2006, 416(1-2): 98-103.
[5] Si J, Fan G, Han P, et al. Simulation on extrusion process of TiAl alloy[J]. Intermetallics, 2011, 19(2): 169-174.
[6] Prasad Y, Gegel H L, Doraivelu S M, et al. Modeling of dynamic material behavior in hot deformation: Forging of Ti-6242[J]. Metallurgical Transactions A, 1984, 15(10): 1883-1892.
[7] Liang X P, Liu Y, Li H Z, et al. Constitutive relationship for high temperature deformation of powder metallurgy Ti–47Al–2Cr–2Nb–0.2W alloy[J]. Materials & Design, 2012, 37(5): 40-47.
[8] Wang G, Lei X, Tian Y, et al. Flow behavior and microstructure evolution of a P/M TiAl alloy during high temperature deformation[J]. Materials Science and Engineering: A, 2011, 528(23): 6754-6763.
[9] Rao K P, Prasad Y, Suresh K. Hot working behavior and processing map of a γ-TiAl alloy synthesized by powder metallurgy[J]. Materials & Design, 2011, 32(10): 4874-4881.
[10] Gupta R K, Murty S, Pant B, et al. Hot workability of titanium aluminide: Development of processing map and constitutive equations[J]. Materials Science & Engineering A, 2012, 551(15): 169-186.
[11] Wang C, Li J, Tang B, et al. Numerical Analysis of Superplastic Bulging Process of TiAl Sheet[J]. Materials Science Forum, 2013, 748: 50-56.
[12] Li J. Dynamic recrystallization behavior of an as-cast TiAl alloy during hot compression[J]. Materials Characterization, 2014, 97: 169-177.

Effect of Alloying Elements and Liquid Film in Late Solidification Stage on Hot Tearing of Al-Cu Alloy

Chunyu Yue[1], Bowen Zheng[1], Ming Su[2], Xiaojiao Zuo[1], Xiaoguang Yuan[1*]

1. School of Materials Science and Engineering, Shenyang University of Technology, No.111, Shenliao West Road, Economic & Technological Development Zone, Shenyang 110870, P.R.China
2. Institute of Metal Research, Chinese Academy of Sciences, No. 72 Wenhua Road, Shenhe District, Shenyang, 110016, P.R.China
*Corresponding address:yuanxg@sut.edu.cn (X.G. Yuan)

Abstract: Hot tearing is an important defect of casting Al alloy, which seriously affects the mechanical properties of alloy. However, there are still many unclear points in the research on the mechanism of hot tearing. The influence of alloying elements on the hot tearing of dendritic Al-Cu alloy is analyzed by using the "cross" hot tearing test system, and the influence of elements on the hot tearing mechanism is discussed. The influence mechanism of liquid film on hot tearing in the late stage of solidification needs to be further studied. Therefore, the semi-solid method is used to prepare the sample, and the near-spherical crystal and liquid film were reproduced by secondary remelting, and then cooled. The hot tearing tendency under different initial solidification states was measured by using the liquid film shrinkage force test system, and the influence of the liquid film on the hot tearing was analyzed, and a new hot tearing evaluation criterion was proposed.

Keywords: Al-Cu alloy, Alloy elements, Liquid film, Hot tearing sensitivity, Hot tearing criterion

1 Introduction

Al-Cu-Mg alloys are commonly applied in aircraft, automobiles, ships and other structural parts owing to their high strength, high heat resistance and excellent forming properties, which can reduce the weight of these structures and improve their comprehensive performance [1]. However, the solidification temperature range of Al-Cu-Mg alloys is fairly wide and dendrite structures are comparatively coarse, which are prone to serious hot tearing on sophisticated large-size components and thin-walled components [2]. Hot tearing is considered to be the most harmful and non-reversible defect [3], which is one of the important difficulties in the industrial production of large-scale uniform defect-free ingots [4]. Therefore, to improve the properties and broaden applications, it is essential to reduce the hot tearing tendency. The effects of alloying elements and liquid film characteristics on the hot tearing of Al-Cu alloys with dendrites and globular crystals were studied. It is helpful to expand the application, and provide theoretical support for research on hot tearing.

2 Experimental procedure

A series of Al-Cu alloys were prepared using high-purity Al, pure Mg, Al-5Zr master alloy and so on. For dendritic structures, the molten metal was poured into the a multi-channel "cross" shaped hot tearing measurement system at 25 ℃ [5]. For non-dendritic structures, the samples were prepared by semi-solid method, and the solidification process of the liquid film was reproduced by secondary remelting. The hot tearing behavior of the alloy was measured under different initial solidification states, and the shrinkage force is analyzed.

3 Result and discussion

1. Effect of alloying elements on hot tearing of dendritic structure

Mg element: With the increase of Mg content, the grain size of the alloy decreases first and then increases, the cracking sensitivity coefficient (*CSC*) decreases first and then increases, and the hot tearing tendency decreases first and then increases. When the Mg content is 1.5wt. %, the average grain size of the alloy is the smallest, and the microstructure is uniform and fine. At the same time, the *CSC* value is 0.583, and the hot tearing tendency of the alloy is the lowest. The effect of Mg element is to improve the hot tearing resistance of the alloy by refining the grain, changing the dendrite morphology and increasing the low melting point phase fraction.

Rare earth element (Sm, Ce, Pr): As the RE content increases, the hot tearing sensitivity (HTS_1) of the alloy first decreases and then increases, as shown in Fig. 1. When the content of Pr is 0.3 wt.%, the hot tearing sensitivity of the alloy is the lowest, and the HTS_1 value is 60. By analyzing the contraction force and solid phase rate curves, with the increase of RE content, the hot tearing initiation temperature first decreases and then increases, and the corresponding solid phase first increases and then decreases. RE elements refine the microstructure, narrow the solidification temperature range, form more intergranular bridges, and reduce hot tearing sensitivity.

2. Research on Hot Tearing of Non-dendritic Structure

with the increase of secondary remelting temperature (t_0) and time (t_0), the liquid film fraction and grain

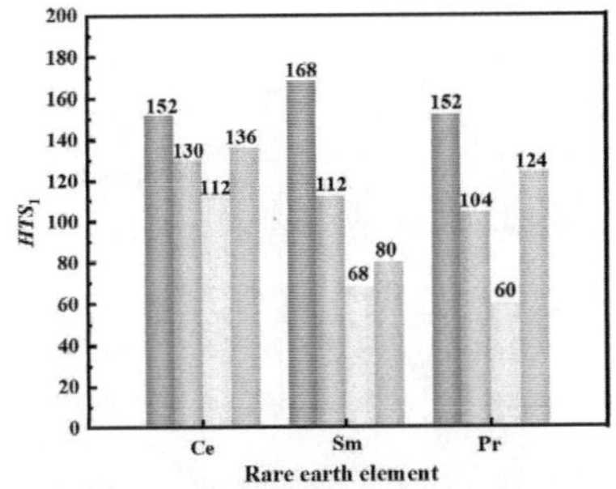

Fig. 1 Effect of RE elements on the hot tearing sensitivity of Al-Cu-Mg alloy

roundness of Al-4Cu alloy increase. As t_0 increases, the temperature of hot tearing initiation decreases, indicating that the hot tearing tendency of the alloy is smaller. When t_0 is 60min, the shrinkage stress curve does not show flattening or attenuation, indicating that no hot tearing have occurred. Increasing the liquid film fraction can change the characteristics of the liquid film, thereby increasing the feeding capacity of the liquid film, healing the hot cracks that have been formed, and effectively reducing the hot tearing tendency.

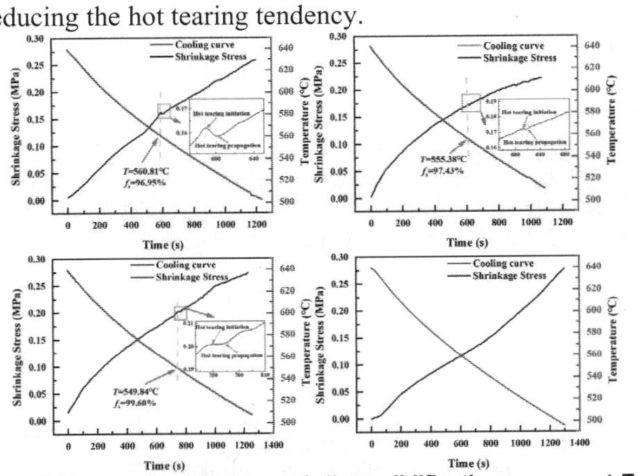

Fig. 2 Shrinkage stress curve of alloy solidification process at T_0 of 640 ℃ under different t_0: (a) t_0=30min; (b) t_0=40min; (c) t_0=50min; (d) t_0=60min

The fundamental reason for the formation of hot tearing is the different volume shrinkage of solid-liquid phases during solidification. However, current criteria simplify volume shrinkage and feeding into two-dimensional problems in unit area, ignoring the volume changes in three-dimensional space. Therefore, based on the characteristics of liquid film and microstructure, the three-dimensional shrinkage volume of solid-liquid two-phase during solidification process is theoretically derived. The expression for the criterion of hot tearing initiation in two adjacent solid grain units:

$$
Criterion\begin{cases}\frac{4}{3}\pi a_0b_0^2\left\{1-\left[(1-\alpha)\Delta T\right]^3\right\}+\left(4a_0b_0L_0-\frac{4}{3}\pi a_0b_0^2\right)-\left(4a_1b_1L_0-\frac{4}{3}\pi a_1b_1^2\right) \\ \quad >-n\frac{\pi r^{*4}}{8\eta l^*}\Delta P^*t \quad \textit{hot cracks will occur} \\ \frac{4}{3}\pi a_0b_0^2\left\{1-\left[(1-\alpha)\Delta T\right]^3\right\}+\left(4a_0b_0L_0-\frac{4}{3}\pi a_0b_0^2\right)-\left(4a_1b_1L_0-\frac{4}{3}\pi a_1b_1^2\right) \\ \quad \le -n\frac{\pi r^{*4}}{8\eta l^*}\Delta P^*t \quad \textit{hot cracks will not occur}\end{cases}
$$

The prediction results based on RDG criterion and Kou criterion are shown in Fig. 3. The criterion proposed in this paper provides a certain guiding significance for the alloy composition, microstructure, mold design and process conditions during the solidification process of the alloy. Compared with other criteria, the biggest advantage of the criterion proposed in this paper is that it can quantitatively calculate whether hot tearing occur in a certain state during solidification.

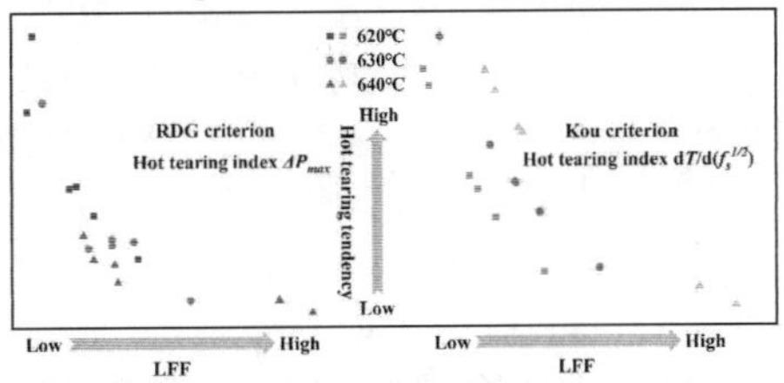

Fig. 3 Hot tearing tendency of Al-4Cu alloy obtained by RDG criterion and Kou criterion

4 Conclusion

When the content of Pr is 0.3 wt.%, the hot tearing sensitivity of the alloy is the lowest, and the HTS_1 value is 60. The feeding effect of liquid film helps to reduce the possibility of hot tearing initiation, and then a criterion of hot tearing initiation based on liquid film characteristics and microstructure is proposed.

Acknowledgments

This work was financially supported by the National Natural Science Foundation of China (No. 51875365).

References

[1] Q.Z. Wang, N. Kang, X. Lin, et al. High strength Al-Cu-Mg based alloy with synchronous improved tensile properties and hot-cracking resistance suitable for laser powder bed fusion, J. Mater. Sci. Technol. 141 (2023) 155-170.

[2] J.W. Liu, S. Kou. Susceptibility of ternary aluminum alloys to cracking during solidification, Acta Mater. 125 (2017) 513-523.

[3] Z.J. Fan, C. Li, H.L Yang, et al. Effects of TiC nanoparticle inoculation on the hot-tearing cracks and grain refinement of additively-manufactured AA2024 Al alloys, J. Mater. Res. Technol. 19 (2022) 194-207.

[4] Y.J. Guo, L. Zhang, G.H Wu, et al. Influence of Cu addition on hot tearing susceptibility and fluidity of Al-Li-Cu alloys: Experimental investigation, criterion prediction and simulation assessment, J. Alloys Compd. 969 (2023) 172301.

[5] C. Y. Yue, X. G. Yuan, M. Su, et al. Effect of adding Pr on the microstructure and hot tearing sensitivity of as-cast Al-Cu-Mg alloys, Mater. Charact., 191 (2022) 112141.

Magnesium on the Brink or Beginning of a New Era?

Christoph Schendera[1]*, Gerhard Geishüttner[2], Andreas Kraly[3]
1. European Research Association for Magnesium, Aalen, Germany
2. GG-Giesstechnik GmbH & Co. KG, Nümbrecht, Germany
3.High Performance Industrietechnik GmbH, Braunau-Ranshofen, Austria
*Corresponding address: e-mail: c.schendera@efm-aalen.de

Abstract: lightweight metalMagnesium, European CO2 targets,uniqueprocesses, Atomizing&Horizontal Continuous Casting of liquid Magnesium

Introduction

Magnesium is a unique lightweight metal in terms of specific mechanical properties and also based on its unlimited eco-friendly raw material sources. However, Magnesium faces enormous challenges in Europe due to (1) self-limiting European CO2 targets, resulting in European regulations, like e.g. „Fit-for-55“ or CBAM, (2) the one-sided dependence on China coupled with unpredictable price-jumps and (3) pressure from the Aluminum industry.

As part of the European Climate Law„Fit-for-55“, the EU has committed to reducing its net CO2 emissions by at least 55% by 2030. This results that with the new CO2 standards, all new passenger cars and light commercial vehicles registered in Europe should be emission-free by 2035.

The increased focus on E-mobility and the associated reorientation of R&D have stalled the development of new magnesium components (like e.g. automatic transmissions) at some European OEMs.

Even worse is the fact thatthe strict CO2 targets mean that proven and long-established Magnesium components are at risk of being replaced by recycled European Aluminumwith lower CO2 emissions.

The European Magnesium Association for Magnesium, a strong supporter of Magnesium, welcomes the growing interest and developments in Magnesium in China and is for their part also working with its members on continuous improvement of the furnace and melting technology as well as unique processes for a better, safer and more environmentally friendly future for Magnesium.

In the following 2 unique processes from liquid Magnesium to semi-finished products are described in more detail:

(1) Horizontal Continuous Casting Line for the production of Magnesium billets

High Performance Industrietechnik GmbH [1] has developed a new continuous Horizontal Casting Line (HPI ForgeMaster Line), which brings the process of horizontal casting of forging stock for Aluminum into a new millennium.

The new ForgeMaster Line has recently been installed at Hydro Rackwitz, Germany and enables the production forging stock in diameters from 42 mm to 110 mm with high surface quality.

Once the starting conditions are met, the furnace is tilted to the desired height in the casting trough and the dams for filling are automatically opened and closed while the system checks that proper temperatures have been reached.

The ForgeMaster system optimizes solidification conditions and ensures a more uniform distribution of material properties (due to improved microstructural isotropy), resulting in a minimal edge zone and limiting losses due to peeling. The casting line is capable of producing forging stock in diameters from 42 mm to 110 mm with high surface quality.

Besides its Hot Top Vertical Continuous Casting (Magnumcast ®), HPI also developed the first Horizontal Casting Line for the production of Magnesium billets or ingots for further processing to profiles, sheet and forgings or as well as high quality Mg-alloy ingots for its use in the Magnesium die casting industry and
A game changer for the Magnesium industry!

Fig. 1. HPI ForgeMaster Line

(2) Atomizing of liquid Magnesium for the Production of Powder and globular Granules

GG-Giesstechnik GmbH & Co. KG (GGG) develops and produces the world's largest and most modern &efficient casting lines, with melting capacities between 80- 1.500 kg/h. [3]

Fig.2 shows one of the largest casting lines in Europe with a melting capacity of 1.400 kg/h, based on customized multi-chamber melting &dosing furnaces as well as cover gas mixing units for N2/Dry Air/SO2/CO2.

Two robots stack the incoming ingots into 4 pre-heating units, which are directly connected to 4 melting furnaces. The molten Magnesium is transferred into 8 dosing furnaces to simultaneously feed 8 fully automated die casting cells. The solidified components are removed by a robot. The gating system is automatically separated and fed back into the melting furnace.

Fig.2. GGG Casting Line with a Melting Capacity of 1.400 kg Mg/h

GGG's and his partnersalso developed novel industrial equipment, such as a novel Atomizer for the direct production of Powder and globular Granules from liquid Magnesium (Fig.3).

The novel Atomizer offers a high degree of flexibility on various Magnesium alloys with different particle size distributions for the use in the Chemical, Steel/Titanium, additive manufacturing as well as for the Mg-Thixomolding industry.

The melting furnace sits on the Atomizer unit, from which the liquid Magnesium is fed into the atomizer chamber flooded with protective gas. Here the liquid Magnesium hits the rotating disk under high pressure (Fig.4).

The actual particle size varies between 0,2 – 1,0mm, as shown in Fig.5. In a further process step, the solid particles are fed into a cyclone, where they are separated into the different particle classes for their respected use.

Fig.3. Novel Mg Atomizer developed by GGG

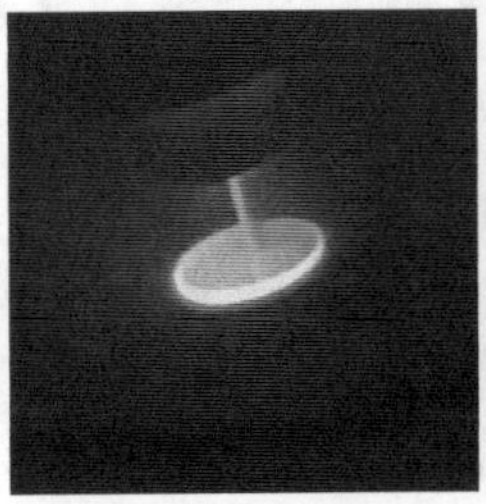

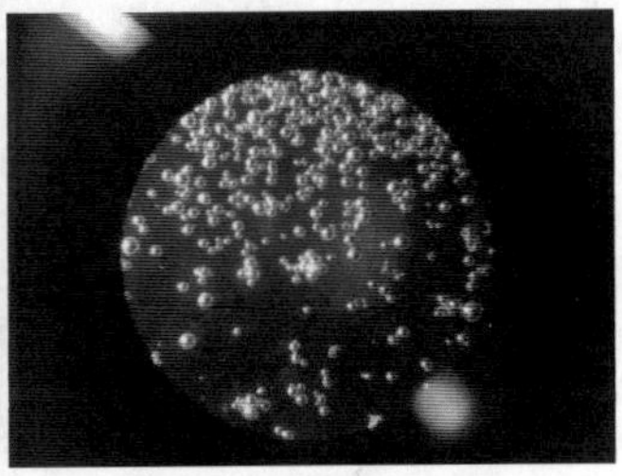

Fig.4. Rotating Disc Fig.5. Particles 0,2-1,0mm

References

[1] https://www.hpi.at
[2] https://www.lightmetalage.com/news/industry-news/forging/hydro-starts-producing-high-quality-automotive-forging-stock-in-germany/
[3] https://giesstechnik.de

Analysis and Solution of Macroscopic Segregation of Mn-containing Phase in Sand Casting of AZ91 Magnesium Alloy

Shiwei Xin[1*], Zehua Li[1], Zhijun Feng[1], Yufei Li[1],Ming Ruan[1],Licheng Zhou[1],Xiao Li[1]
1. National Key Laboratory of Advanced Casting Technologies,Company,Shenyang 110022 China
*Corresponding address: e-mail: 15041225197@163.com (Shiwei Xin)

Abstract: The AZ91 magnesium alloy exhibits excellent properties, yet the sand casting process utilized in its production may give rise to challenges associated with macrosegregation related to manganese-containing phases. Understanding the mechanisms and factors influencing Mn phase segregation is crucial for optimizing the casting process of large magnesium alloy components and enhancing the performance of such components. This paper elucidates the composition and formation mechanisms of segregation defects in AZ91 (ZM5) magnesium alloy through chemical, non-destructive testing, metallography, and scanning electron microscope micrographs, proposing measures to prevent their occurrence. Initially, macrosegregation defects in magnesium alloy sand castings were detected through non-destructive testing, the morphology of the segregation defects was characterized using metallographic analysis, and the morphology and elemental composition of the segregated phase were analyzed using SEM. Through comparative experiments incorporating different element contents and cooling rates combined with phase diagram analysis, the mechanism and solutions for Mn phase segregation were identified.The outcomes of this study hold significant implications for addressing the Mn phase segregation issue in AZ91 (ZM5) alloy sand casting applications and improving the performance of large castings.

Keywords:AZ91 Magnesium Alloy, Macroscopic Segregation, Al8Mn5 phase

1 Introduction

Presently, research on AZ91 magnesium alloy segregation predominantly focuses on die casting or permanent mold casting processes, with limited emphasis on segregation in sand casting. It was observed that for commercial alloys, solidified manganese undergoes redistribution within the primary dendrites during the solidification process, primarily influencing precipitates within non-uniformly distributed matrix grains.

The results of SEM analysis on the alloying elements segregation in the cast, heat-treated, and aged states of AZ91E alloy indicate that manganese is enriched in the dendritic core structure of the casting structure, and even after 24 hours of heat treatment at 410 °C, these manganese-rich regions remain unchanged.[1]

In some magnesium alloys, Mn-containing Phase are closely related to overheating effects. If the melt is cooled from a high temperature, manganese particles crystallize from the melt during the cooling process.[2]

Although existing literature provides valuable insights into segregation phenomena in magnesium alloys, there are still some knowledge gaps that require further research. The specific interactions between the content of manganese in AZ91 magnesium alloy and other alloy elements, and their influence on segregation behavior, remain of interest for research. In addition, there is a need for in-depth exploration of the influence of specific parameters of sand casting (such as mold material, pouring temperature, cooling speed) on segregation patterns. Addressing these knowledge gaps will help to gain a more comprehensive understanding of the segregation mechanisms in magnesium alloys and facilitate the development of effective strategies to mitigate the effects of segregation during casting processes. The segregation behavior of magnesium alloys is influenced by a variety of factors, including alloy composition, particularly the content of alloy elements such as manganese, which can have a significant impact on segregation tendencies. In addition, pouring parameters (such as pouring temperature and mold material) play a crucial role in controlling the segregation patterns within the structure of the castings. The aim of this research is to address these shortcomings and, building on existing literature on magnesium alloy segregation phenomena, to provide new insights for analyzing and solving the macrosegregation issues related to Mn-containing phases in AZ91 magnesium alloy sand casting. Specifically, the research aims to: 1. Investigate the segregation patterns and distribution of Mn-containing phases in casting samples. 2. Identify the root causes of macrosegregation during large-scale casting processes using sand molds. 3. Propose effective solutions or strategies to minimize segregation defects in AZ91 (ZM5) magnesium alloy sand castings.

2 Experimental procedure

To Conducting a comprehensive investigation into the performance of alloy materials and optimizing their manufacturing processes required the execution of a series of casting experiments involving varied pouring temperatures and manganese (Mn) contents. These parameters were intentionally chosen to delve into their influence on the formation of alloy segregation defects. The experimental procedure initially involved preparing the specific alloy compositions in accordance with the designated pouring temperatures and Mn concentrations.

Subsequent to meticulous blending, the molten alloy were cast into meticulously crafted molds. Post-casting, a thorough anatomical examination of the alloy specimens was carried out to scrutinize the internal morphology and distribution of segregation defects. Employing sophisticated flaw detection methodologies, such as ultrasonic and X-ray testing, enabled a comprehensive assessment of the samples, facilitating precise identification of defect locations and sizes. Complementary metallographic analyses were conducted to delve deeper into the microstructural characteristics and properties of the alloy materials.

Table 1. Composition of Alloy .(wt%)

alloy	Al	Mn	Zn
A	8.2	0.15	0.5
B	8.1	0.25	0.49

3 Result and discussion

1. Non-destructive testing results for segregation defects:

Figure 1 shows the results of X-ray detection, indicating the presence of numerous point-like segregations.

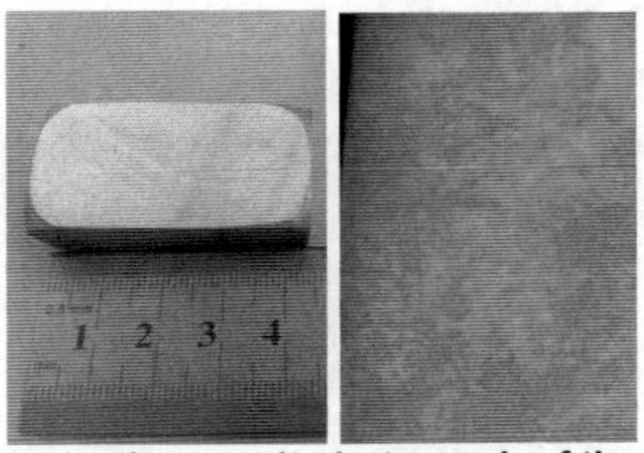

Figure 1 the X-ray testing result photograph of the casting and its corresponding actual location on the sample.

2.Metallographic testing of segregation defects

Figure 2 illustrates the outcomes of metallographic testing conducted on the segregation defects. From the metallographic images in the as-cast, T4, and T6 states, it is evident that numerous short rod-like precipitates are distributed both within and along the grain boundaries, and these precipitates cannot be dissolved through solution treatment.

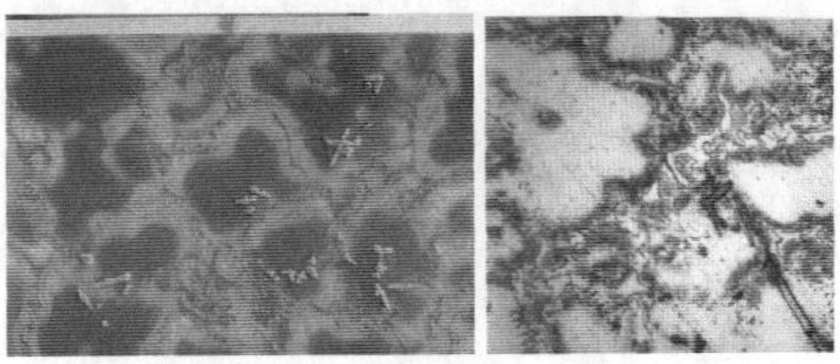

Figure 2 Optical image of the segregation defects area of the cast pattern

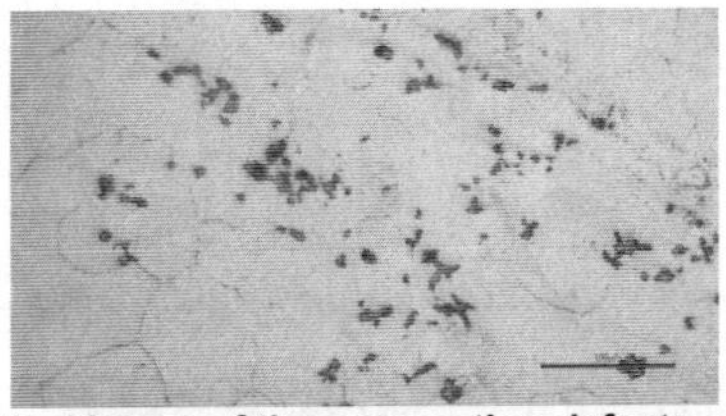

Figure 3 Optical image of the segregation defects area of the T4 pattern

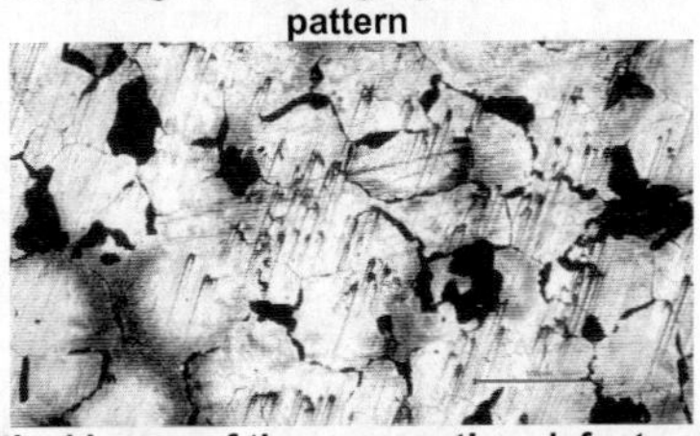

Figure 4 Optical image of the segregation defects area of the T6 pattern

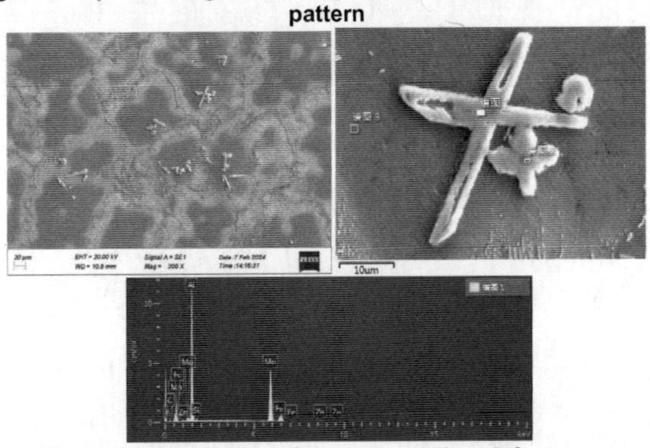

Figure 5 SEM images of the segregation defects area

4 Conclusion

By combining phase diagrams and experimental analysis, it is concluded that during the cooling process of the alloy from 600 °C to 510 °C, the Al8Mn5 phase precipitates from the solid phase. Measures to address macrosegregation of manganese-containing phases include:

1. Reducing the Mn content to the lower limit permitted by the alloy composition range.
2. Shortening the solidification time of the alloy at defect locations in the temperature range of 680-510 °C, increasing the cooling rate to minimize the precipitation of Al8Mn5 phase.
3. Appropriately raising the casting temperature; if the casting temperature is excessively low, close to 680-700 °C, the alloy may prematurely form Al8Mn5 phase locally due to compositional variations.

References

[1] Tamura Y , Kida Y , Tamehiro H ,et al.The effect of manganese on the precipitation of Mg17Al12 phase in magnesium alloy AZ 91.J. Journal of Materials Science, 2008, 43(4):1249-1258.

[2] Tamura Y , Yagi J , Haitani T ,et al.Observation of Manganese-Bearing Particles in Molten AZ91 Magnesium Alloy by Rapid Solidification.J. MATERIALS TRANSACTIONS, 2003, 44(4):552-557

[3] .

Increasing Mechanical Properties of Mg-Y alloy by Adding TiB_2-Al_2Y Particles

Kun Zhang, Yuyang Gao*, Ang Zhang*, Zhihua Dong, Yan Yang, Jiangfeng Song, Bin Jiang, Fusheng Pan
National Engineering Research Center for Magnesium Alloys, National Key Laboratory of Advanced Casting Technologies, College of Materials Science and Engineering, Chongqing University, Chongqing 400044, China
*Corresponding address: e-mail: gaoyuyang@cqu.edu.cn; angzhang@cqu.edu.cn

Abstract: In this study, the effects of TiB_2-Al_2Y particles on microstructures and mechanical properties of Mg-9Y composites were analyzed. The TiB_2-Al_2Y particles refined the α-Mg grains. The yield strength and tensile strength of Mg-9Y alloy were increased by 73 MPa and 57 MPa through adding 4.31 wt.% TiB_2-Al_2Y. The main strengthening mechanisms of the composites were fine grain strengthening and Orowan strengthening.

Keywords: TiB_2-Al_2Y particles, Mg-Y, microstructures, mechanical properties,

1 Introduction

Mg-Y alloys are widely studied due to strong solid-solution strengthening and precipitation strengthening effects [1]. Grain refinement and second-phase strengthening are effective means to improve the mechanical properties of Mg alloys. Adding Al to Mg-Y alloys can generate in-situ Al_2Y particles [2]. As heterogeneous nucleation cores of α-Mg, Al_2Y particles can refine the microstructures. The ceramic nanoparticles (TiB_2, TiC, SiC, etc.) can assist in Orowan strengthening to improve the strength of Mg alloys [3]. Due to high melting point and high thermal stability, TiB_2 particles can effectively hinder the dislocation movement and improve the mechanical properties of Mg alloys.

In this study, by adding in-situ nano TiB_2/Al master alloys into Mg-Y alloys, nano-TiB_2 and micron-Al_2Y particles were successfully introduced to prepare (TiB_2-Al_2Y)/Mg-9Y composites. The effects of TiB_2-Al_2Y particles on the microstructures and mechanical properties of extruded composites were systematically studied.

2 Experimental procedure

Pure Mg (99.99 wt.%), Mg-30Y (wt.%), and TiB_2/Al (30 wt.% TiB_2 + 70 wt.% Al) master alloy were used to prepare (TiB_2-Al_2Y)/Mg-9Y composite material ingots. The ingots were preheated at 440 °C for 30 min and extruded at 440 °C at a speed of 0.5 mm/s to obtain bars with a diameter of 20 mm.

Optical microscopy (OM) (Zeiss Axio Vert.A1) and scanning electron microscopy (SEM) (JEOL JSM-7800F) were used for microstructure characterization, and X-ray diffractometry (XRD) (PANalytical X'Pert Powder) was used for phase identification. A tensile testing machine (CMT5105-100KN) was used to test the mechanical properties of x(TiB_2-Al_2Y)/Mg-9Y composites at room temperature at a strain rate of 1×10^{-3} s^{-1}

3 Result and discussion

As shown in Fig.1, after the TiB_2-Al_2Y particles were introduced into the alloy, the microstructures of the (TiB_2-Al_2Y)/Mg-9Y composites were significantly smaller than those of the matrix alloy. Fig. 2 showed all samples were composed of fine recrystallized grains and elongated non-recrystallized grains with typical bimodal microstructural characteristics. The addition of TiB_2-Al_2Y particles promoted the dynamic recrystallization in the composites by the particle-stimulated nucleation mechanism. Meanwhile, TiB_2 and Al_2Y particles with high thermal stability inhibited grain boundary migration and hindered the growth of recrystallized grains during the hot extrusion process, thus refining the microstructures of composites.

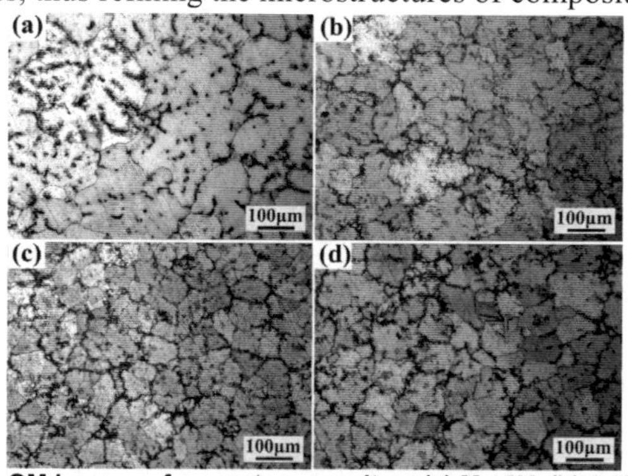

Fig. 1. OM images of as-cast composites: (a) Mg-9Y, (b) 2.15 wt.% TiB2-Al2Y/Mg-9Y, (c) 4.31 wt.% TiB2-Al2Y/Mg-9Y, and (d) 6.46 wt.% TiB2-Al2Y/Mg-9Y.

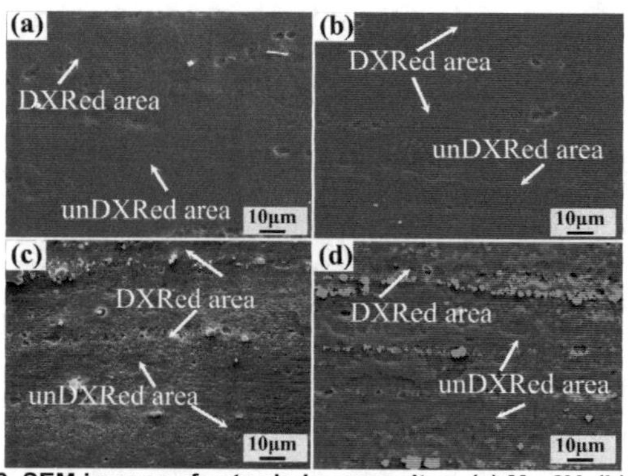

Fig. 2. SEM images of extruded composites: (a) Mg-9Y, (b) 2.15 wt.% TiB2-Al2Y/Mg-9Y, (c) 4.31 wt.% TiB2-Al2Y/Mg-9Y, and (d) 6.46 wt.% TiB2-Al2Y/Mg-9Y.

As indicated by Fig. 2, the Mg-9Y matrix alloy mainly consisted of α-Mg and $Mg_{24}Y_5$ phases, while Al_2Y and TiB_2 phases were detected in the (TiB_2-Al_2Y)/Mg-9Y composites.

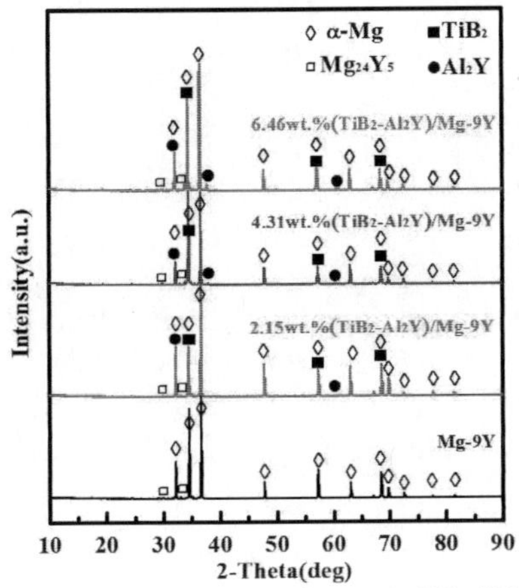

Fig. 3. XRD diffraction patterns of extruded Mg-9Y alloy and composites.

Fig.4 shows the tensile stress–strain curves of the Mg-9Y alloy and composites. The addition of TiB_2-Al_2Y particles increased the YS and UTS of the Mg-9Y matrix alloy but decreased the fracture strain. The YS and UTS of the 4.31 wt.% (TiB_2-Al_2Y)/Mg-9Y composite were 298 MPa and 335 MPa, which were 73 MPa and 57 MPa higher than those of Mg-9Y alloy (225 MPa and 278 MPa), respectively.

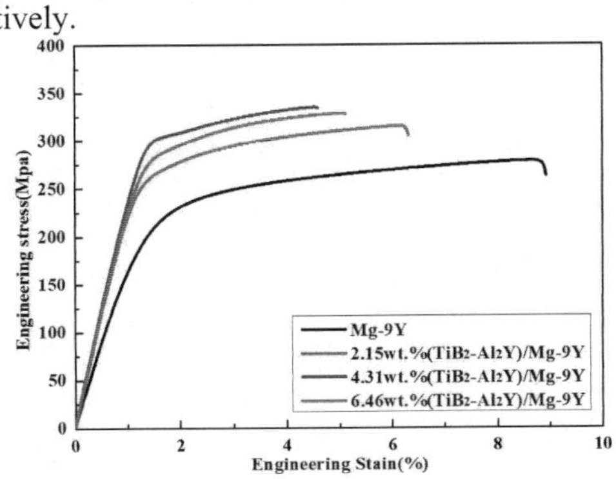

Fig. 4. Tensile engineering stress–strain curves of the Mg-9Y alloy and (TiB2-Al2Y)/Mg-9Y composites.

4 Conclusion

1. The addition of TiB_2-Al_2Y particles significantly refined the as-cast microstructures. After extrusion, the Mg-9Y matrix alloy mainly consisted of α-Mg and $Mg_{24}Y_5$ phases, while Al_2Y and TiB_2 phases were detected in the (TiB_2-Al_2Y)/Mg-9Y composites.
2. The addition of TiB2-Al2Y particles effectively increased the YS and UTS of the composites but decreased the fracture strain. The main strengthening mechanisms of the composites were fine grain strengthening and Orowan strengthening caused by the TiB_2-Al_2Y particles.

Acknowledgments

This work is financially supported by the National Key Research and Development Program of China (Grant No. 2021YFB3701000), the National Natural Science Foundation of China (Grant Nos. 52201106 and U21A2048), and the Natural Science Foundation of Chongqing China (Grant No. CSTB2022NSCQ-MSX1289).

Reference

[1] J.A. Liu, S.J. Liu, B. Wang, W.B. Sun, X.J. Liu, Z.W. Han, Compressive property and energy absorption characteristic of interconnected porous Mg-Zn-Y alloys with adjusting Y addition, Journal of Magnesium and Alloys 12(1) (2024) 171-185.

[2] D. Qiu, M.-X. Zhang, The nucleation crystallography and wettability of Mg grains on active Al2Y inoculants in an Mg–10wt% Y Alloy, Journal of Alloys and Compounds 586 (2014) 39-44.

[3] S.K. Sahoo, B.N. Sahoo, S.K. Panigrahi, Effect of in-situ sub-micron sized TiB2 reinforcement on microstructure and mechanical properties in ZE41 magnesium matrix composites, Materials Science and Engineering: A 773 (2020) 138883.

Microstructural Modification in Aluminum Alloys by Means of a Low Frequency Electromagnetic Field Applied During the Solidification

D. Abedul[1], I. Hurtado[1], N. Herrero[1], G. Arruebarrena[1], P. Garcia[1], Z Azpilgain[1], P. Rodriguez[2]
1. Faculty of Engineering, Mondragon Unibertsitatea, Loramendi 4, 20500 Mondragon, Spain
2. EIPC RESARCH CENTER, Torrekua Kalea, 3, 20600 Eibar, Spain
*Corresponding address: e-mail: dabedul@mondragon.edu

Abstract: This work focuses on the refining of grain size and control morphology by electromagnetic agitation, which offers a clean process without physical contact. The research aims to develop an electromagnetic agitation method to enhance mechanical properties in aluminum components produced through investment casting. A new low-frequency electromagnetic pulsed wave device was created for this purpose, generating variable shape pulses at low frequencies. Results showed significant grain refinement and improved eutectic morphology.

Keywords: Aluminum alloys, microstructural modification, electromagnetic stirring

1 Introduction

The main aluminum alloys used as structural components in the aerospace sector are the high-strength 7xxx series alloys or the 2xxx series alloys, which are known for their improved toughness. The industrial applications of these alloys are explained by their microstructure and physical-mechanical properties, which is why they are commonly used in frames and fuselage parts of aircraft components, where specific strength is also a critical factor. Both the 7xxx series alloys and the 2xxx series alloys are wrought alloys, which limits the complexity of the geometries that can be achieved. If more complex shapes are desired, casting processes must be used. The commonly employed casting process for aeronautics is investment casting, and the A356/A357 alloy is primarily used. To ensure that components made with these alloys achieve the highest properties, they must undergo a T6 treatment, and the size and morphology of the solidification grains must be controlled. A smaller grain size with equiaxed morphology helps improve mechanical properties, allowing the A356/A357 alloys to approach the properties of the previously mentioned wrought alloys. This fact could allow for the transformation of parts typically made by forging into parts made by casting, thus expanding the market for the casting sector.

One way to reduce grain size and control its morphology is by refining the grain of the microstructure. The traditional methods used to achieve this effect include rapid solidification, the addition of alloying elements, or the addition of inoculants. Alternatively, more advanced techniques based on applying vibration to the metal during the solidification phase have been employed. This agitation can be achieved through various methods (mechanically, ultrasonically, or electromagnetically) [1]. However, electromagnetic agitation offers certain advantages over the other mentioned processes. Specifically, the main advantage of this technique is the absence of any physical contact with the metal during processing, resulting in a consistently clean process.

The primary objective of this work is to advance in the development of an electromagnetic agitation method that allows the production of aluminum components through the investment casting process, thereby improving their mechanical properties.

2 Experimental procedure (Bold, 10 pt., Arial)

A new device for Low Frequency EM Pulsed waves (figure 1) have been manufactured in this work. This equipment generates a pulsed EMF wave of a variable shape at a low frequency. The equipment was designed using the values and pulse shapes reported by Wang et al. [2] for microstructural modification. The characteristics of this equipment is:

- Low frequency: 200 Hz
- Variable voltage: range 100 – 280 V
- Pulsed EMF (electromagnetic field): range 1.5 – 10 Hz.
- Variable wave shape: damped wave or single pulse (figure 1)

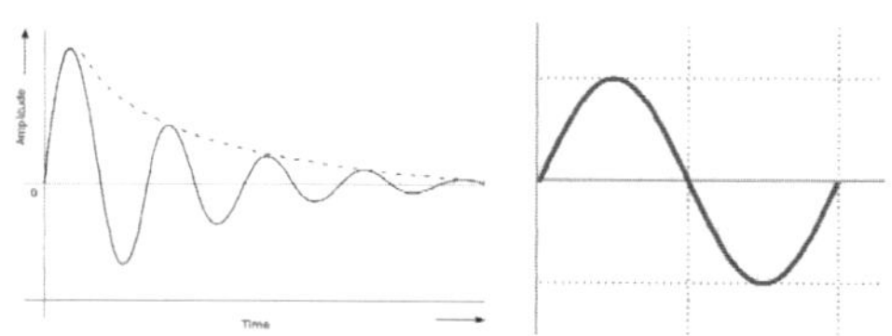

Figure 1. Variable wave shapes used in this work: damped wave (left) and single pulse (right).

Pure aluminum was melted in a electrical melting furnace at 720ºC and degassing tablets were used for avoiding gas porosity. A 60 mm crucible was used as mold and preheated at 650ºC. Grain refiners or alloying elements were then added to pure aluminum. The alloys investigated are the following ones:

- Aluminum
- Aluminum + TiBAl

- Aluminum + 7%Si + Sr

In the results presented in this work a single pulse of 3 Hz was generated at 200 Hz.

3 Result and discussion

Figure 2, 3 and 4 show the results obtained in the mentioned experiments.

Figure 2. Pure aluminum, pulses 3Hz. At high voltage (left) and low voltage (right). The windings for LFEM are placed in the bottom, first 30 mm.

Figure 3. Pure aluminum + TiBAl, pulses 3Hz. At high voltage (left) and low voltage (right). The windings for LFEM are placed in the bottom, first 30 mm.

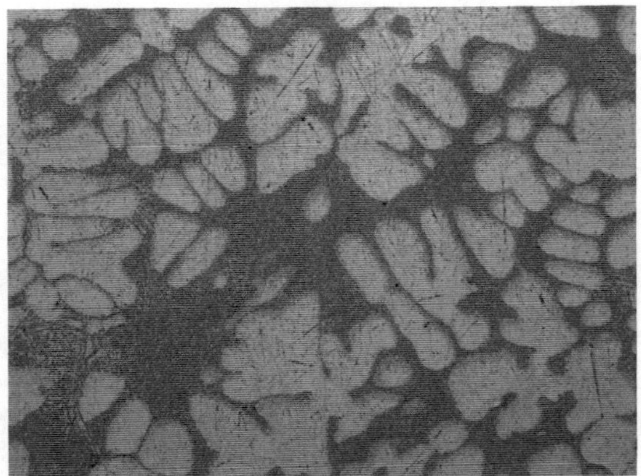

Figure 4. Pure Aluminum + 7%Si + Sr, pulses: 3 Hz and 200 V.

In the case of pure aluminum (figure 2), a single pulse of 3 Hz was able to refine the grain, and the highest voltage shows to have a more pronounced effect in the refinement in a greater area. Even when using TiBAl as grain refiner (figure 3), the effect of the stirring can be appreciated, having a finer grain size in the bottom of the casting comparing with the top. Finally, Si alloying and modification was successfully applied as shown in figure 4, where a fine eutectic morphology can be seen in the interdendritic region.

4 Conclusion

The following conclusions are obtained from this work:

- A device has been built in order to apply controlled LFEM pulses
- Pulsed LFEMP are able to modify the microstructure of aluminium alloys, refining the grain size in aluminium alloys.
- In the presence of an eutectic modifier, i.e. Sr, the silicon structure is very well refined.

Acknowledgments (Bold, 10 pt., Arial)

The authors wish to thank the Basque Government, ELKARTEK projects KK-2024/00061, KK-2016/00010 for their financial support.

References

[1] N. Balasubramani, J. Venezuela, D. StJohn, G. Wang, M. Dargusch, J. Mater. Sci. Technol., 144 (2023), pp. 243-265

[2] B. Wang, Y.S. Yang, X.P. Ma, W.H. Tong, Trans. Nonferrous Metal. Soc., 20 (2) (2010), pp. 283-288

Effect of Cr Element on Microstructure and Mechanical Properties of $Ti_8Zr_6Nb_4V_5Cr_x$ Lightweight High Entropy Alloys

Qin Xu [1], Chengyuan Guo [1], Qi Wang [2], Yajun Yin[3], Ruirun Chen[2*]
1. School of Mechanic and Electrical Engineering, Henan University of Technology, Zhengzhou 450001, China
2. National Key Laboratory for Precision Hot Processing of Metals, Harbin Institute of Technology, Harbin 150001, China
3. State Key Laboratory of Materials Processing and Die & Mould Technology, Huazhong University of Science and Technology, Wuhan 430074, China
*Corresponding address: e-mail: ruirunchen@hit.edu.cn

Abstract: $Ti_8Zr_6Nb_4V_5Cr_x$ (x = 0, 1, 2, 3, 4 at.%) lightweight high entropy alloys with different additions of Cr were prepared and effects of Cr on phases, microstructure and mechanical properties were investigated. Results show that $Ti_8Zr_6Nb_4V_5$ lightweight high entropy alloy is composed of BCC phase and BCC2 phase, while $Ti_8Zr_6Nb_4V_5Cr_x$ alloys by addition of Cr are composed of BCC phase and C15 Laves phase. Dimensional size of BCC phase first increase and then decrease with increasing Cr content, while volume fraction of C15 Laves phase first decreases and then increases with increasing Cr content. It is noteworthy that dimensional size of BCC and C15 Laves phases are refined when the Cr content is 4 at.%. Moreover, density of $Ti_8Zr_6Nb_4V_5Cr_x$ alloys by adding varying Cr content increases slightly, and strength of the alloys increases significantly. Compressive strength of LW-Cr2 alloy with Cr content of 2 at.% reaches the highest of 2326.3 MPa. LW-Cr1 alloy shows excellent comprehensive mechanical properties with compressive strength of 1963.8 MPa and fracture strain of 23.51%. Addition of Cr is playing a significant role in enhancing the mechanical properties of $Ti_8Zr_6Nb_4V_5Cr_x$ lightweight high entropy alloys.

Keywords: lightweight high entropy alloys; phases; microstructure; refinement; mechanical properties

1 Introduction

High entropy alloys exhibit excellent properties for its distinctive structure, such as high strength [1], high hardness [2], thermal stability [3], oxidation resistance [4], etc. With the development of green economy, there is a great demand for lightweight materials in the fields of aviation, aerospace and transportation. Lightweight high-entropy alloys (LHEAs) have been given a special position. Researchers have gradually developed a strong interest in research of LHEAs. Generally, density of LHEAs is less than 7 g/cm^3, and such relatively low density provides important significance for research. Compared with traditional alloys, LHEAs possess excellent mechanical properties like higher specific strength/hardness, excellent corrosion resistance, higher temperature oxidation resistance, etc.

2 Experimental procedure

$Ti_8Zr_6Nb_4V_5Cr_x$ button type ingots were prepared by vacuum arc melting in high purity argon atmosphere. X-ray diffractometer (XRD) of type D8 ADVANCE was utilized to test phase composition and structure for LW-Crx alloys. Microstructure and morphology of LW-Crx alloy were characterized by scanning electron microscopy (SEM) with model Zeiss SUPRA55. Compression test of the specimens was conducted by MTS 370 universal mechanical property testing machine at room temperature. Three measurements were made to avoid accidental errors.

3 Result and discussion

Figure 1 shows X-ray diffraction analysis for $Ti_8Zr_6Nb_4V_5Cr_x$ LHEAs (x = 0, 1, 2, 3, 4 at.%). Results show that $Ti_8Zr_6Nb_4V_5$ (LW-Cr0) alloy is solid solution alloy consisting of two BCC phases, which are BCC+BCC2 dual-phase structure. Notably, BCC2 phase disappears in the alloy with addition of 1 at.% Cr, and new diffraction peaks appear at 2θ of 34.85° and 72.92°, which are characteristic diffraction peaks of C15 Laves phase. It indicates that addition of Cr changes phase composition of LW-Cr0 alloy and leads to phase transformation from BCC2 phase to C15 Laves phase. LW-Cr1, LW-Cr2, LW-Cr3 and LW-Cr4 Cr-containing alloys consist of the main BCC phase and C15 Laves phase.

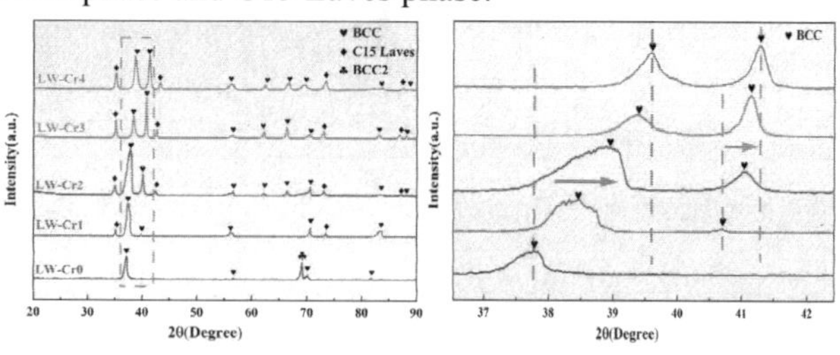

Fig. 1 X-ray diffraction analysis of Ti8Zr6Nb4V5Crx LHEAs
(a) XRD patterns; (b) Diffraction peaks of BCC phase

Figure 2 shows SEM-BSE images of $Ti_8Zr_6Nb_4V_5Cr_x$ LHEAs (x = 0, 1, 2, 3, 4 at.%). Results show that LW-Cr0 alloy is composed of two phases, and the two phases of LW-Cr0 alloy without addition of Cr are a bright phase enriched with both Ti and Nb (much lower than Ti) elements and dark phase enriched with Zr and depleted with V. Both the bright and dark phases are BCC structures (called BCC and BCC2, respectively), which is consistent with results of XRD analyses. Figures 2(b)-(e) show microstructures of Cr-containing alloys LW-Cr1, LW-Cr2, LW-Cr3 and LW-Cr4, respectively. It is obviously that the four alloys are composed of two phases, they are bright phase enriched with both Ti and Nb (much lower than Ti) elements and dark phase enriched with Zr, Cr, and V (Zr at.% > V at.% > Cr at.%) elements. The bright phase is the main phase with disordered BCC (same as LW-Cr0) structure and the dark phase is the ordered CrZrV-type C15 Laves phase. Addition of Cr into $Ti_8Zr_6Nb_4V_5$ LHEAs leads to phase transformation from BCC2 phase to C15 Laves phase, which is consistent with XRD analysis.

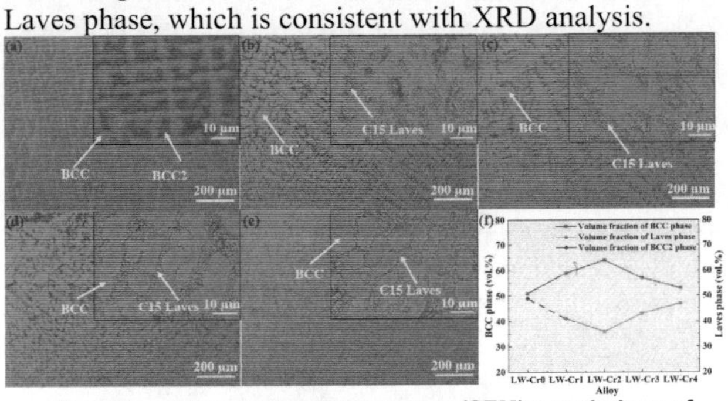

Fig. 2 Scanning electron microscopy (SEM) morphology of Ti8Zr6Nb4V5Crx LHEAs
(a) LW-Cr0; (b) LW-Cr1; (c) LW-Cr2; (d) LW-Cr3; (e) LW-Cr4; (f) Volume fraction of BCC phase and Laves phase

Compressive tests show that compressive strength and fracture strain of the alloys is significantly affected by Cr. Compressive strength of LW-Cr0, LW-Cr1 and LW-Cr2 alloys are 1743.4MPa, 1963.8MPa and 2326.3MPa, respectively. Fracture strain of alloys decreases and then increases with addition of Cr. LW-Cr1 alloy exhibits excellent overall properties with compressive strength of 1963.8MPa and fracture strain of 23.5%. Density of $Ti_8Zr_6Nb_4V_5Cr_x$ LHEAs is quite close to the theoretical results, and both of them decrease slightly with increase of Cr content. Actual density of LW-Cr0 alloy is 6.06 g/cm^3, and that of LW-Cr4 alloy is increased to 6.21 g/cm^3.

Table 1 Comprehensive properties and density of Ti8Zr6Nb4V5Crx LHEAs

LHEAs	Compressive strength (MPa)	Compressive strain (%)	Density (g/cm^3)
LW-Cr0	1743.4	29.6	6.06
LW-Cr1	1963.8	23.5	6.13
LW-Cr2	2326.3	14.8	6.15
LW-Cr3	1975.2	15.4	6.19
LW-Cr4	1787.3	16.2	6.21

Conclusion

(1) $Ti_8Zr_6Nb_4V_5$ (LW-Cr0) LHEA is composed of BCC and BCC2 with volume fraction of 51% and 49%, respectively. After addition of Cr, BCC2 is transformed into C15 Laves.
(2) Compressive strength and yield strength of LW-Cr*x* alloy first increase and then decrease with addition of Cr. Compressive strength and yield strength of the LW-Cr2 alloy reaches maximum of 2326.3 MPa and 2046.1 MPa, respectively.
(3) Fracture strain of LW-Cr*x* alloys decreases and then increases with addition of Cr. Ductility decrease of alloys is related to formation of brittle C15 Laves phase. Fracture strain of LW-Cr4 alloy is greatly increased to 16.2% for the grain refinement.

Acknowledgments

Program for Science & Technology Innovation Talents in Henan Province (23HASTIT022), Scientific Research Fund of State Key Laboratory of Materials Processing and Die & Mould Technology (Grant No. P2023-005).

References

[1] X.M. Wei, L.J. Zhang, C.Z. Zhang, G. Li, A sunflower-like eutectic microstructure in the $AlCrFeMnNi_2$ high-entropy alloy with excellent compressive mechanical properties, Materials Letters, 2023, 339: 134107.
[2] Q.W. Guo, Y. Pan, H. Hou, Y.H. Zhao, Predicting the hardness of high-entropy alloys based on compositions, International Journal of Refractory Metals and Hard Materials, 2023, 112: 106116.
[3] N.X. Zhou, T. Hu, J.J. Huang, J. Luo, Stabilization of nanocrystalline alloys at high temperatures via utilizing high-entropy grain boundary complexions, Scripta Materialia, 2016, 124: 160-163.
[4] T. Xu, Q.J. Chen, L. Ji, H. Wang, S.J. Qu, A.H. Feng, Thermal stability and oxidation resistance of non-equimolar ratio AlTiVCrNb refractory high-entropy alloys, Intermetallics, 2023, 160: 107942.
[5] Y. Qiu, Y.J. Hu, A. Taylor, M.J. Styles, R.K.W. Marceau, A.V. Ceguerra, M.A. Gibson, Z.K. Liu, H.L. Fraser, N. Birbilis, A lightweight single-phase AlTiVCr compositionally complex alloy, Acta Materialia, 2017, 123: 115-124

Sanxingdui Bronze Civilization and Bronze Casting Technology System

Jianbo Guo[1*], Qiutong Cai[2], Jianjun Mei [3], Marcos Martinón-Torres[4]
1. Sichuan Provincial Institute of Cultural Relics and Archaeology, No. 5 Section 4 Renmin South Road Chengdu, China
2. Sanxingdui Museum, No. 133 Xiangxin Road Guanghan City Sichuan Province, China
3. The Needham Research Institute, 8 Sylvester Road Cambridge CB3 9AF, United Kingdom
4. Department of Archaeology, University of Cambridge, Downing Street Cambridge CB2 3ER, United Kingdom
*Corresponding address: No. 5 Section 4 Renmin South Road Chengdu, China. e-mail: 834682583@qq.com

Abstract: The casting process of Sanxingdui bronzes has always been a hot topic in academic circles. This technique has some obvious connections with the production of bronzes in the Central Plains region and the middle and lower reaches of the Yangtze River, but important differences also exist. In the recent years, by using analytical techniques such as X-ray CT scanning and non-destructive testing, scholars have gained a deeper understanding of the internal structure of Sanxingdui bronzes, especially making progress in the application of core rod and strip shim techniques. This article aims to delve into the application of core rod and strip shim techniques in Sanxingdui bronzes, and through comparative studies, it points out that this technique represents a new development in bronze casting technology at the time, highlighting the differences between the Sanxingdui bronze artifacts and other contemporary bronze-making techniques, which is of great significance for a deeper understanding of the technical and cultural background for the rise of the Sanxingdui bronze artifacts.

Keywords: Sanxingdui; Bronze; Sectional casting; Core rod; Strip shim; Cultural exchange

1 Introduction

The Sanxingdui site is located to the north of Nanxing Town in Guanghan City, Sichuan province. The main area of the site is situated within the natural villages of Sanxing, Zhenwu, and Huilong under the jurisdiction of Nanxing Town, as well as the natural villages of Rensheng and Dayan under the jurisdiction of Sanxing Township. The site is distributed across terraces on both banks of Yazi River and Mamu River, covering an area of about 12 square kilometers. Radiocarbon dating results of artifacts unearthed from Pit No. 4 (K4) in the sacrificial area of the Sanxingdui site indicate that the artifacts date back approximately 3,000 years, belonging to the late Shang Dynasty. Since the excavation of Pits No. 1 and No. 2 in the sacrificial area of the Sanxingdui began in 1986, the study of Sanxingdui bronzes has been a focal point in academic circles. Numerous scholars have conducted research on the bronzes from five aspects: base composition, preservation and restoration, casting techniques, origin, and source of minerals, particularly focusing on non-container bronzes. In recent years, with a new round of excavations in the sacrificial area of Sanxingdui, academic interest in the archaeological discoveries of Sanxingdui has surged again, including new research on the craftsmanship and technology of Sanxingdui bronzes.

This article examines the Sanxingdui sectional casting method within the framework of the social and cultural system, based on scientific characterization of bronze production and casting technology. Sectional casting is a prominent technique that, through skilled application of technical details, enabled the Sanxingdui culture to achieve a balance between precision and efficiency in bronze production, showcasing unique innovations in craftsmanship and artistic pursuit. The sectional casting method at Sanxingdui is not merely a technical means but also an expression of art and cultural heritage. This unique production method differs significantly from traditional technological development in other regions, endowing Sanxingdui bronzes with distinctive characteristics. Through empirical evidence, the ubiquity of sectional casting at Sanxingdui is demonstrated. From the dimensions of "commission," "design," and "manufacture," the necessity of choosing sectional casting is explained, with "manufacture" understood as a reflection of the other two dimensions, potentially inspiring creativity and invention to meet cultural demands. By comparative methods, the similarities and differences between the solid casting method and sectional casting are explored, revealing that the primary distinction between solid casting and sectional casting lies not in the size or volume of the cast object but in the nature of the casting goal. The interplay between culture, society, and technology is emphasized. Cultural backgrounds determine the forms of artifacts, thereby stimulating technological innovation and shaping the unique technological manufacturing system of Sanxingdui. Conversely, technological differences gradually reinforce cultural distinctions. We aim to gain deeper insights into ancient bronze technology and its social context while contributing to the understanding of the development and uniqueness of Sanxingdui culture. Moreover, in addition to studying the "input" of Sanxingdui, we should also focus on its "output," namely the materials and ideas that may have flowed out of this region, as well as the modes of exchange.

2 Result and discussion

Sectional Casting

This article presents a new contribution to this area of work, using scientific methods to characterize the bronze crafting and casting techniques, with a view to contextualizing them more broadly in their sociocultural system. The bronze artifacts of Sanxingdui bear witnesses to the broader social and ideological structure that commissioned, created and deposited them. In identifying and discussing idiosyncratic traits of the production system, we offer new pointers for studies of invention, knowledge transfer, and the links between technology and culture. A major focus is placed on sectional casting as a prominent technique that potentially stands as a distinctive feature that sets Sanxingdui apart from the contemporaneous bronze casting cultures in the Central Plains and the middle and lower reaches of the Yangtze River. Through the skillful application of technical details, the Sanxingdui culture achieved a balance between precision and efficiency in bronze production, demonstrating unique innovation in craftsmanship and artistic pursuit. We argue that the sectional casting method in Sanxingdui is not merely a technical means; it is also an art expression, and a form of cultural inheritance. This unique production method evolved in a way that was markedly different from traditional techniques in other regions, imparting a special character to the Sanxingdui bronzes that requires an archaeological explanation. Through a comparative approach, and trying to address the reasons for its widespread use, we hope to gain insight into ancient bronze technology and in its social context, but also contribute to the understanding of the development and uniqueness of the Sanxingdui culture.

Core rod and Strip Shim

In the bronzes from Sanxingdui, we have discovered the use of core rods and strip shims(figure 1), which bear some resemblance to findings in other parts of the world, yet also exhibit unique characteristics. Besides these comparative features, we have, for the first time, identified the combined use of core rods and strip shims. This technique, currently observed only in the bronzes unearthed from Sanxingdui, indicates that although the casting technology of Sanxingdui bronzes broadly belongs to the composite clay mold casting system, it has its own technical innovations in the details of the craftsmanship. These innovations are closely related to the unique cultural characteristics of the artifacts being cast. These casting techniques and features fully demonstrate the exceptional ingenuity of Sanxingdui bronze casting craftsmen in their technical concepts. Additionally, these technical characteristics are applied not only in the lost-wax method but also in the model casting method. At present, there is no archaeological evidence suggesting that this technological innovation originated from other regions, including the Central Plains and the middle and lower reaches of the Yangtze River. It is entirely possible that this innovation originated locally in Sanxingdui.

Figure 1: Bronze statue unearthed from Pit No. 3 of the Sanxingdui and its X-ray photograph (K3QW:100)

3 Conclusion

The bronzes of Sanxingdui have been revealed as an essential new chapter in the history of bronze technology in China, and of technology and art more generally. The widespread use of sequential casting and cast-on joins, in combination with rod cores and strip shims, materializes the mastery of a distinctive technological repertoire that built on traditional materials and technologies but took them in unprecedented directions of sophistication and scale. These developments are reflections of an idiosyncratic cultural system that both enabled and required these creative developments through specific ideas, values and rituals that were translated into specific designs and materials. These unique production methods materialize both inherited tradition and innovation, at the same time reflecting the embeddedness of technology in sociocultural demands that spanned the ritual and the aesthetic. We cannot understand the Sanxingdui bronzes without referring to a broader ideological system, and vice versa, and hence these bronzes provide a precious resource for us to gain a deeper understanding of ancient China's technological and cultural traditions in areas outside the well-known state structures of the Shang.

Research on High-performance Precision Casting Technology for Large and Complex Titanium Alloys

Y. C. Lou[1, 2, *] , J. Zhao[1, 2] , S. B. Liu[1, 2] , T. Y. Liu[1, 2]
1. National Key Laboratory of Advanced Casting Technologies, Shenyang, Liaoning, 110022, China
2. Shenyang Research Institute of Foundry Co., Ltd, Shenyang, Liaoning, 110022, China
*Corresponding address: e-mail:louych@chinasrif.com

Abstract: With the continuous development in the field of high-end equipment, lightweight, ultra-pure, large-scale, complex, high-performance has become the basic requirements of structural materials applied in the field of cutting-edge engineering, and the traditional production and preparation processes have been challenged. In this paper, we will start from the large melting volume of titanium alloy purity melting and precise control of composition, large complex thin-walled precision complete forming and high-performance dense solidification organization control, high-strength heat-resistant and highly stable type / core preparation and deformation control of thin-walled complex cavities and large-capacity vacuum melting and casting equipment, the development of four key technologies, and finally, on the development trend of titanium alloy casting technology (such as: large-scale, digitalization, intelligence and low-cost, etc.) to carry out Summarize.

Keywords: Titanium alloys, Precision Casting, Large-scale, Complex

1 Introduction

Titanium alloys have become crucial structural materials for high-end equipment in aerospace, aviation, and marine fields due to their superior properties such as high specific strength, low density, low thermal conductivity, low coefficient of thermal expansion, excellent corrosion resistance, and high/low-temperature performance[1]. In the aviation field, titanium alloys are mainly used in aircraft structural parts, aircraft engine structural parts, and aviation fasteners, etc. In the aerospace field, titanium alloys are mainly used in carrier rocket engines, satellite antenna brackets, spacecraft cabin frames, etc., with their usage continuing to rise[2, 3].Currently, the global demand for lightweight titanium alloy castings exhibits a diversified and high-standard trend, particularly focusing on requirements for ultra-large sizes, extreme weights, intricately complex structures, extreme thin-wall designs, and exceptional service performance. Specifically, there is an urgent market need for titanium alloy castings capable of producing large outer contour structures exceeding 4 meters in size while meeting load-bearing requirements of over 2 tons. These castings must also possess extremely complex structures featuring multiple flow channels and narrow, multi-layered twisted cavities to cater to demanding operating conditions. Furthermore, challenges arise from designing large-scale extreme thin-wall structures with thicknesses approaching 1mm, necessitating rigorous technological advancements. In terms of performance, these titanium alloy castings must exhibit exceptional service characteristics, including high strength, resistance to high and low temperatures, and wear resistance, ensuring stable operation in various extreme environments.

2 Key technology

The key technologies of high-performance precision casting of large-scale complex titanium alloys face four major engineering challenges: ① large-capacity titanium alloy purification melting; ② high-precision forming of large-scale complex structures; ③ metallurgical defect control of large-scale complex castings; ④ development of ultra-large-capacity vacuum melting equipment. The following methods are adopted to solve the above problems:

1) Large-capacity titanium alloy purification melting and precise composition control: For Ti-6Al-4V and Ti-5Al-2.5Sn alloys, by studying the influence of fine-tuning the main components (Al, Sn, V, etc.) and interstitial impurity elements (C, N, H, O) on the macro- and micro-structures, cryogenic tensile properties, and impact properties of titanium alloys, the mechanism affecting the cryogenic mechanical properties is clarified. An ultra-low interstitial phase cast titanium alloy is invented to achieve a reasonable match between strength and toughness at -253°C. Meanwhile, a large-capacity titanium alloy vacuum levitation + consumable arc hybrid melting technology is developed. By studying the influence of levitation melting process on alloy elements and the effect of levitation melting times on composition purification, the change mechanism is clarified. By adjusting the melting process, compensating for element burnout, and adding trace elements, precise control of the composition is achieved.

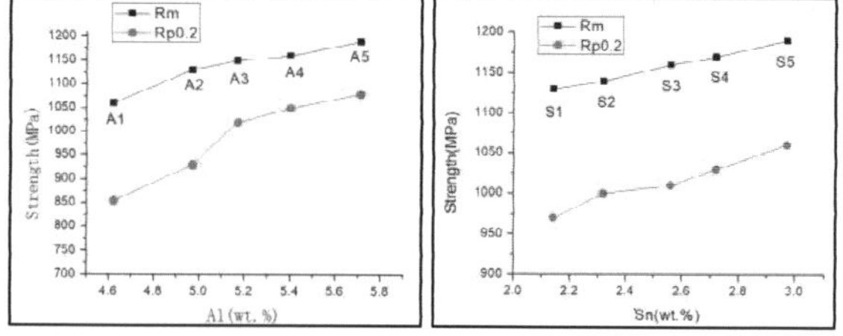

(a) (b)

Figure 1:(a)The Effect of Al Element on the Properties of Ti-5Al-2.5Sn Alloy, (b) The Effect of Sn Element on the Properties of Ti-5Al-2.5Sn Alloy

2) Precision integral forming of large-scale complex thin-walled structures and control of high-performance dense solidification structures: High-precision investment casting patterns are prepared through photosensitive resin laser forming technology. Through research on the preparation technology of composite rare earth oxide ceramic shells, the thermophysical and chemical interaction laws between the shell surface and titanium melt are discovered. New processes such as yttrium rare earth composite surface layers, anti-back-dissolution binders, and flexible reinforcement are developed to solve problems such as interface contamination, mold shell deformation, and thermal shock cracking during pouring. This enables high-precision, high-stability, and high-cleanliness preparation of the mold shell. Applying this technology, large-scale thin-walled frame titanium alloy castings such as aircraft wings and carrier rocket grid frames have been developed, with the largest dimension exceeding 2700mm. The pollution layer thickness is ≤ 6μm, the dimensional accuracy is CT6-7 grade, and the roughness Ra is ≤3.2μm.

3) Preparation of high-strength, heat-resistant cores/molds and deformation control of thin-walled complex cavities:By studying the influence of mold thermal properties on the filling and solidification processes of titanium alloys, a low-thermal-conductivity, high-stability coating for graphite molds is invented. This addresses the issues of insufficient filling capacity caused by graphite chilling and graphite carbon contamination, effectively improving the forming quality of complex thin-walled titanium alloy structures. It solves the challenges of controlling the solidification rate in structures with large wall thickness differences and variable cross-section flow channels, overcomes the difficulties in forming and cleaning extremely narrow structures, and achieves precise forming of complex narrow cavities with a maximum size of 3.6m and a maximum single-piece weight exceeding 2000kg.

4) Development of Large-Capacity Vacuum Melting and Casting Equipment: A mathematical-physical model for high-power vacuum arc melting is established, and components such as a double-chamber U-shaped furnace body, flexible centrifugal mechanism, high-conductivity electrodes, and a coaxial symmetric potential trap arc stabilization device are developed. This achieves arc fluctuation suppression and automatic adjustment, breaking through the challenges of pressure resistance, vibration resistance, and explosion-proof capabilities of large vacuum chambers. A 3-ton titanium melting capacity, 4-meter internal cavity ultra-large vacuum consumable crust melting and casting furnace is developed and manufactured. Through technological breakthroughs in high-power/high-frequency induction power supplies and the development of large-capacity cold crucibles, issues such as multi-station power interference and suspension matching effects are resolved. The world's first 100kg suspension furnace is developed, achieving high-homogeneity, high-purity melting and pouring of titanium alloys.

(a) (b)

Figure 2:(a)3tTitaniumArc Self-consuming Skull Melting Furnace, (b)100kg Titanium Vacuum Magnetic Suspension Melting Furnace

3 Development trend

Titanium alloys hold significant application prospects in high-end equipment fields such as aviation, aerospace, and national defense weapons. The author believes that the development trends of titanium alloy casting technology mainly focus on: ① pushing the limits of casting equipment technology; ② integrated manufacturing through multi-process and multi-technology combinations; ③ digitization and intelligentization of the casting process; and ④ green and low-cost manufacturing

Study on the Thermal Deformation Behavior and Collaborative Recrystallization Mechanism for a Coarse-Grained Near-β Titanium Alloy Ti6455x

Jiaqi Hao, Jichang Yu, Hongze Fang, Ruirun Chen
School of Materials Science and Engineering, Harbin Institute of Technology, Harbin 150001, PR China

*Corresponding address: Heilongjiang Province, Harbin City, Nangang District, 92 West Dazhi Street
*Corresponding e-mail: hjq_hit@163.com

Abstract: To study the effect of hot working on the toughening of titanium alloys, a near-β titanium alloy T4655x was designed, with room temperature tissues having only large-size β grains and a deformation activation energy of only 128.63 kJ/mol, which is lower than that of the pure β-Ti. The thermal deformation behavior of the alloy was also analyzed under different hot working parameters. It was observed that the dissipation efficiency coefficient η value was higher at low strain rates, and the deformed tissues were sufficiently softened with good properties. A synergistic process of CDRX and DDRX exists when machined at 1223 K, $0.001s^{-1}$.

Keywords: Coarse-grained titanium alloy, Dynamic recrystallization (DRX), Low activation energy

1 Introduction

Near-β titanium alloy has high tensile, shear, yield strength and has a very good hot processing performance and strength-plasticity matching development potential[1], suitable for hot processing as a raw material for further process design and development[2]. To realize the enhancement of strength and plasticity of titanium alloy, the influencing of thermal parameters on microstructure are of great significance to explore.

Hu et al.[3] studied the thermal deformation behavior and microstructure evolution of Ti-3.5Al-5Mo-4V alloy in the single-phase region and found that, in as-cast near-β titanium alloys, with large β grain size, long and smooth grain boundaries, few DRX nucleation sites and unfavorable aggregation of dislocations and substructures, the alloys are mainly susceptible to DRV during thermal processing. However, the deformed grains remain elongated and fibrous along the deformation direction during DRV[4].

Therefore, the Ti-4Al-6Cr-5Mo-5Nb-1.6Ta alloy with lower deformation activation energy was designed in this experiment. Isothermal thermal compression of the designed alloy was carried out in the β-single phase region and the cross-sectional microstructure was observed and the microstructure evolution under different parameters were analyzed.

3 Experimental procedure

In this experiment, the alloy matrix composition Ti-4Al-6Cr-5Mo-5Nb-1.6Ta, which meets the category of substable β titanium alloys. The thermal simulation experiments were carried out in Gleeble-1500 thermal simulation testing machine, intercepted Φ6×9mm cylindrical compression specimens. These orthogonal experiments were conducted with four temperature samples at 1123K, 1173K, 1223K and 1273K, and four strain rate samples at 0.001, 0.01, 0.1 and $1s^{-1}$.

The samples after isothermal compression tests were sectioned parallel to the compression direction (CD) for microstructure characterization including optical microscopy (OM), scanning electron microscopy (SEM). The electron backscatter diffraction (EBSD, SUPRA55) was used to analyze the hot processing microstructure.

4 Result and discussion

Microstructure analysis

When the strain rate is between $0.1s^{-1}$ and $1s^{-1}$, the grains are mostly elongated along the direction perpendicular to compression. As temperature reaches 1273K, the grains are refined, but the same elongation along the vertical compression direction. The grain boundaries are completely serrated, which is consistent with the characteristics of DRX grains.

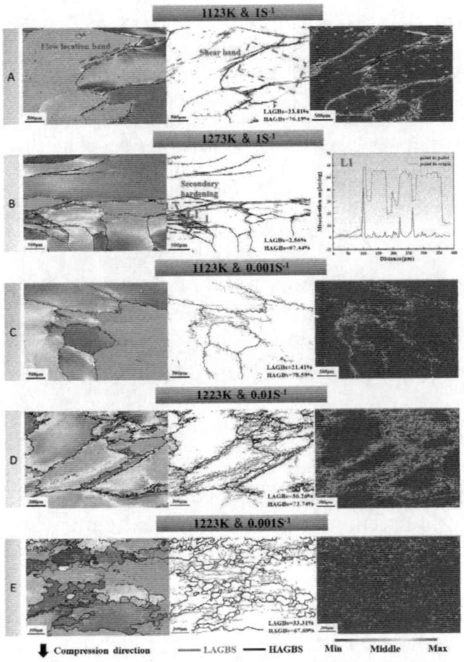

Figure.1 Deformation microstructure of Ti4655x at different thermal parameters

When the strain rate is between 0.001 and 0.01s^{-1}, at 1123K, the grains were fibrillary elongated and larger in size, and the grain boundaries were still very smooth. As the temperature increases to 1173K, the local grain boundaries begin to shift like jagged, and new isometric grains with smaller sizes have appeared within the grain. When the temperature reaches 1223K, fine equiaxed grains appear in the microstructure and some of them adhere to the original grain boundaries.

Deformation kinetic analysis

Combining the calculations, the rheological stress principal equation model for the alloy under normal stress conditions is shown in Equation(1)

$$\dot{\varepsilon} = 3.7\times10^5[\sinh(0.0095\sigma)]^{4.324}exp\left(-\frac{128630}{RT}\right) \qquad (1)$$

The activation energy map of Ti6455x alloy is plotted as Figure.2. The all Q values are lower than that of pure β-Ti, indicating the existence of different softening mechanisms.

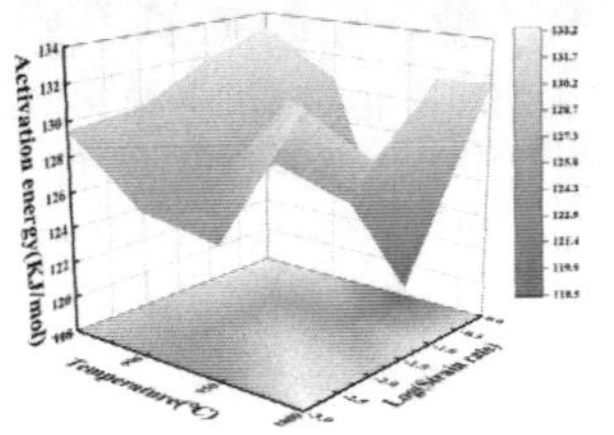

Figure.2 The 3D deformation activation energy map of Ti6455x alloy under different parameters

Combined with organizational and computational data, the machining diagrams can accurately point out the thermal deformation behavior and machining window of the alloy under different thermal processing parameters.

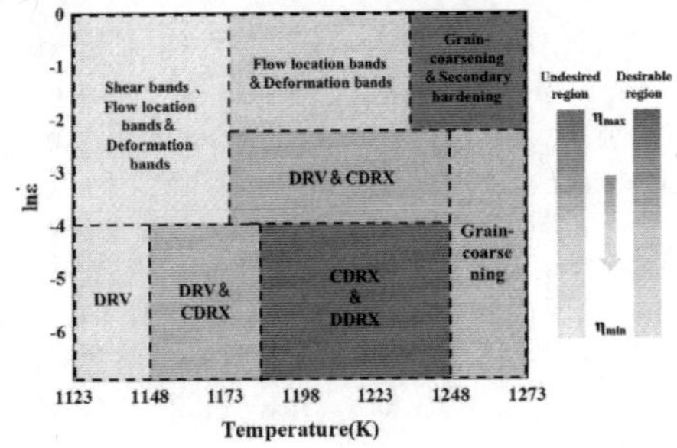

Figure.3 Hot processing map of Ti4655x alloy

5 Conclusion

Under low strain rates 0.001-0.01s^{-1}, with increasing η value, the microstructure is softening by a mechanism of DRV → DRV + CDRX → CDRX + DDRX. The microstructure with highest η is at 1223 K and 0.001s^{-1}, where CDRX and DDRX fine grains coexist.

Acknowledgments

The authors are grateful to the financial support by National Natural Science Foundation of China (Grant No. U21A2042), the National Nature Fund Youth Fund Project of China (No. 52101038), and Young Elite Scientists Sponsorship Program by CAST (2021QNRC001).

References

[1] P. La Roca, M. Sade, Designing a wider superelastic window, Science, 2020, 369: 773–774.
[2] A. S. Chauhan, J. S. Jha, S. Telrandhe, S. V, Laser surface treatment of α - β titanium alloy to develop a β -rich phase with very high hardness, J. Mater. Process. Technol., 2021, 288: 116873.
[3] S. Hu, Y. B. Tan, W. Shi, Microstructure, texture evolution and mechanical behavior of Ti-3.5Al-5Mo-4V titanium alloy during hot rolling in a β field, Mater. Today 2022, 31: 103506.
[4] S. Shi, J. Ge, High-temperature deformation behavior and recrystallization mechanism of a near beta titanium alloy Ti-55511 in β phase region, Mater. Sci. Eng. A, 2022, 847: 143335.

A Novel β-Ti Alloy with Both Strength and Toughness Designed Based on Mo Equivalent and d-Electron Theory

Yili Li[1,*], Shichen Sun[1], Xiang Xue[1], Hongze Fang[1], Ruirun Chen[1,2]
1. National Key Laboratory for Precision Hot Processing of Metals, Harbin Institute of Technology, Harbin 150001, PR China
2. School of Materials Science and Engineering, Harbin Institute of Technology, Harbin 150001, PR China
*Corresponding address: Heilongjiang Province, Harbin City, Nangang District, 92 West Dazhi Street, E-mail: lyllyl3333@163.com

Abstract: Ti-xMo-4Al-4Zr-3Nb-2Cr-1Fe alloys were prepared according to Mo[eq] and d electron theory. Results show that the relative content of β phase increases by adjusting Mo content. Length-width ratios of αp and αs phases decrease from 8.8 to 6 and 10.8 to 9.1 as Mo increases from 5 to 6 wt.%. The dislocation density reaches its maximum at 6Mo. The tensile strength first increases and then decreases, reaching a maximum of 1336 MPa at 6Mo. The toughness of 6Mo alloy is 85 MPa·m$^{1/2}$. The strength increases by 12% while the toughness only decreases by 4%. The precipitation strengthening caused by the optimization of the (α+β) microstructure and the dislocation strengthening caused by the increased dislocations are the determining mechanism of the trade-off between strength and toughness.

Keywords: Titanium alloy; Ultra-high strength and toughness; Microstructure evolution; Mo[eq]; d electron theory

1 Introduction

With the development of the new generation of aircraft in the direction of high speed, large size, complex structure, and low fuel consumption, titanium alloys are required to have high toughness and high damage tolerance while meeting the design requirements of high specific strength [1,2]. The development of ultra-high strength (tensile strength ≥ 1300 MPa) and high toughness (fracture toughness ≥ 55 MPa·m$^{1/2}$) titanium alloys is an inevitable trend in the titanium industry in the 21st century [3,4].

The α phase is one of the most basic and important phases in titanium alloys. The αp grains with larger size can bear more plastic deformation to ensure plasticity [5], while the αs lamellae with smaller size can reduce the dislocation slip length in the residual β matrix, which is conducive to improving the strength [6,7]. The final properties of the alloys are determined by the size and volume fractions of the αp and αs phases. The molybdenum equivalent (Mo[eq]) is a key parameter in the design of the composition of β titanium alloys. It is reported that an alloy with Mo[eq] of 9~15 wt.% has proper α and β phase ratio after aging, therefore it has excellent mechanical properties [8]. The element Mo is one of the most direct elements affecting Mo[eq].

In this study, five alloys (Ti-xMo-4Al-4Zr-3Nb-2Cr-1Fe alloy) with different Mo contents were prepared according to Mo[eq] and the d electron theory, and an alloy with high matching degree of strength-toughness can be obtained by controlling the Mo content to optimize the (α+β) phase, so as to provide alloy ingots with better performance for subsequent deformation.

2 Experimental procedure

Ti-*x*Mo-4Al-4Zr-3Nb-2Cr-1Fe (*x*=5, 6, 7, 8, and 9 wt.%) alloys were made by a high-vacuum arc melting furnace. Solid solution treatment and aging treatment were carried out in a tubular vacuum heat treatment furnace. Optical Microscope and Scanning electron microscopy were used to observe the microstructures of the designed alloys. An electronic universal materials testing machine was used to test the tensile properties and the fracture toughness of the designed alloys.

3 Result and discussion

Microstructural characteristics of different Mo contents

The change of β grain size of designed alloys was observed and the OM images are shown in Fig. 1. The original β grain is relatively coarse and exhibited a characteristic columnar structure. The size of the β grain first decreases and then increases with increasing Mo content. The grains also gradually change from columnar to equiaxed and then to columnar. The 5Mo alloy has the largest size, 0.9 mm, and the 7Mo alloy has the smallest, 0.4 mm.

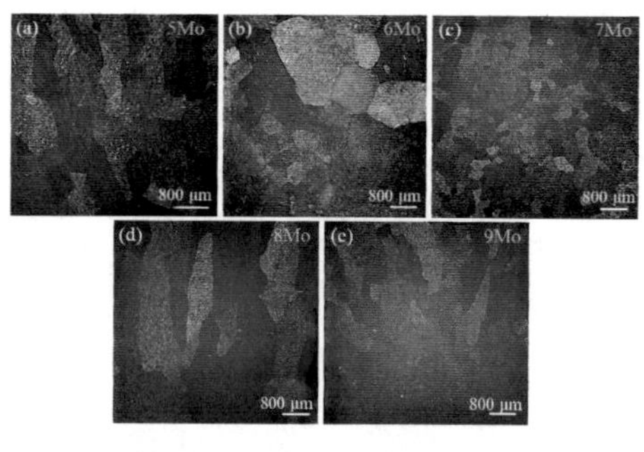

Fig. 1 OM images of Ti-xMo-4Al-4Zr-3Nb-2Cr-1Fe alloys.

Fig. 2 shows the SEM pictures of the designed alloys. The αp phase presents a strip shape and its distribution is uneven. The αs phase is irregularly cross distributed with different orientations. The length of the αp phase first decreases and then increases with increasing Mo content. When the Mo content is 6 wt.%, the length reaches its minimum. However, the volume fraction of the αp phase decreases with increasing Mo content.

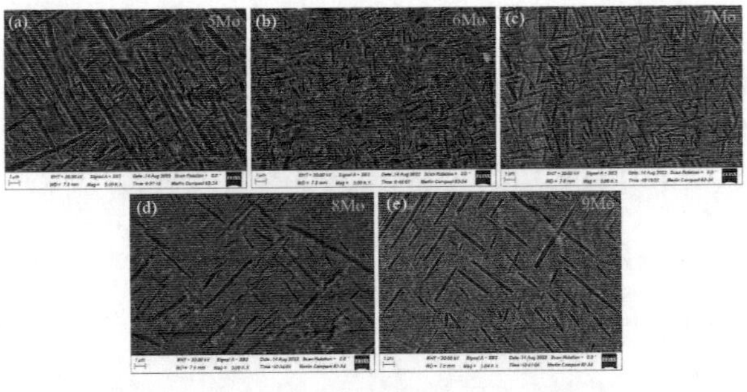

Fig. 2 SEM images of Ti-xMo-4Al-4Zr-3Nb-2Cr-1Fe alloys under the high magnification.

Tensile strength and fracture toughness of Ti-xMo-4Al-4Zr-3Nb-2Cr-1Fe alloys

Table 1 shows the results of the engineering tensile properties and fracture toughness. The strength of 5Mo alloy is 1178 MPa and its elongation is 5.6%. With increasing Mo content, the strength and elongation reach the highest point when the Mo content is 6 wt.%. The strength for 6Mo alloy is 1336MPa and its elongation is 6.9%. The fracture toughness decreases linearly with an increase in Mo content. The value is the highest for 90 MPa·m$^{1/2}$ in 5Mo alloy, while the lowest is 70 MPa·m$^{1/2}$ of 9Mo alloy.

Table 1. Mechanical properties of Ti-xMo-4Al-4Zr-3Nb-2Cr-1Fe alloys.

Alloy	UST (MPa)	YST (MPa)	El (%)	K_{IC} (MPa·m$^{1/2}$)
5Mo	1178	1012	5.6	90
6Mo	1336	1233	6.9	85
7Mo	1314	1198	6.5	80
8Mo	1288	1135	6.1	76
9Mo	1255	1101	5.8	70

Effect of different Mo contents on a trade-off between strength and toughness

The test results show that the strength is improved while the toughness is reduced a little after adding Mo. the increase in strength is mainly due to the contribution of the fineness of the αs phase and the high density of dislocations. Therefore, the precipitation strengthening caused by the optimization of the (α+β) microstructure and the dislocation strengthening caused by the increased dislocations are the determining mechanisms influencing the trade-off between strength and toughness. The content and size of αp phase in the microstructure significantly affect the fracture toughness of β titanium alloy. When passing through the αp phase, the cracks will branch or produce secondary cracks due to turning, making the growth path more tortuous, thus consuming more energy and improving the fracture toughness.

4 Conclusion

In summary, β titanium alloys with good strength and toughness matching can be obtained by regulating the (α+β) microstructure. The optimization of the (α+β) microstructure and the increased dislocations are the determining mechanisms influencing the trade-off between strength and toughness. The alloy design strategy shown in this study can effectively regulate the microstructure of β-Ti alloy, laying the foundation for subsequent rolling, heat treatment and other process methods.

Acknowledgments

The authors are grateful to the financial support by National Natural Science Foundation of China (Grant No. U21A2042), the National Nature Fund Youth Fund Project of China (No. 52101038), and Young Elite Scientists Sponsorship Program by CAST (2021QNRC001).

References

[1] X.Q. Wang, Y.S. Zhang, W.Z. Han. Design of high strength and wear-resistance β-Ti alloy via oxygen-charging, Acta Mater. 2022, 227: 117686.
[2] T.T. Yao, Y.G. Zhang, L. Yang, et al. A metastable Ti-Zr-Nb-Al multi-principal-element alloy with high tensile strength and ductility, Mater. Sci. Eng. A. 2022, 851: 143646.
[3] J. Lv, K. Luo, H. Lu, et al. Achieving high strength and ductility in selective laser melting Ti-6Al-4V alloy by laser shock peening, J. Alloys Compd. 2022, 899: 163335.
[4] N.L. Church, C.E.P. Talbot, G.J. Wise et al. Development of ultra-high strength Ti-Nb-Sn alloys through grain-refinement by in situ Zener pinning, Scr. Mater. 2022, 218: 114809.
[5] Y. Cao, N. Li, Y. Luo, et al. A novel ultra-high strength titanium alloy via hierarchical α/α′ precipitation strengthening, Mater. Sci. Eng. A. 2022, 840: 142878.
[6] S.W. Lee, J.M. Oh, J.H. Kim, et al. Demonstration of martensite reorientation-induced plasticity by ultra-high strength titanium alloys, Mater. Sci. Eng. A. 2021, 807: 140878.
[7] F. Gao, Y. Guo, S. Qiu, et al. Fracture toughness of friction stir welded TA5 titanium alloy joint, Mater. Sci. Eng. A. 2020, 776: 138962.
[8] C. Zhu, G. Peng, Y.C. Lin, et al. Effects of Mo and Cr contents on microstructures and mechanical properties of near β-Ti alloy, Mater. Sci. Eng. A. 2021, 825: 141882.

Influence of Tantalum on the Microstructure and Properties of Ti-46Al-8Nb-1.5Cr Alloy

Jiangshan Liang, Liao Mi, Lingyan Zhou, Hongze Fang, Ruirun Chen
National Key Laboratory for Precision Hot Processing of Metals, Harbin Institute of Technology, Harbin, Heilongjiang, 150001, PR China
*Corresponding address: Heilongjiang Province, Harbin City, Nangang District, 92 West Dazhi Street
*Corresponding e-mail: fanghongze@hit.edu.cn

Abstract: In order to study the effects of different Ta contents on the microstructure and mechanical properties of Ti-46Al-8Nb-1.5Cr-*x*Ta alloy, Ti-46Al-8Nb-1.5Cr-*x*Ta alloy was prepared by vacuum non-self-consumable arc melting method. The results show that the Ti-46Al-8Nb-1.5Cr-*x*Ta alloy is composed of γ phase, α2 phase and B2 phase. With the increase of Ta content and the increase of B2 phase, the microstructure of the microstructure showed a trend of refinement. When the Ta content is 0.2at.%, the compressive strain of the alloy reaches the maximum, which is 39%, and the compressive strength is 1956MPa, when the Ta content is 0.4at.%, the compressive strength of the alloy reaches the maximum, which is 2316MPa, and the compressive strain is 31%. Ta mainly affects the properties of alloys by influencing the microstructure of alloys. The refinement of the microstructure is the most important strengthening method to improve the mechanical properties of the alloy.

Keywords: TiAl alloy ; Ta ; Microstructure evolution ; Mechanical properties

1 Introduction

TiAl alloys have attracted extensive attention due to their high specific strength, low density, and good creep resistance[1], but the intrinsic brittleness and high temperature properties of TiAl alloys have seriously limited their development in more fields. In order to further improve the availability of TiAl alloys, researchers have adopted alloying methods to improve their structure and properties. The fourth-generation TiAl alloy developed by the European Union contains a high content of Ta element, which has good creep resistance at high temperatures above 800°C, and has good application value at high temperatures as a new generation of alloys[2].Vojtnotch et al. compared and analyzed the high-temperature cyclic oxidation properties of Ti-45.2Al-7.2Nb and Ti-44.8Al-6.6Ta (at.%) alloys. The results show that the high-temperature oxidation resistance of Ta is better than that of Nb at 1000°C[3]. However, there are relatively few reports on the effect of Ta on the microstructure and properties of TiAl alloys, so it is necessary to systematically study the effects of Ta as an alloying element on the microstructure and properties of TiAl alloys.

In this paper, Ti-46Al-8Nb-1.5Cr-*x*Ta (x=0.2, 0.4, 0.6, 0.8, 1.0, 1.2at.%)alloy was prepared by vacuum non-self-consumable arc melting method. The microstructure and compression performance of the alloy at room temperature were systematically studied, and the influence of Ta content on the microstructure and properties of Ti-46Al-8Nb-1.5Cr alloy was revealed.

2 Experimental procedure

In this paper, the Ti-46Al-8Nb-1.5Cr-*x*Ta (x=0.2, 0.4, 0.6, 0.8, 1.0, 1.2at.%) alloy was studied. Raw materials such as titanium sponge, aluminum particles, tantalum particles, Al-Nb master alloys are used, and the button ingots are prepared by vacuum non-self-consumable arc melting method, and smelted four times in an argon atmosphere to ensure uniform composition.

A solution of 10 vol% HNO3 + 5 vol% HF + 85 vol% H2O was etched for 10 s for SEM observation and tissue observation was performed using scanning electron microscopy (SEM). The compression performance test is carried out at a tensile rate of 0.3mm/min using an electronic universal testing machine.

3 Result and discussion

Microstructure of Ti-46Al-8Nb-1.5Cr-xTa alloy

Fig. 1 shows the microstructure of Ti-46Al-8Nb-1.5Cr-*x*Ta alloy, and the grain boundaries of Ti-46Al-8Nb-1.5Cr-*x*Ta alloy are very obvious, and its phases are composed of γ phase, α2 phase and B2 phase. With the increase of Ta content, the microstructure underwent two changes.

on the one hand, the precipitation of B2 phase in tissues showed an increasing trend, These B2 phases are distributed at the junction of lamellar and lamellar clusters, and the size is larger, this change is mainly due to the fact that Ta is a strong β stable element, and the increase of Ta content is conducive to promoting the formation of B2 phase, but when the B2 phase is formed, the properties of the alloy will be adversely affected.

On the other hand, with the increase of Ta content, the size of the lamellar was refined, and the refinement effect was most obvious when the Ta content was 0.8at.%. This is due to the fact that when Ta is added to the TiAl alloy, the content of β stable elements in the liquid phase is high, and the composition will be supercooled at the solidification front where the β phase is formed, resulting in the formation of new grains at the solidification front, and there is no directional relationship between the grains and any phase. With the higher the content of β stable elements, the more grain content without orientation relationship in the alloy increases, which increases the heterogeneous nucleation rate during the solidification process, and finally refines the microstructure of the alloy[4].

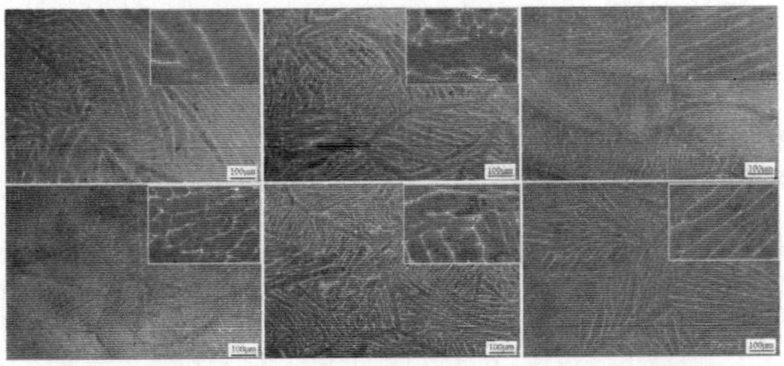

Fig.1 Microstructure of Ti-46Al-8Nb-1.5Cr-xTa alloy (SEM) : (a) 0.2Ta, (b) 0.4Ta, (c) 0.6Ta, (d) 0.8Ta, (e) 1.0Ta, (f) 1.2Ta. Mechanical performances of Ti-46Al-8Nb-1.5Cr-xTa alloy

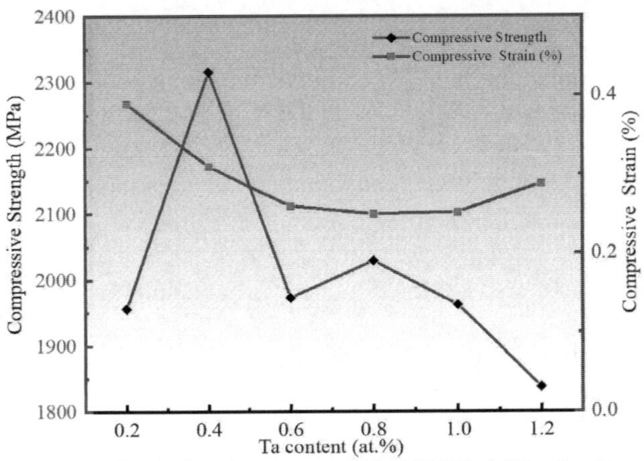

Fig.2 Mechanical performances of Ti-46Al-8Nb-1.5Cr-xTa alloy.

The compressive properties of Ti-46Al-8Nb-1.5Cr-*x*Ta alloy at room temperature are shown in Figure 2. It can be seen from the figure that with the increase of Ta content, the compressive strain of Ti-46Al-8Nb-1.5Cr-*x*Ta alloy decreases first and then increases at room temperature, and its compressive strength increases first and then decreases. When the Ta content is 0.2at.%, the compressive strain of the alloy is 39% and its compressive strength is 1956MPa, and when the Ta content is 0.4at.%, the compressive strength of the alloy is 2316MPa and its compressive strain is 31%. The main reason for the change in its properties is due to the refinement of the microstructure, according to the Hall-Petch equation, the smaller the grain size, the better its properties, but with the increase of B2 phase, the properties of the alloy will deteriorate, and the properties of the alloy will decrease.

4 Conclusion

In summary, improving the mechanical properties of TiAl alloys by adding Ta mainly affects the properties of the alloys by affecting the microstructure of the alloys. The effect of Ta on the microstructure and mechanical properties of Ti-46Al-8Nb-1.5Cr-*x*Ta alloy was studied, and the main conclusions are as follows:

1.The Ti-46Al-8Nb-1.5Cr-*x*Ta alloy is composed of γ phase, $\alpha 2$ phase and B2 phase, and the B2 phase is distributed at the junction of lamellar and lamellar clusters, and the size is larger.

2.With the increase of Ta content, the microstructure of Ti-46Al-8Nb-1.5Cr-*x*Ta alloy was refined, and the refinement effect was most obvious when the Ta content was 0.8at.%, but the content of B2 phase gradually increased with the increase of Ta content.

3.When the Ta content is 0.2at.%, the compressive strain of the alloy is 39% and its compressive strength is 1956MPa, and when the Ta content is 0.4at.%, the compressive strength of the alloy is 2316MPa and the compressive strain is 31%. Its strengthening mechanism is mainly fine-grained strengthening.

Acknowledgments

The authors are grateful to the financial support by National Natural Science Foundation of China (Grant No. U21A2042), the National Nature Fund Youth Fund Project of China (No.52101038), Young Elite Scientists Sponsorship Program by CAST (2021QNRC001).

References

[1] B.G. Liu, L.H. Liu, W.D. Xing, R.C. Liu, R. Yang, P.A. Withey, J. Zhu, R. Yu, Structural stability and the alloying effect of TiB polymorphs in TiAl alloys, Intermetallics 90 (2017) 97–102.

[2] W. J, Zhang, et al. On the origin of superior high strength of Ti–45Al–10Nb alloys[J]. Intermetallics, 2002, 10(5):403-406.

[3] D. Vojtˇech, T. Popela, J. Kub´asek, J. Maixner, P. Nov´ak, Comparison of Nb- and Taeffectiveness for improvement of the cyclic oxidation resistance of TiAl-based intermetallics, Intermetallics 19 (2011) 493–501.

[4] T.T. Cheng. The mechanism of grain refinement in TiAl alloys by boron addition-an alternative hypothesis [J]. Intermetallics, 2000, 8(1): 29- 37

Influence of Nb Modification on the Superplastic Behavior and Mechanical Properties of Titanium Alloy at 650 ℃

Shichen Sun, Hongze Fang*, Yili Li, Ruirun Chen*
National Key Laboratory for Precision Hot Processing of Metals, Harbin Institute of Technology, Harbin 150001, China
*Corresponding address: Heilongjiang Province, Harbin City, Nangang District, 92 West Dazhi Street
*Corresponding e-mail: sunshichen001@126.com

Abstract: To study the superplastic behavior and mechanical properties below recrystallization temperature, titanium alloy with different Nb content were stretched at 650 ℃. The results show that Ti-9Nb titanium alloy shows the best tensile strength (307.2 MPa) and superplasticity (106%). The thermal effect in the process of high temperature drawing leads to the precipitation of α phase. Due to the modification of Nb element, its content decreased gradually. Therefore, the tensile strength and plasticity first increase and then decrease. The superplastic mechanism of Ti-9Nb alloy is mainly caused by dislocation motion.

Keywords: Titanium alloy; Niobium element; Superplastic; Mechanical properties

1 Introduction

Cast titanium alloy needs hot deformation modification because of its poor mechanical properties[4, 5]. Titanium alloy shows excellent superplasticity above 1073K.[6]. In contrast, the deformation of low temperature plays an important role in energy consumption and shortening the production process[7, 8]. Among the current methods of reducing superplastic temperature, alloying is one of the most promising methods. Nb is an important alloying element in titanium alloy. In the previous research of the research group, the addition of Nb element to the room temperature strength is gratifying[10]. Researchers have found that the ratio of α/β phase directly affects the superplasticity at 1023 K[6]. At present, the research on superplasticity is still at a higher temperature, and the mechanism of the effect of Nb on superplasticity at lower temperature still needs to be revealed. Therefore, in this study, by adding different contents of Nb elements, the influence mechanism of elements on superplasticity in the cold deformation temperature range was studied, and the effects of elements on mechanical properties were revealed.

2 Experimental procedure

The raw materials with composition Ti-5Al-5Mo-5Cr-2Zr-*x*Nb (wt. %, *x*= 3, 6, 9, 12, abbreviated as Ti-*x*Nb) were melted by vacuum induction melting, and the specific method was reported by our previous work[10]. After holding 5 min at 923 K, the tensile test was completed at a speed of 0.5 mm/min on the universal testing machine, and was heated continuously during the stretching process. The sample processing process is also shown in previous work[10].

3 Result and discussion

The Ti-*x*Nb alloy is stretched at 923 K. The stress-strain curve and morphology are shown in Fig. 2. The stress-strain curve shows that with the increase of Nb content in the range of 3-9 wt.%, the stress of Ti-*x*Nb alloy increases from 158.6 MPa to 307.2 MPa, and the elongation increases from 75% to 106%. The Ti-9Nb alloy not only shows higher mechanical properties, but also shows as high as 106% superplasticity. When the Nb content increases to 12 wt.%, the stress is 212.1 MPa and the strain is 100.1%, which is slightly lower than that of Ti-9Nb alloy.

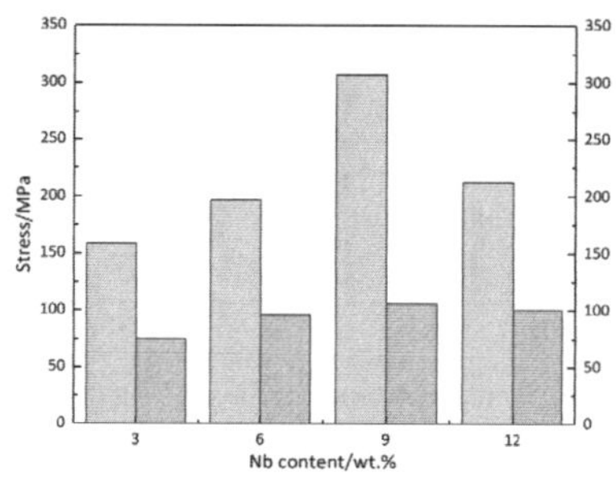

Fig. 2. The stress-strain curve and morphology

The heat preservation in the high temperature drawing process is equivalent to the heat treatment of the sample. Therefore, the microstructure of the unstressed area of the sample is observed as shown in Fig. 2. In Fig. 2(a), the microstructure of Ti-3Nb alloy after high temperature tension is composed of intragranular lamellar α phase and grain boundary α phase. In Fig. 2(b), α phase content and aspect ratio decreased due to the increase of Nb content. In Ti-9Nb alloy, the morphology of intragranular α phase is mainly equiaxed, and the grain boundary α phase is still continuous. According to statistics, the content of α phase is about 15%. As shown in Fig. 2(d), as the Nb content increases to 12 wt.%, the α phase content further decreases.

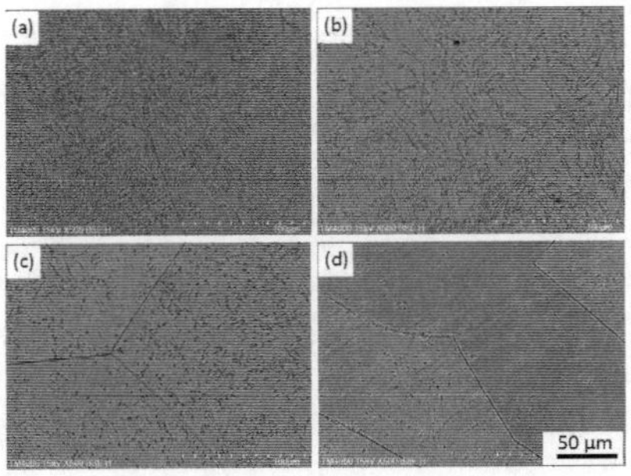

Fig. 2. The microstructure of Ti-xNb alloy: (a) 3 Nb; (b) 6 Nb; (c) 9 Nb; (d) 12 Nb.

Fig. 3(a and b) and 3(c and d) show the microstructure near the fracture surface after high temperature tension_of Ti-3Nb and Ti-9Nb alloys, respectively. As shown in Fig. 3(a), the higher content of α phase in the microstructure captures most of the dislocations and the dislocation density in the β matrix is lower. In Fig. 3(b), the microstructure near the α phase at high resolution shows that the dislocations are arranged near the lamellar α phase to form a dislocation wall, and the dislocation density in the α phase is low. The content of α phase in Ti-9Nb alloy is low, and dispersed dislocations are observed in the β phase shown in Fig. 3(c). The dispersed dislocations cancel each other out with other dislocations during slip, which is the main reason why the superplasticity of Ti-9Nb alloy is as high as 106%. In addition, as shown in Fig. 3(d), a large number of dislocations are deposited near the α phase, and the dislocation density is obviously higher than that of Ti-3Nb alloy.

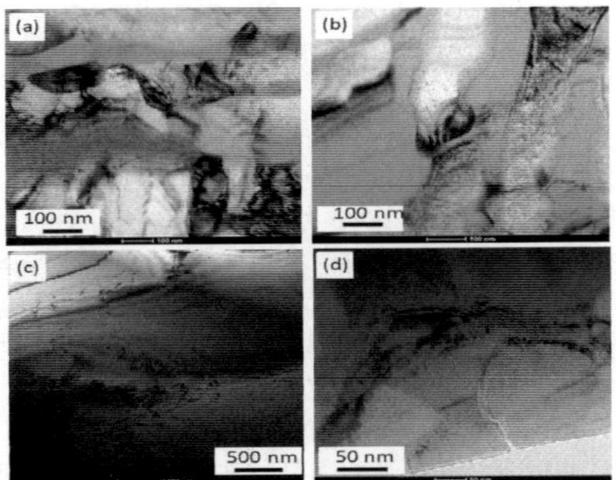

Fig. 3. The microstructure near the fracture surface after high temperature tension: (a, b) Ti-3Nb alloy; (c, d) Ti-9Nb alloy.

4 Conclusion

(1) The as-cast microstructure of Ti-xNb alloy is a single β phase. When the Nb content increases from 3 wt.% to 9 wt.%, the tensile strength increases from 158.6 MPa to 307.2 MPa, and the elongation increases from 75% to 106%.

(2) The dispersed dislocations in β phase and the stacking effect of α relative dislocations in Ti-9Nb alloy are the main reasons for high superplasticity.

5 References

[1] MENG M, FAN X G, YANG H, et al. Precipitation of secondary alpha in competition with epitaxial growth of primary alpha in two-phase titanium alloys [J]. Journal of Alloys and Compounds, 2017, 714: 294-302.

[2] ZHU X, FAN Q, WANG D, et al. A new dynamic recrystallization mechanism in adiabatic shear band of an α/β dual phase titanium alloy: Composition redistribution [J]. Scripta Materialia, 2022, 206.

[3] LIU C, HU X, QI L, et al. Unusual precipitation induced by solute segregation in coherent twin boundary in titanium alloys [J]. Acta Materialia, 2023, 242.

[4] SHI S, GE J, LIN Y C, et al. High-temperature deformation behavior and recrystallization mechanism of a near beta titanium alloy Ti-55511 in β phase region [J]. Materials Science and Engineering: A, 2022, 847.

[5] ZHAO J, GUO H, LUO T, et al. Microstructure evolution and grain refinement mechanism of fine-grained Mg-Gd-Y-Zn-Zr alloy during multi-directional forging [J]. Journal of Alloys and Compounds, 2022, 928.

[6] AKULA S P, OJHA M, RAO K L, et al. A review on superplastic forming of Ti-6Al-4V and other titanium alloys [J]. Materials Today Communications, 2023, 34.

[7] JACKSON M. Superplastic forming and diffusion bonding of titanium alloys [M]. Superplastic Forming of Advanced Metallic Materials. 2011: 227-46.

[8] MIKHAYLOVSKAYA A V, MOSLEH A O, KOTOV A D, et al. Superplastic deformation behaviour and microstructure evolution of near-α Ti-Al-Mn alloy [J]. Materials Science and Engineering: A, 2017, 708: 469-77.

[9] ALABORT E, BARBA D, SHAGIEV M R, et al. Alloys-by-design: Application to titanium alloys for optimal superplasticity [J]. Acta Materialia, 2019, 178: 275-87.

[10] SUN S, FANG H, LI Y, et al. Formation mechanism of titanium solid solution and its influence on equiaxed behavior of α phase of Ti–5Al–5Mo–5Cr–2Zr-xNb alloys [J]. Journal of Materials Research and Technology, 2023, 26: 434-44.

Introduction of Surface Defects in the γ Phase Improves the Mechanical Properties of TiAl Alloys

Xiaokang Yang*, Linyan Zhou, Jiangshan Liang, Hongze Fang, Ruirun Chen
National Key Laboratory for Precision Hot Processing of Metals, Harbin Institute of Technology, Harbin, Heilongjiang 150001, China
*Corresponding address: Heilongjiang Province, Harbin City, Nangang District, 92 West Dazhi Street, Corresponding e-mail: yangxiaokanghit@163.com

Abstract: Actively introducing different types of defects into γ phase is an effective means to improve the mechanical properties of TiAl alloys at high temperatures. Ta was added in Ti-47Al-6Nb-0.1C alloy, and defect types, microstructure, mechanical properties at high temperatures, and related mechanisms of the alloy were studied. Results show that stacking faults (SFs), twin boundaries (TB) and 120° rotation boundaries (RB) are formed in the γ phase between lamellar colonies after adding Ta. Ta promote the generation of SFs and RB by reducing the stacking fault energy and RB formation energy. SF continuously generates and forms nanotwins. The tensile strength and elongation of the alloy at 900 ℃ reach 520.2 MPa and 9.4 %, respectively. TB and RB hinder dislocation slip. SFs decompose dislocations to consume their energy.

Keywords: TiAl alloy; Defect engineering; γ phase; Growth twins; Tensile properties

1 Introduction

Defect engineering aims to artificially change the shape, distribution and quantity of defects in materials. So as to optimize the electrical properties, oxidation resistance, catalytic properties and mechanical properties of the materials [1-3]. Chen et al. [4] regulated the interface gradient distribution from nano-scale to macro-scale to significantly enhance the fatigue resistance of materials. Lu et al. [5] used defect engineering to introduce engineering nanoscale coherent internal interfaces to strengthen nickel-based superalloy. These structures have been proved to be beneficial to the room temperature mechanical properties of the alloy [6-9].

At present, there are still some potential problems in the application of defect engineering in regulating material properties: mainly reflected in the stability of a large number of defects introduced during manufacturing and service, such as loading and special environments [10]. Pippan [11] found that the defects formed by plastic deformation can easily reach submicron saturation. The introduction of defects by heat treatment will inevitably change the original structure of the alloy.

Hence, Ta element is added to TiAl alloy by casting to form nano-scale SFs, growth twins, RBs and LAGBs in γ grains in this study. The high temperature mechanical properties of the alloy are significantly improved. The formation mechanism and strengthening mechanism of surface defects in TiAl alloys were revealed. This result provides a new theoretical basis for the application of defect engineering in TiAl alloys.

2 Experimental procedure

The raw materials for the synthesis of Ti-47Al-6Nb-0.1C-xTa alloy were prepared by vacuum arc melting furnace in the same conditions. The microstructure were observed by scanning electron microscopy (TM 4000 and Merlin Compact). The stacking fault, dislocation and twin were further studied by transmission electron microscopy (JEM 2100 and Talos F200x). The tensile specimen was processed into a plate with a gauge length of 15, a width of 5 and a thickness of 2, and polished. High temperature tensile properties were tested by high temperature electronic universal testing machine 50kN/250kN. The overall temperature of the sample was increased to 900 ℃ and then held for 10 minutes, and then the tensile test was carried out at a tensile speed of 0.5 mm/min.

3 Result and discussion

Fig.1 shows the morphology of γ phase in Ti-47Al-6Nb-0.1C-1.6Ta alloy. The BF image shows that there are a large number of twins in the γ phase, and it is speculated that there should be different γ grains according to the different growth directions of the twins, and the boundary of γ grain corresponds to the end point of the growth twins. Fig. 1(a) confirms that the twins with different growth directions belong to different γ grains. And the twin plane of [1-10]γ_1 is (111). In addition, the upper lattice of the {111} plane rotates at different angles relative to the lower lattice to form 60° pseudo-twins, RB and 180° true twins. The addition of Ta element and the substitution of Al occupancy have the greatest reduction in RB formation energy. Therefore, it is considered that the addition of Ta element is more likely to form a 120° rotation boundary.

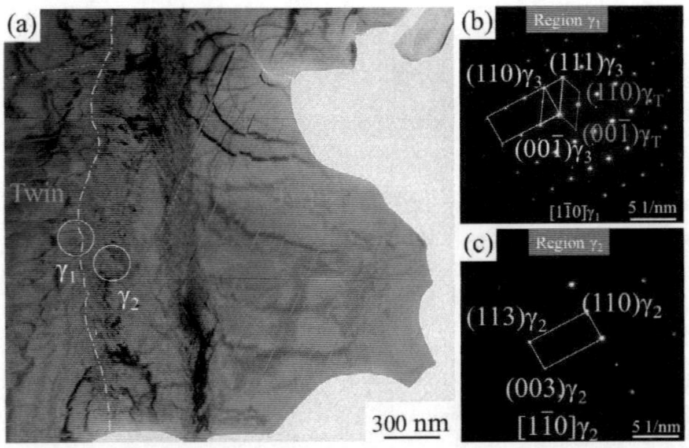

Fig.2 The characteristics of the γ phases in Ti-47Al-6Nb-0.1C-1.6Ta alloys:(a) bright field (BF) image of γ phase; (b-c) SAED patterns of region γ1 and γ2 in (a).

As shown in Fig. 3, when Ta is added, the tensile properties of the alloy at 900 °C were significantly improved after Ta doping. The tensile strength and elongation of the alloy at 900 ℃ reached 520.2 MPa and 9.4 %, respectively. The tensile strength is 16 % higher than that of the alloy without Ta addition. Combined with the above analysis, TB and RB have shear stress field and lattice distortion at the interface. It can hinder dislocation slip and increase the critical shear stress of dislocation activation to improve the deformation resistance of the alloy. SFs can decompose dislocations to consume their energy. Finally, the high temperature mechanical properties of the alloy are improved.

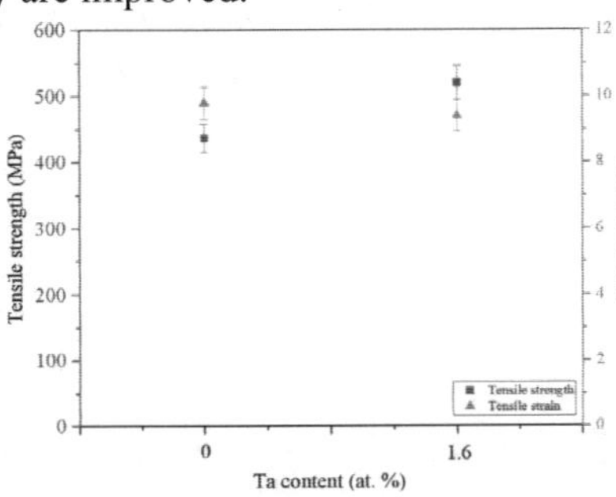

Fig. 3 Stress-strain curves of Ti-47Al-6Nb-0.1C-(0, 1.6) Ta alloys at 900 ℃℃

4 Conclusion

In summary, the formation mechanism and strengthening mechanism of Ta induced surface defects are clarified, which provides a theoretical basis for the application of defect engineering in TiAl alloys. After adding Ta, the stacking fault energy and RB formation energy are reduced, which promotes the formation of SFs and RB. Stacking faults will form twins of atomic thickness and broaden them to the nanoscale. SFs will decompose dislocations, and other interfaces will hinder dislocation slip, ultimately improving the high temperature performance of the alloy.

Acknowledgments

The authors are grateful to the financial support by National Natural Science Foundation of China (Grant No. U21A2042), the National Nature Fund Youth Fund Project of China (No. 52101038), and Young Elite Scientists Sponsorship Program by CAST (2021QNRC001).

References

[1] Yang G, Ni Z, Yu Z J, Larson B W, Yu Z, Chen B, Alasfour A, Xiao X, Luther J M, Holman Z C and Huang J. Defect engineering in wide-bandgap perovskites for efficient perovskite–silicon tandem solar cells, Nat. Photonics 16(8) (2022) 588-594.

[2] Li X, Lu K. Playing with defects in metals, Nat. Mater. 16(7) (2017) 700-701.

[3] Zhu Q, Pan Z, Zhao Z, Cao G, Luo L, Ni C, Wei H, Zhang Z, Sansoz F, Wang J. Defect-driven selective metal oxidation at atomic scale, Nat. Commun. 12(1) (2021) 558.

[4] Chen X, Han Z, Li X, Lu K. Lowering coefficient of friction in Cu alloys with stable gradient nanostructures, Sci. Adv. 2(12) e1601942.

[5] Lu K, Lu L, Suresh S. Strengthening Materials by Engineering Coherent Internal Boundaries at the Nanoscale, Science 324(5925) (2009) 349-352.

[6] Kothari K, Radhakrishnan R, Wereley N M. Advances in gamma titanium aluminides and their manufacturing techniques, Prog. Aerosp. Sci. 55 (2012) 1-16.

[7] Chen G, Peng Y, Zheng G, Qi Z, Wang M, Yu H, Dong C, Liu CT. Polysynthetic twinned TiAl single crystals for high-temperature applications, Nat. Mater. 15(8) (2016) 876-881.

[8] Chen S H, Schumacher G. The application of transformation matrices on the determination of γ/γ interface types in a γ-TiAl alloy, Scr. Mater. 50(1) (2004) 31-34.

[9] Palomares-García A J, Pérez-Prado M T, Molina-Aldareguia J M. Effect of lamellar orientation on the strength and operating deformation mechanisms of fully lamellar TiAl alloys determined by micropillar compression, Acta Mater. 123 (2017) 102-114.

[10] Tao H, Li H, Wang L, Che Y, He W, Li H, Zhou R, Liang X. Improvement on high-temperature tensile strength and ductility of selective electron beam melting TiAl alloys via small-strain forging, Mater. Sci. Eng., A 873 (2023).

[11]Pippan R, Scheriau S, Taylor A, Hafok M, Hohenwarter A, Bachmaier A. Saturation of Fragmentation During Severe Plastic Deformation, Annu. Rev. Mater. Res. 40(1) (2010) 319-343.

Preparation of In-Situ TiB Reinforced Titanium Alloys with Fine TiB Phase Via Ultrasound-Assisted Fusion Casting

JichangYu, Yili Li, Shichen Sun, HongzeFang, RuirunChen
National Key Laboratory for Precision Hot Processing of Metals, Harbin Institute of Technology,Harbin, Heilongjiang, 150001, PR China
*Corresponding address: Heilongjiang Province, Harbin City, Nangang District, 92 West Dazhi Street
*Corresponding e-mail: yujchang2000@163.com

Abstract: The synergistic effects of an ultrasonic field and in situ autogenous TiB on the microstructure and mechanical properties of titanium alloys were examined. The alloy Ti-4Al-6Cr-5Mo-8Nb-1.6B was synthesized via ultrasound-assisted fusion casting. This study analyzed the changes in microstructure and mechanical properties, along with the underlying strengthening mechanisms. The primary β grain size was significantly refined from 1070.8 to 86.9 μm, representing a 91.9% reduction. Additionally, the TiB phase achieved an aspect ratio of approximately 10.3. This refinement is attributed to the effects of ultrasonic cavitation, acoustic streaming, and constitutional supercooling caused by boron. The yield strength of the alloy increased from 809.5 to 1070.4 MPa, which constitutes a 32.2% improvement. This enhancement in yield strength is credited to the refinement of the primary β grains and the supportive role of the TiB phase in load-bearing.

Keywords: Titanium alloy; Ultrasound-assisted; Fine-grain strengthening; Load-bearing strengthening

1 Introduction

Numerous studies have focused on the TiB phase-reinforced titanium alloys due to their potential in strength improvement [1].

TiB phases typically exhibit selective growth, commonly forming whiskers along the [010] direction [2]. The orientation of these in situ TiB phases is usually random, which can be problematic. Specifically, if the stress loading direction does not align with the TiB phase orientation, the elongated TiB phases may adversely impact the alloy's matrix [3]. Thus, mitigating this adverse effect is crucial for enhancing the alloy's strength, achieved by both shortening the TiB whiskers and controlling their orientation.

Furthermore, cast titanium alloys often feature large primary β-grain sizes that negatively affect their mechanical properties [4]. Modulating the microstructure through appropriate heat treatment is vital. However, it invariably raises the production costs of high-strength titanium alloys.

In this study, ultrasound-assisted fusion casting was employed to optimize both performance and cost-effectiveness. This method aimed to directly produce fine primary β-grains and shorten the length of TiB phases using casting techniques.

2 Experimental procedure

Ti-4Al-6Cr-5Mo-8Nb-*x*B (x = 0 and 1.6 wt.%, referred to as 0-1.6B) alloys were prepared under a high purity argon atmosphere. All ingots were melted five times by vacuum arc melting. The frequency of ultrasonic processing used is 20.5 kHz and the processing time is 120 s.

Samples were subjected to etching using Kroll's reagent. Phase analysis was conducted using X-ray diffraction (XRD). Backscattered electron (BSE) was used to analyze the microstructure.

Samples for mechanical performance tests should were tested at least three times (0.5 mm/min tensile rate).

3 Result and discussion

The microstructure and its evolutionary mechanism

As shown in Figure 1(e), the alloy consists solely of a single β phase. With the incorporation of boron (B), the TiB phase emerges within the structure, reducing the β phase content to 72.9%. The main β grain size was significantly reduced from 1070.8 to 86.9 μm after the application of the ultrasonic field and the addition of B. Fig. 1(d) illustrates that the TiB phase achieves an aspect ratio of about 10.3, resembling a short bar.

During solidification, the low solubility of boron results in its expulsion to the solid-liquid interface, leading to constitutional subcooling. This process diminishes the melt's degree of supercooling. The formula for constitutional subcooling is provided below[1]:

$$\Delta T_{CS}=\frac{m_L C_0(1-k_0)}{k_0} \quad (1)$$

where k_0 is the solute partition coefficient (0.013), m_l is the slope of the liquidus line (-66), and C_0 represents the initial solute concentration. Substituting these values into Eq. 1, the constitutional subcooling is calculated to be approximately 82.4 K.

Ultrasonic treatment of the melt enhances the solidification process. The acoustic streaming effect of ultrasonic waves expedites the release of latent heat, thereby accelerating the nucleation and solidification of incipient β-crystals and the TiB phase. The relationship between the critical nucleation radius and the degree of subcooling is given by the following equation [4]:

$$R_g=\frac{2\Gamma}{\Delta T_r}(2)$$

where Γ is the Gibbs-Thompson coefficient and ΔTr is the degree of subcooling required for nucleation. Within a certain range, the increase of the supercooling degree leads to easier nucleation of β-crystals and TiB in the melt, and the β-crystals and TiB phases then undergo refinement.
In addition, the vacuolar effect of ultrasound leads to the rupture of TiB phase and β crystals.

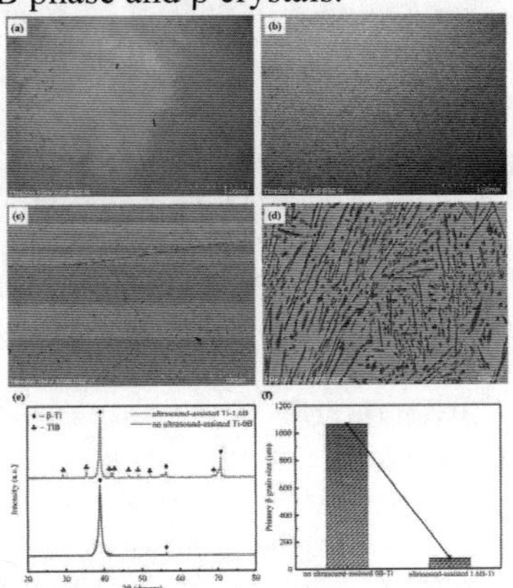

Fig. 1 Microstructure and phase composition: (a, c) Ti-0B; (b, d) Ti-1.6B; (e, f) phase composition and phase ratio.

Strengthening mechanism of alloys

Fig. 2 illustrates the strength, toughness, and elongation properties of the alloy. In the absence of an ultrasonic field and the addition of boron, the yield strength of the alloy registers at 809.5 MPa. Conversely, applying an ultrasonic field and adding boron increases the yield strength to 1070.4 MPa. The enhancement in yield strength is primarily due to the refinement of incipient β-grains and the load-bearing capabilities of the TiB phase.

According to the Hall-Petch formula, the increase in yield strength is quantified as follows[3]:

$$\Delta\sigma_{Matrix}=k_y(d_2^{\frac{1}{2}}-d_1^{\frac{1}{2}})(3)$$

where k_y is the Hall-Petch parameter and d_2 and d_1 denote the dimensions before and after grain refinement. Substituting the data into Eq. 3 yields an increment in yield strength of 156 MPa.

The increment of yield strength due to load bearing strengthening of TiB phase can be expressed by Eq. 4 [2]:

$$\Delta\sigma_{TiB}=\sigma_{Matrix}0.5V_wC_0\frac{l}{d}(4)$$

where V_w is the volume fraction of the TiB phase, l/d is the aspect ratio of the TiB phase, and C_0 is the orientation factor of the TiB phase. Substituting the data into Eq. 4 yields an increment in yield strength of approximately 181.4 MPa due to the TiB phase.

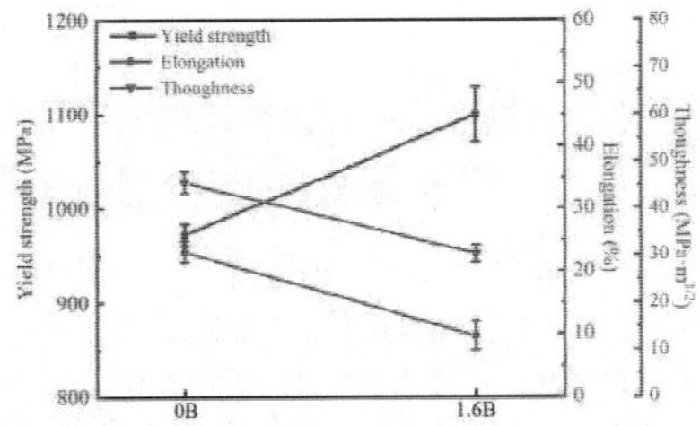

Fig. 3 Yield strength, fracture toughness and elongation of alloys

4 Conclusion

In this work, the effect of ultrasound-assisted fusion casting on the microstructure and mechanical properties of in-situ TiB reinforced titanium alloys was investigated. The mechanism of primary β-grain and TiB phase refinement was revealed, and the strengthening mechanism of the alloy was elucidated. The main conclusions are as follows:

(1) The size of primary β grains is refined from 1070.8 μm to 86.9 μm, and the aspect ratio of TiB phase is 10.3.

(2) The acoustic flow and vacuole effects of ultrasound and B-induced constitutional supercooling led to the refinement of the primary β grains and the reduction of the TiB phase length.

(3) The yield strength of the alloy increased from 809.5 MPa to 1070.4 MPa.

(4) The increase in yield strength is attributed to the refinement of the β-grains and the strengthening of the TiB phase.

Acknowledgments

Theauthors are grateful to the financial support byNationalNatural ScienceFoundation of China·(Grant·No. U21A2042), the National Nature Fund Youth FundProjectofChina(No.52101038),YoungEliteScientistsSponsorshipProgrambyCAST(2021QNRC001).

References

[1] X. Liu, P. Cheng. Dropwise condensation theory revisited: Part I. Droplet nucleation radius. International Journal of Heat and Mass Transfer, 2015, 83: 833-841.
[2] C. Boehlert, S. Tamirisakandala, W. Curtin, et al. Assessment of in situ TiB whisker tensile strength and optimization of TiB-reinforced titanium alloy design. Scripta Materialia, 2009, 61(3): 245-248.
[3] I. Sen, S. Tamirisakandala, D. B. Miracle, et al. Microstructural effects on the mechanical behavior of B-modified Ti–6Al–4V alloys. Acta Materialia, 2007, 55(15): 4983-4993.
[4] K. Kashyap, T. Chandrashekar. Effects and mechanisms of grain refinement in aluminum alloys. Bulletin of Materials Science, 2001, 24: 345-353.

Study on the Microstructure and Properties of Cast High-Temperature Titanium Alloy Based On Cluster Model Design

Tianyu Liu [1,*], Xinqi Li [1], Weichen Qiu [1], Hongyu Liu [1], Hongyu Liu [1], Shibing Liu [1], Kun Shi [1]
1 National Key Laboratory of Advanced Casting Technologies, China Academy of Machinery Shenyang Research Institute of Foundry Co., Ltd. Shenyang 110022, China
*Corresponding address: e-mail: liutianyusrif@163.com.com

Abstract: High-temperature titanium alloys have the characteristics of light weight, high strength and high temperature resistance, and are widely used in compressor discs, blades dan casing of aeroengines. With the further improvement of aero-engine performance, the service temperature of high-temperature titanium alloys parts is increasing, and it is urgent to develop novel high-temperature titanium alloy with resistance above 600℃. In this paper, we designed a series of high-temperature titanium alloys based on the cluster-plus-glue-atom model which describe the short-range order, and studied the influence of alloy elements Mo, Nb, Ta and W on the microstructure, mechanical properties and casting properties of cast high-temperature titanium alloys by OM, SEM, tensile and spiral test methods. The results show that with the increase of [Mo]eq, the alloy strength increases at room temperature, but the alloy strength gradually decreases at 650 °C. Among the four alloy elements of Mo, Nb, Ta and W, the increase of Mo content significantly broadens the solidification range of the alloy, resulting in the greatest influence of Mo on the fluidity of high-temperature titanium alloys, followed by Ta, Nb and W.

Keywords: cast; high-temperature titanium alloy; composition design; the cluster-plus-glue-atom model

1 Introduction

Titanium alloys are widely used due to their good properties in strength, toughness, formability, weld-ability, corrosion resistance and biocompatibility [1]. However, the solid solution strengthening and aging precipitation strengthening that existing high-temperature titanium alloys rely on will reach their limit at 600℃, and further increasing the temperature of the alloy will make it difficult to meet the requirements of thermal strength indicators. At present, high-temperature titanium alloys are mainly characterized by "Ti-Al-Sn-Zr-Mo-Si" multi-component solid solution strengthening. By optimizing the alloy elements and composition, it is still possible to further improve the high-temperature performance of the alloy[2]. In our preliminary work, Mo, Nb, Ta, W, and Si occupied the positions of β--phase cluster type connecting atoms, Sn occupied the positions of α-phase cluster type connecting atoms, Al occupied the positions of α - phase cluster type center and connecting atoms, as well as the positions of β-phase cluster type center atoms, and Zr occupied the shell positions of both α-and β-phase cluster types[3].This article is based on the cluster and connected atom model, and obtains the composition of high-performance cast high-temperature titanium alloy by regulating the matching relationship of alloy elements.

2 Experiment Procedure

In the preliminary work, based on the α and β phase compositions of Ti-6Al-4V alloy, we determined the ideal cluster structural units and their ratios for the two phases, namely α - [Al-Ti12] (AlTi2): β - [Al-Ti14] (V2Ti) ≈ 2.34:1, where the square and small brackets represent clusters and connecting atoms, respectively. Then, the two structural units are regarded as two hard spheres with different radii, and a hard sphere stacking model is constructed. Analogous to atomic resonance theory, it is calculated that one stacking unit contains 17 hard spheres, and the ratio of the two structural units is uniquely determined to be 12:5. Therefore, the cluster structure of Ti-6Al-4V alloy is α-[Al-Ti_{12}]($AlTi_2$): β-[Al-Ti_{14}](V_2Ti), and the ideal alloy composition is $Al_{10.28}V_{3.55}Ti_{86.18}$ (at.%)=Ti-6.05Al-3.94V (wt.%)[4].

The basic cluster structure of the alloy designed in this article is 12 [Al-$Ti_{11.5}Zr_{0.5}$] ($Al_{1.2}5Sn_{0.25}Ti_{1.5}$)+5 [$Al_{0.8}Si_{0.2}$-$Ti_{13.2}Zr_{0.8}$] $(MoNbTaWTi)_3$. Using the cluster structure as a limiting condition, a joint orthogonal experiment was conducted to investigate the effects of alloying elements Mo, Nb, Ta, and W on the microstructure, mechanical properties, and casting performance of high-temperature titanium alloys.

Table 1 Design of alloy composition based on the cluster formula of 5 [Al0.8Si0.2-Ti13.2Zr0.8] (MoNbTaWTi)3

number	Mo	Nb	Ta	W
1	0.2	0.2	0	0
9	1	1	0.2	0

3 Mechanical properties of designed titanium alloys

Figures 1 shows the metallographic structures of alloys 1# and 9# after HIP, respectively. As shown in the figure, the microstructures of both alloys are composed of β-equiaxed crystals, with a Weibull structure within the β-crystals.

(a)

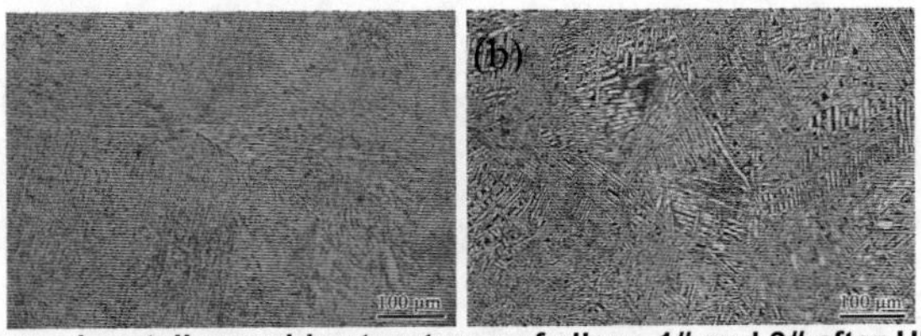

Figure 1 metallographic structures of alloys 1# and 9# after HIP: (a)1#; (b)2#

Figures 2 shows the SEM of 1# and 9# alloys after HIP, respectively. It can be seen from the figure that the β equiaxed crystal of 1# alloy is composed of acicular α phase, and the β equiaxed crystal of 9# alloy is composed of lath α phase. Compared to alloy 1#, alloy 9# exhibits a significant coarsening of the alpha phase.

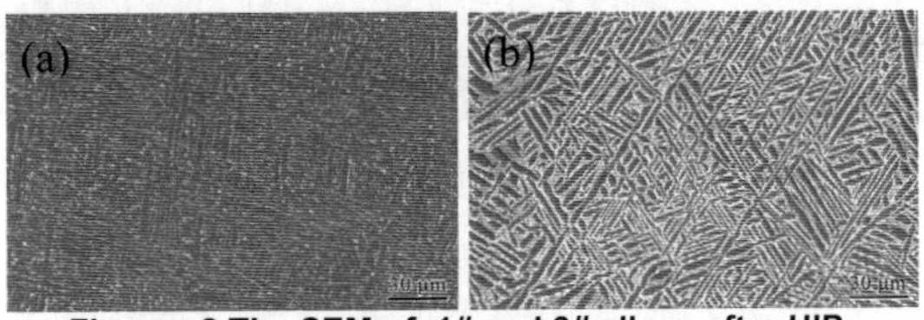

Figures 2 The SEM of 1# and 9# alloys after HIP: (a)1#; (b)2#

According to the formula of Bania's [Mo]eq[5], The stability ability of the β-phase is Mo, W, Nb, and Ta in sequence. Furthermore, according to Bania's [Mo] eq formula, the [Mo] eq of alloy 1# is -5.26, and the [Mo] eq of alloy 9# is -1.32.

Table 2 shows Mechanical properties of different alloy composition at room temperature and 650℃.As shown in the table, with the increase of [Mo] eq, the strength of the alloy increases in room temperature but decrease in 650℃. The elongation rate is on the contrary.. This is because with the increase of [Mo] eq, the stability ability of the alloy's β-phase is enhanced, that is, the content of β phase in the alloy increases, resulting in an increase in the interface between α- and β-phases, which is conducive to hindering dislocation movement, thereby improving the strength of the alloy and reducing its elongation. However, as shown in Figure 4, the alpha phase of alloy 9# is significantly coarsened. This is beneficial for improving the average free path of dislocation movement and increasing the elongation of the alloy.

Table 2 Mechanical properties of different alloy composition at room temperature and 650℃℃

temperature	number	Rm/MPa	$R_{p0.2}$/MPa	A/%
room	1#	958±2	843±1	10±1
	9#	1004±2	888±2	9±2
650℃	1#	582±2	419±3	5.6±1
	9#	537±1	307±2	12.5±1

4 Conclusion

After HIP, compared to the intragranular α-phase of alloy 1#, the α-phase of alloy 9# significantly coarsens;

At room temperature, as [Mo] eq increases, the alloy strength increases; At 650 ° C, as [Mo] eq increases, the alloy strength gradually decreases;

Acknowledgments

This work was supported financially by the Natural Science Foundation of Shenyang, China (Grant No. 22315605).

References

[1] Banerje, D., Williams, J. C. Acta Mater, 61, 2013, 844-879.
[2] YUE K, LIU J, ZHU S, et al. Origins of different tensile behaviors induced by cooling rate in a near alpha titanium alloy Ti65[J]. Materialia, 2018, 1: 128-138..
[3] LIU T Y, LOU Y C, ZHANG S, et al. A novel Ti-5.55Al-6.70Zr-1.50V-0.70Mo-3.41Nb-0.21Si alloy designed using cluster-plus-glue-atom model for laser additive manufacturing[J]. China Foundry, 2023, 20(5): 414-422.
[4] LIU T Y, ZHANG S, WANG Q, et al. Composition formulas of Ti alloys derived by interpreting Ti-6Al-4V[J]. Science China Technological Sciences, 2021, 64: 1732-1740.
[5] MEHJABEEN A, XU W, QIU D, et al. Redefining the β-Phase Stability in Ti-Nb-Zr Alloys for Alloy Design and Microstructural Prediction[J]. JOM, 2018, 70(10): 1-6.

On the Characterization of Microstructure of High-Pressure Die-Cast Alloys with High Thermal Conductivity

Yixian Liu[1], Yongchang Liu[2, 3], Sujing Ge[2, 3], Shoumei Xiong[1, 4*]
1.School of Materials Science, Tsinghua University, Beijing 100084, China
2.Hebei New Lizhong Nonferrous Metals Group Co., Ltd., Baoding071100, Hebei, China
3.Hebei Light Metal Alloy Material Technology Innovation Center, Baoding071100, Hebei, China
4.Key Laboratory for Advanced Materials Processing Technology, Ministry of Education, Beijing 100084, China
*Corresponding address: e-mail: smxiong@tsinghua.edu.cn

Abstract: The high-pressure die-cast Al-1.5Fe alloy was prepared to achieve high thermal conductivity. The microstructure was characterized using various techniques. Die-cast Al-1.5Fe alloy was hypoeutectic and its microstructure mainly consisted of externally solidified crystals primary α-Al (ESCs-α) formed in the shot sleeve, primary α-Al formed in the die cavity (α) and (α-Al+$Al_{13}Fe_4$) eutectics. The microstructure distribution of die-cast Al-1.5Fe alloy was heterogeneous. ESCs-α was enriched in the center of the bar samples because of the melt force and the equivalent diameter of ESCs-α in the center was larger than that in the surface. Moreover, a eutectic-rich band with numerous $Al_{13}Fe_4$ phases was found in the middle of the sample. $Al_{13}Fe_4$ eutectic phases in the eutectic band were more continuous than those in other regions. The volume fraction of porosity increased from surface to center. Due to the low Fe concentration and the large number of aluminum substrates, the die-cast Al-1.5Fe alloy achieved a high thermal conductivity of over 190 W m^{-1} K^{-1} and a high elongation of over 10%.

Keywords: high-pressure die-cast, Al-Fe alloy, microstructure, thermal conductivity

1 Introduction

High pressure die casting is widely used to produce aluminum castings for heat dissipation components in the fields of electrical vehicles, communications and consumer electronics because of its near-net-shape forming characteristics[1,2]. Nowadays, with the development of electronic and renewable energy, the heat generation of components increases rapidly. Traditional Al-Si die-cast alloys with limited thermal conductivity fail to meet the demand of high thermal conductivity. Recently, because of the low solute concentration of element Fe in Al matrix and the necessity of element Fe for the die solder resistance [3], Al-Fe alloy was developed to obtain higher thermal conductivity. Numerous studies have investigated the microstructural characteristics and the effect of elements on microstructure and thermal conductivity of Al-Fe gravity-cast alloys [4]. However, the study on microstructure and thermal conductivity of die-cast Al-Fe alloys is still lacking. In this work, a high-pressure die-cast Al-1.5Fe alloy was prepared, and its microstructure distribution and thermal conductivity of the die-cast Al-1.5Fe alloy were discussed.

2 Experimental procedure

The Al-1.5Fe die-cast sample was produced using a TOYO 350V5 die casting machine. A φ 6.4 mm*10 mm sample was cut from the standard tensile bar, and the sample was used for optical microscope (OM), scanning electron microscope (SEM) equipped with energy dispersive spectrometer (EDS), and electron probe X-ray microanalysis (EPMA). To study the three-dimension morphology and distribution of eutectic $Al_{13}Fe_4$ phases and pores, a φ 1 mm*6.4 mm sample was prepared for the characterization in BL13HB beamline at Shanghai Synchrotron Radiation Facility (SSRF).

3 Result and discussion

1. Al13Fe4 phase in the Al-Fe die-cast alloy

According to the Al-Fe binary phase diagram, the eutectic point is nearly 1.75 wt.%Fe. To obtain the high thermal conductivity and castability, the chemical composition should be hypoeutectic, near the eutectic point. Then, the Fe content is chosen to be 1.5 wt.%. Figs. 1a-a_2 show the eutectics in Al-1.5Fe alloy, which are fully composed of element Fe and Al. It indicates the eutectic consists of α-Al and $Al_{13}Fe_4$ phases. In Fig. 1b, the alloy primarily contains large-size ESCs-α and lamellar eutectics. Few needle-like primary $Al_{13}Fe_4$ phases are also found in Figs. 1c and d.

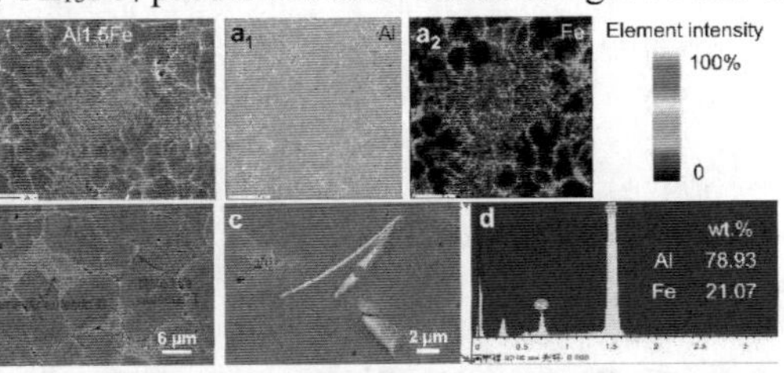

Figure 1 (a-a2) EPMA results of eutectics in Al-1.5Fe die-cast alloy. (b-d) Morphology of eutectics and the primary phase with the corresponding EDS result.

2. The distribution of microstructure in the Al-Fe die-cast alloy

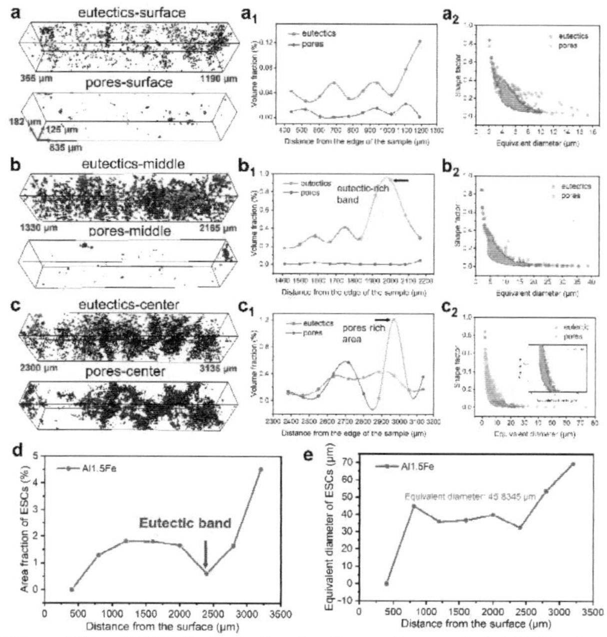

Figure 2 (a-c2) Distribution of eutectics and pores in Al-1.5Fe die-cast alloy. (d-e) Distribution of ESC-α in Al-1.5Fe die-cast alloy.

Figs. 2a-c show the distribution of eutectics and porosity in the surface, middle and center. Because of the limited resolution of the beamline, the reconstructed results can only show the morphology of eutectic cells instead of the individual lamellar phase. In Fig. 2b, eutectics in the middle are coarser and more continuous. The quantitative result in Fig. $2b_1$ indicates that the volume fraction of eutectics increases sharply to ~1% at the position of 2000 μm. There are more eutectics with larger sizes in the middle from Fig. $2b_2$. Figs. 2c and c_1 reveal that the volume fraction of porosity in the center increases dramatically and the most large-size pore is over 70 μm. In Figs. 2d and e, the area fraction and equivalent diameter of ESCs-α both increase in the center because of the Saffman force produced by the different velocity between ESCs-α particles. Notably, in the range of 2000 to 2500 μm, the area fraction of ESCs-α decreases, marking the eutectic-rich band. Because of the gathering of ESCs-α in the center, the size of eutectic cells here is smaller than that in the middle, although the low cooling rate causes the $Al_{13}Fe_4$ eutectic phase becoming larger in size. Additionally, the enrichment of ESCs-α in the center also causes the remaining liquid with a high Fe concentration in the middle. Thus, the eutectic-rich band is formed in this region.

3. Thermal conductivity of the Al-Fe die-cast alloy

Table 1 shows the properties of the Al-1.5Fe die-cast alloy. The thermal conductivity (TC) is 197.1 W m^{-1} K^{-1}. The elongation (EL), yield strength (YS) and ultimate tensile strength (UTS) are 15.7%, 131.7 MPa and 63.4 MPa, respectively. The die-cast alloy has higher thermal conductivity and elongation as well as lower strength compared to commercial die-cast Al-Si alloys. The TC near 200 W m^{-1} K^{-1} is attributed to the low concentration of Fe in Al from the binary phase diagram (~0.05%) and a large number of Al substrate (only 1 vol.% in the eutectic-rich band), which significantly decreases the negative effect of the impurity (solutes and precipitates) scattering on the electron movement in the alloy.

Table 1. Properties of Al-1.5Fe die-cast alloy

YS (MPa)	UTS (MPa)	EL (%)	TC (W m^{-1} K^{-1})
63.4±1.8	131.7±1.9	15.7±6.7	197.1±1.8

4 Conclusion

(1) The die-cast Al-1.5Fe alloy mainly consisted of ESCs-α, αand (α-Al+$Al_{13}Fe_4$) eutectics.

(2) ESCs-α were enriched in the center of the bar and the equivalent diameter of ESCs-α in the center was larger than that in the surface.

(3) An $Al_{13}Fe_4$ eutectic-rich band was in the middle of the sample. $Al_{13}Fe_4$ eutectic phases in the eutectic band were more continuous than those in other regions.

(4) The volume fraction of porosity increased from surface to center.

(5) Due to the low Fe concentration and large number of aluminum substrates, the die-cast Al-Fe alloy had a high thermal conductivity of over 190 W m^{-1} K^{-1} and a high elongation of over 10%.

Acknowledgments

This work was supported by the National Natural Science Foundation of China (Grant No. 52175335) and the beam time support by BL13HB beamline in SSRF.

References

[1] J. Gan, Y. Huang, C. Wen, et al., Effect of Sr modification on microstructure and thermal conductivity of hypoeutectic Al–Si alloys, Transactions of Nonferrous Metals Society of China, 2020, 30: 2879–2890.

[2] L. Hu, S. Chen, Y. Miao, et al., Die-casting effect on surface characteristics of thin-walled AZ91D magnesium components, Applied Surface Science, 2012, 261: 851–856.

[3] H. Zhu, J. Guo, J. Jia, Experimental study and theoretical analysis on die soldering in aluminum die casting, Journal of Materials Processing Technology, 2002, 123: 229–235.

[4] H. Jiang, S. Li, Q. Zheng, et al., Effect of minor lanthanum on the microstructures, tensile and electrical properties of Al-Fe alloys, Materials & Design, 2020, 195: 108991.

In-Situ Observation the Growth of Fe-Rich Intermetallic Particles During Solidification by Synchrotron X-Ray Imaging

Yuliang Zhao[1], Dongfu Song[2], Weiwen Zhang[3]
1. School of Mechanical Engineering, Dongguan University of Technology, Dongguan 523808, P.R. China
2. Institute of New Materials, Guangdong Academy of Science, Guangzhou 510650, P.R. China
3. School of Mechanical and Automobile Engineering, South China University of Technology, Guangzhou 510641, P.R. China
*Corresponding address: e-mail: zhaoyl@dgut.edu.cn

Abstract: Plate-like iron-rich intermetallic phases (Fe-rich phases) greatly decrease the elongation and fatigue strength of recycled Al alloys. The alloy composition and the manufacturing process parameters influence the type, and shape of Fe-rich phases. In this study, the influence of the cooling rate on the nucleation and growth of primary Fe-rich phases during solidification of Al-Fe(-Cu) alloy has been quantified by synchrotron X-ray radiography. The number and number density of primary Fe-rich phases in Al-Fe alloys. The nucleation rate of Al_3Fe in Al-Fe alloys at three cooling rates are 15.5, 44.0, 25.0 No./s, respectively. The maximum length of typical Al_3Fe phases in Al-Fe alloys decrease from 845 μm to 222 μm as the cooling rate increased. The combined Cu addition and higher cooling rate resulting in the constitutional undercooling, and finally promote the formation of refined primary Al_3Fe particles.

Keywords: Al alloy; solidification; Fe-rich intermetallic; synchrotron X-ray imaging.

1 Introduction (Bold, 10 pt., Arial)

Aluminium (Al) alloys are broadly used as engineering materials in the field of transportation, packing, and aerospace industry due to the advantage of lightweight, high-specific-strength, and good corrosion-resistant properties [1]. With increasing use of Al alloys parts in the past years, many of them come to end of service life and resulting in an increase of solid waste [2]. These Al scraps can be easily recycled due to their excellent recyclability [3]. However, recycled Al alloys have a high Fe content, which easily form sharp-edge iron-rich intermetallic phases (shorten as Fe-rich phases), such as, Al_3Fe, β-Al_5FeSi, and β-Al_7Cu_2Fe. They significantly deteriorate the ductility and fatigue properties of the alloys [4-5]. Hence, a lot of trails had been conducted to modify the Fe-rich phases by altering the cooling rates and adding alloying elements. In present study, the effect of cooling rate (0.1, 0.5, and 1.0 °C/s) on the growth behaviour of Al_3Fe in Al-Fe(-Cu) alloys was studied via radiography sequences.

2 Experimental procedure

The in-situ synchrotron X-ray radiography experiment was performed at the BL13HB beamline, Shanghai Synchrotron Radiation Facility (SSRF), P.R. China. The specific information about the furnace and beamline parameters used in this study is reported in [6].

3 Result and discussion

The growth behaviour of Fe-rich phases was captured by radiography sequences, as shown in Fig. 1. The Al-Fe clusters segregate and form the dark Al_3Fe phases shown in the radiograph at 650 °C, Fig. 1a. The Al + Al_3Fe eutectics were observed at the solid/liquid solidification front at 630 °C. When the cooling rate changes from 0.5 °C/s to 1.0 °C/s, the size of Al_3Fe and their nucleation temperature significantly decrease. Therefore, the higher the cooling rate, the higher the number density of Fe-rich particles and the smaller their size. Fig. 2 shows the X-ray radiographies taken during solidification of Al-Cu-Fe alloys. The presence of Cu significantly changes the morphology, amount and shape of primary Fe-rich phases. Similarly, the cooling rates reduced their size and increase their number. The nucleation temperature of the primary Fe-rich phase in the Al-Cu-Fe alloy is 3-4 °C higher than in Al-Fe alloys. In addition, the branch possibility of $Al_3(CuFe)$ phases in the ternary alloys is higher due to the different solute concentrations.

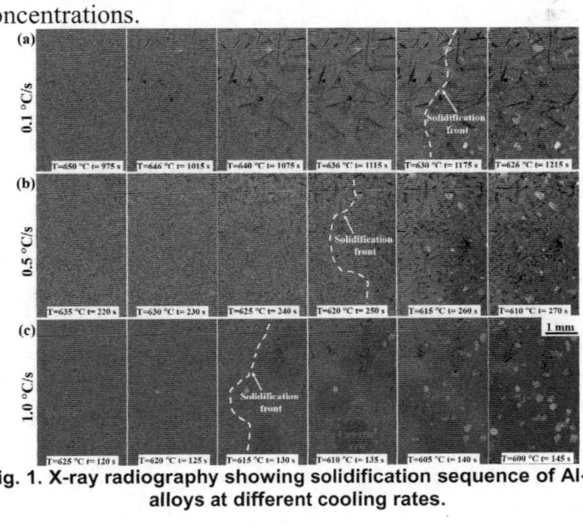

Fig. 1. X-ray radiography showing solidification sequence of Al-Fe alloys at different cooling rates.

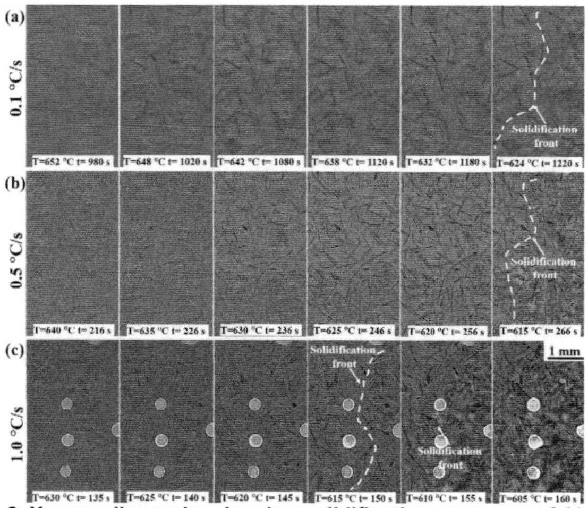

Fig. 2. X-ray radiography showing solidification sequence of Al-Cu-Fe alloys at different cooling rates.

The nucleation evolution of the Al_3Fe in the studied alloys under different cooling rates are illustrated in Fig. 3. The final number of the Al_3Fe in the FOV for the binary alloy firstly increases from 123 to 360 (increment: 192 %) and then decreases to 135 with increasing cooling rate from 0.1 to 1.0 °C/s. The addition of 6% Cu to the binary alloy increases the number of Al_3Fe nucleus in the cooling rate range 0.1 – 1 °C/s. The nucleation rate of the Al_3Fe in the cooling rate range 0.1 – 1 °C /s is plotted in Fig. 3b. In the present study, the area of FOV is 2.78 × 4.46 mm^2, the thickness of the sample is 0.3 mm, hence the FOV volume is 3.72 mm^3. The undercooling in the Al-Fe alloy is approximately 8, 24, and 34 °C with the cooling rates of 0.1, 0.5 and 1.0 °C/s, respectively. The undercooling in Al-Cu-Fe alloys is approximately 37, 47, and 55 °C with increasing cooling rate.

The effect of Cu content and cooling rate on the growth of Fe-rich phases are plotted in Fig. 4. The addition of Cu in the Al–Fe alloys, Cu can solubility in Al matrix and Al_3Fe particles. Cu addition changes the formation of primary Fe-rich phases from Al_3Fe to $Al_3(CuFe)$, the Fe atoms in Al_3Fe are partially substituted by the Cu atoms. Also, the Cu addition block the Fe atoms diffusion in the Al melt at solid/liquid interface. The addition of Cu in the Al–Fe alloys resulting in the constitutional undercooling and increasing the nucleation rate of Al_3Fe particles. As the cooling rate is increased, the temperature gradient is tendency to increased. This resulting in the reduction in size of constitutionally undercooling zone (Fig. 10b). Under the higher constitutional undercooling, which promotes the nucleation of Al3Fe particles and refine their size. At low cooling rate (0.1 ◦C/s), there is a plenty of time for these atoms to diffusion to the equilibrium condition, and vice versa. Thus, the combined Cu addition and higher cooling rate resulting in the constitutional undercooling, and finally promote the formation of refined primary Al3Fe particles

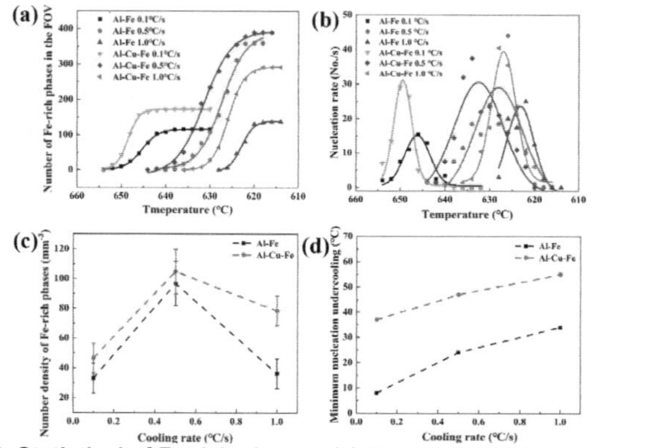

Fig. 3. Statistical of Fe-rich phases: (a) the number; (b) nucleation rate; (c) number density; (d) minimum nucleation undercooling.

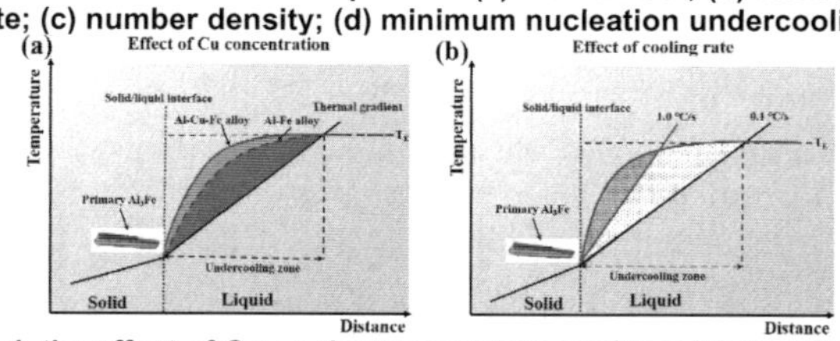

Fig. 4. the effect of Cu on the temperature profiles; (b) the effect of cooling rate on the temperature profiles.

Conclusion

1). With increased cooling rate, the number density of Al_3Fe in both alloys increases and their sizes decreases. The nucleation rate and number density of Fe-rich phases firstly increases and then decreased.

2). The increasing cooling rate and Cu addition increase the minimum nucleation undercooling of Al_3Fe in Al-Fe alloys.

Acknowledgments

This work was supported by the Natural Science Foundation of China (52104373).

References

1. Belov N., Aksenov A., Eskin D., Multicomponent Phase Diagrams: Applications for Commercial Aluminum Alloys, 2005, Elsevier.
2. Green J.A.S., Aluminum Recycling and Processing for Energy Conservation and Sustainability, ASM International, Materials Park, OH, 2007.
3. Schlesinger M.E., Aluminum Recycling, CRC Press, Boca Raton, 2007.
4. Song D.F., Zhao Y.L., Jia Y.W., et al., Effect of B addition on the formation of Fe-rich phases in Al-Si-Fe alloys, J. Alloy. Compd. 2023, 930: 167426.
5. Y.L. Zhao, W.X. He, D.F. Song, et al., Effect of ultrasonic melt processing and Al-Ti-B on the microstructural refinement of recycled Al alloys, Ultrason. Sonochem. 2022, 89: 106139.
6. Zhao Y.L., Song D.F., Wang H.L., Revealing the influence of Fe on Fe-rich phases formation and mechanical properties of cast Al-Mg-Mn-Fe alloys, J. Alloy. Compd. 2022, 901: 163666.

Effects of Ni Contents on Microstructure and Mechanical Properties of Al-8Ce Alloy

Chong Lin[1,2*], Binteng Huang[1], Yanwei Wang[1], Shulin Lü[2], Shusen Wu[2]
1. Hubei Provincial Key Laboratory of Chemical Equipment Intensification and Intrinsic Safety, School of Mechanical and Electrical Engineering, Wuhan Institute of Technology, 206 Guanggu 1st road, Wuhan, China
2. State Key Laboratory of Materials Processing and Die and Mould Technology, School of Materials Science and Engineering, Huazhong University of Science and Technology, 1037 Luoyu Road, Wuhan, China
*Corresponding address: e-mail: chonglin@wit.edu.cn

Abstract: The microstructures and properties of Al-8Ce-yNi alloys were investigated. The results show that with the increase of Ni content, the range of solid-liquid phase zone of Al-8Ce-yNi alloy was narrowed. The ternary eutectic group network structure composed of Al_3Ni and $Al_{11}Ce_3$ was formed. The ultimate tensile strength (UTS) of Al-8Ce-yNi alloys at room temperature and elevated temperature (350°C) are increased with the increase of Ni content. When the Ni content is 6%, the UTS of the alloy at room temperature and elevated temperature reach the maximum value, which are 189 MPa and 72 MPa, respectively.

Keywords: Ni Contents; Al-8Ce alloy; Microstructure; Mechanical Properties

1 Introduction

Ce is an excellent alloying element that can purify melt and improve alloy fluidity [1-2]. Existing studies have shown that Al-Ce alloy has fluidity and casting properties comparable to Al-Si alloy [3]. The eutectic $Al_{11}Ce_3$ phase produced by the eutectic reaction of Ce in Al has excellent high temperature stability. And the higher the Ce content, the better the high temperature performance of the alloy [4-5]. Therefore, Al-Ce alloy has great development potential due to its excellent casting performance and high temperature stability, but there are few relevant studies on Al-Ce alloy at present. In this paper, the effect of Ni content on the microstructure and mechanical properties of Al-8Ce alloy was studied.

2 Experimental procedure

Al-8Ce-yNi alloys with different Ni contents were prepared by pure Al (99.9%, mass fraction, the same below), Al-20Ce and Al-10Ni master alloys. The nominal chemical composition of the alloys is shown in Table 1. First, the prepared materials were fully melted by electrical resistance furnace with graphite crucible. Then the melt was degassed with high-purity argon gas and was cooled down to a temperature of 710 °C after degassing. After that, the melt was poured into a permanent casting mold preheated to about 200 °C, and the standard tensile samples with diameter of 8 mm were obtained. The phase compositions were evaluated by D8 ADVANCE XRD analyzer. Microstructural characterization was revealed and analyzed by ZEISS GeminiSEM300 field emission scanning electron microscope attached with EDS. DSC experiments were performed on a differential scanning calorimeter (NETZSCH STA449F3) at the cooling rate of 10 °C/min. The room and elevated temperature mechanical properties were examined on SHIMADZU AG-IC100KN universal material testing machine and the tensile rate is 1 mm/min.

Table 1. Nominal chemical composition of the alloys (wt.%)

Alloys	Ce	Ni	Al
Al-8Ce	8	-	Bal.
Al-8Ce-2Ni	8	2	Bal.
Al-8Ce-3Ni	8	3	Bal.
Al-8Ce-4Ni	8	4	Bal.
Al-8Ce-5Ni	8	5	Bal.
Al-8Ce-6Ni	8	6	Bal.

3 Result and discussion

As-cast microstructure of Al-8Ce-yNi alloy

Figure 1 shows the XRD pattern of as-cast Al-8Ce-yNi alloy with different Ni content. It can be seen that Al-8Ce alloy is mainly composed of α-Al and Al11Ce3 phases.

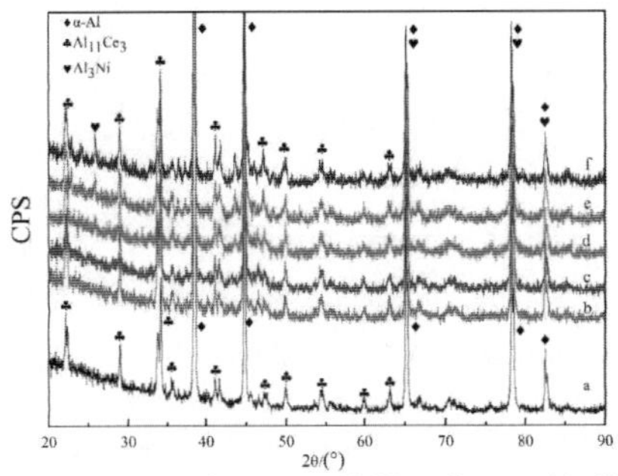

Fig.1 XRD spectrums of as-cast Al-8Ce alloys with different Ni contents: (a) Al-8Ce, (b) Al-8Ce-2Ni, (c) Al-8Ce-3Ni, (d) Al-8Ce-4Ni, (e) Al-8Ce-5Ni, (f) Al-8Ce-6Ni

Al-8Ce-yNi alloy is mainly composed of α-Al, $Al_{11}Ce_3$ and Al_3Ni phases. Fig. 2 shows the BSE micrographs of the microstructure of the as-cast Al-8Ce-6Ni alloy. Combined with the results of XRD and energy spectrum analysis shown in Table 2, it can be seen that in addition to the bright white lamellar $Al_{11}Ce_3$ phase (orange dashed line frame indicated by arrow 1 in Fig. 2) and the light gray fine irregular multilateral Al_3Ni phase (yellow dashed line frame indicated by arrow 2 in Fig. 2), a new network structure composed of the eutectic Al_3Ni phase and $Al_{11}Ce_3$ phase appears in the alloy (red dashed line frame indicated by arrow 3 in Fig. 2).

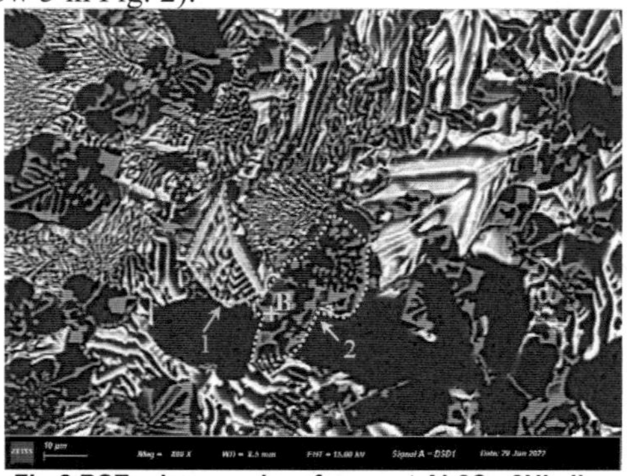

Fig.2 BSE micrographs of as-cast Al-8Ce-6Ni alloy

Table 2 EDS results of corresponding points in Fig. 2 (at%)

Points	Al	Ce	Ni
A	83.97	16.03	0
B	64.15	0	35.85
C	94.01	4.23	1.76

Fig. 3 shows the DSC curve of the temperature rise process of Al-8Ce-yNi alloy, and it can be seen that an obvious heat absorption peak appears on the DSC curve. It can also be seen that with the increase of Ni content, the melting endothermic peak of the primary α-Al in Al-8Ce-(2,3,4,5,6)Ni alloy shifts to the left. This means that with the increase of Ni addition, the liquidus temperature of Al-8Ce-(2,3,4,5,6)Ni alloy is decreased and the solid-liquid two-phase region is narrowed. Therefore, the addition of Ni changes the solidification characteristics of the alloy.

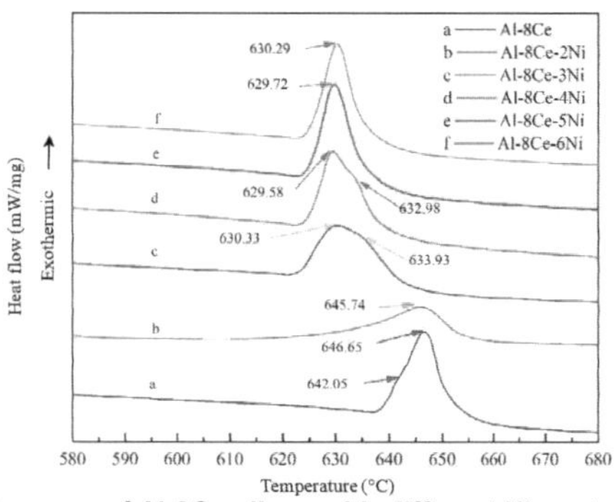

Fig.3 DSC curves of Al-8Ce alloys with different Ni contents during heating process

Mechanical properties of Al-8Ce-yNi alloy

With the increase of Ni content, the ultimate tensile strength (UTS) and yield strength (YS) of the alloy at room temperature are both increased gradually. When the Ni content is 6%, the UTS and YS at room temperature reach the maximum values, which are 189MPa and 59MPa, respectively. The elevated temperature (350°C) UTS and YS of the alloy are also increased with the increase of Ni content. When the Ni content is 6%, the elevated temperature UTS and YS of the alloy reach the maximum value, which are 72MPa and 53MPa, respectively. However, the elongation is decreased with the increase of Ni content either at room temperature or elevated temperature. The very low diffusion rates of Ce and Ni make $Al_{11}Ce_3$ and Al_3Ni phases less prone to dissolution or coarsening at high temperature, and these phases have excellent high temperature stability. Thus, the compact ternary eutectic structure can strengthen the alloy matrix and improve the heat resistance of the alloy.

4 Conclusions

(1) With the increase of Ni content, the liquidus temperature of Al-8Ce-yNi (y=2,3,4,5,6%) alloy is decreased, the solid-liquid two-phase zone is narrowed. And the addition of Ni elements changes the solidification characteristics of the alloy.

(2) The UTS of Al-8Ce-yNi (y=2,3,4,5,6%) alloy at room temperature and elevated temperature (350°C) is increased with the increase of Ni content. When the Ni content is 6%, the UTS of the alloy at room temperature and elevated temperature reaches the maximum value, which are 189MPa and 72MPa respectively, and the elongation decreases with the increase of Ni content.

Acknowledgments

This work was funded by Project (P2022-025) supported by State Key Laboratory of Materials Processing and Die and Mould Technology, Huazhong University of Science and Technology.

References

[1] Weiss D. Improved high-temperature aluminum alloys containing cerium. Journal of Materials Engineering and Performance, 2019, 28: 1903-1908.
[2] Stromme E T, Henderson H B, Sims Z C, et al. Ageless aluminum-cerium-based alloys in high-volume die casting for improved energy efficiency. JOM: the journal of the Minerals, Metals & Materials Society, 2018, 70(6): 866-871.
[3] Weiss D, Rois O, Sims Z, et al. Casting characteristics of high cerium content aluminum alloys. Light Metals 2017, Springer, 2017.
[4] Czrwinski F. Thermal stability of aluminum-cerium binary alloys containing the Al-$Al_{11}Ce_3$ eutectic. Materials Science & Engineering A, 2021, 809: 1-11.
[5] Sims Z C, Rios O R, Weiss D, et al. High performance aluminum-cerium alloys for high-temperature applications. Materials Horizons, 2017, 4: 1070-1078.

NumericalStudy on Dynamic Evolution of a Dual-Frequency Driven Cavitation Bubble in Magnesium Alloy Melt

Zhaoyang Yin[1], Qichi Le[1,*], Qiyu Liao[2], Tong Wang[1] Lei Bao[1], Dazhi Zhao[2]
1. Key Lab of Electromagnetic Processing of Materials, Ministry of Education, Northeastern University, Shenyang 110819, PR China
2.School of Materials Science and Engineering, Northeastern University, Shenyang 110819, PR China
*Corresponding address:qichil@mail.neu.edu.cn

Abstract: The enhancement and control of acoustic cavitation in metal preparation by dual-frequency ultrasonic have been verified experimentally. However, the research is still in infancy and the theoretical mechanism needs to be further explored. Therefore, the dynamic model of a dual-frequency driven cavitation bubble in magnesium alloy established for the first time. The effect of equilibrium radius, driving frequency, acoustic pressure amplitude and ratio were discussed in detail in the f_1-f_2 space. An excessively large equilibrium radius resulted in a long bubble collapse-time, and too small an equilibrium radius resulted in a low collapse strength. In the frequency domain of low-low or low-high frequency coupling, the cavitation bubble collapse strength was high and sensitive to variations in the driving frequency. At both driving frequencies above 100 kHz, the collapse strength reduced to a stable interval and no longer responded significantly to variations in frequency. Increasing the total acoustic pressure amplitude enhanced the synergistic effect between the two distinct driving frequencies. The collapse strength for equal driving frequencies was independent of the acoustic pressure ratio. For two distinct drive frequencies, increasing the acoustic pressure share at a lower frequency or decreasing the acoustic pressure share at a higher frequency increased the collapse strength. The analysis of bubble collapse time indicated that obtaining higher collapse strength often comes at the cost of a longer collapse time.

Keywords: Dual-frequency ultrasonic, Cavitation bubble, Dynamic evolution, Collapse strength

1 Introduction

In recent years, ultrasonic technology has received increasing attention in the field of solidification behavior regulation of metal melts. Dual-frequency ultrasonic has been applied in metal casting due to the excellent ability to refine microstructures[1-3].Although the effective control and enhancement of acoustic cavitation by dual-frequency or multi-frequency ultrasound have been experimentally verified, the study is still in its infancy. The universality needs to be improved, and the theoretical mechanisms need to be further explored. The cavitation dynamic is the most effective way to understand cavitation effects from a theoretical perspective.

In this paper, a dual-frequency driven cavitation bubble dynamic model is established based on the Rayleigh-Plesset equation. The dynamic evolution of cavitation bubble was systematically analyzed in a wide parameter range with AZ80 magnesium alloy melt chosen as a simulation system. The effects of cavitation nucleus equilibrium radius, driving frequency, acoustic pressure amplitude, and acoustic pressure ratio in cavitation strength are discussed in detail in the f_1-f_2 space. This is the first numerical analysis concerning on the cavitation bubble dynamic under dual-frequency driven in a metal melt.

2 Mathematical model

The motion equation of the cavitation bubble radius$R$under the action of the dual-frequency ultrasonic fieldwith various parameters can be expressed as follows:

$$R\left(\frac{d^2R}{dt^2}\right)+\frac{3}{2}\left(\frac{dR}{dt}\right)^2=\frac{1}{\rho}\left[\begin{array}{l}\left(P_0+\frac{2\sigma}{R_0}\right)\left(\frac{R_0}{R}\right)^{3K}-\frac{2\sigma}{R}-\frac{4\mu}{R}\left(\frac{dR}{dt}\right)-P_0+P_V\\+P_{A1}\sin(2\pi f_1t)+P_{A2}\sin(2\pi f_2t)\end{array}\right] \quad (1)$$

where ρ is the density of melt, P_0 is the hydrostatic pressure acting on the bubble wall, σ is the surface tension, R_0 is the initial radius, K is the polytropic exponent, μ is the liquid viscosity, P_V is the vapor pressure inside bubble,P_{A1} and P_{A2} represent the acoustic pressure amplitudeof the two ultrasounds, and f_1, f_2 represent the ultrasonic frequencies.

3 Result and discussion

3.1 Influence of equilibrium radius

The bubble radii increased hundreds of times within the microsecond time variation, implying that the cavitation bubble went through an extremely exaggerated expansion and contraction in the dual-frequency ultrasonic treatment. The calculations indicate that the cavitation bubbles with larger maximum radius have longer collapse times.

3.2 Influence of driving frequency

The simulation results show that a shorter collapse time can be obtained in the frequency domain of 30-500 kHz. In the frequency domain of low-low or low-high frequency coupling, the cavitation bubble collapse strength is high and sensitive to variations in the driving frequency. At both drive frequencies above 100 kHz, the collapse strength reduces to a stable interval and no longer responds significantly to variations in frequency.

3.3Influence of acoustic pressure amplitude

Larger acoustic pressure amplitudes created more intense bubble oscillation, but also caused the bubble to collapse more slowly.The projection of the collapse strength on the f_1-f_2 plane indicates that high collapse strength always occurs on the low-low and low-high coupling frequency

sides in the frequency domain.Furthermore,increasing the acoustic pressure amplitude enhances the synergistic effect between two distinct driving frequencies, and thus increases the cavitation intensity. In contrast, the enhancement does not seem to exist between equal driving frequencies by increasing the acoustic pressure amplitude.
For equal driving frequencies, the collapse strength is independent of the acoustic pressure ratio, but is related to the driving frequency. For distinct driving frequencies, the collapse strength is closely related to the acoustic pressure ratio. Increasing the acoustic pressure ratio at a lower frequency or decreasing the acoustic pressure ratio at a higher frequency will significantly enhance the collapse strength under dual-frequency driven.

4 Conclusion

In this paper, based on the derived Rayleigh-Plesset equation, a cavitation bubble dynamics model under dual-frequency driven is established. The numerical solution is obtained by the 4th-order Runge-Kutta method to clarify the dynamic evolution of a single cavitation bubble radius in the molten medium. The collapse strength was defined based on the cavitation bubble expansion ratio and the influencing factors such as the equilibrium radius of the cavitation nucleus, the driving frequency, and the acoustic pressure amplitude were investigated in the f_1-f_2 space.
The cavitation bubble goes through several exaggerated and irregular expansion-contractions under dual-frequency driven. The bubble radius increases hundreds of times in microseconds. The equilibrium radius of the initial cavitation nucleus present in the liquid is crucial to the cavitation effect. Too large an equilibrium radius leads to a long bubble collapse-time; too small an equilibrium radius results in a low collapse strength. Therefore, there is an optimum interval for the equilibrium radius size of the cavitation nucleus that favors the cavitation efficiency, which is related to the liquid physical properties and the acoustic driving parameters. The driving frequency is also critical for the cavitation effect. In the frequency domain of low-low or low-high frequency coupling, the cavitation bubble collapse strength is high and sensitive to variations in the driving frequency. At both driving frequencies above 100 kHz, the collapse strength reduces to a stable interval and no longer responds significantly to variations in frequency. Increasing the total acoustic pressure amplitude will enhance the synergistic effect between the two distinct driving frequencies. The collapse strength for equal driving frequencies is independent of the acoustic pressure ratio, but the collapse strength for distinct driving frequencies is closely related to the acoustic pressure ratio. Increasing the acoustic pressure share at a lower frequency or decreasing the acoustic pressure share at a higher frequency can effectively increase the collapse strength. The analysis of bubble collapse-time shows that obtaining higher collapse strength often comes at the cost of a longer collapse-time.
The pursuit of the highest collapse strength is not always an ultimate goal. For ultrasonic treatment of magnesium alloy melt, shorter cavitation bubble collapse-time and higher collapse strengths are desired. Therefore, the optimal cavitation effect is always obtained after a comprehensive consideration of collapse strength and collapse time. In addition, the dynamic evolution of a single cavitation bubble existing inside a cavitation cloud is necessarily affected by the pressure released by the surrounding bubble clusters. In the present work, the phenomenon of multiple cavitation bubbles is not considered. As a final remark, an exhaustive analysis of the cavitation bubble dynamic evolution or bubble-collapse modeling requires orders of magnitude higher computational resources.

Acknowledgments

This research was financially supported by the National Natural Science Foundation of China (Grant No.52274377 and 52304391) and the Natural Science Foundation of Liaoning Province, China [No. 2023-MSBA-133].

References

[1] Chen X R, Ning F K, Hou J, Le Q C, Tang Y, Dual-frequency ultrasonic treatment on microstructure and mechanical properties of ZK60 magnesium alloy, UltrasonicsSonochemistry. 2018, (40): 433–441.
[2] Yin Z Y, Le Q C, Chen X R, Jia Y H, The grain refinement of Mg alloy subjected to dual-frequency ultrasonic melt treatment: A physical and numerical simulation, Journal of Materials Research and Technology. 2022, (21): 1554–1569.
[3] Yin Z Y, Le Q C, Zou Q, Ren L, Zhou W Y, Jiang Y C, Chen X R, Atrens A, Liao Q Y, Enhancement of the corrosion resistance of the Mg alloy ZW61 bydual-frequency ultrasonic vibration, Journal of Magnesium and Alloys, 2024, January, 17.

Effect of Two-Stage Aging on the Mechanical Properties of Al-Li Alloy

Youjie Guo, Liang Zhang, Guohua Wu *
School of Materials Science and Engineering, Shanghai Jiao Tong University, Shanghai 200240, China
*Corresponding address: e-mail: ghwu@sjtu.edu.cn

Abstract: In this work, the effect of two-stage aging with low temperature followed by high temperature on the mechanical properties of Al-3.8Cu-1.2Li-0.5Mg-0.15Zr alloy was investigated. The results show that with the increase of the first aging temperature (75-125℃), the time for the alloy to reach the peak hardness in the second aging is shortened, and the peak aging hardness first increases and then decreases. When the first aging temperature is 100℃, with the prolongation of the first aging time (0-32 h), the time for the alloy to reach the peak hardness in the second aging time is also gradually shortened, and the peak aging hardness first increases and then decreases. The strength of the alloy increases significantly after the two-stage aging treatment. With the increase of the first aging temperature, the strength of the alloy after aging first increases and then decreases. The highest room temperature strength of the alloy was obtained when the aging process is 100℃ × 8 h + 175℃ × 32 h, with YS = 685MPa, UTS = 737MPa and EL = 5.58%, respectively. The main reason for the increase in strength and plasticity of double-stage aging alloys is that the first stage of low-temperature aging facilitates the formation of clusters of solute atoms and GP zones in the alloys, which provides more nucleation cores for the formation of the T_1 phase during the subsequent second stage of high-temperature aging. This results in a higher density, smaller size and more uniform distribution of T_1 phase precipitation after the second stage of high-temperature aging.

Keywords: Al-Li alloy; Microstructure evolution; Mechanical properties; Aging treatment

1 Introduction

Al-Li alloys are characterized by low density, high elastic modulus and high specific strength [1]. Studies have shown that the addition of 1% Li can reduce the density of the alloy by 3%, increase the elastic modulus by 6%, and help to improve the strength, fatigue properties and resistance to fatigue crack propagation [2].

However, Al-Li alloys face the problem of strength and plasticity incompatibility, and there is still room for further improvement of their mechanical properties. At present, most of the research focuses on regulating the alloy composition [3], and there is less literature on the optimization of the heat treatment process. Therefore, this paper takes Al-3.8Cu-1.2Li-0.5Mg-0.15Zr alloy as an example to investigate the effect of two-stage aging treatment on the mechanical properties of the alloy and optimize the aging process, to lay a theoretical foundation for the further application of Al-Li alloy.

2 Experimental procedure

The tested alloys were prepared from pure Al, Li, Mg and Al-Cu, Al-Zr master alloys in graphite crucible under the protection of Ar. The melt temperature was maintained at 720°C before being poured into the steel mold.

Harnesses of alloys were tested using a Vickers hardness tester with a load of 5 kg and a dwelling time of 15 s. The tensile properties were conducted on a Z100 testing machine with a constant displacement rate of 1 mm/min. Preparation of specimens for transmission electron microscopy (TEM) observation involves twin-jet electropolishing in a 30% HNO_3 and 70% CH_3OH (volume fraction) solution at -25℃. TEM images were obtained using a transmission electron microscope working at 200 kV.

3 Result and discussion

Fig. 1a shows the hardening curves of the alloys after first-stage aging at different temperatures in second-stage aging at 175°C. The time for the alloy to reach the peak aging at the second-stage of aging decreases with the increase of the first-stage aging temperature when the first-stage aging time is constant. The peak hardness of the alloy in the second-stage of aging increases and then decreases with the increase of the first-stage aging temperature. Fig. 1b shows the hardening curves of the alloys after first-stage aging treatment at the optimized temperature of 100°C for different times of second-stage aging at 175°C. The time to reach peak aging for second-stage aging of the alloy after first-stage aging treatment is gradually shortened with the extension of the first-stage aging time, while the peak hardness first increases and then decreases.

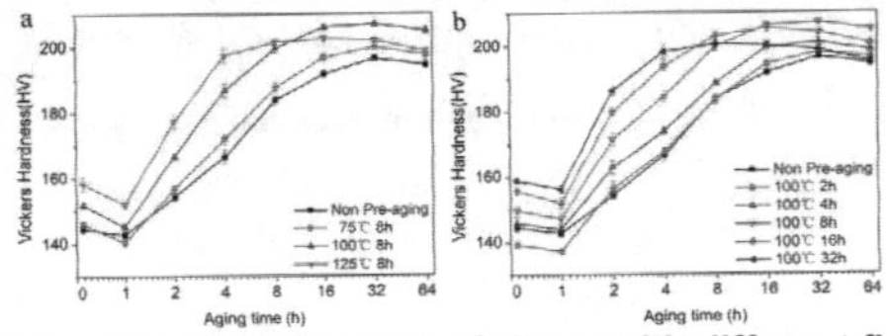

Fig. 1. (a) Age-hardening curves of alloys with different first-stage temperatures at 175°C; (b) Age-hardening curves of alloys with different time aged at 175°C.

Fig. 2a shows that the strength of the alloys increases and then decreases when the second-stage aging time is increased to 32 h. The strength of the alloys with a first-stage aging temperature of 100°C is the highest. The alloy with first-stage aging temperature of 100°C has the highest strength after second-stage aging at 175°C × 32 h as YS = 685 MPa, UTS = 737 MPa and EL = 5.58%. Fig. 2b shows that the strength of the alloy increased when the secondary aging time was increased to 32 h. The strength of the alloy was higher when the second-stage aging time was increased to 8 h. The alloy with first-stage aging time of 8 h (temperature 100°C) has the highest strength after second-stage aging at 175°C × 32 h. The alloy has the highest strength after second-stage aging.

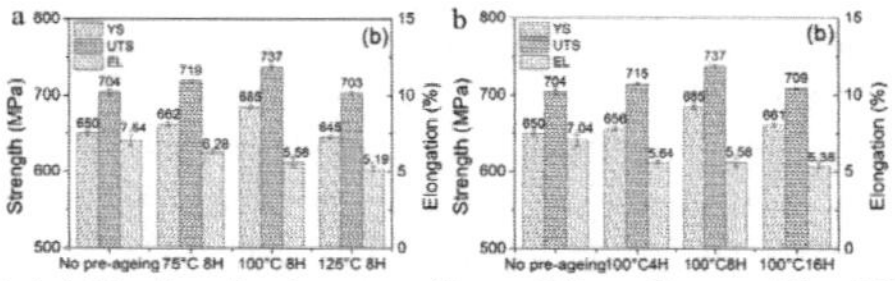

Fig. 2. (a) Mechanical properties of the alloys with different temperature aged at 175°C for 32 h; (b) Mechanical properties of the alloys with different time aged at 175°C for 32 h.

Fig. 3 shows the bright field (BF) image and dark field (DF) image of the alloy after two-stage aging (100°C × 8 h + 175°C × 32 h) and T6 (175°C × 32 h) aging treatment. Obvious T_1 phase diffraction fringes are observed. From the BF images, the main strengthening phases in the alloy are all T_1 phase. From the DF images, the density of T_1 phase is significantly higher than that of the T6 aging treatment after the two-stage aging treatment, and the size of T_1 phase is smaller and more uniformly distributed. The main reason is that first-stage low-temperature aging is conducive to the formation of more solute atom clusters with GP zones, which provides more nucleation cores for the precipitation of T_1 phase.

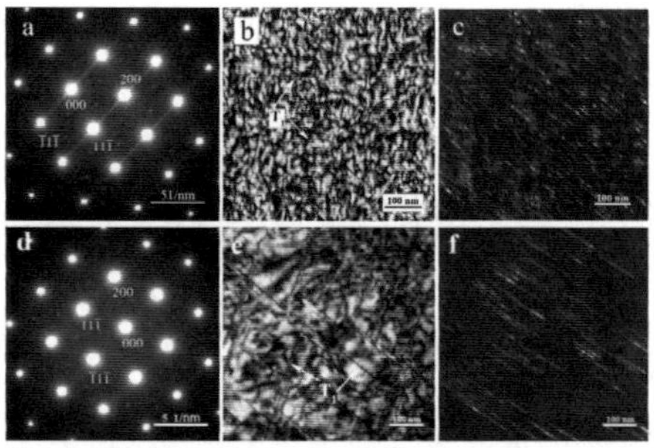

Fig. 3. The SAED patterns with corresponding BF and DF images of the alloys with different aging conditions: (a-c) Two-stage aged; (d-f) Conventional T6 aged.

4 Conclusion

(1) With the increase of the first aging temperature (75-125℃), the time for the alloy to reach the peak hardness in the second aging is shortened, and the peak aging hardness increases first and then decreases. When the first aging temperature is fixed, when the aging time is prolonged (0-32 h), the time for the alloy to reach the peak hardness during the second aging is also gradually shortened, and the peak aging hardness first increases and then decreases.

(2) The strength of the alloy after double-stage aging treatment was significantly increased compared with that of T6 single-stage aging. With the increase of the first aging temperature, the strength of the alloy in the double-stage aging state firstly increased and then decreased, and the highest strength of the alloy was found when the aging process was 100℃ × 8 h + 175℃ × 32 h, which were YS = 685MPa, UTS = 737 MPa, and EL = 5.58%.

(3) Double-stage aging alloy strength and plasticity improvement is due to: the first level of low-temperature aging is conducive to the formation of solute atomic clusters and GP zone, for the subsequent second level of high-temperature aging T_1 phase formation provides more nucleation core. After the second stage of high-temperature aging, the T_1 phase has a higher density, smaller size, and more uniform distribution, which effectively improves the toughness of the alloy.

Acknowledgments

This work was financially sponsored by the National Natural Science Foundation of China (Nos. 51821001 and 51871148).

References

[1] Zhang J S, Wu G H, Zhang L, Zhang X L, Shi C C and Tong X. Addressing the strength-ductility trade-off in a cast Al-Li-Cu alloy——synergistic effect of Sc-alloying and optimized artificial ageing scheme. Journal of Materials Science & Technology, 2021, 96: 212-225.

[2] Wu L, Li X F and Wang H W. The effect of major constituents on microstructure and mechanical properties of cast Al-Li-Cu-Zr alloy, 2021, 171: 110800.

[3] Zhang X L, Zhang L and Wu G H. Influences of Mg content on the microstructures and mechanical properties of cast Al-2Li-2Cu-0.2Zr alloy, Journal of Materials Science, 2019, 54: 791-811.

This page intentionally left blank.

The Study of Extrusion Casting on the Microstructure and Properties of Hypereutectic Al-Si Alloy

Yunguan Zhang[1,2], Zhiyong You[1,2,*], Mengjun Zhang[1,2], Libo Wang[1,2], Chunlian Liu[1,2]
1. School of Materials Science and Engineering, Taiyuan University of Technology, Taiyuan 030024,China
2. Key Laboratory of Interface Science and Engineering in Advanced Materials Ministry of Education, Taiyuan University of Technology, Taiyuan 030024, China
*E-mail:13603511486@163.com

Abstract: In this study, P and Sr composite modified hypereutectic Al-Si alloy samples were prepared by gravity casting and extrusion casting, with extrusion ratios of 400 MPa, 500 MPa, and 600 MPa. The microstructure and properties of hypereutectic aluminum-silicon alloy under gravity casting and different extrusion ratios were studied. The results show that the properties of extrusion cast alloys are better than those of gravity cast alloys, and as the extrusion ratio increases, the refinement effect on the silicon phase becomes better. Under an extrusion ratio of 600 MPa, the refinement effect on primary silicon and eutectic silicon is the best, with the primary silicon partially becoming blunt, decreasing in quantity, and having small dimensional differences with an average size of 29 μm. The density of eutectic silicon increases, and the alloy has good mechanical properties under this condition, with a tensile strength of 316 MPa, an elongation of 1.94%, and a hardness of 115.1 HV.

Keywords: Extrusion Casting; Microstructure; Properties; Hypereutectic Al-Si Alloy

1 Introduction

The hypereutectic Al-Si alloy produced by traditional gravity casting typically exhibits coarse primary silicon and elongated eutectic silicon, both of which, as hard phases, influence the alloy's ductility and wear resistance. The addition of P and Sr composite modification has certain effects: P forms AlP compounds with Al, acting as heterogeneous nucleation sites and dispersing to refine the structure; Sr addition induces an adsorption poisoning mechanism, transforming the elongated eutectic silicon into a coral-like structure. The extrusion casting process significantly reduces defects, increases the density of the alloy structure, further refines the silicon phases, and consequently enhances the alloy's comprehensive mechanical properties. Yong-fei WANG [1]fabricated ZL104 connecting rods using semi-solid extrusion casting and liquid extrusion casting. Comparing with liquid extrusion casting, semi-solid extrusion casting results in higher improvements in tensile strength and elongation of the connecting rods.

2 Experimental procedure

The raw materials used in this study are ZL109 alloy, Al-30Si intermediate alloy, and modifiers including Al-4.5P and Al-10Sr intermediate alloys to melt Al-15Si alloy. The composition of the experimental materials is shown in Table 1.

Table 1. The composition of test alloys

Si	Cu	Mg	Ni	Fe	Al
15	0.9	0.8	0.9	0.9	excess

3 Result and discussion

In the microstructure of the alloy modified by P and Sr composite under gravity casting (Figure 1 (a)), there is a significant difference in the size of primary silicon, with an average size of 24μm. The morphology ofprimary

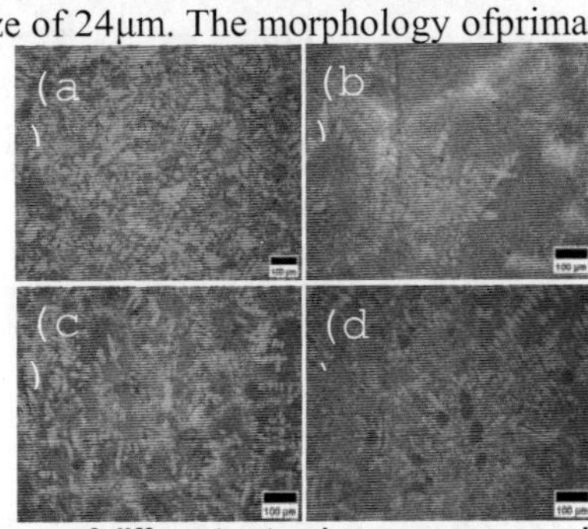

Fig. 1 Influence of different extrusion pressure on microstructure of hypereutectic Al-15Si alloy: ;(a) unmetamorphism (b) 400MPa (c) 500MPa (d) 600MPa

silicon is mostly sharp-edged plate-like with severe agglomeration, while the eutectic silicon consists of long needle-like and small short rod-like structures. After extrusion, the primary silicon exhibits obvious extrusion cracks and more burrs, undergoing a significant change in morphology, size enlargement, and a substantial decrease in density. Meanwhile, the eutectic silicon noticeably refines into small short rod-like structures with a significant increase in compactness. This is because the eutectic point in the Al-Si binary phase diagram at room temperature shifts to the upper right with an increase in

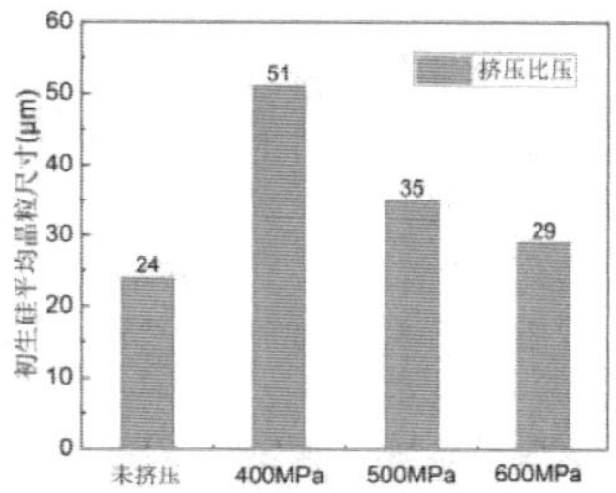

Fig. 2 Average grain size of primary silicon in hypereutectic Al-15Si alloy under different extrusion ratios

extrusion pressure. This leads to an increase in the silicon content in the solid solution at room temperature, an increase in the eutectic temperature of the Al-Si alloy, and the likelihood of generating eutectic structure under pressure, even forming a small amount of subeutectic structure, thereby reducing the quantity of precipitated primary silicon. The eutectic silicon significantly refines into small short rod-like structures with increased compactness, and there are clearly fine α-Al dendrites present in the structure. With the increase in extrusion pressure, the size of primary silicon gradually decreases, attributed to the increase in extrusion force that can enhance the alloy's solidification rate, increase the undercooling of the alloy, and refine the grains.

The extruded alloy exhibits enhanced comprehensive mechanical properties compared to the unextruded alloy. With the increase in extrusion pressure, the tensile strength and elongation of the hypereutectic Al-15Si alloy gradually increase. This is due to the partial improvement in the morphology of primary silicon in the alloy, with reduced size and increased distribution of eutectic silicon, leading to an enhancement in tensile strength through fine crystal reinforcement and an increase in the number of fine α-Al dendrites improving the toughness of the alloy. With the increase in extrusion pressure, the hardness of the alloy initially increases and then decreases, attributed to the increase in compactness of the primary silicon as a hard phase followed by a reduction in compactness.

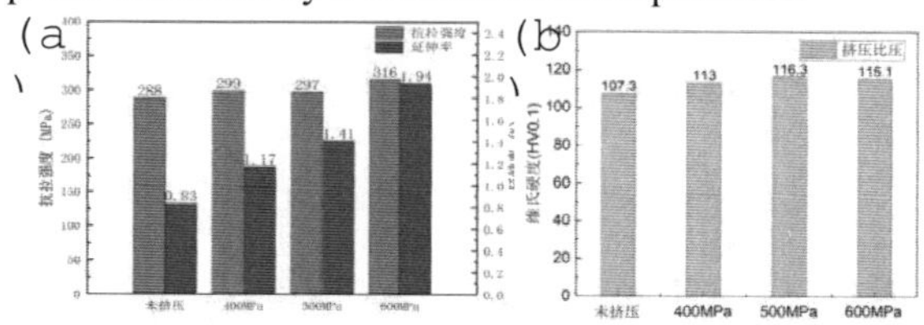

Fig. 3 Mechanical properties of hypereutectic Al-15Si alloy under different extrusion ratios：(a) tensile properties(b) vickers hardness

4 Conclusion

(1) With the increase of extrusion pressure ratio, the average size of primary silicon in Al-15Si alloy decreased, the quantity initially increased and then decreased.

(2) With the increase of extrusion pressure ratio, the tensile strength and elongation of Al-15Si alloy gradually increased, while the hardness initially increased and then decreased.

Acknowledgments

This work is supported by the 2021 Central Government Guide local science and technology development Fund (YDZJSX2021A008)

References

[1] Wang Y F,Zhao S D,Zhang C Y.Microstructures and mechanical properties of semi-solid squeeze casting ZL104 connecting rod[J].Transactions of Nonferrous Metals Society of China,2018,28(2):235-243.

Enhanced Mechanical and Damping Properties of Mg-8Zn-1.5Al-0.5Y Alloys By Rheo-Squeeze Casting

Xiaogang Fang[1,2], Yeliang Zhu[1,2], ZiyunYuan[1,2]

1. School of Materials Science and Engineering, Hefei University of Technology, Hefei 230009, China
2. Engineering Research Center of Ministry of Education for High Performance Copper Alloy Materials and Forming, Hefei 230009, China
*Corresponding address: fangxg@hfut.edu.cn

Abstract: Due to their low density, high specific strength, and stiffness, Mg alloys have significant potential for applications in the automobile, aerospace, and other industries. However, the balance between mechanical properties and damping of Mg alloys has always been a challenge for its applications. The effects of ultrasonic and rheo-squeeze casting on the microstructure and mechanical properties of Mg-8Zn-1.5Al-0.5Y (ZAY8150) alloy were investigated. The results show that the alloy mainly consists of α-Mg, $Mg_{32}(Al, Zn)_{49}$ phase, and rich Y phase. Compared with conventional casting, ultrasonic and rheo-squeeze casting promote the formation and refinement of second phase. Notably, when both ultrasonic treatment and rheo-squeeze casting were applied simultaneously to the alloy, the volume fraction of second phase reaches 10.4%, with axial length decreasing to 2.1 μm and width decreasing to 1.3 μm respectively. At room temperature, the alloys subjected to ultrasound-assisted rheo-squeeze casting retained high damping properties (tanδ is found to be 0.04, and E" is measured as 1100MPa) at high mechanical properties. Tensile property is 260.4 MPa with an elongation of 20.2%.

Keywords: Ultrasonic, Rheo-squeeze casting casting, Quasicrystal, Mechanical, Damping

1 Introduction

Mg alloys exhibit significant potential for application in the aerospace, automotive, and 3C electronics industries due to their advantages of low density, high specific strength and excellent electromagnetic shielding. Meanwhile, Mg alloys are high damping material (tanδ>0.01) because of its good shock absorption and noise reduction properties, making it a candidate material for structural material. However, the balance between mechanical and damping properties of Mg alloys is also a bottleneck restricting its practical application. Therefore, improving the mechanical and damping properties of Mg alloys is the key to further improve the application field of Mg alloys.In this study, the four different treatments (semi-solid casting, ultrasonic vibration (USV), rheo-squeeze casting (SSE), and ultrasonic assisted rheo-squeeze casting casting(USV-SSE)) and conventional casting for Mg-8Zn-1.5Al-0.5Y (ZAY8150) were investigated. And the effect of solidification structure and mechanical properties of alloys under different conditions were compared.

2 Experimental procedure

Pure Mg, pure Zn, pure Al, and Mg-30Y (wt.%) were used to prepare ZAY8150 alloys. The metal ingot were placed in a stainless steel crucible according to its chemical composition and then heated in a resistance furnace. To prevent oxidation a gas mixture of $1\%SF_6+99\%N_2$ was continuously injected into the crucible during melting. Different parameters were employed to the melt.The microstructures of the alloys were analyzed using an X-ray diffractometer, and a scanning electron microscope. Room temperature tensile/compression tests were performed using a high-temperature electronic universal materials testing machine. Damping tests were performed using a dynamic thermo-mechanical analyzer. The frequency scan is a segmented scan: 0.1-5-100 Hz.

3 Result and discussion

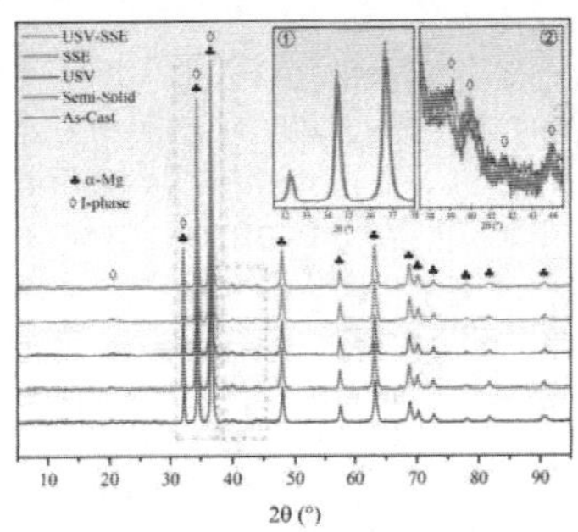

Fig. 1. XRD patterns of alloys under different processes

The XRD patterns in Fig. 1 reveal that the second phase of the alloy primarily consists of α-Mg, $Mg_{32}(Al, Zn)_{49}$, and rich Y phase. Compared to conventionally cast alloys, other alloys exhibit a leftward shift at 32, 34, and 36° (area ①), which can be attributed to the solid dissolution of elements with larger atomic radius into the matrix, resulting in an increase in lattice constant. Notably, quasicrystals show similarity in peak positions; however, their peak intensity slightly surpasses that of other alloys after ultrasonic-assisted rheo-squeeze casting (area ②). Due to the low Y content in the alloys, no Y-rich phase was detected via XRD.

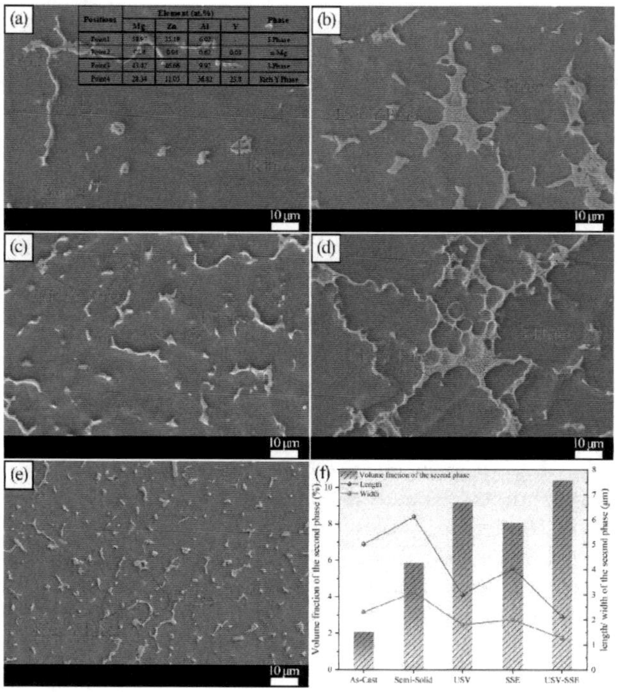

Fig. 2(a-e). SEM of alloys under different treatment processes. (a) As-cast, (b) Semi-Solid, (c) USV, (d) SSE, (e) USV-SSE, Fig. 2(f). Volume fraction and size of second phase

The SEM images of ZAY8150 alloysundergoing different treatment processeswere shown in Fig. 2(a-e). The morphology of the second phase of alloys mainly characterized as blocky, continuous, or granular structures. EDS indicates that the continuous and massive structures are $Mg_{32}(Al,Zn)_{49}$phase (I-Phase), while the granular structures are Y-rich. Both ultrasonic and rheo-squeeze casting promoted the formation and refinement of the second phase. From the Fig. 2(f), it is evident that in conventional casting, the volume fraction of the second phase is small and its size is large When both ultrasonic and rheo-squeeze casting were applied to the alloy, it resulted in the largest volume fraction of the second phase at approximately 10.4%. The average axial length and width of this second phase were reduced to 2.1 and 1.3 μm, respectively.

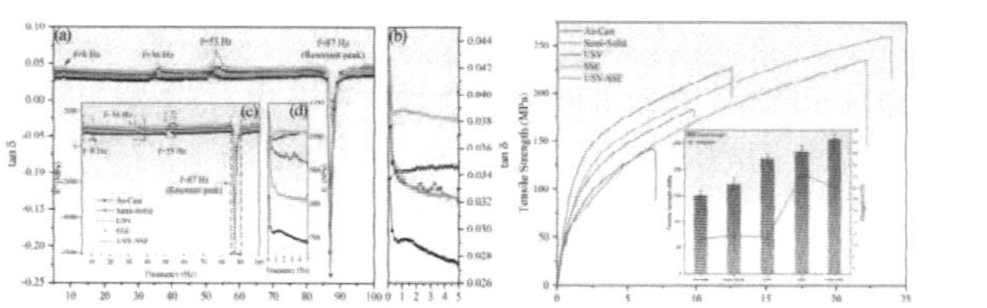

Fig. 3(a-d). tanδ and E" of alloys in frequency scanning mode; (a) tanδ of alloy at scanning frequency of 5-100 Hz, (b) tanδ of alloy at scanning frequency of 0.1-5 Hz, (c) E" of alloy at scanning frequency of 5-100 Hz, (d) E" of alloy at scanning frequency of 0.1-5 Hz, Fig. 3(e). Tensile properties of the alloys

The damping properties of ZAY8150 alloy at different frequencies are shown in Fig. 3(a-d). At low applied vibration frequencies (0.1-5 Hz), the tanδ and E"(Loss Modulus) of the alloys show a sharp decrease pattern as the frequency increases. Compared to the other alloys, alloys with ultrasound-assisted rheo-squeeze casting demonstrate higher damping properties. The tanδ and E" values for this alloy are 0.042 and 1070 MPa, respectively (At an applied vibration frequency of 0.1 Hz,); when the frequency is increased to 5 Hz, these values decrease to 0.038 and 1050 MPa, which are significantly larger than those of conventionally cast alloys (tanδ=0.0275, E"=680 MPa). As the vibration frequency increases further, peaks appear at 8, 36, 53, and 87 Hz. Notably, ultrasonic-assisted rheo-squeeze casting alloys exhibit higher tanδ and E" compared to the other alloys indicating superior damping properties (tanδ is found to be 0.04, and E" is measured as 1100MPa).

From Fig. 3(e), when ultrasound-assisted rheo-squeeze casting was applied to the alloys, their tensile properties increased to 260.4 MPa with an elongation of 20.2%. This represents a significant improvement of 72.7% in tensile strength and 82.0% in elongation compared to conventional casting.

4 Conclusion

In this study, the effects of ultrasonic and rheo-squeeze casting on the microstructure and mechanical properties of the ZAY8150 alloy were systematically investigated. Ultrasonic-assisted rheo-squeeze casting promoted the formation and refinement of the second phase of the alloy, compared with the conventionally cast alloy, the volume fraction of the second phase of alloy (USV-SSE) was 10.4%, and the average length and width were 2.1 and 1.3 μm, respectively. At room temperature, the alloys subjected to ultrasound-assisted rheo-squeeze casting retained high damping properties at high mechanical properties (Tensile strength of 260.4 MPa with elongation of 20.2%).

Acknowledgments

The Project was supported by Anhui Provincial Natural Science Foundation (Grant No. 2308085ME167).

References

[1] Tsai M H, Chen M S, Lin L H, et.al. Effect of heat treatment on the microstructures and damping properties of biomedical Mg–Zr alloy. J. Journal of Alloys and Compounds, 2011, 509(3): 813-819.

[2] Ashraf M N, Guo Z, Wu R, et.al. Historical progress in electromagnetic interference shielding effectiveness of conventional Mg alloys leading to Mg-Li-based alloys: A Review. J. Advanced Engineering Materials, 2023, 25(21).

[3] Wang J, Zou Y, Dang C, et.al. Research progress and the prospect of damping magnesium alloys. Materials. 2024; 17(6):1285.

Thermoelectric Properties Study of ZrCoSb-Based Medium- and High-Entropy Half-Heusler Alloys

Rongchun Chen[1], Huijun Kang[*1,2], Tongmin Wang[*1,2]
1. Key Laboratory of Solidification Control and Digital Preparation Technology (Liaoning Province), School of Materials Science and Engineering, Dalian University of Technology, Dalian, 116024, China.
2. Ningbo Institute of Dalian University of Technology, Ningbo 315000, China
*Corresponding address: e-mail: kanghuijun@dlut.edu.cn; tmwang@dlut.edu.cn

ABSTRACT: As a novel design concept, the design idea of medium- and high-entropy alloy has been successfully applied in various TE systems, which provides a new path for the further development of TE materials. However, it is absent of systematic research on the design and TE performance regulation of medium- and high-entropy HH alloy. Moreover, the research on the TE transport mechanism is relatively shallow, and the mapping relationships between configuration entropy and key TE parameters such as electrical conductivity, Seebeck coefficient, power factor, lattice thermal conductivity, thermal conductivity, and peak ZT are fuzzy. Therefore, based on the iterative design concept among configurational entropy, chemical composition, microstructure, and transport properties, three N-type ZrCoSb-based medium- and high-entropy HH alloys were designed and prepared via multi-site alloying, Zr-site high configurational entropy, and analogous triple crystal structure design strategies. The effects of configurational entropy, V, Nb, Ta, and Ni donor-doping elements on the electrical transport mechanism are elucidated by band structure theory calculation and single parabolic band model. The corresponding phonon transport mechanism is revealed by combining in situ Raman spectroscopy, spherical aberration-corrected transmission electron microscope, with sound velocity, Grüneisen parameter, and other intrinsic physical parameters.

Keywords: Entropy engineering; Triple half-Heusler alloys; Low thermal conductivity

1 Introduction

Thermoelectric performance is directly related to the figure of merit (ZT), ZT = $S^2\sigma T/\kappa$, where S is Seebeck coefficient, σ is electrical conductivity, κ is thermal conductivity, and T is absolute temperature. A lower κ or higher power factor ($PF = S^2\sigma$) enhances ZT. However, high κ prevents further improvement of ZT in n-type ZrCoSb-based HHs. Consequently, phonon engineering and entropy engineering have focused on reducing κ to improve ZT. First, phonon engineering regulates the multiscale microstructure for reducing lattice thermal conductivity (κ_L) by scattering phonons of different frequencies. For instance, owing to differences in the mass and atomic radius between Zr and Hf, the room temperature κ of n-type $ZrCo_{0.9}Ni_{0.1}Sb$ half-Heusler (HH) compound was ~12 W m^{-1}K^{-1}, whereas that of $Zr_{0.5}Hf_{0.5}Co_{0.9}Ni_{0.1}Sb$ HH compound was only ~5.5 Wm^{-1}K^{-1} [1]. In p-type $Zr_{1-x}Hf_xCoSb_{0.9}Sn_{0.1}$ HH compound, Hf substitution considerably reduced phonon mean free path compared to the average crystallite size, affording a very low thermal conductivity of ~2.2 W m^{-1}K^{-1} at 873 K [2]. Besides, the decrease in κ was attributed to the Hf–Ti dual-doping in the FeVSb HH compound by El-Khouly et al. [3]. Second, entropy engineering usually comprises five or more elements with nearly equimolar ratios of 5%–35%. The high mixing entropy, chemical complexity, and large differences in atomic sizes cause severe lattice distortion and high-temperature stability for medium- and high-entropy HHs. Previously, Yan et al. realized a κ_L of ~2.5 Wm^{-1}K^{-1} at 873 K for $Nb_{1-x}M_xFeSb$ (M = Hf, Zr, Mo, V, and Ti, equimolar) HHs developed via the entropy engineering [4], which shows an excellent potential of the entropy engineering in reducing the κ_L. Finally, it includes double HHs except for the 17 and -19-electron systems in the newly developed HH systems. Anand et al. reported that the $Ti_2FeNiSb_2$ double HH compound comprising TiNiSb and TiFeSb shows a considerably lower κ_L than the TiCoSb HH compound [5]. Notably, there is an intrinsic low κ_L in double HHs and high-entropy alloys, substantially inspiring our work. We tried to manipulate the Zr-site composition in the n-type ZrCoSb HH compound to develop low thermal conductivity triple HHs via entropy engineering.

In this study, n-type $M_{0.8}N_{0.2}CoSb$ (M = Ti, Zr, Hf; N = V, Ta, equimolar) triple HHs with a low κ_L of ~3.3 Wm^{-1}K^{-1} at 323 K and ~3.0 Wm^{-1}K^{-1} at 923 K is prepared to trisect the TiCoSb-, ZrCoSb-, and HfCoSb-based HHs. The study focuses on the successful application of the Zr-site entropy engineering strategy in triple HHs with a low κ_L to optimize ZT.

2 Experimental procedure

High-purity Zr (99.5%), Hf (99.95%), Ti (99.99%), V (99.95%), Ta (99.99%), Co (99.95%), and Sb (99.99%) elements are measured according to the nominal stoichiometric ratios of $M_{1-x}N_xCoSb$ (M = Ti, Zr, Hf; N = V, Ta, equimolar). Ingots of $M_{1-x}N_xCoSb$ triple HHs (x = 0.05, 0.1, 0.15, and 0.2) are prepared via levitation melting using a levitation melting furnace (LMF, FM–40, Kejing, China), and the broken ingots are ball milled in a planetary ball mill machine (Pulverisette 4, Fritsch, Germany). 5% excess Sb is added to avoid the loss of Sb during the levitation melting process. The ball-milled powders are sintered at 1173 K for 2 min at 50 MPa using an SPS furnace (SPS, SPS–30, Chen Hua, China).

3 Result and discussion

Fig. 1 shows the temperature-dependent thermal performance of medium-entropy triple HHs. As shown in Fig. 1(a), the thermal diffusivity decreases approximately linearly with the increasing temperature. The entropy engineering indicates a sharp decrease in the thermal diffusion coefficient at 323 K from 4.95 mm^2 S^{-1} for the ZrCoSb HH compound to 1.49 mm^2 S^{-1} for the $M_{0.8}N_{0.2}$CoSb triple HH compound. Cp first increases and becomes stable with the increasing temperature in Fig. 1(b). The overall trend of C_p is downward with the increasing doping content, which may be attributed to the increasing average atomic mass of $M_{1-x}N_x$CoSb triple HHs. Fig. 1(c) demonstrates the decrease in κ for the ZrCoSb HH compound with the increasing temperature. The κ of the ZrCoSb HH compound is higher than that of triple HHs over the studied temperature range. Fig. 1(d) shows that the κ_L is determined by subtracting the electronic contribution ($\kappa_e=L\sigma T$) from κ, where the Lorentz factor (L) is derived from the single parabolic band model.Furthermore, the increasing doping content decreases the κ_L gradually. For instance, the κ_L of the ZrCoSb HH compound decreases from 12.6 Wm^{-1}K^{-1} at 323 K to 7.5 Wm^{-1}K^{-1} at 923 K and the κ_L of $M_{0.95}N_{0.05}$CoSb triple HH compound decreases from 4.0 Wm^{-1}K^{-1} at 323 K to 3.6 Wm^{-1}K^{-1} at 923 K. The obtained results are attributed to entropy engineering, i.e., the strong lattice distortion is beneficial to enhance the scattering of phonons. Moreover, the difference in atomic radius and relative atomic mass causes the local mass and stress field fluctuations, further enhancing the alloy scattering for phonons [6]. Although the κ_e increases with the increasing carrier concentration, the decrease in κ_L caused by lattice distortion offsets the increase in κ_e caused by carrier concentration. Therefore, the κ of $M_{0.8}N_{0.2}$CoSb triple HH compound decreases by 71.4% at 323 K and 50.7% at 923 K.

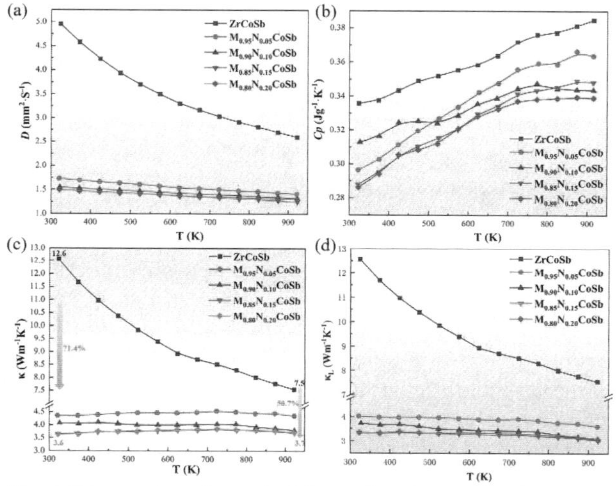

Fig. 1. Temperature-dependent thermal performance of medium-entropy triple HHs: (a) Thermal diffusion coefficient, (b) Cp, (c) κ, (d) κL.

4 Conclusion

In summary, this study proposes a new concept (triple HHs) for designing high-performance HHs. Compared with the ZrCoSb HH compound, the κ of the $M_{0.9}N_{0.1}$CoSb triple HH compound decreases by 67.5% at 323 K and 49.3% at 923 K. The results obtained from this study demonstrate that Zr-site entropy engineering holds great promise for lowering κ_L.

Acknowledgments

This work is financially supported by the National Key Research and Development Program of China (No. 2017YFA0403803), the National Natural Science Foundation of China (Nos. 51971052, 51927801, and 51834009), the Liaoning Revitalization Talents Program (No. XLYC2007183), the Innovation Foundation of Science and Technology of Dalian (Nos. 2020JJ25CY002 and 2020JJ26GX045).

References

[1] R. He, H. Zhu, J. Sun, J. Mao, H. Reith, S. Chen, G. Schierning, K. Nielsch, Z. Ren, Mater. Today Phys. 1 (2017) 24-30.
[2] N.S. Chauhan, S. Bathula, A. Vishwakarma, R. Bhardwaj, K.K. Johari, B. Gahtori, A. Dhar, J. Materiomics. 5(1) (2019) 94-102.
[3] A. El-Khouly, A. Novitskii, I. Serhiienko, A. Kalugina, A. Sedegov, D. Karpenkov, A. Voronin, V. Khovaylo, A.M. Adam, J. Power Sources. 477 (2020) 228768.
[4] J. Yan, F. Liu, G. Ma, B. Gong, J. Zhu, X. Wang, W. Ao, C. Zhang, Y. Li, J. Li, Scr. Mater. 157 (2018) 129–134.
[5] S. Anand, M. Wood, Y. Xia, C. Wolverton, G.J. Snyder, Joule 3(5) (2019) 1226–1238.
[6] S. Chen, K.C. Lukas, W. Liu, C.P. Opeil, G. Chen, Z. Ren, Adv. Energy. Mater. 3(9) (2013) 1210–1214.

Effect of Extrusion Casting and Refining Process on Strength and Elongation of 4032 Aluminum Alloy

Wanneng Liao[1]* Chenxing Zhang[1] Zhongde Shan[1] Wenzhe Song[2] Xixi Dong[1]
1 College of Mechanical and Electrical Engineering, Nanjing University of Aeronautics and Astronautics, Nanjing, 210016, China
2College of Material Science and Technology, Nanjing University of Aeronautics and Astronautics, Nanjing, 210016, China

*Corresponding address: e-mail: liaowanneng@sina.com

ABSTRACT: The 4032 aluminum alloy is a high-Si deformation aluminum alloy, that is commonly generated through forging due to the difficulties involved in its direct casting. In this work, the 4032 aluminum alloy is innovatively prepared via a squeeze casting process, and the Al-6Nb-1.5Ti-0.5B refining agent is incorporated to enhance its mechanical properties. A range of analytical techniques are subsequently employed to analyze the grain refinement mechanism of Al-6Nb-1.5Ti-0.5B under squeeze casting conditions. It was found that the microstructure of the aged 4032 aluminum alloy after refinement treatment was mainly composed of α-Al grains measuring <100 μm in size, in addition to a eutectic Si phase, that is accompanied by stacked dislocations and diffusely distributed Q′-AlCuMgSi and θ′-AlCu phases. The yield strength of the alloy reached 340 MPa, its tensile strength was 404 MPa, and the elongation was 4.1 %. In the presence of the Al-6Nb-1.5Ti-0.5B refiner, the generation of nucleation cores (e.g., MAl(M = Ti, Nb)) in the melt led to a significant reduction (~50 %) in the α-Al grain size of the aged 4032 aluminum alloy (i.e., 95.1 ± 21.0 μm), compared with the standard alloy. After the T6 heat treatment of the refined 4032 aluminum alloy, the size of the precipitation phase was reduced by 36.8 %, and fine silicon particles with a spherical morphology and a more random distribution were observed.

Keywords: 4032 aluminum alloy; Squeeze casting; Al-Ti-Nb-B refiner; Strength; Elongation

2 Introduction

Due to their excellent casting, mechanical, and thermophysical properties, Al-Si alloys have been widely used in the aerospace and automotive industries to prepare components such as engine parts, and compressor scroll discs, among others [1,2]. The α-Al grains present in Al-Si alloys prepared via conventional casting processes tend to be extremely coarse, resulting in poor mechanical properties, especially in the context of elongation [3,4]. To address this issue, elements such as Sr, Sb, and Na are commonly added to the Al-Si alloys to promote melt metamorphism and alter the Si morphology [[5], [6], [7]]. In addition, a range of processes have been investigated to prepare Al-Si alloy castings and to study their mechanical properties following melt metamorphism. For example, Dong et al. [8] reported a high-strength and high-toughness Al-Si alloy fabricated by high-pressure die casting (HPDC); however, the extremely fast filling rate, associated with high-pressure die casting, led to porosity defects in the alloy. In another study, Raju et al. [9] prepared Al-12Si alloys by spray casting; however, this process was expensive and produced only small batches. Moreover, the ultimate tensile strength of the produced Al-12Si alloy reached only 148 MPa.

Thus, we herein report the combinat

ion of squeeze casting with Al-Ti-Nb-B refiners to enhance the mechanical properties of eutectic Al-Si alloys. More specifically, 4032 aluminum alloys were prepared via the squeeze casting in the presence and absence of the Al-6Nb-1.5Ti-0.5B refiners (5.96 wt% Nb, 1.38 wt% Ti, 0.584 wt% B, and a residual amount of Al). The microstructure and mechanical properties of the two alloys in the as-cast and T6 heat-treated states are systematically studied and compared. In addition, the mechanism responsible for the observed microstructure refinement and property strengthening in the 4032 eutectic alloy containing the Al-Ti-Nb-B refining agent is revealed.

1 Experimental procedure

The 4032 aluminum alloy (12.74 wt% Si, 0.33 wt% Fe, 1.29 wt% Cu, 1.00 wt% Mg, 1.15 wt% Ni, 0.059 wt% Ti, residual Al) was melted in a 60 kg capacity crucible using a resistance furnace. During the melting process, the temperature of the resistance furnace was controlled at 720 °C. A total of 30 kg of the alloy melt was prepared, and this melt was homogenized for 1 h. Subsequently, the Al-10Sr intermediate alloy was added, and the alloy melt was subjected to outgassing and smelting for 15 min. After resting for 15 min, the desired alloy melt was obtained (see Fig. 1). In addition, A specimen also added the 1 % weight percentage Al-6Nb-1.5Ti-0.5B refiner (5.96 wt% Nb, 1.38 wt% Ti, 0.584 wt% B, residual Al) when adding Al-10Sr intermediate alloy.

3 Result and discussion

The 4032 aluminum alloy was prepared by squeeze casting both in the presence and absence of a refiner. For this purpose, the following parameters were employed: injection speed of 0.1 m/s, squeeze pressure of 150 MPa, pouring temperature of 710 °C, and mold temperature of

220 °C, respectively. The yield strength of the refiner-free 4032 aluminum alloy (hereinafter referred to as alloy CS1) in the as-cast state was determined to be 169 MPa, the tensile strength was 249 MPa, and the elongation was 1.55 %. Subsequently, alloy CS1 was subjected to T6 heat treatment, according to the following parameters: solid-solution temperature of 525 °C, solid-solution heating insulation time of 2.5 h, aging temperature of 175 °C, aging insulation time of 10 h. After the T6 heat treatment (hereinafter referred to as alloy AS1), the tensile strength of the alloy increased to 364 MPa, the yield strength increased to 334 MPa, and the elongation remained relatively constant at 1.56 %. The above process was then repeated in the presence of the Al-6Nb-1.5Ti-0.5B refiner. As a result, the yield strength of the obtained alloy (hereinafter referred to as alloy CS2) increased to 185 MPa, the tensile strength increased to 285 MPa, and the elongation increased to 2.37 %. After the subsequent T6 heat treatment (hereinafter referred to as alloy AS2), a yield strength of 340 MPa was achieved, along with a tensile strength of 404 MPa, and an elongation of 4.10 %. These results are summarized in Table 1. Considering the process outlined above, it appeared that the use of a high pressure combined with the Al-6Nb-1.5Ti-0.5B refiner synergistically improved the strength and toughness of the alloy. Indeed, the tensile strengths of the CS2 and AS2 alloys were increased by 14.4 and 11.0 %, respectively, compared to the CS1 and AS1 alloys, while the elongation increased by 52.9 and 162.8 %, respectively. In addition, the stress-strain curves for the various alloys are shown in Fig. 2(a), wherein it can be seen that the prepared alloys exhibit excellent mechanical properties. A comparison of their mechanical properties with those of previously reported high-silicon aluminum alloy castings is presented in Fig. 2(b).

4 Conclusion

A 4032 deformed aluminum alloy was prepared by squeeze casting both in the presence and absence of the Al-6Nb-1.5Ti-0.5B refining agent, and the synergistic effect between this process and the incorporation of Al-6Nb-1.5Ti-0.5B was investigated in terms of the alloy's microstructure and mechanical properties. Following the squeeze casting, the refined and unrefined alloys were subjected to T6 heat treatment. It was found that the yield strength of the aged and refined alloy reached 340 MPa, while its tensile strength was 404 MPa, and its elongation was 4.1 %. These results imply that the final obtained alloy was imparted with excellent strength and toughness properties simultaneously.

Acknowledgments

This work is supported by the National Key R&D Program of China (No.2022YFB3404204), the National Natural Science Foundation of China (52105361), the Natural Science Foundation of Jiangsu (BK20210314), China Postdoctoral Science Foundation (2023T160314).

References

[1] Y. Bai, L. Zhao, Y. Wang, et al.Fragmentation of in-flight particles and its influence on the microstructure and mechanical property of YSZ coating deposited by supersonic atmospheric plasma spraying. J. Alloys Compd., 632 (2015), pp. 794-799, 10.1016/j.jallcom.2015.01.265

[2] M.Y. Murashkin Nanostructured Al and Cu alloys with superior strength and electrical conductivity . Mater. Sci., 51 (2016), pp. 33-49, 10.1007/s10853-015-9354-9

Current-Carrying Wear Characteristics of Cast Al-Cu Alloys

Junwen Zhao [1], Yongxin Zhang [1], Hanzhi Feng [1], Lei Xu [2], Jing Han [1]

1. Key Lab of Advanced Technologies of Materials (Ministry of Education), School of Materials Science and Engineering, Southwest Jiaotong University, Chengdu, 610031, China

2. Key Laboratory of materials and surface technology (Ministry of Education), School of Materials Science and Engineering, Xihua University, Chengdu, 610039, Chengdu, China

*Corresponding address: e-mail: swjtuzjw@swjtu.edu.cn

Abstract: The influence of electric current and Cu content on the wear behavior of cast Al-Cu alloys was studied. The results show that the wear rate of cast Al-Cu alloys decreased with the application of electric current, and showed an increasing trend with increasing current. The wear rate of Al-15Cu is higher than that of Al-5Cu and Al-10Cu. Under the condition of current carrying, the difference between Al-15Cu and Al-5Cu/Al-10Cu increases with increasing current. When the current reaches 180 A, the wear rate of Al-15Cu is about three times that of other Al-Cu alloys. In addition, the current carrying wear mechanism of cast Al-Cu alloys was discussed.

Keywords: Cast Al-Cu alloy; Current-carrying wear; Wear rate; Wear mechanism

1 Introduction

Light-weighting is one of the crucial techniques to meet the requirements of higher speed and lower energy consumption for trains. Current-carrying friction pairs are the vital components of trains for effective power transmission [1-3]. Al-Cu alloys are able to be alternative lightweight materials for current-carrying friction pair due to their excellent properties such as high specific strength, good conductivity, thermal conductivity, high temperature performance, and wear resistance [4-6].

2 Experimental procedure

Al-Cu alloys were prepared using electrical resistance furnace with raw materials of pure Al (99.99 wt.%), pure copper (99.7 wt.%) and formed into bars by squeeze casting under a pressure of 106 MPa. Finally, the bars were machined into disc specimens (Φ60×5 mm) by wire-cutting. Alloys were tested on the self-made pin-disc current-carrying wear tester with Cu/C brush as friction pairs (Fig 1). The test was carried out with a load of 24N for 240min, a sliding speed of 60r/s and a current density varying between 0 and 180 A.

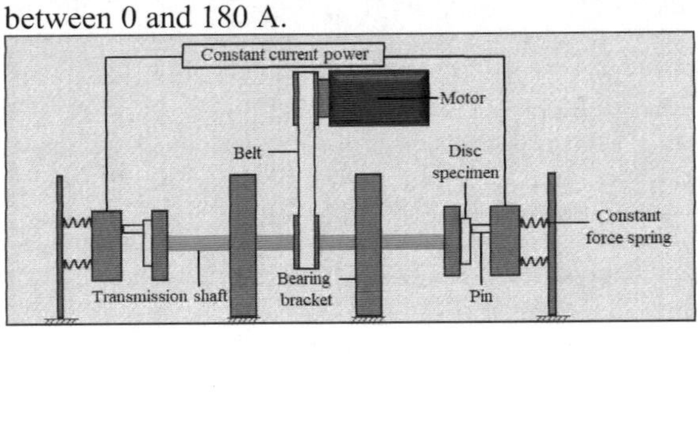

Fig 1 Schematic diagram of the current-carrying wear tester.

Microstructure characterization was carried out by optical microscopy on the samples etched with a Keller reagent. The wear specimens were immersed in alcohol for ultrasonic cleaning, and the mass of each group of wear specimens before and after the test was weighed using the HZK-JA1000 analytical balance to determine wear loss. The micro-morphology of the alloy wear surface and debris were observed using a scanning electron microscope (JSM-IT300).

3 Result and discussion

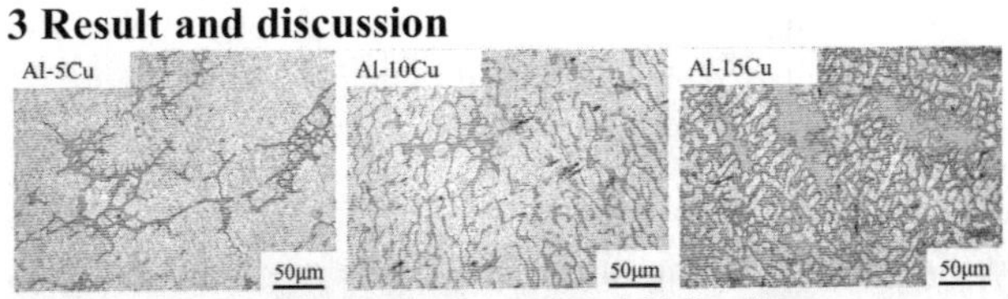

Fig 2 Microstructure of cast Al-Cu alloys

As presented in ***Fig 2***, with the increase of Cu content, the content of eutectic structure increases. The eutectic structure transforms from discontinuous particles and chains into a network-like structure, and even larger block-like eutectic structures emerge in the Al-15Cu alloy.

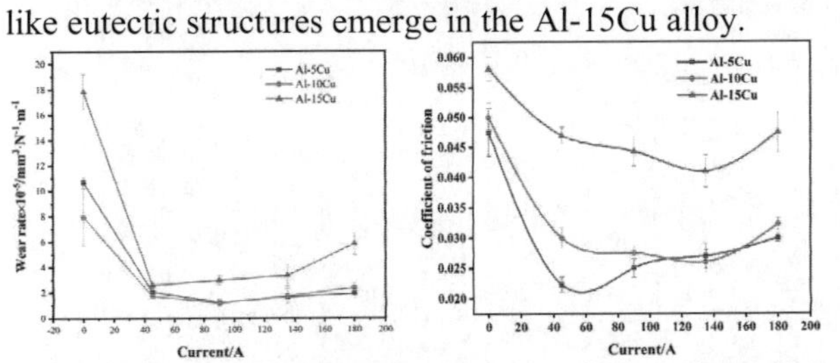

Fig 3 Wear rate and steady-state friction coefficient of cast Al-Cu alloys under different currents

Fig 3 suggests the Cu content has significant effect on the wear rate of Al-Cu alloys. As the current increases from 0 A to 180 A, the wear rate of Al-Cu alloys first decreases and then increases. When the current is 0A, the alloys have the highest wear rate. With the introduction of current, the wear rates decreased sharply, a reduction of approximately 80% with an electric current of 45 A. As the current increases from 45 A to 180 A, the wear rate of the alloys generally shows an increasing trend. The wear rate of Al-15Cu alloy is higher than that of Al-5Cu and Al-10Cu. When no current was applied, the wear rate of Al-15Cu alloy can reach 17.8×10^{-5} $mm^3\cdot N^{-1}\cdot m^{-1}$, which is over twice that of the Al-5Cu alloy. Under the condition of current carrying, the difference in wear rate between Al-5Cu and Al-10Cu is insignificant. However, the difference between Al-15Cu alloy and the two alloys increases with increasing current. When the current reaches 180 A, the wear rate of Al-15Cu is 5.8×10^{-5} $mm^3\cdot N^{-1}\cdot m^{-1}$, which is about three times that of Al-5Cu or Al-10Cu alloy.

As for the steady-state friction coefficient, the friction coefficient of Al-Cu alloys generally exhibits a trend of first decreasing and then increasing as the current increases from 0 to 180 A. Initially, all studied Al-Cu alloys have the maximum friction coefficient at 0 A. After the introduction of current, the friction coefficient decreases significantly. The friction coefficient of Al-15Cu alloy increases significantly from 0.041 to 0.048 when the current is between 135 A and 180 A, representing an increase of approximately 15%. For Al-10Cu and Al-5Cu alloys, the friction coefficient generally shows an upward trend but with a small increase. Comparing the differences in friction coefficients between Al-Cu alloys with different Cu contents, it can be found that the friction coefficient of Al-15Cu alloy is much higher than that of Al-10Cu and Al-5Cu. The change curve of the friction coefficient basically shows a trend of increasing with the increase in Cu content. It is also noted that under current-carrying conditions, as the current increases from 45 A to 180 A, the steady-state wear temperatures of the three alloys show a roughly linear upward trend. Among them, the temperature of the Al-15Cu alloy has the maximum rise, increasing from 47.9°C to 81.9°C, an increase of approximately 120%.

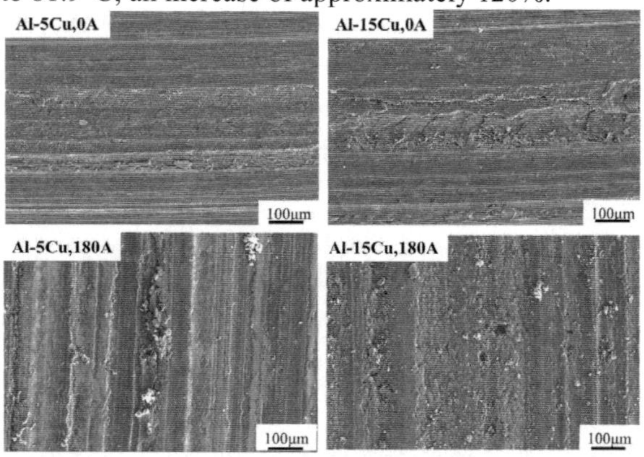

Fig 4 Microscopic wear morphology of cast Al-Cu alloys under different currents

As indicated in Fig 4, when the current is 0 A, the matrix was torn and peeled severely. In the torn and peeled areas of the Al-15Cu wear surface, more large-scale flaky wear debris and peeling can be observed, indicating a poor surface condition. Both alloys have distinct furrow-like groove traces on the wear surface, indicating an abrasive wear mechanism.

The intervention of current promoted graphite transfer and the formation of an oxide mixed layer on the worn surface, which played a lubricating and protective role in reducing the wear rate. When the current is 180 A, for the Al-5Cu alloy, a large amount of adhesion can be observed on the surface, with graphite and Cu adhesion transferred along the wear sliding direction distributed on the wear surface. The surface along the sliding direction is relatively smooth, but furrow-like morphology can also be found. Aggregated spherical granular substances are found in some areas of the wear surface. In contrast, the Al-15Cu alloy exhibits severer tearing and peeling of the matrix, caused by the hard, brittle and low conductivity network eutectic structure. Furthermore, a large amount of flaky wear debris adhesion can be seen on the surface, which is from material transfer under the high-temperature effect of the arc during wear. In addition, after wear with 180 A current, the Al-15Cu alloy exhibits more obvious arc erosion pits on the surface, resulting in a severe degradation of the surface condition and high wear rate.

4 Conclusion

The wear rate of cast Al-Cu alloys decreased with the application of electric current, and showed an increasing trend with increasing current. The wear rate of Al-15Cu is higher than that of Al-5Cu and Al-10Cu. Under the condition of current carrying, the difference between Al-15Cu and Al-5Cu/Al-10Cu increases with increasing current. When the current reaches 180 A, the wear rate of Al-15Cu is about three times that of Al-5Cu or Al-10Cu alloy. The intervention of current promoted graphite transfer and the formation of an oxide mixed layer on the worn surface, leading to a reduction of wear. The increasing current promotes adhesive wear and arc erosion, resulting in high wear rate. The high wear of Al-15Cu alloy can be attributed to the action of eutectic structure during current carrying wear.

Acknowledgments

This work was supported by the Sichuan Science and Technology Program (Nos. 2023YFG0195) and the Key Laboratory of Materials and Surface Technology, Ministry of Education (No. xxx-2023-zd003)

References

[1] Song, Y., et al., Contact Wire Irregularity Stochastics and Effect on High-Speed Railway Pantograph–Catenary Interactions. IEEE Transactions on Instrumentation and Measurement, 2020. 69(10): p. 8196-8206.
[2] Liu, W., et al., Critical life cycle inventory for aluminum die casting: A lightweight-vehicle manufacturing enabling technology. Applied Energy, 2021. 304: p. 117814.
[3] Liu, X., et al., Effect of copper contents on the current-carrying wear properties of carbon brush under different temperatures conditions. Journal of materials research and technology, 2021. 15: p. 3110-3121.
[4] Milligan, B.K., et al., Impact of microstructural stability on the creep behavior of cast Al–Cu alloys. Materials Science and Engineering: A, 2020. 772: p. 138697
[5] Lee, S., et al., Strength and electrical conductivity of Cu-Al alloy sheets by cryogenic high-speed rolling. Materials Science and Engineering: A, 2021. 799: p. 139815.
[6] Yang, C., et al., Nanoparticle additions promote outstanding fracture toughness and fatigue strength in a cast Al–Cu alloy. Materials & Design, 2020. 186: p. 108221

Process Optimization and Wearing Characterization of As-Cast and Heat-Treated Cocrfemnni(Alx) High Entropy Alloy

I. Vicario[1*], E. Villanueva[1], J. Albizuri[2], T. Guraya[2], L. Armendariz[2], O. Gordo[3], P. Garcia-Michelena[3].

1. Metal processing platform, TECNALIA, Basque Research and Technology Alliance (BRTA), Derio, Spain
2. Faculty of Engineering of Bilbao (UPV/EHU), Bilbao, Spain
3. Mechanical and Manufacturing Department, Mondragon University, Arrasate-Mondragon, Spain

*Corresponding address: iban.vicario@tecnalia.com

Abstract: This study aims to study a CoCrFeMnNi(Alx) High Entropy Alloy (HEA) with high wear resistance produced by standard induction melting from scrap and ferroalloys.

To the HEA base alloy, Al was added to promote a decrease in grain size and improve the final wearing properties The thermal treatment (TT) was optimized to increase the solubility of alloying elements at 1,100ºC for 36 hours and immediately water cooling.

The coefficient of friction (COF) and wear rate values are like those obtained from pure metals and with multiple remelting phases by vacuum arc melting. Adding Al in the alloy promotes a decrease in hardness and COF, and an increase in wear rate values, which is anomalous in metals. TT increases hardness and decreases COF and wear rate. The study of samples explains the abnormal behavior by the finer microstructure in Al as-cast samples and the presence of more precipitates in TT samples. The obtained processes and alloys can have future use in automotive or energy applications.

Keywords: HEAs; Wear; Induction melting, Scrap.

1 Introduction

HEAs are very promising because of their excellent mechanical and wear properties. The Cantor alloy [1] is obtained by the combination of five molar metallic elements (5-35%), with a single-phase solid solution structure. These alloys exhibit improved wear [2]. Wear properties information is still scarce, especially for Cantor alloys alloyed with different amounts of Al.

This study focuses on preparing a Cantor alloy by an industrial standard process and using low-cost and CO_2 emissions raw materials. Al addition has been added to increase the wear properties. Al can modify the microstructure of Cantor alloy, promoting smaller grains and precipitates with an increase in wear properties [3].

2 Experimental procedure

30 Kg of casting samples were produced in an induction furnace with no atmospheric protection. Steel scrap, ferroalloys, and an aluminum alloy were employed as raw materials. First, a base of iron scrap melt was obtained, and the highest melting point ferroalloys were added sequentially. The melt was maintained for 5 minutes at 1,650ºC, cleaned, and poured into a sand mold.

Hardness was measured using the Vickers diamond indentation test with a load of 10 kg for 10 s according to UNE EN ISO 6507-1. Dry sliding-wearing friction tests were carried out in a sphere-on-plate reciprocating configuration using a ball-on-disk mode tribometer according to the ASTM G99-05 standard without any lubricant, with a load of 10 N, 0.1 m/s, and 1,000 m.

3 Result and discussion

In Table 1, we can observe the obtained compositions.

Table 1: Compositions of studied alloys (wt.%).

Alloy	Co	Cr	Fe	Mn	Ni	Al
CoCrFeMnNi	20.4	17.7	20.5	18.9	21.8	-
$CoCrFeMnNiAl_{0.03}$	20.5	17.9	20.5	19.4	21.0	0.3
$CoCrFeMnNiAl_{0.3}$	19.8	17.4	20.0	18.9	20.5	2.8

In Table 2, we can observe the obtained hardness values.

Table 2: HV10 Hardness values of studied alloys.

Alloy	As-cast	TT
CoCrFeMnNi	148.5±2.6	149.6±4.2
$CoCrFeMnNiAl_{0.03}$	136.0±4.7	140.4±6.6
$CoCrFeMnNiAl_{0.3}$	132.9±1.0	122.0±4.8

Al addition decreases the hardness and treatment slightly increases the hardness, but not significantly, except in $CoCrFeMnNiAl_{0.3}$. TT increases hardness in samples without Al or with low aluminium.

Microstructure shows a dendrite structure with an interdendritic area and precipitates, with FCC#1 the solid solution of the alloy and FCC#2 the interdendritic phase richer in Mn and Ti. The as-cast sample with 0.03% Al has more precipitates but 0.3% samples less. Cubic precipitates are slightly richer in Cr-Mn [4] and Al dark phase precipitates in TT Al samples alloyed (Approx. Al68CoCrFeNi32 wt.%).

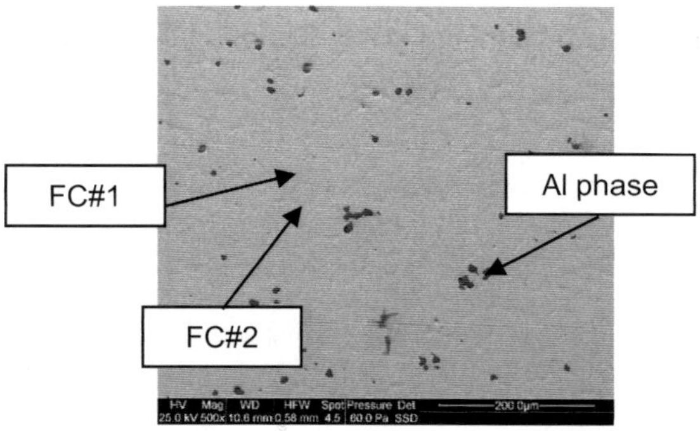

Figure 1 Microstructure of TT CoCrFeMnNiAl0.3 alloy

In Table 3, we can observe the ball-on-disk results.

Table 3: COF and wear rate values (mm3/N.m).

Alloy	μ, COF	Wear rate
CoCrFeMnNi	0.74±0.00	8,72E-04±0.11
$CoCrFeMnNiAl_{0.03}$	0.75±0.01	8,32E-04±0.13
$CoCrFeMnNiAl_{0.3}$	0.71±0.02	1,14E-03±0.14
TT-CoCrFeMnNi	0.73±0.00	8,28E-04±0.06
TT-$CoCrFeMnNiAl_{0.03}$	0.72±0.01	8,94E-04±0.02
TT- $CoCrFeMnNiAl_{0.3}$	0.68±0.02	1,06E-03±0.02

The addition of Al in the alloy promotes in general a decrease in hardness and COF (in disagreement with the decrease in hardness) and an increase in wear rate values, which is not common in HEAs [5,6]. TT samples showed smaller COF (in agreement with the increase in hardness values in TT samples except in $CoCrFeMnNiAl_{0.3}$) and smaller wear rates than as-cast samples (except in $CoCrFeMnNiAl_{0.03}$).

The wear analysis shows the transition from abrasive and adhesive wear to delamination that induces large plastic strains and strain gradients near the worn surface, with smaller wear in the samples with more precipitates, reducing the delamination effect.

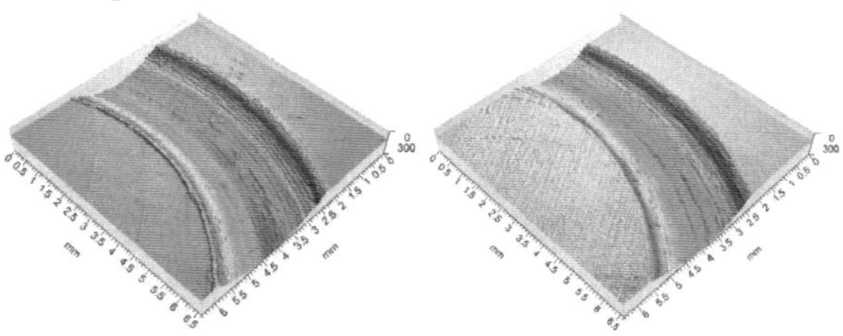

Figure 2 As-cast and TT CoCrFeMnNiAl0.3 alloy 3D profile.

Conclusions

The main results of the study of the new alloy are:

- Obtained wearing values are like high-purity multi-melted vacuum Cantor alloys.
- Adding Al decreases the COF and increases the wearing rate by having a higher matrix area.
- TT samples had smaller COF and smaller wearing rates by presenting more hard precipitated phases.
- Low Al addition seems to act similar to a grain refiner.
- The obtained processes and alloys can have future use in automotive or energy applications.

Acknowledgments

The authors acknowledge the financial support of the Basque Government through the project Elkartek KK-2022/00064 (H2MAT), KK-2024/00021 (H2MAT+), HORIZON-JTI-CLEANH2-2023-1, 101137610 (H2AL) and KK-2023/00020 (DESGAS).

References

[1] [1] B. Cantor, I. T. H. Chang, P. Knight, and A. J. B. Vincent, "Microstructural development in equiatomic multicomponent alloys" *Materials Science and Engineering A*, vol. 375–377, no. 1-2 SPEC. ISS., pp. 213–218, Jul. 2004.

[2] Z. Cheng, S. Wang, G. Wu, J. Gao, X. Yang, and H. Wu, "Tribological properties of high-entropy alloys: A review" *International Journal of Minerals, Metallurgy and Materials,* vol. 29, no 3, pp. 389, Mar 2022.

[3] J.M. Wu, S.J. Lin, J.W. Yeh, S.K. Chen, Y.S. Huang, and H.C. Chen, "Adhesive wear behavior of AlxCoCrCuFeNi high-entropy alloys as a function of aluminum content", *Wear*, Vol. 261, no. 5-6, pp. 513-519, Sep.2006.

[4] X.B. Guo, I. Baker, F.E. Kennedy, S.P. Ringer, H.S. Chen, W.D. Zhang, Y. Liu, and M. Song, "A comparison of the dry sliding wear of single-phase f.c.c. carbon-doped Fe40.4Ni11.3Mn34.8Al7.5Cr6 and CoCrFeMnNi high entropy alloys with 316 stainless steel", *Mater. Charact.*,vol. 170, pp. 1-14, Dec.2020.

[5] L. Yang, Z. Cheng, W.W. Zhu, C.C. Zhao, and F.Z. Ren, "Significant reduction in friction and wear of a high-entropy alloy via the formation of self-organized nanolayered structure", *J. Mater. Sci. Technol.*, vol. 73, pp.1-8, May 2021.

[6] J. Joseph, N. Haghdadi, K. Shamlaye, P. Hodgson, M. Barnett,and D. Fabijanic, "The sliding wear behaviour of CoCrFeMnNi and AlxCoCrFeNi high entropy alloys at elevated temperatures", *Wear*, vol. 428-429, pp. 32-44, Jun. 2019.

Analysis of Nucleation Characteristics and Grain Growth Dynamics in Aluminum and Its Titanium Alloys: A Molecular Dynamics Simulation Study on Critical Nucleus Size, Nucleation Rate, and the Effects of Growth Restriction Factors Based on the MFPT Method

Run Wang, Haijun Huang
School of Materials and Chemistry, University of Shanghai for Science and Technology, No. 516 Jungong Road, Yangpu District, Shanghai, China
*Corresponding address: e-mail: haijun.huang@usst.edu.cn

ABSTRACT: This study utilized LAMMPS molecular dynamics simulation software and the Mean First-Passage Time (MFPT) method to systematically analyze the dynamic behavior of nucleation and grain growth in aluminum (Al) and its titanium (AlTi) alloys. Initially, the research calculated the critical nucleus size, nucleation rate, and grain growth rate for Al. Subsequently, the impact of Ti addition on these parameters was investigated, particularly under various growth restriction factor conditions. The results indicated that the addition of Ti significantly reduced the nucleation rate, critical nucleus size, and grain growth rate of the aluminum alloy. Furthermore, an increase in the growth restriction factors further decreased the nucleation rate, critical nucleus size, and grain growth rate.

Keywords: Growth restriction factors ; Molecular dynamics simulation ; Critical nucleus size ; Nucleation rate

1 Introduction

The properties of metal alloys, such as aluminum and its composites, are largely determined by their nucleation and subsequent grain growth during solidification. These properties in turn critically influence the mechanical behavior and durability of the resulting materials.

Molecular dynamics (MD) simulations have emerged as a powerful tool for exploring these phenomena on an atomic scale, providing detailed insights that are often inaccessible through conventional experimental approaches. This study employs the Large-scale Atomic/Molecular Massively Parallel Simulator (LAMMPS), a sophisticated MD simulation software, to investigate the dynamics of nucleation and grain growth within pure aluminum and its titanium alloys (AlTi).

The inclusion of titanium in aluminum alloys is widely recognized for its potential to enhance material properties through microstructural refinement. However, the mechanisms by which Ti influences nucleation and grain growth in aluminum alloys are not fully understood and require detailed investigation. To this end, our research utilized the Mean First-Passage Time (MFPT) method to systematically quantify the critical nucleus size, nucleation rate, and grain growth rate in these alloys.

Initial simulations focused on pure aluminum to establish baseline measurements of nucleation and grain growth dynamics. Subsequent simulations introduced titanium into the aluminum matrix, and the effects of Ti on these parameters were thoroughly analyzed. The study specifically examined the role of growth restriction factors, which are crucial in controlling the microstructural evolution during solidification.

Our findings revealed that the addition of titanium significantly reduces the nucleation rate, critical nucleus size, and grain growth rate in aluminum alloys. Moreover, increasing the growth restriction factors further diminished these parameters, suggesting that Ti addition and the manipulation of growth restriction conditions could be strategically used to enhance the performance of aluminum alloys. This study not only contributes to a deeper understanding of the solidification processes in Al-Ti alloys but also offers practical insights for the development of advanced materials with improved mechanical properties.

2 Experimental procedure

The MD simulations were conducted using LAMMPS (Large-scale Atomic/Molecular Massively Parallel Simulator) software (Sandia National Laboratories, Livermore, California, USA). In this work, the pure aluminum simulations were performed using a box containing 171,500 FCC lattice atoms in a 141.75 Å×141.75 Å×141.75 Å box; the aluminum-titanium alloys were also simulated using a box with 171,500 FCC lattice atoms of the same dimensions. The models were divided into two groups with different growth restriction factors, containing 48,123 and 39,565 titanium atoms, respectively. The study used the EAM potential. Periodic boundary conditions were applied in all three dimensions, with an integration timestep of 5 fs. The simulations applied an isothermal-isobaric ensemble (NPT), with temperature controlled via a Nosé-Hoover thermostat and pressure controlled by volume scaling. Due to the limited spatial and temporal scales of MD simulations, and because nucleation in metal melts is a random event, independent runs are necessary to obtain statistically meaningful results. In this study, each simulation was repeated five times.

Initially, the perfect single crystal of aluminum was heated from the solid state at 300 K to the liquid state at 2000 K. The liquid was then cooled to 300 K at a rate of 2.0×10^{11} K/s, and nucleation was observed during the cooling process.

For the aluminum-titanium model, the system was heated from 300 K in the solid state to 2000 K in the liquid state, and then relaxed at 2000 K for a period. The relaxed liquid alloy served as the initial state of the model and was cooled from 2000 K to 300 K at a rate of 2.0×10^{11} K/s. All simulations were conducted at 0 MPa pressure.

The microstructural evolution during solidification was observed using the visualization software OVITO. The local environment of atoms was determined using the Adaptive Common Neighbor Analysis (CNA) method. CNA determines the atomic structure by calculating the topology of each atom's nearest neighbors. By comparing the computed values to known characteristics of standard crystals, the local atomic crystal structure can be distinguished. In OVITO, atoms not identified as FCC, HCP, BCC, or any other crystalline type are classified as ICO liquid or ICO solid atoms.

Cluster analysis in OVITO was used to export plots showing the change in the internal atomic count of clusters over time, and the MFPT method was employed to calculate the critical nucleus size and nucleation rate for aluminum and aluminum-titanium.

The growth restriction factor was calculated with reference to the article "Growth Restriction Factor in Al-Si-Mg-Cu Alloys" by Robert-Koch-Str.

3 Result and discussion

The molecular dynamics simulations provided comprehensive insights into the effects of titanium addition and growth restriction factors on the nucleation and grain growth dynamics in aluminum and its titanium alloys.

(1) Nucleation Rate and Critical Nucleus Size: Initial results indicated that titanium addition significantly reduces the nucleation rate and the critical nucleus size in aluminum alloys compared to pure aluminum. The titanium addition enhanced the growth restriction factor, thereby creating a solute-rich environment that hinders nucleation. The Mean First-Passage Time (MFPT) method was used to quantify the nucleation rates and critical nucleus sizes, confirming that as titanium content increases, the nucleation rate decreases due to higher solute concentrations impeding nucleation. The critical nucleus size also reduced, implying that smaller nuclei become stable with titanium addition.

(2) Grain Growth Rate: Titanium's presence in the aluminum matrix also had a marked effect on the grain growth rate. With the increased concentration of titanium, the grain growth rate decreased. This is attributed to the growth restriction factors created by titanium, which inhibited the growth of existing grains. The results demonstrated that higher concentrations of titanium significantly slow the grain growth rate.

(3) Growth Restriction Factors: By analyzing models with varying growth restriction factors, the results showed a clear correlation between an increase in growth restriction factors and a further reduction in nucleation rates, critical nucleus sizes, and grain growth rates. This reinforces the concept that manipulating growth restriction factors via alloying elements like titanium provides a strategic approach to refine microstructure and improve the properties of aluminum alloys.

(4) Implications: These findings underscore the importance of growth restriction factors in the solidification behavior of aluminum-titanium alloys. The strategic addition of titanium, along with optimized growth restriction factors, offers a pathway to designing advanced aluminum alloys with refined microstructures and superior mechanical properties. Further studies should focus on validating these findings through experimental work and exploring the applicability of different alloying elements in other metal matrices.

4 Conclusion

The addition of Ti significantly reduced the nucleation rate, critical nucleus size, and grain growth rate of the aluminum alloy. An increase in the growth restriction factors further decreased the nucleation rate, critical nucleus size, and grain growth rate.

Investigation on Heat Treatment Process Optimization of Super-Slow-Speed Die Casting A356.2 Alloy

Lu Zhang[1,2], Hengcheng Liao[1], Jiang Li[2], Liangchao Tang[2]
1.School of Materials Science and Engineering, Southeast University, Nanjing, China.
2. Nanjing Chervon Auto Precision Technology Co., Ltd. Nanjing, China.
*Corresponding address: e-mail: hengchengliao@seu.edu.cn

Abstract: A356.2 alloy die casting is produced by the Super-Slow-Speed die casting process. The phase composition of the alloy under the Scheil model and DB model (modified Scheil model) was analyzed by the CALPHAD-based method. Combined with DSC, conductivity meter, universal testing machine, and thermomechanical analyzer, the heat treatment process, microstructure, and mechanical properties of the material were studied, and the range of heat treatment process parameters required for actual production was determined.

Keywords: A356.2 alloy, Electrical conductivity,Solution temperature, CALPHAD, Mechanical properties;

1 Introduction

High pressure die casting (HPDC) is widely used but can have defects that affect strength and plasticity. Super-Slow-Speed HPDC (SSS HPDC) reduces these issues, enabling heat treatment for property enhancement. A356.2 alloy, known for toughness, is now used in HPDC. Aging treatments can strengthen but may decrease elongation. This study seeks to develop a heat treatment process for A356.2 in SSS HPDC that balances tensile strength, yield strength, and elongation, utilizing CALPHAD, electrical conductivity, and thermal expansion tests.

2 Experimental procedure

The A356.2 alloy was evaluated via SSS HPDC on an 840Ton Büler Ecoline S machine, with specific process parameters and mold temperatures. The casting, used in a vehicle's toolkit, had a 4mm wall thickness. Analysis included chemical composition with a Bruker spectrometer, thermal properties by DSC, electrical conductivity, microstructure via optical microscopy, and tensile tests with an average of three samples. Dimensional changes were measured with a thermomechanical analyzer across various temperatures.

3 Result and discussion

CALPHAD analysis in solidification

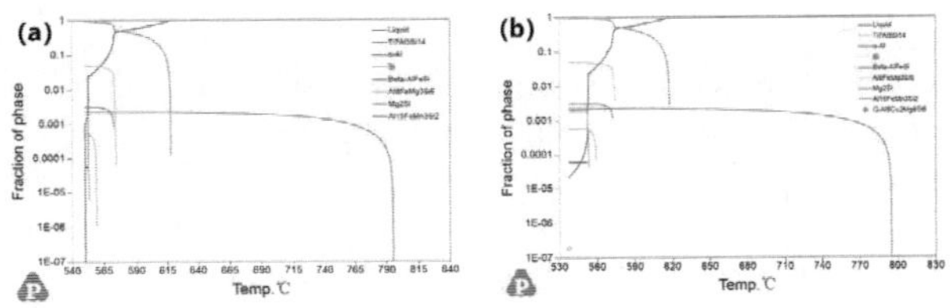

Figure1 Fraction of phase in the Scheil model (a); Fraction of phase in the DB model (b).

The difference is that the Scheil model results in a small amount of Al15FeMn3Si2 phase at 95% solid phase fraction and 550℃ final solidification temperature. BD model results in a termination temperature of 536℃ for multicomponent eutectic reaction due to a very small amount of Q-Al5Cu2Mg8Si6 phase.

Solution heat-treatment process

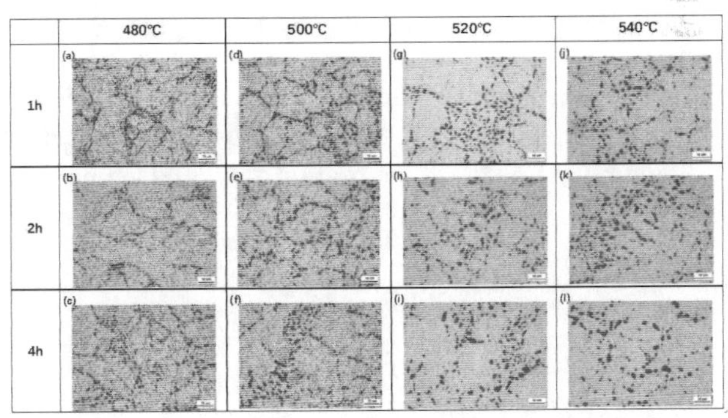

Figure 2 Microstructure after solution heat treatment on 480℃ for 1, 2, 4 h (a, b, c); 500℃ for 1, 2, 4 h (d, e, f); 520℃ for 1, 2, 4 h (g, h, i); and 540℃ for1, 2, 4 h (j, k, l).

Metallographic observations reveal that at 480℃ for 1-hour, eutectic silicon breaks and spreads into the aluminum matrix. Spheroidization occurs at 500℃ for 1 hour and becomes more pronounced with longer holding times. At 520℃ and 540℃ for 1-2 hours, Si particles spheroidize significantly, but after 4 hours, Si particles coarsen again.

Electrical conductivity

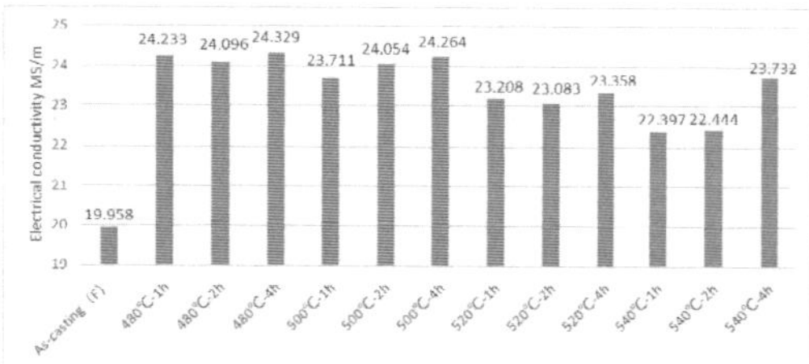

Figure 3 displays the electrical conductivity of an Al-Si alloy that underwent solution treatment at varying temperatures and holding times. The solubility of Si and other alloying elements in the α-Al matrix and Si phase morphology evolution affect the alloy's conductivity.

Mechanical property

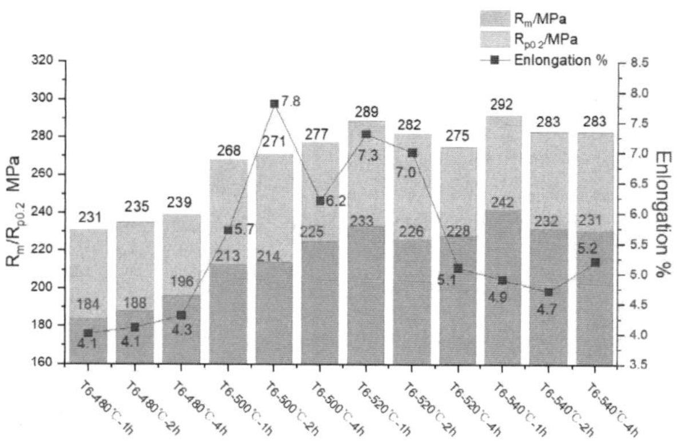

Figure 4 Mechanical properties of different solution treatments followed by aging at 180℃for 6 h, the highest QI should be 418.5MPa when heating at 520℃ and holding for 1 h.

Thermal expansion/length change

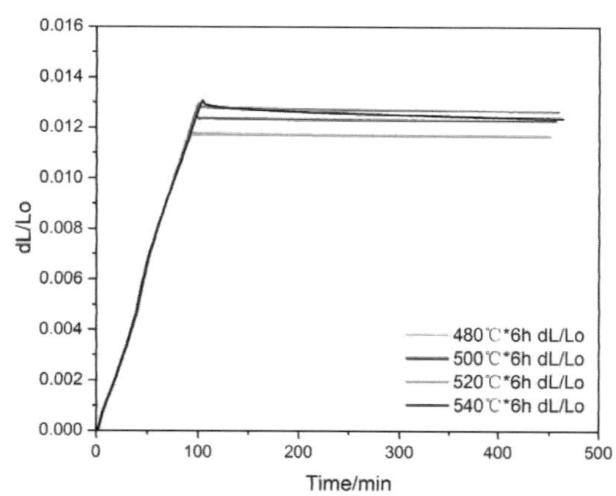

Figure 5 displays the effect of heating as-cast samples to different temperatures ranging from 480℃ to 540℃ for 6 hours. Local remelting of the material at 540℃ caused size deformation, but no significant changes were observed in the temperature range of 480℃-520℃. Therefore, this temperature range is suitable for castings' solid solution heat treatment.

4 Conclusion

The effect of solution treatment on the electrical conductivity of A356.2 alloy is influenced by the solubility of Si and other alloying elements in the α-Al matrix and the morphology evolution of Si phase.

Using the DB model of the solidification module of thermodynamic phase diagram software, the final solidification precipitation temperature was obtained at about 536℃; Combined with the thermal expansion curve at different holding temperatures, the eutectic liquid phase is produced at 540℃, which indicates that the DB model is suitable for the solidification analysis of the material in the super-slow-speed die casting process.

At temperatures of 500℃, 520℃, and 540℃, it is possible to obtain the necessary mechanical properties for the heat treatment process parameters for 1, 2, and 4 hours. The highest QI value of 418.5MPa can be achieved at 520℃ for 1 hour. However, it is not recommended to use the temperature of 540℃ for solution treatment, as it results in a negative size change. In practical production, the appropriate heat treatment process should be chosen according to the specific situation.

Acknowledgments

The financial support from Nanjing Chervon Auto Precision Technology Co., Ltd. (Grant No. 322040)

Thanks to Mr. Wang Xiangyu. Mr. Wang Yuxiao. Mr. Yong Pengling, Mr. Zhao Nan, and Mr. Yao Shaojun for their test work.

Self-Cooling Hollow Mold With Containing Water and Its Effect on Microstructure and Mechanical Properties

Jingying Xu1, Jinwu Kang2,*, Weimin Mao1
1. School of Materials Science and Engineering, University of Science and Technology Beijing, Beijing 100083, China
2. School of Materials Science and Engineering, Key Laboratory for Advanced Materials Processing Technology, Tsinghua University, Beijing 100084, China
*Corresponding address: e-mail: kangjw@tsinghua.edu.cn.

ABSTRACT:The solidification process of castings directly determines the intrinsic quality and performance of castings, so controlling the cooling process during solidification of castings can eliminate or reduce defects such as shrinkage and porosity, cracks, deformation, etc., and then improve the performance of castings. 3D printing sand mold technology provides more options for the production of complex and high-quality castings, and improves the flexibility and innovation of the design of the sand structure. In order to control the sequential solidification of the castings, a new process of self-cooling hollow mold with containing water is proposed. The hollow mold is of a multi-shell structure which can contain water in the gaps between shells. Thus, it is convenient than the water dipping method or the bottom water cooling chill plate method. After pouring the molten metal, cooling water is pumped into the gaps between the shell layers, and the water level of the cooling water can be controlled to rise gradually to drive the solidification front per request, so as to realize the forced cooling of the castings from bottom to top, i.e., sequential solidification. By the casting experiment of barrels of aluminum alloy A357, it was found that the water cooling of the multi-layer shell hollow sand mold accelerates the cooling rate of the castings, the cooling rate increased by approximately 190% compared to natural cooling. The tensile strength of castings is 160 MPa, representing an increase of 14.3%, and the microhardness is 72.3 HV, showing an increase of about 43.5%.

Keywords: Hollow sand mold; Multi-layer shell structure; Controlled cooling; Casting; 3D printing

1 Introduction

Controlling the cooling process of castings can eliminate or reduce the defects and improve the performance of castings. In the solidification process of castings, external media such as cooling water and wind can be used to enforce the cooling of castings. Wladysiak [1, 2] used a rotating nozzle to spray waterfall on the metal mold surface for cooling, and found that an open multipoint waterfall cooling system effectively controlled the cooling process, and the uniformity of the microstructure of the casting was improved. John Robert Grassi [3] proposed the Ablation casting cooling technique, where a water-soluble inorganic binder is used for the sand mold, and a stream of water is sprayed on the outer surface of the sand mold while the casting is solidifying. In addition, the design of special cooling channels inside the sand mold, through the cooling medium for forced cooling, can also control the local temperature of the casting, shorten the local solidification time of a casting. However, the above methods are very limited into certain conditions, and the more complex and larger the casting structure, the more difficult to use these methods.

The emergence of 3D printing sand mold technology for the complex structure of the production of sand mold provides more options, and improves the flexibility and innovation of the design of the sand structure. Jinwu Kang et al [4] proposed hollow sand mold structure which can be adjusted by local spraying cooling water or blowing air to regulate the local cooling of the casting, and hollow sand shell surface and the external air heat transfer is more effective, the casting heat dissipation rate is greater than traditional sand mold. In this paper, the self-cooling device of multi-layer shell hollow sand mold is developed, and the influence of multi-layer shell hollow sand mold on the cooling rate, microstructure and mechanical properties of barrel castings is discussed.

2 Experimental procedure

The self-cooling method of multi-layer shell hollow sand mold was designed, as shown in Fig. 1.

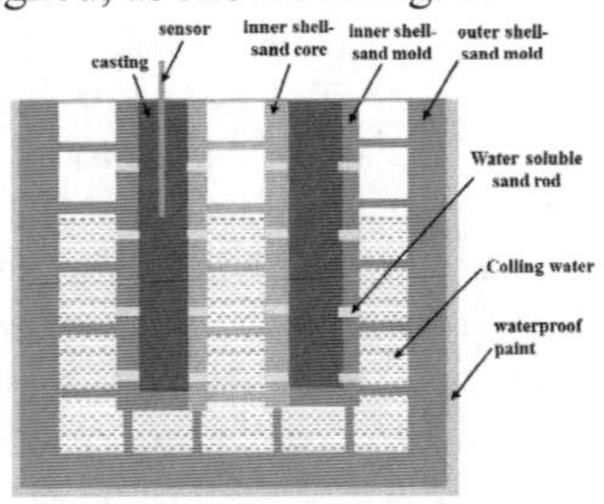

Fig. 1 Self-cooling device with multi-layer shells hollow sand mold.

In this new method, the sand mold/core are designed as an integrated hollow structure, which mainly includes the inner shell, outer shell and reinforcement structure between the shells; the inner shell is used for pouring the liquid alloy, and the cavity between the outer shell and the inner shell is for the cooling water. When water cooling is required for the casting, the cooling water is gradually pumped into the cavity between the shells, thus, the rising cooling water promotes the rapid sequential cooling of the casting from bottom to top. In order to prevent cooling water from leaking out of the sand mold, a waterproof coating is therefore applied to the outer surface of the sand mold.

A barrel casting is designed as shown in Fig. 2(a), the corresponding dense sand mold, double-shells and triple-shells hollow sand molds were designed and generated by using FT-HollowMold software self-developed by Tsinghua University, as shown in Fig. 2(b-d). The alloy used for casting is A357 aluminum alloy. During the solidification of the castings, natural cooling was adopted for the dense sand mold, and water cooling was adopted for the double-shells and triple-shells hollow sand molds, and the rise rate of the cooling water was 0.8 mm/s. When the temperature of castings dropped to below 473K, shakeout was carried out.

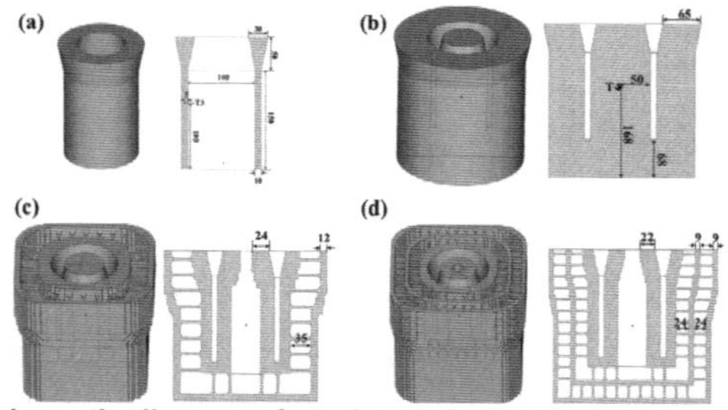

Fig. 2 Schematic diagram of casting and sand molds (unit: mm): (a) Casting; (b) Dense sand mold (P2DN); (c) Double-shells hollow sand mold (P2TN); (d) Triple-shells hollow sand mold (P2MN).

3 Result and discussion

Fig. 3(a) shows the temperature fields of the three groups of sand surface acquired by a infrared imaging camera during casting process. Under natural cooling conditions, the surface temperature distribution of dense sand mold is more uniform, and most of the surface temperature reaches 330K at 800s. For the multi-layer shell hollow sand mold under water cooling, the temperature of water at 100s is low, the heat transferred to the outer shell is low, and the temperature of outer surface is low, while the temperature of cooling water at 800s is significantly increased, and was observedto boil, so the heat transferred by the cooling water to the outer shell increases, then the temperature of outer surface also increases correspondingly. Fig. 3(b) shows the cooling curves of the castings. The castings of multi-layer hollow sand mold were forced to cool immediately after pouring for 200s. When the casting temperature drops to 573K, the casting of dense sand mold needs 1600s, and the castings of multi-layer shell hollow sand molds under water cooling only need 500s, which is 69% shorter than the casting of dense sand mold under natural cooling. The multi-layer shell hollow sand mold under forced water cooling have stronger heat transfer capacity, which significantly shortens the cooling time.

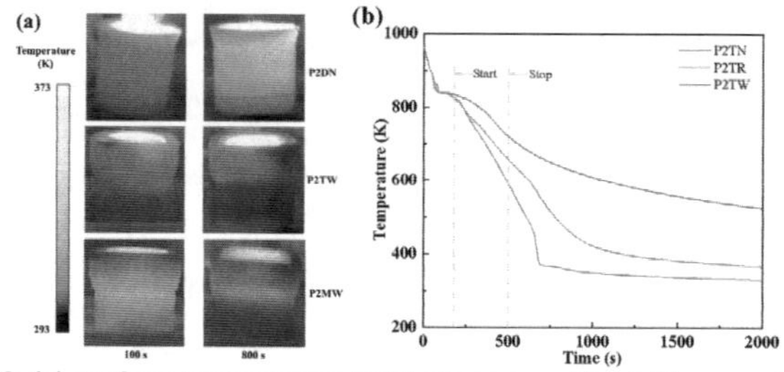

Fig 3. (a) surface temperature field of sand molds; (b) temperature cooling curve of the castings.

Fig. 4 shows the tensile properties of the castings. The tensile strength of casting of dense sand mold is 140 MPa, and the yield strength is 80 MPa. Under the condition of water cooling, the strength is obviously improved, the tensile strength and yield strength are 160 MPa and 116 MPa, respectively, with an increase of 14.3% and 45.0%, but the elongation of castings is reduced. Under the condition of water cooling, the content of strengthening elements in Al matrix phase is improved, thus, the strength properties of castings are improved.

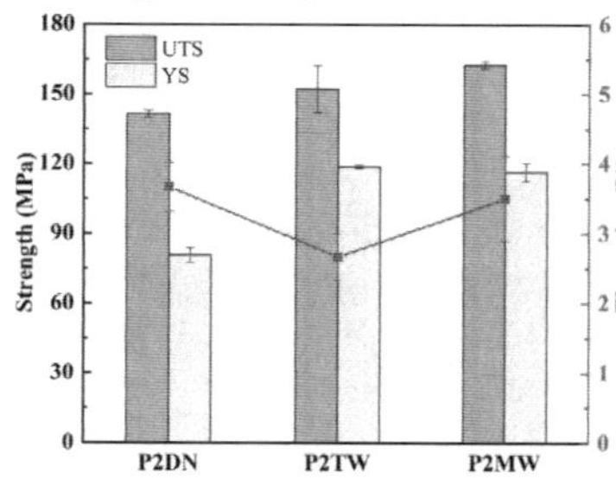

Fig. 4 The tensile properties of the castings.

4 Conclusion

A self-cooling with multi-layer shell hollow sand mold containing water was developed, the water was pumped into the cavity between the shells for forced cooling during casting process, which can enfore the cooling ofcastings and improve the mechanical properties of the castings.

Acknowledgments

The project was funded by the Beijing Nature Sciences Fund—Haidian Originality Cooperation Project (No. L212002).

References

[1] Władysiak R, Pacyniak T, Kozuń A. Effect of Casting Die Cooling on Solidification Process and Microstructure of Hypereutectic Al-Si Alloy[J]. J. Archives of foundry engineering, 2016, 16(4): 175-180.

[2] Władysiak R, Kozuń A, Dębowska K, et al. Analysis of crystallization process of intensive cooled AlSi20CuNiCoMg alloy[J]. J. Archives of foundry engineering, 2017, 17(2): 137-144.

[3] Grassi J, Campbell J, Hartlieb M, et al. The ablation casting process. J. Materials Science Forum, 2009, 16: 591-594.

[4] Kang J, Shangguan H, Deng C, et al., Additive manufacturing-driven mold design for castings[J]. J. Additive manufacturing, 2018, 22: 472-478.

Study on the Microstructure and Mechanical Properties of High Strength and Toughness Al-9Si-Mn-Mg-xCu Casting Alloy

Dongxue Yang, Dezhi Zhu *
School of Mechanical and Automotive Engineering, South China University of Technology, Guangzhou 510640, China
* Building 38, Wushan Campus, South China University of Technology: e-mail: mezhudz@scut.edu.cn

Abstract: With the development of automobile lightweight, lightweight structures and lightweight materials are increasingly used in the automobile manufacturing industry. Aluminum alloy has become the first choice for lightweight materials in automobiles due to its many advantages such as low density, high specific strength, good corrosion resistance, and easy forming. At present, cast aluminum alloy accounts for the highest proportion of automotive aluminum, among which, the largest amount is Al-Si cast alloy. Ordinary Al-Si cast aluminum alloy has good casting performance, but its strength is low and toughness is poor, which is not suitable for the manufacture of load-bearing structural parts such as automobile shock absorbing towers, suspension arms, engine mounts, subframes (it requires the tensile strength of cast aluminum alloy parts≥300 MPa, the yield strength≥200 MPa, and the elongation≥12%).

In order to obtain high-strength and high-toughness cast aluminum alloy materials, methods and means such as solution strengthening (adding Cu, Mg, Mn, Ti and other elements), fine-grain strengthening and heat treatment strengthening are usually used. Among them, heat treatment strengthening requires aluminum alloy castings to have a very low porosity. Obviously, the ordinary die casting process cannot meet the requirements. Squeeze casting and high vacuum die-casting technology are very suitable for the forming and processing of high-strength and tough, heat-treatable strengthened cast aluminum alloys. Therefore, three alloys of Al-9Si-0.4Mn-0.25Mg-xCu (x=1.0 wt.%, 2.0 wt.%, 3.0 wt.%) with different Cu contents were designed. The microstructure and mechanical properties of Al-9Si-0.4Mn-0.25Mg-xCu alloy prepared by squeeze casting and high vacuum die-casting were studied, and the effects of copper content and T6 treatment (solution treatment and peak aging) on the microstructure, precipitated phase and mechanical properties of Al-9Si-0.4Mn-0.25Mg-xCu alloy were analyzed, and the similarities and differences between the two processes were compared and analyzed.

Keywords: Squeeze casting, High vacuum die casting, High strength and toughness AlSiMgMnCu alloy, Heat treatment, Microstructure, Mechanical property

1 Introduction

At present, the development and exploration of lightweight materials is an important development direction to realize the lightweight of automobiles. The density of aluminum alloy is 1/3 that of steel, the specific strength is high, the electrical and thermal conductivity is good, the corrosion resistance is good, the impact toughness is better, about twice that of steel, and it has excellent performance in the fuel consumption, acceleration, braking, and handling of automobiles[1]. Therefore, aluminum alloy is the material of choice for lightweight automobiles. At present, cast aluminum alloy accounts for the highest proportion of aluminum used in automobiles. Cast aluminum alloys are mainly divided into Al-Si series, Al-Cu series, Al-Mg series and Al-Zn series alloys. Among them, Al-Si alloys have been studied as structural materials in the automotive and aerospace industries due to their excellent castability, low coefficient of thermal expansion, and good wear resistance and corrosion resistance[2-5]. However, Al-Si alloys generally have low strength and toughness, which is difficult to meet the requirements of the automotive industry for high-strength and high-toughness structural parts, so some strengthening elements such as Cu are usually added to further improve the mechanical properties of the alloys by heat treatment[6, 7].

Casting forming technology is the main technology for aluminum alloy forming for automobiles, and the most widely used technology is die-casting forming technology. However, ordinary die castings are prone to entrapment during the forming process and reduce elongation. On the contrary, high vacuum die casting technology and squeeze casting technology can effectively reduce porosity and inclusions, and the formed castings not only have high density, but can also be heat treated to improve their mechanical properties. Therefore, it is of great significance to explore the effects of squeeze casting and high vacuum die-casting on the microstructure and mechanical properties of high-strength and toughness Al-Si casting alloys.

缺少 Experimental procedure， Result and discussion 等文章主体内容

3 Conclusion

(1) Squeeze casting Al-9Si-0.4Mn-0.25Mg-xCu alloys mainly include α-Al, Al-Si eutectic, α-Fe, θ phase (Al_2Cu) and Q phase ($Al_5Cu_2Mg_8Si_6$). With the increase of Cu content, the contents of θ phase and Q phase in Al-9Si-0.4Mn-0.25Mg-xCu alloy increased, and the change of θ phase was the most obvious. The tensile strength and yield strength of Al-9Si-0.4Mn-0.25Mg-xCu alloy increased with the increase of Cu content, and the elongation decreased gradually. Among the three alloys, Al-9Si-0.4Mn-0.25Mg-1.0Cu alloy has the best comprehensive mechanical properties, with tensile strength, yield strength and elongation of 252 MPa, 137 MPa and 8.5%, respectively. With the increase of Cu content, the number of dimples and

tear ridges on the fracture surface decreased, the number of microcracks increased, and the porosity increased.

(2) When the solution temperature is constant, with the extension of the solution time, the Si phase in the squeeze casting Al-9Si-0.4Mn-0.25Mg-xCu alloy will coarse and grow, the content of θ phase and Q phase will decrease rapidly, the tensile strength and yield strength of the alloy will gradually increase, and the elongation will decrease. When the solution time is constant, with the increase of solution temperature, the contents of θ phase and Q phase in the alloy decrease, and the tensile strength and yield strength of the alloy increase, and the elongation does not change significantly. The tensile strength, yield strength and elongation of squeeze-cast Al-9Si-0.4Mn-0.25Mg-1.0Cu alloy at 530°C for 4h were 292 MPa, 156 MPa and 13.2%, respectively.

(3) With the increase of Cu content, the average length and size of Q' and θ' precipitated phases in the squeeze-cast Al-9Si-0.4Mn-0.25Mg-xCu alloy decreased, and the area fraction and number density of Q' and θ' phases increased. When the Cu content is 1.0 wt.%, the precipitate phase is dominated by the Q' phase, and when the Cu content increases to 3.0 wt.%, the precipitate phase is dominated by the θ' phase. Among the three alloys, Al-9Si-0.4Mn-0.25Mg-1.0Cu alloy T6 (530°C×3 h+180°C×4 h) has the highest elongation, and its tensile strength, yield strength and elongation are 340 MPa, 255 MPa and 9.8%, respectively. The strength of Al-9Si-0.4Mn-0.25Mg-3.0Cu alloy T6 (530°C×3 h+180°C×5 h) is the highest, and its tensile strength, yield strength and elongation are 404 MPa, 317 MPa and 7.7%, respectively. The tensile fractures of Al-9Si-0.4Mn-0.25Mg-xCu alloy in the peak aging state after different solution temperatures are basically mixed fracture forms, and the brittle fracture characteristics of Al-9Si-0.4Mn-0.25Mg-xCu alloy become more obvious with the increase of Cu content under the same solution treatment process, and the ductile fracture characteristics of Al-9Si-0.4Mn-0.25Mg-xCu alloy become more obvious with the increase of solution treatment temperature under the same alloy treatment process.

(4) The microstructure of Al-9Si-0.4Mn-0.25Mg-1.0Cu alloy is plum-shaped, mainly composed of α-Al, Al-Si eutectic, α-Fe, θ phase and Q phase, and its tensile strength, yield strength and elongation are 296 MPa, 130 MPa and [8] . 11.4%, respectively. In the T6 state (530°C×3 h+180°C×4 h), the precipitated phases in the alloy are mainly Q' phase and θ' phase, and the alloy has the best comprehensive mechanical properties, with tensile strength, yield strength and elongation of 346 MPa, 245 MPa and 14.2%, respectively. The high-vacuum die-cast Al-9Si-0.4Mn-0.25Mg-1.0Cu alloy exhibited ductile fracture.

(5) Compared with the squeeze-cast Al-9Si-0.4Mn-0.25Mg-1.0Cu alloy, the microstructure of the high-vacuum die-cast Al-9Si-0.4Mn-0.25Mg-1.0Cu alloy is similar, but the grain size of α-Al is smaller and the porosity is lower. Under the same heat treatment state (T6: 530°C×3 h+180°C×4 h), compared with the squeeze casting state, the Q' phase size in the high-vacuum die-casting Al-9Si-0.4Mn-0.25Mg-1.0Cu alloy was larger, while the θ' phase was smaller, and the content of the two precipitated phases increased significantly. There is little difference in the strength of the alloys formed by the two processes, but the elongation of the high-vacuum die-cast Al-9Si-0.4Mn-0.25Mg-1.0Cu alloy specimens before and after heat treatment is higher, and the as-cast and heat-treated specimens show ductile fracture.

References

[1] Armao F G. Design & fabrication of aluminum automobiles[J]. Welding Innovation, 2002, 19(2): 2-6.
[2] Wu W, Gong M, Wei B, et al. Atomistic modeling of interface strengthening in Al-Si eutectic alloys[J]. Acta Materialia, 2022, 225: 117586.
[3] Cho Y H, Kim H W, Lee J M, et al. A new approach to the design of a low Si-added Al-Si casting alloy for optimising thermal conductivity and fluidity[J]. Journal of Materials Science, 2015, 50: 7271-7281.
[4] Liu Y Q, Wei S H, Fan J Z, et al. Mechanical properties of a low-thermal-expansion aluminum/silicon composite produced by powder metallurgy[J]. Journal of Materials Science & Technology, 2014, 30(4): 417-422.
[5] Liang Z X, Ye B, Zhang L, et al. A new high-strength and corrosion-resistant Al-Si based casting alloy[J]. Materials Letters, 2013, 97: 104-107.
[6] Niu G, Wang Y, Zhu L, et al. Fluidity of casting Al-Si series alloys for automotive light-weighting: a systematic review[J]. Materials Science and Technology, 2022, 38(13): 902-911.
[7] Ravi K R, Pillai R M, Amaranathan K R, et al. Fluidity of aluminum alloys and composites: A review[J]. Journal of Alloys and Compounds, 2008, 456(1-2): 201-210

Effect of T6 Treatment on the Microstructure and Properties of Mg-Zn-Ce with Different Zn Contents

Aoxue Jiang [1], Zhiyong You [1], Shuaishuai Jin, Chunle Sun [1], Kaiying Wang [1]
Institute of Energy Innovation, Key Laboratory of Interface Science and Engineering in Advanced Materials Ministry of Education, College of Materials Science and Engineering, Taiyuan University of Technology, 030024, China
*Corresponding address: e-mail: 13603511486l@163.com

ABSTRACT:The alloy compositions and heat treatment processes of Mg-Zn-Ce were designed and formulated based on phase diagram calculations. Mg-5Zn-0.2Ce and Mg-7Zn-0.3Ce alloys with different Zn contents were prepared through rapid solidification and T6 treatment. The study showed that when the Zn content is less than its maximum solubility in Mg (6.2 wt.%), T6 heat treatment results in a coarse microstructure and the occurrence of secondary recrystallization. When the Zn content is greater than 6.2 wt.%, T6 heat treatment refines the grains and significantly improves the mechanical properties. However, T6 treatments will generally produce a large number of MgZn and $Mg_{12}Ce$ precipitates distributed within the Mg-Zn-Ce alloy microstructure. These precipitates greatly enhance the comprehensive mechanical properties of the alloy.

Keywords: Mg-Zn-Ce, microstructure, properties, T6 treatment, Zn contents

1 Introduction

Mg-Zn alloys possess advantages such as heat resistance and good biocompatibility, presenting a broad application prospect [1]. Although Mg-Zn alloys have high strength, their low ductility makes them prone to fracture under high-stress conditions, limiting their use in certain high-stress applications. Therefore, it is necessary to modify the material through alloying and process optimization. A series of Mg-Zn-based alloys with good overall mechanical properties have been developed, such as Mg-Zn-Al alloys, Mg-Zn-Cu alloys, Mg-Zn-Sn alloys, and Mg-Zn-RE (rare earth) alloys [2-5].

Among RE elements, Ce has abundant reserves, is inexpensive, has good ductility, and possesses a hexagonal crystal structure, which energetically favors co-segregation of Zn and Ce at grain boundaries. Therefore, adding Ce as a microalloying element to Mg-Zn alloys can significantly improve the strength and elongation of the alloys. However, Mg-Zn alloys exhibit significant grain boundary precipitation during aging, which is detrimental to the plasticity and fracture behavior of the alloy. Given that the maximum solubility of Zn in the Mg matrix is 6.2wt.%, it is necessary to explore the effects of appropriate solution-aging treatment processes on the microstructure and properties of Mg alloys with different Zn contents (up to about 6.2wt.% maximum solubility).

This study designed the optimal compositions of Mg-Zn-Ce alloys through phase diagram calculations and obtained Mg-5Zn-0.2Ce and Mg-7Zn-0.3Ce alloy materials using rapid solidification and T6 solution aging treatment processes. By observing the alloy microstructure and analyzing the elemental composition, the microscopic mechanisms of microstructure and property changes under the influence of different Zn contents were investigated. This provides theoretical guidance and experimental methods for the development and application of Mg-Zn-RE alloys.

2 Experimental procedure

The experimental alloy compositions selected based on phase diagram calculations were Mg-5Zn-0.2Ce and Mg-7Zn-0.3Ce. Additionally, to promote the precipitation of Ce-containing phases in the microstructure, Mn was added at a content of 0.04wt.%. After the alloy elements were uniformly mixed through melting, the melt was poured into a 200 ℃ preheated iron mold and rapidly cooled with cooling water to obtain the cast Mg-Zn-Ce alloy. Subsequently, the alloy samples underwent T6 heat treatment, which consisted of solutionizing at 450℃ for 6 hours and aging at 180℃ for 36 hours, resulting in four sets of alloy samples: Mg-5Zn-0.2Ce, Mg-5Zn-0.2Ce+T6, Mg-7Zn-0.3Ce, and Mg-5Zn-0.2Ce+T6, labeled as alloys 1#, 1-1#, 2#, and 2-1#, respectively.

3 Result and discussion

1. Microstructure Analysis

Figure 1 shows the microstructures of Mg-Zn-Ce alloys with different compositions before and after T6 heat treatment. From Figures 1(a) and 1(b), it can be seen that the average grain size of the Mg-5Zn-0.2Ce (1#) alloy is approximately 107 μm, and there is a network-like second phase at the grain boundaries. EDS analysis identified this second phase at the grain boundaries as the stable MgZn phase, with some Ce atoms adsorbed and Mn elements also gathered near the grain boundaries. After T6 treatment, the second phase at the grain boundaries in the 1-1# alloy disappeared, and the microstructure became significantly coarser (highlighted by the red dashed circles in Figure 1(b)), with an average size of about 310 μm. Additionally, remelting occurred at the grain boundaries, forming small-sized subgrains (highlighted by the black dashed circles in Figure 1(b)), with an average size of about 55 μm. T6 heat treatment also caused the precipitation of lamellar precipitates within the grains along a specific orientation (composition not identified), with a measured precipitation angle of approximately 130° (indicated by the white arrows in Figure 1(b), angles of 137.8° and 131.4°). Related studies in the reference [6] suggest that these phases might be LPSO phases. This could be due to Ce, as a rare earth element, forming LPSO phases with a long-period structure in the microstructure after aging treatment. These LPSO phases can impede dislocation motion, suppress basal plane slip, and increase the strength of the magnesium alloy.

From Figures 1(c) and 1(d), the grain size of the Mg-7Zn-0.3Ce (2#) alloy is about 106 μm, and the grain boundaries still feature a network-like MgZn phase, with rod-like $Mg_{12}Ce$ phases nearby. EDS atomic distribution analysis revealed that this phase nucleates on a Mn substrate. After T6 heat treatment, the grain size of the 2-1# alloy is about 110 μm, with minimal grain coarsening. The microstructure still contains lamellar precipitates, but their size is greatly reduced, and the precipitation angle remains around 130° (as indicated by the white arrows in Figure 1(d), angle of 131.3°).SEM and EDS analyses of the microstructures of the T6-treated 2-1# alloys (as shown in Figure 1(d)) indicate that the grain boundaries consist of island-like MgZn phases, and within the grains, there are point-like MgZn phases. Additionally, the rod-like $Mg_{12}Ce$ phase nucleates near the MgZn phase within the microstructure, both at grain boundaries and within the grains.

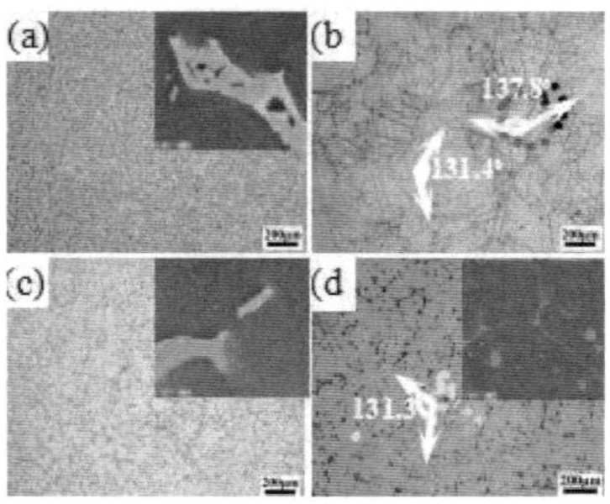

Figure 1. The microstructures of Mg-Zn-Ce alloys with different compositions before and after T6 heat treatment: (a)1#; (b)1-1#; (c)2#;(d)2-1#.

2.Mechanical Properties Analysis

Figure 2 shows the stress-strain curves for Mg-Zn-Ce alloys with different compositions before and after T6 heat treatment, and Table 1 provides the corresponding specific performance values. It can be observed that the strength and elongation of the Mg-5Zn-0.2Ce (1#) alloy deteriorate after heat treatment. This is due to the significant grain coarsening caused by the solution treatment in the T6 process. Additionally, the unstable interface connection between the subgrains at the grain boundaries and the large grains further reduces the overall mechanical properties of the alloy.

In contrast, the strength of the Mg-7Zn-0.3Ce (2#) alloy significantly increases after T6 treatment, reaching 215.95 MPa, while its plasticity slightly decreases to 4.04%. The minimal change in grain size after T6 treatment indicates that the aging treatment in the T6 process has produced an age-hardening effect. The substantial increase in alloy strength is primarily attributed to a significant amount of precipitates in the microstructure. However, the Mg12Ce precipitate phase formed in the microstructure is brittle, and an increased quantity of this phase can reduce the alloy's plasticity. This is the main reason for the reduced elongation observed in the 2-1# alloy.

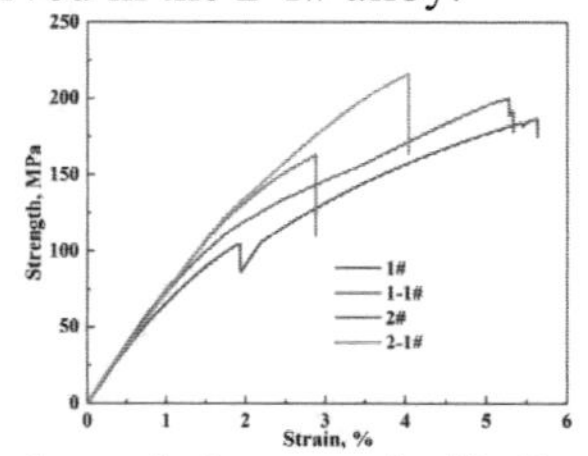

Figure 2. The stress-strain curves for Mg-Zn-Ce alloys with different compositions before and after T6 heat treatment.

Table 1. The strength and strain for Mg-Zn-Ce alloys with different compositions before and after T6 heat treatment

Properties	1#	2#	3#	4#
Strength(MPa)	183.78	162.98	198.60	215.95
Strain(%)	5.43	2.88	5.20	4.04

4 Conclusion

1) T6 solution and aging treatment significantly coarsens the grain size of low-zinc magnesium alloys and deteriorates their overall mechanical properties. For high-zinc magnesium alloys, T6 treatment greatly increases their tensile strength. Therefore, for Mg-Zn alloys, it is necessary to formulate an appropriate heat treatment process based on the Zn content.

2) Due to the higher Zn content and increased Ce content, the strength and elongation of Mg-7Zn-0.3Ce alloy are generally higher than those of Mg-5Zn-0.2Ce alloy before and after T6 treatment. This is because the substantial amount of secondary phases and precipitates in the microstructure has a strengthening effect on the alloy.

References

[1] Zhang J B, Tong L B, Xu C, et al. Influence of Ca-Ce/La synergistic alloying on the microstructure and mechanical properties of extruded Mg-Zn alloy[J]. Materials Science and Engineering: A, 2017, 708: 11-20.

[2] Lv S H, Yang Q, Lv X L, et al. Intermetallic phases and mechanical properties of a Mg-8Zn-6Al-1Sm (wt %) casting alloy[J]. Materials Science and Engineering: A,2022,852: 143719.

[3] Mehrab L, Changiz D, Massoud E, et al. In-vitro corrosion behavior of the cast and extruded biodegradable Mg-Zn-Cu alloys in simulated body fluid[J]. Journal of Magnesium and Alloys, 2021, 9(6): 2078-2096.

[4] Hou C H, Qi F G, Ye Z S, et al. Effects of Mn addition on the microstructure and mechanical properties of Mg-Zn-Sn alloys[J]. Materials Science and Engineering: A, 2020, 774: 138933.

[5] Wang X, Du Y Z, Liu D J, et al. Significant improvement in mechanical properties of Mg-Zn-La alloy by minor Ca addition[J]. Materials Characterization, 2020, 160: 110130.

[6] Gao Z Y, Hu L S, Li J F, et al. Achieving high strength and good ductility in as-extruded Mg-Gd-Y-Zn alloys by Ce micro-alloying[J]. Materials, 2018, 11(1): 102.

Comparative Analysis of Chinese and Foreign Die Casting Aluminum Alloy Standards

Guanghua Qin [1], Jiahui Zhu [1, 2], Yin Zhang [1], Shifang Su [1]

1. National Technical Committee 54 on Foundry of Standardization Administration of China, Shenyang 110022, Liaoning, China;
2. Shenyang Research Institute of Foundry Co., Ltd., CAM, National Key Laboratory of Advanced Casting Technologies, Shenyang 110022, Liaoning, China
*Corresponding address: e-mail:qgh@chinasrif.com

Abstract: Material standards can better guide composition design, improve material properties, and promote the application of engineering materials. This paper provides an overview of the current status of die casting aluminum alloys at home and abroad. By comparing the domestic and international standards of die-casting aluminum alloys, it introduces the material characteristics of die-casting aluminum alloys, and looks forward to the key focus areas and analyzes the urgent problems to be solved in the future of die casting aluminum alloys.

Keywords: Die casting aluminum alloy; High pressure vacuum die cast; Standard interpretation

1 overall

The United Nations Sustainable Development Goals (SDGs) and its commits achieving carbon neutrality by 2050, driven by the target, the automotive industry is striving to implement energy conservation and emission reduction (ECER). Aluminum alloy, an effective ECER solution, has become the primary lightweight material for automobiles. Nearly 60% of automotive aluminum components are produced through die casting (DC), a technology that can achieve a weight reduction of 30%-50%, significantly improving fuel economy during the usage phase of the automotive life cycle, making it an essential method for green manufacturing. The main approaches to vehicle weight reduction involve optimizing the body structure and developing lightweight alloy materials. Aluminum alloy die castings have a particularly widespread application in automobiles, the aluminum alloy die casting industry has vast development potential. This paper comprehensively elaborates on the standards and applications of die casting aluminum alloys at home and abroad, aiming to provide a reference for the subsequent development of low-cost, high-performance die-casting aluminum alloys and advanced die casting

technologies, and promoting the global development of the aluminum die-casting industry.

2 Overview of die-casting aluminum alloy

Take AlSi10Mg series alloy as example, like Y L101, A360, AD3.1 and EN AB-43400 alloy have similar design ideas. At present, the use of Silafont® -36 alloy or AlSi10Mg alloy materials, combined with high pressure vacuum die cast and subsequent T6 and T7 heat treatment processes, are the most used technical route. The following will review the development of the die casting aluminium alloy compared with Chinese national standard, ASTM, JIS and EU standard

2.1 Chinese standards

In the past 20 years, the market for electric vehicles is rapidly expanding, and we see more applications like large-scale structural components and body plate. The application of integrated giga casting technology also promotes the development and application of non-heat treatment materials. The non-heat treatment die casting aluminum alloy Al-Si-X (JDA 1x) and Al-Mg-X (JDA 2x) are developed by Shanghai Jiao Tong University, such as Al-Si-Cu-Mn-RE, Al-Si-Mn-Mg-RE, Al-Si-Cu-Mg-Mn, with high mechanical properties. Al-Si-Cu-Zn-Mg for casting [1] (Hereinafter referred to as HKA-1 aluminum alloy) Al-Mg-Mn-RE, Al-Mg-Si-Mn-RE alloy, has been successfully used in the rear cockpit of automobile. The as-cast alloy of JDA 2b has a yield strength of 200 MPa to 240 MPa, a tensile strength of 360 MPa to 400 MPa, and an elongation rate of 10% to 12%, exhibiting excellent strong toughness.

The newly edition Chinese national standard, GB/T 15115—2024, has introduced new alloy types, YL 118, YL 118, YL 119 and YL 120. YL 118 reduced the content of Cu and Zn on the basis of $AlSi_{12}(Cu)$ specified in ISO 3522:2007 standard, there are more phases in microstructure such as Q-$Al_5Cu_2Mg_8Si_6$ and θ-Al_2Cu with more Cu addition in the alloy. When the amount of Cu exceeds 0.3 wt. %, the θ phase begins to form and its content increases accordingly. However, Si is consumed in the formation of the Q phase, which decreases the total volume of the α-Al (Fe, Mn) Si phase slightly, and the volume of the Q phase was inverse to that of the α-Al (Fe, Mn) Si phase [2]. A small amount of Cu can play the role of precipitation enhancement, improve the elongation, mechanical properties and fatigue properties. Compared with the heat-treated Al-Si alloys, YL 118 requires no Mg. For example, YL 119, is similar to the 365.1 alloy, the upper limit of Fe and Ti is slightly widened. The actual production of Al-Si alloy contains iron at a relatively low level (0.1~0.2%), which is difficult to remove from the melt in a cost-effective way, so the upper limit of Fe is stipulated to be 0.2%. The Mn content is 0.40%~0.80%, which has a benefit effect on the elongation. The Ti content is limited to 0.20%. According to the research, after the addition of 0.0137% Ti, the α(Al) grains can be significantly refined [3], Reine the alloy grain to improve the castability.

2.2 The United States Standard

As early as 1996, the American Aluminum Association accepted the Silafont®-36 alloy, and registered it as 365.0. Currently, roughly 85% of aluminum alloy die castings in the American standard are produced by 380 alloy family, like 383 or 384, which is Al-Si-Cu alloy series. The remaining 15% of aluminum alloy die casts are mainly manufactured using alloys 360, 390, 413, along with smaller volumes of 443 and 518.

2.3 The Japanese Standard

Compare to JIS H 2118:1968, the alloy types in JIS H 2118:2006 were reduced from 13 classes to 7 classes. AD2, AD4, AD11 and AD13 alloys with only a different percentage of iron while other elements remained constant were removed. Additionally, considering practical usage demands, the less used ADC 8 and ADC 9 alloys were removed. During the revision process, reference was made to IS, German standards and American standards, allowing for minor adjustments and eliminations across certain alloy categories. The tolerance for the impurity element iron (Fe) was broadened to make better use of existing domestic resources and decrease reliance on scarce elements in Japan. For example, AD12.1 alloy was adjusted based on the A380.1, and the upper limit of Sn content is modified from 0.35 wt.% to 0.2 wt.%.

2.4 European Union standards

As early as 1994, German company Aluminum Rheinfelden GmbH developed high pressure vacuum die casting Silafont®-36 alloy, which corresponds to EN AB Al Si10MnMg alloy, also known as EN AB-43500. EN AB-43400 has a composition based on EN AB-43500 but with reduced iron (Fe) content, and low Fe content to improve the manufacturability of alloy during die casting. Furthermore, 0.4%~0.8% Mn were added to improve the harmful effects of Fe, and the Mn/Fe ratio was controlled to change the morphology of iron-rich intermetal compounds by extracting Chinese characters, stellate or polyhedral α-$Al_{15}(Fe, Mn)_3Si_2$ [2] phase, reducing the formation of needle or sheet β-Al_5FeSi phase, thus obtaining good strong toughness [3]. The additive of Mg, under the appropriate heat treatment conditions, forms precipitates with Si, to improve the alloy strength. With minimal amounts of Cu, Zn and impurities, this alloy exhibits superior casting properties compared to other die casting alloys.

3 Development trend of die casting aluminum alloy

(1) Develop new aluminum alloy series and alloy designation. Al-Ce and Al-Li alloys were studied to expand new fields of alloy materials. (2) Advanced R&D of non-Heat-Treated high-strength and toughness alloys. Understanding the material design method, mastering the material strong toughness design principle, developing high strength and high toughness, heat and corrosion resistance, easily processable die casting aluminum alloys remain a future development trajectory. (3) Unleash the full potential of secondary alloys. To achieve and primary alloy level with compromising properties, or keep higher impurity tolerance in alloy design, make full use of high recycling alloy can reduce carbon dioxide emissions, development of multi-purpose cross alloy, etc. (4) The development and optimization of manufacturing processes to produce die casting components with exceptional mechanical properties. With the introduction of the 'unbox' concept, ongoing development of innovative processes remains a crucial step in addressing the economics of vehicle manufacturing.

References

[1] Jia Z H, Zhou G W, Zhou H Y, et al. Effects of Cu content and heat treatment process on microstructures and mechanical properties of Al-Si-Mg-Mn-xCu cast aluminum alloys[J]. Transactions of Nonferrous Metals Society of China, 2024, 34(03): 737-754.

[2] Ivana Š, Eva T, Lenka K, Vidžaja K. Possibilities of predicting undesirable iron intermetallic phases in secondary Al-alloys, Transportation Research Procedia, 55(2021): 797-804.

[3] Zhang L L, Jiang H X, He J, et al. Kinetic behaviour of TiB_2 particles in Al melt and their effect on grain refinement of aluminium alloys[J]. Transactions of Nonferrous Metals Society of China,2020, 30(8): 2 035-2 044

Microstructure and Mechanical Properties of Shock Tower Manufactured Using Non Heat-Treated Aluminum Alloy

Yongfei Li[1,2], Chaohang Jia[1], Jianzhong Li[2]*, Pan Fan[1], Haibo Qiao[1], Chao Ma[1], Haifeng Liu[1]
1. CITIC Dicastal Co., Ltd., Qinhuangdao 066011, PR China
2. School of metallurgy, Northeastern University, Shenyang 110819, PR China
*Corresponding address: e-mail: lijz@smm.neu.edu.cn

ABSTRACT: The tensile testing, Kahn tear testing, optical microscopy, and scanning electron microscopy were used to investigate the microstructure and mechanical properties of the shock tower manufactured using non heat-treated aluminum alloy (DK-NHT-1). The results show that the microstructure of shock tower using non heat-treated alloy is uniform and fine, and its density is high. The as cast DK-NHT-1 alloy exhibits high strength and toughness with a yield strength of 125 MPa, a tensile strength of 275 MPa, and an elongation of 14.6%. Additionally, the alloy demonstrates high fracture toughness with tear strength and unit initiation energy measured at 272 MPa and 34 N/mm respectively. Both tensile fracture and tear fracture exhibit typical ductile fracture characteristics.

Keywords: Non heat-treated alloy, Kahn tear testing, Mechanical properties.

1 Introduction

The shock tower is a key load-bearing component that connects car body and chassis, mainly used to absorb the impact load generated during vehicle motion. Due to a harsh service environment, the material of shock tower needs to fulfill high strength and toughness requirements. The shock tower is mostly manufactured by high-pressure die casting process and made of AlSi10MnMg alloy. The AlSi10MnMg alloy exhibits excellent casting fluidity and high tensile strength, while its elongation is low due to a high Si content. Thus, it usually requires T7 heat treatment to improve the toughness to meet the property requirements [1]. The casting is prone to serious deformation due to stress during the solution treatment and the subsequent quenching processes, owing to its complex shape and thin thickness (2-3 mm). Therefore, the correction treatment is often required afterward. In the meanwhile, the casting is prone to surface blistering during solution treatment at elevated temperature, which seriously reduces product yield and increases production cost.

To effectively solve the problems such as the low strength and toughness of the as cast AlSi10MnMg alloy and the deformation and deterioration of casting surface during heat treatment, we successfully developed DK-NHT-1, a non heat-treated aluminum alloy, which exhibits high strength and toughness in the as cast product. In this study, the microstructure, tensile mechanical properties, and fracture toughness of the shock tower cast were investigated, aiming to provide reference for its production and application.

2 Experimental procedure

The samples were taken from the shock tower of an automobile. The DK-NHT-1 alloy was self-developed, and its chemical composition is shown in Table 1.

Table 1. Chemical composition of DK-NHT-1 Alloy (wt.%)

Si	Fe	Mn	Mg	Ti
7.5-8.5	≤0.20	0.40-0.90	0.10-0.40	0.05-0.15
Sr	**Nb**	**V**	**Ca**	**Al**
0.015-0.025	≤0.15	≤0.15	≤0.001	Bal.

The die-casting process parameters of non heat-treated alloy shock tower are as follows: the pouring temperature is 670 ℃, the mould temperature is 150 ℃, the low plunger velocity is 0.2 m/s, the high plunger velocity is 4.2 m/s, the vacuum degree is 80 mbar, and the casting pressure is 40 MPa. Samples are cut from the shock tower body and machined into tensile and Kahn specimens. The metallographic specimens were ground sequentially with 80-2000 grit SiC sandpapers, and then etched using 0.5% aqueous HF solution for 10 s. Microstructure was observed using a fully automated inverted microscope Zeiss Axio Observer 7 m. The tensile test and Kahn tear test are performed according to ASTM 557 and ASTM B871 standards, respectively. The tests were carried out at room temperature using a Zwick/Roell Z100 testing machine. Five specimens were used for each test and then averaged. The fracture was observed using SEM with Zeiss EVO MA 15.

3 Result and discussion

Microstructure

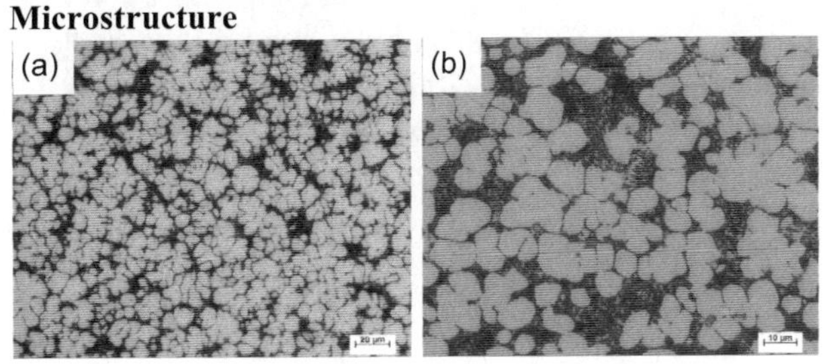

Fig 1. Microstructure of as cast DK-NHT-1 alloy

Figure 1 shows the microstructure of the as cast non heat-treated DK-NHT-1 alloy. It can be seen that the microstructure shows a typical dendrite morphology, consisting of α-Al primary phase, Al-Si eutectic phase and a small amount of Al(Fe, Mn)Si phase. The equiaxed primary α-Al phase is fine and uniform with an average grain size of 19.58 μm, as shown in Fig. 1(a). The Si particles in the Al-Si eutectic phase are extremely fine, with a size at the submicron scale. The Al(Fe, Mn)Si phase exhibits a polygonal morphology and is dispersed in the Al-Si eutectic structure, and no obvious needle-like iron phase is found, as shown in Fig. 1(b).

Tensile testing

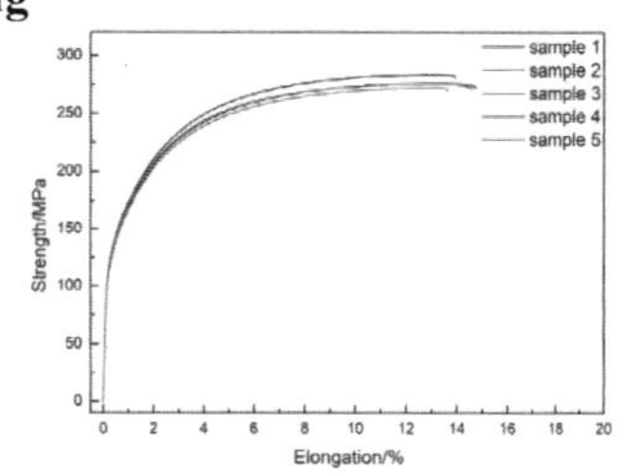

Fig 2. Load-elongation curve from a tensile test

Figure 2 shows the mechanical property curves of the as cast non heat-treated DK-NHT-1 alloy at different locations. It can be seen that the as cast DK-NHT-1 alloy exhibits excellent mechanical properties with an average yield strength of 125 MPa, an average tensile strength of 275 MPa, and an average elongation of 14.6%, meeting the product performance requirements. In DK-NHT-1 alloy, α-Al primary phase is relatively soft, which serves as a barrier to rapid crack propagation to ensure the high toughness of the product (shock tower). The ultra-fine (sub-micron) Si particles in the Al-Si eutectic phase and dispersed Al(Fe, Mn)Si particles can ensure the high strength of the alloy.

Kahn tear testing

The Kahn tear testing is widely used to evaluate the fracture toughness of aluminum alloys [2,3]. The Kahn tear testing can measure the alloy tear strength (TS) and unit initiation energy (UIE). As reported in the literature [4], there is a positive proportional relationship between UIE and fracture toughness KQ^2 (KQ^2 is the fracture toughness). The TS and UIE were measured to be 272 MPa and 34 N/mm, respectively, indicating that the DK-NHT-1 alloy exhibits an excellent fracture toughness.

Fractography analysis

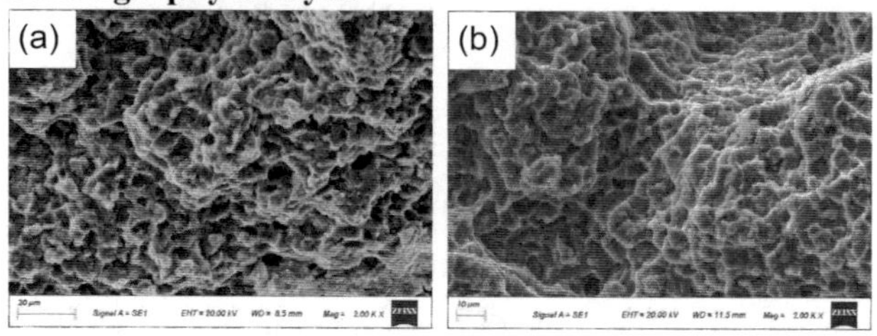

Fig 3. Tensile testing fractography(a) and Kahn tear testing fractography(b) of as cast DK-NHT-1 alloy

The fracture morphology of the tensile testing and the Kahn tear testing are shown in Figure 3(a) and Figure 3(b), respectively. It can be seen that both tensile fracture and tear fracture show typical ductile fracture characteristics, and the fracture contains a large number of small dimples, indicating that the alloy has high toughness.

4 Conclusion

The microstructure of the as cast non heat-treated alloy mainly consists of α-Al primary phase, Al-Si eutectic phase and a small amount of Al(Fe, Mn)Si phase. The α-Al structure is fine and evenly distributed with an average grain size of 19.58 μm. The Al(Fe, Mn)Si phase is polygonal in morphology and dispersed.

The as cast non heat-treated DK-NHT-1 alloy has high strength and toughness with a yield strength of 125 MPa, a tensile strength of 275 MPa, and an elongation of 14.6%. The alloy has a high fracture toughness with tear strength of 272 MPa and unit initiation energy of 34 N/mm.

Both tensile fracture and tear fracture show typical ductile fracture characteristics, containing a large number of small dimples.

References

[1] Rowe J. Advanced materials in automotive engineering. Woodhead Publishing Limited, 2012: 116.

[2] Kaufman J G, Knoll A H. Kahn-type tear tests and crack toughness of aluminum alloy sheet[J]. Materials Research & Standards, 1964, 4: 151-155.

[3] Kumai S, Tanaka T, Zhu H, Sato A. Tear Toughness of Permanent Mold Cast and DC Cast A356 Aluminum Alloys[J]. Materials Transactions, 2004,45(5): 1706-1713.

[4] Dumont D, Deschamps A, Brechet Y. On the relationship between microstructure, strength and toughness in AA7050 aluminum alloy[J]. Materials Science and Engineering A, 2003, 356: 26-336.

Research Progress of Modeling and Casting Technology of Zirconium Alloy

Chunling Bao [1], Dehua Jin [1], Youwei Zhang [1], Ning Zhao [1], Ye Yue [1], Jiawei Li [1], Haitao Yang [1]
1. China Academy of Machinery Shenyang Research Institute of Foundry Co.,Ltd, No.17, Yunfeng South Street, Tiexi District, Shenyang City, Liaoning Province, China
*Corresponding address: e-mail:baochunling0323@126.com

Abstract: Zirconium and zirconium alloy are very important strategic metallic materials in the 21st century. With the advantages of low neutron absorption rate, corrosion resistance and high temperature resistance, they are widely used in chemical equipment manufacturing and nuclear industry. In recent years, zirconium and zirconium alloy have attracted more and more attention from researchers at home and abroad, and the research on modeling and casting technology has been developed rapidly. This paper introduces the main application scenarios and development of zirconium and zirconium alloys, presents the research status and progress of the casting processes by comparison, then points out the problems and difficulties and finally the development prospect of zirconium and zirconium alloy casting technologies.

Keywords: Zirconium; Zirconium alloy; Casting process; Overview;

1 Introduction

This paper is a review paper, which mainly discusses the research progress of zirconium and zirconium alloy casting technology. This paper successively introduces the characteristics and applications, the R&D history and the research progress on modeling technology of zirconium and zirconium alloys. In addition, the smelting technology of zirconium and zirconium alloys such as vacuum self-consumable electrode arc condensation shell technology and vacuum induction condensation shell technology, as well as the modeling technology of zirconium and zirconium alloys such as hard mold casting, sand casting and investment casting are systematically described. Finally, the problems existing in the casting process and forming technology of zirconium and zirconium alloys are summarized, and the future development trend is proposed.

2 Experimental procedure

The smelting technology and formative technology of zirconium and zirconium alloy are mainly introduced.

Smelting technology includes vacuum consumable arc melting and vacuum induction melting. Both smelting techniques have their own advantages and disadvantages, and are widely used in the field of zirconium and zirconium alloy casting.

The molding process includes hard die casting, sand type casting and investment casting. Hard die casting is the use of metal shells and graphite shells for casting. In contrast, the metal type is high cost and short life, is not often used. But graphitic shells are often used to produce large castings. Sand casting has a low frequency of use in industrial production. The investment casting has low modeling cost, and the casting quality is high, which is widely used in industrial production, and more related experimental research. Oxides are often used as a surface refractory material.

3 Result and discussion

Vacuum consumable electric arc furnace melts the electrode generated by the arc discharge between the self-consuming electrode and the crystallization holder. Vacuum induction smelting is a method of heating metal using the magnetic field generated by the induction coil in a sheet water-cooled copper container.

In metal casting, cast steel, cast iron and other materials are mainly used as the type shell. Graphite casting uses high purity, fully graphitized artificial graphite electrode blocks. Sand casting uses bauxite as type sand, type sand surface is coated with a layer of oxide coating, used as zirconium and zirconium alloy casting sand type. The materials used as the mold casting layer are graphite powder, refractory metals, refractory compounds and oxides. The most commonly used materials are oxides, such as yttria, zirconia, etc.

4 Conclusion

1) Compared with other common metals, the research of zirconium and zirconium alloy started late and developed slowly, with only more than 70 years of development history, but zirconium and zirconium alloy are an indispensable material today, especially in the field of nuclear industry, and there is still a long way to go for the research on zirconium and zirconium alloy casting process.

2) At present, there are three main methods for zirconium and zirconium alloy casting: hard mold casting, sand casting and investment casting, among which sand casting is almost not used in industrial production; Hard mold casting is mainly used in large castings with simple structure and low requirements for surface quality, and machined graphite type is generally used; Investment casting is suitable for small and medium-sized castings

with complex structure and high dimensional accuracy requirements, and oxide is generally selected as the shell surface layer material. At present, the commonly used refractories still cannot meet the quality requirements of zirconium alloy castings, and further research is needed to find new refractories with excellent performance, high matching degree with zirconium alloy castings and reasonable cost. $BaZrO_3$ is a new type of refractory material under development, which has lower cost and unique properties than Y_2O_3 and ZrO_2, and has a good application prospect in zirconium and zirconium alloy casting.

3) Vacuum self-consumable electrode condensation shell technology is the main smelting method of zirconium and zirconium alloys, but its energy consumption is high and the process is complex. Vacuum induction shell technology is less costly than it, and the temperature of the molten metal can be controlled more uniformly, the filling capacity is improved, and the quality of zirconium and zirconium alloy castings can be effectively improved.

Acknowledgments

This work was financially sponsored by Liaoning Provincial Science and Technology Plan Joint Plan Project in 2023(2023020930-JH2/1017).

References

[1] Zhang X Q. Rare Metal Materials and Engineering[J]. 1983, (02):84-85.
[2] Hu J, Li G, Zhang T G, et al. Research status of properties of industrial-grade zirconium and zirconium alloys[J]. Metal World, 2020, (03):23-26.
[3] Li X J. Overview of zirconium and zirconium alloys[J]. Titanium Industry Progress, 2011, 28(1):38.
[4] Jia Y j, Lin X H, Zou X W, et al. Research and development history, current situation and development trend of zirconium alloy[J]. China Materials Progress, 2022, 41(05):354-370.
[5] Warr B D, Der H V, Maguier A. O. Characteristics and Corrosion and Hydrogen Uptake in Zr-2.5 Nb CANDU Pressure Tubes[C]//Zirconium in the Nuclear Industry, Eleventh International Symposium. Pennsylvania：ASTM，1996，1295：265-291.
[6] Shcbaldov P V, Perregud M M. E110 Alloy Cladding Tube Properties and Their Interrelation with Alloy Structure-Phase Condition and Impurity Content[C]//Zirconium in the Nuclear Industry，Twelfth International Symposium. Pennsylvania：American Society for Testing and Materials (ASTM), 2000, 1354(324):545-559.
[7] Comstock R J，Schoenberger G，SABLE G. Influence of Processing Variables and Alloy Chemistry on the Corrosion Behavior of ZIRLO Nuclear Fuel Cladding[C]//Zirconium in the Nuclear Industry，Eleventh International Symposium. Pennsylvania：ASTM，1996，1295：710-725.
[8] Wang X F，Li Z K，Zhou J，et al. Material & Heat Treatment[J]. 2012, 41(2): 71-74.
[9] Niku L A. Metal Science and Heat Treatment[J]. 2004，46(11/12)：458-462.
[10] Mardon J P，Charquet D，Senevat J. Influence of Composition and Fabrication Process on Out-of-Pile and In-Pile Properties of M5 Alloy[C]//Zirconium in the Nuclear Industry, Twelfth International Symposium. Pennsylvania:ASTM，2000，1354(125):505-524.

Calculation of Volatilization Behavior of Alloying Elements During Electron Beam Melting of Near-A Titanium Alloy

J. J. Mao[2, 3], J. Zhao[1], Y. Tan[2, 3]S. B. Liu[1], K. Shi[1], Z.Y. Zhang[1], T. Y. Liu[1]
1. Shenyang Research Institute of Foundry Co., Ltd, Shenyang, Liaoning, 110022, China
2. School of Materials Science and Engineering, Dalian University of Technology, Dalian 116024, China
3. Key Laboratory For Energy Beam Metallurgy and Advanced Materials Preparation of Liaoning Province, Dalian University of Technology, Dalian 116024, China
*Corresponding address: e-mail: Msco1027@163.com

Abstract: In this paper, near-α titanium alloy were prepared by electron beam melting. The composition and volatilization behavior of alloying elements were studied. The results show that the mass loss rate of the alloy increases with the increase of melting power. After melting, the mass fractions of Al and Sn decreased, while the mass fractions of other elements increased. The melt was simplified into Ti-Al-Sn ternary alloy model to study the volatilization behavior. The activity coefficient and activity of alloying elements could be predicted by Miedema model. The theoretical volatilization rate of other elements was very small, and the volatilization loss was basically negligible. The volatilization rate of Al and Sn elements is calculated by introducing the activity coefficient compensation factor.

Keywords:near-α titanium alloy, Electron beam cold hearth melting,Volatilization of element

1 Introduction

Titanium alloy has excellent comprehensive properties such as high specific strength, high specific stiffness and high temperature resistance, and the application demand in aerospace is increasing day by day [1,2]. In industrial production, the traditional smelting method of titanium alloy is difficult to fundamentally eliminate the most serious metallurgical defects in high and low density inclusion of titanium alloy. However, the introduction of Electron beam cold hearth melting (EBCHM) technology be a good solution to the above problems. However, in the process of electron beam melting, the alloying elements will be volatilized.

The volatilization behavior of alloying elements during electron beam melting of near α type titanium alloys was studied in this work, which can provide theoretical guidance for element control during electron beam melting of near α type high temperature titanium alloys.

2 Experimental procedure

The measured composition of the raw material in this experiment is Ti-5.88Al-3.87Sn-3.32Zr-1Mo-0.37Si-0.39Nb-0.32Ta-1.11W. The electron beam melting furnace model SEBM-60A is used for melting, and the melting parameters are: 6kW/10min(1), 8kW/10min(2), 15kW/5min(3) and 20kW/5min(4). The mass of the alloy before and after the experiment was measured by an electronic balance(ACS-D II) with a precision of 0.1g, and the composition of the alloy was analyzed by inductively coupled plasma atomic emission spectrometry(ICAPPRO X ICP-OES Duo).

3 Result and discussion

Component analysis

After the experiment, the test results of alloy composition are shown in Table 1. After electron beam melting, the mass fraction of aluminum and tin decreases, while the mass fraction of other elements increases. The mass loss rate of experimental raw materials is shown in Table 2. It can be seen that the mass loss rate increases with the increase of melting power. The reason is that at a given scanning speed, the volatilization loss depends on the line energy, and as the line energy increases, the volatilization loss also increases.

Table1. Chemical Composition Test Results of Raw Materials after experiment (mass fraction/%)

Al	Sn	Zr	Mo	Si	Nb	Ta	W	Ti
3.62	3.28	3.47	1.08	0.42	0.41	0.34	1.18	Bal.
2.58	2.84	3.67	1.13	0.45	0.43	0.36	1.23	Bal.
3.11	3.06	3.6	1.11	0.45	0.42	0.34	1.2	Bal.
2.3	2.83	3.65	1.13	0.44	0.44	0.36	1.24	Bal.

Table2. Mass loss and mass loss rate alloys experiment

Power	Raw material quality	Quality after melting	Mass loss	Mass loss rate
6kW	165g	158g	7g	4.2%
8kW	162g	147g	15g	9.2%
15kW	160g	145g	15g	9.4%
20kW	160g	141g	19g	11.9%

Volatilization behavior of element

According to literature [3], the saturated vapor pressure of pure elements depends on temperature and can be calculated by the Clausius-Capeyron equation:

$$\lg P_i^0 = \frac{A}{T} + B\lg T + CT + D \qquad (1)$$

The volatilization behavior of alloying elements is mainly affected by the interfacial volatilization process.

According to Langmuir equation, the volatilization loss of alloying element i during electron beam refining can be expressed as:

$$\Delta m_i = S \cdot texp\left(\frac{\overline{G}_i^{ex}}{RT}\right) \times \chi_i \times P_i^0 \times \sqrt{\frac{M_i}{2\pi RT}} \quad (2)$$

According to the actual volatilizing loss of Ti element during electron beam melting, the average surface temperature of molten pool under different melting power can be calculated by equation (2) as 1669.4K, 174.4k, 1792.9k and K1807.9K, respectively.The main volatile elements in near-α titanium alloy are Ti, Al and Sn. The solution system can be simplified as a Ti-Al-Sn ternary alloy model to study the volatilization behavior of alloying elements in the alloy. In addition, Zr, Mo, Si, Nb, Ta, and W are relatively low in content, so the solution is first regarded as a dilute solution of these elements when calculating the activity coefficient, which is considered to be 1. According to Miedema model, the activity coefficient and volatilization rate of elements at different temperatures are shown in Fig 1((b) (c)).It can be seen from the figure that the activity coefficient and volatilization rate of elements increase with the increase of temperature.The theoretical volatilization rates of Zr, Mo, Si, Nb, Ta, and W are very small, so the volatilization losses in electron beam melting are basically negligible.Since Al and Sn have a large saturated vapor pressure, their activity coefficient and activity have a great influence on the theoretical volatilization rate, resulting in inaccurate calculation of volatilization loss of Al and Sn through Langmuir equation. It is necessary to introduce the activity coefficient compensation factor ω to optimize the volatilization rate of Al and Sn. After optimization, the theoretical volatilization rate can be expressed as:

$$V_i = \chi_i \times \gamma_i \times \omega_i \times P_i^0 \times \sqrt{\frac{M_i}{2\pi RT}} \quad (3)$$

The approximate value of the activity coefficient compensation factor of Al and Sn in this experiment is 0.0035 and 1.26. The volatilization rate of Al and Sn after optimization and the theoretical volatilization rate are shown in Fig 1(d). It can be seen from the figure that the theoretical volatilization rate of Ti, Al and Sn in the electron beam melting process is in good agreement with the actual volatilization rate.

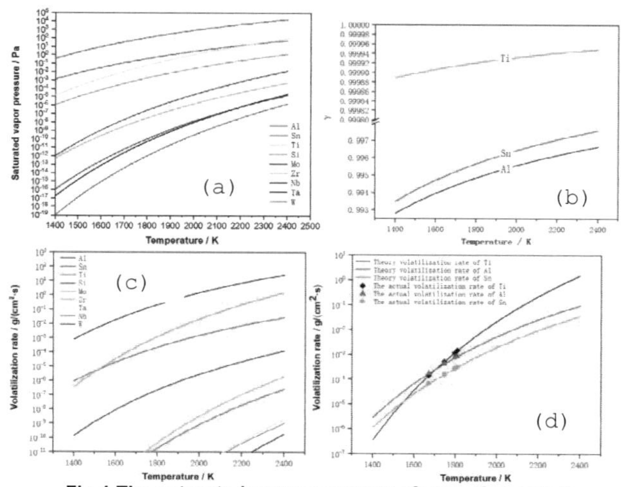

Fig.1 The saturated vapor pressure of pure elements in relationship with temperature for near-α titanium alloy(a); Activity coefficient(b) and volatilization rate(c) of each element at different temperatures;Actual mass loss and theoretical mass loss of volatile elements(d)

4 Conclusion

It is reasonable to simplify the alloy into Ti-Al-Sn ternary system and use Miedema model to calculate the activity coefficients and volatilization rates of Ti, Al and Sn. When calculating the volatilization rate of Al and Sn, it is necessary to introduce the activity coefficient compensation factor to optimize the theoretical volatilization rate.

5 Acknowledgments

The Acknowledgment section is an optional section that uses and “body” text style (10 pt., Times New Roman).

References

[1] Sun Q J, Xie X. Microstructure and mechanical properties of TA15 alloy after thermos-mechanical processing[J]. Mater Sci Eng A, 2018, 724: 493-501

[2] Singh P, Pungotra H, Kalsi N S. On the characteristics of titanium alloys for the aircraft applications[J]. Materials today: proceedings, 2017, 4(8): 8971-8982.

[3] Kvande R, Mjos O, Ryningen B. Materials Science and Engineering A, 2005, 413, 545.

Microstructures and Mechanical Properties of Cast Mg-Zn-Y-Zr Mg Alloys

Ming Sun, Dequan Shi, Guangkui Cheng
School of Materials and Chemistry, University of Shanghai for Science and Technology, 516 Jungong Road, Shanghai, China.
*Corresponding address: e-mail: ssm@usst.edu.cn

Abstract: The as-cast microstructure and mechanical properties of Mg-xZn-yY-0.5Zr (ZWK) alloy were studied, and the effects of Zn and Y elements were analyzed. The results showed that the equilibrium phase of the as-cast alloy is mainly composed of α-Mg matrix and network eutectic phase, and its specific composition is mainly affected by the Zn/Y atomic ratio. For ZW105K, ZW205K, and ZW305K alloys, the equilibrium phase gradually evolves from W-Mg3Zn3Y2 phase to W phase + I phase (I-Mg3Zn6Y), I phase with the increase of Zn/Y atomic ratio (i.e. 2.69, 5.25, 11.89). For ZW202K, ZW205K, ZW215K Alloy, the equilibrium phase gradually evolved from I-Mg3Zn6Y phase to W phase + I phase, W phase with the decrease of Zn/Y atomic ratio (i.e. 10.87, 5.25, 1.71). With the increased of Zn or Y content, the number of eutectic phases gradually increased, the strength (YS, UTS) gradually increased, but the plasticity (EL) gradually decreased. Based on the comprehensive analysis, ZW205K alloy shows the best comprehensive mechanical properties among the two groups of alloys. The YS, UTS and EL was 88.3±0.2 MPa, 187.1±3.7 MPa and 14.1±2.1 %, respectively.

Keywords: Mg alloy; Mg-Zn-Y; microstructures; Mechanical properties

1 Introduction

As one kind of lightweight green structural material, Mg alloy shows great application prospect due to low density, high specific strength and good damping performance. Among many Mg alloys, the medium-high strength Mg-Zn-Y-Zr (ZWK) alloy has attracted widespread attention [1]. However, at present, most ZWK series alloys employ high contents of Zn and Y elements. The high content of rare earth Y greatly increases the alloy cost, which becomes a key bottleneck for engineering applications. Thus, the development of low-alloyed ZWK alloys has become the latest research focus.

In this paper, the microstructures and mechanical properties of low-alloyed ZWK alloys were investigated. The relationships among compositions, microstructures and properties were revealed. The mechanisms for strengthening and toughening was clarified. This paper will help promote the applications of Mg alloys.

2 Experimental procedure

Mg-*x*Zn-0.5Y-0.5Zr (nominal x=1.0, 2.0, 3.0wt.%, i.e. ZW105K, ZW205K, and ZW305K) and Mg-2.0Zn-*y*Y-0.5Zr (nominal y=0.2, 0.5, 1.5wt.%, i.e. ZW202K, ZW205K, ZW215K) alloys were prepared with traditional metal mold gravity casting. The raw materials for melting are pure Mg (99.wt.%), pure Zn, master alloy of Mg-25wt.%Y and Mg-30 wt.%Zr, which were melt in crucible resistance furnace (SG2-3-10). During the whole process of melting and casting, the protective mixed gas 1vol.%SF_6 + 99 vol.%CO_2 was used. After melting pure Mg and pure Zn at 660~700°C, the Mg-25Y master alloy was added at and melt at 700~720°. Then, the grain refiner Mg-30Zr master alloy was added and melt at 730°C. 1.5 wt.% JDMJ flux was used for melt purification. After 15min settling, melt was poured into a metallic mold preheated at 200°C. The real compositions were determined by ICP-AES.

Phase identifications were conducted with a D8 Advance X-ray diffractometer (XRD). Microstructure specimens were prepared through standard metallographic procedures including grounding, polishing and etching (solution: 4.2 g picric acid, 10 ml acetic acid, 80 ml ethyl alcohol and 10 ml water). Microstructure was observed with optical microscopy (OM, Leica DMi8) and scanning electron microscope (SEM, Nova NanoSEM 450) equipped with energy-dispersive spectrum (EDS).

Tensile specimens (Fig. 2) were prepared with electric-spark wire-cutting. Tensile tests were conducted on Zwick/Roell-20KN material test machine with test speed 0.5 mm/min. For each alloy, 7 parallel specimens were tested, and the average values and standard deviations for yield strength (YS, MPa), ultimate tensile strength (UTS, MPa) and elongation (EL. %) were calculated.

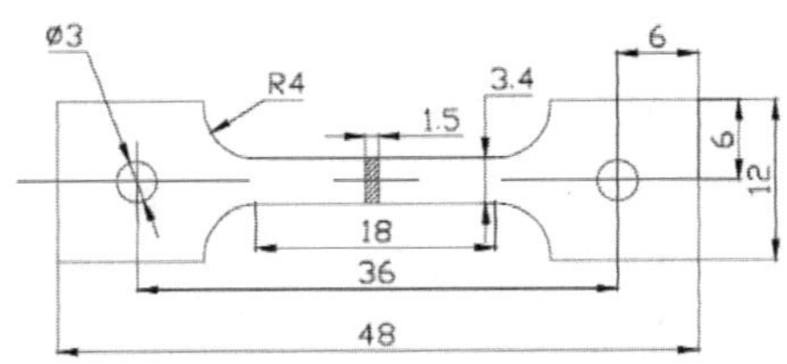

Figure 1. Dimension of tensile test sample ([1]).

3 Result and discussion

Microstructure

Figure 2 shows the XRD patterns of as-cast Mg-Zn-Y-Zr alloys. It can be seen that only the α-Mg matrix phase can be observed. According to [2, 3], the W-$Mg_3Zn_3Y_2$ and I-Mg_3Zn_6Y phases can be presented in Mg-Zn-Y alloys. However, they were not detected with XRD in this study due to a lower content of Zn and Y elements.

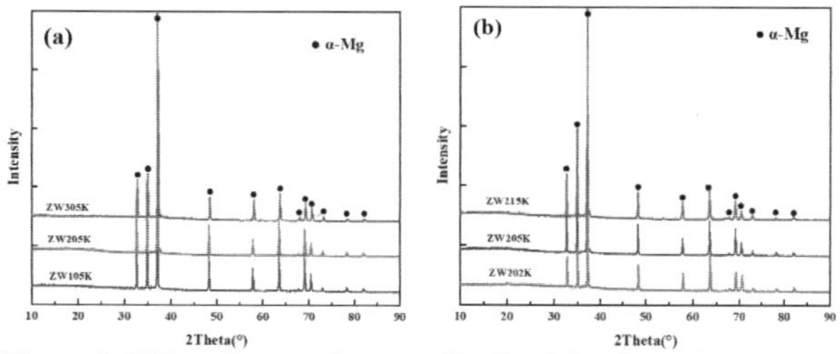

Figure 2. XRD patterns of as-cast Mg_3Zn-Y-Zr alloys: (a) Mg-*x*Zn-0.5Y-0.5Zr, (b) Mg-2.0Zn-*y*Y-0.5Zr.

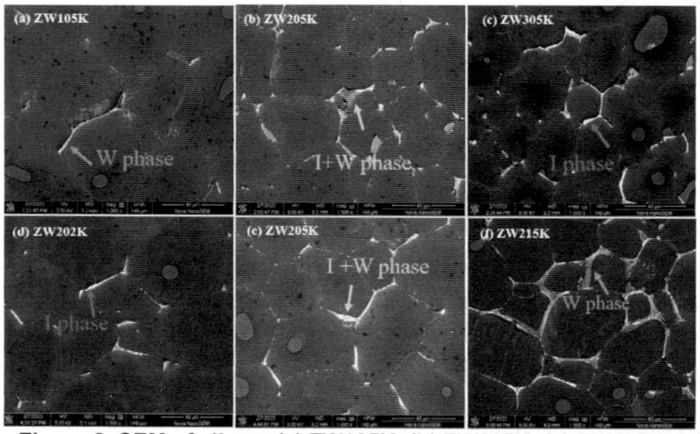

Figure 3. SEM of alloys: (a) ZW105K, (b) ZW205K, (c) ZW305K, (d) ZW202K, (e) ZW205K, (f) ZW215K.

Figure 3 shows the SEM of Mg-Zn-Y-Zr alloys, where EDS of some micro-areas were also analyzed. Firstly, the microstructures of Mg-*x*Zn-0.5Y-0.5Zr alloys were analyzed. It can be seen that the secondary phases in ZW105K alloy are mainly W-$Mg_3Zn_3Y_2$. With the increase in Zn content to 2%, I-Mg_3Zn_6Y phase also presents in ZW205K alloy. When the Zn content increases to 3%, W-phase disappears while the secondary phase is only I-phase. Secondly, the microstructures of Mg-2.0Zn-*y*Y-0.5Zr alloys were also analyzed, the secondary phase in ZW202K, ZW205K, ZW215K alloys is believed to be I, I+W and W phase, respectively. In addition, it has been reported that, when the long-period stacking order (LPSO) phase can present in Mg-Zn-Y-Zr alloy [2, 3]. But in Figure 3, LPSO phase was not observed, which is mainly because of its small amount.

Mechanical properties

Figure 4 shows the mechanical properties. The strength gradually increases with increasing the Zn or Y content. The elongation gradually decreases with an increase in Zn content, but does not change linearly with an increase in Y content, whose reason will be analyzed later. The ZW215K alloy shows the highest UTS (193.1±7.9 MPa) and YS (103.7±1.5 MPa). The ZW205K alloy shows the highest elongation (14.1±2.1%). If the strength and elongation are considered together, the ZW205K alloy shows the best comprehensive mechanical properties.

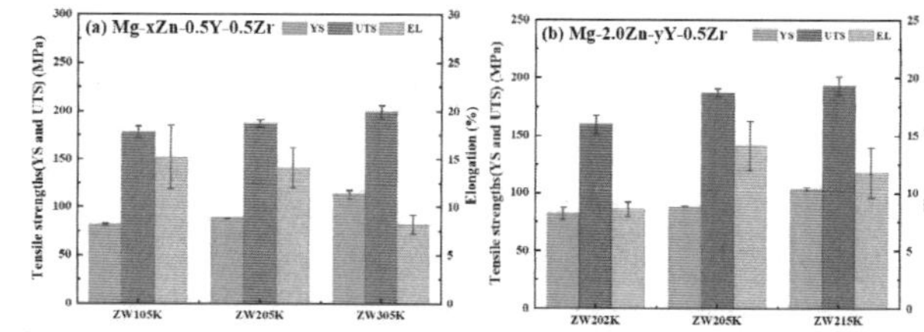

Figure 4. Mechanical properties of Mg-Zn-Y-Zr alloys: (a) Mg-xZn-0.5Y-0.5Zr, (b) Mg-2.0Zn-yY-0.5Zr.

4 Conclusion

The phase constitutions are greatly influenced by the content of Zn or Y content, where the phases can be I, I+W or W. If the strength and elongation are considered in compromise, the ZW205K alloy shows the best comprehensive mechanical properties.

References

[1] Xu D K, Tang W N, Liu L, Y.B. Xu and E.H. Han. Effect of Y concentration on the microstructure and mechanical properties of as-cast Mg-Zn-Y-Zr alloys[J]. J. Alloy Compd., 2007, 432(1-2): 129-134.

[2] Gröbner J, Kozlov A, Fang X Y, Geng J, Nie J F, Schmid-Fetzer R. Phase equilibria and transformations in ternary Mg-rich Mg-Y-Zn alloys[J]. Acta Mater., 2012, 60(17): 5948-5962.

[3] Yamasaki M, Hashimoto K, Hagihara K, Yoshihito K. Effect of multimodal microstructure evolution on mechanical properties of Mg-Zn-Y extruded alloy[J]. Acta Mater., 2011, 59(9): 3646-3658.

In Situ Determination of Nonmetallic Inclusions in AlMg6Si2MnZr Alloy

Primož Mrvar, Mitja Petrič, Sebastjan Kastelic, Matic Žbontar
Faculty of Natural Sciences and Engineering, University of Ljubljana, Aškerčeva 12, 1000 Ljubljana
*Corresponding address: e-mail: primoz.mrvar@ntf.uni-lj.si

Abstract: The study investigates the influence of impurities on the metallurgical quality of the AlMg6Si2MnZr alloy, a relatively new alloy for pressure die casting. Experimental techniques included in situ thermal analysis, electrical resistivity measurement, microscopy, and calorimetry. Findings suggest that improper handling during melting significantly impacts chemical composition and microstructure, necessitating careful process control. From electrical resistivity measurements in liquid, during solidification and in solid it is possible to estimate the quantity of nonmetallic inclusions.

Keywords: AlMg6Si2MnZr, Thermal analysis, Microstructure, Impurities, Pressure die casting.

1 Introduction

The AlMg6Si2MnZr alloy, noted for its enhanced mechanical properties, is less explored metallurgically. High magnesium content increases the likelihood of non-metallic inclusions, affecting the alloy's performance. This study aims to characterize these effects and provide insights for industrial application.

2 Experimental procedure

Methods included thermal analysis, electrical resistance measurements, optical and electron microscopy, and differential scanning calorimetry. A database for virtual casting process calculations was developed and calibrated using cooling curves.

Thermal analysis provided insights into phase transformations and solidification behavior. Electrical resistance measurements evaluated changes in the material's conductivity, indicating its purity and structural characteristics. Optical microscopy observed the overall microstructure, while electron microscopy examined finer structural details and phase distribution. Differential scanning calorimetry measured the heat flow associated with melting and solidification, providing crucial thermal property data.

Additionally, a database for virtual casting process calculations was established and calibrated using experimental cooling curves. This enabled accurate simulation of microstructural evolution and defect prediction during solidification, aiding in the optimization of the alloy's casting process. These combined methods provided a comprehensive understanding of the material's behavior under various processing conditions.

3 Result and discussion

Thermal analysis revealed that there are differences in cooling rates for various melt preparations, though these differences are not significant. In particular, the double 'in situ' thermal analysis indicated that the alloy's tendency for non-metallic inclusions to immerse in the melt and remain is higher compared to traditional aluminum alloys. The experimental series demonstrated that variations in melt preparation influenced the cooling curves to some extent, though significant differences were not observed.

Metallographic analysis provided insights into the microstructural characteristics of the alloy. Optical and electron microscopy revealed a dendritic structure of primary α_{Al} crystals, along with a binary eutectic mixture of α_{Al} and Mg_2Si, and a ternary eutectic of α_{Al}, Mg_2Si, and $Al_6(MnFe)$. The presence of polygonal particles based on Al, Mg, and Si, containing minor concentrations of Mo and V, and Chinese script particles based on Al, Si, Fe, and Mn, were noted. Fine intermetallic compounds such as $(AlSi)_3Zr$ were also observed.

The influence of impurities was investigated through differential scanning calorimetry and thermodynamic calculations, which examined slight variations in element concentrations. These analyses corroborated the presence of non-metallic inclusions, which were found to affect the quality of the alloy. The measured enthalpies of melting were lower for samples containing inclusions, confirming the detrimental impact of impurities on the alloy's quality.

The study underscores the critical need for stringent process control to mitigate the inclusion of impurities during the melting process. Proper handling and careful monitoring are essential to preserve the desired properties of the alloy, ensuring its suitability for industrial applications.

In Figure 1 a comparison of specific electrical resistivities of the investigated alloys is shown depending on the edge sample temperatures for measuring specific electrical resistivity.

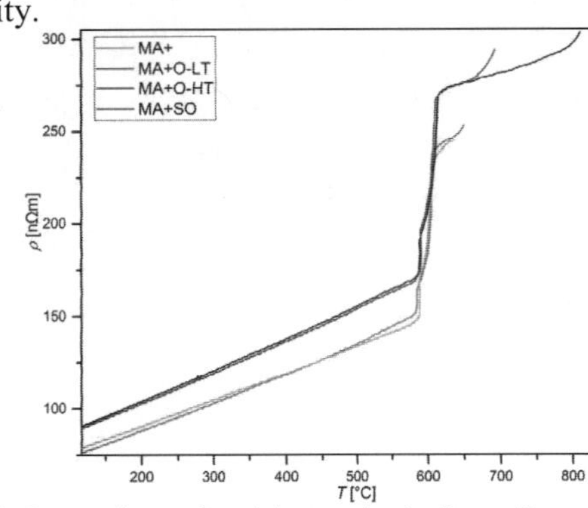

Figure 1: Comparison of resistance charts depending on edge cell temperatures for all investigated series.

The following diagram (Fig. 1) shows, depending on the investigated series, certain average areas between the derivative of the cooling curve and the linear baseline (A), sample masses (m), area-to-mass ratios (A/m), correlation factors of area-to-mass ratios with measured average DSC melting enthalpies of samples (A/m-DSCcor), and average measured DSC melting enthalpies (DSCavg). The correlation factors were determined as the ratio between A/m and DSCavg and calibrated according to the value MA+. Thus, the MA+ sample has a value of 1, while the others show a relative deviation from 1.

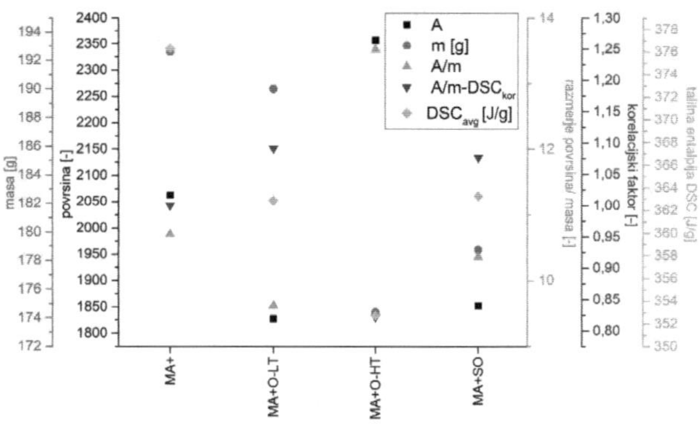

Figure 2: Crucial results from simple thermal analysis and enthalpies for all investigated series.

Conclusion

The study underscores the importance of strict process control in maintaining alloy quality. The presence of impurities significantly affects both microstructural characteristics and mechanical properties, emphasizing the need for precise melting practices.

References

[1] Boyko, V., et al. (2019). "The Effect of Alloying Additions on the Structure and Properties of Al-Mg-Si-Mn Casting Alloy – A Review." Journal of Applied Materials Engineering, 59(2), 89-102.

[2] Feikus, F.F. (2019). "In Situ Detection of Non-metallic Inclusions in Aluminium Melt (1xxx) - Comparison Between a Newly Developed Ultrasonic Technique and LiMCA and PoDFA Method." In Light Metals 2019. Springer, Cham, pp. 1623-1629.

[3] Lipshaw, J. (2020). "Looking at Castings' Environmental Impact Across Their Life Cycles." Modern Casting, 110(10), 34-37.

[4] Salonitis, K.J. (2019). "Life-Cycle and Energy Assessment of Automotive Component Manufacturing: The Dilemma Between Aluminum and Cast Iron." Energies, 12(13), 2557.

[5] Hartlieb, A. & Hartlieb, M. (2023). "The Impact of Giga-Castings on Car Manufacturing and Aluminum Content." In Light Metal Age. Available online: Light Metal Age.

Research on the Tight Fit Technology of New Fuel Element Based on Spinning Necking Forming

Zeming Wang[1,2], Jialong Chen[1], Xianmeng Tu[1], Xiaoyuan Ji[1,*], Jianxin Zhou[1]
1.State Key Laboratory of Materials Processing and Die & Mould Technology, School of Materials Science and Engineering, Huazhong University of Science and Technology, Wuhan 430074, PR China
2.Nuclear Power Institute of China, Chengdu, Sichuan, 610005, China
*Corresponding address: e-mail: jixiaoyuan@hust.edu.cn

Abstract: At present, based on the analysis of the failure characteristics of existing fuel elements, it is found that the failure of existing fuel elements focuses on the oxidation of the inner and outer sides of the fuel cladding, the bulging and bursting of the fuel cladding, the repositioning and diffusion of the fuel, and the melting of the fuel rod. The specific solutions are mainly through two aspects: first, enhancing the cladding of the fuel cladding, and second, improving the thermal conductivity uniformity of the fuel rod. The specific solutions include increasing the functional coating of the inner and outer walls Research and develop new high temperature resistant cladding, increase the height of fuel rod gas cavity, change from solid fuel pellet to annular fuel pellet, especially by enhancing the bonding quality of fuel cladding and fuel pellet interface, reducing the gap between them, so as to improve the cladding ability of fuel cladding, avoid the heat accumulation of fuel pellet and inner wall of cladding, and effectively improve the high sufficiency of single fuel rod High reliability is one of the most feasible solutions in engineering at present, and the close fitting technology between fuel cladding and fuel core is the ultimate success or failure of this process. Based on finite element simulation, this paper analyzes the forming process of the three roller spinning compression diameter, analyzes the radial and axial feed forming process of the roller, metal deformation behavior, metal flow and stress strain state during the forming process, and studies the control and optimization of the forming quality. On this basis, combined with the single fuel rod as the solid object, the spinning compression diameter tight fitting test research is carried out. Based on the test results, the development of special tight fitting equipment and process finalization are carried out. The results show that the best effect is to use the single pass small feed spinning compression diameter to coat the inner wall of the clad tube closely with the coating.

Keywords: Nuclear reactor, fuel element, cladding tube

1 Introduction

According to China's nuclear power mid-to-long term strategy, 4th generation reactors emphasize sustainability, non-proliferation, enhanced safety, and cost-effectiveness. Key designs include sodium, lead, and gas-cooled fast reactors, supercritical water ,ultra-high temp gas, and molten salt reactors, using coolants like liquid sodium, lead, helium, carbon dioxide, molten salts, and water. Despite varied reactor types and coolants, most employ rod fuel elements needing compatibility with liquid metals, high thermal conductivity, closed fuel cycles, and ample fission gas retention [1]. Modern fuel design trends towards optimized coolant flow, compact assemblies, and high-density fuels are pushing the envelope on existing fuel element technology[2].

Therefore, optimizing the fuel core-cladding bond boosts heat transfer, curbs radiation swelling, and bolsters safety in 4th generation reactor fuel, and has notable scientific and practical value.

2 Experimental procedure

The China Nuclear Power Research and Design Institute has independently developed CLA16, a low-activation FM stainless steel, matches T91, T92, CLAM, CLF-1 performance. Cladding OD: 12.02-12.06 mm, ID: 10.39-10.43 mm. Inner Cr coating via CVD (~15 μm). U-Zr alloy test fuel; Mo-W simulant pellets for research (OD: 10.340-10.355 mm) ensure ≤ 0.04 mm gap.

Based on this, taking the research on the tight fitting of new rod fuel elements as the main line, the numerical simulation re-search on the spinning necking tight fitting assembly was carried out. Investigation into cladding-pellet dynamics via spinning compression tests optimized parameters and facilitated specialized equipment development for single fuel rods, aiding full-scale product development and mass production.

3 Result and discussion

3.1 Numerical simulation and analysis of three roller spinning shrink fit assembly

At the beginning of the whole forming process of spinning and reducing, when the three spinning wheels extrude the clad tube radially, they exert a force on the clad tube, thus producing stress and strain. As a whole, only the relatively narrow necking area has large stress, and the equivalent stress is mainly distributed in the necking area, as shown in Fig. 1.

As can be seen in Fig. 2, the clad tube is deformed due to the radial extrusion of the spinning wheel, so as to achieve the effect of reducing. In addition, the area near the necking of the clad tube has almost no axial deformation, and the wall thickness of the clad tube is almost unchanged.

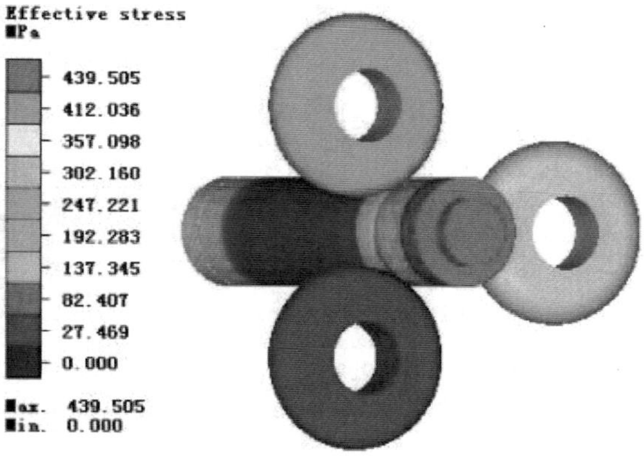

Fig. 1 Equivalent stress distribution in radial feed forming process of spinning wheel

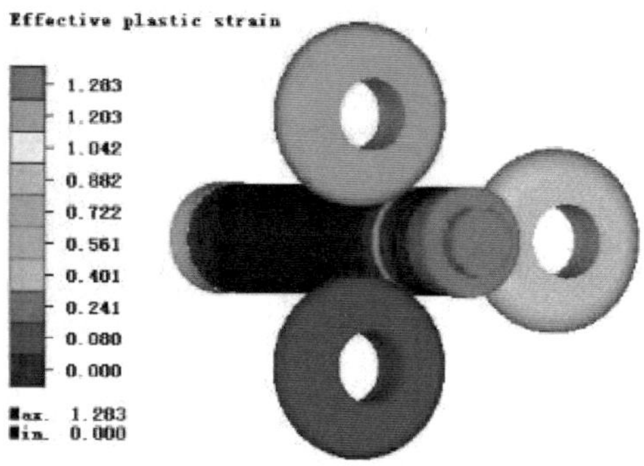

Fig. 2 Equivalent strain distribution in radial feed forming process of spinning roller

3.2 Experimental study on rotary spinning reducing process of single fuel rod

In order to better analyze the influence of different pass feed rates on the forming quality of this product and determine a reasonable feed rate distribution process experiment, this study conducted a total feed rate of 0.09mm, which was simulated in five groups, two groups for two passes, two groups for three passes, and one group for four passes.

Fig.3 shows the product forming effect under different feed rates. High feed rates prevent cracks but cause uneven gap distribution and minor radial bending.

3.3 Research and development of spinning reducing pro-cess equipment and sizing process for single fuel rod

Leveraging simulations and past tests, we'll optimize the spinning necking process for single fuel rods using a specialized roller rotation device, standardizing the tri-roller compression method, and statistically evaluating critical quality parameters to aid in single rod development.

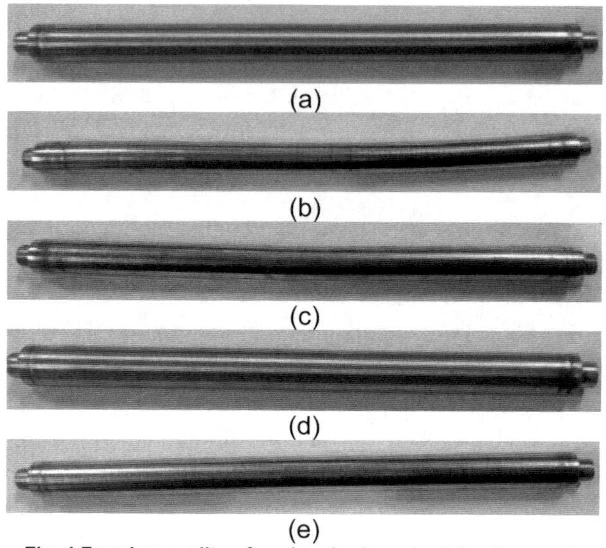

Fig. 4 Forming quality of each spinning wheel feeding mode

4 Conclusions

(1) Optimal settings (250 rpm, 0.03 mm spinning feed, 0.2 mm/s axial feed) in 1-2 passes ensure smooth rods, minimal deformation, and <15 μm max cladding-core clearance.

(2) Small-feed single-pass spinning effectively fits coated cladding tubes, preserving coating integrity without defects like peeling or cracking.

(3) Enhancements needed: NDT for clearance, batch production reliability, stress analysis, and precision straightening in new fuel element development.

5 Acknowledgments

This work is financially supported by National Key R&D Program (2020YFB1710100), Science and Industry Commission Project (2019XXX.XX4007Tm), and National Natural Science Foundation of China (52275337, 51905188, 52090042).

References

[1] Music O, Allwood J M, Kawai K. A review of the mechanics of metal spinning[J]. Journal of materials processing technology, 2010, 210(1): 3-23.

[2] Kwiatkowski L, Tekkaya A E, Kleiner M. Fundamentals for controlling thickness and surface quality during dieless necking-in of tubes by spinning[J]. CIRP Annals, 2013, 62(1): 299-302.

Effects of Squeeze Casting and Heat Treatment on Properties of Low-Cost Mg-Si-Zn-Cu Alloy

Lu Chen, Jianyu Li, Shulin Lü, Wei Guo, Shusen Wu*
State Key Lab of Materials Processing and Die & Mould Technology, School of Materials Science and Engineering, Huazhong University of Science and Technology, Wuhan 430074, P. R. China
* Corresponding Author, ssw636@hust.edu.cn

Abstract: Magnesium alloys possess several advantages such as low density, high thermal conductivity, and excellent seismic resistance, making them ideal materials for laptop frames. However, current research on thermal conductive magnesium alloys focuses mainly on incorporating costly metal elements such as RE and Sn. This not only increases the production costs but also raises the density of magnesium alloys, hindering the commercialization of thermal conductive magnesium alloys. Therefore, the present study aims to develop an Mg-Si-Zn-Cu alloy that offers desirable mechanical properties and thermal conductivity while considering cost and density control. The impact of squeeze casting and heat treatment on the overall properties of this alloy was investigated. The research findings demonstrate that squeeze casting enhances the mechanical properties of the alloy without compromising its heat dissipation performance. Furthermore, heat treatment improves the thermal conductivity of the alloy effectively, primarily attributed to the spheroidization effect on secondary phases resulting from isothermal heat treatment.

Keywords: magnesium, thermal conductivity, Mg_2Si, heat treatment

1 Introduction

Rare-earth elements have shown significant potential in enhancing mechanical properties of magnesium alloys. These elements are commonly employed as alloying agents in the development of magnesium alloys with superior strength and conductivity [1]. Nevertheless, it is important to acknowledge that rare earth elements are generally associated with high cost, which may pose challenges in the commercialization of magnesium alloys. Additionally, the introduction of rare earth elements elevates density of magnesium alloy, which is not desirable for mobile devices. Mg-Si alloys offer advantages in terms of lower density and reduced cost when compared to rare earth element-containing alloys. Mg_2Si is the main strengthening phase in Mg-Si alloys and is an ideal in-situ strengthening phase in magnesium alloys [2]. Mg_2Si possesses high hardness and high modulus, which contribute to overall improved mechanical properties of magnesium. At the same time, the solid solubility of Si in Mg is very small, which avoids decrease in thermal conductivity of magnesium alloys caused by solid solution atoms.

This study aims to develop a new high-strength and high-thermal conductivity magnesium alloy by incorporating small amounts of Zn and Cu elements into the Mg-Si eutectic alloy. The performance of this new alloy undergoing squeeze casting and gravity casting was explored, followed by utilization of heat treatment to enhance its thermal conductivity.

2 Experimental procedure

The raw materials are industrial pure magnesium(99.95%), pure zinc(99.99%), pure copper(99.95%), and Mg-30Si master alloys. After removing the oxide skin from the surface of pure magnesium, it was placed in a resistance furnace, and 0.1vol.%SF_6+99.9vol.%N_2 mixed gas was inlet to prevent melt combustion. When the temperature of melt was raised to 800 ℃, the preheated Mg-30Si master alloy was added and melted completely after holding for 2 hours. Then, the melt was cooled to 750 ℃ and Zn and Cu were added. After 30 min, the melt was cooled to 700 ℃ for pouring. Before pouring, the melt was stirred to prevent Zn and Cu from sinking. The casting mold was metal mold with preheating temperature of 200 ℃ and ingot size of D30mm×90mm. The actual composition and of Mg-Si-Zn-Cu alloy was shown in Table 1.

Table 1 Composition of Mg-Si-Zn-Cu alloys

Mg	Si	Zn	Cu
Bal.	1.34	0.62	0.57

3 Result and discussion

3.1 Effects of squeeze casting on Mg-Si-Zn-Cu alloy

Fig. 1 displays the microstructures of the Mg-Si-Zn-Cu alloy under gravity casting and squeeze casting. After gravity casting, the alloy primarily consists of coarse α-Mg and eutectic Mg_2Si, interspersed with minor worm-like eutectic MgCuZn. Following squeeze casting, the size of α-Mg is significantly refined, resulting in the segmentation of large eutectic Mg_2Si blocks into a network-like framework encircling the α-Mg.

Following squeeze casting, the microstructure of the Mg-Si-Zn-Cu alloy exhibits significant refinement, leading to an improvement in its mechanical properties. The room temperature tensile curves of the Mg-Si-Zn-Cu alloy prepared by gravity casting and squeeze casting are shown in Fig. 2. For gravity cast(GC) alloy, the ultimate tensile strength (UTS) is 165 MPa and the elongation (El.) is 6.2%.

Due to the refined α-Mg segmenting the large eutectic Mg_2Si blocks, the elongation of the squeeze cast(SC) alloy increased to 8.6%. The network-like Mg_2Si framework, compared to large eutectic Mg_2Si blocks, provides a better stress-bearing capability, resulting in an increased UTS of 192 MPa for the SC alloy.

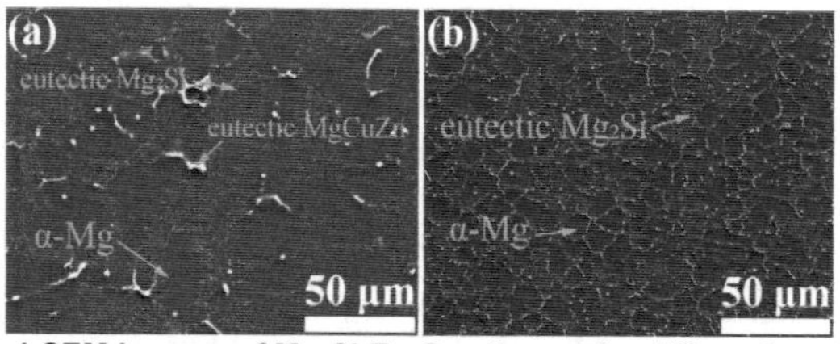

Fig. 1 SEM images of Mg-Si-Zn-Cu alloy under different casting methods: (a) gravity casting; (b) squeeze casting.

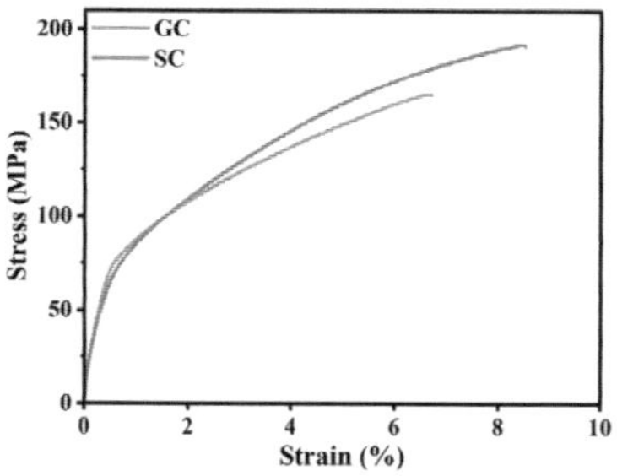

Fig. 2 Room temperature tensile curves of Mg-Si-Zn-Cu alloy prepared by gravity casting and squeeze casting

3.2 Effects of heat treatment on Mg-Si-Zn-Cu alloy

In this study, T1, T4, and T6 heat treatments were applied to the Mg-Si-Zn-Cu alloy to investigate the variations in its thermal conductivity under different conditions. Fig. 3 presents the microstructure of the alloy after the three heat treatments, with Fig. 8(d) focusing on a high-magnification image of the spheroidized Mg_2Si shown in Fig. 8(c). The morphology of Mg_2Si changes significantly with different heat treatments. The T4 and T6 treated samples underwent isothermal heat treatment at 450℃ for 24 hours, resulting in the transformation of Mg_2Si morphology from lamellar to short rod and spherical forms. The T1 treated samples were held at 180 ℃ for 6 hours, maintaining the lamellar morphology of Mg_2Si.

Fig. 4 illustrates the curves of thermal conductivity and thermal diffusivity of the alloy after heat treatment. Compared to the as-cast alloy, the thermal conductivity of the alloy decreases slightly after T1 treatment, but increases after T4 and T6 treatments. T4 heat treatment increases the concentration of solute atoms in α-Mg matrix, but the thermal conductivity exceeds that of as-cast sample. This result indicates that the spheroidization of Mg_2Si enhances the thermal conductivity of the alloy. The above heat treatment experiments confirm that the spheroidization of Mg_2Si leads to an increase in the thermal conductivity of the Mg-Si-Zn-Cu alloy, with the thermal conductivity reaching 142.6 W/(m·K) after T6 treatment.

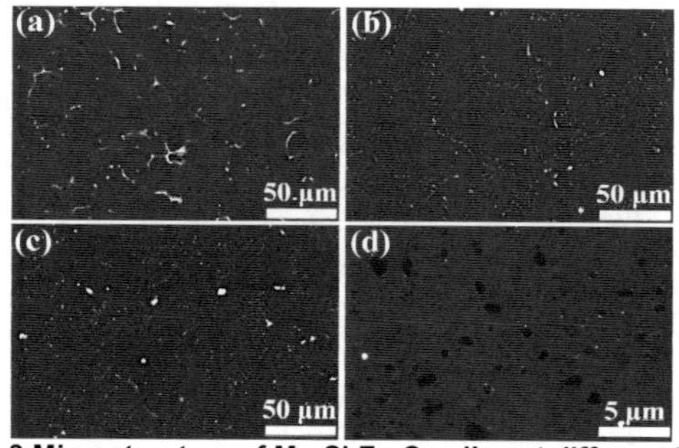

Fig. 3 Microstructure of Mg-Si-Zn-Cu alloy at different heat treatment states: (a)T1 heat treatment; (b) T4 heat treatment; (c)T6 heat treatment; (d) high magnification microstructure after T6 heat treatment

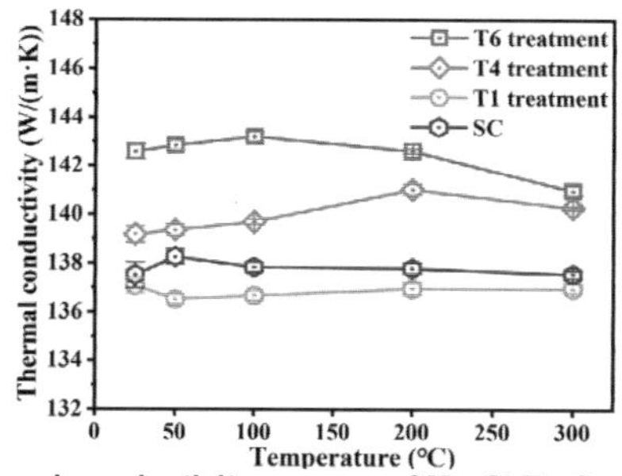

Fig. 4 thermal conductivity cureves of Mg-Si-Zn-Cu alloy after different heat treatments

Conclusion

(1) Squeeze casting refines the primary α-Mg and fragments the eutectic structure, thereby enhancing the strength and elongation of Mg-Si-Zn-Cu.

(2) The isothermal treatment causes the eutectic Mg_2Si to undergo spheroidization, thereby improving the thermal conductivity of Mg-Si-Zn-Cu.

References

[1] Rong J, Zhu J, Xiao W, Zhao X and Ma C. A high pressure die cast magnesium alloy with superior thermal conductivity and high strength. Intermetallics, 2021, 139: 107350.

[2] Dong Y, Xiao P, Gao Y, Zhao Q and Yang H. Microstructure refinement and mechanical properties of eutectic Mg2Si reinforced Mg matrix composites containing Sr element. J Mater Res Technol, 2022, 17: 2614–2623.

Effect of Al_2Y on the Microstructure and Properties of Continuous Squeeze Casting-Extrusion Mg-Y-Al Alloy

Shilong Li, Dijia Zhao, Jianyu Li, Shusen Wu, Shulin Lü*
State Key Lab of Materials Processing and Die & Mould Technology, School of Materials Science and Engineering, Huazhong University of Science and Technology, Wuhan 430074, China
* Corresponding Author, shulin317@hust.edu.cn

Abstract: The evolution of mechanical properties and microstructure of continuous squeeze casting-extrusion Mg-3nY-2nAl (n=0.25, 0.5, 0.75, and 1.0, in wt.%) series alloys was investigated. As the alloy content increases, the microstructure of as-cast state changes from a single Al2Y eutectic to a mixed of primary Al2Y and eutectic Al2Y. In the extruded alloy, the eutectic Al2Y is completely crushed into dispersed small particles, while the larger primary phase is not crushed. As the alloy content increases, the strength first increases and then decreases, reaching a maximum at n=0.75, with ultimate tensile strength and elongation of 314 MPa and 12%, respectively. The study of EBSD results indicates that the reason for the simultaneous increase in strength and plasticity is the heterostructure strengthening brought about by the primary Al2Y particles.

Keywords: magnesium, mechanical property, Al_2Y, continuous squeeze casting-extrusion

1 Introduction

The fine and uniform microstructure of the billet is beneficial for the recrystallization behavior of hot extruded magnesium alloys due to its higher grain boundary density. Zr is the most effective refining element for magnesium alloys. Due to the formation of heat-resistant and stable AlZr phase, Zr cannot refine Al containing magnesium alloys. Therefore, researchers attempted to refine magnesium alloys containing Al by adding heterogeneous nucleating agents such as SiC and ZnO. Al-RE compounds are a very promising class of nucleating agents. The common Al-RE phases in magnesium alloys include two forms: $Al_{11}RE_3$ and Al_2RE. Among them, Al_2RE, such as Al_2Y, Al_2Gd, etc., has been proven to have a good coherent relationship with magnesium matrix, therefore Al_2RE is considered a promising nucleating agent for magnesium alloys [1]. In addition, Al_2Y phase, as a high hardness and large-sized second phase, is generally considered to have a PSN effect, which can effectively promote recrystallization behavior and obtain a uniform hot extruded microstructure. However, in existing research, few researchers have discussed the influence of different Al_2Y forms on the microstructure and properties of deformed alloys, and have not ruled out the influence of other second phases and solid solution elements on the deformation process behavior of alloys. Therefore, the Mg-Y-Al series alloy was designed to control the ratio of yttrium and aluminum elements, so that only Al-Y phase exists in the alloy and reduce the influence of solid solution

2 Experimental procedure

The raw materials were composed of industrial pure magnesium, pure aluminum, and Mg-70Y intermediate alloy. Under the protection of N_2-SF_6(1vol.%), after pure magnesium was melted using a resistance furnace, Mg-70Y intermediate alloy was added. After the intermediate alloy was completely melted, stir the melt evenly, cool it to 720 ℃, and then add pure aluminum preheated to 500 ℃. High-purity argon gas was injected for 10 minutes to refine the melt. The melt was cooled to 705 ℃ and held for 20 minutes before being ready for casting. After the melt was poured into a continuous squeeze casting-extrusion mold preheated to 150℃, a compression pressure of 100MPa was applied to completely solidify the melt under the pressure [2]. Subsequently, the billet was extruded from the mold mouth with an extrusion ratio of 16:1.

Table 1 Composition of Mg-Y-Al alloys

Code name	Mg	Y	Al
WA0.5	Bal.	0.75	0.5
WA1.0	Bal.	1.5	1.0
WA1.5	Bal.	2.25	1.5
WA2.0	Bal.	3.0	2.0

3 Result and discussion

3.1 Effect of alloy content on the microstructure of WA series alloy

Fig.1 shows the as-cast microstructures of WA series alloys with different alloy contents. When the alloy content is low, the microstructure is composed of α-Mg and eutectic Al_2Y, and the area fraction of eutectic structure increases with the increase of alloy content. When the alloy content increases above WA1.5, massive primary Al_2Y phases begin to appear in the microstructure. Further increase in alloy content does not increase the content of primary phases, but increases the size of primary phases and decreases the number density. The results of EDS analysis indicate that there are no intermetallic compounds of other components in the WA series alloys, and no solid solution of aluminum or yttrium elements are detected in the matrix magnesium.

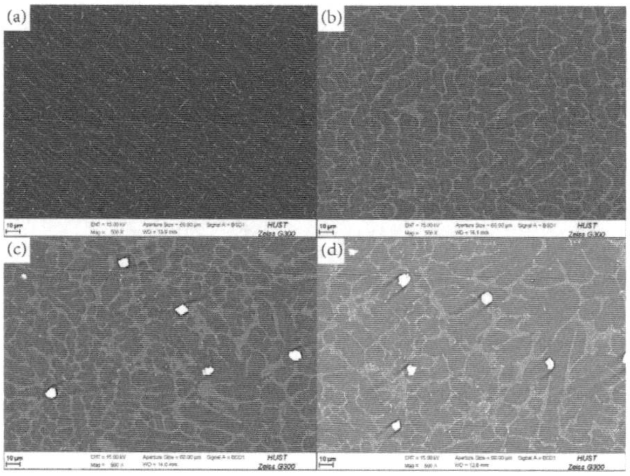

Fig. 1 SEM images of WA series alloy: (a) WA0.5; (b) WA1.0; (c) WA1.5; (d) WA2.0.

Fig.2 shows the IPF diagram of the extruded alloy. After hot extrusion, the microstructure is significantly refined, and the average grain size is refined to about 2μm. The average grain size of the WA0.5 alloy is the smallest, reaching 1.8μm. It is worth noting that although the average grain size is similar, the grain size distribution and recrystallization degree of the alloy are not the same. As the alloy content increases, the recrystallization fraction first increases and then decreases. The WA1.0 alloy has the highest recrystallization fraction (62.5%), while the WA1.5 and WA2.0 alloys have similar recrystallization fractions, both around 31%, which are lower than the 42% of WA0.5. The formation of primary phases does not significantly promote the recrystallization of the alloy, but rather reduces the recrystallization fraction, resulting in a mixed structure of coarse elongated grains + fine recrystallized grains.

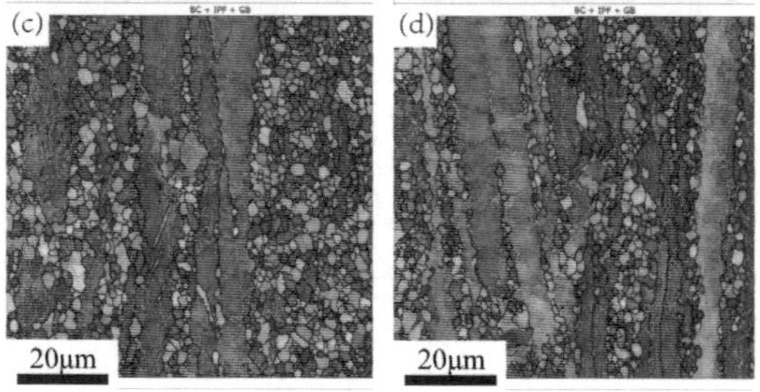

Fig. 2 IPF of WA series alloy: (a) WA0.5; (b) WA1.0; (c) WA1.5; (d) WA2.0.

3.2 Effect of alloy content on the mechanical properties of WA series alloy

The mechanical properties of extruded alloys reach their best at WA1.5. When the alloy content increases from WA0.5 to WA1.0, the strength increases and the plasticity decreases. But when the alloy content further increased to WA1.5, not only did the yield strength increase by 17MPa, but the plasticity increased by more than 60% compared to WA1.0, exceeding the lowest alloy content of WA0.5 alloy. However, higher alloy content did not lead to higher mechanical properties, but instead slightly decreased alloy strength and plasticity. WA1.5 and WA2.0 have similar strain hardening rates and mechanical properties. The HDI strengthening ability of WA1.5 alloy was tested through cyclic tensile tests, and the results showed that the HDI strengthening caused by heterostructure was about 180MPa, accounting for about 70% of the yield strength.

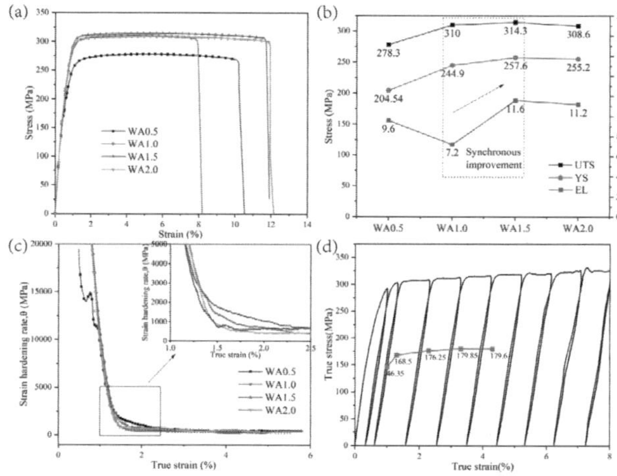

Fig. 3 (a)stress-strain curve of WA alloy; (b) mechanical properties of WA alloy; (c) strain hardening rates of WA alloy; (d)HDI strengthening of WA alloy.

4 Conclusion

(1) The formation of primary Al_2Y phase reduces the recrystallization fraction of the deformed alloy, resulting in a bimodal heterostructure of the alloy.

(2) The formation of primary Al_2Y enhances both the strength and properties of extruded alloys, while excessive alloy content leads to a decrease in mechanical properties of the alloy.

References

[1] Qiu D, Zhang M-X, Taylor JA, Kelly PM. A new approach to designing a grain refiner for Mg casting alloys and its use in Mg–Y-based alloys. Acta Materialia, 2009, 57: 3052–3059.

[2] Zhao D, Lü S, Guo W *et al.* Effects of Mg17Al12 phase on microstructure evolution and ductility in the AZ91 magnesium alloy during the continuous rheo-squeeze casting-extrusion process. Journal of Materials Science & Technology, 2024, 191: 63–79.

Rapid Preheating to Preserve the Initial Microstructure of the Squeeze Casting AZ31 Billet and Effect on Subsequent Hot Extrusion

Dijia Zhao, Shulin Lü*, Zhaoxiang Yan, Wenbo Guo, Shusen Wu, WeiGuo
School of Materials Science and Engineering, Huazhong University of Science and Technology, Wuhan,China.
*Corresponding address: e-mail: shulin317@hust.edu.cn (S. Lü).

Abstract: In this investigation, the squeeze casting process was employed to fabricate AZ31 magnesium alloy billets instead of the traditional semi-continuous casting method. The high-temperature melt was rapidly solidified under pressure. A rapid preheating method was utilized to quickly adjust the temperature of the newly solidified billet to the corresponding extrusion temperature within 10 minutes, bypassing the need for cooling to room temperature. Without homogenization treatment, the fine secondary phase particles were preserved, grain coarsening was avoided, and the microstructure of the AZ31 extruded magnesium alloy was refined, enhancing its mechanical properties. This study helps broaden thecasting process's application range and provides a new strategy for designing thermoplastic deformation processes for different alloys.

Keywords:AZ31 magnesium alloy; Squeeze casting; Hot extrusion; Homogenization; Rapid preheating.

1 Introduction

In addition to producing practical castings, various casting methods are used to create semi-finished materials for manufacturing processes. For example, billets of extruded magnesium alloys are typically made via the semi-continuous casting (SCC) process. This method often results in severe segregation and coarse secondary phases due to non-equilibrium solidification. Therefore, an additional homogenization treatment is performed before hot extrusion to eliminate defects and improve extrudability, though this can lead to grain coarsening and reduced effectiveness of secondary-phase particles in recrystallization nucleation. The low plastic deformation capability of magnesium alloys requires elevated extrusion temperatures, necessitating a preheating process between homogenization and hot extrusion to reach the desired temperature. Both thermal processes affect the billets' microstructure.

The squeeze casting (SC) process, an advanced casting technology, involves the rapid solidification of liquid or semi-solid melt under high pressure, yielding a higher cooling rate that refines matrix grains and produces high-density fine precipitates. This makes SC an ideal method for preparing high-quality billets for hot extrusion. However, the required homogenization and preheating processes in traditional hot extrusion can diminish the benefits of squeeze casting.

This study investigated the effects of homogenization and preheating treatments on the microstructure of various billets. A novel rapid preheating method was developed to adjust the temperature of the squeeze casting billet to the extrusion temperature within 10 minutes. By avoiding homogenization treatment, fine secondary phase particles were preserved, grain coarsening was prevented, and the microstructure of the AZ31 extruded magnesium alloy was refined, enhancing its mechanical properties.

2 Experimental procedure

Figure 1 illustrates the rapid preheating process for squeeze casting billets. The squeeze casting was performed on a hydraulic machine at a pressure of 100 MPa. AZ31 melt at 690 °C was poured into a permanent mold that matches the extruded billet size. After holding the pressure for 10 seconds, the billet was ejected and air-cooled for about one minute. It was then placed in a heat-holding furnace with graphite particles as a heat transfer medium to improve heat transfer. After 10 minutes, the SC billet was hot extruded with an extrusion ratio of 16 and a ram speed of 7 mm/s to produce an extruded rod.

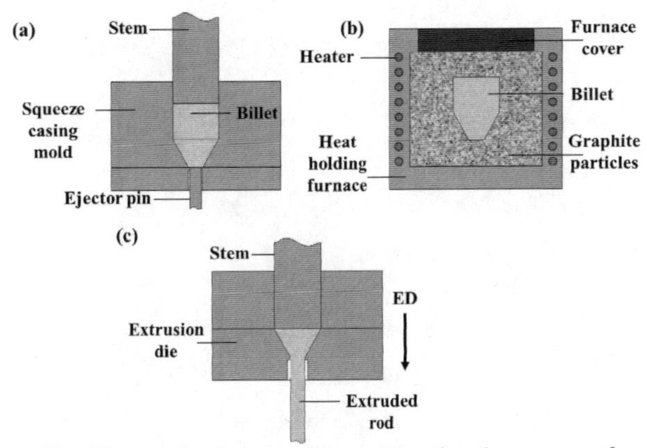

Fig. 1Schematic diagram of therapid preheating process for squeeze casting billets: (a) squeeze casting; (b) rapid preheating; (c) hot extrusion.

The extrusion temperature was set at 380 °C, with the heat-holding furnace and extrusion mold preheated to the same temperature. Homogenization (420 °C for 10 h) was performed on both SC and SSC billets, the latter cut from a commercial AZ31 SSC ingot. Additionally, traditional preheating (380 °C for 1 h) was conducted for the SC billet.

3 Result and discussion

3.1 Microstructure of the billets before and after various heat treatments.

Figure 2 shows the microstructure of the SCC and SC billets. The element distribution maps in Fig. 2(b) indicate that the secondary phase particles are Mg-(Al, Zn) and Al-(Mn, Fe). The SC billet displays denser and finer particles.

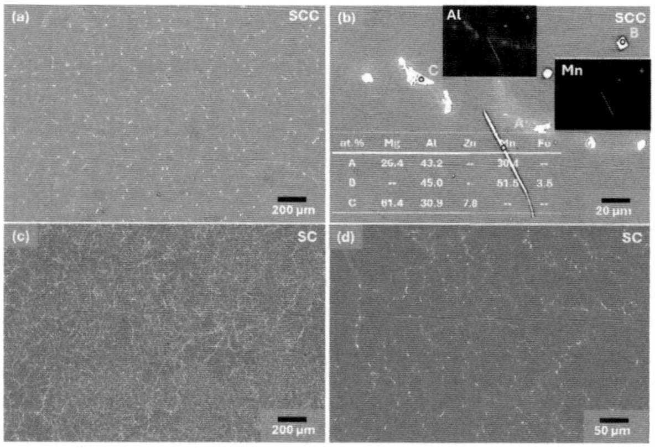

Fig. 2 Microstructure and elements distribution maps of various billets: (a and b) SCC, (c and d) SC.

After homogenization at 420 ℃ for 10 hours, the low-melting-point Mg-(Al, Zn) phase disappeared from both billets, as depicted in Fig. 3(a and b). After preheated at 380℃ for one hour, the Mg-(Al, Zn) phase also vanished in Fig. 3(c). But for the rapid preheating process at 380℃ for only 10 minutes, most of the them remained (Fig. 3(d)).

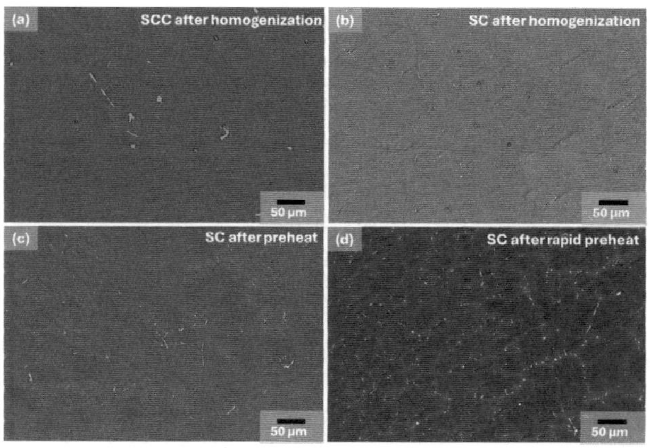

Fig. 3 Microstructure of various billets after different heat treatments: (a) SCC billet after homogenization, (b) SC billet after homogenization, (c) SC billet after preheating, and (d) SC billet after rapid preheating.

Figure 4 analyzes grain size evolution using EBSD. The SCC billet exhibits very coarse grains, with average grain size increasing by 31.4% after homogenization. The SC billet has finer matrix grains. After homogenization and preheating, itsgrain size coarsened by 20.8% and 6.2%, respectively. High-melting-point Al-Mn particles helped suppress coarsening after 10 hours at 420 °C, though it was still significant. Figure 4(f) shows that the grain size of the SC billet after rapid preheating barely increased.

Fig. 4 EBSD maps of various billets after different heat treatments: (a) SCC billet, (b) SCC billet after homogenization, (c) SC billet, (d) SC billet after homogenization, (e) SC billet after preheating, and (f) SC billet after rapid preheating.

3.2 Microstructure and mechanical properties

EBSD maps and tensile mechanical properties along the extrusion direction of extruded rods extruded from various billets are shown in Fig. 5 and Table 1. The rod extruded from the SC billet after rapid preheating shows thefinest microstructure with an average diameter of 5.1 μm.

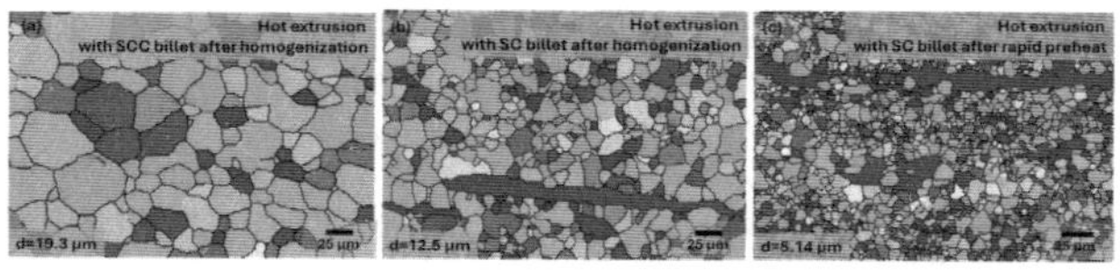

Fig. 5 EBSD maps along the extrusion direction of extruded rods extruded from various billets: (a) SCC billets after homogenization, (b) SC billet after homogenization, and (c) SC billet after rapid preheating.

Table 1 Tensile mechanical properties of extruded rods.

Billet	Heat treatment	UTS/MPa	YS/MPa	EL/%
SCC	Homogenization	259.9±4.4	168.6±2.8	13.8±1.2
SC	Homogenization	274.6±7.2	192.4±7.6	15.2±1.4
SC	Rapid preheating	296.4±5.7	220.8±8.4	15.8±0.7

Conclusion

(1) Both homogenization and preheating processes lead to grain size roughening and dissolution of low-melting-point phases, which is detrimental to subsequent hot extrusion.

(2) A novel rapid preheating method was developed to quickly raise the squeeze billet's temperature to the extrusion level within 10 minutes, preserving its initial microstructure. The solidified squeeze billet was buried in graphite particles at the extrusion temperature, eliminating the need to cool to room temperature. This significantly refined the microstructure of the AZ31 extruded magnesium alloy, enhancing its mechanical properties.

Understanding the Bronze Casting Techniques in the Late Shang Dynasty Through the Solidification Process Simulation of Anyang Niuzun

Huan Yang[1]*, Minghui Fang[2], Yuling He[3], Junchang Yang[1]
1. Institute of Cultural and Heritage, Northwestern Polytechnical University, 127 West Youyi Road, Xi'an, P.R.China
2. School of Materials Science and Engineering, 127 West Youyi Road, Xi'an, P.R.China
3. Institute of Archaeology Chinese Academy of Social Sciences, No.1 North National Stadium Road, Beijing, P.R.China
*Corresponding address: e-mail: yanghuan2019@nwpu.edu.cn

Abstract: This study investigates the advanced bronze casting techniques employed in the Shang Dynasty's capital city, focusing particularly on the cattle-shaped wine vessel (Niuzun) from Anyang. To elucidate the casting techniques used in the creation of the Niuzun, and thereby assess the bronze casting technologies of the Shang Dynasty, this paper integrated advanced solidification theory with cutting-edge research methodologies from materials science. A suite of techniques, including three-dimensional modeling, mesh division, and software simulation, was employed to reconstruct the dynamic solidification process of the vessel. Through this study, we not only elucidated the dynamic solidification process of the bronze Niuzun for the academic community and obtained precise simulation results, but also uncovered numerous previously unrecognized technological features, such as the bronze patches and chaplets. These findings offer a novel research perspective and methodology for understanding the bronze casting techniques of the late Shang period.

Keywords: Yazhangniuzun, Solidification Theory, Gating and Risers, Piece-mold Casting

1 Introduction

Approximately 3,000 years ago, Chinese artisans adopted the distinctive piece-mold casting to produce thousands of bronze vessels, thereby marking the advent of China's unique Bronze Age. The artifacts unearthed from sites such as the Yin Ruins (the capital city of the late Shang Dynasty) and Sanxingdui are particularly noted for their exquisite craftsmanship. Zun is a highly representative type of bronze vessels of this period[1]. In 2000, a cattle-shaped Zun(*Niuzun*) was unearthed from a high- ranking noble tomb at the Yin Ruins, the capital of the late Shang Dynasty (FIG. 1). The vessel measures 40 cm in length, 22.5 cm in height and weighs 7.1 kg [2]. The shape of the cattle statue is accurate and vivid, reflecting the exceptional bronze casting technology of the late Shang Dynasty. In order to accurately understand its casting process and further observe the bronze technology of the late Shang Dynasty, this study employed dynamic simulation methods of solidification process, accurately simulating the filling and solidification process of *Niuzun*, and obtained many understandings about the casting technique of it.

Fig. 1 Niuzun of Yazhang Tomb

2 Experimental procedure

The main research methods of this paper include the observation of the casting traces, the establishment of a 3-D model, meshing, parameter setting, solidification process simulation , etc., to study the casting process of the *Niuzun*. The main task of the observation step is to identify characteristics of casting process, such as chaplets, mold join lines, welding and other technical features, so as to determine the mold sectioning and other casting details of the object. Another major task in this stage is to locate the pouring gate and riser of the mold cavity. The setting of the gate and riser will have a significant impact on the quality and defects of the bronze[3][4].

The three-dimensional model of the bronze was obtained by scanning modeling of the *Niuzun*. After extracting the geometric information of the *Niuzun* , UG software was used for reverse modeling, resulting in a CAD model for simulation. The alloy composition of the *Niuzun* was set as $Cu_{89}Sn_{10}Pb_1$, with the thermophysical property parameters of the alloy materials automatically obtained by ProCAST's built-in material database. The thermophysical property parameters of the cavity and the interfacial heat transfer coefficient between the mold and metal were obtained through actual measurements and calculations of clay mold unearthed at the Zhuangli foundry in Zhouyuan of Shaanxi province (early Western Zhou Dynasty). After completing these steps, the parameters and models were imported into the simulation software(ProCAST) for dynamic simulation of the solidification process.

3 Result and discussion

3.1 THE OBSERVATION OF THE BRONZE

Through observation, it was found that the head and the horn are hollow. The head and the abdomen of the the vessel are separated by a thin bronze plate. The abdominal bottom is closed, and the foot cavities are not communicated. There is a square chaplet in the center of the two front feet, and a square-shaped inlaid bronze sheet can be seen on the wall below the head. Through the X-ray image, we found that there is another patch in the center of the bottom. The thickness of the patch is larger than the surrounding metal wall, with a clear edge, marking the first discovery of such a phenomenon in the bronzes of the late Shang Dynasty.

The four legs of the *Niuzun* are hollow, the edges of right forefoot and left hind foot are smooth, and the edges of left forefoot and right hind foot are thicker than the other two feet. The edge of the right hind foot is not smooth, showing traces of broken stubble, and the thickness of the wall connected with the body cavity is larger than the other three feet, which is speculated to play the role of the indirectly, so the right hind foot should be the pouring gate. A ring of copper liquid traces can be seen on the left front foot, which is seriously rusted and presumed to be a riser.

3.2 THE RESTORATION OF THE SOLIDIFICATION PROCESS OF NUNZUN AND THE OBSERVATION OF DEFECTS

In order to verify the previous inferred results of the sprue and riser, this paper simulated the solidification process of the *Niuzun* with the left front foot and the right rear foot as the pouring gate (shown by the arrow), and observed the differences in solidification time and defects between the two (Figure 2). The comparison revealed that when the right rear foot is used as the gate, the solidification time is shorter and the hole defects volume is minimized, which realizes the better performance of the object. This indicates that the craftsmen of the Shang Dynasty obviously made a better choice in the setting of the pouring gate of the *Niuzun*.

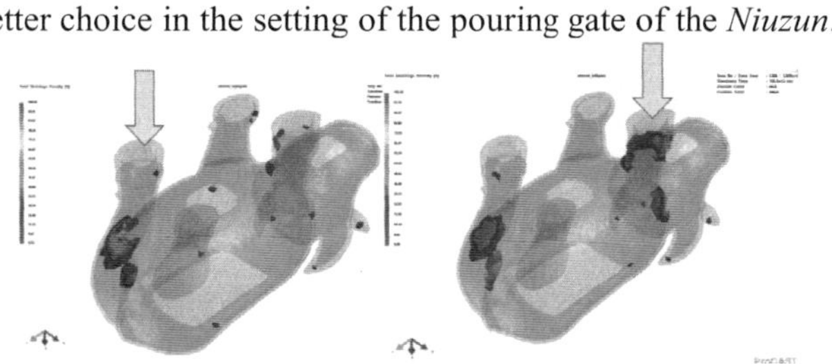

Fig. 2 Comparison of casting defects with different pouring gate settings

4 Conclusion

In this paper, the solidification research in materials science is employed to complete the reverse modeling based on the 3D scanning data of the objects. Through material assignment and finite element simulation of the bronzes and casts, the dynamic restoration of the solidification process of the *Niuzun* unearthed from Yazhang Tomb in the Yin Ruins is realized. The core data such as the solidification time of the objects are obtained, which provides an important reference for the follow-up research. The simulation results of the defects align well with the actual defect locations in the castings, which once again verifies the reliability and accuracy of the solidification simulation method.

It is found that the whole vessel of *Niuzun* adopts an equal wall thickness design, which significantly reduces defects and improves the quality of the objects[5]. After the observation of the object and the comparison of the casting simulation results, it is confirmed that the previous judgment that the pouring gate is the left rear foot is correct. The setting of the sprue and riser for specific objects by craftsmen in the Shang Dynasty achieved an optimal solution from the perspective of modern material science.
Combined with the observation of the casting process and solidification simulation of the artifacts, this research found that the craftsmen of the late Shang Dynasty skillfully employed casting techniques such as metal chaplets, bronze patches, equal wall thickness design, and casting riser settings when casting large-scale artifacts. These techniques ensured the complete filling of the bronze *Niuzun* and improved the quality of the artifacts. It is with these technologies that a large number of bronze vessels of the late Shang Dynasty were cast and their exquisite appearance made a unique bronze civilization in ancient China.

References

[1] Ronggeng. Rong Geng's comprehensive discussion on bronzes. Zhonghua Book Company, 2012:47.
[2] Institute of Archaeology, Chinese Academy of Social Sciences. Shang Dynasty Tombs in Yin Ruins Garden Zhuang in Anyang. Beijing: Science Press, 2007:278.
[3] Jie Wanqi et al. Casting Technology. Higher Education Press, 2013:300-302.
[4] Fan Jinhui, Wei Bing, Wang Feng. Systematic study on hot spot of castings. Foundry Technology, 2000 (03): 5-28.
[5] Yang Huan et al. Observation and research on the phenomenon of equal wall thickness of ancient Chinese bronzes under the view of solidification. Jianghan Archaeology,2022,(05):108-116+144.

Multiscale Simulation of the Effect of Axial Magnetic Field on Macrosegregation and Solidification Structure During Vacuum Arc Remelting of TC17

Junjie Li[1], Panyue Li[1], Jie Guo[1], Chenbao Xu[1], Jincheng Wang[1], Liqing Huang[2], Kai Fan[2]
1. State Key Laboratory of Solidification Processing, Northwestern Polytechnical University, Xi'an, China
2. Hunan Goldsky Titanium Industry Technology Co., Ltd., Changde, China
*Corresponding address: e-mail: lijunjie@nwpu.edu.cn

Abstract: The effects of axial magnetic field on the macrosegregation and microstructure in vacuum arc remelted ingots of titanium alloy TC17 are investigated by a multiscale modelling approach which combines macroscale transport model with a mesoscale cellular automation model. The results show that with the increase in the intensity or reversal time of the axial magnetic field, the enhanced melt swirl drives a stronger second flow in the vertical plane, which is downward along the sidewall of melting pool and upward in the centerline. When such flow in vertical plane evenly matches with the flow driven by the self-induced Lorentz force, the macrosegregation in the ingot is the weak. Moreover, it is found that the columnar grain structure changes to fine equiaxed grain with the increase in the in the intensity or reversal time of the axial magnetic field.

Keywords: Vacuum Arc Remelting, macrosegregation, solidification, grain structure.

1 Introduction

The Vacuum Arc Remelting (VAR) process is extensively used to produce high quality alloys such as titanium, zirconium and nickel-based alloys. During the VAR process an axial magnetic field (AMF) generated by external induction coils is deliberately introduced to confine the arc under the electrode. The interaction of AMF with the melting current gives rise to a stirring Lorentz force, which induces melt swirl in azimuthal direction. It is usually recognized that a unidirectional stirring will deteriorate the ingot quality. So the AMF is reversed periodically. The reversal time and intensity of AMF influence the melt flow and solidification, thus determine the quality of ingot. In recent years such topic has been explored by several numerical studies [1-5]. Spitans et. al. [1] and Karimi-Sibaki et. al. [2] performed simulations to reveal the complex flow behavior induced by the competition between the self-induced magnetic field and time-varying AMF. But the solute transport is absent in these works and thus there is no results about macrosegregation. Cui et. al. [3] and Han et. al. [4] detected the intensity of AMF on the macrosegregation during the VAR of Ni-based alloy and Ti_2AlNb alloy, respectively. They found that the application of AMF can reduce the macrosegregation. But with further increase in the intensity of AMF this effect becomes limited [3] or even results in more significant segregation [4]. It should be mentioned that the reversal time of AMF in these two studies are fixed. By using cellular automation (CA) simulation, the difference in grain structure between situations with and without AMF were revealed by Atwood [5]. Although these fantastic studies have given some insights into the effects of AMF, present understanding of the relations between the parameters of AMF and the ingot properties, such as macrosegregation and grain structure, are still limited. Here we investigate how the intensity and reversal time of AMF affect VAR ingot qualities by a multiscale modelling approach.

2 Experimental procedure

The model for macroscale transport was based on the continuum mixture model for alloy solidification, extended to incorporate electromagnetic field. Details could be found in our previous paper [6]. A 2D axisymmetric model was developed for the TC17 (Ti-1.97Zr-3.85Mo-3.9Cr) VAR ingot with diameter of 680mm and height of 1920mm. The temperature predicted by the macromodel was passed into the CA model for the grain growth simulation. The grid size and time step used in macroscale were 10mm and 0.1s, both of which were ten times larger than those in mesoscale. Therefore, the macroscale temperature field were bi-linear interpolated in space and linear interpolated in time to obtained the mesoscale distribution.

3 Result and discussion

The final distribution of Zr in the ingot in different AMF conditions are shown in Fig. 1. It can be seen from Fig. 1 (a)-(c) that the positive segregation in the center region decreases slightly as the intensity of AMF increases from 10 GS to 30 GS, and then with the further increase of AMF to 90G both the positive segregation in the center part and negative segregation in the periphery of the bottom region enhance. This founding is consistent with the reported results in reference [4]. The effect of reversal time can be found by comparing the results in Fig. 1(a), (d) and (e). It can be seen that the macrosegregation become seriously as the reversal time increases from 16s to infinite (no reversal of AFM). The segregation of Mo and Cr exhibits similar variation trend with the intensity and reversal time of AMF.

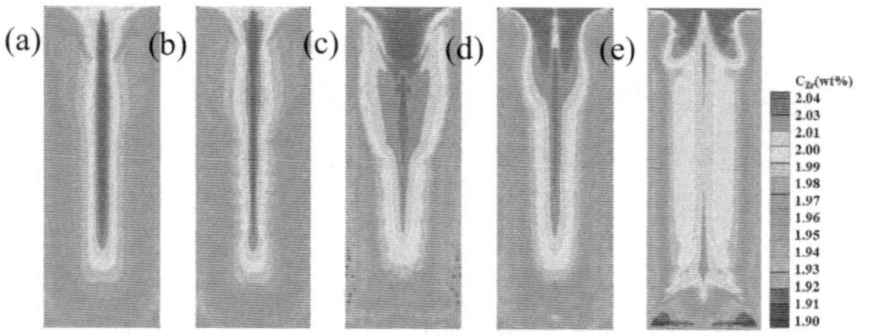

Figure1 Thel distribution of Zr in different AMF conditions (a)10GS-8s, (b) 30GS-8s , (c) 90GS-8s , (d) 30GS-32s , (e) 30GS-infinite

The macrosegregation is determined by melt flow, which is shown in Fig. 2. The left side of each figure is the velocity vector in the vertical plane and the right side is the distribution of swirl velocity. It can be found from Fig. 2(a) that the swirl velocity increases with the intensity of AFM. The enhanced melt swirl drives a stronger second flow in the vertical plane (the so-called Ekman pumping), which is downward along the sidewall of melting pool and upward in the centerline. When AFM gets to 30 GS, Ekman pumping is evenly match with the flow driven by the self-induced Lorentz force, thus the macrosegregation in the ingot is weak. The effects of AFM reversal time on the melting pool shape and flow can be seen in Fig. 2(b). The melt flow along sidewall is enhanced as the reversal time increases to 64s, thus more solute enriched liquid transports to the bottom of melting pool, which results in an enhanced segregation. When the reversal of AFM is stopped, the mushy zone of melt pool enlarges significantly, which results in more serious segregation.

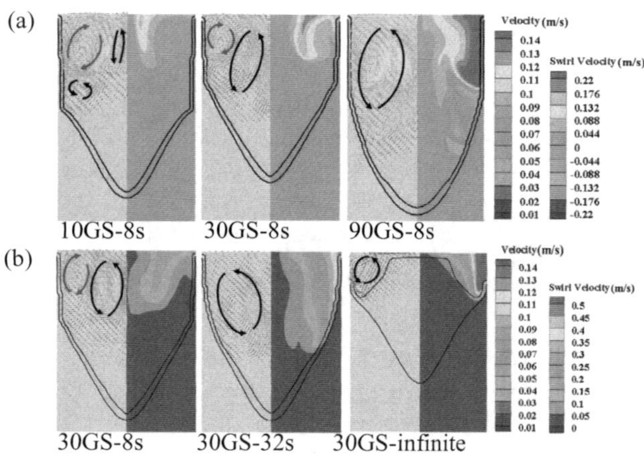

Figure2 Velocity vector(left) and swirl magnitude (right) for various (a) AFM intensity and (b) AFM reversal time.

The final grain structures in different AMF conditions are shown in Fig. 3. As can be seen, both the increase of AFM intensity and lengthening of reversal time can enlarge the equiaxed grain zone and refine grain structure.

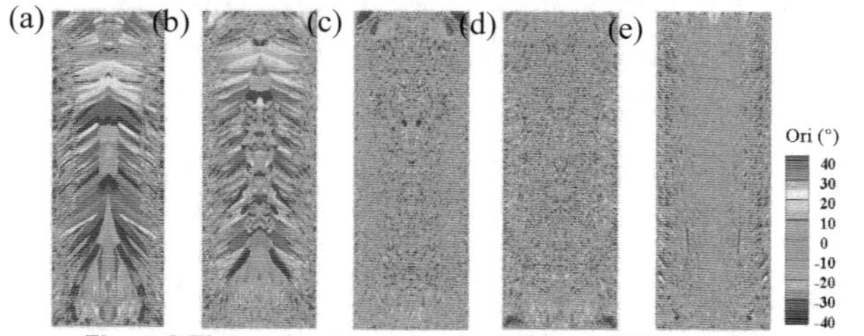

Figure3 The grain structures in different AMF conditions. (a)10GS-8s, (b) 30GS-8s , (c) 90GS-8s , (d) 30GS-32s , (e) 30GS-infinite

4 Conclusion

With the increasing in the intensity or reversal time of AFM, the swirl and Ekman pumping flow is enhanced, which can weaken the flow dirven by the self-induced Lorentz force. The macrosegregation is weak in moderate intensity of AFM, but is enhanced without reversal of AFM. The increase of Lorentz force and lengthening of reversal time can enlarge the equiaxed grain zone and refine grain structure.

5 Acknowledgments

This work was supported by the Research Fund of the State Key Laboratory of Solidification Processing (NPU), China (Grant No. 2023-TS-01).

References

[1] Spitans S, Franz H, Scholtz H, et al. Numerical simulation of the ingot growth during the Vacuum Arc Remelting (VAR) process. Magnetohydrodynamics, 2017,53:557-569.
[2] Karimi-Sibaki E, Kharicha A, Abdi M, et al. A Numerical study on the influence of an axial magnetic field (AMF) on vacuum arc remelting (VAR) process. Metall. Mater. Trans. B, 2021(5):3354-3362.
[3] Cui J, Li B, Liu Z, et al. Numerical investigation of segregation evolution during the vacuum arc remelting process of Ni-based superalloy ingots. Metals, 2021, 11(12): 2046.
[4] Han J, Ren N, Zhou Y, et al. Melt convection and macrosegregation in the vacuum arc remelted Ti2AlNb ingot: Numerical methods and experimental verification. J. Mater. Process Tech., 2022(308).
[5] Atwood R C, Lee P D, Minisandram R S, et al. Multiscale modelling of microstructure formation during vacuum arc remelting of titanium 6-4. J. Mater. Sci., 2004, 39 (24): 7193-7197.
[7] Guo J, Huang L, Wu J, et al. Evolution of macrosegregation during three-Stage vacuum arc remelting of titanium alloys. Acta Metall. Sin., DOI: 10.11900/0412.1961.2022.00544.

The Role of Minor Oxygen on Tensile Propertiesin Cast Ti-6Al-4V Alloy

Yuqing Song[1], Guodong Wang[1],Sisi Xie[1], Mingxiang Zhu[1], Chunhua Yue[2], Hongchao Kou[1*]
1State Key Laboratory of Solidification Processing, Northwestern Polytechnical University, Xi'an, Shaanxi 710072, China
2 Beijing Xinghang Electromechanical Equipment Co. Ltd，Beijing，100071
*Corresponding address: e-mail: hchkou@nwpu.edu.cn

Abstract: In this paper, the effect of oxygen content in the range of 0.15 ~ 0.44 wt.% on the mechanical properties at room temperature of cast Ti-6Al-4V alloy was studied, and the mechanism of oxygen effect on its tensile deformation behavior was discussed. The results show that with the increasing of oxygen content, the strength of the alloy increases,while the elongation increases first and then decreases.It is found that oxygen solution in α-Ti leads to obvious change of lattice parameter c/aratio.The activation of prismatic slip of α-Ti will bring oxygen atoms from octahedral site to hexahedral site to reduce stacking fault energy. The reason why the strength of cast Ti-6Al-4V alloy increases and the ductility decreases with the increasing of oxygen content is that the lattice distortion caused by the solid solution of oxygen inα-Ti prevents the dislocation motion, while the abnormal ductility increasing at 0.22 wt.% oxygen content is due to the dislocation shuffling oxygen atoms into the hexahedral site.

Keywords: Cast Ti-6Al-4V Alloy,Oxygen Content, Mechanical Properties, Deformation Mechanism.

1 Introduction

Ti-6Al-4V alloy has good comprehensivemechanical properties and casting performance, is the most widely used cast (α + β) two-phase titanium alloy[1], oxygen is generally as interstitial atoms into theα-Ti lattice[2], play a hindering role to dislocation motion, with solid solution strengthening effect[3].But the mechanism by which the oxygen affects the mechanical properties is still lack of complete understanding. For this reason, this paper investigates the effect of oxygen content on the mechanical properties of Ti-6Al-4V alloy, and explores the essential mechanism of the effect of oxygen on the tensile properties of cast Ti-6Al-4Valloys.

2 Experimental procedure

In order to study how oxygen affected the mechanical properties of Ti-6Al-4V alloy, by changing the proportion of TiO_2 powder in the raw materials, 16mm Ti-6Al-4V casting rods with four oxygen contents were cast by cold crucible suspension smelting equipment, and then were subjected to hot isostatic pressing (HIP) at 920℃/130MPa/3h, named 0.15O, 0.22O, 0.26O and 0.44O respectively, according to the measured oxygen content of 0.15 wt.%, 0.22 wt.%, 0.26 wt.% and 0.44 wt.% . Tensile mechanical properties at room temperature were tested on an electronic universal material testing machine with a tensile rate of 0.45mm/min. The lattice constants of the samples were determined by selected area electron diffraction (SAED) using FEI TALOS F200X transmission electron microscope. Calculation result was based on the calculation principles of density functional theory and usedthe Vienna Ab initio Simulation Package (VASP) for calculations. The exchange correlation energy was represented by the PBE correlation functional potential in the generalized gradient approximation (GGA).

3 Result and discussion

Effect of oxygen on mechanical properties

Fig. 1(a) shows the stress-strain relationships of Ti-6Al-4V alloys with four oxygen contents at room temperature. It can be seen that there is almost no difference in yield and hardening conditions, which indicates that oxygen does not change its deformation mechanism in essence.

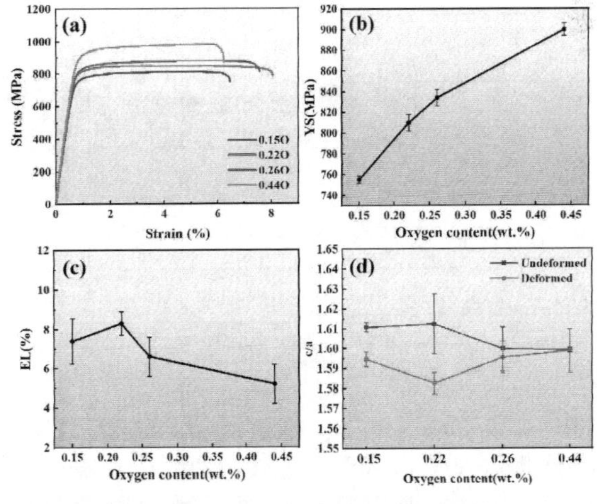

Fig. 1(a) Stress-strain relationships of Ti-6Al-4V alloys with four oxygen contents; (b, c) Changes of average yield strength and elongation with oxygen content; (d) The c/a ratio of the α phase before and after tensile deformation

Fig. 1(b) and (c) show the changes of average yield strength and elongation with oxygen content. With the increasing of oxygen content, thestrength of the alloys are linearly and positively correlated with oxygen content, which is obviously attributed to solution strengthening. The elongation decreases with the increasing of oxygen content, but the linear correlation is not big. It can be seen that the elongation does not decrease in accordance with the law in the range of 0.15 ~ 0.22 wt.% oxygen content, but increases slightly. This is different from the effect of single solution strengthening mechanism, so there must be another mechanism on ductility besides solution strengthening.We calculate the c/a ratio of theα phase before and after deformation by analyzingselected area electron diffraction,in Fig. 1(d).It can be seen that the c/a ratio of the α phase after deformation is smaller than before, and the decreasing degree is positively related to the ductility. It is speculated that the change of c/a ratio is related to the change of oxygen atom position during deformation and acts on ductility.

The role of oxygen in deformation mechanism

Interstitial energies at different sites of solid solution of oxygen in titanium were calculated by first-principles calculation through formula (1).The results are shown in Table 1. When oxygen is at octahedral site, the energy is the lowest, which is the most stable occupying site for oxygen.Energy of oxygen in hexahedral site is greater than that in octahedral site, which means that the hexahedral site is metastable occupying site.By relaxingthe α-Ti unit cells without oxygen and with oxygen at octahedral and hexahedral sites respectively, we find that oxygen increases the c/a ratioat octahedral site and decreases the c/a ratio at hexahedral site.

$$E_I = E_{Ti+O} - E_{Ti} - E_O \quad (1)$$

Table 1. Interstitial energy of oxygen in α-Ti

Position of oxygen	octahedral site	hexahedral site
Interstitial energy/eV	-8.85061	-7.6163

The prism plane of the α-Ti unit cells without oxygen and with oxygen at octahedral and hexahedral site is cut and the vacuum layer is established, and the slip is carried out with $\frac{1}{3}[11\bar{2}0]$ as Berger vector as shown in Figure 2(a). Through first-principles calculation, the stacking fault energy of prismatic slip is calculated at five points in a complete slip. The relationship between stacking fault energy and slip displacement is drawn in Fig. 2(b). It can be seen that the stacking fault energy of prismatic slip increases when oxygen at the stable octahedral site and decreases when oxygen at the metastable hexahedral site. Oxygen mainly occupies the octahedral site before deformation. The c/a ratio should increase gradually with the oxygen content, but the internal stress produced in casting and hot isostatic pressing produces a small number of dislocations in α phase,which shuffle oxygen from stable octahedral site to metastable hexahedral site with lower prismatic slip stacking fault energy. So that c/a ratio does not increase strictly with that increasing of oxygen content. After the deformation began, a large number of dislocations were produced in the lattice. Oxygen atoms were shuffled to hexahedral site, c/a ratio decreases.

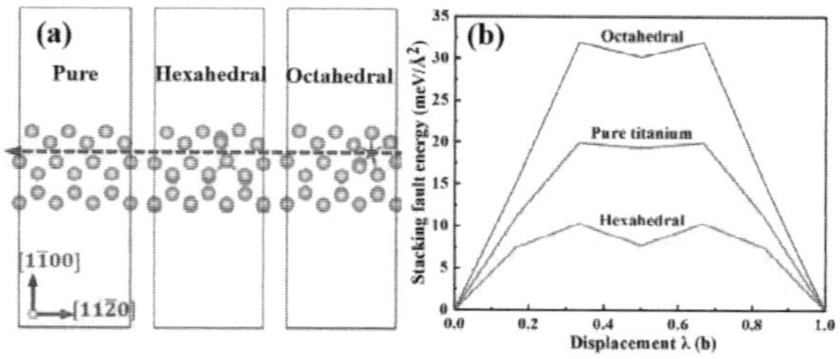

Fig. 2(a) prismatic slip of hexagonal α-Ti unit cells, without oxygen and with oxygen in hexahedral site and octahedral site; (b) The relationship between stacking fault energy and slip displacement.

The prismatic slip stacking fault energy decreases, the ductility problem caused by solid solution strengthening is alleviated.However, with the increasing of solid solubility of oxygen, the lattice distortion becomes larger, which hinders the dislocation activation and the dislocation shuffling oxygen atoms into the hexahedralsite. The c/a ratio decreasing degree decreases, and the ductility alleviation is inhibited.

Conclusion

The activation of prismatic slip shuffles theoxygen atoms from the octahedral to the hexahedral sites in order to reduce the stacking fault energy, exhibiting a decrease in c/a ratio and an alleviation of the damage to ductility.

Acknowledgments

This project was financially supported by Shaanxi Provincial Innovation Capacity Support Plan (2023-CX-TD-4), the State Key Laboratory of Solidification Processing(SKLSP202201) and ND Basic Research Funds, NPU (G2022WD).

References

[1] Leyens C, Peters M. Titanium and Titanium Alloys: Fundamentals and Applications[M]. Wiley-VCH Verlag GmbH & Co KGaA: Weinheim, 2003. 1-36.

[2] Liu C, Cui JZ, Cheng ZY, et al. Direct Observation of Oxygen Atoms Taking Tetrahedral Interstitial Sites in Medium-Entropy Body-Centered-Cubic Solutions[J]. Advanced materials, 2023, 35(13): e2209941.

[3] Shi XZ, Wang XX, Chen B, et al. Precision control of oxygen content in CP-Ti for ultra-high strength through titanium oxide decomposition: An in-situ study[J]. Materials & Design, 2023, 227: 111797.

Characteristics of α Precipitated Phase and Tensile Properties of Cast Ti5553 Alloy

Guodong Wang[1], Yuqing Song[1], Mingxiang Zhu[1], Chunhua Yue[2], Xiangyi Xue[1], Hongchao Kou[1]*
1State Key Laboratory of Solidification Processing, Northwestern Polytechnical University, Xi'an, Shaanxi 710072, China
2 Beijing Xinghang Electromechanical Equipment Co. Ltd，Beijing，100071
*Corresponding address: e-mail: hchkou@nwpu.edu.cn

Abstract: Ti5553 (Ti-5Al-5V-5Mo-3Cr) not only has high strength, but also has good mould filling fluidity, which is a typical high strength casting titanium alloy. The mechanical properties and deformation behavior of Ti5553 alloy mainly depend on the coupling effect of α phase characteristics and residual β matrix. The current research mostly focuses on the deformed alloy and the dependence of the mechanical properties of cast Ti5553 alloy on α phase characteristics is still unclear. In this paper, the room temperature tensile deformation behaviour of cast Ti5553 alloy and its relationship with α precipitated phase characteristics under different heat treatment process are investigated. The results show that: the HCP+FCC structure within the α lamellae of the alloy in the hot isostatic state activates numerous slip systems and the coordinated deformation of α twins and FCC phases leads to yield strength, tensile strength, and elongation of 955 MPa, 1020 MPa, and 10%, respectively; After solid solution aging, a large number of fine acicular α phases precipitate on the β matrix, and the alloy has high strength but very low ductility; the double annealed state alloy has the best strength-ductility match, and the yield strength, tensile strength, and elongation reach 1020 MPa, 1080 MPa, and 8%, respectively. The multiscale α lamellae intralaminar slip shear and α twins are the structural roots of their good strength-ductility.

Keywords: Cast Ti5553 alloy, α phase characteristics, Tensile properties, Deformation behavior

1 Introduction

Metastable β titanium alloys have great potential for application in fields such as aerospace and biomedicine due to their high specific strength, good fatigue resistance, and excellent processing performance[1,2]. Ti5553 alloy is expected to become a typical high-strength cast titanium alloy. The mechanical properties and deformation behavior of Ti5533 alloy actually depends on the characteristics of the α precipitated phase and the coupling of residual β phase. They are highly sensitive to heat treatment processes. In recent years, the strength-ductility of metastable β alloys is mostly improved by heat treatment to obtain the multiscale α phase[3]. Currently, most of the related studies focus on deformed alloys, and the dependence of the mechanical properties of cast Ti5553 alloy on the characteristics of α precipitated phases is not clear. Therefore, in this paper, the room temperature tensile deformation behaviour of cast Ti5553 alloy under different heat treatment process and its relationship with α precipitated phase characteristics are investigated.

2 Experimental procedure

The cast Ti5553(Ti-5.27Al-5.15V-5.25Mo-3.16Cr-0.22O) alloy used in this study, was prepared through investment casting. To address the hole defects in the casting process, a hot isostatic pressing treatment (HIP,820 °C/130 MPa/2 h) was applied to the Ti5553 alloy cast ingot. Three kinds of heat treatment processes, annealing (700 ℃/2 h/AC), double annealing (865 ℃/30min→600 ℃/4 h/AC) and solid solution ageing (880 ℃/30 min/AC+600 ℃/4 h/AC), were carried out on the HIP-Ti5553 alloy.

3 Result and discussion

Microstructure and mechanical properties

The average thickness of the α lamellae in the HIP-Ti5553 alloy is 0.74±0.20 μm, and there is no fine acicular α phase on the β matrix. The α lamellae within the β matrix are slightly coarsened after annealing. After double annealing, β matrix precipitated lamellar α with "fibrous root morphology"[4] and fine acicular α phases. The average thickness of the α lamellae reaches 1.16±0.20 μm, and the thickness of the fine acicular α-phase is about 150~250 nm, and the length is 0.75~1.25 μm. After solid solution aging, a large number of V-shaped or triangular fine acicular α_s phases precipitated in the β matrix The average thickness of the fine acicular α_s is below 100 nm, and the length is about 0.1-0.3 μm. The mechanical properties of the alloys in different states are shown in Fig.1. Annealed alloys have the highest elongation but low strength; the alloys after solid solution ageing have the highest strength but almost brittle fracture; the alloys in the double-annealed state had the best strength-ductility match, with yield strength, tensile strength and elongation up to 1020 MPa, 1080 MPa and 8%, respectively.

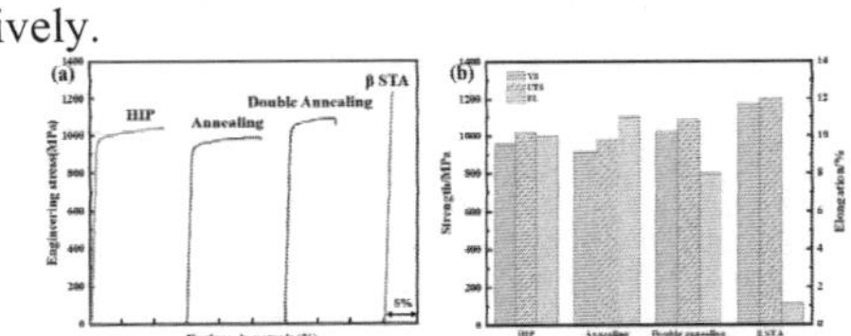

Fig. 1 Comparison of tensile stress-strain curves (a) and properties (b) for cast Ti5553 alloy after different

heat treatment process

4 Deformation behaviour

In order to further analyze the reasons for the good matching of strength-ductility of the HIP-Ti5553 and double annealed Ti5553, the near-fracture microstructure was analyzed by TEM. Fig.2 shows the microstructural characterization of the HIP-Ti5553 with different strains. Due to the high temperature and high pressure environment provided by the hot isostatic pressure, there are a small number of α twins and FCC phases in the α lamellae of the HIP-Ti5553 before tensile deformation. A large number of dislocations and slip bands can be seen in both the α phase and the β matrix when the tensile strain is 2%, in which the α phase is surrounded by neatly arranged misfit dislocations. When the tensile strain is increased to 6%, not only a large number of dislocations are found within the α phase of HIP-Ti5553 alloy, but also parallel distribution of α twins are observed. When the alloy is stretched to fracture, tiny twins of interconnected and zigzag shapes appear within the α phase, and there is a high density of stacking fault within some of the α twins and the α matrix between α twins. The α twins and stacking fault can form the Lomer- Cottrell lock[5]. The initially existing FCC phase deformed during tensile process, resulting in FCC twins. The HCP+FCC structure within the α lamellae activates numerous slip systems and the coordinated deformation of α twins and FCC phases resulting in an ideal strength- ductility match.

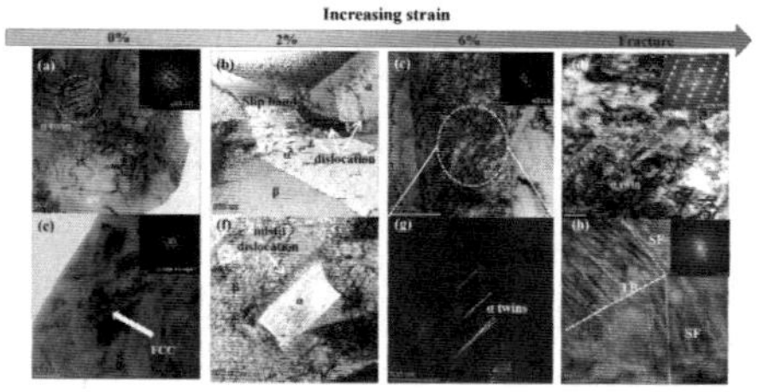

Fig. 2 TEM images of HIP-Ti5553 with different tensile strains:(a,e) 0%;(b,f) 2%;(c,g) 6%;(d,h) Fracture

Fig.3 shows the near-fracture microstructure of the double annealed alloy, which can be clearly seen that there are a small number of twins in the fine acicular α phase, and the coarse α lamellae has obvious slip shear bands and dislocations, which also verifies that the α phase with the "fibrous root morphology"[4] has good deformation ability, and there are high-density dislocations near the fine acicular α. The "fibrous root morphology" α and a small number of α twins ensure the ductility of the alloy, and the distribution of fine acicular α to dislocations in a vertical or parallel manner can better impede the alloy's strength improvement.

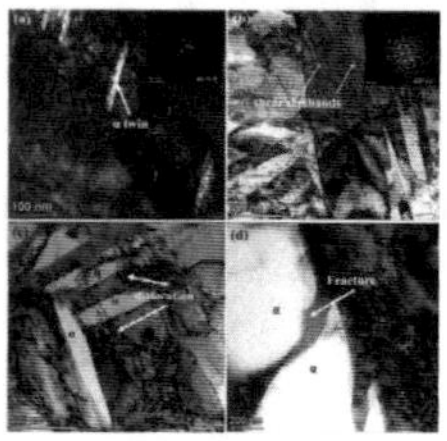

Fig. 3 TEM images of near fracture surface of cast Ti5553 alloy after double annealing: (a) α twins; (b) shear slip bands (c) fine acicular dislocations near α phase; (d) Fracture of α phase

5 Conclusion

After double annealing, the cast Ti5553 alloy has a best strength- ductility match by precipitating multi-scale α phase and inducing α twins with room-temperature tensile yield strength and elongation up to 1020 MPa and 8%.

6 Acknowledgments

This project was financially supported by the National Natural Science Foundation of China (U21A2050), Shaanxi Provincial Innovation Capacity Support Plan (2023-CX-TD-4), the State Key Laboratory of Solidification Processing (SKLSP202201) and ND Basic Research Funds, NPU (G2022WD).

References

[1] P. Singh, H. Pungotra, N. Kalsi. On the characteristics of titanium alloys for the aircraft applications[J]. Materials Today: Proceedings, 2017, 4(8):8971-8982.
[2] C. Cui, B.M. Hu, L. Zhao, et al. Titanium alloy production technology, market prospects and industry development[J]. Materials and Design, 2011, 32(3):1684-1691.
[3] W. Zhu, J. Lei, C. Tan, et al. A novel high-strength β-Ti alloy with hierarchical distribution of α-phase: The superior combination of strength and ductility[J]. Materials and Design, 2019, 168: 107640.
[4] T. Wang, H. Tang, Y. Zhu, et al. Laser additive manufacturing of new α+β titanium alloy with high strength and ductility[J]. Journal of Materials Research and Technology, 2023, 26:7566-7582
[5] Z. Lu, Z.H. Hu, L. Zhang, et al. Enhancing the strength-ductility trade-off in a NiCoCr-based medium-entropy alloy with the synergetic effect of ultra fine precipitates, stacking faults, dislocation locks and twins[J]. Scripta Materialia, 2022: 211114497.

Corrosion Behavior of Ti-6Al-*x*V-2Sn-0.5Cu-*y*Fe Alloys

Chaowen Zheng, Qi Wang*, Ruirun Chen
National Key Laboratory for Precision Hot Processing of Metals, School of Materials Science and Engineering, Harbin Institute of Technology, Harbin 150001, China
*Corresponding address: e-mail:wangqi_hit@hit.edu.cn

Abstract: Due to the high cost of smelting Ti alloys, a low-cost titanium alloy for marine engineering was designed in this study. The expensive V element in Ti-6Al-6V-2Sn-0.5Cu-0.5Fe (TC10) was replaced by the low cost Fe element which was the same as the V element as the β stable element. The corrosion behavior of Ti-6Al-xV-2Sn-0.5Cu-yFe alloys in strong reducing acid (5M HCl) was studied. It was found that although the corrosion resistance of titanium alloy decreased after replacing V with Fe, the corrosion resistance decreased little and the price of alloy raw materials was greatly reduced when only 1.5 wt.% Fe was used to replace 1.5 wt.% V in TC10.

Keywords: Ti alloy; Corrosion behavior; Low cost

1 Introduction

Titanium alloys have the characteristics of high corrosion resistance and high specific strength [1-3]. Therefore, it is also called marine metal. With the rapid development of marine engineering technology, the demand for improving the corrosion resistance of titanium alloys is becoming more and more urgent. However, due to the activity of titanium element, titanium alloy can only be melted in vacuum environment. Moreover, the preparation of titanium sponge is difficult. These two reasons lead to the high melting cost of titanium alloy. V element and Fe element are β-stable elements for Ti alloy. If the expensive V element can be replaced by cheap Fe element, the cost of Ti alloys will be greatly reduced. In this study, the corrosion resisting property of Ti-6Al-*x*V-2Sn-0.5Cu-*y*Fe ($x + y$=6.5, x=6, 4.5, 3, 1.5, 0) alloys in strong reducing acid (5M HCl) was studied. This study provides a method for designing Ti alloys with high corrosion resistance and low cost for offshore engineering.

2 Experimental procedure

The purity of melting raw materials used in the preparation of titanium alloy was more than 99%. Ti alloy was melted by vacuum non-consumable arc furnace. The nominal composition of the melted Ti alloy was Ti-6Al-*x*V-2Sn-0.5Cu-*y*Fe ($x + y = 6.5$, $x = 6$, 4.5, 3, 1.5, 0). They were referred to as 0.5Fe, 2Fe, 3.5Fe, 4Fe, 6.5Fe. The electrochemical experiment was carried out in a three-electrode system with the help of an electrochemical workstation. The test solution was 5M HCl.

3 Result and discussion

Fig. 1 shows the polarization curves of five alloys. From Fig. 1, the anodic branches and the cathodic branches of the polarization curves of the five alloys in 5M HCl solution are similar. It shows that five alloys have similar reactions in electrolyte solution. The anode branches of the five alloys are divided into three stages: active dissolution region, active to passive transition region and passive region. In active dissolution region, Ti-6Al-*x*V-2Sn-0.5Cu-*y*Fe alloys were active dissolved. The main chemical process involved was that Ti element changed to Ti^{3+} ion. In the active to passive transition region, the following reactions are dominant [4]. In the passivation region, an oxide film was formed on the surface of the five titanium alloys. In this region, the current density remained basically unchanged as the potential increased.

$$(TiOH)+\ ads \leftrightarrow (TiOH)2+\ ads+e^-$$

$$H_2O+(TiOH)2+\ ads \leftrightarrow [Ti(OH)_2]2+\ ads+H^++e^-$$

$$H_2O+(TiOH)2+\ ads \leftrightarrow [Ti(OH)_2]^{2+}+H^++e^-$$

$$[Ti(OH)_2]2+\ ads \rightarrow [Ti(OH)_2]^{2+}$$

$$[Ti(OH)_2]2+\ ads \leftrightarrow TiO_2+2H^+$$

It can be seen from 6.5Fe that a certain polarization voltage is required to change from active to passive transition region to passive region. This shows that the passivation film formed on the surface of 6.5 Fe alloy is not stable. When the polarization voltage is about 1V, it can be seen that the perforation corrosion potential is observed in 5Fe alloy but no perforation corrosion potential was observed in other alloys. These indicate that the passivation films of 5Fe and 6.5Fe alloys have a weak ability to protect the surface.

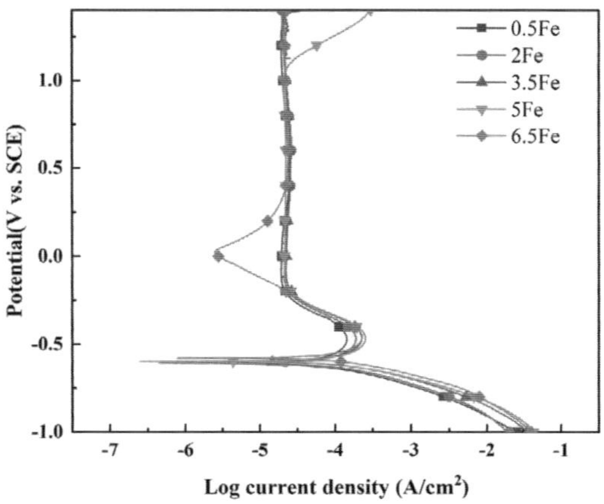

Figure 1. Polarization curves of Ti-6Al-xV-2Sn-0.5Cu-yFe alloys

Because Ti alloy is a common passivation metal. Therefore, the Tafel polarization curves of the five alloys show the characteristics of passive metals, so we use the Tafel extrapolation method to calculate the corrosion potential and corrosion current from the cathode branch.

As shown in Tab. 1, As the Fe element gradually replaces the V element, the corrosion current of the alloy in HCl gradually increases, and the corrosion potential generally shows a gradual increase trend. In general, from the perspective of corrosion kinetics, with the increase of Fe content, the corrosion resistance of the alloys in HCl is getting worse and worse. However, if only 1.5wt. % Fe element is used to replace the V element in the alloy, the corrosion resistance does not decrease much, but its composition decreases a lot. The corrosion current increased by about 36% when only 1.5 wt.% Fe was used instead of V. However, if the Fe content was increased by more than 3.5 wt.%, the corrosion current increased by about 68%.

Table 1. E_{corr} and i_{corr} of Ti-6Al-*x*V-2Sn-0.5Cu-*y*Fe in 5M HCl

	0.5Fe	2Fe	3.5Fe	5Fe	6.5Fe
E_{corr} (mV)	-0.601	-0.605	-5.79	-5.88	-5.78
i_{corr} (μA×cm^{-2})	0.072	0.098	0.121	0.119	0.123

In general, the Fe element is used to replace the V element in the titanium alloy, which reduces the corrosion resistance of the alloys. The reason for this phenomenon is mainly related to the segregation of Fe element in Ti alloys [5]. In the region where Fe element is enriched, the potential difference between the potential of this region and the potential of the alloy matrix increases. This phenomenon greatly increases the galvanic corrosion in the five alloys and increases the corrosion rate of the five alloys.

4 Conclusion

The use of cheap Fe element to replace the expensive V element in TC10 titanium alloy makes the corrosion resistance of titanium alloy worse. However, if only 1.5wt.% Fe element is used instead of V element, the corrosion performance of the alloy will not decrease much. Replacing the V element with 1.5 wt.% Fe element only increases the corrosion current of the alloy in 5M HCl by 36%. if the Fe content was increased by more than 3.5 wt.%, the corrosion current increased by about 68%.

5 Acknowledgments

References

[1] Y.Geng, Y. Ban, X. Li, et al. Excellent mechanical properties and high electrical conductivity of Cu-Co-Si-Ti alloy due to multiple strengthening. J. Material Science and Engineering: A, 2021, 821: 141639.

[2] C. Yi, Y. Yuan, L. Zhang, et al. Antibacterial Ti-35Nb-7Zr-xCu alloy with excellent mechanical properties generated with a spark plasma sintering method for biological applications. J. Journal of Alloys and Compounds. 2021, 879: 160473.

[3] R.W. Schutz and H.B. Watkins, Recent developments in titanium alloy application in the energy industry. J. Material Science and Engineering: A, 1998, 243: 305–315.

[4] Utomo W B and Donne S W. Electrochemical behaviour of titanium in H_2SO_4-$MnSO_4$ electrolytes. J. Electrochimica Acta. 2006, 51(16): 3338-3345.

[5] P. Zhao, Y. Song, K. Dong, et al. Corrosion behavior of dual-phase Ti–6Al–4V alloys: a discussion on the impact of Fe content. J. Journal of Alloys and Compounds. 2021, 858: 157708.

Effect of Heat Treatment on the Microstructure and Mechanical Properties of Cast Mg-5Y-3Nd-3Zn-0.6Zr Alloy

Haojie Wang[1], Yicheng Feng[1,2*],Tao Wang[1],Yuanke Fu[1], Lei Wang[1], Erjun Guo[1]
1. School of Materials Science and Chemical Engineering, Harbin University of Science and Technology, Harbin 150040, China
2. Key Laboratory of Advanced Manufacturing and Intelligent Technology, Ministry ofEducation, Harbin University of Science and Technology, Harbin 150080, China
*Corresponding address: e-mail:fyc7806067@163.com

Abstract: The microstructure evolution and mechanical performance of cast,solid solution, and aging Mg-5Y-3Nd-3Zn-0.6Zralloywere systematically investigated by the scanning electron microscope (SEM), X-ray diffractometer (XRD), transmission electron microscope (TEM), differential scanning calorimeter (DSC) and electronic universal testing machine.in this work. The microstructure analyses show that the cast Mg-5Y-3Nd-3Zn-0.6Zr alloy consist of α-Mg, LPSO phase and W phase. After the solid solution of 510 ℃ × 12h,the reticular W phase can be transformed into spherical W phase. The precipitation sequences of the alloy aged at 210℃ is β''(DO19)→β'(BCO)→β(FCC), and the streaks sequences of the alloy is SF → 14H-LPSO. The peak-agingoccurred at 16h due to the high volume fraction of β' phase. The mechanical properties test shows that the tensile strength of cast and solid solution alloys are 196.3MPa and 209.5MPa, the elongation of cast and solid solution alloys are 5.8% and 7.1%.

Keywords: Mg alloy, microstructure, mechanical properties, heat treatment, heat resistant property

1 Introduction

Magnesium alloy is one of the lightest metal structural materials, so the replace of aluminum alloy and steel by magnesium alloy in aircrafts and cars can greatly reduce weight to save fuel. However, the application of magnesium alloy in engine shell, gearbox shell and other high temperature components puts forward higher requirements for the high temperature performance of magnesium alloy.

Alloying is the general method for optimizing the phase composition to enhance the mechanical performance at elevated temperature of magnesium alloys. The Mg-Y-Zn alloy contains LPSO phase has aexcellent mechanical properties because it can hinder the dislocation movement and rise the ductility by kink deformation. The addition of Nd in Mg-Y-Zn alloy can further improve the strength because it can reduce the solid solubility of Y and Zn to enhance the precipitationstrengthening effect. So it is essential to clarify the effect mechanism of Nd content on the microstructure evolution, aging precipitationbehaviour, and mechanical performance of Mg-Y-Zn alloy.

2 Experimental procedure

The Mg-5Y-3Nd-3Zn-0.6Zr alloy was prepared by pure Mg (Mg ≥ 99.9 wt.%), pure Zn (Zn≥99.9 wt.%), Mg-30Y(29.5wt.%Y) master alloy, Mg-30Nd (29.3wt.%Nd) master alloy, and Mg-30Zr (30.8wt.%Zr) master alloy in well-type resistance furnace. The solid solution of the alloy is 510℃ for 12h and the aging treatment was 210℃ for 1-100h.

The metallographic specimens were ground, polished, and etched, and then were observed by SEMand tested by XRD. The specimens with a diameter of 3mm were polished by ion beam and observed by TEM. The mechanical performance of alloy wastested by a universal tensile testing machine, and the diagram of tensile specimenwas showed in Fig. 1.

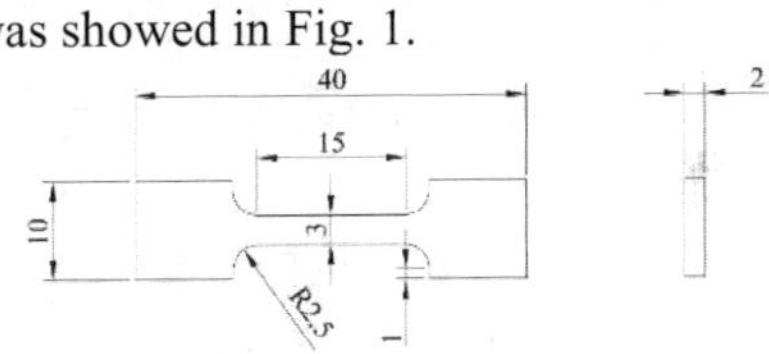

Fig.1 The diagram of tensile specimen

3 Result and discussion

Microstructure of cast WEZK5330 alloy

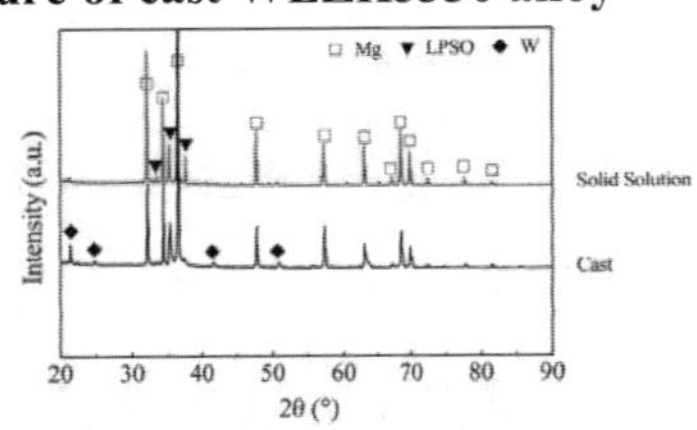

Fig. 2 The XRD of WEZK5330 alloy

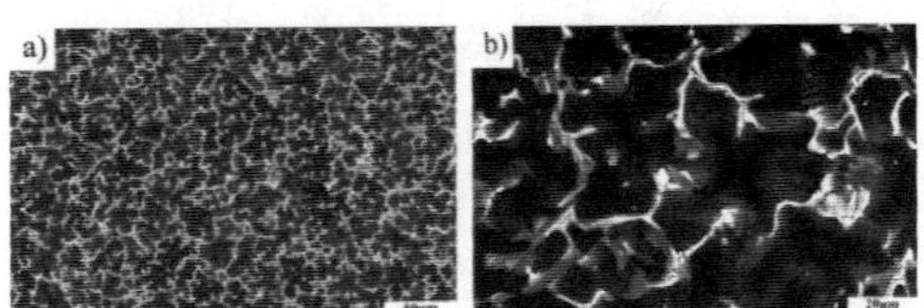

Fig. 3 The SEM image of cast WEZK5330 alloy

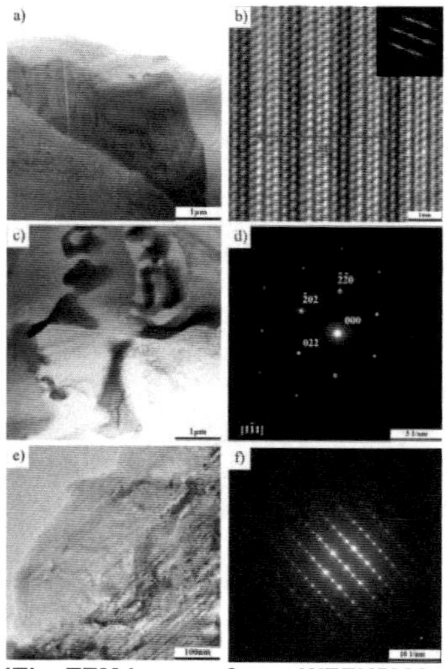

Fig.4The TEM image of cast WEZK5330 alloy

Microstructure of solid solution WEZK5330 alloy

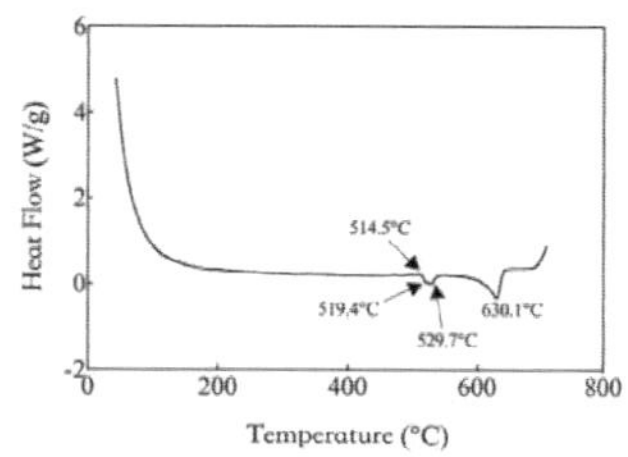

Fig. 5The DSC of cast WEZK5330 alloy

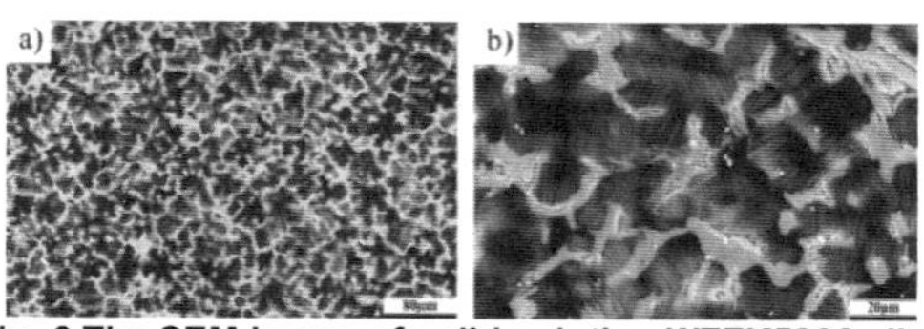

Fig. 6 The SEM image of solid solution WEZK5330 alloy

Microstructure of aging WEZK5330 alloy

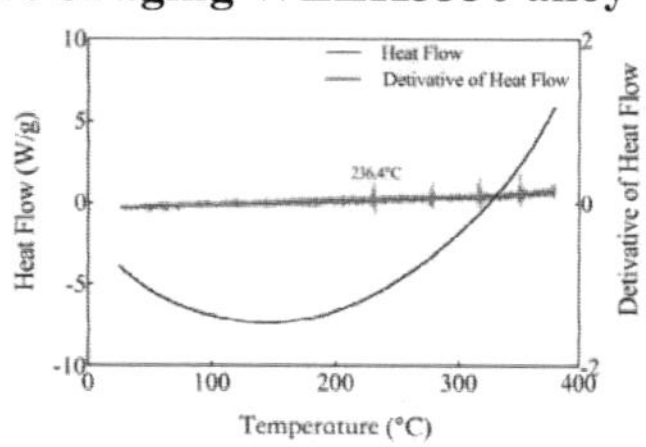

Fig.7The DSC of solid solution WEZK5330 alloy

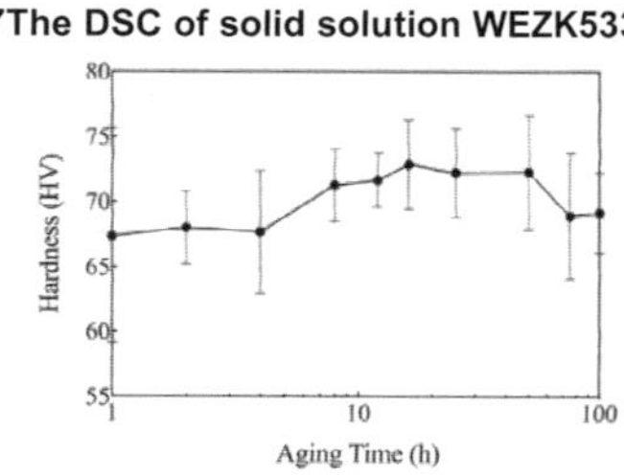

Fig. 8 The aging hardness of WEZK5330 alloy

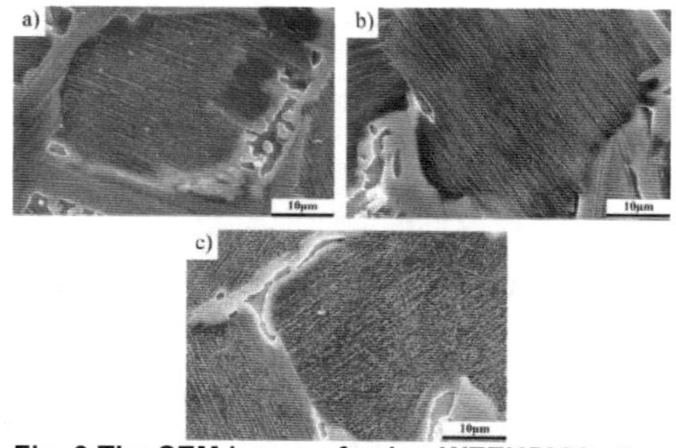

Fig. 9 The SEM image of aging WEZK5330 alloy

Mechanical Properties of WEZK5330 alloy

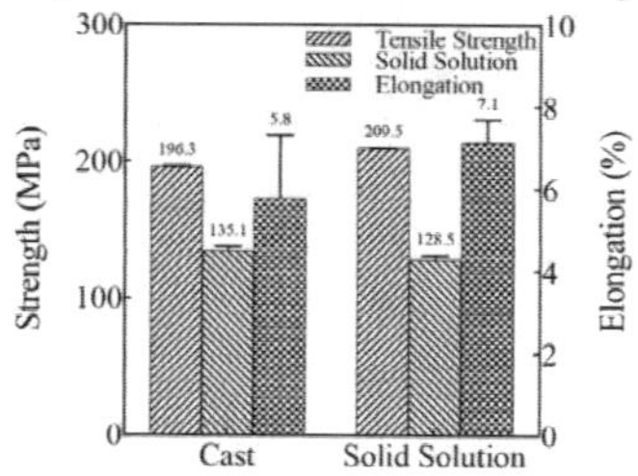

Fig. 10Themechanical propertiesof WEZK5330 alloy

4 Conclusion

1. The cast Mg-5Y-3Nd-3Zn-0.6Zr alloy consist of α-Mg, LPSO phase and W phase.

2. After the solid solution of 510℃ × 12h, the reticular W phase can be transformed into spherical W phase.

3. The precipitation sequences of the alloy aged at 210℃ is β''(DO19)→β'(BCO)→β(FCC)

4. The mechanical properties test shows that the tensile strength of cast and solid solution alloys are 196.3MPa and 209.5MPa, the elongation of cast and solid solution alloys are 5.8% and 7.1%.

References

[1] LUO A A. Recent magnesium alloy development for elevated temperature applications [J]. International Materials Reviews, 2004, 49(1): 13-30.

[2] HORT N, HUANG Y-D, KAINER K U. Intermetallics in magnesium alloys [J]. Advanced Engineering Materials, 2006, 8(4): 235-40.

[3] BAZHENOV V, SAIDOV S, TSELOVALNIK Y V, et al. Comparison of castability, mechanical, and corrosion properties of Mg-Zn-Y-Zr alloys containing LPSO and W phases [J]. Transactions of Nonferrous Metals Society of China, 2021, 31(5): 1276-90.

Effect of Cu Addition and Heat Treatment on the Microstructure and Properties of ZL114A

Zhao Na, Fuwei Kang*, Enhao Wang, Linlin Zhang, Bingpeng Zhang
School of Materials Science and Engineering ,Harbin University of Science and Technology, Building #6, Lin-Yuan ST.4 Harbin, China
*Corresponding address: e-mail: fuwei_kang@163.com

Abstract: This research investigated the impact of varying Cu content and heat treatment on the microstructure and mechanical properties of ZL114A using optical microscope, scanning electron microscope (SEM), transmission electron microscope (TEM), differential scanning calorimeter (DSC), Vickers hardness tester, and room temperature tensioning machine. The findings revealed that the mechanical properties of ZL114A could be enhanced by incorporating different Cu content, with the optimal effect observed at 1.2% Cu content. The hardness post heat treatment was measured at 143.1HV, marking a 13.9% increase, while the tensile strength reached 326MPa, reflecting a 33.4% improvement. The introduction of Cu did not alter the microstructure of ZL114A. Furthermore, heat treatment transformed the eutectic Si morphology from sharp long rod shape to elliptic shape, fractured the dendrites, eliminated stress concentration at the eutectic Si edge, and ultimately enhanced the mechanical properties.

Keywords: Cast aluminum; Eutectic silicon; Grain size; Heat treatment

1 Introduction

ZL114A is a cast Al-Si alloy, which is a significant series o f cast aluminum alloys. Many Al-Si alloys have a relatively low crystallization temperature and contain numerous eutec tic components in their structure. This series of aluminum al loys offers several advantages, such as low density and high strength. Additionally, Al-Si aluminum alloys exhibit stron g casting and mechanical properties, making them suitable f or various casting methods and the production of products with diverse characteristics like different shapes, thin walls, high corrosion resistance, strong air tightness, or high temp erature resistance, making them widely utilized. Aluminum-silicon casting alloys with high toughness generally demons trate excellent performance. After undergoing metamorphic treatment and heat treatment, they exhibit favorable physica l properties, mechanical properties, and processing properti es. These alloys are primarily used for intermediate parts' sh ell structures and load-bearing components with complex sh apes, finding extensive applications in aerospace, transporta tion, and other industries. Copper serves as one of the prima ry strengthening elements in aluminum alloys. The addition of copper significantly enhances the tensile strength and har dness of the alloy by forming Al_2Cu through a reaction with aluminum. This second phase contributes mainly to the soli d solution strengthening and dispersion strengthening of the matrix. Therefore, incorporating copper and subjecting the alloy to solution and aging treatment can notably enhance it s tensile strength and hardness.

2 Experimental procedure

ZL114A alloy was prepared using metal mold casting and s melted in a crucible furnace at 750℃. Once the alloy was c ompletely melted, an Al-50Cu intermediate alloy was added. Various amounts of Cu ranging from 0% to 1.5% were add ed. Following the dissolution of the intermediate alloy, 1% Al-5Ti-B was added for refining and held for 20 minutes. S ubsequently, 1% C_2Cl_6 was used for refining and degassing, and the ingot was obtained through slagging after a 10-min ute hold. The chemical composition of the alloy is detailed i n Table 1. The alloy underwent solution and aging treatmen ts consecutively, with a quenching temperature of 60℃ and a transfer time of less than 10 seconds. The solution and agi ng treatments were conducted in the SX2-2.5-10A Muffle f urnace and the 101A-1 aging furnace, respectively. The hea t treatment process parameters were set at 540℃×12h+16 0℃×8h (T6 treatment). Mechanical properties of each samp le were assessed using a tensile test machine, while the micr ostructure was examined using an FEI Quanta scanning elec tron microscope and analyzed via energy spectrum.

Table 1. Composition of Alloy ZL114A %%

Mg	Si	Cu	Ti	Be	Zn	Al
0.7	6.25	0~1.5	0.18	0.07	0.08	Bal.

3 Result and discussion

Subheads 1

The mechanical properties of ZL114A alloy in T6 state with varying Cu content are presented in Table 2. The data illust rates that the addition of Cu enhances the alloy's mechanical properties. Specifically, with a Cu content of 0.3%, there is minimal impact on the tensile strength of the ZL114A alloy. However, as the Cu content increases to 0.9%, the tensile st rength significantly rises to 325 MPa. At 1.2% Cu content, t he alloy exhibits peak mechanical properties, with a Vickers hardness of 143.1 HV. In comparison to no Cu presence, th e tensile strength increases by 13.9% to 326 MPa, a 33.4% i mprovement, with negligible changes in elongation. Beyon d 1.2%, at 1.5% Cu content, a decrease in mechanical prope rties occurs, likely due to the solubility limit of Cu in the all oy.

Table 2. Mechanical properties of ZL114A alloy T6 state without Cu concent

Cu%	Tensile strength MPa	Hardness HV
0	293	124.3
0.3	301	130.1
0.6	310	132.8
0.9	325	133.6
1.2	326	143.1
1.5	321	133.7

Subheads 2

The cast ZL114A exhibits a coarse dendritic microstructure, with the eutectic Si appearing as long rods with sharp edge s. The addition of Cu has minimal impact on the morpholog y of the eutectic Si. Through heat treatment, the long rod Si can be transformed into fine ellipses, effectively reducing st ress concentrations at the edges of the eutectic Si and enhan cing the mechanical properties of the alloy.

4 Conclusion

With increasing Cu content, the mechanical properties of th e alloy improve. At a Cu content of 1.2%, the alloy exhibits maximum Vickers hardness of 143.1HV, a 13.9% increase i n hardness, and a tensile strength of 326MPa, representing a 33.4% increase. Additionally, the elongation remains relati vely unchanged. Furthermore, the addition of Cu shows min imal impact on the morphology of eutectic Si.

5 Acknowledgments

The authors gratefully acknowledge the financial supports o f this study by the Major Project of Heilongjiang Province (2020ZX03A03).

References

[1] Liang J Y , Hou X W .Effect of Si Content of Matrix Alloy on the Microstructure and Properties of Sip/Al-Si-Mg Composites Prepared by Pressure Infiltration Method[J].Materials science forum, 2017, 898(pt.1):992-999.

[2] Kobayashi T .Strength and Fracture of Aluminium Alloys[J].Materials Science Forum, 2007, 426-432:67-74.

[3] J. H ,Li,M,et al.Solute adsorption and entrapment during eutectic Si growth in A–Si-based alloys[J].Acta Materialia, 2015.

[4] Li J H , Barrirero J , Engstler M ,et al.Nucleation and Growth of Eutectic Si in Al-Si Alloys with Na Addition[J].Metallurgical & Materials Transactions A, 2015, 46(3):1300-1311.

[5] Dahle A K , Nogita K , Zindel J W ,et al.Eutectic nucleation and growth in hypoeutectic Al-SI alloys at different strontium levels[J].Metallurgical and Materials Transactions A, 2001, 32(4):949-960

Microstructure Evolution of the Al-Zn-Mg-Cu Alloy During Solidification under Electric Pulse

Kaijiao Kang[1], Shengsong Wang[1], Shunde Jiang[1], Dayong Li[1,2], Yicheng Feng[1,2*]
1. School of Materials Science and Chemical Engineering, Harbin University of Science and Technology, Harbin 150040, China
2. Key Laboratory of Advanced Manufacturing and Intelligent Technology, Ministry of Education, Harbin University of Science and Technology, Harbin 150080, China
*Corresponding address: e-mail: fyc7806067@163.com

Abstract: In the non-equilibrium solidification process of Al-Zn-Mg-Cu alloy, the distribution of temperature field and flow field is not uniform, which leads to coarse solidification structure and serious component segregation. It is an effective means to further improve the properties of the alloy by refining the solidification structure and improving the segregation degree of the alloy with electric pulse. The effects of pulse current, pulse frequency and pulse application stage on solidification microstructure and properties of Al-Zn-Mg-Cu alloy were studied. The results show that the grain refinement effect and properties of Al-Zn-Mg-Cu alloy solidified under the effect of electric pulse are improved to varying degrees. With the increase of pulse current and pulse frequency, the columnar crystals gradually change to fine equiaxed crystals. The average grain size decreased from 140.3 μm to 102.9 μm after treatment with pulse current of 700 A and pulse frequency of 50 Hz. The tensile strength of the alloy was increased to 210.5 MPa, which is 20.6 % higher than that of conventional solidification.

Keywords: Al-Zn-Mg-Cu alloy, Electric pulse, Microstructure, Mechanical properties

1 Introduction

In the non-equilibrium solidification process of Al-Zn-Mg-Cu alloy, the distribution of temperature field and flow field is not uniform, which leads to coarse solidification structure and serious component segregation. It limits its application. Due to the unique advantages of physical fields such as electric pulse on the microstructure evolution of alloys, it has shown broad application prospects in the field of materials science and engineering, and has attracted extensive attention of foundry workers.

A.K. Misra [1] applied a small pulse current to the solidification process of Pb-15Sb-7Sn alloy with low melting point for the first time, and the microstructure of the alloy was significantly refined after solidification. Cui [2] introduced high density electric pulse during the solidification of LY12 alloy, and found that the larger the current applied, the finer the grain of the alloy after solidification, and the degree of segregation at the grain boundary of the alloy was improved. The optimal electrical pulse treatment parameters and mechanism of different kinds of alloys are not clear. In this paper, the effects of pulse current and pulse frequency on the solidification microstructure and mechanical properties of Al-Zn-Mg-Cu alloy were studied by using control variable method.

2 Experimental procedure

The size of the metal mold is Φ86×150 mm, and two copper electrodes were embedded on the top of the mold. One end of the electrode was inserted into the mold, and the other end was connected with the positive and negative electrodes of the pulse power supply. The molten Al-Zn-Mg-Cu metal liquid was poured into the mold of the same size, and the pulse power supply was turned on. The specific experimental parameters of the electric pulse were determined: the pulse current of 200 A, 700 A, 1200 A, the pulse frequency of 20 Hz, 30 Hz, 50 Hz.

The solidification microstructure and fracture morphology of Al-Zn-Mg-Cu alloy were analyzed by scanning electron microscope (FEI Quanta 200F) and energy dispersive spectrometer. The enthalpy of the alloy after different treatments was measured by a differential scanning calorimeter (METTLER DSC3+), and the dissolution of the phase was analyzed. In the DSC test, the weight of the sample was 15 mg, which was protected by argon. The test temperature ranges from room temperature to 700°C at a rate of 10°C/min. The tensile test was carried out by the MTS-E44304 electronic universal testing machine and the tensile speed of 1 mm/min. In order to ensure the accuracy of tensile test, the average value of each sample was measured four times.

3 Result and discussion

The microstructure of Al-Zn-Mg-Cu alloy under conventional solidification is mainly composed of columnar crystals and a few equiaxed crystals, as shown in Figure 1. A large number of non-equilibrium eutectic structures are distributed on the α-Al matrix in a continuous dendrite network structure. The non-equilibrium eutectic structure composed of Al, Zn, Mg and Cu elements has severe segregation at the grain boundaries. The average grain size of Al-Zn-Mg-Cu alloy under conventional solidification is 140.3 μm.

When the pulse frequency is 20 Hz, the number of equiaxed crystals increases with the increase of pulse current, and the average grain size decreases to 107.2 μm when the pulse current is 700 A. Similarly, when the pulse current is 700 A, the coarse columnar crystals become fine equiaxed crystals with the increase of pulse frequency, and the refinement effect is remarkable. The average grain size decreased from 140.3μm to 102.9μm after the treatment with pulse current of 700 A and pulse frequency of 50 Hz.

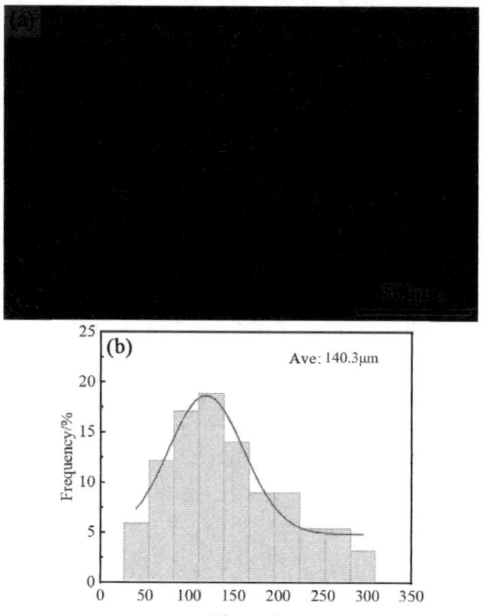

Figure 1. Microstructure of Al-Zn-Mg-Cu alloy under conventional solidification: (a) SEM, (b) average grain size

When the electrical pulse applied during the solidification of Al-Zn-Mg-Cu alloy, the rapidly changing pulsed electric field will generate a pulsed magnetic field in the melt. Due to the interaction of pulsed current and pulsed magnetic field, the metal melt moves in the direction perpendicular to the current, and the columnar crystal will be broken during the movement, and the broken columnar crystal will be used as a new crystal nucleus, which improves the nucleation rate of the alloy. In addition, since the resistivity of liquid Al-Zn-Mg-Cu alloy is higher than that of solid alloy, the current will preferentially select the solid to pass through, and the Joule heat generated by the current segregation at the solid-liquid interface hinders the growth of columnar crystals [3]. The schematic diagram is shown in Figure 2.

The tensile strength and elongation of the alloy after conventional solidification are 174.5 MPa and 4.7%, respectively. When the pulse frequency is 20 Hz and the pulse current is 700 A, the tensile strength and elongation of the alloy are 204.6 MPa and 8.1%, respectively. Compared with the untreated alloy, the tensile strength increased by 17.2%. The tensile strength of the alloy increases to 210.5 MPa when the pulse current is 700 A and the pulse frequency is 50 Hz. Compared with conventional solidification, the tensile strength of the alloy is significantly increased by 20.6 %. In summary, the application of electric pulse during the solidification process of the alloy can significantly improve the mechanical properties of the alloy and improve the microstructure uniformity of the alloy.

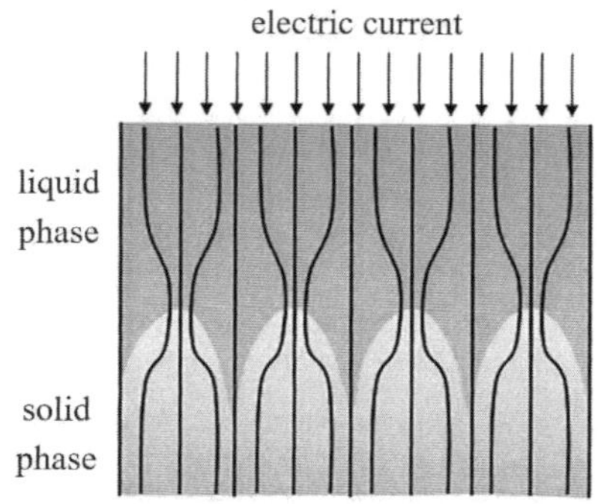

Figure 2. The schematic diagram of current segregation at the solid-liquid interface

4 Conclusion

The microstructure of Al-Zn-Mg-Cu alloy under conventional solidification is mainly composed of columnar crystals and a few equiaxed crystals. The tensile strength and elongation of the alloy after conventional solidification are 174.5 MPa and 4.7%, respectively. The grain refinement effect and mechanical properties of Al-Zn-Mg-Cu alloy solidified under electric pulse are improved to varying degrees. With the increase of pulse current and pulse frequency, the columnar crystals gradually change to fine equiaxed crystals. After electric pulse treatment with pulse current of 700 A and pulse frequency of 50 Hz, the average grain size is reduced from 140.3 μm to 102.9 μm, and the tensile strength of the alloy is increased to 210.5 MPa, which is 20.6 % higher than that of conventional solidification.

5 Acknowledgments

The authors acknowledge the financial support from the Key R&D Projects in Heilongjiang Province (2022ZX04A02).

References

[1] MISRA A K. A Novel solidification technique of metals and alloys: under the influence of applied potential[J]. Metallurgical Transactions A, 1985, 16(7): 1354-1355.
[2] Zi B T, Cui J Z, Ba Q X, et al. Solidification structure of LY12 aluminum alloy under high density pulse current [J]. Special Casting and Nonferrous Alloys, 2000, (4): 4-6.
[3] KOSTANOVSKIY A V, KOSTANOVSKAYA M E. Relationship between force and heat flux in the electric pulse heating of a metal conductor[J]. High Temperature, 2020, 58(5): 747-749.

Effect of Cu and Mg Content on Microstructure and Mechanical Properties of Cast Al-Si-Cu-Mg Alloy

Ning Zhang[1], Yicheng Feng[1,2], Yuanke Fu[1], Lei Wang[1], Erjun Guo[1]
1. School of Materials Science and Chemical Engineering, Harbin University of Science and Technology, Harbin 150040, China;
2. Key Laboratory of Advanced Manufacturing and Intelligent Technology, Ministry of Education, Harbin University of Science and Technology, Harbin 150080, China
*Corresponding address: e-mail: 18800449534@163.com

Abstract: The wide application of traditional cast Al-Si alloy in related industries is seriously restricted because of its low room temperature strength and high temperature performance attenuation. By adding Cu and Mg elements, the effect of Cu and Mg element content on the type and morphology of the second phase in cast Al-8.5Si-xCu-yMg alloy was systematically studied, and its mechanical properties were tested. The results show that when the content of Mg is constant, with the addition of Cu, blocky Q-Al5Cu2Mg8Si6 phase and reticular θ-Al2Cu phase appear in the alloy. When the Cu content is constant, with the increase of Mg content, the size and number of reticular and blocky θ-Al2Cu phases in the alloy decrease, while the blocky Q-Al5Cu2Mg8Si6 phase changes from a small blocky phase to a large flake phase, and the number increases. The tensile strength of Al-8.5Si-2Cu-0.9Mg at room temperature and 250 ℃ is 335.3 MPa and 170.3 MPa, respectively. The yield strength is 254.6 MPa and 147.2 MPa, respectively. The elongation after breaking is 4.0% and 8.8%, respectively.

Keywords: Cast Al-Si alloy, Cu, Mg, Aging treatment, Microstructure, Mechanical properties

1 Introduction

Cast Al-Si-Cu-Mg alloy is an alloy developed by adding Cu and Mg elements on the basis of cast Al-Si alloy. Besides eutectic Si phase, there may exist the second phases such as θ-Al_2Cu and Q-$Al_5Cu_2Mg_8Si_6$ in the alloy. The types, morphology and distribution of these second phases will have an impact on the mechanical properties of the alloy at room temperature and high temperature [1]. The addition of Cu element in Al-Si alloy can improve the microstructure of the alloy, thereby increasing its strength and contributing to its high temperature mechanical properties. Therefore, the existence forms of these two Cu-rich phases in alloys can be effectively regulated by adjusting the content of Cu element, which is of great significance for the research of high-performance Al-Si-Cu-Mg alloys [2].

The content of Mg element determines the content, morphology and distribution of Q-$Al_5Cu_2Mg_8Si_6$ phase in the alloy. Compared with the θ phase, the Q phase can improve the strengthening effect of the alloy more significantly [3]. Therefore, based on the optimal Cu content, the content of Mg element in the alloy can be adjusted to further improve the quantity, morphology and distribution of the second phase in the alloy [4]. Furthermore, the mechanical properties of the alloy at room temperature and high temperature can be improved, which is of great significance for the expansion of the practical application scenarios of casting Al-Si-Cu-Mg alloy.

2 Experimental procedure

The test alloy was prepared by metal type gravity casting method in the paper. The well resistance furnace was used to melt the alloy. The mold used for smelting was metal mold, the crucible was graphite crucible, and the total mass of the raw material was 1000 g. When melting, it was necessary to put the slag scoop, bell jar, crucible tongs, metal mold and other experimental equipment and raw materials into the oven and heat up to 200℃ for 2 hours. First, when the well resistance furnace was heated to 730℃, industrial pure aluminum and Al-Si intermediate alloy were placed in the graphite crucible. After being kept warm until completely melted, the temperature of the resistance furnace was raised to 750 ℃, and Al-40Cu intermediate alloy, industrial pure magnesium, Al-10Sr modifier and Al-5Ti-1B refiner were added to the graphite crucible in turn. After complete melting, the temperature of the resistance furnace was lowered to 730℃, and the refining was carried out with C2Cl6. After the refining was completed, the slag was scraped. After standing for 10 minutes, the metal liquid was poured into a metal mold preheated to 200℃. After the ingot was completely cooled, the test alloy was obtained in the same position and a part of it was subjected to T6 heat treatment. The nominal composition of the alloy was shown in Table 1.

Table 1. Nominal composition of alloy

Alloy	Si	Cu	Mg	Al
1#	8.5	0	0.7	Bal
2#	8.5	2	0.7	Bal
3#	8.5	2	0.9	Bal

3 Result and discussion

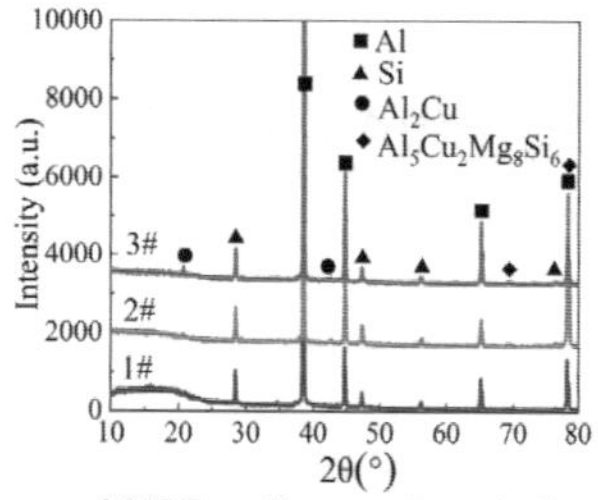

Figure 1 XRD patterns of cast alloys.

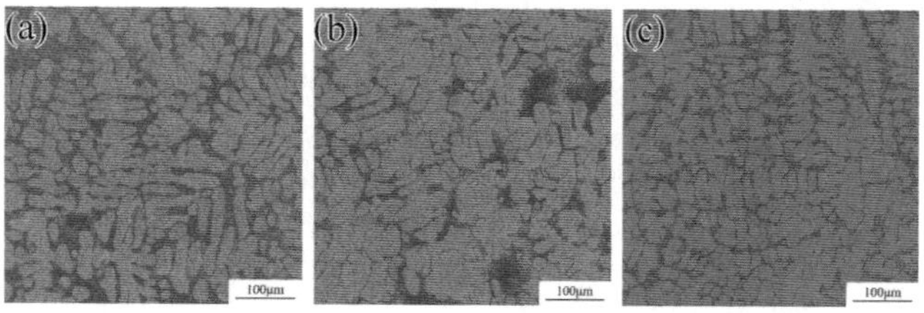

Figure 2. Optical micrographs of alloys in as-cast state (a) 1# (b) 2# (3) 3#.

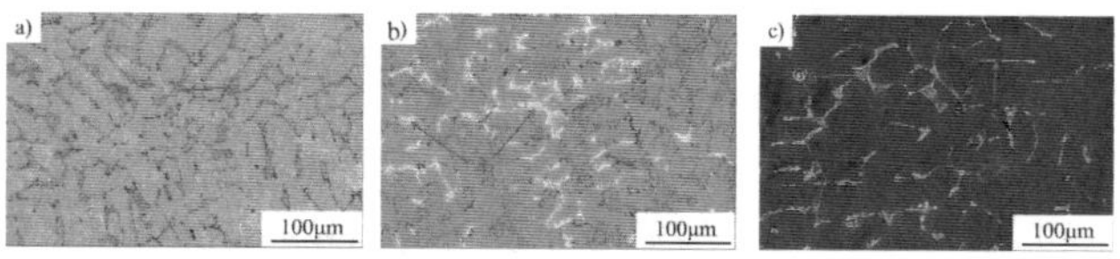

Figure 3. SEM images of as-cast alloys (a) 1# (b) 2# (3) 3#.

Table 2. Mechanical Properties of Al–Si–Cu–Mg Alloy After T6 Heat Treatment

Alloy	Room temperature			250℃		
	UST/ MPa	UST/ MPa	Elongation/%	UST/ MPa	UST/ MPa	Elongation/%
1#	154.6	72.6	14.5	79.3	53.3	20.4
2#	320.3	248.7	4.4	168.1	144.4	9.1
3#	335.3	254.6	4.0	170.	147.2	8.8

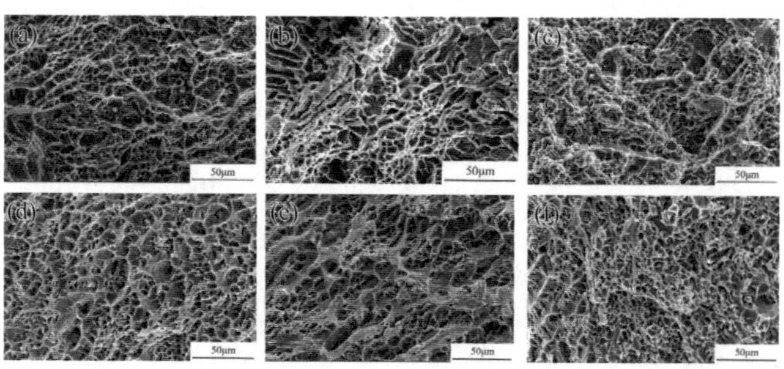

Figure 4. SEM micrographs of fracture surface of alloys tensile specimens (a) 1#, (b) 2#, (c) 3#, (d) 1# 250℃, (e) 2# 250℃, (f) 3# 250℃,

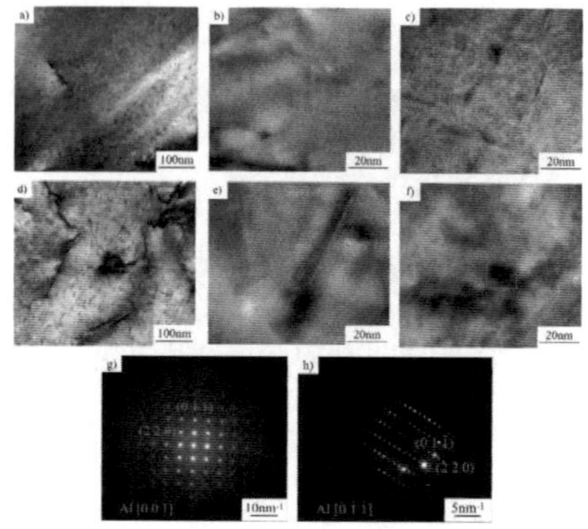

Figure 1TEM images and SAED pattern (a)-(c) 175℃ 5h, (d)-(h) 175℃ 8h.

4 Conclusion

(1)The microstructure of cast Al-Si-Cu-Mg alloy is mainly composed of α-Al grains, eutectic Si phase, θ-Al_2Cu phase and Q-$Al_5Cu_2Mg_8Si_6$ phase.

(2)With the addition of Cu, the bulk Q-$Al_5Cu_2Mg_8Si_6$ phase and the network θ-Al_2Cu phase appear in the alloy. With the increase of Mg content, the size and number of θ-Al_2Cu phase in the alloy decrease, and the morphology of the light-colored Q-$Al_5Cu_2Mg_8Si_6$ phase changes from a small block to a large sheet, and the number increases.

(3)When the content of Cu and Mg elements is 2% and 0.9% respectively, the alloy achieves the best mechanical properties. At room temperature, the tensile strength, yield strength and elongation of the alloy are 335.3 MPa, 254.6 MPa and 4%, respectively. At 250 °C, the tensile strength, yield strength and elongation of the alloy are 170.3 MPa, 147.2 MPa and 8.8%, respectively.

5 Acknowledgments

This work was supported by Key R&D Projects in Heilongjiang Province (2022ZX04A02), Reserve leader of the leading talent team in Heilongjiang Province.

References

[1] Bi J, Liu L and Zhang D S. Research progress of casting, rapid solidification and additive heat-resistant aluminum alloys. J. Transactions of Nonferrous Metals Society of China, 2023, 33(04): 969-996.

[2] Qbau N, Nam N D and Hien N T. Development of Light Weight High Strength Aluminum Alloy for Selective Laser Melting. J. Journal of Materials Research and Technology, 2020, 9(6): 14075-14081.

[3] Zuo L J, Ye B and Feng J. Phases Formation and Evolution at Elevated Temperatures of Al-12Si-3.8Cu-2Ni-1Mg Alloy. J. Advanced Engineering Materials, 2017, 19(3): 1600623.

[4] Zheng Y, Xiao W L and Ge S J. Effects of Cu Content and Cu/Mg Ratio on the Microstructure and Mechanical Properties of Al-Si-Cu-Mg alloys. J. Journal of Alloys and Compounds, 2015, 649: 291–296.

Effect of Directional Solidification on the Thermal and Mechanical Properties of Al-Cu Alloy

Xu Liu, Wentao Jiang, Hongyu Xu, Bo Jiang, Ye Wang*, MaoliangHu, Bo Yu
School of Materials Science and Chemical Engineering, Harbin University of Science and Technology, Harbin 150040, Heilongjiang, China
* Corresponding authors: ohenry1980@163.com (Y Wang)

Abstract: With the increasingly serious global resource and environmental problems and the continuous growth of energy-saving demand, the heat dissipation performance, mechanical properties and lightweight degree of thermal conductive materials directly determine the integration degree and working stability of devices. Aluminum alloy not only has ideal thermal conductivity and mechanical properties, but also has low density. This paper focuses on the influence of directional solidification on the microstructure and thermal conductivity of ZL205 A alloy.

By studying the effect of directional solidification on the microstructure and phase composition of ZL205A alloy, the results show that the phase composition of ZL205A alloy does not change significantly under normal pressure, and the microstructure is still a typical divorced eutectic of α-Al phase + θ-Al_2Cu phase. After the lateral heat preservation measures are applied, the lateral heat dissipation is inhibited. Due to the inhibition of the nucleation of the grains in the direction perpendicular to the directional solidification direction, the size of the α-Al matrix phase increases, the chilling zone becomes narrower and the columnar crystal zone is wider.

By testing the mechanical properties of the alloy under different directional solidification conditions, the results show that the Vickers hardness of the directional solidified ZL205 A alloy increases from 62.66 HV under conventional solidification to 105.62 HV and 127.94 HV under directional solidification without heat preservation measures because of grain refinement, and the performance improvement is very obvious. Similarly, the tensile strength increased from 92.64 MPa to 154.05 MPa, and the elongation increased from 1.61 % to 3.18 %.

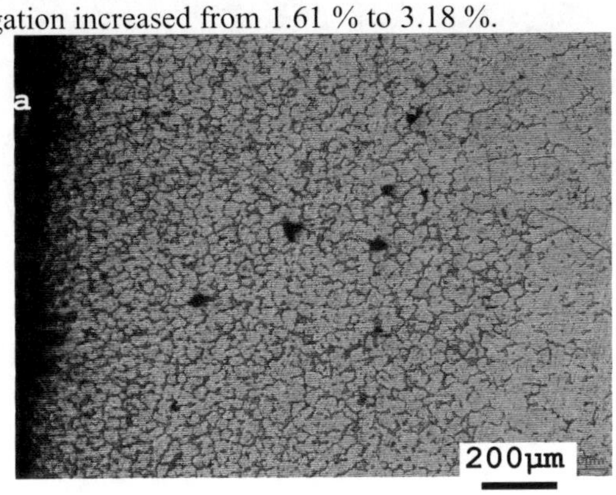

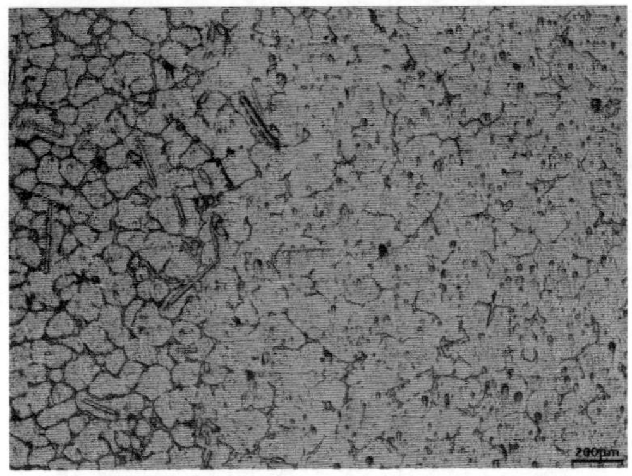

Figure 1 Microstructure of ZL205A alloy along the direction of directional solidification under directional solidification conditions (a) without asbestos (d) with asbestos

Table 1 Test results of physical parameters of ZL205A alloy under different solidification conditions at room temperature

Simples	specific heat J/(g·K)	density g/mm^3	thermal diffusion mm^2/s	heat conductivity W/(m·K)
Conventional solidification	0.896	2.60	36.898	85.958
Uninsulated horizontal direction	0.896	2.60	37.575	95.605
Uninsulated vertical direction	0.896	2.60	39.044	107.005
Insulated horizontal direction	0.896	2.60	38.581	89.904
Insulated vertical direction	0.896	2.60	37.223	89.264

By testing the thermal conductivity of the alloy under different directional solidification conditions, the results show that the average free path of electrons can be effectively improved due to the rapid transport of electrons along the directional growth of columnar crystals formed in the alloy structure, which is no longer affected by the scattering of the original network-like distribution of the θ-Al2Cu phase. The thermal conductivity of the sample obtained by directional solidification without lateral insulation is 107.005 W/(m·K), which is 24.5 % higher than that of the conventional solidified sample.

Keywords: ZL205A alloy, Directional solidification, Microstructure, Mechanical Properties, Thermal conductivity

Study of the Microstructure and Mechanical Properties of Novel As-Cast Titanium Alloys Containing Ta

Chuntong Wang[1], Binguo Fu[1,2*], Yufeng Wang[2], Tianshun Dong[1], Jingkun Li[1], Guolu Li[1], Jinhai Liu[1]

1. Key Laboratory for New Type of Functional Materials in Hebei Province, School of Materials Science and Engineering, Hebei University of Technology, Tianjin 300401, China;

2. Aerospace Product Center, Tianjin Institute of Aerospace Mechanical and Electrical Equipment, Tianjin 300301, China

*Corresponding address: e-mail: fubinguohao@163.com

Abstract: The effect of Ta content on the microstructures, precipitated phases and mechanical properties of novel as-cast titanium alloy Ti-6.5Al-4Sn-5.5Zr-2.7Mo-1.7Nb-3.5W-0.4Si-0.1Y-xTa (x = 0, 1.0, 2.0 wt.%) was systematically studied. The results showed that the microstructure of the as-cast alloys is basket-weave microstructure, mainly composed of α-Ti, β-Ti, Y2O3 and silicides, belonging to α+β type titanium alloy. The addition of Ta element can inhibit the precipitation of Y2O3 at the grain boundary and promote the precipitation of silicides. The orientation relationship between the S1-type silicide and α-Ti matrix of 2.0 wt.% Ta alloy can be identified as $(01\bar{1}0)$S1 // $(01\bar{1}\bar{1})$α and $[\bar{2}11\bar{6}]$S1 // $[2\bar{1}\bar{1}0]$α; the orientation relationship between the S1-type silicide and β-Ti matrix can be identified as $(01\bar{1}0)$S1 // $(01\bar{1})$β and $[1\bar{2}1\bar{3}]$S1 // [100]β; $(01\bar{1}0)$S1 // $(01\bar{1})$β, $(2\bar{1}\bar{1}\bar{1})$S1 // (200)β and $[\bar{2}11\bar{6}]$S1 // $[0\bar{1}\bar{1}]$β. The tensile strength of the as-cast alloys increases with the increase of Ta content, whereas the elongation change is not significant, which is due to the solid solution strengthening of Ta atom and the second phase strengthening of silicides.

Keywords: high-temperature titanium alloy; Ta element; precipitate phases; microstructure; mechanical property

1 Introduction

Titanium and titanium alloys are widely used in aerospace, military, petrochemical and other fields due to their low density, high specific strength and excellent corrosion resistance [1, 2]. In order to adapt to the rapid development of modern aerospace industry, the development of new high-temperature and high-strength titanium alloys has attracted much attention. In this present work, the effects of Ta content (0, 1.0, 2.0 wt.%) on the microstructure precipitation behavior, mechanical properties of the self-designed short-term cast high-temperature titanium alloys (Ti-6.5Al-4Sn-5.5Zr-2.7Mo-1.7Nb-3.5W-0.4Si-0.1Y) were studied. In addition, the corresponding room and high-temperature fracture mechanisms were also discussed.

2 Experimental procedure

Ti-6.5Al-4Sn-5.5Zr-2.7Mo-1.7Nb-3.5W-0.4Si-0.1Y-xTa (*x* = 0, 1.0, 2.0 wt.%) alloys were prepared by the vacuum levitation melting furnace equipped with the water-cooled copper crucible of 5 kg capacity by cooling with the furnace. The microstructures and precipitated phases were characterized by XRD, SEM and TEM. Mechanical properties were studied by tensile test at room-temperature and high-temperature of 700 ℃ and 750 ℃.

3 Result and discussion

1. Phase identification and microstructure

Fig. 1 shows the phase composition of Ti-6.5Al-4Sn-5.5Zr-2.7Mo-1.7Nb-3.5W-0.4Si-0.1Y-xTa (x = 0, 1.0, 2.0 wt.%) analyzed by XRD. The XRD diffraction peak of the alloy indicated that the designed alloys were all composed of α+β phases composition, with Ta element completely dissolved in the matrix.

Fig. 2 shows the OM and SEM microstructures of the titanium alloys with different Ta contents. It can be seen that there was a staggered arrangement between α-Ti phase and β-Ti phase which belongs to the basket-weave structure. With the increased of Ta content, the width of α lamella in the alloy decreased; the content of the α-Ti phase decreased while the β-Ti phase increased in the alloy. This was due to the fact that Ta element as the high-melting point β-Ti phase stabilizing element, promoted the formation of β-Ti phase. It also found that the bright white precipitated phase at local β grain boundary was Y_2O_3 phase by the following TEM analysis.

An ellipsoidal precipitated phase was present in the β-Ti phase with a size of about 130 nm in Fig. 3(a). In order to determine the precipitate phase, the precipitate phase was analyzed by SAED from three different directions, as shown in Fig. 3(b), (c) and (d). The results showed that the precipitated phase on the β-Ti phase of 2.0 wt.% Ta alloy was $(TiZr)_5Si_3$, (S1-type silicide), and the S1-type silicide had the specific orientation relationship with α-Ti and β-Ti phase: $(01\bar{1}0)_{S1}$ // $(01\bar{1}\bar{1})_{\alpha}$ and $[\bar{2}11\bar{6}]_{S1}$ // $[2\bar{1}\bar{1}0]_{\alpha}$; $(10\bar{1}0)_{S1}$ // $(01\bar{1})_{\beta}$ and $[1\bar{2}1\bar{3}]_{S1}$ // $[100]_{\beta}$; $(01\bar{1}0)_{S1}$ // $(01\bar{1})_{\beta}$, $(2\bar{1}\bar{1}\bar{1})_{S1}$ // $(200)_{\beta}$ and $[\bar{2}11\bar{6}]_{S1}$ // $[0\bar{1}\bar{1}]_{\beta}$.

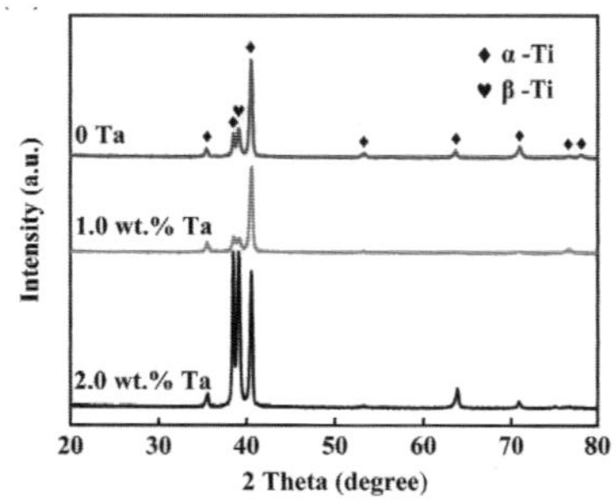

Fig. 1: XRD spectra of alloys with different Ta contents

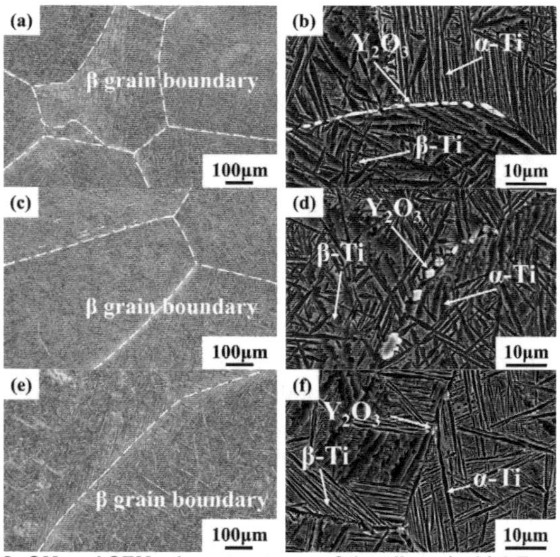

Fig. 2: OM and SEM microstructures of the alloys (a, b) 0 Ta alloy; (c, d) 1.0 wt.% Ta alloy;(e, f) 2.0 wt.% Ta alloy

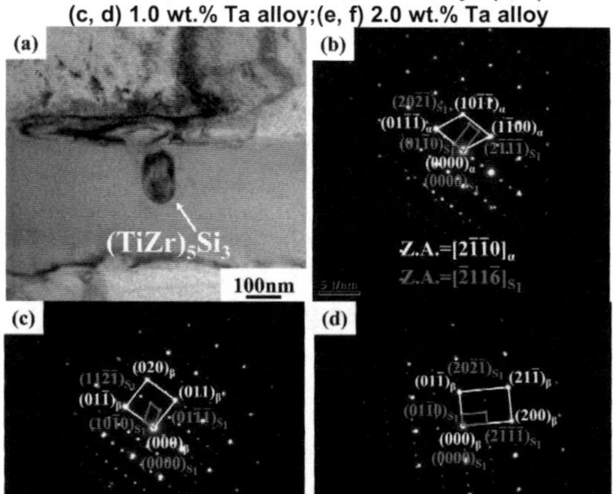

Fig. 3: (a) TEM image and SAED pattern of $(TiZr)_5Si_3$ in 2.0 wt.% Ta alloy;(b, c, d) The complex SAED pattern of $(TiZr)_5Si_3$ in three different directions in (a)

2. Mechanical properties

Table 1 shows the UTS and EL of the alloy at room-temperature and high-temperature. The results showed that the UTS of the three alloys at room temperature and high temperature increased as the Ta content increased, but the EL of the three alloys at the same temperature was not much different. It also can be seen that the tensile properties of the alloy at the same temperature increase with the increase of Ta content. This was due to the different atomic radius of Ta atoms (0.148 nm) and Ti atoms (0.145 nm). When Ta atoms were dissolved in the alloy, the lattice distortion of the matrix alloy occurred, which hindered the movement of dislocations and strengthened the matrix alloy.

Table 1: Mechanical properties of the as-cast alloys

Temperature	Alloy	UTS/MPa	EL/%
25 ℃	0 Ta	1125	0.6
	1.0 Ta	1157	0.5
	2.0 Ta	1175	1.0
700 ℃	0 Ta	707	1.6
	1.0 Ta	708	1.8
	2.0 Ta	874	2.4
750 ℃	0 Ta	606	4.3
	1.0 Ta	612	5.3
	2.0 Ta	749	3.8

4 Conclusion

(1) The as-cast alloys are mainly composed of α-Ti, β-Ti, Y_2O_3 and silicides, belonging to α+β type titanium alloy.
(2) The orientation relationship between the S1-type silicide and α-Ti matrix of 2.0 wt.% Ta alloy can be identified as $(01\bar{1}0)_{S1} // (01\bar{1}\bar{1})_{\alpha}$ and $[\bar{2}11\bar{6}]_{S1} // [2\bar{1}\bar{1}0]_{\alpha}$; the orientation relationship between the S1-type silicide and β-Ti matrix can be identified as $(01\bar{1}0)_{S1} // (01\bar{1})_{\beta}$ and $[1\bar{2}1\bar{3}]_{S1} // [100]_{\beta}$; $(01\bar{1}0)_{S1} // (01\bar{1})_{\beta}$, $(2\bar{1}\bar{1}\bar{1})_{S1} // (200)_{\beta}$ and $[\bar{2}11\bar{6}]_{S1} // [0\bar{1}\bar{1}]_{\beta}$.
(3) The UTS of the as-cast alloys increase with the increase of Ta content, whereas the elongation change is not significant.

5 Acknowledgments

This research was funded by the S&T Program of Tianjin under Grant No. 22YDTPJC00150.

References

[1] Y. N. Sun, H. C. Gong, J. Y. Ying, S. Xu, L. Yang, J. J. Zhang, Q.B. Fan, Quantitative relationship between impact toughness and quasi-static tensile properties of Ti-652 titanium alloy, J. Alloys. Compd. 971 (2024) 172678.
[2] M. D. Hayat, H. Singh, Z. He, P. Cao, Titanium metal matrix composites: An overview, Compos. Part. A. Appl. Sci. Manuf. 121 (2019) 418-438.

Effect of Cooling Rate on the Structure of Cast Ti-20.5al-15Nb-1Mo Alloy

LiangliangLiu, Dong Liu*, Yuyou Cui, Rui Yang
Institute of Metal Research, Chinese Academy of Sciences, 72 Wenhua Road, Shenyang, Liaoning, 110016, China
*E-mail: dliu@imr.ac.con

Abstract: The structure changes of cast Ti-20.5Al-15Nb-1Mo alloy at different cooling rates heated from the B2 field were analyzed. At the cooling rates of 10, 5 and 2.5 ℃/s, the α 2colony precipitated with a subboundary morphology in the retained B2 phase through metallographic microscope observation. The subboundary morphology precipitation differed from real subboundary structure and consisted of a series of acicular α2 phase with approximately identical orientations through scanning electron observation. At decreased cooling rates of 1, 0.5 and 0.1 ℃/s, the subboundary morphology precipitation disappeared, and the α 2 changes from acicular shape to plate. Reheating the alloy with subboundary morphology precipitation at α2 + B2 field, a kind of fine microstructure with homogeneous α2/O plates and few acicular second α 2/O was obtained.

Keywords:cast microstructure, Ti3Al alloy, cooling rate, subboundary

1 Introduction

With low density, high specific strength, good flame retardancy, excellent oxidation resistance, and creep resistance, Ti3Al alloys are being considered as potential lightweight alloys for aerospace structural parts such as pressurized airplane magazines applied at 600~700 ℃. Althoughthe Ti3Al alloys have been showing excellent mechanical properties after thermomechanical processing, the poor welding property limits their application[1, 2]. The precision cast process is another way to manufacture Ti3Al components without welding, while alloys with the cast process show much weaker mechanical properties compared to thermomechanical processing.

Interestingly, a network of subboundaries characteristic of a polygonization phenomenon occurs in Ti3Al alloy during cooling[3-5]. As fine grain size contributes to both strength and plastic, the fine subboundaries may achieve the same effect. But firstly, we should know the characteristics of subboundaries. In this paper, the structure changes, especially the subboundaries phenomenon, of cast Ti-20.5Al-15Nb-1Mo alloy were identified during cooling.

2 Experimental procedure

The Ti-20.5Al-15Nb-1Mo alloy (hereafter designated as 20.5-15-1)casting rods with a diameter of about 32 mm were hot isostatically pressed (HIPed) at 1020 ℃ with an applied pressure of 150 MPa for 2 hours.Table 1. shows the composition of the alloy.The specimen forthe thermal simulated test had a diameter of 6mm and a length of 76mm. The specimens were first heated at a rate of 5 ℃/s to 1100 ℃, holding 10 minutes, and then cooled to room temperature at different cooling rates varying from 0.1 to 100 ℃/s inthe GLEEBLE 3800machine. The microstructure was observed on a Zeiss AXIOVERT 200MAT optical microscope (OM) and an FEI Apreo scanning electron microscope (SEM).

Table 1. Composition of Ti3Al Alloy

Al / wt. %	Nb / wt. %	Mo / wt. %	O / ppm	N / ppm	H / ppm	Ti / wt. %
10.7	27.8	1.9	720	90	9	Bal.

3 Result and discussion

3.1 HIPed microstructure

The HIPed microstructure, shown in Figure 1, consists of coarse α_2 plates and acicular second α_2/O phase in retained ordered B2 phase.

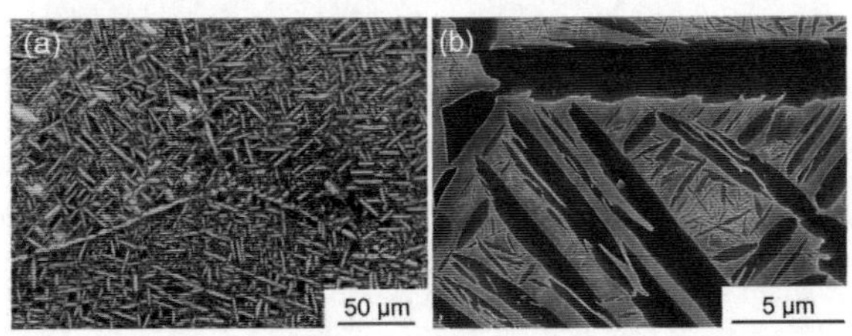

Figure 1. HIPed microstructure of Ti3Al alloy (a) Metallograph, (b) SEM image.

3.2 Effect of cooling rates on metallographic structure

The microstructure changes with different cooling rates from the B2 phase field are shown in Figure 2. The subboundaries phenomenon occurs at cooling rates of 10, 5 and 2.5 ℃/s. The main characteristic of the subboundaries phenomenon is the network precipitation with subboundary morphology in the retained B2 phase. As the cooling rate decreases, the widths of subboundaries increase and more punctiform precipitations appear in the retained B2 phase. When the cooling rate decreases further, the subboundaries phenomenon disappears and acicular α2 appears with no evidence of retained B2.

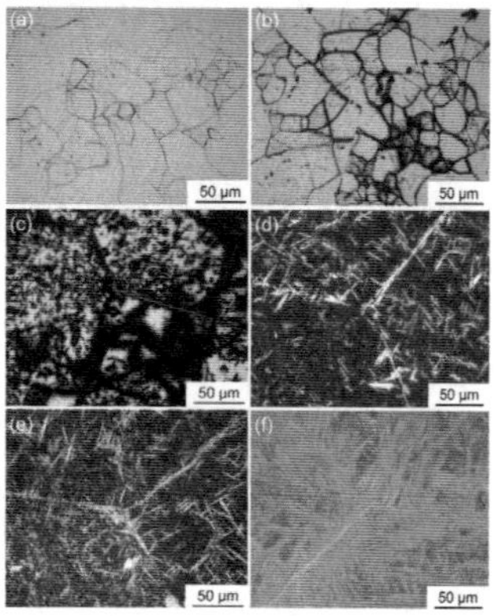

Figure 2. Metallographs of HIPed 20.5-15-1 specimens heated to 1100℃/10min, then cooled at rates of (a) 10℃/s, (b) 5℃/s, (c) 2.5℃ /s, (d) 1℃/s, (e) 0.5℃/s, (f) 0.1℃/s.

3.3 Structure of subboundary morphology precipitation

As shown in Figure 3, the precipitation with subboundary morphology consists of a series of acicular α2 phase with approximately identical orientations.And the precipitation colonies are not real subboundaries. When the cooling rate decreases to 5 ℃ /s, more α_2 colonies precipitate intermixedly and grow up. Although the reason for precipitation showed subboundary morphology is unclear, it indeed stimulates intragranular nucleation instead of the grain boundary.

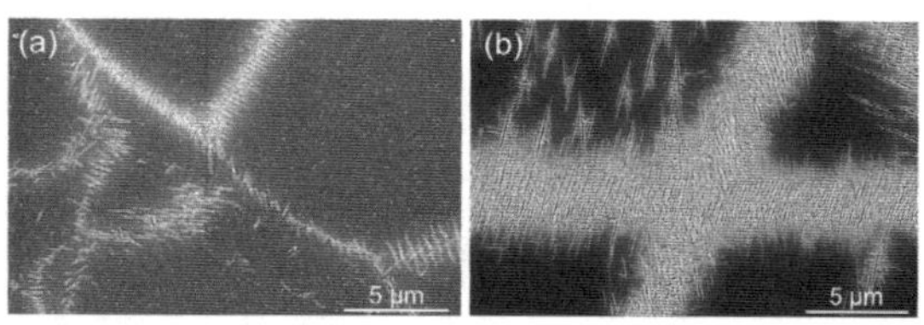

Figure 3. SEM images of HIPed 20.5-15-1 specimens heated to 1100℃/10min, then cooled at rates of (a) 10℃/s, (b) 5℃/s.

3.4 Structure of post heat treatment

Due to the intragranular nucleation of subboundary morphology precipitation, the microstructure of HIPed 20.5-15-1 alloy with 10℃/s cooling rate reheated in 1020℃ consists of fine α_2/O plates and few acicular second α_2/O phase.

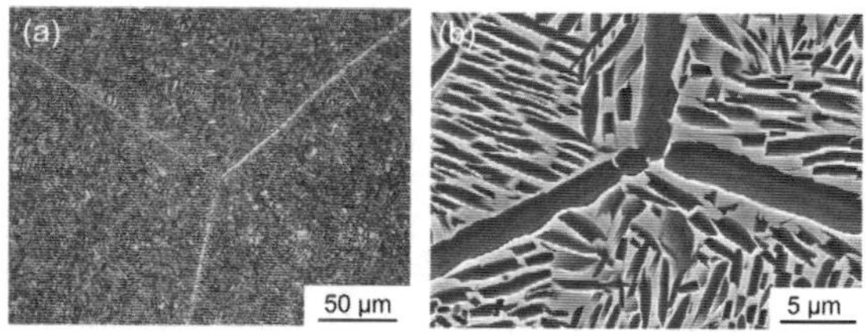

Figure 4. Microstructure of HIPed Ti3Al alloy with heat treatment: 1100℃/10min/10℃/s + 1020℃/1h/furnace cooling (a) Metallograph, (b) SEM image

4 Conclusion

The subboundary morphology precipitation occurs at Ti-20.5Al-10Nb-1Mo alloy at cooling rates of 10, 5 and 2.5 ℃ /s heated at 1100℃. Consisting of a series of acicular α_2 phase with approximately identical orientations, the subboundary morphology precipitation has a different structure from the subboundary. The subboundary morphology precipitationstimulates intragranular nucleation and helps to refine the post heat treatment microstructure of Ti-20.5Al-15Nb-1Mo alloy.

References

[1] W. Baeslack, T. Mascorella, T. Kelly. Weldability of a titanium aluminide [J]. Weld. J., 1989, 68(12): 483.

[2] X. Cai, H. Li, L. Wan , et al. Microstructural characterization and mechanical properties of Ti3Al-Nb/Ni-based superalloy laser-welded joints with different type of V/Cu composite interlayers [J]. Optics and Laser Technology, 2023, 157.

[3] W. Cho, A. W. Thompson, J. C. Williams. CREEP-BEHAVIOR OF TI-25AL-10NB-3V-1MO [J]. Metallurgical Transactions a-Physical Metallurgy and Materials Science, 1990, 21(3): 641-657.

[4] S. Djanarthany, C. Servant, R. Penelle. PHASE-TRANSFORMATIONS IN TI3AL AND TI3AL + MO ALUMINIDES [J]. Journal of Materials Research, 1991, 6(5): 969-986.

[5] W. A. Baeslack,T. Broderick. Effect of cooling rate on the structure and hardness of a Ti-26 at% Al-10 at% Nb-3 at% V-1 at% Mo titanium aluminide [J]. Scripta Metallurgica et Materialia, 1990, 24(2): 319-324.

Effect of Alloy Contents on Shrinkage Porosity Defects in Mg-*x*Y Cast Alloys

Chuangming Li[1], Ang Zhang[1,*], Jiangfeng Song[1], Zhihua Dong[1], Bin Jiang[1], Fusheng Pan[1]
1. National Engineering Research Center for Magnesium Alloys, National Key Laboratory of Advanced Casting Technologies, College of Materials Science and Engineering, Chongqing University, Chongqing 400044, China
*Corresponding address: e-mail: angzhang@cqu.edu.cn

Abstract: As a common defect in cast magnesium alloys, shrinkage porosities can greatly deteriorate material properties. In this study, Mg-xY binary alloys with x ranging from 0.5 wt.% to 19.0 wt.% are investigated. Results show that the porosity distribution is uneven and the morphologies are diverse. As the Y content increases, the porosity tendency of Mg-xY alloys initially rises then declines. The increased Y content refines the grain size and increases the eutectic content, which bolsters resistance to solidification shrinkage and promotes liquid feeding. The shrinkage porosity tendency is lowered after reaching the extreme. Besides, raising Y content reduces the liquidus temperature, while the solidus temperature remains constant, resulting in decreasing solidification temperature range.

Keywords: Mg-Y alloy; Shrinkage porosity; Solidification behavior; Microstructure

1 Introduction

Most magnesium alloy products are shaped through casting processes [1]. Because of wide solidification range and significant volume shrinkage, magnesium alloys are highly prone to shrinkage porosities [2]. Forming at late solidification, shrinkage porosities result from insufficient liquid feeding, which greatly deteriorates material properties [3].

Formation tendency of shrinkage porosities varies with alloy compositions. This paper focuses on the influence of alloy contents on shrinkage porosities by exploring microstructure evolution and solidification behavior in Mg-*x*Y alloys.

2 Experimental procedure

Pure Mg (99.99 wt.%) and Mg-30Y (wt.%) master alloy were used to prepare Mg-*x*Y binary alloys. The actual chemical compositions were listed in Table 1.

Table 1. Compositions of Mg-xY cast alloys (wt.%)

x =	0.5	2	5	7
Mg	Bal.	Bal.	Bal.	Bal.
Y	0.51	2.03	4.80	6.92
x =	10	13	16	19
Mg	Bal.	Bal.	Bal.	Bal.
Y	10.23	13.11	16.07	19.13

Phase analysis was conducted using a Rigaku D/MAX-2500PC X-ray diffractometer (XRD). Micro-Computed Tomography (Micro-CT) analysis was performed using a Phoenix Nanotom m device. Phase diagrams and solidification curves of Mg-*x*Y alloys were calculated using Pandat thermodynamic software. The solidification behavior was investigated using a Differential Scanning Calorimeter (DSC) with a STA 449F3 synchronous thermal analyzer produced by NETZSCH. Computer-aided cooling curve analysis (CA-CCA) was performed using a dual thermocouple thermal analysis device.

3 Result and discussion

Fig. 1 shows the number of shrinkage porosities in Mg-*x*Y alloys. As the Y content increases, the shrinkage tendency of Mg-*x*Y alloys initially rises and then declines.

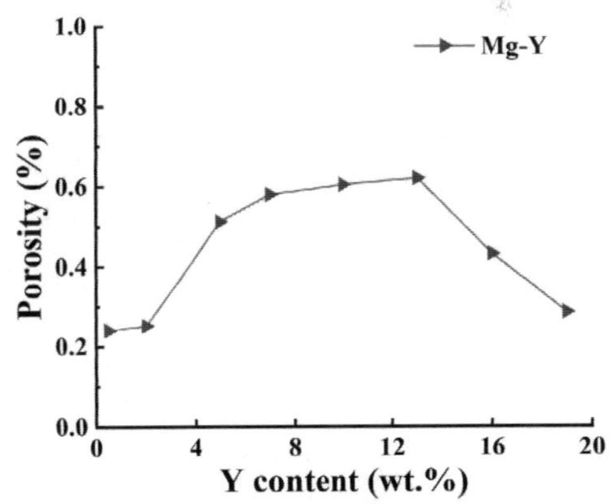

Fig. 1. The number of shrinkage porosities vs Y content.

Fig. 2 shows the micro-CT slice images of Mg-13.0Y alloy. Fig. 3 presents the 3D characteristics of shrinkage porosities. Larger porosities have complex and irregular shapes, while smaller ones tend to be spherical. Fig. 3(b) illustrates the size distribution of porosities. Accordingly, shrinkage porosities exhibit large size and complex morphology characteristics.

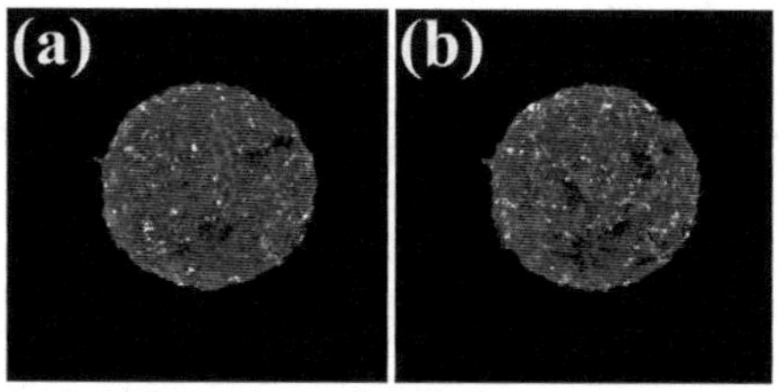

Fig. 2. Selected CT sections of shrinkage porosities in Mg-13.0Y alloy. (a)-(b) Different sections.

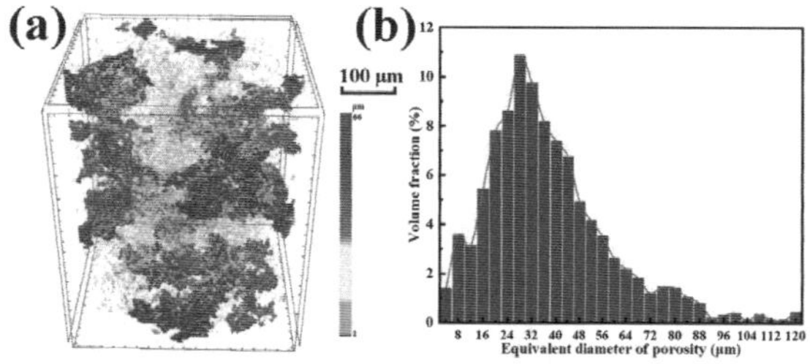

Fig. 3. Shrinkage porosity morphologies in Mg-13.0Y alloy. (a) 3D reconfiguration; (b) Quantitative statistics.

Table 2 lists the volume fraction of the secondary phase in Mg-xY alloys, showing a positive correlation between eutectic content and Y content.

The Y addition promotes formation of the secondary phase, reduces dendrite spacing, and effectively refines the grain. With the increase of Y content, T_l, T_s, and ΔT show similar trends. As ΔT decreases, the coexistence time of solid and liquid phases is shortened, resulting in reduced solidification shrinkage and decreased formation tendency of shrinkage porosities.

Table 2. Volume fraction of the secondary phase in Mg-xY alloys (%)

x =	0.5	2	5	7
f (%)	0.048	0.234	0.334	1.174
x =	10	13	16	19
f (%)	5.224	10.050	16.384	31.096

The XRD patterns of Mg-xY alloys are shown in Fig. 4. Table 3 summarizes the liquidus temperature (T_l), the solidus temperature (T_s), and the solidification temperature range (ΔT) of Mg-xY alloys based on the thermodynamic calculation, DSC, and CA-CCA.

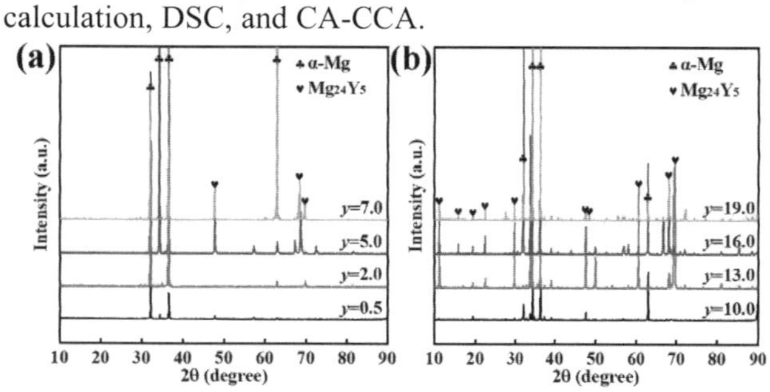

Fig. 4. XRD patterns of Mg-xY alloys: (a) x=0.5-7.0; (b) x=10.0-19.0.

Table 3. Liquidus temperature, solidus temperature, and solidification range of Mg-xY alloys calculated by Scheil model, DSC, and CA-CCA (℃)

Alloy	T_l	T_s	ΔT
Mg-0.5Y	648	574	74
Mg-2.0Y	645	574	71
Mg-5.0Y	639	574	64
Mg-7.0Y	634	574	59
Mg-10.0Y	626	574	51
Mg-13.0Y	617	574	42
Mg-16.0Y	607	574	32
Mg-19.0Y	596	574	21

Conclusion

(1) Shrinkage tendency in Mg-xY alloys follows a "Λ" pattern with increasing Y content, initially rising and then declining. Increasing Y content results in grain refinement and increased eutectic phase. This enhances resistance against solidification shrinkage.

(2) Increasing Y content decreases T_l while T_s remains constant, reducing ΔT and shrinkage tendency. The Y addition promotes formation of secondary phase, promotes liquid feeding and reducing porosity defects.

Acknowledgments

This work is financially supported by the National Natural Science Foundation of China (grant numbers 52101125 and U2037601) and Natural Science Foundation of Chongqing China (grant number CSTB2022NSCQ-MSX1289).

References

[1] Pan Fusheng, Yang Mingbo, Chen Xianhua, A Review on Casting Magnesium Alloys: Modification of Commercial Alloys and Development of New Alloys [J]. Journal of Materials Science & Technology, 2016, 32: 1211-1221.

[2] Li Jilin, Chen Rongshi, Ma Yuequn, Ke Wei, Characterization and Prediction of Microporosity Defect in Sand Cast WE54 Alloy Castings [J]. Journal of Materials Science & Technology, 2014, 30: 991-997.

[3] Khalajzadeh Vahid, D.Carlson Kent, G.Backman Daniel, Beckermann Christoph, A Pore-Centric model for combined shrinkage and gas porosity in alloy solidification [J]. Metallurgical and Materials Transactions A, 2017, 418A: 1797-1816.

Enhancing Mechanical Property Consistency in AE81 Alloy Low-Pressure Die-Casting Wheels Through Ca Addition

Ziwen Li[1], Tian Li[1,2,3*], Bing Jiang[1,2*], JiangFeng Song[1,2], Fusheng Pan[1,2]
1National Engineering Research Centre for Magnesium Alloys, Chongqing University, Chongqing, P.R. China, 400044,
2National Key Laboratory of Advanced Casting Technologies, Chongqing University, Chongqing, P.R. China, 400044,
3Wanfeng Technology Development Co. Ltd, Zhejiang Province, P.R. China, 312432.
*Corresponding author: Tian Li, email: TianLi89@cqu.edu.cn ; Bin Jiang, email: jiangbinrong@cqu.edu.cn

Abstract: AE81M and AE81M+Ca magnesium alloy were employed in the production of 16-inch wheels through a low-pressure die casting (LPDC) process. Defects in the castings were inspected and analysed, revealing entrained oxide films that composed of fluorides and oxides. The mechanical property consistency of the casting wheels was assessed using the Weibull distribution. The study found that while the addition of Ca lead to a slight reduction in the tensile strength of AE81 alloy, it enhances the overall mechanical property consistency of the cast wheels.

Keywords: Mg alloy, low pressure die casting, oxides

1 Introduction

Molten Mg alloys are prone to ignition upon exposure to air, leading to the formation of oxide-related defects, such as oxides or bifilms, within the castings. These defects can detrimentally affect the mechanical properties and its consistency of Mg-alloy castings, significantly impacting the yield rate of casting products. Therefore, in gravity casting processes, the mould chamber is normally flushed with a cover gas before pouring the melt[1]. However, in die casting processes, which are commonly used for producing Mg-alloy automotive products, the mould chamber is typically filled with air rather than a cover gas, exacerbating the oxide defect issue. Researchers are diligently working to reduce the oxide content in Mg-alloy castings.

The addition of Ca has been shown to effectively enhance the oxidation resistance of Mg alloys[2], rendering the oxide film protective and thinner, thus potentially resulting in fewer oxides during the mould filling process. In this study, AE81M Mg alloy and AME81+Ca alloy were utilized to produce 16-inch wheels using a low pressure die casting (LPDC) process. The defects present in the castings were inspected and analysed. The mechanical property consistency of the casting wheels was estimated using Weibull distribution, which is a widely utilized method for reflecting the defect content in castings.

2 Experimental procedure

The casting process was conducted using an LPDC machine provided by Wanfeng Technology Development Co. Ltd. Specially designed melting and holding furnaces for Mg-alloys were employed. During each round of the casting process, 500 kilograms of AE81M or AE81M+Ca alloys were melted and refined, and subsequently transferred to the holding furnace for casting 16-inch wheels.

Table 1. The composition (wt.%) of the AE81M and AE81M+Ca alloys utilized in this investigation, with the residual element (i.e., Mg content) omitted.

Alloy	Al	Ca	Ce	Zn	Mn	Sb	Sr
AE81M	7.6	-	1.15	0.53	0.21	0.18	0.11
AE81M+Ca	7.5	0.93	1.18	0.85	0.28	0.26	0.10

Figure 1 illustrates the dimensions of the wheels produced in this study. For each alloy, a total of 40 castings were manufactured. Given the different property requirements for the rims and spokes of a standard wheel, specific areas of the as-cast wheels were selected for further tensile strength testing. Each casting wheel yielded 40 specimens from the rim area and 10 specimens from the spoke area. The fracture surfaces of the fractured test bars underwent examination using a Scanning Electron Microscope (SEM, Philips JEOL7000) with an accelerating voltage of 5-15 kV. Additionally, the fractured test bars and residual castings were sectioned, polished, and inspected using the same SEM for further analysis.

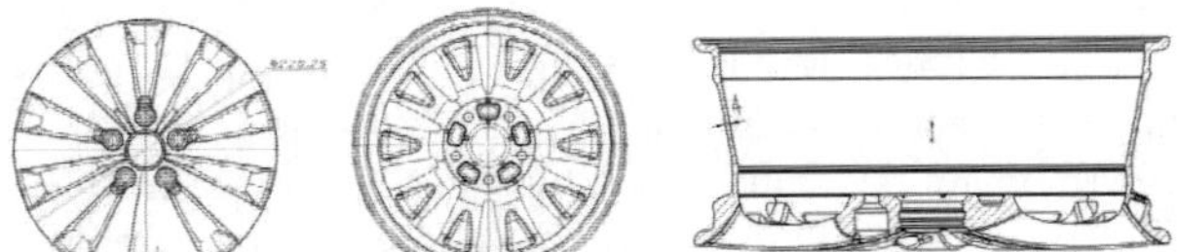

Figure 1. The dimensions of the 16-inch wheels that cast in this study. The wheel's diameter is about 400mm.

3 Result and discussion

Table 2 presents the Weibull modulus of the different sampling areas of the as-cast wheels, obtained using a traditional 2-p linear least square method. The estimator used is P = (i - 0.5) / N.

In the rims area of the AE81M alloy wheel, the UTS Weibull modulus is 3.9, and the elongation Weibull modulus is 1.33. In contrast, the UTS and elongation Weibull moduli of the AE81M+Ca alloy wheel's rims are 10.2 and 1.77, respectively, which are higher than those of AE81M alloy castings. Similarly, in the spoke area, both the UTS and elongation Weibull moduli of the AE81M+Ca castings (36.4 and 7.33, respectively) are higher than those of AE81M alloy wheels (19.5 and 5.65, respectively). These findings suggest that the tensile property consistency of the AE81M+Ca alloy castings is higher than that of AE81M alloy castings.

Table 2. Weibull Modulus of the different sections of the as-cast wheels.

Sampling area	Alloy	Weibull modulus of UTS	Weibull modulus of Elongation
Rim	AE81M	3.9	1.33
	AE81M+Ca	10.2	1.77
Spoke	AE81M	19.5	5.65
	AE81M+Ca	36.4	7.33

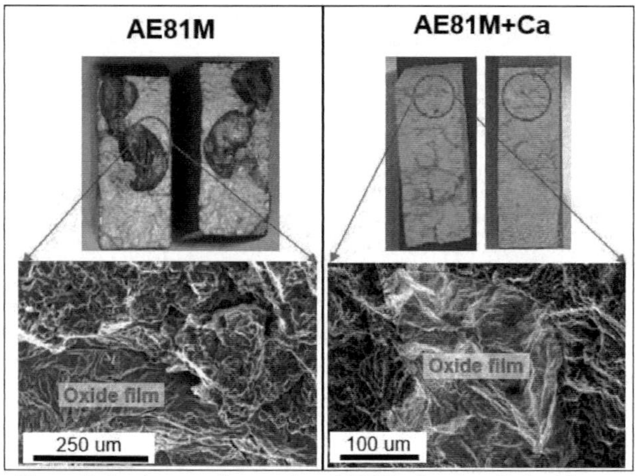

Figure 2. fracture surfaces of an AE81M alloy tensile test bar and an AE81M+Ca alloy test bar, indicating symmetrical oxide films on the fracture surfaces. The dimension of the fracture surface is 4 mm × 10 mm.

Figure 2 shows the fracture surfaces of an AE81M alloy tensile test bar and an AE81M+Ca alloy test bar, both sampled from the rim of the wheels. Symmetrical dark oxide films are evident on both sides of the fracture surfaces. The corresponding EDS results (Table 3 and 4) reveal the presence of fluorine, sulphur, and nitrogen exclusively in the dark regions, indicating that these regions comprise surface protective films entrained into the melt. Therefore, it is plausible to suggest that the dark regions are bifilms, given their symmetrical nature.

Furthermore, the oxide film formed in the AE81M+Ca alloy does not exhibit a higher Ca content compared to the surrounding alloy area. However, it does contain significantly higher cerium content compared to the oxide film formed in the AE81M alloy. This suggests that the addition of Ca may promote the formation of cerium compounds in the oxide film, although further experiments are required to confirm this hypothesis.

Table 3. EDS results corresponding to the AE81M alloy sample shown in Figure 2 (wt%)

	C	O	N	F	Al	Ce	Sr	Sb
Alloy area	2.25	0.5	-	-	7.3	-	0.06	0.08
Oxide film	14.75	16.53	0.80	0.68	6.46	1.00	1.84	-

Table 4. EDS results corresponding to the AE81M+Ca alloy sample shown in Figure 2 (wt%).

	C	N	O	F	Al	Ca	Ce	Sr	Sb
Alloy area	3.76	-	0.67	-	15.43	4.89	0.08	0.17	-
Oxide film	16.46	5.03	16.70	0.73	9.94	3.87	7.67	0.12	0.06

4 Conclusion

The addition of Ca in AE81 alloy led to an enhancement in the mechanical property consistency of the corresponding LPDC wheels. This may be attributed to Ca's role in promoting the formation of a more protective film.

References

[1] K. Strobel, M.A. Easton, V. Tyagi, M. Murray, M.A. Gibson, G. Savage, T.B. Abbott, Evaluation of castability of high pressure die cast magnesium based alloys, International Journal of Cast Metals Research, 23 (2010) 81-91.

[2] Y. Ming, G.Q. You, Q. Li, F.J. Yao, J. Zhang, B. Jiang, Flame resistance and oxidation behaviour of Mg-xCa (x=0.3, 1.0, 2.0 and 3.6 wt%) alloys, Corrosion Sci., 215 (2023).

Low Cost High Performance HPDC AM60 Based Alloys for Super-Sized Integrated Automotive Components

Jing Wang[1,2], Jiangfeng Song[1,2], Bin Jiang[1,2], Fusheng Pan[1,2]
1. National Engineering Research Center for Magnesium Alloys, Chongqing University, Chongqing, Sha Zheng Street 174#, Sha Ping Ba District, Chongqing, China
2.National Key Laboratory of Advanced Casting Technologies, Chongqing University, Chongqing, Sha Zheng Street 174#, Sha Ping Ba District, Chongqing, China
*Corresponding address: e-mail: jiangfeng.song@cqu.edu.cn, fspan@cqu.edu.cn

Abstract: Magnesium alloy super-sized integrated high pressure die casting for automotive structural components has a bright application prospect and attracted much attention. Low cost and high-performance die cast magnesium alloys for super-sized integrated components application are in great need. In this work, microalloying element Mn, Ce and La with an adding content less than 0.2 wt.% were introduced into the commercial AM60 die cast alloys. The ultimate tensile strength, yield strength, elongation of 288.0±1.7 MPa, 158.0±1.0 MPa, and 22.0±3.0 % were achieved in AM60-0.2La alloy. Besides, AM60-0.2La alloy exhibited the best corrosion resistance (0.29 mm/y) and fluidity among the investigated four alloys. The excellent mechanical properties and corrosion resistance are mainly attributed to grain refinement strengthening, low porosity and low content of large shrinkage pores. The current findings offer valuable guidance for the alloy design for magnesium alloy super-sized integrated high pressure die casting application.

Keywords: Magnesium alloys, High pressure die casing, Microstructure, Mechanical properties.

1 Introduction

In this study, small quantities of the alloying elements Mn, Ce, and La were added into the widely used commercial HPDC AM60 alloy. The HPDC alloys, namely AM60, AM61, AM60-0.2Ce, and AM60-0.2La, were subjected to investigate the impact of Mn, Ce, and La on the microstructure, defects, fluidity, mechanical properties, and corrosion resistance of the AM60 alloy. Additionally, it aims to formulate magnesium alloys exhibiting superior comprehensive properties.

2 Experimental procedure

The four AM60 based alloy castings were firstly melted in an electric resistance furnace at 720°C, and then HPDC in a 650 T die caster. The direct die casted tensile specimens with a gauge length of 60 mm and a diameter of 6.4 mm were employed. For corrosion testing, samples measuring 20 mm × 20 mm × 3 mm were cut and immersed for 7 days at room temperature in 3.5 wt.%NaCl solution.

3 Result and discussion

Result

Fig. 1 illustrates the properties of HPDC AM60 based alloys. Fig. 1 (a1) and (a2), depicts the mechanical properties of the AM60 base alloys. The yield strength (YS) of AM60 alloy measures 152 ± 1.0 MPa, the ultimate tensile strength (UTS) is 280.3 ± 1.5 MPa, and elongation (EL) is 19.9 ± 0.8%, respectively. while the addition of 0.2 wt.% La enhances UTS, YS, and EL, yielding optimal mechanical properties. Among commercial and new experimental HPDC magnesium alloys, the developed alloys display the highest UTS, YS, and EL, as shown in Fig. 1 (a2).

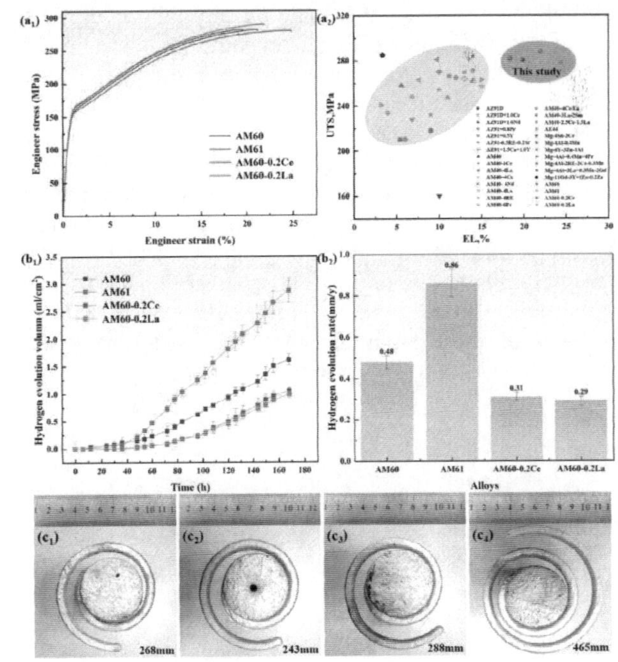

Fig.1. (a) Mechanical properties: (a₁) Tensile stress-strain curve and corresponding mechanical property values of AM based alloys. (a₂) YS, UTS and EL of various die-casting Mg alloys. (b)Corrosion property: (b₁) hydrogen evolution volume-time curves; (b₂) corrosion rate bar chart. (c)Macro-view photographs of fluidity test sample of AM based alloys: (c₁) AM60; (c₂) AM61; (c₃) AM60-0.2Ce; (c₄) AM60-0.2La.

Fig. 1(b1) (b2) depicts the corrosion properties. The corrosion rate in the initial 36 hours of immersion is slow. The cumulative hydrogen evolution for AM60, AM61, AM60-0.2Ce, and AM60-0.2La is 1.26 mL/cm², 2.65 mL/cm², 0.78 mL/cm², and 0.76 mL/cm², respectively, with AM60-0.2La showing the lowest hydrogen evolution.

In addition, the fluidity of four alloys were also investigated and the results were shown in Fig. 1(c1-c4). The fluidity of the alloys shows the trend of AM60-0.2La > AM60-0.2Ce > AM60 > AM61.The increase in Mn content slightly decreases the flow length of the AM60 alloy, while the La addition largely improves the flow length.

Discussion

The primary strengthening mechanism of HPDC magnesium alloys is grain refinement strengthening. Grain refinement significantly enhances the strength of alloys, particularly YS. The studied alloys demonstrate satisfactory grain refinement.Furthermore, AM61 alloy exhibits high amount of ESCs and pore defects, significantly deteriorating mechanical properties. Conversely, AM60-0.2Ce and AM60-0.2La alloys manifest reduced defects. HPDC AM60-0.2La alloy exhibits the best mechanical properties among all.

The fluidity hierarchy of alloys was observed as follows: AM60-0.2La > AM60-0.2Ce > AM60 > AM61. Firstly, precipitation of phases like Al_8Mn_5 before α-Mg increases melt viscosity, thus reducing fluidity. Secondly, the diffusion coefficient of elements in Mg melt, following the order of La > Ce > Al >Mn, contributes to higher fluidity due to faster diffusion rates.

For magnesium alloys, impurity content, second phases, and defects are pivotal factors influencing the corrosion behavior. The addition of Mn, La, Ce will decrease the corrosion rate of AM60 alloy in the aspect of impurity content[1-2].Incorporating RE elements Ce and La results in the precipitation of the $Al_{11}RE_3$ phase [3]. The presence of $Al_{11}RE_3$ phase will decrease the micro-galvanic corrosion of magnesium alloy. In the aspect of casting defects, the addition of Mn to AM60 will largely increase its corrosion rate, while the addition of La and Ce will decrease its corrosion rate. Overall, the corrosion resistance of the four alloys follows: AM60-0.2La > AM60-0.2Ce > AM60 > AM61.

4 Conclusion

In this investigation, we developed AM60-0.2La alloys by incorporating trace amounts of the RE element La into commercial AM60 alloys. This addition not only promotes grain refinement but also reduces the porosity. These microstructural adjustments result in superior fluidity, mechanical properties, and corrosion resistance of the AM60-0.2La alloy. Besides, the minor addition of inexpensive La ensures the low cost of AM60-0.2La alloy, which is very promising for super-sized integrated HPDC automotive part application.

5 Acknowledgments

This work is financially supported by the National Key Research and Development Program of China (grant number 2022YFB3709300, 2021YFB3701000), and the National Natural Science Foundation of China (grant number 52271090, 52071036, U2037601, U21A2048), Chongqing Science and Technology Commission, China (grant number CSTB2022TIAD-KPX0021), and the Fundamental Research Funds for the Central Universities (grant number 2022CDJDX-002).

References

[1] G. Wu, C. Wang, M. Sun, W. Ding, Recent developments and applications on high-performance cast magnesium rare-earth alloys, J. Magnesium Alloys 9(1) (2021) 1-20.

[2] T. Chen, Y. Yuan, T.T. Liu, D.J. Li, A.T. Tang, X.H. Chen, R. Schmid-Fetzer, F.S. Pan, Effect of Mn Addition on Melt Purification and Fe Tolerance in Mg Alloys, Jom 73(3) (2021) 892-902.

[3] M.S. Dargusch, S.M. Zhu, J.F. Nie, G.L. Dunlop, Microstructural analysis of the improved creep resistance of a die-cast magnesium–aluminium–rare earth alloy by strontium additions, Scr. Mater. 60(2) (2009) 116-119.

Multi-Scale Microstructure Regulation and High-Temperature Strengthening Mechanism of Ti_2AlC/TiAl Composites Prepared by Vacuum Melting

Pei Liu*, Jingpei Xie, Aiqin Wang, Zhe Deng, Zhenbo Wang
School of Materials Science and Engineering, Henan University of Science and Technology, Luoyang China
*Corresponding address: e-mail: liupei@haust.edu.cn

Abstract: TiAl alloy is an ideal lightweight high temperature structural material in the aerospace field, however, it could not meet the urgent need of the new generation high thrust-to-weight ratio aero engine blade to withstand the 800℃ high temperature for a long time. In this study, the multi-scale Ti2AlC/TiAl composite is fabricated. The in-situ micro-Ti2AlC is obtained at the grain boundaries of the full lamellar TiAl matrix by vacuum melting, the targeted precipitation of submicro-Ti2AlC at the lamellar TiAl/Ti3Al phase boundary and directional precipitation of nano-Ti2AlC within TiAl crystals are achieved by heat treatment. The composite showed excellent high-temperature strength (the tensile strength of the composite at 850℃ is comparable to that of the TiAl alloy at 750℃). The results show that the micro-Ti2AlC inhibits grain boundary softening and hinders dislocation motion, the submicro-Ti2AlC prevents twin propagation, the nano-Ti2AlC obstructs dislocation motion and refines lamellar microstructure, and the full lamellar TiAl matrix microstructure changes the propagation paths of twins/stacking fault at high temperatures. This study provides theoretical basis and technical support for the preparation technology and application of a new generation of light high temperature resistant TiAl alloy for aeroengines.

Keywords: TiAl Matrix Composite, Multi-scale Ti_2AlC, high-temperature strengthening mechanism

1 Introduction

With the rapid advancement in the aerospace field, the development of a high-temperature, lightweight structural material with exceptional mechanical properties has become an undeniably formidable challenge. TiAl alloy stands as an exemplary lightweight, high-temperature structural material in the aerospace industry, yet it falls short of meeting the stringent demands posed by the next-generation aero engine blades. Multi-scale Ti_2AlC/TiAl composite can effectively improve the high-temperature properties of TiAl alloys. In this study, by optimizing composite components and heat treatment techniques to achieve multi-scale Ti_2AlC/TiAl composite tissue modulation, lightweight and high-strength multi-scale Ti_2AlC/TiAl composite materials can be developed to meet national needs.

2 Experimental procedure

The composite used in this paper are all prepared by vacuum arc melting (VAR) technique. Subsequently, a heat treatment procedure (underwent a 24h solid solution treatment at 1573K, followed by a 0.5h aging treatment at 1173K.) was employed to modulate the microstructural characteristics of the material. The raw materials used for vacuum arc melting were sponge titanium (99.7%), aluminum rod (99.99%), chromium particles (99.98%), aluminum-niobium master alloy (with Nb content of 55%) and graphite powder (99.98%).

Test for tensile properties using the MTS 370.10 hydraulic servo test system. At least three tensile experiments were carried out on each sample to ensure the accuracy of the experimental results. The high-temperature tensile experiments were conducted at 800 °C with a tensile rate of 0.0001 S^{-1}. In order to preserve the microstructure after high temperature tensile deformation, the samples were quenched immediately after the tensile experiments. The microstructure and morphology of the samples were analyzed by SEM. The fracture surface was extensively characterized using a TEM to gain a more detailed understanding of the morphology following the stretching process.

3 Result and discussion

3.1 Effect of graphite powder content on the microstructure of composite

For multi-scale Ti_2AlC/TiAl composites, by controlling the graphite powder content, the micro-Ti_2AlC phase length-diameter ratio can be controlled. Adopting a suitable heat treatment process (1573K solid solution for 24 hours and 1173K aging for 0.5 hours) can precipitate multi-scale Ti_2AlC.

3.2 Effect of micro-Ti2AlC on the high temperature mechanical properties of composite

For multi-scale Ti_2AlC composites, the in-situ micro-Ti_2AlC is obtained at the grain boundaries of the full lamellar TiAl matrix. The micro-Ti_2AlC effectively inhibits grain boundary softening and hinders dislocation motion to enhance the high-temperature performance of composites.

3.3 Effect of submicro-Ti2AlC on the high temperature mechanical properties of composite

The targeted precipitation of submicro-Ti_2AlC at the lamellae $TiAl/Ti_3Al$ phase boundary is achieved by heat treatment. When the multi-scale $Ti_2AlC/TiAl$ composite is subjected to an external tensile load, a large number of deformation twins would be generated along the γ-TiAl movable slip system, and the submicro-Ti_2AlC could hinder the deformation twins further propagation, which is beneficial to improve the strength of the composite.

3.4 Effect of nano-Ti2AlC on the high temperature mechanical properties of composite

The directional precipitation of nano-Ti_2AlC within TiAl crystals are achieved by heat treatment. During high-temperature tensile, the nano-Ti_2 AlC exerts a dual beneficial effect on mechanical properties. Firstly, it acts as an effective barrier, hindering the motion of dislocations. Secondly, the nano-Ti_2 AlC refines the lamellar microstructure, leading to a more homogeneous that further bolsters the mechanical attributes.

3.5 Effect of the full lamellar microstructure on the high temperature mechanical properties of composite

The multi-scale $Ti_2AlC/TiAl$ composite prepared by vacuum arc melting furnace and heat-treated have a full lamellar microstructure. When the multi-scale $Ti_2AlC/TiAl$ composite is subjected to an external tensile load, the full lamellar TiAl matrix microstructure can change the propagation paths of twins/stacking fault at high temperatures and thus impede the further extension of the deformation twins, which is conducive to the enhancement of the strength of the composite.

4 Conclusion

(1)The multi-scale $Ti_2AlC/TiAl$ composite with a full lamellar microstructure can be obtained by vacuum arc melting furnace and heat treatment. It can effectively improve the high-temperature performance of TiAl alloy.

(2)The micro Ti_2AlC is precipitated at grain boundaries in a fully lamellar TiAl matrix. It inhibits the softening of grain boundaries and hinders dislocation motion. The Submicro-Ti_2AlC precipitates on the boundary of the lamellar $TiAl/Ti_3Al$ phase, effectively hindering the propagation of twins during high-temperature stretching.

(3)The nano Ti_2AlC, directedly precipitated within TiAl crystals, effectively hinders dislocation motion and refines lamellar microstructure. Additionally the full lamellar TiAl matrix microstructure changes the propagation paths of twins/stacking fault at high temperatures.

5 Acknowledgments

This work was financially supported by the National Natural Science Foundation of China (52101174).

References

[1] Wang Z B, Liu P, Hou B, Ye F, Wang A Q and Xie J P. Investigation on the in-situ reaction mechanism of $Ti_2AlC/TiAl$ composite prepared by spark plasma sintering. Mater. Charact, 2022, 194: 112417.

[2] Liu P, Liu Z W, Wang Z B, Wang A Q, Xie J P and Hou B. Investigation on the effect of alloying elements on the $Ti_2AlC/TiAl$ interfacial properties: First-principles prediction and experimental verification. Mater. Today. Commun, 2023, 35: 106314.

[3] Liu P, Liu Z W, Hou B, Wang A Q, Xie J P and Wang Z B. A systematic investigation on the surface properties of Ti_2AlC via first-principles calculations. Surf. Sci, 2023, 735:122337.

[4] Hou B, Wang A Q, Liu P and Xie J P. Investigation of the nucleation and growth behavior of Ti_2AlC and Ti_3AlC nano-precipitates in TiAl alloys. Nanotechnol. Rev, 2023, 12(1): 20220510.

[5] Liu P, Hou B, Wang A Q, Xie J P and Wang Z B. Balancing the strength and ductility of $Ti_2AlC/TiAl$ composite with a bioinspired micro-nano laminated architecture. Mater. Design, 2022, 220: 110851.

[6] Liu P, Hou B, Wang A Q, Xie J P, Wang Z B and Ye F. Superior strength-plasticity synergy in a heterogeneous lamellar $Ti_2AlC/TiAl$ composite with unique interfacial microstructure. J. Mater. Sci. Technol, 2023, 159: 21-32.

[7] Wang Z B, Liu P, Wang A Q, Xie J P and Hou B. Effect of spark plasma sintering temperature on the multi-scale microstructure evolution and mechanical properties of $Ti_2AlC/TiAl$ composites with network architecture. J. Mater. Res. Technol, 2023, 25: 6209-6223.

[8] Wang Z B, Liu P, Wang A Q, Xie J P and Hou B. Fabrication, microstructure and mechanical properties of TiAl matrix composite reinforced by submicro/nano-Ti_2AlC. Mater. Charact, 2023, 203: 113141.

[9] Liu P, Wang Z B, Ye F, Hou B, Wang A Q and Xie J P. Hierarchically heterogeneous strategy for $Ti_2AlC/TiAl$ composite with superior mechanical properties. Compos. Part. B-Eng, 2024, 273: 111259.

Effects of Heat Treatment on the Thermal Properties of A356 Aluminum Alloys in Different Casting Processes

Xiaomei Luo [1], Yuan-dong Li[1,2*], Wenjing Liu, Jin Qiu[1], Lu Sun[1], Hongwei Zhou [1,2], Guang-li Bi [1,2]
1. State Key Laboratory of Advanced Processing and Recycling of Non-ferrous Metals, Lanzhou University of Technology, Lanzhou 730050, China;
2. Key Lab. of Non-ferrous Metal Alloys and Processing, Ministry of Education, Lanzhou University of Technology, Lanzhou 730050, China
*Corresponding address: Yuan-dong Li; Tel: +86-931-2976795; E-mail: liyd_lut@163.com

Abstract: The thermal diffusivity and specific heat capacity of continuous casting (SC) and high-pressure diecasting (HPDC) aluminum alloys before and after T4 and T6 heat treatments were measured at room temperature, and their thermal conductivity was calculated. The effects of heat treatments on the thermal properties of A356 alloys in different casting processes were investigated, and the mechanism of the effects was also discussed. Before heat treatment, the SC alloys had higher thermal conductivity than that of the HPDC alloys. Moreover, heat treatment after forming had a remarkable influence on the thermal properties of the SC and HPDC alloys. T4 treatment had positive effects on the thermal conductivity of the SC and HPDC A356 alloys, and the T6 heat-treatment SC and HPDC alloys had higher thermal conductivity than that of the T4 treatment alloys.

Keywords: Thermal properties, A356 alloys, Heat treatment, Casting process

1 Introduction

Aluminum alloys have been used for the production of high thermal or electrical conductor. Due to the excellent strength, low density and high thermal conductivity, the alloying aluminum has been applied in the industry of 3C products and electronic package [1-2]. It has been reported that the thermal conductivity of pure aluminum is 237 W • (m • K)$^{-1}$, which is the fourth highest thermal conductivity among metals next to that of pure Cu, pure Au and pure Ag. Hence, aluminum alloys have attracted immense attention as potential heat sink materials. However, alloying elements, such as Si, Mg and Cu, which strengthen the mechanical properties of Al alloys, significantly reduce the thermal conductivity. For instance, the thermal conductivity of Die-casting ADC12 alloy at room temperature is 92~96 W (m K)$^{-1}$, not even 1/2 that of pure Al. To solve the problem, a series of studies focused on the effects of heat treatments and forming processes on the thermal properties of aluminum alloys.

This work aimed to study the effects of heat treatment on the thermal properties of A356 alloys in different casting processes at room temperature. The thermal conductivity of A356 at different states was calculated. The grain size was analyzed by the electron back scattered diffraction (EBSD) technique, and Electron Probe Micro-Analysis (EPMA) results were used to calculate the atomic fraction of solute Si element in the α-Al matrix, which was aimed at revealing the mechanism of the effects of the heat treatment and casting process on the thermal conductivity of A356.

2 Experimental procedure

The material for the present study was commercial A356 aluminum alloy. The samples of A356 were prepared by continuous casting (SC) and high-pressure die casting (HPDC). The chemical compositions of the samples were measured using Electron Probe Micro-Analysis (EPMA). The samples were machined in the shape of round disks with the size of Φ12.7mm×3mm after T4 or T6 heat treatment. X-Ray Diffraction (XRD) analysis was carried out under CuKα radiation with an incidence beam angle of 2°. Jade 6.0 software was used to identify phases, refine the crystal structure and then calculate the lattice strains by the function of size & strain plot.

The thermal diffusivity and specific heat capacity of the samples were measured at room temperature using the flash method in an argon atmosphere with a NETZSCH LFA 457 Analyzer. Three points were detected for each sample and the average value was obtained. The thermal conductivity was calculated using the following relation:

$$\lambda = \alpha \cdot \rho \cdot C_p \quad (1)$$

where λ is thermal conductivity (W (m K)$^{-1}$), α is thermal diffusivity (mm^2 • s^{-1}), ρ is the density of alloy (g • mm^{-3}) and C_p is the specific heat capacity (J • (g • k)$^{-1}$) at constant pressure.

3 Result and discussion

Fig. 1(a) shows the specific heat capacity of SC and HPDC A356 alloys in different states at 25℃. T4 and T6 treatment reduced the heat capacity of the SC and HPDC A356 alloys. The thermal diffusivity of SC and HPDC A356 alloys in different states at 25 ℃ is shown in Fig. 1(b). The thermal diffusivity of SC and HPDC A356 alloys increased after T4 treatment and rose slightly after T6 treatment. The thermal diffusivity of alloys after T6 treatment was higher than that of as-prepared samples. Fig. 1(c) displays the thermal conductivity of A356 alloys at different states. SC A356 alloys had higher thermal conductivity than the HPDC alloys. Moreover, the changes in the thermal conductivity of SC and HPDC A356 alloys after heat treatments presented the same trends as mentioned in the thermal diffusivity.

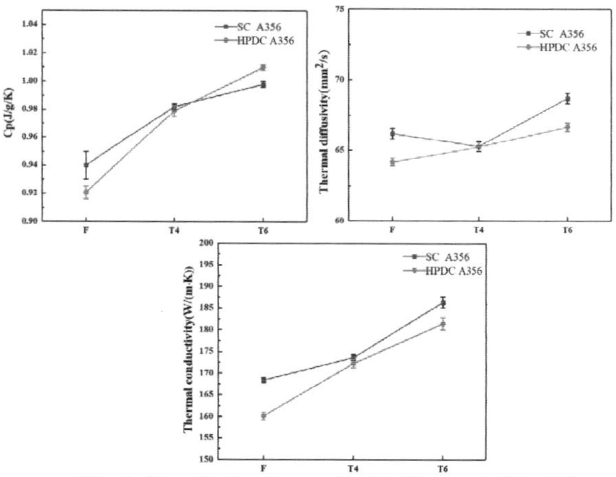

Fig. 1. Specific heat capacity (a), Thermal diffusivity (b) and Thermal conductivity (c) of SC and HPDC A356 alloys in different states at 25℃℃

According to Fig. 1(c), thermal conductivity increased after T4 treatment for the SC and HPDC A356 alloys, this outcome was attributed to dissolution of the block-like eutectic silicon phases and the change of morphology of eutectic silicon from block to fine spherical, the secondary phase volume fraction of both the SC and HPDC A356 alloys decreased. Eutectic silicon phases along grain boundaries were largely dissolved during the solution treatment. It has been shown that thermal conductivity is usually related to the symmetry of a crystal lattice and reduced by the scatter of foreigner atoms. For A356 alloys, the atomic radii of Si and Al are 0.117 nm and 0.143 nm, respectively, and the variance in atomic radii could cause massive amounts of lattice distortion in Al-Si solid solution. The lattice strain of both the SC and HPDC A356 alloys increased after T4 treatment, proving that T4 treatment caused lattice distortion.

The lattice distortion caused by a solute alloying element was much more serious than that of a metallic compound. Hence, the lattice distortion caused by the increased solute Al acted as scattering centres, which limited the mean free path of electrons and phonons and reduced the thermal conductivity of the as-SC and as-HPDC A356 alloy. However, the T4 and T6 heat-treatment SC and HPDC alloys presented higher thermal conductivity than that of the without treatment alloys.

During artificial ageing treatment, the Si phases precipitated near the grain boundaries. Compared with T4 alloys, T6 SC and HPDC alloys had a higher volume fraction of secondary phase and lower solute Al in the α-Al matrix. Moreover, the lattice strain of T6 SC and HPDC alloys was also lower than that of T4 alloys, which accounted for the increase in thermal conductivity of these alloys. Moreover, the thermal conductivity of both T6 and T4 SC and HPDC alloys was obviously higher than that of as-prepared alloys. This is mainly due to the change of the morphology of eutectic silicon.

The thermal conductivity of the SC A356 alloys was higher than that of the HPDC alloys mainly because of the lower solute Si atoms in the Al matrix and larger and more uniform grain size. The results of microscopic observations showed that the eutectic silicon phases in the HPDC A356 alloys were much more dispersed and finer compared with those in the SC A356 alloys. The volume fraction of the secondary phases in the HPDC A356 alloys was higher than that in the SC alloys, the lattice strain of the as-HPDC alloys was lower than that of the as-SC alloys; this outcome meant that less lattice distortion was formed after high-pressure die-casting. Although the grains of the HPDC alloys were finer than those of the SC alloys, which meant that the volume fraction of grain boundary (that could act as the scattering sources blocking the free movement of electrons and decreasing thermal conductivity) was higher It was believed that the lattice distortion had greater influence on the heat transfer than the grain boundaries, therefore, the HPDC alloys had higher thermal conductivity.

4 onclusion

The thermal conductivity of the SC and HPDC A356 alloys before and after T4 and T6 treatment at room temperature were measured and analyzed. The conclusions are summarized as follows:

(1)The thermal conductivity of the SC and HPDC alloys increased notably during T4 treatment and T6 treatment.

(2)The thermal conductivity of the as-SC alloys was higher than that of the as-HPDC alloys.

5 Acknowledgments

The authors gratefully acknowledge the financial support from the Gansu Provincial Basic Research Program-Natural Science Fund, PR China [Project number: 22JR5RA251].

References

[1] R.N. Lumley, I.J. Polmear, H. Groot, J. Ferrier, Thermal characteristics of heat-treated aluminum high-pressure die-castings. Scripta Mater. 58, 1006-1009 (2008)

[2] M.C.S. Jr, A.R. Machado, W.F. Sales, M.A.S. Barrozo, E.O. Ezugwu, Machining of aluminum alloys: a review. Int. J. Adv. Manuf. Technol. 86, 1-14 (2016)

Study of the Interaction Between Tensile Properties and Thermal Conductivity of Al-X (X=Si, Mn, Mg) Alloys

Wenjing Liu[1], Yuandong Li[1,2,*], Zichen Wang[1], Xiaomei Luo[1], Hongwei Zhou[1,2], Guangli Bi[1,2]
1. State Key Laboratory of Advanced Processing and Recycling of Nonferrous Metals, Lanzhou University of Technology, Lanzhou, 730050, PR China;
2. Key Laboratory of Non-ferrous Metal Alloys and Processing, Ministry of Education, Lanzhou University of Technology, Lanzhou, 730050, PR China;
*** Corresponding author:** Yuan-dong Li, email: liyd_lut@163.com, Tel: +86-931-2976706

Abstract: In this study, the Al-1-5wt.% X (X=Si, Mn, Mg) alloys were prepared by permanent mold casting, in order to study the relationship between the mechanical properties and thermal conductivity of the alloys. The micro-structures and the phases compositions of the alloys were characterized by OM, SEM, EDS, and XRD. In addition, the density of electronic states (DOS), elastic constants, and other parameters of several alloys were deeply analyzed by first-principles, which assisted in analyzing the relationship between the mechanical properties of the alloys and their thermal conductivity. The results showed that the microstructure of the alloy was changed with the increase of the content of Si, Mn and Mg elements, producing the effects on the mechanical properties and thermal conductivity of the alloy. Al-Mn alloys showed better mechanical properties than Al-Si and Al-Mg alloys. The thermal conductivity is Al-Mg > Al-Si > Al-Mn. This study reveals an inverse relationship between the reinforcing elements on the mechanical and thermal conductivity of aluminum. It provides a theoretical foundation and experimental basis for the development of aluminum alloy materials with both excellent mechanical and thermal conductivity properties.

Keywords: Aluminum, Alloying elements, Mechanical properties, Thermal conductivity

Introduction

Aluminum alloys have the advantages of low density, high specific strength, and good electrical and thermal conductivity [1, 2]. In recent years, they have gradually replaced copper in power transmission lines and electric vehicle manufacturing. With the continuous advancement of technology and increasing industrial demands, the requirements for the performance of aluminum alloys have become increasingly diverse and stringent [3]. In-depth research into the interaction between the mechanical and thermal conductive properties of aluminum alloys, especially through the regulation of their properties by doping elements, has become a significant topic in the field of materials science.

Existing studies have shown that the impact of solute atoms on the thermal conductivity of aluminum alloys is higher than that of element concentration or secondary phases by one to two orders of magnitude [4]. The fundamental reason is that the addition of solute atoms causes lattice distortion in the aluminum matrix, leading to increased free electron scattering, with different alloying elements causing varying degrees of scattering.

However, the mechanisms by which these doping elements affect the thermal conductivity of aluminum alloys are quite complex, and there has not yet been a systematic study that comprehensively reveals the relationship between mechanical and thermal conductive properties of the alloys. Therefore, this study will utilize a combination of experimental methods and first-principles calculations to investigate the relationship between the thermal conductivity and mechanical properties of aluminum alloys in detail.

Experimental procedure

Al-Si, Al-Mn and Al-Mg alloys were prepared by permanent mold casting. The pure Al and Al-10X (X=Si, Mn, Mg) master alloys are initially placed in a resistance furnace and melted at 790 °C. Once completely melted, the melt was stirred using a graphite rod, and a 1 wt.% covering agent (0.5 wt.% NaCl + 0.5 wt.% KCl) was evenly dispersed over the melt surface. The melt temperature is then adjusted to 760 °C and maintained for 30 min. After insulation, the melt is poured into a preheated metal-type mold at 250 °C to obtain an alloy test bar of Φ18 mm×150 mm. The Al-1-5 wt.% X (X=Si, Mn, Mg) alloys were obtained respectively.

The DK7–725 EDM CNC cutting machine was used to sample the center of the alloy test bar (12.5 mm×12.5 mm× 3mm) for microstructure observation and thermal conductivity testing. The mechanical properties were tested on a WDW-100D electronic universal material testing machine with a tensile rate of 0.5 mm/min and a measurement length of 20 mm. three specimens for each experimental parameter are selected, and the average value is calculated as the final result data. Calculation of material properties by using Materials studio 2023.

Result and discussion

Fig. 1 employs a three-dimensional stress tensor representation to analyze the anisotropy of Al alloys with different doping elements and concentrations. The colors in the figure indicate stress values (GPa), with red representing high-stress areas and blue representing low-stress areas. The study found that anisotropy significantly increases with the concentration of doped elements. The impact of different doping elements on the mechanical properties of the alloys varies: Si and Mn doping results in high-stress areas mainly concentrated in the Z direction, whereas Mg doping makes the anisotropy more complex, with high-stress areas distributed in multiple directions.

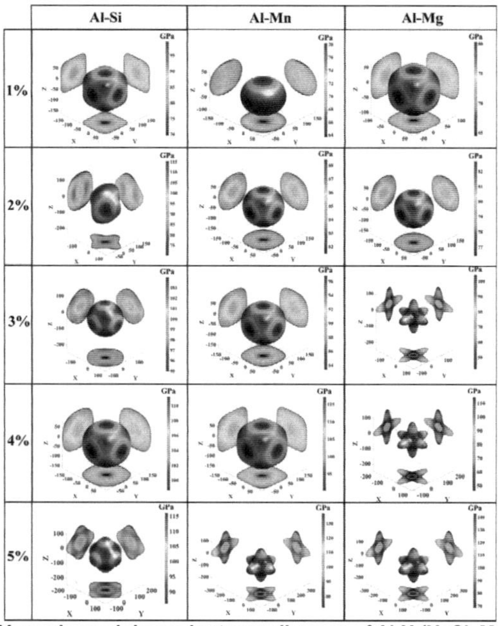

Fig.1 Young's modulus anisotropy diagram of Al-X (X=Si, Mn, Mg)

For the thermal conductivity of alloys, the influence of doping elements is minor at low concentrations, but at high concentrations, due to lattice distortion, formation of secondary phases, and increased phonon scattering, the thermal conductivity significantly decreases. The doping mechanisms of different elements vary, but the main influencing factors include changes in lattice structure, increased phonon scattering centers, and formation of precipitates. In Al-Si alloys, Si phases mainly exist in the form of eutectic Si, while high concentrations of Mn promote the formation of Al_6Mn phases. These second-phase particles not only affect the mechanical properties of the alloy but also significantly reduce the thermal conductivity through interface scattering. Under high Mg doping concentrations, precipitates such as Mg_2Si form within the alloy, reducing its thermal conductivity.

Conclusion

(1)The three elements have varying degrees of impact on the thermal conductivity and mechanical properties of the alloy.

(2)In terms of mechanical properties, Al-Mn alloy exhibits higher mechanical performance than Al-Si alloy, while Al-Mg alloy performs the worst. This experimental result preliminarily demonstrates the inverse relationship between the alloying elements' effect on aluminum's thermal conductivity and mechanical properties.

Acknowledgments

This research is supported by Gansu Provincial Basic Research Program-Natural Science Fund, PR China [Project number: 22JR5RA251].

References

[1] Hagen A B, Wenner S, Bjørge R, et al. Local mechanical properties and precipitation inhomogeneity in large-grained Al−Mg−Si alloy[J]. Materials Science and Engineering: A, 2023,872: 144222.

[2] Lee K, Song Y, Kim S, et al. Genetic design of new aluminum alloys to overcome strength-ductility trade-off dilemma[J]. Journal of Alloys and Compounds, 2023,947: 169546.

[3] Hou H, Dong R, Tan Y, et al. Microstructural characteristics and enhanced mechanical properties of 2024 aluminum alloy resulting from shot-peening treatment[J]. Materials Characterization, 2023,206: 113412.

[4] Li Z, Jarfors A E W, Jansson P. Sustainable choices of alloying element for aluminium for thermal conductivity in circular manufacturing[J]. Sustainable Materials and Technologies, 2024,40:e854.

A Heterogeneous Magnesium Extrusion Alloy with Ultra-High Plasticity

Chenhui Li[1], Hua Hou[1,3], Ruifeng Dong[1], Yuhong Zhao[1,2,4*]

1 School of Materials Science and Engineering, Collaborative Innovation Center of Ministry of Education and Shanxi Province for High-performance Al/Mg Alloy Materials, North University of China, Taiyuan, 030051, China

2 Beijing Advanced Innovation Center for Materials Genome Engineering, University of Science and Technology Beijing, Beijing, 100083, China

3 School of Materials Science and Engineering, Taiyuan University of Science and Technology, Taiyuan, 030024, China

4 Institute of Materials Intelligent Technology, Liaoning Academy of Materials, Shenyang, 110004, China

*Corresponding address: e-mail: zhaoyuhong@nuc.edu.cn

Abstract: Understanding the relationship between the actuate of the slip system and dynamic recrystallization (DRX) during thermo-mechanical processing is vital for developing high-performance Mg-based alloys. In this work, we investigate the effect of slip initiation on DRX behavior of 250 °C hot-extruded. It results that at a temperature of 250 °C, larger sized non-DRX grains undergo continuous dynamic recrystallization (sub-grain growth) with the activation of basal <a> slip. Small-size non-DRX grains incur discontinuous dynamic recrystallization (stress-induced grain boundary migration (SIBM)) with activation of the pyramidal < c + a > slip. In addition, heterostructures contributes to the improvement of the work-hardening of materials and to eliminate localized stress concentrations in small-sized grains. Achieving the synergistic improvement of alloy strength-plasticity.

Keywords: dynamic recrystallization, slip system, heterogeneous magnesium alloy, hot-extruded.

1 Introduction

Mg alloys as the lightest structural metal, which has a wide range of applications [1]. Despite the emerging and critical significance of Mg alloys as a class of engineering materials, magnesium alloys as engineering structural components still have a lot of problems that need to be solved [2]. Thus, how to improve the plasticity of magnesium alloys at room temperature has become a popular research topic today. Hot mechanical methods such as rolling, extruding, forging, etc. are often used to improve the room temperature formability and plasticity of magnesium alloys [3]. Therefore, how to maximize the activation of slip during deformation is the key to preparing highly plastic magnesium alloys. In this work, a heterogeneous structural magnesium alloy is successfully prepared using the extruded process with artificial cooling. The highest plasticity recorded to date has been achieved due to the successful activation of the pyramidal < c + a > slip.

2 Experimental procedure

The original AZ31B alloys are prepared by using pure Mg (99.99%), Al (99.98%), and Zn (99.98%). The as-cast alloys are subjected to a homogenization treatment at 420°C for 12 hours. The extrusion angle is set at 30°. The extrusion ratio is set to 25:1 and the extrusion speed is set to 0.2 mm/s.

3 Result and discussion

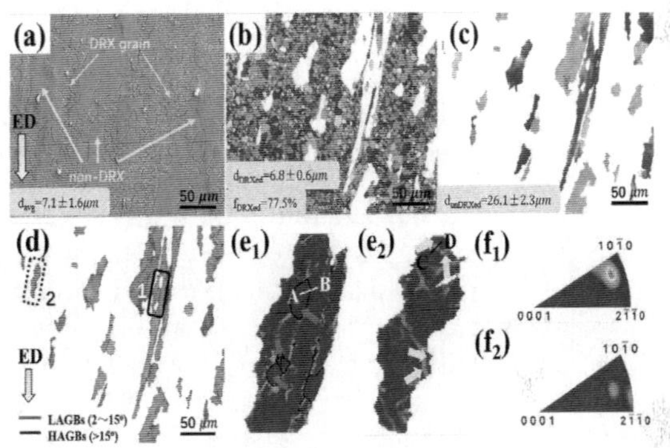

Fig.1 (a) SEM map; (b, c) EBSD maps of DRXed and non-DRXed grains, respectively; (d) the distribution of OIM pattern quality (IQ) facets with superimposed grain boundary orientation differences; (e) KAM map; (f) IGMA analyses of the corresponding magnification of (b).

Fig.1 (a) shows that the average grain sizes of the total are 7.1 ± 1.6 μm. Fig.1 (b, c) show that the average grain sizes of the DRXed grains and the non-DRXed are 6.8 ± 0.6 μm, and 26.1 ± 2.3 μm, respectively. Simultaneously, the volume fractions of non-DRXed grains are statistically 22.5%. Fig.1 (d) shows the corresponding distribution of grain boundaries of non-DRXed grains. It should be noted that some of LAGBs are mainly distributed inside the non-DRX grains with large size, while some of LAGBs are concentrated nearby the grain boundaries of the grains with small size. Therefore, different recrystallization mechanisms exist in non-DRX gains with different sizes. Fig.1 (e) shows the blue grains (strain-free) formed inside the large-size non-DRX grains (red arrows in Fig.1 (e_1)). Similarly, blue grains (strain-free) formed nearby the small-size non-DRX grain boundary are observed (yellow arrows in Fig.1 (e_2)). This indicates the formation of recrystallized grains located inside the grains or nearby the grain boundary. Fig.1 (f_1, f_2) shows that in the early stages of deformation, the activation of the basal < a > slip in large-sized non-DRX grains leads to new grain formation from within their grains. This suggests that a sub-grain growth mechanism (belonging to continuous dynamic recrystallization (CDRX)) is responsible for the recrystallization process in large-sized non-DRX grains. Differently, activation of pyramidal < c + a > slip in small-sized non-DRX grains leads to the formation of new grains at grain boundaries. This shows that the SIBM mechanism (belonging to discontinuous dynamic recrystallization (DDRX)) is responsible for the recrystallization process in small-sized non-DRX grains.

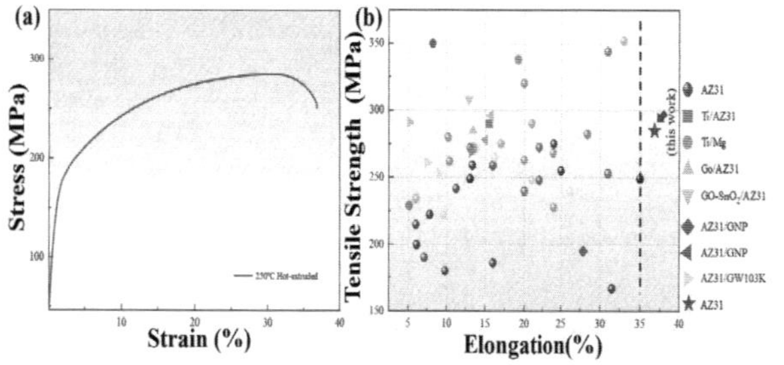

Fig.2 (a) Engineering stress-strain curves of the hot-extruded AZ31B at 250 °C; (b) Comparison of the UTS and El of the hot-extruded AZ31 alloys in this work and other AZ31 alloys.

The unique recrystallization mechanism in the heterogeneous structure leads to excellent mechanical properties. Fig.2 (a) show that the hot-extruded temperature is 250 °C and the ram speed is kept at 0.2 mm/s, the ductility is significantly increased to 36.8 ± 0.3 % in the tensile test, which is a record of extruded AZ31B (Fig.2 (b)). Moreover, the high-ductility AZ31B has a balanced strength of about 285 ± 2.5 MPa.

4 Conclusion

(1)The hot-extruded AZ31B alloys are composed typical bimodal microstructure containing fine DRX grains and coarse non-DRX grains. Significant size differences are also present in the coarse non-DRX grains at a hot-extruded temperature of 250 °C.

(2)When the temperature of 250 °C, larger sized non-DRX grains undergo continuous dynamic recrystallization (sub-grain growth) with the activation of basal <a> slip. Small-size non-DRX grains incur discontinuous dynamic recrystallization (stress-induced grain boundary migration (SIBM)) with activation of the pyramidal < c + a > slip.

5 Acknowledgments

This work is supported by National Natural Science Foundation of China (Nos. 52375394, 52074246, 52275390, 52205429, 52201146, 52305429), National Defense Basic Scientific Research Program of China (JCKY2020408B002), Shanxi Scholarship Council of China (No. 2021-125), Natural Science Foundation of Shanxi Province (No. 20210302124631), and Key Research and Development Program of Shanxi Province (Nos. 202102050201011, 202202050201014).

References

[1] Shuai C, Liu W, Li H Q, Wang K L, Zhang Y T, Xie T Z, Chen L W, Hou H, Zhao Y H. Dual influences of deformation-induced W precipitates on dynamic recrystallization and fracture mechanism of the hot-extruded Mg-Y-Zn alloys: An experimental and phase field study. Int. J. Plasticity, 2023, 170: 103772.

[2] Liu W, Zhao Y H, Zhang Y T, Shuai C, Huang Z Q, Hou H. Deformation-induced dynamic precipitation of 14H-LPSO structure and its effect on dynamic recrystallization in hot-extruded Mg-Y-Zn alloys. Int. J. Plasticity, 2023, 16:103573.

[3] Li C H, Hou H, Dong R F, Zhang X Y, Xu X L, Zhao Y H. Recrystallization behavior and mechanical properties of AZ31B alloy during the hot-rolling process. J Mater Res Technol 2023,23: 1005-1014.

Variations in Recrystallization Kinetics of Magnesium Influenced by <11-20>Twinning During the Hot Rolling Process

Huanqing Li[1], Zhuo Song[1], Yuntao Zhang[1], Hua Hou[1,3], Yuhong Zhao[1,2,4*]
1 School of Materials Science and Engineering, Collaborative Innovation Center of Ministry of Education and Shanxi Province for High-performance Al/Mg Alloy Materials, North University of China, Taiyuan, 030051, China
2 Beijing Advanced Innovation Center for Materials Genome Engineering, University of Science and Technology Beijing, Beijing, 100083, China
3 School of Materials Science and Engineering, Taiyuan University of Science and Technology, Taiyuan, 030024, China
4 Institute of Materials Intelligent Technology, Liaoning Academy of Materials, Shenyang, 110004, China
*Corresponding address: e-mail: zhaoyuhong@nuc.edu.cn

Abstract: In this work, we proposed a spatialized temporal method to reveal the structural evolution of Mg during hot rolling process laterally, and then the recrystallization kinetic behavior of Mg affected by twinning during different passes of rolling was reported.The results show that recrystallization occurs preferentially close to the sample boundary, and multiple rolling passes result in a grain size gradient distribution along the sample boundary at an angle, which is related to the roll radius and the strain. In addition, high-angle $\langle 11\bar{2}0\rangle$ tensile twins impede twinning of recrystallized grains, whereas grains without high-angle tensile twins are prone to form twins within the system when DRX occurred, reducing the stressgradient between the recrystallized grains and the parent grains. It is well established that a clear linear correlation exists between the stress gradient and the strain hardening rate, leading to non-uniform deformation-induced strengthening.Through this work, we are hopeful to provide insightsforstress gradient or homogenization of rolled samples by modulating the initial twinning state.

Keywords: Dynamic recrystallization, Magnesium, Hot rolling, Grain refinement, Stress gradient.

1 Introduction

Dynamic recrystallization in Mg usually affected by tensile twinning due to c-axis or perpendicular-to-c-axis forces during plastic deformation[1]. The twins provide the driving force for recrystallization through lattice strain as well as energy stored in grain boundary defects[2]. It is generally believed that a lower dislocation density prevails within the recrystallized grains, which tends to form a strain gradient with the parent crystal, and this strain distribution leads to non-uniform deformation of the recrystallized and non-recrystallized regions in tensile tests, which creates a source of hetero-deformation induced stresses and enhances the strong plasticity of the material[3].However, we find a strain homogenization phenomenon during pure Mg rolling that differs from the above conclusions, and propose a spatialized time continuum approach to analyze the contribution of high-angle $\langle 11\bar{2}0\rangle$ twins to this stress phenomenon.

2 Experimental procedure

Commercially pure magnesium (purity > 99.9%) was selected as the initial material for rolling to investigate recrystallization behavior independently of the second phase influence.The hot rolling temperature was 673K, with the samples preheated at this temperature for 1 hour before rolling, followed by a five minutes re-hold at 673K after each pass. The work roll temperature was 423 K, operating at a speed of 5 r/min with a roll diameter of 130 mm.
Fig.1(a) illustratesthe dimensional shape of a multi-channel hot-rolled pure Mg sample, segmented into four stages according to appliedthe rolling pressure, with each stage corresponding to the thickness $\boldsymbol{h_i}(\boldsymbol{i}$=0,1,2,3). Additionally, three transitions A, B and C are retained, and the ND directional strain is $\Delta\boldsymbol{\varepsilon_A} = |\boldsymbol{h}_1 - \boldsymbol{h}_0|/\boldsymbol{h}_0 = 30.38\%$, and similarly, $\Delta\boldsymbol{\varepsilon_B} = 30.38\%, \Delta\boldsymbol{\varepsilon_C} = 65.07\%$. For the transitional region of the rolled samples, spatially continuous observations can be used to approximate the temporal evolution of the Mg samples during rolling, known as the spatialized temporal approach, in order to correlate in-situ observations of the recrystallization behavior.Microstructural characterization was conducted using a JEOL JSM-IT800 scanning electron microscopy and an OXFORD electron back scatter diffraction (EBSD) system operatingat an accelerating voltage of 20 kV.

3 Result and discussion

Recrystallization statistics in rolling processes

The spatialized time method combined with the EBSD technique unveils the proportions and evolution of recrystallized grains, substructured grains, and deformed grains in Mg during hot rolling, as depicted in Figs.1 (b1-b8, c1-c5).

The volume fraction of recrystallization increases with the rising downward rolling strain. In the initial stages of rolling, grains deform under stress, generating defects or displacing them, facilitating recrystallization nucleation. However, a significant number of finely deformed grains persisted at the specimen boundaries, attributed to shear forces between the sample boundary and the work roll, along with the strain gradient.

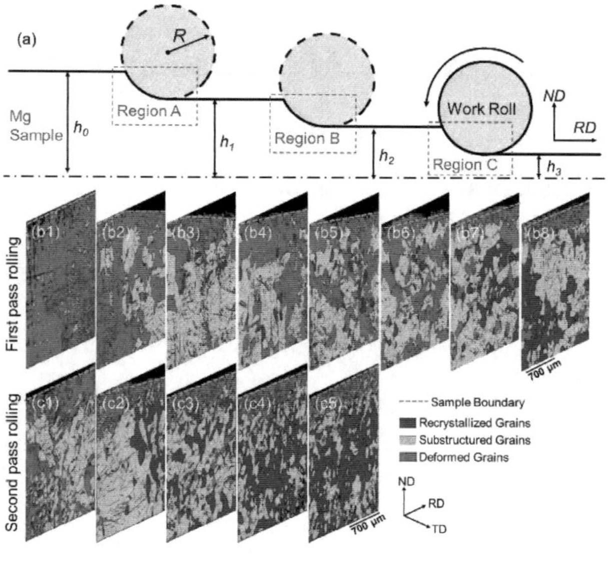

Fig.1 (a) Schematicdiagram of hot rolling and EBSD results show the recrystallization behavior of the first rolling transition zone (b1-b8) and the second rolling transition zone (c1-c5) of Mg samples.

Paradoxical relationship between twins

Further analysis of the recrystallization behavior during the hot rolling process revealed the presence of high-angle tensile twins within the parent grains,as shown in Fig.2, influencing the morphology of the recrystallized grains.

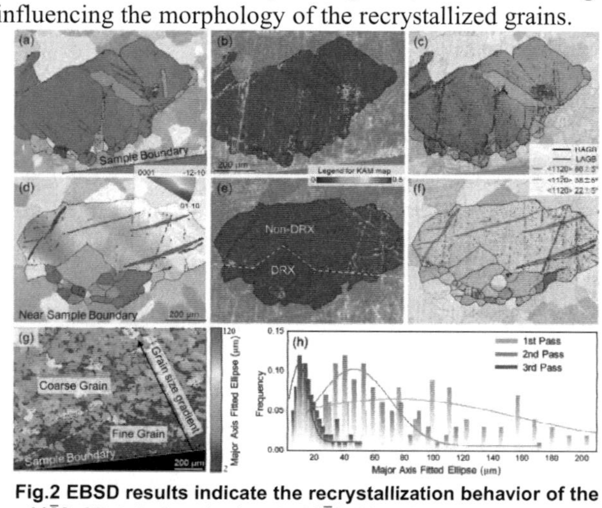

Fig.2 EBSD results indicate the recrystallization behavior of the <11$\bar{2}$0>86° twin-free (a-c) and <11$\bar{2}$0>86° twin systems (d-f), and Figs.(g) and (h) show the grain scale gradient distribution and the grain size statistics during the three rolling passes, respectively.

Within grains with minimal or no <11$\bar{2}$0> 86° high-angle twins, recrystallized grains tend to exhibit a higher density of twins. Conversely, in the recrystallization behavior of grains containing <11$\bar{2}$0>86° grains, there are scarcely any twinswithin the recrystallized grains.Consequently, these grains exhibit lower dislocation densities, creating stress disparities with the parent grains matrix.Additionally, the grain size statistics indicate that recrystallization occurring during the rolling process substantially promotes grain refinement. This refinement effect exhibited non-uniformity, with grains tending to be finer near the sample boundaries,as depicted in Fig.2(g). The grain sizes exhibited a gradient distributionof approximately 30° along the sample boundaries.

4 Conclusion

The spatialized temporal approach proposedin this work offers an avenue for approximatein-situ observation of recrystallization behavior.The existence of <11$\bar{2}$0> twins inhibit the formation oftwins within the recrystallized grains, resulting in an interfacial stress gradient between the recrystallized and non-recrystallized regions.Conversely, parent grains devoid of high-angle twins tend to homogenize interfacial stresses during the recrystallization process.

5 Acknowledgments

This work is supported by the National Natural Science Foundation of China (No. 52375394, 52074246, 52275390, 52205429, 52201146), National Defense Basic Scientific Research Program of China (JCKY2020408B002), Key Research and Development Program of Shanxi Province (202102050201011, 202202050201014), and Graduate Student Research Innovation Program in Shanxi Province (2023KY580).

References

[1] Tiamiyu A A, Pang E L, Chen X, Lebeau J M, Nelson K A, Schuh C A. Nanotwinning-assisted dynamic recrystallization at high strains and strain rates[J]. Nature materials, 2022, 21(7): 786-794.

[2] Li L M, Li H Q, Lei P Y, Liu W, Chen L W, Hou H, Zhao Y H. Microstructure and mechanical properties with different sintering temperature of AZ91D alloy[J]. Journal of Magnesium and Alloys, 2024.

[3] Yang Z S, Liu X, Zhao JJ, et al. Discontinuous coarsening leads to unchanged tensile properties in high-entropy alloys with different recrystallization volume fractions[J]. International Journal of Plasticity, 2024, 176: 103963.

Changing the Shape of the Second Phase Distribution Improves the Mechanical Properties of Mg-6.7Y-2.5Zn Alloy by Adding Trace Ta

Limin Li[1], Pengya Lei[1], Zhuo Song[1,3], Huanqing Li[1], Liwen Chen[1], Hua Hou[1,3], Yuhong Zhao[1,2,4*]
1 School of Materials Science and Engineering, Collaborative Innovation Center of Ministry of Education and Shanxi Province for High-performance Al/Mg Alloy Materials, North University of China, Taiyuan, 030051, China
2 Beijing Advanced Innovation Center for Materials Genome Engineering, University of Science and Technology Beijing, Beijing, 100083, China
3 School of Materials Science and Engineering, Taiyuan University of Science and Technology, Taiyuan, 030024, China
4 Institute of Materials Intelligent Technology, Liaoning Academy of Materials, Shenyang, 110004, China
*Corresponding address: e-mail: zhaoyuhong@nuc.edu.cn

Abstract: The addition of tantalum (Ta) to magnesium alloys, as an alloying element, has not been widely studied. This research employed optical microscopy, scanning electron microscope, and compression tests to investigate the influence of Ta addition on the microstructural evolution and mechanical properties of Mg-6.7Y-2.5Zn alloy. The results indicated that the inclusion of Ta leads to an increase in texture strength and a reduction in the network-like morphology of the second phase. Multiple strengthening mechanisms synergistically contribute to the superior mechanical performance of the Mg-6.7Y-2.5Zn-0.1Ta alloy, with compressive strength and ultimate compressive strength values of 98.6 MPa and 253.4 MPa, respectively. The phase-field simulation unveiled the microscopic mechanisms involved in the sintering process. The synergy between experimentation and simulation enhances our comprehension of the intricate relationship between microstructure evolution and mechanical properties.

Keywords: Mg-6.7Y-2.5Zn-0.1Ta alloy, Phase-field, Second phase, Texture strength, Mechanical properties.

1 Introduction

The tensile yield strength of Mg-Y-Zn alloys produced by sand casting or permanent casting methods is significantly lower [1]. In order to enhance its strength, this study attempts to employ spark plasma sintering to fabricate the alloy. Furthermore, the exploration of the microstructural and macrostructural alloying elements has been undertaken to understand the evolution of the Mg-Y-Zn alloy. Recent research on Mg-Y-Zn alloys has primarily focused on their composition ratios, and the influence of adding Ta on their mechanical properties has not been extensively studied [2]. The impact of Ta addition on the mechanical behavior of Mg-Y-Zn alloys and how it affects the microstructure evolution remain less understood and explained. Therefore, this paper reports on the influence of Ta addition on the microstructural evolution and mechanical properties of Mg-6.7Y-2.5Zn alloy. By combining traditional compression tests with scanning electron microscopy techniques, a deeper understanding of the microstructural evolution of Mg-Y-Zn alloys after Ta addition can be achieved. Additionally, this study aims to explore how microstructural changes affect the mechanical properties of the alloy. In addition, not only did the phase-field simulation study the temporal evolution of porosity at different particle locations, but it also investigated the grain coarsening process through the phase-field simulation. The combination of experimentation and simulation facilitates a better understanding of the relationship between microstructure evolution and mechanical properties.

2 Experimental procedure

High-purity Mg powder (99.99wt.%), high-purity Zn powder (99.9999wt.%), high-purity Y powder (99.99wt.%), and high-purity Ta powder (99.99wt.%) with particle sizes ranging from 75 to 150 μm were ball-milled for half an hour to prepare the alloy powders. The microstructure of the powders is shown in Fig. 1. Using a Spark Plasma Sintering furnace (SPS-20T-10-III), cylindrical samples were prepared with a diameter of 30 mm. The heating rate was set at 20 °C/min, the temperature reached 360 °C, and the samples were held at this temperature for 5 minutes.

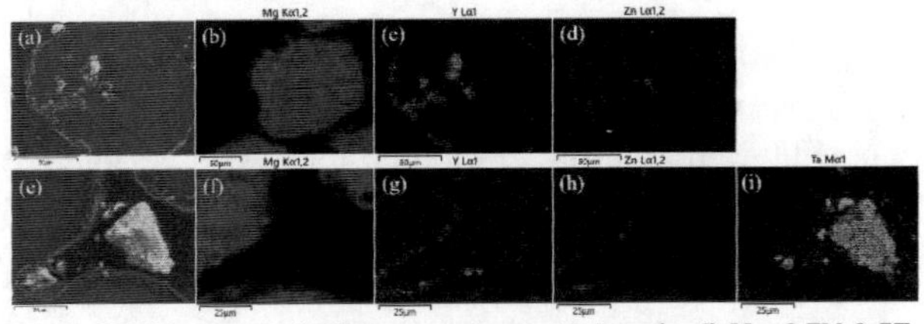

Fig. 1. SEM images of alloys powder particles: (a-d) Mg-6.7Y-2.5Zn, (e-i) Mg-6.7Y-2.5Zn-0.1Ta.

Prior to hardness measurement, the sintered alloy was ground to 1500 grit SiC. Vickers hardness tests were performed on the sintered samples using a TMHVS-1000 microhardness tester under a load of 0.5 kgf. Hardness values for each sample were obtained from 20 indents made along the diameter. In the compression tests, the compressive mechanical properties of the sintered samples were measured using a universal testing machine (AG-X plus). The average mechanical properties were determined from three measurements.

3 Result and discussion

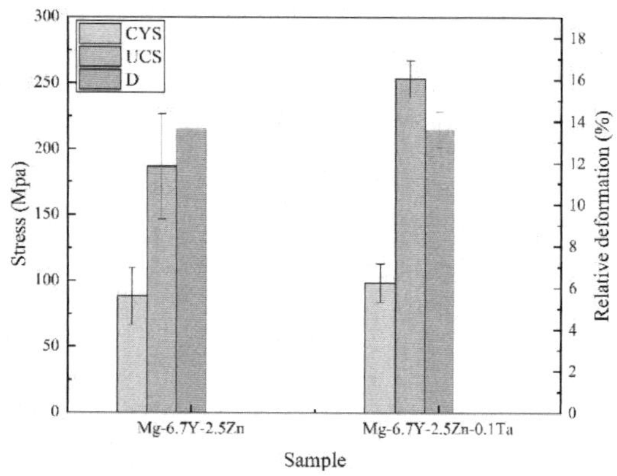

Fig. 2. Compressive properties of sintered alloy.

A comparison of the typical mechanical properties of the two sintered alloys is shown in Fig. 2. The yield compressive strength of Mg-6.7Y-2.5Zn-0.1Ta alloy is 98.6±14.3 MPa, which is approximately 1.1 times higher compared to Mg-6.7Y-2.5Zn alloy. The ultimate compressive strength of Mg-6.7Y-2.5Zn-0.1Ta alloy is 253.4 MPa, approximately 1.4 times higher than that of Mg-6.7Y-2.5Zn alloy. However, relative strain slightly decreased with increasing CYS and UCS. The addition of Ta enhanced both the UCS and CYS of Mg-6.7Y-2.5Zn alloy.The increase in strength in Mg-6.7Y-2.5Zn-0.1Ta alloy can be attributed to multiple strengthening mechanisms working in tandem [3]. The main strengthening mechanisms in different alloys include dislocation strengthening, texture strengthening, solid-solution strengthening, and precipitation strengthening. In summary, the enhanced strength of the Mg-6.7Y-2.5Zn-0.1Ta alloy results from the synergistic effects of these strengthening mechanisms working together. Dislocation strengthening restricts dislocation motion, texture strengthening increases the energy required for deformation, solid-solution strengthening involves lattice distortions, and precipitation strengthening occurs due to interactions with precipitate particles. Additionally, the unique coarse-grained and twin-containing microstructure of the Mg-6.7Y-2.5Zn-0.1Ta alloy may also contribute to its strength. The alloy's elongation did not significantly decrease after the addition of Ta, likely because twin boundaries and dislocation accumulation in grain-dense regions effectively inhibit crack initiation and propagation, ensuring consistent ductility between the two alloys.

4 Conclusion

(1)The addition of Ta enhances the mechanical properties of Mg-6.7Y-2.5Zn alloy, resulting in higher strength in the Ta-containing magnesium alloy.

(2)The increase in compressive strength in Mg-6.7Y-2.5Zn-0.1Ta alloy is complex and may be attributed to the combined action of multiple strengthening mechanisms.

(3)imulation of the phase-field method provides a better understanding of the relationship between microstructure evolution and mechanical properties.

5 Acknowledgments

This work was supported by National Natural Science Foundation of China (Nos. 52375394, 52074246, 52275390, 52205429, 52201146, 52305429), National Defense Basic Scientific Research Program of China (JCKY2020408B002), Shanxi Scholarship Council of China (No. 2021-125), Natural Science Foundation of Shanxi Province (No. 20210302124631), and Key Research and Development Program of Shanxi Province (Nos. 202102050201011, 202202050201014).

References

[1] Li D, Zhang J S, Que Z P, et al. Effects of Mn on the microstructure and mechanical properties of long period stacking ordered Mg95Zn2.5Y2.5 alloy[J]. Materials Letters, 2013, 109: 46-50.

[2] Zhang E, He W W, Du H, et al. Microstructure, mechanical properties and corrosion properties of Mg-Zn-Y alloys with low Zn content[J]. Materials science & engineering A, 2008, 488: 102-111.

[3] Xin T Z, Zhao Y H, Mahjoub R, et al. Ultrahigh specific strength in a magnesium alloy strengthened by spinodal decomposition[J]. Science Advances, 2023, 7: 1-9.

Influence of Heat Treatment Process on Microstructure and Properties of Indirect Squeeze Casting ZM5 Magnesium Alloy baffle

Hua Hou[1,3], Jianhui Jing[1], Liwen Chen[1], Yuhong Zhao[1,2,4*]
1 School of Materials Science and Engineering, Collaborative Innovation Center of Ministry of Education and Shanxi Province for High-performance Al/Mg Alloy Materials, North University of China, Taiyuan, 030051, China
2 Beijing Advanced Innovation Center for Materials Genome Engineering, University of Science and Technology Beijing, Beijing, 100083, China
3 School of Materials Science and Engineering, Taiyuan University of Science and Technology, Taiyuan, 030024, China
4 Institute of Materials Intelligent Technology, Liaoning Academy of Materials, Shenyang, 110004, China
*Corresponding address: e-mail: zhaoyuhong@nuc.edu.cn

Abstract: This work is based on the self-developed ZBYT500 multifunctional liquid molding equipment to prepare ZM5 magnesium alloy baffle by squeeze casting process. The effects of two different heat treatment processes, T4 (solution treatment) and T6 (solution treatment + artificial ageing), on the mechanical properties of castings were investigated. The average tensile strength of the castings after solid solution treatment was 228.37 MPa, the average yield strength was 91.49 MPa and the average elongation was 7.3%. The average tensile strength of the best mechanical properties after aging treatment is 242.49 MPa, the average yield strength is 120.8 MPa, and the average elongation is 3.9%. Considering the comprehensive properties, T4 treated ZM5 magnesium alloy baffle will be more suitable for the application.

Keywords: Squeeze casting; Magnesium alloy; Solution treatment; Aging treatment; Mechanical properties.

1 Introduction

Magnesium alloys as the lightest structural metal materials in current engineering applications, are mainly used in large-scale and complex structural components [1]. However, the absolute strength of magnesium alloys molded by the traditional casting process is relatively low due to the presence of harmful phases. In Mg-Al alloys, β-$Mg_{17}Al_{12}$ dissociated eutectic organizations are usually in the form of coarse network or lamellar distribution, which is prone to stress concentration and fracture, thus reducing the mechanical properties of the alloys and restricting their further development in engineering applications [2,3]. In this study, ZM5 baffles were produced based on ZBYT500 multifunctional liquid molding equipment using indirect squeeze casting process. And the heat treatment research on magnesium alloy baffle castings were expected to provide reference for the promotion and application of heat treatment process technology of squeeze casting magnesium alloy castings.

2 Experimental procedure

The baffle castings were produced through the processes of material preparation, melting, pouring under protective gas and indirect squeeze casting. Baffle casting refers to HB5462-1990 standard for heat treatment process. ZM5 magnesium alloy heat treatment process parameters is shown in Table 1.

Table 1. Heat treatment process parameters

Solid solution temperature/℃	Solid solution time/h	Quenching medium
410	20	Water
Ageing temperature/℃	Ageing time/h	Cooling medium
200	8	Air

3 Result and discussion

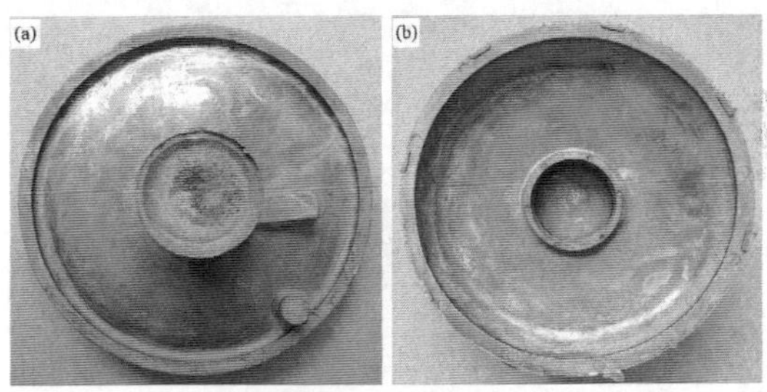

Fig. 1 Photographs showing the macroscopic morphology of ZM5 magnesium alloy baffle.

The ZM5 magnesium alloy baffle prepared by indirect squeeze casting process are shown in Fig. 1(a)(b). It can be found that the casting filling is complete; the surface of the casting is free from defects.

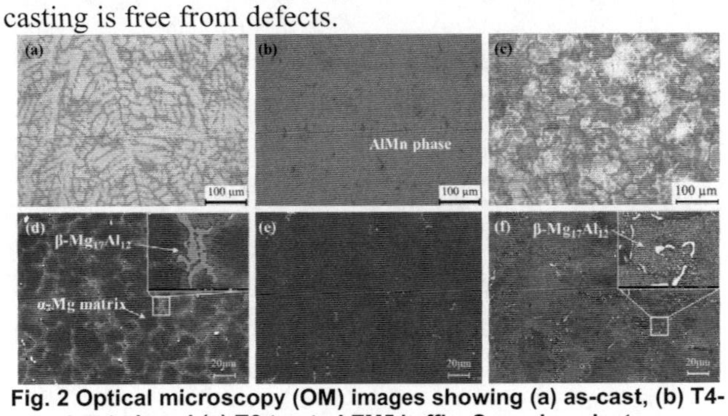

Fig. 2 Optical microscopy (OM) images showing (a) as-cast, (b) T4-treated, and (c) T6-treated ZM5 baffle. Scanning electron microscope (SEM) images showing the microscopic morphology of castings under different heat treatment processes: (a) as-cast, (b) T4-treated, and (c) T6-treated.

The microstructure of the baffle in different heat treatment states is shown in Figs. 2(a)-(f). The coarse eutectic phase $Mg_{17}Al_{12}$ of the as-cast castings is distributed on the grain boundaries in a network structure. After solid solution treatment, the second phase $Mg_{17}Al_{12}$ at the grain boundaries and inside the grains was basically dissolved into the matrix, forming α-Mg oversaturated solid solution. The lattice distortion occurred in the solid solution treated grains, and the grain size increased compared with the as-cast state. In Fig. 2(b), the red box shows the incompletely dissolved AlMn phase inside the α-Mg grains. Figs. 2(c) and (f) are the microstructure of castings after T6 treatment, and it can be found that the castings after aging treatment precipitate the second phase inside the grains in the form of continuous precipitation, and precipitate at the grain boundaries in the form of discontinuous precipitation.

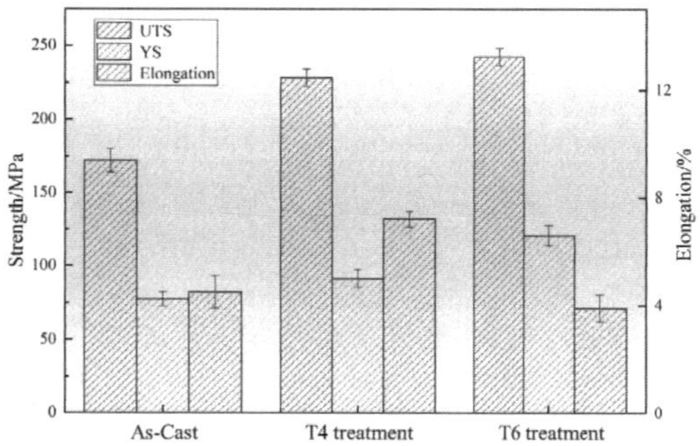

Fig. 3 Mechanical properties of castings under different heat treatment processes.

The test results of mechanical properties of castings under different heat treatment processes are shown in Fig. 3. It can be found that the mechanical properties of the castings in the T4 heat treatment state are improved compared with the as-cast state. While the tensile strength and yield strength of castings after T6 treatment are slightly improved, the elongation decreases and is even lower than the as-cast state. The lattice distortion caused by the solid solution treatment after the dissolution of the coarse second phase into the matrix and the undissolved AlMn particle plays a synergistic effect to enhance the mechanical properties of the castings, while the second phase particles re-precipitated after artificial aging become an effective nailing point to hinder the dislocation slippage, which further improves the mechanical properties, but a large number of brittle second-phase precipitation also reduces the ductility of the castings.

4 Conclusion

ZM5 magnesium alloy baffle plate was successfully prepared by indirect squeeze casting technology, and the casting is complete with no obvious defects on the surface. In terms of comprehensive strength and elongation, the baffle casting after solid solution treatment at 410℃×20h is more suitable for actual industrial production.

5 Acknowledgments

This work was supported by National Natural Science Foundation of China (Nos. 52375394, 52074246, 52275390), National Defense Basic Scientific Research Program of China (JCKY2020408B002), Key Research and Development Program of Shanxi Province (Nos. 202102050201011).

References

[1] Chen L W, Zhao Y H, Hou H, Zhang T, Liang J Q, Li M X and Li J. Development of AZ91D magnesium alloy-graphene nanoplatelets composites using thixomolding process. J. Alloys Compd., 2019, 778: 359-374.

[2] Chen L W, Jing J H, Zhang L L, Li J, Chen W P, Li L M, Zhao Y, Hou H and Zhao Y H. Corrosion behavior of graphene nanosheets reinforced magnesium matrix composites in simulated body fluids. Acta Metall. Sin. (Engl. Lett.), 2024, 37: 525–536.

[3] Song J, She J, Chen D and Pan F S. Latest research advances on magnesium and magnesium alloys worldwide. J. Magnes. Alloy, 2020, 8(1): 1-41.

A Novel Type of Mg-Zn-Cu Alloys with High Thermal Conductivity

Wei Liu[1], Yuntao Zhang[1], Hou Hua[1,2], Yuhong Zhao[1*]
1.School of Materials Science and Engineering, Collaborative Innovation Center of Ministry of Education and Shanxi Province for High-performance Al/Mg Alloy Materials, North University of China, Taiyuan, Shanxi 030051, PR China.
2.School of Materials Science and Engineering, Taiyuan University of Science and Technology, Taiyuan, Shanxi 030024, PR China.
*Corresponding address: e-mail: zhaoyuhong@nuc.edu.cn

Abstract: In this work, we investigate microstructure and thermal conductivity of Mg-3Zn-xCu alloys (x=1, 2, 3, wt. %). Firstly, Mg-3Zn-Cu alloys are composed of α-Mg matrix and MgZnCu phase. Secondly, with Cu content increase, the grain size of α-Mg matrix is refining and the volume fraction of MgZnCu phase is increase. Finally, the Cu addition gradually improves thermal conductivity of Mg-3Zn-Cu alloys. The Mg-3Zn-3Cu alloy shows the best thermal conductivity of 138.5 W/(m·K), which are due to reduction of casting defects, formation of continuous MgZnCu network structure, and the introduction of a large number of free electrons. This work proves that the development of Mg-Zn-Cu based magnesium alloys can provide theoretical basis for applications in the engineering field.

Keywords: Mg-Zn-Cu alloy, MgZnCu phase, Thermal conductivity, Mirostructure organization.

1 Introduction

Mg alloys as the lightest structural metal, whichhas a wide range of applications [1-2]. However, with the number of high-power devices is continuously increasing, and the heat generated needs to be dissipated. How to further enhance thermal conductivity of Mg alloy has become a hot topic in research [3-4].The thermalconductivityofpure Mgis 156 W/(m·K). Butthe thermal conductivity of traditional castMg alloys can hardly exceed 100 W/(m·K). Wu et al. [5] has developed a Mg-5Zn-5Cu alloy with a high thermal conductivity of 136.5 W/(m·K). Therefore, it is reasonable to believe that Mg Zn Cu is a new type of high thermal conductivity alloy. In this work, in order to clarify the microstructural changes and thermal conductivity mechanism of Mg-Zn-Cu alloys, three Mg-3Zn-xCu (x=1, 2, 3, wt. %) casting alloys are designed. The results indicate that the thermal conductivity of Mg-3Zn alloy increases with the increase of Cu content. The result provided a theoretical basis for the development of new high thermal conductivity Mg alloys

2 Experimental procedure

The original ingots with nominal composition of Mg-3Zn-Cu (wt. %) were fabricated in an electric-resistant furnace under argon atmosphere by melting high purity Mg, Zn and Cu at 740 ℃ for 30 min. Then, the melts were cast into a preheated mold at 180 ℃ when the temperature was decreased to 720 ℃. Finally, the cast alloys with a diameter of 40 mm and a height of 180 mm were obtained and their measured chemical compositions were examined by inductively coupled plasma analyzer, in Table 1.

Table 1. actual compositions of Mg-3Zn-Cu sample

Alloy	Mg	Zn	Cu
Mg-3Zn-1Cu	Bal.	3.02	0.98
Mg-3Zn-2Cu	Bal.	2.98	2.03
Mg-3Zn-3Cu	Bal.	3.01	3.02

The thermal diffusivity (α) and the specific heat(Cρ) of the sample was measured by LFA-500 laser thermal conductivity instrument at 25℃. The sample with a diameter of 12.7 mm and a height of 3 mm to ensure the one-dimensional heat transfer state. The density (ρ) was determined by the Archimedes method. The thermal conductivity (λ) was calculated by the following Eq. (1):

$$\lambda = \alpha \rho C\rho \#(1)$$

where λ indicates the thermal conductivity, α is thermal diffusivity, ρ is the density, and Cρ is the specific heat capacity.

3 Result and discussion

Figures. 1(a)-1(c) are the OM images showing microstructure characteristic of these as-cast Mg-3Zn-Cu alloys with different Cu content.Meanwhile, their average grain size of α-Mg phase is measured by the linear intercept methodand related results are listed in Figs. 1(g)-1(i). It can be found that the grain size of α-Mg phases in the Mg-3Zn-1Cu, Mg-3Zn-2Cu, and Mg-3Zn-3Cu alloys are measuredtobe 73.24±12.33 μm, 69.74±10.82 μm, and 63.23±9.12 μm, respectively. In addition, it can be found that the as-cast Mg-3Zn-Cu alloys are mainly consisted of α-Mg matrix, and MgZnCu phase in Figs. 1(d)-1(f). the MgZnCu phases distributed along grain boundaries exhibit a discontinuous network shape in the Mg-3Zn-1Cu alloy. At the same time, a semi-continuous network shape in the Mg-3Zn-2Cu alloy and a continuous network shape Mg-3Zn-3Cu alloys, respectively.

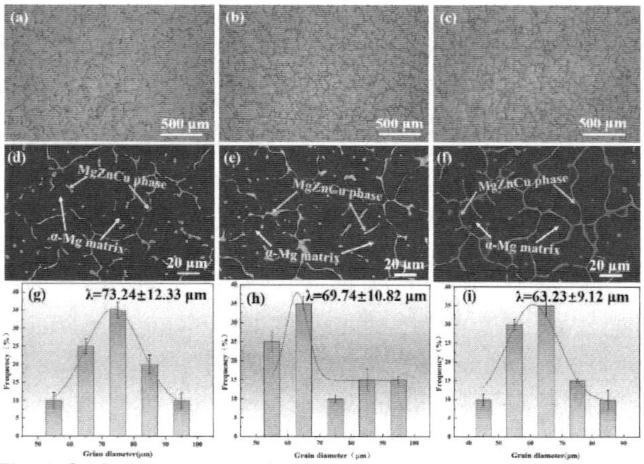
Fig. 1. Optical microscopy (OM) images showing (a) Mg-3Zn-1Cu, (b) Mg-3Zn-2Cu, and (c) Mg-3Zn-3Cu alloys. (e)-(f) Scanning electron microscopy (SEM) images showing secondary phases under different Cu contents: (e) Mg-3Zn-3Cu, (f) Mg-3Zn-2Cu, and (g) Mg-3Zn-3Cu alloys. (g)-(i) correspondingly grain size of α-Mg phases in (a)-(c).

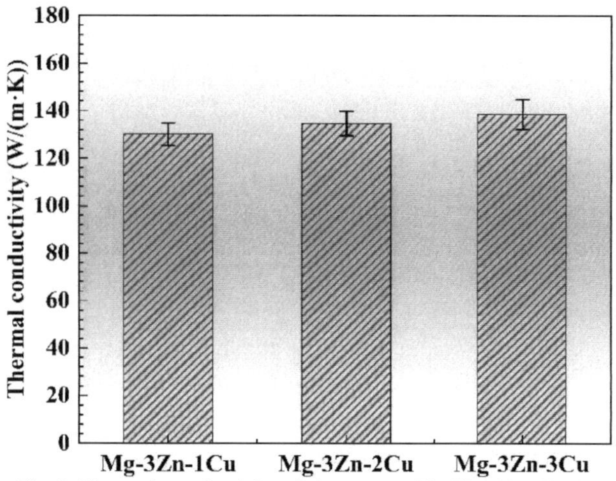

Fig. 2. Thermal conductivity of the as-cast Mg-3Zn-1Cu, Mg-3Zn-2Cu, and Mg-3Zn-3Cu alloys at 25°C temperature, respectively.

Thethermalconductivityofthethreesetsofalloysaremeasuredt obe 130.1 W/(W·K), 134.7 W/(W·K), and 138.5 W/(W·K) in Fig. 2. It can be concluded that the thermal conductivity of Mg-3Zn-Cu alloy increases with the increase of Cu content.The reasons for Cu improving the thermal conductivity of Mg-3Zn-Cu alloys, which could be explained by the following reasons. Firstly, Cu can increase the undercooling of Mg-3Zn-Cu alloy, which can refine grain size of α-Mg and eliminate casting defects. The reduction of casting defects decrease the scattering of electrons and phonons, which improving the thermal conductivity of the alloy. Secondly, with the increase of Cu content, the shape of MgZnCu phase changes from a discontinuous to a continuous. The continuous network shape provides favorable channels for heat conduction. Finally, Cu atoms contain a large number of free electrons, which is beneficial for the thermal conductivity of the Mg-3Zn-Cu alloy.

4 Conclusion

(1)The Cu addition can refine the grain size of α-Mg phase and increase the volume fraction of MgZnCu phase.

(2)The Cu addition can improve the thermal conductivity of Mg-3Zn-Cu alloys. Due to Cu addition can reduce casting defects and Cu contains a large of free electrons, which improve the thermal conductivity of Mg-3Zn-Cu alloys.

(3)The Cu addition forming a continuous network structure, which is beneficial for heat transfer and further improved the thermal conductivity of the Mg-3Zn-Cu alloys.

5 Acknowledgments

This work was supported by National Natural Science Foundation of China (Nos. 52375394, 52074246, 52275390, 52205429, 52201146, 52305429), National Defense Basic Scientific Research Program of China (JCKY2020408B002), Shanxi Scholarship Council of China (No. 2021-125), Natural Science Foundation of Shanxi Province (No. 20210302124631), and Key Research and Development Program of Shanxi Province (Nos. 202102050201011, 202202050201014)

References

[1] LiuW, Zhao Y H, Zhang Y T, Shuai C, Huang Z Q, Hou H. Deformation-induced dynamic precipitation of 14H-LPSO structure and its effect on dynamic recrystallization in hot-extruded Mg-Y-Zn alloys. Int. J. Plasticity, 2023, 16:103573.

[2] LiuW, Su Y, Zhang Y T, Chen L W, Hou H, Zhao Y H. Dissolution and reprecipitation of 14H-LPSO structure accompanied by dynamic recrystallization in hot-extruded Mg89Y4Zn2Li5 alloy. J. Magnes. Alloy., 2023, 11(4):1408-1421.

[3] Zhang W P, Ma M L, Yuan J W, Shi G L, Li Y J, Li X G, Zhang K. Microstructure and thermophysical properties of Mg–2Zn–xCu alloys. T. Nonferr. Metal. Soc., 2020, 30(7): 1803-1815.

[4] Li S B, Yang X Y, HOU J T, Du W B.A review on thermal conductivity of magnesium and its alloys. J. Magnes. Alloy., 2020, 8(1): 78-90.

[5] Chen L, Lü S, Guo W, Li J Y, Wu S S. High thermal conductivity of highly alloyed Mg-Zn-Cu alloy and its mechanism. J. Alloys Compd., 2022, 918: 165614.

Effects of Zn Content and Heat Treatment Process on Microstructures and Mechanical Properties of Al-2Li-1Cu-3Mg-xZn-0.2Zr Cast Aluminum Alloys

Jinzhong Tian[1], Boling Li[1], Zhenghao Bai[1], Shengjie Ma[1], Hou Hua[1,3], Yuhong Zhao[1,2,4*]
1 School of Materials Science and Engineering, Collaborative Innovation Center of Ministry of Education and Shanxi Province for High-performance Al/Mg Alloy Materials, North University of China, Taiyuan, 030051, China
2 Beijing Advanced Innovation Center for Materials Genome Engineering, University of Science and Technology Beijing, Beijing, 100083, China
3 School of Materials Science and Engineering, Taiyuan University of Science and Technology, Taiyuan, 030024, China
4 Institute of Materials Intelligent Technology, Liaoning Academy of Materials, Shenyang, 110004, China
*Corresponding address e-mail: zhaoyuhong@nuc.edu.cn

Abstract: In this work, the effect of Zn addition and heat treatment schemes on the microstructure and mechanical properties of cast Al-2Li-1Cu-3Mg-0.2Zr alloy were investigated. The results indicate that with the increase of Zn element will refine the grain of cast microstructure and improve its mechanical properties. The increase in the aging time hardness showed a trend of first increase and then decrease, and the tensile strength with the increase in the content of Zn increased. The hardness of Al-2Li-1Cu-3Mg-xZn-0.2Zr alloy shows a tendency to increase and then decrease during the aging treatment. The optimal comprehensive mechanical properties are obtained in Al-2Li-1Cu-3Mg-0.9Zn-0.2Zr alloy, which exhibits 404 MPa in ultimate tensile strength (UTS) and 12.1% in elongation (EL).

Keywords: Cast Al-Li alloys, Mechanical properties, Heat treatment.

1 Introduction

Al-Li alloys are age-hardening materials known for their low density, high strength, high modulus of elasticity, and fatigue resistance, making them ideal for aerospace applications where lightweight structural materials are crucial [1]. Microalloying with Al-Li-Mg alloys refines the grain structure, thereby improving strength and ductility [2-3]. In this paper, the effect of Zn and heat treatment schemes on the microstructural evolution and mechanical properties of cast Al-Li-Mg alloy were investigated.

2 Experimental procedure

The alloys were prepared in graphite crucible using resistance furnace. Table 1 lists the chemical compositions and densities of the alloys. The alloys underwent a two-stage solution heat treatment (460°C×24h+530°C×24h), followed by quenching into room-temperature water immediately and isothermal artificial aging at 175 °C for various times in an oil bath.

Table 1 ICP-OES results of the alloys compositions, wt.%

Composition	Li	Cu	Mg	Zr	Zn	Al
0.3Zn	2.1	0.97	2.4	0.14	0.31	Bal.
0.6Zn	1.8	0.99	3.1	0.13	0.59	Bal
0.9Zn	1.8	0.92	2.9	0.15	0.86	Bal

3 Result and discussion

3.1 Microstructure and mechanical properties of as-cast Al-2Li-1Cu-3Mg-xZn-0.2Zr alloy

Figure 1 shows the XRD diagram of Al-2Li-1Cu-3Mg-xZn-0.2Zr with different Zn content. The results show that Al-2Li-1Cu-3Mg-xZn-0.2Zr alloy is mainly composed of α-Al matrix, AlLi, T_1 (Al_2CuLi), T_2 (Al_6CuLi_3), θ'(Al_2Cu) and δ'(Al_3Li) phases.

Figure 2 shows the metallographic diagram of the as-cast structure of the alloy. The average grain sizes of 0.3Zn, 0.6Zn and 0.9Zn alloys are 254 μ m, 161 μ m and 99 μ m, respectively.

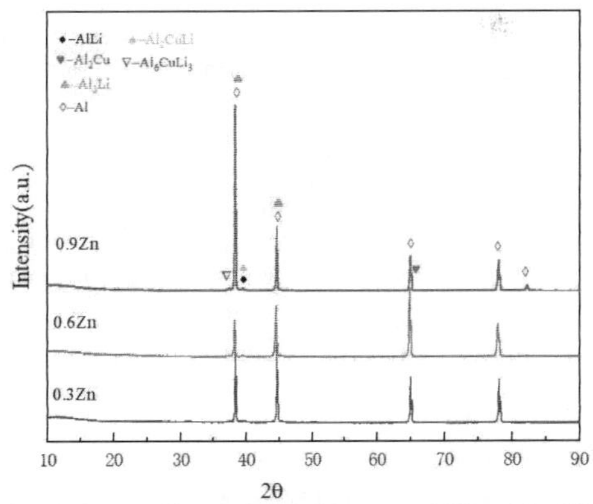

Fig.1 XRD patterns of Al-2Li-1Cu-3Mg-xZn-0.2Zr with different Zn contents

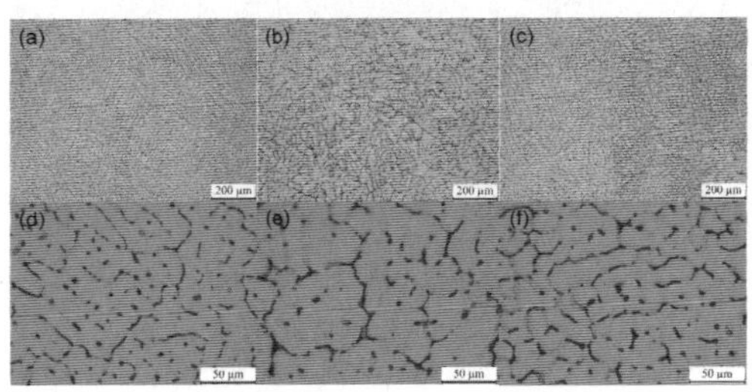

Fig.2 Microstructure of as-cast alloys with different Mg content (a) (d) 0.3Zn, (b) (e) 0.6Zn, (c) (f) 0.9Zn

3.2 Aging Microstructure and Mechanical Properties of Al-2Li-1Cu-3Mg-xZn-0.2Zr Alloy

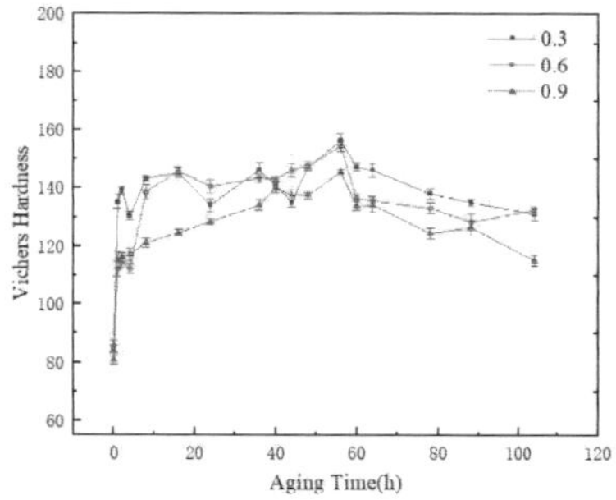

Fig.3 Age-hardening curve of Al-2Li-1Cu-3Mg-yZn-0.2Zr alloy during aging at 175°C.

With the increase of aging time, the microhardness increases at first and then decreases. After aging for 56 h, 0.3Zn, 0.6Zn and 0.9Zn alloys all reached the peak aging, and the microhardness is 156.3HV, 154.3HV and 145.6HV, respectively, indicating that the peak aging hardness did not increase with the increase of Zn content.

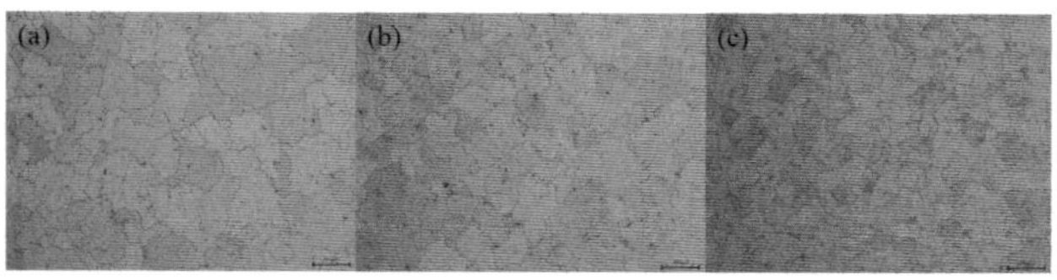

Fig. 4 Optical microstructure of aged alloys (a) 0.3Zn, (b) 0.6Zn, (c) 0.9Zn

Figure 4 shows the optical structure of the aging structure of Al-2Li-1Cu-3Mg-xZn-0.2Zr alloy. With the increase of Zn content, the grains are refined.

3.3 Mechanical Properties of Al-2Li-1Cu-3Mg-xZn-0.2Zr Alloy in the Aging State

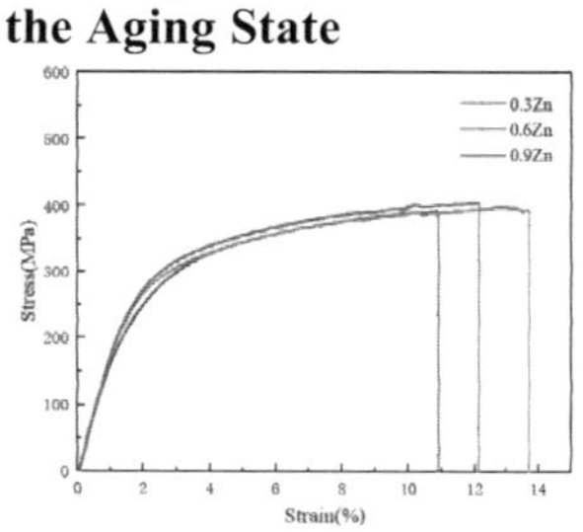

Fig.5 Stress-strain curves of Al-2Li-1Cu-3Mg-xZn-0.2Zr alloy in the aged state

With the increase of Zn content, the tensile strength of 0.6Zn alloy increases to 398MPa, and the elongation (EL) increases to 13.7%. When the Zn content is 0.9wt.%. The tensile strength of 0.9Zn alloy is increased to 404MPa, and the elongation is 12.1%.

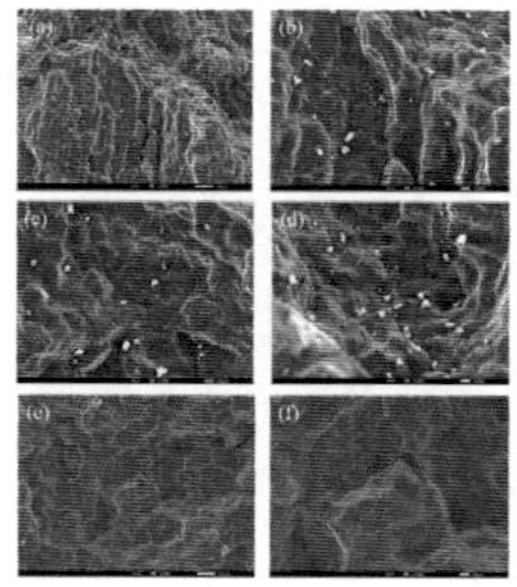

Fig.6 Fracture morphology of as-aged (a)(b)0.3Zn; (c)(d) 0.6Zn; (e)(f) 0.9Zn

The fracture surface of 0.3Zn alloy has obvious long step structure and a large number of granular second phases above the fracture surface. There is no obvious dimple on the fracture surface of 0.9Zn alloy, and the plane is smooth and intergranular fracture, but there are coarse spherical particles on the crystal surface. These precipitates effectively promote the stress concentration of smaller alloys and improve the strength and toughness of the alloys.

4 Conclusion

(1) When the Zn content increases from 0.3wt.% to 0.9wt.%, the average grain size of the as-cast alloy decreases from 254 μm to 161 μm, and the strength and plasticity of the as-cast alloy increase with the increase of zinc content.

(2) The optimal comprehensive mechanical properties are obtained in Al-2Li-1Cu-3Mg-0.9Zn-0.2Zr alloy, which exhibits 404 MPa in ultimate tensile strength (UTS) and 12.1% in elongation (EL).

References

[1] Williams J C, Starke Jr E A. Progress in structural materials for aerospace systems[J]. Acta materialia, 2003, 51(19): 5775-5799.

[2] Lavernia E, Srivatsan T S, Mohamed F A. Strength, deformation, fracture behaviour and ductility of aluminium-lithium alloys[J]. Journal of Materials Science, 1990, 25: 1137-1158.

[3] Abd El-Aty A, Xu Y, Guo X, et al. Strengthening mechanisms, deformation behavior, and anisotropic mechanical properties of Al-Li alloys: A review[J]. Journal of advanced research, 2018, 10: 49-67.

A Novel High-Strength and High Modulus $Mg_{90}Y_4Zn_2Ni_4$ Alloy

Wenyu Liu[1], Wei Liu[1], Hua Hou[1,2], Yuhong Zhao[1*]
1.School of Materials Science and Engineering, Collaborative Innovation Center of Ministry of Education and Shanxi Province for High-performance Al/Mg Alloy Materials, North University of China, Taiyuan, Shanxi 030051, PR China.
2. School of Materials Science and Engineering, Taiyuan University of Science and Technology, Taiyuan, Shanxi 030024, PR China.
*Corresponding address: e-mail: zhaoyuhong@nuc.edu.cn

Abstract: A novel high strength, high modulus hot extruded Mg90Y4Zn2Ni4 alloy is reported here with an ultimate yield strength of 380MPa, elongation at break of 10% and modulus of elasticity of 51.3 GPa. The excellent strength of the hot-extruded alloy is primarily attributed to the combination of grain refinement, long-period stacking ordered (LPSO) strengthening, and precipitation strengthening induced by stacking fault and 14H LPSO phase, among which the LPSO strengthening is dominant. The significant increase in elastic modulus is then attributed to the high-volume fraction of the high modulus 14H LPSO phase. This study can enhance specific theoretical references for the development of high modulus, high strength, and tough deformed magnesium alloys needed for engineering applications.

Keywords: Magnesium, LPSO, Plastic deformation, Elasticity modulus, Mechanical testing

1 Introduction

Elastic modulus is an important measure of a material's ability to resist deformation and is one of the most critical properties of structural materials. The value of E for Mg alloys is ~45 GPa, which is about 3/5 of Al alloys and less than 1/5 of steels[1]. In practice, this means that magnesium alloy components have a shorter service life and are more prone to failure. Because elastic modulus is an intrinsic property for Mg-based materials, which is insensitive to heat treatment and manufacturing process[2]. Therefore, enhancing the elastic modulus of Mg-based materials has been considered a critical challenge in recent decades for the development of new high-performance Mg-based materials.

The LPSO phase is a novel strengthening phase in Mg alloys, with a modulus of elasticity of 70 GPa, surpassing the Mg matrix's 40 GPa. The coherent phase boundary between the LPSO phase and the Mg matrix facilitates perfect load transfer. Thus, Therefore, the introduction of the LPSO phase in Mg alloys can increase the overall modulus of elasticity. In addition, through plasticity processes such as hot extrusion, rolling and equal channel angles, LPSO short-fiber strengthening, LPSO kink band strengthening and refinement strengthening by recrystallisation (DRX) can be introduced into Mg alloys, which significantly improves the strong plasticity of the Mg alloy[3].

2 Experimental procedure

Ingots of $Mg_{90}Y_4Zn_2Ni_4$ alloy (atom percent, at. %) were prepared in a resistance furnace under an Ar_2 protective atmosphere. Then, the as-cast was subsequently homogenized at 500 ℃ for 12 h and subsequently water quenched. Finally, the specimen was extruded at 400 ℃ with extrusion ratio of 25 under extrusion speed of 0.4mm/s. After hot extrusion, the sample is immediately cooled with water. The microstructure of the hot-extruded alloy was observed using scanning electron microscopy (SEM). The metallographic specimens were etched with a magnesium alloy etching solution. Determination of the modulus of elasticity using the Buzz-o-sonic 5.6 Modulus of Elasticity Tester. Tensile results were calculated from the average values of three repeated tests to ensure reliability.

3 Result and discussion (Bold, 10 pt., Arial)

3.1 Mechanical performance of the $Mg_{90}Y_4Zn_2Ni_4$ alloys

Table. 1. Elastic modulus and tensile property of the $Mg_{90}Y_4Zn_2Ni_4$ alloys at different conditions.

Alloy	E/GPa	TS/MPa	YS/MPa	ε/%
Cast	50.85±0.5	211±8	104±6	6.3±0.4
Extruded	51.32±0.7	434±11	362±10	10.2±0.8

The elastic modulus (E) values of the $Mg_{90}Y_4Zn_2Ni_4$ alloys were 50.36 GPa and 51.32 GPa at the as-cast and hot-extruded states, respectively. Figure. 1 depicts the tensile stress-strain curves of as-cast and hot-extruded $Mg_{94}Y_4Zn_2Ni_4$ alloys at room temperature. Table. 1 lists their tensile strength (TS), yield strength (YS) and elongation to failure (ε). The results show that the TS, YS and ε of the as-cast $Mg_{90}Y_4Zn_2Ni_4$ alloy are only 219 MPa, 104 MPa and 6.3%, respectively. After hot extruded, the TS, YS and of the $Mg_{90}Y_4Zn_2Ni_4$ alloy significant increase. The hot-extruded $Mg_{90}Y_4Zn_2Ni_4$ alloy has an outstanding strength-ductility balance with TS=434 MPa, YS=362 MPa and ε=10.2%.

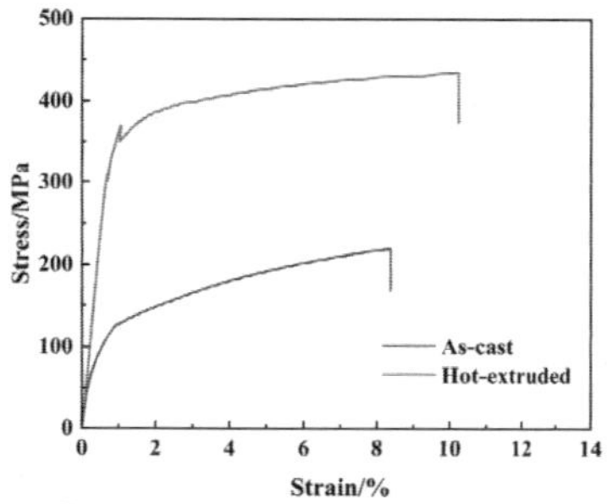

Fig. 1. Tensile properties of the as-cast and hot-extruded Mg94Y4Zn2Ni4 alloy.

3.2 Microscopic structure of the hot-extruded $Mg_{90}Y_4Zn_2Ni_4$ alloy

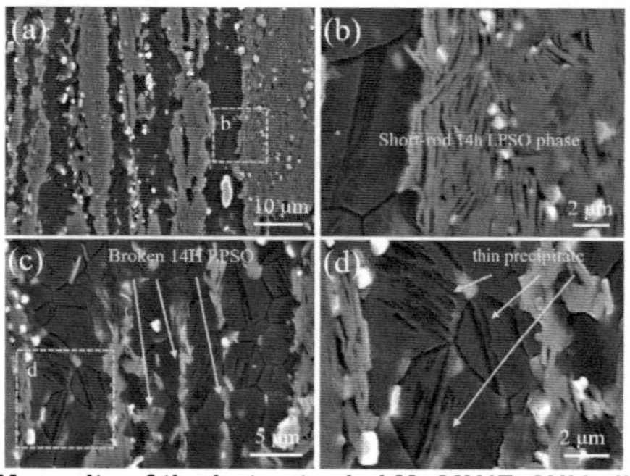

Fig. 2. SEM results of the hot-extruded Mg90Y4Zn2Ni4 alloy along the extrusion direction (ED).

The SEM image of hot-extruded Mg94Y4Zn2Ni4 alloy along the ED direction is presented in Fig. 2. It can be found that, the hot-extruded Mg90Y4Zn2Ni4 alloy exhibit an elongated microstructure in the ED. The hot extruded $Mg_{90}Y_4Zn_2Ni_4$ alloy comprises the 14H LPSO phase, diffusely distributed white granular phase and dynamically recrystallized grains, among which the 14H LPSO phase is dominant. Its volume fraction is about 65 %. The relatively high modulus of elasticity of 14H LPSO results in a significant increase in the overall modulus of elasticity of the $Mg_{94}Y_4Zn_2Ni_4$ alloy[4]. In addition, the LPSO phases display two morphologies: a block LPSO phase elongated along the extrusion direction and a fine, short-rod LPSO phase interspersed with each other. Fig. 2(b) shows a region with a dense distribution of short-rod LPSO phases, which are randomly distributed in different directions and are pinned together, resulting in better staple fibre reinforcement. It is observed in Fig. 2(c), a significant amount of completely fragmented LPSO phases were found to be distributed along the DRX grain boundaries in the region with a higher number of DRX grains. The broken LPSO has a better Particle Stimulated Nucleation (PSN) effect due to the increased number of nucleation sites, thus promoting DRX behavior. Furthermore, the precipitation of the SF and 14H LPSO phases was observed within partially recrystallised grains, which resulted in a strengthening effect on the precipitation.

4 Conclusion

The microstructure and mechanical properties of a novel high-strength and high modulus hot-extruded $Mg_{90}Y_4Zn_2Ni_4$ alloy. The main conclusions drawn from the results are as follows:

(1) The hot-extruded $Mg_{90}Y_4Zn_2Ni_4$ alloy has an outstanding modulus-strength-ductility balance with YS=362MPa, ε=10.2% and E=51.32GPa.

(2) The excellent mechanical properties of the hot-extruded $Mg_{90}Y_4Zn_2Ni_4$ alloy mainly originate from the combination of the combination of grain refinement, LPSO strengthening, and precipitation strengthening induced by stacking fault and 14H LPSO phase, among which the LPSO strengthening is dominant.

(3) The high-volume fraction (65%) of the high modulus 14H LPSO phase significantly increases the overall elastic modulus of the alloy.

5 Acknowledgments

This work was supported by National Natural Science Foundation of China (Nos. 52375394, 52074246, 52275390, 52205429, 52201146, 52305429), National Defense Basic Scientific Research Program of China (JCKY2020408B002), Shanxi Scholarship Council of China (No. 2021-125), Natural Science Foundation of Shanxi Province (No. 20210302124631), and Key Research and Development Program of Shanxi Province (Nos. 202102050201011, 202202050201014)

References

[1] Gu Z, Zhou Y, Dong Q, et al. Designing lightweight multicomponent magnesium alloys with exceptional strength and high stiffness[J]. Materials Science and Engineering: A, 2022, 855: 143901.

[2] Shi H, Xu C, Hu X, et al. Improving the Young's modulus of Mg via alloying and compositing–a short review[J]. Journal of Magnesium and Alloys, 2022, 10(8): 2009-2024.

[3] Liu W, Zhao Y, Zhang Y, et al. Deformation-induced dynamic precipitation of 14H-LPSO structure and its effect on dynamic recrystallization in hot-extruded Mg-Y-Zn alloys[J]. International Journal of Plasticity, 2023, 164: 103573.

[4] Tane M, Suzuki S, Yamasaki M, et al. Insignificant elastic-modulus mismatch and stress partitioning in two-phase Mg–Zn–Y alloys comprised of α-Mg and long-period stacking ordered phases[J]. Materials Science and Engineering: A, 2018, 710: 227-239.

Reinforcing Mechanical Properties of AZ31 By in Situ Autogenous Mg_2Si

Chuan Shuai[1], Wei Liu[1], Hua Hou[1,2], Yuhong Zhao[1*]
1.School of Materials Science and Engineering, Collaborative Innovation Center of Ministry of Education and Shanxi Province for High-performance Al/Mg Alloy Materials, North University of China, Taiyuan, Shanxi 030051, PR China.
2. School of Materials Science and Engineering, Taiyuan University of Science and Technology, Taiyuan, Shanxi 030024, PR China.

Abstract: AZ31 commercial magnesium alloy is unable to meet the requirements of load-bearing structural components due to its limited strengthening effect as its strengthening phase is only Mg17Al12 phase. In this work, the mechanical properties of AZ31 alloys are improved by in situ autogenous Mg2Si phases. The results show that the Mg2Si phase transforms from a short rod-like to a Chinese structure-like to a coarse bulk-like as the Si content increases from 0 to 3%. After extrusion, the Chinese structure Mg2Si phase is refined, but the coarse massive Mg2Si phase is not refined. In addition, with the increase of Si content, the ultimate tensile strength and yield strength of AZ31 alloys increase and then decrease, and the elongation first decrease slowly and then sharply. The best overall performance of the alloy is obtained at 2% Si.

Keywords: AZ31, Mg_2Si, extrusion.

1 Introduction

AZ31 alloy is widely used in automotive components, engine housings and communication equipment due to its good mechanical properties and abundant commercial availability[1]. As aerospace has become more demanding in terms of lightweight and performance, AZ31 is not able to meet the performance requirements and is mainly used in non-structural applications. AZ31 consists mainly of an α-Mg matrix and the $Mg_{17}Al_{12}$ phase[2], which has a limited reinforcing effect. The Mg_2Si phase is used as a reinforcing phase due to its inherent advantages, such as low density, high hardness, and high modulus of elasticity. However, the Mg_2Si phase formed during the casting process has a coarse crystalline structure, and these large particles are highly brittle, leading to deterioration of the mechanical properties[3]. Recently, many researchers have modified Mg_2Si by adding alloying elements[4], but the effect is not obvious. In this study, we improved the mechanical properties of AZ31 alloy by in situ autogenous Mg_2Si phase in AZ31 alloy, followed by refinement of the Mg_2Si phase with Chinese character structure and homogeneous distribution of the Mg_2Si phase in the matrix by extrusion, which significantly improved the yield strengths (YS) and ultimate tensile strengths (UTS) of AZ31 alloy with virtually no reduction in the elongation (E) of AZ31.

Experimental procedure

The raw materials for the experiments are pure Mg, pure Al, pure Zn, pure Mn and pure Si. AZ31 is melted at 750°C using a box resistance furnace and Si powder is added at 580°C to obtain AZ31-xSi (x=0, 1, 2, 3) alloy ingots. The ingots are solution treated at 420°C for 12 hours and then hot extrude, holding for 1 hour before extrusion. The extrusion temperature is 240°C, the extrusion ratio is 25:1, the extrusion angle is 30°, and the extrusion speed is 0.4 mm/s.

2 Result and discussion

Microstructure of alloys

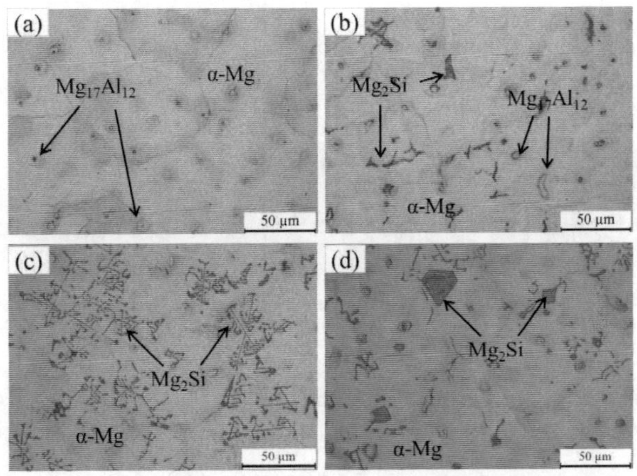

Fig. 1. Optical microstructure (OM) of as-cast AZ31-xSi: (a) AZ31, (b) AZ31-1Si, (c) AZ31-2Si, (d) AZ31-3Si.

Figure 1 shows the microstructure of as-cast AZ31 with different Si contents. The as-cast AZ31 mainly consists of α-Mg matrix and uniformly distributed dotted $Mg_{17}Al_{12}$ phases. As shown in Fig. 1(b), when 1% Si is added, some short rod-like Mg_2Si phases appear. With further addition of Si up to 2%, a large number of Mg_2Si phases with Chinese character structure are clearly observed (Fig. 1(c)). Interestingly, when the Si content is increased to 3%, the Mg_2Si phases with Chinese character disappear and is replaced by a massive Mg_2Si phase. When the Si content is low, the Mg_2Si phase exists mainly in eutectic form, and when the Si content is high, the Mg_2Si phase exists mainly in primary massive form. Figure 2 shows the microstructure of as-extruded AZ31 with different Si contents. From Fig. 2, it can be clearly observed that the microstructure of the as-extruded alloys undergoes significant changes, in addition to the significant grain refinement, the Mg_2Si phase shows a band-like distribution along the extrusion direction. It is noteworthy that the Mg_2Si phase with Chinese character is crushed into small particles during the extrusion process, while the bulk Mg_2Si phase did not undergo any significant change in morphology during the extrusion process.

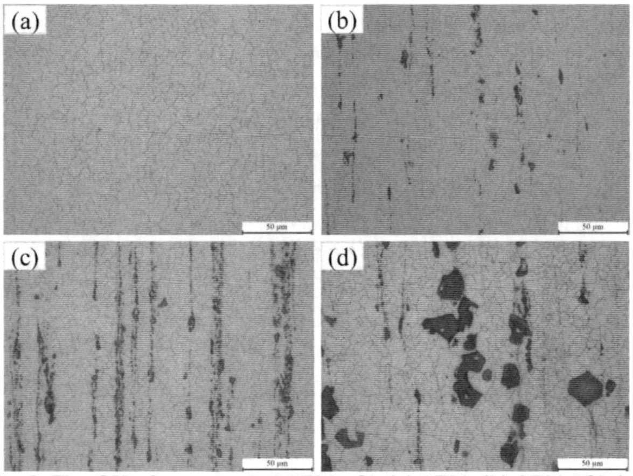

Fig. 2. The OM of as-extruded AZ31-xSi: (a) AZ31, (b) AZ31-1Si, (c) AZ31-2Si, (d) AZ31-3Si.

Mechanical properties of alloys

Table 1. Mechanical properties of as-extruded AZ31 with different Si content

Alloy	UTS	YS	E
0Si	218.6	110.1	28.0
1Si	231.5	128.6	27.2
2Si	289.8	167.8	26.0
3Si	279.1	172.5	19.3

Table 1 summarizes the mechanical properties of as-extruded AZ31 with different Si contents. As the Si content increases from 0 to 2%, the UTS and YS of as-extruded AZ31 alloys increase significantly, while E decreases slightly, especially 2% Si. This indicates that the Mg_2Si phase with Chinese character improves the mechanical properties of AZ31 by extrusion refinement. The refined Mg_2Si phase acts as a reinforcing phase to increase the UTS and YS of the alloys by hindering the movement of dislocations. However, when the Si content is 3%, the E of alloy decreases drastically, and there is a slight decrease in UTS. This result is closely related to the massive Mg_2Si phase, and the coarse massive Mg_2Si tends to lead to early crack initiation, which significantly reduces the E and also has a detrimental effect on the UTS. In conclusion, the Mg_2Si with Chinese character can significantly improve the mechanical properties of AZ31 alloy after extrusion refinement, and the coarse massive Mg_2Si phase is not favorable to the mechanical properties of AZ31 alloy.

3 Conclusion

The Mg_2Si phase with Chinese character is significantly refined after extrusion and deformation, which significantly improves the UTS and YS of the AZ31 alloy with a slight decrease in E. The mechanical properties of the AZ31 alloy are deteriorated due to the fact that the massive Mg_2Si phase is not refined during the extrusion process. The best overall alloy properties are obtained when the Si content is 2%.

4 Acknowledgments

The authors sincerely acknowledge the Provincial and Ministry Collaborative Innovation Center of Development and Application of High-Performance Aluminum/Magnesium Alloy Materials for providing technical support for the research, and financial supports from the Research Project Supported by National Natural Science Foundation of China (No. 52305429, 52375394, 52074246, 52275390, 52205429, 52201146), Shanxi Scholarship Council of China (No. 2021-125), Natural Science Foundation of Shanxi Province (No. 20210302124631), National Defense Basic Scientific Research Program of China (JCKY2020408B002, WDZC2022-12) and Key Research and Development Program of Shanxi Province (202102050201011, 202202050201014).

References

[1] Gupta M, Ling S N M. Magnesium, magnesium alloys, and magnesium composites. John Wiley & Sons, 2011.

[2] Rakshith M, Seenuvasaperumal P. Review on the effect of different processing techniques on the microstructure and mechanical behaviour of AZ31 Magnesium alloy[J]. Journal of Magnesium and Alloys, 2021, 9(5): 1692-1714.

[3] Seth P P, Parkash O, Kumar D. Structure and mechanical behavior of in situ developed Mg_2Si phase in magnesium and aluminum alloys-a review. RSC advances, 2020, 10(61): 37327-37345.

[4] Shin H C, Son J, Min B K, et al. The effect of Ce on the modification of Mg_2Si phases of as-cast eutectic Mg-Si alloys. Journal of Alloys and Compounds, 2019, 792: 59-68.

Effects of thermal compression on thermal conductivity and mechanics of Mg-Zn-Cu alloys

Taoze Xie[1], Wei Liu[1], Hua Hou[1,2], Yuhong Zhao[1*]
1.School of Materials Science and Engineering, Collaborative Innovation Center of Ministry of Education and Shanxi Province for High-performance Al/Mg Alloy Materials, North University of China, Taiyuan, Shanxi 030051, PR China.
2. School of Materials Science and Engineering, Taiyuan University of Science and Technology, Taiyuan, Shanxi 030024, PR China.

Abstract: In order to meet the requirements of thermal conductivity materials, the Mg-Zn-Cu magnesium alloy with better thermal conductivity was selected, and the thermal conductivity and mechanical properties were improved by reducing the alloying degree and combining with hot extrusion deformation. After hot extrusion deformation, the grain is obviously refined, and the morphology of the second phase changes from a reticular structure to a granular distribution along the extrusion direction. After hot extrusion deformation, Mg-2Zn-2Cu alloy obtains the highest thermal conductivity and mechanical properties, thermal conductivity is 153.5 (W/(M·K)), yield strength is 162.9 MPa, tensile strength is 277.8 MPa, and elongation is 37.7 %.

Keywords: Mg-Zn-Cu alloy; hot extrusion; thermal conductivity; mechanical properties

1 Introduction

In the 5G communication era, all kinds of intelligent terminal products are constantly upgrading, and the power consumption is getting higher and higher, which also put forward higher requirements for the thermal conductivity of materials[1]. Magnesium, as the lightest density engineering structural material, has high specific strength, specific stiffness and good heat dissipation performance, which has the potential to meet the demand of thermal conductive materials in the 5G era[2].

Mg-Zn-Cu alloy due to the precipitation of MgZnCu phase reduces the solid solution atoms in the matrix, because the negative effect of the second relative thermal conductivity is much less than that of the solid solution atoms, so Mg-Zn-Cu alloy has excellent thermal conductivity, but its mechanical properties in the as-cast state are low and cannot meet the application requirements[3]. Therefore, this study uses hot extrusion deformation to enhance the thermal conductivity and mechanical properties of the alloy while reducing the degree of alloying[4].

2 vExperimental procedure

The original ingot was prepared by melting high-purity Mg, Zn, and Cu at 750 °C in a resistance furnace under an argon atmosphere, and the nominal composition of the ingot was Mg-1Zn-1Cu, Mg-2Zn-Cu, and Mg-3Zn-3Cu. Then, the molten temperature was reduced to 730°C, and it was poured into the mold preheated to 200°C.

Before undergoing hot extrusion, the cast alloy was homogenized at 430 °C for 60 hours. During the hot extrusion experiment, the sample was preheated at 240°C for 1 hour to achieve uniform temperature. Following this, it was extruded at 240 °C with an extrusion speed of 0.4mm/s and an extrusion angle of 30°, utilizing a die with an extrusion ratio (Re) of 25.

3 Result and discussion

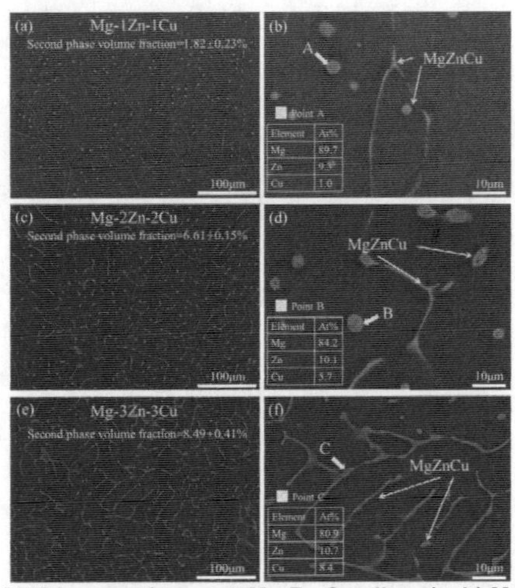

Fig 1. The SEM image of as-cast Mg-Zn-Cu alloy. (a, b) Mg-1Zn-1Cu, (c, d) Mg-2Zn-2Cu, (e, f) Mg-3Zn-3Cu. EDS results are annotated in the lower left corner of the picture.

As shown in Fig.1, in the as-cast alloy, the alloy is composed of α-Mg and eutectic structure. With the increase of Zn and Cu content, the volume fraction of eutectic structure gradually increases, and gradually changes from a semi-continuous network structure to a continuous network structure. In the extruded alloy, it can be observed that the microstructure of as-extruded alloys α-Mg and a kind of second phases which positioned along the extrusion direction in Fig. 2. The second phase changes with the content of Zn and Cu in the extruded alloy have the same trend as that in the cast alloy. Fig.2 show that the grains of the alloy are obviously equiaxial, and second phases are distributed on the grain boundaries. Combined with EDS analysis, the second phase in the alloy is MgZnCu.

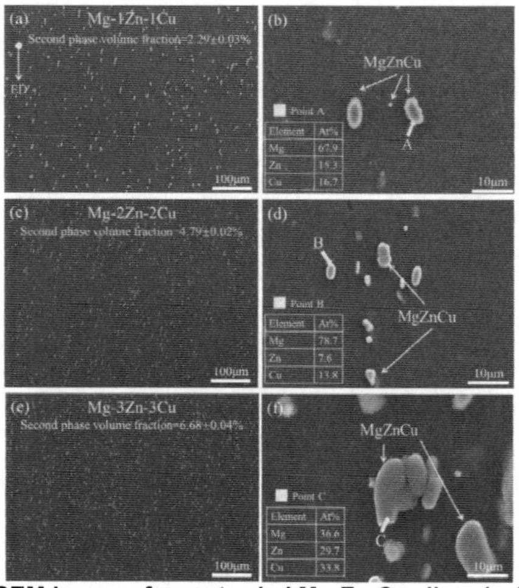

Fig 2. The SEM image of as-extruded Mg-Zn-Cu alloy. (a, b) Mg-1Zn-1Cu, (c, d) Mg-2Zn-2Cu, (e, f) Mg-3Zn-3Cu.

Table 1. Thermal conductivity and mechanical properties of extruded alloys

Alloys	Thermal conductivity (W/(m·K))	YS (MPa)	UTS (MPa)	EL (%)
Mg-1Zn-1Cu	115.3	124.8	237.3	40.2
Mg-2Zn-2Cu	153.5	162.9	277.8	37.7
Mg-3Zn-3Cu	136.4	148.2	248.3	34.2

Table 1 shows the thermal conductivity and mechanical properties of the extruded alloy, and the highest thermal conductivity of the Mg-2Zn-2Cu alloy is 153.5 W/(m·K). At the same time, the mechanical properties of the alloy increase firstly and then decrease with the increase of Zn and Cu content, and on the whole, the mechanical properties gradually increase with the increase of alloy content., and the Mg-2Zn-2Cu alloy achieves the best combination of mechanical properties, with yield strength of 162.9 MPa, ultimate tensile strength of 277.8 MPa and elongation of 37.7%.

According to the microstructure, after hot extrusion deformation, the grain size of the alloy is greatly reduced, and the number of grain boundaries is increased. The resistance of grain boundaries to dislocation movement increases and the yield strength of the alloy is improved. At the same time, during the hot extrusion process, the second phase changes from a reticular structure to a dispersed granular structure in the matrix, which effectively hinders the dislocation movement and improves the tensile strength of the alloy. As for the thermal conductivity, casting defects have a great influence on the thermal conductivity. Hot extrusion can reduce the defects in the as-cast alloy and improve the thermal conductivity of the alloy. On the other hand, during the extrusion deformation process, the solute atoms are dynamically precipitated from the magnesium matrix to form the second phase, which reduces the content of alloying elements in the matrix and improves the thermal conductivity of the alloy.

4 Conclusion

(1) After hot extrusion, the MgZnCu phase in the Mg-Zn-Cu alloy changes from a network structure in the as-cast state to an island distribution at the grain boundaries along the extrusion direction

(2) After extrusion deformation, the casting defects and the content of solute atoms in the matrix are reduced, and the thermal conductivity of the alloy is improved. The highest thermal conductivity of the Mg-2Zn-2Cu alloy is 153.5 W/(m·K)

(3) Mg-2Zn-2Cu alloy obtained the best combination of strength and plasticity after hot extrusion deformation. The yield strength was 162.9 MPa, the tensile strength was 277.8 MPa, and the elongation was 37.7 %.

5 Acknowledgments

The authors sincerely acknowledge the Provincial and Ministry Collaborative Innovation Center of Development and Application of High-Performance Aluminum/Magnesium Alloy Materials for providing technical support for the research, and financial supports from the Research Project Supported by National Natural Science Foundation of China (No. 52305429, 52375394, 52074246, 52275390, 52205429, 52201146), Shanxi Scholarship Council of China (No. 2021-125), Natural Science Foundation of Shanxi Province (No. 20210302124631), National Defense Basic Scientific Research Program of China (JCKY2020408B002, WDZC2022-12) and Key Research and Development Program of Shanxi Province (202102050201011, 202202050201014).

References

[1] Lv H, Tan J, Yuan Q, et al. Recent advancements in thermal conductivity of magnesium alloys. Journal of Magnesium and Alloys, 2024.

[2] Li S, Yang X, Hou J, et al. A review on thermal conductivity of magnesium and its alloys. Journal of Magnesium and Alloys, 2020, 8(1): 78-90.

[3] Zhang Y, Liu W, Chen W, et al. Simultaneously improving thermal conductivity, mechanical properties and metal fluidity through Cu alloying in Mg-Zn-based alloys. Journal of Magnesium and Alloys, 2024.

[4] Zeng Z, Stanford N, Davies C H J, et al. Magnesium extrusion alloys: a review of developments and prospects. International Materials Reviews, 2019, 64(1): 27-62.

Study on Prediction Method of Mechanical Properties of Casting Al-9Si-2Cu-0.4Mg Alloy

Jiayi Liu[1,2], Zhijun Feng[1,2], Yufei Li[1,2], Junjie Xiong[1,2], Zehua Li[1,2], Shiwei Xin[1,2]
1. Shenyang Research Institute of Foundry Co., Ltd. CAM, Shenyang 110022, Liaoning, China
2. National Key Laboratory of Advanced Casting Technologies, Shenyang 110022, Liaoning, China
*Corresponding address: e-mail: liujiayi518@foxmail.com

Abstract: In this paper, we designed stepped samples with different solidification times to investigate the effect of solidification time on the mechanical properties of cast Al-9Si-2Cu-0.4Mg alloy. Based on the mechanical properties of alloys with different solidification times, we used linear fitting to process the data and obtained a predictive model for the relationship between solidification time and mechanical properties, which can be used to predict the mechanical properties of different parts of cast Al-9Si-2Cu-0.4Mg alloy components. The results indicate that with increasing solidification time, the strength and impact of the cast Al-9Si-2Cu-0.4Mg alloy decrease linearly. The predictive model obtained through linear fitting can effectively predict the mechanical properties of alloys with different solidification times. The accuracy of the prediction method for components was verified, and the results show that this method can accurately predict the strength and hardness of Al-9Si-2Cu-0.4Mg alloy with small prediction errors.

Keywords: cast Al-9Si-2Cu-0.4Mg alloy, solidification time, mechanical properties prediction

1 Introduction

The casting of Al-9Si-2Cu-0.4Mg alloy has promising mechanical properties and castability, with wide application prospects in high-tech fields such as aerospace, aviation, and automotive industries [1]. In the early stages of material development, it was found that the mechanical properties of this alloy are sensitive to solidification time. However, most castings have variations in wall thickness, and when the alloy is applied to castings, there may be issues of localized mechanical properties not meeting the required standards due to excessively long solidification times [2]. Therefore, in this study, stepped samples with different solidification times were designed. The solidification times were calculated using Procast simulation, and the mechanical properties of cast Al-9Si-2Cu-0.4Mg alloy with different solidification times were obtained. A predictive model for the relationship between solidification time and mechanical properties was developed using linear fitting. This model was then utilized to predict the mechanical properties of Al-9Si-2Cu-0.4Mg alloy castings.

2 Experimental procedure

In order to study the mechanical properties of cast Al-9Si-2Cu-0.4Mg alloy with different solidification times, a combination of two variables, wall thickness, and chill was used to design stepped samples with different solidification times. The wall thicknesses were set at 7, 10, 13, and 16mm. Using Procast simulation, the solidification times at the ends and in the middle of the tensile test bars were calculated, resulting in a solidification time range of 8 to 356 seconds for the stepped samples. At room temperature, according to the requirements of the GBT 228.1 standard, tensile tests were conducted using a DDL-300 universal testing machine at a speed of 1mm/min. Additionally, hardness was tested using an HB-3000 Brinell hardness tester, with an experimental load (F) set at 0.2kg and a load application time of 10 seconds, following the GB/T 231.1 standard. Based on the mechanical properties of cast Al-9Si-2Cu-0.4Mg alloy with different solidification times, data was processed using linear fitting to obtain a model for the mechanical properties of the alloy with different solidification times. This model was then used to predict the mechanical properties of cast Al-9Si-2Cu-0.4Mg alloy components.

3 Result and discussion

Mechanical properties of cast Al-9Si-2Cu-0.4Mg alloy with different solidification times

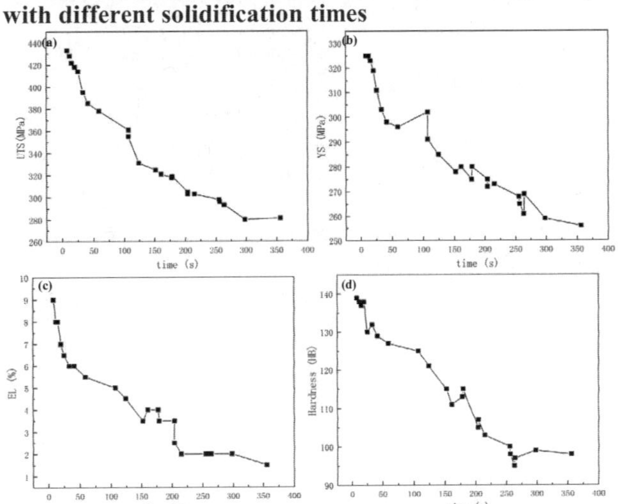

Fig. 1 (a) tensile strength of cast Al-9Si-2Cu-0.4Mg alloy with different solidification times; (b) yield strength of cast Al-9Si-2Cu-0.4Mg alloy with different solidification times; (c) elongation of cast Al-9Si-2Cu-0.4Mg alloy with different solidification times; (d) hardness of cast Al-9Si-2Cu-0.4Mg alloy with different solidification times

Figure 1 illustrates the mechanical properties of cast Al-9Si-2Cu-0.4Mg alloy at different solidification times. It can be observed from the figure that with an increase in solidification time, the tensile strength, yield strength, elongation, and hardness of the alloy linearly decrease. As the solidification time increases from 8 seconds to 356 seconds, the tensile strength of the cast Al-9Si-2Cu-0.4Mg alloy decreases by 35.1%, the yield strength decreases by 21.2%, the elongation decreases by 83.3%, and the hardness decreases by 29.5%.

4 The linear regression model of the mechanical properties of cast Al-9Si-2Cu-0.4Mg alloy as a function of solidification time

Using statistical analysis tools, the mechanical properties of cast Al-9Si-2Cu-0.4Mg alloy at different solidification times were processed to obtain the following predictive models for the mechanical properties: UTS=414.2-0.4766× time; YS=316.3-0.1989 × time; EL=7.36-0.02026 × time; Hardness=137.2-0.1402×time.

The coefficient of determination for all models is greater than 0.7, indicating that the linear models can effectively express the relationship between mechanical performance indicators and solidification time. Specifically, the coefficient of determination for tensile strength is 92.1%, for yield strength is 91%, for elongation is 90.3%, and for hardness is 94.2%.

5 Prediction of the mechanical properties of Al-9Si-2Cu-0.4Mg alloy castings

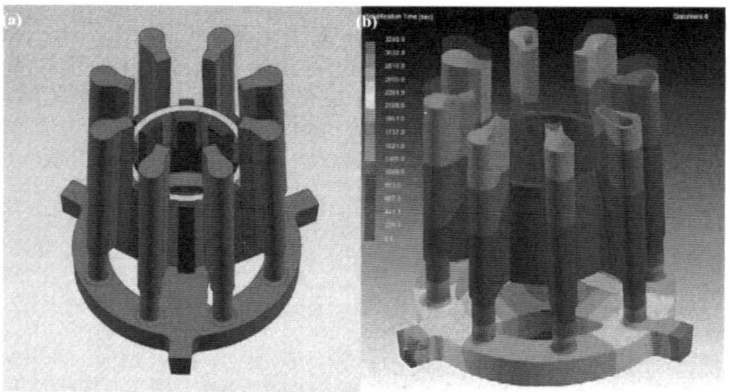

Fig. 2 (a)Casting process for cylindrical castings; (b) Simulation results of the temperature field in Procast

Figure 2 depicts the casting process and Procast temperature field simulation for the cylindrical castings, with a solidification time range of 11 to 209 seconds. Table 1 presents the measured mechanical property values of the cylindrical castings, compared with the predicted performance. The average predicted percentage errors for tensile strength, yield strength, elongation, and hardness are 1.9%, 2.8%, 7.9%, and 1.9% respectively. The predictive models can be effectively utilized for strength and hardness prediction of Al-9Si-2Cu-0.4Mg castings.

Table 1. Measured mechanical properties of cylindrical castings

Time/s	Measured values			
	UTS /MPa	YS /MPa	EL /%	Hardness/ HB
11	401	314	7.0	136
15	402	313	8.0	138
20	395	312	7.0	135
24	399	311	6.5	133
36	383	309	6.0	132
45	385	307	6.0	132
60	382	304	6.0	129
68	373	302	6.0	131
77	375	300	5.5	124
83	367	299	5.5	128
92	361	298	5.0	129
136	355	289	4.5	123
148	351	286	4.0	118
174	341	281	3.5	115
188	321	279	3.0	112
209	321	275	2.5	104

Conclusion

The mechanical properties of cast Al-9Si-2Cu-0.4Mg alloy are sensitive to solidification time. Based on the data of solidification time and mechanical properties, a linear fitting model can be used to predict the strength and hardness of the alloy castings.

Acknowledgments

This project was supported by the National Key Laboratory of Advanced Casting Technology

References

[1] Toschi S . Optimization of A354 Al-Si-Cu-Mg Alloy Heat Treatment: Effect on Microstructure, Hardness, and Tensile Properties of Peak Aged and Overaged Alloy[J]. Metals, 2018, 8(11): 961-961.

[2] Yu W ,Minqiang G ,Bowei Y , et al. Microstructural evolution and mechanical property of Al−Mg−Mn alloys with various solidification cooling rates[J]. Materials Characterization, 2021, (prepublish): 111709.

Analysis of Heating Condition of Copper Crucible in Vacuum Suspension Furnace Melting Process

Jingyan Liu, Zhizhong Liu, Pengcheng Zhang, Yongliang He, Fuli Zhang, Xiaoming Kang, Wanqing Li, Bo Dou, Shaowei Wen
Shenyang Research Institute of Foundry/State Key Laboratory of Advanced Casting Technologies, Shenyang, Liaoning, China
*Corresponding address: e-mail: liujingyan83@126.com

Abstract: The copper crucible is one of the most important parts in the process of vacuum suspension furnace melting, which must work at a certain temperature to ensure its stable physical and chemical properties. The study of its temperature is crucial because it directly affects the quality of the smelting process.
Keywords: Copper Crucible, Suspension smelting,Crucible temperature, The crucible cools

1 Introduction

The temperature of copper crucible is changed by the Heat absorbed and released by the Crucible. The heat absorbed by the crucible mainly includes the heat transfer of burden to it and the induction of melting magnetic field to it. It releases heat in the way that it conducts heat to cooling water.

2 Heat transfer of charge to crucible

At the initial stage of smelting, the charge is not melted, and the heat is transferred from the charge to the crucible in the form of heat transfer, the contact area between liquid metal and crucible will increase gradually, and the heat transfer mode will become convective conduction.

(1) Heat transfer from solid charge tocrucible

The initial stage of smelting. According to Fourier's law, it can be calculated using the following formula:

$$Q_1 = \lambda F(t_1 - t_2)$$

Explain:

Q_1—Heat conduction of charge to crucible (Unit W);

λ —Thermal conductivity between charge and crucible （W/m²*℃）;

F —The contact area between solid charge and crucible is different due to the different particle size of charge;

t_1—Contact side charge temperature;

t_2—Crucible temperature;

3 Heat transfer of liquid metal to crucible

After the metal has melted， Under the action of gravity and levitation， It rolled in the Crucible， But only the bottom touches the crucible

The heat transfer can be calculated using the following formula:

$$Q_2 = \alpha F_2(t_3 - t_2)$$

Explain:

Q_2—Convection heat transfer of liquid charge to crucible;

α — Convective heat transfer coefficient between furnace charge and crucible;

F_2—The contact area between the liquid charge and the crucible;

t_3—Melting charge temperature;

t_2—Crucible temperature;

4 The work of electromagnetic field on the input of Copper Crucible

According to the principle of electromagnetic induction, the work entered can be calculated using the following formula:

$$P = KH^2 q_f \sqrt{\rho\mu f}$$

Explain:

P—The power absorbed by a copper crucible(kW/m²);

K—Constant;

H—Magnetic field strength （A/m);

q_f — The ratio of crucible cross-section diameter to penetration depth;

ρ—The resistivity of copper（1.75*10-5Ω*cm）;

μ—The relative permeability of copper;

f—Electromagnetic frequency;

5 Heat transfer between the crucible and the cooling water

The heat transfer between cooling water and crucible is convective heat transfer.it can be calculated using the following formula:

$$Q_3 = hA\Delta t$$

Explain:

Q_3 —Heat transfer between crucible and cooling water;

h—Convective thermal conductivity between water and copper;

Δt—Temperature difference between crucible and cooling water;

A—Contact area between cooling water and crucible;

From the formula,The larger the contact area A, the larger the temperature difference between crucible and cooling water, and the higher the heat transfer efficiency.

6 For example

The process of titanium suspension melting in vacuum was recorded.

Materials	Weight (kg)	Power supply power (kW)
Titanium	5	200

At the beginning of smelting, the power supply is 50kW, the charge begins to heat up, and when the power is raised to 100kW by 2 minutes, part of the titanium metal begins to melt. As the power increases, the melting speed increases, 3 minutes and 50 seconds, titanium liquid has filled the bottom of the crucible, increase the power to 150 kW, 5 minutes 30 seconds, titanium all melting, into suspension.

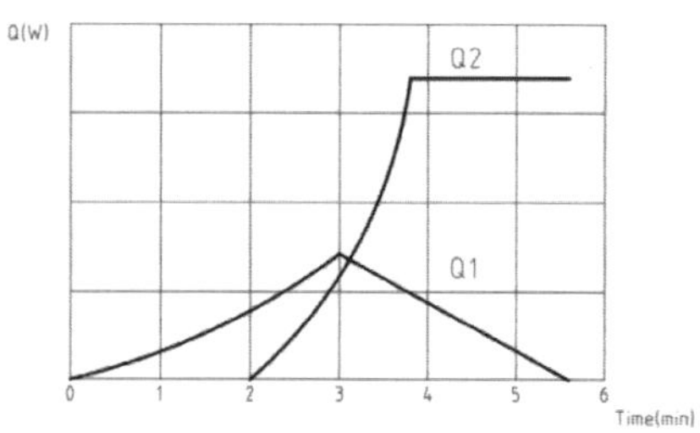

Q1-heat conduction
Q2-convection

Figure.1 Heat transfer of crucibles

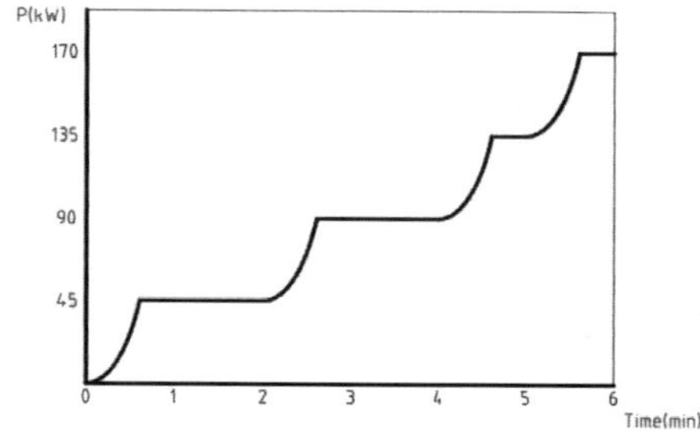

Figure. 2 The heat of the crucible

7 Result and discussion

According to the heating condition of the copper crucible, the section of the water channel of the Crucible and the requirement of the cooling water flow are designed to improve the production efficiency, reduce production costs while ensuring product quality and safe production.

References

[1] Yang shiming，Tao wenquan. 《Heat transfer》2006,8.
[2] Li pei, Zhang sen. METALLURGICAL EQUIPMENT 2025.10.TOTAL NO.222.
Vacuum Suspension Furnace Technology
[3] Zhang yingming, Sun xiangyun, Zhou lian. 《INDUSTRIAL HEATING》1998.
Design of Cold Crucible for Casting Suspension Melting

Lightweight Single-Phase Al-based Complex Concentrated Alloy with High Specific Strength

Yuan Wu[1,*], Mingliang Han[1], Qiaoshi Zeng[2], Zhaoping Lu[1]
1. State Key Laboratory for Advanced Metals and Materials, University of Science and Technology Beijing, Beijing 100083, China
2. Center for High Pressure Science and Technology Advanced Research, Shanghai 201203, China
*Corresponding address: e-mail: wuyuan@ustb.edu.cn

Abstract: Developing light yet strong aluminum (Al)-based alloys has been attracting unremitting efforts due to the soaring demand for energy-efficient structural materials. However, this endeavor is impeded by the limited solubility of other lighter components in Al. Here, we propose to surmount this challenge by converting multiple brittle phases into a ductile solid solution in Al-based complex concentrated alloys (CCA) by applying high pressure and temperature. We successfully develop a face-centered-cubic single-phase Al-based CCA, Al55Mg35Li5Zn5, with a low density of 2.40 g/cm3 and a high specific yield strength of 344×103 N·m/kg (typically ~ 200×103 N·m/kg in conventional Al-based alloys). Our analysis reveals that formation of the single-phase CCA can be attributed to the decreased difference in atomic size and electronegativity between the solute elements and Al under high pressure, as well as the synergistic high entropy effect caused by high temperature and high pressure. The increase in strength originates mainly from high solid solution and nanoscale chemical fluctuations. Our findings could offer a viable route to explore lightweight single-phase CCAs in a vast composition-temperature-pressure space with enhanced mechanical properties.

Keywords: Lightweight ; Al-based ; Complex Concentrated Alloy; High Specific Strength

Nano Particles Reinforced and Directional Solidification in High Entropy Alloys

Ruirun Chen, Gang Qin*, Xu Yang
Harbin Institute of Technology, Harbin 150001, China.
*Corresponding address: E-mail: 249310025@qq.com

Abstract: Nano-precipitation strengthening is an extremely effective method for enhancing the mechanical properties of metallic materials, significantly improving their strength, toughness, and wear resistance. Directional solidification techniques provide a valuable means to control the crystal growth rate and orientation of high-entropy alloys, making them an essential tool for studying the solidification process of these alloys and enriching related theories. This paper systematically presents the recent research progress and key achievements of our research team in the areas of composition design, material preparation, phase formation mechanisms, crystal growth patterns, and strengthening mechanisms of high-entropy alloys. Through the exploration of various compositional combinations, we propose a composition design strategy based on valence electron concentration, revealing the relationship between the microstructure and macroscopic properties of the alloys. Furthermore, we delve into the mechanisms by which nano-precipitates influence the mechanical properties of the materials. These studies not only provide a theoretical foundation for the application of high-entropy alloys but also offer guidance for future material design and optimization.

Keywords: 10 pt., Times New Roman,

1 Introduction

Nanoparticle is an efficient approach for enhancing alloys' strength [8-12] because of the interaction of dislocations with nanoparticles based on the report of Beyerlein et. al [11], which is readily implementable in 3D printed alloys. Nevertheless, the mechanical heterogeneity between the precipitates and surrounding matrix generally causes stress-strain concentration at their interfaces and triggers microcrack nucleation, resulting in ductility loss [13-15]. To overcome the strength-ductility trade-off dilemma in precipitation strengthened alloys, coherent precipitates with a crystal lattice almost the same as that of the surrounding matrix, can be introduced through alloy design and heat treatment [14-16]. This strategy has been demonstrated in certain alloys, e.g., maraging steels [13] and complex concentrated alloys (CCAs) [17], to achieve a gigapascal strength with a large tensile ductility, yet its application is limited due to the delicate alloy design and thermo-mechanical processing.

2 Experimental procedure

The casted samples were prepared by induction melting of a mixture of pure metals (purity>99%) in a high-purity argon atmosphere and then casting into a water-cooled copper crucible. The microstructure was characterized using a Zeiss Sigma 300 field emission scanning electron microscope (SEM) operated at 20 kV with an energy dispersive spectrometer (EDS)

3 Result and discussion

Fig. 1a displays the SEM images of the CoCrCuMnNi HEA. It was found that the produced CoCrCuMnNi HEA has a typical dendrite crystal structure. Fig. 1b shows an image of the dendrite and inter-dendrite zones. Some plate-strip precipitates were observed in the inter-dendrite zone.
Fig. 1c shows an image of the inter-dendrite zone with a high magnification. A mass of nanometer precipitates was found in the inter-dendrite zone. An image of the dendrite zone with high magnification is presented in Fig. 1d. In the dendrite zone, many nanometer precipitates were also observed.

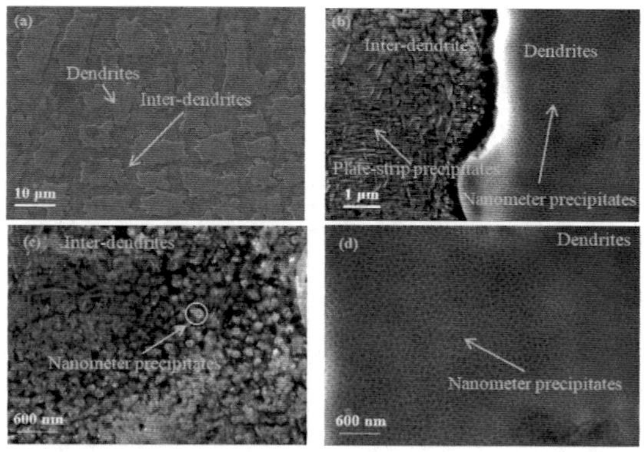

Fig. 1 Microstructure of the CoCrCuMnNi HEA by SEM

4 Conclusion

It should be emphasized that we have not only developed a CoCrCuMnNi HEA in the current work, but our study also sheds light on developing HEAs with remarkable mechanical properties. Furthermore, some main conclusions are drawn.
1. The CoCrCuMnNi HEA is composed of two FCC phases. A large amount of nanometer precipitates (5–50 nm in size) and domains (5–10 nm in size) are found in the inter-dendrite and dendrite zones.

2. This CoCrCuMnNi HEA prepared by vacuum arc melting exhibits excellent mechanical properties. The obtained tensile yield and the ultimate tensile strengths are about 458 MPa and 742 MPa, respectively, and the corresponding elongation is approximately 40%. This HEA has a mechanical response almost identical to those of the single-phase Fe20Mn20Ni20Co20Cr20 HEA (grain-refined) and the dual-phase Fe50Mn30Co10Cr10 HEA (as-homogenized), the two most suc cessful HEAs to date.
3. The enthalpy of mixing between Cu and Co, Cr, Mn, or Ni is higher than that between any two of Cr, Co, Ni, and Mn elements, which is the key factor for the separation of Cu from the CoCrCuMnNi HEA.

5 Acknowledgments

This work was supported by the National Natural Science Foundation of China (No. 52425401) and the Interdisciplinary Research Foundation of HIT.

References

[1] L. Liu, Y. Zhang, J. Han, et al. Nanoprecipitate-Strengthened High-Entropy Alloys. Adv. Sci. 8 (2021), 2100870.
[2] B. Z. Long, Y. Zhang, C. H. Guo, et al. Enhanced dynamic mechanical properties and resistance to the formation of adiabatic shear band by Cu-rich nano-precipitates in high-strength steels. Int. J. Plasticity. 138 (2021) 102924.
[3] S. S. Xu, Y. Zhao, D. Chen, et al. Nanoscale precipitation and its influence on strengthening mechanisms in an ultra-high strength low-carbon steel. Int. J. Plasticity. 113 (2019) 99-110.
[4] A. M. Roach, S. Xu, D. J. Luscher, et al. Interaction of extended dislocations with nanovoid clusters. Int. J. Plasticity. 168 (2023) 103684.
[5] P. Kürnsteiner, M. B. Wilms, A. Weisheit, et al. High-strength Damascus steel by additive manufacturing. Nature 582 (2020), 515–519.
[6] S. Jiang, H. Wang, Y. Wu, et al. Ultrastrong steel via minimal lattice misfit and high-density nanoprecipitation. Nature 544 (2017), 460–464.
[7] T. Yang, Y. L. Zhao, Y. Tong, et al. Multicomponent intermetallic nanoparticles and superb mechanical behaviors of complex alloys. Science 362 (2018), 933–937.
[8] H. Xue, C. Yang, F. D. Geuser, et al. Highly stable coherent nanoprecipitates via diffusion-dominated solute uptake and interstitial ordering. Nat. mater, 22(2023), 434-441.
[9] T. Li, T. Liu, S. Zhao, et al. Ultra-strong tungsten refractory high-entropy alloy via stepwise controllable coherent nanoprecipitations. Nat Commun. 14 (2023) 3006.
[10] Z. Fu, L. Jiang, J. L. Wardini, et al. A high-entropy alloy with hierarchical nanoprecipitates and ultrahigh strength. Sci. Adv. 4 (2018), 8712.

Innovative Design and Future Development of Chemically Complex Intermetallic Alloy

T. Yang[1,*], C. T. Liu[1], B. Xiao[1], Y. L. Zhao[2]
1. Department of Materials Science and Engineering, City University of Hong Kong, Hong Kong, China
2. School of Materials Science and Engineering, Harbin Institute of Technology (Shenzhen), Shenzhen, China
*Corresponding address: e-mail: taoyang6@cityu.edu.hk

Abstract: Intermetallic materials are bestowed by diverse ordered superlattice structures together with many unusual properties. In particular, the advent of chemically complex intermetallic alloys (CCIMAs), which are also called as the high-entropy intermetallic alloys (HEIMAs), has been received considerable attention in recent years and offers a new paradigm to develop novel metallic materials for advanced structural applications [1-5]. These newly emerged CCIMAs exhibit synergistic modulations of structural and chemical features, such as self-assembled long-range close-packed ordering, complex sublattice occupancy, and interfacial disordered nanoscale layer, potentially allowing for superb physical and mechanical properties that are unmatched in conventional metallic materials[1,2]. In this paper, we critically review the historical developments and recent advances in ordered intermetallic materials from the simple binary to chemically complex alloy systems. We are focused on the unique multicomponent superlattice microstructures, nanoscale grain-boundary segregation, and disordering, as well as the various extraordinary mechanical and functional properties of these newly developed CCIMAs. Finally, perspectives on the future research orientation, challenges, and opportunities of this new frontier are provided. This conceptual design of CCIMAs may lead to families of high-temperature structural materials that might avoid some of the drawbacks of high-temperature alloys currently in use, which will be of great interest for a broad range of aerospace, automotive, nuclear power, chemical engineering, and other applications.

Keywords: Ordered superlattice, high-entropy effect, Chemically complex intermetallic, Sublattice occupancy, Grain-boundary engineering

1 Representative results and discussions

Fig. 1. Unusual nanoscale interfacially disordered structure of the superlattice materials.

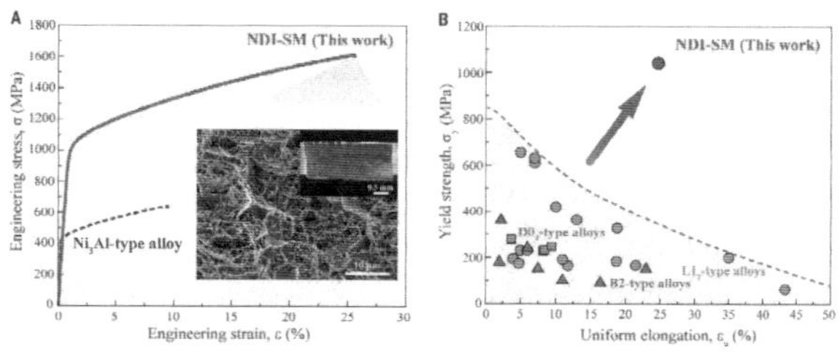

Fig. 2. Exceptional mechanical properties and thermal stability of the newly designed CCIMA.

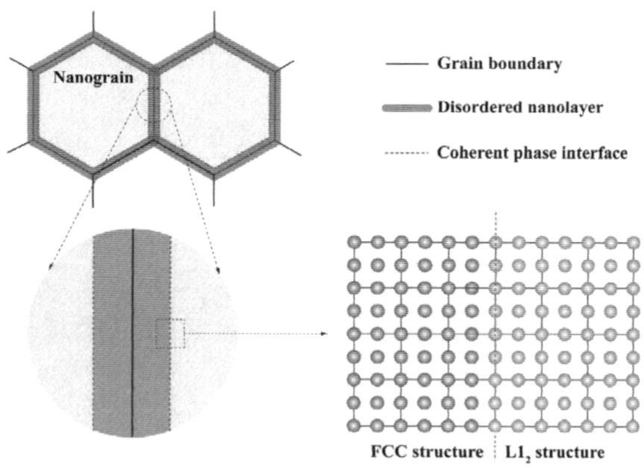

Figure 3. Schematic illustration of the conceptual design of stable nanograined CCIMAs in the future.

We discovered that nanoscale disordered interfaces can effectively overcome the embrittlement problem[1,4,6]. Interfacial disordering is driven by multielement co-segregation that creates a distinctive nanolayer between adjacent micrometer-scale superlattice grains. This nanolayer acts as a sustainable ductilizing source, which prevents brittle intergranular fractures by enhancing dislocation mobilities. Our superlattice materials have ultrahigh strengths of 1.6 gigapascals with tensile ductilities of 25% at ambient temperature. Simultaneously, we achieved negligible grain coarsening with exceptional softening resistance at elevated temperatures. Designing similar nanolayers may open a pathway for further optimization of alloy properties. 3D printing may also greatly promote the large-scale applications of CCIMAs[7].

2 Conclusion

The concept of CCIMAs represents a new and promising frontier for the innovative design of novel materials and devices. As a newly emerged metallic material, several CCIMAs with long-range ordered superlattice structures composed of various multiple alloying elements have been demonstrated with many novel and unprecedented properties and thus they are being actively explored for numerous promising applications, in particular for advanced structural applications. The combinatorial approaches including the state-of-the-art microanalysis techniques (such as the 3D-APT) and theoretical simulations (such as the DFT first-principles calculation) play crucial roles in revealing their intrinsic metallurgical behaviors. Inspired by these attractive findings, we believe that more and more novel high-performance metallic materials with superb structural/functional properties can possibly be discovered based on chemically complex intermetallic systems. In the near future, to further speed up the well-targeted quantitative design and controllable fabrication of CCIMAs with optimized properties, significant effort should be devoted to fundamentally elucidating the alloying behaviors and associated microstructural evolutions of CCIMAs, as well as the atomistic mechanisms behind them. Moreover, to promote their large-scale industrialization, more attention should be paid to the new high-throughput experimental work to systematically evaluate their microstructure-sensitive properties under various service conditions, especially under harsh environments like high temperatures, cryogenic temperatures, high-energy irradiation, etc.

3 Acknowledgments

The authors from City University of Hong Kong acknowledge the financial supports from the National Natural Science Foundation of China (grant no. 52222112 and 52101151), the Hong Kong Research Grant Council (RGC) (grant no. C1020-21G and C1017-21G).

References

[1] T. Yang, Y. L. Zhao, W. P. Li, et al. Ultrahigh strength and ductile superlattice alloys with nanoscale disordered interfaces. Science, 362, (2020): 427-432.

[2] T. Yang, Y. L. Zhao, Y. Tong, et al. "Multicomponent intermetallic nanoparticles and superb mechanical behaviors of complex alloys", Science, 362, (2018): 933-937.

[3] T. Yang, B. X. Cao, T. L. Zhang, et al. Chemically complex intermetallic alloys: A new frontier for innovative structural materials. Materials Today, 52, (2022): 161-174.

[4] F. H. Duan, Q. Li, Z. H. Jiang, et al. "An order-disorder core-shell strategy for enhanced work-hardening capability and ductility in nanostructured alloys." Nature Communications 15.1 (2024): 6832.

[5] B. Xiao, J. Zhang, S. Liu, et al. "Ultrahigh intermediate-temperature strength and good tensile plasticity in chemically complex intermetallic alloys via lamellar architectures." Acta Materialia 262 (2024): 119459.

[6] B. Xiao, J. H. Luan, S. J. Zhao, et al. "Achieving thermally stable nanoparticles in chemically complex alloys via controllable sluggish lattice diffusion", Nature Communications, 13 (1), (2022): 1-8.

[7] T. L. Zhang, Z. H. Huang, T. Yang, et al. In situ design of advanced titanium alloy with concentration modulations by additive manufacturing. Science, 374.6566, (2021): 478-482.

Tuning Microstructure and Mechanical Properties of Additively Manufactured CoCrFeMnNi High Entropy Alloy by Deep Cryogenic Treatment

Yongjiang Huang
School of Materials Science and Engineering, Harbin Institute of Technology, China
*Corresponding address: e-mail: yjhuang@hit.edu.cn

Abstract: Here, deep cryogenic treatment (DCT) is proposed to be an easy and effective approach to tune the microstructure and thus enhance the mechanical performance at room temperature of the laser melting deposited CoCrFeMnNi high entropy alloy (HEA). DCT is found to induce large compressive residual stress and crystalline defects including dislocations, stacking faults, nanotwins and nanograins into the additively manufactured CoCrFeMnNi HEA. Nano-twinning plays a critical role in the strengthening, contributing 51.0–246.8 MPa to the DCT-processed CoCrFeMnNi HEA, while the compressive residual stress is thought to be responsible for the improved ductility by suppressing crack formation. The crystalline defects are evidently generated during the cryogenic soaking, which also result in the compressive residual stress upon reheating at the end of the DCT. The present work establishes DCT as an effective method to simultaneously enhance the strength and plasticity in alloys fabricated by additive manufacturing.

Keywords: High entropy alloy, Deep cryogenic treatment, Laser directed energy deposition

1 Introduction

Laser directed energy deposition (LDED) has been established as a versatile additive manufacturing (AM) technology for the direct fabrication of complex shaped alloy products from powder feedstocks. However, repeated melting and cooling occurred in a molten pool during LDED can lead to a high temperature gradient and thus unevenly distributed residual stress within the as-built components, which will inevitably influence the mechanical performance of the AM components, especially their strength and ductility. DCT has recently been proposed as a nondestructive method for effectively tuning the residual stress state and microstructure for improving the mechanical properties of Ti and Ni alloys such as the TC6 alloy, IN718 super-alloy, and Ti-6Al-4V alloy. In this work, using the CoCrFeMnNi HEA as a prototypic alloy system, we demonstrate that DCT can produce beneficial effects in LDED-fabricated samples. The influence of DCT on the residual stress, microstructural evolution, and mechanical properties of LDED-fabricated CoCrFeMnNi HEA was elucidated. Specific emphasis is placed on revealing the formation of deformation microstructures at cryogenic temperature via microscopy observation to understand the contributions from different strengthening mechanisms to the yield strength of the DCT-processed CoCrFeMnNi HEA samples, and using finite element modeling (FEM) to quantitatively rationalize the formation of the residual stress.

2 Experimental procedure

All the HEA samples were fabricated using an additive manufacturing system. The feedstock material here is the spherical or near spherical gas-atomized CoCrFeMnNi HEA powders. After LDED, all of the samples were cooled to room temperature in the operating chamber, and DCT were immediately carried out to cool the LDED-built samples in the liquid nitrogen chamber. When the samples reached 77 K, they were soaked at this temperature for different durations. The samples subjected to DCT by soaking for 12 h, 24 h, 48 h and 120 h at liquid nitrogen temperature are hereafter labeled as DCT12h, DCT24h, DCT48h and DCT120h, respectively. Details of the DCT process are shown in Fig. 1.

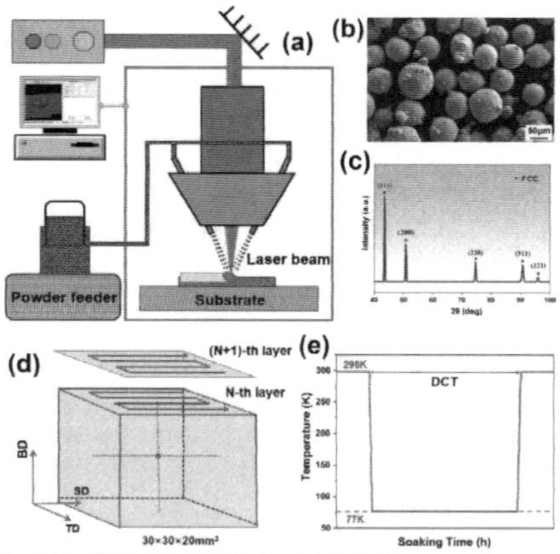

Fig. 1. LDED of CoCrFeMnNi HEA: (a) LDED experimental setup, (b) and (c) SEM image and XRD pattern of the raw powders, and (d) the scanning strategy during the LDED process, and (e) DCT process.

The microstructure was observed by using SEM equipped with EBSD, and TEM with EDS. Room temperature tensile tests were carried out using a universal testing machine at 10^3 s^{-1}. A nonlinear FEM model was established using the commercial software ABAQUS for the coupled thermo-mechanical simulation of the LDED process, and the stress development during the DCT process was simulated using a visco-elastic-plastic model.

3 Result and discussion

All the XRD spectra display only a typical FCC structure, consistent with that of the raw powders, indicating that no new phases were formed and the FCC structure did not change during the DCT process. The compressive residual stress shows a gradual rising trend with the soaking duration, with the DCT120h sample exhibiting the highest compressive residual stress of ~415 MPa measured among all the samples. The variation trend of residual stress in the DCT-processed samples is opposite to that of the LDED-built samples treated via annealing.

The microstructure of the as-built CoCrFeMnNi HEA sample consists mainly of dominant columnar grains, while fine equiaxed grains appear only near the fusion lines indicated between the adjacent deposited layers. There is a high density of dislocations in the as-built sample, which is attributed to the high thermal stress and accumulated micro strain caused by a complex thermal history of the LDED process. Interestingly, in addition to a high density of dislocations located in the matrix, striped nanotwins are observed in the form of bundles in the DCT-processed samples. When the soaking duration is increased to 120 h, a high-density of dislocations, stacking faults and nanotwins exist in the matrix of sample. Homogeneously distributed nanograins (grain size of ~25 nm) can be observed in local regions of the DCT120h sample.

We have conducted the uniaxial tensile tests at ambient temperature and the representative engineering stress-strain curves are plotted in Fig. 2(a). The as-built CoCrFeMnNi HEA possesses a yield strength (σ_y) of 290 ± 9 MPa, ultimate tensile strength (σ_{UTS}) of 456 ± 17 MPa and elongation to failure (ε_f) of 34.8 ± 2.1%. After subjected to DCT for different soaking durations, all of the samples exhibit higher σ_y, σ_{UTS} and ε_f values than the as-built sample, indicating that DCT can promote a significant improvement in both strength and ductility.

How are the deformation microstructures responsible for the strengthening of the DCT-processed samples produced during the DCT process? In this connection, the following should be noted:

(I) The crystalline defects that contribute to strengthening are evidently produced during the cryogenic soaking of the DCT process. The nanotwin density increases on prolonging the DCT duration, and strengthening due to dislocation density and twinning also increases with the prolonged DCT duration.

(II) A state of compressive residual stress is also observed from the DCT-processed samples, with increasing magnitude on increasing DCT duration.

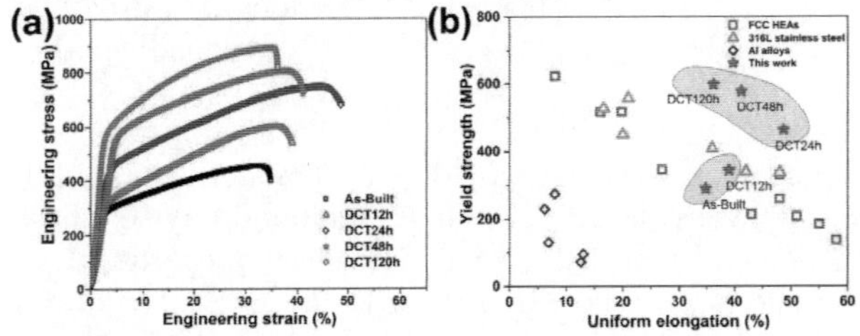

Fig. 2(a) Engineering stress-strain curves of the as-built and DCT-processed HEA samples. (b) Comparison of the yield strength (σ_y) and elongation to failure (ε_f) obtained in this work with FCC HEAs fabricated by various preparation and post-processing methods.

Conclusion

1. Both the as-built and DCT-processed CoCrFeMnNi HEA samples exhibit a single FCC phase, demonstrating their excellent phase stability.
2. DCT causes compressive residual stresses and introduces many defects composed of dislocations, stacking faults, deformation nanotwins and nanograins into the as-built samples, which lead to a significant enhancement in both strength and plasticity.
3. The defects are evidently formed during the prolonged cryogenic soaking of the DCT process under the thermal stress state due to the initial cryogenic cooling.

Acknowledgments

The authors would like to acknowledge the funding from NSFC under Grant nos. 52171154 and 52371106.

References

[1] Li HG, Huang YJ, Zhao WJ, Chen T, Sun JF, Wei DQ, et al. Addit Manuf 50 (2022) 102546.
[2] Li HG, Zhao WJ, Chen T, Huang YJ, Sun JF, Zhu P, et al. J Mater Sci Technol 115 (2022) 40.
[3] Li HG, Huang YJ, Sun JF, Lu YZ. J Mater Sci Technol 77 (2021) 187.
[4] Li HG, Huang YJ, Jiang SS, Lu YZ, Gao XY, Lu X, et al. Mater Design 197 (2021) 109262.

Improvement in the Mechanical Properties of $AlNbTiVCrNi_{1.5}$ High Entropy Alloys Containing Alloyed C

Fangdong Xu, Dezhi Chen*, Ruirun Chen
National Key Laboratory for Precision Hot Processing of Metals, Harbin Institute of Technology, 150001, China
*Corresponding address: Heilongjiang Province, Harbin City, Nangang District, 92 West Dazhi Street
*Corresponding e-mail: chendezhi383@163.com

Abstract: Lightweight high entropy alloys with various Ni and C contents were prepared by vacuum arc melting. The cooperative effects of Ni and C additions on the phase formation, microstructure and mechanical properties. The results shows that The Ni1.5 (AlNbTiVCr) 98.5-xCx (at. %) high entropy alloys are composed of BCC matrix phase, TiC phase, C14 Laves phase, and a small amount of C15 Laves phase. The microstructure of alloys is a typical dendritic structure, with TiC distributed between the dendrites and needle like C14 Laves phase dispersed within the dendrites. The addition of C increases the mechanical properties of alloys first and then decreases. Ni1.5AlNbTiVCrC1 alloy shows the best mechanical properties, and its compressive strength and fracture strain are 1968.1 MPa and 13.7 %, respectively. The improvement in mechanical properties of lightweight high entropy alloys is attributed to solid solution strengthening and second phase strengthening.

Keywords: Lightweight high entropy alloys, Laves phase, Microstructure, Mechanical properties

1 Introduction

The lightweighting of materials is a research focus, as lightweight materials can reduce carbon emissions [1]. Traditional lightweight alloys have reached their development limit due to their poor room temperature strength, low melting point, and high cost. In high entropy alloys, multiple alloying elements tend to form solid solution structures rather than intermetallic compounds [2]. Therefore, the excellent performance of high entropy alloys has attracted a large number of researchers' favor. The emergence of the concept of high entropy alloys has brought new ideas to the design of lightweight alloys. Lightweight high entropy alloys have achieved rich research results in composition design [3]. For example, $AlNbTiVCrNi_{1.5}$ high entropy alloy has the low density and good strength [4]. In order to further improve its room temperature mechanical properties, the $Ni_{1.5}$ (AlNbTiVCr) $_{98.5-x}C_x$ alloys were designed and the phase formation, microstructure and mechanical properties were investigated.

Experimental procedure

Five lightweight high entropy alloys were prepared by vacuum arc melting. Each ingot is repeatedly melted 6 times to ensure the uniformity of chemical composition. Preparation of block shaped samples with dimensions of 8mm × 8mm × 8mm using wire cutting for subsequent phase identification and microstructure observation. Scanning electron microscope (SEM, TM4000) with the energy dispersive spectrometry (EDS). X-ray diffraction (XRD, Empyrean) was used to identify crystal structure. The compression samples dimensions are φ4 mm×6 mm. The compression test was conducted on an AGXplus test equipment with the compression rate of 0.5 mm/min.

2 Result and discussion

Fig. 1 shows the XRD results of Ni1.5 (AlNbTiVCr) $_{98.5-x}C_x$ alloys, where TiC diffraction peaks gradually appear after the addition of C element. Analysis shows that the phase structure after the simultaneous addition of Ni and C is B2, TiC, C14 Laves phase. The increase in C content leads to an increase in the diffraction intensity of the (110) main peak, while the diffraction intensity of TiC and C14 Laves phases decreases. The diffraction intensity of B2 phase gradually increases. Fig. 2 shows the microstructure of the alloy, from which three different contrasts can be observed. Based on the XRD results and previous work analysis, the gray matrix phase is B2 solid solution, the white phase is C14 Laves phase, and the black phase is TiC phase. The increase in C content leads to an increase in the area fraction of TiC. The white needle like C14 Laves phase and the black TiC phase are distributed between the dendrites. There is the most negative mixing enthalpy between C and Ti, which preferentially combines to form TiC.

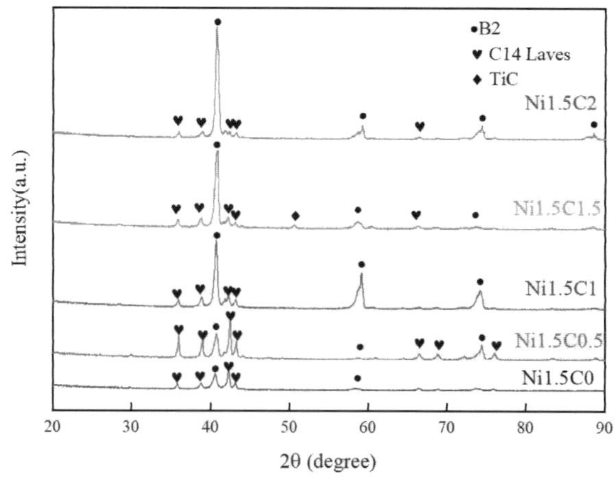

Fig. 1 XRD patterns of Ni1.5 (AlNbTiVCr) 98.5-xCx alloys.

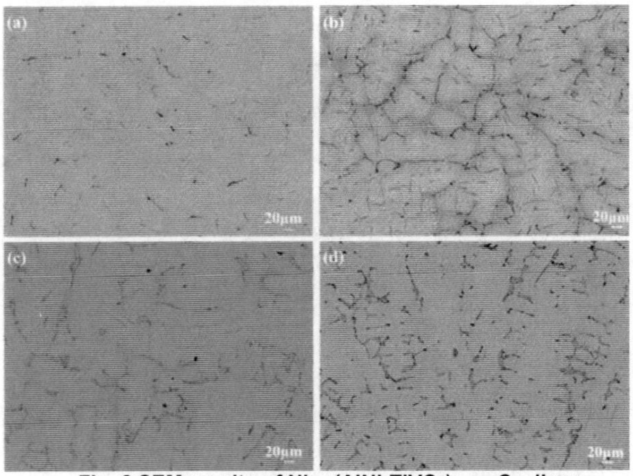

Fig. 2 SEM results of $Ni_{1.5}$ (AlNbTiVCr) $_{98.5\text{-}x}C_x$ alloys.

The distribution of phases is first controlled by melting point and composition, and then the distribution of white needle like C14 Laves phase and black TiC phase depends on grain boundary wetting. The distribution of phases during solidification and cooling mainly depends on grain boundary wetting. The maximum solid solubility of C in Ni is 2.7mol%, and the maximum solid solubility of Ti in Ni is 13.9mol%. Therefore, after the formation of Ni Ti and C-Ti metal compounds, the remaining C will exist in free form within the microstructure of the alloys.

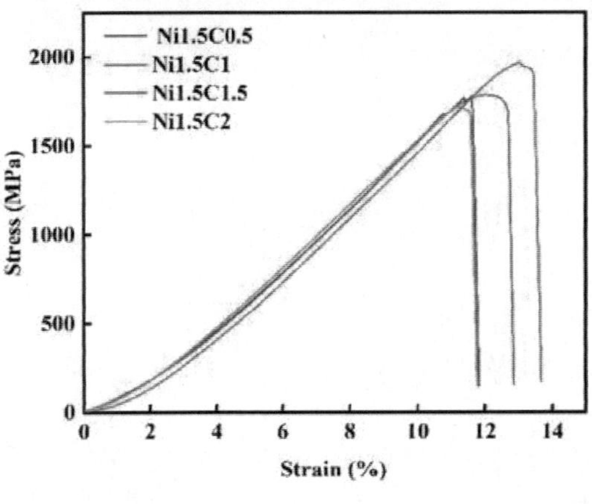

Fig. 3 Compressive stress–strain curve of Ni1.5(AlNbTiVCr)98.5-xCx HEAs.

As shown in Fig. 3, the compressive strength of the alloy gradually increases with the increase of C content. When the C content increases from 0.5 to 1, the compressive strength of the alloy increases from 1776.2 MPa to 1968.1 MPa, and the fracture strain increases from 11.8% to 13.7%. At this point, the mechanical properties of the alloy are optimal. The addition of C causes a slight decrease in the density of the alloy. The maximum density in this combination of gold is 5.75 g/cm^3, and the minimum density is 5.71 g/cm3. As the C content continues to increase, the compressive strength and fracture strain of the alloy decrease. The compressive strength decreases from 1787.3 MPa in Ni1.5C1.5 alloy to 1720.4 MPa in Ni1.5C2 alloy, and the fracture strain decreases from 12.9% in Ni1.5C1.5 alloy to 11.8%.

3 Conclusion

The $Ni_{1.5}$(AlNbTiVCr) $_{98.5\text{-}x}C_x$ high entropy alloys are composed of BCC matrix phase, TiC phase, and C14 Laves phase. The microstructure is a typical dendritic structure, with TiC distributed between the dendrites and needle like C14 Laves phase dispersed within the dendrites. The Ni1.5C1 alloy with the best mechanical properties in this combination of gold has a compressive strength of 1968.1 MPa and a fracture strain of 13.7%.

4 Acknowledgments

This work was supported by the National Key Laboratory for Precision Hot Processing of Metals, Harbin Institute of Technology (6142909230101), the Heilongjiang Postdoctoral Fund (LBH-Z23132) and the China Postdoctoral Science Foundation (2023TQ0099).

References

[1] Buluc, G, Florea, I. Investigation of FeNiCrWMn - a new high entropy alloy. 3rd International Conference on Modern Technologies in Industrial Engineering (ModTech), 2015:6.
[2] Hirotada Hashimoto. Synthesis of Li-Mg-Al-Ti based lightweight high entropy alloys by mechanical alloying and investigation of conditions for solid solution formation. Journal of Alloys and Metallurgical Systems, 4 (2023):100037.
[3] Abayomi Adewale Akinwande. Tribological performance of a novel 7068-aluminium/lightweight-high-entropy-alloy fabricated via powder metallurgy. Materials Chemistry and Physics, 15 (2023) 128207.
[4] Hirth J P and Loathe J. Theory of dislocations, 2nd ed. John Wiley & Sons, New York, 1982: 764.
[5] Fangdong Xu. Lightweight and high hardness $(AlNbTiVCr)_{100\text{-}x}Ni_x$ high entropy alloys reinforced by Laves phase. Vacuum, 213 (2023) 112115.

Effect of Directional Solidification on Microstructure Evolution and Mechanical Properties of Eutectic High Entropy Alloy

Xu Yang, Dezhi Chen, Shu Wang, Gang Qin, Ruirun Chen*, Jingjie Guo
National Key Laboratory for Precision Hot Processing of Metals, School of Materials Science and Engineering, Harbin Institute of Technology, Harbin 150001, P. R. China
*Corresponding address: e-mail: Ruirun Chen, Tel./Fax: +86-0451-86412394, E-mail: ruirunchen@hit.edu.cn

Abstract: Directional solidification, as an advanced method for preparing metal materials, can regulate the microstructure and improve the performance of high entropy alloys by controlling the temperature gradient and solidification rate. This study prepared CoCrFeNi3Al1.25 eutectic high entropy alloy (EHEA) at a directional solidification rate of 150 μm/s and investigated the effect of directional solidification rate on microstructure and mechanical properties. Results indicate that directional solidification significantly changes the microstructure, transforming the microstructure from randomly growing microstructure under as cast conditions to eutectic layer structure with a unified growth direction. The tensile strength and elongation of directionally solidified CoCrFeNi3Al1.25 EHEA are 1125.8 MPa and 22.4%, respectively, which are 14.7% and 83.6% higher than the as cast alloy, achieving a synergistic improvement in strength and plasticity.

Keywords: Eutectic high entropy alloy, Directional solidification, Microstructure, Tensile property

1 Introduction

Eutectic high entropy alloy (EHEA) is a new advanced structural and functional material that has attracted widespread attention because of its excellent strength, high corrosion resistance and wear resistance [1-4]. By changing the solidification parameters, directional solidification can prepare microstructures different from traditional castings, and can simultaneously improve strength and plasticity [5-6].

2 Experimental procedure

$CoCrFeNi_3Al_{1.25}$ EHEA are prepared using a directional solidification furnace at a pulling speed of 150 μm/s. The size of the microstructure observation sample is 8 mm × 8 mm × 8 mm. The gauge length of the tensile specimen is 15 mm, the thickness is 1.8 mm, and the width is 2 mm. The microstructure was observed using scanning electron microscopy (SEM). The mechanical properties were tested using a universal testing machine at a tensile speed of 1 mm/min.

3 Result and discussion (Bold, 10 pt., Arial)

Fig. 1 shows the microstructure of as cast and directionally solidified $CoCrFeNi_3Al_{1.25}$ EHEA. Results show that both as-cast and directionally solidified $CoCrFeNi_3Al_{1.25}$ EHEA are composed of FCC phase and BCC phase. The microstructure of the as-cast EHEA shows a disorderly growth direction, and the directionally solidified EHEA shows a significant single growth direction.

Fig.2 shows the stress-strain curves of as cast and directionally solidified $CoCrFeNi_3Al_{1.25}$ EHEA. The results show that the directional solidification significantly improves the strength and plasticity of EHEA. The tensile strength and elongation of the as-cast EHEA are 981.9MPa and 12.2%. The strength of the directionally solidified EHEA is 1125.9 and 22.4%, respectively, which is 14.7% and 83.6% % higher than that of the as-cast EHEA.

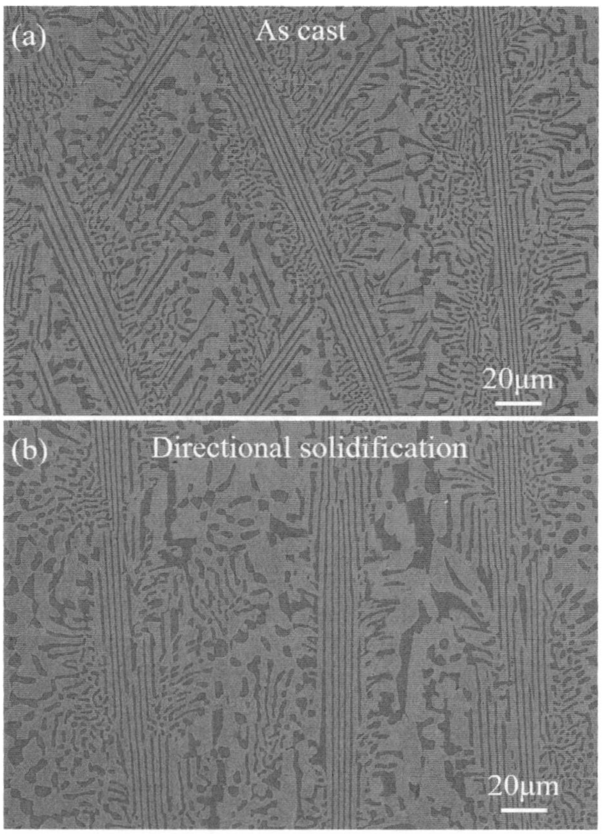

Fig. 1 Microstructure of as-cast and directionally solidified $CoCrFeNi_3Al_{1.25}$ EHEA: (a) As-cast EHEA; (b) Directionally solidified EHEA.

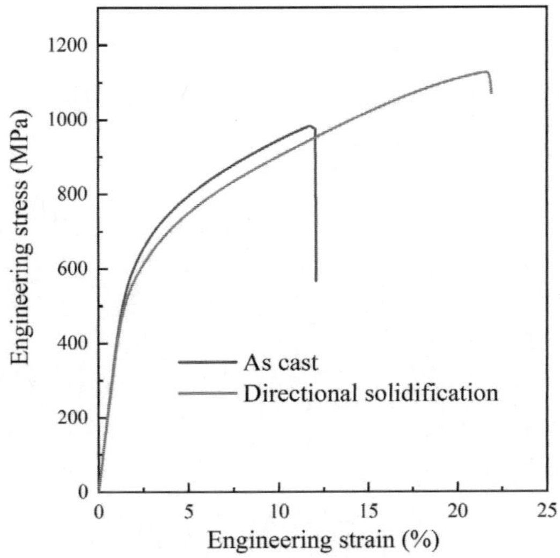

Fig. 2 Tensile stress-strain curves as-cast and directionally solidified CoCrFeNi3Al1.25 EHEA.

4 Conclusion

(1)Directional solidification significantly changed the microstructure of CoCrFeNi3Al1.25 EHEA and generated a layer structure with a single growth direction.

(2)Directional solidification improves the tensile properties. The tensile strength and elongation of CoCrFeNi3Al1.25 EHEA are 1125.8 MPa and 22.4%, respectively, realizing the synergistic increase of strength and plasticity.

5 Acknowledgments

This work was supported by the National Natural Science Foundation of China (No. 51825401), the China Postdoctoral Science Foundation (2023TO0099) and Interdisciplinary Research Foundation of HIT.

References

[1] Mao Z Z, Jin X, Xue Z, Zhang M and Qiao J W, Understanding the yield strength difference in dual-phase eutectic high-entropy alloys, Mater. Sci. Eng. A 2023, 867: 144725.

[2] Tiwary C S, Pandey P, Sarkar S, Das R, Samal S, Biswas K and Chattopadhyay K, Five decades of research on the development of eutectic as engineering materials, Prog. Mater. Sci. 2021, 123: 100793.

[3] Wang M L, Lu Y P, Lan J G, Wang T M, Zhang C, Cao Z Q, Li T J and Liaw P K, Lightweight, ultrastrong and high thermal-stable eutectic high-entropy alloys for elevated-temperature applications, Acta Mater. 2023, 248: 118806.

[4] Chung D H, Lee J, He Q F, Kim Y K, Lim K R, Kim H S, Yang Y and Na Y S, Hetero-deformation promoted strengthening and toughening in BCC rich eutectic and near eutectic high entropy alloys, J. Mater. Sci. Technol. 2023, 146: 1-9.

[5] Zheng H T, Chen R R, Qin G, Li X Z, Su Y Q, Ding H S, Guo J J and Fu H Z, Phase separation of $AlCoCrFeNi_{2.1}$ eutectic high-entropy alloy during directional solidification and their effect on tensile properties, Intermetallics 2019, 113: 106569.

[6] Yang X, Chen R R, Liu T, Gao X F, Fang H Z, Qin G and Su Y Q, Formation of dendrites and strengthening mechanism of dual-phase $Ni_{36}Co_{30}Fe_{11}Cr_{11}Al_6Ti_6$ HEA by directional solidification, J. Alloys Compd. 2023, 948: 169806.

Effect of the Casting Methods on Microstructure and Room Temperature Mechanical Properties of K4800 Superalloy

Wanqi Jie*, Guangyu Yang, Zhiyong Kan, Hao Wu
State Key Laboratory of Solidification Processing, Northwestern Polytechnical University, Xi'an 710072, China
*Corresponding address: e-mail: jwq@nwpu.edu.cn

Abstract: The microstructure and room temperature mechanical properties of K4800 alloy test pieces cast by conventional gravity casting, centrifugal casting and adjusted pressure casting were comparatively studied. The results showed that both centrifugal casting and adjusted pressure casting could refine grain size, the typical grain size was 4.52 mm for normal gravity casting sample, but refined to about 0.28 mm for the sample of adjusted pressure casting, and about 2.22 mm for the sample of centrifugal casting. Meanwhile, the area fraction of shrinkage porosity was decreased from 1.75% of gravity casting sample to 0.27% and 0.25% of centrifugal and adjusted pressure casting samples, respectively. As-cast K4800 alloy by using different casting methods had similar microstructures, namely rod-shaped MC carbides and lamellar borides at grain boundaries, as well as the fine γ' phase within grains, but the sizes of the precipitates were refined for the centrifugal and adjusted pressure casting samples. Due to the reduction of casting defects, as well as the refinement of grain size and γ' phase, K4800 alloy samples prepared by adjusted pressure casting exhibited the much higher room temperature mechanical properties, with yield strength, ultimate tensile strength and plasticity of 798 MPa, 1152 MPa and 11.6%, respectively. Compared with gravity and centrifugal casting, its ultimate tensile strength has increased by 39.9% and 21.8%, and plasticity has increased by 70.4% and 52.5%.

Keywords: K4800; Centrifugal casting; Adjusted pressure casting; Microstructure; Mechanical properties

1 Introduction

With the thrust to weight ratio of aircraft engines further increase, the temperature bearing capacity and mechanics performance of the rear casing components need further improved[1]. The service temperature of the new Ni-based K4800 superalloy significantly increased from 620 ℃-725 ℃ of the existing K4169 superalloy to 800 ℃-850 ℃, exhibiting excellent application potential[2]. The complex structure of the rear casing components restricts the application of traditional welding forming and directional solidification, and most of them are currently prepared by using precision casting methods[3]. Among precision casting methods, there are mainly conventional gravity casting, centrifugal casting and adjusted pressure casting, each with its own advantages[4].

In this paper, As-cast K4800 superalloy test pieces cast by gravity casting, centrifugal casting and adjusted pressure casting, and its microstructure and room temperature mechanical properties were comparatively studied.

2 Experimental procedure

K4800 superalloy was the raw material of rear casing test pieces, and its chemical composition (wt.%) measured by inductively coupled plasma analyzer (ICP, Plasma 400) was Cr 18.0, Co 10.0, Al 2.0, Ti 3.7, Nb 1.9, W 1.4, Mo 2.0, C 0.1, Ni balance. The mold shells of three casting methods were all made of silica sol and zircon sand, and the pouring temperature was all 1530 ℃. The centrifugal speed and holding time of centrifugal casting were 100 rpm and 300 s, respectively. The filling pressure and holding time of adjusted pressure casting were 42 KPa and 1200 s, respectively.

The samples used for microstructure characterization and mechanical property test were prepared by wire electrical discharge machining, and the sampling positions of the three casting methods were all outer ring flanges. The microstructures were characterized by OLYMPUS-GX71 optical microscopy, TESCAN VEGA3 scanning electron microscopy equipped with energy dispersive spectrometer. The dog-bone samples used for mechanical testing were 12.5 mm in gauge length and 3 mm × 2 mm in cross-sectional area. These samples were performed by the Instron 3382 material test machine with a constant speed of 0.75 mm/min, and repeated three times to ensure the accuracy of mechanical properties.

3 Result and discussion

As shown in Fig. 1(a-c), the average grain size of gravity casting sample was about 4.52 mm, while the average grain size was refined to about 2.22 mm of centrifugal casting sample, and about 0.28 mm of adjusted pressure casting sample, respectively. Compared to the adjusted pressure casting, the grain size refinement of centrifugal casting was not significant. Meanwhile, the area fraction of shrinkage porosity in gravity casting sample was about 1.75%, while the proportions of centrifugal casting sample and adjusted pressure casting sample were decreased to about 0.27% and about 0.25%, respectively (Fig. 1(d-e)). Both centrifugal casting and adjusted pressure casting increased the fluidity of the melt, which was conducive to shrinkage and led to a reduction in shrinkage porosity. The adjusted pressure casting not only increased the fluidity of the melt, but also caused the overlapped dendrite skeleton to break, refining the grain size by increasing nucleation sites. In addition, the melting subcooling increased under pressure, inhibiting the grains growth. Therefore, adjusted pressure casting exhibited excellent grain size refinement effects.

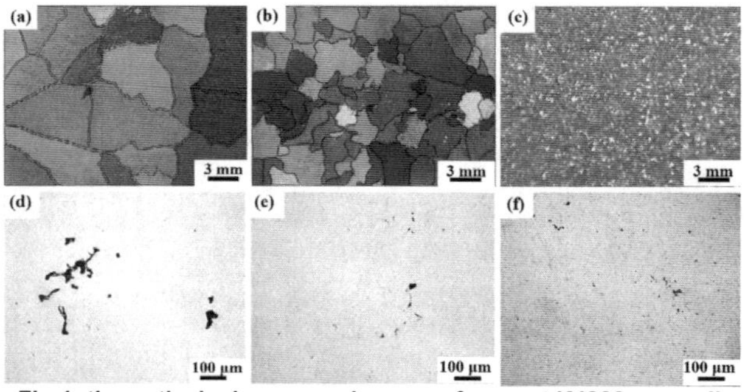

Fig.1. the optical microscopy images of as-cast K4800 superalloy under different casting methods: (a, d) gravity casting, (b, e) centrifugal casting, (c, f) adjusted pressure casting.

It could be clearly observed from Fig. 2 that as-cast K4800 superalloy under different casting methods had similar microstructures, namely rod-shaped MC carbides and lamellar borides at grain boundaries, as well as the γ' phase within grains.就 Meanwhile, compared to the gravity casting sample, both centrifugal and adjusted pressure casting samples refined the size of these precipitates, the size of γ' phase was refined from about 457 nm to about 287 nm and about 105 nm, respectively. In addition, the γ' phase of centrifugal casting sample still presented clusters similar to gravity casting sample, while it was uniformly distributed in adjusted pressure casting sample.

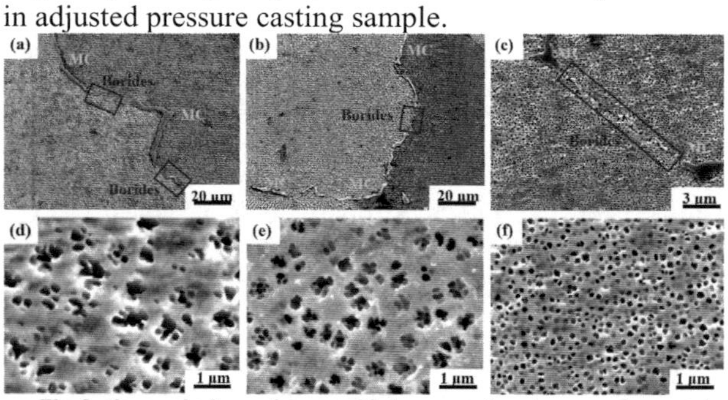

Fig.2. the grain boundary and intragranular phases of as-cast K4800 superalloy under different casting methods: (a, d) gravity casting, (b, e) centrifugal casting, (c, f) adjusted pressure casting.

As shown in Table 1, the as-cast K4800 alloy prepared by adjusted pressure casting exhibited much higher room temperature mechanical properties among three casting methods, with yield strength, ultimate tensile strength and plasticity of 798 MPa, 1152 MPa and 11.6%, respectively. Compared with the gravity casting and centrifugal casting, its ultimate tensile strength has increased by 39.9% and 21.8%, as well as plasticity has increased by 70.4% and 52.5%. The excellent room temperature mechanical properties of adjusted pressure casting sample was mainly attributed to the reduction of casting defects, as well as the refinement of grain size and γ' phase.

Table 1. The room temperature mechanical properties of as-cast K4800 superalloy under different casting methods

Casting methods	Yield strength/MPa	Ultimate tensile strength/MPa	Plasticity /%
gravity	619±7	823±9	6.8±0.5
centrifugal	699±10	946±15	7.6±0.7
adjusted pressure	798±8	1152±18	11.6±0.8

4 Conclusion

(1)The average grain size was 4.52 mm for gravity casting sample, both centrifugal casting and adjusted pressure casting could refine grain size, with average grain size of 2.22 mm and 0.28 mm, respectively.

(2)The area fraction of shrinkage porosity was decreased from 1.75% of gravity casting sample to 0.27% and 0.25% of centrifugal and adjusted pressure casting samples, respectively.

(3)Although the size of precipitates in centrifugal and adjusted pressure casting samples was refined, the microstructure of as-cast K4800 alloy prepared by different casting methods was all composed of the rod-shaped MC carbides and lamellar borides at grain boundaries, as well as the fine γ' phase within grains.

(4)K4800 alloy sample prepared by adjusted pressure casting exhibited the much higher room temperature mechanical properties, with yield strength, ultimate tensile strength and plasticity of 798 MPa, 1152 MPa and 11.6%, respectively. Compared with gravity and centrifugal casting, its ultimate tensile strength has increased by 39.9% and 21.8%, and plasticity has increased by 70.4% and 52.5%.

5 Acknowledgments

This work was supported by the National Science and Technology Major Project (grant number J2019-VI-0004-0118)

References

[1] Reed R C. The superalloys: fundamentals and applications. Cambridge University Press, 2008: 8-50.

[2] Zhao P F, Zhu H Y, Hou K L, et al. Temperature dependence of deformation mechanisms of new Ni-based superalloy and high-temperature property optimization. Journal of Materials Research and Technology, 2023, 27: 1214-1222.

[3] Mishra S, Ranjana R. Reverse solidification path methodology for dewaxing ceramic shells in investment casting process. Materilas and Manufacturing Processes, 2010, 25(12): 1385-1388.

[4] Qin J W, Chen X H, Chen K X, et al. Microstructure and deformation workability of Ni-based superalloy Udimet720Li fabricated by composite shear flow casting. Journal of Alloys and Compounds, 2023, 969: 172150.

As-Cast Microstructure of the Heat-Resistant René N5 Nickel-Based Superalloy

Łukasz Rakoczy[1*], Małgorzata Grudzień-Rakoczy[2], and Rafał Cygan[3]
1. Faculty of Metals Engineering and Industrial Computer Science, AGH University of Krakow, av. Mickiewicza 30, 30-059 Kraków, Poland
2. Łukasiewicz Research Network-Kraków Institute of Technology, Zakopiańska 73, 30-418, Kraków, Poland
3. Rzeszów University of Technology, Research and Development Laboratory for Aerospace Materials, Żwirki i Wigury 4, 35-959 Rzeszów
*Corresponding address: lrakoczy@agh.edu.pl

Abstract: This work focused on the microstructure and mechanical properties of nickel-based superalloy René N5 prepared by investment casting. The microstructure of casting was systematically investigated by X-ray diffraction, light microscopy, energy-dispersive X-ray spectroscopy, and scanning electron microscopy.

Keywords: superalloy; René N5; aerospace; heat-resistant, microstructure

1 Introduction

Nickel-based superalloys are widely used in the turbine blades of modern aero-engines and gas turbines because of their excellent high-temperature mechanical properties, high creep and fatigue resistance, and very good corrosion resistance. The usefulness of Ni-based superalloys during long-term service at harsh conditions depends strongly on the alloying elements, their concentrations, and the morphology of the strengthening phases. In industrial practice, the Ni-based superalloy René N5 is used in a fully heat-treated condition. The solution treatment partially homogenizes the microstructure, and subsequent ageing allows a high-volume fraction of cube-shaped γ' precipitates to be obtained. Therefore, getting more information on the superalloy microstructure and properties in the as-cast state is essential to design and control subsequent heat treatment properly. Solidification segregation of the elements between dendritic and interdendritic spaces induces the formation of non-equilibrium phases, such as carbides, eutectic phases, or other low melting point phases, which should be dissolved during homogenization [1-3].

2 Experimental procedure

René N5 is a second-generation SX superalloy developed by General Electric (Table 1). The presence of Re delays the coarsening of the γ' phase and increases the misfit between γ matrix and γ' precipitates. The superalloy is also strengthened by the effects of chromium, tantalum, and tungsten in solid solution. Additionally, aluminium and titanium contribute to its strength due to the precipitation-strengthening effects of the intermetallic γ' phase. Although the γ' phase is mostly responsible for improvements in thermo-mechanical properties, it also poses challenges during processing, making these alloys prone to crack formation and, consequently, expensive to cast and repair welding. The multi-scale microstructure characterizations were conducted to reveal the primary-dendritic structure, microsegregation, and morphology of γ' phase and MC carbides. The samples for LM and SEM observations, cut from the casting, were ground, polished (diamond suspensions: 3 μm, 1 μm; colloidal silica: 0.25 μm), and chemically etched in No. 17 (25 mL water, 25 mL HCL, 25 mL HNO_3, 1,5 g H_2MoO_4) for 5 seconds. Initial characterization of morphology and distribution of precipitates was carried out on a Leica light microscope. For SEM-EDX analyses, 20 kV accelerating voltage was set on a Phenom XL apparatus equipped with Energy Dispersive X-ray Spectroscopy detector.

Table 1. Nominal chemical composition of René N5 superalloy, wt%

Ni	Cr	Co	Mo	Ta	Re
Bal.	7.0	8.0	2.0	7.0	3.0
W	Al	C	B	Hf	-
5.0	6.0	0.05	0.004	0.15	-

3 Result and discussion

The unetched cross-sections display the distribution of microporosity in the casting and the distribution of carbides (Fig. 1). A dendritic structure was revealed on chemically etched metallographic sections of the castings. The features of the primary structure of the superalloy are very clear. Dendritic areas (primary cores and secondary dendrite arms) are characterized by a relatively uniform two-phase $\gamma+\gamma'$ microstructure, and the precipitates of the γ' intermetallic phase may vary in size. In the interdendritic spaces, in addition to carbides, there are also numerous γ-γ' eutectic islands.

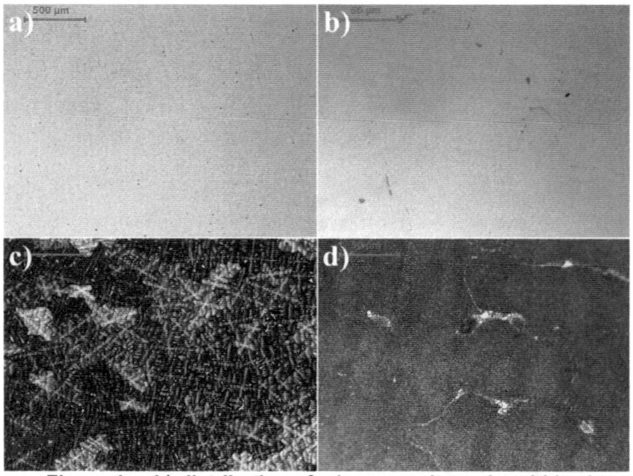

Figure 1. a-b) distribution of microporosity and carbides in unetched sample; c-d) dendritic structure in etched sample, LM

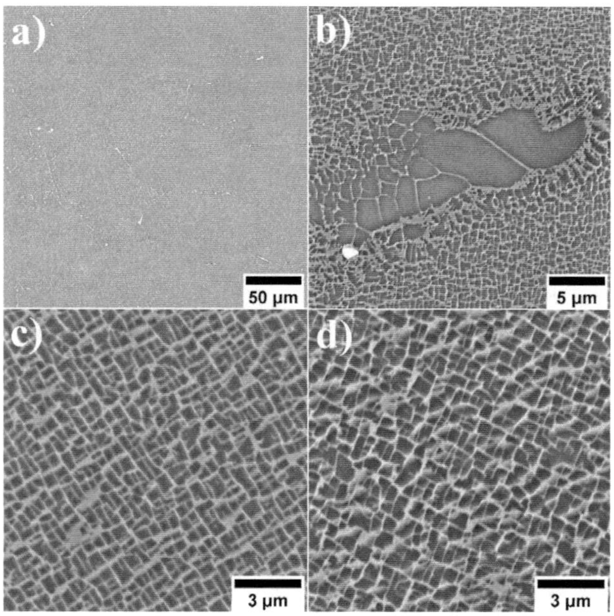

Figure 2. Morphology of precipitates in as-cast René N5: a) carbides; b) primary γ' phase; c) secondary γ' phase in dendritic region; d) secondary γ' phase in interdendritic space, SEM-BSE

Casting solidification is associated with alloying element segregation. This favors the formation of MC carbides near eutectic $\gamma - \gamma'$ islands (Fig. 2a-b). The observed carbides are characterized by complex morphologies, including blocky-shaped, sharp-edged parallelograms and Chinese script-like. The dominant component in the microstructure is the secondary γ' precipitates, which have a cube-like morphology (Fig. 2c-d).

4 Conclusions

This work focused on characterizing the as-cast microstructure and resulting properties of René N5. The obtained results allowed to draw the following conclusions:

-Local microporosity in interdendritic spaces was revealed in unetched samples.

-The types of precipitated phases mainly contain γ'-Ni_3Al phase, (Ta, Hf)C carbides, and eutectic γ+γ' phase.

-The cube-like γ' precipitates in dendritic regions are finer compared to interdendritic regions.

Acknowledgments

The authors gratefully acknowledge the funding by the National Centre for Research and Development Poland under grant FENG.01.01-IP.01-000E/23.

References

[1] Basak A, Das S. Additive Manufacturing of Nickel-Base Superalloy René N5 through Scanning Laser Epitaxy (SLE) – Material Processing, Microstructures, and Microhardness Properties. Adv Eng Mat. 2017; 19(3):1600690. https://doi.org/10.1002/adem.201600690

[2] Wang R, Wang J, Cao T, Zhao R, Lu X, Guan W, Tao H, Shuai S, Songzhe X, Xuan W, Panwisawas C, Chen C, Ren Z. Microstructure characteristics of a René N5 Ni-based single-crystal superalloy prepared by laser-directed energy deposition. Additive Manuf. 2023;61: 103363. https://doi.org/10.1016/j.addma.2022.103363

[3] Cardoso J, Almedia A, Vilar R. Microstructure of coated single crystalline René N5 part repaired by epitaxial laser deposition. Add Manuf. 2022; 49: 102515.

Data-Driven Approach in Ni-Based Superalloy Investment Casting Component Optimization

Zhengyi Ding, Donghong Wang*, Da Shu, Baode Sun, Bang Guan, Jiayu Cui
Shanghai Key Lab of Advanced High-Temperature Materials and Precision Forming, School of Materials Science and Engineering, Shanghai Jiao Tong University, Shanghai, China
*Corresponding address: e-mail: wangdh2009@sjtu.edu.cn

Abstract: Investment casting has been widely applied to fabricate aircraft due to its high dimensional accuracy and excellent surface finish with complex geometries. However, defects such as shrinkage porosity and misrun often occur in these castings. Traditional trial-and-error procedure have hindered the application of investment castings in high-end applications. In this research, a novel data-driven approach for predictive modeling is proposed to address these challenges. With integrated computation of thermodynamic and finite element method (FEM), a variety of components within the boundary of Ni-based superalloy K4169 were evaluated for both fluidity and shrinkage with two patterns. A numerical model incorporating machine learning algorithms was developed based on the integrated computation. For K4169 superalloy castings, the shrinkage porosity is primarily influenced by Al element, while misrun is mainly correlated with Al, Ti, and Mo in a decreasing order. This data-driven model establishes a quantitative relationship between composition and final product defects. The method proposed in this study shows promise for optimizing the investment casting process and can be applied to a wide range of alloys and models.

Keywords: Integrated computation, Data-driven approach, Ni-based superalloy, Investment casting

1 Introduction

Investment casting is a precise forming technique applied in advanced fields such as aerospace, for its intricate design capabilities and high-quality finish.[1] However, the reliance on traditional trial-and-error approaches in the research and development phase often results in huge waste.[2] The introduction of data modeling and big data algorithms has significantly improved the efficiency of the investment casting process.[3] By incorporating advanced computational techniques such as machine learning algorithms and predictive modeling, manufacturers can analyze extensive datasets to identify optimal casting parameters and predict casting defects with enhanced precision.[4]

With great tensile strength, hardness and fatigue life, Ni-based superalloy has been widely applied in aerospace castings.[5] However, the composition of certain superalloy varies in a large bound, which leads to significant uncertainty in properties such as liquidus, solidus and fluidity. These properties contribute to the formation of casting defects, so this uncertainty can lead to misprediction.[6]

In this research, to address the challenge of composition induced casting defects, a data-driven framework is proposed. The schematic is shown in Figure 1.

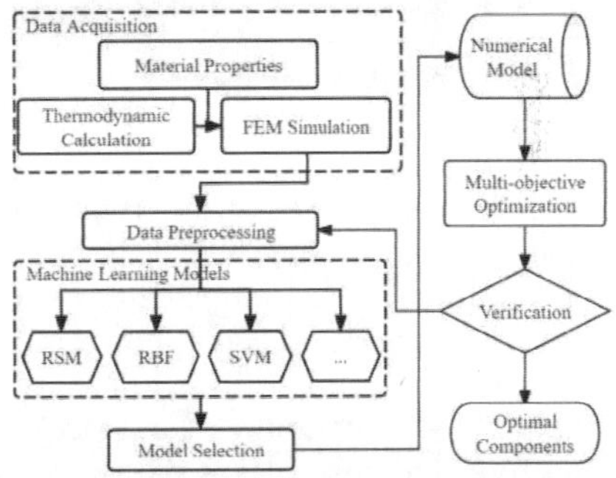

Figure 1. Data-driven framework of this research

2 Data acquisition

Thermodynamic Calculation

In this research, K4169 Ni-based superalloy is taken as the experimental alloy. Its nominal composition is shown in Table 1.

Table 1. Nominal composition of K4169 superalloy

Elements	Al	Cr	Ni	Fe
Lower bound (wt%)	0.30	17.00	50.00	Bal.
Upper bound (wt%)	0.70	21.00	55.00	
Elements	Mo	Nb	Ti	
Lower bound (wt%)	2.80	4.4	0.65	
Upper bound (wt%)	3.30	5.4	1.15	

The performances of different alloy compositions within the limit of K4169 is calculated with Pandat Ni-database. Optimal Latin hypercube sampling was utilized to develop a more effective numerical model. 100 sets of compositions are designed with Latin hypercube sampling, and their performance are calculated with high-throughput calculation in Pandat software.

Finite element integrated calculation

Compositions and their performance are imported into database. Featured patterns in Figure 2 are designed to address shrinkage porosity and misrun defects. FEM

calculations are carried out with ProCAST software. Results are automatically extracted from calculated files.
To establish the relationship model between components and performance, different machine-learning algorithms are applied to the result. In this research, RBF, RSM, linear regression and SVR are applied. The data is divided into training set, validation set and test with a ratio of 2:1:1, and then approximated with the models above.

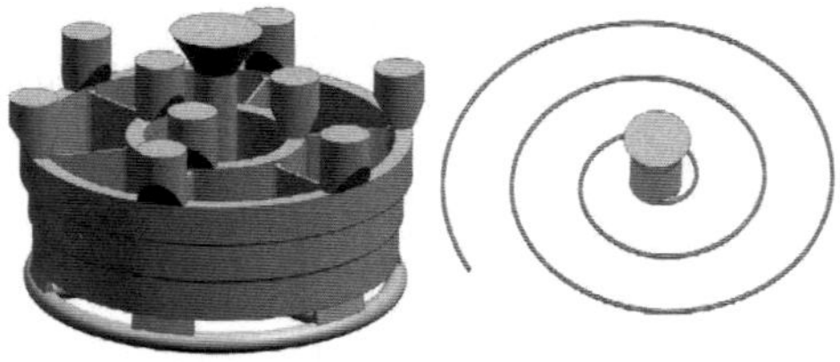

(a) Shrinkage (b) Misrun

Figure 2. Featured patterns

3 Result and discussion

Composition-performance-defect numerical model

The comparison of different machine learning models is shown in Figure 3. In almost all properties in performance, microstructure and defect, RSM algorithm shows the best coefficient of determination. Numerical model is established with RSM algorithm. The correlation table between composition and properties is shown in Table 2.

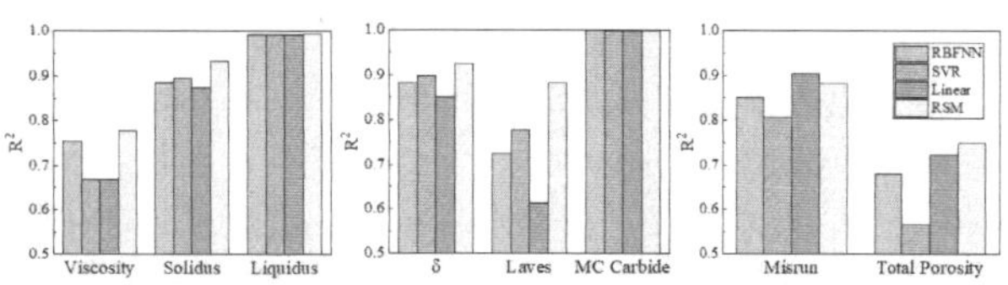

Figure 3. Coefficient of determination among different algorithms

Table 2. Composition-performance-defect correlation table

Elements	Al	Cr	Fe	Mo	Nb	Ti
Misrun	-0.54	0.07	0.03	-0.32	-0.09	-0.45
Porosity	0.79	-0.05	0.05	-0.28	-0.26	-0.39

Component optimization

3rd generation non-dominated sorting genetic algorithm (NSGA-III) is applied to the component optimization. After 40 generations of iteration, the optimal component is: Al 0.79%, Cr 18.1%, Fe 20.1%, Mo 3.08%, Nb 5.25%, Ti 1.09%. Simulation verification in Figure 4 indicates that the optimal composition is enhanced with 23% less internal shrinkage and 13% increased feed length, compared to median composition of K4169 superalloy.

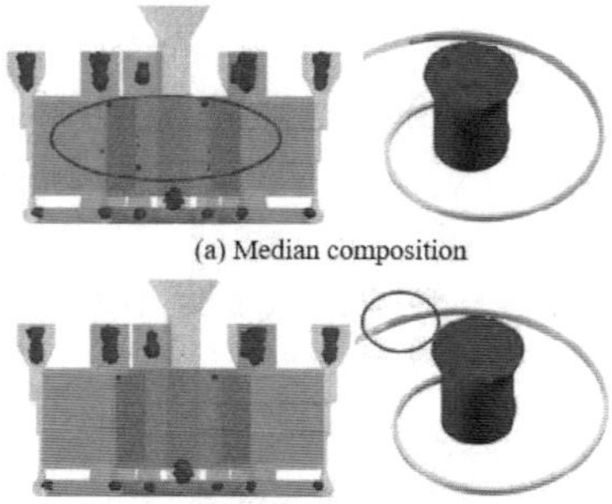

(a) Median composition

(b) Optimal composition

Figure 4. Numerical verification of optimal composition

4 Conclusion

In this research, machine learning algorithms is combined with thermodynamic-FEM integrated computation to develop the numerical model. The NSGA-III algorithm is then applied to optimize the components based on the model. The following conclusions can be made:

a)Within the boundary of K4169 superalloy, shrinkage porosity is primarily influenced by Al, while misrun is negatively correlated with Al, Ti and Mo.

b)The optimal composition within K4169 limit has 23% less internal shrinkage and 13% increased feed length in featured patterns, compared to median.

c)The data-driven approach proposed in this research shows promise in deviation analysis and composition optimization.

References

[1] L. Liu, The progress of investment casting of nickel-based superalloys, Foundry 11 (2012) 1273-1285.

[2] Wang D, Sun F, Shu D, et al. Data-driven design of nickel-based casting superalloys and precise forming of complex castings. Acta Metallurgica Sinica, 2022,58(1):89-102.

[3] Yu J, Wang D, Li D, et al. Engineering computing and data-driven for gating system design in investment casting, Int. J. of Adv. Manuf. Tech., 2020, 111(3-4): 829-837.

[4] Wang W, Li J, Liu W, et al. Integrated computational materials engineering for advanced materials: A brief review, Comput. Mater. Sci., 2019, 158: 42-48.

[5] Selvaraj, S K, Sundaramali G, Dev S J et al. Recent Advancements in the Field of Ni-Based Superalloys, Adv. Mater. Sci. Eng., 2021

[6] Skryabin M L. The study of influence of chemical composition of steel 35HGSL on the characteristics of shrinkage, casting defects and microstructure, J. Phys., Conf. Ser., 2020, 1515: 042107-042112.

This page intentionally left blank.

Study on Coarsening Behavior of γ' Precipitates and Compression Properties in a Novel Co-Ni-Al-W Superalloy

Cheng Zhou[1,2]*, Lei Jin[1,2], Gaoyang Jing[1,2], Boyan Yu[1,2]
1. Shenyang Research Institute of Foundry Co., Ltd.,No. 17 Yunfeng South Street,Shenyang, China
2. State Key Laboratory of Casting Technology for High-end Equipment, No. 17 Yunfeng South Street,Shenyang, China
*Corresponding address: e-mail: qilongzhuone@1163.com

Abstract: In the present study, the coarsening behavior of γ′ phase at different temperatures and the compression property of a novel Co–Ni–Al–W superalloy was investigated. The evolution of the mean radius and volume fraction of the γ′ phase indicates that the coarsening behavior follows the classicalLSW model. The coarsening rate of the γ′ phase exhibits a significant dependence on the aging temperature, increasing from 1.30×10-27 m3/s at 800 ℃ to 9.56×10-27 m3/s at 900 ℃. The activation energy of γ′ phase, influenced mainly by W diffusion in the γ matrix, is found to be 210 kJ/mol. In particular, the Co-Ni-Al-W alloy possesses a good combination of a high γ′ solvus temperature of 1221 ℃ and a low density of 8.7 g/cm³. Besides, compression yield strength of the developed Co-Ni-Al-W alloy at ambient and high temperatures are higher than those of other recently γ′-strengthened Co-based superalloys.

Keywords: Co-Ni-Al-W superalloy, Microstructural evolution, Coarsening behavior, Compression property

1 Introduction

Recent studies have discovered the γ′-Co_3(Al, W) phase with an $L1_2$ structure in Co-Al-W-based superalloys. The γ′ phase is stable at high temperatures and is coherent with the γ matrix, exhibiting superior high-temperature mechanical properties compared to traditional Co-based superalloys. This discovery makes it feasible to develop novel γ′ phase-strengthened Co-based superalloys.

This paper focuses on the microstructural evolution of Co-Ni-Al-W-based alloys under different aging temperatures and times. Specifically, it focuses on the morphological, size, and volume changes of γ′ phase during aging, exploring factors affecting its coarsening. Furthermore, the microstructure stability of the alloy at high temperatures is evaluated, and its compressive performance is analyzed.

2 Experimental procedure

The Co-Ni-Al-W alloy was melted at 1480 ℃, then treated at 1230 ℃ for 24 h, and aged at 800, 850, and 900 ℃ for 24-216 h. The SEM was used to study the size and distribution of the γ′ phase, and compression tests were conducted using a Gleeble-3800 thermo-simulator system.

3 Result and discussion

Microstructural evolution during aging

The size and shape of the γ′ phase change as the aging temperature and time increase. The γ′ phase grows and becomes coarser, with the fastest coarsening occurring at 900 ℃ (Fig.1(b)) and the slowest at 800 ℃ (Fig.1(a)). Higher aging temperatures result in faster growth. The phase changes from near-spherical to cuboidal shape due to the competition between elastic strain energy and interface energy. Changes in composition and element diffusion during aging treatment cause variations in latticeparameters and speed up the change in phase morphology.

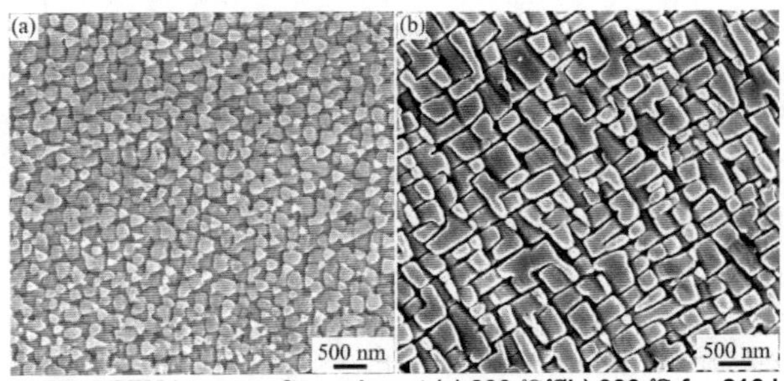

Fig.1 SEM images after aging at (a) 800 ℃;(b) 900 ℃ for 216 h.

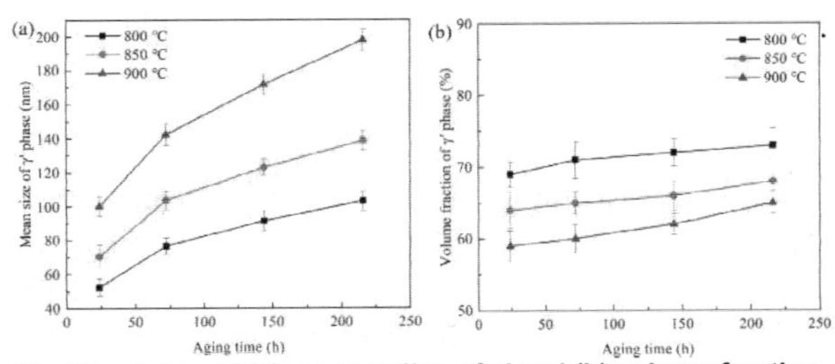

Fig.2Evolution of (a) mean radius of γ' and (b) volume fraction of γ' during aging treatment.

The growth of the γ′ phase results in a gradual increase of its volume fraction as the aging time progresses, as shown in Fig. 2. However, increasing the aging temperature leads to a decrease in the precipitation of the γ′ phase, resulting in a lower volume fraction of the γ′ phase. Additionally, the merging of γ′ phases, with larger ones consuming smaller ones, ultimately results in a decrease in the volume fraction of the γ′ phase. Aging at lower temperatures leads to a higher volume fraction of the precipitated γ′ phase due to higher solute supersaturation than aging at 900 ℃.

4 Coarsening behavior of γ′ phase during aging

The aging time exponent fits with the LSW model theory, indicating that the coarsening of the γ′ phase in the Co-Ni-Al-W alloy follows the LSW model.Theequation for the coarsening kinetics of γ′ phase canbe obtainedthrough linear regression analysis. The equation takes the following form:

$$r_t^3 - r_0^3 = K\cdot(t - t_0) \quad (1)$$

The coarsening rate coefficient K of γ′ phase is determined through the slope of the curve obtained from linear fitting (Fig. 3). The results show that the coarsening rate of γ′ phase is mainly influenced by aging temperature. Moreover, increasing the aging temperature enhances the coarsening rate of γ′ phase. Furthermore, the coarsening of the γ′ phase in the alloy conforms to the LSW model, suggesting that the diffusion of solute elements regulates the coarsening process of the γ′ phase.

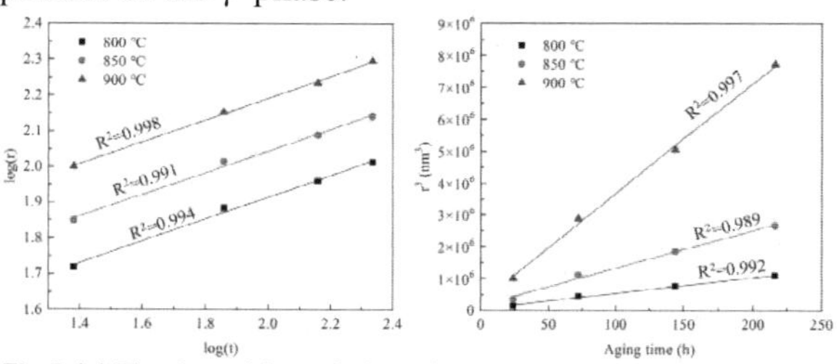

Fig.3 (a) The logarithm of size of γ′ phase vs. the logarithm of aging time, (b) A linear fit of the size of γ′ phase (r^3) versusaging time.

The diffusion activation energy of the solute elements in the Co-Ni-Al-W alloy during aging at 800-900 ℃ is determined to be 210 kJ/mol through linear fitting computations.In Co-Al-Wbased alloys, the coarsening process of γ′ phase is mainly influenced by the diffusion of W element in the alloy. Hence, it inferred that the lower W content in the Co-Ni-Al-W alloy is the main reason for its relatively lower diffusion activation energy.

5 Solvus temperature and density

The studied alloy stands out with an exceptionally advantageous combination of solvus temperature (1221 ℃) and density (8.7 g/cm³).These exceptional characteristics establish the Co-Ni-Al-W alloy as a superior alternative to traditional Co-based alloys, particularly in applications where weight reduction is crucial, as shown in Fig. 4.

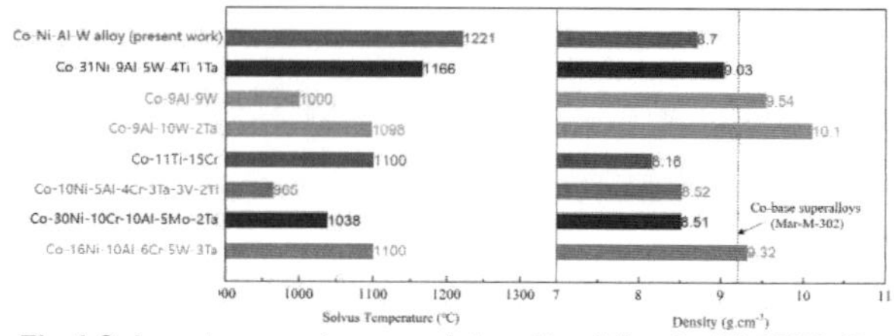

Fig.4 Solvus temperatures and density of the Co-Ni-Al-W alloy, and other Co-based superalloys from literature.

compression properties

The curve of the compression yield strength of the Co-Ni-Al-W alloy with temperature is divided into three stages, as shown in Fig. 5. At temperatures between 25 to 650 ℃, yield strength decreases from 761 to 714 MPa. It exhibits a positive temperature dependence from 650 to 850 ℃. At 850 ℃, it reaches a maximum of 774 MPa. Above 850 ℃, it shows negative temperature dependence. The alloy shows an anomaly behavior in yield strength between 650 to 850 ℃. The phenomenon is associated with the Kear-Wilsdorf mechanism, which locks dislocations at high temperatures.The Co-Ni-Al-W alloy has better mechanical properties at high temperatures than other recently developed W-free/ Cr-containing γ/γ′ Co-based superalloys.

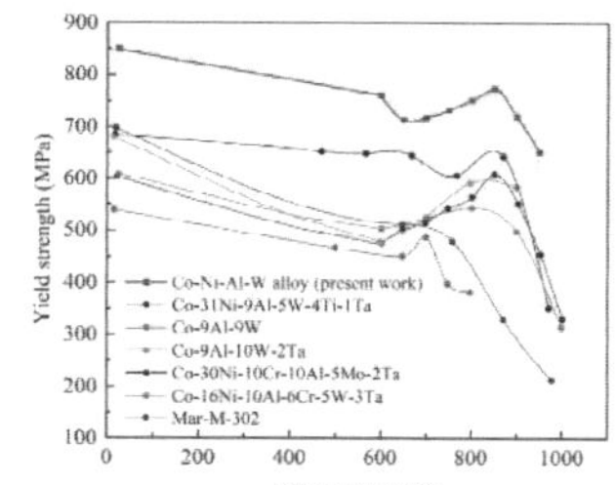

Fig. 5 The temperature dependence of compression yield strength of the Co-Ni-Al-W alloy and other Co-based superalloys.

6 Conclusion

1) The Co-Ni-Al-W alloy has a high γ′ solvus temperature of 1221 ℃ and a low density of 8.7 g/cm³.

2) The Co-Ni-Al-W alloy ages at 800-900 ℃, with γ′ phase morphology changing from spherical to cubic. Its size grows with higher aging temperatures and longer time. The γ′ phase volume fraction decreases with increasing temperature while increases with longer aging time.

3) The coarsening behavior of γ′ phase in the Co-Ni-Al-W alloy follows the diffusion-controlled LSW coarsening mode. The γ′ phase coarsening rates were increased from 1.30×10^{-27} m³/s at 800 ℃ to 9.56×10^{-27} m³/s at 900 ℃.

4) The compression yield strength curve of the Co–Ni-Al-W alloy comprises threestages. From 650 to 850℃, theyield strength exhibits a positive temperature dependence and reaches a maximum of 774 MPa at 850℃.

As-Cast Microstructure of the Alloy 700 Nickel-Based Superalloys Fabricated Via Lost-Wax Casting

Rafał Cygan[1], Konrad Chrzan[2], Krzysztof Jaśkowiec[2], and Małgorzata Grudzień-Rakoczy[2*]
1. Rzeszow University of Technology, Research and Development Laboratory for Aerospace Materials, Żwirki i Wigury 4, 35-959 Rzeszów
2. Łukasiewicz Research Network-Kraków Institute of Technology, Zakopiańska 73, 30-418, Kraków, Poland
*Corresponding address: e-mail: malgorzata.grudzien@kit.lukasiewicz.gov.pl

Abstract: The presented research investigates Ni-based superalloy Alloy 700 castings produced via lost-wax casting. Casting analyses were carried out via thermodynamic simulations, differential scanning calorimetry (DSC), X-ray diffraction (XRD), light microscopy (LM), scanning electron microscopy (SEM), energy dispersive X-ray spectroscopy (EDX), scanning transmission electron microscopy (STEM), and tensile testing. The dendritic regions of the castings are characterized by a relatively homogenous microstructure, consisting of cubic-like γ' precipitates surrounded by the matrix. In the interdendritic spaces, $\gamma-\gamma'$ eutectic, MC carbides and M_3B_2, phases were identified.

Keywords: superalloy; Udimet 700; aerospace; investment casting

1 Introduction

Nickel-based superalloys are commonly used in high-homologous temperature applications due to their ability to maintain high strength and excellent resistance to hot corrosion and oxidation during long-term service [1-2]. The binary Ni-Al phase diagram includes various ordered phases characterized by high ordering enthalpy (e.g., NiAl, $NiAl_3$, Ni_2Al_3, $NiAl_9$, and the most important, Ni_3Al). The γ matrix of superalloys is strengthened by the coherent γ' phase, which possesses an ordered $L1_2$ structure [3]. Considering operation at high temperatures, the volume fraction, misfit, coarsening rate, and antiphase boundary energy (related to the chemical composition) of the γ' precipitates play an important role [4]. The ordering of the γ' phase depends on the strictly defined locations of Al and Ni atoms in its crystal lattice, which influences its behavior during high-temperature exposure and deformation. Ni_3Al is very rigid and impedes dislocation movement from the γ matrix, ensuring the required strength under service conditions [5].

2 Experimental procedure

The lost-wax castings of Alloy 700 nickel-based superalloy were fabricated in vacuum condition. The nominal chemical composition is presented in Table 1.

Table 1. Composition of Udimet 700, wt%

Cr	Co	Ti	Mo	Fe
15.0	18.5	3.4	5.0	<1.0
Al	C	B	Ni	-
4.3	0.07	3.4	Bal.	-

The as-cast specimens (10 × 10 × 10 mm) were cut from the middle section of casting. For microscopic examinations, they were mounted in resin, ground, polished, and chemically etched in No. 17 reagent (25 mL H_2O, 25 mL HCL, 25 mL HNO_3, H_2MoO_4) for light microscopy (LM). LM images were captured by microscope Leica DM/LM. SEM observations were conducted using a Desktop XL instrument with a CeB6 electron source. The accelerating voltage during SEM-BSE (backscattered electrons) observations and X-ray energy dispersive spectroscopy (SEM-EDX) was 20 kV. The chemical compositions of primary precipitates were also measured by SEM-EDX. Quantitative chemical analysis was performed using the ZAF correction.

3 Result and discussion

The chemically etched specimens were analyzed via LM and revealed a typical dendritic structure (Figure 1). The dendritic areas (primary dendrites and secondary dendrite arms) are characterized by a relatively homogenous microstructure, while numerous eutectic $\gamma-\gamma'$ islands and carbide-looking precipitates are located in interdendritic spaces. The morphology of the secondary γ' precipitates resembles a cube (Fig. 2a). The deviation from the cubic shape occurs with increasing precipitate size. It should be noted that the morphology of γ' depends on the local thermodynamical conditions determined by the chemical composition, state of elastic stresses, and the mutual interaction between the precipitates [6-7]. These factors change as a function of temperature and are conditioned by the system's attempt to achieve thermodynamic equilibrium [8]. The morphology of the primary γ' precipitates is shown in Figure 2b. They were formed through the $L \rightarrow \gamma + \gamma'$ phase transformation. The amount of these precipitates is significant, which indicates a strong enrichment of the residual liquid phase in γ'-former. The precipitates are characterized by a diversified distribution and complex morphologies. Casting solidification is associated with alloying element segregation, verified by the calculated partitioning coefficient values. This favors the formation of minor phase precipitates in close proximity to eutectic $\gamma-\gamma'$ islands.

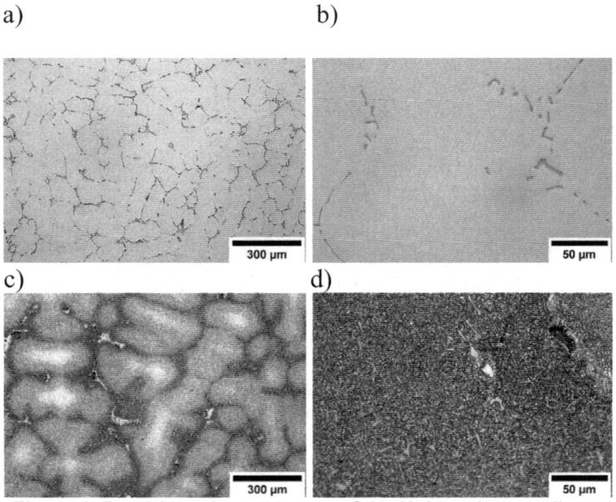

Figure 1. Dendritic microstructure of the Ni-based superalloy: a-b) unetched sample; c-d) etched sample, LM

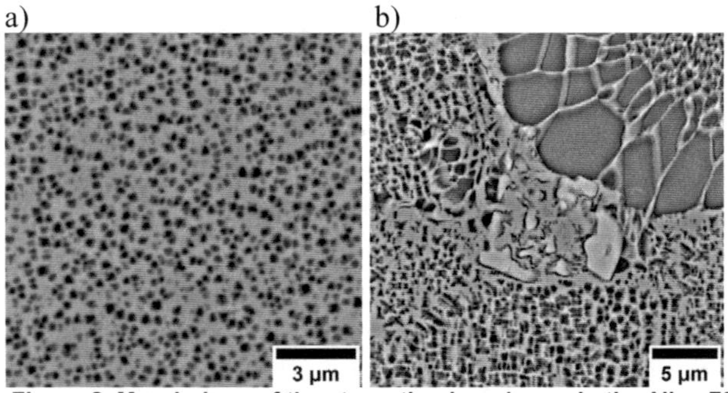

Figure 2. Morphology of the strengthening phases in the Alloy 700, Ni-based superalloy: a) secondary γ' precipitates in dendritic region; b) primary γ' precipitates and borides, SEM-BSE

The MC (enriched in Ti) carbides are characterized by complex morphologies, including blocks, sharp-edged parallelograms and Chinese script-like (feather-, herringbone-like). In the final stage of solidification, the eutectic reaction $L\rightarrow\gamma+\gamma'$ takes place in areas of Al- and Ti-enriched interdendritic liquid. As the solubility of B in γ and γ' is extremely low, it segregates into the boundary (γ–γ') eutectic–residual liquid interface and provide favorable conditions for formation of borides (gray-contrasted precipitates).

4 Conclusions

This work focused on characterizing the as-cast microstructure and resulting properties of Ni-based superalloy Alloy 700, produced via lost-wax casting. The obtained results allowed to draw the following conclusions:

-Casting characterize by typical dendritic microstructure with irregular distribution of strengthening phases.

-Secondary γ' precipitates posses a cubic-like morphology and volume fraction above 40%.

-In the close vicinity of eutectic $\gamma-\gamma'$ islands, MC carbides and borides were detected.

5 Acknowledgments

This work was supported by the National Center for Research and Development (Poland) project LIDER 0147/L-13/2022 (Development of a technology for the regeneration of precise castings of nickel superalloys intended for use in the aviation industry).

References

[1] Cygan R, Rakoczy Ł. The effect of CoAl2O4 as a nucleating agent and pouring temperature on the microstructure and properties of Inconel 713c® nickel-based superalloy castings. Materials, 2023, 16(16):5588.
[2] Pollock T. Tin, S. Nickel-Based Superalloys for advanced turbine engines: Chemistry, microstructure, and properties. J. Prop. Pow, 2006, 22: 361–374.
[3] Rakoczy Ł, Cempura G, Kruk A, Czyrska-Filemonowicz A, Zielińska-Lipiec A. Evolution of γ' morphology a γ/γ'lattice parameter misfit in a nickel-based superalloy during non-equilibrium cooling. Inter. J. of Mat. Research, 2019, 110(1): 66–69
[4] Long H, Mao S, Liu Y, Zhang Z, Han X. Microstructural and compositional design of Ni-based single crystalline superalloys. J. All. Comp. 2018, 743: 203–220.
[5] Cormier J, Milhet X, Mendez J. J. Mater. Sci., 2007, 42:7780–7786.
[6] Doi M, Miyazaki T. Mater. Sci. Eng., 1986, 78:87–94.
[7] Doi M. Prog. Mater. Sci,1996, 40: 79–180.
[8] Rakoczy Ł, Grudzień-Rakoczy M, Hanning F, Cempura G, Cygan R, Andersson J, Zielińska-Lipiec A. Investigation of the γ' Precipitates Dissolution in a Ni-Based Superalloy During Stress-Free Short-Term Annealing at High Homologous Temperatures. Metall Mater Trans A, 2021, 52: 4767–4784.

Microstructural Changes of Ni-Based Superalloy After GTA Welding with Induction Preheating

Rafał Cygan[1], Konrad Chrzan[2], Krzysztof Jaśkowiec[2], and Małgorzata Grudzień-Rakoczy[2*],
1. Rzeszow University of Technology, Research and Development Laboratory for Aerospace Materials, Żwirki i Wigury 4, 35-959 Rzeszów
2. Łukasiewicz Research Network-Kraków Institute of Technology, Zakopiańska 73, 30-418, Kraków, Poland
*Corresponding address: e-mail: malgorzata.grudzien@kit.lukasiewicz.gov.pl

Abstract: This work presents the influence of high-frequency induction preheating on the weldability of Alloy 700 Ni-based superalloy. The investigation was divided into the characterization of the material's microstructure after preheating and determining the influence of preheating on constitutional liquation during gas tungsten arc welding. Microstructure investigation confirmed that the induction preheating allows to prevent liquation cracking formation in HAZ. The eutectic γ–γ' re-solidification products were visible on the liquated edges and allowed for a self-healing effect.

Keywords: superalloy, Alloy 700, aerospace, investment casting

1 Introduction

The use of welding or cladding is desirable for enhancing manufacturing process efficiency and repairing serviced components to extend their lifetime. However, this is very challenging due to the high susceptibility of Ni-based superalloys to liquation cracking during welding [1-2]. Although solidification cracks in the fusion zone are also observed, they are not as problematic as cracks in the heat-affected zone (HAZ). Cracking in the fusion zone can be mitigated by selecting the correct filler metal. However, the chemical composition and microstructure of the base material, in most cases, cannot be arbitrarily altered. Liquation cracks are significantly more challenging because numerous strengthening phases, characterized by a complex chemical composition that decreases weldability, contribute to their occurrence. Nevertheless, these phases are necessary to achieve stability and heat resistance during long-term service. Ni-based superalloys containing substantial amounts of Al and Ti (more than 6 wt%) are generally considered very difficult to weld due to their high susceptibility to cracking. This criterion does not consider other microstructural constituents like carbides and borides [3-5]. Consequently, superalloys with theoretically good weldability unexpectedly crack during welding or repairing. A characteristic feature of cast Ni-based superalloys is their complex chemical composition, resulting from more than 10 alloying elements with significantly different properties. This leads to the formation of various phases, usually with complex morphologies and a high tendency to segregation [6-8]. Therefore, the weldability of each superalloy should be considered individually. This work aims to reveal the role of induction preheat in reducing the susceptibility to liquation cracking formation in the heat-affected zone.

2 Experimental procedure

The lost-wax castings of Alloy 700 nickel-based superalloy were fabricated in vacuum condition. The nominal chemical composition is presented in Table 1.

Table 1. Composition of Alloy 700, wt%

Cr	Co	Ti	Mo	Fe
15.0	18.5	3.4	5.0	<1.0
Al	C	B	Ni	-
4.3	0.07	3.4	Bal.	-

The superalloy was subjected to the gas tungsten arc welding with high-temperature preheating. For microscopic examinations, they were mounted in resin, ground, polished, and chemically etched in Kalling reagent for light microscopy (LM) and in HNO_3, H_3PO_4 and H_2SO_4 for SEM. LM images were captured by microscope Leica. SEM observations were conducted using a Phenom Desktop XL instrument with a CeB6 electron source. The accelerating voltage during SEM-BSE (backscattered electrons) observations and X-ray energy dispersive spectroscopy (SEM-EDX) was 20 kV.

3 Result and discussion

Fig. 1 shows the cross-section microstructure of the preheated Alloy 700 superalloy after GTA welding. The fusion zone and heat-affected zone (HAZ) are free from hot solidification and liquation cracking.

Preheating effectively eliminated liquation cracks in the heat-affected zone (Fig. 2). The preheating temperature facilitated the dissolution of additional γ' precipitates and the generation of a substantial volume of non-equilibrium liquid. Constitutional liquation of the γ' precipitates occurred at the grain boundaries and in areas adjacent to these boundaries. This resulted in the formation of a new eutectic γ–γ' layer at the grain boundaries, with varying thickness likely due to local changes in chemical composition. There was no observed correlation between the thickness or morphology of the newly formed eutectic γ–γ' layer and the distance from the fusion line, where the welding temperature was the highest in the HAZ.

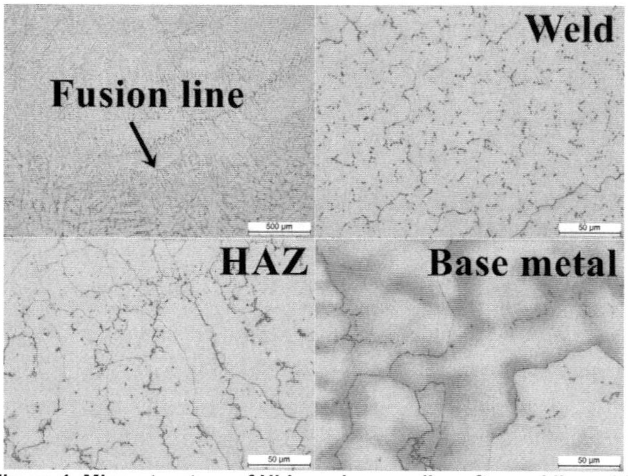

Figure 1. Microstructure of Ni-based superalloy after welding with induction preheating, LM.

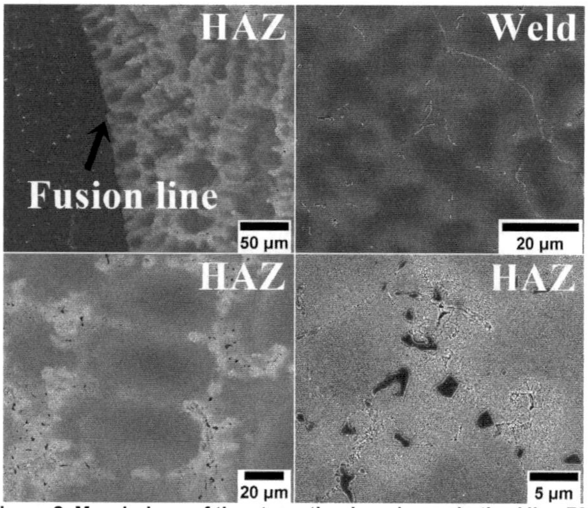

Figure 2. Morphology of the strengthening phases in the Alloy 700 Ni-based superalloy after welding with induction preheating, SEM-BSE.

4 Conclusions

This work focused on analyzing the effects of induction preheating on the improvement of the Alloy 700 superalloy resistance to liquation cracking in HAZ during gas tungsten arc welding. The main conclusions are as followed:

- The partial dissolution of γ' phase precipitates occurred during preheating.
- Rapid heating above the γ' solvus temperature caused the formation of eutectic liquid due to the reaction between partially dissolved γ' precipitates and the matrix.
- A sufficiently high preheating temperature could contribute to eliminating liquation cracks through intensified γ' precipitate dissolution, thermal stress reduction, and microcrack self-healing with a liquid eutectic.

5 Acknowledgments

This work was supported by the National Center for Research and Development (Poland) project LIDER 0147/L-13/2022 (Development of a technology for the regeneration of precise castings of nickel superalloys intended for use in the aviation industry).

References

[1] Rakoczy Ł, Grudzień-Rakoczy M, Cygan R. The Influence of Shell Mold Composition on the As-cast Macro- and Microstructure of Thin-Walled IN713C Superalloy Castings. J. Mat. Eng. Perf, 2019, 28(7): 3974-3985.
[2] Pollock T. Tin, S. Nickel-Based Superalloys for advanced turbine engines: Chemistry, microstructure, and properties. J. Prop. Pow, 2006, 22: 361–374.
[3] Rakoczy Ł, Cempura G, Kruk A, Czyrska-Filemonowicz A, Zielińska-Lipiec A. Evolution of γ' morphology a γ/γ'lattice parameter misfit in a nickel-based superalloy during non-equilibrium cooling. Inter. J. of Mat. Research, 2019, 110(1): 66–69.
[4] Long H, Mao S, Liu Y, Zhang Z, Han X. Microstructural and compositional design of Ni-based single crystalline superalloys. J. All. Comp. 2018, 743: 203–220.
[5] Cormier J, Milhet X, Mendez J. J. Mater. Sci., 2007, 42:7780–7786.
[6] Doi M, Miyazaki T. Mater. Sci. Eng., 1986, 78:87–94.
[7] Doi M. Prog. Mater. Sci,1996, 40: 79–180.
[8] Rakoczy Ł, Grudzień-Rakoczy M, Hanning F, Cempura G, Cygan R, Andersson J, Zielińska-Lipiec A. Investigation of the γ' Precipitates Dissolution in a Ni-Based Superalloy During Stress-Free Short-Term Annealing at High Homologous Temperatures. Metall Mater Trans A, 2021, 52: 4767–4784.

Study on Microstructure and Properties of a New High Temperature, High Strength and High Plastic Titanium Alloy

H. J. Li[1, 2*], S. B. Liu [1, 2], T. Y. Liu [1, 2], J. Zhao[1, 2], K. Shi [1, 2], C.Y.Li[1,2], X. Q.Li[1, 2]
1. National Key Laboratory of Advanced Casting Technologies, Shenyang, Liaoning, 110022, China
2. Shenyang Research Institute of Foundry Co., Ltd, Shenyang, Liaoning, 110022, China
*Corresponding address: e-mail: lihongju545china@163.com.com

Abstract: With the continuous development of aerospace technology, the demand for high temperature and high strength titanium alloys continues to increase, and the comprehensive mechanical properties of high temperature and high strength titanium alloys continue to improve. In this paper, a new high temperature, high strength and high plastic titanium alloy Ti-23Nb-6.7Co-5Cr-4Al(wt.%) was designed based on d-electron alloy design theory, and the alloy was prepared by vacuum induction magnetic levitation melting. The alloy has good high temperature (550℃) performance in the cast state, the tensile strength is 785±5 MPa, and the elongation reaches 35%±1%.

Keywords: Titanium Alloy, High Temperature, High Strength, High Plasticity

1 Introduction

Titanium and titanium alloy, because of their high specific strength, good corrosion resistance, good processing performance, good fracture toughness, good thermal stability and thermal strength and many other advantages, have been widely used in aviation, aerospace, navigation, petrochemical and other fields, so it is also known as "modern metal", "space metal", "Marine metal" and "strategic metal" [1].High-temperature titanium alloy can replace steel to achieve weight reduction effect, and can also be applied to engineering fields where the strength and service temperature of aluminum alloy cannot meet the requirements. As an excellent lightweight structural material, it is often used in the production of high-temperature parts of aerospace engines, such as engine blades, casing and roulette [2].In this paper, a new type of titanium alloy is designed based on the d-electron theory. The as cast titanium alloy has excellent strength and plasticity matching ability at room temperature, and has considerable tensile strength and excellent elongation at high temperature (550 ℃). It is worth noting that, unlike the previous high temperature titanium alloys, Sn, Zr, Si and other elements were not added in the design of the titanium alloy, which showed good high temperature performance.

2 Composition design

The design method of d-electron theory alloy can effectively study the electronic states in the local regions of various crystal structures, and give a correct description of various alloying effects [3]. The theory uses Bo and Md parameters to control the phase stability and properties of the alloy. B_o(Bond order) is used to characterize the strength of covalent bonds between titanium and alloying elements, and M_d(the metal d-orbital energy level) is a parameter closely related to the electronegativity of elements and the radius of metal bonds. The calculation formula is as follows [3]:

$$B_o = \Sigma x_i (B_o)_i \quad (1)$$

$$M_d = \Sigma x_i (M_d)_i \quad (2)$$

Where x_i is the atomic percentage of alloy element i. $(B_o)_i$ and $(M_d)_i$ are the B_o and M_d values of element i, respectively.

After consulting the data, four alloying elements, Nb, Co, Cr and Al, were determined [4]. The B_o and M_d values of the four elements in β-Ti are shown in Table 1. According to alloy design experience [3], β titanium alloys with high strength, easy processing and good plasticity generally have higher B_o value and smaller M_d value, because the larger the B_o value, the better the solid solution strengthening effect, and the smaller of M_d value, the more stable of β phase.

Table 1. B_o and M_d values of each elements in β-Ti

elements	B_o	M_d
Ti	2.790	2.447
Nb	3.099	2.424
Co	2.529	0.807
Cr	2.779	1.478
Al	2.426	2.200

In order to obtain stronger solid solution strengthening effect and more stable β phase, the B_o value and M_d value are set to be greater than 2.76 and less than 2.35, respectively. A new titanium alloy Ti-23Nb-6.7Co-5Cr-4Al(wt.%) was designed by referring to the research of researchers and the calculation of the range of B_o and M_d values [4].

3 Microstructure and mechanical properties of alloys

Figure. 1 shows the backscattering image of Ti-23Nb-6.7Co-5Cr-4Al(wt.%) alloy. It can be seen from the figure that the as-cast structure of the alloy is dendritic, and the secondary arm spacing is about 20 μm. The distribution of elements is investigated by energy spectrometer. The bright part is enriched. The results show that Co is segregated towards the dendrite edge, Nb is segregated in the dendrite center, and Ti, Al and Cr are evenly distributed.

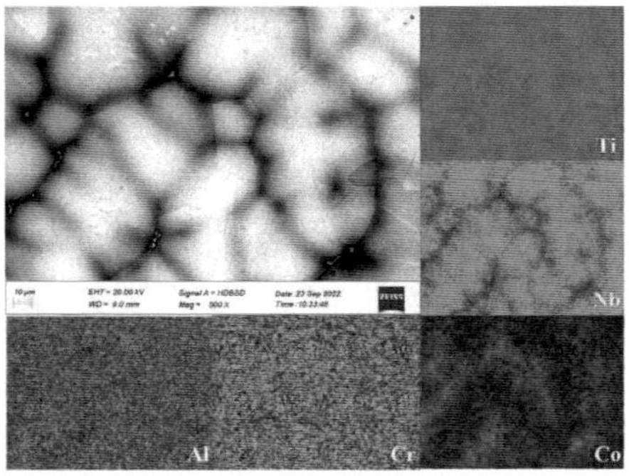

Figure 1. As cast microstructure and EDS analysis of alloys.

Figure. 2 shows the stress-strain curves at room temperature and high temperature (550℃) of Ti-23Nb-6.7Co-5Cr-4Al(wt.%) alloy in as-cast state. The tensile strength of the alloy at room temperature is 1083±5 MPa, the yield strength is 1025±6 MPa, the elongation is 18%±2%, the section shrinkage is 51%±3%, and the alloy has good comprehensive mechanical properties. At high temperature (550℃), the tensile strength is 785±5 MPa, the elongation is 35%±1%, and the section shrinkage is 87%±2%. The alloy has good elongation and section shrinkage at high temperature, and has the potential for superplastic development. The fracture at high temperature is shown in the figure, and the maximum dimples reach 20 μm. This is because the material underwent plastic deformation before fracture, and dimples were gradually formed with the expansion of fracture, which reflects the high toughness of the alloy.

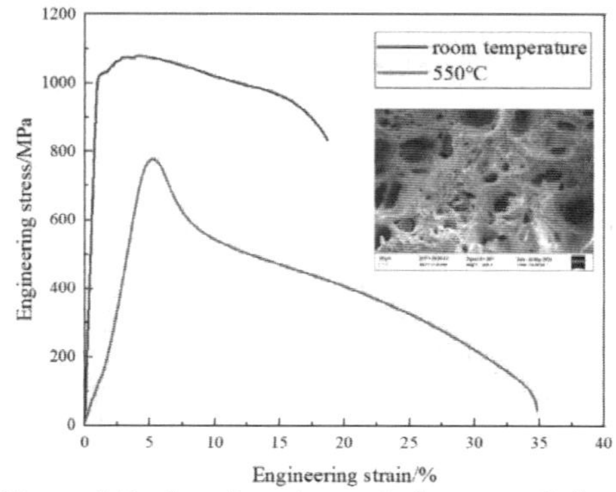

Figure. 2 Engineering stress-strain curve of alloy.

Conclusion

The as-cast Ti-23Nb-6.7Co-5Cr-4Al(wt.%) alloy has excellent properties at high temperature (550℃) (tensile strength of 785±5 MPa, elongation of 35%±1%, and high strength and plasticity), which has considerable development prospects. If Si and other alloy elements are added to the alloy, the high temperature performance will be more excellent, and it has an excellent research and development prospect.

References

[1] ZHAO Q Y, SUN Q Y, XIN S W, et al. High-strength titanium alloys for aerospace engineering applications: A review on melting-forging process[J]. Materials Science Engineering A, 2022, 845: 143260.

[2] KANG L M, YANG C. A review on high-strength titanium alloys: microstructure, strengthening, and properties[J]. Advanced Engineering Materials, 2019, 21(8): 1801359.

[3] WANG G R, GAO Q, LIU J X, et al.Composition Design of Beta-titanium Alloys: Theoretical,Methodological and Practical Advances[J]. MATERIALS REPORTS, 2017, 31(03): 44-51.

[4] OKULOV I V, WENDROCK H, VOLEGOV A S, et al. High strength beta titanium alloys: New design approach[J]. Materials Science Engineering A, 2015, 628: 297-302.

Effect of Cold Crucible Levitation Melting Process on the Solidification Microstructure and Tensile Properties of K4169 Casting

Yipeng Li, Aoqi Li, Guoyan Shui, Xun Sun
1.National Key Laboratory of Advanced Casting Technologies, China Academy of Machinery Shenyang Research Institute of Foundry Co., Ltd, Shenyang 110022 , P. R. China
*Corresponding address: No.17 Yun feng South Street, Tie xi District, Shenyang City, Liaoning Province, China.
e-mail: 945536084@qq.com

Abstract: Aiming at the problems such as the crucible impurities and uneven grain size distribution existing in the Vacuum Induction Melting (VIM) technology, the influences of single-chamber Cold Crucible Levitation Melting (CCLM) technology on the mold filling capacity, purity, grain structure and tensile properties of K4169 superalloy have been investigated. The results show that a 200mm×50mm×3mm thin-wall plate can be completely filled by using of the CCLM technology. Moreover, Compared with the VIM technology, the CCLM technology can be used to improve the purity of liquid metal, which significantly reduces the inclusion defects in the casting and refine the grain structure and improve the tensile properties of the casting obviously.

Keywords: K4169 superalloy, CCLM, grain structure, tensile properties

1 Introduction

It is important to control the purity of the liquid metal and the grain size in the solidification microstructure of cast superalloy. Numerous studies have shown that the presence of inclusions will significantly reduce the service performance of superalloy, and the grain refinement and homogenization contribute to improve the mechanical properties of medium temperature conditions[1].

CCLM technology uses electromagnetic force in the smelting process to make the molten metal material in a suspended or quasi-suspended state, and its most outstanding advantage is that the melt contamination of crucible material can be excluded at high temperature. Previous experimental investigation already proved that the purity and casting surface quality can be significantly improved by the using of this technology [1]. However, due to the use of water-cooled copper crucible, the molten metal overheat is relatively low and the casting temperature is difficult to accurately control, limiting its application in superalloy thin-wall castings. In this project, large-area thin-wall plates were used to investigate the influences of single-chamber CCLM process and VIM process on the mold filling, grain structures and tensile properties of K4169 alloy, providing reference for the application and development of CCLM process in the superalloy thin-wall castings.

2 Experimental procedure

The K4169 alloy bars were selected as the raw materials . The main chemical composition of K4169 alloy was 0.058C, 19.07Cr, 5.06Nb, 3.10Mo, 0.95Ti, 0.55Al, 0.31Co, and Ni others. In order to investigate the application feasibility of CCLM process in the K4169 alloy investment castings, thin-wall plates with a height of 50mm, a length of 200mm and the thicknesses of 3mm, 5mm, and 7mm were designed. The single-chamber CCLM furnace and three-chamber VIM furnace were used to smelt and cast the alloy. The W-Re thermocouples with a diameter of 0.5mm and temperature acquisition device of MV2000 were used to record the temperature variation histories of the metal liquid and the mold shell during the pouring. The thin-wall plate mold shells and the thermocouple positions are shown in Fig. 1.

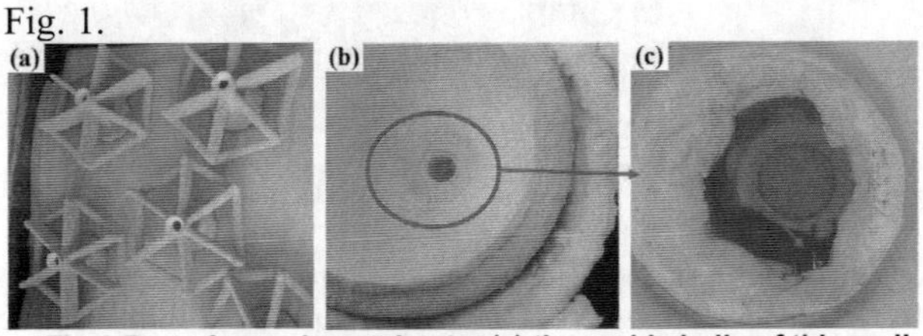

Fig.1 Part of experimental sets: (a) the mold shells of thin-wall plates, and (b),(c) the positions of thermocouples

The main process parameters of the experiment are shown in Tab. 1, in which C1, C2 and C3 denote the CCLM processes, and CC denotes the VIM process.

Tab. 1 Parameters of the main processes

Process	C1	C2	C3	CC
Molding methods	Single mold shell with thermal insulation cotton	Buried sand	Buried sand	Buried sand
Preheating Temperature of mold shell /℃	1000	1000	1000	1000
Melting time/min	30	30	40	30
Measured pouring temperature/℃	1413	1409	1458	1450
Measured mold shell	642	812	782	/

temperature/℃				

3 Results and discussion

The thin-wall plate castings

Fig. 2 shows the experimental mold filling results of the thin-wall plate castings under the conditions of different processes and plate thicknesses. As can be seen from Fig. 2, the thin-wall plates can be completely filled for the CC process, but appears serious misrun situations for the 3mm and 5mm plates manufactured by using of C1 and C2 processes, and under the condition of C3 process, the 3mm and 5mm plates can be basically completely filled. The mold filling ability of the casting is related to the pouring temperature, the mold shell temperature and the characteristics of the casting alloy and mold material. Because the CCLM process adopts a single-chamber furnace, the smelting process and the mold shell insulation process cannot be carried out at the same time, so the cooling time of the mold shell is long, leading to a low mold shell temperature when the pouring is carried out, which results in serious misrun. In the C3 process, the melting time of metal liquid is extended, although the mold shell temperature decreases slightly, the pouring temperature is increased and the mold filling effect is obviously improved. So under the condition of C3 process, the 3 mm plate has been basically completely filled.

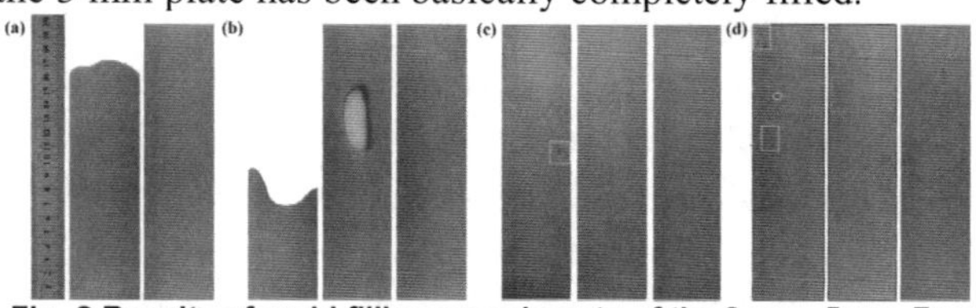

Fig. 2 Results of mold filling experiments of the 3 mm, 5mm, 7mm plates under the conditions of (a) C1, (b) C2, (c) C3, (d) CC, and the surface defects.

4 Grain structures

Fig. 3 shows the grain structures of different thickness thin-wall plates under the conditions of C3 and CC processes. It can be seen from the Fig. 3 (d) ~ (f), when the CC process is applied, the grain structures of the plates are coarse equiaxial crystals with uneven distributions, and with the decrease of the thickness, the distribution of grain is more uneven. However, when the C3 process is applied, the grain structures of the plates become fine and uniform equiaxial crystals. The nucleation rate is an important factor that determines the grain size.

Fig. 3 Grain structures of the K4169 superalloy plates under the conditions of different processes and thicknesses: (a) C3, 3 mm, (b) C3, 5 mm, (c) C3, 7 mm, (d) CC, 3 mm, (e) CC, 5 mm, (f) CC, 7 mm.

5 Tensile properties

Tab. 2 shows the tensile properties of the K4169 superalloy under the conditions of different processes and temperatures. Compared with the CC process, the ultimate tensile strength (UTS) and yield strength (YS) at 20 ℃ and 600 ℃ of the test rods manufactured by the C3 process increased by 4.2%, 3.9% and 4.8%, 3.2%, respectively, which is ascribed to significant inclusion reduction and fine grain structures when the C3 process is applied.

Tab.2 Tensile properties of the K4169 superalloy under different conditions

Process	T/°C	UTS/MPa	YS/MPa	EL/%
C3	20	1119	925	9.5
CC		1072	889	7.0
C3	600	934	770	15.5
CC		889	745	13.5

6 Conclusions

By using the single-chamber CCLM process and the C3 process parameters, excellent mold filling and surface quality of thin-wall plate with thickness of 3 mm can be obtained. Compared with the VIM technology, when the CCLM technology is applied, the purity of the metal liquid can be improved, and the grain structure can be significantly refined, leading to the increasing of tensile properties .

References

[1] Okumura T, Yamamoto K, Shibata M. Large scale cold crucible levitation melting furnace with bottom taping nozzle[C].Proceeding of the 6th International conference. Electromagnetic Proceeding of Materials EMP2009, Oct. 19-23, 2009, Dresden, Germany, 521-524.

Solidification Characteristics and Defect Formation Mechanism of the Fourth Generation Ni-Based Single Crystal Superalloy

Wenchao Yang*, Jun Zhang, Lin Liu
State Key Laboratory of Solidification Processing, Northwestern Polytechnical University, Xi'an 710072, China
*Corresponding address: e-mail: wenchaoyang@nwpu.edu.cn

Abstract: Solidification characteristics and defect formation mechanism of the fourth generation Ni-based single crystal superalloy were systematically investigated. The results indicated that the solidification path of the experimental superalloy during directional solidification was determined as followed: L → γ → β-NiAl → γ′→ (γ+γ′) → TCP. The phase transformation process of the peritecticreaction L + β-NiAl → γ′ of the experimental superalloy during directional solidification had been revealed using the planar interface solidification, and the relaxation of some atoms on the matching planes (110)β and (111)γ′ maximized the atomic matching at the interface, increasing the possibility of peritectic γ′ phase nucleated on primary β-NiAl phase in the peritectic reaction. Finally, it was proposed that it was beneficial to reduce the formation tendency of large-angle grain boundary defects at the platform and reduce the orientation difference of low-angle grain boundaries by controlling secondary dendrite orientation.

Keywords: Ni-based single crystal superalloys; solidification characteristics; secondary dendrite orientation

1 Introduction

Nickel-based single crystal (SX) superalloys are a class of materials extensively used for hot components of turbine engines due to their excellent creep strength, oxidation and hot-corrosion resistance at elevated temperature. The need to continuously improve the performance and efficiency of aircraft and power turbines has driven the development of the successive generations of the nickel-based SX superalloys. Consequently, a large number of refractory elements, such as Re, W, Mo, Ta, and Ru, have been added to further improve the high-temperature mechanical properties of the nickel-based SX superalloy. In particular, as a symbol element in the fourth- and fifth-generation nickel-based SX superalloys, the addition of Ru can greatly improve the microstructural stability and the creep resistance. However, the segregation of high-generation Ru-containing SX superalloys is extremely serious due to the increase of refractory elements, which is more likely to cause casting defects. Therefore, it brings challenges to the composition design and microstructure control of high-generation Ru-containing superalloys [1].

In this paper, the solidification characteristics and solidification microstructures of a fourth-generation Ni-based single crystal superalloy were studied, and the effect of the composition characteristic on the solidification characteristics and solidification microstructures was investigated. Finally, the effects of secondary dendrite orientation on the dendrite growth and orientation evolution within the platforms of simplified blades and solid turbine blades were investigated through a combination of experiment and numerical simulation, and the methods of controlling the stray grain and low-angle grain boundary based on dendrite orientation were further proposed.

2 Experimental procedure

In this work, the nominal composition of the experimental superalloy was as follows (wt.%): 14.0 Co, 6.0 Al, 19.5 (Cr+Mo+W+Ta), 0.1 Hf, 5.4 Re, 3 Ru, and the balance Ni. The single crystal bars with orientation [001] were prepared by liquid metal cooling directionalsolidification furnace. Besides, Singlecrystal blades with different secondary dendrite orientations were prepared byhigh rate solidification furnace. The schematic diagrams of the directional solidification deviceswere shown in Fig. 1.

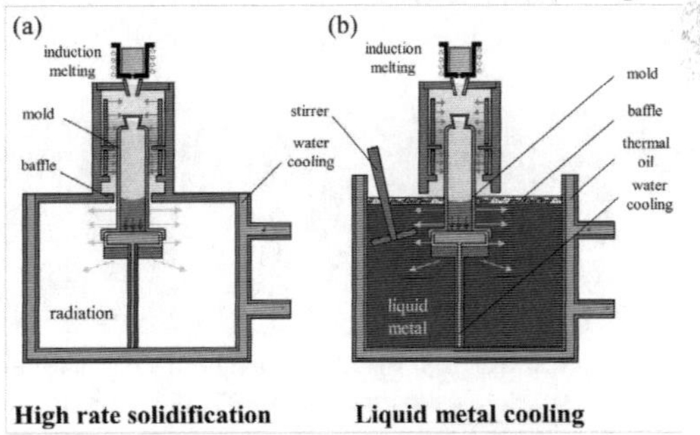

Fig. 1. The schematic of the high rate solidification furnace (a) and liquid metal cooling furnace (b).

3 Result and discussion

Solidification characteristics

Fig. 2 showed the microstructure evolution as a function of the *fs* for the alloy during the planar interface solidification. It could be seen that there were four different microstructure morphologies along the growth direction.The solidification path of the experimental superalloy during directional solidification was determined as followed: L → γ → β-NiAl → γ′→ (γ+γ′) →TCP.

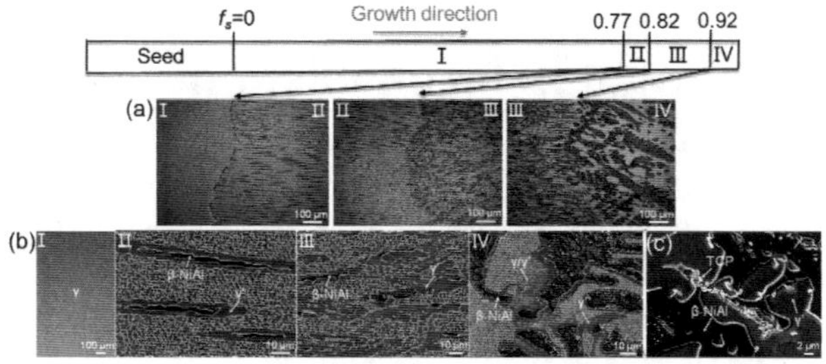

Fig. 2 The microstructure evolution of the alloy at the withdrawal rate of 1 μm/s (a) OM images of the solidification sequence solidified with a planar solid-liquid interface. (b) Higher magnification SEM micrographs of the areas marked as I, II, III, and IV respectively. (c) SEM micrograph showing the area in (b)IV marked by the red dotted rectangle at higher magnification.

The solidification sequence after the solidification of the γ phase was as followed during directional solidification of the experimental superalloy: the primary β-NiAl phase was first precipitated from the liquid directly, and then the coarse γ′ phase was formed by an incomplete peritectic reaction of L + β-NiAl→ γ′ + β-NiAl$_{(Residual)}$.To further reveal the phase transformation mechanism of theperitectic reaction from the crystallographic point of view, TEM observations were performed and the results were shown in Fig. 3. The OR between β-NiAl phase and γ′ phase might be identified as $[001]_\beta$ // $[110]_{\gamma'}$ and $(110)_\beta$ // $(111)_{\gamma'}$. The relaxation of some atoms on the matching planes $(110)_\beta$ and $(111)_{\gamma'}$ maximized the atomic matching at the interface, increasing the possibility of peritectic γ′ phase nucleated on primary β-NiAl phase in the peritectic reaction

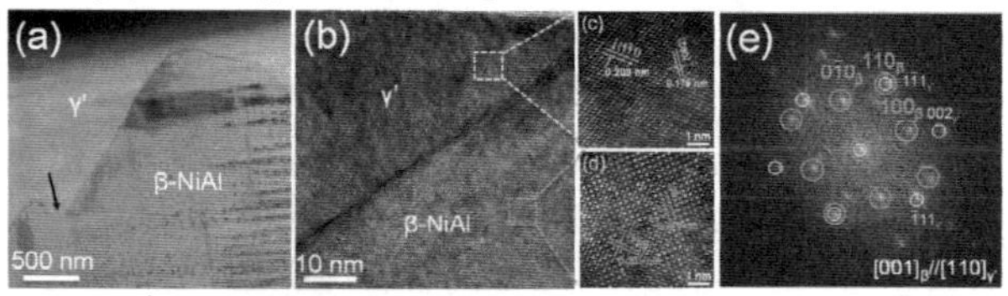

Fig. 3 TEM micrograph and HRTEM analysis of the β-NiAl and γ′ phases. (a) TEM bright-field image. (b) HRTEM image of the two phases and Enlarged HRTEM images showing (c) the γ′ phase and (d) the β-NiAl phase. (e) corresponding FFT image of (b).

Defect formation mechanism

Fig. 4 showed theschematic diagrams of dendrite growth path within the platform base of the blades. The edge B on the bottom surface of the blade platform had the property of being the first to advance to the liquid-line temperature, and the matrix dendrite branching in the leaf body region grew farther into the path of this edge, and the edge B might be the most susceptible to supercooling nucleation at the edge plate to produce stray grains. The formation of stray grainswas determined by the competition between the supercooling time and the growth time of the branching crystals. Changing θ_2 affected the path of the matrix branching crystals in the leaf body region to the edge corner B, which had the greatest tendency to form stray grains in the platform, and this was important for the suppression of the formation of stray grains defects in the platform.

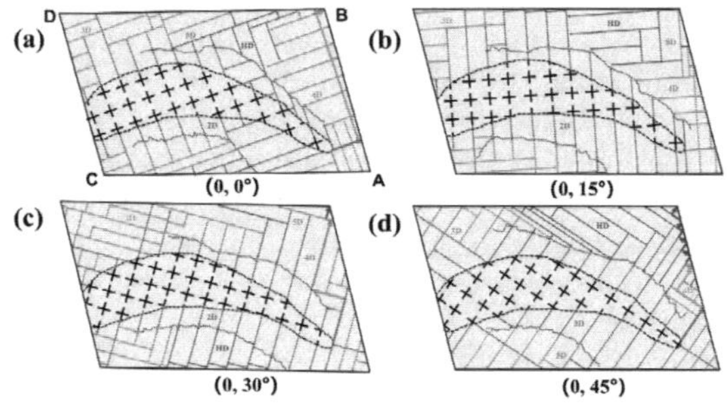

Fig. 4 Schematic diagrams of dendrite growth path within the platform base of blades: (a) θ1=0° andθ2=-2°; (b) θ1=0° and θ2=13°; (c) θ1=0° and θ2=28°; (d) θ1=0° and θ2=45° (The bold arrows showed the paths of dendrite branching from the blade zone growing to the extremity B within the platform base).

Conclusion

1.The results indicated that the solidification path of the experimental superalloy during directional solidification was determined as followed: L → γ → β-NiAl → γ′→ (γ+γ′) →TCP.

2. Controlling the secondary dendrite orientations could effectively inhibit the stray grains and low-angle grain boundaries defects at the blade platform.

Acknowledgments

This work was funded by National Natural Science Foundation of China (51771148, 52071263, 52031012, 52322410), Science Center for Gas Turbine Project (P2021-A-IV-001-001), The Key Research and Development Program of Shaanxi Province (2023-YBGY-432), Natural Science Basic Research Plan in Shaanxi Province of China (2021JC-13) and Research Fund of the State Key Laboratory of Solidification Processing (NPU), China (2021-QZ-03).

References

[1] Z. Shang, X. P. Wei, D. Z. Song, et. al. Microstructure and mechanical properties of a new nickel-based single crystal superalloy. Journal of Materials Research and Technology, 2020, 9(5): 11641-11649.

Nucleation and Transition Sequences of TCP Phases During Heat-Exposure in a Re-Containing Ni-Based Single Crystal Superalloy

Chen Liu, Wenchao Yang*, Jun Zhang, Lin Liu
State Key Laboratory of Solidification Processing, Northwestern Polytechnical University, Xi'an 710072, China
*Corresponding address: e-mail: wenchaoyang@nwpu.edu.cn

Abstract: The nucleation and transition sequences of topologically close-packed (TCP) phases in a Re-containing Ni-based single crystal superalloy were systematically investigated using in-situ transmission electron microscopy (TEM) and three-dimensional atom probe technology (3D-APT). During the initial stage of heat-exposure at 1100 °C, the TCP phase forming elements (Re, Co, Cr, etc.) segregated at the γ/γ' interface near the γ matrix side to provide the concentration undulations for the nucleation sites of TCP phases, following which the σ and P phase coherently nucleated along the $(1\overline{1}\overline{1})\gamma$ and $(022)\gamma$ planes from the γ/γ' interface near the γ matrix side, respectively.With prolonged heat-exposure time, transitions from σ phase to P phase, σ phase to μ phase, and P phase to μ phase occurred. Besides, the orientation relationshipsof TCP phase intergrowth structures indicated that the P phase grew along the $(\overline{1}01)\sigma$plane of the σ phase by co-lattice precipitation, meanwhile the μ phase grew with smaller lattice misfits along the $(0\overline{4}0)\sigma$ plane of the σ phase and the (400)P plane of the P phase. Additionally, the result by first-principles calculation evidenced that the μ phase had the lowest system energy to make the transition of σ phase and P phase to μ phases inevitable, therefore, the TCP phase ultimately existed as the most stable μ phase. Finally, the transitionsequences of TCP phase during heat-exposure could be summarized into three types: γ matrix$\rightarrow\sigma\rightarrow\mu$,$\gamma$ matrix$\rightarrow$P$\rightarrow\mu$, and γ matrix$\rightarrow\sigma\rightarrow$P$\rightarrow\mu$.

Keywords: Ni-based single crystal superalloys; TCP phase; Nucleation; Intergrowth

1 Introduction

The topologically close-packed (TCP) phase precipitated in Ni-based single crystal superalloys during high-temperature service, which was not only likely to be the crack source of alloy fracture, but also greatly weaken the solid solution strengthening effect of the γ matrix [1-4]. Therefore, it was necessary to study the nucleation and transition mechanism of TCP phases to provide theoretical support for the composition design and microstructure regulation of advance Ni-based single crystal superalloys.

In this paper, the nucleation and intergrowth behavior of TCP phases in a Re-containing Ni-based single crystal superalloy was systematically investigated by in-situ TEM and 3D-APT to clarify the nucleation site, the orientation relationship between the TCP phase and γ matrix, and the transition sequences of TCP phases during heat-exposure at 1100 °C. Furthermore, the first-principles calculation based on the density functional theory (DFT) about the system free energies of different types TCP phases were also used to further elucidate the transition mechanisms. The results provided insight to the nucleation and growth of TCP phases, and would also help to offer the theoretical support for the composition design and microstructure regulation of advanced Ni-based single crystal superalloys.

2 Experimental procedure

In this work, the nominal composition of the experimental superalloy was as follows (wt.%): 14.0 Co, 6.0 Al, 19.5 (Cr+Mo+W+Ta), 0.1 Hf, 5.4 Re, and the balance Ni. The single crystal bars with orientation [001] were prepared by liquid metal cooling directionally solidification furnace, as shown in Fig. 1a.The heat treatment process curve of the experimental alloy was shown in Fig. 1b.

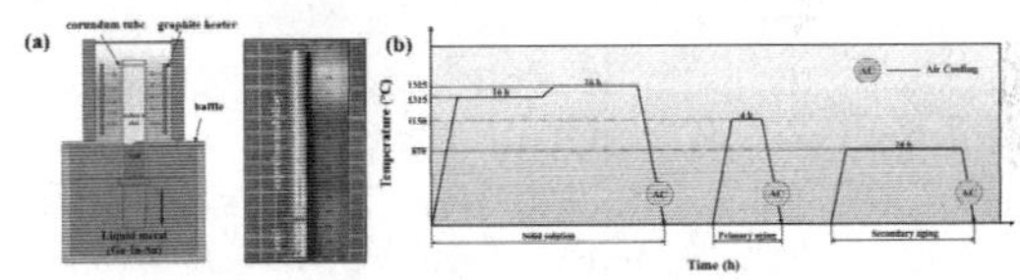

Fig. 1. The schematic of the liquid metal cooling (LMC) furnace and single crystal bar diagram (a). The heat treatment process curve of the experimental alloy (b).

3 Result and discussion

Nucleation of TCP phases

The microstructureand component distributionof the experimental alloys after heat-exposure at 1100°C was analyzed by TEM and 3D-APT, and the results were shown in Fig. 2.After a short heat-exposure, it could be found that the TCP phase preferentially nucleated from the side near the γ/γ' interface near the γmatrix (Fig. 2a). By characterizing the TCP phases, it was demonstrated that the preferentially precipitated TCP phases were the $\sigma$and Pphases, which exhibitedthe specific co-lattice relationships with the γmatrix.In addition, the results of 3D-APT (Fig. 2b) indicated that the TCP phase-forming elements, such as Re, Co, and Cr, were segregated at the γ/γ'interface near the γmatrix side during heat-exposure. Therefore, it could be concluded that the segregation of the $\sigma$and Pphases forming-elements at the γ/γ'interface near the side of the γmatrixprovided concentration undulation for the nucleation of the σand Pphases, and the γ matrix could provide nucleation sites for the σ and P phases to stabilize the nucleus and reduce the critical nucleation size, thus promoting the nucleation of the σ and P phases on the γ matrix near the γ/γ′interface.

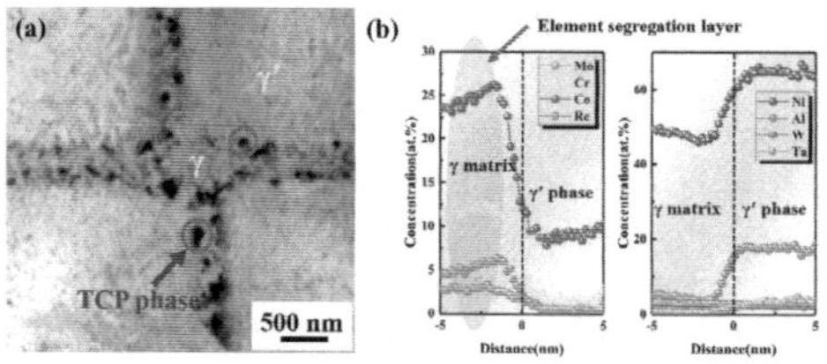

Fig. 2 The bright field image of the alloy (a) after heat-exposure at 1100 °C for 2 h was investigated by TEM. The plots of the variation of the concentrations (b) as a function of distance away from the γ/γ' interface in the experimental alloys during heat-exposure at 1100 °C for 100 h.

Transition behavior of TCP phases

It was known that different types of TCP phases could transform during the growth during heat-exposure at 1100 °C, thus further analysis was done by TEM to target the transition sequences of the TCP phase, and the result was displayed in Fig. 3. As shown in Figs. $3a_1$-c_1, σ/P, σ/μ, and P/μ intergrowth structures were present during heat-exposure for different times, and computational comparison and calibration of their SAED patterns (shown in Figs. $3a_2$-c_2) showed that the intergrowth structures of TCP phases had a certain orientation relationship.After calculating the misfit latticesof the specific interface, the σ/P, σ/μ, and P/μ interfaces might be a coherent lattice matching with the misfit δ of 1.70%, 7.41%, and 4.23%.Besides, the lowest system energy of the μ phase calculated by DFT (as shown in Fig. 4) caused both σ phase and P phase to eventually transform to the μ phase. With the further extension of the heat-exposure time, the TCP phase ended up in the most stable μ phase.

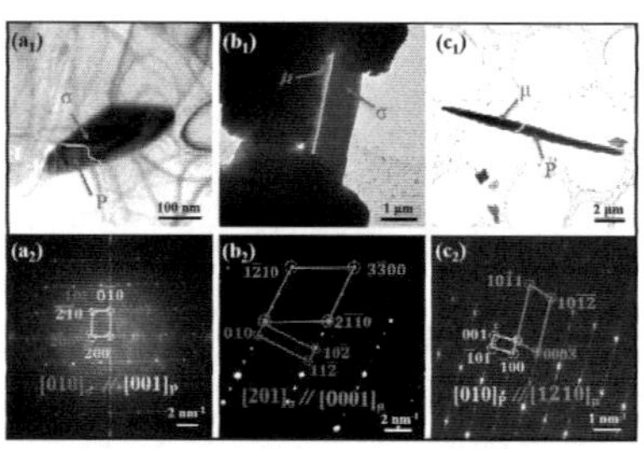

Fig. 3 The Bright field TEM micrograph of σ /P (a1), μ/σ (b1), and μ/P (c1) intergrowth structures. (a2-c2): the SAED patterns of σ /P, μ/σ, and μ/P intergrowth structures.

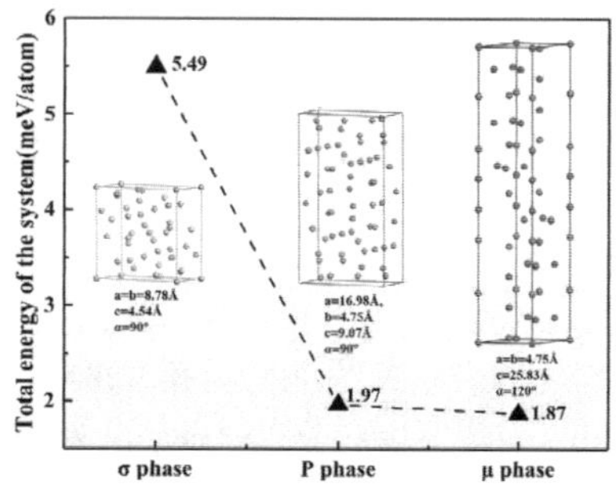

Fig. 4 The system energy of σ phase, P phase, and μ phase calculated based on the first principles calculations.

Conclusion

1.During heat-exposure at 1100 °C, TCP phases preferentially nucleated from the γ/γ′ interface and close to the γ matrix side due to the concentration undulations provided by the segregation behavior of TCP phase forming elements (Re, Co, Cr, etc.) at the γ/γ′ interface close to the γ matrix.

2. The transition sequences of the TCP phase during heat-exposure at 1100 °C were classified into three types: γ matrix→σ→μ, γ matrix→P→μ, and γ matrix→σ→P→μ.

Acknowledgments

This work was funded by National Natural Science Foundation of China (51771148, 52071263, 52031012, 52322410), Science Center for Gas Turbine Project (P2021-A-IV-001-001), The Key Research and Development Program of Shaanxi Province (2023-YBGY-432), Natural Science Basic Research Plan in Shaanxi Province of China (2021JC-13) and Research Fund of the State Key Laboratory of Solidification Processing (NPU), China (2021-QZ-03).

References

[1] M. Simonetti, P. Caron. Role and behaviour of μ phase during deformation of a nickel-based single crystal superalloy, Materials Science and Engineering: A, 254 (1998): 1-12.
[2] C.M. Rae, R.C. Reed. The precipitation of topologically close-packed phases in rhenium-containing superalloys, Acta Materialia, 49 (2001): 4113-4125.
[3] J.X. Yang, Q. Zheng, X.F. Sun, et al. Formation of μ phase during thermal exposure and its effect on the properties of K465 superalloy, Scripta Materialia, 55 (2006): 331-334.
[4]K. Matuszewski, R. Rettig, H. Matysiak, et al. Effect of ruthenium on the precipitation of topologically close packed phases in Ni-based superalloys of 3rd and 4th generation, Acta Materialia, 95 (2015): 274-283.

Effect of Low angle Grain Boundary on the Microstructure and Mechanical Properties in a Fourth Generation Nickel Based Single Crystal Superalloy

Qiang Wang, Wenchao Yang*, Jun Zhang, Lin Liu
State Key Laboratory of Solidification Processing, Northwestern Polytechnical University, Xi'an 710072, China
*Corresponding address: e-mail: wenchaoyang@nwpu.edu.cn; Tel/Fax: +86-29-88492228

Abstract: The effect of different low angle grain boundaries (LAGBs) on the microstructure and mechanical properties in a fourth generation Nickel-based single crystal superalloy was studied. Double seed crystals techniques were used to obtain the specimens with LAGBs. After thermal exposure at 1100 °C for 100 h, the microstructure stability deteriorated when the angle exceeded 12.6°, where the discontinuous precipitation (DP) colonies region occurred along the grain boundary because of high interfacial energy and mobility at LAGBs. With the grain boundary angle increasing, the ultimate tensile strength of the sample rose slightly and then fell, while the elongation and creep rupture life decreased significantly exceeded 4.9°. Under 1100 °C service condition, the carbides and topological close-packed (TCP) phases were precipitated at LAGBs, which seriously impaired the grain boundary bonding strength. On the other hand, the misorientation at LAGBs led to the stress concentration, which increased the difficulty of grain boundary coordinated deformation. Finally, the damage tolerance of the LAGBs in this fourth generation Nickel-based single crystal superalloy was approximately estimated as about 7° at 1100 °C/150 MPa creep rupture condition based on a 70% creep rupture life standard.

Keywords: Nickle-based single crystal superalloy; Low angle grain boundaries; Microstructure; Mechanical properties; Damage tolerance

1 Introduction

In order to meet the need of the increasing thrust-to-weight ratio of aero-engines, the content of refractory elements in single crystal superalloys increased, the turbine blade shape and the internal air-cooled structure became increasingly complex, and the instability factors in the directional solidification process of single crystal turbine blades increased, inevitably resulting in the formation of LAGBs defects [1,2]. LAGBs could provide some preferential nucleation sites for precipitated phases as a high diffusion channel, which destroyed the γ/γ' microstructure and deteriorated the properties of the alloy [3,4]. Therefore, it was significant to reveal the effects of LAGBs on the mechanical properties of nickel-based single crystal superalloys. The aim of the present work was to investigate the effects of LAGBs on the microstructure stability, tensile properties and creep rupture properties of a fourth generation Nickel-based single crystal superalloy. This research could also accelerate the application process of the fourth generation Nickel-based single crystal superalloy.

2 Experimental procedure

The nominal composition of the experimental Ni-based single crystal superalloy is as follows (wt.%): 21.5Co+Mo+W, 14.5Al+Ta, 3.5Cr, 0.1Hf, 5.4Re, 3.0Ru, 0.1C, and the balance Ni. The samples with different LAGBs (4°, 8° and 12°) could be prepared by a bi-crystals method, where the grain boundary angle was controlled by two seeds as shown in Fig. 1(a). The HCl + H_2O_2 etchant was used to slightly corrode the samples for the observation of the grain boundary as shown in Fig. 1(b)

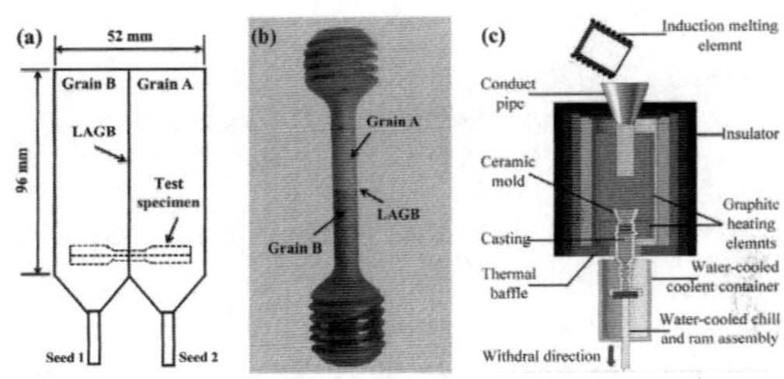

Fig. 1. The schematic diagram of the bi-crystal slab casting (a), the transverse test specimen after macro etching (b) and schematic illustration of the Bridgeman high rate solidification (HRS) furnace (c).

The bi-crystals samples were directionally solidified by high-rate solidification (HRS) process as shown in Fig. 1(c). The standard heat treatment process was adopted as follows: 1315 °C/16 h+1325 °C/16 h/air cooling (AC); 1150 °C /4 h/AC; 870 °C/24 h/AC.

3 Result and discussion

Effect of LAGBs on the microstructure stability

After the thermal exposure for different times at 1100 °C, the γ' phase in matrix gradually coarsened and connected to form a labyrinthine microstructure, and the microstructure at LAGBs evolved differently ascribed to differences in grain boundary angles, as revealed in Fig. 2. A γ-phase film was formed along LAGBs when the grain boundary angle was small as shown in Fig. 2(a) and (b). With the extension of thermal exposure time, the thickness of the γ-phase film increased and was equivalent to the size of the channel in matrix, controlled by the element diffusion behavior. In Fig. 2(c), a DP colonies region occurred at LAGBs when the grain boundary angle was large, such as 12.6°. The regular and ordered γ/γ′ phase microstructure was transformed into a coarse (γ ′ +γ+TCP) microstructure. The DP colonies region altered the orientation of the single crystals within the zone to be consistent with the orientation of the other crystal and precipitated the TCP phases. Simultaneously, the boundary advance into the other crystal to become a new grain boundary. The width of the DP colonies region increased with the thermal exposure time extension, forming a diffusion-controlled solid phase transition. Moreover, the higher mobility at the grain boundary provided the potential for a DP colonies region.

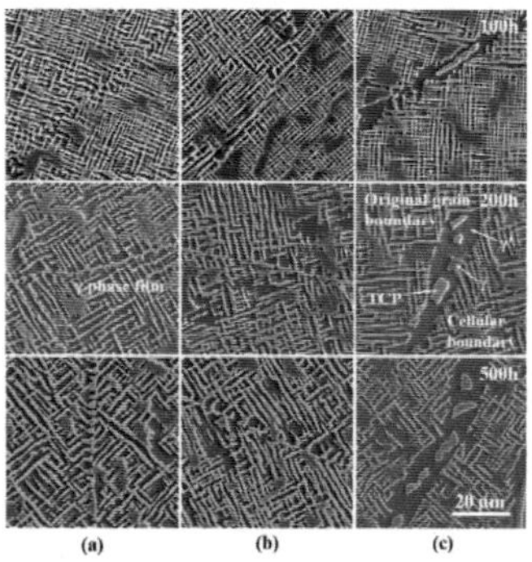

Fig. 2. SEM images of LAGBs microstructure of the alloy with (a) 4.9°; (b) 7.5° and (c) 12.6° after thermal exposure at 1100 °C for 100 h, 200 h and 500 h.

4 Effect of LAGBs on the mechanical properties

The variation of the tensile strength and elongation of the specimen with the grain angle was presented in Fig. 3(a). As the grain boundary angle increased, the strength of the alloy increased slightly at first and then decreased gradually, while the plasticity rapidly decreased. At 1100 °C/150 MPa creep rupture condition, the creep rupture life of specimens containing LAGBs significantly decreased as the grain boundary angle increased, as shown in Fig. 3(b). The effect of LAGBs on mechanical properties was primarily the deterioration of grain boundary microstructure and stress concentration at the grain boundary. The elastic modulus of crystals on both sides was inconsistent along load direction. Then, stress concentration occurred at the grain boundary, and the degree varied with the grain boundary angle, so as to assure the deformation coordination at the grain boundary. From another perspective, LAGBs, with a high energy and element diffusion rate, were easy to form γ phase file, DP colonies region and carbides, which deteriorated the microstructure and lowered the bonding strength of the grain boundary.

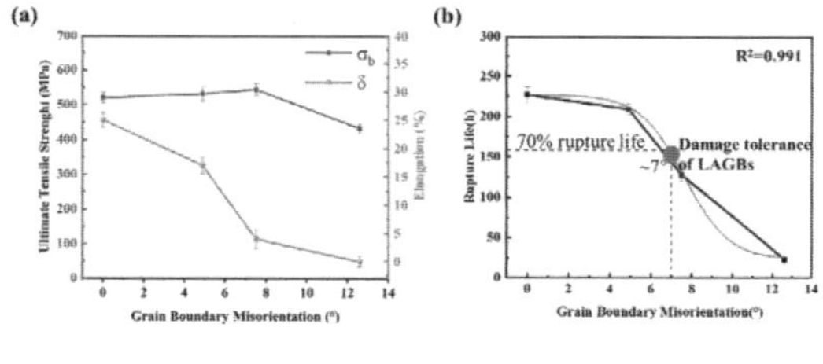

Fig. 3. Transverse tensile properties of the alloy with LAGBs specimens at 1100 °C (a) and transverse rupture properties at 1100 °C/150 MPa (b).

5 Conclusion

1. After thermal exposed at 1100 °C for different times, a γ phase film formed along LAGBs or a DP colonies region occurred, the width of which increased with the extension of the thermal exposure time.

2. Under 1100 °C/150 MPa creep rupture conditions, the alloy has a LAGBs damage tolerance of approximately 7° based on a 70% creep rupture life standard.

6 Acknowledgments

This work was funded by National Natural Science Foundation of China (51771148, 52071263, 52031012, 52322410), Science Center for Gas Turbine Project (P2021-A-IV-001-001), The Key Research and Development Program of Shaanxi Province (2023-YBGY-432), Natural Science Basic Research Plan in Shaanxi Province of China (2021JC-13) and Research Fund of the State Key Laboratory of Solidification Processing (NPU), China (2021-QZ-03).

References

[1] M. Newell, K. Devendra, P.A. Jennings, et al. Role of dendrite branching and growth kinetics in the formation of low angle boundaries in Ni–base superalloys. Materials Science and Engineering: A, 412(1) (2005): 307-315.
[2] G. Liu, L. Liu, X. Zhao, et al. Effects of Re and Ru on the solidification characteristics of nickel-base single-crystal superalloys. Metallurgical and Materials Transactions A, 42(2011): 2733-2741.
[3] Effect of low angle boundaries on the mechanical properties of single crystal superalloy DD6. Superalloy, 2008(2008):443-451.
[4] T.M. Pollock. The growth and elevated temperature stability of high refractory nickel-base single crystals. Materials Science and Engineering: B, 32(3) (1995):255-266.

Investigation on Orientation Control of Single Crystal Blades Through a Using of Heterogeneous Seeds

Jiarun Qin, Wenchao Yang*, Jun Zhang, Lin Liu
State Key Laboratory of Solidification Processing, Northwestern Polytechnical University, Xi'an 710072, China
*Corresponding address: e-mail: wenchaoyang@nwpu.edu.cn

Abstract: As the increasing demand for high temperature performance of the Ni-based single crystal blades, the orientation consistency of the blades has become more and more important. In this work, Ni-W heterogeneous seed was used to prepare the Ni-based single crystal superalloys and the microstructure at remelting interface under different conditions was analyzed. The results showed that lots of fine cellular grains with the basically consistent orientations and compositions were existed in front of the remelting interface under the high-rate solidification (HRS) method, while was not happened under the liquid metal cooling (LMC) condition. Based on the calculation of the constitutional undercooling, it was proved that the HRS method with a small temperature gradient could produce an enough constitutional undercooling at the front of remelting interface to form fine cellular grains. At last, using the heterogeneous seeds, four single crystal blades with the same three-dimensional orientation were successfully produced, which fully improved the preparation efficiencies and orientation quality of the blades, proving the possibility of industrial application for heterogeneous seeds.

Keywords: Directional solidification; Single crystal; Orientation control; Seeds method

1 Introduction

Ni-based single crystal superalloys have been widely used in aero engine and turbine blades due to their excellent mechanical properties, creep and fatigue resistance [1]. However, with the increase of the aeroengine inlet temperature, Ni-based single crystal superalloys with random secondary orientation are no longer sufficient to meet the needs in some circumstances where the accurate orientation is strictly required for properties assurance [2].

However, traditional seeds have usually a complex alloying composition and a large crystallization temperature interval, therefore, it is difficult to avoid the heterogeneous nucleation of stray grains because of dendrite fragmentations in pasty zone. For this reason, Toloraiya and Azhazha et al. [3] found that Ni-W heterogeneous seed could obtain a completely cellular structure at a large withdraw rate and significantly reduced the formation of stray grains to improve the success rate of preparing single crystals. In addition, Zhao et al. [4] studied the planar-cell-dendritic instability transition at the remelting interface of Ni-W heterogeneous seeds and observed the existence of fusion zone. And then, Guo et al. [5] studied the formation mechanism of the fusion zone for Ni-W heterogeneous seed under the high-rate solidification (HRS) condition and found that Ni-W heterogeneous seed effectively avoided the formation of stray grains at remelting interface compared to traditional homogeneous seeds and lots of fine cellular grains were existed in front of the fusion zone.

Based on the investigations above, it can be seen that some important results have been achieved in the preparation of the single crystal blades by heterogeneous seeds. However, the researches above are mainly focused on the primary orientation control and the composition analysis on the remelting interface. The investigations on the three-dimensional orientation control, the microstructural evolution at the remelting interface, the preparation of the single crystal blades, are still severally lacking.

2 Experimental procedure

Material used in this work were first-generation single crystal superalloy DD3 and Ni-20 wt.% W alloy. All seeds were prepared by liquid metal cooling (LMC) directional solidification at 1550 °C and 10 $\mu m \cdot s^{-1}$ and were polished into a cylinder with a diameter of 7 mm and a height of 30 mm. All samples were cut parallelly and vertically from the seed axis, and were polished, etched with the solution of 50% H_2O, 25% HNO_3,25% HF. JXA-8100 and ZEIS SUPRA 55SEM equipped with Electron Probe Micro-analyzer (EPMA) and EBSD were used to observe the microstructures of the samples.

3 Result and discussion

When under the HRS and LMC condition, it was seen that lots of fine cellular grains competed with each other were present above the fusion zone under HRS condition (**Fig. 1a**), while these fine cellular grains were not observed under the LMC condition (**Fig. 1b**).

Aimed to understand the formation mechanism of these fine cellular grains, the constitutional undercooling was calculated as shown in **Table 1**. It was clear that when under the HRS condition, G/R was less than $\Delta T/D_{eff}$, showing a great constitutional undercooling existed in front of the remelting interface, while under the LMC condition, it was in contrast.

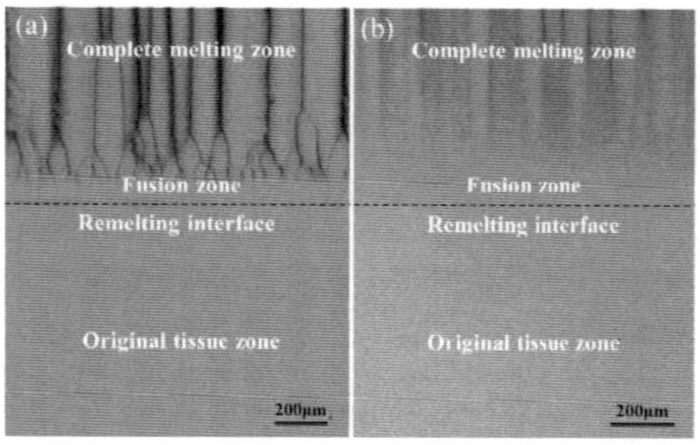

Fig. 1. Vertical microstructures of remelting interface prepared by HRS (a) and LMC (b) conditions.

Table 1. Constitutional undercooling under HRS and LMC methods

	G/R (k·s·m^{-2})	$\Delta T/D_{eff}$ (k·s·m^{-2})
HRS	2-6×10^7	174×10^7
LMC	>200×10^7	174×10^7

At last, a novel high-throughput mold capable of producing four single crystal blades via seeding-grain selection technique was designed. The mold structure and the casting process was simulated by the finite element analysis in the ProCAST software as shown in **Fig. 2**. It was clear that the 3D-orientation of the four blades were all controlled by one seed.

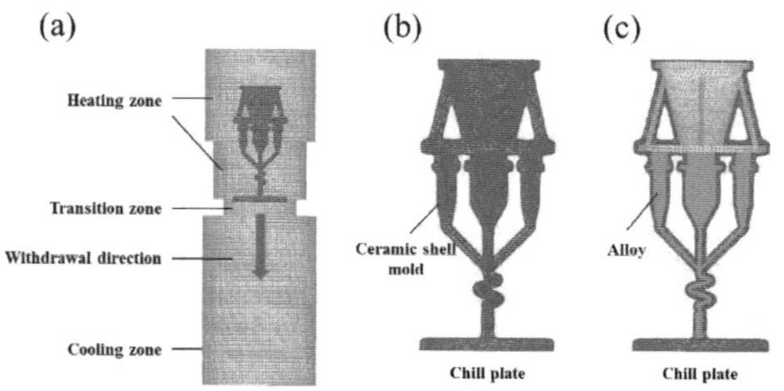

Fig. 2. Casting system (a), ceramic shell (b), and alloy (c) in the grid model.

Using this high-throughput mold, four single crystal blades with the same 3D orientation were successfully produced using this high-throughput mold via heterogeneous seed. There were no stray grains formed during the crystal growth, which fully proved the feasibility of the industrial application for heterogeneous seeds.

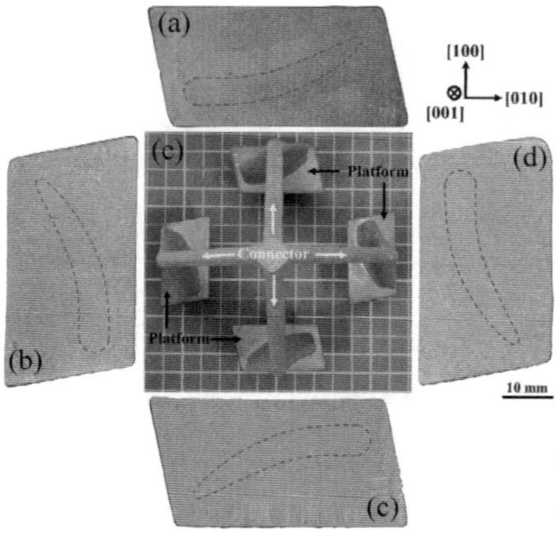

Fig. 3. The main structure of the casting (e) and the platform morphology (a)-(d).

4 Conclusion

1. The cause of the fine cellular grains which existed in front of the growth interface is verified as constitutional undercooling.

2. Four single crystal blades with the same 3D-orientation were successfully produced by one heterogeneous seed.

5 Acknowledgments

This work was funded by National Natural Science Foundation of China (52071263, 52031012, 52322410), Science Center for Gas Turbine Project (P2021-A-IV-001-001), Self-innovation Special Fund Project of Aero Engine Corporation of China (ZZCX-2022-040), The Key Research and Development Program of Shaanxi Province (2023-YBGY-432).

References

[1] GL Erickson, K Harris, RE Schwer. Directionally Solidified and Single-Crystal Superalloys, Metals Han, ASM International, Ohio, 1990.
[2] N Matan, DC Cox, MA Rist, et al. Creep of CMSX-4 superalloy single crystals: effects of misorientation and temperature, Acta Mater 47 (1999) 1549-1563.
[3] VM Azhazha, VY Sverdlov, AN Ladygin, et al. Self-Organization Processes in the Cellular Structure Formation of Ni-W Alloys, J Surf Invest-X-Ray+ 2 (2008) 156-165.
[4] NR Zhao, JG Li, JL Liu et al. Solidification interface in growth of single crystal superalloys by seeding technique, J Mater Eng Perform 11 (2007) 24-27.
[5] J Guo, JG Li, JD Liu, et al. Formation Mechanism of Fusion Zone in Growth of Single Crystal Superalloys with Low-Segregated Heterogeneous Seed, Acta Metall Sin 54 (2018) 419-427.

The Effect of B_4C on Microstructure and Mechanical Properties of Nb-Si Based Superalloy

Xiaowei Wang[1], Qi Wang[1,*], Ruirun Chen[1,*]
1.National Key Laboratory for Precision Hot Processing of Metals, School of Materials Science and Engineering, Harbin Institute of Technology, Harbin 150001, P. R. China
*Corresponding address: Qi Wang, wangqi_hit@hit.edu.cn; Ruirun Chen, ruirunchen@hit.edu.cn

Abstract: The details of microstructure and mechanical properties of Nb-Si based alloys have been shown with the addition of B4C. Results show that the addition of B4C promotes the formation of silicide phases, Moreover, the fracture toughness has been improved with doping of trace B4C, and the maximum value of the investigated alloys is 14.26 MPa·m1/2, which improves by 41.05%.

Keywords: Nb-Si based alloy, B_4C, Fracture toughness,

1 Introduction

The Nb–Si based alloys are one of the most potential candidate materials for the next generation aerospace turbine blades, because of its higher melting point (>1700 ◦C), lower density (6.6–7.2 g/cm^3) and excellent high-temperature strength [1-3]. However, the poor fracture toughness at room temperature has limited its practical application of Nb-Si alloys[4-6].

2 Experimental procedure

All the ingots have been prepared by the non-consumable electric arc melting, and the nominal chemical compositions are Nb-16Si-20Ti-20Zr, Nb-16Si-20Ti-20Zr-1B_4C, Nb-16Si-20Ti-20Zr-2B_4C, and Nb-16Si-20Ti-20Zr-4B_4C. The-X-ray diffraction (XRD) is used to examine the phase structure and crystalline structures. The room-temperature fracture toughness is measured by conducting three-point bending specimens with the size of 20*4*2 mm. The room-temperature compressive test is conducted using the Instron 5569 electronic testing machine with the size of φ4*6.

3 Result and discussion

Mcrostructure

The XRD results can be seen in Fig. 1. It can be seen that the constituent phases no change with the addition of trace B_4C.All the Nbss and γ(Nb, X)$_5$Si$_3$ phases exist in the 0-2at.% B_4C alloys. As the content of B_4C exceed 4 at.%, the new silicide phases ((Nb, X)$_3$Si) appear in the matrix.

Fig. 2 shows the backscattered electron (BSE) images of investigated alloys. The Nbss/γ(Nb, X)$_5$Si$_3$ eutectic and small proportion of γ(Nb, X)$_5$Si$_3$ phases exist in 0B_4C alloys. And It is obvious that the proportion of γ(Nb, X)$_5$Si$_3$ increases as doping of B_4C, suggesting that B_4C can change the solidification of Nb-Si based alloys. In 4B_4C alloys, the gray (Nb, X)$_3$Si phases appear in the matrix.

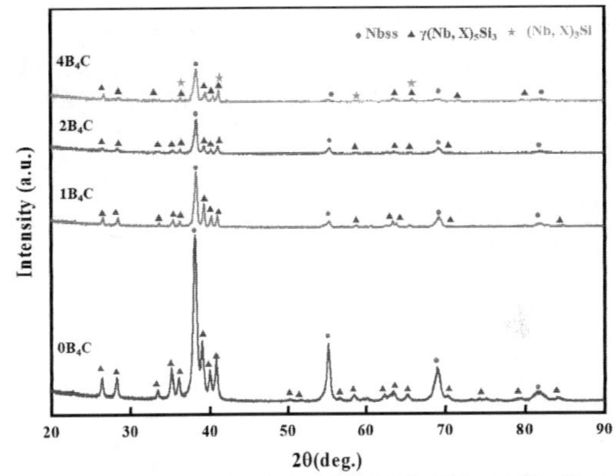

Fig. 1.XRD images of Nb-16Si-20Ti-20Zr-xB4C alloys

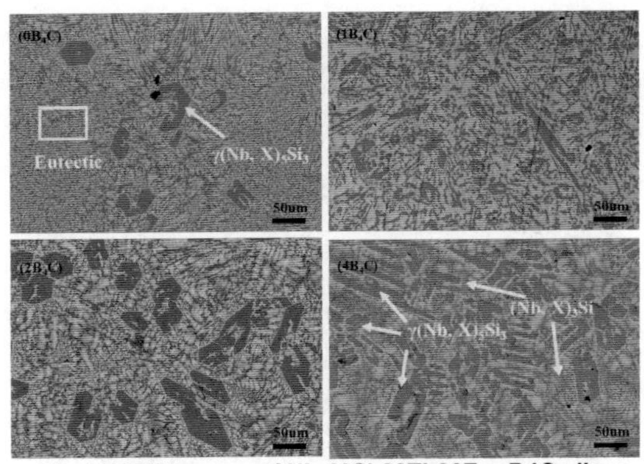

Fig. 2. BSE images of Nb-16Si-20Ti-20Zr-xB4C alloys

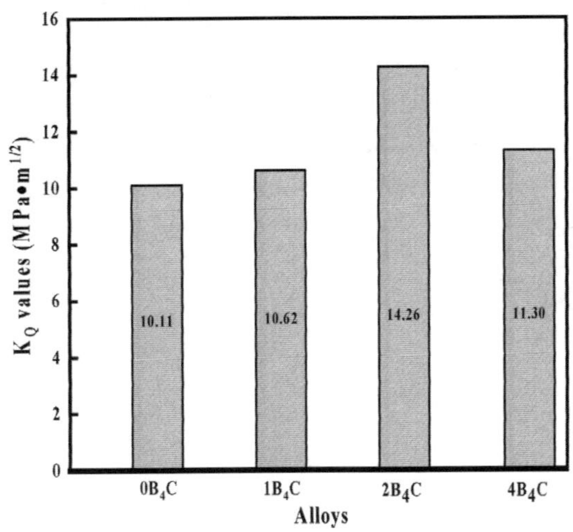

Fig. 3. The K_Q values of Nb-16Si-20Ti-20Zr-xB4C alloys

Room temperature fracture toughness

Fig. 3 shows the room-temperature fracture toughness of investigated alloys. The K_Q valus of based alloys is 10.11 MPa·m$^{1/2}$. The fracture toughness increase with doping of B_4C, and the fracture toughness of $1B_4C$ and $2B_4C$ are 10.62 and 14.26 MPa · m$^{1/2}$,which increase 5.04% and 41.05%. However, the fracture toughness decreases as the content of B_4C exceed 4at.%.

4 Conclusion

(1)The addition of B_4C changes the microstructure of Nb-Si based alloys. The proportion of primary γ(Nb, X)$_5$Si$_3$ phases increase with doping of B4C. In $4B_4C$ alloys, (Nb, X)$_3$Si phases appear in the matrix

(2)The room-temperature fracture toughness increases signally with doping of B4C. The KQ value of $2B_4C$ alloys is 14.26 MPa·m$^{1/2}$, which improves by 41.05%.

5 Acknowledgments

The authors gratefully acknowledge support from the National Natural Science Foundation of China (51825401)

References

[1] K.S. Chan, 13 - Intermetallic composites toughened with ductile reinforcements, in: R. Mitra (Ed.), Intermetallic Matrix Composites, Woodhead Publishing2018, pp. 359-407.

[2] Y. Qiao, X. Guo, Y. Zeng, Study of the effects of Zr addition on the microstructure and properties of Nb-Ti-Si based ultrahigh temperature alloys, Intermetallics 88 (2017) 19-27.

[3] Z. Li, L. Luo, B. Wang, B. Su, L. Luo, L. Wang, Y. Su, J. Guo, H. Fu, Cooperative effects of Mo and B additions on the microstructure and mechanical properties of multi-elemental Nb-Si-Ti based alloys, Mater. Charact. 206 (2023) 113421.

[4] S. Zhu, G. Zhang, Y. Bao, D. Sun, Q. Zhang, X. Meng, Y. Hu, L. Yan, Progress in preparation and ablation resistance of ultra-high-temperature ceramics modified C/C composites for extreme environment, Reviews on Advanced Materials Science 62(1) (2023) 20220276.

[5] A. Annerino, M. Lawson, P.I. Gouma, Future insights on high temperature ceramics and composites for extreme environments, International Journal of Ceramic Engineering & Science 4(5) (2022) 296-301.

[6] X. Yin, J. Liang, X. Zhang, C. Wang, S. Chen, S. Shang, C. Liu, Improving the mechanical properties and oxidation resistance of Nb-16Si-20Ti alloys with B4C addition, Int. J. Refract. Met. Hard Mater. 119 (2024) 106543.

Microstructure and Room-Temperature Fracture Toughness of Nb-Si Alloy with Addition of Mn

Tianyu Zhao[1], Qi Wang[1], Ruirun Chen[1]
1.National Key Laboratory for Precision Hot Processing of Metals, School of Materials Science and Engineering, Harbin Institute of Technology, Harbin 150001, P. R. China
*Corresponding address: e-mail: Ruirun Chen, Tel./Fax: +86-0451-86412394, E-mail: ruirunchen@hit.edu.cn

Abstract: The effect of Mn element on the microstructure and mechanical properties of Nb-Si based alloys is studied, and the investigated alloys compositions are Nb-16Si-22Ti-xMn (x = 0, 2, 4, 8 at.%). Solubility of Mn in Nbss phase is higher than that in silicide. This dissolution of Mn in Nbss phase and silicide causes solid solution strengthening. Tiss phase and Mn2Ti phase eutectic structure precipitates between dendrites of 8Mn alloy, which affect room-temperature fracture toughness adversely.

Keywords: Nb-Si alloy; Mn; Fracture toughness; Microstructure

1 Introduction

Nb-Si alloys have attracted many attentions due to their high melting temperature (~1800 ℃), relatively small density (6.6~7.2 g/cm^3) and excellent mechanical performance in high-temperature environments [1-4]. However, the room-temperature fracture toughness and oxidation resistance of Nb-Si alloys at high temperature is poor, these shortcomings limit its application greatly.

2 Experimental procedure

All the alloys mentioned above were prepared by a vacuum non-comsumable electric arc furnace under an argon atmosphere of 0.05 MPa below standard atmospheric pressure, equipped with a water-cooled copper crucible. Each ingot was melted 5 times repeatedly to ensure that the composition of alloys was uniform. Phase composition was analyzed by XRD. Observation of microstructure morphology adopt Quanta 200FEG scanning electron microscope with backscatter mode. Chemical composition of constituent phases was test by the energy disperse spectroscopy which based on the SEM. Electron diffraction image and the analysis of distribution of chemical elements used transmission electron microscope (Talos F200X). As shown in the Fig. 1. The room-temperature fracture toughness test adopted the electronic universal mechanical testing machine (Instron5569). The room-temperature fracture toughness was characterized by K_Q, which can be obtained from the formula shown below:

$$K_Q=(P\cdot S/B\cdot W^{3/2})\cdot F(A/W) \quad (1)$$

$$F(A/W)=3(\alpha/W)^{1/2}\cdot\frac{1.99-(\alpha/W)(1-\alpha/W)[2.15-3.93(\alpha/W)+2.7(\alpha/W)^2]}{2(1+/2\alpha W)(1-\alpha/W)^{3/2}} \quad (2)$$

where P is the corresponding load at the 95% slope of the linear elastic part of the load-displacement curve, S stands for the load span (S = 4W), W stands for the width of sample, B is the thickness and a is the length including the slit notch of the pre-crack.

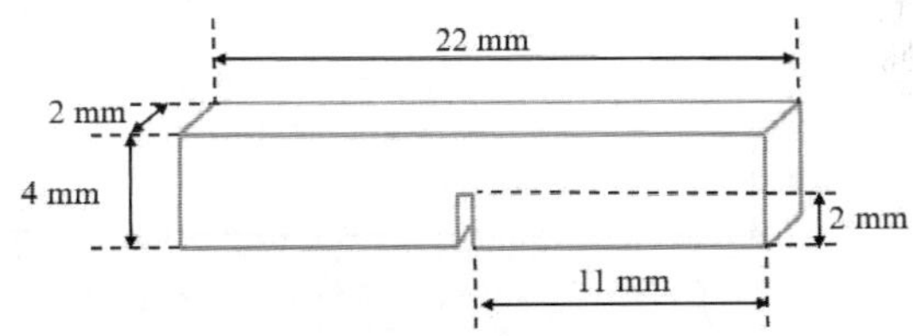

Fig. 1. The size of sample for fracture toughness testing

3 Result and discussion

Phase constitutions and microstructure of Nb-Si based alloys

Fig. 2 shows the microstructure of as-cast Nb-16Si-22Ti-xMn (x = 0, 2, 4, 8 at.%), and the microstructure of the alloys is identified. All the alloys maintain microstructure characteristics of hypoeutectic because of the primary Nbss phase distributed in (Nb, Ti)$_3$Si matrix. Primary Nbss phase exhibits dendritic feature and Nbss phase in eutectic structure remains the morphology of fine block, as shown in Fig. 3(a)~(d). Previous studies have pointed out that the intensity (I_j) of the diffraction peak is proportional to the weight percentage of phase content (ω_j, wt.%).

When atomic ratio of Mn reaches 8 at.%, a fine eutectic structure appears between dendrites as shown in Fig. 2(d). XRD analysis of phases constituent and previous research, the darker gray phase has a lot of Ti, it can be certain as Tiss phase. The black phase is identified by Mn_2Ti phase.

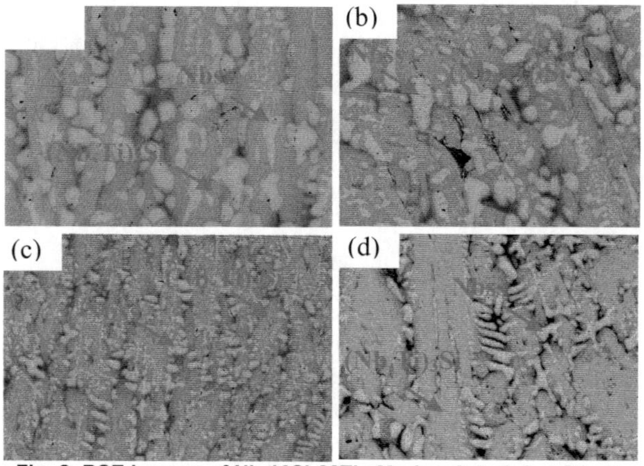

Fig. 2. BSE images of Nb-16Si-22Ti-xMn (x = 0, 2, 4, 8 at.%) alloys: (a) Nb-16Si-22Ti, (b) Nb-16Si-22Ti-2Mn, (c) Nb-16Si-22Ti-4Mn, (d) Nb-16Si-22Ti-8Mn

4 Room temperature mechanical properties

Fig. 6 shows the K_Q value which represents room-temperature fracture toughness of Nb-Si based alloys. As the content of Mn in alloys increases (atomic ratio from 2 to 8), the room-temperature fracture toughness shows a trend of increasing at first and then decreasing. When the atomic ratio of Mn reaches 4 at.%, the fracture toughness of the as-cast alloys get a peak (8.07 MPa • $m^{1/2}$). When the atomic ratio exceeds 4 at.%, the fracture toughness is severely degraded. Compared with Nb-16Si-22Ti alloy, the fracture toughness of the 4 at.% Mn alloy is increased by 17%.

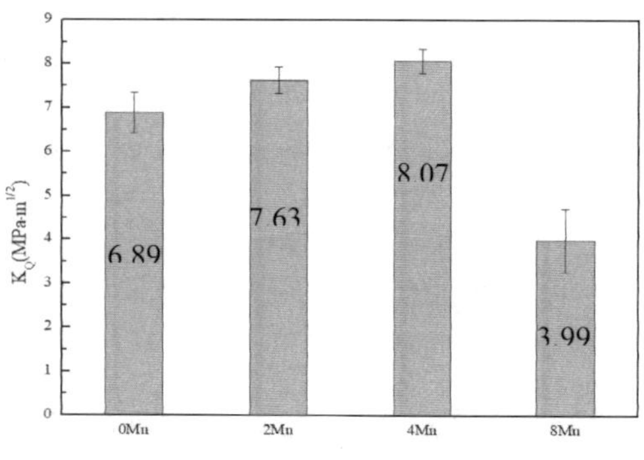

Fig. 6. Average room-temperature fracture toughness K_Q value of Nb-16Si-22Ti-xMn (x = 0, 2, 4, 8) alloys

5 Conclusion

(1)With the increase of Mn, the volume fraction of Nbss decreases. Solubility of Mn in Nbss phase is higher than that in silicide. When the content of Mn reaches 8 at%, Tiss and Mn_2Ti eutectic structure precipitates at the grain boundary.

(2)Solid solution strengthening effect of Mn in Nbss contributes to the improvement of room-temperature fracture toughness greatly. However, the Mn_2Ti at the interface in the 8Mn alloy is harmful to fracture toughness. Due to the increase of Mn content in Nbss, K_Q value of Nb-16Si-22Ti-8Mn alloy has increased by 17% compared to Nb-16Si-22Ti alloy.

6 Acknowledgments

The authors gratefully acknowledge support from the National Natural Science Foundation of China (51825401)

Reference

[1] B. Wan, H. Zhang, M. Gao, P. Bai, H. Zhang, High-temperature wettability and interactions between Hf-containing NbSi-based alloys and Y2O3 ceramics with various microstructures, Mater Design 2018;138(1):103-110.

[2] B. Kong, L. Jia, H. Zhang, J. Sha, S.Shi, K.Guan, Microstructure, mechanical properties and fracture behavior of Nb with minor Si addition, Int J Refract Met H 2016;58:84-91.

[3] B.P. Bewlay, M.R. Jackson, J.C. Zhao, P.R. Subramanian, A review of very-high-temperature Nb-silicide-based composites, Metall Mater Trans A 2003; 34A(10) 2043-2052.

[4] B.P. Bewlay, M.R. Jackson, H.A. Lipsitt, The balance of mechanical and environmental properties of a multielement niobium-niobium silicide-based In Situ composite, Metall Mater Trans A 1996; A27(12):3801-3808.

Optimization of Microstructure and Fracture Toughness of Nb_4FeSi-Containing Nb-Si Based Alloy Using V, Zr and C Elements

Qibin Wang[1], Qi Wang[1,*], Ruirun Chen[1,*]
1. National Key Laboratory for Precision Hot Processing of Metals, School of Materials Science and Engineering, Harbin Institute of Technology, Harbin 150001, China
*Corresponding address: Qi Wang, wangqi_hit@hit.edu.cn; Ruirun Chen, ruirunchen@hit.edu.cn

Abstract: Nb-Si alloy is considered an ideal material to replace Ni-based alloys, but toughness is the key limiting factor for its application. In this work, the microstructure and room temperature fracture toughness of the alloy by adding V, Zr and C elements were investigated. After doping with 3 at.% V, the microstructure is similar to the base alloy, consisting of Nbss and β-Nb5Si3. After doping with 3 at.% Zr, partial β-Nb5Si3 phase evolves into hexagonal γ-Nb5Si3 blocks, while obtaining fine eutectic structures. After doping with 3 at.% C, dispersed TiC particles appear, and the silicides are refined and spheroidized. The toughness of the alloy has been improved after alloying, and the toughness of the C-doped alloy is the best, reaching 13.61 MPa·m1/2, which is 64% higher than the base alloy.

Keywords: Nb-Si, alloying, microstructure, toughness

1 Introduction

With the development of aerospace, the demand for temperature bearing capacity of the hot-section components of engine has been increased [1-3]. Nb-Si in-situ composites can withstand service temperatures above 1200 °C and are considered as alternative materials for Ni-based alloys [2]. The main constituent phases of this alloy are niobium solid solution (Nbss) and silicides [3]. However, the insufficient room temperature toughness limits the industrial application of the alloys.

Extensive attempts have been made to improve the properties of the Nb-Si based alloy via suitable alloying using Ti, Hf and B, etc. [4, 5]. By adjusting the morphology and proportion of the constituent phases, the optimized properties maybe meet the requirements of industrial applications [5]. Up to now, the effect of V, Zr, and C elements on microstructure and fracture toughness of Nb_4FeSi-containing Nb-Si based alloy is still unclear.

In this work, the constituent phase and structural evolution of the alloy after adding V, Zr, and C were investigated. In addition, the room temperature fracture toughness of the optimized alloy was tested.

2 Experimental procedure

The investigated alloys with the nominal compositions of Nb-24Ti-16Si-2Al-4Fe-X (X=3V, 3Zr and 3C, at.%) were prepared by vacuum non-consumable arc melting in a water-cooled copper crucible.

The microstructure of the alloy was analyzed using the BSE-SEM (Merlin Compact). The single-sided notch bending test was conducted on the Instron 5569 universal electric testing machine to obtain the fracture toughness. The size of the test sample is 20 × 4 × 2 mm^3, with a notch depth of 2 mm. The fracture toughness K_Q value is calculated using the formula mentioned in reference [6]. The fracture toughness test of each alloy shall be conducted at least 3 times at a loading rate of 0.2 mm/min.

3 Result and discussion

Microstructure

Fig. 1 shows the microstructure characteristic of the alloy after adding alloying elements. As shown in Fig. 1(a), the constituent phase of the matrix alloy is Nbss, β-Nb_5Si_3 and Nb_4FeSi. Nbss is the light gray contrast, β-Nb_5Si_3 is the dark gray, and Nb_4FeSi is the darker contrast in BSE-SEM mode. Nbss and β-Nb_5Si_3 exist in a eutectic structure, while Nb_4FeSi appears near β-Nb_5Si_3 block. The microstructure of 3V alloy (Fig. 1(b)) is mainly composed of three phases: Nbss, β-Nb_5Si_3 and Nb_4FeSi phases. Element V is mainly enriched in the Nbss phase. Thus, the addition of V element has little effect on the microstructure and the constituent phase. The microstructure of 3Zr alloy (Fig. 1(c)) is composed of Nbss, β-Nb_5Si_3, γ-Nb_5Si_3 and Nb_4FeSi phases. The addition of Zr element causes the formation of hypereutectic structure and the appearance of primary hexagonal γ-Nb_5Si_3 blocks. During the solidification process, the leading precipitation of primary γ-Nb_5Si_3 causes a decrease in the surrounding Si content, thereby forming a fine eutectic structure around it. The microstructure of 3C alloy (Fig. 1(d)) is composed of Nbss, β- Nb_5Si_3, Nb_4FeSi, and TiC phases. The addition of C element is mainly enriched in fine TiC particles, leading to their formation. In addition, compared to the initial microstructure, the silicides are refined and spheroidized after adding 3 at.% C.

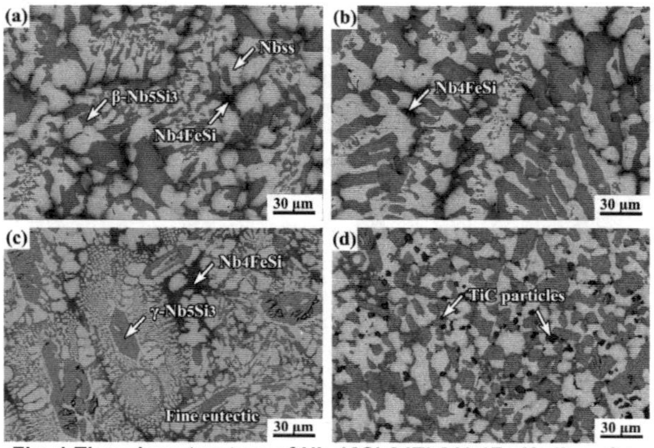

Fig. 1 The microstructure of Nb-16Si-24Ti-2Al-4Fe-X alloy after alloying: (a) base alloy; (b) 3V; (c) 3Zr; (d) 3C.

4 Room temperature fracture toughness

The room temperature fracture toughness of the alloys by the notched three-point bending tests is shown in Fig. 2. The fracture toughness K_Q value of the base alloy is 8.29 $MPa \cdot m^{1/2}$. After adding 3 at.% V and 3 at.% Zr, the fracture toughness K_Q value increased to 10.46 $MPa \cdot m^{1/2}$ and 12.78 $MPa \cdot m^{1/2}$, respectively. The fracture toughness K_Q value of the C-doped alloy is the best, reaching 13.61 $MPa \cdot m^{1/2}$, which is 64% higher than the base alloy. The increase of K_Q value after V doping is mainly due to the solid solution of V in phase Nbss, which promotes the toughening of the alloy. The fine eutectic structure obtained by doping Zr is beneficial for hindering crack propagation and improving toughness. After adding C element, it is attributed to the refinement of the silicides and the presence of small TiC particles, which played a role in toughening the alloy.

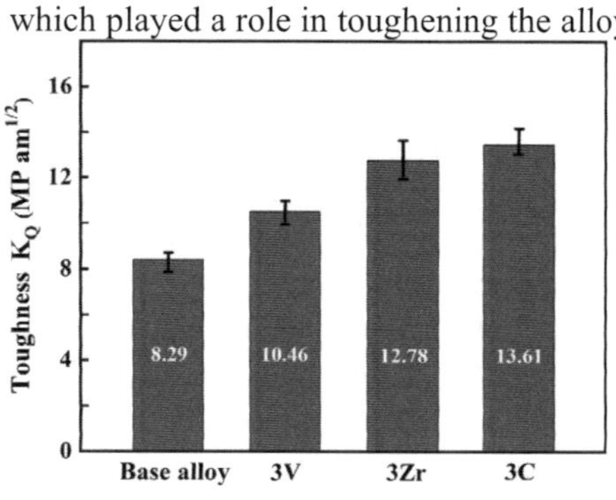

Fig. 2 The room temperature fracture toughness of the alloys.

5 Conclusion

(1)The microstructure doped with 3 at.% V is composed of Nbss, β-Nb_5Si_3 and Nb_4FeSi. The γ-Nb_5Si_3 appears and fine eutectic is obtained by doping 3 at.% Zr. Dispersed TiC particles appear and the silicides are refined and spheroidized by doping 3 at.% C.

(2)After adding 3 at.% V and 3 at.% Zr, the K_Q value increases to 10.46 and 12.78 $MPa \cdot m^{1/2}$, respectively. The K_Q value of the C-doped alloy reaches 13.61 $MPa \cdot m^{1/2}$, which is 64% higher than the base alloy.

6 Acknowledgments

This work was supported by National Natural Science Foundation of China (No. 51825401 and 52004077).

References

[1] Bewlay B P, Jackson M R, Zhao J C, et al. A review of very high temperature Nb silicide-based composites, Metall. Mater. Trans. A 2003, 34 (10): 2043-2052.
[2] Chen D Z, Wang Q, Chen R R, et al. An as-cast Nb-Si-based alloy with fine-grains and remarkable fracture toughness by minor Sc addition, J. Alloy. Compd. 2022, 895: 162707.
[3] Tang Y, Guo X. Flow softening behavior during hot compression of a Nb-Si based ultrahigh temperature alloy, J. Alloys Compd. 2018, 731: 985-994.
[4] Wang Q, Zhao T Y, Chen R R, et al. Collaborative optimization of microstructure, phase composition and room-temperature fracture toughness of Nb-Si based alloys using Ti, Zr and Hf elements, Trans. Nonferrous Met. Soc. China, 2024, 34 (1): 194-202.
[5] Ma R, Guo X P. Cooperative effects of Mo, V and Zr additions on the microstructure and properties of multi-elemental Nb-Si based alloys, J. Mater. Sci. Technol. 2023, 132: 27-41.
[6] S. Zhang, X.P. Guo, Alloying effects on the microstructure and properties of Nb-Si based ultrahigh temperature alloys, Intermetallics 2016. 70: 33-44.

Effect of Ultrasonic Assisted Solidification on Microstructure and Mechanical Properties of Nb-Si Alloy

Qi Wang
National Key Laboratory for Precision Hot Processing of Metals, School of Materials Science and Engineering, Harbin Institute of Technology, Harbin 150001, China
*Corresponding address: Qi Wang, wangqi_hit@hit.edu.cn; Ruirun Chen, ruirunchen@hit.edu.cn

Abstract: Nb-Si alloy is an ideal material to replace Ni-based superalloy as the hot end component of aerospace engine. In this study, the microstructure and the room temperature fracture toughness of the Nb-16Si-20Zr-2C alloys before and after ultrasonic treatment were studied. It is found that the application of ultrasonic wave changes the phase content in the alloy. Under the action of ultrasonic wave, a large number of primary silicides phases are precipitated, and the shape of the primary silicide is a regular polygon with sharp corners. Therefore, after ultrasonic treatment, the room temperature fracture toughness of the alloy decreased by 24.1 %, mainly due to the increase of the brittle phase and the sharp edge of the brittle phase splitting the continuous Nbss phase matrix, resulting in a decrease in the room temperature fracture toughness of the material.

Keywords: Nb-Si, ultrasonic, microstructure, toughness

1 Introduction

With the continuous development of aerospace technology, the requirements for aircraft structural materials are getting higher and higher [1-3]. Ni-based superalloy has been difficult to meet the temperature requirements of new aero-engine turbine blades due to its high density and temperature limitation. Nb-Si superalloy with lower density and higher melting point is a strong contender to replace the Ni-based superalloy. The basic phases of Nb-Si based alloys include ductile Nbss phase and intermetallic compound Nb_5Si_3 phase which is thermodynamically stable and maintains high strength at 1600-1800 ℃.

Because Nb-Si superalloy has the characteristics of poor room temperature fracture toughness and insufficient oxidation resistance at high temperature. In the research of Nb-Si alloy, alloying and improving the preparation process are mainly used to improve its comprehensive mechanical properties [4, 5].

In recent years, the research institutions led by GE and AFRL have conducted in-depth research on Nb-Si superalloys by alloying [2, 6]. Ultrasonic assisted solidification process is a method that can improve the mechanical properties of the prepared materials. Because of the limitation of ultrasonic probe material, the most studied ultrasonic assisted solidification materials are alloys with medium and low melting points. The effect of ultrasonic wave on the solidification process of Nb-Si alloy with high melting point is rarely studied.

In this study, the effect of ultrasonic wave was introduced into the high temperature melt of Nb-Si alloy by indirect introduction. And we studied the effect of ultrasonic assisted solidification on the microstructure and mechanical properties of Nb-Si ultra-high temperature alloy.

2 Experimental procedure

The investigated alloys with the nominal compositions of Nb-16Si-20Zr-2C were prepared by vacuum non-consumable arc melting in a water-cooled copper crucible and an arc-melting device which installs an ultrasonic generator under the water-cooled copper crucible. In order to facilitate the analysis later, the as-cast Nb-16Si-20Zr-2C alloy and the Nb-16Si-20Zr-2C alloy which was ultrasonically treated are referred to 20Zr-2C(AC) and 20Zr-2C(UST). The current of the arc gun is reduced to 350A and the ultrasonic wave is applied. The ultrasonic treatment time is 100s.

The microstructure of the 20Zr-2C(AC) and 20Zr-2C(UST) were analyzed using the BSE-SEM (Merlin Compact). The three-point bending test was carried out on the Instron 5569 universal electric testing machine to calculate the fracture toughness. The size of three-point bending specimen is $20 \times 4 \times 2$ mm^3, with a notch depth of 2 mm. The fracture toughness K_Q value is calculated using the formula mentioned in reference [7]. The fracture toughness test of each alloy shall be conducted at least 3 times at a loading rate of 0.2 mm/min.

3 Result and discussion

Fig. 1 shows the microstructure characteristic of the 20Zr-2C(AC) and 20Zr-2C(UST). As shown in Fig. 1(a), it can be seen that in the 20Zr-2C(AC) alloy obtained by arc-melting process, the size of silicide phase is small, the edges of Nbss phase and silicide phase are smooth and round, and the Nbss phase is large and continuous. As shown in Fig. 1(b), after ultrasonic treatment of Nb-16Si-20Zr-2C alloy for 100s, more primary silicide phases are produced in the microstructure of the Nb-Si superalloy, and the edges of the primary silicide phases show a regular polygonal shape. There are many fine eutectoid Nbss/γNb_5Si_3 phases between the primary silicides, which are the products of final solidification.

It can be clearly seen from Fig. 1 that when 100s ultrasonic wave is applied during the solidification of the alloy, the phase content in the final formed structure changes. The content of Nbss-based solid solution decreases significantly, while the content of silicide phase increases. This shows that the phase diagram of Nb-Si alloy is changed because of the effect of ultrasonic wave, which in turn changes the content of Nbss phase and the silicide phase in the final solidified microstructure.

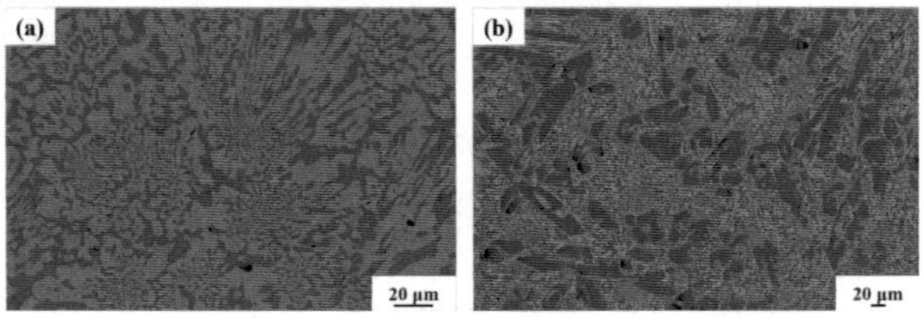

Fig. 1 The microstructure of Nb-16Si-20Zr-2C (a) 20Zr-2C(AC) (b) 20Zr-2C(UST)

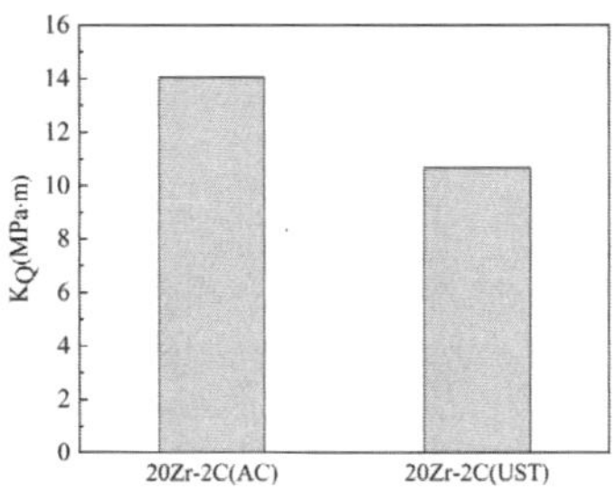

Fig. 2 The room temperature fracture toughness of Nb-16Si-20Zr-2C (a) 20Zr-2C(AC) (b) 20Zr-2C(UST)

The room temperature fracture toughness of the alloys by the notched three-point bending tests is shown in Fig. 2. The room temperature fracture toughness K_Q of the 20Zr-2C(AC) is 14.05 MPa·m$^{1/2}$. And the fracture toughness K_Q of the 20Zr-2C(UST) is 10.67 MPa·m$^{1/2}$. After ultrasonic treatment, the room temperature fracture toughness of the alloy decreased by 24.1%. This is related to the increase of the number and the change of the shape of primary silicides in 20Zr-2C(UST) alloy. As a brittle phase, the silicide phase is unfavorable to the room temperature fracture toughness of the Nb-Si alloy. Therefore, the increase of silicide phase will reduce the room temperature fracture toughness of the Nb-Si alloy. Secondly, there are sharp corners in the polygonal silicide phase in the alloy, which have obvious splitting effect on the Nb-Si alloy matrix, which is also the reason for the decrease of the toughness of the alloy.

4 Conclusion

(1) After 100s ultrasonic treatment, a large number of primary silicide phases appear in the 20Zr-2C alloy, and the shape of the primary silicide phase is regular polygon. The Nbss phase in the 20Zr-2C(AC) alloy is more continuous.

(2) After 100 s ultrasonic treatment, the room temperature fracture toughness of the alloy decreased by about 24.1 %, which was mainly due to the increase of the brittle phase of the primary silicide content and the splitting effect of the sharp edge on the alloy matrix.

5 Acknowledgments

The authors gratefully acknowledge support from the National Natural Science Foundation of China (51825401).

References

[1] Bewlay B P, Jackson M R, Zhao J C, et al. A review of very high temperature Nb silicide-based composites, Metall. Mater. Trans. A 2003, 34 (10): 2043-2052.
[2] Chen D Z, Wang Q, Chen R R, et al. An as-cast Nb-Si-based alloy with fine-grains and remarkable fracture toughness by minor Sc addition, J. Alloy Compd. 2022, 895: 1627070.
[3] Tang Y, Guo X. Flow softening behavior during hot compression of a Nb-Si based ultrahigh temperature alloy, J. Alloy. Compd. 2018, 731: 985-994.
[4] Zhou Z C, Wang Q, Chen R R, et al. Ultra-fine Nbss/Nb_5Si_3 in situ composites with remarkable properties prepared by ultrasonic melt treatment, J. Alloy. Compd. 2023, 940: 168940.
[5] Zhou Z C, Wang Q, Chen R R, et al. Microstructure evolution and mechanical properties of as-cast and ultrasonic treated Nb-16Si-*x*Cr alloys, J. Mater. Res. Technol. 2022, 17: 2856-2863.
[6] Bewlay B P, Jackson M R, Zhao P R, et al. Ultrahigh-Temperature Nb-Silicide-Based Composites. MRS Bull. 2003, 28(9): 646-653.
[7] Zheng C W, Wang Q, Chen R R, et al. Mechanical properties and microstructure evolution of NbSiZr based alloy, J. Alloy Compd. 2023, 963: 171155.

Optimization of Solid-Liquid Bimetallic Casting Process Parameters for Dual-Performance Blade

Jinlin Li, Tiantian Wang, Hongyu Xu, Bo Jiang, Ye Wang*, Maoliang Hu
School of Materials Science and Chemical Engineering, Harbin University of Science and Technology, Harbin, 150040, China
* Corresponding authors: ohenry1980@163.com (Y Wang)

Abstract: In this paper, the solid-liquid bimetal investment casting of turbine blade for nickel-based superalloys DD412 and K447A has been investigated by numerical simulation methods. The effect of investment casting process parameters on the depth of bimetallic bonding and the grain size of the disc casting has also been studied by combining Taguchi method and ANOVA (Analysis of Variance) method, and the solid-liquid bimetal investment casting parameters were optimized to obtain the best combination of process parameters, this work can provide a theoretical basis and experimental foundation for the preparation of high-quality bimetallic turbine blisk castings.

A quarter of the original casting was retained to improve the calculation accuracy on the premise of ensuring the calculation accuracy by the Soildworks software, due to the casting had the axisymmetric characteristics. The finite element model of turbine blisk casting for solid-liquid bimetallic casting process and the corresponding casting parameters for simulation were both established in ProCAST software, as shown in Fig. 1. The simulation parameters of the body shape nucleus boundary conditions, employed for CAFE advanced module, were obtained by inverse computation from the actual experiments under the conditions of pouring temperature of 1350 ℃ and preheating temperature of 1100 ℃, with the measured casting grain size of 0.2 mm , which were DTm = 1.5 K, DTs = 2 K, and Nmax = 5×10^{11} 1/m^3.

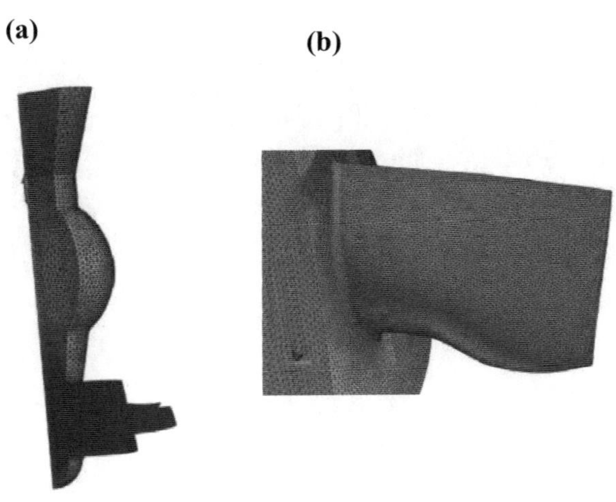

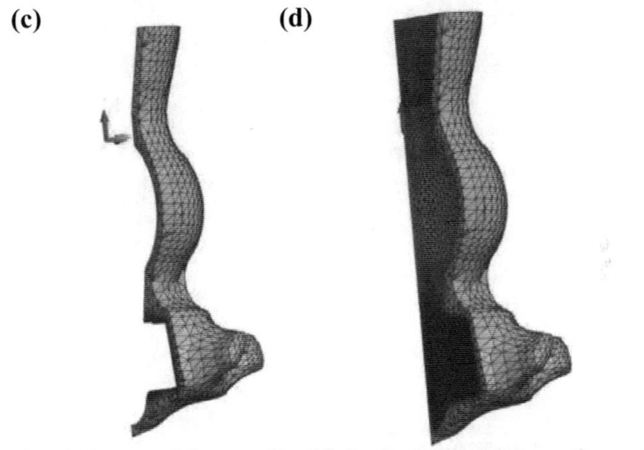

Fig. 1 The generation results of finite element mesh in casting and mold shell, (a) mesh of the casting, (b) mesh at the blade, (c) mesh of the mold shell, (d) overall finite element mesh

According to the analysis, three process parameters of solid-liquid bimetallic investment casting process, including pouring temperature, preheating temperature of mold shell, and pouring time, with great influence on the quality of dual-performance turbine blisk castings were selected, and a three-level, five-factor orthogonal test was designed, then 25 groups of numerical simulation experiments were carried out with the quality evaluation indexes of grain size and the depth of the bimetallic bonding. The simulation results were analyzed by Taguchi method and ANOVA method, the best combination of process parameters for the turbine blisk casting was finally optimized as follows: pouring temperature of 1350 ℃, preheating temperature of the mold shell of 1200 ℃, and pouring time of 5 s.

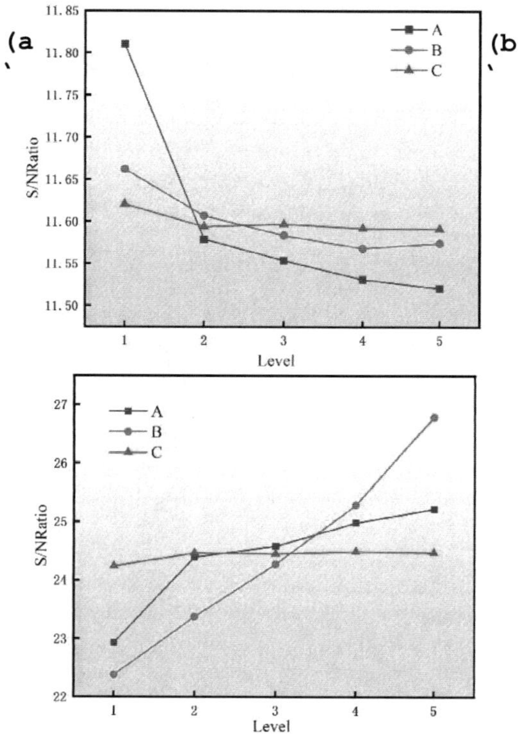

Fig.2 S/N response curve: (a) Grain size S/N response curve (b) Bimetallic bonding depth S/N response curve

Finally, the temperature field and microstructure under the best combination of process parameters and other combinations in the orthogonal design experiment were compared and analyzed, which verifying the accuracy and rationality of Taguchi method.

Key words: nickel-based superalloy; bimetallic casting; bimetallic bonding; grain size; Taguchi method.

The 75th World Foundry Congress
25~30 October 2024, Deyang, Sichuan, China

Effect of Al, Ti and W on microstructure and properties of high tungsten K416B superalloy

Xiang Fei[1, 2], Naicheng Sheng[1, *], Yuping Zhu[1], Jun Xie[1], Jinjiang Yu[1], Guichen Hou[1], Jinguo Li[1], Yizhou Zhou[1], Xiaofeng Sun[1]
1. Shi-changxu Innovation Center for Advanced Materials, Institute of Metal Research, Chinese Academy of Sciences, Shenyang 110016, China
2. School of Materials Science and Engineering, University of Science and Technology of China, Shenyang 110016, China
*Corresponding address: e-mail: nccheng@imr.ac.cn

Abstract

The macro-segregation of W in the K416B alloy ingot has been investigated. It was observed that the distribution of tungsten-rich α-W phases in the ingot is not uniform. And the bottom of the ingot enriched in various shapes and sizes of tungsten-rich α-W phases, consistent with the higher mass fraction of W at the bottom of the ingot. Thermo-Calc thermodynamic software was employed to analyze the influence of major alloying elements on the microstructure of the alloy. The results show that when the content of Al, W and Ti elements exceed certain thresholds, the tungsten-rich α-W phases would be precipitated, and the precipitation temperature and peak content of the tungsten-rich α-W phases increases with the raising content of Al, W, and Ti element. The alloy test rods with different Al and Ti contents were cast at 1550 °C. When the Al content reaches 7 wt. % and the Ti content reaches 2.5 wt. %, the tungsten-rich α-W phase was precipitated in the alloy, and the stress rupture life of the alloy decreases significantly. By adjusting the Al, W, and Ti element contents and casting the alloy ingot of K416B, the macro-segregation of W and Al elements was suppressed. The stress rupture life of the alloy is more than 40 hours, and the fluctuation is small.

Keywords: K416B, α-W, M_6C, the stress rupture life

Introduction

Nickel-based superalloys are the key materials for hot-end components of advanced aero-engines due to their excellent high-temperature mechanical properties, oxidation resistance and hot corrosion resistance [1]. High-W superalloys are strengthened by adding a large amount of solid solution strengthening element W to enhance their high-temperature mechanical properties. The addition of W increases the bonding energy between atoms and affects the distribution of alloy elements between the γ matrix and γ' phase, altering the lattice constants and mismatch of the two phases. W also reduces the stacking fault energy of the alloy and improves its high-temperature strength [2]. K416B alloy is one of the typical high-W superalloys [3], with a W element content of 16 wt. %, which is one of the equiaxed crystals with the highest temperature bearing capacity at present. The addition of a large amount of W element not only strengthens the alloy, but also causes the macro-segregation of W element during the solidification process of the alloy. It is difficult to obtain the alloy with uniform and optimal compositions, which significantly reduces the performance and stability of the alloy. Therefore, studying the phase of tungsten-rich and performance regulation of K416B alloy is crucial for enhancing the performance level of the alloy to ensure the safe and reliable operation of aero-engines.

Experimental procedure

Thermo-Calc software was employed for thermodynamic simulation to investigate the influence of major alloying elements such as Al, Ti and W on the precipitation of the W-rich phase. The test rods of alloy with different contents of Al and Ti were prepared, and the influence of Al and Ti elements on the microstructure and mechanical properties of the alloy were analyzed. By comparing the microstructures and compositions of the alloy ingots at different heights before and after compositional adjustment, the relationship between the tungsten-rich phase and macro-segregation was investigated, and the effect of alloy composition regulation on macro-segregation was verified. The main composition range of K416B alloy is shown in Table 1.

Table 1. Main composition of Alloy K416B (wt. %)

	Al	W	Ti	Cr	Co	Nb	Hf	C	Ni
Upper	5.6	15.3	0.7	4.6	6.0	1.6	0.7	0.08	Bal
Lower	6.2	16.5	1.2	5.2	8.0	2.1	1.2	0.14	Bal

Result and discussion

1. Thermo-Calc calculations

The distribution of element in the ingot would be affected by the precipitation of tungsten-rich phase. Therefore, the solidification process of K416B alloy was simulated by Thermo-Calc, and the effects of Al, W and Ti on the formation of tungsten-rich α-W phase were compared. The results show that when the content of Al, W and Ti exceeds the threshold, the tungsten-rich α-W phase would be precipitated, as shown in Fig. 1. The calculated results are shown in Fig. 1(b-d) for a mere change of Al (W, Ti) elements content while the content of other alloying elements fixed and unchanged. The results indicate that the precipitation of the α-W phase begins when the mass fractions of Al, W, and Ti reach 6.1 %, 15.7 %, and 1.1%, respectively. As the contents of Al, W, and Ti elements increase, the precipitation temperature and peak content of the α-W phase increase.

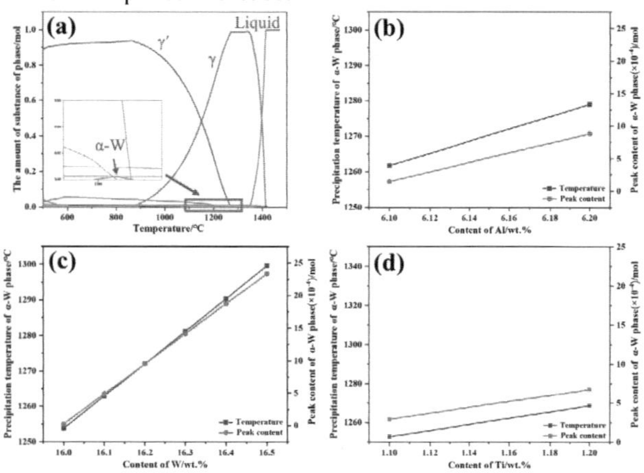

Fig. 1 Thermodynamic calculation results: (a) equilibrium phase diagram of K416B alloy; (b) (c) (d) precipitation temperature and peak content of tungsten-rich α-W phase with different Al, W and Ti contents

2. Distribution of element and microstructure

According to the results of thermodynamic calculations, the alloying elements Al, W and Ti were mainly regulated, and the K416B alloy ingot was recast. The content of elements at different positions of the ingot before and after composition control of the alloy was analyzed and the microstructure was studied, the results are shown in Fig. 2. Before the composition control, the macro-segregation of W and Al elements in the alloy ingot is more significant, and the content of W element at the lower end is about 5wt. % higher than that at the upper end. In addition, the Al element shows an opposite trend, and the content of Al in the lower part of the alloy is significantly lower than that in the upper part (Fig.2a. More spherical tungsten-rich α-W phases are found in the lower part of the alloy ingot than in the upper part, which is also the reason for the higher W content in the lower part (Fig.2(b, c After the composition control of the alloy, the mass fractions of the upper and lower ends of the W element are 15.6 % and 15.9 %, respectively. And the mass fractions of the upper and lower ends of the Al element are 5.97 % and 5.93 %, respectively. Nearly no tungsten-rich α-W phases are observed in the alloy. The macro-segregation of W and Al elements has been significantly suppressed (Fig.2(d-f)).

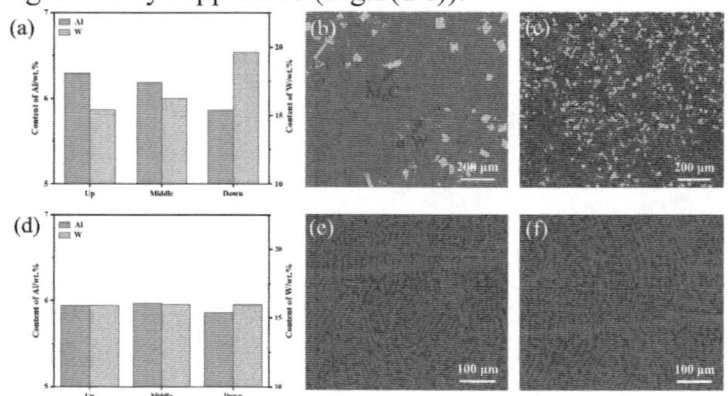

Fig. 2 The content of W and Al elements at different positions of ingot before and after element regulation: (a) before, (b) after; (b, c, e, f) microstructure at different positions

Conclusion

After controlling the alloying elements of Al, W and Ti, no tungsten-rich α-W phase precipitated. The macro-segregation of W and Al elements in the alloy ingot is suppressed, and the composition of the alloy ingot is uniform and the microstructure stability is enhanced.

References

[1] R.C. Reed, The superalloys: fundamentals and applications, 2006, Cambridge University Press.
[2] D. L. Shu, J. Xie, F. J. Zhang, G. C. Hou, Z. J. Wang, S. L. Xun, J. J. Yu, X. F. Sun, Y. Z. Zhou, Room temperature tensile deformation behavior of a Ni-based superalloy with high W content, China Foundry 18 (2021) 192–198.
[3] Q. Li, J. Xie, J. J. Yu, D. L. Shu, G. C. Hou, X. F. Sun, Y. Z. Zhou, Solidification behavior and segregation characteristics of high W-content cast Ni-Based superalloy K416B, J. Alloys Compd. 854 (2021) 156027.

Carbon Addition and Temperature-Dependent Tensile Properties of a Low-Cost 3rd-Generation Ni-Based Single Crystal Superalloy

Yongmei Li, Xinguang Wang*, Zihao Tan, Jinguo Li, Yizhou Zhou, Xiaofeng Sun
1. Shi-changxu Innovation Center for Advanced Materials, Institute of Metal Research, Chinese Academy of Sciences, Shenyang 110016, China
*Corresponding address: e-mail: xgwang11b@imr.ac.cn

Abstract

Doping with an appropriate carbon is anticipated to balance the mechanical properties and castability of Ni-based single crystal (SX) superalloys. This work examined the effects of carbon content on tensile properties of a novel Ni-based low-cost third-generation SX superalloy, at 760 °C and 1120 °C. The results show that increasing carbon content improved the strength and plasticity at 760 °C. However, the elevated temperature to 1120 °C caused a loss in strength and plasticity. With the Vickers hardness and high-resolution X-ray diffraction tests, it was confirmed that a minor addition of carbon could improve the solid solution strength and interface strength of SX superalloy. However, the high-temperature strength of high-carbon alloys will be compromised for the large consumption of refractory elements by interdendritic MC carbides, although the blocky M_6C carbides could provide an additional hindrance to dislocation motion. Regarding the macroscopic deformability, EBSD confirms that MC carbides could apply as an obstacle to impede the rapid shear decohesion at 760 °C, nevertheless, the severe stress concentration and deformation incompatibility near MC carbides would promote microcrack initiation and propagation behaviors at 1120 °C. Overall, future compositional optimization should focus on reducing the formation of Ta-enriched MC carbides to minimize refractory element depletion and deformability reduction at high temperatures.

Keywords: Ni-base single crystal superalloys, carbon, tensile properties, carbides

Introduction

In Ni-based SX superalloys, the boundary strengthener carbon had been completely removed to obtain superior mechanical properties. Practical applications require SX superalloys with favorable castability to boost the qualification rate of turbine blades during manufacturing. Appropriate inclusion of carbon in advanced Ni-based SX superalloys is critical to limit the formation tendency of stray grain, recrystallization, and high-angle grains (HAG) [1, 2]. The effect of carbon addition on the mechanical properties of Ni-based SX superalloys has been extensively studied. However, these studies concentrated on the effects of carbon on the creep and fatigue properties of first and second-generation SX superalloys [3, 4]. Limited studies on the tensile properties of Ni-based SX superalloys as a function of carbon content can be found. Therefore, it is necessary to ascertain the effect of carbon addition on the tensile properties of the novel alloy to indicate its' further creep and fatigue properties.

In this work, we investigated the impact of temperature and carbon content on the tensile characteristics of a novel low-cost 3rd-generation Ni-based SX superalloy. The synergistic *carbon content & temperature* effects on tensile properties were highlighted, which could guide future carbon modifications and practical applications of this newly developed SX superalloy.

Experimental procedure

The chemical composition of the novel low-cost 3rd-generation Ni-based SX superalloys with three different carbon content is given in Table 1. Master ingots were prepared from pure metals in a vacuum induction melting (ZG-0.01). Single bars with dimensions ϕ16×220 mm were then produced using the selecting crystal method in a vacuum directional solidification furnace (HRS, ZGD-15), at a withdrawal rate of 6 mm/min. From these bars, only single rods with an orientation deviation [001] within 10° were wire-cut into ϕ16×60 mm rods for subsequent heat treatment. Here, the same high-temperature solution and two-step aging heat treatments were carried out for the three experimental alloys as follows:
1320~1325℃/1h + 1325~1330℃/3h + 1335~1340℃/4h (A.C.) → 1100℃/6h (A.C.) + 870℃/24 h (A.C.).

Table 1. Composition of Alloy 0C, 0.045C and 0.85C, wt.%

Alloys	C	Re	Cr	Co	Mo+W+Ta	Al	Ni
0C	0	2.93	4.56	7.95	14.76	5.70	Bal.
0.045C	0.048	2.93	4.53	7.92	14.80	5.68	Bal.
0.085C	0.083	2.91	4.51	8.03	14.73	5.72	Bal.

Result and discussion

Tensile properties

The engineering stress-strain (ε-σ) curves of the three alloys at 760 °C and 1120 °C were displayed in Fig.1. At 760 °C, the ε-σ curves were almost indistinguishable from each other, and all of them exhibited insignificant yielding phenomena. While differences appeared at 1120 °C, especially for the ultimate strength and rupture elongation. Table 2 compares the strength and plasticity of three alloys at 760 °C and 1120 °C. It could be noticed that the yield strength (YS) and ultimate tensile strength (UTS) at 760 °C increased slightly with increasing carbon content. Particularly, the plasticity of alloy 0.085C outperformed about 60% than that of alloy 0C and 0.045C. At 1120 °C, YS gradually decreased from 313 MPa of alloy 0C to 285 MPa of alloy 0.085C. Simultaneously, plasticity decreased significantly with increasing carbon addition.

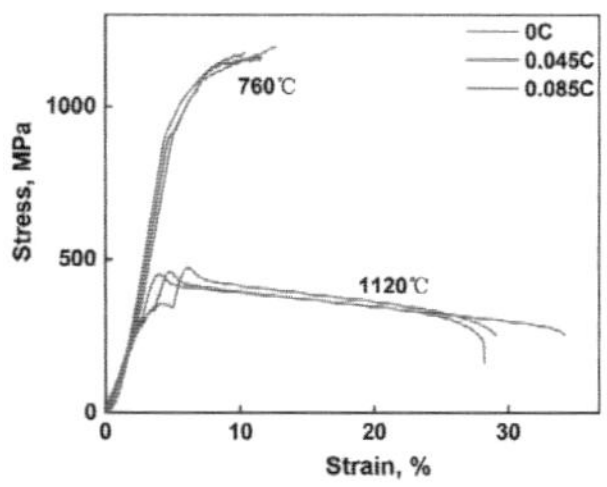

Fig. 1 Stress-strain curves of three alloys at 760 ℃ and 1120℃

Table 2 Strength (YS and UTS, MPa) and plasticity (δ and A, %) of Alloy 0C, 0.045C and 0.085C.

Alloys	760 ℃				1120 ℃			
	YS	UTS	δ	Z	YS	UTS	δ	Z
0C	921	1165	4	5.5	313	459	34	50
0.045C	938	1174	4.8	4.5	290	475	24	42
0.085C	944	1196	8	9.5	285	452	27	36

Carbide effects

The influences of carbon addition on tensile properties were discussed in terms of the intrinsic microscopic resistance to dislocation movements and deformability to macroscopic damage accumulation. It was found that the solid solution strength and interface strength increased with increasing carbon content, as confirmed by HV and HR-XRD tests. With TEM characterization, the precipitation of blocky M_6C carbides was found to apply as a hindrance to dislocation movement. However, the high-temperature strength was severely weakened due to the massive consumption of refractory elements caused by the formation of interdendritic Ta-enriched MC carbides. Here, the effects of MC carbides on macroscopic deformability were examined via electronic backscatter detection technology (EBSD). At 760 ℃, extensive secondary slip bands were observed upon MC carbides, nevertheless, only a modest amount of slip bands were detected at locations below those carbides (Fig.2(a)). At 1120℃, severe recrystallization occurred in regions containing large MC carbides, while the areas containing small carbide particles still showed severe crystal rotation (Fig.2(b)). Summarily, we could assume that the improved plasticity at 760 °C originated from the attenuating slip band extension, whilst the reduced plasticity at 1120 °C was generated by collectively promoting microcracks initiation and propagation near interdendritic MC carbides.

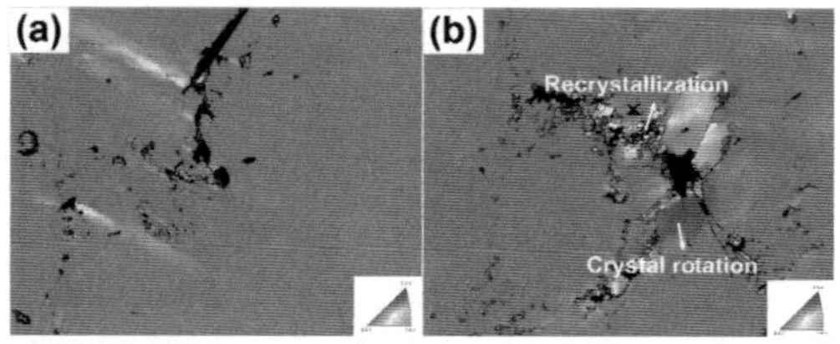

Fig. 2 EBSD-IPF maps in interdendritic region of alloy 0.085C at (a) 760 ℃, (b) 1120℃.

Conclusion

Minor carbon addition could improve the strength of the novel low-cost third-generation SX superalloy by enhancing its solid solution strength and interface strength. However, the formation of interdendritic MC carbides would cause massive consumption of refractory elements, thus leading to the strength slump at high temperatures, even though blocky M_6C carbides provided additional hindrance for dislocation movement. As for plasticity, the existence of MC carbides could hinder the extension of slip bands at 760 ℃, leading to an increase in plasticity. Nevertheless, those MC carbides would facilitate microcracks initiation and propagation behavior, which resulted in a drastic decrease in plasticity at 1120 ℃. Therefore, future compositional optimization should focus on reducing the formation of MC carbides to diminish the strength loss and deformation incompatibility at high temperatures.

References

[1] K.A. Al-Jarba, G.E. Fuchs, Carbon-containing single-crystal nickel-based superalloys: Segregation behavior and carbide formation, JOM 56 (2004) 50–55.
[2] J. Krawczyk, R. Paszkowski, W. Bogdanowicz, et. al, Defect Creation in the Root of Single-Crystalline Turbine Blades Made of Ni-Based Superalloy, Materials 12 (2019) 870.
[3] A. Iqbal, J. King, The role of primary carbides in fatigue crack propagation in aero-engine bearing steels, Int. J. Fatigue 12 (1990) 234–244.
[4] L. Wang, D. Wang, T. Liu, et. al, Effect of minor carbon additions on the high-temperature creep behavior of a single-crystal nickel-based superalloy, Mater. Charact. 104 (2015) 81–85.

The 75th World Foundry Congress
25~30 October 2024, Deyang, Sichuan, China

Micro-Scale Damage Mechanisms and Life Prediction Method of the Fourth-Generation Single Crystal Superalloy During Thermal-Mechanical Fatigue

Zihao Tan, Chenglu Zou, Xinguang Wang*, Jianchao Pang, Jinguo Li, Zhefeng Zhang, Xiaofeng Sun, Yizhou Zhou
1. Shi-changxu Innovation Center for Advanced Materials, Institute of Metal Research, Chinese Academy of Sciences, 72 Wenhua Road, Shenyang 110016, China
*Corresponding address: e-mail: xgwang11b@imr.ac.cn

Abstract
Thermal-mechanical fatigue (TMF) is considered as an important damage mode for single crystal superalloys during real service. It is of great significance to uncover the damage mechanisms of the alloy and establish fatigue life prediction method for enhancing the service reliability of the alloys. Therefore, in this work, the TMF behavior and damage mechanisms of the single crystal superalloy DD91 has been systematically clarified via advanced analysis methods. The results showed that during the TMF deformation, the high-level of stress concentration existed in the vicinity of micro-pores and fatigue cracks, which led to the presence of deformation bands and brought about micro-scale plastic damage to the alloy, thereby affecting the overall fatigue life of the alloy. In addition, the process of γ ′-degradation would be accelerated by the introduction of local micro-defects. Under in-phase cycling condition, plenty of superdislocations would shear into the γ ′ particles, while for out-of-phase cycling condition, the deformation twins would extend through both γ and γ ′ phases. Based on the micro-scale plastic deformation mechanisms, a TMF life prediction method has been proposed through considering both plastic strain energy and elastic strain energy. These findings have provided useful guidance for the safe service of the fourth-generation single crystal superalloys. **Keywords:** Fourth-generation single crystal superalloy; Thermal-mechanical fatigue; Local micro-scale damage mechanism; Life prediction method.

Introduction
During the real service of single crystal (SX) turbine blades of aircrafts, they would definitely undergo the cyclic mechanical stress and thermal stress during the stages of starting-up, accelerating, decelerating and shutting-down. Hence the potential failure mode of thermal-mechanical fatigue (TMF), which poses a threat to safe service of SX superalloys, would be introduced [1]. During TMF process, the presence of SF bands is common in the first- and second-generation SX superalloys under out-of-phase (OP) cycling [2]. When it was transformed into in-phase (IP) cycling, the creep damage played a vital role during the TMF deformation [3]. Due to the reduced stacking fault energy of the fourth-generation SX superalloy, it might be thus considered that the primary mechanisms during fatigue deformation would be different. In addition, another pivotal factor which would promote the fatigue failure and might even control the fatigue life is the local micro-defects [4]. Moreover, the elevated service temperature and long-term solution treatments would undoubtedly lead to the formation of excessive oxidation cracks and micro-pores in alloys, respectively. Therefore, the TMF experiments concerning the fourth-generation SX superalloy and the systematic clarification of damage mechanisms based on local micro-defects are in demand.
In the present work, therefore, systematical fatigue experiments had been carried out on a fourth-generation SX superalloy. The fatigue behaviors, overall deformation mechanisms and local micro-scale damage mechanisms during TMF deformation were characterized and clarified. Subsequently, based on the local micro-scale plastic damage mechanisms, a novel TMF life prediction method has been proposed and verified.

Experimental procedure
The fourth-generation SX superalloy DD91 with the excellent high-temperature properties was selected. For TMF experiments, the temperature range was determined ranging from 600 °C to 1000 °C and the frequency of 1/200 Hz were adopted. Both IP- and OP-TMF tests were carried out at different strain amplitudes. In-depth dislocation configurations and deformation mechanisms were analyzed via transmission electron microscopy of Talos F200X and aberration-corrected microscopy.

Result and discussion
1. TMF damage mechanisms based on local defects
As shown in figure 1, during IP-TMF process of the DD91 alloy, the common dislocation configuration was γ/γ ′ interfacial dislocation networks. The matrix dislocations would overcome the γ ′ particles via the method of Orowan by-passing. While in regions near pores or surface oxidation cracks, plenty of long and curved superdislocations were observed to shear into γ ′ phase directly. While for OP-TMF process, the primary deformation mechanism was still Orowan by-passing. Nevertheless, when the local defects were introduced, the local stress concentration had on the one hand facilitate the decomposition of a/2<110> matrix dislocations, hence

a large number of a/6<111> twinning dislocations would be generated in the matrix. On the other hand, the viscous slipping of {111}<112> systems would also be activated. Consequently, the deformation twins would gradually take shape, as shown in figure 1. To sum up, the introduction of local micro-defects would promote the processes of dislocations entering γ′ particles, thereby leading to severe local micro-scale damage to the alloy.

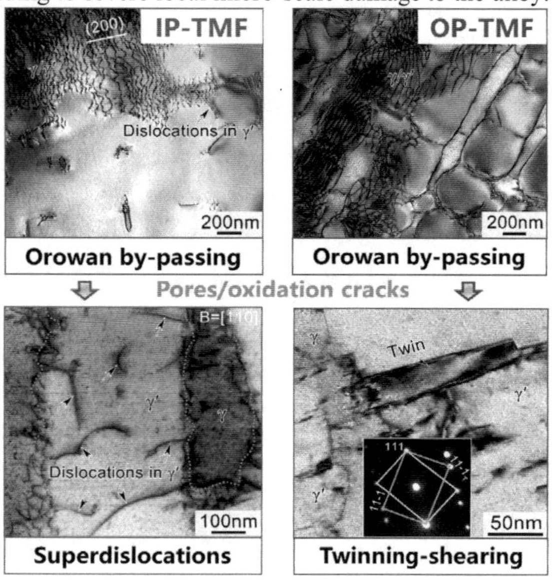

Figure 1. Evolution of dislocation configurations when pores or oxidation cracks were introduced during TMF.

2. Life prediction method based on local defects

In this work, the energy model was used to achieve TMF life prediction. The key point lies in correlate the different energy part to the damage modes. Based on the micro-scale plastic damage mechanisms, the schematic diagram of the total strain energy model is demonstrated in Figure 2. It could be seen that the first part was hysteresis strain energy (Wp), which was reflected by the widespread deformation bands or twinning bands (not much dependent on the local micro-defects). Apart from that, the energy compositions of We^{+} and We^{-} are corresponded to the micro-scale plastic damages generated in tensile and compressive half-cycles, respectively. On this basis, a high-accuracy fatigue life prediction results were obtained through data calculations.

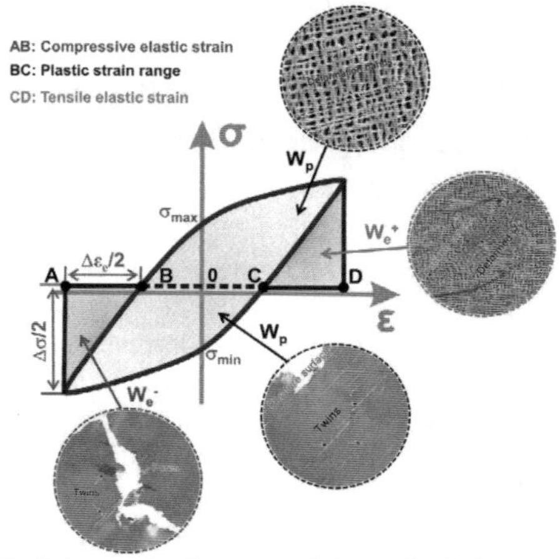

Figure 2. Schematic diagram of the calculation method of total strain energy as well as the relationship between the energy composition and typical deformed microstructures.

Conclusion

During the TMF deformation of fourth-generation SX superalloy, the local micro-scale damage was verified to exist near micro-pores and oxidation crack. The process of γ′-degradation would be accelerated by the introduction of local micro-defects while the entire TMF life would be affected. At high strain amplitude conditions, the alloy was subjected to the severe macroscopic plastic deformation damage. While for low strain range conditions, the alloy sustained oxidation damage and considerable micro-scale plastic deformation damage, which was induced by the excessive superdislocations (IP-TMF) as well as stacking faults and deformation twins (OP-TMF) shearing γ′ precipitates. Based on the micro-scale plastic damage mechanisms, an optimized energy model (total strain energy model) considering both plastic and elastic strain energy had been proposed for fatigue life prediction. In comparison to conventional model, a much higher predicting accuracy was obtained when the optimized model was applied.

References

[1] Moverare J J, Johansson S, Reed R C, Deformation and damage mechanisms during thermal-mechanical fatigue of a single-crystal superalloy, Acta Mater. 2009, 57: 2266-2276.
[2] Han G M, Yu J J, Sun X F, et al., Thermo-mechanical fatigue behavior of a single crystal nickel-based superalloy, Mater. Sci. Eng. 2011,A 528 :6217-6224.
[3] Yu J J, Han G M, Chu Z K, et al., High temperature thermo-mechanical and low cycle fatigue behaviors of DD32 single crystal superalloy, Mater. Sci. Eng. (2014, A 592:164-172.
[4] Tan Z, Wang X, Pang J, et al., Pore-induced defects during thermo-mechanical fatigue of a fourth-generation single crystal superalloy, Mater. Res. Lett. 2023,11(8) :678-687.

Effects of Powder Forming Method and Heat Treatment Process on the Microstructure and Mechanical Properties of a Ni-Based Superalloy by Laser Additive Manufacturing

Zhipeng Zhang[1, 2, 3], Jide Liu[3, *], Jianjun Wang[1, 2], Jinguo Li[3, **]
1. Key Laboratory for Anisotropy and Texture of Materials (MOE), School of Materials Science and Engineering, Northeastern University, Shenyang 110819, PR China
2. Research Center for Metallic Wires, Northeastern University, Shenyang 110819, PR China
3. Shi-changxu Innovation Center for Advanced Materials, Institute of Metal Research, Chinese Academy of Sciences, Shenyang, 110016, PR China
*Corresponding address: e-mail: jdliu@imr.ac.cn (Jide Liu), jgli@imr.ac.cn (Jinguo Li)

Abstract

In this study, the effects of powder forming method and heat treatment process on microstructure and mechanical properties of laser additive manufactured ATI 718Plus alloy were investigated. It was found that the porosity of the samples fabricated using plasma rotation electrode preparation (PREP) powders was significantly lower compared to vacuum induction melting gas atomization (VIGA). The porosity of the samples fabricated with PREP powder was controlled below 0.05% by adjusting the laser process parameters. In addition, after direct aging (DA) treatment, a large number of nano-sized γ′ phases were precipitated in the interdendritic and dendritic trunk regions, but the Laves phase was not significantly affected. After homogenization + solution + aging (HSA) treatment, the Laves phase in the alloy was basically eliminated, and a large number of needle-like η phases precipitated along the grain boundary positions, but the grains underwent obvious coarsening. The tensile results show that the ultimate tensile strength and elongation of the samples after DA treatment were significantly increased at the same time. Compared with DA samples, the tensile strength of HSA samples decreased, but the elongation significantly increased.

Keywords: laser additive manufacturing, microstructure, mechanical properties, Ni-based superalloys

1 Introduction

ATI 718Plus alloy (hereafter referred as 718Plus alloy) is a precipitation strengthened Ni-based superalloys, which is widely used in aero-engine turbine disks, compressor blades and other key components due to its excellent mechanical properties and high temperature stability [1]. Damage defects will inevitably occur during the long-term service of aero-engine, and repair technology is crucial. Laser additive manufacturing (LAM) technology can accurately manufacture and repair superalloy parts. Excellent performance of additive manufacturing alloys cannot be separated from the control of raw materials, manufacturing process and heat treatment process. Alloy powder, as a raw material for laser additive manufacturing, directly affects the strength of the repaired parts. High cooling rate can lead to significant residual stress and the formation of a large amount of Laves phase inside the parts, so it is necessary to perform heat treatment to improve the microstructure of the alloy. In addition, different heat treatment processes can obtain specific microstructures and mechanical properties under different application conditions. In this work, the effects of powder forming method and heat treatment process on the microstructure and mechanical properties of LAM 718Plus alloy were investigated.

2 Experimental procedure

The alloy powders were prepared by vacuum induction melting gas atomization (VIGA) method and plasma rotation electrode preparation (PREP) method. A forged 718Plus alloy was used as a substrate and two different alloy powders were printed using a coaxial powder feeding device with different laser process parameters. The samples prepared with the optimal process parameters were subjected to different heat treatment processes, respectively. (1) direct aging (DA) treatment, 788 °C/8 h, 55 °C/h cooling to 704 °C /8 h, followed the air cooling; (2) homogenization + solid solution + aging (HSA) treatment, 1080 °C /1.5 h air cooling +980 °C /1 h air cooling +788 °C/8 h, 55 °C/h cooling to 704 °C /8 h, followed the air cooling.

3 Result and discussion

3.1 Effect of powder forming method on microstructure

The levels of porosity in the as-deposited samples fabricated by different types of powders under the same laser process parameters are shown in Figure 1. A large number of near spherical pores exist in the samples fabricated with VIGA powders, with the maximum pore diameter reaching 200 μm. According to literatures [2], there are two mechanisms for the formation of spherical pores in the LAM process: (1) the presence of hollow powder in the original powder, which results in inert gases being trapped in the original powder and remaining inside the alloy at the end of printing. (2) A mismatch in the input laser process parameters results in the formation of spherical holes due to gases trapped in the molten pool during deposition or gases above the molten pool being caught in the solidified tissue. However, the samples fabricated with PERP powders at the same laser process parameters have lower levels of porosity that can reach below 0.05%. This indicates that hollow powder leads to the formation of large-sized pores, which is independent of the laser process parameters. It should be noted that samples prepared from PREP powder are used in the following studies.

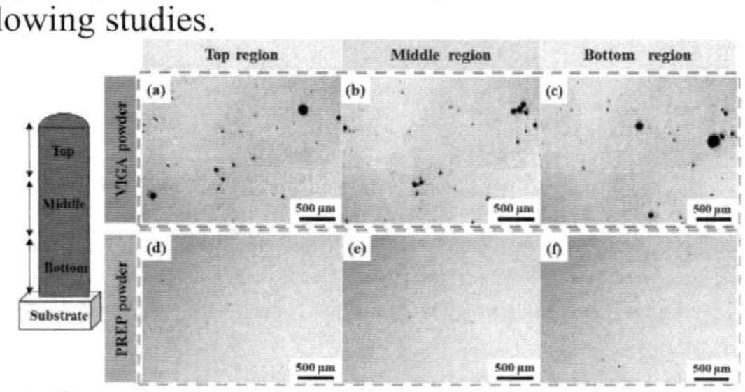

Fig. 1. Porosity levels of as-deposition samples fabricated with two different types of powders: (a, b and c) VIGA powders, (d, e and f) PREP powders.

3.2 Influence of heat treatments on the microstructure and mechanical properties

Due to the low DA treatment temperature, a large amount of Laves phase still exists in the interdendritic region, while the precipitation of short bar-like η phases at grain boundaries and a large number of γ' phases precipitated within the grains are observed, as shown in Figure 2(a)-(d). After HSA treatment, the Laves phase is completely eliminated, and only some blocky carbides remain, but the grains have undergone significant coarsening. A large number of needle-like η phases precipitated along the grain boundaries can be observed in Figure 2(f). In addition, the γ' phase is precipitated sufficiently by aging treatment.The room temperature tensile properties of the samples after different heat treatments are shown in Figure 3. After DA treatment, the strength of the samples increases significantly, but the plasticity remains basically unchanged. After HSA treatment, the strength and plasticity of the sample are comprehensively improved.

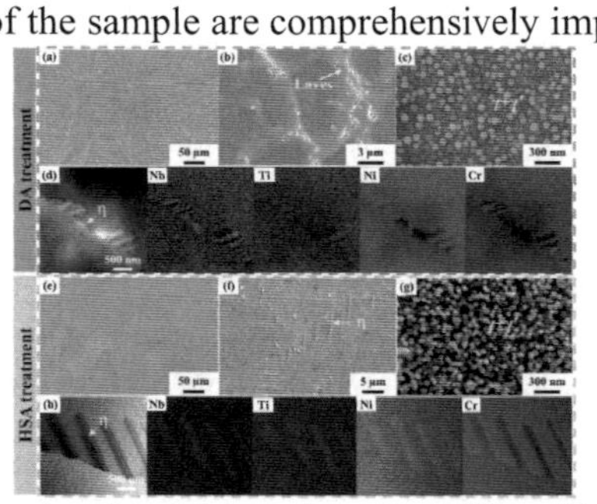

Fig. 2. Microstructure of samples after different heat treatments: (a-d) DA treatment, (e-h) HSA treatment

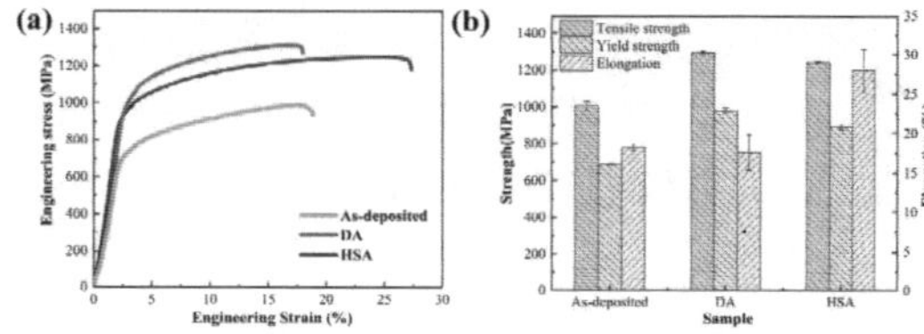

Fig. 3. Tensile properties at room temperature of the as-deposited samples and different heat treatment samples: (a) Engineering stress-strain curves, (b) Bar chart of tensile properties.

4 Conclusion

(1) Compared to PREP powder, VIGA powder has a higher percentage of hollow powder, which is the main source of porosity in LAM 718Plus alloy. The porosity of the samples fabricated with PREP powder was controlled below 0.05%.

(2) After DA treatment, the strength of the alloy increases significantly and the elongation remains essentially unchanged, which is mainly attributed to the precipitation of a large number of homogeneous nano-sized γ' phases in the interdendritic and dendritic trunk regions. After HSA treatment, the Laves phase in the alloy was completely dissolved, leading to a significant increase in elongation. However, the coarsening of the grains and the precipitation of the needle-like η phase limited the strength of the alloy.

Acknowledgments

This work was supported by the financial support from the National Science and Technology Major Project (J2019-VI-0004-0144).

References

[1] Pickering E J, Mathur H, Bhowmik A, et al. Grain-boundary precipitation in Allvac 718Plus[J]. Acta Materialia, 2012, 60(6–7): 2757–2769.

[2] Panwisawas C, Gong Y, Tang Y T, et al. Additive manufacturability of superalloys: Process-induced porosity, cooling rate and metal vapour[J]. Additive Manufacturing, 2021, 47: 102339.

The 75th World Foundry Congress
25~30 October 2024, Deyang, Sichuan, China

Investigate Tensile and Stress Rupture Behavior of Crystallographic Lamellar Microstructure in Ni Superalloy Prepared by Laser Powder Bed Fusion

Peng Wang[1,2], Wei Song[1,2], Yi Qiu[3], Jingjing Liang[2,*], Yuping Zhu[2], Yingju Li[2], Jinguo Li[2,*]
1.School of Materials Science and Engineering, University of Science and Technology of China, Shenyang, 110016, China
2. Shi-changxu Innovation Center for Advanced Materials, Institute of Metal Research, Chinese Academy of Sciences, 72 Wenhua Road, Shenyang 110016, China
3.School of Materials Science and Engineering, Beijing Institute of Technology, Beijing, China.
*Corresponding authors:E-mail address: jjliang@imr.ac.cn (J.J. Liang), jgli@imr.ac.cn (J.G. Li)

Abstract

As a unique solidification microstructure of Ni superalloys produced by laser powder bed fusion (L-PBF), the crystallographic lamellar microstructure (CLM) exhibits a strong texture with <001> and <110> orientation, but in-depth mechanical properties studies remain lacking. In this study, comparative studies are conducted to reveal the unique deformation and strengthening mechanism of the CLM under different tensile and stress rupture conditions. The mechanical properties results showed that the specimen with CLM exhibited an outstanding strength-ductility synergy and exceptional stress rupture plasticity due to the presence of high-proportioned <110> texture and unique <001>/<110> boundary. The regularly distributed <001> grains can serve as strong obstacles for slip bands and cracks propagating in <110> grains. Moreover, the <110> texture together with <001>/<110> boundary can elevate the local flow stresses and result in the activation of profuse deformation mechanisms, such as Planar slip bands, Stacking faults (SFs), and deformation twins. This work provides cross-scale insights into the deformation mechanisms and underscores the potential of CLM for developing high-temperature structural applications in Ni superalloy.

Keywords: laser powder bed fusion; crystallographic lamellar microstructure; Ni superalloys; mechanical property; scanning strategy.

Introduction

Laser powder bed fusion (L-PBF), one of the mainstream additive manufacturing technologies, possesses the unprecedented capacity to fabricate high-precision superalloy parts (resembling multi-cavity hollow structure blades) attributed to its ultrafine spot size and digit control technique [1, 2]. In the L-PBF process, the solidification microstructure is governed by the heat input and melt pool geometry, which can be manipulated by various printing parameters, such as laser power, scanning speed, laser spot size, scanning strategy, etc [3].

In 2021 [4], a unique crystallographic lamellar microstructure (CLM) in Inconel 718 alloys was first designed through variation laser power and scanning speed in the L-PBF process. This CLM has a strong texture with a regular alternant <110>-oriented layer and <100>-oriented layer. In the original, the CLM exhibited exceptional strength-ductility combination at room temperature compared to single-crystal and polycrystalline-like microstructure [4]. However, the CLM reports are scarce, and some particular areas need to be clarified: First, it remains to be seen whether such a CLM with a strong texture is suitable for high-temperature conditions. At present, there are relatively fewer reports of L-PBF fabricated superalloy that apply service temperatures up to 1000℃. Second, the stress rupture behavior of CLM at different loading conditions has not been reported. To investigate the stress rupture feature of ZGH451 with CLM is crucial, especially under prolonged exposure to high temperature and stress.

Therefore, this study aims to comprehensively understand the observed mechanical differences and microstructural evolution from L-PBFed CLM superalloy. The tensile and stress rupture tests with different conditions were conducted to evaluate applicability in different service environments. A suite of complementary characterization techniques is carried out to reveal the strengthening mechanism and deformation behavior. This work provides insights into the deformation mechanisms of CLM superalloy for high-temperature applications.

Experimental procedure

The L-PBF experiments were conducted in a sealed working chamber filled with argon gas to keep the oxygen level below 10 ppm. A Concept Laser M2 machine manufactured all specimens (25 × 25 × 60 mm^3) with the same powder feedstock, scan strategy, and substrate preheating (100 ℃) to minimize the effects of other factors.

The as-built microstructure and fracture morphologies were investigated by a suite of complementary characterization techniques including optical microscopy (OM), scanning electron microscopy (SEM), electron backscatter diffraction (EBSD) and transmission electron microscopy (TEM) et.al,

Uniaxial tensile samples were tested in the universal testing machine with a constant strain rate of 10^{-3} s^{-1} at room temperature and 10^{-2} s^{-1} at 1000℃. The stress rupture test was conducted at a constant condition of 760 °C/780 MPa and 980 °C/260 MPa. The mechanical properties tests were conducted three times under each condition.

Result and discussion

The CLM specimens exhibit a higher yield strength (YS) and uniform elongation (UE) compared to non-CLM at both different tensile conditions, indicating synergistically enhanced strength-ductility. Meanwhile, the non-CLM samples exhibits the shorter stress rupture life, whereas the CLM sample demonstrates the longer stress rupture life. Regarding stress rupture plasticity, the non-CLM exhibit relatively lower plasticity compared to CLM alloy. Upon comparing of the life and plasticity, it is evident that the CLM sample demonstrates superior stress rupture life while maintaining a relatively high level of plasticity.

In the sub-grain scale, both CLM and non-CLM alloys exhibited typical cellular structures with dense dislocation accumulation at their boundaries, and relatively few were distributed at cellular interiors. The CLM specimens possess more multiple ultrafine planar slip bands than non-CLM samples. These planar slip bands and dislocation cells strengthen the dislocation pinning and enhance the strain hardening during plasticity deformation. The <110> texture is relatively easy to initiate planar slip band associated with rearrangement of dislocations due to its low yield stress and pronounced back stress[5]. The proportion of the <110> and <001> texture layers is approximately 3.6:1 in CLM specimens. Hence, large amounts of planar slip bands are generated within <110> texture layer and increased global plasticity. Simultaneously, these dispersed slip bands facilitate homogeneous plastic deformation and avoid stress concentration at the local region. In addition, the sufficient dislocation pile-ups and theoretically higher effective stacking faults energy in CLM can be enough to slide partials a long distance [6]. Hence, the PLBs and SFs behavior in the CLM specimen is expected during the high strain stage.

Conclusion

The texture-related strengthening plays a dominant role in CLM alloys. The asynchronous deformation of <001> and <110> oriented layer induced strain gradient and produced HDI strengthening, which enhanced yield stress and strain-hardening ability. In addition, the abundant deformation mechanism can be activated in CLM alloys under room temperature tensile, including Planar slip bands (PLBs), Stacking faults (SFs), and deformation twins (DTs).

References

[1] O. Andreau, I. Koutiri, P. Peyre, J.-D. Penot, N. Saintier, E. Pessard, T. De Terris, C. Dupuy, T. Baudin, Texture control of 316L parts by modulation of the melt pool morphology in selective laser melting, Journal of Materials Processing Technology,2019,264:21-31. https://doi.org/10.1016/j.jmatprotec.2018.08.049.

[2] W. Song, J. Yang, J. Liang, N. Lu, Y. Zhou, X. Sun, J. Li, A new approach to design advanced superalloys for additive manufacturing, Additive Manufacturing 84. (2024). https://doi.org/10.1016/j.addma.2024.104098

[3] M.-Y. Jiang, X.-Y. Meng, J. Zhou, Z.-Y. Wei, H.-F. Zhang, P. Shen, Design and manufacture of near-net-shape metal-ceramic composites with gradient-layered structure and optimized damage tolerance, Additive Manufacturing ,2024,84. https://doi.org/10.1016/j.addma.2024,104112.

[4] O. Gokcekaya, T. Ishimoto, S. Hibino, J. Yasutomi, T. Narushima, T. Nakano, Unique crystallographic texture formation in Inconel 718 by laser powder bed fusion and its effect on mechanical anisotropy, Acta Materialia,2021, 212. https://doi.org/10.1016/j.actamat.2021.116876.

[5] Z. Sun, X. Tan, S.B. Tor, C.K. Chua, Simultaneously enhanced strength and ductility for 3D-printed stainless steel 316L by selective laser melting, NPG Asia Materials , 2018,10(4):127-136. https://doi.org/10.1038/s41427-018-0018-5

[6] Y. Liu, J. Ren, S. Guan, C. Li, Y. Zhang, S. Muskeri, Z. Liu, D. Yu, Y. Chen, K. An, Y. Cao, W. Liu, Y. Zhu, W. Chen, S. Mukherjee, T. Zhu, W. Chen, Microstructure and mechanical behavior of additively manufactured CoCrFeMnNi high-entropy alloys: Laser directed energy deposition versus powder bed fusion, Acta Materialia, ,2023, 250. https://doi.org/10.1016/j.actamat.2023.118884

Effect of Temperature on High-Cycle Fatigue Behavior of A W-rich Nickel-Based Superalloy

Jun Xie*, Maokai Chen, Wu Hao, Li Qi, Fengjiang Zhang, Hongyu Chai, Yuejiao Ma, Jinjiang Yu, Jinguo Li
Institute of Metal Research, Chinese Academy of Sciences, China
*Corresponding address: Jun Xie, e-mail: JunXie@imr.ac.cn

Abstract: The influence of temperature on the high-cycle fatigue behavior of a W-rich nickel-based superalloy has been comprehensively investigated. Results indicate that dislocations predominantly slip within the matrix channels forming dislocation networks and deformation dislocations shear the γ' phase. With an increase in the applied temperature, the dislocation density within the matrix channels escalates, and dislocation pairs shearing the γ' phase were evident at 1000°C. Additionally, fatigue initiation primarily occurs at micropores located on the alloy's surface and subsurface. This leads to the formation of slip lines with varying orientations in the crack propagation area. These slip lines tend to converge at locations such as eutectics, carbides, and grain boundaries, which facilitates crack initiation. As fatigue advances, cracks propagate through the interdendritic region, culminating in cleavage fractures within the instantaneous fracture zone.

Keywords: W-rich nickel-based superalloy; High-cycle fatigue; Deformation mechanism; Fracture characteristics; Microstructure

1 Introduction

Nickel-based superalloys are pivotal in the fabrication of high-temperature components, notably in advanced aero-engines and gas turbines [1]. In challenging high-temperature environments, the interplay between temperature and oscillating loads significantly contributes to high-cycle fatigue damage, leading to the deterioration of these materials [2]. Recent studies indicate that the high-cycle fatigue properties of alloys are intrinsically linked to their chemical composition, microstructural characteristics, and metallurgical defects [3]. Typically, fatigue cracks in nickel-based superalloys initiate and propagate predominantly on the alloy surface, subsurface regions, and at carbides [4]. The fatigue fracture is conventionally characterized by a fatigue source, a fatigue zone, and a rapid failure zone [5]. Owing to the unforeseen nature of fatigue fractures that can lead to severe mishaps, studying the fatigue behavior of blade materials holds profound implications for alloy design and life expectancy predictions [6]. The alloy under investigation is a high-tungsten nickel-based superalloy, distinguished by its superior thermal corrosion resistance and thermal fatigue resilience. This study focuses on evaluating the impact of temperature on the high-cycle fatigue behavior of the high-tungsten Ni-based alloy.

2 Experimental procedure

Using a 10 kg vacuum induction melting furnace, a W-rich master ingot was remelted and subsequently cast into an equiaxed crystal test bar. The alloy's nominal composition (by mass fraction) was as follows: C 0.13%, Cr 4.90%, Co 6.82%, Nb 2.06%, Al 5.75%, W 16.3%, Ti 1.00%, Hf 1.00%, with the remainder being Ni. The specimens underwent high-frequency axial fatigue testing under controlled loads using the PLG-100C fatigue testing machine. The testing parameters included experimental temperatures of 800 ℃ , 900 ℃ and 1000 ℃ , an applied frequency of approximately 140 Hz, a stress ratio (R) of -1, and a sinusoidal loading wave.

3 Result and discussion

The relationship between high-cycle fatigue stress and fatigue life for the alloy under varying temperature conditions is illustrated in Fig. 1. The fatigue limits of the alloy at 800°C, 900°C, and 1000°C were calculated to be 196 MPa, 207 MPa, and 186 MPa, respectively.

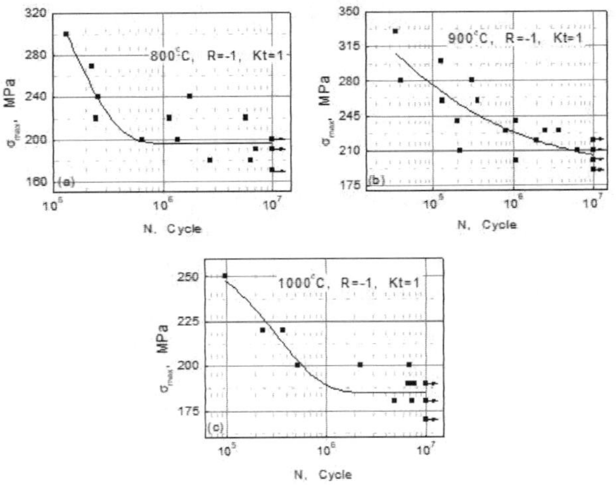

Fig. 1 High-cycle fatigue S-N curve of the alloy at different temperature: (a)800 °C, (b)900 °C and (c)1000 °C

The SEM morphology of high-cycle fatigue-fractured specimens at different temperature is presented in Fig. 2. The fracture of the alloy is relatively flat, and there is loose aggregation at the fatigue source. The massive M_6C carbide remain with the micro-cracks at the fracture. On the other hand, more slip traces remain in the eutectic and the cracks extend along the eutectic interface.

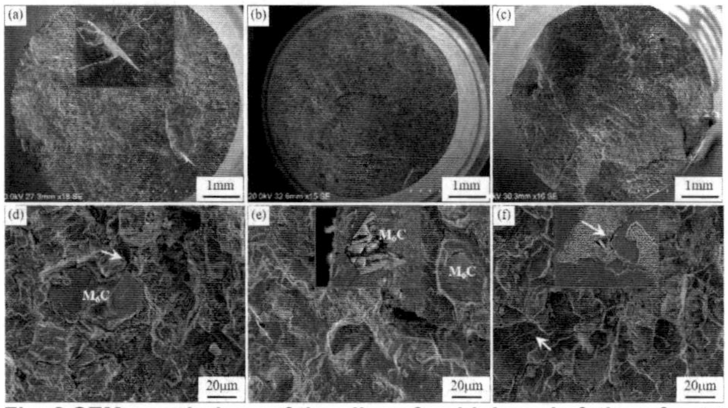

Fig. 2 SEM morphology of the alloy after high-cycle fatigue fracture at different temperature: (a)(d)800 °C, (b)(e)900 °C and (c)(f)1000 °C

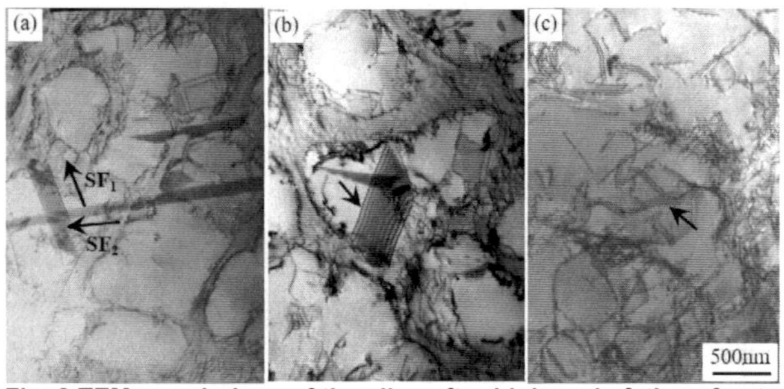

Fig. 3 TEM morphology of the alloy after high-cycle fatigue fracture at different temperature: (a)800 °C, (b)900 °C and (c)1000 °C

The TEM morphology of high-cycle fatigue-fractured specimens at different temperature is presented in Fig. 3. The results indicate that deformation dislocations predominantly slipped within the γ matrix channels forming dislocation networks or tangles and shear the γ' phase forming stacking faults. At a temperature elevation to 1000°C, the deformation mechanism transitioned from stacking fault formation to dislocation pairs.

4 Conclusion

1. At temperatures of 800°C, 900°C, and 1000°C, and a stress ratio of R=-1, the high-cycle fatigue limit of the W-rich Ni-based alloy was measured at 196 MPa, 207 MPa, and 186 MPa, respectively.

2. Deformation characteristics during high-cycle fatigue revealed that, deformation dislocations predominantly slipped within the γ matrix channels forming dislocation networks or tangles and shear the γ' phase forming stacking faults. At a temperature elevation to 1000°C, the deformation mechanism transitioned from stacking fault formation to the generation of dislocation pairs.

3. The principal fatigue source originated from micro-pores located on and beneath the alloy's surface. Slip lines of diverse orientations were identified within the crack propagation zone, converging notably at the eutectic, carbide, and grain boundaries.

Acknowledgments

This work was financially supported by National Key Research and Development Program of China (2023YFB3712003); National Science and Technology Major Project (J2019-VI-0018-0133); Youth Innovation Promotion Association Project, Chinese Academy of Sciences (2020)

References

[1] Fu C, Chen Y D, Li L F, et al. Evaluation of service conditions of high pressure turbine blades made of DS Ni-base superalloy by artificial neural networks[J]. Mater. Today Commun., 2020, 22: 100838.

[2] Tanaka Y, Okazaki S, Ogawa Y, et al. Inability of precipitation-hardening to improve the fatigue limit of Ni-based superalloy 718 through a perspective of shear-mode cracking threshold[J]. Mater. Lett., 2020, 277: 128377.

[3] Li X L, Zhang Y C, Li W, et al. High-cycle and very-high-cycle fatigue behavior and life prediction of Ni-based superalloy at elevated temperature[J]. Fatigue Fract. Eng. M., 2021: 1-17.

[4] Xie J, Yu J J, Sun X F, et al. High-cycle fatigue behaviors of K416B Ni-based casting superalloy at 700°C[J]. Acta Metall. Sin., 2016, 52(3): 257-263.

[5] Evans W J, Screech J E and Wliilams S J. Thermo-Mechanical Fatigue and Fracture of INCO718[J]. Int. J. Fatigue, 2008, 30(2): 257-267.

[6] Suzuki S and Sakaguchi M. Fatigue crack retardation associate d with creep deformation induce d by a tension hold in a single crystal Ni-base superalloy[J]. Scripta Mater., 2020, 178: 346-350.

Study on the Counter Gravity Adjusted Pressure of Large and Complex Thin-Wall Ni-Based Superalloy Components

Baode Sun, Anping Dong*, Dafan Du*, Lianghua Xiong, Lin He, Dashan Sui
Shanghai Key Lab of Advanced High-temperature Materials and Precision Forming and State Key Lab of Metal Matrix Composites, School of Materials Science and Engineering, Shanghai Jiao Tong University, Shanghai 200240, China
* Corresponding author: apdong@sjtu.edu.cn; dafand@sjtu.edu.cn.

Abstract: Counter-gravity casting (CGC) is a widely adopted material processing technique in metals due to its notable benefits, including enhanced filling behavior, reduced defect occurrence, and elevated mechanical properties. It plays a pivotal role in fabricating intricate, high-quality components. Casting experiments and multi-scale numerical simulations were conducted to examine the microstructure characteristics of K4169 nickel-based superalloy casting with varying cross-sections during the investment casting process. First, microstructure analysis was conducted on the casting using scanning electron microscopy (SEM) and electron backscatter diffraction (EBSD). Moreover, the Cellular Automaton Finite Element method was employed to study the pressure on the dendrite growth and filling behavior in the thin-wall structure. The experimental results revealed that the samples fabricated by counter gravity adjusted pressure investment casting perform finer microstructure and superior mechanical properties. Recently, the CGC was applied to producing large, thin-wall Ni-based superalloy components. Compared to traditional gravity casting, the filling speed, holding pressure, and solidification rate during CGC can be controlled. With a high filling pressure, the defects in the solidification microstructure decreased, and the strength and ductility improved. The large-size CGC facility for large and complex thin-wall Ni-based superalloy components was designed and fabricated in our group; moreover, some representative productions with high quality were successfully made by the large-size CGC facility.

Keywords: Conter gravity adjusted pressure investment casting, K4169 alloys, Microstructure, mechanical properties

1 Introduction

Counter-gravity Nickel-based superalloys have a good combination of performance at high temperatures [1-2], making them widely used in hot-end components, such as aero-engines and gas turbines [3]. Most of these castings are Al, and Mg alloys, which have lower densities, casting temperatures, and narrower solidification ranges than superalloys. Due to these differences in physical properties, the fabrication of superalloy castings is more challenging because of the back pressure brought by the Laplace effect and higher requirements for equipment caused by the high casting temperature. The study studied the filling behavior, microstructure and mechanical properties of K4169 fabricated by counter gravity adjusted pressure investment casting (CG-APC). The results indicated that the K4169 alloy fabricated by CGAPC exhibited superior mechanical properties compared to the K4169 alloy fabricated by gravity casting. These results could be attributed to the forced convection imposed on the dendrites during filling and solidification processes, which was also numerically studied. The purpose of this investigation is two-fold: (1) to investigate how the forces convection affects the microstructure and mechanical properties; (2) to gain a deeper understanding of the dynamics of the melt pool, which can lead to improvements in CG-APC fabricated Ni-based superalloys in industry.

2 Experimental procedure

The raw material used in the casting experiments was the as-cast K4169 nickel-based superalloy. The investment casting experiments were conducted. The mold shell was preheated to 850 ℃ and held for 3hours in a resistance furnace. 5kg of K4169 alloy was melted in a vacuum induction furnace. The preheated mold shell was placed in a vacuum chamber, and the pouring temperature was approximately 1500 ℃. The specimens were machined, ground, and corroded using a ratio of $CuSO_2$: HCl: H_2SO_4: = 30: 100: 7 and a corrosion time of 10 seconds. The microstructures of the specimens were analyzed using optical microscopy and electron microscopy.

3 Result and discussion

Numerical Simulation

In order to study the pressure on the dendritic growth during the thin-wall structure, the simulation on different pressure on the convection and dendritic growth was performed, as shown in Fig. 1. The results show that the dendrite growth from crucible wall to the center due to the temperature gradient. Meanwhile, with the melt flow up which may break the dendrite and even remelt it. For a 5 kPa/s pressure, the dendrite tips lap joint together at 0.8 s. For a higher 7 kPa/s pressure, the dendrite tips lap joint together at 1.1 s. The simulation results demonstrated that the dendrite tips lap joint can be restrained for a higher-pressure during counter-gravity casting, which can decrease the porosity at the end of solidification period.

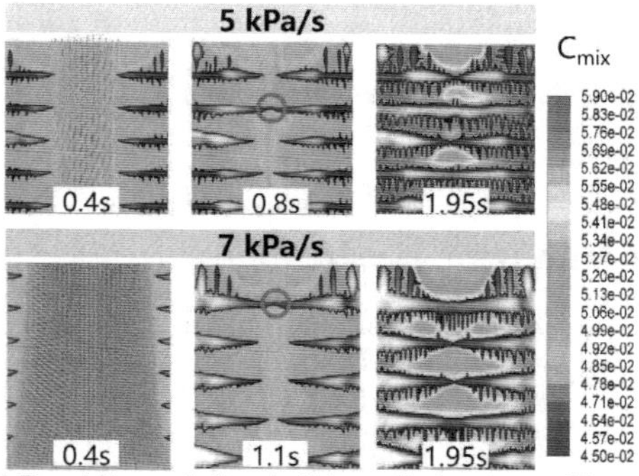

Fig. 1 Dendrite evolution in thin-wall structure with various pressure during counter-gravity casting.

Moreover, the simulation to predict the solid phase fraction and defects was performed by ProCAST, as shown in Fig. 2. Comparing two methods, the final solidification area of the casting is located near the upper inner wall surface connected to the main runner. Due to insufficient shrinkage, there are certain defects in this area. However, by using the counter gravity adjusted pressure investment casting, the shrinkage defects in the thin-walled area have been significantly decreased. Due to the effect of pressure, the shrinkage defects in the lower thin-walled area have been significantly eliminated, and the effect of pressure closer to the gate is more apparent.

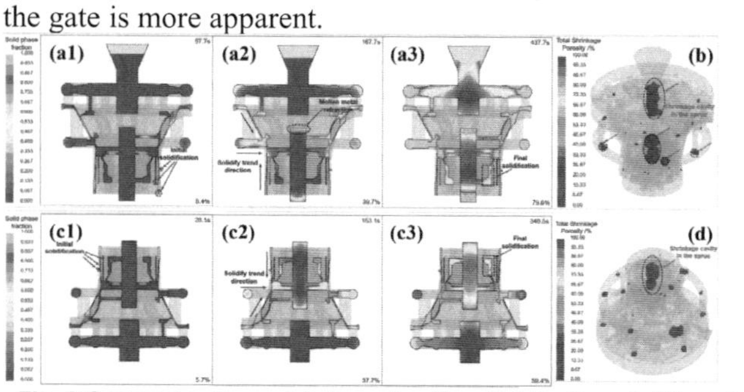

Fig. 2 Comparing the solid phase fraction distribution at different time points (a and c) and defect distribution (b and d) during traditional gravity casting (a-b) and counter gravity casting (c-d) by numerical simulation.

Microstructure and mechanical property

Moreover, the microstructure of K4169 alloys for two casting methods was analyzed as shown in Fig. 3. It is obviously that the grain size is smaller by using counter gravity adjusted pressure investment casting. This may be attributed to the remelt of dendrite due to the melt flow during forced convection, the fragments would a new nucleation and refine the microstructure. For this reason, it is easy to associate with that the samples fabricated by counter gravity adjusted pressure investment casting can perform higher mechanical property. Compared with the gravity process, the room temperature tensile yield strength of the alloy has increased from 487.5MPa to 533.4MPa, and the ultimate tensile strength has increased from 678.7MPa to 858.7MPa. The plasticity has decreased to a certain extent, from 26.1% to 13.3%.

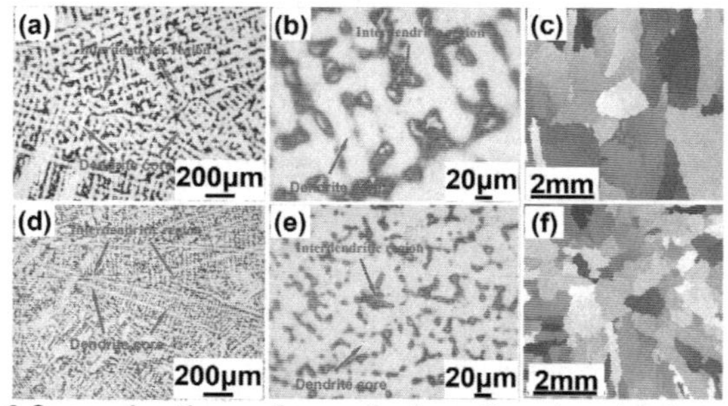

Fig. 3 Comparing the casting methods on the microstructure and grain morphology of as cast K4169 alloys: (a)-(c) Gravity investment casting; (d)-(f) Counter gravity adjusted pressure investment casting.

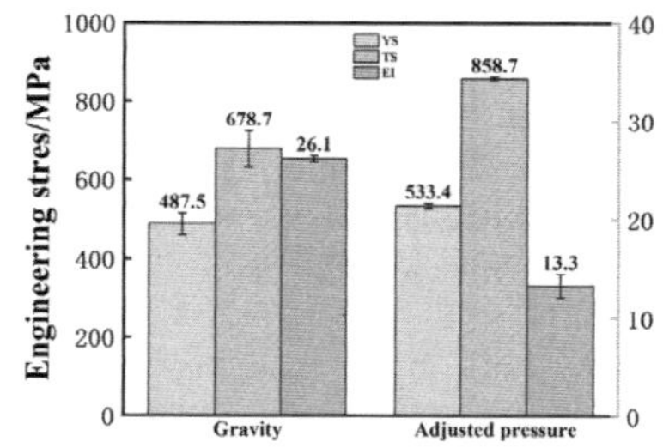

Fig.4 Room temperature tensile properties of as-cast K4169 alloy under different casting processes

4 Conclusions

The filling solidification process of the complex thin-walled casting of K4169 alloy was analyzed by numerical simulation for gravity precision casting and pressure regulating precision casting. Present results indicate that pressure-controlled precision casting has the characteristics of smoother filling speed, better cooling effect, and fewer pore defects for complex thin-walled parts. In addition, the use of pressure regulating precision casting technology can reduce the secondary dendrite spacing of the alloy. The average grain size and improved element segregation of the alloy result in excellent tensile properties at room temperature.

References

[1] JIANG Ju-fu, XIAO Guan-fei, WANG Ying, et al. High temperature deformation behavior and microstructure evolution of wrought nickel-based superalloy GH4037 in solid and semi-solid states I. Transactions of Nonferrous Metals Society of China, 2020, 30: 710-726.

[2] HAO Xin, LIU Guohuai, WANG Ye, et al. Optimization of investment casting process for K477 superalloy aero-engine turbine nozzle by simulation and experiment [J]. China Foundry, 2022,19(4): 351-358.

[3] WHITMORE L, AHMADIM R, GUETAZ L, et al. The microstructure of heat-treated nickel-based superalloy 718Plus. Materials Science and Engineering A 2014,610: 39-45.

The Formation Mechanism of Abnormal Grains of Superalloys Used for the Turbine Blade

Shijie Sun, Naicheng Sheng*, Guichen Hou, Jinguo Li, Yizhou Zhou, Xiaofeng Sun
Institute of Metal Research, Chinese Academy of Sciences, 72 Wenhua Road, Shenyang 110016, China
*Corresponding address: e-mail: ncsheng@imr.ac.cn

Abstract： The change of blade size has a significant impact on the solidification microstructures, resulting in the fluctuation mechanical properties of the turbine blade, which would affect its service performance. It is found that the abnormal grains with special grain boundaries were easy to occur in the airfoil of the turbine blade made of K465 alloy, while no abnormal grain was found in the root of the blade. The size of the abnormal grain is in micron scale, which is much smaller than the equiaxed grain with millimeter scale in cast K465 superalloy. The stability of stress rupture properties of the alloys with the abnormal grains decreased. The samples with different thickness were designed to simulate the microstructures at different positions of the turbine blade. The nucleation and growth mechanism of abnormal grains with special grain boundaries were revealed. In addition, the optimization the alloy composition was conducted to reduce the tendency of abnormal grains in cast superalloys and the adverse influence of the size effect on the microstructural stability and mechanical properties, which could provide ideas for improving the stability of mechanical properties of turbine blades.

Keywords: nickel based superalloy, size effect, residual stress, abnormal grains

1 Introduction

In recent years, with the continuous increase of thrust to weight ratio in aeroengines, the service temperature of turbine blades has become higher and higher. The internal cooling structure of turbine blades would be sophisticated, which results in the significant differences in wall thickness at different positions of a turbine blade [1]. Generally, the changes in blade cross-sectional dimensions would affect the heat and cooling rate during the solidification process, and the key parameters have a significant impact on the solidification structure. In addition, during the alloy solidification process, the residual stress in the thin-walled site is relatively high, which can easily lead to significant plastic strain [2]. Some researchers have found that the microstructure with high residual stress are prone to recrystallization during the heat treatment process, resulting in the formation of abnormal grains [3]. While the abnormal grains would cause the decrease of the mechanical properties of blade alloys [4]. It is found that the abnormal grains at the thin-walled part of a K465 alloy blade, which decrease the safety of the blade during service. This study focuses on the formation mechanism of abnormal grain in the K465 alloy. Specimens with different thicknesses were designed to simulate different positions of the blade, revealing the fundamental mechanism of the formation of abnormal grain and providing guidance for the optimization of the alloy.

2 Experimental procedure

The samples with different thicknesses (3mm, 5mm, 10mm, and 20mm) were designed to simulate different positions of the blades, with a width of 15mm and a height of 80mm. The K465 master alloy was melted using the VIM-F25 vacuum induction furnace, and the tested alloy is as follow in weight percent: Al 5.41, Co 10.0, Cr 8.6, Mo 2.0, Nb 1.1, Ti 2.5, W 10.3, Zr, 0.03, B 0.024, C 0.17, and the balance of nickel. The liquid alloy was cast into the alumina-silicate mold buried in the sand box to obtain the thin-walled specimens. The specimens were taken out until the mold and sand box were cooled to room temperature. The above specimen preparation method was similar as the method of the cast blade.

3 Result and discussion

Abnormal grains

Fig. 1a shows the typical microstructure of a K465 alloy specimen with a thickness of 20 mm. There are coarse equiaxed grains in the alloy structure, with a grain size of about a few millimeters. The microstructure includes γ', γ matrix, eutectic and carbide; In addition, the abnormal grain was observed in the alloy, as indicated by the red arrow in Fig. 1a. As the thickness of the specimens decreases, the number of abnormal grains increases (Fig. 1b, c, and d). In addition, abnormal grains are more likely to appear near carbides between dendrites (as indicated by the blue arrow in Fig. 1a). The size of the abnormal grain is about tens of micrometers, much smaller than the equiaxed grains, and contain a large amount of precipitated phases. The abnormal grain is high-angle grain boundaries (HAGBs), as indicated by the inverse pole plot (IPF) and corresponding misorientation.

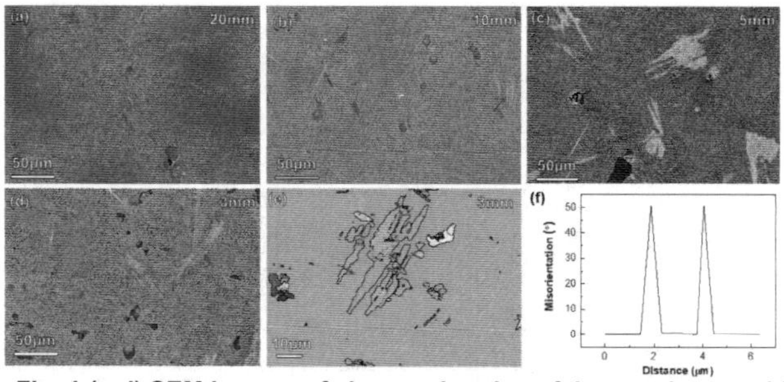

Fig. 1 (a-d) SEM images of abnormal grains of the specimens with different thickness; (e) IPF images of the 3 mm-specimen; (f) corresponding misorientation of the arrow in (e)

Residual stress

The GND values of regions containing several carbides in samples of different thicknesses were calculated, and the results are shown in Fig. 2. It can be clearly seen that the GND density in the sample increases with the decrease of specimen thickness, indicating that the residual stress in the specimen increases with the decrease of sample thickness. The thermal stress that occurs in the casting during alloy solidification can lead to residual stress, and numerical simulation of thermal stress during solidification is an important method for predicting residual stress in castings. This article conducted residual stress simulations on samples of different thicknesses using ProCAST commercial software, and the results are shown in Fig. 2b. It can be seen that residual stress increases with the decrease of sample thickness. The residual stress of the 20mm thick specimen was about 50 MPa, and as the thickness decreases to 3mm, the stress increased to about 250 MPa. The trend of increasing macroscopic residual stress of the sample with decreasing sample thickness is consistent with the trend of increasing GND density with decreasing sample thickness.

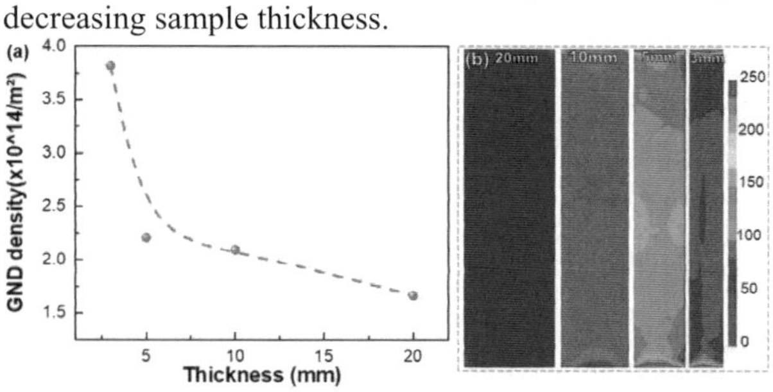

Fig.2 GND density distribution of in the specimens with different thickness; (b) the residual stress distribution of the specimens with different thickness

Size effect induced abnormal grains

During solidification and subsequent cooling process, residual plastic strain was induced in the microstructures of the K465 superalloy, which was due to the different thermal expansion coefficients of the metal, mold and core. In general, the non-deforming particle (such as a carbide) could be as a site for the development of a recrystallized grain due to that the region surrounding the particle contains a high dislocation density [5]. In this work, the carbides with high dislocation density in the vicinity could serve as the nucleation site, which induce the onset of the recrystallized grain. Hence, it is found that abnormal grains exist around block carbides in the interdendritic region. In addition, the macroscopic residual stress increases with the decreases of the specimen thickness. Therefore, the quantity of the feather-like grains increases with the decrease of the specimen thickness. Moreover, it is reported that the thin-walled sites in single crystal superalloys have an increased propensity for recrystallization [2]. In addition, due to the thin-walled samples prepared by the sand embedding process, the sand box has a high temperature after the solidification, and the sand box with a high temperature in the subsequent cooling process would have a heat treatment effect on the samples. It is speculated that recrystallization would occur near carbides in thin-walled samples due to the heat treatment effect of high-temperature sandbox.

4 Conclusion

The abnormal grains with HAGBs were detected in the thin-walled specimens. The experimental results show the cooling rate and the residual stress increases with the decrease of the specimen thickness. The appearance of feather-like recrystallized grains is due to the residual stress generated during the solidification process and the subsequent heat treatment resulted from the sand box with high temperature.

5 Acknowledgments

This work was financially supported by National Natural Science Foundation of China (No. 52303394).

References

[1] G. Liang. Thin turbine rotor blade with sinusoidal flow cooling channels. United States: 2010.

[2] Z. Li, J. Xiong, Q. Xu, et al. Deformation and recrystallization of single crystal nickel-based superalloys during investment casting[J]. J. Mater. Process. Tech. 2015, 217: 1-12.

[3] Y.T. Tang, N. D'Souza, B. Roebuck, et al. Ultra-high temperature deformation in a single crystal superalloy: Mesoscale process simulation and micromechanisms[J]. Acta Mater. 2021, 203: 116468.

[4] L. Wang, G. Xie, J. Zhang, et al. On the role of carbides during the recrystallization of a directionally solidified nickel-base superalloy[J]. Scr. Mater., 2006, 55(5): 457-460.

[5] L.H. Rettberg, Localized recrystallization during creep in nickel- based superalloys GTD444 and René N5[J].Acta Mater,2014,73: 287-297.

Microstructural and Mechanical Properties of Extremely Long-Term Thermal Exposed Superalloy Mar-M247After Rejuvenation Heat Treatment

Song Wu[1,2],Xiufang Gong[1,2], Zhenhuan Gao[1,2], Nieliping[1,2], Changshuai Wang[1,2], Lexin Yang[1,2], Chao Luo[1,2]
1. State key Laboratory of Clean and Efficient Turbomachniery Power Equipment, Deyang 618000, Sichuan, China;
2. Donfang Electric Corporation Dongfang Turbine Co., Ltd. Deyang 618000, China)
*Corresponding address: e-mail:Song Wu, Tel/ 086+18782014700, song.wu2022@outlook.com

Abstract: Superalloy Mar-M247 is a polycrystalline nickel-based high-temperature alloy utilized for the hot section blades of heavy-duty gas turbines. However, the degradation of the microstructural and mechanical properties of Mar-M247 occurred after ultra-long-term thermal exposure. In this paper, rejuvenation heat treatments were first performed on alloy samples after 16,000 and 24,000 equivalate operating hour thermal exposures. The differences in microstructural and mechanical properties of the above-rejuvenated alloys were investigated.

Meanwhile, different heat treatment procedures and their re-service stability have been further discussed. The results have shown that when the solution temperature is low, the rejuvenated alloy has more eutectic/carbide, and the size of the γ' phase gets larger; but after the solution temperature is high, the γ' film gets wider, and the hardness value is higher. The above conclusions indicate that the proper choice of rejuvenation heat treatment can restore the damaged alloys. The rejuvenated alloy has the same mechanical properties as the original alloy; meanwhile, it also has a higher stability.

Keywords: Long-term thermal exposure; rejuvenation heat treatment; gas turbine blade; microstructural and mechanical properties; stability.

1 Introduction

Gas turbine blades operate in high-temperature and high-stress environments and withstand temperatures reaching 800℃-1000℃ for a total service life of up to 50,000 hours. It is reported that the proper examining and repairing interval of blades is 16000 and 24000 EOH due to different maintenance plans. Between each repair interval, the blade's metallurgical organization and performance degradation are inevitably caused by the long-term thermal exposure of the blade. Microstructure degradation includes the strengthened phase getting γ' dissolved, coarsening growing, and volume fraction reducing. The microstructure changes ultimately seriously affect the alloy's high-temperature mechanical properties, significantly shortening the operation life of the gas turbine.

Steven and Flewitt [1] have investigated the effect of heat treatments on the morphology of superalloy IN738. The distribution and area fraction of γ' phases have been analyzed and related to changes in the hardness.

Yao et al. [2] have discussed the microstructural degradation of Ni-superalloy CMSX-4 material after high temperatures and a series of rejuvenation heat treatments. The observed microstructures have been investigated for the effectiveness of the heat treatment procedures and their effect on the subsequent mechanical performance. The tensile test results showed that, when the degraded CMSX-4 material was restored using the RHT, these were subject to more rapid decline than those of the as-received alloy during further subsequent degradation treatment. However, the process parameters affecting microstructure and mechanical properties during RHT have not been discussed in detail.

This paper aims to provide a greater understanding of the suitability of rejuvenation heat treatments and their role in extending component life in power plant applications. It focuses on the recovery of microstructure and mechanical properties after ultra-long service damage and discusses in detail the effect of solid solution temperature in the rejuvenation heat treatment procedure on the stability of microstructure and properties after re-service.

2 Experimental procedure

The material used in this work was a conventional polycrystalline nickel-based superalloy Mar-M247. Experimental samples had a diameter of 5 mm and a thickness of 4 mm and were machined by EDM from an unserved turbine blade. These samples were first exposed to high temperature and long-term thermal exposure and then restored through rejuvenation heat treatment. The detailed experimental conditions are listed in Table 1.

Table 1. Damaged alloys and their RHT procedures.

Thermal Exposure Temperature/℃-Time/h	RHT No.	Solid Solution Treatment Condition
1000℃/16000h-24000h	RHT-1	1190°C/4h/AC
	RHT-2	1220°C/4h/AC
	RHT-3	1250°C/4h/AC

3 Result and discussion

To investigate the effect of solid solution temperature on the microstructure of thermally exposed alloys, rejuvenated alloys with different thermal exposure periods from 16000h to 24000h are compared and analyzed from the aspects of precipitation-strengthened phases γ' and mechanical properties in Fig. 1-Fig. 2.

Fig.1 presented the area(or volume) fractions of precipitation-strengthened phases γ' . It indicated that the γ' phases of thermally exposed alloys for both 16000h and 24000h can be restored into uniformly distributed γ' phases after RHT. However, with a higher solid solution temperature, the area fractions of γ' phases in the rejuvenated alloys in Fig. 1(c) and Fig. 1(f) tend to have a higher fraction value and hardness value (in Fig. 2) as compared to those in Fig. 1(b) and Fig. 1(e).

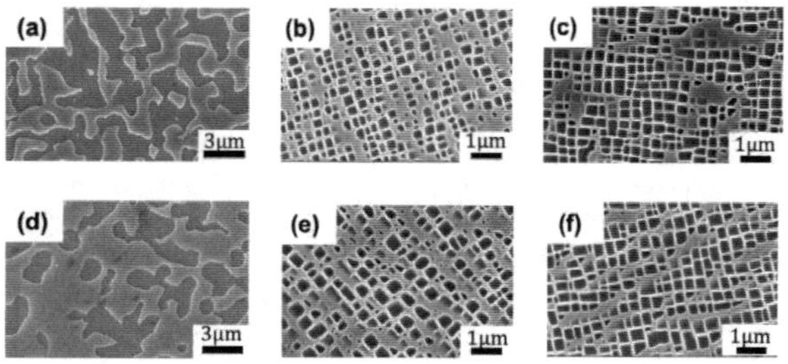

Fig.1 Morphology of γ' phases for thermal exposed alloys after (a)-(c) 16000h and (d)-(f) 24000h. (a), (d) thermal exposed alloys; (b), (e) rejuvenated alloys by RHT-1(1190 ˚C); (c), (f) rejuvenated alloys by RHT-3(1250 ˚C).

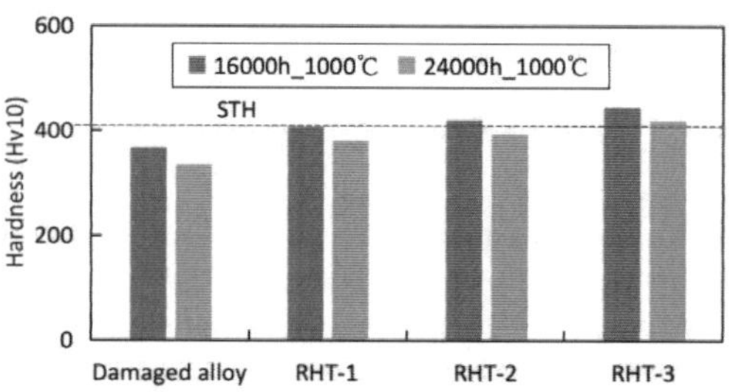

Fig. 2 Mechanical properties comparison between different rejuvenation heat treatments.

Fig. 3 described the coarsening behavior of γ' phases during the short-term re-aging at a temperature of 1000 ℃ using the coarsening equation of Lifshitz-Slyozov-Wagner(LSW) [3]. It is indicated that when the solid solution temperature of RHT is 1250℃, the coarsening rate of γ' phase rejuvenated by high solid solution temperature (the slope value of the coarsening curves) tends to be lower as compared to those of the other RHT procedures, and the SHT alloy.

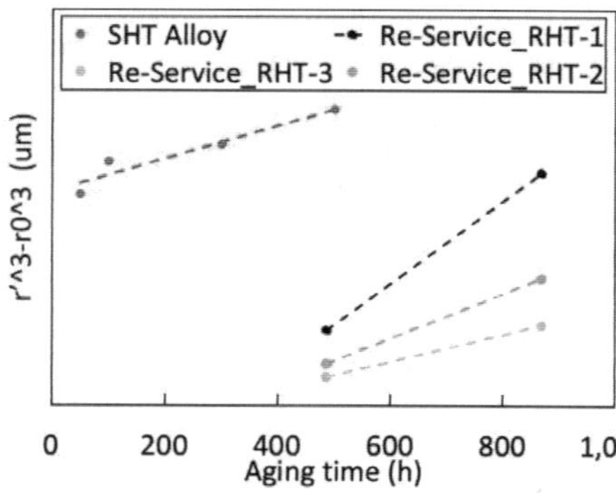

Fig. 3 The coarsening behavior of the γ' phase for the rejuveated alloys at the aging temperature of 1000℃℃

4 Conclusion

1) The microstructure of alloy Mar-M247 was significantly degraded after 24,000h of long-time thermal exposure compared to 16,000h, where it showed a more obvious growth of the γ' precipitation-strengthened phases.

2) Appropriate selection of solid solution temperature in RHT can effectively restore the degraded microstructure and properties. The microstructure stability after rejuvenation is high, and the hardness and durability are comparable to the as-received alloy.

References

[1]Steven, R. A., Flewitt, P. E. J. (1978). Microstructural changes which occur during isochronal heat treatment of the nickel-base superalloy IN-738. Journal of Materials Science, 13, 367-376.

[2] Yao, Z., Degnan, C. C., Jepson, M. A. E.,Thomson, R. C. (2013). Effect of rejuvenation heat treatments on gamma prime distributions in a Ni based superalloy for power plant applications. Materials Science and Technology, 29(7), 775-780.

[3]Zhang, J., Liu, L., Huang, T., Chen, J., Cao, K., Liu, X., Fu, H. (2021). Coarsening kinetics of γ' precipitates in a Re-containing Ni-based single crystal superalloy during long-term aging. Journal of Materials Science & Technology, 62, 1-10.

Freckle Defect Prediction for Single Crystal solidification of Superalloys

Xiaoshan Liu [1,2], Ma Dexin[1,2], Guo Jianzheng[1,2*]
1. State Key Laboratory of Powder Metallurgy, Central South University, Changsha 410083, Hunan, China
2. Wedge Central South Research Institute Co Ltd, Shenzhen 518045, Guangdong, China
*Corresponding address: e-mail: 218062@csu.edu.cn

Abstract: Defects in high performance aerospace investment castings can reduce mechanical properties and consequently degrade both component life and durability. It is critical to understand the mechanism of defects formation so that an effective tool can be developed to prevent defects. Those defects are dependent on the chemistry of alloys, casting design, and casting processes. Casting engineers must be able to both accurate predict and reduce or even completely remove casting defects. In the present study, qualitative models were applied to predict the location-specific freckles during single crystal (SC) solidification of superalloy castings. These predictions are validated by comparing predicted results against several special designed superalloy castings as well as production castings. The simulation results agree well with experimental measurements.

Keywords: Freckle; Superalloy; Directional solidification; Single crystal casting

1 Introduction

There are many kinds of casting defects in SC components, such spray grains, freckles and slivers. Freckle is a detrimental defect in superalloy castings, especially in high generation SC superalloy blades. Researchers have found that freckle formation mainly depends on the chemistry of the alloy, casting processing parameters, and the geometry variation of the castings[1-4]. Special designed simple shape castings have been performed experimentally to assess the formation and severity of freckles previously to study the effect of the geometric variation comprehensively [1-2]. In addition to those extensive experimental works, some researchers[3-4] built fluid dynamics models to deeply analyze the convection phenomenon of liquid in the mushy zone during directional solidification, and proposed Ra criteria for the formation of freckles. However, those criteria are mostly based on simplified and ideal conditions, and the simulation results are often inconsistent with the actual castings. Recently, a new mathematical model based on Ra was proposed [5], which takes into account the geometric effect of castings, and the prediction results of freckle defects in complex castings are relatively consistent with the experiments. In this paper, the directional solidification process of a third-generation nickel-based superalloy SC casting is simulated based on an inhouse code which takes into account of the chemistry, casting processing parameters as well as the geometry effect. In order to evaluate the accuracy of the prediction, the area of the freckle on the casting surface was measured and compared with the simulation. The prediction is rather accurate with the error of less than 10%, which proves the accuracy of the current prediction method.

2 Experimental procedure

A third-generation SC superalloy WZ30 was used in the experiment, and its chemical composition is shown in Table 1.

Table 1. Nominal composition of the alloy WZ30 (wt.%)

Cr	Co	W	Mo	Al	Ti	Ta	Re	Nb	Ni
3.5	6.0	6.5	0.4	5.8	0.15	8.0	4.95	0.1	base

The castings designed in this experiment have 3 different cross-section shapes, and the casting wax molds were arranged around the center to assemble a circular wax mold model. Figure 1 shows the casting cluster arrangement and detail of one zigzag casting bar. This paper presents the detail experimental results and simulation results of the zigzag casting only.

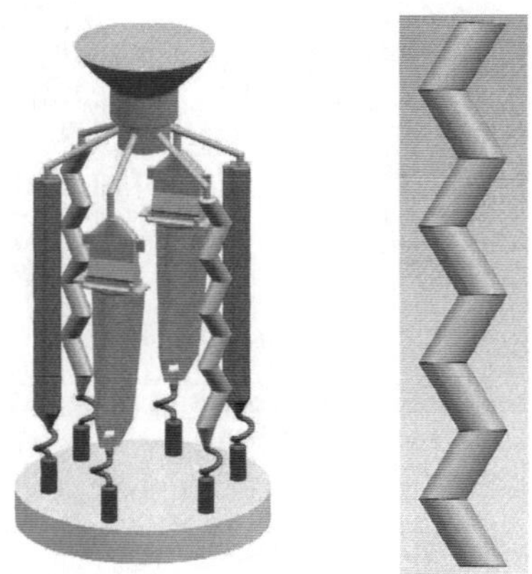

(a) (b)

Fig. 1 Casting cluster arrangement (a) and zigzag casting (b)

The pouring temperature was 1500 ° C, and the withdraw rate was 1.5 mm/min. The location of freckles in the casting was identified by visually inspection on the macroscopically etched casting surface. The freckle defect area of the actual casting was marked and measured using Image Pro Plus, to characterize the freckling severity in the investigated casting.

3 Result and discussion

Fig. 2 shows the comparison between the simulated and experimental observed freckle results of the studied casting. In order to quantitatively assesses the prediction, the castings were segmented, and the freckle areas in each section of castings were measured and analyzed in detail along the height of the casting.

(a) (b)

Fig. 2 Comparison of the whole zigzag casting freckles between simulation and experiment (a), and local zoom view of top part of the casting (b)

The comparison between the measured freckle defect and the simulated results in one section of the casting is shown in Figure 3. The predicted freckle area was defined by the area of Ra greater than a critical value which can be determined by experiments or theory calculation.

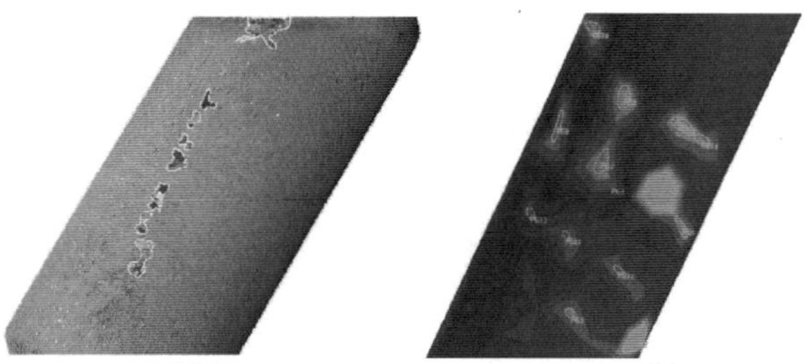

(a) (b)

Fig. 3 Freckle defect area map of one Section of the casting from experiment (a) and simulation (b)

The freckle defect in the zigzag casting from experiment and simulation along the height of the casting were statistically analyzed, and the comparison results were shown in Figure 4. The error between simulation and measurement is less than 10% .

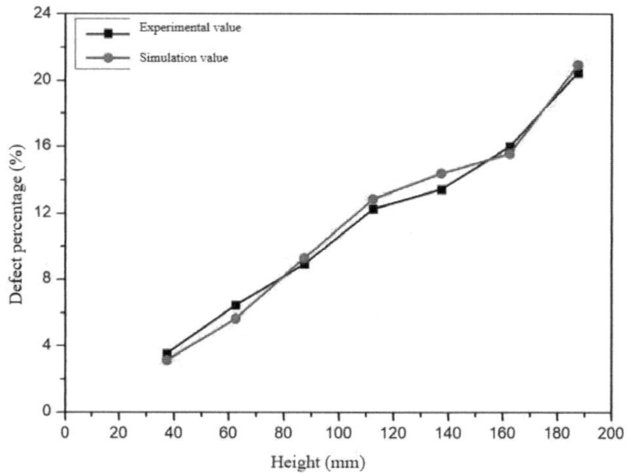

Fig. 4 The freckle area with the height of the casting by experiment and simulation

4 Conclusion

With the increase in casting height, both simulated and measured freckle area on the casting surface shows an increasing trend. This is because during conventional directional solidification process, the temperature gradient becomes normally weaker, leading to the continually coarsening of dendrite structure and the increasing tendency of freckle formation. Simulation results agree well with the experimental observation.

References

[1] Ma Dexin, Zhao Yunxing, Xu Weitai, et al. Single crystal superalloy castings of freckles study [J]. Journal of special casting and nonferrous alloys, 2021, 41 (11): 5. DOI: 10.15980 / j.t. ZZZ. 2021.11.005.

[2] Dexin MA. Freckle formation during directional solidification of complex castings of superalloys [J]. Acta Metallurgica Sinica, 2016, 52(4): 426-436. 2016.DOI:10.11900/0412.1961.2015.00379.

[3] Beckermann C, Gu J P, Boettinger W J, Development of a freckle predictor via Rayleigh number method for single-crystal nickel-base superalloy castings[J]. Metallurgical and Materials Transactions A, 2000.DOI:10.1007/s11661-000-0199-7.

[4] Zhang Haijie, Liu Xiaoshan, Ma Dexin, et al. Digital twin for directional solidification of a single-crystal turbine blade[J]. Acta Materialia,Volume 244, 2023, 118579, ISSN 1359-6454, https://doi.org/10.1016/j.actamat.2022.118579.

[5] Liu Yang, Wang Fu, Ma Dexin, et al. Freckle prediction model incorporating geometrical effects for Ni-based single-crystal superalloy components[J]. Acta Materialia, Volume 266,2024,119702, ISSN 1359-6454, https://doi.org/10.1016/j.actamat.2024.119702.

This page intentionally left blank.

Study on Ceramic Cores for Directional and Single Crystal Castings

Xinli Guo[1*], Zhongliang Lu[2], Lei Jin[1], Dichen Li[2], Ling Zhang[1], Guoyan Shui[1], Kai Miao[2]
1. China Academy of Machinery Shenyang Research Institute of Foundry Co., Ltd., 17 South Yunfeng Street Tiexi District, Shenyang, Liaoning 110022, P.R. China
2. Xi'an Jiaotong University, No.28 Xianning West Road, Xi'an, Shaanxi 710049, P.R. China
*Corresponding address: E-mail: guoxl@chinasrif.com (10 pt. Times New Roman)

Abstract: ccording to the requirements of high strength and low deformation of ceramic cores for large-sized directional blades of gas turbines, silica was used as the matrix material. Based on the study of the influence of powder size gradation, the amount of mineralizer and sintering process on the core high temperature performance, the ceramic cores with length over 400mm for heavy-duty gas turbine were manufactured and applied in the directional solidified blades. According to the requirement of precisely dimensional accuracy control of ceramic cores for Single crystal blade of aero-engine, a new gel casting process was used to manufacture ceramic cores rapidly. By the gel casting process, the influence of structure on the deformation of ceramic cores was studied, and the deformation law was obtained in advance. Then the manufacturing of ceramic cores by hot injection moulding process was guided, and the material and die design was optimized. The dimensional stable ceramic cores with the profile of the core surface controlled in ± 0.1 mm were manufactured and applied to the single crystal blades and vanes.

Keywords: ceramic cores, strength, shrinkage

1 Introduction

With the increase of front inlet temperature of aero-engine and gas turbine, the working temperature of turbine blade, which is the key hot-end part of engine, is also increasing. Limited by the melting point of the alloy, the heat-bearing capacity of the alloy is close to the limit. Therefore, improving the air-cooling structure and the cooling efficiency of the blade has become the key to improve the heat-bearing capacity of the blades. The key to high efficiency of air-cooling lies in the forming of complex inner cooling channel inside the turbine blades, and the ceramic cores are the key to form complex inner cooling channel.

The performance and quality of ceramic cores directly affect the pass rate, dimensional precision, casting cost and service performance of hollow blades. The ceramic core is required to meet the following conditions, sufficient high temperature strength and thermal stability, the linear expansion contraction is small, and easy to remove from castings.

In this paper, the properties of the core were studied from the particle size gradation of powder, amount of mineralizer addition and sintering process, and the formulation and preparation process of the core were optimized under the guidance of gel casting process.

3 Experimental procedure

High purity silica powder was selected as the matrix material for ceramic core, and zircon powder was selected as mineralizer. 280 # , 400 # , 800 # quartz glass powder were selected for gradation test. The gradation test scheme is shown in Table 1. The slurry was made up of 85% silica powder, 15% zircon powder and 16% plasticizer by weight. The preparation process of performance sample was as follows, the hot injection pressure was 45 bar, the mold temperature was 40 ℃, the slurry temperature was 100 ℃, and the final sintering temperature was 1160℃.

Table 1. Gradation scheme of silica powder

Gradation scheme	280 # silica powder	400 # silica powder	800 # silica powder
ZG-1	45	30	15
ZG-2	40	35	15
ZG-3	35	40	15

4 Result and discussion

(1)The effect of powder size on the properties of ceramic core is shown in Table 2 from ZG-1 to ZG-3. From scheme ZG-1 to ZG-3, the bending strength of ceramic core at room and high temperature increased first and then decreased, the sintering shrinkage and high temperature deflection increased in turn.

Table 2. Properties of ceramic cores

Gradation scheme	Bending strength at room temperature (MPa)	Bending strength at 1550℃(MPa)	Deflection at 1550℃ (mm)	Shrinkage after sintering (%)
ZG-1	18.67	20.33	0.48	0.33
ZG-2	21.45	25.58	0.55	0.36
ZG-3	14.13	19.17	0.62	0.40

The ceramic core samples were manufactured by using ZG-02 graded silica powder as matrix material, and the addition of zircon powder was 5%, 10%, 15%, 20% respectively. The properties of the samples are shown in table 3. The bending strength and thermal deformation were the best when the content of mineralizer was 15%. The addition of mineralizer improved the sintering of ceramic powder, which resulted in the increase of core strength and the decrease of thermal deformation.

Table 3. Properties of ceramic cores

Zircon powder content (%)	Shrinkage after sintering (%)	Bending strength at room temperature (MPa)	Bending strength at 1550℃(MPa)	Deformation at 1550℃(mm)
5	0.21	8.51	18.64	0.68
10	0.37	16.87	20.01	0.42
15	0.38	17.32	21.17	0.41
20	0.47	16.00	18.64	0.52

(2)According to the requirement of precisely dimensional accuracy control of ceramic cores for single crystal blade of aero-engine, the ceramic cores were rapidly manufactured by gel casting process.

The shrinkage and strength of ZG -2 ceramic cores were improved by adding Al powder(5%,10%,15%) and mullite fiber(2%,4%,6%). With the result, the following scheme was chosen to manufacture the core, the Al content was 10%, and the mullite fiber content was 4%. The properties of the core samples obtained were as follows, the high temperature strength was 23.7 MPa, the room temperature strength was 17.2 MPa, and the shrinkage was 0.3%.

The influence of structure on the deformation of ceramic cores was also studied. The sintering deformation of ceramic cores tends to hug shrink. The deformation of the leading edge and trailing edge was bigger. And the tip of the edge deformed most.

Through optimized material , new design of the die, and pre-deformation treatment of green core, The dimensional stable ceramic cores with the profile of the core surface controlled in ± 0.1 mm were manufactured.

5 Conclusion

(1)When the silica powder was graded according to the following proportion, 40% of 280#, 35% of 400# and 15% of 800#, the core strength at room temperature and high temperature both reached the maximum 21.45 MPa and 25.58 MPa respectively, and the shrinkage after sintering reached the minimum 0.36%.

(2)The shrinkage of the core was improved by adding 10% of Al powder and 4% of mullite fiber to ZG-2 ceramic core scheme, when the properties of the core samples obtained were as follows, the room temperature strength was 17.2 MPa, the high temperature strength was 23.7 MPa, and the shrinkage was 0.3%.

6 Acknowledgments

Thanks for the support and guidance from the Institute of Metal Research, Northwestern Polytechnical University, Tsinghua University, Beijing Institute of Aeronautical Materials and other partners.

References

[1] Weizhe Tang, Tong Zhao, Rui Dou, et al. Additive manufacturing of low-shrinkage alumina cores for single-crystal nickel-based superalloy turbine blade casting. J. Ceramics International, 2022, 48: 15218-15226.
[2] Gye Seok An, Soo Wan Choi, Tae Gyun Kim, et al. Amino-functionalization of colloidal alumina particles for enhancement of the infifiltration behavior in a silica-based ceramic core. J. Ceramics International, 2017, 43: 157-161.
[3] Fuchu Liu, Zitian Fan, Xinwang Liu, et al. Aqueous gel casting of water-soluble calcia-based ceramic core for investment casting using epoxy resin as a binder. J. Int J Adv Manuf Technol, 2016, 86: 1235-1242.
[4] Magdalena Gromada, Adam Swieca, Marek Kostercki, et al. Ceramic cores for turbine blades via injection moulding. J. Journal of Materials Processing Technology, 2015, 220: 107-112.
[5] Gye Seok An, Soo Wan Choi, Tae Gyun Kim, et al. Amino-functionalization of colloidal alumina particles for enhancement of the infifiltration behavior in a silica-based ceramic core. J. Ceramics International, 2017, 43: 157-161.
[6] A A Wereszczak, K Breder, M K Ferber, et al. Dimensional changes and creep of silica core ceramics used in investment casting of superalloys. J. Journal of materials science, 2002, 37: 4235-4245.
[7] Weigou Jiang, Kaiwen Li, Jiuhan Xiao, et al. Effect of silica fifiber on the mechanical and chemical behavior of alumina-based ceramic core material. J. Journal of Asian Ceramic Societies, 2017, 5: 410-417.

Effect of Mo Addition on the Microstructure and Creep Property of Ni-Based Single Crystal Superalloys

Yuan Cheng[1], Xinbao Zhao[1, 2*], Yuefeng Gu[1, 2*], Quanzhao Yue[1], Wanshun Xia[1], Ze Zhang[1, 2]
1. Institute of Superalloys Science and Technology, School of Materials Science and Engineering, Zhejiang University, Yuhangtang Road, No. 866, Hangzhou, China
2. State Key Laboratory of Silicon and Advanced Semiconductor Materials, School of Materials Science and Engineering, Zhejiang University,Yuhangtang Road, No. 866, Hangzhou, China
*Corresponding address: e-mail: guyf@zju.edu.cn

Abstract: A self-developed 4th generation Ni-based single crystal superalloy with different Mo contents has been developed to investigate the microstructure and creep property. The Mo addition did not change the dendrite spacings significantly. The segregation of refractory elements, such as W and Re, was significantly restrained. With the relief of segregation, the content of eutectic islands was decreased by Mo addition. After heat treatments, Mo addition decreased the size of γ' phase and γ channel width. Mo addition mainly affected the creep life by influencing the formation of creep pores, dislocation networks, elemental inhomogeneity, TCP phases and γ channel widening. a suitable Mo content range was proposed.

Keywords: Mo addition, superalloys, solidification characteristics, microstructure

1 Introduction

As a key strengthening element in Ni-based single crystal superalloys (SCs), Mo has attracted much attention in recent yearsdue to its low-cost and low-density[1-3]. Previous studies have researched the effects of Mo addition on the microstructures and creep properties of SCs[2, 4]. However, previous studies on the effects of Mo were carried out mainly in low-generation alloy systems[1, 2]. The behavior of Mo might act differently in the new-generation SCs due to the complex composition. Therefore, it was crucial to investigate the effects of Mo addition on the microstructure and creep property in new-generation SCs for further alloy design. In this paper, this goal was achieved by developing a 4th generation SC with different Mo content.

2 Experimental procedure

The 4th generation SCs were directionally solidified by the Bridgeman method in our lab. The composition of the investigated alloys was listed in Table 1. The solid solution treatment procedure was set mainly based on the results of incipient tests, the highest temperature of which was set as 1350 ℃. The creep tests were carried out at 1100 ℃ and 137 MPa.

Table 1. Composition of the investigated alloys. (wt.%)

	Al+Ta	W+Re+Cr+Co	Ru	Hf	Mo	Ni
2Mo	12	19.8	2.5	0.1	2	Bal.
3Mo	12	19.8	2.5	0.1	3	Bal.
4Mo	12	19.8	2.5	0.1	4	Bal.

The images of microstructure were mainly gained by scanning electron microscope (SEM). The samples were mechanically grinded and polished. The polished samples were then etched by a solution consisted of HF, HNO3 and glycerin with the volume ratio of 1: 2: 3.

3 Result and discussion

The as-cast structure

The primary and secondary dendrite structure of the as-cast sample was exhibited in Fig.1a and 1b. Mo addition did not significantly influence the dendrite arm spacings. Fig. 1c showed a typical eutectic morphology, which could be classified as the fine γ/γ' structure and coarse γ' phase. Fig. 1d indicated that Mo addition significantly decreased the segregation ratio of refractory elements. Such alleviation of segregation behavior led to the decrease of the eutectic content, as the statistical results in Fig.1e showed.

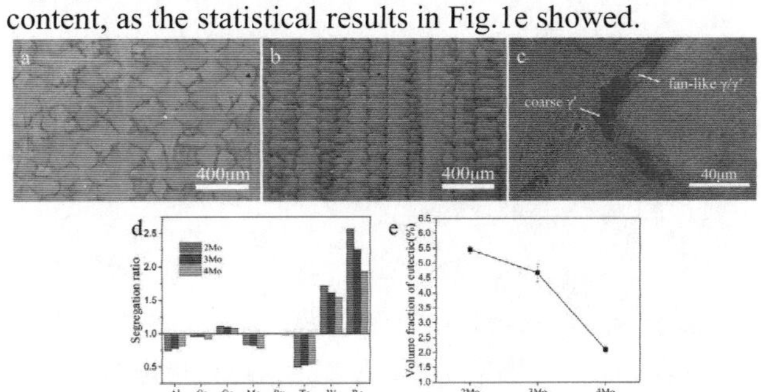

Fig.1 The as-cast microstructure of 3Mo SC, including (a) the primary dendrite, (b) the secondary dendrite, (c) eutectic island, (d) the elemental segregation ratio and (e) the volume fraction of eutectic islands[3].

Microstructure after heat treatments

To explore the upper limit temperature of the solid solution treatment, the incipient melting tests were carried out. The incipient melting stopped to occur near 1350 ℃, thus the upper limit of solid solution temperature was set as 1350 ℃. During the solid solution treatment procedure, the temperature rose in multiple steps from 1320 ℃ to 1350 ℃. The samples were held at 1350 ℃ for 10 h, followed by air cooling. The two-step aging treatments were carried out at 1100 ℃ and 870 ℃, respectively. Fig.2 showed the microstructure in the dendrite cores after heat treatments. Due to the residual segregation, the size of γ' in the interdendritic regions was slightly larger than in the dendrite cores. Moreover, Mo addition decreased the size of γ' phase and the width of γ channel.

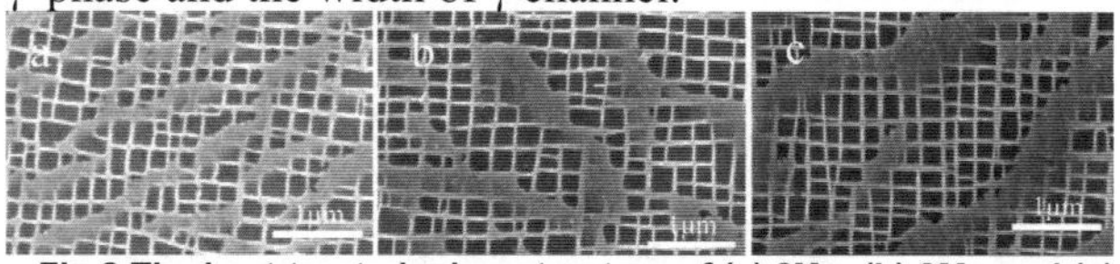

Fig.2 The heat-treated microstructure of (a) 2Mo, (b) 3Mo and (c) 4Mo in the dendrite cores[3].

Creep property at 1100 ℃/137 MPa

The creep life at 1100 ℃/137 MPa was exhibited in Fig.3. 2Mo and 3Mo alloy possessed similar creep life reaching 500 h. However, the creep life of 4Mo alloy decreased to less than 300 h. The influence mechanisms of Mo addition on the creep properties were analyzed. Mo influenced the creep property in both positive and negative ways, which were summarized in Fig.3. Mo addition inhibited the formation of creep pores, delayed the elemental inhomogeneity and promoted the formation of dense interfacial dislocations, which were beneficial factors enhancing creep resistance. However, Mo also promoted the precipitation of topologically close-packed phases (TCP phases) and accelerated the widening of the γ channel, which weakened the creep strength. Based on the analysis of the creep property and microstructure, the suitable range of Mo content in the 4th generation Ni-based SCs was proposed.

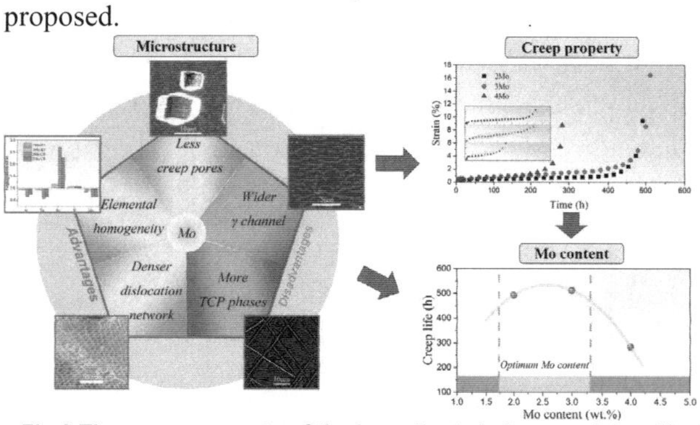

Fig.3 The creep property of the investigated alloys and the effect mechanisms of Mo addition[5]. The suitable range of Mo content was proposed based on the analysis.

4 Conclusion

(1)Mo decreased the segregation ratio of refractory elements, decreasing the content of eutectic islands.

(2)After heat treatments, the γ/γ' became typical cubic patterns. Mo addition slightly decreased the size of γ' phase and the width of γ channel.

(3)Mo enhanced the creep property by affecting the formation of creep pores, interfacial dislocation networks and elemental inhomogeneity. Meanwhile, the γ channel widening and TCP phase induced by Mo addition weakened the creep strength.

5 Acknowledgments

This work was jointly supported by the Zhejiang Provincial Natural Science Foundation of China (LR22E010003, LQ21E010005), the National Natural Science Foundation of China (91960201, 52101155), the National Science and Technology Major Project of China (J2019-III-0008-0051)

References

[1] Mackay R A, Gabb T P, Smialek J L, et all. A new approach of designing superalloys for low density[J]. JOM,2010,62(1): 48-54.

[2] Zhang L F, Huang Z W, Jiang L, et al. Effect of Mo:W ratio on segregation behavior and creep strength of nickel-based single crystal superalloys[J].. Mater Sci Eng, A,2019:744481-9.

[3] Cheng Y, Zhao X, Xia W, et al. Effects of Mo addition on microstructure of a 4th generation Ni-based single crystal superalloy[J].Progress in Natural Science: Materials International. 2022,32(6): 745-51.

[4] Ai C, Liu L, Zhang J, et al. Influence of substituting Mo for W on solidification characteristics of Re-containing Ni based single crystal superalloy[J]. J Alloys Compd. 2018:75485-92.

[5] Cheng Y, Zhao X, Yue Q, et al. The role of Mo on the creep deformation of a 4th generation Ni-based single crystal superalloy at 1100 °C[J]. Materials Today Nano, 2023,24.

Study on the Effect of Static Magnetic Field on the Nickel-Based Superalloy Fabricated by Additive Manufacturing

Dafan Du, Anping Dong*, Baode Sun
Shanghai Key Lab of Advanced High-temperature Materials and Precision Forming and State Key Lab of Metal Matrix Composites, School of Materials Science and Engineering, Shanghai Jiao Tong University, Shanghai 200240, China
* Corresponding author: apdong@sjtu.edu.cn;

Abstract: The influence of a static magnetic field on microstructure evolution during laser direct energy deposition was studied. Our results reveal that dendrite spacing increases with increasing magnetic field flux density (MFFD). Moreover, electron backscatter diffraction results reveal that the epitaxial growth was preferred with increasing MFFD. We discuss these findings in terms of the influence of a magnetic field on melt convection and propose that an applied magnetic field effectively limits Marangoni convection.

Keywords: Ceramic mold, laser powder bed fusion, infiltration, turbine blade

1 Introduction

Directed energy deposition (DED) processes have lower production rates when compared to those of other additive manufacturing (AM) processes, such as selective laser melting (SLM) and selective electron beam melting (S-EBM). DED processes are capable of producing components with complex geometries and are particularly well suited for repairing or fabricating high value parts at low production volumes [1-3]. Hence, it is not surprising to find numerous published studies on the influence of laser energy density [4, laser source type [5], scan strategy [6] and build direction on the microstructure of AM parts by DED. Inspection of the published literature reveals some research on the solidification behavior of the melt pool during DED; however many questions remain unanswered. For example, Lee et al. [7] investigated the influence of fluid convection on dendrite arm spacing during solidification and reported an inverse relationship between temperature gradient G and growth speed R. Moreover, the relationship of simulated weld pool convection patterns to predict solidification boundary shape, cooling rate distribution, and dendrite arm spacing (DAS) was also examined. Wei et al. numerically studied heat transfer and liquid metal flow calculations in an effort to provide insight into the evolution of solidification morphology and texture during multi-layer AM processes. Further, O. Nenadl et al[8]. studied the role of solid/liquid interface on the grain orientation structure and texture based on both experimental findings and theoretical analyses.

The above discussion underscores the critical influence that solidification mechanisms in the melt pool exert on the final microstructure during DED. Inspired by results from the casting literature, researchers have proposed the idea of using a magnetic field to effect melt convection during directional solidification. For example, studies on the influence of a weak static magnetic field (in the order of 0.1 T) on the liquid melt during directional solidification suggest that a static magnetic field promotes thermoelectric magnetic convection (TEMC), and thereby ultimately influences the thermal and solute fields. Chen et al.[9] applied a 0.32 T static magnetic field during laser beam welding of austenitic stainless steel and reported a reduction in austenite grain size, and crack and thickness of brittle intermetallic compounds in the interface. Moreover, research described in reference showed that by applying a 0.3 T horizontal magnetic field the "humping" welding defect, which is analogous to "balling" in AM that leads to beading, could be suppressed and they concluded that this effect could be attributed to electromagnetic forces. In view of the above discussion, and the apparent lack of information on melt pool solidification during DED, in this research we investigated the influence of a static magnetic field on the microstructure of IN718 alloys during DED, paying particular attention to the evolution of microstructure during solidification in the melt pool. The selection of this particular material was prompted by the availability of numerous AM published studies using IN718.

2 Experimental procedure

Gas atomized IN718 powder (Staffordshire, United Kingdom) with a particle size 53 μm-150μm was used as the feed material, and laser aided direct metal tooling (DMT) was used to fabricate the specimens. A laser-aided DMT machine (InssTek, MX-450, South Korea) was used to fabricateIN718 cuboid samples with dimensions 8× 8 × 12mm. The laser scan speed used in our work was 0.85 m/min. The slicing layer depth was 0.3 mm and totally 40 layers for the sample. The scanning direction rotated 90° after each layer. It is worth pointing out that as an Ni based superalloy, the IN 718 platform has a low magnetic susceptibility ($\chi \approx 6\times10^{-4}$ SI) and hence its effect on the applied magnetic field is likely to be negligible.

The microstructures were characterized by optical microscopy (Olympus, Japan), SEM (Tescan, GAIA3, Czechia) and electron backscattered diffraction (EBSD) (Oxford, Nordlys Max2, United Kingdom) along the longitudinal cross sections (XZ plane).

3 Result and discussion

Fig. 1 show the optical microscopy of the longitudinal plane of the samples deposited at various laser energy density with various MFFDs. In all samples, black spherical pores can be observed in the microstructure. The results also showed that the porosity decreases with increasing SMF for samples fabricated at different laser energy density.

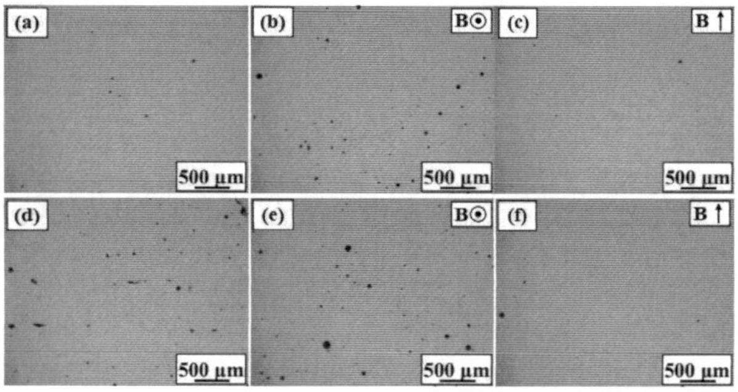

Fig. 1 Optical microscope of Inconel 718 alloys processed using different laser energy density and SMF intensities.

Fig. 2 show the EBSD maps of the longitudinal section of deposited samples with various MFFD. Fig. 2(a) and (b) are the IPF-Z maps for the samples at a constant laser energy density. EBSD map shows the grains oriented at different angles due to the variation of the thermal gradient in each layers with fine equiaxed grains with random orientations distributed along the edge of the melt pool. The pole figures also show that the texture increases with increasing MFFD from 0 T to 0.1 T.

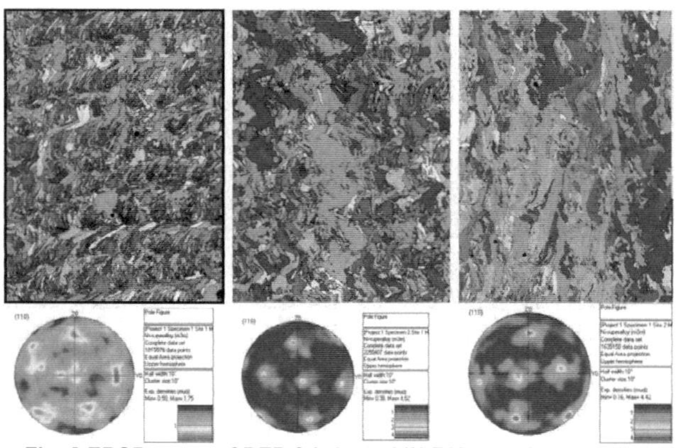

Fig. 2 EBSD maps of DED fabricated IN 718 samples with varied magnetic field intensities.

The above results demonstrate that by employing a static magnetic field (up to 0.1 T) during the DED process, the microstructure and crystallographic texture of IN718 can be modified. In the discussion that follows, it is important to note that the diameter of IN718 samples is much smaller than the substrate; therefore, the influence of the magnet on the temperature field of the substrate can be safely ignored. The change of liquid metal flow in the melt pool can easily affect the thermal distribution in and near the melt pool and alter the final microstructure. Previous studies also show that a static magnetic field can modify the melt convection and thereby alter the solidification process. During the AM process, the solidification in the melt pool mainly consists of nucleation and growth of γ cellular dendrites; the underlying phenomena have been widely studied during DED, selective laser melting (SLM) and selective electron beam melting (SEBM). On the basis of these and other studies, a superimposed magnetic field has two primary effects on melt convection. The first effect involves causing thermoelectric magnetic convection (TEMC) in the melt pool; the second one is the damping effect that a magnetic field can exert on fluid flow [10].

4 Conclusion

Influence of static magnetic field on the microstructure of IN718 alloys during DED process was studied. The following conclusions can be drawn from this work: The applied magnetic field modifies the microstructure of IN718 during DED process. With increasing MFFD, the primary dendrite spacing increases. Meanwhile, epitaxial growth leads to the formation of columnar grains through multi layers with increasing MFFD. In addition, the <110>-texture dominates the build direction and the fraction of LAGBs increases with increasing MFFD. The above results should attribute to the effect of the magnetic field on restraining the liquid convection by electromagnetic braking forces.

5 Acknowledgments

This work was supported by the National Natural Science Foundation of China (Nos. 51901135, 51771118, U1760110 and 51821001).

References

[1] W.R. Morrow, H. Qi, I. Kim, et al. Clean Prod., 2007, 15 : 932-943.
[2] G.F. Sun, S. Bhattacharya, G.P. Dinda, et al, Scripta Mater., 2011, 64:454-457.
[3] Y.J. Liang, J. Li, A. Li, et al, Scripta Mater., 2017, 127:8-62.
[4] Y.W. Zhai, D.A. Lados, E.J. Brown, et al. Manuf., 2019, 27:334-344.
[5] M. Mohammadi, H Asgari, Addit. Manuf., 2018, 20:23-32.615–622.
[6] M.Y. Mendoza, P. Samimi, D.A. Brice,et al.Metall. Mater. Trans. A 2017, 48(7):3594-3605.
[7] Y. Chen, Y.B. Guo, M.J. Xu, C.F. Ma, Q.L. Zhang, L. Wang, J.H. Yao, Z.G. Li, Mater. Sci. Eng. A 2019, 754:339-347.
[8] G.P. Dinda, A.K. Dasgupta, J. Mazumder, Scripta Mater., 2012, 67: 503-506.
[9] L.L. Parimi, G.A. Ravi, D. Clark, M.M. Attallah. Mater. Charact., 2014, 89:102-111.
[10] Y.L. Kuo, S. Horikawa, K. Kakehi, Scripta Mater.,2017, 129:74-78.

This page intentionally left blank.

Progress in Cast Magnesium-Matrix Composites with Low Expansion and High Thermal Conductivity

Shusen Wu*, Lu Chen, Shulin Lü, Wei Guo, Jianyu Li
State Key Lab of Materials Processing and Die & Mould Technology, School of Materials Science and Engineering, Huazhong University of Science and Technology, Wuhan 430074, P. R. China
*Corresponding address: e-mail: ssw636@hust.edu.cn

Abstract: A type of magnesium alloy that urgently needs to be developed is one that has both high thermal conductivity and low thermal expansion, in order to ensure that the thermal physical properties of magnesium alloy components match with other components, and to ensure the dimensional stability of the components. However, in general, the high thermal conductivity and low expansion of magnesium alloys are contradictory, as the fewer alloying elements in magnesium metal, the higher the thermal conductivity (TC), but when magnesium alloys are close to pure magnesium, the thermal expansion coefficient (CTE) is large. The author's research group innovatively designed a Mg-Zn-Cu cast alloy with high TC and good mechanical properties. Due to the Laves structure phase of AB2 type C15 structure, the content of alloy elements in the solid solution is reduced, resulting in a high TC of the alloy, close to pure magnesium. On this basis, in order to further reduce the CTE, ceramic particles with low expansion and high thermal conductivity such as AlN, SiC and BN could be added to prepare cast Mg-matrix composites. The results show that the 10 wt.% AlN /Mg-Zn-Cu composite can reach a room temperature TC of 149.1 W/(m·K), and CTE of around 17.3×10^{-6} K^{-1} at 20 ~ 100℃, which can meet requirements for the thermal physical properties of magnesium alloys in many applications.

Key words: thermal expansion; thermal conductivity; magnesium-matrix composites; Mg-Zn-Cu alloy

1. Introduction

At room temperature, pure Mg has a thermal conductivity of 156 W/(m·K), which is higher than most metals and organic polymer materials, but it is lower than Al (237 W/(m·K)) and Cu, especially when it is alloyed by many elements, such as AZ91 alloy (only 45.1 W/(m·K)) [1].

Less research has been done on the CTE of magnesium alloys. Numerous studies show that the alloying elements are typically low in order to maintain high TC. But low-alloyed Mg alloys have high CTE, usually higher than many metallic materials, such as steels. The average CTE (20-100°C) of pure magnesium is 26×10^{-6} K^{-1}. By combining those components of high TC and low CTE, such as AlN and SiC, with magnesium-based materials, composites with high TC and minimal CTE may be produced. For example, the CTE of AlN is only 4.5×10^{-6} K^{-1} and with high TC of about 320 W/(m·K).

2. Experimental methods

The Mg-xZn-xCu (x=0.5, 1, 3, 5 wt.%) alloys were obtained by melting the raw materials of pure Mg, Zn and Cu in a resistance furnace. The mold used in this study was a cylindrical mold made of steel, the preheating temperature was 200°C, and the ingot size was Φ35mm×120mm. Through semi-solid stirring method, AlN particles with an average particle size of 10 μm are introduced to the melt to form AlN/Mg-Zn-Cu composites, with 10 wt.%, 20 wt.%, 30 wt.% respectively, and process can be found in reference [2]. After the synthetic process, the melt is quickly poured into the same preheated steel mold and solidified under a pressure of 50 MPa.

Metallography and SEM methods can be found in reference [3]. A LFA-427 laser thermal conductivity meter was used to measure the TC and the specimen size was Φ12.7 mm×2.5 mm. The CTE of materials was measured using a NETZSCH DIL402 thermal expansion meter at a temperature range of 20-300 ℃.

3. Results and discussions

3.1 Composition design, microstructure and properties of Mg-Zn-Cu alloys with high TC

The Mg-*x*Zn-*x*Cu (*x*=0.5, 1, 3, 5) alloys were designed to study the effects of elemental amount on TC. The as-cast alloys consist of α-Mg (dark parts) and second-phase (white parts) distributed along the grain boundary, as shown in Fig. 1. Combined with the XRD spectrum results, the second phase is identified as CuMgZn ternary phase. The CuMgZn phase precipitates in the form of eutectic in the post process of solidification, and the reason for the grain refinement should be that solute atoms aggregated at the solidification front, resulting in composition overcooling.

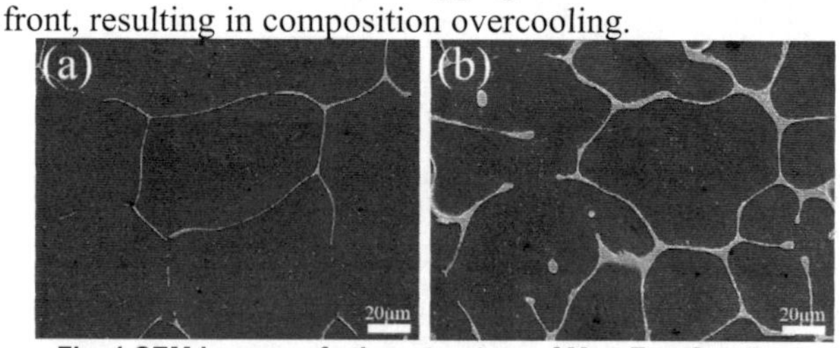

Fig. 1 SEM images of microstructure of Mg-xZn-xCu alloys: (a)x=1; (b)x=3

Fig.2 shows the TC of the alloys. The TC shows a trend of increasing with the increase of temperature. The Mg-0.5Zn-0.5Cu alloy has highest TC at room temperature of 153.0 W/(m·K), and with the increment of Zn and Cu, the Mg-5Zn-5Cu alloy has the lowest thermal conductivity, which is 136.5 W/(m·K). The TC of Mg-1Zn-1Cu is also high of 148.0 W/(m·K). With the increment of the total content of alloying elements, the content of Zn and Cu dissolved in the α-Mg matrix increases greatly, which is the main reason for the change of TC of the alloys.

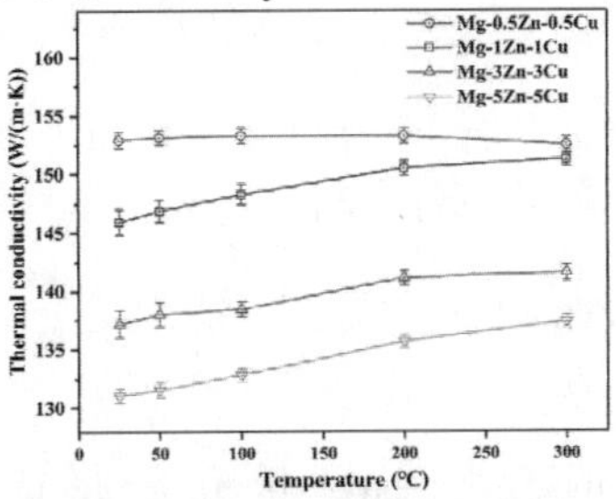

Fig. 2 TC variation of Mg-Zn-Cu alloy with temperature

Fig. 3 shows two presentative CTE of Mg-xZn-xCu alloys. Contrary to TC, the CTE increases with the increase of alloying element content. The average CTE of Mg-0.5Zn-0.5Cu alloy at 20-100°C is 22.8×10^{-6} K^{-1}, while the CTE of Mg-5Zn-5Cu alloy is 24.7×10^{-6} K^{-1}, which is 8.3% higher. Moreover, the CTEs of all Mg-Zn-Cu alloys are still very high. Therefore, although the designed and developed Mg-Zn-Cu alloy has a high TC, its CTE is also relatively high and does not meet the ideal requirements. Other measures need to be taken for regulation, such as adding a second phase with low CTE to form composites.

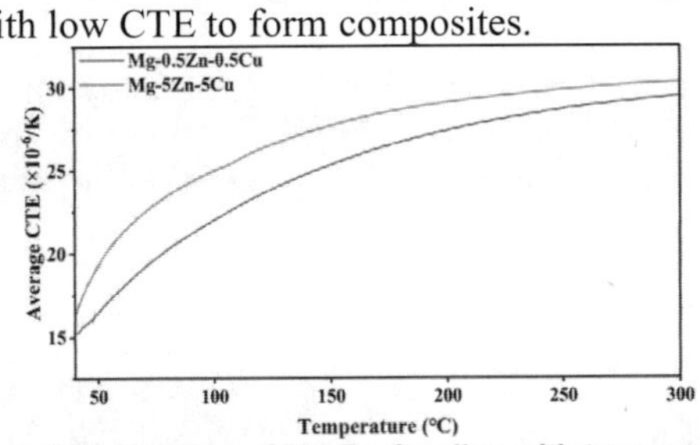

Fig. 3 CTE variation of Mg-Zn-Cu alloy with temperature

3.2 Structure and properties of low expansion and high thermal conductivity AlN/Mg-Zn-Cu composites

Fig.4 is representative microstructure of AlN/Mg-0.5Zn-0.5Cu composites, and the semi-solid stirring technique allows AlN to enter the magnesium matrix and distribute uniformly. The AlN particles are distributed inside the grains and at the grain boundaries, surrounded by the white eutectic structures α-Mg+CuMgZn.

Fig. 5 shows the temperature dependence of TC of AlN/ZC0505(wt.=0, 10%, 20%, 30%) composites. The TC of the composites decline with increasing AlN concentration, and AlN does not have the expected strengthening impact on TC. Adding 10 wt.% AlN decreases the TC at room temperature by 2.5%, from 153.0 W/(m·K) to 149.1 W/(m·K) in this study. But overall, TC of the composites is still very high. Fig. 6 shows the temperature dependence of CTE. The average CTE of 10% AlN/ZC0505 composites is 19.0×10^{-6} K^{-1}, and it is 16.7% smaller than matrix alloy. AlN

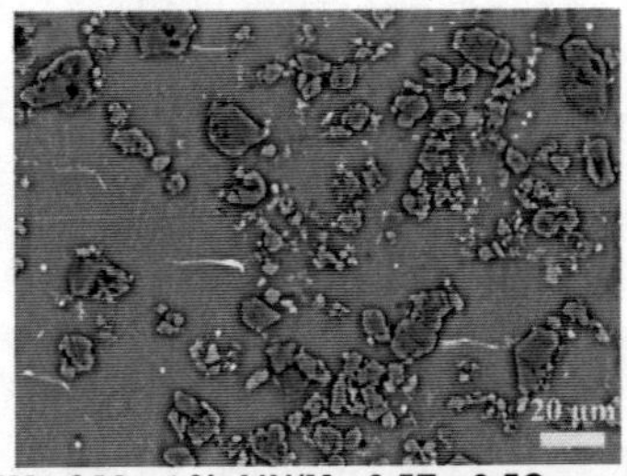

Fig. 4 SEM of 20 wt.% AlN/Mg-0.5Zn-0.5Cu composites

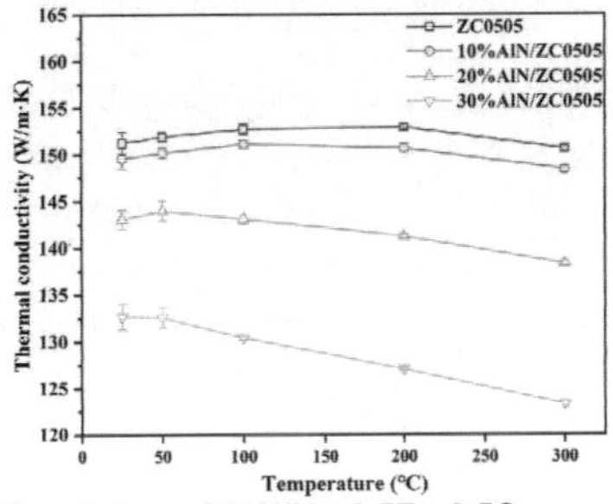

Fig. 5 TC variation of AlN/Mg-0.5Zn-0.5Cu composites

effectively reduces the CTE of the composites due to its low CTE. With the addition of 20 wt.% and 30 wt.% AlN, the CTE of the composites is only 17.3×10^{-6} K^{-1} and 15.6×10^{-6} K^{-1}, which is 24.1% and 31.6% less than that of the matrix alloy respectively.

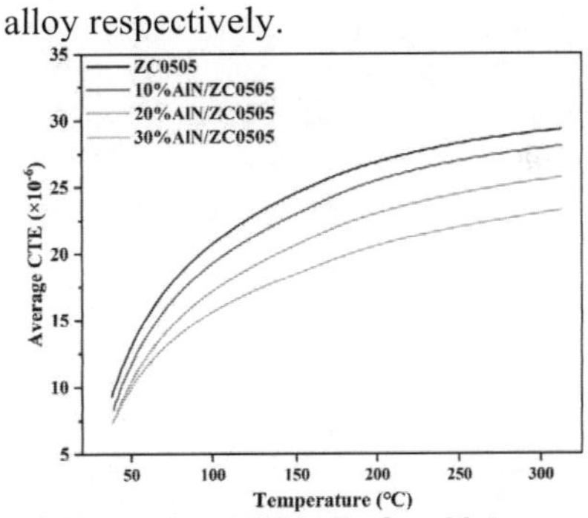

Fig. 6 CTE variation AlN/Mg-Zn-Cu with temperature

4. Conclusions

The TC of newly designed Mg-0.5Zn-0.5Cu alloy is as high as 153.0 W/(m·K). For every 1wt.% increment in element content, the TC decreases by about 1.95W/(m·K). The TC of AlN/ZC0505 composites decreased slightly with the increase of AlN content, TC of 10% AlN/ZC0505 decreased 2.5% from 153.0 W/(m·K) to 149.1 W/(m·K).
The AlN limit effectively the thermal expansion of the composites. The CTE of 20% AlN/ZC0505 is only 17.3×10^{-6} K^{-1}, which is 24.1% lower than that of the matrix alloy.

References

[1] Cheng L, Shuai Z, Yu Z, et al. Alloys Compd,2014, 597: 119–123
[2] Chen L, Qin J, Li J,et al. Solid State Phenomena, 2023, 347: 91-96
[3] Chen L, Lü S, Guo W, et al. Alloys Compd., 2022, 918, 165614

Development of Al/Mg Bimetal Prepared by Ultrasonic Vibration-Assisted Compound Casting

Qingqing Li, Wenming Jiang*, Yuancai Xu, Linghui Yu, Zitian Fan
State Key Lab of Materials Processing and Die & Mould Technology, Huazhong University of Science and Technology, Wuhan 430074, PR China
*Corresponding address: e-mail: wmjiang@hust.edu.cn

Abstract: The coarse Al-Mg intermetallic compounds and oxide film at the Al/Mg bimetal interface strongly limit and weaken the interfacial bonding performance. The introduction of ultrasonic vibration into the liquid medium could provide high-frequency oscillations and induce unique acoustic effects in liquid melt, which have significant impacts on the flow field and temperature field of the liquid medium, and may be aided for the enhancement of the Al/Mg bimetal interface. In this work, the Al/Mg bimetals were fabricated by ultrasonic vibration-assisted solid-liquid compound lost foam casting. The ultrasonic vibration treatment (UVT) was introduced to the Al/Mg interfaces, and the effects of UVT on the interfacial microstructures and mechanical properties were investigated. The results revealed that the microstructures of the Al/Mg interface with UVT were refined and got more homogeneous, and the oxide film at the Al/Mg interface was fractured and eliminated. The microhardness was more uniform, and the bonding strength of the Al/Mg bimetal with UVT was enhanced to 61.4 MPa, with an increase of 89.5%, compared with that of the Al/Mg bimetal without UVT.

Keywords: Ultrasonic vibration; Lost foam casting; Al/Mg bimetal; Interfacial microstructures; Mechanical properties

1 Introduction

The Al/Mg bimetal composites combines the advantages of Al alloys and Mg alloys [1, 2], which have prominent comprehensive properties on the basis of ensuring light weight, thus expanding the application range of aluminum, magnesium and their alloys in industrial fields [2].

The compound casting is suitable for manufacturing the Al/Mg bimetals with complex shapes and large sizes. However, Al and Mg could dissolve infinitely and form brittle Al-Mg intermetallic compounds (IMCs). These brittle IMCs are with much higher microhardness than the Al and Mg matrix. Cracks initiate at the IMCs firstly, which are detrimental to the bonding performance of Al/Mg bimetals. The oxide inclusions easily formed at the Al/Mg interface, which inhibit the interdiffusion of elements and weaken the metallurgical reaction between Al alloy and Mg alloy. Besides, the Al-Mg IMCs are coarse.

Majorities of efforts have been made to enhance the Al/Mg interface, such as setting interlayers, microalloying of melt and vibration-assisted compound casting [1-3]. The vibration-assisted compound casting is a pollution-free method to enhance castings. Compared with low-frequency mechanical vibration, the ultrasonic vibration introduced to the melt could induce acoustic cavitation and acoustic streaming flow effects, which shows potential application in strengthening Al/Mg interface.

In this study, the effects of ultrasonic powers on the interfacial microstructures and mechanical properties of the Al/Mg bimetals were investigated.

2 Experimental procedure

The commercial AZ91D alloy and A356 alloy ingots were used as the raw materials to fabricate the Al/Mg bimetals. The chemical compositions of the AZ91D alloy and A356 alloy ingots are listed in Table 1.

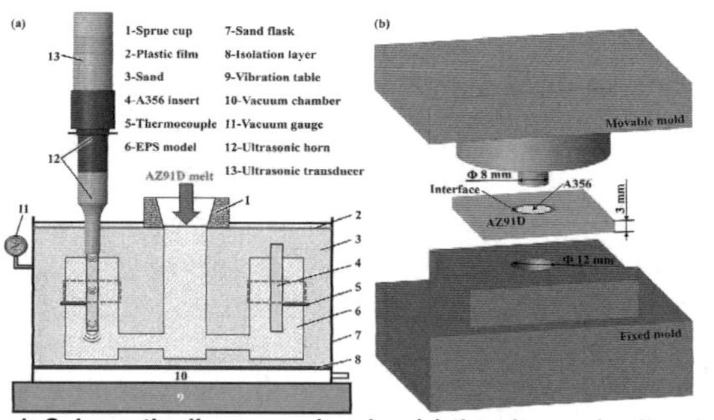

Fig. 1. Schematic diagrams showing (a) the ultrasonic vibration treatment assisted lost foam casting system, the red boxes marked in the image illustrate the position of the Al/Mg bimetal samples cut for microstructure observation and mechanical measurement, (b) the shear strength measurement.

Fig. 1(a) shows the schematic diagram of the UVT-assisted LFC system. The frequency of the ultrasonic device is 20 kHz. After the pouring process, the ultrasonic device was activated, with duration time of 5 s, and ultrasonic power of 75, 150 and 225 W respectively.

Table 1. Compositions of A356 and AZ91D Alloy

Alloy	Mg	Al	Si	Ti	Mn	Fe	Zn
AZ91D	Bal.	9.11	0.02	-	0.24	-	0.61
A356	0.31	Bal.	7.08	0.11	-	0.10	-

The microstructure characteristics and the mechanical properties of the Al/Mg interfaces were investigated. Fig. 1(b) shows the schematic diagram of the shear strength measurement. The loading rate was 0.5 mm/min.

3 Result and discussion

1. Interfacial microstructure

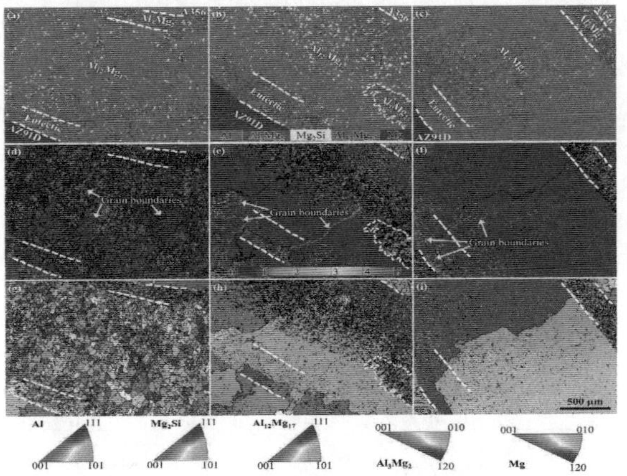

Fig. 2 EBSD results of the Al/Mg interfaces of the UVTed-75 W, UVTed-150 W and UVTed-225 W specimens. (a-c) Phase map, (d-f) grain boundaries and stress map, and (g-i) grain boundaries and IPF map of (a, d, g) the UVTed-75 W specimen, (b, e, h) the UVTed-150 W specimen, (c, f, i) the UVTed-225 W specimen. The white dashed lines illustrate the boundaries among A356, Al3Mg2, Al12Mg17, eutectic structure and AZ91D. IPF: inverse pole figure.

Fig. 2 shows the EBSD results of the Al/Mg interfaces of the UVTed-75 W, UVTed-150 W and UVTed-225 W specimens. The Al/Mg interfaces of the UVTed specimens are composed of Al_3Mg_2, Mg_2Si, $Al_{12}Mg_{17}$ and δ-Mg. Compared with the phase map of the specimen without UVT, the compositions of the Al/Mg interfaces with UVT are not changed. However, the distributions of phases have been significantly changed. Evidently, the Mg_2Si particles are dispersed and uniformly distributed at the whole Al/Mg interface, rather than only distributed at the IMCs area and gathered at the $Al_{12}Mg_{17}$ layer.

2. Mechanical properties

Fig. 3(a) and (b) exhibits the results of shear strength measurement of the Al/Mg bimetals. The shear strength of the specimen without UVT reaches 32.4 MPa. The shear strengths of the UVTed-75 W, UVTed-150 W and UVTed-225 W specimens are 61.4, 51.7 and 50.8 MPa respectively, which are increased by 89.5%, 59.6% and 56.8% respectively, compared with that of the specimen without UVT.

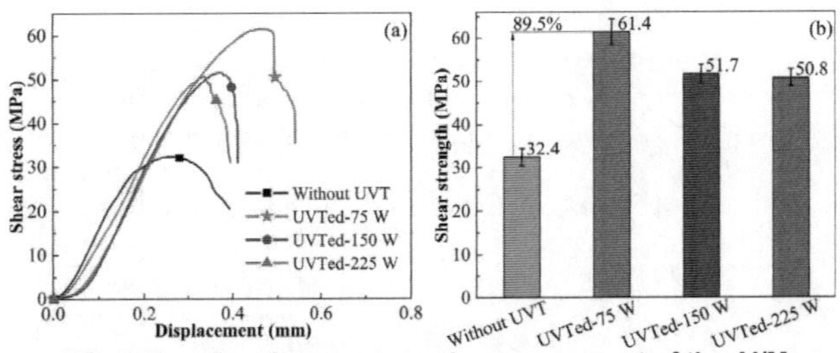

Fig. 3. Results of shear strength measurement of the Al/Mg bimetals: (a) the stress-displacement curves and (b) the shear strengths of the Al/Mg bimetals.

4 Conclusion

(1) With UVT, the oxide film was removed by UVT, and the Mg_2Si particles were refined and dispersed at the whole Al/Mg interface by the acoustic cavitation effect induced by UVT. Grain refinement of the Al/Mg interface with UVT only occurred when the ultrasonic power was 75 W.

(2) The microhardness of the Al/Mg interfaces was increased and got more uniform by UVT. The shear strength of the Al/Mg bimetal with UVT under the ultrasonic power of 75 W was enhanced to 61.4 MPa, increased by 89.5%, compared with that of the Al/Mg bimetal without UVT. It could be ascribed to the more homogeneous microstructure, the refined and uniformly dispersed Mg_2Si particles and the removal of the oxide film.

Acknowledgments

This work was financially supported by the National Natural Science Foundation of China (Nos. 52271102 and 52075198), and the Analytical and Testing Center, HUST.

References

[1] Li Q, Jiang W, Fan Z. Development of prominent bonding strength in Al/Mg bimetal composites prepared by ultrasonic vibration-assisted compound casting: Effects of ultrasonic powers. J. Mater. Sci. Technol. 2024,197:78-93.
[2] Li Q, Fan Z, Jiang W. Development of Al/Mg bimetal prepared by ultrasonic vibration-assisted compound casting: Effects of interface treatment temperatures. Mater. Sci. Eng. A 2024,890:145911.
[3] Li Q, Jiang W, Fan Z. Development of Al/Mg bimetal processed by ultrasonic vibration-assisted compound casting: Effects of ultrasonic vibration treatment duration time. Materials 2023,16:5009.

Fabrication and Microstructure Characterization of Diamond/Aluminum Composites by Gas Pressure Infiltration Technology

Qiang Zhang[1, 2*], Ping Zhu[1], Yifu Ma[1], Gaohui Wu[1, 2], Yinyin Pei[3], Sujuan Zhong[3]
1. School of Materials Science and Engineering, Harbin Institute of Technology, Harbin 150001, China
2. State Key Laboratory of Precision Welding & Joining of Materials and Structures, Harbin Institute of
3. Technology, Harbin 150001, China
4. State Key Laboratory of Advanced Brazing Filler Metals & Technology, Zhengzhou Research Institute of
5. Mechanical Engineering Co., Ltd., Zhengzhou 450001, China
*Corresponding address: e-mail: zhang_tsiang@hit.edu.cn

Abstract
The necessity to remove the local heat accumulation of electronic devices has brought great challenges to the thermal management aluminum matrix composites designed to effectively dissipate heat in electronic applications. In this work, diamond/aluminum composites have been produced by gas pressure infiltration of the continuous aluminum matrix into the diamond particles preforms at 800℃. The interfacial product of Al_4C_3 was found mostly on diamond (100) surfaces. Nanoscale W coating was produced on diamond particles by magnetron sputtering method. Then the formation of hydrophilic Al_4C_3 was supressed, with reduced content and size of Al_4C_3 and newly formed Al_5W in the composite. The effect of these interfacial phases on the phonon scattering and consequent thermal conductivity of diamond/aluminum composites was also discussed. This research helps to understand the microstructure features in diamond/aluminum composites.

Keywords: Diamond, Aluminum matrix composites, Interfacial microstructure

1 Introduction
The miniaturization, high-performance and multifunction of electronic devices have caused huge local heat accumulation which needs to be removed to avoid functional failure of devices. The traditional thermal management materials, such as Cu, Mo-Cu or W-Cu composites, cannot satisfy the current requirements of high thermal conductivity and low quality [1].

Diamond, with an ultra-high thermal conductivity (~2200 W/(m·K)), is a prominent filler material to produce aluminum matrix composites for thermal management applications [2]. Diamond has poor wettability with most metals because of its inert feature. Therefore, it is a major challenge to tailor the interfacial microstructure and corresponding properties. In this work, pristine and W-coated diamond particles were added into the continuous aluminum matrix by the gas pressure infiltration process. The interfacial microstructure was characterized and the corresponding effect on the thermal conductivity was discussed.

2 Experimental procedure
The diamond particles were deposited with 50 nm tungsten coating by the magnetron sputtering method. The pure aluminum was chosen as the matrix. Then the diamond/aluminum composites were fabricated by the gas pressure infiltration process.
The composite surfaces were polished by IB-09020CP cross-section ion polishing instrument. The interfacial microstructure was observed by Quanta 200FEG field emission environmental scanning electron microscope (SEM).

3 Result and discussion
Fabrication and distribution of diamond particles
The fabrication process of diamond/aluminum composites is illustrated in Fig.1. Firstly, the pristine or W-coated diamond particles were filled into a graphite mold to produce the preform according to the given volume fraction. Then the mold, with the aluminum ingot on top of the diamond preform, was put into the vacuum furnace. When the mold was heated to 800 °C and maintained for 10 min, a gas pressure of 15MPa was applied to force the molten aluminum to infiltrate into the diamond preform completely. When the furnace was cooled down to the room temperature, the diamond/aluminum composites were obtained.

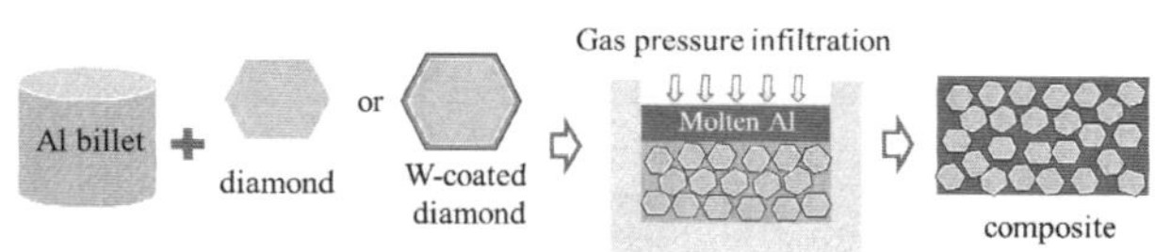

Fig.1 Fabrication route of diamond/aluminum composites

Fig.2 shows the optical microstructure of as-fabricated diamond/aluminum composite. The diamond particles distributed uniformly in the composite. The composite was dense and macroscopically homogeneous, and seldom particles cluster was observed. A dense microstructure was beneficial to thermal management applications because of improvement in heat conduction.

Fig.2 Optical micrograph of diamond/aluminum composite

4 Interfacial microstructure

Fig. 3 is the SEM morphology showing the interfacial microstructure of uncoated diamond/aluminum composite after cross-section ion polishing. The interfacial reaction product of Al_4C_3, was found at the interface of uncoated diamond/aluminum composite, mostly on diamond (100) surfaces.

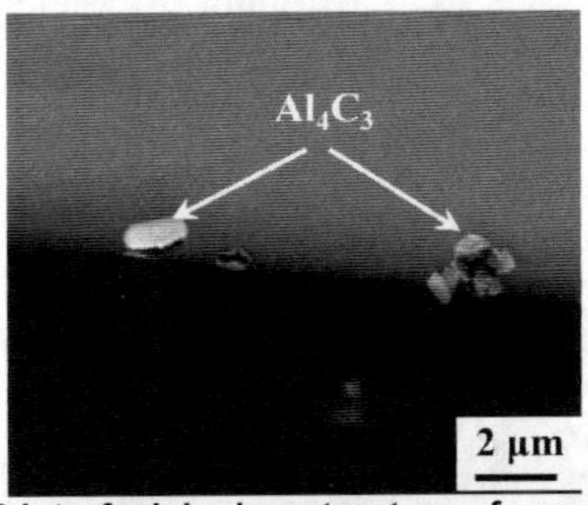

Fig.3 Interfacial microstructure of uncoated diamond/aluminum composite

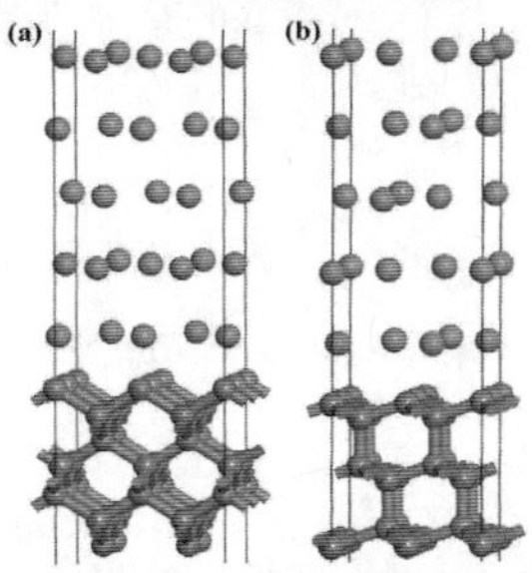

Fig.4 Model of diamond (100)/aluminum (111) (a) and diamond (111)/aluminum (111) (b)

The first-principles calculation was used to study the interfacial properties and reaction of diamond/aluminum composites. Fig. 4 is the model for calculation. The calculated interfacial adhesion work of diamond (111)/aluminum (111) was 4.14 J/m^2, while that of diamond (100)/aluminum (111) was 5.85 J/m^2. It was 41% higher than that of diamond (111)/aluminum (111), suggesting a stronger interfacial bonding of diamond(100)/aluminum(111). More charge transfer was also found at the diamond(100)/aluminum (111) interface during the differential charge density analysis. As a result, the diamond (100) was prone to react with aluminum, as was found in Fig.3.

For the W-coated diamond/aluminum composite, the interfacial microstructure is shown in Fig.5. The Al_4C_3 and Al_5W were both found, with a decrease in the content and size of Al_4C_3. The Al_4C_3 had an acoustic impedance between diamond and aluminum, which was beneficial to the thermal conductivity of composites, while the Al_5W would lead to additional phonon scattering. But the hydrophilic feature of Al_4C_3 would deteriorate the long-term reliability in humid environment. Therefore, it is important balance the interfacial microstructure and the final properties.

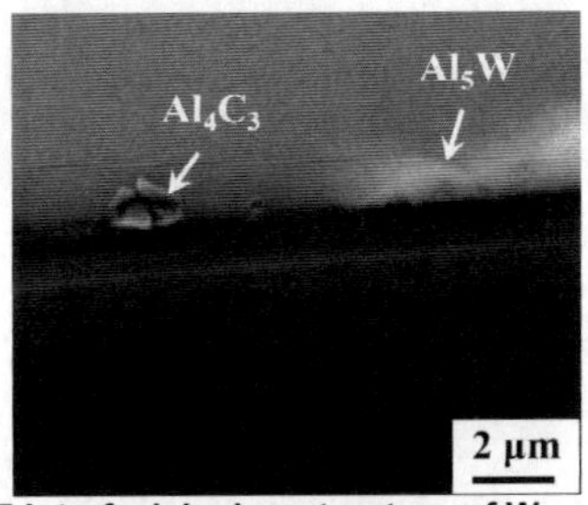

Fig.5 Interfacial microstructure of W-coated diamond/aluminum composite

5 Conclusion

The interfacial microstructure of diamond/aluminum composites were modified using nanoscale W coating on diamond particles. The Al_4C_3 was found on the diamond (100) of diamond/aluminum composite. The first-principles calculations suggested higher interfacial adhesion work and more charge transfer at the diamond (100)/aluminum (111) interface. The nanoscale W coating on diamond particles suppressed the formation of Al_4C_3, then to tailor the final properties of composites.

6 Acknowledgments

This work was funded by the National Key Research and Development Program of China (No. 2022YFE0121400), and the National Natural Science Foundation of China (No. 52071117, No. 52111530297).

References

[1] Qian X, Zhou JW, Chen G. Phonon-engineered extreme thermal conductivity materials. Nature Mater., 2021, 20(9): 1188-1202.
[2] Kondakci E, Nuri S. Enhanced thermal conductivity and long-term stability of diamond/aluminum composites using SiC-coated diamond particles. J. Mater. Sci., 2022, 57: 3430-3440.

Ultrasonic Bonding of Bare Steel Inserts to A354 Aluminum Alloy in a Sand-Casting Process

Dashan Sui[1], Qingyou Han[2*]
1. School of Materials Science and Engineering, Shanghai Jiao Tong University, Shanghai 200030, China
2. School of Mechanical Engineering, Southeast University, Nanjing 211891, China
*Corresponding address: e-mail: hanq@seu.edu.cn

Abstract: We explored the feasibility of using ultrasound-assisted cast-on (UACO) methods for producing bimetal sand castings using as-received bare steel inserts and un-degassed molten A354 alloy. The ultrasound-assisted methods included a direct ultrasound-assisted cast-on (DUACO) method, and an indirect ultrasound-assisted cast-on (IUACO) method. Conventional cast-on (CCO) method was unable to metallurgically bond bare steel inserts to the un-degassed A354 alloy. The IUACO method was capable of producing metallurgical bonding between bare inserts and un-degassed A354 alloy with bonding quality identical to that obtainable using the Cummins process associated with using coated inserts following stringent procedures for insert and molten metal processing. The DUACO method was much effective than the IUACO method in producing high-quality metallurgical bonding between bare steel insert and un-degassed A354 alloy. Push-out test revealed that the shear strength of the bond produced by the DUACO method approached the shear strength of the as-cast A354 alloy. Mechanisms under which defect-free metallurgical bonding were formed were discussed.

Keywords: Aluminum alloy, steel insert, bimetal casting, cast-on method, high-intensity ultrasonic vibration.

1 Introduction

Cast-on method is a cost-effective method for joining iron or steel insert to a lightweight alloy using a casting process [1]. To obtain a high-quality metallurgical bond using the cast-on method, the insert surface has to be absolutely clean and coated with a sacrificial metallic coating, the bond region has to be free from defects such as oxides and porosity, and the molten metal has to be thoroughly degassed and poured under a protection atmosphere [2]. Even with such stringent procedures for insert and molten metal preparation, the bonding is still defective [3]. Methods that do not require the use of such stringent procedures would be of great importance in cost savings in producing bimetal castings of high bonding quality. The aim of this work was to study the feasibility of using UACO methods in bonding bare steel inserts with un-degassed molten A354 alloy in a sand-casting process. The ultrasound-assisted methods included the DUACO method and the IUACO method.

2 Experimental procedure

Fig. 1 is a schematic of the casting configuration. It consisted of a downsprue, a thin runner (6.35×38×203 mm), two thin in-gates (6.35×19×29 mm), a riser (45.7×63.5×244 mm) set on the top of the casting (107×50.8×193 mm), and three inserts. The casting was bottom fed. The middle insert shown in Fig. 1 was not connected to an ultrasonic system. The other two side inserts were each bolted to an ultrasound transducer vibrating at a resonant frequency of 20 kHz. Thus, each side insert was under DUACO conditions while the middle insert was under the IUACO conditions.

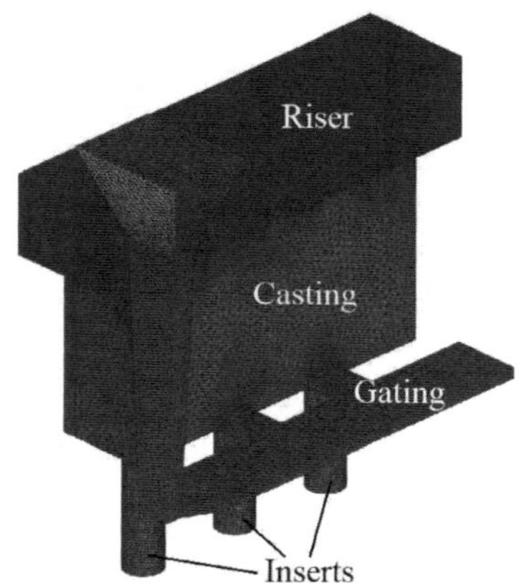

Figure 1. Schematic of the casting configuration.

5 kg of A354 alloy was melted in an electrical resistant furnace, held at 760ºC for 30 min., cooled to a selected pouring temperature and then poured into sand molds. Five disks, about 5 mm thick, were cut near the tip of the insert for the measurement of shear strength using a push-out test carried out on an ATS 10000 tensile tester [4].

3 Result and discussion

There was a large gap between the steel insert and A354 alloy solidified on the insert using the CCO method, indicating that there was no metallurgical bonding between the steel insert and the aluminum casting. Such a result is well known as a bare steel insert is difficult to be bonded to aluminum alloy, especially when the molten alloy is not thoroughly degassed, and the insert is not thoroughly cleansed.

The bonding quality was improved when ultrasonic vibrations were injected into the molten alloy after the alloy was poured over the inserts. However, the bonding quality varied with the way by which ultrasonic vibrations were injected.

Under the IUACO conditions, the bond on the vertical surface of the insert seems acceptable but a large gap existed on the top surface of the insert. This gap resulted from bubbles that adhered to the top surface of the insert. Ultrasonic vibrations transmitted from the neighboring side inserts were not strong enough to shake those bubbles off the top of the middle insert. Small gaps and isolated pores existed at the vertical insert/aluminum interface but the majority of the vertical surface of the insert was metallurgically bonded to the aluminum alloy. A thin and discontinuous layer of intermetallic phases was formed on the bonding regions of the insert surface. Beyond this layer was the dendritic aluminum structure which contained large branches of aluminum dendrites over 1000 μm long.

Under the DUACO conditions, small and isolated pores still existed at the insert/aluminum interfaces including the top and vertical surfaces of the insert. A thin and continuous layer of intermetallic phases, about 40 μm thick, was formed on bonded regions of the insert surface. Beyond this intermetallic layer was equiaxed aluminum grains of about 100 μm. These grains had less dendritic feature as their secondary dendrite arms were less clearly defined.

Fig. 2 provides experimental data of the push-out test on samples. Data on the Cummins process was taken from literature [2-3].

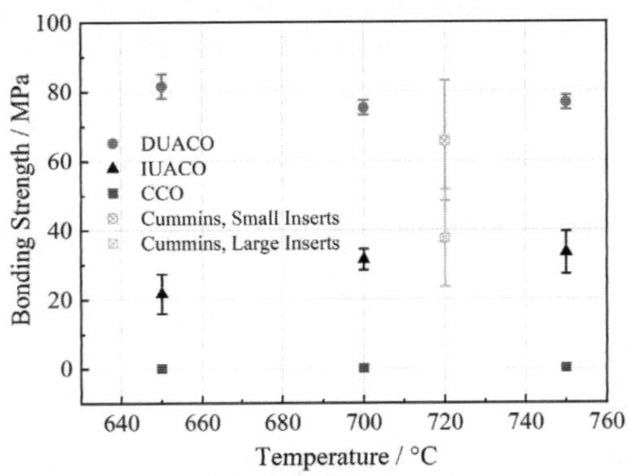

Figure 2. Bonding strength vs. pouring temperature under various conditions.

Without subject to a thorough cleaning and coating procedure, the as-received bare steel insert did not exhibit any significant bonding between the insert and the un-degassed A354 aluminum alloy using the CCO method. The measured bonding strength was between 20 to 40 MPa depending on the pouring temperatures using the IUACO method Such levels of bonding strength were comparable to that obtainable on similar sized inserts in castings made by the Cummins process which has stringent requirements for insert and molten metal preparations. The measured shear strength was about 80 MPa, insensitive to the pouring temperatures, using the DUACO method. Such a bonding strength was comparable to the shear strength of the as-cast A354 alloy. The application of high-intensity ultrasonic vibrations shook off oxides and bubbles that adhered on the surface of the insert and enhanced the steel/aluminum interfacial temperatures, resulted in the formation of a metallurgical bond between the steel insert and the aluminum alloy.

4 Conclusion

High-intensity ultrasonic vibrations are effective in forming a metallurgical bond between a bare steel insert and an un-degassed A354 aluminum alloy. The IUACO method is capable of producing bonding strength comparable to that achievable using the Cummins process. The DUACO method could be used for making a bimetal casting with the bonding strength comparable to that of the shear strength of the A354 alloy.

5 Acknowledgments

This research was partly sponsored by the North American Diecasting Association (NADCA) project #145 and partly by National Science Foundation STTR Project # IIP-0637532.

References

[1] Jorstad, J.L., Morley, R.A., Overbagh, W.H., Steele, G.W., Process for Creation of Metallurgical Bonded Inserts Cast-in-Place in a Cast Aluminum Article, U.S. Patent No. 5,333,668 (1991).

[2] Han, Q., A Modified Cast-on Method for the Reinforcement of Aluminum Castings with Dissimilar Metals, Metall. Mater. Trans. B, 47B, 3266-3273 (2016).

[3] Sui, D., Chen, Y., Han, Q., Reducing Macroscopic Defects at the Aluminum/Steel Interface in Bimetal Casting Made by the Cummins Process, Metall. Mater. Trans. B., s11663-023-02774-9 (2023).

[4] Sui, D., Han, Q., Ultrasound-Assisted Cast-On Method: Obtaining High-Quality Metallurgical Bonds between a Bare Steel Insert and A354 Aluminum Alloy, Journal of Materials Processing Technology, 2023, 117783.

Processing of Porous AC-AlSi11 Silumin Matrix Composites by Casting

Joanna Borowiecka-Jamrozek
Faculty of Mechatronics and Machine Design
Kielce, University of Technology, 25-314 Kielce, Al. Tysiąclecia PP 7, Poland
*Corresponding address: e-mail: jamrozek@tu.kielce.pl

Abstract: The paper presents the application of the casting method to produce porous composites, called syntactic foams, of the casting alloy-solid particles type. Using this method, composites were produced based on Al alloys reinforced with clinoptilolite particles, a natural mineral from the zeolite group. Before the casting process, tests were carried out on the morphology, physicochemical properties and chemical composition of zeolite, which was obtained from a rock called zeolite tuff, mined in a quarry in Kucin (VSK PRO-ZEO s.r.o., Slovakia). Observations of the microstructure of the produced composites were also carried out using a scanning electron microscope, diffractometric studies of the zeolite rock as delivered for testing and the produced samples reinforced with zeolite particles. The suitability of the presented composite production method was assessed on the basis of structural tests, with particular emphasis on the distribution of particles in the alloy matrix.

Keywords: composite materials, syntactic foam, zeolite, aluminum alloy, foaming granules, matrix, microstructure, mechanical properties, porosity

1 Introduction

Porous metal materials are materials with a high degree of structural discontinuity. Their structure can be described as a geometrically disordered arrangement of pores in the metal matrix. They are mainly made of light metals or their alloys. One of the technologies for their production is adding a blowing agent directly to the molten metal in a liquid-solid state. Titanium hydride is used as a gas-forming agent, which decomposes into titanium and hydrogen gas [1].

An interesting method of generating the cellular structure are foams reinforced with particles by mixing liquid systems of aluminum alloy - ceramics. The presented work proposes an innovative method of producing metal foam using the method in which aluminum alloy was used as a matrix, and as foaming granules - clinoptilite, a natural mineral from the zeolite group [2]. Klinoptilites are microporous crystalline aluminosilicates, used as industrial adsorbents and catalysts containing a large amount of silicon, calcium, sodium, potassium and admixtures of various metals, such as: barium, strontium, potassium, magnesium, and manganese [3].

Aluminum alloy composites reinforced with zeolite particles give a material with good properties, lower density, higher strength, and a lower coefficient of thermal expansion compared to the solid material obtained by casting.

2 Experimental procedure

The innovative technology of casting metal foam with an aluminum alloy matrix, developed in the presented work, using clinoptilolite as foaming granules, can provide the ability to suppress sounds and vibrations [5].

The process of manufacturing a porous structure in the conducted experiment consisted in adding to the cast aluminum alloy AC-AlSi11 at a temperature of 790°C, zeolite granules with a grain size of 4-6 mm in the amount of 20% by weight. The foaming agent in the form of zeolite granules was mixed with the liquid metal while pouring the mold - die. After solidification and cooling down to room temperature, the ready foamed element with a porosity of about 55% was removed from the mold. Upon contact with metal at elevated temperatures, water vapor is released from the hydrated aluminosilicates, leading to the formation of pores in the liquid or solidifying metal. The zeolite grains that have not decomposed due to the intense evolution of water vapor are the syntactic part of the material. The melting process was carried out in a laboratory muffle furnace at 760°C for zeolite granulate and aluminum alloy. All processes were carried out without a protective atmosphere. Cylindrical samples with a diameter of φ29 were produced for further research.

Before the casting process, the physicochemical properties and chemical composition of the zeolite were tested using the JSM-7100f scanning electron microscope. The content of the elements was determined by x-ray microanalysis in the EDS analyzer by Oxford Instruments. The spectrum of x-rays taken from the surface of the zeolite particles and the content of individual elements are shown in figure 1.

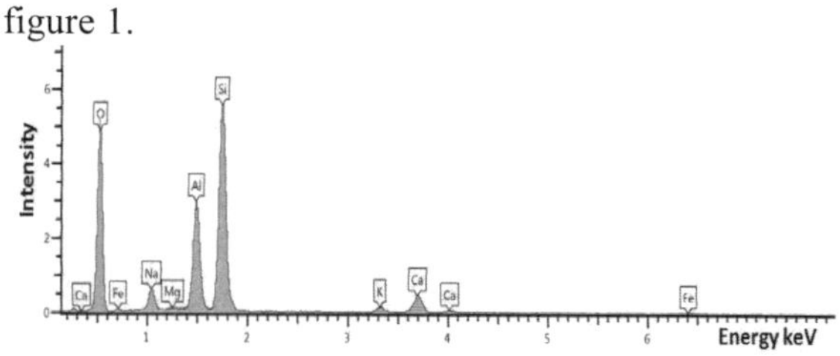

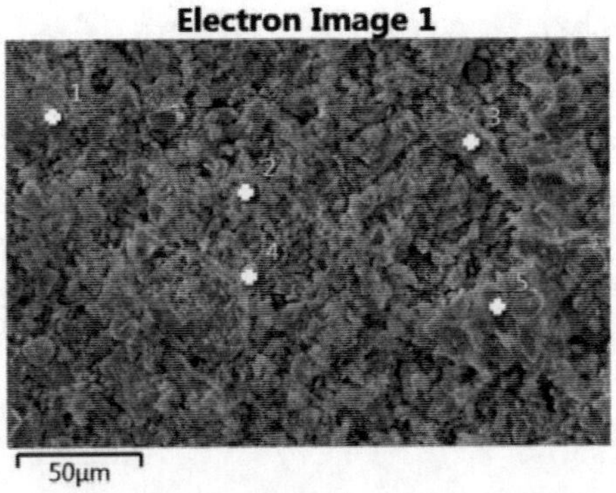

Fig.1. The energy spectrum for emitted X-rays for zeolite particles

Studies of the chemical composition of zeolite as delivered have shown that the main elements present in zeolite are aluminum, silicon, potassium, calcium and iron. This study confirmed that the tested rock is an aluminosilicate mineral.

Fig. 2. Photo of the manufactured composite

3 Conclusion

Optimal parameters of the sintered production technology were selected on the basis of previously conducted own research and it is satisfactory to repeatedly produce a porous structure in the experiment. As a result of the experiment, a casting with a density lower than that of a casting made in a solid mold was obtained. The highest porosity obtained for the composite was 58%. The results of the total porosity test for the produced foams made of aluminum alloy with zeolite allow for the conclusion that the introduction of zeolite particles increases the porosity of composites. The macrostructure of composites based on an aluminum alloy with the participation of zeolite particles are presented in figures 2.

As a result of the observations, no discontinuities were found at the interface between the matrix and zeolite particles. A very good combination of zeolite particles with an aluminum matrix was obtained, without the occurrence of voids, only the pores present in the composite are visible in the photomicrographs. The zeolite particles are clearly visible in the fractures in the form of irregular precipitations.

On the basis of the conducted research, it was assessed that the optimal parameters of the technology for the production of the porous structure on the matrix of the AC-AlSi11 aluminum alloy with the participation of zeolite particles were correctly selected and it is satisfactory to repeatedly obtain the above-mentioned porous structure in accordance with the proposed technology. Acoustic tests planned at a later date will confirm whether the produced porous composite can be used for elements intended to dampen sounds and vibrations.

References

[1] M. Taherishargh, M.A. Sulong, I.V. Belova,et al. On the particle size effect in expanded perlite aluminium syntactic foam, Materials and Design,2015,66: 294–303.

[2] Joanna Borowiecka-Jamrozek, Wojciech Depczyński; The effect of the addition of zeolite on the properties of a sintered copper-matrix composite; In Metal 2017: 26rd international conference on metallurgy and materials. ostrava: tanger, 2017:1652–1657.

[3] G. Gottardi, E. Galli; Natural zeolites, mineral and rocks. 18. Springer-verlag, Berlin Heidelberg, 1985: 256-284.

[4] H. Nanbin, G. Dianyue, H. Bekkum,et al. Introduction to Zeolite Science and Practice. 2nd Completely revised and expanded edition, Elsevier, New York 2001, 137:54-59.

[5] N. Gupta, A Functionally Graded Syntactic Foam Material for High Energy Absorption under Compression, Materials Letters, 2007, 61(4-5):979-982.

Analysis of the Microstructure and Selected Properties of the MMC Composites Produced Via Suction Casting

Łukasz Rakoczy[1*], Krzysztof Pajor[1], Dawid Kozień[2], Tomasz Kargul[1], Aldric Plusch[3], and Rafał Cygan[4]

1. Faculty of Metals Engineering and Industrial Computer Science, AGH University of Krakow, av. Mickiewicza 30, 30-059 Kraków, Poland
2. Faculty of Materials Science and Ceramics, AGH University of Krakow, av. Mickiewicza 30, 30-059 Kraków, Poland
3. University of Dijon, Polytech Dijon (ex-Esirem), 9 avenue Alain Savary, 21078 Dijon, France
4. Rzeszów University of Technology, Research and Development Laboratory for Aerospace Materials, Żwirki i Wigury 4, 35-959 Rzeszów

*Corresponding address: lrakoczy@agh.edu.pl

Abstrac
t: Over the years, metal matrix composites (MMCs) have been developed for various industries. In this work, the 718-TiB2 metal-matrix composites were fabricated through suction casting. The composites' microstructure, chemical composition, phase constitution, phase transformation temperatures, and hardness were investigated by scanning and energy-dispersive X-ray spectroscopy, thermodynamic simulation, and Vickers microhardness measurements. The results indicated that the MMC composites fabricated through suction casting exhibit a dendritic microstructure with the segregation of strengthening precipitates into interdendritic spaces. The addition of TiB2 particles increased the hardness and ultimate tensile strength.

Keywords: MMC, superalloy, composite, nanoparticles

1 Introduction

Composites can be manufactured using two main methods based on the temperature at which the metallic matrix is processed: liquid-state and solid-state processes. Creating MMCs aims to optimize the properties of both the matrix and the reinforcing precipitates, resulting in a unique combination of characteristics. MMCs integrate the strength and rigidity of non-metallic reinforcing particles with the ductility of a metal matrix, often resulting in a lower density that reduces the weight of the components. Ni-based alloys are particularly favored as a matrix for refractory composites in high-temperature applications due to their high strength, flexibility during service, and excellent resistance to hot corrosion and oxidation. The selection of ceramic particles is based on criteria such as their reactivity with the matrix, wetting ability, coefficient of thermal expansion, and density. For optimal performance, the chosen reinforcing particles should have a thermal expansion coefficient closely matching that of the matrix, minimizing the variance and promoting the generation of compressive or tensile stresses during service [1-3].

2 Experimental procedure

Alloy 718 and TiB_2 powders were used to produce the composites (Table 1). Its morphology is presented in Figure 1.

Table 1 Chemical composition of Alloy 718 powder, wt%

Ni	Fe	Cr	Mo
Bal.	18.55	19.8	2.98
Ti	**C**	**Al**	**Nb**
0.80	0.005	0.56	5.2

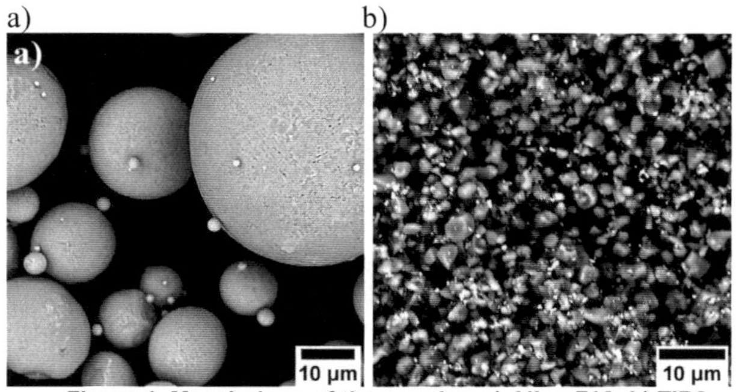

Figure 1. Morphology of the powder: a) Alloy 718; b) TiB2

The samples for microstructure analysis were taken and hot-embedded in Struers Polyfast conductive resin. Then, the grinding process was carried out on sandpaper of increasing grit levels, starting from #100 to #4000. After grinding, a three-stage polishing was conducted. Firstly, on Struers DiaDuo diamond suspension with a 3 μm particle size, followed by Struers DiaDuo diamond suspension with a 1 μm particle size, and finally using Struers OP-S reagent with a silica particle size of 0.06 μm. The unetched samples were then observed using a Leica DM/LM light microscope for preliminary observations. Subsequently, observations were carried out using a BSE (back-scattered electrons) detector to identify microstructural constituents and characterize the precipitates morphology. Point semi-quantitative analysis of the chemical composition, linear analysis of the distribution of selected elements, and distribution maps of selected alloying elements were performed using the EDX detector. The accelerating voltage for imaging and analyses was set at 20 kV.

3 Result and discussion

The microstructure of the 718-TiB_2 composites is presented in Fig. 2 and 3. The images of unetched microstructure confirm lack of the hot cracks and unacceptable discontinuities. Only a local microporosity has been observed.

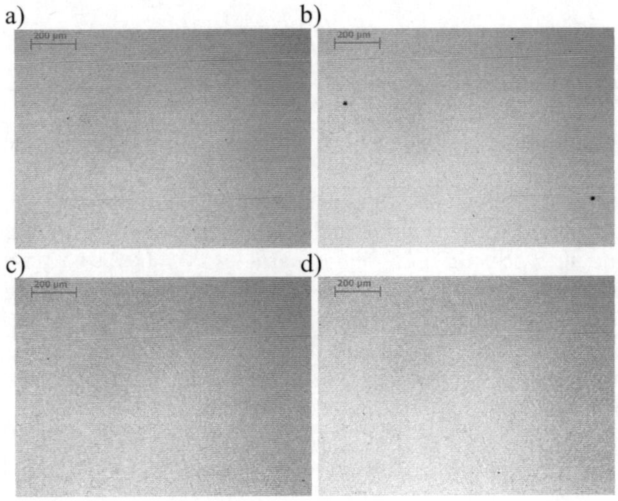

Figure 2. Microporosity in the 718-xTiB2 composite: a) x=1.25%; b) x=2.5%; c) x=3.75%; d) x=5.0%, LM

On the etched samples a dendritic structure is revealed (Fig. 3). A relatively homogeneous microstructure characterizes the dendritic regions, while in the interdendritic spaces the primary strengthening precipitates has been revealed.

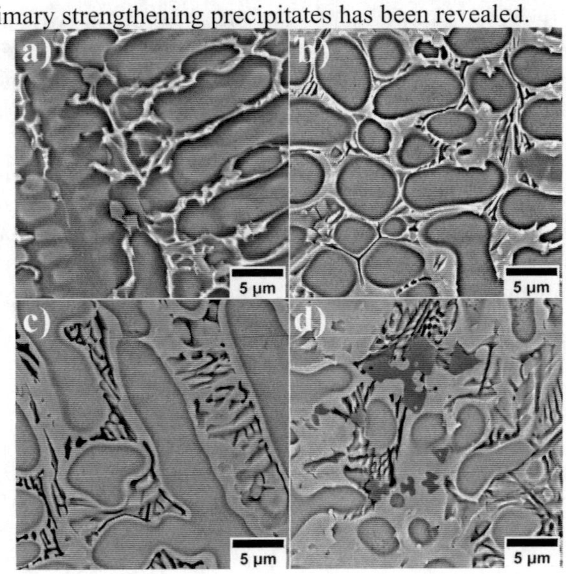

Figure 3. Morphology of precipitates in as-cast 718-xTiB2 composite: a) x=1.25%; b) x=2.5%; c) x=3.75%; d) x=5.0%, LM

4 Conclusions

The microstructure and selected properties of the MMC with different TiB_2 particles produced via suction casting were investigated. The following key results were obtained:

(1) Microporosity is the investigated composites occurring only locally and not exceeding 0.1%.

(2) The addition of TiB_2 particles led to an increase of the Alloy 718 hardness.

(3) Composites characterized by a dendritic morphology with segregation of alloying elements.

5 Acknowledgments

The authors gratefully acknowledge the funding by National Centre for Research and Development Poland under grant LIDER13/0036/2022 "Development of the manufacturing and deposition technology of metal ceramic nanocomposite coatings for the structural reconstruction of heat-resistant nickel-based superalloys".

References

[1] Lemos G, Márcio C, Pyczak F, et al. Development of a TiCp reinforced Ni-based superalloy MMC, with high creep resistance and reduced weight[J]. Key Eng Mater. 2017, 742:189–96.

[2] Olejnik E, Szymanski Ł, Tokarski T, et al. Local composite reinforcements of TiC/FeMn type obtained in situ in steel castings[J]. Arch Civ Mech Eng, 2019, 19:997–1005.

[3] Arsenault R, Fishman S, Taya M. Deformation and fracture behavior of metal-ceramic matrix composite materials[J]. Progress Mater Sci. 1994; 38:1–157.

On the Preparation and Properties of Al-7Si-0.4Mg Matrix Composites Reinforced by In-situ Intragranular Al_3Ti and Intergranular TiB_2

Cuicui Yang[1], Zhiwu Dong[1], Zhiwei Liu[1,*]
1. State Key Laboratory for Mechanical Behavior of Materials, School of Materials Science and Engineering, Xi'an Jiaotong University, No. 28, Xianning West Road, Xi'an, Shaanxi Province 710049, P.R. China
*Corresponding address: e-mail: liuzhiwei@xjtu.edu.cn

Abstract: In particulate reinforced Al matrix composites (PRAMCs), the micro-sized ceramic particles tend to segregate at the α-Al grain boundary, resulting in a sharp drop in the ductility of the composite and thus restricting its application. It is promising to improve the ductility of the TiB_2-containing composites while maintaining high strength by introducing the intragranular Al_3Ti particles to replace part of TiB_2 particles. In the present work, ($Al_3Ti+TiB_2$)/Al-7Si-0.4Mg composites were prepared by using the ultrasound assisted in-situ casting. The results showed that most of modified Al_3Ti particles were inside the α-Al grains while TiB_2 particles were mainly located at the α-Al grain boundaries. Under the ultrasonic stirring, both particles were uniformly distributed in the Al alloy matrix. Compared with Al-7Si-0.4Mg alloy, the ($3Al_3Ti+2TiB_2$)/Al-7Si-0.4Mg composite exhibited the best comprehensive mechanical properties, and its yield strength, ultimate tensile strength, and elongation (254MPa, 310MPa, and 3.8%) were improved by 21.0%, 21.6%, and 111%, respectively. This work provides a novel composite design approach to achieve a strength-ductility synergetic improvement.

Keywords: Al_3Ti; TiB_2; Hybrid reinforced particles; Microstructure; Mechanical properties

1 Introduction

PRAMCs have been widely used in the automotive and aerospace industries due to its specific lightweight and high strength characteristics [1]. The reinforced particles of PRAMCs include ceramic (TiB_2 [2], TiC [3], etc) and intermetallic compound (Al_3Ti [4], Al_3Zr [5], etc). TiB_2 particles in as-cast PRAMCs tend to aggregate together in the Al melt, resulting in the microstructure with the agglomerated TiB_2 particles at the α-Al grain boundary, which is detrimental to the mechanical properties of PRAMCs and thus restricts their application [6]. In contrast, during the cooling and solidification process of Al_3Ti-containing Al melt, the peritectic reaction occurs, in which Al_3Ti particles are used as the heterogeneous nucleation sites of α-Al grains [7]. In this case, Al_3Ti particles are located within the α-Al grains, which will not significantly deteriorate the ductility of PRAMCs.

Hence, in order to achieve the superior strength-ductility synergy of TiB_2-containing PRAMCs, it is necessary to replace part of TiB_2 particles with Al_3Ti particles. Moreover, ultrasound treatment contributes to promote the uniform dispersion of small-sized particles [4,8]. In the present work, the dual-phase particles ($Al_3Ti+TiB_2$) with different rations were introduced into the Al-7Si-0.4Mg matrix (wt.% unless otherwise noted) to prepare the composites by the ultrasound assisted in-situ casting. Furthermore, the phase, microstructure and tensile properties of T6-composites were investigated.

2 Experimental procedure

The ($xAl_3Ti+(5-x)\ TiB_2$)/Al-7Si-0.4Mg composites (Table 1) were fabricated by using Al-7Si-0.4Mg matrix (pure Al (99.7%), Si (99.9%) and Mg (99.9%)), Al-10Al_3Ti (Al-K_2TiF_6 system) and Al-10TiB_2 (Al-K_2TiF_6-KBF_4 system) master composites as raw materials. The detailed fabricating process of the master composites can be referred to our previous research [4,8]. Al-10TiB_2, pure Al and Si were initially melted at 750℃ for 30min within a graphite crucible in a high-frequency induction furnace. Then Al-10Al_3Ti and pure Mg were added into the melt when the temperature was stabilized at 700℃. After that, the ultrasound probe (1.6kW, 20kHz) was immersed into the melt for 5min with the aim of making Al_3Ti and TiB_2 particles distribute uniformly in the melt. Finally, the composite melt was poured into a graphite mold to obtain an ingot. Then all ingots were T6-heat treated: solution treatment (540℃ 4 h, 60℃ water quenching) + aging treatment (170℃ 7 h, air cooling).

Table 1. Composition of the composites (wt.%)

Samples	Al_3Ti	TiB_2	Matrix
0-0#	0	0	Si: 7 Mg: 0.4 Al: Bal.
5-0#	5	0	
4-1#	4	1	
3-2#	3	2	
2-3#	2	3	
1-4#	1	4	
0-5#	0	5	

The phases and microstructure of samples were examined by X-ray diffractometer (XRD, Bruker D8 ADVANCE) and scanning electron microscopy (SEM, SU3500). Flat specimens (gauge length×width×thickness: 20×4×3.5mm) were tensile tested by using a testing machine (CMT5305, MTS, USA) equipped with an extensometer at a constant cross-head speed of 0.5mm/min.

3 Result and discussion

1. Phase analysis and microstructure

Figure 1 shows the XRD patterns of T6-samples. As shown in Figure 1, Al, Si, in-situ formed Al_3Ti and TiB_2 phases were found in the composites. Owning to the low Mg content (0.4%), the diffraction peak of Mg_2Si cannot be detected in the matrix. The SEM images of the composites are shown in Figure 2(a-f). It can be clearly seen that in Al_3Ti-containing composites, the in-situ formed Al_3Ti particles were distributed uniformly in the Al matrix and most of them were located inside the α-Al grains, which showed they could act as heterogeneous nucleating sites [7]. Furthermore, most of Al_3Ti particles were blocky in shape and a few of them had a short rod-like morphology. It should be noted that no large-sized flaky Al_3Ti particles were found in the composites. Here, the formation of modified Al_3Ti particles with uniform distribution in the matrix was attributed to the fragmentation and dispersion of Al_3Ti particles by high-intensity ultrasound [4]. In TiB_2-containing composites, small-sized TiB_2 particles were uniformly dispersed along the grain boundaries of α-Al grains, which is a result of the particles pushing effect during the cooling and solidification process of the composite melt. Similarly, ultrasonic treatment promoted uniform dispersion of TiB_2 particles [8].

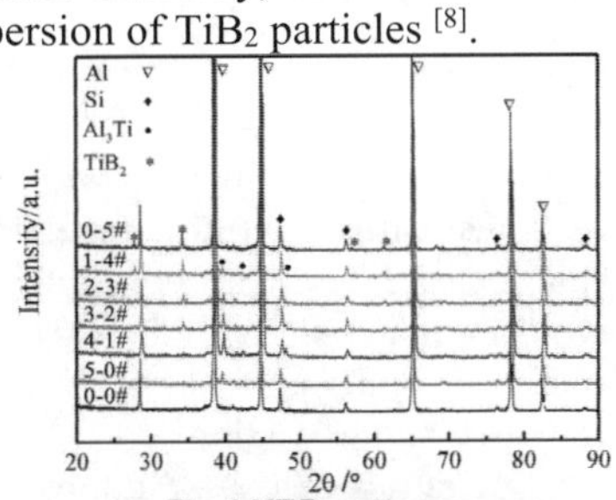

Fig.1 XRD patterns

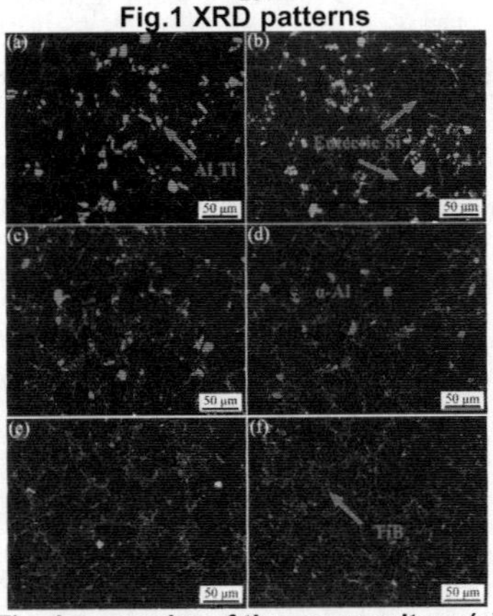

Fig.2 SEM (BSE) micrographs of the composites: (a) 5-0#; (b) 4-1#; (c) 3-2#; (d) 2-3#; (e) 1-4#; (f) 0-5#

2. Tensile mechanical properties

Figure 3 presents the typical tensile results of T6-samples. As displayed in Figure 3, with the increase of TiB_2 content, the values of yield strength (0.2% YS), ultimate tensile strength (UTS) and elongation (El) of T6-samples first increased and then decreased. The mean values of 0.2% YS, UTS and El of the 3-2# composite was 254MPa/310MPa/3.8% (compared to 0-0# matrix alloy, improved by 21.0%/21.6%/111%), which had the best mechanical properties. The improvement of comprehensive mechanical properties of dual-phases reinforced composites was attributed to grain refinement, load-bearing effect, coefficient of thermal expansion mismatch and Orowan strengthening. Furthermore, it can be obviously observed that Al_3Ti contributed to the improvement of the ductility while TiB_2 was beneficial for enhancing the yield strength of composites.

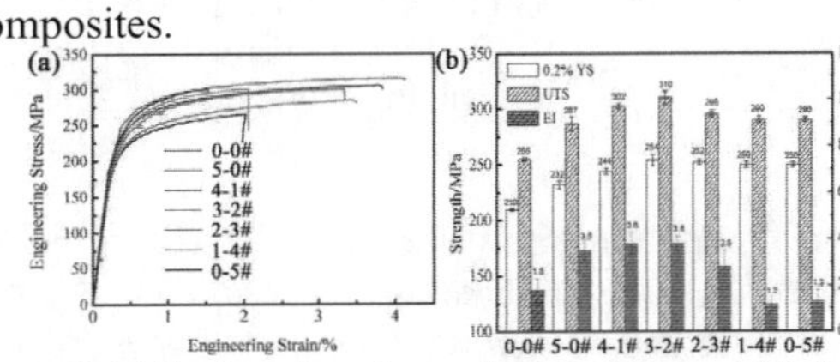

Figure 3 (a) Engineering stress-strain curves and (b) detailed mechanical properties of T6-samples.

4 Conclusion

In the present work, the dual-phase particles (intragranular-Al_3Ti+intergranular-TiB_2) reinforced Al-7Si-0.4Mg matrix composites with high strength and good ductility were successfully prepared using the ultrasound assisted in-situ casting. Most of modified Al_3Ti particles were inside the α-Al grains, while TiB_2 particles were mainly located at the grain boundaries of α-Al grains. Both particles were uniformly distributed in the Al alloy matrix as a result of the ultrasonic dispersion. Comparing with the alloy matrix, the $(3Al_3Ti+2TiB_2)$/Al-7Si-0.4Mg composite exhibited the best comprehensive mechanical properties, and its 0.2% YS/UTS/El (254MPa/310MPa/ 3.8%) were improved by 21.0%/21.6%/111%. This work provides a novel composite design approach to achieve a strength-ductility synergetic improvement.

Acknowledgments

The authors thank the National Natural Science Foundation of China (Grant Nos. 52174372, 51974224 and 51604211) for supporting the work.

References

[1] V. Chak, H. Chattopadhyay, T. L. Dora. Journal of Manufacturing Processes, 2020, 56: 1059-74.
[2] K. Zhao, M.M. Liu, H.J. Kang, et al. Journal of Alloys and Compounds, 2022, 916: 165461.
[3] S.H. Pan, J. Yuan, K.Y. Jin, et al. Materials Science and Engineering: A, 2022, 840: 142992.
[4] C.C. Yang, Z.W. Liu, Q.L. Zheng, et al. Journal of Alloys and Compounds, 2018, 747: 580-90.
[5] P. Pandee, P. Sankanit, V. Uthaisangsuk. Materials Science and Engineering: A, 2023, 866: 144673.
[6] Y.H. Wu, L.W. Li, H.J. Kang, et al. Materials Characterization, 2024, 208: 113652.
[7] R. Gupta, G.P. Chaudhari, B.S.S. Daniel. Composites Part B: Engineering, 2018, 140: 27-34.
[8] J. Liu, Z.W. Liu, Z.W. Dong, et al. Journal of Alloys and Compounds, 2018, 765: 1008-17.

Combination of Superior Strength and Excellent Ductility of Al-3%Cu Alloy with Homogeneous Layer and Lamella Structure

Ruirui Wu [1], Fang Wang [1], Huiqin Chen [1]
1. School of Materials Science and Eng ineering, Taiyuan University Of Science and Technology, No.66 Waliu Road, WanBailin District, Taiyuan city, Shanxi Province, China
*Corresponding address: No.66 Waliu Road, WanBailin District, Taiyuan city, Shanxi Province, China
e-mail: 2018016@tyust.edu.cn

Abstract: A non-uniform layered Al-3%Cu composites was prepared by diffusion bonding process. First, Al-3%Cu alloy sheets with diameter of 15 mm and thickness of 1 mm prepared by cold rolling (CR) and cold rolling + annealing (AR) was alternately superimposed into a cylindrical specimen. The cylinder was then thermally compressed by a thermal simulator, and finally hydraulically compressed into 1 mm Al-3%Cu alloy composites with non-uniform layered structure. The composites were annealed at 200 ℃, 250 ℃, and 300 ℃ respectively for 3h. The results show that the coarse crystal in the soft zone (AR) of Al-3%Cu alloy with non-uniform layered structure is the basis of plasticity. After annealing at 200 ℃ for 3h, the proportion of hard region (CR) in the non-uniform region is the highest, and the hard region can limit the geometric necessary dislocation (GND) generated in the soft region to the greatest extent, resulting in large strain gradient and non-uniform deformation induced stress. Simultaneously, the additional work hardening and the precipitated θ′ and other secondary phases resist dislocation slip and increase the strength. The uniform elongation increases by 97.4%, from 3.8% to 7.5%, compared with that of the unannealed specimens. Meanwhiles, the strength remains almost unchanged, from 311.07 MPa to 308 MPa, means that the comprehensive mechanical properties are improved.

Keywords: Al-3%Cu alloy; anneal; geometric necessary dislocation; mechanical properties

1 Introduction

As an excellent lightweight material, aluminum-copper alloy has become an irreplaceable material in the aerospace and automotive industries because of its high strength and high stiffness [1-4]. However, with the improvement of industrial development level and innovation awareness, from the previous single pursuit of high strength to plasticity, strength, fatigue resistance and other aspects of excellent characteristics, the traditional structure of aluminum-copper alloy has gradually not met people's needs.

2 Experimental procedure

The as-cast Al-3%Cu alloy was homogenized at 500 ℃ for 24h, and then cut into 10 ×8 ×5 mm sheets. A sample of 1 mm thickness was obtained by 80% cold rolling (CR), and the sample was annealed at 575 ℃ for 1h after cold rolling (AR). The CR and AR discs are alternately laminated into a cylinder, the total number of layers in the cylinder is odd, and the middle layer is CR. The preparing processes of heterogeneous laminated structure Al-3%Cu alloy in Fig.1.

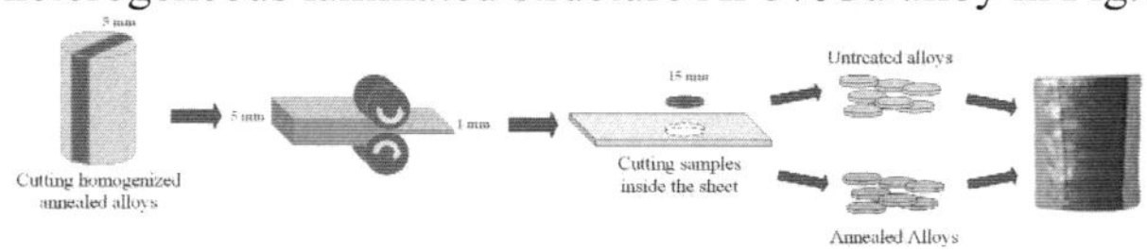

3 Result and discussion

microstructure

The dislocation density of Al-3%Cu alloy with non-uniform layered structure decreases during annealing without changing the grain morphology. The subsequent nucleation of the grains is not simultaneous, and the growth rate is different, resulting in a significant change in the size of the new grains, and two types of regions, soft and hard. The soft zone is the aggregation of coarse grains with a grain size greater than 1 μm after annealing. Meanwhile, the hard area is the aggregation of ultra-fine grains due to severe deformation [28].

Fig. 2 is a metallographic microstructure of the non-uniform layers under different annealing conditions of Al-3%Cu alloy. The visibility of the interface layer with a uniform contrast expression as smooth connection without micro-holes and cracks. It shows that during the preparation and annealing process of non-uniform layered structure of Al-3%Cu alloy, the integrity of the interface is maintained reasonable, and there is the continuous interface layer forming between the soft and hard area, thus, the boundary is obvious. Fig.2(a) shows the non-uniform layered of Al-3%Cu alloy without annealing. The surface of CR and AR layers is rough, and the second phaseAl2Cu on the surface is less precipitated. Fig.2(b) shows the non-uniform layered of Al-3%Cu alloy annealed at 200 ℃ for 3h. The bonding rate of the interface is higher, the surface of the AR layer becomes smoother than that of the non-annealed alloy, and the second phase is precipitated on the Al matrix. The defects on the surface of Al2Cu gradually disappear, the number of holes is reduced, and the grain size is basically unchanged, but it still keeps fine grain. Fig.2(c) shows the non-uniform layered Al-3%Cu alloy annealed at 250℃ for 3h. The surface holes appear, the defects increase, the second phase decreases, and the interfacial bonding rate decreases. Fig.2(d) shows the non-uniform layered Al-3%Cu alloy annealed at 300 ℃ for 3h. Under the annealing condition of high temperature and long time, the grains of the CR and AR layers are coarsened completely. The strength is significantly reduced, and minor damage occurs at the interface. After low temperature annealing at 200 ℃ for 3h, 250 ℃ for 3h and 300 ℃ for 3h, the α phase in the non-uniform layered Al-3%Cu alloy produces solid solution strengthening, and the precipitation of the second phase in the CR layer is much higher than that in the AR layer. The strengthening effect of Cu diffusion results in the formation of θ′(Al2Cu) phases, which leads to the increase of the hardness of the alloy [29,30]. When annealed at 200 ℃ for 3h, the metallurgical bonding rate of the interface increases.

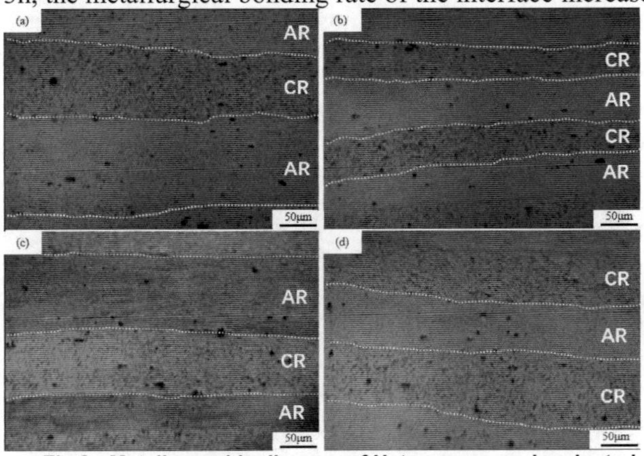

Fig.2 - Metallographic diagram of Heterogeneous Laminated Structure Al-3%Cu Alloy under different annealing conditions ((a) untreated (b) at 200 ℃ for 3h (c) at 250 ℃ for 3h (d) at 300 ℃ for 3h)

4 Conclusion

Starting from the improvement of the ductility and strength of Al-3%Cu alloy, this paper adopts the diffusion connection + hydraulic forming composite process to prepare a non-uniform layered structure Al-3%Cu alloy, performs different annealing treatments for non-uniform layered structure Al-3%Cu alloy, and optimizes the mechanical properties through microstructural design and annealing process adjustment.

(1) Through the comparison of the three annealing processes, it was found that the interfaces of untreated and annealed were free of defects such as micropores and cracks, and the interfaces were metallurgically combined.

(2) At 200 °C annealing for 3h, the metallurgical binding rate of the interface is the highest, the distribution of oxygen elements in the CR layer, AR layer and interface is uniform and the alloy precipitates (Al2Cu, etc.) are generated at the same time.

(3) At the different annealing temperatures, the soft zones are not much different, ensuring the basic plasticity. At 200 °C for 3h annealing process, the hard zone has the highest proportion and the highest strength, and the uniform elongation increases by 97.4%, from 3.8% to 7.5%. In addition, the strength is basically unchanged (311.07 MPa to 308 MPa), and the strength-plasticity match is achieved.

(4) Non-uniform layered structure Al-3% Cu alloy needs to optimize the microstructure through precipitation strengthening, and its hard zone can be maximum constraint on the GND, resulting in a large strain gradient. The strain gradient will produce non-uniform deformation induction stress, which is a kind of back stress strengthening, resulting in additional working hardening and strength. Simultaneously, due to the large density of the hard area, there are a large number of second phase precipitations such as Al2Cu in the non-uniform area, which can resist dislocation slip and increase the strength.

5 Acknowledgments

This work was supported by Shanxi Province Science Foundation for Youths(No. 202203021222198).

References

[1] Chen GQ, Liu JP, Dong ZB, et al. Underlying reasons of poor mechanical performance of thick plate aluminum-copper alloy vacuum electron beam welded joints[J]. Vacuum 2020,182:109667.

[2] Matvienko YI, Polishchuk SS, Rud AD, et al. Effect of graphite additives on microstructure and mechanical properties of Al–Cu composites prepared by mechanical alloying and sintering[J]. Mater Chem Phys 2020,254:123437.

[3] Jiang L, Rouxel B, Langan T, et al. Coupled segregation mechanisms of Sc, Zr and Mn atθ′interfaces enhances the strength and thermal stability of Al-Cu alloys[J]. Acta Mater 2021, 206:116634.

[4] Perrin A, Bahl S, Leonard D N, et al. Phase stability in cast and additively manufactured Al-rich Al-Cu-Ce alloys[J]. J Alloys Compd,2022, 926:166984.

[5] Wang RF, Guo WG, Liu LT, et al. Simultaneously improved strength and ductility in aluminum matrix composite with heterogeneous structures under impact loadings[J]. J Mater Res Technol 2023,23:191-208.

Metal Mesh Reinforced Metallic-Glass Composites Prepared by High-Pressure Die Casting

S. H. Hu, W. J. Gao, Lehua Liu*
School of Mechanical and Automotive Engineering, South China University of Technology, Guangzhou, 510640, China
*Corresponding address: e-mail: liulh@scut.edu.cn

Abstract： Using high-pressure die casting, metallic-glass composites (MGC) with improved plasticity and strength were successfully fabricated by introducing different metal mesh. Experimental results demonstrate that the prepared interpenetrating structure composites exhibit a fracture plastic strain of 6.7% and a fracture strength of 1900 MPa in compression tests. This outstanding performance is attributed to the effective suppression of shear band propagation by the high hardness Mo woven toughening agents. The discussion emphasizes that the hardness of the toughening agent material is the dominant factor influencing the mechanical properties of MGC.

Keywords: Bulk metallic glass composite; HPDC; plasticity; toughening agent

1 Introduction

High-pressure die casting (HPDC) is a widely utilized metal forming technique [1]. Recent advancements in HPDC have expanded its application in the production of bulk metallic glass (BMG) engineering components, such as amorphous alloy hinges and middle frames for 5G folding mobile phones, allowing the fabrication of metallic glass parts with intricate structures [2]. Despite the exceptional properties of BMGs, such as high strength and superb surface finish, their brittleness at room temperature remains a significant limitation [3].

To address these limitations, this study employs HPDC to inject a glass-forming liquid into a metal mesh under high filling ratios and pressures, resulting in a dual continuous interpenetrating phase composite structure. The study focuses on investigating the impact of the metal mesh hardness on the performance of the metallic-glass composite (MGC).

2 Experimental procedure

This study fabricates interpenetrating MGCs using HPDC. Reinforcement is achieved using Mo, Ni, and Ti materials with a 200-mesh size. The matrix consists of a commercial $Zr_{41.2}Ti_{13.8}Cu_{12.5}Ni_{10}Be_{22.5}$ (at. %) bulk metallic glass (Vit1) known for its high glass-forming ability. The phase constitution of the composites is characterized using X-ray diffraction (XRD). The microstructure and fracture morphology are investigated and analyzed using scanning electron microscopy (SEM). Additionally, the Vickers hardness of the different material wire meshes and the room temperature compression properties of the MGCs are tested and evaluated.

3 Result and discussion

Microstructure characterization

The die-casting process, illustrated in Fig. 1(a-c), is conducted in a high vacuum environment. Fig. 1(d) shows the SEM image of the 200-mesh Mo material used as a 3D reinforcement. The uniformly distributed woven mesh effectively suppresses the propagation of shear bands, enhancing the plasticity of the MGC. Fig. 1(e) presents the XRD patterns of the toughened MGC with various woven mesh materials and their corresponding prepared meshes. From the normalized XRD in Fig. 1(f), typical halo diffraction peaks of the metallic glass, superimposed with crystallization peaks of Mo, Ni, and Ti, can be observed, with no additional crystallization peaks appearing in the diffraction patterns of the prepared MGC. The SEM morphology of the MGC is shown in Fig. 1(g-i), demonstrating that the woven meshes of different materials have achieved good bonding with the metallic glass matrix, with no obvious pores or cracks around the bonding interfaces.

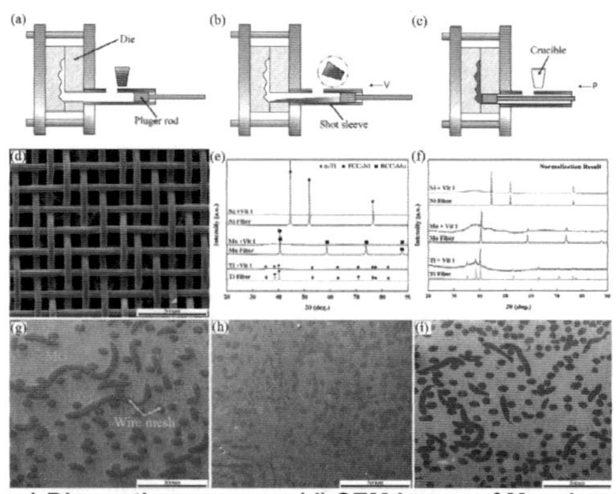

Fig. 1 (a-c) Die casting process; (d) SEM image of Mo wire meshes; (e-f) XRD spectrum of MGC; (g-i) SEM morphology of MGC prepared using different material (In sequence: Mo, Ni, Ti).

Compression deformation behavior

The hardness of metal meshes made from different materials is shown in Fig. 2(a). Mo displays the highest hardness. Fig. 2(b) presents the room temperature compression curves of the toughened MGC and Vit 1. It can be observed that all composite materials exhibit a significant brittle-to-ductile transition. Notably, the MGC reinforced with Mo wire mesh demonstrates the best plasticity under room temperature compression conditions, with a fracture strain of approximately 6.74% and a fracture strength of around 1900 MPa.

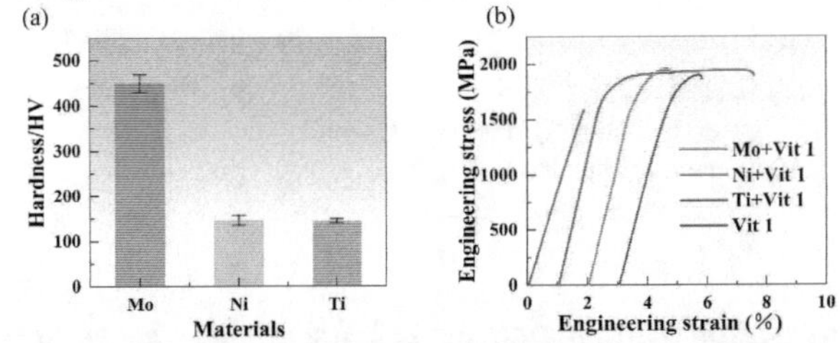

Fig. 2 Mechanical properties of metallic-glass composites materials; (a) Hardness of wire meshes; (b) Room temperature compressive performance of MGC.

Fig. 3 shows the compressive fracture surface morphology of the MGC, where the metallic glass exhibits a typical river-like pattern. Observations from Fig. 3(a-c) reveal that debonding occurred at the interfaces between the Mo, Ti, and Ni wire meshes and the amorphous matrix after plastic deformation, leading to crack formation at these interfaces. Fig. 3(d-f) illustrates the fracture modes of different material metal frameworks. The SEM image in Fig. 3(d) shows clear cleavage steps on the fracture surface of the Mo wire, indicating cleavage fracture during compression. In Fig. 3(e), the fracture surface of the Ni wire appears relatively smooth, suggesting a ductile fracture mode similar to brittle shear fracture. In Fig. 3(f), the fracture surface of the Ti wire shows a certain degree of necking, indicating significant plastic deformation before fracture.

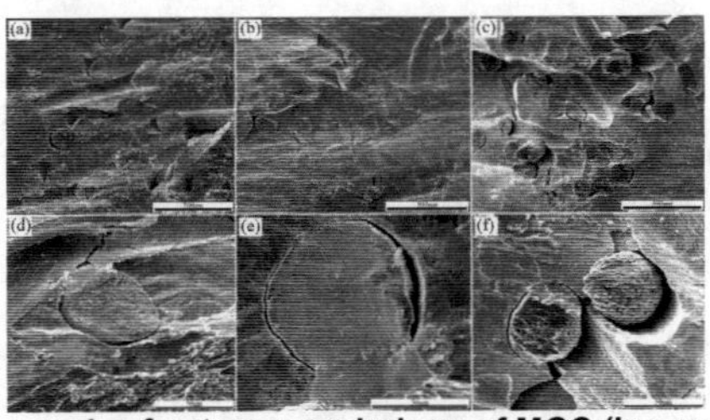

Fig. 3 Compressive fracture morphology of MGC (In sequence: Mo, Ni, Ti); (a) wire mesh and metallic glass; (b) Metallic scaffold fracture surface.

As shown in Fig. 4, the process of shear propagation in interpenetrating structure MGC is illustrated. The extension and propagation of shear bands will be hindered by the metallic mesh, thereby enhancing the plasticity of the material. Furthermore, the higher the hardness of the metallic mesh, the more pronounced the hindrance effect on shear propagation, which is beneficial for enhancing the plasticity of the MGC.

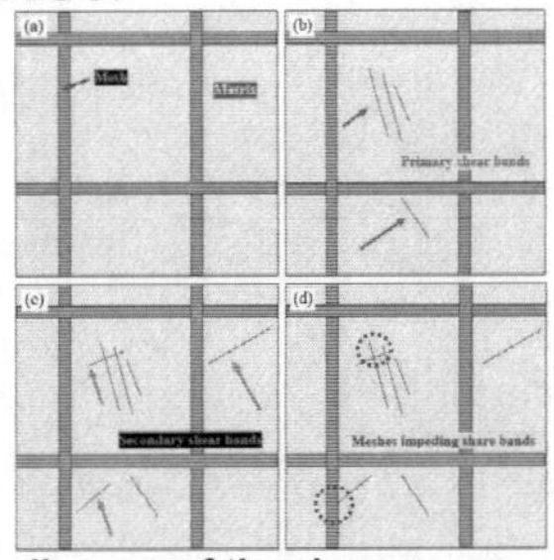

Fig. 4 Schematic diagram of the shear process in metallic-glass composites materials.

4 Conclusion

A novel HPDC method has been successfully employed to prepare MGC with a greater tensile elongation. The outstanding mechanical performance mainly stems from the effective suppression of shear band extension and propagation by the high-hardness Mo woven mesh. The hardness of the toughening agent material is the dominant factor influencing the mechanical properties of MGC. This work provides insights for the design of metallic-glass composites.

5 Acknowledgments

The work is sponsored by the Guangdong Basic and Applied Basic Research Foundation (No.2022A1515012627) and Guangzhou Science and Technology Plan Project (No.2024A04JB668).

References

[1] Vinarcik E J. High integrity die casting processes[M]. John Wiley & Sons, 2002.

[2] Liu L H, Ma J, Yu C Y, et al. Determination of forming ability of high pressure die casting for Zr-based metallic glass[J]. Journal of Materials Processing Technology, 2017, 244: 87-96.

[3] Liu L, Zhang T, Liu Z, et al. Near-net forming complex shaped Zr-based bulk metallic glasses by high pressure die casting[J]. Materials, 2018, 11(11): 2338.

Effects of Silica-Modified Aluminum on the Microstructure and Properties of Diamond/Al-Si Composites

Jiaping Fu [1, 2], Canxu Zhou [1, 2], Yuan Liu [1, 2,*]
1. School of Materials Science and Engineering, Tsinghua University, Beijing 100084, China
2. Key Laboratory for Advanced Materials Processing Technology, Ministry of Education, Beijing 100084, China
*Corresponding address: e-mail: yuanliu@tsinghua.edu.cn

Abstract : To improve the wettability of the Diamond/Aluminum (Dia/Al) composites interface and prevent the excessive formation of Al4C3, this study used silicon-modified aluminum as the matrix. The composites with different silicon content in the matrix were fabricated by the vacuum-assisted pressure infiltration method, and the effects of silicon content on the microstructure and properties of the composites was systematically studied. The results showed that the Dia/Al-Si composites had a uniform and dense microstructure with good interfacial bonding. With the increase in silicon content, both the thermal conductivity and bending strength of the composites first increased and then decreased, while the coefficient of thermal expansion decreased monotonically. When the silicon content was 12 wt.%, the composites prepared with 180 μm diamond particles exhibited the highest thermal conductivity and bending strength, reaching 654 $W \cdot m^{-1} \cdot K^{-1}$ and 296 MPa, respectively, and the coefficient of thermal expansion was as low as $6.13 \times 10^{-6} \cdot K^{-1}$. This study indicates that the silicon-modified aluminum matrix can effectively suppress the excessive formation of Al4C3, and the coexisting of appropriate amount of Al4C3 or SiC at the interface enhances interfacial bonding, thereby improving the performance of the composites.

Keywords: thermal conductivity; electronic packaging; Silicon-modified aluminum; diamond/aluminum.

1 Introduction

Dia/Al composites with excellent performance have broad application prospects in the field of electronic packaging and heat dissipation. However, due to the poor wettability of diamond with aluminum, it is difficult to significantly increase the thermal conductivity of Dia/Al composites[1-3]. At sustained high temperatures, diamond and aluminum liquids are also prone to generate excessive amounts of the brittle and easily deliquescent phase Al_4C_3, which greatly deteriorates the interface and reduces the stability of the properties[2]. Aluminum matrix modification is an important method commonly used for interfacial modification of Dia/Al composites. Mizuuchi et al.[4] concluded that wettability can be promoted and interfacial bonding can be strengthened by aluminum matrix modification Beffort et al.[5] obtained composites with 7 wt.% Si-modified aluminum as a matrix with a thermal conductivity of only 130 $W \cdot m^{-1} \cdot K^{-1}$. Subsequently, Ruch[6] concluded that Si precipitated on the diamond surface hindered the generation of Al_4C_3 such that the interfacial bonding became weaker, and the thermal conductivity was reduced. In order to further enhance the thermal conductivity of Dia/Al composites, this paper systematically investigated the effect of Si content on the microstructure and properties of the composites by using silicon-modified aluminum as the matrix.

2 Experimental procedure

The Pure aluminum and silicon (Tianjin Zuoyuan New Materials Technology Co., China) and synthetic diamond single-crystal particles (Henan Huanghe Whirlwind Co., China) were used as the raw materials. The average particle sizes of the diamond are 180 μm. The Dia/Al composites were prepared by the vacuum-assisted pressure infiltration (VAPI) process. More information concerning preparation details of VAPIed Dia/Al composites can be found in paper[1].The microstructure was examined by field emission scanning electron microscopy (SEM, ZEISS Merlin Compact, Germany). Electrochemical etching was performed by using a direct current (DC) power supply in a 15vol.% HNO3 solution at an etching current of 2A. The thermal conductivities of composites were given by the equation: $\lambda = \alpha \times \rho \times Cp$. The thermal diffusivity (α) was measured using a laser flash apparatus (LFA467, Netzsch, Selb, Germany). Density (ρ) is given by the Archimedes method. The specific heat capacity Cp was derived by the mixture rule based on the mass fraction of each component. The CTE was test by using a dilatometer (DIL 402, Netzsch, Germany). The test was conducted over a temperature range from 25 °C to 200 °C. Bending strength was evaluated using a universal testing machine (MTS E45, MTS Systems Corporation, USA).

3 Result and discussion

Figure 1 shows the microstructure of Dia/Al-Si composite. Fig 2 shows the interfacial morphology of the composites. The microstructures of different composites are uniform and dense, with good interfacial bonding and a certain number of interfacial products distributed on the interface. With the increase of Si content, the interfacial products gradually increased and tended to be uniformly distributed, and their shapes gradually changed from point-like or rod-like to dendritic or coral-like. Fig 3 shows the EDS-SEM analysis of the interfacial products, indicating that the enrichment of silicon occurred on the surface of diamond particles during the preparation of the composites[5]. When the Si content in the aluminum matrix is low, the concentration of silicon atoms in the aluminum liquid is low and the driving force for diffusion is insufficient, so only a small amount of silicon is enriched at the interface. Whereas, when the Si content increases, the concentration of Si atoms in the aluminum liquid is higher, the driving force for diffusion is larger, and a large number of Si atoms are enriched at the interface and form dendrites.

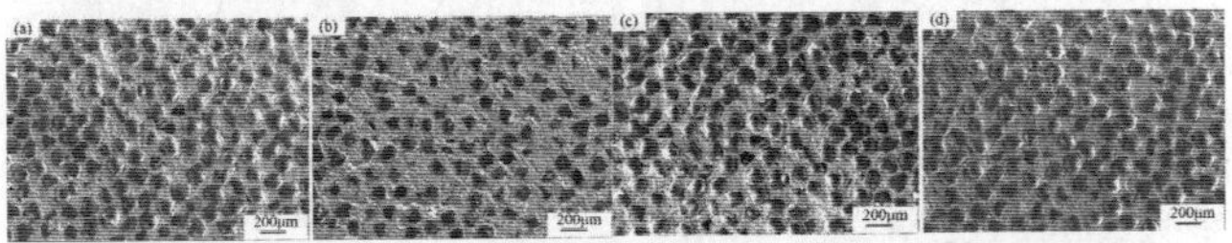

Fig. 1: Microstructure of Dia/Al-Si with different Si contents (a) 4 wt.% ;(b) 8 wt.% ;(c) 12 wt.% ;(d) 16 wt.%
(b)

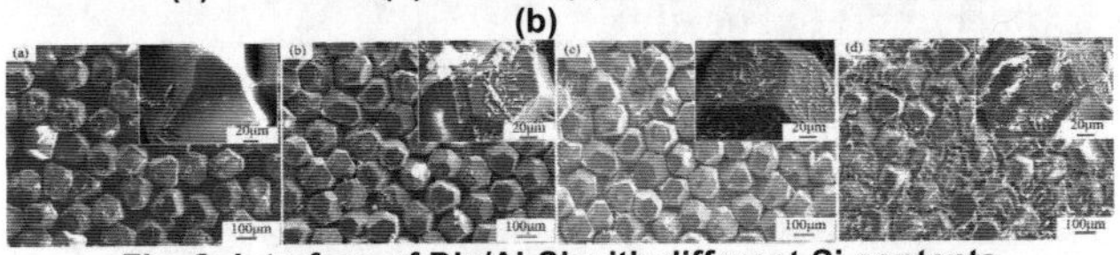

Fig. 2: Interface of Dia/Al-Si with different Si contents (a) 4 wt.% ;(b) 8 wt.% ;(c) 12 wt.% ;(d) 16 wt.%
(b)

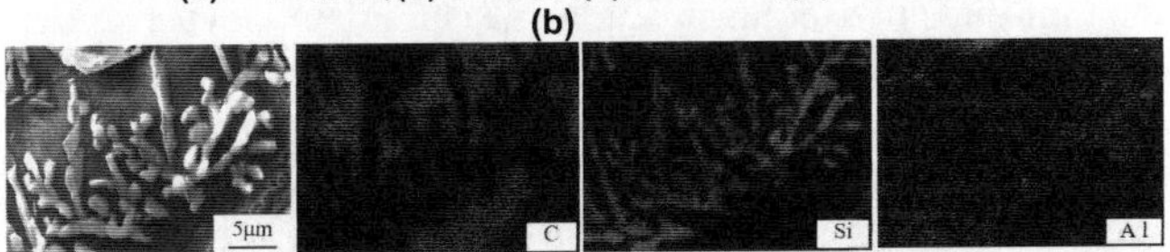

Fig. 3: EDS-SEM analysis of composites interfaces

Table 1 shows the properties of Dia/Al-Si composites. As the Si content in the aluminum matrix increases, both the bending strength (BS) and thermal conductivity (TC) of the composites tend to increase and then decrease, while the coefficient of thermal expansion (CTE) monotonically decreases. The former can be attributed to the changes in the interfacial structure in the composites[7]. When the Si content is low, the concentration of silicon atoms in the aluminum liquid is low and the driving force for diffusion is insufficient, the interfacial reaction is more difficult to carry out. The interfacial bonding is weaker, and at the same time, cracks are easier to germinate and extend, so the TC and BS of the composites are lower. With the increase of Si content, the concentration of silicon atoms increases, and it is easier to diffuse, the interfacial reaction is driven to form a metallurgical bond between the interfaces, and the interfacial strength increases, so the TC and BS are improved. However, when the Si content is too high, it leads to the formation of a thick brittle interfacial layer at the interface, which greatly reduces the interfacial bonding strength and increases the interfacial thermal resistance, so the TC and BS are reduced. The CTE of Al-Si alloys is lower with increasing Si content due to the decrease in the CTE of Al-Si alloys after alloying Al with Si and the transformation of tensile stresses into compressive stresses within the aluminum matrix[3].

Table 1. Properties of Diamond/Al-Si composites

Silicon content (*wt.*%)	TC ($W \cdot m^{-1} \cdot K^{-1}$)	BS (MPa)	CTE ($\times 10^{-6} \cdot K^{-1}$)
4.0	476	268	7.5
8.0	532	273	6.87
12.0	654	296	6.13
16.0	589	276	5.96

4 Conclusion

In summary, the silicon modified aluminum matrix can effectively promote interfacial wetting, reduce the generation of the side reaction Al_4C_3, and enhance the performance of Dia/Al-Si composites. The change of Si content leads to changes in the interfacial structure of the composites, but hardly affects their microstructure. With the increase of Si content in the aluminum matrix, the TC and BS of Dia/Al-Si composites show a tendency of increasing and then decreasing, while the CTE only monotonically decreases. When the Si content is 12 wt.%, the composites with 180 μm diamond exhibit excellent overall performance and meet the electronic packaging needs of typical semiconductor devices. In practical applications, composites with different properties can be prepared by regulating the alloy content to meet different application requirements.

5 Acknowledgments

The authors have no other competing interests to disclose.

References

[1] Fu J, Zhou C, Mi G, et al. Effects of diamond particle size on microstructure and properties of diamond/Al-12Si composites prepared by vacuum-assisted pressure infiltration[J]. CHINA FOUNDRY, 2024, 21(4): 405-413.

[2] Li N, Zhang Y, Zhang Y, et al. Realizing ultrahigh thermal conductivity in bimodal-diamond/al composites via interface engineering[J]. Materials Today Physics, 2022, 28: 100901.

[3] Sang J, Chen Q, Yang W, et al. Architecting micron SiC particles on diamond surface to improve thermal conductivity and stability of Al/diamond composites[J]. Surfaces and Interfaces, 2022, 31: 102019.

[4] Mizuuchi K, Inoue K, Agari Y, et al. Thermal conductivity of diamond particle dispersed aluminum matrix composites fabricated in solid–liquid co-existent state by SPS[J]. Composites Part B: Engineering, 2011, 42(5): 1029-1034.

[5] Beffort O, Khalid F A, Weber L, et al. Interface formation in infiltrated Al(Si)/diamond composites[J]. Diamond and Related Materials, 2006, 15(9): 1250-1260.

[6] Ruch P W, Beffort O, Kleiner S, et al. Selective interfacial bonding in Al(Si)–diamond composites and its effect on thermal conductivity[J]. Composites Science and Technology, 2006, 66(15): 2677-2685.

[7] Zhang H, Wu J, Zhang Y, et al. Mechanical properties of diamond/Al composites with Ti-coated diamond particles produced by gas-assisted pressure infiltration[J]. Materials Science and Engineering: A, 2015, 626: 362-368.

The Wettability of Molten Al and Sn on Solid Substrates Under High Magnetic Fields

Tie Liu[1]*, Peng Miao[1, 2], Yubao Xiao[1], Shuang Yuan[2], Qiang Wang[1]
1. Key Laboratory of Electromagnetic Processing of Materials (Ministry of Education), Northeastern University, Shenyang 110819, China.
2. School of Metallurgy, Northeastern University, Shenyang 110819, China.
*Corresponding address: e-mail: liutie@epm.neu.edu.cn

Abstract: This study investigated the wetting behavior of molten Al (paramagnetic) and Sn (antimagnetic) on Al2O3 and SiO2 substrates under high magnetic fields. It was observed that the contact angles of the Sn/Al2O3 and Al/Al2O3 systems both significantly decreased under high magnetic fields. The contact angle of the Sn/Al2O3 system exhibited a linear decrease with the increase in magnetic flux density, while the Al/Al2O3 system showed a non-linear decrease. With or without a high magnetic field, the contact angles of the Al/Al2O3 system both decreased with the increase in temperature. Without a high magnetic field, the contact angles in the Sn/Al2O3 system decreased with temperature, whereas under high magnetic fields, they increased with temperature. Additionally, at different isothermal temperatures, as the magnetic flux density increased, the contact angle of the Al/SiO2 system gradually decreased. Whereas, there was no clear pattern of change in the contact angle of the Sn/SiO2 system. Specifically, at an isothermal temperature of 263 °C, as the magnetic flux density gradually increased, the contact angle of the Sn/SiO2 system gradually increased. Furthermore, high magnetic fields, in conjunction with material characteristics and temperature, couples to influence the wetting behavior of molten metal on a solid substrate.

Keywords: Wettability; High magnetic fields; Molten metal; Ceramics; Contact angle

1 Introduction

Wettability is a major parameter in the fields of materials, metallurgy, biology, and manufacturing. The contact angle (θ), which is a measure of wettability, can be derived from Young's equation [1]. Wettability can be modified by increasing the temperature [2], modifying the ambient atmosphere [3] or substrate surface properties [4], or by adding active elements to molten metal [5]. However, these techniques may alter the intrinsic characteristics of a system or fall short of meeting practical usage requirements. Therefore, there is an urgent need to develop new methods of modifying wettability, possibly based on novel mechanisms.

Recently, high magnetic fields, considered a form of high-energy physics field, have been shown to change solid-liquid and liquid-vapor interfacial energy [6, 7]. Therefore, utilizing a high magnetic field to regulate wetting behavior is highly feasible.

The present study assessed variations in the contact angle of molten Al and Sn on Al_2O_3 and SiO_2 substrates in conjunction with high magnetic fields. This work examined the effects of magnetic flux density and isothermal temperature on wettability.

2 Experimental procedure

In this experiment, Al and Sn cylinders with dimensions of $\varphi 3$ mm × 3 mm were selected as the molten metal. The substrate materials chosen were single-sided polished Al_2O_3 and SiO_2 substrates with a purity of 99.99 wt.% and dimensions of 20 × 20 × 3 mm³. The experiment utilized a custom-designed contact angle measurement device under HMFs [8]. Prior to each trial, the substrates and metals were cleaned by ultrasonication in alcoholic, following which the specimens were placed in the vacuum chamber of the apparatus and the chamber was evacuated to a pressure of 4.0×10^{-4} Pa. Subsequently, the superconducting magnet was initiated, and the required magnetic flux density was applied to the apparatus. Finally, the chamber was heated to the required isothermal temperature after which this temperature was maintained for 30 min. Real-time observations were performed during this period while photographic images were captured using a charge coupled device camera. The equipment was then allowed to naturally cool and, when the furnace temperature dropped below the melting point of the molten metal, the superconducting magnet was turned off. Upon further cooling to room temperature, the vacuum system was vented and the sample was retrieved.

3 Result and discussion

Figure 1 shows that after utilizing a 6 T magnetic field,

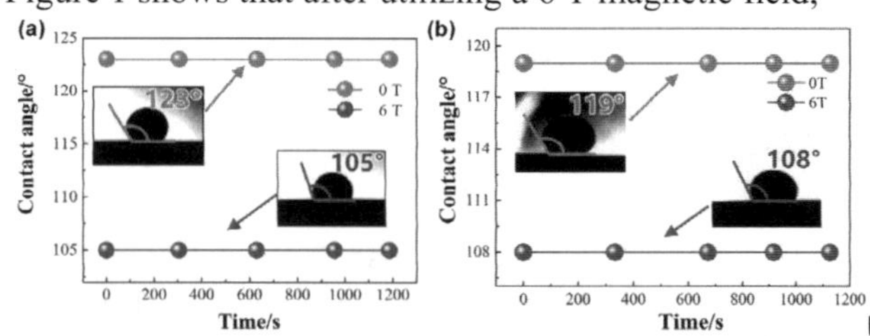

Figure 1. The changes in contact angles of molten (a) Al and (b) Sn on an Al_2O_3 substrate with increasing isothermal time in the presence or absence of a high magnetic field.

the equilibrium contact angles of molten Al and Sn on the Al_2O_3 substrate significantly decreased, reducing by 18°

and 11° respectively. Additionally, the morphology of molten Al and Sn transformed from spherical when no magnetic field was present to flattened spheres under a 6 T high magnetic field.

Figure 2 shows that the contact angle of the Sn/Al_2O_3 system exhibited a linear decrease with the increase in magnetic flux density, while the Al/Al_2O_3 system showed a non-linear decrease.

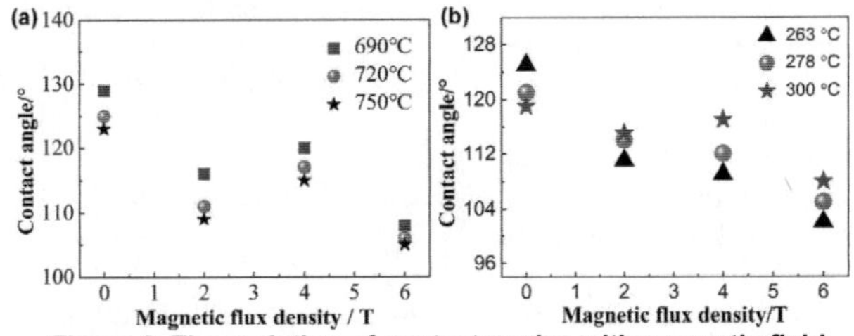

Figure 2. The variation of contact angles with magnetic field intensity for (a) Al/Al_2O_3 and (b) Sn/Al_2O_3 systems at different isothermal temperatures.

Figure 3 shows that the contact angle of the Al/Al_2O_3 system decreased with increasing isothermal temperature regardless of the presence of a high magnetic field. Conversely, in the absence of a high magnetic field, the contact angles in the Sn/Al_2O_3 system decreased with temperature, while under high magnetic fields, they increased with isothermal temperature.

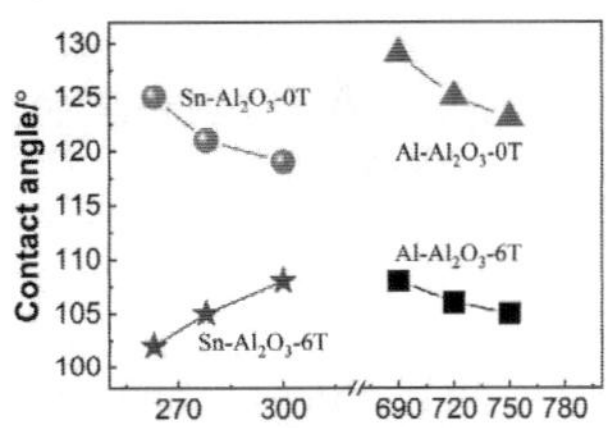

Figure 3. The variation of contact angles of molten Al and Sn on Al2O3 substrates with increasing isothermal temperature in the presence or absence of a high magnetic field.

Figure 4 shows that at different isothermal temperatures, as the magnetic flux density increased, the contact angle of the Al/SiO_2 system gradually decreased. In contrast, there was no clear pattern of change in the contact angle of the Sn/SiO_2 system.

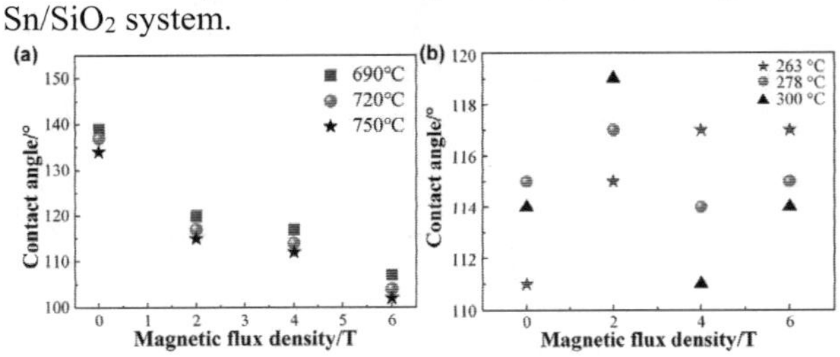

Figure 4. The variation of contact angles with magnetic field intensity for (a) Al/SiO_2 and (b) Sn/SiO_2 systems at different isothermal temperatures.

Specifically, Figure 5 shows that at an isothermal temperature of 263 °C, as the magnetic flux density gradually increased, the contact angle of the Sn/SiO_2 system gradually increased.

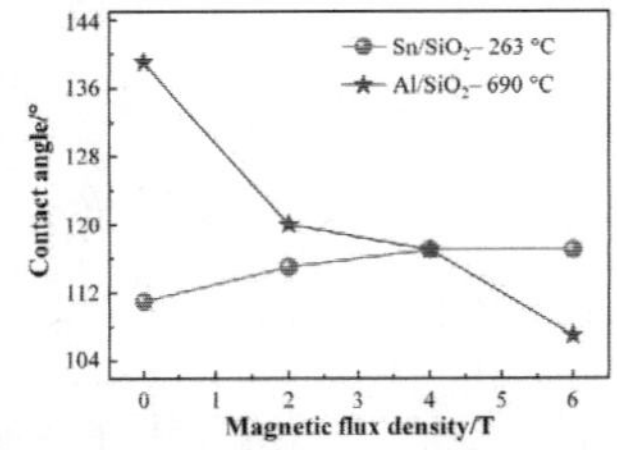

Figure 5. The variation of contact angles with magnetic field intensity for Al/SiO_2 at 690 °C and Sn/Al_2O_3 at 263 °C.

4 Conclusion

The utilization of high magnetic fields significantly affected the wettability of two molten metals with opposite magnetic properties on different substrates, generally leading to a decrease in contact angle. Additionally, magnetic flux density, material characteristics, and temperature collectively affected wettability.

5 Acknowledgments

This work was supported by the National Natural Science Foundation of China (NNSFC) (Grant Nos. 52127807 and U2241230) and the fund of the State Key Laboratory of Solidification Processing in NWPU (Grant No. SKLSP202101).

References

[1] T. Young, III. An essay on the cohesion of fluids, Philosophical transactions of the royal society of London, (1805) 65-87.

[2] S. Bao, K. Tang, A. Kvithyld, et al. Wettability of aluminum on alumina[J]. Metallurgical and Materials Transactions B, 2011,42:1358-1366.

[3] P. Shen, H. Fujii, T. Matsumoto, K. Nogi, The influence of surface structure on wetting of α-Al_2O_3 by aluminum in a reduced atmosphere[J]. Acta Mater., 51 (2003) 4897-4906.

[4] E. Bormashenko, Apparent contact angles for reactive wetting of smooth, rough, and heterogeneous surfaces calculated from the variational principles[J]. J. Colloid Interface Sci., 2019,537 :597-603.

[5] Q. Lin, L. Wang, R. Sui, Wetting of AlN by molten Cu-8.6 Zr-*x*Ti ternary alloys at 1373 K[J]. Acta Mater., 2021,203:116488.

[6] C. Huang, S. Shuai, P. Wang, et al, The effect of static magnetic field on solid-liquid interfacial free energy of Al-Cu alloy system[J]. Scripta Mater., 2020,187:232-236.

[7] J. Yang, X. Bian, C. Yang, et al. Viscous features of single-component melt under horizontal magnetic field[J]. Physica B Condens. Matter, 2013,415: 18-22.

[8] Y. Xiao, T. Liu, Z. Lu, et al. Instrument to characterize the wetting behavior of molten metal on a solid substrate under high magnetic field[J]. Rev. Sci. Instrum., 2019,90:063902.

The Role of Wettability in Liquid-Assisted Processing of Ceramic-Reinforced Iron-Based Metal Matrix Composites

Ł. Szymański[1,2*], J. Sobczak[1], N. Sobczak[3], K. Peddeti[4], E. Olejnik[1,5], A. Bigos[3], K. Żak[1], S. Sobula[1], K. Biegun[5], B. Karakkunnummal[6]

1. Faculty of Foundry Engineering, AGH University of Krakow, 23 Reymonta St., 30-059 Krakow, Poland
2. FLSmidth Mining Technology GmBH, 12 Schleebergstrasse, 59320 Ennigerloh, Germany
3. Institute of Metallurgy and Materials Science, Polish Academy of Sciences, 25 Reymonta St., 30-059 Krakow, Poland
4. FLSmidth Inc., 1100 American Parkway, Allentown, PA, 18109, USA
5. Innerco sp. z o.o., 17 Władysława Siwka St., 31-588 Krakow, Poland
6. FLSmidth Private Limited, 34 Egatoor, Kalembakkam, Old Mahabalipuram Road, 603103 Chennai, Tamil Nadu, India
*Corresponding address: e-mail: lukaszszymanski14@gmail.com

Abstract：The high-temperature wettability tests were conducted for various types of ceramics commonly used as reinforcements in fabrication of metal matrix composites (MMCs) by liquid-assisted processes. Significant differences during wettability tests in the course of kinetic curves were identified with regards to ex-situ and in-situ techniques used in manufacturing process of MMCs. Two different types of MMC castings were manufactured, i.e. reinforced by TiC particles (in-situ) and preforms made from oxide ceramics (ex-situ). Compared to monolithic alloys, the presence of the ceramic phases (TiC, Al2O3, ZrO2) in the both cases significantly improved its hardness (at least two-fold increase) and wear resistance (weight loss decrease at the level from 30 to 300%).

Keywords: MMC, wettability, ceramic, wear resistance

1 Introduction

Literature data indicate that the wear resistance of commercial monolithic iron alloys can be significantly increased by introducing hard ceramic particles into the metal matrix [1]. The most common ceramic reinforcements for iron alloys are oxides and carbides, including Al_2O_3, ZrO_2, B_4C, WC, and TiC [2]. Independently on manufacturing process of metal matrix composites (MMCs), their utility properties are conditioned by obtaining the relevant bonding at the reinforcing phase/matrix interface. For liquid-assisted processes, the wettability between a molten metal matrix and a reinforcement is a key factor affecting interface structure and properties that are responsible for high quality and reliability of final MMC products [3,4]. This work aims at experimental study of wetting behavior of molten cast iron in contact with different ceramic reinforcements followed by the trials to manufacture castings made from either cast steel or cast iron, both locally reinforced with TiC particles and oxide preforms.

2 Experimental procedure

In this study, a sessile drop method combined with non-contact heating of an alloy/substrate couple to the test temperature was adopted for the real-time observation of wetting and infiltration phenomena taking place between molten cast iron and (Ti + C_{gr}) substrate. The detailed experimental procedure has been presented in the previous work of authors [4]. On the contrary, wettability of ceramic oxides by molten high chromium cast iron was investigated by contact heating procedure.

For the *in-situ* production of TiC/Fe composite layers on the surface of castings an aqueous solution with the addition of TiC-forming (Ti + C_{gr}) mixture was used. Next, it was deposited in the cavities of the sand molds as a reactive coating in a contact with molten cast steel or cast iron. Because after pouring, the high temperature of molten alloy initiated the synthesis reaction of TiC.

The second type of composite castings, reinforced by ceramic preforms of a porous spatial structure made from ZrO_2 and Al_2O_3 powders, were produced by pressure-less infiltration. The manufactured composite castings were subjected to microstructure characterization (LM, SEM/EDS) as well as measurements of hardness (Vickers test) and wear resistance (ball on disc test, Miller test and cavitation erosion in slurry).

3 Result and discussion

High-temperature wettability test

The wetting kinetic curve for molten cast iron on reactive (Ti + C_{gr}) substrate is shown in Fig. 1a. The course of the wetting process can be divided into four stages. Stage I corresponds to non-wetting behavior with the contact angle values of $\theta>90^o$. A threshold value of $\theta=90^o$ is reached in only 1.5 s and subsequent continuous decrease in the contact angle value up to 72° (Stage II – wetting) and 18° (Stage III – good wetting). In the last stage (Stage IV), θ approached zero as a result of the complete wetting and infiltration of the molten alloy into the substrate. Such behavior is characteristic of spontaneous reactive wetting in the systems with the very fast kinetics of the synthesis reaction. Fig. 1b shows the top-view image of substrate surface after the wettability test. The presence of the three zones could be observed: Zone I (complete infiltration area), Zone II (partial infiltration area), and Zone III (limited infiltration area). In contrast, oxide ceramics showed with the same alloy non-wetting behavior and a lack of permanent bonding between them (the drops broke away from substrates after cooling process).

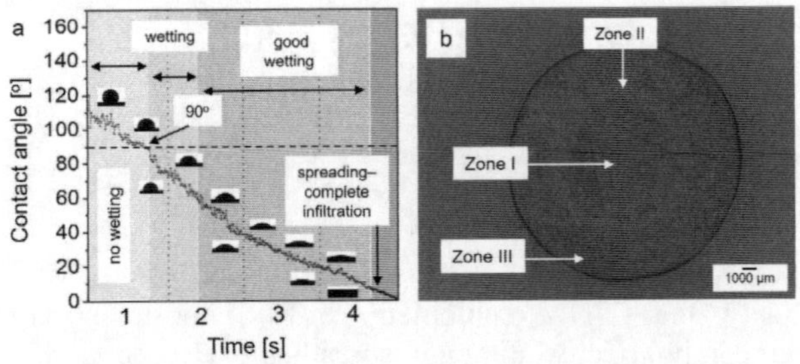

Fig. 1. a) wetting kinetics recorded in the sessile drop test of molten cast iron on (Ti + Cgr) substrate; b) LM micrograph showing the top-view of the substrate after its interaction with molten cast iron drop

Manufacturing process of composite castings

The microstructure of the TiC/Fe-type composite layer *in-situ* formed in locally reinforced cast steel casting is shown in Fig. 2a. The SEM/EDS analysis indicated its non-uniform structure and thickness varying from 1.0 to 2.0 mm. The maximum hardness values within this layer were above 1000HV1 which is more than a double increase as compared to the monolithic alloy. Similar to the sessile drop sample of the same system, the transition area was formed between the TiC-reinforced layer and the base alloy showing a lack of discontinuities and a good bonding between them related to the phenomenon of the reactive infiltration of the casting coating by liquid metal and high exothermic reaction synthesis of TiC carbide. Wear resistance tests showed significant decrease in the weight loss of composite samples ranged from 2 to 3.5 times as compared to the reference samples. On the contrary, the Fe-based MMCs reinforced with TiC exhibited lower resistance to cavitation erosion than monolithic materials.

The illustrative microstructure of composite zone reinforced by a ceramic preform is presented in Fig. 2b. The observations evidenced that the open porosity of preforms was fully infiltrated by molten Fe-based alloys while the occasional occurrence of discontinuities was related to the closed porosity formed in the preforms themselves during their fabrication. The chemical analysis (EDS) of the composite produced with high-Cr alloys and the ZrO_2 reinforcement showed the presence of complex Cr-rich oxide within the ceramic/matrix transition area. These results point to the positive role of Cr, which promoted a physicochemical interaction between the matrix and the reinforcement despite the lack of wetting found in the sessile drop tests. The hardness measurements indicated that the presence of the ceramic phase caused a two-fold increase in the hardness and around a 250 HV0.1 more increase for the cast steel and high chromium cast iron respectively, while their wear resistance was increased up to 20%.

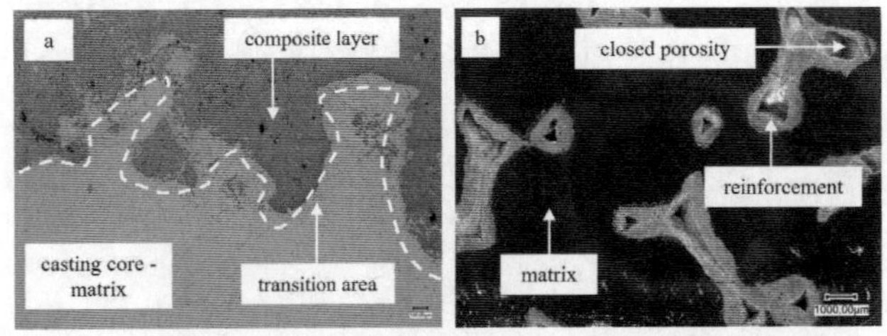

Fig. 2. a) SEM image of composite layer of steel casting *in situ* reinforced by TiC; b) LM image showing the microstructure of composite zone reinforced by ceramic preform

4 Conclusion

The mechanism of *in-situ* synthesis of TiC in ferrous alloys was explained based on high-temperature wettability test. Two types of composite castings were produced by *in-situ* and *ex-situ* methods. The use of reactive coatings is an effective method for *in situ* production of composite layers reinforced with TiC particles on the surface of ferrous castings while introduction of porous ceramic preforms allowed to produce MMC zones in Fe-based castings. The usage of ceramic preforms as well as TiC particles significantly increased the hardness and wear resistance, as compared to monolithic cast steel, and cast iron.

5 Acknowledgments

Scientific work financed by The National Centre for Research and Development of Poland within Project No. LIDER/49/0200/L-11/19/NCBR/2020.

References

[1] Luo ZC, Ning JP, Wang J, et al. Microstructure and wear properties of TiC-strengthened high-manganese steel matrix composites fabricated by hypereutectic solidification[J]. Wear, 2019, (202970): 432-433.

[2] Szymanski L, Sobczak J, Peddeti K, et al. Production of metal matrix composites reinforced by TiC by reactive infiltration of cast iron into Ti + C preforms[J]. Ceram. Int. 2024 (50): 17452-17464.

[3] Chao L, Jun W Y, Ke Q, et al. The research status of the wetting of metal/ceramic in the high temperature self-lubricating materials[J]. Int. J. Mater. Sci. Appl., 2016, 5(2): 108-112.

[4] Sobczak N, Singh M, Ashtana R. High-temperature wettability measurements in metal/ceramic system – some methodologies issues[J]. Curr. Opin. Solid. State Mater. Sci., 2005 (9): 241-253.

High-Temperature Interaction Between Liquid Pure Al and Al-7Si Alloy with Ti+C Compacted Substrate

K. Janus[1], S. Terlicka[2], J.J. Sobczak[1], N. Sobczak[2], W. Maziarz[2], E. Olejnik[1]
1. Faculty of Foundry Engineering, AGH University of Krakow, 23 Reymonta St., 30-059 Krakow, Poland
2. Institute of Metallurgy and Materials Science of Polish Academy of Sciences, 25 Reymonta St., 30-059 Krakow, Poland
*Corresponding address: kjanus@agh.edu.pl

Abstract: The interaction of aluminum and its alloys with titanium carbide (TiC) is crucial in producing TiC-reinforced aluminum matrix composites. In this study, the high-temperature wettability tests were conducted using the sessile drop method combined with a capillarity purification procedure (CP) to investigate the phenomena at the Al/Ti+C and Al-7Si/Ti+C interfaces. The tests were performed in a high vacuum environment under isothermal conditions at 800°C using compacted substrates produced from Ti+graphite powder mixture. The results evidenced a good wetting of Ti+C substrates both by liquid pure Al and Al-7Si alloy (a contact angle $\theta << 90°$). A comprehensive structural characterization of the solidified couples conducted with a scanning electron microscope (SEM) equipped with energy-dispersive x-ray spectroscope (EDS) revealed the mass transfer from drop to substrate and the formation of the new phases at the drops/substrate interface during wettability tests.

Keywords: aluminum matrix composites, wettability, microstructural characterization

1 Introduction

Aluminum matrix composites are highly valued in the automotive and aerospace industries due to their excellent mechanical properties, thermal stability, and high strength-to-density ratio [1]. The interaction of aluminum (Al) and its alloys with titanium carbide (TiC) is of critical importance in the production of TiC-reinforced aluminum matrix composites. TiC is a highly valuable reinforcing material due to its exceptional mechanical and physicochemical properties [2]. A comprehensive understanding of the high-temperature wetting behavior and reactivity between liquid aluminum (or its alloys) and TiC is essential for the successful production of metal matrix composites. Recently, the wetting of TiC by liquid Al has been extensively studied [3]. Research indicates that at temperatures below 1000°C, the oxide film on the Al surface results in a large contact angle, and as the temperature increases, the wetting behavior changes from non-wetting to wetting [4]. In [5], alloying Al with 7 wt% Si was found to improve its spreading on the TiC substrate, while another study [6] have shown better wetting of TiC by pure Al, as compared to Al-7Si alloy. For both Al/TiC and Al(-Si)/TiC systems, the porosity and C/Ti stoichiometry of TiC substrate may significantly affect wetting and spreading [7]. Hence, in this study, the real-time observation of wetting behavior of oxide-free Al and Al-7Si drops on compacted Ti+C substrates was made in a high vacuum using a sessile drop method combined with capillary purification procedure in order to elucidate the role of Si in the wettability and reactivity in the Al/TiC system.

2 Experimental procedure

Powdered Ti (55%) and graphite (45%) were mechanically mixed for 2 h, dried at 150°C for 2 h and next the mixture was cold pressed at 500 MPa to produce the Ti+C compacted substrates. The ingots of pure Al and Al-7Si were machined to a cylindrical shape with a diameter of 10 mm. Before the wettability tests, the pure Al and Al-7Si alloy cylinders and Ti+C substrates were ultrasonically cleaned in isopropanol for 15 min to remove contaminants. The high-temperature wettability tests were conducted using the sessile drop method. The experiments were performed in a high vacuum of 10^{-6} mbar at $T_{exp} = 800°C$. To eliminate the effect of native oxide films on Al and Al-7Si drops and to avoid the heating history of Al/Ti+C and Al-7Si/Ti+C couples the capillary purification (CP) procedure coupled with non-contact heating was employed. [8-9]. For this, Al and Al-7Si samples were placed in an alumina capillary above the substrates and then squeezed onto the substrate once the test temperature was reached. To observe the wetting behavior of Al/Ti+C and Al-7Si/Ti+C couples, high-speed cameras were used to record side-view images from two different directions at a speed of 57 frames per second. These images were later used to determine the contact angle values between the liquid Al and Al-7Si and the substrate under investigation.

Microstructural observations and chemical composition analyses of Al/Ti+C and Al-7Si/Ti+C couples were made using an FEI E-SEM XL30 scanning electron microscope (SEM) equipped with an EDAX GEMINI 4000 energy-dispersive X-ray spectrometer (EDS).

3 Result and discussion

Wetting kinetics

Fig. 1 shows the curve of wetting kinetics with time recorded for a liquid Al/Ti+C couple during 240 s upon isothermal heating at T_{exp}. The average contact angle (θ_{av}) values were calculated for 2 s intervals. Analysis of θ_{av} shows that the first contact formed by the oxide-free Al drop after its deposition on Ti+C substate was around 120°, thus showing a non-wetting phenomenon. Further holding the couple at T_{exp} resulted in a fast decrease in the contact angle resulting in wetting phenomenon ($\theta_{av} < 90°$) in 75th second and a very good wetting (θ_{av} ~45°) in 145th, second followed by a small decrease to a final contact angle value of 41° in 240th second. Under the conditions of this study, nonwetting-to-wetting transition was also observed for the Al-7Si/Ti+C couple.

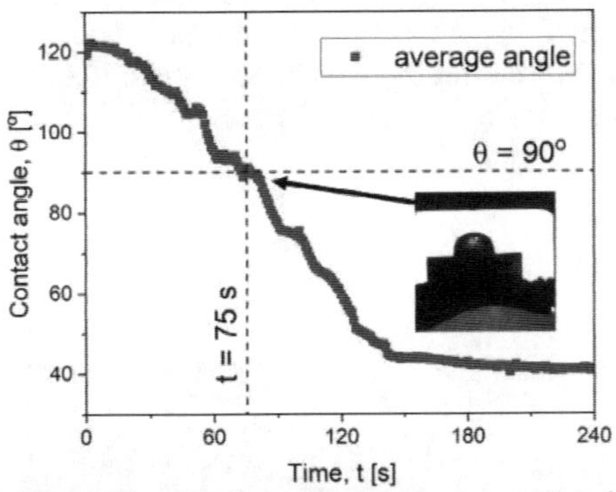

Fig. 1. The wetting kinetics of liquid Al on compacted Ti+C substrate recorded during the sessile drop tests by the CP procedure at 800℃ for 240 seconds in a high vacuum.

Microstructural characterization

Side-view observations of the couples provide extensive information on the wettability behavior of pure Al and Al-7Si alloy on the compacted Ti+C substrate. Initially, the liquid drops were perfectly spherical, but in time they changed their geometry, indicating good wetting. For example, in the case of the Al/Ti+C couple, it was visible that during the wettability test, the drop of pure Al transformed from a spherical shape to the shape of a cone, which is presented in Fig. 2. In addition, the real-time videos reveal that the mass transfer occurred from compacted Ti+C substrate to the Al and Al-7Si drop direction. The detailed SEM/EDS observations of the cross-sectioned sessile drop couples evidenced the presence of reactively formed phases. Similar findings were reported in [6] for wettability tests of oxide-free Al and Al-Si alloys on bulk TiC substrates. The mechanisms of the formation of Al_4C_3 and TiC in the Al/Ti+C couple and TiC and $TiSi_2$ in the Al-7Si/Ti+C couple will be discussed.

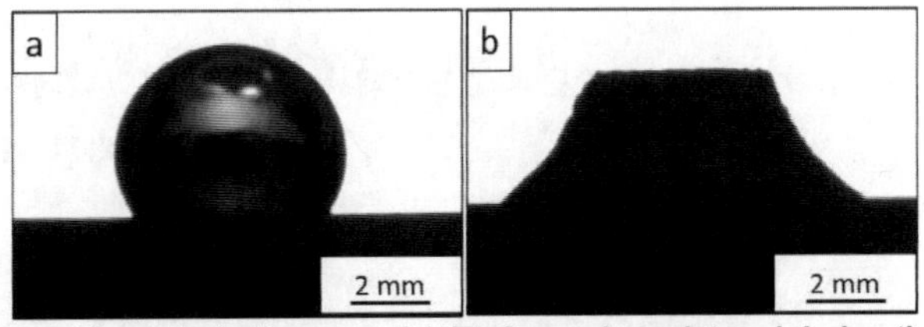

Fig. 2. The selected images of Al/Ti+C couple registered during the sessile drop test at 800°C with the CP procedure in a high vacuum: (a) after the drop deposition; (b) after 240 s contact interaction.

4 Conclusion

The study investigated the interaction of liquid Al and Al-7Si alloy with the compacted Ti+C substrate using the sessile drop method in combination with non-contact heating to a test temperature of 800 °C and mechanical cleaning the drop from its native oxide film directly in a vacuum chamber by the capillary purification procedure. After deposition on the substrate, the Al and Al-7Si drops reacted, wetted, and infiltrated a substrate consisting of titanium and graphite powders. Structural characterization of the solidified couples by SEM with EDS analysis confirmed the presence of new phases reactively formed by dissimilar mechanisms.

Acknowledgments

This research was supported by the National Science Centre of Poland within OPUS 22 project no. 2021/43/B/ST8/03271.

References

[1] Zhou M.Y., et al., Journal of Alloys and Compounds (2020), 838: 155274.
[2] Anandkumar R., et al., Wear ,2012, 282:31–39.
[3] Contreras A., et. al., Scripta Materialia ,2003, 48(12): 1625–1630.
[4] Laurent V., et al., Acta Metallurgica,1998, 36(7): 1797–1803.
[5] López V.H., et al., Journal of Materials Science .2010,45:4321–4331.
[6] Guo R.F., et al., Journal of Materials Science,2021, 56: 7791–7798.
[7] Froumin N., et al., Scripta Materialia,1997, 37:1263–1267.
[8] Terlicka S., et al., Materials ,2022, 15(4):9024.
[9] Terlicka S., et al., Journal of Materials Engineering and Performance, 2023, 32: 5689–5696.

Fabrication of Lightweight Aluminium Metal Matrix Composites and Validation in Green Vehicles

Christian Kneissl
Austrian Foundry Research Institute (OGI), Parkstrasse 21, 8700 Leoben, Austria
*Corresponding address: e-mail: christian.kneissl@ogi.at

The FLAMINGo project, funded under the Horizon 2020 program (LC-GV-06-2020 theme), represents a pioneering effort to enhance the development of high-performance aluminium composites specifically designed for automotive applications. The core objective of the project is to engineer reinforced aluminium metal matrix nanocomposites (Al-MMnCs) that surpass the strength and stiffness of conventional aluminium alloys. These advanced materials are crucial in the burgeoning battery electric vehicle (BEV) industry, where weight reduction is directly linked to improved energy efficiency and extended range.

1 Introduction

As electric vehicles (EVs) gain momentum, driven by both market demands and regulatory pressures aimed at reducing carbon emissions, automakers are seeking innovative lightweight materials to replace heavier components like steel. Aluminium has long been favoured for its impressive strength-to-weight ratio, and FLAMINGo aims to push these advantages further. By integrating nanotechnology into aluminium composites, the project aims to create materials that can outperform existing solutions, enabling the substitution of steel with lighter Al-MMnC parts. This innovation will help the automotive industry meet weight reduction targets and contribute to the production of greener, zero-emission vehicles.

2 Experimental procedure

One key application in development is the aluminium steering knuckle, which replaces steel (S355) components that typically require welding. By casting and optimizing the design of the aluminium part using topology optimization techniques, FLAMINGo is able to achieve the same mechanical performance as steel while significantly reducing weight. The project envisions producing right and left configurations for a full front suspension system, further contributing to vehicle weight reduction in BEVs.

The demand for composite materials in BEVs continues to rise as manufacturers strive to make vehicles lighter and more efficient. FLAMINGO's approach focuses on producing reinforced metal-matrix castings or extruded parts (Al-MMnCs) with superior properties in terms of weight, strength, and stiffness compared to the steel currently used in electric vehicles. These parts will be produced using advanced manufacturing techniques such as high-energy ball milling (HEBM) by MBN, ensuring improved mechanical properties while maintaining cost-efficiency.

Figure 1 highlights the differences in hardness between samples containing nanoparticles and those without. The results indicate that the hardness increase observed in the samples without nanoparticles can be attributed to the formation of thermally activated precipitations. These precipitates form under heat treatment and contribute to an increase in hardness.

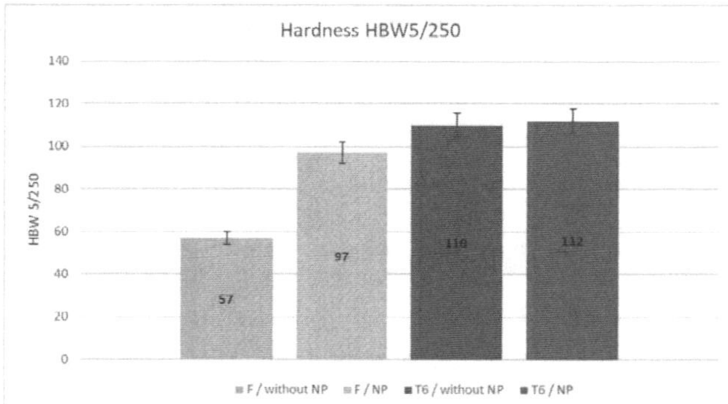

Figures 1: Differences in hardness between samples containing nanoparticles and those without

On the other hand, the samples with nanoparticles exhibit a distinct behaviour. The increase in hardness in these samples is primarily due to the presence of non-thermally activated dispersoids, introduced by the nanoparticles. These nanoparticles are temperature-stable, meaning they do not undergo changes at elevated temperatures, and as a result, the hardness of the samples remains relatively constant across different temperature ranges.

This sustained hardness, irrespective of thermal influence, serves as indirect evidence of the existence and stability of nanoparticles within the casting. Their presence effectively reinforces the material and provides consistent mechanical properties, even under thermal conditions that would typically alter the hardness in nanoparticle-free samples.

3 Project Objectives

The overarching goal of the FLAMINGo project is to significantly reduce vehicle weight by improving the mechanical properties of aluminium alloys. By developing Al-MMnCs with enhanced strength and stiffness, the project enables manufacturers to use less material without sacrificing performance. This is expected to result in a 50%

reduction in weight for certain vehicle components, even those that have already undergone previous weight optimization (Fig 2.).

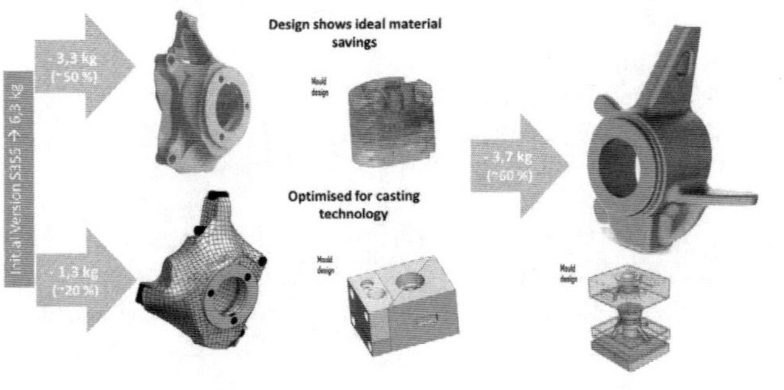

Figure 2: Weight was reduced by 58% from the initial geometry to a first topology-optimized casting.

In line with the growing trend toward lightweight materials in electric vehicles, FLAMINGo focuses on producing reinforced metal-matrix castings or extruded parts that outperform existing aluminium alloys. One of the key tasks is optimizing the casting and extrusion processes by adapting nanoparticle inclusions within the material matrix. Masterbatches, produced using HEBM, introduce nanoparticles into the molten metal, enhancing the material's properties. Rigorous testing and evaluation of these components will ensure they meet the stringent mechanical property requirements for use in electric vehicles.

4 Alignment with Market Trends

The electric vehicle market is experiencing rapid growth, driven by increasing environmental awareness and government initiatives promoting zero-emission technologies. The FLAMINGo project is strategically positioned within this market, responding to the need for lighter, more efficient materials in the automotive sector. Aluminium, a key material in this transformation, plays a vital role in reducing vehicle weight while maintaining high performance. Through its focus on advanced aluminium nanocomposites, FLAMINGo is well-aligned with market trends and is poised to make a significant impact on the development of next-generation vehicles.

5 Impact

Composite materials will be increasingly used in battery electric vehicles (BEV) to make them lighter. By replacing heavier steel parts with lightweight Al-MMCs, BEVs can achieve better energy efficiency, longer driving range, and improved performance. Additionally, the enhanced mechanical properties of these composites contribute to higher structural integrity, ensuring that the reduction in weight does not compromise vehicle safety or durability. This shift towards Al-MMCs in BEVs aligns with industry trends aimed at creating more sustainable, efficient, and high-performing electric vehicle.

The FLAMINGo project is expected to deliver significant benefits across several key areas of the automotive industries:

Vehicle Weight Reduction: The development of reinforced aluminium alloys enables a reduction in the weight of vehicle components, leading to better fuel efficiency and improved overall performance.

Optimized Manufacturing Processes and Lead Times: By integrating advanced material modelling techniques, FLAMINGo streamlines product development, reduces lead times, and simplifies the adoption of new materials by manufacturers.

Improved Structural Integrity and Safety: The project emphasizes stringent monitoring and control of the structural integrity of FLAMINGo components, ensuring they meet the safety standards required for automotive applications.

Compliance with Circular Economy and Zero-Emission Legislation: In line with EU goals for sustainability and circularity, FLAMINGo supports zero-emission technologies, helping the automotive industry meet environmental regulations.

6 Summary

The FLAMINGo project represents a transformative step forward in the development of lightweight materials for the automotive sectors. Through innovative metallurgical processes and a focus on high-performance aluminium nanocomposites, the project is well-positioned to support the global shift toward more efficient, sustainable transportation solutions.

References

[1] Li, J., Kneissl, C., Tose, P., et al. Effect of SiC additions on solidification microstructure and mechanical properties of Al-7Si-0.3Mg based alloys. Materials Science and Engineering: A (in preparation).

Insights Into the Interface Properties of Mg/Ti Via Experiments and First-Principles Calculations

Wenbo Guo[a], Shulin Lü[a*], Shusen Wu[a],Wei Guo[a]
[a] State Key Lab of Materials ProcWenbo Guoessing and Die & Mould Technology, School of Materials Science and Engineering, Huazhong University of Science and Technology, Wuhan 430074, P. R. China
* Corresponding Author, shulin317@hust.edu.cn

Abstract: In order to understand in depth the interface characteristics between Ti particles and Mg matrix, and reveal the strengthening mechanism of Ti particles reinforced Mg matrix composites, the interface structure, work of adhesion and the interface electronic structure of the Mg(0001)/Ti(0001) interface have been studied by experiments combined with first-principles methods. The calculation results of the work of adhesion show that the Ti-HCP interface is more stable than the Ti-OT interface. The electronic structure analysis results show that the Ti-Mg bond formed at the Ti-OT interface has a weak covalence and exhibits the characteristics of a Ti-Mg ionic bond. The electronic characteristics of the Ti-HCP interface are metallic bonds.

Keywords: First-principles calculation; Mg/Ti interface; Work of adhesion; Electronic structure

1 Introduction

Compared to conventional Mg alloys, particle reinforced Mg matrix composites (PMMCs) have higher strength, better wear resistance and better fatigue and creep properties, which has attracted a lot of research attention in recent years [1, 2].

Reinforcement particles are divided into ceramic particles and metal particles. Compared to ceramic particles, metal particles such as Cu, Ni, Fe and Ti have high ductility and excellent elastic modulus and can be used as candidate materials for strengthening Mg matrix composites [3, 4]. In particular, Ti particles can improve the strength of composites while maintaining good ductility, and experimental studies have been reported.

For Ti particle reinforced Mg matrix composites, the interface properties between Ti particles and the Mg matrix have an important effect on the ductility, strength and fracture mechanism of the composites, so it is necessary to have a basic understanding of these interface properties. However, it is very limited to systematically investigate the interface properties of composites by experiments alone. In recent years, what is recognised as a first-principles method for the study of microscopic surfaces and interfaces has been widely used. This method can reveal the interfacial properties of Ti/Mg from the atomic scale, such as atomic configuration, interfacial adhesion and interfacial bonding, which will help to understand the strengthening mechanism of Ti particle reinforced Mg matrix composites and improve the overall mechanical properties of the composites. In this work, the interfacial properties of Mg(0001)/Ti(0001) are calculated by the first-principles method in order to clarify the interfacial structure, work of adhesion and interfacial fracture mechanism, and to guide the design of novel Mg/Ti composites.

2 Experimental procedure

In this work, high purity magnesium ingot was used as the matrix material of the composite. Spherical Ti metal particles with a particle size range of 1-20μm were used as the reinforcing phase. The Mg-10wt%Ti composite was prepared by the stirring casting method.

In this work, based on the density functional theory (DFT) plane-wave pseudopotential method, generalized gradient approximation (GGA) of Perdew-Burke-Ernzerhof (PBE) approach was employed to describe the exchange-correlation functional. The plane-wave cutoff energy was set as 400eV. The k-point sampling meshes obtained by using the Monkhorst-Pack scheme were set to 9×9×9 for bulk Mg and bulk Ti, and 9×9×1 for all the surfaces and interfaces.

3 Result and discussion

1.Microstructure of Ti/Mg composites

Fig.1 shows the microstructure of Ti/Mg composites, and it can be seen that the particles are uniformly dispersed.

Fig. 1. Microstructure of Ti/Mg composites

2. Surface properties of Mg and Ti

In order to ensure the accuracy of the calculation, 7-layer Mg(0001) and Ti(0001) slabs were selected to establish the Interface model for the subsequent calculations in this work.

3. Interface properties of Mg and Ti

Two different Mg(0001)/Ti(0001) interface models were established, as shown in Fig. 2. Two different stacking sequences are "OT" and "HCP".

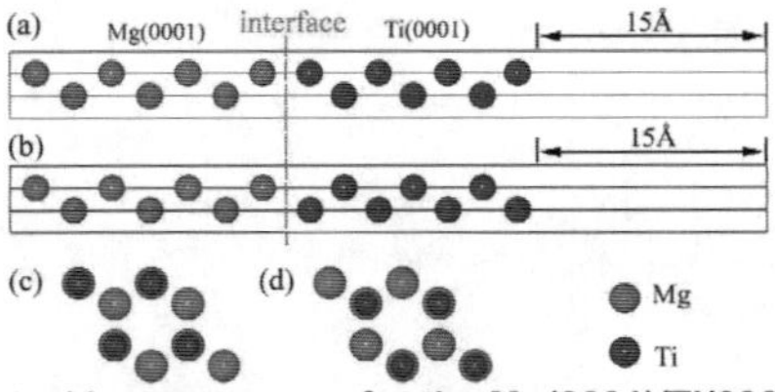

Fig. 2. Two stacking sequences for the Mg(0001)/Ti(0001) interface. (a) (c): "OT", (b) (d): "HCP"

The Work of adhesion (W_{ad}) of the Ti-HCP interface after relaxation is the largest, increasing from 0.961J/m^2 before relaxation to 1.430J/m^2 after relaxation. The W_{ad} at the Ti-OT interface is 0.107J/m^2 before relaxation and 0.673J/m^2 after relaxation.

Table 1 The calculated Wad and interfacial distance and that after relaxation

Stacking sites	Unrelaxed		Relaxed	
	d_0(Å)	W_{ad}(J/m^2)	d_{int}(Å)	W_{ad}(J/m^2)
Ti-OT	2.8	0.107	2.904	0.673
Ti-HCP	2.4	0.961	2.520	1.430

4. Electronic structure and bonding

Fig. 3 shows the charge density of the Mg(0001)/Ti(0001) interface. A change in color represents a change in charge concentration, with blue indicating less charge and red indicating more charge. The formation of charge sharing regions at the interface can be clearly observed (see red arrows). However, the distribution of charge sharing region on the Ti-HCP interface is large, which indicates that the interface interaction is strong and the interface bonding strength is high.

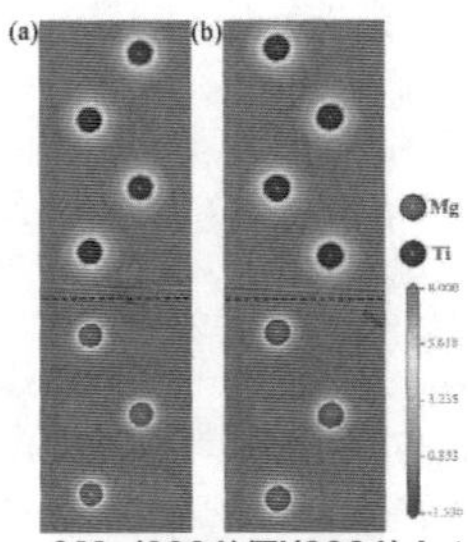

Fig. 3. Charge density of Mg(0001)/Ti(0001) interfaces: (a) OT model, (b) HCP model.

Fig. 4 shows the charge density difference of the Mg(0001)/Ti(0001) interface. The red regions represent charge accumulation and the blue regions represent charge dissipation. For the Ti-OT interface, the charge distribution between the Ti atom and the Mg atom has some directionality, but the amount of charge is relatively small, indicating that the Ti-Mg bond formed at the interface has a weak covalence. For the Ti-HCP interface, the interfacial Ti atom and Mg atom form an charge accumulation region at the interface, showing the characteristics of non-directional metal bonding. It is also observed that the charge aggregation on the Ti atom side of the interface is significant, showing the characteristics of Ti-Mg ionic bonding.

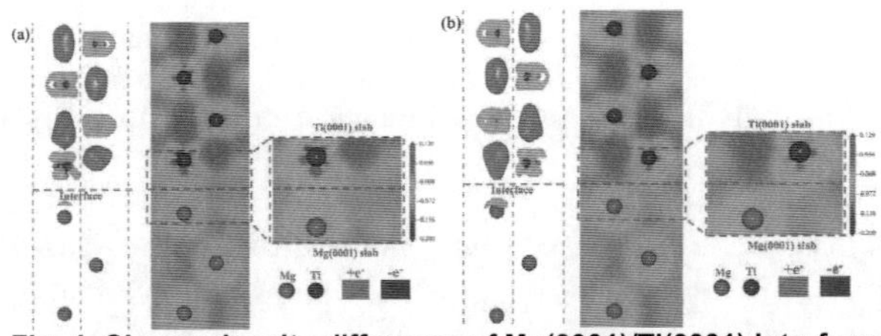

Fig. 4. Charge density difference of Mg(0001)/Ti(0001) interfaces: (a) OT model, (b) HCP model.

Fig. 5a shows the PDOS of the Ti-OT interface. Comparing the electron orbital curve of the interfacial atoms, it can be seen that the trough of the 2p electron orbital curve of the interfacial Mg atoms near the Fermi level is lower than that of the inner layer, indicating that the metallic property of the interfacial Mg atoms is reduced and the covalence is improved. Fig. 5b shows the PDOS of the Ti-HCP interface. The interface Mg atom shows a new DOS peak at -2.26 eV compared to the internal Mg atom, while the peak at 0.258 eV is significantly enhanced.

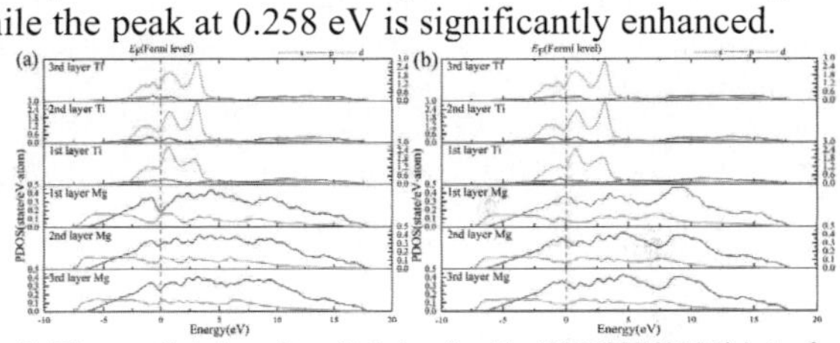

Fig. 5. The partial density of states for Mg(0001)/Ti(0001) interfaces: (a) OT model, (b) HCP model.

4 Conclusion

(1) In the Ti/Mg interface, the Wad of the Ti-HCP interface is 1.430J/m^2, and that of the Ti-OT interface is 0.673J/m2, which indicates that the binding strength of the Ti-HCP interface is higher.

(2) For the Ti-OT interface, the Ti-Mg bond formed at the interface has weak covalency. At the Ti-HCP interface, Ti atoms and Mg atoms form an electron accumulation region on the interface, showing the characteristics of non-directional metal bonding.

References

[1] Synthesize of AZ31/TiC magnesium matrix composites using friction stir processing[J].J. Magnes. Alloy,2015,3:76–78.
[2] Y.S.A. B, C.Y.A. B, B.Z.A. B, et al, The aging behavior, microstructure and mechanical properties of AlN/AZ91 composite[J].J. Magnes. Alloy,2022.
[3] M.J. Shen, X.J. Wang, M.F. Zhang, et al. Fabrication of bimodal size SiCp reinforced AZ31B magnesium matrix composites, Mater[J].Sci. Eng. A. 2014,601:58–64.
[4] Jie Chen, Chonggao Bao, Wenhui Chen, et al. Mechanical Properties and Fracture Behavior of Mg-Al/AlN Composites with Different Particle Contents[J].J. Mater. Sci. Technol,2017.

Design of In-Situ Mg-Based Amorphous Alloy Matrix Composites

Wei Guo[1,2,*], Mi Zhao[3], Hidemi Kato[2], Shusen Wu[1]
1. School of Materials Science and Engineering, State Key Laboratory of Materials Processing and Die & Mould Technology, Huazhong University of Science and Technology, Wuhan 430074, China
2. Institute for Materials Research, Tohoku University, Sendai 980-8577, Japan
3. School of Aerospace Engineering, Huazhong University of Science and Technology, Wuhan 430074, China
*Corresponding address: e-mail: weiguo@hust.edu.cn

Abstract: The Mg-based amorphous alloys possess low density, high specific strength, and high anti-corrosion property, which makes them have great application potentials in the 3C industry, biomaterials, etc. However, the room-temperature brittleness of Mg-based amorphous alloys, which results from the localized shear flow, limits their applications. The approaches to improve the ductility of these materials limit in ex-situ direct adding method. The present study designed a novel fabrication process based on the dealloying in metallic melt reaction and has successfully induced a variety of in-situ reinforcing phases. We further optimized the composite structure and mechanical property via process parameter. The present study provides a theoretical basis and practical experiences in the following research related with the composite fabrication via dealloying reaction.

Keywords: Mg-based metallic glass; composites; dealloying reaction; mechanical property

1 Introduction

Due to the high strength and good corrosion resistance compared with the corresponding crystalline counterparts, bulk metallic glasses (BMGs) have attracted great attention in the last decades. Mg-based BMGs exhibit considerable application prospects in lightweight structural components. However, Mg-based BMGs tend to break into small pieces under uniaxial compression before their elastic limit at ambient temperature, which greatly limits their further application. Introducing crystalline particles as secondary phases into the BMG matrix has been proven to be an effective method to improve the plasticity of the amorphous alloys. Considering the introducing approaches of the secondary phase, the preparation process of the Mg-based BMG matrix composites (BMGCs) can be classified as the ex-situ method and the in-situ method. In fact, most Mg-based BMGCs are synthesized by ex-situ method with adding the fibers or crystalline particles. However, the poor interfacial bonding between the added phase and the matrix in the ex-situ BMGCs degrades the reinforcing effects in mechanical properties. In contrast, the in-situ BMGCs, fabricated by precipitation during solidification or partial nanocrystallization in BMGs via thermal treatment, always exhibit better mechanical property than the ex-situ ones as a result of the lack of interfacial oxidation layer or other crystalline layers. Recently, a novel dealloying reaction in metallic melt has been proposed by Wada et al. in fabrication porous alloys. The reaction is based on the differences in the enthalpy of mixing between pre-alloy element and the metallic melt, i.e., negative value in the enthalpy of mixing indicates miscibility or reactivity and visa versa. In the present study, we have applied the above dealloying reaction in metallic melt to synthesis several kinds of Mg-based BMGCs with various reinforcing phases. Furthermore, the composite structure has been optimized to improve the mechanical properties of such composites.

2 Experimental procedure

The fabrication process has been firstly designed based on dealloying reaction in the metallic melt. With different composition of the materials, the experimental procedure is similar. Firstly, the precursors were fabricated by arc-melting, then the precursors were immersed in Mg-based melt to induce dealloying reaction. The melt temperature and time were set in different values. After the dealloying reaction finished, the melt were poured into mould to prepare master alloys. Finally, the BMGCs were fabricated by injection copper mould casting. The phase constitutions of the specimen were characterized by X-ray diffraction (XRD, Shimadzu XRD-6100) with monochromatic Cu Kα radiation. Thermal properties of the glassy transition temperature, the onset first crystallization temperature and the endothermic enthalpy of crystallization were measured by differential scanning calorimeter (DSC, NETZSCH STA 449C DSC) under ultra-high Ar gas atmosphere with a constant heating rate of 20K/min. Microstructures of the specimen were examined by scanning electron microscope (SEM, Carl Zeiss GeminiSEM300) equipped with energy-dispersive x-ray spectroscopy (EDS, AZtec X-Max 50). The chemical composition of the submicron crystalline phases was analyzed by electron probe micro analysis (EPMA). The specimens were tested by uniaxial compression at ambient temperature at a strain rate of 1×10^{-4} s^{-1} with a ZwickRoell Z020 machine. Five samples were tested for each specimen to ensure the reproducibility.

3 Result and discussion

All BMGCs showed the diffraction hump from the amorphous matrix and the sharp crystalline peaks from the reinforcing phases, which indicates that the dealloying reaction has been successfully induced. The DSC curves of all specimens show an obvious glassy transition, supercooled liquid region, and exothermic peak related with the crystallization process. The SEM images of the composites show the homogenously distributed porous Ti /Fe particles in the glassy matrix. According to the image analysis, the average size and volume fraction of such phases are given. The pores in the bicontinuous porous Ti particles are fully filled with the amorphous matrix, agreeing well with our design. To further analyze the pore structure, we used nitric acid solution to separate the particles from the composites, the surface morphology of the porous particle and the detailed pore characteristics, including porosity and the distribution of pore size, is measured by a mercury porosimeter. As the processing zone size of Mg-based BMGs is about 0.1–1 μm, thus the present pore size or the filled matrix size in the pores is very close to such value and optimum reinforcing effects should be realized in the local area inside porous particles.

The uniaxial compression tests are performed for all samples. A sudden fracture with almost no plastic strain can be found for the monolithic amorphous sample. With the introduction of porous Ti/Fe particles, the BMGCs exhibit the obvious yield phenomenon due to the avoidance of premature fracture, and the fracture strength is improved when compared with Base alloy. In addition, the plastic strain of the composites is also increased. The porous Ti/Fe particle provides effective obstacle for the rapid propagation of main shear bands, prohibiting the propagation of shear bands and promoting the multiplication of shear bands, therefore improving the mechanical properties of BMGCs. The size of glassy matrix inside one porous particles should lead to a more homogeneous deformation instead of localized shear flowing in shear bands, thus avoiding the temperature rise and softening in shear bands as well as transformation from shear bands to cracks, which should further contribute to the improved mechanical properties of the present porous particle reinforced BMGCs. The obtained composite structure can be further optimized by adjusting the composite composition and the process parameters. The present results have shown the feasibility of applying the dealloying reaction in fabricating metal matrix composites.

4 Conclusion

A novel fabrication process of in-situ porous Ti/Fe particle reinforced Mg-based BMGCs have been proposed based on the dealloying reaction in metallic melt, and the composites have been successfully synthesized. The introduction of porous Ti/Fe particle into the glassy matrix improves the fracture strength and the plasticity. The pores are fulfilled with the glassy matrix, and the ligament of porous particle has a good bonding with the matrix. The average pore size diameter is close to the processing zone size of the present glassy matrix. The high interface area is in favor of arresting the propagation of main shear bands, and also facilitating the multiplication of shear bands. The present study offers a feasible way to synthesis the bicontinuous porous particle reinforced BMGCs, which should contribute to the breakthrough in designing novel metal matrix composites.

Study of Phase Separation Theory of Alloys and its Applications

Jiuzhou Zhao[1, 2*], Hongxiang Jiang[1, 2], Lili Zhang[1, 2]
1 Shi-changxu Innovation Center for Advanced Materials, Institute of Metal Research, Chinese Academy of Sciences, Shenyang 110016, PR China;
2 School of Materials Science and Engineering, University of Science and Technology of China, Shenyang 110016, PR China
*Corresponding address: e-mail: jzzhao@imr.ac.cn

Abstract: The in-situ metal matrix composite materials have become one of the hot topics in the field of materials science and engineering due to their advantages of high thermodynamic stability, excellent performance, and low production cost. Phase separations, which include the liquid-liquid phase separation, liquid-solid phase separation, and solid-solid phase separation, are the main approaches for the development of the in-situ composite materials. The thermodynamic and kinetic characteristics of all these phase separation processes are similar. Among them, the liquid-liquid phase separation is the most complicated one. The theories on the microstructure formation during the liquid-liquid phase separation can be used to describe the other two kinds of phase separations. We investigated the phase separation process in depth. This report elaborates on the microstructure formation, key affecting factors and controlling methods of liquid-liquid phase separation process. It makes a simple introduction to the application of the theoretical research results in the development of the materials needed nowadays.

Keywords: In-situ composite, phase separation alloy, solidification, magnetic field, electric pulses

1 Introduction

Phase separations, which includes the liquid-liquid phase separation, liquid-solid phase separation, and solid-solid phase separation, are the main approaches for the development of the in-situ composite materials. The thermodynamic and kinetic characteristics of all these phase separation processes are similar. Among them, the liquid-liquid phase separation is the most complicated one. The theories on the microstructure formation during the liquid-liquid separation can be used to describe the other two kinds of phase separations. We investigated the solidification of the liquid-liquid phase separation alloys. This paper will elaborate the microstructure evolution during the liquid-liquid phase transformation, its key affecting factors and the possibility to control the microstructure formation by using the static magnetic field and electric pulses (ECPs).

2 Experimental procedure

Directional solidification experiments were carried out with Al-5t%Pb alloy and Bi-10%Cu-10%Sn (wt.%) by using a Bridgman type solidification setup with a magnetic field producer and an ECPs producer. The Al-5t%Pb alloy sample was prepared by melting pure Al and Pb in a graphite crucible, heating the melt to 950°C, holding temperature for 30 minutes to form a homogeneous liquid and then withdrawing the crucible into a bath of a Ga-In-Sn liquid alloy. The Bi-10%Cu-10%Sn alloy sample was prepared by melting pure Cu, Bi, and Sn in an alumina crucible, heating the melt to 900°C, holding temperature for 30 minutes to form a homogeneous liquid and then withdrawing the crucible into the Ga-In-Sn liquid alloy bath. The temperature profile in the samples was measured by the thermocouples locating inside the melt.

3 Result and discussion

Fig. 1 shows the microstructures of the sample solidified at the rate of 5mm/s in a static magnetic field 0T and 0.3T. The black phase is α-Al and the white phase is the Pb-rich particles. The microstructures demonstrate that the magnetic field causes a decrease in the size of the Pb-rich particles. The magnetic field causes a refinement of the in-situ Pb particles. Fig. 2 shows the microstructures of the Bi-10%Cu-10%Sn samples solidified at the rate of 8mm/s with or without the effect of ECPs. The results indicate that ECPs promote this alloy to form a well dispersed microstructure [1].

The microstructure formations were calculated. The results demonstrate that there exists an undercooling region in front of the Solid/Liquid interface and the nucleation of the Minority Mhase Droplets (MPDs) occurs around the peak of the undercooling, as shown in Fig. 3. The nucleation and growth of the droplets lead to a decrease in the supersaturation and thus the nucleation occurs only in a small region. After nucleation the droplets grow and coarsen as they move towards the solidification interface until they are caught by the S/L interface.

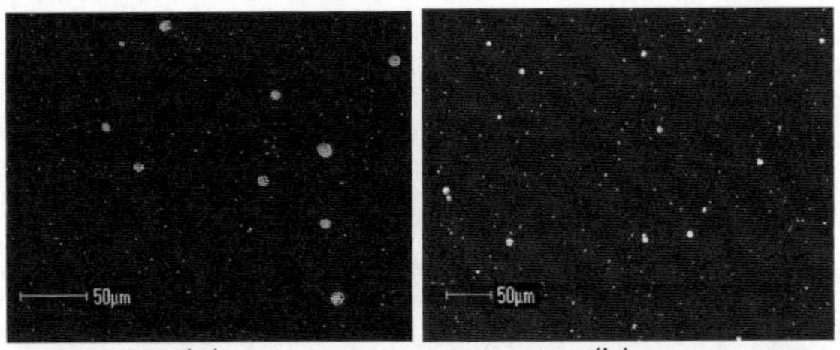

(a) (b)

Fig.1 Microstructures of Al-5%Pb alloy solidified at the rate of 5 mm/s in magnetic field of 0 T (a) and 0.3 T.

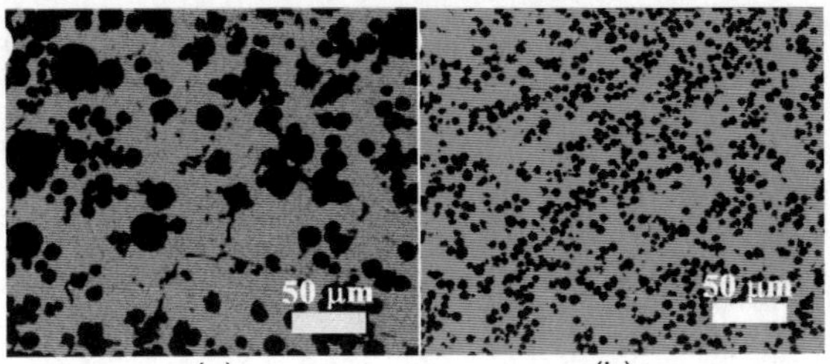

(a) (b)

Fig.2 Microstructures of Bi-10%Cu-10%Sn alloy solidified at the rate of 10 mm/s without ECPs (a), and with ECPs of 50 Hz and 30,000 A/cm2.

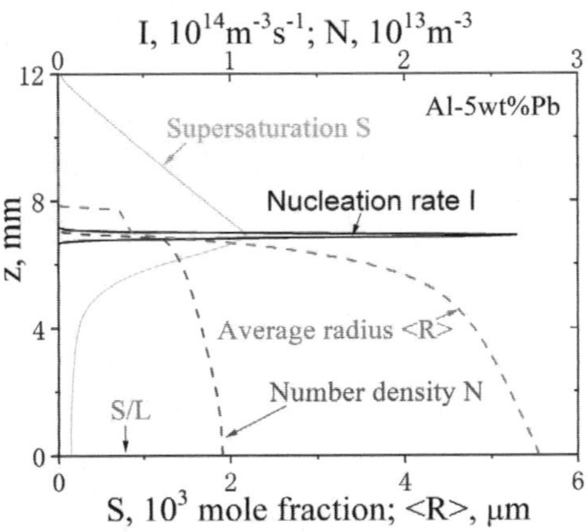

Fig.3 Supersaturation, nucleation, number density and average radius of the MPDs in front of the solidification interface along the central longitudinal axes of the sample.

The static magnetic field affects the microstructure formation mainly through the suppression of convections. It causes a more uniform distribution of the nucleation rate and the number density of the MPDs along the sample's radial direction, as shown in Fig. 4. It also causes a decrease in the size of the maximum droplet in front of the solidification interface. All these are favorable for the formation of a well dispersed microstructure. ECPs mainly affect the microstructure formation through changing the nucleation behaviors of the MPDs. They cause an increase in the nucleation rate of the Cu-rich droplets during the liquid-liquid phase transformation and promote the formation of a well dispersed solidification microstructure, as shown in Fig. 5.

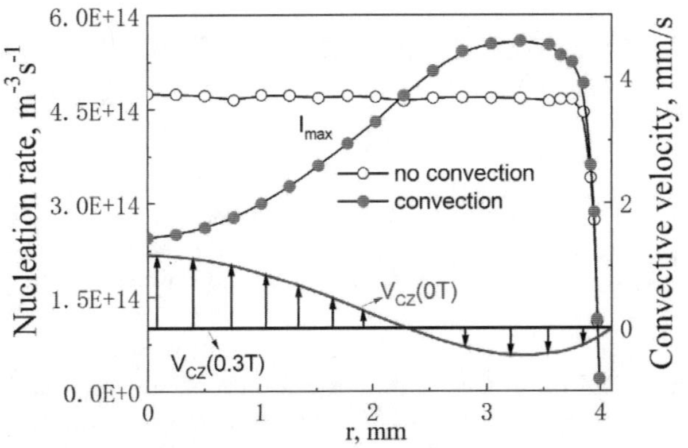

Fig. 4 Variation of the maximum nucleation rates of the Pb-rich droplets and the z vector of the flow velocities at the corresponding position with the sample radial direction.

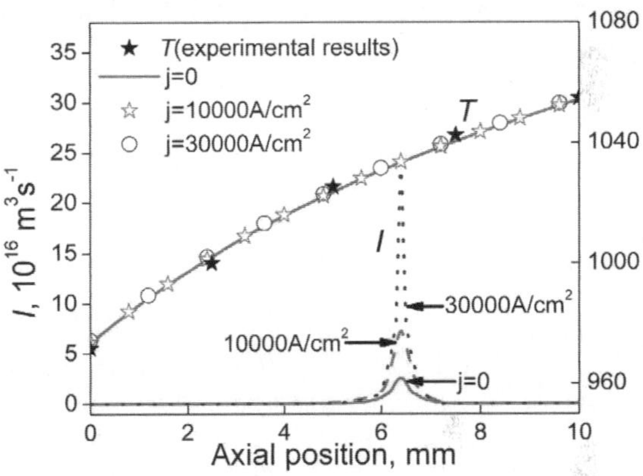

Fig. 5 Temperature profile in front of the solidification interface and nucleation rate of precipitated phase droplets (I) during solidifying Bi-10%Cu-10%Sn alloy.

4 Conclusion

The microstructure formation during the liquid-liquid phase transformation is a result of the concurrent actions of the nucleation, growth, Ostwald ripening, motions and collision-and-coagulations of the MPDs. Both a static magnetic field and ECPs can be applied to control solidification process and microstructure of the liquid-liquid phase separation alloys.

5 Acknowledgments

This work was supported by the National Key Research and Development Program of China [2021YFA0716303], the China's Manned Space Station Project.

References

[1] Zhao J Z, Ahmed T, Jiang H X, et al. Solidification of immiscible alloys: A review, Acta Metall. Sin. (English Letters) , 2017,30:1-28.

The Effect of Sm on the Microstructure and Properties of Nb-Si Based in-Situ Composites

Wei Wei, Qi Wang*, Ruirun Chen*
National Key Laboratory for Precision Hot Processing of Metals, School of Materials Science and Engineering, Harbin Institute of Technology, Harbin 150001, China
*Corresponding address: Qi Wang, wangqi_hit@hit.edu.cn; Ruirun Chen, ruirunchen@hit.edu.cn

Abstract: This study examines the microstructure and mechanical properties (with a focus on room-temperature toughness), of Sm-doped Nb-Si based in-situ composites. The base alloy consists of the coarse primary Nb5Si3 phase and the Nb3Si + Nbss (Nb solid solution) eutectic cells. The microstructure significantly refines with increasing Sm content. In addition, Sm forms a stable Sm oxide phase, which alleviates the oxygen contamination problem to some extent. Thus, with 0.2 at% Sm doping, the KQ value is 18.10 MPa·m1/2, which is 31.6% higher than that of the base alloy.

Keywords: Nb-Si based composites; rare earth elements; microstructure; properties

1 Introduction

The TET of the next-generation aero-engines is predicted to be approximately 2400K [2]. The blade material must survive temperatures of 1600 K even with advanced cooling systems and thermal barrier coatings [3]. However, nickel-based single crystal superalloys have a temperature bearing capacity of only 1450 K [4]. Thus, the search for new-generation ultra-high temperature structural materials has begun.

Nb-Si based in-situ composites become the most expected candidate materials because of its low density (7.3–8.6 g/cm^3) and quite high-temperature strength (270–320 MPa at 1673 K) [7]. However, issues with their poor room-temperature toughness and oxidation resistance must be resolved [10].

In this study, the highly active rare-earth element Sm was doped into the classical Nb-Ti-Si-Hf-Cr-Al system. It is expected that some progress will be made in the application of Nb-Si based in-situ composites by Sm doping.

2 Experimental procedure

The base alloy had the standard composition of Nb-24.7Ti-16.0Si-8.2Hf-2.0Cr-1.9Al (MASC alloy, at.%) [29]. The Sm doping amounts used were 0.1, 0.2, 0.4, and 0.8 at%, hereinafter referred to as *x*Sm alloys. Button ingots were prepared using a non-consumable vacuum arc furnace. To ensure the accuracy of the alloy composition, the alloy was melted six times. Electrical discharge machining (EDM) was used to prepare the specimens.

X-ray diffractometry (XRD) was used to examine the phase composition. Field emission scanning electron microscopy equipped with an energy dispersive spectrometer (FE-SEM; Hitachi TM4000) was used to examine the microstructure, fracture morphologies and composition. Wave dispersive spectrometry (WDS), wherein a wave dispersive spectrometer equipped with an electron probe X-ray microanalyzer (EPMA), was used to study the light element O. Quantitative analysis of the phase or microstructure was performed using Image Pro Plus 6.0. In-situ transmission electron microscopy (TEM; Talos F200X) was used to determine the phase structure. An AGXplus electronic universal tensile machine was used to measure the mechanical properties.

3 Result and discussion

Phases and microstructure

The phase compositions of all the alloys are Nbss phase, Nb_3Si phase and γ-Nb_5Si_3 phase. Sm doping causes changes in the number, intensity, and position of the diffraction peaks. The diffraction peak intensity of the Nb_3Si phase decreases with the increasing of Sm content, and the diffraction peaks of the Nb_3Si phase at 39° disappear in the Sm-doping alloys. The phase content is directly related to changes in the number and intensity of the diffraction peaks. This demonstrates that Sm doping results in a a relative decrease in the Nb_3Si phase content.

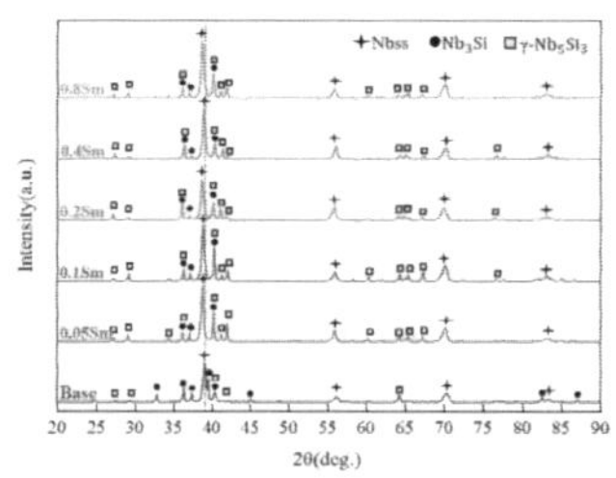

Figure 1 XRD patterns of base-xSm (x = 0, 0.05, 0.1, 0.2, 0.4, and 0.8) alloys

Combined with the XRD results and phase compositions, the dark grey phase is the Nbss phase. The thin hexagonal phase with a darker color is the γ-Nb_5Si_3 phase, which typically forms a eutectic structure with the secondary Nbss phase. The Nb_3Si phase is the light gray phase with a large area percentage. It was discovered that microstructure especially the Nb_3Si phase significantly refined as the Sm content increased.

Figure 2 Microstructure morphology: (a, b) base alloy; (c, d) 0.05Sm alloy; (e, f) 0.1Sm alloy;(g, h) 0.2Sm alloy; (i, j) 0.4Sm alloy; (k, l) 0.8Sm alloy.

Mechanical properties

The K_Q value of base alloy is 13.75 $MPa{\cdot}m^{1/2}$, and the K_Q value is greatly increased by the doping of an appropriate amount of Sm. The K_Q value of 0.2Sm alloy reaches 18.10 $MPa{\cdot}m^{1/2}$, which is 31.6% higher than that of the base alloy. The reduction in the size of brittle silicides is directly responsible for the increase in toughness. In addition, Sm doping with strong oxygenophilicity reduces the oxygen content in the base alloy. However, the microstructure is too fine to resist cracking through when the Sm content exceeds 0.4 at.%.

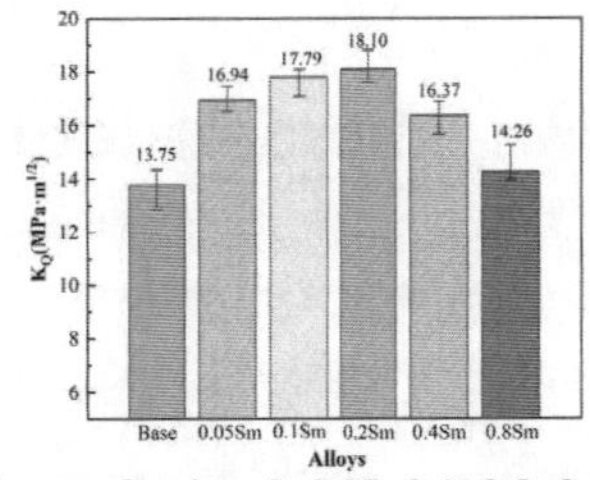

igure 3 KQ values of base-xSm (x = 0, 0.05, 0.1, 0.2, 0.4, and 0.8) alloys

4 Conclusion

(1) The phase compositions are Nbss phase, Nb_3Si phase and γ-Nb_5Si_3 phase. microstructure especially the Nb_3Si phase significantly refine as the Sm content increased.

(2) With 0.2 at% Sm doping, the KQ value is 18.10 $MPa{\cdot}m^{1/2,}$ 31.6% higher than that of the base alloy.

Acknowledgments

The authors gratefully acknowledge support from the National Natural Science Foundation of China (51825401).

References

[1] Zeng Q, Chen X. Combustor technology of high temperature rise for aero engine[J]. Prog Aerosp Sci. 2023, 140: 100927.

[2] Zhang G, Zhu R, Xie G, Li S, Sundén B. Optimization of cooling structures in gas turbines: A review[J]. Chin J Aeronaut. 2022, 35: 18–46.

[3] Unnikrishnan U, Yang V. A review of cooling technologies for high temperature rotating components in gas turbine[J]. Propuls Power Res. 2022, 11: 293–310.

[4] Bewlay BP, Jackson MR. A review of very-high-temperature Nb-silicide-based composites[J]. Metall Mater Trans A. 2003, 34: 2043–52.

[5] Knittel S, Mathieu S, Portebois L, et al. Effect of tin addition on Nb–Si-based in situ composites[J]. Part II: Oxidation behaviour, Intermetallics. 2014, 47: 43-52.

[6] Bewlay BP, Jackson MR, Lipsitt HA. The balance of mechanical and environmental properties of a multielement niobium-niobium silicide-based in situ composite[J]. Metall Mater Trans A,1996, 27: 3801–08.

Preparation, Microstructure and Mechanical Properties of AlN / Mg-Al-RE Composites

Yuyang Gao[1,2,*], Yuanlin Li[1,2], Zhihua Dong[1,2], Ang Zhang[1,2], Yan Yang[1,2], Jiangfeng Song[1,2], Bin Jiang[1,2,*], Fusheng Pan[1,2]

1. National Engineering Research Center for Magnesium Alloys, College of Materials Science and Engineering, Chongqing University, Chongqing, 400044, China
2. National Key Laboratory of Advanced Casting Technologies, Chongqing University, Chongqing, 400044, China
*Corresponding address: gaoyuyang@cqu.edu.cn (Yuyang Gao); jiangbinrong@cqu.edu.cn (Bin Jiang)

Abstract: In the process of high temperature service, the Al11RE3 phase in Mg-Al-RE alloy will soften and even decomposition into Al2RE phase, while the Mg17Al12 phase is easy to grow, coarsen, and dissolve when the service temperature is higher than 150 ℃ , which is not conducive to the service temperature and high temperature performance improvement of Mg-Al-RE alloy. The high temperature performance of Mg-Al-RE alloy can be improved by adding ceramic particles with high temperature stability. In this paper, the in-situ AlN/Mg master alloy were prepared by Mg-Al-Mg3N2 reaction system. The effects of nano-AlN particles on the microstructure and mechanical properties of Mg-Al-RE composites were studied.

Keywords: in-situ AlN particles, Mg-Al-RE composites, mechanical properties

1 Introduction

Magnesium (Mg) alloys are the most promising lightweight materials for engineering applications, mainly because of their low density. However, the potential engineering applications of Mg alloys are limited due to their relatively low absolute strength. Adding ceramic particles to Mg alloys to obtain Mg matrix composites is an effective way to further improve its high temperature mechanical properties. It is found that AlN particle has excellent physical properties such as high melting point (2200 °C), high elastic modulus (310 GPa), low thermal expansion coefficient (4.4×10^{-6}/°C), and low density (3.2 g/cm^3), which can be used as a reinforcing phase to improve the properties of Mg alloys. In this paper, the in-situ AlN/Mg master alloy was prepared and the effects of nano-AlN particles on the microstructure and mechanical properties of Mg-Al-RE composites were discussed and analyzed.

2 Experimental procedure

In this study, 40wt.% AlN/Mg master alloy was prepared by Mg-Al-Mg_3N_2 reaction system. The Mg-4Al-4La-0.3Mn alloy was used as the matrix alloy. The 0.5-3.0wt. % AlN/AE44 composites were prepared by cast method. The microstructure was observed by OM (Zeiss Axio Vert.A1), SEM (JEOL JSM-7800F), and TEM. A tensile testing machine (CMT6305-300KN) was used to test the mechanical properties of composites at room temperature at a strain rate of 1×10^{-3} s^{-1}.

3 Result and discussion

Microstructure of in-situ AlN master alloy

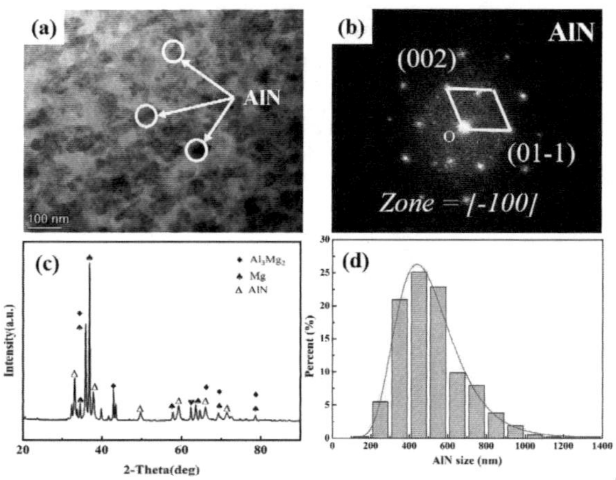

Figure 1. In-situ AlN/Mg master alloy:(a) TEM image, (b) SAED image, (c) XRD results; (d) size statistics diagram of AlN particles.

Fig. 1(a) shows the TEM image of in-situ AlN/Mg master alloy. It is found that AlN particles are uniformly distributed in the Mg matrix, with good dispersion and no obvious agglomeration. Fig. 1(b) is the selected area electron diffraction pattern (SAED) pattern of AlN particles. Combined with the XRD results (Fig. 1(c)), the in-situ AlN/Mg master alloy is mainly composed of AlN, Al_3Mg_2, and α-Mg phases. As shown in Fig. 1(d), the size of AlN particles generated by in-situ reaction varies from 100 to 1200 nm, and the size is mainly distributed in the range of 400-600 nm.

Microstructure of AlN/AE44 composites

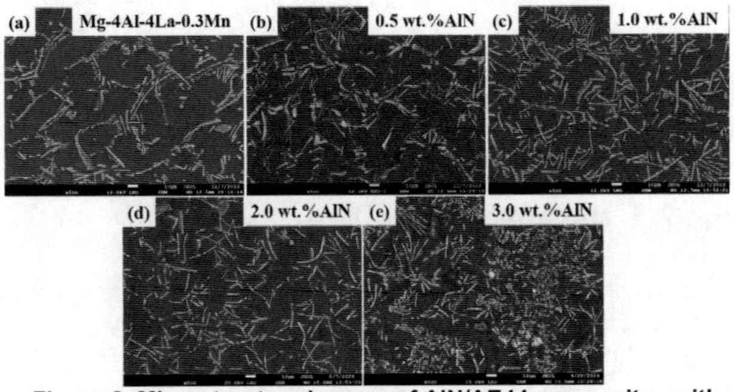

Figure 2. Microstructure images of AlN/AE44 composites with different AlN content

The in-situ nano-AlN particles were added into the AE44 alloy to obtain AlN/AE44 composites. Fig. 2 shows the microstructure of cast AE44 alloy and AlN/AE44 composites. As shown in Fig. 2(a), the microstructure of AE44 alloy is mainly composed of dendritic grains. The morphology of the second phase is mainly needle-like and rod-like, and there are also a small number of bulk shaped phases. As indicated by Figs. 2(b-e), compared with the matrix alloy, the addition of nano-AlN particles significantly improves the distribution of needle-like $Al_{11}La_3$ phase and bulk Al_2La phase. The size of the second phases was firstly reduced and then increased with increasing AlN content. The 2.0 wt.% AlN/AE44 composite shows the most uniform distribution of the second phases.

Mechanical properties of AlN/AE44 composites

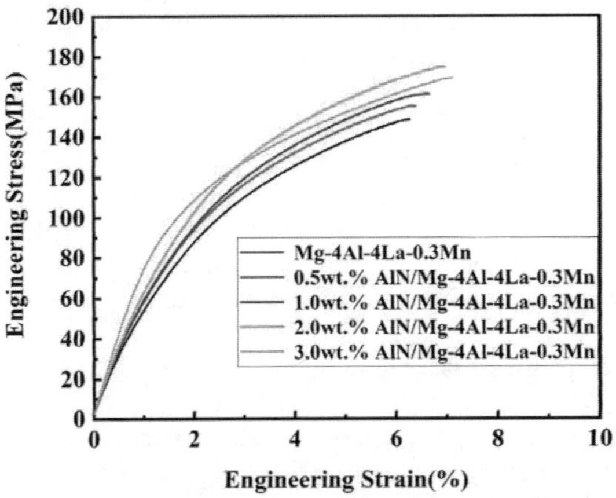

Figure 3. Room temperature tensile engineering stress-strain curves of AlN/AE44 composites.

Fig.3 shows the tensile engineering stress-strain curves of AlN/AE44 composites, and the corresponding mechanical properties are shown in Table 1. The addition of nano-AlN particles simutanously increased the yield strength, tensile strength, and ductility of AE44 matrix alloy. The 2.0 wt. % AlN/AE44 composite shows the best comprehensive mehcanical properties with the yield strength, tensile strength, and ductility of 96 Mpa, 175 MPa, and 7.0%, respectively.

Table 1. Tensile properties of AlN/AE44 composites.

Composites	$\sigma_{0.2}$/MPa	σ_{UTS}/MPa	ε/%
AE44	70	148	6.3
0.5 wt.% AlN	82	155	6.4
1.0 wt.% AlN	90	161	6.7
2.0 wt.% AlN	96	175	7.0
3.0 wt.% AlN	99	169	7.1

4 Conclusion

The addition of AlN particles significantly reduces the sizes of the second phases, and improves the distribution of the second phases. The nano-AlN/AE44 composites show improved comprehensive mechanical properties compared to matrix alloy. The increased mechanical properties of the composites are attributed to the strengthening effects of nano-AlN particles and finer sized AlRE phases.

5 Acknowledgments

This work is financially supported by the National Key Research and Development Program of China (Grant No. 2021YFB3701000), the National Natural Science Foundation of China (Grant Nos. 52201106 and U21A2048), and the Natural Science Foundation of Chongqing China (Grant No. CSTB2022NSCQ-MSX1289).

References

[1] Joost W J, Krajewski P E. Towards magnesium alloys for high-volume automotive applications[J]. Scripta Materialia, 2017, 128: 107-112.

[2] Mehra D, Mahapatr M.M., Harsha S P. Processing of RZ5-10wt%TiC in-situ magnesium matrix composite[J]. Journal of Magnesium and Alloys, 2018, 6: 100-105.

[3] Yang C L, Zhang B, Zhao D C, et al. Microstructure and mechanical properties of AlN particles in situ reinforced Mg matrix composites[J]. Materials Science & Engineering A, 2016, 674:158-163.

[4] Chen J, Bao C G, Chen F L. Evolutions of microstructure and mechanical properties for Mg-Al/AlN composites under hot extrusion[J]. Materials Science & Engineering A, 2016, 667:426-434.

[5] Chen Q, Chen G, Han L N, et al. Microstructure evolution of SiC_p/ZM6 (Mg–Nd–Zn) magnesium matrix composite in the semi-solid state[J]. Journal of Alloys and Compounds, 2016, 656:67-76.

Study on Strength and Toughness Regulation of SiCp/Al Composite for Aerospace Vehicles

Jingpei Xie [1*], AiqinWang[1], Shiming Hao [2], Yahu Song [3], Pei Liu [1], Pengru Liu [2], Ying Liu [1]
1. School of Materials Science and Engineering, Henan University of Science and Technology, Luoyang China
2. School of Physics and Engineering, Henan University of Science and Technology, Luoyang China
3. School of Intelligent Manufacturing, Luoyang Institute of Science and Technology, Luoyang China
*Corresponding address: e-mail:xiejp@haust.edu.com

Abstract: Aerospace vehicles face harsh environments such as extreme temperature changes and various load coupling effects during their service in space, which puts forward more stringent requirements for the dimensional stability, strength, plasticity and toughness of their key structural components. Improving the strength and toughness of Aluminum Matrix Composite, breaking the contradiction between strength, plasticity, toughness, and thermal properties, is a key scientific problem in the development of composite. This work focuses on the preparation of medium high volume fraction SiCp/Al composite by powder metallurgy method, and focuses on the strength and toughness regulation strategies of Aluminum Matrix Composite. The regulation methods and influencing laws of factors such as reinforcement types and sizes, aluminum alloy matrix, interface bonding, and heat treatment process are explored, and the interface structure characteristics, matrix microstructure, and second phase evolution law of composite are revealed. The strength and toughness mechanism of composite under the synergistic effect of interface effects and multi-scale second phases is elucidated. The research provides theoretical basis and technical support for the engineering application of lightweight, low expansion, and high-strength Aluminum Matrix Composite.

Keywords: Aluminum Matrix Composite, Interface Bonding, Strength and Toughness Regulation

1 Introduction

In order to meet the special requirements of key structural components of aerospace vehicles in terms of dimensional stability, strength, and plasticity and toughness. Medium to high volume fraction SiCp/Al composite are expected to become substitutes due to their integrated structural and functional properties. However, the addition of SiC improves strength while sacrificing plasticity, which limits its wider application. Finding the optimal balance of strength and plasticity inversion has been a challenging problem in the development of SiCp/Al composites. By optimizing composite components, composite processes, and heat treatment techniques to achieve multi-scale second phase and interface structure control, lightweight, high-strength, and low expansion SiCp/Al composite materials can be developed to meet national needs.

2 Experimental procedure

The composite used in this paper are all prepared by powder metallurgy method. Spherical aluminum alloy powders and irregular 6H–SiC powders were used as raw materials. Pretreated SiC particles (SiCp) were mixed with aluminum alloy powders at different fraction ratio. The mixed powders were ball milled and sintered by vacuum hot pressing sintering furnace. The phase compositions, micro-morphology and elemental distribution were analyzed by XRD, SEM, HAADF-STEM snd EDS. The tensile and compressive properties were tested using a universal testing machine.

The CASTEP quantum mechanics program in Material Studio software was used for all of the first-principles calculations conducted in this study. Different interfacial configurations of $MgAl_2O_4$/Al and $MgAl_2O_4$/SiC low index interface models were constructed. The adhesion properties of the interface were predicted by calculating the interfacial spacing, adhesion work, and interfacial energy for different $MgAl_2O_4$/Al and $MgAl_2O_4$/SiC interface models. The uniaxial tensile tests were conducted on eight different 6H-SiC/Al interface models through the molecular dynamics method(LAMMPS). The deformation of the interface was monitored to obtain the main mechanical property parameters and to explore the dynamic process of atomic migration, stress-strain characteristics, and dislocation interactions.

3 Result and discussion

3.1 Effect of preparation process on the microstructure and properties of composite

For SiC/Al-Si composites, by controlling the sintering temperature, extrusion temperature, and solid solution temperature, the Si phase size can be controlled. Adopting a suitable heat treatment process (500 ℃ solid solution and 180 ℃ aging for 6 hours) can achieve synergistic improvement of strength and plasticity.

3.2 Effect of adding rare earth elements on the microstructure and properties of composite

For SiC/Al-Si composites,the appropriate addition of CeO_2 can achieve a synergistic improvement in strength and plasticity. The addition of rare earth elements inhibits the dissolution, precipitation and growth of Si element, refines the precipitated phase, and forms a rare earth compound strengthening phase.

3.3 Interface characteristics of composite

The SiC/Al-Si and SiC/Al composite prepared by powder metallurgy method have good interfacial bonding. The interfaces of SiC/Al, Si/Al, and SiC/Si are clear, with no obvious interface reactants or defects. The SiC/Al interface is mainly a clean interface, and the interface atoms formed between Si/Al and SiC/Si are matched as a semi coherent interface.

3.4 Effect of interface reaction layer regulation on the microstructure and properties of composite

Moderate interface reaction(Generating $MgAl_2O_4$ phase) is beneficial for improving interface bonding strength. By controlling SiC oxidation and the content of Mg in the matrix, the thickness and coating effect of the $MgAl_2O_4$ layer can be effectively adjusted to change the fracture mode during deformation and achieved a comprehensive improvement in strength and ductility.

3.5 Intrinsic computational simulation and in-situ experimental research on interface properties of composite

First principles calculations indicate that the interfacial bonding strength between SiC and Al is higher than that between atomic layers within the Al matrix. Molecular dynamics calculations indicate that the tensile fracture position at the SiC/Al interface is more likely to occur within the Al matrix near the interface. The TEM in-situ tensile test of the SiC/Al interface micro zone shows that fracture first occurs in the aluminum matrix, and the bonding strength of the SiC/Al interface is higher than that of the aluminum matrix.

4 Conclusion

（1）By regulating the multiple scales second phase, the synergistic improvement strength and plasticity of composite can be achieved. The control technical parameters vary depending on the aluminum alloy matrix.

（2）Moderate interface reaction is beneficial for improving interface bonding strength and enhancing load transfer effect; The combination of SiC oxidation pretreatment and appropriate control of Mg element in the matrix is an effective method for enhancing toughness.

（3）Using the concept of material genetic engineering to carry out material research and development, using first principles, molecular dynamics and other methods, the intrinsic characteristics of composite can be studied in depth at the atomic scale, which plays an important role in the development of composite.

5 Acknowledgments

This work was financially supported by the National Natural Science Foundation of China (51371077, 51771070, 52171138).

References

[1] Liu P R, Hao S M, Wu H Z, et al. Synergistic enhancement of strength and plasticity of SiCp/Al composites by designing and controlling SiCp@$MgAl_2O_4$ core-shell micro-interfaces[J]. Ceram. Int. 2023, 49(24PA): 41016-41021.

[2] Liu Y, Wang A Q, Liu P,et al. First-principles study of structural stability and electronic properties of polar $MgAl_2O_4$/Al interface[J]. J. Phys. Chem. Solids, 2023, 182: 111602.

[3] Song Y H, Wang A Q, Ma D Q, et al. Dynamic recrystallization behavior and nucleation mechanism of dual-scale SiCp/A356 composites processed by P/M method[J]. Nanotechnol. Rev. 2023, 12: 20220506.

[4] Dong X D, Wang A Q, Xie J P, et al. Optimization of volume fraction and microstructure evolution during thermal deformation of nano-SiCp/Al-7Si composites. Nanotechnol[J]. Rev. 2021, 10(1): 1306-1317.

[5] Dong X D, Wang A Q, Xie J P, et al. Effect of CeO_2 Content on Microstructure and Properties of SiCp/Al-Si Composites Prepared by Powder Metallurgy[J]. Materials, 2021, 14(16): 4685-4685.

[6] Wang C Q, Chen W G, Xie J P. Effects of transition element additions on the interfacial interaction and electronic structure of Al(111)/6H-SiC(0001) interface: a first-principles study[J]. Materials, 2021, 14(3): 630-630.

[7] Yang B, Wang A Q, Liu K D, et al. Effects of CeO_2 on the Si precipitation mechanism of SiCp/Al-Si composite prepared by powder metallurgy[J]. Materials, 2020, 13(19): 4365-4365.

[8] Song Y H, Wang A Q, Ma D Q, et al. Hot-deformation behavior and microstructure evolution of the dual-scale SiCp/A356 composites based on optimal hot-processing parameters[J]. Materials, 2020, 13(12): 2825.

[9] Wang Z, Wang A Q, Han H H, et al. HRTEM study of interfacial structure in SiCp/A390 composites[J]. Mater. Res. Express, 2020, 7(4): 046514.

The Microstructure and Mechanical Properties of Aluminum/Steel Composite Plates Controlled by Rheological Cast-Rolling and the Rolling

Jin Qiu[1], Yuandong Li[1,2,*], Xiaoyu Liang[1], Hongwei Zhou[1], Chi Cao[1,3], Guangli Bi[1,2]

1. State Key Laboratory of Advanced Processing and Recycling of Nonferrous Metals, Lanzhou University of Technology, Lanzhou 730050, China;
2. Key Laboratory of Nonferrous Metal Alloys and Processing, Ministry of Education, Lanzhou University of Technology, Lanzhou 730050, China;
3. Wenzhou Pump and Valve Engineering Research Institute of Lanzhou University of Technology, Wenzhou 325105, China

*Corresponding author. E-mail address: liyd_lut@163.com (Y.D. Li)

Abstract: In this study, a new idea of prefabricated diffusion layer to improve the mechanical properties of composite plates is proposed. The results indicate that the metallurgical mechanical bonding formed by rheological cast-rolling + rolling can enhance the mechanical properties of the composite plate, and the shear strength and tensile strength under 50% reduction are as high as 105.0 MPa and 342.7 MPa, respectively. The mechanism analysis shows that cracks are formed in the diffusion layer along the normal direction during rolling, and the diffusion layer and the matrix always maintain good metallurgical bonding. The crack gaps of the diffusion layer are small when the reduction is 20%, and the soft matrix cannot enter the gaps to form secondary metallurgical bonding. The residual compressive stress is the main reason for the increase in shear strength. The aluminum matrix is recrystallized and refined when the reduction is 50%, the crack gaps of the diffusion layer increase, the fresh metal is squeezed into the gaps, the matrix re-forms metallurgical bonding. The broken intermetallic compounds and the matrix simultaneously form a mechanical lock and a metallurgical bonding, and the shear strength and tensile strength increase. The thickness of the diffusion layer increases, the unfilled crack gaps increase, the residual compressive stress of the matrix around the diffusion layer decreases, the shear strength decreases significantly, but the tensile property changes little.

Keywords: Prefabricated diffusion layer; Rheological cast-rolling; Rolling; Aluminum/steel composite plate; Shear strength; Metallurgical mechanical bonding

1 Introduction

Aluminum/steel bimetal is a typical bimetal composite material that combines aluminum alloy and steel through a certain mechanical or metallurgical method [1]. However, the preparation of high performance aluminum/steel bimetals is widely recognized as a challenge.

Semi-solid processing temperature is low, low solidification shrinkage, low slurry flow, low requirements for equipment cooling capacity [2]. This is very beneficial for controlling diffusion layer growth and stabilizing production. In this paper, the aluminum/steel composite plates with different thicknesses diffusion layers are prepared by rheological cast-rolling technology.

2 Experimental procedure

The quantitative aluminum alloy ingots were melted in the SG2-7.5-10 well crucible melting furnace. When the temperature reached 730 ℃ then refining, degassing, slag removal. Following the aluminum alloy temperature being controlled to 680 ℃, flowed through the inclined plate to prepare semi-solid slurry, the temperature of inclined plate outlet was 645 ℃ (50% solid phase temperature). Before preparing the slurry, the cleaned steel plate was placed in a mold preheated at 645 ℃. After all the slurry was poured on the surface of the steel plate, the isothermal diffusion was carried out at 645 ℃ for 30 s, 60 s, 120 s, 180 s and 300 s. Isothermal diffusion was followed by rheological cast-rolling, followed by air cooling. The process is shown in Fig.1 (a). The aluminum/steel composite plate prepared by rheological cast-rolling was cut into 60 mm×60 mm×5.5 mm small squares and then rolling (Fig.1 (a)), The reduction was 20% and 50% respectively.

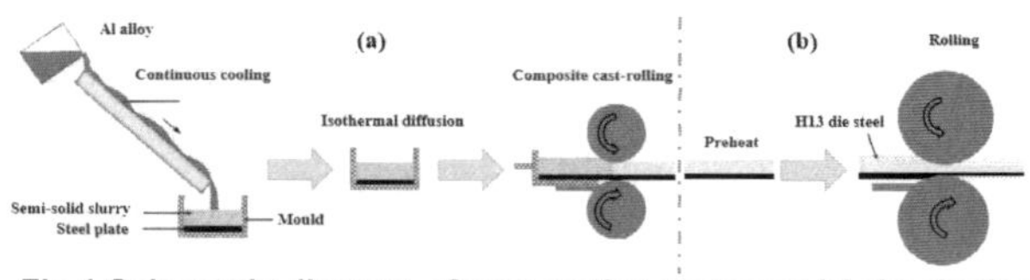

Fig.1 Schematic diagram of preparation process: (a) rheological cast-rolling; (b) rolling.

3 Result and discussion

Mechanical Properties

It can be seen that the ultimate tensile strength and uniform elongation of composite plates increased first and then decreased with the extension of isothermal diffusion time, but the change was little (Fig.2). This was mainly related to the aluminum matrix microstructure, aluminum matrix microstructure was fine and roundness at isothermal diffusion 60 s. The transverse direction ultimate tensile strength and uniform elongation of composite plates were higher than the rolling direction under different isothermal diffusion times. There may be two reasons: there were more cracks in the diffusion layer in the rolling direction than in the transverse direction, so the rolling direction ultimate tensile strength was lower than that in the transverse direction; In the rolling process, the deformation in the rolling direction was greater than that in the transverse direction, and the degree of work hardening of rolling direction was higher, so the transverse direction elongation was higher than that in the rolling direction. Fig.2(b) shows the ultimate tensile strength and uniform elongation of composite plates at 50% reduction under different isothermal diffusion times. The rolling and transverse direction ultimate tensile strength and uniform elongation reached the maximum, were 337.5 MPa, 342.7 MPa and 1.7% and 2.0%, respectively.

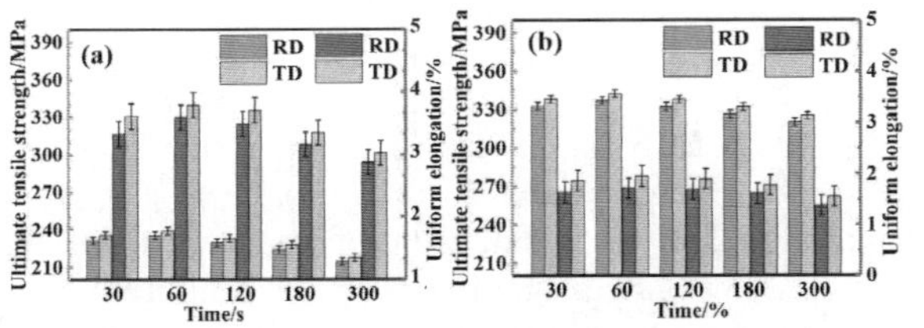

Fig.2 Tensile properties of composite plates under different parameters: (a) ultimate tensile strength and uniform elongation of composite plates at 20% reduction under different isothermal diffusion times; (b) ultimate tensile strength and uniform elongation of composite plates at 50% reduction under different isothermal diffusion times.

It can be seen that the shear strength decreased continuously with the extension of isothermal diffusion time (Fig.3). The rolling and transverse direction shear strength of composite plates decreased from 78.2 MPa and 81.8 MPa at isothermal diffusion 30 s to 52.6 MPa and 54.0 MPa at isothermal diffusion 300 s, respectively (Fig.3(a)). This shows that the thickness of the prefabricated diffusion layer had an important effect on the subsequent rolling treatment. It is found that the transverse direction shear strength was slightly higher than the rolling direction shear strength under different isothermal diffusion times, which may be related to the uneven deformation in the transverse and rolling directions. Fig.3(b) shows that the shear strength of the composite plate at 50% reduction under different isothermal diffusion times. The rolling and transverse direction shear strength decreased from 100.3 MPa and 105.0 MPa at isothermal diffusion 30 s to 73.3 MPa and 74.4 MPa at isothermal diffusion 300 s.

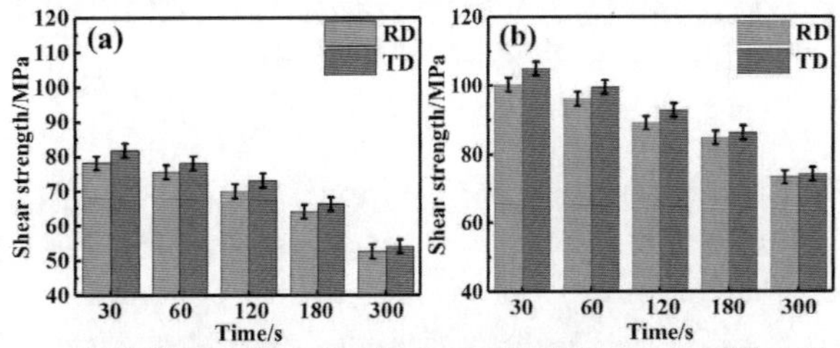

Fig.3 Shear strength of composite plates under different parameters: (a) shear strength of composite plates at 20% reduction under different isothermal diffusion times; (b) shear strength of the composite plates at 50% reduction under different isothermal diffusion times.

4 Conclusion

The prefabricated hard and brittle diffusion layer significantly improves the bonding properties of composite plates. The transverse and rolling direction shear strength reach 105.0 MPa and 100.3 MPa, respectively.

The tensile properties of the composite plate are mainly determined by the matrixes and affected by the bonding properties of the diffusion layer. The rolling and transverse direction ultimate tensile strength and uniform elongation are 337.5 MPa, 342.7 MPa and 1.7% and 2.0%, respectively.

5 Acknowledgments

The authors would like to acknowledge the financial supports from the National Natural Science Foundation of China (Fund name, Study of Semi-Solid Printing Fabrication of Aluminum Laminate Composites and its Interface Regulation. Fund number, 52261010). Major Science and Technology Project of Gansu Province (Fund number, 22ZD6GA008).

References

[1] Qiu J, Li Y D, Luo X M, et al. Silicon regulates the interface microstructure and shear strength of rheological cast-rolling aluminum/steel composite plates[J]. J. Journal of Materials Research and Technology, 2024(30): 6812-6828.

[2] Li G, Lu H X, Hu X, et al. Current Progress in Rheoforming of Wrought Aluminum Alloys: A Review[J]. J. Metals (Basel), 2020, 10(2): 238.

Synergistic Crack Inhibition in 2024 Al-GNPs/TC4 Laminated Composites: Experimental, Phase-Field Modeling, and Molecular Dynamics Investigation

Muxi Li[1], Kaile Wang[1], Qingwei Guo[1], Xiaolin Tian[1], Yihong Liu[1], Kangan Wang[1], Hua Hou[1,3], Yuhong Zhao[1,2,4,*]

1. School of Materials Science and Engineering, Collaborative Innovation Center of Ministry of Education and Shanxi Province for High-performance Al/Mg Alloy Materials, North University of China, Taiyuan 030051, P. R. China
2. Beijing Advanced Innovation Center for Materials Genome Engineering, University of Science and Technology Beijing, Beijing 100083, P. R. China
3. School of Materials Science and Engineering, Taiyuan University of Science and Technology, Taiyuan 030024, China
4. Institute for Materials Intelligent Technology, Liaoning Academy of Materials, Shenyang 110004, P. R. China

*Corresponding address: e-mail: zhaoyuhong@nuc.edu.cn

Abstract: The Al layer always cracks first in Al/Ti LMCs due to its low strength. In this study, 2024 Al-GNPs/TC4 LMCs were prepared by replacing the conventional Al alloy with 2024 Al-GNPs composites. Its ultimate tensile strength of 757 MPa surpasses that of almost all other Al/Ti LMCs. The lamellar structure and GNPs were found to have a synergistic effect in enhancing the mechanical properties by inhibiting crack extension across and within the layers, respectively. The hindrance of crack extension by the lamellar structure is primarily due to strain delocalization in TC4, as confirmed by phase-field-crystal simulation. On the other hand, GNPs impede crack extension through passivation and deflection, which was revealed by molecular dynamics to be attributed to dislocation hindrance and load transfer.

Keywords: Laminated composites, Graphene, Phase-field simulation, Mechanical properties.

1 Introduction

Although the lamellar structure provides a significant strengthening effect, the performance limit of Al/Ti LMCs is constrained by the lower strength of the Al layer. Under stress, cracks tend to initiate in the Al layer first, which results in stress concentration in the Ti layer and ultimately leads to fracture failure of the Al/Ti LMCs.Compared with Al, aluminum matrix composites (AMCs) exhibit superior strength. It is thus speculated that replacing Al with AMCs may be an effective way to further enhance the performance of LMCs. Numerous studies have confirmed the feasibility of replacing Al with AMCs to enhance the strength of LMCs. However, the incorporation of granular reinforcing materials makes AMCs suffer from disadvantages such as transient damage, limited ductility, and poor processability. As a new type of reinforcing material, graphene nanoplatelets (GNPs) with excellent mechanical properties and unique lamellar structure are expected to improve the strength and ductility simultaneously [1]. This work reported 2024 Al-GNPs/TC4 LMCs with an ultimate tensile strength (UTS) surpassing that of almost all other Al/Ti LMCs. The correlation between the mechanical properties, microstructure, and crack extension of 2024 Al-GNPs/TC4 LMCs was investigated. The reason for the different crack evolution in Ti and Al was explained by the phase-field-crystal (PFC) method. The molecular dynamics (MD) method was used to analyze the mechanism of crack blocking by GNPs. In addition, the interfacial atomic diffusion behavior was investigated by the phase-field method. This work finds that GNPs and lamellar structure synergistically block the crack extension to break the performance limit of Al/Ti LMCs and provides a new idea for the design of LMCs.

2 Experimental procedure

The TC4 and 2024 Al-GNPs composites foils were surface-treated and stacked as shown in Fig. 1. After sintering in SPS, the 2024 Al-GNPs/TC4 LMCs were obtained.

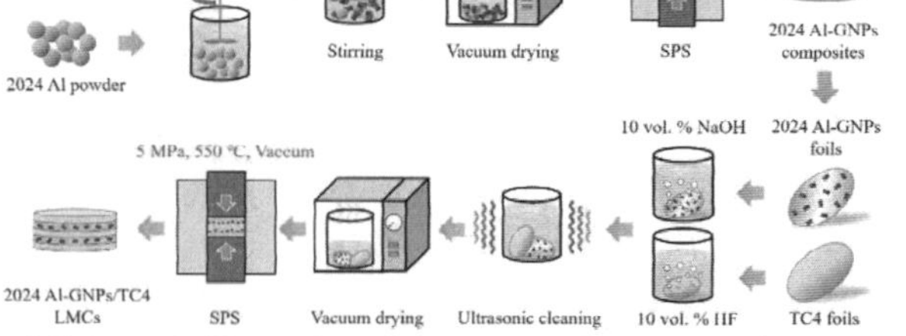

Fig. 1 Schematic illustration of the preparation process for 2024 Al-GNPs/TC4 LMCs.

3 Result and discussion

The morphology of the 2024 Al-GNPs/TC4 LMCs is shown in Fig. 2a. The grains of the 2024 Al-GNPs composites are significantly larger than those of TC4, while the Al_3Ti layer has the smallest grains. This formed a special structure with alternating coarse-fine grain distribution.

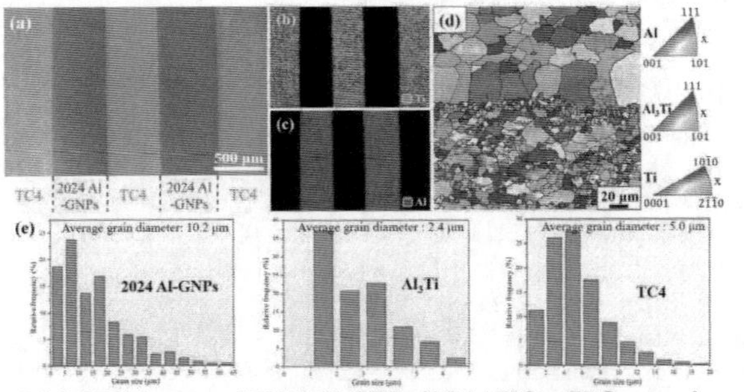

Fig. 2 (a) SEM image of 2024 Al-GNPs/TC4 LMCs, EDS mapping of (b) Ti, (c) Al, (d) IPF map of the 2024 Al-GNPs/TC4 LMCs interface, (e) Grain size statistics of 2024-GNPs composites layer, Al3Ti layer, and TC4 layer.

Figs. $3a_1$-c_1 show the crack evolution without dislocation by phase-field simulation [2, 3]. The main crack propagates along the direction of the maximum shear stress and ultimately penetrates the whole layer. As the dislocation density increases, the dislocations act as sources of cracks and sprout microcracks, as shown in Figs. $3a_2$-c_2. Microcracks disperse strain to prevent excessive crack extension caused by strain concentration, as shown in Fig. $3b_2$ and its inset. This finding reveals that TC4 disperses local stress concentrations through uniformly distributed dislocations, thus avoiding the continuous extension of individual cracks.

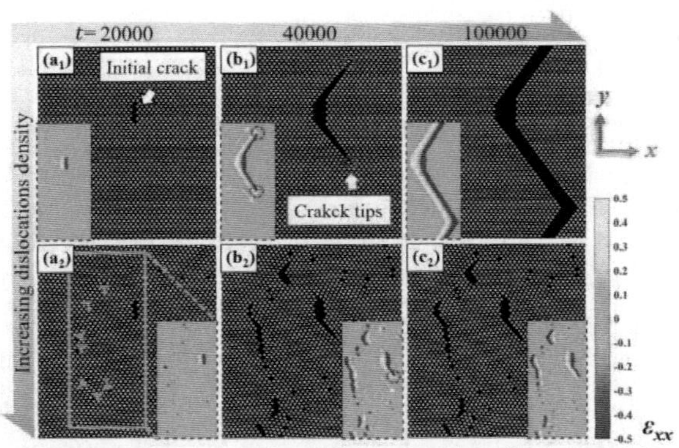

Fig. 3 Crack evolution by Phase-field-crystal method. (a_1-c_1) Crack evolution without dislocation, (a_2-c_2) Crack evolution with high-density dislocations, Insets show the strain field distribution: blue and yellow represent compressive strain and tensile strain, respectively.

A high-density dislocation region was observed near the GNPs, which contains a large number of dislocations and dislocation tangles. This is attributed to the obstruction of dislocations by GNPs. Molecular dynamics confirms that GNPs hinder the connection between dislocations and crack tips.

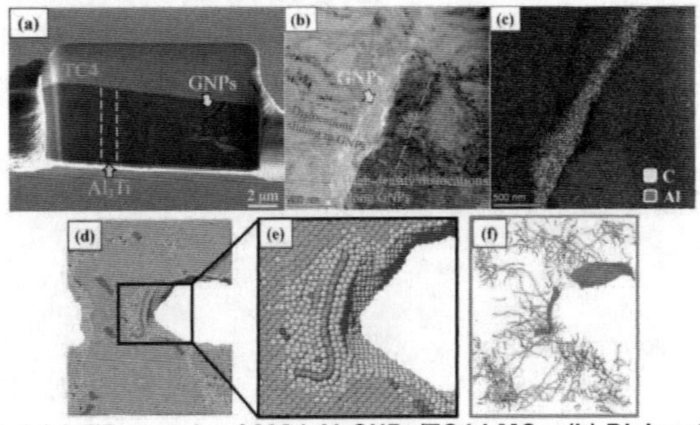

Fig. 4 (a) FIB sample of 2024 Al-GNPs/TC4 LMCs, (b) Dislocation distribution around GNPs, (c) distribution of Al and C elements, and (d)-(f) Molecular dynamics study of GNPs hindering dislocation motion

4 Conclusion

(1) Dense microcracks are generated in the TC4 layers to compensate for the strain difference.
(2) GNPs were found to prevent crack extension within the 2024 Al layer by passivating and deflecting.
(3) Stress is transferred from the matrix to GNPs, which disperses the stress concentration.
(4) GNP prevents dislocations from connecting with the crack tip.

5 Acknowledgments

This work was supported by the National Natural Science Foundation of China (No. 52074246, 52275390, 52205429, 52201146); National Defense Basic Scientific Research Program of China (JCKY2020408B002, WDZC2022-12); Key Research and Development Program of Shanxi Province (202102050201011, 202202050201014); Science and Technology Major Project of Shanxi Province (20191102008, 20191102007); Guiding Local Science and Technology Development Projects by the Central Government (YDZJSX2022A025, YDZJSX2021A027).

References

[1] Li M X, Yang W K, Tian X L, et al. Precipitation and refining of Al_2Cu in graphene nanoplatelets reinforced 2024 Al composites[J]. Mater. Charact,2023, 200: 112854.
[2] Zhao Y H, Understanding and design of metallic alloys guided by phase-field simulations[J]. npj Computational Materials,2023, 9(1): 94.
[3] Xin T Z, Zhao Y H, Mahjoub R, et al. Ultrahigh specific strength in a magnesium alloy strengthened by spinodal decomposition[J]. Sci. Adv, 2021, 7(23): eabf3039.

Molecular Dynamics Simulation of Mechanical Properties of Graphene/Magnesium Matrix Composites

Dunwei Peng [1,2], Qingwei Guo[1], Xiaopan Wang[1], Hua Hou[1,3], Yuhong Zhao[1,4,5*]
1 School of Materials Science and Engineering, Collaborative Innovation Center of Ministry of Education and Shanxi Province for High-performance Al/Mg Alloy Materials, North University of China, Taiyuan, 030051, China
2 School of Materials Science and Engineering, Xi'an University of Technology, Xi'an, Shaanxi 710048, China
3 School of Materials Science and Engineering, Taiyuan University of Science and Technology, Taiyuan, 030024, China
4 Beijing Advanced Innovation Center for Materials Genome Engineering, University of Science and Technology Beijing, Beijing, 100083, China
5 Institute of Materials Intelligent Technology, Liaoning Academy of Materials, Shenyang, 110004, China
*Corresponding address: e-mail: zhaoyuhong@nuc.edu.cn

Abstract: In this work, molecular dynamics simulations were used to investigate the effects of monolayer graphene (GR) with varying orientation angles on Mg-9Al-1Zn (AZ91 wt.%) on mechanical properties of magnesium alloy. The simulation results show that Young's modulus and tensile strength of AZ91/GR composites decrease gradually with the increase of the orientation angle of the 1LG. The Young's modulus and tensile strength of AZ91/GR composites can be improved by the 1LG orientation angle of 0°~10°, where the two are enhanced by 21.7% and 19. 7% respectively, at an orientation angle of 0°. However, the Young's modulus and tensile strength of 1LG are decreased for orientation angles of 20°~90°. It can provide some technical guidance for the experimental process design of AZ91/GR composites.

Keywords: AZ91 Magnesium matrix composites, Graphene, Molecular Dynamics, Mechanical property

1 Introduction

The research and development of graphene magnesium matrix composites can provide new solutions for the application of lightweight materials. In recent years, researcher found that the GNPs were aligned along the tensile direction, enhancing the strength of the graphene/magnesium matrix composites [1, 2]. However, the contribution of GNPs perpendicular to the tensile direction was minimal, limiting their ability to enhance the overall mechanical properties. The impact of graphene nanosheets at different angles to the stretching direction was not specifically addressed with current experimental equipment in detail. This paper focuses on modeling AZ91/GR composites with graphene nanosheets of varying orientation angles using molecular dynamics simulations to understand the tensile deformation mechanism.

2 Experimental procedure

This work is carried out using the Large-scale Atomic/Molecular Massively Parallel Simulator (LAMMPS) [3]. Ten groups of tensile models of AZ91/GR magnesium alloy composites with different orientation angles (θ = 0°, 10°, 20°, 30°, 40°, 50°, 60°, 70°, 80°, 90°) were designed in this study. The Mixed Element Atomistic Method (MEAM) [4] potential function is used in MD simulations to describe the interatomic interactions in Mg-Al-Zn alloys. At the same time, the interatomic interaction of GR is described by the Adaptive Intermolecular Reaction Empirical Bond Order (AIREBO) potential function [5]. In addition, the interactions between Mg-Al-Zn alloys and GR were described using the Lennard Jones (L-J) potential function [6] with the cutoff radius set to 1 nm. The specific energy and distance constants are as follows:
$\varepsilon_{Mg\text{-}C} = 0.0027eV$, $\sigma_{Mg\text{-}C} = 3.5015$ Å; $\varepsilon_{Al\text{-}C} = 0.035078eV$, $\sigma_{Al\text{-}C} = 3.0135$; $\varepsilon_{Zn\text{-}C} = 0.00265699eV$, $\sigma_{Zn\text{-}C} = 3.9735$ Å.

3 Result and discussion

Fig.1(a) shows stress-strain curves for AZ91/GR composites and AZ91 magnesium alloy with graphene monolayers with different orientation angles. The results show that the tensile strength of GR increases to 4.66 GPa at orientation angles of 0°and 10°, while it decreases when the angles range from 20° to 90°. In all models, the stress initially increases linearly with strain, following Hooke's law. The slope of the stress-strain curve during the initial elastic stage represents Young's modulus of the AZ91/GR composites. The results presented in Fig.1(b) indicate that Young's modulus of AZ91/GR magnesium composites is higher and the likelihood of deformation is reduced when the orientation angle of GR is at 0°,10°and 20°. Conversely, as the orientation angle is 30°~90°, Young's modulus decreases, leading to an increased likelihood of deformation. The Young's modulus of AZ91/GR magnesium alloy composites decreases as the orientation angle increases and tends to increase after surpassing 60°, but remains lower than Young's modulus of the AZ91 magnesium alloy. The results presented in Fig. 1(c) demonstrate that incorporating GR can improve the tensile strength of AZ91 magnesium alloy to 4.66 GPa. Furthermore, it is observed that as the orientation angle increases, there is a gradual decrease in the tensile strength.

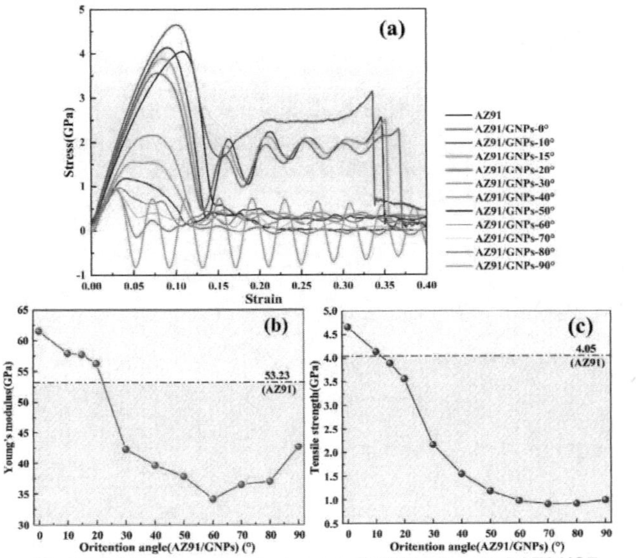

Fig.1 (a) The stress-strain curves of AZ91 alloy and AZ91/GR composites were measured at different orientation angles of GR; (b) The Young's modulus of the composite material depicted in the fig. (a); (c) The tensile strength of the composite material depicted in the fig. (a)

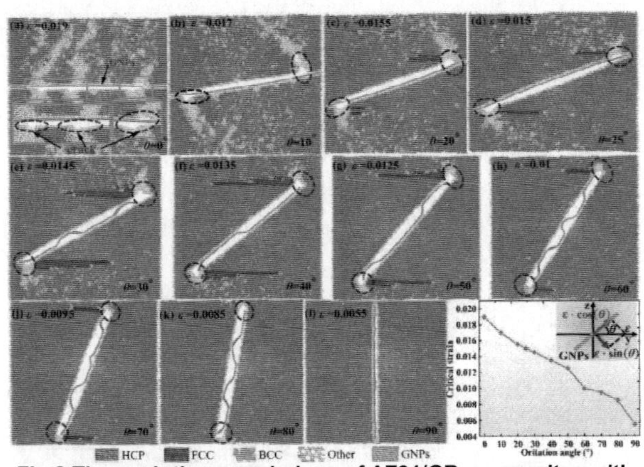

Fig.2 The evolution morphology of AZ91/GR composites with different orientation angles of GR under critical strain.

The evolution of morphology in AZ91/GR composites with varying orientation angles of GR under critical strain is depicted in Fig.2. The critical strain of AZ91/GR composites shows a distinct correlation with the orientation angle of GR. Specifically, a larger orientation angle of GR results in lower stress at the ends of the GR and a reduced critical strain. A simplified model is employed to elucidate the impact of orientation angle on the behavior of GNPs. The forces acting on the GR are resolved into components of $\varepsilon \cdot sin(\theta)$ and $\varepsilon \cdot \cos(\theta)$, where $0^{\circ} \ll \theta < 30^{\circ}$, causing the GR to extend along the $\cos(\theta)$ direction. As $30^{\circ} < \theta \ll 90^{\circ}$, $sin(\theta)$ becomes dominant, leading to the formation of local wrinkles in the graphene structure. This results in the bending and debonding of the GR in AZ91/GR composites. The simulation reveals that the critical orientation angle for GR is 30°. Furthermore, the critical strain of the AZ91/GR composite matrix decreases as the orientation angle of the GR increases, consequently reducing the load-bearing capacity of the GR.

4 Conclusion

(1)The strength of AZ91/GR-0° and AZ91/GR-10° single crystal composites surpass that of the AZ91 matrix magnesium alloy. As the orientation angle of GR increases (30°~90°), debonding occurs, leading to reduced tensile strength and increased susceptibility to deformation in AZ91/GR composites.

(2)The fracture time of AZ91/GR single crystal composites advances with higher orientation angles of GR. The impact of GR on the strength of AZ91/GR single crystal composites is primarily determined by the orientation angle of GR.

5 Acknowledgments

This work was supported by the National Natural Science Foundation of China (Nos. 52275390, 52375394), Key Research and Development Program of Shanxi Province (202102050201011, 2022ZDYF035).

References

[1] Rashad M, Pan F S, Hu H H, et al. Enhanced tensile properties of magnesium composites reinforced with graphene nanoplatelets[J]. Materials Science & Engineering A, 2015, 630: 36-44.

[2] Yuan Q H, Zeng X S, Wu J B, et al. Preparation and mechanical properties of graphene nanosheets reinforced AZ91 alloy composites[J]. Special Casting Nonferrous Alloys, 2016, 36: 282-286.

[3] Plimpton S, Fast parallel algorithms for short-range molecular dynamics[J]. Journal of Computational Physics. 1995, 117: 1-19.

[4] Dickel D E, Baskes M I, Aslam I, et al. New interatomic potential for Mg–Al–Zn alloys with specific application to dilute Mg-based alloys[J]. Modelling and Simulation in Materials Science and Engineering, 2018, 26: 045010.

[5] Stuart S J, Tutein A B, Harrison J A, A reactive potential for hydrocarbons with intermolecular interactions[J]. Journal of Chemical Physics, 2000, 112: 6472-6486.

[6] Zhou X, Liu X X, Mechanical Properties and Strengthening Mechanism of Graphene Nanoplatelets Reinforced Magnesium Matrix Composites[J]. Acta Metallurgica Sinica, 2020, 56: 240-248.

Semi-Solid Casting Preparation Technology for Graphene-Reinforced Magnesium Matrix Composites

Liwen Chen[1], Jianhui Jing[1], Yuan Zhao[1], Hua Hou[3, 1], Yuhong Zhao[1, 2, 4*]
1. School of Materials Science and Engineering, Collaborative Innovation Center of Ministry of Education and Shanxi Province for High-performance Al/Mg Alloy Materials, North University of China, Taiyuan, 030051, China
2. Beijing Advanced Innovation Center for Materials Genome Engineering, University of Science and Technology Beijing, Beijing, 100083, China
3. School of Materials Science and Engineering, Taiyuan University of Science and Technology, Taiyuan, 030024, China
4. Institute of Materials Intelligent Technology, Liaoning Academy of Materials, Shenyang, 110004, China
*Corresponding address: zhaoyuhong@nuc.edu.cn

Abstract: Traditional magnesium alloy materials have problems such as low strength and elongation, low elastic modulus and poor wear resistance, which limit their further application. How to effectively improve the comprehensive properties of magnesium alloy to meet the development needs has become the development goal of new materials. Graphene with excellent mechanical, thermal and electrical properties, has been considered as an ideal reinforcement of magnesium matrix composites. However, due to the uniform dispersion of graphene in the metal matrix, the interface bonding and complex preparation processes, the progress of graphene-reinforced magnesium matrix composites has not achieved the expected results. In this paper, semi-solid thixomolding process was used to prepare graphene nanoplatelets (GNPs) reinforced magnesium matrix composites. The preparation technology was studied to reveal the GNPs uniform dispersion mechanism, fracture mechanism and strengthening mechanism. The effects of process parameters on the properties of the composites were studied. The developed technology is expected to realize mass production and application of graphene-reinforced magnesium matrix composites.

Keywords: Graphene nanoplatelets, Magnesium matrix Composites, Semi-solid thixomolding process

1 Introduction

Semi-solid processing (SSP) has received considerable interests as a promising technology for processing of magnesium matrix composites (MMCs) relies on the thixotropic behavior of alloys in the semi-solid state [1]. It is expected that the low processing temperature of SSP can prevent the problem of Mg oxidation and GNPs structural damage effectively. Furthermore, the semisolid-state mixing might facilitate the engulfment of GNPs into the semi-solid slurry by reducing their buoyancy. Several recent reports have described the synthesis of MMCs base on SSP [2]. However, there is limited data in the literature on the SSP preparing methods of GNPs/Mg composites in addition to thixomolding process. Even more important, although most of work focused on the preparing methods and the effects of GNPs on the microstructure, little work has been reported on the microstructure evolution of MMCs in the semi-solid state.

2 Experimental procedure

The AZ91D-chips and GNPs (0.6 wt% content) were mixed and stirred in the V-pattern mixer to obtain AZ91D-chips fixed GNPs (AZ91D-chips-GNPs). After extruding and shearing by the screw of Mg thixomolding machine (JLM220-MG), the AZ91D-chips-GNPs were processed into semi-solid mixed slurry in the end of barrel. The barrel temperature and the screw rotate rate were 600℃ and 140 r/min, respectively. When the quantity reaches a fixed value, the semi-solid mixed slurry was injected into the mold with injection velocity of 1.5 m/s to mold for AZ91D-GNPs composites. The semi-solid mixed slurry and AZ91D-GNPs composites, respectively named as SSM@0.6GNPs and AZ@0.6GNPs, were collected individually to assess microstructural evolution in AZ91D-GNPs composites fabricated by thixomolding process, as shown in Fig. 1 [3-5].

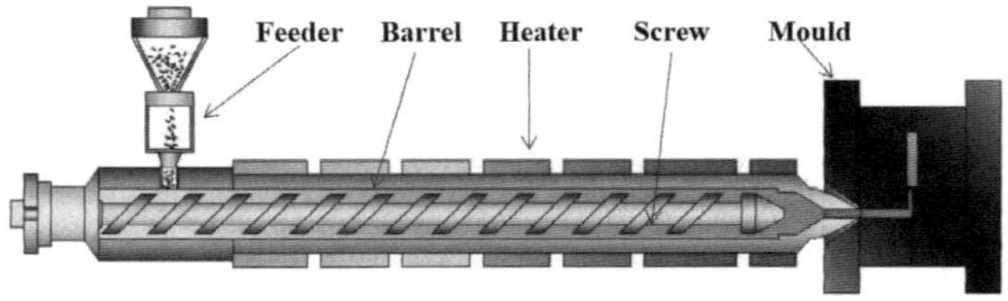

Fig. 1 Semi-solid Thixomolding Process

3 Result and discussion

As can be observed in Fig. 2, the *UTS* and δ of AZ@0GNPs were dropped slightly compared with SS@0GNPs. Although grain structure became much finer after injection process, the high porosity would lead to a significant decline of the performance for AZ@0GNPs. However, the *UTS* and δ of AZ@0.6GNPs had achieved 247.1 MPa and 4.3%, showing a significant increase in mechanical properties by adding GNPs to AZ91D alloy. According to the above results, the strength improvement should be attributed to grain refinement, porosity decline and special GNPs distribution characteristic. More importantly, the uniformly dispersed intragranular GNPs with anchoring MgO particles were supposed to be the main reason for the higher performance of AZ@0.6GNPs.

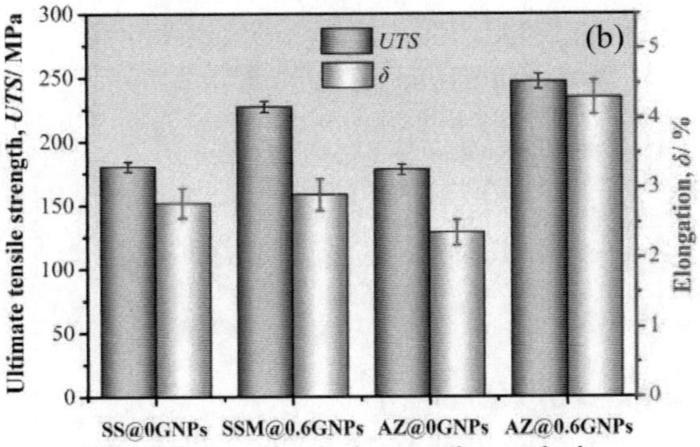

Fig. 2 The mechanical properties analysis

The mechanical properties of the composites strongly depend on the interfacial load transfer efficiency between the matrix and reinforcement. The GNPs/MgO interface was intimate and thus the compressive stress might be initiated at the interface, leading to the formation of a stress contrast. The appropriate amounts of MgO particles could contribute to enhance the anchor effect of GNPs from the AZ91D alloy matrix and enhance the load transfer efficiency at the interface. The MgO particles could be bonded to the outer layer of GNPs and share the applied load from the AZ91D alloy matrix to all GNPs layers, which is expected to increase the load-bearing capacity of GNPs. Furthermore, the anchor effect of GNPs would contribute to the improvement of effective interfacial shear strength. Consequently, due to the increased load-bearing capability and the effective interfacial shear strength, the appropriate amounts of MgO particles could effectively improve the interfacial load transfer, leading to the enhancement of mechanical properties for AZ@0.6GNPs. However, what would the best size, morphology and content of MgO particles in strengthening the interfacial bonding is still unclear, which will be highlighted in the following work.

4 Conclusion

The GNPs reinforced MMCs were fabricated by Semi-solid thixomolding process, which could enable reference for the practical application of SSP in GNPs reinforced MMCs. The strength improvement of AZ@0.6GNPs can be attributed to grain refinement, porosity decline and special GNPs distribution characteristic. The appropriate amounts of MgO particles were anchored around the GNPs, which could effectively improve the interfacial load transfer.

5 Acknowledgments

This work was supported by National Natural Science Foundation of China (Nos. 52375394, 52275390, 52205429, 52201146, 52305429), National Defense Basic Scientific Research Program of China (JCKY2020408B002), Key Research and Development Program of Shanxi Province (Nos. 202102050201011).

References

[1] Pan S, Li Q, Yu B ,et al. Research progress of Mg alloy semisolid forming[J]. Rare Met. Mater. Eng,2019, 48: 2379-2385.
[2] Chen Q, Chen G, Han L, et al. Microstructure evolution of SiCp/ZM6 (Mg-Nd-Zn) magnesium matrix composite in the semi-solid state[J]. J. Alloys Compd,2016, 656: 67-76.
[3] Chen L, Zhao Y, Hou H, et al. Development of AZ91D magnesium alloy-graphene nanoplatelets composites using thixomolding process[J]. J. Alloys Compd, 2019,778: 359-374.
[4] Chen L, Zhao Y, Li M, et al. Reinforced AZ91D magnesium alloy with thixomolding process facilitated dispersion of graphene nanoplatelets and enhanced interfacial interactions[J]. Mater. Sci. Eng. A, 2021, 804: 140793.
[5] Chen L, Zhao Y, Li J, et al. Microstructural evolution in graphene nanoplatelets reinforced magnesium matrix composites fabricated through thixomolding process[J]. J. Alloys Compd, 2023, 940: 168824.

Investigation on Tensile Strength and Interface of Fiber-Metal Laminates Based on Carbon Fiber Reinforced Zn-Al Alloy

Xinwei Hao[1], Liwen Chen[1], Hua Hou[3, 1], Yuhong Zhao[1, 2, 4*]
1. School of Materials Science and Engineering, Collaborative Innovation Center of Ministry of Education and Shanxi Province for High-performance Al/Mg Alloy Materials, North University of China, Taiyuan, 030051, China
2. Beijing Advanced Innovation Center for Materials Genome Engineering, University of Science and Technology Beijing, Beijing, 100083, China
3. School of Materials Science and Engineering, Taiyuan University of Science and Technology, Taiyuan, 030024, China
4. Institute of Materials Intelligent Technology, Liaoning Academy of Materials, Shenyang, 110004, China
*Corresponding address: No. 3, Xueyuan Road, Taiyuan, Shanxi Province e-mail: zhaoyuhong@nuc.edu.cn

Abstract: Fiber metal laminates (FMLs) which consist of AZ31 magnesium layers and continuous carbon fiber reinforced Zn-8%Al alloy (mass fraction) composite layer was fabricated by hot-pressing technology. The reaction between magnesium and Zn-Al alloy leaded to a well-bonded interface. Tensile test was carried out and the result indicated an improvement of tensile strength by 102% compared with original AZ31 magnesium sheets. Both the interface of Cf-Zn-Al and the interface of Mg-Zn-Al were characterized by SEM equipped with EDS. This research proves the feasibility of joining metal sheets with carbon fiber reinforced low-melting-point alloy composites sheets to fabricate metallurgically bonded FMLs.

Keywords: Fiber Metal Laminates; Carbon fiber reinforced Metal Matrix Composites; Tensile strength; Interface; Zn-Al alloy.

1 Introduction

Fiber metal laminates (FMLs) are hybrid composites which bond alternative metal sheets and fiber reinforced composites sheets [1, 2]. FMLs have excellent performance because of the combination of advantages of both metal alloys and composites. Due to their excellent properties, FMLs are being applied to many structural parts in aerospace and aeronautical industry, such as fuselage skin structures of the next generation commercial aircrafts [3, 4].

This work is stimulated by the lack of researches in the MMCs-based FMLs. The aim is to investigate the feasibility of producing metallurgically bonded FMLs.

（缺关键词）

2 Experimental procedure

Zn-8Al matrix composite reinforced with continuous carbon fiber was made using pressure infiltration method. At this point, a hydraulic machine was used to press the semi-solid slurry to infiltrate into the carbon fiber at 1 MPa.

AZ31 Mg alloy sheets with a thickness of 0.8mm were used to fabricate FMLs. Covered by Mg alloy sheets on both sides, the composites sheet was put back into the mould and heated to 340℃. The pressure was 0.5MPa and maintaining time was 3 minutes.

3 Result and discussion

Fig. 1 shows SEM images of sections of FMLs which is perpendicular to fiber orientation. As shown in Fig. 1, the pressure during casting process does not change the original configuration. The Zn-8Al alloy sheets between fiber layers function as barriers and no agglomeration of fibers occurs. Fig. 2 (a) shows magnified images of composites. It can be seen that all the interstices among fibers were filled with Zn-8Al alloy semi-solid slurries, which manifests a good infiltration during the fabrication process of Cf/Zn-8Al composites. Besides, no serious casting defects, such as porosity could be found. The alternate configuration and semi-solid processing technology make the pressure for slurries to infiltrate carbon fiber much less than that used in common pressure infiltration process. The magnified image of interface is shown in Fig. 2 (b). An obvious and uniform interface area (the dark area) without any micro-cracks or debonding is observed between carbon fiber and Zn-8Al alloy. The thickness of the interface is about 100nm.

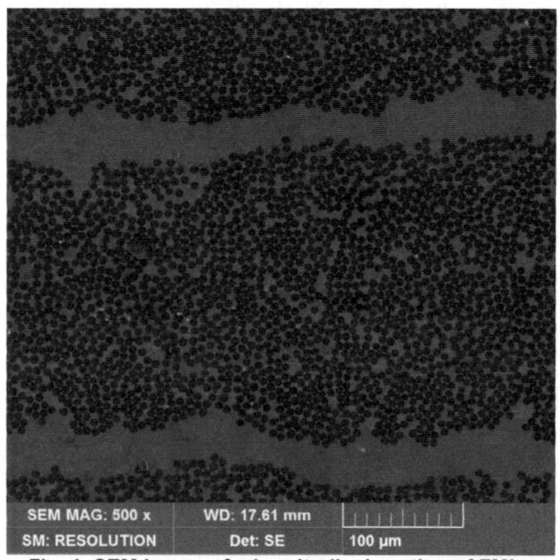

Fig. 1. SEM image of a longitudinal section of FMLs

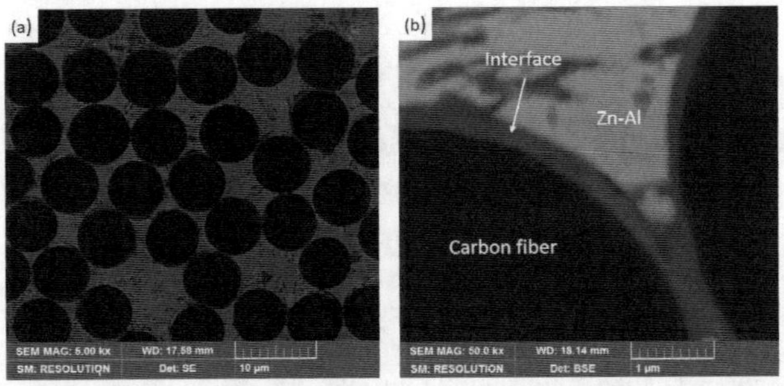

Fig. 2. SEM image and magnified Back-scatter electron (BSE) image at Cf/Zn-Al interface. (a) SEM image at 5kx magnified times, (b) BSE image at 50kx magnified times.

The mechanical properties of FMLs and AZ31 Mg alloys sheets used in this research are shown in Fig. 3. The ultimate tensile strength of FMLs reached 495MPa, which increased by 102% compared with that of AZ31 Mg alloy. However, the elastic modulus decreased from 45GPa to 37GPa. This may be the result of overreaction between Mg and a certain amount of carbon fibers, as shown in Fig. 4. The extremely low modulus of damaged carbon fibers decreased the modulus of FMLs at macro level to a great extent. The tensile curve shows no obvious yield point or plastic deformation stage, which represents typical deformation mechanism of fiber metal laminates.

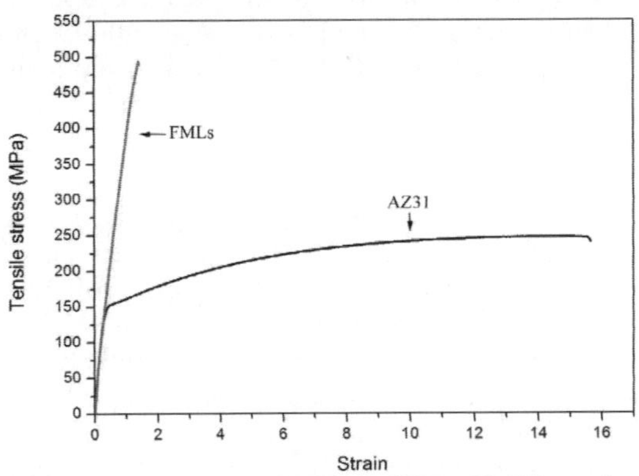

Fig. 3. Tensile curve of AZ31 Mg/Cf-Zn-Al FMLs and AZ31 Mg alloy sheet

4 Conclusion

In this work, the mechanical properties and interfacial characteristics of a new kind of metallurgically bonded FMLs were investigated. The results are as follows:

1)The Zn-Al alloy reacted with carbon fiber and formed a chemical bonding interface. The formation of brittle interfacial phases was prevented.

2) The tensile strength of AZ31 Mg/Cf-Zn-Al FMLs reached 495MPa, which increased by 102% compared with that of AZ31 Mg alloy.

5 Acknowledgments

This work was supported by This work was supported by National Natural Science Foundation of China (Nos. 52375394, 52074246, 52275390, 52205429, 52201146,52305429).

References

[1] Y.B. Hu, H.G. Li, L. Cai, et al. Preparation and properties of Fibre-Metal Laminates based on carbon fibre reinforced PMR polyimide[J]. Compos Part B-Eng,2015,69:587-591.

[2] T. Sinmazcelik, E. Avcu, M.O. Bora, et al. A review: Fibre metal laminates, background, bonding types and applied test methods[J]. Mater Design, 2011, 32(7): 3671-3685.

[3] L. Ferrante, F. Sarasini, J. Tirillò, et al. Low velocity impact response of basalt-aluminium fibre metal laminates[J]. Mater Design, 2016, 98:98-107.

[4] A. Asundi, A.Y.N. Choi, Fiber metal laminates: An advanced material for future aircraft[J]. J Mater Process Tech,1997, 63(1):384-394.

Effect of Si content of Hot-Dip Plating on the Interface Structure and Properties of Solid-Liquid Cast and Rolled 6061/Q345 Composite Plate

Xuyuan Zhang[1], Yuandong Li[1,2,*], Tao Ju[1], Guangyu Liu[1], Hongwei Zhou[1], Sheng Wang[1,2], Chi Cao[1,3], Guangli Bi[1,2]

1. State Key Laboratory of Advanced Processing and Recycling of Nonferrous Metals, Lanzhou University of Technology, Lanzhou, 730050, PR China;
2. Key Laboratory of Non-ferrous Metal Alloys and Processing, Ministry of Education, Lanzhou University of Technology, Lanzhou, 730050, PR China;

3.Wenzhou pump and Valve Engineering Research Institute of Lanzhou University of Technology, Wenzhou 325105, China)

* **Corresponding author:** Yuan-dong Li, e-mail: liyd_lut@163.com, Tel: +86-931-2976795

Abstract: Employing hot-dip coating with Al-XSi alloys, combined with phase diagram analysis and first-principles calculations, the influence of Si elements on the interfacial microstructure, phase composition, and mechanical properties in the cast-rolled 6061/Q345 composite sheets was investigated. Results demonstrated that the addition of Si in the hot-dip coating led to the formation of novel ternary phases, Al-Fe-Si, namely Al3FeSi2, Al9Fe2Si2, and Al0.5Fe3Si0.5, at the aluminum-steel interface, which reduced the brittle phases Al5Fe2 and Al3Fe, thereby enhancing the deformation resistance and toughness of the composite. Compared to the composite sheet with pure Al coating (yielding a tensile strength of 402.4 MPa and shear strength of 51.3 MPa), the maximum values of 438.5 MPa and 120.8 MPa for tensile and shear strengths were achieved with an Al-1%Si (mass fraction) coating, representing increases of 8.9% and 135.4%, respectively. Further increments in the Si content in the hot-dip bath to 7.5% resulted in a decline in both the tensile strength (395.7 MPa) and shear strength (78.6 MPa) of the composite sheet. The tensile and shear strengths exhibited a trend of initially increasing and then decreasing with the rising Si content in the hot-dip coating.

Keywords: Composite board; Solid/liquid casting and rolling; Q345 steel; 6061 aluminum; Interface organization; mechanical property

1 Introduction

Bimetallic composite materials compensate for the shortcomings of single-metal materials in terms of performance, possessing extensive application prospects [1]. However, the formation of brittle intermetallic compounds (IMCs) at the interfaces of these composites significantly deteriorates their mechanical properties, thereby severely restricting their applications [2]. Jiang W [3] prepared aluminum-steel bimetals via composite casting, resulting in an irregular tongue-shaped reaction layer with an average thickness of 30 μm at the aluminum-steel interface. This layer was composed of Fe_2Al_5, $FeAl_3$, Al_8Fe_2Si, and $Al_2Fe_3Si_3$ IMCs. The use of surface modifiers and aluminum plating promoted good interface and metallurgical bonding, increasing the shear strength of the castings by 40% compared to untreated ones.

So far, when preparing composite materials, most studies have focused on the effects of adding elements such as Ni, Cu, Zn. as intermediate layers on the interface organization and properties of composite plates. This study uses aluminum-silicon alloy as an intermediate layer to connect different aluminum and steel materials. On the one hand, it improves the wettability of the interface between aluminum and steel, promotes metallurgical bonding, and enhances the interface bonding strength. On the other hand, by adding Si to suppress the growth of Al-Fe brittle intermetallic compounds, the relationship between the microstructure and mechanical properties of composite plates with different Si contents is analyzed, providing reference for the preparation of aluminum/steel composite plates.

2 Experimental procedure

Alloys of various silicon contents were placed in an SG2-7.5-10 pit-type resistance furnace and heated to 730°C. Refining was conducted with 1% C_2Cl_6, followed by skimming to remove impurities. A 1% cover flux (50% KCL and 50% NaCl) was uniformly sprinkled on the molten surface. The temperature was then adjusted to 720°C for holding. During the holding process, the melt was stirred thoroughly every 10 minutes to ensure the accuracy of the alloy composition. Treated Q345 steel plates were vertically immersed in the 720°C aluminum-silicon alloy melt for immersion coating, with a dipping time of 30 seconds. The immersion coating process was performed with melts of pure Al, Al-0.5%Si, Al-1%Si, Al-2.5%Si, Al-5%Si, and Al-7.5%Si, respectively. Upon completion of the immersion coating, solid-liquid casting rolling was immediately executed.

During hot dipping, the cut 6061 aluminum ingots are placed in another SG2-7.5-10 pit-type resistance furnace and heated to 730°C. 1% (mass fraction) of C_2Cl_6 is added for refining and slag removal. After that, 1% (mass fraction) of cover agent (50% KCL and 50% NaCl) is uniformly sprinkled on the surface of the melt, and the temperature is adjusted to 680°C for holding. The roll gap of the rolling mill is set to 4mm, and the 680°C 6061 aluminum liquid is quantitatively poured onto the Q345 substrate treated by hot dipping. The rolling speed is 42mm/s, and the maximum rolling force is 170kN. After starting the rolling mill, the two alloys are sent between the two rolls for solid/liquid casting and rolling, followed by air cooling.

3 Result and discussion

As depicted in Fig 1, the schematic illustration of phase formation in the aluminum-iron composite bonding layer , with the progression of diffusion (Fig 1b), intermetallic

and Fe atoms enter the intermediate layer from both sides. Due to the continuous supply of energy from the 6061 melt, the bonding layer has a slower cooling rate, allowing for sufficient mixing of Al, Fe, and Si elements during the cooling process. with the addition of Si atoms, the Si atoms tend to segregate in the intermediate layer, hindering the diffusion of Fe and Al atoms (Fig 1d). Additionally, the Si atoms can occupy vacancies in the Al_5Fe_2 phase, inhibiting the growth of the Al_5Fe_2 phase (Fig 1e, Fig 2). Furthermore, as shown, when the Si content increases (exceeding the solid solubility of Si in Al_5Fe_2), new Al-Fe-Si phases (Al_3FeSi_2, $Al_9Fe_2Si_2$, and $Al_{0.5}Fe_3Si_{0.5}$) are formed to suppress the growth of the original Al-Fe intermetallic compounds, resulting in a smoother appearance on the steel side and a reduction in the thickness of the intermediate layer (Fig 1f).

When the C-axis of the Al_5Fe_2 phase is oriented at 90° to the initial aluminum-steel interface, it exhibits the fastest growth rate, suppressing the growth at other angles, as demonstrated in Fig 1b and c. Al_5Fe_2 is the predominant phase in the aluminum/steel intermetallic compounds, while the thermodynamically most stable Al_3Fe phase constitutes a relatively small proportion in the actual phase composition, which deviates significantly from thermodynamic predictions. Research indicates that the growth of intermetallic compounds is not solely governed by thermodynamic driving forces but also involves the state of diffusing atoms in dissimilar metals and interface conditions.

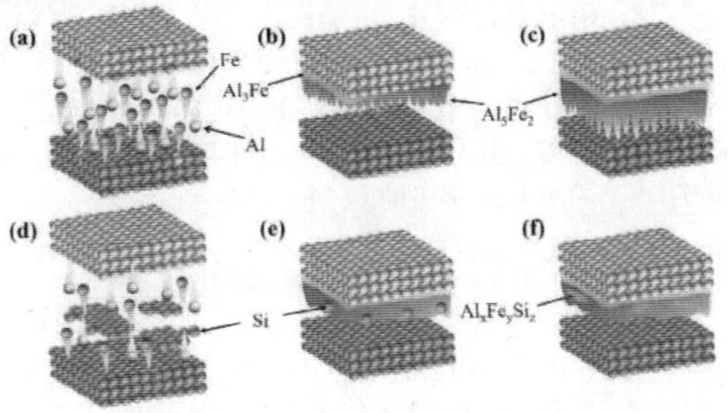

Fig. 1 Diagram of the formation of the interlayer phase (a) Diffusion of Al and Fe atoms; (b) Initial stage of interlayer phase formation; (c) Termination stage of interlayer phase formation;(d) Si atom bias and enrichment; (e) Si atom occupying vacancies; (f) formation of AlxFeySiz phase

compounds will initially appear on the aluminum side. The thermodynamically most stable Al_3Fe phase is the first to emerge. As aluminum atoms diffuse further, the Al_5Fe_2 phase starts to grow at the interface between the Al3Fe phase and the Fe(Al) solid solution. The Al_5Fe_2 phase has an orthorhombic lattice structure with a high concentration of vacancies (30%) along the C-axis, as shown in Fig 2 [4]. During the bonding process of al/steel composite materials with an Al-Si alloy intermediate layer, the Al-Si intermediate layer that is first in contact with the molten 6061 aluminum liquid melts first. Subsequently, Al atoms

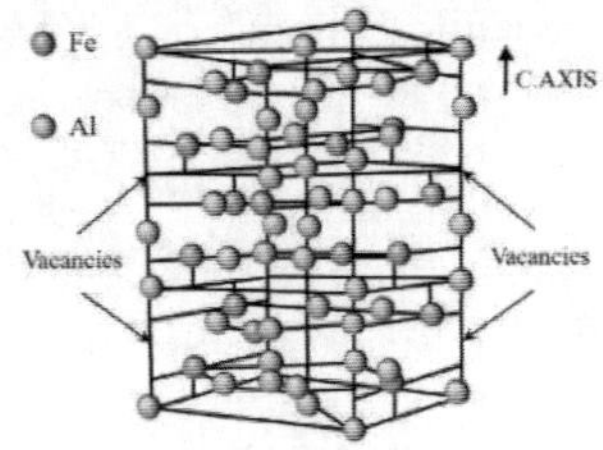

Fig. 2 Diagrammatic representation of the structure of the Al5Fe2 phase[4]

4 Conclusion

(1) The mechanical properties of the composite plate containing Si elements in the middle layer are higher than those of the composite plate without Si elements.

(2) The introduction of Si element in the hot-dip plating solution leads to the formation of new phases such as $Al_9Fe_2Si_2$ phase, Al_3FeSi_2 phase and $Al_{0.5}Fe_3Si_{0.5}$ phase at the composite interface, which changes the phase composition at the interface and improves the mechanical properties of the composite plate.

5 Acknowledgments

The authors would like to acknowledge the financial supports from the National Natural Science Foundation of China (Fund name, Study of Semi-Solid Printing Fabrication of Aluminum Laminate Composites and its Interface Regulation. Fund number, 52261010). Major Science and Technology Project of Gansu Province (Fund number, 22ZD6GA008).

References

[1] Chinababu M, Rao E B, Sivaprasad K. Fabrication, microstructure, and mechanical properties of Al-based metal matrix-TiB2-HEA hybrid composite[J]. Journal of Alloys and Compounds, 2023, 947: 169700.

[2] Joshani E, Beidokhti B, Davodi A, et al. Evaluation of dissimilar 7075 aluminum/AISI 304 stainless steel joints using friction stir welding[J]. Journal of Alloys and Metallurgical Systems, 2023, 3: 100017.

[3] Jiang W, Fan Z, Li C. Improved steel/aluminum bonding in bimetallic castings by a compound casting process[J]. Journal of Materials Processing Technology, 2015, 226: 25-31.

[4] Grushko B. A study of the Al-rich region of the Al–Fe–Ir alloy system[J]. Journal of Alloys and Metallurgical Systems, 2023, 4: 100040.

Modeling and Experimental Study of Cu/Al Clad Sheet by Twin-Roll Casting

Zhiping Mao[1,2,*], Jingpei Xie[1,2], Aiqin Wang[1] , Pei Liu[1], Douqin Ma[1,2]
1. School of Material Science and Engineering, Henan University of Science and Technology, Luoyang, China
2. Collaborative Innovation Center of Nonferrous Metals, Henan University of Science and Technology, Luoyang, China, 471003
*Corresponding address: e-mail: zpmao@haust.edu.cn

Abstract: A mathematical model of Cu/Al clad sheet by twin-roll casting was developed to study the influences of casting parameters on temperature field of cast-rolling process. The effect of the casting speed, casting temperature of melt, and roll gap on the flow and temperature field was predicted. Meanwhile, the microstructure of Cu/Al clad sheet interface was studied.

Keywords: Mathematical model, Cu/Al clad sheet, Twin-roll casting.

1 Introduction

The Cu/Al clad sheet can use aluminium to instead of most copper, which not only reduce the total material cost, but also has good electrical and thermal conductivity. It is more wildly applied to the fields of building, electric-power, chemical and so on [1-3]. Many kinds of developed techniques can fabricate the clad sheets, such as rolling bonding [4], explosive cladding [5], cast cladding [6] and friction-stir welding [7]. Among these composite techniques, twin-roll casting (TRC) has many advantages such as shorter production routines, lower capital investment, low production cost, etc. [8]. It can be modified for the TRC process to produce many clad metals. The interface of metal layered composite materials is particularly critical to their performance [9]. Mechanical properties of the clad sheet depend on the quality of interface layer, which is influenced by temperature at the interface and compressive stresses from the roll force [10].

For the complex situation of TRC, mathematical and numerical simulation methods are widely used to describe the distribution of temperature and flow inside the TRC area. Plenty of researchers use simulation software to study the relationships between the main process parameters and field of temperature and flow in the TRC process [11,12]. However, further research is needed on the influence of key process parameters on the heat flux field of horizontal cast-rolling of copper aluminum composite plates. In this paper, a mathematical model of Cu/Al clad sheet by twin-roll casting was developed to study the influences of casting parameters on temperature field of cast-rolling process. The effect of the casting speed, casting temperature of melt, and roll gap on the flow and temperature field was predicted. The Cu/Al clad sheets with different thicknesses were produced by horizontal twin roll caster. Simultaneously analyzed the influence of different TRC process parameters on the interface layer structure of copper aluminum composite plates.

2 Experimental procedure

The description of the TRC composite process is as follows: after melting, the liquid aluminium is fed to the region between the rotating rolls. Copper sheet was inserted in between the melt stream and the roll, then heat transfer from the molten metal to the clad/roll interface and solidification of the melt takes place resulting in a Cu/Al clad sheet.

3 Result and discussion

3.1 Optimization and Impact of Heat transfer coefficient of copper and roller (HTC_1)

During the casting and rolling process, HTC_1 is mainly related to the contact state of copper/roller. The variation of outlet temperature in the casting and rolling zone under different HTC_1 conditions was studied. The results indicate that when the HTC1 value is set to 10000 $wm^{-2}k^{-1}$, the error between the simulation results of the outlet temperature and the experimental results is the smallest.

3.2 The influence of casting speed

The effect of casting and rolling speed on the interface melt length and centerline melt pool depth indicated that with the increase of casting and rolling speed, the depth of interface molten state and the depth of molten pool at the centerline increase. The casting and rolling speed is fast, and the heat transfer time from aluminum to copper strip and from copper strip to roller is short.

3.3 The influence of casting temperature

Figure 1 shows the effect of casting temperature on the interface melt length and centerline melt pool depth. The results indicate that as the casting temperature increases, the amount of heat entering the casting and rolling zone from the inlet increases rapidly, but the heat transfer rate between copper/aluminum and aluminum/roller remains unchanged, resulting in an increase in the depth of the interface molten state and the depth of the centerline molten pool.

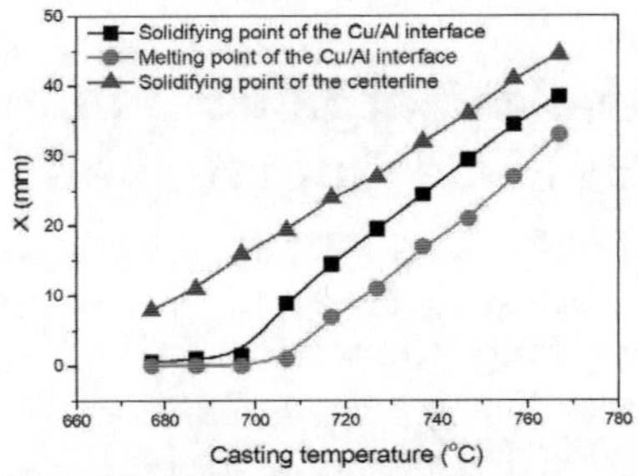

Fig. 1 The relationship between casting temperature and sump depths of the Cu/Al interface and centerline.

4 The microstructure of Cu/Al clad sheet interface

Using transmission electron microscopy to perform diffraction spot analysis on the interface characteristic area and determine the phase of the interface, as shown in Figures 2. The area with darker contrast is the copper matrix, exhibiting twinning characteristics, while the area with brighter contrast is the aluminum matrix, whose grain size is much larger than that of the interfacial compound. Through the analysis of diffraction spots in the characteristic region, it can be concluded that the compound layer near aluminum is Al_2Cu, while the layer near copper is Al_4Cu_9. The composition distribution of the interface layer is uniform, and the grain size is mostly within 500 nm, exhibiting certain equiaxed crystal characteristics.

Fig. 2 BF TEM image of the interface for the as-cast clad sheet

4 Conclusion

In this paper, a thermal flow coupling model for the horizontal casting and rolling process of copper aluminum composite plates was constructed. The influence of casting and rolling parameters on the temperature field of the casting and rolling zone was studied. The microstructure of Cu/Al clad sheet interface was revealed. Two intermetallic compound layers (Al_2Cu and Al_4Cu_9) with the total thickness of around 500 nm were formed at the interface.

5 Acknowledgments

The authors gratefully acknowledge the support from National Key R&D Program of China (2021YFB3701304) for their kind financial support of this work.

References

[1] Mai T A, Spowage A C. Characterisation of dissimilar joints in laser welding of steel–kovar, copper–steel and copper–aluminium[J]. Mater Sci Eng, 2004, 374:224–233.

[2] Tseng H, Hung C, et al. An analysis of the formability of aluminum/copper clad metals with different thicknesses by the finite element method and experiment[J]. Int J Adv Manuf Technol, 2010, 49:1029–1036.

[3] Eslami P, Karimi Taheri A. An investigation on diffusion bonding of aluminum to copper using equal channel angular extrusion process[J]. Mater Lett, 2011,65:1862–1864.

[4] Sheng L Y, Yang F, Xi T F, et al. Influence of heat treatment on interface of Cu/Al bimetal composite fabricated by cold rolling[J]. composite：part B, 2011,42:1468-1473.

[5] Doh J M, Byun J Y, Yoon J K, et al. Microstructure and mechanical properties of Ti/Al explosive cladding[J]. Materials and Design, 2014,56:1014–1019.

[6] Liu T, Wang Q, Sui Y, et al. Microstructure and mechanical properties of overcast 6101–6101 wrought al alloy joint by squeeze casting[J]. Journal of Materials Science & Technology, 2015,32:298-304.

[7] Ivan Galvão, Loureiro A, Verdera D, et al. Influence of tool offsetting on the structure and morphology of dissimilar aluminum to copper friction-stir welds[J]. Metallurgical and Materials Transactions A, 2012,43(13):5096-5105.

[8] Barekar N S, Dhindaw B K, Twin roll casting of Al alloys – an overview[J]. Mater. Manuf. Process. 2014,29: 1661–1665.

[9] Grydin O, Gerstein G, Nuernberger F, et al. Twin-roll casting of aluminum–steel clad strips[J]. Journal of Manufacturing Processes. 2016,15:501–507

[10] Hua G H, Ce J I, Yi K D, et al. Thermal-flow coupled numerical simulation and experimental research on bonding mechanism of Cu/Al composite strip by solid-liquid cast-rolling[J]. The Chinese Journal of Nonferrous Metals， 2016,3:623-629.

[11] Stolbchenko M， Grydin O， Samsonenko A, et al. Numerical analysis of the twin-roll casting of thin aluminium-steel clad strips [J]. Forschung im Ingenieurwesen, 2014, 78(3):1-10.

[12] Hadadzadeh A, Wells M A, Jayakrishnan V. Development of a mathematical model to study the feasibility of creating a clad AZ31 magnesium sheet via twin roll casting[J]. Int J Adv Manuf Technol. 2014,73:449-463.

Dual-Scale Near-Net Structure Design to Achieve Synergistic Enhancement of Strength and Plasticity of $(TiC+Ti_5Si_3)/TC4$ Composites

Tiantao He[1], Chunyu Yue[1], Bowen Zheng[1*], Feng Gu[1], Fuyu Dong[1], Xuejian Lin[1], Xiaojiao Zuo[1], Yinxiao Wang[2], Hongjun Huang[1], Xiaoguang Yuan[1],
1 School of Materials Science and Engineering, Shenyang University of Technology, No. 111, Shenliao West Road, Economic & Technological Development Zone, Shenyang 110870, P. R. China
2 School of metallurgy, Northeastern university, NO. 3-11, Wenhua Road, Heping District, Shenyang 110819, P. R. China
*Corresponding address: e-mail: zhengbowen89@163.com (B.W. Zheng)

**A
bstract:** To solve the problem of low plasticity in homogeneous particle-reinforced composites, micro-nano dual-scale near-network (TiC+Ti5Si3)/TC4 composites were prepared by C powders+Si powders and in-situ melting-casting technology. The microstructural evolution, strengthening mechanisms of micro-nanoscale composites were systematically investigated. The results show that with the increase of the reinforced phase content, the morphology of micron-sized TiC changed significantly from the initial short plume to granular gradually, and the nano-sized Ti5Si3 particles changed from small particles to ellipsoidal gradually precipitated at the β -Ti and α / β interfaces. The (3vol.%TiC+3vol.%Ti5Si3)/TC4 (TMCs-2) sample exhibits excellent mechanical properties with hardness of 52.8HRC, compressive strength of 1774MPa and plasticity of 24.5%. The increase in strength of the composites is attributed to the synergistic effect of matrix (fine grain strengthening, solid solution strengthening) and particle strengthening.

Keywords: $(TiC+Ti_5Si_3)/TC4$ composites, Multiphase multi-scale, Mechanical properties, Strengthening mechanism,

1 Introduction

Discontinuous particle reinforced titanium matrix composites (DRTMCs) are considered to be one of the most promising candidate structural materials in high-tech fields such as automobiles and aerospace, with low density, high specific strength, good corrosion resistance and excellent high temperature performance [1,2]. TiC particles (TiCp) are considered an ideal reinforced phase for TMCs due to their excellent physical properties (high modulus, high hardness, excellent thermal stability, and good chemical compatibility) [3,4]. However, the poor interfacial bonding strength and the trade-off between strength and toughness are still the main problems in the research of TMCs [5,6]. Inspired by the concept of natural structures, many researchers have constructed various bionic structures to obtain better strength-ductility synergistic effects [7,8]. In this study, based on the design of network and multi-scale structure, the dual-scale near-network $(TiC+Ti_5Si_3)/TC4$ composites were prepared by in-situ melting-casting technology using nano-Si powder and nano-C powder. The effects of different proportions of mixed micron TiC and nano Ti_5Si_3 particles on the microstructure evolution and mechanical properties of the composites were systematically studied.

2 Experimental procedure

The experimental raw materials include commercial Ti-6Al-4V (TC4) alloy, Si powder (purity 99.9%, average particle size 800nm), and C powder (purity 99.8%, average particle size 100nm). The experimental raw materials were melted using a vacuum non consumable arc furnace, and $(TiC+Ti_5Si_3)/TC4$ composites were prepared based on in-situ self generated reactions between materials. The microstructure of the samples was characterized using a scanning electron microscope (SEM, ZEISS, Gemini 300) equipped with an energy dispersive spectrometer (EDS). The SEM samples were etched with Kroll reagent for 7-10s. The hardness of the samples was measured using HR-150A Rockwell hardness tester, and the average value was taken after 5 tests on each sample. The room temperature compression performance test was conducted using an MTS810 electronic universal testing machine with a strain rate of $1.38\times10^{-4}s^{-1}$.

3 Result and discussion

3.1 Analysis of microstructure and mechanical properties

When a small amount of reinforced phase content is added, TiC is rod-like and uniformly distributed in the matrix grains or grain boundaries, and a small amount of spherical Ti_5Si_3 particles are precipitated in the β-Ti phase and the α/β interface. With the increase of the content of the reinforced phase, the morphology of TiC changed significantly. The TiC phase gradually evolved from rod-like to short pinnate distribution, and the number of Ti_5Si_3 particles increased significantly. When excessive reinforced phase content is added, a large number of equiaxed TiC particles are distributed at the grain boundaries of the matrix, and a small amount of spherical Ti_5Si_3 particles are precipitated at the β-Ti grain boundaries. More spherical

Ti_5Si_3 particles are attached to the TiC particles, showing obvious aggregation of the reinforced phases. The coarse Ti_5Si_3 particles are easily connected to each other at the interface (TiC interface, grain boundary, etc.) to produce local aggregation, which adversely affects the mechanical properties of the TMCs [9]. The TMCs-2 sample exhibits excellent mechanical properties with hardness of 52.8HRC, compressive strength of 1774MPa and plasticity of 24.5%.

3.2 Strengthening mechanism

During the deformation process, with the continuous loading of stress, cracks begin to form and propagate along the interface between the reinforced phase and the matrix. The reinforced phase can interact with dislocations and inhibit dislocation motion. When the crack propagates along the interface between the reinforced phase and the matrix, the reinforced phase strongly hinders the crack propagation, effectively passivates the crack tip and hinders the further propagation of the crack. In addition, the deflection due to the near-network structure during crack propagation effectively consumes the propagation energy of the crack. The fine α-Ti grain sheets with high strength in ductile β-Ti grains inhibit the deformation of β-Ti grains, which is the reason for the enhancement of TMCs [10]. In addition, due to the multi-scale structure of TMCs. Micron-sized TiC particles can be used as an additional heterogeneous nucleation site to promote the formation of equiaxed α-Ti particles to refine grains and strengthen grain boundaries, which is conducive to improving the mechanical properties of TMCs. Due to the special distribution and size characteristics of Ti_5Si_3 reinforcement, sufficient space can be provided for dislocation movement to ensure the ductility of TMCs. Therefore, the design of dual-scale near-net structure can effectively improve the deformation coordination of the composites.

4 Conclusion

The in-situ formation of TiC and Ti_5Si_3 phases have a significant influence on the growth of grains. The designed dual-scale near-net structure $(TiC+Ti_5Si_3)/TC4$ composites have excellent strength and plasticity. The increase of strength is attributed to the synergistic effect of solid solution strengthening, fine grain strengthening and reinforced phase strengthening. The improvement of plasticity is attributed to the design of the two-scale near-net structure, which can effectively hinder the crack propagation, resulting in crack deflection and passivation, thereby increasing the energy required for crack propagation and delaying the fracture of the composite.

5 Acknowledgments

This work is supported by the Liaoning Province Science and Technology Plan Joint Fund Project (2023-BSBA-248) and the Scientific Research Fund of Liaoning Provincial Education Department (No. LJKZ0122).

References

[1] S. Wang, L.J. Huang, L. Geng, et al. Microstructure evolution and damage mechanism of layered titanium matrix composites under tensile loading[J]. Materials Science and Engineering: A, 2020, 777:139067.

[2] Q. Zhang , S.F. Li, Y. Cao, et al. Nanostructure evolution of reticular nano-TiB whiskers reinforced titanium matrix composite subjected to ultrasonic shot peening[J]. Journal of Alloys and Compounds, 2023, 948:169704.

[3] X.J. Zhang, W.J. Yu, J. Wang, et al. Rapid in-situ synthesis, microstructure and mechanical properties of titanium matrix composites with micro/nano-sized TiB/TiC hybrid structures[J]. Vacuum, 2023, 207:111635.

[4] B.W. Zheng ,T.T. He, C.Y. Yue, et al. Realization of synergistic enhancement for strength, wear resistance and ductility via adding La_2O_3 particles in (TiB+TiC)/Ti6Al4V composite[J]. Wear, 2023, 526-527:204945.

[5] H.J. Wang, Z.Y. Huang, X. Li, et al. Microstructure and properties of TiCx/TC4 composites with a quasi-network structure prepared by the decomposition of the Ti_2AlC precursor[J]. Journal of Alloys and Compounds, 2022, 920:166011.

[6] X.Y. Huang, Y.M. Gao, Z.P. Wang, et al. Microstructure, mechanical properties and strengthening mechanisms of in-situ prepared $(Ti_5Si_3+TiC_{0.67})/TC4$ composites[J]. Journal of Alloys and Compounds, 2019, 792:907-917.

[7] C.X. Lei, Y. Pan, F. Kuang, et al. A novel configuration design of strong and ductile $(TiC+Ti_5Si_3)/Ti$ laminated composites[J]. Materials Science and Engineering: A, 2024, 892:145961.

[8] K. Feng, H.M. Zhang, X.W. Cheng, et al. Breaking through the strength-ductility trade-off in graphene nanoplatelets reinforced titanium matrix composites via two-scale laminated architecture design[J]. Materials Characterization, 2023, 205:113290.

[9] F.S. Sun, F.H.S. Froes. Precipitation of Ti_5Si_3 phase in TiAl Alloys[J]. Materials Science and Engineering: A, 2002, 328(1-2):113-121.

[10] Z.G. Zhou, Y.Z. Liu, X.H. Liu. Constructing targeted bi-lamellar microstructure via heat treatment for high compressive strength and plasticity in selective laser melted Ti6Al4V-5vol% TiB composite[J]. Materials Science and Engineering: A, 2022, 844:143173.

Refining Mechanism of Ti_2AlC Particles and Improvement Mechanism of Mechanical Properties by Er Addition

Kexuan Li, Hongyuan Zhai, Hongze Fang, Ruirun Chen
National Key Laboratory for Precision Hot Processing of Metals, Harbin Institute of Technology, Harbin, Heilongjiang 150001, China
*Corresponding address: Heilongjiang Province, Harbin City, Nangang District, 92 West Dazhi Street
*Corresponding e-mail: 23B909099@stu.hit.edu.cn

Abstract: Revealing the refinement mechanism of Er on Ti_2AlC particles and the strengthening mechanism of TiAl composites is of great significance for promoting the development prospects of TiAl composites. As-cast Ti42Al6Nb2.6CxEr alloys are prepared by vacuum non consumable arc melting. Results show that as Er increases from 0.04 to 0.08 at. %, the relative content of γ phase, $\alpha2$ phase, and Ti_2AlC particle remain stable. As Er increases to 0.4 at. %, the relative content of γ phase decreases, and that of $\alpha2$ phase increases. Er consumes Al to form $ErAl_3$ phase during the $\alpha2 \rightarrow \alpha2+\gamma$ transformation, resulting in a decrease in Al content involved in the formation of γ lamellae. Aspect ratio of Ti2AlC particle decreases from 7.97 to 3.92 and the morphology changes from slender rod to short rod as Er increases to 0.08 at. %, which is due to the Er_2O_3 preferring to be adjacent to Ti_2AlC particle, inhibiting its growth of length and radial direction. Alloying with Er results in refinement in terms of lamellar colony size. Dispersed Er_2O_3 and refined Ti_2AlC both serve as nucleation sites, resulting in refinement of the lamellar colony. $Ti_{42}Al_6Nb2.6C0.08Er$ alloy exhibits optimum compressive properties with the strength of 2057 MPa and the strain of 22.44%, which is due to the refinement of Ti_2AlC particle and lamellar colony, and the precipitation of Er_2O_3 particle.

Keywords: TiAl alloy; Er element; Refinement; Microstructure evolution; Mechanical properties

1 Introduction

The comprehensive performance of a single intermetallic compound is ill-equipped to meet the overall requirements of aircraft engines [1]. The in-situ self-generation composite technology further expands the application prospects of TiAl alloys. The segregation distribution and type of reinforcing particle have a significant effect on the mechanical properties of composite materials [2]. For carbides, uniformly distributed Ti_2AlC and Ti_3AlC are ideal reinforcement particles. Therefore, obtaining great cast microstructure is inseparable from effective control of the matrix and reinforcing particle. Rare earth (RE) element has played a valuable role in this field , which has great solid solution strengthening effect, and tends to accumulate at the front of the solid-liquid interface, increasing the undercooling and inhibiting grain growth [3]. Its characteristic of easily reacting with oxygen to form stable oxides not only improves the purity of TiAl melt, but also generates significant precipitation strengthening. Research has shown that RE elements exhibit positive regulation on the microstructure of TiAl alloys, and they also have a positive effect on the reinforcing particles [4]. Different from previous studies, we have specifically investigated the effect of RE element on the morphology and content of MAX type ceramic particles, and revealed the mechanisms of Er element promoting the refinement of Ti_2AlC particles in β-solidified TiAl alloy.

2 Experimental procedure

The vacuum non-consumable electric arc furnace was used to prepare Ti42Al6Nb2.6C*x*Er (*x*=0.04, 0.06, 0.08, 0.1, 0.4 at. %) alloys. Each ingot was solidified in a copper crucible chilled by water. Vacuum degree was maintained below 3×10^{-3} Pa until the completion of melting to prevent oxidation. All materials were remelted five times under argon environment to guarantee composition consistency. To evaluate the microstructures, the ingots were grinded, polished, and chemically etched with Kroll reagent (5 HF+10 HNO_3+85 H_2O (vol. %)) for 7s. Quanta 200F (SEM) with EDS device were used to analyze the microstructure evolution. Nanometer scale of the microstructure was observed by Talos F200X (TEM). For compressive properties, on the Instron 5569, a cylinder with the size of Ø 4mm×6 mm is deformed at a loading rate of 0.5 mm/min under a pressure head of 250kN/50kN. To assure the reliability of the experiment, the compression test requires at least three samples.

3 Result and discussion

Figure 1 shows the constituent phases are α_2, γ, Ti2AlC , Er_2O_3 with different Er contents. When Er content is below 0.08 at. %, it exists in the form of Er_2O_3 particle, the relative contents of the three constituent phases are stable. When Er content exceeds 0.08 at. %, the content of γ phase decreases, while that of $\alpha2$ phase increases, which due to Er consumes Al element to form ErAl3 phase during the $\alpha_2 \rightarrow \alpha_2+\gamma$ transformation, resulting in a decrease in Al content involved in the formation of γ lamellae. It can also be seen that appropriate Er content is beneficial for reducing the aspect ratio of Ti_2AlC particles. Ti_2AlC particles exhibit the morphology with small size and uniformly distribution as Er content increases to 0.08 at. %.

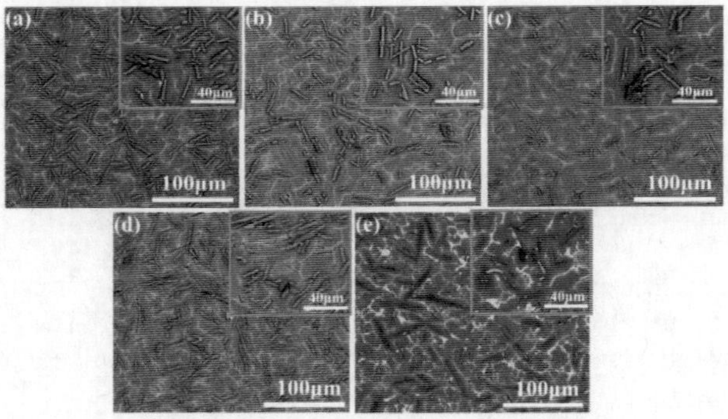

Figure 1 Microstructures of Ti42Al6Nb2.6CxEr alloys: (a) 0.04Er; (b) 0.06Er; (c) 0.08Er; (d) 0.1Er; (e) 0.4Er.

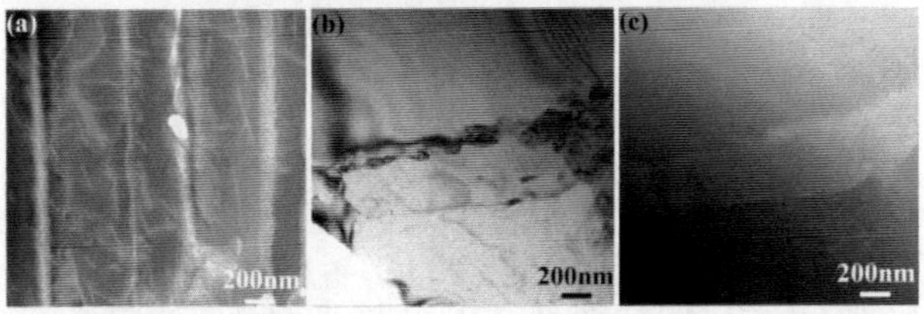

Figure 2 TEM analysis of the Er2O3/Ti_2AlC phase interface of Ti42Al6Nb2.6C0.08Er alloy.

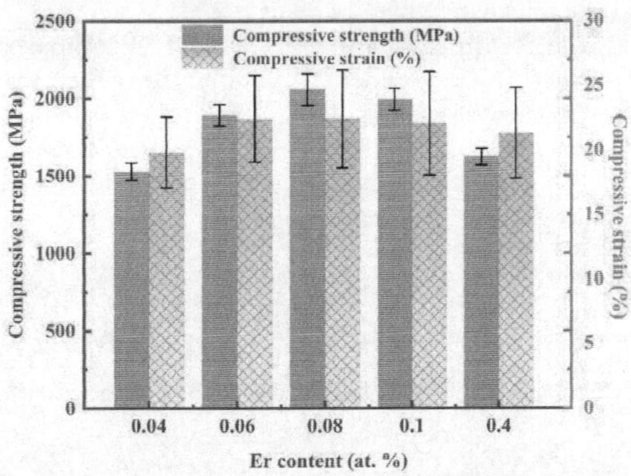

Figure 3 Compressive properties of Ti42Al6Nb2.6C*x*Er alloys.

Er_2O_3/Ti_2AlC phase interface is shown in Figure 2. Er_2O_3 particles are not only distributed inside the lamellar colony to hinder the dislocation slipping, but also prefer to be adjacent to Ti_2AlC particles. Er_2O_3 particles are connected sequentially, hindering the growth of Ti_2AlC particles in terms of length and radial direction. Therefore, Ti_2AlC particles transform from slender-needle shaped to short-rod shaped. The fine Er_2O_3 and Ti2AlC particles jointly refine the lamellar colony. Based on the refinement of Ti_2AlC particles and lamellar colony, the compressive properties at room temperature increases from 1528 to 2057 MPa without sacrificing the compressive strain (Figure 3). Therefore, Er element strengthens TiAl alloy by refining Ti_2AlC particles and the precipitation of Er_2O_3 particles.

4 Conclusion

1. The aspect ratio of Ti_2AlC particle decreases from 7.97 to 3.92 as Er content increases from 0.04 to 0.08 at. %, which is due to the Er_2O_3 preferring to be adjacent to Ti_2AlC particle, inhibiting its growth of length and radial direction.
2. The dispersed Er_2O_3 and refined Ti2AlC particles both serve as nucleation sites, resulting in significant refinement of the lamellar colony.
3. The TiAl alloy exhibits optimum compressive properties, which is due to the refinement of Ti2AlC particle and lamellar colony, and the precipitation of Er_2O_3 particle.

5 Acknowledgments

The authors are grateful to the financial support by National Natural Science Foundation of China (Grant No. U21A2042), the National Nature Fund Youth Fund Project of China (No. 52101038), and Young Elite Scientists Sponsorship Program by CAST (2021QNRC001).

References

[1] Li D X, Zhang G Y, Lu G, et al. Precipitation of Ti_2Al phases at lamellae interfaces in a high-Nb-containing TiAl alloy during thermal exposure[J]. Journal of Materials Science & Technology, 2022, 126: 132-140.
[2] Chen C G, Luo J, Guo Z M, et al. Microstructural evolution, and mechanical properties of in situ TiB_2/Al composites under high-intensity ultrasound[J].Rare Metals, 2015, 34(3): 168-172.
[3] Wang X, Luo R, Liu F, et al. Characterization of Gd-rich precipitates in a fully lamellae TiAl alloy[J]. Scripta Materialia, 2017,137: 50-54.
[4] Baek K, Kim H, Shin H, et al. Multiscale study to investigate nanoparticle agglomeration effect on electrical conductivity of nano-SiC reinforced polypropylene matrix composites[J].Mechanics of Advanced Materials and Structures, 2023, 30(12): 1-11.

Microstructural Evolution and Silicide Precipitation Behavior of (TiBw+TiCp)/ Titanium Matrix Composite During Multi-Directional Forging

Zhenbo Yang, Dezhi Chen*, Ruirun Chen
National Key Laboratory for Precision Hot Processing of Metals, Harbin Institute of Technology, 150001, P R China
*Corresponding address: Heilongjiang Province, Harbin City, Nangang District, 92 West Dazhi Street
*Corresponding e-mail: chendezhi383@163.com

Abstract: In this study, (TiBw+TiCp)/ titanium matrix composite underwent multi-directional forging deformation to investigate the microstructure evolution and silicide precipitation behavior. The results shown that the microstructure is mainly composed of fragmented TiBw and TiCp, equiaxed α phases, and silicides after deformation. The accumulation of strain during hot deformation and the formation of high-density dislocations near the reinforcement induce and promote the precipitation of silicides. Silicides are mainly distributed in the reinforcement and grain boundaries of low nucleation energy.

Keywords: Titanium matrix composite, Multi-directional forging, Microstructural evolution, Precipitation behavior of silicide;

1 Introduction

Titanium matrix composites (TMCs) have broad application prospects in the aerospace field due to their high specific strength, specific stiffness, and excellent high-temperature resistance [1]. Among the numerous reinforcements in TMCs, TiBw and TiCp reinforcement prepared by in-situ melt casting method have attracted the attention of many researchers. The in-situ casting method has the advantages of simple process and low cost, and is widely used in the preparation of TMCs. However, the obtained composite materials have problems such as reinforcement segregation and room temperature brittleness. Multi-directional forging(MDF) can effectively eliminate the defects and greatly improve the mechanical properties of the TMCs [2]. Silicon is an important matrix alloy element in TMCs, which can significantly improve their strength and creep resistance, generally exists in the form of solid solutions and silicides. The silicides precipitated during the deformation process can significantly affect the microstructure, thermal stability, and creep performance of TMCs [3]. Therefore, it is necessary and feasible to study the influence of multi-directional forging on microstructure and the precipitation of silicides during deformation.

2 Experimental procedure

In this study, the matrix alloy is a new near α titanium alloy Ti-6.5Al-2.5Sn-9Zr-0.5Mo-1W-1Nb-0.25Si (wt.%). The reinforcements are TiCp and TiBw. Firstly, a water-cooled copper crucible induction melting furnace is used to prepare composite material ingots, and The β transformation temperature of the ingot determined by metallographic method is approximately 1085 ± 10 ℃. Then, decreasing MDF was carried out sequentially at three temperatures: 1100 ℃, 1050 ℃, and 1035 ℃.

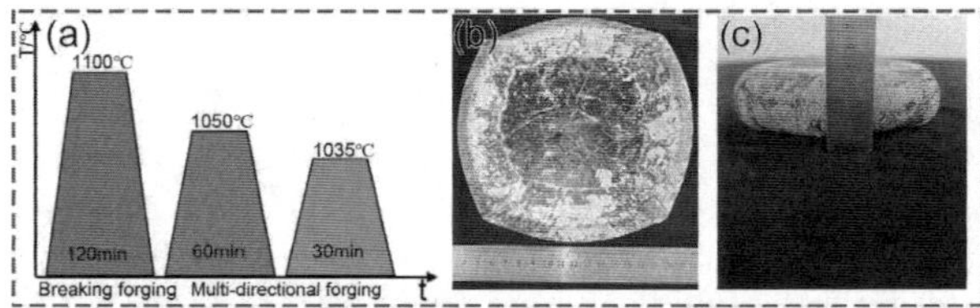

Fig. 1 Schematic diagram of the process of MDF

3 Result and discussion (Bold, 10 pt., Arial)

3.1. The microstructure of as-cast TMCs

As shown in Fig. 2, shows the OM and SEM images of the as-cast (TiBw+TiCp)/Ti composite, the matrix microstructure of as-cast (TiBw+TiCp)/Ti composite is a typical widmanstatten structure. Moreover, TiBw and TiCp have a high aspect ratio, which is not conducive to the improvement of titanium alloy properties.

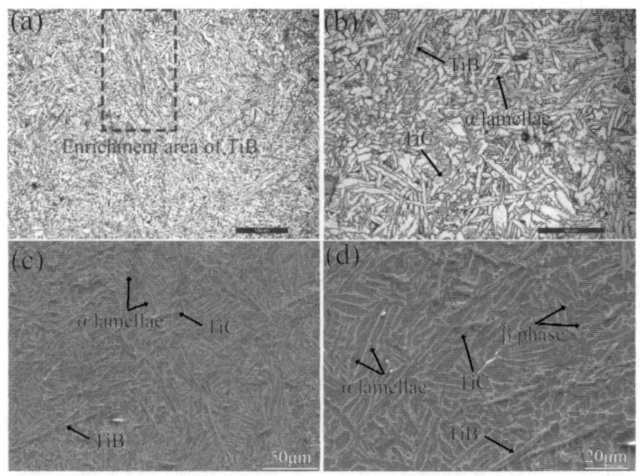

Fig. 2 The microstructures of as-cast (TiBw+TiCp)/Ti composite: (a), (b) OM images; (c) (d) SEM images
The microstructure of as-forged TMCs

Fig. 3 shows the microstructure of (TiBw+TiCp)/Ti composite after MDF. According to the Fig. 3a and 3b, after forging deformation, the microstructure is changed from widmanstatten structure to near equiaxed α structure.

The reinforcements with high aspect ratio fractures are broken, resulting in a decrease in the aspect ratio and they are uniformly distributed in the matrix. In addition, a large amount of white precipitates were precipitated during the forging process (as shown in Fig. 3c and 3d).

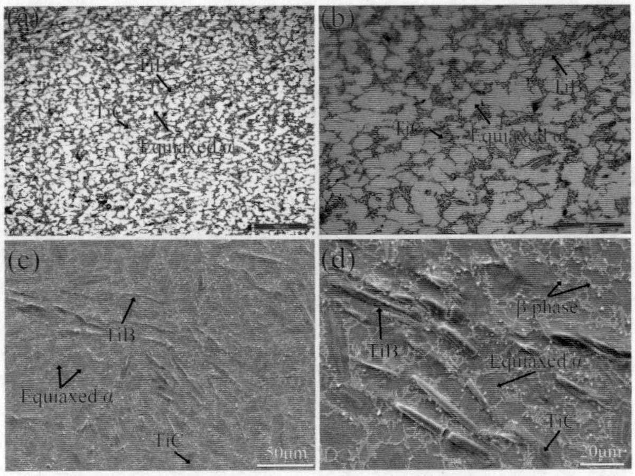

Fig. 3 The microstructure of as-forged (TiBw+TiCp)/Ti composite: (a), (b) OM images; (c) (d) SEM images

As shown in Fig. 4a, it can be further observed that the precipitates are mainly distributed at the reinforcements and grain boundaries. Fig. 4b shows the EDS point scan analysis results of the precipitated phase. The diffraction peaks of Ti, Si, and Zr elements dominate, indicating that the precipitate is mainly composed of these three elements. Therefore, it can be inferred that the precipitate is a silicide. Notably, silicides are mainly distributed at grain boundaries and reinforcements, as the formation of silicides during deformation mainly depends on the diffusion channels provided by dislocations and grain boundarie. During the deformation process, the reinforcements will act as a pinning dislocation to promote the precipitation of silicides. Due to the low nucleation energy at crystal defects, silicides always precipitate in these places [4].

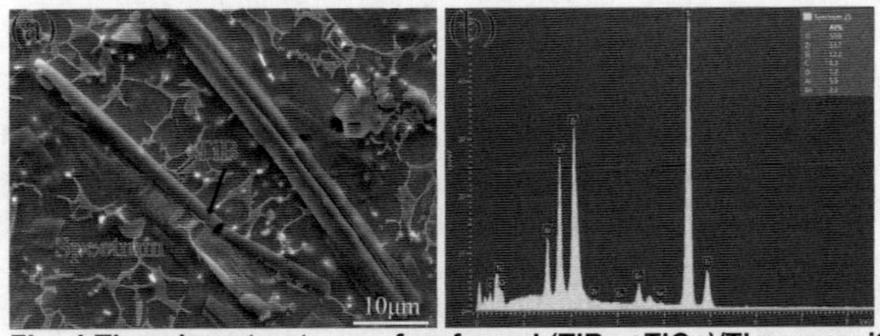

Fig. 4 The microstructures of as-forged (TiBw+TiCp)/Ti composite: (a) SEM image; (b) EDS analysis results

4 Conclusion

(TiBw+TiCp)/Ti composite was prepared, and the as cast matrix structure is a widmanstatten structure with high aspect ratio reinforcements. After multi-directional forging, the grain size and reinforcements are significantly refined, and the microstructure is changed to near equiaxed α structure. The precipitation of silicides during the forging process is mainly located in the aggregation area of the reinforcement and grain boundaries of low nucleation energy.

5 Acknowledgments

This work was supported by the National Key Laboratory for Precision Hot Processing of Metals, Harbin Institute of Technology (6142909230101), the Heilongjiang Postdoctoral Fund (LBH-Z23132) and the China Postdoctoral Science Foundation (2023TQ0099).

References

[1] Huang L J, An Q, Geng L, et al. Multiscale architecture and superior high-temperature performance of discontinuously reinforced titanium matrix composites[J].Advanced Materials, 2021,33: 2000688.

[2] Zhang R, Wang D J , Yuan S J. Effect of multi-directional forging on the microstructure and mechanical properties of TiBw/TA15 composite with network architecture[J].Materials and Design, 2017, 134: 250-258.

[3] Wang T B, Li B L, Wang Z Q , et al. A microstructure with improved thermal stability and creep resistance in a novel near-alpha titanium alloy[J]. Materials Science and Engineering A, 2018,731:12-20.

[4] Sun F, Li J S, Kou H C, Tang B and Cai J M. Nano-precipitation and tensile properties of Ti60 alloy after exposure at 550°C and 650°C[J].Materials Science and Engineering A, 2015,626: 247-253.

High Mechanical Properties TiAl Composite Materials by Introducing In-Situ Nano-Borides Under Ultrasound

Lingyan Zhou, Xiaokang Yang, Hongze Fang, Ruirun Chen
National Key Laboratory for Precision Hot Processing of Metals, Harbin Institute of Technology, Harbin, Heilongjiang 150001, China
*Corresponding address: Heilongjiang Province, Harbin City, Nangang District, 92 West Dazhi Street
*Corresponding e-mail: ruirunchen@hit.edu.cn

Abstract: To form in-situ nano borides in the solid-liquid two-phase region, Ti46Al8Nb2.6C0.8TaxB alloys were prepared under ultrasound to improve the microstructure and mechanical properties. The morphology and distribution of borides, and characteristics of lamellar colonies were observed through SEM, TEM, and XRD, revealing the influence of borides on microstructure evolution and the improving mechanisms of strength and plasticity. Results indicate that the morphology of borides is elongated and spherical. As the content of B increases, borides gradually spheroidized, and nanoscale borides precipitate on the lamellar phase. There are a large number of dislocations around the borides to form the SFs and dislocation wall. The size of the lamellar colony has decreased from 26.9 to 19.3 μm. The increasing nucleation site of borides with increasing content of B and primary borides being broken by ultrasound are the reasons for forming B27-type borides. And the increase of nucleation particles refines the lamellar colony. The compressive strength increased from 2026 to 2224 MPa, and the strain increased from 26.3 to 29.6% when the addition of boron increased from 0 to 1.2%. Uniformly distributed nano borides hinder the movement of dislocations to form SFs and dislocation walls and the lamellar colonies are refined are the main reasons for the significant improvement of mechanical properties.

Keywords: TiAl, Nano borides, Compressive property, Dislocation.

1 Introduction

High Nb-TiAl alloy has received widespread attention as the most high-temperature-resistant lightweight material to prepare aerospace turbine blades [1-2]. However, the insufficient high-temperature strength and unreasonable matching of strength and plasticity of TiAl alloy severely limit its development. The formation of in-situ self-generated particles is expected to enhance the high-temperature usability of alloys, providing prospects for TiAl alloys to meet more stringent service conditions [3]. This work focuses on improving the unbalanced strength plasticity and the microstructure containing multiple reinforcing phases of TiAl alloy. Under UT conditions, boron is introduced into TiAl alloy to form borides, refining the microstructure and improving the strength of TiAl alloy.

2 Experimental procedure

The designed compositions are Ti46Al8Nb2.6C0.8Ta-*x*B (x = 0, 0.4, 0.8, 1.2, 1.6, 2.0 at. %). In the following research, it will be named TANCT-*x*B. The button ingot is completed through arc melting and UT. The five remelting treatments performed during melting are to ensure that the alloy material is completely melted. Then perform UT for 100s under the same ultrasound process.

3 Result and discussion

3.1 Microstructure characteristics

As the boron content increases, most boride particles appear in small spherical shapes between the layers, while a small amount of boride particles appear in a white curved shape in the matrix. When the content of boron increases from 0 to 1.2 at.%, the size of the lamellar colonies decreases from 26.9 to 19.3 μm.

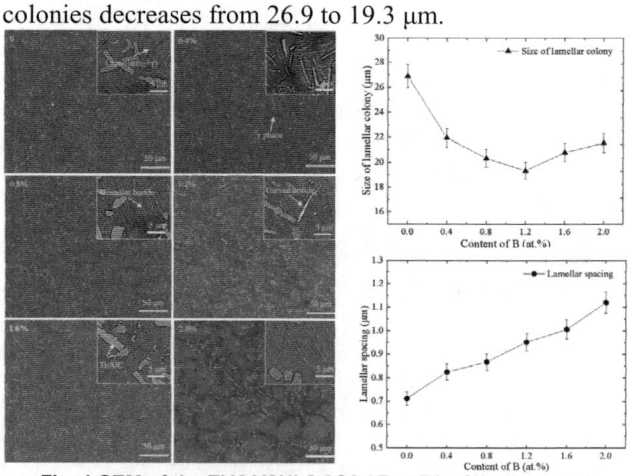

Fig. 1 SEM of the Ti46Al8Nb2.6C0.8Ta-xB(at.%) alloys and schematic diagram of statistical results

3.2 Element distribution and phase composition

In Ti46Al8Nb2.6C0.8Ta-*x*B alloy, the main phase of the alloy is γ, α_2. Ti_2AlC and (Ti, Nb, Ta) B phases. Among them, (Ti, Nb, Ta) B phase is a TiB phase that solidly dissolves some Nb and Ta. man).

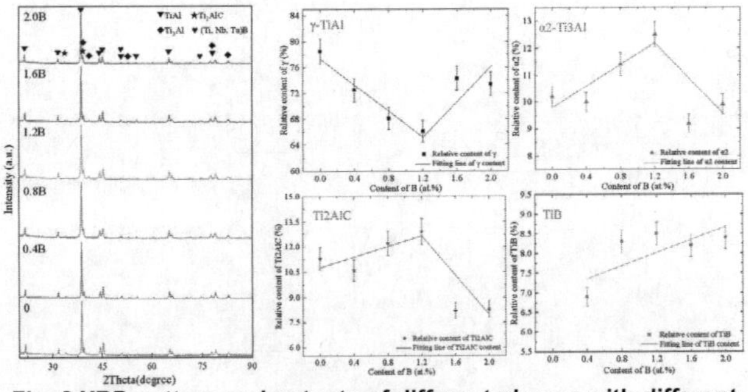

Fig. 2 XRD pattern and cntents of different phases with different B content

3.3 Dislocation characteristics around precipitates

Micron and nanoscale Ti_2AlC precipitated in the matrix, a large number of dislocations with interlaced and entangled distributions can be observed around the Ti_2AlC phase.

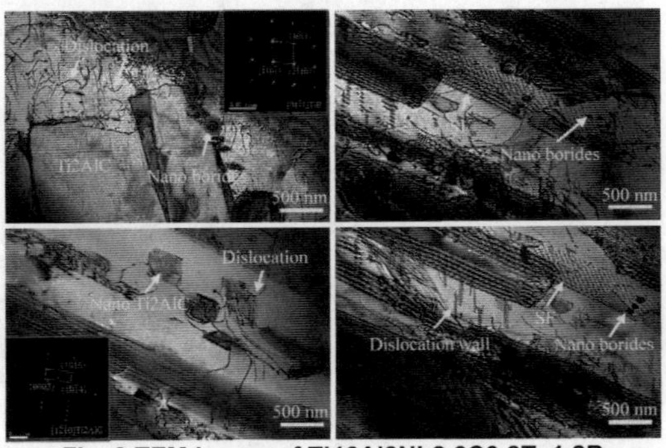

Fig. 3 TEM image of Ti46Al8Nb2.6C0.8Ta1.2B

In addition to the nano borides attached to the Ti_2AlC phase, many nano TiB particles are also precipitated between the γ/α_2 lamellar. The stacking faults (SFs) caused by these nano borides are obvious, the dislocation wall formed around the boride due to dislocation entanglement. The intricate and intertwined dislocations around the precipitates are one of the main reasons that significantly affect the mechanical properties of the alloy.

3.4 Mechanical properties

When the boron content increased from 0 to 1.2 at.%, the compressive strength increased from 2026 to 2224 MPa, the compressive strain increased from 26.3 to 29.6%. The improvement of compression performance is mainly influenced by grain refinement strengthening, dislocation strengthening, and boride precipitation strengthening.

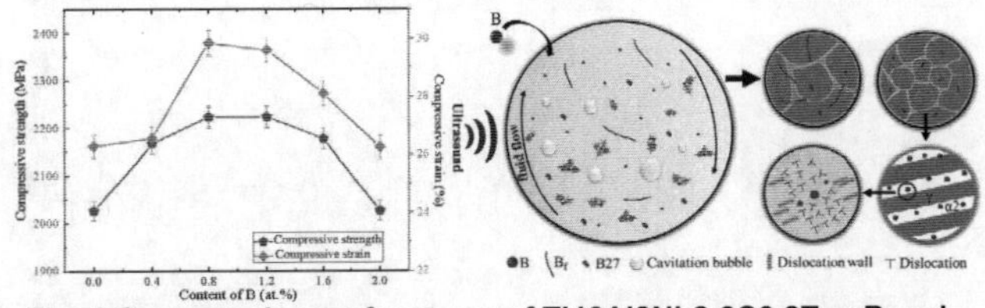

Fig. 4 Compression performance of Ti46Al8Nb2.6C0.8Ta-xB and the schematic diagram on microstructure evolution

4 Conclusion

By forming in-situ generated nano borides, a large number of dislocations are generated, leading to the formation of SFs and dislocation walls. The mechanical properties of the high-Nb Ti_2AlC/TiB TiAl alloy alloy have been improved, with a maximum strength of 2224 MPa and a maximum strain of 29.6%.

5 Acknowledgments

The authors are grateful to the financial support by National Natural Science Foundation of China (Grant No. U21A2042), the National Nature Fund Youth Fund Project of China (No. 52101038), and Young Elite Scientists Sponsorship Program by CAST (2021QNRC001).

References

[1] Salvador C A F, Maia E L, Costa F H, et al. A compilation of experimental data on the mechanical properties and microstructural features of Ti-alloys[J]. Scientific Data, 2022, 9: 188.

[2] Huang L, Liu B, Wang Z, et al. Effect of carbon nanotube content on the microstructure and mechanical properties of CNTs/TiAl alloys[J]. Ceramics International, 2023, 49: 8196-8203.

[3] Edwards T E J, Di Gioacchino F, Goodfellow A J, et al. Transverse deformation of a lamellar TiAl alloy at high temperature by in situ microcompression[J]. Acta Materialia, 2019, 166: 85-99.

AUTHOR INDEX

World Foundry Organization Ltd
Winton House, Lyonshall
Kington, Herefordshire
HR5 3JP, United Kingdom

ISBN 979-8-3313-1254-1

CURRAN ASSOCIATES INC.
proceedings.com

75th World Foundry Congress (WFC 2024)

Deyang, China
25-30 October 2024

Volume 2 of 2

ISBN: 979-8-3313-1254-1

Printed from e-media with permission by:

Curran Associates, Inc.
57 Morehouse Lane
Red Hook, NY 12571

Some format issues inherent in the e-media version may also appear in this print version.

Printed with permission by Curran Associates, Inc. (2025)

For permission requests, please contact World Foundry Organization Ltd
at the address below.

World Foundry Organization Ltd
Winton House, Lyonshall
Kington, Herefordshire
HR5 3JP, United Kingdom

Phone: +44 (0) 1544 340332
Fax: +44 (0) 1544 340332

www.thewfo.com

Additional copies of this publication are available from:

Curran Associates, Inc.
57 Morehouse Lane
Red Hook, NY 12571 USA
Phone: 845-758-0400
Fax: 845-758-2633
Email: curran@proceedings.com
Web: www.proceedings.com

TABLE OF CONTENTS

VOLUME 1

PART 1 : CAST IRON AND CAST STEEL TECHNOLOGY

PART 2 : NON-FERROUS ALLOY

PART 3 : HIGH ENTROPY ALLOY

PART 4 : HIGH-TEMPERATURE ALLOY

PART 5 : METAL MATRIX COMPOSITE MATERIALS

VOLUME 2

PART 6 : RARE EARTH MATERIALS

PART 7 : FUNCTIONAL MATERIALS

PART 8 : MATERIALS GENOME ENGINEERING

PART 9 : KEY CASTINGS MANUFACTURING FOR RAIL TRANSPORTATION

PART 10 : DIE-CASTING TECHNOLOGY

PART 11 : LOST FOAM CASTING TECHNOLOGY

PART 12 : PRECISION CASTING TECHNOLOGY

PART 13 : ADDITIVE MANUFACTURING

PART 15 : MOULDING MATERIAL AND TECHNOLOGY

PART 16 : GREEN AND LOW-CARBON DEVELOPMENT

Effect of Trace La on Microstructure and Mechanical Properties of Al-Si-Mg alloy

Zaijun Su*, Qiuxian Liang, Shengqiang Liu, Kexin Shen, Zhufu Chen
New materials Research and development Department, Guangdong brilliant metal products Co., LTD, E17, Zhaoqing Asia Metal Renewable Resource Industry Base, Sihui City, Guangdong Province, China;
*Corresponding address: e-mail: suzjy@hnsyu.edu.cn

Abstract: The effects of La elements (0, 0.02, 0.04, 0.08 wt.%) on the microstructure and mechanical properties of Al-Si-Mg sand casting alloy were studied by means of optical microscopy(OM), X-ray diffractometry (XRD) and scanning electron microscopy (SEM), the mechanical properties test. The results show that the size of coarse eutectic silicon and secondary dendrites in the as-cast alloy can be significantly refined by the addition of trace La element. The as-cast alloy without La element is mainly composed of Al phase, Si phase, and a small amount of Mg2Si phase, Al3Ti phase and AlFeSi phase. When the content of La element exceeds 0.08%, Al3La phase appears in the alloy. With the increase of La content, the tensile strength of the alloy increases gradually, the elongation increases first and then decreases. The alloy containing 0.04%La has the best mechanical properties, the tensile strength and elongation of the alloy are 160 MPa, 4.52% and 350 MPa, 2.35%, respectively, in the as-cast and aged states. Continue to increase La content, the tensile strength and elongation of alloy decreased rapidly. The re-coarsening of eutectic silicon and secondary dendrites, and the appearance of Al3La phase results in rapidly decreased in tensile strength and elongation of alloy.

Keywords: Al-Si alloy; La; microstructure; mechanical properties,

1 Introduction

A357 alloy is a hypoeutectic Al-Si alloy with excellent casting and mechanical properties, which is suitable for casting high-strength parts with complex shapes[1,2]. It has been widely used in aviation, aerospace, military, automotive and other industrial fields[3,4]. At home and abroad, the tensile strength of the sand mold casting alloy is usually required to be more than 320 MPa and the elongation is more than 5%, but in the traditional sand mold casting alloy production process, it has the large grain size and precipitated Si-phase size due to the slow cooling rate, which is difficult to achieve the required performance indicators[5]. Modifying Si is the most effective way to increase tensile strength, especially elongation. The commonly modification method is strontium modification because it has a high absorption rate and a long effective period, however, the burn-off rate is high and the absorbs gas in alloying solution is serious during the holding process, which affect the quality and mechanical properties of the castings[6]. Due to the fact that rare earth elements not only have good deoxidation and slag removal effects, but also have long-term modification of mono-silicon and refinement of grain size, they have become a research hot spot for alloying elements of aluminum alloy. Many studies have shown that adding La, Ce, Sc and Y elements to the Al-Si alloy can refine the α-Al grains and modify the eutectic Si, thereby improving the mechanical properties of the alloy[7-10]. However, addition of a single rare earth element needs a relatively high content to achieve a better modification effect, and the modification mechanism is different from that of the commonly used industrial modification agents, resulting in a large gap in modification effect. Therefore, the present study takes the Al-7Si-0.5Mg alloy (A357) as the research object and studies the effects of trace Sr (0.015 wt%) and La (0-0.08 wt%) elements on the microstructure and mechanical properties of the alloy. It aims to find the optimal addition level of rare earth metal, provide experimental basis for the development of new low-cost Al-Si alloys with better performance.

2 Experimental procedure

Gravity casting is performed in Al-7Si-0.5Mg-0.2Ti-0.015Sr-xLa (x=0, 0.02, 0.04, 0.06, 0.08wt.%, mass fraction, the same below) alloys. The main experimental raw materials are industrial pure aluminum ingot (99.95wt.%), pure Mg ingot (99.95wt.%), Al-10wt.%Si intermediate alloy, Al-10wt.%Sr intermediate alloy, Al-10wt.%La intermediate alloy and Al-5wt.%Ti intermediate alloy. The alloy is smelted by resistance furnace. After melting, the alloy is refined and degassed at 730 °C and kept for 15 minutes before casting into a gravity sand mold at 720 °C. The solid solution treatment of alloys were conducted at 535 °C for 12h, and then the alloys containing different La contents were aged at 170℃ for 10h. The alloy samples were etched with 0.5% hydrofluoric acid solution, and the microstructure was observed on OLYMPUS microscope. Phase analysis was carried out by DPMax 2500 X-ray diffractometer. The microstructure and energy spectrum were analyzed by Sirion-200 scanning electron microscope. The tensile test was carried out on the CSS-44100 universal electronic experimental machine.

3 Result and discussion

1. Influence of trace La on microstructure of A357 alloy

The X-ray diffraction patterns, the SEM microstructure and EDS analysis show that the as-cast A357 alloy without La mainly consists of Al phase, Si phase, a small amount of Mg_2Si phase and AlFeSi phase. After adding 0.08%La, a small peak of Al_3La phase appears in the alloy. The optical microstructures of the as-cast A357-xLa (x=0, 0.02, 0.04, 0.08 wt.%) alloys shows that the content of long bar-shaped silicon phase gradually decreases and it transforms into cubic and short rod-shaped phases. The size of eutectic Al dendrite also gradually decreases, with the La content increases to 0.04%. However, when the content of La increases further, the eutectic Si phase appears with more long bar-shaped particles and rapidly coarsens, and the eutectic Al dendrite also re-coarsens.

2. Effect of trace La on the mechanical properties of A357 alloy

It can be seen that the tensile strength and elongation of the cast alloy without La element were 157 MPa and 4.38%, respectively. As the La content increases, the tensile strength of the cast alloy gradually increases, while the elongation first increased and then decreased. The tensile strength and elongation of the alloy with 0.04% La element were improved to 160 MPa and 4.52%, respectively. Continuing to increase the La content, the elongation rapidly decreased. After aging, the tensile strength of the alloy was significantly improved, but the elongation sharply decreased. The tensile strength and elongation of the peak-age alloy without La element were 334 MPa and 2.18%, respectively. With the increase of La content, the tensile strength and elongation of the peak-age alloy were improved. The tensile strength and elongation of the alloy with 0.04% La element were improved to 350 MPa and 2.35%, respectively. The tensile strength and elongation decrease rapidly with further increase of La content.

4 Conclusion

1. With the addition of La element, the coarse, long rod-shaped and square-shaped eutectic silicon at the grain boundary of the cast alloy is significantly refined, and the secondary dendrite arm spacing also gradually decreases. When the addition of La exceeds 0.04%, the eutectic silicon and secondary dendrite arm spacing become coarse again.

2. The cast alloy without La mainly consists of Al phase, Si phase, and a small amount of Mg_2Si phase, Al_3Ti phase, and AlFeSi phase. After the addition of La exceeds 0.08%, the alloy contains Al_3La phase.

3. With the increase of La content, the tensile strength of the alloy gradually increases, while the elongation first increases and then decreases. The alloy with 0.04% La content has the best mechanical properties, with a tensile strength of 160 MPa and an elongation of 4.52% in the cast state and 350 MPa and 2.35% in the aged state. Continuing to increase the La content results in the re-coarsening of the eutectic silicon and the secondary dendrite arm spacing, as well as the appearance of the Al_3La phase, which lead to a rapidly decrease in tensile strength and elongation.

References

[1] Li S S, et al. Development and applications of aluminum alloys for aerospace industry. Journal of Materials Research and Technology, 2023, 27: 944-983.
[2] Wang T, et al. Mechanisms related to the influence of natural aging on the precipitation behavior of the A357 aluminum alloy during artificial aging. Vacuum, 2023, 217: 112581.
[3] Lorella C, et al. Effect of Cu addition on overaging behaviour, room and high temperature tensile and fatigue properties of A357 alloy. Transactions of Nonferrous Metals Society of China, 2020, 30(11): 2861-2878.
[4] Zhu K, et al. Study on Aging Characteristics of Cast A357 Alloy. Procedia Manufacturing, 2019, 37: 247-252.
[5] Sathiyaseelan G, et al. Enhancing the strength and structure of Al–Si alloys for biomedical applications through the addition of Sr and Zr grain refiners on secondary AlSi7Mg aluminum alloys. Chemical Physics Impact, 2024, 8: 100557.
[6] Lee J Y, et al. A study on the interaction between a Sr modifier and an Al-5Ti-1B grain refiner in an Al-7Si-0.35Mg casting alloy. Journal of Alloys and Compounds, 2023, 938: 168598.
[7] Li Z D, et al. Influence of Sr, Fe and Mn content and casting process on the microstructures and mechanical properties of AlSi7Cu3 alloy. Materials Science and Engineering: A, 2017, 689: 286-297.
[8] Pourbahari B, et al. Effects of La intermetallics on the structure and tensile properties of thin section gravity die-cast A357 Al alloy. Materials & Design, 2016, 94: 111-120.
[9] Sezgin C, et al. Effect of Ce addition on microstructure, thermal and mechanical properties of Al-Si alloys. Materials Today Communications, 2023, 34: 105518.
[10] Wei Q, et al. Effect of Sc on microstructure and properties of A357 alloy under different casting conditions. Journal of Materials Research and Technology, 2022,20: 2051-2059.

Influence of Rare Earth Elements Erbium and Europium Addition on Microstructure and Mechanical Properties of Al-Cu-Mg-(Ag) Alloys

Hayati Sahin, Derya Dispinar
Non-ferrous Metal Treatment, Vesuvius Foundry R&D Center, Enschede, The Netherlands
*Corresponding address: e-mail: hayati.sahin@vesuvius.com

Abstract: The alloying behavior of rare earth Erbium (Er) and Europium (Eu) at different ratios in Al–Cu–Mg (A206) and Al-Cu-Mg-Ag (A201) alloys with/without T7 heat treatment on mechanical properties were investigated. The potential effect of Er and Eu on the tensile and hardness of the alloy was characterized by optical metallographic examination, SEM and EDS examinations. The mechanical properties first increased with the addition of alloying element, however decreased with the increase in the amount of alloying element. Even after 0.2 wt%, the mechanical properties have decreased considerably. In SEM and EDS analyses, the increase in the number and size of intermetallic with the addition of Er and Eu is associated with negative mechanical properties.

Keywords: Al-Cu-Mg-(Ag), Rare earth, Er, Eu

1 Introduction

The precipitation of fine Ω phase in Al–Cu–Mg (A206) and Al-Cu-Mg-Ag (A201) alloys provides attractive strength and excellent creep resistance at elevated temperatures. These excellent properties make Al-Cu alloys very suitable for aerospace applications [1-4]. The microstructure and mechanical properties of Al-Cu alloys can be changed with different amount of Mg addition. These changes are achieved by the precipitation of Ω phases in the Al-Cu alloy with the addition of Mg [1]. The amount of precipitation of Ω phases is significantly increased by addition of Ag to Al-Cu-Mg alloys. On the other hand, adding Ag to Al-Cu alloys does not provide this effect. As a result, adding Mg and Ag together to Al-Cu alloys can shape Mg-Ag co-clusters, precipitating a considerable amount of Ω phases [2]. In addition to this, the small addition of Ag to Al-Cu-Mg alloys improves the aging hardening response by the precipitation of the fine and uniform Ω phase on the $\{111\}_\alpha$ planes [3].

Lately, it has been shown that trace additions of rare earth elements can greatly improve the mechanical properties of aluminum alloys and provide enhanced capabilities [1-4]. The present work investigates the effect of the addition of erbium (Er) and europium (Eu) to Al-Cu-Mg and Al-Cu-Mg-Ag alloys on the microstructure and mechanical properties of the as-cast and T7 heat-treated condition.

2 Experimental procedure

In the experimental work, Al-Cu4-Mg commercial purity alloy was used, and the chemical composition is given in Table 1.

Table 1. Composition of Al-Cu4-Mg base alloy

Cu	Fe	Si	Mn	Mg	Ti	Al
4.79	0.12	0.14	0.36	0.23	0.25	Bal

15 kg base alloy (Al-Cu-Mg) were melted in SiC crucible at 750°C in induction furnace. After the alloy was melted, an appropriate amount of Ag was added so that the content was 0.6 wt% Ag. Before the addition of master alloys, it was held for 15 minutes for the added Ag to be dispersed homogeneously. Al-3 wt% Er and Al-10 wt% Eu master alloys were used for alloying in appropriate proportions (0.1 and 0.3 wt% combinations). Degassing was carried out with N_2 gas for 300 seconds. Sand molds, which provide 10 cylindrical test bars were prepared with 60 AFS sand, 2 wt% resin and 0.5 wt% hardener. 7 out of 10 bars are heat treated. Regarding T7 state, base alloy and alloyed bars were solution treated for 22 hours at 540 °C at resistance furnace. Following by the cold water quenched at room temperature, the alloys were natural aged at room temperature for 24 hours then artificial aged at 160 °C for 13 hours and cooled in the air. Tensile test sample was machined according to the ASTM E8. The hardness test was employed. Sanding and polishing were applied to the samples for metallographic and SEM examination.

3 Result and discussion

Tensile test was employed for each parameter, with 3 samples in as-cast and 7 samples in T7 state. Yield Stress (YS), Ultimate Tensile Stress (UTS) and Elongation at fracture results were given at Figure 1.

In the case of as-cast condition, although the elongation decreases with the addition of alloying elements, there is a small fluctuation from the YS and UTS values. On the other hand, YS and UTS values have generally doubled with T7 heat treatment, while percentage elongations have decreased by more than half. The best mechanical properties were achieved in T7 heat treated samples without rare earth elements in the base alloy. YS was 381 MPa, UTS was 450 MPa and elongation was 4%. The mechanical properties decreased with the increase of the alloying elements.

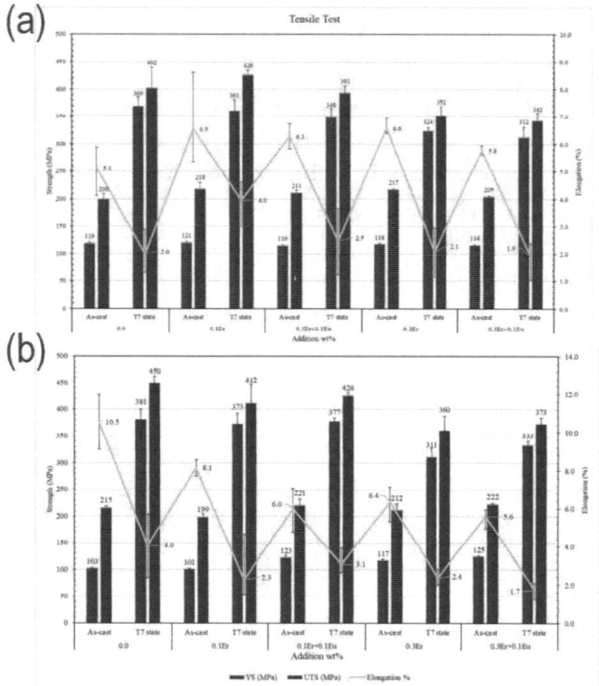

Figure 1. Mechanical properties of (a) Al-Cu-Mg and Al-Cu-Mg-Ag alloys with different master alloy addition ratios.

Figure 2 showed SEM micrographs and EDS analysis of as-cast Al–Cu-Mg-Ag alloys. In the SEM photograph of the Al-Cu-Mg-Ag alloy with 0.1 wt% Er and 0.1 wt% Eu added, it was seen in the Figure 2a that the dendritic intermetallic surround the structure. It was determined by EDS analysis that the intermetallic formed were Al-Cu-Er-Eu. (Figure 2b). In the SEM photograph of the Al-Cu-Mg-Ag alloy with 0.3 wt% Er and 0.1 wt% Eu added, it was seen in the Figure 2c that with the increase of alloying element, intermetallic structures such as spine skeleton or fish bone-like and a structure resembling a flower petal are seen. It was determined by EDS analysis that the intermetallic formed were Al-Cu-Er-Eu. (Figure 2d). The weight ratios of Er and Eu added are in agreement with the EDS analysis results.

The number and shape of intermetallic grew with the increase of alloying elements. As a result of the incompatibility between the main structure and intermetallic and the increase in the amount and size of the intermetallic, the mechanical properties decreased with the addition of the alloying element.

Figure 2. SEM micrographs of Al-Cu-Mg-Ag alloys a) 0.1 wt% Er and 0.1 wt% Eu, b) EDS analysis of marked area in (a), c) 0.3 wt% Er and 0.1 wt% Eu, d) EDS analysis of marked area in (c)

4 Conclusion

The mechanical properties of Al-Cu-Mg alloy with T7 heat treatment were greatly improved compared to the as-cast state. On the other hand, the mechanical properties decreased with the increase in the amount of rare earth elements Er and Eu added as alloying elements in T7 heat-treated samples. The mechanical properties decrease with the increase of alloy addition. The reason for this decrease was observed by SEM and EDS analyzes that the number and size of noncoherent needle-like intermetallic increased with the increase of alloy element addition.

References

[1] Bai S, Huang T, Xu H, Liu Z, Wang J, Yi X. Effects of small Er addition on the microstructural evolution and strength properties of an Al–Cu–Mg–Ag alloy aged at 200° C. Materials Science and Engineering: A. 2019 Oct 24;766:138351.

[2] Lee SL, Wu CT, Chen YD. Effects of minor Sc and Zr on the microstructure and mechanical properties of Al-4.6 Cu-0.3 Mg-0.6 Ag alloys. Journal of Materials Engineering and Performance. 2015 Mar;24:1165-72.

[3] Bai S, Liu Z, Li Y, Hou Y, Chen X. Microstructures and fatigue fracture behavior of an Al–Cu–Mg–Ag alloy with addition of rare earth Er. Materials Science and Engineering: A. 2010 Mar 25;527(7-8):1806-14.

[4] Chen X, Liu Z, Bai S, Li Y, Lin L. Alloying behavior of erbium in an Al–Cu–Mg alloy. Journal of alloys and compounds. 2010 Aug 27;505(1):201-5.

Sr Study on the Metamorphic Mechanism of Eutectic Silicon in Hypereutectic Al-Si Alloys

Mengjun Zhang[1,2], Zhiyong You[1,2,*], Yunguan Zhang[1,2], Libo Wang[1,2]
1. School of Materials Science and Engineering, Taiyuan University of Technology, Taiyuan 030024, China
2. Key Laboratory of Interface Science and Engineering in Advanced Materials Ministry of Education, Taiyuan University of Technology, Taiyuan 030024, China
*E-mail:13603511486@163.com

Abstract: The addition of a Sr modifier during the forming process of a hypereutectic Al-Si alloy results in the transformation of the acicular eutectic silicon within the alloy structure into rod or particle form. This process leads to an improvement in the comprehensive mechanical properties of the alloy. However, there is a lack of research on the mechanism of eutectic silicon metamorphism, especially on the interaction between the modifier and silicon phase. In this paper, the structural stability of Si(100) clean surface and adsorption model surface is studied by using the first-principle calculation method based on density functional theory. The results show that Sr is easily adsorbed on the surface of silicon phase and tends to fall at the Vacancy site. Then, Al-15%Si alloys with different Sr content are prepared by metal mold casting. By observing the SEM scans of the microstructure of alloy samples, it is found that Sr atoms are completely adsorbed and enriched on the surface of eutectic silicon.

Keywords: Hypereutectic Al-Si alloy; microstructure; first-principle; surface property

1 Introduction

Hypereutectic Al-Si alloys have been widely used in the manufacture of engine components such as cylinders or pistons because of high specific strength, low coefficient of thermal expansion and wear resistance[1]. However, the microstructure of the hypereutectic Al-Si alloy comprises long acicular eutectic silicon which can result in the matrix being cut or cause stress concentration at its angular tip. Modification treatment is an effective way to modify the silicon phase morphology and improve the mechanical properties of hypereutectic Al-Si alloys. As a traditional eutectic silicon modifier, Sr can not only modify the morphology of eutectic silicon, but also act as a green and environmentally friendly modifier. Furthermore, it has long-term effectiveness. Yan et al[2] explored the nucleation and growth mechanism of silicon phase through morphology analysis and non-isothermal analysis kinetics, and the results showed that Sr modified eutectic silicon mainly in three aspects: (1) toxic nucleation sites; (2) the interface energy between Si phase and liquid is reduced; (3) improve the activation energy of solid-liquid interface diffusion.

2 Experimental procedure

In this study, the CASTEP calculation program is employed to perform first-principle calculations. The exchange correlation energy parameter selected in the geometric optimization process is the PW91 model. The plane wave energy of 300 eV and the Monkhorst-Pack of 6×6×6 are selected as the truncation energy and k-point grid of the volume phase calculation, respectively.

Then ZL109 material was selected as the master alloy, and Al-15%Si alloys were prepared by adding Al-30%Si. Sr modifier is added in the form of Al-10Sr alloy. The Al-15%Si alloys containing 0.03%, 0.06% and 0.09% (wt.%) Sr were prepared by metal mold casting. The microstructure of the alloys were analyzed by SEM scanning.

3 Results and discussion

The Si(100) clean surface and the surface of the adsorption model corresponding to three different adsorption sites were constructed, as shown in Figure 1.

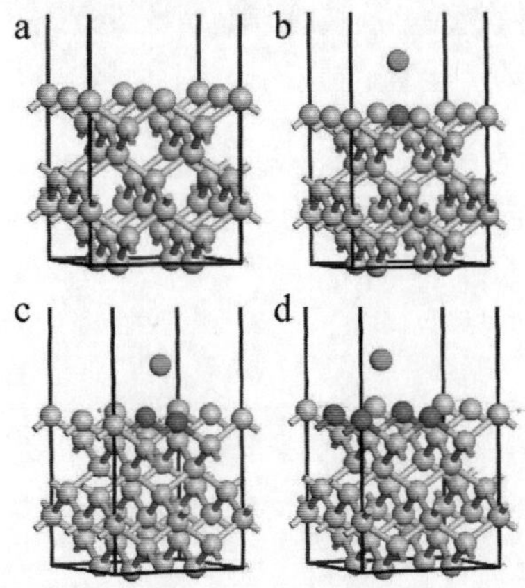

Fig. 1 Si(100) clean surface and adsorption surface model: (a) clean surface; (b) Top site; (c) Bridge site; (d) Vacancy site(The green atom is Sr atom and the rest are Si atoms)

The adsorption energy and surface energy of different surface models are calculated to determine the effect of Sr on the surface stability of silicon phase. The calculation formulas are shown in equation (1) and equation (2).

$$E_{ad} = E_{Sr/Si(1\,0\,0)} - E_{Sr} - E_{Si(1\,0\,0)} \quad (1)$$

$$E_{surf} = \frac{E_{Sr/Si(1\,0\,0)} - E_{Sr} - N_{slab}E_{Si}}{2A} \quad (2)$$

E_{ad} is adsorption energy; $E_{Sr/Si(100)}$ is the total energy of adsorption model; E_{Sr} is the energy of a single Sr atom; $E_{Si(100)}$ is the total energy of clean surface; E_{surf} is the surface energy; N is the number of Si atoms; A is the surface area of the surface model.

The adsorption energy and surface energy of different surface models are shown in Table 1. As shown in Table 1, the adsorption energy and surface energy of the vacancy site adsorption surface model are both the smallest, which are -1.98 eV and 0.04 eV, indicating that Sr atoms are easily adsorbed on the surface of silicon phase and are more inclined to fall at the vacancy site of silicon phase.

Table 1. Calculation results of adsorption energy and surface energy of different surface models

Model	a	b	c	d
E_{ad}/eV	-	-0.76	-1.32	-1.98
E_{surf} /eV	0.06	0.06	0.05	0.04

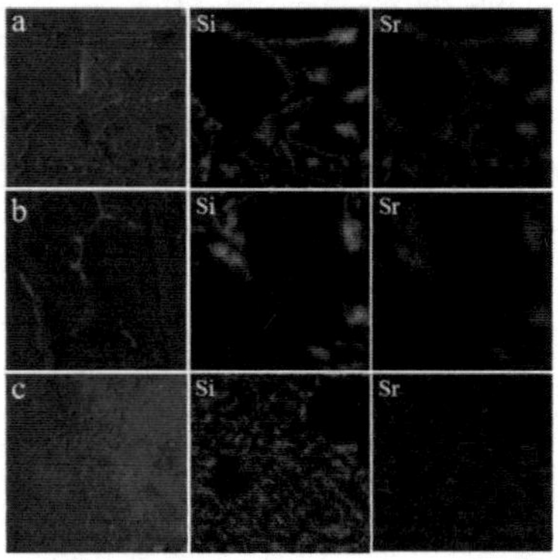

Fig.2 SEM scanning images of alloys after modification with different content of Sr: (a)0.03% (b)0.06% (c)0.09%

The SEM images of the three Sr modified alloy are shown in Figure 2. It can be seen that the positions and distributions of Si and Sr atoms in the alloy completely overlap, which indicates that after adding Sr atoms in the forming process of hypereutectic Al-Si alloy, all Sr atoms will be absorbed and enriched on the surface of silicon phase.

4 Conclusion

(1) Sr atoms easily adsorb on the surface of the silicon phase and tend to fall at the vacancy site of the silicon phase.

(2) The SEM images confirm the above view and show that Sr atoms are fully adsorbed and enriched on the surface of the silicon phase.

5 Acknowledgments

This work is supported by the 2021 Central Government Guide local science and technology development Fund (YDZJSX2021A008) and key Laboratory of Interface Science and Engineering in Advanced Materials Ministry of Education

References

[1] Slavica M, Blaža S, Sandra G, et al. Hypereutectic Aluminum Alloys and Composites: A Review. J. Silicon, 2022, 15(6).

[2] Yan P, Mao W, Fan J, et al. Simultaneous Refinement of Primary Si and Modification of Eutectic Si in A390 Alloy Assisting by Sr-Modifier and Serpentine Pouring Channel Process. J. Materials, 2019, 12(19).

Effect of Sc Content on Microstructure and Mechanical Properties of Cast Al-2.5Li-1.5Cu Alloy

Zhaoxiang Yan, Jianyu Li, Shulin Lü*,Wei Guo, Shusen Wu
State Key Lab of Materials Processing and Die & Mould Technology, School of Materials Science and Engineering, Huazhong University of Science and Technology, Wuhan 430074, P. R. China
* Corresponding Author, shulin317@hust.edu.cn

Abstract: Casting Al-Li alloys are widely used in the aerospace industry due to their advantages such as low density, high stiffness, and high elastic modulus. However, poor toughness is a significant bottleneck for the application of cast Al-Li alloys. In this study, a high-strength and tough Al-2.5Li-1.5Cu-xSc alloy was successfully prepared by ultrasonic-assisted squeeze casting, and the influence of different Sc contents (x=0, 0.15, 0.2, 0.25, 0.3wt.%) on the microstructure and mechanical properties of cast Al-2.5Li-1.5Cu alloy was systematically investigated. The research indicates that with the continuous increase of Sc content, the grain size of the as-cast alloy significantly refines, and the grains transform from coarse dendrites to equiaxed grains. Meanwhile, as Sc content increases, the variation trend of the mechanical properties is consistent with the microstructural evolution. The Al-2.5Li-1.5Cu-0.3Sc alloy exhibits the best mechanical properties after ultrasonic treatment, with tensile strength, yield strength, and elongation of 338.2MPa, 182.4MPa, and 12.2%, respectively. Compared to the 0Sc alloy, its performance has been improved by 45.5%, 62.3%, and 161.1%, respectively.

Keywords: Casting Al-Li alloy, Sc element, Squeeze casting, Microstructure, Mechanical property

1 Introduction

Al-Li alloys are widely used in aerospace applications due to their many advantages such as low density, high stiffness and high modulus of elasticity [1]. In the past few decades, the successful development and commercialization of wrought Al-Li alloys (such as AA2099 and AA2195) have achieved great success. However, the high anisotropy caused by the deformation process and the inability to form complex components at one time have been plaguing researchers. Therefore, researchers have gradually turned their attention to the preparation of Al-Li alloys by casting method to ensure the isotropy of the alloys. Al-Li alloys allow higher Li content, and the advantages of lightweight and high stiffness of the alloys are more obvious. So it is of far-reaching significance to develop light-weight, high-strength and high-toughness cast Al-Li alloys and expand their application prospects.

As a common strengthening element in Al alloys, the addition of Sc in minor quantities can produce considerable strengthening effects and significant grain refinement [2]. However, there are fewer studies on the strengthening mechanism of Sc element in cast Al-Li alloys.

The study aims to develop a new type of high strength casting Al-Li alloy by ultrasonic-assisted squeeze casting forming process by adding Sc elements. The effects of different Sc contents (x=0, 0.15, 0.2, 0.25, 0.3 wt.%) on the microstructure and mechanical properties of cast Al-2.5Li-1.5Cu alloy were also investigated.

2 Experimental procedure

The nominal composition of the Al-Li alloy in this paper is Al-2.5Li-1.5Cu-xSc (x=0, 0.15, 0.2, 0.25, 0.3wt.%).The actual composition of the Al-Li alloy was measured by ICP-OES and the results are shown in Table 1. Pure Al, Cu, Li, and Al-2Sc intermediate alloys were smelted in a vacuum furnace as follows: a graphite crucible with pure Al was placed in the vacuum smelting furnace when the furnace temperature was raised to a certain temperature, pure Al was melted and Cu and Al-2Sc were added to the vacuum smelting furnace, and finally pure Li grains were added under argon atmosphere. After complete melting of the raw materials, the Al-Li alloy slurry was poured into the metal mould cavity preheated to 200 ℃, and the Φ30 mm×90 mm alloy ingots were obtained by squeeze casting. In addition, 0.3Sc alloy was selected and the alloy slurry was prepared by ultrasonic treatment before squeeze casting. The temperature and time of ultrasonic treatment were 660~700 ℃ and 1~2min, respectively.

Table 1 Composition of Al-2.5Li-1.5Cu-xSc alloys

Alloys	Measured compositions (wt%)			
	Li	Cu	Sc	Al
Base	2.46	0.48	—	Bal.
0.15Sc	2.48	0.50	0.12	Bal.
0.2Sc	2.55	0.53	0.19	Bal.
0.25Sc	2.42	0.46	0.23	Bal.
0.3Sc	2.52	0.52	0.30	Bal.

3 Result and discussion

1.The effect of Sc content on the microstructure of cast Al-2.5Li-1.5Cu alloy

Fig.1 shows the as-cast microstructure of Al-2.5Li-1.5Cu alloys with different Sc contents(x=0, 0.15, 0.2, 0.25, 0.3 wt.%). As shown in Fig.1a, the matrix alloy exhibits obvious dendrites and coarse grains. With the increase of Sc content, the grain refinement of the as-cast Al-Li alloy was obvious, especially when 0.3 wt% Sc was added, the grains became more homogeneous and close to equiaxed grains. Furthermore, when 0.3Sc alloy was treated with ultrasound, the grains of the alloy became smaller and more uniformly distributed. Therefore, the addition of Sc and ultrasonic treatment have a positive effect on the microstructure of Al-2.5Li-1.5Cu-xSc alloy, which can significantly refine the grains and improve their uniformity.

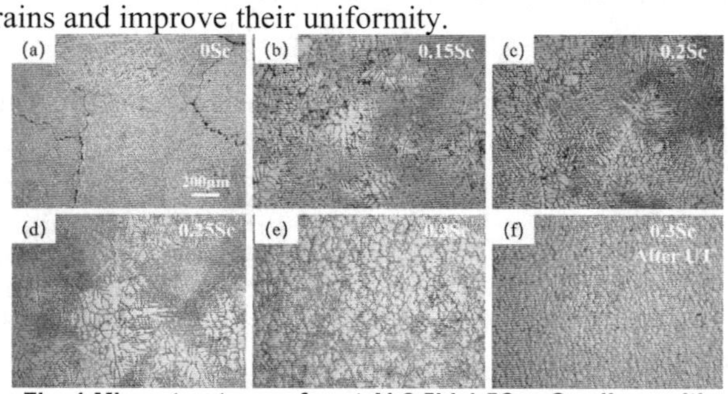

Fig. 1 Microstructures of cast Al-2.5Li-1.5Cu-xSc alloys with different Sc contents: (a-e) Squeeze casting alloy; (f)

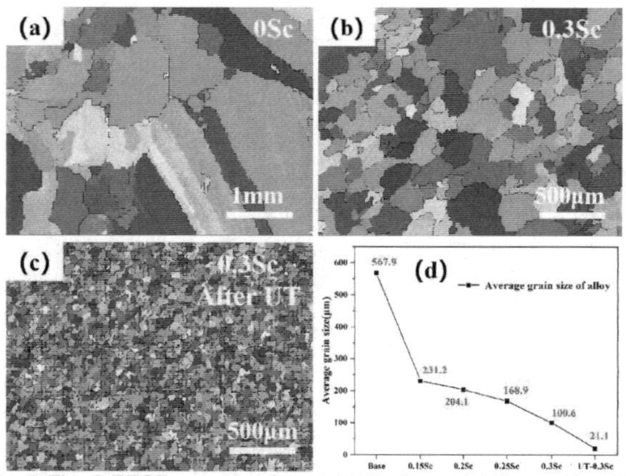

Fig. 2 Fig. 2 EBSD results of cast Al-Li alloys: (a) Base alloy, (b) 0.3Sc alloy, (c) 0.3Sc alloy after ultrasonic treatment; (d) grain size of the alloys at different Sc contents.

2.The effect of Sc content on the mechanical properties of cast Al-2.5Li-1.5Cu alloy.

Fig.3 shows the stress-strain curves and mechanical properties of cast Al-2.5Li-1.5Cu-xSc alloys with different Sc contents. It can be found that when Sc content reaches 0.3%, the ultimate tensile strength reaches 339.4 MPa, the yield strength is 192.9 MPa, and the elongation reaches 10.7%, which is increased by 46.6%, 40.3%, and 214.2%, respectively, compared with the base alloy. The elongation of 0.3Sc alloy after ultrasonic treatment further reaches 12.2%, an increase of 13.6% Ultrasonic treatment-Squeeze casting of 0.3Sc alloy

Fig.2 shows the EBSD results of cast Al-Li alloys with different Sc contents and the grain size of the alloys. The EBSD results show that the grain size of the base alloy is 567.9μm, and the distribution is not uniform. With the increase of Sc content, the grain size of cast Al-Li alloy decreases. When 0.3% Sc is added, the grain size is refined to 100.6μm, which is 82.3% lower than that of the base alloy, and the grain size distribution is more uniform. After ultrasonic treatment, 0.3Sc alloy is further refined, the grain size is 21.1μm, and the grain distribution is more uniform. It is reduced by 96.3% compared with the base alloy and 79.0% compared with the non-ultrasonic alloy. The addition of Sc element and ultrasonic treatment can effectively refine the grain size of cast Al-Li alloy and improve the microstructure of the alloy.
compared to the non-ultrasonicated alloy. The addition of Sc element and ultrasonic treatment have a great positive impact on improving the mechanical properties of cast Al-Li alloy.

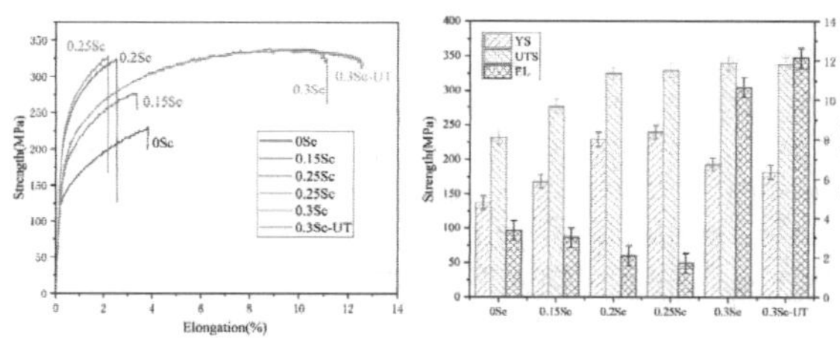

Fig.3 Stress-strain curves and mechanical properties of cast Al-2.5Li-1.5Cu-xSc alloys with different Sc contents.

4 Conclusion

(1) The addition of Sc and ultrasonic treatment have a significant effect on the microstructure of Al-2.5Li-1.5Cu alloy, especially in the refinement of grains and the reduction of dendrite formation. When Sc content reaches 0.3 % and after ultrasonic treatment, the grain size of the alloy is further refined to 21.1μm, which is 96.3 % smaller than that of the base alloy.

(2) When the Sc content reaches 0.3%, the ultimate tensile strength reaches 339.4 MPa, the yield strength is 192.9 MPa, and the elongation reaches 10.7%, which is increased by 46.6%, 40.3%, and 214.2%, respectively, compared to the base alloy. The elongation of the 0.3Sc alloy after ultrasonic treatment further reached 12.16%.

References

[1] J. Li, Y. Pan, D. Zhao, S. Lü, S. Wu, W. Guo, Development of a novel high strength and toughness Al-Cu-Li alloy casting billet with a new process, Materials Science and Engineering: A, 854 (2022).

[2] S. Lü, Z. Yan, Y. Pan, J. Li, S. Wu, W. Guo, Enhancement of Strength–Ductility Synergy of Al-Li Cast Alloy via New Forming Processes and Sc Addition, Materials, 17 (2024) 1558.

Effect of La Content on Microstructure, Mechanical Properties and Electrical Conductivity Of Cast Al-Mg-Si Alloys

Mingliang Li, Hongyu Xu*, Han Yu, Bo Jiang, Ye Wang, Maoliang Hu, Zesheng Ji
School of Materials Science and Chemical Engineering, Harbin University of Science and Technology, Harbin 150040, China
*Corresponding address: e-mail: xuhongyu@hrbust.edu.cn

Abstract: The effect of rare earth addition on the microstructure, tensile properties and electrical conductivity of Al-Mg-Si alloy have drawn increasing attention recently. La is introduced to modify Al-Mg-Si alloy by the melting method in this study. The microstructural evolution and phase constitution are examined using optical microscopy (OM), scanning electron microscopy (SEM), X-ray diffraction (XRD) and transmission electron microscopy (TEM). Results indicate that when La content is 0.2 wt.%, the alloy has the optimal tensile properties and electrical conductivity with ultimate tensile strength of 170 MPa, yield strength of 88 MPa, elongation of 18.9% and electrical conductivity of 44.0%IACS, which are 9.0%, 15.8%, 70.3% and 17.3% higher than that of Al-Mg-Si alloy, respectively. The main reasons for enhancing the tensile properties of Al-Mg-Si-xLa alloys are the grain refinement and the change of second phases. The second phases include Mg2Si phase, AlFeSi phase and Al11La3 phase. The main reason for improving electrical property is that Mg and Si solute atoms in Al matrix are reduced by the formation of Mg2Si phase. Moreover, excessive La will deteriorate the properties of Al-Mg-Si-xLa alloys. Therefore, the appropriate La content is helpful to improve the strength and electrical conductivity of Al-Mg-Si alloy. This study may provide a theoretical basis for simultaneously improving the strength and electrical conductivity of Al-Mg-Si alloy.

Keywords: Al-Mg-Si alloy, rare earth La, microstructure, tensile properties, electrical conductivity

1 Introduction

6101 aluminum alloys, as a typical conductor material, are widely applied to transportation, medical, power electronics and other fields due to their low density, high specific strength, good conductivity, formability as well as low cost. With the developing modern industry, there is an urgent requirement for Al-Mg-Si alloys with excellent comprehensive properties, especially high strength and excellent electrical conductivity [1]. However, the strength and conductivity properties of Al-Mg-Si alloy are contradictory. It has been shown that micro-alloying with rare earth elements is one of the important ways to modify the microstructure of Al-Mg-Si alloy, thus affecting their comprehensive performance [2]. The most commonly used rare earth element is La. Zhao et al. [3] added La into Al-0.5Mg-0.24Si alloy to form the $LaSi_2$ phase, which promoted the transformation of Fe-rich phase from β-Al_5FeSi to α-Al_8Fe_2Si, and the appropriate La content improved the tensile properties and electrical conductivity of the alloy. Jiang H X et al. [4] added trace La could reduce the grain size of Al-Mg-Si alloys and promote the precipitation of Mg_2Si phases. Moreover, it also improved the tensile strength of Al-Mg-Si alloys from 310 MPa to 361 MPa. But it is not consistently reported about the effect of La on the microstructure and properties of Al-Mg-Si alloy, which needs to be further studied. The effects of La contents on the microstructure, tensile properties and electrical conductivity of cast Al-Mg-Si alloy are investigated in this study. The reason for improving the comprehensive properties of cast Al-Mg-Si alloy is discussed.

2 Experimental procedure

Al-Mg-Si-*x*La alloys (*x*=0, 0.1, 0.2, 0.3, 0.4, 0.5 wt.%) were prepared from pure aluminum (99.9%), pure magnesium (99.9%), Al-20Si master alloy and Al-20La master alloy. Al-Mg-Si-*x*La samples for microstructure observation were ground, polished and corroded in sequence. The etchant is hydrofluoric acid solution with a volume fraction of 1%. Microstructure observation of the alloys was conducted by using an optical microscope (OM, Leica DM ILM) and a scanning electron microscope (SEM, FEI Sirion 200). The phase was analyzed by an X-ray diffractometer (XRD, X'Pert Pro) with scanning speed of 9°/min and scanning angle of 10°-90°. The phase in the alloys was characterized by a transmission electron microscope (TEM, JEM-2100). The tensile properties of the alloys were tested on an Instron 569/50K universal tensile testing machine at room temperature, where the load speed is 1mm/min. The electrical conductivity of the alloys was performed by an RTS-11 metal four-probe tester.

3 Result and discussion

Fig. 1 presents the microstructures of Al-Mg-Si-xLa alloys. The average grain size of Al-Mg-Si-xLa alloys with increasing La content is about 271.9 μm, 201.6 μm, 141.7 μm, 146.8 μm, 175.5 μm and 180.1 μm, respectively. It can be seen that the grains are refined after the addition of La. The grain refinement effect of Al-Mg-Si-0.2La alloy is optimal and the average grain size is reduced by about 47.7% compared with that of the cast Al-Mg-Si alloy. However, when the La content exceeds 0.2 wt.%, the average grain size of the alloys gradually increases compared with that of Al-Mg-Si-0.2La alloy. Furthermore, after adding La, the distribution of the grains is more uniform in Al-Mg-Si-xLa alloys.

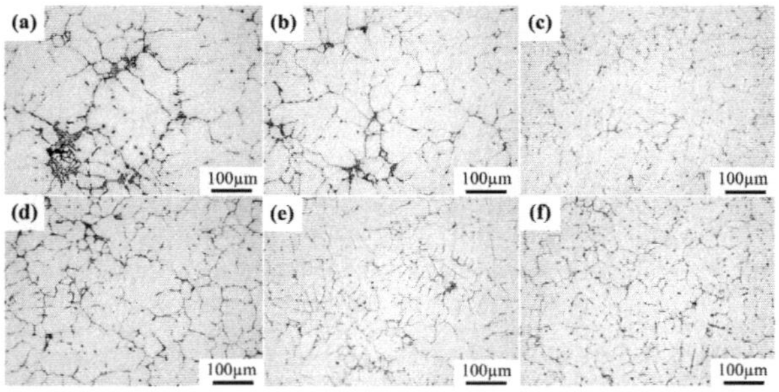

Figure 1. Microstructures of Al-Mg-Si-xLa alloys: (a) Al-Mg-Si alloy, (b) Al-Mg-Si-0.1La alloy, (c) Al-Mg-Si-0.2La alloy, (d) Al-Mg-Si-0.3La alloy, (e) Al-Mg-Si-0.4La alloy, (f) Al-Mg-Si-0.5La alloy.

Fig. 2 shows the engineering stress-strain curves and the tensile properties of the cast Al-Mg-Si-xLa alloys. The ultimate tensile strength (UTS), yield strength (YS) and elongation (EL) of Al-Mg-Si-xLa alloys present a tendency of the first increasing and then declining with La content. Compared with the Al-Mg-Si alloy (UTS: 156 MPa, YS: 76 MPa and EL: 11.1%), the Al-Mg-Si-0.2La alloy has the best tensile properties with UTS of 170 MPa, YS of 88 MPa and EL of 18.9, which are improved by 9.0%, 15.8% and 70.3%, respectively. However, the tensile properties of the Al-Mg-Si-xLa alloys gradually decrease when the content of La exceeds 0.2 wt.%.

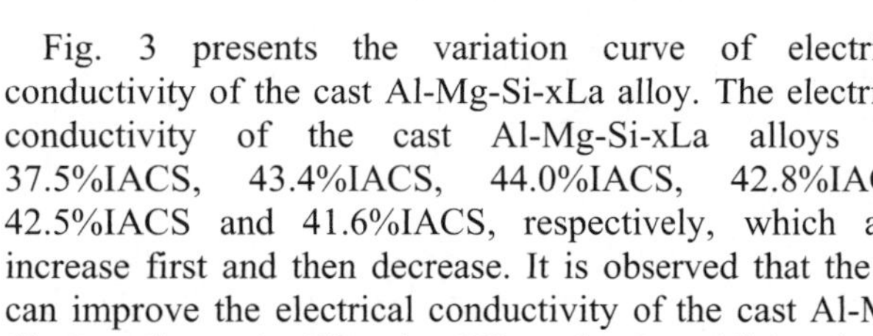

Figure 2. Tensile properties of the cast Al-Mg-Si-xLa alloys: (a) engineering stress-strain curve; (b) tensile strength and elongation.

Fig. 3 presents the variation curve of electrical conductivity of the cast Al-Mg-Si-xLa alloy. The electrical conductivity of the cast Al-Mg-Si-xLa alloys are 37.5%IACS, 43.4%IACS, 44.0%IACS, 42.8%IACS, 42.5%IACS and 41.6%IACS, respectively, which also increase first and then decrease. It is observed that the La can improve the electrical conductivity of the cast Al-Mg-Si-xLa alloys significantly. When the La addition is 0.2 wt.%, the electrical conductivity of the Al-Mg-Si-0.2La alloy achieves the maximum value, which is about 17.3% higher than that of the Al-Mg-Si alloy. When the La content exceeds 0.2 wt.%, the electrical conductivity declines gradually, and it decreases to 41.6%IACS of Al-Mg-Si-0.5La alloy, while which is still 10.9% higher than that of the Al-Mg-Si alloy.

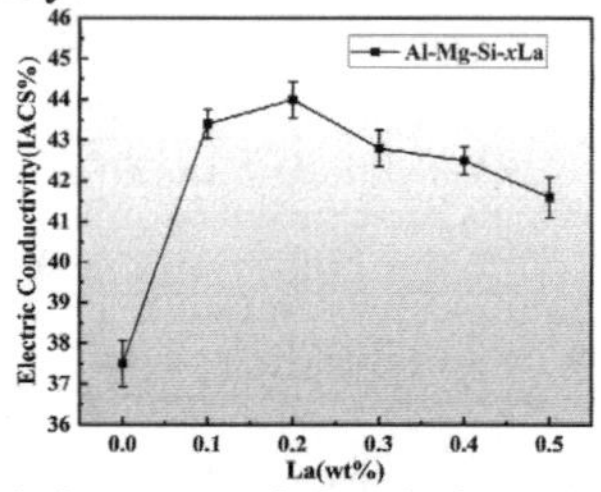

Figure 3. The variation curve of electrical conductivity of cast Al-Mg-Si-xLa alloys.

4 Conclusion

(1) La can refine the grain of the cast Al-Mg-Si-*x*La alloys by reducing the wetting angle. The optimum La content is 0.2 wt.%, the average grain size of the Al-Mg-Si-0.2La alloy is reduced by 44.7% compared with that of the cast Al-Mg-Si alloy.

(2) La promotes the formation of Mg_2Si phase and the morphology transformation of the AlFeSi phase from continuous state to discontinuous state. But the excessive La will form many needle-like $Al_{11}La_3$ phase.

(3) The Al-Mg-Si-0.2La alloy has the best tensile properties and electrical conductivity with ultimate tensile strength of 170 MPa, yield strength of 88 MPa, elongation of 18.9% and electrical conductivity of 44.0%IACS, which are 9.0%, 15.8%, 70.3% and 17.3% higher than that of the Al-Mg-Si alloy, respectively.

(4) The enhancement of tensile properties of Al-Mg-Si-0.2La alloy is mainly attributed to the fine grain strengthening and the second phase strengthening. The improvement of electrical conductivity is mainly ascribed to the lower Mg and Si solute atoms in the Al-Mg-Si-0.2La alloy.

5 Acknowledgments

This work was financially supported by the National Natural Science Foundation of China (No. 51704087).

References

[1] Jiang S Y, Wang R H. Grain size-dependent Mg/Si ratio effect on the microstructure and tensile/electrical properties of Al-Mg-Si-Sc alloys. Journal of Materials Science Technology, 2019, 35(7): 1354-1363.

[2] Ussadawut P, Julathep K, Chaowalit L. Grain refinement mechanism in an Al-Si-Mg alloy with scandium. Journal of Alloys and Compounds, 2012, 524(25): 177-186.

[3] Zhao Y, Zhang Y B, Li T J, et al. Effects of rare earth element La on the microtructure and properties of Al-Mg-Si alloy. Special Casting Nonferrous Alloys, 2021, 41(10): 1219-1223.

[4] Jiang H X, Zheng Q J, Song Y, et al. Influence of minor La addition on the solidification, aging behaviors and the tensile properties of Al-Mg-Si alloys. Materials Characterization, 2022, 185: 111750.

Study on the Interfacial Reaction Mechanism Between Superalloy Melt and Alkaline Earth-Rare Earth Composite Oxide System

Shigang Fan [1,*], Naicheng Sheng [1], Chenyu Lin [1,2], Shijie Sun [1], Guichen Hou [1], Jinjiang Yu [1], Jinguo Li [1], Yizhou Zhou [1], Xiaofeng Sun [1]
1. Shi-Changxu Innovation Center for Advanced Materials, Institute of Metal Research, Chinese Academy of Sciences, Shenyang 110016, China
2. School of Materials Science and Engineering, University of Science and Technology of China, Shenyang 110016, China
*Corresponding address: e-mail: sgfan@imr.ac.cn

Abstract: Superalloys are the most important materials for aeroengine hot section. The improvement of its trace element control level has always been a hot topic in the field. The precise control of sulfur (S) and rare earth (RE) elements, which are the most harmful and beneficial trace elements affecting the properties of superalloys, has been increasingly valued by the academic and engineering fields. The Alkaline earth-Rare earth ceramic system has the application prospects of both desulfurization and increasing rare earth elements. In this paper, the phase and element distribution of interfacial reaction products between MgO-Y_2O_3, CaO-Y_2O_3, MgO-CaO-Y_2O_3 ceramics and typical high activity superalloys were systematically studied by innovative experimental methods combined with XRD and SEM-EDS characterization methods. Furthermore, the reaction mechanism equations of different refractories and superalloy melts were summarized. The CaO-Y_2O_3 and MgO-CaO-Y_2O_3 systems have been confirmed to have the ability of desulfurization and rare earth addition by reacting with active elements such as Al in the melt. The interface reaction mechanism is revealed.

Keywords: Alkaline earth-Rare earth composite oxide; superalloy; interface reaction; ultra-low sulfur; rare earth addition

1 Introduction

Nickel based superalloys are the most important materials for the aeroengine hot section, and their usage in advanced turbine engines has exceeded 50% [1,2]. With the background of its temperature bearing capacity approaching the theoretical upper limit, improving the purity of the melt and accurately controlling the content of trace beneficial elements are important means to maximize the alloy performance[3,4].

Sulfur(S) has a significant impact on the elevated temperature service performance of casting superalloys and is recognized as one of the harmful elements. Smialek[5,6] reported that S content in superalloy is less than 1ppm, and its oxidation resistance is greatly improved. Rare earth (RE) elements (La, Ce, Y, etc.) are recognized by researchers as the most significant beneficial trace elements affecting superalloys[7]. The addition of a small amount of REs can significantly improve the properties of superalloys[7-8]. Alkaline earth-Rare earth oxide refractory materials can be used as a promising refractory system for superalloys melt in desulfurization and rare earth element control.

In this study, typical Alkaline earth-Rare earth oxide systems (MgO-Y_2O_3, CaO-Y_2O_3, MgO-CaO-Y_2O_3) are taken as the research object. The interface reaction behavior between highly active K417G superalloy melt and ceramics has been systematically studied. By analyzing the morphology, phase, and elemental composition of the reaction layer before and after the reaction, the interfacial reaction mechanism is revealed.

2 Experimental procedure

The immersion method was used to observe the interface reaction between ceramics and superalloys melt. The mixed Rare earth-Alkaline earth oxide was pressed into a rod shape (8 *10*100mm). According to specific heating gradients and insulation stages, the final product was kept at 1600 ℃ for 1 hour and cooled in the furnace to produce a dense ceramic rod. The ceramic samples were inserted into the alloy melt and held for 5 minutes before being directly taken out and cooled to room temperature for further characterization.

3 Result and discussion

The morphology and element distribution after the reaction at 1500 ℃ are shown in Figure 1. There are two continuous dense reaction layers with coexisting Ca and S, as well as coexisting Ca, Al, Y, and O (Figure 1 (a)). Further microscopic elemental analysis indicates the presence of a small amount of elongated channels of $CaAlYO_4$ composite oxide phase in the region dominated by solid CaS in the outer reaction layer. At the same time, there is a clear trend of internal infiltration in the oxide layer dominated by $CaAlYO_4$, which exists between unreacted CaO and Y_2O_3 particles (Figure 1 (b)).

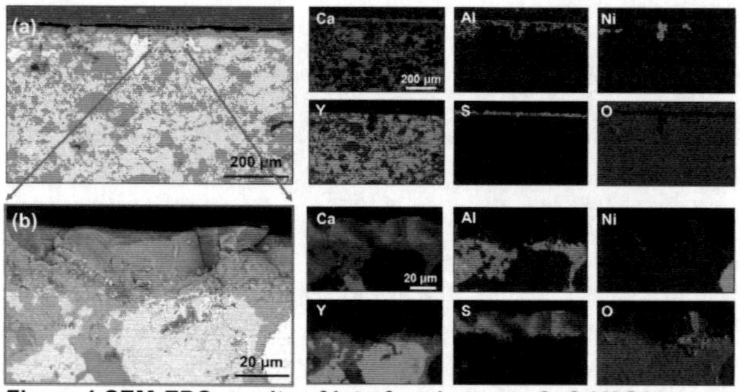

Figure 1 SEM-EDS results of interface between CaO-Y2O3 ceramic and superalloys after 1500 ℃ reaction

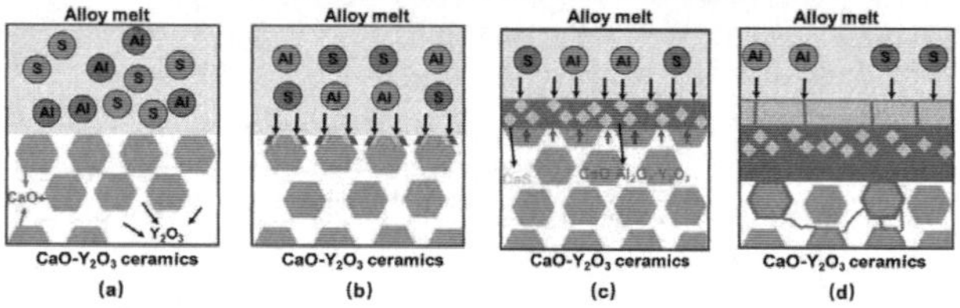

Figure 2 Interfacial reaction mechanism between CaO-Y_2O_3 and superalloy melt

The interfacial reaction can be divided into four stages. The first stage is the contact between the high temperature alloy melt (containing Al and S elements) and the uniformly distributed dense CaO-Y_2O_3 ceramics. The second stage is the initial stage of the reaction. The reaction occurs in the area where the melt contacts with CaO and Y_2O_3 particles, and the reaction products are solid CaS and CaO-Al_2O_3-Y_2O_3 ternary oxides. The third stage is the development stage of the reaction. Subsequently, the two products generated in the second stage form a continuous reaction layer. With the progress of the reaction, the solid CaS accumulates and occupies most of the physical space in the external desulfurization reaction layer containing S. At the same time, in order to ensure the continuous reaction, the $CaAlYO_4$ composite oxide medium layer presents a network channel to ensure the transmission of Al and S in the melt and CaO and Y_2O_3 phases in the ceramics (Fig.2).

4 Conclusion

In this study, the interfacial reaction behavior between highly active K417G superalloy melt and CaO-Y_2O_3 ceramics was studied.

(1) By in situ reaction with superalloy melt, MgO-Y_2O_3 system has the ability of rare earth addition, CaO-Y_2O_3 and CaO-MgO-Y_2O_3 systems have the ability of both desulfurization and rare earth addition.

(2) The interface reaction occurs between CaO-Y_2O_3 ceramics and Al, S elements in the superalloy melt. The reaction equation is CaO-Al_2O_3+[Al]+[S] → CaS+ $xCaO\cdot yAl_2O_3\cdot zY_2O_3$ + [Y];

(3) $CaO\cdot Al_2O_3\cdot Y_2O_3$ multi-oxides play a decisive role as the interface reaction of the medium layer. The internal CaO, Y_2O_3 and Al, S in the melt maintain the continuity of the interfacial reaction through its transport channel.

5 Acknowledgments (Bold, 10 pt., Arial)

This work was financially supported by the National Key R&D Program of China (No. 2022YFB3708100) and the National Natural Science Foundation of China (Nos. 52104352 and U2341267), and the National Natural Science Foundation of Liaoning province (No 2023-MS-015).

References

[1] Tang Y T, Panwisawas C and Ghoussoub J N. Alloys-by-design: Application to new superalloys for additive manufacturing. Acta Mater, 2021, 202: 417-436.
[2] Xu H, Zhang Y and Fu H. Effects of boron or carbon on solidification behavior of Co-Ni-Al-W-based superalloys, J. Alloy. Compd, 2022,891: 161965.
[3] Petrushin N V, Elyutin E S and Visik E M. Development of a Single- Crystal Fifth-Generation Nickel Superalloy, Russian Metall. Metally, 2017,2017: 936 -947.
[4] Reed R C. The Superalloys: Fundamentals and Applications, New York, 2006 :1-5.
[5] Smialek JL, Jayne DT and Schaeffer JC. Effects of hydrogen annealing, sulfur segregation and diffusion on the cyclic oxidation resistance of superalloys: a review. Thin Solid Films, 1994, 253(1): 285-92.
[6] Smialek J. Origins of a low-sulfur superalloy Al2O3 scale adhesion map. Crystals. Crystals, 2021, 11(1):60.
[7] Zhou P J, Yu J J and Sun X F. Influence of Y on stress rupture property of a Ni-based superalloy. Mater. Sci. Eng., 2012, A551: 236.
[8] Wang R M, Song Y G and Han Y F. Effect of rare earth on the microstructures and properties of a low expansion superalloy. J. Alloys Compd., 2000, 311: 60.

Insights into the Effects of Sm on Microstructure Evolution Mechanism and Mechanical Properties of Mg-8Gd-2Y-1Zn-0.5Zr Cast Alloy at High Temperature

Jianfei Xiao[1,2], Yuyang Gao[1,2,*], Jiangfeng Song[1,2,*], Yuling Xu[3], Jianxin Zhou[1,2], Xiaojun Luo[1,2], Xin Guo[1,2], Jing Wang[1,2], Bin Jiang[1,2], Fusheng Pan[1,2]

1. National Engineering Research Center for Magnesium Alloys, College of Materials Science and Engineering, Chongqing University, Chongqing, 400044, China
2. National Key Laboratory of Advanced Casting Technologies, Chongqing University, Chongqing, 400044, China
3. Baomarc (Hefei) Technology Co., Ltd., Hefei, 238000, China

* gaoyuyang@cqu.edu.cn (Yuyang Gao); * jiangfeng.song@cqu.edu.cn (Jiangfeng Song)

Abstract: In this study, the effects of 1% Samarium (Sm) on the microstructure and room and high temperature mechanical properties of Mg-8Gd-2Y-1Zn-0.5Zr alloy are discussed and analyzed. The addition of Sm increases the volume fraction of lamellar and blocky long-period stacking ordered (LPSO) phases and promotes the precipitation of lamellar β' and β phases. The high yield strength and ultimate tensile strength of Mg-8Gd-2Y-1Sm-1Zn-0.5Zr alloy at room and high temperatures are mainly attributed to the strengthening effects of LPSO, β, and β' phases. The LPSO, β, and β' phases effectively hinder the movement of dislocations and inhabit the migration of grain boundaries. The Mg-8Gd-2Y-1Sm-1Zn-0.5Zr alloy shows superior yield strength (203.3 MPa) and ultimate tensile strength (258.9 MPa) at 350 ℃, which are respectively 67.2 MPa and 23.8 MPa higher than those of Mg-8Gd-2Y-1Zn-0.5Zr alloy (136.1 MPa and 235.1 MPa).

Keywords: Mg-Gd-Y-Zn-Zr, Sm addition, high-temperature mechanical properties, strengthening mechanism

1 Introduction

In the aerospace field, the use of Mg alloys in key military equipment such as airplanes, spaceships and satellites can enhance the maneuverability of aircraft by reducing weight [1]. In some aerospace components, Mg alloys need to work at high temperatures above 300 ℃, such as rocket cabins, aircraft engine casings, and helicopter gearbox housings [2]. However, the commercial Mg-Al-Zn and Mg-Al-Mn series alloys tend to rapidly lose their strength when exposed to temperatures above 125 ℃ [3].

With the incorporation of rare earth (RE) [4], Mg alloys can simultaneously achieve solid solution strengthening and precipitation strengthening, and subsequently exhibit favorable high-temperature mechanical properties. Among RE-containing Mg alloys, the Mg-Gd-Y-Zn-Zr series alloys have gained widespread attention for the high strength and excellent creep resistance [2-4]. It is reported that Mg-10Gd-3Y-1Zn-0.5Zr alloy exhibits excellent yield strength (199 MPa) at 250 ℃, which is 25% higher than the commercial WE54 alloy [1, 5]. However, its yield strength rapidly decreases to 142 MPa at 300℃. Additionally, the studies on the deformation mechanisms and microstructural evolution of Mg alloys at high temperatures are mainly focused on the range of 150 ℃ to 250 ℃. When the temperature is further increased to 350 ℃, the strength of the alloys decreased significantly. The microstructure evolution, deformation mechanism and strengthening mechanism of Mg alloys during tensile deformation above 300 ℃ need to be further studied.

In this study, the effects of Sm on the microstructure and mechanical properties of Mg-8Gd-2Y-1Zn-0.5Zr at room and high temperatures (ranging from room temperature to 350 °C) are discussed. The fracture modes and strengthening mechanisms of the cast alloys at high temperatures are systematically discussed and analyzed. This investigation intends to provide some guidance for the design and preparation of high-strength heat-resistant cast Mg-RE alloys.

2 Experimental procedure

Actual chemical compositions of Mg-8Gd-2Y-1Zn-0.5Zr (VW82) and Mg-8Gd-2Y-1Sm-1Zn-0.5Zr (VW82-1Sm) alloys were tested by the inductively coupled plasma (ICP-OES) and the results were listed in Table 1. The samples for tensile testing were cut from the bottom part of the ingot (as shown in Fig. 1a). The samples for phase analysis and characterization were shown in Fig. 1b. The heat treatment process parameters are 510 ℃×12 h+225 ℃×9 h.

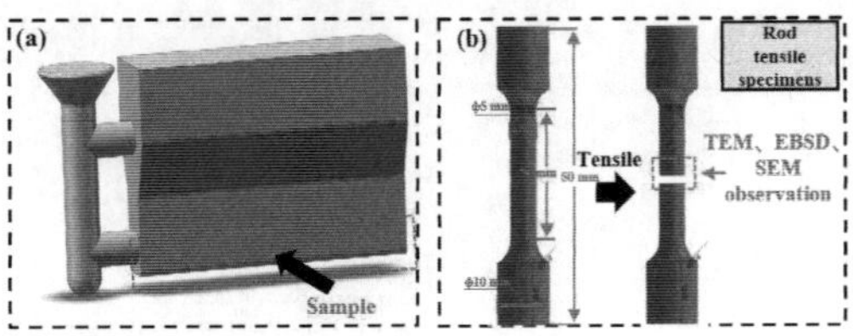

Figure 1. (a) casting mould, (b) tensile sample and observation sample.

Table 1. Composition of Alloy

Alloy	Mg	Gd	Sm	Y	Zn	Zr
VW82	Bal.	8.04	—	1.94	1.09	0.52
VW82-1Sm	Bal.	8.13	1.10	2.25	1.07	0.47

Tensile tests at 25 ℃, 250 ℃, 300 ℃, and 350 ℃ were carried out on a CMT 6305-300 KN testing machine. During the high-temperature tensile test, the tensile specimen was firstly heated to the set temperature and maintained for 30 minutes, and then the test began. The tensile test was carried out at a strain rate of 0.001 m/s. The tensile test was repeated three times to ensure the accuracy and reliability of the data.

3 Result and discussion

1. Mechanical properties and microstructure at aged state

The introduction of Sm results in a reduction in the average grain size of α-Mg from 40.3 μm in VW82 alloy to 35.4 μm in VW82-1Sm alloy. The volume fraction of LPSO phases increases from 12.6% to 21.8%. At room temperature, the YTS (235.3 MPa) and UTS (325.7 MPa) of VW82-1Sm alloy are 62.9 MPa and 43.3 MPa higher than those of VW82 alloy (172.4 MPa and 282.4 MPa), respectively. At room temperature, although the density of the β' phase is not high, the co-precipitation of β' precipitates and γ' precipitates can prevent basal slip more effectively.

2. Mechanical properties and microstructure at high temperature

When the temperature is further increased to 350 °C, the YTS (203.3 MPa) and UTS (258.9 MPa) of VW82-1Sm alloy are 67.2 MPa and 23.8 MPa higher than those of VW82 alloy (136.1 MPa and 235.1 MPa), respectively. The continuous precipitation of strengthening phase β' within the grains at high temperatures, along with the continuous growth and densification of β phase, which can pin the grain boundaries. This enables VW82-1Sm to maintain a high yield strength even as non-basal slip initiation and grain boundary sliding tendency increase with temperature.

4 Conclusion

1. The addition of 1 wt.% Sm effectively refine the grain sizes of the cast alloys. In addition, the introduction of Sm element increases the volume fractions of β', β, β1, and LPSO phases in VW82-1Sm alloy.

2. The introduction of Sm enhances the high temperature strength of VW82-1Sm alloy. The high strength of VW82-1Sm alloy can be attributed to the continuous precipitation of β and β' phases, as well as the increase in volume fraction of LPSO phase.

5 Acknowledgments

This work is financially supported by the National Natural Science Foundation of China (Grant Nos. U2037601, 52271090, 52071036, U21A2048, and 52201106), the National Key Research and Development Program of China (Grant Nos. 2021YFB3701000 and 2022YFB3709300), and the Fundamental Research Funds for the Central Universities (Grant Nos. SKLMT-ZZKT-2022Z01 and SKLMT-ZZKT-2022M12).

References

[1] G. Wu, C. Wang, M. Sun, W. Ding, Recent developments and applications on high-performance cast magnesium rare-earth alloys, J. Magnes. Alloy 9(1) (2021) 1-20.

[2] J. Song, J. She, D. Chen, et al. Latest research advances on magnesium and magnesium alloys worldwide, J. Magnes. Alloy, 8(1) (2020) 1-41.

[3] X. Dong, L. Feng, S. Wang, et al. A new die-cast magnesium alloy for applications at higher elevated temperatures of 200–300 °C, J. Magnes. Alloy, 9(1) (2021) 90-101.

[4] H. Pan, G. Qin, Y. Huang, et al. Development of low-alloyed and rare-earth-free magnesium alloys having ultra-high strength, Acta. Mater, 149 (2018) 350-363.

[5] Z. Chang, Y. Wu, N. Su, et al. Microstructural evolution of Mg-10Gd-3Y-1Zn-0.4Zr (wt%) alloy prepared by strain-induced melt activation process, Mater. Charact, 171 (2021) 110831-110831.

Effect of Nano Refiners on the Microstructure and Mechanical Properties of Cast Al-Zn-Mg-Cu Alloy

Yongxiang Xu, Jialin Gong, Baoyao Zhang, Xiangping Li, Juncheng Ma
CRRC Qishuyan Institute Co.,Ltd,Jiangsu Changzhou 213011, China
*Corresponding address: e-mail: 76787209@qq.com

Abstrac： This article aims to develop high-performance casting alloys. The effect of trace amounts of TiB2+TiC nano refiners on the microstructure and mechanical properties of Al-Zn-Mg-Cu alloy was studied through observation under an optical microscope, combined with detection by a tensile testing machine and a brinell hardness tester. The results indicate that the addition of trace amounts of TiB2+TiC nano refiners can significantly refine the alloy grains and improve the mechanical properties of the alloy. The grain size decreased from 66μm to 20.9μm, and the tensile strength of the cast alloy increased from 175MPa to 217.3MPa. After heat treatment, the tensile strength, yield strength, and elongation of Al-Zn-Mg-Cu alloy containing trace amounts of TiB2+TiC nano refiner were 602.3MPa, 582.3MPa, and 2%, respectively, with an increase of 34.2%, 34.8%, and50%.

Keywords: TiB_2+TiC nano refiners, Al-Zn-Mg-Cu alloy, microstructure, mechanical properties

1 Introduction

Al-Zn-Mg-Cu alloy has low density, high strength, and good cutting performance, and is often used in aerospace, automotive, and shipbuilding fields[1]. The preparation process of this alloy usually includes melting and casting, extrusion deformation, and heat treatment. Among them, extrusion deformation is a key process to improve the mechanical properties of the alloy, but it also limits the size and application range of the product. In recent years, with the upgrading and replacement of products involved in the casting field, higher requirements have been placed on the mechanical properties of materials, so there is an urgent need to develop a high-performance aluminum alloy material.

Alloying is considered to be the most effective method to improve the microstructure and properties of Al-Zn-Mg-Cu alloys. Studies have shown that elements such as Sc, Zr, Er, and Yb can form dispersed strengthening phases of Al_3X (X=Sc, Zr, Er, Yb), but these elements can also form coarse phases such as Al_8Cu_4X (X=Sc, Zr, Er, Yb), which is detrimental to the alloy properties. However, the nano TiC and TiB_2 ceramic particles have small size, relatively stable performance, and good interface bonding with the aluminum matrix, making them ideal strengthening agents for Al-Zn-Mg-Cu alloys[2]. However, there is relatively little research on adding this nano refining agent to Al-Zn-Mg-Cu alloys.

2 Experimental procedure

The raw materials are pure Al, pure Zn, pure Mg, Al-50Cu intermediate alloys, and nano refining agent rods, which were melted using a resistance furnace to obtain the two alloys shown in Table 1. The specific process is as follows: place pure A1 and A1-50Cu intermediate alloys in a crucible furnace, heat up to 750℃ to melt them, then add pure Zn, stir evenly, wrap the pure Mg block in aluminum foil, and use a bell shaped tool to press it into the aluminum liquid. For the 2# alloy in Table 1, place the nanoparticle refining agent rod in the aluminum liquid, heat up to 730℃, stir evenly and remove the slag, and then perform refining treatment. After refining, heat up to 750℃ and keep it warm for 10 min. Then, the furnace will naturally cool down to 700℃ before casting. The casting mold is a preheated iron mold.

The microstructure of two alloys was observed using a optical microscope. The sanding paper used was 320#, 500#, and 800# water sandpaper, the polishing agent was 0.25μm diamond particles, and the corrosive agent was Keller reagent. The mechanical properties of two alloys were tested using a tensile testing machine and a Brinell hardness tester, with a tensile rate of 1mm/min.

Table 1. Composition of Alloy

Alloy	Zn	Mg	Cu	TiB_2 +TiC	Fe	Si	Al
1#	6.14	2.05	1.82	/	0.11	0.06	Bal.
2#	6.25	2.08	1.76	0.35	0.08	0.05	Bal.

3 Result and discussion

Cast microstructure

Figures 1a and 1b show the metallographic structure of 1# alloy without the addition of nano refining agents. It can be seen from the figures that the grains are dendritic and there is a phenomenon of dendritic segregation. The grain size is relatively dispersed, with an average grain size of 66μm. After the addition of nano additives to TiB_2 and TiC, as shown in Figures 1c and 1d, the grains of 2# alloy are equiaxed and significantly refined, with an average grain size reduced to 20.9μm. This indicates that the addition of nano refining agents can effectively refine the cast Al-Zn-Mg-Cu alloy.

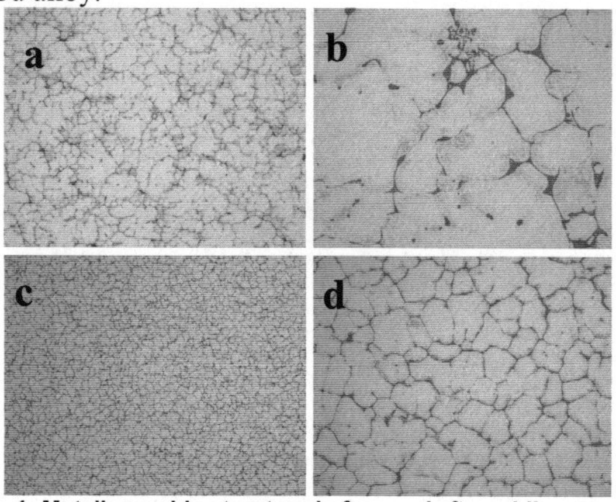

Figure 1: Metallographic structure before and after adding nano refining agent
(a)(b)Metallographic structure of the tested alloys (100x);
(c)(d)Metallographic structure of the tested alloys (500x).

Mechanical property

Based on previous research, 470℃, 24h was selected

Table 2.mechanical properties of the tested alloys

Alloy	State	Tensile strength/MPa	Yield strength/MPa	Elongation/%	Hardness/HB
1#	As-cast	175.5	/	/	114
		177	/	/	116
		172.5	/	/	115
	T6	474	459	1.5	165
		434	420	1	165
		438	417	1.5	164
2#	As-cast	216	/	/	115
		221	/	0.5	116
		215	/	/	115
	T6	588	566	2	181
		624	601	1.5	185
		595	580	2.5	184

as solid solution processes, and 130℃, 24h was selected as aging processes. Table 2 shows the mechanical properties of alloy 1# and 2# in as-cast and T6 states. As shown in the table, the addition of nano refining agents has little effect on the hardness of the alloy under as cast conditions, while the tensile strength of the alloy has increased from 175MPa to 217.3MPa, with an increase of 24.2%. In the T6 state, the addition of nano refining agents significantly improved the mechanical properties of Al-Zn-Mg-Cu alloy. The tensile strength, yield strength, elongation, and hardness of the alloy increased from 448.7MPa, 432MPa, 1.3%, and 164.7HB to 602.3MPa, 582.3MPa, 2%, and 183.3HB, with increases of 34.2%, 34.8%, 50%, and 11.3%, respectively.

4 Conclusion

(1) The addition of TiB_2 and TiC nano refining agents can refine the as-cast Al-Zn-Mg-Cu alloy, with grains changing from dendritic to equiaxed. The average grain size is reduced from 66μm to 20.9μm, and the mechanical properties are improved. The tensile strength is increased from 175MPa to 217.3MPa.

(2) After solid solution treatment at 470℃, 24 h, and aging heat treatment at 130℃ and 24 h, Al-Zn-Mg-Cu alloy with excellent mechanical properties was obtained. The tensile strength, yield strength, and elongation were 602.3MPa, 582.3MPa, and 2%, respectively.

References

[1] Peng Y H, Liu C Y, Wei L L, et al. Quench sensitivity and microstructures of high-Zn-content Al-Zn-Mg-Cu alloys with different Cu contents and Sc addition. Transactions of Nonferrous Metals Society of China, 2021, 31: 24-35.

[2] Xu Y X, Zhu M Z, Fang H Cet al. Effect of Yb microalloying on microstructure and properties of Al-Zn-Mg-Cu alloy. Rare Metal Materials and engineering, 2023, 52(10): 3556-3562.

[3] Liu Q. Microstructure construction and strengthening mechanisms of Al-Zn-Mg-Cu alloy reinforced by trace TiB2+TiC nanoparticles. Jilin: Jilin University, 2021.

Effect of Rare Earth Aluminum Titanium Boron Refiner on the Microstructure and Mechanical Properties of Aluminum Alloys

Wenxin Hu[1,2,*], Zhiqiang Chen[1,2]

1. State Key Laboratory of Baiyunobo Rare Earth Resource Research and Comprehensive Utilization, Baotou 014030, China;
2. Baotou Research Institute of Rare Earths, Baotou 014030, China
*Corresponding address: e-mail: brirehuwenxin@126.com

Abstract: The effect of different proportions of mixed rare earth LaCe and Al-Ti-B composite addition on the microstructure and grain refinement of 6061 aluminum alloy, the existence form of rare earth and its effect on the second phase of the alloy were discussed. The synergistic effect of rare earth and Al-Ti-B on the tensile properties, fracture morphology, and thermal conductivity of the alloy was analyzed. After adding rare earth and Al-Ti-B intermediate alloy, the grain size of the alloy decreased; Rare earths are mainly distributed at grain boundaries in the form of AlFeSiREMg phase and AlSiTiMgRE. In addition, the addition of rare earths promotes the transformation of β - AlFeSi phase to α - AlFeSi phase, reduces the size of Mg2Si phase, forms various complex compounds such as AlFeSi and AlFeSiREMg, and reduces impurities in the iron rich phase at grain boundaries. Compared with 6061 alloy without rare earth elements, 6061 aluminum alloy with 0.05% LaCe and 0.2% Al-Ti-B (mass fraction) intermediate alloy showed an increase in tensile strength, elongation, and thermal conductivity of 15.3%, 80%, and 9%, respectively; At the same time, after the combination of rare earth and Al-Ti-B intermediate alloy, the rough and irregular dimples in the fracture morphology transform into small dimples, and the fracture form is ductile fracture.

Keywords: Al-Ti-B-RE refining agent; refining effect; mechanical properties;

1 Introduction

Addition of grain refiners has become a common industrial practice to obtain microstructure with fine equiaxed grains and thus enhances the quality of a casting . The most widely used grain refiner for Al alloys is the Al-5Ti-1B master alloy, which contains both soluble TiAl3 intermetallic compound and insoluble TiB2 particles in an aluminum matrix. Grain size and morphology of TiAl3 and TiB2 phases have an important influence on the grain refinement efficiency . But, there are some problems for Al-5Ti-1B as a refiner, such as agglomeration of boride, low refining efficiency, and poisoned by certain elements like Zr, V and Cr. All these are related to the morphology and distribution ofTiB2 phase .

2 Experimental procedure

The experimental raw materials are 6061 aluminum alloy semi continuous cast bars produced by Baotou Huize Aluminum Industry and its provided Al-Ti-B intermediate alloy, as well as Al-20LaCe, Al-20La, Al-20Ce intermediate alloys produced by Baotou Rare Earth Research Institute. The effect of rare earth elements is evaluated through grain refinement experiments and SEM testing

3 Result and discussion

The Effect of Rare Earth Elements on the Microstructure of 6061 Alloy

The XRD analysis results of 6061 alloy indicate that there are mainly phases such as α - Al, Mg2Si, AlFeSi in the alloy. After adding LaCe (1:1) composite rare earth, AlSiMgREFe phase was added to the alloy. From the SEM image of the alloy, it can be seen that there are large coarse and thin straight stripes at the grain boundaries of the alloy. The larger coarse strip-shaped phase at the grain boundary is Mg2Si phase, and the elongated straight strip-shaped phase is AlFeSi phase.

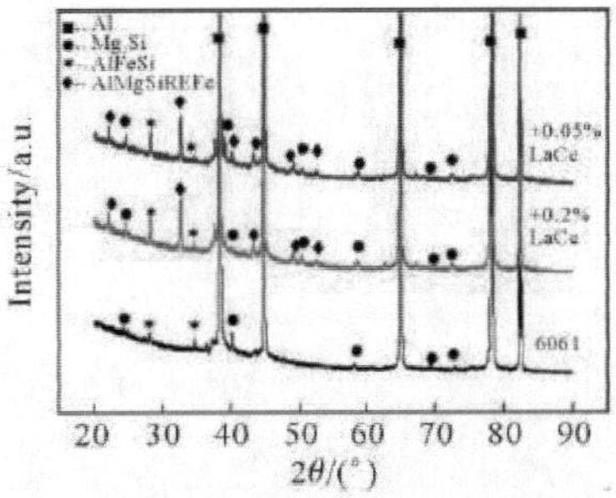

Fig. 1 XRD analysis results of 6061 alloy

The Effect of Rare Earth Elements on the Tensile Strength and Elongation of 6061 Alloy

Many small dispersed phases are formed in the alloy, mainly due to the addition of rare earth elements, which form more high melting point compounds, increase the strengthening effect, and lead to an increase in strength after heat treatment. But with the increase of rare earth content, the number of strengthened Mg2Si particles decreased, forming complex rare earth rich compounds. These compounds are non coherent with the matrix, have high interfacial energy, and are in an unstable state, resulting in a slight decrease in strength.

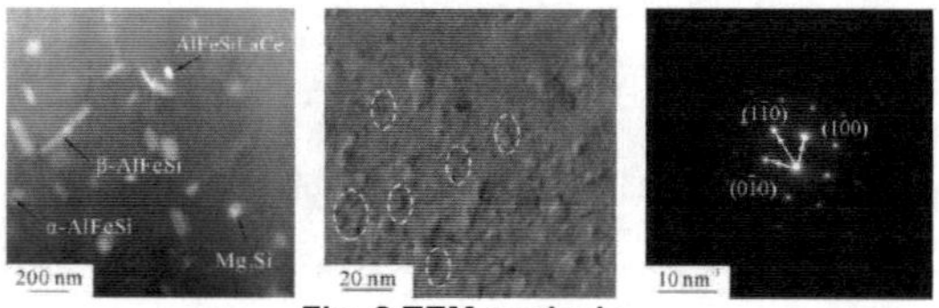

Fig. 2 TEM analysis

The Effect of Rare Earth Elements on the Thermal Conductivity of 6061 Alloy

Aluminum alloy wheels have a great demand for alloy heat dissipation. The frictional heat generated by car braking and the frictional heat between the tire and the road surface can cause the wheel temperature to become too high, increasing the risk of tire blowouts. Therefore, in addition to utilizing structural characteristics in surface design to improve heat dissipation capacity, another way for aluminum alloy wheels is to enhance their own heat dissipation capacity. By improving the heat dissipation capacity of the wheel hub, even in the event of continuous braking during long-distance driving, the wheel system can be ensured to be within a safe range, thereby reducing the risk of tire blowouts and improving safety..

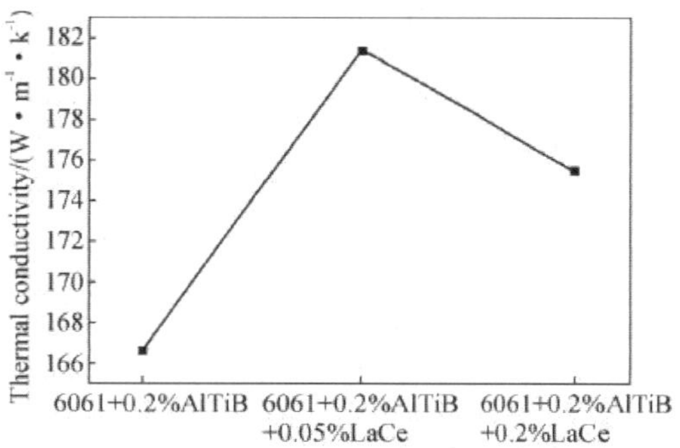

Fig. 3 The Effect of Rare Earth Elements on the Thermal Conductivity of T6 6061 Alloy

4 Conclusion

(1) The effect of rare earths on the thermal conductivity of T6 6061 alloy (1) Different ratios of La/Ce rare earths added separately or mixed with Al-Ti-B have a refining effect on the grain size of 6061 aluminum alloy. But the best refining effect is achieved by the composite of La/Ce=1:1 mixed rare earth and Al-Ti-B intermediate alloy.

(2) Rare earth and Al-Ti-B composite are added to refine 6061 aluminum alloy, and the rare earth is distributed in the form of AlFeSiREMg phase and AlSiTiMgRE at the grain boundaries. In addition, the addition of rare earths promotes the transformation of β - AlFeSi phase to α - AlFeSi phase, reduces the size of Mg2Si, forms various complex compounds such as AlFeSi and AlFe SiREMg, and reduces impurities in the iron rich phase at grain boundaries.

(3) The refinement of rare earth and Al-Ti-B composite improves the tensile strength, elongation, and thermal conductivity of the alloy. The tensile strength of 6061 aluminum alloy cast rod with 0.05% LaCe (1:1)+0.2% Al-Ti-B added reached 198.5 MPa, and the elongation rate reached 23.74%. After extrusion heat treatment, the tensile strength reached 369 MPa and the elongation rate was 18.3%.

(4) The addition of rare earth elements plays an important role in improving the fracture surface. At the same time, after adding rare earth and Al-Ti-B intermediate alloy to 6061 aluminum alloy, the rough and irregular dimples appearing in the fracture morphology transformed into small dimples, and the fracture form was ductile fracture.

Development and Application of Rare Earth in Magnesium Alloy

Qiyu Liao[1,2], Qichi Le[1,2,*]
1. School of Materials Science and Engineering, Northeastern University, Shenyang 110819, China
2. Key Laboratory of Electromagnetic Processing of Materials, Ministry of Education, Northeastern University, Shenyang 110819, China
*Corresponding address: e-mail: mail@mail.com

Abstract： There are currently a series of "bottleneck" difficulties and pain points in magnesium alloy, such as the high chemical activity and poor plastic deformation ability, which greatly affect their melting and casting preparation and stable forming. In addition, the insufficient strength and poor heat resistance of magnesium alloys also limit their application in specific service environments. Rare earth elements, as alloying elements, have the reputation of "industrial MSG" in magnesium alloys and play an extremely important role in improving the performance of magnesium alloys. The development of high-performance magnesium alloys mostly relies on rare earth additives. The development and application of magnesium alloy materials is also a highly valued field for the comprehensive utilization of rare earth resources. The close combination and mutual promotion of magnesium and rare earths are the inevitable path for the development of the industry. This article focuses on the research and application of rare earth magnesium alloys, and further explores new ideas for the comprehensive utilization of rare earths.

Keywords: rare earth; magnesium alloy; high performance; research and development; application

1 Introduction

Currently, China's rare earth industry is facing the problem of low utilization rate of light rare earths and serious imbalance in the amount of light and heavy rare earths used. Magnesium and rare earth are both important advantageous minerals in China and valuable resources with significant strategic value. However, there are currently a series of "bottleneck" difficulties and pain points in the preparation and application of magnesium alloy materials, such as the high chemical activity and poor plastic deformation ability of magnesium alloys, which greatly affect their melting and casting preparation as well as stable forming; In addition, the insufficient strength and poor heat resistance of magnesium alloys also limit their application in specific service environments.

Rare earth elements, as alloying elements, play an extremely important role in improving the properties of magnesium alloys. Rare earth addition has significant effects on inhibiting high-temperature oxidation and combustion of magnesium alloy melt, improving the purity of magnesium melt, refining solidification structure, weakening deformation texture, etc. It can greatly improve the formability, mechanical, corrosion resistance, and high-temperature performance of magnesium alloy. The development and application of magnesium alloy materials is also a highly valued field for the comprehensive utilization of rare earth resources. The close combination and mutual promotion of magnesium and rare earths are the inevitable path for the development of the industry. This article focuses on the work of our research group in the development of rare earth magnesium alloys.

2 Experimental procedure

This article aims to improve the comprehensive properties of magnesium alloys, including mechanical properties, corrosion resistance, and heat resistance, by adding rare earth elements such as La, Ce, Y, Nd and Gd to different magnesium alloys. The preparation of casting samples is first carried out by melting the alloy in a small resistance furnace in the laboratory, and then poured into a 3kg metal circular mold to form a casting billet. The preparation of homogenized samples involves subjecting the ingot to high-temperature and long-term heat treatment in a muffle furnace. The billet size for the extrusion experiment is Φ46×10mm, the extrusion speed is 1mm/s, and the extrusion ratio is 17.

3 Result and discussion

The influence of La and Ce on the microstructure and mechanical properties of AZ alloy

After hot extrusion deformation, the tensile strength UTS of AZ110-1.5LC alloy is greater than 370MPa, the yield strength YS is greater than 260MPa, and the elongation at break δ is greater than 15% [1]. Meanwhile, due to its high m value, small and equiaxed grains, weak basal texture, and GBS deformation mechanism, the alloy can achieve low-temperature superplasticity.

The influence of Ce on the microstructure and mechanical properties of Mg-Zn alloy

Based on the coordinated deformation of basal texture grains, a high plasticity and high energy absorption Mg-2Zn-0.4Ce-0.4Mn alloy was designed, which significantly refined the microstructure and weakened the texture [2]. This alloy has good plasticity, and after extrusion, the elongation can reach 33%, exhibiting superior cold bending performance compared to AZ31.

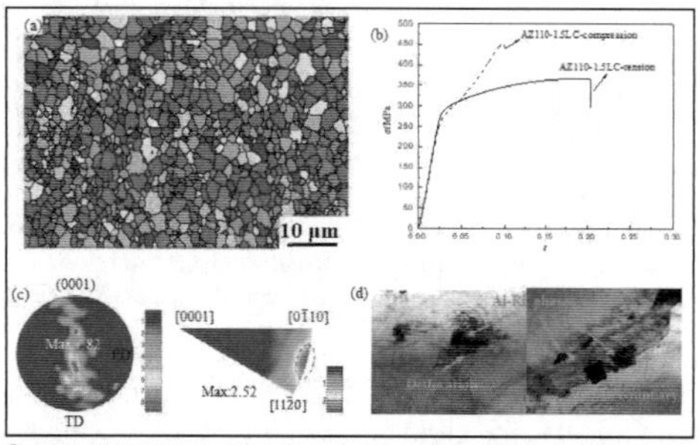

Fig. 1 Grain structure, mechanical properties, texture, and rare earth phases of extruded AZ110-1.5LC alloy [1]

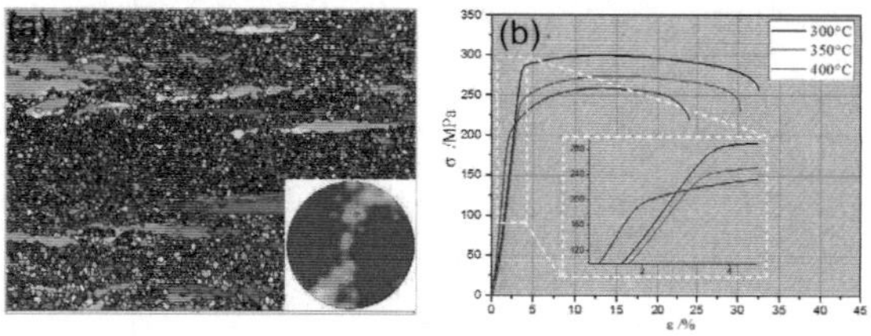

Fig. 2 Microstructure and stress-strain curves of Mg-Zn-Mn-La-Ce alloy[2]

Developed Mg-2Zn-1Al-Ca-Gd magnesium alloy capable of low-temperature extrusion

ZA21EX has excellent comprehensive mechanical properties, with a wide temperature range for thermal/hot extrusion deformation. This is because the decrease in extrusion temperature refines and disperses the Al_2Ca phase, promotes DRX occurrence, weakens the basal texture, and hinders grain growth [3].

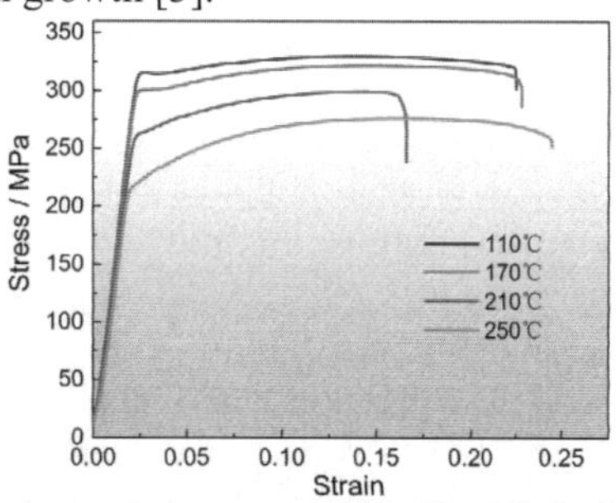

Fig. 3 The stress-strain curves of Mg-2Zn-1Al-Ca-Gd alloy[3]

The influence of Y/Nd on the high-temperature oxidation kinetics behavior of AZ alloy

Both Y and Nd can reduce and refine β-$Mg_{17}Al_{12}$ in AZ80, and precipitate Al_2Y phase and Al_2Nd, increasing the oxidation activation energy, significantly reducing the oxidation rate of the alloy, and improving the oxidation film-forming properties of the alloy [4]. The oxidized surface becomes smooth and dense.

Fig. 4 Macroscopic morphology after 6 hours of high-temperature oxidation in air[4]

4 Conclusion

Rare earth elements can greatly improve the comprehensive performance of magnesium alloys and provide strong support for their application in structural components. At present, there has been in-depth research on the design and development of rare earth magnesium alloys. However, the large-scale application of rare earths in magnesium alloys in China is severely limited by the high material and preparation costs of rare earth magnesium alloys. Therefore, it is necessary to focus on the entire industry chain from rare earth extraction to alloy preparation and processing, and develop new processes and methods aimed at achieving efficient and low-cost preparation of high-performance rare earth magnesium alloys.

5 Acknowledgments

This research was financially supported by the National Natural Science Foundation of China (No.52274377 and No.52304391).

References

[1] Liao Q Y, Hu W X, Le Q C, et al. Improvement of yield asymmetry and enhancement of mechanical properties of extruded AZ110 alloy with La-rich Misch Metal addition. Met. Mater. In. 2022, 5: 1143-1156.

[2] Hu K, Liao Q Y, Li C Y, et al. High ductility induced by un-DRXed grains in a Mg-Zn-Mn-La-Ce alloy, J. Mater. Sci., 2019, 54: 10902-10917.

[3] Niu Y X, Song Z T, Le Q C, et al. Excellent mechanical properties obtained by low temperature extrusion based on Mg-2Zn-1Al-0.2Mn-0.2Ca-0.2Gd alloy, J alloy. Compd., 2019,801: 415-427

[4] Cheng C L, Li X Q, Le Q C, et al. Effect of REs (Y, Nd) addition on high temperature oxidation kinetics, oxide layer characteristic and activation energy of AZ80 alloy, J Magnes. Alloy., 2020, 8: 1281-1295.

Effect of Gd on Interfacial Microstructure and Mechanical Properties of Mg/Al Bimetallic Composites by Compound Casting

Wenming Jiang*
State Key Lab of Materials Processing and Die & Mould Technology, Huazhong University of Science and Technology, Wuhan 430074, PR China
*Corresponding address: e-mail: wmjiang@hust.edu.cn

Abstract: To strengthen the bonding properties of Mg/Al bimetallic composites, rare earth Gd was added to the Mg/Al bimetallic composites manufactured by a novel compound casting, and the interfacial microstructure, mechanical properties and strengthening mechanism of the Mg/Al bimetallic composites with Gd addition were studied. Without Gd addition, there was a thin oxide film consisting of an alumina phase in the Mg/Al bimetallic interface, which hindered the diffusion of Mg2Si phase. After adding the Gd, the oxide film in the interface was broken and dissolved, significantly increasing the diffusion distance of the Mg2Si phase. The Al2Gd and Al8Mn4Gd precipitated in the interface. The primary Al12Mg17 phase and eutectic structure in the interface were greatly refined due to the heterogeneous nucleation and the inhibition of growth caused by the Mg2Si phase and Gd-rich precipitates. The interface was strengthened under the comprehensive effect of the fine grain strengthening and precipitation strengthening of the Mg2Si and Gd-rich precipitates.

Keywords: Gd; Mg/Al bimetallic composites; Compound casting; Interfacial microstructure; Mechanical properties; Strengthening mechanism

1 Introduction

Rare earth elements have strong metallicity, so they can easily combine with other metal elements to form RE precipitates. These precipitates have the functions of precipitation strengthening and can serve as the heterogeneous nucleation substrate to refine the microstructure of the metals. Besides, the rare earth elements can react with oxygen, sulfur, and other impurities in the molten metal to purify the melt and improve the properties of the materials. So, the rare earth elements are often used to modify various metal materials.

Heavy rare earth elements, such as Gd, have a high melting point and are often used to strengthen the mechanical properties and high-temperature resistance of the magnesium alloys. The application and research of the Gd mainly focus on its effect on the microstructure and properties of the single alloy. The investigation of the impact of the Gd on the bimetallic interface has not been reported yet. [1, 2].

2 Experimental procedure

The A356 aluminum alloy (Al-6.81Si-0.44Mg-0.21Fe-0.02Ti wt.%) was selected as the solid insert. AZ91D-xGd (x=0, 0.6) were prepared as the melt to cast around the solid insert. Foam patterns were prepared using expanded polystyrene (EPS) material with a density of 0.012g/cm^3. The schematic illustration of the lost foam compound casting experimental equipment in this work is illustrated in Fig. 1.

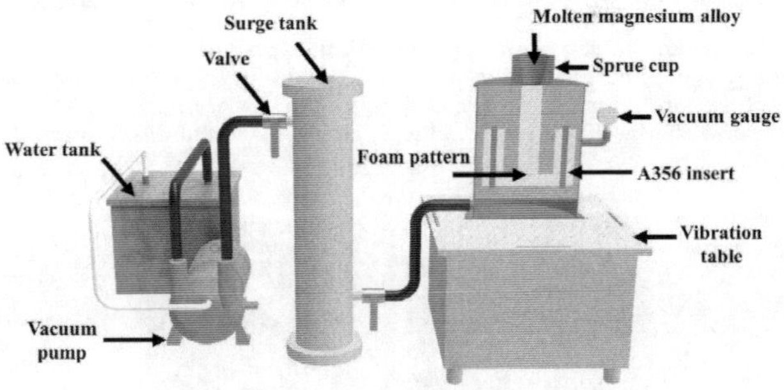

Fig. 1. The schematic diagram of the experiment.

The AZ91D magnesium alloy (Mg-9.08Al-0.62Zn-0.23Mn wt.%) was melted in a resistance furnace under CO_2 and SF_6 mixture gas protection. When the AZ91D melt reached the temperature of 740°C, the Mg-30Gd master alloy was added to the melt. After the Mg-30Gd master alloy was completely melted, the melt was thoroughly stirred. The molten metal was poured when it reached the pouring temperature of 720 °C, and the vacuum value in the sand flask was adjusted to 0.03 MPa. The microstructure and the mechanical properties of the Al/Mg interfaces were investigated.

3 Result and discussion

3.1 Microstructure of the Mg/Al bimetallic composites

Fig. 2 (a) shows the interfacial morphology of the Mg/Al bimetal composite without Gd addition. There are many black precipitate phases in the intermetallic compounds (IMCs) region. Plenty of coarse dendrites that grow perpendicular to the interface can be observed in the Eutectic region. In the E-IMCs region, a thin black boundary is found. Fig. 2(b-d) shows the distributions of O, Si, and Al elements in the Mg/Al bimetallic interface.

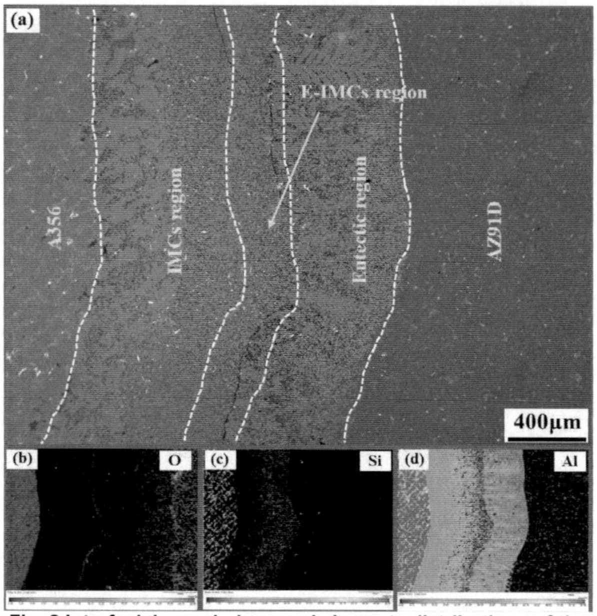

Fig. 2 Interfacial morphology and elements distributions of the Mg/Al bimetallic composite without Gd addition:(a) microstructure of the Mg/Al bimetallic composite; (b) the distribution of the O element; (b) the distribution of the Si element; (d) the distribution of the Al element.

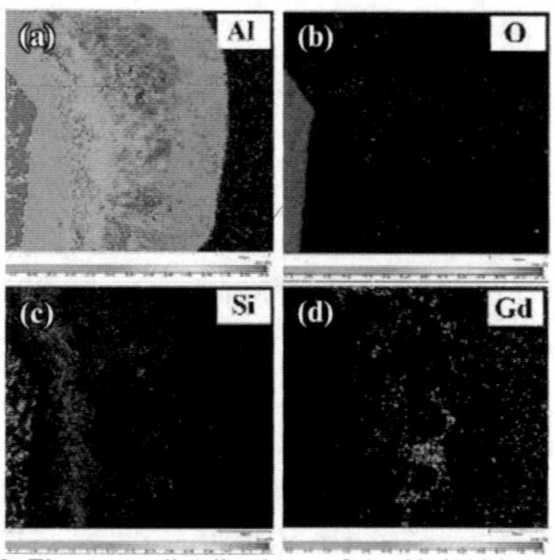

Fig. 3. Elements distributions of the Mg/Al bimetallic composites with Gd.

Fig. 2(b) indicates a region of oxygen enrichment in the middle of the interface. It can be found out that the morphology of that region is similar to the black boundary in the transition zone. The Si element in the interface is mainly concentrated in the range of the IMCs. In addition, a transparent edge is detected in the interface region with the Si element, and the position of this edge is highly coincident with the oxygen element enrichment film in Fig. 2(b). Fig. 2(d) displays that the distribution of the Al element in the IMCs is not uniform. As shown in Fig. 3, the diffusion distance of the Si element significantly increases. The apparent oxygen element enrichment film disappears.

3.2 Mechanical properties

Fig. 4 indicates the shear test results for the Mg/Al bimetallic composites without and with Gd addition. With Gd addition, the shear strength of the Mg/Al bimetallic composites reaches 46 MPa, which is 39% higher than that of without Gd.

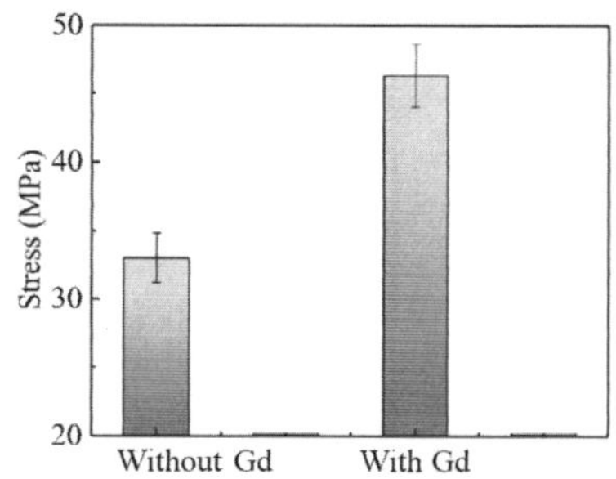

Fig. 4. Shear test results of the Mg/Al bimetallic composites without and with Gd.

Conclusion

1.Under the condition of no Gd, there was a thin oxide film in the Mg/Al bimetallic interface. The alumina phase was confirmed in the oxide film. With Gd, the oxide film was removed, and the primary $Al_{12}Mg_{17}$ in the E-IMCs region and the eutectic groups in the Eutectic region were refined.

2.The average shear strength of the Mg/Al bimetallic composites with Gd reached 46 MPa, which was 39% higher than that of the group without Gd addition

Acknowledgments

This work was financially supported by the National Natural Science Foundation of China (Nos. 52271102 and 52075198), and the Analytical and Testing Center, HUST.

References

[1] Gaun F, Jiang W, Fan Z. Interfacial microstructure, mechanical properties and strengthening mechanism of Mg/Al bimetallic composites produced by a novel compound casting with the addition of Gd. Mater. Character. 2023,200:112898.

[2] Gaun F, Jiang W, Fan Z. Significantly Enhanced Mg/Al Bimetallic Interface by Compound Casting via Combination of Gd Addition and Vibration. Metall. Mater. Trans. A 2023,54A:3389.

This page intentionally left blank.

Effect of Mg on Microstructure and Mechanical Properties of the Novel Al-Based Composite for Thermal Neutron Shielding

Yigeng Ding[1], Zhiwei Liu[1,*]
1. State Key Laboratory for Mechanical Behavior of Materials, School of Materials Science and Engineering, Xi'an Jiaotong University, No. 28, Xianning West Road, Xi'an, Shaanxi Province 710049, P.R. China
2. *Corresponding address: e-mail: liuzhiwei@xjtu.edu.cn

Abstract: The structural materials used for neutron shielding are typically fabricated as thin-walled grillage structures, which require a combination of superior neutron shielding capacity and acceptable mechanical properties. A novel (5 wt.% B_4C_p+4.5 wt.% Gd)/Al-5Cu neutron shielding composite which exhibited promising commercial potential was designed based on equivalent B areal density (EBAD) calculations. Additionally, the influence of Mg on microstructure and mechanical properties was investigated. The primary phase of the composite was τ-Al_8Cu_4Gd, which altered from a lamellar to a massive structure along with the addition of Mg, and the Mg was completely dissolved in the Al matrix and Al_3Gd. After hot rolling, the large-size τ-Al_8Cu_4Gd phase was broken into smaller particles, resulting in a more homogeneous dispersion within the matrix. This fragmentation process proved advantageous for enhancing both the strength and ductility of the composites. The addition of Mg significantly increased both the yield strength (61.7%↑) and ultimate tensile strength (42.9%↑) after hot rolling, which was due to a synergistic effect of refinement of second phase and solution strengthening. The study presents a high-performance Al-based composite with outstanding casting ability containing biphasic neutron-absorbing elements.

Keywords: Aluminum alloy; Mechanical properties; Casting; Neutron shielding materials

1 Introduction

High abundance of B^{10} makes B_4C_p/Al composite suitable for the ideal neutron radiation shielding materials [1]. However, B_4C_p is a brittle phase, which introduced excessively led to the decline of the deformability and plastic toughness of the material [2]. Therefore, there are many researchers have introduced the Gd element, which has a higher neutron absorption cross-section (48,800 b), to partially replace B_4C_p. Chen et al. [3] developed 6TiB_2/Al-6Mg-1Gd composite wich the UTS and El were 464 MPa and 15.6%, respectively. The composite had outstanding mechanical properties, but its absolute neutron shielding property was below the standard. Chen et al. [4] fabricated (10B_4C_p+3.6Gd)/6061Al by ultrasound-assisted casting, the UTS and El of as-cast composite were only 117.8 MPa and 0.8%, respectively, which was still lack of ductility.

In consideration of the trade-off between shielding phase content and properties, adding elements for alloying is an undeniably effective method. Mg has a notable solution-strengthening effect in Al, because of high solid solubility in Al and 13% larger than Al at atomic radius, which introduced lattice distortion [5]. This work designed and fabricated high Gd content (4.5 wt.%) composites and investigated the influence of Mg element on the microstructure and corresponding mechanical properties of as-cast and rolled 5B_4C_p/Al-5Cu-4.5Gd composites.

2 Experimental procedure

According to the formula of B_{Eq} [6] ,when the B_4C content was determined to be 5%, the Gd content can be calculated to be 4.5%. The B_4C_p/Al-5Cu-4.5Gd-xMg (x=0, 2) composites were fabricated by using pure Al (99.7%), Cu (99.9%), Mg (99.9%), Al-30Gd master alloys and Al-10B_4C_p (Al-K_2TiF_6 system [7]) master composites as raw materials. Al-30Gd master alloys, Al-10B_4C_p master composites, pure Al, and pure Cu were initially melted at 750 ℃ for 20 min within a graphite crucible in a high-frequency induction furnace. After that, Mg was added to the melt, and mechanical stirring was introduced into the melt with a rotating speed of 380 r/s for 5 min. Finally, the composite melt was poured into a graphite mold to obtain an ingot. For simplicity, the B_4C_p/Al-5Cu-4.5Gd composite and the B_4C_p/Al-5Cu-4.5Gd-2Mg composite were referred to as the 0Mg composite and the 2Mg composite, respectively.

The phases and microstructure of samples were examined by X-ray diffractometer (XRD, Bruker D8 ADVANCE) and scanning electron microscopy (SEM, Sigma300). Flat specimens (gauge length×width×thickness: 20×4×3.5 mm) were tensile tested by using a testing machine (CMT5305, MTS, USA) equipped with an extensometer at a constant cross-head speed of 0.5 mm/min.

3 Result and discussion

Phase analysis and microstructure

The XRD patterns of as-cast 0Mg and 2Mg composites were shown in Figure 1. The primary peaks except for α-Al correspond to $τ_1$-Al_8Cu_4Gd (marked by orange verticals) and Al_3Gd phase. No peaks of Mg-contain were found, which indicated Mg existed as a solid solution, it was also suggested by the leftward shift of the peak of α-Al. The continuous net-like phase was proved to be α-Al+$τ_1$ eutectic structure by the EDS analysis, as shown in Table 1. Mg solubilized mainly in α-Al matrix and $τ_1$, consistent with the XRD results. Perceptibly, Mg addition altered the precipitation behavior of τ_1, due to Gd can reduce the diffusion coefficient of Cu and Mg in Al, the Mg and Cu elements were segregated on the surface of the τ_1 phase at the later period of precipitation of eutectic τ_1, resulted in overcooling of the composition, which inhibited the growth of the τ_1 phase, resulting in the transformation of the τ_1 phase from a lamellar to a massive structure (Figure 2(a-b)). The continuous net-like τ_1 phase was broken into fine particles aligned along rolling direction after hot rolling, as shown in Figure 2(c-d).

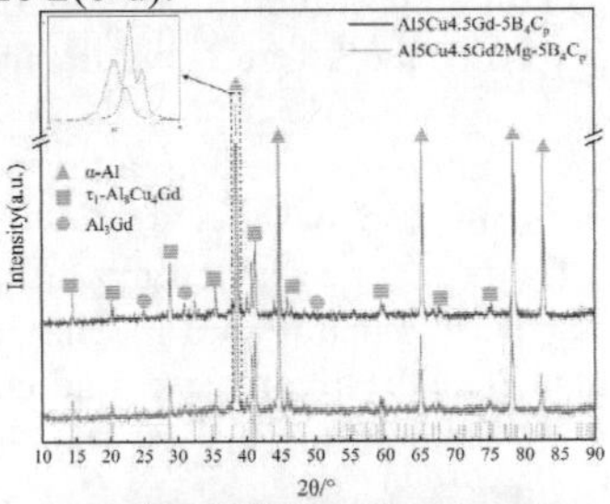

Fig.1 XRD patterns

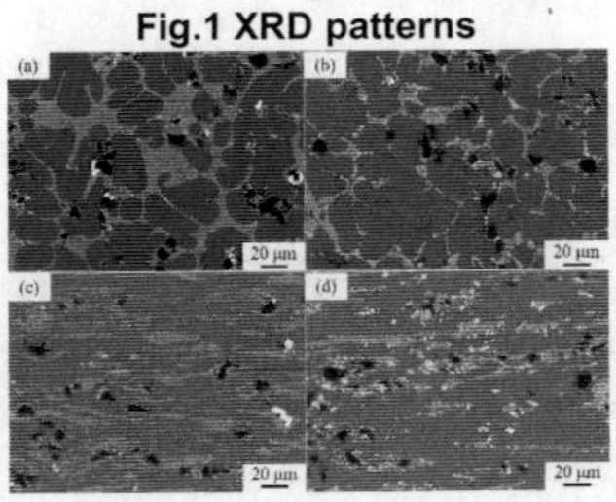

Fig.2 BSE micrographs of the 0Mg ((a)as-cast, (c)as-rolled) and 2Mg ((b)as-cast, (d)as-rolled) composites

Table1 EDS results of point in Figure 2 (a-b)

Point		Al	Cu	Gd	Mg
Fig. 2a	A	99.27	0.73	0.00	0.00
	B	71.31	0.00	28.69	0.00
	C	88.31	9.29	2.41	0.00
	D	60.45	31.65	7.90	0.00
Fig. 2b	A	98.09	0.40	0.00	1.51
	B	70.8	0.00	29.20	0.00
	C	88.89	8.92	2.19	0.89
	D	62.13	30.55	7.32	0.00

Tensile properties and fracture fractography

Figure 3 exhibited the typical strain-stress curves and corresponding tensile properties of as-cast and as-rolled B_4C_p/Al composites with and without Mg. For all curves, significant Portevin-Le Chatelier (PLC) effects were observed, indicating a high content of solute atoms. With the addition of Mg, YS and the UTS increased while the El decreased. The improvement of YS and UTS was attributed to solution strengthening and refinement of τ phase. According to the fracture fractography, large and shallow dimples with lamellar τ_1 phase inside were observed in as-cast 0Mg composite (Figure 4(a-b)), indicating a mixed ductile-brittle fracture. While the cleavage planes were observed (Figure 4(c)) and demonstrated to be a massive τ_1 phase by BSE diagrams (Fig. 4(d)), in as-cast 2Mg composite, which led to brittle fracture. The improvement of comprehensive mechanical properties of the composites with and without Mg after hot rolling was attributed to the broken of τ_1 phase (Figure 4(e-h)), which diminished stress concentration. The as-rolled 2Mg composite had excellent comprehensive tensile properties, the YS and UTS were 217 MPa and 284 MPa, respectively, which increased by ~76 MPa and ~74 MPa over as-rolled 0Mg composite.

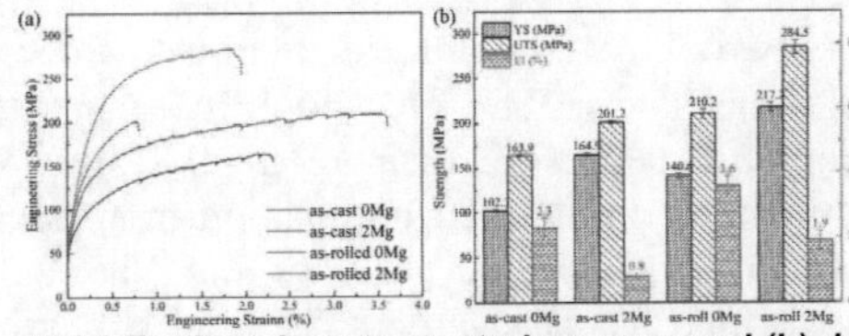

Figure 3 (a) Engineering stress-strain curves and (b) detailed mechanical properties.

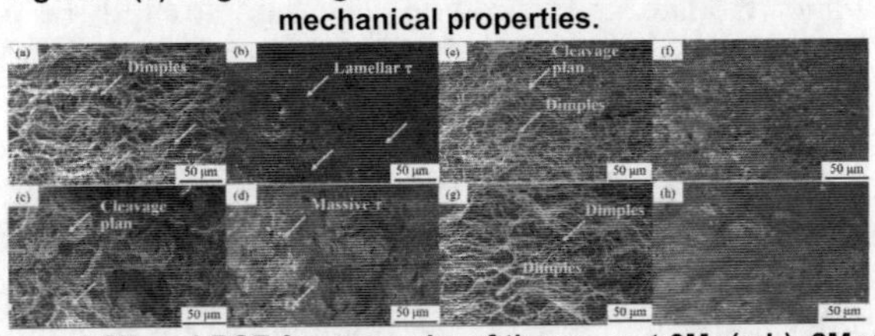

Figure 4 SE and BSE fractography of the as-cast 0Mg(a-b), 2Mg(c-d), as-rolled 0Mg(e-f), and 2Mg(g-h) composites.

4 Conclusion

In the present work, a B_4C_p/Al-5Cu-4.5Gd-xMg (x=0, 2) composite material with high neutron shielding ability and desirable mechanical properties was designed and prepared by the stir casting method, and the effect of the addition of Mg on the microstructure and properties of the composites was investigated. The addition of Mg did not form a new second phase, instead, it was dissolved in the Al matrix and the τ_1 phase. The τ_1 phase altered from reticular to massive because Mg reduced the diffusion coefficient and inhibited the growth of the τ_1 phase. The addition of Mg improves the YS and UTS due to solid solution strengthening and the refinement of the second phase, while hot rolling improves the properties as a result of the broken second phase which reduces the stress concentration.

Acknowledgments

The authors thank the National Natural Science Foundation of China (Grant Nos. 52174372, 51974224 and 51604211) for supporting the work.

References

[1] Zhang. P, et al. J. Nucl. Mater, 2013, 437: 350–358.
[2] Chen H.S, et al. Mater. Des. 2016, 94: 360–367.
[3] Chen. Y, et al. Intermetallics 2022,148: 107630.
[4] Chen. M, et al. Mater. Sci. Eng. A 2022, 861: 144376.
[5] Xue. L.W, et al. Mater. Sci. Eng. A 2024, 908: 146775.
[6] Xu. Z.G, et al. Mater. Des. 2016, 111: 375–381.
[7] Chen. M, et al. J. Alloys Compd. 2021, 858: 157659 .

Composition Design, Property Optimization and Additive Manufacturing of High-Strength Invar Alloy

Fang Lin, Jie Wan*, Yuzeng Chen*
State Key Laboratory of Solidification Processing, Northwestern Polytechnical University, Xi'an, Shaanxi 710072, People's Republic of China
*Correspondence: wan@nwpu.edu.cn (Jie Wan); yzchen@nwpu.edu.cn (Yuzeng Chen)

Abstract: Invar alloys, which have an ultra-low coefficient of thermal expansion (CTE), are widely used in precision instrumentations. However, their relatively low strength poor machinability limits their application in wider fields. Introducing diffused carbide particles into the Fe-Ni-Co Invar alloy, which has a negative CTE, can strengthen the alloy and regulate its CTE value. The high-strength Fe-Ni-Co Invar alloy was designed and fabricated via selective laser melting (SLM). Mechanical properties and CTE for designed alloy were measured using electronic universal tensile testing machine and thermal expansion analyzer, respectively. Results showed that the designed alloy demonstrates a low CTE and significantly improved strength compared with the traditional Invar alloys. Strengthening mechanism is being investigated utilizing SEM and TEM.

Keywords: Invar alloys; Selective laser melting; Strengthening mechanism

1 Introduction

Inva36 is widely used in precision components such as space telescopes, reference scales and arms of precision balances for its low CTE under Curie temperature[1]. However, its low yield strength (300MPa) limits its application in wider fields[2]. The excellent plasticity of Invar36 also make it difficult and inefficient to form complex parts by conventional manufacturing methods.

Selective laser melting (SLM) can realize the near-net forming of metal parts, which provides a new favorable way for the efficient forming of Invar36 parts. The process parameters, mechanical properties and CTE of Invar36 fabricated via SLM have been deeply studied. It was reported that the as-built Invar36 samples show a high relative density when the laser energy density is between 90-110J/mm^3[3, 4]. SLM-Invar36 shows a higher yield strength (YS, 300-400MPa) than conventionally manufactured specimens [4, 5]. While the CTE of SLM-Invar36 is lower than that of conventionally manufactured counterpart[6].

Conventionally, carbides were introduced into Invar36 to enhance its YS and ultimate tensile strength (UTS)[7, 8]. But the addition of these elements causes an increase of the CTE. It is reported that replacing Ni with Co in Invar36 can make its CTE lower, showing obvious negative expansion [9]. For high-precision parts, neither expansion nor negative expansion is expected. Therefore, the simultaneous introduction of Co and strengthening elements into Invar36 is expected to improve the strength while maintaining its low CTE.

In this work, Fe-Ni-Co-V Invar alloy was designed, with V and C elements added and with 4 wt.% Co instead of Ni element in Invar36. The dense samples were obtained by SLM. The thermal expansion and mechanical properties of the as-built Fe-Ni-Co-V Invar were studied. The effects of heat treatment on the properties of the designed alloy were also investigated.

2 Experimental procedure

Nitrogen atomized Fe-Ni-Co-V Invar powder were used in this study. Table 1 presents the composition of the alloy powder. The specimens were fabricated using a BLT-S320. In this study a laser power of 300W, a scanning speed of 700mm/s, a hatch spacing of 0.15mm and a powder layer thickness of 30μm were used to fabricate specimens. In order to obtain nano-carbides, some specimens were aged at 650℃ for 3h.

Table 1 Composition of Fe-Ni-Co-V Invar powders (wt.%).

Elements	Ni	Co	V	C	Fe
Fe-Ni-Co-V Invar	32.10	3.47	1.80	0.08	Bal.

3 Result and discussion

Microstructure

Fig.1(a) and (b) display the metallography of as-built and aging treatment (AT) Fe-Ni-Co-V Invar, respectively. It can be observed that the angle between scanning tracks in different layers was about 67°, consistent with the scanning strategy (Fig. 1(a)). After aging treatment, the grains arranged along the scanning tracks in the specimens (Fig. 1(b)). There is no obvious change in grains size after aging treatment (30μm on average). It is obvious in Fig. 1(c) that the as-built sample is full of cellular structures with a high density of dislocations accumulated along the cell boundaries. Diffused carbide particles can be observed in the specimen after aging treatment, with dislocation cells existing (Fig. 1(d)).

Fig. 2 shows tensile stress-strain curves of as-built and AT Fe-Ni-Co-V Invar specimens. Obviously, AT can improve the strength of the alloy: as-built specimen possessed the YS of 495 MPa, while AT specimens demonstrated a higher value of 594MPa.

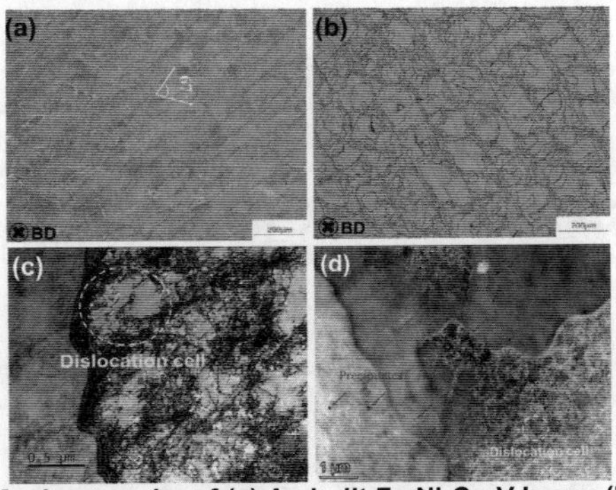

Fig. 1 OM micrographs of (a) As-built Fe-Ni-Co-V Invar; (b) AT Fe-Ni-Co-V Invar and TEM micrographs of (c) As-built Fe-Ni-Co-V Invar; (d) AT Fe-Ni-Co-V Invar.

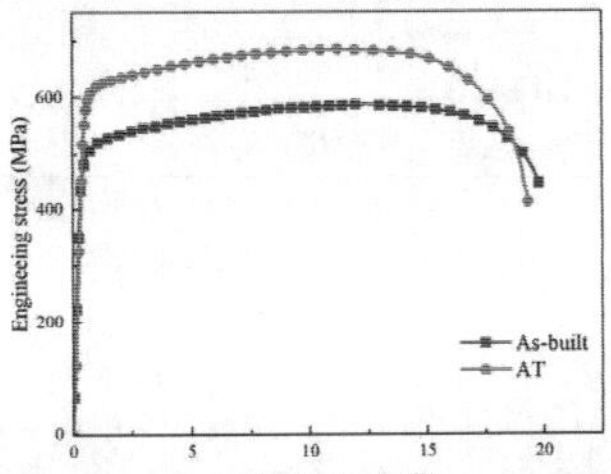

Fig. 2 Tensile engineering stress-strain curves of As-built and AT Fe-Ni-Co-V Invar specimens.

Tensile behavior

The YS of as-built Fe-Ni-Co-V Invar is significantly higher than that of Invar36 fabricated via SLM, which may be due to the solid solution strengthening caused by the addition of V and C. After AT, the strength is further enhanced because of the precipitation of carbide.

Thermal expansion properties

Fig. 3(a) displays the change in length of as-built and AT specimens as a function of temperature. AT has no obvious effect on the CTE of Fe-Ni-Co-V Invar fabricated via SLM. This may due to the release of residual stress and precipitation of carbide. Fig. 3(b) shows the absolute values of CTE (|CTE|) of different alloys at different temperature ranges. In the low temperature range (30-100℃), Fe-Ni-Co-V Invar shows a low |CTE|, which is similar to Invar36. While Invar36 has a lower |CTE| at the high temperature range (30-200℃).

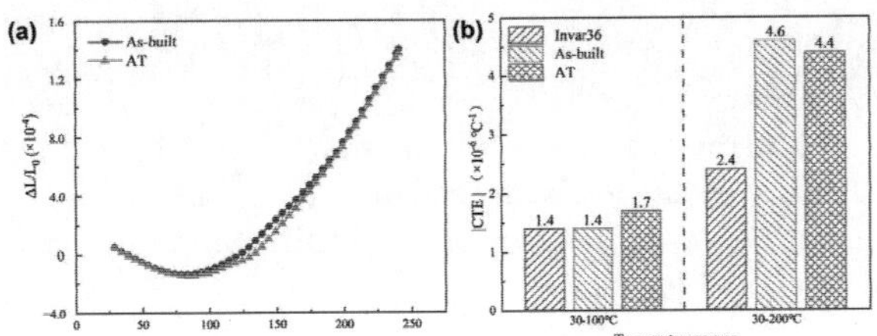

Fig. 3 (a) Thermal expansion curves of As-built and AT Fe-Ni-Co-V Invar; (b) absolute value in different temperature ranges of Invar36, As-built and AT Fe-Ni-Co-V Invar.

4 Conclusion

The as-built Fe-Ni-Co-V Invar has significantly higher strength than Invar36, and aging treatment can further enhance its strength without affecting its thermal expansion properties. Fe-Ni-Co-V Invar demonstrates a low absolute value of CTE at low temperature.

5 Acknowledgments

This work was supported by the National Key Research and Development Program of China (Nos. 2022YFB3404203) and the Natural Science Basic Research Program of Shaanxi Province, China (No. 2023-JC-QN0421).

References

[1] M. Yakout, A. Cadamuro, M.A. Elbestawi, S.C. Veldhuis, The selection of process parameters in additive manufacturing for aerospace alloys, The International Journal of Advanced Manufacturing Technology 92(5-8) (2017) 2081-2098.

[2] H. Liu, Z. Sun, G. Wang, X. Sun, J. Li, F. Xue, H. Peng, Y. Zhang, Effect of aging on microstructures and properties of Mo-alloyed Fe–36Ni invar alloy, Materials Science and Engineering: A 654 (2016) 107-112.

[3] M. Yakout, M.A. Elbestawi, S.C. Veldhuis, Density and mechanical properties in selective laser melting of Invar 36 and stainless steel 316L, Journal of Materials Processing Technology 266 (2019) 397-420.

[4] K. Wei, Q. Yang, B. Ling, X. Yang, H. Xie, Z. Qu, D. Fang, Mechanical properties of Invar 36 alloy additively manufactured by selective laser melting, Materials Science and Engineering: A 772 (2020).

[5] Q. Yang, S. Yang, S. Ma, J. Zhou, Y. Zhou, R. Huang, K. Wei, Z. Qu, X. Yang, In-situ X-ray computed tomography tensile tests and analysis of damage mechanism and mechanical properties in laser powder bed fused Invar 36 alloy, Journal of Materials Science & Technology 175 (2024) 29-46.

[6] N.J. Harrison, I. Todd, K. Mumtaz, Thermal expansion coefficients in Invar processed by selective laser melting, J Mater Sci 52(17) (2017) 10517-10525.

[7] K. Nakama, K. Sugita, Y. Shirai, Effect of MC Type Carbides on Age Hardness and Thermal Expansion of Fe–36 wt%Ni–0.2 wt%C Alloy, Metallography, Microstructure, and Analysis 2(6) (2013) 383-387.

[8] Q.S. Sui, J.X. Li, Y.Z. Zhai, Z.H. Sun, Y.F. Wu, H.T. Zhao, J.H. Feng, M.C. Sun, C.L. Yang, B.A. Chen, H.F. Peng, Effect of alloying with V and Ti on microstructures and properties in Fe–Ni–Mo–C invar alloys, Materialia 8 (2019).

[9] X. Rao, Y. Ru, F. Guo, Y. Ren, L. Cui, K. Yu, Abnormal strain evolution of austenite upon martensitic transformation in FeNiCo alloy with invar effect: An in situ study, Scripta Materialia 220 (2022).

Competition Improvement of Energy Storage Between In-Plane Domains and Domain Walls in $K_{1-x}Na_xNbO_3$ (KNN) Thin Films Using High-Throughput Phase-Field Methods

Jinming Cao[1], Hua Hou[1,3], Yuhong Zhao[1,2,4*]
1 School of Materials Science and Engineering, Collaborative Innovation Center of Ministry of Education and Shanxi Province for High-performance Al/Mg Alloy Materials, North University of China, Taiyuan, 030051, China
2 Beijing Advanced Innovation Center for Materials Genome Engineering, University of Science and Technology Beijing, Beijing, 100083, China
3 School of Materials Science and Engineering, Taiyuan University of Science and Technology, Taiyuan, 030024, China
4 Institute of Materials Intelligent Technology, Liaoning Academy of Materials, Shenyang, 110004, China
*Corresponding address: e-mail: zhaoyuhong@nuc.edu.cn

Abstract: In this paper, the effect of domains and domain walls on energy storage properties was discussed, and the influence of elements doping and misfit strain were also studied. The results show that with the enhancement of external fields, the energy storage efficiency increases, while the energy storage density increases first and then decreases. Besides, and the in-plane domains and domain walls were proved to be the essential element of energy storage,where energy storage density and efficiency increased by 2.3 and 1.7 times, respectively. Finally, the optimal energy storage location, where "misfit strain = 1.9%" and "Na ratio = 0.3", was obtained. The energy density and efficiency reached 29.97kJ/cm3 and 99.99%, respectively.

Keywords: $K_{1-x}Na_xNbO_3$ (KNN) thin film, In-plane phase, Energy storage, Phase-field methods.

1 Introduction

Ferroelectric nano-electronic capacitors have attracted intensive interest over the past two decades [1]. The improvement in energy storage properties is essentially characterized by an increase in maximum polarization and a decrease in residual polarization [2]. Numerous works have discussed the benefit of the in-plane phase in experiments and simulations [3, 4]. Koukhar proved that a positive misfit strain would transform the orthorhombic domain phases [5]. Liu et.al produce in-plane domains via voltage control to multiple non-volatile resistance states, thus demonstrating the key functional property of being a memristor a few unit cells thick [6]. Sánchez-Santolino controls twist angles to induce in-plane vortex domains [7]. However, Some works proved that the presence of domain walls will hinder the energy storage of ferroelectric materials [8, 9].

In this work, a high-throughput phase-field method (more than 4000 calculations) was performed here to optimize energy storage density and efficiency. Results show that in-plane vortex domains are the equilibrium positions of doping and misfit strain in energy storage.

2 Result and discussion

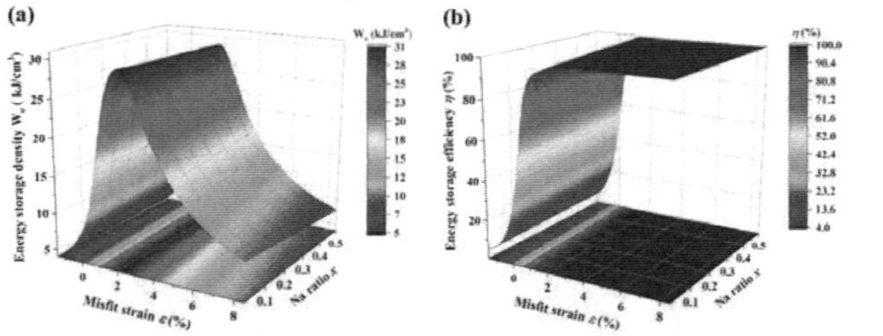

Fig. 1. The results of the high-throughput phase-field method. The influence of misfit strain, whose range is -1.5% (the minus sign responding to compressive strain) to 8.5%, and Na ratio on (a) energy storage density and (b) energy storage efficiency.

Considering a plane parallel to the YOZ plane, taking $x = 0.5$ as an example, the energy storage density shows a parabolic-like change, reaching a maximum $\sim 29.6\ kJ/cm^3$ when $\varepsilon = 1.7\%$. In contrast, energy storage is positively correlated with the misfit strain. It increases rapidly from the initial 5.114%to 99.318%in a small misfit strain ($-1.5\% \sim 1.7\%$) while slowly in a large misfit strain ($1.7\% \sim 8.5\%$) and gradually becomes stable to $\sim 99.858\%$ .For a plane parallel to the XOZ plane, the energy storage density increases with the Na ratio in a small misfit strain ($-1.5\% \sim 1.7\%$) and decreases in a large misfit strain ($1.7\% \sim 8.5\%$). Specifically, the growth rate is first slow and then fast when $\varepsilon < 0.5\%$, which fits well with the existing experimental data [10, 11]. As the tensile strain continues to increase, the growth rate becomes fast first and then slow. Finally, the maximum energy storage density with different Na ratios is reached at $1.7\% \sim 2.1\%$. When the misfit strain exceeds this range, the energy storage density will decrease with the Na ratio. The change in energy storage efficiency with the Na ratio is similar to the misfit strain. It increases obviously with the Na ratio when the efficiency is 99%.

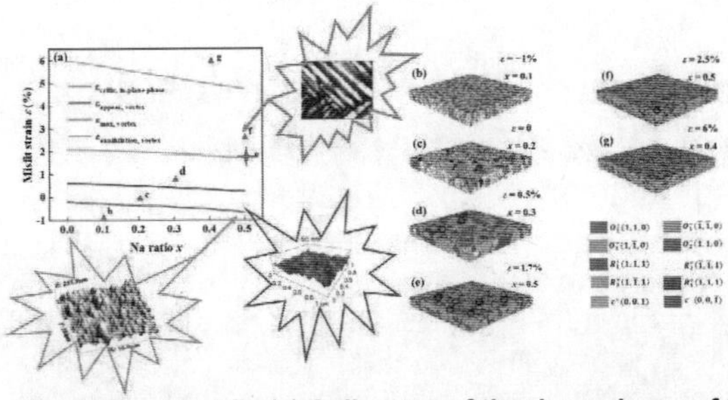

Fig. 2. Empirical fit. (a) A diagram of the dependence of the phase state and domain structures on misfit strain ε and Na ratio x of the KNN thin film. (b)-(g) correspond to the red triangles and stars in (a) which have been labelled as b-g.

3 Conclusion

(1)The energy storage properties depend on the competition between in-plane domains and domain walls and the vortex domains are the equilibrium positions.

(2)The emergence of exotic ferroelectric vortex domain increases the energy storage density of KNN thin film (e.g., $x = 0.2$for $\varepsilon = 0.5\%$);

(3)Both tensile strain and Na ratio promote the creation of in-plane phase, which is achieved at the cost of decreasing elastic free energy but has the effect of accelerating and then inhibiting the generation of vortexes (e.g., $x = 0.2$for $\varepsilon = 5.5\%$).

4 Acknowledgments

This work was supported by National Natural Science Foundation of China (Nos. 52375394, 52074246, 52275390, 52205429, 52201146, 52305429), National Defense Basic Scientific Research Program of China (JCKY2020408B002), Shanxi Scholarship Council of China (No. 2021-125), Natural Science Foundation of Shanxi Province (No. 20210302124631), and Key Research and Development Program of Shanxi Province (Nos. 202102050201011, 202202050201014).

References

[1] Liu C., Wang J., Strain engineering of ferroelectric negative capacitance in PbZr(1-x)TixO3 thin films, Acta Mater. 2021,206:116607.

[2] Kuang W., Wang H., Li X., Zhang J., Zhou Q., Zhao Y., Application of the thermodynamic extremal principle to diffusion-controlled phase transformations in Fe-C-X alloys: Modeling and applications, Acta Mater. 2018, 159:16-30.

[3] Aspe B., Castel X., Demange V., Passerieux D., Pinault-Thaury M.A., Jomard F., Députier S., Cros D., Madrangeas V., Bouquet V., Sauleau R., Guilloux-Viry M., Enhanced tunability and temperature-dependent dielectric characteristics at microwaves of $K_{0.5}Na_{0.5}NbO_3$ thin films epitaxially grown on (100)MgO substrates, J. Alloys Compd. 2021, 856.

[4] Liu D., Wang J., Wang J.-S., Huang H.-B., Phase field simulation of misfit strain manipulating domain structure and ferroelectric properties in $PbZr_{(1-x)}Ti_xO_3$ thin films, Acta Phys. Sin. 2020, 69(12):127801-1-127801-10.

[5] Koukhar V.G., Pertsev N.A., Waser R., In-plane polarization states and their instabilities in polydomain epitaxial ferroelectric thin films, Appl. Phys. Lett. 2001, 78(4):530-532.

[6] Liu, Z., Wang, H., Li, M. et al. In-plane charged domain walls with memristive behaviour in a ferroelectric film. Nature 2023, 613:656–661.

[7] Sánchez-Santolino G., Rouco V., Puebla S., Aramberri H., Zamora V., Cabero M., Cuellar F.A., C., Mompean F., Garcia-Hernandez M., Castellanos-Gomez A., Íñiguez J., Leon C., Santamaria J., A 2D ferroelectric vortex pattern in twisted BaTiO3 freestanding layers, Nature 2024, 626(7999):529-534.

[8] Bassiri-Gharb N., Fujii I., Hong E., Trolier-McKinstry S., Taylor D.V., Damjanovic D., Domain wall contributions to the properties of piezoelectric thin films, J. Electroceram. 2007, 19(1):49-67.

[9] Porokhonskyy V., Damjanovic D., Domain wall contributions in $Pb(Zr,Ti)O_3$ ceramics at morphotropic phase boundary: A study of dielectric dispersion, Appl. Phys. Lett. 2010, 96(24).

[10] Lai F., Li J.-F., Sol-gel processing and characterization of (Na, K)NbO_3 lead-free ferroelectric films, Ferroelectr. 2007, 358(1):181-187.

[11] Yu Q., Li J.-F., Sun W., Zhou Z., Xu Y., Xie Z.-K., Lai F.-P., Wang Q.-M., Electrical properties of $K_{0.5}Na_{0.5}NbO_3$ thin films grown on Nb: $SrTiO_3$ single-crystalline substrates with different crystallographic orientations, J. Appl. Phys. 2013, 113(2):024101.

Selective Foaming Agents for Preparation of Foam Glass Ceramics as Insulation Materials from High-Titanium Blast Furnace Slag

Qiaoling Jiang, Liangping Cai, Keqin Feng*
School of Mechanical Engineering, Sichuan University, Chengdu 610065, China
*Corresponding address: e-mail: kqfeng@scu.edu.cn

Abstract: Foam glass ceramics as insulation materials were fabricated using high-titanium blast furnace slag and waste glass through a one-step sintering. Notably, selecting a suitable foaming agent is crucial during preparation. In order to select a suitable foaming agent, foaming agents CaCO3, SiC, and AlN are mainly compared in terms of usage effectiveness and cost in this paper. Based on the results of Differential Thermal Analysis, the foaming temperatures of SiC and AlN are more suitable. Further experiments reveal that the foaming effects of 0.1 wt% SiC and 1 wt% AlN are similar. And the foam glass ceramics prepared with SiC have predominantly round, closed pores, leading to superior thermal insulation and compressive strength compared to those made with AlN. Therefore, SiC is relatively the most appropriate foaming agent.

Keywords: Foam glass ceramics; High titanium blast furnace slag; Foaming agent; Insulation material

1 Introduction

As urbanization progresses, the demand for energy-efficient building insulation is rising. Foam glass ceramics, made from industrial waste, offer a superior choice due to their light weight, high strength, and good thermal insulation properties [1]. High-titanium blast furnace slag (HTBFS) is a kind of solid waste produced by blast furnace smelting of vanadium-titanium magnetite [2]. Our research group has prepared foam glass ceramics by one-step sintering using HTBFS and waste glass (WG) as the main raw material successfully [3, 4]. In this preparation, the foaming and crystallization processes can be completed simultaneously, where the two processes would affect each other. Notably, a suitable foaming agent is significant for producing foam glass ceramics with good comprehensive properties. Selection on foaming agent mainly considers from two aspects: (1) Impurities cannot be brought to the existing glass system; (2) Lower cost while ensuring good foaming effect. Therefore, This study investigated common foaming agents $CaCO_3$, SiC, and AlN to select the more suitable one for fabricating foam glass ceramics as insulation materials via one-step sintering to utilize HTBFS and WG.

2 Experimental procedure

In search of the most suitable foaming agent, $CaCO_3$, SiC, and AlN were analyzed by using a Hitachi STA 7000 instrument to identify reaction temperature of foaming agents.

In this study, 42 wt% HTBFS and 58 wt% WG were used as the main raw materials, with 7 wt % $Na_2B_4O_7$ as a fluxing agent and 5 wt % H_3BO_3 as a stabilizer added according to previous exploration. Then the appropriate amount of selected foaming agent were to be added. These powders were ball milled and pressed into cylindrical samples with dimensions of Φ 18 mm × 13 mm and Φ 45 mm × 15 mm and sintered at 1050 ℃。

The sintered samples were etched with a 5 wt% HF solution and then observed for microstructure using SEM. The compressive strength was measured using a universal testing machine. Porosity was measured with a 3H-2000TD true density analyzer using the Archimedes principle. Thermal conductivity was measured with a DRE-2C meter on samples sized Φ40×13 mm.

3 Result and discussion

In this study, using software of Inorganic Glass Engineer System, the characteristic temperature of different mass ratios of HTBFS and WG were obtained in Table 1 (H: G means HTBFS: WG). It can be found that crystallization temperatures of these systems are range from 886 ℃ to 1061℃. For foaming agents, the foaming temperature should be similar to the crystallization temperature.

Table 1 Characteristic temperature of different combinations of HTBFS and WG

H:G	Softening Temperature (℃)	Melting Temperature (℃)	Crystallization Temperature (℃)
42:58	801	1390	886-1056
44:56	806	1384	890-1058
46:54	811	1378	895-1059
48:52	817	1372	899-1061

From the DTA-TG curve of $CaCO_3$ in Fig. 1(a), $CaCO_3$ decomposes between 680 ℃ and 814 ℃ (see reaction equation (1)).

$$CaCO_3 = CaO + CO_2 \quad (1)$$

The decomposition reaction of $CaCO_3$ is intense and gas production rate is fast. More importantly, the highest reaction temperature of $CaCO_3$, 814℃ is much lower than crystallization temperatures as shown in Table 1. Thus, it is not considered despite its low cost in this study.

AlN achieves foaming by reacting with O_2 to produce N_2, as shown in reaction equation (2). According to the curve in Fig. 1(b), AlN undergoes an oxidation reaction between 800 ℃ and 1200 ℃, close to the range of crystallization temperature. It is an appropriate foaming agent.

$$4AlN + 3O_2 = 2Al_2O_3 + 2N_2 \quad (2)$$

Fig. 1(c) shows the DTA-TG curve of SiC, which lacks significant exothermic or endothermic peaks. The mass first decreases and then increases due to the reaction of SiC with O_2 to form SiO_2 (see reaction equation (3)). According to this curve, as the temperature exceeds 850℃, the mass increases without notable exothermic peaks. This is because a dense SiO_2 layer forms on the SiC surface, requiring O_2 to diffuse inward for further reaction [5]. Thus, the oxidation of SiC is continuous and stable, with no distinct exothermic peaks in the DTA curve. So, SiC is also a suitable foaming agent as AlN.

$$SiC + 2O_2 = SiO_2 + CO_2 \tag{3}$$

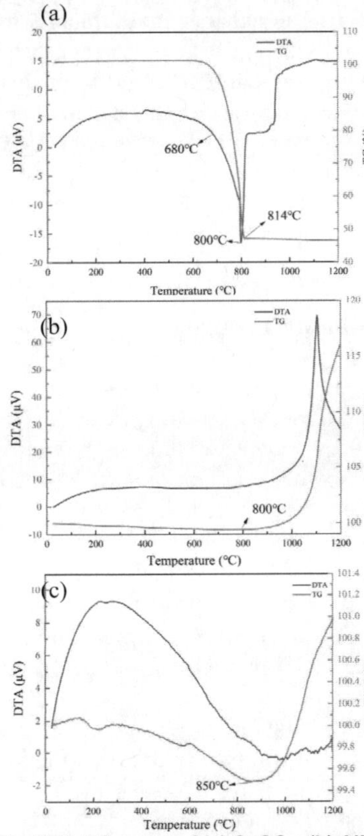

Fig. 1 The DTA-TG curve of (a) $CaCO_3$; (b) AlN; (c) SiC

To select the better foaming agent, 0.1 wt% SiC and 1 wt% AlN were separately added into basic components, and final foam glass ceramics were produced. Fig. 2. shows the microstructure of prepared foam glass ceramics.

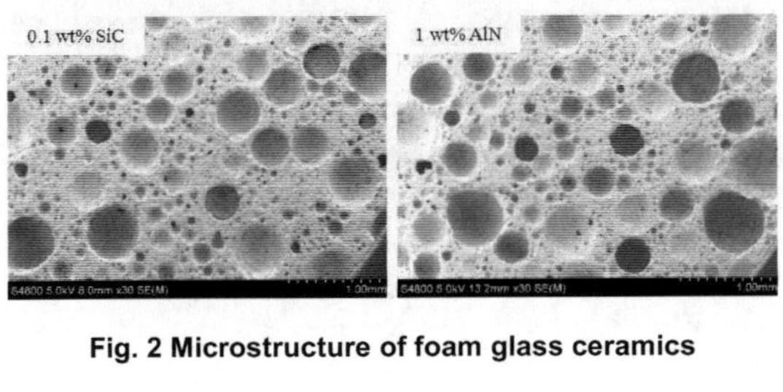

Fig. 2 Microstructure of foam glass ceramics

The foam glass ceramics prepared with SiC has more uniform and predominantly round, closed pores, with fewer interconnected structures. This can reduce air turbulence during heat transfer, inhibiting convective heat exchange and enhancing thermal insulation. Performance of foam glass ceramics with SiC and AlN are shown in Table 1. From a porosity difference of only 0.8%, SiC produces similar foaming effect as AlN at lower addition. Moreover, the foam glass ceramics prepared with SiC have a better compressive strength, 20 MPa and a lower thermal conductivity, 0.380 W/m·K. Additionally, SiC is more cost-effective than AlN. Therefore, the foaming agent SiC is more suitable for preparation of foam glass ceramics as insulation materials.

Table 2 Performance of foam glass ceramics with 0.1 wt SiC and 1 wt% AlN

	Bulk density (g/cm^3)	Porosity (%)	Compressive strength (MPa)	Thermal Conductivity (W/m·K)
SiC	1.42	54.2	20	0.380
AlN	1.40	55	15	0.415

4 Conclusion

Based on curves of DTA-TG, it is found that, compared with $CaCO_3$, SiC and AlN have more appropriate foaming temperature. Further experiments show that only 0.1 wt% SiC achieves the similar foaming effect of 1 wt% AlN, with a porosity difference of 0.8%. The foam glass ceramics prepared with SiC has more uniform and mainly round, closed pores. Moreover, its compressive strength increases, and the thermal conductivity decreases significantly. By adjusting the addition of SiC, the thermal insulation of foam glass ceramics can be further improved.

References

[1] Jiang C, Huang S, Li G, et al. Formation of closed foam ceramics from granite scraps. Ceramics international, 2018, 44(3): 161-167.

[2] C. Gan, J. Yang, X. Du, et al. A. Nikitin, Vanadium mobilization and redistribution during mineral transformation of vanadium-titanium magnetite tailings with different weathering degrees. Science of The Total Environment, 2023, 894: 165068.

[3] H. Zhou, K. Feng, Y. Liu, et al. Preparation and characterization of foamed glass-ceramics based on waste glass and slow-cooled high-titanium blast furnace slag using borax as a flux agent. Journal of Non-Crystal Solids, 2022, 590: 121703.

[4] Wang H, Feng K, Sun Q. Effect of calcium carbonate on the preparation of glass ceramics foams from water-quenched titanium-bearing blast furnace slag and waste glass. Advances in applied ceramics, 2018, 117(5): 312-318.

[5] Wang H, Chen Z, Liu L, et al. Synthesis of a foam ceramic based on ceramic title polishing wasted using SiC as foaming agent. Ceramics international, 2018, 44(9): 10078-10086.

Mg-Ca-Based Hydrolysis Hydrogen Generation Technology

Chaoling Wu [1,2,3*], Yao Wang [3,4], Yigang Yan [3,4], Yungui Chen [3,4]
1. College of Materials Science and Engineering, Sichuan University, Chengdu, PR China
2. State Key Laboratory of Intelligent Construction and Healthy Operation and Maintenance of Deep Underground
3. Engineering, Sichuan University, Chengdu, PR China
4. Engineering Research Center of Alternative Energy Materials & Devices, Ministry of Education,
5. Sichuan University, Chengdu, PR China
6. Institute of New Energy and Low-Carbon Technology, Sichuan University, Chengdu, PR China
*Corresponding address: e-mail: wuchaoling@scu.edu.cn

Abstract: Real-time hydrogen generation from hydrolysis of low-cost Mg-based materials is a perfect combination of hydrogen generation, storage and transportation. The theoretical hydrogen storage densities of Mg and MgH2 are 8.2wt% and 15.2wt% (excluding H2O), respectively. However, the main by-product Mg(OH)2 covers on the surface of the materials during hydrolysis process, thus preventing the reaction from proceeding. We have developed the Mg-Ca-based hydrolysis materials composed of MgH2 and Ca4Mg3H14 phases, which display better hydrolysis performance than pure Mg/MgH2 because Ca4Mg3H14 is prone to hydrolyze, as the byproduct Ca(OH)2 is more easily soluble than Mg(OH)2, providing more gates for water penetration deep into the particles. By adjusting the content of Ca, the mass transfer was enhanced and the conversion rate was improved. By further adding additives (salts, expanded graphite, etc.) and adjusting the reaction solutions (chloride solutions), the pulverization efficiency and hydrolysis performances at low temperatures were improved. Combined with pressing technology, the safety and air stability of the materials were improved. Our work is promising to promote the engineering application of Mg-Ca based hydrolysis materials.

Keywords: Mg-Ca alloy hydrides, Hydrolysis, Hydrogen generation

1 Introduction

The H_2 generation through the hydrolysis of cost-effective Mg-based materials is a perfect combination of hydrogen generation, storage and transportation, namely "integration of H_2 generation-storage technology". The theoretical H_2 storage densities of Mg and its hydride (MgH_2) are 8.2 wt% and 15.2 wt% (excluding H_2O), respectively.

However, during the hydrolysis process of both Mg and MgH_2, the main by-product $Mg(OH)_2$ is produced and deposited on the surfaces of the materials, thereby impeding the progression of the reaction. To break down this barrier, researchers have proposed a series of methods, for example, refining particles to increase active sites, or breaking $Mg(OH)_2$ to enhance mass transfer by modifying the materials or adjusting the reaction accelerants[1].

In our work, the Mg-Ca-based hydrolysis materials with better hydrolysis performance than pure Mg/MgH_2 were developed. Mass transfer was enhanced to improve the conversion rate by regulating the content of Ca. By further adding proper additives, the pulverizing efficiency, the hydrolysis performance at low temperature and the air-stability were also improved.

2 Experimental procedure

First, the Mg-Ca alloys were prepared in a resistance furnace with high purity Mg (99.9%) and Ca (99.5%) under SF_6. Then, the Mg-Ca alloys were milled under dry argon atmosphere by SPEX 8000 high energy ball mixer with a ball-to-powder mass ratio of 20:1 for 2 h. Later, the alloys powders were placed in the Sieverts type device and hydrogenated under a hydrogen pressure of 4-5 MPa at 400 ℃. It is worth noting that various additives were added during the ball milling process. The generated Mg-Ca-based hydrolysis materials were stored in an argon glove box with <0.01 ppm O_2 and <0.01 ppm H_2O.

3 Result and discussion

3.1 Effect of Ca content on hydrolysis performance

The x wt.% Ca-Mg alloy hydrides (x = 10, 20 and 30) were prepared by hydrogenating the brittle Mg-Ca alloys[2]. The 30 wt.% Ca-Mg alloy hydride (MCH) shows the best hydrolysis properties, which releases 1419.8 mL g^{-1} H_2 within 1 h at 70 ℃ and its conversion rate is about 95%. The superior performances of Mg-Ca alloy hydrides is attributed to an easy hydrolysis nature of $Ca_4Mg_3H_{14}$ and a high solubility of the by-product $Ca(OH)_2$, which provides gates to make water penetrate deeply inside the particles. Increasing the content of Ca enhances the hydrolysis performances of Mg-Ca alloy hydrides due to the formation of more $Ca_4Mg_3H_{14}$ and larger specific surface areas.

3.2 Common ion effect in the hydrolysis reaction of Mg-Ca alloy hydride-salt composites

The effects of two salts, $MgCl_2$ and $CaCl_2$, on microstructures and the hydrolysis performances of MCH were investigated[3]. Due to the common ion effect of Mg^{2+} in the hydrolysis processes, the xMCH-$MgCl_2$ (x = 5, 10, 15 and 20) composites generate less H_2 than MCH. The 10MCH-$MgCl_2$ composite milled for 0.5 h displays the lowest H_2 yield of 962 mL g^{-1} of the composite at 25 ℃, and has slower H_2 generation rate (692 mL g^{-1} in 10 min) than MCH (722 mL g^{-1} in 10 min). At the same conditions, compared with MCH, the xMCH-$CaCl_2$ (x = 5, 10, 15 and 20) composites exhibit better H_2 generation performances. Especially, 15MCH-$CaCl_2$ has a H_2 yield of 1141 mL g^{-1}.

3.3 Low-temperature hydrolysis performances and mechanisms of Mg-Ca-based hydride

The influence of different cations (Na^+, Ca^{2+}, Mg^{2+} and NH_4^+) in chloride solutions on low-temperature (-20, -10, 0, 10 ℃) hydrolysis performances of MCH and mechanisms of interaction are investigated[4]. The results demonstrate that Mg^{2+} and NH_4^+ in chloride solutions significantly enhance the rate of hydrolysis reaction between MCH and water. In NH_4Cl solution, MCH exhibits the best initial hydrolysis performance and the highest H_2 yield, 1198 mL g^{-1}(MCH) of H_2 is generated within the first 5 min and the final conversion rate reaches 89% at 0 ℃. MCH displays the best hydrolysis dynamics in $MgCl_2$ solution, where the activation energy is low, only 3.56 kJ mol^{-1}. This can be attributed to the fact that the types of hydroxide products and the corresponding physical states with respect to the hydrolysis of MCH in various chloride solutions differ with the variations of OH^- affinity toward M^{n+} cations (i.e. pK/(n+1)) when pH values of chloride solutions are altered. Thus, different low-temperature hydrolysis performances can be observed.

3.4 Enhanced air-stability of Mg-Ca hydride-chloride composites

The hydrolysis properties and air-stability of MCH-based material improved by ball-milling with NH_4Cl are investigated[5]. The addition of NH_4Cl by ball-milling significantly enhances the hydrolysis performance. The MCH-5%NH_4Cl composite (powders) milled for 0.5 h displays the highest H_2 yield of 1006 mL g^{-1} at 25 ℃ without stirring, and has the fastest H_2 generation rate (872 mL g^{-1} min^{-1} in the initial 1 min). The air-stability of the MCH-based material significantly increases when being pressed into plate with NH_4Cl due to the reduction of the contacting areas to the air and hygroscopicity of NH_4Cl. The H_2 yield of the MCH-5%NH_4Cl composite plate still reaches 726 mL g^{-1} at 25 ℃ without stirring when being exposed to the air for 5 h with 80% humidity at 30 ℃.

3.5 Hydrogen generation behaviors from hydrolysis of cold-welding free Mg-Ca hydride-expanded graphite composites

The composites including MCH and 0-20 wt% expanded graphite (EG) were explored to examine the impact and underlying mechanisms of EG as additive in mitigating cold-welding and enhancing hydrolysis[6]. EG coats on the material surface as a dispersant to prevent cold-welding, optimize ball milling efficiency and reduce particle size, further facilitating hydrogenation and hydrolysis. In deionized water, MCH with 10 wt% EG (MCH-10EG) displays the best hydrolysis performance, with a low activation energy of 11.26 kJ mol^{-1} and a complete conversion within 8 min at 55 ℃. Moreover, at room temperature (25℃), combined with the corrosion effects of Cl^- and H^+ to break the formed $Mg(OH)_2$ passivation layer, MCH-10EG achieves a complete conversion within only 5 min in 0.3 mol L^{-1} $AlCl_3$ solution.

4 Conclusion

In our work, the Mg-Ca-based hydrolysis materials composed of MgH_2 and $Ca_4Mg_3H_{14}$ phases were developed. By adjusting the content of Ca, adding proper additives and adjusting the reaction solutions, the pulverization efficiency, air stability and hydrolysis performances were all improved. Our work is promising to promote the engineering application of Mg-Ca based hydrolysis materials.

5 Acknowledgments

This research was supported by the National Key Research & Development Project (2022YFB3803705-1).

References

[1] Wu C L, Huang Q, Gong B S, Zhou J, Xia G H, He F N, Wang Y, Yan Y G, Chen Y G, J. Mater. Sci. Techno., 2023, 163, 1-16.
[2] Liu P P, Wu H W, Wu C L, Chen Y G, Xu Y M, Wang X L, Zhang Y B, Int. J. Hydrogen Energ., 2015, 40, 3806.
[3] Li J F, Liu P P, Wu C L, Chen Y G, Int. J. Hydrogen Energ., 2017, 42, 1429-1435.
[4] Zhong S, Wu C L, Li J F, Chen Y G, Wang Y, Yan Y G, J. Alloy. Compd., 2021, 851, 156762.
[5] Zhong S, Wu C L, Chen Y G, Feng Z, Zhao Y, Xia YT, J. Alloy. Compd., 2019, 792, 869.
[6] Huang Q, Gong B S, Habibullah, Wang Z Y, Wang Y, Yan Y G, Chen Y G, Wu C L, J. Power Sources, 2024, 595, 234004.

Preparation and Hydrogen Absorption/Desorption Mechanism of High-Performance and Low-Cost BCC Type Hydrogen Storage Alloys

Qingjun Chen*, Huazhou Hu, Luocai Yi
Ganjiang Innovation Academy, Chinese Academy of Sciences, Ganzhou 341000, China
*Corresponding address: e-mail: qjchen@gia.cas.cn

Abstract: Efficient, safe and economical hydrogen storage technologies are essential for the widespread application of hydrogen as an energy carrier, and Vanadium (V)-based BCC alloys stand out due to their high theoretical hydrogen storage capacity. However, the high cost of V limits their practical application. In this work, three types of BCC hydrogen storage materials were developed. Firstly, a novel high entropy BCC hydrogen storage alloy was constructed by co-doping Mo and Ce; secondly, the BCC structure of low-V alloy was maintained based on melt-spining; finally, V-free BCC hydrogen storage alloy was constructed by completely replacing V with Mo. The effective hydrogen desorption capacity of the alloys reaches 2.4 wt% above 0.1 MPa, which provides guidance for the application of BCC-type alloys in gas-solid hydrogen storage. The hydrogen absorption/desorption mechanism of the BCC alloys were studied.

Keywords: Hydrogen storage alloys; BCC phase; High entropy alloys; Low-V; V-free.

1 Introduction

Hydrogen energy vital for energy security and dual-carbon goals faces storage/transportation challenges due to its physicochemical properties. Solid-state storage, offering safety, and low pressure, is a promising alternative attracting research focus. Alloy-based hydrogen storage materials are mature, used in gas-solid storage, compression, recovery, and purification. Rare earth alloys have rapid kinetics but Ni cost limits gas-solid use [1]. Ti-based alloys, low-cost and stable, favor large-scale systems but have low capacity and production issues [2]. Mg-based materials, high-capacity and cost-effective, need advancements in material prep, tank design, and heat integration for industrialization [3]. V-based BCC alloys [4-6], like V-Ti-Cr, show impressive hydrogen storage capacities, outperforming rare earth and Ti-based alloys. However, high V costs and impurity issues hinder their use. Researchers aim to reduce costs, enhance stability, and improve performance through element substitution, innovative preparation methods, and seeking cheaper V sources, to develop cost-effective BCC alloys for practical hydrogen storage.

Inspired by the research described above, three distinct categories of BCC-type hydrogen storage alloys (high-entropy, low-V, V-free alloys) were prepared in this work. This comprehensive research endeavor underpins the foundation for the practical application of cost-effective and efficient hydrogen storage solutions.

2 Experimental procedure

The alloy was prepared by arc melting Ti, Cr, V, Mn, Mo, and Ce in a vacuum furnace. The alloys were crushed, sieved, and analyzed via XRD, SEM-EDS, TEM, and XPS. After vacuum activation, their hydrogen storage performance was tested, including kinetics, PCT curves, cycling performance. Effective hydrogen desorption capacity was measured at pressures above 0.1 MPa.

3 Result and discussion

1 High entropy BCC type alloy

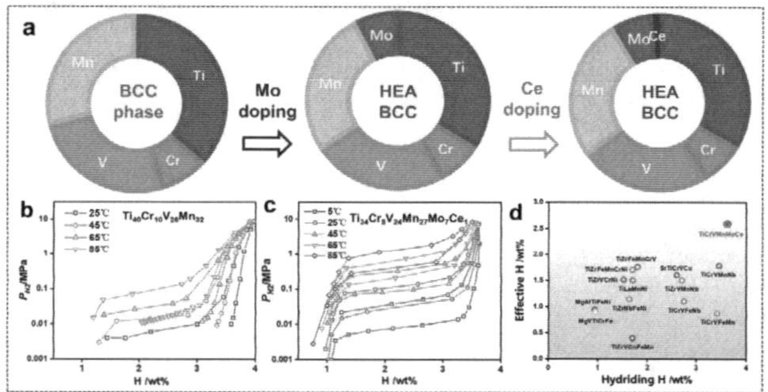

Fig. 1 (a) Diagram of preparation of high entropy alloy, PCT curves of HEA (b) before and (c) after doping, (d) Comparison of hydrogen storage capacity with HEA.

To overcome these challenges, a BCC-type TiVCrMnMoCe HEA with an effectively dehydriding capacity of 2.5 wt% above 0.1 MPa was prepared (Fig. 1). By introduction of Mo and conducting heat treatment, the precipitation of Ti-rich phase in HEA was successfully suppressed, resulting in improved compositional uniformity and dehydriding capacity. Consequently, the effective dehydriding capacity increased significantly from 0.60 wt% to 2.50 wt% at 65 °C, surpassing that of other types of hydrogen storage alloys under the same conditions. Moreover, the addition of 1 wt% Ce enabled initial hydrogen absorption at 25 °C without the need for activation at 400 °C. Furthermore, Ce doping reduced the dehydriding activation energy of the HEA from 52.71 to 42.82 $kJ \cdot mol^{-1}$. Additionally, the enthalpy value of dehydrogenation decreased from 46.89 to 17.96 $kJ \cdot mol^{-1}$, attributed to a decrease in the hysteresis factor from 0.68 to 0.52.

2 Low-V BCC type alloy

Herein, a low-V BCC alloy ($Ti_{41}Cr_{50}V_5Mo_4$) with an excellent dehydriding capacity (2.4 wt%) is developed via Mo partial substitution and coupling with the melt-spin process (Fig. 2). The alloys' crystal structures, micromorphology, and de-/hydrogenation mechanisms are investigated systematically. It is found that the Mo-substitution and the melt-spin process can keep and extend the BCC zone in the low-V portion of Ti-Cr-V phase diagram, and Mo atom is uniformly distributed in the BCC alloy. The dehydrogenation mechanism of the low-V BCC alloy was determined to follow a diffusion-controlled model with an activation energy of 52.38 kJ/mol. The cycle test also proves that the synergistic effect of Mo substitution and melt spin improved the durability of the low-V alloy. The retention rate of hydrogen capacity can achieve 94 % after 15 cycles.

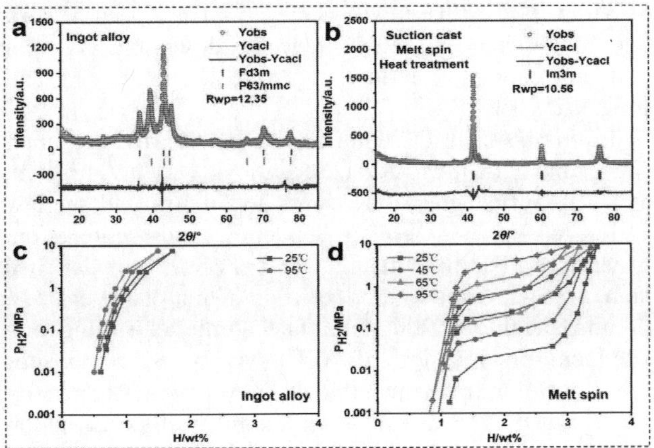

Fig. 2 Crystal structure of low-V alloy (a) before and (b) after melt spin, PCT curves of low-V alloy (c) before and (d) after melt spin.

3 V-free BCC type alloy

An easily activated and high-performance V-free BCC type alloy ($Ti_{40}Cr_{50}Mo_{10}Ce_1$) was successfully synthesized by heat treatment and Ce doping (Fig. 3). The heat treatment significantly increased its dehydriding capacity to 2.5 wt%, attributed to a reduced slope factor (0.76−0.17), making it comparable to V-based BCC-type alloys. And after Ce doping, enabled room temperature hydrogen absorption without high-temperature activation. Ce doping did not significantly affect the activation energy, enthalpy change, or hysteresis factor of the alloy during de/hydrogenation. Additionally, the V-free BCC-type alloy, after heat treatment and Ce doping, exhibited excellent cyclic stability, with a capacity retention rate of 90% after 100 cycles.

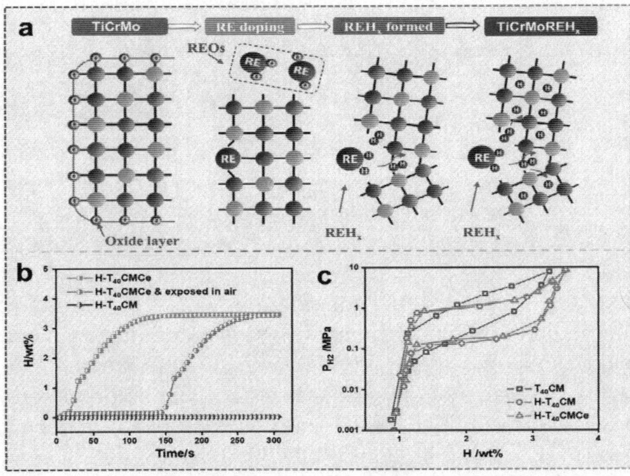

Fig. 3 (a) Activation diagram, (b) activation properties of rare earth-doped alloys and (c) PCT curves of V-free alloys before and after Ce doping and heat treatment.

Conclusion

In this work, three high-performance BCC type hydrogen storage alloys were successfully synthesized. A BCC-type Ti–V–Cr–Mn–Mo–Ce HEA with an effectively dehydriding capacity of 2.5 wt% above 0.1 MPa was prepared. a low-V BCC alloy ($Ti_{41}Cr_{50}V_5Mo_4$) with an excellent dehydriding capacity (2.4 wt%) is successfully developed via Mo partial substitution and melt-spin process. An easily activated and high-performance V-free BCC-type alloy ($Ti_{40}Cr_{50}Mo_{10}Ce_1$) was successfully synthesized by heat treatment and Ce doping.

Acknowledgments

This work was supported by the National Key R&D Program of China (2022YFB3504700).

References

[1] Zhang YX, et al. A_2B_7-type La-Mg–Ni alloys prepared by Mg thermal diffusion for improved hydrogen storage performance. Rare Met. 2024,43:3260-3272.

[2] Qiao WF, et al. Effects of Cu doping on the hydrogen storage performance of Ti-Mn-based, AB_2-type alloys. Chem. Eng. J. 2023,465:142837.

[3] Shang Y, et al. Mg-based materials for hydrogen storage. J. Magnesium Alloys. 2021,9(6):1837-1860.

[4] Hu HZ, et al. Development of Ti-V-Cr-Mn-Mo-Ce high-entropy alloys for high-density hydrogen storage in water bath environments. Rare Met. 2024.

[5] Hu HZ, et al. Hydrogen storage in Mo substituted low-V alloys treated by melt-spin process. Chem. Eng. J. 2023, 455:140970.

[6] Hu HZ, et al. Development of V-free BCC structured alloys for hydrogen storage. ACS Appl. Energy, 2023, 6(21):11108-11117.

Advancements in the Development of High-Performance Al-Cu-Mg Based Composites

Jiehua Li
Chair of Casting Research, Department of Metallurgy, Montanuniversität Leoben, A-8700, Leoben, Austria
*Corresponding address: e-mail: jiehua.li@unileoben.ac.at

Abstract: In this talk, the addition of 3 wt.% TiB_2 (as grain refiner) into Al-Cu-Mg based alloys was investigated, which can refine grain size down to 50 μm and thereby greatly reduce hot tearing tendency and segregation. Furthermore, additive manufacturing (laser powder bed fusion) was used to produce rods with a diameter of 16 mm and height of 200 mm. The solute clustering and subsequent precipitation of Al-Cu-Mg based composites produced by additive manufacturing was investigated by using high-angle annular dark-field scanning transmission electron microscopy (HAADF-STEM) and atom probe tomography (APT). A significant solute clustering (Mg-Ag-rich) was observed after the solution treatment (T4). After further over-ageing treatment (T7), three different types of precipitates (Al-Cu (θ) phase, Al-Cu-Mg-Ag (Ω) phase and Mg-Si (β) phase) were observed, which is fully consistent with the case of samples produced by conventional gravity die casting condition. More interestingly, one Mg, Ag-rich layer was observed in the vicinity of Al-Cu (θ) phase. These careful HAADF-STEM characterizations and quantitative atom probe data analyses give us some hints to design high performance Al-Cu-Mg based composites.

Keywords: Al-Cu-Mg based composites; Solute clustering; Precipitation; Investment casting; Additive manufacturing; TEM; APT

1 Introduction

High-performance precipitation hardening Al-Cu-Mg based alloys have been widely used in aerospace and automotive industries, especially at elevated temperatures (i.e. up to 250 °C). During the heat treatments, cluster clustering and precipitation very often occurs to provide strengthening effects for the high-performance Al alloys. Tailoring the solute clustering and precipitation is of great necessity to improve the performance of the Al alloys. However, the precipitates are not stable, in particular at elevated temperatures. Their strengthening effect becomes less or lost. Improving their thermal stability is therefore of great necessity.

Alloying is one of the most important methodologies to improve their thermal stability. Indeed, Al-Cu-Mg-Ag alloy system, as an important Al alloy system, has been used to develop high strength Al alloys currently with plate-like Ω (AlCuMgAg) precipitate on the {111} planes and θ (Al2Cu) precipitate on the {001} planes. The high strength and thermal stability of these Al-Cu-Mg-Ag alloys has been attributed to the presence of the Ω precipitate and its propensity for interfacial segregation. Despite of its high strength of Al-Cu-Mg-Ag alloy system, it should be noted that Al-Cu-Mg-Ag alloy system has a significant hot tearing tendency and a serious segregation issue due to their large solidification range. An effective grain refinement can be used to solve the hot tearing and segregation issues. Although the addition of 0.1 wt.% TiB2 (as grain refiner) into Al-Cu-Mg-Ag based alloys (regarded as Al-Cu-Mg-Ag based composites) can refine grain size down to 200 μm, it is still not enough to reduce or avoid the hot tearing and segregation issues.

In this abstract, the addition of 3 wt.% TiB2 (as grain refiner) into Al-Cu-Mg-Ag based alloys was investigated, which can refine grain size down to 50 μm and thereby greatly reduce hot tearing tendency and segregation. Furthermore, additive manufacturing (laser powder bed fusion) was also used to produce rods with a diameter of 16 mm and height of 200 mm. The solute clustering and subsequent precipitation of Al-Cu-Mg based composites produced by additive manufacturing was investigated by using HAADF-STEM and APT. A significant precipitation was observed in as built condition, which can be attributed to the repeated heating and cooling during additive manufacturing. However, it should be no significant solute clustering (Mg-Ag-rich) was observed after the solution treatment (T4), which is in contrast to the case of samples produced by conventional gravity die casting. After further over-ageing treatment (T7), three different types of precipitates (Al-Cu (θ) phase, Al-Cu-Mg-Ag (Ω) phase and Mg-Si (β) phase) were observed, which is fully consistent with the case of samples produced by conventional gravity die casting condition. More interestingly, one Mg, Ag-rich layer was observed in the vicinity of Al-Cu (θ) phase. These careful HAADF-STEM characterizations and quantitative APT data analyses give us some hints to design high performance Al-Cu-Mg based composites.

Experimental procedure

Conventional gravity die casting was performed by using Diez mold (produced by cast iron). Additive manufacturing was done by using laser powder bed fusion. HAADF-STEM was done by using Cs-corrected JEOL-2100F microscope operated at 200 kV, FEI Titan^{3TM} G2 60–300 (S/TEM) monochromated and probe corrected microscope operated at 300 kV, and Nion UltraSTEM100 operated at 300 kV. APT was done by using LEAP 5000XR.

Result and discussion

Grain size under conventional gravity die casting condition

Figure 1 shows the grain size under conventional gravity die casting condition by using optical microscopy, SEM and EBSD. It was found that grain size is less than 50 μm, TiB_2 particles are located along grain boundaries, and pores are also observed.

Figure 1 shows the grain size under conventional gravity die casting condition by using optical microscopy, SEM and EBSD.

Precipitates under conventional gravity die casting condition

Figure 2 shows the HAADF-STEM imaging and EELS maps under conventional gravity die casting condition. It was found that Al-Cu-Mg-Ag (Ω) phase are dominant after ageing. Similar precipitated were observed after additive manufacturing and/or ageing.

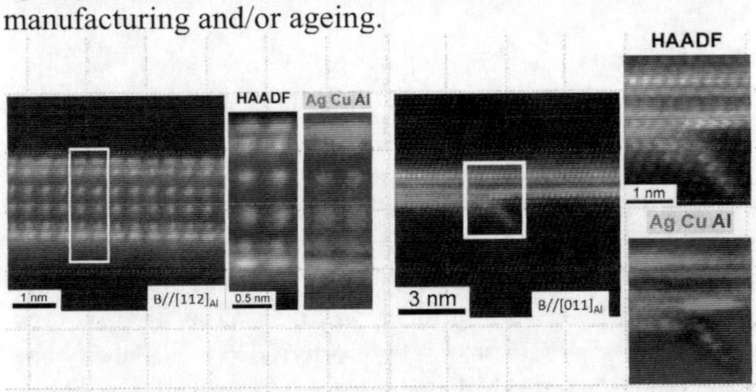

Figure 2 shows the HAADF-STEM imaging and EELS maps under conventional gravity die casting condition.

Grain size under additive manufacturing condition

Figure 3 shows the grain size under additive manufacturing condition by using SEM EBSD. It was found that grain size is less than 3 μm.

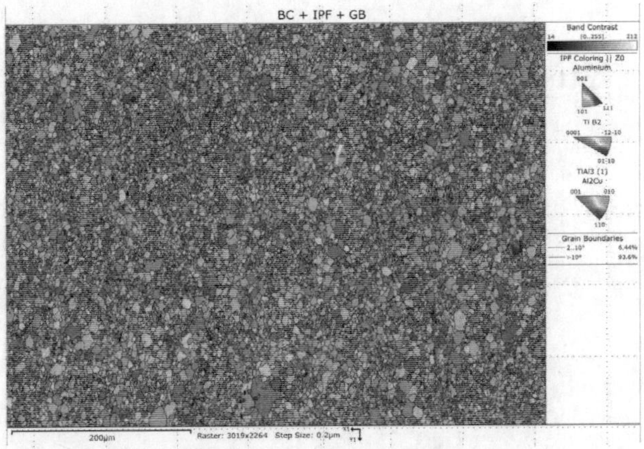

Figure 3 shows the grain size under additive manufacturing condition by using SEM and EBSD.

Conclusion

1. The addition of 3 wt.% TiB_2 (as grain refiner) into Al-Cu-Mg based alloys can refine grain size down to 50 μm and thereby greatly reduce hot tearing tendency and segregation.
2. Under conventional gravity die casting condition, a significant solute clustering (Mg-Ag-rich) was observed after the solution treatment (T4). After further over-ageing treatment (T7), three different types of precipitates (Al-Cu (θ) phase, Al-Cu-Mg-Ag (Ω) phase and Mg-Si (β) phase) were observed.
3. Under additive manufacturing condition, after further over-ageing treatment (T7), three different types of precipitates (Al-Cu (θ) phase, Al-Cu-Mg-Ag (Ω) phase and Mg-Si (β) phase) were also observed, which is fully consistent with the case of samples produced by conventional gravity die casting condition.
4. These careful HAADF-STEM characterizations and quantitative atom probe data analyses give us some hints to design high performance Al-Cu-Mg based composites.

Suppression of Detrimental Fe-Related Phases in Secondary Aluminum Alloys with CALPHAD-Guided Strategies

Chuan Zhang[1,*], Fan Zhang[1], Alan A Luo[2]
1. Computherm LLC, 8401 Greenway Blvd. STE248, Middleton WI-53562, USA
2. The Ohio State University, 140 W 19th Ave, Columbus, OH 43210, USA
*Corresponding address: e-mail: chuan.zhang@computherm.com

Abstract: The recycling of secondary aluminum alloys has gained increasing attention due to its economic and environmental benefits. However, the presence of Fe-related intermetallic phases, particularly the β-AlFeSi phase, poses significant challenges to the mechanical properties of these alloys. This study explores the routes of suppressing the detrimental phases in Al-Fe-Si alloys by adding Cr and/or Mn through both experimental investigation and CALPHAD simulations. The results reveal that CALPHAD simulations can accurately predict the microstructures observed in as-cast alloys, and the simulated processing map offers valuable guidance for optimizing the recycling process of secondary aluminum alloys.

Keywords: Casting, Aluminum alloy, Recycling, CALPHAD, Solidification.

1 Introduction

The recycling of secondary aluminum alloys, primarily from post-consumer scrap, has emerged as a critical area of research in response to growing environmental concerns and the need for resource conservation [1]. One of the primary challenges in recycling secondary aluminum alloys is the presence of impurity elements, particularly iron. The higher concentration of Fe will then enable the formation of Fe-related intermetallic phases, such as the β-AlFeSi phase (platelet-type), which is known to be brittle and detrimental to the mechanical properties of the alloy. Various strategies have been developed to mitigate the detrimental effects of Fe-related phases in secondary aluminum alloys. The addition of alloying elements such as Cr and Mn has been shown to stabilize the less harmful α-$Al_{15}(Cr,Fe,Mn)_3Si_2$ phase (with script or polygonal morphologies), which not only enhance the mechanical properties of the alloy but also improve its resistance to fracture and wear [2-4]. Studies have also demonstrated that rapid cooling can limit the formation of detrimental β-AlFeSi phases by reducing the time available for their growth, thereby enhancing the mechanical properties of the alloy [5]. The precise control of the cooling rate, in conjunction with the addition of alloying elements like Cr and Mn, can therefore be leveraged to optimize the microstructure and properties of secondary aluminum alloys. Despite significant advancements in these alloying and processing techniques, fully and efficiently suppressing the formation of detrimental phases in recycled secondary aluminum alloys remains a challenging goal. The complexity of multicomponent alloy systems and the variability in scrap composition necessitate a more systematic approach. The CALPHAD (CALculation of PHAse Diagrams) simulations offer a powerful tool to predict phase stability and guide the design of alloy compositions and processing conditions [6]. The integration of CALPHAD simulations into this process allows for the prediction of phase formation of multicomponent alloys under varying cooling conditions. This study aims to demonstrate the value of CALPHAD simulations in the recycling process of secondary aluminum alloys.

2 Methodology

A series of Al-Fe-Si alloys with varying Cr and Mn content were prepared using high-purity aluminum (99.7%) pure Si, Al-25 %Fe, Al-25 %Mn and Al-20 % Cr master alloys. The alloys were melted in the induction furnace and cast into the wedge molds under a wide range of cooling rates (1.5~65°C/s). The as-cast microstructures were characterized by scanning electron microscopy (SEM), and energy-dispersive X-ray spectroscopy (EDS) to identify the phases present and their distribution.

CALPHAD simulations were conducted using the Pandat™ software and PanAl database developed by Computherm LLC [7] to predict solidification paths in the Al-Fe-Si system with added Cr and Mn. The simulations considered a range of compositions and cooling rates to replicate the experimental conditions. The resulting phase diagrams and microstructure predictions were compared with experimental data to validate the accuracy of the simulations.

3 Result and discussion

The chemical compositions of all alloys prepared in this work were analyzed using a Spectromaxx (AMETEK) optical emission spectrometer and listed in Table 1.

Table 1. Composition of Alloy (wt.%)

	Si	Fe	Cr	Mn	Al
#1	8.71	1.01	0.21	0.26	Bal.
#2	8.79	1.05	0.23	0.52	Bal.
#3	8.98	0.99	0.21	0.78	Bal.

Figure 1 shows the BSE images of alloy #2 at the cooling rate ~1.5°C/s. Both script-type α-$Al_{15}(Cr,Fe,Mn)_3Si_2$ and needle-type β-AlFeSi intermetallic phases were observed. The EDS measured composition of the script-type phase is about Al-12Si-8Fe-4.7Mn-2.2Cr (at. %).

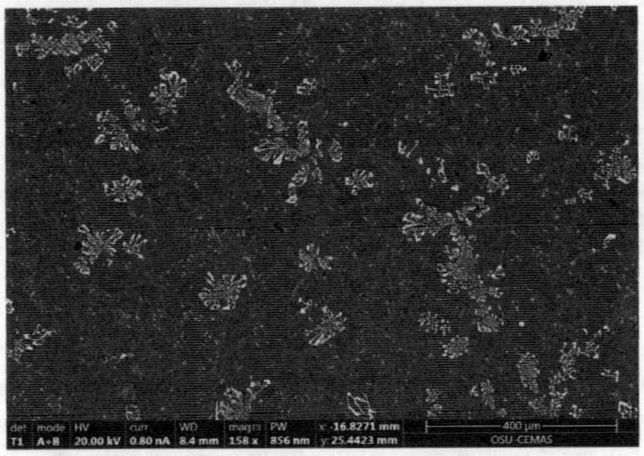

Figure 1: BSE image of alloy #2

Figure 2(a) shows calculated solidification path of alloy #2 using the Scheil model, which shows that the primary α-$Al_{15}(Cr,Fe,Mn)_3Si_2$ phase forms at about 665.6°C. The coexistence of the α-$Al_{15}(Cr,Fe,Mn)_3Si_2$ and β-AlFeSi at later solidification stage agrees well with the experimental observations. The elemental distribution of each alloying component within the α-$Al_{15}(Cr,Fe,Mn)_3Si_2$ phase is shown in Figure 2(b), which reasonably describes the EDS measurement.

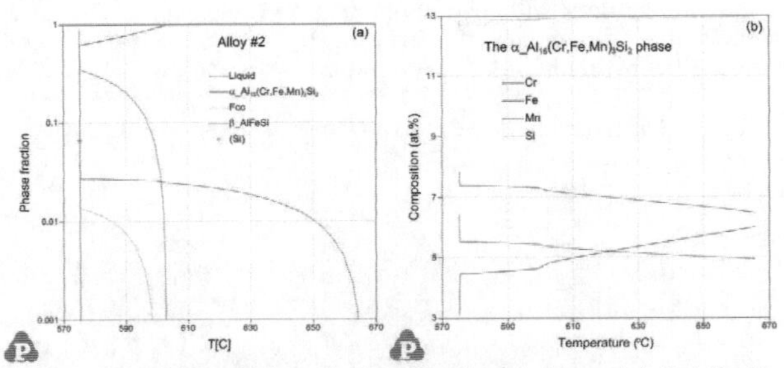

Figure 2: Scheil simulation of alloy #2 (a) and elemental distribution within the α-Al15(Cr,Fe,Mn)3Si2 phase (b)

4 Conclusion

This study demonstrates the effectiveness of CALPHAD simulations in predicting the microstructure of Al-Fe-Si alloys with Cr and Mn additions. The simulated results closely match experimental observations. This study demonstrated that CALPHAD method can be used as a feasible approach to optimize the recycling use of secondary aluminum alloys. By integrating CALPHAD simulations with key experiments, we can address the challenges posed by Fe-related intermetallic phases in the secondary aluminum alloys.

5 Acknowledgments

Thanks for the financial support by the U.S. Department of Energy 's Office of Energy Efficiency and Renewable Energy (EERE) under the Advanced Manufacturing Office Award Number DE-EE0 0 07897, awarded to The REMADE Institute, a division of Sustainable Manufacturing Innovation Alliance Corp.

References

[1] H. Nunes, O. Emadinia, R. Soares, M.F. Vieira, and A. Reis, Adding value to secondary aluminum casting alloys: a review on trends and achievements. Materials, 2023. **16**(3): p. 895.

[2] S. Ferraro, A. Bjurenstedt, and S. Seifeddine, On the formation of sludge intermetallic particles in secondary aluminum alloys. Metallurgical and Materials Transactions A, 2015. **46**: p. 3713-3722.

[3] E. Cinkilic, C. Ridgeway, X. Yan, and A. Luo, A formation map of iron-containing intermetallic phases in recycled cast aluminum alloys. Metallurgical and Materials Transactions A, 2019. **50**: p. 5945-5956.

[4] N. Balasubramani, M. Moodispaw, and A.A. Luo, Controlling the Fe-intermetallic phases and mechanical properties of secondary Al-9Si-1Fe alloy with Cr and Mn additions. Journal of Materials Science & Technology, 2025. **206**: p. 135-152.

[5] D.V. Malakhov, D. Panahi, and M. Gallerneault, On the formation of intermetallics in rapidly solidifying Al–Fe–Si alloys. Calphad, 2010. **34**(2): p. 159-166.

[6] F. Zhang, C. Zhang, S.-M. Liang, D. Lv, S. Chen, and W. Cao, Simulation of the composition and cooling rate effects on the solidification path of casting aluminum alloys. Journal of Phase Equilibria and Diffusion, 2020. **41**: p. 793-803.

[7] CompuTherm LLC, www.computherm.com.

Ni Coarsening in Ni-Yttria Stabilized Zirconia Electrodes: Three-Dimensional Quantitative Phase-Field Simulations Supported by Ex-Situ Ptychographic Nano-Tomography

Ming Chen[1,*], Shenglan Yang[1, 2], Lijun Zhang[2]
1. Department of Energy Conversion and Storage, Technical University of Denmark, Kgs. Lyngby 2800, Denmark
2. State Key Laboratory of Powder Metallurgy, Central South University, Changsha 410083, China
*Corresponding address: e-mail: minc@dtu.dk

Abstract: Solid oxide cells (SOCs) are electrochemical devices for efficiently converting chemical energy to electrical energy when operated as solid oxide fuel cells (SOFCs), or vice versa when operated as solid oxide electrolysis cells (SOECs). The durability of SOCs during long-term operation is essential for a cost-effective and efficient energy storage [1]. The porous cermet of nickel (Ni) and yttria-stabilized zirconia (YSZ) composite electrode is the current state-of-the-art SOC fuel electrode [2, 3]. However, during the long-term operation at high temperature (600-1000 oC), the Ni particles in the active Ni-YSZ electrode coarsen, resulting in both a reduction of the active three-phase-boundary (TPB) density and loss of Ni percolation [4, 5]. Therefore, in-depth understanding of nickel (Ni) coarsening is helpful for improving the service life of Ni-yttria stabilized zirconia (YSZ) electrodes in solid oxide cells. Unfortunately, very few quantitative experimental/ theoretical descriptions of Ni coarsening in Ni-YSZ electrodes during long-term operation exist.

In this work, quantitative modelling of Ni coarsening in Ni-YSZ electrodes was achieved through three-dimensional (3D) phase-field simulation supported by ex-situ ptychographic nano-tomography and input of reliable thermophysical parameters. A pragmatic procedure was proposed to refine and verify the materials parameters for the simulations. Moreover, the microstructures of the Ni-YSZ electrode in the pristine and annealed states obtained via the ex-situ ptychographic nano-tomography were used as the initial input and experimental validation for the phase-field simulations. After that, comprehensive comparison between the simulated and the experimental 3D microstructures was conducted, indicating the successful quantitative phase-field simulation of Ni coarsening in Ni-YSZ electrodes presented here. The success of this work is expected to pave the way for accurate prediction of the service life and even design of high-performance Ni-YSZ electrodes.

Keywords: Ni coarsening; Ni-YSZ electrode; Phase-field modelling; Ex-situ ptychographic nano-tomography

1 Result and discussion

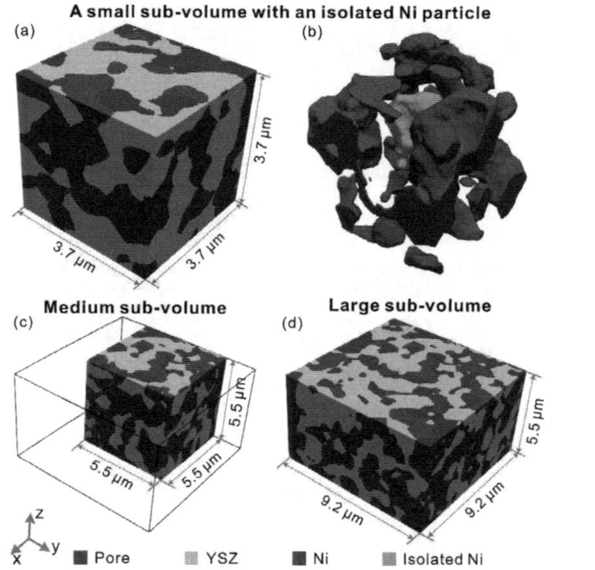

Fig. 1. Representative 3D microstructures taken from a real microstructure of the Ni-YSZ electrode. (a) a small sub-volume with a domain of 3.7×3.7×3.7 um3 containing an isolated Ni particle; (b) the same volume as in (a) showing only the Ni phase and the isolated Ni particle (i.e. the pore and YSZ phases are transparent); (c) a medium sub-volume with a domain of 5.5×5.5×5.5 um3, and (d) a large sub-volume with a domain of 9.2×9.2×5.5 um3. The blue region is the pore phase, the gray region is YSZ phase, the red region is Ni phase, and the orange region is the isolated Ni particle.

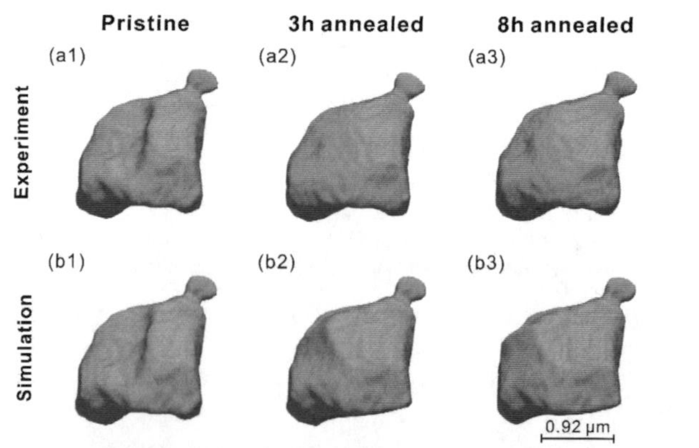

Fig. 2. 3D Morphology evolution of the isolated Ni particle in the microstructure shown in Fig. 1a and 1b. Experimental: (a1) Pristine state, (a2) after annealing at 850 oC for 3 h, and (a3) for 8 h; Phase-field simulation: (b1) Pristine state, (b2) after annealing for 3 h, and (b3) for 8 h.

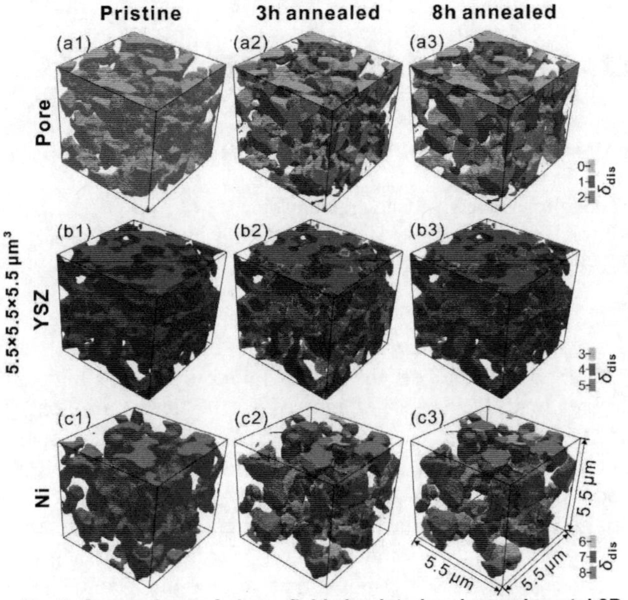

Fig. 3. Comparison of phase-field simulated and experimental 3D microstructures. $\delta_{dis}(t,x)$ for (a1 – a3) pore, (b1 – b3) YSZ, and (c1 – c3) Ni phases in phase-field simulated 3D microstructures with a medium domain of (a1 – a3, b1 – b3 and c1 – c3) 5.5×5.5×5.5 μm³ in the pristine state (a1, b1, c1), after annealing for 3 h (a2, b2, c2) and for 8 h (a3, b3, c3).

2 Conclusion

• A strategy of 3D phase-field simulation using the MPF model coupled with reliable thermophysical parameters and ex-situ experiment was proposed for the quantitative description of Ni coarsening in Ni-YSZ electrodes. With this strategy, the quantitative phase-field simulation of Ni coarsening in Ni-YSZ electrodes was realized.

• Based on the phase-field simulation of a small 3D sub-volume (i.e., the domain of 3.7×3.7×3.7 um³) with an isolated Ni particle calibrated with the experimental data, a set of material parameters were obtained. Their reliability was further confirmed through a comprehensive comparison between the phase-field simulation of twelve 3D small sub-volumes taken from different locations in the experimental microstructure of the Ni-YSZ electrode and the corresponding experimental data.

• A direct comparison of the simulated 3D microstructures with the subsequent experimental ones at annealing time steps was implemented, and their consistency was quantified. Moreover, the various characteristic information of 3D microstructures including microstructure parameters, PSDs and TPB tortuosity, were statistically analyzed and comprehensively compared. A good agreement between phase-field simulation and experimental results was achieved, and the phase-field simulation results performed in the large sub-volume were more consistent with experimental ones.

• The reliable time evolution of the TPB density of Ni-YSZ electrodes obtained by phase-field simulations can be used as input of an electrochemical model, and thus the time evolution of the electrochemical property of the Ni-YSZ electrode can be predicted. Combining phase-field-electrochemical simulations with artificial structure generators can be used in the future to model the performance of hypothetical structures to predict optimum electrode designs.

3 Acknowledgments

This work was supported by the project "Efficient Power2Gas Combining SOEC and Biomass Gasification" (EUDP no. 64017–0011). Shenglan Yang acknowledges the China Scholarship Council (CSC No: 201906370120). Lijun Zhang acknowledges the financial support from the Natural Science Foundation of Hunan Province for Distinguished Young Scholars (Grant No. 2021JJ10062).

References

[1] C. Graves, S.D. Ebbesen, S.H. Jensen, S.B. Simonsen, M.B. Mogensen, Eliminating degradation in solid oxide electrochemical cells by reversible operation, Nat. Mater. 2015, 14(2): 239-244.

[2] M.S. Khan, S.-B. Lee, R.-H. Song, J.-W. Lee, T.-H. Lim, S.-J. Park, Fundamental mechanisms involved in the degradation of nickel–yttria stabilized zirconia (Ni–YSZ) anode during solid oxide fuel cells operation: a review, Ceramics International, 2016, 42(1): 35-48.

[3] T.L. Skafte, J. Hjelm, P. Blennow, C. Graves, Quantitative review of degradation and lifetime of solid oxide cells and stacks, Proceedings of 12th European SOFC & SOE Forum, 2016, pp. 8-27.

[4] S. Koch, P.V. Hendriksen, M. Mogensen, Y.L. Liu, N. Dekker, B. Rietveld, B. De Haart, F. Tietz, Solid oxide fuel cell performance under severe operating conditions, Fuel Cells, 2006, 6(2): 130-136.

[5] D. Simwonis, F. Tietz, D. Stöver, Nickel coarsening in annealed Ni/8YSZ anode substrates for solid oxide fuel cells, Solid State Ionics, 2000, 132(3-4): 241-251.

Air Batteries Discharge Behavior of Mg-Al-Zn and Mg-Li-Al-Zn Alloys

Liang Wu[1,2,*], Huabao Yang[1], Yuantai He[1], Wenhui Yao[1,2], Jingfeng Wang[1,2], Bin Jiang[1,2], Fusheng Pan[1,2]
1. College of Materials Science and Engineering, Chongqing University, Chongqing, China
2. National Engineering Research Center for Magnesium Alloys, Chongqing University, Chongqing, China
*Corresponding address: e-mail: wuliang@cqu.edu.cn

Mg-air batteries have attracted increasing attention as promising electrochemical energy storage and conversion equipment because they have the advantages of having high theoretical voltage (3.09V) and high specific capacity density (2230Ah kg^{-1}), being safe to use, using abundant materials. However, some issues still limit the application of Mg-air batteries, and these issues include low anode efficiency and the actual discharge voltage being far less than the theoretical value. The former results from high self-corrosion of the magnesium alloy, and the latter results because discharge products (e.g., $Mg(OH)_2$) adhere to the magnesium anode surface. Thus, the magnesium anode, which has a high corrosion resistance and high discharge performance, tends to be used in the research of Mg-air batteries. One way to improve the corrosion resistance and discharge properties of Mg alloys is to alloy Mg with other elements.

AZ series (Mg-Al-Zn) magnesium alloys have been widely used and studied as anode materials. The Mg-Al-Zn and Mg-Li-Al-Zn alloys were used as the anode materials for magnesium-air batteries. The effects of texture, second phase, grain size, twins, and discharge products on the discharge performance of Mg-Al-Zn and Mg-Li-Al-Zn alloys were investigated by controlling the microstructure of the above alloys. Firstly, four representative Mg-3Al-1Zn anodes with different texture characteristics, i.e., ET0°, ET30°, ET60°, and ET90°, are assembled as the anodes for Mg-air batteries, and the correlation between texture and discharge products and the influences of texture and discharge products on the discharge performance are studied. The results show that the discharge performance of the four anodes follows the order: ET90° > ET60° > ET30° > ET0°. The discharge performance of AZ31 anode for Mg-air battery is affected by the texture and discharge products. It can be seen that the discharge products formed on the ET90° anode surface are more easily to fall off than those formed on the ET0° anode surface. The prismatic oriented grains such as the (10-10) and (11-20) oriented grains have higher discharge activities than the basal oriented (0002) grains. The ET90° anode provides higher and steadier discharge voltage and higher anodic efficiency than the ET0° anode. Although the ET90° anode discharged at 40 mA cm^{-2} still suffers a more severe self-corrosion. However, the ET90° anode exhibits more uniform dissolution without the existence of large pits, resulting in its higher anodic efficiency. Considering the influences of chunk effect and self-corrosion on the anodic efficiency, the ET90° anode still has higher anodic efficiency than the ET0° anode. Therefore, the ET90° anode possesses highest and steadiest discharge voltages of 1.46 V at 2.5 mA cm^{-2} and 0.71 V at 40 mA cm^{-2}. The corrosion behavior at open circuit potential and discharge properties under applied anodic currents of Mg-1Li-3Al-1Zn (LAZ131) and Mg-5Li-3Al-1Zn (LAZ531) alloys are investigated to analyze the effect of microstructural features for primary Mg-air batteries. The results show that the grain boundaries contribute equally to the corrosion and discharge processes, which are attacked preferentially than the grain interiors and accelerate the dissolution processes of α-Mg based Mg-Li alloys. The (10−10)/ (11−20) orientated grains are attacked preferentially than the (0002) orientated grains on the corrosion and discharge processes. The increased corrosion rate and improved discharge properties of α-Mg based Mg-Li alloys are attributed to the refinement of grain size, the decreased content of (0002) orientated grains and the increased content of (10−10)/ (11−20) orientated grains. The LAZ531 alloy possesses high and steady discharge voltage at small discharge current density for long time, with the values of 1.4801 V at 2.5 mA cm^{-2} and 1.3742 V at 10 mA cm^{-2}. Additionally, its anodic efficiency is higher than the LAZ131 alloy. Further, in order to improve their anodic efficiencies of the LAZ alloys, the Mg–5Li–3Al–1Zn (LAZ531) alloy sheets with different rolling reductions are investigated using hydrogen evolution, potentiodynamic polarization, electrochemical impedance spectroscopy, and galvanostatic discharge measurements. The results show that rolling changes the microstructure of LAZ531, such as the texture (grain orientation) and the presence of twins, and rolling increases the corrosion resistance and strengthens the discharge ability of LAZ531 alloy, which is attributed to the increased density of twins and the change of grain orientation. The grain boundaries and twins play the same roles on the corrosion and discharge processes. After application of the discharge current, the grain boundaries dissolve preferentially, meanwhile the twins are attacked. With increasing discharge time, more grain boundaries and twins are attacked. With the discharge time further increasing, the grain interiors are attacked. The uniform distribution of twins is helpful to the uniform dissolution of the anode material during the discharge process. Eventually, the LAZ531%-20% alloy has high and steady discharge voltage at small discharge current density for long time, with the values of 1.490 V at 2.5 mA cm^{-2} and 1.411 V at10 mA cm^{-2}. Besides, the LAZ531%-20% alloy has the highest anodic efficiency at each discharge current density, with the values of 36.9% at 2.5 mA cm^{-2} and 45.2% at 10 mA cm^{-2}.

In conclusion, the corrosion and discharge behavior of AZ alloy (LAZ131 and LAZ531) is studied. Based on the Li elemental content, extrusion and rolling, the different microstructure characteristics of alloys is prepared and the suitable microstructure characteristics of Mg alloys for the anodes of magnesium-air batteries were summarized as follows: fine grain size, a high content of (11-20)/ (10-10) orientated grains, introduction of twins and uniform distribution of fine second phases. These microstructure characteristics can effectively improve Mg-air battery anode performance.

Keywords: Magnesium alloys, Microstructure, Mg-air batteries, Discharge properties

Atomic Mobilities of Li-X-O (X=Co, Ni, Mn) Cathodes and Their Applications

Fangzhou Xing[1,2], Lijun Zhang[1,*]
1. State Key Laboratory of Powder Metallurgy, Central South University, Changsha, Hunan, China
2. Thermo-Calc Software AB, Stockholm, Sweden
*Corresponding address: e-mail: lijun.zhang@csu.edu.cn

Abstract: In this paper, the Li-X-O (X=Co, Mn, Ni) cathodes are chosen as the target materials to be investigated. The CALPHAD (CALculation of PHAse Diagram) method, combined with the models of anisotropic diffusion and the diffusion under electrochemical potential gradients, as well as the first-principles calculations, is utilized to develop the atomic mobility descriptions and quantitatively simulate the phase transitions of cathodes during the charging and discharging process, which may provide the important theoretical guidance for the design as well as the property optimization of cathode materials of lithium batteries.

Keywords: Li-X-O (X=Co, Ni, Mn) Cathodes, Atomic mobility, CALPHAD, first-principles calculation

1 Introduction

The cathode materials of lithium batteries determine the electrochemical properties of lithium batteries such as the capacity, voltage and cyclability. The phenomenological study of diffusion kinetics of cathodes is essential for understanding the mechanisms of cathode cyclability affected by the service parameters. So far, it is difficult to utilize a single phenomenological method or first-principles calculation to predict the dynamic relations of "element - composition - structure - electrochemical property" of cathodes due to the limitation of each method.

In this study, the Li-X-O (X=Co, Mn, Ni) cathodes are chosen as the target materials to be investigated. The CALPHAD (CALculation of PHAse Diagram) method [1,2], combined with the models of anisotropic diffusion and the diffusion under electrochemical potential gradients, as well as the first-principles calculations [3], is utilized to develop the atomic mobility descriptions and quantitatively simulate the phase transitions of cathodes during the charging and discharging process.

2 Modeling

The VASP (Vienna Ab-initio simulation package) software package and the NEB (Nudged Elastic Band) method were employed to explore the diffusion barrier of Li ions in cubic spinel $Li_xMn_2O_4$, while the CALPHAD approach was utilized to describe atomic mobilities of different phases in Li-X-O (X=Co, Ni, Mn) cathodes.

3 Result and discussion

The integrated computational approach of materials based on the CALPHAD method and the first-principles calculations has been established. The diffusion model considering the effects of electrochemical potentials and anisotropy and the moving boundary model are built, and the programming is implemented. The codes can obtain atomic mobility descriptions, coupling the thermodynamic database, as well as quantitatively simulate the phase transition process during the charging and discharging process.

The concept of anisotropic atomic mobility [4] based on the CALPHAD method is firstly proposed, and the anisotropic diffusion model is developed. Then, the atomic mobility database of O3 Li-Co-O cathodes is established according to the critically assessed experimental data. After that, the starting times of O3-O3' phase transition changing with the current and crystal orientation during the charging process are simulated. The diffusivity of layered cathodes as a function of crystal orientation, composition and temperature is model-predicted. The simulated results show that the Li-Co-O cathodes have better battery performance as the charging currents are in the range of 7.5~12.5 mA and the crystal orientation θ approaches to zero.

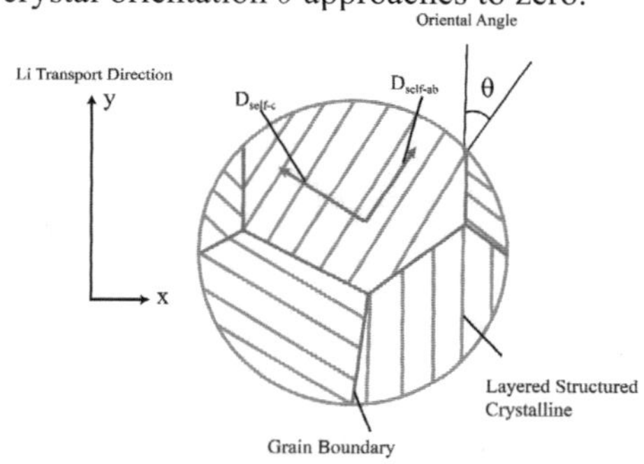

Fig. 1 Schematic diagram of microstructure of O3-layered hexagonal LixCoO2

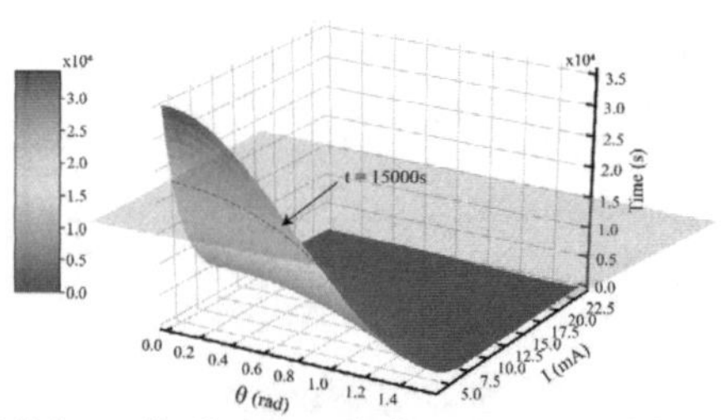

Fig. 2 Schematic diagram of microstructure of O_3-layered hexagonal LixCoO2

The atomic mobility database of O3, O3' and H1-3 phases of Li-Ni-O cathodes are established, and the diffusion kinetics of cathodes is quantitatively simulated. The concentration profiles, phase fractions and moving boundary process during the charging and discharging process are model-predicted. The simulated results show that reducing charging currents within the range of 4.5~15 A/m^2 may avoid the formation of H1-3 phase; when the H1-3 phase formation cannot be prevented considering to realize fast charging process, increasing the charging current slightly can lower the maximum content of H1-3 and O1 phases, and then relieve the capacity face of Li-Ni-O cathodes.

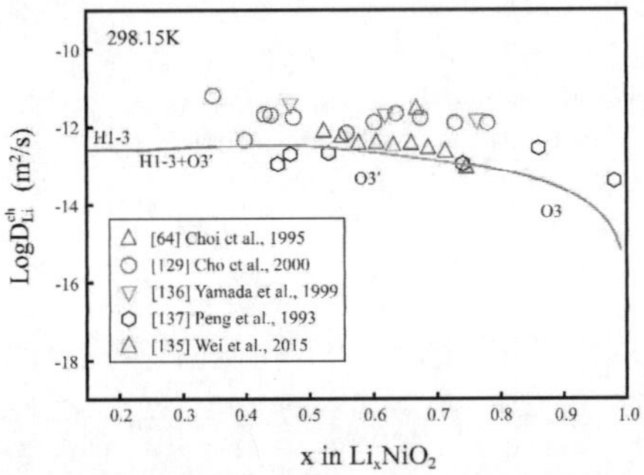

Fig. 3 Calculated chemical diffusion coefficients as a function of compositions at 298.15K based on the Li-Ni-O thermodynamic descriptions and the obtained atomic mobility

Furthermore, the reliable experimental data of diffusivity is firstly filtered, and the atomic mobility database of spinel Li-Mn-O cathodes are established by means of the CALPHAD method and the first-principles calculations, considering the effect of the grain boundaries. Then, the evolutions of $LiMnO_2$ phase fraction, capacity as well as the cell voltage under different current densities and cycles during the discharging process are quantitatively simulated. The electrochemical properties changing with the currents are model-predicted. The simulated results show that larger discharging currents may cause more serious capacity fade, which can be avoided when the current density is in the range of 0~0.367mA/cm^2.

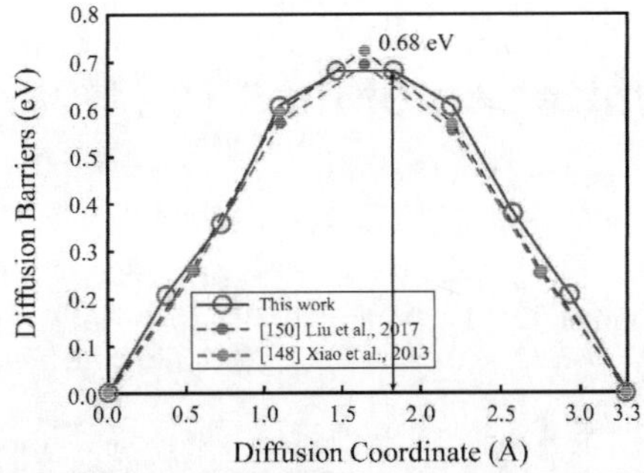

Fig. 4 First-principles calculated diffusion barriers of Li in the spinel $Li_xMn_2O_4$ along the migration pathway of 8a-16c-8a sites in the present work, compared with those in the literature

Conclusion

Accurate atomic mobility descriptions of Li-X-O (X=Co, Mn, Ni) cathodes were established by using the CALPHAD approach together with first-principles calculations. Moreover, the phase transitions of cathodes during the charging and discharging process Li-X-O (X=Co, Mn, Ni) cathodes were quantitatively simulated.

Acknowledgments

The financial support from the National MCF Energy R&D Program of China (Grant No. 2018YFE0306100) and the Natural Science Foundation of Hunan Province for Distinguished Young Scholars (Grant No. 2021JJ10062) is acknowledged.

References

[1] Wang, Q., Zhang, G., Li, Y. Application of phase-field method in rechargeable batteries. npj Computational Materials, 2020, 6, 176.
[2] Ma, S., Xing, F. Z., Deng, C. et al. A novel analytical approach to describe the simultaneous diffusional growth of multilayer stoichiometric compounds in binary reactive diffusion couples. 2021, 191, 111-115.
[3] Haldar, P., Chatterjee, A. Nudged-elastic band study of lithium diffusion in bulk silicon in the presence of strain. Energy Procedia, 2014, 54, 310-319.
[4] Tang, Q., Ma, S., Xing, F.Z., Zhang, L.J. Anisotropic atomic mobilities of hcp Zr (O) solid solution and their application in description of early-stage oxidation process of pure Zr. Corrosion Science, 2021, 186, 109445.
[5] Xing, F.Z., Ma, S., Zhang, L.J. Anisotropic atomic mobilities of layer-structured O3 Li_xCoO_2 cathodes and their applications in optimization of battery performance. The Journal of Physical Chemistry C, 2020, 124, 10832-10844.

Optimization of the First Hydrogenation Reaction Via Introducing Active Sites Through Y-Doping

Shiyang Hu, Xin Ding*, Ruirun Chen**, Xiangfeng Ma
National Key Laboratory for Precision Hot Processing of Metals, Harbin Institute of Technology, Harbin 150001, China
*dingxin_hit@hit.edu.cn; ** ruirunchen@hit.edu.cn

Abstract: Mg-based hydrogen storage alloys are considered as hydrogen storage materials with great development potential, but their sluggish kinetics restrict their applications. Activation property is also considered to be one of the important indicators for evaluating the performance of hydrogen storage alloys. In this study, as the doping amount of Y element increases, the grain size decreases and the initial hydrogenation performance gradually improves. The first hydrogenation time of Y1 sample is only 8554 s. Due to the strong affinity between Y and H atoms, the number of active sites increases, which shortens the "incubation period" and accelerates the adsorption and dissociation process of H molecules.

Keywords: Hydrogen storage, Mg-based alloy, Activation performance

1 Introduction

Mg-based alloys are considered to be one of the most promising hydrogen storage materials with important practical value and application prospects due to their widespread source, low price and high capacity(7.6 wt.% and 110 g/L) [1]. Nevertheless, the practical application of MgH_2 is limited by its poor thermodynamic and kinetic properties, which are attributed to the difficult dissociation of H molecules, the slow diffusion rate of H atoms and the high energy of Mg-H bonds [2]. Alloying is an effective method to improve the performance of Mg-based hydrogen storage alloys [3].It is found that the addition of transition metal (TM) elements and rare earth (RE) elements can effectively weaken the bond energy between Mg and H atoms and accelerate the recombination and dissociation of H molecules [4].

2 Experimental procedure

Materials preparation

The $Mg_{97}Ni_{2-x}Y_x$ (x=0, 0.25, 0.5, 1) as-cast alloys were prepared in an electric resistance-heated furnace under the protection of SF_6 shielding gas atmosphere.The as-cast alloys were synthesized by the industrial pure magnesium (99.9%), Mg-70wt.%Ni and Mg-20wt.%Y intermediate alloys. An excess of 3 wt.% Mg was added in order to compensate for Mg evaporation losses from the melting process. To simplify the description, $Mg_{97}Ni_{2-x}Y_x$ (x=0, 0.25, 0.5, 1) samples are abbreviated as Y_0, $Y_{0.25}$, $Y_{0.5}$ and Y_1, respectively.

The microstructures of the as-cast alloys were characterized by scanning electron microscopy (SEM). Hydrogen storage properties of the as-cast alloy powder samples were measured with a precision Sieverts-type apparatus.

3 Result and discussion

Microstructure characteristics

The microstructures of the as-cast alloys are exhibited in Fig. 1. Two phases are distinguished in Y_0 alloy, the primary phase is a blossom-shaped Mg phase, and the eutectic is $Mg+Mg_2Ni$. Doping with Y element in $Y_{0.25}$, $Y_{0.5}$ and Y_1 alloys, blocky NiY_3 phases are preferentially precipitated, which act as nucleation sites for Mg phases and hinder the growth of Mg grains. As the amount of Y doping increases, the sizes of the Mg and NiY_3 phases decrease, and the distribution of NiY_3 phases becomes more diffuse.

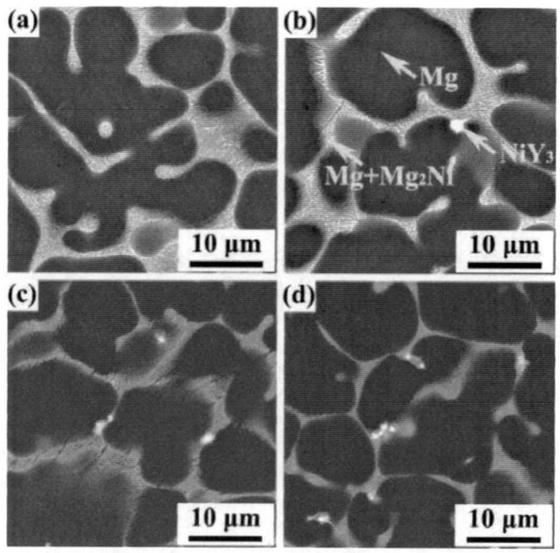

Fig.1 Microstructures of the as-cast alloys: (a)Y_0, (b)$Y_{0.25}$, (c)$Y_{0.5}$, (d) Y_1.

4 Activation performance

Due to the active chemical properties of Mg, a dense MgO film is easily formed in oxygen-containing environments, which hinders the reaction between Mg and H atoms. During the activation process, cracks gradually appear on the oxide film and expose the fresh surface owing to the expansion or contraction of lattice volume. Therefore, activation property is considered to be one of the important indicators for evaluating the performance of hydrogen storage alloys.

In this experiment, each sample is carried out three cycles of hydrogen absorption and desorption activation at 400 ℃ under 3 MPa and 0.1 MPa). Fig.2 (a-d) shows the hydrogen absorption curves, revealing a great difference in the first cycle. The first hydrogenation process of the Y_1 sample is only 8554 s. However, the first hydrogenation process of Y_0 sample is 13581s, which is 1.5 times that of the Y_1 sample.The initial stage of activation with almost no hydrogen absorption is called the "incubation period". There is no "incubation period" in Y_1 alloy, but there is a brief incubation period in $Y_{0.25}$, $Y_{0.5}$ and Y_1 alloys.

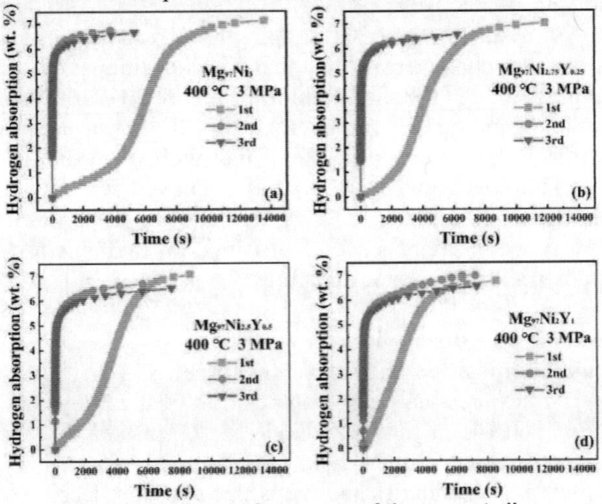

Fig.2 Activation performances of the as-cast alloys: (a)Y_0, (b)$Y_{0.25}$, (c)$Y_{0.5}$, (d) Y_1.

The hydrogen absorption process can be summarized as follows: (1) Adsorption and dissociation of H_2 molecules on the surface, (2) diffusion of H atoms, (3) formation of α-solid solution, (4) formation of β-metal hydrides. Due to the strong affinity between Y and H atoms, NiY_3 phase first attracts H molecules and acts as an active site, which accelerates the adsorption and dissociation process of H molecules. The decrease in grain size after doping with element Y accelerates the diffusion process of hydrogen atoms, which is because grain boundaries are fast diffusion channels for H atoms. Therefore, doping with Y element can significantly improve the initial hydrogenation performance.

5 Conclusion

(1) As the amount of Y doping increases, the sizes of the Mg and NiY_3 phases decrease.

(2) The first hydrogenation process of the Y_1 sample is only 8554 s, which is 0.63 times that of the Y_0 sample.

(3) The number of active sites increases with increasing Y element content, which shortens the "incubation period" and accelerates the adsorption and dissociation process of H molecules.

6 Acknowledgments

This work is financially supported by the Natural Science Foundation of Heilongjiang Province, China (LH2021E51).

References

[1] Yang X, Zhang J, Hou Q and Guo X. Regulation of kinetic properties of chemical hydrogen absorption and desorption by cubic K_2MoO_4 on magnesium hydride. Nanomaterials, 2022, 12(14): 2468.

[2] Bogdanović B, Bohmhammel K, Christ B, Reiser A, Schlichte K, Vehlen R and Wolf U. Thermodynamic investigation of the magnesium-hydrogen system. J. Alloy Compd., 1999, 282(1-2): 84-92.

[3] Cao W, Ding X, Zhang Y, Zhang J, Chen R, Su Y, Guo J and Fu H. Enhanced de-/hydrogenation kinetics of a hyper-eutectic $Mg_{85}Ni_{15-x}Ag_x$ alloy facilitated by Ag dissolving in Mg_2Ni. J. Alloy Compd., 2022, 917: 165457.

[4] Zhao D L and Zhang Y H. Research progress in Mg-based hydrogen storage alloys. Rare Metals, 2014, 33: 499-510.

Study on Microstructure and the Hydrogen Storage Behavior of $Ti_{25}V_xCr_{25}Nb_{50-x}$ Multi-Principal Elements Alloys with Customized Parameters

Xiangfeng Ma, Xin Ding*, Ruirun Chen*, Shiyang Hu
National Key Laboratory for Precision Hot Processing of Metals, Harbin Institute of Technology, Harbin 150001, China
*Corresponding authors: Xin Ding, dingxin_hit@hit.edu.cn; Ruirun Chen, ruirunchen@hit.edu.cn

Abstract: Multi-principal elements alloys show great potential in hydrogen storage due to their severe lattice distortion and wide range of compositional designs, but their highly stable hydrides limit further applications. In this study, the Ti25VxCr25Nb50-x alloys with tailored parameters (VEC = 5) were designed and the microstructure and hydrogen storage behavior of the alloys were investigated. The results show that all alloys consist of a single BCC phase with a dendritic microstructure. Compared to V25 and V35, V15 has the largest lattice constant and the finest dendritic crystals, which are responsible for the excellent hydrogen storage properties of V15. The reversible hydrogen storage capacity of V15 is 1.4 wt.% at 120 °C.

Keywords: Hydrogen storage, MPEAs, BCC phase

1 Introduction

As a newly emerging alloy system in recent years, multi-principal elements alloys (MPEAs) have been widely concerned as hydrogen storage materials due to their significant lattice distortion and the ability to tailor a range of interesting properties by changing the alloy composition [1].However, the development of new things is not smooth sailing. The general disadvantage of MPEAs is that its hydride phase has high stability, and it can only be completely desorbed after long-term desorption in dynamic vacuum at high temperature [2].Predicting alloy properties from calculated parameters to aid in alloy composition design is important for MPEAs with far greater numbers of alloys than conventional alloys [3]. Nygård et al. found that the VEC value is positively correlated with the expansion of the lattice volume, and that larger VECs may lead to instability of the hydride [4]. Based on these insights, $Ti_{25}V_xCr_{25}Nb_{50-x}$ (x = 15, 25, 35) alloys with customized parameters (VEC = 5) were designed.

2 Experimental procedure

Ingots of $Ti_{25}V_xCr_{25}Nb_{50-x}$ were prepared in water-cooled copper crucibles by the arc-melting technology. The entire preparation process was carried out in an argon (purity 99.999%) atmosphere. The sample used for phase characterization and hydrogenation test was taken from the core of each ingot. r. Prior to all tests, removal of oil and oxide from the surface of the specimens with sandpaper. For the convenience, the $Ti_{25}V_xCr_{25}Nb_{50-x}$ alloys with x = 15, 25 and 35 are designated as V15, V25 and V35, respectively.

X-ray diffraction (XRD) using Cu-Kα radiation was utilized to characterize the phase compositions of the samples, and the test angle and test step are 20 ~ 100° and 4°/min. Scanning electron microscopy (SEM) was used to identify the microstructural morphology. Energy-dispersive spectrometer (EDS) was employed to study the elemental distribution. The hydrogen storage properties of alloy powder samples were measured by a homemade volumetric Sieverts-type apparatus.

3 Result and discussion

Phase composition and microstructure

Fig. 1 illustrates the phase composition of the alloys. All alloys consist of only a single BCC phase. As the elemental content of V rises, the diffraction peaks are shifted to higher angles and the lattice constant decreases, which is attributed to the fact that the atomic radius of Nb is larger than that of V.

Fig. 2 shows backscattered electron images andEDS mappings for V15, V25 and V35. All alloys show typical dendritic microstructure. All alloys show a typical dendritic microstructure with slightly different chemical compositions in the dendritic and interdendritic regions. Dendritic crystal coarsening was found with increasing V element content.

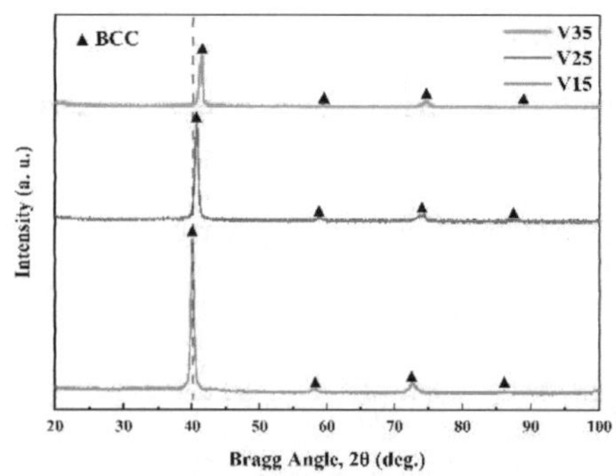

Fig.1 XRD patterns of V15, V25 and V35.

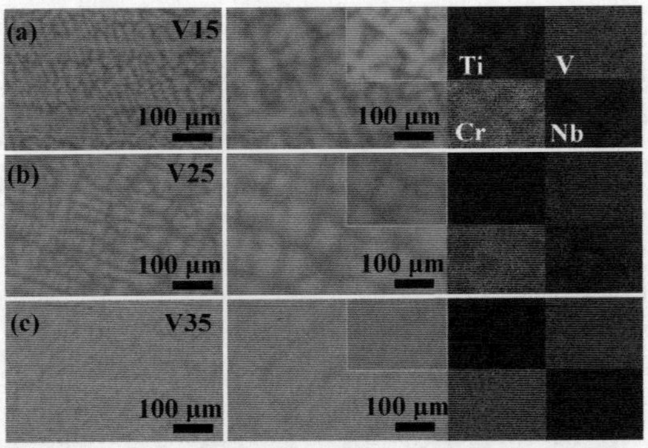

Fig.2 Backscattered electron images and EDS maps of as cast (a)V15, (b)V25 and (c)V35.

4 Hydrogen storage behavior

Hydrogenation tests were performed with about 2 g of alloy powder, from 100 mesh to 200 mesh. The alloys require activation prior to hydrogenation due to breakage of the alloy in air and oxidation on the powder surface. The activation conditions used in this experiment were holding at 350 °C for 2 hours, followed by the first hydrogenation at 120 °C and 3 MPa.

It is worth noting that the V25 and V35 alloys were not able to absorb hydrogen in the first hydrogenation, which is attributed to the fact that the alloys could not be activated by holding at 350 °C for 2 hours. A comprehensive hydrogen storage test was performed on the activated V15 alloy and the results are shown in Fig. 3.

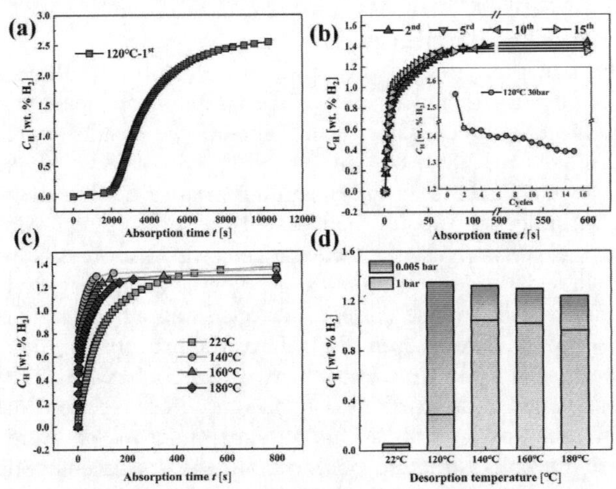

Fig.3 Hydrogen storage behavior of V15: (a) first hydrogenation (b) cycling performance (c) hydrogenation behavior at different temperatures and (d) Hydrogen Desorption at Different Temperatures

V15 Alloy reached 2.54 wt.% hydrogen absorption capacity and 1.4 wt.% reversible hydrogen absorption capacity in the first hydrogenation. After 15 cycles of hydrogen absorption and desorption at 120 °C, the hydrogen absorption capacity decayed to 1.35 wt.%. Following this, hydrogenation tests were carried out at different temperatures. The hydrogen storage capacity of V15 alloy decreases with increasing temperature. V15 alloy has the highest hydrogen storage capacity, 1.37 wt.%, at room temperature, requiring only 800 seconds for essentially complete hydrogenation.

5 Conclusion

(1) All alloys consist of a single BCC phase with a dendritic microstructure.
(2) The BCC phase lattice constant decreases and dendrites coarsen with increasing V element content.
(3) The V15 alloy has a reversible hydrogen absorption capacity of 1.4 wt.% at 120°C.

6 Acknowledgments

This work is financially supported by the National Key Research and Development Program of China (2023YFB4005401).

References

[1] Yang F, Wang J, Zhang Y, Zhen W, Zhang Z, Zhao F, Huot J, Novaković J and Novaković N. Recent progress on the development of high entropy alloys (HEAs) for solid hydrogen storage: a review. Int. J. Hydrogen Energy, 2022, 47(21): 11236-11249.

[2] Chen S, Lee P, Lee H and Su H. Hydrogen storage of C14-$Cr_uFe_vMn_wTi_xV_yZr_z$ alloys. Mater. Chem. Phys., 2018, 210: 336-347.

[3] Zhang Y, Zhou Y J, Lin J P, Chen G L and Liaw P K. Solid-solution phase formation rules for multi-component alloys. Adv. Eng. Mater., 2008, 10(6): 534-538.

[4] Nygård M, Ek G, Karlsson D, Sørby M, Sahlberg M and Hauback B. Counting electrons - a new approach to tailor the hydrogen sorption properties of high-entropy alloys. Acta Mater., 2019, 175: 121-129.

Crack Initiation Mechanism in Casting AC4B Aluminum Alloy Parts with Complex Structure

Daliang Yu[1], Wen Yang[2], Wanqing Deng[2], Songzhu Zhu[2], Qingwei Dai[1, 3*], Dingfei Zhang[3]
1 School of Metallurgy and Materials Engineering, Chongqing University of Science and Technology, Chongqing, 401331, China.
2 Chongqing Zhicheng Machinery Co., LTD, Chongqing, 400039, China.
3 College of Materials Science and Engineering, Chongqing University, Chongqing, 400045, China
*Corresponding address: e-mail: daiqingwei@cqust.edu.cn (Q. Dai)

Abstract: Hot tearing is a common defect observed in aluminum (Al) components with intricate geometries. In this study, optical microscopy (OM), scanning electron microscopy (SEM), energy-dispersive X-ray spectroscopy (EDS), and the finite difference method (FDM) were employed to investigate the mechanisms underlying hot tearing in cast AC4B aluminum alloy engines. Shrinkage porosities, ranging from 300 to 500 μm in size, were identified near the surface of the cast specimens. Additionally, numerous fractured brittle Fe-bearing phases were detected on the crack surfaces of the castings. Analysis of the secondary dendrite arm spacing (SDAS) across five typical locations revealed that the cooling rate at position 5 was significantly higher than at position 1. The improper solidification sequence between these positions disrupted the feeding of liquid metal to position 1 during solidification, potentially resulting in casting defects or crack formation. Numerical simulations further corroborated these findings, demonstrating that the solidification sequence influenced defect formation, consistent with the microstructural observations. Moreover, impurity accumulation was observed at position 5 during the filling process. The study concludes that the presence of large β-Fe intermetallics, coupled with an improper solidification sequence, jointly contributed to crack initiation in the aluminum engine components.

Keywords: Aluminum alloy, casting defects, casting sequence, microstructure

1 Introduction

To enhance vehicle fuel efficiency, aluminum (Al) alloys have been increasingly employed as engineering materials in the automotive industry. Among them, Al-Si-Cu series cast alloys are widely utilized for engine components and engine blocks due to their superior castability and excellent mechanical properties at both ambient and elevated temperatures. However, the performance of cast Al alloy components is influenced by various factors, including alloy composition, casting techniques, and processing parameters. Common casting defects such as hot tearing, brittle secondary phases, porosity, and solidification shrinkage can significantly degrade the mechanical performance of these components in service.

The newly developed cast AC4B Al alloy engine, characterized by its complex structural design, has frequently exhibited hot cracking at specific locations. Understanding the mechanisms of hot crack initiation in these engines is crucial for enhancing product quality and provides valuable insights into the interplay between component geometry and casting defects.

In this study, the microstructural evolution in five representative regions of the engine components was examined to elucidate the changes occurring during the casting process. Detailed observations of crack surfaces were performed to investigate the mechanisms of crack initiation during manufacturing. Numerical modeling techniques were also applied to analyze the temperature distribution and solidification sequence during casting. The relationship between microstructure, solidification behavior, and crack initiation mechanisms in the cast alloys was systematically evaluated.

2 Experimental procedure

The materials used in this study were AC4B Al alloys. The chemical compositions of the alloys were measured by the optical emission spectrometer, the results were 9.26 wt.% Si, 2.89 wt.% Cu, 0.65 wt.% Mg, 0.78 wt.% Zn, 0.86 wt.% Fe, 0.12 wt.% Mn and the balance Al. The melt was refined at 700 °C then casted by low pressure die casting. After casted, the engines were successively processed by decore, solution treatment, quenching and aging treatment. The casted, quenched and aging treated specimens were chosen for microstructure observation. These specimens were cut into two parts, microstructure observation was carried out on the cross section in different positions. SEM and optical microscope were used for microstructure observation. The specimens and positions that selected for microstructure were illustrated in Fig.1. Five typical positions were chose. The position 3 is in the button of engine and position 4 is in the top of engine. The casted, quenched and aged engine were tested by eddy current inspection. If macro cracks were detected, then the crack area were cut off and specimens were break into two parts along cracks for crack surface observation.

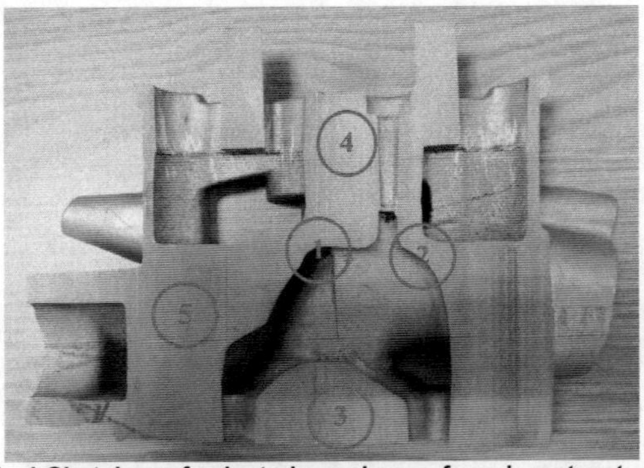

Fig.1 Sketches of selected specimens for microstructure observation

3 Result and discussion

Solidification sequence

The secondary arm spacing and solidification sequence of eligible specimens were shown in Table 1. In order to obtain precise secondary arm spacing, hundreds of secondary dendrite were measured for each position. It can be seen that the specimens in different positions had different secondary arm spacing.

Table 1 Secondary arm spacing and solidification sequence of rejected specimens

Position	Statistical magnitude	Secondary Arm Spacing	Solidification sequence
1	795	48.8	
2	504	39.8	
3	413	52.2	4 2 5 1 3
4	308	22.2	
5	830	42.2	

The cooling rate and solidification sequence has a great influence on casting quality. Since the engine had a complex structure, the cooling rate was various in different positions that result in the unpredictable solidification sequence, impertinent solidification sequence could lead to casting defects.

Crack initiation mechanisms

During casting process in AC4B alloy, casting defects were closely related with casting techniques especially cooling rate. As discussed above, since the engine had complex structure, the unexpected solidification sequence resulted in casting defects. What was worse, these casting defects might promote crack initiation. In casting Al alloys, the plate-shaped β-Fe (β-AlxFeySiz) was always considered to be the most detrimental intermetallic in crack initiation. The crack initiation mechanisms in the engine could be summarized in Fig.2.

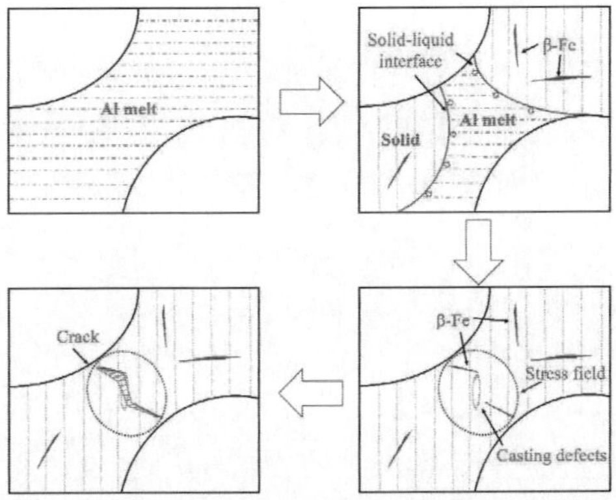

Fig.2 Sketches of crack initiation during casting in position 1

Fig.2 showed the sketches of crack initiation in position 1. During casting process, the solid-liquid interface converged at position 1. The improper solidification sequence impeded solidification feeding in position 1, and two interfaces' interaction could cause air entrainment. As the result, casting defects emerged in position 1. Since the solidification feeding was hindered, the solidification contraction induced stress field in this region. Plate-shaped β-Fe precipitated in Al alloy. In stress field, the elasticity and plasticity incompatibility between β-Fe and Al matrix lead to crack initiation in these region. The presence of casting defects facilitated crack coalescence with each other in the stress field. Therefore, cracks were frequently found in position 1.

4 Conclusion

- Fe-bearing bulk was found in crack surface, the cracked big compound with many cleavage-liked features. There brittle Fe-bearing phase would promote crack initiation in service, and lead to deteriorative mechanical properties.
- The secondary dendrite arm spacing in five typical positions were significantly different. The secondary dendrite arm spacing in position 5 is smaller than that in position 1. The variation of secondary dendrite arm spacing is related to cooling rate during casting.

5 Acknowledgments

This work was sponsored by Zhicheng Company, Chongqing Science and Technology Commission (cstc2018jcyjAX0607, cstc2019jcyj-msxmX0284), Scientific and Technological Research Program of Chongqing Municipal Education Commission (KJQN201801523), Project funded by China Postdoctoral Science Foundation (2019M653317, 2016T90831, 2015M572447), Chongqing Talent Project (CQYC201905100).

Prediction of Printing Performance of Additively Manufactured M2052 Alloy by Simulation Coupled with Machine Learning

Haoran Zhou, Huaiqian Li, Yingchun Ma, Jian Shi, Zhiquan Miao, Yanpeng Wei*,
National Key Laboratory of Advanced Casting Technologies, Shenyang Research Institute of Foundry Co., Ltd. CAM, Shenyang 110022, China
*Corresponding address: e-mail: weiyp1609@163.com

Abstract: M2052 alloy is a twinned damping alloy with damping and shape memory effects. This paper employs a simulation-machine learning-experimentation approach to establish a connection between the different dampings and mechanical properties of M2052 alloy produced under various printing process parameters. Utilizing the Single Bead Parametric Simulation and Microstructure Simulation modules in ANSYS simulation software, datasets of melt pool morphology and microstructure were generated. Five machine learning models are selected and ultimately optimized to establish the relationship between process parameters and microstructure. The use of machine learning effectively predicts the performance of specimens under different process parameters.

Keywords: M2052; damping alloy; simulation; machine learning.

1 Introduction

MnCu alloy is a typical twin-shaped damping alloy, and its excellent damping performance mainly comes from the high migration rate of {101} twin boundaries caused by the martensitic phase transformation, accompanied by low critical twin/de-twinning stress and tetragonal lattice distortion (1-c/a). Due to its outstanding damping and mechanical properties, it has a wide range of applications in fields such as radio frequency communication, sensors and actuators, energy harvesters, and biomedical devices. The development of the Mn-20Cu-5Ni-2Fe (at. % M2052) alloy is relatively mature. It is characterized by its ability to simultaneously achieve excellent mechanical and damping properties, and has a broad potential for application in complex service conditions such as deep space and deep sea. Typically, M2052 alloy is manufactured using traditional methods such as vacuum induction melting, forging, and rolling techniques to produce finished components. However, in recent years, the precision requirements for MnCu alloy components have been increasing, along with demands for complex spatial gyration structures and even lattice structures. Traditional processing methods are unable to fully meet the economic and time cost requirements for these types of applications, to a certain extent restricting the prospects for the use of MnCu alloys.

Selective laser melting (SLM) is a typical additive manufacturing process capable of integrally forming complex spatial structures while maintaining extremely high heat input (10^{5-7} W/cm^3), a large temperature gradient in the molten pool (10^6 K/m), and a high cooling rate (10^{3-8} K/s). The rapid heating and cooling characteristics of the thermal source, along with the thermal cycling effect of layer-by-layer melting and printing, facilitate the formation of fine and dispersed equiaxed crystals as well as dense and tough complex components. The properties of SLM can also effectively assist in the forming of MnCu alloy components, increase the solubility limit, refine the grain structure, and promote the formation of metastable phases. Therefore, M2052 high-damping alloy prepared using SLM technology holds broad research and application prospects.

2 Experimental procedure

Materials and equipment

Preparation of M2052 alloy powder with a particle size of 15-53 μm using electrode induction melting atomization method. The powder composition is listed in Table 1. The morphology and size of M2052 powder are shown in Figure 1.

Table 1 Chemical composition of the M2052 power.

Elements	Mn	Cu	Ni	Fe	O
wt. %	Bal.	21.66	4.69	1.65	0.11

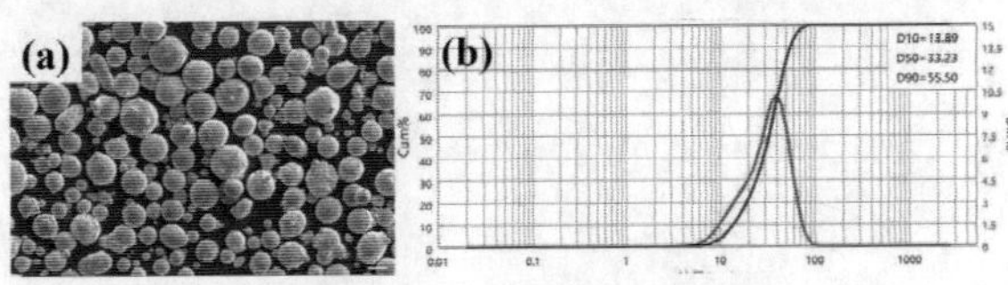

Fig. 1 (a)SEM morphology of M2052 powders (b) particle size distribution of alloy powder

The specimens were fabricated using a Pro X200 laser powder bed fusion (LPBF) system. The standard for setting the printing process parameters was based on the laser volumetric energy density (Ev), which is determined by the following factors: laser power (P), scan speed (v), hatch spacing (h), and powder layer thickness (t). The volumetric energy density can be calculated using the formula(1):

$$E_v = P/(v \times h \times t) \quad (1)$$

Simulation calculation and machine learning prediction algorithms

By designing simulated experimental data effectively, rich training and testing sets can be provided for machine learning models, thereby improving the prediction accuracy and generalization ability of the models. In this study, the machine learning algorithm's training and testing data were mainly obtained using the Single Bead Parametric Simulation and Microstructure Simulation modules within the finite element simulation and cellular automata.

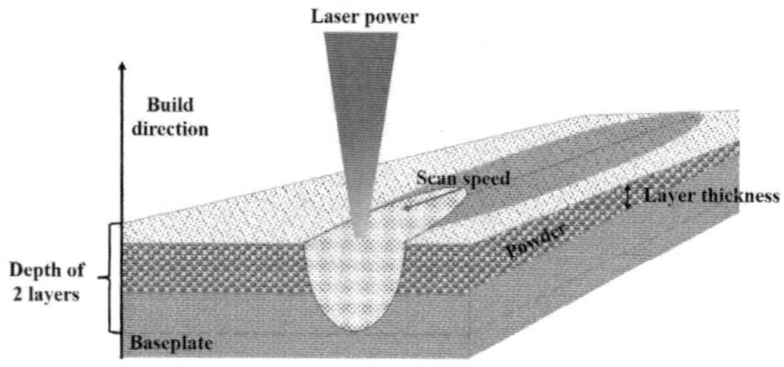

Fig.2 Schematic diagram of SLM process

3 Result and discussion

Machine learning based prediction of molten pool morphology

For the as-printed M2052 alloy, the damping performance originates from the relaxation effect of twin grain boundaries, as well as the energy dissipation caused by the fcc→fct phase transformation and the paramagnetic→ferromagnetic transition-induced lattice distortion. According to Cochardt's magnetomechanical damping model, under the same strain amplitude, the damping capacity and the saturation magnetostriction constant (λ) of the twinned damping alloy exhibit a positive correlation, and the value of λ can be established as a functional relationship with the degree of tetragonal lattice distortion.

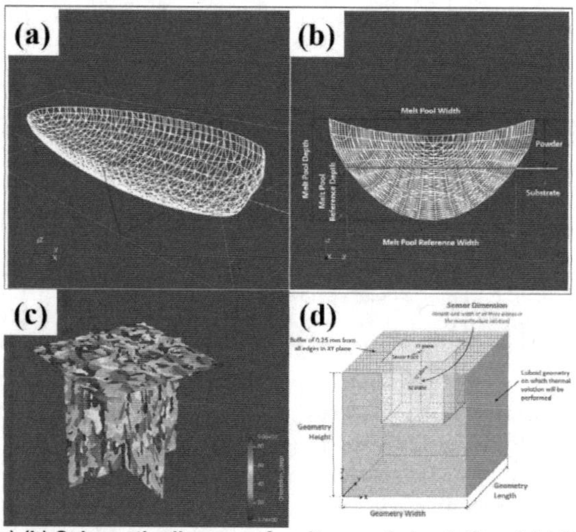

Fig.3 (a) (b) Schematic diagram of molten pool simulation, (c) (d)

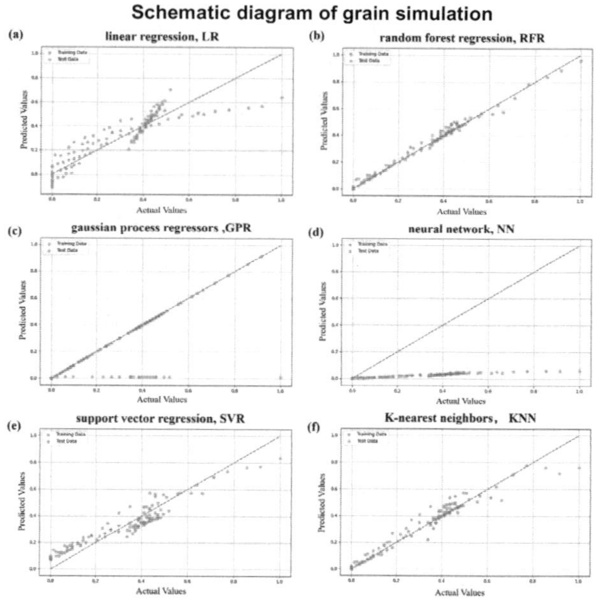

Fig.4 The depth fitting effect of different machine learning models on the melt pool (a)LR (b)RFR (c)GPR (d)NN (e)SVR (f)KNN

4 Conclusion

The experimental results of this study demonstrate that Random Forest Regression (RFR) and Support Vector Regression (SVR) exhibit desirable generalization capabilities and predictive accuracy when dealing with small sample learning processes. Based on the simulation calculation results and machine learning model predictions, the relationship between process parameters and performance is ultimately established. The model developed in this work can effectively predict the mechanical and damping performance of printed samples under different printing parameters.

5 Acknowledgments

This research works was supported by the Liaoning Province Applied Fundamental Research Program (No. 2023JH2/101700039) and Liaoning Province Natural Science Foundation (No. 2023-MSLH-328).

References

[1] J. Zhao, L. Sun, P. Ji, et al, The effect of scanning strategies on the microstructure and mechanical properties of M2052 alloy manufactured by selective laser melting, Journal of Materials Research and Technology 27 (2023) 7084-7093.

[2] J. Yang, C. Zhao, H. Liang, et al. Spinodal decomposition and martensitic transformation of the high manganese Mn–*x*Cu alloys fabricated by additive manufacturing, Applied Materials Today 25 (2021).

[3] F.X. Yin, S. Iwasaki, T. Sakaguchi, et al. Susceptibility of damping behavior to the solidification condition in the ss-Cast M2052 high-damping alloy, Key Engineering Materials 319 (2006): 67-72.

A Cluster-Formula Composition Design Approach in β-Ti Alloys

T. Y. Liu[1,2*], J. Zhao[1,2], S. B. Liu[1,2], X. Q. Li[1,2], K. Shi[1,2]
1. National Key Laboratory of Advanced Casting Technologies, Shenyang, Liaoning, 110022, China
2. Shenyang Research Institute of Foundry Co., Ltd, Shenyang, Liaoning, 110022, China
*Corresponding address: e-mail: liutianyusrif@163.com.com

Abstract: Cluster-plus-glue-atom model was proposed to deal with short-range-order structures in quasicrystals, metallic glasses, and solid solution alloys. In this model, the characteristic short-range-order structure is simplified into a local structural unit, where the atomic interactions are best satisfied locally, covering only the nearest neighbors (or cluster for short) and a few outer-shell atoms (called glue atoms). The unit is then formulated as [cluster](glue atoms), indicating that one cluster is matched with a few glue atoms. In this paper, a novel Ti-4.13Al-9.36V (wt. %) alloy with cluster-formula of 4[Al-Ti_{12}]($AlTi_2$)+12[Al-Ti_{14}](V_2Ti) was designed based on the cluster-formula of Ti-6Al-4V. Compared to Ti-6Al-4V, the tensile strength of Ti-4.13Al-9.36V alloy increased to 1045MPa, by 13%, and the ductility increased to 13.2%, by 164%.

Keywords: Cluster-plus-glue-atom model, β-Ti alloys, composition design

1 Introduction

Titanium alloys are widely used due to their good properties in strength, toughness, formability, weld-ability, corrosion resistance and biocompatibility [1]. However, limited by the missing model for chemical short-range-ordering, their composition rule is unresolved for long which has hindered the development of high-performance Titanium alloys. Cluster-plus-glue-atom model is an effective method to describe chemical short-range-ordering [2]. In this model, the chemical short-range-ordering is simplified into a local structural unit, where the atomic interactions are the best satisfied locally, covering only the nearest neighbors (or cluster for short) and a few outer-shell atoms (called glue atoms). The unit is formulated as [cluster](glue atoms)x, indicating that one cluster is matched with a few glue atoms. The model has been successfully used as an effective alloy composition design method in Titanium alloys [3, 4]. In this paper, according to the cluster formula of Ti-6Al-4V alloy, the β-Ti alloys have been designed.

2 Composition design

In our previous work, the chemical units of α and β phases in Ti-6Al-4V were revealed which are [Al-Ti_{12}]($AlTi_2$) and [Al-Ti_{14}](V_2Ti), respectively. Being analogous to the atomic resonance theory, the stacking unit of Ti-6Al-4V alloy can be calculated to contain 17 hard spheres, and the ratio of the α phase and β phase structural units only meets 12: 5. Therefore, the cluster-plus-glue-atom model of Ti-6Al-4V is 12[Al-Ti_{12}]($AlTi_2$)+5[Al-Ti_{14}](V_2Ti) [5].

Considered the β cluster formula of Ti-6Al-4V is taken as solvent, and α cluster formula is taken as solute, the cluster formula of novel alloy can be expressed as $[\{\beta\}\text{-}\{\alpha\}_{12}]\{\alpha\}_x\{\beta\}_y$. x and y are the number of glue clusters of {α} and {β}, respectively. The value of x and y can be calculated by the following equation:

$$x\cdot R_{\alpha/\beta}^3 + y = 2.03\cdot(R_{\alpha/\beta} + 1)^3 - R\alpha/\beta^3 - 12 \quad (1)$$

where, $R_{\alpha/\beta}$ is the radius ratio of the hard sphere of α and β phases cluster formula. Using the hard sphere radius of α and β phases of 0.367 nm and 0.379 nm, respectively, the above equation goes to:

$$1.154x + y \approx 4.309 \quad (2)$$

The solutions of Eq. (2), the closest to integers, are x=3 and y=0, meaning that the novel alloy cluster formula is 4[Al-Ti_{12}]($AlTi_2$)+12[Al-Ti_{14}](V_2Ti), and the composition is Ti-4.13Al-9.36V (wt.%).

3 Mechanical properties of designed Titanium alloys

Figure1 displays the microstructure of both alloys contained prior-β equiaxial grains. The width of the prior-β equiaxial grains decreases as the cluster ratio changes from 12:5 to 4:12. The microstructure within the prior-β equiaxial grains consists of Widmanstätten acicular α phase in both alloys, but as the cluster ratio changes from 12:5 to 4:12, α phase content decrease.

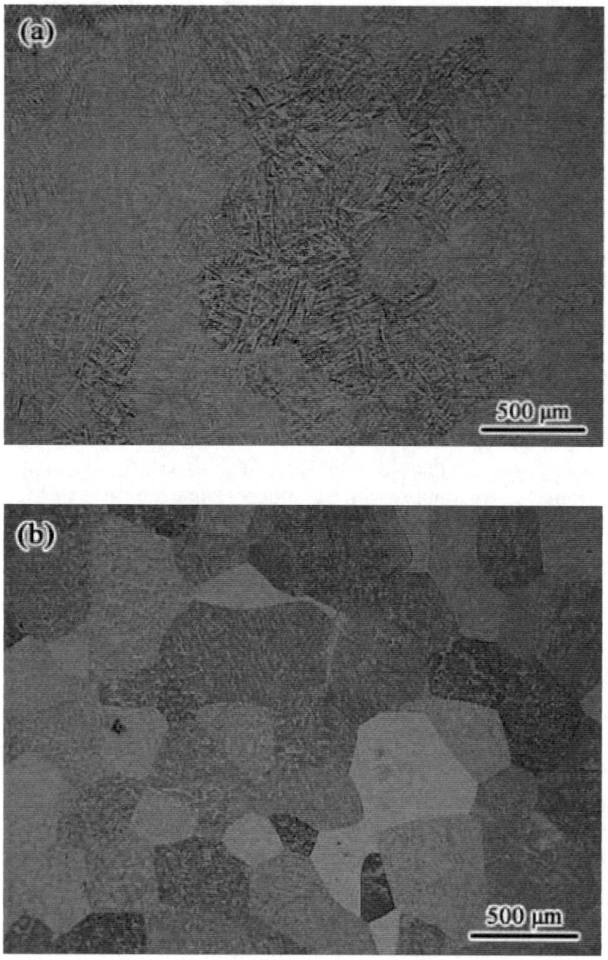

Figure 1 OM images of 12[Al-Ti12](AlTi2)+5[Al-Ti14](V2Ti) (a), and 4[Al-Ti12](AlTi2)+12[Al-Ti14](V2Ti)

Fig. 2 shows the engineering stress-strain curves of 12[Al-Ti_{12}]($AlTi_2$)+5[Al-Ti_{14}](V_2Ti), and 4[Al-Ti_{12}]($AlTi_2$)+12[Al-Ti_{14}](V_2Ti) at room temperature. The UTS and YS with the cluster ratio of 12:5 alloy is 925±3 MPa and 793±2 MPa , respectively, and the ductility is 5±1%. Compared to the cluster ratio of 12:5 alloy, the cluster ratio of 4:12 alloy has a significantly increase of strength and ductility.The UTS and YS are 1045±5 MPa and 819±2 MPa, respectively, and the ductility is 13±2%.

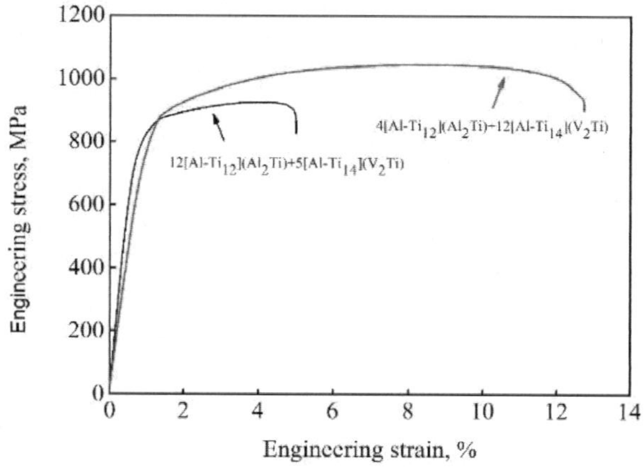

Figure 2 Engineering stress-strain curves at room temperature

4 Conclusion

Compared to Ti-6Al-4V, the tensile strength of Ti-4.13Al-9.36V alloy increased to 1045MPa, by 13%, and the ductility increased to 13.2%, by 164%.

5 Acknowledgments

This work was supported financially by the Natural Science Foundation of Shenyang, China (Grant No. 22315605).

References

[1] Banerje, D., Williams, J. C. Acta Mater, 61, 2013, 844-879.
[2] Dong, C., Wang, C., Qiang, J. B., Wang, Y. M., Jiang, N., Han, G., Li, Y. H., Wu, J., Xia, J. H. J Phys D: Appl Phys, 40, 2007, R273-R291.
[3] Liu, T. Y., Lou, Y. C., Zhang, S., Zhu, Z. H., Zhao, J., Liu, S. B., Shi, K., Zhao, N. China Foundry, 20, 2023, 414-422.
[4] Liu, T. Y., Zhu, Z. H., Zhang, S., Min, X. H., Dong, C. China Foundry, 18, 2021, 424-432.
[5] Liu, T. Y., Zhang, S., Wang, Q., Min, X. H., Dong, C. Sci China Tech Sci, 64, 2021, 1732-1740.

The Couple of Advanced Thermal Analysis and Casting Simulation to Predict the Component's Soundness Directly on the Pouring Line and Before Shaking Out. The Kasandra® Method

A. Areitioaurtena, J. M. Guitierrez[1], E. Martinez[1], G. Zarrabeitia[1], R. Suárez[1, 2]
1. AZTERLAN, Basque Research and Technology Alliance (BRTA), Durango, Bizkaia, Spain
2. Veigalan Studio 2010 SLU, Askatasun Etorbidea 16, 48200 Durango, Bizkaia, Spain
*Corresponding address: aareitio@azterlan.es

Abstract: The use of powerful computers has led to develop complex simulation solidification software's that go beyond the use of Chorinov rules. Nevertheless, this huge advance in the comprehension of solidification phenomena the problem has not been solved completely. Some remaining residual soundness scrap appear randomly after production, given rise to an expensive unitary control to detect the castings that do not fulfill the quality requirements. This extra control generates a lack of productivity, extra cost, etc.

The use of commercial solidification software's cannot predict the variations on metallurgical quality during the production, and this is the reason why unexpected micro shrinkages appear in some castings given rise to the extra unitary control.

With the aim to solve this important problem a new tool coupling the thermal evolution are and the measurement of metallurgical quality in the production line has been developed given an instantaneous picture about castings soundness.

Keywords: Simulation software, Thermal analysis, LFA (last to freeze areas), Micro shrinkage prediction, WIP (working in process)

1 Introduction

Cast Iron is a unique alloy that depending on the nucleation capacity can solidify following the stable Fe-C-Si diagram with graphite formation as well in an intermediate state with presence of graphite and cementite or with fully presence of carbides and austenite. These three-solidification behaviours determine the final properties of the castings.

Nowadays the classical casting foundry design in ductile, compact or gray iron where the thermal evolution goes to the risers as a final solidification element has not place, due to the highest competitiveness introduced by globalization. It is imperative design models that reduce the number of risers, increase the pattern yield, lower the energy consumption, reduce the amount of master alloys and the number of final controls on the castings and avoiding operations that do not add any value, such as degating, trimming, fleeting, producing castings in a lean way without WIP.

The only way to obtain soundness castings with high yield is combining thermal evolution with the capacity to form graphite. The isolated thermal pockets detected by the study of thermal evolution in the LFA and the corresponding contraction due to the residual liquid must be counteract by the graphite formation in these areas. The Kasandra® software combine the study of thermal evolution of the casting pointing out the isolated pockets and the capacity of graphite formation during solidification to predict directly in the production line the soundness of the poured components avoiding the further triage of the casting to scrap parts with porosity because of micro-shrinkage presence.

3 Experimental procedure

Caliper castings for automotive brakes systems were produced in a vertical molding machine (Disamatic 230 type Z) with six cavities per mold and a pattern yield of 70% in D.I quality GGG 50-7. The melting shop of the foundry is composed by three induction medium frequency furnaces 25 ton of capacity each and the material of charge is 50% steel scrap from automotive stamping, 40% foundry returns and 10% of pig iron.

Some graphite and CSi was added to adjust the chemical composition. Spheroidization treatment is performed by 1.2% of Mg master alloy with 5-6% Mg 1.2% of RE following the sandwich method in a ladle of 2-ton capacity. The metal is introduced in a pouring unit with heating of 15-ton capacity. Every 20 minutes a new ladle is introduced into the pouring unit. The metal is poured into the mold by means a rod-stopper system and the flow rate is guided by automated laser system.

The chemical composition is highlighted in table 1.

Table 1. Composition of Alloy

C	SI	Mg	P	S	Cu
3.65	2.30	0.040	0.002	0.006	0.35

The inoculation is performed in the liquid stream during the pouring in 0.2%. Grain size 0.2-0.6 mm. The composition of inoculant is the following.

Table 2. Composition of Inoculant

Si	Ca	Al	RE	Fe	
70-75	0.7-1.2	0.7-1.2	1.5-2	5.01	6.01

The range of pouring temperatures is 1390-1410ºC and the pouring time is around 8 sec.
The metallurgical quality of the poured melt is measured by Thermolan® thermal analysis system
The lay out shows that the casting is fill and fed only for one riser and the thermal evolution is pointed out in Figure 1. It is of interest to remark the different liquid fractions inside of the isolated pockets

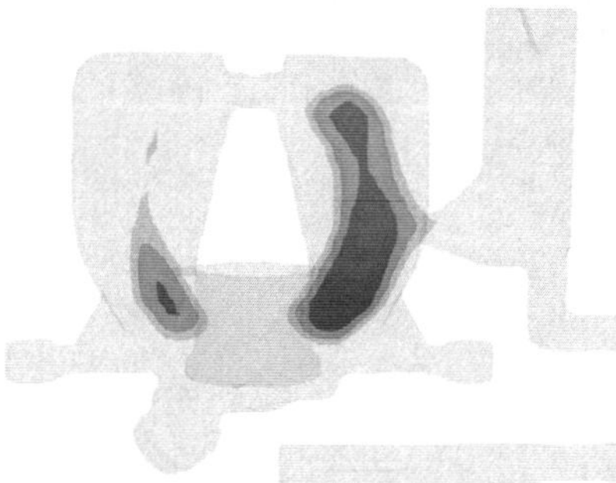

Figure 1. Lay out of filling and feeding system.

4 Result and discussion

The prediction of Kasandra® software who integrate in its calculations the metallurgical quality of the metal treated and inoculated in terms of capacity of graphite formation along all the solidification period paying special attention at the final end (latest fraction liquid to solidify) show that it is possible to produce this casting with appropriate quality in terms of soundness and mechanical properties if the metallurgical quality is classified as "premium" with good capacity to form graphite till the latest states of solidification otherwise the prediction forecast defects in both sides of the isolated area.

Figure 2 shows how, during the production, it is possible to produce good castings with the appropriate metallurgical quality as it is shown by Thermolan® and Kasandra® prediction.

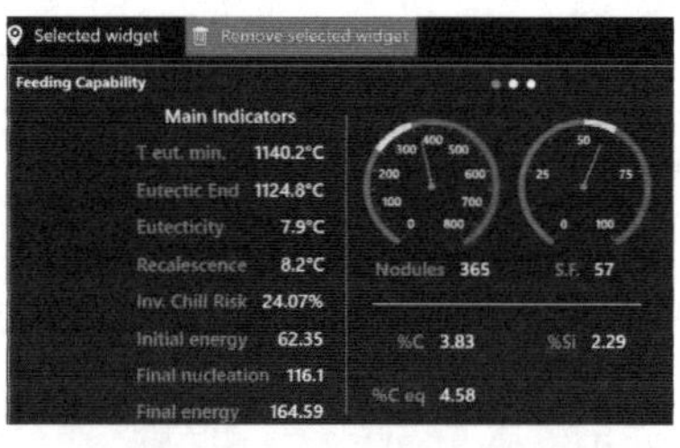

Figure 2. Picture taken from Thermolan® showing the main features of the cooling curves as well as the prediction knowledge

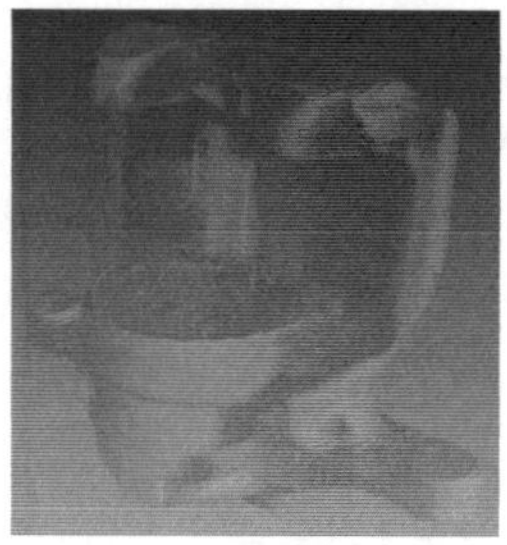

Figure 3. Image of the casting analysed by tomography.

Conclusion

1.A new software which integrates filling, thermal evolution of the castings as well as the implementation of metallurgical quality of the metal to be poured into the mold have been developed.

2.This software can be used well in the technical department to design the lay out the new project as well as to calculate the final cost for RFQ proposal and also directly in the shop floor together with the thermal analysis to predict the soundness of the products that are producing. In such a way it is possible well reduce the xR control of the parts or eliminate this costly control.

Acknowledgments

The authors would like to acknowledge Diputación Foral de Bizkaia for supporting this research.

References

[1] R.Suarez, J. Guitierrez, G.Zarrabeitia, A.Areitio. Science behind of Kasandra. Rev. Metalurgia.
[2] Lee J K, Choi J K and Hong C P. Thermal parameter-based quality criteria for the prediction of shrinkage defects in steel casting[J]. J. Korea Foundry Society, 1998, 18(2): 77-84.
[3] Jae-Hoon Nam, Sang-Mok Lee, Sang-Hwan Lee. Guaranteed soundness of heavy section spheroidal graphite cast iron cast iron 2 based on a reliable C and Si ranges design.
[4] R. Suarez, J.M. Gutierrez, M. Ferrer. Model of prediction of graphite growth in cast iron. Cours of Foundry Technology. Azterlan 2020
[5] N. K. Vedel-Smith, N. S. Tiedje Technical University of Denmark, Department of Mechanical Engineering, Kgs. Lyngby, Denmark Review on Numerical Simulation of Cast Iron Castings A Guide to Mold, Mesh, Metallurgy, Filling, Feeding and Stresses

Coupled Modeling of Gas and Shrinkage Microporosity with a Gas-Liquid-Solid Multi-Phase-Field Model

Yu Chen, Lei Wang*, Junjie Li, Jincheng Wang
State Key Laboratory of Solidification Processing, Northwestern Polytechnical University, Xi'an, Shanxi, China
*Corresponding address: Northwestern Polytechnical University e-mail: lei.wang@nwpu.edu.cn

Abstract: One of the most important types of defects that form during solidification is microporosity. There are two types of microporosity, gas porosity and shrinkage porosity. Gas porosity is caused by the gas element supersaturation originating from the difference in solubility between liquid and solid while shrinkage porosity is due to the volume difference between liquid and solid combined with restricted feeding during the last stage of solidification. Due to the difference in formation mechanism, there is a convention that gas and shrinkage porosities are examined separately. However, the two types of porosity can be coupled, i.e., both gas supersaturation and shrinkage flow can contribute simultaneously to the formation of porosity. As a model to incorporate both mechanisms is lacking, we developed a gas-liquid-solid multi-phase-field model to fill the gap. This model contains two core components: a quantitative multi-phase-field model describing the motion of different interfaces, which can reduce to the sharp interface model in the thin-interface limit; and a diffuse interface model of two-phase flows to introduce the shrinkage flow and the motion and deformation of the gas bubbles. A lattice Boltzmann method is used to solve the phase-field model and calculate the shrinkage flow. It is expected our model can provide a unified approach to study gas and shrinkage porosity and shed a new sight on the porosity formation.

Keywords: Phase Field; Lattice Boltzmann; Solidification microporosity; Shrinkage flow.

1 Introduction

Microporosity is one of the most serious solidification defects that can considerably reduce mechanical properties of casting parts during solidification. Thus, understanding the formation and growth of porosity is vital to optimize the materials performance [1].

The microporosity defects can be summarized into two types based on the formation mechanism: gas porosity and shrinkage porosity. Gas porosity is caused by the gas element supersaturation originating from the difference in solubility between liquid and solid while shrinkage porosity is due to the volume difference between liquid and solid combined with restricted feeding during the last stage of solidification [2]. Due to the difference in formation mechanism, gas and shrinkage porosities are investigated separately in numerical modeling. Carslon and Beckermann [3] proposed a dimensionless Niyama criterion to predict shrinkage pore volume fraction. The key of the criterion is the critical pressure drop combined with continuity equation and Dracy's law. Sun [4] developed a multiphase-field model for the solid-liquid-gas system present in gas porosity formation during solidification. Gas supersaturation acts as the driving force for bubble growth, Hele-Shaw flows are adopted to study the evolution of microstructure and the bubble motion. However, it is difficult to solve the Hele-Shaw flows equations. Recently, the lattice Boltzmann (LB) models, having the merits of simple algorithm and high parallel computational efficiency, have received great attention in the fields related to solidification. Zhang et al. [1] used a multiphase-field lattice-Boltzmann model to describe the complex multiphase interaction during solidification. The dendrite growth, bubble growth, motion and deformation, melt flow, and partition of both alloy solute and dissolved gas species are taken into account. The driving force for bubble growth is pressure difference between bubble, liquid and gas-liquid interface. Melt flow, e.g., shrinkage flow can decrease the liquid pressure. However, a constant pressure in liquid is used in their model.

Indeed, both gas supersaturation and shrinkage flow have influence on simultaneously the formation and growth of porosity [5]. Therefore, it is necessary to develop a coupled modeling of gas and shrinkage porosities to describe the growth, motion, and deformation of porosity in the solidification process of real metals.

2 Modeling procedure

Sharp-interface equations

The system is assumed that there are three components: a gas atom from which the bubble is composed, the other two are metal atoms, e.g., hydrogen dissolved in an Al-Cu melt. The sharp-interface equations for solid-liquid-gas system are given by

Liquid:

$$\frac{\partial C_l^{Cu}}{\partial t} = D^{Cu}\nabla^2 C_l^{Cu} \tag{1}$$

$$\frac{\partial C_l^{H}}{\partial t} = D^{H}\nabla^2 C_l^{H} \tag{2}$$

Solid-liquid interface:

$$C_l^{Cu}\left(1-k^{Cu}\right)\boldsymbol{u}_i\cdot\boldsymbol{n}_{sl} = -D^{Cu}\boldsymbol{n}_{sl}\cdot\nabla C_l^{Cu}\big|_l \tag{3}$$

$$C_l^{H}\left(1-k^{H}\right)\boldsymbol{u}_i\cdot\boldsymbol{n}_{sl} = -D^{H}\boldsymbol{n}_{sl}\cdot\nabla C_l^{H}\big|_l \tag{4}$$

$$T = T_m + m^{Cu}C_l^{Cu} - \Gamma\kappa_{sl} - \frac{\boldsymbol{u}_i \cdot \boldsymbol{n}_{sl}}{\mu_k} \tag{5}$$

Gas-liquid interface:

$$C_l^{Cu}\boldsymbol{u}_i \cdot \boldsymbol{n}_{gl} = -D^{Cu}\boldsymbol{n}_{gl} \cdot \nabla C_l^{Cu}\big|_l \tag{6}$$

$$\left(1 - C_l^H\right)\boldsymbol{u}_i \cdot \boldsymbol{n}_{gl} = -D^H\boldsymbol{n}_{gl} \cdot \nabla C_l^H\big|_l \tag{7}$$

$$C_{l,i}^H = \beta\boldsymbol{u}_i \cdot \boldsymbol{n}_{gl} + C_{eq}^H\left(1 + \frac{\sigma\kappa_{gl}}{\rho RT}\right) \tag{8}$$

where subscripts g, l, and s denote the gas, liquid, and solid, respectively, and subscripts sl and gl denote the solid-liquid and gas-liquid interfaces, respectively. D is the mass diffusivity, $\boldsymbol{u}_i$ is the interface velocity, $\boldsymbol{n}$ is the unit normal to the interface and pointing into the liquid, k is the partition coefficient, T is the interface temperature, m is the liquidus slope, Γ is the Gibbs-Thomson coefficient, κ is the local curvature of the interface, μ_k is the kinetic coefficient at the solid-liquid interface, β is the chemical kinetic at the gas-liquid interface, $C_{l,i}$ is the liquid concentration at the interface, C_{eq}^H is the equilibrium solubility of the gas species in the liquid and calculated by Sievert's law, ρ is the density, and R is the universal gas constant.

Phase-field model

A quantitative multi-phase-field model is constructed, which can reduce to the sharp interface model in the thin-interface limit and a diffuse interface model of two-phase flows to introduce the shrinkage flow and the motion and deformation of the gas bubbles.

Assuming equal densities and surface tensions among phases, a complete model for porosity formation in alloy solidification with flow in liquid can be written by [4]

Phase-field:

$$\begin{aligned}\tau(\phi_g,\phi_l,\phi_s)\frac{\Gamma_g}{\rho_l} = {} & \nabla^2\phi_g + \frac{2}{3}\left[-2\phi_g(1-\phi_g)(1-2\phi_g) + \phi_l(1-\phi_l)(1-2\phi_l) + \phi_s(1-\phi_s)(1-2\phi_s)\right] \\ & +8\lambda\left[\frac{\partial g_g}{\partial\phi_g}\left(\frac{C_l^H - (C_{eq}^H)^0}{1-(C_{eq}^H)^0} - \theta_g\right) - \frac{\partial g_s}{\partial\phi_g}\left(\frac{C_l^{Cu} - (C_l^{Cu})^0}{(1-k)(C_l^{Cu})^0} + \theta_s\right)\right]\end{aligned} \tag{9}$$

$$\begin{aligned}\tau(\phi_g,\phi_l,\phi_s)\frac{\partial\phi}{\partial t} = {} & \nabla^2\phi_s + \frac{2}{3}\left[-2\phi_s(1-\phi_s)(1-2\phi_s) + \phi_g(1-\phi_g)(1-2\phi_g) + \phi_l(1-\phi_l)(1-2\phi_l)\right] \\ & +8\lambda\left[\frac{\partial g_g}{\partial\phi_s}\left(\frac{C_l^H - (C_{eq}^H)^0}{1-(C_{eq}^H)^0} - \theta_g\right) - \frac{\partial g_s}{\partial\phi_s}\left(\frac{C_l^{Cu} - (C_l^{Cu})^0}{(1-k)(C_l^{Cu})^0} + \theta_s\right)\right]\end{aligned} \tag{10}$$

$$\tau(\phi_g,\phi_l,\phi_s) = \begin{cases}\frac{1}{2}\left(\frac{1-C_l^H}{1-(C_{eq}^H)^0} + \frac{C_l^{Cu}}{(C_l^{Cu})^0}\right) + \frac{1}{2}\left(\frac{C_l^{Cu}}{(C_l^{Cu})^0} - \frac{1-C_l^H}{1-(C_{eq}^H)^0}\right)\frac{\phi_s-\phi_g}{\phi_s+\phi_g} & \text{if } \phi_l \neq 1 \\ \frac{1}{2}\left(\frac{1-C_l^H}{1-(C_{eq}^H)^0} + \frac{C_l^{Cu}}{(C_l^{Cu})^0}\right) & \text{if } \phi_l = 1\end{cases} \tag{11}$$

$$\phi_s + \phi_l + \phi_g = 1 \tag{12}$$

Species:

$$\begin{aligned}\frac{\partial}{\partial t}\left[(\phi_l + k\phi_s)C_l^H\right] + \phi_l\boldsymbol{u}_l \cdot \nabla C_l^H = {} & \nabla\cdot\left(D_l^H\phi_l\nabla C_l^H\right) - (1-r_p)C_l^H\left(\frac{\Gamma_g}{\rho_l}\right) - r_pC_g^H\left(\frac{\Gamma_g}{\rho_l}\right) \\ & +\nabla\cdot\left[\frac{W}{\sqrt{2}}(C_g^H - C_l^H)\left(\frac{\Gamma_g}{\rho_l}\right)\frac{\nabla\phi_l}{|\nabla\phi_l|}\left(-\frac{\nabla\phi_l}{|\nabla\phi_l|}\cdot\frac{\nabla\phi_g}{|\nabla\phi_g|}\right)\right] \\ & -\nabla\cdot\left[\frac{W}{\sqrt{2}}(1-k^H)C_l^H\frac{\partial\phi_s}{\partial t}\frac{\nabla\phi_l}{|\nabla\phi_l|}\left(-\frac{\nabla\phi_l}{|\nabla\phi_l|}\cdot\frac{\nabla\phi_s}{|\nabla\phi_s|}\right)\right]\end{aligned} \tag{13}$$

$$\begin{aligned}\frac{\partial}{\partial t}\left[(\phi_l + k\phi_s)C_l^{Cu}\right] + \phi_l\boldsymbol{u}_l \cdot \nabla C_l^{Cu} = {} & \nabla\cdot\left(D_l^{Cu}\phi_l\nabla C_l^{Cu}\right) - (1-r_p)C_l^{Cu}\left(\frac{\Gamma_g}{\rho_l}\right) \\ & -\nabla\cdot\left[\frac{W}{\sqrt{2}}(1-k^{Cu})C_l^{Cu}\frac{\partial\phi_s}{\partial t}\frac{\nabla\phi_l}{|\nabla\phi_l|}\left(-\frac{\nabla\phi_l}{|\nabla\phi_l|}\cdot\frac{\nabla\phi_s}{|\nabla\phi_s|}\right)\right] \\ & -\nabla\cdot\left[\frac{W}{\sqrt{2}}C_l^{Cu}\frac{\Gamma_g}{\rho_l}\frac{\nabla\phi_l}{|\nabla\phi_l|}\left(-\frac{\nabla\phi_l}{|\nabla\phi_l|}\cdot\frac{\nabla\phi_g}{|\nabla\phi_g|}\right)\right]\end{aligned} \tag{14}$$

where ϕ is order parameter to distinguish the phases. The superscript 0 denote equilibrium values. τ is the relaxation time. W is width of interface. $\Gamma_g = \frac{\partial\phi_g\rho_g}{\partial t} + \nabla\cdot(\phi_g\rho_g\boldsymbol{u}_g)$ is mass transfer rate per unit volume.

Lattice Boltzmann method

The multi-relaxation-time (MRT) scheme is implemented in the LB model to simulate the liquid-gas flows. The MRT LB equation is written as [6]

$$f_i(\boldsymbol{x}+\boldsymbol{e}_i\Delta t, t+\Delta t) - f_i(\boldsymbol{x},t) = -\left(\mathbf{M}^{-1}\mathbf{\Lambda}\mathbf{M}\right)_{ij}\left(f_j - f_j^{eq}\right) \tag{15}$$

$$i,j = 0,1,\ldots,8$$

where f_i is the density distribution function and f_i^{eq} represents its equilibrium distribution function. $\boldsymbol{x}$ is the spatial position $\boldsymbol{e}_i$ represents the discrete velocities along the ith direction and Δt represents the time step in the LB model. $\mathbf{M}$ and $\mathbf{M}^{-1}$ are orthogonal transformation matrices and the corresponding inverse matrix.

The density distribution function $\boldsymbol{f}$ and its equilibrium function $\boldsymbol{f}^{eq}$ can be transformed from the velocity space into the moment space via $\boldsymbol{m}=\mathbf{M}\boldsymbol{f}$ and $\boldsymbol{m}^{eq}=\mathbf{M}\boldsymbol{f}^{eq}$, respectively. The MRT LB equation becomes

$$\boldsymbol{m}^* = \boldsymbol{m} - \mathbf{\Lambda}\left(\boldsymbol{m} - \boldsymbol{m}^{eq}\right) + \Delta t\left(\mathbf{I} - 0.5\mathbf{\Lambda}\right)\boldsymbol{S} \tag{16}$$

The streaming process of the liquid-gas particles is given as

$$f_i(\boldsymbol{x}+\boldsymbol{e}_i\Delta t, t+\Delta t) = f_i^*(\boldsymbol{x},t) \tag{17}$$

where f^* can be obtained by $\boldsymbol{f}^* = \mathbf{M}^{-1}\boldsymbol{m}^*$.

References

[1] A. Zhang, Z. Guo, B. Jiang, et al., Multiphase and multiphysics modeling of dendrite growth and gas porosity evolution during solidification, Acta Materialia, 2021, 214: 117005.

[2] P.D. Lee, A. Chirazi, D. See, Modeling microporosity in aluminum-silicon alloys: a review, Journal of Light Metals, 2001, 1: 15-30.

[3] K.D. Carlson, C. Beckermann, Prediction of Shrinkage Pore Volume Fraction Using a Dimensionless Niyama Criterion, Metallurgical and Materials Transactions A, 2008, 40(1): 163-175.

[4] S. Ying, Phase-field simulation of microporosity formation in solidification, The University o f Iowa, 2006.

[5] V. Khalajzadeh, K.D. Carlson, D.G. Backman, et al., A Pore-Centric Model for Combined Shrinkage and Gas Porosity in Alloy Solidification, Metallurgical and Materials Transactions A, 2017, 48(4): 1797-1816.

[6] Y. Chen, X. Chen, Q. Zhang, M. Zhu, F. Liu, X. Wang, H. Nagaumi, Z. Yao, Numerical Modeling of Welding Porosity Formation and Dendrite Growth of 6xxx Aluminum Alloys, Jom, 2022, 74(6): 2428-2436.

Composition design of Nickel-Based Superalloy Based on Data-Driven Methods in Investment Casting

Zhengyi Ding[1,2], Donghong Wang[1,2*], Bang Guan[1,2], Jiayu Cui[1,2], Da Shu[1,2], Baode Sun[1,2]
1. Shanghai Key Lab of Advanced High-Temperature Materials and Precision Forming, School of Materials Science and Engineering, Shanghai Jiao Tong University, Shanghai, China
2. State Key Laboratory of Clean and Efficient Turbomachinery Power Equipment, School of Materials Science and Engineering, Shanghai Jiao Tong University, Shanghai 200240, China
*Corresponding address: e-mail: wangdh2009@sjtu.edu.cn

Abstract: Investment casting has been widely applied to fabricate aircraft due to its high dimensional accuracy and excellent surface finish with complex geometries. However, defects such as shrinkage porosity and misrun often occur in these castings. Traditional trial-and-error procedure have hindered the application of investment castings in high-end applications. In this research, a novel data-driven approach for predictive modeling is proposed to address these challenges. With integrated computation of thermodynamic and finite element method (FEM), a variety of components within the boundary of Ni-based superalloy K4169 were evaluated for both fluidity and shrinkage with two patterns. A numerical model incorporating machine learning algorithms was developed based on the integrated computation. For K4169 superalloy castings, the shrinkage porosity is primarily influenced by Al element, while misrun is mainly correlated with Al, Ti, and Mo in a decreasing order. This data-driven model establishes a quantitative relationship between composition and final product defects. The method proposed in this study shows promise for optimizing the investment casting process and can be applied to a wide range of alloys and models.

Keywords: Integrated computation, Data-driven approach, Ni-based superalloy, Investment casting

1 Introduction

Investment casting is a precise forming technique applied in advanced fields such as aerospace, for its intricate design capabilities and high-quality finish.[1] However, the reliance on traditional trial-and-error approaches in the research and development phase often results in huge waste.[2] The introduction of data modeling and big data algorithms has significantly improved the efficiency of the investment casting process. By incorporating advanced computational techniques such as machine learning algorithms and predictive modeling, manufacturers can analyze extensive datasets to identify optimal casting parameters and predict casting defects with enhanced precision.

With great tensile strength, hardness and fatigue life, Ni-based superalloy has been widely applied in aerospace castings. However, the composition of certain superalloy varies in a large bound, which leads to significant uncertainty in properties such as liquidus, solidus and fluidity. These properties contribute to the formation of casting defects, so this uncertainty can lead to misprediction. In this research, to address the challenge of composition induced casting defects, a data-driven framework is proposed. The schematic is shown in Figure 1.

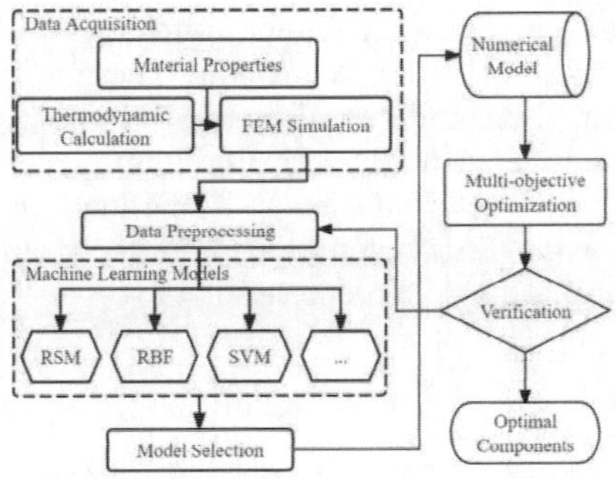

Figure 1. Data-driven framework of this research

2 Experimental procedure

In this research, K4169 Ni-based superalloy is taken as the experimental alloy. Its nominal composition is shown in Table 1.

Table 1. Nominal composition of K4169 superalloy

Elements	Al	Cr	Ni	Fe
Lower bound (wt%)	0.30	17.00	50.00	Bal.
Upper bound (wt%)	0.70	21.00	55.00	
Elements	Mo	Nb	Ti	
Lower bound (wt%)	2.80	4.4	0.65	
Upper bound (wt%)	3.30	5.4	1.15	

The performances of different alloy compositions within the limit of K4169 is calculated with Pandat Ni-database. Optimal Latin hypercube sampling was utilized to develop a more effective numerical model. 100 sets of compositions are designed with Latin hypercube sampling, and their performance are calculated with high-throughput calculation in Pandat software.

Compositions and their performance are imported into database. Featured patterns in Figure 2 are designed to address shrinkage porosity and misrun defects. FEM calculations are carried out with ProCAST software. Results are automatically extracted from calculated files.

To establish the relationship model between components and performance, different machine-learning algorithms are applied to the result. In this research, RBF, RSM, linear regression and SVR are applied. The data is divided into training set, validation set and test with a ratio of 2:1:1, and then approximated with the models above.

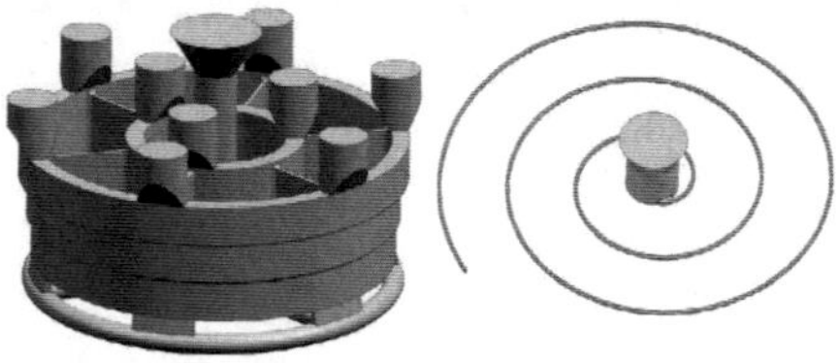

(a) Shrinkage (b) Misrun

Figure 2. Featured patterns

3 Result and discussion

Composition-performance-defect numerical model

The comparison of different machine learning models is shown in Figure 3. In almost all properties in performance, microstructure and defect, RSM algorithm shows the best coefficient of determination. Numerical model is established with RSM algorithm. The correlation table between composition and properties is shown in Table 2.

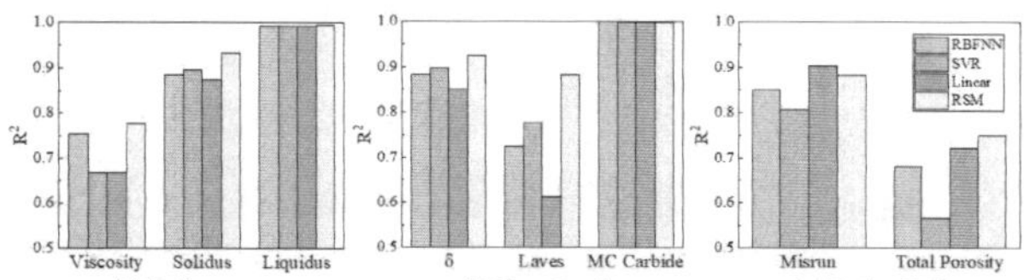

(a) Performance (b) Phase Fraction (c) Casting Defect

Figure 3. Coefficient of determination among different algorithms

Table 2. Composition-performance-defect correlation table

Elements	Al	Cr	Fe	Mo	Nb	Ti
Misrun	-0.54	0.07	0.03	-0.32	-0.09	-0.45
Porosity	0.79	-0.05	0.05	-0.28	-0.26	-0.39

3rd generation non-dominated sorting genetic algorithm (NSGA-III) is applied to the component optimization. After 40 generations of iteration, the optimal component is: Al 0.79%, Cr 18.1%, Fe 20.1%, Mo 3.08%, Nb 5.25%, Ti 1.09%. Simulation verification in Figure 4 indicates that the optimal compositions is enhanced with 23% less internal shrinkage and 13% increased feed length, compared to median composition of K4169 superalloy.

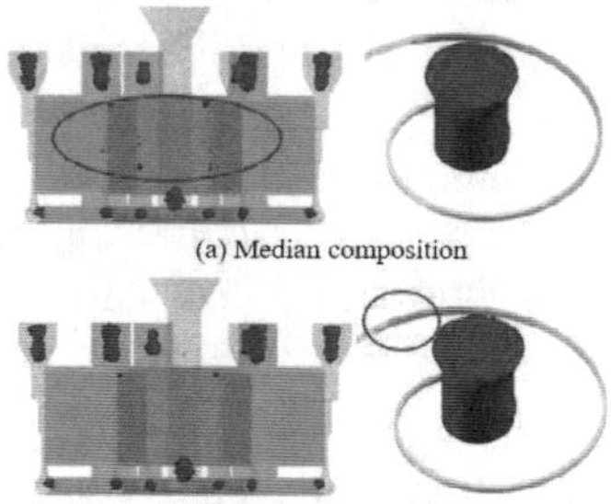

(a) Median composition

(b) Optimal composition

Figure 4. Numerical verification of optimal composition

4 Acknowledgments

This work was financially supported by the National Science and Technology Major Projects of China (J2019-VI-0004-0117), the National Key Research and Development Program of China (2022YFB3706800), the National Natural Science Foundation of China (52090042).

5 Conclusion

In this research, machine learning algorithms is combined with thermodynamic-FEM integrated computation to develop the numerical model. The NSGA-III algorithm is then applied to optimize the components based on the model. The following conclusions can be made:

a) Within the boundary of K4169 superalloy, shrinkage porosity is primarily influenced by Al, while misrun is negatively correlated with Al, Ti and Mo.
b) The optimal composition within K4169 limit has 23% less internal shrinkage and 13% increased feed length in featured patterns, compared to median.
c) The data-driven approach proposed in this research shows promise in deviation analysis and composition optimization.

References

[1] L. Liu, The progress of investment casting of nickel-based superalloys. J. Foundry, 2012, 12: 1273-1285.
[2] Wang D, Sun F, Shu D, et al. Data-driven design of nickel-based casting superalloys and precise forming of complex castings. J. Acta Metallurgica Sinica, 2022, 58(1): 89-102.

Rapid Simulation Platform and Application in Investment Casting Process

Pan Zhang[1], Renchen Cao[2], Donghong Wang[1*], Zhen Huang[1], Baode Sun[1]
1.Shanghai Key Laboratory of Advanced High Temperature Materials and Precision Forming; State Key Laboratory of Clean and Efficient Turbomachinery Power Equipment; School of Materials Science and Engineering, Shanghai Jiao Tong University, Shanghai 200240, China
2.School of Materials Science and Engineering, Shanghai University of Engineering and Technology, Shanghai 200240, China
*Corresponding address: E-mail: wangdh2009@sjtu.edu.cn

Abstract: A rapid simulation platform for high-temperature alloy investment casting process was constructed to address the issues of cumbersome process parameter operation and low simulation efficiency in numerical simulation of investment casting. This platform integrates the process from CAD design to CAE simulation, functions such as parameterization of pouring system, design of casting process parameters, and dynamic display of simulation results. It has established 10 types of simulation templates and 237 sets of model libraries, achieving automated and rapid simulation. Taking a certain power turbine casing as an example, using this platform to construct a simulation model, parameter design variables, and result prediction optimization indicators for simulation calculation; The simulation prediction analysis results are consistent with the actual production defects; Simulation efficiency increased by 89.8%.

Keywords: Investment casting; Rapid simulation of casting process; Template Library

1 Introduction

With the real arrival of big data and the "Internet plus" era, the traditional casting production mode has been seriously challenged, and intelligent casting came into being [1]. In recent years, numerical simulation technology has been flourishing in the investment casting process, and software such as ProCAST, MAGMA have been widely used [2]. Among them, ProCAST can effectively and accurately predict the defects produced by castings [3]. The use of ProCAST simulation in the above research requires a high level of simulation technology, and the simulation operation is cumbersome, resulting in low efficiency and inability to achieve rapid process prediction. CHENG H et al. [4] proposed an automatic optimization design method based on Isight platform digital modeling, mesh partitioning, numerical simulation for the design of multi-objective and multi parameter cavity structures under the influence of flow fields. However, the above research lacks direct calls to various modules of ProCAST, and requires researchers to write corresponding script files.

Based on the simulation of investment casting process and analysis of simulation results, this article aims to design a rapid simulation platform for casting process with the goal of automated rapid simulation. This platform achieves automated rapid simulation and efficient prediction of casting process defects.

2 Experimental procedure

In order to achieve the automation of casting simulation process and the efficient prediction of defects. The rapid simulation platform for casting process will be divided into two modules: pre-processing, post-processing.

Pre-processing module: This module establishes 10 types of simulation templates, and through parameterized design of the pouring system in 237 sets of model libraries, the adjustable parameters of 15 simulation design variables and 3 simulation results for predicting optimization indicators are uniformly encapsulated, such as geometric models, 2D meshes, and boundary conditions. After submitting the calculation, the ProCAST software is automatically called for simulation calculation using batch processing, ultimately achieving the associative storage of simulation results.

Post-processing module: This module extracts simulation results and dynamically displays temperature field, flow velocity field, and contour maps through visualization. The shrinkage porosity map can be generated based on the specified porosity; At the same time, simulation reports can also be downloaded to view the process parameters and result cloud maps of castings, achieving efficient defect prediction and analysis.

3 Result and discussion

Taking the filling and solidification of a certain power turbine casing of an aircraft engine as an example, this casing serves as the core component of a certain helicopter engine model, as shown in Figure 1. In the 237 sets of pouring system model library, select the parameterized design and assembly of the casing pouring system. The material of the casing casting is nickel based high-temperature alloy K4169, and the shell material is mullite. The shell roasting temperature is 950 ℃, the pouring temperature is 1450 ℃, and the pouring time is 5 s. The casing template library is called for simulation. The simulation results are shown in Figure 2.

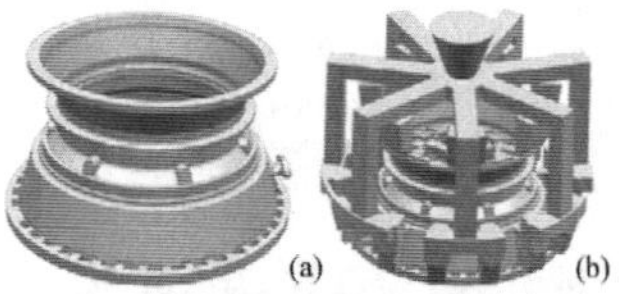

Fig.1 (a)Power turbine casing model(b)pouring system model

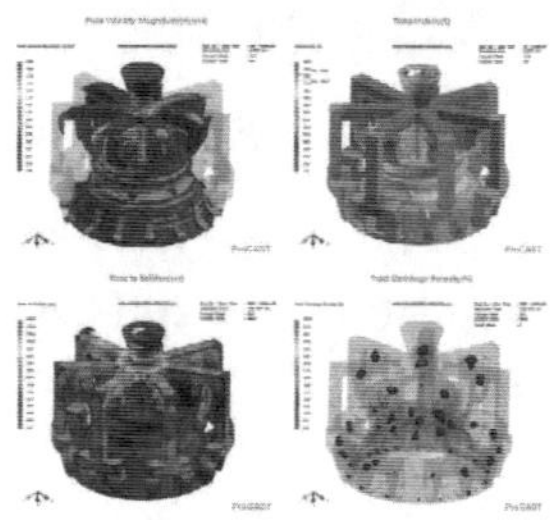

Fig.2 Post processing result interface

The shrinkage porosity diagram of the casing is shown in Figure 2, with shrinkage porosity appearing in the bottom end face and the lower thin-walled area. As shown in Figure 3, not all thin-walled areas between the risers show shrinkage and looseness. This is because the feeding distance of the riser is not long enough, resulting in the formation of isolated liquid phase zones between the thin-walls, with a high shrinkage porosity (as shown in Figure 3(b)). The casing was put into actual production and it was found that the defect results predicted and analyzed by the casting process rapid platform were consistent, as shown in Figure 3(c).

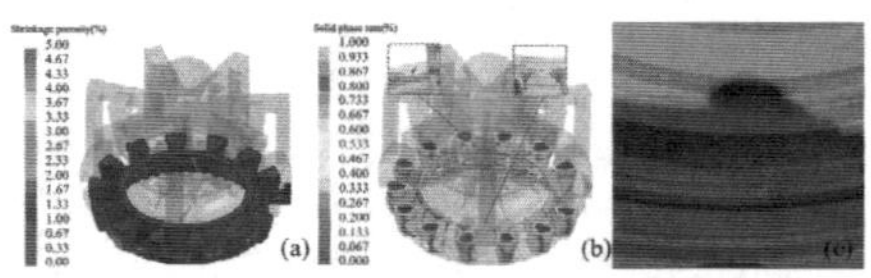

Fig.3 (a)Casting profile solidification defect distribution (b)Solid phase fraction distribution in casting profile (c)Defects in power turbine casing casting

As shown in Table 1, during the numerical simulation process, it takes about 31 minutes to use general finite element simulation software for the power turbine casing, while using the casting process rapid simulation platform only takes about 3 minutes, saving nearly 28 minutes and improving efficiency by 89.8%, greatly shortening the simulation time.

Tab.1 Comparison of simulation time

Category/time	Pre-process	Post-process
General simulation	17min28s	14min41s
Rapid simulation platform	<1min	2min13s

4 Conclusion

Based on the simulation of investment casting process and analysis of simulation results, a rapid simulation platform for casting process has been successfully constructed by integrating the process from CAD design to CAE simulation. Taking a certain power turbine casing as an example, a simulation model, parameter design variables, and result prediction optimization indicators are constructed through a fast simulation platform for simulation calculation; Compared with general simulation software simulation, the simulation time was saved by 28 minutes and the efficiency was improved by 89.8%; It is found that the simulation results are consistent with actual production defects.

5 Acknowledgments

This work was financially supported by the National Science and Technology Major Projects of China (J2019-VI-0004-0117), the National Key Research and Development Program of China (2022YFB3706800), the National Natural Science Foundation of China (52090042).

References

[1] Xu Q Y. Foundry Technology Roadmap: Digitalization, Networking, and Intelligent Foundry. J. Foundry, 2017, 66 (12): 1243-1250.

[2] Li B. Numerical simulation and process optimization of bracket die-casting based on Magma software. J. Special Casting and Nonferrous Alloy, 2019, 39 (09): 952-955.

[3] Li F, Cui X P, Nan H. Process analysis and optimization of a circular titanium alloy casting based on ProCAST. J. Special Casting and Nonferrous Alloy, 2021, 41 (12): 1469-1471.

[4] Cheng H, Zhao N, Li J, et al. Integrated design method for a multi-objective and multi-parameter cavity structure under the flow field effect. J. Journal of the Chinese Institute of Engineers, 2020, 43(8): 752-762.

Detection and Segmentation of Cracks and Cold Shut Defects Based on YOLOv8 Algorithm

Zhiqiang Duan[1], Hua Hou[1,3], Yuhong Zhao[1,2,4*]
1. School of Materials Science and Engineering, Collaborative Innovation Center of Ministry of Education and Shanxi Province for High-performance Al/Mg Alloy Materials, North University of China, Taiyuan, 030051, China
2. Beijing Advanced Innovation Center for Materials Genome Engineering, University of Science and Technology Beijing, Beijing, 100083, China
3. School of Materials Science and Engineering, Taiyuan University of Science and Technology, Taiyuan, 030024, China
4. Institute of Materials Intelligent Technology, Liaoning Academy of Materials, Shenyang, 110004, China
*Corresponding address: e-mail: zhaoyuhong@nuc.edu.cn

Abstract: The detection of casting defects is crucial to ensure the quality and performance of the final product. This paper aims to comprehensively explore the advantages and disadvantages of various detection technologies and methods in the field of casting defect detection. On the premise of improving the image quality of aluminum and magnesium alloy casting defects through data enhancement algorithms, a casting defect detection method based on the YOLOv8 instance segmentation algorithm is proposed. The method can effectively detect and segment crack and cold shut defects in castings, and is experimentally verified to have a segmentation confidence level of up to 0.9 and 0.8 for detecting cracks and cold shut defects, respectively. This proves that the method provides an adoptable machine-learning solution for defect detection in castings.

Keywords: Casting Defects, Defect Detection, Data Enhancement, Instance Segmentation, YOLOv8 Algorithm.

1 Introduction

As a traditional metal processing technology, casting is widely used in automobile manufacturing, aerospace, energy and many other fields. The quality of casting products is directly related to the stability and cost-effectiveness of the entire manufacturing process. However, despite the continuous development and improvement of the casting process, casting products are still inevitably plagued by various defects, which may be derived from raw materials, process parameters, operator skills and other factors. These defects not only pose a threat to the quality and performance of the product, but also may cause damage to the equipment, affect production efficiency, and even endanger the safety of personnel.

In this context, casting defect detection has become a key link to ensure product quality and improve production efficiency. The timely detection and solution of casting defects can not only reduce the scrap rate, improve the reliability and durability of products, but also reduce the production cost and improve the competitiveness of enterprises. Therefore, the in-depth research and continuous innovation of casting defect detection technology are of great significance for promoting the development of manufacturing industry and realizing intelligent, green and efficient production.

2 Experimental procedure

Instance Segmentation is one of the most difficult tasks in the four classic visual tasks. It not only has the characteristics of Semantic Segmentation, but also needs to be classified at the pixel level. It also has some characteristics of Object Detection, and it needs to locate different instances, even if they are the same kind. The real challenge of instance segmentation is that it needs to correctly detect the target in the image, determining the contour of the target, and accurately segment each instance at the pixel level.

As a powerful and flexible object detection and image segmentation tool, YOLOv8 introduces new features based on YOLOv5 model to further improve the efficiency and accuracy of the model. One of its key features is scalability, designed as an architecture that supports all previous versions of YOLO, enabling users to easily switch and compare their performance [1]. In addition to scalability, YOLOv8 also introduces a number of innovations, including a new backbone network, anchor-free detection head, and loss function, making it the preferred model for object detection and image segmentation tasks.

3 Result and discussion

Through the statistics of 240 defect data images marked in the data set, the results shown in Fig.1 are obtained. Fig.1(a) represents the size and distribution of the bounding box marked on the defect site, the relative distribution of the defect size in the data set and the center-aligned bounding box are shown in Fig.1(b). The relative width and relative height represent the ratio of the width and height of the target to the width and height of the whole image, respectively. The deeper the color, the more the number of targets corresponding to the size. From Fig.1, it can be observed that the relative size of the target area is mostly concentrated in the range of 0 ~ 0.40, and the small and medium target defects and their boundary frames account for the vast majority.

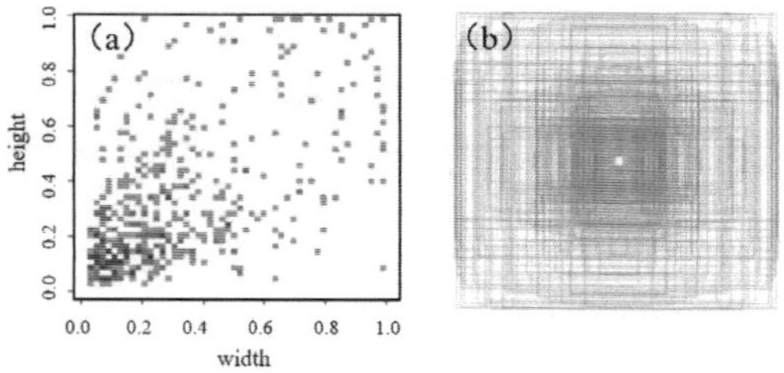

Fig.1 Relative size distribution map of target site.

Subsequently, the YOLOv8 algorithm model was used for training after the enhancement preprocessing of the casting surface defect image. After 200 epochs, the model training set and the verification set reached the lowest and stabilized with the increase of the number of iterations. As shown in Fig.2, the model achieves the best detection state.

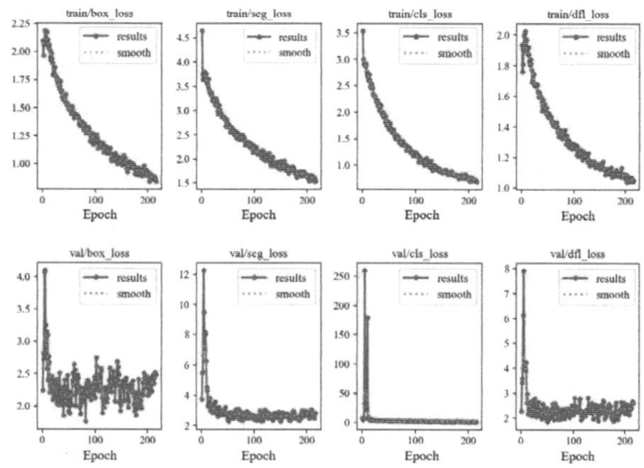

Fig.2 YOLOv8 model dataset is trained on various loss results.

In order to evaluate the detection accuracy of the model, we select the accuracy rate, recall rate, average detection accuracy value (IOU threshold is 0.5) and average detection accuracy value (IOU threshold is in the range of 0.5-0.95) as performance indicators. According to the confusion matrix, the performance index fitting is obtained as shown in Fig. 3. With the increase of the number of iterations, the average precision rate is 0.7375, the average recall rate is 0.581, and the average detection accuracy is close to 0.6.

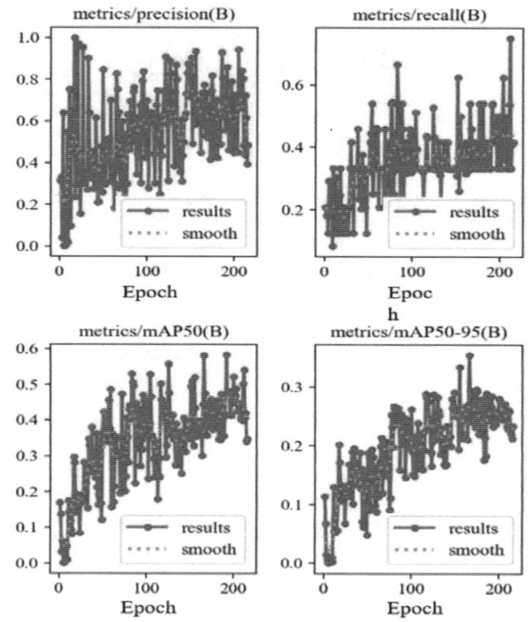

Fig.3 YOLOv8 model performance index training curve.

4 Conclusion

In summary, under the premise of improving the defect image quality of aluminum-magnesium alloy castings by data enhancement algorithm, this paper proposes a casting defect detection method based on YOLOv8 instance segmentation algorithm, which can effectively detect and segment the cracks and cold shut defects of castings. The experimental results show that the segmentation algorithm model based on this example can accurately detect and segment the cold shut and crack defects in the casting defects, and determine the confidence range of the best detection segmentation effect between 0.6-0.8. After subsequent comparison, it is found that the contrast of the defect image after image enhancement and grayscale processing is more obvious, and the defect contour is clearer, which greatly improves the efficiency of the instance segmentation and avoids the phenomenon of missed detection or false detection.

5 Acknowledgments

This work was supported by the National Natural Science Foundation of China (Nos. 52275390, 52375394), Key Research and Development Program of Shanxi Province (202102050201011, 2022ZDYF035)

References

[1] Zhou F, Chao Y, Wang C, Zhang X, Li H, Song X, A small sample nonstandard gear surface defect detection method. Measurement 2023,221:113472.

A Modified Phase-Field Model for Non-Isothermal Sintering Induced by Local High-Temperature

Pengya Lei[1], Limin Li [1], Wenkui Yang [1], Hua Hou[1,3],Yuhong Zhao[1,2,4*]
1. School of Materials Science and Engineering, Collaborative Innovation Center of Ministry of Education and Shanxi Province for High-performance Al/Mg Alloy Materials, North University of China, Taiyuan, 030051, China
2. Beijing Advanced Innovation Center for Materials Genome Engineering, University of Science and Technology Beijing, Beijing, 100083, China
3. School of Materials Science and Engineering, Taiyuan University of Science and Technology, Taiyuan, 030024, China
4. Institute of Materials Intelligent Technology, Liaoning Academy of Materials, Shenyang, 110004, China
*Corresponding address: e-mail: zhaoyuhong@nuc.edu.cn

Abstract: The current isothermal sintering model cannot adequately describe the effect of local high-temperature on particle aggregation during spark plasma sintering (SPS), and a new non-isothermal sintering phase-field model suitable for spark plasma sintering is developed in this work. The model couples the heat conduction equation to the free energy to explore the impact of temperature on particle growth and observe its evolution mechanism. Using this model, the temperature not only affects the contribution of body energy by changing the temperature-dependent parameters but also significantly changes the gradient terms. When using this model to simulate non-isothermal sintering, we found that two identical particles grow asymmetrically under non-isothermal conditions, which is caused by the gradients of surface and grain boundary energies induced by local high-temperature. In addition, local high-temperature induces faster grain growth near the grain boundary, while the grain size in the center is smaller, which is confirmed in the experiment. The simulation results provide valuable insights for investigating particle growth under non-isothermal conditions.

Keywords:phase-field model, particle sintering, local high-temperature, temperature gradient, spark plasma sintering

1 Introduction

Solid sintering is widely used in powder metallurgy involving the applied heat and pressure to coalesce powder particles into a solid structure [1].The temperature is a critical parameter in the sintering process because it can act as a thermodynamic indicator and significantly affect the particles growth.

Although several studies on spark plasma sintering have been published in recent years, most of them regard the particles as a whole with uniform temperature without considering the influence of local high-temperature and temperature gradient on the particle evolution during sintering, which may lead to uneven temperature distribution inside the material, affecting the consistency and properties of the material. A new phase-field model for non-isothermal sintering is presented in this paper to observe the evolution of particle morphology under different temperature distributions, which is helpful in understanding the effect of temperature in the sintering process.

2 Phase field method

In a sintering system with multiple particles and pores, the total free energy F of the subdomain Ω within the system can be expressed as [2]:

$$F(T,\rho,\{\eta_\alpha\}) = \int_\Omega \begin{bmatrix} f(T,\rho,\{\eta_\alpha\}) + \frac{1}{2}T\kappa_\rho|\nabla\rho|^2 \\ +\frac{1}{2}T\kappa_\eta \sum_\alpha |\nabla\eta_\alpha|^2 \end{bmatrix} d\Omega \#(1)$$

Where the free energy density is:

$$f(T,\rho,\{\eta_\alpha\}) = f_{ht}(T)\left(A\rho + B\sum_\alpha \eta_\alpha\right) + C[\rho^2(1-\rho)^2] +$$

$$D\left[[\rho^2+6(1-\rho)]\sum_\alpha \eta_\alpha^2 - 4(2-\rho)\sum_\alpha \eta_\alpha^3 + \left(3\sum_\alpha \eta_\alpha\right)^2\right] \#(2)$$

$$f_{ht}(T) =- c_r T ln\frac{T}{T_0} + c_r(T-T_0) \#(3)$$

The modified temperature evolution equation is[3]

$$\frac{\partial T}{\partial t} = \mu\left(\frac{\partial^2 T}{\partial x^2} + \frac{\partial^2 T}{\partial y^2} + \frac{\partial^2 T}{\partial z^2}\right) \#(4)$$

3 Result and discussion

Particle agglomeration of two identical particles with temperature gradients

As shown in Fig 1(a), we apply a temperature gradient to two equal particles, which causes gradient changes in surface energy and grain boundary energy. The gradient between low and high temperatures creates a force that drives mass transfer, resulting in an uneven concentration distribution when non-isothermal particles gather. Thus, in this process, an asymmetric shape appears between two identical particles. Comparing the isothermal sintering process in Fig. 1(c), it can be found that temperature plays a dominant role in the process.

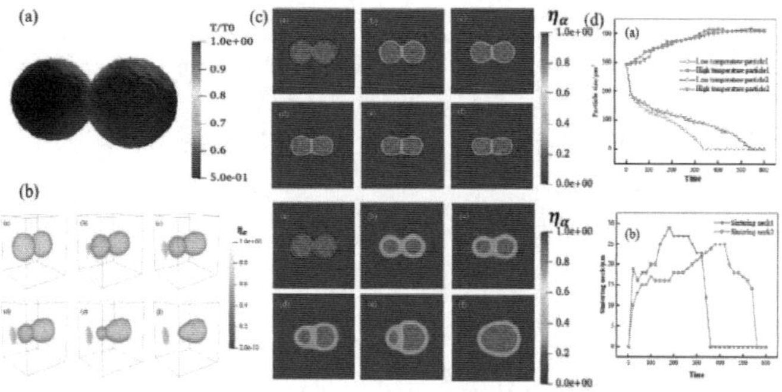

Fig. 1. (a) A three-dimensional diagram of two particles with a temperature gradient; (b) Three-dimensional results of particles evolution under temperature gradient; (c) Isothermal sintering and non-isothermal sintering particles evolution; (d) Under different temperature gradients, the particle area and the sintering neck on the cold side and the hot side evolves with time

Fig 1(d) shows the influence of different temperature gradients on the sintering process. The results show that the cold side particles and sintering neck disappear at the same time in the higher temperature gradient, which indicates that the higher temperature gradient will accelerate the sintering process.

Effect of local high-temperature on sintered grain size In the sintered particles, the particle size of each position in the initial stage is close, and there is no obvious difference in particle size. Due to the high local temperature at the grain boundary, the grain diffusion ability is strong and the annexation speed is fast. Therefore, at the later stage of sintering, the grains around the grain boundaries are significantly larger, while the particles in the center of the particles are smaller, as shown in Fig 2. The simulation results are in good agreement with the experimental results of AZ91D spark plasma sintering.

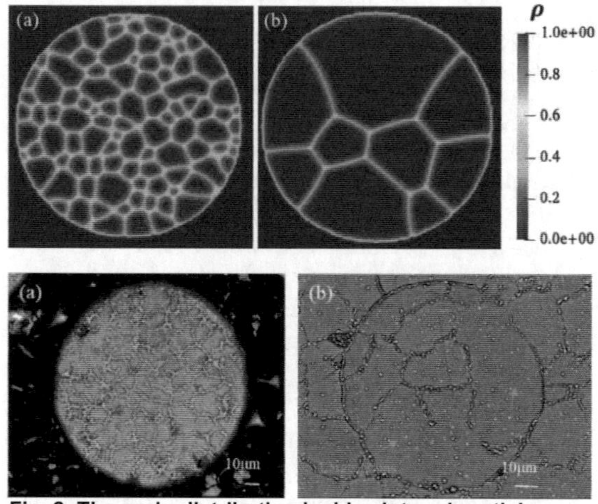

Fig. 2. The grain distribution inside sintered particles was observed by phase field simulation and experiment

4 Conclusion

In the sintering process of non-isothermal particles, the local high temperature of the gap causes the temperature gradient, and the gradient change of surface energy and grain boundary energy leads to the uneven distribution of the concentration of non-isothermal particles. At the same time, high temperature makes the grain at the grain boundary grow preferentially, so that the grain at the grain boundary is obviously larger than the grain at the core.

5 Acknowledgments

This work was supported by the National Natural Science Foundation of China (Grant No. 52074246). Shanxi Graduate Innovation Project, China (Grant No. 2021Y592).

References

[1] Li X Y, Zhang Z H, Cheng X W, etal. The Development and Application of Spark Plasma Sintering Technique in Advanced Metal Structure Materials: A Review, Powder Metallurgy and Metal Ceramics, 2021,60:410-438.

[2] Yi M, Wang W, Xue M, etal. Modeling and Simulation of Sintering Process Across Scales, Archives of Computational Methods in Engineering, 2023,30:3325-3358.

[3] Miranville A, Quintanilla R. A Phase-Field Model Based on a Three-Phase-Lag Heat Conduction, Applied Mathematics & Optimization, 2011,63133-150.

Phase Field Simulation for Grain Refinement in Dendrite Growth of A356 Aluminum Alloy

Jiaqi Pei[1], Weipeng Chen[1], Wenda Zhang[1], Hua Hou[1,3], Yuhong Zhao[1,2,4*]
1 School of Materials Science and Engineering, Collaborative Innovation Center of Ministry of Education and Shanxi Province for High-performance Al/Mg Alloy Materials, North University of China, Taiyuan, 030051, China
2 Beijing Advanced Innovation Center for Materials Genome Engineering, University of Science and Technology Beijing, Beijing, 100083, China
3 School of Materials Science and Engineering, Taiyuan University of Science and Technology, Taiyuan, 030024, China
4 Institute of Materials Intelligent Technology, Liaoning Academy of Materials, Shenyang, 110004, China
*Corresponding address: e-mail:zhaoyuhong@nuc.edu.cn

Abstract: Although A356 aluminum alloy possesses poor ductility and strength, the refinement of the initial α-Al phase has shown promise in improving these properties. The impact of grain refinement mechanisms on the organization and morphology of A356 aluminum alloys has not been thoroughly investigated. In this paper, we constructed a Phase-field lattice-Boltzmann model to explore the influence of the refinement mechanism on the growth of α-Al dendrites during the non-isothermal solidification process of A356 aluminum alloy. The results demonstrate that the approach of alloying by adding elements, introducing forced convection, and increasing the degree of subcooling has proven to be effective in refining the dendrites of A356 aluminum alloy. The presence of Fe element further promotes the thinning of α-Al dendrites, and this effect is also found after melting ultrasonic treatment at a lower initial casting temperature. These findings are expected to provide theoretical support for effectively improving the hardness, ductility, and overall performance of Al-based alloys.

Keywords: Dendrites,Grain refinement, Phase-field method, Aluminum alloys

1 Introduction

A356 aluminum alloy is frequently used in automobile engine components and wheel rims. Its excellent castability, high strength-to-weight ratio, and exceptional corrosion resistance are what make it attractive[1], but its ductility and strength are somewhat limited. The refining of the α-Al phase in the alloy can significantly improve the strength and ductility of the alloy.The methods can be broadly categorized as physical refinement, thermodynamic refinement, and chemical refinement [2,3].Nevertheless, there is a lack of comprehensive research on the microstructure refinement mechanism and the growth of polycrystalline structures in Al-based multicomponent alloys.

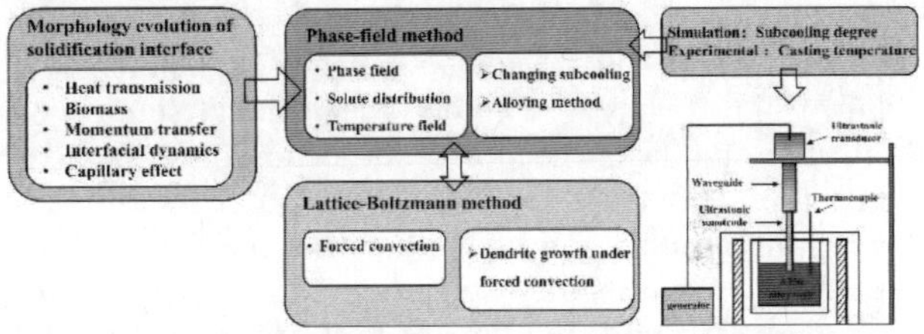

Figure 1. Schematic correlation between phase-fieldsimulation and melt ultrasonic treatment.

For the first time, we utilize the phase-field lattice-Boltzmannmodel to explore the grain refinement mechanisms on thedendritic morphology in A356 aluminum alloyin this paper. The simulation results were also correlated with the experimental results after ultrasonic melt treatment, the schematic is shown in Figure 1. The results may provide research insights for improving the ductility and hardness of other Al-based alloys.

2 Phase-field Method

In this study, the solidification phase-field model in the EasyPhase software package of Prof. Yuhong Zhao's group was used to describe the α-Al dendrite growth of A356 aluminum alloy under non-isothermal solidification conditions. The phase field order parameter is denoted by ϕ, with solid phase $\phi =+1$ and liquid phase $\phi =-1$ .To simulate the movement of the liquid phase and its convection effect on the concentration evolution equation, we employed a lattice-Boltzmann method.

In our calculations, we discretize the phase and solute field equations using the finite volume method, the temperature field equations using the alternating direction implicit (ADI) algorithm, and the flow field using the lattice Boltzmann method.

3 Result and discussion

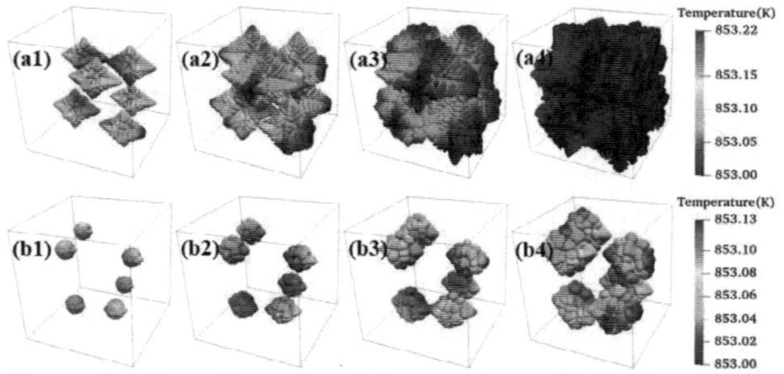

Figure 2. Three-dimensional temperature distribution of A356 aluminum alloy without (a1-a4) and with (b1-b4) Mg solute element. (a1-c1) t = 3000Δt, (a2-c2) t = 5000Δt, (a3-c3) t = 7000Δt, (a4-c4) t = 10000Δt

Temperature distribution of A356 aluminum alloy before (a1-a4) and after (b1-b4) the addition of Mg solute elements is shown in Figure2. During the initial stages of solidification, the latent heat of solidification diffuses from the center towards the surrounding areas. At $t = 5000\Delta t$, following the introduction of Mg element, the temperature of the dendrite tip near the boundary is higher than before. Consequently, the collision time of heatbetween individual dendrites is delayed, thereby influencing the growth rate of the dendrites. Moreover, the diffusion rate of latent heat during solidification is decreased. This extended diffusion time allows for enhanced grain refinement and serves to impede the growth of the dendrites.

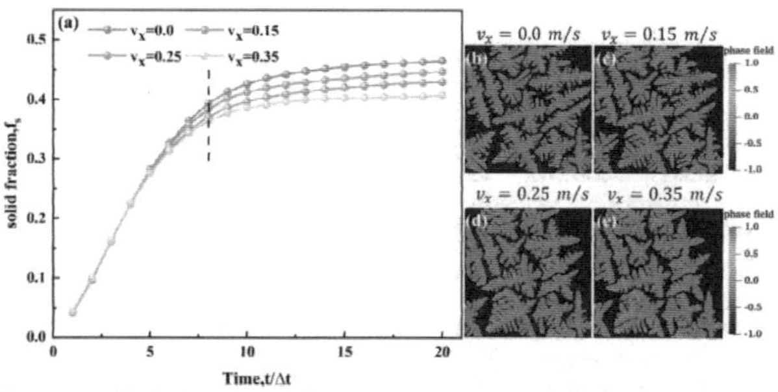

Figure 3. (a) Solid phase fraction curves of A356 aluminum alloy at different flow rates and dendritic morphology corresponding to the black dashed line, the flow rates are (b) $v_x = 0.0$ m/s, (c) $v_x = 0.15$ m/s, (d)$v_x = 0.25$ m/s, (e)$v_x = 0.35$ m/s, respectively.

The solid phase fraction of A356 aluminum alloy with time at different flow rates as well as the corresponding dendrite morphology at $t = 7000\Delta t$ (black dashed line) as shown in Figure 3. As the flow rate increases from 0.0 m/s to 0.35 m/s, the rate of solidification in the dendrites decelerates,resulting in a reduced in the solid phase fraction. This phenomenon can be attributed to the increased transport of heat and solute from the countercurrent side to the tip of the dendrite arm on the downstream side at higher flow rates. As a result, the growth rate of the dendrite arm on the countercurrent side accelerates, causing the dendrite tip to become more susceptible to interactions with boundaries or neighboring dendrites. Additionally, minor protrusions emerged on the countercurrent side of the horizontal dendrite arms, but they did not fully develop into well-formed secondary dendrite. Consequently, these factors contributed to a decrease in the solid phase fraction.

4 Conclusion

(1) The addition of the Mg element, implementation of forced convection, and increase in subcooling in A356 aluminum alloy can effectively achieve grain refinement and improve its ductility and hardness. The increase of Mg content can provide more effective nucleation sites for α-Al nucleation and promote grain refinement. The introduction of forced convection increases the thermal and compositional subcooling in the downstream zone, which hinders dendrite growth on the downstream side. Increasing the degree of subcooling accelerates the secondary dendrite differentiation and the primary dendrite arms are finer, forming finer dendrites.

(2) The application of melt ultrasonication during casting lowers the initial casting temperature, consequently augmenting the degree of subcooling. This increase in subcooling leads to the refinement of α-Al dendrites. Both simulations and experiments have corroboratedthis trend of morphological changes.

(3) Elemental Fe has been observed to reduce the driving force during the solidification and further increases the refinement of α-Al dendrites.

5 Acknowledgments

This work was supported by the National Natural Science Foundation of China (52074246, 52275390, 52205429, 52201146), National Defense Basic Scientific Research Program of China (JCKY2020408B002, WDZC2022-12), Key Research and Development Program of Shanxi Province (202102050201011, 202202050201014).

References

[1] Reyes R V, Bello T S, Kakitani R, etal. Tensile properties and relatedmicrostructural aspects of hypereutectic Al-Si alloys directionally solidified under different melt superheats and transient heat flow conditions. Mater Sci Eng, A 2017, 685:235-43.

[2] Chen ZN, Kang HJ, Fan GH, etal. Grain refinement of hypoeutectic Al-Si alloys with B. Acta Mater 2016, 120: 168-78.

[3] Campo KN, Proni CT, Zoqui EJ. Influence of the processing route on the microstructure of aluminum alloy A356 for thixoforming. Mater Char 2013, 85: 26-37.

A Prediction of Atomic Occupation Probability and Order Parameter Changes in the Precipitation Process of $Ni_{75}Al_{2.5}V_{22.5}$Alloy was Carried Out Using a BP Neural Network

XiaonaWang[1],Zeyu Wang[1],Hua Hou[1,3],Yuhong Zhao[1,2,4*]

1 School of Materials Science and Engineering, Collaborative Innovation Center of Ministry of Education and Shanxi Province for High-performance Al/Mg Alloy Materials, North University of China, Taiyuan, 030051, China
2 Beijing Advanced Innovation Center for Materials Genome Engineering, University of Science and Technology Beijing, Beijing, 100083, China
3 School of Materials Science and Engineering, Taiyuan University of Science and Technology, Taiyuan, 030024, China
4 Institute of Materials Intelligent Technology, Liaoning Academy of Materials, Shenyang, 110004, China
*Corresponding address: e-mail: zhaoyuhong@nuc.edu.cn

Abstract: The Ni75Al2.5V22.5 alloy was used as the research object, and the precipitation process of the alloy at different temperatures was simulated by the microscopic phase-field method. We found that it takes a long time to change the temperature conditions many times. Therefore, for the first time, this paper proposes to use the simulated data of the microscopic phase-field method as a data set to construct an artificial neural network model to directly and quickly predict the results: The back propagation (BP) neural network model is established. When the total prediction step is 50000, the result is about 970 times shorter than the micro-phase field simulation calculation time. Under the premise of ensuring that the maximum error is 0.00672, the calculation efficiency is greatly improved.

Keywords: $Ni_{75}Al_{2.5}V_{22.5}$ alloy,Back propagationneural network,Microscopic phase-field, Precipitation.

1 Introduction

Microstructure determines macroscopic properties. Computer simulation technology and artificial intelligence technology are widely used in materials science [1]. We can observe the microstructure evolution process of alloy materials at different time and space scales, and can also directly predict the microstructure under different process conditions[2].The machine learning method has been applied to find and identify phase transition models in material physics [3], extracting properties based on material microstructure [4] and accelerate the prediction of material properties [5], which provides ideas for our research. Therefore, we try to combine BP neural network technology with the microscopic phase-field phase field method.

2 Experimental procedure

Firstly, a large amount of data is established by microscopic phase-field simulation, and then the data set is randomly divided into training set and test set. The training set is used to construct the BP neural network model, and the test set is used to evaluate the model. Finally, the fitting results of BP neural network are used to optimize.

The microscopic phase-field model

The phase field method mainly describes the microstructure evolution by obtaining the changes of field variables in time and space. The microscopic phase-fieldequation

$$\frac{dP(r,t)}{dt}=\frac{C_0(1-C_0)}{k_BT}\sum_{r'}L(r-r')\frac{\partial F}{\partial P(r',t)}\#(1)$$

The Back propagationneural network model

The most significant feature of BP neural network is the function of error back propagation [8]. The calculation method of BP neural network algorithm is:
training data input value:

$$X=(x_1,x_2,\dots,x_m)$$

the actual output value of the model :

$$Y=(y_1,y_2,\dots,y_n)$$

the expected output value:

$$T=(t_1,t_2,\dots,t_n)$$

For the kth pair of calculated data $(X_k,T_k)$of the BP neural network, the error of the kth pair of calculated data, that is, the error between the actual output and the expected output, is defined as:

$$E_k=\frac{1}{2}\|Y_k-T_k\|^2=\frac{1}{2}\sum_{z=1}^{n}(y_z-t_z)^2\#(2)$$

Simulation process

The atomic evolution images of $Ni_{75}Al_{2.5}V_{22.5}$ at 1050K and 1350K were obtained by microscopic phase-field model simulation, Fig.1, asa reference.Each la (50,33),(52,34),(54, 35)…

*Data collection.*The input variable is temperature 1000K~1500K, and the output variable is atomic occupation probability.Select a complete θ phase, a total of 15 lattice points,A total of 101 sets of data were collected.
*Algorithm selection.*The BP neural network algorithm in the artificial neural network system is selected.

*Modeling.*Training data 70%, validation data 15%, test data 15%.

*Model analysis.*Themean square error

$$E(W,b) = \frac{1}{m}\sum_{i=1}^{m}\frac{1}{k}\sum_{j=1}^{k}\left(\hat{y}_{ij} - y_{ij}\right)^2 \#(3)$$

$m$the number of nodes in the input layer.

Finally, the goodness of fit index R is used to evaluate the model.

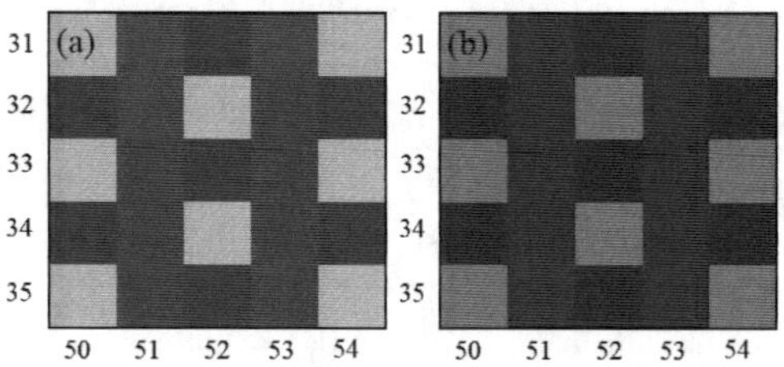

Fig. 1Local the simulation area.(a)1050K(b)1350K,there is aθ precipitated phase structures, completly.

3 Result and discussion

After the training is completed, the mean square error MSE < 6.72E-03, the average error of the training set is 1.33%, the maximum error of the training set is 5.88%, the average error of the verification set is 1.23%, and the maximum error of the verification set is 5.96 %, the error of R value is not more than 0.9%,indicating that the BP neural network of the structure can achieve a better prediction effect.

In order to objectively evaluate the generalization ability of the model, 10 sets of data are used to predict and compare with the reference value.$R^2 = 0.9832$, the fitting degree is good.

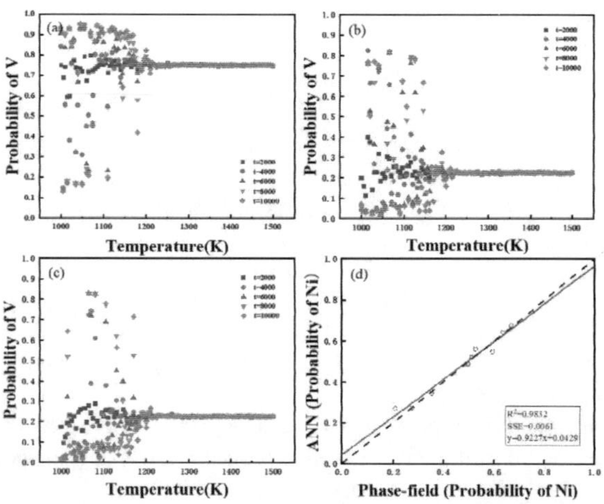

Fig. 2Lattice data sets and prediction results. (a)lattice (50,33) (b) lattice (52,34) (c) lattice(54,35) (d)There is a linear relationship between the atomic occupation probability predicted by BP neural network and the atomic occupation probability calculated by microscopic phase- field.

Predict the order parameters

We use a similar method, only need to change the output value.The R value is concentrated near 0.85, and the prediction effect of the model is general, so more data need to be collected.

Conclusion

The BP neural network is a feed-forward supervised network model, which can better replace the microscopic phase-field model.The BP neural network model takes about 0.6 seconds to perform a prediction, and the calculation time is about 970 times shorter than the phase field model.

Acknowledgments

This project was financially supported by Shaanxi Provincial Innovation Capacity Support Plan (2023-CXTD-4), the State Key Laboratory of Solidification Processing (SKLSP202201) and ND Basic Research Funds, NPU (G2022WD).

References

[1] Gregory H， Krishna G. Machine learning materials physics: Surrogate optimization and multi-fidelity algorithms predict precipitate morphology in an alternative to phase field dynamics[J].Computer Methods in Applied Mechanics and Engineering, 2019, 344:666-693.

[2] Evert PL, Liu Y H.Learning phase transitions by confusion[J].Nature Physics, 2016,435-440.

[3] Carrasquilla J and Melko M G. Machine learning phases of matter. NaturePhysics,2017, 13:431–434.

[4] Steinmetz P, YabansuY C, et al Analytics for microstructure datasets produced by phase-fieldsimulations[J]. Acta Materialia,2016, 103:192–203.

Dislocation Controlled Segregation-Assisted Precipitation and Pinning Grain Growth: A Phase-Field Study

Wenkui Yang[1], Kaile Wang[1], Pengya Lei[1], Hailong Nie[1], Hua Hou[1,3], Yuhong Zhao[1,2,4,*]

1. School of Materials Science and Engineering, Collaborative Innovation Center of Ministry of Education and Shanxi Province for High-performance Al/Mg Alloy Materials, North University of China, Taiyuan, 030051, China
2. Beijing Advanced Innovation Center for Materials Genome Engineering, University of Science and Technology Beijing, Beijing, 100083, China
3. School of Materials Science and Engineering, Taiyuan University of Science and Technology, Taiyuan, 030024, China
4. Institute of Materials Intelligent Technology, Liaoning Academy of Materials, Shenyang, 110004, China
*Corresponding address: e-mail: hchkou@nwpu.edu.cnn

Abstract: Based on the grain growth model and dislocation stress field model, we investigated the dynamic processes of grain growth and nucleation and growth of CRPs, with the combined effect of dislocations and grain boundaries. Meanwhile, CRPs can pin grain growth to maintain smaller grain sizes. Simulation results reveal that both grain boundaries and dislocations promote the nucleation and growth of Cu phase, hindering grain growth. The precipitate is a dual core-shell structure of Mn/NiAl/Cu at the grain boundary in the middle stage of aging. The precipitate at the grain boundary is a co precipitate of MnNiAl and Cu rich, while the intragranular precipitate is a dual core-shell structure of Mn/NiAl/Cu in the later stage of aging. The precipitation at the grain boundary undergoes a process of strip→spherical→strip with increasing dislocation density. On the whole, this modeling framework provides an avenue to explore the role of dislocation on the grain growth and microstructural evolution of Cu-rich precipitates.

Keywords: Grain growth, Dislocation, Phase-field method, Cu-rich precipitates.

1 Introduction

Fu-Cu alloy—as a typical nanocrystalline metal, its high-density grain boundaries exhibit a unique combination of characteristics, which also constitute a huge driving force for grain growth [1]. The ability to stagnate grain growth and retain the nano-crystalinity of Fu-Cu alloy under high temperature and for extended time scales is arguably one of the enormous challenges to the RPVs [2]. Several studies, such as recrystallization, pore or bubble drag, precipitate pinning and solute segregation have been shown to mitigate grain growth [3]. In recent years, GBs segregation-assisted precipitates has been the subject of active research as it has been proposed as important route to stagnate grain growth in Fu-Cu alloy. However, coarse precipitates and strip/needle shaped precipitates at GBs lead to stress concentration, increasing the risk of catastrophic failure. Ensuring the pinning of GBs segregation-assisted precipitates and the fine distribution of precipitation dispersion is one of the important issues that still need to be addressed. Although the precipitation process of multi-component Fe-Cu-based alloys has been studied, the effect of dislocations on the morphology of precipitates at grain boundaries and precipitates at grain boundaries pinned grains growth have not been reported. On this basis, a phase field model was established by coupling grain growth, multi-component field, and dislocation elastic field. The simulation model and results in this work were integrated into the EasyPhase software package developed by Professor Zhao. The purpose is to study the process of grain boundary precipitation, dislocation-controlled grain boundary precipitation, and precipitation pinning grain growth.

2 Phase field model

Our phase field model is built upon the incorporation of structural order parameters (OPs) $\phi_p(\boldsymbol{r},t)(p=1,\ldots,m)$ ($\phi_p(\boldsymbol{r},t)$ characterizes the grains present in the alloy, each possessing distinct crystal orientations) and compositional field $c_i(\boldsymbol{r},t)$ ($i=2,3,4,5$ are Cu, Mn, Ni and Al atoms, respectively). We propose the expression for the overall free energy F, which can be written as

$$F = \int \left[\begin{array}{c} f_{ch}(c_i,T) + \frac{1}{2}\sum k_{c_i}^2(\nabla c_i)^2 + \frac{1}{2}\sum k_{\phi_p}^2(\nabla\phi_p)^2 \\ + f_{\text{grain}}(\phi_p) + f_{inter}(c_i,\phi_p) + f_{el}(c_i,\phi_p) \end{array} \right] dV \quad (1)$$

According to Khachaturyan' micromechanical theory, the elastic stress $\sigma_{ij}^{el}(\boldsymbol{r})$ can be expressed as:

$$\begin{aligned}\sigma_{ij}^{el}(r) &= C_{ijkl}\varepsilon_{kl}^{el}(r) \\ &= \left(C_{ijkl}^0 + \Delta C_{ijkl}\Delta c\right)\left(\delta\varepsilon_{ij} - \varepsilon_{ij}^0 - \varepsilon_{ij}^d\right)\end{aligned} \#(2)$$

The governing equations for the evolution of the concentration field c and OPs ϕ_i, can be stated respectively as

$$\frac{\partial c_i(\boldsymbol{r},t)}{\partial t} = \nabla \cdot M \cdot \nabla \frac{\delta F}{\delta c_i(\boldsymbol{r},t)} + \xi_{c_i}(\boldsymbol{r},t)$$
$$\frac{\partial \phi_i}{\partial t} = - L_i \frac{\delta F}{\delta \phi} \quad \#(3)$$

3 Result and discussion

The simulation results in Fig. 1 show that solute atoms preferentially aggregate at the triple intersections. Subsequently, solute atoms aggregate at grain boundaries and gradually form granular CRPs. Meanwhile, granular CRPs absorb solute atoms near grain boundaries to connect as strips of CRPs (a Cu/Mn/NiAl double core-shell structure), leading to the formation of solute-poor and solute-rich regions. Mn, Ni and Al atoms continue to aggregate at the shells of the CRPs gradually forming co-precipitates of Cu-rich and MnNiAl.

THE GRAPHIC ON THIS PAGE WAS NOT AVAILABLE ON THE E-MEDIA VERSION OF THIS PAPER.

Fig. 1. 2D element distribution diagram of polycrystalline grain boundaries and intra particle precipitation

While the intracrystalline randomly formed spherical CRPs of double core-shell structure. There is a competitive mechanism between the nucleation and growth of CRPs within the grains and the growth of CRPs at the grain boundaries, that is, the nucleation and growth of CRPs within the grains leads to the refinement of CRPs at the grain boundaries. The equivalent stress result in Fig. 1 (b) shows that the strip-shaped CRPs lead to stress concentration at grain boundaries. The 2D results in Fig. 2 indicate that solute atoms preferentially aggregate at dislocations of grain boundaries. Solute atoms continue to aggregate at dislocation to form spherical CRPs, leading to an increase in solute poor regions around grain boundaries. Subsequently, a small portion of spherical CRPs at the grain boundaries gradually connects to form short rod-shaped CRPs, while spherical CRPs form gradually within the grain. Competition between precipitation within the grain and at the grain boundaries leads to the refinement of spherical and short rod-shaped Cu-rich cores at the grain boundaries. The equivalent stress in Fig. 2(b) shows almost no stress concentration. This indicates that a $3.43 \times 10^{13}/m^2$ dislocation density can significantly regulate the morphology of CRPs at the grain boundaries, thereby resulting in a more uniform stress distribution.

THE GRAPHIC ON THIS PAGE WAS NOT AVAILABLE ON THE E-MEDIA VERSION OF THIS PAPER.

Fig. 2 Dislocation of $3.43 \times 10^{13}/\mathrm{m}^2$ regulates the process of polycrystalline precipitation

4 Conclusion

This study investigates the process of grain growth, grain boundary precipitation and pinning, as well as the regulation of grain boundary precipitation and pinning of grains by dislocations of grains in Fe-Cu-Mn-Ni-Al alloys. The segregation of solute atoms near grain boundaries forms CRPs that can effectively pin grain growth. But the strip-shaped CRPs lead to significant stress concentration. A uniformly distributed dislocation with a density of $3.43 \times 10^{13}/m^2$ at the grain boundary can effectively regulate the morphology of precipitated phases at the grain boundary. The precipitates at the grain boundaries are double core-shell structures of Cu/NiAl/Mn in the middle stage of aging, which are co-precipitates of Cu-rich and NiAl/Mn in the late stage of aging.

5 Acknowledgments

This project was financially supported by Shaanxi Provincial Innovation Capacity Support Plan (2023-CXTD-4), the State Key Laboratory of Solidification Processing (SKLSP202201) and ND Basic Research Funds, NPU (G2022WD).

References

[1] Darvishi Kamachali R, Kwiatkowski da Silva A, McEniry E, et al. Segregation-assisted spinodal and transient spinodal phase separation at grain boundaries[J]. npj Computational Materials, 2020, 6(1): 191.

[2] Zhou X, Kamachali R D, Boyce B L, et al. Spinodal decomposition in nanocrystalline alloys[J]. Acta Materialia, 2021, 215: 117054.

[3] Rehn V, Hötzer J, Rheinheimer W, et al. Phase-field study of grain growth in porous polycrystals[J]. Acta Materialia, 2019, 174: 439-449.

Effect of Natural Convection on Dendrite-Bubble Interaction

Jing Cheng, Ang Zhang *, Lang Qin, Minghang Yang, Bin Jiang, Fusheng Pan
National Engineering Research Center for Magnesium Alloys, National Key Laboratory of Advanced Casting Technologies, College of Materials Science and Engineering, Chongqing University, Chongqing 400044, China
*Corresponding address: e-mail: angzhang@cqu.edu.cn

Abstract: The interaction between solidifying dendrite and rising bubble is investigated by adopting a multiphase-field lattice Boltzmann method. The bubble changes the dendrite skeleton by hindering the solid growth, including entrapped in the dendrite root, tilting growth towards the neighboring bubble, forming the knot structure, and growing along the radial direction of the bubble. The bubble in the columnar dendrite array with random orientation distribution has larger effects on the average primary dendrite arm spacing. A phase diagram is proposed to distinguish different interaction regimes. The size of the closed liquid phase region is further measured to evaluate the porosities tendency and guide how to reduce the porosities.

Keywords: Solidification; Bubble; Magnesium; Phase-field simulation

Introduction

Prediction and control of gas porosity defect during solidification is challenging but has attracted increasing interest for decades. The gas porosity deteriorates material properties by reducing loading area, causing stress concentration, and increasing fracture sensibility [1, 2].

Based on the employed multiphase-field lattice-Boltzmann approach, the objective of this work is to discuss the multiphase competitive growth under convection.

Experimental procedure

The multiphase-field model introduces three order parameters with the sum of one to denote the solid, liquid, and gas phases [3]. The Mg-6wt.%Gd alloy with sixfold symmetry pattern is simulated. The nucleation process of both dendrite and bubble is ignored. A circle seed is planted at the top side and the dendrite solidifies directionally from top to bottom. A circular bubble is released at the lower part.

Result and discussion

Figure 1 shows the typical multiphase simulation result and the solute concentration distribution along the direction indicated by the horizontal arrow in Figure 1a at different times. The dendrite presents semi-sixfold pattern and the bubble rises to the interdendritic root. Local concentration extreme is observed at the solid-liquid interface due to solute repartition. The solute concentration at the top closed region reaches local extreme, because the bubble impedes solute transport towards elsewhere.

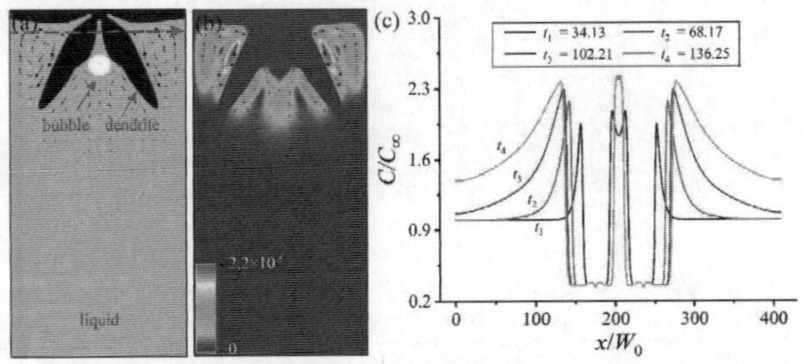

Figure 1. Typical multiphase simulation result. (a)-(b) Multiphase-field and flow field. The arrows denote the flow velocity vectors. Solute concentration distribution along the direction indicated by the horizontal arrow in Fig. 1(a) at different times.

Figure 2 shows the multiphase contours under different combinations of the growth orientation and the undercooling. When the flow intensity is fixed, the larger the undercooling, the more complex the dendrite morphology and also the larger the distance between the bubble-dendrite striking position and the seed center. The developed dendrite arms entrap the bubble in the solid skeleton. Different interaction behaviors, including the tilting growth towards the neighboring bubble, the formation of the knot structure, and the radial growth along the bubble boundary, can be observed as the growth orientation increases.

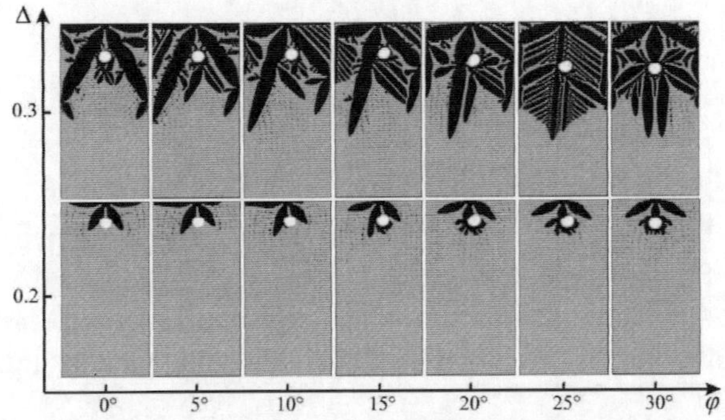

Figure 2. Multiphase contours under different growth orientations and different undercoolings. The flow intensity g/g0 = 1. The undercoolings Δ = 0.2 and 0.3, and the growth orientations φ change from 0° to 30° with the interval of 5°. The arrows denote the flow velocity vectors.

The relative position between the bubble and the dendrite seed center along the *x* axis affects the multiphase interaction by changing the striking position. The change of the relative position is similar to the change of the growth orientation, i.e., changing the contact point between the rising bubble and the dendrite.

Figure 3 shows the change of primary dendrite arm spacing (PDAS). Both the change of the dendrite orientations and the change of the bubble position can cause different PDAS, further illustrating the important influence of the striking position on the interaction between the growing dendrite and the rising bubble.

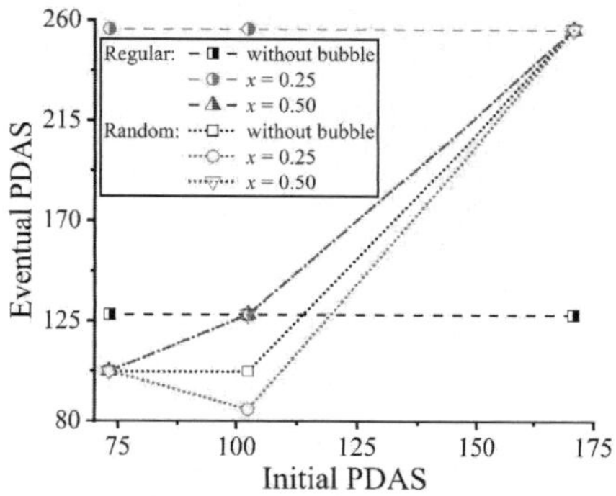

Figure 3. Change of the average primary dendrite arm spacing (PDAS) under different settings.

Figure 4 shows the dendrite-bubble interaction cases. The increase of Δ, g/g_0, and φ results in increasing the number of the primary dendrite arms, including forming the knot structure and growing along the radial direction. By providing a summary of different interaction results identified, the phase diagram shown in Figure 4 can be depicted to distinguish different cases.

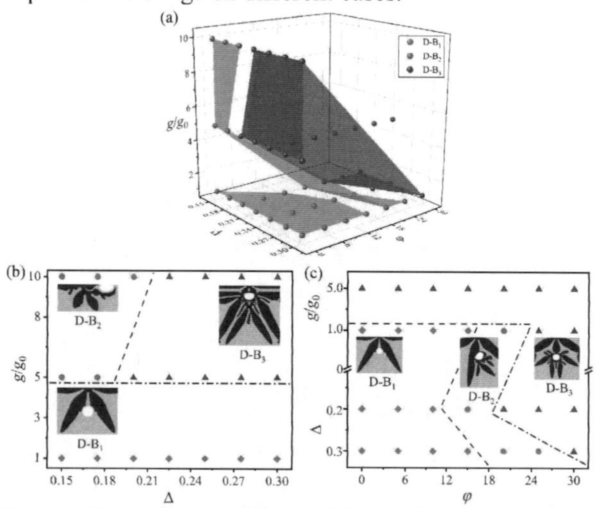

Figure 4. Phase diagram of the dendrite-bubble interaction regimes. (a) 3D diagram; (b)-(c) 2D projection.

Conclusion

(1) The bubble rises to the interdendritic region and deforms to adapt the arm surface as well as changes the solid skeleton. As the dendrite grows, the bubble is entrapped in the dendrite skeleton.

(2) The growth orientation affects the multiphase interaction by changing the striking position between the bubble and the dendrite, which is similar to that caused by the change of the relative position between the bubble and the dendrite.

(3) A phase diagram is proposed to distinguish different interaction regimes based on the summary of different interaction results identified.

Acknowledgments

This work is financially supported by National Natural Science Foundation of China (grant numbers 52101125 and U2037601), and Natural Science Foundation of Chongqing China (grant number CSTB2022NSCQ-MSX1289).

References

[1] Dantzig J., Rappaz M., 2016. Solidification. 2nd. EPFL Press, Lausanne.
[2] Kurz W., Rappaz M., Trivedi R., 2020. Progress in modelling solidification microstructures in metals and alloys. Part II: dendrites from 2001 to 2018. Int. Mater. Rev. 72, 1–47.
[3] Zhang A., Guo Z., Jiang B., Du J., Wang C., Huang G., Zhang D., Liu F., Xiong S., Pan F., 2021. Multiphase and multiphysics modeling of dendrite growth and gas porosity evolution during solidification. Acta Mater. 214, 117005.

Advances in Macro- and Micro-Modeling of Squeeze Casting Process

Zhiqiang Han[1,2,*]
1. School of Materials Science and Engineering, Tsinghua University, Beijing 100084, China
2. Key Laboratory for Advanced Materials Processing Technology (Ministry of Education), Tsinghua University, Beijing 100084, China
*Corresponding address: e-mail: zqhan@tsinghua.edu.cn

Abstract: Squeeze casting is an advanced manufacturing process. The physics involved in squeeze casting is pressurized solidification, and it is important to understand the fundamental knowledge of pressurized solidification and to develop numerical models at macro and micro scales. This review presents the major challenge and novel research dedicated to macro- and micro-modeling on squeeze casting of aluminum and magnesium alloys, including metal displacement and free surface tracking, thermal-mechanical coupled simulation, casting-mold interfacial heat transfer model, pressurized solidification and microstructure modeling.

Keywords: squeeze casting, thermal-mechanical coupling, casting-mold interfacial heat transfer model, pressurized solidification, microstructure simulation

1 Introduction

Squeeze casting is an advanced near-net-shape forming process that offersmany remarkable advantages over conventional casting process. The combination of squeeze casting process with light alloys is an effective approach to produce lightweight components with excellent performance and precise geometry.

Squeeze casting process involves complicated physics at both macro and micro scales. Fig. 1 illustrates the essential physics at macro- and micro-scales and the major challenge in modeling and simulation of the squeeze casting process.

This article addresses the challenge in macro- and micro-modeling and simulation of squeeze casting process and reviews the major progress made in the past two decades.

2 The progress in macro-modeling and simulation of squeeze casting

Simulation of mold filling process

The major challenge in the simulation of mold filling process of squeeze casting is to deal with the movement of the punch, which means to solvethe fluid flow and track the free surface in a computational domain subject to a continuous change.Lewis and Han proposed a quasi-static Eulerian finite element method, where a velocity boundary condition was imposed at the interface between the punch and the molten metal to simulate the effect of the punch action (Fig. 2).

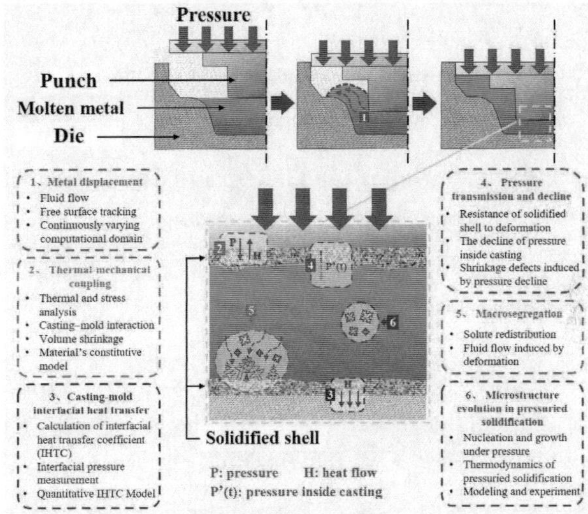

Fig. 1 The essential physics at macro- and micro-scales and the challenge in modeling and simulation of squeeze casting process.

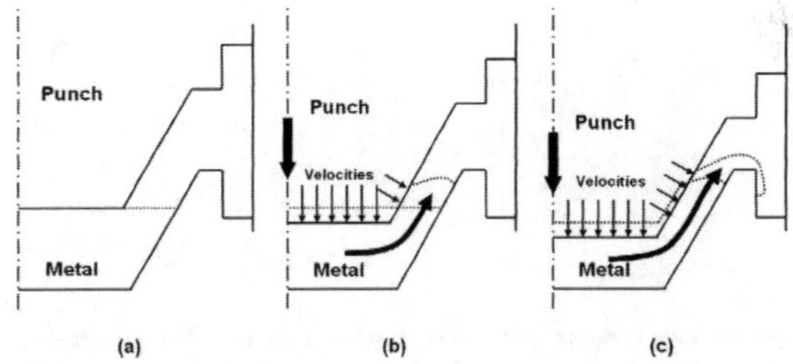

Fig. 2 The schematics of the metal displacement in direct squeeze casting process.

3 Thermal-mechanical coupled simulation

In squeeze casting, thermal and mechanical are inter-dependent, resulting in some important and complicated issues in the solidification process. In the work of Zhu and Tang et al., a three-dimensional thermal-mechanical coupled simulation method for squeeze casting components with complex geometry has been developed.

4 Casting-mold interfacial heat transfer model

In squeeze casting, the interfacial heat transfer coefficient (IHTC) depends upon the casting-mold interfacial contact status varying with time and location of the interface. Wang et al. established a quantitative model (Fig. 3) describing the intrinsic dependence of the IHTC on the interfacial pressure and temperature, providing a way to couple the thermal-mechanical inter-dependence.

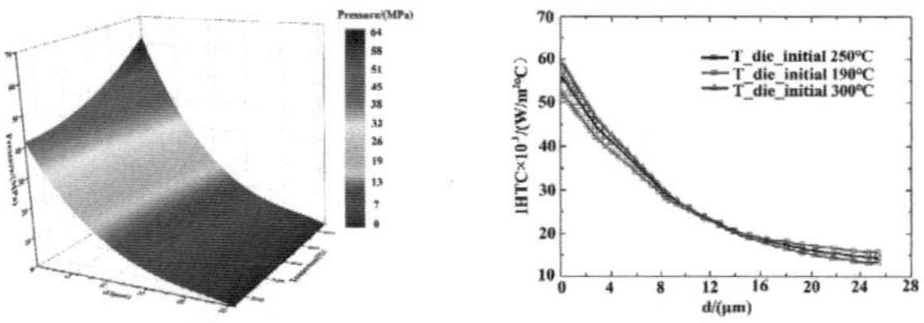

(a) (b)

Fig. 3 (a) The dependence of mean plane separationon pressure and temperature; (b) The relationship between IHTC and the mean plane separation.

5 The progress in micro-modeling and simulation of squeeze casting

Nucleation model of pressurized solidification

Squeeze casting is essentially a pressurized solidification process. Pressuresignificantly affects the nucleation in solidification.Huang et al. developed a quantitative model (Eq. 1) for describing the heterogeneous nucleation in pressurized solidification.Fig. 4 shows the dependence of nucleation rate on the undercooling and pressure.

$$ {}^{v}I_{\mathrm{he}} = 10^{42}K_{\mathrm{he}}\exp\left[-\frac{\mathrm{b}f(\theta)\frac{\sigma^3}{(\Delta G_V+\varepsilon P)^2}+\Delta G_{\mathrm{d0}}+\Delta VP}{\mathrm{k_B}T}\right] \quad (1) $$

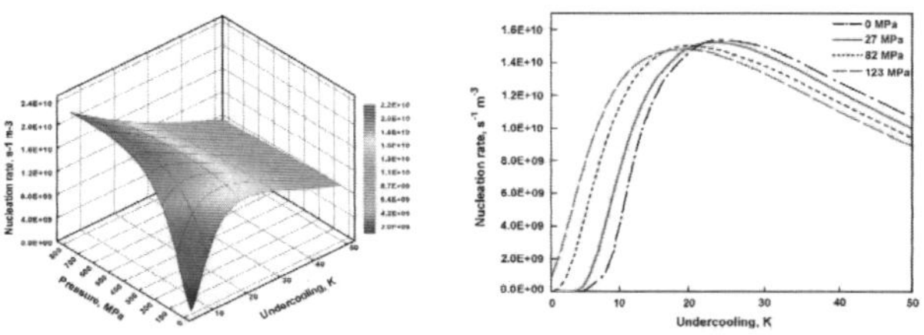

(a) (b)

Fig. 4 The effects of pressure and undercooling on the heterogeneous nucleation in squeeze casting.

Microstructure simulation of squeeze casting

For the simulation of the microstructure evolution in squeeze casting, it is essential to appropriately incorporate the effects of pressure on the thermodynamics and kinetics of growth. Huang et al. developed a mathematical expression of the solid-liquid interface equilibrium of a binary system during its pressurized solidification:

$$ C^*-C_0=\left(\frac{\partial C}{\partial T}\right)_{\mathrm{P}}(T^*-T_{\mathrm{L}}^{\mathrm{eq}})+\left(\frac{\partial C}{\partial P}\right)_{\mathrm{T}}(P^*-P_0)+\left(\frac{\partial C}{\partial T}\right)_{\mathrm{P}}\Gamma K f(\varphi,\theta) \quad (2) $$

This work provided an approach to calculate the growth dynamics in cellular automaton simulation of the pressurized solidification of binary systems.

Shang et al. developed a phase field model incorporating the effect of pressure on solidification of multi-component system. As an example, Fig. 5 demonstrates a pressure-effect-embedded calculation of phase diagram, which was used to simulate the dendritic growth of a Mg-Al-Sn ternary alloy solidified under pressure (Fig. 6).

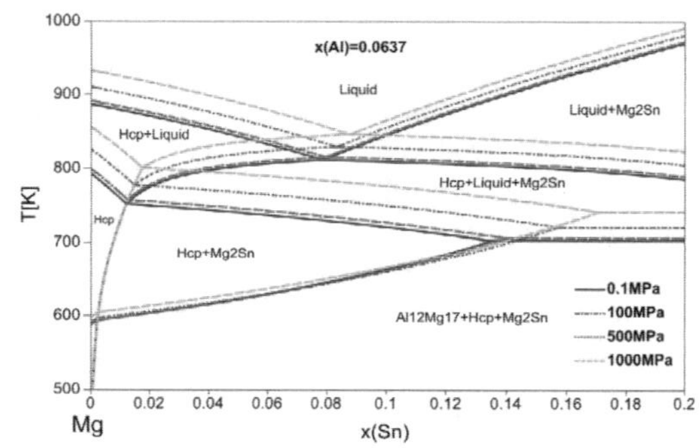

Fig. 5 The vertical section of the Mg-Al-Sn ternary phase diagram at 6.37 at.% Al.

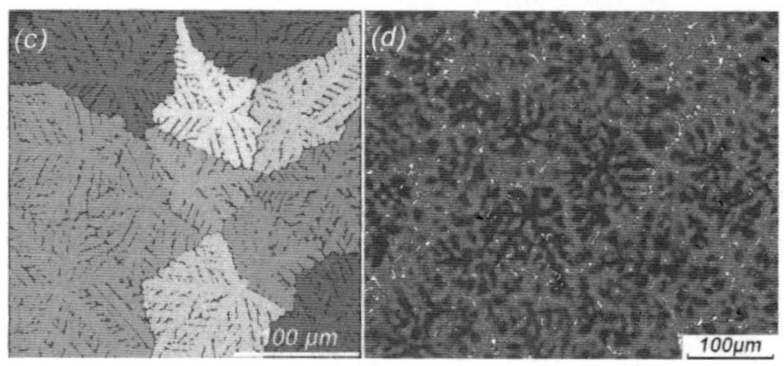

Fig. 6 The simulated and observed microstructure of a Mg-Al-Sn alloy under 85 MPa.

Summary and prospect

The major challenge and progress in modeling and simulation of metal displacement, thermal-mechanical coupled phenomenon, casting-mold interfacial heat transfer, and microstructure formation during pressurized solidification of squeeze casting are addressed, and a prospect for future work is provided.

Acknowledgments

The author would like to acknowledge the support of Beijing Natural Science Foundation (L223001).

Intelligent Digital Simulation of Casting Process in Context of Materials Genome Engineering

Hua Hou[1,2*],XiaolongPei[1], Yuhong Zhao[1,3,4]

1. School of Materials Science and Engineering, Collaborative Innovation Center of Ministry of Education and Shanxi Province for High-performance Al/Mg Alloy Materials, North University of China, Taiyuan, 030051, China

2. School of Materials Science and Engineering, Taiyuan University of Science and Technology, Taiyuan, 030024, China

3. Beijing Advanced Innovation Center for Materials Genome Engineering, University of Science and Technology Beijing, Beijing, 100083, China

4 Institute of Materials Intelligent Technology, Liaoning Academy of Materials, Shenyang, 110004, China

*Corresponding address: e-mail: houhua@nuc.edu.cn

Abstract: Casting is a fundamental and irreplaceable method in advanced manufacturing. The design and optimization of casting processes are crucial for producing high-performance, complex metal structural components. Numerical simulation is the primary tool for achieving this optimization. The Materials Genome Engineering (MGE) aims to accelerate material design and development and has shown significant potential in enhancing numerical simulations of casting processes. This article reviews the impact of material genome engineering on numerical simulation technology in casting and forecasts future trends in intelligent digital simulation for casting processes.

Keywords: Casting,Numerical simulation,Materials Genome Engineering, Machine Learning

1 Introduction

In recent years, the MGE has significantly transformed materials design and manufacturing, particularly in developing new materials and optimizing processes. MGI aims to accelerate material development by integrating high-throughput computing, experimental databases, and materials informatics. This innovative approach has profoundly influenced traditional manufacturing.Casting involves complex multi-stage processes, including the flow and solidification of molten metal, which present significant challenges due to the interplay of multiple scales and physical fields. The introduction of material genetic engineering has opened new opportunities for enhancing numerical simulation in casting.Traditional simulation methods, such as Finite Element Analysis (FEA) and Computational Fluid Dynamics (CFD), can predict macroscopic results but are computationally intensive and often struggle with complex microstructure evolution and multi-scale coupling. Material genetic engineering has improved the accuracy and efficiency of numerical simulations by leveraging big data and machine learning technologies.A key advancement of MGI is the introduction of data-driven models that allow machine learning to capture complex physical relationships in casting processes. The material genome database offers extensive data on material composition, processing techniques, and performance relationships, providing a solid foundation for simulations. By employing machine learning and deep learning techniques, researchers can quickly develop accurate predictive models, enhancing simulation precision. Overall, material genetic engineering has accelerated the optimization of casting processes and advanced intelligent digital simulation technology. This article will explore the latest developments in numerical simulation for casting within the context of material genetic engineering and provide insights into future advancements.

2 Employing MGEto expedite numerical simulation

Material genetic engineering reverses the traditional product development process by starting with application requirements and deriving the necessary material composition and structure to meet those functional needs. This approach significantly bridges the gap between academic research and industrial production, driving innovation in new materials.

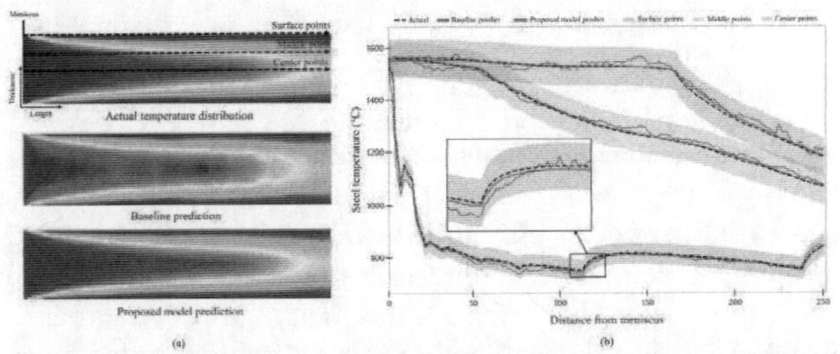

Figure 1 An example of temperature predictions of the solidifying steel [1]

Traditional physics-based computational simulation methods face limitations in efficiency and accuracy due to the precision of thermal property parameters, model complexity, and high-performance computing devices. In contrast, data-driven models, characterized by "data + artificial intelligence," focus on data generation and processing, representing the core direction of material genetic engineering.The predictive power of machine learning can enhance traditional numerical simulations by significantly reducing computation time while maintaining accuracy. Current data-driven methods to accelerate numerical simulations include solving equations more efficiently, rapidly predicting thermal properties parameters, and establishing alternative models for metallurgical changes. Additionally, machine learning models can replace traditional simulations, exemplified by deep learning techniques for predicting shrinkage porosity, the temperature distribution (as shown in Figure 1) [1]and generating microstructures.

3 Integrated Computational Materials Engineering (ICME)

The connection between ICME and MGE is evident in their data sharing and model complementarity: MGE provides extensive material data, while ICME leverages computational models to optimize manufacturing and processing. Additionally, MGE can refine ICME models. MGE has expanded ICME's philosophy across the entire materials science, technology, and engineering chain. A notable example of ICME in the casting process is the Virtual Aluminum Casting (VAC) manufacturing process developed by Allison et al. from Ford Motor Company in 2006. Furthermore, the casting ICME framework proposed by Alan A. Luo et al. [2] (shown in Figure 2) parallels the industrial product development process that utilizes various CAD-CAE-CAM tools, including microstructure and performance simulations for local casting positions.

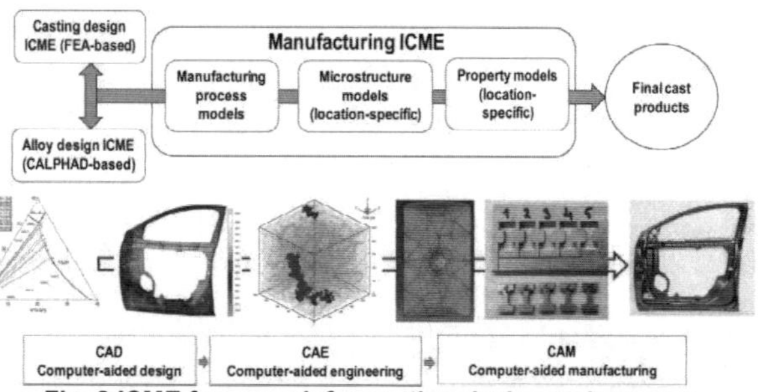

Fig. 2 ICME framework for casting design and process development as compared to industrial CAD-CAE-CAM approach [2]

4 Establishment of relevant databases

Currently, there is extensive research on material databases, such as the Thermocalc thermodynamic database, AFLOWLIB, and the Materials Project calculation database. However, studies on casting-related databases are relatively scarce and typically limited to specific categories, like casting process databases, generative process databases, or material databases. Cho et al. [3] proposed an integrated database system for casting production, which includes shape databases, material physical properties (melting point, solidus line, liquidus line, etc.), process databases, and expert systems. Despite this, the high flexibility, complexity, and variability of casting production processes, along with organizational challenges and isolated departmental data, result in difficulties in data storage, retrieval, and application. To advance digital casting technology, integrate process design, ensure casting quality, and predict lifecycle, it is imperative to establish a standardized key material performance database. This database should store information about the relationships between processing technology and microstructure, microstructure and material properties, and thermal properties parameters to support the development and validation of computational models.

5 Conclusion

The Materials Genome Initiative (MGI) has introduced significant innovations to the numerical simulation of casting processes. Material genetic engineering has enhanced the efficiency of traditional numerical simulations. Moreover, the integration of Integrated Computational Materials Engineering (ICME) with material genetic engineering has further improved the multi-scale coupling capabilities of these simulations. This advancement allows for more accurate predictions from microstructure to macroscopic performance, offering comprehensive optimization solutions for casting processes. Future development trends in numerical simulation of casting within the framework of material genetic engineering include:

1) A rich and perfect database of material thermophysical parameters used to improve the calculation accuracy, as well as a technology to synergistically improve the accuracy and efficiency of numerical simulation.
2) The establishment of special database for casting and the improvement of process data quality evaluation and control technology.
3) The deep integration of data drive, simulation software and ICME, and the construction of digital platform based on digital twins throughout the life cycle.

6 Acknowledgments

This research project was supported by the National Natural Science Foundation of China (Nos. 52074246), National Defense Basic Scientific Research Program of China (Nos. JCKY2020408B002), Key R&D Program of Shanxi Province (Nos.202102050201011).

References

[1] Lee S Y, Tama B A, Choi C, Hwang J, Bang J and Lee S. Spatial and sequential deep learning approach for predicting temperature distribution in a steel-making continuous casting process. IEEE Access, 2020, 8: 21953-21965.
[2] Luo A A, Sachdev A K and Apelian D. Alloy development and process innovations for light metals casting. J. Mater. Process. Technol., 2022, 306: 117606.
[3] Cho I, Yoo S, Lim C and Choi J.Application of integrated database to the casting design. J. Mater. Sci. Technol., 2008, 24(03):351-355.

Design of Dual-Phase Steel Based on Active Learning

Jincheng Wang[1]*, Xiaobing Hu[2], Junjie Li[1]
1. State Key Laboratory of Solidification Processing, Northwestern Polytechnical University, Xi'an 710072, China
2. Xi'an Rare Metal Materials Institute Co. Ltd, Xi'an 710016, China
*Corresponding address: e-mail: jchwang@nwpu.edu.cn

Abstract: Dual-phase (DP) steels are an important family of steel grades widely used in the automotive industry, aerospace, ultra-supercritical generating units, etc.Reducing costs throughout the process from raw material preparation to experimental design is a critical challenge that needs to be addressed urgently. This paper develops an effective active machine learning (AL) method to explore and exploit new DP steels with excellent mechanical properties. A simple case of hardness optimization is first reported to validate the reliabilityand efficiency of the AL method. Simultaneous enhancement of strength and plasticityis then realized by fast learning in a vast design space free of Co, finding several desired low-cost DP steels. More importantly, convenient application software has beensuccessfully developed, which has practical significance for the engineering application of the AL method.

Keywords: Dual-phase steels； active learning, mechanical properties.

1 Introduction

DP steels have potentialmechanical properties, making them widely applied in the automotive industry, aerospace, ultra-supercritical generating units, etc. Uncovering the complicated interaction between compositions/processing, microstructures, mechanical properties is extravagant in terms of time andmoney by traditional trial-and-error. Therefore, it is necessary to develop a new method to accelerate the design of DP steels.

A potential solution to the problem above is machine learning (ML),which can quickly explore the complex composition space to improve the microstructure and mechanical properties. However, the prediction was only induciblefor the samples near the training data. To solvethis problem, an adaptive active learning (AL)method is a wise direction for the iterative experiments.To the design of DP steels, one is bound to meet a sparse and small dataset. The efforts should beperformed in early iterative experiments before adaptive sampling to balance data distribution as much as possible (reducethe overall σ of ML prediction), which may be beneficial for the subsequent search for the global optimum.

In this study, an improved AL method is proposed, and validated for its ability in finding global optimum by only several iterations, then used to design new strength-plasticityenhanced DP steels free of Co.

2 Experimental procedure and AL method

Experimental procedure

The DP steelswere prepared using a mold of a WKII model vacuum arc melting furnace. Homogenization, hot-rolling, and heat treatment were executed sequentially to ensure appropriate microstructure.

3 AL method

We trained a Gaussian process regression (GPR) model. Employing the model, we then ranked the samples by their σ of prediction. Three top-ordered candidates were recommended to experiment and then feedback to the initial set, which ensures a rapid reduction of σ on the design space.This procedure was repeated several times until GPR generated a stable prediction for the unexplored alloys. Subsequently, EI was used to explore the region wherethe alloys have a superior property than the best-so-far until anoptimal candidate was found.To enhance the strength and plasticitysimultaneously, some expert experience is brought in ML.

4 Result and discussion

Validation of the AL method in finding global optimum

Executing the AL method, we explored a total of 27alloys in the designspace. As shown in Fig.1a, we first extracted the candidates with large σ in the 1–6iterations, their hardness varies over a widerange (≈10–55 HRC). Fig.1b reveals that the predicted σ ofthe ensembles in the virtual set rapidly decreases and convergesto the position where the mean is ≈5 during the 1–4iterations. Interestingly, the curves seem to change slightly in the 4–6iterationsand a stable plateau of σ occurs, demonstrating negligible benefits from the sequentially recommended alloys according to exploration.At the first exploration stage, the variation of average EI with iteration is ruleless, as shown in Fig. 1(c). When the extremely sparse region in the design space is filled, the new recommended alloys tend to minimize σ, and EI is therefore reduced, just as displayed in the 4-6 iterations in Fig. 1(c). Once the improvement of data distribution is completed, exploitation starts, the reduction of EI decreases, and the new alloy approaches the global optimal, as depicted after the sixth iteration in Fig. 1(c). Notsurprisingly, the mean absolute error (MAE) between the predictions and the measurements quickly drops in iterations 1–4, asshown in Fig.1(d) but fluctuates slowly in iterations 4–6, revealing that the alloys recommended by ranking σ haveinsignificant contribution to the improvement of the predictionprecision on the design space after the fourth iteration, as shown in Fig. 1(f). MAEs in the last three iterations are thus close to 0.

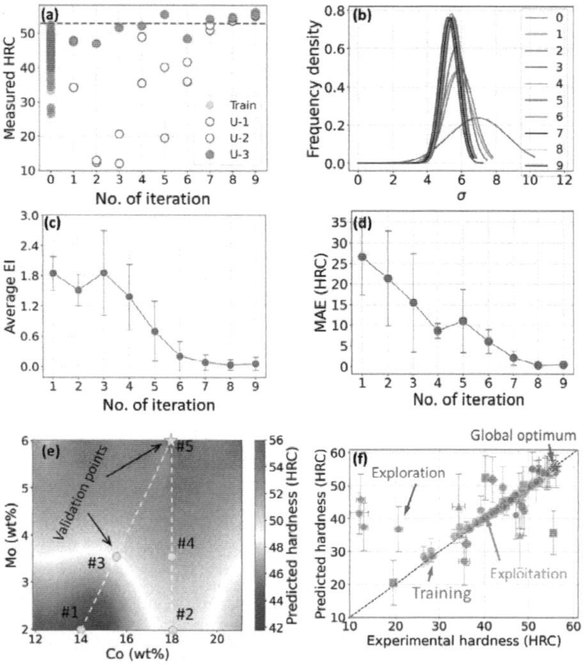

Fig.1 Design process of DP steels with the highest HRC. (a) Experimental HRC *vs* iterations. (b) Frequency density of σ in 9 iterations. (c) El *vs* iterations. (d) MAE *vs* iterations.(e) Predicted HRC of Fe-0.05C-12Cr-1Ni-Co-Mo DP steels.(f) Experimental validation of the designed alloys.

5 Application of the AL method to design strength-plasticity enhanced DP steels free of Co

The experimental dataset with the variates of Ti, Mo and W was constructed, and GPR model was used to learn the relationship between chemical compositions and the properties (UTS and elongation at 650 ℃), as shown in Fig. 2(a)-(f). Our designed steels show the highest product as shown in Fig. 2(h), followed by the 15Cr steelsmarked by the gray arrow. In addition, the product of steel D3 is156%, 31%, and 62% higher than the best ones in 9Cr, 15Cr, and20Cr steels, respectively. The product of V1 shows an increase ofapproximately 33% compared to that of the steel marked by thecyan arrow. Moreover, our designed steelsare free of Co and contain low Ni content (5 wt%–8 wt%), and the exploitation cost is thus lower thanthat of traditional DP steels.

6 Conclusion

The design, validation and application of the AL method are successfully realized by finding new DP steels free of Co. The main conclusions are summarized as follows:

1. An accurate and high-efficiency AL method is developed, which achieves both data distribution optimization and high-property alloy exploitation.
2. The DP steel with the highest hardness was successfully designed from a given composition space,

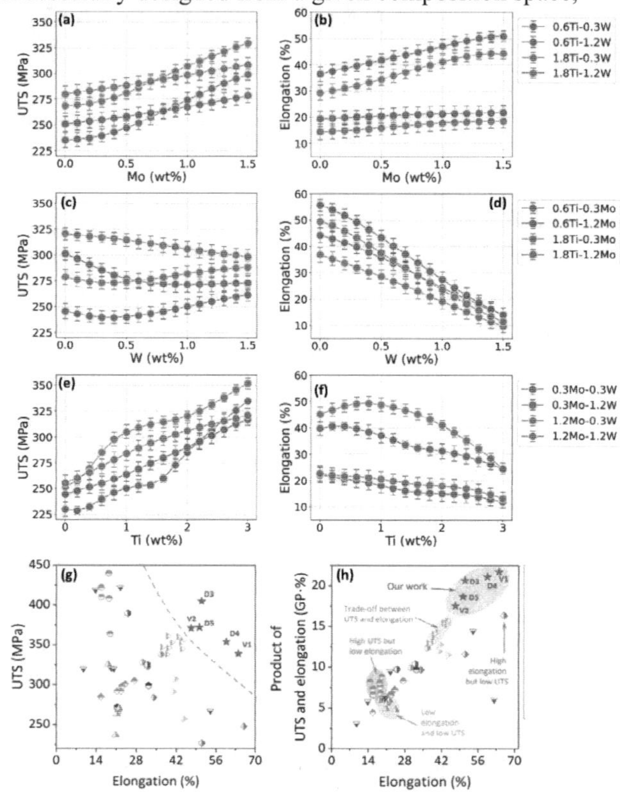

Fig.2 Design process of DP steels free of Co with enhanced strength and plasticity. (a)-(f) Predicted effect of Ti, Mo and W contents on UTS and elongation. (g) UTS vs elongation and (h) Product of UTS and elongation vs elongation for the typical 9-12Cr, 14Cr, 15Cr and our designed steels.

which uncovers the capability of global optimization of the AL method.

3. Several new low-cost 15Cr DP steels are designed.

7 Acknowledgments

This research works was supported by the National Natural Science Foundation of China (No. 52474423) and the National Key Research and Development Program (No: 2023YFC3010902).

References

[1] Q. Wu, Z. Wang, X. Hu, et al., Uncovering the eutectics design by machine learning in the Al–Co–Cr–Fe–Ni high entropy system, Acta Materialia 182 (2020) 278-286.

[2] X. Hu, J. Wang, Y. Wang, et al., Two-way design of alloys for advanced ultra supercritical plants based on machine learning, Computational Materials Science 155 (2018)331-339.

[3] C. Guo, X. Hu, X. Han, et al., Laser Precise Synthesis of Oxidation-Free High-Entropy Alloy Nanoparticle Libraries, Journal of the American Chemical Society 146 (2024) 18407-18417.

Gradient-Enhanced Plasticity Theory for Size Effects and Mesh-Independent Numerical Results

Jaeheum Yeon[1], Yooseob Song[2*],
1. Department of Regional Infrastructure Engineering, Kangwon National University, Chuncheon 24341, Republic of Korea
2. Department of Civil and Environmental Engineering, University of Alabama in Huntsville, Alabama, 35899 USA
*Corresponding address: e-mail: yooseob.song@uah.edu

Abstract: In this work, a thermodynamically consistent constitutive formulation for the coupled thermomechanical strain gradient plasticity theory is developed in the context of the finite deformation framework. A corresponding finite element solution is presented to investigate the micro-structural features of metallic volumes. Size effects and mesh-sensitivity are simulated using the developed numerical framework to test the feasibility of the proposed gradient-enhanced plasticity theory.

Keywords: strain gradient plasticity; size effects; mesh-sensitivity; finite element method.

1 Introduction

In the authors' previous works [1-9], the coupled thermo-mechanical strain gradient plasticity model was proposed to investigate the microstructural features of metallic materials and solved numerically in the context of one-dimensional small deformation finite element implementation. In the current work, a two-dimensional finite element solution for finite deformation considering the temperature effect is proposed. The developed model is established based on an extra Helmholtz-type partial differential equation, and the nonlocal quantity is calculated in a coupled method based on the equilibrium conditions. This approach is well known for its computational strength, however, it is also commonly accepted that it cannot capture the size effect phenomenon observed in the micro-/nanoscale experiments during hardening. In order to resolve this issue, a modified strain gradient approach which can capture the size effects under the finite deformation is constructed in this work. The simple shear problem is then solved to carry out the feasibility study of the developed model on the size effect phenomenon. Lastly, the uniaxial plane strain tension problem with shear bands is examined to perform the mesh sensitivity tests of the model during softening.

2 Implicit Gradient-enhanced Plasticity Theory

The implicit gradient-enhanced plasticity theory (IGT) is introduced as an alternative method with only a C^0-continuous requirement to overcome a computational inherent difficulty in the implicit gradient-enhanced plasticity theory. In the IGT, the Helmholtz-type partial differential equation (PDE) is obtained, and the nonlocal quantities are calculated in a coupled method based on the equilibrium conditions. The nonlocal effective plastic strain $\bar{\varepsilon}^p$, which indicates a weighted average of the local effective plastic strain ε^p, is introduced and obtained by the following PDE with the material length scale parameter l_p:

$$\bar{\varepsilon}^p - l_p^2 \nabla^2 \bar{\varepsilon}^p = \varepsilon^p \quad (1)$$

3 Result and discussion

Size effects

The schematic illustration of an infinitely long plate geometry, the initial conditions, and the macroscopic and microscopic boundary conditions under shear is shown in Figure 1. The direct comparison between the non-gradient plasticity theory and the proposed IGT is given in Figure 2 on the size effects in terms of the true stress versus true strain behaviors.

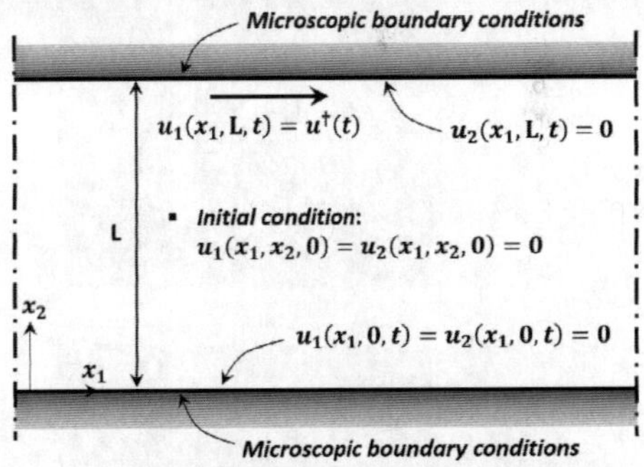

Fig.1 Shearing of a layer problem: macro-/microscopic boundary conditions, initial conditions, and geometry.

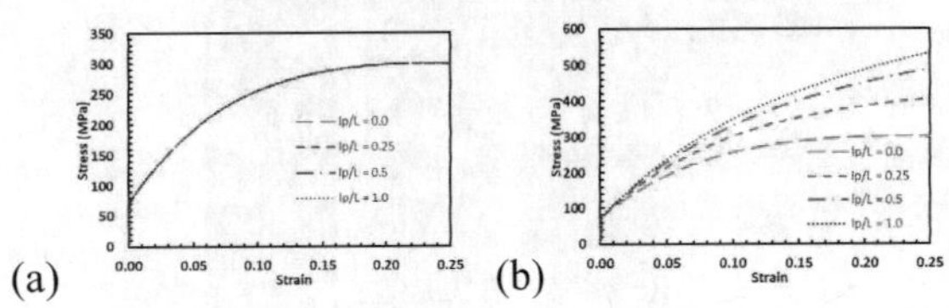

Fig.2 Direct comparison between (a) non-gradient plasticity theory and (b) the proposed IGT on the size effects.

4 Mesh-sensitivity tests

The schematic illustration of a plane strain problem under uniaxial tensile loading to explore the mesh-sensitivity characteristics of the proposed model during strain softening is presented in Figure 3. To do this, the three different mesh sizes, i.e. Mesh 1: 10×50=500, Mesh 2: 15×75=1125, and Mesh 3: 20×100=2000 elements are used. Figure 4 shows the comparison between non-gradient plasticity theory and the proposed IGT. As expected, the non-gradient theory results in strong mesh-dependent behaviors, whereas, the proposed IGT results in mesh-insensitive results.

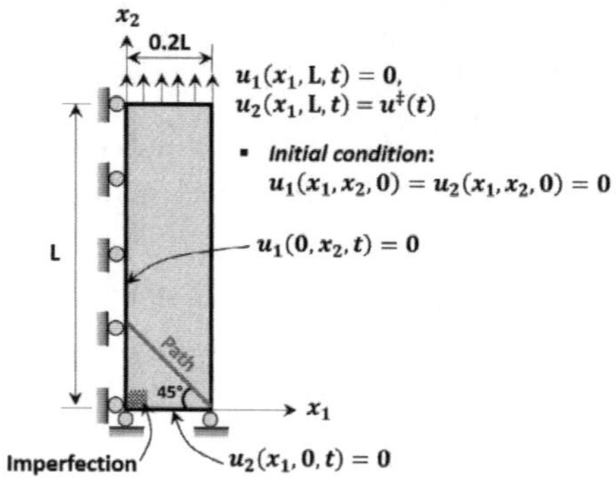

Fig.3 Plane strain problem under uniaxial tensile loading: boundary conditions, initial conditions, and geometry.

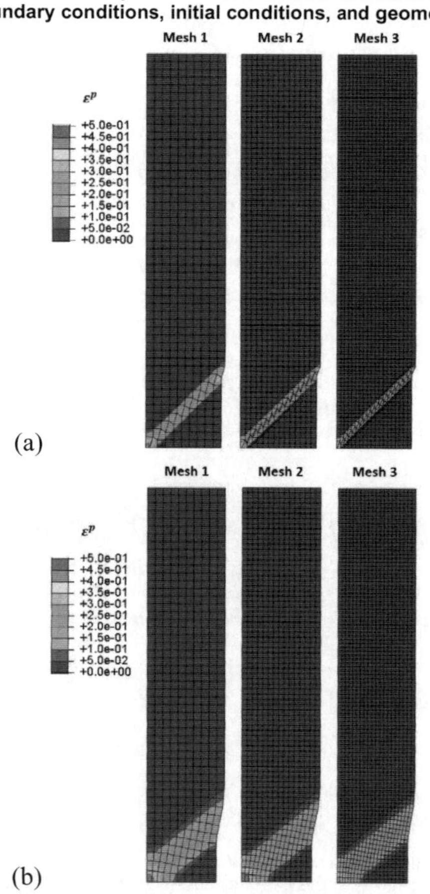

Fig.4 Direct comparison between (a) non-gradient plasticity theory and (b) the proposed IGT on the mesh-sensitivity tests.

5 Conclusion

The thermodynamically consistent constitutive model for the coupled thermo-mechanical strain gradient plasticity theory is constructed in the finite deformation context. In the IGT, it was well observed that the material responses during hardening were stiffer as the sizes of the sample got smaller, viz. the size effect. The main cause of this difference was the existence of the nonlocal effective plastic strain in the yield criterion. In addition, the mesh-sensitivity of the proposed model on the strain localization during strain softening was examined in this work. In the classical non-gradient plasticity model with no gradient regularization, it was well observed that the widths of the shear band drastically vary according to the mesh size, whereas the mesh-independent material responses are obtained in the proposed model.

References

[1] Voyiadjis GZ and Song Y. Effect of passivation on higher order gradient plasticity models for non-proportional loading: energetic and dissipative gradient components. Philos Mag 2017; 97: 318-345.

[2] Voyiadjis GZ, Song Y and Park T. Higher-Order Thermomechanical Gradient Plasticity Model With Energetic and Dissipative Components. J Eng Mater-T ASME 2017; 139.

[3] Voyiadjis GZ and Song Y. Finite Element Analysis of Thermodynamically Consistent Strain Gradient Plasticity Theory and Applications. Handbook of Nonlocal Continuum Mechanics for Materials and Structures. 2018, pp.1-58.

[4] Song Y. A coupled thermo-mechanical theory of strain gradient plasticity for small and finite deformations. Louisiana State University and Agricultural & Mechanical College, 2018.

[5] Song Y and Voyiadjis GZ. Small scale volume formulation based on coupled thermo-mechanical gradient enhanced plasticity theory. Int J Solids Struct 2018; 134: 195-215.

[6] Song Y and Voyiadjis GZ. A two-dimensional finite element model of the grain boundary based on thermo-mechanical strain gradient plasticity. Journal of Theoretical and Applied Mechanics 2018; 56: 377-391.

[7] Song Y, Yeon J and Na B. Numerical Simulations of the Hall–Petch Relationship in Aluminium Using Gradient - Enhanced Plasticity Model. Advances in civil engineering 2019; 2019: 7356581.

[8] Voyiadjis G and Song Y. Finite element analysis of thermodynamically consistent strain gradient plasticity theory and applications. Handbook of nonlocal continuum mechanics for materials and structures 2018: 1-58.

[9] Voyiadjis GZ and Song Y. Strain gradient continuum plasticity theories: theoretical, numerical and experimental investigations. Int J Plasticity 2019; 121: 21-75.

This page intentionally left blank.

Prediction of Microstructure and Thermophysical Properties of Steel for Brake Discs

Shiting Zhu*, Fang Wang, Huiqin Chen, Shimeng Guo

1. School of Materials Science and Engineering, Taiyuan University Of Science and Technology, No.66 Waliu Road, WanBailin District, Taiyuan city, Shanxi Province, China

*Corresponding address: No.66 Waliu Road, WanBailin District, Taiyuan city, Shanxi Province, China

e-mail: 382717435@qq.com

Abstract: With the increasing speed of high-speed trains, the brake disc heat load has also been increasing, particularly during emergency braking. Therefore, to address such issues, strict requirements for brake disc materials are suggested. The microstructure and mechanical properties of steel for brake discs was investigated. Based on the material thermodynamic calculation software JMatPro, the equilibrium phase composition, hardenability and other thermophysical parameters of steel for brake discs are calculated and predicted. The paper also calculates and draws the continuous heating austenitizing (TTA) curve and continuous cooling transformation (CCT) curve of the steel for brake discs, and predicts the martensite transformation temperature and other relevant material parameters, in order to provide theoretical basis for the heat treatment process.

Keywords: high-speed train, brake disc, cast steel, microstructure, thermophysical properties.

1 Introduction

The brake disc is the core component of the brake device, which plays a vital role in the safe running of the high-speed trains[1-3]. However, with the increase of train speed, the brake disc needs to have sufficient strength, such as room temperature/low temperature plastic toughness, high temperature strength and plasticity, thermal fatigue performance, high temperature wear performance, thermal conductivity, etc., to meet the use of high-speed trains[4-8].

2 Experimental procedure

The element composition of low alloy cast steel for brake disc of high-speed train is shown in Table 1.

Table 1. Chemical composition of low-alloy cast steel for high-speed iron brake disc (mass fraction /%)

C	Si	Mn	Cr	Ni	Mo	Al	Cu
0.23	0.44	1.07	0.87	0.94	0.60	0.03	0.09

The test material is heat treated to change its matrix structure, and then the relationship between its structure and properties is investigated. The heat treatment process parameters of the material are shown in Fig.1 The microstructure of the sample after heat treatment is observed.

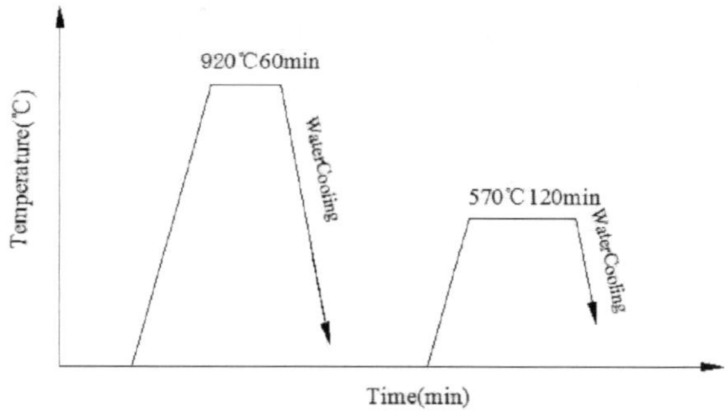

Fig.1 - Heat treatment process parameters of steel used for high-speed rail brake disc.

3 Result and discussion

Tempered sorbite was formed after tempering the quenched structure of the experimental steel at high temperature. The quenched lath martensite recrystallized to form equiaxed grains. Carbide dispersion distribution. The volume fractions of ferrite, cementite and carbide were 61%, 38.8% and 0.2%, respectively. Jmatpro software was used to calculate the equilibrium phase composition of the cast steel from room temperature to 1600℃.The equilibrium precipitated phase in brake disc steel is: M_2 (C,N), M (C,N), M_7C_3, MNS, M_6C, $M_{23}C_6$. At room temperature, the structure content of the low-alloy cast steel used for the brake disc is respectively: ferrite (F) 95.51%, M_2 (C,N)phase 1.34%, G phase (a special intermetallic compound containing Ni and Si elements) 1.56%, M_7C_3 phase 1.15% and about 0.24% cementite. The Young's modulus and density of low alloy cast steel used for high iron brake disc are negatively correlated with temperature. A maximum of 212.66GPa and 7.83g/cm^3 occurred at room temperature. When the cast steel melts and the solid-liquid phase changes, both values decrease significantly. Poisson's ratio of cast steel is positively correlated with temperature and has a minimum value of 0.289 at room temperature. When the specific heat capacity of cast steel increases from room temperature to melting point, it decreases first and then increases. It has a maximum value of 62.42W/(m·K) at 20℃ and a minimum value of 25.89W/(m·K) at 793.41℃In addition, the analysis of equilibrium phase diagram, with the increase of temperature, the G phase in the low-alloy cast steel material used in the brake disc dissolves rapidly, resulting in a rapid decline in specific heat capacity. At

793.41℃, the dissolution of ferrite and the formation of austenite gradually increase the specific heat capacity of cast steel. TTA diagram of low alloy cast steel for high-speed rail brake disc is shown in Fig 2. According to the TTA diagram, Ac1, Ac3 and austenite homogenization temperatures all increase with increasing heating rate. The CCT curve of cast steel is drawn according to thermodynamic simulation, as shown in Fig 2. It can be seen that the initial martensite transition temperature (Ms) is 352.45℃, the temperature is 317.81℃ when the transformation to 50% martensite, and the temperature decreases to 237.37℃ when the transformation fraction is 90%. The initial transition temperature of bainite (B) is 542.7℃. The perlite (P) transition temperature is 668.39℃.

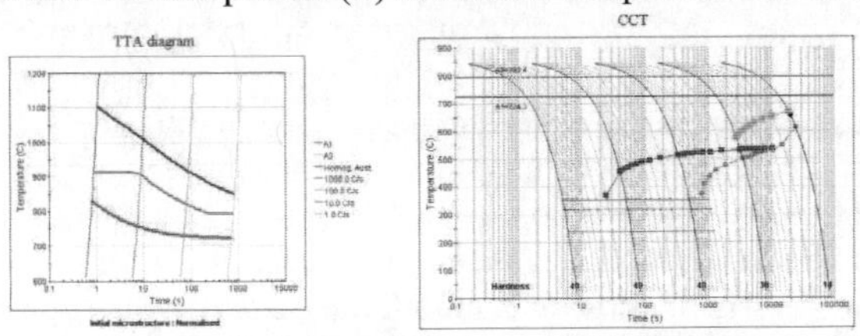

Fig.2 - TTA diagram and CCT curve of low alloy cast steel for brake disc of high speed train

4 Conclusion

In this paper, the microstructure and properties of steel used for high-speed train brake discs are preliminarily analyzed, and the thermophysical properties of low-alloy cast steel used for high-speed train brake discs are predicted and analyzed based on thermodynamic simulation software. According to the calculation results, the equilibrium phase composition diagram and CCT curve of this steel at room temperature are obtained. In order to provide guidance for the casting of the cast steel and the formulation of subsequent heat treatment parameters, the conclusions are as follows:

(1) Tempered sorbite is formed in low alloy cast steel for high-speed train brake disc after quenching structure is tempered at high temperature.

(2) At room temperature, the equilibrium structure and content of low-alloy cast steel used for high-speed train brake disc are: ferrite (F) 95.51%, M_2(C,N) phase 1.34%, G phase 1.56%, M_7C_3 phase 1.15% and about 0.24% cementite.

(3) Density and Young's modulus were negatively correlated with temperature, and Poisson's ratio was positively correlated with temperature. The specific heat capacity of low-alloy cast steel used for brake discs of high-speed trains decreases first and then increases with the increase of temperature.

(4) The calculation of CCT curve shows that when the cooling rate of cast steel is greater than 0.5℃/s, martensite phase precipitation, the initial martensite transformation temperature (Ms) is 352.45℃, the temperature is 317.81℃ when the transformation to 50% martensite, and the temperature decreases to 237.37℃ when the transformation fraction is 90%. The initial transition temperature of bainite (B) is 542.7℃. The perlite (P) transition temperature is 668.39℃.

References

[1] Sheng Huan, Wang Zehua, Shao Jia, et al. Research Status and Prospect of Brake Disc Materials for high-speed Trains [J]. Mechanical Engineering Materials,2016, 40(326): 5-9+46.

[2] Tang Cengjiang, Chen Yunbo, Zuo Lingli, et al. Application Status and Research Progress of brake disc Materials for high-speed Trains [J]. Materials Review, 2018, 32(s1): 443-448.

[3] Naveen Kumar1, Ajaya Bharti1, etc. The evolution of brake friction materials:A Review[J]. Materials Physics and Mechanics, 2021(47): 796-815.

[4] Shao Lin, Zhang Xi, Yang Xin. Application Research on Non-adhesive Braking of higher speed EMUs [J]. Railway Rolling Stock,2016, 36(3): 37-42+87.

[5] Guo Hongwei. Study on Friction and Surface Damage of Alloy Steel Brake Disc Materials [D]. Beijing: Beijing Jiaotong University,2018.

[6] Niu Yuecheng, Li Fu, Li Xinrong, et al. Current Situation and Prospect of Thermal Analysis of Train Brake Disc [J]. Locomotive Electric Transmission, 2019(6): 5-9.

[7] Li Z, Han J, Yang Z, et al. Analyzing the mechanisms of thermal fatigue and phase change of steel used in brake discs[J]. Engineering Failure Analysis, 2015(57): 202-218.

[8] WU S C, ZHANG S Q, XU Z W. Thermal crack growth-based fatigue life prediction due to braking for a high-speed railway brake disc[J]. International Journal of Fatigue, 2016, （87）: 359-369.

Development of an Innovative Carbon-Free Ecological Process as an Alternative to Explosion for Manganese Steel Hardening

P. Caballero*, A. Abuin
TECNALIA, Basque Research and Technology Alliance (BRTA), Derio, Spain
*Corresponding address: e-mail: patricia.caballero@tecnalia.com

Abstract: Explosion hardening is currently the go-to method for increasing manganese steel hardness in critical applications such as railway turnouts. While effective, environmental considerations, sustainability, and competitiveness, are now more critical in industry, exposing this costly method's weaknesses; transport to external specialists, explosions release greenhouse gases, lack precision, and disrupt the casting's internal crystallographic network. Follow-up work is also required prior to installation. Consequently, there is a need for a more technological and cost-effective eco-hardening process, better matching today's ethos and industrial decarbonization goals.

This project presents an alternative to explosion treatment, 'Single Process Hardening' (SPH), which innovatively combines three established technologies for the first time in a single process: percussion hammering, ultrasonic waves, and shotblasting. Unlike the explosion method, SPH castings would not require follow-up work, e.g. crack repair, prior to installation.

Following promising results using simulation software, several trial batches of 5m turnout frogs were cast, and divided between explosion hardening and SPH, prototype equipment being used for the latter. The following were analyzed; contact surface and cross-sectional hardness, permanent and residual stress, and tribological aspects. Empirical data was fed back into the software to improve future results through machine learning (AI). For analysis of all the castings, x-ray diffractometry, USM, hole ring and ring core methods were used to obtain an in-depth picture of the relevant metallurgical characteristics.

With the new process, approximately one hour of treatment obtained the 400+/- HB obtained by explosion. Potentially, by tweaking the process, +/- 420HB could be obtained as the internal crystallographic integrity is not disrupted. The estimated reduction in the total cost with this new method is 75% less than using the explosion method. Regarding decarbonization goals, the elimination of road transport, and the use of green electricity, would make this method a valuable contributor.

Keywords: Decarbonization, manganese steel hardening, railway turnout, explosion hardening, eco-hardening process, percussion hammering, shotblasting, ultrasonic waves Single Process Hardening (SPH).

1 Introduction

Major railway operators have established standards and criteria that include surface hardness and depth requirements required for crossings that have undergone a hardening process.

Currently, hardening with explosives is the process used to increase the surface hardness of rail track turnouts, specifically the crossing turnout "frog" (austenitic manganese steel) prior to installation. Originally, surface hardening occurred as a result of impact through use, which required follow up inspection and maintenance. The benefit of pre-hardened turnouts was mainly a reduction in this maintenance. The explosion method was patented in 1955, becoming extensively used in the mid-eighties. [1]

The method relies on the plastic deformation that steel undergoes when subjected to the pressure wave produced by the detonation of an explosive material, typically Pentaerythritol Tetranitrate (PETN), [2] which happens to increase surface hardness. The process consists of adhering a specific charge to the section of the piece that requires hardening, running areas and then detonating it, usually one cycle, but sometimes more [3,4]. See table 1 below. And to locate area hardened see figure 1.

Table 1: Hardness of studied turnouts (HBW)

Nº explosions	Area D1/D2	AreaD3/ D4	Area D5
Explosion 1	310	330	350
Explosion 2	350	390	383

With a single detonation, Brinell hardness values greater than 310 HB (initial 190 HB) can be reached up to a depth of about 2 mm below the rolling surface. [5]. To achieve greater hardness (>350 HBW), this process can be repeated once.

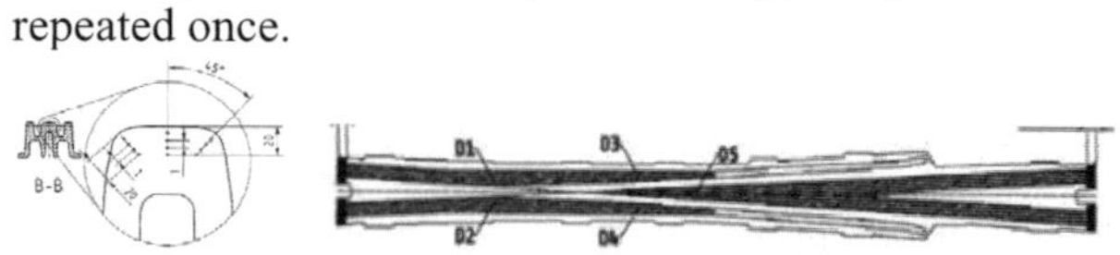

Figure 1. Essays areas in turnouts and section essay.

2 Experimental procedure

For experimental testing purposes, 100 units of UIC60 profile rails were cast.

Firstly, individual hardness tests were carried out using ultrasonic hardening, hammering and shotblasting. The premise was that ultrasonic waves, [6,7] would energize the surface of the steel at an atomic level, preparing it for hammering then shotblasting. [8]. Hammering would harden the surface, shotblasting would homogenize the surface and reduce residual tension, while ultrasonic waves would increase the depth of the hardened layer.

These three methods [9,10,11] were then combined into a single process (SPH). Batches of ten were then compared.
In all processes the surface of austenite is transformed into martensite, but the depth of the latter is greater with the SPH process. In table 2 we can observe the median hardness values obtained.

Table 2: Hardness median values

T	Hardening process	D1/D2 Area	D3/D4 Area	D5 Area
A	1.Ultrasonic	330	320	321
B	5.Explosion (x2)	365	397	390
C	6. SPH 2+3+4	410	412	420
D	8.Hammering	293	287	305
E	9.Shotblasting	345	330	340

Following these promising results with the rails and using Abacus simulation software, [12] five trial 5m crossings were cast for the new SPH, using prototype equipment. Three lines of post-treatment analysis were carried out. [13] Firstly, contact surface and cross-sectional hardness, secondly, permanent and residual stress,[14] and thirdly, tribological aspects. Empirical data was fed back into the software to improve future results through machine learning (AI). For analysis of the treated castings, x-ray diffractometry, USM, and ring core and hole-ring methods, were used to obtain an in-depth picture of the relevant metallurgical characteristics. [15,16,17].

The prototype equipment used for the new SPH is currently patent applied for.

3 Result and discussion

In Table 3, we can observe the obtained hardness values in the 5m SPH crossings.

Table 3: HB Hardness values of studied alloys

Crossing SPH hardened	1	2	3	4	5
D1/D2	408	403	410	398	405
D3/D4	400	405	418	400	406
D5	399	410	407	420	410

Unlike explosion hardened crossings, these SPH crossings were ready to use without follow-up processes.

The following table (4) shows that crossings treated using the new method suffered no internal structural alteration. The diffractogram obtained clearly indicated the presence of austenite (Fe-γ) with an FCC (face-centered cubic).

Table 4: Diffractrogram results

Lattice	Interplane distance(Å)	*Diffraction angle 2θ (copper anode)*
(111)	2.088	43.278
(200)	1.809	50.403

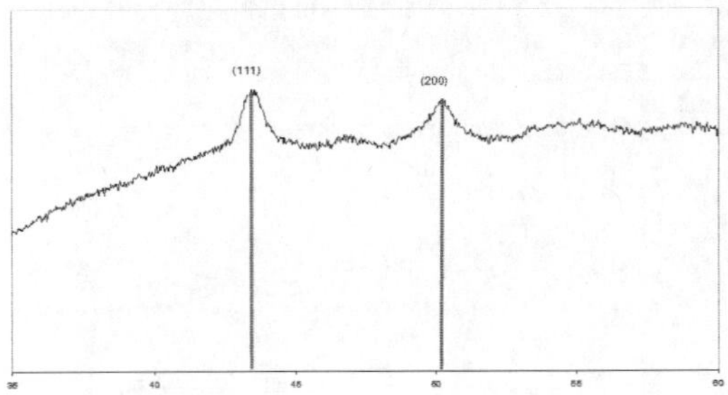

Fig. 2. Diffractogram of base Mn material sample.

The results present that:

- Hardness obtained from the new SPH can be higher than that of explosion method.
- The depth of the hardened area exceeds 2mm.
- There is no damage to the crystallographic network.
- Tribological trials demonstrate superior wear characteristics.
- There is no need for follow-up work due to cracks etc.

In Figure 3, we can observe the obtained hardness.

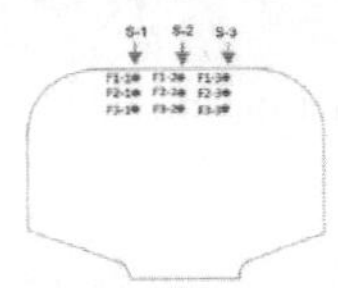

Figure 3. Hardened surface depth

To measure the residual tension in the studied samples, we used firstly the hole drilling strain gauge method as shown in figure 4 below (norm ASTM E837-13). At high speed, the drill bits, both 1mm and 2mm diameter were used, release residual stresses which can be measured. This is a NDT technique. The results shown in graphic in figure 4, which mirror the sequence of Table 2 (five hardening process), compare residual tension by hardening method. We can observe that the highest tension results from the explosion method, and the lowest tension results from the new SPH. In terms of load capacity, the SPH gives optimal results, maintaining mechanical properties such as tenacity. The tension resulting from the explosion method is five times greater than that of the new SPH, which implies a significant reduction in the stability of mechanical properties. With this test we demonstrate that the explosion method increases residual tension times 5, while the new SPH increases tension times 1. Ring core method being used also as a complement to hole drilling for comparison purposes and results were found to be similar.

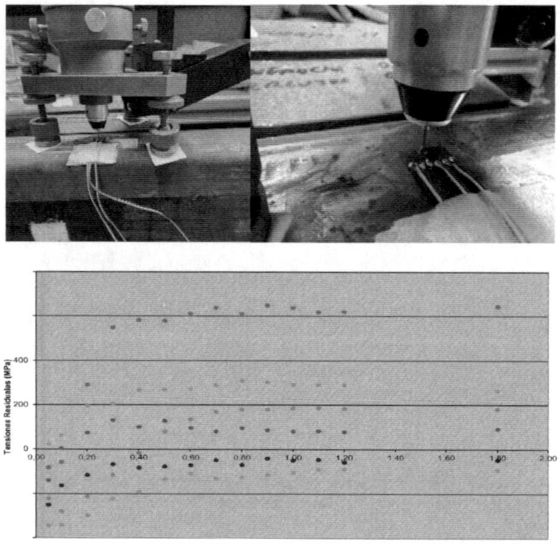

Figure 4. These figures show the ring core method being used as a complement to hole drilling. And graphic showing residual tensions tested in the five-hardening process.

Tests were also carried out to analyse the tribological behaviour of the hardened surfaces with the 5 processes according to table 2: A. Ultrasound (1), B. Explosion (5), C.SPH (6), Hammering (8) and E. Shotblasting (9) to identify differences and the best option.

Therefore, 3 tests are carried out for each type of hardening process under the same conditions; PE-05-MO-110 with a MICROTEST Tribometer (05-168.03) PoD humidity control, under the following conditions: Normal Load (50N), Lineal speed (36.65), angular speed (500rpm), radius 7mm, time (60 minutes) temperature (rt), relative humidity (50%), ball/pin (al2o3), pin dimensions (∅ 6mm).

The test, known as Pin on Disk, measures wear by volume and consists of placing a "pin" in contact with the stationary test material with the pin in motion, applying a constant load in accordance with the above criteria for a given period of time. The equipment records the value of the Dynamic Friction coefficient in real time. See table 5.

Table 5. Wear results for each pin on disk test

Referencia Muestra	Repetición	Referencia Ensayo	Desgaste Muestra Velocidad Desgaste (mm³/Nm)	Desgaste Muestra ▯(Velocidad Desgaste) (mm³/Nm)	Desgaste Pin Velocidad Desgaste (mm³/Nm)	Desgaste Pin ▯Velocidad Desgaste (mm³/Nm)
Sample 1	A	1_A	9,74E-06	5,59E-07	7,37E-08	1,09E-07
	B	1_B	9,67E-06	1,90E-06	9,77E-08	9,41E-08
	C	1_C	1,06E-05	3,62E-07	6,38E-08	1,01E-07
Sample 5	A	5_A	7,46E-06	7,41E-07	1,00E-07	6,12E-08
	B	5_B	7,02E-06	1,29E-06	1,18E-07	8,38E-08
	C	5_C	7,47E-06	1,51E-06	1,22E-07	7,03E-08
Sample 6	A	6_A	7,42E-06	5,53E-07	4,10E-08	7,10E-08
	B	6_B	6,56E-06	5,78E-07	4,17E-08	4,76E-08
	C	6_C	6,49E-06	2,54E-07	2,55E-08	4,21E-08
Sample 8	A	8_A	7,01E-06	5,31E-07	8,25E-08	6,31E-08
	B	8_B	8,50E-06	3,39E-07	7,69E-08	5,33E-08
	C	8_C	7,67E-06	3,59E-07	5,54E-08	5,84E-09
Sample 9	A	9_A	7,61E-06	5,98E-07	7,25E-08	1,13E-07
	B	9_B	9,31E-06	1,29E-06	1,04E-07	1,46E-07
	C	9_C	8,48E-06	9,82E-07	8,01E-08	9,59E-08

To compare the results between the test samples, the volume of wear in each case has been normalized according to the simulated distance travelled.

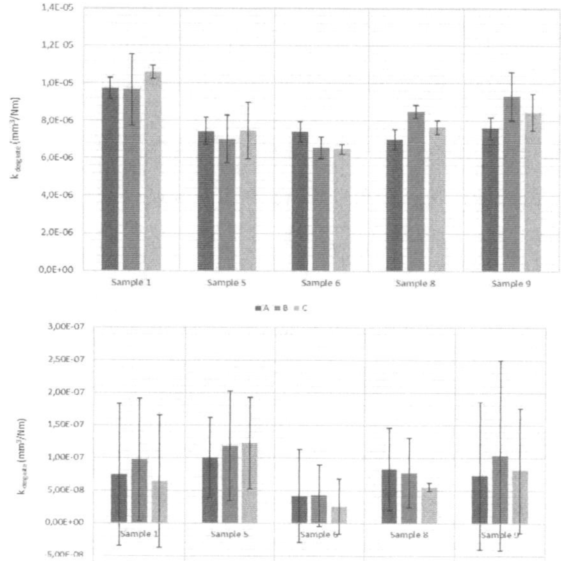

Figure 5. These graphs show the wear on the steel and the pins of the five samples (3 essays each).

Once the test was carried out and the Friction Coefficient was obtained, the volume of wear was evaluated, providing a series of wear profiles.

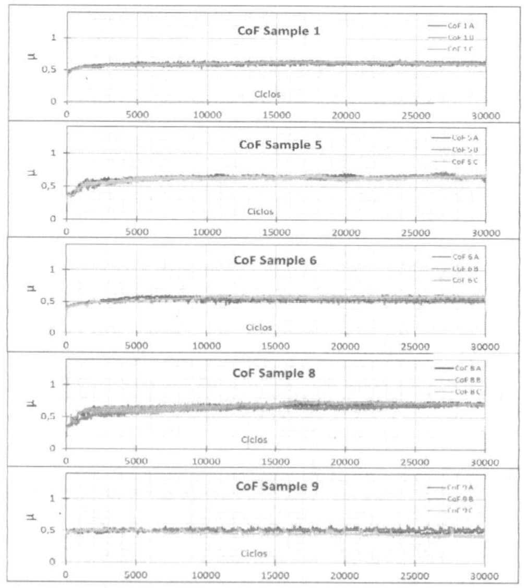

Figure 6. These figures show Friction Coefficients.

Material C, test 6, which is the material hardened by SPH, shows the least wear under test conditions, notably coefficient of friction.

Another diffractometry analysis was carried out with the Bruker D8 Advanced equipment on 5 samples hardened with the 5 different process and similar results were obtained as with the hole drilling method.

4 Conclusion

This project on developing an alternative to explosion hardening has resulted in the discovery of a more technological and non-destructive method, which better fits today's industrial values and decarbonisation goals. As an indirect consequence, for the first time, we are able to show the extent to which explosion disrupts the crystallographic network of castings, both internally and externally. The use of our innovative single process hardening, SPH; combing three technologies, ultrasonic waves, shotblasting and percussion hammering is not only structurally non-destructive but is also superior in all respects. Namely, our process is carbon free, and can be carried out on-site, implying a significant cost and time saving, resulting in an overall reduction in cost of around 75%. Investment would be quickly recoverable. The process is currently patent pending, and a demonstration model will be in operation in a foundry in Spain in 2025.

5 Acknowledgments

The authors acknowledge and appreciate the support of the Basque Government through the projects HAZITEK (DES2) ZL2024/00368 and (H2MAT+) KK-2024/1122.

References

[1] G.Calboreanu "Increased Wear Performance of Abrasive Particies", AFS Transactions, V. 98,1990.
[2] SergejToedtll"Development Work on Blast CleaningPlants", Stahl und Eisen 98 (1978) . 6April03.
[3] H.J. PLASTER, "Blast Cleaning and Allied Processes"
[4] Eugene Tarabek - "Principles of Centrifugal Blast Machines", Modern Casting, Feb 198305.
[5] AE J-444 - "Cast Shot and Grit Size Specifications for Peening and Cleaning", 1984.
[6] A. Pineda, A.M. Balu, J.M. Campelo, A.A. Romero, D.Carmona, F. Balas, J. Santamaria, R. Luque, ChemSusChem. (2011), doi:10.1002/cssc.201100265.
[7] P. Kalita, N.M. Gupta, R. Kumar, J. Catal. 245 (2007) 338–345.
[8] K. Yoshida, C. Gonzalez-Arellano, R. Luque, P.L. Gai, Appl. Catal. A 379 (2010) 38-44.
[9] R. Luque, J.M. Campelo, D. Luna, J.M. Marinas, A.A. Romero, Microporous Mesoporous Mater. 84 (2005) 11–20. [21] E.P. Parry, J. Catal. 2 (1963) 371–379.
[10] I.S. Pieta, M. Ishaq, R.P.K. Wells, J.A. Anderson, Appl. Catal. A 390 (2010) 127–134.
[11] Laser in Manugfacturing and material processing effect laser sock peening on the microstructure, tensil and heat transport properties. 28May 2020. Volum7. Pag 259-277.
[12] Efim,S: Statnikov, OlegV. Physics and mechanism of ultrasonic impact. Ultrasonics 44-e538 (2006).
[13] Michael A. Sutton J. Image correlation for shape, motion and deformation measurements. Springer-Verglag, Boston (2009).
[14] Metal Handbook. Vol2. Head treatment, cleaning and finishing.
[15] Fatigue and fracture mechanics 29thvolume Tina Panontin, STP1332.
[16] The physical metallurgy of microalloyed steels. T.Gladman. The institute of materials.1997.
[17] Petr Havlicek, Katerina Busova.Metal,23-25.5.2012.

Precision Solidification for Creating Multifunctional Material Systems

Daniel Liang
Commonwealth Scientific and Industrial Research Organisation (CSIRO)
*Corresponding address: e-mail: daniel.liang@csiro.au

Abstract: Transformation of our societies in the age of electrification and climate change increasingly demands metal manufacturers, including casting industries, to supply metallic-based materials, structures, and systems capable of integrating mechanical strength with other functionalities. This brief will focus on the pivotal roles that solidification-based technologies can play in meeting this challenge. By highlighting examples of relevant progress using near-net-shape and extreme-condition solidification processes, their potential to create multifunctional material systems through multi-materials and multi-technology approaches, will be explored.

Keywords: Foundry, casting, solidification, multifunctional materials, multi-processing,

1 Introduction

The 20th century witnessed significant technical advances in fluid dynamics, solidification sciences, and quality control through X-ray technologies. These advancements enabled the foundry industries to produce near-net-shape as-cast components, minimizing the need for extensive machining, thus optimizing both cost and energy efficiency. Recently, the imperatives of climate change and environmental preservation have accelerated transformative megatrends aimed at reducing carbon footprints. The shift towards miniaturization, electrification, renewable energy, and materials recycling across various domains requires the development of novel metallic-based components. These new components not only uphold traditional roles in providing mechanical strength but also integrate functionalities such as heat transfer, electron transport, energy conversion and harvesting, sensing, and adaptive responses to environmental variations.

These evolving demands require multilocational materials systems, characterized by complex chemistries, hierarchical microstructures at the nano-micro scale, and near-net-shape geometries. However, the manufacturing of these sophisticated materials poses significant challenges for traditional foundry processes. This presentation offers a brief overview of the challenges confronting the foundry industry and explores strategic approaches to advance casting processes to overcome these obstacles. The focus is on creating multifunctional materials systems that align with the rapidly changing landscape of technological, environmental, and societal demands.

2 Technologies

Solidification technologies highlighted in the presentation include planar flow foil casting, twin roll strip casting and stir casting. These processes are representative examples of non-equilibrium solidification conditions, near-net-shape features and multi-materials fabrication. The as-cast materials produced through the solidification routes can serve as feeds for other processing technologies, such as solid-state friction stir welding and friction stir additive manufacturing, realising sizes, geometries, structures and functionalities that solidification technologies alone cannot achieve. Furthermore, the role of digital technology, including artificial intelligence (AI) and machine leaning (ML), is highlighted for optimising either the operations of dynamic solidification processes or designing complex materials systems.

3 Discussion

The abilities of non-equilibrium solidification processes to generate a variety of atomic structures offers opportunities to induce functional properties while enhancing original structural properties. It is well established that near-rapid solidification of metal alloys at higher cooling rates, such as two roll casting (TRC) aluminium (Al) and magnesium (Mg) alloys [1][2], increases, for example, the concentration of alloying elements in the Al phase. The supersaturated Al phase alters atomic arrangement within ordered lattice structures, i.e. the unit cells repeating in 3D, resulting in significant increase in functional corrosion resistance [2]. With an extremely high cooling rate during rapid solidification, such as planar flow casting (PFC) as shown in Figure 1, this atomic orderly arrangement within the unit cell can be completely replaced by a random arrangement of atoms, leading to the formation of an amorphous metal [4]. This dramatic transformation of the unit cell from an orderly to a random atomic arrangement creates new functionalities [4] - [6]. For example, amorphous steel foils produced by planar flow casting exhibit superior soft magnetic properties [4] - [5], which have become the preferred materials choice for manufacturing modern electrical transformers and electric motors with high energy efficiency.

Resembling repeating atomic unit cells in a single-phase metal, metal matrix composites (MMCs) can be considered as consisting of large unit cells at the microscale with multiple phases, characterized by pseudo repeats of uniformly distributed phases. By carefully selection of these phases from different materials, MMCs with such pseudo unit cells can demonstrate unique functionalities that are not possible to achieve with the constituent materials alone. Stir casting technology has been demonstrated the most cost-effective technology to produce MMCs, including Al matrix composites (AMCs) with carbide (SiC) particles as enhancement. This multi-materials AMCs indeed exhibit advanced braking function in brake discs that neither Al nor SiC alone would possess [7].

Braking functionality of AMC-based brake discs, for example, involves converting the kinetic energy of moving vehicles into thermal energy through friction, followed by the subsequent removal of heat from the braking interfaces. This functionality replies on a complex interaction between various mechanical and functional materials properties, such as coefficient of friction, thermal conductivity, hardness and toughness. The key to optimising these interactions is the precise control of the multi-material phase distribution within the AMC pseudo unit cells. However, achieving this microstructural precision, particularly a uniform and hierarchical distribution of a high-fraction SiC particles, has been challenging for the stir casting process.

Microstructural optimisation of this SiC distribution in AMC multi-materials systems is achieved only when a multi-technology manufacturing approach is deployed. In this approach, Stir-solidified Al and MMC semi-finish parts are integrated using friction stir additive manufacturing [7]. This multi-materials and multi-technology method allows for precision control of the microstructures enhancing materials functionality in the final product. This increasing sophistication of performance requirement, materials and manufacturing methods predicts emerging development of digital technologies, such as artificial intelligence (AI) optimisation and machine learning (ML) modelling [8], as an integral part of future solidification-based materials technology.

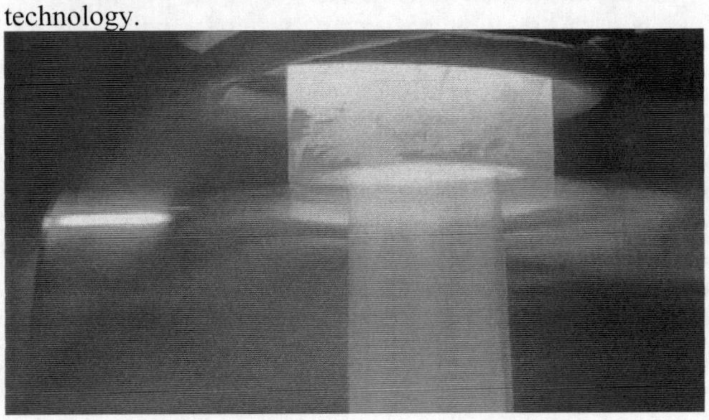

Fig.1 Planar flow casting

4 Conclusion

Solidification technology plays a pivotal role in emerging advanced manufacturing by integrating multiple materials through a multi-technology fabrication approach to meet increasing industrial demands. Precision control of solidification and microstructure, aided by AI optimization and ML modelling, is anticipated to be key for this technology to fulfill its critical role in the future.

5 Acknowledgments

The author would like to acknowledge the support of CSIRO Manufacturing and Data61 Research Units in Australia.

References

[1] Liang, D., Cowley, C.B. The twin-roll strip casting of magnesium. JOM 56, 26–28 (2004).

[2] Birol, F., Birol, Y., Corrosion Behavior of Twin-Roll Cast Al-Mg and Al-Mg-Si Alloys, Materials Forum Volume 28, 2004, Edited by J. F. Nie, A. J. Morton and B. C. Muddles.

[3] Theisen, E.A., Weinstein, S.J. An overview of planar flow casting of thin metallic glasses and its relation to slot coating of liquid films. J Coat Technol Res 19, 49–60 (2022).

[4] Parsons, R., Suzuki, K., Nanocrystalline soft magnetic materials produced by continuous ultra-rapid annealing (CURA), AIP Advances 12, 035316 (2022)

[5] Guo, Y, Liu, L, Ba, X, Lu, H, Lei, G, Sarker, P., Zhu, J., Characterization of Rotational Magnetic Properties of Amorphous Metal Materials for Advanced Electrical Machine Design and Analysis, Energies 2022, 15(20), 7798.

[6] Oropeza, D., McMahan, C., Hofmann, D.C., Self-deploying amorphous metal origami structures for environmental protection of lunar payloads, International Journal of Space Structures, 2023, Vol. 38(3) 204–213.

[7] Dey, D., Chintada, S. K., Bhowmik, A., Biswas, A., Evaluation of wear performance of Al2024-SiC ex-situ composites, Materials Today: Proceedings, Volume 26, Part 2, 2020, Pages 2996-2999, ISSN 2214-7853.

[8] Tan, D., Xia, S., Yob, A., Yang, K., Yan, S., Givord, M., Liang*, D., Evaluation of the wear resistance of aluminium-based hybrid composite brake discs under relevant city rail environments, Materials & Design, 215 (2022) 11054.

Effect of Tempering on ADI Microstructure and Properties

Rui Wang, Han Deng, Yanqing Mu, Bin Zhang
Materials and Processes R&D Center, CRRC Qishuyan Institute Co. Ltd, No.258 Wuyi Road Qishuyan, Changzhou, Jiangsu, China
*Corresponding address: e-mail: wangrui.qs@crrcgc.cc

Abstract: The transformation of mechanical properties occurs in ADI at different tempering temperatures and times. The microstructure of ADI is analyzed using the univariate comparison method, which involved observing metallography, scanning electron microscopy (SEM), and backscattering electron diffraction technique (EBSD) methods after the material is treated with different tempering temperatures and times. The mechanical properties of the material were also examined. The tensile test and Charpy impact test are employed to examine the mechanical properties of the material, namely strength, plasticity, and toughness. The results indicated that the transformation temperature of the ADI is in the range of 310 to 330°C. When the tempering time is certain, the strength increases with an increase in tempering temperature, while hardness remains almost unchanged. And plasticity is reduced by 45%, and impact toughness is reduced by 67%. These changes affect the normal use of the material. The reason for the transformation is that the structure of the residual austenite transformed into ferrite and carbide, reducing the plasticity and toughness of the material. When the tempering temperature is certain and higher than the transformation temperature, the tempering time will only reduce the plasticity, strength, hardness, and impact toughness to a negligible extent, while the matrix structure remains largely unaltered.

Keywords: ADI; ausferrite; temper; residual austenite

1 Introduction

Austempered ductile iron (ADI) is a novel type of ductile iron material that has been developed over the past four decades. It is produced by isothermal quenching of ductile iron through austenitization. ADI exhibits enhanced tensile strength, fatigue strength, fracture toughness, and superior wear resistance compared to conventional ductile iron[1]. The structure of austenite in ADI materials remains stable at room temperature. If the material is subjected to prolonged stress or elevated temperatures, the matrix structure may undergo a transformation[2], resulting in a significant decline in the conventional mechanical properties and limiting the applicability of ADI components. In this study, tempering tests were conducted at varying temperatures or times to observe the tissue changes and detect the mechanical property alterations, thereby determining the tempering transformation temperature and transformation time.

2 Experimental procedure

The experiment employs QTD1050-6 grade ADI material. The raw materials employed in this experiment are pig iron, scrap, and furnace material. The casting temperature is 1340℃~1360 ℃, with the flow of pregnancy. The final composition is C: 3.3 ~ 3.8% , Si: 2.2 ~ 2.9%, Mn: 0.4-0.9%, P: <0.03%, S: <0.02, Mo: <0.2%, Cu: 0.4- 0.9%, Mg: 0.03-0.05%. Molten iron is poured into a 25-millimeter-thick Y-shaped test piece. Rectangular blocks measuring 25 mm × 25 mm × 220 mm should be extracted from the Y-shaped test block below for use as an isothermal quenching test object. The equal quenching process is as follows: the austenite temperature is 900°C, the austenitization time is 2.5 hs, and the isothermal quenching temperature is set at 345°C, with a quenching time of 2.5 hours. The bars should be cut into 20 mm × 220 mm pieces..

3 Result and discussion

3.1 Effect of tempering temperature on properties and microstructure

6 specimens were prepared. Heat each specimen by 5℃ per minute from room temperature to 310℃, 330℃, 350℃, 370℃ and 390℃. After being insulated for 5 hs, the specimens were removed using a sample clamp and air-cooled to room temperature.

From the metallographic test results, it was observed that the base structure consisted of ausferrite, with carbides appearing after tempering at 330°C. In terms of mechanical properties, both tensile strength and yield strength exhibited a decreasing trend with increasing environmental temperature. However, there was no significant change in hardness. The post-fracture elongation and impact performance drastically decreased when tempered at 330°C. The post-fracture elongation decreased from 11% to 6%, representing a reduction of 45%. And the impact performance dropped from12 J to 4 J, with the decrease of 67%.

The morphology of specimens 310℃,330℃ and 390℃ is observed under the scanning electron microscope. The tempering temperature of 310℃ is associated with a matrix structure of ausferrite (AF) that is arranged in an orderly manner. The ferrite needles are fine and uniform. Additionally, islands of residual austenite (RA) are observed between the ausferrite grains. At a tempering temperature of 330 ℃, the residual austenite in the matrix structure has undergone transformation into ferrite and carbides, accompanied by an increase in the thickness and length of the ferrite needles in the ausferrite. At a tempering temperature of 390 ℃, the ferrite needle coarsening and elongation is significant, and the grain direction is no longer ordered.

Residual austenite is an unstable phase that transforms at higher temperatures. In order to measure the specific content of residual austenite in the specimens, EBSD (electron backscatter diffraction) tests are performed on specimens 310℃, 330℃ and 390℃. The volume fractions of residual austenite are 27.3%, 13.5%, and 0%, respectively. The data indicates that when the tempering temperature is higher than the tissue transformation temperature, the residual austenite in the ADI material transforms into ferrite and carbide. The higher the tempering temperature, the more the residual austenite transformed until it is completely transformed.

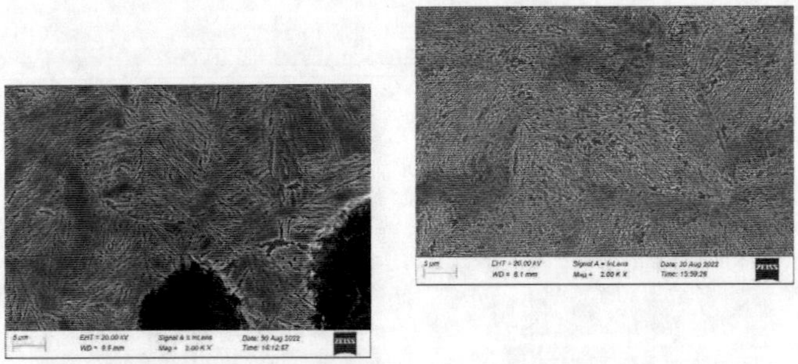

Fig 1 SEM morphology of specimen in 310℃ & 390℃

The aforementioned results demonstrate that in the tempering process, the ADI material microstructure transformation temperature is between 310℃ and 330℃. When the tempering temperature is below this range, the microstructure and properties of the material remain largely unaltered. Island-shaped residual austenite plasticity exerts a second-phase plasticizing effect. High-carbon austenite in ausferrite can also enhance the plastic deformation resistance. However, when the tempering temperature exceeds 330 ℃, the microstructure begins to transform. Residual austenite into ferrite, which no longer serves the role of a second-phase plasticizing agent. The ferrite needle becomes longer and coarser, and the grain transitions from an ordered to a disordered state, resulting in a sharp decline in plasticity and toughness, ultimately leading to material failure.

3.2 Effect of tempering time on properties and microstructure

The tempering process involves the preparation of five specimens. Each specimen is heated at a rate of 5 ℃/min from room temperature to 330 ℃, held for 0.5hs, 1h, 2hs, 3hs, or 4hs. Cool them to room temperature by air.

The metallographic test results indicated that the matrix structure is predominantly ausferrite, with no discernible differences. The mechanical properties demonstrated a tendency for the post-break elongation to decrease with the extension of the tempering time, while there is no significant change in tensile strength, yield strength, hardness and impact toughness. The post-break elongation exhibits a notable decline, decreasing from 12% to 5.5%, representing a 54% reduction.

To assess the microstructure of the specimens, EBSD tests conducts directly on specimens of 0.5hs, 2hs and 4hs, resulting in residual austenite volume fractions of 33.7%, 31.8%, and 35.1%, respectively. The results indicate that when the tempering temperature is higher than the tissue transformation temperature, the residual austenite in the ADI material doesn't decompose with the extension of tempering time, and the morphology of ferrite needles neither change. This implies that as long as the tempering temperature is higher than the transformation temperature, the tempering time only affects the plasticity of the material, and the strength, toughness, and microstructure don't transform significantly. In comparison to time, temperature exerts a more pronounced influence on the microstructure and properties of ADI materials.

4 Conclusion

(1)The transformation temperature of the ADI is in the range of 310 to 330°C.

(2)When the tempering time is certain, the strength increases with an increase in tempering temperature, hardness remains, and plasticity and impact toughness reduced.

(3) When the tempering temperature is certain and higher than the transformation temperature, the tempering time will only reduce the plasticity, strength, hardness, and impact toughness to a negligible extent.

References

[1] Liu Jincheng, Gong Wenbang. Some recent progress of foreign ADI R&D[J]. Modern cast iron, March 2020.

[2] S.K. Putatunda, S. Kesani, R. Tackett, and G. Lawes, Development of Austenite Free ADI (Austempered Ductile Cast Iron), Mater. Sci. Eng. A, 2006,435–436,p112–122. doi:10.1016/j.msea.2006.07.051.

Microstructure, Mechanical Properties and Wear Resistance of Aluminum Matrix Composites Brake Disc Prepared by Vacuum Pressure Infiltration

Kangxi Fu, Shiqi Zhang, Dong Tan, Baowang Wang
CRRC Qishuyan Institute, Changzhou, Jiangsu 213011, China;
*Corresponding address: e-mail: 15603335884@163.com

Abstract: In this study, lightweight hybrid composite brake discs consisting of a SiCp/A356 aluminium matrix composite top layer and an A356 Al alloy bottom base are fabricated by vacuum pressure infiltration process. The SiC preform is solely utilized for a 5mm friction layer on the friction surface, which significantly reducing the manufacturing cost, processing and recycling difficulty of brake discs. The friction and wear properties of the brake disc sliding against semi-metallic brake pads were investigated using MM3000 brake dynamometer. The testing conditions included a braking pressure of 0.57~0.72 MPa and an initial braking speed of 40~220 km/h. It is found that the hybrid composite brake discs demonstrate excellent wear and friction performance under relevant environments for operating a city rail vehicle. The excellent performance of the lightweight hybrid composite brake discs confirms its high technology readiness level to replace heavy iron and steel brake discs in city rail vehicles.

Keywords: Brake disc; Lightweight; Wear; SiCp/Al; Friction film.

1 Introduction

SiC particle-reinforced aluminum matrix composites (AMC) offer numerous advantages, such as low density, high specific strength, high specific stiffness, and exceptional wear resistance, making them a highly promising choice for brake materials [1]. Currently, AMC brake discs are primarily manufactured using the stir casting method and powder metallurgy process [2]. The size of rail transit brake discs is large, requiring a significant amount of AMC material. The cost of AMC material is expensive, typically 3-5 times that of aluminum alloy [3]. The high cost of preparation hinders the widespread commercial application of AMC brake discs. In this study, we fabricated AMC brake discs by vacuum pressure infiltration. The SiC preform is solely utilized for a 5mm friction layer on the friction surface, which significantly reducing the manufacturing cost, processing and recycling difficulty of brake discs. More importantly, the wear performance of the hybrid composite brake disc was evaluated under the brake environment relevant for operating a city-rail vehicle at 220 km/h.

2 Experimental procedure

Fig. 1 shows the manufacturing process of the AMC brake disc developed in this study. The friction and wear properties of the brake disc sliding against semi-metallic brake pads were investigated using MM3000 brake dynamometer. The microstructures of these samples were analysed using optical microscopy and a scanning electron microscope (SEM). The roughness analysis and 3D morphology were observed using a confocal laser scanning microscope.

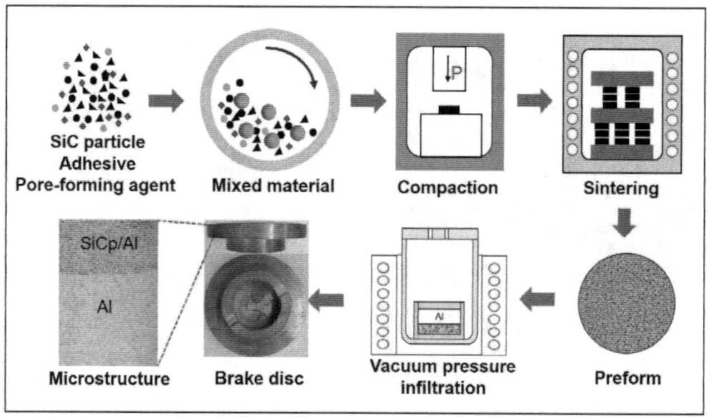

Fig.1 Fabrication stages of AMC brake disc

3 Result and discussion

Microstructure and physical properties

Fig. 2 shows the metallographic structure of the AMC brake disc. The SiCp/A356 composites has a dense structure without any noticeable holes or crack defects. The SiC particles are uniformly dispersed and no particle agglomeration phenomenon is observed. The interface between the SiCp/A356 composites and the A356 Al alloy bottom base is well-bonded, and no pores or microcracks are present at the interface.

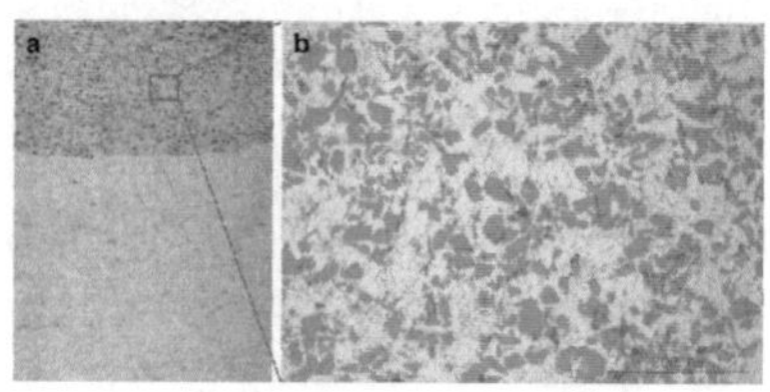

Fig.2 Microstructure of AMC brake disc

Table 1 shows the mechanical properties of the AMC brake disc. The SiCp/A356 composites exhibits higher tensile strength and hardness than A356 alloy. The density of the AMC brake disc is only 38% of that of a cast iron brake disc, which offers clear advantages in terms of lightweight design.

Table 1. Mechanical properties of brake disc

Brake disc	Tensile strength (MPa)	Elongation (%)	Hardness (HBW)	Density (g/cm^3)
SiCp/A356	162	1.2	112	2.82
A356 alloy	154	2.0	58	2.70

Friction and wear behaviour

Fig. 3 shows the shows the mean coefficient (μm) under different braking conditions. Under different braking pressures, the μm shows a trend of first increasing and then decreasing with the increase of braking speed. The measured values of all the μm for the entire test within a narrow band between 0.3 and 0.5. After a series of repeated braking, the wear loss of the brake disc and brake pads was only 0.36 g and 1.62 g.

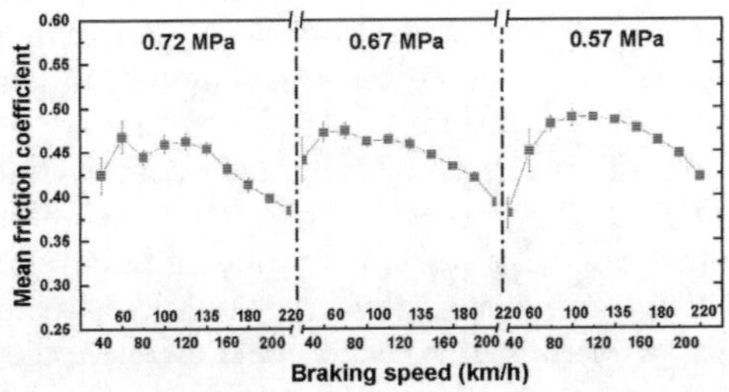

Fig.3 The mean coefficient under different braking conditions

Worn surface

Fig. 4 shows the 3D morphology analysis of worn surface, the depth of the circular scratch was only approximately 16 μm.

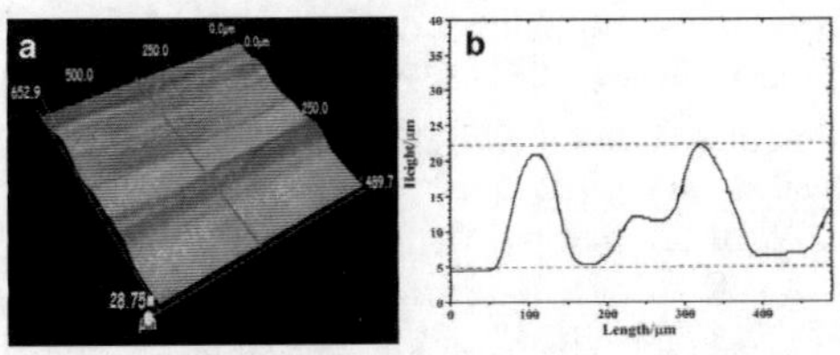

Fig. 4 The 3D morphology analysis of worn surface

Fig. 5 shows the SEM image of worn surface, numerous ploughing grooves can be observed on the surface, a friction film formed by the bonding of granular debris can be observed at higher magnification.

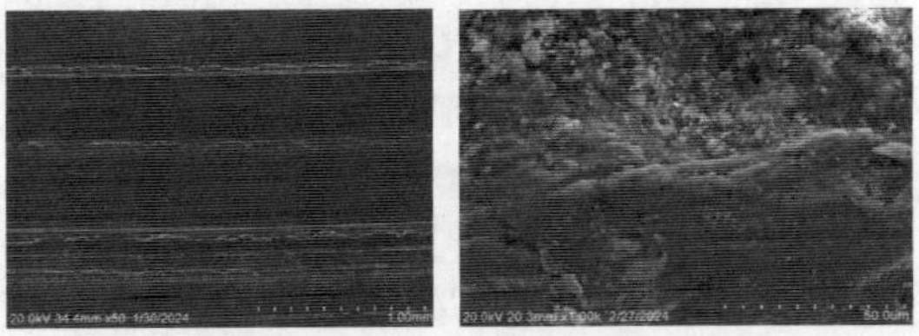

Fig. 5 The SEM image of worn surface

4 Conclusion

Hybrid composite brake discs, consisting of a A357/SiC composite top layer and an A356 Al alloy base, were fabricated by vacuum pressure infiltration process. The brake discs have successfully passed 150 braking test cycles under relevant environments for operating a city rail vehicle. It is found that the hybrid composite brake discs demonstrate excellent wear and friction performance, providing a stable friction coefficient within a range of 0.35~0.50, and maintaining the integrity and stability of the hybrid structure. The wear loss of the brake disc was only 0.36 g, the depth of the circular scratch was only approximately 16 μm. The excellent performance of the lightweight hybrid composite brake discs confirms its high technology readiness level to replace heavy iron and steel brake discs in city rail vehicles.

5 Acknowledgments

This work is financially supported by National Key R&D Program of China (2023YFB3710600).

References (Bold, 10 pt., Arial)

[1] Ghandvar H, Farahany S and Idris M H. Effect of Wettability Enhancement of SiC Particles on Impact Toughness and Dry Sliding Wear Behavior of Compocasted A356/20SiCp Composites[J]. Tribology Transactions, 2017, 61(1): 88-99.

[2] Laden K, Guerin J, Watremez M and Bricout J. Frictional characteristics of Al–SiC composite brake discs[J]. Tribology Letters, 2000, 8: 237-247.

[3] Tan D, Xia S, Yob A, Yang K, Yan S, Givord M and Liang D. Evaluation of the wear resistance of aluminium-based hybrid composite brake discs under relevant city rail environments[J]. Materials & Design, 2022: 215.

Composition Analysis and Micro-Alloying Study of ZG25MnCrNi Casting Steel for Railway Wagon Bolsters and Side Frames

Cunlei Zou[1]*, Junhe Wang[1], Shuang Zhang[1], Sheng-tian Liu[2], Changliang Tian[2], Hao Wang[2], Chuang Dong[1]
1. School of Materials Science and Engineering, Engineering Research Center of Continuous Extrusion, Ministry of Education, Dalian Jiaotong University, Dalian, Liaoning, China
2. CRRC Qiqihar Vehicle Co, Limited, Qiqihar, Heilongjiang
*Corresponding address: e-mail: clzou@djtu.edu.cn

Abstract: Due to the demand for high speed and heavy load, railroad wagons have increasingly stringent requirements for the reliability of vehicle structure materials. However, at present, there is a need to control the cost while obtaining enhanced performance of the material to meet the requirements of the service. This is an urgent problem in this field.Furthermore, there is a critical challenge in controlling the cost of the precursor in order to obtain a more excellent performance of the material to meet the service requirements. In this paper, we address the issue of the unclear composition of ZG25MnCrNi steel for railroad wagon rocker sidestands and the insufficient yield strength redundancy of this material. We employ a cluster composition analysis method to obtain the cluster formula of ZG25MnCrNi steel. This enables us to determine the precise composition of ZG25MnCrNi steel, which we then optimize through alloy reinforcement. This results in the production of new materials with enhanced performance. The new material with excellent performance was obtained through material preparation and analysis. The following results were obtained.

Keywords: ZG25MnCrNi; Cluster-plus-glue-atom Model; Mechanical properties; Microalloying; Microstructure

1 Introduction

The railroad wagon is a crucial component of cargo transportation. To meet the demands of high-speed and heavy loads, the standards for rolling stock reliability are becoming increasingly rigorous. The ZG25MnCrNi steel is utilized in railroad wagon bogies for the rocking pillow and sidestand, and it is essential to meet the heightened performance and reliability expectations. The domestic active rocker and sidestand are primarily constructed with B + steel, which is ZG25MnCrNi steel. This steel is required to have a yield strength greater than 345 MPa, and an impact absorption energy of at least 20 J when tested at -7 ℃. The main alloying elements for this steel are C, Si, Mn, Cr, and Ni. However, there is a serious shortage of domestic B + steel with the required yield strength, which presents a significant challenge for the active rocker and sidestand industry. The composition and organization of a material are closely related, with a change in composition resulting in a change in organization and performance. The current AAR M-201-05 "cast steel parts specification" of the standard only provides the range and carbon equivalent of the five major elements composing the material, namely C, Si, Cr, Mn, and Ni. The composition of CE and other alloying elements, such as Cr, Ni, Mo, and V, is based on the requirements of manufacturers. There is no calibration of the specific composition range, which leads to an inaccurate theory to guide the production of the target composition and the development of internal control standards. This results in fluctuations in material performance. It is therefore imperative to clarify the target composition of materials and internal control standards. Furthermore, it is well established that the addition of elements that can make steel microalloying in B+ alloy steel will result in significant changes to the microstructure and mechanical properties of B+ steel. One effective method to enhance the mechanical properties of ZG25MnCrNi for sidestand is to employ microalloying techniques.

ZG25MnCrNi steel is suitable for use in railroad wagon bogie rocker sidestands, with a carbon content of approximately 0.25 wt.%, and a high-temperature single-phase austenitic structure. This composition will satisfy the specific composition formula. Consequently, this paper will apply the cluster composition design method to analyze the compositional law of alloyed ZG25MnCrNi steel from the perspective of the chemical proximity program. This will entail the summary of its cluster formula and the sorting out of the role of elements Ti, Mo, V and other alloying elements in the steel. The microalloyed composition of ZG25MnCrNi steel will be analyzed using the "cluster plus connected atom model." The microalloying composition analysis involves the replacement of other elements in the cluster with Ti, Mo, and V microalloying elements. This process optimizes the composition of ZG25MnCrNi steel, resulting in enhanced grain refinement and superior performance.

2 Experimental procedure

The optimized and microalloyed compositions of ZG25MnCrNi steel, were prepared as specimens and subjected to testing and analysis. The specimens were prepared by melting in an electric arc furnace and a vacuum induction melting furnace. The as-cast samples were subjected to a normalizing heat treatment with a holding time of 900 °C for 3.5 hours. The microstructure of the samples was observed using a Leica DMi8A inverted metallographic microscope, a Zeiss SUPRA 55 field emission scanning electron microscope, and a Hitachi H-800 transmission electron microscope, respectively. The hardness of the cast and heat-treated specimens was determined using a HBE-3000 micro-Vickers hardness tester with a pressure of 500 g and a holding time of 15 s. The stress-strain curves were measured using an MTS series C43 electronic universal testing machine. The experiment used a JB-300s Charpy impact test pendulum machine to measure the low temperature impact work at -7 ℃.

3 Result and discussion (Bold, 10 pt., Arial)

1. Compositional analysis of the ZG25MnCrNi steel

The optimized ZG25MnCrNi steel class molecular composition formula for the C3Fe250Ni0.5Mn2.5Cr1Al0.25Si1.75, The composition of the alloy is 0.25C-97.82Fe-0.21Ni-0.96Mn-0.36Cr-0.05Al-0.35Si. The footnote of the element symbols indicates the corresponding atomic number, while the number before the element symbol indicates the corresponding mass percentage.

2.Result of Experimental

The refinement of microalloying elements, including grain refinement and pearlite lamellar refinement, has led to a significant improvement in the material's performance. The yield strength of tensile parts has increased significantly, up to 380 MPa, compared to the standard parts, which have been enhanced with Ti, Mo, and V elements. The tensile ratio of the parts has also been improved, reaching up to 28% and 29% with Ti and V elements, respectively, while the addition of Mo elements has resulted in a lower tensile ratio of 19%. Furthermore, the impact work at -7 ℃ has been enhanced, reaching up to 80 J, which is much higher than the standard. The tensile ratio of the parts after the addition of Ti and V elements is up to 28% and 29%, respectively, while that of the Mo elements is only 19%, which is below the standard of 24%. Furthermore, the impact work at -7 ℃ is up to 80 J, which is considerably higher than that of the standard. Following the addition of Ti and V elements, the tensile ratio can reach 28% and 29%, respectively. However, the addition of Mo elements results in a tensile ratio of only 19%, which is below the standard of 24%. Furthermore, the impact work at -7 ℃ can reach 80 J, which is significantly higher than the standard of 20 J. The tensile ratio at a temperature of -7 ℃ can reach 80 J, which is considerably higher than the standard of 20 J. Similarly, the sectional shrinkage can be as high as 73%, which is much higher than the standard of 36%..

Table 1 Comparison of properties of ZG25MnCrNi(B+) steel

Steels	R_m/ MPa	R_{sl}MPa	A_{50}/%	Z/%	R_{el}/MPa
ZG25	≥450	≥230	≥22	≥32	≥25
B	≥485	≥260	≥24	≥36	≥20(-7℃)
B+	≥550	≥345	≥24	≥36	≥20(-7℃)
C	≥620	≥415	≥22	≥32	≥20(-18℃)
Sample of B+	613±4	358±12	27.3±1.1	64.5±0.9	27.5±3.5
optimized	573±5	359±10	28.1±1.0	67.7±1.2	39.5±7.7
With Ti	609±3	380±9	28.6±0.9	69.8±3.6	77.5±2.1
With Mo	590±10	388±12	19.1±4.5	60.9±3.7	41.5±10.6
With V	620±8	379±10	29.4±0.5	73.7±3.5	82.5±3.5

ADI Spoke Wheels

Chengqi WANG
National Innovation Center of High Speed Train, Qingdao 266111, Shandong, China
*Corresponding address: e-mail: 520053200073@crrcgc.cc

Abstract: Railway companies need quieter wheels, lighter wheels and higher wear-resistant wheels. National Innovation Center of High Speed Train is carrying out research on new ADI spoke wheels. The research optimizes the chemical composition and heat treatment process parameters of wheel ADI materials, and shows that the key properties of the stereotyped ADI materials, such as wear performance, damping performance and thermal stability, meet the requirements of railway wheel application. A new wheel with spoke structure is designed by topology optimization method, and simulation studies on static/fatigue strength, dynamics, vibration characteristics, and acoustic radiation performance are carried out. The casting process, machining and non-destructive testing methods of the new wheel were studied, and a series of bench tests were conducted.

The study shows that the strength and fatigue strength of the new ADI spoke wheels have reached the standard of rolled steel wheels, and compared with the traditional rolled steel wheels, the new wheels have a weight reduction of more than 10% and a noise reduction of more than 2dB, which is expected to improve the wheel-rail abrasion and enhance the reliability and comfort of the vehicle operation.

Keywords: ADI material; spoke wheel; casting process; performance testing; quieter, lighter, wear resistance wheel

1 Introduction

Wheel is the key component of wheelsets and bogies, plays an important role of bearing, guiding and driving, and bears complex loads. The plate rolled steel wheels are the main choice for metros and high speed train today thanks to their prices, excellent mechanical properties and easy approval processes. However, rolled steel wheels are not perfect, and their performance breakthrough is difficult to make, and a wheel with lighter quality, better vibration and noise reduction performance, and better wear resistance is needed.

In this report, through the application development of ADI materials and new topology optimized structure design, a new type of ADI spoke wheel is developed, the casting processing of the new wheel is completed, and the test verification is carried out.

2 Experimental procedure

The report researches and develops a new type of ADI spoke wheel, and the research mainly includes 4 aspects:

Research on high-performance ADI materials for wheels: In order to develop ADI materials for wheels, orthogonal tests on chemical composition and heat treatment of ADI were carried out, and the chemical composition and heat treatment process with optimal comprehensive performance were obtained, and the wear performance, damping performance, and thermal stability of ADI materials were studied.

Structural design and simulation analysis of new wheel: Based on the railway applications and boundary conditions, using topology optimization method, we carry out the optimal design of wheel structure, taking into account the strength, dynamics, acoustic radiation and other properties, as well as material properties and molding process, and carry out the static/fatigue strength, dynamics, vibration characteristics, acoustic radiation performance and other systematic simulation analysis of the designed wheel structure.

Wheel casting process: carry out new wheel casting process design, prototype trial production, non-destructive testing, machining process design, optimize the spoke wheel casting process, processing and testing methods.

Experimental procedure: The examinations were carried out with reference to UIC 510-5,13979-1 standards and other relevant standards.

3 Result and discussion

Research on high-performance ADI materials for wheels

Through the orthogonal test to explore the alloy composition content and isothermal quenching heat treatment of ADI, we got the ADI material chemical composition and heat treatment process route for the wheel . Through the damping test and friction wear test, the superiority of ADI material compared with rolled steel material in terms of noise reduction and wear resistance is verified. The feasibility of the new wheel tread braking was verified by conducting stability tests on the microstructure transformation and mechanical properties changes of ADI materials at different temperatures.

New wheel structure design and simulation analysis

The topology optimization design of wheel structure is carried out to optimize the wheel structure of multi-spoke scheme and performance simulation analysis, and the optimal structure of spoke wheel is determined. The static and fatigue strength of the new wheel is evaluated, and the vibration characteristics of the new wheel are studied. The new wheel is better than the rolled steel wheel in terms of the vertical force of the wheel-rail, modal frequency and contact stress of the wheel-rail. The sound radiation performance and dynamic performance of the new wheel were simulated and analyzed, and they were better than the rolled steel wheel.

Wheel casting process

Multiple casting process designs and optimizations have been carried out for the new wheel. Based on the design and simulation results, the trial production of the iron mold with sand coating for the new wheel has been completed, and the prototype of the wheel has been completed The average weight of the new wheel is reduced by more than 10% compared with the weight of the rolled steel wheel, and the flaw detection shows that there are no defects on the surface of the wheel and inside the spokes.

Verification test of new wheel

The wheel axle assembly test, braking bench test, fatigue bench test and acoustic behaviour test of the new wheel have been carried out, and the performance of the wheel meets the design expectation. Rolling vibration test of the new wheel is being carried out to further study the dynamic characteristics of the new wheel, vibration and noise reduction, wear resistance and other aspects of performance.

4 Conclusion

The National Innovation Center of High Speed Train is developing a new ADI spoke-type wheel. ADI materials for the wheel was developed, and the new spoke-type structure was also developed by topology optimization method, and the casting of wheel samples was completed, as well as the verification tests. The results of the research show that the ADI spoke wheels have certain advantages in lightweight, vibration and noise reduction, wear resistance and other aspects compared with rolled steel wheels. The research of ADI spoke wheels will promote the lightweight work of trains and metros, especially in the higher requirements for vibration and noise reduction in metros. The next step will be to continue to assess the performance of the new wheel, and to promote the railway applications of the new wheel.

5 Acknowledgments (Bold, 10 pt., Arial)

This research was funded by Shandong Province Major Science and Technology Innovation Project Funding (grant number: 2021ZDPT02)， National Key R&D Program of China (grant number: 2022YFB4300100).

Sand CastingProcess for HeavyIron Castings:Front &Rear AxleHousings for Trucks

Gorka PáramoArgote
LoramendiS.Coop.Alibarra 26,01010,Vitoria-Gasteiz,Spain
gparamo@loramendi.com

Abstract: Thisdocumentexploresthebenefitsandvarious applicationsof the CorePackage Systemintheproductionofironcastings. Initiallyrenownedforitsapplicationinaluminumcastingsincethe 1990s,particularlyintheautomotivesectorfor producingengineblocks,thisinnovativeprocesshasnowexpandedtoironcastings,includingcylinderblocksand heads.The latest advancements haveextendeditscapabilitiestoheavycastingssuchasaxlehousingsfor trucks.Furthermore,thetechnologyisevolvingtowards sustainability,shifting from thickercastingwallstothinner,moreefficientones.

Keywords:CPS,core package,sand core,iron casting, rearaxle,frontaxle,housings.

1 Introduction

TheCorePackageSystementailscreatinganentiremoldfromapackageof individualsandcores,includingallthe running and feeding systems.This method is especiallysuitedfor partswithhighly complexshapesandfor medium to high production volumes.

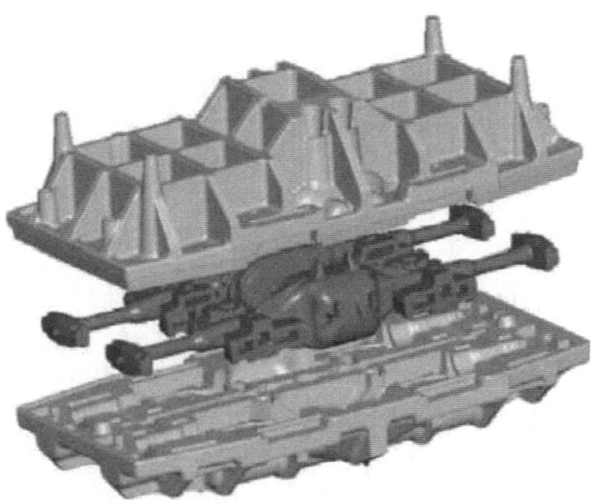

Figure 1 Example of aCore Package System to produce arear axlehousings.

2 Processdescription

ThecastingmanufacturingprocessofCPSforaxlehousingsconsistsof thefollowingstages:
Core manufacturing:
Coreproductionutilizes an automated coldboxtechnologyprocess,operatingonacycletimeof120seconds(upto 1.000kgof sand core).Post-production,the cores are automatically unloaded, painted,dried,andstored,adheringstrictly toLEANmanufacturingprinciples.

Core assembly:
Corepackages are predominantly assembledautomatically.Current productionlinesarecapableofassemblingonecoreorcorepackageevery 120 seconds.Theassemblyprocessmaintainsprecisetolerancesof +/-0.15mm,facilitatingthestraightforwardintegrationof chillersandexothermic materials into thecoreorcore package.

Corepackageinpouringboxes:
Thecorepackageisplacedintoapouringboxand surrounded byloosesilicasand.Through the application ofvibration,the corepackageissecured and compressed,ensuringstabilityand readinesstowithstandferrostatic pressure duringthemetal pouring process.
Pouring:
The metalispouredintothecorepackage,which includesthecompletedesignofthepouringcup, downsprue,runners,and risers.

3 Design characteristics

The characteristics ofthe CorePackageSysteminclude:
1)Complexityof designs:
-Thesystemallowsthecreationofcomplexgeometries, offering significant flexibilitytointegratecastingsandvarioustypesofchannels.

- Itfacilitatestheincorporationoffunctionalparts,therebyreducingthenumberofcomponentsinthefinalproduct.
- Theweightof thecastpartcanbeminimizedthrough optimizeddesign. Additionally, thelowthermalconductivityofsandenablestheachievementof thinnerwall thicknesses compared to other casting methods.

2) Mechanical properties:

- Thesystempermitslocalizedcoolinginareasof highmechanicalstressthroughtheuseofchills,resultinginsuperior materialpropertiessuchastensileandyieldstrength, along with enhanced elongation characteristics.
- Castingscanbeorientedaccordingtothefinalworkingconditions, significantly extendingthecomponent'slifespanbyupto300%, from2million toover6million cycles.

Due to these characteristics, theSandCastingProcessoffersclearcostadvantagesforparts withcomplexgeometries.

4 Advantages of the Sand CastingProcess

Introduction

The sandcastingprocessoffersnumerousbenefitsintermsofcastingquality,flexibility,efficiency,andcost-effectiveness.

Casting quality/ Flexibility:

- Facilitates the integrationof complexgeometries,suchascooling jacketsandoilpassages,alongwithadditional castings.
- Enablestheproductionofcomponentswiththinnerwalls.
- Ensureshighprecision,allowingforcastingorientationaccordingtospecificoperationalrequirements.
- Offerslow-costprototypedevelopmentandhighflexibility in production lines.
- Expandstherangeof applicationsforcastpartswithin the industry.

Casting process:

- Reducescycletime.
- Provideseaseandprecisioninassemblingcoresandintegrating inserts made from various materials..

Investment/ Cost:

- Requires lower investment compared to traditionalgreen sand processes.
- Reducesoperationalandlaborcoststhroughsimplified process automation.
- Presentsaflexibleandcost-effectivesolution.
- Ensuressimpletoolingmaintenance.

Conclusion

Whiletheprocesshas beenestablishedforover30years,evolvingtechnicalrequirements,thedevelopmentof newproducts,andstricterenvironmentalregulationshaveopened up new opportunitiesfor thesandcastingprocess. Theoutstandingdimensionalqualityand possiblecomplexityofthecastingsmakethisprocesssuitedfor themanufacturingof heavyironcastingswithhighlevelofintegratedparts and thinner walls. Additionally, thelevel ofinvestment compared to the traditionalgreensandprocess is lowso it makes thesystemsustainableand withenough flexibility for the new production challenges.

References

[1] AAM, TheAluminiumAutomotiveManual, 2002,p.18-28
[2]Smetan,H.:KernpaketverfahrenimAluminiummotorenguss, MTZ 61, 2000, No.10,P.712-715
[3] Feikus, F.J.: OptimierungeinerAlSi-Gusslegierung und anwendungsorientierte Entwicklung derGießtechnik zur Herstellung hochbelasteterMotorblöcke. Giesserei 88, 2001, No.11, p.25-32
[4]John Campbell, inCompleteCastingHandbook(SecondEdition), 2015
[5]Walton, C. F.,Opar,T. J.:IronCastingHandbook. IronCastingSociety, NewYork,1981.

The Next Generation of Carbon Bonded Filters

Ian Andrews*
Capital Refractories Ltd
*ian.andrews@capital-refractories.com

Abstract: Filtration of molten metal is established practice for many foundries around the world. Filters are used to reduce the risk of non-metallic inclusions being present in the casting and to modify the metal flow – essentially to reduce metal turbulence in the casting cavity.

Traditional filters are generally either foam, pressed, or extruded refractory ceramic products. Each of these filter types offer some advantages and disadvantages. This paper describes a next generation filter that Capital Refractories has recently introduced that offers most of the advantages with less of the disadvantages of traditional filters.

Key Words: Metal Filtration; High Capacity Filters; Steel Filtration; Heavy Iron Filtration; Next Generation Foundry Filters

1 Introduction

There are several types of molten metal filter available to foundries. The most popular "traditional" filters include foam, pressed and extruded. More recently, "Next Generation" filters have emerged thanks to new manufacturing technologies: such as additive manufacture.

2 Different Types of Molten Metal Filters

Foam Filters

Foam filters are manufactured using a ceramic slurry. The slurry is usually an aqueous mixture of the main refractory powders This slurry is then used to coat reticulated polyurethane foam. After coating, the filter is dried and fired. During firing, the polyurethane foam pre-curser will burn out and the strands of the foam filter will be hollow. This process lends itself to utilising a wide range of materials for different alloys types: such as zirconia/alumina or carbon-bonded alumina for steel, silicon carbide for iron and non-ferrous metals.

Foam filters are characterised by their sponge like reticulated structure; this makes for a tortuous path through the filter. Their porosity is expressed as "ppi" (pores per inch) and will range from about 10ppi for coarse filters to about 30ppi for fine filters. Finer porosities are available for more specialised applications.

3 Pressed Filters

Pressed filters are manufactured by pressing refractory powders into a complex mould. This mould contains many pins that form the holes in the filter. The powders are formed around the pins, and these are then extracted to form the filter. The pressed filters are then dried and fired. They are usually made using mullite or alumina with a clay bond.

Pressed filters are characterised by a honeycomb design with circular holes. Their porosity is usually expressed as a cell diameter: e.g., 2.5 mm cell diameter. Unlike foam filters, pressed filters do not create a tortuous path but a simple linear one.

They are generally not as refractory as foam filters and are rarely used for steel but do find considerable use for grey and ductile iron.

4 Extruded Filters

Extruded filters are manufactured by extruding a refractory dough through a die. The dough is usually an alumina-based composition with rheological additives. The extrudate emerges from the die as a continuous "log". This log is then sliced into individual pieces after which they are dried and fired.

Extruded filters are again characterised by a honeycomb design with square holes. They are commonly made from alumina and their porosity is conventionally expressed in CSI (cells per square inch).

Like pressed filters, they are not normally as refractory as foam filters and are mainly used for melting iron and rarely used for steel. The lack of refractoriness also gives a limit to the time of exposure to hot metal.

5 Next Generation Filters

New manufacturing techniques have made new filter structures possible. An example of a new manufacturing technique is additive manufacturing which uses a process in which layers of material are laid down upon the previous layer in such a manner as to create an engineered structure that can be designed to form a tortuous path of differing complexities. They are then dried and fired as per other filter types.

A next generation filter can generally be characterised as having a consistent and repeatable structure (like pressed and extruded filters) but combined with a tortuous flow network (like foam). They are made from various materials such as zirconia, alumina and can be carbon bonded.

6 Desirable Properties in a Molten Metal Filter

There are many desirable traits that an ideal molten metal filter should have. Some of these properties are in conflict with each other. For example, a filter with a very large capacity may not have a very high filtration efficiency. Filters with a very consistent structure (e.g. pressed cellular filters), may not have very good flow modification properties. Ideally a filter should have the following attributes:

Good filtration efficiency
High pour rate
High capacity
Consistent structure
Good flow modification abilities
High strength (hot and cold)
Be cost effective.

7 Filtration Mechanisms

Filters prevent inclusions from occurring in the finished casting by a variety of mechanisms. These include screening, deep bed filtration and flow modification.

Screening is simply the prevention of large pieces of slag or dross passing through the filter due to their physical size compared to the size of the filter cells/pores.

Particles that are smaller than the lattice gap that are able to pass through the initial layers may eventually stopped due to the tortuosity of the internal structure. This is deep bed filtration

This sort of filtration process is realistically only offered by filters that force the metal to take a tortuous path through them such as foam filters and next generation filters.

Flow modification is essentially the reduction of turbulence in the molten metal stream and is arguably the most important attribute of a filter. A reduction in turbulence will reduce the chances of inclusions being generated from re-oxidation, the effects of mould erosion, and other flow related defects such as caused by mould erosion.

8 Attributes of Foam Filters

Foam filters have very good filtration efficiency by virtue of their structure; both screening and deep bed filtration mechanisms will be effective. Additionally, foam filters have a very good flow modification effect due to their tortuous structure. This reduces the kinetic energy of the metal stream and essentially calms the flow.

The capacity of foam filters is limited by the porosity. One of the biggest issues with foam filters is their consistency. Every foam filter will have a unique structure. This results in variation in both capacity and flow rates.

Another issue with foam filters is spalling or friability. This is when small strands of the filter break off; it is possible that these ceramic pieces can end up in the casting.

In terms of hot strength, foam filters are generally good. However, as the filters get larger (200 mm and above) they are more susceptible to breakage.

Fig. 1 – A zirconia foam filter

9 Attributes of Pressed Filters

Pressed filters usually have round holes. Unlike foam filters, they have a very repeatable and consistent structure. However, they do not have as good a filtration efficiency or flow modification effects. They are very robust and have good cold strength. Larger sizes can suffer from hot strength and creep issues.

Fig. 2 – A Pressed Ceramic Filter

10 Attributes of Extruded Filters

Extruded filters are similar to pressed filters except they usually have square holes and a higher open area. They also have similar filtration efficiency and flow modification effects to pressed filters. Again, larger filters can suffer from strength issues at high temperatures.

Fig. 3 – An Extruded Ceramic Filter

11 Attributes of MetCon C – Next Generation Filter

There are now several types of "next generation" molten metal filter available. These filters generally use a form of additive manufacture to produce a more consistent, tortuous structure that can be more customisable in terms of pore size etc. Even though these filters offer an "ideal" combination of physical properties, they are generally very expensive and can often be uneconomic to use.

MetCon C filters are manufactured using a unique additive manufacturing type of process with the aim of offering a cost-effective next generation filter. Whilst they may be slightly more expensive than traditional filters, the improvement in physical properties make them cost effective. Like foam filters, they have very good filtration efficiency due to the tortuous path, good screening and deep bed filtration, and good flow modification effects. However, unlike foam filters, their structure is consistent and repeatable. It is also possible to offer very coarse porosities so very high capacity/flow rate filters are possible. The exceptional hot strength of MetCon C filters makes it possible to make larger filters so that very large castings can be poured directly through a single filter.

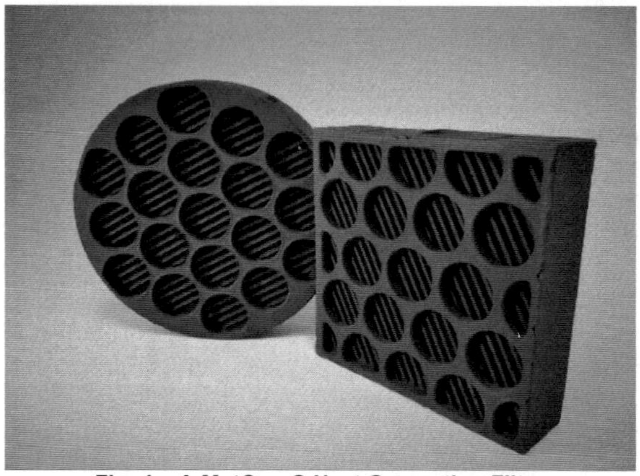

Fig. 4 – A MetCon C Next Generation Filter

High Capacity/High Strength Filters

Foam filters can be limited in their capacity by their porosity. Generally, the coarsest porosity available is 10ppi. There are occasions when a coarser porosity would be desirable such as when pouring manganese steel.

Pressed and extruded filters can have larger hole sizes to increase capacity, but the larger the hole, the more they resemble standard strainer cores – giving very little flow modification and filtration efficiency. Furthermore, large cellular filters can suffer from high temperature creep and lose strength during the pour. This gives limitations on the sizes that can be produced.

MetCon C filters are available in coarse porosities designed for very large castings and/or metals that are difficult to filter such as manganese steel. Because the hot strength is so high, they are available as large filters that can withstand very long pour times and high flow rates. The porosities that are currently available are 3G2S (3 mm gap and 2 mm strand), 4G2S (4 mm gap and 2 mm strand), and 5G2S (5mm gap and 2mm strand).

12 Direct Pour Applications

It is generally advised that direct impingement of the metal onto the filter should be avoided and the filter should be positioned in the runner system so that the metal passes through the filter in a more gentle manner.

However, a pouring/gating method known as direct pour is becoming more and more common. Direct Pour is when a filter/sleeve combination is used to replace a standard gating system. The metal is poured directly on to the filter positioned at the base of a sleeve. This means that the standard runner system can be eliminated. This has the advantage of improving yield and simplifying the gating system. It is not suitable for every type of casting but is good for castings where the sleeve is positioned on a heavy section. Direct pour applications can however be quite limited. For example, it is usual that only one filter is used. If the casting is too large to be filtered with only one filter, then direct pour would likely be unsuitable.

When there is direct impingement, the impact shock on the filter can be very severe. This is especially true when large, bottom pour ladles are used with a large diameter nozzle – or when large lip pour ladles are poured from a great height. The strength of traditional filters is somewhat lacking in applications such as this. Additionally, the capacity of traditional filters can cause issues.

Currently castings of 3.5 tonnes are being poured through a 200 mm 4G2S MetCon filter. Flow rates of 50 kg s^{-1} are being achieved.

New porosities are being produced all the time and it is envisaged that castings over 5 tonnes through a single filter will be achieved in the future with flow rates up to 60-70 kg s^{-1} with the soon to be available 300mm diameter 5G2S porosity.

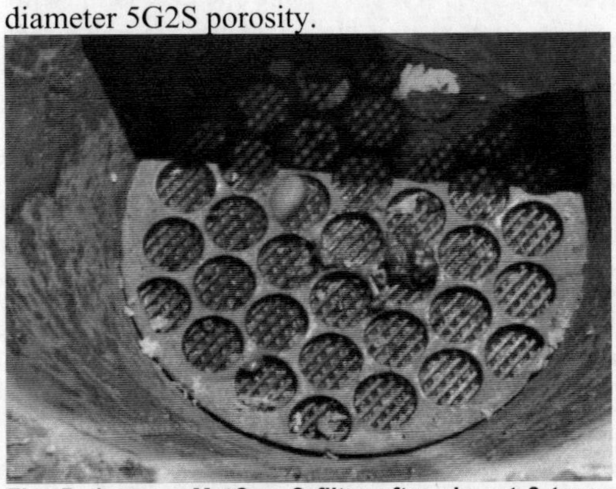

Fig. 5 shows a MetCon C filter after almost 3 tonnes of molten steel have been poured through it. The filter is a 200 mm tapered 3G2S porosity and the pour rate is 50 kg s-1. In this particular application, zirconia foam filters did not have sufficient strength.

Conclusions

Traditional molten metal filters can be effective but have some limitations on their properties. Foam filters are limited by their pore size and have issues related to consistency. Cellular filters are consistent but have issues related to filtration efficiency and flow modification. Many filters on the market suffer from hot strength issues especially for larger sizes. New "next generation" filters can offer the ideal combination of physical properties but, are often expensive and uneconomical to use.

MetCon C filters have been designed to offer the best combination of properties in a cost effective way. They have a consistent and repeatable structure that is engineered to give the best possible capacity and flow rate whilst retaining a good level of filtration efficiency and flow modification. They have excellent cold strength and superior hot strength so they can withstand long pour times with very high flow rates. They are suitable for a variety of metals including carbon steels, manganese steels, chrome irons, and ductile iron.

MetCon filters are increasing the type of castings that can be produced using the "direct pour" filter/sleeve combination pouring technique. Their high strength and high capacity means that they can be used on larger and larger castings even when bottom pour ladles are used

Research on Intelligent Casting Process Design of Rail Transit Steel Castings

Guoxiang Wang[1, 3],*,, Jun Wang[1, 3], Jianzhong Li[1, 3], Xianzhi Li[2], Song Cao[2]
1. School of Materials Science and Engineering, Shanghai Jiao Tong University, No. 800, Dongchuan Road, Minhang District, Shanghai 200240, China
2. CRRC QiShuYan Institute Co.,Ltd., No.258 Wuyi Road,Changzhou City, Jiangsu Province 213011, China
3. Jiangsu Zhongchao Aerospace Precision Casting Technology Co., Ltd., No.16, Changxing Road, Yixing City, Jiangsu Province 214241, China
*Corresponding address: gxwang@sjtu.edu.cn (Guoxiang Wang)

Abstract: Due to the complex structure, large size, lightweight design and economic requirements of key steel castings in rail transit, high-strength casting materials are accompanied by a decline in casting processability, however the casting quality requirement is much higher, which greatly enhances the difficulty of casting process design. On the other hand, there is a relative shortage of experienced foundry process designers. Therefore, the intelligent design of casting process of rail transit parts is the only way to solve this problem. CRRC QiShuYan Institute Co.,Ltd has gradually applied digital casting technology since the 1990s,The integrated application of CAD/CAE/CAM greatly improves the accuracy and efficiency of casting process design. We study the application of AI on steel casting process design, step-by-step, we use while building, on the basis of expert system, we have built a core knowledge database. We study intelligent defect detection of castings. Then tracing the defect results to the corresponding process parameters used through the MES system and AVI system. The next step is to study the combined technology of AI+CAE.

Keywords: AI; rail transit castings; steel castings; casting process design

1 Introduction

As a safe, green, efficient and large-volume mode of transportation, rail transit has been booming in recent years, and the scale of the global rail transit market continues to expand, especially in China, where the rapid growth of the rail transit market is maintained due to the government's strong investment in infrastructure construction. With the implementation of "One Belt, One Road" strategy, rail transit equipment has gradually been applied globally. Rail transit is characterized by safety, reliability, comfort and high-speed. With the demand for higher speed, the safety and reliability of rail transit equipment are facing more difficult challenges. Rail transit casting products, especially key steel castings, are generally more complex in structure, complex and variable in use conditions, and play a key role in equipment, so their castings have more stringent technical requirements.

Due to the complex structure, large size, lightweight design and economic requirements of key steel castings in rail transit, high-strength casting materials are accompanied by a decline in casting processability, however the casting quality requirement is much higher, which greatly enhances the difficulty of casting process design. On the other hand, there is a relative shortage of experienced foundry process designers. The working environment of foundry enterprises is poor, and the salary is relatively low, resulting in fewer employees; Since the 1990s, China's colleges and universities have moved from professional education to general education, and most colleges and universities have canceled casting, and casting is only a part of professional courses. Limited by school hours, casting knowledge cannot be taught systematically and comprehensively, so that graduates have shortcomings in solving technical problems. Therefore, the intelligent design of casting process of rail transit parts is the only way to solve this problem.

2 Result and discussion

CRRC QiShuYan Institute has gradually applied digital casting technology since the 1990s,the application of casting simulation and optimization improves the production efficiency and casting quality. The integrated application of CAD/CAE/CAM greatly improves the accuracy and efficiency of casting process design. With the development of network technology, digital equipment, information management, such technologies improve the automation degree of casting production, improve efficiency and quality. But casting is a complex process, it include the casting process design, modeling, melting, pouring and other processes, involving the process, alloy, equipment, testing and many other aspects. Casting process need to consider metal metallurgy characteristics, mold composition, the heat and mass transfer of the high temperature liquid metal in the cavity, involving metallurgy, mechanics, fluid, thermodynamics, etc., also you need to avoid casting deformation and cracks caused by the complexity of the casting structure. Therefore, the casting process design still has to rely on semi-empirical mode, and the design quality depends on the designer’s experience, so the efficiency is low, and it is more often to have errors.

Large Language models (LLMS) can already use deep learning models trained on large amounts of text data to generate natural language text or understand the meaning of language text. Processing multiple natural language tasks, such as text classification, question and answer, conversation, etc., is an important path to artificial intelligence. We can introduce artificial intelligence technology into the casting process design, carry out casting process identification, understanding, thinking, learning, reasoning, rationality judgment, and finally achieve the best process plan given by artificial intelligence.

CRRC QiShuYan Institute, together with Shanghai Jiao Tong University, carry out intelligent exploration of steel casting process design, take a step-by-step approach, has initially built the core knowledge data base, and classify the products according to the material and structural characteristics. The data include the drawings, technical elements, process design and corresponding defects. All these works help to optimize the subsequent process designing.

Due to the rapid development of the information industry, big data has become a major subversive technological. Among them, Data is the foundation, Data is the food of AI. We collect data from the following three aspects, collect it into the enterprise big data center, organize it, and add it into the enterprise core database of rail transit steel casting parts.

1. Casting defect results, we carry out research on AI identification of casting defects, we have a large number of X-ray, CT, ultrasound, metallography, fluorescence, coloring method detection results, the use of large models to accurately identify the defects.

2. The application of casting tracking technology, MES system (manufacturing execution system) and AVI system (casting automatic identification system) are adopted at the workshop to accurately correspond the casting defects and the process parameters used.

3. Casting simulation, we study the causes of casting defects via the casting simulation; contrast the defects of castings，than the simulation boundary conditions are fine-tuned to make the simulation results more accurate and closer to the reality. Through DOE technology, multiple gating systems are automatically generated to find out the optimal process method. The next step is to study the combined technology of AI+CAE.

3 Conclusion

We study the application of AI on steel casting process design, step-by-step, we use while building, on the basis of expert system, we have built a core knowledge database. We study intelligent defect detection of castings. Then tracing the defect results to the corresponding process parameters used through the MES system and AVI system.

References

[1] Ye Maolin, Yan Dengkun, The Latest research and Exploration on the application of AI in Foundry industry [J]. Foundry Equipment and Processes, 3, 2024.

[2] Xu Jianlin, Wang Zhiping. Artificial Intelligence Technology in Foundry production [J].China Mechanical Engineering, 2002 (12):4.

[3] Liu Dongmei, Sun Runchao, Wang Yunxia, Zhang Jin, Zhang Chunyan, Current Situation and Prospeect of Casting Profssional Talent training in China [J]. Casting, 4/V68,2019

[4] Guojun macro: artificial intelligence to assist people, replace people, become "people". Finance [citation 2023-04-13]

Use of Grain Refiner Flux in Wheel Castings

Hayati Sahin[1], Kerem Can Dizdar[2], Derya Dispinar[1*]
1. Vesuvius Foseco R&D, Enschede, The Netherlands
2. NTNU, Trondheim, Norway
*Corresponding address: e-mail: derya.dispinar@vesuvius.com

Abstract: Al5TiB is the most common grain refiner master alloy used in aluminum castings. The low efficiency, heterogeneity and fading of Ti rods have been a great concern. In this work, 25 wheels were cast (Al11Si alloy) with two grain refiners: Al5Ti1B in the form of rod, and Nucleant 1582 in the form of flux. The wheels were sectioned into three parts: bore, spoke and rim. Samples were machined to check the mechanical properties and grain size measurements. It was found that both Al5Ti1B rods and Nucleant 1582 had revealed quite similar results, although the addition ratio of flux was half of the Ti rod addition. In the case of use of Ti rod, a cleaning flux was also used. By the use of Nucleant 1582, reduction of consumables from 2 to 1 was achieved, because the flux had both cleaning and grain refinement features at the same time.

Keywords: Al11Si; Grain refinement; Flux

1 Introduction

Achievement of finer grains in casting of aluminum alloys known to have positive effects such as increased fluidity, increased feedability, decreased shrinkage, decreased porosity and increased mechanical properties. The process of grain refinement is the heterogeneous nucleation initiation by Al_3Ti, TiB_2 and AlB_2 intermetallic phases peritectically nucleates the α-dendrites. Due to the fundamentals of a peritectic reaction, the solid particles present in the liquid melt have the tendency to settle at the bottom of the crucible during casting operations known as the fading effect [1-2]. In foundries, the economical way is to melt large quantities and cast as many parts as possible. Therefore, the liquid metal is kept for long durations until the castings are complete. It was shown that 20 minutes was enough to start the settlement process of TiB_2 and $TiAl_3$ [3]. Thus, the grain refinement efficiency decreases.

Began [4] on the other hand introduced chemical grain refiner. It was in the form of granulates which was added to the melt during degassing operations. TiB_2 phases were formed in-situ which eliminated the requirement of surface energy and wettability. The risk of presence of possible oxides and impurities as in the case rods or waffles (i.e. quality) were no longer an issue in these fluxes. Additionally, since it was added in the form of flux (Nucleant 1582), it had the advantage of both cleaning the melt and grain refine simultaneously.

2 Experimental procedure

In this work, the effect of two different grain refiners on the microstructure and mechanical properties of Al11Si alloy was investigated. Al5Ti1B rod were added in the ratio of 1 kg/1 ton melt. On the other hand, from the family of Nucleant 158X grain refiner fluxes, Nucleant 1582 was selected for these trials and 0.55 kg/1 ton was added to melt during treatment. Degassing was carried out for 10 minutes with Foseco MTS 1500 system in 900 kg capacity transfer crucible. The chemical composition of the starting melt is given in Table 1. 25 wheels were collected in the sequence of 5 parts. Baker test samples were collected for grain size measurements. Wheels were heat treated for the conditions of 540 °C solutionizing for 6 hours, quenched in water and aged at 145°C for 4 hours. Samples were collected from three regions: bore, spoke and rim; and subjected to tensile testing.

Table 1. Composition of Alloy

Si	Fe	Cu	Mn	Mg	Sr	Ti
10.8	0.13	0.003	0.003	0.14	0.03	0.09

3 Result and discussion

The grain size measurements showed that Al5Ti1B rod added and Nucleant 1582 flux treated melt had quite similar grain size which was 291 and 276 μm, respectively (Fig 1).

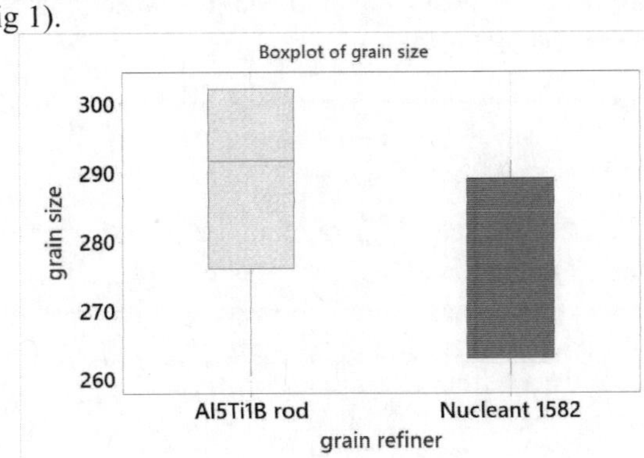

Figure 1. Grain size change with different grain refiners

The strength of Ti-rod and Nucleant 1582 flux treated melts revealed similar values in the different sections of the wheel as seen in Figure 2a. Same scenario and findings were observed for the elongation at fracture values which can be seen in Figure 2b.

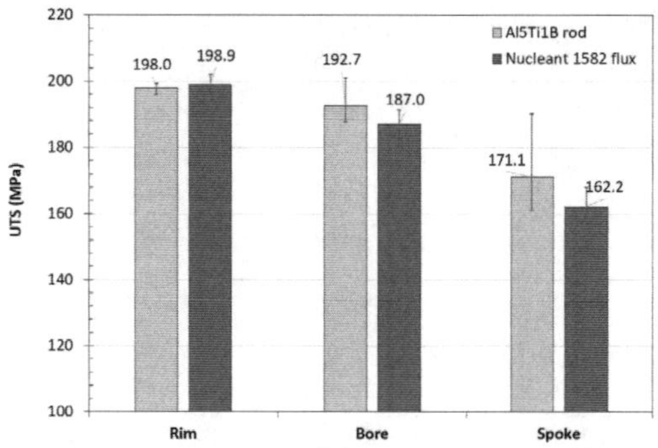

(a)

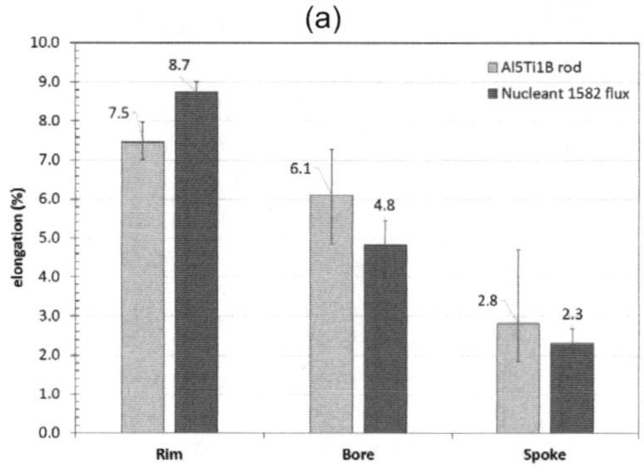

(b)

Figure 2. Mechanical property change by section (a) UTS, (b) elongation at fracture

One of the interesting observations was the change in Ti content of the melt by the casting sequence. At every 5th wheel, an OES sample was collected from the furnace and the results are presented for all three grain refiners in Figure 3. It can be seen that Ti content was decreasing towards the end of the melt when Ti rod was added to melt. On the other hand, when Nucleant 1582 flux was used, the Ti level was not changing and had remained constant through the casting process.

Figure 3. Ti wt% change by duration

4 Conclusion

There is no significant difference between Ti rod and Nucleant 1582 flux treatment of Al11Si alloy on the microstructure and mechanical properties of cast wheels. The addition ratio of the flux is half of the Ti rod addition, and the consumption of products was decreased to 1, instead of two when Ti rod was used (because no extra cleaning flux was added to melt as in Ti rod use). No fading effect was observed when flux was used compared to Ti rod added trials.

Acknowledgments

The authors would like to thank Doktas, Manisa Wheels, Turkey for their support.

References

[1] Easton M, Stjohn D, “Grain refinement of aluminum alloys: Part I. the nucleant and solute paradigms—a review of the literature,” Metall. Mater. Trans. A, 30-6 (1999) 1613–1623.

[2] Easton M, StJohn D, “Grain refinement of aluminum alloys: Part II. Confirmation of, and a mechanism for, the solute paradigm,” Metall. Mater. Trans. A, 30-6 (1999) 1625–1633.

[3] Schaffer PL, Dahle AK, "Settling behaviour of different grain refiners in aluminium." Materials Science and Engineering: A 413 (2005): 373-378.

[4] Began B, Careil P, “Theory and Practice of Grain Refining for Aluminium Alloys - Utilizing COVERAL MTS 1582”, Foundry Practice, 268 (2019) 4-12.

Designing a New High-Pressure Die Casting Al-Si Based Alloy with High Strength for Cylinder Block

Pengyue Wang*, Tianhua Wang, Chenggang Wang, Feng Li, Jingyao Sun, Yongquan Yang, Linghai Hu
FAW FOUNDRY CO.,LTD., Crossing of Hexie Street & Bingwu Road, Changchun 130013, Jilin, China
*Corresponding address: e-mail: wangpy0105@163.com

Abstract: With the improvement of automobile engine power, the strength of engine cylinder block is increased. Herein, a new Al-Si based alloy designed by micro-alloying of Zr and Ti subjected to high-pressure die casting method. The as-cast cylinder block using the new alloy has excellent tensile strength at room, i.e. yield strength of 180 MPa, ultimate tensile strength of 261MPa. The role of Zr and Ti on microstructure and tensile strength was systematically investigated in Al-Si based alloy. The introduced Zr and Ti element can lead to formation of AlSiTiZr phase, which can hinder grain growth and result in a fine grain size. Meanwhile, numerous phases promote the effect of the second phase strengthening. The refined grains as well as AlSiTiZr particles result in the high strength. This work will contribute to improve the strength of engine cylinder block, which makes this alloy attractive for the automotive industry.

Keywords: Al-Si based alloy; Engine cylinder block; High performance

1 Introduction

With the proposal of environmental protection requirements, low-carbon has become one of the main directions for the future development of the automotive industry. Weight reducing of car can reduce carbon emissions. As the most important lightweight material, aluminum alloy has been widely used in the automotive field. At present, the automobile engine block is mostly die-cast cylinder block using $AlSi_9Cu_3$ alloy, which can reduce the weight. However,with the improvement of automobile engine power, the strength of engine cylinder block is increased. Herein, it is urgent to develop new high-performance die casting aluminum alloy for the engine cylinder block.

2 Experimental procedure

The experimental engine cylinder block was manufactured by a 3500-ton die casting machine in accordance with the preset die casting process. The tensile sample was prepared for the cylinder body. The diameter of the sample is 8mm and the length is 70mm.

3 Result and discussion

By adding Zr and Ti elements, the development of new alloys is completed. It can be found from the table that with the addition of Zr and Ti elements, the mechanical properties of the cylinder block have been significantly improved. The introduced Zr and Ti element can lead to formation of AlSiTiZr phase, which can hinder grain growth and result in a fine grain size. Meanwhile, numerous phases promote the effect of the second phase strengthening. The refined grains as well as AlSiTiZr particles result in the high strength.

Table 1. The mechanical properties of cylinder block

Sample	Yield strength/MPa	Ultimate strength/MPa	Elongation/%
sample	180	261	1.2
blank sample	118	205	1.2

4 Conclusion

(1) A new Al-Si based alloy is designed by micro-alloying of Zr and Ti subjected to high-pressure die casting method, which has excellent tensile strength at room, i.e. yield strength of 180 MPa, ultimate tensile strength of 261MPa on the as-cast cylinder block.

(2) The high performance of the new alloy is due to the refined grains and numerous AlSiTiZr particles.

Design of Low-Pressure Casting Process for Aluminum Alloy Reducer Housing

Xiaofang Zhang[1], Mei Yin[2,*], Lan Jiang[2], Yanping Yang[2], Xinyu Zeng[2], Ke Feng[1], Yanran Fu[2]
1.Deyang Yinghe Machinery Manufacturing Co., Ltd., Deyang, Sichuan, 618000;
2. Sichuan Polytechinic University, Deyang, Sichuan, 618000
*Corresponding address: e-mail: bakongbk@qq.com

Abstract: In recent years, with the changing environment and the demand for energy conservation, emission reduction, and environmental protection, the research and development focus of automobile manufacturing enterprises has shifted from traditional fuel vehicles to new energy vehicles. This article takes the middle shell of the HD reducer housing as the research object, explores the typical casting process scheme of the motor housing using low-pressure technology, and evaluates the feasibility of the process scheme using MAGMA simulation software. The results indicate that the produced motor casing fully meets the requirements of the relevant technical conditions.

Key words: reducer shell; low-pressure; casting process for middle shell; CAE

1 Introduction

By simulating and analyzing the casting process, casting defects can be predicted and the feasibility of the process can be evaluated. Using the casting simulation software to simulate and analyze the casting process, as well as validate the process.

1.Technical requirements for reducer housing - middle housing

1.1 Product overview

The product size is 982×545×370 mm, with a main wall thickness of 8mm and a rib plate wall thickness of 4 mm in the middle position. The casting weighs 74.3 kg. The material is ZL101A aluminum alloy.

1.2 Technical requirements

1) After T6 heat treatment, the specified position of the casting requires a tensile strength of ≥ 295MPa and a hardness of ≥90HB.

2 Design of casting process for 3 reducer shell middle shell

2.1 Casting scheme selection

Low pressure casting is used, and sand cores are used to form the sprue and non-moldable positions. The thick parts are cooled by a water circulation system.

2.2 Riser design

Multiple risers are installed at the top and thick parts: the riser function (1) is to supplement and shrink; (2) Beneficial for exhaust and slag removal; (3) Avoid shrinkage and looseness. The diameter and height should be selected between 60-100mm.

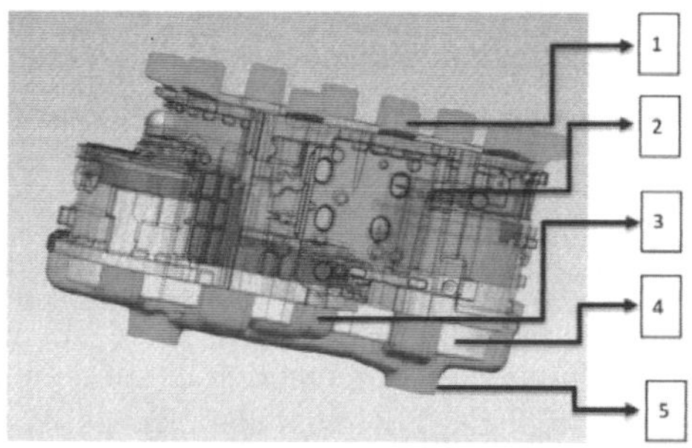

1. Riser; 2. Product; 3. Sprue; 4. Sand core; 5 water inlet

Fig.1 Process Diagram

2.3 Design of sprue

The pouring system adopts a vertical sprue to separate the internal sprue and feed the material, with a sprue diameter within the range of 80-100mm.

2.4 Water cooling design

The water-cooling effect (1) Adjusting the solidification time; (2) Shorten the molding cycle (3) to avoid blockage of the lifting pipe; use a water circulation system for cooling thick parts.

3 Simulation and Analysis of Casting Process

The analysis of the metal liquid flow state and temperature field during the filling process is shown in

Figure 2. The metal liquid flows smoothly and the cooling state is also ideal (Figure 5a-b). The temperature of the metal liquid is also above the liquidus temperature, and the risk of cold shut off or porosity is relatively low.

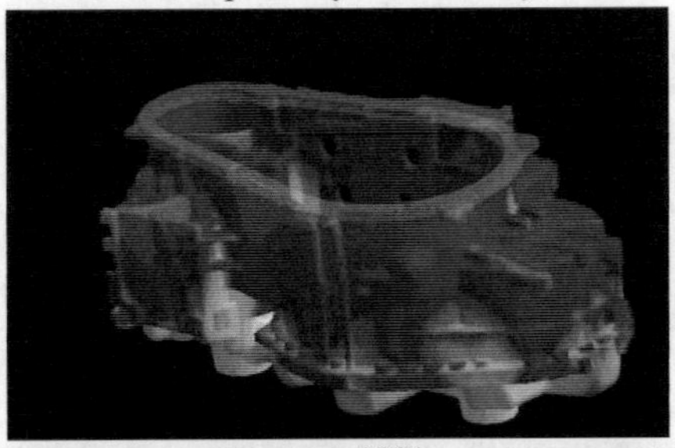

a.10%

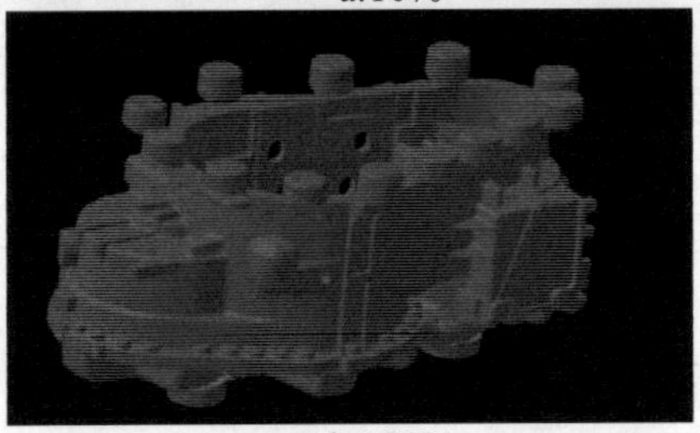

b.100%

Fig.2 Solidification Liquid Phase Ratio during Solidification Process

4 Production validation

The sand core adopts a core assembly method; 70-140 mesh coated sand, metal hot box core making. Rotary degassing method is used for pre furnace treatment.

Pouring parameters: pouring temperature of 700 ℃, liquid lifting time parameters are shown in Table 1, and filling time of 500s. The boosting pressure during the pressure holding process is 150mbar, Open the mold and take out the parts after 600 seconds.

Table 1 Shell filling parameters in

Filling stage	Filling time (s)	Filling pressure (Mbar)
first phase	12	160
phase 2	7	230
Third stage	13	390
Stage 4	3	460
Stage 5	3	550

The actual castings produced have passed the leak test after inspection and processing.

Fig.3 Pouring Products

5 Conclusion

(1) Low pressure process is a shell casting metho d for producing aluminum alloy water-cooled motor shells. If the riser, water-cooled position, and pourin g system are designed reasonably, qualified motor s hells can be produced.

(2) The casting simulation software provides conv enience for product process design, and through risk analysis, casting process design can be evaluated. S hortened the trial production time of the product

References

[1] Design of Control System for Electric Vehicle Motor Spiral Sand Core Shooting Machine by Chen Yuxua n, Han Wei, Liu Chusheng, et al. [J]. Foundry (201 9): 1261-1265

[2] Design of Low Voltage Casting Motor Spiral Water way Sand Core Cleaning Machine by Yan Jian, Dua n Haifeng, Han Wei, et al. [J]. Casting (2019): 291-294

[3] Casting process design and optimization of aluminum alloy motor shells by Shang Xianguang, Hu Yukun, Li Changyun, and others [J]. Hot working process (2012): 64-67/70

[4] Mao Hongkui and Xu Hong's Casting Process Simul ation and Process Design [M].Beijing: National Defe nse Industry Press, 2011

Discussion on Improvement of PT Series Rear Swing Arm Low Pressure Casting Pouring System

Ke Feng[1], Shunshan Wang[2], Tingjia Wu [2], Wang Dai[2], Jiaqing Gan[2], Jun Fu[2], Xiaofang Zhang[1]
1. Deyang Yinghe Manufacturing Co., Ltd., Deyang 618000, Sichuan , China
2. Sichuan Polytechnic University, Deyang 618000, Sichuan , China

Abstract: The PT series rear swing arm low-pressure casting pouring system scheme adopts a metal mold casting for the outer mold and a sand core forming for the inner cavity. It adopts a dual riser and dual gate, and an open pouring system. During small-scale production, after 100% X-ray inspection, it was found that shrinkage defects were concentrated in the overlapping area. The group increased the thickness of the transverse pouring channel of this building from 31mm to 35mm; Fully apply 7029 insulation coating to this pouring channel to reduce mold cooling and ensure smooth filling and contraction channels. Afterwards, the shrinkage ratio significantly decreased, and the 100% pass rate of product X-ray inspection increased to over 95%, proving the effectiveness of this plan.

Keywords: rear swing arm, low pressure casting, gating system, defect analysis

1 Question raising

The PT series rear swing arm products produced by our company require 100% X-ray inspection to be defect free in key areas. During the small-scale production stage, there were defects in the inspection, with a qualification rate of only 75-80%, which could not meet the delivery conditions for batch production.

2 Product conditions

2.1 Product pouring system

This product is produced by aluminum alloy die-casting method, as shown in Figure 1. The outer mold is cast with a metal mold, and the inner cavity is formed with a sand core. It adopts a double riser and double gate, and an open pouring system [1].

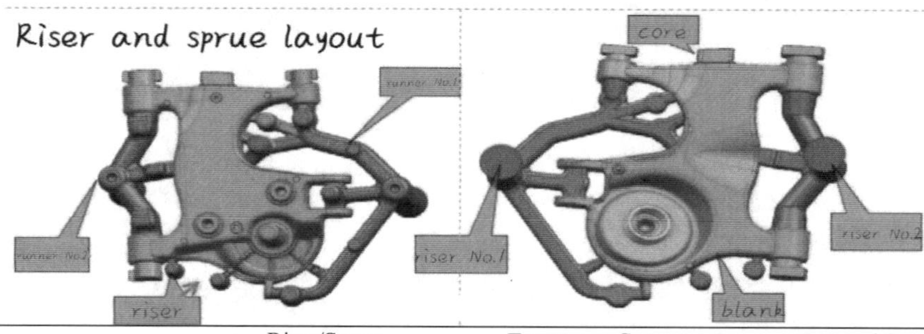

	Riser/Sprue	Transverse Sprue	Internal Sprue
quantity	2	6	11
Total cross-sectional area	2770(choke area)	7250	10250
Design Formula---sprue: transverse sprue: internal sprue=1 : 2.6 : 3.7 (open type)			

Figure 1 Process plan for PT series rear swing arm

2.2 Defects that occur during small-scale production processes

During small-scale production, after 100% X-ray inspection, it was found that shrinkage defects were concentrated in the overlapping areas. The distribution of product waste is shown in Table 1.

The inspection results of the loosening position of the PT series rear swing arm are shown in Figure 2.

3 Analysis and improvement of the causes of defects

3.1 Analysis of shrinkage and looseness at the junction

A typical hot spot is formed at the junction of this joint and the circular wall. During pouring and filling, it is at the end of the runner. After the casting is filled, the temperature at the end of the runner is low, and the filling ability of the runner to this joint is poor, resulting in shrinkage and looseness of the product at this point [2].

Table 1 Inspection Record of PT Series Rear Swing Arm

The Critical area						
Tie Zi Shrink Loose	Bearing hole 1 air hole	Bearing hole 2air hole	Shrinkage and looseness of bearing hole 2	Bearing hole 3 air hole	Large disk	the remaining
55	4	1	1	1	31	24

Non-Critical areas		The Others		Inspection quantity	Qualified quantity	Qualification rate
Disk R porosity	the remaining	Rotten sand	Lack of meat/deformation/cold insulation			
1	2	1	1	621	499	80.4%

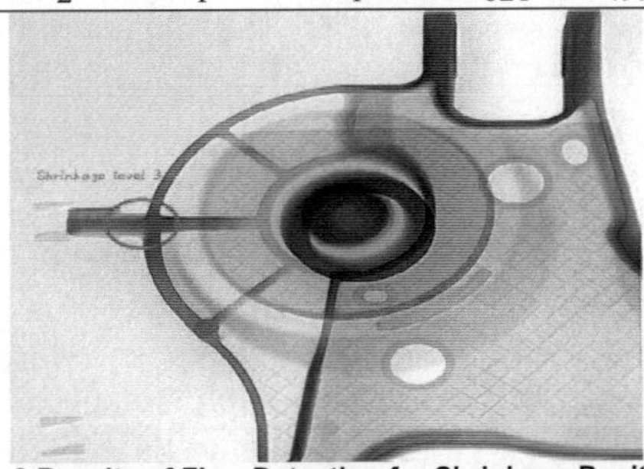

Fig.2 Results of Flaw Detection for Shrinkage Position

3.2 Rectification plan

Determine improvement plan after group discussion:

1) The thickness of the transverse sprue for this scaffold has been increased from 31mm to 35mm, as shown in Figure 4.

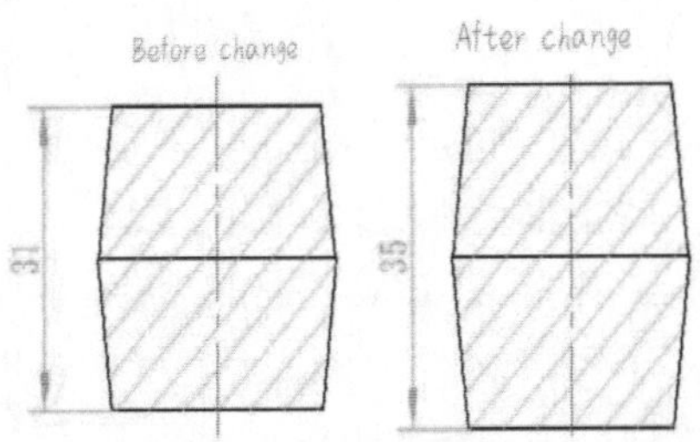

Fig.3: Increasing the thickness of the transverse runner

2) Fully apply 7029 insulation coating to this pouring channel to reduce mold cooling and ensure smooth filling and contraction channels.

Fig.4 Horizontal pouring channel coated with insulation paint

3.3 Verification of Improved Results

After improvement according to the analyzed plan, production verification was carried out, and the shrinkage and looseness ratio of this product was significantly reduced. The 100% pass rate of X-ray inspection of the product was increased to over 95%. After verification, it has been proven that this plan is effective and can meet the batch production conditions of the product [3].

Therefore, improvements have been made to the same series of products according to this plan.

4 Conclusion

(1) Regarding the hot spot shrinkage at the far end of t he casting, the cooling effect of the shrinkage channel ne eds to be considered.

(2) Properly thickening the sprue is beneficial for incre asing the shrinkage effect.

(3) Reasonable use of insulation coatings can effective ly slow down the cooling of the mold runner during the s olidification process.

References

[1] Editorial Board. Casting Handbook (Volume 6, Special Cast ing) [M]. Beijing: Machinery Industry Press, 2021

[2] He Y, Lu D, Li Z, et al. Multi-Objective Optimization of Th e Low-Pressure Casting of Large-Size Aluminum Alloy Wheels through a Systematic Optimization Idea[J]. Materials,2023,16 (18).

[3] Liu B, Kang J, Yang X, et al. Effects of hollow sand mold o n the microstructure and mechanical properties of a low-pressur e aluminum alloy casting[J]. Journal of Materials Research and Technology,2024,28

Advancing Die Casting: Exploring the Benefits and Techniques of Rheocasting

Maria Pammer[1,*], Florian Roeper[2]
1. Chair of Casting Research, Montanuniversität Leoben, Franz-Josef Str. 18, 8700 Leoben, Austria
2. Austrian Foundry Research Institute, Parkstraße 21, 8700 Leoben, Austria
*Corresponding address: e-mail: maria.pammer@unileoben.ac.at

Abstract: Rheocasting is becoming increasingly important in the die casting industry as it offers innovative solutions for the production of die cast parts. One advantage of rheocasting over conventionally produced die-cast parts is the improvement in the material properties of those parts. The semi-solid molten metal allows finer microstructures to be produced, resulting in increased strength, rigidity and durability of the parts.

The difference between conventional high pressure die casting and rheocasting is discussed in this paper. An Al7SiMg alloy was used to produce the cast parts using the RheoMetal process. The casting parameters, particularly in the second phase, were varied. Afterwards, the components were examined using computed tomography and a porosity analysis was carried out. The areas with the largest accumulation of pores were metallographically prepared to determine the differences between metallographic porosity analysis and that from CT.

Keywords: AlSi, rheocasting; RheoMetal; SSM; CT; computed tomography; porosity; microstructure

1 Introduction

Sustainability and the pursuit of a low carbon footprint are central to modern industrial development, influencing various sectors. The potential for lightweight component processing is particularly significant in this context [1]. Therefore, aluminium has become increasingly important in the automotive industry in recent years [2]. To meet the industry's high standards, there is a significant demand for sustainable technologies.

Processing aluminum in a semi-solid state presents additional opportunities and effective alternative methods. These methods generally fall into two categories: thixocasting and rheocasting [3]. Thixocasting involves heating the material from a solid state to a semi-solid state, whereas rheocasting achieves the semi-solid state through controlled cooling and treatment directly from the liquid melt [4]. Rheo processes currently in use include the GISS, RheoMetal, and SEED processes [5]. The RheoMetal process, utilized in this study, stands out because it does not require external cooling of the melt. Instead, the semi-solid state is achieved through enthalpy exchange between the melt and a solidified enthalpy exchange material (EEM) made of the same alloy as the melt.

To produce defect-free parts with good mechanical properties, as demanded bei e.g. automotive industry, minimizing porosity is crucial [6]. Venting to the atmosphere is the most common method for HPDC, but increasing demands necessitate advanced methods like vacuum application [7]. Evacuating gas from the mold cavity can prevent gas pores during mold filling and compress small pores to microscopic sizes during intensification [6]. As Jarfors [8]] indicates, semi-solid processing introduces new possibilities, requiring novel approaches and adaptations to these developing technologies. For this reason, casting parameters must also be adapted to the new technologies, as these have a major influence on the distribution of porosities in the component. It is this aspect that is examined in this paper.

2 Experimental procedure

A test mold was used for this experiment, which is shown in Figure 1. Due to the large gate, it is well suited for rheocasting but is not adapted to this technology. The components are cooled with two different cores, one made of tungsten and one of hot-work tool steel, resulting in different cooling rates. The focus of this work was particularly on different casting parameters in the filling phase and the comparison between conventional high pressure die casting and rheocasting. In detail, only the filling phase of rheocasting was carried out with different parameters. The greatest deviation was shown by the use of conventional high pressure die casting parameters with a piston speed of 2.5 m/s and a filling speed of 0.4 m/s in the filling phase.

After casting, the casting components were separated from the gate and venting system and examined using computed tomography. A small overflow was then prepared metallographically, the CT scan was oriented according to the cutting plane in order to compare the same planes from the polished section and the CT scan. Different regions of interest were also created for different areas of the casting part. However, the analysis parameters per region of interest remained constant between the individual casting parts.

In addition, the chemical composition was analyzed using spark spectrometry. The alloy's chemical composition is shown in Table 1.

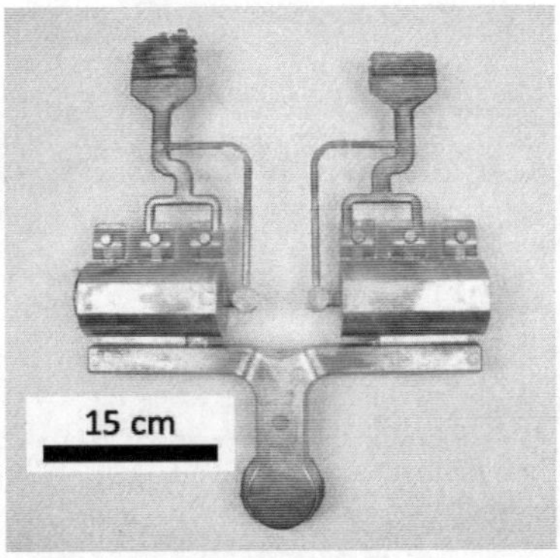

Figure 1: Entire casting used for the tests.

Table 1. Chemical composition of the used alloy

Mg	Fe	Mn	Cu	Al
0.33	0.11	0.024	0.01	Bal.

3 Result and discussion

Computed Tomography

Figure 2 (from left to right) shows the difference in the CT scans between conventional die casting, rheocasting with conventional die casting parameters of 2.5 m/s during the mold filling phase and rheocasting with adapted shot parameters of 0.4 m/s during the filling phase. Rheocasting already changes the morphology of the porosities which become significantly smaller and more homogeneously distributed in the component. However, this also leads to deceptive results, as the largest proportion of porosities, 4.9 % of the total volume, is present in the sample using rheocasting and conventional HPDC parameters, followed by conventional casting. By far, the rheo sample with adjusted HPDC parameters of 0.4 m/s and 1.9% porosity of the total volume performed best.

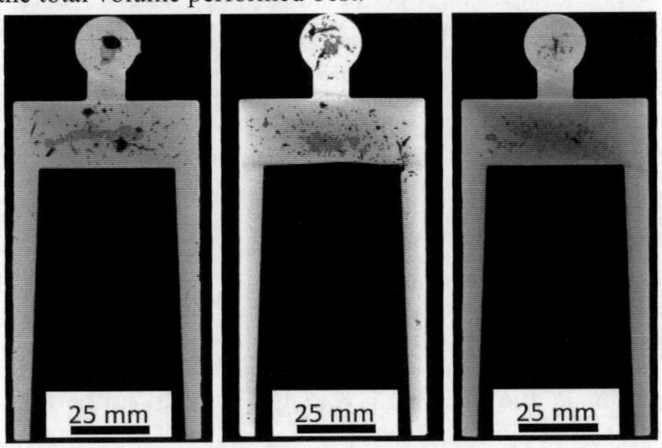

Figure 2: CT images with porosity analyses of (left) conventional die casting, (middle) rheocasting with conventional die casting parameters in the filling phase and (right) rheocasting with adapted filling parameters.

Metallography

The comparison between the microsection plane from metallography and the plane in the CT scan revealed a further deviation between the porosity analyses. While the sharp-edged porosities from metallography showed a porosity percentage of 3.3 %, only 1.14 % could be determined in the same region of interest using the CT image, see Figure 3.

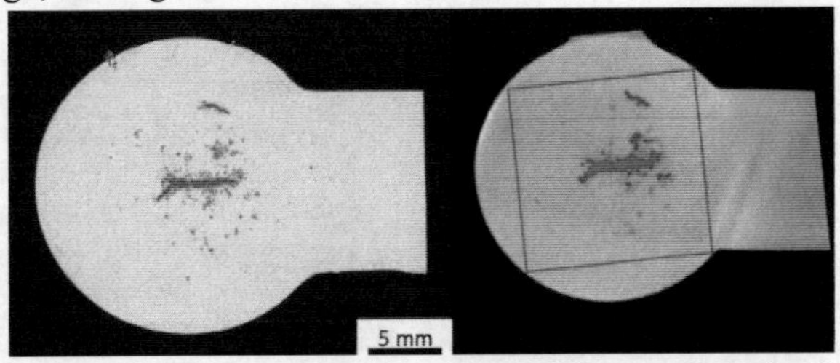

Figure 3: Comparison of the porosity of the same region in (left) metallographic micrograph and (right) CT scan.

4 Conclusion

Rheocasting is an extremely promising technology for the production of complex cast parts. However, many areas of the previously mentioned new technologies have not yet been researched well enough to utilize their full potential. An important aspect in this context is the possibility of adjusting the shot curve in high pressure die casting, which is discussed in this paper. It is clearly shown that an adjustment of the shooting curve is essential in order to use rheocasting appropriately.

Acknowledgments

This project is funded by the Innovation Foundation for Education and is carried out as part of the Innovation Labs for Education program. The authors gratefully acknowledge the financial support from Österreichische Forschungsförderungsgesellschaft FFG (CircRheo project, No. F0999891118).

References

[1] A.C. Serrenho, J.B. Norman, J.M. Allwood, Philosophical transactions. Series A, Mathematical, physical, and engineering sciences 375 (2017).
[2] S. Cecchel, D. Chindamo, M. Collotta, G. Cornacchia, A. Panvini, G. Tomasoni, M. Gadola, Int J Life Cycle Assess 23 (2018) 2043–2054.
[3] K. Anderson, J. Weritz, J.G. Kaufman (Eds.), Aluminum science and technology, ASM International, Materials Park, OH, 2018.
[4] S. Ji, K. Wang, X. Dong, Crystals 12 (2022) 1044.
[5] A.E. Jarfors, A.D. Di, G.G. Yu, J.C. Zheng, K.K. Wang, J. Wannasin, Semi-Solid Processing of Alloys and Composites XII 327 (2022) 287–292.
[6] B.F. Nogovicyn, Theorie und Praxis des Druckgusses, Schiele & Schön GmbH, Berlin, 2011.
[7] Genick Bar–Meir, Fundamentals of Die Casting Design, Free Software Foundation, Inc., Boston, 2012.
[8] A.E.W. Jarfors, Metals 10 (2020) 1368.

Integrated Evolution Process of Large Integrated Castings Based on Digital Simulation Analysis

Moritz Weidt[1], Jinsheng Xu[2], Hang Gao[3], Yibing Fan[4]
1. MAGMA Engineering (Suzhou) Co., Ltd, Suzhou, Jiangsu, China
2. Shenzhen Precisioner Diecasting Mold Co..Ltd, Shenzhen 518110, Guangdong, China.
3. Bohler Special Steels（Shanghai）Co., Ltd, Shanghai, China.
4. Hangzhou Fuxian New Materials Co.,Ltd, Hangzhou, Zhejiang, China
*j.liu@magmasoft.cn

Abstract: This paper explores the application of integrated die casting technology in the automotive industry, focusing on the development and application of heat-treatment-free, high-strength, and high-toughness aluminum alloys to overcome the limitations of traditional materials. These alloys can prevent deformation and surface defects caused by heat treatment during high vacuum die casting, reducing part scrap rates and production costs. The study emphasizes the importance of controlling the eutectic structure and improving impurity tolerance to enhance alloy performance. Using MAGMASOFT® simulation analysis, the design schemes for components such as the front cabin and lower floor were optimized, addressing molding defects and gas porosity issues. Additionally, the paper examines the selection of mold steel and its heat treatment process, which significantly impact mold performance and lifespan. Experimental and numerical simulation results confirmed the adverse effects of bainite content on toughness. Overall, this paper highlights the importance of material development, design optimization, and process control in advancing the industrial application of ultra-large integrated die-cast vehicle body structures.

Keywords: MAGMA; Simulation; Material; GIGA Part

1 Introduction

Since the integrated die casting technology proposed, it has quickly become a focal point in the automotive industry, widely applied in the integrated production of vehicle body structures. The automotive industry's shift towards lightweight and integrated vehicle designs has highlighted the importance of advanced manufacturing technologies. Integrated die casting has quickly become a key focus, offering significant advantages for producing large, complex vehicle body structures. Traditional materials like Silafont-36 alloy, used with high vacuum die casting and heat treatment, face challenges such as deformation and surface defects in large, thin-walled components, leading to increased scrap rates and costs.

To overcome these issues, developing heat-treatment-free, high-strength, and high-toughness aluminum alloys is essential. These alloys need excellent eutectic structure control, high impurity tolerance, strong toughness, good fluidity, and low shrinkage to meet the demands of large integrated castings. This paper explores the development and application of such alloys, optimizing their performance and design through MAGMASOFT® simulation and experimental analysis. The study aims to enhance the efficiency and effectiveness of integrated die casting technology in automotive manufacturing.

2 Experimental procedure

This paper studies and optimizes large integrated die-cast aluminum alloys and mold steels through several key experiments. At first, the heat-treatment-free, high-strength, and high-toughness aluminum alloys were developing, testing various compositions for mechanical properties, fluidity, and shrinkage rates. Using MAGMASOFT® simulation, it refines the design of components like the front cabin and lower floor to address potential defects. Then, different mold steels and their heat treatment processes, assessing their hardness, toughness, and thermal stability. By analyzing molding defects with optical and SEM methods and optimizing process parameters, it aims to reduce defects. Lastly, mechanical performance tests using DEFORM software simulations confirming the alloys' and steels' suitability for large-scale integrated die casting in the automotive industry.

3 Result and discussion

The development of heat-treatment-free aluminum alloys

The experiments successfully developed high-strength, high-toughness heat-treatment-free aluminum alloys, addressing traditional materials' limitations in large integrated die casting. Using efficient modifiers improved impurity tolerance, with tests confirming superior mechanical properties, fluidity, and shrinkage rates suitable for large structures.

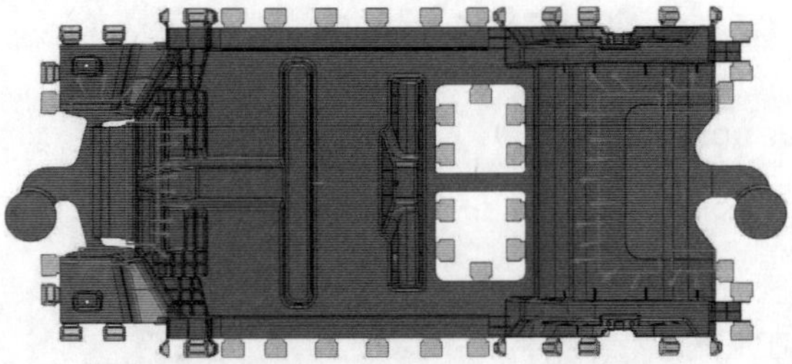

Fig.1 Design of Integrated lower floor

Design optimization

Design optimization using MAGMASOFT® simulations refines the front cabin and lower floor components. Adjusting thickness, filling time, and gate velocity to resolve defects and gas porosity issues, significantly enhancing molding quality and efficiency.

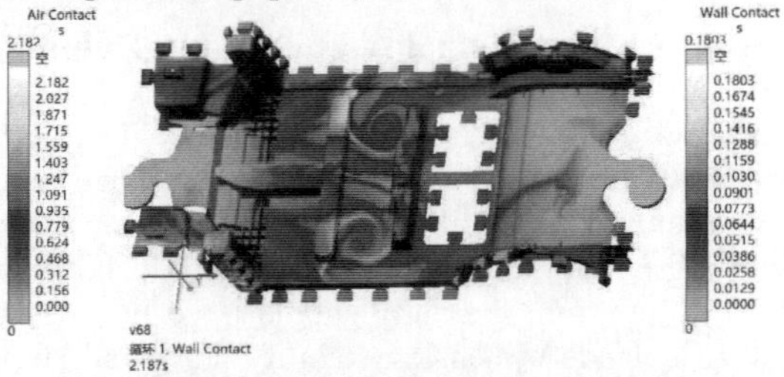

Fig.2 Simulation of the casting process

Mold steel selection and heat treatment

Mold steel selection and heat treatment identified suitable hot work mold steels, with experiments showing that proper quenching and tempering improved hardness, toughness, and thermal stability, extending mold lifespan.

Molding defect analysis

The problem areas were identified by molding defect analysis using optical and SEM methods to optimize process parameters and reduce defects. Mechanical performance tests on mold steel samples showed that cooling rates affected phase composition and material performance. Higher bainite content reduces toughness.

Mechanical performance testing

The problem areas were identified by molding defect analysis using optical and SEM methods to optimize process parameters and reduce defects. Mechanical performance tests on mold steel samples showed that cooling rates affected phase composition and material performance. When the bainite content is higher, the toughness decreases.

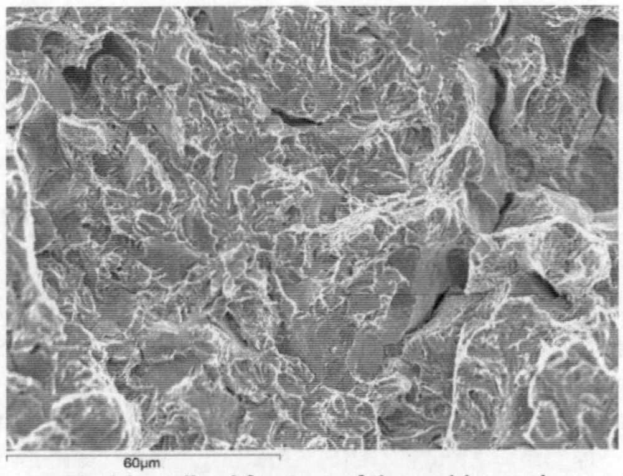

Fig.3 Localized fracture of the mold sample

4 Conclusion

The development of heat-treatment-free aluminum alloys successfully addresses the limitations of traditional materials in large integrated die casting, offering superior mechanical properties, fluidity, and shrinkage rates. Design optimization using MAGMASOFT® simulations significantly improves the quality and efficiency of components like the front cabin and lower floor. The selection and heat treatment of suitable hot work mold steels extend the lifespan of the molds by enhancing hardness, toughness, and thermal stability. Molding defect analysis and mechanical performance testing, including the impact of cooling rates on phase composition, provide critical insights for reducing defects and ensuring the reliability of die-cast components. Overall, these findings support the application of large-scale integrated die casting technology in the automotive industry, enhancing component quality and laying the foundation for future advancements.

Acknowledgments

Thank you for shenzhen precisionerdiecasting mold co..ltd, bohler special steels（shanghai）co., ltd, hangzhou fuxian new materials co.,ltd cooperation and support.

References

[1] www.magmasoft.cn.

[2] Conley J G, Huang J, Asada J, et al., et al .Modeling the effects of cooling rate, hydrogen content, grain refiner and modified on microporosity formation in Al A356 alloys [J]. Materials Science and Engineering，2000，A285：49-55.

Ultra-Large Aluminum Castings in Automobiles

Qigui Wang*, Andy Wang, Jason Coryell
Advanced Materials Technology & Virtualization
GM Global Technical Center, Warren, MI USA
*Corresponding address: e-mail: qigui.wang@gm.com

Abstract: Ultra-large aluminum castings are increasingly used in automobiles, particularly in electric vehicles for light-weighting and manufacturing cost reduction. As most of them are structural components subject to both quasi-static, dynamic and cyclic loading, the quality and quantifiable performance of the ultra-large aluminum castings is critical to their success in both design and manufacturing. This paper reviews applications of ultra-large aluminum castings in automotive industry and outlines their advantages and benefits as well as challenges. Factors affecting quality, microstructure and mechanical properties of the ultra-large aluminum castings are evaluated and discussed.

Keywords: Ultra-Large Castings; Aluminum; Light-weighting; Quality; Microstructure; Materials Properties

1 Introduction

According to Ducker [1], applications of aluminum castings in automobiles have grown up by more than 300 pounds per vehicle (100%) in past twenty years. Recently, automakers have introduced ultra-large aluminum castings in critical structures to integrate tens or hundreds of different parts into a single piece casting. For instance, Tesla uses front and rear aluminum giga castings in the model Y vehicles [2]. GM has used 6 mega aluminum castings to form the entire lower body structure for the Cadillac Celestiq vehicle, Fig. 1 [3]. Toyota plans to implement giga castings to significantly reduce the number of parts used in its front and rear body frames [4]. Many other automakers are taking the similar approach.

Ultra-large aluminum castings require good quality, high properties, and predictable performance. These goals are made more challenging by the complexity of ultra-large casting geometry and subsequent processing. The aluminum casting quality and final product performance is determined by alloy composition, melt treatment, casting and gating system design, and particularly casting process. Therefore, it is critical and economically important to design and manufacture the ultra-large castings correctly.

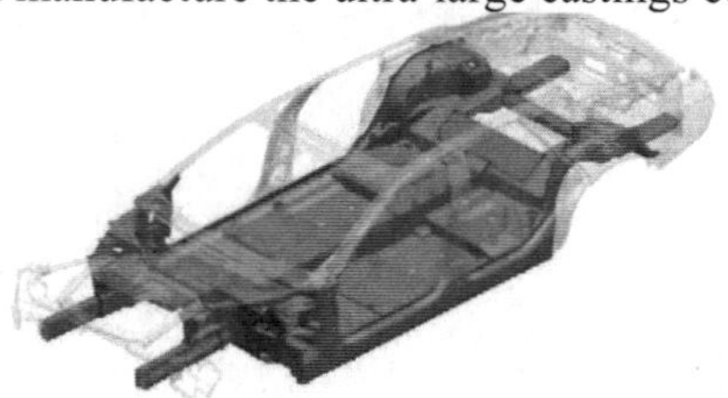

Fig.1. Six mega Al castings in Cadillac Celestiq [3].

2 Opportunities and Challenges

Ultra-large aluminum castings offer great opportunities and benefits including reducing number of parts (60+), saving tooling cost (>40%), saving energy (>30%), saving mass (>30%), reducing lead time, and broadening applications. There are also challenges associated with the ultra-large aluminum castings such as dimensional stability, sustainability, repairability and serviceability, and particularly difficulty in making quality casting.

3 Key factors affecting quality of ultra-large aluminum castings

Cast aluminum alloy

With the long metal flow length in the ultra-large casting, cast aluminum alloy shall have much better castability including high fluidity and low shrinkage and hot tearing tendency. In addition, the alloy should have limited aging response to balance dimension stability and yield strength requirements. Currently, C611 and Aural 5 alloys are mainly used in high pressure die casting (HPDC) giga castings. Both alloys have about 7% Si and low Fe (<0.2%). Tesla uses AA386 alloy which has high Fe content (<0.5%) and small amount of Cu (up to 0.8%).

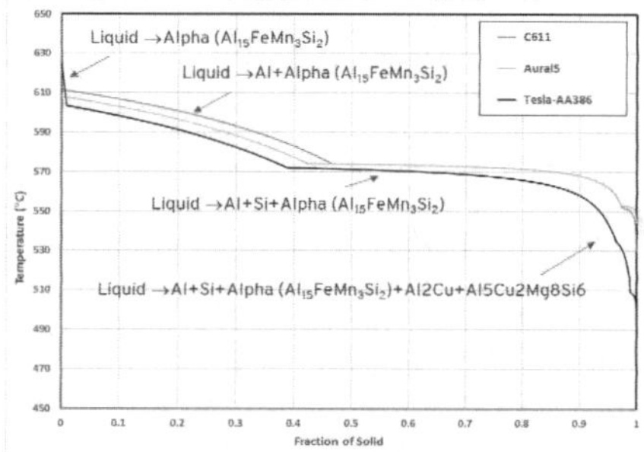

Fig.2. Calculated fraction of solid and solidification sequences of three alloys.

The Cu in the alloy increases freezing range (Fig. 2), porosity and crack susceptibility coefficient. It also decreases corrosion resistance. GM developed a sustainable alloy called UniCAST which is good for structure casting [5].

Melt cleanliness control

Liquid aluminum cleanliness is vital to produce high-quality castings since most defects in final casting are usually related to inclusions and gases coming from the liquid metal. The liquid metal should be cleaned to the highest level possible before it is introduced into mold cavity. One technology was reported from GM to reduce liquid metal contamination with scrap return charge [6]. The method includes coating the scrap charge with a layer of flux that can help remove oxides from the scrap charge surfaces and meanwhile provide a cover flux to protect the melt bath from oxidation.

Gating system design

The optimal design of gating system can significantly reduce the turbulent flow of the molten metal, minimizing the amount of gas and trapped impurities. For production of quality aluminum castings, the design of a naturally pressurized gating system shall be used. Melt velocity at any time during mold filling should be kept below the critical velocity of 0.5 m/s. For HPDC, however, mold filling needs to be completed within 100-200 milliseconds to be able to fill up the thin walls. The minimum melt velocity at ingates is usually controlled around 40 m/s, which is 80 times greater than the critical velocity. Therefore, it is vital to correctly design HPDC gating system and particularly ingate locations and sizes to avoid converging melt fronts.

Mold surface treatment

Mold surface condition determines not only the quality of the surface finish but also the quality of the casting skin layer that can significantly affect casting durability performance as fatigue cracks usually form from the casting surfaces. Castings with relatively large flat surfaces can experience excess surface appearance and quality issues due to the formation and entrapment of young oxides in the melt front during filling. Mold/die surface should be treated with unique surface patterns or texture so that the oxide film in the metal front can be peeled off during mold filling, improving surface quality.

Casting process

Casting process optimization and control is so important to the casting quality particularly for the ultra-large aluminum castings. Casting process parameters such as mold/die temperature, melt pouring temperature, and shot profile should be aligned with the gating system design for optimal results as a given gating design literally determines process parameters.

For a given size of casting, there is an ideal/optimal mold temperature to achieve a good combination of best casting quality and prolonged die life although it is generally accepted that the ideal die temperature should be kept about one third of the melt pouring temperature. The actual ideal mold temperature should be determined and optimized using a casting process simulation code incorporating with a thermal imaging technique.

Pouring temperature influences both metal flow and casting quality and mechanical properties. Although typical pouring temperature is adapted to be around 50-100 °C higher than the alloy liquidus, comprehensive casting process simulations are needed to determine the optimal pouring temperature, particularly for the ultra-large castings due to longer metal flow distance.

In HPDC, shot profile is very critical as it determines final casting quality such as trapped air and oxide entrainment. Fig. 3 compares the calculated accumulation of trapped air for four combinations (two gating designs and two shot profiles). The delayed plunger fast speed stage and pressurization (shot profile 2) significantly reduces the trapped air with gating design 2.

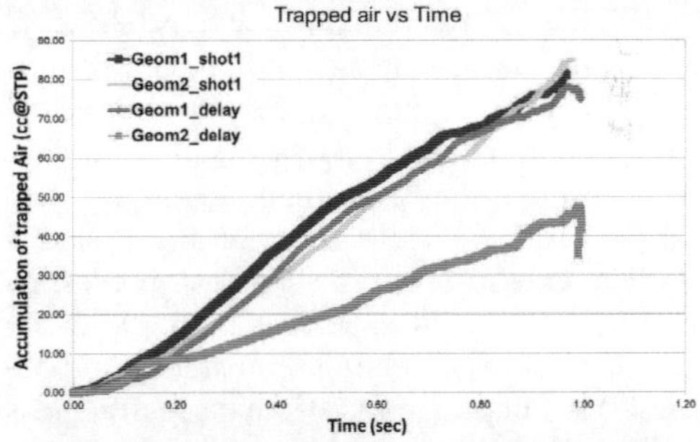

Fig.3. Accumulation of the trapped air calculated for different shot profiles.

4 Conclusion

Application of ultra-large aluminum castings in automobiles has brought great opportunities and posed challenges that can be addressed with the advanced metal casting technologies and virtual casting tools.

References

[1] Ducker Frontier, North America Light Vehicle Al Content and Outlook. Final Report Summary: July 2020.
[2] https://www.theautopian.com/how-tesla-made-gigacasting-the-most-important-word-in-the-car-industry/ (Sept 14, 2023)
[3] https://insideevs.com/news/617108/cadillac-takes-page-from-tesla-book-uses-mega-castings-on-celestiq/ (Oct 18, 2022)
[4] https://www.linkedin.com/pulse/what-do-toyotas-announcements-battery-technologies-giga-castings/ (Aug. 2023)
[5] Zhan, H., G. Zeng, Q. Wang, C. Wang, P. Wang, Z. Wang, Y. Xu, D. Hess, P. Crepeau, J. Wang, Unified Cast (UniCast) Aluminum Alloy Journal of Materials Science & Technology, 154 (2023) 251–268
[6] Wang, Q., D. Gerard, US Patent 9574252 (Feb 21, 2017)

Research on Local-Loading and Forced-Feeding Squeeze Casting of Aluminum Alloy Flywheel Housing with Large Size and Complex Shape

Jufu Jiang*
School of Materials Science and Engineering, Harbin Institute of Technology, Harbin 150001, China;
*Corresponding address: e-mail: jiangjufu@hit.edu.cn

Abstract: In this paper, in order to meet the forming and manufacturing requirements of large and complex aluminum alloy automotive load-bearing components with large wall thickness difference, 12500 kN low-cost and high-efficiency large aluminum alloy squeeze casting equipment was developed by combining direct injection technology, general hydraulic technology and quantitative pouring technology. The equipment reduced 60% cost than foreign equipment. Accordingly, the squeeze casting technology with local-loading and forced-feeding was proposed. The finite simulation results showed that the forming and solidification process could be divided into four stages: low resistance channel flow, high resistance horizontal area filling, height direction filling and the last difficult filled area filling. The optimal process scheme determined by orthogonal experiments of 18 kg flywheel housing component was pouring temperature of 655 °C, B scheme mold, holding time of 20 s and specific pressure of 34 MPa. The orthogonal experiments of 20 kg flywheel housing showed that the decreasing order of influence on mechanical properties was mold temperature, extrusion pressure, pressure delay, pouring temperature and local pressure. The tensile strength and elongation of 35 kg flywheel housing components with or without local pressure application were 205 MPa, 215 MPa and 6.02%, 14.2%, respectively, significantly increasing elongation by 135.9%. The ultimate tensile strength, yield strength and elongation of 35 kg flywheel housing component after heat treatment reached 306.1 MPa, 237.5 MPa and 13.29%, respectively. Under the optimum heat treatment parameters, numerous fine dispersed acicular β'' phase particles were precipitated. Moreover, β'' phase had significant precipitation strengthening effect, which caused lattice distortion, hindered dislocation movement, and improved mechanical properties.

Keywords: Local-loading; Force-feeding; Squeeze casting; Aluminum alloy; Flywheel housing.

1 Introduction

With the rapid development of advanced rail transit, new energy vehicle and aerospace, high performance and quality, and related lightweight component manufacturing technology of aluminum alloy is becoming a major strategic and urgent need of the country. For example, large aluminum alloy flywheel housing with large wall thickness difference and complex shape is attracting widespread attention [1]. Accordingly, squeeze casting with local-loading and forced-feeding was proposed. As shown in Fig. 1, 12500 kN large-scale, low cost and high-efficiency squeeze casting equipment was developed including key components designing, main frame designing, hydraulic designing, equipment and molds designing and manufacturing.

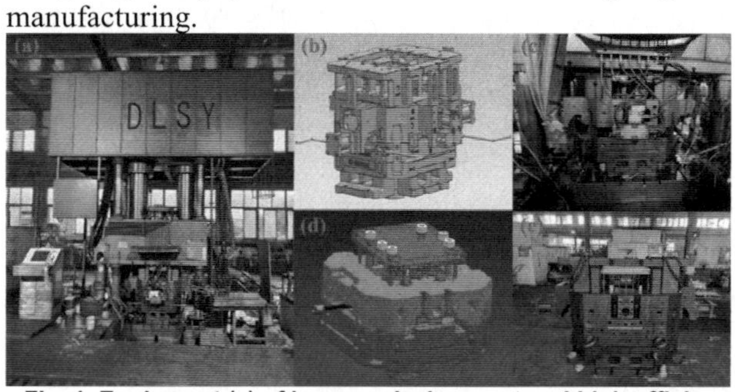

Fig. 1. Equipment (a) of large-scale, low cost and high-efficiency squeeze casting, squeeze casting design drawing (b) and mold (c) of 18 kg aluminum alloy flywheel housing, squeeze casting design drawing (d) and mold (e) of 35 kg aluminum alloy flywheel housing.

2 Experimental procedure

In this paper, the squeeze casting technology of local loading and forced feeding was put forward by combining finite element simulation technology with hydraulic loading technology. Microstructures and fracture morphology were characterized by optical microscope and scanning electron microscope. Elements distribution and the precipitation phases were characterized by Talos F200X transmission electron microscope. In addition, the mechanical properties were measured by Instron 5569 testing machine with the rate of 1 mm/min.

3 Result and discussion

As shown in Fig. 2, simulation and forming processes were mentioned and the mechanical properties were displayed. The finite simulation results showed that the forming and solidification process of flywheel housing castings could be divided into four stages: low resistance channel flow, high resistance horizontal area filling, height direction filling and the last difficult filled area filling. Moreover, temperature field, locations and distributions of shrinkage and porosity defects, effect stress zones and fraction solid were explained. Fig. 2(b) and (c) illustrated that there was no obvious macroscopic cracks and other serious defects, and the silvery white surface has further proved the outstanding quality during forming process. In addition, Fig. 2(d) shows the average mechanical properties of 18 experimental samples under different parameters. The optimal process scheme determined by orthogonal squeeze casting forming experiments of 18 kg flywheel housing component. The macro morphology of cross section at the same position with local loading and without local loading was observed, at the same time. Heat treatment (including solution and aging treatment) was applied to improve the mechanical properties.

Fig. 2. Results of simulation (a), macroscopic morphology of insufficient filling process (b) and complete filling process (c) and mechanical properties of forming samples under different parameters (d)

The orthogonal squeeze casting experiments of 20 kg flywheel housing showed that the mold temperature was the most significant technological factor affecting the mechanical properties, and the primary and secondary order of influence on mechanical properties was mold temperature, extrusion pressure, pressure delay, pouring temperature, local pressure. The tensile strength and elongation of 35 kg flywheel housing component with or without local pressure application were 205 MPa, 215 MPa and 6.02%, 14.2% respectively. In addition, the elongation was significantly increased by 135.9%. The ultimate tensile strength, yield strength and elongation of 35 kg flywheel housing component after heat treatment reached 306.1 MPa, 237.5 MPa and 13.29%, respectively.

Fig. 3(a) illustrated the example and range of parameters exploration under heat treatment. Fig. 3(b) and (c) shows the distributions of elements and precipitation phases to explain the distribution and strengthening mechanism. A large number of fine-dispersed acicular β″ phase [2] particles were precipitated. At this condition, β″ phase was the main strengthening phase, which caused lattice distortion, hindered dislocation movement, improved mechanical properties, and has significant precipitation strengthening effect [3]. Furthermore, Fig. 3(d) shows that the mechanical properties of 35kg flywheel housing with local loading position for ultimate tensile strength, yield strength and elongation after T6 heat treatment are 306.1MPa, 237.5MPa, and 13.29%, increased by 93.0MPa and 138.0MPa and 1.6%, respectively.

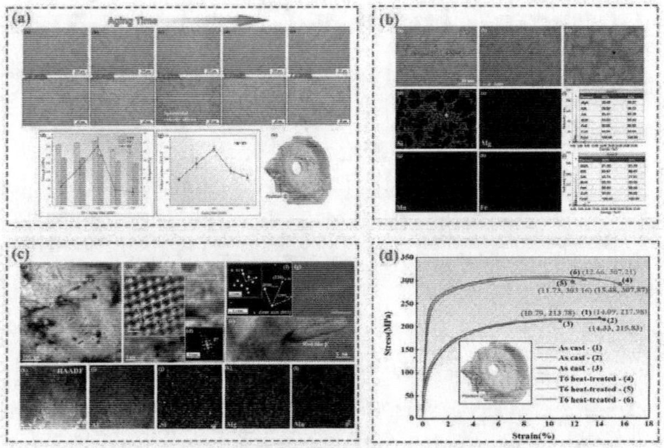

Fig. 3. Results of 35kg flywheel housing performances microstructure under different aging treatment parameters (a), element maps (b), distribution and characterization of precipitation phases (c) and stress-strain relationships (d) stress-strain curve with or without heat treatment.

4 Conclusion

(1) A technology of squeeze casting by local loading and forced feeding to meet the forming and manufacturing requirements of large and complex aluminum alloy components with large wall thickness differences was proposed.

(2) A large-scale aluminum alloy squeeze casting equipment, achieving a cost reduction of over 60% compared to foreign equipment was developed.

(3) The squeeze casting technology by local loading and forced feeding was applied in 18, 20, and 35kg aluminum alloy automobile flywheel housing with large size

Acknowledgments

This work is supported by the National Natural Science Foundation of China (NSFC) under Grant Nos.U2241232, U2341253 and 52375317 and the National Key R&D Program of China (No.2022YFB3404204).

References

[1] P. X. Qi, Survey of squeeze casting technology development home and abroad, Spec. Cast. Nonferrous Alloy. 2 (2002) 20-23.

[2] J. H. Peng, H. X. Chen, S. C. Wang, K. H. Zheng, W. J. Qi, Effect of heat treatment on microstructure and tensile properties of A356 alloys, Trans. Nonferrous Met. Soc. China, 2011, 21: 1950-1956.

[3] X. H. Zhou, Y. P. Sun, Y. Y. Zhang, J. Fu, Effect of solution and aging on microstructure and properties of 7055 aluminum alloy, J. Guangxi Univ. Sci. Technol. 27 (2016) 81-86.

Impact of HPDC's Core Temperature on Internal Homogeneity of the EC Electromotor Housing Casting

Marek Matejka[1,*], Dana Bolibruchová[1], Peter Oslanec[2], Radka Podprocká[3]
1. University of Žilina, Department of Technological Engineering, Univerzitná 1, 010 26 Žilina, Slovakia
2. Institute of Materials and Machine Mechanics, Dúbravská cesta 9/6319, 845 13 Bratislava, Slovakia
3. Rosenberg-slovakia S.r.o., Kováčska 442, 044 25 Nižný Medzev, Slovakia
*Corresponding address: e-mail: marek.matejka@fstroj.uniza.sk

Abstract: High pressure die casting (HPDC) is one of the most productive casting methods to produce a wide range of aluminum components with high dimensional accuracy and complex geometry. The process parameters of high-pressure casting generally directly affect the resulting quality of the castings, such as the presence of pores in the casting or the microstructure. These defects can lead to a significant reduction in mechanical properties. The aim of the presented article is to describe the effect of the temperature the high-pressure mold on the presence and distribution of porosity and the microstructure of the aluminum casting in two geometric variants. The temperature of the core was changed due to the use of two flowing media in the thermoregulation circuit, i.e. demineralized water and heat transfer oil. The change in core temperature fundamentally influenced the presence of gas porosity, microporosity and microstructure changes in the evaluated area of the high-pressure die casting.

Keywords: HPDC; Al-Si-Cu alloy; Porosity; Microstructure

1 Introduction

High-pressure die casting is a complex process that considers a number of factors that significantly affect the final quality of the cast part. The basic parameters of the high-pressure die casting process are mainly the maintenance temperature, casting and molds temperature, hydrostatic pressure, pressing pressure, and alloy flow rate in the inlet channel. The mechanical properties of a HPDC product are primarily related to the temperature of the mold, the speed of the metal at the inlet and the applied casting pressure [1].

The combination of mold temperature, the fluidity of the molten metal, the complexity of the part geometry, and the cooling rate during high-pressure die casting all affect the integrity of the cast component. If these factors are not properly controlled, we can expect the appearance of various defects in the final casting. The thermal profile of the tool during operation is another important aspect of producing high quality components. Mold temperature is a key factor affecting heat removal from the molten metal, as well as for filling the mold and for correct setting of the casting properties [2,3].

The key factors for effective temperature control of the die are the temperature control unit, suitable heat transfer medium and tempering circuits in the mold die. The heat-transfer medium plays an important role in ensuring optimal temperature regulation. The better the heat transfer properties of the used medium, the more efficiently it is possible to transfer a large amount of heat [5].

2 Experimental procedure

In the experimental part, an aluminum alloy AlSi12Cu1(Fe) was used, the chemical composition of which is shown in Table 1. For experimental purposes, a casting marked Statorbuchse EC75 and EC55 was used in two height variants of the tube part (Figure 1). EC 75 and EC 55 castings are part of the EC electromotor.

Table 1. Chemical composition of the AlSi12Cu1(Fe) alloy

	Si	Fe	Cu	Mn	Mg	Cr	Ni	Zn
wt.%	11.4	0.69	0.89	0.32	0.07	0.01	0.07	0.34

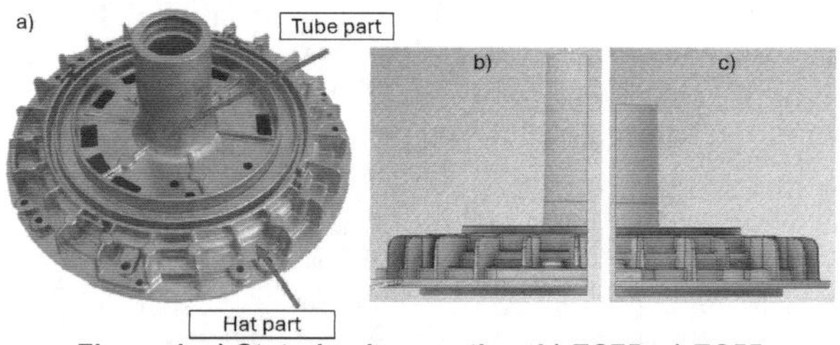

Figure 1. a) Statorbuchse casting; b) EC75; c) EC55

Both variants of castings were cast with a change of the medium flowing in the thermoregulating circuit of the mold in the core part, namely water and heat transfer oil - ITERM 6MB. Other machine parameters did not change during the casting process.

3 Result and discussion

Simulation ProCast software

From the images of mold cycling, it can be observed that oil-tempered cores reach higher temperatures than water-cooled cores (Figure 2). The use of oil also caused a higher temperature of the subsurface layers to a temperature of approximately 380°C at the time of opening the mold, which reduced the cooling rate of the solidification melt.

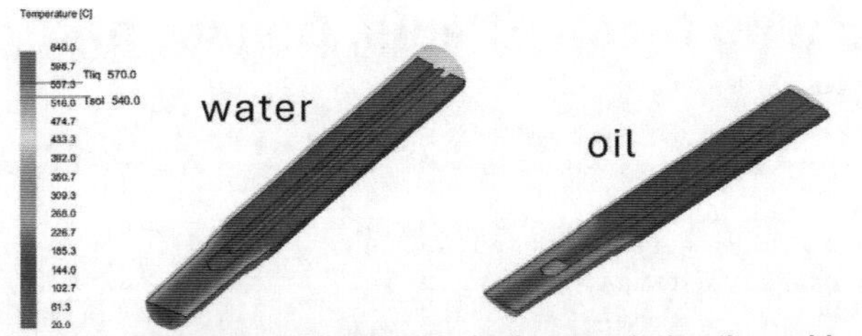

Figure 2. Temperature distribution in the core during the mold opening phase within the tenth cycle for EC 55

Porosity evaluation

The presence of pores in the castings was evaluated by X-ray and computed tomography (CT). The tube part of the casting, which represents the critical area for the formation of pores, was evaluated. When comparing EC 55 castings, we see that the number of pores when using tempering oil is smaller compared to water-cooled castings. By observing castings with a larger tube (EC 75) that have been tempered with oil, we see that the porosity presents in the volume of the tube. In castings that were cooled by water, the number and size of pores increased significantly. The range of pores found in EC 75 castings is significantly larger than in EC 55 castings. In EC 75 castings that were water tempered, the pore size also increased (Figure 3).

Figure 3. Porosity evaluation: a) CT; b) X-ray

Microstructure analysis

The microstructure is formed by the eutectic (dark gray color) distributed in the α-Al phase (pale gray color) and intermetallic phases based on Fe and Cu. When taking a closer look at the morphology of eutectic Si, a clearly visible change is the effect of the used medium in the thermoregulation system of the core (Figure 4). The morphology of the eutectic in the water-cooled casting is observed in the cutting plane as various round grains and lamellae, which in the 3D morphology are observed as rods. The morphology of the eutectic in the tempered casting is in the form of sharp-edged lamellae (needles) with an uneven surface (protruding steps), which are related to the twinning of the basic Si crystalin 3D morphologies observed as hexagonal plates.

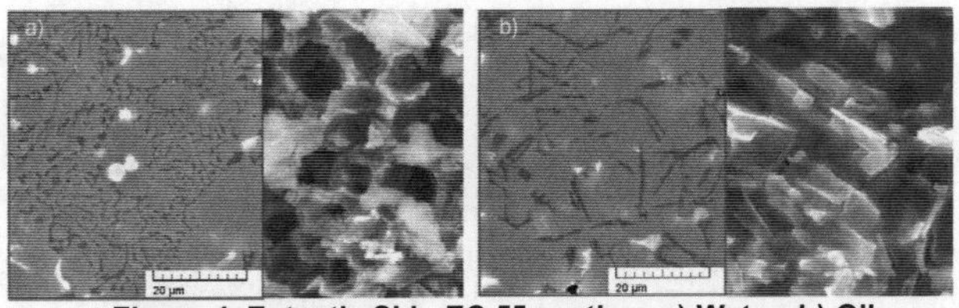

Figure 4. Eutectic Si in EC 55 casting; a) Water, b) Oil

4 Conclusion

The use of heat transfer oil caused an increase in the temperature of the core during the work cycle and a decrease in the cooling rate of the aluminum melt. This caused the presence of pores in a smaller number and dimensions in the volume of the EC 55 and EC75 cast tube. By applying water to the thermoregulation circuit, the initial temperature of the core during the work cycle was reduced, which resulted in the appearance of a larger number of pores in the form of closed air in the entire volume of the tube. The effect of tube growth confirmed the trend of increasing the presence of pores in the evaluated area of the EC 75 casting.

A better microstructure was achieved by the influence of water flow in the core. The microstructure of the water-cooled castings was characterized by finer grains of the α phase (Al) and eutectic silicon in the form of round grains and lamellae. Coarse grains of the α phase (Al) and Si lamellae in the form of sharp-edged lamellae (needles) with an uneven surface were characteristic of the oil-tempered castings.

Acknowledgements

The article was created as part of the VEGA 1/0160/22. The authors thank the agency for their support.

References

[1] Sadeghi M and Mahmoudi J. Experimental and Theoretical Studies on the Effect of Die Temperature on the Quality of the Products in High-Pressure Die-Casting Process. Advances in Materials Science and Engineering, 2012, 9.

[2] Jarfors A, Sevastopol R, Karamchedu S, Zhang Q, Steggo J and Stolt R. On the Use of Conformal Cooling in High-Pressure Die-Casting and Semisolid Casting. Technologies, 2021, 9(2):39.

[3] Shin S S. Lee S K, Kim D, Lee B. Enhanced cooling channel efficiency of high-pressure die-casting molds with pure copper linings in cooling channels via explosive bonding. Journal of Materials Processing Technology. 2021, 297.

[4] Hu H, Chen F, Chen X, Chu Y and Cheng P. Effect of cooling water flow rates on local temperatures and heat transfer of casting dies. Journal of Materials Processing Technology, 2004, 148(1): 57-67.

Heat Checking of High Pressure Die Casting Dies - Effects, Causes and Solutions

Yang You *
Huaxinglong Mould & Die Material Technology Co., Ltd. Dongguan, Guangdong, China
*Corresponding address: e-mail: youyang@hxlsteel.com.cn

Abstract: Heat checking is the main failure type of high pressure die casting (HPDC) dies, which have significant impact on the production, efficiency, cost, and other aspects of the new energy vehicle industry. This article discusses the impact of heat checking on die-casting production, the causes of heat checking, and targeted prevention and control of heat checking from several perspectives, attempting to propose more effective solutions for enterprises.

Keywords: Heat checking; HPDC; casting; die steel

1 The Influence of Heat checking on Die Casting production

HPDC Dies

The impact of heat checking on die-casting production is profound and diver When heat checking occurs on the dies, it will cause imprints at the corresponding positions of the casting. As the degree of heat checking increases and the cracks become deeper, the imprints on the casting will gradually become burrs. When the cracks become more severe and the burrs become unacceptable, it often declares the final life of the die.

Casting

Heat checking brings burrs, particles, and excess flesh to castings, corresponding to cracks, slag, and flesh on the dies. At this point, in order to ensure the delivery quality of castings, it is necessary to arrange deburring or even additional processing procedures. How to do deburring more economically and effectively is not only related to the customer's equipment conditions, but also to the cracking characteristics.

Die casting production

When the problem of burrs on castings becomes unacceptable, it is necessary to request a shutdown for die repair, which will obviously cause delays in delivery. In the situation of extremely tight delivery time, the severely cracked dies cannot be taken off the machine due to production capacity maintenance, and there is no window for welding repair/maintenance work. Only by deploying backup dies to replace the old dies can delivery be guaranteed. On the one hand, before the preparation of the mold is in place, the delivery of castings must bear multiple pressures such as deburring costs, yield/delivery time reductions, etc; On the other hand, the cost of die preparation itself is also high, and both customers and suppliers bear losses.

In order to increase production capacity, many die-casting factories invest heavily to shorten the cycle time, such as improving the pouring arrangement, adding die cooling/warming machines, adopting 3D printing inserts, and so on. If a production capacity bottleneck is formed in the deburring step at this time, the effect of the initial investment cannot be realized. Therefore, it becomes more practical to correctly solve the heat checking problem corresponding to burrs.

2 Cause of heat checking

The temperature gradient provides conditions for the inconsistent expansion and contraction at both ends of the gradient field, and the contraction determines the location of heat checking, which occurs at the cold end of the gradient field. the hot end of the gradient field expands while the cold end contracts. The hot end tends to "pull back" the contracted cold end to the length before contraction, which generates tensile stress at the cold end (while the corresponding hot end receives compressive stress). This tensile stress repeatedly occurs with the repeated punching of the die and accumulates back to cause surface cracks on the die, which originate from the repeated temperature gradients and are also known as thermal fatigue.

3 Factors affecting heat checking

Furthermore, we will explore the causes of more practical forms of heat checking, namely how different forms of thermal cracking such as dense thermal cracking, sparse thermal cracking, and local deep cracking are formed. As shown in formula 1, N is the shots before heat checking initiation, S is the high-temperature creep strength of the material, proportional to the die temper-back resistance and room temperature hardness, and has about 80% ratio, D is the material plasticity, proportional to the die elongation and Charpy-V impact energy, and has about 20% ratio, R is the creep scale, proportional to the die temperature and thermal stress

$$N = \frac{S \times D}{R} \quad (1)$$

Assuming R remains constant and external conditions such as die temperature and thermal stress are unified, there are four possible scenarios:

1. S is low, and D is high. Due to the high weight of S, the initiation time of heat checking is advanced, and the appearance of heat checking is early. Due to the high D, the crack propagation distance is short, so the heat checking presents a dense form, and it is easy to develop into slag dropping after increasing the die shots.

2. S is low, D is low, the initiation time of heat checking is advanced, and heat checking appears early. Due to the low D, the crack propagation distance is long, so the crack presents a sparse and "large grid" shape, and it is easy to develop into chipping after increasing the die shots.

3. S is high, and D is high, the initiation time of heat checking is delayed, and the appearance of heat checking is late. Due to the high D, the crack propagation distance is short, so the thermal cracking also presents a dense form. After increasing the die shots, it will also develop into slag, but the degree is lighter later.

4. S is high, and D is low. Due to the high weight of S, the initiation time of heat checking will still be delayed, and the appearance of heat checking will be delayed. Due to the low D, the crack propagation distance is long, so the crack also presents a sparse and "large grid" shape. After increasing the die shots, it is easy to develop into chipping, and the degree is more severe.

4 Heat checking response plan

Understanding the formation mechanism of different heat checking forms is essential to better generate response plans for different situations.

Firstly, from the perspective of mold materials, it is always correct to improve toughness without damaging other properties under any circumstances. High toughness can improve D, delay the initiation of heat checking, and inhibit crack propagation, avoiding the development of heat checking into large chippings.

If it is possible to improve toughness or not reduce toughness while increasing hardness, it is a more ideal situation, which is equivalent to simultaneously improving S and D. To achieve such effects, high toughness reserve mold steel materials such as EX55 are needed. Due to their metallurgical properties and excellent heat treatment processes, this type of material is more likely to achieve high toughness at high hardness.

The performance of die steel belongs to the basic preparation for dealing with heat checking, and on this basis, it also requires the support of supporting heat treatment equipment and technology. The hardness scheme that adapts to the specific structural characteristics of the die is the most important and the core direction of subsequent work.

High Pressure Die Casting Process Optimization Using Artificial Intelligence

Malinowski Pawel*
Faculty of Foundry Engineering, AGH University of Krakow, Al. Mickiewicza 30, 30-059 Krakow, Poland
*Corresponding address: e-mail: pamalino@agh.edu.pl

Abstract: High Pressure Die Casting technology makes it possible to make precision castings of light metal alloys such as Aluminum. The PLC of the pressure machine monitors dozens of process parameters such as temperature, pressure, doping curve, vacuum level, etc. The number of parameters that affect casting quality is very large. Using the artificial intelligence system, it is possible to build a model that can predict the quality of a casting in real time, suggest to the operator what steps he should take to achieve high casting quality. By using the artificial intelligence system, it is possible to improve the quality of castings, reduce the number of defects, and increase production efficiency. The large number of production parameters makes it impossible for humans to analyze the process. Artificial intelligence algorithms including machine learning algorithms or neural networks perform these tasks very well. Improving the quality of castings is one of the most essential elements of optimizing the production process.

Keywords: Artificial Intelligence; machine learning; production process optimization; High Pressure Die Casting process optimization

1 Introduction

The HPDC process is a complex process that generates more than 100 parameters during one process cycle, which are stored in a PLC - Programmable Logic Controller. Such a large amount of data makes the process very complex and artificial intelligence can optimize it. The entire process is divided into several steps that make up the entire manufacturing process cycle.

2 Experimental procedure

By collecting data from the process combined with the quality of the casting as measured by e.g. leak test, a dataset was created. This data underwent a preparation process for use in machine learning. Outliers were eliminated, checked for blank data, were rescaled and normalized so that the artificial intelligence model created would be better. The data was then subjected to various machine learning algorithms. Separate models were created for each algorithm, which were compared with each other using appropriate metrics to measure model performance.

The entire machine learning process used the regression method, as there was a continuous value as the objective function.

3 Result and discussion

As a result of research using machine learning algorithms, a model was developed with which the optimum parameters for the HPDC process were determined. Such a model should be periodically trained so that its accuracy is even better. With the optimum process parameters set, they should be implemented.

4 Conclusion

The use of artificial intelligence to optimize the HPDC process yields very good results. There is an improvement in casting quality, there are fewer defects, higher profits for the foundry and an increase in the foundry's competitiveness.

The High Pressure Die Casting process consists of many process parameters, alloy and mold, which have a significant impact on the result of the process - casting. Optimization of this process, narrowing the range of parameters, selection of optimal parameters will significantly reduce the number of defects, improve the quality of products, competitiveness of the company and, above all, profits. The use of the machine learning process will not only allow us to explore the potential of the data, but above all thanks to it, based on production data, we can forecast the quality of the casting, measured in this case by leak test.

Various regression models were used to optimize the HPDC process. Appropriate metrics were selected to measure the performance of the model and based on them the best algorithm was selected. Before the data entered the algorithm, it had to be properly prepared. Carry out the appropriate transformations to obtain a normal distribution, scale the data, clean it, carry out the feature selection process to select the most important features (process parameters).

Acknowledgments

The research is financed within GREEN CASTING LIFE project. GREEN CASTING LIFE project is co-funded by European Union's LIFE program under grant agreement LIFE21-ENV-FI 101074439. Views and opinions expressed are however those of the author(s) only and do not necessarily reflect those of the European union or CINEA. Neither the European union nor the granting authority can be held responsible for them. GREEN CASTING LIFE project (LIFE21-ENV-FI-101074439) is co-funded by the National Fund for Environmental

Protection and Water Management (NFOŚiGW) under grant agreement 276/2023/Wn-06/OZ-PO-LF/D.

References

[1] Rahul Kapoor, Vishal S Sharma, Harwinder Singh, Optimization of machining process parameters of high pressure die casting for improve the quality of manufactured product, Materials Today: Proceedings, 2022, 57(2): 793-800.

[2] Andres Perez, Javier Nieves, Jorge Angulo. Artificial Intelligence Based Foundry Processes Optimization Technology. 2nd Foundry Young Researchers and Early Professional Careers Conference.

Effect of Different Pressures on Microstructure and Properties of Squeeze Casting Al_5Mg_2SiMn Alloy

Yongsheng Li, Dongyang Yang, Jingsi Chen, Chuanyao Tang, Weiwen Zhang*
School of Mechanical and Automotive Engineering, South China University of Technology, Guangzhou, 510640, China
*Corresponding address: e-mail: mewzhang@scut.edu.cn

Abstract: In this paper, the mechanical properties and microstructure of Al_5Mg_2SiMn alloys formed under different pressures are investigated. It is found that the yield strength and elongation of the alloys show an increasing trend with the increase of pressure, but the increase gradually decreases. Meanwhile, the Mg_2Si interlamellar space (ILS) decreases with increasing pressure, and the magnitude of the decrease decreases with increasing pressure. The pressure of squeeze casting increases the melting point of the alloy, and the solidification of the alloy provides the subcooling degree, which increases with the increase of pressure. And the subcooling degree provides the interfacial free energy for the formation of Mg_2Si, so the interlamellar space of Mg_2Si decreases with the increase of subcooling degree. The mechanical properties of Al_5Mg_2SiMn alloys are mainly affected by the interlamellar space of Mg_2Si. The smaller the Mg_2Si interlamellar space is, the better the tensile strength and plasticity of the alloy is.

Keywords: Al_5Mg_2SiMn alloys; pressures; Mg_2Si; subcooling degree; interlamellar space

1 Introduction

Under the general trend of automotive lightweighting, aluminum alloys with low density and high specific strength can reduce carbon emissions under the premise of ensuring automotive performance and safety [1]. Heat-treatment-free aluminum alloys can further reduce the cost and energy consumption in the production process. Al_5Mg_2SiMn alloy relies on the Mg element to form the second phase, possessing heat-treatment-free characteristics, and having better stress corrosion resistance, which is used in the inner door panels, vibration-damping towers, and other automotive components [2]. In this paper, the heat treatment-free Al_5Mg_2SiMn alloy is taken as the object. The most suitable production process of heat treatment-free aluminum alloy Al_5Mg_2SiMn is determined by regulating the extrusion casting pressure, which reveals the mechanism of the pressure's influence on the microstructure and mechanical properties of the alloy.

2 Experimental procedure

The alloy used in the experiment was Al-5Mg-2Si-0.6Mn-0.3Fe-0.3Cu. It was solidified under a pressure of 0,30,40,50,60,80,100,125,150,200 MPa and then cooled in air. The microstructure and composition of the samples were characterized and analyzed using OM, SEM, and EDS. Mechanical properties are determined by averaging the results of at least three valid measurements.

3 Result and discussion

Microstructure

Fig. 1 shows the microstructure of the alloys. In the figure, the white phase is detected as $Al_{15}(Fe,Mn)_3Si_2$ phase by composition, the black phase is Mg_2Si, and the gray phase is Al matrix. From the Fig.1, it can be seen that with the increase of extrusion pressure, the ILS of Mg_2Si is decreases gradually, and the ILS of Mg_2Si with pressure is shown in Fig. 2（b）.

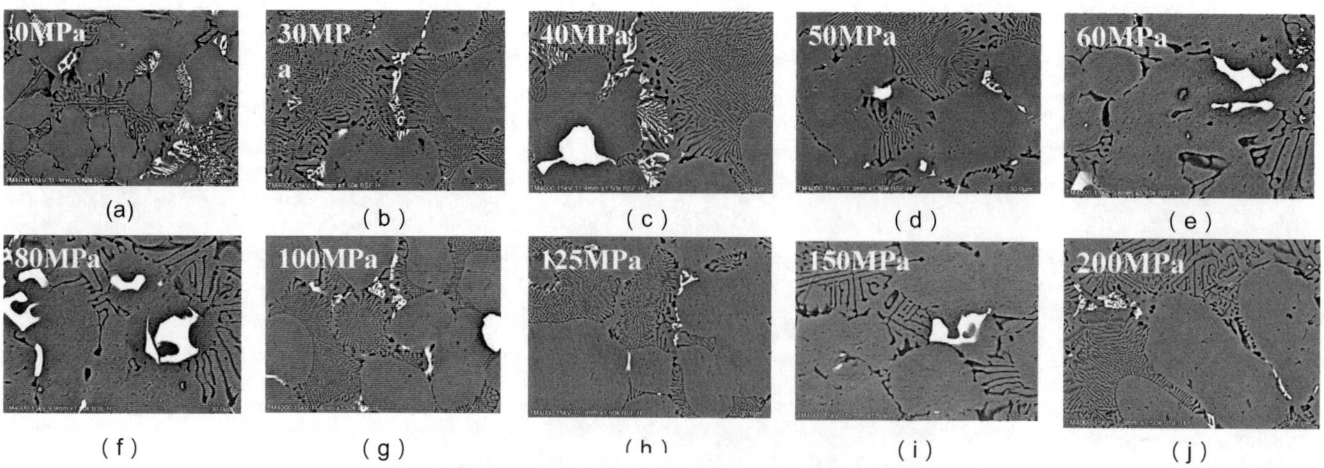

Fig. 1 SEM images of of alloys with different Fe pressures

Mechanical properties

The mechanical properties of the alloy at different pressures are shown in Fig. With the increase oh pressure, the tensile strength of the alloy increases from a minimum of 194 MPa to a maximum of 263 MPa, the yield strength increases from a minimum of 131 MPa to a maximum of 154 MPa, and the elongation increases from a minimum of 1.6% to a maximum of 5.1%.

The ILS of the eutectic organization has a significant effect on the properties of the alloy. The smaller the ILS of the eutectic organization, the higher the strength of the alloy, which can be described by the Hall-Petch (Hall-Petch) formula [3]. Let be the yield strength and σ^*, m is constant related to the material, then we have

$$\sigma = \sigma^* + m\lambda^{-\frac{1}{2}} \tag{1}$$

Effect of pressure on Mg_2Si

Al-Mg_2Si eutectic belongs to lamellar eutectic. the growth rate of eutectic R determines the speed of diffusion. Diffusion rate depends on the two-phase laminar sheet spacing λ [3]. Minimum spacing of the eutectic lamellar sheet is determined by the thermodynamic conditions. The change of Gibbs free energy in the formation of eutectic consists of two parts, one is the bulk Gibbs free energy and the other is the change of interfacial Gibbs free energy, which is related to the eutectic lamellae spacing λ.

$$\lambda * = -\frac{2\gamma_{\alpha\beta} \cdot T_E}{\Delta H \cdot \Delta T_0} \tag{2}$$

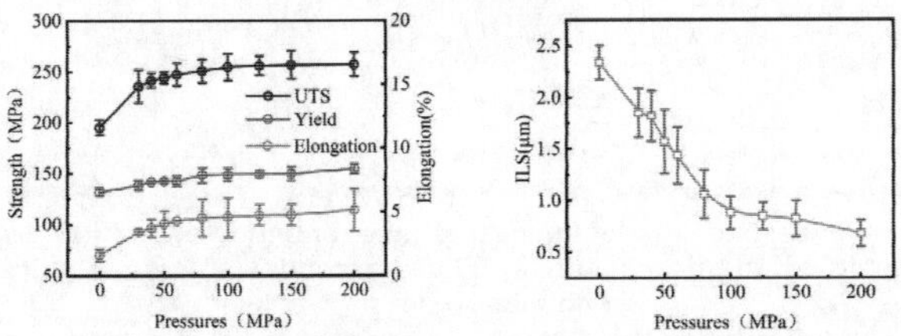

(a) Mechanical property (b) ILS of Mg_2Si

Figure 2 Mechanical properties of alloys at different pressures with Mg_2Si lamella spacing

$\gamma_{\alpha\beta}$ is the interfacial energy, ΔT_0 is the subcooling degree, T_E is the eutectic temperature.

4 Conclusion

The pressure of squeeze casting raises the melting point of the alloy, and the solidification of the alloy provides subcooling, which increases with the increase of pressure. The subcooling provides interfacial free energy for the formation of Mg_2Si, therefore the ILS of Mg_2Si decreases with the increase of subcooling. The mechanical properties of Al_5Mg_2SiMn alloys are mainly affected by the ILS of Mg_2Si, and the smaller the ILS of Mg_2Si is, the better the tensile strength and plasticity of the alloy.

Acknowledgments

We would like to acknowledge the financial support from the National Natural Science Foundation of China (No. 52074131)

References

[1] Chapuis M P, Estoup A. Microsatellite null alleles and estimation of population differentiation[J]. Molecular biology and evolution, 2007, 24 (3) : 621-631.

[2] Jason Rowe. Advanced materials in automotive engineering[M]. Germany:WOODHEAD, 2012. 126-138.

[3] Hu Gengxiang, Cai Xun, Rong Yonghua. Fundamentals of materials science. Shanghai: Shanghai Jiao Tong University Press, 2010. 310-312.

Damage of HVDC AlSiMgMn Alloys Using In-Situ Tensile with X-ray Tomography

Xueling Wang, Haidong Zhao*
National Engineering Research Center of Near-net Shape Forming for Metallic Materials, South China University of Technology, Guangzhou 510641, China
*Corresponding address: e-mail: hdzhao@scut.edu.cn; zhaohaidong@tsinghua.org.cn

Abstract: AlSiMgMn alloys with different primary phases were prepared by high vacuum die-casting (HVDC) process. To evaluate the role of Fe-rich intermetallic characteristics on crack initiation, three-dimensional visualization of tensile damage process in these alloys was investigated by in-situ X-ray computed tomography (XCT). The results show that cracks in the alloy with primary α-Al were initiated at connection of shrinkage pore and damaged eutectic intermetallics while clustered primary intermetallics in the alloy with primary α-Fe led to crack formation.

Keywords: HVDC AlSiMgMn alloy; in-situ XCT; Fe-rich intermetallics; crack initiation

1 Introduction

In automotive industry, AlSiMgMn alloys are widely used in high vacuum die castings for their excellent castability and strength [1]. With increasing demands for economics and environmental sustainability, higher Fe content in AlSiMgMn alloys, which is detrimental to mechanical properties [2], has recently drawn much attention. With the change of Fe contents, Fe-rich intermetallics would precipitate in different solidifications, affecting mechanical behavior. Therefore, understanding the effects of different Fe-rich intermetallic characteristics and distributions on damage process is important for application of high Fe content alloys.

Benefiting from development of high-resolution X-ray computed tomography (CT) techniques, 3D in-situ analysis by X-rays implemented with mechanical test systems has become a powerful characterization option in materials science [3]. In this study, tensile damage processes of HVDC AlSiMgMn alloys with different primary phases were present, and the effects of eutectic and primary intermetallics on crack initiation were studied.

2 Experimental procedure

AlSi10MgMn alloys with different Mn contents were prepared by HVDC process (high vacuum of about 48 mbar), and their chemical compositions measured by ARL4460 are presented in Table 1. In our previous work [4], the Fe-rich intermetallics of alloy AE formed in eutectic while α-Fe phase first precipitated in alloy AP.

Table 1. Compositions of HVDC alloys (wt.%)

No.	Si	Mn	Fe	Mg	Al
AE	10.2	0.61	0.16	0.40	Bal.
AP	10.2	0.82	0.15	0.40	Bal.

The equipment of in-situ X-ray micro-CT experiments was ZEISS Xradia 520 versa coupled with tensile stage, as shown in Fig. 1a. The standard zone of specimens was 2 mm × 1 mm × 3.5 mm (Fig.1b) to obtain resolution of 2.531 μm for well reconstructions of Fe-rich intermetallics and microcracks. A constant displacement rate of 0.001 mm/min was applied to the tensile specimens and each alloy was tested three times.

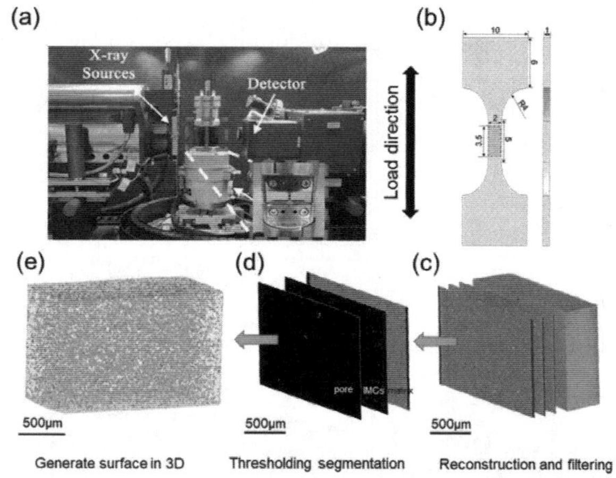

Fig. 1. (a) In-situ 3D tensile test set-up, (b) size of the specimens for tensile test, (c) original images volume rendering and slices, (d) segmentation of phases, and (e) 3D visualizations of intermetallics and microcracks.

3 Result and discussion

Tensile curves of alloys with different primary phases

The eutectic intermetallics in alloy AE had polyhedron shape and were with dispersed distribution, while lots of hexahedron or octagonal dendrite shapes primary intermetallics in alloy AP clustered distributed.

In Fig. 2, each specimen has five scanning points (indicated by red dots), in which point A represents the time before loading. Point B represents the time when the microcracks were first observed. It can be seen that damage occurred at higher strain in alloy AE. Point D represents the time start of crack propagation, which is determined by a sudden drop in force. Point E represents the time of specimen complete fracture. Comparing stress-strain curve of the alloys, lower elongation in alloy AP indicated that clustered intermetallics would be detrimental to ductile.

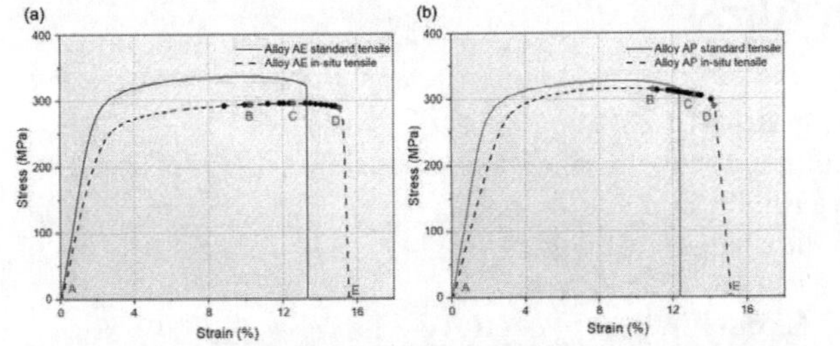

Fig. 2. Stress-strain curves in standard tensile and in-situ tensile of alloys AE (a) and AP (b).

Crack initiation in alloy with primary α -Al

Fig. 3a shows the 3D morphologies of shrinkage pores and cracks in alloy AE. It can be seen that pores in the alloy were isolated in distribution. They enlarged under the load and didn't connect before fracture. As shown in Fig. 3b, six polyhedral intermetallics were around the shrinkage pore, which is near the specimen surface. With the strain increased, the shrinkage pore further enlarged, and many intermetallics were damaged at high strain. The connection of shrinkage pore and microcracks resulted in the formation of crack in alloy AE.

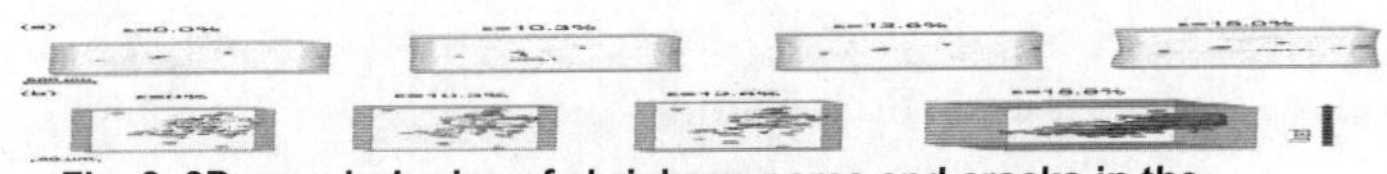

Fig. 3. 3D morphologies of shrinkage pores and cracks in the tensile process (a) and crack initiation (crack-2) in alloy AE (cracks indicated by red, intermetallics indicated by green and shrikage pores indicated by grey).

Crack initiation in alloy with primary α -Fe

Different in alloy AP, with the strain increased, new cracks formed and connected, resulting in a significant crack propagation, as shown in Fig. 4a. From Fig. 4b and c, clustered primary intermetallics fractured and formed larger microcracks at low strain. These microcracks connected with each other, becoming the crack initiation sites.

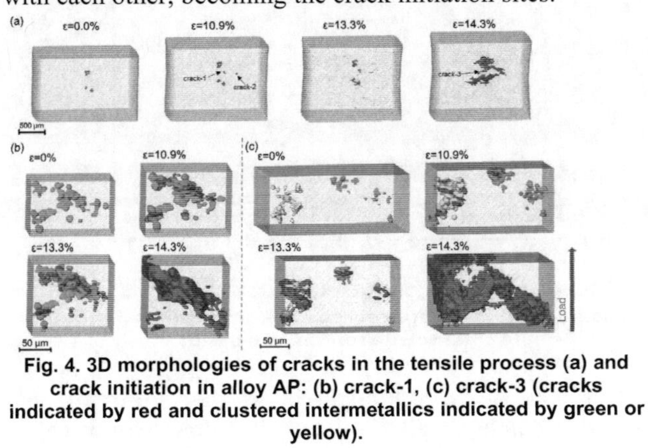

Fig. 4. 3D morphologies of cracks in the tensile process (a) and crack initiation in alloy AP: (b) crack-1, (c) crack-3 (cracks indicated by red and clustered intermetallics indicated by green or yellow).

4 Conclusion

3D in-situ tensile of HVDC AlSiMgMn alloys was conducted and the initiation of cracks was investigated. For alloy with primary α-Al, eutectic Fe-rich intermetallics were damaged at high strain, and the crack was initiated at the connection of shrinkage pore and microcracks. For alloy with primary α-Fe, clustered primary intermetallics fractured at low strain, and cracks initiated at the connection of microcracks.

Acknowledgments

The authors thank for the support of the Beijing Natural Science Foundation (No. L223001) and the Natural Science Foundation of Guangdong Province (No. 2023A1515012730).

References

[1] Niu Z C, Liu G Y, Li T and Ji S X. Effect of high pressure die casting on the castability, defects and mechanical properties of aluminium alloys in extra-large thin-wall castings. J. Mater. Pro. Tech., 2022, 303: 117525.

[2] Emre C, Michael, M, Zhang J Y, Miao J S and Luo A A. A new recycled Al–Si–Mg alloy for sustainable structural die casting applications. Metall. Mater. Trans. A, 2022, 53: 2861–2873.

[3] Kong X, Helfen L, Hurst M, Hänschke D, Morgeneyer T F, et al. 3D in situ study of damage during a 'shear to tension' load path changes in an aluminium alloy. Acta Mater., 2022, 231: 117842.

[4] Wang X L, Zhao H D, Xu Q Y, Han Z Q. Clustering characteristics of Fe-rich intermetallics in high vacuum die cast AlSiMgMn alloys with high resolution μ-CT inspection. Mater. Char., 2024, 207: 1044-5803.

Investigation of Quench Sensitivity of High Vacuum Die Casting AlSi10MgMn Alloy

Wenhui Bai, Yueqin Lei, Haidong Zhao*
National Engineering Research Center of Near-Net-Shape Forming for Metallic Materials, South China University of Technology, Guangzhou 510640, China
*Corresponding address: e-mail: hdzhao@scut.edu.cn

Abstract: In this study, the quench sensitivity of high vacuum die casting (HVDC) AlSi10MgMn alloy was investigated through time-temperature-transformation (TTT) and time-temperature-property (TTP) curves. The mechanical properties during different quenching rates were predicted by quench factor analysis (QFA). Phase transformation kinetics was analyzed by Avrami equation. The results show that the sensitivity temperature range is 295~445°C and the nose temperature is 370°C. During the isothermal treatment, a number of rod-shape β-Mg_2Si phases precipitate and grow up with the extension of time. The number density of the age-induced precipitates β″reduce after aging treatment. Based on simulated results of TTT and continuous cooling transformation (CCT) curves, the accuracy of QFA methods was validated. To obtain good mechanical properties, the critical cooling rate should be higher than 7.7 °C/s.

Keywords: AlSi10MgMn alloy; high vacuum die casting; quench sensitivity; quench factor analysis; microstructure

1 Introduction

Al-Si alloy is a typical age hardening alloy [1]. Solution, quenching, and aging treatment are usually used to improve alloy mechanical properties. However, the quenching cooling rate is essential for the supersaturation of the solid solution [2]. A low cooling rate may precipitate coarse phases from the supersaturation solid solution, leading to a decrease of strength [3]. This phenomenon is called quench sensitivity. When the quenching cooling rate is relatively high, residual stress would generate in the die castings, which is harmful to the mechanical behavior. Therefore, the quenching cooling rate should be controlled and well-studied to balance the mechanical properties and residual stress [4]. Unfortunately, there are little literature about the evolution of mechanical properties and microstructure of HVDC AlSi10MgMn during the quenching.

In this paper, TTT and TTP curves of HVDC AlSi10MgMn alloy were obtained by interrupted quenching method. QFA methods was verified by the simulated TTT and CCT curves. Microstructure evolution was studied, and the microstructure transformation during isothermal treatment was analyzed by phase transformation kinetics.

2 Experimental procedure

In this study, commercial AlSi10MgMn alloy and Al-50%Mg mater alloy were used to prepare the experimental alloy. The compositions of this alloy are shown in Table 1. A TOYO BD-350V5 cold chamber HVDC machine was used in this experiment. As for interrupt quenching experiment, the solution treatment was performed at 530°C for 3h and the isothermal temperature was 280-460°C with isothermal time of 0-1200 s. In addition, the aging treatment was performed at 185°C for 7 h.

Table 1. Compositions of the experimental alloy

Si	Mg	Fe	Mn	Ti	Zn	Al
10.58	0.42	0.07	0.61	0.08	0.04	Bal.

3 Result and discussion

Performance prediction by QFA

As can be seen in Fig.1, the quench factor decreases, and alloy hardness increases with the cooling rate increases. When the cooling rate is higher than 7.7 °C/s, the alloy hardness can reach 95% maximum value (102HB). Below this cooling rate, the hardness decreases rapidly with the cooling rate decreasing.

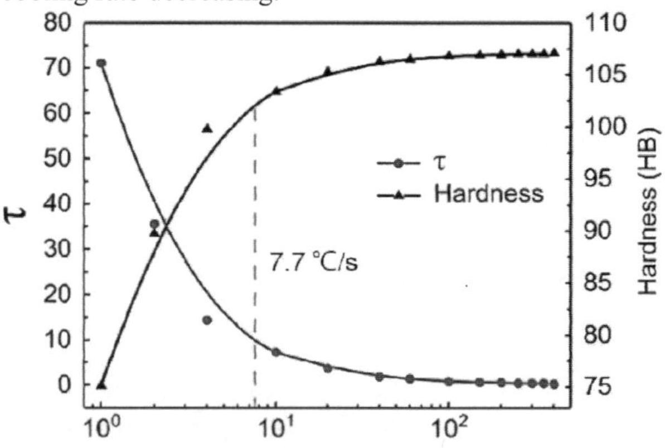

Fig. 1. Effect of average cooling rate on the quench factor and predicted hardness of the alloy.

Microstructure evolution during isothermal treatment

As shown in Fig.2, after isothermal treatment for different time at 370°C, the number density of point-like β″ phase is obviously different. As the isothermal treatment progresses, solute atoms precipitate from the solute solution and the rod-shaped β phase forms. Especially for 1200 s, there are almost no β″ phase, only the β phase.

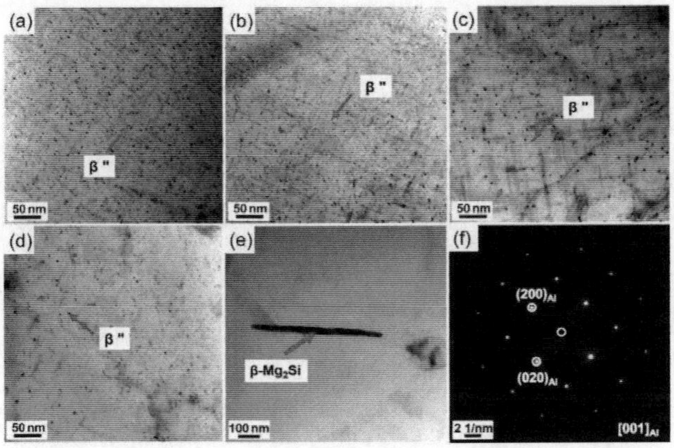

Fig. 2. Effects of different isothermal time at 370°C on β″: (a) 0 s; (b) 20 s; (c) 240 s; (d) 480 s; (e) 1200 s; (f) the corresponding SAED pattern of (a).

Simulated TTT and CCT curves

Fig.3a shows the simulated TTT and CCT curves of HVDC AlSi10MgMn alloy. The simulated TTT curves are also "C" shape with nose temperature of about 360°C. This is close to the experimental result. In addition, as can be seen in Fig. 3b, the critical cooling rate of non-precipitation of β phase is about 0.1 °C/s, and the critical cooling rate of non-precipitation of β' and β″ phase is about 7 °C/s. To hinder the formation of quench-induced precipitates and obtain a comprehensive mechanical property, the cooling rate should be higher than 7 °C/s at the high quench sensitivity zone. At higher/lower temperature, the cooling rate can be reduced appropriately. This result is also agreement with the QFA method.

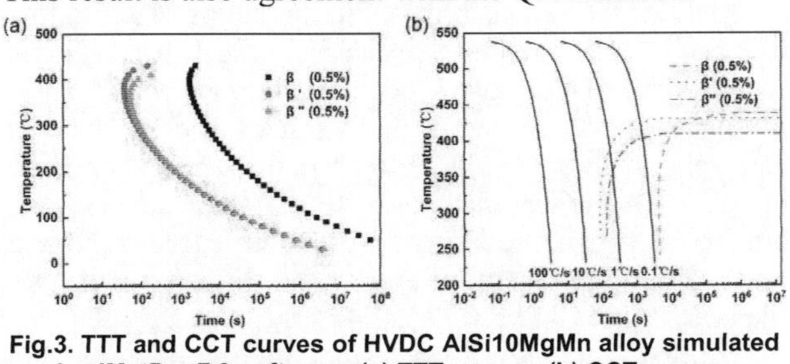

Fig.3. TTT and CCT curves of HVDC AlSi10MgMn alloy simulated by JMatPro 7.0 software: (a) TTT curves; (b) CCT curves.

Phase transformation kinetic analysis

Fig. 4 shows the phase transformation kinetic curves at different temperature. The curves are "S" shape, indicating that there is an incubation period during the phase transformation process. The curve of 370°C is at the far left, indicating the highest transformation rate. As for the curves of 280°C and 460°C, the curve of 280°C is at the far right, indicating that the phase transformation rate at the low temperature zone is lower. Therefore, the quench sensitivity is lower at the low temperature zone.

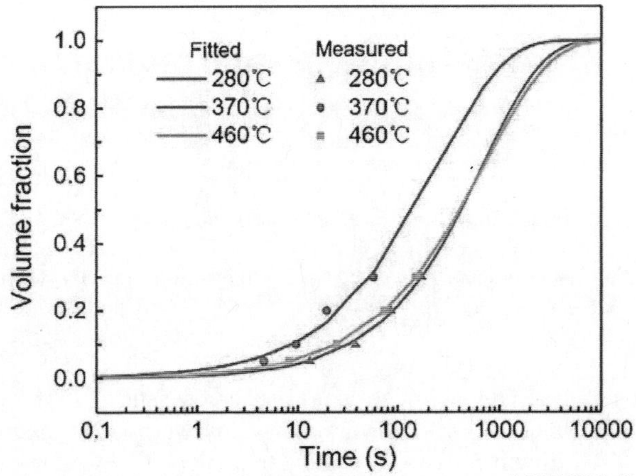

Fig. 4. The phase transformation kinetics curves of AlSi10MgMn alloy.

4 Conclusion

1) The high quench sensitivity range of HVDC AlSi10MgMn alloy is 295-445°C with a nose temperature of 370°C. Quench rate should be higher than 7.7 °C/s to obtain excellent mechanical properties at the high quench sensitivity zone.

2) At the nose temperature of 370°C, the average size of the β phase increases and the number density of β″ phase decreases with the extension of the time, which is detrimental to the mechanical properties.

Acknowledgments

The research was financially supported the support of Beijing Natural Science Foundation-Millet Innovation Joint Fund (No. L223001), Natural Science Foundation of Guangdong Province (No. 2023A1515012730) and the Key Area Research and Development Program of Guangdong Province (No. 2020B010186002).

References

[1] Timelli G, Lohne O, Arnberg L and Laukli H I. Effect of solution heat treatments on the microstructure and mechanical properties of a die-cast AlSi7MgMn alloy, Metall. Mater. Trans. A, 2008, 39: 1747-1758.

[2] Zhang Y X, Yi Y P, Huang S Q, Dong F and Wang H M. Investigation of the quench sensitivity of forged 2A14 aluminum alloy by time-temperature-tensile properties diagrams, J. Alloys Compd., 2017, 728: 1239-1247.

[3] Wang H M, Yi Y P and Huang S Q. Investigation of quench sensitivity of high strength 2219 aluminum alloy by TTP and TTT diagrams, J. Alloys Compd., 2017, 690: 446-452.

[4] Liu S D, Zhong Q M, Zhang Y, Liu W J, Zhang X M and Deng Y L. Investigation of quench sensitivity of high strength Al-Zn-Mg-Cu alloys by time-temperature-properties diagrams, Mater. Des., 2010, 31: 3116-3120.

Characteristics of Intermetallics of High Vacuum Die-Cast AlSiMgMnCu Alloys with 3D Electron Microscopy

Fei Liu [1], Huiting Zheng [1], Yuanhang Jiang [1], Bo Chen [2], Haidong Zhao [1*]
1. National Engineering Research Center of Near-net Shape Forming for Metallic Materials, South China University of Technology, Guangzhou 510641, China
2. Key Laboratory of Advanced Civil Engineering Materials (Ministry of Education), School of Materials Science and Engineering, Tongji University, Shanghai 201804, China
*Corresponding address: e-mail: hdzhao@scut.edu.cn

Abstract: The microstructural characteristics of high vacuum die casting AlSiMgMnCu alloy were investigated in three dimensions (3D) using serial block-face scanning electron microscopy (SBFSEM). The characteristics of α-Fe, β-Mg_2Si, Q-$Al_5Cu_2Mg_8Si_6$ and θ-Al_2Cu phase in the skin layer, segregation band and core were quantitatively investigated. In addition, the morphology of the Si particles and the intermetallics in the alloy after T6 heat treatment were compared with those in the as-cast alloys. The dissolution of the phases during the solution treatment was quantified and analyzed.

Keywords: high vacuum die casting; AlSiMgMnCu alloy; intermetallics; 3D characteristics; SBFSEM

1 Introduction

High vacuum die casting (HVDC) process has been widely in Al alloy automotive components. AlSi based alloys have excellent castability and good comprehensive mechanical properties. Microalloying is an economical method for strengthening the alloys to meet requirement of increasing properties. Mg and Cu elements are usually added into the alloys forming strengthening intermetallic phases such as β-Mg_2Si, Q-$Al_5Cu_2Mg_8Si_6$ and θ-Al_2Cu phases [1]. The microalloying not only form new phases but also change the phase morphology and characteristics. In addition, die casting Al alloys usually are of three typical regions, namely skin layer, segregation band, and core, in which microstructures show serious heterogeneity [2]. In this study, the microstructural characteristics in 3D of HVDC AlSiMgMnCu alloy were investigated with SBFSEM.

2 Experimental procedure

The tensile samples with 4 mm thickness were produced with a TOYO cold-chamber die casting machine attached to a high vacuum unit. The alloy compositions were: Si: 10.26 wt.%; Mg: 0.37 wt.%; Cu: 0.67 wt.%; Mn: 0.54 wt.%; Fe: 0.07 wt.%; Ti: 0.08 wt.%; Al rem. The specimens for SBFSEM were cut from the tensile sample in three regions of skin layer, segregation band, and core. Another specimen was taken from the core in the T6-treated sample. The specimens were machined with sizes of 1×1×10 mm. The morphology and distribution of different intermetallics in 3D were measured by a Gatan 3View device, a SBFSEM system, equipped with a Zeiss Sigma 300 VP scanning electron microscope (SEM). Raw images of 2048×2048 pixels with 30 nm during serial sectioning were obtained to render the intermetallic 3D morphology in 61×61×21 μm^3 with Avizo software. The SBFSEM procedures are shown in Fig. 1.

Fig. 1. Procedures of SBFSEM measurement

3 Result and discussion

Intermetallics in skin layer, segregation band and core

Fig. 2 shows the reconstructed intermetallics of the alloy, and their volume fractions are list in Table 1. The volume fractions of the α-Fe phase in the three regions were 1.53%, 1.37%, and 1.94%, respectively. The sizes of α-Fe phase in skin layer and core were larger than in the segregation band zone. A similar phenomenon for β and Q phases existed. On the contrary, the fraction of θ phase in segregation band was 0.78%, significantly higher than in the other two regions.

According to the SBFSEM results, the seaweed-like and plate-like α-Fe phase, rod-like β phase, seaweed-like Q and θ phases were found in skin layer, formation of which was determined by the heat transfer there and their formation competition. In segregation band and core, the seaweed-like and near-spherical α-Fe phases, smaller and spherical β phase, and irregular θ phase formed. The small Q phase was in uniform distribution in segregation band whereas the phase was with complex shape in core. The formation of the different morphology and segregation of the phases are related to cooling, phase formation, solute concentration and convection in the regions [3].

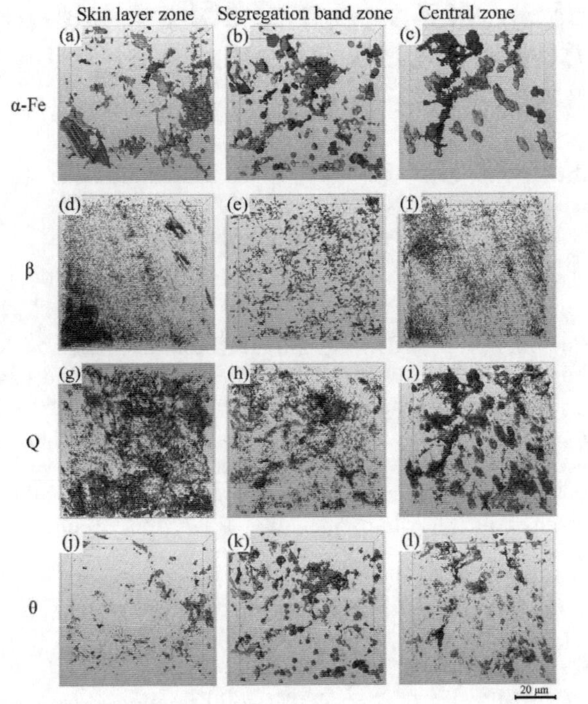

Fig. 2 The reconstructed intermetallics of the alloys in skin layer (a, d, g, j), segregation band (b, e, h, k) and core (c, f, i, l).

Table 1 Volume fractions of intermetallics in the different regions

region	α-Fe,%	β,%	Q,%	θ,%
Skin	1.53	0.28	1.11	0.18
Band	1.37	0.22	0.25	0.78
Core	1.94	0.30	1.04	0.38

Change of Intermetallics in T6-treated alloys

The Si and intermetallics of the T6-treated alloys were quantitively characterized and compared with those in the as-cast alloy. The dissolution of Si, Mg, Cu elements were quantified and compared with theorical calculation at nonequilibrium solidification. The actual solute of Si, Mg and Cu elements in the as-cast alloy are much supersaturated than the calculation, and in the T6-treated alloy, those are lower than the equilibrium condition [4]. In the treatment, only 3% eutectic Si particles were dissolved, and the remained particles spheroidized, increasing the particle number; the α-Fe phase was hardly dissolved due to its higher melting temperature than the solution temperature; due to the β phase with high mean curvature contacting with the matrix, the phase dissolution was sufficient, and it was not detected after the treatment. As shown in Fig. 3, the Q phase was micro-sized and nano-sized particles in the as-cast alloys. After the treatment only micro-sized phase of ellipsoid-like shape with high positive mean curvature was found. Similar change was also found for the θ phase. 83% of the phase were dissolved, and the isolated θ phase easily dissolved into α-Al matrix, while the dissolution of larger interconnectivity θ phase was limited.

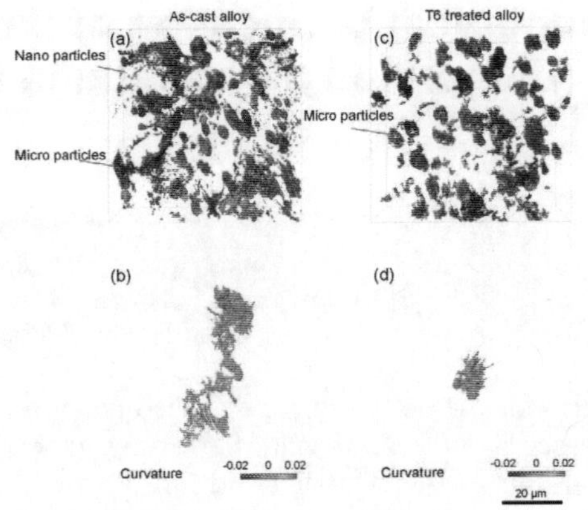

Fig. 3 3D morphology of the Q phase in as-cast (a and b) and T6 treated alloy (c and d).

4 Conclusion

(1) The fractions of α-Fe, β, and Q phases in skin layer and core were higher than in segregation band. On contrary, the fraction of θ phase in segregation band was higher. In skin layer, α-Fe phase showed the preferential growth orientation parallel to heat flux, while the other phases had a complex seaweed-like morphology. In segregation band and core, the α-Fe and β phases were regular whereas the Q and θ phases showed irregular shape.

(2) During T6 heat treatment, the eutectic Si phase become spheroidization, and only 3% of the phase were dissolved; the α-Fe phase underwent partial fragmentation with no obvious dissolution; about all β phase, 33% of Q phase and 83% of θ phase were dissolved. The isolated θ phase was easily dissolved in the matrix whereas the dissolution of larger interconnected θ phase was limited.

Acknowledgments

The study was supported by National Natural Science Foundation of China (51875211, 51971160 and 52142305), Beijing Natural Science Foundation (L223001), and Natural Science Foundation of Guangdong Province (2023A1515012730).

References

[1] Yang H, Ji J, Yang W, Wang Y and Fan Z. Effect of Mg level on the microstructure and mechanical properties of die-cast Al-Si-Cu alloys, Mater. Sci. Eng. A, 2015, 642: 340-350.

[2] Gourlay C M, Laukli H I and Dahle A K. Segregation Band Formation in Al-Si Die Castings Metall. Mater. Trans, 2004, 35A:2881-2891.

[3] Liu F, Zhao H D, Chen B and Zheng H T. Investigation on microstructure heterogeneity of the HPDC AlSiMgMnCu alloy through 3D electron microscopy, Mater Des., 2022, 218:110679.

[4] Liu F, Qin J, Zhao H D, Jiang Y H, Zheng H T and Chen B. Influence of intermetallics and element concentrations on the performance of the heat treated HVDC AlSiMgMnCu alloys at nano-scale, Mater Des., 2023, 229: 111902.

Numerical Verification of Feasibility of Small Amount of solid phase dispersion die casting method by CONTROLLING Molten Metal Temperature

Eitaro Koya[1,*], Atsushi Kawauchi[2]
1. R&S Cast Solutions, Honcho, Shiki-shi, Saitama, Japan
2. Kyushu Yanagawa Seiki Co., Ltd,1106-1 Kawabe Kyokushi, Kikuchi, Kumamoto, Japan
*Corresponding address: e-mail: koya.rscastsolutions@gmail.com

Abstract: One method of improving the quality of the die casting process is the semi-solid die casting method using semi-solid slurry. The Partial Solid Die Casting (PSDC) method was developed to address the issue of the semi-solid die casting method, which is an increase in cost due to the increased slurry production process. By reducing the temperature of the molten metal, the proportion of solid fraction in the sleeve immediately before injection is controlled to 0.05 or below, which disperses α-Al crystal nuclei and suppresses the growth of coarse porosity in thick-walled parts, thereby achieving high quality. Furthermore, a significant reduction in the manufacturing stage energy is achieved. The distribution of the solid fraction in the sleeve was precisely verified by numerical analysis of the ladle and the full model of the injection mechanism. This analysis was conducted to determine whether molten metal with the prescribed solid fraction could be produced stably under the manufacturing conditions of the PSDC method.

Keywords: semi-solid die-casting; solid fraction; aluminum; eutectic alloy

1 Introduction

One method of improving the quality of the die casting process is the semi-solid die casting method using semi-molten slurry. One disadvantage of the semi-solid die casting method is the increased cost associated with the production of the slurry. Additionally, the method is not easily applicable to JIS ADC12, a eutectic alloy with a narrow solidification range. To address these issues, Itamura et al. reported the development of a semi-solid die casting method using a hypoeutectic alloy, JIS AC4CH, by controlling the temperature in the sleeve [1]. Furthermore, Koide et al. recently reported the practical application of a semi-solid manufacturing method using slurry of JIS ADC12 alloy [2].

In contrast to these developments, the Partial Solid Die Casting (PSDC) method proposed in this study aims to suppress the coarse growth of porosity in the thick wall of the product at the completion of filling by generating and dispersing a small amount of crystal nuclei in the sleeve through molten metal temperature control. For the mass production of PSDC, it is essential to accurately predict the sleeve temperature at the time of ladle feeding in order to control the solid fraction in the sleeve by the molten metal temperature. In order to validate the PSDC method, the temperature distribution of the sleeve was obtained based on a mold temperature cycle calculation that considers the operating position of the plunger tip when filling the sleeve, and the flow and solidification calculations were verified by coupling the molten metal temperature calculation during ladle feeding with the flow and solidification calculations.

2 Overview of PSDC Method and Experimental Methods

Overview of PSDC Method

The PSDC method controls the solid fraction in the sleeve by lowering the molten metal temperature. In ADC12, which is a eutectic alloy, there is concern about the formation of cold flakes on the inner wall of the sleeve. In contrast to a solid fraction of 0.1, which increases the possibility of cold flakes, the PSDC method sets the solid fraction at about 0.05. The reduction in the molten metal temperature confers an advantage upon the injection mechanism. The decreased occurrence of galling between the sleeve and injection tip is anticipated to result in the stable operation of the injection mechanism. Furthermore, the cycle time is shortened by a reduction in the quantity of molten metal and an acceleration of the solidification of the stamped part. The lower molten metal temperature reduces the amount of heat input to the mold, but the shorter cycle time allows the mold temperature to be set at a higher level. This prevented the occurrence of flow marks in the product section.

Experimental Conditions and Results

Table 1 shows the experimental conditions used to confirm the effectiveness of the PSDC method. Tests were conducted to compare the PSDC method (PSDC) with the conventional condition of holding the molten metal temperature high (general DC). Casting experiments were conducted using a casting machine with a clamping force of 350 tons and a mold for a product mass of 850 g and a minimum wall thickness of 5 mm. JIS ADC12 alloy was used as the material. Figure 2 shows a comparison of X-ray transmission images under each experimental condition. Porosity was observed in the general DC, but not in the PSDC.

Table 1 Comparison of conditions for casting experiments

Casting parameter	General DC	PSDC
Molten metal temperature (℃)	640	605
Low injection speed (m/s)	0.15	0.27
High injection speed (m/s)	2.7	3.2
Casting pressure (MPa)	50	65
Thickness of stamp (mm)	45	20
Delay time of injection (s)	1.0	0
Cycle time (s)	53.9	41.0

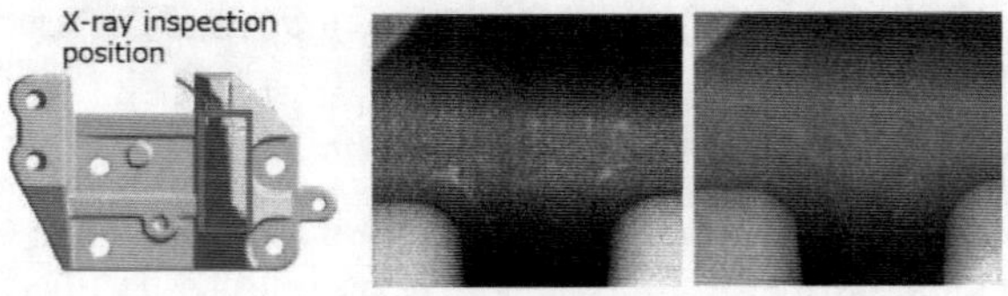

(a) General DC (b) PSDC

Figure 1 Comparison of X-ray transmission images of thick wall

3 Verification by simulation

Temperature Distribution Verification of Injection Mechanism Section

The software utilized for the calculations was CAPCAST®. Figure 2 shows the mold model for flow and solidification calculations and the ladle pouring model. Figure 3 shows the temperature distribution in the sleeve cross section just before the molten metal is poured. It is evident that there is a significant temperature gradient in the sleeve from the pouring section to the stamping section just before the pouring.

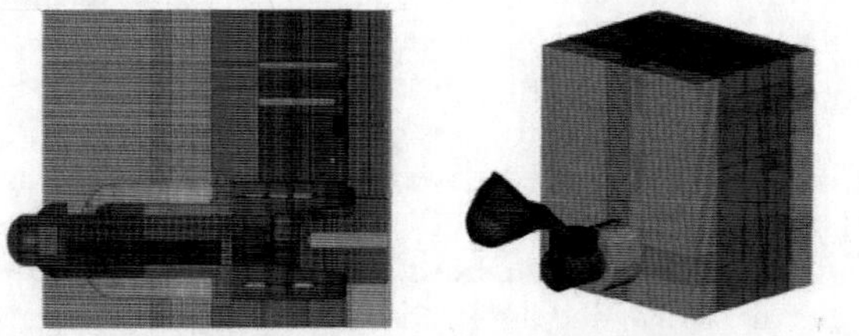

(a) Solidification, Flow calculation model (b) Ladle molten metal pouring model

Figure 2 Full mold model including ladle and injection section

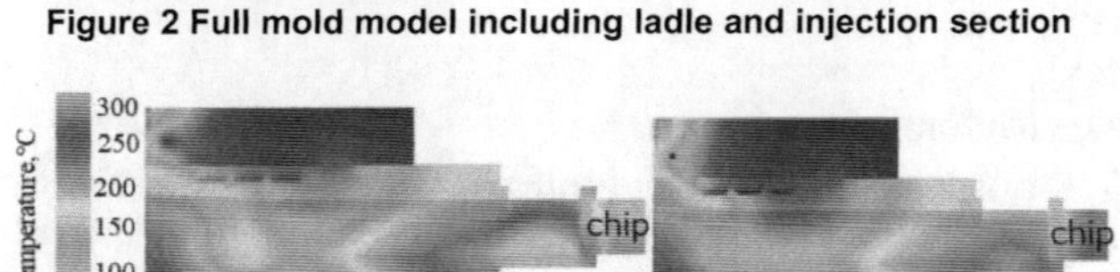

(a) General DC (b) PSDC

Figure 3 Comparison of sleeve cross-sectional temperature distribution during filling

Verification of solid fraction distribution in the sleeve

Figure 4 shows the isosurface distribution of the solid fraction (fs=0.01) of the molten metal in the sleeve immediately after the completion of molten metal injection. In general DC, the solid phase is distributed on the bottom of the sleeve in the form of a skin. In contrast, in PSDC, almost the entire molten metal is covered by a small amount of solid fraction distribution. Figure 5 shows the isosurface volume for each fs as V_{fs}, the total molten metal volume V_{all}, and the isosurface volume ratio $R_{atio} = Vfs / V_{all}$. The molten metal with a large solid fraction at fs=0.5 is almost gone.

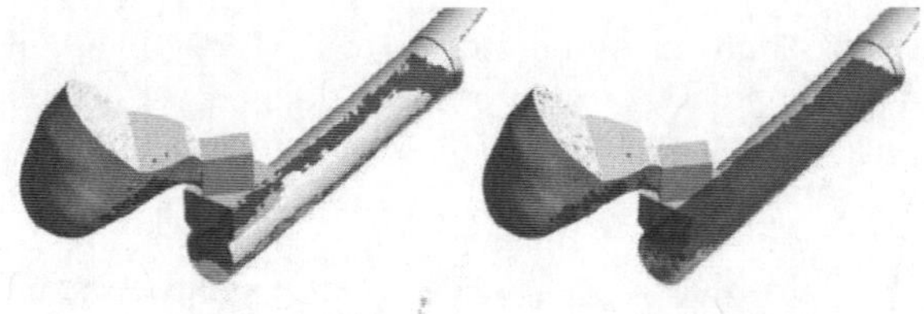

(a) General DC (b) PSDC

Figure 4 Comparison of solid phase ratio isosurface in the sleeve (fs=0.01)

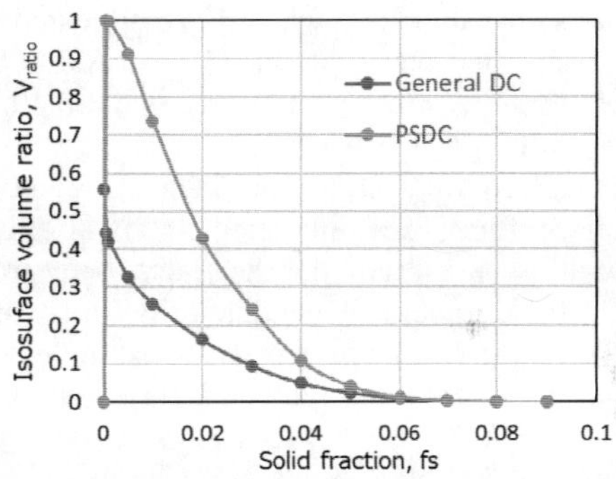

Figure 5 Variation of isosurface solid fraction fs and isosurface volume fraction

4 Conclusion

The solid phase distribution of the molten metal was verified by full model simulation of the PSDC die casting with molten metal temperature control, including the injection and ladle pouring sections.

1) In the PSDC method, a small amount of solid fraction fs=0.0005 is uniformly distributed in the molten metal at the completion of pouring.
2) The decrease in porosity is thought to be due to the small amount of solid phase in the molten metal of the PSDC method, which acts as a crystal nucleus, dispersing the formation of porosity and promoting solidification around the porosity.

References

[1] M. Itamura, K.Anzai, N,Hirata: J-DEC JD12-27 (2012)

[2] R. Koide, M. Kito, J. Shinoda, et al: Report of the 182nd JFS Meeting11(2023):11.

Hot-Work Steels for Additive Manufacture and Additive Manufacture of High Pressure Die Casting Die Inserts Containing Conformal Cooling Channels

Yuzeng Chen[1], Jie Wan[1], Fang Lin[1], Jinyue Wang[2], Zhiqiang Liu[2], Yuanyuan Wang[2]
1. State Key Lab. of Solidification Processing, Northwestern Polytechnical University, Xi'an 710072, P.R. China
2. Suzhou Yunjing Metal Technologies Co., Ltd., Suzhou 215024, P.R. China
*Corresponding address: e-mail: yzchen@nwpu.edu.cn

Abstract: High pressure die casting (HPDC) is an important approach to fabricate Al-based structural components. The structures of HPDC Al-based castings are usually complex. Therefore, to ensure the precision forming of the HPDC castings, it's important to realize an accurate and efficient temperature control of the casting process. As a carrier of the high pressure die casting process, the temperature field of the HPDC die is mainly controlled via cooling channels. To achieve an efficient temperature control, conformal cooling channels embedded in mold core inserts are fabricated, which is complicated in geometry and can be only produced via additive manufacture. However, the additive manufacture processibility of traditional hot-work tool steels are poor and can't meet the requirement for producing the mold inserts containing conformal cooling channels via additive manufacture. In this presentation, the latest research progress of our team on additive manufacturing hot-work tool steels will be shared. Meanwhile, relevant examples on additive manufacture of the mold core inserts containing conformal cooling channels and its advantages in reducing the casting defects of Al die castings and improving the producibility will be shared.

Keywords: Hot-work steels, Additive manufacture, High pressure die casting, Mold inserts

1 Introduction

High pressure die casting (HPDC) is one of the most efficient ways in producing Al, Mg, Zn, and Cu alloys castings. The geometries of the HPDC castings are usually complex. Temperature control via the HPDC die is essentially important for controlling casting defects and production rate of these castings. The reasons are as follows. It is known that solidification of metals is a volume shrinkage process. The finally solidified parts of a casting inevitably subject to significant volume shrinkage which leads to the formation of shrinkage induced defects, such as pores and hot cracks. If the solidification sequence of the castings is manipulated by temperature control via the HPDC die, it will be possible to keep the shrinkage induced defects in the regions that will be removed from the castings. On the other hand, the solidification of the castings releases latent head of the metals which leads to increase of the die temperature. For a scale production, if the latent heat of the die can be conducted out, the production rate of the castings will be inevitably decreased, leading to an increased cycle time of the castings. Further, the over-heat of the die may cause pre-failure of the die. The traditional way in temperature control of the HPDC die is to machine channels in backside of the die where the liquid cooling medium, e.g. water and oil, can be filled in and used to control the temperature of the die. However, due to the restriction of the machining techniques, only straight channels can be usually machined. This will reduce the efficiency and precision of the temperature control.

The concept of conformal channels was proposed to enable a more efficient and precise control of die temperature [1]. Apparently, due the complexity of the conformal channels, these channels are not able to be prepared by traditional machining method.

An additive manufacture technique i.e., selective laser melting (SLM) is a most widely used method to produce the HPDC die inserts. The mostly investigated SLM hot-work steels include medium carbon hot-work steels (e.g., H13, Dievar, a trademark of Assab, and W360, a trademark of Böhler) and maraging steels (e.g., 18Ni300 and 18Ni350). The investigations focus on development of the SLM parameters, failure analyses of the as-produced inserts, and effects of the SLMed inserts containing conformal channels on the casting effects. A comprehensive investigation on evaluation of the properties of the hot-work steels for SLM, and applications of the SLMed inserts in solving casting defects and production rates is still limited. In this work, a systematic study on these issues will be reported.

2 Experimental procedure

A hot-work steel with a trademark of YJ01 which was developed by Suzhou Yunjing Metal Technologies Co., Ltd was used to prepare the specimens for mechanical property tests and produce the HPDC inserts. The hardness and Charpy V notched impact energy of the as-prepared specimens were performed. The as-produced inserts was installed in a HPDC die for tests.

3 Result and discussion

Figure 1 shows an SEM image of the YJ01 steel manufactured by SLM method and subjected to heat treatment. It is seen that the microstructure shows a typical cellular morphology of the alloys manufactured by SLM method. Formation of the cellular structure of the SLMed alloy can be ascribed to high positive temperature gradient up to 10^7-10^8 K/m, extremely high cooling rate upon solidification (10^6-10^7K/s), and ultrahigh solidification rate (around 1m/s) which suppress the development of secondary dendrite arms.

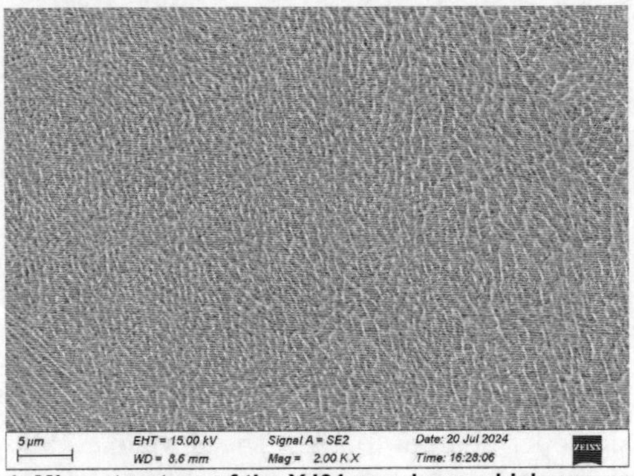

Figure 1. Microstructure of the YJ01 specimen which was prepared by SLM method and subjected to heat treatment.

Table 1 shows the mechanical properties of the SLMed and heat treated YJ01 steel where it can be seen that the YJ01 exhibits excellent mechanical performance as a hot work steel. Especially, the Charpy V notched impact energy is as high as 39 J which is higher than commonly used hot work steels, e.g., H11, H13, and Dievar, with a value usually less than 30 J at a similar hardness. Also, the yield strength and elongation to failure are also comparable to the normal commercial hot work steels.

Table 1. Mechanical properties of the SLMed and heat treated YJ01 steel.

Properties	Hardness, HRC	Charpy V notched impact energy, J	Yield strength, MPa	Elongation to failure, %
Values	45	39	1433	11

Figure 2 shows the fading of the hardness against heating at 550 ℃ and 600 ℃, which are two typical surface temperatures of HPDC dies. The data for a well-known commercial hot work steel, Dievar, were also added in the plot for comparison. It is seen that after 100 h heating at 550 ℃ and 600 ℃, the hardness of YJ01 faded by 2 HRC and 6 HRC, respectively, which are much lower than those of Dievar. This indicate that the SLMed YJ01 exhibits an excellent thermal resistance during service.

Figure 3 shows the results of a comparison test for HPDC castings, which is highly susceptible to solidification cracks in deep cavity of the castings (see Figure 3a). It is shown in Figure 3b, by replacing the machined insert with a straight cooling channel by the SLMed one, the casting defects are eliminated, showing the enhanced efficiency of temperature control of the inserts with conformal cooling channels.

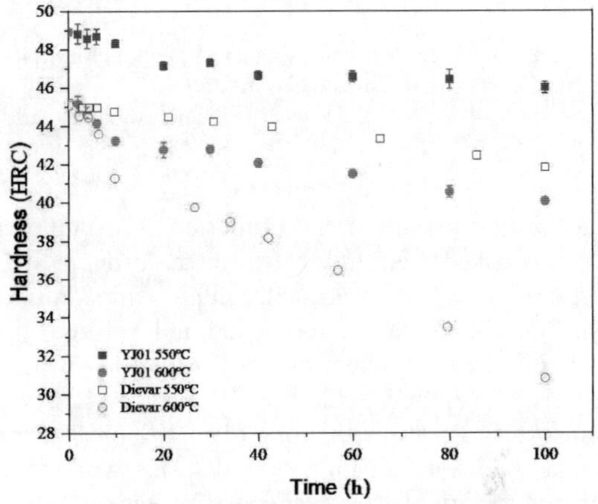

Figure 2. Evolution of hardness of the YJ01 specimen at 550 ℃ and 600 ℃ for a period up to 100 h. Data of Dievar were cited from Ref. [2]

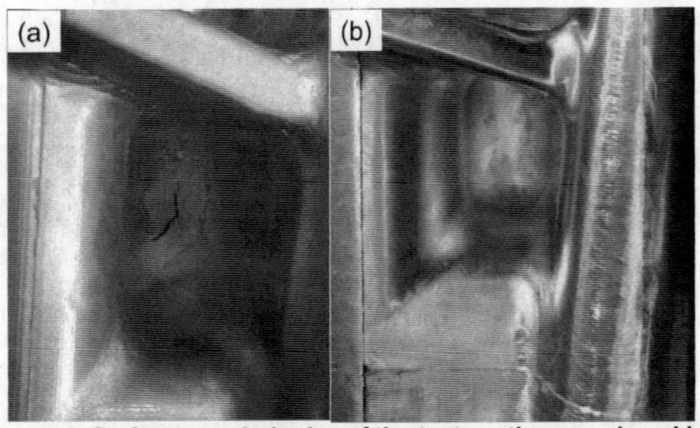

Figure 3. Surface morphologies of the test castings produced by using (a) machined insert with straight cooling channel, and (2) SLMed insert with conformal cooling channel.

4 Conclusion

The YJ01 hot work steel for SLM method exhibits excellent performance when used as a material to fabricate HPDC die inserts with conformal cooling channels. The application of the SLMed insert is highly helpful in controlling the casting defects.

Acknowledgments

This work was supported by the National Key R&D Program of China (No. 2022YFB3404203).

References

[1] Fiorentini F, Curcio P, Armentani E, Rosso C, Baldissera P. Study of two alternative cooling systems of a mold insert used in die casting process of light alloy components. Procedia Structural Integrity 2019, 24: 569–582.

[2] https://www.assab.com/app/uploads/sites/133/2019/05/Dievar_EN.pdf.

Development of High Temperature Die-Cast Magnesium Alloys for Small IC Engine Application

Xixi Dong[1,*], Shouxun Ji[2], Zhongde Shan[1]
1. College of Mechanical and Electrical Engineering, Nanjing University of Aeronautics and Astronautics, Nanjing 210016, China
2. Brunel Centre for Advanced Solidification Technology (BCAST), Brunel University London, Uxbridge UB8 3PH, United Kingdom
*Corresponding address: e-mail: dongxixi@nuaa.edu.cn

Abstract: An innovative and superior coarsening/creep resistant Al-based ternary short-range order (SRO,0–2nm)/cluster(2–10nm) at critically high temperature and stress of 300°C/50MPa was proposed and verified in Mg, and a highly heat-resistant die-cast Mg alloy was thus successfully developed with the super low steady-state creep rate ($1.35\times10^{-10}s^{-1}$) and long creep life (>600h) under the critical condition of 300°C/50MPa. This work changes the common belief of the detrimental effect of Al on the high temperature creep of Mg alloys over the past few decades, and it provides key evidence for the long-term and massive existence of SRO/clusters in Mg alloys under critically high temperatures and stresses. It greatly increases the working temperature of die-cast Mg alloys from 120-200°C to 300-350°C. The developed alloy has been transferred into manufacturing high temperature (300-350°C) Mg components used in small internal combustion (IC) engines.

Keywords: Magnesium alloys; Die casting; High temperature; Engine application

1 Introduction

Magnesium (Mg) alloys are excellent candidate materials for lightweight structures in the automobile, aerospace, rail and other industries when compared to steel and aluminum (Al) alloys [1,2]. Over the past few decades, one of the beneficial applications of magnesium alloys is the manufacturing of powertrain components, operating at the elevated temperatures of 150–200 °C [1,2]. Recently, this advancement has focused on the die-castable components in small IC engines, used in machinery such as powered tools, drones and unmanned aerial vehicles, with a working temperature capable of reaching 200–350 °C, and where the creep resistance is a major challenge [1,2].

In the past, for Mg alloys, Al was considered beneficial for castability, but it was found to be detrimental for creep resistance at elevated temperatures [1,3]. One effective approach to improve elevated creep resistance is to develop Al-free die-cast Mg-RE alloys. However, the RE content in die-cast Mg–RE alloys was usually lower than 4 wt.% as a high RE resulted in brittleness, and this restricted the die-cast Mg-RE alloys from working at higher temperatures, that is above 200 °C [1,2,4].

In this work, an innovative and superior coarsening/creep resistant Al-based ternary short-range order (SRO,0–2nm)/cluster(2–10nm) at critically high temperature and stress of 300°C/50MPa was proposed and verified in Mg, and a highly heat-resistant die-cast Mg alloy was developed with the proper addition of Al. The developed alloy has been transferred into manufacturing high temperature (300-350°C) Mg components used in small IC engines.

2 Experimental procedure

Cylindrical creep samples designed with ISO 204:2018 were prepared by high pressure die casting (HPDC) on a 4500 kN cold chamber machine; the gauge size of the creep samples were Φ6.35×20 mm, respectively [1]. The developed die-cast Mg3.5RE1.5GdMnAl alloy was melted in a steel liner and held at 720 °C, under the mixed protection gas of N_2 (6 L/min) and SF_6 (0.025 L/min). The die and pour temperatures were controlled at 225 °C and 715 °C during HPDC [2].

Tensile creep tests were conducted on a lever-type Instron machine following ISO 204:2018, and the temperature of during creep was controlled by thermocouples with an accuracy of ±1 °C [1]. Transmission electron microscopy (TEM) analysis was carried out on a Thermofisher Titan Themis 60-300 microscope operated at 300 kV.

3 Result and discussion

High temperature creep properties

Fig. 1 shows the representative tensile creep curve of the die-cast Mg3.5RE1.5GdMnAl alloy tested under the critical temperature of 300 °C and the applied stress of 50 MPa. The alloy went into the tertiary state after creep for a considerable long time of 460 h and still did not reach the fracture point after 620 h, and the alloy had a super low steady state creep rate of $1.35\times10^{-10}s^{-1}$.

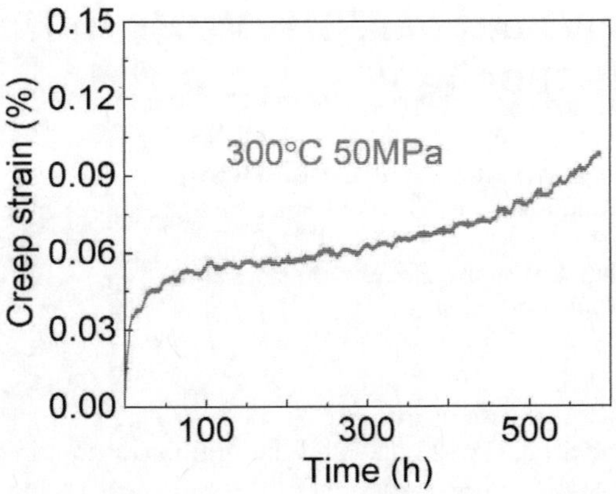

Fig 1. Tensile creep curve of the die-cast Mg3.5RE1.5GdMnAl alloy tested under the critical temperature of 300 °C and the applied stress of 50 MPa.

Microstructure of SRO and cluster

Fig. 2 presents the TEM image showing the Al-based ternary heat-resistant SRO and cluster in the Mg matrix of the die-cast Mg3.5RE1.5GdMnAl alloy, and the SRO and clusters were still smaller than 10 nm after creep at 300 °C/50MPa for 400 h.

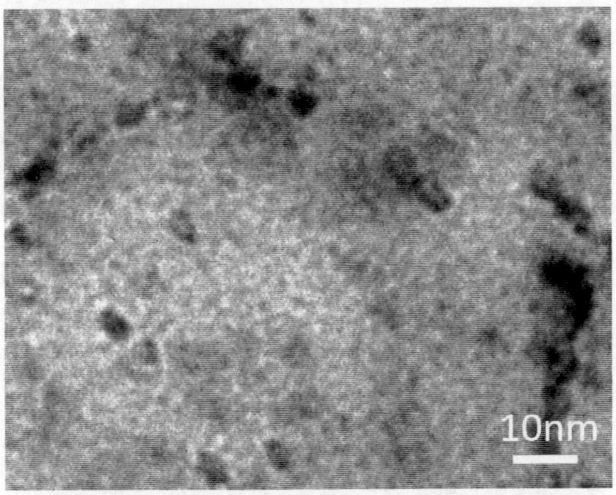

Fig 2. TEM morphology of the Al-based heat-resistant SRO and cluster in the Mg matrix of the die-cast Mg3.5RE1.5GdMnAl alloy after creep at 300 °C/50MPa for 400 h.

Application in small IC engine

The developed die-cast Mg3.5RE1.5GdMnAl alloy has been transferred into manufacturing high temperature (300-350°C) Mg pistons used in small IC engines, as shown in Fig. 3.

Fig 3. Application of the die-cast Mg3.5RE1.5GdMnAl alloy for the manufacturing of the piston used in small IC engines.

4 Conclusion

(1) An innovative and superior coarsening/creep resistant Al-based ternary SRO(0–2nm)/cluster(2–10nm) at critically high temperature and stress of 300°C/50MPa was proposed and verified in Mg.

(2) A highly heat-resistant die-cast Mg3.5RE1.5GdMnAl alloy was developed with the super low steady-state creep rate ($1.35\times10^{-10}s^{-1}$) and long creep life (>600h) under the critical condition of 300ºC/50MPa.

(3) The developed die-cast Mg3.5RE1.5GdMnAl alloy was transferred into manufacturing high temperature (300-350°C) Mg components used in small internal IC engines.

Acknowledgments

This work has been supported by the National Natural Science Foundation of China.

References

[1] Dong X X, Feng L Y, Wang S H et al. On the exceptional creep resistance in a die-cast Gd-containing Mg alloy with Al addition. Acta Mater., 2022, 232: 117957.

[2] Dong X X, Feng L Y, Wang S H et al. A new die-cast magnesium alloy for applications at higher elevated temperatures of 200–300° C. J. Magnes. Alloy., 2021, 9: 90–101.

[3] Mo N, Tan Q Y, Bermingham M et al. Current development of creep-resistant magnesium cast alloys: A review. Mater. Des., 2018, 155: 422–442.

[4] Easton M, Gibson M A, Zhu S M et al. Development of magnesium-rare earth die-casting alloys. Magnes. Technol., 2018, 329–336.

Effect of Mold Wall Thickness on the Defect Band of AlSi10MgMn in High-Pressure Die Casting Technology

Zhenyu Sun[1], Wenbo Yu[*,1], Junjie Li[1], Weichen Zheng[1], Shoumei Xiong[2]
1. Center of Materials Science and Engineering, School of Mechanical and Electronic Control Engineering, Beijing Jiaotong University, Beijing 100044, China
2. School of Materials Science and Engineering, Tsinghua University, Beijing 100084, China
*Corresponding address: e-mail: wbyu@bjtu.edu.cn

Abstract: In order to investigate the defect band within the casting, the design of a stepped mold with a thickness gradient of 1mm (the thickness of its five stepped areas were 5mm, 4mm, 3mm, 2mm and 1mm), the selection of AlSi10MgMn alloys to carry out high pressure die casting experiments. Statistical analysis reveals that the band thicknesses of 2mm、3mm and 4mm die castings were measured to be in the range 7 - 18 mean grains wide, which is substantial evidence on that defect bands form due to strain localization in partially solidified alloys during HPDC. In addition, the defect band of 1mm, 2mm and 5mm areas are relatively indistinct. For 1mm and 2mm area, the filling and solidification process of castings plays a deterministic role. But for 5mm area, the distribution of ESCs in band is the key reason. At the same, the distribution of ESCs also makes the defect bands in the 3mm and 4mm regions well defined. Finally, no additional porosity occurs in the AlSi10MgMn alloys, which display defect bands of positive macrosegregation.

Keywords: AlSi10MgMn; High Pressure Die Casting; Defect Band; Externally Solidified Crystals; Porosity

1 Introduction

At present, the ultra-large integrated die casting technology equipped with large and complex metal molds is the first choice for achieving efficient near-net forming of complex lightweight aluminum alloy structural parts. Logically, the ultra-large integrated die is inevitably composed different wall thicknesses. This study focused on the microstructure and defects of different wall thickness areas of stepped castings under the same die casting process, and to analyze the influence of ESCs on defect bands. In addition, 3D characterization was conducted on different wall thickness areas within the stepped casting, and their porosity was statistically analyzed. The reasons for the abnormal porosity in the 1mm, 2mm, and 5mm wall thickness areas were explained.

2 Experimental procedure

The stepped sample with dimensions of 5mm, 4mm, 3mm, 2mm and 1mm was prepared by cold chamber HPDC TOYO-BD-350V5. The die casting molten metal liquid temperature is 695°C and the mold preheating temperature is 150°C.

3 Result and discussion

The defect band width (w) and the grain size (d_{sb}) inside of defect band were extracted and their ratio w/d_{sb} was plotted in Fig.1. The w/d_{sb} for the defect bands in the steps of 2mm, 3mm and 4mm were in the range of 7-18. This result is consistent with that for dilatant shear bands in both classical granular materials and from controlled rheology experiments on partially solid alloys[2]. This provides substantial evidence that defect bands in HPDCs are dilatant shear bands. However, the w/d_{sb} in the step of 5mm is about 24.47, which is beyond the range of 7-18.

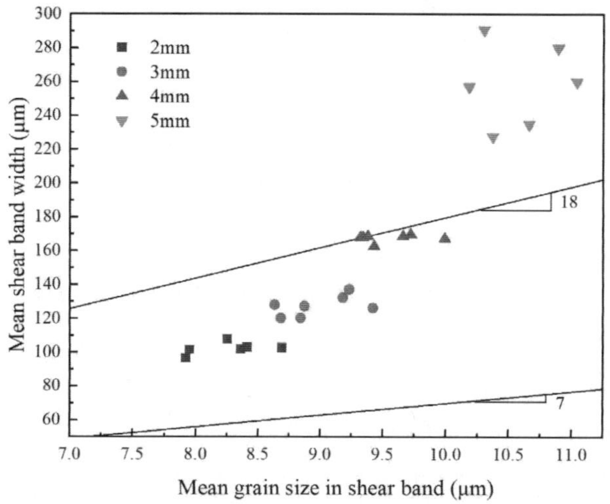

Fig. 1 The relationship between band thickness and average grain size in the defect band (w/d_{sb}) for different wall thickness area

The f_s^{coh} marks the point of impingement of the dendrites and where strength development begins, and the f_s^{pk} corresponds to the maximum packing solid fraction. When the solid fraction is higher than f_s^{pk}, the shear strength of the grain network increases rapidly[2]. The value of f_s^{coh} and f_s^{pk} is closely related to the grain size and morphology, as shown in Fig. 2. So, the distribution of ESCs in the defect band will have a serious impact on them.

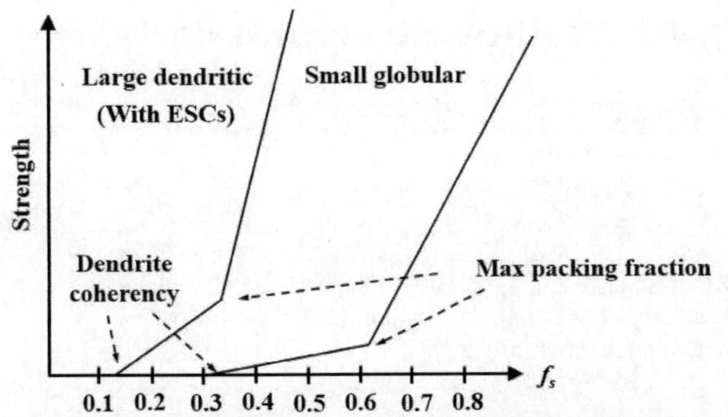

Fig. 2 General behavior of the increase of shear strength vs solid fraction in the mushy zone for two extremes in microstructures: large dendritic grains and small globular grains [2]

4 Conclusion

In order to investigated the influence of wall thickness on the defect band, AlSi10MgMn alloy was casted by HPDC equipped with one stepped die containing different thicknesses (1mm, 2mm, 3mm, 4mm and 5mm). This study has revealed the following.

The w/d_{sb} of defect bands in the steps of 2mm-4mm are in the range of 7-18. This supports the hypothesis that partially solid HPDC microstructures deform as granular materials and that strain localizes into dilatants shear bands during processing. However, the w/d_{sb} in the step of 5mm is beyond the range of 7-18. The abnormal ESCs distribution of defect bands and the lowest shear rate in the step of 5mm were considered to be the key factors.

By statistically analyzing the ESCs distribution of different wall thickness areas with different locations, filling and solidification processes, the influence of ESCs and shear rate on defect band formation was elucidated. The extensive distribution of ESCs within the defect band is not conducive to the formation of the defect band; excessive or insufficient filling speed is detrimental to the formation of defect bands.

References

[1] Rodrigo P D D, Ahuja V. Effect of casting parameters on the formation of 'pore/segregation bands' in magnesium die castings[J]. 2000.

[2] Laukli, H.I., Gourlay, C.M. & Dahle, A.K. Migration of crystals during the filling of semi-solid castings. Metall Mater Trans A 36, 805-818 (2005).

Squeeze Casting Technology for Mg-Al-Ca Alloy Structural Parts

Shengyue Cui [1], Hua Hou[1,3], Liwen Chen[1], Jianhui Jing[1], Yuan Zhao[1], Yuhong Zhao[1,2,4*]
1 School of Materials Science and Engineering, Collaborative Innovation Center of Ministry of Education and Shanxi Province for High-performance Al/Mg Alloy Materials, North University of China, Taiyuan, 030051, China
2 Beijing Advanced Innovation Center for Materials Genome Engineering, University of Science and Technology Beijing, Beijing, 100083, China
3 School of Materials Science and Engineering, Taiyuan University of Science and Technology, Taiyuan, 030024, China
4 Institute of Materials Intelligent Technology, Liaoning Academy of Materials, Shenyang, 110004, China
*Corresponding address: e-mail: zhaoyuhong@nuc.edu.cn

Abstract: Four kinds of Mg-Al-Ca-Zn-Mn alloys with different Ca and Al mass ratios of 0.3, 0.6, 0.9, and 1.2 were studied. They are Mg-6.2Al-1.8Ca-0.7Zn-0.4Mn, Mg-5Al-3Ca-0.7Zn-0.4Mn, Mg-4.2Al-3.8Ca-0.7Zn-0.4Mn, and Mg-3.6Al-4.4Ca-0.7Zn-0.4Mn, respectively. By squeeze casting, alloys with high strength, good thermal conductivity, and acceptable heat resistance are obtained. The effect of the Ca/Al ratio on the microstructure and mechanical properties of the squeeze casting alloy was investigated. The Ca/Al ratio can be used to manipulate the mechanical properties of the alloy system, and the volume fraction, type, and morphology of the phases can also be controlled by the Ca/Al ratio. With the increase of the Ca/Al ratio, the yield strength increases, and the tensile strength decreases; the precipitated phase changes from Al_2Ca phase to $(Mg, Al)_2Ca$ phase and finally to Mg_2Ca phase.

Keywords: Mg-Al-Ca-Zn-Mn alloys; Squeeze casting; Organizational evolution; Mechanical property

1 Introduction

At present, most magnesium components are usually processed by high-pressure die casting (HPDC), and this process can produce magnesium alloy products with a certain complex structure at a low cost. However, in the production of complex thin-wall castings, it is easy to produce defects such as shrinkage holes and pores; as a result, the performance of magnesium alloys produced by die casting is low, and they often fail to meet the requirements of complex mechanical parts [1-3]. Therefore, the application of Mg-Al-Ca alloy in squeeze casting is studied in this paper. The squeeze casting process can produce castings with a certain complex structure and effectively eliminate defects such as porosity, shrinkage, and blowholes caused by casting during the solidification process [4]. The addition of Ca helps to improve thermal conductivity, refine the structure, and enhance the fluidity of the liquid metal, playing a beneficial role in both the casting and squeeze casting processes of mold filling and solidification [5]. In this study, C15-Al_2Ca, C36-$(Mg, Al)_2Ca$, and C14-Mg_2Ca phases in squeeze-cast Mg-Al-Ca alloys with four components were systematically studied. The four alloys showed different mechanical properties in the as-cast states, and the phenomenon was reasonably explained by a comprehensive analysis of their microstructures. Therefore, we systematically changed the Ca/Al ratio in Mg-Al-Ca alloys and studied the microstructure formation and mechanical properties of Mg-Al-Ca alloys in different squeeze-cast states.

2 Experimental procedure

Four kinds of Mg-Al-Ca-Zn-Mn alloys, with Ca and Al mass ratios of 0.3, 0.6, 0.9, and 1.2, were prepared. Their compositions are Mg-6.2Al-1.8Ca-0.7Zn-0.4Mn, Mg-5Al-3Ca-0.7Zn-0.4Mn, Mg-4.2Al-3.8Ca-0.7Zn-0.4Mn, and Mg-3.6Al-4.4Ca-0.7Zn-0.4Mn (they are expressed as C/A0.3, C/A0.6, C/A0.9, and C/A1.2, respectively).The alloys were fabricated in an electric-resistant furnace under an argon atmosphere by melting high purity Mg, Al, Zn, Mn, and Mg-30Ca intermediate alloys at 750℃ for 30 min. Magnesium alloy partition casting was produced by a SHP34-500A multifunctional liquid forming machine. The detailed parameters used in this study are as follows: pouring temperature, ~750℃; punch pressure, 125 MPa; punch speed, 0.63 m/s; mold holding time, ~55 s. Through scanning electron microscope (SEM) tests, we observe the analysis of the sample's microstructure and phase distribution. Through point-scan energy spectrum analysis, we examine the element distribution in the sample's microstructure to determine the composition of the phases.

3 Result and discussion

The SEM images of the four alloys in their cast states are shown in Fig. 1. EDS analysis was performed for different microstructure components at the points marked by characters in Fig. 1 (b, d, f, and h). EDS analysis shows that the main intermetallic phase in C/A0.3 alloy is $Mg_{17}Al_{12}$ and Al_2Ca; the main intermetallic phase in C/A0.6 alloy is Al_2Ca and $(Mg,Al)_2Ca$; the main intermetallic phase in C/A0.9 alloy is $(Mg,Al)_2Ca$; and the main intermetallic phase in C/A1.2 alloy is $(Mg,Al)_2Ca$ and Mg_2Ca.

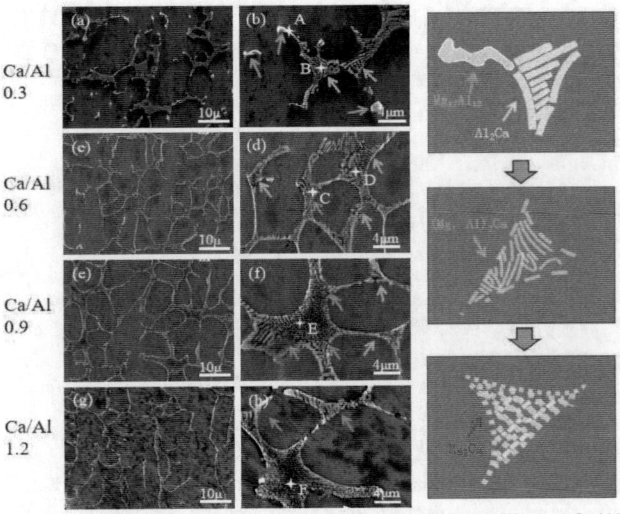

Figure 1. SEM images of as-cast alloys containing different Ca/Al ratios and a schematic diagram of the second phase changes, in which the green arrow marks the $Mg_{17}Al_{12}$ phase, the yellow arrow marks the Al_2Ca phase, the blue arrow marks the $(Mg,Al)_2Ca$ phase, and the red arrow marks the Mg_2Ca phase.

As the Ca/Al ratio increases, the YS increases, and the elongation decreases. Alloy C/A0.3 has a tensile strength of 152 MPa and 4.1% uniform elongation at room temperature, while alloy C/A1.2 shows a tensile strength of 134 MPa and 1.3% uniform elongation at room temperature. Alloys C/A0.6 and C/A0.9 exhibit intermediate YS, UTS, and UE values between alloys C/A0.3 and C/A1.2; alloy C/A0.6 has a tensile strength of 147 MPa and 3% uniform elongation.

4 Conclusion

(1) The Ca/Al ratio is an effective tool for controlling the type of second phase in Mg-Al-Ca alloys: from Al_2Ca and $(Mg,Al)_2Ca$ Laves phase at low Ca/Al ratios to $(Mg,Al)_2Ca$ and Mg_2Ca Laves phase at high Ca/Al ratios.

(2) As the Ca/Al ratio increases, the volume fraction of the second phase increases, along with their morphological changes: from a disconnected network of coarse laminae (Ca/Al=0.3) to an interconnected skeleton network of fine particles (Ca/Al=1.2).

(3) The second phase is the key factor affecting the strength of the alloy. The hardness, elastic modulus, and yield strength of the alloy increase with the increase of the volume fraction of the second phase, but the elongation and tensile strength decrease with the increase of the Ca/Al ratio.

Acknowledgments

This work was supported by National Natural Science Foundation of China (Nos. 52375394, 52275390, 52205429, 52201146, 52305429), National Defense Basic Scientific Research Program of China (JCKY2020408B002), Key Research and Development Program of Shanxi Province (Nos. 202102050201011).

References

[1] Wang S L, Nie K B, Yang A and Deng K K. Effect of Ca/Al ratio on microstructure and mechanical properties of slow extrusion high-strength Mg-Al-Ca-Mn alloys. Mat. Sci. Eng. A-Struct., 2023, 32(7): 3129-3141.

[2] Koltygin A V, Bazhenov V E, Belova E A and Nikitina A A. Development of a magnesium alloy with good casting characteristics on the basis of Mg–Al–Ca–Mn system, having Mg–Al_2Ca structure. J. Magnes. Alloy, 2013, 1(3): 224–229.

[3] Kondori B and Mahmudi R. Effect of Ca additions on the microstructure, thermal stability and mechanical properties of a cast AM60 magnesium alloy. Mat. Sci. Eng. A-Struct., 2010, 527.7-8: 2014-2021.

[4] Zubair M, Sandlöbes S, Wollenweber M A, Kusche C F, Hildebrandt W and Broeckmann C. On the role of Laves phases on the mechanical properties of Mg-Al-Ca alloys. Mat. Sci. Eng. A-Struct., 2019, 756: 272-283.

Characterization of Externally Solidified Crystals (ESCs) and Porosity of HPDC AlSi9MnVZr Alloy

Saria Akhtar[1], Shoumei Xiong[1, 2*]
1. School of Materials Science, Tsinghua University, Beijing 100084, China
2. Key Laboratory for Advanced Materials Processing Technology, Ministry of Education, Beijing 100084, China
*Corresponding address: e-mail: smxiong@tsinghua.edu.cn

Abstract: An investigation was conducted to characterize externally solidified crystals (ESCs) and porosities in a high-pressure die-cast AlSi9MnVZr alloy using an optical microscope (OM), SEM, and computed tomography (CT). Particularly two sets of HPDC process parameters were designed. The microstructure of HPDC AlSi9MnVZr alloy mainly comprised of primary α-Al, ESCs, Al-Si eutectic and iron-rich phase. ESCs nucleate in the shot sleeve whereas α-Al form in die cavity. In sample produced without fast shot speed and intensification pressure revealed the aggregation of larger-size dendritic ESCs in the casting. Moreover, big gas pores and shrinkage pores along with many small dispersed pores with high porosity fraction were observed. When both slow shot speed and fast shot speed were induced, the diameter and the area fraction of ESCs in casting were significantly reduced. In addition, the value of porosity substantially lowered and obvious reduction in the porosity fraction percent was observed as its value decreased from 1.07 % to 0.1 %.

Keywords: high-pressure die-cast, Al-Si alloy, ESCs, porosity

1 Introduction

High pressure die casting process is extensively used to manufacture aluminum castings due to its ability to produce complex shapes with excellent dimensional accuracy and surface finish [1-2]. However, due to the characteristics of slow shot speed in the shot sleeve, higher speed filling and high intensification pressure in the die cavity result in the formation of ESCs and porosities in die-cast parts, which limits their potential applications [3]. Currently, heat-treated Al-Si alloys are most commonly used alloys, heat treatment consumes energy which leads to the deformation of large die-casting parts and increases the difficulty of subsequent machining and the cost. Therefore a non-heat treated AlSi9MnVZr alloy was developed to obtain highly efficient properties with less energy consumption. Previous studies have reported that several process parameters influence the extent of defects in die-cast parts such as die temperature, melt temperature, low shot speed, fast shot speed, intensification pressure, and casting geometry. However, the study on the influence of process parameters on ESCs and porosities in Al-Si alloys is insufficient. In this study, HPDC AlSi9MnVZr alloy was developed and the influence of two sets of process parameters on the microstructures was studied.

2 Experimental procedure

A TOYO BD-350V5 cold chamber die-casting machine was used to produce AlSi9MnVZr alloy. The samples with dimensions of φ6.4 mm*10 mm were cut from the center of the standard tensile test bar. The samples were ground using 400# to 2000# SiC paper, polished with diamond past, and etched using Murakami reagent (10 g of NaOH, 5 g of $K_3(Fe (CN)_6$, and 60 ml of H_2O). The microstructure was examined using a Zeiss scope A1 optical microscope (OM) and laboratory CT scan (nanotom m CT). The working electric current, voltage, and the voxel in CT were set as 110 μA, 100 kV and 3 μm, respectively.

Table 1. Key parameters of HPDC process

Sample	Slow shot speed	Fast shot speed	Intensification pressure
S-F	0.1	1.04	-
S	0.1	-	-

3 Result and discussion

Fig.1 (a-b) shows the microstructure of AlSi9MnVZr alloy. The microstructure comprises of primary α-Al, ESCs, Al-Si eutectic, pores, iron-rich phase and Zr-rich phase. ESCs nucleate in the shot sleeve whereas α-Al form in die cavity.

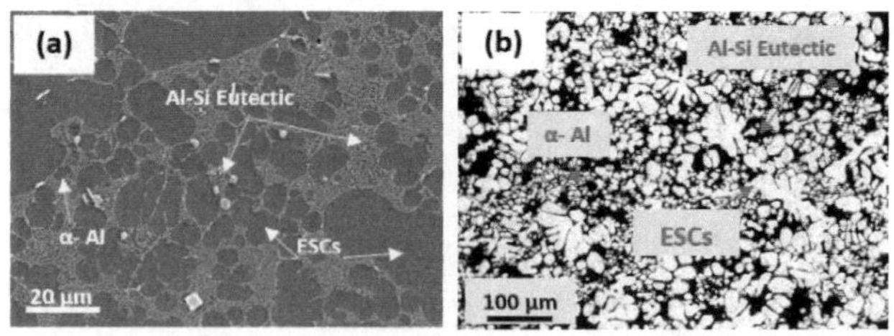

Figure 1 Microstructure of AlSi9MnVZr alloy (a) OM; (b) SEM

Fig.2 shows distribution of ESCs at surface and center of bar samples at two different process parameters. The equivalent diameter and the number of ESCs increased from the surface to the center for both sample.

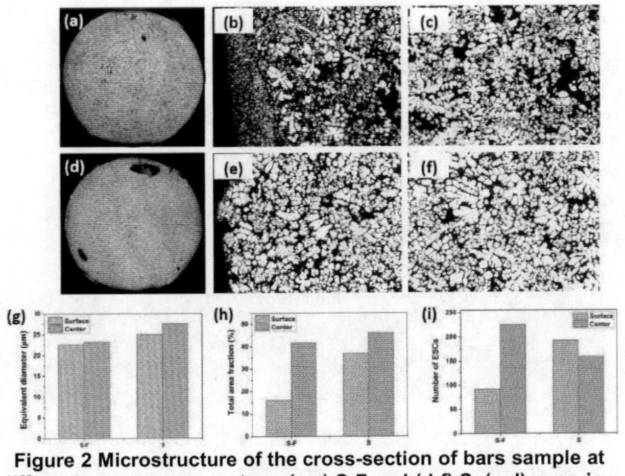

Figure 2 Microstructure of the cross-section of bars sample at different process parameters; (a-c) S-F and (d-f) S. (a,d) overview of the microstructure across the cross-section; (b,e) and (c,d) magnified view of surface and center; (g-i) statistical distribution of ESCs.

In Fig.2 (b-c), from the surface to the center, the size of ESCs was smaller in the bar sample with slow shot speed and fast shot speed (S-F), whereas both the size and the total area fraction of ESCs increased from surface to center in the bar sample without fast shot speed and intensification pressure (S), as shown in Fig.2 (e-f). Fig 2(g-h) shows that the average equivalent diameter increased from 23 μm to 27 μm and the total area fraction increased from 28.8% to 41.5%. During the filling process in HPDC, higher melt flow speed caused a turbulent state, which created considerable thermal shock and convection, resulting in ESC remelting and fragmentation.

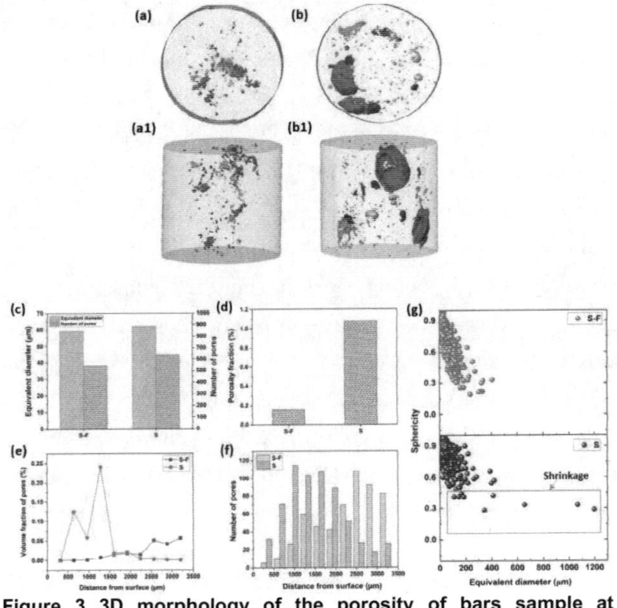

Figure 3 3D morphology of the porosity of bars sample at different process parameters; (a-a1) S-F and (b-b1) S; (c-g) statistical distribution of porosity.

Fig. 3 shows 3D morphology of the porosity of two different process parameters. It is evident from the sample S-F that the fast shot speed could significantly reduce porosity. As shown in Figs.3 (a, a1), the size of pores dramatically decreases in the specimens produced with slow shot speed and fast shot speed. As shown in Figs. 3(b-b1), without fast shot speed and intensification pressure, larger pores are distributed in the bar cross-section accompanied with a lot of dispersed small pores. According to the statistics in Fig. 3(c-d), the average diameter of the pores increased from 59 μm to 62 μm and the porosity fraction percent exhibited an obvious declined trend, and its value decreased from 1.07 % to 0.1 % in S-F sample, compared to S sample. Fig. 3(e-f) shows the volume fraction percent of pores and the number of pores aggregated. The ESCs had continued growth and showed an increase in size in the S bar sample. As a result, additional and larger ESCs would be linked together, creating greater spaces between them. This makes it more challenging to fill the remaining molten material, leading to the formation of more and larger pores.

4 Conclusion

(1) The AlSi9MnVZr alloy comprised of α-Al, ESCs, Al-Si eutectic, iron-rich phase and Zr-rich phase.

(2) The absence of fast shot speed and intensification pressure resulted in larger dendritic ESCs aggregation and the formation of larger pores with high porosity fraction percent.

(3) When slow shot speed and fast shot speed were introduced, the equivalent diameter and area fraction of ESCs decreased. Moreover, the porosity fraction percent exhibited an obvious decline, and its value decreased from 1.07 % to 0.1 %.

Acknowledgments

This work was financially supported by the National Key Research and Development Program of China (2022YFB3404201), and the Major Science and Technology Project of Changchun City, Jilin Province (Grant No. 20210301024GX).

References

[1] Liu F, Zhao H, Chen B, Zheng H. Investigation on microstructure heterogeneity of the HPDC AlSiMgMnCu alloy through 3D electron microscopy. Materials and Design. 2022;218

[2] Dong X, Yang H, Zhu X, Ji S. High strength and ductility aluminium alloy processed by high pressure die casting. Journal of Alloys and Compounds. 2019;773:86-96

[3] Soares G, Neto R, Madureira R, et al. Characterization of Al Alloys Injected through Vacuum-Assisted HPDC and Influence of T6 Heat Treatment. Metals. 2023;13(2)

High Thermal Conductivity Tooling Steel Fastcool 55 Solving Soldering Problems in Die Casting Applications

Shi Li[1,*], Mingchu Xu[2]
1. Shanghai Excellence Mould materials Co.. Ltd. Shanghai, China
2. Ningbo Beilun Excellence Mould Technology Co. Ltd. Ningbo, Zhejiang, China
Corresponding address: Room 916, Shanghai GongheXin Lu 2449; 200072
*E-mail:semmpeter@163.com

Abstract: Fastcool 55 material has a thermal conductivity of 47 W/mC which doubles the figure of commonly used hot work tooling steel H 13 material. The result of this leading to a higher and faster solidification of liquid Aluminum when in contact with Fastcool 55 materials in die casting dies.

With typical application of insert pins and square inserts, Fastcool 55 material increased the insert tooling life 3 times and more.

Also, Fastcool 55 material has much higher tempering resistance. This property increased the thermal fatigue resistance of the material. Leading to longer tooling life.

Applicatiion case studies are cited which illustrate that Fastcool 55 material has much better performance in resisting soldering problems when used as die casting die steels.

Key words: Fastcool 55; soldering; tooling life

1 Introduction

High Pressure Die casting is an application technology which produces near net shape Al/Mg/Cu alloy products at a rather fast speed as against normal low pressure or gravity castings. The process of die casting is designed with a runner and gating system which sprays the alloys into die cavities to form the demanded products. With different designs of the runner and gates the liquid metal has certain designed speed to inject into the cavity.

In Al die casting, soldering may happen when the Al temperature is too high or gating speed is over 60m/s. With normal Din 1.2344 (H13) tooling steels or modified Din1.2367 materials such as NADCA C category soldering occur when injected liquid Al temperature is over 690°C or gating speed over 60m/s[(1)]. This is due to the fact that Al has solubility to iron.

When tooling steel changed to Fastcool 55 [(2)] material which has a higher thermal conductivity than Din 1.2344 (H13) materials, the material has superior tooling failure resistance leading to longer tooling life.

2 Mechanism

The Fastcool 55 material has a chemical composition which has high Mo content which yields more ferrite formation leading to high thermal conductivity of the material. The Si and Cr content lowered down to< 0.05%wt.

Typically, Din1.2344 (H13) tooling steels have a thermal conductivity of 25W/mC; whereas Fastcool 55 material has a thermal conductivity of 47W/mC. The higher thermal conductivity of Fastcool 55 transfers the liquid Al solidification heat faster into the tooling steel materials than conventional Din 1.2344 (H13) steels or NADCA C category steels.

In Al high pressure die casing, when liquid Al in contact with Fastcool 55 tooling steel it solidifies immediately which over comes the soldering problem between tooling steels and Al alloy.

Figure 1 shows Fastcool 55 material being used as die casting dies which have no water cooling lines. The die surface temperature is between 215-236°C in the high temperature zones and 122-188°C in the low temperature zones.

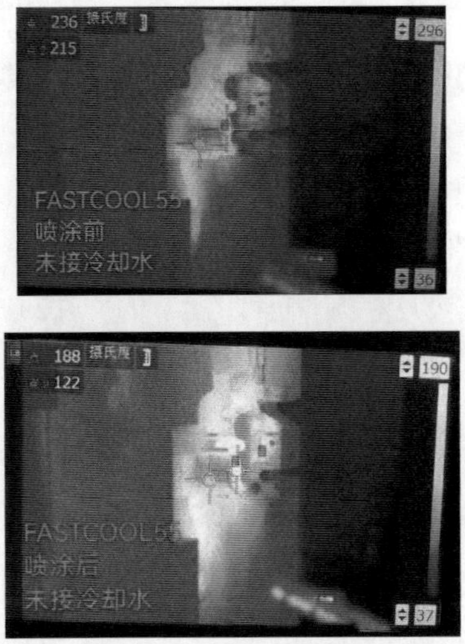

Figure1 Surface temperature of Fastcool 55 material in die casting dies

3 High temperature tempering resistance

Figure2 shows the tempering curve of Fastcool 55. The tempering resistance is exceptionally high. Conventional hot work tooling steel such as modified Din 1.2367 materials i.e. typically NADCA C category materials have tempering resistance of 600°C to reach hardness of 46HRC hardness. Whereas, Fastcool 55 needs 640°C temperature to temper the material in order to reach 46HRC which has a high temperature margin of 40°C than the conventional hot work tooling steels.

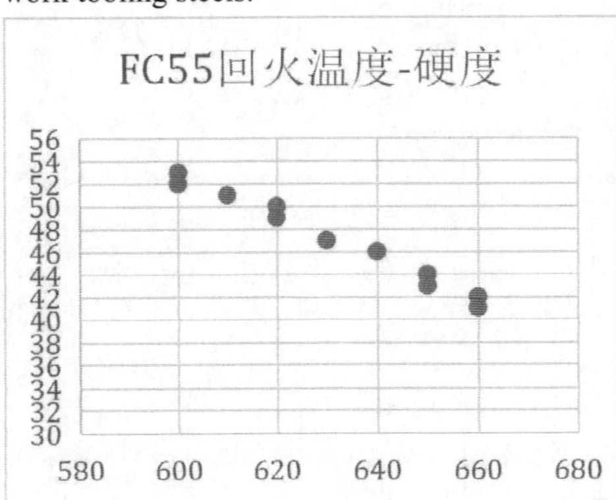

Figure2 Tempering curve of Fastcool 55 material.

4 Thermal fatigue resistance

When performing thermal fatigue test, Fastcool 55 material showed maximum crack length of 120μm whereas NADCA C category materials showed 500μm maximum crack length. Also, much less cracks are found with Fastcool55 material. Figure 3 shows the cracks distribution of Fastcool 55 material with maximum crack length of 120μm. By comparison NADCA C category material which has maximum thermal fatigue crack length of 500μm. The tests are conducted with a V notch specimen by induction heating up the samples to 600°C with water spray cooling to room temperatures after 2400 cycles.

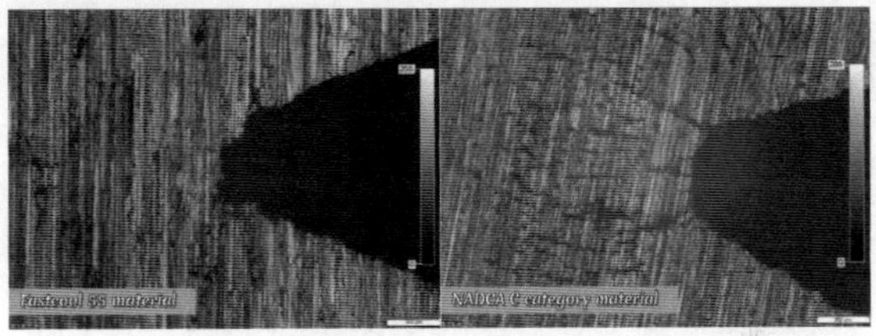

Figure 3 Thermal fatigue tests showed maximum crack length of 120 μm with Fastcool 55 material; and crack length of 500μm with NADCA C category materials.

5 Conclusion

1. Fastcool 55 material has high thermal conductivity、which over comes soldering problems in many die casting applications when used as die steels.
2. Fastcool 55 material has high tempering resistance which leads to better thermal fatigue resistance
3. Thermal fatigue tests demonstrates that Fastcool 55 material has more than 4 times of better thermal fatigue resistance than NADCA C category materials.
4. With many die casting applications when Fastcool 55 material is used the die casting cycle time is being reduced over 20%.
5. Fastcool 55 material yields longer tooling life in Al high pressure die casting.

Reference

(1) Morris Murray, Defects- how you have changed, Die casting Engineering, July 2012.

(2) Isaac Valls, Extremely high conductivity low cost steel, US Patent: 20200291496.

Investigation on Heat Treatment Process Optimization of A356.2 Alloy Prepared by Super-Slow-Speed Die Casting

Lu Zhang[1,2], Hengcheng Liao[1], Jiang Li[2], Liangchao Tang [2]
1.School of Materials Science and Engineering, Southeast University, Nanjing, China.
2. Nanjing Chervon Auto Precision Technology Co., Ltd. Nanjing, China.
*Corresponding address: e-mail: hengchengliao@seu.edu.cn

Abstract: A356.2 alloy die casting is produced by the Super-Slow-Speed die casting process. The phase composition of the alloy under the Scheil model and DB model (modified Scheil model) was analyzed by the CALPHAD-based method. Combined with DSC, conductivity meter, universal testing machine, and thermomechanical analyzer, the heat treatment process, microstructure, and mechanical properties of the material were studied. The optimized heat treatment process for the prepared A356.2 alloy was determined as solution at 520℃ for 1 h and aging at 180℃ for 6 h.

Keywords: A356.2 alloy; electrical conductivity; solution temperature; CALPHAD; mechanical properties

1 Introduction

High pressure die casting (HPDC) is widely used but always causes porosity defect that greatly affect strength and plasticity [1,2]. The castings manufactured by Super-Slow-Speed HPDC (SSS HPDC) can be heat-treated (T6) for property enhancement [3]. A356.2 alloy, usually gravity-cast, is now used in HPDC. Aging treatments without corresponding solution treatment can increase the strength but decreases the elongation [4,5]. This study seeks to develop a heat treatment process for A356.2 in SSS HPDC that balances strength and elongation, utilizing CALPHAD, tensile test and electrical conductivity and thermal expansion measurements.

2 Experimental procedure

A356.2 alloy casting was manufactured via SSS HPDC on an 840Ton Büler Ecoline S machine, with specific process parameters and mold temperatures. The casting, used in a vehicle's toolkit, had a 4mm wall thickness. Analysis included chemical composition with a Bruker spectrometer, thermal properties by DSC, electrical conductivity, microstructure via optical microscopy, and mechanical properties with an average of three samples. Dimensional changes were measured with a thermomechanical analyzer across various temperatures.

3 Result and discussion

CALPHAD analysis in solidification

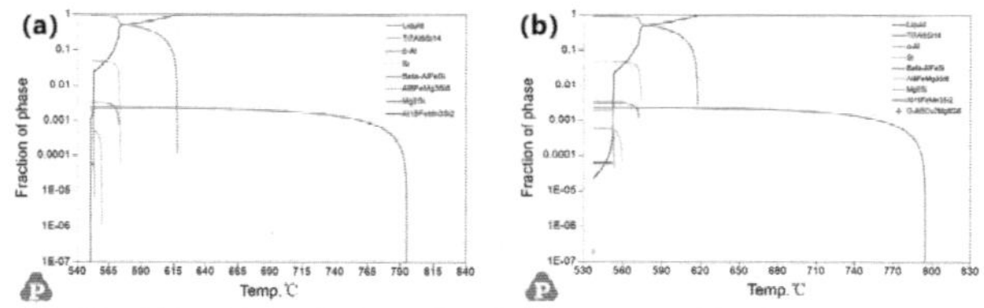

Fig.1 Fraction of phase by Scheil model (a) DB model (b)

Fig.1 show the fraction of phases obtained by the Scheil model (a) DB model (b). The difference is that: in Scheil model, a very small amount of Al15FeMn3Si2 phase is formed at 95% solid phase fraction and the final solidification ends at 550℃; in BD model the end of solidification is at 536℃, where a multicomponent eutectic reaction occurs, containing a very small amount of Q-Al5Cu2Mg8Si6 phase.

Solution heat-treatment process

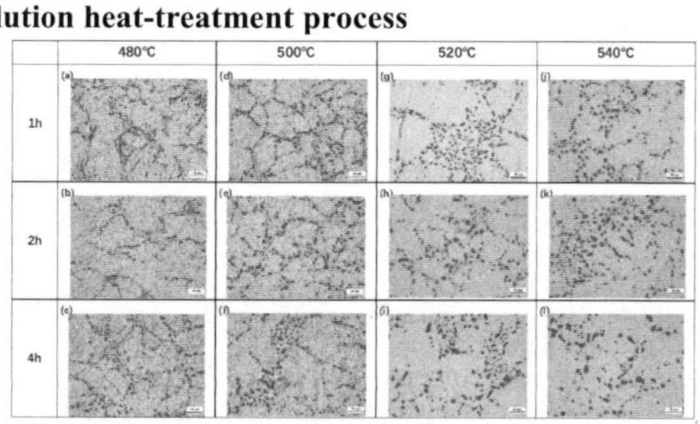

Fig.2 Microstructure after solution heat treatment at 480℃ for 1, 2, 4 h (a, b, c); 500℃ for 1, 2, 4 h (d, e, f); 520℃ for 1, 2, 4 h (g, h, i); and 540℃ for1, 2, 4 h (j, k, l)

Metallographic observation in Fig. 2 reveals that at 480℃ for 1-hour, eutectic silicon becomes segments. Spheroidization occurs at 500℃ for 1 hour and becomes more pronounced with longer holding times. At 520℃ and 540℃ for 1-2 hours, Si particles are spheroidized significantly, but after 4 hours, Si particles become coarsened again.

Electrical conductivity

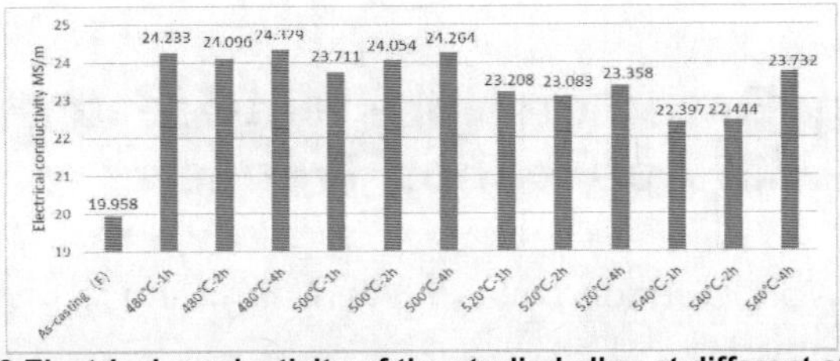

Fig. 3 Electrical conductivity of the studied alloy at different states

Fig. 3 displays the electrical conductivity of the studied alloy in different states. The changes in solution temperature and time results in the variations of Si and Mg in the α-Al matrix and Si phase morphology, thus affecting the alloy's conductivity. The highest conductivity is obtained by solution treatment at 480℃-1h.

Mechanical property

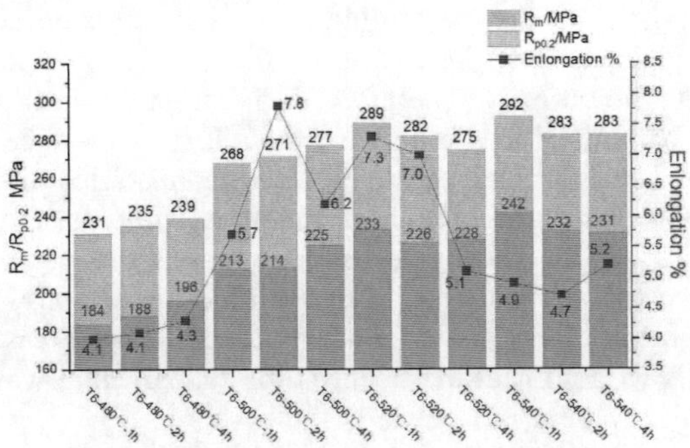

Fig. 4 Mechanical properties of the alloy in different solution states followed by aging at 180℃ for 6 h

It is seen that the highest strength is obtained by solution-treatment at 540℃-1h, the highest elongation is at 500℃-2h, the highest QI is at 520℃-2h from Fig.4.

Thermal expansion/length change

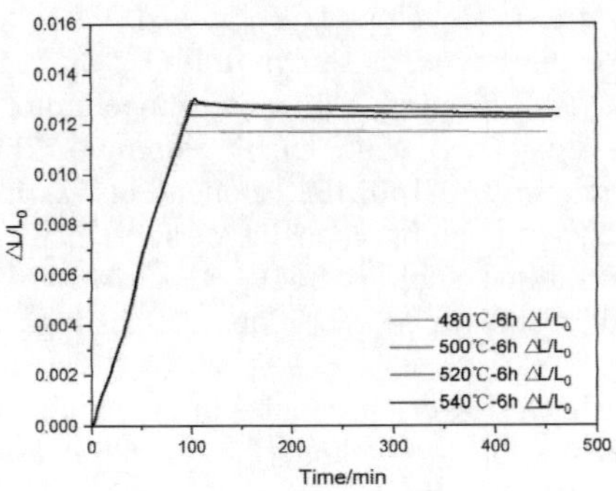

Fig.5 Thermal expansion/ length change at 480℃ to 540℃ for 6 hours

Fig. 5 displays the heating effect on the length of as-cast samples to different temperatures ranging from 480℃ to 540℃ for 6 hours. Local remelting of the material at 540℃ caused a considerable change in size, but in the temperature range of 480℃-520℃, the change of length is very small. Therefore, this temperature range is suitable for castings' solid solution heat treatment.

4 Conclusion

The electrical conductivity of A356.2 alloy is influenced by the solubility of Si and other alloying elements in the α-Al matrix and the morphology evolution of Si phase during solution treatment. The DB model of thermodynamic phase diagram software is suitable for analyzing the solidification process in super-slow-speed die casting. The highest QI value of 418.5MPa is achieved at 520°C for 1 hour. It is not recommended to use 540°C for solution treatment due to negative size changes. Therefore, the appropriate heat treatment process should be chosen based on specific production requirements.

Reference

[1] F. Bonollo, N. Gramegna, G. Timelli, JOM 67 (2015) 901–908. DOI: 10.1007/s11837-015-1333-8

[2] Adamane, A.R., Arnberg, L.. et al.. Inter Metalcast 9, 43–53 (2015). https://doi.org/10.1007/BF03355601

[3] Yanfei Bai, Haidong Zhao, Materials & Design,Vol. 31,9,2010 ,https://doi.org/10.1016/j.matdes.2010.04.005.

[4] Zhang J, Cinkilic E, Huang X, et al. Materials Science and Engineering: A, 2023, 865: 144604.

[5] Emi Yanagihara, Shin Orii, Takuya Iketani, Seiji Saikawa, et al., MATERIALS TRANSACTIONS, 2015, https://doi.org/10.2320/matertrans.

Integrated HPDC Gigacasting Technology Based on Non-Heat-Treated (NHT)High-Ductility Magnesium and Its Application Trends

Liming Peng*, Penghuai Fu, Dejiang Li , Jiangchen Wang, Qudong Wang, Wenjiang Ding
School of Materials Sci. and Eng., Shanghai Jiao-Tong University, P.R.C
Address: No.800 Dongchuan Rd., Minhang District, Shanghai, China
*Corresponding address: e-mail: plm616@sjtu.edu.cn

Abstract: In recent years, there has been a growing demand for lightweight automotive components, such as magnesium alloy/aluminum alloy integrated die casting, which is different from traditional automotive castings. New body or chassis structural components require higher requirements for alloy materials, casting processes, and mold design. This report provides a detailed overview of the design and development of heat-treatment-free high-strength and ductile magnesium alloys, research on high-density Giga-casting processes, design and manufacture of die casting molds, die casting experiments and evaluation of microstructure and mechanical properties, and finally reports the application and trend of magnesium alloy integrated die casting technology in automotive structures.

Keywords: High-strength-ductility Magnesium alloy; HPDC Giga-casting; Automotive structural components

1 Introduction

With the world's energy becoming increasingly tight and stricter requirements for low-carbon and environmental protection, the full electrification of cars has become an irreversible trend. However, the problem of "range anxiety" has always existed in new energy vehicles, and light weighting is an important way to solve this problem. In recent years, the body of new energy vehicles has been developing rapidly towards a "diversified material system" and "replacing steel with aluminum/magnesium integrating design, and casting instead of forging." Especially after Tesla introduced an aluminum alloy rear floor part with Giga-casting in 2019, it further accelerated the above light weighting technology.

As the lightest metal structural material, magnesium alloys have excellent die-castability and can achieve higher lightweight effect than aluminum alloy material. Therefore, the development of new high-strength and high-toughness die-casting magnesium alloys and large-scale complex thin-wall die-casting technology to meet the requirements of vehicle body service conditions is of great significance to help the lightweight of new energy vehicles. This report systematically introduces the research and development of high strength-and-toughness die-casting magnesium alloys without heat treatment, integrated design of large thin-walled structural parts, die manufacturing, and high density die-casting technology and its application in automotive integrated die-casting.

Material development must be considered in conjunction with the casting process. As well-know, the inclusion of a certain amount of gas in die casting is the inherent characteristics of the die casting process itself, even if the use of high vacuum die casting is difficult to completely eliminate the porosity. Therefore, if the large thin-wall die-casting parts need to use heat-treated alloys to produce, on the one hand, it may cause the bubble blistering of the entrapped gas during the heat treatment process and be scrapped, and on the other hand, the parts are seriously deformed by heat resulting in unqualified dimensions. The development of high strength-and- toughness die casting materials without heat treatment is the fundamental way to solve this problem.

2 Result and discussion

In this project, based on the academic idea of regulating the precipitated phases during solidification, potential multiple-element alloy systems and ranges were screened by using high-output thermodynamic calculation in combination with the rapid solidification conditions in the die casting process, and then a new die-casting alloy of Mg-RE-Al-X series (EA54V) and one of Mg-Y-Zn-X series (WZ42AM) have been successfully developed, combined with the verification of die casting tests, as shown in FIG. 1 and FIG. 2. Their typical tensile properties at room temperature in the die-casting state can reach: tensile strength 260-300MPa, yield strength 140-160MPa, elongation 12-16% excellent performance, the comprehensive performance has exceeded aluminum alloys such as C611, BFA etc., which are currently used in integrated rear floor die-castings of auto-bodies.

Secondly, the shock tower and subframe, which are typical representatives of the vehicle body and chassis system, were selected as the research objects, and the integration design technology for large and complex thin-walled magnesium alloy structural parts was studied based on the characteristics parameters of the new magnesium alloy material, the decomposition of vehicle

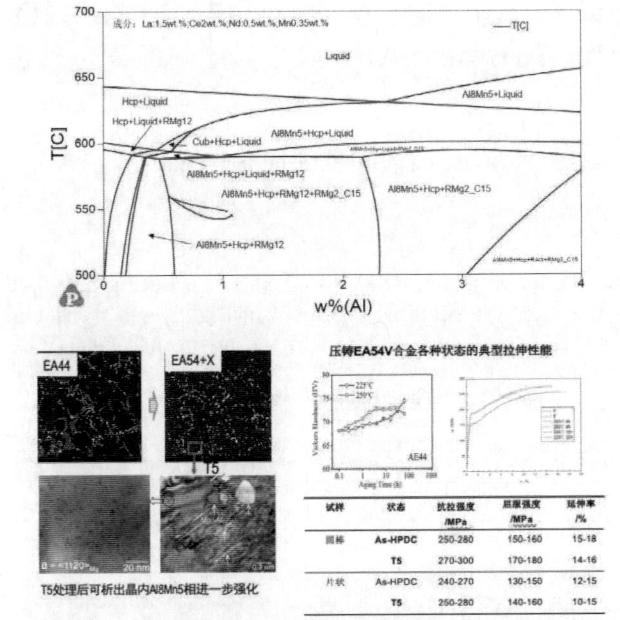

试样	状态	抗拉强度/MPa	屈服强度/MPa	延伸率/%
圆棒	As-HPDC	250-280	150-160	15-18
	T5	270-300	170-180	14-16
片状	As-HPDC	240-270	130-150	12-15
	T5	250-280	140-160	10-15

Fig.1 Design scheme of new high-strength-ductility HPDC Mg alloy EA54V and its typical microstructure and mechanical properties

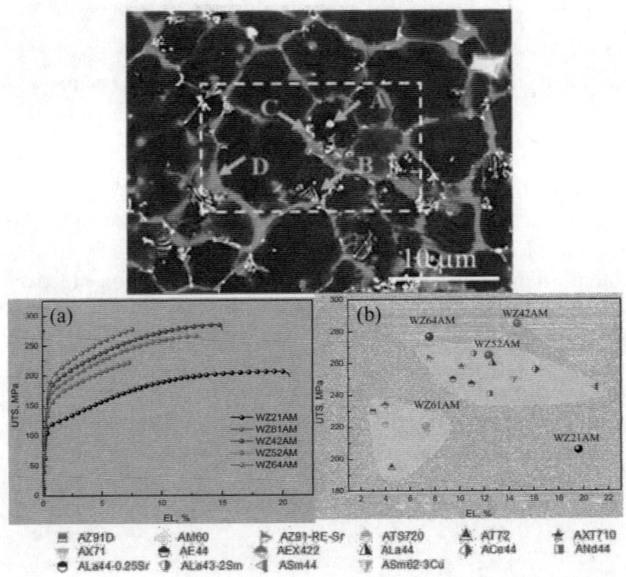

Fig.2 The microstructure and mechanical properties of new high-strength-ductility HPDC Mg alloy WZ42AM

load and performance indicators, FEM finite element strength analysis, and casting process simulation. "Five-in-one" and "seven-in-one" highly integrated magnesium alloy casting structural parts were designed, achieving a maximum weight reduction rate of 62% compared to the original steel parts. As shown in Figure 3.

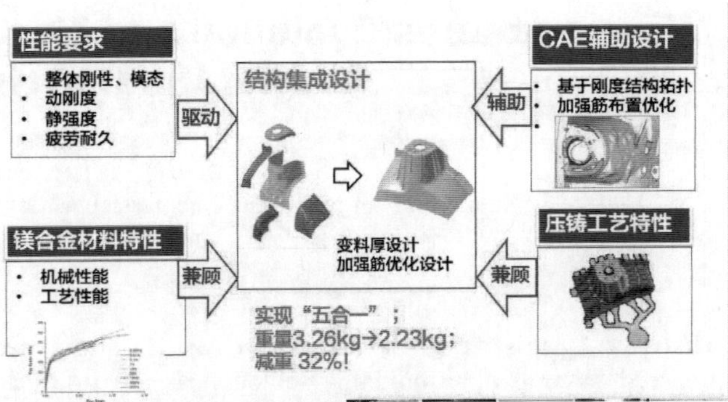

Fig.3 Integrated design scheme of auto-body structural components of Mg alloys

Meanwhile, to solve the manufacturing and defect control problems of large thin-wall magnesium alloy castings, this project redefined the technical connotation of high-density HPDC, and we closely combined "moderate vacuum technology, reliable process simulation, pre-crystallization control, and precise mold thermal balance technology" to develop a genuine high-density HPDC technology for magnesium alloys. We achieved casting and machining of magnesium alloy shock tower and subframe, and solved the dissimilar-material jointing technology between magnesium alloy castings and steel body parts, becoming the first in the world to pass bench and on-vehicle tests for magnesium alloy body and chassis structural parts. In 2023, the team successfully developed the world's largest and more functionally integrated giga casting of magnesium alloy for body structural component - the rear floor of FAW Co. As shown in Figure 4.

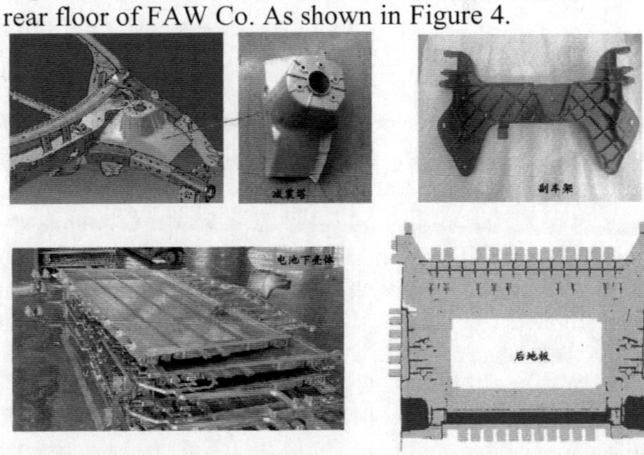

Fig.4 Partial auto structural parts made by new HPDC Mg alloys

3 Conclusion

The large integrated casting technology based on non-heat-treated magnesium alloys shows a very promising application prospect.

Microstructures Evolution and Mechanical Properties of Semi-Solid AZ91D Alloy at Different Injection Temperatures

Jun Zhou[1], Tiejun Sui[1], Xiang Li[1], Larry Wang[1]
Key Laboratory of Intelligent Compression Molding Equipment, YIZUMI Holdings Co., Ltd, Foshan, China
*Corresponding address: e-mail: j.zhou@yizumi.com

Abstract: In this work, the semi-solid AZ91D alloy was prepared by injection molding method near the liquidus temperature. The microstructure evolution and mechanical properties of the alloy were analyzed by optical microscopy (OM), scanning electron microscopy (SEM/EDS) and X-ray diffraction (XRD). The results indicated that the semi-solid AZ91D alloy was mainly composed of α-Mg matrix and β-$Mg_{17}Al_{12}$ phase, and the precipitates were small and dispersed when the injection temperature is 595 ℃. At this time, the tensile strength (R_m), yield strength ($R_{p0.2}$) and elongation (A) of the semi-solid AZ91D alloy was 273MPa, 160MPa and 7.3%, respectively, which are higher than the lower limit of liquid die-casting specified in the national standard.

Keywords: AZ91D magnesium alloy; injection molding; semi solid state; liquidus temperature.

1 Introduction

Magnesium alloy is the lightest metal structural material in engineering applications, and the environmental friendly material in the 21st century[1-2]. At present, the near liquidus semi-solid injection molding technology for aluminum alloys has becoming increasingly mature, while the technology applying in the magnesium alloy, which is also belongs to the light alloy, is also gradually attracting attention. The temperature range of the solid-liquid phase line of AZ91D alloy is 470 ℃ to 595 ℃ [3-5]. This study integrated the near liquidus injection molding technology with semi-solid injection molding equipment, and investigated the microstructures evolution and mechanical properties of semi-solid AZ91D alloy under different near liquidus temperatures. The research results are helpful for providing basis for the formation of semi-solid magnesium alloys.

2 Experimental procedure

Adopted the UNI300MG II magnesium alloy injection molding machine produced by YIZUMI company, this study prepared the semi-solid AZ91D alloy with the commercial AZ91D alloy. Transfer and shear the alloy particles with a size of 4mm × 1.5mm × 1.5mm through a rotating screw to obtain slurry. The injection molding was performed separately, when the slurry temperature was 580 ℃, 595 ℃ and 610 ℃. Among them, the mold temperature was 280 ℃, the injection speed was 2.3m/s, and the locking force was 3000kN. The working principle of the injection molding is shown in Fig. 1, and the actual chemical composition of the alloy is shown in Table 1.

Table 1. Composition of semi-solid AZ91D alloy (wt.%)

Al	Zn	Mn	Si	Cu	Ni	Fe	Mg
9.13	0.73	0.25	0.035	0.0025	0.0004	0.005	Bal

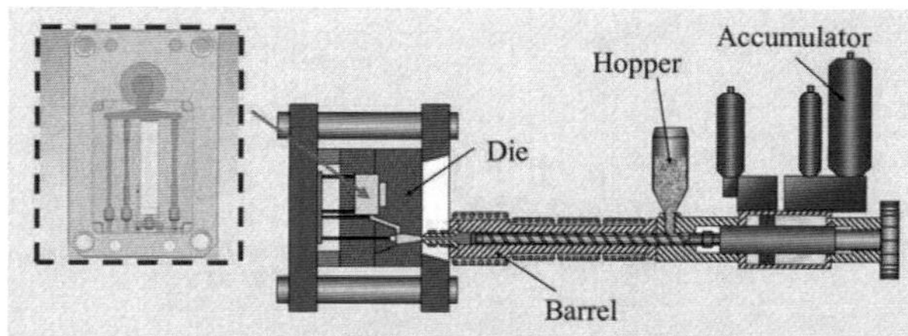

Fig. 1 Schematic diagram of the preparation process for semi-solid AZ91D alloy

The mechanical properties of semi-solid AZ91D alloy were tested by the electronic universal material testing machine with a tensile rate of 2.5mm/min. The microstructure of the semi-solid AZ91D alloy were characterized by Zeiss optical microscopy (Smart-zoom 5), field emission electron scanning electron microscopy (S4800), and X-ray diffraction (Smart Lab 9kw XRD). Using Image Pro Plus software to statistic and calculate the primary phase α-Mg. The solid phase fraction (S) and shape factor (F) of Mg were calculated using the following formulas[3]:

$$S=NA_0/A_s \quad (1)$$

$$F = [\Sigma p_0^2 \,/\, (4\pi A_0)] \,/\, N \quad (2)$$

3 Result and discussion

Microstructures evolution of semi-solid AZ91D alloy at different injection temperatures

Fig. 2 shows the OM images of AZ91D alloy. It can be seen that as the injection temperature increases, both of the solid phase fraction *S* and shape factor *F* decrease. When injected at 580 ℃ and 595 ℃, the alloy exhibits a semi-solid structure with similar shape factors. Among them, the *S* reaches to 30.4% at 580 ℃, which is the highest, and the α-Mg primary phase has formed and grown, while the *S* at 595 ℃ is 3.1%, and the grain sizes are smaller and uniform.

When injected at 610 ℃, the *S* is 0, but there are some pores in the matrix.

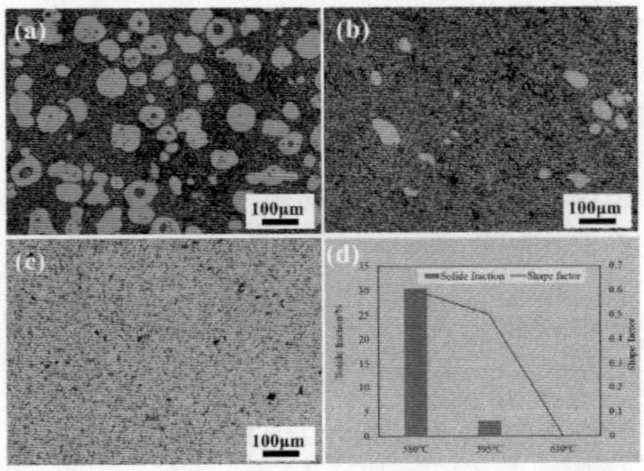

Fig.2 OM images of the semi-solid AZ91D alloy: (a) 580 °C; (b) 595 °C; (c) 610 °C; (d) S and F

Fig. 3 shows the SEM images and XRD patterns of the semi-solid AZ91D alloy. It can be seen that the three injection temperatures have no effect on the phase composition, which are α-Mg matrix and β-$Mg_{17}Al_{12}$ phase, but affect the phase morphology significantly. At 580 ℃, the second phases are rounded, distributing along the grain boundaries with continuous network structures; At 595 ℃, the second phases are mostly pinned to grain boundaries, small and dispersed; At 610 ℃, the second phases are mostly distributed in the matrix, forming a semi-continuous network structure.

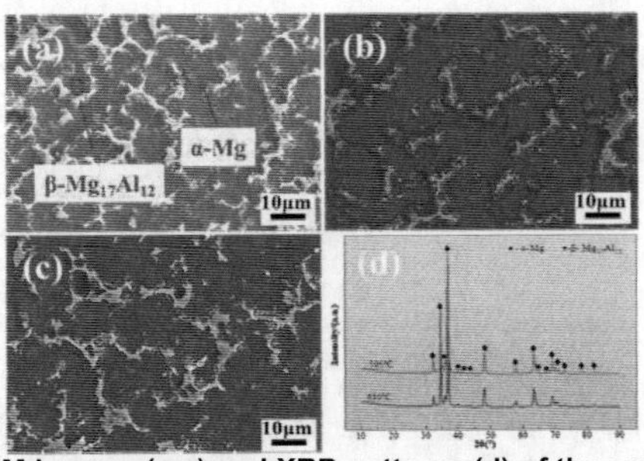

Fig.3 SEM images (a-c) and XRD patterns (d) of the semi-solid AZ91D alloy: (a) 580 °C; (b) 595 °C; (c) 610 °C

Mechanical properties of semi-solid AZ91D alloy at different injection temperatures

Mechanical properties and fracture surfaces at room temperature are shown in Fig. 4. It can be seen that as the injection temperature increases, the mechanical properties of the alloy first increase and then decrease, besides, the fracture surfaces are composed of cleavage planes, tearing edges and a small amount of ductile dimples, showing typical brittle fracture. When injection molding is carried out at 595 ℃, the obstruction to the dislocations movement by the small and dispersed second phases is the strongest[4-5], and the mechanical properties of the alloy are the best, with R_m of 273MPa, $R_{p0.2}$ of 160MPa, and *A* of 7.3%, respectively, which are higher than the lower limit values specified in the national standard GB/T 25747-2022 for liquid die-casting parts.

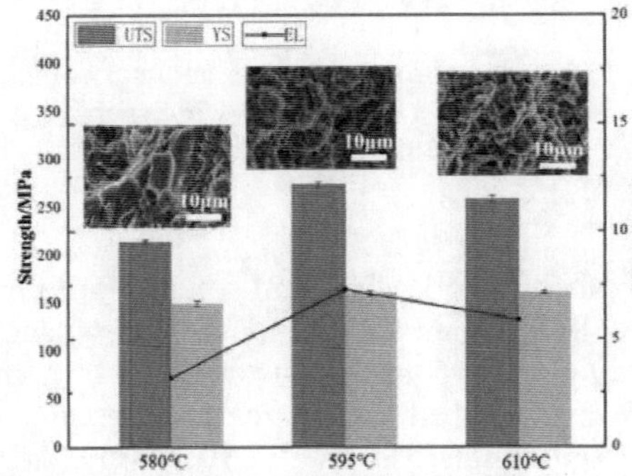

Fig. 4 Mechanical Properties and fracture surface of semi-solid AZ91D alloy at room temperature

4 Conclusion

(1) The semi-solid AZ91D alloy is mainly composed of α-Mg matrix and β-$Mg_{17}Al_{12}$ phases. The phase type is not affected by injection temperature, but when injection molding at 595 ℃, the second phase is the smallest and most dispersed.

(2) The semi-solid AZ91D alloy, formed by injection molding at 595 ℃, has the best mechanical properties, with R_m of 273MPa, $R_{p0.2}$ of 160MPa, and *A* of 7.3%, respectively, which are higher than the lower limit values specified by the national standard for liquid die-casting parts.

References

[1] Ramalingam V V, Ramasamy P, Kovukkal M D and et al. Research and development in magnesium alloys for industrial and biomedical applications: a review[J]. Metals and Materials International, 2020, 26: 409-430.

[2] Tan J and Ramakrishna S. Applications of magnesium and its alloys: A review[J]. Applied Sciences, 2021, 11(15): 6861.

[3] Liu Z, Mao W, Wang W and et al. Preparation of semi-solid A380 aluminum alloy slurry by serpentine channel [J]. Transactions of Nonferrous Metals Society of China, 2015, 25(5): 1419-1426.

[4] Shuai S S, Guo E Y, Phillion A B and et.al. Fast synchrotron X-ray tomographic quantification of dendrite evolution during the solidification of Mg-Sn alloys[J]. Acta Materialia, 2016, 118: 260-269.

[5] Dargusch M S, Balasubramani N, Venezuela J and et al. Improved biodegradable magnesium alloys through advanced solidification processing[J]. Scripta Materialia, 2020, 177: 234-240.

Effect of T6 Treatment on Microstructures and Mechanical Properties of Semi-Solid A356 Alloy

Jun Zhou, Caihua Wang*, Larry Wang
Guangdong Provincial Key Laboratory of Intelligent Molding Equipment Enterprises, Yizumi Holdings Co. Ltd.
*Corresponding address: e-mail: larry.wang@yizumi.com

Abstract: In this work, the semi-solid A356 alloy was prepared by SEED semi-solid pulping process, and effect of T6 treatment on microstructures and mechanical properties were studied by optical microscopy (OM), scanning electron microscopy (SEM), X-ray diffraction (XRD) and other methods. The results indicated that the phase composition of A356 alloy remained unchanged after the T6 treatment, and included α-Al matrix, eutectic Si, and Mg_2Si phases, while, the solid phase fraction (S) of the alloy increased from 66% to 75%, and the shape factor (F) decreased from 0.7 to 0.6, and the mechanical properties significantly improved. The tensile strength (UTS), yield strength (YS), and elongation (EL) of the T6-A356 alloy was 310MPa, 245MPa, and 12.1%, respectively.

Keywords: A356 alloy; rheological forming;microstructure; mechanical properties; high solid faction.

1 Introduction

A356 aluminum alloy is a hypoeutectic alloy with Si and Mg as the main alloying elements, which has good fluidity and a wide temperature range in the two-phase zone, and is commonly prepared by the semi-solid metal forming. Due to its lightweight, excellent mechanical strength, ductility, fatigue performance, compactness and workability, A356 aluminum alloy is becoming more and more attractive in the automotive and aerospace industries [1-2]. The present study prepared the semi-solid A356 alloy by SEED semi-solid die-casting, and studied the influence of T6 treatment on microstructure evolutions and mechanical property changes.

2 Experimental procedure

The A356 commercial alloy ingot was melted at 760 ℃ and kept for 30 minutes. Then, 15 minutes of rotational degassing was performed under argon gas protection. The hydrogen content of aluminum melt was measured, and kept the qualified melt insulated at 630 ℃. The chemical composition of the qualified melt is shown in Table 1.

Pouring the above melt into the SEED crucible by the robotic arm for slurry production. The principle of SEED pulping process is shown in Fig. 1[3]. The LEAP-840T die-casting machine from YIZUMI company was used to cast the slurry. The slow and fast injection speeds was 0.5m/s and 1.5m/s, respectively. The boost pressure was 100MPa. The T6 treatment condition was 540 ℃ × 2h + 170 ℃ × 4h.

Table 1. Composition of Aluminum Melt (wt.%)

Si	Cu	Mg	Mn	Fe	Ti	Al
6.46	< 0.0002	0.34	< 0.0005	0.082	0.09	Bal

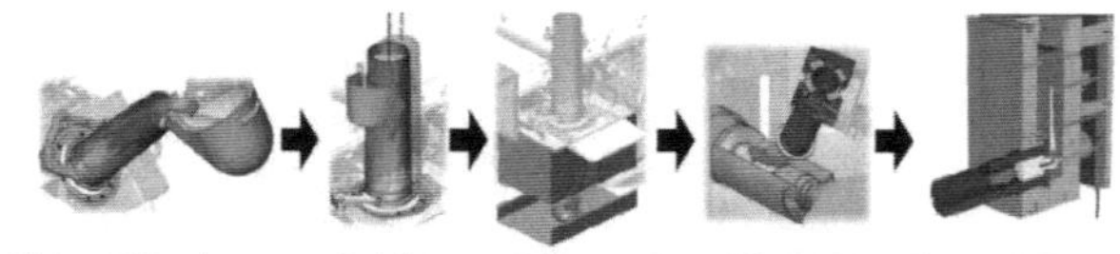

Fig. 1 SEED pulping process principle [3]

The microstructure of the semi-solid A356 alloy were characterized by Zeiss optical microscopy (Smart-zoom 5), field emission electron scanning electron microscopy (S4800), and X-ray diffraction (Smart Lab 9kw XRD). Using Metallographic analysis software to statistic and calculate the primary phase α-Al The solid phase fraction (S) and shape factor (F) of Al are calculated using formulas (1) and (2) [4].

$$S = NA_0/A_S \quad (1)$$
$$F = \left[\Sigma p_0^2 \,/\, (4\pi A_0)\right] / \,N \quad (2)$$

3 Result and discussion

Effect of T6 treatment on Microstructures of A356

Figure 2 shows OM images, SEM-EDS images, and XRD patterns of as-cast and T6-treated A356 alloys. It can be seen that after SEED and semi-solid rheological die-casting, both as-cast and T6-treated A356 alloys have non-dendritic structures with uniform distribution of primary phases, while T6 treatment has minor impact on grain sizes, and the solid phase fraction (S) and shape factor (F) of the T6-treated and as-cast alloys are 75% and 66%, 0.6 and 0.7, respectively.

Among them, the as-cast A356 alloy mainly contains nearly spherical α-Al, black fishbone Mg_2Si eutectic phase (point A) and lath eutectic Si. After T6 treatment, the Mg_2Si phase between the original grains dissolves and enters in the α-Al matrix, the original lath eutectic Si breaks while short rod or near spherical eutectic Si (point B) appears.

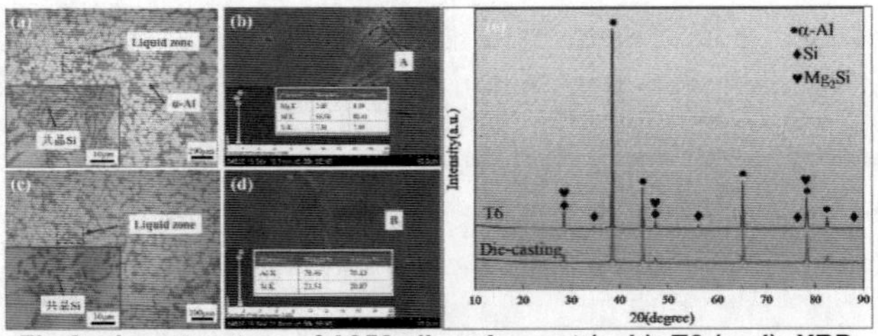

Fig.2 microstructure of A356 alloy: As-cast (a, b); T6 (c, d); XRD patterns (e)

Effect of T6 treatment on Mechanical Properties of A356

Table 2 shows mechanical properties of semi-solid A356 al loys at the room temperature. It can be seen that after T6 tre atment, there's significant increase in the tensile strength (R_m) and yield strength ($R_{p0.2}$) of the alloy, which increases fro m 210 MPa and 115 MPa to 310 MPa and 245 MPa, respec tively, while the elongation (A) decreases from 14.4% to 12. 1%. That is to say, the performance indicators of semi-solid T6-A356 alloy are generally higher than the lower limit of the current standard 《GB/T40809-2021 Casting aluminum alloys-Process specification forsemisolid rheo-diecasting f orming》.

Before T6 heat treatment, fishbone Mg_2Si is mainly distributed between α-Al grains. However, after T6 heat treatment, most of Mg diffuses in the matrix and distributed more uniformly, besides, a small portion of Si dissolves into the α-Al matrix. The above changes formed solid solution strengthening effect, improving the R_m and $R_{p0.2}$ of the alloy significantly. Meanwhile, most of the eutectic Si between Al grains transforms into eutectic Si. It should be noted, the small and dispersed short rod-shaped or nearly spherical eutectic Si in the α-Al matrix, hindering the dislocation movement more seriously, and resulting in a significant increase in the $R_{p0.2}$ of the alloy. This is consistent with the research results of PENG[5].

Table 2 Mechanical Properties of semi-solid A356

Alloys	UTS/MPa	YS/MPa	EL%
A356	210	115	14.4
T6-A356	310	245	12.1

4 Conclusion

(1) Microstructures of both as-cast and T6-treated A356 alloys are composed of α-Al matrix, eutectic Si, and Mg_2Si phases. The solid phase ratio and shape factor of α-Al is 66% and 75%, 0.7 and 0.6, respectively.

(2) T6-treated A356 alloy exhibits superior mechanical properties with R_m, $R_{p0.2}$, and A of 310MPa, 245MPa and 12.1%, which is overall higher than the lower limit values specified by the national standard for semi-solid rheological die castings.

References

[1] Colombo M, Gariboldi E and Morri A. Er addition to Al-Si-Mg based casting alloy[J]. Journal of Alloys and Compounds, 2017, 708: 1234-1244.

[2] Mao F, Yan G, Xuan Zand et al. Effect of Eu addition on t he microstructures and mechanical properties of A356 alum inum alloys[J]. Journal of Alloys and Compounds, 2015, 65 0: 896-906.

[3] Côté P, Larouche M E and Chen X G. New developments w ith the SEED technology[J]. Solid State Phenomena, 2013, 192: 373-378.

[4] Liu Z, Mao W, Wang W and et al. Preparation of semi-soli d A380 aluminum alloy slurry by serpentine channel[J]. Tra nsactions of Nonferrous Metals Society of China, 2015, 25 (5): 1419-1426.

[5] Peng J, Tang X, He J and et al. Effect of heat treatment on microstructure and tensile properties of A356 alloys[J]. Tra nsactions of Nonferrous Metals Society of China, 2011, 21 (9): 1950-1956.

This page intentionally left blank.

Influence of Super Large Single Piece Die Casting Production to Automotive Industry

Liu Siong Song*, Chan Kwok Keung, Qiu Ge
L.K.Machinery International Limited, Hong Kong/China
*general@lk.world

Abstract: Under the motivation of the innovation idea from a new energy vehicle company in US, and the great jobs and cooperation of several parties, a new type of Super Large Single Piece die-casting production was established, which not only reduced production costs, but also accelerated production speed and further reduced the weight of the car body. In new energy vehicles, with continued explosive growth and the demand for energy conservation and emission reduction, lightweighting of new energy vehicles has become an inevitable trend, which has also promoted the faster development of Super Large Single Piece die-casting development.

Large aluminum alloy die-casting parts such as car front underbody, rear underbody, and battery trays used to be composed of up to 70+ small parts. With this new technology, they can be produced by die-casting in one shot. It is not just to amplify all the equipment can be done, but there involves many difficulties and challenges. This article starts from the perspective of the die-casting unit, through the design and process optimization of each step, and through on-site verification, to improve the performance and production efficiency of die-casting products.

The production process of the die-casting unit includes metal ladling, injection, die cooling, parts removal, spraying, parts cooling, trimming, marking, and output.

Keywords: Super Large Single Piece die-casting, new energy vehicles new production, underbody casting

1 Introduction

The automotive industry is constantly evolving, with technological advancements driving improvements in vehicle design, manufacturing efficiency, and overall performance. One such groundbreaking innovation is the introduction of super large single-piece die casting production. This technology has the potential to revolutionize the way vehicles are manufactured, offering numerous benefits in terms of cost, weight, strength, and production speed.

In the traditional method, the underbody of the vehicle was produced by jointing 70~100 pieces of small parts [1] together to form the whole underbody. This method has involved hundreds of robots and large space for building the assembly line. The time used is in hours basis.

On using the super large die casting machine, the front underbody, or the rear underbody can be produced in a single piece which take around 100 seconds. Space required is much less, only around 800~1000 square meter per die casting cell (around the size of two basket ball court).

Die casting is a manufacturing process that involves injecting molten metal into a die cavity under high pressure. Traditionally, this process has been used to create smaller components that are then assembled into larger structures. However, recent advancements have enabled the production of much larger components in a single piece, a process known as super large single-piece die casting. This is also called the Giga casting.

In 2019, there is 6000 ton die casting machine, and then in 2020, the 9000 ton die casting machine was announced. In 2021, the 12000 ton die casting machine was shown and in 2023 the 16000 ton die casting machine was displayed.

When the clamping force of the die casting machine become larger and larger, this has indicated that the size of a single piece casting can be bigger and bigger.

2 Experiments in using Single Piece casting

On using the Super Large Single piece die casting, there is a lot of advantages:

(1) Cost Efficiency: One of the most significant advantages of this technology is the reduction in manufacturing costs. By producing large components in a single piece, the need for multiple smaller parts and the associated assembly processes is eliminated. This leads to a decrease in labor costs, tooling costs, and overall production expenses.

(2) Weight Reduction: Super large single-piece die casting allows for the production of components with complex geometries and optimized material distribution. This results in lighter components without compromising strength or durability. Weight reduction is a critical factor in improving vehicle fuel efficiency and performance, making this technology highly attractive to automotive manufacturers.

(3) Enhanced Structural Integrity: Traditional methods of joining multiple smaller components can introduce points of weakness due to welds, bolts, or adhesives. By

producing a single large piece, these potential weak points are eliminated, leading to improved structural integrity and overall safety of the vehicle.

(4) Increased Production Speed: The ability to produce large components in a single casting cycle significantly reduces production time. This increased speed not only enhances manufacturing efficiency but also allows for quicker turnaround times, enabling manufacturers to respond more rapidly to market demands.

(5) Design Flexibility: Super large single-piece die casting offers greater design flexibility, allowing engineers to create more complex and innovative vehicle structures. This can lead to the development of new vehicle designs that were previously not feasible with traditional manufacturing methods.

3 Result and discussion

The adoption of super large single-piece die casting is poised to have a profound impact on the automotive industry. Some of the key areas of influence include:

(1) Electric Vehicles (EVs): The demand for electric vehicles is growing rapidly, driven by environmental concerns and government regulations [2]. Super large single-piece die casting is particularly beneficial for EVs, as it allows for the production of lightweight and structurally sound components that can accommodate battery packs and other electrical systems more efficiently.

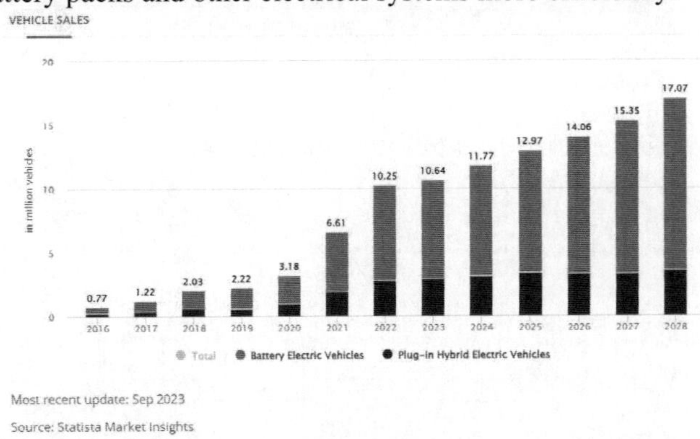

(2) Manufacturing Efficiency: Automotive manufacturers are constantly seeking ways to improve efficiency and reduce costs. The integration of super large single-piece die casting into production lines can streamline manufacturing processes, reduce the number of required parts, and minimize assembly times. This, in turn, can lead to significant cost savings and higher profit margins.

(3) Supply Chain Simplification: By reducing the number of components needed for vehicle assembly, manufacturers can simplify their supply chains. This reduction in complexity can lead to fewer supply chain disruptions, improved inventory management, and a more streamlined production process.

(4) Sustainability: The automotive industry is under increasing pressure to adopt more sustainable practices. Super large single-piece die casting contributes to this goal by reducing material waste, energy consumption, and emissions associated with traditional manufacturing processes. The production of lighter vehicles also contributes to lower fuel consumption and reduced greenhouse gas emissions.

4 Conclusion

Super large single-piece die casting represents a significant advancement in automotive manufacturing technology. Its potential to reduce costs, improve efficiency, enhance vehicle performance, and contribute to sustainability makes it a highly attractive option for automotive manufacturers. As the industry continues to evolve, the adoption of this technology is likely to play a pivotal role in shaping the future of vehicle production and design.

Since the Super Large Single Piece die casting is not only used in EV, but also possible to used for producing traditional IC vehicles. It is believed that the technology can lower the production cost of the automotive industry and thus lower the vehicle price in the coming future.

References

[1] https://x.com/tesla/status/1513886756923445254?lang=th

[2] https://www.statista.com/outlook/mmo/electric-vehicles/worldwide#unit-sales

Application and Notes of Super Large Single Piece Die Casting Production in Automotive Industry

Liu Siong Song*, Chan Kwok Keung, Chen Yuk Lin
L. K. Machinery International Limited, Hong Kong, China
*general@lk.world

Abstract: Under the motivation of the innovation idea from a new energy vehicle company in US, and the great jobs and cooperation of several parties, a new type of Super Large Single Piece die-casting production was established, which not only reduced production costs, but also accelerated production speed and further reduced the weight of the car body. In new energy vehicles, with continued explosive growth and the demand for energy conservation and emission reduction, lightweighting of new energy vehicles has become an inevitable trend, which has also promoted the faster development of Super Large Single Piece die-casting (Giga Casting) development.

Large aluminum alloy die-casting parts such as car front underbody, rear underbody, and battery trays used to be composed of up to 70+ small parts. With this new technology, they can be produced by die-casting in one shot. It is not just to amplify all the equipment can be done, but there involves many difficulties and challenges. This article starts from the perspective of the die-casting unit, through the design and process optimization of each step, and through on-site verification, to improve the performance and production efficiency of die-casting products.

More and more motive OEM has applied the production process of the Super Large One-piece die-casting in manufacturing cars. A case study of battery tray production was carried out. Some conclusions were made in implementing the Super Large Die Casting Cell.

Keywords: Super Large Single Piece die-casting, Giga casting, battery tray, new production process for automotive.

1 Introduction

High-pressure die casting is a production method in which molten metal materials are filled into the die cavity at high speed and high pressure, and then cooled and solidified to form castings. According to the different design characteristics of the product, it is necessary to choose the appropriate die-casting production process.

The normal production process of integrated products:
Pick up → Cool → Remove the runner/overflow → Output
Now we will introduce a case of a large battery tray product (Figure 1) to demonstrate how to choose a suitable die-casting cell production process. This project verified the characteristics of the die-casting product through multiple die trials in the early stage, followed the principle of high efficiency and stability, selected the most appropriate die-casting cell production plan, and successfully achieved the expected results after implementation.

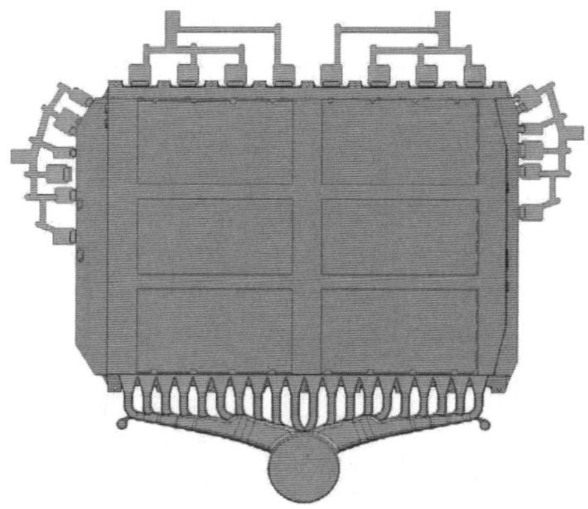

Figure 1 Large battery tray product

Product size 1800×1500×100; product basic average wall thickness 3mm; product net weight 50kg; total casting weight 100kg; heat treatment-free material C611; product deformation ≤1mm. It was produced on a LK 12000 ton die casting machine cell.

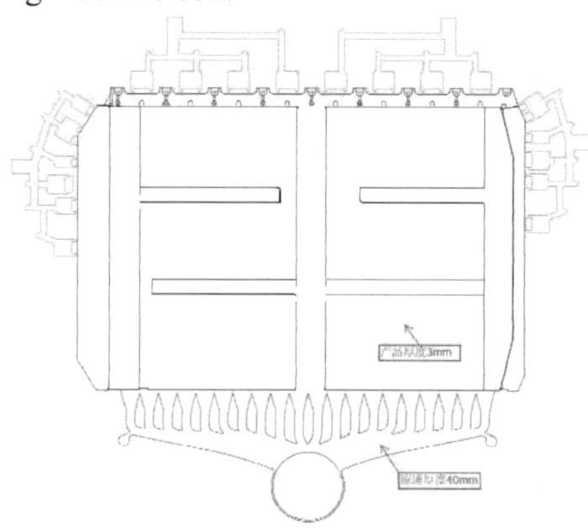

Figure 2 Size of product and runner

2 Experimental procedure

An obvious feature of battery tray die-casting parts is that the product is thin and the runner is thick (as shown in Figure 2). This feature has a great impact on the product during cooling, because the shrinkage of the runner and the product during the cooling process is inconsistent. The thick runners shrinkage is large, while the shrinkage of thin products is small, resulting in serious deformation of the product on the runner area (≥5mm). Through a CMM (Coordinate Measuring Machine), 9 positions of the product is detected, and the deformation results are obtained (Figure 3).

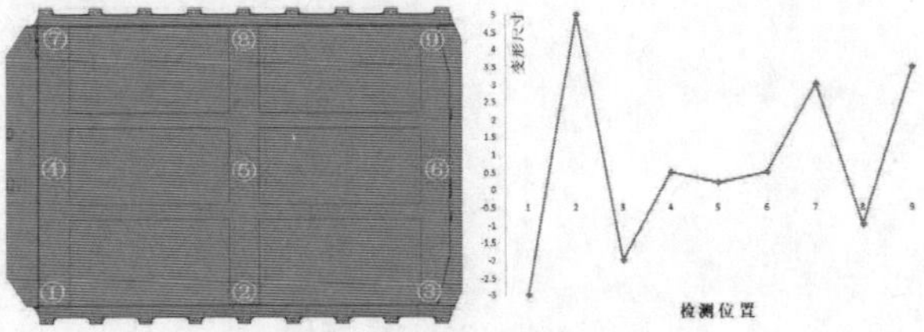

Figure 3 Deformation analysis

3 Result and discussion

In order to reduce product deformation caused by runner shrinkage, the thick runner needs to be separated from the product as soon as the casting is taken out of the die-casting die. There are currently three separation methods: Trimming by hydraulic press, Laser cutting, and Plasma cutting (see Table 1). After extensive evaluation, this project finally chose the Trim Press method to remove the runners and overflow because this method has far more advantages than other methods. This kind of production efficiency best meets our production needs.

The deformation of the product caused by the thermal shrinkage of the runner is eliminated, and the deformation is reduced from 5mm to 3mm, with the ultimate goal of controlling it within 1mm. The product itself has reinforcing rib structures of different thicknesses. During its own cooling process, the deformation caused by shrinkage is still large (Figure 4).

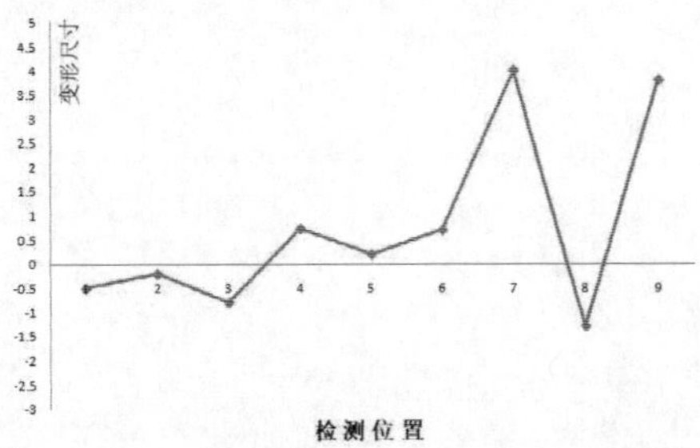

Figure 4 Deformation analysis after 1st optimization

At the beginning, it is tried to manually hammering to rectify the shape of products outside the die-casting cell, but found that it was very difficult to rectify the shape of the cooled products. The efficiency was extremely low and the workload was very large, making it impossible to achieve stable mass production. Through research, it is considered to add a shape rectify station to the die-casting cell, to carry out shape rectification when the product is still under warm status, and to perform cooling and shape rectifying at the same time. Although this solution is ideal, its specific implementation is a very big challenge. After repeated evaluation and discussion, the die-casting cell finally adopted the following production process.

picking up → punching runner/overflow → shape rectifying + cooling → output (Figure 5)

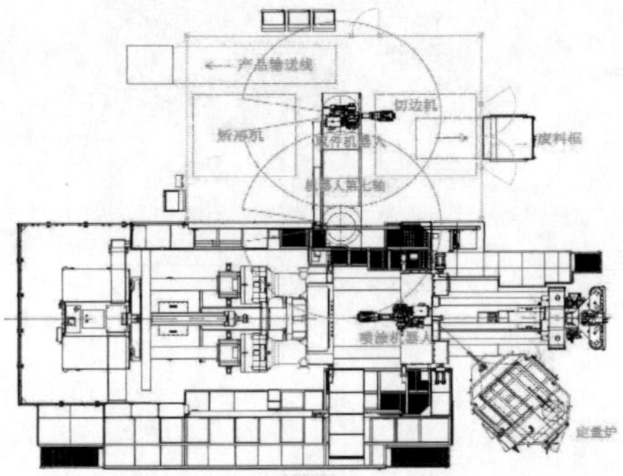

Figure 5 Final process

This solution has been verified through continuous production, and the product deformation is ≤1mm (Figure 6), which meets the processing and assembly requirements of the product.

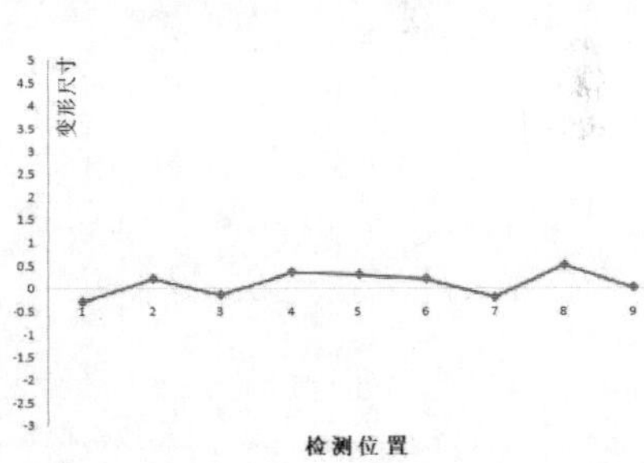

Figure 6 Final deformation

4 Conclusion

Super-large die-casting cell equipment have high investment costs (Compare to traditional assembly line with robots, the cost maybe lower). The design of the die-casting cell should follow the principle of efficiency and stability. Non-essential workstations, such as product polishing and X-ray inspection, should be placed outside the die-casting cell. Reduce the failure rate of die-casting cell production and thereby increase the equipment operating rate and thus raise the total OEE, bringing higher economic benefits to the enterprise.

Research Progress on New Technologies of Lost Foam Casting for Aluminum and Magnesium Alloys

Wenming Jiang*, Zitian Fan
State Key Lab of Materials Processing and Die & Mould Technology, Huazhong University of Science and Technology, Wuhan 430074, PR China
*Corresponding address: e-mail: wmjiang@hust.edu.cn

Abstract: Lost foam casting (LFC) is a precision green casting technology with some special advantages. However, there are some problems in aluminum and magnesium alloys lost foam casting, such as poor filling ability, serious hole defects, coarse solidification microstructure and low mechanical properties. In view of the above problems, this paper mainly introduces the latest research progress of several new technologies of aluminum and magnesium alloy lost foam casting. Vacuum and low pressure lost foam casting technology can ensure the stable filling of liquid metal, improve the aluminum and magnesium liquid filling and filling capacity, and protect the filling process. The vibration solidification and pressure solidification increased the compactness of the castings and refined the microstructure, thereby significantly improves the mechanical properties of the castings. The expendable shell casting technology could solve the pore, carburization as well as inclusion defects of the traditional LFC method, obtaining the castings with the good surface quality. Magnesium/aluminum bimetal lost foam compound casting technology can directly form complex magnesium/aluminum bimetal castings at low cost. These proposed novel LFC technologies will have positive effect for solving the current technological issues and promoting the technological progress for the LFC process.

Keywords: LFC under vacuum and low pressure; Vibration solidification; Pressure solidification; Expendable shell casting; Bimetallic castings

1 Introduction

The lost foam casting (LFC) process is a near net forming technology that represents 21st century casting technology. Currently, LFC technology is widely employed in the cast iron and cast steel materials. The applications of Al and Mg alloys on the LFC technology are relatively few, especially in the Mg alloy[1]. The technology and control of the LFC process encounter much difficulty because of the material properties of the Mg alloys; thus, the Mg alloy LFC remains on the test stage. The technical difficulties for the Al and Mg alloy LFC are as follows. Firstly, the loss of heat is largely due to the decomposition of the foam materials, leading to poor filling ability of the molten metal. On the other hand, the pouring temperature of the Al and Mg alloy LFC is higher (i.e., 30 °C to 50 °C) compared with that in the common casting process. As a result, the oxidation and burning of the Al and Mg alloys lead to a coarse microstructure, pore and inclusion defects, and poor mechanical properties. Basic studies on the foam pattern materials for the Al and Mg alloys are still lacking[1-3].

2 Experimental procedure

Some novel LFC technologies for the Al and Mg alloys are proposed and investigated to solve the aforementioned technical problems, including porosity defect, poor filling ability, and low mechanical properties for the Al and Mg alloy LFC technology.

3 Result and discussion

1) LFC under vacuum and low pressure

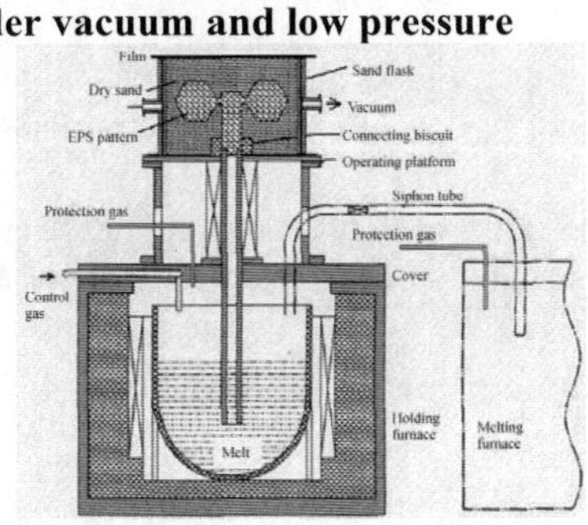

Fig. 1 Schematic of the LFC under low pressure and vacuum.

The LFC under vacuum and low-pressure technology combines the LFC process and low pressure casting (LPC). It possesses the advantages of the LFC and LPC processes. The schematic of this new casting technology is presented in Fig. 1. The filling ability of the melt is increased as the melt is filled with a control pressure.

2) LFC under vibration solidification

The mechanical vibration of the LFC under vibration solidification technology has been applied on the solidification process of the casting. The relevant schematic is shown in Fig. 2. During the vibration solidification process, the forced convection of the melt is caused by the vibration.

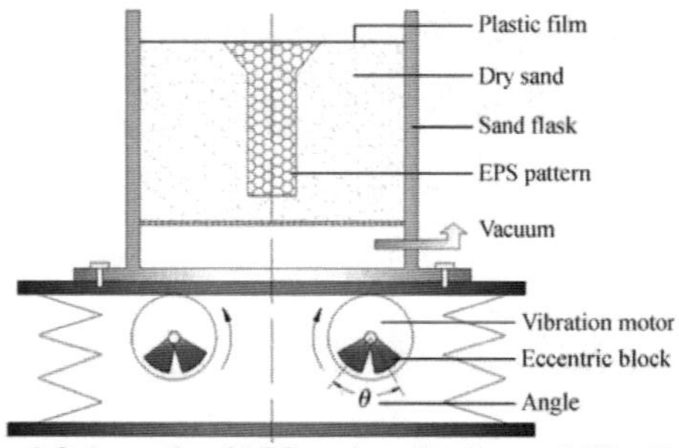

Fig. 2 Schematic of LFC under vibration solidification.

3) LFC under pressure solidification

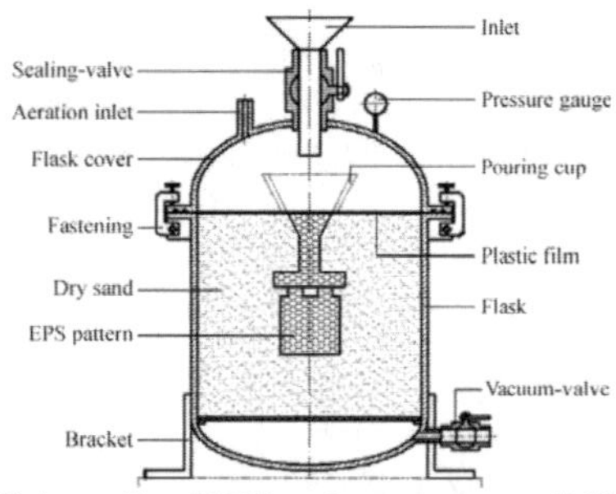

Fig. 3 Schematic of LFC under pressure solidification

The LFC under pressure solidification technology combines LFC and pressure solidification technology, which can increase the density and improve the mechanical properties of the castings. Fig. 3 illustrates the equipment and method of the LFC under pressure solidification for technology.

4) Expendable shell casting technology under vacuum and low pressure

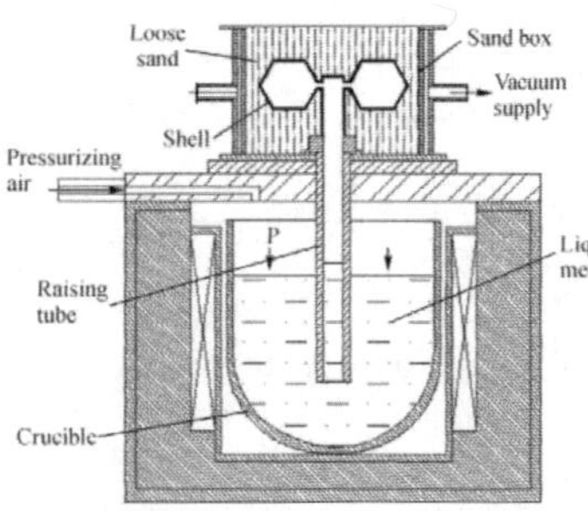

Fig. 4 Schematic of the expendable shell casting process under vacuum and low pressure

The expendable shell casting process combines the LFC and the investment casting technologies. The expendable shell casting technology under vacuum and low pressure simultaneously adopts the preparation of the foam pattern of the LFC process, the fabrication of the ceramic shell of the investment casting, and the vacuum and low-pressure casting based on the expendable shell casting. Fig. 4 presents the schematic of this casting technology.

5) Preparation technology of bimetallic castings based on the LFC method

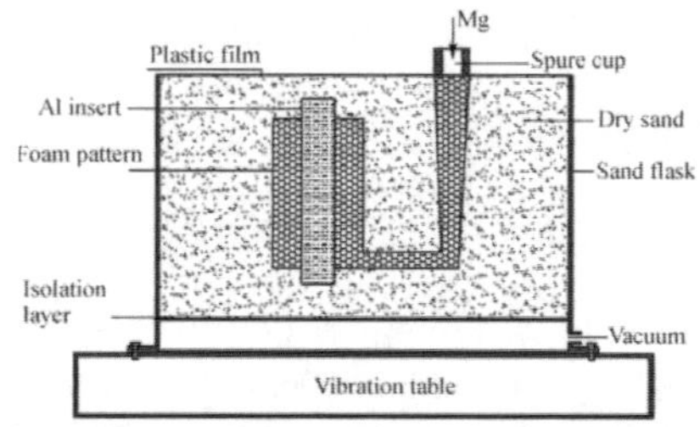

Fig. 5 Schematic of the preparation technology of bimetalliccastings based on the LFC method

Currently, a large number of methods to produce the bimetallic castings, such as casting, welding, and rolling, are available. In particular, the LFC method can be used to directly fabricate the bimetallic castings. The schematic of the preparation technology of bimetallic castings based on the LFC method is illustrated in Fig. 5.

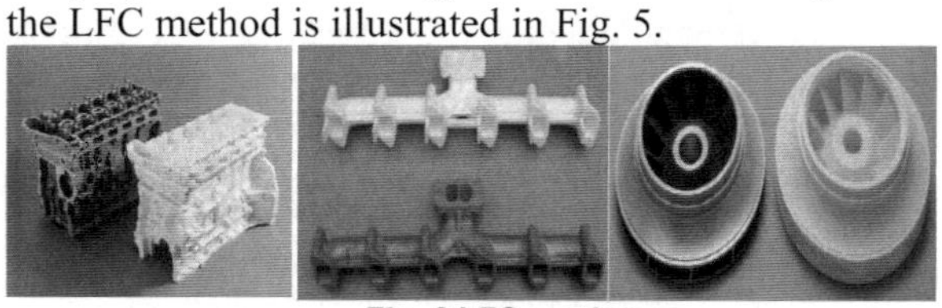
Fig. 6 LFC castings

4 Conclusion

(1) The LFC under vacuum and low pressure, the vibration and pressure solidifications, the expendable shell casting technology under vacuum and low pressure, can reduce the pouring temperature, improve the filling ability, refine the microstructure and which is suitable for producing the complex and thin-wall castings.

(2) The Al/Mg and Al/Al bimetallic castings with acceptable metallurgical bonding can be successfully prepared by the LFC method.

(3) These proposed novel LFC technologies can solve the current technological problems and promote the technological progress of the LFC process. Finally, other important development directions for the LFC technology include foam materials, coating, and purification technology of the tail gas

References

[1] Jiang WM, Fan ZT. Novel technologies for the lost foam casting process. Front. Mech. Eng. 2018,13(1):37–47.

[2] Li GY, Jiang WM, Fan ZT. Preparation, interfacial regulation and strengthening of Mg/Al bimetal fabricated by compound casting: A review. J. Magnes. Alloy. 2023,11(9):3059-3098.

[3] Jiang WM, Fan ZT. Combined effects of mechanical vibration and wall thickness on microstructure and mechanical properties of A356 aluminum alloy produced by expendable pattern shell casting. Mater. Sci. Eng. A 2014,619(12):228-237.

Lost Foam Casting Technology and Defects Control of Box Shell Castings

Zhongchao WU [1], Yusong Liu[2,*], Guangbin Zou[1]
1. Liugong Liuzhou Foundry Co., Ltd., Liuzhou 545007, Guangxi China
2. Foundry Branch of Sichuan Province Mechanical Engineering Society, Chengdu 610065, Sichuan China
*Corresponding address: e-mail:wuzc@liugong.com

Abstract: The production of loading case shells using the lost foam process has advantages such as low molding cost, surface smoothness, and light casting weight. However, for box shell castings with complex inner cavity structures and significant wall thickness differences, there are often defects such as shrinkage porosity, sintering, and carbon defects.The box shell adopts the "two big die pieces" EPC slicing process, selects appropriate gating system, pouring temperature, foam bead density and other control measures, and the produced box shell iron castings are qualified in size and performance, and the defects such as embedded tube sintering and porosity of castings are effectively solved and mass produced.

Keywords: Lost Foam Casting; porosity; Sintering; box shell castings

1 Introduction

The lost foam casting process, also known as gasification casting or full mold casting.Compared with traditional sand casting, the lost foam casting process has the advantages of lower cost and lighter blank weight. However, lost foam casting also has limitations, as not all casting structures are suitable. Below, the author introduces a lost foam casting process for a box , as well as control measures for defects such as looseness, sintering, and carbon defects.

2 Process scheme

The material of this casting is HT200, with a basic wall thickness of 12 mm and a maximum wall thickness of 75 mm. The contour dimensions are 1040mm x 600mm x 606mm (Figure 1). Design and manufacture the lost foam casting process based on the geometric shape of the casting.

The traditional "three major molds" process for the disappearing mold of the box body, Scheme 1 (Figure 2),. Each box body is formed by three major molds, and the requirements for the molding machine and mold design are not high. However, there are many adhesive molds with large cumulative errors and low adhesive efficiency. Therefore, the characteristics of the "three piece structure" process determine that the forming process has high costs and low production efficiency of white molds.

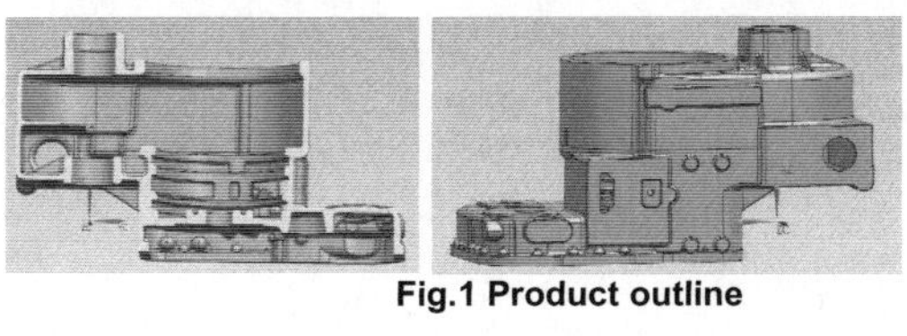

Fig.1 Product outline

The traditional "three major molds" process for the disappearing mold of the box body, Scheme 1 (Figure 2), includes molds from the end cover surface to the middle cover surface of the inner cavity, between the middle cover surface and the cylinder bottom surface, and between the cylinder bottom surface and the mating surface. Each box body is formed by three major molds, and the requirements for the molding machine and mold design are not high. However, there are many adhesive molds with large cumulative errors and low adhesive efficiency. Therefore, the characteristics of the "three piece structure" process determine that the forming process has high costs and low production efficiency of white molds.

Fig.2 Process scheme 1

The author designed two process plans for the mold, Plan 2 (Figure 3): two large molds, seven small mold processes, with "cross" ribs applied to the inner hole, serving as a sprue and preventing deformation by pulling ribs. The pouring system is made as a whole mold, with stepped water inlet, smooth casting punching, and overflow risers set at the top to reduce the risk of carbon slag defects.

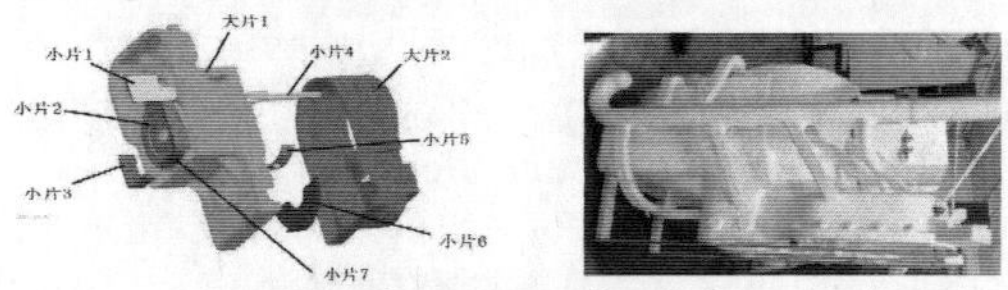

Fig.3 Process scheme 2

Weigh and compare the advantages and disadvantages of the two typical process options mentioned above: Option 2, two large mold pieces are formed, with high adhesive efficiency, improved dimensional accuracy, less occupation of tooling racks, and less space in the drying room; Considering the comprehensive quality and cost of Scheme 1, Scheme 2 is adopted for mold opening testing.

The specific parameters are: the density of pre produced foam beads in the mold is controlled (19.5 ± 0.5) g/L, and the cross-sectional area ratio of the pouring system is controlled according to the ratio of straight runner: horizontal runner: inner runner=1:5: 1.2. A section of overflow slag collection is set at the bottom of the straight runner.

The main difficulty in the process lies in the deep grooves in the inner cavity of the casting, as well as the presence of grooves on the flange surface. Both sides should be filled with sand to prevent defects such as bulging and carbon slag on the machining surface caused by loose sand filling.

3 Process testing

After determining the process plan, proceed with mold production and experimental verification, including the following steps:

(1) The mold is formed and produced by a well-known specialized mold factory in China. EPS gray iron beads are used, and the density of pre produced foam beads is controlled at 19.5 ± 0.5g/L.

(2) Two large mold pieces are glued together, and the pouring system is made as a whole. The glued efficiency is high, and the mold frame and drying room are occupied less.

(3) The thickness of the coating is controlled within (1.2-1.4) mm. While meeting the requirements of coating thickness and weight, the three coat coating process is improved to a two coat coating process, saving 30% of coating time and reducing the occupation of the drying room.

(4) Due to the deep oil pipe of the box and the bending of the valve surface oil port structure, it is necessary to use Φ 17 (inner hole Φ 12) Cold drawn steel pipes are pre embedded to create oil ducts, and coated sand is pre filled inside the steel pipes.

(5) Shaped buried boxes, are produced using ceramic sand instead of pearl sand. The density of ceramic sand is relatively lighter than that of pearl sand, and the amount of molding sand material used is reduced by more than 30%. when the box is buried and filled with sand, the foam inclines (2-3) ° to the back of the cylinder bottom to prevent the inner cavity groove from being filled with sand loosely, which may cause bulge defects.

(6) Before pouring, check to ensure that the negative pressure value reaches (-0.05) - (-0.06) Mpa.

(7) The composition before the furnace is controlled within the range of C%=3.1~3.3, Si%=1.7~1.9, Mn%=0.65~1.0, P% ≤ 0.07, and S% ≤ 0.1 to meet the mechanical properties and processing requirements.

(8) Pouring temperature control, first pouring temperature: 1480 ± 5 ℃, pouring time (30-40) seconds.

(9) The unboxing time after pouring is ≥ 3 hours.

(10) By adjusting the key cylinder bottom size, center hole size and 3D scanning of castings in the foam bonding process of the control box, the casting tolerance meets the drawing requirements.

4 Defect improvement measures

(1) To address the looseness of the pump surface machining surface of castings, the sprue on the pump surface machining surface where the looseness defect occurs is moved to the non machining surface with a thin wall next to it, in order to eliminate the looseness defect caused by prolonged heating of the sprue (Figure 4).

(2) In response to the sintering defects of buried pipes, the inner hole of the embedded steel pipe remains unchanged, and the wall thickness of the buried pipe increases from 2.5mm to 4mm. Baozhu chromite coated sand is pre filled to eliminate high-temperature deformation and internal sintering defects of the buried pipe..

Fig.4 porosity defects and improvement

5 Conclusion

For the lost foam casting process of box shell castings, two large molds are used for molding, pre foaming bead density control, integral pouring system molding, and coating secondary pouring process to control pouring temperature and pouring time. Significant results have been achieved in solving the problems of loose pump surface, buried pipe sintering, and carbon slag defects in castings. The product quality meets the requirements, and mass production can achieve a comprehensive waste rate of less than 5%.

Ceramic Particles Reinforced Ductile Iron Prepared by Lost Foam Casting and Its Properties

W. Zhang[1,2], Y. Zhao[1], Y.D. Qu[1,2 *], G.L. Li[1,2], S.L. Dong[1,2], S.R. Zhang[1,2]
1. School of Materials Science and Engineering, Shenyang University of Technology, Shenyang 110870, Liaoning, China
2. Technical Innovation Center for Lightweight and High Performance Metal Materials of Liaoning Province, Shenyang University of Technology, Shenyang, 110870, China
*Corresponding address: quydong@sut.edu.cn

Abstract: Based on the integrated composite addition technology of ceramic particles and Fe powder foam preform, ceramic particle reinforced ductile iron materials were prepared by lost foam casting, and the effect of TiC ceramic particles on the microstructure, mechanical properties and friction and wear behavior of ductile cast iron were studied. The results show that the heterogeneous nucleation of cementite can be enhanced by using TiC particles (1~5 μm) as the reinforcement phase, and the content of pearlite in ductile iron increases. The tensile strength and hardness of ductile iron increase with the increase of TiC content, while the elongation decreases. The wear mechanism changes from adhesive wear to abrasive wear. When the addition of TiC is 1wt. %, the strength of ductile iron reaches 498MPa and the elongation is 18.1 %.

Keywords: ductile iron composites; ceramic particles; mechanical properties

1 Introduction

Ductile iron is widely used in structural parts due to its excellent mechanical properties such as strength and plasticity. For some ductile iron castings, e.g., hubs, gears and crankshafts, etc., in addition to the need for good strength and toughness, the wear resistance is often their assessment index as well [1]. Adding hard insoluble ceramic particles to alloy materials, that is, based on the dispersion strengthening method, can generally improve the strength, hardness and wear resistance of the alloy simultaneously [2], which provides a new way for the development of wear-resistant ductile iron. Research indicates that the addition of 0.5 wt.% CeO_2 can make the austenitic ductile iron have 452 MPa ultimate tensile strength, 14.4 % elongation and 85 % wear volume reduction [3]. However, the inherent density difference, wettability and reaction problems between molten iron and ceramic particles make it difficult to achieve a good combination between the two by using traditional casting methods to prepare composite materials, and it is also difficult to achieve uniform distribution of ceramic particles [4]. Therefore, it is imperative to develop new technologies that can maturely prepare ductile iron / ceramic particles composites. Hence, in this study, a method for preparing ductile iron / ceramic particle composites by Fe powder / foam preform and lost foam casting technology was proposed, and the TiC particles was used as the reinforced phase to add into ductile iron. The variation trend of microstructure and mechanical caused by TiC ceramic particles are evaluated.

2 Experimental procedure

The preparation process of ceramic particles reinforced ductile iron is shown in Figure 1. The iron powder with a size of about 5 μm and the TiC (1~5μm in size) ceramic particles with the expected addition amount were ground together to form a mechanical mixture by high-energy ball milling. Then the mechanically mixed Fe/TiC powder was mixed with EPS beads and foamed together to prepare foam preform for lost foam casting. After which the foam preform was embedded in the molding sand and awaited pouring under negative pressure. Pig iron, scrap steel, carburant and ferrosilicon were used as raw materials for smelting molten iron, and they were melted using a induction melting furnace. Spheroidization and inoculation were carried out in ladle by punching method, and the smelting temperature of molten iron is about 1500 °C. Then the molten iron in the ladle is poured into a sand mold containing prefabricated foam, and the pouring temperature is about 1380°C. After solidification and cooling, the sanding was carried out and the ductile iron ingot was obtained.

The metallographic samples, tensile performance samples and wear test samples were cut from the ingot. The 4% nitric acid alcohol solution was used to corrode the metallographic sample. Microstructure of the TiC ceramic particles reinforced ductile iron were analyzed using the OLS4100 laser scanning confocal microscope (LSCM) and the TM3030 scanning electron microscope (SEM). The tensile properties of the samples were tested using an E45-305 microcomputer-controlled electronic universal testing machine (MTS) at the speed of 1 mm/min. The wear resistance of the ductile iron was analyzed by reciprocating friction of WC balls at a speed of 5 mm/s for 90m under a load of 50 N on an MFT-5000 multi-function tribometer. The surface morphology of wear test samples after wear was observed using the OLS4100 LSCM.

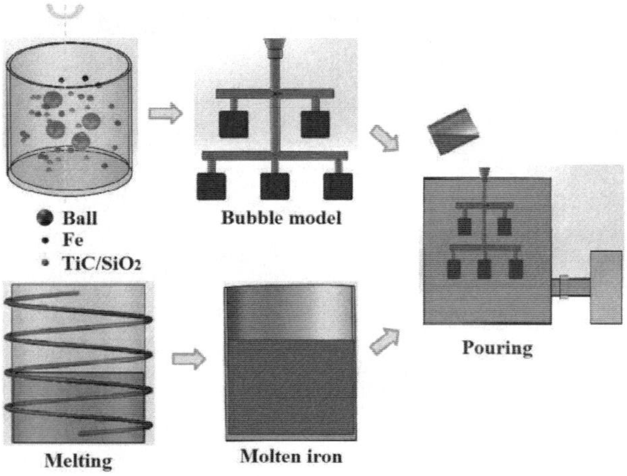

Fig. 1. Process diagram of preparing ceramic particle reinforced ductile iron by lost foam casting method.

3 Result and discussion

Microstructural characterization

Fig.2 shows the effect of TiC ceramic particles on the microstructure of ductile iron. With the addition of TiC, the pearlite content in the matrix increases significantly, and the value for ductile iron without TiC addition is 0.63%, while 10.75% for ductile iron with the TiC addition of 1%. In addition, the addition of TiC will lead to a decrease in the spheroidization rate of graphite. The pearlite exhibits a typical lamellar structure with cubic TiC particles distributed in it, and the TiC particles are well combined with the cementite in the pearlite. However, no obvious TiC particles were found in the ferrite matrix. The phenomenon indicates that TiC can be used as the heterogeneous nucleation of cementite, which promotes the increase of pearlite in ductile iron [5].

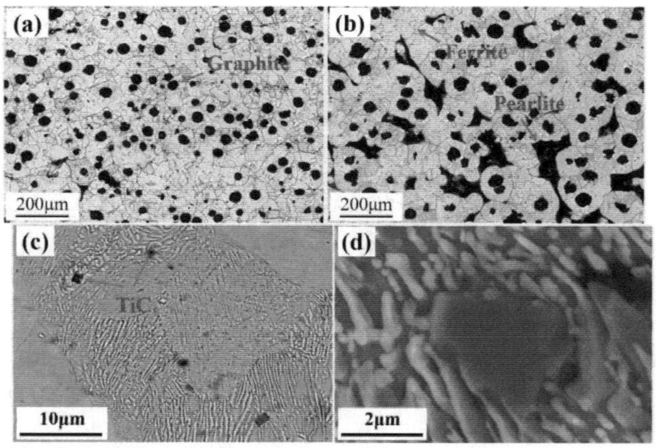

Fig. 2. the effect of TiC ceramic particles on the microstructure of ductile iron, in which (a) is the ductile iron without TiC, (b) is the ductile iron with the TiC addition of 1.0wt.%, (c) is the distribution of TiC in the ductile and (d) is the enlarged TiC particle.

Microstructural characterization

Fig.3 shows the effect of TiC addition on strength and elongation of ductile cast iron. With the increase of TiC content, the strength of ductile iron tends to increase gradually, while the elongation decreases gradually, which are mainly related to the increase of pearlite proportion caused by TiC addition. When the addition of TiC is 1 wt.%, the tensile strength of ductile iron is 498MPa, and the elongation is 18.1%.

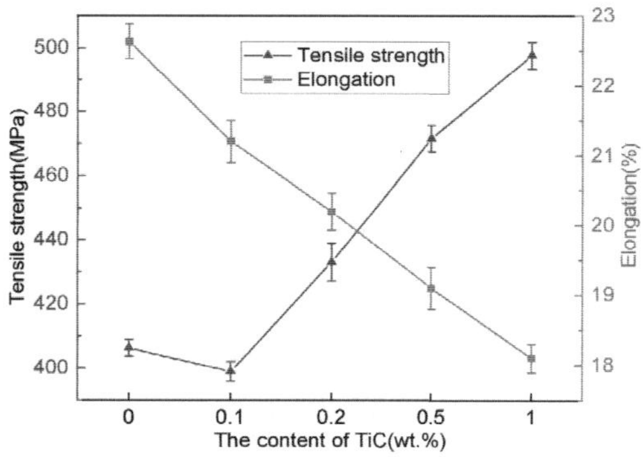

Fig. 3. Effect of TiC addition on strength and elongation of ductile cast iron

4 Conclusion

In this work, prefabricated foams were prepared by co-foaming of Fe powder / TiC ceramic particles obtained by high-energy ball milling and EPS beads. Then TiC / ductile iron composites were prepared by lost foam casting technology. TiC ceramic particles are uniformly dispersed in the pearlite of the matrix structure of ductile iron, and increases the proportion of pearlite in the ductile iron, which is related to the fact that TiC can be used as the core of heterogeneous nucleation of cementite. The increase of TIC content leads to the increase of pearlite content in ductile iron, resulting in the increase of strength and the decrease of elongation of ductile iron.

Acknowledgments

This research was supported by the Program for Natural Science Foundation of Liaoning Province (No. 2022-BS-181).

References

[1] Fontanari V., et al., Wear (2016), 350, 68-73.
[2] Qiu F., et al., China Foundry (2020), 17(2), 111-126.
[3] Zhao Y., et al., International Journal of Metalcasting, (2024).
[4] Kalogeropoulou S., et al., Acta Metallurgica et Materialia (1995), 43(3), 907-912.
[5] Zhao Y., et al., Tribology Letter (2023), 71, 27.

Study on Interfacial Strengthening of Al/Mg Bimetal Prepared by the Lost Foam Compound Casting

Guangyu Li[1,2,*], Wenming Jiang[3]

1. School of Materials Science and Engineering, Dalian University of Technology, Dalian, China
2. Ningbo Institute of Dalian University of Technology, Ningbo, China
3. State Key Laboratory of Materials Processing and Die & Mould Technology, School of Materials Science and Engineering, Huazhong University of Science and Technology, Wuhan, China

*Corresponding address: e-mail: gyli@dlut.edu.cn

Abstract: Al/Mg bimetal combines the advantages of both aluminum and magnesium and has broad application prospects in automotive, aerospace, weapons, digital products and so on. The compound casting has the characteristics of low cost, easy to achieve metallurgical combination and suitable for the preparation of complex Al/Mg bimetallic parts. However, bimetallic joint strength is low due to the differences of physical properties between Al and Mg, oxide film on the metallic surface and interfacial Al-Mg IMCs, which is closely related to the interfacial microstructure and properties. Therefore, how to control the interface of the bimetal to achieve the performance enhancement is the focus and difficulty in this field. To solve the above problems, we put forward the following solutions. Firstly, the undesirable Al-Mg IMCs were reduced or eliminated by adding the interlayers (Zn, Ni and Ni-Cu). Secondly, the evolution process of interfacial microstructure was changed and fine strengthening phases were formed by adding Si element to Al alloy or rare earth element to Mg alloy. Thirdly, mechanical vibration and ultrasonic vibration were applied in the process of the filling and solidification to refine and homogenize the interfacial structure. Finally, some other methods, including heat treatment and constructing exterior 3D morphology, also can be used to regulate the interfacial microstructure and compositions. The above strengthening methods can be used alone or in combination to achieve bimetallic strengthening. The highest shear strength (69.8 MPa) was obtained by adding plasma sprayed Ni coating on the surface of A356 alloy in the lost foam compound casting so far. Finally, the future development direction of the Al/Mg bimetal is prospected, which provides some new ideas for the development and application of the Al/Mg bimetal.

Keywords: Al/Mg bimetal; Compound casting; Interfacial regulation; Interface strengthening

1 Introduction

Al/Mg bimetal combines the performance characteristics of aluminum alloy and magnesium alloy, and maintains excellent comprehensive performance while achieving lightweight, which has broad application prospects in aerospace, automotive, weapons and other fields [1-2]. According to the state of bimetal preparation, the preparation methods can be divided into solid-solid composite, solid-liquid composite and liquid-liquid composite, each of which has its own characteristics and application fields. Among them, solid-liquid composite is also known as composite casting, which has the advantage of low-cost preparation of complex parts. However, the use of composite casting to prepare bimetals also has problems such as too thick interfacial layer, oxidation inclusion, cracks, etc. These problems lead to the bimetal performance cannot meet the requirements of use, therefore, limiting the development of bimetals. These problems are mainly related to the interface, so how to regulate the interface of bimetal is the focus and difficulty of this field. At present, researchers have proposed many methods to regulate the Al-Mg bimetallic interface, including removing oxide film, adding intermediate layer, applying external field, alloying and so on [3-4].

Our research team has more than ten years of research experience in the field of composite casting Al/Mg bimetal. In this paper, the research results on the interface regulation and strengthening of composite casting bimetal are reviewed, and the future development direction is prospected to provide theoretical basis and practical experience for the application of bimetal.

2 Experimental procedure

Al/Mg bimetal is prepared by the lost foam compound casting, the principle of which is shown in Fig. 1. The microstructures and properties of the bimetals were analyzed by SEM, EDS, TEM, EBSD, XRD and material property testing machine.

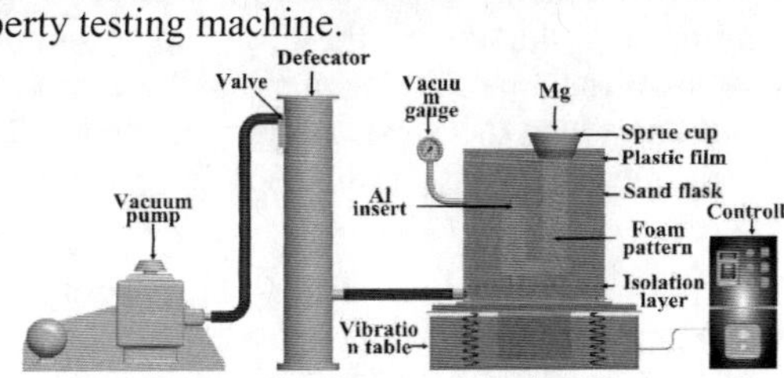

Fig. 1. Schematic diagram of the lost foam casting

3 Result and discussion

ADDING INTERLAYER

The effects of the Ni coatings prepared by different methods, namely, electro nickeling (EN), electroless nickel plating (ENP), plasma sprayed nickel (PSN), on the microstructure and properties of the Al/Mg bimetal were compared, as exhibited in Fig. 2. There is metallurgical bonding between Ni coating and Al matrix but no metallurgical bonding between Ni coating and Mg matrix at the bimetallic interface of EN and ENP, and the interfacial layers are mainly composed of Ni solid solution (Ni SS) and Al_3Ni, as exhibited in Fig. 2(a) and (b). For the bimetal containing PSN coating, metallurgical reaction layers occur between the Ni coating with Al substrate and Mg substrate, and the interfacial layer is mainly composed of Mg_2Ni, Ni SS and Al_3Ni, because the rough coating is more conducive to the metallurgical bonding, as displayed in Fig. 2(c). EBSD results verify the existence of these phases, and these phases have small grain size and random orientation, as exhibited in Fig. 2(d) and (e). Only the shear strength of the bimetal with PSN coating is improved, with a shear strength of 69.8 MPa, which is 69% higher than that of the bimetal without coatings.

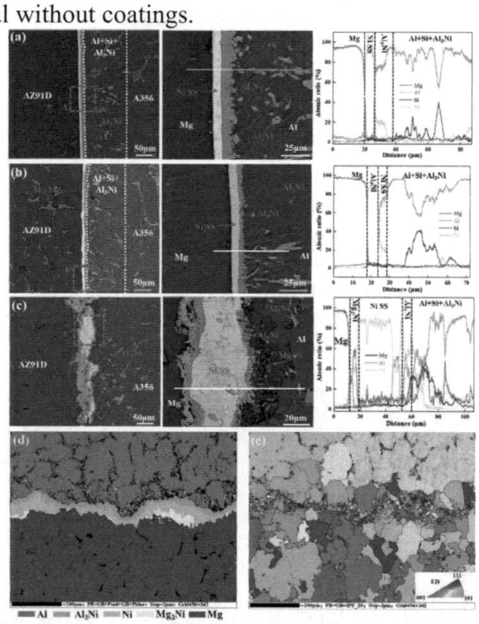

Fig. 2 Microstructure and EBSD results of the bimetal with different Ni coatings

Alloying

It is well known that if the bimetallic matrix is pure magnesium and pure aluminum, the phase composition of the interface layer must be Al-Mg intermetallic compound. However, we know that Al-Mg IMCs are brittle and hard phase, accompanied by coarse grains, uneven composition, stress concentration, preferred orientation and other problems. In practice, the matrix alloy usually contains other elements, and the presence of these elements will also affect the interface structure of the bimetal. Therefore, alloying is also an effective method to improve the structure and properties of the interface by adjusting the element type or content of the matrix alloy. At present, the elements used in the study of alloying in compound casting can be divided into two categories: one is Si element; the other is rare earth alloy, including La, Ce, Nd, Y and Gd, as listed in Table 1.

Table 1. Alloying element used in compound casting

Material	Alloying element	Improvement of shear strength/MPa (Content)
Al-Si/AZ91D	Si	6.0→58.6 (15 wt.%)
A356/AZ91D	La	39.36→51.54 (1 wt. %)
	Ce	37.22→53.29 (1 wt. %)
	Nd	36.8→52.2 (0.7 wt. %)
	Y	36.51→50.98 (0.2 wt. %)
	Gd	33.1→46.0 (0.6 wt. %)

4 Conclusion

The above strengthening methods can be used alone or in combination to achieve bimetallic strengthening. The highest shear strength (69.8 MPa) was obtained by adding plasma sprayed Ni coating on the surface of A356 alloy in the lost foam compound casting so far.

Acknowledgments

The authors gratefully acknowledge the supports provided by the National Natural Science Foundation of China (Nos. 52271102, 52075198 and 52205359), the China Postdoctoral Science Foundation (No. 2021M691112), and the Analytical and Testing Center, HUST.

References

[1] Lei J, Ma L, Cai Z. Interfacial fracture characteristics of Mg/Al composite plates with different thickness ratios by asymmetrical rolling with differential temperature rolls. Materials Science and Engineering. A, 2023, 869: 144764.

[2] Liu B, Yang J, Zhang X. Development and application of magnesium alloy parts for automotive OEMs: A review. Journal of Magnesium and Alloys, 2023, 11(1): 15-47.

[3] Siengchin S. A review on lightweight materials for defence applications: Present and future developments. Defence Technology, 2023, 24: 1-17.

[4] Yu H, Liu S, Cao L. Preface to the special issue on aluminum alloys for transportation. Journal of Central South University, 2022, 29(3): 741-743.

Solidification Microstructure and Properties of Gray Cast Iron Made by Lost Foam Casting with Vibration

Botao Xiao [1,*], Wenming Jiang [2], Zitian Fan [2]
1. School of Materials and Engergy, Jiangxi Science and Techonlogy Normal University, Nanchang 330036, China
2. State Key Laboratory of Materials Processing and Die & Mould Technology, School of Materials Science and Engineering, Huazhong University of Science and Technology, Wuhan 430074, China
*Corresponding address: e-mail: xltao_2005@163.com

Abstract: The effects of sample wall thickness, vibration frequency and amplitude on graphite morphology, primary austenite and properties of gray cast iron (GCI) made by lost foam casting (LFC) with vibration were studied. The results show that when the thickness of the sample is relatively thin (20 mm), type B and type C graphite are formed in the GCI, but no type B and type C graphite are observed in the sample thickness of 40 and 60 mm. The results of GCIwith a thickness of 70 mm made by LFC with mechanical vibration show that when the vibration frequency is 35 Hz and the amplitude is 2 mm, the flake graphite length of the GCIis small, the austenitic dendrite arms are thick and evenly distributed, and an effective network skeleton is formed in the GCIwith large tensile strength and elongation. This can be attributed to the effect of vibration on heat and mass transfer during solidification of GCImelt.

Keywords: Solidification; Vibration; Gray cast iron; Microstructure; Property; Lost foam casting

1 Introduction

Lost foam casting (LFC) is a near-net forming method, known as "the new casting technology in the 21st century" and "green engineering in casting", is widely used in the production of complex GCIcastings, such as cylinder block, cylinder head and so on. However, the dry sand molding is used in the LFC, and the pouring temperature is 30~50 ℃ higher than that of the traditional casting, resulting in the solidified structure of the iron base alloy castings is coarse and the mechanical properties are low, which is very unfavorable to the service life of the castings. For this reason, refining the microstructure and improving the performance of the alloy castings made by LFC are particularly key to improving the service life of the castings, among which mechanical vibration can refine the alloy microstructure and improve the alloy properties without adding additional equipment. Therefore, in this paper, mechanical vibration is applied to GCIin the LFC, and the effects of sample wall thickness, vibration frequency and amplitude on graphite morphology, primary austenite and properties of GCIin the LFC are studied to lay a foundation for improving the structure and properties of GCImade by the LFC.

2 Experimental procedure

Scrap steel, pig iron, ferromanganese, carburant and 75 ferrosilicon was used as raw materials for preparing HT 100 iron liquid. Melting equipment was a one-ton medium-frequency induction furnace. The scrap steel, pig iron and carburant were firstly added into the furnace. When the temperature increased to 1,500 °C, the scrap steel, ferromanganese and ferrosilicon were added into the furnace to adjust the component of the molten metal. Finally, the melt was modified with 75 ferrosilicon and then poured into the sand box. The Y-type specimen of GCI was produced and their dimensions are shown in Figure 1. During pouring, vibration solidification parameters are shown in Table 1. The cube specimen for the density measurement was cut using wire cutting and its dimensions were 10.4 mm x 10.4 mm x 10.4 mm. It was then polished to the dimensions of 10 mm x 10 mm x 10 mm using 100, 400 and 800 grit abrasive papers in sequence and the surface was cleaned using ultrasonic cleaning in alcohol. Then the cube mass was measured using a balance, and the density of the GCI was calculated. The average density value was obtained from three specimens for each specified condition. The flake graphite of the GCI before corrosion and the morphology of the primary austenite after corrosion were observed by an optical microscope (OM), and the length of the flake graphite was measured by the optical microscope software system. The etching solution was 4% nitric acid and alcohol solution, and Alkaline sodium caustic acid solution. The Brinell hardness of the GCI was measured by a HBE-3000A-type Brinell hardness tester with a load of 1,000 kgf and a loading time of 15 s. The average value of the Brinell hardness was taken from three measurements. The mechanical properties of the casting were tested by a SHIMADZU AG-IC universal testing machine. The tensile test was carried out at room temperature at the tensile speed of 1 $mm \cdot min^{-1}$. Then, the fracture morphology was observed by Quanta 200 environmental scanning electron microscopy (ESEM). Finally, the temperature field of the columnar zone during solidification of the GCI fabricated by the LFC with vibration frequency of 0 and 35 Hz was measured by a thermocouple with a diameter of 0.3 mm. The thermocouple wire was welded and embedded into the ceramic tube, and embedded in the foam pattern. The embedded depth of the thermocouple wire was 12.5 mm.

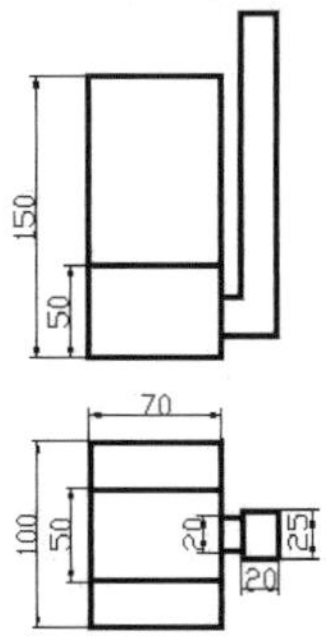

Figure 1. The size of EPS shape

Table 1. Vibration solidification parameters

Amplitude/ mm	3			
Frequency/ Hz	0	35	50	100
Frequency/ Hz	35			
Amplitude/ mm	0.75	2	3	4

3 Result and discussion

1.Effect of vibration frequency on microstructure

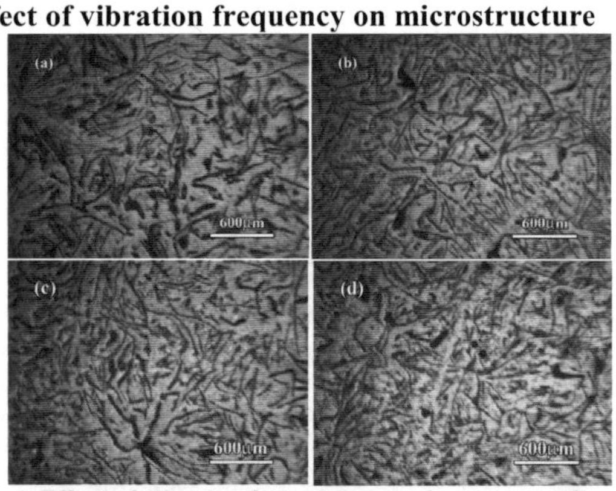

Figure 2. Effect of vibration frequency on primary austenite of GCI: (a) 0 Hz, (b) 35 Hz, (c) 50 Hz, (d) 100 Hz

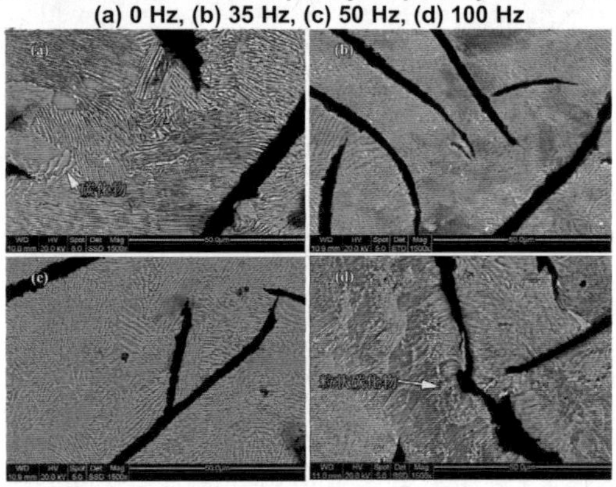

Figure 3. Effect of vibration frequency on matrix structure of GCI: (a) 0 Hz, (b) 35 Hz, (c) 50 Hz, (d) 100 Hz

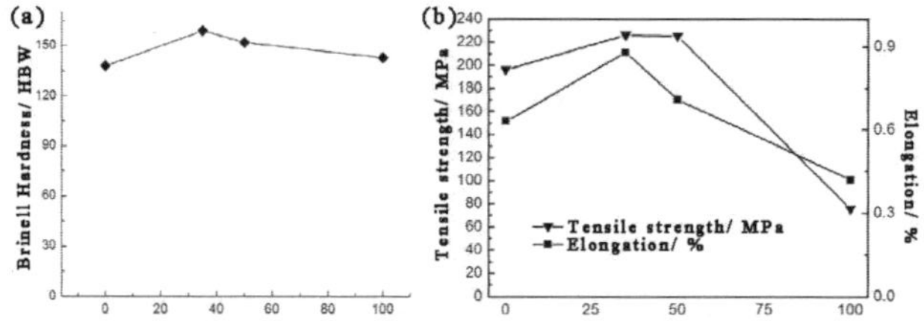

Figure 4. Effect of vibration frequency on mechanical properties of GCI: (a) Brinell Hardness, (b) Mechanical property

2.Effect of amplitude on matrix structure of GCI

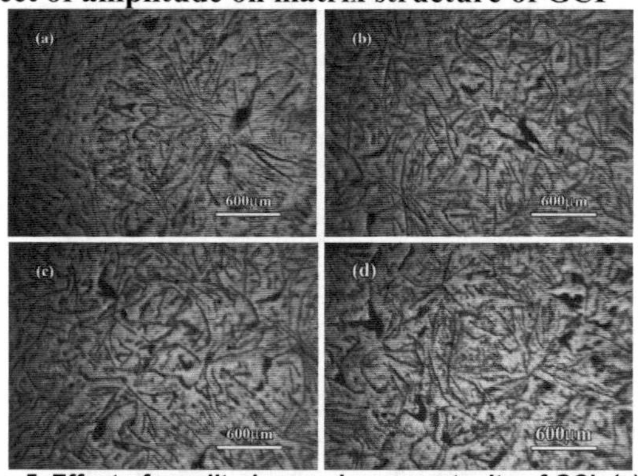

Figure 5. Effect of amplitude on primary austenite of GCI: (a) 0.75 mm, (b) 2 mm, (c) 3 mm, (d) 4 mm

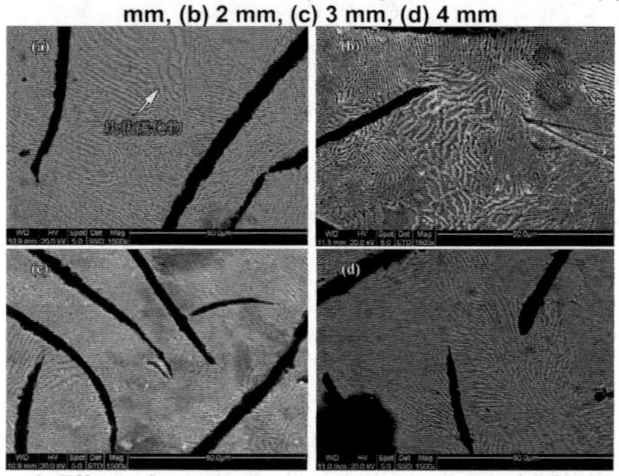

Figure 6. Effect of amplitude on matrix structure of GCI: (a) 0.75 mm, (b) 2 mm, (c) 3 mm, (d) 4 mm

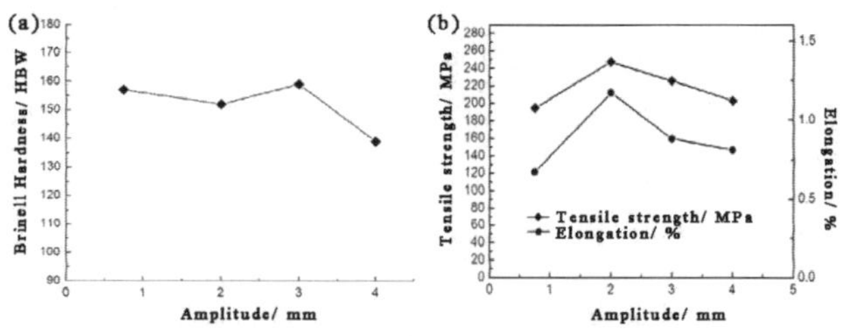

Figure 7. Effect of amplitude on mechanical properties of GCI: (a) 0.75 mm, (b) 2 mm, (c) 3 mm, (d) 4 mm

Casting Technology and Application Prospect of Titanium Alloy EPS Lost Mold Cladding

Bobo Li*, Bing Sun, Huifa Tao, Huaqiang Chang, Hongzhe Sun, Fei Wang, Haibin Qiao
Luoyang Shuangrui Precision Casting Titanium Co., LTD., Luoyang 471000, Henan, China
*Corresponding address: e-mail: libobo0816@126.com

Abstract: Domestic titanium and titanium alloy castings are mainly produced by two process methods: machined graphite type and investment precision casting, but subject to technical limitations, the actual production process can not fully meet the development needs of customers for high quality, low cost and short delivery, so it is necessary to promote the research and development and application of new casting technology, and complement the existing technology. Domestic titanium and titanium alloy castings are mainly produced by two process methods: machined graphite type and investment precision casting, but subject to technical limitations, the actual production process can not fully meet the development needs of customers for high quality, low cost and short delivery, so it is necessary to promote the research and development and application of new casting technology, and complement the existing technology. Lost die casting has advantages that other casting methods do not have, and it has been widely used in steel, aluminum and other metal castings, but there are few reports on its application in titanium alloy. This paper mainly introduces the technical advantages and application prospects of titanium alloy EPS lost mold cladding casting. Lost die casting has advantages that other casting methods do not have, and it has been widely used in steel, aluminum and other metal castings, but there are few reports on its application in titanium alloy. This paper mainly introduces the technical advantages and application prospects of titanium alloy EPS lost mold cladding casting.

Keywords: Titanium alloy; Casting; Evanescent mode

1 Introduction

Titanium alloy has the advantages of low density, high specific strength, corrosion resistance, non-magnetic, good biocompatibility, etc. It is an excellent structural and functional metal material, which can significantly reduce the weight of equipment, improve the life and performance of equipment, and is widely used in aerospace, petrochemical, ship, ocean engineering and other fields [1].

2 Titanium alloy EPS lost mold shell casting technology production technology

This part introduces the casting process of EPS lost mold shell, foam model preparation process and oxidized ceramic shell preparation process. EPS lost mold shell casting process includes "foam model preparation → coating preparation → shell preparation → shell roasting → vacuum casting". EPS vanishing mold preparation method can be divided into processing molding and foam molding, which can be flexibly selected according to the shape and structure of the casting, production batch and quality requirements. Processing molding is the foam billet,

However, due to the high temperature reactivity of titanium alloy, titanium alloy casting is difficult, the production process is long, the equipment and process are complex, and the casting material is expensive, making the price of titanium castings 10 to 50 times higher than that of steel castings or more, limiting the further promotion and application of titanium castings, resulting in a small market size. Therefore, it is very necessary to develop new titanium alloy casting technology, complement existing technology, and promote the development of titanium alloy casting technology and the application of titanium castings [2-5].

3 Development history and characteristics of lost die casting technology

The development history of lost die casting technology is introduced. The process is simple and easy to produce. After decades of technological innovation and development, advanced countries such as the United States, Germany and Japan have adopted this technology to produce cast iron, cast steel, aluminum alloy and other metal parts. It is widely used in the production of engine cylinder head, cylinder block, motor shell, box, intake manifold and other complex parts in the automotive machinery industry [6-7]. Lost mold casting is the use of foam board processing into the same size as the casting model, the surface is coated with refractory paint and dried, dry sand modeling, vibration compaction and pouring, so that the model gasification, metal liquid filling, solidified to get consistent with the model metal parts casting method, is a near zero margin, accurate forming casting process. Compared with other casting technologies, it has many advantages [8].

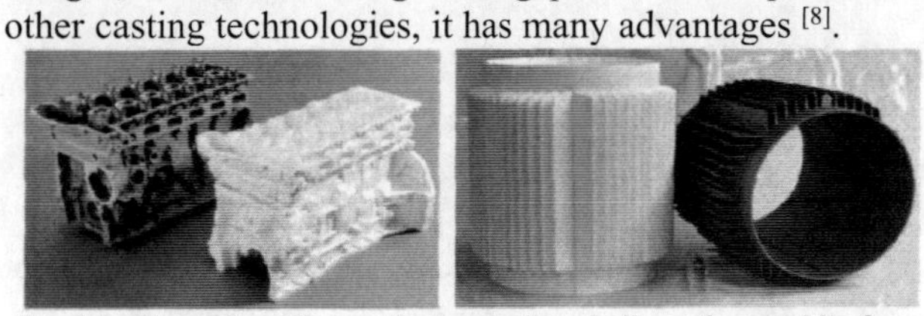

Fig. 1 Lost mold casting cylinder, motor shell casting and its foam model

prepared by CNC machine tool processing, the surface roughness is large (as shown in Fig 2), suitable for simple structure, small production quantity of medium and large castings; The foam type is to prepare the metal mold first, and then foam the model inside the metal mold, and the surface of the model is smooth and delicate (as shown in Fig 3), which is suitable for castings with a certain batch and relatively high precision and roughness requirements. In the preparation process of oxidized ceramic shell, the wax dipping process, the preparation process of ceramic shell surface and back layer, and the control of process defects are introduced.

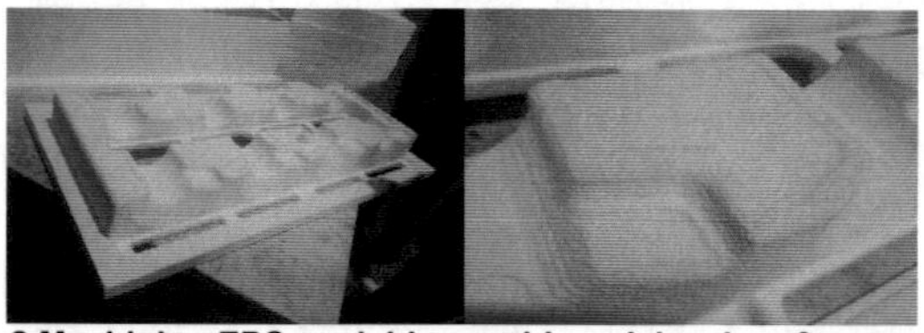

Fig. 2 Machining EPS vanishing mold model and surface quality

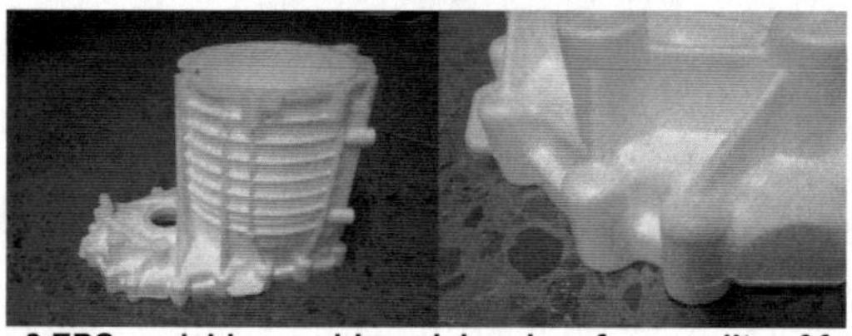

Fig. 3 EPS vanishing mold model and surface quality of foam forming

4 Application prospect and advantages of titanium alloy EPS lost die cladding casting technology

This part introduces the difference between the lost mold cladding casting and machining graphite mold casting, metal mold injection wax investment mold precision casting, and states the advantages of titanium alloy EPS lost mold cladding casting technology, which is complementary to other casting methods and has greater development potential

5 Conclusion

Titanium alloy EPS lost die cover casting technology has the advantages of simple production operation, high dimensional accuracy, good surface quality, and can form highly complex castings, which solves the problem of flow marks, cracks, cold insulation, poor filling performance of machined graphite castings, as well as large wax injection machine and mold investment, significantly reducing the production cost of titanium alloy castings and significant economic benefits. Therefore, combined with the production characteristics of titanium castings, fully tap the potential of EPS lost mold shell casting, technological innovation and application promotion, and have a positive role in promoting the low-cost production of titanium castings and expanding the market application scale.

References

[1] Huo D, Liang J, Li H et al. Research and Application Progress of Titanium Alloys [J]. Casting Technology, 2016,37 (10) : 2065-2066.

[2] Jia H, Lu F, Hao B et al. 2017 China Titanium Industry Development Report [J]. China Titanium Industry, 2018 (2) : 3-11.

[3] Jia H, Lu F, Hao B, et al. China Titanium Industry Development Report 2018 [J]. Titanium Industry Progress, 2019,36 (3) : 42-48.

[4] Shen X, He T, Luo G et al. Investment precision casting of Titanium and its Alloys [J]. Foundry Technology, 2015,36 (7) : 1876-1878.

[5] Shi X, Zheng Z. Fabrication of Graphite mold for Titanium and titanium alloy casting [J]. Foundry, 1998,47 (3) : 37-39. (in Chinese)

[6] Liu J, Ramsay C, Askeland D. Effects of foam density and Density gradients on metal fill in the LFC process [J]. AFS Transactions, 1997(105): 435-442.

[7] Brown J. The lost foam casting process [J]. Metals Mater. , 1992(8):550-555.

[8] Fan Z, Jiang W. Current status and Development trend of lost die casting technology [J]. Casting, 2012(6): 1072-1075.

This page intentionally left blank.

Tensile Bar Castings

Mark A. White
Impro Aerotek USA, Inc. 21680 Gateway Center Drive, Suite 368A, Diamond Bar, CA 91765, USA
*E-mail: mark_white@improusa.com

Keywords: Mechanical Properties; Metallurgical Effects on Cast Shapes; Process Approval; Inimitable Cast Shapes

1 Introduction

In new casting development projects, there exists some form of an agreement between the casting purchaser and the casting manufacturer which outlines the technical requirements for qualification to production. These arrangements may range from more comprehensive specific dimensional and metallurgical requirements associated with aerospace fixed processes to a lone AS9102 dimensional requirement, which may apply to an industrial part application. There exists a gap, in the middle, where casting designer/purchaser assume metallurgical requirements including the notion that machined from casting (MFC) test bars are intended to and must meet ASTM requirements. Oftentimes, the designer/purchaser requires MFC expecting ASTM properties. Many times, casting houses agree, because they think there is no other choice and the designer must be correct. This can cost the casting producer in terms of excessive mechanical testing and lost production.

2 Experimental procedure

The paper thoroughly discusses the metallurgical differences and the inimitability of cast shapes including case studies, mechanical and process data, a metallurgical summary, photos and a practical guide to manage this topic.

3 Result and discussion

The paper represents collaboration with and data from our foundries where we successfully worked through these customer situations.

4 Conclusion

This paper is offered to the 75th World Foundry Congress as a service to both sides of this topic- casting purchaser and casting manufacturer. My hope in offering this work to the 75th World Foundry Congress, is that the audience gains valuable information to go on and successfully handle this topic in their own companies where the situation is turned from "production stopping" to smooth production.

References

[1] Monroe, R., "Mechanical Properties of Heavy Section Castings: An Overview," Steel Foundry Facts 354 :19-23 (February 1983).
[2] Shank, M.E., "Control of Steel Construction to Avoid Brittle Fracture", Welding Research Council, pp. 27-31, 35-39, 55 (1957).
[3] Campbell, J., "Complete Casting Handbook" 2nd ed., pp. 447-451. Butterworth-Heinemann, London (2015).
[4] Roes, H.L., and Wittie, W., "Effects of Wall Thickness on Primary Structure and Mechanical Properties of Plain Carbon and Alloy Steels," AFS Cast Metals Research Journal, pp. 80-85 (June 1968).
[5] See, A., and Hollandsworth, R., "Mechanical Properties of Test Bars Compared to Castings," SFSA Technical & Operating Conference, pp. 2-12 (2001).
[6] "Standard Specification for Castings, Steel and Alloy, Common Requirements, for General Industry Use," ASTM A781/A 781/M-03A (2003) ASTM Subcommittee A01.18.
[7] "Standard Test Methods and Definitions for Mechanical Testing of Steel Products," ASTM A 370 – 03a (October 2003) ASTM Subcommittee A01.13."Standard Specification for Steel Castings, General Requirements, for Pressure-Containing Parts," ASTM A 703/A 703M – 07 (March 2007) ASTM Subcommittee A01.18.
[8] "Steel Castings Specification M-201-00", AAR Manual of Standards and Recommended Practices Couplers and Freight Car Draft Components, pp. 7-8 (2000).
[9] Reed-Hill, R. E., "Physical Metallurgy Principles", 3rd ed., pp. 172, 447-459, 679, 785-796. PWS Publishing, Boston (1992).
[10] "Steel Castings Handbook", 5th ed., General Properties of Steel Castings, pp. 3-23 Steel Founders Society of America, Chicago (1980).
[11] Morris, J. W., "The Effects of Grain Size on the Mechanical Properties of Steel.", Department of Material Science and Engineering, University of California, Berkeley Center for Advanced Materials, Lawrence Berkeley Laboratory.
[12] "Specification for Steel Castings for General Engineering Purposes", British Standard BS-3100-1991 Committee reference ISM/6. pp. 12-14, Fig 5, Fig 6. (January 1991) (London).

Elephant Caput Mould Reliable Sampling for Quality Control

Hayati Sahin*, Derya Dispinar
Non-ferrous Metal Treatment, Vesuvius Foundry R&D Center, Enschede, The Netherlands
*Corresponding address: e-mail: hayati.sahin@vesuvius.com

Abstract: Defect free casting begins with starting melt cleanliness, casting speed and well-designed sprue and runner systems. Failure of one of these three main components has a significant negative impact on the quality of the final product. ASTM B108/B 108M Tensile Test Specimen Casting mold (ASTM mold), which produces only 2 tensile test specimens in a single casting, is widely used to determine mechanical properties in aluminum alloys. The idea of statistical work is fundamental to science. Therefore, the more test samples are produced and analyzed, the higher the reliability of the results are. A new tensile test mold was needed to be designed instead of ASTM mold due to its negative aspects such as turbulence in filling the mold, taking only two samples in one casting, increase operator effect, time loss and excessive metal loss in feeder and runners. In our previous study, different mold designs and different casting (gravity and tilt casting) methods were examined to produce 10 samples in a single casting without turbulence in sand mold. The statistically enough samples were produced in a single casting to give a reliable mechanical property result, eliminating the extrinsic effects. It is also critical to observe the microstructural changes with regard to cooling rate. Therefore, a step mould was also incorporated into the new design together with cylindrical tensile bars.

Keywords: Gravity Casting; Quality Control; ASTM B108/B 108M; Permanent Mold

1 Introduction

In order to obtain reliable results from experiments, a minimum of 3 experiment results are typically required. ASTM B108/B 108M Tensile Test Specimen Casting mold is widely used to determine mechanical values in aluminum alloys as it is determined by ASTM standards. ASTM B108/B 108M Tensile Test Specimen Casting mold produced only 2 tensile test specimens in a single casting. In order to provide sufficient data, it is necessary to make more than one casting. Turbulent filling of the mould is one of the ways to introduce extrinsic defects that could deteriorate the mechanical properties of any casting [1]. In this work, the different pouring basin, sprue, and runner systems were designed to optimize a mould in an aim to achieve defect free castings. The filling and solidification behavior were analyzed with simulation software. The inclined sprue has been designed to prevent turbulent flow resulting from free fall in the sprue with proper pouring basin. The location and thickness of the 5 bars and step mold were optimized.

2 Experimental procedure

The experimental study consists of two stages. The first stage is the simulation for optimized mould design, and the other is the comparison of these analyzes with the actual casting results.

Table 1. Composition of Aluminium alloys

	Si	Fe	Cu	Mn	Mg	Ti	Al
CP	0.18	0.21	0.05	0.02	0.08	0.01	Bal
A360	9.86	0.45	0.05	0.01	0.09	0.1	Bal

Different mold designs were drawn in 3D (Figure 1) in SolidWorks software and were subjected to filing and solidification analysis in Anycast software. A360 alloy was selected and simulated in permanent mold with a casting temperature of 750ºC and mold was preheated at 250°C for all different designs. Tensile test sample was machined according to the ASTM E8.

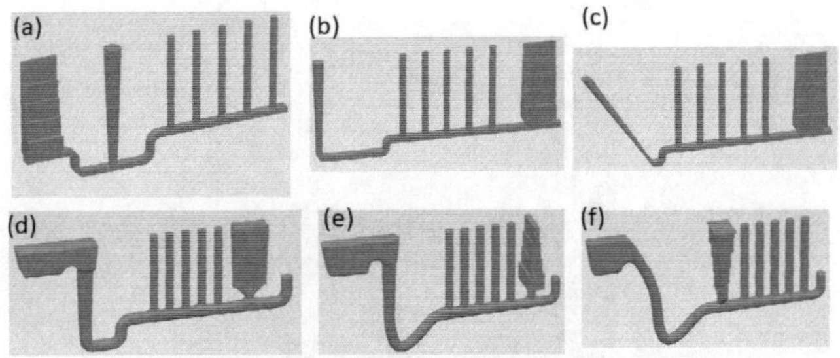

Figure 1. Different mould designs.

3 Result and discussion

Due to the taper sprue design, the liquid metal does not have enough space to provide turbulence. Provides a 50% reduction in its speed for each turn in the runner. When the turn does not have radius, flow is unstable and provides turbulence. There are two turns with 90-degree angle that decreases the velocity without providing turbulence. There are so many turbulences as the velocity of the liquid increases. The main reason for the speed increase is due to the metallostatic pressure caused by the mold height. The turns have not enough radius to decrease the velocity of the molted metal. Even though pouring basin and stopper was used, the turbulence at the entrance of the liquid metal to the bars is reduced, however it was not completely prevented [2].

Although different designs were tried, the most stable runner design and smooth filling were achieved with the taper runer and trapper in Figure 3 (d). Despite the pouring basin, narrowing runner and 3 ninety-degree turns, the liquid metal velocity, turbulence and uncontrol movement caused by free fall could not be controlled. These issues were eliminated by the slope of the runner that guides the molten metal movement, smoother turns rather than ninety-degree turns and the pouring basin design.
The new mould design was named as "Elephant Caput" mould which could produces 5 cylindrical bars and a step mould with thickness ranging from 5 to 30 mm (Figure 2).

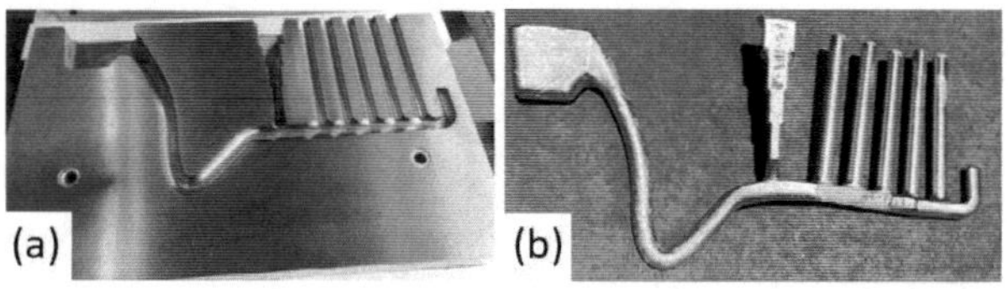

Figure 2. New design (a) permanent mold and (b) casting part.

The tensile test conducted with CP and A360 alloys. The average Yield Stress (YS), Ultimate Tensile Stress (UTS) and Elongation at fracture results were given at Figure 3.

YS was 90 MPa, UTS was 142 MPa and 2.1 elongation% with A360 alloy. YS was 53 MPa, UTS was 96 MPa and 22.1 elongation% with CP alloy. In the tensile test results showed that the standard deviation was ±10 MPa, whereas it was 0.8% for elongation.

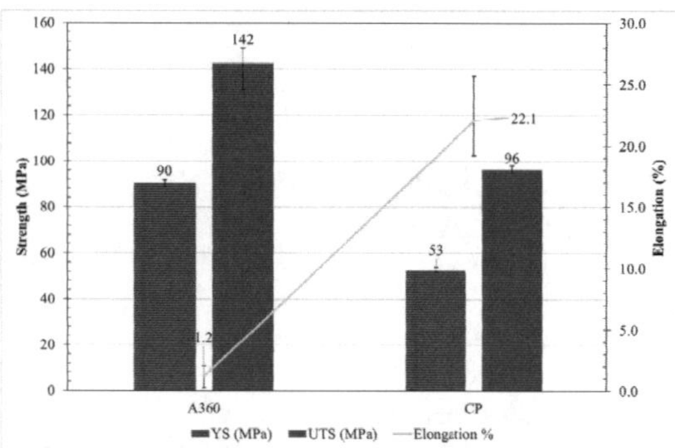

Figure 3. Elephant Caput mould tensile test result with A360 and CP alloys.

4 Conclusion

In this work, the different pouring basin, sprue, and runner systems were designed to optimize a mould in an aim to achieve defect free castings. The filling and solidification behavior were analyzed with simulation software. The inclined sprue has been designed to prevent turbulent flow resulting from free fall in the sprue with proper pouring basin. The location and thickness of the 5 bars and step mold were optimized. A series of tests were conducted with different alloys. Statistical analysis had shown that for UTS, the standard deviation was ±10 MPa, whereas it was 0.8% for elongation.

The new mould design was named as elephant caput mould which could produces 5 cylindrical bars and a step mould with thickness ranging from 5 to 30 mm. The reproducibility of the test results was proven to be reliable with a very low scatter. This permanent mould was proposed as a simple quality control measure that can be used easily in foundry floor together with reduced pressure test.

References

[1] Campbell, J., 2004. Castings practice: the ten rules of castings. Elsevier.

[2] Birsan, G., Ashtari, P. and Shankar, S., 2011. Valid mould and process design to cast tensile and fatigue test bars in tilt pour casting process. International Journal of Cast Metals Research, 24(6):378-384.

The Influences of Residual Heat Quenching and Applied Pressure on the Mechanical and Thermal Conductivity of Cast Al-Zn-Mg-Mn-Fe Alloy

Jingsi Chen[1], Yuliang Zhao[2], Chuanyao Tang[1], Le-hua Liu[1], Weiwen Zhang[1,*]
1. School of Mechanical and Automotive Engineering, South China University of Technology, Guangzhou, 510640, China
2. School of Materials Science and Engineering, Dongguan University of Technology, Dongguan, 523808, China
*Corresponding address: e-mail: mewzhang@scut.edu.cn

Abstract: The microstructure, mechanical properties, and thermal conductivity of Al-3.5Mg-3.5Zn-0.8Mn-0.5Fe alloy were analyzed. Applied pressure and residual heat quenching refined the grain and significantly reduced the volume fraction of the Al_6(FeMn) phase, enhancing the alloy's mechanical and thermal properties. The AP+RQ alloy exhibited the best overall performance, with a tensile strength of 310 MPa, a yield strength of 144 MPa, an elongation of 6.9%, and a thermal conductivity of 90.0 $W·m^{-1}·K^{-1}$.

Keywords: Applied pressure; Residual heat quenching; Fe-rich phase; Mechanical and thermal properties

1 Introduction

The complex digital driving system has led to the high integration of automotive electronic control units, significantly increasing power density and causing heat dissipation issues [1]. At the same time, to align with the trend of energy saving, emission reduction, and green travel, higher demands have been placed on new-generation automotive materials. Aluminum alloys, with their light weight and excellent thermal conductivity, are the preferred materials for reducing weight and improving heat dissipation in vehicles. Among them, Al-Zn-Mg alloys, with their high specific strength, hold promise for replacing forging with casting, thereby shortening the preparation cycle of high-performance aluminum alloys, and substantially reducing costs [2].

This paper investigates the effects of squeeze casting and residual heat quenching processes on the microstructure and properties of Al-Mg-Zn alloys formed by squeeze casting.

2 Experimental procedure

The alloy used in the experiment was Al-8Zn-3.5Mg-0.8Mn-0.5Fe. It was solidified under a pressure of 100 MPa and then cooled in air and room temperature water. The microstructure and composition of the samples were characterized and analyzed using OM, SEM, and EDS. The mechanical and thermal properties were determined by averaging the results of at least three valid measurements.

3 Result and discussion

Microstructure

Fig. 1 shows the microstructure of the alloys which display the grain size. The grain size was significantly refined by both applied pressure (AP) and residual heat quenching (RQ). In gravity cast (GC) alloy, the average grain size of 225.3 μm, while in AP+RQ alloy, the grain size is reduced to about 69.6 μm.

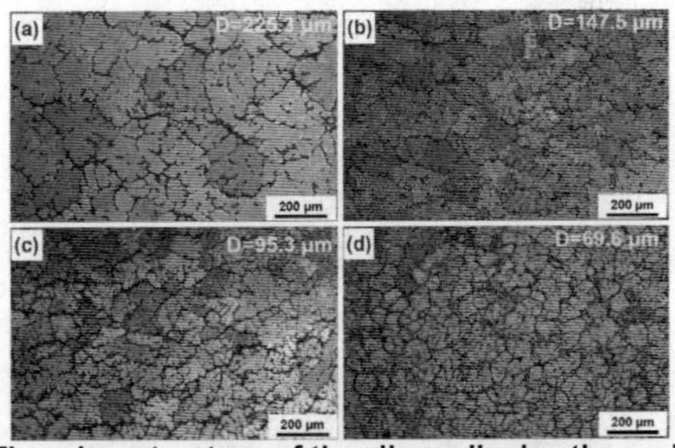

Figure 1 The microstructure of the alloys display the grain size. (a) GC; (b) GC+RQ; (c) AP; (d) AP+RQ.

Figs. 2 shows SEM-BSE microstructures of different samples. EDS-point scanning results indicated that the alloy microstructure consists of $Mg_{32}(Al,Zn)_{49}$ and Al_6(Fe,Mn). AP significantly refined the second-phase particles and improved the morphology of Al_6(Fe,Mn). In the GC and GC+RQ alloys, there is a significant amount of lath-like Al_6(Fe,Mn) phases, which is widely reported to be a primary phase formed before α-Al [3]. In contrast, in the AP and AP+RQ alloys, the Al_6(Fe,Mn) phase mainly exists in a Chinese script-like morphology. Additionally, the AP resulted in more numerous and clustered distribution of the $Mg_{32}(Al,Zn)_{49}$ phase.

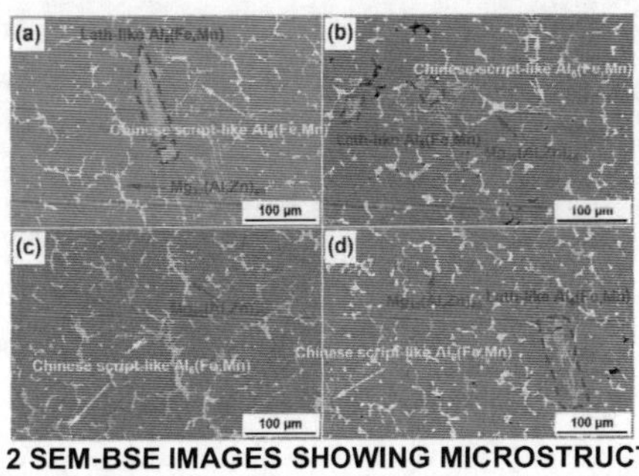

FIGURE 2 SEM-BSE IMAGES SHOWING MICROSTRUCTURE OF DIFFERENT SAMPLES: (A) GC; (B) GC+RQ; (C) AP; (D) AP+RQ.

Mechanical properties

The mechanical properties of the alloys are shown in Fig. 3. Relative to the GC alloy, under the applied pressure, the AP alloy improved the UTS from 197 MPa to 289 MPa, which was a 46.7% increase. The YS increased from 132 MPa to 138 MPa, a 4.6% increase, and the elongation increased from 3.8% to 6.1%, a 60.5% increase. The GC+RQ alloy achieved optimal mechanical properties, with a UTS of 310 MPa, a YS of 144 MPa, and an elongation of 6.9%.

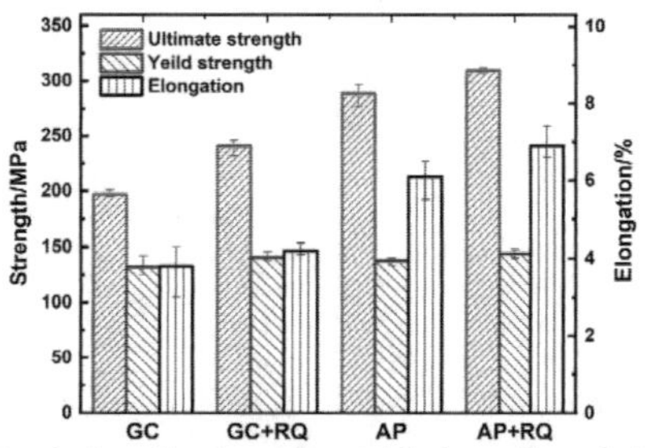

Figure 3 Ultimate tensile strength, yield strength and elongation of various samples.

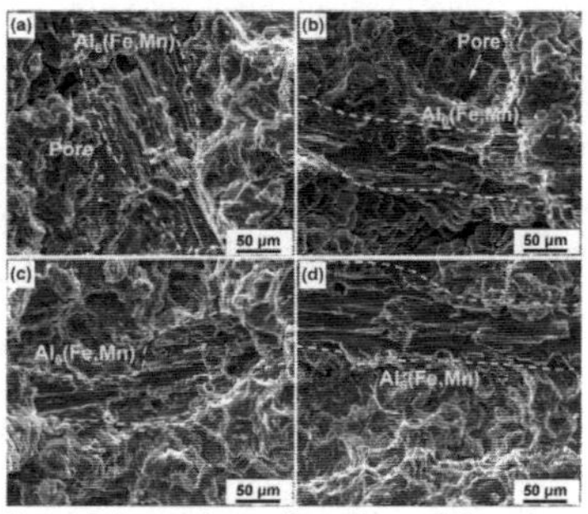

Figure 4 Statistics of the volume fraction of the phases and defects in different samples.

As shown in Fig. 4, the long hollow diamond shaped $Al_6(Fe,Mn)$ phases easily cause stress concentration during deformation, becoming crack sources that significantly reduce the alloy's plasticity. Additionally, some pores were observed on the fracture surfaces of the GC and GC+RQ alloys.

Thermal properties

Tab. 1 shows the thermal conductivity of different samples. The AP significantly increased the thermal conductivity, from 81.1 $W \cdot m^{-1} \cdot K^{-1}$ in the GC sample to 89.5 $W \cdot m^{-1} \cdot K^{-1}$ in the AP sample, an increase of 10.4%. Additionally, RQ had no significant effect on the alloy's thermal conductivity.

Table 1. Thermal conductivity of the samples ($W \cdot m^{-1} \cdot k^{-1}$)

Samples	GC	GC+RQ	AP	AP+RQ
Thermal conductivity	81.1	81	89.5	90.0

The increase in the thermal conductivity of the samples is due to two main factors. Firstly, AP promotes the precipitation of the $Mg_{32}(Al,Zn)_{49}$ phase, reducing the solid solubility of Zn and Mg atoms in the alloy, thereby decreasing lattice distortion. Secondly, AP significantly refines the second-phase particles in the alloy, reducing the likelihood of electron scattering.

4 Conclusion

AP and RQ refined the grain and significantly reduced the volume fraction of the $Al_6(FeMn)$ phase, enhancing the alloy's mechanical and thermal properties. The AP+RQ alloy exhibited the best overall performance.

Acknowledgments

We would like to acknowledge the financial support from the National Natural Science Foundation of China (No. 52074131)

References

[1] Mallik S, Ekere N, Best C, et al. Investigation of thermal management materials for automotive electronic control units[J]. J. Applied Thermal Engineering, 2011, 31(2-3): 355-362.

[2] Zhilin A S, Jianguo L, Yalunina V R, et al. Influence of Silicon on Temperature Dependence of Thermal Conductivity of Al–Si–Fe Alloys[J]. J. KnE Engineering, 2018: 298–301.

[3] Zhu X, Blake P, Ji S. The formation mechanism of $Al_6(Fe, Mn)$ in die-cast Al–Mg alloys[J]. J. Cryst Eng Comm, 2018, 20(27): 3839-3848.

Minimal Sintering Shrinkage Aluminum-Based Ceramic Cores Fabricated by Stereolithography

Haoyuan Wang[1,2], Kehui Hu[1,2,*], Zhigang Lu[2,3,4]
1. State Key Laboratory of Tribology in Advanced Equipment, Tsinghua University, Beijing 100084, China
2. Department of Mechanical Engineering, Tsinghua University, Beijing 100084, China
3. Key Laboratory of Advanced Materials Processing Technology, Ministry of Education, Tsinghua University, Beijing 100084, China
4. State Key Laboratory of Clean and Efficient Turbomachinery Power Equipment, Tsinghua University, Beijing, 100084, China
*Corresponding address: Department of Mechanical Engineering, Tsinghua University, Beijing 100084, China
e-mail: hukehui@tsinghua.edu.cn

Abstract: This study explores the stereolithography-based fabrication of aluminum-based ceramic cores, aimed at minimizing sintering shrinkage. Stereolithography-based ceramic cores often exhibit significant shrinkage during sintering, limiting their applications. In this work, by adding silica and kyanite into the alumina, high mechanical performance while minimizing shrinkage were achieved. Results demonstrate that cores with 22% silica and 8% kyanite show nearly zero shrinkage, making them suitable for complex, high-precision investment casting applications.

Keywords: stereolithography; ceramic cores; alumina; sintering shrinkage

1 Introduction

Ceramic cores play a crucial role in forming the intricate cooling channels within turbine blades used in gas turbine engines. These cores must exhibit high mechanical strength, thermochemical stability, and precise dimensional accuracy [1]. Ceramic stereolithography can achieve high-precision, high-performance, and complex core manufacturing. However, a significant challenge with stereolithography-based cores is their tendency to experience significant sintering shrinkage due to the high binder content [2,3]. This research addresses this challenge by modifying the ceramic composition to control shrinkage while maintaining mechanical integrity.

2 Experimental procedure

Aluminum-based samples were prepared with varying contents of silica and kyanite to evaluate their individual and effects on sintering shrinkage and mechanical properties. The contents are shown in Table 1. The samples were firstly solidified with a digital light processing stereolithography device under a UV light intensity of 15 mW/cm^2 and a layer thickness of 100 μm and sintered at 1500°C or 1600°C for 2 hours. Room-temperature flexure strength tests and high-temperature creep deformation analysis were then conducted.

Table 1. Contents of samples

Samples	Al_2O_3 (wt.%)	SiO_2 (wt.%)	Kyanite (wt.%)
K5S25	70	5	25
K6S24	70	6	24
K7S23	70	7	23
K8S22	70	8	22
K9S21	70	9	21
K10S20	70	10	20

3 Result and discussion

Microstructural analysis

SEM images (Figure 1) and XRD results showed that silica addition increased particle contact areas, promoting liquid-phase sintering and forming mullite through reaction with alumina. Kyanite decomposed into primary mullite and silica at lower temperatures, which further reacted with alumina to form secondary mullite.

Figure 1. Microstructures of the samples: (a) original alumina; (b) silica added; (c) and (d): kyanite added.

Thermogravimetric analysis.

As shown in Figure 2, the DTA curve for the sample with 30 wt.% kyanite shows a 15% mass loss between 200°C and 600°C due to the oxidation and removal of the photosensitive resin. Above 600°C, an endothermic peak at 1221°C indicates kyanite decomposition into primary mullite. An exothermic peak at 1321°C signifies the reaction between resultant silica and alumina, forming secondary mullite.

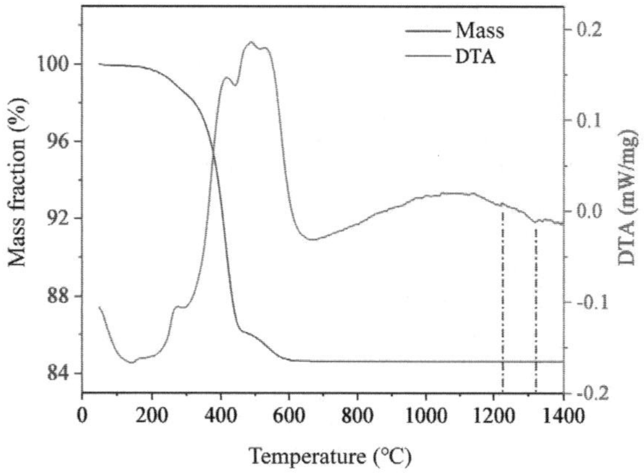

Figure 2. TG-DTA curve of binder removal and sintering.

Sintering shrinkage and mechanical properties

By comparing the effects of silica and kyanite on aluminum-based ceramic cores, it was found that adding silica increases sintering shrinkage and reduces deflection. Initially, flexure strength improves due to enhanced bonding between alumina grains and the formation of mullite, but it declines with further silica addition due to micro-cracks from cristobalite's phase transformation. On the other hand, kyanite reduces sintering shrinkage, even causing volume expansion. This initially decreases flexure strength due to cracks formed during kyanite decomposition, but strength improves with additional kyanite as more mullite forms, enhancing mechanical properties.

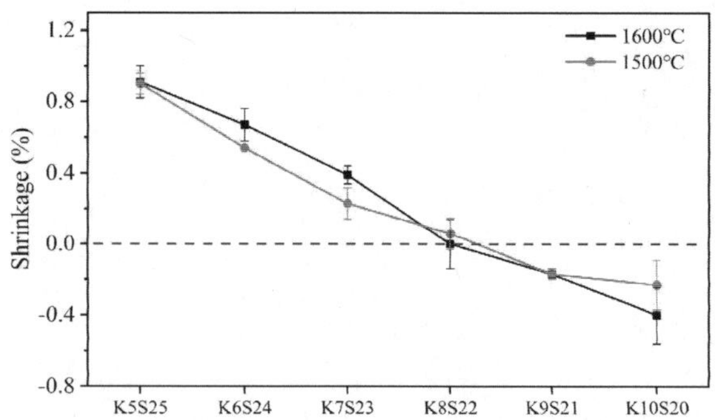

Figure 3. Sintering shrinkage of the samples.

Combining silica and kyanite leverages their complementary effects: while silica increases shrinkage, kyanite reduces it. A composition of 22% silica and 8% kyanite achieves minimal sintering shrinkage, with a deflection of 0.04 mm and a flexure strength of 29.4 MPa. This optimal powder system compensates for shrinkage while maintaining mechanical strength, enabling the production of high-precision ceramic cores suitable for practical applications.

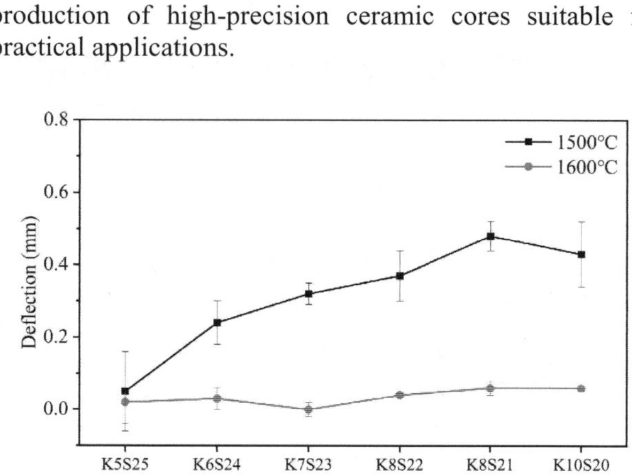

Figure 4. Deflection of the samples.

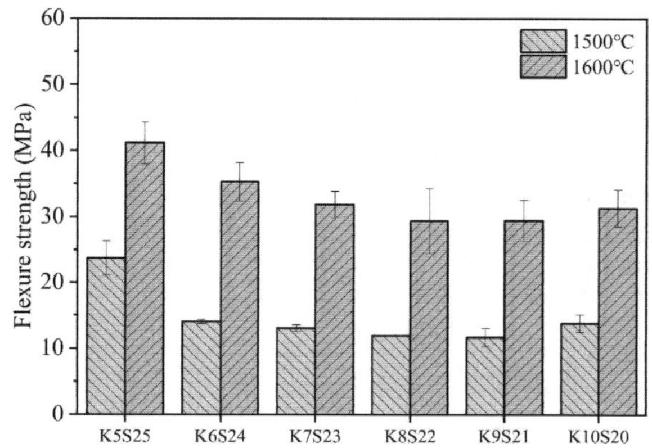

Figure 5. Flexure strength of the samples.

4 Conclusion

By optimizing the ceramic composition with silica and kyanite, we successfully minimized sintering shrinkage while maintaining high mechanical performance. This innovation addresses a key limitation in stereolithography-based ceramic core manufacturing, enabling the production of complex, high-precision turbine blades. Further work will focus on practical casting verifications and scaling up to mass production.

References

[1] Hu K, Wang H, Lu K, et al. Fabrication of silica-based ceramic cores with internal lattice structures by stereolithography. China Foundry, 2022, 19 (5): 369-379.

[2] Hu K, Lu Z, Lu K, et al. Additive manufacturing of complex ceramic cores and verification of casting process. J. Mech. Eng., 2021, 57 (03): 227-234.

[3] Li H, Hu K, Liu Y, et al. Improved mechanical properties of silica ceramic cores prepared by 3d printing and sintering processes. Scr. Mater., 2021, 194: 113665-113669.

Visualization and Monitoring of Shell-making Process in Investment Casting

Qian Feng[1,2], Kehui Hu[1,2,*], Zhigang Lu[2,3,4]
1. State Key Laboratory of Tribology in Advanced Equipment, Tsinghua University, Beijing 100084, China
2. Department of Mechanical Engineering, Tsinghua University, Beijing 100084, China
3. Key Laboratory of Advanced Materials Processing Technology, Ministry of Education, Tsinghua University, Beijing 100084, China
4. State Key Laboratory of Clean and Efficient Turbomachinery Power Equipment, Tsinghua University, Beijing, 100084, China
*Corresponding address: Department of Mechanical Engineering, Tsinghua University, Beijing 100084, China
e-mail: hukehui@tsinghua.edu.com

Abstract: Shell-making process is a critical step in investment casting for shape control, and breaking through the "black box" of this process remains a challenging issue in its research. This paper introduces blue light scanning technology into the study of investment casting shell processes, using samples with cantilever beam structures as research objects. It explores the influence of cantilever structure inclination angles on shell wall thickness and deformation. The feasibility and effectiveness of this method are demonstrated through numerical simulation results. This approach achieves visualization and digitalization of the layer-by-layer shell preparation process, providing a new technical pathway for investigating shell formation mechanisms.

Keywords: investment casting; ceramic shell; shell quality monitoring; dimensional deformation

1 Introduction

The fabrication of high-quality ceramic shells with dimensional stability and uniform wall thickness is a critical prerequisite for achieving near-net-shape investment casting. To ensure shell quality, real-time monitoring and analysis of key parameters, such as shell wall thickness, are essential during the preparation phase. However, current research methodologies predominantly rely on post-fabrication or post-dewaxing evaluations using techniques like industrial CT[1], which precludes real-time data acquisition of crucial factors such as layer thickness uniformity and deformation during the shell preparation process. This limitation impedes the timely detection and correction of potential quality issues, thereby increasing the risk of shell rejection. To address this challenge, there is an urgent need to develop novel technologies and methodologies capable of real-time monitoring of key quality metrics during the layer-by-layer fabrication process of ceramic shells.

2 Experimental procedure

A cantilever beam sample model was prepared using additive manufacturing technology, referencing the local features of castings, as shown in Figure 1. The upper part of the model is the shell-making area. Cantilever beam structures with inclination angles ranging from 30° to 150° are evenly distributed on both sides of the model at 15° intervals, forming narrow spaces of varying degrees at the roots of the cantilever beams. Referencing the actual investment casting shell-making process, the shell-making area of the model was immersed in slurry and then evenly sprinkled with refractory powder (sand). This process was repeated after each layer was completely dry to complete the shell preparation work. After each shell layer dried, blue light scanning equipment was used to collect surface data of the model.

To achieve precise fitting of data with different layer thicknesses, a calibration area that does not participate in the shell-making process was designed on the model. The calibration area consists of a truncated cone and a plane, which can achieve complete alignment of six degrees of freedom in the spatial coordinate system. Feature structures for evaluation were also designed in the calibration area. When all feature structures show good positional coincidence, it indicates that the two models have achieved complete alignment, allowing for further extraction and analysis of surface data.

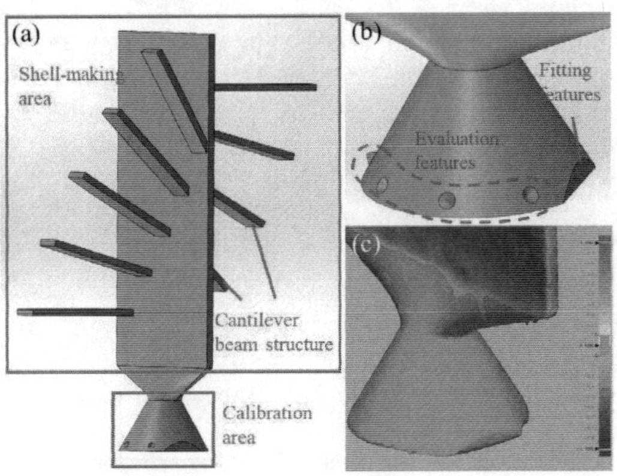

Figure 1. Experimental sample with cantilever beam structure. (a) Whole sample, (b) Calibration area, (c) Alignment status.

3 Result and discussion

Tracking the shell-making process

Figure 2 shows the thickness cloud map of the sample's surface layer shell. It can be observed that the wall thickness uniformity on both sides of the shell is relatively good, with the shell thickness at corresponding positions on both sides being essentially at the same level. However, the shell thickness at the bottom is greater than at the top, which is due to the slow downward flow of the slurry under the influence of gravity before it is completely dry. The cantilever beam structure significantly reduces the uniformity of the shell thickness. There is a noticeable increase in shell thickness at the root of the cantilever beam, reflecting the downward flow process of the slurry. In contrast, as shown in Figure 2, the shell thickness in the bottom area of the cantilever beam is significantly lower.

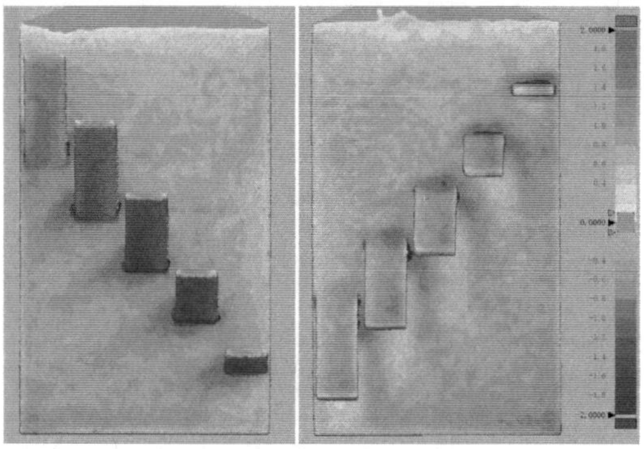

Figure 2. Thickness cloud map of the surface layer shell.

Shell deformation status

In addition to visually tracking the shell preparation process, the deformation during the shell-building process can be characterized using external calibration markers. Figure 3 illustrates the displacement of calibration markers at various positions on the cantilever beam sample, revealing three types of deformation behavior: (1) All cantilever beams move downward, indicating downward bending deformation of the cantilever beams under the gravitational force of the coating material. (2) Both calibration markers 1 and 2 move downward, indicating that the entire sample undergoes clockwise bending deformation along the X-axis. (3) Compared to calibration marker 1, calibration marker 2 shows more severe displacement with a degree of leftward offset, suggesting torsional deformation along the Z-axis at the bottom of the sample.

Numerical simulation verification results demonstrate that the displacement changes of external calibration markers can effectively reflect the deformation of the sample during the layer-by-layer shell preparation process. This method has been applied to the shell preparation process of actual casing castings, revealing the inward shrinkage deformation of its outer ring during the initial shell-making and burnout stages.

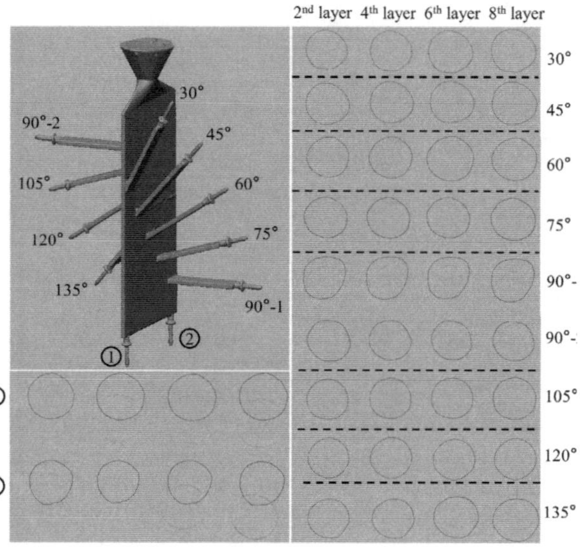

Figure 3. Displacement of calibration markers at different positions on the cantilever beam sample.

4 Conclusion

This study introduces blue light scanning technology to investment casting shell preparation, addressing challenges in analyzing layer thickness uniformity and deformation during shell-making. By aligning surface data from different stages using calibrated regions, it enables visualization and digital analysis of the shell preparation process, providing reliable evidence for quality assessment. Using a cantilever beam sample, the study investigates the impact of beam inclination on shell thickness uniformity and deformation, validated through numerical simulations. This method is then applied to actual production, analyzing radius and roundness data of large casing outer rings during shell-making. This reveals the inward shrinkage pattern of the casing outer ring during initial shell preparation and burnout stages.

References

[1] Liu C. Study on stream of variation modeling and robust control method for complex thin-walled parts in the investment casting process. (Doctoral dissertation). Shanghai Jiao Tong University, China.

Interface Reaction of ZTA15 Titanium Alloy and $CaZrO_3$ Investment Casting

X. Q. Li[1, 2, *], S. B. Liu[1,2], T. Y. Liu[1, 2], W. L. Wang[1, 2], W. C. Qiu.[1, 2], H. J. Li[1, 2], C. Y. Li[1,2]
1. National Key Laboratory of Advanced Casting Technologies, Shenyang, Liaoning, 110022, China
2. Shenyang Research Institute of Foundry Co., Ltd, Shenyang, Liaoning, 110022, China
*E-mail:2018300940@mail.nwpu.edu.cn

Abstract: Molten titanium alloys have extremely high chemical activity and are prone to interface reactions with precision casting shells during the casting process, resulting in a decrease of the surface quality of the castings. Now Y_2O_3 is widely used as casting shells, but it's so expensive that there is an urgent need for a low-cost shell surface layer material to replace Y_2O_3 to reduce the cost of titanium alloys precision castings. In this paper, $CaZrO_3$ was used to prepare two types of titanium alloy precision casting shells, namely neutral binder surface layer and acidic binder surface layer, for TA15 titanium alloy precision casting. The results indicate that the use of neutral binders can lead to the presence of a certain thickness of Zr diffusion layer and O diffusion layer on the surface of castings, while the use of acidic binders only results in an O concentration zone on the surface of castings, without a Zr diffusion layer. At the same time, there is a certain thickness of tissue transition zone on the surface of the casting, among which the interface structure of the acidic binder surface layer casting is much better than that of the neutral binder surface layer casting.

Keywords: Interface reaction; Investment casting

1 Introduction

Titanium alloys are widely used due to their good properties in strength, toughness, formability, weld-ability, corrosion resistance and biocompatibility[1]. Investment casting is the most important casting process for titanium alloys. Currently, more than 90% of small and medium-sized titanium alloy castings use investment casting technology[2]. However, Limited by the high cost of surface material such as Y_2O_3, there is an urgent need for a low-cost shell surface layer material to replace Y_2O_3 for large-scale of titanium alloy. According to Ellingham diagram[3], $CaZrO_3$ is a promising refractory material for titanium alloy investment casting.

In this paper, The interface reaction between the surface layer materials prepared by $CaZrO_3$ combined with different binders and TA15 was mainly studied.

2 Experimental procedure

In the experiment, silica sol and zirconium acetate were used as binders, and 325 mesh $CaZrO_3$ was used for the surface layer. Neutral surface layer shells with a powder to liquid mass ratio of 3.5:1, 3.25:1, and 3:1 were prepared using silica sol, as well as acidic surface layer coatings with a powder to liquid mass ratio of 1.75:1. 100 mesh, 60 mesh, and 30 mesh alumina coarse sand were used for the back layer sanding. Use this shell to cast TA15 titanium alloy, and then perform SEM, XRD and metallographic analysis on the obtained titanium alloy and shell.

3 Result and discussion

Microscopic organization

The microstructure of the casting is shown in Figure 1. It can be seen that the microstructure of the sample is a layered Weinstein structure, but there is a clear boundary transition region in the surface area. The transition zone structure is consisting of columnar crystal structures arranged along the direction of heat flow. This is mainly due to the chilling action of the shell and the presence of a large number of heterogeneous nucleation at the edges.

By comparing Figures 1(a), (b), and (c), it can be observed that as the powder to liquid ratio gradually decreases, the thickness of the edge tissue transition region increases from 320 μm in 1# to 216 μm in 2#, and then to 290 μm in 3#. The columnar crystal structure at the edge also changes accordingly, with columnar crystal grain widths of 12.18 μm, 11.59 μm, and 11.96 μm, respectively. From Figure 1(d), it can be observed that the interface quality of the alloy with an acidic surface layer is higher than that with a neutral surface layer. The transition region of the microstructure is only about 218 μm, and the structure is uniform without the appearance of obvious coarse columnar crystals.

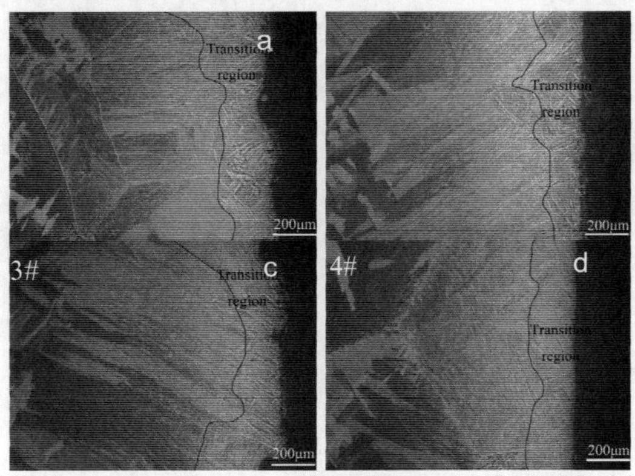

Fig.1 Surface microstructures of specimens:(a)1#, (b)2#, (c)3#, (d)4#.

The scanning electron microscope at the edge of the

sample is shown in Figure 2. Different interface reaction layers were found on the surfaces of each sample, with a reaction layer of about 50 μm in 1#, a thicker reaction layer of about 60-90 μm in 2#, and a relatively thicker reaction layer of about 30 μm in 3#. However, the reaction layer of the acidic surface layer is thicker, higher than 50 μm. The results observed in the metallographic structure of the sample with a neutral surface layer are consistent, while the acidic surface layer structure is relatively thick, indicating that the acidic binder reacts with $CaZrO_3$ during the shell hanging and calcination processes, producing substances that can dissolve with the titanium liquid and promote the crystallization of the titanium alloy.

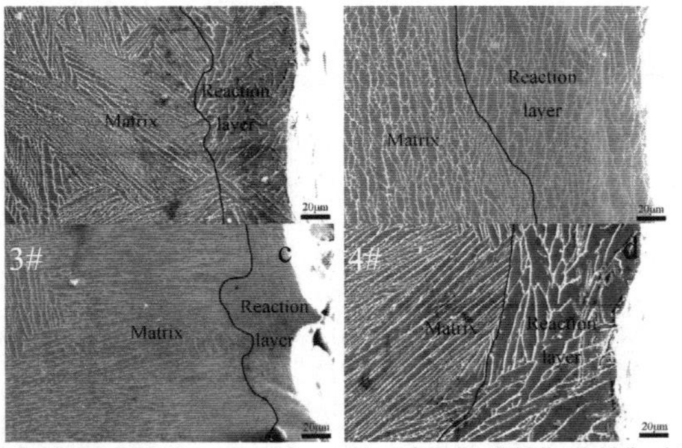

Fig.2 Surface SEM images of samples:(a)1#, (b)2#, (c)3#, (d)4#.

Phase analysis

The XRD results of four sets of samples are shown in Figure 3. Except for the α - Ti group, the surface of 1#, 2#, 3#, and 4# mainly consists of various oxides of Ti and V, as well as $CaCO_3$ and $CaZrO_3$ attached to the surface. The surface layer material O diffused into the sample and reacted with the matrix, while the oxidation of various elements in 4# using an acidic surface layer was relatively severe, and even the Ti peak was lower. The various oxide peaks of titanium were higher than Ti.

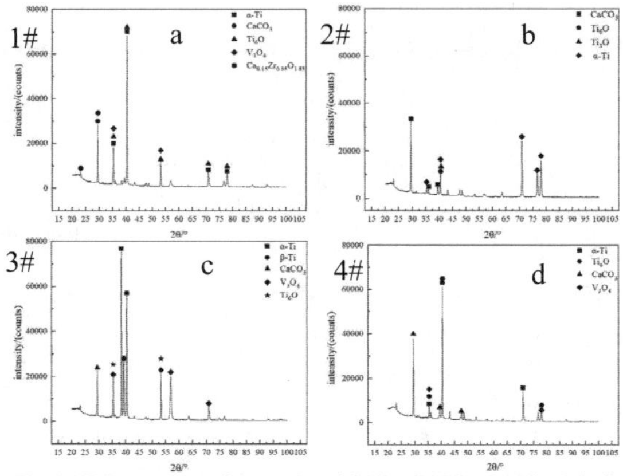

Fig.8 XRD pattern of samples:(a)1#, (b)2#, (c)3#, (d)4#.

4 Conclusion

There is a transition region of 200-320 μm on the surface of the casting, with the grains mainly being columnar. The thickness of the transition zone of the neutral surface layer is 216-320 μm, while the transition zone of the acidic surface layer is only about 218 μm. The structure obtained by the acidic surface layer is significantly better than that of the neutral surface layer.

References

[1] Zhao Q Y, Sun Q Y, Xin S W, et al. High-strength titanium alloys for aerospace engineering applications: A review on melting-forging process[J]. Materials Science & Engineering A, 2022, 845.

[2] Lin K F, Lin C C. Interfacial reactions between Ti6Al4V alloy and zirconia mold during casting[J]. J Mater Sci, 1999, 34(23): 5899–5906.

[3] Kim M G, Kim S K, Kim, Y J. Effect mold material binder on metal-mold interfacial reaction for investment casting of Titanium alloys[J]. Mater Trans, 2002, 43(4): 745-750.

Effect of Thermal Controlled Solidification Process on Microporosity of GH3230 Alloy Thin-wall plate

Aoqi Li*, Xun Sun, Lei Jin, Yipeng Li
National Key Laboratory of Advanced Casting Technologies, China Academy of Machinery Shenyang Research Institute of Foundry Co., Ltd, Shenyang 110022 , P. R. China
*Corresponding address: No.17 Yun feng South Street, Tie xi District, Shenyang City, Liaoning Province, China.
e-mail: 2629978503@qq.com

Abstract: Aiming at the problems such as the shrinkage porosities existing in the large superalloy complex thin-wall castings under the conditions of traditional investment casting conditions, the influence of Thermal Controlled Solidification (TCS) process on the microporosity of thin-wall plate of GH3230 alloy was investigated. The results show that the microporosity of the thin-wall plate can be reduced significantly by TCS process and the reduction degree of microporosity of different height and thickness plates is different. The microporosity content below the upper-middle part of the plate is controlled below 0.5%, and the microporosity content in the lower part and middle part is controlled below 0.25%. The microporosity content in the upper part of 3mm and 5mm plates are higher, which is 1.65% and 1.44%, respectively, the microporosity content in the 7mm plate is only 0.49%.

Keywords: TCS; GH3230 alloy; microporosity

1 Introduction

With the rapid development of high-end equipment, the demand for weight reduction, high performance and high reliability of high-end equipment structures has promoted the development of superalloy structural parts in the direction of thin-wall lightweight, structural complexity, low defects and high dimensional accuracy [1]. The use of integral casting technology to form large complex thin-wall castings instead of welding by multiple parts is one of the current development directions. However, the problems of uneven grain distribution, misrun, cold shut and porosities of large and complex thin wall castings formed by traditional investment casting process are easy to occur.

In order to solve these problems of mold filling and solidification defects of large and complex thin-walled castings under the condition of traditional investment casting process, Precision Castparts Corp of the USA proposed a new equiaxed crystal casting process, which is developed from the directional solidification technology. By controlling the temperature gradient G and solidification rate R at the front edge of the solidification interface, the shape and refinement of the solidification structure are controlled [2]. TCS technology can be used to create strong sequential solidification conditions, which has obvious advantages for improving the mold filling ability of complex thin-walled castings. In this project, taking GH3230 alloy thin-wall plates as example, the influence of TCS process on the microporosity of the thin-wall plates with different height and thickness was investigated.

2 Experimental procedure

GH3230 alloy bars were selected as raw materials in the experiments. The main chemical composition of GH3230 alloy bar was 0.10C, 21.84Cr, 1.97Mo, 2.10Fe, 13.84W, 2.96Co, 0.46Al, 0.55Si, 0.76Mn and Ni others. In order to study the effect of TCS process on the microporosity of large size thin-walled castings, thin-wall plates with a height of 300mm, a width of 50mm and the thicknesses of 3mm, 5mm and 7mm were designed, as shown in Fig 1.In the TCS experiment, ZGD-25BYF directional solidification furnace was used to melt and cast the alloy, and the main process parameters were shown in Tab 1.

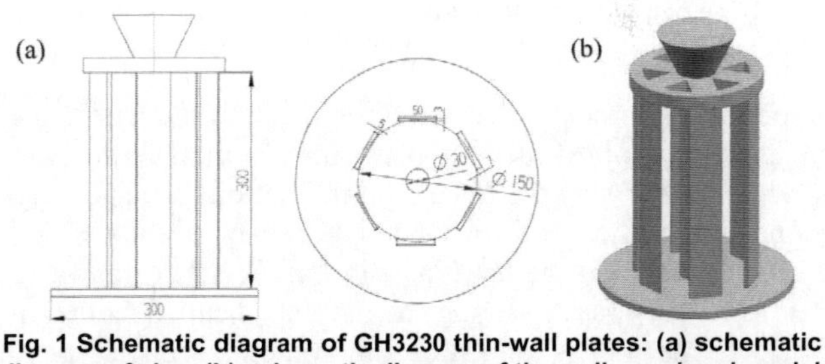

Fig. 1 Schematic diagram of GH3230 thin-wall plates: (a) schematic diagram of size; (b) schematic diagram of three-dimensional model.

Tab.1 Parameters of the main process

Upper mold temperature /℃	Lower mold temperature /℃	Refining temperature /℃
1400	1350	1550
Refining time /s	pouring temperature /℃	withdrawal rate (μm/s)
600	1400	500

3 Result and discussion

Fig 2 and Fig3 show the diagram of GH3230 plate microporosity and the content of microporosity with different thickness and height. As can be seen from the shrinkage diagram in Fig 2 and Fig 3, the shrinkage of 3mm, 5mm and 7mm plates below the upper-middle parts is less than 0.5%, and the shrinkage of the lower and middle parts is less than 0.25%, while the shrinkage of the upper parts of 3mm and 5mm plates is higher, which is 1.65% and 1.44%.

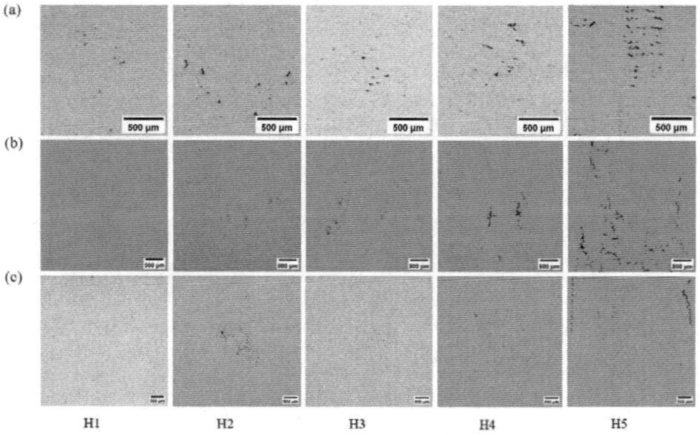

Fig. 2 Schematic diagram of the plate microporosities with different thickness and height: (a) 3mm plate, (b) 5mm plate, (c) 7mm plate; H1: lower, H2: lower-middle, H3: middle, H4: upper-middle, H5: upper.

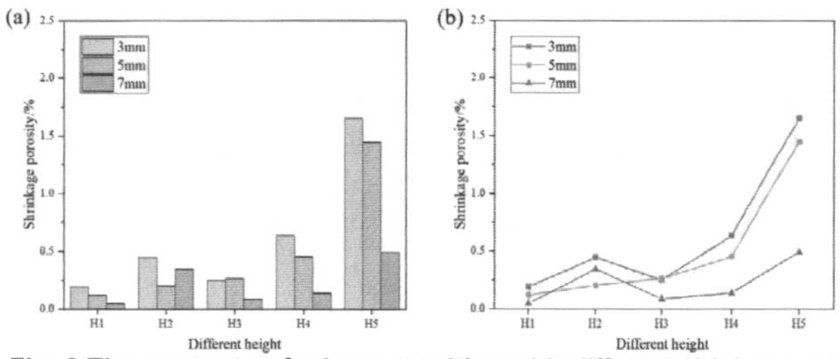

Fig. 3 The contents of microporosities with different thicknesses and heights of the plates.

In the TCS process, the heat transportation models of the plates during solidification process mainly include the heat conduction between the plates and the bottom water-cooled copper plate, the heat conduction between the plates and the shell wall, and the heat radiation between the shell and the furnace body. Among them, the heat conduction between the plates and the water-cooled copper plate accounts for the main part, which is also the premise of ensuring the sequential solidification. In the early stage of the plate solidification process, heat dissipation is mainly through the heat conduction between the bottom of the plates and the water-cooled copper plate, and the heat conduction between the plates and the shell wall and the heat radiation between the shell and the furnace body can be ignored. As the solidification process proceeding, the distance between the water-cooled copper plates and the solidification interface increases, the longitudinal heat conduction efficiency decreases, while the heat radiation from the shell to the furnace body does not decrease obviously. At this time, the transverse heat transfer between the shell and the furnace body cannot be ignored, and the conditions for sequential solidification are limited, and the solidification pattern turns to intermediate solidification dominated by dendrite growth. This increases the possibility of the formation of isolated liquid phase in the late solidification period, which increase the possibility of the formation of shrinkage porosity defects.

When the plates are extracted from the holding furnace, the thin-wall plates solidify quickly due to its rapid cooling and the micro-molten pools formed between the dendrites have solidified before the upper liquid can be supplemented. Therefore, the microporosity content of thin-wall plate is relatively high. However, the solidification of thick plates is relatively slow, and the micro-molten pools formed between dendrites can obtain the feeding of the upper liquid, therefore, there is little increase in microporosities. The above test result is also according with the Niyama criterion proposed by Niyama et al. That is, reducing the temperature gradient or increasing the cooling rate will increase the microporosity.

4 Conclusion

In this paper, the effect of TCS process on the micro-porosity of GH3230 alloy thin-wall plate was studied. The results show that the microporosity of the thin-wall plate can be significantly reduce by the TCS process, but for 3mm and 5mm plates, the upper part of the microporosities is higher. The main reason is that the longitudinal temperature gradient of the upper part of the plate is not enough, and the transverse cooling rate is increased, leading to the increase tendency of microporosity formation.

References

[1] Zhang Jun, Jie Ziqi, Huang Taiwen, et al. Research and Development of Equiaxed Grain Solidification and Forming Technology for Nickel-Based Cast Superalloys[J]. Acta Metall Sin, 2019, 55(9): 1145-1159.

[2] Ronald R. Brookes, North Canton, Ohio, et al. Thin wall casting[P]. US, US4724891A. 1988.

Inclusion Defect Prediction and Multi-process Process Control of Outlet Pipe Casting

Tianyou Li, Yu WANG, Peng Yu, Le Mao, Xuzheng Pan ,
Xiaoyuan Ji*, Yajun Yin, Jianxin Zhou**
State Key Laboratory of Materials Processing and Die & Mould Technology, Huazhong University of Science and Technology, Wuhan 430074, China; 2. Xi'an Aerospace Engine Co., Ltd, Xian 710025, China
*E-mail: jixiaoyuan@hust.edu.cn ; **zhoujianxin@hust.edu.cn

Abstract: Inclusion defects have a great influence on the performance of investment casting. When optimizing the process for inclusion defects, due to the production fluctuation in the casting site, the actual value of the process parameters will deviate from the set value. Therefore, this paper proposes a defect prediction and multi-process process control strategy combining BP neural network and improved particle swarm optimization algorithm. Firstly, data mining is carried out based on Huazhu ERP, and data cleaning is carried out on the collected process parameters. Secondly, an inclusion defect prediction model based on particle swarm optimization and BP neural network (PSO-BP) is established. Compared with the ordinary BP neural network, the accuracy is improved from 92.1% to 94.7%. Thirdly, a multi-process process control strategy combining K-nearest neighbor interpolation method and improved particle swarm optimization algorithm (KNN-IPSO) is proposed to control the process parameters of the subsequent process according to the deviation of the process parameters of the previous process, so as to ensure the stability of the casting quality. Finally, the simulation experiment was carried out. Compared with the ordinary pre-production process optimization method, the defect rate was reduced by 23%.

Key words : investment casting ; BP neural network ; improved particle swarm optimization algorithm ; k nearest neighbor interpolation method.

1 Introduction

Inclusion defects seriously affect the performance of castings, and the formation process is complex and difficult to eliminate. Therefore, it is very important to take preventive measures against inclusion defects. In recent years, researchers have made some progress in defect prediction using deep learning network models[1], and used meta-heuristic algorithms[2][3][4] to optimize process parameters to ensure casting quality. However, in the production process of investment casting, the actual value of the process parameters often deviates from the set value, and the deviation is accumulated, which is likely to cause huge product quality fluctuations. Therefore, it is urgent to control the process of the subsequent process in time for the production fluctuation. In this paper, a defect prediction and multi-process process control strategy combining BP neural network and improved particle swarm optimization algorithm is proposed. The technical route is as follows.

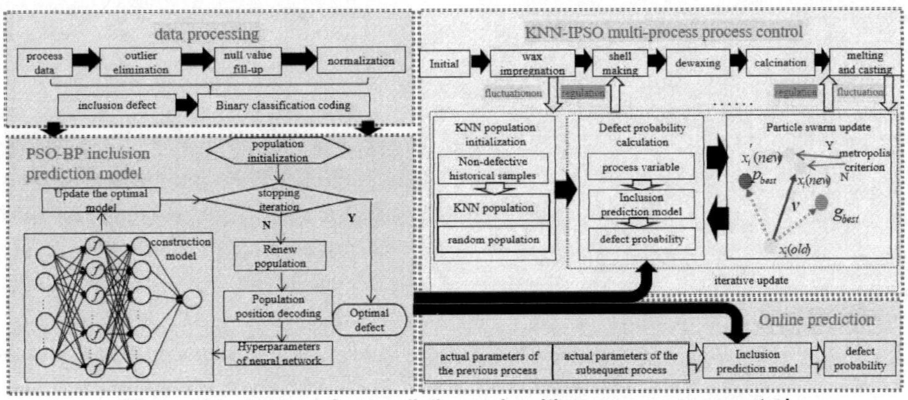

Figure 11 Inclusion defect prediction and multi-process process control

2 Experimental procedure

(1) Based on the Huazhu ERP system of a factory, through the process parameters of the original mold cleaning, wax leaching, shell making, dewaxing,

roasting, vacuum melting and pouring process of the single-piece number-related outlet pipe casting, the abnormal value is detected and eliminated based on the quartile distance criterion. The null value is filled with the mode, and the shell-making parameters with similar data distribution are reduced in dimension. After removing the unchanged parameters in the data, a total of 24 process parameters in the data set are obtained, and 251 single-piece records of the complete casting are obtained.

(2) The training set and test set are divided according to the ratio of about 7 : 3. The inclusion defect results are binary coded, and the process parameters are normalized by Z-score. At the same time, the sigmod ( ) function is used to process the output layer data and obtain the probability of defect occurrence, and then the threshold judgment is converted into a binary classification result. The particle swarm optimization algorithm is used to optimize the neural network hyperparameters, and the number of hidden layer nodes, the type of activation function and other hyperparameters are determined. Combining the F_1 score and the number of nodes, the objective function of neural network hyperparameter optimization is established as follows :

$$fitness\_net = F_{\beta} + 0.01\prod_{i=1}^{n} hidden\_size_i / (50\hat{\ } n) \qquad (1)$$

Here n is the number of layers of the neural network, and $hidden\_size_i$ is the number of nodes in the i-th layer.

(3) After each step of single-piece production, the improved particle swarm optimization algorithm combined with KNN nearest neighbor interpolation method can be used to search the hyper-parameter combination of subsequent processes, and the KNN-IPSO multi-process process control model can be established. When the population is initialized, the KNN nearest neighbor interpolation method is used to search for records similar to the process parameters of the previous process in the historical defect-free samples, and the parameters of the subsequent process in the first ten records are selected, and then 10 corresponding records are randomly generated to form the KNN initial population. In the update stage of the particle swarm, because the population initialization refers to the historical solution set, the particles are more likely to converge quickly and fall into the local optimal solution. In order to reduce the impact of this problem, the Metropolis criterion is used to consider whether to accept a worse solution set.

3 Result and discussion

In order to test the effect of KNN-IPSO solution strategy, the following simulation experiment is designed : after the planned parameters of the previous process are determined, Gaussian white noise is added to simulate the production fluctuation, and then the KNN-IPSO control process is carried out to change the planned value of the process parameters of the subsequent process. At the same time, 100 single-piece pre-production optimization experiments were carried out as a control group. The results are shown in Table 1. The results show that the KNN-IPSO multi-process process control scheme can effectively improve the inclusion defects of castings, and the inclusion rate is reduced by 23 % compared with the pre-optimized inclusion rate.

Table 2 Comparison of the number of inclusions in simulated experimental castings

Control group	Total number	Number of inclusions	Inclusion rate
Historical data	251	18	86.1%
Pre-production optimization	100	44	44%
Multi-process control	100	21	21%

4 Conclusion

In this paper, the particle swarm optimization algorithm is used to optimize the hyperparameters of BP neural network. Compared with the ordinary BP neural network, the accuracy of PSO-BP inclusion defect prediction model is improved from 92.1% to 94.7%. The KNN-IPSO multi-process process control strategy is proposed. Compared with the pre-production optimization strategy, the inclusion rate is reduced from 44% to 21%. Although the simulation experiment can not fully represent the effect of engineering application, it shows the advancement and effectiveness of the method to a certain extent.

Acknowledgments

This work is financially supported by the National Key Research and Development Program of China (No. 2020YFB1710100) and National Natural Science Foundation of China (Nos. 52275337, 52090042, 51905188)

References

[1] Sata, B. Ravi. Bayesian inference-based investment-casting defect analysis system for industrial application. The International Journal of Advanced Manufacturing Technology, 2017, 90(9): 3301-3315.

[2] J. Kim, J. Y. Lee. Development of a cost analysis-based defect-prediction system with a type error-weighted deep neural network algorithm. Journal of Computational Design and Engineering, 2022, 9(2): 380-392.

[3] Beheshti Z, Shamsuddin S M H. A review of population-based meta-heuristic algorithms[J]. Int. j. adv. soft comput. appl, 2013, 5(1): 1-35.

[4] Kennedy J, Eberhart R. Particle swarm optimization[C]//Proceedings of ICNN'95-international conference on neural networks. ieee, 1995, 4: 1942-1948.

Study on Microstructural Uniformity Control of Support Plate Castings of ZTC4 Alloy During Investment Casting and Heat Treatment

Xiwang Qie[1,2], Yajun Yin[1], Xu Shen[1], Jianxin Zhou[1,*]
1. School of Materials Science and Engineering, Huazhong University of Science and Technology, Luoyu Road 1037, Wuhan, China
2. Baimtec Material Co., Ltd., Yongxiang North Road 5, Beijing, China
*Corresponding address: e-mail: zhoujianxin@hust.edu.cn

Abstract: The structure of support plate castings of ZTC4 titanium alloy is complex. And the wall thickness varies greatly from 3.4 mm to 22.8 mm. Therefore, the mircostructure of the casting is ununiform. In order to realize microstructure homogeneity of support plate castings during the whole processes of investment casting and heat treatment, the solidification nucleation model and grain growth model of ZTC4 alloy were established based on the cooling curves and solidification microstructure at different positions of the step casting and reasonable nucleation hypothesis. The simulation results are in agreement with the experimental results. The numerical simulation analysis of support plate castings under different casting processes were carried out using this model. The results show that when the conditions of preheating temperature of shell is 200 ℃, the rotating speed is 150 r/min and the heat treatment temperature is 730 ℃, the microstructure of support plate castings is more uniform.

Keywords: Support plate, Titanium alloys, Investment casting and heat treatment, Modeling, Microstructural uniformity

1 Introduction

The final mechanical properties of titanium alloy castings are determined by their microstructure, which is affected by solidification and heat treatment conditions [1,2]. At present, most of the research work [3,4] is limited to the casting process or heat treatment process, without considering the correlation and inheritance between the processes.

In this paper, numerical simulation technology was used to predict the microstructure of the whole process of solidification and heat treatment of titanium alloy castings, and the influence of different casting processes and heat treatment processes on the microstructure of support plates was analyzed. And the optimal processes to realize the uniform control of microstructure were finally obtained.

2 Experimental procedure

In order to obtain different solidification conditions of ZTC4 alloy, the molten metal was poured into a cavity with a step shape. A thermocouple was placed in the center of each step to record the solidification process of different steps (as shown in Figure 1), and the cooling curve of each temperature measurement point was finally obtained. In order to analyze the grain morphology and size under different solidification conditions, samples of the same size at each temperature measuring point were made.

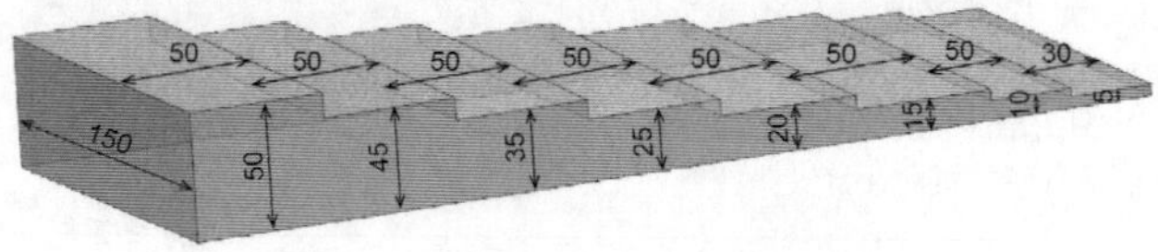

Figure 1 The size of step casting

Numerical simulation procedure

Grain Nucleation model

The continuous nucleation model proposed by Rappaz and Gandin et al. [5] is used to describe the grain nucleation. The total grain density n(ΔT) is calculated by:

$$n(\Delta T)=\int_0^{\Delta T}\frac{dn}{d\Delta T'}d(\Delta T') \tag{1}$$

Where ΔT is the undercooling. The continuous nucleation distribution dn/dΔT′ is given by:

$$\frac{dn}{d\Delta T'}=\frac{n_{\max}}{\sqrt{2\pi}\Delta T_\sigma}\exp\left[-\frac{1}{2}\left(\frac{\Delta T-\Delta T_N}{\Delta T_\sigma}\right)\right] \tag{2}$$

where n_{max} is the total density of grains, ΔT_σ is the standard deviation of undercooling, and ΔT_N is the mean undercooling in nucleation, respectively.

Grain growth model

The expression of mesoscopic grain growth rate is obtained by linear polynomial fitting of the relationship curve between tip growth rate V and subcooling degree:

$$V(\Delta T)=a_1(\Delta T)^2+a_2(\Delta T)^3 \tag{3}$$

where a_1 and a_2 correspond to the fitting coefficients of the polynomial curves respectively.

3 Result and discussion

Simulation and experimental verification of equiaxed dendrite structure of cast titanium alloys

The final microstructure of step castings under different solidification conditions obtained by experiment and numerical simulation was compared. It can be seen that with the increase of cooling rate, the number of nucleation increases, the grain size decreases. The simulation results are in agreement with the experimental results.

Effects of preheating temperatures of shell on grain sizes

The numerical simulation analysis of support plate castings under different casting processes were carried out using the established model. The results show that the preheating temperature of mold shell has little effect on the microstructure of casting. While when the conditions of preheating temperature of shell is 200 ℃, the microstructure of support plate castings is more uniform.

Effects of rotating speeds on grain size

The numerical simulation analysis of support plate castings under different rotating speeds were carried out using the established model. The results show that the microstructure of support plate castings is more uniform under the rotating speed of 150 r/min.

Effects of heat treatment temperatures on grain size

The numerical simulation analysis of support plate castings under different rotating speeds were carried out using the established model. The results show that the microstructure of support plate castings is more uniform under the heat treatment temperature of 730 ℃.

4 Conclusion

The model established in this paper is in good agreement with the experimental results, and it is proved that it can accurately simulate the microstructure evolution during whole process of investment casting and heat treatment of support plate. When the conditions of preheating temperature of shell is 200 ℃, the rotating speed is 150 r/min and the heat treatment temperature is 730 ℃, the microstructure of support plate castings is more uniform.

Acknowledgments

This work was financially supported by the National Key Research and Development Program of China (Grant No. 2022YFB3706800), National Natural Science Foundation of China (Grant No. 52201042 and 52275337). Thanks to Baimtec Material Co., Ltd. for raw materials and temperature measuring experiments. Thanks to the Analytical and Testing Center of HUST for materials characterization.

References

[1] Warnken N, Ma D, Drevermann A, et al. Phase-field modelling of as-cast microstructure evolution in nickel-based superalloys[J]. Acta Materialia, 2009, 57(19): 5862–5875.

[2] Su B, Han Z, Zhao Y, et al. Numerical Simulation of Microstructure Evolution of Heavy Steel Casting in Casting and Heat Treatment Processes[J]. ISIJ International, 2014, 54(2): 408–414.

[3] Guo Z, Zhou J, Yin Y, et al. Numerical Simulation of Three-Dimensional Mesoscopic Grain Evolution: Model Development, Validation, and Application to Nickel-Based Superalloys[J]. Metals, Multidisciplinary Digital Publishing Institute, 2019, 9(1): 57.

[4] Gu C, Lu Y, Cinkilic E, et al. Predicting grain structure in high pressure die casting of aluminum alloys: A coupled cellular automaton and process model[J]. Computational Materials Science, 2019, 161: 64–75.

[5] Rappaz M, Gandin Ch-A. Probabilistic modelling of microstructure formation in solidification processes[J]. Acta Metallurgica et Materialia, 1993, 41(2): 345–360.

Enhanced Properties of Ceramic Shell for Investment Casting by Introduction of Mullite

Weichen Bai[1], Nana Guan[1,2], Chaowei Zhang[1], Jie Meng[1,*], Jinguo Li[1], Yizhou Zhou[1], Xiaofeng Sun[1]
1. Institute of Metal Research, Chinese Academy of Sciences, Wenhua Road No 72, Shenyang, China
2. University of Science and Technology of China, School of Materials Science and Engineering, Shenyang 110016, China
*Corresponding address: E-mail: jmeng@imr.ac.cn

Abstract: Ceramic shells are important for investment casting of superalloy castings. Alumina based ceramic shells faced up with insufficient mechanical properties at high temperature during casting process. In this research, a high stable performance of ceramic shell was performed by introducing mullite powders in alumina refractory. XRD results showed that more mullite phases were characterized in ceramic shells with mullite powders addition after sintering and high-temperature heating treatment. By binding strength and thermal expansion measurement, ceramic shells with alumina and mullite powders showed better performance at both room and high temperature compared with traditional alumina based shells, which was because alumina and mullite powders mutually embedded in each interval, forming more compact ceramic slurry part in the shell.

Keywords: Ceramic shell, mullite, binding strength, thermal expansion

1 Introduction

Alumina based ceramic shells are widely used for investment casting, which is a popular method for the fabrication of superalloy castings with high dimension accuracy and surface smoothness. However, it was reported that alumina based ceramic shells faced up with problems that mechanical properties of shells decreased at the temperature around 1300 ℃, which would affect the dimensional accuracy of the blade castings[1].

Based on the demand on the properties of ceramic shells, including binding strength and thermal expansion coefficient at high temperature up to 1500 ℃ of ceramic shells, additives have been produced to the ceramic shell system to facilitating formation of mullite phase in the shell at high temperature for casting[2]. Besides, an ideal refractory fused EC95 with mullite and alumina phase as principal phases has been proposed by Zhang for the preparation of ceramic shells[3]. The increased mullite phase in the refractory significantly enhanced the mechanical performance of ceramic shell at high temperature. However, the mullite phase content in refractory is hard to control, and further research is still in need for enhancing properties of ceramic shells by the introduce of mullite phase needs further research.

In this research, we proposed a new method to realize stabilized mullite phase content in the refractory and an optimized structure of refractory cluster utilizing pure fused mullite with only mullite phase. Ceramic shell with different amount of fused alumina and mullite powders were prepared, the content of mullite phase was measured by quantitative XRD analysis, and the influence of mullite on the binding strength and thermal expansion behavior of ceramic shell were evaluated.

2 Experimental procedure

Ceramic slurry used for fabricating ceramic shells were prepared with pure alumina (marked as Al-Shell) and 75 wt.% alumina with 25 wt.% mullite powders (marked as AlM-Shell). Ceramic shells were sintered at 980 ℃ for 2 hours. XRD measurements were conducted to quantitatively analyze the phase content in shells. The 3-point binding tests were conducted to evaluate the binding strength, while thermal expansion behavior of ceramic shells was measured with thermomechanical analyzer.

3 Result and discussion

Phase characterization results

The XRD results of ceramic shells with and without mullite powders were shown in Fig. 1. Only phases of alumina and mullite are shown in XRD results of each shell. Only alumina and mullite phase are shown in XRD results.

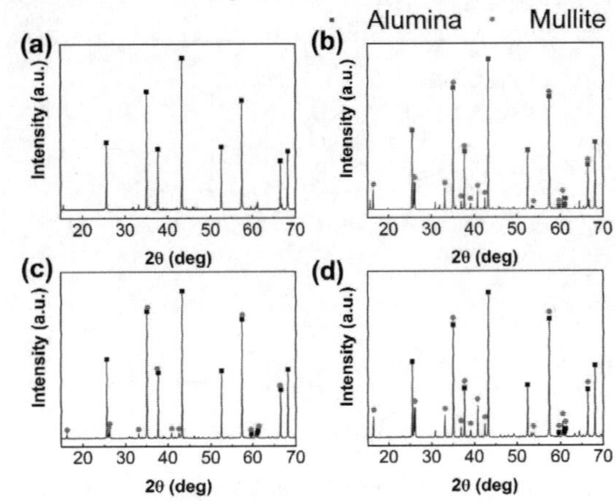

Fig. 1 XRD results for sintered Al-Shell(a) and AlM-Shell (b), and shells after 1500 ℃ heating treatment, (c) Al-Shell (d) AlM-Shell.

Further quantitative analysis results were conducted by WRF refinement method. For Al-Shell, mullite phase content increased from 0 to over 10 wt.% after 1500 ℃ heating treatment. While for ceramic shell with 25 wt.% mullite powders, mullite phase content increased from around 25 wt.% to 32 wt.%. The results show that for ceramic shells both before and after 1500 ℃ heating treatment, the mullite phase content in ceramic shells with 25 wt.% mullite powders are more than that in ceramic shells with only alumina powders.

Binding strength measurements

Binding strength results as shown in Fig. 2 indicate that the addition of mullite powders can help improve the binding strength at room temperature and high temperature. The enhanced binding structure was related to a more compact structure in ceramic slurry parts, with columnar mullite and globular alumina powders mutually embedded in each interval.

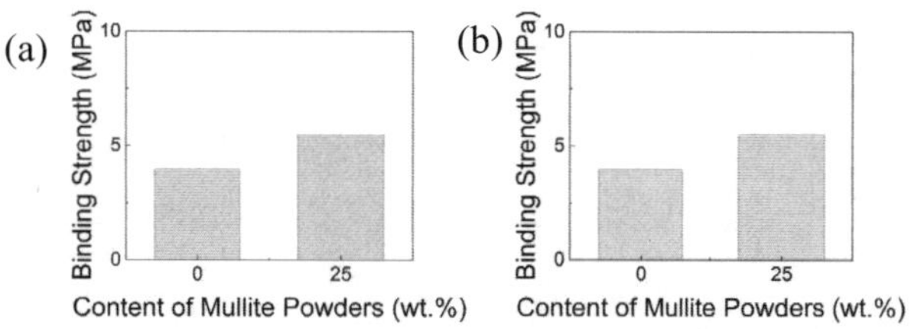

Fig. 2 Binding strength for ceramic shells with and without mullite powders before sintering (a) and at 1500 ℃(b).

Thermal expansion coefficient measurements

The curves of linear thermal expansion increment as temperature increased are shown in Fig. 3. AlM-Shell showed a continuous increasing in the expansion, while Al-Shell faced with change after 1300 ℃. Therefore, the thermal expansion performance of cereamic shells with fused mullite powders is more stable ,which benefits investment casting process.

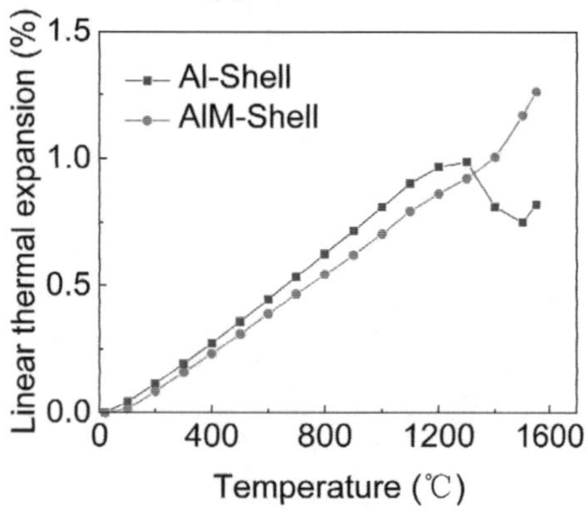

Fig. 3 Linear thermal expansion curve with temperature increasing for Al-Shell and AlM-Shell.

4 Conclusion

By introducing mullite powders as refractory in alumina based ceramic shell, the mullite phase content in the shell was increased. By this means, the binding strength for unsintered shell and sintered shell at 1500 ℃ was increased. Besides, the thermal expansion behaviour of the shell was more regular, which benefits the wax mold design for investment casting.

Acknowledgments

This work was financially supported by the National Key R&D Program of China (grant No.2017YFA0700704 and 2019YFA0705300) The authors are grateful for those supports.

References

[1] LI M, DU H, ZHANG L, et al. Study on High-Temperature Properties of a Ceramic Shell [J]. Foundry, 2012, 61(6): 610-612.

[2] SACKS M D, PASK J A. SINTERING OF MULLITE-CONTAINING MATERIALS .1. EFFECT OF COMPOSITION [J]. Journal of the American Ceramic Society, 1982, 65(2): 65-70.

[3] ZHANG S, LIU Y, ZHANG M, et al. Study on Heat Creep of the A1_2O_3 Shell Mold for Directional Solidification Casting [J]. Foundry, 2013, 62(1): 55-58.

Surface Quality Control of Wax Pattern for Ring Plate Structure

Qingyuan Fang[1,2], Zixiang Jiao[3], Donghong Wang[1*], Guoliang Zhu[1], Zhen Huang[1], Baode Sun [1]

1. Shanghai Key Laboratory of Advanced High Temperature Materials and Precision Forming; State Key Laboratory of Clean and Efficient Turbomachinery Power Equipment; School of Materials Science and Engineering, Shanghai Jiao Tong University, Shanghai, China
2. School of Materials Science and Engineering, Shanghai University of Engineering and Technology, Shanghai, China
3. Xi'an Aerospace Engine Company Limited, Xi'an, China

*Corresponding address: E-mail：wangdh2009@sjtu.edu.cn

Abstract: The thermal properties, rheological properties and melt compressible PVT characteristics of KC-2656L wax were investigated. Numerical simulation of the wax pattern process of rotary body structure was carried out based on Moldflow software. The calculation results show that the melt confluence exhibited by the wax pattern filling process matches with the location of fusion marks appearing in the actual injected wax pattern with the same process parameters, which verifies the accuracy of the numerical simulation results. After increasing the wax injection rate, reducing the wax pattern filling time, lowering the temperature difference between the flow front of the wax mold and the freezing layer factor at the end of filling, the wax melt flow resistance was effectively reduced. Correspondingly, the surface flow state and melt marks of the wax pattern were minimized, and the surface quality was improved.

Keywords: Investment casting; Numerical simulation; Mold filling process; Rheology of wax

1 Introduction

Investment casting is a recent near-net forming production process, suitable for difficult to machine parts and large complex thin-walled parts of precision molding, widely used in aerospace, military equipment and other fields [1]. The castings made by investment casting do not require much processing, and the castings are characterized by low surface roughness and high dimensional accuracy. As the first production process, the surface quality and dimensional characteristics of the wax pattern will be transferred to the ceramic shell, which will ultimately affect the surface quality and dimensions of the castings [2].

Problems such as surface quality and dimensional accuracy of the wax pattern produced during the wax pattern pressing process will directly affect the wax pattern repair workload of the subsequent workers, which will waste massive manpower and material resources. The traditional trial and error method is used to redesign the wax pressing process to try to adjust the quality of the wax pattern molding. With the continuous development of the numerical simulation technology in recent years, it has provided a useful reference to improve the quality of the wax pattern molding. Cui et al [3] simulated the wax injection molding process of a large aircraft magazine wax pattern, and found that the numerical simulation and the actual wax pattern in the same location are underfilled holes, which verifies the accuracy of the numerical simulation. This research utilized Moldflow simulation software to numerically simulate the wax pattern forming process and addressed surface flow marks and weld lines issues by optimizing the wax injection parameter.

2 Experimental procedure

The experiment utilized a capillary viscometer to measure the viscosity of wax at different temperatures and shear rates. Additionally, the wax volume change was examined using the PVT module. The specific heat capacity of the wax was tested with a differential scanning calorimeter, while the thermal conductivity was evaluated using the plane heat source method. Parameters were obtained through multi-variable nonlinear regression fitting using the Cross-WLF equation and Tait equation, and then were imported into Moldflow software for numerical simulation. In the simulation, the wax injection rate ranged from 100 to 220 cm^3/s, the injection temperature varied from 62 to 68°C, and the holding pressure was set at 4 MPa. The wax injection experiments were conducted under the same process parameters for comparison.

3 Result and discussion

Figures 1 presented the PVT characteristics curve, rheological curve, specific heat capacity curve, and thermal conductivity curve of the wax.

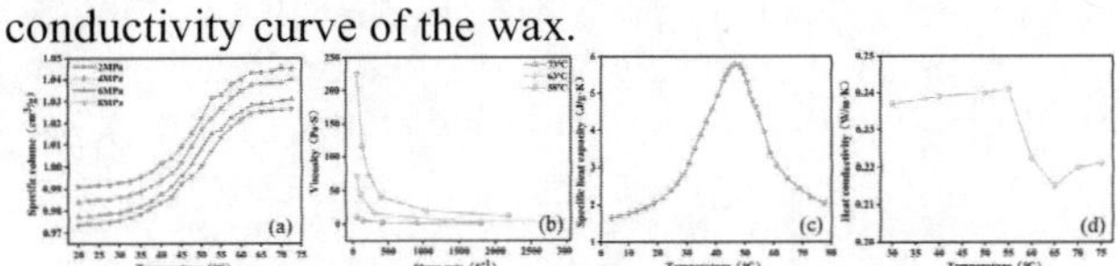

Fig.1 (a)PVT characteristic curve (b)rheological curve (c)specific heat capacity curve (d)thermal conductivity

Figures 2 depicted the schematic of the filling process, temperature field, and solidification field under the condition of a wax injection rate of 100 cm^3/s and an injection temperature of 65°C through numerical simulation. As illustrated, with the increase in filling time, the temperature at the leading edge of the wax pattern decreased, the solidification layer factor at the filling end decreased, and the flow resistance of the wax melt decreased. Figure 3 presented the physical wax pattern obtained under the same process parameters, where the weld marks on the opposite side of the wax mold injection port matched with the convergence point of the wax melt during the numerical simulation filling process, validating the accuracy of the numerical simulation results. Additionally, a large number of flow lines were observed on the surface of the wax pattern, which affected the surface quality of the wax pattern.

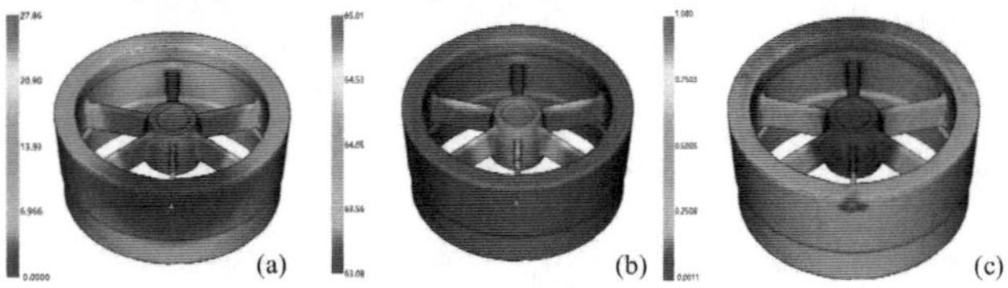

Fig.2 (a)Schematic diagram of filling process (b)Schematic diagram of temperature field distribution (c)Distribution of solidification field

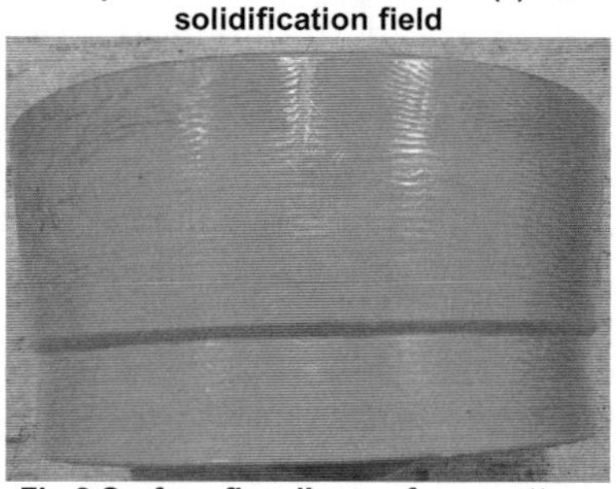

Fig.3 Surface flow lines of wax pattern

This study optimized the surface quality of the wax pattern by increasing the wax injection rate. Figures 4 depicted the schematic of wax pattern filling, temperature field, and solidification field under the condition of a wax injection rate of 220 cm^3/s. The filling time was reduced from 27.86s to 12.67s. The temperature difference at the leading edge of the flow decreased from 1.93°C to 0.54°C. The solidification layer factor at the filling end also decreased correspondingly. Increasing the wax injection rate reduced the filling time and heat loss during the filling process, optimized the uniformity of the melt temperature during wax mold filling, and reduced the flow resistance of the wax melt. Figure 5 presented the actual wax pattern after optimization, showing almost no flow lines on the surface and significantly reduced weld marks.

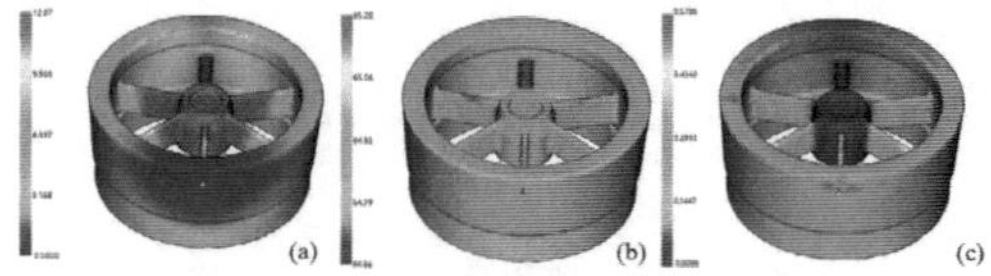

Fig.4 (a)Schematic diagram of filling process (b)Schematic diagram of temperature field distribution (c)Distribution of solidification field

Fig.5 The optimized wax pattern

4 Conclusion

By elevating the wax injection rate, the filling time of the wax pattern was decreased, as well as the temperature differential at the leading edge of the flow and the solidification layer factor at the filling end. The temperature distribution of the melt during the filling process became more uniform, resulting in a reduction of the flow resistance of the wax melt during filling and an optimization of the flow lines and weld marks on the surface of the wax pattern.

Acknowledgments

This work was financially supported by the National Science and Technology Major Projects of China (J2019-VI-0004-0117), the National Key Research and Development Program of China (2022YFB3706800), the National Natural Science Foundation of China (52090042).

References

[1] Song H, Han D, Zhao J, et al. Development status of titanium alloy investment casting technology. J. Foundry, 2020, 69(12): 1304-1311.

[2] Zhang Z Y, Liang B N. Factors affecting the dimensional accuracy of castings in the mold making process. J. Mechanical Research and Application, 2002 (2):15-16.

[3] Cui X P, Li F, Cai S Y. Numerical simulation of large-scale wax mold forming process in precision casting. J. Foundry, 2022, 71(9): 1150-1153.

Effect of Shell Dwell Time and Cooling Conditions on Shrinkage Defects of Superalloy Turbine Blade in Investment Casting

Pan Zhang[1,2], Yizhou Xiao[3], Donghong Wang[1,2*], Zhen Huang[1,2], Guoliang Zhu[1,2], Baode Sun[1,2]
1.Shanghai Key Laboratory of Advanced High Temperature Materials and Precision Forming, School of Materials Science and Engineering, Shanghai Jiao Tong University, Shanghai 200240, China;
2.State Key Laboratory of Clean and Efficient Turbomachinery Power Equipment, School of Materials Science and Engineering, Shanghai Jiao Tong University, Shanghai 200240, China
3.School of Materials Science and Engineering, Shanghai University of Engineering and Technology, Shanghai 200240, China
*Corresponding address: e-mail: wangdh2009@sjtu.edu.cn.

Abstract: This study employed finite element simulation to investigate the effects of the mold dwell time and cooling conditions before melt pouring process on the shrinkage defects of Inconel 718 superalloy turbine blade. The simulation results suggest that with the mold dwell time increasing from 0s to 180s, the shell temperature drops from 1000 °C to 680 °C and the total shrinkage volume increases from 5.19 cm^3 to 6.14 cm^3, however, the two types of mold shell insulation measures, addition of insulation cotton and sand filling, reduced the shrinkage volume to 3.26 cm^3 and 3.76 cm^3 respectively. This phenomenon was mainly caused by the mold shell temperature dropping and rising in the superalloy casting.

Keywords: Numerical simulation; Investment mold casting; mold shell dwell time; cooling conditions

1 Introduction

Gas turbine blades were one of the most typical superalloy investment castings, which have high requirements for their microstructure and mechanical properties [1]. Shrinkage stood out as a primary defect in these castings, leading to decreased fatigue life of the blade castings [2]. Numerical simulations emerged as the most effective method for unveiling shrinkage formation during the investment casting of turbine blades. Otherwise, the model's accuracy may be insufficient to align with the real casting process. The shell mold temperature served as a crucial input parameter for the finite element simulation of superalloy investment casting. It significantly influenced the flow and solidification of the high-temperature melt exceeding 1400 ℃. In the practical application of superalloy investment casting, ceramic shells are typically heated to approximately 1000 ℃ to minimize the temperature differential between the molten metal and the shell. Nonetheless, with the high-temperature shell undergoes the transition from the preheating furnace to the vacuum induction furnace, its temperature inevitably diminishes. Notably, the longer the interval between the mold shell's transfer and the alloy casting, the greater the reduction in shell temperature. Additionally, experimental data indicated that the temperature decrease rate in the shell is neither linear nor uniform. Consequently, in determining the crucial mold dwell time, it was imperative to investigate the impact of mold dwell time before pouring the melt on shrinkage pores, through numerical simulations in investment casting. This study utilized numerical simulation to investigate the impact of mold dwell time and cooling conditions before pouring the molten metal on shrinkage defects in investment castings of Inconel 718 superalloy turbine blades.

2 Experimental procedure

Various parameters, such as the melt pouring temperature set at 1460 ℃, mass flow rate of 6 kg/s, mold shell preheating temperature of 1000 ℃, and air cooling for cooling conditions. Change the mass flow rate curves to alter the mold shell dwell time.

3 Result and discussion

The predicted results of the mold shell temperature during the investment casting of the turbine blade were illustrated in Fig. 1(a1-a4) and mold shell addition of insulation cotton and sand filling in Fig. 2(a1-a2). The minimum mold shell temperature decreased with longer mold dwell times, and insulation cotton and sand filling reduced the decrease in mold temperature.

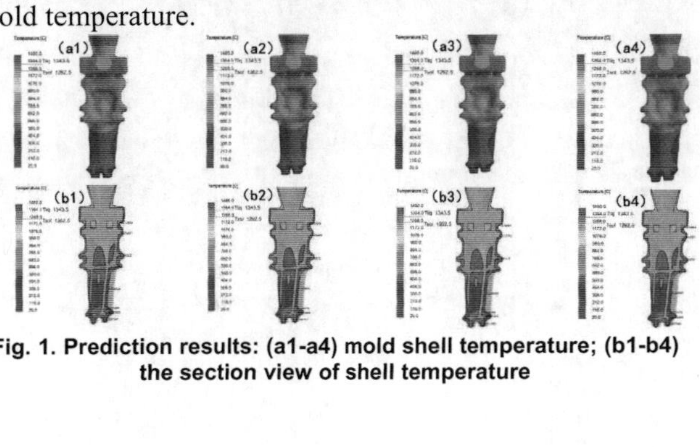

Fig. 1. Prediction results: (a1-a4) mold shell temperature; (b1-b4) the section view of shell temperature

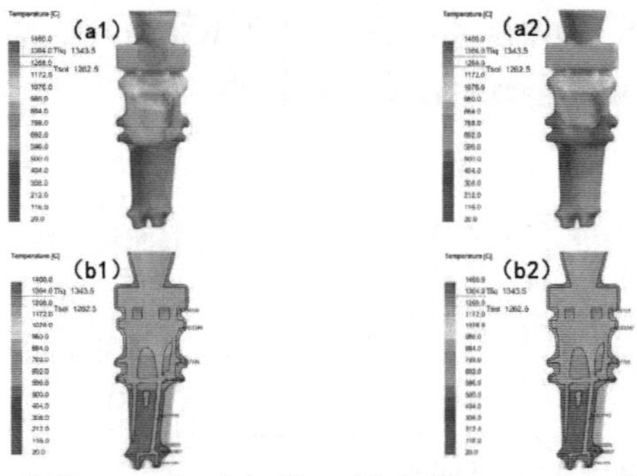

Fig. 2. Prediction results: (a1-a2) mold shell temperature; (b1-b2) the section view of shell temperature

The anticipated hot spots of the superalloy turbine blade are depicted in Fig. 3(a1-a4) and Fig. 4(a1-a2). Clearly, the volume of these hot spots decreased with the mold shell dwell time increasing, but mold shell addition of insulation cotton and sand filling significantly reduced the hot spot area and volume.

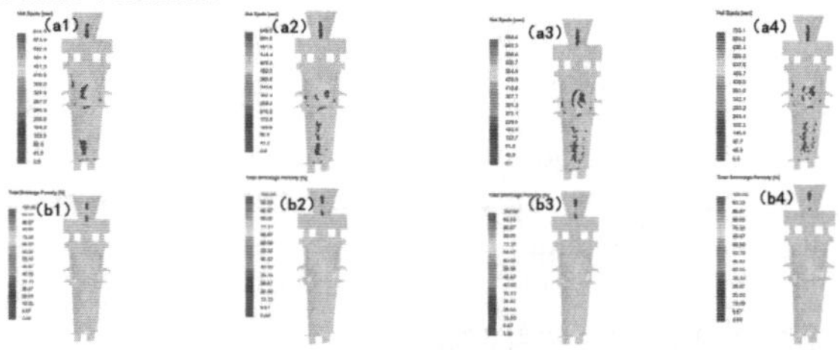

Fig. 3. Prediction results: (a1-a4) hot spots; (b1-b4) total shrinkage porosity

This trend suggested that an excessively low shell temperature would markedly augment the overall volume of hot spots. The insulation cotton and composite sand filling provided thermal insulation to the mold, thereby significantly reducing the volume of hot spots. The total predicted shrinkage porosity of the superalloy turbine blade is depicted in Fig.3(b1-b4) and Fig.4(b1-b2). It was evident that total shrinkage increase notably as the mold shell dwell time escalates from 0 to 120 s. This observation underscored the substantial impact of shell temperature on the generation of shrinkage porosity in superalloy castings.

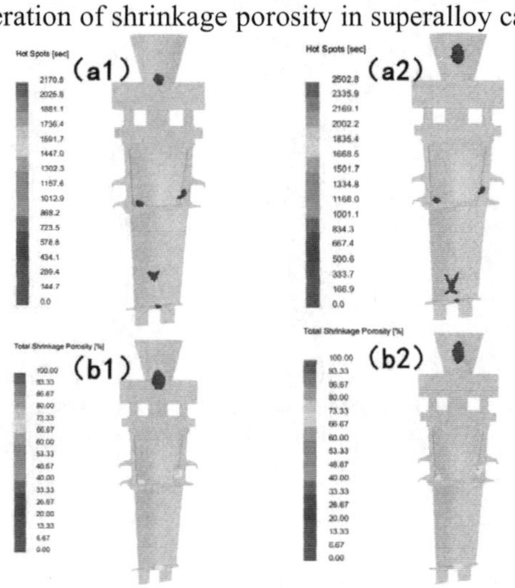

Fig. 4. Prediction results: (a1-a2) hot spots; (b1-b2) total shrinkage porosity

4 Conclusion

With the mold shell dell time increased, the temperature of the mold shell decreased and the total shrinkage volume increased significantly. But the heat preservation measures of the shell, the addition of insulation cotton and sand filling, prevented the decrease of mold temperature, reduced the hot spot area and volume, and played a heat preservation effect on the shell. Notably, the optimal shell temperature for mitigating shrinkage porosity in the turbine blade castings is 950 ℃ rather than 1000 ℃.

Acknowledgments

This work was financially supported by the National Science and Technology Major Projects of China (J2019-VI-0004-0117), the National Key Research and Development Program of China (2022YFB3706800), the National Natural Science Foundation of China (52090042).

References

[1] Ma D. Novel casting processes for single-crystal turbine blades of superalloys. J. Frontiers of Mechanical Engineering, 2017, 13: 3-16.

[2] Liu X f, Lou Y c, Yu B, Su G q, Li C c, Guo X l, Li B, Shui G y. Directional solidification casting technology of heavy-duty gas turbine blade with liquid metal cooling (LMC) process. J. China Foundry, 2019, 16: 23-30.

Ceramic Stereolithography in Investment Casting: Recent Progress and Applications

Kehui Hu[1,2,*], Haoyuan Wang[1,2], Zhigang Lu[2,3,4]
1. State Key Laboratory of Tribology in Advanced Equipment, Tsinghua University, Beijing 100084, China
2. Department of Mechanical Engineering, Tsinghua University, Beijing 100084, China
3. Key Laboratory of Advanced Materials Processing Technology, Ministry of Education, Tsinghua University, Beijing 100084, China
4. State Key Laboratory of Clean and Efficient Turbomachinery Power Equipment, Tsinghua University, Beijing, 100084, China
*Corresponding address: Department of Mechanical Engineering, Tsinghua University, Beijing 100084, China
e-mail: hukehui@tsinghua.edu.com

Abstract: Stereolithography-based ceramic additive manufacturing represents a promising technology for creating complex ceramic components with high precision. The technology's ability to form intricate structures and its compatibility with various ceramics make it particularly advantageous for rapid prototyping and low-volume production. This work discusses the recent progress and applications of stereolithography-based ceramic parts in the investment casting industry, particularly in the production of ceramic cores, shells, integrated core-shells, and filters, highlighting its potential to significantly accelerate production and enhance complexity.

Keywords: ceramic stereolithography; investment casting; ceramic core; integrated core-shell; ceramic filter

1 Introduction

Stereolithography-based ceramic additive manufacturing is a promising technology for producing complex ceramic components with high precision. This process involves the use of photo-curable resin mixed with ceramic powders to create suspensions, which are then cured with UV light to form green bodies. Post-processing involves burning out the resin and sintering the ceramic powders to achieve the desired mechanical properties. This technology is advantageous for rapid prototyping and low-volume production due to compatibility with various ceramic materials and ability to form complex structures. The application of stereolithography-based ceramic parts in the investment casting industry can greatly accelerate the production and further increase the complexity.

2 Ceramic stereolithography

There are multiple types of stereolithography devices being used to form green bodies with different characteristics. The mask-projection devices cure an entire layer simultaneously, achieving high efficiency with layer formation times between 10-30 seconds and a resolution of 35 micrometers. However, its maximum forming dimensions are limited by the light source. On the other hand, point-by-point scanning devices use point light sources, with layer formation taking 2-5 minutes. It allows for larger dimensions but has a lower resolution compared to the mask-projection type. The mask-projection scanning type combines the advantages of the previous two, using a movable mask-projection light source to achieve larger dimensions while maintaining high efficiency. It can produce parts up to 600 mm in size with layer formation times of 1-2 minutes.

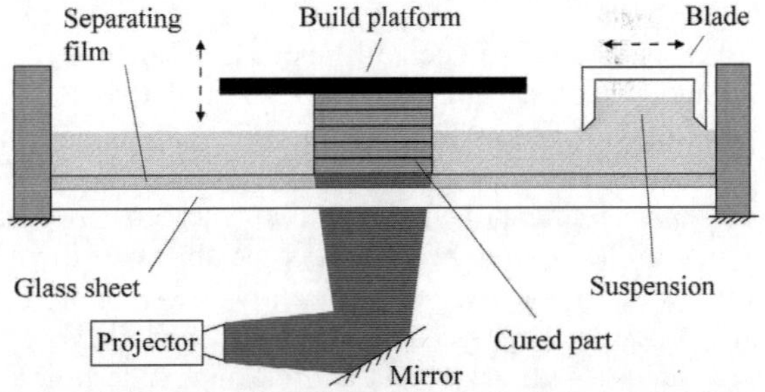

Figure 1. Principle of mask-projection stereolithography device.

3 Application in investment casting

Ceramic cores

Ceramic cores must withstand high temperatures and mechanical stresses. Advances in ceramic stereolithography technology have led to the production of complex double-layer ceramic cores, which are essential for the next generation of aero-engines operating at intake temperatures up to 2500 K. Research has shown that these cores can achieve high dimensional accuracy and surface quality, with examples including successful casting of single crystal blades with complete and high-quality inner cavities [1]. Multi-layer ceramic cores have also been successfully produced with casting verifications of single crystal vane blades [2]. These advancements also allowed the mass production of ceramic cores for industrial applications, significantly reducing costs and production times.

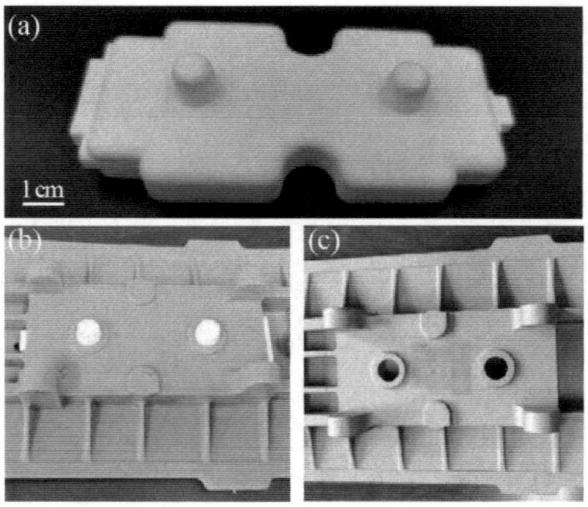

Figure 2. Stereolithography-based ceramic core and the corresponding casting verification.

Ceramic shells

Additively manufactured ceramic shells are designed based on 3D models of castings and have a wall thickness of approximately 1 mm. These shells are assembled into traditional clusters for subsequent processing. The advantages include the elimination of wax costs, higher efficiency in the coating process, and the ability to flexibly control wall thickness, significantly shortening the production cycle. High-accuracy ceramic shells have been demonstrated with systematic cleaning techniques, achieving a dimensional accuracy of 0.1 mm.

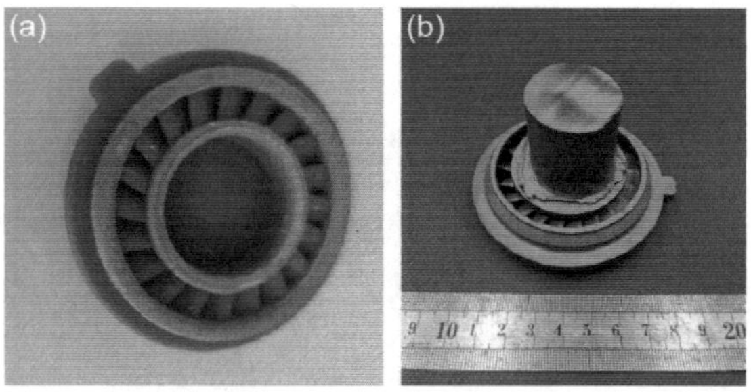

Figure 3. Stereolithography-based ceramic shell green body and the corresponding casting.

Integrated core-shells

The ceramic stereolithography technology allows to produce integrated core-shells. Combining the core and shell into one part reduces the number of molds required, shortening the production cycle from several months to just weeks. Integrated core-shells have shown high accuracy in both external shapes and internal cavities, with no defects detected in the cast parts. Integrated core-shells with pouring cups, cores, and shells have been developed. Permeation tests showed no cracks, and X-ray images confirmed complete and defect-free inner structures. The castings produced with these core-shells exhibited high surface roughness and dimensional accuracy within 0.1 mm.

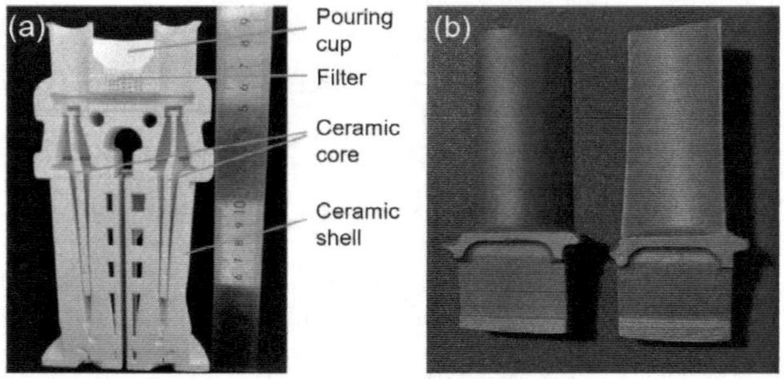

Figure 4. Stereolithography-based integrated core-shells and corresponding casted blades.

Ceramic filters

Traditional ceramic foam filters are prone to defects, while 3D printed filters with regular geometric structures provide better control over melt flow and are less likely to break. These filters enable repeatable pouring conditions and feasible simulations and structural calculations. Gradient filters have been designed and manufactured with 20 PPI in the upper region for impact resistance and 40 PPI in the lower region for inclusion control. This design enhances local temperature control and promotes single crystal growth with fewer stray grains.

4 Summary

Rapid advancements in ceramic stereolithography technology have expanded its applications in investment casting, including the production of ceramic cores, shells, integrated core-shells, and filters. Stereolithography-based ceramic cores have been demonstrated to meet the requirements for single crystal blade casting, with ongoing improvements needed for high-temperature performance. Shells and integrated core-shells have been successfully validated in practical castings, though further development in detection and evaluation techniques is necessary for industrial applications. Lastly, additively manufactured ceramic filters require verification and optimization for industrial use, emphasizing the importance of reducing device costs for widespread adoption in the investment casting industry.

References

[1] Hu K, Wang H, Lu K, et al. Fabrication of silica-based ceramic cores with internal lattice structures by stereolithography. China Foundry, 2022, 19 (5): 369-379.

[2] Hu K, Lu Z, Lu K, et al. Additive manufacturing of complex ceramic cores and verification of casting process. J. Mech. Eng., 2021, 57 (3): 227-234.

Precision Casting Technology for Superalloy Transfinite Components

Jun Wang[1, 2], Yahui Liu[1, 2], Ying Ning[1, 2], Guoxiang Wang[1, 2], Jianzhong Li[1, 2]

1. School of Materials Science and Engineering, Shanghai Jiao Tong University, No. 800, Dongchuan Road, Minhang District, Shanghai 200240, China

2. Jiangsu Zhongchao Aerospace Precision Casting Technology Co., Ltd., No. 9, Changxing Road, Yixing City, Jiangsu Province 214242, China

*Corresponding address: junwang@sjtu.edu.cn (Jun Wang)

Abstract: High performance aircraft engines are one of the domestically produced large-scale equipment. Precision superalloy castings (such as turbine casings) formed by investment casting are key load-bearing components of aircraft engines, and their dimensional accuracy and metallurgical quality determine multiple service performance of aircraft engines. For example, dimensional accuracy can affect the starting performance of the turbine casing, while metallurgical defects can reduce the fatigue life of castings. Due to the fact that controlling the dimensional accuracy and metallurgical quality of precision superalloy castings has always been an important way to improve the service life of engines, it has attracted the attention of the aviation manufacturing industry.

With the structural integration and lightweight design of aircraft engines, the profile dimensions of casing structural components continue to increase, the wall thickness decreases, and the quality requirements are significantly improved, which poses a huge challenge to traditional investment casting processes. This article focuses on the technical bottlenecks of casing castings, including micropore defects, under casting, and dimensional deviations. Relevant research has been conducted and four key technologies have been achieved, including:

(1) Melt processing and melt state control technology: Specifically, by utilizing the structural transformation and genetic characteristics of the melt, a high-temperature treatment process for the melt has been established, which reduces segregation and TCP phase precipitation. During the high-temperature treatment process, there was no severe element burnout, the grain size of the casting was refined, and the performance was significantly improved. The comprehensive use of melt treatment and new thermal control processes can reduce casting defects and comprehensively improve microstructure and properties.

(2) Design technology for transfinite precision casting process: Based on hydraulic models, a shape following distributed pouring system has been developed to achieve rapid and stable filling, ensuring the rapid transportation of molten metal and the uniformity of overall temperature. A design criterion for transfinite casting process of large complex thin-walled castings has been proposed, breaking through traditional casting design methods and forming a new casting process system.

(3) Pressure regulating precision casting technology: We have conducted research on pressure regulating precision casting equipment and processes for 1.0mm superalloy castings, achieving complete filling of large-area thin-walled castings with a wall thickness of 1mm.

(4) Full process dimensional accuracy and surface quality control technology. A wax material flow and pressure holding calculation model has been constructed, realizing simulation calculation of wax material flow and shrinkage deformation, breaking through the long-term dependence on the experience + "trial and error" method in wax mold forming. We have established a reverse compensation deformation design method based on the full process displacement field and a reverse compensation deformation design method for node displacement transfer, and invented a new method for controlling the dimensional accuracy of various parts of castings.

These technologies have successfully solved the manufacturing problem of casing castings and played an important role in the independent development of commercial aviation engines.

Keywords: superalloy; investment casting; rear casing; metallurgy defects; dimensional deviations

Refinement of the Secondary Phase Structure in Molybdenum-Based Refractory Alloys Achieved Through Mechanical Milling

Haolong Liu[1], Ning Li[1]
State Key Laboratory of Materials Processing and Die & Mould Technology, and School of Materials Science and Engineering, Huazhong University of Science and Technology, Wuhan, 430074, People's Republic of China
*Corresponding address: e-mail: d202180439@hust.edu.cn

Abstract: This study investigates the impact of mechanical milling on the ~~microstructural properties~~ microstructure of TZM alloy prepared via powder metallurgy. Increasing milling time significantly reduces the size of doped particles in the green bodies, due to the breakdown of larger particles during the milling process. The size of doped particles in the green body milled for 8 hours was reduced by 38% compared to that in the green body pressed from unmilled powder. Sintered samples show a significant decrease in the size of second-phase particles and grain size with increased milling time. The grain size decreased from 9.45 μm in the sample prepared from unmilled powder to 6.89 μm in the sample prepared from powder milled for 8 hours. This grain refinement is attributed to the overall reduction in powder particle size and the pinning effect of refined second-phase particles.

Keywords: TZM alloy ; Powder metallurgy ; Ball milling

1 Introduction

TZM molybdenum alloy exhibits superior specific strength at temperatures above 1000K. It possesses high mechanical strength and creep resistance at high temperatures (1373K), and its low thermal expansion coefficient combined with high thermal conductivity minimizes thermal stress during rapid heating and cooling, ensuring the working dimensions of the mold remain unchanged. Therefore, molybdenum alloy molds can meet the long-term, high-temperature (1373K), and high-load usage requirements that traditional mold materials cannot satisfy. [1,2].

TZM alloy, composed of Mo-Ti(0.4~0.55)-Zr(0.06~0.12)-C(0.01~0.04), benefits from the addition of alloying elements titanium (Ti) and zirconium (Zr), which not only contribute to solid solution strengthening but also help to purify the grain boundaries. These elements form fine second-phase particles within the molybdenum (Mo) matrix, which pin dislocations and grain boundaries, thereby refining the grains. This allows TZM ~~molybdenum~~ alloy to achieve solid solution strengthening, grain refinement strengthening, and second-phase particle strengthening simultaneously[3-6].

In this work, the second-phase particles in TZM alloy prepared by powder metallurgy were significantly refined through mechanical milling.

2 Experimental procedure

The material preparation method is powder metallurgy. The raw powders used are TiH_2 powder, ZrH_2 powder, graphite powder, and molybdenum powder. First, the raw powders are mixed in a three-dimensional mixer for 10 hours at a speed of 30 revolutions per minute (rpm). Then, the mixed powder is ball milled for 2, 4, and 8 hours, with a ball-to-material ratio of 1:1, at a speed of 200 rpm. The milling medium consists of 3 mm ZrO_2 balls, and the milling jar is made of stainless steel. To minimize the ~~increase in~~ oxygen content in the powder, the entire milling process is carried out under argon protection. The mixed powder was compacted into green bodies using cold isostatic pressing at 280 MPa. The green bodies were then sintered at 1900°C for 4 hours.

The microstructural analysis was performed using a Leica DM2700 M optical microscope and a Phenom XL G2 desktop scanning electron microscope. The powder particle size was measured using a Mastersizer 3000 laser particle size analyzer.

3 Result and discussion

Microstructural analysis of green bodies

The SEM images shown in Figure 1 illustrate the distribution of doped particles (black) and molybdenum powder particles (white) in the green bodies of TZM alloy prepared with different milling times. The size of doped particles in the green bodies significantly decreases with increased milling time. This reduction is attributed to the further breakdown of large particles in the mixed powder during the milling process. As shown in Table 1, the D_{50} of powder particles decreases from 13.3 μm to 6.8 μm. Due to the varying degrees of fragmentation of the doped powders, different sizes and distributions are observed. The specific particle size distribution of the doped particles in the green body is depicted in Figure 2. The doped particle size in the green body prepared by 8 hours of milling is reduced by 38% compared to that in the green body pressed from unmilled powder.

Microstructural analysis of sintered samples

Figure 3 presents the optical micrographs of the microstructure of sintered TZM alloy prepared with powders milled for different durations. The black particles within the white molybdenum matrix are second-phase particles formed during sintering. It is clear that with increasing milling time, the size of the second-phase particles in the sintered alloy significantly decreases. Furthermore, the grain size also shows a decreasing trend, from 9.45 μm in the sample prepared from unmilled powder to 6.89 μm in the sample prepared from powder milled for 8 hours. This grain refinement is attributed to the overall reduction in powder particle size due to milling, as each powder particle forms an initial grain during sintering. The finer the powder, the finer the initial grains. Additionally, the refined second-phase particles hinder grain growth during the sintering process. The combined effect of these factors results in smaller grain sizes.

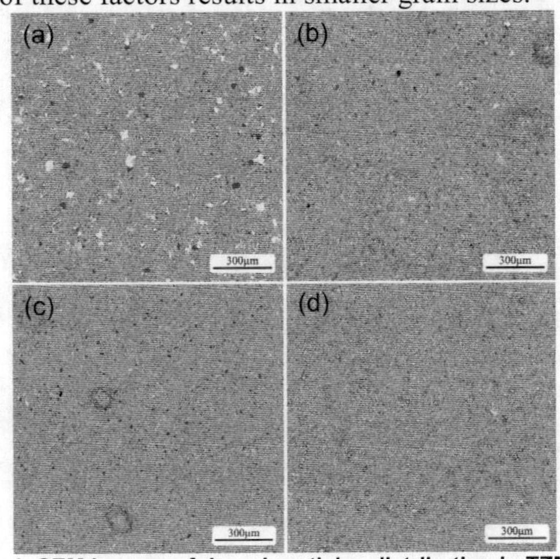

Figure 1. SEM images of doped particles distribution in TZM alloy green bodies with different milling times:(a) 0h.(b) 2h.(c) 4h.(d) 8h.

Table 1. Laser particle size analysis of powders with different milling times

Ball milling time(h)	0	2	4	8
D_{50}(μm)	$13.3_{\pm0.6}$	$8.3_{\pm0.7}$	$7.8_{\pm0.4}$	$6.8_{\pm0.5}$

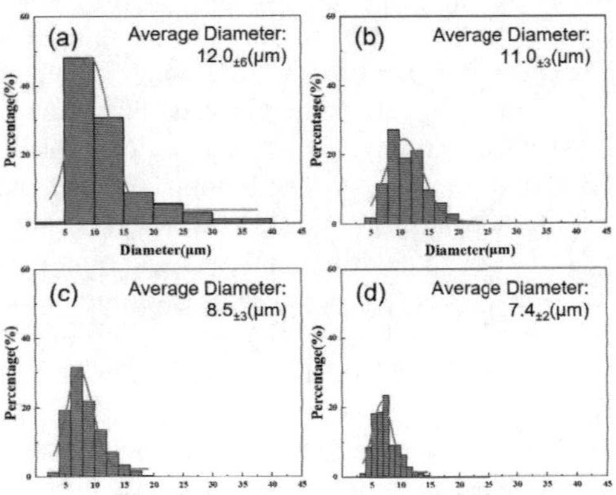

Figure 2.Particle size distribution of the doped particles in TZM alloy green bodies with different milling times:(a) 0h.(b) 2h.(c) 4h.(d) 8h.

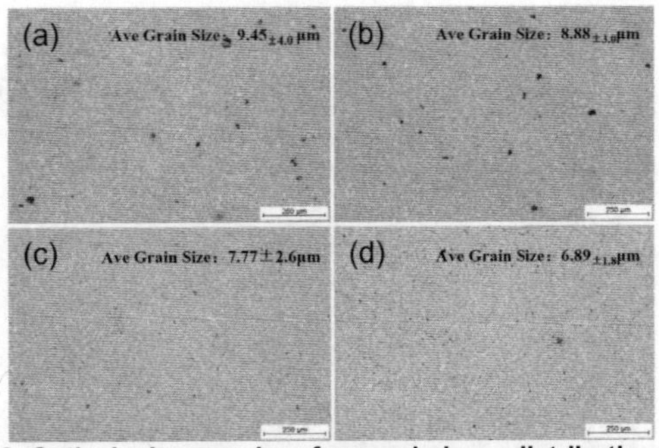

Figure 3. Optical micrographs of second phase distribution in TZM alloy as-sintered samples with different milling times:(a) 0h.(b) 2h.(c) 4h.(d) 8h.

4 Conclusion

The microstructural analysis of TZM alloy green bodies and sintered samples demonstrates that mechanical milling effectively refines the size and distribution of doped particles. With increased milling time, the D50 of powder particles significantly decreases, resulting in finer doped particles in the green bodies. This refinement continues through the sintering process, leading to smaller second-phase particles and reduced grain size. The combined effects of reduced powder particle size and the pinning action of refined second-phase particles hinder grain growth, resulting in smaller grains. These finer grains enhance the material's strength through grain refinement strengthening. Thus, mechanical milling is a critical process for optimizing the microstructure and overall performance of TZM alloy.

References

[1] El-Genk M S. and Tournier J-M. A review of refractory metal alloys and mechanically alloyed-oxide dispersion strengthened steels for space nuclear power systems[J]. Journal of Nuclear materials, 2005, 340(1): 93-112.

[2] Walser H and Shields J A. Traditional and emerging applications of molybdenum metal and its alloys[J].. IMOA Newsletter, 2007.

[3] Chaudhuri A, et al. Hot deformation behaviour of Mo-TZM and understanding the restoration processes involved[J].Acta Materialia, 2019, 164: 153-164.

[4] Li S-L, et al. The formation mechanism of micro-nano secondary phase in solid-liquid doped TZM alloy[J].Materials Characterization, 2022, 186: 111800.

[5] Li S-L, et al. Formation mechanism of coronal composite secondary phase in titanium-zirconium-molybdenum alloy[J].Materials Characterization, 2022, 193: 112235.

[6] Li S-L, et al. Micro-nano secondary phase greatly increases the plasticity of titanium-zirconium-molybdenum alloy[J].International Journal of Refractory Metals and Hard Materials, 2023, 112: 106152.

New Development of Intelligent Detection Technology for Complex Aerospace Castings Ray Flaw Detection

Xiaoyuan Ji*, Mingjun Hou, Hao Dong, Haozhe Duan, Chuhao Wu, Qinyang Li, Shuohong Li, Yajun Yin, Jianxin Zhou

State Key Laboratory of Materials Processing and Die & Mould Technology, Huazhong University of Science and Technology, Wuhan 430074, China

*Corresponding address: e-mail: jixiaoyuan@hust.edu.cn

Abstract: Complex castings of light alloys such as aluminum, titanium and magnesium are widely used in high-end national defense equipment such as aerospace. The main means of internal defect detection are "X-ray flaw detection two-dimensional DR Manual division parameter adjustment and film shooting + manual visual experience evaluation + manual trial and error three-dimensional positioning". The three links are highly dependent on manual labor, and defects are easy to miss and mis-detect, and the efficiency is low, resulting in potential safety risks for equipment. Therefore, a new idea of intelligent casting inspection in the whole process of path planning, automatic machine assessment and 3D defect positioning is proposed. A defect recognition and evaluation model based on casting knowledge and deep learning is established, an inspection path optimization and planning algorithm based on quality key points, and a 3D defect reconstruction and position technology based on multi-angle 2D images. The results of engineering application or test verification show that the accuracy of casting defect assessment exceeds 98% (miss detection is close to 0), full-size filming without blind area, and 3D positioning error is less than 2%, which is expected to transform manual inspection into intelligent inspection of the whole process of filming, scoring and positioning.

Keywords: Defect 3D positioning; Internal defect detection; Light alloy castings; X-ray inspection images

1 Introduction

Light alloy castings, such as titanium, aluminum, and magnesium, are widely used in critical aerospace and defense applications, including engine casings, air intake ducts, and ribbed plates, which are typical titanium alloy aerospace components. The precision casting process for light alloys involves multiple stages, including wax removal, baking, melting, and pouring, with any anomaly in these steps potentially leading to internal defects. To ensure the reliability and safety of critical light alloy castings for aerospace equipment, radiographic inspection and 3D defect positioning have become essential procedures before installation. Current detection methods rely heavily on manual adjustments in image acquisition parameters and visual inspection, resulting in lengthy processes with a high likelihood of missed or misidentified defects. Furthermore, existing methods for internal defect positioning depend on manual analysis of X-ray images, which makes it difficult to intuitively obtain information about defect depth and size, leading to low-quality and inefficient defect positioning and repair. Therefore, there is an urgent need to shift from manual inspection to a fully intelligent detection process encompassing imaging, evaluation, and defect positioning.

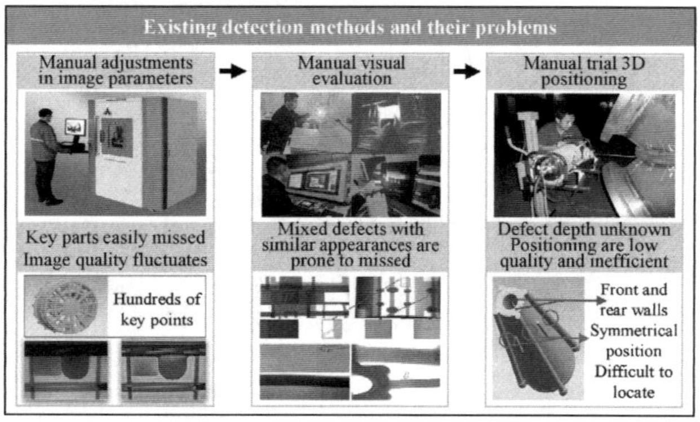

Figure 1. Existing detection methods and their problems

2 Experimental procedure

(1) To improve the efficiency of radiographic path planning, we first introduced the concept of quality control points[1], which were identified based on a feature point extraction algorithm applied to the casting. Next, we generated a cylindrical surface of motion around the 3D model of the casting, representing the potential positions of the radiation source, and map the quality control points onto this surface. Finally, using a genetic algorithm, we generated the optimal movement paths for the radiation source, the workpiece platform, and the imaging plate, drawing 3D curves to complete the path planning process.

(2) To achieve automated online radiographic image evaluation, we first developed the ASCUnet (Ameliorated Skip Connection Unet) model. We embedded an efficient, lightweight attention module[2] at appropriate positions within the Unet architecture to build an attention-enhanced multiscale feature pyramid. Additionally, we proposed a data augmentation strategy tailored for X-ray defect segmentation in light alloy castings and used transfer learning to train the ASCUnet model. Finally, comparative tests and validation showed that the model achieved a segmentation accuracy of 98% for multiscale defect detection.

(3) To achieve intelligent 3D defect positioning, we first integrated filtering and morphological processing techniques to extract defect contours from the images. Next, we constructed a coordinate dimensionality elevation model to transform 2D image defect coordinates into 3D defect coordinates on the casting[3]. Finally, we calculated the intersections between the radiographic rays and the casting model, established a boundary feature polyhedron, and developed an algorithm for defect morphology reconstruction. The results demonstrated that the relative error in 3D defect positioning was less than 2%.

3 Result and discussion

To validate the performance of ASCUnet, we created a titanium alloy dataset and trained seven semantic segmentation models, including Res50DeepLabV3+, for comparison with ASCUnet, with the results shown in Figure 1. Define the defect detection accuracy:

$$\eta = \frac{T-M}{T} \times \frac{D-F}{D} \qquad (1)$$

Data analysis indicated that the total number of actual defects(T) was 259, with 260 defects detected(D), 4 false positives(F), and 0 missed defects(M). The defect detection accuracy(η) was 98.46%.

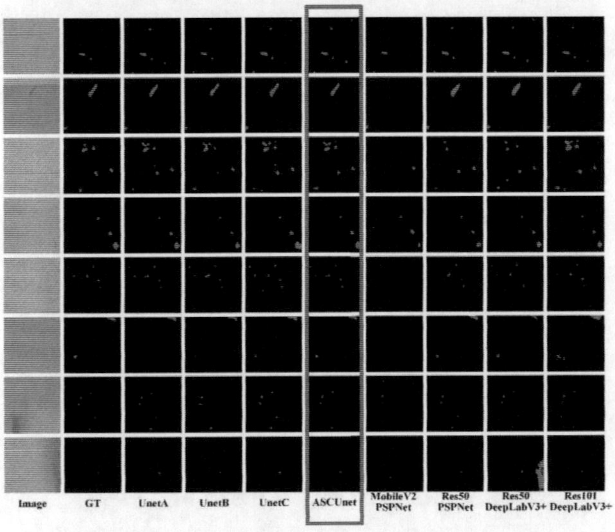

Figure 2. Comparison results of the eight models

To verify the accuracy of 3D defect positioning, we simulated 3D positioning for hole-type and crack-type defects, calculating the centroid coordinates of the feature polyhedra obtained from positioning, the centroid coordinates of the simulated defects in the model, the model dimensions, and the relative positioning error. The results are presented in Table 1.

Table 1. Defect positioning result quantification table (CG means the center of gravity)

Object	X	Y	Z
Model size	81.77	81.77	43.81
CG of Defect A	-2.145	-8.776	19.264
CG of Result A	-1.858	-9.085	20.014
Relative error	0.35%	0.38%	1.71%
CG of Defect B	-1.040	10.392	32.471
CG of Result B	-0.828	10.216	32.679
Relative error	0.26%	0.22%	0.47%

4 Conclusion

We propose a new approach for intelligent casting inspection that integrates radiographic path planning, automated image evaluation, and 3D defect positioning. This includes developing a defect recognition and evaluation model that combines casting expertise with deep learning, an optimized inspection path planning algorithm based on quality control points, and a technique for 3D defect reconstruction and positioning using multi-angle 2D images. Engineering applications and test validation demonstrate that our method can achieve an image evaluation accuracy of over 98%, full-size radiographic inspection with no blind spots, and a 3D positioning error of less than 2%. This approach is expected to transform manual inspection into a fully intelligent process encompassing imaging, evaluation, and positioning.

Acknowledgments

This research was funded by the National Natural Science Foundation of China (Nos. 52275337, 52090042, 51905188), and the National Key R&D Program of China (2020YFB1710100).

References

[1] Zhang Z, Ji X, Zhou J, et al. A method for modeling and extracting 3D structural features of castings considering size[J]. Procedia Manufacturing, 2019, 37: 563-570.

[2] Wang Q, Wu B, Zhu P, et al. ECA-Net: Efficient channel attention for deep convolutional neural networks[C]//Proceedings of the IEEE/CVF conference on computer vision and pattern recognition. 2020: 11534-11542.

[3] Chen L, Li B, Zhang L, and Shang Z. 3D positioning of defects for gas turbine blades based on digital radiographic projective imaging. J. NDT&E International, 2023, 133: 102751.

Study on the Performance and Precision of 3YSZ-based Ceramic via Layered Extrusion Forming

Li Yang, Lei Qian, Zhiyuan Yang, Zitian Fan*
State Key Laboratory of Materials Processing and Die & Mould Technology, School of Materials Science and Engineering Huazhong University of Science and Technology, Wuhan 430074, China
*Corresponding address: fanzt@hust.edu.cn

Abstract: Low-cost layered extrusion forming (LEF) technique is often considered to fabricate complex ceramic parts. The 3 mol% yttria-stabilized zirconia (3YSZ) ceramic has been widely used in industrial field because of its excellent comprehensive properties such as high strength, high wear resistance, high temperature resistance and corrosion resistance. In this work, we adopted LEF to directly prepare 3YSZ ceramic green bodies, the effects of silica contents on the properties of 3YSZ-based ceramic samples were studied, and the forming process parameters were optimized by orthogonal test. When the SiO_2 content was 8.4wt.%, the linear shrinkage rate of the 3YSZ-based ceramic sample was 15.66% and the bending strength was approximately 208 MPa after sintering at 1300 ℃ for 60 min. Besides, the forming parameters were optimized, and the best upper and side surface roughness (Ra) of the ceramic sample were 3.05 μm and 20.01 μm under the optimal process parameters, respectively.

Keywords: Layered extrusion forming; 3YSZ ceramic; SiO_2; Mechanical properties; Precision

1 Introduction

The 3YSZ ceramic has been widely used in modern industrial and practical engineering because of its excellent mechanical properties[1]. With the increasing demand for ceramic parts with complex structure, especially in aerospace, traditional ceramic processing methods were no longer adequate. The rapid development of AM technologies provides a new method to prepare complex ceramic parts[2-3]. Among AM techniques, LEF that relied on computer-controlled robotic deposition of highly dispersed ceramic slurries through a fine nozzle to form three-dimensional structure was often considered due to its simple process, low-cost and the flexibility to print a wide variety of materials[4]. The dental structures produced by direct ink writing present very promising properties regarding porosity, surface finish and mechanical performance, however, the corresponding contraction of 3YSZ ceramic is still large (>30%). Therefore, it is necessary to reduce the sintering shrinkage rate and maintain high strength of 3YSZ ceramics formed by layered extrusion forming. In this work, the effects of silica contents on the linear shrinkage, bending strength, open porosity properties of 3YSZ-based ceramic samples were studied to reduce shrinkage while maintaining high ceramic strength on the basis of previous research[5]. In addition, the forming process parameters were optimized by orthogonal test. Finally, 3YSZ-based ceramic samples with good surface quality were obtained.

2 Experimental procedure

3YSZ (D_{50}=1.47 μm), SiO_2 (D_{50}=29.6 μm) powders, Methylcellulose (MC), Nano-TiO_2 dispersion solution, ammonium salt of polyacrylic acid (NH4PAA) and Polyethylene glycol (PEG-20000) were used as raw materials. The preparation process of 3YSZ-based ceramic shown in Fig.1.

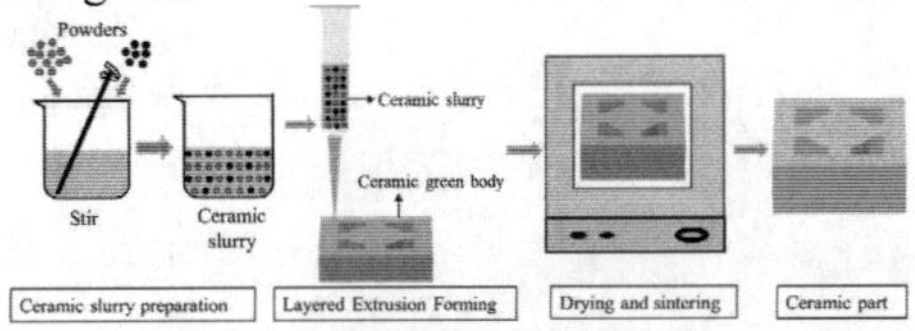

Fig. 1 preparation process of 3YSZ-based ceramic

$$\varphi = \frac{L_0 - L}{L_0} \times 100\% \qquad (1)$$

$$P = \frac{M_2 - M_0}{M_2 - M_1} \times 100\% \qquad (2)$$

Where Φ is the linear shrinkage, L_0 is the length of the original state, L is the length after sintering, P is the open porosity, M_0 is the weight of the dry ceramic samples, and M_1 and M_2 are the weight of water-saturated samples in water and air, respectively.

3 Result and discussion

As shown in Fig. 2, with the SiO_2 additions increasing from 0 wt.% to 16.8 wt.% after sintering at 1300 °C for 60 min, the shrinkage of 3YSZ-based ceramic sample decreases from 18.33 % to 11.68 %, the bending strength of 3YSZ-based ceramic sample decreases from 230 MPa to 113 MPa. And correspondingly, the open porosity of increases from 3.28 % to 10.98 %, and the bulk density decreases from 5.8 g/cm^3 to 4.2 g/cm^3, respectively. The influence mechanism of SiO_2 on reducing sintering shrinkage could be attributed to the phase transformation of SiO_2, changing its crystal structure from di-trigonal to hexagonal when temperature is above 573 °C, which would result in volume expansion change of around 10% at the transformation temperature. As shown in Fig 3(a~e), SiO_2 with large particle size is introduced into the 3YSZ-based ceramic as a brittle phase. Although the interface reacts with the ceramic, it still exists in the large particle and weakens the strength of the 3YSZ ceramic. In the corresponding XRD analysis results, shown in Fig. 3(f), there are mainly ZrO_2 and $Zr_{0.92}Y_{0.08}O_{1.96}$ phases. The intensity of the $ZrSiO_4$ diffraction peak increased with the increase of SiO_2 powder addition amount and consequently the intensity of the ZrO_2 diffraction peak decreased with the increase of silica powder additions.

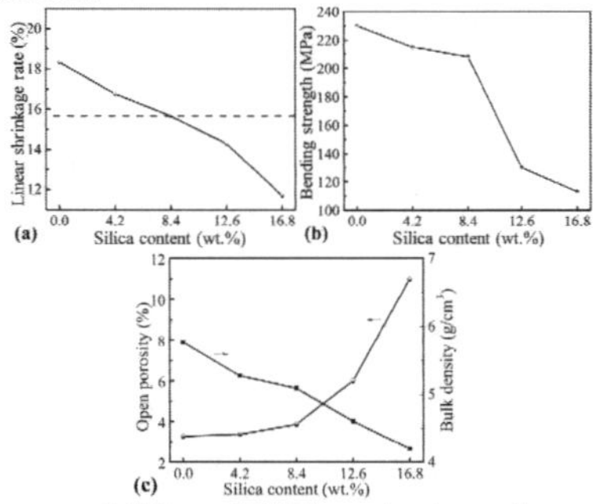

Fig. 2 Influence of different SiO2 contents on the properties of 3YSZ-based ceramic samples: (a) linear shrinkage rate; (b) bending strength; (c) open porosity

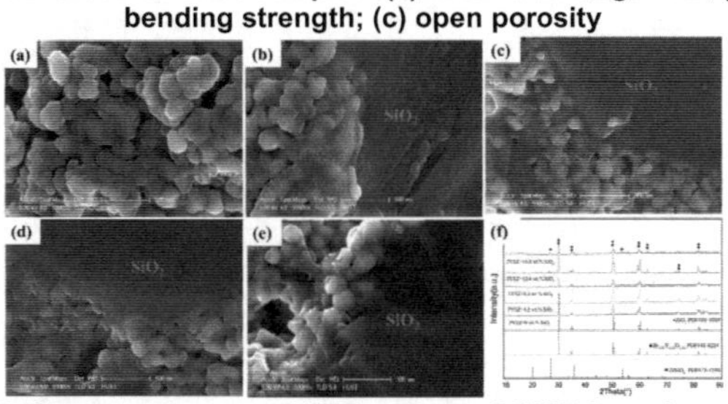

Fig. 3 Microstructures and XRD pattern of 3YSZ-based ceramic with different SiO2 contents after sintering at 1300 ℃ for 60 min: (a) 0 wt.%; (b) 4.2 wt.%; (c) 8.4 wt.%; (d) 12.6 wt.%; (e)16.8 wt.%; (f) XRD

The important factors affecting the forming accuracy of ceramic in the layered extrusion forming technology were investigated, including print speed (Vs), nozzle diameter (d), and layer height/Nozzle diameter (h/d) were investigated. A three-factor and three-level orthogonal test was designed, and the orthogonal test results were shown in Table 1. The parameters influenced the fabrication process in the order as: d> h/d > Vs.

Table 1 Results of orthogonal experiment

Number	Influence factors			
	Vs (mm/s)	d (mm)	(h/d)	Ra/μm
1	20	0.34	75%	35.331
2	20	0.41	85%	26.004
3	20	0.51	95%	34.466
4	25	0.34	85%	20.012
5	25	0.41	95%	34.166
6	25	0.51	75%	36.413
7	30	0.34	95%	31.595
8	30	0.41	75%	27.902
9	30	0.51	85%	35.988
K1	95.801	86.938	99.646	
K2	90.592	88.072	82.005	
K3	95.485	106.867	100.226	
k1	31.934	28.979	33.215	
k2	30.197	29.357	27.335	
k3	31.828	35.622	33.409	
R	1.736	6.643	6.074	
Primary and secondary order	d>h/d>Vs			
Optimization level	Vs-k2	d-k1	h/d-k2	
Optimal combination	Vs=25mm/s, d=0.34mm, h/d=85%			

When Vs is 25mm/s, D is 0.34mm, and the h/d is 85%, the optimal surface accuracy Ra=20.012μm of 3YSZ-based ceramic after sintering at 1300 ℃ for 60 min. Fig. 4 shows the 3YSZ-based ceramic sample with optimal process, the surface morphology and ultra-depth 3D microscope images.

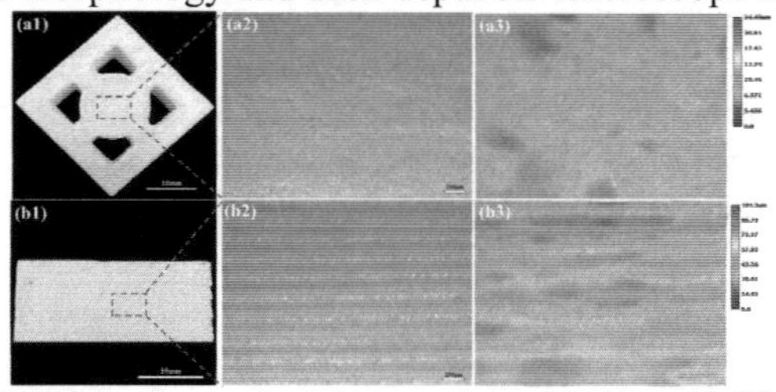

Fig. 4 3YSZ-based ceramic: (a) top surface; (b) side wall; (1) picture; (2) morphology; (3) ultra-depth 3D microscope image

4 Conclusion

When the SiO_2 content was 8.4wt.%, the linear shrinkage rate of the 3YSZ-based ceramic sample was 15.66% and the bending strength was 208 MPa after sintering at 1300 ℃. The parameters influenced the fabrication process in the order as: nozzle diameter > layer height / Nozzle diameter > print speed. The 3YSZ-based ceramic with good quality were obtained.

Acknowledgments

This project has been financially supported by the National Natural Science Foundation of China (No.52205365, No.52275334 and No.51775204).

References

[1] Fabris D, Mesquita-Guimarães J, Pinto P, et al. Mechanical properties of zirconia periodic open cellular structures. Ceramics International, 2019,45(13):15799-15806.

[2] Yang Z, Yang L, Fan Z, et al. Effect of sintering aid combined vacuum infiltration on the properties of Al_2O_3-based ceramics via binder jetting. Additive Manufacturing, 2024, 79: 103898.

[3] Yan S, Chen Y, Li D, et al. Mechanically robust, transparent, conductive hydrogels based on hydrogen bonding, ionic coordination interactions and electrostatic interactions for light-curing 3D printing. Chemical Engineering Journal, 2024, 486: 150289

[4] Mooraj S, Welborn S S, Jiang S, et al. Three-dimensional hierarchical nanoporous copper via direct ink writing and dealloying. Scripta Materialia, 2020, 177: 146-150.

[5] Cao Y, Li C, Ma Y, et al. Mechanical properties and thermal conductivities of 3YSZ-toughened fully stabilized HfO_2 ceramics. Ceramics International. 2019, 45(10): 12851-12859.

Quasi-static Compression Behavior of Bio-inspired Hybrid Lattices with Non-periodicity Prepared by a Rapid Casting Method

Jiawen Li[1], Yu Bai[1], Endian Liu[1], Hai Hao[1,2*]
1. Key Laboratory of Solidification Control and Digital Preparation Technology (Liaoning Province), School of Materials Science and Engineering, Dalian University of Technology, Dalian, 116024, China
2. Ningbo Research Institute of Dalian University of Technology, Ningbo 315016, China
*Corresponding address: e-mail: haohai@dlut.edu.cn

Abstract: Inspired by the cross-lamellar structures commonly found in mollusks, a strengthening concept by adding crossed struts to the adjacent units in a cubic lattice structure was proposed, for the purpose of achieving strength gain by simple structure design. In this study, lattice structures were prepared by a rapid casting combing additive manufacturing and infiltration casting Quasi-static compression experiments were conducted to investigate the crushing behavior and mechanical performance of the non-periodic hybrid lattice (NPHL)with crossed struts. In addition, numerical simulations were employed to illustrate the stress distribution inside the structure under different strains. The anisotropy of NPHL in the elastic deformation range was investigated using the asymptotic homogenization method and validated by numerical simulations. Compared with the typical cubic lattices, the NPHL structures have a reduced tendency of loaded struts to buckle, altered crack propagation paths in the cubic lattice, and crushing behavior. At the same time, the NPHL structure has higher compressive strength, specific strength, and specific energy absorption.

Keywords: Non-periodic Hybrid Lattice, Rapid casting, Crossed struts, Energy absorption

1 Introduction

Porous materials are capable of possessing customizable properties due to their internal topology, and many works have investigated the structure-property relationship from microscopic to macroscopic perspectives with the help of experimental, numerical, and simulation tools. Many complex topologies are difficult to construct due to the limitations of traditional preparation processes, which limits the applications and research on porous materials[1]. The emergence of additive manufacturing technology provides great design space for selecting substrate materials and preparing complex configurations. Porous materials can be categorized into ordered and disordered based on their structure, and ordered porous materials have become the focus of research because of their superior mechanical properties. Porous materials are called lattice materials because of their periodic arrangement and structural characteristics in three-dimensional space [2]. Lattice materials are often put to use as mechanical metamaterials due to their excellent specific strength and counter-intuitive special properties that can be realized by topological design. The properties of lattice materials depend on the structural design of the representative volume element (RVE), and the initial design strategy was the body-centered cubic, face-centered cubic, and simple cubic crystal structures common in crystals[3]. Subsequently, the design strategy shifted to hybridizing multiple lattice structures or making simple topological changes to optimize performance further. With the development of bionics, new ideas have been given to the design strategy of lattice structures. In recent years, RVEs can be composed of pillars, plates, or shells [4], which are then periodically and repeatedly arranged in three dimensions to build lattice metamaterials. It is worth noting that the scale of RVEs can reach the micrometer level, which makes lattice materials have special properties that cannot be found in solid materials [5]. However, existing design strategies limit the topological transformation to a single unit cell, which greatly restricts the enhancement of the mechanical properties of lattice materials. In this study, three crossing patterns and two pillar placement strategies were designed based on the common cross-laminated structures in mollusk shells, three hybrid lattice structures were designed and samples were prepared and evaluated for their mechanical properties using a combination of selective sintering and percolation casting processes.

2 Experimental procedure

In this study, geometric modeling was carried out using Creo Parametric 3D modeling software. The sample preparation process is shown in Fig. 1: the samples were prepared using an indirect additive manufacturing technique that combines selective sintering (SLM) and percolation casting processes. The raw material was ZL111 alloy (2.71 g·cm^{3,} elemental composition: 8.0-10.0 Si, 1.3-1.8 Cu, 0.4-0.6 Mg, 0.1-0.35 Mn,0.1-0.35 Ti, wt%. Bal. Al). The raw material used for additive manufacturing was coated resin sand (75-250 μm). The prefabricated body required for the casting process was first prepared using additive manufacturing technology, and then molten aluminum at 750°C was poured into the prefabricated body that had been preheated at 200°C.

Subsequently, in order to eliminate casting defects, the samples were heat-treated at 400 °C for two hours. Finally, the samples were desanded to obtain the designed lattice material.

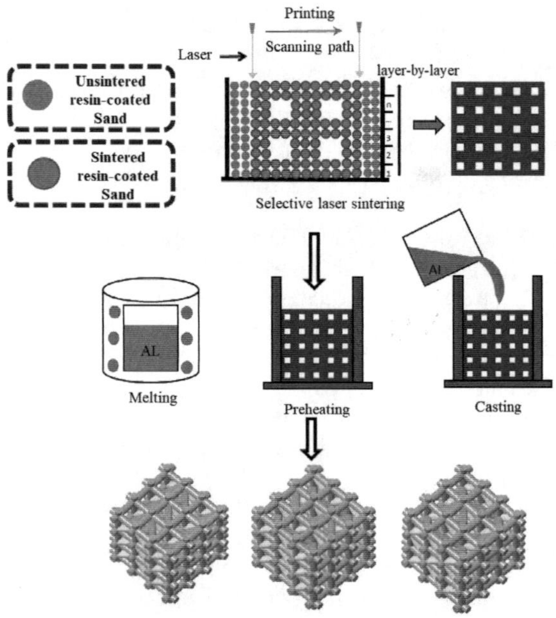

Figure 1 Schematic diagram of preparation process

Table 1 The elastic modulus E, yield stress σ_y, density ρ of specimens

Samples	E (MPa)	σ_y (MPa)	ρ ($g \cdot cm^{-3}$)
A	671.87	17.17	0.75
B	649.46	16.12	0.77
C	729.8	21.17	0.76
SC	444.62	10.54	0.48

3 Result and discussion

The stress-strain curves of the three samples are shown in Fig. 2. It can be seen that the compressive strength of the structure can be effectively enhanced by placing intersecting pillars within and on the lattice body. The three crossing strategies have similar enhancement of the compressive strength of the samples, but the enhancement is more than 70%.

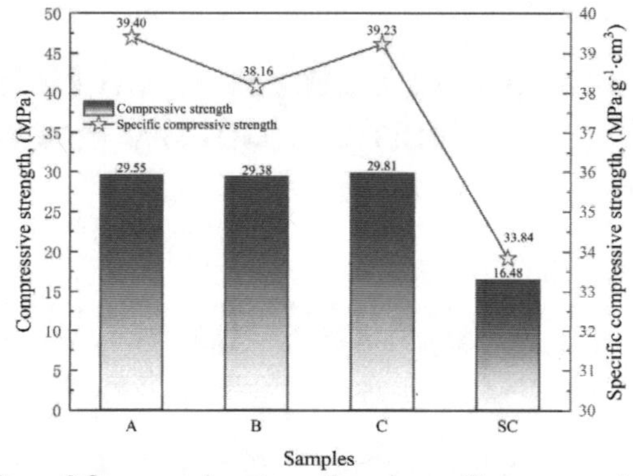

Figure 3 Compressive strength and specific compressive strength of samples

4 Conclusion

The non-periodic lattice structures formed by placing cross-pillars inside the simple cubic lattice have excellent compressive strength, and the strength enhancement effect is significant, with all three cross-strategies achieving strength enhancement of more than 70% for the SC lattice and specific compressive strength enhancement of more than 10%, with a maximum of 16%. In addition, the yield strength of Sample C increased by 68.82% and the modulus of elasticity increased by 64.14% compared to SC.

Acknowledgments

The work was supported by the National Natural Science Foundation of China [52171030], and the National Key Research and Development Program of China [2018YFA0702903]

References

[1] Wu, Y., et al., Additively manufactured materials and structures: A state-of-the-art review on their mechanical characteristics and energy absorption. International journal of mechanical sciences, 2023. 246: p. 108102.

[2] Wang, P., et al., Bio-inspired vertex modified lattice with enhanced mechanical properties. International journal of mechanical sciences, 2023. 244: p. 108081.

[3] Li, L., et al., A new hybrid lattice structure with improved modulus, strength and energy absorption properties. Science China. Technological sciences, 2023. 66(7): p. 2119-2133.

[4] Zhong, H.Z., et al., Ultralight, ductile metal mechanical metamaterials with super elastic admissible strain (0.1). Journal of Materials Science & Technology, 2023. 162: p. 227-233.

[5] Yu, X., et al., Mechanical metamaterials associated with stiffness, rigidity and compressibility: A brief review. Progress in Materials Science, 2018. 94: p. 114-173.

Applications of 3D Printing Technology in Sand Casting

Qingwen Chen*, Garrison Zong
3D Printing Technology, Inc., Beijing, China
*Corresponding address: e-mail: qingwen.chen@3dptek.com

Abstract: Compared with the traditional casting process, 3D printing technology aided sand casting has advantages in terms of efficiency, costand quality in high-end complex structural parts casting. This paper details the use of a 3D printing composite process that combines an SLS resin coated sand core with 3DP resin sand shell to create a combined sand mold for casting key components of the aero-engine, and produced casting products of complex structural parts with large thin-walled, complex small pipelines and other complex structures. And the use of 3DP sand mold combined with low-pressure casting technology and advanced solidification control process, produced the casting of key components of an underwater vehicle power unit with large-size thin-walled structure and complex runners. It further discusses the impact of 3D printing on the cost, quality, environmental protection, and other aspects in the application of sand casting. It was found that printing process parameters affect the cost of sand molds, for example, increasing the thickness of the printing layer can significantly reduce the printing cost.

Keywords: 3D Printing, Sand Casting, Precision Casting, High-end Complex Structure Casting, SLS, 3DP, EP, Intelligent

1 Introduction

3D printing technology is a multidisciplinary technique based on digital models, integrating laser technology, inkjet technology, and more. It uses filament or powder materials to construct objects directly through the layer-by-layer stacking of materials without the need for molds. In the field of high-end complex structural parts casting, the integration of 3D printing technology with sand casting technology has led to the development of rapid sand mold casting technology, which has shortened the production cycle of products, with saving development time and costs, and enhancing the performance and quality of products, and thus promoting the transformation and upgrading of the high-end complex product manufacturing industry.

1. The Current State of 3D Printing Technology

High-end complex structural parts have a wide range of application demands across various fields. In the field of high-end complex structural part casting, traditional casting processes often face many challenges, such as long mold production cycles, high costs, and difficulty in adapting to the production of increasingly complex structural parts. 3D printing sand mold casting technology provides effective solutions to these problems and shows significant advantages. Therefore, this technology has been widely applied in the fields of new energy vehicles, energy power, aerospace, and so on. In recent years, the key technologies of 3D printed sand molds, Selective Laser Sintering (SLS) and Three Dimension Printing (3DP) have begun to be applied on a large scale in sand mold casting production services.

2. Introduction to 3DP Technology and Application Cases

This papaer elaborates on the principles of 3DP technology and the sand mold casting process workflow, comparing to the conventional sand mold casting process, to demonstrate the efficiency and cost benefits of 3D printed sand mold technology in the casting of high-end complex structural parts. It showcases an application example of 3DP casting technology in the propulsion devices of modern underwater vehicles, with large dimensions, thin walls, ribs and protrusions, and intricate curved flow channels, which present challenges to traditional casting techniques. This paper delves into the use of 3DP sand printer to construct casting sand molds, allowing the flow channels to conform to more complex curved surface designs. Utilizing low-pressure precision casting technology, a gap-style casting system was thoughtfully designed, and state-of-the-art solidification control techniques were employed, enabling the casting of large-sized thin-walled structures and complex flow channels with an overall dimension exceeding 2 meters and a wall thickness of 5-7mm.

3. Introduction to SLS Technology and the Application of SLS-3DP Composite Process

This papaer outlines the working principle of SLS technology and the process flow of sand mold casting, followed by an introduction to the composite sand mold casting process that integrates both SLS and 3DP technologies. It then elaborates on the application of this composite technique in the casting of casing components for aero-engines. Traditional casting methods encounter considerable challenges when manufacturing large, thin-walled, and intricate structures with fine channels. The adoption of a composite process involving SLS and 3DP printing has become an innovative and efficient solution. Utilizing SLS technology, the laser sintering of specialized coated sand forms sand cores with dimensional precision of ±0.1mm, and the tensile strength in specific areas exceeds 4MPa, capable of creating complex structures with fine channels. For sand shells that are less complex but large scale, the 3DP technology uses inkjet bonding of silica sand, facilitating lightweight design, and ensuring high molding efficiency and reduced costs while maintaining quality.

5. Analysis and Summary

In the traditional sand mold casting process, especially in the fields of high-end complex structural parts and large structural parts, the mold-making cycle is long and the costs are high. 3D printing sand mold technology provides a completely new solution for the manufacturing of casting sand molds and demonstrates significant advantages. The advantages of 3D printing technology in the application of sand mold casting in terms of cost, efficiency, quality, intelligence, and environmental protection were also discussed.

6. Conclusion and Prospects

The application of 3D printing sand mold technology in the casting industry has propelled the development and innovation of traditional casting techniques, bringing more innovation and breakthroughs to the sand mold casting field. It has facilitated the transformation and upgrading of the traditional casting industry and plays a significant role in the intelligent, green, and high-end development of the casting industry. With the continuous research and development of new 3D printing technologies, materials, and processes, as well as the integration of advanced manufacturing technologies such as digitalization and intelligentization, rapid sand mold casting technology will continue to reduce costs, improve quality, and increase efficiency. The application prospects of 3D printing technology in the sand mold casting field will be even broader.

Research on the Development of Melting and Continuous Casting Technology of Feedstock Materials based on Modified Copper Alloys, Intended for the Use in 3D WAAM Additive Manufacturing Technology

J. Kulasa*, M. Maleta, R. Drajewicz, B. Cwolek, A. Brudny
Łukasiewicz Research Network – Institute of Non-Ferrous Metals, Sowińskiego 5, Gliwice, Poland
*Corresponding address: joanna.kulasa@imn.lukasiewicz.gov.pl

Abstract: Due to the continuous development of Additive Manufacturing Technologies and the search for new material solutions in this area, research was undertaken to develop a technology for producing feedstock materials based on copper alloys for the use in the WAAM (Wire Arc Additive Manufacturing) Additive Manufacturing Technology. The research in this work focused on modifying the chemical composition of commercial copper alloys, i.e. aluminum-nickel bronzes and cupronickel bronzes intended for operation in corrosive conditions and showing good mechanical and welding properties. The modification of the chemical composition of the selected alloys was aimed at producing feedstock materials in the form of rods enabling the production of wires with a target diameter of 1 mm, as well as the possibility of obtaining 3D printed products with improved mechanical properties while maintaining an appropriate level of corrosion resistance. Modification of the chemical composition of selected copper alloys resulted in the need to develop melting and continuous casting technology for them, as well as wire drawing technology with inter-operational heat treatment. This work presents selected test results of rods cast using variable conditions of the continuous casting process for aluminum bronzes and cupronickel with modified chemical composition.

Keywords: melting and casting, continuous casting, copper alloys, feedstock material for 3D printing, Metal 3D Printing, Additive Manufacturing Technology, Wire Arc Additive Manufacturing

1 Introduction

Additive manufacturing offers building of components or products layer-by-layer using materials in powder or wire forms and depositing them over a defined toolpath of the given 3D design data. Wire Arc Additive Manufacturing is emerging as a new promising economical technique for metal additive manufacturing. Large volume custom-built metallic structures can be produced in relatively less time using WAAM techniques [1]. Additive Manufacturing (AM) technologies are developing dynamically and are becoming increasingly important in industry. The advantages of WAAM are rapid manufacturing of three-dimensional objects, low price of feedstock materials, and the possibility of implementation for commercial production. [2]. WAAM technology has proven that it can produce medium to large components because of its high deposition rate and potentially unlimited build size. Like all additive manufacturing technologies, however, an optimized process planning that provides uniform, defect-free deposition is key for the production of parts. Moreover, AM, particularly WAAM, is no longer just a prototyping technology, and most of today's attention is on its transformation to a viable and cost-effective production [3]. Therefore, WAAM technology still requires the search for new material solutions, including continuous development and modification of currently used feedstock materials.

2 Experimental procedure

Pure components were used for producing the alloys: cathode Cu, Al, electrolytic Ni and the master alloys: CuMn30, CuMn50, CuFe20, and CuTi20. The alloys being the input material for the continuous casting process were cast statically in the form of bars using the PIT10 and PIT100 crucible induction furnaces. The melting process was done in graphite crucibles and chamotte-graphite crucibles (PIT100). Gravity casting was carried out into a pre-heated cast iron ingot mould, obtaining an input material in the form of bars for the continuous casting process and in the form of Ø 50 mm ingots for further tests. Produced alloys were the input material for the continuous casting process. The process was carried out using the ConCasTech MINI SCC station for horizontal continuous casting for aluminum bronze alloys. The continuous casting process was carried out with the use of graphite double-strand crystallizer (2x Ø 8 mm). The loaded material in the form of bars was melted, and then, after stabilizing the bath temperature, the continuous casting process was started. The parametres of the process were: feed 10-12 mm, feed speed 20 mm/s with a reverse feed equal to 1 mm. During the process, the temperature of the casting surface at the exit from the crystallizer was monitored using a contact temperature meter and the ThermaCAM SC640 thermal imaging system. Rods with a diameter of 8 mm made of the CuNi11Fe2Mn1.5Ti0.5 alloy were cast on a vertical VCC3000 continuous melting and casting system from Indutherm. The melting and casting system uses elements made of boron nitride.

The melting of the charge material was carried out after initially obtaining a vacuum and then under an argon flow. The continuous casting process of the alloy was conducted in the temperature range of 1280°C-1300°C, using a strand stroke in the range of 3 to 5 mm and a strand stop in the range of 1-1.5 s.

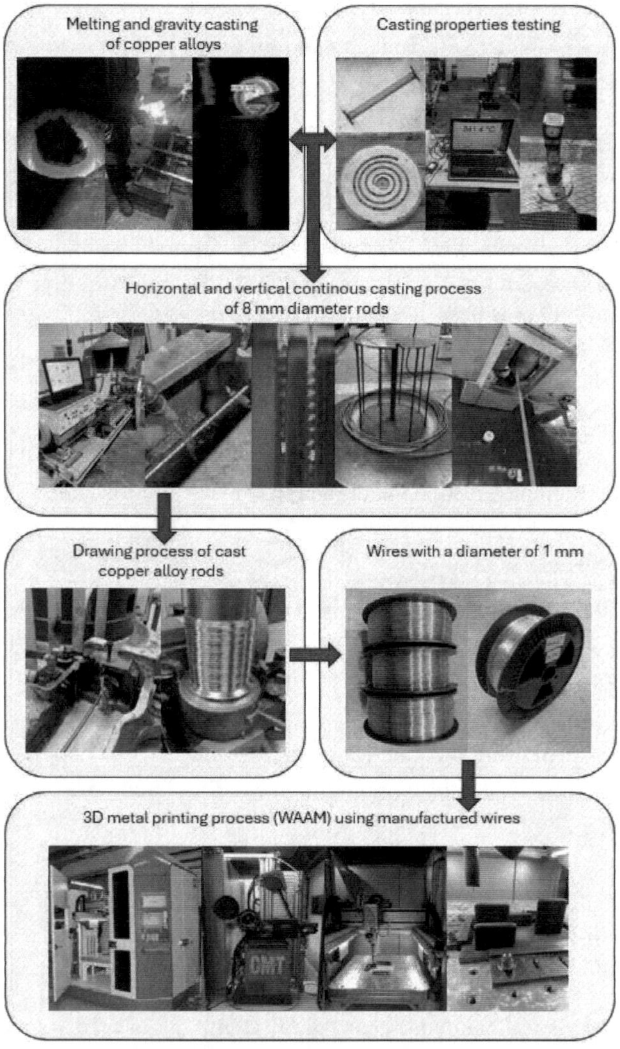

Fig.1. Production technology of the wires (feedstock materials intended for 3D Metal Printing process) and their verification by means of WAAM

3 Result and discussion

The developed technology of continuous casting of copper alloys enabled the production of high-quality, defect-free rods with a diameter of 8 mm, intended for further processing. The as-cast hardness of the CuNi11Fe2Mn1.5Ti0.5 alloy ranged from 153 to 160 HV10. The cast rods material was coiled and intended for further processing: plastic working by drawing with inter-operational heat treatment in a protective gas atmosphere

.

4 Conclusion

The production technology of wires from multi-component alloys from aluminum bronzes and cupronickel group includes the process of melting and gravity casting of bars, continuous casting of semi-finished products in the form of rods in a horizontal or vertical arrangement and plastic working with inter-operational heat treatment.

The conducted technological tests of continuous casting have shown that the use of the process parameters: feed 10-12 mm, feed speed 20 mm/s with a reverse feed equal to 1 mm for alloys from the aluminum bronze group (horizontal continuous casting) and strand stroke in the range from 3 to 5 mm and strand stop in the range from 1-1.5 s in the case of cupronickel (vertical continuous casting) enable the production of high-quality castings being input material for further processing, in this case, the production of high-quality wires as a feedstock material for WAAM 3D printing technology.

Acknowledgments

The research was carried out as part of the projects: The project 3DMPWire. Material-efficient Cu wire-based 3D printing technology, 2019-2021 (Upscaling) co-funded by EIT RawMaterials - project number: 18072, and the Implementation Doctorate Program in the Łukasiewicz Research Network - Institute of Non-Ferrous Metals financed by a subsidy from the Ministry of Education and Science of the Republic of Poland.

Supported by
eit RawMaterials Connecting matters
Co-funded by the European Union

References

[1] Singh S., Sharma S., Rathod D.W. A review on process planning strategies and challenges of WAAM. Materials Today: Proceedings 47 (2021), 6564-6575.

[2] Maleta M., Kulasa J. Kowalski A., Kwaśniewski P., Boczkal S., Nowak M. Microstructure, Mechanical and Corrosion Properties of Copper-Nickel 90/10 Alloy Produced by CMT-WAAM Method. Materials 2024, 17, 50. https://doi.org/10.3390/ma17010050

[3] Jafari D., Vaneker T.H.J., Gibson I. Wire and arc additive manufacturing: Opportunities and challenges to control the quality and accuracy of manufactured parts. Materials and Design 202 (2021) 109471.

Micro-Extrusion 3D Forming and Anti-hydration Strengthening of Water-Soluble Calcia Ceramic Cores for Manufacturing Complex Hollow Castings

Yingpeng Mu[1,2], Bolin Hu[1,2], Hao Liu[1,2], Fuchu Liu[1,2*]
1. China University of Geosciences, Wuhan, China
2. Shenzhen Research Institute, China University of Geosciences, Shenzhen, China
*Corresponding address: e-mail: liufuchu@cug.edu.cn

Abstract: Poor forming accuracy and susceptibility to hydration swelling were bottlenecks in the casting application of micro-extruded 3D printed CaO ceramic cores. In this manuscript, three key process parameters, namely, the inner diameter of the needle (d), the layer height (h) and the printing speed (v), were optimized through response surface experiments, and the regression equations for surface roughness were established and analysis of variance (ANOVA) was carried out. The results showed that the main factor affecting the forming accuracy of the printed green bodies was the inner diameter of the needle, followed by the layer height, and the printing speed was the least affected. When d=0.41mm, h=0.31 mm, v=29.87 mm/s, the response value was the smallest and the billet indicated the best quality. The moisture absorption rate of CaO ceramic cores was tested by introducing Nano-ZrO_2 powder, and the results showed that when the addition of Nano-ZrO_2 was 10 wt.%, the moisture absorption rate of CaO ceramic cores decreased by 43.9%, and at the same time the shrinkage rate decreased by 21.5%, the flexural strength increased by 63.8%, and the water solubility rate in 60 ℃ water was 4.21 g/s·m^2, which was important for the rapid moldless. The SEM microscopic morphology and XRD physical phase analysis showed that the introduction of the additive generated a new phase substance $CaZrO_3$, which was uniformly distributed around the CaO experience to effectively reduced the moisture absorption rate of CaO. The research results were of great theoretical and practical significance for the rapid moldless accurate forming of hydration water-soluble CaO ceramic cores.

Keywords: CaO ceramic core,Extrusion 3D printing technology,Formability,Hydration resistance

1 Introduction

The structure of new generation aerospace, high-speed rail and new energy vehicle castings develops in the direction of complexity and integrality, the casting cores became very complex, and it is difficult to clean the inner cavity of the casting, the traditional core preparation process needed to form a number of small cores through the mold, and then obtained the whole core from the group, which was a complex process, high cost, and low precision[1,2]. The preparation of water-soluble CaO ceramic cores by micro-extrusion 3D printing technology can simultaneously solved the forming and cleaning problems of complex cores. However, the disadvantages of CaO ceramic cores, such as reduced strength due to moisture absorption and low extrusion precision, had seriously hindered their practical application[3]. In this manuscript, a three-factor, three-level Box-Behnken test was firstly conducted based on the response surface method with the inner diameter of the needle, layer height/inner diameter and printing speed. A quadratic regression model was established using the surface roughness of the printed ceramic green body, and the relationship between the three factors and the response value surface roughness was discussed by analysis of variance (ANOVA), and the interaction between the three factors was analysed to determine the optimal process parameter combinations. Then, additives were introduced to improve the anti-hydration properties of CaO, and the effects of different Nano-ZrO_2 additions on the linear shrinkage, flexural strength, moisture absorption and water dissolution rate of ceramic cores were investigated. The microstructural evolution of CaO samples containing different amounts of Nano-ZrO_2 additives was described by SEM microscopic morphology analysis and XRD physical phase analysis, and the optimal amount of Nano-ZrO_2 was determined based on the comprehensive evaluation of the anti-hydration and water solubility rate, which realized the unity of the anti-hydration and water solubility rate of the CaO ceramic cores. This study provided a new way to obtain high-performance, high-precision, anti-hydration water-soluble ceramic cores, which was of great theoretical and practical significance for industrial development.

2 Experimental procedure

1250 mesh $CaCO_3$ powder was used as the matrix material and Nano-ZrO_2 powder with an average particle size of 30 nm was used as the additive. A 30 wt.% solution of polyethylene glycol (Mw=4000 g/mol) was configured as a binder using deionised water as a solvent. Ceramic green bodies were micro-extruded and formed using an indigenously built printing device. Printing parameters were controlled and adjusted using Simplify 3D software. Green bodies printed first print substrate on the completion of natural drying and then transferred to a constant temperature drying oven to remove excess moisture on the surface. The fully dried green bodies were transferred to a high temperature sintering furnace, and the CaO ceramic cores are obtained after stepwise sintering to 1400°C. The CaO ceramic cores were then sintered at a constant temperature to remove excess moisture from the surface.

3 Result and discussion

According to the Box-Behnken design principle, a three-factor, three-level response surface experimental table was designed using the inner diameter of the needle (A), layer height/inner diameter (B) and printing speed (C) as the independent variables, and the surface roughness of the ceramic core green body as the response value Y.

By measuring the surface roughness of green bodies printed with different parameters, the experimental results were subjected to quadratic regression analysis to establish the regression equation for the surface roughness of ceramic core green bodies:

$$Y = 55.05 + 12.42A + 4.57B - 2.32C + 2.01AB + 0.69AC - 2.94BC + 3.16A^2 - 3.35B^2 - 4.75C^2 \quad (1)$$

After analysis of variance, factors A, B and C had a significant effect on the surface roughness of the green body in the order of A > B > C. The optimal parameter combination was obtained by analysing and solving the regression equation, and the surface roughness of the ceramic core green body was 35.41 μm when the A was 0.41 mm, the B was 0.75, and the C was 29.87 mm/s.

Due to the fact that CaO was highly susceptible to hydration expansion and pulverisation during storage, its mechanical properties and dimensional accuracy were affected. Therefore, its hygroscopic resistance was improved by introducing additives, and five groups of pastes were formulated with Nano-ZrO_2 content from 0-20% at 5% intervals. After micro-extrusion 3D printing, drying, and sintering, CaO ceramic cores were obtained and tested for performance, and the results showed that the moisture absorption rate of all five groups of ceramic cores increased with the increase of storage time and decreased with the increase of Nano-ZrO_2 content. In addition, the linear shrinkage of the ceramic cores decreased with the increase of the Nano-ZrO_2 content, while the bending strength showed a gradual increase. As can be seen from the XRD patterns, the diffraction peaks of the ceramic cores with the addition of Nano-ZrO_2 generate $CaZrO_3$ phase. From the SEM microscopic morphology, it can be seen that the CaO grains were surrounded by the formed $CaZrO_3$, which effectively inhibits the moisture absorption. The water solubility rate of the moisture-resistant reinforced cores was tested by placing them in different water temperatures, and the results showed that the water solubility could still be faster at water temperatures ranging from 60℃ to 80℃. In addition, the higher residual temperature during the water-soluble cleaning creates a good condition for the rapid removal of the ceramic core.

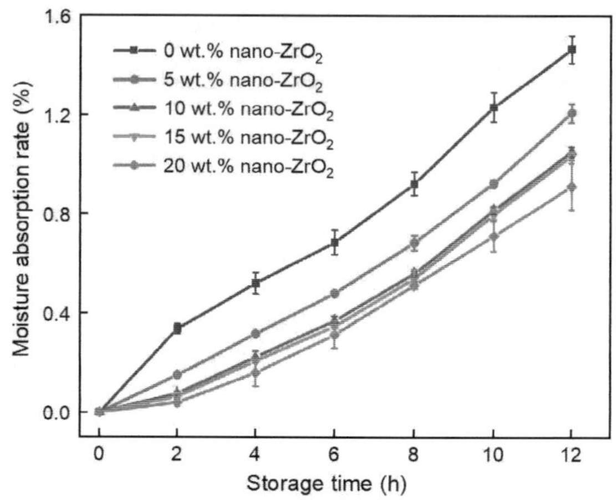

Figure 1 Moisture absorption rate of CaO core

4 Conclusion

In this manuscript, the process parameters of micro-extrusion 3D printing forming ceramic cores were optimized by response surface experiments, and the best surface quality of ceramic core green bodies with a roughness of 35.41 μm was achieved when the inner diameter of the needle was 0.41 mm, the layer height/inner diameter ratio was 0.75, and the printing speed was 29.87 mm/s. When the amount of NanoZrO_2 was added at 10%, the hygroscopicity of the ceramic cores with CaO decreased by 43.9%, which was due to the generation of hydrophobic phase $CaZrO_3$ after the introduction of the additive.

Acknowledgments

This work was supported by Guangdong Basic and Applied Basic Research Foundation [No.2024 A1515013258] and Free Exploration Basic Research Project of Local Science and Technology Development Funds Guided by the Central Government [No. 2021Szvup158].

References

[1] Fan J, Li Q and Jin F. High solid loading, low viscosity stereolithography 3D printing ceramic cores slurry.Ceramics International, 2023, 49(6): 7681-7689..

[2] Li Q, Hou W and Liang J. Controlling the anisotropy behaviour of 3D printed ceramic cores: From intralayer particle distribution to interlayer pore evolution. Additive Manufacturing, 2022, 58(6): 103055.

[3] Li X, Su H and Dong D. Enhanced comprehensive properties of stereolithography 3D printed alumina ceramic cores with high porosities by a powder gradation design. Journal of Materials Science & Technology, 2022, 131(6): 264-275.

Enhancing the Characteristics of 3D Printed Polymer Patterns in the Rapid Investment Casting Process

Bethala Sai Sujith[1], D. Benny Karunakar[1*], Thomas Mathew[2]
1. Mechanical & Industrial Engineering Department, Indian Institute of Technology Roorkee, Roorkee – 247667, India
2. Mechanical Engineering Department, COER University, Roorkee – 247667, India
*Corresponding address: e-mail: benny.karunakar@me.iitr.ac.in

Abstract: In the rapid investment casting process, the polymer patterns replace the conventional wax patterns, which are produced on 3D printing machines within a shorter period. Though the strength of the polymer pattern is good enough, they undergo larger expansion during melting compared to wax patterns, which would break the ceramic shell. Different infill patterns were considered while developing the 3D printed patterns, out of which Gyroid and Cubic infill patterns were found to have better compressive strength, compared to the other infill patterns.

Further, the surface roughness of a polymer pattern is always inferior to that of a wax pattern. Hence, an attempt was made to enhance the surface roughness of the polymer patterns by creating a thin wax coating around them by dipping them in a paraffin molten wax bath. Several experiments were conducted by varying wax bath temperatures, dipping time, etc. It was found that the wax bath's temperature of 69°C and a dipping time of 2 seconds could improve the surface roughness of the polymer pattern to a significant level.

Keywords: Rapid Investment Casting, Fused Deposition Modelling, Polymer Patterns, Surface Roughness, Wax Coatings

1 Introduction

In Rapid Investment Casting (RIC), the wax pattern is replaced with a 3D-printed pattern, which allows for more complex shapes and faster production times. The 3D printed pattern is then used to create a ceramic mould, which is then used to cast the metal part. This process is often used for prototyping and low-volume production runs, as it allows for rapidly producing complex metal parts without needing tooling. Though the strength of the polymer pattern is good enough, they undergo larger expansion during melting compared to wax patterns, which would break the ceramic shell. Further, the surface roughness of a polymer pattern is always inferior to that of a wax pattern. Hence, in the present work, some investigations were conducted to overcome these challenges.

2 Experimental procedure

Initially, polymer patterns were developed as per the geometry shown in Fig 1, using an Fused Deposition Modelling (FDM) machine (Make: Ultimaker).

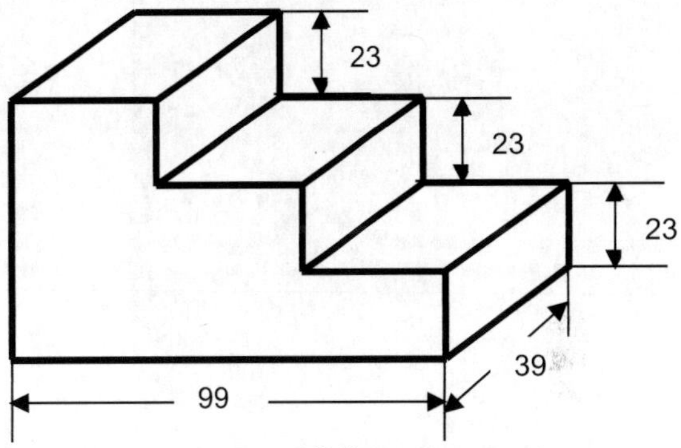

Fig. 1 Dimensions of the pattern

The raw material used in FDM is solid in the form of a wire roll. This chamber is surrounded by a heating coil which maintains the temperature of the chamber at a required level. At the entry of this chamber is a pair of wheels through which the raw material wire passes through into the chamber. At the exit, a nozzle of appropriate diameter is fixed. The nozzle diameters are varied at 0.4mm and 0.8mm.

Polymer patterns (using the materials PLA and ABS) were developed using FDM, varying the parameters like layer height, line width, infill line width, wall line width, top-bottom line width, print speed, infill speed, wall speed, and top-bottom speed travel speed. Cooling and support have been selected appropriately.

The ceramic slurry was prepared by mixing aluminum silicate with the binder (Ethyl silicate), along with some wetting agents and anti-foaming agents. The ingredients and the process parameters of the ceramic slurry are shown in Table 1.

Table 1 Process parameters of ceramic slurry

Coat	Slurry type	Stucco	Dip time (s)	Drain time (s)	Dry time (h)
1	Primary	Zircon 80/120	30	60	24
2-5	Secondary	Zircon 16/30	30	60	4
6	Outer	N/A	30	60	24

The ceramic shells developed are shown in Fig. 2. These shells were fired inside an oven and the polymer pattern material was removed from the ceramic shell. Molten metal was then poured into the dried ceramic shell and the casting was taken out after the solidification.

Fig. 2 Ceramic shells during drying

Influence of Pattern Interior

The problems with the solid pattern are wastage of material and cracking of ceramic shell due to excessive thermal expansion of polymer while melting. Hence, hollow patterns were printed initially at 90% infill and gradually the infill was reduced to 10%.

Influence of wax coating

The 3D printed polymer patterns have an inadequate surface finish compared to the wax patterns. Hence, an attempt was made to enhance the surface roughness of the polymer patterns by creating a thin wax coating around them by dipping them in a paraffin molten wax bath. Several experiments were conducted by varying wax bath temperatures, dipping time, etc.

3 Results and discussion

Influence of pattern interior:

Even at 10% infill, the patterns were found to be suitable for foundry requirements without compromising factors like strength, surface finish, etc. The different hollow pattern interiors are shown in Fig. 3.

Influence of wax coating:

A thin coating with wax could successfully improve the surface roughness of the polymer pattern.

Fig. 3 Pattern interiors used in the present study

4 Conclusions

By reducing the infill of the pattern material, the thickness of the ceramic shell could also be reduced by reducing the number of coatings around the pattern, which in turn would optimize the strength and stability of the ceramic shell.

A thin wax coating around the polymer pattern could successfully improve its surface roughness. A wax bath's temperature of 69°c and a dipping time of 2 seconds could improve the surface roughness of the polymer pattern to a significant level.

References

[1] Cheah C M, Chua C K, Lee C W, Feng C and Totong K. Rapid prototyping and tooling techniques: A review of applications for rapid investment casting. Int J Adv Manuf Technol, 2005, 25: 308–320.

[2] Chua C K, Feng C, Lee CW and Ang G Q. Rapid investment casting: direct and indirect approaches via model maker II. Int J Adv Manuf Technol. 2005, 25: 26–32.

[3] Kumar P and Singh R. Application of fusion deposition modelling for rapid investment casting—A review. Int J Mater Eng Innov, 2012, 3: 204–27.

[4] Volpato N, Foggiatto J A and Schwarz D C. The influence of support base on FDM accuracy. Rapid Prototyping J, 2014, 20: 182–191.

[5] Tong K, Joshi S and Lehtihet E A. Error compensation for fused deposition modeling (FDM) machine by correcting slice files. Rapid Prototyping J, 2008, 14: 4–14.

Research on Slurry Extrusion-based Additive Manufacturing Process for Bauxite Shell Mould

Hongjun Liu*, Yamin Li, Guoling Su, Junhong Wang, Jiao Liu
State Key Laboratory of Advanced Processing and Recycling of Non-ferrous Metals, Lanzhou University of Technology, 287 Langongping Road, Lanzhou, 730050 Gansu, China
*Corresponding address: hongjun_liu@hotmail.com

Abstract: Directly fabricating casting shell mould through additive manufacturing technologies can simplify the process, shorten the manufacturing time, and enhance the rapid manufacturing and trial production capabilities of castings. A slurry extrusion-based additive manufacturing process for casting shell mould was proposed in this paper, which can fabricate green shell parts by extruding and stacking bauxite slurry. An additive manufacturing equipment with Delta mechanism was designed and manufactured, with a recommended extrusion speed of no more than 100 mm/s, a ratio of feeding speed to extrusion speed of 1.0, and a layer height of 80% of the nozzle diameter. By analyzing the rheological properties and extrusion formability of the slurry, it is recommended to use 42.5 vol.% bauxite and 51 vol.% silica sol in the slurry, with a yield stress range of 300-400 Pa for the best forming effect. When the infill pattern is the honeycomb pattern and the filling density is 80%, the bending strength, porosity, and shrinkage are 11.62 MPa, 41.6%, and 1.9%, respectively, and the thermal conductivity is also good. Finally, the feasibility of this additive manufacturing process was verified through the preparation of sample shell mould and their trial casting.

Keywords: slurry extrusion; additive manufacturing; shell mould; bauxite

1 Introduction

Investment casting is a precision casting process used to create complex and intricate metal parts. It is used in industries such as aerospace, automotive, jewelry, and medical devices. Additive manufacturing technologies can directly fabricate shell mould for investment casting, eliminating the wax pattern preparation, repeated wet dipping and stuccoing, and wax removal processes in traditional investment casting [1]. It can greatly simplify the process, shorten the manufacturing time, and enhance the rapid manufacturing and trial production capabilities of castings. The additive manufacturing technologies used for shell mould mainly include SLS (Selective Laser Sintering), SLA (Stereolithography), 3DP (Three Dimensional Printing), and slurry extrusion-based additive manufacturing. SLS uses a mixture of organic binder and ceramic powder; SLA uses a mixture of UV laser cured photosensitive resin and ceramic powder; 3DP selectively sprays organic binders onto ceramic powders. There are many organic binders in the moulds prepared by these methods, and post-treatment may have a negative impact on the environment. The slurry extrusion-based additive manufacturing uses aqueous ceramic slurry with shear thinning characteristics to prepare shell mould [2], which is simple in process, suitable for a variety of materials, and environmentally friendly. The main problems faced are the specialized equipment, materials and supporting processes. It is necessary to further research on the equipment, ceramic slurry, preparation process, and shell mould post-treatment, to promote the development and application of the slurry extrusion-based additive manufacturing process.

2 Experimental procedure

Firstly, the CAD model of the shell mould is designed and then the STL file is output. G codes are generated through the slicing software. The bauxite slurry is extruded from a nozzle that moves according to the G codes, and the extruded filaments are stacked layer by layer to obtain a green part, which has a wall structure composed of the outline layer and the internal infill. Finally, the green part is vacuum dried for 24 hours in a freeze dryer, and then sintered at 1100 ℃ for 4 hours in a resistance furnace.

The composition of ceramic slurry includes 40-50 vol.% bauxite powders, 45-55 vol.% silica sol, 3.5-4.5 vol.% sodium polycarboxylate, 3 vol.% glycerol, and 1 vol.% PEG400. The mixture of silica sol, sodium polycarboxylate, glycerol, PEG400 is prepared by magnetic stirring for 20 minutes. Then the bauxite powders are added to the mixture in batches during the ball milling process. After ball milling for 12 hours, the viscosity of the bauxite slurry is adjusted with an acidic thickener, and then subsequent experimental tests and fabrication of shell mould can be carried out.

3 Result and discussion

Additive manufacturing equipment

A slurry extrusion-based additive manufacturing equipment for shell mould preparation was developed at Lanzhou University of Technology, as shown in Figure 1. It adopts a Delta mechanism composed of a fixing plate and three groups of guide pillars, pulleys and connecting rods, and has a relatively small footprint but a large motion range. The maximum workspace is 700 mm × 700 mm × 950 mm. A stepping motor in the push-rod feeding device provides pressure to push the slurry in the barrel to be transported to the extrusion head. The screw in the extrusion head rotates to extrude the slurry from the nozzle and the extruded filaments deposit layer by layer.

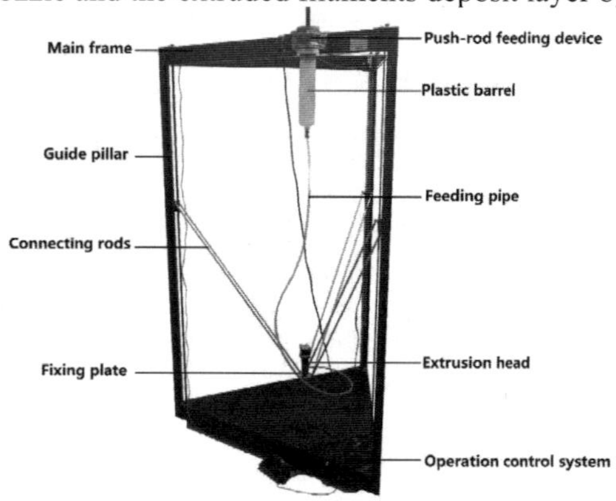

Figure 1 Slurry extrusion-based additive manufacturing equipment for shell mould

To balance preparation efficiency and part quality, the printing speed should not exceed 100 mm/s and not be less than 20 mm/s. When the feeding speed does not match the printing speed, the extrudate will accumulate or become too thin. Generally, the ratio of feeding speed to printing speed is set to 1.0. 80% nozzle diameter as the layer height can achieve a finer surface while ensuring strong interlayer bonding.

Bauxite slurry

The effects of composition on slurry properties and the extrusion process were analyzed. With the increase of silica sol content, the viscosity of the slurry after ball milling gradually decreases, and the average width of the extruded filament increases. Based on the principles of moderate viscosity and yield strength, as well as uniform and smooth extrusion, a silica sol content of 51 vol.% is selected. The optimized slurry composition also contains 42.5 vol.% bauxite and 4.1 vol.% sodium polycarboxylate. The rheological properties of bauxite slurry are important indicators determining the quality of additive manufactured shell mould. A suitable slurry should have shear thinning characteristics. The viscosity is very high at low shear rate, and the slurry can quickly form after extrusion; Viscosity rapidly decreases at high shear rate, reducing extrusion resistance. Through adjusting by the thickener, the yield stress of the slurry is within the range of 300-400 Pa. The slurry shows a very obvious shear thinning characteristic and can have good extrusion fluency and self-supporting ability.

Infill pattern

The porous shell mould with three infill patterns (rectilinear, grid, and honeycomb) was prepared using bauxite slurry, and the effects of the infill pattern on properties were evaluated. The different filling paths of the infill patterns lead to changes in the stacking form and connections of the extruded filaments, resulting in differences in the number and shape of pores formed in the wall, which affects the properties of the shell moulds. The samples prepared with the honeycomb pattern have the highest bending strength (11.62 MPa) and porosity (41.6%), as well as good heat-transfer ability, with an average shrinkage rate of 1.9%. The honeycomb pattern has more comprehensive advantages compared to the other two infill patterns.
A trial casting was carried out for shell mould prepared by the slurry extrusion-based additive manufacturing process. Figure 2 shows an example of additive manufactured shell mould and trial cast part.

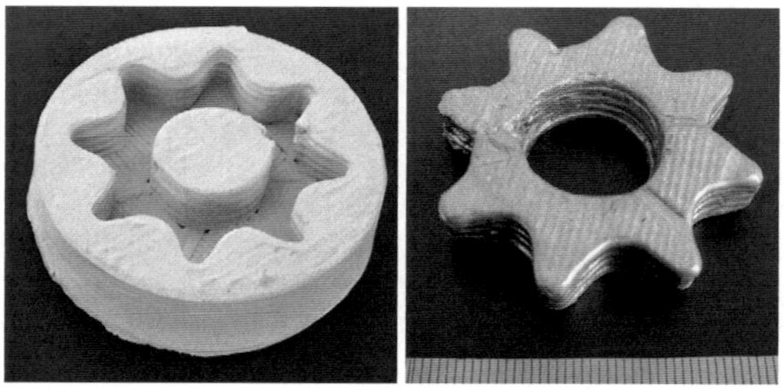

Figure 2 Prepared shell mould and trial cast part

4 Conclusion

The shell mould can be innovatively designed and fabricated using the slurry extrusion-based additive manufacturing developed in this paper. This work provides an attractive feasibility for fabricating shell mould with complex shapes, lightweight, high-strength and controllable hierarchical porous walls.

Acknowledgments

The authors acknowledge the financial support by the Key Science and Technology Project of Gansu Province (18YF1GA064).

References

[1] Tang S, Yang L., Fan Z., et al. A review of additive manufacturing technology and its application to foundry in China. China Foundry, 2022, 18(4): 249-264.
[2] Liu H, Su G and Li Y. Effect of Wall Structure on the Dimensional Accuracy of Shell Mould Prepared by Slurry Extrusion-based Additive Manufacturing Process. International Journal of Cast Metals Research, 2022, 35(4): 102-110.

Impact of Gas Pressure on Atomization Process and Particle Size of the Fe-Based Powder

Shuwen Guan, Shichang Liu, Zhen Liu, Yanpeng Wei, Zhiquan Miao, Bo Yu *
State Key Laboratory of Advanced Casting Technologies, Shenyang Research Institute of Foundry Co., Ltd. CAM, Shenyang 110022, China
*Corresponding address: e-mail: yub@chinasrif.com

Abstract: Gas atomization is a commonly employed method for producing metal powder in additive manufacturing. This paper utilizes computational fluid dynamics to simulate the flow field and analyze the gas-melt two-phase mechanism in two-stage atomization, based on experimental data. The study explores the influence of gas pressure on the gas flow field, the gas-liquid field, and particle size distribution. The findings suggest that the powder required for Laser Melting Deposition can be produced at an atomization pressure of 4 MPa, yielding the highest powder output. Additionally, an increase in gas pressure results in a decrease in the particle size of the powder. The Volume of Fluid model and Discrete Phase Model effectively simulate the gas atomization process, with the simulated particle size distribution closely matching the experimental results. The experimental Fe-based alloy powder exhibits good sphericity with minimal satellites, rendering it suitable for laser additive manufacturing with metallic materials.

Keywords: gas atomization, metal powder, numerical simulation, gas pressure, atomization process

1 Introduction

High-quality, low-cost metal powder is the fundamental material for rapidly producing parts. Gas atomization is the predominant method for producing metal powders in industrial settings [1]. The entire atomization process encompasses the supersonic flow of inert gas and the breakup of high-temperature melt, and the atomization mechanism is not yet fully understood [2]. Therefore, this study aims to examine the impact of gas pressure on the atomization process and the particle size distribution of the powder through a combination of simulation and experimental methods.

2 Experimental procedure

Fig. 1 shows the computational domain and boundary conditions. Argon was used as the atomizing gas and was defined as an isentropic, compressible, ideal gas. Fe-based alloy was chosen as the atomizing material. The VOF (Volume of Fluid) multiphase flow model was used to simulate the primary atomization, and the DPM (Discrete Phase Model) was used to simulate the secondary atomization. The experiments were conducted using a close-coupled discrete-jet atomizer within a Vacuum Induction Melting Inert Gas Atomization (VIGA) system.

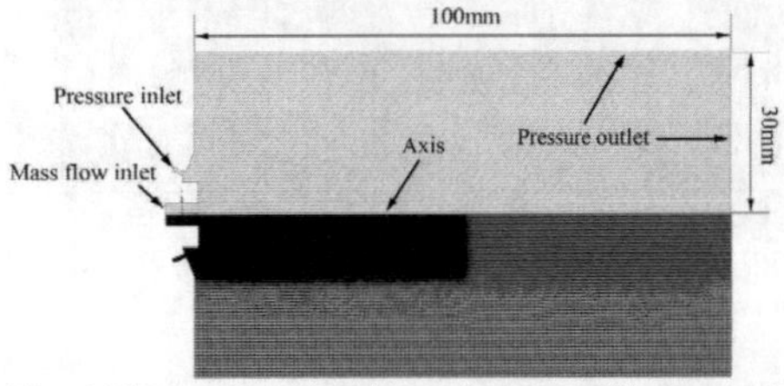

Fig. 1 The model size, boundary conditions, and grid for simulating the atomization process.

3 Result and discussion

Impact of gas pressure on the gas flow field

Fig. 2 illustrates the velocity field, directional vector, axial velocity, and static pressure at different gas pressures. The position of the Mach disk in the flow field gradually shifts to the right with increasing atomization pressure, leading to an increase in the length and width of the supersonic expanding gas flow.

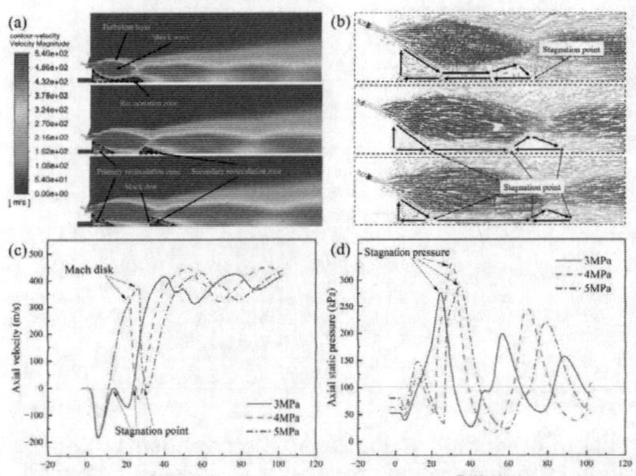

Fig. 2 The velocity field contour (a) and directional vector (b) at different gas pressures, while figures (c) and (d) illustrate the effects of gas pressure on axial velocity and static pressure.

Primary atomization and secondary atomization

Fig. 3 illustrates the primary atomization process under varying gas pressures. Notably, at 4 MPa, the melt requires less time to break under the influence of high-speed airflow compared to 3 MPa, thus contributing to the primary breaking process. At 5 MPa, the melt moves upward due to reverse airflow, leading to radial flow and subsequent breakage into smaller droplets, albeit with a risk of backflow.

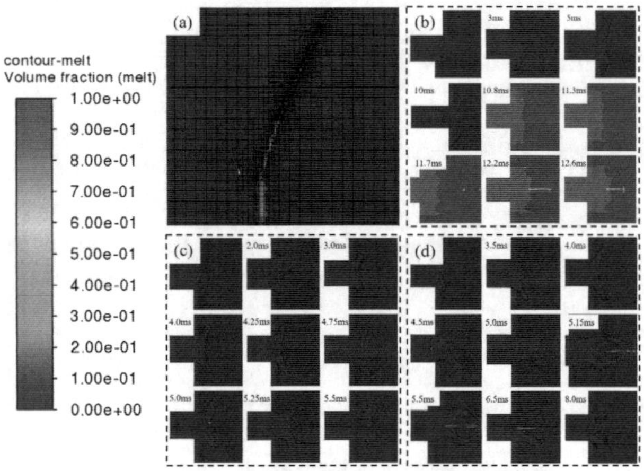

Fig. 3 Adaptive grid schematic at the gas-liquid interface (a) and the impact of gas pressure on primary atomization. (b) 3 MPa (c) 4 MPa (d) 5 MPa

Fig. 4(a) illustrates the process of secondary atomization, while Fig. 4(b) to (d) depicts the particle size distribution at varying gas pressures. With an increase in atomization gas pressure from 3 MPa to 4 MPa and 5 MPa, the median particle size decreases to 76.79 μm, 71.12 μm, and 57.88 μm, respectively.

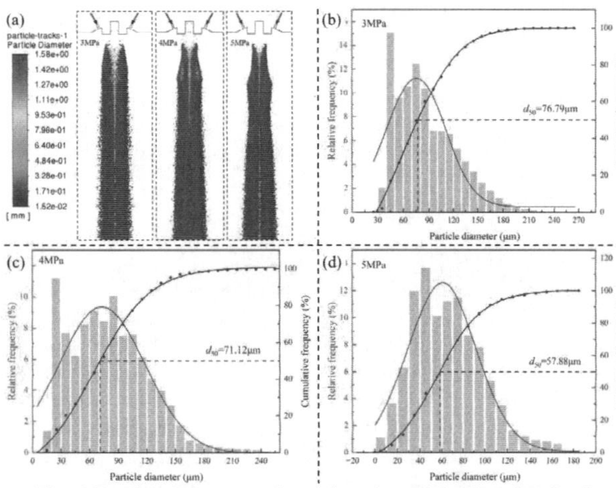

Fig. 4 Particle fragmentation (a) and particle size distribution during the secondary atomization process. (b) 3 MPa (c) 4 MPa (d) 5 MPa

Powder characteristics

The particle size distribution of the Fe-based powder obtained at various atomizing gas pressures is depicted in Fig. 5(a) to (c). As the gas pressure increases, the median particle size of the powder gradually decreases to 87.6 μm, 70.1 μm, and 53.9 μm, respectively. In comparison with the simulation results, the errors are 12.3%, 1.5%, and 7.4% respectively, indicating that the simulated particle size distribution aligns with the experimental results. The surface morphology of the powder is illustrated in Fig. 5(d). Overall, the powder exhibits good sphericity, with a small number of irregular and defective particles, such as coated particles and satellites.

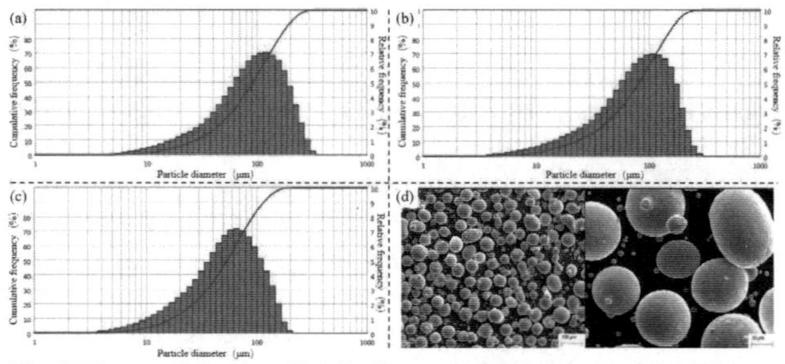

Fig. 5 The particle size distribution and SEM image of the Fe-based powder (d) produced by gas atomization. (a) 3 MPa (b) 4 MPa (c) 5 MPa

4 Conclusion

The simulation and experimental findings suggest that the powder necessary for Laser Melting Deposition can be generated at a pressure of 4 MPa, resulting in the highest yield. Furthermore, elevating the gas pressure leads to a decrease in the size of powder particles. The VOF and DPM models proficiently replicate the gas atomization process, and the simulated particle size distribution closely corresponds to the experimental results. The Fe-based powder produced by gas atomization demonstrates favorable sphericity with minimal satellite particles, rendering it suitable for laser additive manufacturing.

Acknowledgments

The authors gratefully acknowledge the financial support from Shenyang Research Institute of Foundry Co., Ltd. CAM. Technology Development Foundation (170222ZS03) and China Academy of Machinery Science and Technology Group Co., Ltd. Technology Development Foundation (170222ZY02).

References

[1] Anderson I E, White E M H, Dehoff R. Feedstock powder processing research needs for additive manufacturing development[J]. Current Opinion in Solid State and Materials Science, 2018, 22(1): 8-15.

[2] Shi Y, Lu W, Sun W, et al. Impact of gas pressure on particle feature in Fe-based amorphous alloy powders via gas atomization: Simulation and experiment[J]. Journal of Materials Science & Technology, 2022, 105: 203-213.

Rectification Effect of 3D-Printed Filters with Deflectors

Chaoke Li[1,2], Haoyuan Wang[1,2], Qian Feng[1,2], Kehui Hu[1,2,*], Zhigang Lu[2,3,4]
1. State Key Laboratory of Tribology in Advanced Equipment, Tsinghua University, Beijing 100084, China
2. Department of Mechanical Engineering, Tsinghua University, Beijing 100084, China
3. Key Laboratory of Advanced Materials Processing Technology, Ministry of Education, Tsinghua University, Beijing 100084, China
4. State Key Laboratory of Clean and Efficient Turbomachinery Power Equipment, Tsinghua University, Beijing, 100084, China
*Corresponding address: Department of Mechanical Engineering, Tsinghua University, Beijing 100084, China e-mail: hukehui@tsinghua.edu.cn

Abstract: A novel 3D-printed filter with integrated deflectors is proposed to enhance the state of metal liquid outflow. The rectification effect of filters with varying structural sizes was investigated through physical simulations. It was found that filters with smaller structural sizes allow the outflowing liquid to interact more readily under disturbance, promoting fluid convergence, but at the cost of reduced flow rates. With 3D printing, deflectors were integrated into a larger structural-size filter to guide fluid convergence, significantly enhancing the rectification effect and flow stability without reducing the flow rate. The filters proposed in this work allow for the decoupling of rectification and filtration functions, providing greater design flexibility for improving filter strength, filtration efficiency, and other critical performance attributes.

Keywords: ceramic filter; 3D printing, rectification; deflector

1 Introduction

Extreme application fields such as aerospace engines, gas turbines, and nuclear power plants present significant challenges to the performance of castings[1]. Ceramic filters play an important role in investment casting by filtering impurities, improving the flow state of molten metal, and controlling the flow rate, significantly impacting the performance of the final castings[2-4]. However, the rectification effect of ceramic filters has not received sufficient attention, as the structure of traditional foam ceramic filters can't be designed freely. The chaotic state of the outflowing molten metal could increase the contact area between the metal and air, raising the risk of oxidation and air entrapment, which would increase the number of defects in the final casting[5]. In this work, a novel 3D-printed filter with integrated deflectors is proposed to enhance the rectification effect, making the outflowing molten metal stable and uniform.

2 Experimental procedure

The physical simulation device is shown in Figure 1a. The basic structure of the filter is formed by periodically translating and stacking simple grids, as shown in Figures 1b and 1c. Water was used in place of molten metal to analyze the effect of the filter structure on flow, with blue dye added to make the flow contour visible.

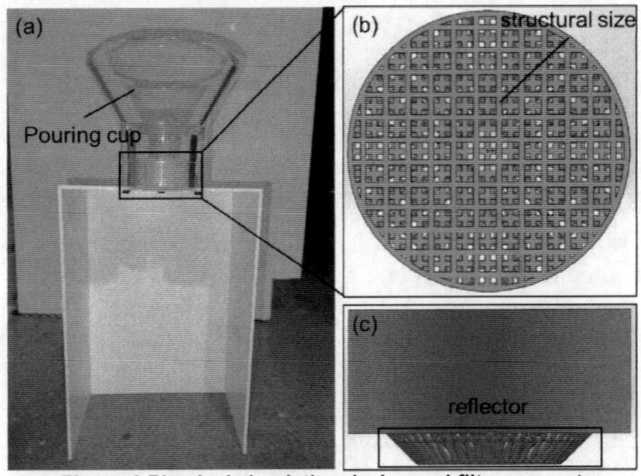

Figure 1 Physical simulation device and filter geometry.

3 Result and discussion

The effect of structural sizes

Filters with 1.5, 1, and 0.5 mm structural sizes were studied As the size decreases, the outflowing metal no longer forms separate liquid columns but shows some degree of convergence, as shown in Figure 2. The improved rectification effect is due to the smaller structural size allowing the outflowing liquid to more easily contact each other under perturbations, leading to convergence under surface tension, as visually recorded in Figure 3. However, because the outflowing liquid is not subjected to the force directing it toward the center, the convergence process primarily relying on perturbations is probabilistic and unstable. Figure 2b shows the convergence in the center, but a significant portion of the separated liquid columns, illustrating the instability of the convergence phenomenon. Additionally, in Figure 2c, the marked reduction in the diameter of the outflowing liquid indicates that fluid flow is significantly obstructed in smaller-sized structures.

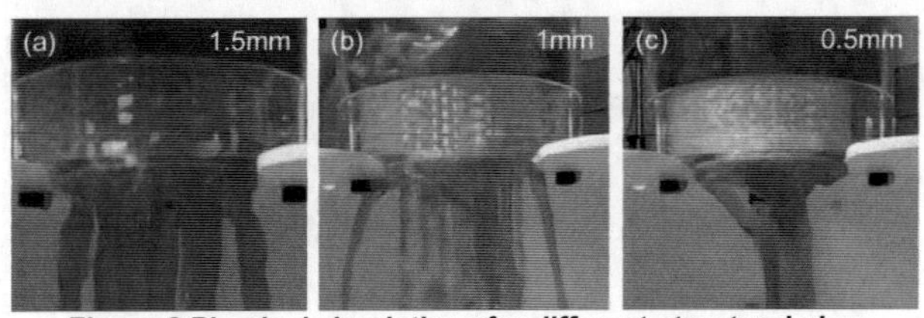

Figure 2 Physical simulations for different structural sizes.

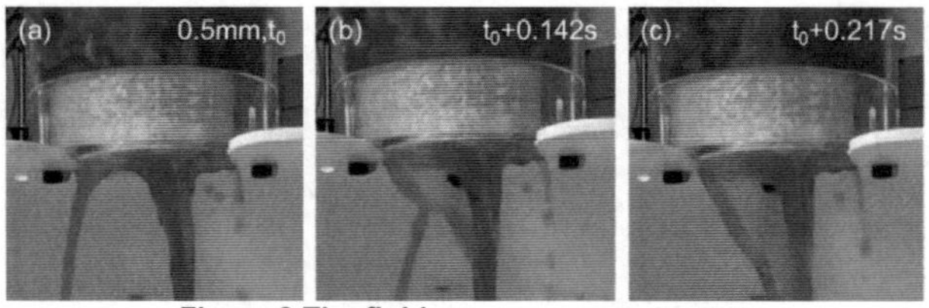

Figure 3 The fluid convergence process.

The effect of deflectors

Various deflectors were integrated into a filter with a 1 mm structural size. As shown in Figure 4, with the increased number and coverage area of the deflectors, the outflow liquid exhibited significant convergence. In Figure 4f, the outflow liquid is uniform, stable, and converged, without impact on the flow rate. Notably, the diameter of the converged outflow corresponds to the external diameter of the deflectors, which indicates that the coverage area of external deflectors determines the affected region. It is attributed to the filter's internal structure, which consists of rod-like elements similar to deflectors, allowing internal deflectors to affect only a limited region of the fluid flow. The deflectors provide the fluid with a centripetal force, enabling stable contact and subsequent convergence under surface tension, as shown in Figure 5. Through 3D printing, deflectors were seamlessly integrated into the filters, achieving a highly effective rectification outcome.

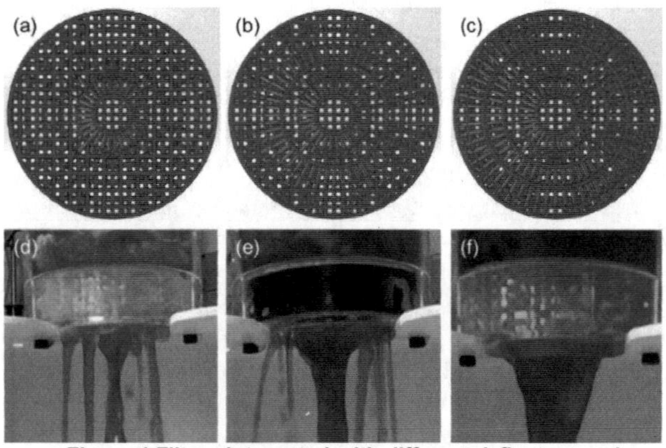

Figure 4 Filters integrated with different deflectors and corresponding physical simulations.

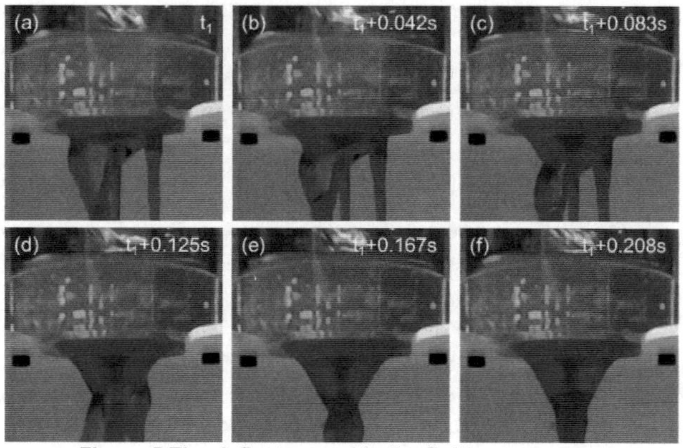

Figure 5 The deflectors guide the fluid convergence.

4 Conclusion

- The rectification effect of the filter was investigated through physical simulations, which was found to be an effective method for further studies.
- Filters with smaller structural sizes exhibit a superior rectification effect but at the cost of reduced flow rates.
- The integration of deflectors significantly improves the rectification effect of the filters and the external coverage area determines the affected region.
- The deflectors through 3D printing decouples the filtration and rectification functions of the filters, enabling designers to prioritize the filters' strength and filtration performance while maintaining rectification.

References

[1] Selvaraj S K, Sundaramali G, Jithin Dev S, et al. Recent Advancements in the Field of Ni-Based Superalloys. Advances in Materials Science and Engineering, 2021: 1-60.

[2] Demchenko A I, Ryabtsev A I, Korovin V A, et al. Effect of Filtration on the Structure and Properties of an EP-648-VI Nickel Superalloy. Russian Metallurgy (Metally), 2020: 1468-1471.

[3] Yan W, Chen Z, Li G, et al. Preparation and enhanced mechanical properties of novel Al_2O_3 - C ceramic filter reinforced by microporous powder and SiC whiskers. Journal of the American Ceramic Society, 2024, 107(4): 2725-2737.

[4] Wetzig T, Baaske A, Karrasch S, et al. Application of exchangeable carbon-bonded alumina foam filters in an industrial tundish for the continuous casting of steel. Ceramics International, 2018, 44(18): 23024-23034.

[5] Brůna M, Galcik M, Pastircak R, et al. Effect of Gating System Design on the Quality of Aluminum Alloy Castings. Metals, 2024, 14(3): 312.

A Study of Microstructure, Mechanical Properties and Thermal Expansion in Super Invar Alloy Fabricated Via Selective Laser Melting

Mingke Yang, Haoran Zhou, Huaiqian Li, Yingchun Ma, Zhiquan Miao, Yanpeng Wei, Bo Yu*
National Key Laboratory of Advanced Casting Technologies, Shenyang Research Institute of Foundry Co., Ltd. CAM, Shenyang 110022, China
*Corresponding address: e-mail:yub@chinasrif.com

Abstract: In this paper, the mechanical properties of the low-expansion alloy 4J32(super-Invar alloy) manufactured by selective laser melting(SLM) were studied, and the differences of the thermal expansion properties caused by the process were discussed. The results show that the appropriate selection of process parameters makes the additively manufactured super-Invar alloy exhibit higher tensile properties, which also shows a certain anisotropy. This development is mainly coming from the grain refinement due to rapid solidification and the grain selection due to process selection. The coefficient of thermal expansion of the alloy has also remained at a lower level than the existing research, showing the great advantages of additively manufactured super-Invar alloys.

Keywords: super Invar alloy; selective laser melting; tensile properties; thermal expansion properties

1 Introduction

Super Invar alloy is an iron-nickel-based alloy with excellent low expansion performance, which can even reach half of that of Invar 36 alloy (4J36) in the temperature range of 20-100℃[1]. It also has good plasticity. These properties make it one of the research hotspots in aerospace, precision instruments and other fields. However, its poor machinability and low mechanical properties limit its production and application[2]. For these reasons, this paper investigated the tensile properties and thermal expansion properties of super-Invar alloy via SLM.

2 Experimental procedure

To study the properties of additively manufactured super-Invar alloys, super-Invar powders (average particle size of 15-53μm) were used, Table 1 shown the elemental percentages. The test was carried out on the BLT210 laser printing equipment, and 15 sets of process parameters with the printing power of 200-300W and the scanning speed of 600-1000mm/s were set for experimental comparison. The tensile mechanical properties of the alloy at room temperature and 230℃ were tested at a tensile rate of 0.55mm/min on an electronic universal materials testing machine, and the experimental group with the best performance was selected to test the linear thermal expansion properties of the alloy at a cooling rate of 3K/min on dilatometer.

Table 1. Chemical composition of powders in the study(wt.%)

Elements	C	Si	Mn	P	S	Ni	Co	Fe
wt.%	0.05	0.10	0.23	0.005	0.003	31.88	5.14	Bal.

3 Result and discussion

ANALYSIS OF MICROSCOPIC TOPOGRAPHYThe microstructure of each group of test samples was characterized. The results showed that all samples have high compactness degree, and no obvious defects and cracks. The super-Invar alloy fabricated by SLM is mainly composed of austenite phases, so it exhibits a typical Invar effect on the macroscopic level. At the same time, the grain is mainly columnar grains oriented along the build direction and the average size is smaller than as-cast alloy. The reason for the above situation is that the rapid condensation of additive manufacturing technology largely avoids the growth of crystallization, and the large temperature gradient inside the melt pool leads to the growth of grains mainly along the construction direction. It's also the reason why the mechanical properties of additively manufactured super-Invar alloys show a certain anisotropy.

Analysis of tensile performance

Additively manufactured metals typically have higher strength than as-cast due to grain refinement, which is a effective way for invar alloys to improve the mechanical properties. Fig. 1(a) showed the stress-strain curve of the optimal experimental alloy at room temperature compared with the standard curve of the as-cast super-Invar alloy[3], and it is not difficult to find that the super-Invar alloy made by SLM shows higher tensile mechanical properties than the as-cast state. It also showed the difference in performance between the horizontal and build directions for the same print parameter. The tensile strength of the alloy is lower in the build direction, but the elongation is slightly higher, which corresponds to the grain preference orientation exhibited under this printing parameter.

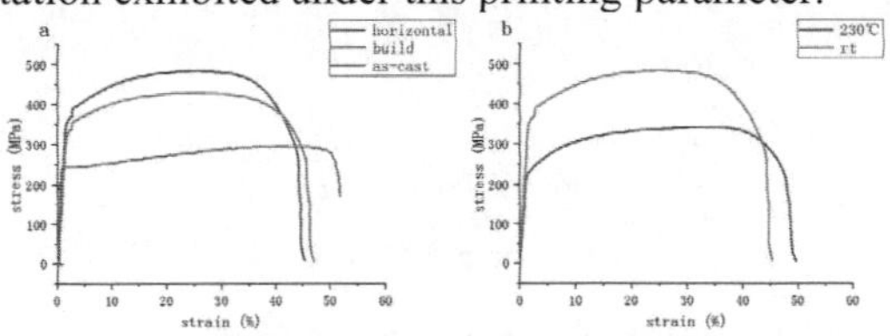

Fig. 1(a) The test stress-strain curve of horizontal and build orientation samples and the standard curve of as-cast super-Invar alloy at room temperature; (b) The test stress-strain curve of horizontal sample at room temperature and 230℃

Fig. 1(b) showed the test stress-strain curve of horizontal sample at room temperature and 230℃. The stress-strain curve in the high-temperature environment showed a significant decrease in tensile strength, which is mainly caused by the weakening of the strengthening mechanisms such as grain boundary strengthening and dislocation strengthening in the high-temperature environment. It also showed that the mechanical strengthening brought about by the SLM process cannot maintain- at high temperature.

Analysis of thermal expansion performance

Figure 2 and Table 2 shows the analysis curve of the thermal expansion test of the selected test alloy. The thermal expansion coefficient(CTE) of the super-Invar alloy used in this test has reached the level of 0.5×10^{-6}/K, showing a low thermal expansion performance far exceeding the as-cast alloy[3]. In addition, the thermal expansion curves of heating and cooling are in good agreement, indicating that the low thermal expansion characteristics brought by additive manufacturing are relatively stable and will not be affected by the temperature variation.

At the same time, it can be observed from the experimental curve that there is a rapid decline of the low thermal expansion performance of the sample over 100℃, and the average thermal expansion coefficient of the alloy in the temperature range of 20-230℃ has been consistent with the as-cast standard, which is consistent with the poor low-temperature stability of the as-cast super-Invar alloy[3].

Table 2 CTE within 20-100 ℃/20-230 ℃(10^{-6}/K)

	Heating	Cooling
20-100℃	0.4135	0.5590
20-230℃	2.7572	3.1131

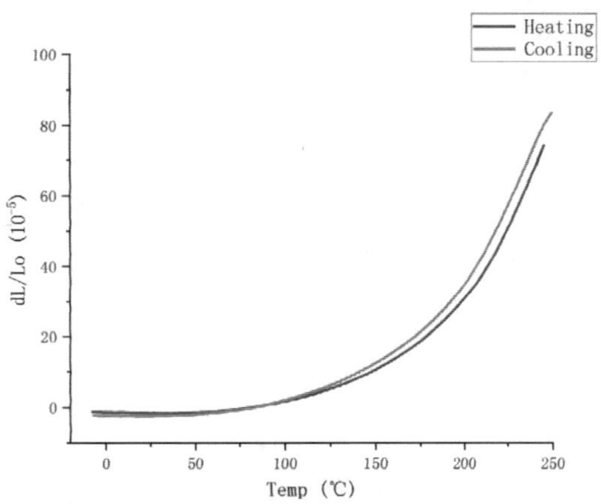

Fig 2: Test the thermal expansion test curve of sample

4 Conclusion

The super-Invar alloy fabricated by SLM has better tensile mechanical properties and exhibits certain anisotropy, which is related to grain refinement and preferential grain orientation. At the same time, the test samples showed excellent low thermal expansion performance, and reached the level of 0.5×10^{-6}/K within 20-100℃, which showed that additive manufacturing has great potential in the research direction of further reducing the CTE of super-Invar alloys.

Acknowledgments

This research works was supported by the Liaoning Province Applied Fundamental Research Program (No. 2023JH2/101700039) and Liaoning Province Natural Science Foundation(No. 2023-MSLH-328).

References

[1] Cao Y, et al. Ultrawide Temperature Range Super-Invar Behavior of $R_2(Fe, Co)_{17}$ Materials (R = Rare Earth). Phys. Rev. Lett., vol. 127, no. 5, p. 055501, Jul. 2021.

[2] Arora A and Mula S. Effect of Ti addition on thermal stability and phase evolution of super-invar based yttria added ODS alloys developed by mechanical alloying and spark plasma sintering. J. Alloys Compd., vol. 899, p. 163336, Apr. 2022.

[3] Sakaguchi N, Ohno H, and Nakada N. Simultaneous Optimization of Rigidity and Strength of Super Invar Cast Steel using by Martensitic Reversion. ISIJ Int., vol. 62, no. 3, 2022.

Study on the Effect of Transition Layers on the Interfacial Microstructure Between BJ Ceramic Shells and Iron-Based Metals

Zhiyuan Yang, Li Yang*, Jingfei Liu, Zitian Fan*****
State Key Laboratory of Materials Processing and Die & Mould Technology, School of Materials Science and Engineering Huazhong University of Science and Technology, Wuhan 430074, China
*Corresponding address: lyangcl@hust.edu.cn, smartliu1206@163.com, fanzt@hust.edu.cn,

Abstract: Ceramic coatings are commonly employed to enhance the wear and corrosion resistance of metal components. Among them, metal ceramic components with Al_2O_3 coatings have been widely utilized in industrial applications. However, existing ceramic coating processes typically result in thin coatings with limited wear and corrosion resistance. Moreover, achieving efficient and high-quality coating on the surfaces of complex metal components remains challenging. In this study, complex Al_2O_3-based ceramic shells were prepared using Binder Jetting technology based on additive manufacturing principles. Subsequently, transition layers including Cr, Ni, and 316 stainless steel were coated on the surface of the shells, followed by liquid pouring of iron-based metal, and controlled temperature treatment to achieve metallurgical bonding at the Al_2O_3/Fe interface. The influence of transition layers on the microstructure of the Al_2O_3/Fe interface was explored. Experimental results indicate that coatings using silica sol as a solvent facilitate better bonding between the transition layer aggregate and the Al_2O_3 ceramic matrix. The transition layer effectively suppresses the formation of cracks at the Al_2O_3/Fe pouring interface, with the Cr transition layer demonstrating significant effectiveness.

Keywords: coating,Al_2O_3/Fe Interface; binder jetting; microstructure

1 Introduction

The ceramic/metal (coating/matrix) interface, as a functional coating system, has been widely applied in various fields such as aerospace, energy, electronic devices, and mold steel industries [1–2]. Currently, ceramic coatings on metal surfaces are primarily prepared using surface coating methods. These methods are costly, and it is challenging to achieve efficient, high-quality coatings on complex metal parts. Additionally, the ceramic layer is relatively thin (100–600 μm thick), resulting in limited improvements in wear and corrosion resistance [3]. In recent years, the rapidly developing additive manufacturing (AM) technology has provided new approaches for the rapid fabrication of complex ceramic shells [4]. Binder Jetting (BJ), as one of the main ceramic additive manufacturing technologies, offers low cost, high forming efficiency, and does not require support structures during forming. It has significant potential for development and industrial application [5].

In this study, high-strength Al_2O_3-based ceramic shells fabricated using BJ [6] were used as shell blanks. Through dipping, Cr, Ni, and 316 stainless steel coatings were applied as intermediate layers. High-temperature sintering was performed to tightly bonds the coatings with the ceramic substrate, followed by casting to achieve transitional bonding at the Al_2O_3/Fe interface. The effects of various intermediate layers on the interfacial structure and properties between the BJ ceramic shell and the iron-based metal were explored. The ceramic/metal interface was observed, and its microstructure and interfacial bonding mechanisms were analyzed. This study provides a new method for further enhancing the wear and corrosion resistance of iron-based metals.

2 Experimental procedure

The raw materials used included Al_2O_3 powder (d_{50}=15.1 μm), MnO_2 powder (d_{50}=12.9 μm), chromium, nickel, and 316 stainless steel powders (d_{50}=13.7 μm), phenolic resin binder, titanium dioxide sol, silica sol, sodium carboxymethyl cellulose (CMC-Na), silicone, and 65Mn spring steel. The overall preparation process for the Al_2O_3/Fe ceramic/metal samples is illustrated in Fig. 1, and mainly includes the following steps: BJ fabrication of the ceramic shell, preparation and dipping of coatings, sintering of the coatings, and casting of the metal liquid.

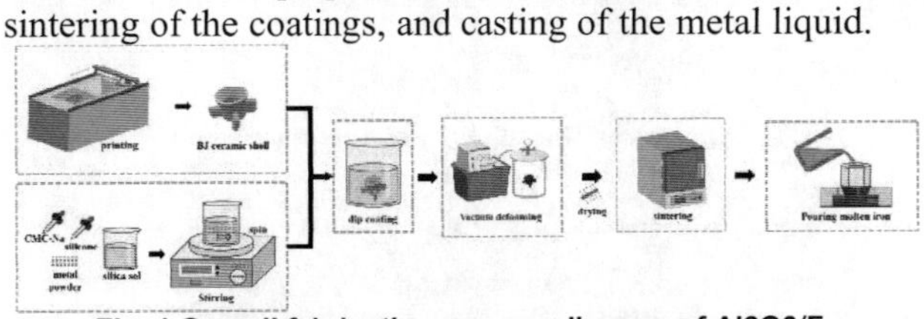

Fig. 1 Overall fabrication process diagram of Al2O3/Fe ceramic/metal specimens and components.

$$\rho_b = \frac{m_1}{m_3 - m_2} \times \rho_1 \quad (1)$$

$$D = \frac{\rho_b}{\rho_t} \times 100\% \quad (2)$$

where ρ_b is the bulk density (g/cm³), ρ_1 is the density of water (g/cm³), m_3 is the weight of the water-saturated sample in air (g), m_1 is the dry weight of the sample (g), m_2 is the weight of the water-saturated sample in water (g), ρ_t is the true density (g/cm³), and D is the relative density (%).

3 Result and discussion

To ensure good coating performance on the ceramic substrate, the coating should exhibit shear-thinning characteristics. Additionally, after application, the coating must maintain high viscosity to prevent cracking and delamination. The coating performance of each metal powder on the Al_2O_3 substrate should remain stable within a certain viscosity range. Using Cr coating as an example, the rheological properties of the coating at various Cr solid contents and their impact on the substrate performance after coating are shown in Fig. 2.

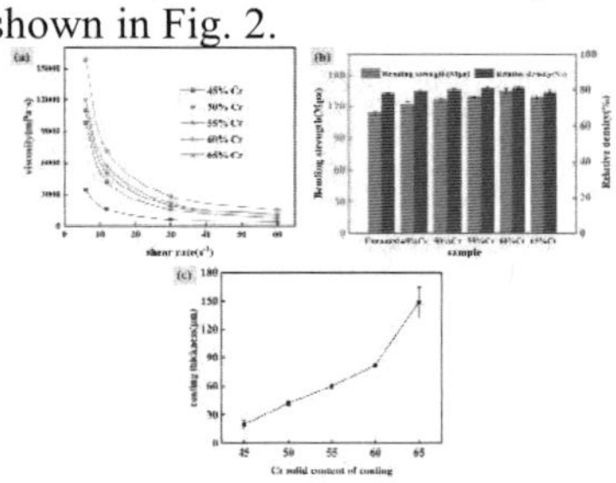

Fig. 2 Effects of different Cr contents on the properties of coatings and ceramic samples: (a) rheological properties of the coatings; (b) properties of the sintered ceramic substrate; (c) coating thickness.

As shown in Fig. 2, the coating exhibits shear-thinning characteristics, with viscosity increasing as the Cr solid content increases. The flexural strength and relative density of the coated samples are both improved compared to the uncoated samples. However, the flexural strength and relative density of the coated samples initially increase and then decrease with the increase in Cr solid content. The adhesion of the Cr coating on the Al_2O_3-based ceramic surface primarily relies on the reaction between SiO_2, produced by the decomposition of silica sol at high temperatures, and the Al_2O_3 substrate, forming mullite ($3Al_2O_3 \cdot 2SiO_2$). A small amount of mullite can enhance the physical properties of Al_2O_3-based ceramic, while an excess amount, due to its brittleness, can lead to a decline in the physical properties of the Al_2O_3 ceramics.

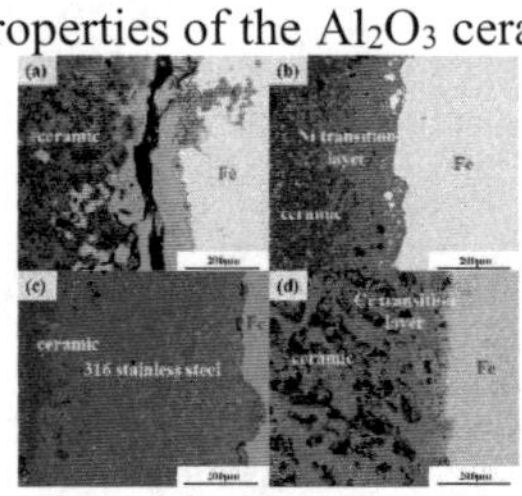

Fig. 3 Al_2O_3/Fe interface: (a) without transition layer; (b) with Ni transition layer; (c) with 316 stainless steel transition layer; (d) with Cr transition layer

As shown in Fig. 3, the bright region on the right is iron, while the black and gray region on the left is Al_2O_3-based ceramic. The Al_2O_3/Fe interface without a transition layer exhibits significant cracks. This is due to the large difference in the coefficient of thermal expansion (CTE) between Fe ($(9.2\text{-}11.8)\times10^{-6}$ K^{-1}) and Al_2O_3 ($(7.85\pm0.02)\times10^{-6}$ K^{-1}), along with the presence of a substantial amount of iron oxides at the interface, leading to interface cracking.

After adding metal transition layers, the Al_2O_3/Fe interface showed significant improvement, with no large cracks observed. When the 316 stainless steel transition layer was added, a few pores were still present at the transition layer and Fe interface boundary, which could lead to cracks under external force. With the addition of the Ni transition layer, no obvious cracks or pores were found at the transition layer and Fe interface boundary. However, there were regions of Ni element segregation and enrichment, as shown in Fig. 3(b). This segregation and enrichment of Ni could cause stress concentration at the interface, reducing interface performance. The Cr transition layer showed the best improvement for the interface, with no significant cracks or elemental segregation observed.

4 Conclusion

The metal transition layer coatings prepared according to the viscosity range of 60% wt. Cr solid content can form transition layers on the Al_2O_3-based ceramic surface and improve the flexural strength and relative density of the ceramic substrate. The transition layers significantly enhance the microstructure of the Al_2O_3/Fe interface, with the Cr transition layer showing the best improvement in the interfacial properties.

Acknowledgments

This project has been financially supported by the National Natural Science Foundation of China (No. 52275334, No. 52205365 and No. 51775204).

References

[1] Xu H, Zhao Y, Yang J, et al. Investigation on wetting behavior and mechanism of AgCu-X wt.%Ti filler metal/AlN ceramic reactive wetting system: Experiments and first-principles calculations. Journal of Alloys and Compounds ,2021; 869: 159323.

[2] Nath S, Manna I, Ray SK, et al. Studies on nanotribological and oxidation resistance properties of yttria stabilized zirconia (YSZ), alumina (Al_2O_3) based thin films developed by pulsed laser deposition. Ceramics International, 2016; 42: 7060–71.

[3] X. Guo, Y.G. Jung, Z. Lu, J. Zhang, Novel lanthanum zirconate-based thermal barrier coatings for energy applications, Novel Lanthanum Zirconate-Based Thermal Barrier Coatings for Energy Applications,2020, 1–85.

[4] Li N, Huang S, Zhang G, et al. Progress in additive manufacturing on new materials: A review. Journal of Materials Science & Technology.2019; 35: 242–69.

[5] Mariani M, Beltrami R, Brusa P, et al. 3D printing of fine alumina powders by binder jetting. Journal of the European Ceramic Society,2021; 41: 5307–15.

[6] Yang Z, Yang L, Wang P, et al. Effect of sintering aid combined vacuum infiltration on the properties of Al_2O_3-based ceramics via binder jetting. Additive Manufacturing, 2024; 79: 103898.

Automated Pore Segmentation in Powders for Additive Manufacturing Based on Convolutional Neural Network

Hao Xue, Xinyue Li, Xingjie Li, Han Yu*
State Key Laboratory of Advanced Casting Technologies, China Academy of Machinery Shenyang Research Institute of Foundry Co., Ltd, Shenyang 110022, Liaoning, China
*Corresponding address: yuhan@chinasrif.com

Abstract: The pores in metal powders have a great impact on the mechanical properties of 3D printed parts. Thus, it is very important to realize automated pore segmentation. However, the tiny size and unclear boundaries limit this task. Considering this, we propose an Efficient-UNet network with the deep-supervised method for pore segmentation. On the one hand, more representative features are extracted by EfficientNet and fused with features in the decoder. On the other hand, deep supervision branches are connected with the decoder to prevent gradient disappearance. Benefiting from the above aspects, our proposed network model can segment small-sized pores with ambiguous boundaries. Experiments on our metal powder dataset demonstrate that our method achieves more precise results with clear advantages than other semantic segmentation methods.

Keywords: pore segmentation; micro-CT; convolutional neural network

1 Introduction

Selective-laser melting (SLM) is a representative technology for additive manufacturing (AM), which employs a high-energy laser beam to melt and solidify the pre-coated alloy powder layer by layer in a short time [1]. Its raw material is generally a single-component metal powder, mainly including stainless steel, nickel-based alloys, titanium alloys, etc.

The current processing technologies for metal powder mainly include the gas-atomization process and the plasma rotating electrode process. Both methods will inevitably produce pores during the manufacturing process, leading to the formation of hollow powder [2]. Cunningham [3] studied that powders influence the porosity rate and mechanical properties of 3D printed parts, and found that pores in the powders would be passed on to the completed components.

To improve the mechanical properties of 3D printed parts, metal powder for SLM requires a lower hollow powder rate. The testing methods of hollow powder mainly include the metallographic method and micro-CT method. Through the metallographic method, a 2D cross-section image of metal powder can be obtained by optical microscope. However, because this method cannot achieve 3D characterization of the powders, it has significant randomness and low measurement efficiency [4]. To overcome its shortcomings, the micro-CT testing technology for metal powder has attracted extensive attention. Plessis [5] provided information on powder internal porosity through high-resolution micro-CT scans and a detailed analysis of Ti6Al4V metal powders. The above research has established the foundation to further study automated pore segmentation in metal powders based on micro-CT. However, due to the unclear boundary of the pore caused by the small scale, it is quite difficult to test and represent the pores for traditional image processing algorithms. Above all, an Efficient-UNet network with a deep-supervised method is proposed to recognize and segment the pores accurately.

2 Experimental procedure

As shown in Figure 1,the powder materials we selected are steel and TC4. To solve the problem of poor dispersion, the metal powders were embedded in the resin. Before using CT scanning, the two samples were sanded into regular cylindrical shapes respectively.

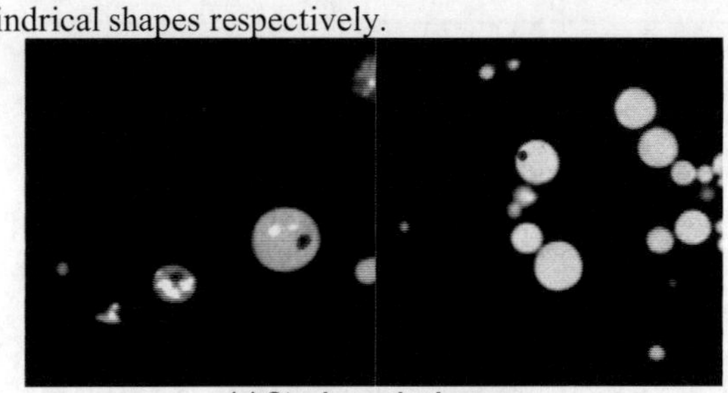

(a) Steel powder image;
(b) TC4 powder image
Figure 1 Metal Powder Image with annotations

Data collection

The equipment for data collection is nan-Voxel 3000 micro-CT produced by Sanying Precision Instruments Co., Ltd. It is equipped with a micro-level focus and transmission radiation source. So the flat panel detector can reach a spatial resolution of 500nm, which can meet the requirements of metal powder defects on high resolution. The sample is placed on a high-precision rotary table and then CT scanning is performed on it. The projection data obtained by scanning is reconstructed using the Feldkamp (FDK) method to obtain a 3D volume data model, so as to obtain 2D tomographic image data.

Detection

We employ sensitivity (sen), specificity (spe), intersection over union (IoU), Dice coefficient to measure the accuracy of the network model in our task.

$$\text{sen}=\frac{TP}{TP+FN} \tag{1}$$

$$\text{spe}=\frac{TN}{TN+FP} \tag{2}$$

$$\text{IoU}=\frac{TP}{TP+FP+FN} \tag{3}$$

$$\text{Dice}=\frac{2TP}{2TP+FP+FN} \tag{4}$$

3 Result and discussion

Comparation with state-of-the-art methods

The quantitative results in Table 2 have proved it. We achieve 0.6691 IoU and 0.7717 Dice by our network. Compared with UNet, UNet++, and Deeplabv3plus, it has higher sensitivity of detection for positive results. But for specificity, the quantitative results of the four networks were almost the same, proving that their ability to judge negative results was almost identical.

Tabel 1 Comparisons with state-of-the-arts on dataset

Method	SEN	SPE	IoU	Dice
UNet	0.8720	0.9996	0.6547	0.7566
UNet++	0.8599	0.9996	0.6552	0.7549
Deeplabv3plus	0.8469	0.9994	0.6242	0.7300
Ours	0.9057	0.9995	0.6691	0.7717

The influence of deep supervised loss

No matter where the deep supervision loss is, it will lead to a better performance than without deep supervised loss. Phase 1+ Phase 3 performs best in Dice score. Adding Phase 2 fails to produce significant improvement. Therefore, Phase 1 is necessary for our task to boost performance.

Table 2 Ablation Studies for deep supervised loss.

Method	SEN	SPE	IoU	Dice
No deep supervised loss	0.8718	0.9996	0.6491	0.7524
Phase 1 + Phase 2 + Phase 3	0.8989	0.9990	0.6696	0.7692
Phase 2 + Phase 3	0.8700	0.9997	0.6581	0.7579
Phase 1 + Phase 3	0.9057	0.9995	0.6691	0.7717

4 Conclusion

In this paper, a pore segmentation method is proposed to achieve automatic testing for powders in SLM. Specifically, the EfficientNet is utilized as our backbone network for multi-level feature extraction. The decoder module with a deep supervision branch fuses the multi-level features and restores them to high resolution. Through them, our method can obtain the pore edge even if the boundary is not clear in an end-to-end manner. Experiments demonstrate that the proposed method achieves 0.6691 IoU and 0.7717 Dice score, which has obvious advantages over other semantic segmentation methods.

Acknowledgments

This research was funded by the Central-guided fund of local development in science and technology (2022JH6/100100011) and the Technology Development Foundation of China Academy of Machinery Science and Technology Group Company, Ltd (200221ZS02).

References

[1] Zhang L C, Attar H. Selective laser melting of titanium alloys and titanium matrix composites for biomedical applications: a review[J]. Advanced engineering materials, 2016, 18(4): 463-475.

[2] Chen G, Zhao S Y, Tan P, et al. A comparative study of Ti-6Al-4V powders for additive manufacturing by gas atomization, plasma rotating electrode process and plasma atomization[J]. Powder technology, 2018, 333: 38-46.

[3] Cunningham R, Nicolas A, Madsen J, et al. Analyzing the effects of powder and post-processing on porosity and properties of electron beam melted Ti-6Al-4V[J]. Materials Research Letters, 2017, 5(7): 516-525.

[4] Murr L E. A metallographic review of 3D printing/additive manufacturing of metal and alloy products and components[J]. Metallography, Microstructure, and Analysis, 2018, 7(2): 103-132

[5] Du Plessis A, Sperling P, Beerlink A, et al. Standard method for microCT-based additive manufacturing quality control 4: Metal powder analysis[J]. MethodsX, 2018, 5: 1336-1345.

Collaborative Optimization of High Strength, Ductility,and Low Porosity for Selective Laser Melting GH4169 Superalloy by Fusion of Data-Driven and Theory

Xianmeng Tu[1], Jialong Chen[1], Xiaoyuan Ji[1*], Huanqing Yang[2], Shengzhi Zhou[1], Dongjian Peng[2], Peng Yu[1], Zeming Wang[1], Wenhao Zhou[1], Yajun Yin[1], Zejun Zhang[1], Jianxin Zhou[1*]

1. State Key Laboratory of Materials Processing and Die & Mould Technology, School of Materials Science and Engineering, Huazhong University of Science and Technology, Wuhan, Hubei, 430074, China
2. Xi'an Space Engine Co., Ltd., Xi'an, 710100, China

*Corresponding address: e-mail: jixiaoyuan@hust.edu.cn (Xiaoyuan Ji); zhoujianxin@hust.edu.cn (Jianxin Zhou)

Abstract: Multi-performance collaborative optimization is a trend for improving the properties of additive manufacturing products and expanding their applications. A fusion of data-driven and theory approach is proposed for multi-performance objective optimization of selective laser melting. Firstly, a global, stepwise, centrally dense hexagonal design of experiments is applied to the additive manufacturing of 65 GH4169 nickel superalloy materials each employing different laser processing conditions. Secondly, 4 multiple stepwise regression models are established to quantify the complex relationship between laser process parameters and relative density, yield strength, tensile strength, and elongation. Thirdly, a stepwise regression non-dominated sorting genetic algorithm with relative density energy criterion constraints is developed to achieve multi-performance objective process parameters collaborative optimization. Finally, a high strength, ductility, and low porosity component is manufactured by this approach. Compared to traditional optimization approaches, this approach can more accurately and efficiently achieve multiple process parameters collaborative recommendations.

Keywords: Additive manufacturing (AM), Selective laser melting (SLM), Fusion of data-driven and theory (FDDT), GH4169 superalloy, Multi-performance objective, Process parameters collaborative optimization

1 Introduction

Additive manufacturing (AM) is one of the fundamental technologies for rapidly moving towards intelligent manufacturing. Selective laser melting (SLM), as a typical metal AM technology, has dozens of parameters[1-2] as well as complex heat and mass transfer phenomena in molten pools [3] which lead to collaborative optimization being extremely difficult[4].

In the paper, a fusion of data-driven and theory approach is proposed for multi-performance objective optimization of selective laser melting GH4169 superalloy. 4 key process parameters, laser power(P), scanning speed(v), hatch space(h), and layer thickness(t), were tuned by the intelligent recommendation systems and the locally conflicting properties, relative density, yield strength, tensile strength, and elongation were optimized.

2 Experimental Procedure

Bone shape GH4169 samples (FIGURE 6FIGURE 6) were fabricated by SLM with a global, stepwise, centrally dense hexagonal design (Table 1).

Table 1 Design of Experiments

No.	*P* (W)	*v* (mm/s)	*h* (μm)	*d* (μm)
Group1				
1	260	700	90	20
2	290	900	100	20
3	320	1100	110	20
4	350	1300	120	20
Group2				
1	100	300	60	20
2	150	600	75	30
3	200	900	90	40
4	250	1200	105	50
5	300	1500	120	60
6	350	1800	135	70
7	400	2100	150	80

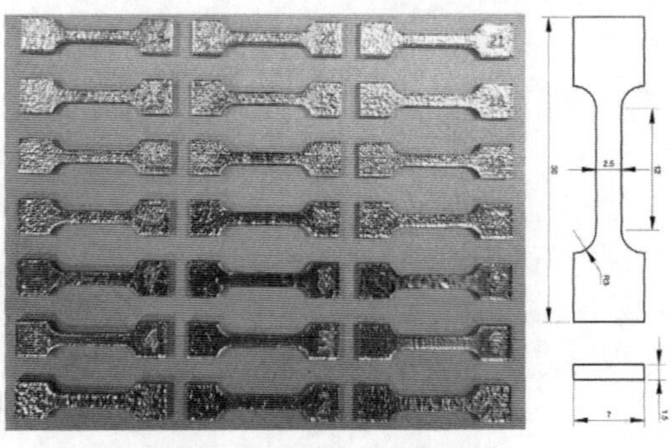

Figure 6 SLM samples with engineering drawing

3 Result and discussion

3.1 Model construction and evaluating

The four properties prediction models for relative density, tensile strength, yield strength, and tensile strain were constructed by stepwise regression modelling (Figure 7). The process parameters recommendation system assembly the prediction model and multi-objective optimization algorithm Non-dominated Sorting Genetic Algorithm (NSGA-Ⅱ) (Figure 8, Figure 9, Figure 10).

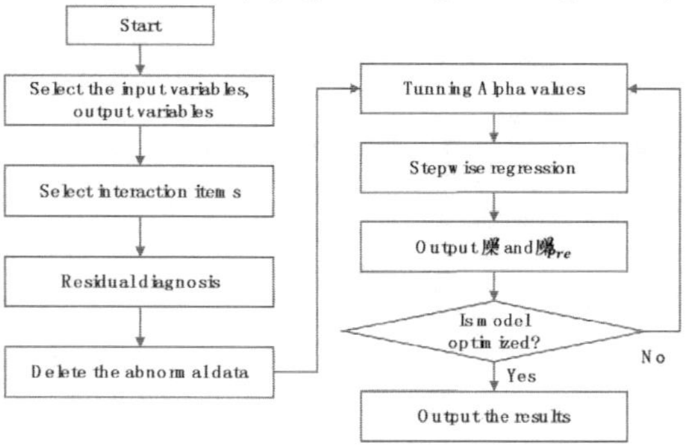

Figure 7 Stepwise regression model construction procedure

The evaluating metrics of predictive models, $RSME$, R^2, $R^2{}_{adj}$, $R^2{}_{pre}$, are the following formula:

$$RSME = \sqrt{\frac{1}{n}\sum_{i=1}^{n}(y_i - \hat{y}_i)^2}\text{ , i=1, 2, ..., } n \quad (1)$$

$$R^2 = 1 - \frac{\sum(y_i - \hat{y})^2}{\sum(y_i - \bar{y})^2} \quad (2)$$

$$R^2{}_{adj} = 1 - \left\{\frac{\sum(y_i - \hat{y})^2}{\sum(y_i - \bar{y})^2}\right\}\left(\frac{n-1}{n-p-1}\right) \quad (3)$$

$$R^2{}_{pre} = 1 - \frac{\sum_1^n\left(\frac{e_i}{1-h_i}\right)^2}{\sum_1^n(y_i - \bar{y})^2} \quad (4)$$

3.2 COLLABORATIVE OPTIMIZATION

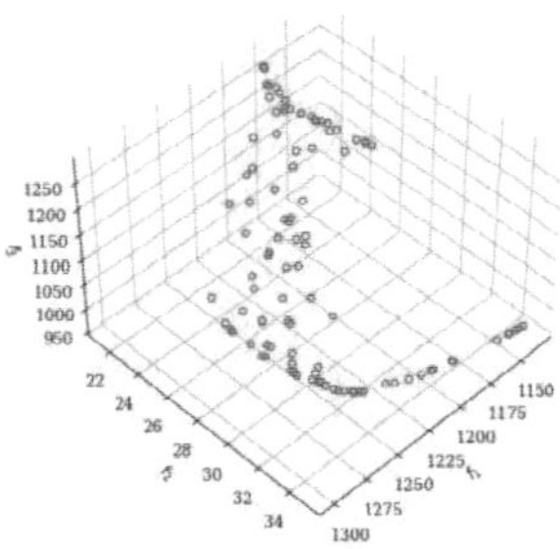

Figure 8 The Pareto front of 3

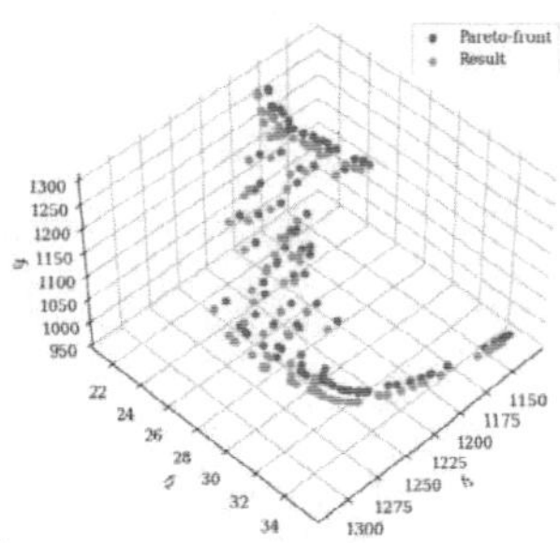

properties

Figure 9 The recommendation points with 3 properties

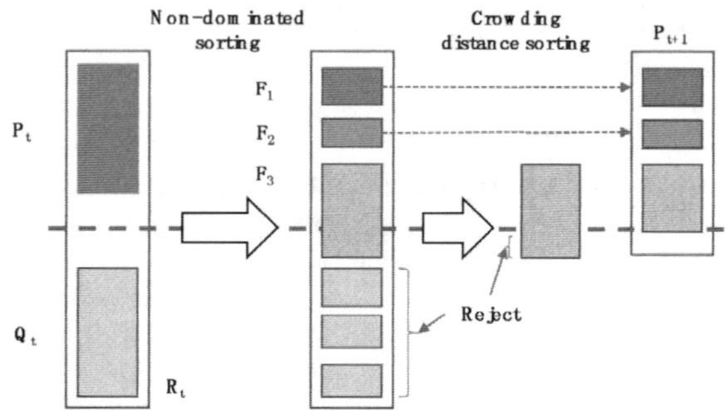

Figure 10 NSGA-Ⅱ algorithm procedure illustration

Conclusions

(1) 4 key properties predictive models were established by stepwise regression model.

(2) 4 key properties were collaboratively optimized as well as process parameters intelligently recommended by NSGA-IIalgorithm.

Acknowledgements

The authors acknowledge financial support from the KGW Program (grant No. 2019XXX.XX4007Tm), the National Key Research and Development Program of China (grant No. 2020YFB1710100), and the National Natural Science Foundation of China (grant No. 52275337, grant No. 52090042, grant No. 51905188).

References

[1] X. Tu, J. Chen, X. Ji, H. Yang, D. Peng, J. Zhou, Y. Yin, X. Shen, Process parameters optimization of selective laser melting SUS316L components based on response surface methodology, the 74th world foundry congress, BEXCO, BUSAN, Republic of Korea, 2022.

[2] Z. J. Zhang, Y. J. Wu, Z. M. Wang, X. Y. Ji, W. Guo, D. J. Peng, X. M. Tu, S. Z. Zhou, H. Q. Yang, J. X. Zhou. Recommendation of SLM Process Parameters Based on Analytic Hierarchy Process and Weighted Particle Swarm Optimization for High-Temperature Alloys, Materials. 16 (2023). https://doi.org/10.3390/ma16165656.

[3] T. DebRoy, T. Mukherjee, H.L. Wei, J.W. Elmer, J.O. Milewski, Metallurgy, mechanistic models and machine learning in metal printing, Nat. Rev. Mater. 6 (2021) 48–68. https://doi.org/10.1038/s41578-020-00236-1.

[4] K. Shi, D. Gu, H. Liu, Y. Chen, K. Lin, Process-structure multi-objective inverse optimisation for additive manufacturing of lattice structures using a physics-enhanced data-driven method, Virtual Phys. Prototyp. 18 (2023)1–20. https://doi.org/10.1080/17452759.2023.2266641.

Performance Prediction and Process Optimization of 316L Stainless Steel Produced by Selective Laser Melting Based on Response Surface Methodology

Jialong Chen[1], Xianmeng Tu[1], Xiaoyuan Ji[1,*], Huanqing Yang[2], Guomin Le[3], Zeming Wang[1,4], Dongjian Peng[2], Yajun Yin[1], Jianxin Zhou[1*]

1. State Key Laboratory of Materials Processing and Die & Mould Technology, School of Materials Science and Engineering, Huazhong University of Science and Technology, Wuhan 430074, China
2. Xi'an Space Engine Co., Ltd., Xi'an, 710100, China
3. Institute of Materials, China Academy of Engineering Physics, Mianyang, Sichuan, 621900, China
4. Nuclear Power Institute of China, Chengdu, Sichuan, 610005, China
*Corresponding address:e-mail: jixiaoyuan@hust.edu.cn (Xiaoyuan Ji); zhoujianxin@hust.edu.cn (Jianxin Zhou)

Abstract: A key aspect in 3D printing is studying how different process settings in Selective Laser Melting (SLM) affect part quality. Optimizing SLM requires a model that predicts part performance based on these settings. Therefore, we conducted a study using 316L stainless steel as an example, manufacturing and testing 31 sets of specimens to investigate the influence of different process parameters on forming performance. Initially, we employed response surface methodology to design and carry out four-factor printing experiments, which considered laser power, powder thickness, scanning speed, and hatch space. Subsequently, the specimens were tested for density and mechanical properties. Based on statistical analysis, we established four predictive models that depicted the relationship between four process parameters and properties like relative density, tensile strength, yield strength, and elongation. We also calculated the optimal response value and its corresponding process parameter combination. The established models exhibited good fitting and prediction capabilities, with all four models achieving R2 values of 85.56%, 82.19%, 90.08% and 91.72% respectively. After process optimization, the formed parts demonstrated a relative density of 99.96%, tensile strength of 834 MPa, yield strength of 685 MPa, and elongation of 43.95%. This method can obtain good optimization results with fewer experiments.

Keywords: Selective laser melting; Process parameters; Properties; Response surface methodology; Predictive model; 316L stainless steel

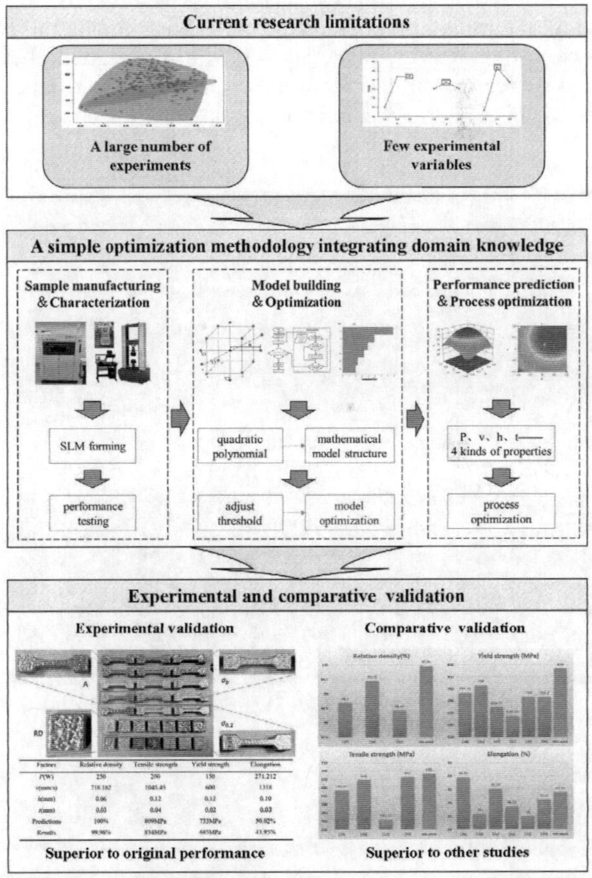

Graphical Abstract

This paper presents a method merging domain knowledge with response surface methodology for optimizing process parameters using fewer, high-quality experiments. It uses 316L stainless steel as a case study, examining laser power, powder layer thickness, scanning speed, and hatch space in printing experiments.

1 Introduction

Selective laser melting (SLM) is one of the most widely used additive manufacturing metal materials methods.

SLM is widely used in aerospace, automotive industry, biomedical and defense fields [1] due to its pollution-free [2], high precision [3], freedom design of shape and short manufacturing cycle.

2 Experimental procedure

The 316L stainless steel samples were formed by SLM with a four-factor central composite design (CCD) in the response surface methodology, and the parameters varied as presented in Table 1. Relative density is tested with drainage.

Table 1 Design of Experiments for CCD

Factors	Axial point1	Lower level	Central point	Upper level	Axial point 2
P (W)	150	200	250	300	350
v (mm/s)	600	825	1050	1275	1500
h (mm)	0.06	0.075	0.09	0.105	0.12
t (mm)	0.02	0.03	0.04	0.05	0.06

3 Result and discussion

3.1 Properties prediction

Four prediction models for material properties (relative density, tensile, yield strengths, tensile strain) were developed via RSM with variables P, v, h, t (Equations 1-4). Stepwise regression optimized these models by adjusting α to maximize R-sq (prediction), employing a hierarchical model structure in the process.

Fig. 1 Samples of SLM SUS 316L

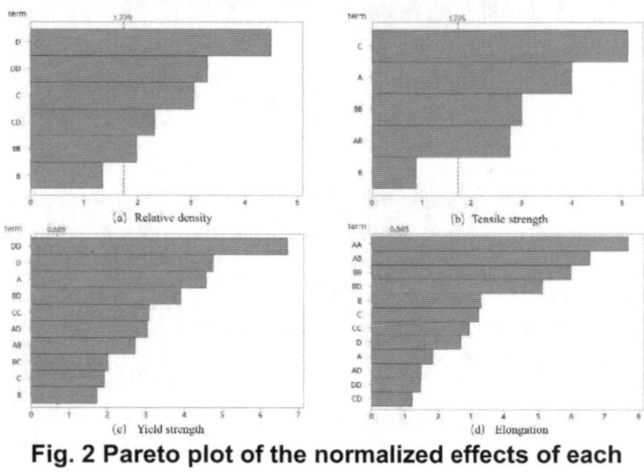

Fig. 2 Pareto plot of the normalized effects of each model

The effect of prediction models' terms on properties is ordered as shown in Fig. 2.

3.2 Parameters optimization

Validation tests were done to verify effectiveness. For the density-response surface model, the optimal value of the response (density) reaches 99.96% , the optimal value of the response (tensile strength) reached 834 MPa.The results of yield strength are up to 685 MPa and the elongation is up to 43.95%.

Fig. 3 Verification samples

4 Conclusions

(1) 4 good fitting ability properties prediction models are constructed based on a domain knowledge-based and response surface methodology fusion methodology.

(2) the relationship prediction model between the four process parameters and the density, tensile strength, yield strength and elongation are established

(3) Process parameters optimization of SUS316L is realized in selective laser melting.

Acknowledgments

This work is financially supported by National Key R&D Program (2020YFB1710100), Science and Industry Commission Project (2019XXX.XX4007Tm), and National Natural Science Foundation of China (52275337, 51905188, 52090042).

References

[1] M. Benedetti, A. du Plessis, R.O. Ritchie, M. Dallago, N. Razavi, F. Berto, Architected cellular materials: A review on their mechanical properties towards fatigue-tolerant design and fabrication, Mater. Sci. Eng., R 144 (2021).

[2] A Majeed, Y. Zhang, S. Ren, J. Lv, T. Peng, S. Waqar, E. Yin, A big data-driven framework for sustainable and smart additive manufacturing, Rob. Comput. Integr. Manuf. 67 (2021).

[3] E. Liverani, S. Toschi, L. Ceschini, A. Fortunato, Effect of selective laser melting (SLM) process parameters on microstructure and mechanical properties of 316L austenitic stainless steel, J. Mater. Process. Technol. 249 (2017) 255-263.

Effect of Deposition Layers on the Microstructure and Mechanical Properties of Cu/Al Bimetal Fabricated by Wire Arc Additive Manufacturing

Fuyuan Liang[1], Pengfei Xing[2], Mengwu Wu[1,*]
1. School of Automotive Engineering, Wuhan University of Technology, Wuhan 430070, China
2. National Key Laboratory of Intelligent Agricultural Power Equipment, Luoyang 471039, China
*Corresponding address: e-mail: wumw@whut.edu.cn

Abstract: Copper/aluminium bimetallic materials were fabricated by CMT wire arc additive manufacturing, and the effect of deposition layers on the microstructure and mechanical properties of the bimetal interface was investigated. Experimental results show that with the increase of deposition layers, the subsequent thermal cycle promotes the growth of copper/aluminium interfacial layer. And when the number of deposition layers exceeds 5, the thickness of the interfacial layer increases sharply. The shear strength of the interface shows a tendency to increase firstly and then decrease later.

Keywords: Cu/Al bimetal; wire arc additive manufacturing; microstructure; mechanical property

1 Introduction

Copper/aluminium bimetallic materials have the advantages of light weight, good electrical/thermal conductivity, etc., and have been widely used in automotive, power electronics, metallurgical machinery and other fields [1-2]. Wire arc additive manufacturing (WAAM) has the advantages of high deposition efficiency and low cost [3]. Applying WAAM process to fabricate bimetallic materials has become a research hotspot in recent years. In the process of copper/aluminium bimetallic fabrication by WAAM, the subsequent depositing metal will have a thermal effect on the deposited metal, so it is important to understand the effect of thermal cycling on the microstructure and mechanical properties of the copper/aluminium interface.

2 Experimental procedure

304 steel with a dimension of 200×200×5mm is used as a substrate. The filler materials are 1.2mm-diameter Al (4043) and Cu (CuSi-3) alloys. Table 1 shows the composition of the filler materials. The wire arc additive manufacturing equipment consists of a Fronius CMT Advanced 4000R welding machine and a ABB 2600 robot. The DC-CMT mode is used for the deposition of Al alloy and Cu alloy. Initially, 10 layers of Cu alloy is deposited on the 304 steel substrate. Then, Al alloy with various layers is deposited on the Cu alloy. During the WAAM process, k-type thermocouple is used to measure the thermal cycle of Cu/Al interface. After obtaining the Cu/Al bimetal, metallographic samples are cut, polished and etched for 15 s using a standard Keller solution. The microstructure of Cu/Al interface is characterized by scanning electron microscopy (SEM) equipped with an energy dispersive spectroscopy (EDS). Shear tests are carried out using an electromechanical universal testing machine.

Table 1 Composition of the filler materials

Wire	Al	Cu	Si	Mn	Fe
4043	Bal.	0.30	6.00	0.15	0.60
CuSi-3	--	Bal.	3.00	0.80	--

3 Result and discussion

Microstructure of the interface layer

Figure 1 shows the macroscopic morphology of Cu/Al bimetal. It can be seen that there is no obvious macro cracks on the surface of the specimen.

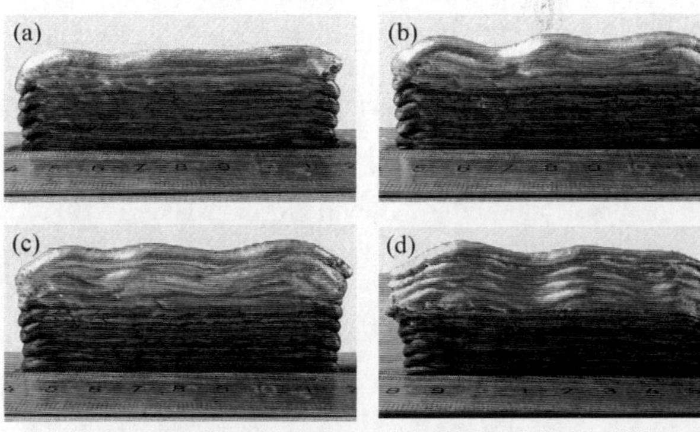

Fig. 1 Macroscopic morphology of Cu/Al bimetal with different deposition layers: (a) 3; (b) 5; (c) 7; (d) 10

Figure 2 shows the microstructure of Cu/Al interface with various Al alloy deposition layer. From Figures 2 (a-b), it can be seen that the thickness of the interface layer is 15 μm and 18 μm, respectively, and the thickness of the interface layer does not change much. As shown in Figure 2 (c-d), the thickness of the interface layer increases dramatically when the aluminium alloy deposition layer is 7 and 10, while the thickness of the interface layer reaches 100 μm and 300 μm, respectively. In addition, it can be seen that when the number of aluminium alloy deposition layer is 10, the intermetallic compounds within the interface layer are obviously broken. This is because the intermetallic compounds within the interface layer are more brittle in the process of depositing aluminium alloy. Affected by the subsequent layer of aluminium alloy depositing, the first layer of aluminium alloy will be remelted and locally uneven melting, the liquid metal flow in the molten pool will impact on the solid-liquid interface. In the process of the liquid metal flow, the brittle intermetallic compounds within the interface layer will be fractured.

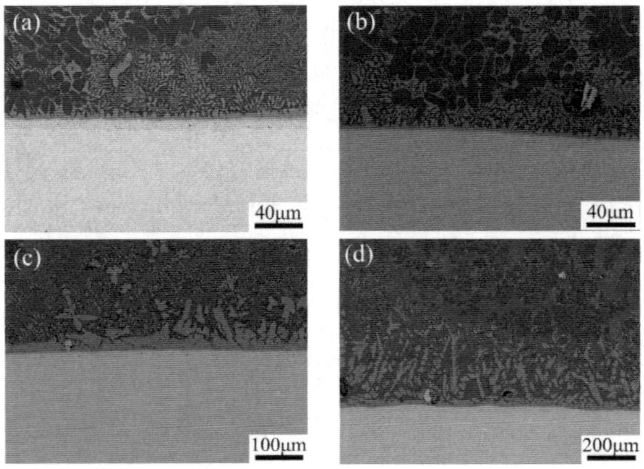

Fig. 2 Microstructure of Cu/Al interface with different deposition layers: (a) 3; (b) 5; (c) 7; (d) 10

Mechanical property

Figure 3 shows the thermal cycle curve at the Cu/Al interface. It can be seen that the layer 1 aluminium alloy is subject to multiple remelting and post-thermal effects. With the increase of the aluminium alloy deposition layer, the distance of the electric arc from the interface gradually increases, the thermal impact of subsequent deposited layer on the interface diminishes, the peak temperature of the interface gradually decreases. When the number of aluminium alloy deposition layer is 10, the peak temperature at the Cu/Al interface still exceeds 400 °C, which indicates that there is a certain thermal accumulation at the Cu/Al interface. The thermal accumulation promotes the growth of intermetallic compounds at the interface.

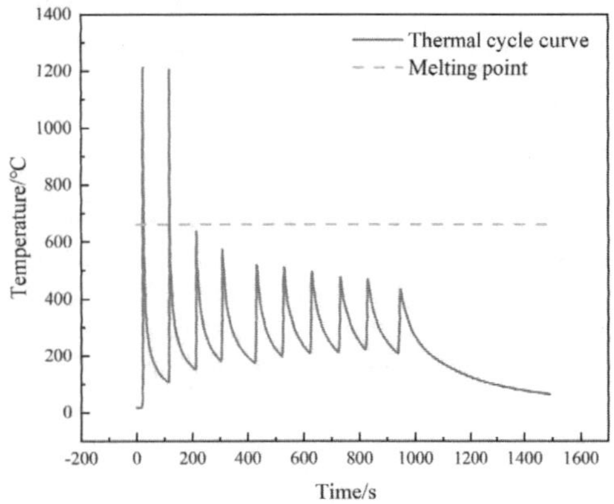

Fig. 3 Thermal cycle curve at the Cu/Al interface

Figure 4 shows the shear strength of the Cu/Al bimetal interface. It can be seen that the shear strength of the composite interface shows a tendency of increasing first and then decreasing. This is due to the fact that the thickness of the interface layer increases dramatically when the number of aluminium alloy deposition layer increases to 7 and 10. Under the action of the load, it is easy to lead to stress concentration and promote the generation of microcracks, which are easy to expand rapidly along the interface layer, and ultimately make the shear strength of the composite interface drop significantly.

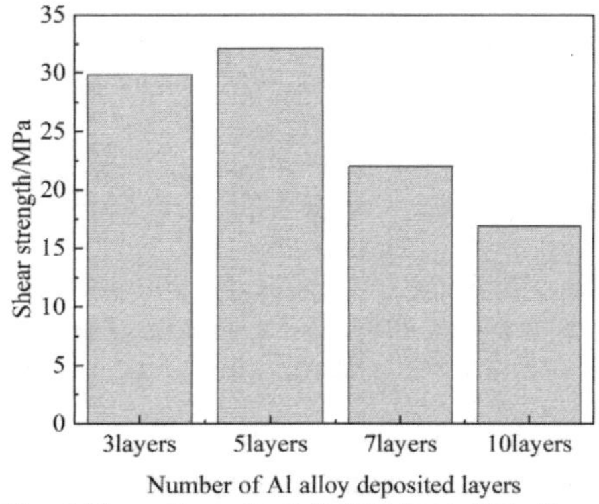

Fig. 4 Shear strength of Cu/Al bimetal interface

4 Conclusion

With the increase in the number of aluminium alloy deposition layer, the Cu/Al bimetal interface undergoes more thermal cycles. The thickness of the copper/aluminium interface layer gradually increases, and the shear strength of the composite interface shows a tendency of increasing first and then decreasing.

Acknowledgments

This work was financially supported by the Key R&D Program of Hubei Province, China (Grant No. 2022BAD100) and the Open Fund of National Key Laboratory of Intelligent Agricultural Power Equipment (Grant No. SKLIAPE2023014).

References

[1] Chen X, Chen Z, Hu B, et al. Strengthening behavior and thermal conductivity of Cu/Al composite with penetration architecture[J]. Transactions of Nonferrous Metals Society of China, 2024, 34: 236-245.
[2] Wu Z, Zuo L, Zhang H, et al. Effect of Liquid-Solid Volume Ratio and Surface Treatment on Microstructure and Properties of Cu/Al Bimetallic Composite[J]. Crystals, 2023, 13: 794.
[3] Gu J, Yang S, Gao M, et al. Influence of deposition strategy of structural interface on microstructures and mechanical properties of additively manufactured Al alloy[J]. Additive Manufacturing, 2020, 34:101370.

Effect of Carbon Content on the Printability and Mechanical Properties of Ni-based Superalloy in Laser Additive Manufacturing

Shaofeng Chen[1,2], Wei Song[2], Nannan Lu[2], Jingjing Liang[2,*], Hao Yu[1] Wei Xu[1], Jinguo Li[2,*]
1. The State Key Laboratory of Rolling and Automation, Northeastern University, Shenyang 110016, China
2. Shi -Changxu Innovation Center for Advanced Materials, Institute of Metal Research, Chinese Academy of Sciences,
3. Shenyang 110016, China
*Corresponding address: jjliang@imr.ac.cn (J. Liang), jgli@imr.ac.cn (J. Li)

Abstract: The role of trace elements such as C, B, Zr, and Hf in additive manufacturing of superalloys remains a subject of debate. This study aims to analyze the impact of carbon content on the printability and mechanical properties of a specialized additive manufacturing superalloy. The findings indicate that the cracking susceptibility of the alloy initially increases and then decreases with rising carbon content. The porosity gradually decreases with the addition of carbon. However, as the carbon content increases, the stress rupture properties of the alloy progressively deteriorate. The mismatch in deformation between the carbides and the matrix triggers crack initiation and propagation at the carbide interface, resulting in diminished property. These results elucidate the influence of carbon content within a specific range on both printability and mechanical properties of the alloys, thus providing valuable guidance for regulating carbon content in future alloy designs.

Keywords: Additive manufacturing; Superalloy; Carbon; Printability; Mechanical properties

1 Introduction

Additive manufacturing technology offers a promising solution for the production of complex superalloy parts with advantages such as short production period, high precision and good formability [1,2]. However, grain boundaries with severe elements segregation and high internal stress are more susceptible to crack initiation. Carbon is the most common grain boundary strengthening elements in superalloys, typically added in the range of 0.06-0.15 wt.%. Carbon plays a complex role in nickel-based superalloys, and properties can be improved if the appropriate proportion of carbon is present [3]. In the field of additive manufacturing, carbides are considered to be one of the causes for crack formation [4]. However, recent studies have shown that even with a high carbon content of 0.6%, good formability can still be achieved in the additive manufacturing process, leading to a controversy surrounding the role of carbon [5]. In addition, in the performance research of additively manufactured superalloys, the main focus is on room temperature tensile properties, and there is little research on high temperature tensile and rupture properties.

In conclusion, further investigations are required to explore the roles and mechanisms of trace elements in AM superalloys. This study aimed to examine the relationship between varying carbon content in additively manufactured printability and properties. The results of this study can serve as a reference for determining the optimal carbon content for additive manufacturing of superalloys in the future.

2 Experimental procedure

In this study, an alloy specifically designed for additive manufacturing was utilized. Four different levels of C content (0 wt.%, 0.05 wt.%, 0.1 wt.%, 0.15 wt.%) in the alloy powders were prepared through gas atomization of the as-cast alloys. Directed energy deposition (DED) equipment was used to prepare the sample. The substrate was GH3536, and the samples size were 16×16×60 mm. Uniform laser power (2000 W), scanning speed (1100 mm/min), layer thickness (0.167 mm), spot diameter (2 mm), and powder feeding volume (11 g/min) were employed for all specimens.

3 Result and discussion

Alloy printability

Table 1 shows the statistics of some microstructure characteristics of as-built samples. The crack density did not show a linear relationship with carbon content. Instead, it increased initially, peaked at 0.06 wt.%, and then decreased significantly. Cracks were classified according to their causes, and the crack density was calculated separately. The statistical results were compared with the results of thermodynamic calculations, which well explained why the crack density first increased and then decreased. Owing to the rapid cooling rate in DED, carbides have limited time for growth following nucleation, resulting in minimal changes in carbide size with increasing carbon content. These finely granular carbides are expected to exert minor impact on liquid phase feeding. The reduction of porosity can be examined from two complementary perspectives. Firstly, during the inherent heat treatment process of additive manufacturing, carbides gradually grow and fill existing pores. Secondly, thermodynamic calculations indicate that the carbon lowers the pressure drop within the mushy zone, promoting the closure of pores.

Table 1 As-built sample microstrcture characteristics statistics

Alloy	Crack density (μm/μm^2)	Porosity (%)	Carbides Size (μm)
0C	2.48×10^{-4}	0.36	—
0.05C	3.27×10^{-4}	0.19	1.12
0.1C	5.32×10^{-4}	0.14	1.58
0.15C	3.22×10^{-4}	0.13	1.96

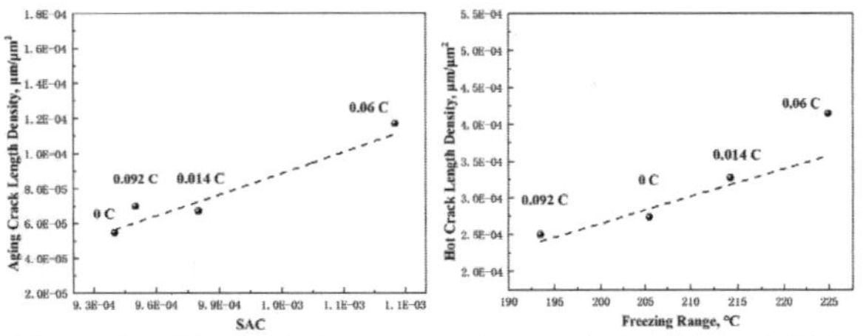

Figure 1 Relationship between aging crack density and SAC and between hot crack and solidification range.

Stress rupture properties

The stress rupture properties of the alloys under different conditions are presented in Table 2. It is evident that the alloys experience a decline in rupture properties as the carbon content increases. Since the size and volume fraction of the γ′ phase are minimally affected by the addition of carbon, the main source of performance degradation is likely the increasing presence of carbides. EBSD analysis reveals a notable stress concentration at the carbide interface, along with observations of dynamic recrystallization in the vicinity of the carbides. The large difference in thermal expansion coefficients and deformation coordination ability between carbides and the matrix lead to high stress levels at the interface. Furthermore, an excessive quantity of carbides tends to deplete solid solution elements, further diminishing the alloy's solid solution strength and facilitating dislocation movement within the matrix.

Table 1. Alloy rupture life under different conditions

Alloy	760°C/800MPa (h)	980°C/260MPa (h)
0.05C	132±15	26±4
0.1C	57±10	21±3
0.15C	15±5	12±4

4 Conclusion

Carbon can influence various factors including the solidification interval of the alloy and the precipitation rate of desired phases, thereby aiding in the prevention of crack initiation and reduction of porosity. However, excessive carbon content can significantly impair the stress rupture properties of the alloy. This deterioration is attributed to the high interface stress between the carbide and the matrix, which promotes microcrack formation. Therefore, when designing the alloy composition, a comprehensive consideration of printability and mechanical properties is essential, with careful control of the carbon content playing a crucial role.

Acknowledgments

This study was supported by the National Key R&D Program of China (Grant No. 2021YFB3702503), National Defense Foundation Strengthening Plan (Grant No. 2021-JCJQ-JJ-0092), Science Center for Gas Turbine Project (Grant No. P2022-C-Ⅳ-002-001) and National Natural Science Foundation of China (Grant No. 52204383).

References

[1] Dinda G P, Dasgupta A K , and Mazumder J. Laser aided direct metal deposition of Inconel 625 superalloy: Microstructural evolution and thermal stability. Materials Science and Engineering: A. 2009, 509: 98-104.

[2] Jia Q B. and Gu D D. Selective laser melting additive manufacturing of Inconel 718 superalloy parts: Densification, microstructure and properties. Journal of Alloys and Compounds. 2014, 585: 713-721.

[3] Wei C N, Bor H Y. and Chang L. The effects of carbon content on the microstructure and elevated temperature tensile strength of a nickel-base superalloy. Materials Science and Engineering: A. 2010, 527: 3741–3747.

[4] Tomus D, Rometsch P A, and Heilmaier M. Effect of minor alloying elements on crack-formation characteristics of Hastelloy-X manufactured by selective laser melting. Additive Manufacturing. 2017, 16: 65-72.

[5] Zhou W Z, Tian Y S, and Tan Q B. Effect of carbon content on the microstructure, tensile properties and cracking susceptibility of IN738 superalloy processed by laser powder bed fusion. Additive Manufacturing. 2022, 58: 103016.

Formation Mechanism of Lamellar Structure and Anisotropy Control of Ceramic Core via Stereolithography Additive Manufacturing

Qiaolei Li*, Chaowei Zhang, Jingjing Liang, Yizhou Zhou, Xiaofeng Sun, Jinguo Li*
Institute of Metal Research, Chinese Academy of Sciences, 72 Wenhua Road, Shenyang, Liaoning, 110016, China
*Corresponding address: E-mail: jgli@imr.ac.cn, qlli@imr.ac.cn

Abstract: The influence mechanism of zircon distribution on sintering shrinkage was analyzed, and the strengthening mechanism of mullite on cores was analyzed from thermodynamics and dynamics. Through the comprehensive evaluation of the properties of ceramic slurry and core, the best choice of alumina ceramic core was the zircon content of 15%. The bending strength of alumina-based ceramic core increased from 14.8 MPa to 61.54 MPa, an increase of 315.8% at 25 ℃. The bending strength of alumina-based ceramic cores increased from 4.91 MPa to 11.59 MPa, an increase of 136% at 1500 ℃. The shrinkage in the height direction was reduced by 21%, which better weakens the anisotropy of the shrinkage of 3D printing alumina-based ceramic cores. Nano zirconia phase and nano mullite phase were formed by strengthening alumina ceramic cores with zircon, which improved the comprehensive properties of alumina-based ceramic cores. The successful preparation of 3D printing ceramic cores by stereolithography has promoted the industrial application of 3D printing ceramic cores technology in the field of preparing ceramic cores with complex structures.

Keywords: additive manufacturing; ceramic cores; anisotropy; flexural strength; sintering shrinkage

1 Introduction

The hollow blade of single crystal turbine is the key component of aeroengine[1]. The temperature resistance of the blade is affected by the complexity of the hollow structure[2]. The cooling channel inside the hollow blade is formed by the ceramic core, so the preparation of the ceramic core is particularly critical[3]. At present, the traditional hot injection molding technology is facing the challenge of difficult preparation of ceramic core with complex structures[4]. The stereolithography 3D printing has become a new choice of ceramic cores because it is apt to prepare ceramic cores with complex structures with low cost and high accuracy[5]. However, the disadvantage of anisotropy is still the key problem that restricts the wide industrial application of 3D printing ceramic cores technology[6].

In order to improve the strength of the ceramic core and improve the anisotropy of the ceramic cores at the same time, this study used stereolithography 3D printing technology to add zircon to the alumina-based ceramic cores, which significantly improved the anisotropic sintering shrinkage of the ceramic cores and improved the flexural strength of the ceramic cores.

2 Experimental procedure

The ceramic powder was composed of sieved alumina and zircon(sieve aperture 75 μm). The volume fraction of the zircon is 0%, 5%, 10%, 15%, 20%. The ceramic slurry was printed vat photopolymerization 3D printer (Beijing Ten Dimensions Technology Co., Ltd) to obtain 40×10×4 mm ceramic bodies. The laser energy was 3.56 mW/cm^2, the exposure time of each layer was 5 s, and the thickness of each layer was 100 μm. The ceramic bodies were embedded in 100 μm corundum sand and sintered in air.

3 Result and discussion

1. Slurry optimization design

Fig. 1 shows the viscosity of alumina ceramic slurry at room temperature with different zircon content. The viscosity of alumina ceramic slurry with different zircon content was less than 1500 mPa/s, and the viscosity first decreased and then increased with the increase of zircon content.

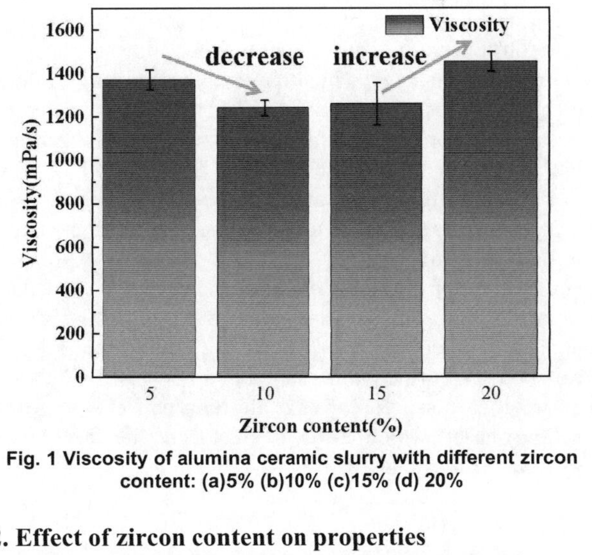

Fig. 1 Viscosity of alumina ceramic slurry with different zircon content: (a)5% (b)10% (c)15% (d) 20%

2. Effect of zircon content on properties

Fig. 2 shows the performance test results of ceramic cores with different zircon contents. After adding zircon, zircon reduced the sintering shrinkage because it was evenly distributed between alumina particles, which hindered the sintering of alumina particles. When the zircon content was less than or equal to 20%, the porosity of the ceramic cores remains above 30%. When the zircon content increased from 0% to 15%, the bending strength of alumina ceramic cores increased sharply at 25 ℃.

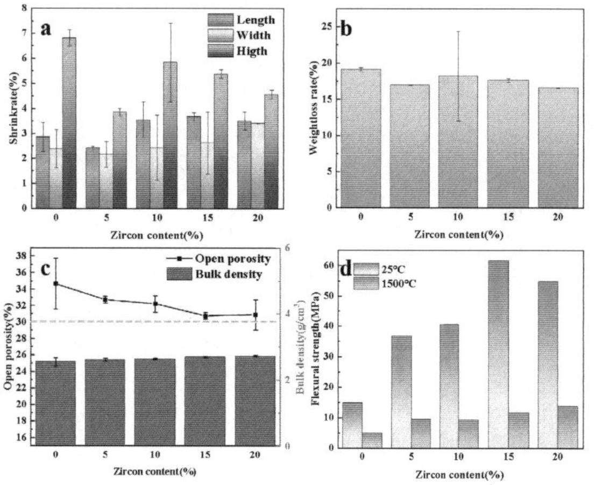

Fig. 2 Properties of alumina ceramic cores with different zircon contents:(a) Shrink rate (b) Weight loss rate (c) Open porosity and bulk density (d) Flexural strength

3. Formation and strengthening mechanism of printing layered interface

Fig. 3 shows the influence of zircon on the interface of alumina core. The distribution of zircon in the interface of ceramic layer. Large zircon particles were mainly distributed in the interior of the lamellar structure, while small ceramic particles were distributed at the edge of the lamellar structure. The decomposition of zircon particles and the diffusion behavior of silicon. The decomposition of zircon was accompanied by the diffusion and migration of silicon to alumina, and viscous flowed in the pores of ceramic core particles, which reduced the porosity of core and hinders crack propagation. Zirconia and silica were precipitated from the surface layer of zircon. Zirconia, which was distributed between alumina and zircon particles as a high-temperature stable phase, migrated to the pores and lamellae of alumina particles and hindered the shrinkage of printing layer. As an unstable phase, silica diffused from zircon to alumina particles, and part of silica filled between alumina particles in viscous flow, which promoted lamellar sintering of alumina and weakened the influence of lamellar on core. In addition, part of silica dissolved around alumina particles to form mullite phase, which further enhanced the interlayer performance of the layer.

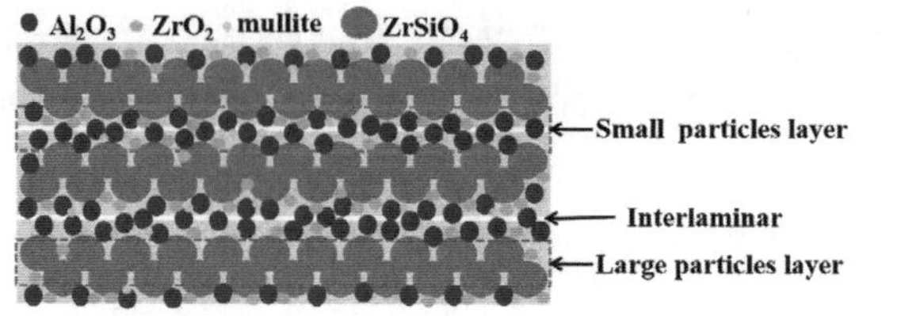

Fig. 3 Formation and strengthening mechanism of printing layer structure.

4 Conclusion

When the zircon content was 15%, the comprehensive performance of alumina ceramic core is the best. By 3D printing zircon to strengthen alumina ceramic core, while maintaining 30% open porosity, the bending strength of alumina ceramic cores at 25 ℃ was increased from 14.80 MPa to 61.54 MPa, and the bending strength at 1500 ℃ was increased from 4.91 MPa to 13.76 MPa.

References

[1] J. Liang, Q. Lin, X. Zhang, et al. Effects of alumina on cristobalite crystallization and properties of silica-based ceramic cores[J]. Journal of Materials Science and Technology, 2017, 33: 204-209.

[2] Q. Li, J. Liang, Y. Zhang, et al. Fused silica ceramic core based on network-structured zircon design via 3D printing[J]. Scripta Material, 2022, 208: 114342.

[3] Padture N., Gell M., Jordan E.H., Thermal barrier coatings for gas-turbine engine applications. Science 2002, 296: 280-284.

[4] C. Bae, K. Daniel, W. Halloran. Mechanical and kinetic studies on the refractory fused silica of integrally cored ceramic mold fabricated by additive manufacturing[J]. Journal of the European Ceramic Society, 2019, 39: 618-623.

[5] Z. Lu, J. Cao, H. Jing, et al. Review of main manufacturing processes of complex hollow turbine blades[J]. Virtual and Physical Prototyping, 2013, 8: 87-95.

[6] A. Kazemi, M. Faghihi-Sani, M. Nayyeri, et al. Effect of zircon content on chemical and mechanical behavior of silica-based ceramic cores[J]. Ceramics International, 2014, 40: 1093-1098.

Defect Control of Ceramic Core via Stereolithography Additive Manufacturing

Xiantian Meng*, Qiaolei Li, Chaowei Zhang, Jingjing Liang, Yizhou Zhou, Xiaofeng Sun, Jingguo Li
Institute of Metal Research, Chinese Academy of Sciences, 72 Wenhua Road, Shenyang, Liaoning, 110016, China
* E-mail: xtmeng@imr.ac.cn, jgli@imr.ac.cn

Abstract: Stereolithography additive manufacturing technology is used to prepare complex double-walled ceramic core. In order to study and control the defects of complex double-walled ceramic core, the slurry flow in printing process is simulated. The defects of double-wall core of complex structure are observed, the causes of defects are explained, and the corresponding solutions are put forward. The effects of placement position and different parameters on complex double-walled ceramic core are studied. The results show that the placement position has a significant effect on the yield of complex ceramic core. The optimization of process parameters such as slurry viscosity, slurry spreading speed, laser moving speed and irradiation time can significantly reduce the defects of core green body. Based on the systematic observation of complex core structure, the defect prediction, evaluation and process optimization of ceramic core green bodies are proposed.

Keywords: additive manufacturing, ceramic core, defect, simulated, control method

1 Introduction

Hollow turbine blades are crucial components of aeroengines[1-3]. In order to improve the thrust-weight ratio of the engine, higher requirements are put forward for the turbine inlet temperature of the aero engine. In the investment casting process, the ceramic core is the key to forming complex cooling channels inside the blade. Traditional hot injection moulding process cannot meet the increasingly complex cooling channel requirements. The emergence of SLA technology makes it possible to prepare ceramic cores with complex structures[4-5]. In this paper, the defect factors and control methods of SLA technology for preparing complex double-wall ceramic cores are studied.

2 Experimental procedure

The experiment will use a silicon-based ceramic slurry consisting of a premix consisting of an active monomer (HDDA+PPTTA), a photoinitiator and a dispersant, and a ceramic fraction. High purity fused quartz powder and zirconia are used as the powder composition for preparing complex ceramic core. The powder is screened by 1000 mesh. The samples are heated to 600℃ at 1℃/min and degreased by holding for 120min. Then the temperature is raised to 1250℃ at 1℃/min, and the heat is kept for 600min for sintering. The complex structure ceramic core is prepared by CeraBuilder100Pro equipment. Fluid simulation analysis is carried out by simulation software. The microstructure of the sample is observed with Zeiss AXIOVERT200MAT optical microscope.

3 Result and discussion

1. Influence of slurry viscosity on defects and solution

Slurry viscosity is one of the key factors to determine the success of core printing. In the SLA equipment printing process, the viscosity is too low will make the slurry backflow, cannot control the thickness of the material, resulting in each layer of printing thickness thicker, accumulated will make the scraper scrape down the part [6-8]. Figure 1 shows this. The appropriate viscosity range can avoid defects and make the equipment operate efficiently.

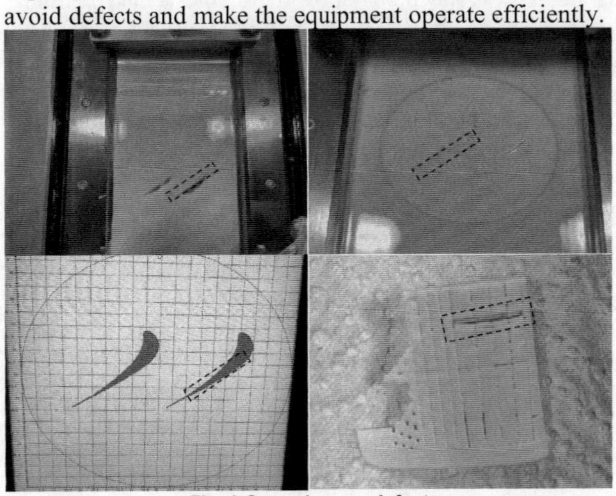

Fig. 1 Ceramic core defects.

2. Influence of model placement on defects and solutions

The mode of model placement is also one of the key factors affecting the success of core printing, and some special structures in complex structural core have an impact on the laying process. The influence of model placement mode can be found in time through the fluid simulation of the laying process by simulation software[9]. The effect can be reduced or eliminated by modifying the display mode. Fig. 2 shows the influence of model placement on defects and solutions.

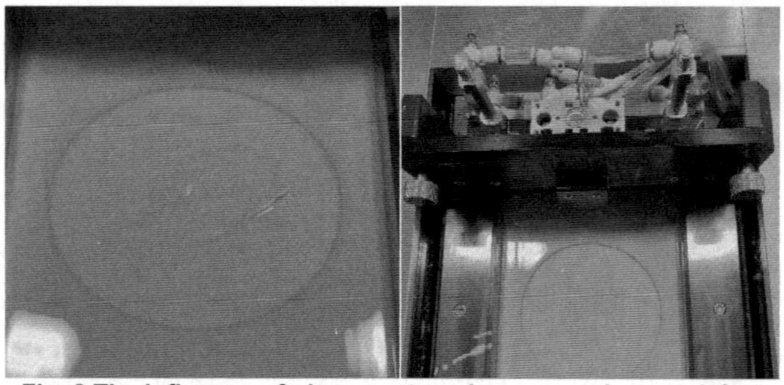

Fig. 2 The influence of placement mode on ceramic core defects.

3. Influence of process parameters on defects and solutions

Parameters such as the moving speed of the scraper and the size of the laser spot are also key factors affecting the success of core printing, and the combination of the appropriate slurry viscosity and the corresponding laying speed can avoid the production of defects in the printing process. The laser parameters not only determine the success of printing but also affect the accuracy, and the appropriate laser parameters will determine the size of the printing window period.

4 Conclusion

The defects in the preparation of double-walled ceramic cores with complex structure by SLA technology may be caused by slurry viscosity, model placement and process parameters. Through simulation and analysis, corresponding measures are adopted to reduce or eliminate the defects and improve the printing success rate.

Acknowledgments

This work was financially supported by the the National Defense Basic Scientific Research Program of China (No. JCKY2022130C005); China Postdoctoral Science Foundation (No. 2023M743571); Postdoctoral Fellowship Program of CPSF (No. GZC20232743); National Natural Science Foundation of China (Nos. U234120139 and U22A20129); National Key Research and Development Program of China (Nos. 2021YFB3702503 and 2018YFB1106600); Innovation Project of IMR (2024-PY11); Open Research Fund of National Key Laboratory of Advanced Casting Technologies (No. CAT2023-006).

References

[1] Q.L Li, W.Q Hou, J.J Liang, et al. Controlling the anisotropy behaviour of 3D printed ceramic cores: From intralayer particle distribution to interlayer pore evolution[J]. Additive Manufacturing,2022,58.

[2] Y.H Mu, J.W Chen, X.LAn, et al. Defect control in digital light processing of high-solid-loading ceramic core[J]. Ceramics International,2022,48(19):28739-28744.

[3] K.H Hu, Z.G Lu, K. L, et al. Additive Manufacturing of Complex Ceramic Cores and Verification of Casting Process[J]. Journal of Mechanical Engineering, 2021, 57(03): 227-234.

[4] K.Q Zhang, Q.Y Meng, Z.L Qu, et al. A review of defects in vat photopolymerization additive-manufactured ceramics: Characterization, control, and challenges[J]. Journal of the European Ceramic Society, 2024,44(3): 1361-1384.

[5] Q.L Li, Y.X Qiu, W.Q Hou, Jet al. Slurry flow characteristics control of 3D printed ceramic core layered structure: Experiment and simulation[J]. Journal of Materials Science & Technology,2023,164: 215-228.

[6] J. Li, X. An, J. Liang, Y. et al. Recent advances in the stereolithographic three-dimensional printing of ceramic cores: challenges and prospects[J].Journal of Materials Science & Technology,2022, 117:79-98.

[7] C. Hinczewski, S. Corbel, T. Chartier. Ceramic suspensions suitable for stereolithography[J]. Journal of the European Ceramic Society,1998,18(6):583-590.

[8] S. Zakeri, M. Vippola, E. Levanen. A comprehensive review of the photopolymerization of ceramic resins used in stereolithography[J]. Additive Manufacturing.2020,35.

[9] Alfred I.Y. Tok, Freddy Y.C. Boey, Y.C. Lam. Non-Newtonian fluid flow model for ceramic tape casting[J]. Materials Science and Engineering: A, 2000, 280(2): 282-288.

Research on the High-Temperature Concession of Internal Topology Structure Sand Mold

Baozhi Li[1, 2], Junhang Xu[1, 2], Zhaowei Song[1, 2], Yunbao Gao[1, 2], Jingming Li[1, 2], Zengrui Wang[1, 2,*]
1. National Key Laboratory of Advanced Casting Technologies, Shenyang 110022, China
2. Shenyang Research Institute of Foundry Co., Ltd. CAM, Shenyang 110022, China
*Corresponding address: e-mail: wangzr_srif@foxmail.com

Abstract: For the existing 3DP sand casting, castings are prone to casting crack defects, which increases their scrap rate and greatly wastes resources. Moreover, if castings have crack defects, there will be significant safety hazards after service. One of the reasons for the occurrence of crack defects is the poor high-temperature concession performance of sand molds. Therefore, this article designs various internal topology structures of 3DP sand molds and studies the influence of different internal topology structures on the high-temperature concession performance of sand molds.

Keywords: 3D printed non-mold sand forming; topological structure; high-temperature concession

1 Introduction

3D printing technology allows for flexible and unrestricted design, enabling the production of more complex structures with high dimensional accuracy [1]. Taking advantage of the strengths of 3D printing, numerous scholars have conducted extensive research on the topology structure to further enhance the performance of 3D printed sand molds. Wang et al. [3] proposed a constrained topology optimization method to redesign 3DP sand· molds, developing a new design framework for complex structural metal castings. Structural analysis simulations proved that despite a 50% reduction in the bracket weight, the mechanical performance of the optimized design was increased by 30%. Sama et al. [4] found that both parabolic and conical helix sprues improved the quality of complex gating systems when using 3D sand printing. Therefore, the research and practical application of the internal topology structure of 3D printed sand molds are of great significance. Previous studies on the use of sand mold topology focused on complex gating systems, improving the ability to fill shrinkage during solidification, and testing the topology of castings. However, there remains a gap in research regarding the improvement of concession properties and the elimination of crack defects in 3D printed sand molds through modifications in internal topology structure.

2 Experimental procedure

Six samples were prepared using a VX2000 3D experimental machine. The high-temperature thermal expansion stress experiment of the sand molds was conducted using a ZGY resin sand high-temperature performance tester, under controlled conditions of room temperature at (25±5) ℃ and relative humidity at (60±5)%. In this experiment, the samples were placed into a furnace of the ZGY tester preheated to 1,000 ℃ with a preload compression value of 10 N. While maintaining the sample length unchanged, the force values for different samples were recorded, and the maximum value was considered as the limited thermal expansion stress. By comparing the peak values of thermal expansion force among the different samples, an analysis of the improvement in high-temperature concession can be conducted.

Table 1: Internal structure and size of specimen

Sample serial number	Internal mesh shape and size
A	Solid sand mold structure
B	Each layer has 3 mm×2 4 mm square grids, a total of 2 layers, and the square spacing of 2 mm
C	Each layer has 3 mm×2 5 mm square grids, a total of 2 layers, and the square spacing of 2 mm
D	Each layer of 3 mm×2 6 mm cube grid, a total of 2 layers, cube interval of 2 mm
E	Each layer of 3 mm×2 6 mm sphere grid, a total of 2 layers, the sphere spacing of 2 mm
F	The internal cavities are connected to each other. The side length of the bottom surface of the skeleton is 4 mm square , and the interval between parallel cuboids is 2 mm

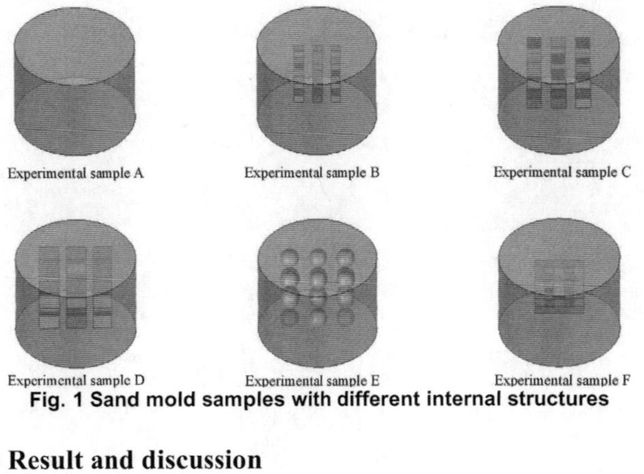

Fig. 1 Sand mold samples with different internal structures

3 Result and discussion

The high-temperature thermal expansion force peak values of different topology grids and skeleton structures are compared, as shown in Fig. 2. The results demonstrate that, in comparison to the original sand mold sample with a peak value of 44.6 N for high-temperature thermal expansion stress, the stress peaks of the modified sand molds with altered internal cavity structures exhibit a reduction of approximately 24.4%-51.3%. Among the various samples, the improvement in thermal expansion stress for the internal structure samples with cubic shapes and side lengths of 4-6 mm is quite similar. It can be seen that as the grid volume increases and the resin content decreases, the thermal stress decreases accordingly. Moreover, the spherical shape with a diameter of 6 mm shows superior retraction performance compared to the internally tangent square with a side length of 4 mm, resulting in a decrease of 4.7 N in thermal expansion stress. The skeleton structure depicted in the figure, which demonstrates the highest reduction of 51.3% in the peak value of high-temperature thermal expansion stress, features interconnected internal topology cavities. This structure ensures sand mold excellent high-temperature concession performance while meeting the required strength for sand backing in casting.

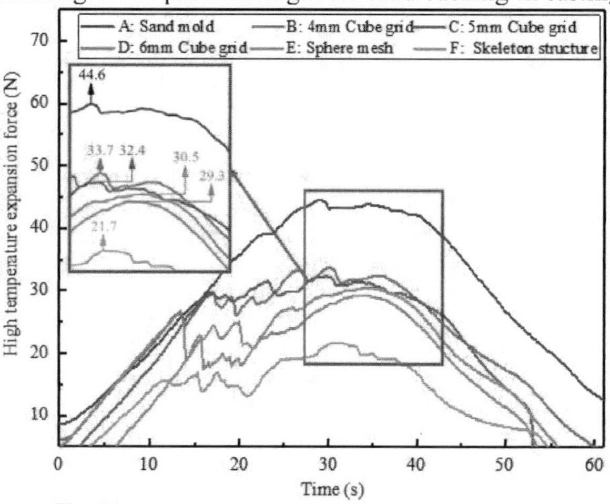

Fig. 2 High-temperature thermal expansion stress curve

4 Conclusion

Due to the improvement of the internal cavity structure, the overall amount of resin added decreased, resulting in a decrease of approximately 24.4%-51.3% in the peak thermal expansion stress compared to the original sand mold sample. The improvement of high-temperature yielding performance of the skeleton structure is excellent. Due to the interconnection of internal topological cavities, the total volume of cavities is larger and can meet the requirements of casting for sand strength. The peak high-temperature thermal expansion stress is reduced by 22.9 N.

References

[1] Shi Y, Zhang J, Wen S, et al. Additive manufacturing and foundry innovation. China Foundry, 2021, 18(4): 286-295.

[2] Wang J, Sama R S, Manogharan G. Re-thinking design methodology for castings: 3D sand-printing and topology optimization. International Journal of Metalcasting, 2019, 13(1): 2-17.

[3] Sama R S, Badamo T, Lynch P, et al. Novel sprue design in metal casting via 3D sand-printing. Additive Manufacturing, 2018, 25: 563–578.

[4] Shangguan H, Kang J, Yi J, et al. The design of the 3D printed lattice reinforced thickness-varying shell mold for casting. Materials, 2018, 11(4): 535-545.

On 3D Printing Spatial Grid Sand Mold

Zhaowei Song, Jingming Li, Rui Chen*, Ling Zhao, Yunbao Gao, Haijun Zhang
National Key Laboratory of Advanced Casting Technologies, China Academy of Machinery Shenyang Research Institute of Foundry Co., Ltd, Shenyang 110022 , P. R. China
*Corresponding address: No.17 Yunfeng South Street, Tiexi District, Shenyang City, Liaoning Province, China.
E-mail: 13998261547@126.com

Abstract: Conventional 3D printing methods face limitations in simultaneously producing a high-strength sand facing layer and a low-strength sand backing layer. The high-strength sand layer typically exhibits poor air permeability and collapsibility, resulting in inefficient use of strength in the central part of the sand mold. In order to address this challenge in 3D printing solid sand molds for casting, a novel approach involving the use of a spatial grid sand mold has been proposed. This method allows for simultaneous printing of high-strength and low-strength sand layers. By employing a spatial grid structure, this approach offers several advantages. It not only optimize the distribution of strength throughout the mold, but also enables the production of sand molds with improved air permeability and collapsibility.

Keywords: spatial meshing, 3D printing, sand mold performance, air permeability

1 Introduction

The origins of 3D printing technology can be traced back to the United States in the 1980s. Over the course of several decades, it has found widespread application in various industries such as automotive, energy, healthcare, and aerospace [1]. Within the realm of casting, the application of 3D printing technology primarily encompasses prototype or pattern printing [2], part printing, as well as sand mold and core printing. There are many advantages associated with 3D printed sand molds. They create a favorable production environment, leverage digital transmission to expedite information exchange, eliminate the need for pattern production, thereby reducing production cycle times, enhancing the dimensional accuracy of sand molds, consequently decreasing the machining allowance for castings, and ensuring product design flexibility.

2 Experimental procedure

As depicted in Fig. 1, when modeling the 3D sand mold for casting, a spatial grid structure is utilized, incorporating a "facing sand layer" whose thickness is contingent upon the sand mold size and the casting specifications. The grid comprises loose sand and is not treated with a binder. Normal printing and spraying operations are conducted within the interstices of the grids. These loose sand grids remain unbonded, with the frames between them constituting the backing sand. Ultimately, the "facing sand" satisfies the casting's strength prerequisites, while the backing sand layer enhances the air permeability of the sand mold, reduces binder consumption, and diminishes gas evolution from the sand mold, thereby lowering costs and enhancing overall mold performance.

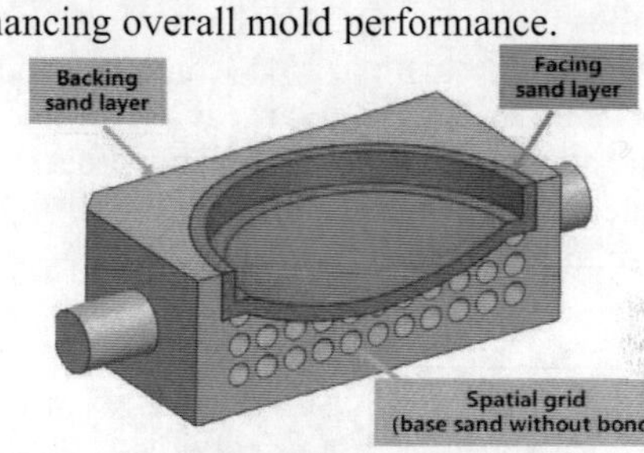

Figure 1 Schematic diagram of cross-section of the spatial meshed sand mold.

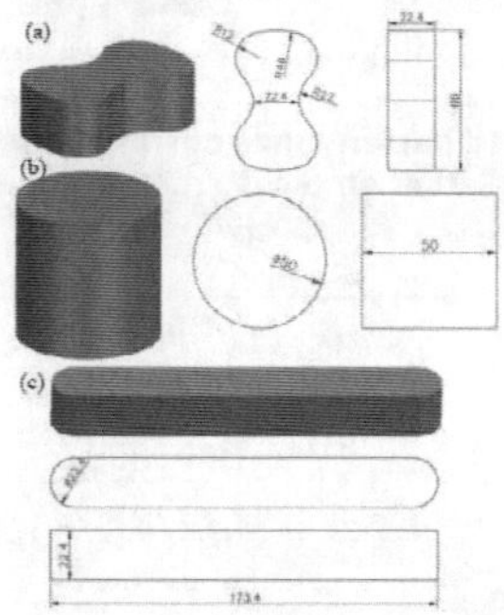

Figure 2 Standard solid sand sample

3 Result and discussion

High surface strength of the mold and good air permeability of the core are desirable during the casting process. In this experiment, cubic grids with a decreasing circumferential sphere diameter from 4 mm to 0.2 mm per layer is employed, arranged concentrically from large to small diameters, thus segmenting the internal space of the sand mold into gradient grids of varying sizes, as shown in Fig. 3. The performances of the gradient mesh sand mold are detailed in Table 1.

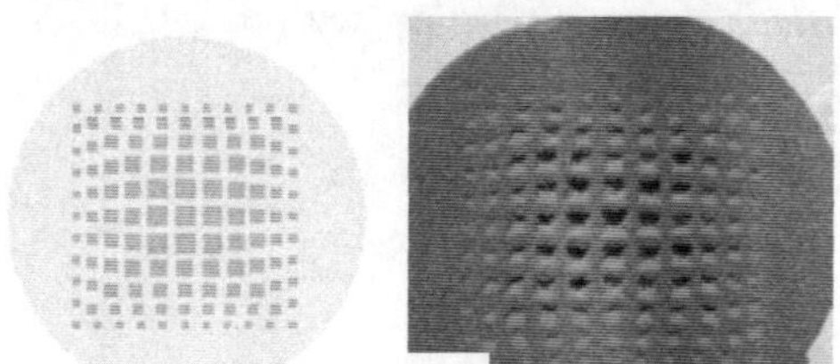

(a) (b)

Figure 3 Gradient mesh division diagram.

Table 1 Sand mold performance corresponding to gradient mesh.

Grid shape	Bending strength (MPa)	Tensile strength (MPa)	Compression strength (MPa)	Breathability	Binder dosage ratio	Binder reduction (%)
0-4 mm gradient mesh	1.395	1.178	5.121	160	0.75	42.2
Cube with side length 2 mm	1.395	1.234	4.832	150	0.96	29.6

A space is configured within the interconnected cavity structure, as illustrated in Fig. 4. The cavity takes on a rectangular shape with a bottom side length of 3 mm. The interspace between the rectangular cavity is filled with parallel sprayed adhesive powder sand. This process will be directly executed within the 3D printing sand mold. The detection outcomes are outlined in Table 2.

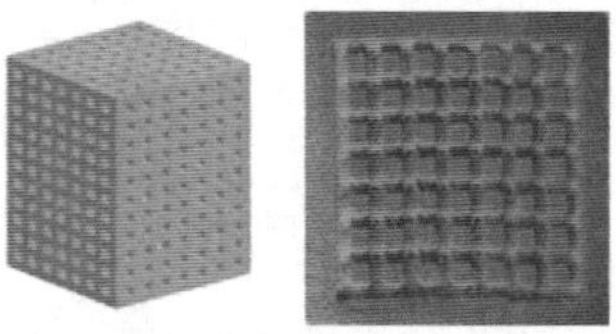

Figure 4 Topological skeleton division diagram.

Table 2 Sand mold performance corresponding to topological frameworkThe simulation employs electrical parameters and input power.

Grid shape	Bending strength (MPa)	Tensile strength (MPa)	Compression strength (MPa)	Breathability	Binder dosage ratio	Binder reduction (%)
Topological skeleton	1.101	0.926	4.114	160	0.65	50%
Cube with side length 3 mm	1.589	1.402	5.402	110	1.14	12.50 %

Combined Table 1 with Table 2, it can be found that the binder dosage is approximately 50% lower than that of the solid sample. The tensile strength, bending strength, and the compressive strength measured 0.925 MPa, 1.101 MPa, and 4.114 MPa, decreased by 42.8%, 0.396%, and 29.1%, respectively, compared to the solid sample, while the air permeability yielded a value of 160, a 190% increase compared to the solid sample.

In comparison to the sand mold comprising 3 mm cubes, the topological skeleton demonstrates an obvously enhancement in air permeability with far less binder dosage. Although the tensile strength is decreased by about 30%, it still meets the casting requirements for the base sand.

The internal penetration of the topological grid's cavity can function as an exhaust channel, leading to a more substantial improvement in air permeability compared to that achieved with a conventional grid.

Furthermore, topology optimization can be employed to further refine the design of the topological skeleton. This involves determining the topological configuration of the connections of the structural elements, as well as the presence or absence of cavities in the structure, along with their number and location, based on the given design domain, constraints, and loading conditions.

4 Conclusion

This study demonstrates that the utilization of spatial grid sand molds can decrease the strength of the sand mold by 10% to 50%, augment air permeability by over 100%, and reduce the amount of binder needed by 10% to 50%. Additionally, the employment of space lattice sand molds improves the deformability, air permeability, and air release of the sand mold. Due to its advantages of low strength, high air permeability, and reduced binder dosage, space lattice sand molds are well-suited for back-sand printing.The strength of a simple cubic packing sand mold is optimal. Hexagonal dense packing and face-centered cubic packing exhibit improved air permeability with reduced binder requirements. The distance mechanism sand between topological cavities enhances air permeability owing to the connectivity between cavities.

Acknowledgments

The authors appreciate the financial support by the National Nature Science Foundation of China(No. U1808216).

References

[1] Druschitz A, Williams C, Snelling D, et al. Additive manufacturing supports the production of complex castings. 143rd Annual Meeting and Exhibition, USA: Shape Casting, 2014: 51-57.

[2] Sun D, Xu X C, Wang D H, Liu K F, et al. Rapid investment casting process for impeller based on 3D printing technology. Special Casting & Nonferrous Alloys, 2016, 36(11): 1172-1174. (In Chinese).

Mega and Giga Casting: A New Technological Paradigm for Die Material and Design

Isaac Valls, Thibault Le Bourdiec*, Anwar Hamasaiid
ROVALMA S.A., C/ Collita 1-3, 08191 Rubí, Spain
*Corresponding address: e-mail: tlebourdiec@rovalma.com

Abstract: The sociotechnological transition to electric mobility requires much lighter, more economic and more sustainable life cycle Electric Vehicles (EVs). To meet with these conditions, the technological ecosystem of the automotive industry needs to develop and shape new light materials, complex and large geometries as well as functionally and compositionally graded components, at reduced cost. Aluminium based alloys are among the choice candidates to meet these requirements considering their high formability and recyclability features, as well as good combination of density, mechanical and physical properties.

According to a study carried out for the European Aluminium Association [1], the aluminium content of passenger vehicles in Europe will increase from 205 kg in 2022 to 256 kg in 2030. Very similar forecasts are also cited for American vehicles [2]. Accordingly, the decrease in demand for castings associated with the combustion engine will be largely counterweighed by the demand for new aluminium based components for EVs, such as for motor housings, BEV and PHEV battery enclosure assemblies and different structural parts. The demand for automotive structural parts, manufactured by die casting, is expected to increase sharply from 8.2 million in 2021 to 25 million in 2030 [3]. The cited studies agree that more than 50% of the projected aluminium based parts will be shaped by die casting methods in particular High Pressure Die Casting (HPDC). In these studies, the fast uptake of Mega and Giga high pressure die casting was not taken into consideration. Hence, yet much higher demand for HPDC parts can be expected than forecasted in the coming years.

The geometries of these new parts, the new aluminium alloys, and the size of the components are significantly changing the technological paradigm of the die casting process in particular die making processes. There are indisputable needs to new die design, cooling strategies, die materials, die machining and die life management technology in order to address the challenges related to the three Key Technological Enablers (KTEs) of Mega and Giga casting which are i) durability of the expensive dies and the related die components, ii) the productivity of the process (scrap rate, cycle time, production stability, etc.) and iii), the part and process quality and the related monitoring and controlling technologies.

In this paper, three innovative technologies, with significant potential to address the challenges related to the above mentioned three KTEs will be presented and discussed.

The first technology is the disruptive high thermal conductivity of hot work tool steels for die casting known as HTCS© grades which features very high thermal conductivities, namely up to 55-62 W/m·K compared to conventional tool steel grades, used as materials for dies that present much lower thermal conductivity, typically between 20-29 W/m·k.

Considering the size for Mega and Giga casting dies, new grades of high thermal conductivity tool steel have been specially optimised and further developed for these new promising casting processes. These grades are named FASTCOOL®-55, FASTCOOL®-20 and FASTCOOL®-35. The table below shows some properties of these grades compared to conventional EN-DIN 1.2344 / H13.

Die Materials	Thermal Conductivity	Wear Resistance	Hardness Max.	Heat Treatment
1.2344 (H13)			54 HRc	Quench + Temper
FASTCOOL®-35			43 HRc	Aging
FASTCOOL®-55			53 HRc	Quench + Temper
FASTCOOL®-20 (Mainly for LPDC and Gravity Casting)			420 HB	Prehardened

FASTCOOL®-35 is particularly promising for big dies as it can be hardened by precipitation, achieving the mechanical properties optimized for dies of die casting by a simple age hardening process. This hardening method allows reducing the die making time often by more than two weeks. Additionally, the high thermal conductivity allows to significantly increase the productivity and decrease the scarp rate through reducing the cycle time and hot spots generation. Several industrial use cases that will be presented in the paper further highlight these advantages.

The second innovative technology is a disruptive additive manufacturing methods that is able to print the above mentioned high productivity die tool steels as well as carbon based conventional hot work too steels. This Metal Additive Manufacturing method, ROVALMA® MAM can print high performant die components with speeds and costs that are competitive to conventional die making by machining, in addition to the possibility to print large sizes up to 600 x 700 x 2200 mm, while offering moreover the flexibility to print conformal cooling ducts beyond the known State of Art (SoA) of MAM technology.

Furthermore, over the past few years, there has been a notable surge in the adoption of MAM technology across diverse sectors including tooling. This is attributed to the newfound design freedom they offer, allowing for the realization of optimized geometries that were once unattainable through conventional die making methods, in particular when it comes to conformal cooling systems inside the dies. Unfortunately, most of SoA of MAM methods, in addition to the prohibitive costs of the process, do not result in the thermo-mechanical properties required from the tool. It is hence not surprising that several companies and institutes have started to invest great efforts in mitigating the present limitations of available MAM technologies, specially through combining MAM and conventional manufacturing technologies.

ROVALMA has charted a distinct course by introducing a groundbreaking, cost-effective MAM technology. The ROVALMA® MAM approach enables the production of large, intricately designed metal components with high performance, while maintaining a competitive production cost along with pushing the potential for environmentally sustainable manufacturing of MAM parts beyond the hereto existing technologic boundaries. This new technology allows manufacturing of dies and die components by combining conformal cooling with advanced materials featuring high thermal conductivity and tribological properties, at a considerably lower production cost and environmental impact, compared to conventionally manufactured tools, and even versus alternative MAM technologies.

Currently, there are MAM technologies that succeed in manufacturing large components, such as WAAM, but have difficulties in terms of yield, design flexibility with cooling channels and material selection (only those with good weldability). ROVALMA® MAM offers a performance that is capable of exceeding forged material counterparts and an extremely wide choice of materials can be used to manufacture large high performance components. In terms of cost, ROVALMA® MAM is often more cost-effective than conventional manufacturing for a large range of geometries, especially when part weight is to be reduced.

Nowadays, a very limited range of die components is produced with conformal cooling and additive manufacturing with disputable success rate in terms of cooling performance and durability. This is because in the majority of case, the maraging steel type EN-DIN 1.2709 is used due to the limitation of the SoA of MAM technology to print other carbon-based tool steels. However, 1.2709 steel presents a very low thermal conductivity, typically between 14-18 W/m.K and very low abrasive wear resistance, at the working temperature range of die casting. Therefore, often the improvement gained by conformal cooling is diminished by the low thermal conductivity and low durability of the die material used.

In this paper, the advantage of the combination of high thermal conductivity tool steels with ROVALMA® MAM and innovative cooling system design will be highlighted through different uses cases.

As Mega castings, casted alloys, MAM, conformal cooling and new steels are pushing the boundaries of the cooling systems and die design, a performant thermal management and monitoring of the dies are also increasingly required to monitor and extend the die life span, reduce scrape and maintain an optimised productivity considering the metallurgical characteristics of both die materials and casting alloys. In this paper, the combination of above-mentioned innovative technologies with advanced embedded sensors will be also presented. These types of sensors are multirole sensors with the capacity of measuring the temperatures of the injected alloy and die together using the most advanced temperature measuring methods. The figure bellow explains the temperature measured by one of that multirole sensors embedded to a die run on a clod chamber die casting machine with clamping force of 1650T. The casted alloy was Al-9Si-3Cu (A380). These measurements are directly related to the die and process working conditions through advanced digital technologies.

References

[1] Ducker Research & Consulting, Aluminium Content in Passenger Vehicles (Europe) 2022 assessment and outlook 2026, 2030, public summary, April 2023.

[2] Svendsen A., Aluminium Continues Unprecedented Growth in Automotive Applications, Light Metal Age Magazine, 2020.

[3] Roos H.J., Lagler M. et Quintana L: Whitepaper: The Future of Structural Components in HPDC. Bühler AG, 2019.

Customized Heat Treatment Process Enabled Excellent Mechanical Properties in Wire Arc Additively Manufactured Mg-RE-Zn-Zr Alloys

Dong Ma[1, 2, 3], Chunjie Xu[1, 2, 3*], Shang Sui[1, 2, 3]

1. School of Materials Science and Engineering, Xi'an University of Technology, Xi'an, Shaanxi, China
2. Xi'an Shechtman Nobel Prize New Materials Institute, Xi'an, Shaanxi, China
3. Xi'an Key Laboratory of Advanced Magnesium Alloy Additive Manufacturing and Precision Forming, Xi'an, Shaanxi, China
*Corresponding address: Beilin District, Xi'an City, Shaanxi Province; e-mail: xuchunjie@xaut.edu.cn

Abstract: Customized heat treatment is essential for enhancing the mechanical properties of additively manufactured metallic materials, especially for the alloys with complex phase constituents and heterogenous microstructure. However, the interrelated evolutions of different microstructure features make it difficult to establish optimal heat treatment processes. Herein, we proposed a method for customized heat treatment process exploration and establishment to overcome this challenge for such kind of alloys, and a wire arc additively manufactured (WAAM) Mg-Gd-Y-Zn-Zr alloy with layered heterostructure was used for feasibility verification. Through this method, an optimal microstructure (fine grain, controllable amount of LPSO structure and nano-scale β' precipitates) and the corresponding customized heat treatment process (520 °C/ 30 min + 200 °C/ 48 h) were obtained to achieve a combination of a high strength of 364 MPa and a considerable elongation of 6.2 %, which surpassed those of other state-of-the-art WAAM-processed Mg alloys. Furthermore, we evidenced that the favorable effect of the undeformed LPSO structure on the mechanical properties was emphasized only when the nano-scale β' precipitates were present. It is believed that the findings promote the application of magnesium alloy workpieces and help to establish customized heat treatment process for additively manufactured materials.

Keywords: wire arc additive manufacturing; heat treatment; Mg-RE-Zn-Zr alloys; LPSO structure; mechanical properties

1 Introduction

The multiple thermal cycles and rapid cooling behavior of additive manufacturing process make the microstructure different from the traditional components, so it is difficult to optimize through the traditional heat treatments to fully tap its mechanical properties potential. The exploration and establishment of customized heat treatment is one of the current research hotspots in the field of additive manufacturing. In addition, the evolution of complex phase characteristics of Mg-RE-Zn alloy during heat treatment is interrelated. For example, the dissolution of grain boundary eutectic phase (detrimental phase) is usually accompanied by grain growth, and the formation of LPSO structure also affects the subsequent β' enhanced phase precipitation. In view of the above problems, this work presents a method to explore and establish a custom heat treatment of additive manufacturing metal materials. Through the design and comparison of the microstructure of different solution treatments, the evolution was clarified, the strengthening and toughening mechanism was explored, and the optimal heat treatment process for additive manufacturing of magnesium alloys was finally determined.

2 Experimental procedure

We employed two criteria to preliminary screen heat treatment processes to prevent the time waste caused by blind heat treatment process optimization. Firstly, the grain-boundary eutectic phase should be eliminated, since it not only occupies the strengthening element and affects the following precipitation of strengthening phase, but also becomes the crack origin site during bear loading process and causes premature failure. Secondly, the grain size should remain almost unchanged after heat treatment. Based on the above two principles and our preliminary experiments, the solution heat treatment processes utilized in this work were set as follow: 460 °C/ 1 h, 490 °C/ 1 h, 520 °C/ 30 min, 520 °C/ 1 h, 520 °C/ 8 h and 580 °C/ 15 min. For the solid solution heat treatment and aging treatment, the samples were put in the quartz tubes and oil bath and then quenched in water.

3 Result and discussion

The detailed phase evolution with temperature and time of heat treatment is shown in Fig. 1. The $(Mg, Zn)_3(Gd, Y)$ phase (yellow arrows) was gradually dissolved and transformed into the block-like LPSO structure (blue arrows) on the grain boundary with increasing the duration at 520 °C. With further prolonging the duration, Fig. 1d shows the transformation of the bulk LPSO structure into the lamellar LPSO structure (green arrow). Some reports thought the block-like LPSO structure had an 18R-type structure, while the lamellar LPSO structure is the 14H-type [1]. The TEM test results further evidenced the above statement by presenting both two type LPSO structures in sample after 520 °C for 1 h.

As the duration was extended from 1 h to 4 h, some lath LPSO structures (orange arrows) were formed long the grain boundary. Previous investigations demonstrated that the lath LPSO structure, also known as the X phase, was easily formed at temperatures over 500 °C. With further increasing the duration to 8 h, more X phase was obviously formed in the microstructure.

Under the same duration, the increase in heat treatment temperature will also have an important effect on the phase constituent evolution. High temperatures can effectively reduce the content of the (Mg, Zn)3(Gd, Y) phase, while the LPSO structures rarely precipitate from the microstructure.

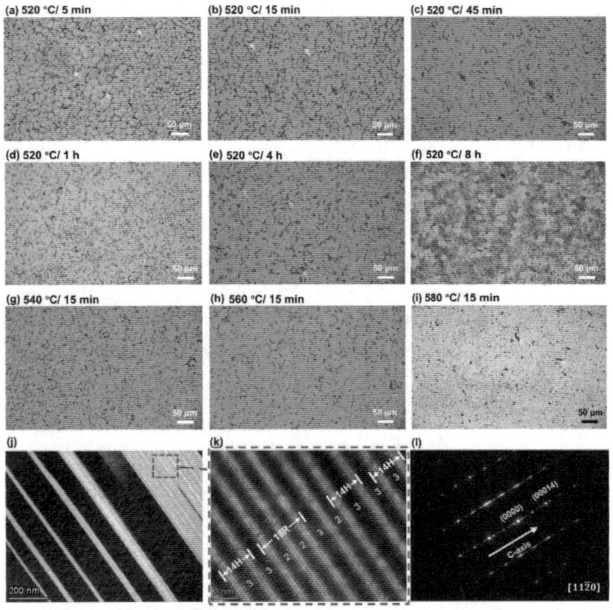

Fig. 1 The OM characterization of the solution treated samples: (a) (520 °C/ 5 min); (b) (520 °C/ 15 min); (c) (520 °C/ 45 min); (d) (520 °C/ 1 h); (e) (520 °C/ 4 h); (f) (520 °C/ 8 h); (g) (540 °C/ 15 min); (h) (560 °C/ 15 min); (i) (580 °C/ 15 min); (j, k) STEM images of the LPSO structures and (l) SAED pattern of 14H-LPSO structure in (520 °C/ 1 h) sample.

Three kinds of heat-treated samples (460 °C/ 1 h, 490 °C/ 1 h and 520 °C/ 1 h) with different combinations of eutectic phase and LPSO structures were selected to determine the peak aging time at 200 °C. It is evident from Fig. 2a that the difference in phase constituent did not influence the peak aging time, which remained 48 h for all the samples. In the following work, the peak aging time of 48 h was used for all the samples to induce the formation of the strengthening phase.

Fig. 2b-2d show the microstructure of the sample after 520 °C/ 30 min + 200 °C/ 48 h. The LPSO structures remained on the grain boundaries as evidence by TEM image. According to the EBSD test, the average grain size was 16.3±2.4 μm. The SAED result shows the β′ precipitates precipitated after aging treatment. The lens-shaped β′ precipitates can significantly increase the tensile strength of the Mg-RE alloy workpieces by hindering the basal dislocation slip.

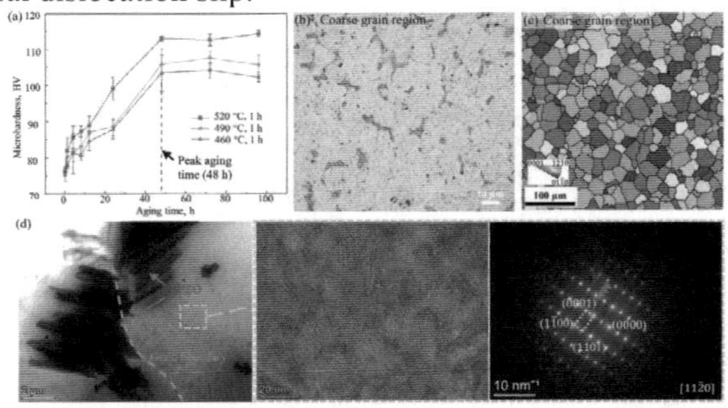

Fig. 2 (a) The aging hardening curve of the solution treated (460 °C/ 1 h), (490 °C/ 1 h) and (520 °C/ 1 h) samples at 200 °C; (b-d) OM image, inverse pole figure and TEM images in (520 °C/ 30 min + 200 °C/ 48 h) sample.

4 Conclusion

The phase constituents of the heterogeneous structure exhibit different evolution behaviors during the heat treatment process. Moreover, with continuously prolonging the duration at 520 °C, the metastable block-like structure is gradually changed into the lamellar LPSO structure, and then into the lath LPSO structures.

An optimal microstructure (fine grain + a certain amount of LPSO structures + nano-scale β' precipitates) and the corresponding customized heat treatment process (520 °C/ 30 min + 200 °C/ 48 h) are obtained to achieve a combination of a high strength of 364 MPa and a considerable elongation of 6.2%.

References

[1] Liu H, Xue F, Bai J and Sun Y S. Effect of heat treatments on the microstructure and mechanical properties of an extruded $Mg_{95.5}Y_3Zn_{1.5}$ alloy. *Mater. Sci. Eng. A*, 2013, 585: 261-267.

New Advances in Printed Sand

Ingo Ederer[1,*], Alfred Grießer[1], Alexander Kudernatsch[1], Dominik Rumschöttel[2]
1. Voxeljet AG, Paul-Lenz-Str. 1a, 86316 Friedberg, Germany
2. Fraunhofer IGCV, Am Technologiezentrum 2, 86159 Augsburg, Germany
*Corresponding address: e-mail: info@voxeljet.de

Abstract: Printed sand-molds and cores have become indispensable in foundries to produce prototypes and small series. But, the process offers far more advantages than just a shorter production time. Complex geometries can be created without additional effort in pattern making and core assembly [1]. These and other advantages have convinced BMW and GE Vernova (beforehand GE Renewable Energies), for example, to produce certain cast components in series using printed sand. The requirements range from several hundred thousand of cores per year in the case of BMW to very large molded parts with lateral dimensions of 7 x 9 m in the case of GE Vernova printed in a single piece. This article presents the two use cases and their respective additive production solutions from the perspective of the 3D printer manufacturer. The possibilities and limitations of the current Binder-Jetting technology are shown and an outlook on future possibilities is given.

1 Introduction

At first glance, the requirements in the following two applications could hardly be more different: BMW produces very delicate cores for Aluminum cylinder head casting in large numbers, several hundred thousand cores per year to be precise.

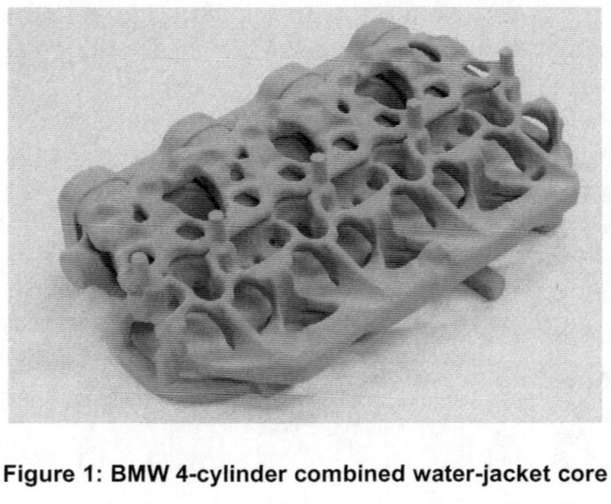

Figure 1: BMW 4-cylinder combined water-jacket core

As the wall thicknesses in the casting are only a few mm, the demands on the accuracy of the cores are immensely high; +/- 0.3 mm is permissible for dimensions of approx. 600 mm. For environmental reasons, the use of an inorganic binder system was required [2]. In addition, manual intervention was to be kept to a minimum. This project was carried out in collaboration with Loramendi - a leading supplier of foundry equipment

For GE Vernova, on the other hand, the focus is on the wind power components to be cast. Here we are typically talking about cast parts with dimensions of several meters and a casting weight of up to 80 tons [3]. For this purpose, the molds should be printed in as few parts as possible and with a maximum duration of 3 days per full mold. The material used is recycled sand from the foundry and furan resin.

2 Solutions

For the BMW application, voxeljet worked with the customer and joined forces with Loramendi to decide on a combination of several printers and an automated pre- and post-processing line. In principle, another variant with fewer or just one printer, but with a larger build area, would also have been possible here. Ultimately, the cost of handling larger boxes and automated post-processing was deemed too high. The core of the system therefore consists of five 3D printers, each with a build volume of 1300 x 600 x 500 mm.

Figure 2: BMW core printing cell, developed by Loramendi & voxeljet

The printers have a special application system that combines two recoaters with a print head in the middle which spans over the entire width of the build area. This allows one layer of sand to be processed per pass. Printing up to 6 water jacket cores in one job box thus takes just a few minutes. The finished job box is then transported via a conveyor system into an industrial microwave, where it is processed for a few minutes. At the same time, an already emptied box moves into the printer fully automated and the printing process starts again. In the microwave, the printed sand is heated to over 100 °C and the inorganic binder hardens immediately. After the microwave, the unbound sand is automatically emptied from the job box and the printed cores are cleaned by a robot. The process concludes with an optical measurement of the component, which can then be fed into the serial casting process.

For the large-scale casting application, voxeljet had to go in a different direction with the printer. GE Vernova Fraunhofer IGCV and voxeljet AG created a joint research partnership to develop the world's largest 3D printer to optimize the production of major components of offshore wind turbines. The desired sand mould size of up to 9 x 7 x 3 m means that a sand mass of several hundred tons must be processed. This is where our existing VX4000 concept comes into play, where the printer works with a stationary platform and lifts its printing-axes layer by layer, printing both job box walls and molds as construction progresses. This has the advantage that only a relatively small weight must be moved for the sand application device, offering less engineering efforts to further reduce cost. In addition, the molding sand for each layer rests during the curing process and therefore experiences little disturbance that could affect accuracy. Furthermore, the consortium decided to assemble the respective mold using individual printed plates of maximum 1 m in height. This makes it easier to clean and to handle the molds. The basic idea in the research partnership is that after printing the respective mold plate, a steel frame is placed over the mold parts, which may need to be backfilled with molding sand. The mold plate can then be lifted off the building platform and cleaned using the steel frame and crane. This is followed by the insertion of chill plates where needed and the application of coating before the mold plate is loaded into the casting pit. Depending on the mold size, several printed mold plates are stacked on top of each other and then cast. The printer itself consists of 8 vertical axis units that lift a frame layer by layer, on which the traversing units for the print head and sand coater are located. The print head consists of several indivi9dual print modules approximately 1 m wide which, depending on the print speed requirement, can be constructed in such a way that the axis only needs two passes over the construction field to print the entire layer.

The recoater is also made up of individual modules to simplify maintenance. In addition, the recoater is designed so that it can apply layers of up to 0.7 mm in height. The maintenance units for the print head are located on one side of the printer and the sand feed for the recoater on the other.

The entire sand feed concept is designed so that circulating sand from the foundry is taken from the silo and fed to a batch mixer. It is also planned that the unprinted sand produced when cleaning the molds will also be fed back into the batch mixer in predetermined quantities. In the mixer itself, the activator is added in the appropriate quantity. The mixed molding sand is then transported directly to the printer. This concept makes it possible to limit the amount of sand to be disposed of to a minimum.

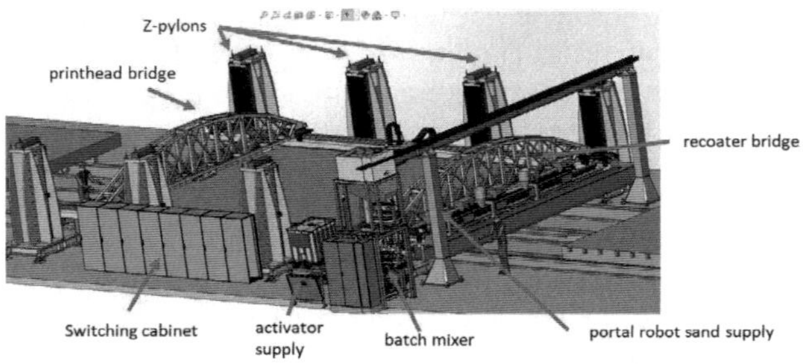

Figure 3: Large Format Printer Layout

3 Conclusion

The BMW application shows that it is technically possible to 3D print and successfully cast sand cores in very large quantities, even in the demanding inorganic material set. The challenges here include the properties of the binder, which requires very precise process control, and the control of many other critical variables that influence the process, such as the sand composition and climatic conditions. The combination with microwave hardening limits the application to core geometries with thin surrounding walls. Nevertheless, this is the first breakthrough in printed sand cores for mass production, which can be transferred to other applications, other geometries and other molding materials as well.

The second application with GE Vernova is still in the development phase. Nevertheless, it is already clear that printing the molds can also bring considerable advantages here and this is technically – despite the huge dimensions - feasible. For the continued success of this technology, it is important to utilize the possibilities that arise with printing, e.g. in terms of freedom of design.

References

[1] Erhard, P., Hartmann, C., Li, R. et al. Advanced Procedures for Series Production with 3D-Printed Core Packages. Inter Metalcast 17, 2572–2583 (2023)

[2] Weissenbek W,, Kautz T., Brotzki J. and Müller J. Tomorrows Cylinder Head Production…. Industry Cylinder Head, 2012

[3] Kang, Jw., Ma, Qx. The role and impact of 3D printing technologies in casting. China Foundry 14, 157–168 (2017).

Application of Rapid Manufacturing Process & Equipment System for Large Complex Light Alloy Structural Components

Haichun Hong*
State Key Laboratory of Advanced Forming Technology and Equipment,
Beijing National lnnovation Institute of Lightweight Ltd., Beijing, China
* E-mail: honghc025@163.com

Abstract: This paper introduces the key technologies of light alloy fire-retardant sand printing and forming processes, including material system optimization, feedstock ratios, mixing methods, and in-house developed additive manufacturing equipment. Through the shape-controllable composite mold technology, the challenges of matching and precise regulation between mold and casting properties have been addressed, enabling rapid manufacturing through mold-less composite forming. The impact of multi-material composite molds on the performance of castings has been investigated, leading to the invention of corresponding control principles and assembly methods. Furthermore, addressing the influence of sand parameters on forming quality and tool wear, formulas comprising materials such as quartz sand, zircon sand, and chromite sand have been devised, achieving synergistic optimization of mold properties. Finally, for future needs, ceramic core additive manufacturing equipment has been developed to further advance material forming technology.

Keywords: Rapid manufacturing, Multi-material, Light alloy

1 Introduction

The advancement of manufacturing technologies has been pivotal in the evolution of industrial production, particularly in the realm of casting processes. This paper delves into the innovative domain of light alloy fire-retardant sand printing and forming, a field that has witnessed significant strides in recent years. The research presented here encapsulates the core technologies that have been instrumental in enhancing the efficiency, precision, and adaptability of casting processes, thereby facilitating the rapid manufacturing of complex components with high dimensional accuracy and material integrity.

Light alloy fire-retardant sand printing and forming

The core of this research lies in the optimization of the sand material system for the printing process. The study delves into the effects of various parameters such as the content of curing agents, the shape function of the wedge-shaped sand spreading mouth, and the particle size distribution of the original sand [1]. These factors are critical in determining the mechanical properties of the sand molds, including their resistance to deformation and flow characteristics.

Material system and additive manufacturing equipment

The paper presents a detailed analysis of the composite flame retardant raw materials, highlighting the importance of pre-dispersion and high-precision addition. The study also emphasizes the need for thorough mixing with the original sand to achieve a homogenous mixture [2]. The optimization of the flame retardant additive quantity is crucial, as it directly impacts the mechanical strength and gas emission properties of the sand molds. The data presented in the paper demonstrates the relationship between the flame retardant content and the resulting tensile and flexural strengths, as well as the gas emission volume and permeability of the sand molds [3]. And a fire-retardant additive manufacturing equipment has been developed, as shown in Figure 1.

Fig.1 Fire-retardant additive manufacturing equipment

Shape-controllable composite mold technology

One of the key contributions of this research is the invention of shape-controllable composite mold technology, as shown in Figure 2. This technology addresses the longstanding challenges of matching and precise regulation between mold and casting properties [4]. By establishing an active control mechanism for different types of sand materials, the technology ensures that the mold's shape and properties are precisely tailored to the requirements of the casting process.

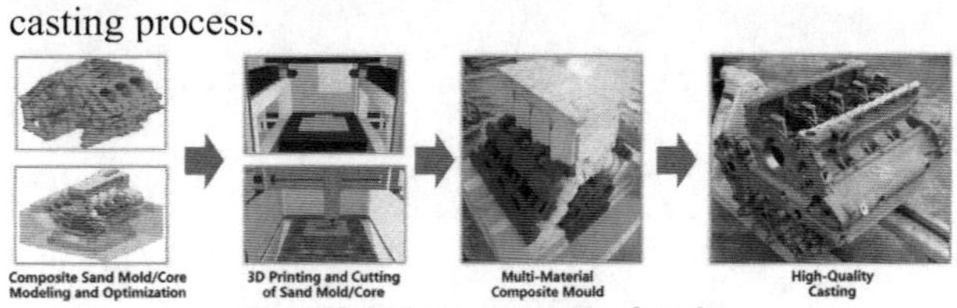

Fig.2 Mold-less composite forming

Impact of multi-material composite molds

As shown in Figure 3, the research has also investigated the impact of multi-material composite molds on the performance of castings. Through a series of experiments, the influence of different types of sand, such as quartz sand, zircon sand, and chromite sand, on the temperature field, microstructure, and mechanical properties of the castings has been studied [5, 6]. This has led to the invention of control principles and assembly methods that are specifically tailored to the needs of high-quality casting production.

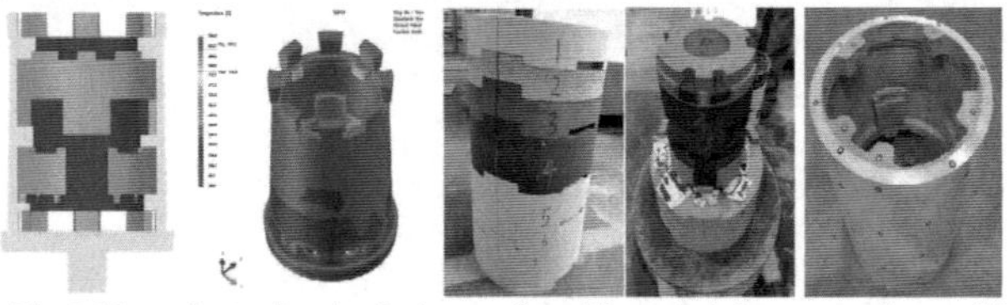

Fig.3 Manufacturing technique of multi-material composite molds

Synergistic optimization of mold properties

Addressing the influence of sand parameters on forming quality and tool wear, this study has formulated materials that are capable of synergistic optimization of mold properties [7]. The formulations have been designed to control the mold's permeability, curing strength, and cutting performance, ensuring that the molds are not only suitable for the casting of complex components but also exhibit high dimensional stability and resistance to wear.

2 Product cases

Casing Category

Casing Sand Mold: Dimensions 542mm x 446mm x 244mm. Using moldless composite precision forming technology, a sand mold with dimensional accuracy of ±0.5mm was obtained within two days, allowing the casting of 3mm thick-wall products that meet usage requirements.

Compartment category

Transition Section Sand Mold: Dimensions 2400 mm × 2400 mm × 1000 mm, weight 5000 kg. Using moldless forming technology, one set was completed within three days, controlling wall thickness error to within 0.5mm, completing high-precision assembly of separate sand molds and preliminary development of the compartment.

Shell category

Shell Part Sand Mold: Dimensions 700 mm × 700 mm× 500 mm, weight 250 kg. Using moldless forming technology, two sets were completed within two days, controlling wall thickness error to within ±0.5 mm, meeting the rapid production requirements of thin-walled parts.

3 Conclusion

This paper presents a comprehensive study on the key technologies that underpin the rapid and efficient manufacturing of light alloy components through fire-retardant sand printing and forming processes. The innovations discussed herein have the potential to revolutionize the casting industry by offering solutions that are not only cost-effective but also environmentally friendly and adaptable to a wide range of applications.

References

[1] Fuchu Liu, Li Yang, Ying Huang, Peng Jiang, Guang Li, Wenming Jiang, Xinwang Liu, Zitian Fan, Performance of resin bonded sand for magnesium alloy casting, Journal of Manufacturing Processes, Volume 30, 2017, Pages 313-319.

[2] Dana H R, El Mansori M. Mechanical characterisation of anisotropic silica sand/furan resin compound induced by binder jet 3D additive manufacturing technology[J]. Ceramics International, 2020, 46(11): 17867-17880.

[3] De Camargo I L, Fortulan C A, Colorado H A. A review on the ceramic additive manufacturing technologies and availability of equipment and materials[J]. Cerâmica, 2022, 68: 329-347.

[4] Gu Zhaoxian. Research on rapid combined forming process of casting. 2012.

[5] Liu Limin, Shan Zhongde, Lan Dun, Liu Feng. Research of Concavo-convex Concerted Structures Based on Precision Casting without pattern. Foundry Technology, 2017:38(2).

[6] YAN Dandan, SHAN Zhongde, Study on Properties of Multi-material Mixed Sand Mold and Its Effect on Solidification Microstructure of A356 Aluminum Alloy, Material Reports, 2023, Vol. 37 No. 1.

[7] LIU Li-min, SHAN Zhong-de, Liu Feng, Numerical Simulation and Experimental Research on Multi-Material Hybrid Sand-Mold Process of Iron Castings, Foundry, Nov.2018, Vol. 67 No.11: 955-960.

Effect of Catalyst on Quality of 3D Printing Sand Mold/Core

Ruilong Yu[1,2,*], Rui Tan[1,2], Shaokui Yin[1,2], Yan Wang[1,2], Yingwei Zhou[1,2], Xin Zhang[1,2]

1. Shenyang Research Institute of Foundry Co., Ltd. CAM, Shenyang 110022, China
2. National Key Laboratory of Advanced Casting Technologies, Shenyang 110022, China

*Corresponding address: e-mail: yuruilong0513@163.com

Abstract: The strength and the dimensional accuracy are two important factors affecting the quality of sand mold/core produced by 3D sand printing (3DSP), which are not only closely related to the reactivity of furan resin and the phase transition of silica sand, but also to the catalyst of furan resin. Taking the furan resin system as an example and using a sand specimen produced by a 3DSP machine, the influence of catalyst on the strength of a 3DSP sand mold/core was studied. Experimental results show that the amount of catalyst of 0.22% to 0.25% by sand weight for the sand mixture is the most suitable for 3D printing core-making. However, it should be noted that the amount of catalyst is less than 0.22%, the tensile strength of a no-bake sand mold/core is higher than that of a 3DP sand mold/core, while the amount of catalyst was more than 0.22%, the result was the opposite.

Keywords: 3D sand printing; catalyst; gel time; sand mold/core.

1 Introduction

3D sand printing is by far one of the most successful technologies applied in foundry[1], which accelerated rapid product development and unit production. 3D sand printing cores differed from traditional sand cores for its lower density, higher porosity and rougher surface quality. During the 3D sand printing process, the substrate will drop 0.2-0.6 mm after recoating sand and jetting binder for one layer. If the curing rate of resin is too low with lower strength, a small amount of relative displacement will occur between the sand layers. Determining the most suitable amount of catalyst of the bonding system for the 3DP process is the goals of this study.

2 Experiment procedures

The sand specimens were prepared with the same silica sand, furan resin binder and sulfonic acids catalyst by no-bake resin sand method and 3DP method, respectively. When preparing no-bake sand samples, 1 kg sand was put into the sand-mixing machine with 0.12%-0.3% catalyst (by sand weight) and mixed for 2 min, then mixed with 1% furan resin for 30 s. After that, the mixed sand was put into the eight-character die to complete the sample preparation within 1 min, and then the tensile strength was measured at 1 h and 24 h.

When preparing the 3DP samples, the 0.12%-0.3% catalyst was mixed with silica sand by the continuous sand mill. 10 kg sand was mixed each time, and then the mixed sand was dried over 12 h in the sack at 25±2 °C, 45±5 RH. Then, the 3DP samples were printed on a Voxeljet VX200 printer using the prepared sand as powders and 1% (by sand weight) furan resin as spray binder. The printed samples were cleaned 2 h after the printing finished, and the tensile strength at 24 h was measured.

3 Results and discussion

A different amount of catalyst was added, and the tensile strength of samples by no-bake sand process and 3DP process were compared. As can be seen from Table 1, as for 24 h strength, it firstly increased with the increasing content of catalyst and went down when the amount of it was more than 0.22%.

Table 1: Effect of catalyst addition amount on strength of no-bake resin sand

Addition amount (%)	Gel time (min)	Stripping time (min)	24h tensile strength (MPa)
0.12	30	100	1.50
0.15	20	55	1.60
0.18	13	51	1.60
0.20	9	36	1.85
0.22	5	28	1.65
0.25	2	18	0.20
0.30	1	12	0.05

Figure 1 compares the tensile strength of 3DP sample and no-bake sand sample as a function of the amount of catalyst. When catalyst was less than 0.22% by sand weight, the tensile strength of no-bake sand was higher than that of 3DP sand, and When catalyst was more than 0.22%, the result was the opposite. The sand core fabricated by 3DP possessed the highest tensile strength when the addition of catalyst was between 0.22% to 0.25% by sand weight. As for the no-bake sand, the tensile strength increased when the addition is less than 0.22%, which decreased significantly when the amount of that was more than 0.25%. As the amount increases 0.22% to 0.3%, the 24h tensile strength decreases from 1.35 to 1.05 MPa, decreased by about 22%. What's more, the best tensile strength of no-bake sand specimen was 37% higher than that of the 3DP specimen.

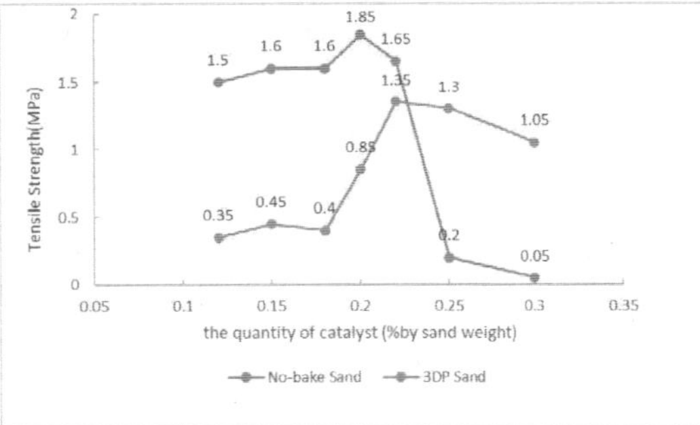

Figure.1: the tensile strength of specimen by no-bake sand and 3D sand printing with different amount of catalyst

Sand compaction

The density of 3DP specimen (1.35-1.45 g/cm^3) is less than that of no-bake sand (1.45-1.55g/cm^3) for the sand spreading pressure in 3DP process is less than that of no-bake sand. The sand compaction was a main technical specification for sand casting and became more and more important in the 3DP process. The porosity of the 3DP specimen reached up to 40% because the raw sand concentration was up to 95% by the two-sieve screening method or the three-sieve screening method [8]. The sand was covered with binder by capillary action without fully stirring as in the no-bake sand[2]. As illustrated in Fig. 2, the 3DP specimen showed fewer and smaller adhesion bridges compared with the no-bake sand specimen, which is the reason why the 3DP specimen had lower strength than no-bake sand under the same amount of binder and catalyst. The liquid binder flowed slightly by gravity and surface tension in large space because of lower compaction, resulting in a longer gel time, which would deteriorate the strength of the specimen.

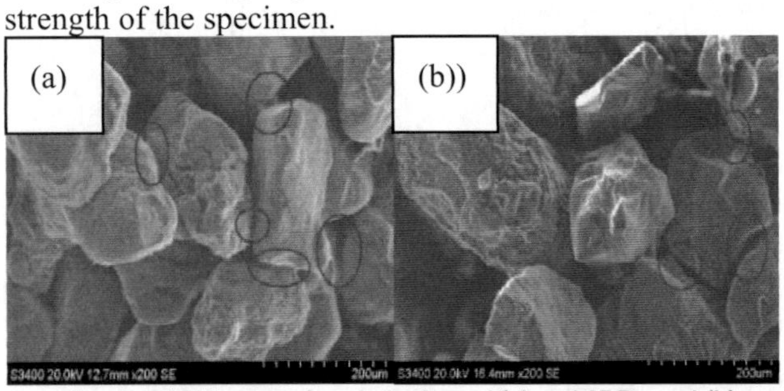

Fig. 2: SEM image of no-baked sand (a) and 3DP sand (b)

Catalyst crystallization

As shown in Fig. 3, the sulfonic acids could crystallize on the sand surface after solvent evaporated in the process of drying mixed sand. For better recoating sand, the sand mixed catalyst was dried for more than 12h, and most of the solvent was lost during drying.

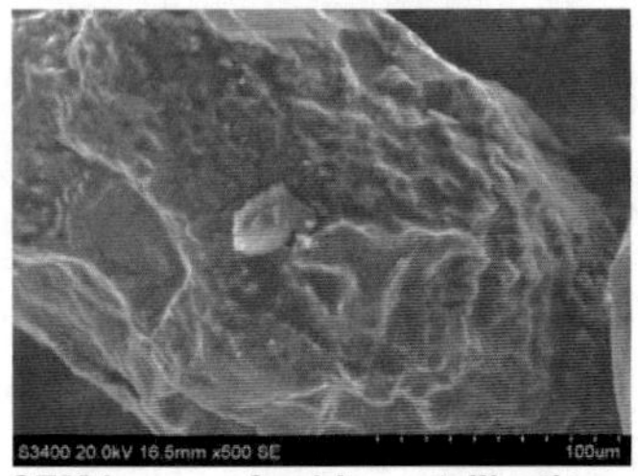

Fig. 3: SEM image of acid crystallization on sand

In the 3DP process, the binder printed on the sand, the sulfonic acid of catalyst needs to be dissolved in the binder, which provides the H^+ required for the curing of the resin. During the process of forming the network from the resin to the body structure, a considerable amount of acid has not been dissolved, which cannot move to the adhesion bridge in the form of ions from liquid, resulting in an insufficient acid amount. Therefore, the catalyst in the 3DP did not fully function and strength was lower compared to the self-hardening mold/core. In this case, when the amount of catalyst was less, the acid of the catalyst reacting with binder was dissolved while the binder was being cured, so that the strength was lower; this is an unavoidable problem during the use of the sulfonic acid-based catalyst in the 3DP process. To reduce the gel time, excessive catalyst was added to increase the sulphur amount which would be corrosive to the recoater. To sum up, the key point to solve this problem was to promote activity of a catalyst which could gel in a short time with less amount, or to develop a new catalyst without crystallization or corrosion. Besides this, wet-recoating is another effective way to solve the above problems.

4 Conclusions

The amount of catalyst had significant impact on sand strength. The most suitable amount of the catalyst in 3D sand printing was 0.22% to 0.25% by sand weight. The compaction rate of the sand mold/core was low, resulting in bond bridges with low number and small area, which caused the strength of the 3DP sample to be lower than that of no-bake sand. Sulfonic acids catalyst would crystallize and redissolve during 3DP process, which required sufficient amount of acid to improve the tensile strength.

Reference

[1] Hartmann C, Silberhorn J, Erhard P, et al. On the mechanism of binder migration in furan binder jetting of sand molds and cores[J]. Additive Manufacturing, 2024, 83: 104073.

[2] Yang W D, Jia P F, Ma Y Y, et al. Modeling and simulation of binder droplet infiltration in 3D printing technology. Nanotechnology and Precision Engineering, 2017, 15(4): 246-253.

Ceramic Mold with Integral Core and Shell for Hollow Turbine Blades Fabricated by Integrated Additive Manufacturing Technology

Chunze Yan[1,2*], Shixiang Zhou[1,2], Shuang Chen[1,2], Dong Sun[1,2], Yusheng Shi[1,2]
1. State Key Laboratory of Materials Processing and Die & Mould Technology, School of Materials Science and Engineering, Huazhong University of Science and Technology, Wuhan 430074, China
2. Engineering Research Center of Ceramic Materials for Additive Manufacturing, Ministry of Education, Wuhan 430074, China
*Corresponding address: e-mail: c_yan@hust.edu.cn

Abstract: Ceramic molds are extensively applied in investment casting processes of hollow turbine blade due to their outstanding high-temperature chemical stability and creep resistance. Herein, laser powder bed fusion (LPBF) method was utilized to fabricate an ceramic mold with integral core and shell (CMCS). The vacuum infiltration process was adopted to improve mechanical properties of ceramic after the sintering, where the solid loading of ceramic slurry was investigated. After the first infiltration and pre-sintering, the separated two parts were bonded by a ceramic binder and then underwent second infiltration and final-sintering to obtain a complete CMCS. The ceramic mold fabricated through LPBF and post-treatment possessed low sintering shrinkage (1.51–2.03%), adequate apparent porosity (30.82 ± 0.01%), and room-temperature strength (13.03 ± 2.90 MPa for unbonded specimen, 9.72 ± 1.34 MPa for bonded specimen). Strikingly, the subsequent casting experiment has proved that the method proposed in this paper is promising to fabricate superior ceramic based CMCSs.

Keywords: ceramic mold, laser powder bed fusion, infiltration, turbine blade

1 Introduction

Superalloy turbine blades with intricate interior cooling passages are the key components of gas turbine engines[1]. The traditional method of fabricating ceramic mold involves a series of procedures including die-casting, shaping wax patterns, stuccoing etc., and all of these steps require long production cycles and high cost but the yield is not satisfactory. Additive manufacturing (AM) technology shows capability of producing parts with complex structures in a layer-by-layer mode, which makes it possible to eliminate these complicated and time-consuming processes in the fabrication of ceramic casting molds[2]. In recent years, researchers have made remarkable progress on the application of AM to fabricate ceramic core and shell molds. With the development of ceramic AM technology and related post-treatment technology, the preparation of core and shell respectively are becoming more mature. But the separated ceramic core and shell must be assembled to form a complete mold before the casting process, and the assembly accuracy is hard to control. Therefore, producing a ceramic mold with integral core and shell (CMCS) is a preferred choice in this aspect[3]. The structures of ceramic molds for large-scale gas turbine blades are sophisticated. The total height of the mold could reach about 700−800 mm, while the narrowest gap in the mold cavity is only about 1−2 mm in width, and there are only two narrow openings on top and bottom of the mold respectively. Such conditions make cleaning the residual powder inside the ceramic mold complicated cavities very difficult. Besides, the low density and high porosity of alumina ceramics made by LPBF usually result in low strength due to low powder bed packing density, which means that the post-treatment techniques are needed to enhance the mechanical properties of the alumina ceramics[4,5]. Infiltration is a method that can fill micro-pores of ceramic green body with ceramic slurry, thus improving the density of ceramic parts. But the solid loading of ceramic slurry has a significant influence on the infiltration effect. The infiltration efficiency would be decreased if the solid loading is too low, while the cavity would be blocked if the solid loading rises too high, so how to design an optimal ceramic slurry and infiltration scheme is another problem.

2 Experimental procedure

Once the printed specimens were cleaned, the first vacuum infiltration was applied to enhance their densities. The specimens were firstly placed in a vacuum container and the pressure was kept under − 0.08 MPa, then the ceramic slurry was pumped into the container until the specimens were fully immersed. The specimens were kept immersed under vacuum for 30 min to achieve uniform infiltration. After the specimens were dried, they were subjected to debinding and pre-sintering in a furnace. The debinding and pre-sintering process was: the specimens were heated up to 600 ℃ with a rate of 1 ℃/min and held for 2 h to remove binder completely, and then were directly heated up to 1200 ℃ with a rate of 2 ℃/min and held for 4 h. Afterward, the specimens were cooled in the furnace to room temperature. In order to further improve the mechanical properties of the ceramic for casting purposes, the specimens needed to undergo the second vacuum infiltration and final-sintering. The second vacuum infiltration followed the same protocol as the first vacuum infiltration. The final sintering process was: the specimens were heated up to 1600 ℃ with a rate of 5 ℃/min and held for 2 h, and then cooled in the furnace to room temperature.

3 Result and discussion

The effects of solid loading of ceramic slurry on properties of infiltrated green bodies and pre-sintered bodies are shown in Fig. 1. The weight increasing ratio increases as the solid loading increases, indicating that introducing fine alumina powder is an efficient way to promote material density. The weight increasing ratio of the 20 vol% ceramic slurry is 22.24 %, which is almost four times as large as that of the counterpart treated with pure silica sol (5.05 %). Different from the properties discussed above, solid loading of the ceramic slurry does not have significant impact on the pre-sintering shrinkage. All the specimens show low shrinkage rates below 1.4 % along different directions, and specimens infiltrated with ceramic slurries which contain fine alumina powder have even lower shrinkage rates around 1 %. It can also be found that the shrinkage rate along the Z-axis direction is just slightly larger than the shrinkage rate along the X/Y direction during this stage, which means the shrinkage anisotropy is not significant. Due to the low pre-sintering temperature (1200 ℃) and low sintering activity of the coarse spherical alumina powder, the thermodynamic conditions of mass transfer and particle rearrangement between the coarse alumina particles cannot be fulfilled.

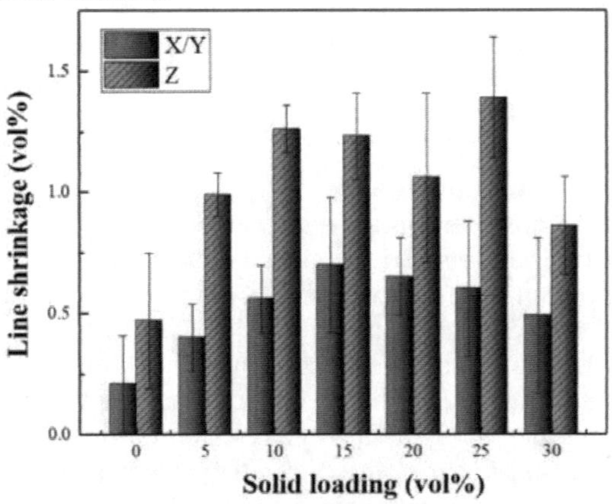

Figure 1 Effects of fumed silica suspension solid loading on line shrinkage of LPBF-fabricated ceramic mold

The overall shape of the casting mold is intact, and there are no obvious cracks and other defects on the outer surfaces. However, cracks can be found inside the ceramic mold, and some trapped bubbles can be seen inside the bonding joint. To verify the feasibility of this process, casting experiment is conducted and a nickel-based superalloy turbine blade is successfully fabricated. The turbine blade is basically intact and the cracks in the ceramic mold did not lead to fatal failures. Also, no noticeable metal leakage can be seen around the bonding joint, proving that the bonding process is feasible for casting superalloy turbine blade.

4 Conclusion

CMCS with low sintering shrinkage and adequate properties was prepared by LPBF and its feasibility for casting superalloy turbine blade was successfully verified. The CMCS model was split into two parts for the convenience of cleaning residual powder. Ceramic slurry with high solid loading (≥25 vol%) is not suitable due to low infiltration efficiency and clogging phenomenon, and the ceramic slurry with 20 vol% is determined to be optimum for both first and second infiltration. After final-sintering, alumina-based material with low sintering shrinkage (1.51–2.03 %), adequate apparent porosity (30.82 ± 0.01 %) and room-temperature strength (13.03 ± 2.90 MPa) is obtained. The two parts of CMCS were bonded by a ceramic binder with adequate strength (9.72 ± 1.34 MPa), and no obvious cracking was found at the interface. Finally, the potential to cast superalloy turbine blades was demonstrated by a cast case using the CMCS.

Acknowledgments

The work was supported by the National Science and Technology Major Project, China [2017-Ⅶ-0008-0102].

References

[1] A. Ehk, A. Hyp, B. Cll, C. Jbp, A. Scy, A. Ygj, Single crystal casting of gas turbine blades using superior ceramic core, J. Mater. Res. Technol. 9 (3) (2020) 3348–3356.
[2] J.E. Kanyo, S. Schaffoner, R.S. Uwanyuze, K.S. Leary, An overview of ceramic molds for investment casting of nickel superalloys, J. Eur. Ceram. Soc. 40 (15) (2020) 4955–4973.
[3] W. Xu, Z. Lu, G. Tian, K. Miao, D. Li, W. Zhu, F. Wang, H. Zhang, Y. Wang, Y. Song, Fabrication of single-crystal superalloy hollow blade based on integral ceramic mold, J. Mater. Process. Technol. 271 (2019) 615–622.
[4] S. Chen, C.S. Wang, W. Zheng, J.M. Wu, C.Z. Yan, Y.S. Shi, Effects of particle size distribution and sintering temperature on properties of alumina mold material prepared by stereolithography, Ceram. Int. 48 (5) (2022) 6069–6077.
[5] N. Travitzky, A. Bonet, B. Dermeik, T. Fey, I. Filbert-Demut, L. Schlier, T. Schlordt, P. Greil, Additive manufacturing of ceramic-based materials, Adv. Eng. Mater. 16 (6) (2014) 729–754.

This page intentionally left blank.

Future Brake Systems in the Automotive Industry

Mustafa Ata*
Continental Automotive Technologies GmbH, Guerickestr. 7, 60488 Frankfurt am Main, Germany
*Corresponding address: mustafa.ata@continental-corporation.com

Abstract: Regardless of the architecture of the vehicle and the architecture of future braking systems, the brake remains the most important element of active safety. Reliability is a must, and "brake manufacturers" do not make compromises. This mentality of the experts and the know-how that has grown over decades will become even more important in the future in order to equip the brake for future requirements and maintain absolute confidence in it: The status and appreciation of the "brake" function will remain untouched.

However, almost everything else is beginning to change.

Vehicles are being rethought. Global megatrends are behind this. With the electrification of the drive and the growing capability for automated driving (AD), vehicle architecture is changing. Braking systems are becoming increasingly intelligent in order to meet future requirements through automated driving and electrification. And this is newly designed vehicles with a modified architecture. This functional expansion requires a deep understanding of the system in order to combine uncompromising safety and sustainability in future braking systems, which will be also modular and distributed in the long term.

Digitalization and networking are bringing about a fundamental reorganization of the electrical and electronic architecture (E/E architecture), which is increasingly geared towards the software. Because it is the software that will determine the character of the car and the driving experience in the future!

Bits are taking the place of horsepower.

Apps and services are expanding the car into an immersive experience that is becoming ever safer and more comfortable. This presentation looks at what this means for future braking systems and for the foundry industry.

Keywords: brake system; automotive industry; autonomous driving; future technology; electric vehicles

1 Introduction

The roadmap ranges from FBS 0 to FBS 3 and shows the development from brake control to systems for controlling the entire vehicle movement. It charts the path to completely "dry" brakes on the rear and front axles.

At the beginning of this roadmap (FBS 0) is the award-winning, second generation of the MK C2 brake-by-wire brake system as an introduction to all Future Brake Systems. The subsequent generations are so called "one-box brake-by-wire" systems.

In the next development stage (FBS 1), it is no longer necessary to mount the system directly on the bulkhead at a specific point in front of the driver in order to enable a mechanical fallback level. Instead, an electronic pedal supports new vehicle concepts with different interior spaces and dimensions, such as the skateboard chassis of electrified vehicles, on which different superstructures can be mounted. Decentralization increases the degree of freedom of vehicle architectures

With FBS 2, the braking system begins to "go dry on one axle". The first step is to stop applying the brakes hydraulically on the rear axle. For vehicle manufacturers, this decentralization means "breaking up" the conventional architecture and further increasing the degree of freedom of vehicle architectures.

With FBS 3, the braking system can ultimately be broken down into modules. In the long term, the hydraulic system could be completely eliminated, including all fluids, which would contribute to sustainability.

To achieve this, all four-wheel-corners could be actuated electromechanically and would be completely dry. Individual functions of a braking system are packaged as stand-alone products in modular, validated and proven software blocks

One of the further strategies is focusing on the future of drum brake system. This brake system technology is reducing CO_2 and particulate emissions and closed system offers additional protection.

The drum brake is almost as old as the automobile itself. For many, it is considered as an old-fashioned product that is used today at most in a low-powered small car. However, many overlook the great potential of the drum brake. Durability, robustness and brake dust reduction are just some of the reasons why car manufacturers should rely on this modified and further developed technology in the future. Old technology in modern cars-closed system- offers additional protection

Drum brakes are a real alternative to disc brakes, especially for electric cars. Due to the high recuperation performance of new electric cars, the brake is rarely used in many everyday situations. What at first sounds like a longer service life for brake linings and brake discs actually leads to the exact opposite with disc brakes. Due to their infrequent use, disc brakes are more susceptible to corrosion, they lose braking power after a long period of inactivity, which must be available 100% and spontaneously in an emergency.

The drum brake is a system that is completely encapsulated in a housing. This means that the mechanics and the braking surfaces inside are well protected against the effects of corrosion such as rain and salt. In addition, the brake shoes are reliably released from the drum surface by the force of a spring, so that residual braking torque is not an issue with the drum brake.

Particle emission reduction is given due to closed design.

While combustion engines have had to meet increasingly stringent exhaust emission limits in recent years, the focus is now also shifting to wheel brakes. This is because, depending on the size of the particles emitted, brake dust can have far-reaching harmful effects on human health. Experts are expecting stricter legal requirements from the EU Commission by 2025 at the latest. Closed design, accumulates brake dust inside the drum brake and can be therefore collected.

However, the use of the drum brake on the front axle is also very practical. The duo-servo principle used achieves a high braking torque with low actuation forces; the self-reinforcing effect of this functional principle can be positively utilized. This makes the drum brake on the front axle an ideal component for the next generation of dry-actuated wheel brakes. Continental is preparing for a wide range of customer requests and is continuing to develop a wide variety of wheel brakes.

In view of this mix of system characteristics, the drum brake is an attractive option not only for the rear axle of electric vehicles, but also for the front axle of vehicles, depending on the vehicle segment and requirements

"Green Caliper" is a further option with advanced features to improve residual resistance and the option of using slide rails with a special surface treatment for low friction. The pad surface is reduced, taking recuperation into account, and the weight of the brake caliper is optimized. Further weight optimization is possible by reducing the pad thickness for BEVs.

Another technology is the "Wheel hub drive with integrated brake

2 Conclusion

The brake **sys**tem architecture is changing.

A lot of researches and investigations are ongoing for the future brake systems, as the electrification of vehicles is progressing rapidly.

Automated driving (AD) and electrification pose new challenges for brakes as does the need for greater sustainability and lower friction losses due to residual resistance.

The mechanical part of brake actuates on the hydraulics can be replaced by electrical control.

Hydraulics may become optional in the long term, depending on the strategy of the individual vehicle manufacturers. In addition, electromechanical brake actuation offers more freedom in defining the vehicle architecture. Both basic types of wheel brake, the disc brake and the drum brake, offer potential for future requirements, regardless of whether with hydraulic or electromechanical actuation.

Integrated Structural Casting Development with Digital Twin Technology

Shouxun Ji[1,*], Zhichao Niu[1], Xixi Dong[1,2], Xiangzhen Zhu[1,3]
1. Brunel Centre for Advanced Solidification Centre (BCAST), Brunel University London, Uxbridge, UB8 3PH, UK
2. College of Mechanical Engineering, Nanjing University of Aeronautics and Astronautics, Nanjing 210016, China
3. Technology Centre (Institute of Mg and Mg Alloys), Baosteel Metal Ltd., Baowu Group, 2003 Baoyang Rd., Baoshan, Shanghai, China
*Corresponding address: e-mail: Shouxun.Ji@brunel.ac.uk

Abstract: In the present paper, we introduced the development of integrated structural castings with digital twin technology for the rear structure of high-end E-sports cars. The topological optimisation, casting structural design, materials verification, melt flow and solidification, mechanical property prediction and the final verification using low pressure die casting (LPDC) for the component manufacturing are described in details for digital twin. The microstructures and mechanical properties of the casting with satisfied quality are described under heat-treated condition.

Keywords: aluminium alloys; casting; microstructures; mechanical properties; digital twin

1 Introduction

Aluminium alloys are being increasingly used as lightweight materials in transport since weight reduction has been proved as one of the most effective and efficient approaches to improve energy consumption and reduce CO_2 emissions [1, 2]. Castings of aluminium alloys are critical and important for geometrically complex parts. Because of the high temperature and complex process during casting, recognition and using potentials for property improvement of lean structures before the components are produced will be extremely significant for product development. As a new development technique, digital twin is now reliable for the smart development of the manufacturing process of how to make a component from scratch and considering all process steps until dispatch, but still rare in the applications in casting manufacturing. In the present study, we introduced the digital twin assisted development of integrated structural castings for the rear structures of high-end E-sports cars.

2 Experimental procedure

A356.2 alloy was used for manufacturing castings using a low pressure die casting (LPDC) machine, with a locking force of 24 tonne and a plate size 1.9 m×1.6 m, made by LPM (Italy). The pressure was set as 0.8 bar during casting. The tensile tests were conducted following ASTM E08/E08M, using an Instron 5500 Universal Electromechanical Testing System equipped with Bluehill software. The gauge length of the extensometer was 25 mm and the ramp rate for extension was 2 mm/min. The microstructures of cast components were checked using a Zeiss optical microscope with quantitative metallography. Digital twin including melt flow, solidification and mechanical properties prediction was conducted using Magmasoft with specially developed modules.

3 Results and discussion

The computational procedures used to create the digital twin included gating development, melt filling and solidification with structural optimization, the prediction of microstructures and the mechanical property under as-cast and as-heat treated conditions, the validation of digital results with casting samples for geometry tolerance and mechanical properties.

Figure 1 shows the full design procedure for topological optimisation of the integrated casting structure for rear chassis structures. This holistic design approach includes the separation of subframe, all the hardpoints for suspension links on the same casting, the connection of upper and lower castings for better integration, the increase of webs for hardpoint stiffness improvement, the incorporation of damper top and the definition of tolerance requirement, the incorporation in primary structure, the socketed longitudinal crush element and improved bond connection at front, the load path improvement for better crash load transfer, the casting manufacturability and primary feedback from flow analysis, the final optimised and refined structure.

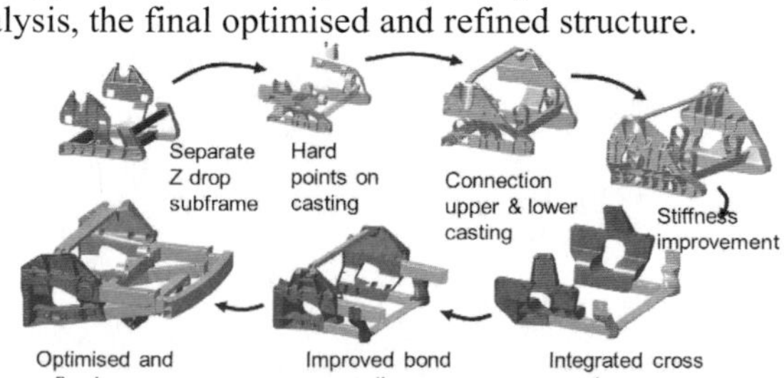

Figure 1 Topological design and optimisation of the integrated casting structure for rear chassis.

Figure 2 shows the materials modelling and verification, in which a standard casting structure widely used in OEMs was adopted. A356.2 alloy was melted using a standard procedure and cast into standard components. The tensile properties were then measured and analysed. From the digital analysis, the middle section of top flanges showed the yield strength from 243 to 246 MPa, UTS from 320 to 330 MPa and elongation from 8.2 to 8.7%. The verification casting in Fig. 2d was sampled at the same position of the middle section of top flanges. T6 heat treatment was conducted at 540 °C/8 h for solution and 180 °C/6 h for ageing. Table 1 shows that the yield strength and UTS were better than the digital prediction, but close each other. However, the elongation was slightly lower in the physical casting. Thus, it is important to satisfy the elongation in the real castings.

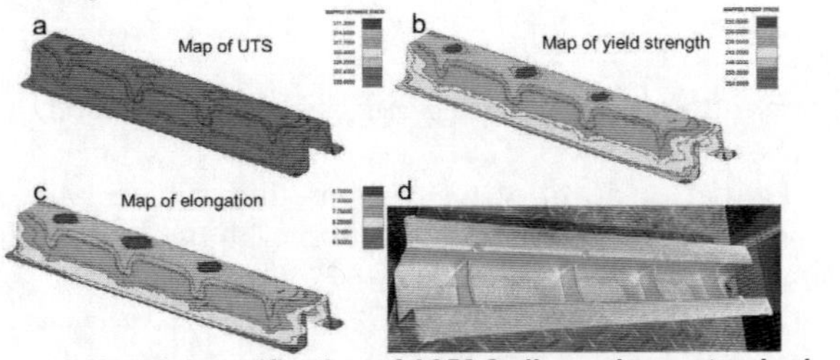

Figure 2 Materials verification of A356.2 alloy using a standard casting structure at auto OEMs.

Table 1. Tensile properties of aluminium alloys under T6.

	Yield (MPa)	UTS (MPa)	Elongation (%)
A356 T6	250±7	330 ±13	7±1.5

Figure 3 shows the casting defects in the digital analysis of the casting with optimised gating system for the melt flow, solidification and defect formation during casting. The results showed the design of gating system was superior, no obvious defects could be found in the key areas of castings and only tiny defects were found at the cross-section of ribs.

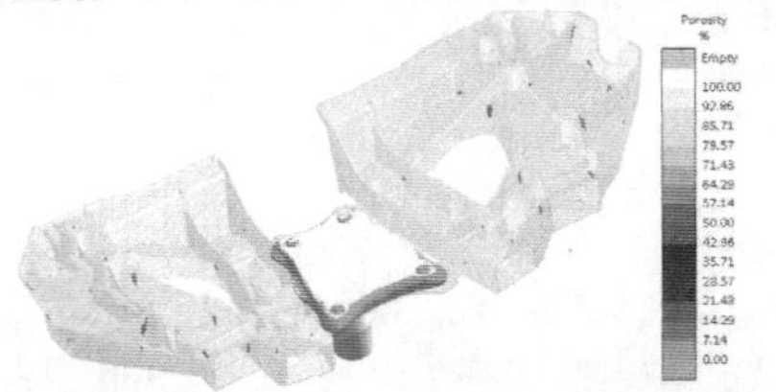

Figure 3 Simulation of casting defects after melt filling and solidification simulation of A356.2 alloy

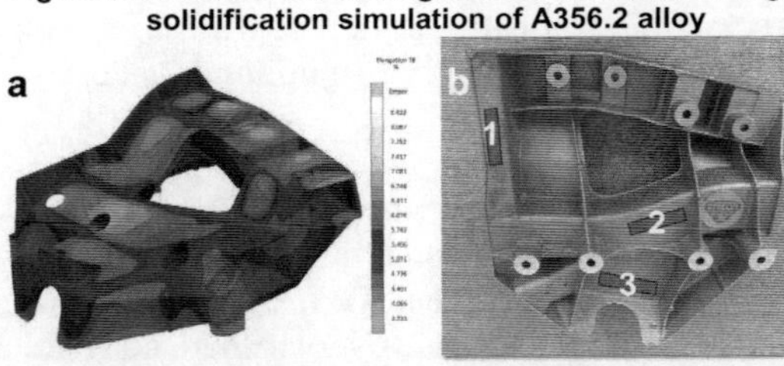

Figure 4 Digital twin of casting made by A356.2 alloy for (a) virtual simulation of elongation under T6, (b) physical casting made by LPDC with sampling for tensile test

Figure 4 shows the elongation prediction from digital analysis and the final verification of castings using LPDC. From Figure 4(a), the predicted elongation of the casting varied from 4.5% to 7.5% under T6 condition. The casting in Figure 4(b) was for the final verification of mechanical properties with the locations of sampling. The measured tensile properties of different locations are shown in Figure 5, in comparison with the elongation data obtained from prediction. Clearly, the elongation of tested sample was better than the prediction. As such, the safety factor should be higher in the castings, which can be used in industry.

The microstructure of the casting can be seen in Fig. 6, in which the dendrites with the size of 300-400 μm were formed under as-cast condition. After T6 heat treatment, the tips of dendrites were blunt and the eutectic Si phase was spheroidized.

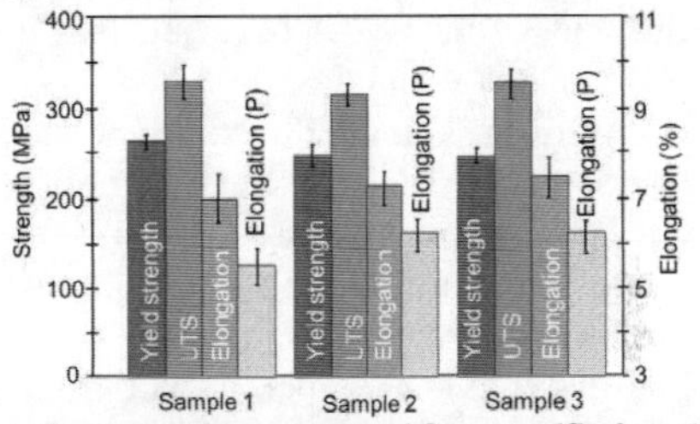

Figure 5 Tensile properties measured from verified casting and the data obtained from prediction (P) of A356.2 alloy

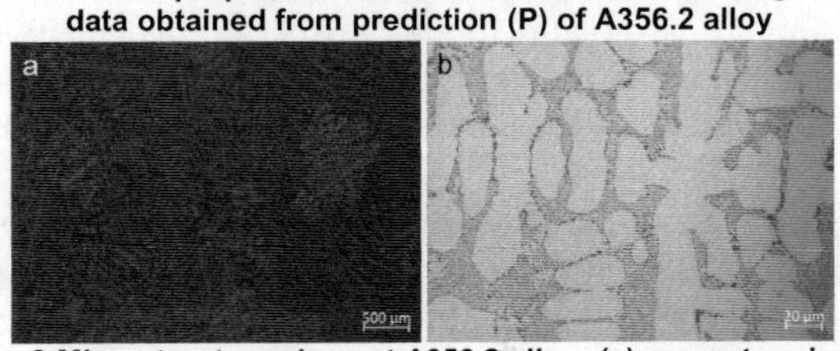

Figure 6 Microstructures in cast A356.2 alloy: (a) as-cast grain size; (b) morphology of primary α-Al phase after T6

4 Conclusion

The digital twin technique is capable of developing casting manufacturing from structural optimisation design to final mechanical properties. The gating system optimisation, melt filling, modification and defect formation, and the mapping of mechanical properties can be obtained from digital analysis with relatively accurate data. The casting verification of A356.2 tensile samples have a yield strength of 250 MPa, a UTS of 330 MPa and an elongation of 7%, which is close to the digital results.

References

[1] Dong X, Zhang Y, Ji S. Enhancement of mechanical properties in high silicon gravity cast AlSi9Mg alloy refined by Al3Ti3B master alloy. Mat. Sci. Eng. A, 2017, 700: 291-300.

[2] Amirkhanlou S, Ji S. Casting lightweight stiff aluminium alloys: A review. Critical Reviews in Solid State and Materials Sciences, 2020, 45(3): 171-186.

Implementing IIoT and AI-Driven Process Optimization for Foundries

Per Larsen*
DISA Industries A/S, Højager 8, 2630 Taastrup, Denmark
per.larsen@noricangroup.com

Abstract: Implementing a proven IIoT solution can quickly deliver a real-time digital picture of a foundry process and build a consistent, trustworthy database for monitoring, reporting and analysis. Adding AI optimization maximizes the value of this process data and delivers performance improvements that are impossible with traditional manual techniques.

This paper will discuss how collecting and storing data via an IIoT system can deliver huge benefits for any foundry, from identifying bottlenecks, faultfinding and measuring energy consumption to scrap reduction and precise metrics for quality improvement.

It will explain how to bring together data from multiple vendors' equipment and other systems like ERP and MES via a single IIoT cloud platform, how a step-by-step cloud implementation minimizes project risk and how to effectively translate insight from the data into process improvements.

With an accurate foundry database available, full process optimization using service-based Artificial Intelligence (AI) and machine learning software is now a mature and proven route to large performance improvements. The presentation will share IIoT and AI learnings from multiple global foundry IIoT and AI implementations including Huaxiang (China), Condals (Spain), Morikawa (Japan) and MAT Group (UK). Deployed with application knowledge, AI has driven an average of 40% scrap reduction at these companies.

The paper will also discuss full casting-level traceability. This technology links every single casting to the process parameters used to produce it. This casting-level tracking enhances IIoT effectiveness in areas like faultfinding and scrap reduction, and improves AI modelling accuracy for faster, more successful optimization.

Keywords: digital foundry; IIoT; Huaxiang; Morikawa; MAT group; AI; process optimization

1 Introduction

The green sand process is complex. Each sub-process involves tens or hundreds of machines, material and other variables, from sand moisture content to pouring temperature. This complexity makes it challenging to manually monitor production and find improvements.

Troubleshooting often relies on experience. As more experienced staff retire, a skills gap appears and it's even harder to control casting quality, cut costs and lower emissions.

Digital is an effective way to find substantial gains. That may be lower scrap levels higher productivity, sustainability or equipment utilization.

2 Digitalization

A single IIoT (Industrial Internet of Things) solution is the best basis for foundry-wide digitalization. A secure in-house network connects sensors, PLCs, gateways and edge devices to a central storage platform. Sensor data is forwarded to the central database where it is stored and combined with other data from the whole process.

The central data platform should be instantly compatible – "plug and play" – with all other system components such as edge devices. The platform should be able to automatically import, check and store all input data while computing any required derived variables in real time.

Dashboards and KPIs should be able to be created and edited easily without any knowledge of SQL or the system's back-end coding or data configuration. This will allow every user to construct their own view of the data available to suit their actual needs.

The platform should include further visualization features such as charting, tabular reporting and color coding. The central platform's data and applications should be cloud-hosted. This makes it easier to link together multiple global sites to feed a single system, though this is also possible with an in-house system.

3 Optimization

Digitalization and analysis are effective in improving single sub-processes like moulding or pouring. But the complexity and volume of the data needed to optimize the entire green sand process overwhelms conventional analytics.

Artificial Intelligence (AI) is the solution, built on top of a mature data platform as described above. An AI-driven Expert Execution System (EES) like Norican's Monitizer | PRESCRIBE can consider all the process parameters from an entire production line in order to maximize one target variable: casting quality.

The AI examines historical data to learn how parameters like sand compressibility or pouring temperature influence each other and affect final casting quality. It calculates which combination of machine and material settings will produce the best results for each pattern.

During production, the EES updates its recommendations for the control plan every 30 minutes in response to AI predictions based on real-time data from the line. That maintains stable, high-quality production, even as factors like air temperature or sand moisture content vary.

4 Casting-level tracking

Process data gathered via IIoT has many applications. But conventional batch-level data does not have the resolution needed to precisely pinpoint the root causes of quality problems.

That's because, even when a parameter like pouring temperature is sampled for every mould or casting, it's impossible to know which data points relate to which individual defective castings. There can be multiple, interwoven reasons for defects like porosity or sand inclusions.

Adding a unique physical ID to each casting allows each one to be tracked through the entire process. Scanning or digitally entering any casting ID calls up its individual process parameters and, if scrapped, its defect type from the process database.

Casting-level data will also reveal when in-tolerance process parameters interact with each other and cause scrap, something that is very hard to diagnose at batch level.

5 Digitalization case study: Grede

Grede is currently implementing IIoT and AI-driven optimization at all 10 of its US foundries. It is connecting its melt, sand and moulding sub-processes in the first phase, then AI in the second.

The main drivers of the project were to create a single standardized process database, replace slow manual data with automated collection, timestamping and data integration, give new staff more support, and to react very quickly to any operational problems. Upon full implementation, Grede's foundries will be able to see historical and real-time production data side by side, helping them to solve quality issues and prevent scrap.

6 Digitalization case study: Morikawa

Leading Japanese foundry Morikawa started implementing IIoT in 2021. It connected its entire casting process and began to use its data to implement AI-driven full process optimization.

The results of the first test showed an uplift of 66.6% and 86.9%, respectively for two different patterns.

7 Digitalization case study: Huaxiang

Huaxiang is China's fifth largest private foundry. It started implementing IIOT in 2021 and was implementing full process optimization within 2 years. Testing produced scrap reduction figures from four automotive castings of 86.1%, 57.3%, 46.6% and 30.3%.

8 Digitalization case study: Condals

Spanish foundry group Condals is a data-driven company; it takes decisions based on the data from its process. It delivered a step-by-step IIoT program that has so far connected three casting lines. During process optimization testing, Condals saw a 40% scrap reduction and that figure has continued to be improved.

9 Conclusions

Traditional manual techniques are insufficient for the needs of a modern green sand foundry. Digitalisation provides the data and tools required to manage and further improve casting production. An IIoT system that can collect and send data to a single, central database is the optimal solution. The central database must be easily accessed and be combined with tools that allow monitoring, visualization, reporting and analytics.

As the database builds over time, it becomes a foundry's most valuable asset. If desired, it can support AI-driven, real-time process control across one or more entire lines to deliver the most efficient and stable possible process.

Numerical Simulation and Process Optimization of Electromotor Shell

Yanzhu Jin[1,*], Linghai Hu[1], Wei Jiang[1], Feng Li[1], Dianfu Zhao[2], Hongjun Liu[2], Yunfeng Zhang[1]
1. FAW foundry Co., Ltd. Changchun 130013, China
2. FAW foundry Co., Ltd., Chengdu Non-Ferrous Branch, Chengdu 610100, China
*Corresponding address: e-mail: jinyanzhu2009@163.com

Abstract: Based on the low pressure sand casting process, aiming at the common defects of electromotor shell such as shrinkage, misrun, cold shut, pores and poor fitting of bushing, the casting process was analyzed by the MAGMA software, and process scheme were determined and verified combined with the DOE design. The results indicate that the gating system scheme of annular runner with multiple inner gates can effectively solve the shrinkage at the bottom of the casting, achieve stable filling and avoid cold isolation. The use of top feeders and side chills can effectively solve the shrinkage and loosening of other parts of the casting, and effectively solve the pores defect.

Keywords: electromotor shell; low pressure sand casting; casting process; numerical Simulation

1 Introduction

The structure of the motor shell is complicated. The side wall of the motor shell surrounds the cooling water jacket. Ensuring the sealing of the water jacket is an important technical requirement of the product, as well as the biggest casting difficulty of the product. At the same time, the shrinkage of the upper and lower end faces and side walls of the motor shell are also casting defects that need to be avoided in the process development.

This paper introduces the structural characteristics and common casting defects of the motor shell. On this basis, the casting process design of motor shell and the application of computer simulation technology in the rapid trial production of motor shell castings are discussed and shared.

2 Experimental procedure

Technical specifications for products: The main technical requirements of motor shell in this paper are as follows: the casting surface and the processed surface are not allowed to have porosity, shrinkage, cold isolation, cracks, slag inclusion and other casting defects; Mechanical properties of castings: hardness of the top flange surface ≥ 90HbW, tensile strength (Rm) ≥ 275 MPa and elongation (Aa) ≥ 2% at the sampling part of the body. Casting internal defects shall be controlled to ASTM E155 Class III; The air tightness shall meet the requirements of water test at 0.6 MPa for 10 minutes after the product is blocked, and there is no gas leakage around the water jacket.

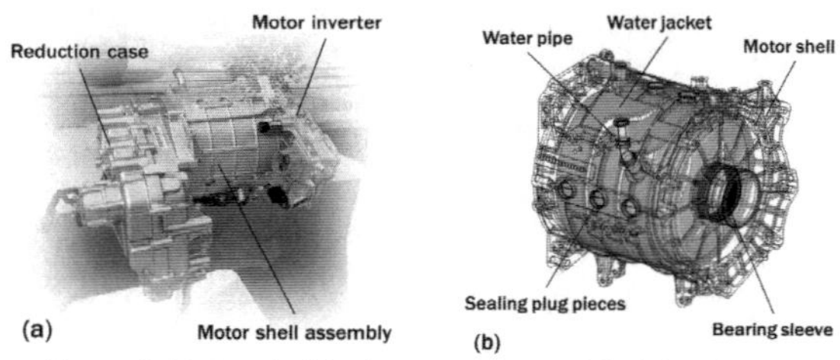

Figure 1 Motor shell in the power assembly (a) and motor shell construction (b)

Gating system design: The casting is filled with the top flange facing up, so that the area around the bearing bushing can be filled earlier during the filling process. The advantage of this is that the metal liquid in the filling is in contact with the bearing bushing, and the temperature loss is small, which is conducive to improving the fitting effect of the bearing bushing and the casting. A riser needs to be placed on the top of the casting to realize the feeding of the top of the casting. The cross runner of the casting is arranged in a ring and connected to the gate at the top of the riser through a cross structure. The inner gate is arranged in the ring runner, and the inner gate corresponds to the position of the bolt hole at the bottom of the casting. The pressure feeding eliminates the shrinkage hole and porosity defect at the bottom of the casting, and the advantage of the ring runner is to make the liquid flow in the filling process more stable and avoid air trapping.

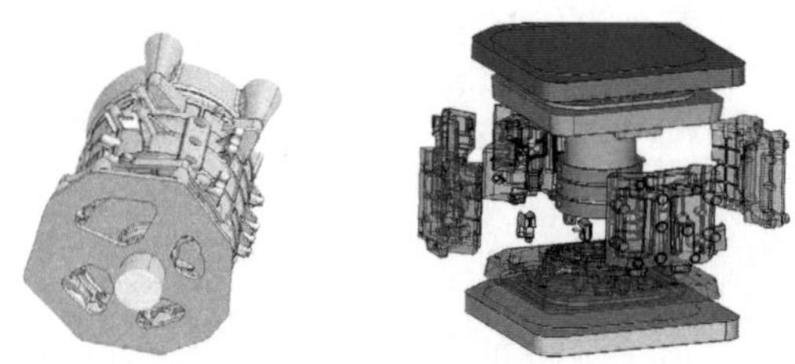

Figure 2 Gating system design and core combination modes of the mass production

For the mass production of motor shell, the water jacket core is made by hot core box, and the outer contour cores are made by cold core boxes, which is shown in Figure 2(b).

In order to avoid the casting defects of cold shut and burn on sand in the filling process, filling temperature is controlled in the ideal range of 720-740 ℃. The pressure-time curve is shown in Figure 6. The filling time is 10 s, and the pressure holding time is 210 s.

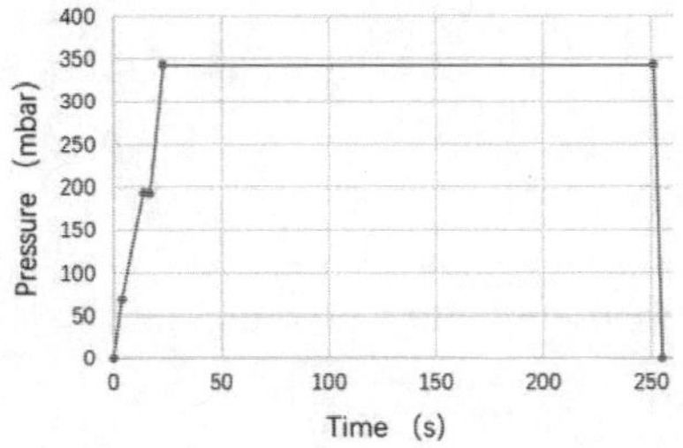

Figure 3 Pressure-time curve for pouring and solidification

3 Result and discussion

Numerical simulation:

The simulation of filling process is shown in Figure 4. During the whole filling process, the liquid metal flows smoothly without turbulence or air-entrapment. And, until the end of filling, there is no serious loss of temperature of metal liquid.

The fraction-solid field during solidification is shown in Figure 5. The metal liquid in the mold cavity is solidified in an ideal solidification sequence, and no isolated liquid region appears.

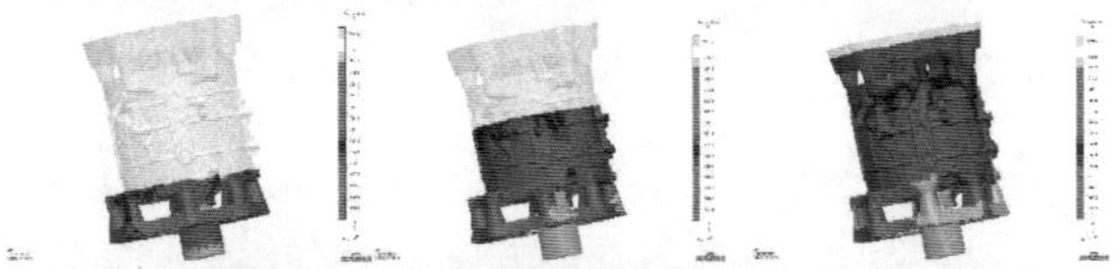

Figure 4 Simulation of filling process

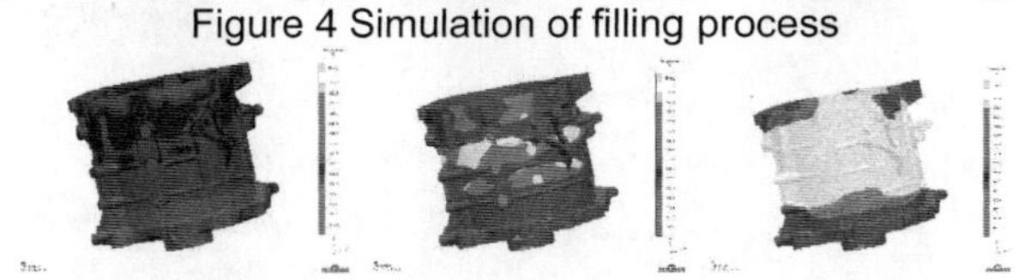

Figure 5 Simulation of solidification process

As shown in Figure 6, the shrinkage and cavity defects exist in the feeders, and there is no serious shrinkage cavity and porosity in the casting.

Figure 6 Simulation of shrinkage

Production: The final motor shell is shown in Figure 7. There is no casting defect in the motor shell. Through pressure leakage test to verify the product sealing to meet the technical requirements of the product.

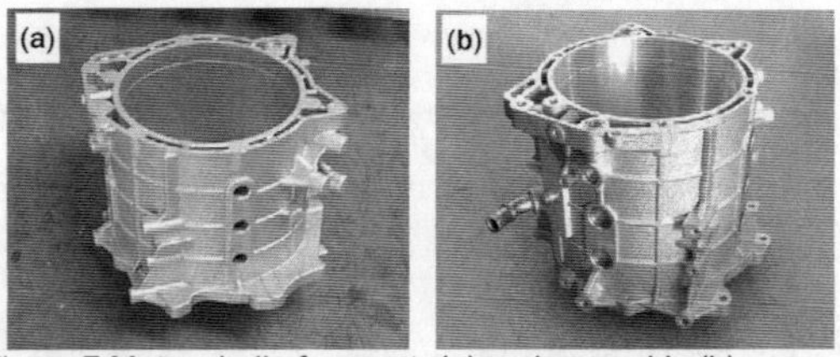

Figure 7 Motor shell of as-cast (a) and assembly (b)

4 Conclusion

(1) For the motor shell, the low-pressure sand casting process can be used for the mass production and small batch production.

(2) For the complicated motor shell, the gating system and feeding system need to be carefully thought when the casting process design.

(3) In order to avoid the appearance of casting defects as much as possible, it is necessary to make full use of simulation methods.

References

[1] Huang M T, Gao X M. Motor shell temperature prediction based on ARIMA. Applied Mechanics and Materials, 2013: 997-1000.

[2] Guang Y M, Li L, Dong J X. Research on quality control of motor shell coating process. Applied Mechanics and Materials, 2014, 699-702.

[3] Han W, Liang Q H, Huang L S, et al. Design and analysis of hot core box mold for new energy vehicle motor shell. International Journal of Metalcasting, 2018, 12: 779-784.

[4] An Z Y, Huang Z Y, Yan F, et al. Defect analysis and solutions for aluminum alloy motor-cover developed by die-casting. Foundry Equipment and Technology, 2020, 2: 25-28.

[5] Li L, Ma Z M. CAE analysis of aluminum alloy casting process for motor shell. China Foundry Machinery & Technology, 2020, 55(6): 95-98.

[6] Han W, Duan H F, Tong Z, et al. Design and optimization of low-pressure casting process for water-cooled casting of electric bus. Special Casting & Nonferrous Alloys, 2020, 40(8): 857-860.

[7] Sigworth G K, Wang C. Mechanisms of porosity formation during solidification: A theoretical analysis. Metallurgical Transactions, 1993, 24(2): 349-364.

[8] Asta M, Beckermann C, Karma A, et al. Solidification microstructures and solid-state parallels: Recent developments, future directions. Acta Materialia, 2009, 57(4): 941-971.

Development Practice of Casting Foundation Parts Database

JianchengFu[1], YuanqiZhou[2], YanqiuWang[1], JunFu[1], JiaqingGan[1],TingjiaWu[1],ShunshanWang[1]
1.College of Materials Engineering, Sichuan Polytechnic University, Deyang, Sichuan, 618000, China
2.College of Engineering and Technology, Xihua University, Chengdu, Sichuan, 610039, China

Abstract: This project is in response to the promotion of enterprises using digital technology to transform traditional process equipment and production lines. So we carried out secondary development of UG, including the implementation of points, lines, surfaces, and bodies, as well as the creation of menus and toolbars. Using Grip programming technology, we can generate three-dimensional entities with similar structures and different sizes by inputting parameters, which can assist designers in quickly designing product shapes, reduce modeling time, and solve the problem of errors in drawing.

Keywords:secondary development; Grip programming technology; digital collaborative platform

1 Introduction

The user interface design of UG includes menus and toolbar icons. The development language uses the built-in Grip [1].

2 Custom menu production process

In the UG installation path, open the custom_dirs.dat file in Notepad and write the path in the last line. Create a new folder and create two new folders under the file: Startup and Application. In the Application folder, place a. grx file, such as a sphere. grx. In the Startup folder, create a new text document. men. Write the code and place an image in the. bmp format.Carry out a task

```
VERSION 139
EDIT UG_GATEWAY_MAIN_MENUBAR
BEFORE UG_HELP
CASCADE_BUTTON MY_MENU1
LABEL basic parts
END_OF_BEFORE
MENU MY_MENU1
BUTTON MENU_ basic parts 1
LABEL basic parts 1
MESSAGE Click to generate basic component 1 and
run basic component 1
BITMAP basic parts 1.bmp
ACTIONS basic parts 1.grx
BUTTON MENU_ basic parts 2
LABEL basic parts 2
MESSAGE Click to run basic component 2
BITMAP basic parts 2.bmp
ACTIONS basic parts 2.grx
END_OF_MENU
```

Open UG and click on the menu.

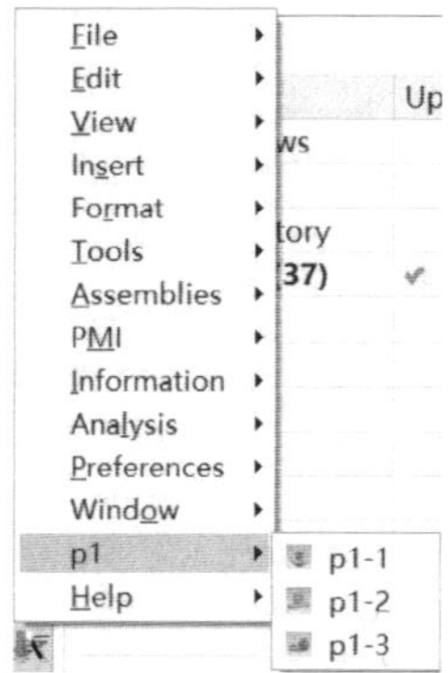

Figure 1 Custom Menu

3 Custom toolbar icon production process

Place images in BMP format in the Startup folder. Create a new text document, input the code and save it as a. tbr file. Switch to the Classic interface, click the Load button in the menu, and select the. tbr file generation toolbar icon.

```
TITLE Customize toolbar icons
VERSION 139
DOCK TOP
BUTTON
FUNA
LABEL Function 1
BITMAP basic parts 1.bmp
ACTION c:\UGgrip\application\ basic parts 1.grx
BUTTON FUNB
LABEL Function 2
BITMAP basic parts 2.bmp
ACTION c:\UGgrip\application\ basic parts 2.grx
```

4 Interactive window creation

Interactive windows allow technicians to generate corresponding 3D entities by simply clicking command buttons.

```
DATA/AA,5,BB,8,CC,10
PARAM/'select','radius1',AA,'diameter',int,BB,'radius2
',CC,rsp
```

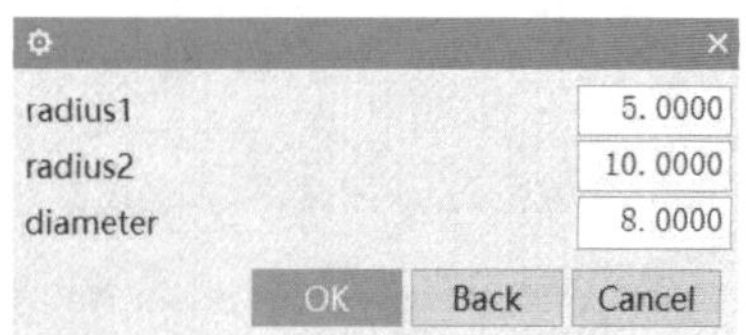

Figure 2 Creating an interactive window

5 Development practice

The complete development process was demonstrated using the A-type base of a new energy vehicle.

Part of the code is as follows:

ln(1)=LINE/0,0,(x120-x90)/2,0
ln(2)=LINE/(x120-x90)/2,0,(x120-x90)/2+z6,z6
ln(3)=LINE/(x120-x90)/2+z6,z6,x120-(x120-x90)/2-z6,z6
ln(4)=LINE/x120-(x120-x90)/2-z6,z6,x120-(x120-x90)/2,0
ln(5)=LINE/ x120-(x120-x90)/2,0,x120,0
ln(6)=LINE/ x120,0,x120,z20
ln(7)=LINE/ x120,z20,0,z20
ln(8)=LINE/0,z20,0,0
enta=SOLEXT/ln(1..8),HEIGHT,y80

Ultimately, you only need to input parameters in the interactive window to generate the corresponding A-shaped base 3D entity.

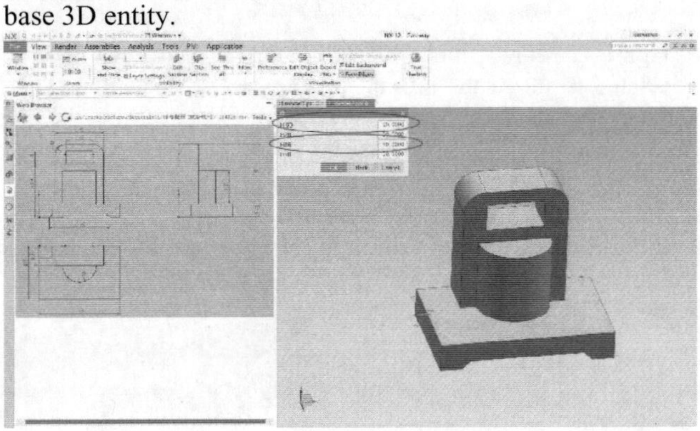

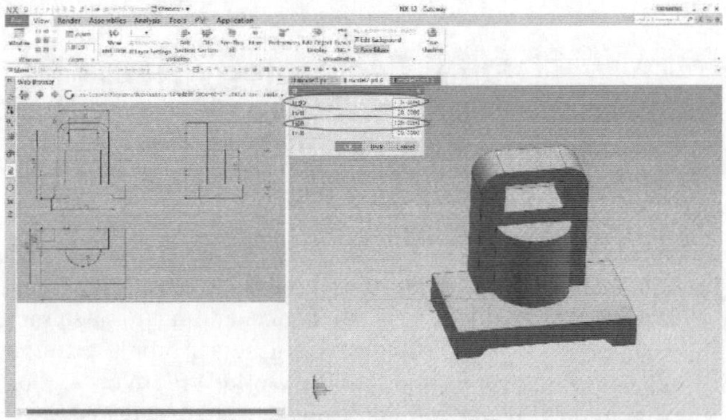

Figure 3 Generating A-type bases with different parameters

6 Conclusion

The "Foundry Basic Parts Library System" developed and designed using Grip language can obtain three-dimensional parts with similar structures but different sizes by inputting parameters, freeing technical personnel from a large amount of tedious work and driving upstream and downstream enterprises to implement intelligent manufacturing synchronously.

Reference

[1] Shi S W. Research on parametric modeling and static meshing simulation of spiral bevel gears based on UG/Open GRIP [D]. Shijiazhuang Railway University, 2023. DOI: 10.27334/d.cnki.gstdy.2023.00104

Digital Technology Helps Promote the Culture of Bronze Ware

Yanping Yang[1], Lingyue Wei[1], Mei Yin[1], Xinyu Zeng[1], Yanqiu Wang[1], Peng Feng[2], Jun Fu[1]
1.Sichuan Polytechnic University, Deyang, Sichuan, 618000, China;
2.Chengdu Technological University, Chengdu, Sichuan, 610031, China)

Abstract: Utilize open-source artificial intelligence platform APIs to obtain access token tokens and achieve artificial intelligence generation and editing of materials. Using Wen Xin AI technology to generate bronze artifacts. The development engine uses the World Wide Web version, which only requires filling out an Excel spreadsheet to achieve development. The Bronze Ware Park is divided into five exhibition halls. AIGC has designed the Bronze Ware Park and designed five halls one by one. The homepage of the works features five buttons, including background music, exhibition introduction, character roaming, camera rotation, appearance and hiding, video playback, highlighting, and knowledge Q&A. Successfully developed VR works to promote the culture of bronze ware.

Keywords artificial intelligence; virtual reality; AIGC; bronze ware; world wide engine national leaders

Introduction

National leaders have repeatedly proposed and emphasized the importance of strengthening cultural confidence, deepening academic research, innovating exhibitions and exhibitions, promoting the revitalization and utilization of cultural relics, promoting cultural exchange and mutual learning, and safeguarding, inheriting, and showcasing the excellent achievements of Chinese civilization. In ancient Chinese etiquette, bronze ware played an important role. The Xia, Shang, and Zhou dynasties were the heyday of bronze ware manufacturing in China. The three generations of bronze ware were loved by people for their unique shapes, exquisite patterns, and elegant inscriptions. Figure 1 shows some exhibits of ancient Chinese bronze artifacts.

Figure 1 China's splendid ancient bronze culture

1 Artificial intelligence solutions for materials

The materials used for developing virtual reality VR works include images, text, voice, video, etc. They are generated using AI generated AIGC technology and are also used for AI based editing.

1.1 Artificial intelligence generating AIGC technology for materials

The generation of work content evolved from professional PGC and user generated UGC to artificial intelligence generated AIGC.

The key code for Wen Xin's painting is,

input_dict={"text":prompt, "style":fengge, "resolution":chicun, "number":shu}
*rst = TextToImage.create(**input_dict)*
print(rst)

Figure 2 shows the artificial intelligence AIGC generated result of the prompt word prompt="ancient Chinese bronze ware, plate".

Figure 2 Ancient Chinese bronzes produced by Baidu Wenxin AI AIGC

2 Development practice of web version bronze vr works

2.1 Selection of development engine

Figure 3 is the logical relationship diagram of the World Wide Web version engine.

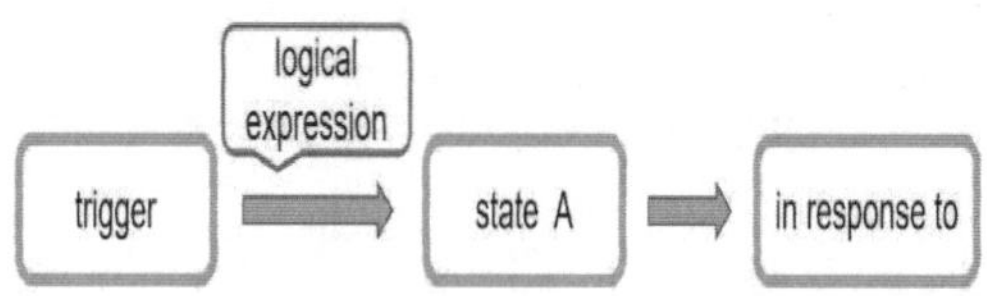

Figure 3 Logic diagram of VeryEngine

The World Wide Engine uses Excel as the development script, where users fill in objects, states, triggers, and responses in a specified format, and through the free combination of triggers and responses, complex program logic is achieved.

2.2 Park design

The park is divided into five exhibition halls: General Education Hall, Theme Hall, Country Hall, Technology Hall, and Information Hall, each of which is further divided into several exhibition rooms.

Figure 4 is the initial draft of the bronze ware park designed by the project team using AIGC technology, and detailed design will be carried out on this basis in the later stage.

Figure 4 Preliminary design draft for bronze ware park (Hugging Face AIGC Technology)

2.3 Functional realization

2.3.1 Process design

1） There are five exhibition hall names on the homepage. Clicking on one of them will take you to the exhibition hall. After entering the exhibition hall, background music will automatically play. Click the "Background Sound" button to play and stop music.

2） Operate the keyboard to achieve character roaming in the venue. When a character approaches an exhibit, the exhibit highlights and automatically plays its introduction. Users can choose one or more methods such as text, voice, images, and videos. When leaving the exhibit, the highlight and introduction will end.

2.3.2 Home page design

There are 5 buttons for venue names on the homepage. When the mouse moves over one of these buttons, the button turns red and highlights the corresponding venue name. Figure 5 is the homepage of VR works.

Figure 5 The Entrance of "Hello, Bronze" VR Works (In Chinese)

2.3.3 Exhibition introduction voice

The code for the introduction voice of the exhibits is , *WindowsVoice, *voiceStr.* Among them, *VoiceStr is the text to be converted into speech.

2.3.4 Role roaming

Switch to a roaming camera in the focus camera state and use the roaming camera code to achieve scene roaming. The code is:*Active_Camera, Main Camera, Camera1.*

2.3.5 Appearance and hiding of text/images

The code is:*GUI_Enable,2D canvas. Container, true.*

2.3.6 Highlight

The object highlighting code is: *OutlineHighlight,* Boolean, greenBox2, (255,0,0), (0,255,0), 0.5, false.*

2.3.7 Knowledge Q&A

The knowledge Q&A code is:GUI_Control, *keyStr,2D canvas,1: the first book,(2: the second book,The third book).

Figure 5 shows the "Knowledge Q&A" interface. When the user moves the mouse over a button for a certain answer, the button will turn red.

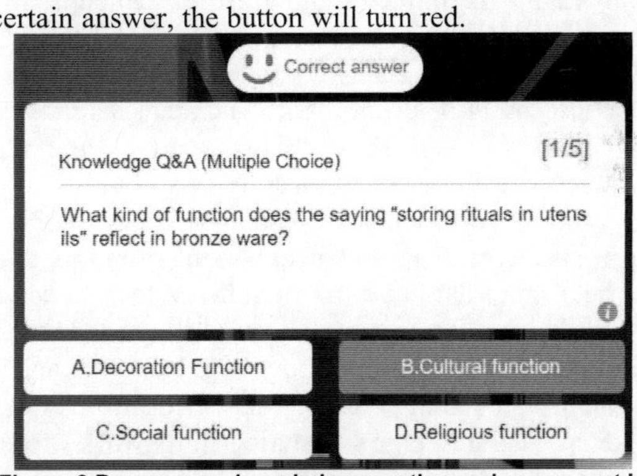

Figure 6 Bronze ware knowledge question and answer set in the World Wide Engine

3 Conclusion

We have prepared VR materials using artificial intelligence technology and developed "Hello, Bronze Ware" VR works using the World Wide Web version engine, which combines education with entertainment and promotes traditional Chinese culture.

Acknowledgments

Thanks to Hangzhou Wanwei Mirror Technology Co., Ltd. for providing technical support.

References

[1] Wang H H. The practice path of integrating excellent traditional culture into strengthening the awareness of the Chinese National Community among college students. Journal of Southwest University for Nationalities (Humanities and Social Sciences Edition), 2023, 44(1): 210-217

[2] Zhai Y, Li J. Reflection on the development path of AIGC: New opportunities for the popularization of big model tools. Internet World, 2022, 11: 22-27.

Data Analytic Framework for University - Industrial Collaboration for Implementing Industry 4.0 in Metal Manufacturing Sector

Hongbiao Dong*, Maryam Khaksar Ghalati, Jianbo Zhang
School of Engineering, University of Leicester, Leicester, UK
*Corresponding e-mail: hd38@leicester.ac.uk

Abstract: The metals industry is facing many challenges, from productivity, capital investment cycle, margin squeeze, and the global challenge of decarbonization. The revolution in digital manufacturing, from the concept of Industry 4.0 is a key enabler in meeting these challenges and ensuring companies remain fit for the future. Recent advances in information technologies and data sciences enable data-driven-decision-making in metal manufacturing, which can significantly improve their operations and profitability. Most of the large manufacturing enterprises now start to benefit from this as they can collect more data that can be utilized to enhance their decision-making processes. Universities, and small and medium enterprises (SMEs) have limited data and resources, thus reducing the possible gains. In this work, we propose a shared data analytic framework for university and industry to collaborate and share data, the data will be then jointly analyzed, feasibility and quality of data analytics and decision-making processes could be significantly enhanced.

Keywords: data sharing framework; digital manufacturing; metal manufacturing

1 Introduction

The metal industry is undergoing digital transformation due to challenges like productivity, capital investment cycles, margin squeeze, and the global push for decarbonization [1, 2]. Industry 4.0, which promotes the seamless integration of digital and physical systems, is key to meeting these challenges by leveraging artificial intelligence (AI) and vast data generated throughout manufacturing processes [3]. In 2023, global spending on digital transformation reached 2.15 trillion USD and is expected to grow to 3.9 trillion USD by 2027 [4]. Large enterprises typically lead in digital transformation due to advanced IT capabilities, optimizing operations through big data analytics. While big data analytics also advance university research and benefit SMEs, these smaller entities often face limitations in data and resources. A shared data analytic framework is needed to help these institutions fully leverage digital transformation.

2 Challenges in industrial-academia collaboration

Figure 1 illustrates the complex intersections between AI, academia, and the metal industry, highlighting the multifaceted challenges and opportunities in fostering effective industrial-academic data collaboration.

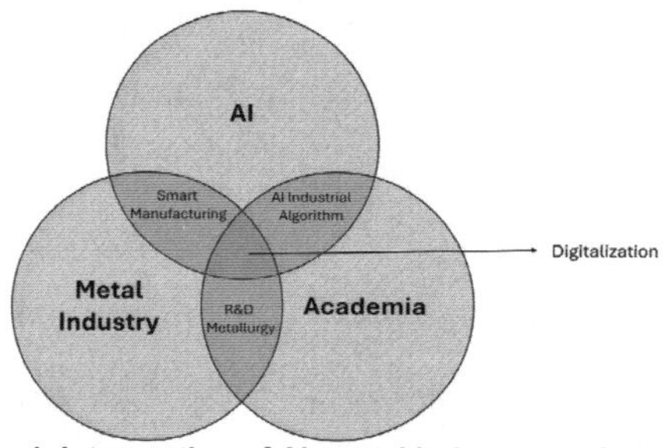

Figure 1 Intersection of AI, metal industry, and academia

Data nature and storage strategy in the metal industry. In the metal industry, data is generated from various sources such as sensors, ERP systems, and automated processes, leading to a mix of structured and unstructured data. This data is characterized by its high volume, velocity, variety, and variability, particularly in processes like BOF steelmaking, where real-time data on temperature, chemical composition, and emissions is crucial for maintaining product quality [5]. However, the fragmented legacy systems typically in place within metal-producing companies limit the accessibility and integration of this data. To address these challenges and fully leverage the insights offered by such diverse data, a comprehensive digital transformation is needed. This involves creating a unified data platform that enhances data storage, accessibility, integration, and analytics capabilities, thereby optimizing operational efficiency and decision-making across the organization.

Data confidentiality. Data confidentiality is a major concern in industry-academia collaborations, as industrial partners fear intellectual property theft or losing their competitive edge (Brettel et al., 2014). Effective data governance, such as those used in healthcare under HIPAA or in banking under Basel III standards, ensures that sensitive data is managed securely. The proposed framework emphasizes data governance solutions like anonymization, secure transfer protocols, and clear usage agreements. Centralizing data management within academic institutions can simplify data access for research while ensuring industrial partners retain control over data usage—critical for protecting proprietary processes in the metals industry.

Standards for AI and industry 4.0. Adopting AI and machine learning in the metal industry can greatly enhance efficiency and innovation, but the implementation is challenging due to the complexity of metallurgical data and the need for specialized expertise. Developing models for predicting equipment failures or optimizing production processes requires deep integration of data science with metallurgical engineering. Effective data governance is crucial to ensure data quality, consistency, and regulatory compliance. Additionally, scalable data infrastructure is needed to manage growing data volumes and support real-time processing, enabling successful Industry 4.0 initiatives.

3 Data sharing and analytics framework

The shared data analytic framework facilitates seamless data integration and analysis across universities, SMEs, and large manufacturing enterprises. Its modular design allows flexible data input and processing, integrating data from various sources into a centralized, secure database. This database is protected with robust security measures, including encryption, access controls, and secure transfer protocols, ensuring confidentiality throughout the process.

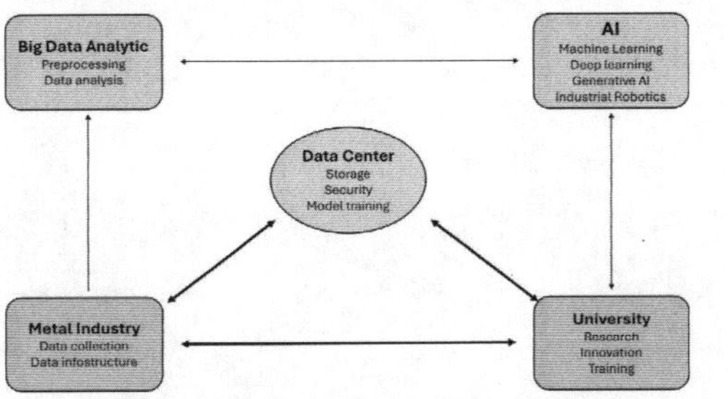

Figure 2 Data sharing and analytics framework

This secure database serves as the foundation for data analysis activities, including training machine learning models to predict outcomes, optimize processes, and generate actionable insights. The framework also incorporates advanced data visualization tools, making complex analyses accessible and particularly valuable in processes like casting, where precision is crucial. By integrating real-time data, the framework allows dynamic production adjustments, enhancing efficiency and reliability.

Additionally, the framework promotes industry-academia collaboration, accelerating technology development and innovation. This exchange benefits both sectors: universities gain access to real-world data for impactful research, while industry partners gain access to cutting-edge research and a skilled talent pipeline. Figure 2 illustrates this framework.

4 Conclusion

This paper discusses the characteristics and challenges of industrial data in the metals industry, focusing on the complexities of analyzing such data and the importance of secure data sharing between universities and industry. Embedding standards for data science and Industry 4.0 in future research is crucial for advancing the sector. The proposed shared data analytic framework facilitates the integration of experimental and modeling data from various laboratories and partners into a unified workflow. This integration enables advanced processes such as training machine learning algorithms to improve data-driven decision-making. The framework holds significant potential to enhance the reliability, availability, and efficiency of manufacturing processes, particularly in areas like casting. By fostering collaboration between academia and industry, the framework not only supports innovation but also strengthens the overall competitiveness and sustainability of the metals industry.

5 Acknowledgment

We acknowledge the support from the NISCO-UK center, School of Engineering, University of Leicester.

References

[1] Izatt R M, Izatt S R, Bruening R L, Izatt N E, Moyer B A. Challenges to achievement of metal sustainability in our high-tech society. Chemical Society Reviews, 2014, 43(8): 2451-2475.

[2] Kermeli K, Crijns-Graus W, Johannsen R M, Mathiesen B V. Energy efficiency potentials in the EU industry: Impacts of deep decarbonization technologies. Energy Efficiency, 2022; 15(8): 68.

[3] Jan Z, Ahamed F, Mayer W, Patel N, Grossmann G, Stumptner M, et al. Artificial intelligence for industry 4.0: Systematic review of applications, challenges, and opportunities. Expert Systems with Applications, 2023, 216: 119456.

[4] https://www.statista.com/statistics/870924/worldwide-digital-transformation-market-size/.

[5] Ghalati M K, Zhang J B, El - Fallah G M A M, Nenchev B, Dong H B. Toward learning steelmaking—A review on machine learning for basic oxygen furnace process. Materials Genome Engineering Advances, 2023, 1(1), https://doi.org/10.1002/mgea.6.

Integrated Computational Materials Engineering and Artificial Intelligence for Sustainable Casting Industry

Alan A. Luo[1], *, Diran Apelian[2]
1. Advanced Casting Research Center, The Ohio State University, Columbus, OH 43210, USA
2. Advanced Casting Research Center, University of California, Irvine, CA 92697, USA
*Corresponding address: e-mail: luo.445@osu.edu

Abstract: The global metal casting industry is facing enormous sustainability and regulatory challenges related to carbon reduction and carbon neutrality. Digital design and manufacturing of metal castings, enabled by integrated computational materials engineering (ICME) technologies and the recent boom in artificial intelligence and machine learning (AI/ML), provide great opportunities for the industry to overcome these challenges. This presentation provides some examples of cast alloy design and process innovations using the ICME approach. The talk will also present application cases of AI/ML tools to support casting quality control and property prediction. We will also discuss future opportunities to combine ICME and AI/ML tools to revitalize and revolutionize the metal casting industry for sustainable growth.

Keywords: metal casting; sustainability; integrated computational materials engineering (ICME); artificial intelligence (AI); machine learning (ML)

1 Introduction

Materials and manufacturing industries including metal casting are the backbone of global economies but also a climate liability, thus making their decarbonization a key priority for carbon mitigation strategies. To achieve the Paris Climate Agreement goal of reaching climate neutrality by 2050, major economies including the United States, China, and European Union have announced ambitious emission reduction commitments. Therefore, the global casting industry is facing enormous regulatory and sustainability challenges related to carbon emission reduction, as its customers in all sectors especially the automotive industry are transitioning to clean energy and sustainable production. Presently, a large portion of the industrial emissions comes from the production of iron and steel (24%) and aluminum (3%). However, the recycling rates for these metals are astonishingly low, 45% for steel and 30% for aluminum. Increasing the circularity of these nonrenewable metals and reducing energy consumption in manufacturing activities would significantly contribute to a carbon-neutral society and a circular economy. Figure 2 shows a vision of material circularity and sustainability which can be achieved by 1) reduced/prolonged usage via better material design and manufacturing/energy efficiency; 2) increased repair, reuse, refurbish, remanufacturing and recycling; 3) limited but clean primary material production; and 4) minimal or no disposal of non-renewable materials.

2 Integrated computational materials engineering

Integrated computational materials engineering (ICME) is defined as the integration of materials information, captured in computational tools, with engineering product performance analysis and manufacturing-process simulation. Figure 2 shows a CALPHAD-based ICME approach for designing secondary (recycled) aluminum alloys for structural die casting applications.

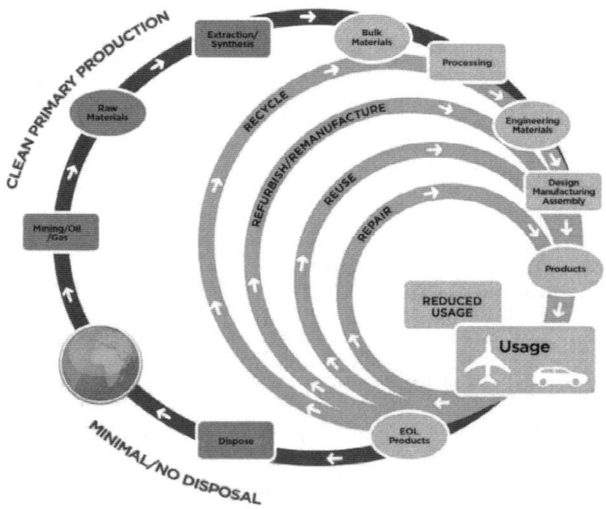

Figure 1 Vision of material circularity and manufacturing sustainability (modified, based on [1])

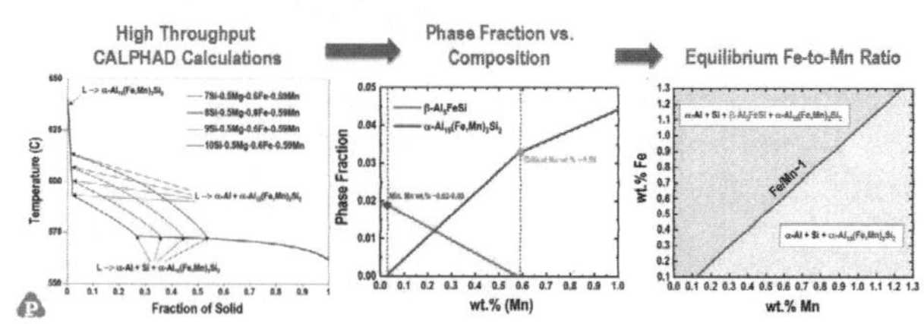

Figure 2 CALPHAD-based approach for alloy design [2]

In the last few decades, casting simulation and digital manufacturing tools have been increasingly used in the industry as part of product development process, based on computer-aided design (CAD), computer-aided engineering (CAE), and computer-aided manufacturing (CAM). As shown in Figure 3, ICME is a relatively new approach based on location-specific microstructure and property models predictions, as opposed to the current CAD/CAE/CAM approach largely using uniform material properties. These ICME models, based on computational thermodynamics, kinetics and process models, can provide further mass and cost savings as well as improved quality/efficiency. The recent development of large thin-wall die casting (giga-casting), using ICME tools, will enhance the sustainability of the metal casting industry.

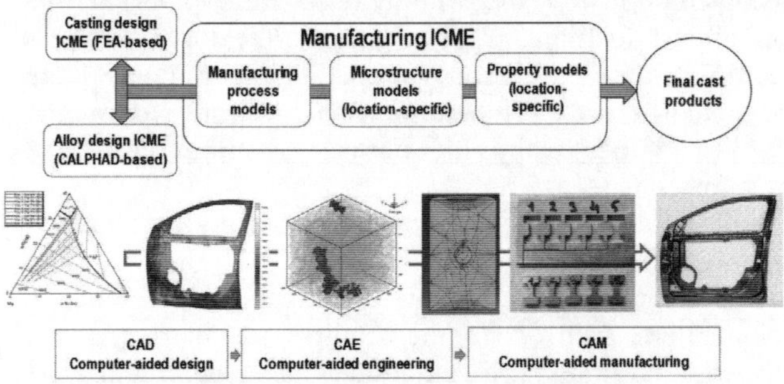

Figure 3 ICME framework for casting design and process development[3]

3 Artificial intelligence and machine learning

Artificial intelligence (AI), including machine learning (ML), is a broad term defined as a "system that can, for a given set of objectives, generate outputs such as predictions, recommendations or decisions influencing real or virtual environments" [4]. These systems use data and human-built algorithms to simulate how humans perceive, learn and respond to questions and prompts. Metal casting processes such as high-pressure die casting can generate extensive datasets, which can be used to train deep ML algorithms to model the prediction of good parts and process scrap as determined by a die-casting machine. Additionally, neural network ML models can be used to predict the ultimate tensile strength (UTS) of die castings (Figure 4). Future applications of AI/ML in metal casting include defect control, predictive maintenance, and supply chain logistics. AI/ML and ICME tools can be combined to co-design of alloys, processes and topology for multi-objective optimization of sustainable castings.

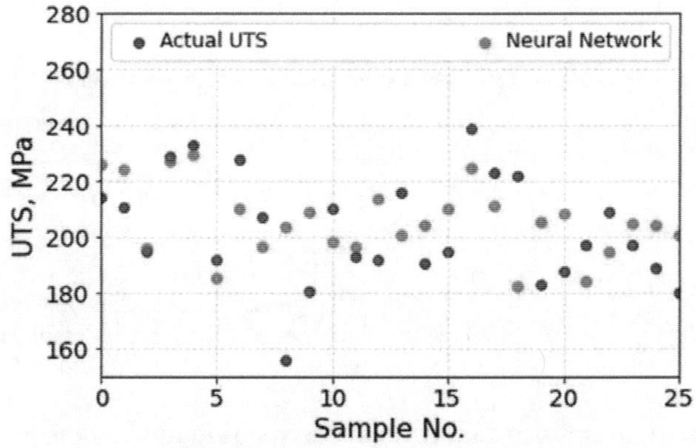

Figure 4 Actual UTS testing data vs. the values predicted by the Neural Network model of die castings[5]

4 Conclusion

Metal casting has a long history but is presently facing technical and social challenges related to carbon footprint. It is imperative to use recycled alloys in the industry, embracing the circular material concept. To overcome these challenges, the casting industry needs to utilize new ICME and AI/ML tools to improve its efficiency and reduce energy consumption. Opportunities also exist in transportation industries as they migrate to clean energy technologies will require lightweight and high-performance castings.

5 Acknowledgments

The authors are grateful to the support and collaboration of numerous colleagues and students from The Ohio State University, University of California at Irvine, and many industrial partners/members of the Advanced Casting Research Center (ACRC). The authors acknowledge the financial support from the United States National Science Foundation and Department of Energy.

References

[1] https://www.brunel.ac.uk/research/Centres/BCAST/About-us.
[2] CinkilicE, RidgewayCD, YanX, Luo AA, Metall. Mater. Trans. A, 2019, 50A: 5945-5956.
[3] LuoAA, SachdevAK, ApelianDJ. Mater. Process.Technol., 2022, 306: 117606.
[4] https://nvlpubs.nist.gov/nistpubs/ai/NIST.AI.100-1.pdf.
[5] KopperA, Karkare, R, Paffenroth, RC, Apelian A. Integr. Mater. Manuf. Innov., 2020, 9: 287–300.

Digital Twin Model of Casting Temperature and Shrinkage Porosity Prediction

Bang Guan[1], Donghong Wang[1, 2],*, Da Shu[1], Baode Sun[1]

1. Shanghai Key Laboratory of Advanced High Temperature Materials and Precision Forming, School of Materials Science and Engineering, Shanghai Jiao Tong University, Shanghai 200240, China
2. State Key Laboratory of Clean and Efficient Turbomachinery Power Equipment, Deyang 618000, China
*Corresponding address: e-mail: wangdh2009@sjtu.edu.cn

Abstract: Shrinkage porosity is one of the main defects of investment castings. In this work, a simulation-based digital twin (DT) methodology of porosity distribution prediction is proposed. The methodology is based on a regression supervised learning approach to predict porosity from thermal history time series. The temperature reduced order model (ROM) based on Proper Orthogonal Decomposition (POD) and Radial Basis Function (RBF) network interpolation is developed. Learning data are generated by casting simulation with different process parameters. Furthermore, the optimal distribution of temperature measurement points is determined and the POD coefficient is updated based on the thermocouple data in casting process, so that the time series of temperature field are update, and then the porosity distribution is predicted.

Keywords: digital twin; reduced order model; casting; shrinkage porosity

1 Introduction

Porosity is one of the major defects that can affect a casting part. Optimizing the porosity formation in casting parts can greatly reduce the time cost of the optimization process through machine learning techniques such as artificial neural networks [1]. This effectiveness has been exemplified in the work of Tsoukalas [2], where the total porosity volume is reduced with a rate of 66% in aluminum alloy pressure die casting. To predict shrinkage porosity in real time, the ROM was used to calculate real-time temperature field [3], and the machine learning models can quickly respond and integrate with actual measurement data to construct real-time digital twin models for shrinkage porosity prediction [4].

2 Experimental procedure

Basic architecture

An overview of proposed digital twin methodology is presented and detailed. As shown in Figure 1, a dataset is created from the results of the simulations, primarily consisting of nodal thermal histories and corresponding porosity distribution values, and then, utilizing a POD approach to create ROM. For a case with fixed process parameters, the modal coefficient is computed through radial basis function (RBF) interpolation, enabling the reconstruction of the temperature field. During the supervised learning process, essential features, such as solidification time and temperature gradients, are extracted from the temperature field. A regression model is then trained using the porosity distribution as the outcome.

In the online stage, by correcting the pod coefficient, the predicted temperature is consistent with the collection of time-series data, thereby establishing the complete temperature field. Then the features are extracted and imported into the regression model. Thus, completing the online prediction of porosity. Take group of parts as casting, presented in Figure 2, consists of parts, runner and sprue cups.

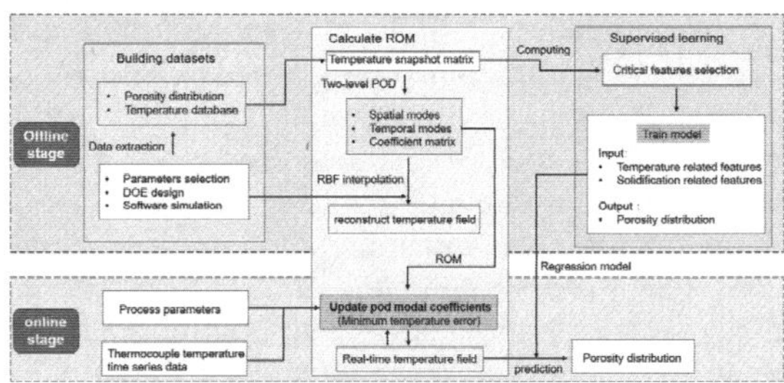

Figure 1 Digital twin implementation architecture of shrinkage porosity prediction

Figure 2 Geometric model of casting

3 Result and discussion

Study of processing parameter

Based on the simulation result, extract the shrinkage and porosity distribution results. Statistically analyze the distribution of shrinkage porosity at the riser. Calculate the volume occupied by shrinkage cavities and their average location. As shown in Figure 3, it can be found that the temperature change of the mold shell has a greater impact on shrinkage porosity when the shell temperature changes from 900 ℃ to 1,200 ℃. The simulation result of other cases are shown in Figure 4.

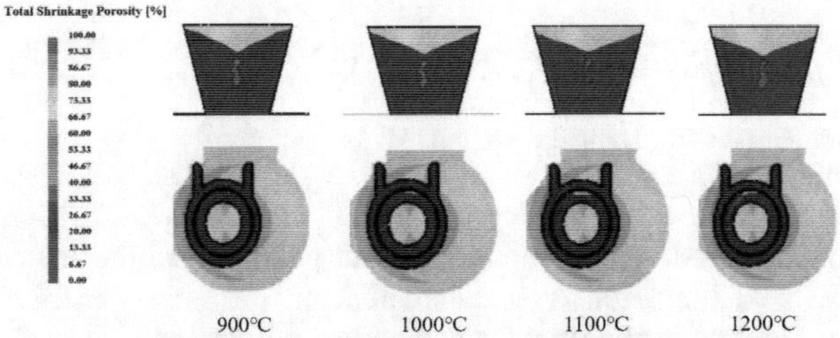

Figure 3 Shrinkage position porosity changed with shell temperature

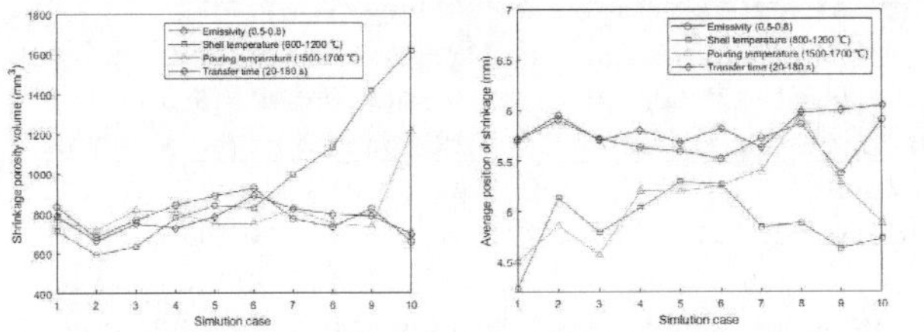

(a) (b)

Figure 4 Statistical distribution of shrinkage porosity position changed with process parameters, porosity in riser (a) and part (b)

Temperature field reconstruction of reduced order model

The transient temperature field is established with numerical simulation to calculate the node temperature value at each discrete time. POD was used to analyze the snapshots and perform an eigenvalue analysis according t in Figure 5, the first two eigen values accounts for 99.82% of the sum of the eigenvalues and the first four eigenvalues account for more than 99.99% of the total. when the solidification fraction reaches 98%, the reconstructed temperature field error is shown in the Figure 5(b).

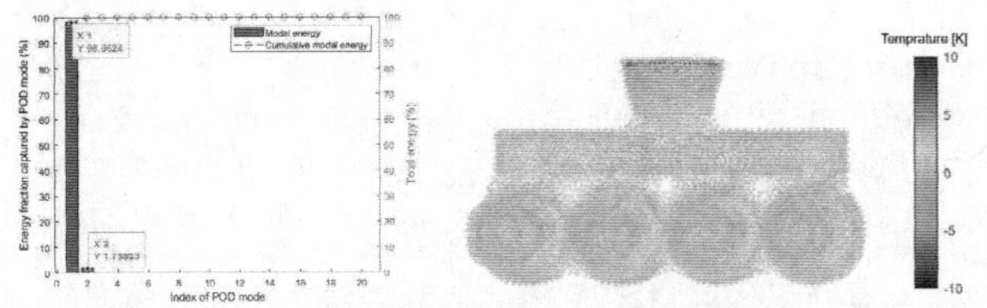

(a) (b)

Figure 5 Shrinkage position porosity changed with shell temperature (a), and reconstructed temperature field error (b)

In the arrangement of temperature measurement points, the correlation of the node temperatures is analyzed, and the greedy algorithm is used to select the eight least relevant node as shown in Figure 6(a). Nodes location is shown in Figure 6(b). The error of temperature field reconstructed by measuring point temperature based on simulation is shown in Figure 6(c).

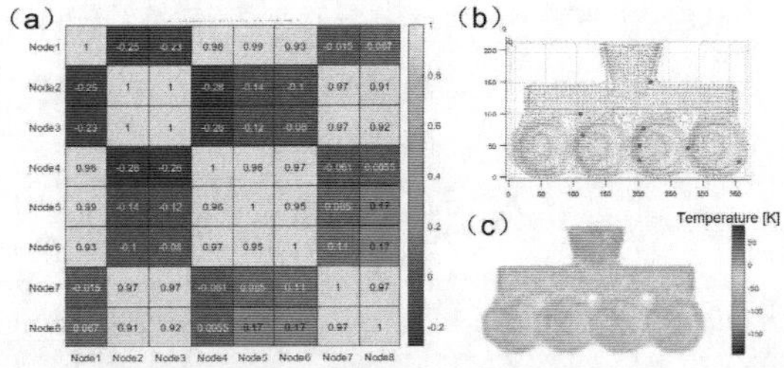

Figure 6 Correlation matrix of vectors at the locations of the eight nodes (a), nodes location (b), and reconstructed temperature field error (c)

Shrinkage porosity prediction model

The characteristics of the temperature field were extracted, including 20 related feature, such as critical solidification moment (initial solidification, solid fraction equal to 0.7, solidification end), temperature gradient, Laplace operator, etc. As shown in Figure 7(a), when the shrinkage porosity value is greater than 1, the prediction model has a better prediction effect. We analyze the importance of the characteristics of temperature field. In this model, temperature gradient and solidification time are important.

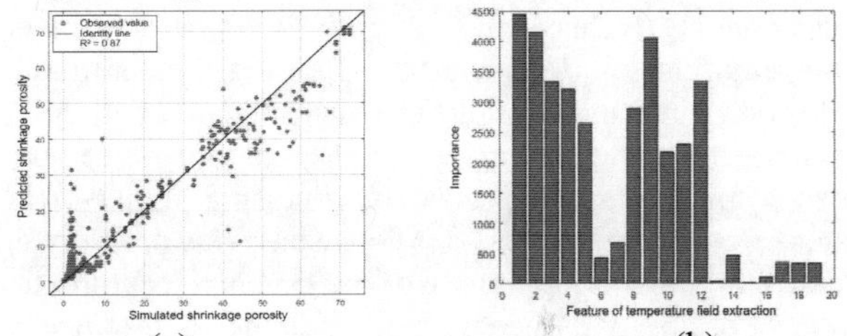

(a) (b)

Figure 7 Porosity prediction of our solution against Procast simulation (a), and importance of temperature feature (b)

4 Conclusion

The digital twin model is constructed by POD and machine learning model, which can real-time predict the temperature field according to the actual measured point temperature. Shrinkage porosity distribution prediction model is based on temperature field feature.

References

[1] Cheng Y W. Optimization of die casting processing parameters based on BP neural network and GA algorithm [J]. Journal of Plasticity Engineering, 2011, 18(3): 105-110.

[2] Tsoukalas V D. Optimization of porosity formation in AlSi9Cu3 pressure die castings using genetic algorithm analysis [J]. Materials & Design, 2008, 29(10): 2027-2033.

[3] Wu T, Yang F, Farooq U, et al. An online learning method for constructing self-update digital twin model of power transformer temperature prediction [J]. Applied Thermal Engineering, 2024, 237: 121-728.

[4] Nouri M, Artozoul J, Caillaud A, et al. Shrinkage porosity prediction empowered by physics-based and data-driven hybrid models [J]. International Journal of Material Forming, 2022, 15(3): 25.

Key Technologies of Automatic Production Line and Digital-Twin System of Mega Low-Pressure Die-Casting

Xuxia Zhang[1], Weipeng Liu[1, 2], *, Zhihui Wang2, Jun Wu[1], Tao Peng[2]
1. Zhejiang Wanfeng Technology Development Corporation Limited, Shengzhou, China
2. School of Mechanical Engineering, Zhejiang University, Hangzhou, China
*Corresponding address: e-mail: lwp@zju.edu.cn

Abstract: Mega casting represents a groundbreaking advancement in the global automobile manufacturing industry, offering enhanced quality and productivity. Low-pressure die-casting, as a significant form of mega casting technology, presents numerous advantages for the production of mega castings utilized in new energy automobiles. However, the low efficiency, high reject rate, and elevated manufacturing costs associated with mega low-pressure die-casting present notable challenges. In response, this study proposes a range of key technologies for the automatic production line and digital-twin system to address these shortcomings. First, critical issues were identified and research requirements were refined. Corresponding technologies were then proposed to meet these requirements, focusing on enhancing the automatic production line and digital-twin system. Building upon these key technologies, a high-quality and cost-effective scheme for automatic production line and digital-twin system for mega low-pressure die-casting was developed. A case study involving the use of the developed automatic production line and management systems to produce a subframe was selected to validate the effectiveness of the technology. The results demonstrated a significant improvement in production efficiency and qualification rate.

Keywords: low-pressure die-casting; automatic production line; digital-twin system; mega casting

1 Introduction

Mega casting, a term that has recently gained significant traction within the automotive sector, epitomizes a revolutionary leap forward in the realm of global automobile manufacturing. This innovative approach to vehicle construction entails the utilization of colossal casting machines capable of producing large, intricate aluminum structural components in a single, seamless piece. By supplanting the traditional multi-part assembly process, mega casting not only streamlines manufacturing workflows but also yields a substantial augmentation in both the structural integrity and overall quality of the final product [1].

Low-pressure die-casting (LPDC) represents a transformative approach within the mega casting domain, particularly for the production of critical components in new energy automobiles. Its ability to produce lightweight, high-strength, and complex castings with minimal defects and reduced environmental impact positions LPDC as a key enabler for the advancement of new energy automobile technology. As the demand for new energy automobiles continues to escalate, LPDC is poised to play an increasingly vital role in the evolution of sustainable and efficient vehicle manufacturing [2].

The implementation of mega LPDC technology, while heralded for its potential to revolutionize the production of large-scale castings, is not without its significant challenges. Chief among these are the low efficiency, high reject rate, and elevated manufacturing costs that currently hamper its widespread adoption in the industry. Compounding these issues are the elevated manufacturing costs intrinsic to LPDC. These challenges necessitate ongoing research and development efforts to optimize the automatic production line and management systems, enhancing its viability for broader industrial application.

2 Key technologies of automatic production line

Figure 1 illustrates the typical workflow of the casting process, which encompasses three primary stages: smelting and refining, transferring and casting, and subsequent post-processing. The pivotal technologies integral to the automated production line are explicated in detail herein.

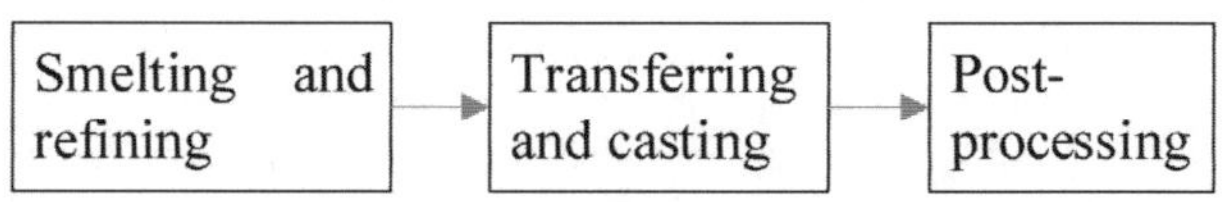

Figure 1 shows common casting process

Smelting and refining

Melting operations constitute the most energy-intensive equipment within the thermal processing workshop and are pivotal for energy conservation and consumption reduction. Manufacturers of melting furnaces spare no effort in technological advancements, such as the enhancement of the internal cavity structure, insulation layer architecture, and strategic placement of burners, to elevate the utilization rate of flue gas heat, thereby striving to minimize energy consumption and reduce aluminum melting loss. Furthermore, the optimization of exhaust gas collection technology facilitates the comprehensive preheating of aluminum ingots using waste heat; the adoption of regenerative burners significantly curtails gas consumption. Additionally, the implementation of aluminum scrap recycling devices allows for the melting of aluminum scraps, risers, and other minor waste materials. This process also drives the circulation of molten aluminum within the water jacket and internal cavity, equalizing their temperatures and substantially reducing melting losses in these areas.

Transferring and casting

The low-pressure die casting machine is the core equipment for obtaining castings, capable of automating the entire process of mold closing, filling, cooling, and part extraction, with a high degree of automation already achieved. The potential for further enhancement of this equipment is concentrated on reducing energy consumption, increasing efficiency, and refining process implementation. For instance, the adoption of servo-hydraulic and accumulator technologies in the hydraulic system, coupled with an optimized control system integrated with the production process, can effectively reduce energy consumption. Proper coordination and speed regulation of various mechanisms, combined with an effective integration of fast and slow operations, can minimize idle cycle time and increase per-shift production. Moreover, the mold temperature control system and cooling system, utilizing temperature-controlled and proportional flow closed-loop cooling control systems, not only prevent excessive cooling but also enhance the quality of the castings.

Post-processing

The castings delivered by the part extraction system can be transferred to the cleaning line via a cooling conveyor. The cleaning line is configured according to the characteristics of the castings, with the most comprehensive processes currently including loading, sand shaking, sand dropping, deburring, cutting, grinding, drilling, and marking, which can be selectively configured. Under the management and scheduling of the central control system, this production line achieves orderly interlinked operations and unmanned production.

3 Digital-twin casting system

In automated production lines, the reliance on equipment and sensing technology is increasingly critical, with manual intervention based on human experience becoming progressively minimal. Consequently, the traditional method of manually recording and collecting on-site data, characterized by poor timeliness and incomplete information, can no longer ensure the continuous operation of automated production lines. This necessitates a matching digital twin management system for comprehensive control, overall scheduling, detailed diagnostics, and quality analysis of the entire field. This digital twin system networks the equipment from the melting furnace to the cleaning line, retrieving real-time data and information from the devices. It displays equipment status and fault information on the system interface, allowing for timely access to equipment information. Simultaneously, it conducts real-time monitoring of the equipment's consumable and critical parts to ensure the continuity of production.

4 Application case

The product delineated in this manuscript has been deployed across more than 20 production lines within leading foundry corporations, primarily for the fabrication of large subframes for new energy vehicles. This deployment has actualized the automation of casting production lines and the digital management and control of operations.

5 Conclusion

This study underscores the transformative potential of mega casting in revolutionizing the global automobile manufacturing landscape, particularly through low-pressure die-casting for new energy vehicles, despite its inherent challenges such as low efficiency, high rejection rates, and increased production costs. It introduces a suite of pivotal technologies tailored for the automatic production line and digital-twin system, aimed at mitigating these drawbacks. The research delineates a comprehensive approach to refine research directives and implement technologies enhancing both the automatic production line and digital-twin system. The culmination of this research is the development of a cost-efficient, high-quality automatic production and digital-twin system framework for mega low-pressure die-casting.

6 Acknowledgments

This study was funded by the Key Research and Development Program of Zhejiang Province (Nos. 2023C01153 and 2024C01139).

References

[1] Fiedler F, Bergweiler G, Burggräf P, Schuh G. Fields of action in mega-casting for the industrialization of automotive car bodies. RWTH Publications, 2022.

[2] Zhang X. Intelligent production technology of hot working workshop for low pressure casting. MW Metal Forming, 2023, 1: 15-18+25.

Analysis of the Possibility of Using Exploration and Learning Algorithms in the Production of Castings

Dorota Wilk-Kołodziejczyk[1,2, *], Mateusz Witkowski[1], Krzysztof Jaśkowiec[1,2], Marcin Małysza[1,2], Adama Bitka[2], Łukasz Marcjan[1]
1. Department of Applied Computer Science and Modelling, Faculty of Metals Engineering and Industrial Computer Science, AGH University of Krakow, Czarnowiejska 66, 30-054 Krakow, Poland
2. Łukasiewicz – Krakow Institute of Technology, Zakopiańska 73, 30-418 Kraków, POLAND
*Corresponding address: e-mail: dwilk@agh.edu.pl,Dorota.wilk@kit.lukasiewicz.gov.pl

The paper is devoted to the study of the effectiveness of selected machine learning algorithms - Linear Regression, Decision Tree, Random Forest, AdaBoost, Gradient Boosting, XGBoost, SVR, K-Nearest Neighbors - in the issue of supporting the production of ADI cast iron. Knowledge and information on the production of cast iron and the effects of its chemical composition and heat treatment parameters on mechanical properties were obtained through an extensive literature review. Articles on topics similar to the subject of this work were also analyzed in the context of practical application of machine learning algorithms. As part of the work, hyperparameter tuning and data augmentation experiments were conducted to test the impact of the optimizations performed on the final results of the generated models. During the evaluation, metrics such as root mean square error and coefficient of determination were used with prior cross-validation allowing for more realistic results. For each of the predicted mechanical parameters, Gradient Boosting proved to be the best algorithm. The work culminated in a web application with a graphical user interface allowing interaction with the best predictive models created during the study.

The input data set consists of 513 records aggregating information on ADI cast iron production parameters:
•Chemical composition of cast iron – percentage of elements:
o C – Carbon
o Si – Silicon
o Mn – Manganese
o Mg – Magnesium
o Cu – Copper
o Ni – Nickel
o Mo – Molybdenum
o S – Sulfur
o V – Vanadium
o Cr – Chromium
o P – Phosphorus
o Ti – Titan
o Sn – Tin about Al-Glin
• Heat treatment process parameters:
o Austenitization temperature – in degrees Celsius
o Austenitization time – in minutes
o Ausferitization temperature – in degrees Celsius
o Ausferitization time – in minutes
• Product thickness – in millimeters:
• Mechanical parameters.

The purpose of this experiment was to check the model results for default parameters in order to obtain a reference point for subsequent tests. One configured parameter in this study was random_state in models using randomness. This was intended to obtain repeatable results. Model dictionary was passed as the input parameter "models_dict".

Brinell hardness models - HB:
• The best results for fitting the model to the training set were achieved by: DecisionTreeRegressor and ExtraTreeRegressor
• The best results for fitting the model to the training set with 5-fold cross-validation were obtained by: XGBRegressor
• The best results for fitting the model to the test set were achieved by: GradientBoostingRegressor
• The worst-fitting model for each set is: SVR

Models for tensile strength - Rm:
• The best results for fitting the model to the training set were achieved by: DecisionTreeRegressor
• The best results for fitting the model to the training set with 5-fold cross-validation were obtained by: ExtraTreesRegressor
• The best results for fitting the model to the test set were achieved by: ExtraTreesRegressor
• The worst-fitting model for each set is: SVR Models for yield strength - Rp02:
• The best results for fitting the model to the training set were achieved by: DecisionTreeRegressor and ExtraTreeRegressor
• The best results for fitting the model to the training set with 5-fold cross-validation were obtained by: ExtraTreesRegressor
• The best results for fitting the model to the test set were achieved by: XGBRegressor
• The worst-fitting model for each set is: SVR.

Elongation models - A5: • The best results for fitting the model to the training set were achieved by: DecisionTreeRegressor and ExtraTreeRegressor

• The best results for fitting the model to the training set with 5-fold cross-validation were obtained by: ExtraTreesRegressor
• The best results for fitting the model to the test set were achieved by: XGBRegressor
•The worst-fitting model for each set is: SVR Impact strength models - K
• The best results for fitting the model to the training set were achieved by: DecisionTreeRegressor and ExtraTreeRegressor
• The best results for fitting the model to the training set with 5-fold cross-validation were obtained by: ExtraTreesRegressor
• The best results for fitting the model to the test set were achieved by: XGBRegressor
• The worst-fitting model for each set is: SVR Summary: Analyzing the results of the basic study, it was noticed that algorithms based on decision trees are significantly overfitted. This is especially noticeable in the DecisionTreeRegressor and ExtraTreeRegressor models. In most studies, the mentioned algorithms matched the training set 100%, but when verifying the results using cross-validation or on the test set, the same algorithms performed much worse. Models using decision trees but based on ensemble learning were also overfitted, but showed significantly better generalization abilities, i.e. operating on new, previously unknown data that were not included in the training set. This is especially visible in the results on the test set. The exception in this case is the AdaBoostRegressor model, which uses weak regression models as base models, which, if the data is highly non-linear and contains many outliers, may negatively affect the results of the entire committee. LinearRegression and KNeighborsRegressor are models that assume a linear or locally linear relationship between the input parameters and the purpose of the calculation. When the actual problem is non-linear, the models mentioned may have difficulty fitting effectively. This is most likely why they performed so poorly in the study. The SVR model performed the worst in the entire study and this is most likely related to the default value of the "C" parameter, which for this algorithm is 1. Such a small value means that the model is very tolerant of prediction errors and tries to find a solution very close to linear regression.

References

[1] KochanskiA, PerzykM, KlebczykM. Knowledge in imperfect data. Advances in Knowledge Representation, 2012, 181-210.
[2] YescasMA. Prediction of the Vickers hardness in austempered ductile irons using neural networks. International Journal of Cast Metals Research, 2003, 15(5): 513-521.
[3] RojekI, Mik D, Kotlarzet P, al. Intelligent system supporting technological process planning for machining and 3D printing. Bulletin of the Polish Academy of SciencesTechnical Sciences, 2021, 69(2): 233178742.
[4] LiashchynskyiP. Grid search, random search, genetic algorithm: A big comparison for NAS. arXiv preprint arXiv, 2019, 1912:06059.
[5] Wu J, Chen X Y, Zhang H, et al. Hyperparameter optimization for machine learning models based on Bayesian optimization. Journal of Electronic Science and Technology, 2019, 17(1): 26-40.
[6] NaftalyU, IntratorN, HornD. Optimal ensemble averaging of neural networks. Network: Computation in Neural Systems, 1997, 8(3): 283.
[7] PavlyshenkoB. Using stacking approaches for machine learning models. In: 2018 IEEE Second International Conference on Data Stream Mining & Processing (DSMP). IEEE, 2018, 255-258.
[8] AwadM, Khanna R. Support vector regression. Efficient learning machines: Theories, concepts, and applications for engineers and system designers. Efficient Learning Machines, Theories, Concepts, and Applications for Engineers and System Designers, 2015, 67-80.
[9] MammoneA, TurchiM, CristianiniN. Support vector machines. Wiley Interdisciplinary Reviews: Computational Statistics, 2009, 1(3): 283-289.
[10] SuX G, YanX, Tsai CL. Linear regression. Wiley Interdisciplinary Reviews: Computational Statistics, 2012, 4(3): 275-294.
[11] SongY S, Liang J Y, Lu J, et al. An efficient instance selection algorithm for *k* nearest neighbor regression. Neurocomputing, 2017, 251: 26-34.
[12] NoriegaL. Multilayer perceptron tutorial. School of Computing. Staffordshire University, 2005, 4(5): 444.

An Attempt to Identify Heat Transfer Coefficients in Casting CAE Using Cooling Curves

Kato Hiroto[1], Yasuhiro Maeda[2], *
1.Graduate Student, Dept. of Mechanical Engineering, Daido University
2.Professor, Dept. of Mechanical Engineering, Daido University
*Corresponding address: e-mail:y-maeda@daido-it.ac.jp

Abstract: The Casting CAE is used practically but is not almighty at all. There are some needs to improve the calculation accuracy of defect prediction, develop new functions, and be an easy operation. An efficient and accurate method for determining heat transfer coefficients (HTC) in heat transfer and solidification simulations is important. HTC is a physical property value that depends on the material properties and conditions at the interface. Still, it cannot be easily determined because HTC is influenced by many factors. Therefore, it is often used as a tuning parameter to control the accuracy of analysis.

Assume that we have the practical cooling curve in a simple experiment. In this case, one way to identify the HTC is to search for the HTC of simulation so that the cooling curves of the experiment and simulation match. Alternatively, there is a method to obtain the actual HTC using the inverse analysis of the measured cooling curve. In this study, as a new attempt, we constructed the inverse model of casting CAE by machine learning, using the cooling curve obtained by casting CAE as the input and the HTC as the output. Next, HTC was calculated by inputting the measured cooling curve into the constructed inverse model. The heat transfer and solidification simulation using the obtained HTC was carried out to confirm the cooling behavior between the experiment and the simulation.

Keywords: casting CAE; aluminum alloy; heat transfer coefficient; cooling curve

1 Introduction

Many researchers have studied how to identify the heat transfer coefficient (HTC) in the casting CAE[1-3]. HTC is influenced by many factors or conditions, such as pouring conditions, mold release agents, molten metal and mold material, repetition times, and so on.Further, HTC depends on time and temperature. Therefore, it is difficult to determine the actual value of HTC.

On the other hand, HTC is used as a tuning parameter in the reconciliation of experimental and simulation results.Some literature has reported how to determine the HTC from the cooling curve measured by experiments to identify the heat transfer coefficient between molten metal and mold [2,3]. Further, the HTC is identified by using data assimilation[4,5].

In this study, we propose a series of procedures, from measuring the cooling curve to determining physical properties other than HTC, and until to identifying HTC. In particular, the identification method using an inverse neural network (NN) model is proposed, and its results are compared to other methods.

2 Experimental setup

Figure 1 shows the schematic illustration of the experiment using the taper mold [3]. Assuming one-dimensional heat transfer, the temperatures of castings and mold(chill) measured in the experiment are stored in a recorder. Thermo-couples are located at 5, 10, and 15mm away from the interface. Their measurement points are named to MP1, MP2, MP3, CP4, CP5, CP6 from top, respectively. The pouring material is an aluminum alloy of JIS-AC4C(A360), and the chill material is a steel of JIS-S45C. As a mold release agent, the boron-nitride spray was applied to the interface between the castings and the chill. An example of the cooling curves measured a few times is shown in Figure 2. In addition to the above experiment, the cooling curve is also measured when the same material was slowly cooled in the crucible.

3 Analysis procedures

The results of two types of experiments identify the liquidus temperature, solidus temperature, and the relationship between temperature and solid fraction. Then, assuming one-dimensional heat transfer, the thermo-physical properties, such as thermal conductivity, etc., are also identified as constant values.

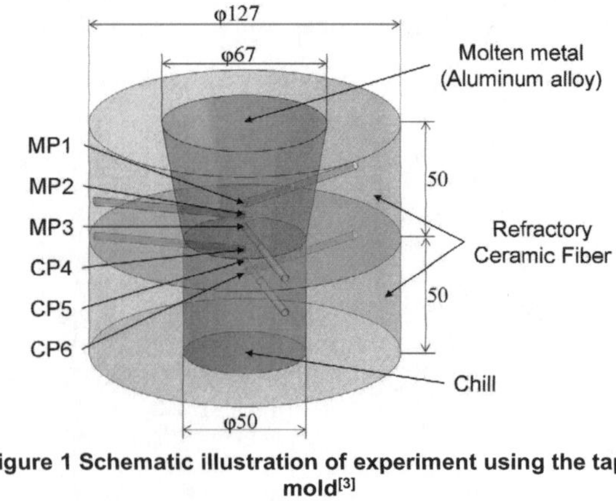

Figure 1 Schematic illustration of experiment using the taper-mold[3]

The above experimental results are simulated using the casting CAE software TopCAST™. We assume that the heat transfer and solidification processes can be constructed forward and inverse models using an NN. In this case, the inputs in the forward model include casting conditions, thermal properties, and heat transfer coefficients, and the outputs are the cooling curves. In this study, the heat transfer coefficient (HTC) is identified from the obtained cooling curve and the inverse model of the NN.

Figure 3 shows an overview of the NN model. It is a general all-coupled NN consisting of an input, hidden, and output layer. The open-source software "Jupyter Notebook" from "Project Jupyter" is used to create the NN. The total number of hidden layers and the number of nodes were considered in this study to determine the number of layers and nodes with higher accuracy.

4 Results and discussion

As a machine learning example, which uses the sigmoid function as the activation function, and Adam as the optimization method, the learning results using 10 data sets and 1 hidden layer are shown in Figure 3 and Figure 4.Figure 4shows the relationship between mean squared error (MSE) and the number of trials. The normalized input and output data are used in machine learning. From the figure, it is clear overlearning occurs when the number of trials is about 300. The NN model constructed in this study can calculate the error with an accuracy of approximately 150W/m²·K.The results of the simulation using the HTC identified by the inverse model are affected by a convergenceerror and did not show a good agreement. Then, in the future study, it may be necessary to modify the NN structure or increase the number of data sets.

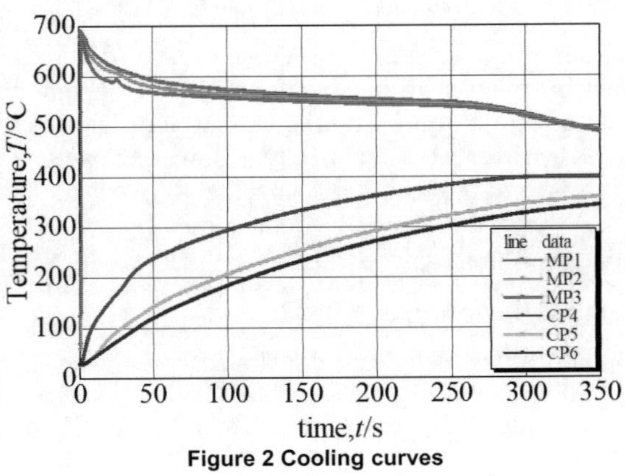

Figure 2 Cooling curves

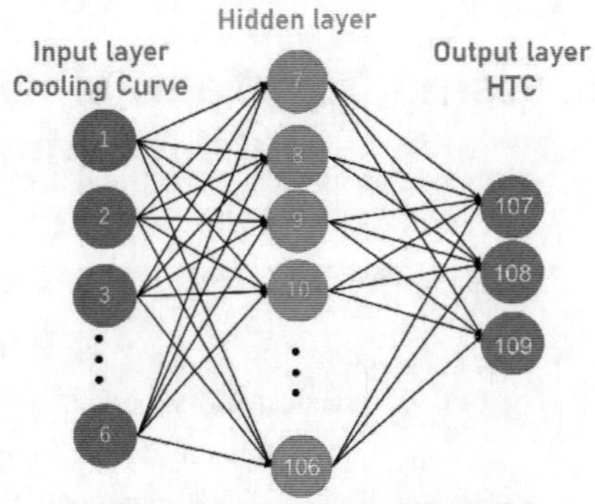

Figure 3 Overview of the NN model

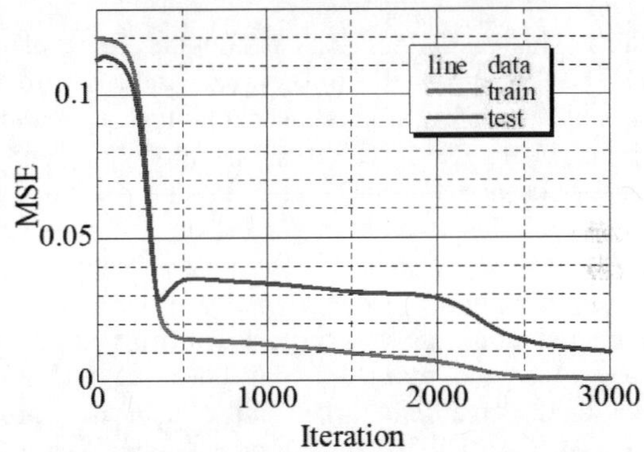

Figure 4 Relationship between mean squared error (MSE) and the number of trials in the case of 10 data sets

5 Conclusion

A method for identifying heat transfer coefficient (HTC) using NNs is investigated in the present study. Although the identification accuracy is still insufficient, it is thought that it can be improved by reconsidering the number of data sets, etc.

References

[1] SunHC, ChaoLS. An investigation into the effective heat transfer coefficient in the casting of aluminum in a green-sand mold.Materials Transactions, 2009, 50(6): 1396-1403.

[2] NakamuraI,KunoT,OkimuraY,KambeH,OkaneT, YoshidaM. Effect of contact conditions mold release agent on heat transfer coefficient between Al-7%Si alloy and permanent mold during solidification process. JFS, 2023, 95(10): 552-560.

[3] MizunoA,NiwaD,OzakiT,MaedaY.Report of the 177th. JFS Meeting, 2021: 58.

[4] OkaY,OhnoS, MatsuuraK. Parameter estimation in heat conduction problem of casting processes based on data assimilation. Tetsu-to-Hagane, 2017, 103(12):755-762.

[5] NatsumeY, OkaY,OgawaJ, OhnoM. Estimation of time-dependent heat transfer coefficient in unidirectional casting using a numerical model coupled with solidification analysis and data assimilation. Int. J. Heat and Mass Transfer, 2020, 150: 119222.

Assessing Fixed and Moving Mesh Methods for Hydrodynamic Free Surface Simulation in Induction Melting

P. Garcia-Michelena[1*], O. Gordo-Burgoa[1], N. Herrero-Dorca[1], I. Vicario[2], I. Crespo[2], G. Arruebarrena[1], X. Chamorro[1]

1. Mechanical and Manufacturing Department, Mondragon University, Arrasate-Mondragon, Spain
2. Manufacturing Processes and Materials Department, Tecnalia, Derio, Spain

*Corresponding address: email: pgarciam@mondragon.edu

Abstract: This study develops a coupled finite element method (FEM) approach to optimize induction-melting processes by accurately simulating multiphysics phenomena. It integrates a magneto-hydrodynamic model to represent magnetic fields and Lorentz forces, crucial for fluid flow and pressure fields. Comparing fixed and moving mesh techniques, it assesses their effectiveness in capturing free surface deformation. Experimental validation in an aluminum induction-melting furnace confirms the model's accuracy. Insights gained contribute to advancing computational methods in metallurgical process optimization, particularly in understanding magneto-hydrodynamic behavior during induction melting.

Keywords: induction melting; multiphysics modelling

1 Introduction

The optimization of casting processes remains a field of study, where melting processes represent one area where the greatest benefit can be obtained by reducing energy consumption, emissions and process times. Induction melting (IM) is one of the processes with the best expectations [1]. This requires the design of advanced numerical models validated against experimental measurements. The interaction between the magnetic field and the liquid metal leads to recirculation of the melt and surface deformation, with the necessity of accurate modeling for calculating flow recirculation, heat transfer, alloy dissolution, and crucible erosion. Traditional fixed mesh methods like Volume of Fluid (VOF) are commonly used [2]. However, these methods require very fine meshing, high computational resources, and the results are highly dependent on the numerical diffusion of the interface. Thus, this study compares the use of mobile mesh as an alternative, by comparing both methods[3].

2 Experimental procedure

For the IM test, an Inductotherm® furnace with a VIP POWER-TRAK+® power generator was employed. The inductor coil was composed of 10 loops having a rated power of 50 kW at 3 kHz. The melted Al 99% ingots underwent gradual heating up to 750 °C and power was maintained. Once the free surface of the metal reached a stable state, contact probes were introduced into the metal. After allowing the aluminum to wet the probes, the device was removed, leaving a visible mark indicating the melt height and the free surface profile. Also, the electrical variables corresponding to current, voltage, phase, and frequency were measured to feed the posterior simulation. Six trials were done for 5, 7.5, and 10 kg respectively, and two power conditions at 25 kW and 40 kW for the filling case (Figure 1).

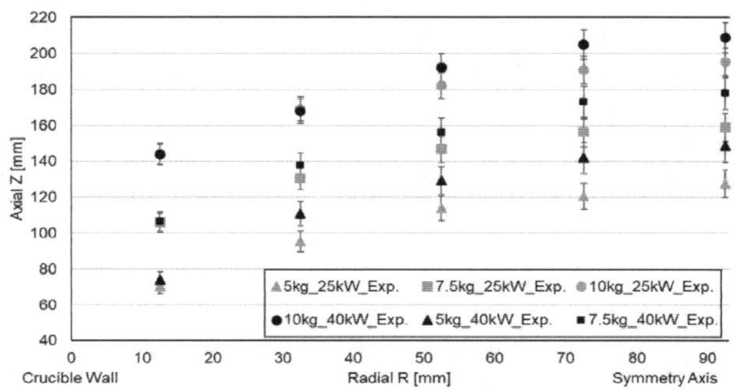

Figure 1 Measured surface profile for IM trials

A meniscus shape surface profile was observed across all the trials. For lower filling levels, the melt's height was higher, leading to more significant surface deformation. This trend was repeated for different power supplies, with the top surface of the melt higher for the 40 kW than the 25 kW case, with the position of the crucible wall probes remaining in a similar place.

3 Numerical Procedure

The numerical model was developed using the finite element tool COMSOL Multiphysics 6.2®. Due to the symmetry of the IM furnace, an axisymmetric 2D model was considered, thus reducing the size of the geometry to be discretized. Based on the dimensions of the IM furnace, the geometry and domains corresponding to the molten aluminum, the crucible of Al_2O_3, the coil of copper and air were defined with their corresponding material properties. To determine the numerical model that rules the coupled magneto-hydrodynamic model, it was first necessary to derive Maxwell's equations in the frequency domain. The resulting equation for the conducting metal is as follows:

$$\mathbf{J_e}=(j\omega_f\sigma)\mathbf{A}+\nabla\times(\mu_r^{-1}\mu_0^{-1}\nabla\times\mathbf{A})+\sigma\mathbf{u}\times(\nabla\times\mathbf{A}) \quad (1)$$

The external current density (J_E) derives from the magnetic field generated by the coil. To model this phenomenon, in the coil solid section the measured power cases have been imposed.

$$P_c=1/2\mathrm{Re}(V_C I_C^{*}\cos(\varphi)) \quad (2)$$

Concerning the fluid dynamics sub-model, the typical Navier-Stokes equations for the conservation of mass and momentum over time have been considered.

$$\nabla\cdot(\rho\mathbf{u})=0 \quad (3)$$

$$\rho(\partial\mathbf{u}/\partial t+(\mathbf{u}\cdot\nabla)\mathbf{u})=\nabla\cdot[-p\mathbf{I}+\mathbf{K}]+\mathbf{F}+\rho\mathbf{g} \quad (4)$$

For the reproduction of turbulence, two additional transport equations are added: the turbulent kinetic energy (k) and dissipation (ε), based on the standard k-ε model and for the model parameters [4].

The necessity for enabling displacement and interaction between phases prompted the adoption of the moving mesh method, utilizing the Arbitrary Lagrangian-Eulerian (ALE) approach. This method tracks the moving boundaries by allowing mesh nodes to follow fluid flow. Free mesh normal direction deformation in the metal domain and slips to the metal-crucible wall boundary for tangential displacement were imposed. To ensure numerical stability during mesh displacement the Winslow smoothening method was employed.

4 Result and discussion

The results of the fixed mesh with the experimental measurements in the case of 25 kW power are represented in Figure 2. The model reproduces the free surface profile in the center of the metal, with an increasing discrepancy towards the walls of the crucible for all three cases of crucible filling. For the 40 kW case, similar trend of the surface profile is achieved, although the discrepancy with the experiment is larger.

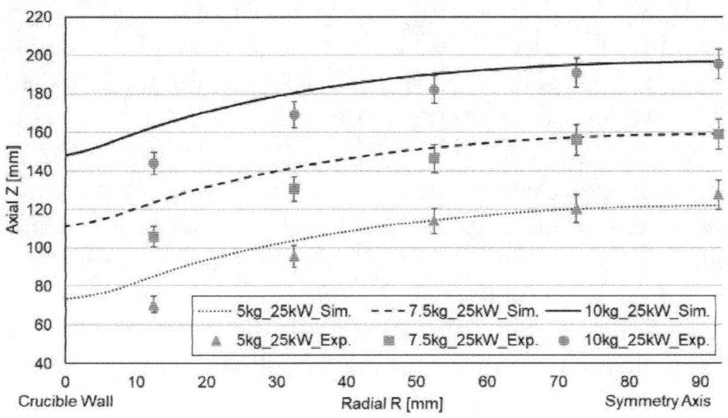

Figure 2 Free surface profile for 25 kW with fixed mesh

In the second case with the moving mesh model, the model also reproduces the free surface profile in a relatively correct way, approaching the experimental measurement, particularly in the region close to the wall (Figure 3). However, in the case of 40 kW and 5 kg, the simulation did not converge, due to the distortion of the grid elements during mesh displacement.

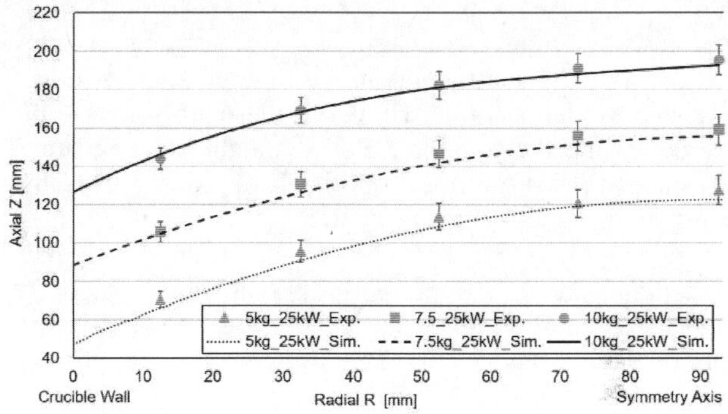

Figure 3 Free surface profile for 25 kW with moving mesh.

5 Conclusion

This analysis has shown the importance of the method employed to define the type of mesh and its impact on the numerical results of magneto-hydrodynamic simulations of the free surface for the IM process. The results obtained indicate that the moving mesh method was more accurate and required less computational time. However, the method lost its validity for large deformations of the free surface, since the stretching and distortion of the mesh elements was excessive leading to computation convergence failures.

6 Acknowledgments

The authors thank the Basque Government and Department of Economic Development, Sustainability and Environment for its financial support by means of the ELKARTEK DESGAS (KK-2023/00020) and the program of the RVCTI.

References

[1] HoltzerM,DankoR, and Zymankowska-KumonS. Foundry Industry. Current State and Future Development, Metalurgija, 2012, 51: 337–340.

[2] SpitansS, JakovicsA, BaakeE, and NackeB. Numerical modeling of free surface dynamics of melt in an alternate electromagnetic field: Part I. Implementation and verification of model. Metall. Mater. Trans. B: Process Metall. Mater. Process. Sci., 2013, 44(3): 593–605.

[3] VollerVR, SwaminathanC R, andThomasBG. Fixed grid techniques for phase change problems: A review. Int. J. Numer. Methods Eng., 1990, 30(4): 875–898.

[4] WilcoxD C. Turbulence modeling for CFD. DCW Industries, 2006.

Supporting Foundry Workers with an AI-Based Multimodal System Tunned Using Foundry Knowledge

Javier Nieves*, Andres Perez
Azterlan member of Basque Research Team Alliance (BRTA), Aliendalde Etxetaldea 6, 48200 Durango (BI), Spain
*Corresponding address: e-mail: jnieves@azterlan.es

Abstract: Currently, society is experiencing a new Artificial Intelligence revolution with the creation of Large Language Models (LLMs). However, these tools missed the specific and needed knowledge of a domain to be accurate. This paper presents a novel approach of using LLMs but including the knowledge generated by a research center. Hence, by providing access through a simple human interaction system, we present a possible ecosystem of models that combining textual and image knowledge will support day-by-day work. The system was tested and evaluated by foundry experts, assuring that this kind of tools could be a fundamental future consulting tool.

Keywords: artificial Intelligence; LLM; RAG; foundry knowledge; intelligent systems

1 Introduction

Although we have passed 4 different industrial revolutions, currently, we are in a new one, Industry 5.0., where the human being is the center of the system. The new trend in Artificial Intelligence, Large Language Models (LLMs) [1], makes possible the introduction of the human being into the Artificial Intelligence and Knowledge Ecosystem, providing and easy way to improve human-machine interactions. These models are a good but, if we are involved in a domain specific problem, we need to improve them.

In this way, when we want to adapt the specific knowledge to the aforementioned AI models, it can be done applying one of the following manners.

Create new models. The most expensive. It needs extensive work to get all data for learning and the energy consumption to generate it (i.e., learning the GPT-3 model in 2020 by OpenAI was close to 4.6 million dollars [2]).

Transfer learning[3]. Reuse the elements of a previously trained machine learning model in a new model. This technique is being widely used especially in machine vision systems [4].

Retrieval-Augmented-Generation (RAG)[5]. Optimizing the output of a model making use of a new knowledge base outside of the training data used for the creation for the model. This avoids the generation of false or unrealistic answers as well as obsolete.

Against this background, the solution proposed in this paper will focus on the generation of a multimodal system based on an LLM extended through the RAG technique with the expert knowledge of a research center such as Azterlan. The system presented introduces bibliographic knowledge through books, scientific knowledge through articles, technical knowledge through user manuals and, finally, visual knowledge about the real foundry analyses.

2 Experimental procedure

Once the desired style of our multimodal system has been determined, a working methodology was designed. More accurately, several models were built and integrated as we describe below.

Firstly, we selected an existing LLM to use it as a base. In this case, Meta's Llama3 was employed, but configuring its personality. The model must work as a researcher from our center, giving answers as it was an academic. The system will always answer any topic, but if it deviates from the defined field, it will indicate that the model is not made to answer those type of questions.

Subsequently, we generated 4 different models. These incorporate the knowledge of a book of our own production explaining different aspects of the sand, scientific articles on digitalization solutions developed by Azterlan and the user's manual of its thermal analysis system. Each one has been processed, extracted and stored in a vector database for RAG using the *Nomic Embed* model.

Then, we have generated a model for classifying metallographies by similarity. This is based on the weighted average of the 10 most similar ones extracted from the comparison of its vectorial representation created with *image2vec* model.

Finally, we have generated a routing agent. This system is responsible for determining to which model(s) the job should be sent. In this way, we have succeeded in creating the generalization of an agentic RAG multimodal system for assists in foundry related topics.

3 Result and discussion

The result of this research has been the generation of a multimodal AI assistant. This system is able to solve doubts related to sand (book added), digital twins and use of Deep Learning in the foundry (some papers included), the use of Thermolan® (manual of the thermal analysis system developed by Azterlan) and make classifications of nodular iron metallographies. Specifically, Figure 1 shows an example of a query related to sand problems.

For evaluating our development, several experts prepared relevant questions about the provided knowledge. These questions have been solved by the tool and they have evaluated the answers. In summary, everyone thinks that results are good, although they can be improved adding more knowledge to the system.

The obtained AI assistant is a proof of concept. This is not the definitive supporting tool. To achieve it, it would need to face other problems:

Multi-Language capability, something that can be solved using another AI like NLLB from Meta.

Ability to manage speech questions. This problem can be solved, also, using Whisper (OpenAI *Speech-2-Text* free model).

Adding more knowledge to cover more tasks.

In addition, this proof of concept can be the basis for, later, increase the ecosystem with new models that our aforementioned router agent can integrate and redirect jobs to them.

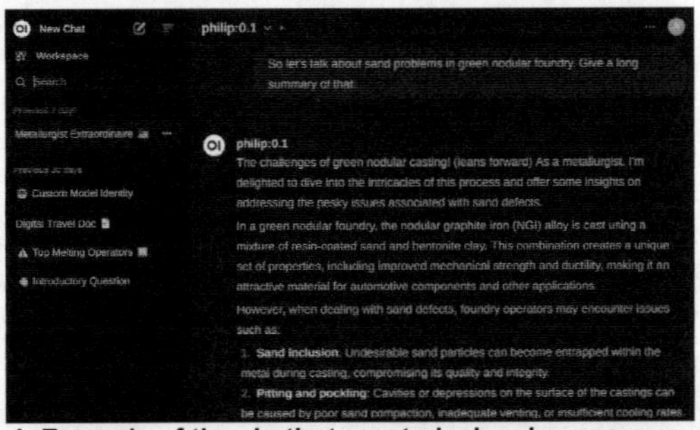

Figure 1 Example of the chatbot created, showing answers related to sand problems in green nodular foundry

4 Conclusion

Industry 5.0 has arrived, and we must facilitate the use of new technologies. Therefore, in this paper we present a multimodal system specialized in foundry that, through a simple communication, gives support to different tasks. The experts who analyzed it indicate that its quality is high and, by providing more data and more models, it will be very useful in the near future.

References

[1] ChangY P, WangX, WangJ D, WuY Y, Yang L Y, ZhuK J, et al. A survey on evaluation of large language models. ACM Transactions on Intelligent Systems and Technology, 2024, 15(3): 1-45.

[2] https://fortune.com/2024/04/04/ai-training-costs-how-much-is-too-much-openai-gpt-anthropic-microsoft/.(29 June 2024).

[3] TorreyL, and ShavlikJ. Transfer learning. In handbook of research on machine learning applications and trends: Algorithms, methods, and techniques. IGI global, 2010: 242-264.

[4] SarrionandiaX, Nieves J, Bravo B, Pastor-López I, and BringasPG. An objective metallographic analysis approach based on advanced image processing techniques. Journal of Manufacturing and Materials Processing, 2023, 7(1): 17.

[5] Lewis P, Perez E, Piktus A, Petroni F, Karpukhin V, Goyal N, et al.Retrieval-augmented generation for knowledge-intensive NLP tasks. Advances in Neural Information Processing Systems, 2020, 33: 9459-9474.

Determination and Study of Heat Transfer Coefficient in Frozen Sand Mold

Shijie Dong[1], Zhongde Shan[2], *, Feng Lin[1], *
1. Department of Mechanical Engineering, Tsinghua University, Beijing100084, China
2. College of Mechanical and Electrical Engineering, Nanjing University of Aeronautics and Astronautics, Nanjing 210016, China
*Corresponding address:e-mail: shanzd@nuaa.edu.cn (Zhongde Shan); linfeng@tsinghua.edu.cn (Feng Lin)

Abstract: Resin and other binders are widely used in traditional sand casting. It has caused serious environmental pollution in the casting process and has affected the physical and mental health of foundry workers. Frozen sand mold casting (FSMC) technology, which has the advantages of low cost, high performance, and green manufacturing, will become an essential means of green manufacturing in the automotive, rail transportation, aerospace, and other fields. However, the interfacial heat transfer behavior of FSMC technology in the casting process is still unclear. As an important parameter to describe the interfacial heat transfer behavior, the interfacial heat transfer coefficient (IHTC) has not been further studied in the field of FSMC. This method provides a reference for the determination of IHTC of different kinds of alloys in the process of FSMC. Moreover, the determination of the IHTC during the casting process of ZL101 aluminum alloy in this work provides important boundary conditions for the numerical simulation of the FSMC process.

Keywords:frozen sand mold casting (FSMC); inverse thermal conductivity problem (IHCP); cast aluminum alloy; interfacial heat transfer coefficient (IHTC); numerical calculation

1 Introduction

Casting aluminum alloy has the advantages of high performance, low cost, easy mass production of precision parts, etc. It is widely used in aircraft manufacturing, new energy vehicles, deep space probes, high-speed railway construction, and other fields [1-4]. With the continuous improvement of human manufacturing level and environmental awareness, the manufacturing of aluminum alloy gradually tends to be highly efficient, high performance, low cost, and green [5-6]. FSMC technology, which has the advantages of low cost, high performance, and green manufacturing, will become an essential means of green manufacturing in the automotive, rail transportation, aerospace, and other fields. The materials used for casting mold mainly include water and sand. The mold which mixed with water is formed in a lower temperature to content with a casting strength. Castings are obtained after pouring the mold [7]. As a result of less pollution, easy of shakeout, high recovery rate of molding sand, and better performance of casting, frozen casting is an advanced casting pathway for greening.

This study establishes a method for determining the IHTC during FSMC by combining the FDM and the CGM. The IHCP method is established by the FDM architecture and the CGM for solving the forward and inverse heat transfer problems, respectively. The accuracy of the IHCP process established based on the 1-D heat transfer assumption is verified experimentally and numerically. The IHTC was determined for different FSMC process conditions, including initial frozen temperature, water contents, and types of molding sand. Moreover, the influence of water content in frozen sand mold on IHTC was analyzed.

2 Numerical modeland experimental procedure

Figure 1 is the flow chart of FSMC technology, which is divided into six parts: sand mixing, vibration compaction, frozen billet making, cutting and processing, deep cooling and melting and pouring.

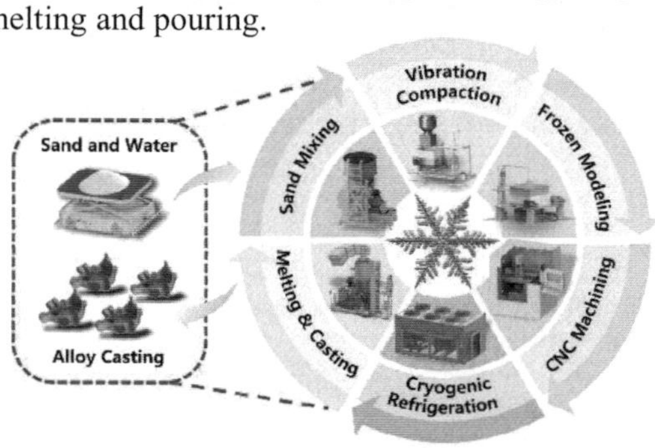

Figure 1 FSMC technology diagrammatic drawing

Based on the 1-D heat transfer assumption, the inversion method of IHCP for FSMC process was established by combining FDM and CGM. The calculation flow of the one-dimensional heat transfer model and the IHCP inversion model is shown in Figure 2. The calculation error of which is less than 1.6°C, which is lower than other related reports. In this work, a frozen casting thermometry experiment was designed, as shown in Figure 3. During the experiment, a data recorder was used to collect the temperature. Table 1 shows the process parameters of frozen casting thermometry experiment.

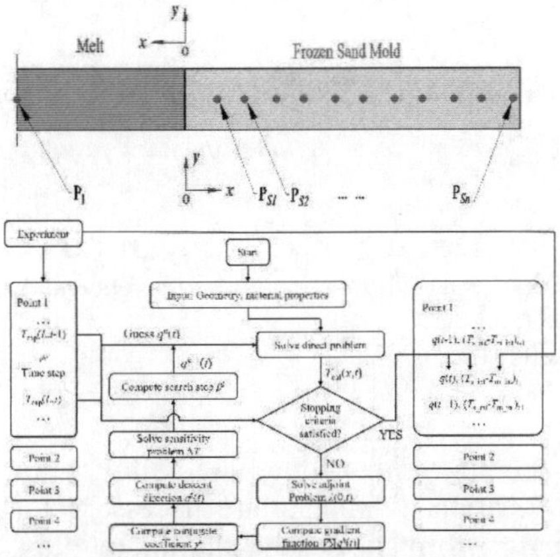

Figure 2 1-D heat transfer assumptions and IHCP inversion modeling calculation procedure

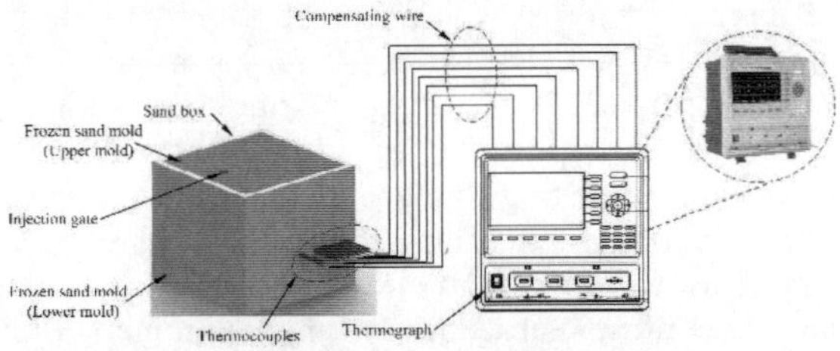

Figure 3 The frozen casting thermometry experiment

Table 1. Process parameters of frozen casting thermometry experiment

Initial frozen temperature (℃)	Water content	Sand type
-34	2wt.%	Silicon and Chromite sand
-32	3wt.%	
-30	4wt.%	
-28	5wt.%	
-26	6wt.%	

3 Result and discussion

The IHTC and temperature fields were calculated for 17 different process conditions by the IHCP process established in this work. For frozen sand mold, although different water content led to two distinct types of IHTC change rules, the essential reason was the same air gap change process. Figure 4 shows the variation of IHTC with 3wt.% and 5wt.% water contents with temperature.In the temperature range of 575 ℃ to 550 ℃ on the casting surface, the IHTC of the frozen sand mold with the water content of 2wt.%, 3wt.%, and 4wt.% appeared as 'Fluctuation'. Meanwhile, the IHTC of the frozen sand mold with the water content of 5wt.% and 6wt.% appeared as 'Turning'.

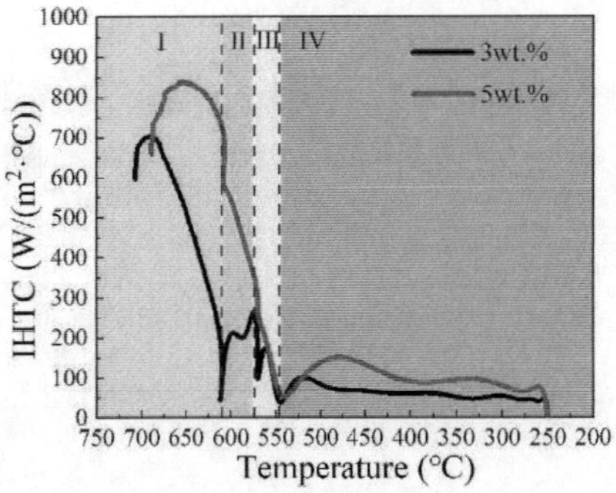

Figure 4 Two types of frozen sand mold IHTC trend with temperature variations

4 Conclusion

The calculated results show that the inverse analytical method can be successfully applied to determining the IHTC of ZL101 alloy during the frozen sand mold casting process. It suggested that the turning point for two IHTC change types was between 4%-5% water content. When the water content was below this value, the IHTC change followed a 'fluctuation type', while when the water content exceeded this value, the IHTC change showed a 'turning type'.

5 Acknowledgment

This work was financially supported by the National Key R&D Program of China (No. 2021YFB3401200).

References

[1] PhinikettouV, PapamichaelI, VoukkaliI, Zorpas AA. Comparison of methodologies for microplastic isolation through multicriteria analysis (AHP). Microplastics, 2024, 3: 184-204.
[2] LiuGB,WeiC H, LiuT,LuoHY,ZhouH,RongHW,ChenD,NgoHH. Electroplating wastewater treatment and resource recovery via a hybrid process of stepwise alkalization, Fenton, and chlorination. Separation and Purification Technology, 339 (2024) 126658.
[3] GeW, LiH, WenX, LiC, CaoH,XingB. Mathematical modelling of carbon emissions and process parameters optimisation for laser welding cell. Int. J. of Prod. Res., 2023, 61: 5009-5028.
[4] LiuM,YiH, CaoH. Additive manufacturing (AM) for advanced materials and structures: Green and Intelligent development trend. Crystals, 2023, 13: 92.
[5]ChenQ,GeN,JiangJ,HuangM,LiM,WangY,DongJ,DingC,ZouD. Effect of local pressurization on microstructure and mechanical properties of aluminum alloy flywheel housing with complex shape. Mater., 2023,17: 106.
[6] Abou El-khairMT. Microstructure characterization and tensile properties of squeeze-cast AlSiMg alloys. Mater. Lett., 2005, 59: 894-900.
[7] ShanZ, GuoZ, DuD, LiuF, LiW. Digital high-efficiency print forming method and device for multi-material casting molds. Front. Mech. Eng., 2020, 15: 328-337.

Enhanced Melt Quality Rapid Analysis of Aluminum Alloy Assisted by Machine Learning

Yunxiang Tan[1], Juhuai Ma[1], Haifeng Liu[2], Jianlei Jia[2], Liguo Yang[2],Qingyan Xu[1], *
1. Key Laboratory for Advanced Materials Processing Technology, Ministry of Education, School of Materials Science and Engineering, Tsinghua University, Beijing 100084, China
2. CITIC Dicastal Co., Ltd., Qinhuangdao066004, China
*Corresponding address: e-mail: scjxqy@tsinghua.edu.cn

Abstract: The quality of aluminum alloy melts decisively influences the molding process and mechanical properties of castings. Rapid and accurate pre-pour inspection of melt quality has significant engineering value and scientific significance. This research conducts temperature measurement experiments on aluminum alloy melts to obtain cooling curves. Algorithms to identify characteristic values during the solidification process of aluminum alloy melts are proposed and developed. A database on the quality of aluminum alloy melts is established. Data including cooling curves and characteristic values obtained during the solidification process of the aluminum alloy melts utilizing algorithms such as random forests, support vector machines, and deep neural networks, combined with results from thermodynamic calculations and numerical simulations of the solidification process of aluminum alloy melts with inclusions. This research identifies factors affecting melt quality and proposes a new criterion for melt quality assessment. Results demonstrate that the melt quality analysis method developed in this study is significantly faster than traditional methods. The proposed melt quality criterion accurately and stably reflects the quality of the melt, facilitating the optimization of production processes and enhancing the quality of castings.

Keywords: aluminum alloy; melt quality;melt quality criterion; machine learning

1 Introduction

Aluminum alloys play a pivotal role in the lightweighting of automobiles. Optimizing the performance of aluminum alloy castings has received widespread attention. The liquid quality of aluminum alloys before casting is a critical factor that directly influences the microstructure and mechanical characteristics of the castings[1]. To enhance the overall performance of aluminum alloy castings, it is essential to first improve the quality of the liquid melt[2]. There are numerous factors affecting the liquid quality of aluminum alloys. Among these, increasing the cooling rate prior to casting can effectively improve shrinkage porosity, dendritic spacing, grain size, and microstructure in aluminum-silicon alloys[3]. However, due to limitations in production conditions and casting processes, the cooling rate cannot be infinitely increased, machine learning should be introduced into melt quality analysis [4]. Typically, refining and degassing of the liquid alloy, grain refinement, and modification treatments are necessary to improve the quality and performance of the aluminum alloy.

2 Experimental procedure

The thermal analysis experiment of aluminum alloy melts conducted at CITIC Dicastal Co., Ltd. primarily takes place in a low pressure die casting melting furnace(Figure 1). Initially, it is imperative to ensure that all equipment, including the melting furnace and the test cup, are in optimal working condition, followed by preheating the furnace to an appropriate temperature for melting the aluminum alloy. Subsequently, the aluminum alloy raw material is placed in the furnace for melting. During this phase, the temperature is regularly monitored to ensure uniform heating, and slag removal agents are employed to purify the melt from impurities. Prior to sampling, an appropriate test cup is selected and preheated to the correct temperature. During the sampling process, the aluminum alloy melt, at an ideal pouring temperature, is carefully extracted from the furnace and slowly and steadily poured into the test cup, taking caution to avoid splashing and oxidation. Following this, temperature data are immediately recorded using an IMC temperature acquisition device and thermocouples to monitor the cooling process of the melt. A cooling curve is then plotted, which illustrates the cooling trajectory of the aluminum alloy from the pouring temperature to room temperature. Analyzing this curve is crucial for understanding the thermal characteristics of the aluminum alloy and is essential for optimizing casting processes and enhancing product quality.

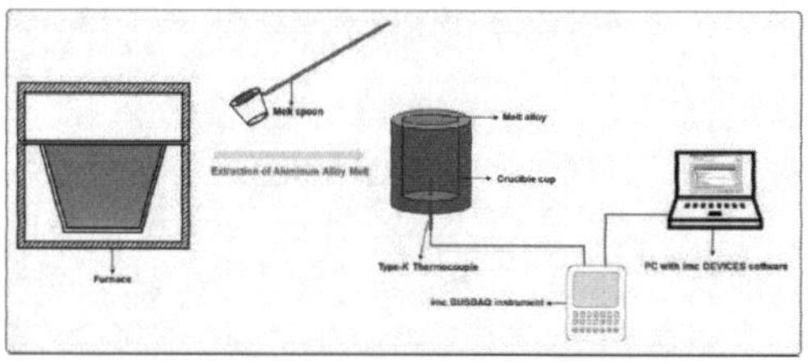

Figure 1 Experimental schematic diagram of aluminum alloy melt quality analysis

3 Result and discussion

Melt quality criteria have been defined by considering the area under the cooling curve within the primary and eutectic phase transformation temperature ranges. The cooling curve and its higher-order derivatives is shown in Figure 2.

Cooling curve slope criteria, such as the slopes of the pure liquid phase and pure solid phase, have been considered. The slope of the cooling curve before and after the phase transformation temperature points can reflect changes in cooling rate during solidification. A larger slope (first derivative) indicates rapid cooling. Therefore, a standard can be established where the cooling curve slope near the primary and eutectic phase transformation temperature points should be within a specific range to ensure ideal solidification characteristics and material quality.

Melt quality analysis programis shown in Figure 3. The criteria for the ductility at the phase transformation temperature points involve observing the behavior of the phase transformation temperature points on the cooling curve, such as the length of time the temperature remains stable. This reflects the stability of the phase transformation process. A longer stable phase may indicate a more uniform phase transformation process, resulting in a more uniform microstructure. A standard can be set where the cooling curve should maintain a stable state for a certain period at the phase transformation temperature point to ensure good solidification characteristics.

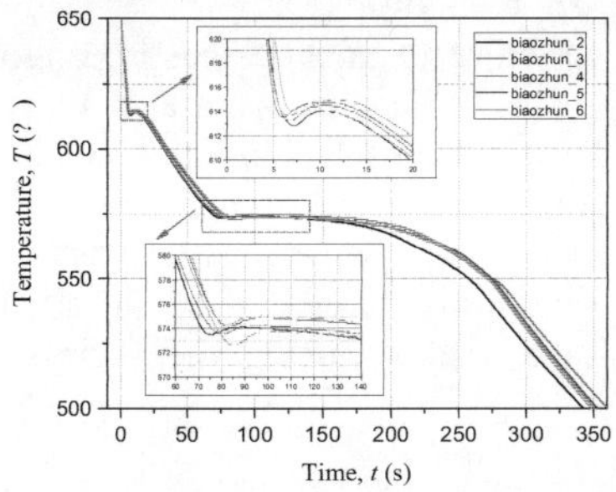

Figure 2 Cooling curve and its higher-order derivatives

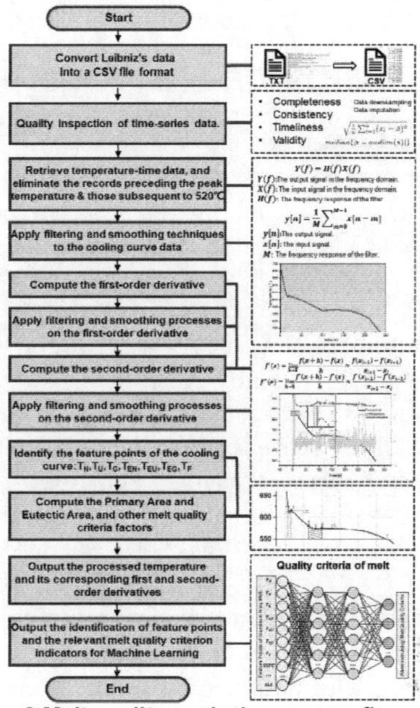

Figure 3 Melt quality analysis program flow chart

4 Conclusion

An experiment on the measurement of aluminum alloy melt temperature was conducted.The raw cooling curve data is processed using methods such as smoothing and filtering. An algorithm for feature point identification is established, which identifies the characteristic values during the primary solidification stage and the eutectic solidification stage, respectively. Based on the extracted characteristic values, criteria for melt quality assessment are developed using approaches like machine learning and linear regression, enabling the scoring of melt quality.

5 Acknowledgments

This work was supported by a grant from the Beijing Natural Science Foundation (No. L223001) and Tsinghua-CITIC Dicastal Cooperation Project.

References

[1] LuoAA,SachdevAK,ApelianD. Alloy development and process innovations for light metals casting. Journal of Materials Processing Technology, 2022,306:117606.

[2] Stefanescu D M, Suarez R, Kim S B.90 years of thermal analysis as a control tool in the melting of cast iron. China Foundry, 2020, 17(2):69-84.

[3] Li Y X, Hu X,Xu X R. Pattern recognition of thermal analysis cooling curves and quality evaluation of melt cast alloys.2001, 17(1): 73-74.

[4] Yi W, Liu G, Gao J,et al.Boosting for concept design of casting aluminum alloys driven by combining computational thermodynamics and machine learning techniques.Journal of Materials Informatics, 2021, 1(2).

Study on Macrosegregation of Squeeze Casting Based on Multiphase Flow Model

Chuanyao Tang, *, Weiwen Zhang
School of Mechanical and Automotive Engineering, South China University of Technology, Guangzhou, 510640, China
*Corresponding address: e-mail: metangcy@mail.scut.edu.cn

Abstract: Multiphase flow model was used to simulate the extrusion casting process of Al5Cu0.4Cu alloy. Pouring temperature, mold temperature, extrusion pressure and delay time before extrusion all have influence on Cu segregation in squeeze casting, and the influence of pressure is particularly important. The variation of Cu macrosegregation in castings was observed. The segregation of Cu is mainly due to the forced filling of Cu rich liquid phase and pressure at the solid-liquid interface. Compared with the results of experiment of casting, it is proved that the model simulation is more accurate.

Keywords: multiphase flow model; macrosegregation; simulation

1 Introduction

The macroscopic segregation of castings is mainly due to the inhomogeneous composition caused by the relative motion of solid and liquid phases during casting solidification [1]. This relative motion is mainly caused by convection of thermal solutes during solidification, flow caused by volume contraction, flow caused by floating or sinking of free crystals, and flow caused by different processes [2]. Although the squeeze casting process can effectively improve the segregation of castings, it is still inevitable to produce segregation. The pressure has a particularly important impact on the macro segregation of castings, and the change of pressure can effectively improve the segregation of castings.

In this paper, the distribution of copper element in squeeze casting $Al_5Cu_{0.4}Mn$ was studied by multiphase flow model and compared with experimental results. By comparing the results of experiment with simulation, It provides a basis for Al-Cu alloy squeeze casting to reduce macro segregation.

2 Experimental procedure

The experimental materials, shown in Table 1, contained very little Si(< 0.04%, mass fraction) and Fe(< 0.1%, mass fraction) and other unavoidable trace impurities by spectral composition analysis. It was squeeze cast to form cylindrical ingots with diameter φ=68mm and height H=57 mm. Because the pouring temperature and mold temperature mainly affect the temperature gradient and cooling rate of the metal, thus affecting the alloy structure. The higher pouring temperature and mold temperature will make the grain coarser, and the larger the grain size, the more serious the macroscopic segregation of Cu. So the extrusion parameters are shown in Table 2.

Table 1. Composition of alloy (wt. %)

Al	Cu	Mn
bal	5	0.4

Table 2. Main parameters of squeeze casting

No.	Pouring temperature /℃	Die temperature /℃	Applied pressure /Mpa
1	730	200	0
2	730	200	25
3	730	200	50

Result and discussion

Macrostructure of casting

The microstructure of squeeze casting ingots 1 to 3 is shown in Figure 1. With the increase of the pressure, the shrinkage and porosity of the ingot gradually weakened. This is because the increase of pressure made the metal liquid continuously flow to the central paste area, and the liquid was successfully forced to fill through the forced drive of pressure, thus reducing the shrinkage hole and porosity.

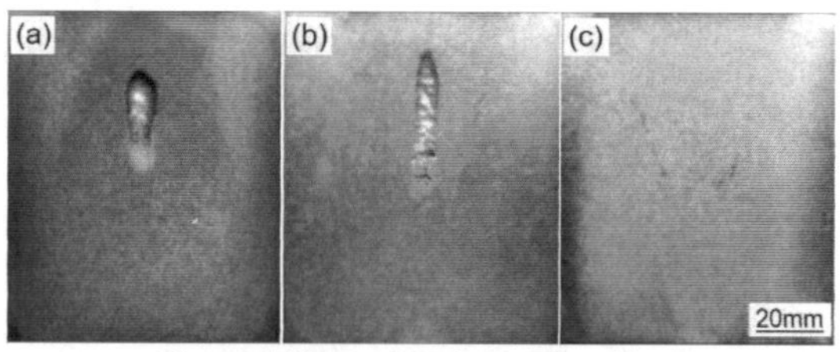

Figure 1 Macrostructure of casting: (a)1; (b) 2; (c) 3

Composition of casting

A multiphase flow model was used to simulate the squeeze casting process of No. 2, and the final local Cu positive segregation was selected as shown in Figure 2. 20 positions were taken from each ingold for composition testing, which were the spectral components of five different heights along the radial direction (R = 4 mm, 12 mm, 20 mm, and 28 mm), and the equal height schematic diagram of Cu element distribution was obtained by fitting, as shown in Figure 2(1/4 ingot). It can be seen that the Cu egation of $Al_5Cu_{0.4}Mn$ alloy after squeeze casting is mainly concentrated in the upper center of the ingot. This is caused by the forced extrusion of the low melting point copper rich liquid phase to the core of the casting.

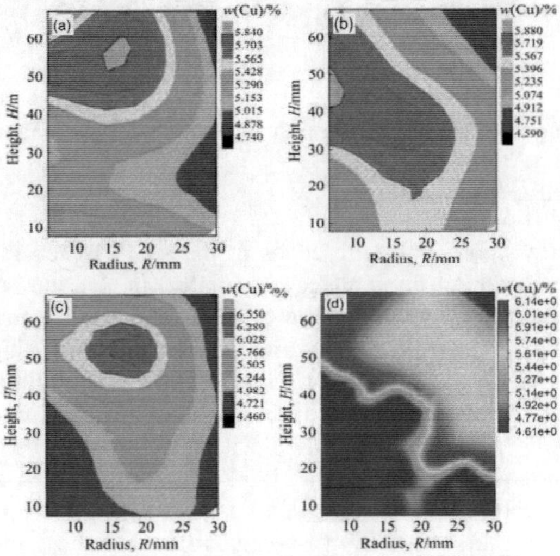

Figure 2 Local schematic of the distribution of Cu elements in an ingot: (a) 1; (b) 2; (c)3; (d) simulation result

Effect of pressure on Cu segregation

Cu segregation mainly depends on the release rate of Cu atoms at the front of the solid-liquid interface, the amount of Cu rich liquid phase and the flow state.

Zhong Yong et al. [3] pointed out that under high pressure, liquid with low melting point will flow into the dendritic gap, which will weaken the segregation, and with the increase of pressure, the segregation will gradually weaken. Bo Shuming et al.[4] pointed out that plastic deformation and metal liquid flow were compensated under pressure in the process of squeeze casting, so that the solid phase and interdendritic liquid phase were flowing toward the solid-liquid two-phase area, liquid phase area and shrinkage hole area with low resistance, so as to carry out forced shrinkage. In squeeze casting, the interdendritic liquid phase is forced to the core of the casting through the dendritic channel under the effect of pressure filling, and the abnormal positive segregation phenomenon is presented.

The comparison between the experimental segregation maximum and the simulated segregation maximum is shown in Table 3, which shows that the simulation results are more reliable.

Table 3. The experimental values are compared with the simulated values

	Experiment value	Value of simulation	Error value
Maximum segregation value of Cu	5.88	6.14	4.4%

4 Conclusion

In squeeze casting with different process parameters, the macrosegregation of the casting can be effectively weakened by changing the pressure. The multiphase flow solidification model has small error between simulation and actual situation under pressure and is more reliable.

Acknowledgments

We would like to acknowledge the financial support from the National Natural Science Foundation of China (No. 52074131)

References

[1] Li J, Wu M, Ludwig A, et al. Simulation of macrosegregation in a 2.45-ton steel ingot using a three phase mixed columnar-equiaxed model [J]. Int. J. Heat Mass Tran., 2014, 72: 668-679.
[2] Ge H H, Ren F L, Li J, Hu Q D, Xia M X, Li J G. Modelling of ingot size effects on macrosegregation in steel castings [J]. Mater. Process. Technol., 2018, 252: 362–369.
[3] Zhong Y, Yan D S, Su G Y, Yang K. Microsegregation and improving method of a squeeze cast LY12 alloy [J]. Acta Metallurgica Sinica, 2001, 37(1): 42−46.
[4] Xing S M, Ma J, Chen W S. Squeeze casting shrinkage predicting process designing [J]. The Chinese Journal of Nonferrous Metals, 1998, 8(S1): 205−209.

Automatic Grinding Difficulties Analysis and Solutions for Dynamic Arm Plate (Movable Arm Plate, Boom Plate)

Wenjun Li
Dalian YuYang Industrial Intelligence Co, Ltd., No.3 Yinggang Road, Shagangzi Village Street, Ganjingzi District, Dalian, Liaoning Province, China
*Corresponding address: e-mail: lidantong@kinhanyan.com

Abstract: As the castings with multiple varieties and specifications, as well as the inconsistencies, it is difficult for the dynamic arm plate to achieve automatic grinding. There are over a hundred different types of castings, and each type of casting undergoes flame and plasma cutting to produce different sizes of deformation, resulting in significant changes in appearance. The traditional automatic fettling method requires continuous programming adjustment and different fixture designs according to the different models of products and the same model of products with different shape variables, which is an extremely huge workload, nearly impossible to complete. Therefore, the traditional automatic grinding method cannot achieve continuous production of dynamic arm plate.

Keywords: dynamic arm plate (movable arm plate, boom plate); automatic grinding; AI learning; mathematical model; castings with multiple varieties and specifications; large deviation of dimensions

1 Automatic grinding difficulties analysis for dynamic arm plate

Engineering machinery castings, especially dynamic arm plates, are characterized by strong rigidity and high hardness. After mixed-line production, there are more than a hundred models, each of which is subject to different dimensional deformations after thermal cutting by flame, plasma and so on.

For example, the deformation generated by overheating processing is not only a variation of 2-4 mm in the product size itself, but also the thermal warping deformation caused by heat in different parts of the steel plate. This thermal deformation is irregular and most of the deformation is between 5-15 mm, with some even reaching 20 mm, and the shape of the casting is extremely variable.

In this situation, how to realize automatic grinding of dynamic arm plate has been a challenge for the whole industry, including common issues of tooling and fixture, massive manual programming problem, high precision issues of chamfering and grinding, and so on.

2 Automatic grinding solutions for dynamic arm plate

2.1 AI automatic recognition of workpiece size

There are significant differences in the size of each model of the dynamic arm plate, especially in length. After thermal cutting, the workpiece of each model will generate significant deformation, which puts forward higher difficulties for the design of fixtures and the positioning and gripping of loading. The working principle of tooling and fixtures is to find a suitable position firstly to push the workpiece to ensure its orientation and position dimensions, and the second step is to select a suitable position to tighten the workpiece so that it will not move during the grinding process.

Due to the wide variety of workpieces, the above two points need to be completed through AI intelligent recognition. The positioning mechanism and fastening mechanism used for large and small workpieces are different, therefore, during the loading process, the 3D vision system scans and identifies the specific model of the workpiece, and the AI mathematical model will call out the matching tooling mechanism to complete the positioning and fastening of the workpiece through self-learning and calculation. This requires the design of tooling to meet all workpiece types, but also to solve the interference and influence of different size workpieces on the same fixture.

2.2 3D scanning to avoid manual programming

In the process of workpiece size detection, there is a significant difference between the actual workpiece shape position and the theoretical size, which is one of the important issues affecting the grinding quality. The traditional automatic fettling method requires continuous shutdown programming and adjustment based on different models of products and different deformation variables of the same model of product, which is a huge workload that is almost impossible to complete, so the traditional automatic grinding method cannot realize continuous production of dynamic arm plate.

3D scanning automatic grinding technology breaks through traditional visual technology, each contour including 3200 data points will be scanned through 3D laser, which can quickly and accurately have production point cloud data for the scanned object. The data will be converted into robot coordinates, which are then converted into grinding trajectories, that is, automatic production grinding program.

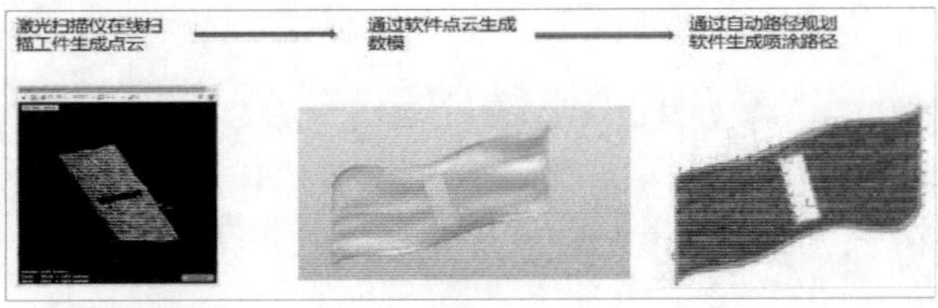

Fig. 1

By collecting drawing data to understand the shape, size, grinding position, and process of the dynamic arm plate, and design the underlying software that meets the processing technology. The scanning parameters are set through a customized software system and based on the actual situation of the workpiece, and input the required types and quantities of toolings into the software in advance for use in subsequent program generation. The visual system (customized software) will generate mathematical models through software to automatically calculate the trajectory points of the robot, convert them into recognizable data, and send to the robot.

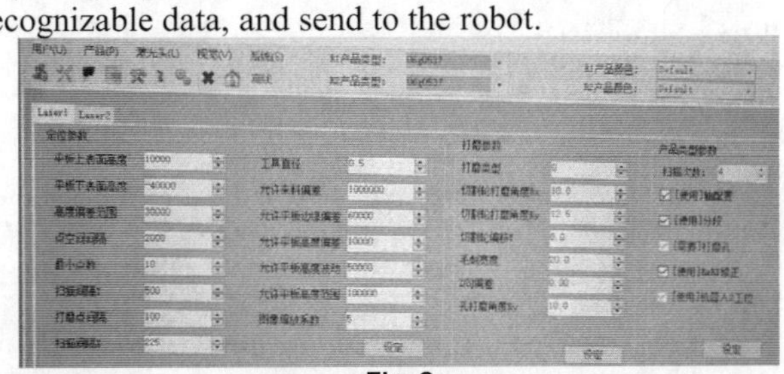

Fig. 2

When a new workpiece enters the grinding area, the system gives a command to the robot to start executing the scanning command. Scanning can finally give two modules, one is the robot point position, the other is the path module, through the visual module, the two modules will be transformed into the final grinding trajectory. The trajectory is given by the software, and the automatic grinding function is realized after simple input of processing parameters. Not only can the trajectory be automatically produced, but robot can also achieve automatic tool changing to meet the requirements of processing technology. After completing the above, the automatic grinding program can be produced such as similar parts are only slightly different in size, the above parameters can be directly copied.

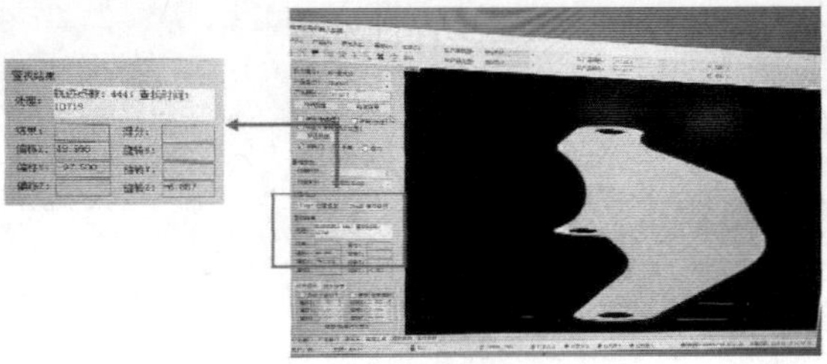

Fig. 3

2.3 Real time force control system ensures the accuracy of chamfering and grinding

During the whole grinding process, strict grinding accuracy control to the dynamic arm plate is required, especially the chamfering of the workpiece. The traditional grinding method adopts airbag and spring type, both of which are not precise enough in controlling the grinding force. When the spindle changes orientation, it is greatly affected by gravity, for example, the spring type of grinding, which is unable to make corresponding adjustments to the force according to different grinding positions and the amount of grinding, and thus fail to satisfy the requirement of high-precision fettling of chamfering.

For the dynamic arm plate grinding, it adopts 3D scanning, and add real-time force control system at the same time, to achieve flexible grinding by detecting and controlling the grinding force. The real-time force control system can control the different grinding forces required by the spindle in all directions and angles to ensure a constant output effect. No matter what tools the robot uses or what position angle the spindle is placed at, the system can accurately calculate the designed grinding force through real-time force control and maintain it in this state to meet grinding quality requirements.

3 Conclusion

From the above, it can be seen that the successful application of the dynamic arm plate grinding technology has overcome many difficulties encountered in the actual production of engineering machinery castings in the field of post-casting fettling, such as a variety of types, models, and large deviations in the external dimensions. It has filled the gap in the AI self-learning system for building mathematical models in the subdivision field of post casting fettling, and achieved full automatic programming of 3D scanning for casting grinding technology. With the continuous development of science and technology, the research on the grinding of such castings will certainly have new breakthroughs. The continuous improvement of grinding technology will also enhance our manufacturing production level.

Melting Efficiency in β-Ga_2O_3 through Induction Skull Melting Technology

Yanshen Wang[1, 2], Mingzhi Zhu[1, 2], Yuan Liu[1, 2], *
1. School of Materials Science and Engineering, Tsinghua University, Beijing 100084, China
2. Key Laboratory for Advanced Materials Processing Technology (Ministry of Education), Beijing 100084, China
*Corresponding address: e-mail: yuanliu@mail.tsinghua.edu.cn

Abstract: Oxides crystal growth using the cold crucible (OCCC) method is a new application of induction skull melting technology (ISM). In this study, a 3D numerical model was proposed by analyzing the ignition efficiency of starting material and the influence of power, frequency, and crucible diameter on the area of steady melting zone during β-Ga_2O_3 melting by ISM. The steady Ga_2O_3 temperature by ISM was calculated under different start-up conditions, which considers the effects of electrical properties, locations, and shaped of starting material. In addition, the minimum power required to hold the melt without the starting material and corresponding melt volume under different frequencies and crucible diameter were also calculated. The results show that starting material with an electrical conductivity of 10^4 to 10^5 S/m, location at the middle of coils, and plate shape exhibited high ignition efficiency. After removing the starting material, the critical power and steady melt volume decrease with increased frequency. Crucibles with 90 mm diameter and power supply with frequency of 200 kHz are suitable for the melting of β-Ga_2O_3 considering the energy loss and economic efficiency.

Keywords: Ga_2O_3; ignition; ISM; numerical simulation

1 Introduction

With the development of semiconductors in the field of microelectronics, researchers have increasingly favored semiconductors with wider bandgaps. Gallium oxide (Ga_2O_3), one of the fourth-generation ultra-wide bandgap (UWBG) semiconductors, has become a competitive candidate due to its excellent properties. With a bandgap of more than 4.5eV, Ga_2O_3 has a higher breakdown electric field and better tolerance to high temperature and radiation than those of the third-generation semiconductors[1-3]. Coupled with a superior Baliga's figure-of-merit (BFOM) and low on-resistance, Ga_2O_3 becomes a promising material for high-power applications in electronic devices[4]. Moreover, the large bandgap leads to its cut-off wavelength located in the deep ultraviolet region ranging from 250nm to 280nm, which meets the requirements of deep ultraviolet light (DUV) detection. Photodetectors based on Ga_2O_3 are insensitive to visible and infrared light exhibiting highly accurate ultraviolet (UV) detection [5]. Besides, Ga_2O_3 offers great potential for nuclear radiation detection with high density (6.44 g·cm^{-3}), excellent scintillation properties, and radiation resistance [6]. However, it is difficult to reduce the cost of β-Ga_2O_3 single crystal growth due to the fact that precious metal crucibles are essential for the oxygen-rich growth environment. In recent years, oxide crystal growth from cold crucible (OCCC), which was a new application of induction skull method, was proposed to reduce the single crystal cost with the substitution of cold copper crucible for precious metal crucibles. As a new process for β-Ga_2O_3 single crystal growth, it is quite effective to reduce the cost of trail by using numerical simulation. In this paper, the ignition efficiency of starting material and the influence of power, frequency, and crucible diameter on the area of steady melting zone during β-Ga_2O_3 melting by ISM were analyzed.

2 Experimental procedure

Figure 1 depicts the ISM melting system. Periodic boundary condition was applied to a portion of the crucible to reduce the simulation calculations.

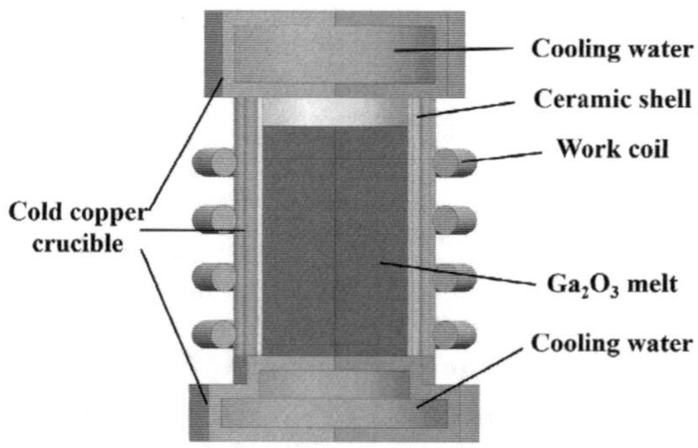

Figure 1 Schematic view of ISM melting system

Result and discussion

Effect of start-up conditions on the steady Ga_2O_3 temperature.

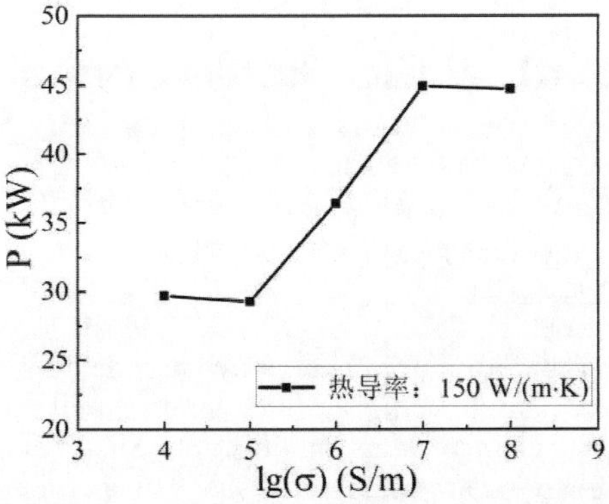

Figure 2 Minimum power required to hold the melt under different electrical conductivity of ignitions

Ignitions with an electrical conductivity of 10^4 to 10^5 S/m, location at the middle of coils, and plate shape exhibited high ignition efficiency.

Effect of frequency on the minimum power required to hold the melt.

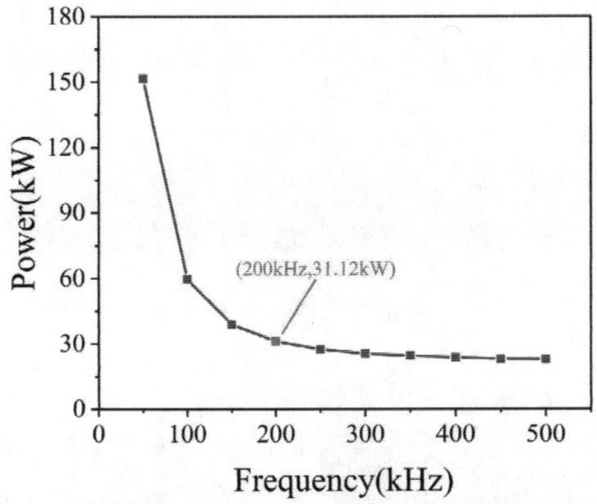

Figure 3 Minimum power required to hold the melt under different frequency

The minimum power required to hold the melt reduced with increased frequency. In addition, the minimum power under 200 kHz is 31.12 kW, which is similar to the experimental power 30 kW. Crucibles with 90 mm diameter and power supply with frequency of 200 kHz are suitable for the melting of β-Ga_2O_3 considering the energy loss and economic efficiency.

3 Conclusion

In this work, we proposed a numerical model for Ga_2O_3 melting with ISM technology. The simulation results showed that ignitions with an electrical conductivity of 10^4 to 10^5 S/m, location at the middle of coils, and plate shape exhibited high ignition efficiency. In addition, the minimum power required under different crucibles diameters and frequencies are calculate. Crucibles with 90 mm diameter and power supply with frequency of 200 kHz are suitable for the melting of β-Ga_2O_3.

References

[1] Xue H W, He Q M, Jian G Z, et al. An overview of the ultrawide bandgap Ga_2O_3 semiconductor-based schottky barrier diode for power electronics application. Nanoscale Res. Lett., 2018, 13(1): 290.

[2] Mastro M A, Kuramata A, Calkins J, et al. Perspective—opportunities and future directions for Ga_2O_3. Ecs J. Solid State Sci. Technol., 2017, 6(5): 356–359.

[3] Galazka Z, Uecker R, Irmscher K, et al. Czochralski growth and characterization of β-Ga_2O_3 single crystals. Cryst. Res. Technol., 2010, 45(12): 1229–1236.

[4] Higashiwaki M, Sasaki K, Kuramata A, et al. Development of gallium oxide power devices. Phys. Status Solidi A, 2014, 211(1): 21–26.

[5] Xu J J, Zheng W, Huang F. Gallium oxide solar-blind ultraviolet photodetectors: A review. J. Mater. Chem. C, 2019, 7(29): 8753–8770.

[6] Bauman D A, Borodkin A I, Petrenko A A, et al. On improving the radiation resistance of gallium oxide for space applications. Acta Astronaut., 2021, 180: 125–129.

Dynamic Simulation for Top-Seeded Solution Growth of SiC Single Crystal

Gangqiang Liang[1], Jie Zhou[1], Yuan Liu[2]
1. School of Materials Science and Engineering, Tsinghua University, Beijing 100084, China
2. Key Laboratory for Advanced Materials Processing Technology (Ministry of Education), Beijing 100084, China
*Corresponding address: e-mail: yuanliu@mail.tsinghua.edu.cn

Abstract: SiC-MOSFETs have partially replaced traditional Si-IGBT devices in high-voltage fields such as new energy vehicles and photovoltaics, but they still face the problem of high cost today. Solution method is a method capable of growing high-quality SiC single crystals at low cost, and numerical simulation is considered an effective way to probe its process parameters. The current research work reports the prediction of temperature, flow rate, and concentration distributions under initial steady-state conditions, however, detailed discussion on the kinetic coupling process of crucible corrosion and seed crystal growth during solution growth is limited. In this work, the prediction of the geometrical change of the interface between crucible corrosion and seed crystal growth is realized by using COMSOL software with the moving mesh module, and the effects of seed crystal rotational speed and temperature distribution on the stability of the growth interface are discussed.

Keywords: SiC single crystals; solution growth; numerical modeling

1 Introduction

Silicon carbide (SiC) power devices have found extensive applications in industrial equipment, including electric vehicles, photovoltaics and railways. As one of the IV-IV compound semiconductors, SiC is capable of applying to scenarios at high temperatures, high voltage and strong radiation attributed to its superior properties. The production of SiC substrate relies on SiC bulk growth technology. Most of the commercial N-type SiC substrates are produced by the physical vapor transport (PVT) method, while it is challenging to prepare P-type SiC ingots since the Al source is released excessively in the initial stage of sublimation, leading to large defect density. As an ideal alternative candidate, top-seeded solution growth (TSSG) was proved to be an effective way to fabricate P-type crystals with a low resistivity of 35mΩ cm [1]. Besides, micropipes can be easily eliminated induced by step flow, and most of the threading dislocation (TDs) can be converted into BPD or Frankel-type stacking faults, realizing the reduction of dislocation density in the grown crystals [2]. In the development of TSSG technology in the past 20 years, the crystal diameter has been enlarged from 4 mm to 15 mm by the meniscus growth [3], the maximum growth rate of 2 mm/h with a FWHM value of 20 arcsec [4] and 4H-SiC with ultra-low dislocation density have been reported [5]. TSSG therefore became a promising method to grow high-quality single crystal SiC ingots and the corresponding finite elements analysis (FEA) of SiC crystal growth process has been well established in previous investigations [6-10].

2 Experimental procedure

Figure 1 depicts the top-seeded solution growth system. A radiofrequency induction coil (1.8 kHz) with a power of 5 kW is employed to heat the graphite hot wall and graphite crucible under argon atmosphere at 50 kPa. The pure Si melt is 40 mm in height and 200 mm in diameter. The SiC seed crystal of 150 mm in diameter is mounted on the graphite seed holder connected by a graphite shaft and then dipped 1mm below the solution surface.

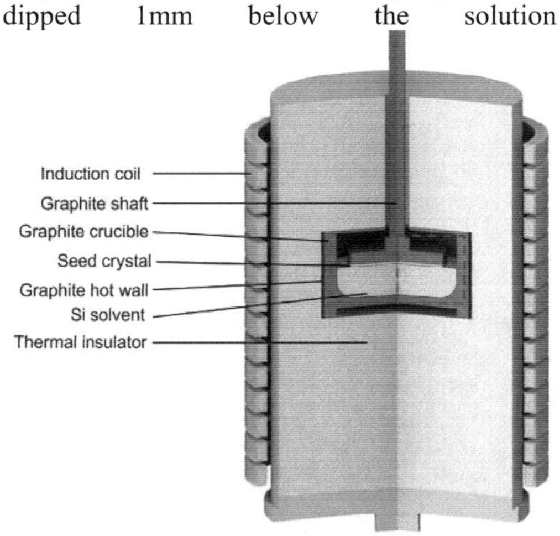

Figure 1 Schematic view of TSSG growth system

3 Result and discussion

Effect of crystal rotation rate on the corrosion of graphite crucible and grown crystal surface

The seed rotation brings the upward flow at the center of Si melt, enhancing the transport of carbon solute towards the growth surface. However, the rapid melt flow placed at the edge of the melt surface induced by seed rotation accelerates the local erosion of graphite wall, which may lead to leakage of the melt once crucible wall is cracked.

Effect of crucible rotation rate on the corrosion of graphite crucible and grown crystal surface

By employing the crucible rotation, the concentration distribution becomes uniform so that more carbon solutes are applied to the growth interface. The shape of the erosion area shifts from local erosion to uniform erosion at the interface between the solution and the graphite crucible wall.

4 Conclusion

In this work, we proposed a numerical model for predicting the geometrical change of the growth surface and graphite crucible during top-seeded solution growth of silicon carbide single crystals. The simulation results imply that seed rotation enhances the growth rate by facilitating the upward flow of carbon solute. Meanwhile, the melt leakage may occur under a long period of growth due to the local erosion of the crucible wall. This phenomenon is reduced by applying the crucible rotation since the erosion area expands. Higher growth rates are therefore obtained as well.

5 Acknowledgment

This work was financially supported by the Beijing Municipal Science and Technology Commission (Grant No. Z231110000270000).

References

[1] Shirai T, Danno K, Seki A, Sakamoto H, Bessho T. Solution growth of p -type 4H-SiC bulk crystals with low resistivity. 2014, 75–78.

[2] Harada S, Yamamoto Y, Xiao S, Koike D, Mutoh T. Dislocation conversion during SiC solution growth for high-quality crystals. Materials Science Forum, 2015, 821-823: 3–8.

[3] Danno K, Yamaguchi S, Kimoto H, Sato K, Bessho T. Trials of solution growth of dislocation-free 4H-SiC bulk crystals. Materials Science Forum, 2016, 85:19–22.

[4] Kado M, Daikoku H, Sakamoto H, Suzuki H, Bessho T, Yashiro N, Kusunoki K, Okada N, Moriguchi K, Kamei K. High-speed growth of 4H-SiC single crystal using Si-Cr based melt. Materials Science Forum, 2013, 73–76.

[5] Murayama K, Hori T, Harada S, Xiao S, Tagawa M, Ujihara T. Two-step SiC solution growth for dislocation reduction. Journal of Crystal Growth, 2017, 468: 874–878.

[6] Dang Y, Kutsukake K, Liu X, Inoue Y, Liu X, Seki S, Zhu C. A transfer learning-based method for facilitating the prediction of unsteady crystal growth. Advanced Theory and Simulations, 2022, 9(5): 2200204.

[7] Yu W, Zhu C, Tsunooka Y, Huang W, Dang Y, et al. Geometrical design of a crystal growth system. CrystEngComm., 2021, 23: 2695–2702.

[8] Sui Z, Xu L, Cui C, Wang R, Pi X, Yang D, Han X. Effects of solution height and crystal rotation on the solution flow behavior in the top-seeded solution growth of SiC single crystals. CrystEngComm., 2024, 2024, 26: 1022–1031.

[9] Liu B, Yu Y, Tang X, Gao B. Optimization of crucible and heating model for large-sized silicon carbide ingot growth in top-seeded solution growth. Journal of Crystal Growth, 2020, 533: 125406.

[10] Li F, He L, Yan Z, Qi X, Ma W, Chen J, Xu Y. Effects of crystal rotation on the carbon transport in the top-seeded solution growth of SiC single crystal. Journal of Crystal Growth, 2023, 607: 127112.

Modeling Macrosegregation in Continuously Cast Round Bloom with Combined Electromagnetic Stirring

Yuwei Yang[1, 2], Sen Luo[1, 2, 3], *, Weiling Wang[1, 2, 3], Miaoyong Zhu[1, 2, 3]
1. Key Laboratory for Ecological Metallurgy of Multimetallic Mineral (Ministry of Education), Northeastern University, Shenyang 110819, Liaoning, China
2. School of Metallurgy, Northeastern University, Shenyang 110819, Liaoning, China
3. Institute of Steel Sustainable Technology, Liaoning Academy of Materials, Shenyang 110000, Liaoning, China
*Corresponding address: e-mail: luos@smm.neu.edu.cn

Abstract: The electromagnetic stirring (EMS) can effectively improve the macrosegregation, central porosity, shrinkage, cracking, and other defects of continuous casting round blooms. Thus, based on the volume-averaged method, a three-phase solidification model is established, which considers the nucleation, growth of the columnar and equiaxed crystals, deposition of equiaxed crystals, and transmission of solutes. The effect of the individual EMS and combined EMS on the temperature, velocity, phase distribution, and macrosegregation distribution in a large vertical CC round bloom was studied. The results show that the CC bloom shell is built by columnar crystals, and equiaxed crystals fabricate the central zone. Individual M-EMS has a notable impact on expediting the dissipation of superheat, enhancing the initiation of grain formation, and enlarging the equiaxed region. The percentage of equiaxed grains increased from 58.56% to 62.17% with the increment in current intensities of M-EMS from 150 to 250 A. Furthermore, the implementation of individual F-EMS demonstrated its effectiveness in reducing center positive segregation. The ratio of center positive segregation decreased from 1.14 to 1.10, when the current intensity increases from 150 to 250 A. The combine EMS has multiple advantages of M-EMS and F-EMS. The notable enhancement in the equiaxed grain ratio up to 60.13% was observed combined EMS models were simultaneously employed during continuous casting of round blooms. And, there was a reduction in center segregation ratio for round blooms to as low as 1.09.

Keywords: continuous casting; EMS; three-phase solidification model; macrosegregation

1 Introduction

Recently, various industries based on large-scale equipment have been developed, including wind power generation, petrochemical industry and machinery manufacturing. In order to meet the production demands, the products of special steel continuous casting round blooms are increasing gradually. However, macrosegregation is a key problem in continuously cast round blooms [1,2]. Numerous scholars [3] have made long-term commitments to the mitigation of macrosegregation. Such investigations will contribute to the development of enhanced production techniques for continuously cast round blooms, facilitating the manufacturing of defect-free rolled products caused by macrosegregation. Presently, electromagnetic stirring (EMS) [4] stands out as one of the most efficient measures to diminish macrosegregation in continuously cast round blooms.

Although there are numerous experimental and theoretical investigation on the effect of EMS on the macrosegregation in as cast steel, the synergistic effect of combined M-EMS and F-EMS are still needed to be further studied in order to achieve the optimum metallurgical function of EMS on elimination of macrosegregation in continuously cast steel. This study is based on a three-phase solidification model. By studying the effects of the EMS, combined EMS electromagnetic parameters, and stirring models on the microscopic structure evolution and solute segregation behavior, the influence of only EMS and combined EMS on the internal quality of slab is further revealed.

2 Experimental procedure

In this paper, a control volume based on the finite difference method was applied to solve the governing equations and define the source terms and exchange terms through the UDF functions in ANSYS FLUENT software. All phases shared the same pressure field and obtained the pressure correction equation by the phase coupled SIMPLE algorithm. The 3D/2D structured grid used in the simulation initially consists of 952000 computational cells. The growth of calculation region was considered by the dynamic layering method under unsteady conditions. For each time step, 60 iterations were adopted to decrease the convergence criterion of the normalized residual of the continuity equations, momentum equations, species transport equations, and user-defined scalar equations below 10^{-5} and the energy equation below 10^{-6}.

3 Result and discussion

Figure 1 shows the area fraction of the equiaxed phase zones at the end of solidification with different EMS processes. The area fraction of the equiaxed phase zones of individual EMS, and combined EMS were 59.26%, 51.59%, and 60.13%, respectively. The results demonstrate that M-EMS significantly improves the extension of the equiaxed grain zone, whereas the F-EMS effect on increasing the equiaxed grain zone is not as pronounced. The main reasons are that the CET transition and solidified equiaxed grain zones are established before the installation position of the F-EMS.

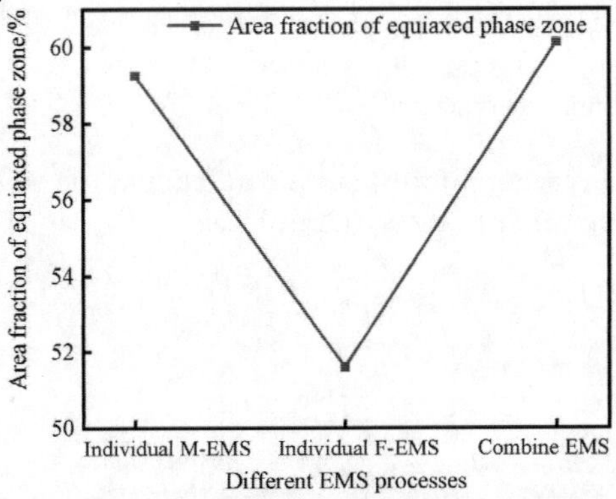

Figure 1 Area fraction of the equiaxed phase zones at the end of solidification with different EMS processes

Figure 2 shows the distribution of macrosegregation ratio with different EMS processes. The ratio of center positive segregation at the center of round bloom was 1.17 with only M-EMS. The ratio of center positive segregation at center position was 1.10 with only F-EMS. In the case of the combined EMS models, the ratio of center positive segregation reached at 1.09. The results indicate that M-EMS has limited improvement on center positive segregation, while F-EMS can effectively reduce center positive segregation. The combined EMS models further decrease the center positive segregation based on the metallurgical efficacy of F-EMS.

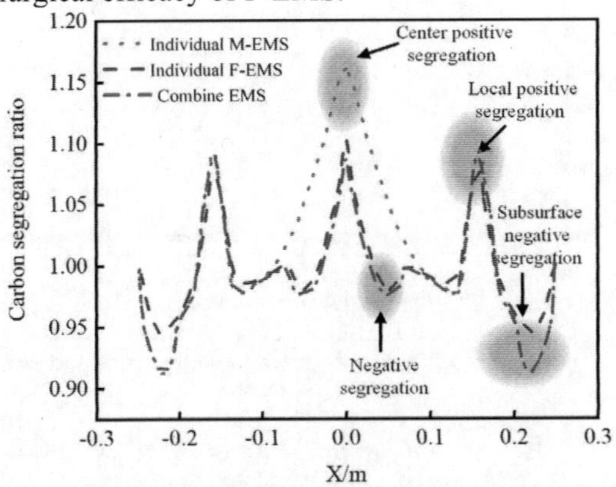

Figure 2 Distribution of macrosegregation ratio with different EMS processes

4 Conclusion

Based on the volume-averaged method, a three-phase solidification model is established. The effect of the individual EMS and combined EMS on the macrosegregation distribution in a large vertical CC round bloom was studied. The main conclusions can be drawn:

(1) Individual M-EMS has a notable impact on expediting the dissipation of superheat, enhancing the initiation of grain formation, and enlarging the equiaxed region. The percentage of equiaxed grains increased from 58.56% to 62.17% with the increment in current intensities of M-EMS from 150 to 250 A.

(2) The implementation of the individual F-EMS demonstrated its effectiveness in reducing center positive marcosegregation.

(3) The combine EMS has multiple advantages of M-EMS and F-EMS.

5 Acknowledgments

This study was financially supported by the National Key Research and Development Plan (No. 2021YFB3702000), National Natural Science of China (Nos. 52074076, 52174306, and U20A20272), and Fundamental Research Funds for the Central Universities No. N2225023.

References

[1] Sun H, Li L. Formation and control of macrosegregation for round bloom continuous casting. Ironmak Steelmak. 2015, 42(9): 683-688.

[2] An H, Bao Y, Wang M, et al. Numerical and experimental investigation of solidification structure evolution and reduction of centre segregation in continuously cast GCr15 bloom. Ironmak Steelmak, 2020, 47(9): 1063-1077.

[3] Li L, Zhang Z, Luo M, et al. Control of shrinkage porosity and spot segregation in Ø195 mm continuously cast round bloom of oil pipe steel by soft reduction. Metals, 2020, 11(1): 9.

[4] Ren B, Chen D, Xia W, et al. Numerical simulation of electromagnetic field in round bloom continuous casting with final electromagnetic stirring. Metals, 2018, 8(11): 903.

Modeling of Macrosegregation with Liquid Core Reduction in Slab Casting

H F Shen[1],*, L Y Xu[2], Y H Dou[1], S Q Wen[1]

1. School of Materials Science and Engineering, Tsinghua University, Beijing 100084, China
2. Hunan Xiangtan Iron and Steel Co., Ltd., Xiangtan 411101, China

*Corresponding address: e-mail: shen@tsinghua.edu.cn

Abstract: During the continuous slab casting process, macrosegregation affects the product quality, and it can be improved by the liquid core reduction in which the strand thickness is reduced whilst its core is still in a liquid state. The formation of macrosegregation is a complex process related to various factors such as thermosolutal convection, mush deformation and grain sedimentation[1-2]. In this paper, a multiphase volume averaging model is developed with the liquid as the first phase and the columnar and equiaxed dendrites as the secondary and the third phases. The transportation conservation equations of each phase as well as the grain growth dynamics are coupled and solved. The results show that the model can be used to simulate the solidification process of continuous slab casting.

Keywords: continuous casting; macrosegregation, liquid core reduction; numerical simulation

1 Modeling of macrosegregation in slab casting

Figure 1 shows the numerically simulated liquid volume fraction f_l during the solidification process of continuous slab casting. It indicates that the high temperature molten steel enters the mold from the submerged entry nozzle (SEN) and a solid shell forms because of heat transfer by the water cooling.

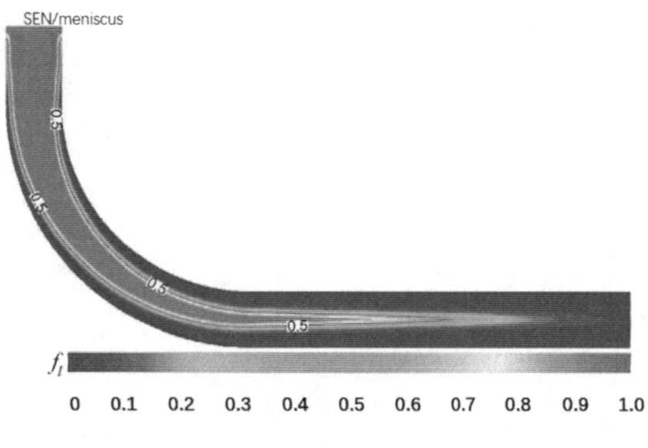

Figure 1 Solidification process of continuous casting slab

Figure 2 shows the carbon segregation index (SI=C/C_0) during the continuous slab casting. Comparing with the solidification front in Figure 1, it can be found that solute is enriched at the solid-liquid interface, and the enrichment increases with the liquid core depth. Finally, an obvious positive segregation forms at the centerline while a negative segregation forms at the surface of slab.

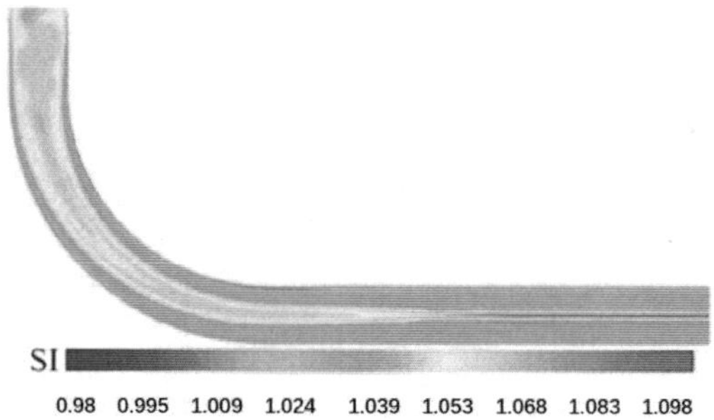

Figure 2 Macrosegregation in continuous slab casting

Figure 3 shows the evolution of liquid fraction and segregation index along the casting centerline. Segregation index is highest after fully solidification.

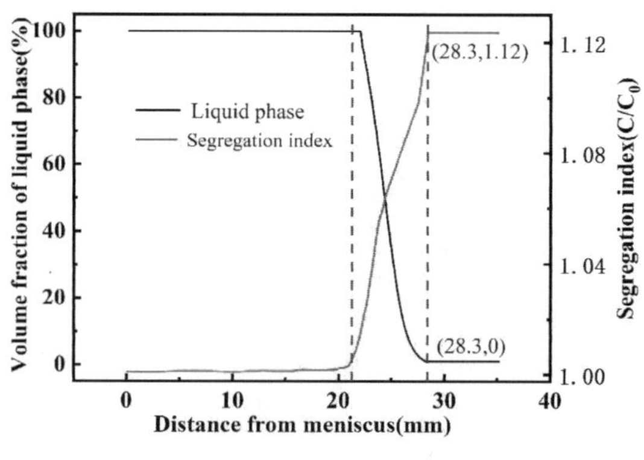

Figure 3 Evolution of liquid fraction and segregation index along casting centerline

Figure 4 shows the distribution of the carbon segregation index along the slab thickness at different distances to the meniscus L. Obvious solute enrichment is also found at the solid-liquid interface (referring Figure 1). When L is less than 20m, the concentration in the liquid pool is relatively homogenous compared to that at the solidification front and it increases with the increasing L. However, when L is larger than 20m, the concentration at the solidification front decreases first and then increases sharply. This decreased concentration at the solidification front may be lower that the melt initial concentration C_0, so the negative segregation forms near the solidification end, which can be explained later.

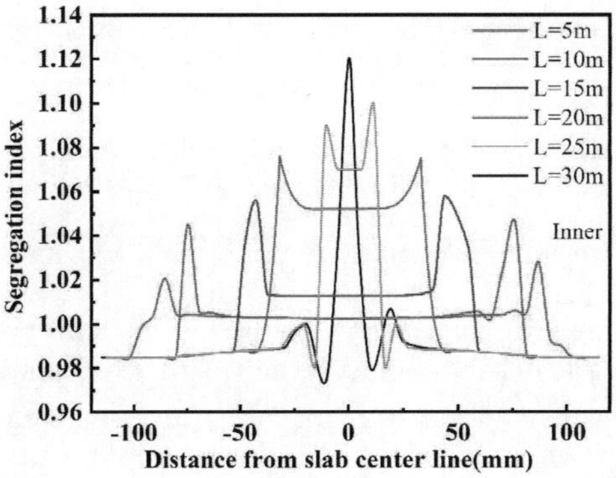

Figure 4 Segregation index at different thickness position to meniscus

Figure 5 shows the distribution of volume fraction of equiaxed phase and liquid, as well as segregation index of equiaxed along the thickness at 24m from the meniscus. Equiaxed nucleates ahead of solidification front because of decreasing temperature. Equiaxed with lower concentration sediments because of its heavier density and negative segregation may forms as shown in Figure 4.

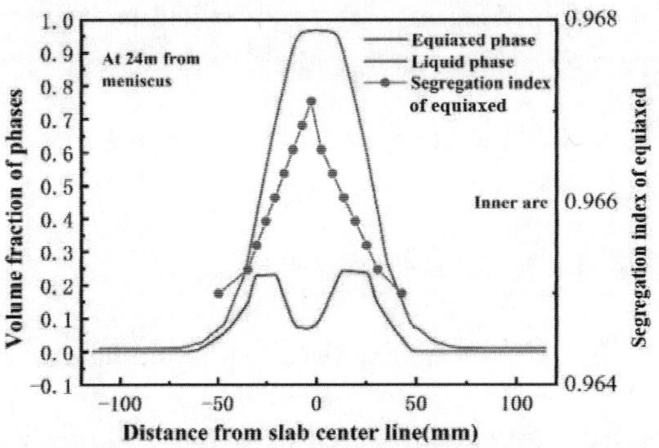

Figure 5 Liquid and equiaxed fractions as well as equiaxed segregation index along thickness

2 Effect of liquid reduction on macrosegregation

Figure 6 shows the distribution of carbon concentration along the thickness at the center of slab with different thickness reductions. It is found that the central segregation is reduced by using liquid core reduction. The segregation index C/C_0 at the center decreases from 1.11 to 1.06 when the thickness reduction is applied from 0 mm to 5 mm. Further more, negative segregation may form if the over large thickness reduction of 10 mm is applied. Concentration beyond the slab center increases because the species is pushed from the center with the reduction.

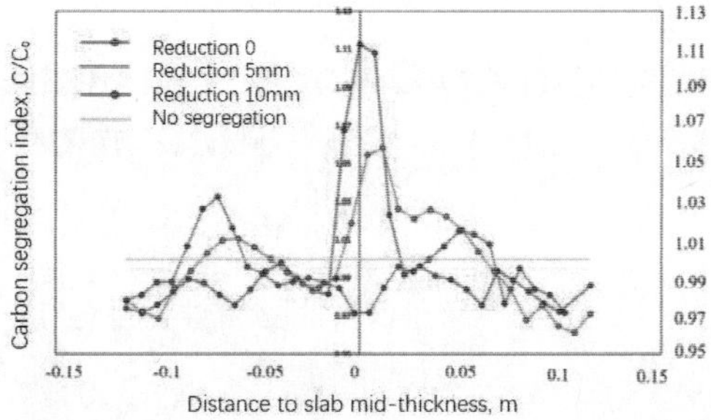

Figure 6 Effect of liquid reduction on macrosegregation

Figure 7 shows the fluid flow in the mush during the solidification of cast slab with the liquid core reduction. It is found that a backflow is pushed at the solidification end because of the enforcement of reduction. It can be explained that the center segregation is reduced because of homogenous of species caused by the backflow.

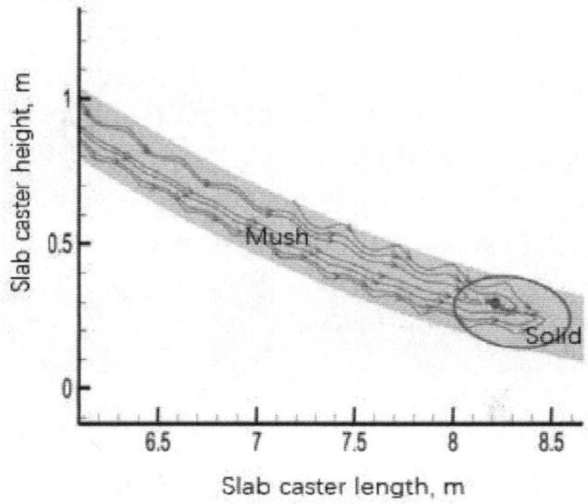

Figure 7 Backflow caused by the mush compression

3 Conclusion

The serious positive segregation is found at the center of continuous casting slab and the negative segregation is found near both sides of the thickness center, which is caused by the equiaxed sediment. The center segregation can be improved by the liquid core reduction because of the homogenous of species and the backflow induced by the mush compression.

References

[1] Flemings M C, Nereo G E. Macrosegregation I [J]. Transactions of the Metallurgical Society of AIME, 1967, 239(9): 1039-1053.

[2] Wu M H, Ludwig A. A three-phase model for mixed columnar-equiaxed solidification [J]. Metallurgical and Materials Transactions A-Physical Metallurgy and Materials Science, 2006, 37A(5): 1613-1631.

Determining Defects Formation on the Castings Surface by Computer Simulation of Casting Processes

Yury Nikalaichyk[1], *, Yevgeny Marukovich[2],*

1. Belarusian national technical university, 220013, Minsk, 65 Nezavisimosty ave., Republic of Belarus
2. Association of Foundrymen and Metallurgists of Belarus, 220013, Minsk, 65 24 Kolasa Str., Office 8M, Republic of Belarus

*Corresponding address: e-mail: yuni@bntu.by, maruko46@mail.ru

Abstract: Foundry production is one of the energies and resource intensive industries. Despite the undeniable advantages of foundry production, one of its disadvantages is the insufficiently high quality of the castings surface due to the possible formation of various surface defects such as burn-on, calcination, metal penetration, wrinkles and vitrification. Prevention of the casts surface defects on the surface of steel and cast-iron castings is carried out by applying special foundry (protective) coatings. Foundry coating is one of the most important parts in the foundry industry. Prediction of the casting surface defects, appearance and internal quality of castings depends on the correct choice of foundry coatings, which can be done due to analysis of the casting technology parameters and computer simulation of casting processes.

Keywords: castings surface; casting processes simulation; defects prediction

1 Introduction

Information about local overheating of the foundry mold during pouring and solidification may be used to predict casting surface imperfections and identify defect formation. The poured melt is cooled due to the foundry mold absorption of its physical heat (about 20%), heat of crystallization (70%) and phase transformations (1%-2%). Temperature of the foundry mold and its boundary layers and temperature in the contact zone "alloy – foundry mold" determine interaction processes between them, such as possible melt penetration, probability and kinetics of chemical reactions between melt oxides and material of the foundry mold and stress-strain state of the casting-mold system, including the occurrence of thermomechanical stresses.

2 Experimental procedure

The intensity of the aforementioned processes is the root cause of a diverse range of casting surface defects, including burn-on burn-in sand, metal penetration with and without chemical reactions. For example, according to opinions about the mechanism of burn-in formation, filtration of the melt is needed when the metallostatic pressure of the melt exceeds the capillary and gas back pressure of the foundry mold. At the same time, temperature in contact zone between alloy and boundary layers of foundry mold must be above a certain temperature or a critical temperature (T_{crit}). T_{crit} may be found from crystallization range of the alloy because it loses its ability to flow after the formation of the solid phase. It is rather close to the zero-fluidity temperature and can take on different meanings for various types of alloys. Alloys which crystallize in the form of a solid solution (austenitic steels) T_{crit} will close equal to the liquidus temperature, but for alloys which crystallize with a high eutectic content it will close to the solidus temperature. For alloys which crystallize with peritectic transformation, T_{crit} is recommended to take as an equal to half the sum of the liquidus temperature and the solidus temperature of the alloy. The critical temperature value for iron-based alloys can be determined based on the iron-cementite diagram.

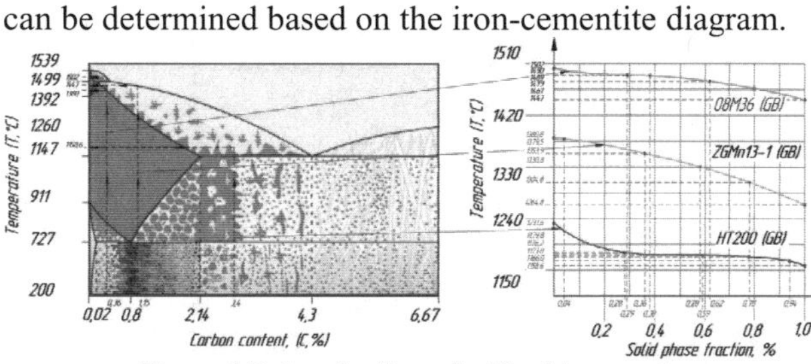

Figure 1 Determination of critical temperature

Figure 1 schematically shows the determination of the T_{crit} value for casting alloys which are widely used in the production of various castings such as steel 08M36 (GB), which has a peritectic transformation, austenitic steel ZGMn13-1 (GB) and cast iron HT200 (GB) which is crystallized with an amount of eutectic.

It is necessary to identify an area of the foundry mold and time intervals where the temperature exceeds the critical temperature in order to determine defect formation on the casting surface due to computer simulation. This approach to predict the formation of surface defects in castings can serve as the basis for the development of effective methods for their prevention, including through foundry coating application.

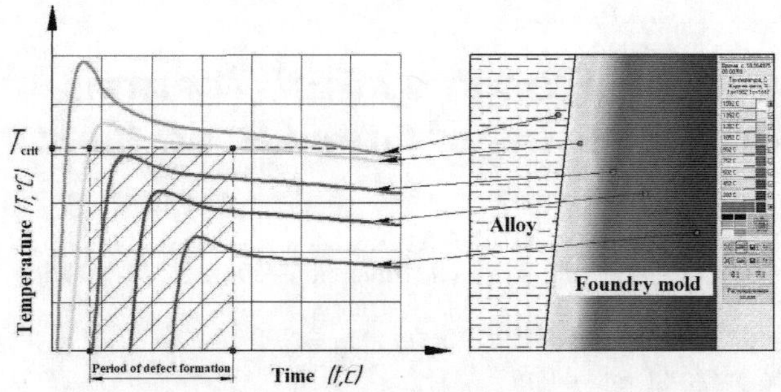

Figure 2 Temperature distribution in the contact zone "alloy-mold" and near-surface layers of the mold

Potential period of defect formation can be found by cooling curves, which describe melt temperature changes near the casting surface (Figure 2). Interval is situated between the points where the mold temperature is higher than T_{crit}.

3 Result and discussion

Computer simulation of casting processes was implemented by Polygon software. V-block [1, 2] were used for simulation, the shape of which is the inclined tetrahedral prism with a V-shaped through channel (Figure 3).

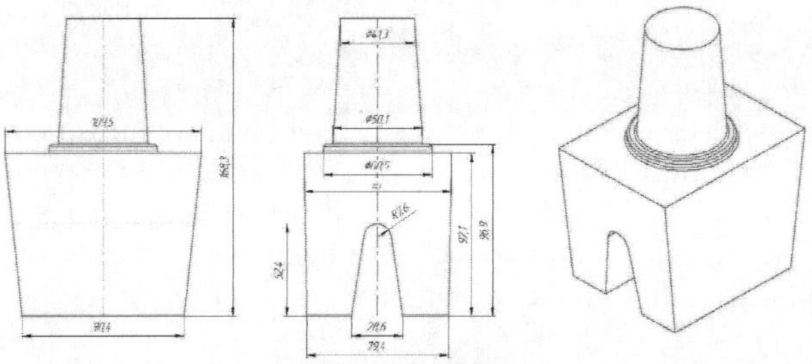
Figure 3 V-block

First of all, we simulated pouring processes with main goal to take into account the heating of the mold during the pouring period (Figure 4).

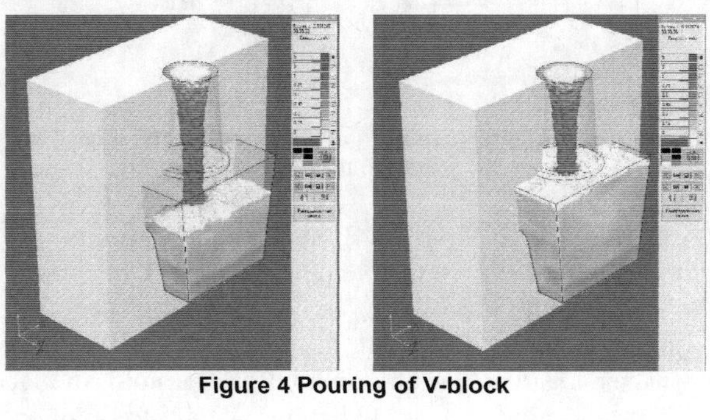
Figure 4 Pouring of V-block

It was established that filling occurs in 9.01 s for steel 08M36 (GB), 8.97 s for steel ZGMn13-1 (GB) and 8.99 s for cast iron HT200 (GB). Maximum temperature of foundry mold during this period is 284 °C, 260 °C and 240 °C.

Simulation of solidification processes made it possible to establish local areas of the casting mold and time intervals where the temperature exceeds T_{crit} (Figure 5).

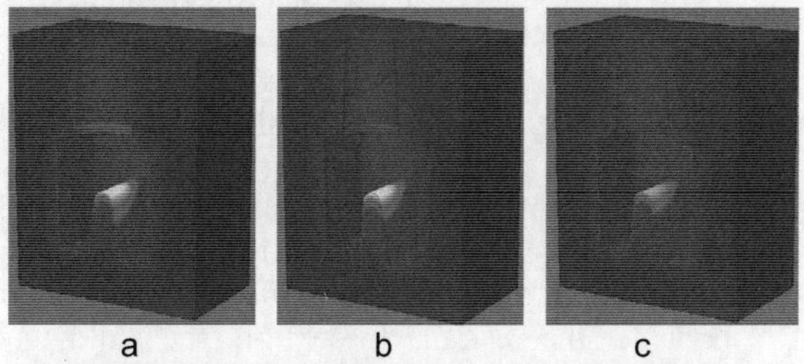
a b c

Figure 5 Temperature distribution in the foundry mold (a- steel 08M36, b - steel ZGMn13-1, c - cast iron HT200)

T_{crit} equal 1,475 °C for steel 08M36, time interval is 962.5 s, and the maximum surface temperature of the casting mold occurs in the area of the V-shaped protrusion and is 1,490 °C. For steel ZGMn13-1 this time interval is 716.6 s (T_{crit} = 1,360 °C), and the maximum surface temperature of the foundry mold also reaches into the area of the V-shaped protrusion and equal 1,369 °C. For cast iron HT200, the time interval for potential defect formation is 3,066 s (T_{crit} = 1,158 °C), and the maximum mold surface temperature (1,230 °C) reaches in the area of the V-shaped protrusion.

4 Conclusion

It is important to note that identified time intervals for defect formation and areas of the mold where the temperature exceeds T_{crit} can be used to perform thermodynamic analysis of chemical reactions between melt oxides and material of the foundry mold. A similar approach could be implemented to analyze the stress-strain state of the casting-mold system in order to determine areas of the casting mold where high-temperature stresses may arise, which could lead to deformations and cracking of the foundry mold.

References

[1] Richards V L, Lekakh S, and Kruse B. Burn-in/Burn-on: Coating Characterization and Designed Experiments. In Proceedings of the Steel Founders' Society of America 57th Annual Technical and Operating Conference, Chicago, IL, November 2004.

[2] Brooks B E, Beckermann C. Prediction in steel castings using simulation. In Proceedings of the SFSA Technical and Operating Conference, Steel Founders Society of America, Chicago 2006.

Prediction of Warpage Deformation of Casting Flange by FEM Thermal Stress Analysis Using Revised Drucker-Prager Model of Green Sand Mold

Yasuhiko Okimura[1], Takashi Hanai[2], Makoto Yoshida[1]
1. Department of Modern Mechanical Engineering,School of Creative Science and Engineering, Waseda University, #59-322, 3-4-1 Okubo, Shinjuku, Tokyo 169-8555, Japan
2. Osaki Works,Sintokogio, Ltd., 1-1, Kado, Osaki-cho, Toyokawa, Aichi 442-8515, Japan
*Corresponding address: e-mail: arco@asagi.waseda.jp

Abstract: This study experimentally measured the warpage deformation of an iron casting flange and tried to predict it by FEM thermal stress analysis during the casting cooling process. In the analysis, the revised Drucker-Prager model was applied as a mechanical model of the green sand mold to consider the restraint force from the mold to the casting undergoing thermal deformation. As a result of the comparison between the experimentally measured warpage and that calculated in the analysis, the calculated transition of the warpage during the cooling process qualitatively agrees with the experimentally measured value. The result suggests that the RDP model will significantly contribute to establishing a technique to predict the casting deformation by FEM thermal stress analysis.

Keywords:mechanical modeling; finite element method; deformation; green sand mold

1 Introduction

For large-scale products made by sand casting such as body components of large construction machines, the dimensional defect due to the warpage deformation during their cooling process is a serious issue. Prediction of the deformation by FEM thermal stress analysis at the designing stage has been attempted, however, the prediction accuracy has yet been enough for practical uses. Previous studies have developed the mechanical models required in the FEM thermal stress analysis to predict the casting deformation [1,2]. For a green sand mold, Okimura et al. [1] experimentally proposed and validated the revised Drucker-Prager (RDP) model to show that the calculated stress-strain curve with the RDP model agreed well with the experimentally obtained curve. The present study aimed to examine the accuracy of predicting the warpage deformation of the casting flange by the FEM thermal stress analysis with the RDP model. The experiment was first conducted to measure the warpage of the flange of the iron casting during its cooling process in the green sand mold. Subsequently, the calculated warpage of the flange in the FEM thermal stress analyses simulating the above experiment was compared to the experimentally measured value. The RDP model and a conventionally employed model were applied as a model of the green sand mold in the analyses to compare the accuracies of predicting the warpage deformation of the casting flange.

2 Procedure of experimental measurement

Figure 1 illustrates the experimental device to measure the warpage deformation of the casting flange. The casting was bar-shaped and had a square flange at one end. The steel grip held the opposite end to fix its displacement against the mold frame. The displacement of the flange was measured as the movement of the quartz rods with the laser displacement meters. The displacement was measured at three points: the center (z_{center}) and both edges (z_{edge}) of the flange, to calculate the warpage of the flange Δz by the following equation:

$$\Delta z = z_{center} - z_{edge} \quad (1)$$

The displacement at each point was continuously measured from the solidification of the casting until the mold removal. The cast alloy was gray cast iron (ASTM A48 class No. 45 equivalent) poured at 1,380 ℃. The mold was a green sand mold using silica sand and bentonite.

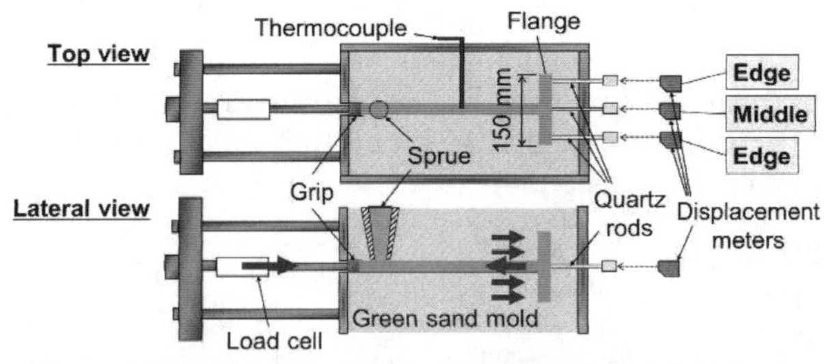

Figure 1 Schematic illustration of the experimental device to measure the warpage of the casting flange

3 Procedure of FEM thermal stress analyses

Figure 2 depicts the finite element model of the casting, mold, and grip for the analyses. The model creation and analyses were performed using Abaqus, the general-purpose FEM solver. As a constitutive model of the mold, the RDP and the modified Cam-Clay (MCC) models were applied to compare the prediction accuracies of the warpage by these two models. The latter model had been conventionally employed for green sand molds [3]. For the cast iron, the elasto-plastic-creep (EPC) model proposed by Motoyama et al. [2] was employed. The values of material constants to construct these models were experimentally determined by mechanical tests for each material.

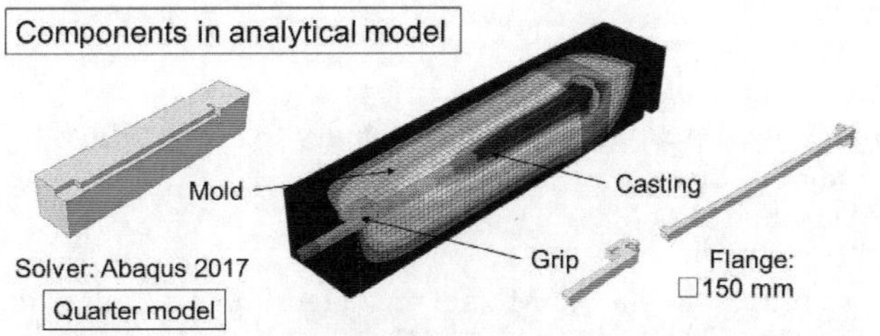

Figure 2 Finite element model of the casting, mold, and grip for the thermal stress analyses

4 Result and discussion

Experimental measurement of the flange warpage

Figure 3 shows the comparison of the flange warpage obtained in the experiment and the FEM analyses. From the transition of the warpage during the cooling process of the experiment, the change in the flange shape is schematically illustrated in Figure 4. The warpage once became negative during solidification and then positiveafter the casting finished solidification. This indicates that the flange once became arrowhead-shaped during solidification and then changed to Y-shaped. The warpage increased until the eutectoid transformation but was almost constant after that. At a temperature lower than the eutectoid point, the strength of the cast iron was probably much higher than the mold, and thus the flange shape was no longer affected by the restraint force from the mold.

Comparison of the experiment and FEM analyses

In Figure 3, comparing the FEM analyses to the experiment, the warpage with the conventional MCC model at the end of the cooling process (at 300 ℃) agreed well with the experimental value. However, the warpage with the MCC model continued increasing after the eutectoid transformation, which did not agree with the experiment. This result suggests that the MCC model failed to predict the strength relationship between the casting and the mold. On the other hand, the RDP model qualitatively predicted the decrease and increase in the warpage including the maintenance of the flange shape after the eutectoid transformation. The calculated warpage was smaller than the experiment presumably because the EPC model of the casting overestimated its strength, especially above the eutectoid temperature.When used with a model of the cast alloy in which the above fault was addressed, The RDP model would significantly contribute to a quantitative prediction of warpage deformation of castings.

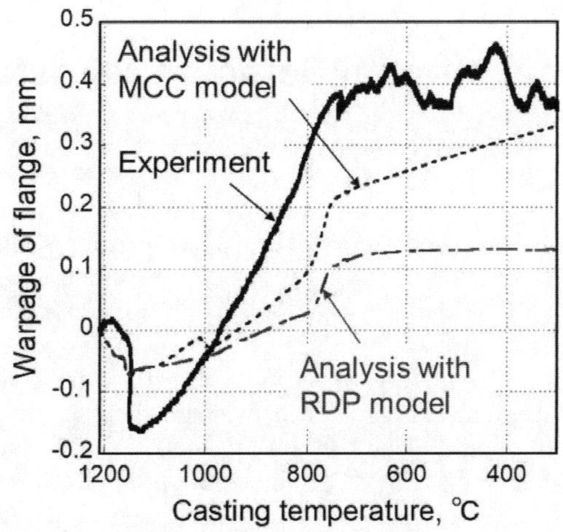

Figure 3 Comparison of the flange warpage obtained in the experiment and the FEM analyses

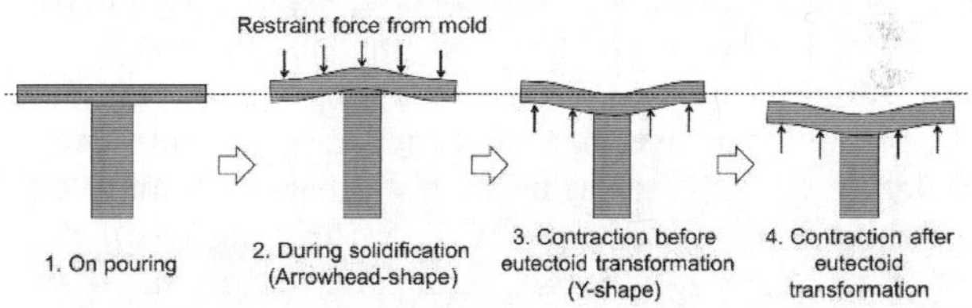

Figure 4 Schematic illustration of the transition in the flange shape during the casting cooling process

5 Conclusion

This study attempted to predict the casting warpage deformation by FEM thermal stress analysis with the RDP model for the green sand mold. The calculated warpage of the flange was compared to the experimentally measured value. As a result, the RDP model enabled a qualitative prediction of the warpage, which was not possible with the conventional model. Though the model for cast alloy would still have to be modified, the RDP model would contribute to realizing a quantitative prediction technique of the casting warpage deformation.

References

[1] Okimura Y, Kato Y, Makino H, Hashimoto K, Maeda Y, Okane T, Faiz M K, Komine H, Miyashita T and Yoshida M.Proposition of new yield criterion for green sand mold and its experimental validation by FEM stress analysis of triaxial compression test[J]. J.Mater. Process. Technol., 2024, 318: 118020.

[2] Motoyama Y, Shiga H, Sato T, Kambe H and Yoshida M. Elasto-plastic-creep constitutive equation of an Al-Si-Cu high-pressure die casting alloy for thermal stress analysis[J]. Metall. Mater. Trans. A, 2016, 47: 5598-5608.

[3]Ami Saada R, Bonnet G and Bouvard D. Thermomechanical behavior of casting sands: Experimental and elastoplastic modeling[J]. Int. J. Plast., 1996, 12(3): 273-294.

Typical Internal Defects Characterization in Castings Based on Rapid Total Focusing Method

Shangyu Liu[1], Han Yu[1], Xinyue Li[1], Zhida Jin[1], Xingjie Li[1,*]
National Key Laboratory of Advanced Casting Technologies, Shenyang Research Institute of Foundry Co., Ltd., CAM, Shenyang 110022, Liaoning, China
*Corresponding address: lixj@chinasrif.com

Abstract: The quality of castings directly affects their service life, and it is crucial to evaluate their quality by means of non-destructive testing (NDT). The total focusing method (TFM) has gained a lot of applications in many fields because of its higher sensitivity and visualisation of results. In order to achieve the application of this technology in casting quality inspection, this paper optimises both imaging speed and imaging quality. Firstly, by training the deep learning model, the complex calculation of high-resolution images is skipped, and high-resolution images are generated directly from low-resolution images to achieve fast characterisation of defects inside the castings by deep learning. And the characterisation results are compared with the DR inspection results to achieve fast and high-quality characterisation of typical defects inside the castings, which provides the basis for the application of this technology in casting quality inspection.

Keywords: castings; ultrasonic testing; TFM; deep learning

1 Introduction

Casting as the basic industry of manufacturing is widely used, with the development of technology, the structure is more complex, the working environment requires more stringent castings have become the focus of research and development. Such as aircraft engines in the magazine, hydraulic turbine runner, etc., the casting process is complex. In the casting process, due to stress concentration and other issues, it is inevitable to produce a variety of defects, affecting the quality of castings, which in turn reduces the service life of castings.

In the field of nondestructive testing, ultrasonic testing is often used in the detection of internal defects in castings because of its fast detection speed, low cost and other advantages. With the development of ultrasound technology, phased array ultrasound technology has a higher sensitivity, but also can achieve the visualisation of the detection results. The total focusing method (TFM) ultrasound detection is the use of a number of array elements containing a phased array ultrasound probe successive emission of ultrasound waves, to be detected in the area of detection, each array element and then successively receive these ultrasound data (FMC), and according to the TFM imaging algorithms through the post-processing of data by the way of the ultrasound detection technology. Compared with conventional phased array ultrasound, TFM ultrasound inspection can be post-processing algorithms to focus on any position (pixel point) of the imaging area, and thus can obtain higher detection sensitivity and resolution, while intuitive imaging can better characterise the defects in the inspection area.

At present, the TFM inspection technology has been widely used in bolt and weld inspection, but there are few applications in castings. The quality of TFM imaging directly depends on the number of pixel points, but each pixel points amplitude needs to be calculated by the TFM, and the more pixel points, the longer the algorithm calculation time. In addition, coarse grains inside the casting, poor surface condition, etc. will affect the effect of focusing imaging. Therefore, if the TFM is to be applied in casting inspection, it is necessary to solve the problems of imaging speed and imaging quality, i.e., to achieve rapid characterisation of typical defects inside the casting.

At present, the optimisation of the imaging time for the confocal algorithm is divided into several aspects. First, the wave number imaging algorithm can avoid the calculation of the ultrasonic propagation path in the algorithm, which can improve the imaging speed to a certain extent, but the calculation needs to be processed for three-dimensional data, and the memory load is too large, which makes it difficult to be applied in portable ultrasonic phased array equipment. Simplification of the FMC is also commonly used to speed up the method, such as triangular matrix, sparse matrix, etc., to reduce the data can improve the imaging speed, but a large number of experiments have also shown that the reduction of data for a specific detection need, the reduction of data will reduce the quality of the imaging, and it is difficult to meet the high quality of the detection needs. With the development of computer hardware, accelerated imaging methods through the use of GPUs, DSPs and FPGAs have been applied to TFM imaging, and powerful hardware can nearly shorten the imaging time of TFM. However, the cost of hardware is difficult to match with the development of increasing detection needs. In summary, many of the current methods have certain limitations for higher resolution TFM images. With the development of artificial intelligence, deep learning is widely used in the fields of data classification and image processing. Homin Song et.al combine simulation data and test data as a database to train deep learning models to achieve super-resolution imaging, but the results are only verified in the test, and are not applied in the actual industrial inspection [1].

In summary, this paper proposes a method based on deep learning to achieve fast high-resolution TFM imaging, for cast steel parts, cast titanium parts and other castings for the FMC acquisition, the construction of low-resolution images and high-resolution image pairs as a database, to achieve fast high-resolution imaging. And in this way, the typical defects inside the castings are characterised, and the results are compared with the DR inspection results, so as to realise the accurate characterisation of the defects inside the castings by TFM. It provides a theoretical basis for the application of total focus technology in casting inspection.

2 Experimental procedure

The research idea of this paper is shown in Fig. 1. Firstly, the FMC containing defects are acquired, and the traditional TFM is used to generate images with different resolutions (from low resolution to high resolution). Then the low resolution image and the high resolution images as a set of image pairs, which is used to train the deep learning model. Finally, the low resolution image is calculated by the traditional focusing algorithm, and then the high resolution image is generated by the trained deep learning model to achieve fast focusing high-resolution imaging. In this paper, a 32-array phased array probe is used for data acquisition of cast steel and titanium parts containing typical defects such as porosity and porosity, with a confocal imaging area of 40 mm × 50 mm. after fast high-resolution imaging is achieved, a phased array probe equipped with an encoder is used to perform a confocal B-scan and C-scan on the casting and the C-scan image is compared with the DR image, so as to realise the characterisation of the casting's typical defects.

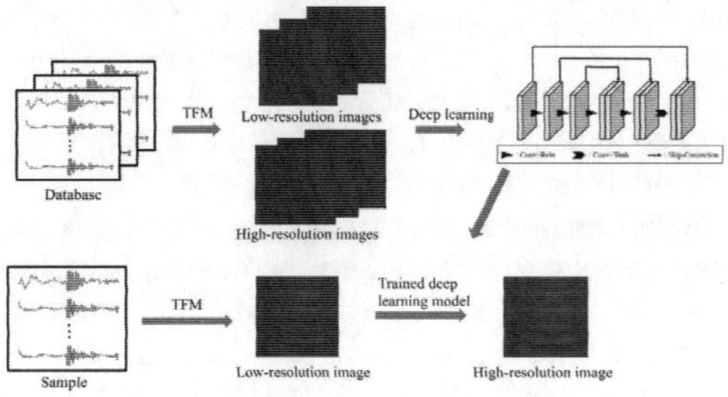

Figure 1 Technical route

3 Result and discussion

(1) Database and model

The same FMC were calculated using the TFM for64×64 pixel, 128×128 pixel, 256×256 pixel, 512×512 pixel, and 1024×1024 pixel, respectively. It can be found that the low-resolution imaging time is only 0.04 s, and when the resolution reaches 512×512 pixel, the single picture imaging time is 7.28 s, which is already difficult to meet the needs of industrial inspection.

In addition, as shown in Table 1 for the similarity between different resolution images, were selected PSNR, SSIM, Lpips three parameters (in order to ensure that the image size is the same, the low-resolution image using linear interpolation with 1024 × 1024 pixel for comparison). It can be found that as the resolution increases, the higher the similarity with the 1024×1024 pixel image, indicating the availability of the database.

Table 1. Comparison of image similarity at different resolutions

	64*64	128*128	256*256	512*512	1024*1024
Time(s)	0.04	0.12	0.62	7.28	28.63
PSNR	33.49	35.25	38.52	42.62	/
SSIM	0.8646	0.8719	0.9194	0.9646	/
Lpips	0.6257	0.5108	0.3228	0.1788	/

(2) Characterization of typical defects inside castings

In order to advance the application of TFM in casting inspection, the technology was used to characterise internal defects in castings. A 64-array probe with a mechanical encoder is used to sweep the casting, showing both the TFM B-scan and the TFM C-scan. As shown in Fig. 2, comparing the detection results with the DR results, typical defects such as shrinkage and porosity can be clearly distinguished in the C-scan. Moreover, the depth of defects can be determined in the B-scan, which cannot be achieved by DR.

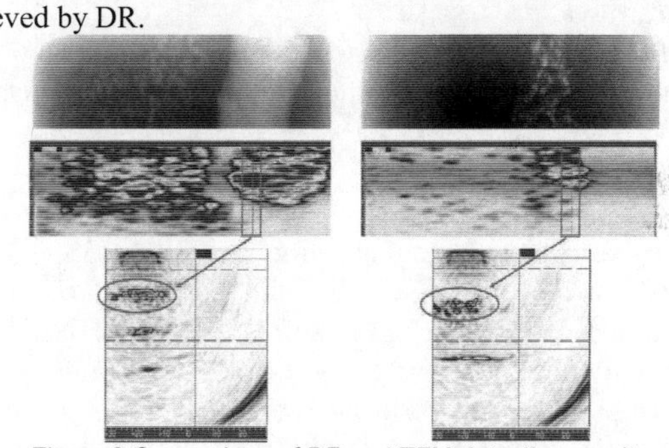

Figure 2 Comparison of DR and TFM detection results

4 Conclusion

In this paper, we construct a database containing images of different resolutions by means of deep learning to optimise the generation time of high-resolution images generated by the traditional TFM, which, compared with the 1024×1024 pixel generated by the TFM, can reach PSNR of about 39, SSIM of about 90, and the imaging time using the deep learning model is only 0.05s, which significantly improves the imaging speed. After realising the fast imaging, TFM B-scan and C-scan inspections were carried out on the castings, and the typical defects such as shrinkage and porosity in the castings were similar to the results of the DR inspection, and the TFM can provide the depth of the defects, which provides a solid theoretical basis for the application of the technology on the castings.

Reference

[1] Song H M, Yang Y C. Super-resolution visualization of subwavelength defects via deep learning-enhanced ultrasonic beamforming: A proof-of-principle study. NDT and E International, 2020, 116: 102344.

A Casting Retrieval Method Based on Cavity and External Shape Features

Shuren Guo, Chuhao Zhou, Beibei Li, Huatang Cao, Xuanpu Dong*
State Key Laboratory of Materials Processing and Die & Mould Technology, School of Materials Science and Engineering, Huazhong University of Science and Technology, Wuhan Hubei 430074, China
*Corresponding address: e-mail: dongxp@hust.edu.cn

Abstract: Model retrieval is one of the key technologies in parts management for the reuse of casting process and design. However, current retrieval schemes usually treat parts as a whole, thereby neglecting the impact of part cavities on functionality and manufacturing. This study has investigated an encoding method based on the cavity features of parts. The external shape feature and cavity features are extracted through cavity filling and connected domain algorithms. These features are integrated with the volume and modulus to enable a formulation of casting feature coding. Then a retrieval system based on model input was created, utilizing a specified coding similarity calculation algorithm. Experiments show that it is possible to obtain the desired results within the top 2 retrieval results. This significant enhancement had substantially improved both the efficiency and accuracy of the retrieval process.

Keywords: casting retrieval; cavity features

1 Introduction

3D CAD models have been the basic form of data for modern advanced manufacturing. Efficient 3D model retrieval is instrumental in the domain of parts management. Moreover, it is also beneficial for the design of new parts and new processes for manufacturing [1]. The content-based model retrieval, which utilizes 3D models as the input for retrieval, is favorably conducive to the parts management. It is a research area of significant interests within the domain of model retrieval. The key of content-based retrieval methods lies in the computation of model features through feature extraction algorithms, utilizing the similarity between features to represent the similarity between models. For instance, content-based retrieval methods represent 3D models by employing techniques such as surface distance statistics [2], reflective symmetry [3], topological structures [4], skeletal structures [5], and maximum inscribed sphere decomposition graphs [6], among others to calculate the similarity of models. These descriptive methods treat the model as an indivisible entity. Nevertheless, in casting production, the internal cavities of components are typically formed by sand cores, and the external shape are formed by the molds. Both the cavities and the shape are of equivalent significance in the function and the manufacturing process of parts. Therefore, in the study, an approach has been introduced for casting retrieval by synthesizing characteristics of both the cavities and external shapes.

2 Experimental procedure

Casting Feature Encoding

Considering various factors that influence the casting process, the encoding structure of the casting's process features is depicted in Figure 1. It contains modulus, volume, and structural code with external shapes and cavities of parts.

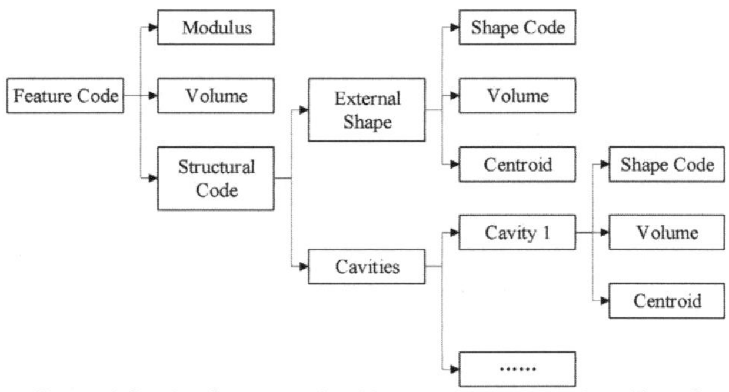

Figure 1 Casting feature code with external shape and cavities of model

To compute the feature encoding of the casting, the initial step involves the extraction of the cavity from the model. In this methodology, the model is normalized by the pose-normalization method [7] and projected onto planes along the three principal axes – X, Y, and Z. Then, a statical analysis is conducted to determine the maximum and minimum values for each point along the principal axes within the projection. Subsequently, the interior of the 3D model is filled with the intermediate content between these maximum and minimum values. Finally, the cavities in the model are obtained by performing a difference operation between the filled model and the original model as shown in Figure 2. Although the cavities extracted is composed of several non-adjacent entities, these cavities are marked as a whole, as shown in Figure 2(c).

Each entity within the cavity and the external shape entity devoid of cavities are subjected to the computation of their respective shape feature coding, volume and centroid position. These computations are then systematically organized in accordance with the structure delineated in Figure 1, thereby yielding the feature encoding of the casting.

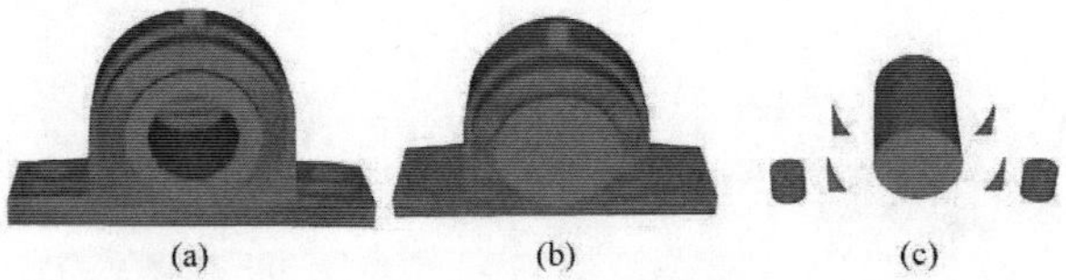

Figure 2 The model and cavities of a casting: (a) original model; (b) the filled model; (c) the cavities

Similarity Calculation

Another key aspect of model retrieval is to compute the similarity among encodings to reflect the similarity between models. In the study, KL divergence was utilized to calculate the similarity between two distinct entities, as shown in Equation (1).

$$s = \frac{1}{1+distance}$$
$$distance = \sum_{i=1}^{N} p(x_i) log\left(\frac{p(x_i)}{q(x_i)}\right) \quad (1)$$

For multiple cavity entities of parts, the similarity of entities is weighted by distance and volume. Here, volume and centroid distance represent the influence of the cavity on the part and the confidence level of the similarity. The structurally similar parts obtained ultimately need to be further screened based on volume and modulus, as these factors greatly influence the casting process design of parts.

Dataset experiments

Based on the Engineering Shape Benchmark dataset [8], the study marked 400 parts with similar levels of each other for the verification of retrieval effectiveness. The performance of a retrieval system is generally evaluated through expected reciprocal rank (ERR) in Equation (2).

$$ERR = \sum_{r=1}^{n} \frac{1}{r} PP_r = \sum_{r=1}^{n} \frac{1}{r}\left(\prod_{i=1}^{r-1}(1-R_i)\right)R_r$$
$$R(g) = \frac{2^g-1}{2_k}, g \in 0, 1, 2, \cdots, n-1 \quad (2)$$

where g is the similarity level of parts, k is the count of similarity level, and n is the number of retrieved parts.

3 Result and discussion

Figure 3 illustrates the impact of different volume and modulus similarities thresholds on ERR. It can be observed that as the similarity threshold increases, the performance metrics of the retrieval system improve because it eliminates more cases. However, an excessively high similarity threshold may lead to the erroneous elimination of reasonable similar parts. A reasonable similarity threshold needs to be further determined. The ERR of the retrieval results is always above 0.68, indicating that an ideal part outcome can be expected in the top 2 retrieval parts.

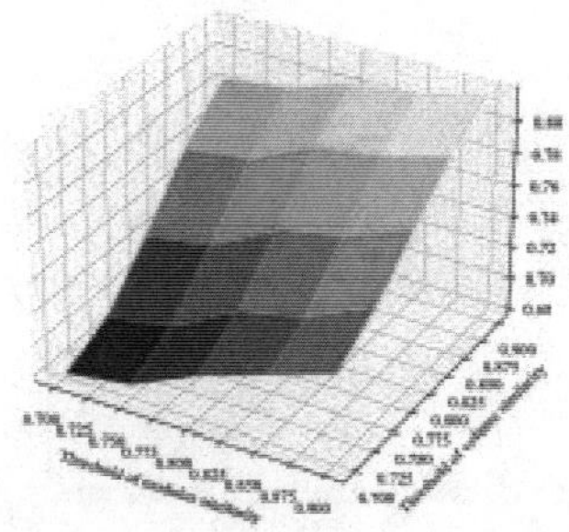
Figure 3 Influence of the similarity threshold of volume and modulus on the ERR

4 Conclusion

The study has constructed a similarity system for castings. Based on the 3D model, a cavity extraction algorithm was designed, and a method for encoding that integrated casting features and external shape features was proposed. A method for calculating the similarity of castings was proposed, and the similarity retrieval of castings was realized, lying a technical foundation for the establishment of a casting process database and intelligent design of casting processes.

References

[1] Guo S R, Zhou C H, Dong X P, et al. A transformation method for gating system designing between similar castings [J]. Journal of Manufacturing Processes, 2024, 112: 82-91.

[2] Osada R, Funkhouser T, Chazelle B, et al. Shape distributions [J]. ACM Transactions on Graphics, 2002, 21(4): 807-832

[3] Kazhdan M, Chazelle B, Dobkin D, et al. A reflective symmetry descriptor for 3D models [J]. Algorithmica, 2004, 38(1): 201-225.

[4] McWherter D, Peabody M, Regli W C, et al. Transformation invariant shape similarity comparison of solid models. In: ASME 2001 International Design Engineering Technical Conferences and Computers and Information in Engineering Conferenc [J]. American Society of Mechanical Engineers Digital Collection, 2020: 303-312.

[5] Sundar H, Silver D, Gagvani N, et al. Skeleton based shape matching and retrieval. In: 2003 Shape Modeling International [J]. 2003: 130-139.

[6] Chen J, Ilieş H T. Maximal disjoint ball decompositions for shape modeling and analysis [J]. Computer-Aided Design, 2020, 126: 102850.

[7] Guo S R, Dong X P, Xiang D, et al. A pose-normalization method for casting voxel models using second-order central moment matrix [J]. IEEE Access, 2023, 11: 36849-36855.

[8] Jayanti S, Kalyanaraman Y, Iyer N, et al. Developing an engineering shape benchmark for CAD models [J]. Computer-Aided Design, 2006, 38(9): 939-953.

Data-Driven Prediction and Process Optimization of Deformation in Ti-alloy Casing

_XiaoyuanJi[1], *,PengYu1,XiwangQie[1,2], Yajun Yin[1], MeijuanZhang[1], XuShen[1], Wen Li1, Hai Nan[2], Jianxin Zhou[1]
1.State Key Laboratory of Materials Processing and Die & Mould Technology, Huazhong University of Science and Technology, Wuhan 430074, China
2. Beijing Aerospace Materials Research Institute Co., Ltd, Beijing 100094, China
*Corresponding address:e-mail: jixiaoyuan@hust.edu.cn

Abstract: With the increasing demands of high-tech weaponry in the field of national defense on the yield and performance of parts, Ti-alloy casings are evolving towards more complex structures, such as thin-walled, large wall thickness differences, multi-crossing, rings, and blind holes. The complex structures make it extremely difficult to accurately predict and control deformation of casings during casting. Therefore, a data-driven deformation prediction and process optimization technique for Ti-alloy castings is proposed. First, a multi-source dataset is constructed. By building InteCast ERP system, the full-process quality big data of 54 casings in single-piece production investment casting, including molding, pouring, and heat treatment process parameters, are collected. The post-pouring and post-heat treatment shape point cloud data are obtained via 3D scanning, constructing a dataset with 21 types and 9,720 key dimensions, including distance, diameter, contour, and wall thickness. Second, data preprocessing is performed. Abnormal fluctuating data are removed, missing data are interpolated using regression and clustering algorithms, parameters are reduced by correlation and principal component analysis, and the processed parameters are normalized. Third, deformation prediction models are built and tuned, using pouring temperature, mold temperature, vacuum degree, heat treatment temperature and other key process parameters as decision variables to establish post-pouring and post-heat treatment dimensional deformation prediction models based on neural networks. Hyperparameters (1,0.1,0.01,0.001, etc.) and optimization algorithms (SGD, Adam, AdaGrad, etc.) are combined to optimize the models. The optimal model with minimum mean squared error is chosen, achieving prediction accuracy over 80%. Fourth, the multi-objective particle swarm algorithm (MOPSO) is used to realize the process parameter recommendation. The aim is to reduce the dimensional deformation of casing. The sum of the positive deviations of the predicted and theoretical values of all key parameters after heat treatment was taken as the objective function, and the upper and lower bounds of the decision variables were set to 105% of the maximum value and 95% of the minimum value of each parameter for actual, respectively. The goal of predicting dimensional deformation with an average value of less than 1 mm for 50 sets of recommended process parameters was achieved. This study proposes a data-driven technology based on accurate prediction of Ti-alloy casings dimensions and process parameter recommendation for multiple critical dimensional deformations, which realizes effective control of casings deformation.

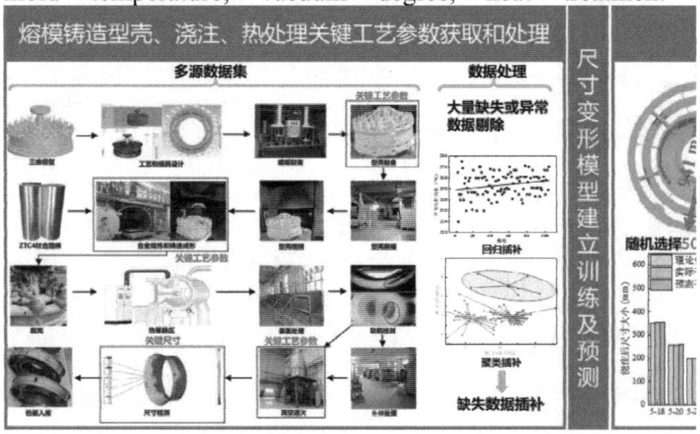

Figure 1 Investment casting parameter acquisition, model training, prediction and optimizing

Keywords: data-drive; Ti-alloy casing; dimensional deformation; prediction model; multi-objective optimization

Funds: National Key Research and Development Program (2020YFB1710100), National Natural Science Foundation (52090042, 52275337, 51905188).

Achievement Awards in 2021 (ranked No.1), the first prize of Hubei Provincial Scientific and Technological Progress Award in 2022 and other first prizes at the provincial and ministerial levels. Guided the Challenge Cup Special Competition in 2021 to win the national special prize, and was personally awarded the Most Beautiful Science and Technology Worker of the National Foundry Industry in 2022 and other prizes. He serves as the executive secretary-general of Hubei Foundry Society, deputy secretary-general of China Foundry Association Mould Branch, senior member of China Mechanical Engineering Society, youth editorial board member of China Foundry, Special Casting Non-ferrous Alloys, and Foundry Technology.

Casting-RCG and GRImage-Net: Generation of Casting Flaw Detection Image and Defect Intelligent Detection

Mingjun Hou, Hao Dong, Xiaoyuan Ji*, Qinyang Li, Chuhao Wu, Shuohong Li, Yajun Yin, Jianxin Zhou
State Key Laboratory of Materials Processing and Die & Mould Technology, Huazhong University of Science and Technology, Wuhan 430074, China
*Corresponding address: e-mail: jixiaoyuan@hust.edu.cn

Abstract: The automatic defect detection of casting X-ray inspection images based on deep learning has the problems of single defect sample topography and poor network generalization. Therefore, the research on casting inspection image generation and multi-source image deep feature fusion defect detection is carried out. Firstly, a defect image simulation generation model based on RCG (Casting-RCG) was constructed, and a quasi-unconditional generation framework was built according to the defect features, which realized the high-quality generation of multiple defects such as shrinkage holes in Casting inspection images. Secondly, a defect localization and classification network based on deep feature fusion of generated images and real images (GRImage-Net) was created, and ablation experiments were carried out to verify the performance of the network. The results show that GRImage-Net based on Casting-RCG can significantly improve the accuracy, precision and recall rate of defect detection. It has better generalization than the single feature input model. The Casting-RCG and GRImage-Net created in this paper provide a new way to enrich the defect morphology and improve the generalization of the network, which has important theoretical guidance and engineering application value.

Keywords: casting; defect detection; flaw detection image; generation; deep feature fusion

1 Introduction

X-ray defect detection of castings is an important means to detect the internal defects of castings [1]. The traditional manual detection method is inefficient and easily affected by subjective factors [2]. In recent years, deep learning technology has made significant progress in the field of image processing, and automatic detection methods based on deep learning are gradually applied to defect detection of X-ray inspection images of castings [3]. However, there are two main problems with existing methods:

Single defect sample topography: Due to the difficulty of obtaining defect samples and the limited types of defect topography in the existing data set, the trained model does not perform well in the face of new types of defects.

Poor network generalization: The model performs well on the training data, but its performance is unstable on the test data or real-world applications, and it is difficult to generalize the application.

In order to solve these problems, this study proposes a defect detection method combining defect image simulation generation and multi-source image deep feature fusion, aiming to improve the detection ability and generalization of the model.

2 Experimental procedure

(1) as shown in Figure 1. The defect image simulation generation model (Casting-RCG) is constructed. In order to generate high-quality defect images with various morphologies, a defect image generation model based on RCG (Random Conditional Generative) is constructed. The model is implemented by the following steps.

Data collection and preprocessing: The real casting X-ray inspection images are collected and preprocessed, such as denoising and normalization.

defect feature extraction: Convolutional Neural Network (CNN) was used to extract the features of different types of defects and construct a feature library.

Simulation generation model construction: Based on the RCG framework, combined with the extracted defect features, the defect images of various morphologies are generated. RCG model achieves high-quality simulation generation of different defect morphologies by introducing random noise and conditional constraints.

(2) Build a fault localization and classification network (GRImage-Net). In the defect detection stage, a defect localization and classification network (GRImage-Net) based on deep feature fusion of generated images and real images is established. The specific steps are as follows.

Feature extraction network: Design a deep convolutional neural network for extracting deep features of real and generated images.

Feature fusion module: The extracted deep features are fused, including feature stitching and attention mechanism, to enhance the feature representation ability.

Defect localization and classification: through multi-layer convolution and pooling operation, the precise localization and classification of defects are realized.

(3) Ablation experiments were carried out for verification. To verify the performance of GRImage-Net, we design the following ablation experiments.

Single feature input model: only real images or generated images are used for training and testing.

Multi-source image feature fusion model: combining the

deep features of real images and generated images for training and testing.

Comparison of feature fusion strategies: The effects of different feature fusion strategies (such as simple concatenation, weighted fusion, and attention mechanism) are compared.

Figure 1 Framework of this paper

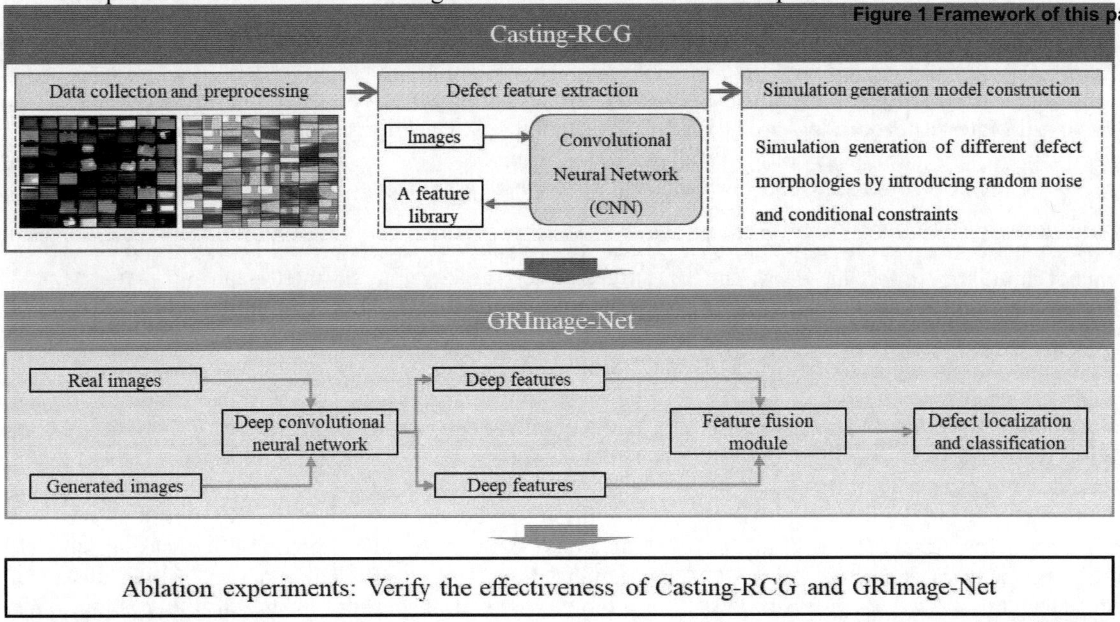

3 Result and discussion

The model is expected to achieve the following:

Defect image generation quality: Casting-RCG is able to generate high-quality defect images of various morphologies, and these generated images are visually very close to the real defect images, providing rich samples for network training.

Defect Detection Performance: GRImage-Net is expected to achieve significant gains in accuracy, precision, and recall for defect detection compared to models trained using only real images.

Generalization: GRImage-Net is expected to perform better than traditional models on different datasets, and the network may have better generalization and be able to maintain high detection performance in different detection tasks.

4 Conclusion

In this paper, a defect image simulation generation model based on RCG (Casting-RCG) and a defect detection network based on deep feature fusion of generated images and real images (GRImage-Net) are proposed, which can effectively solve the problems of single defect sample topography and poor network generalization. The experimental results are expected to be realized: high-quality defect images generated by Casting-RCG provide rich samples for network training, and GRImage-Net significantly improves the performance and generalization of defect detection through multi-source image deep feature fusion. The research in this paper provides a new method and idea for casting X-ray flaw detection image defect detection, which has important theoretical and practical application value.

5 Acknowledgments

This research was funded by the National Natural Science Foundation of China (Nos. 52275337, 52090042, 51905188), and the National Key R&D Program of China (2020YFB1710100).

References

[1] Wu B, Zhou J, Yang H, Huang Z, Ji X, Peng D, Yin Y, and Shen X. An ameliorated deep dense convolutional neural network for accurate recognition of casting defects in X-ray images. Knowl.-Based Syst., 2021, 226: 107096.

[2] Wu B, Zhou J, Ji X, Yin Y, and Shen X. Research on approaches for computer aided detection of casting defects in X-ray images with feature engineering and machine learning. Proc. Manuf., 2019, 37: 394–401.

[3] Ji X, Yan Q, Huang D, Wu B, Xu X, Zhang A, Liao G, Zhou J, and Wu M. Filtered selective search and evenly distributed convolutional neural networks for casting defects recognition. J. Mater. Process. Technol., 2021, 292: 117064.

3D Positioning Method of Defects in Light Alloy Castings Based on Multi-Angle X-Ray Detection Images

Hao Dong, Mingjun Hou, Xiaoyuan Ji*, Jianxin Zhou
State Key Laboratory of Materials Processing and Die & Mould Technology, Huazhong University of Science and Technology, Wuhan 430074, China
*Corresponding address: e-mail: jixiaoyuan@hust.edu.cn

Abstract: The two-dimensional X-ray inspection images of light alloy castings cannot intuitively reflect the depth and size of defects, and the correspondence between image defect coordinates and the three-dimensional body of the casting is unclear. Therefore, manual repeated drilling and inspection are needed to determine defect locations, resulting in low-quality and inefficient welding repairs. To address these problems, we propose a three-dimensional defect positioning method for light alloy castings based on multi-angle X-ray inspection images. First, we collect images of the castings at different inspection angles, and integrate image processing methods such as filtering to extract the contours of defect areas in the images. Next, we establish the relative positions and angles between the radiation source, casting, and imaging plate to construct a coordinate upscaling model from the image to the three-dimensional model, achieving the conversion from two-dimensional image defect coordinates to three-dimensional casting defect coordinates. Finally, we calculate the intersections between the rays and the casting model, establish boundary feature polyhedrons, and develop a defect morphology reconstruction algorithm to achieve three-dimensional defect positioning based on multi-angle inspection images. This method achieves three-dimensional defect positioning of castings by simulating actual DR inspections and has significant engineering application value.

Keywords: defect three-dimensional positioning; multi-angle defect detection; three-dimensional model; X-ray inspection images

1 Introduction

After X-ray inspection imaging of light alloy castings, it is necessary for personnel to manually judge the defect positions on the three-dimensional castings based on the defect information in the DR images. This often requires drilling holes in the castings for positioning and re-imaging. Because the correspondence between the two-dimensional DR images and the three-dimensional castings is unclear, this method often damages the casting

body during defect positioning, requiring subsequent welding repairs. Therefore, relying on multiple manual drillings to position defects not only damages the casting body but also severely restricts the quality and efficiency of casting defect welding repairs. Therefore, there is an urgent need to develop an intelligent three-dimensional defect positioning method for light alloy castings.

2 Experimental procedure

(1) The grayscale value of an X-ray inspection image can be calculated based on the light intensity after the X-rays penetrate the medium [1]. The light intensity is related to the attenuation coefficient of the medium and the penetration thickness. Based on this principle, we develop a virtual imaging system that, after inputting the casting model, casting attenuation coefficient, and inspection distance, outputs a virtual inspection image of the casting. We generate holes in the casting model to simulate actual defects, and then put the model into the virtual imaging system to obtain a virtual inspection image with defects. Next, we process the image with median filtering, binarization, and contour searching to obtain a set of two-dimensional coordinates for the defect contours on the virtual inspection image.

(2) In actual inspection, the three-dimensional positions of the radiation source, casting, and imaging plate are correlated [2]. When simulating actual inspection, the generated virtual inspection images are located on the imaging plate. Therefore, the two-dimensional coordinate set of the defect contours needs to be converted into a three-dimensional coordinate set in the world coordinate system. Based on the angular relationship between the radiation source, casting, and imaging plate, the two-dimensional coordinates can be converted into two components in the world coordinate system. The third component can be calculated based on the distance from the imaging plate to the origin of the world coordinate system.

(3) Connect the radiation source to each defect contour point on the imaging plate, and calculate the intersection points of each line with the casting. Use the Delaunay triangulation method to generate a polyhedron that includes all the intersection points. After changing the inspection direction, repeat the above steps. Intersect the two generated polyhedrons to obtain a characteristic polyhedron that represents the defect location, as shown in Figure 1.

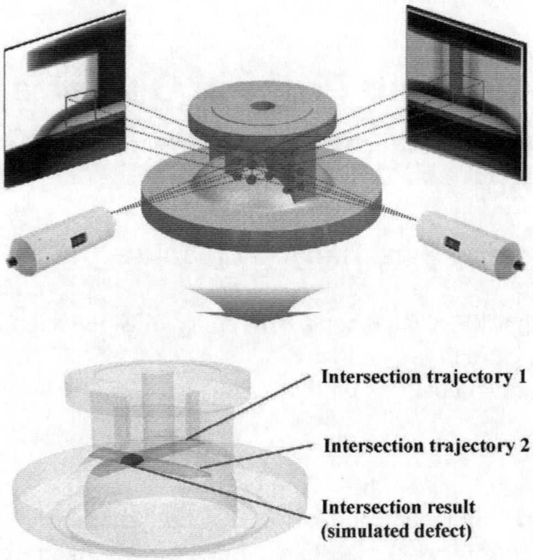

Figure 1 Defect three-dimensional positioning

3 Result and discussion

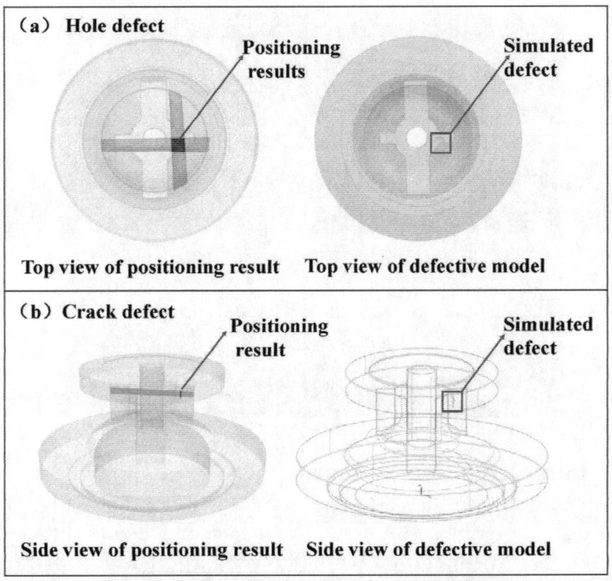

Figure 2 Positioning result comparison

We simulate the three-dimensional positioning of two types of defects: Group A consists of hole defects, and Group B consists of crack defects. The comparison between the defect localization results and the original model is shown in Figure 2. To quantitatively demonstrate the accuracy of defect positioning, we calculate the centroid coordinates of the characteristic polyhedron, the centroid coordinates of the simulated defects in the model, the model dimensions, and the relative positioning error. As shown in Table 1, the three-dimensional defect positioning method based on multi-angle X-ray inspection images achieves a relative positioning error of less than 2% in the *X*, *Y*, and *Z* directions, indicating a high accuracy of defect positioning. In practical applications, the virtual inspection images can be replaced with actual inspection images of the casting, and by inputting the three-dimensional positions of the radiation source, casting, and imaging plate, the three-dimensional positioning results of the defects can be obtained.

Table 1. Defect positioning result quantification table
(CG means the center of gravity)

Object	*X*	*Y*	*Z*
Model size	81.77	81.77	43.81
CG of Defect A	-2.145	-8.776	19.264
CG of Result A	-1.858	-9.085	20.014
Relative error	**0.35%**	**0.38%**	**1.71%**
CG of Defect B	-1.040	10.392	32.471
CG of Result B	-0.828	10.216	32.679
Relative error	**0.26%**	**0.22%**	**0.47%**

4 Conclusion

We propose a three-dimensional defect positioning method for light alloy castings based on multi-angle X-ray inspection images. First, we generate multi-angle virtual inspection images of the casting model using a virtual imaging system, and obtain a set of two-dimensional coordinates for the defect contours on the virtual inspection images. Next, we convert the two-dimensional coordinate set into a three-dimensional coordinate set in the world coordinate system. Finally, we develop a defect morphology reconstruction algorithm to achieve three-dimensional defect positioning based on multi-angle inspection images. The results show that the relative error in three-dimensional defect positioning is less than 2%, indicating that this method has significant engineering application value.

5 Acknowledgments

This research was funded by the National Natural Science Foundation of China (Nos. 52275337, 52090042, 51905188), and the National Key R&D Program of China (2020YFB1710100).

References

[1] Tillack G R, Nockemann C, and Bellon C. X-ray modeling for industrial applications. NDT & E International, 2000, 33: 481-488.

[2] Chen L, Li B, Zhang L, and Shang Z. 3D positioning of defects for gas turbine blades based on digital radiographic projective imaging. NDT & E International, 2023, 133: 102751.

Three-Dimensional Quantitative Characterization and Process Optimization for Simulating Shrinkage Defects of Titanium Alloy Aeroengine Casing in Investment Casting

Tao Sun, Peng Yu, Xiaoyuan Ji *, Wenhao Zhou, Xuzheng Pan, Tianyou Li, Yajun Yin, Jianxin Zhou
State Key Laboratory of Materials Processing and Die & Mould Technology, Huazhong University of Science and Technology, Wuhan 430074, China
*Corresponding address:e-mail: jixiaoyuan@hust.edu.cn

Abstract: A three-dimensional characterization system of shrinkage defects in numerical simulation and a multi-objective optimization method are proposed for titanium alloy aeroengine casing in investment casting using the traditional optimization methods of subjective evaluation and trial-and-error method. The method extracts the three-dimensional quantitative characterization system of shrinkage defects in the criterion of size, location and shape based on the numerical simulation results of titanium alloy aeroengine casing, and establishes a prediction model based on the three-dimensional feature values of shrinkage obtained from multiple sets of numerical simulations through a machine learning algorithm. Using the characteristic prediction model of size and location as the objective function, the second-generation non-dominated sorting genetic algorithm (NSGA-II) is employed to perform multi-objective optimization of the casting process parameters, obtaining a craft window for shrinkage defects in numerical simulation.

Keywords: titanium alloy casing; defect characterization; machine learning; multi-objective optimization

1 Introduction

In the actual production process, the casting and solidification process of titanium alloy castings are in a closed and opaque environment (as shown in Figure 1), making it difficult to observe the solidification process. It relies on the experience of previous production batches and difficult to evaluate shrinkage defects quantitatively. The casting process of titanium alloy castings requires the three-dimensional quantitative characterization system of shrinkage defects of numerical simulation and the optimization of process parameters using intelligent optimization algorithms. In terms of quantitative characterization of shrinkage defects, ASTM (American Society for Testing and Materials) uses the ASTM E155-2005 standard to classify various defects in castings, with the defect level based on the volume, maximum diameter, depth of influence, and cross-sectional thickness of the defect area [1]. Wenbo Yu from Beijing University of Technology used the volume of shrinkage defects, sphericity, defect band width, and average external solidification crystal size to characterize shrinkage defects [2]. Shantanu Shahane from the University of Illinois at Urbana-Champaign used the NSGA-II algorithm to solve multi-objective optimization problems in low-pressure die casting [3].

Figure 1 Solidification process of titanium alloy castings

2 Characterization System

The initial results of numerical simulation of the solidification process of aeroengine casing using InteCast CAE post treatment software show a large number of tiny and clustered shrinkage defects. These defects are analyzed using by DBSCAN (Density-Based Spatial Clustering of Applications with Noise) clustering algorithm.

Firstly, a quantitative characterization system is established for individual shrinkage defects after clustering, based on the criteria of "size, location, and shape". At the size level, the characteristic values are volume and minimum enclosing sphere diameter. For the location level, two characteristic values are used to represent the defect's belonging region and depth, with the casting region divided into internal and external areas. At the shape level, the sphericity of the shrinkage defect is defined, which is the ratio of the diameter of a sphere with the same volume as the defect to the diameter of the minimum enclosing sphere, serving as the characteristic value.

Subsequently, a quantitative characterization system targeting three criteria layers is established for the overall casting defects. On the size criteria layer, four feature values are selected: the total volume of casting shrinkage defects, the sum of the minimum bounding sphere diameters, the number of large shrinkage defects, and the number of large-diameter defects. On the location criteria layer, two feature values are chosen: the proportion of internal region shrinkage volume and the weighted average depth based on volume. On the shape criteria layer, the weighted average sphericity is used as the feature value.

3 Multi-objective optimization

138 groups of different process parameters were numerically simulated. The value ranges of the three key process parameters, namely pouring temperature, pouring time, mold temperature, and mold thickness, were 1,680-1,800 ℃, 4-7 s, 230-290 ℃. Three-dimensional feature values were extracted from the simulation results of porosity defects. Finally, a dataset of 138 groups of process parameters and simulated porosity defect three-dimensional feature values was obtained as sample data for machine learning, with a structure of three input variables and six output variables.

Through the dataset, separate second-order and third-order polynomial stepwise regression models were established for each individual feature value. The root mean square error (RMSE) and coefficient of determination (R^2) were compared between the second-order polynomial model and the third-order polynomial model, and the optimal regression analysis model was obtained for each dataset.

For six different input variables, six single-output neural networks were designed to predict six feature values, respectively. Each neural network has a basic structure of [3-X-1]. The training sample dataset consisted of 138 groups, which is a small sample size. Therefore, for each training, the samples were randomly divided into training, validation, and test sets in a ratio of 60%, 20%, and 20%, respectively. A single hidden layer neural network was chosen. For the selection of hidden layer nodes, an empirical formula was used to determine the range of hidden layer nodes as 3-10. The activation function used was the tanh function.

The overall data's mean squared error (MSE), correlation coefficient R, were selected as evaluation metrics. A comparison was made among 11 training algorithms and different hidden layer network structures to obtain the optimal model. The results showed that the models trained using the Bayesian regularization algorithm outperformed other algorithms in all prediction models.For all models, the fitting accuracy of the BP neural network model was higher than that of the regression model. Therefore, the prediction models used in this paper were all based on neural network prediction models.

Table1. Optimal BP neural network model parameters

Feature values	Hidden layer nodes	MSE	R
total volume	7	0.075	0.989
sphere diameters	6	255.6	0.959
number of large	7	0.282	0.881
number of large-diameter	4	0.521	0.795
proportion of internal region	5	0.0015	0.808
average depth	10	21.5	0.890

The feature values on the same criterion layer were normalized and then unified through linear weighting. The constraint ranges of the three process parameters were the same as the data set ranges. Using the NSGA-II algorithm, the pouring temperature, mold temperature, and pouring time obtained from the Pareto front within the constraint ranges were around 1,680 ℃, 290 ℃, and 5 s, respectively.

3 Conclusion

This paper addresses the optimization of casting process parameters for porosity defects in titanium alloy engine casing castings. A three-dimensional quantitative characterization system is established, and multiple groups of process parameter simulation results are collected. A machine learning prediction model is built, and a genetic multi-objective optimization algorithm is used to search for the Pareto front solution set, resulting in a process window with pouring temperatures around 1,680 °C, mold temperatures around 290 °C, and pouring times around 5 s.

Acknowledgments

This work was financially supported by the National Key R&D Program (2020YFB1710100), the Science and Industry Commission Project (2019XXX.XX4007Tm), and the National Natural Science Foundation of China (52275337, 51905188, 52090042).

References

[1] ASTM E155-2005, ASTM E155-2005 Standard Reference Radiographs for Inspection of Aluminum and Magnesium Castings [S].

[2] Diao Y P, Yan L C, Gao K W. A strategy assisted machine learning to process multi-objective optimization for improving mechanical properties of carbon steels [J]. Journal of Materials Science & Technology, 2022, 109: 86–93

[3] Shahane S, Aluru N, Ferreira P, et al. Optimization of solidification in die casting using numerical simulations and machine learning [J]. Journal of Manufacturing Processes, 2020, 51: 130–141.

Design and Development of a System for Optimizing Boundary Conditions and Thermophysical Parameters in Casting Simulation

XuewenQian, Qi Zhang, XuShen,Xiaoyuan Ji,YajunYin*,Wen Li,JianxinZhou
State Key Laboratory of Materials Processing and Die & Mould Technology, Huazhong University of Science and Technology, Wuhan 430074, China
*Corresponding address:e-mail: yinyajun436@hust.edu.cn

Abstract: Casting process numerical simulation can predict casting defects and performance,improving the efficiency of casting product process design and optimization. However, the accuracy of casting process numerical simulation results is affected by boundary condition parameters and material thermophysical parameters. Unreliable boundary condition parameters and material thermophysical parameters can lead to deviations between the calculated results of casting process numerical simulation and actual situations. To obtain reliable boundary condition parameters and material thermophysical parameters, this paper aims to design and develop a casting simulation boundary condition/ thermophysical parameter optimization system to improve the accuracy of casting process numerical simulation. Firstly, the overall requirements analysis and overall structural design of the parameter optimization system are carried out. Secondly, according to the requirements analysis, the design and construction of the system hardware are carried out, including the design of a standardized temperature acquisition model and the construction of temperature acquisition equipment. Then, according to the requirements analysis, the development of system software is carried out, including temperature acquisition control software and a parameter optimization calculation platform. Finally, experiments verify that the optimized boundary conditions and thermophysical parameters obtained by the system significantly improve the agreement between the simulation results and actual castings, proving the effectiveness of the system.

Keywords: numerical simulation;boundary condition; thermophysical prameter;designs and development of system

1 Introduction

The numerical simulation technology of casting process can achieve the prediction of casting defects and performance in computers, providing strong basis for casting structure design and casting process design, greatly improving the quality of casting products and the efficiency of casting process optimization. However, the boundary condition parameters and material thermal properties parameters used in casting numerical simulation calculations often differ significantly from the actual values, greatly reducing the reliability of casting numerical simulation results. In practical applications, experimental measurement and manual reverse calculation methods are generally used to obtain reliable values of corresponding parameters, but there are drawbacks such as limited applicability and high empirical dependence. Therefore, this article designs and develops a casting simulation boundary condition/thermal property parameter optimization system.

2 Experimental Procedure

The casting simulation boundary condition/thermal property parameter optimization system built in this article aims to achieve fast and simple optimization of relevant target parameters by measuring the temperature curve of the standardized temperature acquisition model, combined with the parameter optimization algorithm based on multi strategy improved particle swarm optimization [1]. The main requirements of the system are as follows:①Design a standardized temperature measurement model to facilitate the conduct of temperature measurement experiments;②Build temperature data collection equipment to achieve stable and accurate collection of temperature data for different types of materials; ③ Implement temperature data analysis function, which can process the collected temperature data and simulated temperature data, analyze and extract valuable information, such as the characteristics of casting cooling curves in each temperature curve;④Implement task setting and optimization calculation functions, capable of setting parameters for casting simulation boundary conditions, thermal properties, optimization algorithm control parameters, and other parameters.

The developed casting simulation boundary condition/ thermal property parameter optimization system includes four parts: ① Standardized temperature acquisition model [2], which is based on the thermal analysis sample cup as a reference, as shown in Figure 1(a). Its upper and lower surfaces are both square, with a height of 60mm. Two temperature acquisition points are set, one at the geometric center of the lower bottom surface of the model and the other at the sand mold 10mm away from the edge of the bottom surface of the model The integrated temperature acquisition device, as shown in Figure 1(b), adopts the cDAO9174 data acquisition system, a temperature data acquisition module of model 9211, an industrial control microcomputer with anti-interference and vibration resistance, an AC-DC power module with a rated power of 150W, and a box made of thickened aluminum alloy sheet Develop temperature acquisition and control software based on LabVIEW platform, as shown in Figure 1(c). ④An intelligent parameter optimization calculation platform has been developed based on the Qt framework, as shown in Figure 1(d). It has the functions of setting target curves, setting calculation tasks, parameter optimization calculation, and displaying calculation results.

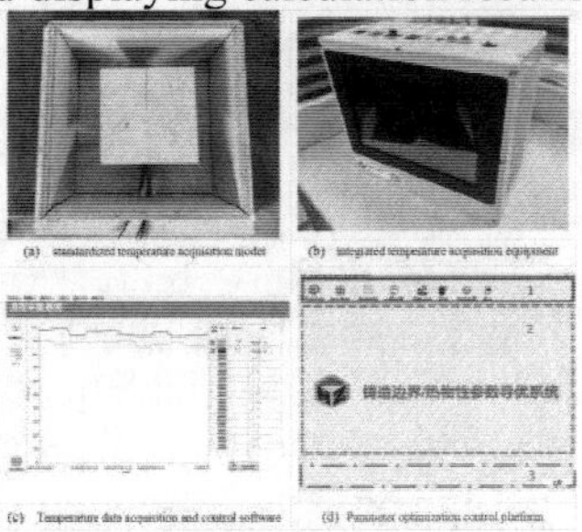

Figure 1Casting simulation boundary condition/thermophysical parameters optimization system

Conduct parameter optimization testing using a standard temperature acquisition model, with the alloy material grade of ZG24Mn6. Based on the calculation results of Huazhu CAE database and JMatPro software, the initial values of simulation boundary conditions/thermal properties parameters used in numerical simulation were determined. The initial parameters were simulated and calculated. The comparison between the simulated temperature curve and the measured temperature curve is shown in Figure 2(a). Using the optimized parameters for further simulation calculations, the results are shown in Figure 2(b).

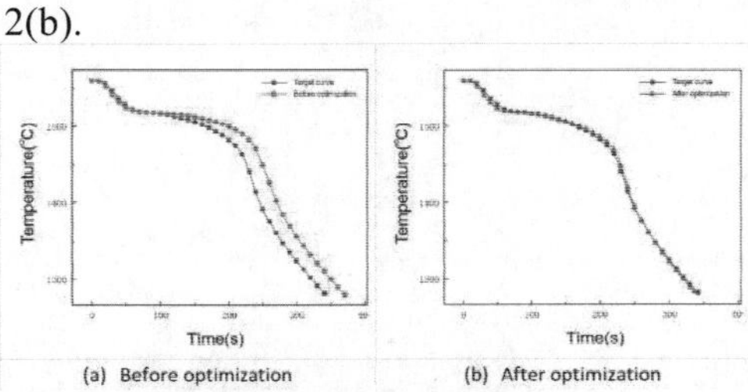

Figure 2 Comparison between simulated temperature curve and measured temperature curve

To verify the application effect of optimized parameters in numerical simulation, a test piece was designed for pouring test, and the pouring results of the test piece are shown in Figure 3.

3 Result and Discussion

Using the parameters before and after optimization, numerical simulation calculations were conducted on the test piece. Comparing the two simulation results with the actual pouring results, it can be concluded that the calculation results using the parameters before optimization were poor, only simulating the defects in the left cylindrical structure of the test piece, and there were significant differences in the overall shape, type, and actual situation of the defects, as shown in Figure 4(a); The calculation results of optimized parameters were used to simulate the defects at the root of the right cylindrical structure, and the shape, type, and actual situation of the defects in the left cylindrical structure are highly similar, as shown in Figure 4(b).

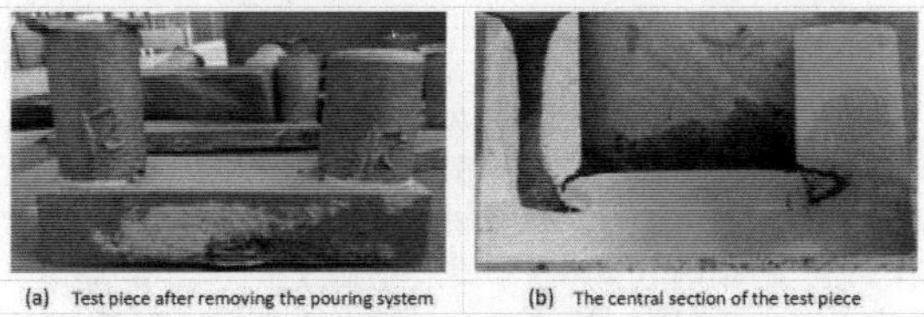

Figure 3Test piece and its defects

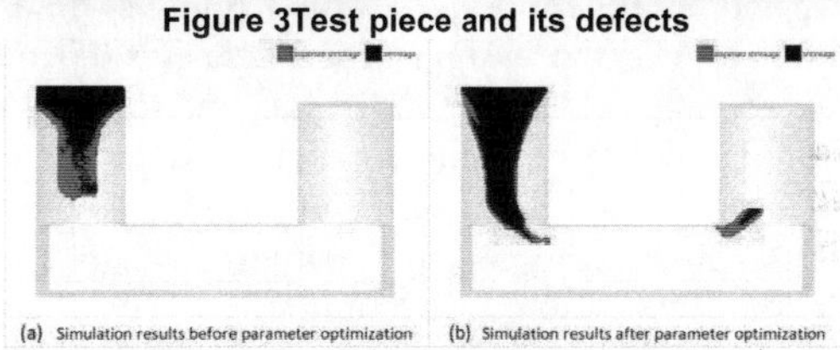

Figure 4Simulation results of test pieces before and after parameter optimization

4 Conclusions

(1) Developed a casting simulation boundary condition/thermal property parameter optimization system that combines software and hardware. The system includes a standardized temperature acquisition model, temperature acquisition equipment, temperature acquisition control software, and parameter optimization calculation platform.

(2) The experimental results show that this system can achieve the optimization calculation of boundary condition parameters and thermal properties parameters in casting simulation, greatly improving the accuracy of casting numerical simulation.

Acknowledgment

This work was financially supported by the National Key Research and Development Program of China (No. 2020YFB1710100).

References

[1] Zhang Q, Shen X, Yin Y J, et al. Optimization of casting thermophysical parameters based on cooling curve characteristics and particle swarm optimization algorithm [J]. Special Casting and Nonferrous Alloy, 2024,44 (2): 165-169. DOI: 10.15980/j.tzzz.2024.02-004.

[2] Foundry Branch of China Mechanical Engineering Society Casting Handbook Volume 5 Casting Process (Third Edition). Beijing: Machinery Industry Press, 2011.

Intelligent Design and Optimization of Feeding System inCastings Based on Reinforcement Learning

LechuanLi, YajunYin*, Xiaoyuan Ji, Xu Shen, Wen Li, Jianxin Zhou
State Key Laboratory of Materials Processing and Die & Mould Technology, Huazhong University of Science and Technology, Wuhan 430074, China
*Corresponding address:e-mail: yinyajun436@hust.edu.cn(Yajun Yin)

Abstract: The feeding system can effectively reduce the shrinkage porosity in the design of casting process and then improve the casting quality and process yield. therefore, the automatic and intelligent design of the feeding system plays an essential role in the casting process design of complex structural castings. In this paper, an intelligent design method of feeding system for complex structural castings based on reinforcement learning is proposed. By combining simulation with reinforcement learning model,a novel design method with high quality and efficiency of feeding system is realized. Firstly, the adaptive design method of the initial process of feeding system based on temperature gradient descent is established to provide the required environment for reinforcement learning model. Secondly, The tabular form Q-learning method in reinforcement learning is adopted, and the initial process design is regarded as the environment, and the return is obtained from the results of the solidification and heat transfer simulation. The specific size of the riser is optimized by selecting the optimal action and state in the optimization process of each step, and the above steps are iterated repeatedly until the termination condition is met. Thirdly, the influence of changes in different model parameters on the optimization results is discussed by modifying the Q-learning model, and the model parameters are updated to determine the optimal reinforcement learning model suitable for this environment. Finally, the model is tested by intelligent design of feeding system for several groups of complex structure castings. The test results show that the model has a good effect on a variety of complex structure castings, which proves the validity and reliability of this model.

Keywords: feeding system; reinforcement learning; solidification simulation; process optimization

1 Introduction

The design of feedingsystem is very essential in the casting process.The feeding system can affect the casting filling solidification process, thus affecting the feeding effect and the distribution of defects inside the casting.Especially for castings with complex structure, the design of the feeding system is extremely difficult, which makes the casting defects difficult to effectively eliminate, thus affecting the quality and service performance of the castings.In the design of feeding system, the key problem to be solved is to find out the internal relationship between feeding process and casting defect distribution.At present, many methods have been devoted to solving this internal relationship,mainly including modular method, geometric heat knot method, and numerical simulation based test design.

2 Experimental procedure

(1)An adaptive generating model for initial process of feeding system based on temperature gradient descent is proposed.Firstly, the solidification heat transfer process of the casting itself was simulated to obtain the temperature change law at different times in the casting and the temperature gradient at different positions.Secondly, all possible locations of the feeding system are obtained by the geometric surface discrimination method, and the effectiveness screening is carried out.Finally, the initial process of the feeding system is generated(as shown inFigure 1).

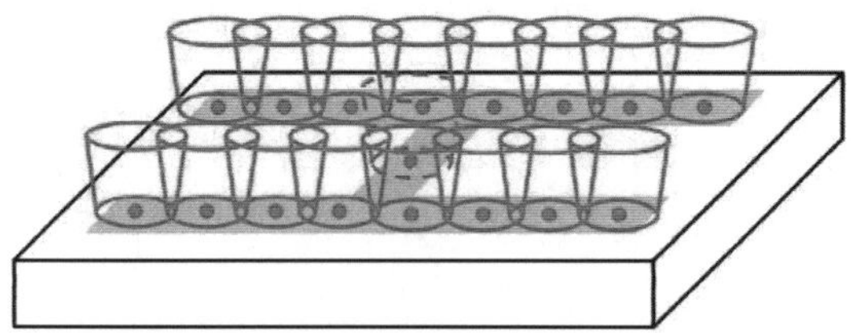

Figure 1 Riser design pointsgenerated randomly

(2)An automatic position optimization method of feeding system based on temperature gradient descent is proposed.In the heat transfer process of casting solidification, the relatively low temperature part of the casting is the last part of solidification.Therefore, the final solidification position can be found by the method of temperature gradient descent, and the feeding system can be set here.The temperature field calculation results generated in the first step can be used to optimize the distribution position of randomly generated feeding system according to the direction of temperature gradient decline, so as to achieve the purpose of automatic design of feeding system(as shown in Figure 2).

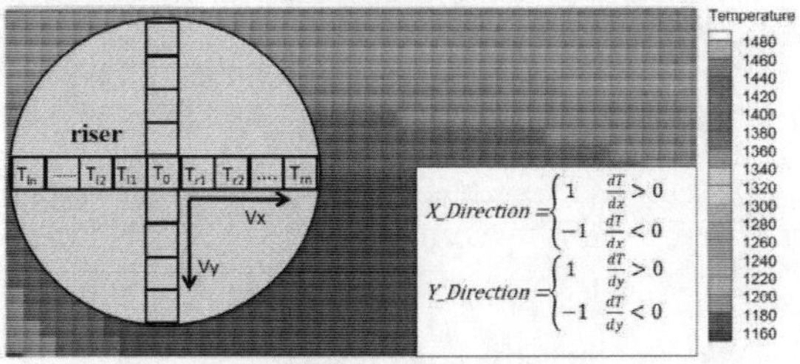

Figure2 Optimization of riser position based on temperature gradientdescent

(3)After completing the initial riser design scheme constructed above, we can use the above design scheme as the environment to build a reinforcement learning model (as shown in Figure 3).The reinforcement learning method adopted in this paper is Q-learning method.Q-learning is a temporal difference method used to solve problems based on Markov decision process (MDP).The core idea of Q-learning method is to optimize the decision-making strategy of the agent by constantly updating the Q-value function.Specifically, Q-Learning uses the Bellman equation to update the Q-value function, which states that the value of Q in the current state is equal to the current reward plus a discount on the maximum value of Q in the future.

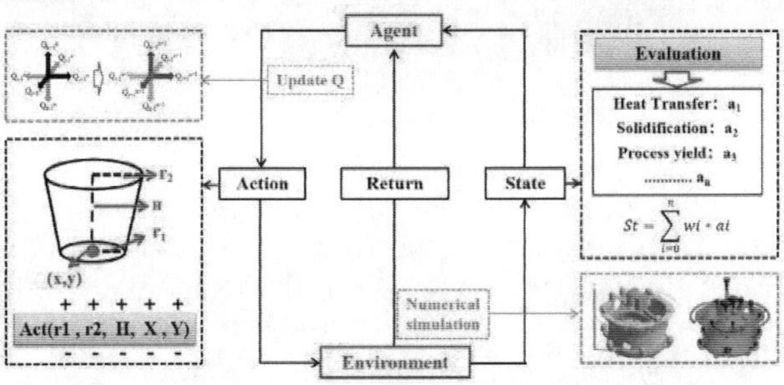

Figure 3 Reinforcement learning methodology framework

3 Result and Discussion

A typical casting with complex structure is selected for testing. Firstly, the surface of the three-dimensional model of the casting is analyzed to find out the area where the riser can be designed. Secondly, the temperature field simulation is carried out for the casting.Finally The designing results are obtained based on reinforcement learning (as shown in Figure 4).As can be seen from Figure 4, risers are concentrated in the region with the highest casting temperature and the final solidification, and the higher the temperature, the larger the risers are, which is conducive to eliminating shrinkage holes. Therefore, the design method of the model is consistent with the experience of the casting process.

In order to test the effect of the feeding system, solidification simulation is used, and the defect results showed that the shrinkage holes could be effectively eliminated(as shown in Figure 5).

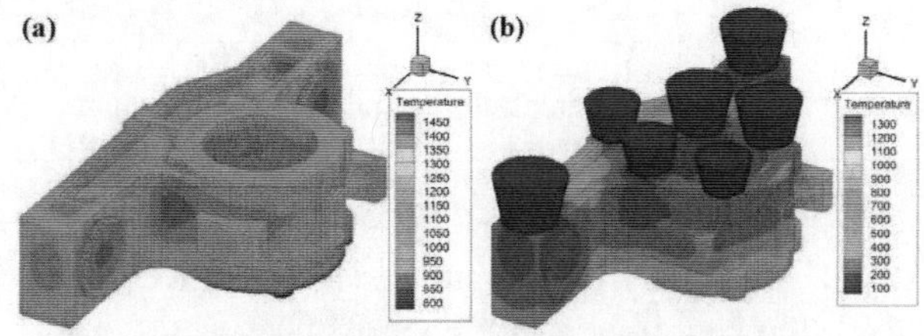

Figure 4 The results ofriser designing: (a) simulation results of temperature field; (b) riser designing results

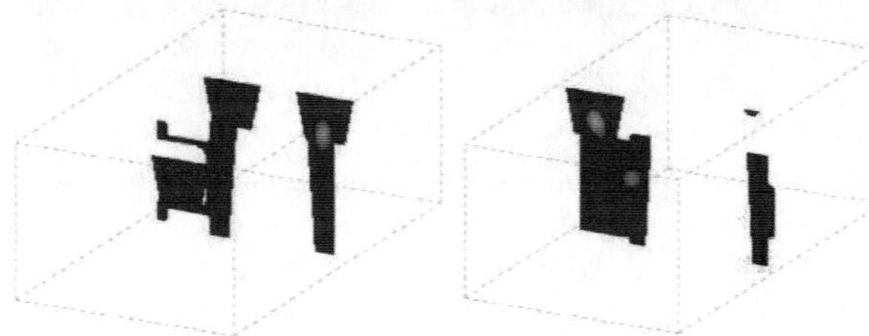

Figure 5 Simulation of shrinkage based on the designed feeding system

4 Conclusions

(1) The design of riser based on structure identification and temperature field simulation is realized, and the feeding system is effectively designed without the assist of manual experience.

(2) The designed feeding system can eliminate the shrinkage and porosity of castings well, which proves that the design model in this paper is effective.

5 Acknowledgment

This work is financially supported by the National Key Research and Development Program of China (No. 2022YFB3706800).

References

[1] Gong X, Liao D, Chen T, et al. Optimization of steel casting feeding system based on BP neural network and genetic algorithm[J]. China Foundry, 2016, 13(3): 182-190.

[2] Zhou J, Wang M, Yin Y, et al. Feed paths and hot spots computation based on a time gradient method in casting[J]. The International Journal of Advanced Manufacturing Technology, 2017, 93: 261-272.

[3] Wang T, Yin Y, Zhou J, et al. Optimal riser design method based on geometric reasoning method and fruit fly optimization algorithm in CAD[J]. The International Journal of Advanced Manufacturing Technology, 2018, 96(1-4): 53-65.

[4] Zhang C, Fu Y, Wang H, et al. Multi-objective optimization of process parameters during low-pressure die casting of AZ91D magnesium alloy wheel castings[J]. China Foundry, 2018, 15(5): 327-332.

Study on the Forming Process of Prefabricated Reinforced Concrete Thin-Walled Molds in Complex Cavities

Yingtao Yang[1, 2], Yajun Yin[1], Xu Shen[1], Xiaoyuan Ji[1], Jianxin Zhou[1], *
1. School of Materials Science and Engineering, Huazhong University of Science and Technology, Wuhan 430074, China
2. CHINA RAILWAY 11TH BUREAU GROUP CORPORATION LIMITED, Wuhan 430061, China
*Corresponding address: e-mail: zhoujianxin@hust.edu.cn

Abstract: This study investigates the pouring process and quality control of PRC thin-walled permanent formworks with complex cavities. Computational Fluid Dynamics (CFD) simulates the pouring process, L-shaped box flowability tests calibrate simulation parameters. Simulations focus on an actual bridge project, analyzing flow characteristics and potential defects. Criteria for pouring PRC thin-walled molds are established, along with auxiliary pouring methods. This study enhances understanding and optimization of PRC formwork, improving pouring uniformity and quality assurance for construction applications.

Keywords: PRC thin-wall molds; concrete forming process, computational fluid dynamics (CFD); L-shaped box test; forming quality prediction.

1 Introduction

The emergence of prefabricated reinforced concrete (PRC) thin-walled permanent molds, utilizing core-shell separation, marks a significant advancement in prefabrication technology. These molds, tailored for high-load components, excel in transportation and lifting. PRC molds serve as formwork for pouring concrete and enhance overall structural load-bearing capacity, gaining attention in civil engineering.

In practical engineering, designing complex cavities in thin-walled structures poses challenges for concrete pouring due to limited flowability and dense reinforcement. To tackle this, the paper utilizes CFD with the Volume of Fluid (VOF) method to simulate Ultra-High Performance Concrete (UHPC) flow. Comparing simulation with L-shaped box tests, it determines H-B model parameters for UHPC flow. Lastly, using a real cantilever cap beam project, the study simulates UHPC pouring into PRC thin-walled molds, examining flow characteristics and predicting quality defects.

2 Experimental procedure

The shape and dimensions of the L-shaped box are illustrated in Figure 1. During the test, the L-shaped box is placed on a level, solid surface, and the movable gate is closed. The inside of the mold is wetted with water and any excess water is wiped away. The vertical section of the apparatus is then filled with a concrete sample. After allowing it to rest for one minute, the movable gate is lifted, and the time it takes for the concrete to flow through the designated points is recorded. During the test, the time for the concrete to reach the furthest point and the flow patterns at different times are recorded. The experiment utilized two types of concrete materials: normal strength concrete (NC) and UHPC.

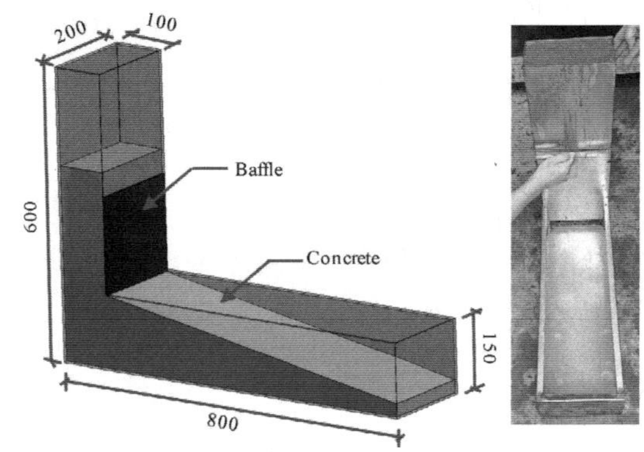

Figure 1 L-shaped box dimensions (unit: mm)

3 Numerical simulation procedure

Research shows UHPC shares shear thickening traits with self-compacting concrete [1, 2]. To simulate this, the Herschel-Bulkley (H-B) rheological model, with yield stress, offers superior accuracy over the Bingham model [3, 4]. Employing a pressure-based, transient, implicit segregated solver, the simulation sets a reference pressure point outside the model at standard atmospheric pressure. Gravity is considered in the y-direction, while thermal exchange is omitted. In the multiphase material, air and concrete are represented, with concrete characterized by the H-B model. Boundary conditions include a pressure inlet with zero relative pressure and a pressure outlet to aid convergence during backflow. Flow equations employ second-order upwind discretization, with pressure-velocity coupling via the SIMPLE algorithm. Momentum equations use first-order upwind discretization, and fluid interface tracking employs the Geo-Reconstruct scheme.

4 Result and discussion

L-shaped box tests

L-shaped box tests were conducted separately for NC and UHPC. The NC could not flow autonomously and required placement on a vibration table to complete the L-box flow test. In contrast, UHPC exhibited excellent flowability, flowing smoothly upon the opening of the movable gate and reaching the marked position (600mm from the gate) within 10.10 seconds. As a result, this study focuses solely on simulating the flow of UHPC, as shown in Figure 2, the number 1 represents air, and the number 0 represents concrete, and any cross-section consists of these two materials. During the test, the time taken for the UHPC to reach the furthest point and the flow patterns at different times were recorded. The results demonstrate that CFD simulation can effectively replicate the pouring process.

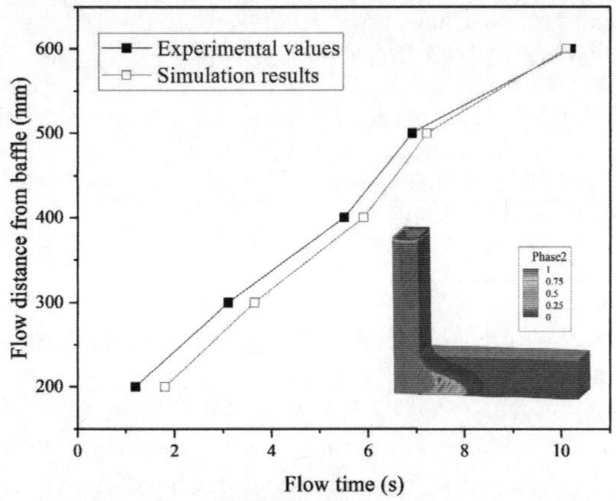

Figure 2 Comparison between UHPC L-shaped box test and simulation

PRC thin-walled molds

The thin-walled cap beam shell utilizes a three-segment prefabricated scheme, as shown in Figure 3(a). Its overall dimensions are 38.2 m in length and 4m in width, with symmetrical sections on both sides. The left and right segments are each 13.7 m in length, while the middle segment is 10.8 m, and the maximum depth is 5.5 m. The thickness of the bottom slab is 0.3 m, and the thickness of the web is 0.28 m.

The CFD model employs simulation parameters consistent with those of the L-shaped box test. The pouring inlet and outlet are illustrated in Figure 3(b), where the blue arrow represents the calculated inlet with a velocity of 0.5 m/s. The top surface, excluding the inlet gate, serves as a pressure outlet indicated by the red arrow, while the remaining surfaces are designated as walls. Due to the structural complexity and to enhance computational efficiency, a tetrahedral unstructured mesh with four nodes is utilized for discretization. The edge sections of the cap beam are partitioned into 490,282 mesh elements, while the central section comprises 498,536 mesh elements.

The computational results are depicted in Figure 3(c). As shown in the figure, the red area represents the completed pouring region, while the blue portion indicates areas where pouring is incomplete, mainly situated above the outlet. These areas can be filled after the pouring process is finished. The yellow and green regions lie between the completed pouring and incomplete areas, indicating zones containing air bubbles. They are predominantly concentrated at the bottom of the model, corners, and the bottom surface farther from the pouring inlet. These areas require auxiliary vibration to ensure pouring quality.

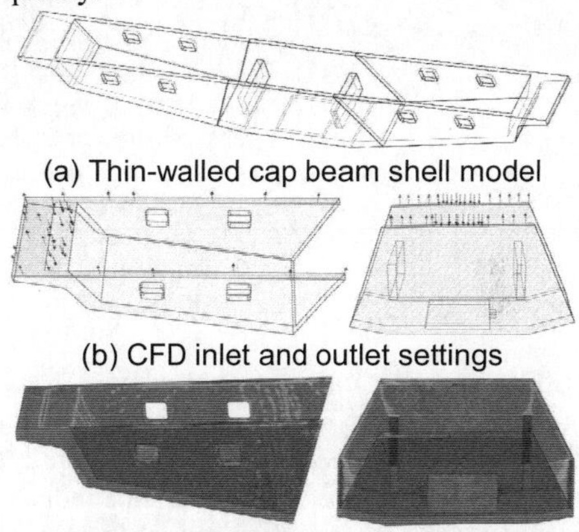

(a) Thin-walled cap beam shell model

(b) CFD inlet and outlet settings

(c) Post processing results of tecplot

Figure 3 Thin-walled cap beam shell CFD model and results

5 Conclusion

This study utilizes CFD with the H-B rheological model to simulate the flow characteristics of UHPC. By comparing numerical simulation results with experimental data, the reliability of the H-B model in simulating the flow behavior of shear-thickening power-law fluid UHPC within molds is validated. The simulation findings provide guidance for predicting the quality of UHPC pouring in thin-walled structures.

References

[1] Wang X, Ma Z, Wang X, Xue S, Shen W, Wu D, et al. Design of self-compacting ultra-high performance concrete (SCUHPC) towards to the cementitious materials packing optimization. Cement and Concrete Composites. 2024;148:105443.

[2] Yu R, Xu L, Li K, Leng Y, Wang Z. Dynamic behaviors assessment of steel fibres in fresh Ultra-High Performance Concrete (UHPC): Experiments and numerical simulations. Journal of Building Engineering. 2022;59:105084.

[3] He J, Cheng C, Zhu X, Li X. Effect of silica fume on the rheological properties of cement paste with ultra-low water binder ratio. Materials. 2022;15(2):554.

[4] Jirout T, Krupica A, Kolomijec A. CFD analysis of ultra-high-performance concrete rheological tests. Fluids, 2024, 9(2): 45.

Software and Application of Big Data Analysis Platform for Complex Casting Manufacturing Process Quality

Jianxin Zhou[1],*, Wenhao Zhou[1], Peng Yu1, Wen Li[1], Xiaoyuan Ji[1], Yajun Yin[1], Xu Shen[1], Meijuan Zhang[2], Xiwang Qie[2], Hai Nan[2], Shouqin Zhu[3], Mingguo Xie[3], Yu Wang[4], Hulin Li[4], Xiao Liu[5], Baoming Zhang[5], Gang Yao[6], Chuansheng Wang[7], Donghong Wang[8], Jianping Hu[9], Xuewen Qian[10]
1. State Key Laboratory of Material Processing and Die & Mould Technology, Huazhong University of Science &
2. Technology, Wuhan, 430074, China
3. Baimtec material Co.,Ltd., Beijing, 100094,China
3. Hefei Casting and Forging Factory of Anhui Heli Co., Ltd. Hefei, 230000, China
4. Xi'an Aerospace Engine Co., Ltd. Xian, 710000, China
5. CRCC High-Tech Equipment Co,. Ltd., Kunming, 650215, China
6.Guizhou Aircraft Engine Precision Casting Co., Ltd. Guiyang, 550014,China
7. Wuxi Branch of FAW Casting Co., Ltd., Wuxi, 214177, China
8. Shanghai Key Laboratory of Advanced High-temperature Materials and Precision Forming, Shanghai Jiao Tong University, Shanghai, 200240, China
9. Wuhan Yimo Technology Co., Ltd, Wuhan, 430074, China
10. Wuhan Huazhu Information Technology Co., Ltd., Wuhan, 430074, China
*Corresponding address: e-mail: zhoujianxin@hust.edu.cn

Abstract: The complex castings used in major equipment in industries such as aerospace, automotive, rail transportation, and engineering machinery are prone to defects such as porosity, deformations, and inclusions due to the coupling effects and fluctuations of key process parameters. This leads to common problems such as "key quality points exceeding standards" and large quality fluctuations, seriously affecting the reliability and stability of major equipment manufacturing products. In this paper, a systematic study on the model-method-technology of typical defect quality big data was conducted. Firstly, relevant models were constructed, such as the single piece quality tracing model, the multi-source heterogeneous information fusion model, the model of production parameters and intrinsic relationship of typical defects, and the model of comprehensive element time-varying disturbance and defect evolution mapping. Then, five key technologies were developed, including high-precision and efficient online monitoring and prediction of typical defects, tracing the causes of quality problems in multiple links, and quality control based on the relationship between production fluctuations and metallurgical defect evolution. Based on the above research, data collection, cleaning, mining, sensitivity analysis, and relationship modeling algorithms were established for the casting production process. Special core algorithms for predicting defects such as shrinkage, looseness, and gas entrapment were proposed, achieving high-precision prediction of typical defects. A "1+N" quality big data analysis software and hardware integrated platform was developed. After applying the integrated platform to the five major industries of aerospace, automotive, rail transportation, engineering machinery, and aerospace industries, the results showed an accuracy of quality prediction of ≥80% and a reduction in defects of ≥30%, leading to the formulation of solutions and key demonstrations in these industries.

Keywords: complex castings; manufacturing process; quality big data model; single-piece model; high-precision defect prediction; quality traceability and root cause analysis; 1+N software platform

1 Introduction

Casting is the foundation of manufacturing industry and the preferred technology for forming complex metal parts of major equipment such as aviation and aerospace. There are some common problems in the manufacturing process of complex castings, such as "the key quality points are out of tolerance and the quality fluctuates greatly", which lead to the reliability and service life of major equipment can not meet the requirements, and seriously affect the establishment of a manufacturing power. With the development of artificial intelligence technology, big data technology can effectively study the influence of process parameter fluctuation on complex casting defects, and become a new way and method to effectively solve the above common problems [1].

To this end, Firstly, this research has studied quality big data model and management method. Then, quality intelligent control and process optimization technology is studied. In addiction, the core algorithm, software and platform for quality big data analysis of complex casting manufacturing process are developed and applied. Finally, the application demonstration was carried out in the five major industries of aviation, aerospace, automotive, rail transit and construction machinery to realize the control of key quality points and quality fluctuations of complex core castings, and achieve quality improvement.

Experimental Procedure

As shown in Figure 1, firstly, data is the basis of this research, so we study the quality big data model and management methods first. For this:

(1) The production equipment was transformed by PLC to automatically collect part of the production data, and then the whole process data of castings from process design to quality detection are collected;

(2) Due to the variety of data forms collected, the flexible fusion method of multi-source heterogeneous data is studied. The semi-structured data is processed by row and column retrieval model, and the three-dimensional features of castings are extracted by deep learning model, so that the unstructured data and semi-structured data can be directly stored in the relational database.

(3) A one-piece quality traceability big data model was established to correlate the design, production and quality data of castings, providing the basis for quality traceability and data mining.

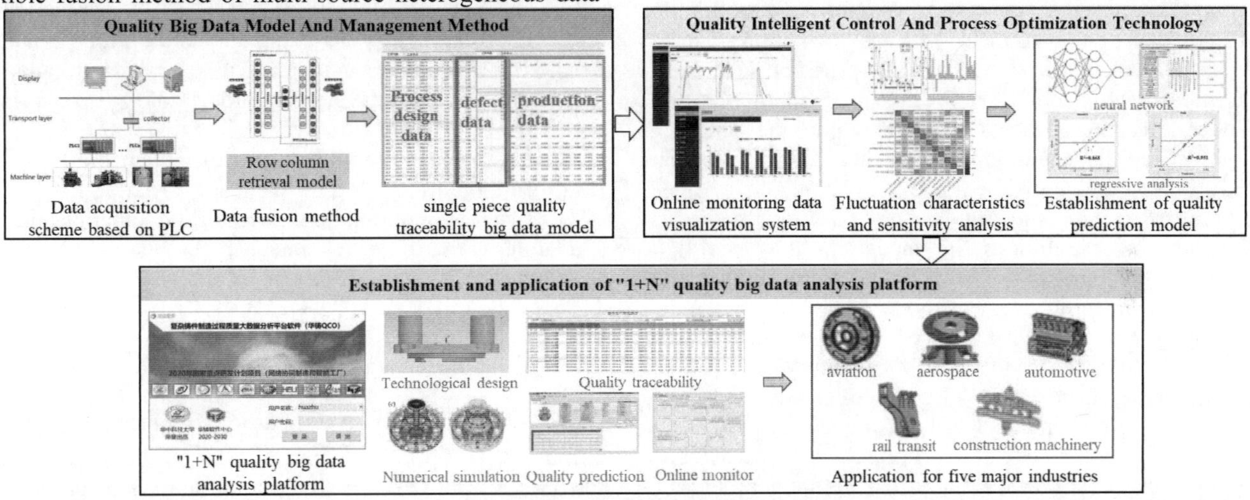

Figure 1 Framework of this paper

Then, In addition to accurately collecting data, the use of artificial intelligence technology to achieve quality control is another major study content. Hence:

(1)Visualization system was developed. And fluctuation characteristics and sensitivity of the collected parameters was analyzed;

(2)The mapping relationship between process parameters and defect evolution was analyzed by regression analysis and neural network method, and the mapping relationship model between time-varying disturbance of all factors and defect evolution was established. Based on this model, a quality prediction method is developed.

In order to improve the quantity and quality of data, this study also developed algorithms related to numerical simulation, data cleaning and sensitivity analysis. Finally, a "1+N" quality big data analysis platform was established based on the above research combined with other software like Huazhu CAE, ERP.

Result and discussion

After verified in typical casting of five major industries, including aviation, aerospace, automotive, rail transit and construction machinery., as shown in Table 1:

(1)The defect prediction accuracy of the "1+N" platform for typical castings in the five industries has exceeded 90%.

(2)According to statistics, after using the "1+N" quality big data analysis platform, the defect rate of typical castings in the five industries has decreased by more than 30%.

Table 1. Verification result

Casting type	Prediction accuracy	Defect rate before use	Defect rate after use
Gearbox	98.97%	20.00%	10.00%
Outlet pipe	99.94%	3.30%	2.10%
cylinder block	95.00%	5.42%	1.83%
pickaxe arm	96.00%	20.00%	3.80%
Steering axle	94.00%	6.00%	3.16%

Acknowledgments

National Key Research and Development Program (2020YFB1710100), National Natural Science Foundation (52275337, 52201042, 52090042, 51905188). The authors would like to thank all the organizations and members of the project team of the National Key Research and Development Program "Software and Application of Big Data Analysis Platform for Quality in Complex Casting Manufacturing Processes".

Reference

[1] Yu P, Ji X, Sun T, Zhou W, Li W, Xu Q, Qie X, Yin Y, Shen X, Zhou J. Data–physics fusion-driven defect predictions for titanium alloy casing using neural network. Materials, 2024, 17: 2226.

Numerical Simulation of Multi-Mathematical Model Coupling in the Squeeze Casting Process

Zhaoxuan Liu, Yingzi Zhang, Taotao Ge, Yan Zhou, Xiaofeng Niu*
School of Materials Science and Engineering, Taiyuan University of Technology, Taiyuan 030024, China
*Corresponding address: e-mail: 519129407@qq.com

Abstract: This paper presents the development of an SPH-FEM coupled computational model, combining the Smoothed Particle Hydrodynamics (SPH) method with the Finite Element Method (FEM), for numerical simulation of the squeeze casting process. Drawing upon the unique strengths of the SPH method in simulating free surfaces, a computational framework for numerical simulation of the filling process in squeeze casting is established, maximizing its capabilities in modeling flow fields. Furthermore, the temperature field results are transmitted as initial data to the finite element model, thereby achieving the coupling between SPH and FEM. This integration leverages the advantages of the FEM in addressing stress-strain problems. Consequently, a finite element model based on the FEM method is constructed to conduct numerical simulation of the solidification process in squeeze casting, enabling the analysis of its thermal stress field.

Keywords: squeeze casting; smoothed particle hydrodynamics; finite element method; coupled model

1 Introduction

Squeeze casting is an advanced metal forming process that combines casting and forging. It produces high-performance castings with few defects by solidifying the molten metal in the cavity under high pressure [1]. However, due to limitations in domestic equipment and process levels, defect issues are difficult to avoid in actual production. Simulation technology can provide solutions covering the entire product lifecycle for the squeeze casting process, thereby improving production efficiency and shortening the production cycle [2].

Smoothed Particle Hydrodynamics (SPH) is a meshless numerical method used to simulate fluid motion. It discretizes continuous media into a series of interacting particles to reflect the changing behaviors of mass, energy, and momentum [3]. Due to the unique advantages of the SPH method in calculating free surfaces, a mathematical model is established using the SPH method to simulate the filling process of squeeze casting.

The Finite Element Method (FEM) is a numerical computation technique for finding approximate solutions to boundary value problems involving partial differential equations [4]. It divides the complex continuum into a finite number of non-overlapping units, lists the equations and solves them. Given the advantages of FEM in calculating stress-strain problems, this paper adopts the Finite Element Method to establish a mathematical model for simulating and calculating the solidification process of squeeze casting.

2 Mathematical modelling and numerical methods

Mathematical modelling of the filling process

The particle approximation for the field function and its spatial derivative at particle i is given as follows:

$$\langle f(x_i)\rangle = \sum_{j=1}^{N}\frac{m_j}{\rho_j}f(x_j)W_{ij}$$

$$\langle \nabla\cdot f(x_i)\rangle = \sum_{j=1}^{N}\frac{m_j}{\rho_j}f(x_j)\cdot\nabla_i W_{ij}$$

where, i denotes the central particle, j represents particles within the support domain, $<f(x_i)>$ is the approximate function for particle i, N is the total number of particles j, W_{ij} is the smoothing kernel function, ∇ is the gradient operator, The kernel function is chosen to be a cubic spline function.

By employing the SPH particle approximation method, the Navier-Stokes equations are discretized to obtain the SPH expressions for the density equation, momentum equation, and energy equation. These equations are presented as follows:

$$\frac{\mathrm{d}\rho_i}{\mathrm{d}t} = \sum_{j=1}^{N} m_j v_{ij}^{\beta}\frac{\partial W_{ij}}{\partial x_i^{\beta}}$$

$$\frac{\mathrm{d}v_i^{\alpha}}{\mathrm{d}t} = \sum_{j=1}^{N} m_j\left(\frac{p_i}{\rho_i^2}+\frac{p_j}{\rho_j^2}\right)\frac{\partial W_{ij}}{\partial x_i^{\beta}}$$

$$\frac{\mathrm{d}e_i}{\mathrm{d}t} = \frac{1}{2}\sum_{j=1}^{N} m_j\frac{p_i+p_j}{\rho_i\rho_j}\cdot\frac{\partial W_{ij}}{\partial x_i^{\beta}}$$

where, ρ represents the density, v denotes the velocity, e is the internal energy of the fluid parcel, p is the pressure, α and β indicate coordinate directions.

Mathematical modelling of the solidification process

The equation of motion for three-dimensional finite element dynamics problems can be expressed as:

$$M\ddot{a}(t)+C\dot{a}(t)+Ka(t)=Q(t)$$

where, $a(t)$ represents the nodal displacement, $\dot{a}(t)$ is the nodal velocity, $\ddot{a}(t)$ is the nodal acceleration, and M, C, K, Q are respectively the mass matrix, damping matrix, stiffness matrix, and load vector.

During the solidification process in squeeze casting, the mechanical behavior of the metal within the mold cavity is highly complex. To address this, this paper employs the thermoelastic-viscoplastic constitutive model established by Zhu Wei [5] et al. for analysis:

$$\{\Delta\varepsilon\} = \{\Delta\varepsilon_{el}\} + \{\Delta\varepsilon_{th}\} + \{\Delta\varepsilon_{in}\}$$

where, $\{\Delta\varepsilon_{el}\}$ represents the increment of elastic strain, $\{\Delta\varepsilon_{th}\}$ denotes the increment of thermal strain, and $\{\Delta\varepsilon_{in}\}$ is the increment of inelastic strain.

3 Numerical simulation results

Filling process

Figure 1 shows the filling status of the molten metal at different time points, with the time taken for the molten metal to completely fill the cavity being 0.83 seconds. As can be seen from the figure, using the vertical counter-gravity squeeze casting process, the molten metal fills the cavity from bottom to top at a low speed, resulting in a smooth filling process that generally aligns with real-world conditions[6].

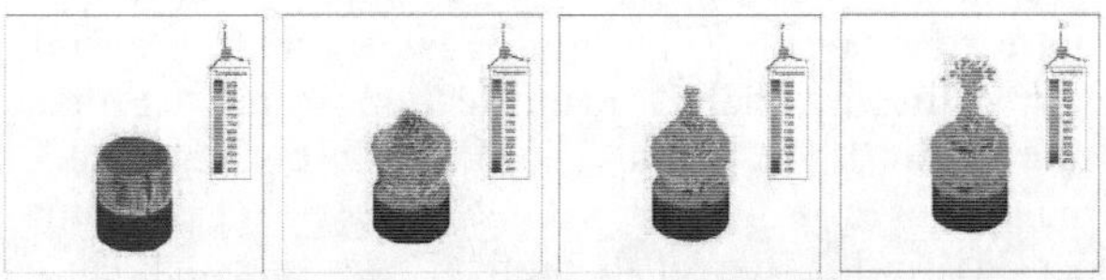

Figure1 The filling status and temperature distribution of the casting at different time points during the filling process: (a) *t*=0.03 s; (b) *t*=0.23 s; (c) *t*=0.33 s; (d) *t*=0.53 s

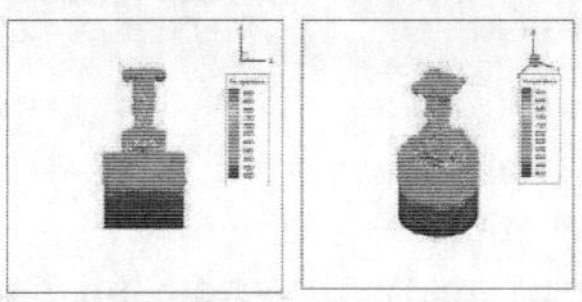

Figure 2 The filling status and temperature distribution of the casting at the end of the filling process

Solidification process

The simulation results of the temperature field during the solidification process of the casting are shown in Figure 3. The time required for the casting to solidify completely is 0.48 s, and the overall solidification order is from outside to inside.

Figure 4 presents the contour plot of equivalent stress during the solidification process of the casting. It can be observed from the figure that at the initial stage of solidification, the overall equivalent stress of the casting is relatively low. As the pressure from the punch continues to be applied, the overall equivalent stress of the casting gradually increases. The maximum equivalent stress is located at the thin-walled sections of the casting, and the stress distribution generally aligns with real-world conditions[3].

Fig. 3. The temperature field of the casting at different time points during the solidification process: (a) t=0s; (b) t=0.04s; (c) t=0.16s; (d) t=0.44s

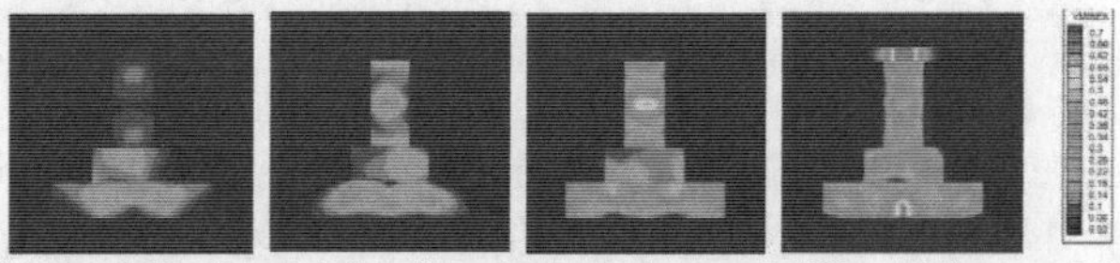

Figure 4 Contour plots of equivalent stress at different time points during the solidification process of the casting: (a) *t*=0.06 s; (b) *t*=0.19 s; (c) *t*=0.36 s; (d) *t*=0.40 s

4 Conclusion

This paper establishes a mathematical model for numerical simulation of the squeeze casting process based on the SPH-FEM coupling method. By analyzing the flow field, temperature field, and stress field during the filling and solidification processes of the I-shaped specimen in squeeze casting, the accuracy of the mathematical model established in this paper is verified, which has certain guiding significance for the actual production process of squeeze casting.

Acknowledgments

This research was financially supported by the National Natural Science Foundation (No.52474418).

References

[1] Jiang J F, et al. The Chinese Journal of Nonferrous Metals, 2021, 31(9): 2313-2329.
[2] Liao Dun-ming, et. al. Acta Metallurgica Sinica (2018), 54(2): 161-173.
[3] Ge Tao-tao, et al. Special Casting & Nonferrous Alloys (2024), 44(7): 909-916.
[4] Sun X L, et al. Journal of Logistical Engineering University, 2006, 22(1): 67-69.
[5] Han Z Q, et al. Acta Metallurgica Sinica, 2009, 45(3): 356-362.
[6] Huang H F, et al. China Foundry Machinery & Technology, 2022, 57(4): 81-85.

Aerospace Magnesium-Aluminum Alloy Casting Defect Diagnostic Expert System

Weizhong Liang*,Xinrui Dong,Guoxin Wang,Jinfeng Liu
Heilongjiang University of Science and Technology, No. 2468, Puyuan Road, Songbei District, Harbin City, Heilongjiang Province, China
*Corresponding address:E-mail: wzliang1966@126.com

Abstract: Sand casting, as a leading traditional casting process, is widely used in various scales of casting production due to its accessible materials and simplified manufacturing. However, frequent occurrences of casting defects such as pores, shrinkage cavities, and cracks not only impact product quality but also increase production costs and scrap rates, posing a severe threat to the economic benefits and reputation of casting enterprises. Therefore, rapidly and accurately determining the causes of casting defects is crucial to enhancing product quality.

This study developed an expert analysis system specifically targeting casting defects in aviation transmission load-bearing gearboxes. By leveraging advanced machine learning technology combined with casting expertise, the system achieves automatic diagnosis, classification, and cause analysis of defects. The system collects relevant textual data on casting defects and interacts with the SQLite database through a user interface designed with PyQt5, facilitating users' queries and browsing of defect information. In the recognition section, the system comprehensively collects defect features, including morphology, size, distribution, and other critical information. After multiple experimental validations, the Bernoulli Naive Bayes algorithm model has demonstrated superior performance on the dataset, enabling rapid and accurate automatic classification and diagnosis of user-selected defect features.

Moreover, the system supports semi-automated updating of domain knowledge, ensuring alignment with the latest casting technologies. The research findings provide a novel approach for quality control in aviation transmission load-bearing gearbox casting and serve as a reference for similar fields. In the future, we will continue to refine and enhance the system to adapt to the analysis needs of a broader range of casting defects, driving the sustained progress of the casting industry.

Keywords:sand casting;bernoulli naive bayes algorithm; expert system

1 Introduction

In the aviation industry, which requires high precision and reliability, the quality of the transmission system load-bearing casing is directly related to flight safety and performance. However, although the traditional sand casting process is mature and widely used, defects such as porosity, shrinkage porosity, and cracks will inevitably appear in the casting process, which not only reduces product quality but also increases production costs and scrap rates, posing a severe challenge to the economic benefits and brand image of aviation enterprises.

The system mainly includes four functional modules: user management, analysis and diagnosis, defect knowledge, and defect acceptance criteria, of which the user management module includes user registration, user login, and user registration code verification functions; The analysis and diagnosis module includes defect reasoning, defect category creation, defect feature creation, defect picture display, defect knowledge graph display, defect identification result formation cause and prevention measures display, defect identification result acceptance standard display functions. The defect knowledge module is mainly used to display defect knowledge, including defect description, defect theoretical causes, actual causes, improvement measures, and what kind of improvement measures will have any effect. The Defect Acceptance Criteria module provides defect acceptance criteria documents.

2 Experimental procedure

In order to meet this challenge, we have successfully developed an expert analysis system for casting defects in the load-bearing casing of the aero drivetrain. The system deeply integrates advanced machine learning technology with deep casting expertise, aiming to achieve fast and accurate identification, classification, and cause analysis of casting defects, and provide strong technical support for casting quality control.

The system adopts a hierarchical architecture, and the front-end interface is designed with Qt to interact with the SQLite database. As an intermediate layer, the Qt program is responsible for passing data between the front-end interface and the database and completing the corresponding logical processing. As the backend of data storage and management, MySQL database is responsible for data persistence and management. The advantage of using a layered architecture is that there is less coupling between the layers, and their respective responsibilities are clear, making it easy to maintain and upgrade. At the same time, the architecture is also scalable, and each layer can be scaled and optimized according to specific needs.

In Qt, we can implement the MVC architecture using Qt's MVC framework and signal slot mechanism. Qt's MVC framework provides a Model-View/Delegate architecture that helps us manage and present data. At the same time, Qt's signal slot mechanism can help us communicate and collaborate between various components, making the logic of the application clearer and easier to maintain.

3 Result and discussion

The system uses PyQt as the front end in the knowledge database of experts on key assembly processes of the engine. PyQt is a cross-platform Python application framework with a variety of features, including GUI, networking, databases, scripting, XML, OpenGL, and more. PyQt provides a number of classes and functions that make it easy for developers to develop cross-platform applications. The advantage of PyQt is that it is cross-platform and efficient, as it supports multiple operating systems including Windows, Linux, macOS, etc. This means that developers can use PyQt to develop cross-platform applications, reducing development costs and time.

In addition, PyQt uses a mechanism called meta-objects that automates signal and slot connections, which makes PyQt's programming model more flexible and efficient. Signals and slots are one of the most important concepts in PyQt, as they are used to communicate between objects, making it easy for developers to implement a variety of interactions and event handling. Through signal and slot connections, developers can extend and modify various functions without changing the code structure. This flexibility and efficiency is what makes PyQt unique and one of the reasons why it is so widely used.

4 Conclusion

In this study, an expert analysis system was successfully developed to solve the frequent defects in the casting process of the load-bearing casing of the aero transmission system. Through the deep integration of advanced machine learning technology and casting expertise, the system realizes automatic diagnosis, classification, and cause analysis of casting defects, providing an efficient and accurate solution for casting quality control.

In the process of system development, we used the combination of SQLite database and PyQt5 user interface to realize convenient data management and visual display and provide users with a good operation experience. At the same time, through several experimental verifications, we choose the Bernoulli naïve Bayes algorithm model as the core classifier, which shows superior performance on the casting defect dataset and can quickly and accurately complete the classification and diagnosis tasks of defect features.

The application of this expert analysis system not only significantly improves the accuracy and efficiency of casting defect identification, reduces the scrap rate and production cost, but also helps enterprises to find and solve problems in the casting process in a timely manner, and improves the overall quality and reliability of products. This is of great significance for the manufacture of key components such as the load-bearing casing of the aviation transmission system, which helps to ensure flight safety and enhance the market competitiveness of aviation enterprises.

Looking ahead, with the continuous development of foundry technology and the continuous innovation of machine learning algorithms, we believe that this expert analysis system will continue to be refined and optimized to meet the needs of a wider range of foundry defect analysis. At the same time, we also expect that the system can be promoted and applied in more fields, and contribute more to the sustainable development of the foundry industry.

Simulation on Dendritic growth of Aluminum Alloy under Magnetic Assisted Solidification

Yinqi Qiu[1], MengwuWu[1,*], ShoumeiXiong[2]
1. School of Automotive Engineering, Wuhan University of Technology, Wuhan 430070, China
2. School of Materials Science and Engineering, Tsinghua University, Beijing 100084, China
*Corresponding address: e-mail: wumw@whut.edu.cn

Abstract: Applying electromagnetic fields during casting or welding processes has been a hot research topic in recent years. In the absence of direct contact with the material, an external magnetic field can affect the flow of the melt, solute distribution, heat transformation, and dendritic growth morphology during the solidification process, which has a decisive impact on the mechanical properties of the product. It is crucial to understand the potential alloy solidification mechanism in order to improve the casting quality.This paper investigates the influence of magnetic damping effect and Seebeck effect on dendritic growth process under magnetic assisted solidification.

Keywords:aluminum alloy; dendritic growth; magnetic field; numerical simulation

1 Introduction

The inhibitory effect of magnetic field on the flow of conductive liquid is called magnetic damping effect, and the melt during solidification is considered as conductive liquid. When the melt is subjected to convection, applying a magnetic field perpendicular to the direction of melt movement will generate an induced current, and the resulting induced current and magnetic field will generate a Lorentz force that is completely opposite to the direction of melt flow. However, it has also been found that applying a stable magnetic field during the solidification process drives the flow of the molten pool through the thermoelectric magnetic force generated by the Seebeck effect. This thermoelectric magnetic force is generated in the paste-like region between dendrites, thereby altering the transfer process of heat and solute along the solid-liquid interface and ultimately affecting the formation of solidification microstructure.

2 Numericalmodel

The mathematical model for simulating aluminum alloy dendrite growth using the phase field method is detailed in Ref. [1], and the magnetic damping effect is described by the following equation [2]:

$$\frac{\partial \rho}{\partial t}+\nabla\cdot\rho u=F_d+F_b+F_{MF} \tag{1}$$

$$\frac{\partial \rho u}{\partial t}+\nabla\cdot\left(pI+\frac{B^2}{2}I+uu-BB\right)= \upsilon\nabla\cdot\left(\rho\left(\nabla u+\nabla^T u\right)\right)+F_d+F_b+F_{MF} \tag{2}$$

The magnetohydrodynamic control equation is obtained by introducing the Ohm's law modified by the Seebeck effect [3]:

$$F_{TEMF}=\sigma_e((-S\cdot G)\times B+(u\times B)\times B) \tag{3}$$

$$\frac{\partial \rho}{\partial t}+\nabla\cdot\rho u=F_d+F_b+F_{TEMF} \tag{4}$$

$$\frac{\partial \rho u}{\partial t}+\nabla\cdot\left(pI+\frac{B^2}{2}I+uu-BB\right)= \upsilon\nabla\cdot\left(\rho\left(\nabla u+\nabla^T u\right)\right)+F_d+F_b+F_{TEMF} \tag{5}$$

Where $S$and $G$are the Seebeck coefficient and temperature gradient, respectively.

3 Result and discussion

Dendrite growth under magnetic damping effect

The lateral flow velocity is set to 0.01, and the magnetic field direction is vertically upward. Magnetic field strength $B=(B_x,B_y)$, where B_x and B_y correspond to the transverse and longitudinal magnetic field strengths, respectively. Fig. 1 shows the dendritic growth, solute distribution, and fluid velocity under four different conditions of forced convection, with longitudinal magnetic field strengths of 0, 0.01, 0.05, and 0.1, respectively. In the absence of a magnetic field, the growth rate of upstream dendrite tips is greater than that of downstream dendrite tips, as shown in Figure 1(a). When a longitudinal magnetic field is added, an induced current perpendicular to the paper surface is generated in the fluid with transverse flow velocity. Subsequently, an orthogonal magnetic field acts on this current and applies a Lorentz force opposite to the flow to suppress the flow. As the magnetic field intensity increases, the flow velocity within the computational domain decreases, the difference in growth rate between upstream and downstream dendrite tips decreases, and the phenomenon of dendrite tilting caused by convection gradually disappears, as shown in Figures 1(b-d).

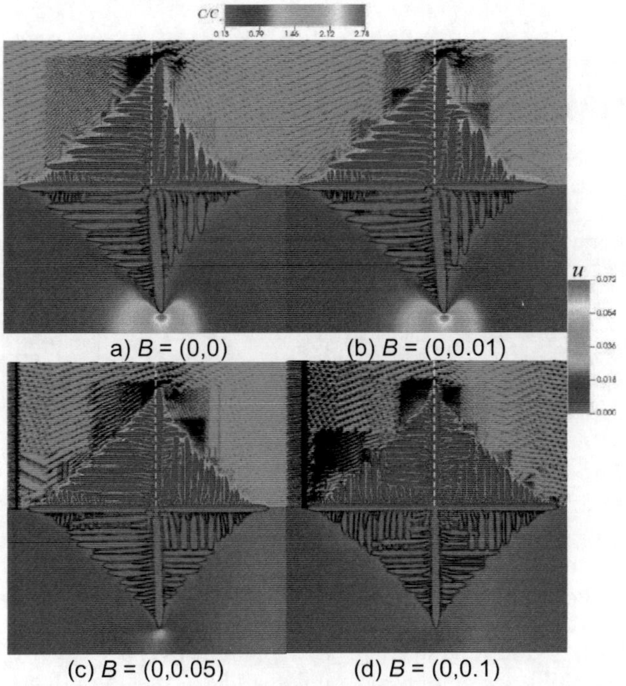

Figure 1 Dendrite growth under different longitudinal magnetic field intensities

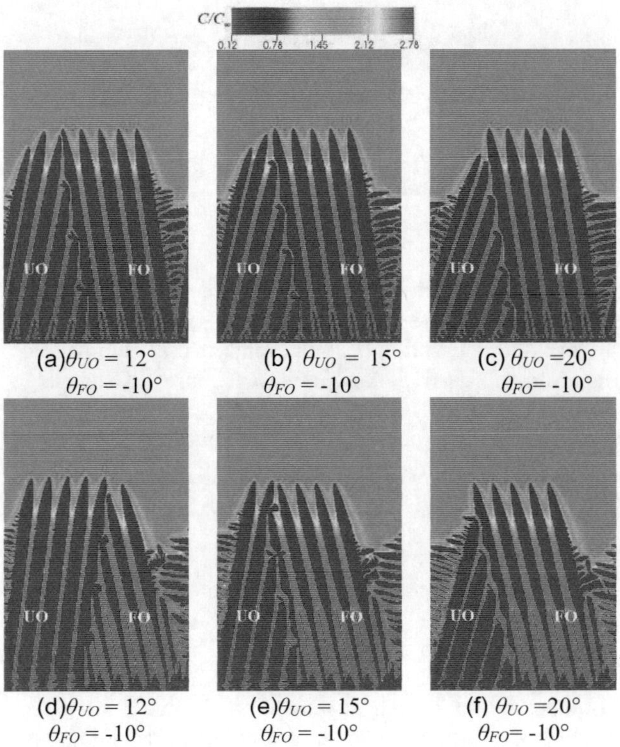

Figure 2 Simulation of columnar dendritic growth: (a-c) B=0; (d-f) B=0.6

Dendritic growth under the Seebeck effect

Figure2 shows the simulated competitive growth of converging columnar dendrites under pure diffusion and thermoelectric convection with different growth orientation angles. Ten seeds with equal distance distribution are arranged at the bottom of the computational domain, and the first half of the seeds have different growth orientations compared to the second half. In Figures 2(a-c), under pure diffusion conditions, UO columnar dendrites are always eliminated by FO columnar dendrites. When a magnetic field is applied, the position between UO and FO columnar dendrites changes, and UO columnar dendrites instead eliminate FO columnar dendrites, as shown in Figure 2(d). In Figure 2(e), when the angle between the growth direction of UO columnar dendrites and the temperature gradient direction increases to 15°, three main branches of UO columnar dendrites are relatively backward and are eliminated by FO columnar dendrites, but FO columnar dendrites can no longer continue to eliminate UO columnar dendrites. In Figure 2(f), When θ_{UO} increases to 20°, even with the addition of a magnetic field, UO columnar dendrites are completely eliminated.

4 Conclusion

Under the action of lateral forced convection, the longitudinal magnetic field can generate a Lorentz force opposite to forced convection, thereby suppressing flow within the computational domain. During the columnar dendritic growth, the competitive behavior between dendrites is influenced by both the intrinsic orientation angle and magnetic field strength. The magnetic field affects the competitive`growth process by changing the growth orientation angle of dendrites.

5 Acknowledgments

This work was financially supported by the National Natural Science Foundation of China (Grant No. 52275370), and the Key R&D Program of Hubei Province, China (Grant No. 2022BAD100).

References

[1] Qiu Y, Wu M, Qin X, et al. Phase-field lattice-Boltzmann study on fully coupled thermal-solute-convection dendrite growth of Al-Cu alloy[J]. China Foundry, 2024, 21(2): 125-136.

[2] Dellar P J. A note on magnetic monopoles and the one-dimensional MHD Riemann problem[J]. Journal of Computational Physics, 2001, 172(1): 392-398.

[3] Cao L, Liu D, Jiang P, et al. Multi-physics simulation of dendritic growth in magnetic field assisted solidification[J]. International Journal of Heat and Mass Transfer, 2019, 144: 118673.

Numerical Simulation and Optimization of the Casting Process for a 6 MW Large-Scale Wind Turbine Main Frame

Fei Su[1, 2], Donghong Wang[1*], Qianhai Gong[3], Da Shu[1], Baode Sun[1]
1. Shanghai Key Laboratory of Advanced High Temperature Materials and Precision Forming; State Key Laboratory of Clean and Efficient Turbomachinery Power Equipment; School of Materials Science and Engineering, Shanghai Jiao Tong University, Shanghai 200240, China
2. School of Materials Science and Engineering, Shanghai University of Engineering Science, Shanghai 200240, China;
3. Zhejiang Jiali Wind Power Technology Co., Ltd., Hangzhou 310051, Zhejiang, China
*Corresponding address: e-mail: E-mail：wangdh2009@sjtu.edu.cn

Abstract: This study focuses on the main frame of a 6 MW large-scale wind turbine generator as the research object, utilizing the ProCAST finite element analysis software for numerical simulation and optimization. Through numerical simulation, potential shrinkage and porosity locations in the casting were successfully predicted, which served as the basis for the simulation and optimization of the gating system. The optimized gating system was then applied to practical production at the factory, resulting in the successful casting of main frame products that met the delivery requirements. This research provides an effective method for optimizing the casting process and predicting defects in large-scale wind turbine ductile iron castings, contributing to resource conservation and quality assurance in industrial production.

Keywords: wind turbine nodular cast iron component; casting process optimization; numerical simulation; wind turbine main frame

1 Introduction

As wind turbines continue to trend toward larger and heavier designs, critical aspects such as production, transportation, installation, and maintenance of wind turbine structural components face significant challenges. To minimize damage to wind turbine structures, reduce maintenance costs and ensure a minimum service life of at least 20 years [1], it's crucial to control the metallurgical quality of castings during manufacturing.

Employing traditional trial-and-error methods [2–3] for producing large castings in small batches significantly increases the cost. However, utilizing computer-aided simulation to analyze the flow field during casting and the temperature field during solidification enables the prediction of defect locations in the castings. This approach improves the efficiency of casting process design and reduces the costs associated with casting development and production. Researchers such as Wei, Chen, You, and Vladimír have effectively used numerical simulations to address casting defects in wind power equipment [4].

This study utilizes ProCAST software to simulate the filling and solidification processes of a 6 MW wind turbine main frame, aiming to enhance casting processes and mitigate issues such as shrinkage and shrinkage holes. The simulation results serve as a guide for production, ensuring the quality of turbine main frames.

2 Experimental procedure

The 6 MW wind turbine main frame, measuring 4,533 mm in length, 4,217 mm in width, and standing at a maximum height of 1,772 mm, with a total weight of around 23 tons, is utilized. The preliminary process design incorporates two sets of symmetric bottom pouring systems to ensure the smooth filling of molten metal. Additionally, 424 equally thick chills are employed to optimize local cooling effects during casting. Moreover, five risers are strategically positioned on the casting flange surface to facilitate shrinkage compensation.

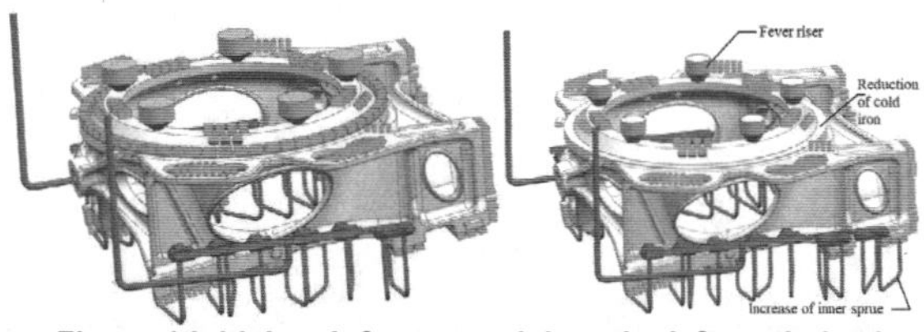

Figure 1 Initial mainframe model on the left, optimized model on the right

In the optimization of the casting process, heated risers were utilized to replace traditional risers, enhancing shrinkage compensation. Concurrently, the removal of chills from the casting flange surface improved shrinkage compensation at the base of the casting. Furthermore, two sets of symmetric inner gating systems were introduced to expedite filling at the casting ends, effectively reducing the formation of shrinkage voids and porosity.

3 Result and discussion

With the chills removed, the solidification rate was slightly slowed on the flange surface of the casting. This led to an extended connection time between the liquid metal at the thick wall and that at the flange surface, indirectly assisting in compensating for the heat-induced shrinkage of the risers. Consequently, unlike the large-volume isolated liquid phase region shown in Figure 2, the volume and density of the isolated liquid phase region formed at the bottom of the casting were reduced in Figure 3(b). Additionally, there was a trend of solidification segregation in the middle of the isolated liquid phase region. As solidification progressed, the larger isolated liquid phase region was divided into multiple smaller ones, providing convenience for the self-compensation of ductile iron.

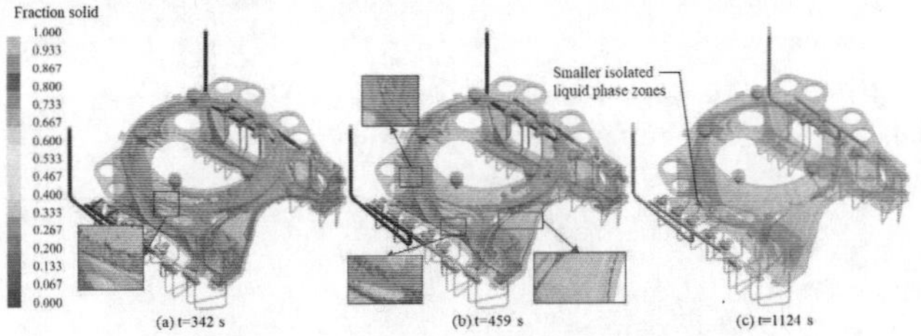

Figure 2 Solid fraction at different time after optimization of the main frame

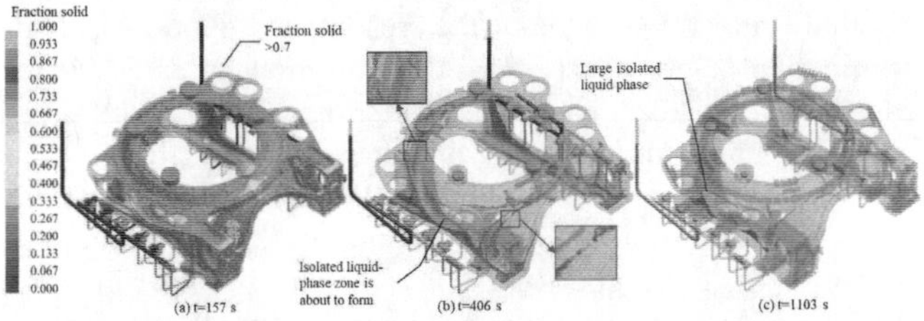

Figure 3 The simulated solid fraction of the main frame at different times

Figure 4 depicts the predicted shrinkage outcomes of the main frame before and after optimization. During the initial casting process, the casting exhibited numerous instances of shrinkage and shrinkage holes, amounting to a total volume of 2,643.402 cm^3. Following optimization, only a few shrinkage holes were present at the casting's end, with a total volume of just 126.137 cm^3. As shown in Figure 5, the wind turbine main frame was successfully produced with numerical simulation as a reference.

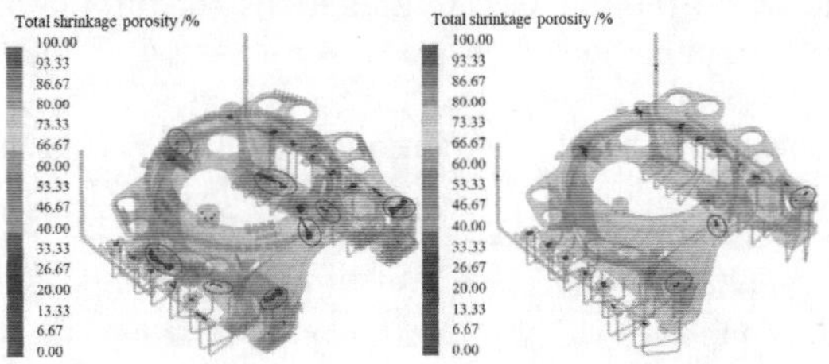

Figure 4 On the left are the results of the initial shrinkage simulation and on the right are the optimized results

Figure 5 The 6 MW wind power main frame

4 Conclusion

Using numerical simulation, the casting defects of the wind turbine main frame were accurately predicted. Rational adjustments to the cold iron, replacement of heated risers, and addition of inner runners led to a 95.2% reduction in shrinkage and shrinkage holes, facilitating the smooth production of the casting.

5 Acknowledgments

This work was financially supported by he National Key Research and Development Program Zhejiang Province (2022C01147), the National Key Research and Development Program of China (2022YFB3706800), and the National Natural Science Foundation of China (52090042).

References

[1] Fu M K, Song X F, Chen Q H, et al. Casting process design of box body in large wind turbine. J. Casting, 2017, 66(1): 16-19.
[2] Deng W, Jiang F Z, Zhou S Y, et al. Optimization of hot forging forming process for offshore wind power flange. J. Metallurgical Equipment, 2022, 4: 1-6.
[3] Zhao S Z. Simulation and parametric study of wind turbine spindle molding process. D. Jinan: Shandong University of Architecture, 2022.
[4] Wei D D, Xu G B, Zhang J, et al. Research on the casting process of ductile iron planetary frame for wind power. J. Casting, 2021, 70(10): 1233-1238.

Integrated Computational Framework for Controlling Dimensional Accuracy of Thin-Walled Turbine Blades During Investment Casting

Leyao Zhou[1, 2], Pang Zhang[1, 2], Daiyin Zhao[2, 3], Donghong Wang[1, 2, *], Da Shu[1, 2], Baode Sun[1, 2]
1.Shanghai Key Laboratory of Advanced High Temperature Materials and Precision Forming, School of Materials Science and Engineering, Shanghai Jiao Tong University, Shanghai 200240, China
2.State Key Laboratory of Clean and Efficient Turbomachinery Power Equipment, School of Materials Science and Engineering, Shanghai Jiao Tong University, Shanghai 200240, China
3.Dongfang Electric Corporation Dongfang Turbine Co., LTD, Deyang 618000, China
*Corresponding address: e-mail: wangdh2009@sjtu.edu.cn

Abstract: To control the dimensional accuracy of the turbine blades, this study proposed a novel integrated computational framework named AICAST, implanted with the response surface methodology optimization model and multi-layer perceptron neural network is proposed. The deformation of the blade casing reaches the minimum via AICAST of 0.1504 mm. The maximum prediction difference of the MLP neural network model is 0.0123 mm, and the mean absolute percentage error is 2.71%, indicating that the deformation of blade casting can be modeled and predicted.

Keywords: turbine blade; dimensional accuracy; ICME; investment casting

1 Introduction

The increasing sophistication of computer hardware and software led to the computer simulation area towards the new stage of ICME (Integrated Computational Materials Engineering), which was first launched by the U. S. National Materials Advisory Board Committee in 2008 [1-3]. ICME is a simulation-driven design approach that employs batch-processing script files to realize multiscale-multiphysics modelling [4]. It is based on the understanding of structure-process-dimension relationships . In this study, the core concept of the ICME approach from gas turbine perspectives is that it employs bottom-up modeling and simulation, calibrated and validated by experiments and measurement, combined with the top-down requirements-driven exploration of deviation transmission during different stages in IC. Many stages are involved in the ICME approach from a gas turbine perspective, such as ceramic core preparation, wax injection, shell roasting, and metal casting. To achieve better dimensional accuracy of turbine blades, the causes and effects of each stage during IC should be analyzed.

2 Experimental procedure

The traditional manual settings of these numerical simulations are a long cycle and cost-inefficient. To solve this problem, an ICME framework named AICAST is proposed to provide a simulation service framework for simulation tools integration and process optimization for IC. The main pages and functional architecture of the AICAST platform are shown in Figure 1. The overall architecture is divided into three layers: the resource, functional, and application layers. Since the ICME framework is designed as a platform for general use, the users can build their workflows and automatically operate the script files on this platform conveniently following the steps (a - c): (a) Firstly, users should build the module and workflow, including name, input, the output of DOE (Design Of Experiments). (b) Secondly, users should upload all the simulation files to the workflow. The platform will automatically replace the process parameters according to DOE. (c) Finally, users can choose appropriate optimization algorithms or neural networks to achieve satisfactory results. In this study, all the numerical simulations for turbine blades can be integrated and optimized via the AICAST platform, including wax pattern deformation simulation and solidification process simulation. The RSM (Response Surface Methodology) model and MLP (Multi-Layer Perceptron) neural network are chosen to realize the optimization of wax injection and alloy solidification processes via the AICAST platform. In this study, all the numerical simulations for turbine blades can be integrated and optimized via the AICAST platform, including wax pattern deformation simulation and solidification process simulation. The RSM (Response Surface Methodology) model and MLP (Multi-Layer Perceptron) neural network are chosen to realize the optimization of wax injection and alloy solidification processes via the AICAST platform. In addition, the RSM and MLP models are encapsulated in the AICAST platform so that users can use these algorithms directly.

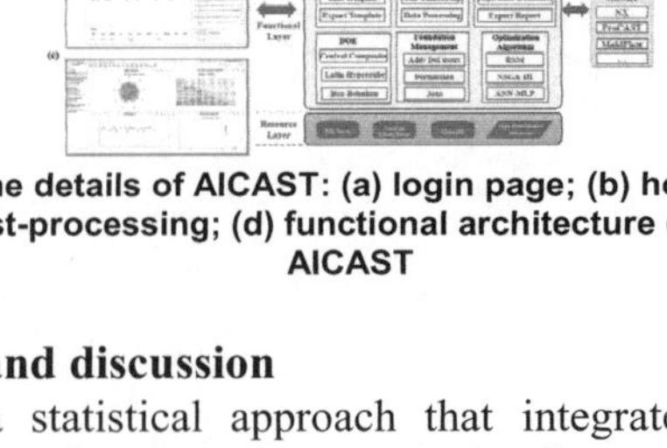

Figure 1 The details of AICAST: (a) login page; (b) home page; (c) data post-processing; (d) functional architecture diagram of AICAST

3 Result and discussion

RSM is a statistical approach that integrates the DOE technique and regression analysis for modeling the relationship between multiple inputs and responses. The response surfaces of the injection velocity, packing pressure, melt temperature, and warpage deformation are modeled. The optimal combination of process parameters for key nodes and X direction is X = [2.0, 64.8, 140], and the warpage deformation of key nodes is 1.79 mm, which is 60.39% lower than that of the original (4.519 mm). The optimal combination of process parameters for Y and Z directions is X = [2.0, 64.8, 108] and X = [2.0, 64.0, 100], respectively. MLP is one of the most widely used neural network architectures in literature for classification or regression problems. The output of the MLP structure is produced by linear combinations of the outputs of hidden layer nodes in which every neuron maps a weighted average of the inputs through a sigmoid function. The MLP neural network performs well in blade casting deformation prediction under the influence of process parameters. The MAPE (Mean Absolute Percentage Error) of the MLP neural network is 2.71%. The max difference between true values and predicted values is 0.0123 mm, indicating that the prediction accuracy is satisfactory in the MLP model when X = [1490, 1220, 8], the deformation of the blade casing reaches the minimum amount, which is 0.1504 mm. The real wax pattern, the shell, and the final blade casting are shown in Figure 2.

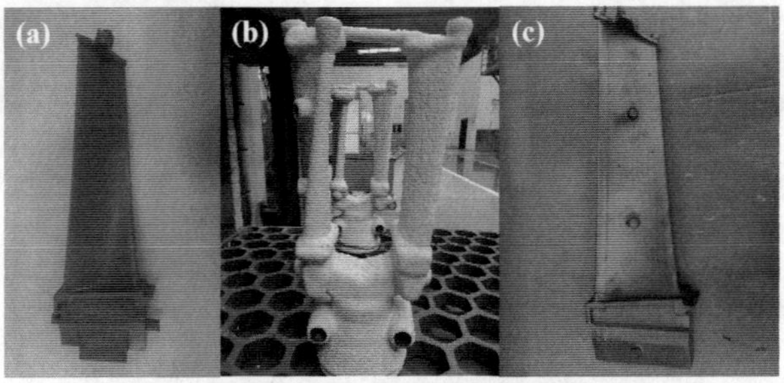

Figure 2 Experimental verification: (a) wax pattern; (b) the shell; (c) blade casting

4 Conclusion

A novel ICME framework (AICAST) coupled with the RSM and MLP model is proposed to investigate the mechanisms of blade deformation and achieve the appropriate process parameters. The optimal combination of process parameters for key nodes is $X1$ = [2.0, 64.8, 140], and the warpage deformation of key nodes is 1.79 mm, which is 60.39% lower than that of the original (4.519 mm). When $X2$ = [1490, 1220, 8], the deformation of the blade casting reaches the minimum amount of 0.1504 mm. The max difference of MLP neural network model is 0.0123 mm and MAPE is 2.71%, indicating that deformation of the blade casting can be modeled and predicted.

5 Acknowledgments

This work was financially supported by the National Science and Technology Major Projects of China (J2019-VI-0004-0117), the National Key Research and Development Program of China (2022YFB3706800), and the National Natural Science Foundation of China (52090042).

References

[1] Pollock T M, Allison J E, Backman D G. Integrated computational materials engineering: A transformational discipline for improved competitiveness and national security. National Materials Advisory Board, NAE, National Academies Press, Washington DC, 2008.
[2] Joost W J. Reducing vehicle weight and improving U.S. energy efficiency using integrated computational materials engineering. JOM, 2012, 64: 1032-1038.
[3] Allison J, Li M, Wolverton C, Su X M. Virtual aluminum castings: An industrial application of ICME. JOM, 2006, 58: 28-35.
[4] Himanen L, Geurts A, Foster A S, Rinke P, Smith D, Brinson C. Data-driven materials science: Status, challenges, and perspectives. Adv. Sci., 2019, 6: 1900808.

Development of Automatic Flame Cutting Equipment for Medium and Large Castings

Mengliang Lu*, Tianying Ji, Kai Lu, Zhilong Chen
Intelligent Manufacturing Division, CRRC Qishuyan Institute Co.,Ltd., 258 Wuyi Road, Wujin District, Jiangsu Province, China
*Corresponding address: e-mail: 011500003384@crrcgc.cc

Abstract: For the pouring and riser cutting process of cast steel, cast iron, and carbon steel parts, CRRC Qishuyan Institute Co.,Ltd. has independently developed an intelligent flame cutting process equipment for castings to promote the greenization, intelligent, and digitalization of customers' casting post-processing procedures. The equipment uses two industrial six-axis robots integrated with two sets of flame cutting devices, integrating key technologies such as laser vision recognition, intelligent control technology, electrical proportional control technology, Dual robot coordination technology, flame recognition and sensing technology, and advanced casting cutting technology. The equipment can autonomously identify the position of the pouring and riser and correct the cutting path,automatically correct the pressure and flow of oxygen, gas or propane. The workpiece is fixed on a dual-position fixture such as a rotary table, and the loading and cutting can be performed simultaneously. It can automatically and adaptively complete the cutting of various sizes, materials, and types of castings pouring and risers. The cutting surface has a smooth and flat surface quality, and it has now formed a demonstration application.

Keywords: dual robot coordination technology; intelligent flame cutting

1 Introduction

The bolster and side frame are crucial components of the railway truck bogie, with a workpiece length of approximately 2,500 mm. The general dimensional tolerance is ±3mm, and the maximum cutting thickness of the pouring gate is around 120mm. These parts are cast using B+ grade steel (ZG25MnCrNi), making it challenging to cut.

Currently, manual hand-held torches are commonly used in the cleaning process for cutting the bolster and side frame, but this approach does not enhance casting quality. The automation level in post-processing remains low, resulting in a heavy workload during cutting operations while demanding high operator skills and quality standards. Moreover, working conditions involve high temperatures, dust, noise, arc pollution, safety risks as well as occupational disease hazards. Therefore, addressing post-processing challenges through automation and intelligent technology represents a key area for future research.

Figure 1 Composition of flame cutting module.

The system utilizes 2 Motoman GP35L robots for coordinated operations (multiple robots can be selected for large workpieces), and is equipped with an equal pressure cutting torch, laser displacement sensor, anti-collision sensor, flow proportional control valve, automatic ignition mechanism, flame monitoring sensor, sheet metal protection room, running mechanism, waste recycling box, etc. The workpiece is secured on the reverse positioner.

The flame cutting task involves a form of target coordination, requiring significant spatial and temporal synchronization between the two robotic arms despite the absence of physical interaction.

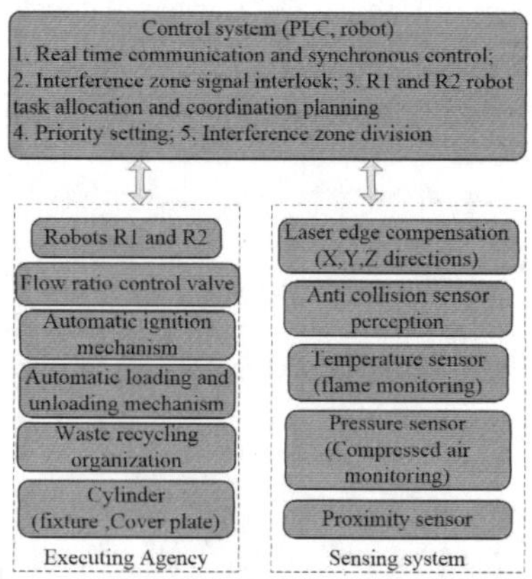

Figure 2 Composition of control module

The robot uses a laser displacement sensor to correct the cutting path. The sensor searches the cutting position in *X*, *Y* and *Z* directions, compares it with the reference position, and automatically corrects the cutting path according to the position deviation.

Table 1. Automatic cutting equipment technical parameters

Type	Configurations
Cutting gas pressure	Efficient natural gas(or propane) ≥0.08 MPa,Oxygen pressure≥0.8 MPa;
Ignition mode	Automatic ignition
Support cutting material	Cast iron, cast steel, carbon steel
Maximum cutting thickness	300 mm
Robot workspace	2,538 mm
Residual height after cutting	Edge cutting: ≤3 mm, Non-edge cutting: ≤5 mm
Cutting speed	160-230 mm/min
Gas control system	Automatic proportional control
Flame detection	The temperature sensor monitors the flame in real time
Collision detection	Anti-collision sensor

2 Experimental procedure

Test preparation: Pipeline natural gas (CH_4 accounted for 94%) and oxygen with 99% purity were selected as the test gas. The test object was the spout of the bolster side frame, equipped with Jierui isobaric cutting gun and 307-6 cutting nozzle; Gas and oxygen are controlled proportionally through the flow control valve, and the pressure and flow data are monitored during the test.

Figure 3 Automatic cutting test platform.

3 Result and discussion

The robot system performs automatic adaptive flame cutting in turn, without manual participation, the root of each pouring gate is cut completely, there is no adhesion after cutting, and the riser can fall automatically. The cutting surface was compared by sample blocks, about 50-500 μm, there was a little hanging slag after cutting, the cutting surface quality was good, the residual height was less than 3 mm, and the consistency was good after cutting.

Figure 4 surface quality of automatic cutting

4 Conclusion

The development of automatic flame cutting equipment for medium and large castings solves the problems such as low degree of automation, poor consistency of manual cutting quality and high labor intensity within the industry for a long time, and provides a solution for speeding up the intelligent upgrading of the cutting process of the head. The cutting surface quality, cutting residual height and cutting precision are better than manual cutting.

References

[1] Li Y J. Cutting technology and application [M]. Beijing: Beijing Industry Press, 2004.

[2] Patricia M E, Maria C P. Analysis and influence of acetylene and propane gas during gas cutting of 1045 carbon steel [J]. Materials Engineering and Perfor - Mance, 2006, 15: 684-692.

Research on Metal Mold Casting Technology of Copper Alloy High Pressure Valve

Ling Zhao[1],*, Huijun Kang[2], Yunbao Gao[1], Haijun Zhang[1], Qiu Du[1], Can Wang[1]
1. National Key Laboratory of Advanced Casting Technologies, Shenyang 110022, China
2. Dalian University of Technology, Dalian 116024, China
*Corresponding address: e-mail: zhaoling1214@126.com

Abstract: This paper introduces the structure and technical requirements of copper alloy high-pressure valve bodies. Based on the structural characteristics and quality requirements of the castings, a precision casting process combining metal mold with 3D printed cores is proposed, and the casting process is designed accordingly. Detailed explanations are provided for the parting surface, gating system, and riser design of the casting. Computer simulation software is employed to analyze the filling and solidification processes of the copper liquid, and the process is optimized based on the simulation results. Various issues such as the integrity of filling for thin-walled complex castings like high-pressure valve bodies, mold venting, structural shrinkage, and solidification cracks are addressed, leading to the development of a stable production technology suitable for complex structured high-pressure valve bodies with high quality.

Keywords: copper alloy; high-pressure valve body; metal casting; numerical simulation

1 Introduction

Shipboard high-pressure valve bodies serve as critical components in high-pressure pipeline systems, with numerous supporting products and stringent quality requirements. These products are structurally complex, with significant differences in wall thickness and numerous thermal junctions, making it difficult to compensate for solidification shrinkage. Traditional sand casting methods, previously used for similar valve bodies, suffer from poor dimensional accuracy, high gas emission rates, and low thermal conductivity and capacity, significantly impacting the solidification quality and dimensional accuracy of copper alloy castings [1]. As a result, the yield rate of the product has been consistently low, and its quality unstable [2].

To address these challenges, a study was conducted on the precision casting process for copper alloy using a combination of metal molds and 3D printed cores, taking into account the characteristics and quality requirements of the product and leveraging advanced production modes for copper alloy castings. This approach maximizes the advantages of metal molds, such as fast heat transfer, dense casting structure, excellent surface quality, and high dimensional accuracy, resulting in superior internal quality of the castings.

2 Experimental procedure

The maximum wall thickness of the copper alloy high-pressure valve body is 39 mm, while the minimum wall thickness is only 6 mm. The material of the casting is ZCuAl10Fe3, with its chemical composition and mechanical properties meeting the requirements of the GB/T 12225-2018 standard. 100% of the casting's interior undergoes radiographic inspection, conducted in accordance with the procedures outlined in NB/T 20003.3, and acceptance criteria align with ASTM-E-272 standards at Level 3. The casting dimensions adhere to the CT10 level tolerance specified in GB/T 6414-97. High-pressure testing involves applying pressures ranging from 10 to 80MPa. Consequently, the quality requirements for high-pressure valve body castings used in QT applications are extremely stringent.

3 Result and discussion

The rationality of the aforementioned casting process scheme was validated through simulation analysis using ProCast software. The simulation process parameters were as follows: mold preheating temperature of 300 °C, pouring temperature of 1,200 °C, ambient temperature of 23 °C, and pouring time of 14 seconds.

Filling Simulation Results

The filling simulation results of the casting at different times are shown in Figure 1. The filling of the alloy liquid is smooth without significant turbulence. This facilitates the overflow of gases, slag, and other impurities, preventing the formation of secondary oxide inclusions.

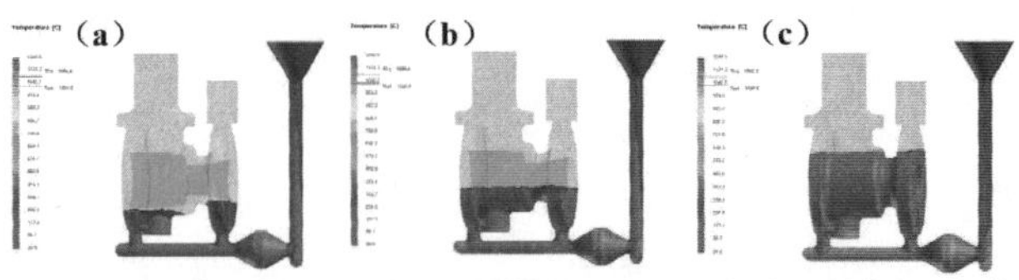

Figure 1 Simulation of mold filling process (a-3 s, b-5 s, c-8 s)

Simulation results of temperature field

The simulation results of the temperature field are shown in Figure 2. From Figure 2(a), it can be observed that at location I, the alloy liquid has the longest solidification time. To avoid the formation of shrinkage cavities, additional measures such as adding chills to accelerate its solidification rate need to be implemented at this location.

Since risers are not suitable to be placed at this location, internal chills are added instead.

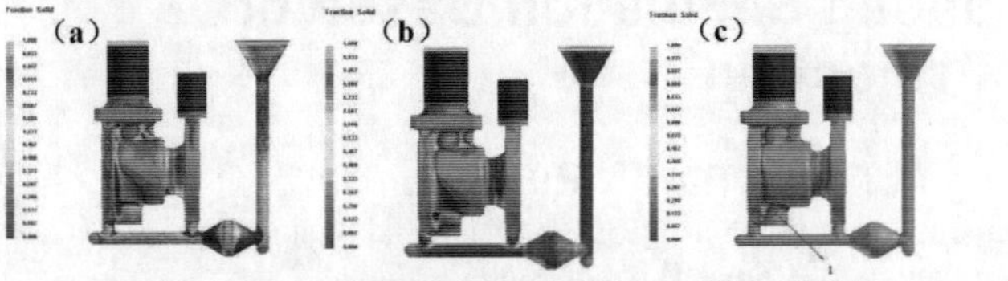

Figure 2 Simulation results of temperature field (a-17 s, b-27 s, c-30 s)

Simulation results of defects

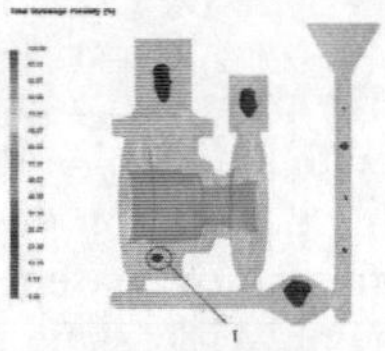

Figure 3 Simulation of defect distribution

The simulation results of shrinkage and shrinkage porosity defects are shown in Figure 3. A relatively severe shrinkage porosity defect has formed at location I, which is consistent with the conclusion drawn from the temperature field simulation, where the longest solidification time is required at this location.

Final process plan and production validation

The final process plan, optimized accordingly, is depicted in Figure 4. Internal chills are placed at position I of the casting to expedite its solidification rate, thereby achieving sequential solidification.

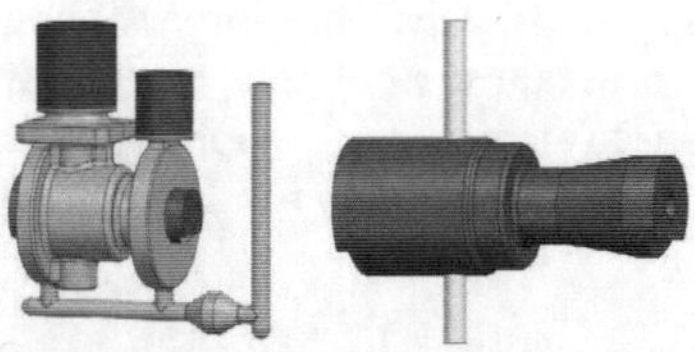

Figure 4 Optimized process scheme

Following the optimization of the casting process, production validation was conducted on the valve. The castings produced after process optimization exhibited good formability and precise dimensional control. The mechanical properties, high-pressure test results, and radiographic inspection results of the castings all meet customer requirements. The mechanical performance test results of the product are shown in Table 1.

Table 1. Product mechanical performance test results

No.	R_m/MPa	$R_{P0.2}$/MPa	A/%	HBW
GB/T12225	≥490	≥180	≥13	≥100
ZCu201914	642	250	21.5	145

4 Conclusion

This study combines advanced copper alloy casting production methods and employs precision casting technology using metal molds with 3D printed cores for copper alloys. Through the design of the process and optimization using computer simulation-assisted techniques, the produced castings have undergone high-pressure testing (import pressure ranging from 10 to 80 MPa) with excellent air tightness. The composition, dimensions, radiographic inspection results, and mechanical properties of the castings all meet the relevant requirements. Various issues such as the integrity of filling for thin-walled complex castings like high-pressure valve bodies, mold venting, structural shrinkage, and solidification cracks have been effectively addressed.

References

[1] Tang B. Research on casting technology of large thin-walled parts for automobiles [J]. Foundry Technology, 2013, 34(11): 1577-1579.

[2] Li S Z. Research on high pressure die casting technology and mechanical properties analysis of aluminum alloy shock tower for a certain automobile [D]. Jilin University, 2020.

Mesh Segmentation Method for Casting Simulation Based on Boundary Mesh Projection

YuweiCui[1], Yuhong Zhao[1,2,4,*], Xiaolong Pei[1], Yue Pan[1,3], Hua Hou[1,3]
1 School of Materials Science and Engineering, Collaborative Innovation Center of Ministry of Education and Shanxi Province for High-performance Al/Mg Alloy Materials, North University of China, Taiyuan 030051, China
2 Beijing Advanced Innovation Center for Materials Genome Engineering, University of Science and Technology Beijing, Beijing100083, China
3 School of Materials Science and Engineering, Taiyuan University of Science and Technology, Taiyuan030024, China
4 Institute of Materials Intelligent Technology, Liaoning Academy of Materials, Shenyang110004, China
*Corresponding address: e-mail: zhaoyuhong@nuc.edu.cn

Abstrac: Mesh segmentation is an important part of the casting numerical simulation process, many of the current casting simulation software in the processing of complex castings with sloped and curved surfaces are still unable to effectively avoid the jagged error at the boundary, resulting in the loss of the casting slope information, thus affecting the accuracy of the subsequent computational simulation. For this reason, this paper proposes a mesh segmentation method for casting simulation based on boundary mesh projection. Firstly, the STL file of the 3D model of the casting is read and the defects are processed, then the casting is structured as a whole, the boundary grid is obtained by differentiating the inner and outer grids of the casting, and finally the boundary grid is projected. Application examples show that this method can get the mesh file that fits the shape of the casting 3D model, which is of great significance for the improvement of casting CAE accuracy.

Keywords:numerical simulation of casting; structured mesh sectioning; boundary meshes; jagged errors; projection processing.

1 Introduction

Casting moulding is a process in which the metal is melted and poured into a casting pattern, cooled, solidified and cleaned to obtain a casting of a predetermined shape, size and properties[1,2]. Traditionally, this complex process relies on a large number of experiments, which is both time-consuming and costly[3,4]. With the development of computer technology, numerical simulation technology in the casting of the increasingly wide range of applications[5]. The numerical simulation system includes three modules: pre-processing, intermediate calculation and post-processing. Pre-processing involves three-dimensional solid modelling and mesh splitting, mesh splitting quality directly affects the calculation accuracy and speed.

Pang et al.[6] and Hou et al.[7] used adaptive Cartesian grid and immersed boundary method to achieve non-uniform mesh dissections, respectively, but these two methods suffer from high computational complexity and numerical errors when dealing with complex geometries.Owen et al.[8] proposed a parallel raster method-Sculpt algorithm, but with a poor overall structure.

Based on this, this paper proposes a mesh sectioning method for casting simulation based on boundary mesh projection based on the pre-processing system of the home-made EasyCast software independently developed by the group, which is based on reading the STL file and performing structured mesh sectioning, dividing the casting interior and boundaries, obtaining the mesh file that fits the shape of the three-dimensional model, and improving the simulation accuracy of the complex castings.

2 Boundary mesh projection processing algorithm

The mesh dissecting method proposed in this paper for casting simulation includes six main steps: firstly, read the STL file and and use the ray method to deal with the defects, then use the slicing method to carry out structured mesh dissections, then divide the interior and boundary of the casting model to obtain the boundary mesh, and project the boundary mesh to match the shape of the casting, and finally, save and output the mesh file for subsequent simulation use.

3 Examples of boundary mesh projection algorithms and analysis and discussion

In order to verify the effectiveness of the proposed mesh sectioning method for casting simulation based on boundary mesh projection, this paper carries out boundary mesh projection processing for a common partition piece (Figure 1), the casting has a diameter of 32 cm, a height of 8 cm, and a sectioning size of 0.5 × 0.5 × 0.5 cm. Processing is carried out in sequence according to the method proposed in this paper, and Figure 2 is the structured mesh sectioning of the casting as a whole, and Figure3 shows the sectioning result map after being processed by the boundary mesh projection algorithm. Comparison of the two results shows that the method has a great advantage in dealing with the castings with rounded and bevelled surfaces, which effectively avoids the "jagged error" and improves the accuracy of the subsequent casting simulation.

Figure 1CAD drawing of partition parts

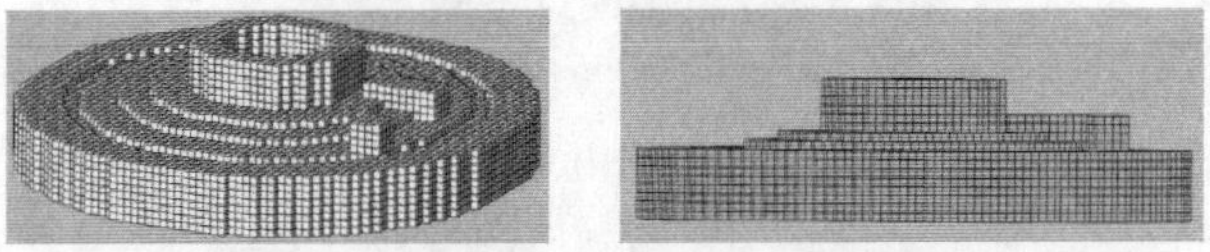

Figure 2 Schematic diagram of the results of structured grid division

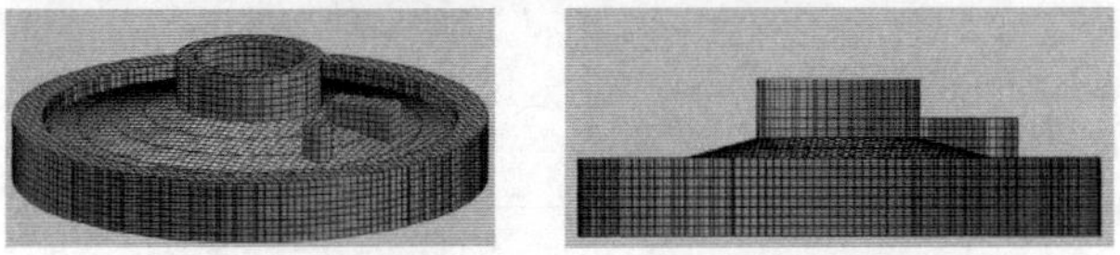

Figure 3 Schematic diagram of partition results afterboundary grid projection processing

4 Conclusion

This paper proposes a boundary mesh projection-based method for casting simulation and simulation grid partitioning to address the issue of jagged edge errors when dealing with complex castings with inclined and curved surfaces. The method involves reading the STL file of the casting's 3D model and performing defect processing, conducting structured grid partitioning, and dividing the internal and boundary of the casting to obtain boundary meshes. Finally, it performs projection processing on the surface boundary mesh. The generated hexahedral mesh maximally preserves the structural information of the casting, effectively avoids the loss of casting slope information, enhances the simulation accuracy of casting CAE, and is of great significance for optimizing casting processes.

5 Acknowledgments

This work was financially supported by the National Natural Science Foundation of China (Nos. 52375394, 52275390, 52205429, 52201146, 52305429), the National Defense Basic Scientific Research Program of China (JCKY2020408B002), and the Key Research and Development Program of Shanxi Province (No. 202102050201011).

References

[1] Hou H. Macro-micro numerical simulation of casting solidification process and engineering applications. Shanxi:North Central University,2005.

[2] Jin Y C, Hou H and Zhao Y H. Numerical simulation of material forming process. Beijing:Weapon Industry Press,2004.

[3] Sun X P, Zhao Y H, Chen L W, Li Z Q, Jing J H, Li M X, Tian J Z and Hou H. Indirect extrusion casting of ZAlSi7Mg aluminium alloy spacer parts for mine shells. Special Casting and Nonferrous Alloys, 2022, 42: 446-450.

[4] Sun Y Y,Zhao Y H, Zhao B J, Guo Z, Tian X L, Yang W K and Hou H. Multi-component phase-field simulation of microstructural evolution and elemental distribution in Fe-Cu-Mn-Ni-Al alloy. Calphad, 2020, 69: 101759.

[5] Zhao Y H, Liu K X, Hou H and Chen LQ. Role of interfacial energy anisotropy in dendrite orientation in Al-Zn alloys: A phase field study. Materials & Design, 2022, 216:110555.

[5] Zhao Y H, Liu K X, Hou H and Chen LQ. Role of interfacial energy anisotropy in dendrite orientation in Al-Zn alloys: a phase field study. Mater. Design, 2022, 216:110555.

[6] Pang S Y, Chen L L, Liu R X andZhou J X. CSDF: A STL-based non-sawtooth differential mesh dissecting algorithm. China Mechanical Engineering, 2006, 10:219-222.

[7] Mao H K, Mao W M and Hou H. Numerical simulation of mould filling process based on boundary cut mesh. Casting Technology,2010,31(9):1237-1240.

[8] Owen S J, Brown J A, Ernst C D, Lim H and Long K N. Hexahedral mesh generation for 3D computational materials modeling. Procedia Engineering, 2017: 167-179.

A Phase-Field Simulation of Easily Switchable Vortex Structure for Multilevel Low-Power Ferroelectric Memory

Zhonglei Liu[1], Jinming Cao[1, 4], Qiaoling Wang[2, 3], Hua Hou[1], Yuhong Zhao[1, 2, 3, *]

1 School of Materials Science and Engineering, Collaborative Innovation Center of Ministry of Education and Shanxi Province for High-performance Al/Mg Alloy Materials, North University of China, Taiyuan, 030051, China

2 Beijing Advanced Innovation Center for Materials Genome Engineering, University of Science and Technology Beijing, Beijing, 100083, China

3 Institute of Materials Intelligent Technology, Liaoning Academy of Materials, Shenyang 110004, P R China

4 School of Materials Science and Engineering, Taiyuan University of Technology, Taiyuan 030024, P R China

*Corresponding address: e-mail: zhaoyuhong@nuc.edu.cn

Abstract: In this work, domain types are simplified from 26 to 9 through virtual polarization vectors. By temperature selection, we accomplished a large easily switchable vortex structure with 20% side length. Under electric-field pulses, diverse storage states were presented through 0/1 measurement of 8 positions.

Keywords:topological vortex structure; charged domain walls; multilevel low-power ferroelectric memory; phase-field method

1 Introduction

Experimental studies have shown that a series of changes in the low-resistance state of DWs can be achieved by directly changing the distance between metal electrodes [1]. Although this method can theoretically achieve multilevel data storage, it is difficult to change the distance between metal electrodes in highly integrated ferroelectric domain walls memory in practice. It is necessary to consider changing the conductive properties of DWs from other angles, especially by constructing functional units such as easily switchable topological vortex structures[2, 3], which can achieve domain wall changes with low-resistance and low power.

2 Phase-filed method

In the ferroelectric phase-field model, the order parameter is spontaneous polarization. The evolution of polarization over time and the spatial distribution of polarization can be obtained by solving the Time-Dependent-Ginzburg-Landau (TDGL) equation. In the model, the total free energy density f including Landau energy density f_{Land}, electrostatic energy density f_{elec}, gradient energy density f_{grad}, and elastic energy density f_{elas}.

3 Result and discussion

Multilevel data ferroelectric memory and design of the conductance state

In KNN, the in-plane vortex topological structure can exhibit only eight polarization directions because of the perovskite-type ferroelectric structure. In this way, the vortex can be represented in crystal structure by 8 in-plane directions. After simplification, the equivalent domain contents at different temperatures can be obtained, as shown in Figure 1(a). From Figure 1(b), we got a normal easily switchable vortex structure at 500K. The width of the topological vortex structure is about 13 nm accounting for about 20 percent of the whole system.

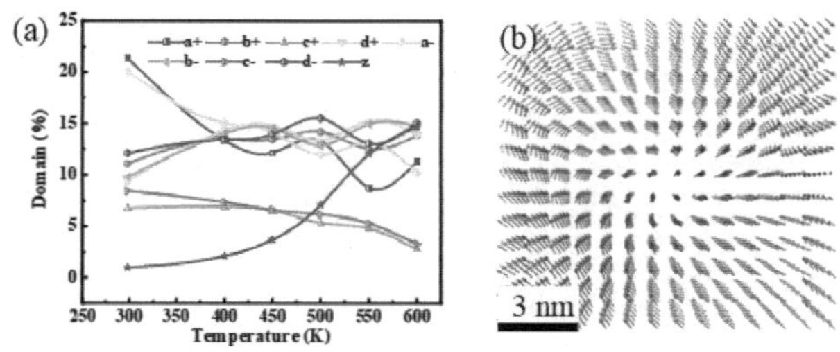

Figure 1(a) Equivalent domain contents at different temperatures; (b) Image of vortex domains dimension

Conductive state changes and data storage

For the application of data storage, only a change in the conductance state of DW is required, including the change in quantity or position. The output state at a given location can be determined by measuring the conductive domain wall state at the local position, as shown in Figure 2(a). In topological structures, the vortex domain wall's conductive can be switched with 0/1 by applying low electric-field pulses. During the measurement process, each voltage corresponds to a special storage state "cloud map", as shown in Figure2(b). Therefore, simply measuring binary conductivity can achieve multilevel data storage with higher recognition accuracy. Different "cloud maps" represent different data maps, thereby providing the possibility for achieving multilevel data ferroelectric memory.

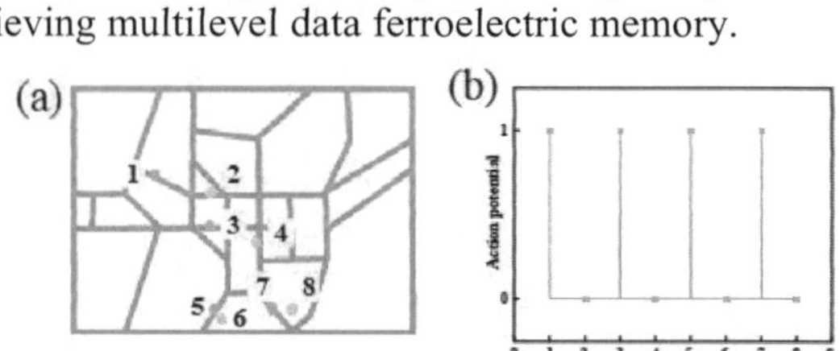

Figure 2 (a) Electrical conduction at vortex structure; (b) Images of storage state "cloud map"

Conclusion

By temperature control to ensure that external low electric-field pulses regulation of topological vortex structures have better diversity and clarity. These easily transformable domain wall conductive states can help identify positive and negative conductive states. Domain state maps were obtained through the measurement of 8 positions under different electric-field pulses.

Acknowledgments

This work was financially supported by the National Natural Science Foundation of China (No. 52074246),the National Defense Basic Scientific Research Program of China (JCKY2020408B002), andthe Key Research and Development Program of Shanxi Province (202102050201011).

References

[1] SharmaP, ZhangQ, SandoD, et al. Nonvolatile ferroelectric domain wall memory. Sci. Adv., 2017, 3(6): e1700512.
[2] Liu D, Shi X, Wang J, et al. Phase-field simulations of surface charge-induced ferroelectric vortex. J. Phys. Appl. Phys., 2021, 54(40):405302.
[3] Wang M, Liu D, Wang J, et al. Phase-field simulations of topological conservation in multi-vortex induced by surface charge in $BiFeO_3$ thin films. J. Adv. Dielectr., 2023:2343002.

Optimizing Casting Process Using a Data-Driven Approach with Tiny Dataset

Yuhong Zhao[1,2,4*], XiaolongPei[1], Jiaqi Pei[1], Hua Hou[1,3], Zhiqiang Duan[1]
1. School of Materials Science and Engineering, Collaborative Innovation Center of Ministry of Education and Shanxi Province for High-performance Al/Mg Alloy Materials, North University of China, Taiyuan030051, China
2. Beijing Advanced Innovation Center for Materials Genome Engineering, University of Science and Technology Beijing, Beijing 100083, China
3. School of Materials Science and Engineering, Taiyuan University of Science and Technology, Taiyuan 030024, China
4 Institute of Materials Intelligent Technology, Liaoning Academy of Materials, Shenyang110004, China
*Corresponding address: e-mail: zhaoyuhong@nuc.edu.cn

Abstract: Due to high cost of experiments and complexity of process parameters, achieving precise and cost-effective casting process optimization has been a persistent challenge. A novel method was proposed that initially employed Latin hypercube sampling that uniformly covering the design range of each process parameter to enhance data quality. Subsequently, it integrates Bayesian optimization accounting for uncertainty, to optimize the casting process for tiny dataset. The optimization of the squeeze casting processes for aluminum alloy partition part was used for experimental verification.Through 25 datasets and 6 experimental iterations, casting properties have been enhanced, with both tensile strength and elongation increasing by approximately 20%.

Keywords:casting; process optimization; machine learning; Bayesian optimization

1 Introduction

Casting stands out as one of primary manufacturing methods for key components in aviation, aerospace, and other sectors. To enhance performance and prevent defects like shrinkage cavities, and cracks, optimizing process parameters becomes crucial and essential once material and casting system structure are determined.The problem that exists with traditional empirical trial-and-error methods, which is limited by time, material and labor costs. Design of experiment (DoE)has been employed to enhance efficiency of process optimization. However, when confronted with multiple factors (process parameters) and levels (process range), the cost required for DoE significantly escalates. More seriously, global optimal process remains elusive. Additionally, casting numerical simulation techniques, especially some casting software such as ProCAST [1], EasyCast [2] etc., can offer valuable process optimization insights prior toproduction. Nevertheless,most simulation results are qualitative. Establishing a quantitative relationship between casting process and properties through numerical simulation is challenging [3]. Consequently, there is an urgent demand for a more efficient strategy that directly optimizes casting performance while avoids local optima as much as possible. In this study, we introduce a method to optimize casting processes with extremely small datasets by integrating Latin hypercube sampling (LHS) and Bayesian optimization (BO). The effectiveness of this method is demonstrated in optimizing squeeze casting processes for aluminum alloy partition part.

2 Experimental procedure

Our method mainly includedthree stages: 1) Data collection: We utilized LHS strategy to select test points (process parameters) and conducted experiments to obtain casting properties data, including ultimate tensile strength (UTS) and elongation; 2) Surrogate model establishment: Lasso and XGBoost were employed to develop predictive models based on collected dataset; 3) BO: Through query strategies, we identified potentially optimal process points in unlabeled datasets, followed by experimental validation. The resulting experimental data was incorporated into the original dataset to update the surrogate model. This iterative stage continued until no further improvement in casting properties was observed. Figure 1 illustrates3Dmodel ofpartition parts.The material of the partition is ZL101.This study focuses on optimizing the key process parameters for squeeze casting, including pouring temperature, die temperature, squeeze pressure, dwell time, and squeeze velocity.

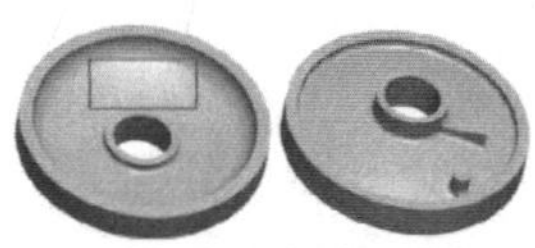

Figure 1 3D model of partition parts

The expected improvement (EI) was selected as the acquisition function due to its efficient global optimization (EGO) [4]. To achieve process optimization for both UTS and elongation improvements, final acquisition function is shown in Eq. (1):

$$EI_{fin} = EI_{UTS} \times EI_{EL} (1)$$

3 Result and discussion

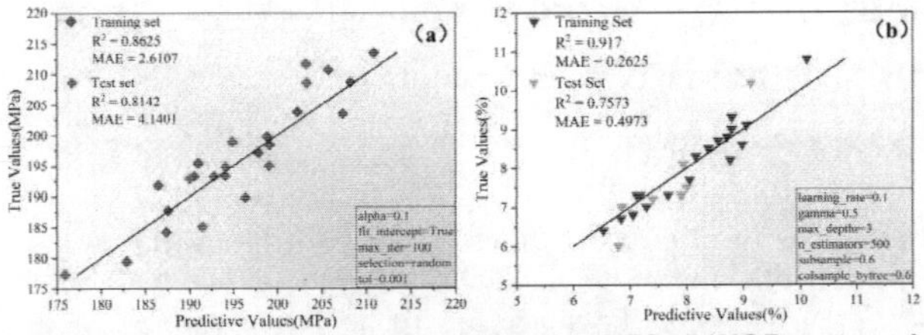

Figure 2 (a)Lasso model for predicting UTS; (b)XGBoost model for predicting elongation

Figure 2 displays the R^2 and MAE results of UTS and elongation prediction models on the training set and test set. Lasso model achieved R^2 value of 0.8142, indicating an undeniable gap in accurately predicting strength values with the initial model. XGBoost achieved R^2 value of 0.7573on the test set.In contrast to UTS prediction model, XGBoost elongation predictionmode had a lower R^2 and higher uncertainty, making it challenging to accurately captureelongation distribution across the entire process parameter set using the initial model.

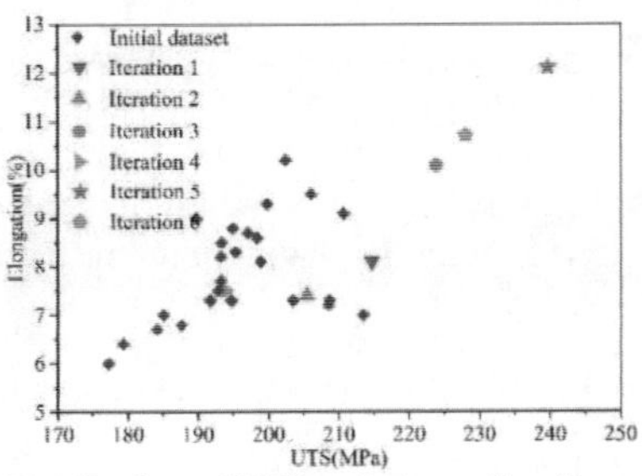

Figure 3 Distribution of UTS and elongation throughout the iteration process

Figure 3 illustrates the distribution of UTS and elongation throughout the iteration process. Remarkably, after only 5 iterations, UTS and elongation had both increased by approximately 20% compared to LHS dataset. However, properties started declining after the 6th iteration, prompting us to conclude iterative process.

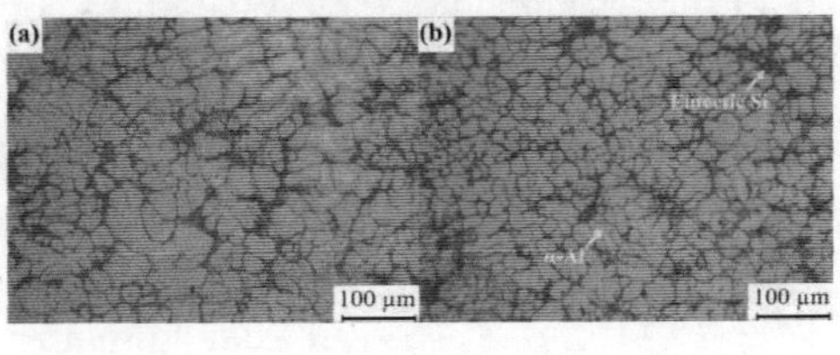

Figure 4 Microstructure of partition parts under different processes: (a) initial dataset; (b) 5th iteration

Figure 4 displays the microstructure of partition parts under various processes. The microstructure of the partition part is composed of primary α-Al and eutectic Si. Significant refinement of α-Al occurs after BO, which is the reason for the improved properties of partition parts.

4 Conclusion

We proposed a process optimization strategy for cast process tiny data set, which integrates Latin hypercube sampling, machine learning, Bayesian optimization, and experimental feedback to adaptively identify the optimal process. We focused on optimizing squeeze casting processes for aluminum alloy partition parts to illustrate this strategy. The optimal process was achieved after six iterations, leading to an increase of approximately 20% in both UTS and elongation.

5 Acknowledgments

This research project was financially supported by the National Natural Science Foundation of China (Nos. 52074246), the National Defense Basic Scientific Research Program of China (Nos. JCKY2020408B002), and the Key R&D Program of Shanxi Province (Nos.202102050201011).

References

[1] Jing J H, Chen L W, Wang K L, Hou H and Zhao Y H. Indirect squeeze casting of ZM5 baffle with linear segregation and its formation mechanism.J. Mater. Res. Technol., 2023, 27: 7024-7036.

[2] Hou H, Zhao Y H and Zhao Y H. Prediction method of shrinkage cavity and shrinkage porosity in solidification process under low-pressure casting.J. Comput. Theor. Nanos., 2012, 9(9): 1447-1452.

[3] Jolly M and Katgerman L. Modelling of defects in aluminium cast products. Prog. Mater. Sci., 2022, 123.

[4] Liang Q, Gongora A, Ren Z, Fisher J and Buonassisi T. Benchmarking the performance of Bayesian optimization across multiple experimental materials science domains. Npj Comput. Mater., 2021, 7(1): 188.

Simulation of Microstructure Phase-Field during Wire Arc Additive Manufacturing of Aluminum Alloys

Fuchen Wang[1], Weipeng Chen[1], Hua Hou[1,3], Yuhong Zhao[1,2,4*]
1. School of Materials Science and Engineering, Collaborative Innovation Center of Ministry of Education and Shanxi Province for High-performance Al/Mg Alloy Materials, North University of China, Taiyuan030051, China
2. Beijing Advanced Innovation Center for Materials Genome Engineering, University of Science and Technology Beijing, Beijing100083, China
3. School of Materials Science and Engineering, Taiyuan University of Science and Technology, Taiyuan030024, China
4. Institute of Materials Intelligent Technology, Liaoning Academy of Materials, Shenyang110004, China
*Corresponding address: e-mail: zhaoyuhong@nuc.edu.cn

Abstract: Wire Arc additive manufacturing (WAAM) technology has attracted wide attention due to its advantages of high deposition rate, low cost, stable product performance and high material utilization. However, the complex thermal history under rapid solidification limits the study of the solidification structure under rapid solidification. In this work, we studied the dendritic growth and solute segregation behavior at different scanning speeds in the WAAM process based on finite element and phase field methods. It is found that the dendrite growth under rapid solidification is strongly dependent on the temperature gradient. With the increase of temperature gradient, the dendrite growth rate is accelerated, and the solute segregation is more serious.

Keywords: WAAM;rapid solidification; dendrite growth; solute segregation

1 Introduction

At present, the integrated molding of complex parts is the trend in the manufacturing industry. Because the traditional manufacturing methods (machining, forging, powder metallurgy, etc.) can't realize the increasingly complex industrial needs. Therefore, Additive Manufacturing (AM) technology has gradually received widespread attention. The complexity of process parameters and the sensitivity of microstructures to processing conditions in WAAM make it difficult to accurately grasp the evolution of microstructures under rapid solidification. However, investigating the evolution of solidified tissues under rapid solidification in additive manufacturing by experimental methods is bound to incur a lot of unnecessary consumption. Based on the finite element, phase field method framework, Keller et al. [1] investigated the microstructure evolution during laser powder bed melting of nickel-based high-temperature alloys and verified the reliability of the simulation results using experiments. However, in the current phase field simulations, researchers tend to ignore the non-equilibrium-dominated solidification effects under rapid solidification. This can lead to biased predictions of the microstructure. In this work, investigates the influence of non-equilibrium effects on the microstructure evolution under rapid solidification. The microstructure evolution at different scanning speeds was investigated. The results show that with the increase ofscanning speed, the temperature gradient during solidification decreases, the growth rate of dendrites decreases, and the solute segregation is weakened. The results provide a general strategy for optimizing the WAAM processing.

2 Experimental procedure

In this work, a 3D finite element model was used to calculate the temperature gradient at different scanning speeds during the WAAM process.
In this work, the governing equation for the phase-field simulation is given by the following equation [2, 3]:

$$\tau a_s^2(\vec{n})\frac{\partial\phi}{\partial t}=\vec{\nabla}\cdot\left[a_s(\vec{n})^2\vec{\nabla}\phi\right]+\sum_{m=x,y}\partial_m\left(|\vec{\nabla}\phi|^2a_s(\vec{n})\frac{\partial a(\vec{n})}{\partial(\partial_m\phi)}\right)+$$

$$\phi-\phi^3-\lambda(1-\phi^2)^2\left(U+\frac{\left(z-\int_0^t V_p(t')dt'\right)}{l_T}\right)$$

$$\left(\frac{1+k-(1-k)\phi}{2}\right)\frac{\partial U}{\partial t}=\vec{\nabla}\cdot\left(D_l q(\phi)\vec{\nabla}U-\vec{j}_{at}\right)+[1+(1-k)U]\frac{1}{2}\frac{\partial\phi}{\partial t}$$

where $a_s(\vec{n})$ is the interface energy anisotropy,$\theta$is the angle between theinterface normal and the coordinate axis， α_0 isthe angle of misorientation of the crystal to the coordinate axis,ε_4 is the strength of the surface tension anisotropy,λ is the coupling constant.V_p is the solidification speed,l_T is the thermal length,m is theliquidus slope,c_l^0 is the concentration of the liquidphase when the temperature is T_0,k is the non-equilibrium partition coefficient, $k=(k_e+V_p/V_D)/(1+V_p/V_D)$, k_e is theequilibrium partition coefficient, V_D is the solute diffusion velocity. D_l is the liquid diffusion coefficient , $\vec{j}_{at}=-a(\phi)[1+(1-k)U](\partial\phi/\partial t)\left(\vec{\nabla}\phi/\left|\vec{\nabla}\phi\right|\right)$ is the antitrapping current, $a(\phi)=c_1(1-[c_2/ln(1+\varepsilon)](1-\phi^2))$, $c_1=1/2\sqrt{2}$, $c_2=[1/4(1-k_e)](V_S/V_D)ln(k_e+1/2k_e)$, $V_S=D_l/d_0$, d_0 is the capillary length, $q(\phi)=[1-\phi+k_e(D_s/D_l)(1+\phi)]/[1+k_e-(1-k_e)\phi]$ is theinterpolation function, D_S is the solid diffusion coefficient.

3 Results and discussion

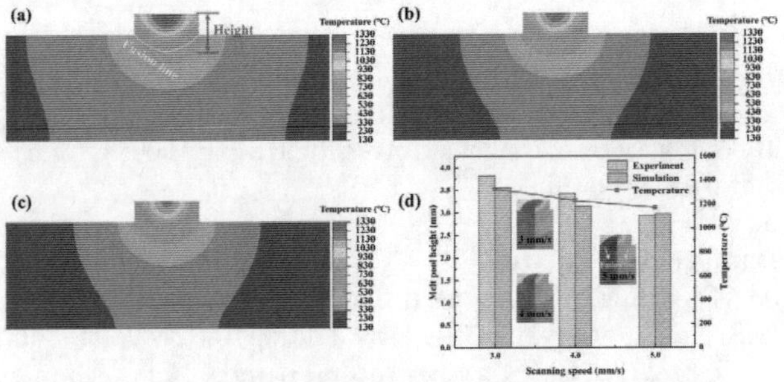

Figure 1 Finite element temperature distribution at different scanning speeds: (a) 3 mm/s; (b) 4 mm/s; (c) 5 mm/s.

Figure 1 shows the finite element analysis results at different scanning speeds. The maximum temperatures in the melt pool at different scanning speeds are 1,330 °C, 1,230 °C, and 1,171 °C, respectively. This is mainly because the effective heat input decreases with the increase of scanning speed. Figure 1(d) shows the variation of the height of the melt pool. The height and maximum temperature of the melt pool are inversely proportional to the scanning speed. Meanwhile, we compared the molten pool morphology of the WAAM experimental results and the finite element simulation results at different scanning speeds, as shown in Figure 1(d). It can be observed that with the increase of scanning speed, the errors between the melt pool depths obtained from the experiments and those from the finite element simulation are 0.257 mm, 0.283 mm, and 0.033 mm, respectively. The melt pool morphology obtained by finite element simulation is in good agreement with the experimental results.

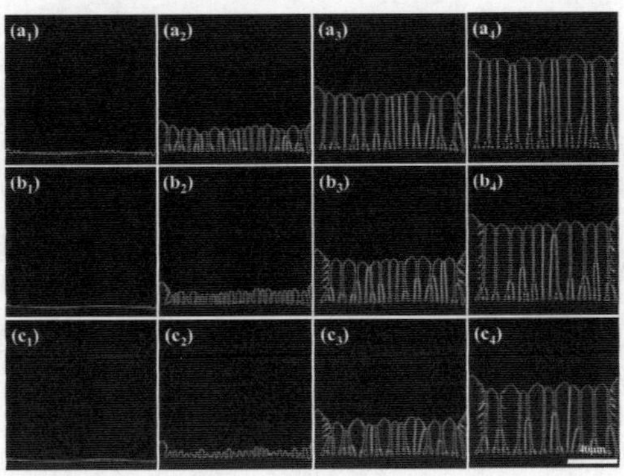

Figure 2 Dendritic morphology at different scanning speeds: (a_1) -(a_4) 3 mm/s; (b_1) -(b_4) 4 mm/s; (c_1) -(c_4) 5 mm/s

As shown in Figures 2(a_1) -(c_1), a lower scanning speed causes the solid-liquid interface to destabilize earlier. This is because the scanning speed has a significant effect on the temperature gradient during solidification in the WAAM process. As the scanning speed increases, the temperature gradient during dendrite growth decreases and the driving force for dendrite growth decreases, so it can be observed that the dendrite growth rate decreases as the scanning speed increases. As shown in Figures 2(a_2) -(c_2), the solid-liquid interface is destabilized, and the planar interface generates dendrites and penetrates deep into the liquid phase. Due to the negative temperature gradient at the solid-liquid interface front, the growth of dendrites is promoted. Therefore, at the junction of the WAAM substrate and the deposited layer, the dendrites first grow in the form of columnar dendrites during solidification. Since there is a preferential orientation in the process of dendrite growth, the growth tendency of dendrites parallel to the direction of heat flow is more intense. Therefore, the phenomenon of competing growth of dendrites occurs in the early stage of solidification, as shown in Figures 2(a_3)-(c_3).

4 Conclusions

(1)As the scanning speed increases the heat input to the WAAM process decreases and the temperature gradient during solidification decreases.

(2)As the temperature boost decreases the driving force for dendrite growth during solidification decreases and the dendrite growth rate decreases. Solute segregation is reduced.

5 Acknowledgments

This work was supported by the National Natural Science Foundation of China (Nos.52074246, 52275390, 52205429, 52201146), the National Defense Basic Scientific ResearchProgram of China (JCKY2020408B002, WDZC2022-12), the Key Research and DevelopmentProgram of Shanxi Province (202102050201011, 202202050201014).

References

[1] Keller T, Lindwall G, Ghosh S, et al. Application of finite element, phase-field, and CALPHAD-based methods to additive manufacturing of Ni-based superalloys. Acta Materialia, 2017, 139: 244-253.

[2] Echebarria B, Folch R, Karma A, Plapp M. Quantitative phase-field model of alloysolidification. Physical Review E, 2004, 70(6 Pt 1):061604.

[3] Kavousi S, Asle Zaeem M. Quantitative phase-field modeling of solute trapping in rapidsolidification. Acta Materialia, 2021, 205:116562.

Study on Machine Learning for the Prediction of Tensile and Impact Properties of Aluminum Alloys by Incorporating Multi-Modal and Multi-Scale Information

Jiale Ma[1, 2], Wenchao Zhang[1, 2], Zhiqiang Han[1, 2, *]
1. School of Materials Science and Engineering, Tsinghua University, Beijing 100084, China
2. Key Laboratory for Advanced Materials Processing Technology (Ministry of Education), Tsinghua University, Beijing 100084, China
*Corresponding address: e-mail: zqhan@tsinghua.edu.cn

Abstract: Machine learning (ML) methods have played an increasingly important role in materials design. In this work, an explainable deep learning (DL) model was developed based on a multi-modal and multi-scale dataset for predicting the tensile properties of aluminum (Al) alloys. In addition, the internal mechanism of the model was analyzed by using a visualization technique. Further, a relationship of tensile property-impact property was established using a deep neural network (DNN) model. The correlation between tensile properties and impact properties was analyzed, providing scientific insight for the design of Al alloys with good comprehensive mechanical properties.

Keywords: machine learning; aluminum alloys; microstructure; tensile property; impact property

1 Introduction

By virtue of their high strength-to-weight ratio and high corrosion resistance, Al alloys are widely used in automotive industries. Recently, ML models have been incorporated to assist Al alloy design [1]. For instance, DL models, based on convolutional neural networks (CNN), can automatically extract features from images and analyze intricate correlations in high-dimensional data, which has been successfully used to predict tensile properties by compositions and microstructures [2].

Certain components in automotive applications are often exposed to dynamic impact loading during service life. Impact property [3], a measure of a material's ability to absorb energy and resist fracture under rapid loading, is a positive correlation both with ultimate tensile strength (UTS) and elongation (El). However, there is often a trade-off between UTS and El. Establishing the relationship model between tensile and impact properties can provide guidance for the design of Al alloys with good comprehensive mechanical properties.

In this work, we developed a DL model based on multi-scale microstructure information to predict the tensile properties. Then, by establishing a DNN model, the intrinsic connection between tensile and impact properties was further analyzed.

2 Experimental procedure

Dataset construction

Al-7Si-0.35Mg alloys modified by adding 0.02wt.% Sr (ASr) and jointly modified by adding 0.02wt.% Sr and 0.1wt.% La (ASrLa) were prepared. The samples underwent solution treatment at 535 ℃ for 0-12 h, and then aging treatment at 170 ℃ for 0-24 h. To obtain the multi-scale microstructure information, optical microscope (OM) and scanning electron microscope (SEM) characterization were performed. Tensile testing and impact testing were conducted at ambient temperature. Finally, a "composition-microstructure-tensile property" dataset with 78 sets of data and a "tensile property-impact property" dataset with 41 sets of data was established.

Construction of ML architectures

For the DL model, an OM image and an SEM image of each sample are combined as input to form the image features. The component information is input into the model to form text features. Then, the image and text features are spliced and trained together to predict the tensile properties. The architecture of the DNN model is "input layer-hidden layer 1-hidden layer 2-output layer". The squared correlation coefficient (R^2) and mean absolute error (MAE) were used to evaluate the models' performance.

3 Results and discussion

The relationship model of composition-microstructure-tensile property

For the UTS results of the DL model, the values of R^2 and MAE are 0.94 and 18.9 MPa, respectively. For the El result, the mean values of R^2 and MAE are 0.96 and 0.01%, respectively. The results indicate that the predicted values are in good agreement with the experimental values, demonstrating the model's high prediction accuracy.

To improve the interpretability of the DL model, the Gradient-weighted Class Activation Mapping (Grad-CAM) technique was applied. Grad-CAM can produce a heatmap in which regions with hotter spots indicate areas of the input image that contribute more to the prediction of the target value. The visual analysis of the DL model is performed using the SEM image of the ASrLa sample as an example (Figure 1). The results show a strong correlation between the UTS and the precipitates, as most of the hotspots coincide with the elliptical particles. This result proves that the visualization technology can reflect the mechanism knowledge learned by the DL model.

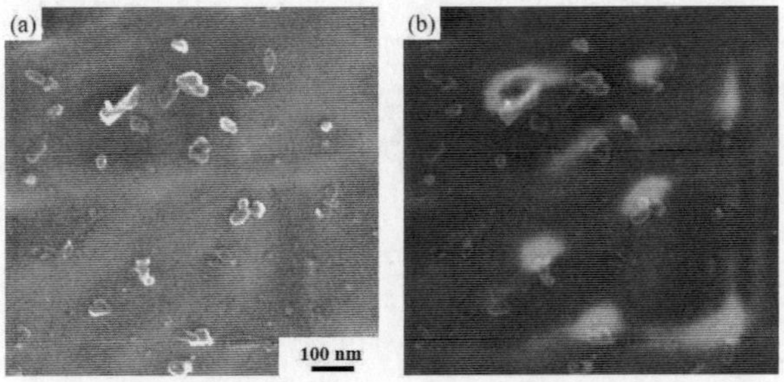

Figure 1 The ASrLa sample: (a) original SEM image and (b) visualized results of UTS in stacked SEM image

The relationship model of tensile property-impact property

For the prediction results of the DNN model, the values of R^2 and MAE are 0.89 and 5.87 J, respectively. Figure 2 shows the 3D surface plot of the "UTS-El-impact property" predicted by the DNN model. Further, an analysis of the correlation between tensile properties and impact properties was carried out by calculating the sensitivity factors, which is the average partial derivative of the output with respect to the input. For solid solution state Al alloys, the sensitivity factors of UTS and El are -4.7 and 16.3, respectively. For aging state Al alloys, the sensitivity factors of UTS and El are 1.1 and 13.1, respectively. The results indicate that El has a stronger positive correlation with the impact properties of Al alloys compared to UTS. The relationship between UTS, El, and impact properties provides valuable insights for the design of impact-resistant Al alloys.

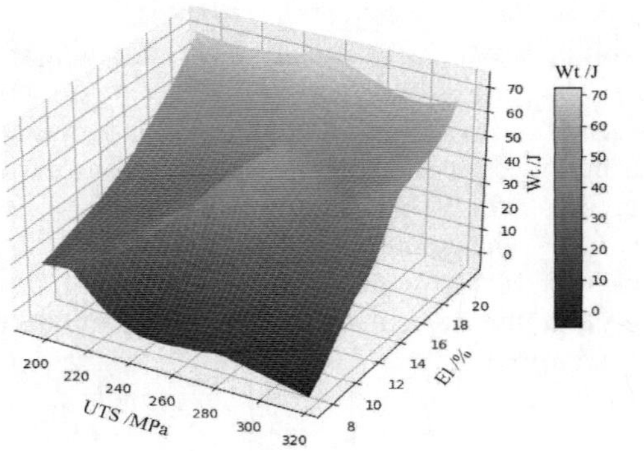

Figure 2 The correlation results of "UTS-El-impact property"

4 Conclusions

(1) The "composition-microstructure-tensile property" and "tensile property-impact property" relationship models were established using machine learning methods.

(2) The DL model can be used to comprehensively analyze the multi-scale microstructure information and the intrinsic mechanism was elucidated by visualization technology.

(3) The correlation between tensile properties and impact properties was analyzed, providing scientific insight for the design of Al alloys with good comprehensive mechanical properties.

5 Acknowledgment

The work was financially supported by Beijing Natural Science Foundation (L223001).

References

[1] Chaudry U M, Hamad K and Abuhmed T. Machine learning-aided design of aluminum alloys with high performance. Mater. Today Commun., 2021, 26: 101897.

[2] Ren D, Wang C C, Wei X L, Lai Q Q and Xu W. Building a quantitative composition-microstructure-property relationship of dual-phase steels via multimodal data mining, Acta Mater., 2023, 252(15): 118954.

[3] Zheng X L, Li S X, Ma J L, Xu Q Y, Zhao H D and Han Z Q. Improvement of impact properties of Al-Si-Mg alloy via solution treatment and joint modification with Sr and La. Rare Met., 2024.

Research on the Effect of Thermal Convection of Steel Liquid on Riser Shrinkage in Steel Castings Based on Numerical Simulation Analysis

Tongtong Han*, Junbao Zhang, Xiaobing Yang, Youqing Lv
Erzhong (Deyang) Heavy Equipment Co., Ltd., Deyang, Sichuan, China
*Corresponding address: e-mail: 954427390@qq.com

Abstract: In order to practice the green and low-carbon development concept, further explore the potential for material saving and cost reduction in large steel castings, and improve the material utilization of steel castings, this paper combines the characteristics of molten steel heat convection, relies on the casting steel products being manufactured, and uses numerical simulation analysis software to carry out the simulation of heat convection in steel castings with different specifications. It compares the shrinkage conditions with and without convection calculations for different specifications of risers, and verifies the accuracy of the simulation results through actual dissection of the risers, mastering the impact of heat convection on the shrinkage height of risers of different specifications. The conclusions of this study can provide a reference for the design of risers for large steel castings.

Keywords: steel castings; riser shrinkage hole; thermal convection of molten steel; numerical simulation

1 Introduction

To achieve the "dual carbon" goal, China is optimizing its energy structure, vigorously developing clean energy such as hydropower, photovoltaics, and wind power, and promoting energy conservation and material reduction in various industries. As the foundation of equipment manufacturing, large-scale steel casting manufacturing is also striving to achieve green and low-carbon transformation. Despite the increasing investment in new processes and materials, the production of large steel castings still relies mainly on traditional manual sand molds and gravity shrinkage solidification methods, which leads to material utilization rates generally below 60% and a high proportion of riser weight. In large steel castings, commonly used types of risers include ordinary sand mold risers, insulation risers, and heating risers, with insulation and heating risers being more widely used. The precise and scientific use of riser feeding to reduce consumption has become a new research direction.

The quality of steel castings largely depends on the design of the riser, and the local temperature difference of the steel liquid is the driving force of thermal convection. Therefore, this study used numerical simulation methods to analyze the effect of thermal convection on the height of riser shrinkage holes in different specifications of risers, with the aim of effectively improving the material utilization rate of steel castings.

2 Experimental procedure

This article relies on the production of steel castings and conducts experiments on three different specifications of risers for low alloy steel (GS20Mn5) castings. The steps are as follows:

(1) Riser selection: Select three different specifications of risers (300 mm, 600 mm, and 1,014 mm) for simulation and verification.

Table 1. Riser size

Riser number	Riser diameter/mm	Riser height/mm
1#	300	450
2#	600	900
3#	1,014	1,100

(2) Thermal convection simulation of molten steel: Using MAGMA software to simulate the thermal convection phenomenon of molten steel in the riser and analyze its impact on the height of the shrinkage cavity.

(3) Physical verification: Verify the accuracy of simulation results by dissecting the riser and observing the shrinkage situation.

3 Results and discussion

Comparison between the actual shrinkage height of the riser and the simulated height

The thermal convection of the insulation riser has a significant impact on the height of the shrinkage cavity, and the degree of influence increases with the increase of the riser diameter. The comparison is shown in Table 2.

Data analysis results

By analyzing the data, the following conclusions can be drawn:

(1)The thermal convection of the insulation riser has a significant impact on the shrinkage height: for the insulation risers with a diameter of 600 mm and a diameter of 1,014 mm, considering thermal convection, the shrinkage height of the riser is increased by 110 mm and 170 mm, respectively. For example, using a 600 mm riser, the simulated and actual riser shrinkage height can be seen in Figures 1 and 2.

Table 2. The actual shrinkage height of the riser and the simulated height

Riser number	Riser diameter/mm	Riser shrinkage height/mm		
		Actual anatomy	Simulate convection calculation	Simulate not calculating convection
1#	300 (No insulation sleeve placed)	20	24	10
1#	300 (Placing insulation sleeves)	250	135	100
2#	600	450	480	370
3#	1,014	450	460	290

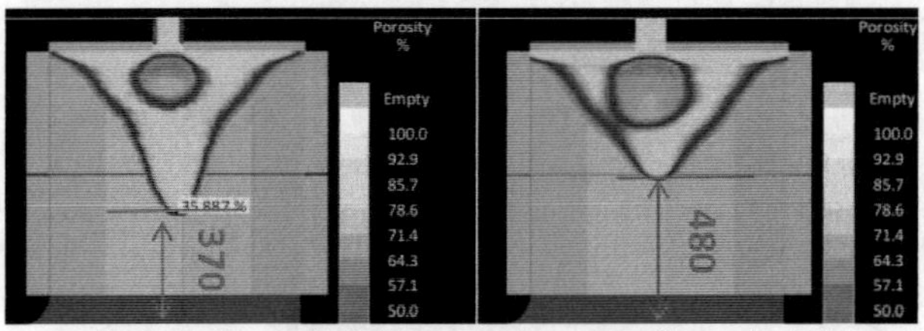

a not convection calculation b convection calculation

Figure 1 Simulated shrinkage of a 600mm riser

Figure 2 Actual shrinkage of a 600mm riser

(2)The degree of influence of thermal convection is directly proportional to the diameter of the insulation riser: as the diameter of the insulation riser increases, the degree of influence of thermal convection on the height of the shrinkage cavity also increases. The small riser with a diameter of 300mm is greatly affected by insulation materials, and the effect of thermal convection is relatively small; The large risers with diameters of 600mm and 1014mm are greatly affected by thermal convection, and the larger diameter, the higher the height of the shrinkage cavity.

(3)The numerical simulation results are consistent with the physical verification results: By dissecting three specifications of risers and analyzing them, it was found that the actual shrinkage height is basically consistent with the numerical simulation results, proving the reliability of the numerical simulation method.

(4)Thermal convection has a significant impact on the cost of large steel castings: Taking the insulation riser with a diameter of 1,014 mm as an example, considering thermal convection, the shrinkage height of the riser is increased by 110 mm and 170 mm respectively. If the height of the riser is reduced by 100 mm, it can save 0.59 t of molten steel, which accounts for about 9% of the total weight of the riser. Based on the annual production of 50,000 tons of molten steel by the foundry, the weight of the riser is approximately 15,000 tons. Taking into account the thermal convection effect of the riser, the weight of the riser molten steel can be reduced by 5%-7%, saving 750-1,000 tons of molten steel per year. Calculated at a cost of 4,000 yuan per ton of molten steel (considering partial recyclability), the cost can be saved by 3-4 million yuan per year.

Conclusions

(1) The design of the insulation riser should consider the influence of thermal convection: When designing the riser, numerical simulation results should be referred to, and the influence of thermal convection on the height of the shrinkage cavity should be considered. The size of the riser should be optimized, the weight of the riser should be reduced, and the material utilization rate should be improved.

(2) Numerical simulation is an effective tool for riser design: With the help of numerical simulation technology, the impact of thermal convection on riser shrinkage can be accurately predicted, providing a scientific basis for riser design.

(3) Green and low-carbon development is the only way for the large-scale steel casting industry: By optimizing the design of the riser and reducing the weight of the riser, production costs can be effectively reduced, promoting the green and low-carbon development of the large-scale steel casting industry.

Reference

[1] Foundry Branch of China Mechanical Engineering Society Foundry Handbook - Volume 5, Foundry Processes.

Research and Application of Intelligent Control Systems for Vacuum Induction Suspension Furnace

Wanqing Li*, Pengcheng Zhang, Yongliang He, Xinfeng Liu
Shenyang Research Institute of Foundry Co., Ltd., CAM, National Key Laboratory of Advanced Casting Technologies, Shenyang 110022, Liaoning, China
*Corresponding address: e-mail: 374333797@qq.com

Abstract: Vacuum induction furnaces mainly is the production of high entropy alloys, high temperature alloys, titanium alloy vacuum smelting equipment high-quality special alloy materials, the vacuum smelting process is realized by using PLC and touch screen the design of the automatic monitoring and control, between the various components of the automatic implementation logic, booster pump, the oil temperature control, and security monitoring, under the premise that meet the requirements of production technology, make the system to achieve the optimal state of control, system maintenance, easy to use, safe, stable and accurate operation, automatic monitoring system for safety production and improve product quality play an important role.

Keywords: vacuum induction furnace; interval temperature; PLC

1 Introduction

Since the 1960s, vacuum induction melting technology has been rapidly developed in China. A vacuum induction suspension furnace is an important vacuum smelting equipment for the production of high entropy alloy, high temperature alloy, titanium alloy and other high quality alloys, precision alloy, ultra-high strength steel and special steel [1].

A vacuum induction suspension furnace is developed on the basis of an ordinary vacuum induction furnace, with the purpose of melting higher-quality alloy materials. Under vacuum conditions, the vacuum induction suspension furnace uses alternating current to act on the induction coil to generate an alternating magnetic field, and the alternating magnetic field induces alternating current - "eddy current" on the charge, and the charge is heated and melted by "eddy current", thus melting metal [2]. Compared with other types of electric furnaces, the functions and system components of each part of the intelligent control system of a vacuum suspension furnace are analyzed.

The PLC control system is the core part of the whole vacuum furnace control, other functions must be calculated by PLC, command control, PLC selected Siemens S7-300, this component belongs to the more advanced products of similar components, with no pollution, fast computing speed, high precision, strong anti-interference ability, high-cost performance advantages. The touch screen is the main operation panel in the automatic control of the vacuum furnace, which is the function of controlling the operation of each valve, displaying the value, recording the curve, displaying the alarm and so on. The function of the encoder is to measure the rotation Angle of the furnace body, and the pulse sent by the encoder is converted into the corresponding furnace body Angle after PLC collection and calculation. The sensor includes temperature, pressure, and flow detection, and all analog and switching quantities are collected and controlled by PLC in real-time.

2 Realization method of intelligent control system of vacuum induction suspension furnace

The vacuum induction suspension furnace smelting starts from the melting of solid raw materials or scrap steel and goes through four stages: melting period, refining period, alloying period and pouring. The whole process is completed in the vacuum chamber, and the process is highly compact and concentrated. The software for programming and application is STEP7-V5.5 and MCGS-7.2. Among them, STEP7-V5.5 is to realize the data acquisition and control function of the PLC of the lower computer, and MCGS-7.2 is the upper configuration interface, which is to realize the upper configuration and data display and operation control function. As shown in Figure 1, this is the touch screen operation interface of the whole vacuum furnace, including the main charging chamber, vacuum chamber, pump control and valve control. The operation interface and the physical object correspond to a clear layout, easy to use, safe and reliable.

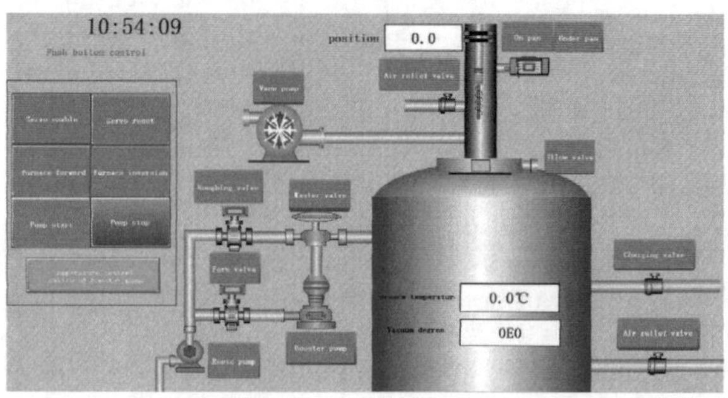

Figure 1 Intelligent operation interface of vacuum induction suspension furnace

Figure 2 shows recording historical data, statistical reports, and saving data to facilitate users to query historical data records. In addition, the most important is that you can use the U disk operation, U disk to check the touch screen in the start time and end time box to enter the period, automatically all records during this period in the form of an Excel spreadsheet exported to the U disk, convenient for leaders to view and test.

	mcgs-time	Electroplate	power	Molten steel temperature	Vacuum degree
1	2018-04-19 09:57:20	0.00 Kwh	0.00 Kw	0.00 ℃	0.00
2	2018-04-19 09:57:30	0.00 Kwh	0.00 Kw	0.00 ℃	0.00
3	2018-04-19 09:57:40	0.00 Kwh	0.00 Kw	0.00 ℃	-2.33
4	2018-04-19 09:57:50	0.00 Kwh	0.00 Kw	0.00 ℃	234543.55
5	2018-04-19 09:58:00	0.00 Kwh	0.00 Kw	0.00 ℃	234543.55
6	2018-04-19 10:02:21	0.00 Kwh	0.00 Kw	0.00 ℃	0.00
7	2018-04-19 10:02:31	0.00 Kwh	0.00 Kw	0.00 ℃	10000000000.00
8	2018-04-19 10:02:41	0.00 Kwh	0.00 Kw	0.00 ℃	10000000000.00
9	2018-04-19 10:02:51	0.00 Kwh	0.00 Kw	0.00 ℃	3553554432.00
10	2018-04-19 10:03:01	0.00 Kwh	0.00 Kw	0.00 ℃	10000000000.00
11	2018-04-19 10:03:11	0.00 Kwh	0.00 Kw	0.00 ℃	10000000000.00
12	2018-04-19 10:03:21	0.00 Kwh	0.00 Kw	0.00 ℃	34344.00
13	2018-04-19 10:03:31	0.00 Kwh	0.00 Kw	0.00 ℃	-244344.00
14	2018-04-19 10:03:41	0.00 Kwh	0.00 Kw	0.00 ℃	-244344.00
15	2018-04-19 10:03:51	0.00 Kwh	0.00 Kw	0.00 ℃	-244344.00

Start time   End time   Data export

Figure 2 Data saving and exporting

3 Field application of vacuum induction suspension furnace

Figure 3 shows the overall structure and the included parts of the vacuum induction suspension furnace. The charging chamber is the function of adding alloy ingredients and thermocouple temperature detection. The vacuum melting chamber contains an induction ring for melting molten steel and melting molten steel in a suspended vacuum state. After the molten steel composition is qualified, the ingot mold is put into the pouring position in the ingot mold chamber for pouring, and the whole process is completed in the vacuum state. The lid has a viewing window and infrared temperature measurement. There is a regulating valve at the side of the furnace to fill the furnace with an argon gas channel. The whole system has a gas path system and a water system. The water system is to cool the furnace body and the medium-frequency induction power supply, and the gas path system is to control the switch of each valve.

Figure 3 Vacuum suspension melting field practical application

4 Conclusion

As a kind of special metallurgical equipment, the safety and intelligence of vacuum induction melting furnaces are more and more important in their design and manufacturing process. Aiming at the application of an automatic monitoring system for vacuum induction furnaces, this paper adopts self-programmed control software to realize automatic monitoring function, improves the control method of vacuum furnaces, and solves the trouble of manual operation and manual writing records. By analyzing the characteristics and application scope of the design idea, the reliability and authenticity of the records are increased. Because PLC and touch screen man-machine interface are used in the system design, the control is more simple, convenient, safe and reliable.

5 Acknowledgments

Thanks to Shenyang Metal Research Institute Co., Ltd. for the intelligent control system of our vacuum suspension melting furnace.

References

[1] Lv Y N. Vacuum metallurgy of non-ferrous Metals [M]. Beijing: Metallurgical Industry Press, 1998.
[2] Sun D J. Vacuum metallurgical equipment (4) -- vacuum induction furnace [J]. Vacuum Science and Technology, 1985, 5(1): 69-80.

Development Trend Analysis of High Purity Vacuum Suspension Melting Casting Equipment

Zhizhong Liu, Yongliang He*, Pengcheng Zhang, Kun Shi, Jianqian Yang, Wanqing Li, Xinfeng Liu
National Key Laboratory of Advanced Casting Technologies, Shenyang Research Institute of Foundry Co., Ltd., CAM, Shenyang 110022, Liaoning, China
*Corresponding address: e-mail: hehe32787152@163.com

Abstract: Aiming at the common problems of high temperature, high purity and refractory alloy materials, which are needed for the development of aviation, aerospace and other future Strategic Emerging industries, the high purity vacuum melting casting technology is put forward. The advantages of vacuum induction suspension melting casting equipment in preparing new materials in high purity vacuum melting casting technology and the development direction of equipment, process and technology are introduced. By analyzing the problems facing the current application market, the development direction of vacuum induction suspension melting equipment in the future is pointed out.

Keywords: high purity vacuum casting technology; floating melting; auto-control

1 Introduction

Materials are the material basis for developing the national economy and people's livelihood, strengthening the modernization of national defense, and improving the country's overall industrial strength. Before forming high-quality alloy materials, melting is the most important link, and the pollution problem in the melting process is almost a common problem faced by hot processing enterprises. With the introduction of the concept of high purity vacuum casting technology, the casting of many high temperature, high purity and refractory alloy materials has become a reality.

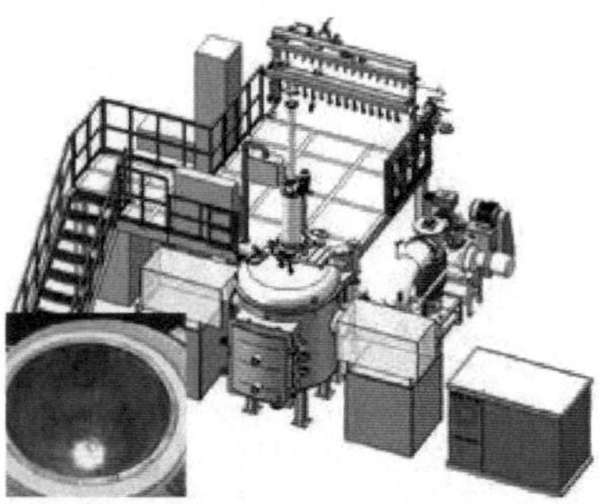

Figure 1 Suspension melting equipment and water-cooled copper crucible diagram

In the high purity vacuum melting casting technology, the vacuum induction suspension melting casting technology is a special metallurgical process which combines the high frequency and high power induction heating method with the slit water-cooled copper crucible to melt the alloy, and is the most ideal melting casting technology for high temperature, high purity and refractory alloy. The technology uses the water-cooled copper crucible as the carrier of the alloy material to completely eliminate the pollution of the alloy material in the melting process. The melt is stirred during the heating process to obtain a uniform superheat and chemical composition.

2 The development status of equipment, technology and technology

Composition and shortcomings of suspension melting equipment

Vacuum induction suspension melting casting equipment is mainly composed of vacuum system, water cooling system, control system, induction melting system and water-cooled copper crucible, among which induction melting system and water-cooled copper crucible are the most critical components. However, with the continuous increase of the capacity of water-cooled copper crucible, it is difficult to promote industry due to the limitation of high-power and high-frequency induction power supply. Only suitable for scientific research institute laboratories, in recent years, the United States, Germany, Russia and other developed countries to promote cold crucible induction melting process, has a certain scale of cold crucible melting capacity has reached 100 kilograms.

Research direction of large-scale special suspension melting equipment

Because the advantages of suspension melting equipment are very obvious, the suspension effect and alloy melting capacity have become the most concerned development direction. At present, the suspension effect is mainly optimized by improving the structural design of the crucible, changing the magnetic field frequency and magnetic field distribution state. We can find an optimal structural design and magnetic field distribution state through software simulation.

The melting capacity is 80 kg at home and 100 kg at abroad. With the diversification of functions and the innovative design of the structure, the maximum capacity can be more than 200 kilograms through multi-power joint casting, which truly meets the needs of actual production. Under the premise of realizing the diversity of functions, the interference problem between induction power supplies is brought. Through the simulation and optimization of the magnetic field of the power supply under different layouts, the interference problem between induction power supplies is reduced, and the functional requirements of large capacity are met.

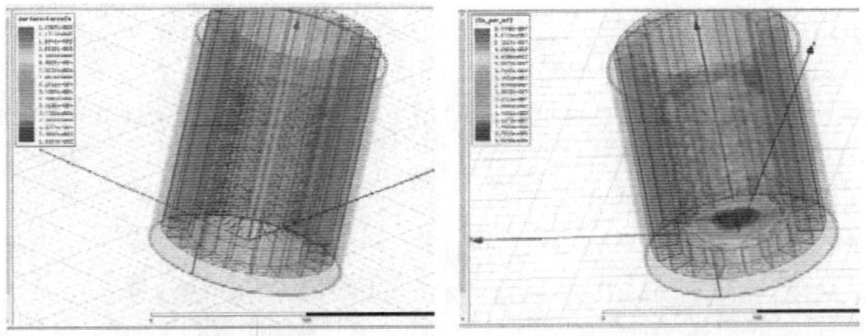

Figure 2 Magnetic field state simulation

Development direction of suspension melting process

Suspension melting in the realization of pure melting, casting functions at the same time, the types of melting alloys are increasing, including: titanium aluminum alloy, aluminum-scandium alloy, high entropy alloy, tungsten alloy, amorphous alloy and many functional materials and so on. Equipment functions are also increasing, including: suspension ingot, suspension centrifugation, suspension continuous casting, etc., the increasing function represents the continuous improvement and innovation of the casting process. This requires engineers to understand the entire casting process and casting equipment, and realize the entire process through equipment.

3 Analysis of future development trend of suspension melting

Functionality tends towards modularity

In the future, it is necessary to consider the establishment of independent modules for different functions to facilitate the iteration of individual functional modules. Customized equipment functions according to process requirements. For example, real-time on-line monitoring of casting alloy temperature is realized by adding temperature measurement module; Equipped with a centrifugal module, it is possible to centrifuge castings that require specific centrifugal casting functions, and the centrifugal speed can be set according to the process requirements to optimize the performance of the alloy. In the future, the suspension melting equipment will meet the needs of the actual process through the combination of different modules, and prepare materials with better performance.

The system tends towards intelligence

The control system of suspension melting equipment is gradually developing in the direction of intelligence, to solve the problem that the system can complete the complex system according to the set procedure in the case of no intervention. At present, most equipment is still in the state of manual adjustment, manual observation and human intervention when melting alloy materials. With the advent of the era of big data, it is entirely possible to combine the advantages of PLC and industrial computers to form an industrial Internet platform, through data accumulation, analysis, optimization, to develop the optimal process curve, to avoid human intervention caused by misoperation and uncertainty and other problems.

More diversified types and processes of melting alloys

The melting process generally refers to the process of melting alloy materials at high temperatures and casting. With the continuous increase of the types of smelting alloys, the casting process is constantly changing, and the process needs to be continuously refined and improved. Different melting processes will directly affect the performance of the final product, melting process will produce a lot of information, including power real-time parameters, electromagnetic field, charge, crucible, water temperature, water pressure and physical properties of alloy materials, casting process will also produce a lot of information, including casting materials, preheating temperature, whether centrifugal, casting speed and so on. Therefore, it is necessary to develop different melting casting processes according to different materials and different casting processes, to form a specific process route, to reduce human intervention in the melting casting process, so as to obtain more excellent alloy materials.

4 Conclusion

With the increasing requirements for cutting-edge alloy materials in national defense and military industry, high purity vacuum melting casting technology has become the most important process technology to meet the requirements, and the emergence of vacuum suspension melting equipment has promoted the development of high purity vacuum melting casting technology in the preparation of high temperature, high purity, refractory alloy materials preparation technology. In the future, with the rapid development of digital, visual and intelligent equipment, suspension melting and casting equipment will also be customized according to the needs of users, the system is intelligent, and the process is diversified to meet the development needs of new materials.

References

[1] Yan J Q, Bai Z Y, Zhang Z Y, et al. Application of vacuum induction suspension melting technology in the preparation of high purity materials [J]. Foundry, 2024, 73(2): 253-256.

[2] Jiang T. Research and application of vacuum induction melting technology in experimental teaching of materials science and engineering [J]. Human Resources Management, 2017, 12: 502-503.

[3] Lan L J, Gu Y Y, Pu T J, et al. Research status of high entropy alloy properties [J]. Metallurgical Engineering, 2018, 5 (1): 17-24.

Accessing the Effects of Composition and Process Fluctuations on the Quality of Investment Castings

Da Shu[1],*, Zhengyi Ding[1], Bang Guan[1], Chengbo Xiao[2], Baode Sun[1]

1. Shanghai Key Laboratory of Advanced High Temperature Materials and Precision Forming, School of Materials Science and Engineering, Shanghai Jiao Tong University, Shanghai 200240, China

2. Science and Technology on Advanced High Temperature Structural Materials Laboratory, AECC Beijing Institute of Aeronautical Materials, Beijing 100095, China.

*Corresponding address: e-mail: dshu@sjtu.edu.cn

Abstract: The composition and process fluctuations cause the uncertainty of the quality of investment casting of Ni-based superalloys. This study assesses the influence of fluctuations in alloy composition and process parameters on shrinkage porosity defects in castings based on integrated computation and data-driven approach. Besides, a dynamic correction method for determining boundary condition changes during the shell transfer process is presented. The results show that more accurate shrinkage porosity predictions are obtained with the dynamic temperature corrections, facilitating the quality control of investment casting.

Keywords: investment casting; fluctuations; shrinkage porosity

1 Introduction

Investment casting is a complex manufacturing process. Different parameters have their impact on the casting defects [1]. On one hand, the composition of specific grades of Ni-based casting superalloys is usually a range rather than a fixed value. The composition fluctuation leads to difference in thermodynamic and physical properties, causing the fluctuation in casting defects [2]. On the other hand, some studies rely on historical data and use inverse algorithms to obtain more accurate parameters [3]. However, the actual production process is not fixed, and there are deviations from historical data. Simulation with fixed processing parameters may cause large error. Besides, in some cases, boundary conditions cannot be accurately set in advance.

2 Method

Composition fluctuations

Thermodynamic-finite element integrated computation is applied to quantify the influence of composition fluctuations of K4169 superalloy on shrinkage porosity and feed length. 100 sets of alloy compositions within the boundary limit are designed with optimal Latin hypercube sampling. Thermodynamic calculations are made with Pandat and incorporated into ProCAST software to predict the casting defects.

Processing parameter fluctuations

Stochastic uncertainty is unavoidable in real production, which may cause large error in traditional numerical simulation with fixed processing parameters. To address this challenge, the data-driven approach should be applied. From 20 groups of real production, the fluctuations in processing parameters are collected, which is shown in Table 1. More simulation experiments are designed within the range of process parameter variations.

Table 1. Uncertainty modelling of parameters

Parameters	Fluctuation range
Gating system assembly error	±3 mm
Shell transfer duration	100±25 s
Pouring temperature	1500±30 ℃
Interface HTC	±30%

Boundary condition fluctuation during shell transfer

The shell transfer process is a process in investment casting during which the preheated ceramic shell is transferred from an oven to a pouring furnace prior to casting. During the transfer process, the shell temperature drops rapidly, forming an uneven temperature field that is different from the preheating temperature. In order to real-time capture the fluctuation in boundary condition (ambient temperature), we present a simplified mathematical model of heat transfer based on the finite volume method and a dynamic correction method based on measured data by a thermocouple as shown in Figure 1.

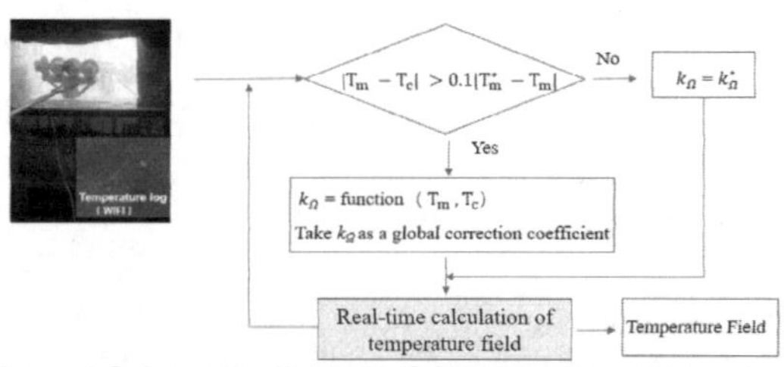

Figure 1 Schematic diagram of dynamic corrections of shell temperature.

Using the dynamically corrected shell temperature at the end of shell transfer as the boundary conditions, the formation of shrinkage porosity defect in the castings is simulated with ProCAST.

Results and discussion

Composition fluctuations

The result of 100 sets of compositions is automatically extracted from integrated computation. Response surfaces shown in Figure 2 are established with machine learning algorithms including RBF and BPNN. The maximum fluctuation due to composition uncertainty is 36% for shrinkage porosity and 21% for feed length.

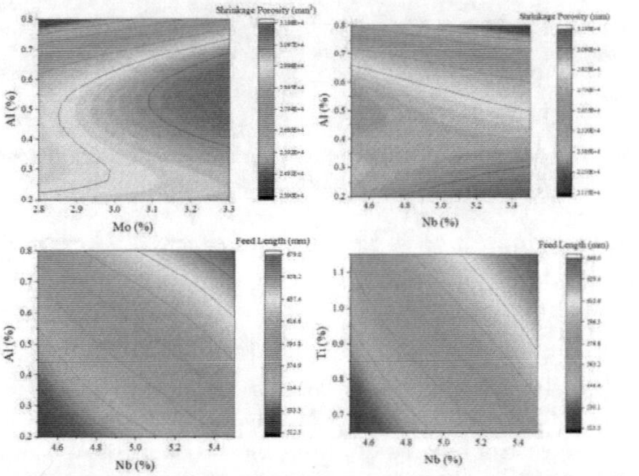

Figure 2 Response surface of shrinkage porosity and feed length.

Processing parameter fluctuations

The impact of parameter fluctuations on the shrinkage porosity is quantified with numerical model. Figure 3 shows the response surface of the shrinkage porosity inside the casting and the safe margin of the porosity above the top of flange for different parameters. Within the limit in Table 1, the contribution of fluctuations is shown in Table 2.

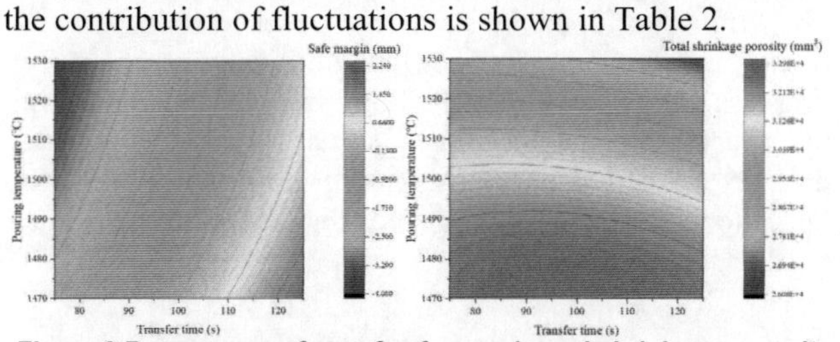

Figure 3 Response surface of safe margin and shrinkage porosity

Table 2. Contribution of process fluctuations

Parameters	Contribution to porosity (%)
Gating system assembly error	32.4
Shell transfer duration	24.5
Alloy composition	22.3
Interface HTC	16.6
Pouring temperature	4.2

Boundary condition fluctuation during shell transfer

Figure 4(a) shows the calculated temperature field of the mold shell after 30 s transfer with dynamic corrections. The trend of the corrected temperature is consistent with the experiment, providing reference values of the process parameters for the subsequent casting process simulation. Castings are more prone to concentrated shrinkage porosity, as indicated in Figure 4(b).

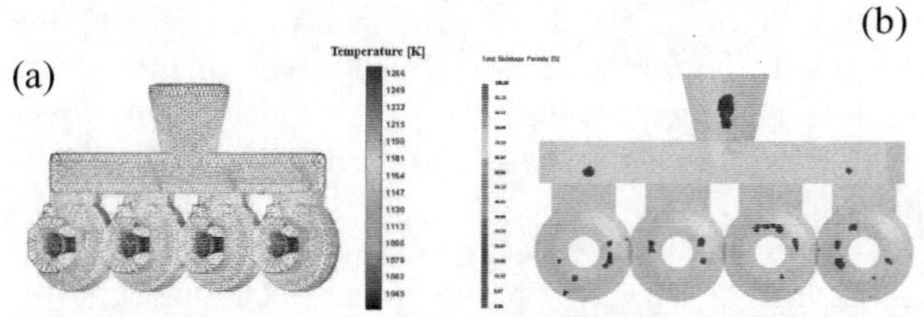

Figure 4 (a) The snapshot of the temperature field of the mold shell with dynamic corrections, (b) Simulation results of the shrinkage porosity in the casting.

Conclusions

In this study, the uncertainty in investment casting from composition and process fluctuations are accessed. The following conclusions can be made:

a) Within the composition boundary limit of K4169 superalloy, the internal shrinkage porosity volume can fluctuate by up to 36%, while the feeding length exhibits a variability of up to 21%.
b) The fluctuations in processing parameters result in different shrinkage porosity defects, among which the gating system assembly error and the shell transfer duration contribute mostly.
c) More accurate shrinkage porosity predictions are obtained with the dynamic temperature corrections during the shell transfer process.

Acknowledgments

This work was financially supported by the National Key Research and Development Program of China (2022YFB3706800), and the National Natural Science Foundation of China (51821001, 52090042).

References

[1] Suthar J, Persis J, Gupta R. Critical parameters influencing the quality of metal castings: A systematic literature review [J]. International Journal of Quality & Reliability Management, 2023, 40(1): 53-82.
[2] Nawrocki J, Szeliga D, Kubiak K. Influence of process parameters on cooling conditions in nickel base superalloy investment casting [J]. Key Engineering Materials, 2015, 641: 124-131.
[3] Nouri M, Artozoul J, Caillaud A, et al. Shrinkage porosity prediction empowered by physics-based and data-driven hybrid models [J]. International Journal of Material Forming, 2022, 15(3): 25.

Inside a Smart Foundry: Digitalized Operations Management

Hongliang Zhang, Jieyuan Wu*, Jian Tong, Quanfeng Li
Suzhou Mingzhi Technology Co., Ltd., No. 1999 West Tongxiao Road Suzhou 215216 Jiangsu, China
*Corresponding e-mail: wu_jie_yuan@mingzhi-tech.com

Abstract: This article seeks to demonstrate some of the principles, features, benefits, and challenges of digitalized operations management inside a “smart foundry”. A three-stage approach was used to achieve this digital transformation.

Specific digital use cases relevant and unique to the sand casting industry will be presented, including: end-to-end traceability and quality control by the MES system’s automated data acquisition; predictive tooling manage-ment that reduces waste and improves OEE; preventive safety measures powered by IoT and computer vision; cross-platform data integration to achieve granular cost transparency and control, etc.

Keywords: smart foundry; digital transformation; mes system; operations management

1 Introduction

Digital transformation has swept across many traditional manufacturing industries in recent years, where advanced technologies including robotics and automation, IoT, and Big Data / AI have been used to elevate decade-old operations management best practices such as Lean, Kaizen, TQM, TPM, JIT and Six-Sigma (See Table 1 for explanations of all acronyms). Playing a fundamental role in this digital transformation is the MES, a software platform that connects the shopfloor to the broader business context [1].

However, the millennium-old sand casting industry has been slow to adapt to the digital transformation overall, despite being under pressure from other competing metal fabrication technologies like forging, die casting and 3D printing [2]. This is due to a combination of factors: the intrinsic long and complex chain of processes in sand casting that involves various types of equipment (e.g., Figure 1), the often high-mix low-volume nature of casting products, and financial constraints – in 2022, all US foundries averaged $26.7 million in revenue but only 57% of them were profitable [3]. It is therefore of paramount importance that foundries develop and adapt digital tools to their specific needs in pursuit of operational excellence.

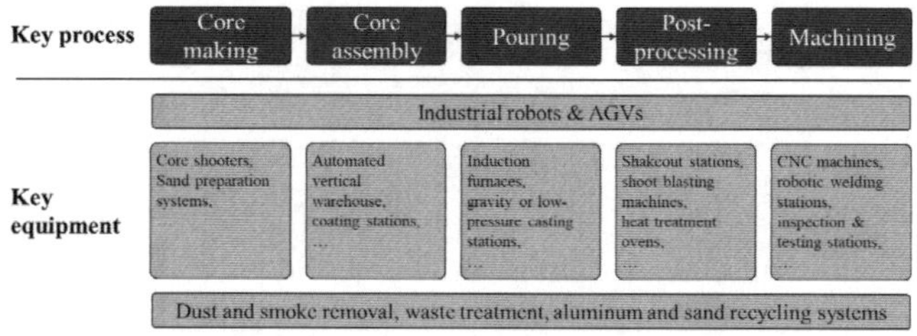

Figure 1 Overview of MZT’s modularized aluminum casting process (non-exhaustive)

2 Methodology

Suzhou Mingzhi Technology Co., Ltd. (MZT) developed the digital technology platforms inside and around its highly automated and modularized aluminum casting factories in three stages:

- Stage 1: Independently developing IT platforms and their data infrastructure, including the MES, QDAS, WMS, ERP, PLM, etc. In the case of the MES system, Figure 2 illustrates how operational data is acquired and aggregated in a number of ways.
- Stage 2: Linking IT with OT in order to analyze, extract insights, hence gain value from the data, by developing specific use cases such as TQM and TPM. Examples of such use cases are described in the following sub-chapters.
- Stage 3: Bridging the information flow across different IT platforms developed in stage 1 to enable holistic data-based operations management beyond just the production lines, as shown in Figure 3. For example, depending on the business context, the optimal tradeoff between delivery, quality, and cost may vary, hence specific operational decisions would differ accordingly across products and over time.

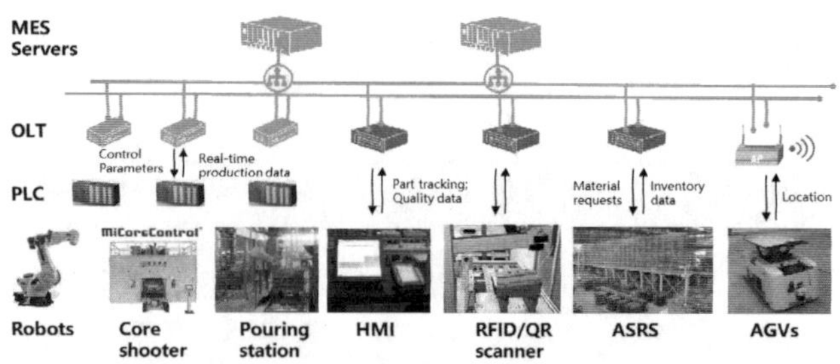

Figure 2 MES data acquisition from various sources

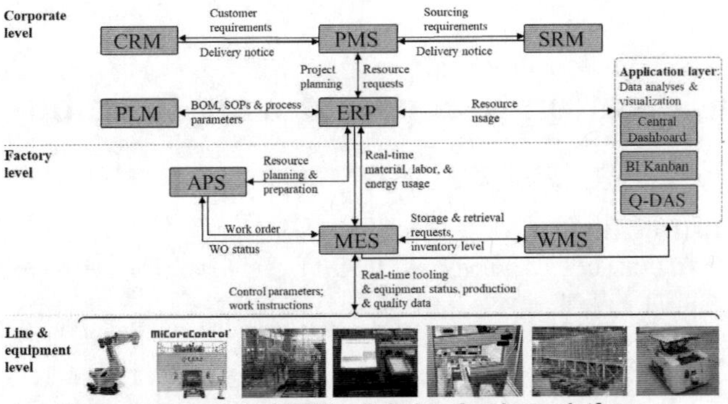

Figure 3 Blueprint of MZT's digital technology platforms

3 Use cases

End-to-end traceability and quality control

The aim of end-to-end traceability is to be able to track each individual casting part and to trace it back to its mold assembly or core package, or even each core, while linking the part to metadata like timestamps, batch number, operators IDs, testing results, and process parameters. At MZT, a combination of QR code printed on the casting, RFID-implanted core & core package trays, and computer vision are used to achieve this. The QDAS software will then analyze all the data to identify patterns and potential origins of defects.

Predictive tooling management

By collecting and storing the production data on the MES server, the spot-checks, maintenance, and replacement of various tooling (including molds, cutting tools, jigs and fixtures) can evolve from fixed-schedule to usage-based to predictive. I.e. prediction algorithms suggest whether, when, and what actions are required, based on the tools' usage and maintenance history. Over time, with more data gathered, the predictions will become more accurate, pushing the tools' utilization closer to limits, reducing the replacement cost and downtime, improving OEE.

Preventive safety features

At MZT, safety is always the highest priority, above delivery, quality and cost. IoT technologies are being used to implement even more preventive safety features. At the pouring station, one of the highest risk areas, only one operator who has been trained and certified can enter the designated area of the station at a time. If violated, all production equipment in the area will halt and alarms will raise. The operators are identified, marked and tracked by RFID chips planted in their helmet, as well as AI image recognition from the camera feed.

4 Challenges

The challenges in digitalized operations management at a foundry are nuanced. For example, with high-mix low-volume production, the processes vary across products, thus complex rules are required to accurately allocate resources and calculate costs (material, labor, energy and equipment) for each product. Another example is the APS - production and resource planning cannot be made fully automatically unless a certain level of process cabaility is achieved, as the optimization algorithm is sensitive to the process variabilities likely found in a foundry.

Table 1. List of acronyms

Acronym	Meaning
AGV	Automated Guided Vehicle
APS	Advanced Planning System
ASRS	Automated Storage & Retrieval System
ERP	Enterprise Resource Planning
HMI	Human-machine Interface
JIT	Just-in-Time
MES	Manufacturing Execution System
MZT	Suzhou Mingzhi Technology Co., Ltd.
OEE	Overall Equipment Effectiveness
OLT	Optical Line Termination
OT	Operational Technology
PLC	Programmable Logic Controller
PLM	Product Lifecycle Management
PMS	Project Management System
QDAS	Quality Data Analysis Software
RFID	Radio Frequency Identification
TPM	Total Productive Maintenance
TQM	Total Quality Management
WMS	Warehouse Management System

5 Result and conclusion

Transforming towards digitalized operations management at MZT has already brought measurable benefits across multiple dimensions. At one of its newer aluminum casting production lines for heat exchanger products, capacity (measured by pieces per shift) has increased by 58% whilest the number of operators has decreased by 70% since the beginning of the digital transformation, more than tripling the output per capita. In the meantime, average scrap rate went down from ~6% to ~2.5%, fastest turnaround time from 7 days to 0.5 days, and unit energy consumption down by 28%.

However, Kaizen is a journey with only beginning and no end. Along the way, there will always be new challenges to overcome and features to be developed. Furthermore, no matter how digitalized or intelligent operations management gets, people are still at the center of it. The data is meaningless unless the engineers and managers can derive the right insights and actions from it.

References

[1] Nurdiyanto H, Kindiasari A. Critical role of manufacturing execution systems in digital trans-formation of manufacturing industry. Journal of Electrical Systems, 2024, 20(7s): 2432-2436.

[2] Banganayi F C, Nel H, Nyembwe K. Benefits of industry 4.0 in foundry engineering's greensand moulding process. University of Johannesburg. https://core.ac.uk/outputs/245881299.

[3] Silhacek B, Wise M. Modern Casting March 2022: 66-67.

"In-situ observation" of melting and solidification process of Cold Crucible UO_2

Shu Wang[1], Jianpeng Tan[1], Dezhi Chen[1],* , Houjun Gong[2], Yang Li[2], Qi Wang[1], Hongze Fang[1], Ruirun Chen[1],*

1. National Key Laboratory for Precision Hot Processing of Metals, Harbin Institute of Technology, No. 92, Xidazhi Street, Huayuan Street, Nangang District, Harbin, Heilongjiang Province, China
2. Nuclear Power Institute of China, No. 328, Section 1, Changshun Avenue, Huayang, Shuangliu County, Chengdu City, Sichuan Province, China

*Corresponding address: e-mail: chendezhi383@163.com (Dezhi Chen), ruirunchen@hit.edu.cn (Ruirun Chen)

Abstract: The electromagnetic cold crucible (EMCC) technology enables to melt and directional solidification of reactive and high-temperature materials with no contamination. The solidification of nuclear waste by cold crucible is beneficial to reduce the environmental risk of radioactive waste. In this paper, a 3-D EMCC model for investiagting the physical fields of UO2 were established by finite element method. The whole process of UO2 solidification was simulated. The results show that the maximum magnetic flux density mode of the charge increases with increasing power, as does the maximum temperature, and the turbulent kinetic energy of the melt. As the power increases, the vortex range expands and the melt is stirred adequately.

Keywords: electromagnetic cold crucible; nuclear waste; magnetic field; numerical simulation

1 Introduction

The electromagnetic cold crucible is a revolutionary, highly efficient, and environmentally friendly melting technique. It is particularly beneficial when melting refractory materials and high temperature active metals [1, 2]. The issues of refractory melting of some wastes can be resolved by using electromagnetic cold crucible technology for the solidification of nuclear waste [3, 4]. Currently, one of the most important aspects of treating nuclear waste is the application of electromagnetic cold crucible solidification technology.

From the current research, for various aspect ratios of the cold crucible, Yang et al. [5] established a 2-D model to study the temperature field for Ti-Al alloy. The magnetic induction intensity inside the cold crucible was approximated by Huang et al. [6]. Due to the lack of research on the visualization of the melting and solidification process of UO_2. In this paper, the coupled calculation of magnetic field, temperature field and flow field under different power input parameters, and the law is systematically analyzed.

2 Experimental procedure

3D Model of EMCC

As shown in Figure 1(a) a circular inner cavity crucible is taken as the research object. Meanwhile, the 1/20 symmetric structure is used for modeling to improve the computational efficiency substantially. Impedance boundary conditions as well as coil excitation are added at the coil in the computational domain, no-slip conditions as well as radiative heat dissipation are added at the wall. The crucible temperature is initially 303.15 K, and the material physical parameters vary with temperature.

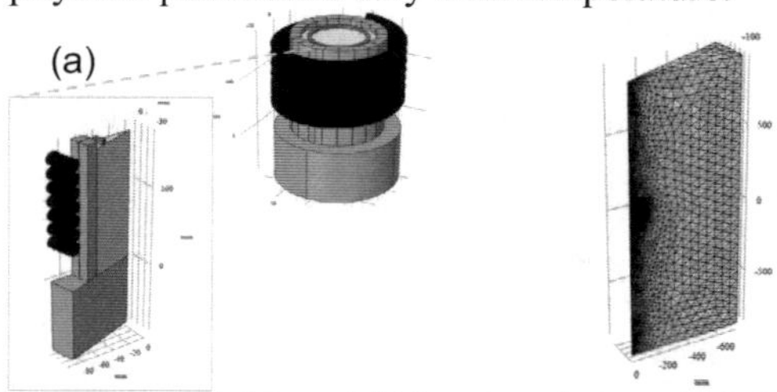

Figure 1 3D model (a) and mesh for calculation (b)

3 Mathematical Mode

The electromagnetic field is described by Maxwell equations [7].

$$\nabla \cdot \vec{J} = 0 \tag{1}$$

$$\vec{J} = -\sigma \frac{\partial \vec{A}}{\partial t} - \sigma \nabla \phi \tag{2}$$

$$\nabla \cdot H = \vec{J} \tag{3}$$

$$B = \nabla \times \vec{A} = \mu_0 \vec{H} \tag{4}$$

where, $\vec{J}$ is the current density; σ is the permeability of the magnetic fluid; $\vec{A}$ is the magnetic potential vector; t is the time; φ is the electric potential; $\vec{H}$ is the magnetic field strength; $\vec{B}$ is the magnetic flux density; μ_0 is the magnetic permeability of the medium.

The temperature change can be described by the classical transient heat transfer equation:

$$\rho C_p \frac{\partial T}{\partial t} = \nabla(\lambda \nabla T) + q_{in} \tag{4}$$

where, ρ is the material density; c_p is the material specific heat capacity; λ is the material thermal conductivity.

The flow in the molten pool in the electromagnetically cooled crucible satisfies the N-S equation:

$$\nabla \cdot \vec{v} = 0 \quad (5)$$

$$\frac{\partial(\rho\vec{v})}{\partial t} + (\vec{v} \cdot \nabla)\vec{v} = \mu_v \nabla^2 \vec{v} - \nabla p + \rho \vec{g} \beta_T \Delta T + \vec{F}_d \quad (6)$$

4 Result and discussion

The ingot is installed in the crucible, and the magnetic flux density is calculated at different powers. The calculation results are shown in Figure 2. The calculation results show that the induced current in the load increases with the increase of power.

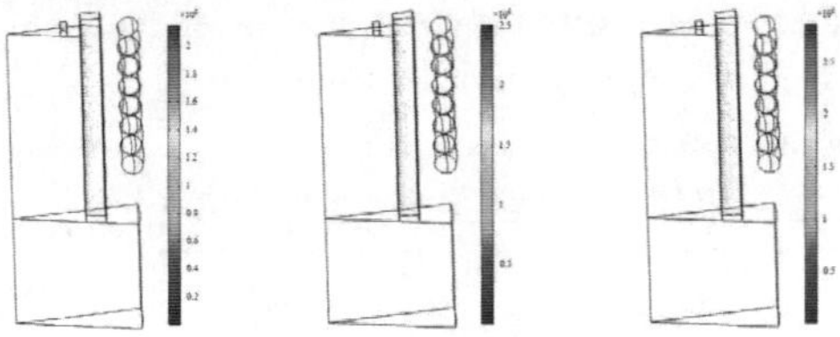

Figure 2 Induced current cloud diagram: (a)300 A; (b) 350 A; (c) 400 A

The temperature field and turbulent energy on the melt is calculated and the results are shown in Figure 3 and Figure 4.

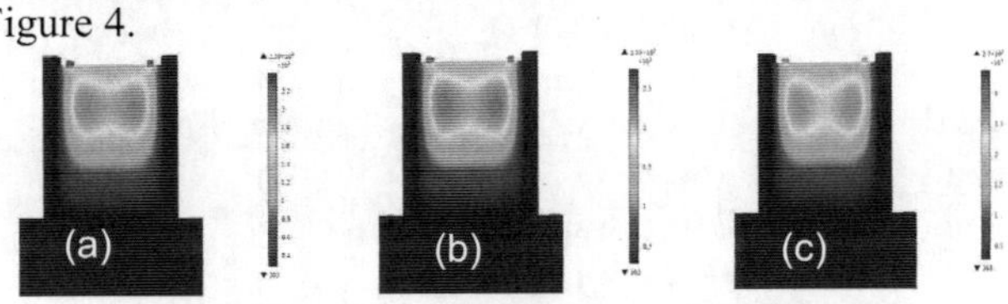

Figure 3 Temperature field variation with power: (a)141.34 kW; (b) 166.18 kW; (c)193.25 kW

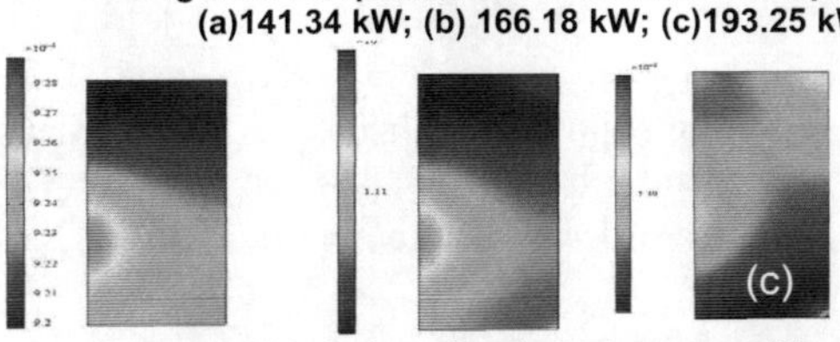

Figure 4 Variation of turbulent energy with power: (a)141.34 kW; (b) 166.18 kW; (c)193.25 kW

5 Conclusion

allics, 2012, 31: 264-273.

In this paper, the main conclusions are summarized as follows: 1. The induced current at the load center increases with the increase of power. 2. Along the height direction, the magnetic field increases first and then decreases. 3. As the power increases, the highest temperature and turbulent kinetic energy of the melt increase. When the power increased to 166.18 kW, a vortex appeared in the core of the melt.

6 Acknowledgments

This work was supported by National Natural Science Foundation of China (No. 52304393), and Heilongjiang Postdoctoral Fund (LBH-Z23173).

References

[1] Wang S, Wang Q, Chen R R, et al. Numerical analysis for solid-liquid interface shape at various temperature gradient in electromagnetic cold crucible directional solidification [J]. International Journal of Heat and Mass Transfer, 2022, 199.

[2] Yang Y H, Chen R R, Wang Q, et al. Dominant dimensionless parameters controlling solute transfer during electromagnetic cold crucible melting and directional solidifying TiAl alloys [J]. International Communications in Heat and Mass Transfer, 2018, 90: 56-66.

[3] Liu J, Wang F, Liao Q, et al. Synthesis and characterization of phosphate-based glass-ceramics for nuclear waste immobilization: Structure, thermal behavior, and chemical stability [J]. Journal of Nuclear Materials, 2019, 513: 251-259.

[4] Nuclear waste management facilities [M]. Elsevier Inc, New York: 2024.

[5] Yang Y H, Chen R R, Guo J J, et al. Experimental and numerical investigation on mass transfer induced by electromagnetic field in cold crucible used for directional solidification [J]. International Journal of Heat and Mass Transfer, 2017, 114: 297-306.

[6] Huang H T, Ding H S, Xu X S, et al. On real-time control of microstructure of Ti Al specimens with varied cross-sections based on numerical calculation and machine learning [J]. Journal of Alloys and Compounds, 2023, 938: 168549.

[7] Ding H, Chen R R, et al. Directional solidification of Ti-Al-W-Si alloy by electromagnetic confinement of melt in cold crucible [J]. Intermet

Research on the Defect Formation Mechanism of Shell Mold Cast Gas Turbine Blades

Chen Xiong[1], Shu Wang[1],*, Qiuju Zhu[2], Xiaoming Wang[2], Yalong Gao[2], Xin Ding[1], Qi Wang[1], Dezhi Chen[1], Ruirun Chen[1], *

1. National Key Laboratory for Precision Hot Processing of Metals, Harbin Institute of Technology, No. 92, Xidazhi Street, Huayuan Street, Nangang District, Harbin, Heilongjiang Province, China

2. Harbin Xinrun Industry Co., Ltd， No. 6 Pingling Road, Acheng Economic Development Zone, Harbin City, 150046, China

* Corresponding address: e-mail: wangshu@hit.edu.cn (Shu Wang), ruirunchen@hit.edu.cn (Ruirun Chen)

Abstract: Gas turbines are thermal mechanical devices that utilize gas power to drive rotating shafts, widely used in applications such as power generation, aircraft propulsion, and industrial processes. The current issues with gas turbines include manufacturing defects, turbine blade damage, operational wear during operation, and challenges in detection and maintenance. In this paper, we have analyzed the physical field distribution during the filling and solidification processes and eliminate defects.

Keywords: shell mold casting; numerical simulation; temperature field; defect

1 Introduction

Shell molding is a precision casting process that involves creating a heat-resistant ceramic shell on the surface of a mold, into which molten metal is poured to produce components with complex shapes and precise dimensions. The defects of gas turbine blades were analyzed in slices and the causes of defects were discussed and eliminate the casting defects [1,2].

Smith and Johnson et al. analyzed and optimized the manufacturing process of aerospace engine turbine blades using simulation and experimental methods. They focused on key manufacturing parameters and proposed methods to reduce manufacturing defects [4,5]. Brown et al. predicted and prevented manufacturing defects in gas turbine blade production using computational methods. They proposed strategies for process optimization and parameter control to minimize production issues [6]. Gupta et al. analyzed thermal stresses in turbine blade manufacturing using finite element simulation and proposed effective solutions to mitigate thermal stress risks [7].

2 Experimental procedure

The blade material is a nickel-based superalloy, and the shell material is Mullite, with thermal property parameters selected from a database. Interface heat transfer includes two parts: between the casting and mold shell, and between the mold shell and thermal insulation cotton. The gravity direction is set according to the actual pouring direction. Relevant boundary condition parameters are specified, and the step length exceeds 50,000. The mathematical model of filling and solidification process is as follows：

Continuity equation for compressible fluids [8]：

compressible fluid：

$$\frac{\partial \rho}{\partial t}+\frac{\partial \rho u}{\partial x}+\frac{\partial \rho v}{\partial y}+\frac{\partial \rho w}{\partial z}=0 \quad (1)$$

Incompressible fluid：

$$\frac{\partial u}{\partial x}+\frac{\partial v}{\partial y}+\frac{\partial w}{\partial z}=0 \quad (2)$$

u, v, w - the fraction of velocity in x and z directions; ρ-Density of liquid gold.

Three-dimensional heat transfer model：

$$\rho C\frac{\partial T}{\partial t}=\lambda\left(\frac{\partial^2 T}{\partial^2 x}+\frac{\partial^2 T}{\partial^2 y}+\frac{\partial^2 T}{\partial^2 z}\right)+Q \quad (3)$$

x, y, z - three-dimensional coordinates; ρ- Object density scale, g/w3; Cg specific heat, J/m3; T - Temperature, K; t – time, s and λ-Calorimetric conductivity, W/m • K; Q-Internal heat source, J

3 Result and discussion

Based on the numerical simulation results conducted during the forming process of gas turbine blade castings, the Niyama criterion implemented in ProCAST was utilized to predict shrinkage and porosity within the castings. The simulation outcomes indicate the following observations:

Shrinkage and porosity were observed in the large edge plate region, particularly away from the pouring riser and in the middle area between the two straight runners. In the blade body section near the intake side of the small edge plate, varying degrees of shrinkage and porosity distribution were identified. The resulting shrinkage and porosity distribution closely matched that observed in the actual pouring process.

Analysis of the stress field from the simulation results revealed a higher flow rate of liquid metal at the base of the blade body near the small edge plate during pouring, coupled with a smaller modulus in this region, which consequently favors quicker cooling compared to other areas. As the solid phase fraction reaches a critical value, thermal stress and strain develop, increasing the likelihood of hot cracks in this region. In practical pouring scenarios, hot cracks often originate from grain boundaries. The surface exhibits oxidation and lacks metallic luster, aligning roughly with the location of predicted hot cracks from the simulation results. The temperature field distribution cloud map and defect distribution are shown in the Figures 1(a) and (b) below:

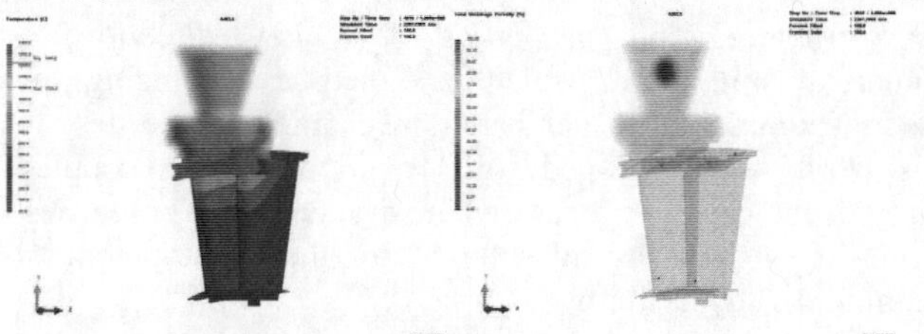

Figure 1 Temperature field distribution cloud map (a) and defect distribution (b)

4 Conclusion

Simulating the filling and solidification processes of castings using ProCAST software involves setting various casting temperatures and process parameters to analyze the filling process, temperature field, and flow field simulation results. This analysis helps determine the appropriate casting temperature and optimize casting process parameters. During the simulation, different casting temperatures and process parameters are tested to observe the flow of molten metal and temperature distribution during the filling process. Analyzing these simulation results, the optimal casting temperature of around 1,490 °C can be found to ensure smooth filling and prevent defects.

Using the HTI module in ProCAST simulation software, predictions were made regarding locations prone to thermal cracking. Through analysis of the flow field and stress field, specific thermal hotspots within the casting were identified， characterized by the highest concentration of thermal stress and strain. These identified areas closely align with actual observations. By optimizing the cooling system, it becomes possible to precisely adjust local temperature distribution, thereby slowing down solidification rates and reducing temperature gradients. These optimizations effectively mitigate the risk of thermal cracking, contributing to improved quality and integrity of the turbine blade castings.

5 Acknowledgments

This work was supported by National Natural Science Foundation of China (No. 52304393), and Heilongjiang Province Science and Technology Plan Project (No. MSSJH20240016), and Major Technology Achievement Transformation Project in Heilongjiang Province, Project Number: ZC2023SH0075.

References

[1] Ahmed M M, Atia A A. Optimization of gas turbine blade casting process using ProCAST software. Journal of Manufacturing Processes, 2019, 41: 53-62.

[2] Bhatia M, Sharma V. Numerical simulation of gas turbine blade casting using ProCAST, 2017.

[3] Smith R, Johnson S, Williams T. Analysis and optimization of aerospace engine turbine blade manufacturing process. Journal of Manufacturing Science and Engineering, 2019, 10(2): 123-135.

[4] Johnson M, Brown P, Wilson L. Optimization of gas turbine blade manufacturing process using numerical simulation. Materials Science and Engineering: A, 2015, 7(4): 289-301.

[5] Brown A, Miller J, Lee K. Prediction and Prevention of Defects in Gas Turbine Blade Manufacturing Process. Journal of Manufacturing Processes, 2017, 5(3): 210-225.

[6] Gupta S, Wilson M, Thompson E. Analysis of thermal stresses in turbine blade manufacturing using finite element simulation. International Journal of Mechanical Sciences, 2016, 12(1): 75-89.

[7] Chen X, Wang Y. Analysis of the continuity equation in computational fluid dynamics. Computers & Fluids, 2021, 202: 105432.

Multi-scale Modeling and Simulation of Low Pressure Die Casting of Aluminum Alloys

Yunxiang Tan, Juhuai Ma, Qingyan Xu*
Key Laboratory for Advanced Materials Processing Technology, Ministry of Education, School of Materials Science and Engineering, Tsinghua University, Beijing 100084, China
*Corresponding address: e-mail: scjxqy@tsinghua.edu.cn

Abstract: Aluminum alloys distinguish themselves through their notable properties, including low weight, robust strength, and versatile formability, which facilitate the production of extensive, intricate, and thin-walled castings using low pressure die casting processes.These distinctive features significantly contribute to the advancement of automotive lightweighting initiatives, which are essential in optimizing fuel consumption and mitigating environmental footprints. This study aims to optimize the low pressure die casting process of Al-Si alloys for automotive components through integrated numerical simulations. A software framework has been developed to couple macro-simulation of mold filling and solidification with microstructural evolution and mechanical property prediction. A novel nucleation model, validated with experimental data, enhances the accuracy of primary phase growth simulations. Models for dendritic and eutectic phase growth are also presented, considering solute diffusion, growth kinetics, and interface anisotropy. This multiscale approach enables the prediction of microstructures and mechanical properties, facilitating the optimization of casting parameters to enhance the final properties of the castings. The results demonstrate the potential to reduce production costs and accelerate the development cycle for aluminum alloy castings.

Keywords: aluminum alloys; simulation; modelling; low pressure die casting

1 Introduction

Al-Si alloys are widely used in the fabrication of complex components in automobiles [1]. Low pressure die casting is a common method for manufacturing aluminum alloy castings. The casting process determines the evolution of microstructures at different scales, which in turn dictates the final mechanical properties of castings (see Figure 1).
Employing numerical simulation to replace the traditional Trial and Error method can optimize the manufacturing process [2].

After decades of development, macro-simulation has matured, with some commercial software being applied in production [3]. However, most commercial software only focuses on mold filling and solidification simulation, and little attention was paid to microstructure and mechanical property prediction [4]. Therefore, through simulation, explaining the relationship between the casting process, microstructure, and properties is a key direction in numerical simulation, which is aimed at optimizing the casting parameters to get better mechanical properties. In this work, we developed software to conduct a coupled study of the casting process, microstructure, and properties, proposing a new casting solution for the development of aluminum alloy castings.

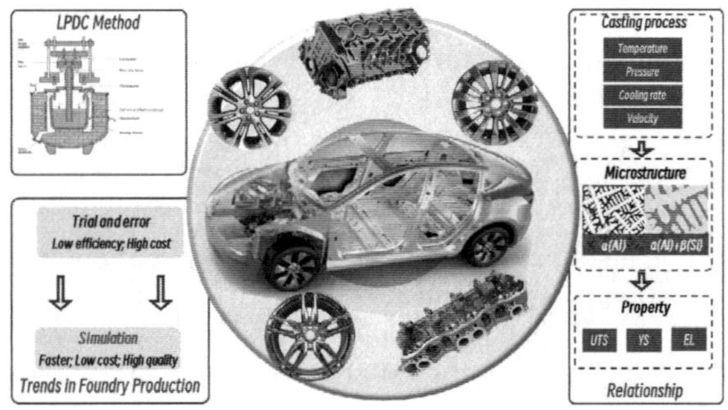

Figure 1 Casting process-structure-property relationship diagram of low pressure die casting of aluminum alloy

2 Mathematical modeling

The physical and mathematical modeling process pertaining to low pressure die casting can be methodically delineated into five interconnected components(see Figure 2). Firstly, the calculation of the solidification path, which lays the foundation for predicting the phase transformation behavior during casting. Secondly, macro-simulation, encompassing the broader-scale phenomena influencing the casting process, such as temperature distributions and fluid flow. Thirdly, the establishment of nucleation models, an important step that governs the initiation of solidification by modeling the formation of crystal nuclei. Fourthly, primary phase growth simulation, focusing on the detailed evolution of the initial solid phase, crucial for understanding microstructure development. Lastly, eutectic phase growth simulation, addressing the simultaneous growth of multiple phases, often occurring at eutectic temperatures, which is instrumental in refining the final microstructure and properties of the casting. This segmented approach ensures a comprehensive and accurate portrayal of the low pressure die casting process, facilitating process optimization and the production of high-quality castings.

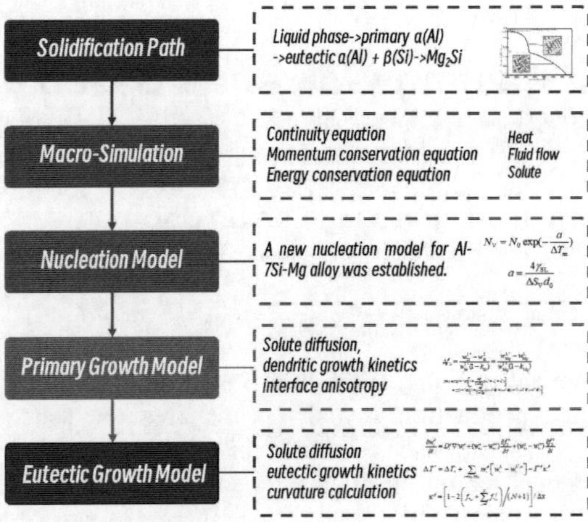

Figure 2 Mathematical modeling of low pressure die casting

3 Result and discussion

The mold filling and solidification processes of step casting were simulated. Temperature changes were acquired by inserting thermocouples at the center of each step. Subsequently, dendrite growth on the step casting, coupled with the macro-simulation, was simulated in both two and three dimensions. Given that different steps correspond to varying solidification conditions, the final dendrite microstructure was derived from these simulations.

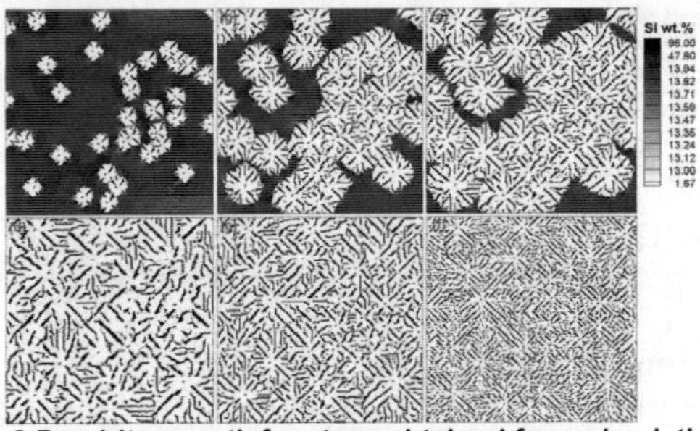

Figure 3 Dendrite growth for steps obtained from simulation and experiment

The evolution of unmodified and Strontium-modified Al-Si eutectic alloys was simulated by us under varying degrees of undercooling (see Figure 3). During the growth process, the eutectic colonies emerged in a nearly spherical form, with the two eutectic phases alternating in their growth patterns.

4 Conclusion

In conclusion, this study presents a novel casting methodology that significantly advances the current practices in the field. By developing a comprehensive nucleation model tailored specifically for Al-Si-Mg alloys and experimentally determining the crucial parameters, we have gained a deeper understanding of the underlying mechanisms governing alloy solidification. This model enabled the simulation of both the primary and eutectic phases of the aluminum alloy, which were seamlessly integrated with macro-scale simulation results, fostering a holistic view of the casting process.

Furthermore, we propose an innovative casting solution that harmoniously incorporates casting process dynamics, microstructural evolution, and material properties, thereby offering a robust framework for optimizing casting procedures and ensuring the production of defect-free castings. The practical application of this research to critical automotive components manufactured by Dicastal Company underscores its potential to revolutionize casting process optimization, leading to enhanced productivity and efficiency in industrial settings. In essence, this work represents a pivotal step forward in the development of advanced casting technologies, paving the way for future advancements in materials science and engineering.

5 Acknowledgments

This work was supported by a grant from the Beijing Natural Science Foundation (No. L223001).

References

[1] Luo A A, Sachdev A K, Apelian D. Alloy development and process innovations for light metals casting. Journal of Materials Processing Technology, 2022, 306: 117606.

[2] Hao Z Y, Ju Y, Chen L X. The use of aluminium and magnesium alloys in automotive lightweight technologies. Journal of Mechanical Science and Technology, 2023, 37(9): 4615-4622.

[3] Chen R, Xu Q Y, Liu B C. Simulation of the dendrite morphology and microsegre-gation in solidification of Al-Cu-Mg aluminum alloys. Acta Metallurgica Sinica (English Letters), 2015, 28(2): 173-181.

[4] Guo J W, Wu M W, Xiong S M, et al. Research progress in numerical simulation technology for die casting process. Special Casting & Nonferrous Alloys, 2023, 43(10): 1366-1375.

Research on the Master Plan Scheduling of Casting Workshop Based on Deep Reinforcement Learning

Wenhao Zhou, Fayuan Chen, Peng Yu, Xiaoyuan Ji*, Yajun Yin, Jianxin Zhou
State Key Laboratory of Material Processing and Die & Mould Technology, Huazhong University of Science & Technology, Wuhan, 430074, China
*Corresponding address: e-mail: jixiaoyuan@hust.edu.cn

Abstract: With the continuous enhancement of productivity and management levels, many foundry enterprises have begun to adopt multi-workshop distributed production, with each workshop capable of independently completing the full-process machining of castings. Consequently, the complexity of production scheduling tasks has further increased, making the master scheduling of distributed workshops the primary challenge in scheduling. This paper investigates a multi-objective optimization method for distributed workshop scheduling. Initially, a multi-objective optimization model is established, targeting the minimization of the maximum completion time, the minimization of order earliness/tardiness penalties, and the minimization of workshop load imbalance. By employing triangular and trapezoidal fuzzy numbers, the production process of castings is abstracted into a black-box model, upon which master scheduling research is conducted. The production status of workshops is conceptualized as agents, with the state of the workshop agents described by order status, current completion times across workshops, and the total completion time of the task queue. The allocation of tasks and the selection of workshops serve as actions, and a reward mechanism is set based on the objective functions. A dual-network deep reinforcement learning algorithm is utilized to solve the master scheduling model, and the results obtained are compared with those from an in-house improved particle swarm algorithm. The findings indicate that the proposed deep reinforcement learning outperforms the improved particle swarm algorithm in terms of performance for smaller-scale problems.

Keywords: distributed production; foundry workshop; master scheduling; deep reinforcement learning; neural network

1 Introduction

Casting is a discrete manufacturing process characterized by a fixed workflow. With the advancement of productivity, foundry enterprises currently consist of multiple parallel casting workshops, each capable of performing the entire production process of castings. This resembles a parallel machine scheduling problem, which has been a subject of considerable interest in both academic and industrial circles for an extended period[1]. In scenarios where there are numerous orders and production times are variable, the manual scheduling method is greatly influenced by the dispatcher's experience, particularly with the introduction of distributed production, which complicates the scheduling process. In response to this, this paper establishes a mathematical model for the master scheduling of distributed workshops based on fuzzy time theory. Compared to traditional metaheuristic algorithms, deep reinforcement learning algorithms, possess superior generalization capabilities, solution quality, and flexibility. Therefore, this paper employs a deep reinforcement learning algorithm to solve the established master scheduling model.

2 Experimental procedure

As shown in Figure 1, Initially, leveraging the principles of fuzzy theory, the processing time and ideal delivery time of castings are estimated through the utilization of triangular and trapezoidal fuzzy numbers. Subsequently, a mathematical model is formulated, focusing on the minimization of three objective functions: the maximum completion time, the penalty for early or late delivery, and the imbalance in workshop load distribution. Next, the workshop agent system is established through the following methods:

(1) State: The agent's state is described by a one-dimensional vector, comprising the scheduling status of each order, the completion time of each workshop, and the required processing time of the task queue.

(2) Action: A pair of data consisting of the task number selected for allocation at the current moment and the workshop number chosen for that allocation is used as the current action.

(3) Reward: Based on the three objective functions, the corresponding reward calculation rules are established.

Finally, the model is solved through following methods:

(1) An initial action selection policy is established, and a trajectory of length 10,000 is generated to create a historical data repository.

(2) An experience replay mechanism is utilized to select a dataset of length 1,000 from the historical data repository.

(3) The action values for each state-action pair are updated through a dual neural network. The first neural network is updated with weights in every iteration, while the second neural network serves as the target network, which is updated every 20 iterations. During the update, the weights from the first neural network are directly assigned to the second neural network.

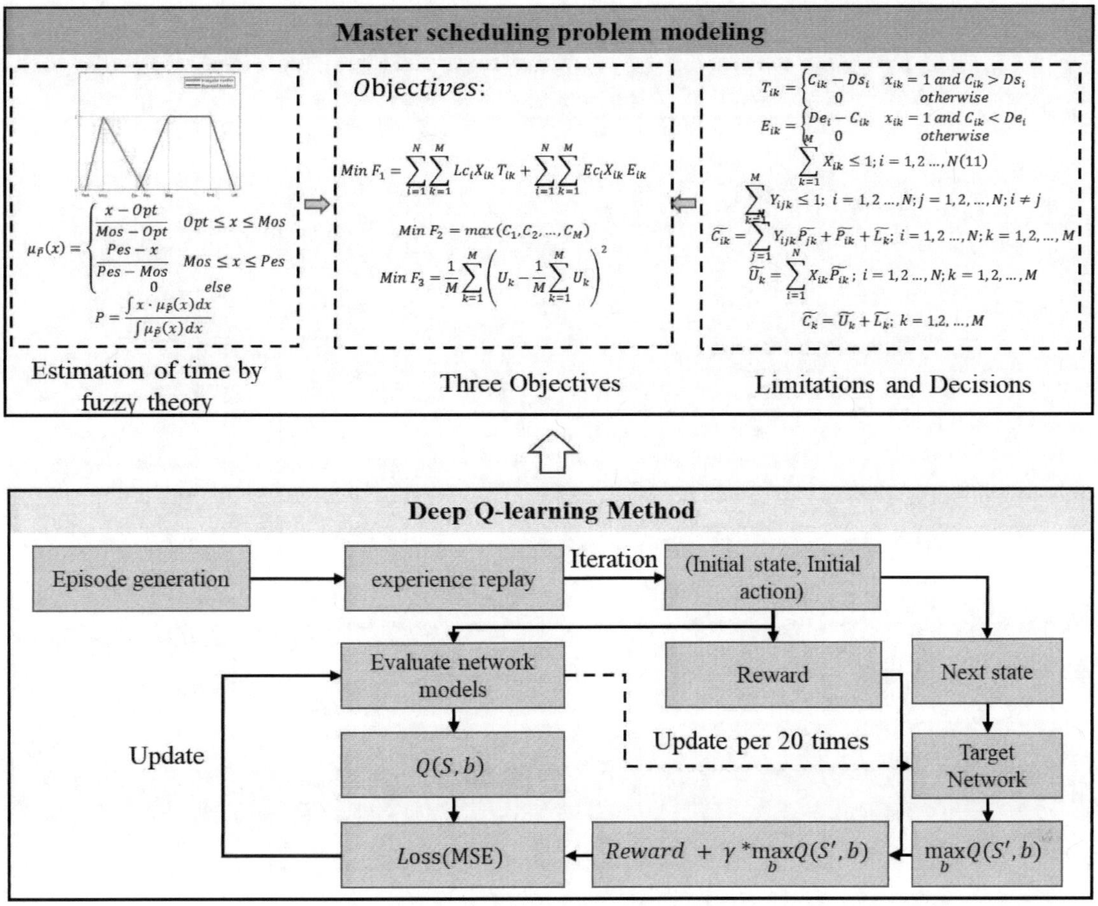

Figure 1 Framework of this paper

3 Result and discussion

Comparing the solution quality of deep reinforcement learning algorithm and improved particle swarm algorithm on this problem through Coverage metric comparison: comparing the dominance of the solutions resolved by two algorithm in the Pareto frontier:

Table 1. Verification result

	(Q-LEARNING, PDPSO)	(PDPSO, Q-LEARNING)
M1N1	**0.50**	0.30
M1N5	**0.40**	0.30
M2N6	0.20	**0.60**
M2N10	0	**0.30**

It can be observed that Q-learning yields better results than PDPSO[2] in M1N1 and M1N5. However, as the scale of the problem increases, Q-learning performs worse than PDPSO. This is because as the scale of the problem increases, Q-learning requires a larger computational scale. Due to limitations in computing power, the number of iterations set during computation does not reach the optimal parameters required for Q-learning.

4 Acknowledgments

National Key Research and Development Program (2020YFB1710100), National Natural Science Foundation (52275337, 52201042, 52090042, 51905188).

References

[1] Cheng T C E, SinC C S. A state-of-the-art review of parallel-machine scheduling research. Eur. J. Oper. Res., 1990: 271-292.

[2] Zhou W H, Chen F Y, Ji X Y, Li H L, Zhou J X. A Pareto-based discrete particle swarm optimization for parallel casting workshop scheduling problem with fuzzy processing time. Knowledge-Based Systems, 2022, 256: 109872.

Study on the Specificity of Artificial Spherical Ceramic Casting Sands

Qingzhou Sun[1], Wei Zhao[2], Jinhuai Wang[2], Shanghe Zhang[2]
1. College of Material Science and Engineering, Shandong University of Construction, Jinan, Shandong 250001, China
2. Shandong Jinpu Advanced Material Ltd. Binzhou, Shandong 256216, China

Abstract: The common production processes for artificial spherical ceramic casting sands include fusion blowing, sintering after spray granulation, and sintering after rolling granulation. Spherical ceramic casting sand produced by the fusion blowing process is referred to as fused ceramic casting sand; those produced by spray granulation and rolling granulation are known as sintered ceramic casting sand. The surface of fused ceramic casting sand can be described as "smooth orange peel", the main crystalline phase being amorphous phase, corundum phase and with a small amount of mullite phase; the surface microstructure of sintered ceramic casting sands is "rough orange peel", the main crystalline phases being mullite phase, amorphous, and a small amount of corundum phase or quartz. Compared to sintered ceramic casting sands, fused ceramic casting sand has lower thermal stability, poorer impact abrasion resistance, and higher density; and for the same mold (core), results in higher sand usage and higher gas emission.

Keywords: spherical; ceramic; foundry sand; specificity

1 Introduction

Artificial spherical ceramic foundry sand has spherical shape, high refractoriness, low acid consumption, and low thermal expansion, making it suitable for various sand casting processes and for various alloy casting materials. In addition to the common properties of ceramic materials such as high strength, hardness, wear resistance, and high temperature resistance [1], ceramic foundry sand also possesses specific characteristics attributed to its preparation process, chemical composition, and microstructure. These specific characteristics lead to subtle differences in the casting performance of artificial spherical ceramic foundry sands. This article takes artificial spherical ceramic foundry sand prepared from Al-Si refractory raw materials as an example to introduce the specificity of artificial spherical ceramic foundry sands and their correlation with casting process performance.

1. Types of artificial spherical ceramic foundry sands

The common production processes for artificial spherical ceramic casting sand include fusion blowing, spray granulation, or rolling granulation, followed by sintering. Fusion blowing involves mixing sand-making materials and fusing them into a liquid in a furnace. When the liquefied material exits the furnace, it is blown with high-speed airflow to disperse it into small droplets. The droplets finally form spherical sand particles. These particles are then sieved and graded to obtain artificial spherical ceramic casting sand. The artificial spherical ceramic casting sand produced by the fusion blowing process is referred to as fused ceramic casting sand, with commercial names such as 'Baozhu Sand', etc.

Spray granulation includes two main processes: spraying and drying. The sand-making materials are mixed, crushed, and pulverized, then made into a uniform slurry with a proper solid content. This slurry is sprayed into mist-like micro-droplets in a hot airflow environment, where the moisture is evaporated by the hot air. The solid phase within the droplets aggregates into dry particles (sand blanks). The sand blanks are then dried, sintered, cooled, sieved, and graded to obtain sintered ceramic casting sand.

Rolling granulation involves first mixing and pulverizing the sand-making materials, then placing the powdered materials into a rotating granulation disk. The sand billet is made by rolling granulation process. The sand blanks are then dried, sintered, cooled, sieved, and grading to obtain sintered ceramic casting sand.

Sintered ceramic casting sands commercial names

include Cerabeads, CPS, etc.

2. The specificity of artificial spherical ceramic foundry sands

The specificity of artificial spherical ceramic foundry sand mainly lies in three aspects: morphology, structure, and density of the sand. Taking the fused ceramic foundry sand produced by a certain company and the sintered ceramic foundry sand produced by Shandong Jinpu New Materials Co., Ltd. as research objects, their morphology and structure were observed under electron microscope and X-ray diffractometer, and their bulk density and true density were measured using the tapping method.

It can be seen that both the fused ceramic foundry sand and the sintered ceramic foundry sand exhibit uniformly spherical grain shapes. When observed under relatively low magnification, both show smooth porcelain-like surfaces. Upon further magnification, it is observed that the surface of the fused ceramic foundry sand appears to have a "smooth orange peel", while the surface of the sintered ceramic foundry sand appears "rough orange peel."

It can be observed that sintered ceramic foundry sand is composed of mullite phase, amorphous structure, and a small amount of quartz phase. The fused ceramic foundry sand, on the other hand, consists of an amorphous structure, corundum phase, and mullite phase. The content of the amorphous structure in fused ceramic foundry sand is much higher than that in sintered ceramic foundry sand. The true density and bulk density of fused ceramic foundry sand are higher than those of sintered ceramic foundry sand.

3. Correlation between the specificity of artificial spherical ceramic sands and performance of casting process

The sintered ceramic foundry sand and the fused ceramic foundry sand were separately placed into a ball mill for grinding. After 30 minutes of milling, the surface of the fused ceramic foundry sand surface was damaged, while it took 60 minutes of milling for the surface of the sintered ceramic foundry sand to exhibit similar damage. This indicates that the sintered ceramic foundry sand has a better resistance to impact, abrasion, and damage compared to the fused ceramic foundry sand.

The average bulk density of fused ceramic foundry sand is 1.31 times greater than that of sintered ceramic foundry sand. When manufacturing the same mold (core), the amount of fused ceramic foundry sand used will be 1.31 times that of sintered ceramic foundry sand.

The ceramic foundry sand samples were heated to 1400°C and then cooled to room temperature. The amorphous content before and after calcination was determined using X-ray diffraction. It was found that before calcination, the fused ceramic foundry sand had a significantly higher amorphous content than the sintered ceramic foundry sand. After calcination at 1400°C, the amorphous content in the sintered ceramic foundry sand remained essentially unchanged, while the amorphous content in the fused ceramic foundry sand decreased significantly.

2 Conclusion

(1) The surface of fused ceramic foundry sand exhibits a "smooth orange peel" appearance, with its matrix composed of amorphous structure, corundum phase, and a small amount of mullite phase. The surface of sintered ceramic foundry sand appears as "rough orange peel" and its matrix comprises mullite phase, amorphous structure, and a small amount of quartz phase.

(2) Compared to sintered ceramic foundry sand, fused ceramic foundry sands demonstrate reduced thermal stability and poorer resistance to impact and abrasion.

(3) The density of fused ceramic foundry sand is higher than that of sintered ceramic foundry sand and for the same mold (core), results in higher sand usage and higher gas emission.

References

[1] Wang Qi, Process Technology of Inorganic Non-metallic Materials. Beijing: China Building Materials Industry Press, October 2005.

[2] Tao Zhendong, Zheng Shaohua. Powder Engineering. Beijing: Chemical Industry Press, January 2015.

Production of Aluminum Castings in Green Sand

Per Larsen
DISA Industries A/S, Højager 8, 2630 Taastrup, Denmark.
per.larsen@noricangroup.com

Abstract: Many Chinese and US foundries have already adopted the green sand process for producing aluminum castings but there are still plenty of other foundries around the world that could benefit. With growing global demand for aluminum castings and the mounting pressure to reduce costs, the green sand alternative fits well with the requirements of key markets like automotive.

This paper will discuss the opportunities and challenges involved in casting aluminum in green sand. The opportunities include significant cost savings driven by factors such as increased: productivity, flexibility, reduced manpower requirements due to increased automation and high outputs, simple core handling, rapid prototyping, and low tooling costs.

Specifically, it will argue that green sand is often a superior choice for producing aluminum components that are, today, produced using die casting. It will consider which types of aluminum components are most suitable and why.

The paper will consider the molding and pouring options available to suit different casting applications and annual casting volumes. It will include the low pressure pouring option that optimises castings' mechanical properties and, at the same time, fully automating the pouring process which further reduces the need for manpower. It will also include a detailed case study of a foundry that pours aluminum using green sand and its success in transferring aluminum components to green sand that were previously produced using die casting.

Keywords: green sand, aluminum, die casting, automotive

1 Introduction

Many automotive parts are already cast in aluminum, and many other components previously made from iron are now being transitioned to aluminum. Aluminum castings also play a key role in the transition to electric vehicles (EVs). As the automotive industry makes this change, new high-volume aluminum parts for EV drivetrains are introduced like e-motor housings and covers.

The green sand process is well known for producing high quality iron parts efficiently and inexpensively. Growing global demand for aluminum castings combined with rising pressure to reduce costs is leading more foundries and manufacturers to explore the green sand process for aluminum applications as well.

Lower costs along with other benefits like rapid prototyping and flexibility are the rewards for successfully transferring the benefits of the green sand process enjoyed by iron foundries to the production of aluminum parts.

This paper investigates the opportunities and challenges of efficiently producing aluminum castings in green sand. It considers both safety-critical aluminum components and those with less stringent demands on mechanical properties, as well as high and low production volumes.

Historically, green-sand-cast aluminum parts have been perceived as having lower dimensional accuracy and less desirable mechanical properties caused by slow cooling. Neither of these is no longer justified, as will be shown below.

Moulding

Green sand molds can be produced both vertically and horizontally, with vertical moulding delivering the highest production speeds.

The best-known vertical molding lines can achieve very high speeds of up to 555 molds per hour without cores and up to 485 moulds per hour with cores. However, at least one manufacturer now supplies vertical molding lines with lower speeds.

Horizontal molding using matchplate technology is an attractive alternative to vertical, especially with lower production volumes and when high flexibility is required. It demands relatively low investment and employs the well-known horizontal principle.

Moulding material composition

Quartz sand mixed with clay and water remain as the moulding material. These are similar to the parameters used for cast iron but will have to suit the specific aluminum application. The use of coal dust is unnecessary for aluminum.

Advantages of green sand process

The green sand process offers several advantages, including very high productivity, a simple and proven core-setting process and rapid tooling changes: production only halts for one or two minutes.

Initial pattern costs are low and pattern plate lifetime is high. Pattern plate maintenance costs are low. Lead times from the first 3D CAD drawing to the first casting prototypes and the start of series production are short.

Pouring the mould

Aluminum forms oxides very quickly; these harm its mechanical properties. Because it lets the melt flow relatively freely, gravity pouring from the top of the mould generates these undesirable oxides. To achieve the best possible mechanical properties, another way of pouring method is needed.

A better alternative is low-pressure pouring through a sprue positioned low down on the bottom of the mold. Filling from the bottom up gives laminar flow and full control over the liquid aluminum's inflow, minimising oxide formation. Mould filling speed is no longer dependent on gravity and the casting profile can be easily optimised for each individual component.

This variant of low-pressure sand casting not only helps to improve the mechanical propertiesbut also fully automates the casting process. The runner bars can be very small which improves yield compared to conventional gravity casting from above. Smaller runner bars allow for more parts to be accommodated per mold on the pattern plate.

For aluminum parts with less stringent demands on their mechanical properties, manual gravity pouring of green sand molds is a good alternative. Hand pouring requires minimal investment and offers high flexibility, making it well suited to shorter series.

Metallurgy

Apart from mold filling, handling the metallurgy correctly is also important when aiming to achieve the best possible mechanical properties for aluminum castings produced in green sand.

Compared to the different die-based processes, casting in green sand moulds gives a lower cooling rate. This makes it more suitable for parts with thinner wall thicknesses from 3 mm up to approximately 10 mm where cooling power is less critical. However, thicker-walled parts can still be suitable for the green sand process.

The modification of the silicon phase is achieved with strontium, making it possible to have a continuous production flow.

Casting accuracy

Traditionally, sand casting was known for its lower casting accuracy. But the accuracy of castings produced in green sand has improved over the years, today achieving as little as 0.1 mm machine-related mismatch. Modern systems for in-line checking of mismatchs or mould gaps just before the moulds are poured give extra protection against possible dimensional problems and probable differences in weight.

Design for the green sand process

Most parts are designed with a production process in mind. The full potential of green sand can be reached by designing castings for the green sand process. This way, further weight reduction, productivity gains and, of course, cost savings can be achieved.

2 Conclusion

The green sand process's historic reputation for inconsistent quality and challenges around the mechanical properties of aluminum castings is unjustified.

Both safety-critical aluminum castings and castings with less stringent demands on their mechanical properties can be feasibly produced in green sand. Systems available today allow the highest production speeds of several hundred moulds per hour, but lower throughputs of less than 100 moulds per hour are also efficiently served by green sand lines.

It does not make sense to produce every aluminum casting in green sand, and this paper does not claim this. Instead, the suitability of each component and application should be judged individually.

The well-known benefits of the green sand process, along with the fact that certain castings are inherently well suited for it, and the increased efforts, especially by automotive manufacturers, to further reduce costs, suggest that aluminum and the green sand process are an increasingly attractive combination.

Coating Application of Heat-resistant Cast Steel Turbine Shell

Min Fang[1], Weidong Fu[1], Chao Deng[1], Hongyun Wan[2]
1. Xiangyang Ju-Li High Technology Material Co. Ltd., Xiangyang, Hu Bei, 441705, China
2. Wescast Industrial Inc., Wuhan 430058, China

Abstract
This article introduces the common casting defects of heat-resistant steel turbine shells. The causes of black spots and slag inclusion defects are analyzed using Sigma 300 scanning electron microscope, AZtecX-Max20 energy spectrometer, and X-ray fluorescence analyzer (XRF). A new zircon water-based coating (105KH coating) is used to solve the defect of coating peeling off in the inner cavity of the flow channel; the new composite water-based thermal insulation coating (857 coating) is used to solve the black spots and cold isolation defects within the inner cavity of the three-channel turbine shell, while ensuring that the inner cavity is free of defects such as sand and pores.

Keywords: turbine shell; heat-resistant cast steel; black spots; zircon coating; thermal insulation coating

1 Introduction
This article discusses common casting defects in heat-resistant steel turbine casings. The pouring temperature of heat-resistant steel is relatively high, sometimes as high as 1650°C (or even higher). Choosing appropriate coatings can effectively eliminate sand sticking, reduce surface pores and wrinkles, improve the surface finish of castings, and reduce inclusions and inclusions caused by coating peeling. Defects such as leakage [2]. At present, the domestic coatings used for heat-resistant steel turbine shells are mainly zircon powder or zircon powder composite coatings. However, in actual production, it is found that zircon powder coatings have problems with cold isolation at the end of the flow channel when producing complex three-channel turbine shell castings, the defect of the inner cavity coating falling off, and a small number of pore defects.

In response to these problems, our company has developed a new type of zircon powder water-based coating (105 KH coating), which solves the problem of coating peeling off in the inner cavity of the flow channel. The new composite water-based thermal insulation coating (857 coating) is used to solve the black spots and cold insulation defects in the inner cavity of the three-flow turbine shell, while ensuring that the inner cavity is free of defects such as sand sticking and pores. After production practice, the appearance and flow channel can meet customer requirements.

2 Experimental procedure
Water-based 105 KH paint trial process

To solve the problem of coating peeling, 105KH coating is used. This paint is white in color and has no pungent smell. After the paint is diluted and stirred, no bubbles are visible in the paint tank, and the defoaming property is very good. The paint is diluted with water to a Baume degree of 63 to 66 Bé and dip-coated. The dip-coating performance is excellent. The coating on the surface of the flow channel core has good coverage without any undesirable problems such as water breaks, bubbles, and flow marks (see Figure 1). After drying, the cross-section of the sand core was observed. The coating penetrated evenly and the coating penetration depth was 1.40-1.60 mm (see Figure 2), which is higher than that of workshop coating B. The wet thickness of the coating is 175~200 μm.

Fig. 1. Surface after drying

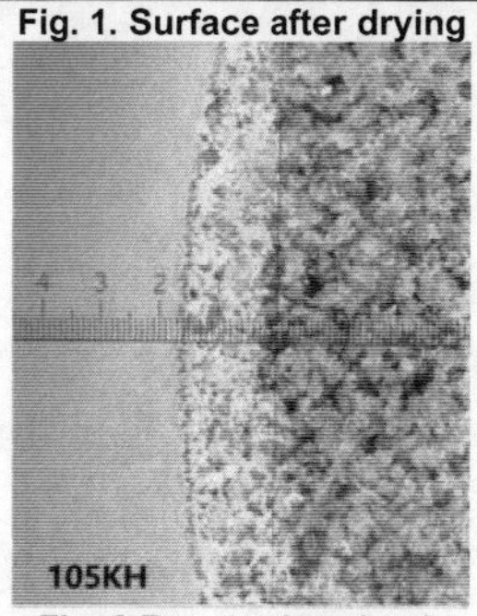

Fig. 2.Penetration depth

Water-based 105 KH paint trial results
The workshop uses B paint for normal production. After shakeout, the castings have a certain proportion of coating peeling defects following shot blasting. Using 105KH paint, 300 pieces of castings were produced after falling sand. The flow channels were inspected after shot blasting. No defects such as sand sticking, sintering, or coating peeling were found (Figure 3). Use a roughness meter was used to measure the surface roughness of the flow channel, and the measured roughness Ra≤6.5. The result meets the

customer's requirements. Afterwards, 3,000 pieces were produced in batches, all of which were qualified.

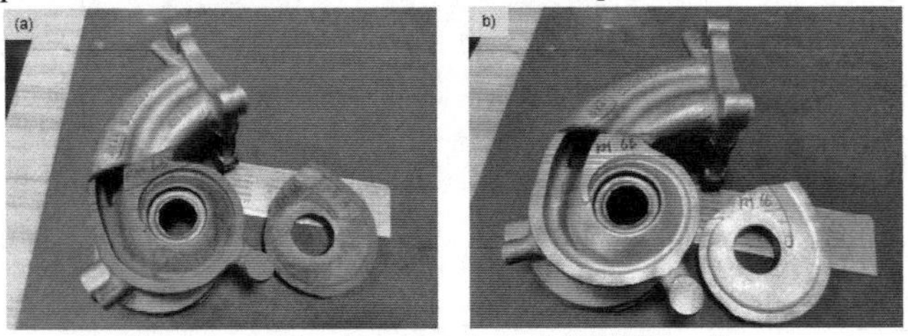

Fig. 3 Samples with water-based 105 KH coatings

3 Conclusion

(1) The black spot defects appearing in the internal flow channel of the heat-resistant steel turbine shell are mainly oxides of Al and Si, and it can be concluded that they are related to the coating.

(2) The weight proportion of Al and Si elements in the slag inclusion defects that appear in the internal flow channels of the heat-resistant steel turbine shell is very small, and they are mainly oxides of Fe, Ni, and Cr. This indicates that the defects come from the oxidation of the matrix, which is related to the oxidation of the matrix. Paint and molding sand have nothing to do with it.

(3) The use of water-based 105 KH zirconium-based paint can effectively solve the coating peeling defects in the internal flow channel of the heat-resistant steel turbine shell. The surface roughness can reach Ra≤6.5, which meets the casting design requirements.

(4) The use of water-based 857 thermal insulation coating can effectively solve defects such as black spots, pores, and cold insulation in the internal flow channels of heat-resistant steel turbine shells, and has a good effect on improving product quality.

References

[1] Xu Jialong. Research on austenitic heat-resistant steel volute casting process [J]. Equipment Machinery, 2014:20.

[2] Chen Shibin, Zhang Shiwei. Application of Italian corefon coating in China [C]. Proceedings of 2019 China Foundry Activity Week, 2019: 11-12.

[3] Zhang Yongfu, Zhao Xinwu. Application of Z889CH coating on heat-resistant cast steel turbine shell [C]. Proceedings of 2020 China Casting Activity Week, 2020:

Revolutionizing Metal Casting: Integrating Inorganic Binders with Refractory Coatings for Structural Applications

Christian Lustig, Ekaterina Potaturina
Hüttenes-Albertus Chemische Werke GmbH, Germany
*Corresponding address: e-mail: christian.lustig@ha-group.com , ekaterina.potaturina@ha-group.com

Abstract: This paper explores advances in metal casting, focusing on inorganic sand cores with specialized developed water-based coatings in HPDC for automotive applications. This technology can be used to produce lightweight, hollow structures especially required for the manufacture of structural components.

Keywords: Inorganic Binder, HPDC, Sand Cores, Sustainability, Refractory Coating

1 Introduction

This paper presents a comprehensive overview of the recent advancements in metal casting, focusing on the integration of inorganic sand cores in aluminum high-pressure die casting (HPDC), with a special emphasis on the automotive industry. The innovation lies in the synergy of traditional sand cores coated with a (water based) core wash and the dynamic HPDC processes, addressing the need to produce large, lightweight, hollow structures for automotive applications.

Key Considerations for the use of Inorganic Sand Core Usage in High-Pressure Die Casting

As the automotive industry shifts from internal combustion engines (ICE) to battery electric vehicles (BEV), the requirements for the body-in-white (BIW) structure must evolve accordingly. This evolution entails a transition from designing simple, single-piece components to more complex structures, offering several benefits such as increased design flexibility, the ability to create undercuts, and the elimination of the need for draft angles in high-pressure die casting (HPDC) tools.

A significant innovation to realize these demands is the use of sand cores in the casting of large, volumetric components. Traditionally, various methods have been employed to create cavities within HPDC parts these include steel cores and salt, or ceramic cores mainly used for smaller cavities.

Inorganic sand cores in HPDC applications represent a new approach, not intended to replace existing technologies but rather to complement them–with focus on bigger and more complex cavities. They provide many advantages, such as greater design freedom and especially a well-established and economically efficient production processes. Sand cores have been a part of foundry technology for many years, with the first machine-made sand cores dating back several decades.

Especially inorganic bonded sand cores show high potential for the use in HPDC applications as they are environmentally friendly not only in their production but especially also in casting and suitable for wet decoring and reclamation process. Especially as inorganic cores are water disposable, these cores can be easily and gently removed from thin-walled castings. This is the basis for the application of this technology in the field of HPDC castings.

2 Experimental procedure and results

Combining Sand Cores and HPDC: The Best of Both Worlds

A demonstrator was designed to develop and evaluate the concept of lost cores in HPDC applications. An existing HPDC casting node was modified with integrated sand cores that have ribs and undercuts. The tests were run at piston speed between 5.5 and 7.0 m/s. The after pressure varing between 300 and 900 bar.

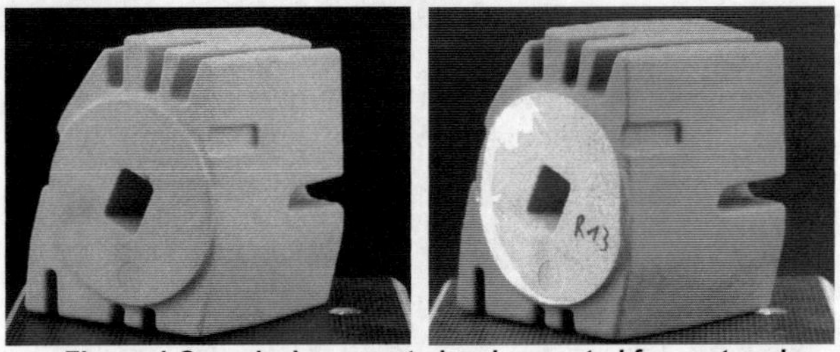

Figure 1 Core design – coated and uncoated for cast node

MAGMA simulations were also carried out to understand the flowability of the demonstrator cast node.

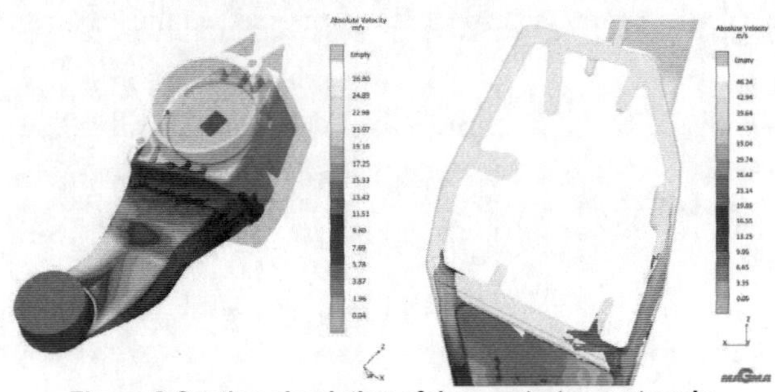

Figure 2 Casting simulation of demonstrator cast node

As a result, most of the metal penetrations and sand adhesion between the ribs were present until the optimum parameters were found.

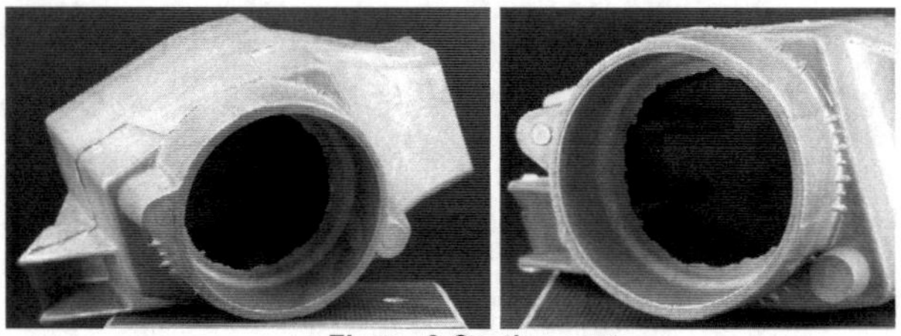

Figure 3 Casting part

Integrating sand cores with HPDC casting technology requires several process steps:

- machine production of sand cores
- surface sealing of the cores
- core removal
- recycling of core sand

While the core-making process for inorganic sand cores is well established, the surface sealing of these cores is a new and challenging topic. The coating process plays a crucial role in this integration. One of the biggest challenges is the development of a water-based coating that is suitable for inorganic and water-disposable sand cores and does not damage them. When developing such coatings, factors such as the composition and rheological properties of the coating must be considered, especially as these cores should have long, deep, narrow ribs to be coated. Among the various application methods, dipping is a viable option. It is crucial that the cores do not absorb too much water and that the coating does not penetrate too deeply. The coating should be applied well to avoid dripping and the formation of runs to ensure a smooth and even coating layer.

When transferring the parameters from the demonstrator casting to the real, various machine-side parameters, such as the flow rate and the casting pressure, were varied. The influence of the sand core type and, in particular, the applied coating was evaluated. It is easy to see that the casting results correlate to a high degree with the applied flow rates and the applied coating. As the sand core is porous by nature, the coating has a dominating influence on the casting result and especially on the surface finish.

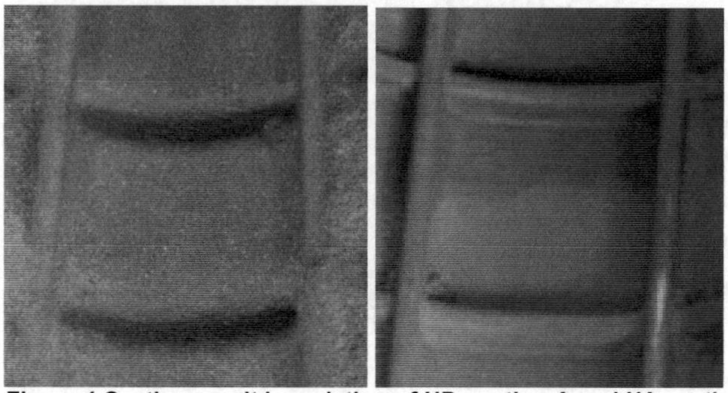

Figure 4 Casting result in variation of HD coating A and HA coating B

3 Conclusion

For HPDC applications, the sand cores used are important but not the only criterion for successful casting. The selection of a suitable coating and its application to the sand core is crucial for the successful use of sand cores in HPDC casting.

The suitability of this new technology was proven based on several demonstrator castings. All process steps were examined and evaluated:

- Core production
- Core finishing - Sealing of the core
- Casting
- Sand removal
- Recovery of the sand

With these findings, a start has been made on further evaluating this technology and bringing it into initial series applications.

References

[1] Kallien L Böhnlein C Dworag A Müller B. Ergebnisse aus dem Forschungsprojekt 3-D-Freiform– medienführende Kanäle im Druckguss Giesserei 2013 (100)12: 36-43.

[2] Zaretskiy, L. Modified Silicate Binders New Developments and Applications. Inter Metalcast 2016 (10): 88–99.

Development and Application of Crushing-shaped Silica Sand with High Silica Content and Roundness from Quartzite Crushing and Shaping

Jianxun Zhu [1,*], Jie Chen[2], Zhenguo Xing[3], Na Li[1]

1. Shengquan Group Share Holding Co.Ltd., Jinan, China
2. Handan Yuexin New materials Technology Co., Ltd. Handan, China
3. Hebei University of Engineering, Handan, China

*Corresponding address: Diaozheng Industrial Development zone, Zhangqiu, Jinan, China
Email: sq007@shengquan.com

Abstract: Sand casting accounts for about 80% of the world's casting production, and more than 50 million tons of silica sand are consumed each year, the vast majority of which are high-quality natural silica sand. Limited by resources, many foundries have to purchase silica sand from natural silica sand plants thousands of miles away, resulting in not only a higher logistics cost than the cost of silica sand itself, but also a great increase of carbon emissions. In this study, a new crushing and shaping technology was used to develop a shaped silica sand with high silicon content and roundness. The silica content of shaped sand is above 98.5%, and the roundness is higher than 0.89. Moreover the sand broken rate is low, and the refractoriness is high. Tests in various processes and foundries show that shaped silica sand can replace completely high-quality natural sand.

KeyWords: quartzite, shaped silica sand, high roundness, high quality silica sand, carbon reduction

1 Introduction

Quartzite is a mineral widely existing in nature, its silica content is greater than 99%. Some foundries use silica sand from quartzite crushing in steel casting, but due to poor grain shape, it can only be used for CO_2 hardening water glass process along with problems of high amount water glass addition, poor collapsibility, and low regeneration rate. Therefore, the development of silica sand with high silica content and round grain shape can help solving the resource constraint, reducing logistics costs and carbon emissions in steel casting.

2 Production process of shaped silica sand

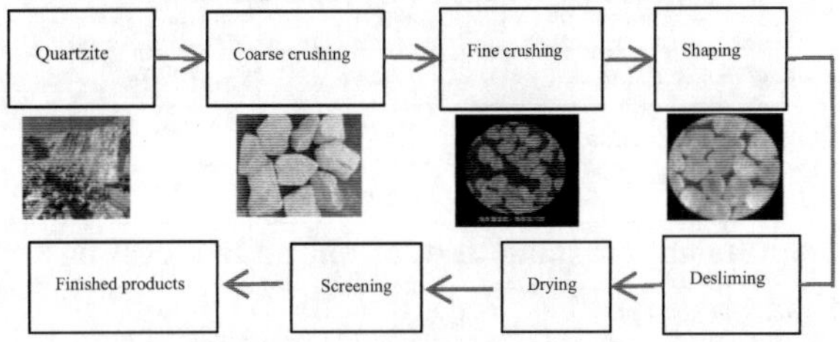

Fig.1 High silica content and roundness shaped silica sand production process

This process is determined after screening and optimisation of various shaping equipments and processes.

3 Comparison of physical and chemical properties

Table 1 Physical and chemical properties of sand

Silica sand	Roundness*	Specific surface area (cm²/g)	Clay content	ADV	refractoriness (℃)	LOI	SiO_2 content
Fujian sand	0.878	89.63	0.41%	4.81	1710	0.19%	>98%
Hainan sand	0.885	82.20	0.33%	1.42	1720	0.11%	>99%
Crushed sand	0.854	93.25	0.56%	15.64	1730	0.18%	>99%
Shaped sand	0.902	75.58	0.30%	4.79	1730	0.15%	>99%

The data in Table 1 shows that the silica content, roundness and specific surface area of shaped silica sand are better than high-quality natural silica sand and crushed silica sand, which is conducive to reducing core binder addition and improving quality of castings.

4 Broken Rate

The formation of micro-cracks on the surface of

shaped sand during the crushing and shaping process can significantly influence sand breakage during casting and sand regeneration, affecting both particle size distribution and reusability. To investigate this, we conducted a series of tests and experiments.

4.1 Broken rate test

The broken rates of cast iron natural silica sand, Hainan natural sand and shaped silica sand were respectively tested, and the results are shown as below in Table 2.

Table 2 Test results of breakage rate

Silica sand	Broken rate (%)	Test pressure
Cast ion natural silica sand	2.74%	35MPa
Hainan natural sand	2.84%	35MPa
Shaped silica sand	0.89%	35MPa

4.2 Simulated crushing test of simulated casting and sand regeneration process

We heated natural silica sand and shaped silica sand with same particle size to 1000℃, and then naturally cooled to room temperature, grinding under the same conditions for 10 hours. The changes of particle size and surface morphology of the two kinds of sand before and after grinding were compared. Results are shown in Figure 2 and 3.

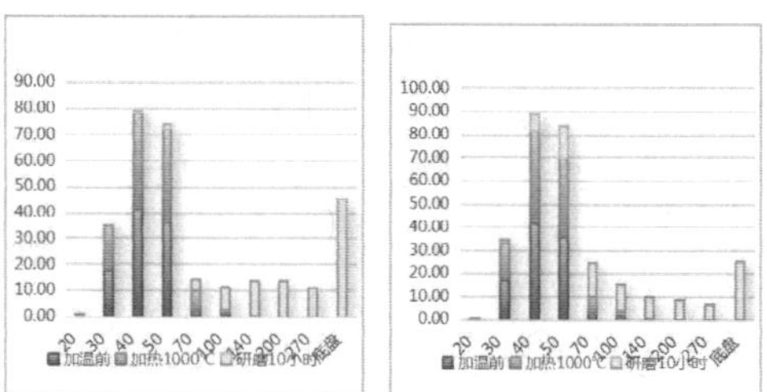

(a) cast steel natural silica sand (b)shaped silica sand

Fig 2. Simulated crushing tests result

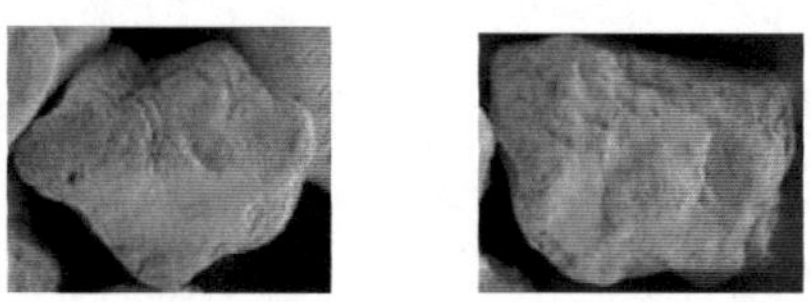

Natrual silica sand before grinding Natrual silica sand after grinding

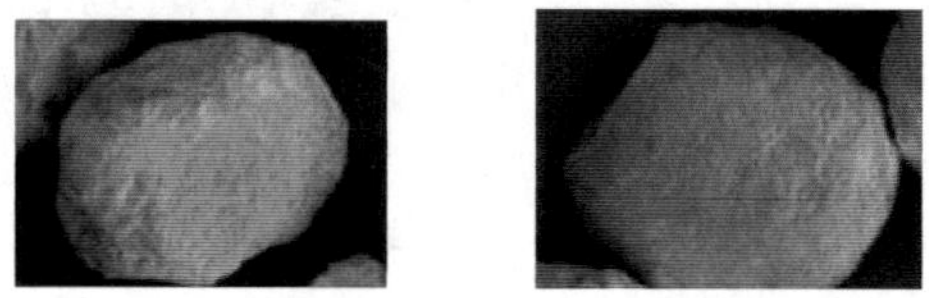

Shaped silica sand before grinding shaped silica sand after grinding

Fig.3 SEM photos of sand images

All the tests and results show that there are no micro-cracks on the surface of shaped silica sand, and its crushing resistance is better than that of natural silica sand.

5 Process properties tests

Sand (sameAFS)	24h pulling strength（MPa）			
	Furan process①	Alkaline phenolic process②	Easter hardening water glass③	RCS④
Fujian Sand		0.70		4.11
Hainan Sand	0.96			
Shaped sand	1.45	0.83	2.55 (compressive)	5.48
Crushed sand			1.08 (compresive)	

①resin addition1.0% , hardener addition 45% (of resin)
②resin addition1.63% , hardener addition 20% (of resin)
③water glass addition 2.7% ,ester addition 18%(of water glass)
④resin addition 2.5% , hexa addition15% (of resin)

The strength test results of several molding and core-making processes show that the strength of shaped silica sand is better than that of natural silica sand under the same amount of binder.

6 Conclusion

The new high roundness and high silicon content shaped silica sand can be used for a variety of core molding processes, it has high SiO_2 content (>99%), roundness (>0.90) and better reusability, especially the binder consumption can be reduced, and can completely replace the high-quality natural silica sand in steel casting.

Cooling Behavior by Water Spray and Air-Blow on Green Sand with High-Temperature

Yasuhiro Maeda [1*], Shingo Yoshida [2], Yuuka Ito [2,3], Takashi Hanai [3], Hiroyasu Makino [3]
1. Dept. of Mechanical Engineering, Daido University, 10-3 Takiharu-cho, Minami-ku, Nagoya 457-8530, Japan
2. Graduate Student, Dept. of Mechanical Engineering, Daido University
3. SINTOKOGIO Ltd., 3-1 Honohara Toyokawa 442-8502, Japan
*Corresponding address: e-mail: y-maeda@daido-it.ac.jp

Abstract: The sand cooling process before the mulling of green sand is fundamental to obtaining the green molding sand with proper conditioning and performance characteristics. In a previous study [1], the water-spay cooling behavior of the high-temperature green sand that exceeds 100℃ by water spray is investigated under static conditions. It is clarified that in the early stage of watering cooling, the evaporation cooling of water mainly occurred by watering. Further, in the later stage of watering cooling, the temperature is decreased under heat transfer behavior the same as air (natural) cooling.

In the present study, both the water spray cooling and air-blow cooling behavior of the high-temperature molding sand with a high temperature that exceeds 100℃, which is mulling, is investigated by the experiment and the numerical analysis. It clarifies the effect of watering on the temperature drop of the hot sand in the early stage of sand cooling. Then the air-blow cooling phenomenon is also clarified in the later stage of cooling. Further, two types of cooling behavior are explained by two equations introduced by the heat transfer and the modified evaporation of water.

Keywords: Green Sand, Hot Sand, Watering, Air-blow, Heat transfer

1 Introduction

The temperature of return sand becomes higher because of the change of the component of green molding sand and the speedup of the molding line [1-3]. The sand cooling process before the mulling of green sand is very important to make the comfortable green sand mold and to obtain comfortable quality molding sand. The green sand before the mulling process requires suitable properties of temperature and moisture content.

In a previous study [4], the high-temperature molding sand that exceeds 100℃ is investigated under static conditions, then the water spray cooling behavior of the molding sand is investigated. It is clarified the cooling phenomenon of molding sand with high temperatures that exceed 100℃.

In the present study, both cooling behavior by water spray and air-blow on the high-temperature molding sand is investigated by the experiment and the calculation. The effect of the addition of air-blow to water spray on colling behavior is analyzed. The temperature of green molding sand is estimated by using the theoretical equations using a heat transfer and an evaporation equation.

2 Experimental procedure

The green molding sand consists of synthetic sand (mullite beads) with 10% sodium bentonite, and water. The properties of molding sand are controlled 40% of the compact ability (CB) index using Sympson-type mulling machine. The molding sands of 500g are pre-heated to the initial temperature of 120, 100, and 80℃ in the heater.

The experimental conditions are detailed in Table 1. In the experiment, the molding sand, which undergoes mulling and air-blowing, is cooled naturally by atmospheric air, with variations in watering amount and air-blow speed. Then, the temperature and moisture content of the molding sand are measured during cooling at 30s of intervals.

3 Result and discussion

The comparison of the cooling behavior varying with the air-blow speed in the case of the initial temperature of 100℃ and the watering amount of 10g is shown in Fig.1. From the figure, the air-blow affects the early stage of temperature drop of molding sand. The temperature gradient slope of molding sand after cooling is nearly the same. These two points show a similar temperature drop even if the initial temperature of casting sand is changed.

Table 1 Experimental conditions

Green molding sand	Sand 90%+Bentonite10%+Water
Sand	Ceramic Sand (Mullite beads)
Bentonite	Na-Type (Volclay)
Compactability	40%
Initial Temp.	80, 100, 120 ℃
Weight of sample	500 g
Cooling type	Air-blow, Air-blow + Water spray
Watering amount	0, 10, 20 30 g
Air-blow speed	0, 0.8, 1.6 m/s.

From the previous study in the case of static conditions, the cooling behavior is analyzed in two mechanisms. In the early stage of watering cooling, the evaporation cooling of water mainly occurs by watering. Later stage of watering cooling, the temperature is decreased in accordance with heat transfer behavior the same as air (natural) cooling. The tendency of cooling behavior obtained in the case of present conditions such as the air-blow and the air-blow + Water spray cooling is the same with the previous ones.

Modifying the theoretical analysis model proposed in the previous study [4], two equations to estimate the cooled molding sand temperature are obtained in the present study. Equation (1) is the water evaporation equation for the early stage of water spray cooling.

$$T_{s2} = T_{s1} + \frac{L_s}{\beta_s \times C_s}\Delta w - \left(T_{s1} - T_1\right)\frac{1-e^{-H(1-\varphi')t_0}}{1-\varphi' e^{-H(1-\varphi')t_0}} \qquad (1)$$

Equation (2) is the heat transfer cooling equation for the later stage of cooling.

$$T_{s2} = T_{s1} - \left(T_{s1} - T_1\right)\frac{1-e^{-H(1-\varphi')t_0}}{1-\varphi' e^{-H(1-\varphi')t_0}} \qquad (2)$$

Where, T: temperature [K], t: time [s], L: latent heat of vaporization [J/kg], C: specific heat [J/kgK], w: moisture content [kg/dry.kg], α: heat transfer coefficient [W/m^{2}K], β: rate of wet sand, A: surface area [m^2], m: Mass [kg], subscript: a: air, s: sand, sw: wet sand, sd: dray sand.

Fig.2 shows the comparison of water spray cooling behavior between experiments and calculations. From Fig.2, it is possible to infer the behavior close to the gradient of temperature decrease of the experiment by setting the specific heat correction coefficient to β_s=0.25 and the heat transfer coefficient correction coefficient to β_c=0.60. The temperature of molding sand is dropped by the evaporation of water in the early stage, and then it is decreased by heat transfer behavior in the later stage.

4 Conclusion

It is clear the cooling behavior of both the watering and air-blow on the green molding sand with high temperatures that exceed 100℃. The air-blow affects the initial stage of temperature decrease of molding sand. Two types of cooling behavior are explained by two equations introduced by the heat transfer and the evaporation of water using two correction coefficients.

References

[1] Japan Foundry Engineering Society: Report of Technical Session "Trend of Sand Mold Technics", No. 164 of JFS Meeting, Kyoto (2014), pp.1-37.

[2] Japan Foundry Engineering Society: Research Report No.90, (2002), pp.71-81.

[3] Japan Foundry Engineering Society: Research Report No.117, (2015), pp.12-40.

[4] Y. Maeda and H. Makino: "Basic Investigation into Water Cooling of the Green Molding Sand with High Temperature", International Journal of Metalcasting, 11(2017), pp27-32.

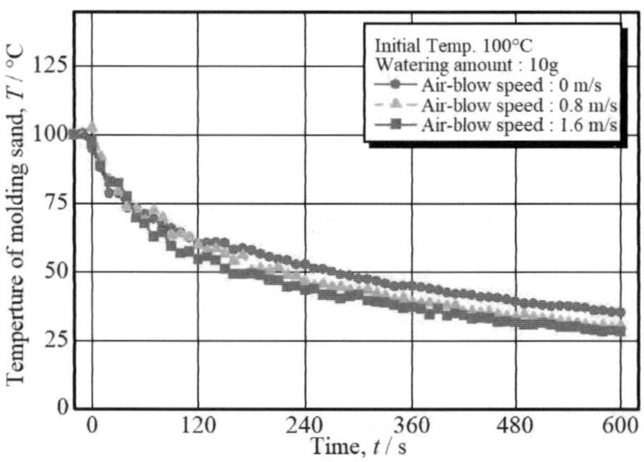

Fig.1 Comparison of the cooling behavior varying with the air-blow speed in the case of the initial temperature of 100℃ and the watering amount of 10g.

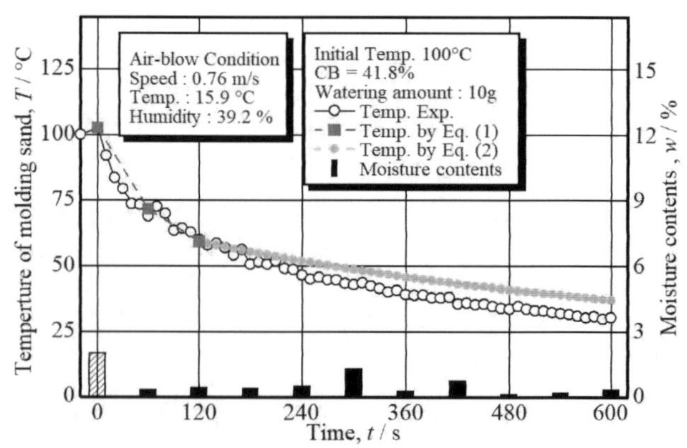

Fig.2 Comparison of cooling behavior between experiments and calculation in the case of initial temp. of 100℃, watering amount of 10g and air-blow speed of 0.8m/s.

Calculation Model of Casting Filter in Mold Filling Simulation using Particle-based SPH Method

Kento Miwa[1], Kousuke Taki[1], Yasuhiro Maeda[2,*]
1. Graduate Student, Dept. of Mechanical Engineering, Daido University, Japan
2. Professor, Dept. of Mechanical Engineering, Daido University, Japan
*Corresponding address: e-mail: y-maeda@daido-it.ac.jp

Abstract: In order to produce good castings, it is necessary to suppress the occurrence of casting defects. One means of reducing the occurrence of casting defects is the use of a casting filter. Casting filters are expected to have the effects of rectification, removal of inclusions, and early filling of the gate. On the other hand, a challenge arises with the particle-based SPH method in simulations, where the number of calculated particles required to analyze mold filling with the casting filter becomes excessively large. We have reported in a previous study, that it was recommended that the particle size *d* in the calculation be smaller than $d=D/16$ concerning the casting filter pore size *D*. Then, it would be possible to observe the molten metal behavior passing through the filter. However, the number of calculated particles is not small even when $d=D/16$ is used.

In this study, a "filter scaled model" is investigated to increase the filter pore size *D* while maintaining the permeability on the surface and the porosity of volume. The permeability on the surface and the porosity of the volume of the filter are closely related to the pore size of *D* and the number of pores of *N*. If the filter pore size of *D* becomes large then the number of pores of *N* decreases. The flow behavior passing through the filter is investigated varying the number of pores of *N* in the simulation. The particle fraction in the upper, middle, and lower areas of the filter is compared to the flow behavior with the visualization experiment. As a result, the number of calculated particles can be significantly reduced while maintaining a certain degree of calculation accuracy.

Keywords: Strainer filter, Honeycomb filter, SPH, Permeability on the surface, Porosity of the volume

1 Introduction

Casting filters are expected to have the effects of rectification, removal of inclusions, and early filling of the sprue and runner. However, the designer's know-how or experience determines the casting plan with the filters. It is not clear the appropriate specifications of filter type, aperture ratio, installation position, and other conditions. Previous studies on molten metal flow through filters are as follows. Yang et al. [1] investigated the distribution trends of inclusions in filters from water model observation and cast-iron casting experiments. Jonas Bäckman et al. [2] also investigated how different filter positions, filter roughness, filter effective cross-sectional area, and filter length affect filling. Further, in the previous studies, Taki et al. [3] have evaluated filter performance using the particle-based SPH method. From the report of Taki et al., the particle size of *d* in the calculation is recommended to be smaller than $d=D/16$ concerning the casting filter pore size of *D*.

According to the previous report [3], the number of particles in the calculation is not few even when $d=D/16$ is used. In addition, it did not sufficiently examine whether $d=D/16$ is appropriate in the terms of calculation accuracy, calculation time, and memory usage. In this study, a "filter scaled model" is investigated to increase the filter pore size of *D* while maintaining the permeability on the surface and the porosity of volume.

2 Analysis object and method

The analysis object in this study is the casting geometry called L-220, which was the visualization experiment in the previous report [5]. The casting filters are installed at the sprue. Since it was difficult to analyze the mold filling behavior for the entire castings due to the huge number of particles, the simulation at a 1/2 scale is proposed. For the filter part, the filter 1/4 model is also proposed. The permeability on the surface and volume porosity of the 1/4 filter model are the same as those of the actual filter. Fig. 1 shows the model used for the calculations. Permeability on the surface and volume porosity of the filter are related by using the filter pore size of *D* and the number of pores of *N*.

In the present study, a filter scaled model is evaluated using particle-based SPH software "COLMINA CAE" [4]. The five filter types, which are Strainer filter of 20 t and 12 t, and Honeycomb filter of 100, 200, and 300cpsi, used in the visualization experiments are investigated. The particle

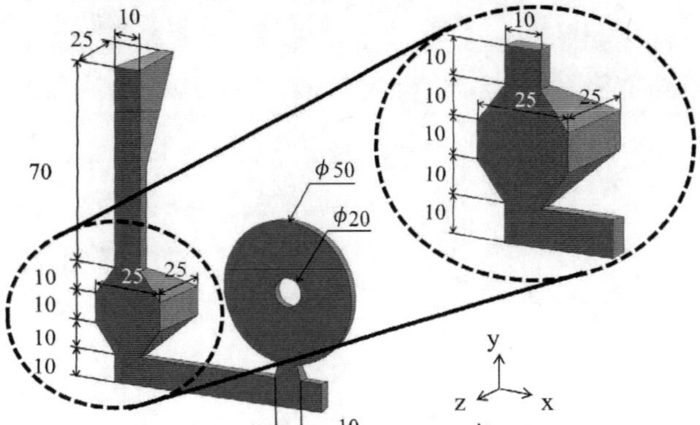

Fig. 1 The calculation model (L220, filter 1/4 model)

size in the calculation is used $d=D/16$. Although the same value of permeability on the surface and volume porosity of the filter, the number of pores of N decreases following the filter pore size of D increases. The number of pores of N is changed from a minimum of 4(=2×2) until a real number.

3 Result and discussion

As an example of calculated results, Fig. 2 shows the comparison of flow behavior of molten aluminum alloy passing through the honeycomb filter 200cpsi by varying the numbers of pores N. Comparisons of flow behavior were made at 0.040 s after the molten metal impacted the top surface of the filter. Regardless of the filter pores conditions, the flow is observed to split into two types of flows at the top of the filter, one splitting to the left/right and one splitting to the inside of the filter, almost the same as in the direct observation results. The results are affected by adopting the condition of $d=D/16$ proposed in previous research. On the other hand, the simulation results for those with a small number of filter pores slightly differ in the actual behavior.

Fig. 3 shows the results for the particle fraction at the upper, inside, and lower the filter. The particles inside and lower the filter are those that have passed through the pores of the filter. From Fig. 3, increasing the number of pores of N will be almost the same percentage as the permeability on the surface and volume porosity of the filter. Conversely, if the number of filter pores of N is decreased to some extent, the filter performance will be varied. Further, it also investigated the relation between the position of the molten metal inlet and filter pores.

4 Conclusion

It is confirmed that the flow behavior of molten metal passing through the filter is highly reproducible if the number of pores N in the filter is sufficient to some extent. It is found that the "filter scaled model" should not only compare particle fractions but also consider the aperture ratio of the molten metal inlet. The number of calculated particles can be significantly reduced while maintaining a certain degree of calculation accuracy.

References

[1] Y. Yang, H. Nomura, M. Takita, Y. Maeda : "Effect of filter on flow behavior and inclusion removal during pouring", J. Japan Foundry Engineering Society, 69 (1977), pp.62–68.

[2] Jonas Bäckman, Ingvar L. Svensson, Y. Maeda: "Influence of filter on the mould filling of aluminium melts in vacuum-sealed moulds. Research report 99:6, ISSN 1404-0018, Jönköping University, (1999)

[3] K. Taki, G. Endo and Y. Maeda: "Molten Aluminum Alloy Flow Through Casting Filtered Installed at Sprue", Int. J. Metalcasting, 17(2023), pp2477-2483.

[4] M. Kazama, T. Suwa, Y. Maeda: "Modeling of Computation of Molten Aluminum Alloy Flow with Oxide Film by Smoothed Particle Hydrodynamics", J. Japan Foundry Engineering Society, 90 (2018), pp.68-74.

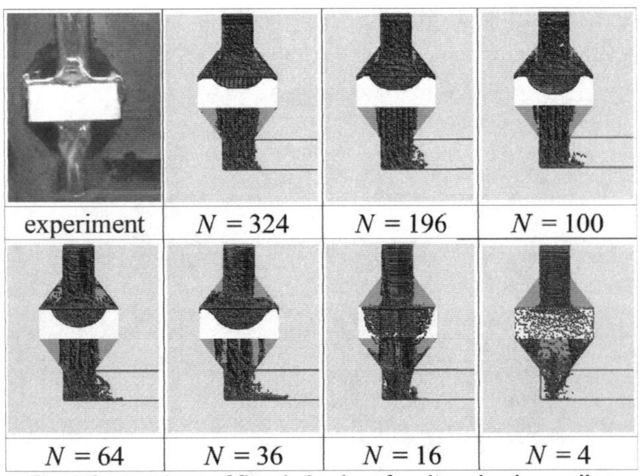

Fig. 2 Comparison of flow behavior of molten aluminum alloy passing through the honeycomb filter of 200cpsi by varying the number of pores N.

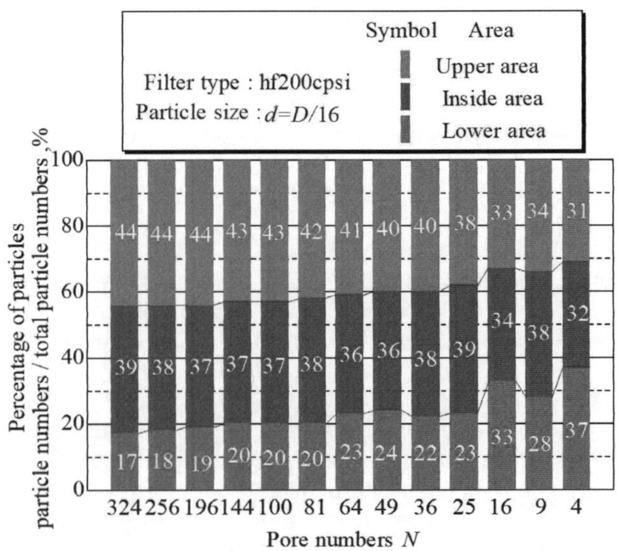

Fig. 3 Comparison of particle fraction at the upper, inside, and lower the filter

Research on High-performance Furan Urea-formaldehyde Phenol-formaldehyde Copolymer/Composite Resin

Shuzhong Xie[1], Chunjing Liu[2], Jiahao Liang[2], Shengshan Feng[3]
1. Guangdong Zhuli Casting Materials Technology Co., Ltd., Zhuhai, Guangdong, China
2. Guangdong Foundry Modeling Materials Engineering Technology Research Center, Zhuhai, Guangdong, China
3. Guangdong Polytechnic of Science and Technology, Zhuhai, Guangdong, China

Abstract: In response to the need for high-quality development, the foundry industry continues to demand higher performance from resin sands with wide-ranging applications. Addressing the shortcomings of furan resins currently available in the market, this study develops a resol resin that possesses both acid and heat-curing characteristics similar to traditional urea-modified furan resins. This resin is then co-polymerized and compounded with traditional urea-formaldehyde-modified furan resin under certain conditions, resulting in a multi-component resin with uniform texture and excellent stability. The advantages of the developed resin combine the high room temperature strength and good collapsibility of traditional urea-formaldehyde modified furan resin with the high-temperature strength， smooth gas evolution and no nitrogen generation of phenol-formaldehyde resin. Additionally, it exhibits low free formaldehyde content and reduces furfuryl alcohol consumption by approximately 20%, thereby significantly reducing resin costs especially during periods of high furfuryl alcohol prices. The advantages of this composite resin are particularly evident in the production of thick and large ductile iron castings.

Keywords: resol resin; copolymerization; composite; cost reduction; performance improvement

1 Introduction

There are two main types of furan resins used in casting: one is urea-formaldehyde modified furan resin, which has the advantages of high ordinary temperature strength, fast curing speed, and good collapsibility, but has high nitrogen and formaldehyde content, produces a large amount of smoke, and makes castings prone to nitrogen porosity. The other type is phenolicmodified furan resin, which does not contain nitrogen, can effectively reduce the risk of nitrogen porosity in castings, and has higher high-temperature strength but has the drawback of low ordinary temperature strength in sand cores. Both types of resins share common shortcomings: their modification components are singular and in small quantities, and the high proportion of furfuryl alcohol leads to abnormally high product costs when the price of furfuryl alcohol is elevated.

2 Experimental procedure

1. Reducing furfuryl alcohol and improving performance is the main direction

Since the main material of furan resin, furfuryl alcohol is derived from furfural extracted from corn cobs and then reduced, it cannot currently be produced through chemical synthesis. Therefore, the supply of furfuryl alcohol is sometimes hard to secure, resulting occasional price spikes. In the long term, its price has an upward trend which increases the cost of furan resins and significantly undermines its application advantages. Hence, to promote better and wider application of furan resin, reducing the consumption of furfuryl alcohol and improving its performance are the keys.

There are three common methods to reduce furfuryl alcohol consumption: urea-formaldehyde modification, phenolic modification, and the addition of other modification materials that are less expensive than furfuryl alcohol. All these methods have shortcomings.

2. Furan-urea-formaldehyde-phenolic copolymer/composite resin

2.1 Resin synthesis mechanism

The synthesis mechanisms of urea-formaldehyde, phenolic, and their modified furan resins are well-known. The mechanism of composite resin is as follows: if two different resins can maintain a uniform texture for a long time after mixing and have the same or similar curing characteristics (such as thermosetting and acid-setting), or even complement each other (e.g. one having lower thermal strength and the other higher thermal strength), then compounding can potentially achieve the goal of complementing strengths and improving performance. If the other resin involved in the compounding is significantly cheaper, it can improve performance while reducing costs. Phenolic resin, with its widely available raw materials and stable prices, is generally cheaper than furfuryl alcohol and is therefore an ideal material for composite resins.

2.2 Study on phenolic resin for compounding

2.2.1 Property requirements for composite phenolic resins

To enable the phenolic resin to be compounded with traditional urea-formaldehydemodified furan resin, it is crucial to ensure that the compounded resin can maintain a uniform texture over time and not separate. Using a weak acid as a catalyst to control the degree of polymerization of the phenolic resin, it kept the resin's average molecule amount small, and its degree of polymerization low. This forms a resole phenolic resin liquid with a good water solubility, a water dilution ratio of about 50%, a solid content of about 70%, numerous active groups, and highly active with a low degree of polymerization. This resin, when copolymerized/ compounded with traditional urea-formaldehyde-modified furan resin under suitable conditions, exhibits good compatibility.

2.2.2 Copolymerization/compounding conditions

Extensive experiments have determined the optimal process conditions for copolymerization/compounding: a pH value of 5.5-6.5, a temperature of 55-60°C, and a duration of 30-40 minutes. If the pH value is too low, the compounding temperature is too high, or the compounding time is too long, the viscosity of the compounded resin will become too high due to excessive copolymerization.-Conversely, if the two resins are mixed only at room temperature, minimal copolymerization occurs, leading to an unstable final product prone to separation. The experiments show that with a phenolic resin content lower than 20%, the viscosity changes very little within three months. However, if the phenolic resin content reaches 30%, the viscosity starts to increase after one month.

2.2.3 Study on curing characteristics

Because the phenolic resin used for compounding has a low degree of polymerization and is rich in active groups such as -OH, it rapidly polymerizes into a highly polymerized resin in an acidic environment, achieving a curing effect. This is similar to the curing characteristics of traditional acid-cured self-hardening furan resin. By controlling the degree of polymerization of the synthesized phenolic resin liquid, the two resins can have almost synchronous curing.

2.3 Performance of composite resin

2.3.1 High-temperature resistance

Numerous research findings have shown that phenolic resin has better high-temperature resistance than furan resin. When phenolic resin constitutes 20-30% of the composite resin, the composite resin's high-temperature resistance is superior to that of urea-formaldehyde-modified furan resin, while also retaining the good collapsibility characteristics of urea-formaldehyde-modified furan resin.

2.3.2 Gas emission performance

Experiments have demonstrated that phenolic resin emits gas more smoothly and in smaller quantities than furan resin, particularly not producing nitrogen. Since the composite resin contains 20-30% phenolic resin, its gas emission is significantly lower than that of traditional urea-formaldehyde-modified furan resin, and emits at a smoother rate.

2.3.3 Reduction of free aldehydes and harmful substances, significant reduction in furfuryl alcohol consumption

Comparing the composite resin ZFP801 with a nitrogen content of ≤3.0% used for large ductile iron parts with the non-composite resin ZF801, the experimental results showed that the consumption of furfuryl alcohol per ton of product decreased by 34.0%, free formaldehyde decreased by 46.4%, and the tensile strength increased by 13.5%. It indicates that the composite resin can significantly reduce the consumption of furfuryl alcohol and free formaldehyde while improving the tensile strength.

3 Result and discussion

(1) Exceeding 20% of phenolic resin in the composite will lead to a rapid increase in its viscosity. Further research is required to enable greater composite proportions.

(2) Composite resin isn't just a simple mixture; it requires copolymerization/compounding to ensure uniformity of the resin after blending. The mechanisms and conditions for copolymerization/compounding require further research.

4 Conclusion

(1) The proportion of phenolic in the composite resin should optimally be around 20%, with a shelf life of up to 4 months. If the proportion exceeds 30%, the viscosity of the resin will increase rapidly, and the shelf life will be limited to one month.

(2) The composite resin leverages the complementary advantages of the two resins, enhancing high-temperature resistance and improving gas emission performance, effectively reducing the occurrence of porosity defects in castings while maintaining good collapsibility.

(3) The composite resin can achieve lower nitrogen content. Urea and other aldehyde removers can be added during compounding, reducing free formaldehyde which is more environmentally friendly.

(4) The composite resin can significantly reduce the dependency on furfuryl alcohol, with a reduction of over 30%. When the price of furfuryl alcohol is significantly higher than that of phenol, it can substantially reduce the costs.

References

[1] Tianyou Huang. Zao Xing Cai Liao[M]. Beijing. China Water&Power Press, 2006.

[2] Yuancai Li. Zhu Zao Zao Xing Cai Liao Shi Yong Shou Ce[M]. China Machine Press, 2009.

[3] Shengshan Feng. Sha Xing Zao Xing Gong Yi De Xuan Ze[J]. Jin Ri Zhu Zao,2008,(10).

Regulation of the Flexural Strength and Porosity of Metakaolin Ceramic Filter Fabricated by Slurry Micro-Extrusion Direct Forming Method

Ming Wu[1, 2], Miao Wang[1, 2], Yi Wang[1, 2], Hao Liu[1, 2], Fuchu Liu[1, 2 *]
1. School of Mechanical Engineering and Electronic Information, China University of Geosciences, Wuhan 430074, China
2. Shenzhen Research Institute, China University of Geosciences, Shenzhen 518057, China
*Corresponding address: e-mail: liufuchu@cug.edu.cn

Abstract: Ceramic filters were widely applied in casting. However, regulating their porosity and strength has been a research difficulty. In this work, high-porosity and high-strength metakaolin-based ceramics were fabricated using the slurry micro-extrusion direct forming method via optimizing four key process parameters, including nozzle internal diameter, height to diameter ratio, filling rate and printing speed. The orthogonal experiments were used to adjust flexural strength and porosity, and the optimal process parameters were obtained by comprehensive scoring and range analysis methods. The predictive models of strengthparameters, porosityparameters and strength-porosity were established and validated. The results showed that the optimal process parameters for high-strength and high-porosity ceramics were 0.51 mm nozzle internal diameter, 60% height-to-diameter ratio, 60% filling rate, and 25 mm/s printing speed. The flexural strength and porosity of the ceramic samples fabricated by optimal process parameters were 17.12 MPa and 48.65%, respectively. The correctness and predictability of three predictive models were proved by two methods, which were the mutual validation and comparison between theoretical and actual values. And the error rates between theoretical and actual results were less than 7%. This work provides guidance for the rapid fabrication of ceramics with adjustable strength and porosity by material extrusion, and the established predictive models can pave the way for its wider application in practice.

Keywords: Material extrusion; Metakaolin-based porous ceramic; Flexural strength; Porosity; Predictive model

1 Introduction

Porous ceramics are widely used in the filtration of molten metal or particulate exhaust gases, catalyst carriers, and gas sensors due to high specific surface area, low bulk density, high-temperature resistance, and corrosion resistance [1]. Traditional processes for preparing porous ceramics include freeze casting and direct foaming [2,3]. However, the traditional processes in the preparation of porous ceramics with complex structures require complex shape molds, and there are some problems such as long manufacturing cycles and low resource utilization. Slurry micro-extrusion direct forming technology, also named direct ink writing (DIW), is flexible in form, simple in operation, and adjustable in pore size, which is suitable for preparing porous ceramics [4]. In practical applications, there are also certain requirements for the strength of porous ceramics in fields such as filtration, catalysis, and insulation. However, there are fewer studies on the correlation between porosity and flexural strength of porous ceramics by DIW, and there is a contradictory relationship between porosity and strength. Therefore, it is important to investigate the correlation between porosity and flexural strength of porous ceramic samples.

2 Experimental procedure

The raw materials used in this study were metakaolin and binder PVA. In the previous experimental research [5], it was found that the four parameters of nozzle internal diameter, height-to-diameter ratio (printing layer height/nozzle internal diameter), filling rate, and printing speed had important impacts on the formability of the green bodies and the property of sintered ceramic, as shown in Fig. 4. Three superior levels of each parameter to well form green bodies were selected. In this work, the four-factor and three-level orthogonal experiments were designed, as shown in Table 1. Comprehensive scoring and range analysis methods were used to study the influence mechanism of each parameter on the flexural strength and porosity of ceramics, and a set of process parameters for fabricating high-strength, high-porosity ceramics were gained [5,6].

Table 1 The orthogonal experimental design

Level	Factors			
	A: Nozzle internal diameter (mm)	B: Height to diameter ratio (%)	C: Filling rate (%)	D: Printing speed (mm/s)
1	0.41	50	60	15
2	0.51	60	80	20
3	0.61	70	100	25

3 Result and discussion

1. Parameter optimization

The results of the orthogonal experiments are shown in Table 2. The optimal process parameters of 0.51 mm nozzle internal diameter, 70% height-to-diameter ratio, 100% filling rate, and 15 mm/s printing speed were obtained by comprehensive scoring and range analysis methods.

Table 2 Results of the orthogonal experiments

Num	Factors A	B	C	D	Flexural strength (MPa)	Porosity (%)	Bulk density (g/cm^3)
1	0.41	50	60	15	24.37±1.48	37.54±2.87	1.64±0.06
2	0.41	60	80	20	23.60±1.54	38.68±3.31	1.61±0.07
3	0.41	70	100	25	31.62±4.34	32.96±0.78	1.71±0.03
4	0.51	60	60	25	16.76±0.74	50.76±5.05	1.28±0.14
5	0.51	70	80	15	26.42±2.62	36.84±2.95	1.66±0.08
6	0.51	50	100	20	32.76±2.38	32.04±1.16	1.73±0.03
7	0.61	70	60	20	16.88±1.17	48.88±2.46	1.36±0.07
8	0.61	50	80	25	21.47±1.06	41.82±1.90	1.53±0.04
9	0.61	60	100	15	30.42±1.45	34.68±1.88	1.68±0.02

The actual flexural strength and porosity of the ceramics fabricated using the optimal parameters were 17.12 MPa and 48.65%, respectively.

2. Predictive models of strength and porosity

Based on the results of orthogonal experiments, Origin software was used to fit multiple linear regression models for strength-parameter, porosity-parameter, and strength-porosity respectively, the fitting results are shown in Eqs. (1), (2) and (3).

$$\sigma=-18.03A-0.06B+0.31C-0.39D+20.85 \quad (1)$$

$$\varphi=27.00A+0.12B-0.31C+0.55D+32.32 \quad (2)$$

$$\sigma = 114.78771e^{-0.03956\varphi} \quad (3)$$

where σ and φ are flexural strength and porosity, respectively

The validation process for Eqs. (1), (2) and (3) can be carried out in two ways. The one approach is the mutual validation between the three models. The other approach is to compare the actual and theoretical values. The specific validation logic is shown in Fig. 1.

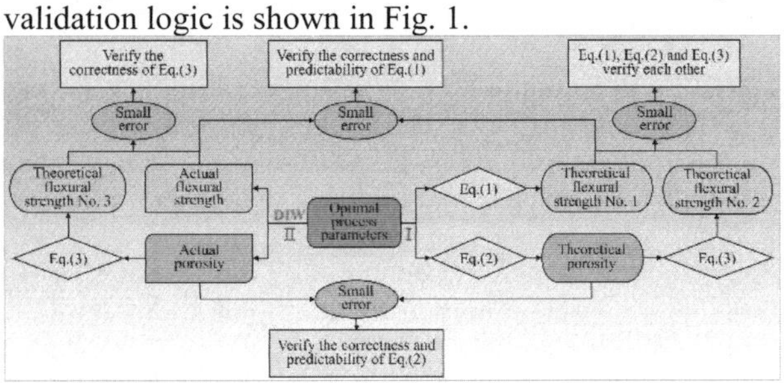

Fig. 1. Validation logic diagram for the fitted models.

3 Conclusion

This work focused on the effect and optimization of four key DIW process parameters on the flexural strength and porosity of metakaolin-based porous ceramics. The orthogonal experiments were designed to investigate the influence law of process parameters on the flexural strength and porosity, and the optimal process parameters were obtained. Models were used to fit the results of orthogonal experiments for achieving the goal of predicting flexural strength and porosity through process parameters. The specific conclusions are drawn as follows:

(1) There was a gradual caramization of metakaolin-based green bodies during the sintering process. The powder particles were tightly bonded to each other, resulting in a decrease in porosity and dimension, and an increase in the strength of the ceramic samples. The volatilization of the organic binder PVA created a large number of pores in the ceramic matrix, which slowed down the reduction of porosity.

(2) The optimal process parameters of 0.51 mm nozzle internal diameter, 70% height to diameter rate, 100% filling rate, and 15 mm/s printing speed for fabricating high-strength and high-porosity ceramics were determined by the comprehensive scoring and range analysis method. Among the process parameters, the filling rate had the most significant effect in changing the flexural strength and porosity.

(3) Predictive models for strength-parameter, porosity-parameter, and strength-porosity were established based on the orthogonal experimental results. The correctness and predictability of the predictive models were proved by two methods of mutual validation among the three models and comparison between the theoretical and actual values. The error rates between the theoretical and actual values did not exceed 7%.

Acknowledgments

This work was supported by Guangdong Basic and Applied Basic Research Foundation [No.2024A1515013258] and Free Exploration Basic Research Project of Local Science and Technology Development Funds Guided by the Central Government [No. 2021Szvup158].

References

[1] Man Y, Ding G and Xudong L. A review on porous ceramics with hierarchical pore structure by 3D printing-based combined route. J. Asian. Cera. Soc, 2021, 9(4): 1377-1389.

[2] Zhou J, Fan J and Sun G. Preparation and properties of porous silicon nitride ceramics with uniform spherical pores by improved pore-forming agent method. J. Alloy. Compd, 2015, 632:655–660.

[3] Yang J, Yu J and Huang Y. Recent developments in gelcasting of ceramics. J. Eur. Ceram. Soc, 2011, 31(14):2569–2591.

[4] Liu G, Zhang X and Chen X. Additive manufacturing of structural materials. Mater. Sci. Eng: R: Reports, 2021, 145: 100596.

[5] Wu M, Liu F and Lin Y. Parameter Optimization via Orthogonal Experiment to Improve Accuracy of Metakaolin Ceramics Fabricated by Direct Ink Writing. Chin. J. Mech. Eng: Addit. Manuf. Front, 2023, 2(4):100098.

[6] Kang P, Zhao Q and Guo S. Optimisation of the spark plasma sintering process for high volume fraction SiCp/Al composites by orthogonal experimental design. Ceram. Int, 2021, 47(3):3816–3825.

Effect of Sintering Temperature on the Precision and Properties of DIW 3D-printed Porous ZrO_2 Ceramic Filter for Casting

Yuxiao Lin[1,2], Yi Wang[1], Miao Wang[1], Liang Zhang[1], Hao Liu[1,2], Fuchu Liu[1,2*]
1. School of Mechanical Engineering and Electronic Information, China University of Geosciences, Wuhan 430074, China
2. Shenzhen Research Institute, China University of Geosciences, Shenzhen 518057, China
*Corresponding address: e-mail: liufuchu@cug.edu.cn (Fuchu Liu)

Abstract: In this manuscrip, porous ZrO_2 ceramic filters were fabricated by direct ink writing (DIW) 3D printing technology, and the effects of sintering temperature on the sintering shrinkage, surface roughness, porosity, flexural strength, and compressive strength of porous ZrO_2 ceramic samples were systematically investigated. Results showed that when the sintering temperature increased from 1400 °C to 1500 °C, the sintering shrinkage of porous ZrO_2 ceramic samples initially increased and then decreased, which can be explained using the variations in ceramic grain size and internal pores caused by different sintering temperatures. When the shrinkage rate was higher, the stepped effect on the side surface of the sintered ceramic samples was smaller, resulting in a higher flatness, and so the surface roughness decreased first and then increased. The main reason for the initial increase followed by a decrease in the flexural and compressive strength of porous ZrO_2 ceramics with increasing sintering temperature was related to the number of grain boundaries and grain boundary bonding properties of the sintered ceramic samples at different sintering temperatures. When the sintering temperature was 1450 ℃, the sintering shrinkage of porous ZrO_2 ceramics in the X, Y, and Z directions were 8.95%, 8.58%, and 9.21% respectively. The roughness of the side surface was 19.18 μm, the flexural strength was 6.01 MPa, the compressive strength reached 22.70 MPa, the porosity was 62.37%, the bulk density was 2.14 g/cm^3, and the comprehensive properties were optimal.

Keywords: Direct ink writing; Porous ZrO_2 ceramic; Strength; Porosity; Sintering temperature

1 Introduction

Porous ZrO_2 ceramics are a high-performance structural material composed of a three-dimensional network skeleton structure, and exhibit advantages such as high specific surface area, low bulk density, excellent high-temperature thermal shock resistance, and high-temperature compressive strength, which are widely used in the field of high-temperature alloy precision casting [1]. The use of porous ZrO_2 ceramics for the filtration and purification of molten metals can efficiently remove large and fine suspended inclusions in the metal melt and promote the precipitation of some gases through mechanisms such as mechanical interception, rectification of floating slag, and surface adsorption. This process effectively improves the mechanical properties and service life of high-temperature alloy precision castings [2]. However, traditional methods for preparing porous ceramics, such as direct foaming, gel casting, and freeze casting, have complex preparation processes, long production cycles, require mold development and are difficult to meet the demand for rapid manufacturing of porous ceramics with excellent comprehensive properties such as high porosity and high compressive strength [3]. Therefore, there is an urgent need to develop a new method for preparing porous ZrO_2 ceramics with excellent comprehensive properties.

Using DIW 3D printing technology, the high solid content ceramic slurry is deposited on a substrate through an extrusion nozzle, and complex structures are formed by layer-by-layer stacking, effectively enabling the preparation of porous ceramics with pore sizes ranging from micrometers to millimeters.

In this manuscript, DIW 3D printing technology was employed to fabricate porous ZrO_2 ceramic filters using ZrO_2 ceramic powder as the raw material, methyl cellulose (MC) as a binder, and ammonium polyacrylate ($C_3H_4O_2$) as a dispersant. On this basis, the influence of different sintering temperatures on the accuracy (sintering shrinkage and surface roughness) and properties (porosity, flexural strength, and compressive strength) of porous ZrO_2 ceramics were investigated. The microstructural evolution of porous ZrO_2 ceramic samples at different sintering temperatures was analyzed using scanning electron microscopy (SEM). Moreover, an attempt was made to establish the relationship between the properties and microstructural evolution of porous ZrO_2 ceramics. Finally, porous ZrO_2 ceramic filters with optimal comprehensive properties were successfully prepared, with a surface roughness of 19.18 μm, porosity of 62.37%, and a compressive strength of 22.70 MPa, at a sintering temperature of 1450 ℃.

2 Experimental procedure

First, a uniformly dispersed ZrO_2 ceramic slurry was prepared. The composition ratio of the ceramic slurry is listed in Table 1.

Subsequently, porous ceramic green bodies were printed with a self-built DIW 3D printing device. The forming process parameters are listed in Table 2.

Table 1. Porous ZrO_2 ceramic slurry composition.

Slurry composition	ZrO_2	2 wt.% MC	$C_3H_4O_2$
Content (wt.%)	79	20.3	0.7

Table 2. DIW printing parameters used in this manuscript

Printing parameter	Parameter value
Nozzle inner diameter (mm)	0.41
Filling rate (%)	50
Layer height (mm)	0.246
Printing speed (mm/min)	700

Based on the test results from the TG-DSC simultaneous thermal analyzer, the sintering temperature curves for porous ZrO_2 ceramic samples were designed. Initially, the temperature was increased to 500 ℃ at a rate of 2 ℃/min, and held for 2 hours to ensure thorough removal of organic binders and complete evaporation of residual moisture from the samples. Subsequently, the temperature was increased to 800 ℃ at a rate of 2 ℃/min, and held for 1 hour to facilitate the decomposition of internal organic impurities. Then, the temperature was increased at a rate of 2.5 ℃/min to 1400 ℃, 1450 ℃, and 1500 ℃, respectively, and maintained for 2 hours at each temperature to promote full migration of ceramic particles and bonding at grain boundaries, enhancing ceramic densification. Finally, the sintered ceramic samples were obtained by furnace cooling to room temperature.

3 Result and discussion

Microstructural evolution

When sintered at 1400 ℃, the grain migration distance was small, and grain boundaries bonding was slow, resulting in a loosely structured overall grains arrangement and larger pores. As the temperature further increased to 1450 ℃, the larger pores gradually shrank, and the bonding between grain boundaries became tighter. Once the sintering temperature was increased to 1500 ℃, the grain migration distance increased, and the grains fused with each other, disrupting grain boundaries bonding and forming larger agglomerates. The microstructural evolution results indicated that high sintering temperatures can lead to abnormal grain growth and weakened grain boundary bonding. Simultaneously, due to the abnormal grain growth, the ceramic samples were prone to microcracks during the sintering process, significantly affecting the ceramics' mechanical properties.

As the sintering temperature increased, the average grain size continuously enlarged. For instance, when the temperature increased from 1400 ℃ to 1450 ℃, the average grain size coarsened from 1.57 ± 0.08 μm to 1.91 ± 0.12 μm, indicating a pronounced trend of grain growth. As the temperature further increased to 1500 ℃, the average grain size continued to coarsen, due to the abnormal growth of some grains, although the increasing trend decelerated, as depicted in Fig. 1.

Sintering property

The shrinkage rate, surface roughness, porosity, density, flexural and compressive strength of porous ZrO_2 ceramic samples at different sintering temperatures are shown in Table 3.

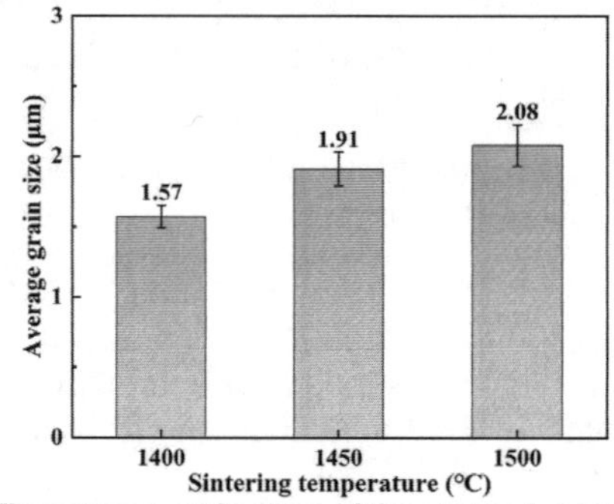

Fig. 1. The average grain sizes of the sintered porous ZrO_2 ceramics

Table 3. Effect of different sintering temperatures on the precision and properties of porous ZrO_2 ceramics

Sintering temperature (℃)	Flexural strength (MPa)	Compressive strength (MPa)	Porosity (%)	Density (g/cm^3)	Roughness (μm)	Shrinkage (%) X	Y	Z
1400	3.9	18.3	66.8	1.9	29.3	7.8	6.5	9.1
1450	6.1	22.7	62.4	2.1	19.2	8.9	8.6	9.2
1500	3.6	17.8	64.3	2.1	26.7	7.9	6.6	9.1

4 Conclusion

(1) The average grain size of porous ZrO_2 ceramics continuously increases with elevated sintering temperature, but the increasing trend gradually slowes down.

(2) The sintering shrinkage rate of porous ZrO2 ceramics initially increases and then decreases with elevated sintering temperature. When the shrinkage was higher, the side surface of the ceramic exhibited higher smoothness, hence the surface roughness decreased initially and then increased with elevated sintering temperature.

(3) Due to the higher number of grain boundaries and higher bulk density of porous ZrO_2 ceramics sintered at 1450 °C, the flexural and compressive strength of the ceramic samples were higher.

Acknowledgments

This work was supported by Guangdong Basic and Applied Basic Research Foundation [No.2024A1515013258]and Free Exploration Basic Research Project of Local Science and Technology Development Funds Guided by the Central Government [No.2021Szvup158].

References

[1] Olson III R A, Martins L C B. Cellular ceramics in metal filtration. J. Advanced Engineering Materials, 2005, 7(4): 187-192.

[2] Li S, Baeyens J, Dewil R, et al. Advances in rigid porous high temperature filters. J. Renewable and Sustainable Energy Reviews, 2021, 139: 110713.

[3] Ohji T, Fukushima M. Macro-porous ceramics: processing and properties. J. International Materials Reviews, 2012, 57(2): 115-131.

Eco-Friendly Casting By Using Innovative Inorganic Binder Systems (IIBs)

Angelika Kmita[1]*, Rafał Dańko[2*], Mariusz Holtzer[2], Józef Dańko[2], Dariusz Drożyński[2], Mateusz Skrzyński[2], Agnieszka Roczniak[2], Daniel Gruszka[2], Jarosław Jakubski[2], Sara Tapola[3]
1. AGH University of Krakow, Academic Centre for Materials and Nanotechnology, Al. A. Mickiewicza 30 St., 30-059 Krakow, Poland
2.AGH University of Krakow, Faculty of Foundry Engineering, Al. A. Mickiewicza 30 St, 30-059 Krakow, Poland
3. Meehanite Technology Oy, Kuokkamaantie 4, 33800, Tampere, Finland
*Corresponding address: e-mail: akmita@agh.edu.pl , rd@agh.edu.pl

Abstract: The article presents the gas emissions results, released during thermal destruction of molding sands with innovative inorganic binder systems (IIBs). These studies were conducted in terms of assessing the harmfulness of these molding sands based on the emissions of compounds such as benzene, toluene, ethylbenzene, xylenes, polycyclic aromatic hydrocarbons, phenol, formaldehyde and others. Moreover, equally important factors studied from the technological point of view were volume of released gases and the gas release rate. It has been shown that the use of innovative inorganic binder systems (IIBs) in molding sands significantly reduce emissions of harmful substances generated during the casting process compared to organic binder systems. These studies were conducted as part of a European Green Casting Life project (LIFE21 ENV/FI/101074439) [1].

Keywords: green casting; BTEX; PAHs emissions; dust; harmful substances; clean production; new inorganic binders; sustainable development.

1 Introduction

European foundries mainly use sands with bentonite (green sand) and sands with organic binders (e.g. no-bake) to produce molds and cores. Only 1% of foundries in Europe use technologies based on inorganic binder systems [2]. The main reason for this situation is technological limitations. When choosing a technology for the production of molds and cores, the following elements should be taken into accounts: efficiency, casting quality and impact on the environment and working conditions, which consequently determines the costs [2]. The advantages of technologies using organic binders are: small resin addition, high reactivity and low binder viscosity, high strength and good reclaimability of molding sands [2]. Their disadvantage is the emission of numerous compounds, including harmful ones (e.g. formaldehyde, phenol, benzene and their derivatives and sulfur compounds), generated during pouring, cooling and knocking out molds. In addition, the high price of components eg. furfuryl alcohol, nitrogen content in furfuryl resins (unfavorable for steel castings) etc.

2 Conclusions

It has been shown that inorganic innovative binder systems (IIBs) significantly contribute to reducing emissions of hazardous compounds (Fig. 1) and improving working conditions compared to organic binder systems [1].

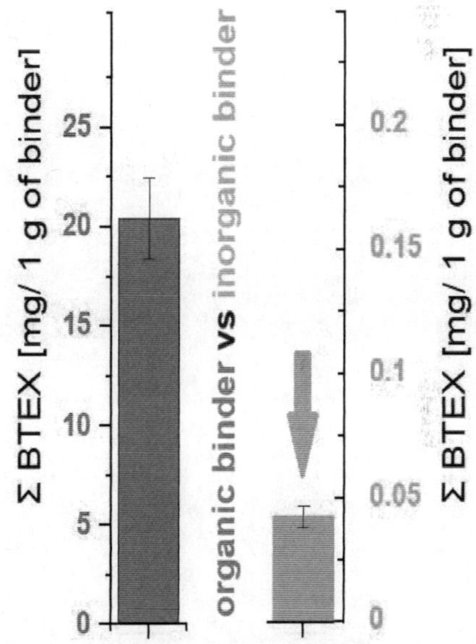

Fig. 1 Exemplary results of BTEX content in emitted gases from molding sands with organic and innovative inorganic binder system (IIBs).

Moreover, replacing organic binders with inorganic binders can potentially expand the possibility of managing used moldings and thus contribute to reducing the amount of waste, which is also consistent with the circular economy. The introduction of innovative inorganic binder systems (IIBs) in the foundry industry is consistent with the sustainable development strategy and will have a positive impact on the development of this industry. This means it is necessary to take into accounts the interdependence between the environment and socio-economic aspects when making strategic decisions and production activities [1, 2].

Acknowledgments

The research is financed within GREEN CASTING LIFE project. GREEN CASTING LIFE project is co-funded by European Union's LIFE program under grant agreement LIFE21-ENV-FI101074439. Views and opinions expressed are however those of the author(s) only and do not necessarily reflect those of the

European Union or CINEA. Neither the European Union nor the granting authority can be held responsible for them. GREEN CASTING LIFE project (LIFE21-ENV-FI-101074439) is co-funded by the National Fund for Environmental Protection and Water Management (NFOŚiGW) under grant agreement 276/2023/Wn-06/OZ-PO-LF/D.

References

[1] Green Casting Life project, grant agreement LIFE21-ENV-FI 101074439,https://webgate.ec.europa.eu/life/publicWebsite/project/details/101074439

[2] Holtzer M and Kmita A. Mold and Core Sands in Metalcasting: Chemistry and Ecology Sustainable Development,Springer Nature Switzerland AG 2020: 359.

Getting Closer To Green Casting Foundry - Sustainable Development

Rafał Dańko[1]*, Angelika Kmita[2*], Mariusz Holtzer[1], Józef Dańko[1], Sara Tapola[3]
1. AGH University of Krakow, Faculty of Foundry Engineering, Al. A. Mickiewicza 30 St, 30-059 Krakow, Poland
2. AGH University of Krakow, Academic Centre for Materials and Nanotechnology, Al. A. Mickiewicza 30 St., 30-059 Krakow, Poland
3. Meehanite Technology Oy, Kuokkamaantie 4, 33800, Tampere, Finland
*Corresponding address: e-mail: rd@agh.edu.pl, akmita@agh.edu.pl

Abstract: The article presents the main assumptions of the activities carried out as part of the European project GREEN CASTING LIFE (LIFE21 ENV/FI/101074439) [1], which is an initiative to assess the potential benefits of inorganic binder systems in European ferrous alloy foundries. GREEN CASTING LIFE project aims to improve the environment and working conditions in European iron alloy foundries by introducing cleaner and ecological production methods using new innovative inorganic binders [1-3]. The Green Casting Consortium [1] includes 16 partners from 8 European countries such as Finland (Project Coordinator – Meehanite company), Spain, Poland, Italy, Czech Republic, Germany, France and Estonia including 6 flagship foundries [1]. The work carried out as part of the GREEN CASTING LIFE project will enable the implementation and validation at the industrial level of the use of inorganic binder systems for the production of molds and cores in iron alloy foundries, achieving the level of Technological Readiness Level TRL8 [1]. In addition, the new knowledge obtained will will be included in the new version of the Reference Document on Best Available Techniques (BAT) in the Smitheries and Foundries Industry (BREF), which is expected to be published in 2024, as a future reference BAT for all EU finds. [4].

Keywords: green casting, innovative technologies, green production, sustainable development,

1 Introduction

LIFE programme is the EU Programme for Environment and Climate Action [1]. It's one of the key contributor programme to the European Green Deal (COM(2019) 640 final) which aims [1]: To transform the EU into a fair and prosperous society, with a modern, resource-efficient and competitive economy where there are no net emissions of greenhouse gases in 2050 and where economic growth is decoupled from resource use; To protect, conserve and enhance the EU's natural capital, and protect the health and well-being of citizens from environment and climate related risks and impact.

GREEN CASTING LIFE project aims to demonstrate the technical and environmental feasibility of using new inorganic binders, instead of traditional organic ones, in ferrous foundries [1]. Casting represents one of the major sources of dust and Hazardous Air Pollutants releasing to the atmosphere hazardous compounds (NOx, SO_2, PM2.5 Polycyclic Aromatic Hydrocarbon, BETX, etc) and thus affecting directly to the occupational risk of the workers and air quality indicators at urban areas [1]. In addition, casting processes produce a huge amount of undesirable waste representing 6 million tons of spend foundry sand in Europe per year, being mostly landfilled [1]. Arising from promising results obtained in a funded ongoing LIFE project: LIFE17 ENV/FI/000173 GREEN FOUNDRY LIFE (2017- 2022), where harmful emission reduction between 70% and 90% was achieved in indoor air quality and sand circuit stack emissions [1, 5]. However, the implementation of the inorganic system in a foundry goes beyond this emission results and needs to be carefully planned, dealing with technical and process challenges [1]. In GREEN CASTING LIFE project, 6 heterogeneous and complementary European flagship ferrous foundries from Spain, Poland, Estonia, Finland and Italy have joined the consortium to implement at industrial scale the inorganic system on their production process implying major changes starting with the selection of the suitable inorganic system, changes in processes and investments in new equipment [1]. In parallel, different sand reclamation methods will be demonstrated and external reusing applications will be studied [1]. Based on the results obtained, the foreseen techniques will represent an outstanding knowledge for BAT technologies participating to tackle environmental problems and to produce greener products in the European market [1, 4].

2 Conclusion

The main objective of GREEN CASTING LIFE is to demonstrate that the implementation of tailor-made inorganic binder systems in Ferrous foundries is a very relevant positive Environmental impact solution for European foundries industries [1]. Using inorganic binder systems can decrease hazardous emissions from casting processes, improve indoor air quality and quality of waste sand and improves waste sand reuse applications [1]. To achieve this main goal, Green Casting Life is divided into the following specific objectives [1]:

Technical objectives [1]:

To develop tailor made inorganic binder systems solutions for iron and steel foundries;

To successfully implement and validate tailor made inorganic binder systems at industrial scale, for the first time in Europe, in 6 flagship foundries from 5 European countries;

To demonstrate three different type of sand reclamation methods for inorganic binder system sands (thermal, mechanical and wet methods);

To develop the mixed inorganic and organic foundry sand regeneration systems by demonstrating different regeneration methods for the flagship foundries to reach an amount of recycled sand of min 75% in the moulding processes;

To enhanced exploitation and replication potential by analysing the status of European foundries, guiding and assisting selected follower European foundries;

Providing the results and experiences for the preparation of the Reference Document on Best Available Techniques in the Smitheries and Foundries Industry (BREF document) [4];

To ensure an efficient communication and dissemination of the project activities and results to the general public and relevant stakeholders at European level [1].

Environmental objectives [1]:

To reduce total hazardous airborne emissions from casting processes by 90-95% by targeting the following figures for the most critical compounds eg. benzene, toluene, ethylbenzene, xylenes (BTEX), SO_2, NOx, formaldehyde etc.;

To create a healthier environment for the workers by improving indoor air quality in foundries by 70-80% for the most critical compound eg. BTEX, formaldehyde, polycyclic aromatic hydrocarbons (PAHs), etc.;

To reduce the harmful substances in waste sand by 80-90% in following compounds: dissolved organic compounds, phenol, fluoride, BTEX;

To reduce total hazardous airborne emissions from casting processes of the 6 flagship foundries by 90-95%.

Acknowledgments

The research is financed within Green Casting Life project. Green Casting Life project is co-funded by European Union's LIFE program under grant agreement LIFE21-ENV-FI101074439. Views and opinions expressed are however those of the author(s) only and do not necessarily reflect those of the European Union or CINEA. Neither the European Union nor the granting authority can be held responsible for them. Green Casting Life project (LIFE21-ENV-FI-101074439) is co-funded by the National Fund for Environmental Protection and Water Management (NFOŚiGW) under grant agreement 276/2023/Wn-06/OZ-PO-LF/D.

References

[1] Green Casting Life project, grant agreement LIFE21-ENV-FI 101074439, https://webgate.ec.europa.eu/life/publicWebsite/project/details/101074439

[2] Dańko R, Kmita A, Dańko J, Holtzer M. Key assumptions for Green Casting Life Project: Towards zero emissions in European ferrous foundries using inorganic binder systems. IV International Conference of Casting and Materials Engineering ICCME 23 Krakow, Poland, November 6 - 7, 2023.

[3] Holtzer M and Kmita A. Mold and Core Sands in Metalcasting: Chemistry and Ecology Sustainable Development, Springer Nature Switzerland AG 2020: 359.

[4] Commission, Best Available Techniques (BAT) Reference Document for the Smitheries and Foundries Industry Draft 1 (February 2022).

[5] Green Foundry Life project https://webgate.ec.europa.eu/life/publicWebsite/project/LIFE17-ENV-FI-000173/inorganic-binder-system-to-minimize-emissions-improve-indoor-air-quality-purify-and-reuse-of-contaminated-foundry-sand

Development of the foundry powder coatings for chill mold

Yabo Hou*, Yury Nikalaichyk
Mechanical and technological faculty, Belarusian national technical university, 220013, Minsk, 65 Nezavisimosty ave., Republic of Belarus
*Corresponding address: e-mail: houyabo@gmail.com, yuni@bntu.by

Abstract: The economic efficiency of producing castings in permanent metal molds (chill molds) is largely determined by their durability. Failure of the chill mold occurs due to the formation of cracks and gas corrosion on its internal surface which is in contacted with liquid alloy many times. During the pouring period of the chill mold by the alloy, its internal surface and nearby layers very quickly heat up to high temperatures, while other layers located further from the internal surface heat up more less. Compressive and tensile stresses may arise in the chill mold as a result. Various plastic deformations of the chill mold material can lead to the formation of small cracks or crack grid. As a result, quality of the casting surface is deteriorated and problems arise associated with removing the castings from the chill mold. Also, its contributes to further destruction of the chill mold from gas corrosion. In recent years there is various trend in foundry coating technology such as developing of anti-defect coating (for example anti-veining coating), developing of ecological friendly coating (water-based coating), developing of non-zircon coating and etc. In our opinion one of the potential and promising area of research in foundry coating development is powder coating technology.

Keywords: chill mold; powder foundry coating; refractory materials; casting quality; obtaining method.

1 Introduction

Various types of coatings have been very spread in different areas such as friction reduction, corrosion protection, wear resistance, self-cleaning, antifouling etc. Implementation Foundry is an industry which is widely implemented refractory coatings or foundry coatings and its allows a number of beneficial effects during casting.

The use of foundry coatings for molds and cores during casting is very necessary as a means of achieving high quality surface finish of castings more especially in complex internal channels created by use of cores [1]. Foundry coating eliminates the direct chemical interaction between casting alloy and material of the mold as well.

However, the main role of the foundry coating layer is the ability to create a certain, predetermined thermal resistance at the contact zone between casting and mold.

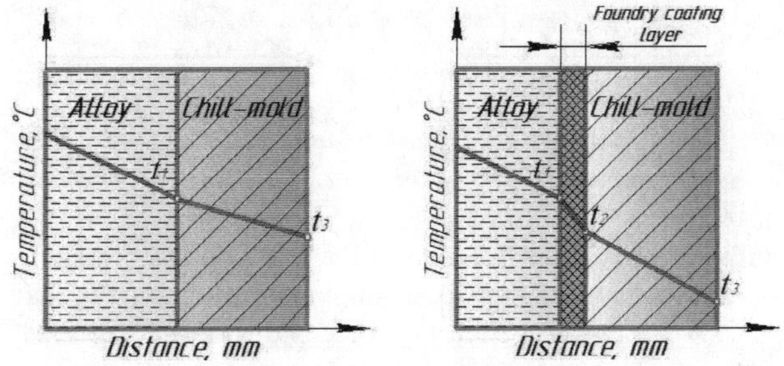

Fig. 1 Temperature distribution in the chill-mold with and without foundry coating

It's an excellent ability to regulate the amount of heat transferred from the casting to the mold (Fig. 1). The thermal resistance of the foundry coating affects to the skin casting formation and the mold as well. On the one hand, the presence of thermal resistance may lead to decrease of the cooling intensity and drop of temperature in the casting cross section. The presence of thermal resistance may lead to decrease of the cooling reduction of the physicochemical reaction at the metal-coating interface.

Research was carried out within the framework of this work and aimed to develop the technology of foundry powder coatings manufacturing, their application to the casting mold is carried out by electrostatic methods and the formation of strength (hardening) occurs due to polymerization when heating the thermosetting resins (binders) included in their composition.

2 Experimental procedure

The main stages of the research consisted of the investigation the coatings ability to form a protective layer of the required thickness on the surface of chill-mold, research and identification of strength models formation of the foundry powder coatings, investigation of the adhesive ability of foundry powder coatings and development of optimal modes and parameters of the application process (electrostatic coating) of chill molds.

The manufacturing of foundry powder coatings was carried due to laboratory extruder according to a pre-calculated recipe (Fig. 2).

Fig. 2 Laboratory extruder for foundry powder coatings manufacturing

Powdered highly refractory fillers such as kyanite-sillimanite and cryptocrystalline graphite were used for manufacturing of coatings. It's traditionally used in the manufacture of liquid refractory coatings, but at the same time its characterized by opposite physical-chemical and electrical properties. During the preparation process, simultaneously with mixing, the highly refractory filler and additives were clad with a thermosetting binder. At the next step, a homogeneous composition of highly refractory filler, additives and binder in the form of plates was ground and dispersed into fractions. The coarsely dispersed fraction (more than 70 μm) was returned to the attritor for repeated grinding. Next, foundry powder coatings were used for testing.

3 Result and discussion

Experiments results of the abrasion strength investigation have been established polyester resin optimal content which is necessary for creating required layer on the surface of chill-mold. It was established (due to mathematical processing of experimental data) that polyester resin content (C) influences on abrasion strength (σ) of foundry powder coatings as a one of the determining parameters according equations (1) and (2).

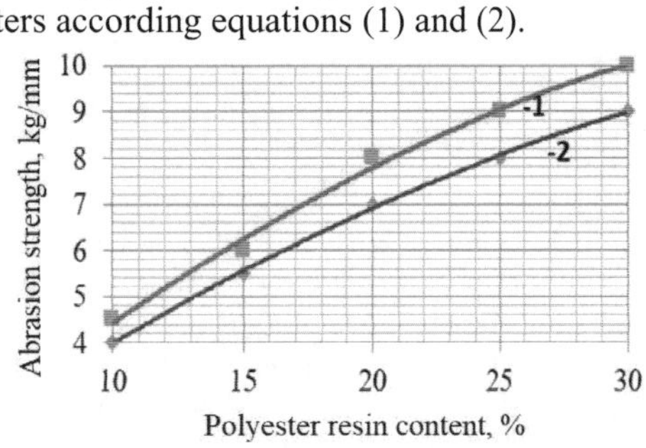

Fig. 3 Abrasion strength of kyanite-sillimanite (1) and cryptocrystalline graphite (2) foundry powder coatings

Polyester resin provides an inextricable connection between particles of highly refractory filler and affects other strength characteristics and when the percentage content of polyester resin increases from 10 to 30%, the abrasion strength increases proportionally from 4 kg/mm to 9 kg/mm for a coating based on kyanite-sillimanite and from 4.5 kg/mm to 10 kg/mm for cryptocrystalline graphite (Fig. 4).

$$\sigma_1 = 0{,}0057C^2 + 0{,}5086C - 0{,}1 \qquad (1)$$

$$\sigma_2 = 0{,}0043C^2 + 0{,}4214C + 0{,}2 \qquad (2)$$

It was also found that with an increase in layer thickness (δ) from 250 μm to 805 μm, the adhesive properties (A) deteriorate from 5 to 3 points for both types of coatings (Fig. 5).

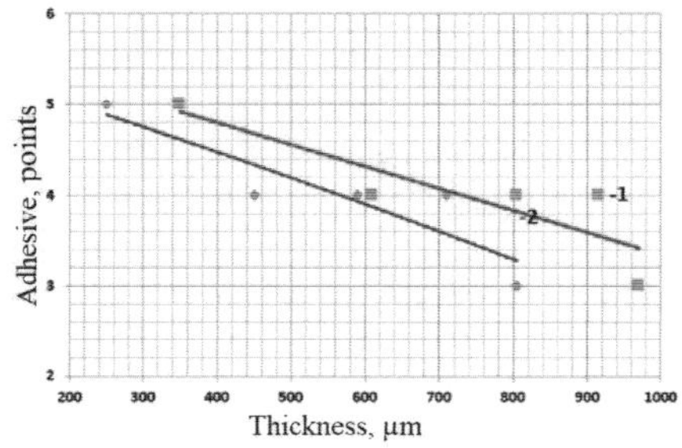

Fig. 4 Adhesive properties of kyanite-sillimanite (1) and cryptocrystalline graphite (2) foundry powder coatings

Mathematical processing of experimental data made it possible to obtain equations (3) and (4) as well.

$$A_1 = 07\,\delta^2 - 0{,}0025\,\delta + 5{,}5267 \qquad (3)$$

$$A_2 = 08\,\delta^2 - 0{,}0024\,\delta + 5{,}7479 \qquad (4)$$

Equations describe the changing of foundry powder coating adhesive properties based on kyanite-sillimanite and cryptocrystalline graphite fillers.

4 Conclusion

Powder coatings technology develops rapidly and has great potential for wide implementation in foundry industry. The foundry powder coatings can greatly reduce the packaging and transportation costs of the coatings because it does not contain solvent. The global safety regulations will be more strict, and the safety requirements for the production, transportation, and storage of flammable and explosive dangerous materials will be increasing, which will also greatly increase the transportation cost of alcohol-based foundry coatings and will greatly encourage the widespread application of foundry powder coatings. Also, it is very important for environmental friendly and safety trends in foundry industry because the foundry powder coatings is free of toxic chemicals and solvents, it's may deemed safe for future use and disposal.

References

[1] Nwaogu U.C., Tiedje N.S. Foundry Coating Technology: A Review. Materials Sciences and Application, 2011, 2(8): 1143-1160.

Viable Inorganic Binder Solutions for Ferrous Applications

A. Rangel[1], V.A.C. Haanappel[2*]
1. Vesuvius USA-Cleveland, 20200 Sheldon Rd, Brook Park, OH 44142, USA
2. Foundry R&D Centre, Foseco Nederland BV, Pantheon 30, 7521 PR Enschede, the Netherlands
*Corresponding address: email:vincent.haanappel@vesuvius.com

Abstract: The application of inorganic binder systems (IOBs) in serial production cores has gained significant traction in the aluminum automotive sector. However, the adoption of IOBs for ferrous applications remains limited. This study investigates the technical enhancements in IOB technology, particularly focusing on gray and ductile iron, to validate its efficacy for ferrous applications. Through semi-serial production of castings for various applications, the study demonstrates the benefits of IOBs. For ductile iron, high-quality castings can be achieved without the need for refractory coatings. Conversely, for gray iron, a waterborne coating is generally required to mitigate metal penetration. The findings underscore the feasibility of IOBs to maintain casting quality and efficiency in ferrous casting applications.

Keywords: Inorganic, Binder, Ferrous, Gray, Ductile.

1 Introduction

Inorganic binder systems (IOBs) have emerged as a well-established technology for serial production cores, particularly in the aluminum automotive sector. This success has bridged the use of these binders for core-making in ferrous applications, including gray iron, ductile iron, and steel castings. Despite their potential, the use of inorganic binder systems in ferrous foundries remains limited.

Sand cores using IOBs in ferrous applications must withstand temperatures around 1400°C, posing stringent demands. At these high temperatures, the lack of hot strength is a critical issue, as the waterglass-based binder softens due to its relatively low glass transition temperature. Additionally, the collapsibility or de-coring after the casting process becomes more challenging, requiring more complex binder system formulations. However, one significant advantage of IOBs in ferrous foundries is their lower sensitivity to veining formation.

The underlying study is dealing with the development of IOB systems for casting processes with ductile and gray iron. For ductile iron, a differential housing and a brake caliper were examined in detail, while for gray iron, a simple brake disc model was reviewed.

2 Experimental procedure

Sand mixtures were prepared using quartz sand and various inorganic binder systems, including liquid binders and powder additives. Recipes were engineered to achieve high-quality castings based on the specific requirements of the casting process and the type of metal. Laboratory-scale tests were conducted using a core shooter Laempe L1, which produces transverse bars (for strength measurements) and other small cores, such as cylinder-type cores and those for small brake discs.

Sand cores for semi-production trials were produced using a Mingzhi core shooter type 20 L, which produced cores for differential housings and brake calipers.

The ductile iron used was GJS600 for the brake calipers and GJS500 for the differential housings, while the gray iron used was GJL250 being used for brake discs.

Result and discussion

Casting trials on lab scale (GJS400-15)

Initial trials using existing IOB systems developed for the aluminum automotive sector on ferrous applications (GJS-400-15) resulted in the formation of amorphous silica deposits (see Figure 1 left IA001 0.0 wt%).

However, further laboratory tests revealed that adding a small concentration of IA001 led to a smooth, silica-free casting surface (see figure 1 middle and right) and improved collapsibility after the casting trials.

Figure 1: Casting surface after removing core residue (left: 0.0 wt% IA001; middle: 0.2 wt% IA001; right: 0.4 wt% IA0001).

Differential housings (GJS500)

Using the optimized IOB system, including a small amount of IA001 for ferrous applications, sand cores were produced using the Mingzhi 20L core shooter. Previous trials with coated sand cores, produced using the cold-box process principles, showed a smooth, sand-free and defect-free casting surface. This coating was necessary to prevent metal penetration and sand adhesion.

Figure 2: Complete differential housing (examples).

With the addition of a small quantity of the IA001, more than 100 non-coated sand cores were produced, and were casted on site. After de-coring, fettling and shot blasting, the inner surface of the castings was in all cases as-serial. Figure 2 shows two examples of differential housings.

Brake calipers (GJS600)

Ductile castings of brake calipers were manufactured without refractory coating application. After optimizing the core shooting parameters, cores were produced with smooth surfaces and no defects. Figure 3 shows a series of sand cores with on the right a macrograph of part of the inner surface of the casting.

Observations of the castings indicated no gas formation during the casting process. Other feature observed during the trials was the functional bench life of the sand mixture even after several hours, no issues occurred with manufacturing these cores: good flowability of the sand mixture, and high-quality cores with sufficient compaction and surface smoothness.

Figure 3: left: series of sand cores to be casted; right: smooth and sand-free casting surface.

Brake discs (GJL250)

Laboratory trials with small sand cores used for a simplified brake disc was coated with a waterborne dip-coating. After applying the wet coating layer, samples were dried in a pre-heated furnace set at 120 °C. No issues occurred with weakening of the cores during dipping and during the drying process. Figure 4 shows on the left 4 coated sand cores ready for casting and on the right the inner surface of the brake disc.

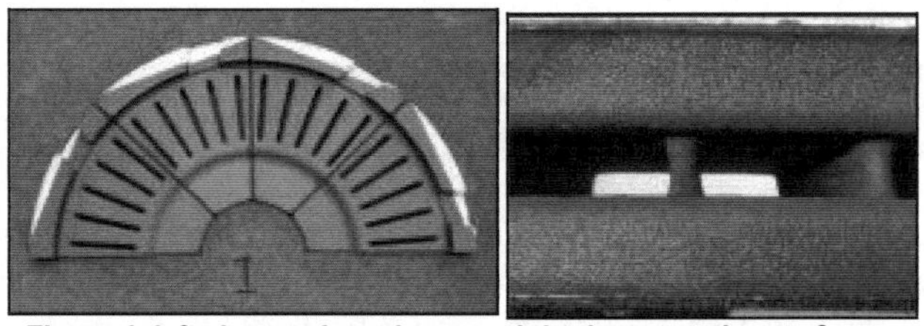

Figure 4: left: 4 coated sand cores; right: inner casting surface. Observations showed a high-quality casting surface: after easy de-coring a smooth, sand- and veining-free surface.

3 Conclusions

The findings from the presented case studies and through engineered formulations of the IOB systems, affirm the viability of IOB binder systems in ferrous applications, particularly in gray and ductile iron production. This study demonstrates that with appropriate adaptations, IOBs can produce ferrous castings with quality, consistency, and efficiency, reducing the need for refractory coatings and improving the overall work environment and carbon footprint associated to this metal casting segment.

Acknowledgments

The authors wish to express their gratitude to the management team Foundry and the complete R&D team of the Foseco R&D Centre in Enschede, the Netherlands.

A Study on the Application of Carbon-free Green Sand Casting Technology

Shigen Zhu
Donghua University, 2999 North Renmin Road, Songjiang District, Shanghai 201620, China
*Corresponding address: e-mail: sgzhu@dhu.edu.cn

Abstract: The casting method can be categorized as sand mold casting and special casting. There are a variety of specific processes. Among these, sand mold casting is used most commonly. It has the largest dosage which accounting for more than 80% of the industry. However, current sand mold casting technology has four worldwide problems: 1). Heavy pollution.2). High energy consumption. 3). Poor casting quality. 4. Poor reuse performance. The author has developed Carbon-free Green sand mould casting technology which featuring lower energy consumption while is environmentally friendly. This has laid a solid technical foundation for solving the above problems and will provide a strong support for the sustainable development of the Foundry industry. Technical features include: 1, No black pollution.2. Low energy consumption.3. Improvement of casting quality. 4. Outstanding reusing and recycling performance.

1 Introduction

Foundry is the foundation of machinery industry. Casting is one of the most economical and convenient forming technologies for metal parts. There are no restrictions on the size, batch or shape of casting parts. There are no restrictions on the material of casting parts. Also, the cost of metal castings is lower.

The casting method can be categorized as sand mold casting and special casting. Among them, sand mold casting is used most widely and has the largest amount of use. More than 80% of castings are produced by sand molds.

The existing sand mold casting technology has long been facing four worldwide issues: High pollution, high energy consumption, poor casting quality, significate resource demand.

1) High pollution. China has the world's largest casting industry, producing 50 million tons of casting every year. Among them, sand mold casting is about 38 million tons （seacoal sand mould casting 30 million tons）. With the existing foundry sand production, will produce waste sand 30 million tons, powder dust 1.5 million tons, exhaust gas 50 billion cubic meters.

Black wasted sand and dust containing seacoal powder and organic matter discharged from casting production, these not only bring serious black pollution to the environment, because of the complete or incomplete combustion of seacoal and resin, also bringing in carbon monoxide (CO), carbon dioxide (CO_2), sulfur dioxide (SO_2), nitrogen dioxide (NO_2), formaldehyde (CH_2CHO) and volatile organic chemicals (VOCs). The worse is which contains formaldehyde, α- benzopyrene, dioxins and other so-called "environmental hormones", these are the primary carcinogens listed on the *Stockholm Convention*. The high pollution of foundry sand has seriously restricted the development of the industry.

From these photos taken on site, we can see how terrible the working condition can be. And it's obvious that it's harmful to workers'health.

Fig. 1 Figure caption

2) High energy consumption. The annual energy consumption of China's foundry production accounts for about 45% of the total energy consumption in the machinery industry, and the energy utilization rate is only about 17%. The energy consumption of castings mainly comes from the smelting, the molding, cleaning and remelting of castings. But a considerable part also comes from the coal powder added in molding sand.

3)Poor casting quality. Burning-on and scab of the casting are still serious, the surface of the casting is not smooth and clean. There are many gas holes in the castings, the repair work is heavy, and the reject rate is high.

4) High resource consumption. Only part of the clay sand in sand mold casting can be directly reused. Sodium silicate sand and resin sand can only be used once, can not be reused directly.

The foundry industry in China consumed very large amount of base sand, binder and bituminous coal in 2017.

Fig. 2 Figure caption

2 Experimental procedure

Dry carbon-free clay sand and water were added to the high-speed rotor sand mixer for 120-150 seconds to make carbon-free clay wet molding sand. Then various casting production lines are used to shape, cast and drop sand, and the castings are inspected and analyzed after shot blasting.

3 Result and discussion

Through long-term efforts, we have developed Innovative Sand Mould Casting Technology Featuring Lower Energy Consumption and Less Pollutions, which has laid a solid technical foundation for solving the mentioned problems and will possibly provide strong support for the sustainable development of the foundry industry.

1) No black pollution

The raw materials are all non-toxic and inorganic. No waste gas is produced during casting. Waste sand is not toxic and can be integrated with the environment.

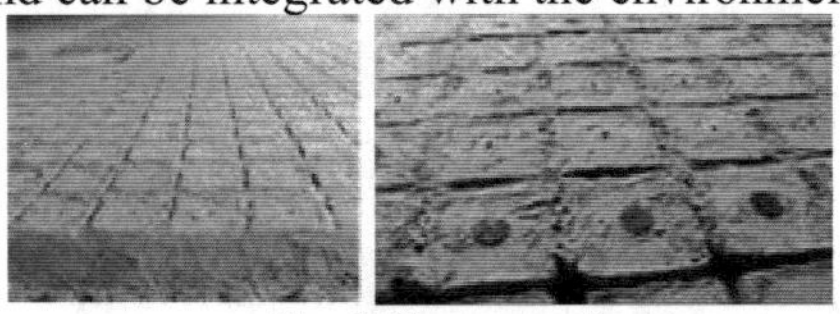

Fig. 3 Figure caption

The picture below is WELL completely automatic vertical moulding site.

Fig. 4 Figure caption

The picture below is SITO completely automatic static pressure moulding line site.

Fig. 5 Figure caption

The picture below is TOKYU horizontal moulding worksite.

Fig. 6 Figure caption

The pictures below show that only a small amount of water vapor is produced, and no pollution gas produced during pouring.

Fig. 7 Figure caption

2) Low energy consumption

The pictures below show that sand mixture does not contain seacoal, which can save energy.

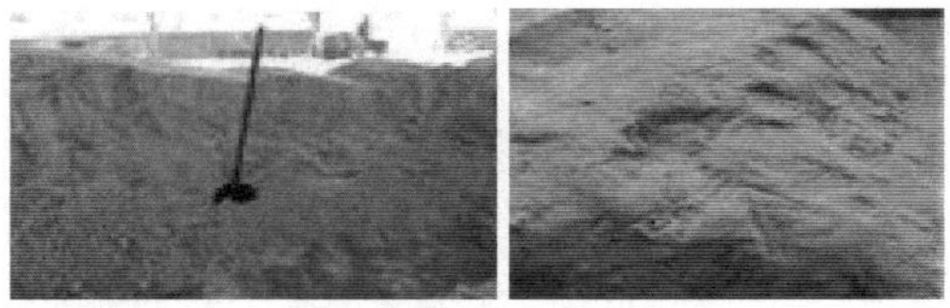

Fig. 8 Figure caption

Burning-on on the surface of casting is improved obviously and the cleaning workload is reduced. The surface roughness and precision of casting are improved, and the subsequent machining is more convenient. Thus cleaning and finishing energy consumption can also be reduced.

Fig. 9 Figure caption

3) Good casting quality

The surface of casting has less burning-on. The defects of scab and wash are lower. Gas holes are rarely found in castings. The rejection rate is lowered.

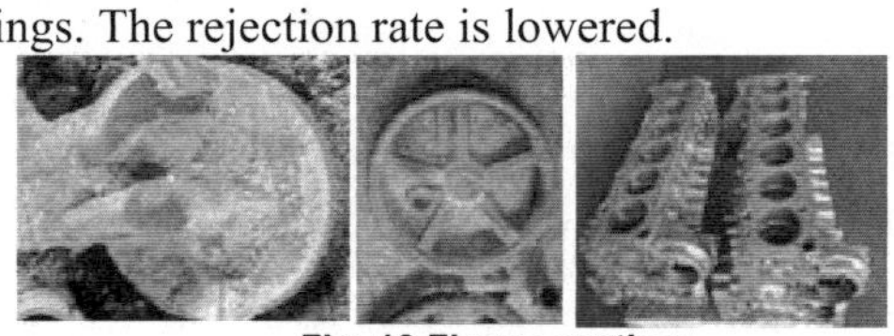

Fig. 10 Figure caption

4)Low resource consumption

Recycling performance of moulding sand is good, and moulding sand can be reused.

The picture is taken after 9 months using without adding in any new sand.

Fig. 11 Figure caption

Up to now, the new technology has been used in more than 100 foundries, and more than 1.2 million tons of castings have been successfully produced.

Fig. 12 Figure captions

4 Conclusion

Compared with the traditional sand mold casting process, the new technology has the following advantages: green and environmentally friendly, low carbon and energy saving, recyclable and better casting quality.

Formation of Oolitic Layer in Green Sand under Heat Effect

Yuyang Qi[1], Zefen He[2], Xin Peng[1], Yajun Yin[1*], Xiaoyuan Ji[1], Yuancai Li[1], Jianxin Zhou[1]
1. State Key Laboratory of Materials Processing and Die & Mould Technology, Huazhong University of Science & Technology, Wuhan, China
2. Jiangsu Sud-Chemie Chemical Materials Co., Ltd. Jiangsu, China
*Corresponding address:e-mail: yinyajun436@hust.edu.cn (Yajun Yin); zhoujianxin@hust.edu.cn (Jianxin Zhou)

Abstract: The green sand is repeatedly heated during casting, and the failed clay binder deposits and accumulates to form an oolitic layer, which affects the performance and regeneration of the molding sand. To study the formation mechanism of the oolitic layer, bentonite was mixed with sands and fully reacted at high temperatures ranging from 500℃ to 1000℃. The transformation process of bentonite forming a oolitic layer on the surface of sands was investigated using analysis methods such as TG/DTG, FTIR, and XPS. The results indicate that at temperatures above 600℃, the crystal structure of bentonite begins to be disrupted, and some form chemical bonds with sands, beginning to deposit and form a oolitic layer. Compared with the used sands, the oolitic sands have the same surface functional groups and similar composition, both formed by high-temperature deposition of bentonite. The oolitic layer is a mixture of glassy material, SiO_2, Al_2O_3, Fe_2O_3, etc.

Keywords: Green sand; Oolitic layer; Heat effect

1 Introduction

The recovery and reuse of foundry green sand has great environmental and economic value [1]. However, in the recycled green sand, the bentonite layer deactivated by high temperature will adhere tightly to the surface of the sand grains and form an oolitic layer in multiple accumulations [2]. It may influence properties of molding sand and quality of casting [3]. In order to control the oolitic rate of the green sand system, stabilize the quality of the sand, and improve the yield of castings, basic research was conducted on the formation mechanism of the oolitic layer. Samples containing an oolitic layer were prepared through high-temperature thermal action. And the transformation and deposition process of bentonite were investigated. By comparing with the used sand from the foundry, it was confirmed that the reaction conditions can simulate the formation process of the oolitic layer and reveal the formation mechanism of the oolitic layer.

2 Experimental procedure

The used sand samples and bentonite are from Huangshi Dongbei Foundry in China. Keep the bentonite at different temperatures for 40 minutes to analyze its reaction when heated. The sand used in the laboratory is Dalin standard sand. Mix the standard sand with 10% bentonite by mass, keep warm at different temperatures for 40 minutes. After cooling, wash off the mud, and dry the sample. In this way, the sample after heating with the oolitic layer is obtained.

The oolitic rate of the sample was measured by a multiple approximation method. The process includes pretreatment, blank control group design, multiple reactions with KOH and final calculation of oolitic rate.

3 Result and discussion

The thermal analysis experiment of bentonite is shown in Figure 1. In the low temperature region, the endothermic dehydration process begins at 61.5℃. At this time, the adsorbed water and interlayer water in the sample are removed, resulting in a 2.99% decrease in mass, while the crystal structure of montmorillonite is not affected. When the temperature rises to 651.5℃, the endothermic dihydroxylation process begins. The two hydroxyl groups of montmorillonites form a water molecule, leaving a chemically bound oxygen as the superoxide in the crystal lattice. And the crystal structure of montmorillonite is destroyed. The sample mass is reduced by 6.07%.

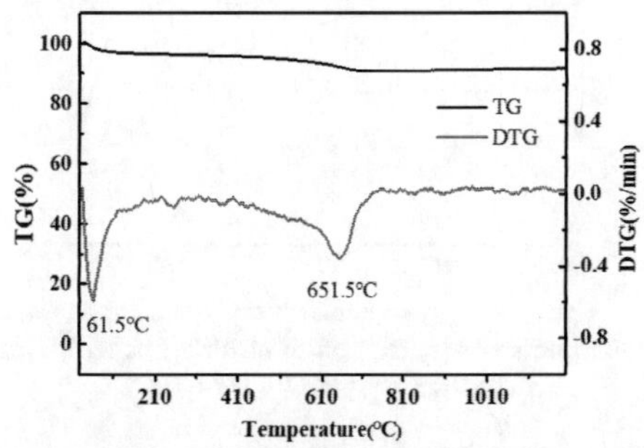

Figure 1. TG/DTG results of bentonite.

The FTIR results of bentonite at different temperatures are shown in Figure 2(a). The absorption bands at 3447 cm^{-1} and 1640 cm^{-1} are related to the stretching vibration of the -OH group and the bending vibration of the H-O-H group in the adsorbed water, respectively. Its strength decreases with increasing temperature, which is due to the dehydration of montmorillonite at high temperatures. The absorption band at 3628 cm^{-1} is related to the -OH groups in montmorillonite, including Al-OH and Si-OH. As the temperature rises above 600℃, montmorillonite loses its structural water, and the strength of the absorption band at this point significantly decreases until it disappears. These all indicate that the crystal structure of montmorillonite is destroyed at high temperatures.

From Figure 2(b), it can be seen that the infrared spectra of the model bentonite mixture and the used sand from the foundry, which contain the oolitic layer, are basically the same as those of the bentonite treated at high temperature. The absorption bands at 1080 cm^{-1}, 778 cm^{-1}, 694 cm^{-1}, and 458 cm^{-1} are derived from SiO_2, indicating the transformation of bentonite into orthorhombic quartz at high temperatures. This also indicates that the model bentonite mixture prepared in the laboratory can restore the surface functional group state of the used sand in the foundry, which is transformed from the deposition of bentonite.

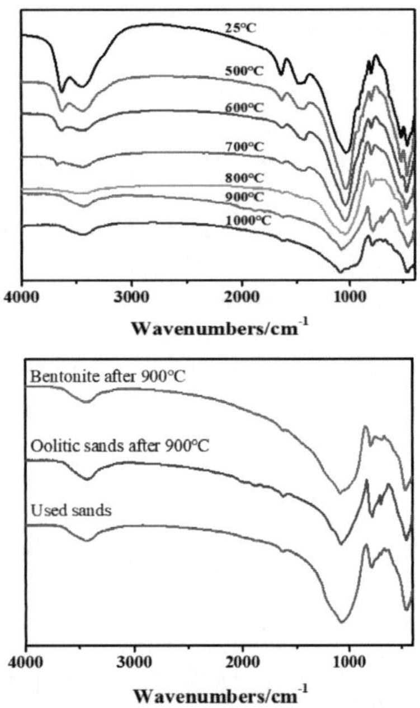

Figure 2. FTIR spectra of (a) Bentonite at different temperatures; (b) samples with oolitic layers.

The content of elements by XPS is shown in Table 1. Compared to the model bentonite mixture prepared in the laboratory, the Si and O content of the oolized layer on the surface of the used sand is higher. This may be due to the fact that components that are the same as the substrate are more prone to bonding during repeated pouring, so they preferentially combine and accumulate. After oxidation, Al and Fe are retained in their original proportions, while Na is lost under thermal action. Despite the presence of coal powder and other additives containing C in the reaction, the C content in the used sand is still the lowest, indicating that the C in the additives does not constitute a component of the oolitic layer.

Table 1. Element content in the surface of bentonite, oolitic sands and used sands.

Element	Atomic/%			
	Bentonite	Bentonite after 900℃	Model bentonite mixture after 900℃	Used sands
Si	9.6	8.5	9	13.2
C	28	41.1	33.2	19.7
O	56.5	45.8	52.7	63.6
Fe	0.7	0.7	0.6	0.7
Na	2.3	1.3	1.8	0.4
Al	2.9	2.7	2.7	2.4

4 Conclusion

The structure of bentonite is significantly affected by temperature. The results of TG-DSC show that the crystal structure of montmorillonite is destroyed when the structure water is removed at 600℃. When bentonite is mixed with sands and the temperature reaches more than 600℃, the crystal structure of bentonite is destroyed, so some of it forms chemical bonds with sands, and deposits into oolitic layer. The results of FTIR show that the groups on the surface of the prepared oolitic sand and the old sand are the same, and both are formed by the deposition of bentonite. XPS results further show that bentonite is deposited on the sand surface at high temperature, and its chemical state is not affected by the matrix. The oolitic layer of oolitic sands and used sands is a mixture of glassy material, SiO_2, Al_2O_3, Fe_2O_3 and so on.

Acknowledgments

This work is financially supported by the National Key Research and Development Program of China (No. 2022YFB3706800), the National Natural Science Foundation of China (Grant No. 51905188 and 51775205).

References

[1] Maria C.Z., Silvia F. Conservation and Recycling, "Foundry processes: the recovery of green moulding sands for core operations", 38 (2003) 243-254.

[2] Khan, M.M., Mahajani, S. M.; Jadhav, G.N. et.al. Applied Clay Science, "Transformation of bentonite used in green sand molds during metal casting process and its relevance in sand reclamation", 206 (2021) 106072.1-106072.10.

[3] Chengyan Z, Xijun Z, Fang Z, et al. Hot Working Technology, "Review of Study on Oolitization of Used-clay-sand", 41 (2012) 34-37. (In Chinese)

Research, Application and Development of Inorganic Binder for Casting Process

Xiaolong Gong, Shengli Hu, Zitian Fan*
State Key Laboratory of Materials Processing and Die & Mould Technology, School of Materials Science and Engineering, Huazhong University of Science and Technology, Wuhan 430074, China
*Corresponding address: e-mail: fanzt@hust.edu.cn

Abstract: Inorganic binder used in casting process has the advantages of low odor, labor-friendly conditions, and relatively low cost, which is one of the main development directions for casting molding materials in the future. However, compared to organic binders (such as resin binders), inorganic binders exhibit lower bonding strength and are more sensitive to environmental humidity. This sensitivity poses challenges, particularly in the reclamation of used sand, thus limiting their broader application. In this paper, the research and application status of inorganic binders (mainly silicate inorganic binders) and their curing methods are summarized. Meanwhile, the development direction of inorganic binders is clarified.

Keywords: inorganic binder; silicate system; used sand reclamation; development trend

1 Introduction

With the increasingly stringent environmental regulations, foundries are under pressure to enhance their development and adopt eco-friendly inorganic binder sand casting technology. There is a particular focus on improving and refining silicate inorganic binders with the aim of replacing organic binder sand casting processes with more environmentally friendly alternatives [1-3]. The development of silicate inorganic binder has generally undergone four stages: CO_2 cured sodium silicate sand process, powder cured sodium silicate sand process, ester cured sodium silicate sand process, and new silicate inorganic binder sand process. Currently, the CO_2 cured sodium silicate sand process and the ester cured sodium silicate sand process are mainly used in the production of steel castings, while the new silicate inorganic binder sand process is mainly applied in aluminum alloy castings [4, 5].

At present, silicate inorganic binders remain the most successful environmentally friendly binders applied in sand casting process. This paper focuses on the research and application status of silicate inorganic binders and their curing methods. Finally, the future development direction of inorganic binders for casting process is discussed.

Silicate system inorganic binder

Currently, the inorganic binders used in the foundry industry mainly refer to sodium silicate binders [6]. According to different curing methods and a wide range of applications, silicate inorganic binders have undergone four development stages, as shown in Fig. 1. Among them, CO_2 cured sodium silicate sand process and ester cured sodium silicate sand process are mainly used in the manufacturing of steel castings in industries such as railway, shipbuilding, and engineering machinery. However, the new silicate inorganic binder sand process cannot withstand high temperatures and is mainly used for aluminum alloy casting.

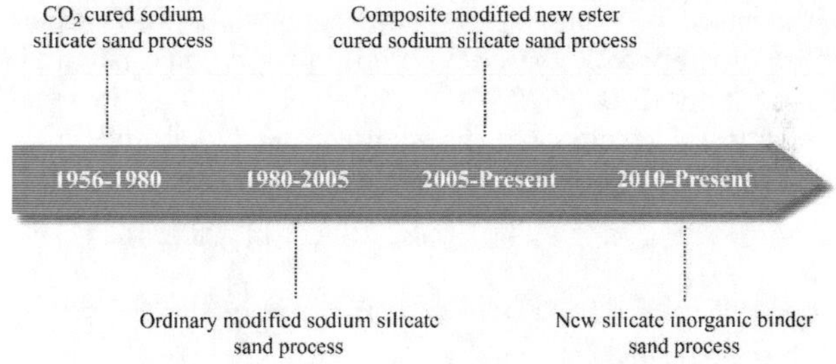

Fig. 1. Development of silicate inorganic binder

CO_2 cured sodium silicate sand process

The CO_2 cured sodium silicate sand process involves blowing CO_2 gas into the sodium silicate sand to harden the sodium silicate binder, thereby achieving the bonding strength. Currently, the ordinary CO_2 cured sodium silicate sand process still has a large application market in the production of steel castings because of its simple process operation, low cost, rapid hardening speed, and strong adaptability. However, with increasingly strict environmental requirements, the adoption of new CO_2 cured sodium silicate sand processes, such as modified sodium silicate binder combined with gas-blowing processes like pulse method and heating method, further reduces the dosage of sodium silicate (<3%), and improves the bonding strength and heat collapsibility. This is expected to be the development trend of the CO_2 cured sodium silicate sand process.

Ester cured sodium silicate sand process

Ester cured sodium silicate sand has the comprehensive advantages of sodium silicate sand and resin self-hardening sand, such as high bonding strength, adjustable curing speed, convenient operation, good working environment. Furthermore, the dosage of sodium silicate binder can be decreased to 3.5%-4.5%. In addition, the ester-cured modified sodium silicate sand process can further decrease the content of sodium silicate binder to less than 3%. At present, the new ester cured sodium silicate sand process is gradually replacing ordinary sodium silicate sand and part of resin self-hardening sand process in the actual production of steel castings for industries such as metallurgy and mining machinery, rail transit, engineering machinery, coal mining machinery, and pump valves, etc. With the continuous improvement of this process, as well as the technical and equipment levels of modified sodium silicate materials, the new ester cured sodium silicate sand process has become the preferred technology for the production of steel castings.

New silicate inorganic binder sand for aluminum alloy

At the beginning of the 21st century, German company HA developed the Cordis system inorganic binder, and the ASK company invented the INOTEC system inorganic binders, which are suitable for aluminum alloy castings [7]. This process uses a modified alkaline silicate binder (1.8%-2.5%, based on sand weight) and a metal oxide powder (approximately 50%, based on binder weight), and is cured under hot air (150 °C-200 °C) conditions. Fig. 2 shows the comparative results of tar content and harmful gas emissions in the core making process [8].

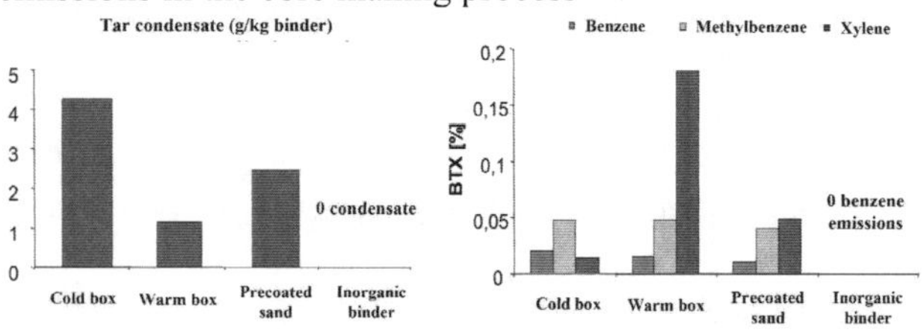

Fig. 2. Comparative results of tar content and harmful gas emissions in the core making process[8]

In the past decade, Chinese foundry material manufacturers and research institutions have successively developed various new silicate inorganic binder materials based on modified sodium silicate and conducted production trials. They have achieved good results in several areas, including the inorganic binder warm box process, composite blow air cold box process, and inorganic coated sand process.

2 Conclusion

CO_2 cured sodium silicate sand and ester cured sodium silicate sand remain the most widely used inorganic binder sand in the production of steel castings, while new silicate inorganic binder sand has seen rapid development and application in the production of aluminum alloy castings. For silicate inorganic binders, whether applied to steel castings or aluminum alloy castings, improving the ambient temperature strength of molding (core) sand, enhancing the collapsibility of used sand, and promoting the reusability of reclaimed sand are the goals pursued in the development of this type of inorganic binder. Furthermore, deeper research is needed into the bonding mechanism and hardening characteristics of silicate inorganic binders, improving the quality of binder and hardener materials, refining traditional curing methods, processes, and equipment, and focusing on solving the recycling and reclamation problem of used sand.

Acknowledgments

This work was financially supported by the National Natural Science Foundation of China (Nos. 52275334, 52205361, 51075163, and 50575085).

References

[1] Zhou J Y. Application of sodium silicate inorganic binder to foundry production at home and abroad and its latest development. Foundry, 2012, 61(3): 237-245. (In Chinese)

[2] Holtzer M, Kmita A. Mold and core sands in metalcasting: chemistry and ecology: sustainable development. Cham: Springer International Publishing, 2020.

[3] Dańko R, Kmita A, Holtzer M, et al. Development of inorganic binder systems to minimise emissions in ferrous foundries. Sustainable Materials and Technologies, 2023, 37, e00666.

[4] Merta V, Beňo J, Obzina T, et al. Innovative inorganic binder systems for the production of cores for non-ferrous metal alloys reflecting the product quality requirements. Metals, 2021, 5: 1-14.

[5] Lafay V. Application of no-bake sodium silicate binder systems. International Journal of Metalcasting, 2012, 6(3): 19-26.

[6] Fan Z T, Zhu Y S, Dong X P, et al. The water-glass sand process principle and application (2nd Edition). Beijing: Machinery Industry Press, 2016.

[7] Li Y C. Foundry handbook volume 4: Molding materials (4th Edition). Beijing: Machinery Industry Press, 2020.

[8] The Foundry Institution of China Mechanical Engineering Society, International Foundry Organization Molding Materials Committee. Research report on the current status and development trends of foundry molding materials in China. 2023.

Research on Water-Based Quick-drying Coatings for Foundry and the Film-Forming Mechanism

Bao Liu[1]*, Weihua Liu[1], Lai Song[1]
1. School of Materials Science and Engineering, Shenyang University of Technology, Shenyang,China
*Corresponding e-mail: 94175562@qq.com

Abstract: In this paper, the influence of refractory aggregates, binders, and other components on the performance of the coating was studied through single-factor tests using water as the carrier liquid. The optimal ratio of the coating was determined using the orthogonal test method: 75% of refractory aggregate (white corundum powder: bauxite powder = 6:4), 4.5% of acrylic emulsion, 3% of sodium bentonite, and the remaining balance was water and additives. Under the conditions of a temperature of 25 ± 5 ℃ and relative humidity of ≤ 50% RH, the drying time was ≤ 50 minutes, gas evolution was 9.3 mL/g, wear loss was 0.0217 g, suspension property was 99%, and high-temperature crack resistance was grade I. The film-forming mechanism of water-based coatings was studied using scanning electron microscopy and Fourier transform infrared spectroscopy. The results indicated that the water-based quick-drying coating prepared in this study has the advantages of a compact and smooth film, fast drying speed, high film strength, and low gas evolution, which can meet the requirements of casting production.

Keywords: water-based quick-drying coating, drying time, acrylic emulsion, orthogonal experiment

1 Introduction

During the casting process, a layer of coating is applied to the surface of the mold (core) to enhance its surface quality and effectively prevents or minimizes casting defects[1]. Water-based coatings have excellent suspension, thixotropy and leveling properties. It is non-toxic and odorless which can meet the criteria for energy conservation, emission reduction, and environmental protection. This coating offers a promising market prospect in the future. Many domestic and international research institutions and foundries had conducted studies on the water-based coatings [2-6].

At present, the primary obstacle which hinders the widespread use of water-based casting coatings is the long drying time, and drying needs specialized drying equipment. This not only increases production costs but also prolongs the production cycle. This study investigates the influence of refractory aggregate, acrylic emulsion, and sodium bentonite on the drying time of water-based coatings through experiments. The coating components and their optimal ratios were determined through orthogonal testing. The film-forming mechanism of water-based coatings was examined using SEM and FTIR spectroscopy.

2 Experimental procedure

The main raw materials used in the test are white corundum powder (100-200 mesh), high alumina powder (200-300 mesh), acrylic emulsion (commercially available), sodium bentonite (commercially available), and water. The coating was prepared using a combination of gentle stirring and high-speed stirring. Initially, the additives and water were added to the mixing tank for gentle dispersion (5-10 minutes), followed by the refractory aggregate for high-speed dispersion (20-40 minutes). Subsequently, the binder suspending agent was added and mixed at low speed (5-10 minutes) before being packed for use.

The evaluation of coating performance primarily depends on drying time, wear loss, and gas evolution. The drying time of the coating was measured using the QGZ-24 automatic coating drying time tester. Gas evolution was analyzed using the SFL type gas evolution tester, and coating wear loss was assessed with the SUM type coating wear resistance tester, following the guidelines of JB/T 9226-2008.

3 Result and discussion

The influence of refractory aggregates on the effectiveness of coatings.

White corundum was mixed with bauxite in a 6:4 ratio, using varying proportions of 60%, 70%, 80%, and 90%. The study aimed to achieve consistent viscosity by adjusting the amount of carrier liquid. The technical specifications can be found in the accompanying figure 1. Analysis of the figure indicates that the most desirable overall properties, including drying time, are achieved when the refractory aggregate content was at 80%.

The influence of acrylic emulsion on the effectiveness of coatings:

The acrylic emulsion comprises 2-6%, and the technical specifications of the coating were delineated in the figure 2. Examination of the graph indicates that the drying time and overall properties demonstrated optimal performance at an acrylic emulsion content of 5%.

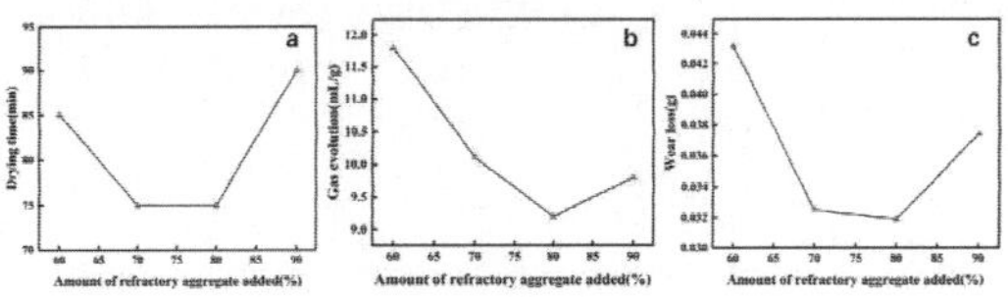

Figure 1 Relationship between refractory aggregates and coating performance

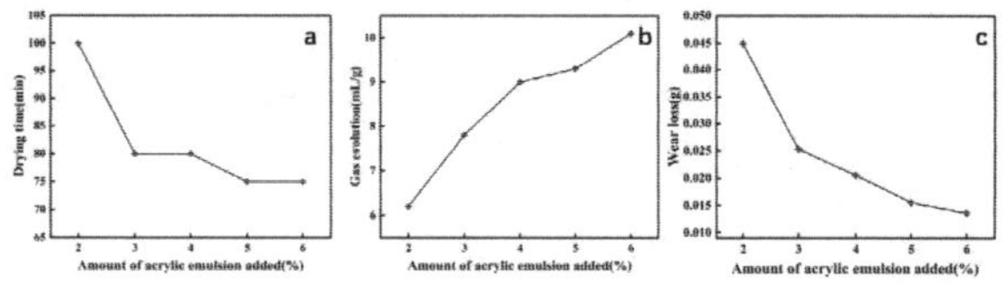

Figure 2 Relationship between acrylic emulsion and coating performance

The influence of sodium bentonite on the effectiveness of coatings:

The sodium bentonite comprises 2-6%, and the technical specifications of the coating were delineated in the figure 3. Examination of the graph indicates that the drying time and overall properties demonstrated optimal performance at an sodium bentonite content of 3%.

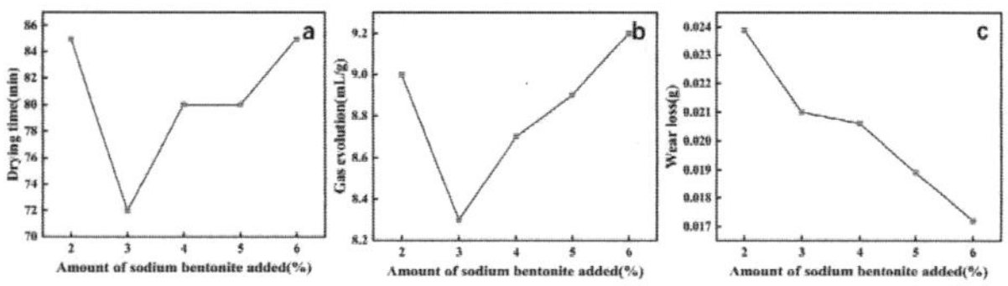

Figure 3 Relationship between sodium bentonite and coating performance

The influence of other components on the drying time of coatings:

The study investigated the impact of quick-drying additives, water-reducing agents, and ethanol on the drying kinetics of coatings. Findings indicated that the combined application of 0.25% quick-drying agent, 0.5% water-reducing agent, and 3% ethanol resulted in a coating drying time of 50 minutes.

Orthogonal experiment to determine the optimal ratio:

A three-factor, three-level orthogonal experiment was conducted to ascertain the optimal composition of the coating. The range analysis revealed that the most efficient blend consisted of 75% refractory aggregate (white fused alumina powder: bauxite powder = 6:4), 4.5% acrylic emulsion, and 3% sodium bentonite. The technical properties of the coating at this optimal proportion were outlined in Table 1.

Table 1. Optimizing the blending ratio of technical indicators

Suspension 48h	Gas evolution ml/g	Wear loss g	Drying time min
99	9.3	0.0217	50

Study on the Film-Forming Mechanism of Coatings:

See Figure 4 for the microscopic morphology, infrared spectrum, and film-forming mechanism of the coating.

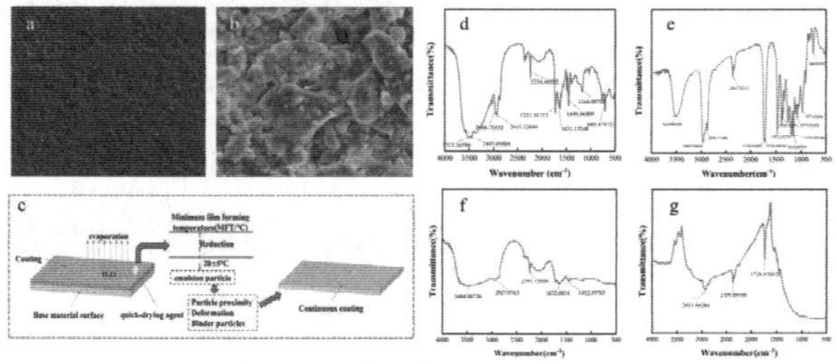

Figure 4 The microscopic morphology, infrared spectrum, and film-forming mechanism of the coating

4 Conclusion

The high glass transition temperature and high minimum film-forming temperature of acrylic emulsion coating bonding system can significantly improve the drying time of water-based coatings. The content of refractory aggregate and sodium bentonite can affect the amount of coating carrier liquid. Through the analysis of coating micro-morphology, infrared spectrum and film-forming mechanism, it was found that the curing speed of the binder, the content of carrier liquid, and volatilization speed were the key factors affecting the drying time of water-based coatings. In this experiment, the room temperature drying time of the water-based quick-drying coating is $\leq$ 50 minutes, which can meet the casting requirements.

References

[1] Li Yuancai. Casting coatings and applications [M]. Beijing: Mechanical Industry Press, 2007.

[2] Gao Tianjiao, Yin Shaokui, Li Ling, et al. A fast-drying water-based coating for casting and its preparation method [P]. CN110586854B, 2021-07-30.

[3] Chen Jianfeng, Deng Bo, Chen Shuai. Preparation method of a fast-drying high suspension water-based casting coating [P]. CN108405797A, 2018-08-17.

[4] Long Dawen. A series of water-based quick-drying coatings for cast steel and cast iron [P] CN107433321A, 2017-12-05.

[5] Feng Shengshan, Xie Shuzhong, Liu Chunjing, et al. A water-based self-drying/rapid-drying sand mold casting coating and its preparation method [P]. CN103878302B, 2016-01-20.

[6] Bian Qingqing, Feng Shengshan, Dong Huang, et al. Process characteristics of water-based self-curing coatings for sand casting[J]. Journal of Physics: Conference Series, 2021, 2109(1).

Development Trends of Eco-friendly Binder for Foundry Molding

Jian Tan, Yanxia Ma
HUETTENES- ALBERTUS China, No.26 Taxiu Road, Jiading District, Shanghai, China
*Corresponding address: e-mail: jian.tan@ha-group.com

Abstract: Reducing emissions and promoting carbon neutrality are inevitable choices for sustainable development in all countries around the world. HA focuses and owns the cutting-edge technology and eco-friendly products for this. Compared to traditional molding binders, the Alkaline Phenolic No-bake reduces VOCs emissions by 30% in the hand molding field, while the Furan system uses a new generation of low sulfur hardener(Without sulfuric acid), reducing emissions of SO_2 and NOx by nearly 70%; In the field of automatic molding lines, HA's the fourth generation eco-friendly Cold-Box resin introduces inorganic components in the production process of base resins, reducing the emissions more than 50%; And the ultimate product for emission reduction, 100% pure inorganic binder, has not only been widely used in non-ferrous castings, but also achieved breakthroughs in the application of iron castings.

At the end of the article, shares successful cases for well-known domestic foundries how to achieve emission reduction, for reference.

Keywords: sustainable development; emission reduction; molding binders; resin

1 Introduction

With the official implementation of the Emission Standards for Air Pollutants in the Foundry Industry, foundry workers have to face the problem of how to implement the standards. However, due to a lack of experience, many Foundries have invested a large amount of funds with little effect. 100 years old international enterprise, HA Group, with the most cutting-edge products and mature solutions. This article introduces the latest development and application of furan, alkaline phenolic, cold box resin, and pure inorganic binders of zero emission, in order to help Foundries avoid detours and promote the Eco-friendly development.

Furan no bake system

Traditional furan system contains nitrogen and sulfur, which is not only increases emissions and also strong odor. HA developed a new material to replace sulfuric acid and sulfonic acid, which can reduce more than 50% S content of recycled sand. At the same time, the N content is controlled ≤ 1%, reducing NOx emissions by nearly 70%

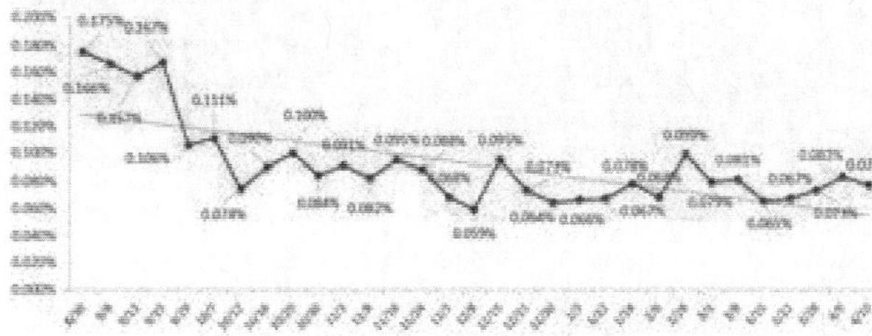

Fig. 1 S content changes in recycled sand

Alkaline phenolic no bake system

Alkaline phenolic no bake has a good high-temperature stability and is commonly used in steel castings, but it is difficult to reclaim. HA has improved the regeneration performance, by using mechanical regeneration method, a minimum of 15% new sand addition can maintain the balance of recycled sand strength, and make 30% reduction in emissions.

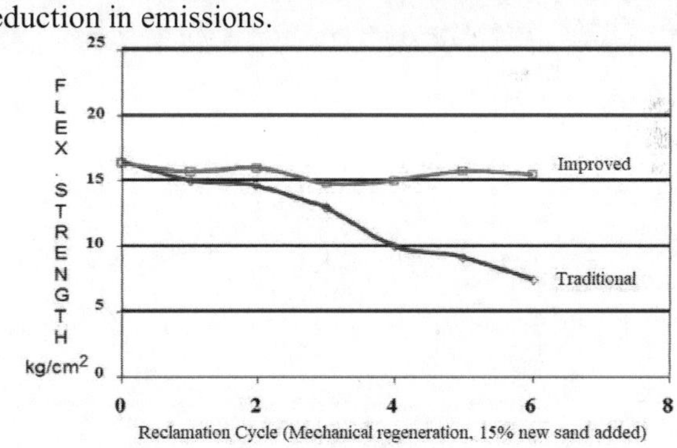

Fig. 2 Figure caption

Cold box resin

HA's the fourth generation eco-friendly Cold-Box resin introduces inorganic components in the production process of base resins, reducing the emissions 50% and C content 23%.

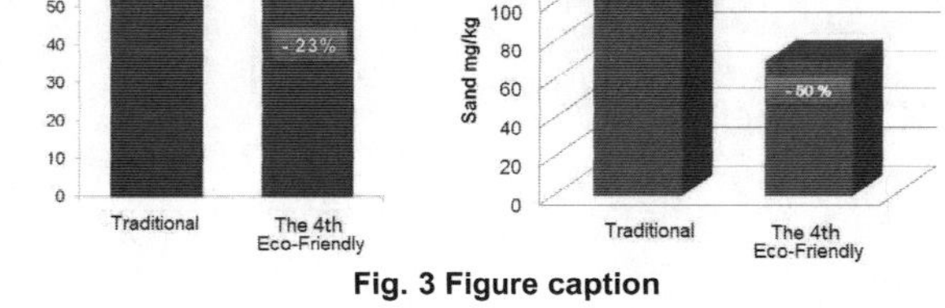

Fig. 3 Figure caption

Pure inorganic binder

During core making and filling, there are no unpleasant smelling gases, liquids and no condensation by using inorganic systems, Emission=0. Now it is not only applied to Al casting, and also iron parts

Fig. 3 Figure caption

Study case

The way to help H customer successfully reduce NMHC from 120 to below 60, meeting GB39726-2020 standard (requirement<100).

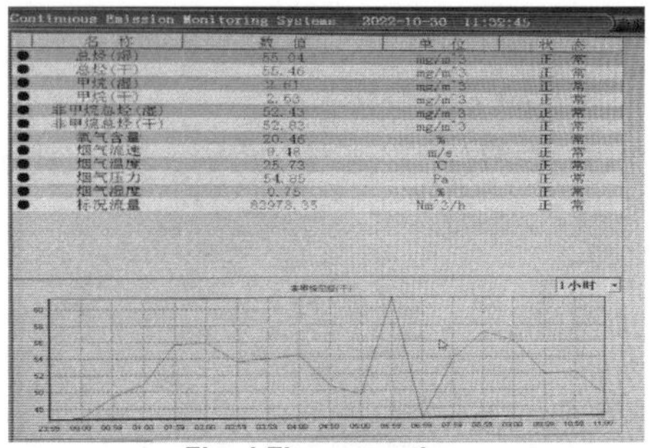

Fig. 4 Figure caption

2 Conclusion

HA product Compared to traditional:

[1] The Alkaline phenolic no bake process reduces emissions by 30% and improved the regeneration performance.

[2] Furan resin system without sulfuric acid, reducing help of SO2 and NOx by nearly 70%;

[3] The 4th eco-friendly Cold-Box introduces inorganic components, reducing the emissions more than 50%.

[4] 100% pure inorganic binder, emission=0, has not only non-ferrous, but also iron castings.

[5] Can reduce emissions, help foundry meet standard.

Boron Anhydride on the Performance of Covering Sand

Bo Tian*, Guozhong Li , Lilin Zheng, Wen-heng Duan , Jin-shengWang
Country CRRC changzhou Auto parts CO., LTD. Changzhou, Jiangsu, china,213011
*Corresponding address: e-mail: 18437953667@163.com

Abstract: This article studies the laws of boron anhydride on the performance of covering sand. The results of the study show that after adding 3% boron anhydride and 7% Fe_3O_4 to the covering sand, after the sand core is solidified by the case of the sand, its room temperature resistance can reach more than 3MPa, and it has good storage stability. In addition, after baking at 550℃/2h and 800℃/2h, it can still maintain the strength of more than 1MPa.

Through analysis, the mechanism of boron anhydride to keep the strength of covering sand after baking at high temperature was discussed, which provided a way to expand the use covering sand.

Keywords: boron anhydride, covering sand core, storage stability, strength.

1 Introduction

Shell molding is a sand molding method in which resin coated sand is contacted with a metal mold preheated to about 230 degrees Celsius to form a hard shell[1-2]. The coated sand used in the production of ordinary cast iron is generally composed of silica sand, thermoplastic phenolic resin, urotropine and calcium stearate, and usually does not contain relevant additives, so it does not have high strength, high temperature resistance, low expansion, low gas and other characteristics.Steel casting has high requirements for dimensional accuracy, surface quality and stability of coated sand, so it is necessary to add some additives to the coated sand used in steel casting to make it have high strength, high temperature resistance and low expansion.

2 Experimental procedure

The core is formed by hot box method for tensile strength sample and core forming effect evaluation.

The strength before roasting, the strength after roasting at 550℃, the strength after roasting at 800℃ and the core forming effect of the sample were used as evaluation indexes. The factors to be considered include the presence or absence of impregnating coating, the amount of boron anhydride added and the amount of Fe_3O_4 added. The performance and application evaluation were carried out by comprehensive performance data and sand core.

The equipment used in this test is SAR-II intelligent temperature controlled coated sand making machine, box furnace, 8615 core making machine, and the raw materials used in this test are covering sand, boron anhydride, Fe_3O_4, 300H paint. The evenly mixed sand is made into a tensile sample. The core structure for tensile sample and sand core forming effect evaluation is shown in Figure 1.

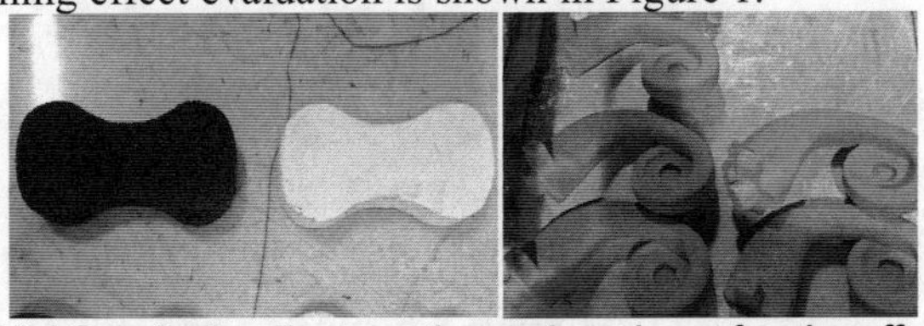

FIG.1 Core for tensile strength sample and core forming effect evaluation

3 Result and discussion

In order to obtain the core with good performance, the following main factors were determined through several tests and screening: the content of boron anhydride, the content of ferric oxide, whether the sand core was immersed in coating, the time of sand placement, etc.

According to the test data of each factor, the relationship curve between the level and intensity of each factor was drawn, as shown in FIG. 2 to FIG. 3.

It can be seen from Figure 2(a) that the strength of the core after sintering at high temperature can be increased with the increase of boron anhydride content. After sintering of the sand core at 800℃/2h, the resin gradually decomposes and burns, so the resin has basically no effect on the strength of the core.

At high temperature, boron anhydride can react with Al_2O_3 and other substances in the coated sand to solidify the sand core. However, it is found that when the content of resin is fixed and the content of boron anhydride is too high, the molding sand is sticky, the forming effect is poor, and it is difficult to remove the core from the mold. And the produced core is easy to damp, easy to break. The content of boron anhydride is set at 2%~3% after comprehensive consideration.

Then considering the characteristics of boron anhydride being easy to absorb moisture and water, the strength changes of sand exposed to air was studied, as shown in Figure 2(b). As can be seen from the figure, with the passage of time, the strength of the sand core at room temperature decreased slightly, and the strength after high temperature sintering did not change much, so it can be concluded that the sand can be used stably for a long time when it is sealed and stored, which has little impact on its overall performance.

In order to further improve the stability of the sand core at high temperature, the influence of whether or not the sand core is impregnated with paint on its strength after sintering at high temperature was studied. As can be seen from the figure, the strength of the sand core impregnated with paint after sintering at high temperature is generally higher than that of the sand core without impregnating paint, so the whole sand core should be impregnated with paint later.

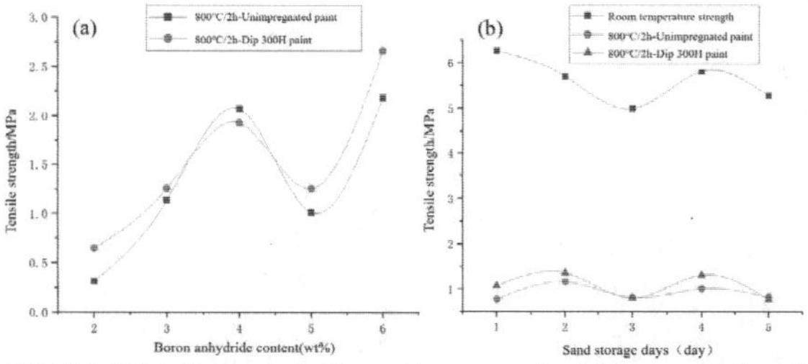

FiG.2 (a)Relationship between boron anhydride content and sample strength. (b)The variation of sand days and tensile strength.

In order to further improve the sintered strength of the sand core and reduce the influence of boron anhydride on moisture, Fe_3O_4 was considered to be added to the sand. As can be seen from Figure 3, with the increase of Fe_3O_4 content, the strength of the core after sintering can be maintained above 1MPa. Therefore, the content of Fe_3O_4 should be controlled at 7%.

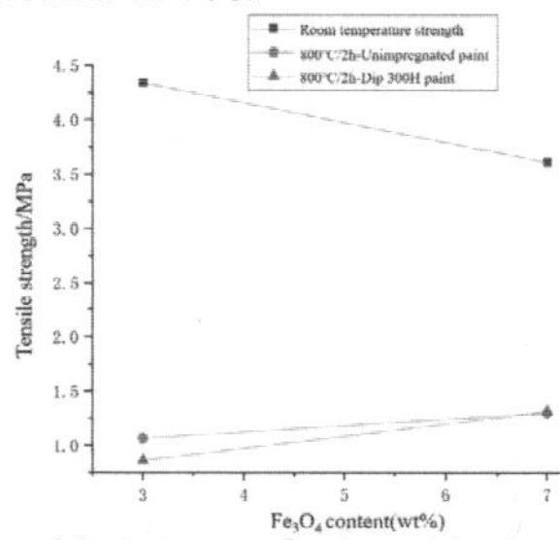

FIG.3 Relationship between ferric oxide content and tensile strength.

As shown in Figure 4, in the process of core making, it is found that the formability of the core of this process formula is not very good, and the core is sticky to the mold, which is not easy to clean; After adjusting the core parameters and changing the mold structure, the core making situation has been improved.

FIG.4 (a)The picture shows the bad situation of sand core making, and (b)the picture shows mass preparation of sand core after modifying core parameters and mold structure.

The formulated core can be used in the lost mold casting process; When casting complex products, the pouring system needs to carry out the process of burning the empty shell at 800℃/0.5h, and the sand core still maintains a high strength and does not break after this process, which provides a guarantee for the successful casting of products.

After the casting is completed, as shown in Figure 5(b), when cleaning the inner cavity of the casting, it is found that although the residual core is still massive, the strength is low and it can be easily removed. Therefore, it can be judged that the collapsibility and cleanability of the residual core are better.

The results of pouring test show that the inner cavity of the casting is easy to clean after pouring, and it has the advantage of mass production.

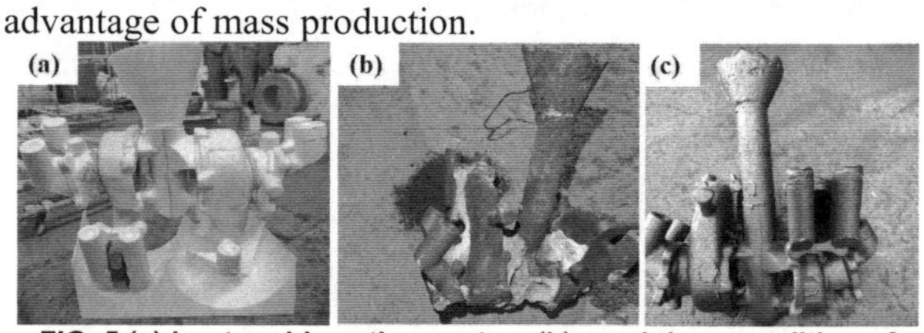

FIG. 5 (a) Lost mold casting system (b) sand drop condition of castings after casting and (c) castings after shot blasting

4 Conclusion

1. Sand mixed with boron anhydride and Fe_3O_4 can be used as a binder to solidify the sand core in the process of lost mold casting.The sand formula has excellent resistance to high temperature deformation. After dipping coating, it can meet the core requirements for investment casting.
2. The result shows that the residual sand core in the casting has good collapsibility and is easy to clean.
3. When the addition amount of boron anhydride is 3% and Fe_3O_4 is 7%, the sand core state is the best.

Acknowledgments

Thank CRRC changzhou Auto parts CO., LTD. platform support.

References

[1] AFS. Mold and core test handbook [M]. Illinois: AFS Des Plaines,1978:1121.

[2] Ross Foti. A look back at the 20th century: part 7 the shell process[J].

A quick method for determining angularity of foundry sand

Na Li[1], Yong Yuan[1], Ye Liu[1] *, Zhiming Wang[1, 2]
1. Jinan Shengquan Group, Jinan 250204, Shandong, China
2. School of Mechanical Engineering, Qilu University of Technology (Shandong Academy of Sciences), Jinan 250353, China
*Corresponding address: e-mail: danny.liu@shengquan.com

Abstract: A method for rapid determination of sand shape based on imaging technology is developed. Combing the calculation formula of circularity and the principle of measuring angularity, the relationship between circularity and angularity is deduced, that is, the angularity can be approximately calculated by the reciprocal of the squared circularity. Based on this relationship, the circularity of base sand is directly measured by imaging method, and the angularity then can be calculated quickly. Comparative analysis between the calculated and measured angularity of the base sand confirms the good accuracy of angularity calculated by the new developed method. This quick method provides a new reference for determining angularity and standard formulation of foundry sand.

Keywords: Circularity; Angularity; Foundry sand

1 Introduction

The shape and particle size of foundry sand determine largely the permeability, mold strength, and surface roughness of castings, and is one of the key parameters for foundries to select sand. Particle size is always measured by grains fineness number (American Foundrymen's Society-Grains Fineness Number [1], AFS-GFN) and average grain size (AGS), but which cannot reflect the particle size distribution and shape of the sand.

Angularity is always used to evaluate the shape of sand. According to the China national standard GB/T49442 "Foundry Sand" [2], angularity is defined as the ratio of the actual specific surface area to its theoretical value. The actual specific surface area of sand is measured by the ventilation method which is always complex and time-consuming, and the accuracy depends largely on the testing operation. In this situation, to develop a quick and accurate measuring method for angularity will have great practical significance. In the present work, a new rapid method for evaluating angularity was developed based on dynamic image method and mathematical derivation and conversion.

2 Experimental procedure

The morphology and particle size of the base sand were analyzed using the BT-2900 particle size and shape analysis system. Its main working principle is to take snapshots of the moving particles by a high-speed camera, and then, information as particle size and circularity can be obtained by processing the snapshots using the built-in analysis software.

Actual angularity was measured according to the method specified in National Standards of China 'GB/T9442-2010 Foundry silica sand' [2].

3 Result and discussion

(1) Derivation of the relation between angularity and circularity

Circularity (R) is a parameter that to characterize the degree of roundness of particles in two-dimensional projection, which is defines as:

$$R = (4\pi * S)/L^2 \qquad (1)$$

Where, S is the projection area of a single particle, L is the circumference, and π is the circumference ratio.

From equation (1), it can be known that, among all the 2D shapes, circle has the largest projection area at the same circumference, corresponding to a circularity of 1. The more irregular the shape is, the smaller the circularity, and the greater the deviation from circle. Such as, the circularity of a circle is 1.0, while the circularity of a square is $\pi/4$, which is approximately 0.79, and the circularity of a triangle is approximately 0.60.

Define an equivalent diameter D_1 which can be calculated from S as $S = \pi(D_1/2)^2$, and define another equivalent diameter D_2 which can be calculated from L as $L=\pi D_2$. Then, equation (1) can be further written as:

$$R = D_1/D_2 \qquad (2)$$

National Standards of china 'Foundry silica sand' [2] specifies the measurement methods of angularity, which is based on the principle that spherical shape among geometric shapes of equal volume has the minimum surface area. Then, angularity (A_n) is written as:

$$A_n = S_w/S_t \qquad (3)$$

Where, S_w is the measured surface area of base sand particles, and St is the theoretical specific surface area of sands. S_t is calculated based on the mesh diameter corresponding to the sand particle size.

Here, S_t and S_w can be approximately calculated by the equivalent diameter D_1 and D_2 respectively as:

$$S_t \approx \pi(D_1/2)^2 \quad (4)$$

$$S_w \approx \pi(D_2/2)^2 \quad (5)$$

Combining equation (2) ~ (5) simultaneously yields:

$$A_n = (D_2/D_1)^2 = (1/R)^2 \quad (6)$$

That is, Angularity is approximately equal to the square of the reciprocal of circularity. Angularity reflects the shape of sand particle in three-dimension. Although circularity is obtained from two-dimensional projection of sand particle, it is based on a large statistical sample size so is enough to reflect the morphology of three-dimensional sand particles. Therefore, equation (6) can be used as a convenient way to calculate angularity, and, even more, circularity R is easier and faster to be measured by the above-mentioned imaging method.

(2) Comparison between the calculated and measured angularity

To verify the accuracy of equation (6), comparisons between the calculated and measured angularity were carried out on base sand from some typical regions and suppliers, as the results shown in Table 1. It indicates that the calculated angularity based on circularity is in good consistency with the measured values, with a maximum deviation of less than 0.16%.

Fig. 1 shows the morphology of the base sand from different suppliers list in Table.1. It can be found that the shape characteristics reflected by the overall morphology are in good consistent with the circularity and angularity listed in the Table 1.

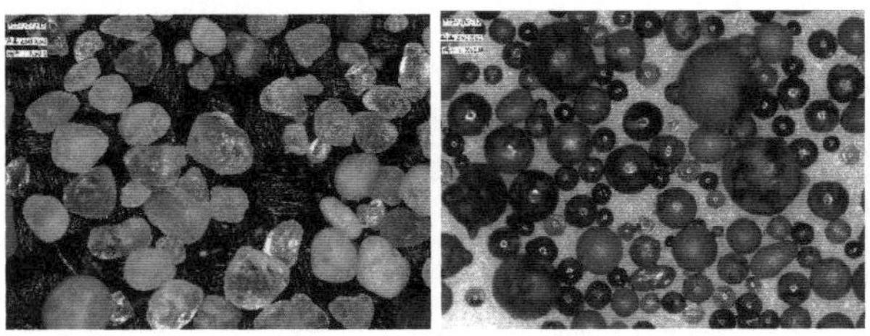

There are no (a)(b)(c)(d) marked in the figures

Fig.1 Morphology of base sand from different suppliers. (a) Hebei sand;(b) Ceramic sand (40/70); (c) Dalin sand (40/70);(d) China Building Material sand

Table.1 Circularity and angularity of sand from some typical suppliers

Base sand	AFS	Circularity	Calculated angularity	Measured angularity
Hebei sand	51.13	0.888	1.27	1.25
Ceramic sand (40/70)	44.63	0.924	1.17	1.18
Dalin sand (40/70)	35.27	0.881	1.29	1.28
China Building Material sand (40/70)	34.39	0.878	1.30	1.30

4 Conclusion

A new method for quickly determining the angularity of base sand was developed based on imaging technology and calculation. Using this method, angularity can be approximately calculated based on the relationship between circularity and angularity ($A_n=(1/R)^2$), which is in good agreement with those measured by method specified in the standard. Due to circularity can be quickly obtained by the two-dimensional imaging approach, the new method to determine angularity has the advantage of being 100 times faster than traditional methods.

References

[1] American Foundry Society. Metal casting training- sand testing. 2010.

[2] GB/T9442-2010 National Standards of the People's Republic of China, Foundry silica sand.

Cavity Forming Mechanism of Green Inorganic Foamed Sand

Yanhai Li, Tianjiao Gao, Zhiyong Yu, Pengwei Wang, Jiajun Liu, Haidong Zhang, Shaokui Yin, Rui Tan, Ruilong Yu, Ling Li
Shenyang Research Institute of Foundry Co., Ltd. CAM, Shenyang 110022, China
National Key Laboratory of Advanced Casting Technologies, Shenyang 110022, China
*Corresponding address: e-mail: 13080840917@163.com

Abstract: Green and environmentally friendly inorganic foamed sand with high strength and excellent sand cleaning performance has been obtained by modify inorganic binder system by adding foaming agent and foaming promoter. The foamed sand is extruded into a heating mold to harden and form a hollow sand mold/core. It has excellent sand cleaning performance, which can remove the sand mold through low-cost and simple equipment such as vibration and air flow, and omit the complicated sand cleaning methods such as crushing treatment, heat treatment and sandblasting treatment that have been used to simplify the casting process.

Keywords: Inorganic foamed sand, inorganic binder, foaming agent, foaming promoter

1 Introduction

For castings with complex structure, high dimensional accuracy and surface quality requirements, it is usually necessary to use complex combined sand molds/cores, such as water jackets, air inlets, and air outlets[1-2].The traditional organic binder sand mold/core has serious pollution to the environment, and it is necessary to adopt a cumbersome and complicated process method when cleaning sand. Inorganic binder has less emissions and low comprehensive cost in the production process, which is a preferred product in line with the development trend of green environmental protection casting. However, the difficulty of sand cleaning of inorganic binder sand limits its further application.

In this paper, the modified WJ-1 was used as the binder, and the modified FT-1 was added during the sand mixing. The green and environmentally friendly inorganic foamed sand mold /core with high strength and excellent sand cleaning performance was obtained by heat hardening[3-7].

2 Experimental procedure

1. Materials

Table 1. Physical and chemical properties of experimental materials

Name	Model	Appearance	Viscosity (mPa·s/ 25℃)	Density (g/cm³/25 ℃)
Inorganic binder	WJ-1	Clear shade	100-300	1.45-1.55
Additive	FT-1	Clear shade	100-300	1.15-1.20
Foaming promoter	YA-230	Yellow transparent	1.2-2.0	1.15-1.18
Foaming agent	JK-50	Clear shade	1.2-2.0	1.15-1.18

2.Preparation and detection of inorganic foamed sand

(1) The modified WJ-1: Add WJ-1 into 3-mouth flask and start the stirrer. Pour some foaming agent into it and blend slowly for 20 minutes.

(2) The modified FT-1: Add FT-1 into 3-mouth flask and start the stirrer. Pour some foaming promoter and pure water into it and blend slowly for 20 minutes.

(3) Flowability detection: The test cylinder was placed in the center of the horizontal glass plate marked with a scale, and the mixed flow sand was poured into the test cylinder and scraped flat. After holding for 90s, the test cylinder was lifted vertically, and the sand flowed naturally along the plate until it lost its flow ability.

(4) Foam duration detection: Put 100g foamed sand (100 mesh) into a glass barrel (250ml, φ60mm), record the volume and start timing. When the volume began to decrease, stop timing, and the time was the foam duration.

(5) Foam multiple detection: The ratio of foamed sand volume to ceramic sand volume.

(6) The addition of WJ-1 inorganic binder is 2.0% of sand weight, and FT-1 is 1.0 % of sand weight.

3 Result and discussion

The effect of foaming agent addition on the properties of foamed sand

Foaming agent is a surface active substance, soluble in water, can significantly reduce the surface tension of the aqueous solution, so that it is easier to wet the surface of the sand. The foaming agent is fully stirred, dispersed and produces a large number of bubbles, which are dispersed between the sand particles and reduce the friction between them[8-11]. After the hydrolysis of the foaming agent, the charge carried is aligned on the surface of the liquid, so that the surface of the bubbles and sand particles has the same charge. Due to the repulsion of the homogeneity, the polymerization force between the sand particles is greatly weakened, and the flow state is further enhanced.

It can be seen from Table 2 that the foaming agent addition has an effect on the flowability, tensile strength, residual strength, foam duration, and foam multiple.

Table 2. The effect of foaming agent addition on the properties of inorganic binders

Foaming agent addition	Tensile strength(2h/MPa)	Tensile strength(4h/MPa)	Residual strength/MPa	Flowability/mm	Foam duration/min	Foam multiple
0.5%	1.85	2.52	0.25	45	20	1.1
1.0%	1.68	2.49	0.24	58	22	1.4
1.5%	1.52	2.33	0.19	60	22	1.5
2.0%	1.66	2.34	0.11	72	28	1.6

2.5%	1.60	2.28	0.10	78	29	1.85
3.0%	1.52	2.10	0.08	80	32	2.12

The addition of foaming promoter during this experiment was 1.0 % of the FT-1. With the increase of the addition of foaming agent, the flowability increases, the volume of foamed sand increases, the foam duration increases, the foam strength increases, and the residual strength decreases. 2-hour and 4-hour tensile strength decrease first and then increase. After reaching a certain peak, they decrease again. Through the results, the optimal addition of foaming agent is 2.0 % of the WJ-1.

The effect of foaming promoter addition on the properties of foamed sand

The foaming promoter can improve the speed of foaming, increase the amount of foam, make the foam fine and uniform, and improve the duration of foam. After shooting the core, the hollow core can be formed in a short time after heating.

Table 3. The effect of foaming promoter addition on the properties of inorganic binders

Foaming agent addition	Tensile strength (2h/MPa)	Tensile strength (4h/MPa)	Residual strength /MPa	Flowability /mm	Foam duration /min	Standard '8'-sand sample weight/g
0.5%	1.66	2.34	0.21	72	27	75
1.0%	1.68	2.34	0.15	73	28	73
1.5%	1.66	2.33	0.11	75	30	70
2.0%	1.60	2.10	0.08	78	30	69
2.5%	1.55	2.00	0.05	80	26	68
3.0%	1.42	1.98	0.03	81	24	66

With the increase of the amount of foaming agent promoter, the flowability increases, the foam duration increases first and then decreases, the residual strength gradually decreases, and 2-hour and 4-hour tensile strength decrease first and then increase to a certain peak. Through the results, the optimum addition of foaming promoter is 1.5 % of the FT-1.

Application

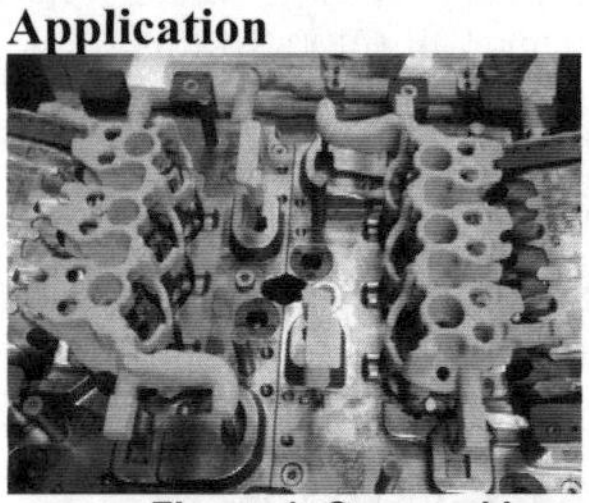

Figure 1. Core-making

Figure 2. Vibration cleaning

The tensile strength was 1.60 MPa for 2 hours, 2.25 MPa for 4 hours, and the residual strength was 0.10 MPa. The surface quality and dimensional accuracy met the design requirements, and there was no residue after vibration cleaning.

4 Conclusion

The green and environmentally friendly inorganic foamed sand with high strength and excellent sand cleaning performance can be obtained by using WJ-1 inorganic binder modified by foaming agent and FT-1 additive modified by foaming promoter. The optimum addition of foaming agent is 2.0 % (of the WJ-1), and that of foaming accelerator is 1.5 % (of the FT-1). Compared with the traditional water-glass sand mold/core, the foamed sand mold/core after casting has excellent sand cleaning performance, which can remove the sand mold by low-cost and simple equipment such as vibrating sand cleaning, so as to simplify the casting process.

References

[1] Lai L D, Zhang T L. Study of foam drainage agent based on g-C3N4 nano sheets reinforced stabilization. 2017 International symposium on resource exploration and environmental science (REES 2017). Volume 657 Part A. DOI: 10.1016/j.colsurfa.2022:130607

[2] Qiao S Y; Liu Q W. Synthesis and analysis of foam drainage agent for gas well in Jilin Oilfield. 2017 International symposium on resource exploration and environmental science (REES 2017). Volume 64. DOI: 10.1088/1755-1315/64/1/012025

[3] Li Z C; Zhang A J.Synthesis of compound foaming agent of lightweight foamed doncrete. 3rd International Conference on Chemical Engineering and Advanced Materials (CEAM 2013).Volume 785-786: 163

[4] Hou L; Li J.Influence of foaming agent on cement and foam concrete. Construction and building materials. Volume 280. DOI: 10.1016/j.conbuildmat.2021.122399

[5] Qiu Y Q;Liu Y. Development of fiber compound foaming agentand experimental study on application performance of foamed lightweight soil. Applied rheology. Volume 33 Issue1. DOI: 10.1515/arh-2023-0108

[6] Sun Y Q; Zhang Y P. Research Progress on New Highly Efficient Foam Drainage Agents for Gas Wells (A Review). Peteroleum chemistry. Volume 63 Issue 9: 1119-1131. DOI: 10.1134/S0965544123080029

[7] Wang J N;Xu K D. Optimal type and ratio of foaming agent foaming stabilizer and fluxing agent of foam glass fabricated with red mud and coal gangue. Journal of ceramic processing research. Volume 23 Issue 1: 79-85. DOI: 10.36410/jcpr.2022.23.1.79

[8] Chen M; Giorgetti M. Effects of temperature on the properties of the new gel foam. Advances in machinery materials science and engineering application. Volume 24: 95-103. DOI: 10.3233/ATDE220423

[9] Falliano D;Restuccia. LA simple optimized foam generator and a study on peculiar aspects concerning foams and foamed concrete.Construction and building materials. Volume 268. DOI: 10.1016/j.conbuildmat. 2020.121101.

[10] Choi J;Lee H. An experimental study on the development of structural lightweight concrete using micro foam agents. Advances in fracture and damage mechanics VI. Volume 348-349: 469. DOI: 10.4028/ www.scientific.net/ KEM.348-349.469.

[11] Denkov N;Tcholakova S. Physicochemical control of foam properties. Current opinion in colloid & interface science. Volume 50. DOI: 10.1016/j.cocis.2020.08.001

Research Progress of Modified Furan Resins

Jincheng Wang[1], Xuyi Tao[2], Ling Yu[1], Wenying Zhu[1], Yangbin Shen[1,2*]
1. Suzhou Sinye Material Technology Co., Ltd, Suzhou, China
2. School of Materials Science and Engineering, Suzhou University of Science and Technology, Suzhou, China
*Corresponding address: e-mail: syb1989@126.com

Abstract: Modified furan resin is a green binder which could widely applied in foundry field. This resin contains furan ring, could replace non-renewable petrochemicals. So modified furan resin has potential to reduce the usage of fossil materials. It is an alternative product of traditional furan resins, exhibits great development potential and socio-economic value. In this paper, we will introduce our recent progress in the synthesis of furan resin by using modified compounds, such as 5-hydroxymethylfurfural. We will also share the achievement of some production process for this resin.

Keywords: modified furan resin; process condensation;

1 Introduction

Traditional molding binding agent of furan resin is made from petroleum-based furfuryl alcohol, formaldehyde, urea and benzaldehyde. Formaldehyde and phenol are non-renewable petrochemicals; Modified polymer materials are the new polymer materials made from natural renewable resources, such as protein, starch, cellulose, vegetable oil, etc. They are green and renewable materials. The modified furan resin is green and environment-friendly materials with excellent performance. It has better biodegradability, and could avoid secondary pollution. There is basically no harm to the environment in the total life cycle that contains raw materials, production, use and final abandonment. Biomass materials have the advantage of abundant functional groups, wide sources and low price. This paper will mainly introduce the research progress and existing problems in production of modified furan resins from 5-hydroxymethylfurfural.

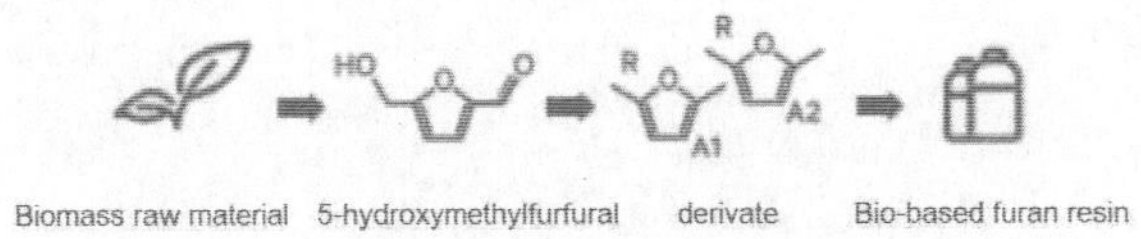

Figure 1 Preparation of modified furan resin.

2 Result and discussion

Preparation of modified furan resin

Our company will share various modification technologies based on furan resin used in the casting, and study the mechanism for producing derivatives of 5-hydroxymethylfurfural. We also optimize the ingredients of modified furan resin, increase the proportion of modified compounds (5-hydroxymethylfurfural, lignin, xylitol, phytophenols, etc.) in resin, which could improve the performance of furan resin in a wide temperature range. The Figure 2 shows the production process of modified furan resin. The performance of modified furan resin prepared by the above process is shown in Table 1.

Table 1. Performance of modified furan resin.

Entry	Index	Performance Data	Modified furan resin	
			Alloy steel molding	Alloy iron molding
1.	Environmental performance	VOC, g/L, ≤	50	50
2.		Char Yield, ≥%	56	56
3.		Free Phenol, ≤%	0.1	0.1
4.		Free Formaldehyde,	≤0.1%	≤0.1%
5.		Nitrogen content,	≤0.3%	<2~5%
6.	Room-Temperature Performance	Viscosity, mPa.s,20℃	50	60
7.		Room-Temperature tensile strength after 24h,	2.5≥Mpa	2.8≥Mpa
8.		The temperature of 5% weight loss℃,	≥260	≥280

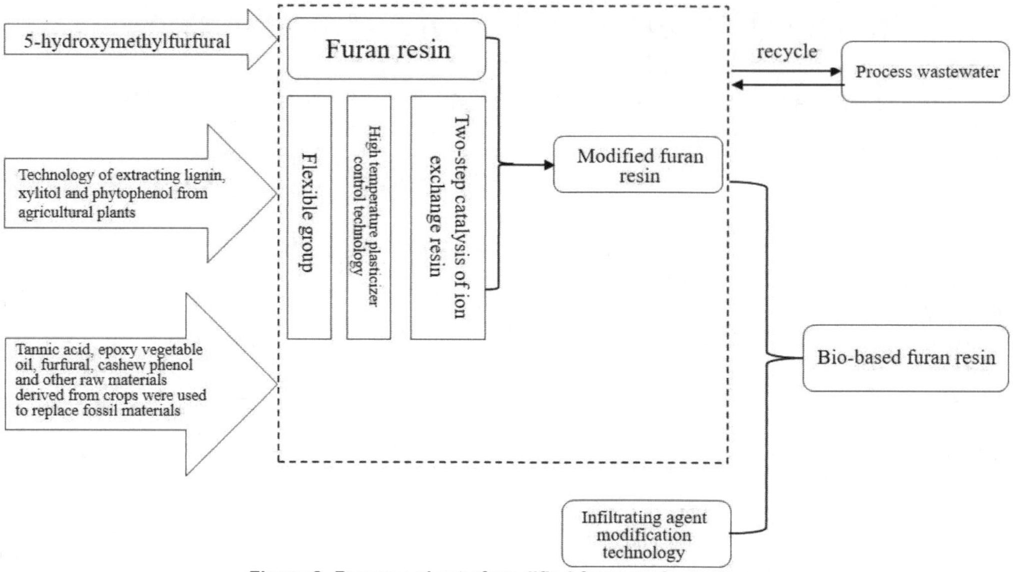

Figure 2. Process chart of modified furan resin system.

Various bio-acids are employed in the production process of modified furan resin. Some natural acids could act as active monomers and catalysts. Moreover, some flexible organic groups and infiltrating agent are added in the polymers which could improve the performance of resin. As exhibited in Table 2, the modified furan resin has higher performance compared with conventional one.

Table 2. Comparison of modified and conventional resin

Entry	Index	Performance Data	Type of Resin	
			Conventional	Modified
1	Environmental Performance	VOC, g/L, ≤	300	50
2		Char Yield, ≥%	50	60
3		Free Phenol,	<0.3%	/
4		Free formaldehyde,	<0.3%	<0.01%
5	Room-Temperature Performance	Room-Temperature Tensile Strength After 24h,	≥1.8Mpa	≥2.8Mpa
6	High-Temperature Performance	Load Thermal Deformation, 500℃/60s, Mm	≤2.5	≤1.0
7		Thermal Deformation Duration, S	100～110	90～100

3 Conclusion

Biomass material is made from sustainable and abundant resources, which has the advantage of low price and bio-degradation. It could produce modified furan resin, then replace petroleum, coal and other fossil resources. It shows us an effective way to solve the current resource crisis and protect the environment. Benefited from its green and environmental resource, modified furan resin can partially, even completely replace petroleum-based furan resin. Thus, it has become a rapidly growing emerging industry, and widely applied in sand casting production. While modified furan resin is a new material, its structure, preparation process and application method are not completely understanded. Therefore, it is urgent to increase the proportion of biomass materials in the modified furan resin, develop high-performance and low-cost furan resin. We share some production process for the novel biomass materials, and we find such new materials could exhibit outstanding performance as casting binder.

References

[1] Chu Fuxiang. New modified polymer materials[D], Internet Resources, 2021.

[2] Yin Shaokui, Zhou Jingyi, Tan Rui et al.Molding material[J].Casting equipment and technology, 2019(05): 60-64.

[3] Gong Yubo. Study on the preparation of furan resin modified by lignin[D]. Shenyang University of Technology, 2022.

[4] A 3D printing furan resin preparation method[P].Chinese patent：CN107778491A.2022.02.15

[5] Xiong Gaohu.Study on the properties of cashew phenol modified furan resin[J]. Cast, 2019, 68(09):1005-1008.

New Solution on Invaded Gas Hole Formation from Green Sand Process

Zefen He[1,*], Jun Ma[1], Tongshun Liu[2], Xiaoli Dai[2]
1. Clariant Chemical Technology(Shanghai) Co., Ltd, No. 388, Chunde Road, Shanghai, China
2. Yinfeng Foundry (Wuhu) Company Ltd, Wanzhi, Wuhu, China
*Corresponding address: e-mail: Zefen.he@clariant.com

Abstract: Gas blister or invaded gas hole is one of the public defects for castings made from green sand process, which forms through the interplay of gas development and permeability [1]. To investigate the new solution on improvement of gas defect, one anti veining additive on green sand properties has been studied in Clariant Zhenjiang Lab. The results indicate, the casting surface finish has been improved even eliminating coating at 0.7% addition by weight of sand for core making, while the gas defect percentage for casting increases as well.

Keywords: Gas blister, green sand, anti-veining additive, green sand properties, invaded gas defect

1 Introduction

Green sand is one of the most economical and environmental molding sands, which has been widely applied to produce kinds of volume production for small and middle size castings, embracing such advantageous as high efficiency, low cost as well as good environmental compatibility.

Gas hole is one of the common defects for ductile iron production with green sand process, which is always forming during pouring and solidification. There are a lot of physical, chemical factors as well as interplay factors on this defect. Normally our foundry engineer shoots this defect by reducing moisture, active clay, total clay, loss on ignition, gas development of green sand, reducing influx of organic core sand. It rarely had articles on how the anti-veining additives from core does influence on gas defect.

The physical properties of one brand additives for Yinfeng application has been tested, the relative effects on green sand properties have also been studied, the relationship between the dosage and rate of casting has been observed as well. defect is investigated in this article, new solution has been founded through matching the solidification time with proper gas evolution and gassing speed.

2 Experimental procedure

The physical properties of anti- veining additive have been tested with the facilities from Clariant Lab, and the returned sand was simulated by the core sand influx into green sand in accordance with the additive dosage implemented at Yinfeng core shop, 0.4% and 0.7% by weight of core sand are studies, 120kg core sand influx into green sand during differentiate case production.

3 Result and discussion

Dosage of additive on Casting defect

Gas defect free for casting with 0.4% additive for core, while burn on occurring without coating, 5% gas defect occurring with core sand with 0.7% additive, no any penetration was founded as fig. 1.

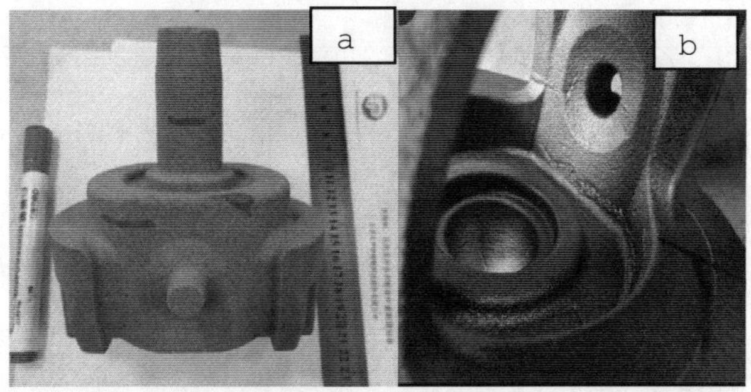

Fig. 1 Defected casting and core(a)Core with 0.7% additive;(b) defected casting.

Effect of additive on green sand properties

Table1. Effect of additive on properties

	Permeability	GCS	WTS	DCS	Gas
	$m^3/m^2\cdot min$	Kpa	Kpa	Kpa	mL/g
Without additive	189	189	4.7	206	9.8
With 0.7% additive	176	192	4.7	214	11

It observed the additive reduces the dry compressive strength and gas evolution, while almost no influence on green compressive strength, wet tensile strength and permeability.

Effect of additive on gas evolution and velocity

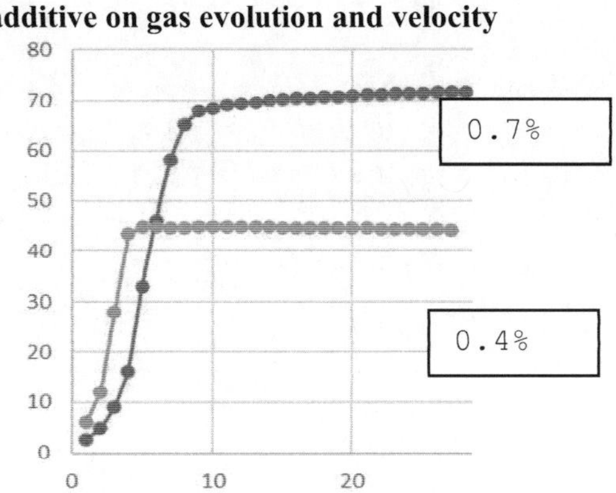

Fig.2 Additive on gas evolution

It understood with the increasing of the additive, the gas evolution is higher, while the gas velocity keeps same.

Effect of additive dosage on casting defect

It reported the gas defect rate is high as 5% at a dosage of 0.7% additive, casting surface is acceptable without coating, while zero gas defect at a dosage of 0.4% while needs coating to avoid penetration.

4 Conclusion

Additive dosage below 0.7% doesn't create gas defect while rough surface finish obtained without coating.

Additive dosage above 0.7% improve surface finish without coating while 5% of gas defect occurring.

Anti veining agent reduces hot compressive strength and no obvious effects on other green sand properties.

Proper additive dosage could balance casting surface finish and gas defect.

Acknowledgments

This study has got fully support from Clariant Bentonite (Jiangsu) Company Lab and Yinfeng (Wuhu) Company Ltd.

References

[1] Werner Tilch, Michael Franke, Handbook on bentonite-bonded moulding materials. P 279.

The Application of CCITEK® Inorganic Technology for Foundry Sand

Ying Xiong*, Weisong Luo
Chongqing Changjiang River Moulding Materials (Group) Co., Ltd., Chongqing, China
*Corresponding address: email: xiongyin@ccrmm.com.cn

Abstract: The CCITEK® Inorganic coated wet sand is a casting (core) sand product based on the CCITEK® inorganic binder technology. The core temperature is low, and it belongs to the warm core box coated sand using inorganic materials in the current casting industry.

In this paper, through the introduction of the technical principle and the application of the technology and the sand process and the application effect of sand mixing, core making and casting, the author shows that this new inorganic coated wet sand technology has the environmental protection characteristics of dust-free, smokeless and tasteless in the whole life cycle.

The application of this technology has no pollution to the environment, zero harm, can reduce the input of foundry equipment, reduce energy consumption, solve the defects of orifice, shrinkage and expansion of casting, improve the yield of casting, and improve the production efficiency.

Keywords: inorganic binder; sand; waste sand regeneration; environmental protection; foundry

1 Introduction

CCITEK® inorganic coated wet sand technology is independently developed by Chongqing Changjiang River Moulding Materials (Group) Co., Ltd（hereafter simply called: CCRMM, with more than 10 related patents, with full intellectual property rights.

Changjiang Materials has raised funds, established internal projects, planning and research, and has been put into practical application since 2012, and completed the new product appraisal and acceptance of Chongqing Municipal Government Economic and Information Technology Commission in the same year.

After more than ten years of development and application optimization, it has formed a mature and reliable original technology system, which is a new green sand technology with unique technical characteristics. It has a broad application prospect.

2 Experimental procedure

The basic technical characteristics of CCITEK® are based on the research of existing inorganic cementitious materials, using other non-homologous inorganic gel materials to modify them, and a new silicate inorganic gel technology route that breaks away from traditional water glass technology has been determined. And this unique composite modification technology is extended to the phosphate inorganic cementing system, achieving a new technological innovation of water-based inorganic binder technology with two major categories of silicate and phosphate inorganic cementing structures.

The binder: component one, liquid, its composition content is shown in Table 1. The binder: component two, powder, it's element content are shown in Table 2.

Table 1. The composition content of component one

Composition	Silicate	Chemical doping components	Modified & additives, water	PH	Viscosity（s, 25℃, 4# flow cup）
Content	24%-27%	6%-17%	59%-67%	13-14	12-17

Table 2. The element content of component two

Element	SiO_2 %	Al_2O_3 %	Others %	Water content %	LOI %	Loose accumulation density g/cm³	Degree of dispersion
Content	70-80	≥12	2-3	≤1	≤6	0.4-0.6	≥5%

CCITEK® inorganic wet coated sand technology is a new sand casting technology developed based on inorganic cementing technology and five key technologies. The essence of the inorganic cementing technology is the modification of composite material technology, five key technologies include the key technology of preparation and evaluation, inorganic wet sand preparation and evaluation, core (type) pouring and evaluation of key technology, inorganic waste sand regeneration and evaluation of key technology, etc.

Recommended dosage of binder: component one, liquid 1.8-3.0% (sand amount basis); component two, powder 40% (component one, liquid amount basis).

Refractory aggregate： Baking silica sand, Dry silica sand, Recycled sand, etc.

The main technical performance of CCITEK® inorganic wet coated sand are shown in Fig.1,2.3,4,5 and 6.

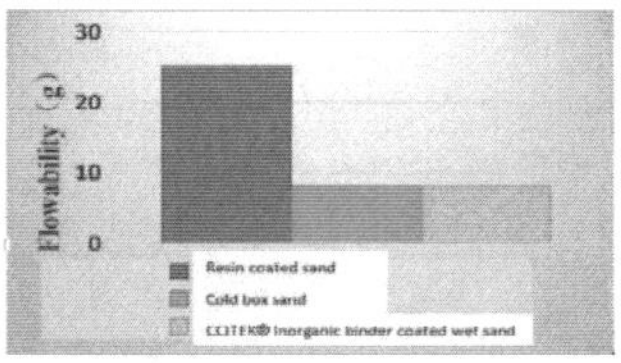

Fig.1 Comparison of fluidity of several kinds of coated wet sands

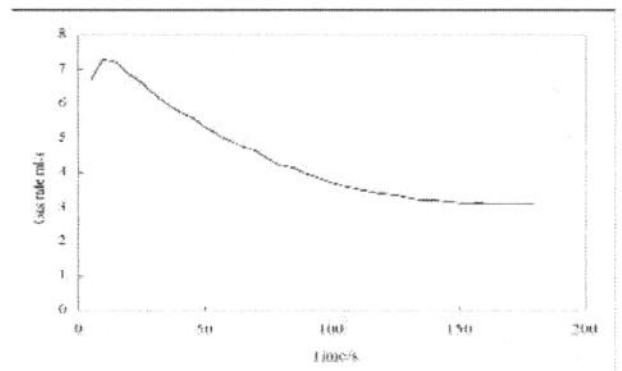

Fig.2 Gas evolution of inorganic wet coated sand

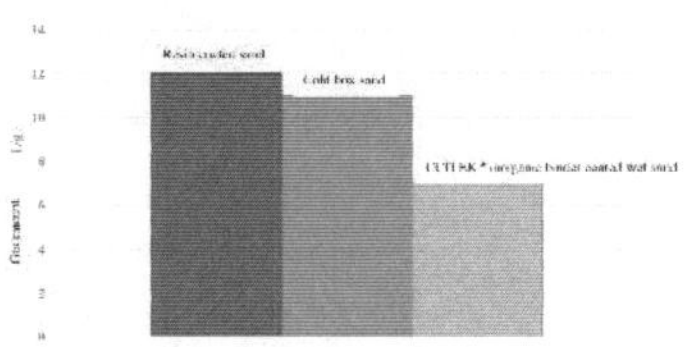

Fig.3 Comparison of the maximum gas output of several molding materials

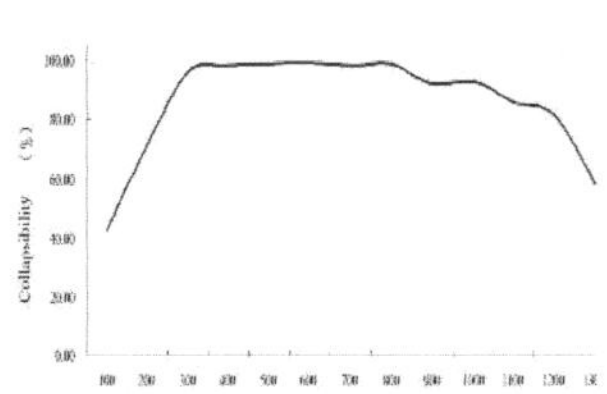

Fig.4 Collapsibility of CCITEK® inorganic wet coated sand at different temperatures

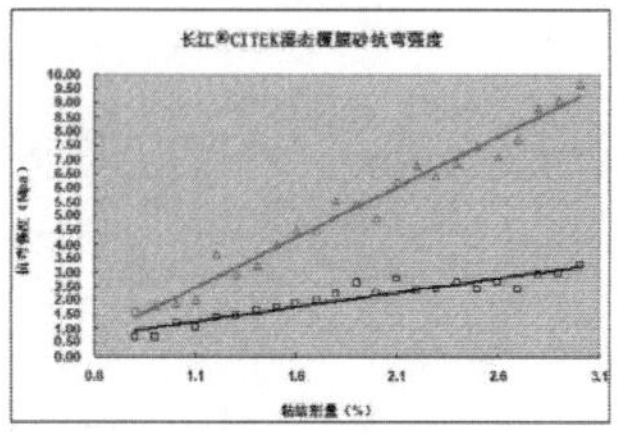

Fig.5 The relationship between bonding dose and strength

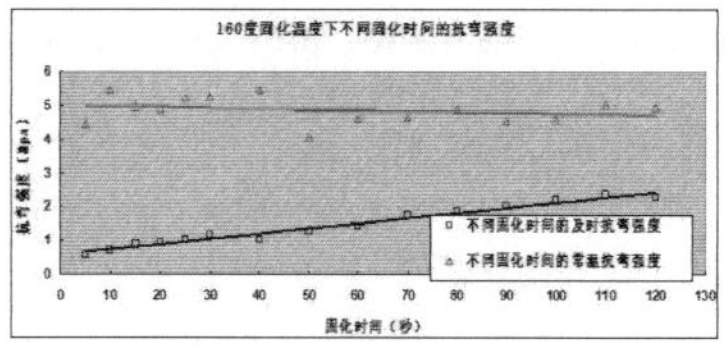

Fig.6 The relationship between curing time and strength

Core curing behavior mainly in the hot sand caused more system of chemical polymerization between inorganic materials, makes the interface of inorganic cement gradually adhesive curing, the formation of inorganic film high shear strength between sand consolidation into sand (core), at the same time in the process of sand curing only inorganic gel solvent phase-water in the form of water vapor, the whole core link only water vapor. The curing process is shown in Fig.7.

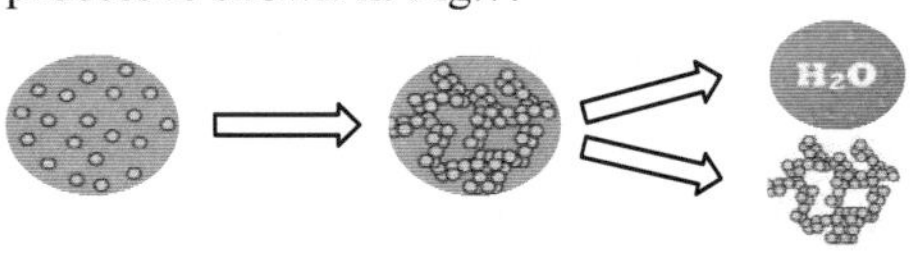

Fig.7 Schematic diagram of curing

The curing of inorganic binder belongs to the chemical curing mechanism triggered in a hot environment, which is different from the sand mixing method and curing conditions of other traditional molding sands. Therefore, the preparation technology of wet sand is completely different from the traditional preparation technology of other molding sands, which is reflected in the following aspects, as shown in Table 3.

3 Result and discussion

CCITEK® coated wet sand, which based on CCITEK ® inorganic binder technology, is a casting material (core) sand product The core making temperature is low (150℃~200℃). The core making and casting are smokeless, tasteless and non-toxic, which is an environmentally friendly product. It is especially suitable for the use of core aluminum casting of non-ferrous metal casting, to solve the pores, shrinkage, veins, holes, expansion defects, improve the strength of casting parts, solve the collapse of sand core, and improve the casting yield, improve production capacity, reduce the comprehensive cost, but also suitable for some simply shell for iron casting core making.CCITEK ® inorganic binder is founded on Five key technology systems, that is , key technologies for the preparation and application evaluation of binder, key technology for wet sand preparation, sand (shell) core engineering and equipment key technology, key technology of casting application and management, inorganic waste sand regeneration process and equipment key technology. In Fig.8 and Fig.9 inorganic waste sand regeneration process and equipment are shown.

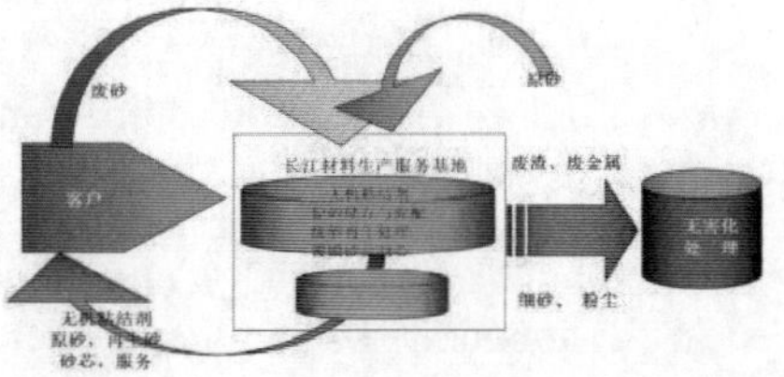

Fig.8 Inorganic waste sand regeneration process

Fig.9 CCRMM's CZS Re generation Equipment

4 Conclusion

CCITEK ® wet-coated sand is a casting (core) sand product based on 长江 ® CCITEK ® inorganic binder technology. It has a low core-making temperature and is currently the warm core box coated sand using inorganic materials in the foundry industry.

This technology product has the characteristics of being smokeless and odorless during core making and casting, does not pollute the environment, has zero harm, can reduce foundry investment, reduce energy consumption, and improve production efficiency; Solved the defects of casting porosity, shrinkage, and expansion, and improved the casting yield.

Acknowledgments

I would like to thank all the leaders and colleagues for their support and help for the preparation of the paper and experiment.

References

[1] Xiong Jie, Technical research and industrial application of inorganic sand, Foundry engineering, 2023(1): 59-63.
[2] Xiong Ying, Xiong Jie, Han Yue, Application of new inorganic wet state coated sand technology, 2013China foundry industry high-level forum.
[3] Luo Weisong, Interpretation of the industry standard "Inlet binder coated wet sand for non-ferrous alloy casting", 2017.
[4] Wang Lifeng, A regeneration method for silicate, 2020 Chongqing Casting Annual Conference.

Table 3 Comparison of inorganic coated wet sand with cold box resin sand and self-hardening sand

	Inorganic coated wet sand	Cold box resin sand	Self-hardening sand
Binder	Water-based low viscosity liquid	Two-component liquid	Viscous liquid
Curing agent	Ultrafine powder	Triethylamine	Liquid
Curing method	Hot (blowing hot air) curing	Blowing triethylamine gas	Room temperature self-curing

Technology for Full Utilization of Foundry Solid Waste and Its Applications

Chunlei Liu
Guangxi Lanco Resource Recycling Co., Ltd., Yulin 537000, Guangxi, China

Abstract: Foundry solid waste has become more and more foundry enterprises must solve the problem, the traditional landfill is not friendly to the environment. In this paper, the full utilization technology and application of foundry solid waste, including foundry sand, dust and slag, are discussed. Foundry waste sand can be divided into organic - type, inorganic- type and mixed-type according to its binder type. According to its type, the foundry waste sand can be recycled by mechanical regeneration, thermal regeneration, wet regeneration, combined thermal and wet regeneration and micro-wet regeneration. Dust and slag can be comprehensively used as building materials and organic fertilizers. There are many kinds of building materials that can be used as building materials, such as permeable bricks, light fireproof boards, mineral rigid plastics. The full utilization technology of foundry solid waste has been applied in large-scale batches in factories, which greatly alleviates the environmental anxiety caused by foundry solid waste.

Keywords: Foundry solid waste; Full utilization technology; Regeneration of foundry waste sand; Dust and slag; Environmental Protection Technology

1 Introduction

Foundry solid waste, mainly including waste sand, dust, and slag, is an inevitable byproduct in the foundry production process. These solid wastes,If not properly handled,not only occupy a lot of land resources,but also may cause long-term environmental pollution.Therefore, it is of great significance to develop efficient and environmentally friendly technology for the full utilization of foundry solid waste to achieve the green transformation of the foundry industry.

2 Full utilization technology for foundry solid waste

The full utilization technology of foundry solid waste involves multiple stages such as waste sand regeneration, dust and slag utilization. In order to achieve full utilization of solid waste, it is necessary to integrate all kinds of regeneration technologies and comprehensive utilization methods to construct a perfect solid waste treatment and utilization system. The specific process includes mechanical regeneration of waste sand, thermal regeneration, wet regeneration, and new building material conversion of dust and slag.

Foundry waste sand is divided into silica sand and non-sand substances according to the different uses of the substances it contains. Silica sand is produced through regeneration technology to produce reclaimed sand, which is recycled in the foundry to achieve this cycle; Non sand materials (inorganic non-metallic materials) are mainly refractory materials, which are produced through advanced molding technology to produce fire-resistant building materials and achieve secondary cycle.

3 Foundry waste sand recycling technology

Foundry waste sand can be divided into inorganic , organic , and mixed types according to the type of binder. Different types of waste sand have different physical and chemical properties, so different regeneration technologies need to be used for treatment.

4 Dust and slag utilization technology

As important components of foundry solid waste, dust and slag have rich chemical composition and unique physical properties. Through comprehensive utilization of technology, dust and slag can be converted into a variety of building materials products, such as permeable bricks, lightweight fireproof boards, mineral rigid plastics, and so on. These building materials products not only have excellent physical properties, but also have fire prevention,environmental protection and other advantages,can be widely used in construction and other fields.

5 Industrial application cases

According to the example of comprehensive utilization and discharge treatment of waste sand which our company has been engaged in for many years, the total reclamation of waste sand must embody high level comprehensive utilization and minimum environmental impact

6 Conclusion

Technical feasibility: foundry waste sand regeneration technology and the comprehensive utilization technology of dust and slag have a high level of technological maturity and can be applied on a large scale in factories.

Environmental benefits: Full use of foundry solid waste can significantly reduce the waste of land resources and environmental pollution, which is of great significance for protecting the ecological environment.

Economic benefits: By recycling foundry solid waste, enterprises can reduce raw material costs, improve product quality, and reduce reject rates, thereby achieving an increase in economic benefits.

References

[1] Hu Pengsheng, Editor in Chief, Sand, Shanghai: Shanghai Science and Technology Press, 1994.

[2] Edited by Guo Jingchun et al., Recycling and Pollution Control of Casting Old Sand, Guangzhou, 2001.

[3] Dai Weiping, Wet regeneration and recycling of foundry waste sand, Pan Pearl River Delta Foundry Forum paper, 2013.

[4] Sandra Bunker et al.: "Regeneration, dust and concretization of clay sand and their effects on sand properties and casting quality". (German) "Casting Practice" 2014 Issue 1.

Green Foundry Technical Comprehensive Solution —The Cross-Border Applications of New Materials and Process

Songhai Xue [1,2*], Wenlong Chen [1,2]
1. Haixi（Fujian）Institute,China Academy of Machinery Science&Technology Group
2. Quanzhou Innovation Center，China Academy of Machinery Science&Technology Group
*Corresponding address: e-mail: xuesh@cam.com.cn

Abstract: "Green Foundry"as a new name which is an environment- friendly Foundry Technical Comprehensive Solution with the meaning of lower emissions, higher casting yield and the best cost-effective molding process with service package. The systematic balance of foundry sand system in foundry is the key point of the "Green Foundry" which covers two issues of new silicon-based materials with the corresponding molding process in a broad sense.From the systematic molding sand recycling equipment to the application with this new silicon-based material additives in the molding process corresponding to the molding sand system in foundry, this paper payed more attention on the technical node of reclamation and sustainability optimization of molding sand system.New materials emphasis this new additives with micron-sized fused quartz powder can improving the application performance of reclaimed sand. Another,New molding process with equipment, this paper expounds the process node of sand recycling equipment combined with new additives in order to achieve the best process balance and cost-effective in foundry.This article is more objective to clarify the foundry industry in the shape of raw/auxiliary materials,sand treatment equipment and process control is an inseparable whole. It need to be recognized in a specific process environment with cross-border ideas to discover the technical commonalities in different industries.

Keywords: Green Foundry; molding sand; silicon-based material; sand mold additives; foundry sand system; micron-sized fused quartz powder; environment- friendly

1 Introduction

The molding process in the foundry industry is developing around the sand system, and the sand molding process is composed of three main molding materials: foundry sand, binder and coating. Sand treatment equipment in the traditional sense is composed of sand recycling equipment and related auxiliary machinery to form a complete set of systems. Therefore, materials and equipment are the two key elements of the sand molding process. In recent years, green foundry mainly lies in the effective utilization of the integration of molding materials and sand treatment equipment to achieve the cost-effective comprehensive indicator.With the promotion of environmental awareness of foundries all around the world,the solution of lower VOC emission and molding cost is the double factor of foundries. Based on the balance of economy and technology while facing a variety of molding process on the dimensions of materials and equipment, the introduction of cross-industry common cognition for foundry enterprises to achieve a better sand system solution for the case and relevant data analysis in order to promote the productivity of foundry industry and reducing costs and increasing efficiency. The foundry used sand recycling equipment is discussed in cross-industry cases related to thermal and wet method combined with Industrial water treatment sewage treatment and thermal cycling. The new silicon-based additives get relevant inspiration from EMC packaging industry and applied in the foundry industry.

2 Experimental procedure

Taking inorganic sodium silicate molding process as an example, wet reclamation is the most suitable method so far, and the practice has proven that the reclaimed sand can meet the needs of the whole sand molding process.The sodium silicate used sand samples that has mixed with 5000 mesh (the average particle size is 2.5 μ m) fused silica powder as a test object. The optimal reclaimed result was achieved under the conditions of distilled water with the corresponding temperature, soaking and scrubbing retention time. Hot water in the whole testing process can accelerate the dissolution of silicate binders to effectively reduce the residual Na_2O index of the reclaimed sand.After the test, the recycled water is simulated by the condenser for MVR water treatment in the sewage treatment industry and the index of recycled water reaches the effluent standard.The heat source required for water heating in wet reclamation and MVR sewage treatment comes from the reuse of waste heat in thermal reclamation system.The heat source of the wet reclamation process in the thermal and wet process linkage line comes from the waste heat of the hot calcinator and water cooling system dedicated for the used sand reclamation of the organic binder sand molding process.The heat source of the wet reclamation simulation testing came from the constant temperature electric heating under the same testing conditions with the process of soaking in hot water at 60 ℃ for 5 minutes and scrubbing at different mortar concentrations for 5 minutes.

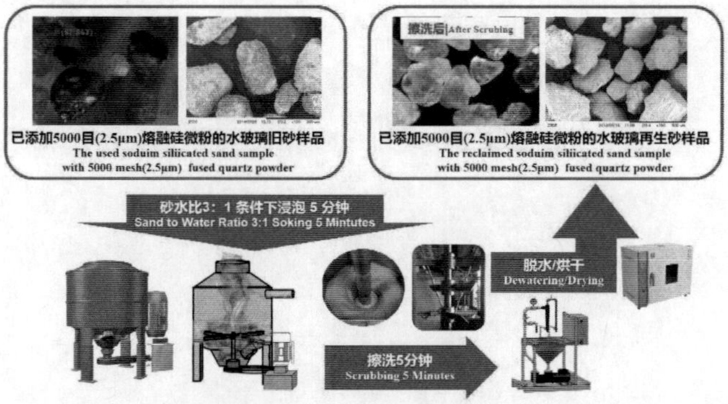

Figure 1 Full Test Process Analysis

3 Result and discussion

The corresponding inorganic sodium silicate used sand reclamation testing process simulated the advanced sewage treatment technology. After the sewage in this new process is treated by server main processes such as precipitation，flocculation and neutralization, the chroma of clean water chroma less than 60, then through the MVR technology to obtain pure circulating water in order to achieve near-zero emission of sewage and Solid.

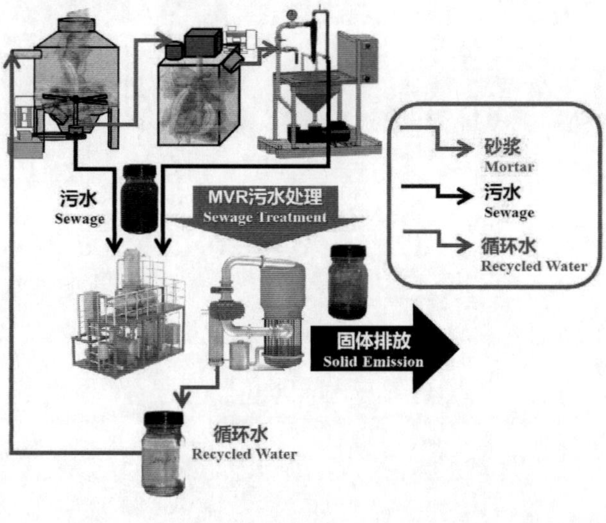

Figure 2 Full Sewage Treatment Process Analysis

Based on the basic strength analysis of four sand samples concluded that the reclaimed sand with new additives of fused silica powder has the best performance under the same testing conditions. It means that fused silica powder can form a dense ceramic sintered layer covered the silica sand grain when meet temperature above 1200℃ in order to improve the molding sand strength.And it can make a perfect bridge between binder and sand particle.The follow Table 1 and Figure 1 shows basic performance of these four initial strength，24h strength and retained strength. So as a new cross-border silicon-based materials can be applied in our foundry industry. It will have a good potential value.

Table 1. Basic Performance Parameters of Sand Sample

Sand Sample	Initial Strength /MPa	24h Strength /MPa	Retained Strength /MPa
New Sand	0.461	1.892	1.822
Used Sand	0.045	0.339	1.576
Reclaimed Sand	0.618	1.529	1.573
Reclaimed Sand（Additives）	0.652	2.601	1.936

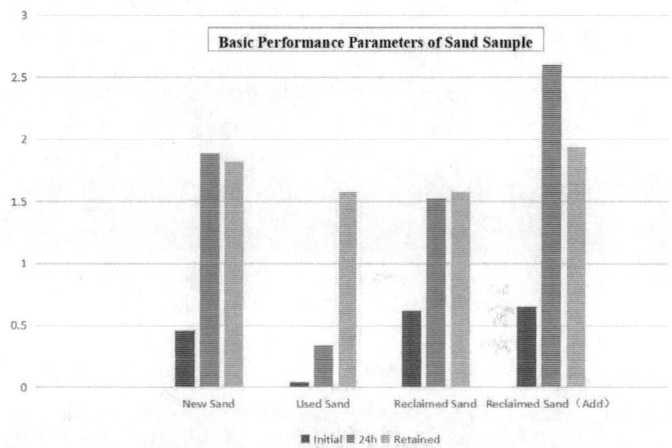

Figure 1 Effects of Reclaimed Sand with New Additives

4 Conclusion

Foundry Sand Treatment System that combined with new process and materials need a cost-effective balance point and Standardized processing environment. Because flexible raw and auxiliary molding materials, we should find a green and universal sand molding solution face different foundries.

Acknowledgments

The work was supported by 2024 75th World Foundry Congress, World Foundry Organization and Foundry Institution of Chinese Mechanical Engineering Society .

Exploration and Practice of Avoiding "Engineering Scientifization" in Undergraduate Vocational Education

Jun Fu*, Xianru Mi, Jiye Wei, Mei Yin, Xinyu Zeng, Lingyue Wei
School of Materials Engineering, Sichuan Polytechnic University, Deyang, JingYang District,Taishan South Road 801, Sichuan, China
*Corresponding address: e-mail: fujun@scetc.edu.cn

Abstract: Engineering has independence. Engineering is not the application of science and technology, nor is it a subsidiary of them. The current engineering education system needs to be reformed. Engineering activities integrate a large number of engineering elements with different properties into a specific functional engineering entity, and must be organized according to the theories and methods of systems science. We need to strengthen the basic training of science subjects such as mathematics, physics, and chemistry for engineering students. Avoid scientificization in engineering evaluations. COMET evaluates technical solutions from eight aspects: intuitiveness, functionality, value orientation, economy, work process orientation, social acceptance, environmental friendliness, and creativity, in order to evaluate the functional abilities of the subjects. All of these provide an implementation path to avoid the "engineering oriented" approach.

Keywords: undergraduate vocational education; engineering ont ology; COMET

Introduction

Based on the epistemology that "technology is the application o f science, and engineering is the application of technology", the current order of the training system for engineering majors at all levels and types of vocational education is: scientific foundatio n - technical foundation - engineering foundation - graduation p roject. People firmly believe in this and rarely reflect on it [1][2].

In fact, engineering has independence. Engineering is not an application of science, nor is it a subsidiary of science. From a h istorical perspective, engineering activities existed in primitive s ocieties, while scientific activities only had a brief history of sev eral thousand years. For a long time in history, engineering acti vities did not rely on the discovery of scientific principles, and e ven the first industrial revolution was not the result of the "push " of scientific theories.

In recent years, some people have suddenly been surprised to find that there is a phenomenon of "detachment from reality to v irtuality" and "engineering and science oriented" in vocational e ducation, and these engineering graduates cannot meet the need s of engineering construction. Personnel from various industries have engaged in a series of discussions and come up with ideas. Vocational undergraduate education, in particular, should overc ome the inertia of "undergraduate education" and take effective measures in the process of talent cultivation to avoid the emerge nce of "engineering and science oriented".

1 Cultivation of science knowledge for engineering students

For engineering students, scientific knowledge is an indispensab le foundation for mastering engineering science, and can only b e strengthened rather than weakened. Science refers to the colle ctive term for formal science and natural science. Mathematics belongs to formal sciences, while natural sciences include physi cs, chemistry, biology, and so on. Engineering science is the sci ence that links scientific knowledge with engineering technolog y [1].

1.1 Content of engineering science

Any project always requires a variety of interdisciplinary knowledge and technologies. Engineering science starts with the overall goal of engineering, comprehensively considers various constraints, and then repeatedly optimizes the process through analysis integration, integration analysis, and so on. A project needs to integrate multiple technical systems and elements, as well as non-technical factors such as resources, capital, land, labor, and market. To pursue the optimal configuration of different engineering elements and processes, with minimal investment in personnel, finance, materials, and information, to achieve maximum economic and social benefits.

1.2 Characteristics of engineering science

Engineering activities integrate a large number of engineering el ements with different properties into a specific functional engin eering entity, and must be organized according to the theories a nd methods of systems science. Engineering activities add socia l and cultural complexity to the complexity of natural things, re quiring the study and handling of various factors, and have com plex scientific characteristics.

Engineering science has interdisciplinary characteristics. The interdisciplinary field in engineering needs to address the inters ection of natural sciences and social sciences, technical sciences and humanities, and mathematics and social sciences. Engineer ing starts from the existing basic scientific theories, integrates th e theoretical provisions and laws of different disciplines accordi ng to human social goals, and objectifies them into an artificial creation.

1.3 The science foundation of engineering science

Natural science is the theoretical foundation of modern engineer ing activities, and both engineering technology and engineering science are based on natural science. Engineering science must f ollow scientific theories and conform to the basic principles and laws of natural science. All engineering activities and all aspect s of engineering activities cannot be separated from or violate n atural science.

Strength analysis and structural optimization in engineering a ll involve advanced mathematics knowledge, including limit thi nking, differential and integral thinking in advanced mathematic s, especially partial differential, to help engineering students bet ter understand and understand the natural world. Linear algebra embodies the connection between geometric concepts and algeb raic methods, and its determinant and matrix tools are widely us ed in mathematics, mechanics, physics, computer-aided design, virtual reality, and other technologies. The concepts embodied i n mathematics, such as axiomatic assumptions, axiomatic metho ds, induction and synthesis, and logical reasoning, are very bene ficial for strengthening people's engineering thinking.

Physics mainly studies topics such as force, heat, sound, light, electricity, atoms, etc. It mainly teaches the basic structure and interactions of matter, as well as the most basic and common for ms of motion and transformation laws of matter. All fields in na tural science and engineering activities cannot do without the la ws of physics. Engineering chemistry is the discipline that studi es the properties, structures, changes, and synthesis of substance s. In fields such as materials science, manufacturing processes, environmental engineering, and energy research, these issues all require chemical knowledge to solve.

2 The path of avoiding "engineering and scientifization"

The management personnel and frontline teachers of vocational education institutions are mostly promoted to the vocational edu cation podium through receiving regular undergraduate educatio n. Due to the inertia of undergraduate education, vocational coll eges attach importance to paper publication and neglect practica l innovation in the production line. The evaluation system for v ocational education teachers is like this, and the conditions for u ndergraduate and master's graduation are formulated accordingl y. Some people believe that there is a serious phenomenon of sc ientization in Chinese engineering education based on this.

The business and academic communities have already taken a ction to explore beneficial paths for cultivating engineering tale nts suitable for social needs. The theoretical framework of the " Five Theses" in engineering philosophy and the evaluation of pr ofessional abilities, COMET, provide powerful guidance for bre aking the "scientific transformation of engineering" [3][4].

2.1 ENGINEERING ONTOLOGY

In ancient times, people always consciously or unconsciously b uilt projects to adapt to nature, and then summarized invention t echniques and techniques, as well as discovered scientific princi ples, to further guide better construction projects. Any project is carried out under certain natural environment (conditions). In o rder for engineering activities to succeed, they must follow natu ral laws and be carried out within the range that the natural ecol ogical environment can bear.

Engineering Examples (Casting).

The "Chinese College Student Mechanical Engineering Innovation and Creativity Competition: Casting Process Design Competition", sponsored by the China Society of Mechanical Engineering and organized by the Casting Branch of the China Society of Mechanical Engineering, requires participants to start from two-dimensional drawings and submit design specifications. The main contents of the works include: part names, material requirements, structural analysis, main production technology requirements (shape, melting, casting, heat treatment, etc.), process plans, and process descriptions Parameters, process diagrams and tables (cards), casting quality control (preventive measures for defects), etc. These are all engineering requirements for a casting enterprise to produce high-quality castings.

2.2 COMET MODEL FOR OCCUPATIONAL COMPETE NCY assessment

The COMET (Competence Measurement) model for assessing p rofessional competence and professional identity is an internatio nal collaborative project based on modern vocational education theory. It establishes a professional competence model and asse ssment model, and conducts rigorous psychological measureme nt technology verification. Not only can it be used to diagnose t he professional ability level of students, but it can also evaluate the overall situation of professional identity and commitment [7].

3 Result and discussion

Engineering should avoid being science oriented. Engineering s hould attach importance to application and practical innovation, while science should attach importance to basic research and the oretical innovation. The evaluation of engineering should focus on the practical application of scientific research achievements and technological breakthroughs and innovations.

The COMET assessment of occupational ability and occupati onal identity can diagnose the level of occupational ability of en gineering students, and can also evaluate the overall situation of occupational identity and commitment. It is a powerful tool for engineering to avoid scientification.

4 Conclusion

Based on the ontology of engineering, the current order of traini ng engineering students in the fields of "Science Fundamentals - Technical Fundamentals - Engineering Fundamentals - Graduat ion Design" is unreasonable. For engineering college students, we should strengthen their knowledge of science rather than we aken it.

References

[1] Qian X S. Engineering and engineering science [J]. Engineering Research - Engineering in an Interdisciplinary Perspective, 2010, 2(4): 282-289
[2] Yin R Y, Wang Y L, Li B C, et al. Philosophy of engineering (Se cond Edition) [M]. Beijing: Higher Education Press, 2013.
[3] Chen B S, Zhou T, Liang J H. Research and practice of professio nal teaching diagnosis and reform mechanism based on COMET ability evaluation [M]. Beijing: Beijing Institute of Technology P ress, 2022.
[4] Cha J Z, He Y S. Three major strategies for China's engineering e ducation reform [M]. Beijing: Beijing Institute of Technology Pr ess, 2010.

Application of Technical Consulting into the Engineering Design in Foundry Projects

Zhipeng Liu*, Bas van Gemert, Kitty Eman
Gemco Metal casting technology B.V., Science Park Eindhoven 5053, 5692 EB Son, the Netherlands
*Corresponding address: e-mail: zhipeng.liu@gemco.nl

Abstract: The engineering design plays a crucial role in ensuring efficient and safe production of castings in a foundry project. Engineering design generally includes factory layout, production equipment, material logistics and storage, safety and environmental consideration, quality control and inspection, utilities and services, and environmental sustainability. Technical consulting is widely applied during casting production. It involves providing expertise and guidance to individuals or organizations in the casting industry. Technical consulting covers a wide range of topics, which are related to casting processes, materials, equipment, quality control, technology application and overall foundry operations. Technical consulting also offers many benefits when applied during the engineering design of foundries. In this paper, the necessity of technical consulting during the foundry design is analyzed, and the related benefits and extra value are presented. By providing specialized expertise and guidance to individuals or organizations, the aspects of implementing technical consulting in engineering design will include: facility layout and planning, equipment selection, environmental and safety compliance, energy efficiency. material handling systems, quality control and inspection facilities, waste management, training programs, technology integration, cost analysis, and environmental sustainability.

Keywords:engineering design; technical consulting; project management; technology application

1 Introduction

Engineering design is crucial in running a successful foundry project. Engineering design has significant influences on the foundry efficiency, casting quality, and cost-effectiveness of process and production. Effective engineering design ensures that the casting production can be maximally optimized for the casting function and performance. State-of-the-art design will minimize the material wastage, and contribute to environmental sustainability. The proper engineering design also minimizes the casting defects, and ensures higher quality and reliability of castings. Meanwhile, the efficient design ensures proper manufacturing, which reduces the production time and required energy consumption. As a result, the cost of castings is reduced, and the competitiveness of the company is increased. The engineering design generally includes factory layout, production equipment, material logistics and storage, safety and environmental consideration, quality control and inspection, Utilities and services, and environmental sustainability. Concept engineering design is seen as the first phase of design for a new foundry project. The concept design involves developing a preliminary plan that outlines the key elements and features of the foundry facility.

Technical consulting generally provides expert guidance and assistance to the operation and management of foundries. By offering extra value, it helps optimize processes, improve efficiency, enhance quality, ensure safety, and address specific challenges. Technical consulting for foundries focuses on many topics, like process optimization, material selection, scrap reduction and quality improvement, equipment evaluation and upgrade, safety and environmental compliance, energy efficiency, quality control and inspection, waste reduction and recycling, worker training and skill development, troubleshooting and problem resolution, cost analysis and financial planning, technology integration, market and industry analysis. Technical consultants need to have expertise or knowledge in metallurgy, materials science, mechanical engineering, operation and management processes.

2 Result and discussion

2.1 Engineering design and technical consulting

Engineering design is a complex and interdisciplinary process that requires expertise in metallurgy, materials science, mechanical and civil engineering, industrial design, and project management. Collaboration with experienced engineers and specialists is essential to create a successful design that meets production goals while adhering to safety and environmental standards. For an engineering design, technical consulting is highly beneficial in a foundry project. Engineering design for a foundry project involves the creation of detailed plans, specifications, and layouts for the establishment of a foundry facility or the improvement and expansion of an existing one. Technical consultants bring extensive knowledge and experience in engineering, metallurgy, materials science, and foundry operation. They can provide insights and solutions that may not be readily available within the client's organization. Consultants can also assess the proposed design and suggest improvements to optimize processes, reduce waste, and increase overall efficiency. Their expertise generally leads to cost savings and higher productivity. Consultants help implement quality control measures and inspection processes to ensure that castings meet the required standards and specifications. This minimizes the risk of defects and rejects.

Foundries must adhere to various safety, environmental, and regulatory standards. Technical consultants can assist in designing facilities and processes that comply with these regulations, reducing the risk of fines and legal issues. Consultants can recommend the most suitable equipment for the project, considering factors such as capacity, energy efficiency, and automation. They can also ensure that the selected equipment is integrated into the facility effectively. Consultants can help identify potential safety hazards within the design and develop safety protocols and systems to mitigate risks, protect workers and prevent accidents. Consultants can advise on energy-efficient practices and technologies, helping to reduce operational costs and environmental impact. In the event of issues or challenges during the construction or operation of the foundry, consultants can provide rapid and effective solutions, minimizing downtime and production unsafety.

In general, technical consulting for engineering design is seen to be essential in a foundry project, to ensure the successful planning, construction, and operation of the facility. The goal of engineering design in a foundry is generally to ensure the efficient and safe production of cast metal components while optimizing resources and meeting quality standards. Technical consulting offers related foundry expertise, quality assurance, compliance, cost-effectiveness, and risk mitigation to the project, ultimately contributing to its overall success.

2.2 Principles of technical consulting for an engineering design

When implementing technical consulting during the engineering design phase, certain principles can enhance the effectiveness of the consulting process and contribute to the success of the engineering project. The key principles include:

• To involve technical consultants early in the engineering design phase to benefit from their expertise during critical decision-making stages. Early engagement helps in identifying potential challenges, aligning strategies, and ensuring a more integrated approach to the design process.

• To ensure that technical consultants possess deep domain knowledge relevant to the specific engineering project. Expertise in the relevant technology, standards, and the best practices is essential for providing valuable insights and recommendations.

• To conduct collaborative design workshops involving both the engineering team and technical consultants. Ideas can be exchanged efficiently, and design challenges can be addressed collectively.

• To ensure that the engineering design follows relevant standards and regulations. Technical consultants should provide guidance on compliance, ensuring that the design meets all requirements.

• To conduct a thorough assessment of available technologies that can enhance the engineering design process. The innovative technologies are evaluated and adopted to improve efficiency and performance.

• To identify potential risks associated with the engineering design. Technical consultants contribute to risk analysis and propose strategies to address potential challenges.

• To Propose solutions with potential future expansions or changes in requirements. Technical consultants can guide the engineering team in making choices that align with long-term business goals.

• To conduct thorough cost-benefit analyses for different design options. Technical consultants can contribute by evaluating the feasibility analysis of various design choices and recommending cost-effective solutions.

The principles of implementing technical consulting during the whole process of engineering design have to be detailed in the different steps, demand evaluation, technology and process selection, facility layout, equipment selection, material logistics, environmental evaluation, safety, energy efficiency, quality control and inspection, cost analysis, and project management.

3 Conclusion

Application of technical consulting should be considered at the early stage of engineering design as much as possible. Our experiences show that it will add many benefits to the foundry project, particularly when introducing new products or updating the old production facilities. The potential risks are identified and the corrections are given. All processes are monitored or conducted by consultants during the engineering phase. By integrating the above principles, technical consulting can ultimately contribute to the successful realization of the foundry project.

Making Sustainability Business as Usual for the World's Foundries

Morten Bechlund
Norican Group, Højager 8, 2630 Taastrup, Denmark
morten.bechlund@noricangroup.com

Abstract: Sustainability and finding a path to net zero is crucial for our industry's future as well as the world's. This paper will discuss why foundries must put sustainability at the heart of their strategy, how they can start measuring emissions and set targets for reducing them, and the partners and programmes available to help them.

It will outline how the new EU legislation CSRD will influence companies globally, and reference relevant global standards like ISO 14001 and Ecovadis. It will cover scope 1, 2, and 3 emissions auditing, and explain how frameworks like the Science Based Targets initiative (SBTi) can help to comply with legislation and focus on carbon reduction efforts.

The paper will share what Norican has learned in the three years since its commitment to SBTi: its measurement initiatives, approved carbon reduction targets, ongoing global sustainability initiatives and partnerships with customers.

It will also tell the story of MAT Group that shows how foundries can exploit the latest digital technologies, efficient equipment and supplier expertise in order to measure and cut emissions across their supply chain.

Keywords: sustainability; emissions; SBTi; scope; energy; CSRD; MAT Group

1 Introduction

The world is facing the existential challenges of depleting resources and rising global temperatures. Climate science tells us we need to slow global warming urgently to prevent runaway global heating and irreversible damage to the planet. The economics of depleting resources means some will become so scarce as to be unaffordable to most.

Since the Paris Agreement in 2015, a global push towards binding emission reduction targets and stricter sustainability regulations has gathered pace, affecting all regions and making its way across supply chains.

2 Foundries in the spotlight

Foundries are of course energy-intensive, largely fossil-fuel-powered operations, and users of several diminishing resources. As such, our industry is part of the problem but, in turn, can be part of the solution.

The industry's most powerful customer segment, Automotive, is under pressure to decarbonize its own production and, increasingly, its supply chains. This is happening fast and is starting to impact foundries, which will be asked to track and declare the carbon footprint of their castings and set emission reduction targets to be invited to tender.

Many people in the foundry industry recognize the transition to more sustainable operations as a moral duty. They are supported in pursuing it by increasing regulatory and customer pressures, but also by the simple fact that, in many cases, sustainable practices make business sense.

3 Regulations and frameworks

The regulations driving the adoption of more sustainable industrial practice and the frameworks helping companies work towards them are plentiful. Regional and local regulations are emerging around the world.

In Europe, the EU's CSR Directive (CSRD) has taken a central role. It directly impacts industrial businesses operating in Europe and those with a significant turnover in European markets. Its influence will ripple along global supply chains, and foundries selling into Europe should be aware of and prepared to meet key requirements.

The main purpose of CSRD is to enhance sustainability reporting as part of corporate reporting of non-financial information. The Directive expressly references existing reporting standards to look to when seeking to comply with the new rules. For companies looking to comply with these emerging rules, existing structures and frameworks are available, offering both a basis and a roadmap for their sustainability transformation.

For example, the **Science-Based Targets initiative (SBTi)** offers an emissions reporting framework and guidance tailored to businesses, including specific reporting advice for certain sectors. It helps businesses establish a baseline and set emission reduction targets rooted in climate science, with a clear pathway to achieving them.

Sustainability assessment and scoring providers help companies assess their sustainability performance, often by awarding an annual score with a progression path. Norican Group has experience using EcoVadis, a French company widely used around the world.

Frameworks like the above help foundries embed emissions tracking and sustainability reporting in their processes. Commonly used standards like ISO 14001:2015 for environmental management and ISO

50001 for energy management provide a good basis for both. So does, as we will see, having an IIoT system in place that already collects sustainability-related data.

4 Sustainability in the foundry process

In foundries of all sizes, we have seen a broad range of successful sustainability initiatives that unlock significant improvements in sustainability and business performance, with knock-on effects that help address the key challenges the industry is facing – from attracting talent to dealing with materials shortages and rising prices.

Important opportunities for sustainability progress we have seen are: radical scrap reduction through automation and use of new technology and data/AI; tight control of consumption ratios (through digital monitoring); increased use of recycled materials (from recycled sand to using less virgin metal); electrification and energy-efficiency upgrades (including lower-emission equipment such as pneumatic instead of thermal sand reclamation). Across all of these, digital monitoring and IIoT are often key, not least to identify opportunities for improvement – as shown in the following case study.

5 Case study: MAT Group

MAT Foundry Group is a specialist manufacturer of cast and machined parts for the automotive sector. The MAT team realized early on that the ability to track emissions and report environmental performance would be key to winning tenders from automotive companies. At their UK site in Poole, UK, MAT currently calculates a per-ton figure for embedded scope 1 and 2 carbon, based on its total casting output and energy consumption. Adding in product-specific metrics like weight and yield produces a carbon footprint for each individual casting.

Over the years, MAT Group has steadily created robust foundations for its progress on sustainability. All its facilities comply with ISO 14001 for environmental management and OE plants are all ISO 50001 certified for energy management. The team introduced comprehensive digital process optimization (using the IIoT platform Monitizer | DISCOVER) across several of its sites more than five years ago, which has made it easier to now get emissions data from the process.

Dashboards were initially set up to report energy consumption per ton in real-time. "What we're aiming for is a real-time carbon metric for every sub-process and a Monitizer digital dashboard that gives us a per-batch emissions figure for each product as it's produced," says Luke Batter, Process Development Manager at MAT Group. Measuring **scope 3 emissions** is the next big challenge. MAT is gathering data as fast as it can but, with data on only 10% of supplier emissions, it still doesn't have quite enough to get an accurate figure. First, it wants to be ready to commit to SBTi.

The group's commercial director, Shaun Lindfield, explains: "We haven't signed up to SBT, we're waiting so we can align our standards with our customers. But we've been measuring scope 1 and 2 emissions across the group for the last two years and already offer our customers a product carbon footprint based on that. We started measuring scope 3 emissions in our own supply chain this year *[2022]*."

The team also realize the **key role of scrap reduction** in improving sustainability performance. Shaun concludes: "Quality and emissions reduction are being linked like never before. When you start to reduce your scrap, there's a significant reduction in scope 1, 2 and 3 emissions. There's the melting energy but also all the other carbon emissions that go into producing scrap – the alloys, the energy to dry the sand, bentonite, the transport emissions and so on. There's a complete shift in mindset coming over the next three to five years. **Getting to zero scrap** is going to be crucial because the effect on emissions will make you so much more competitive. If you are producing 3%-5% scrap, it's going to cost you a whole load more to get to carbon neutrality. Again, it's about process improvement leading to fewer emissions. It saves you money and makes you more profitable – so why isn't everybody doing this?"

6 Conclusion

Fast progress is possible for foundries who want to produce more sustainably and retain access to increasingly sustainability-driven global supply chains. Frameworks are available to guide a foundry's sustainability roadmap, with most foundries farther along on their journey than they might think. IIoT can deliver fast, cost-effective sustainability reporting and improvements, with significant business benefits.

Case Study: From Annual Loss of CNY 10 Million to Annual Profit of More Than 50 Million, in 2 Years

James Jing Zhang[1], Crystal Leung[2],
1. HongKong Foundry Association, No.3 of 10, Road Xingsheng, Tianhe District, Guangzhou, Guangdong, China
2. Wah Fung Metal Co., Ltd., No.1 Fengri Street, Huangsiwei Village, Shijie Town, Dongguan, Guangdong, China
*E-mail: Jameszh@vip.163.com; **E-mail: crystal@hoihing-group.com

Abstract: This paper analyzes, studies and summarizes the real case of an old small- and medium-sized foundry's operation from net loss to greater profit, adopts the method of 'Cost and gross profit analysis, referred to as **Value analysis**, guide profit improvement strategies', lists the corresponding improvement measures with their implementation, which greatly improved the actual operating profit of this foundry.
The case shows that old small- and medium foundries, through the implementation of appropriate improvement measures of management, are entirely possible to significantly boost profits in a short time.

Keywords: small and medium-sized enterprises; operation; profitability; improvement countermeasures

1 Introduction

Old small and medium-sized foundry enterprises sometimes have poor operating efficiency and even losses. This is of course directly related to the external economic environment, customer conditions, product category, process technology level, production conditions and other factors, but in the case of the above conditions are unchanged, how to find their own weak points and improve, to obtain better profits, is also a special concern of small and medium-sized entrepreneurs. This paper describes a practical case of an old small and medium-sized foundry enterprise. Based on the in-depth analysis of cost and gross profit, the weaknesses of the enterprise are found and the corresponding improvement measures are accurately found. Therefore, the profit of the enterprise has been rapidly improved.
This kind of thinking and method has a positive reference value for enterprises to improve their profitability.

2 Case background

Enterprise age: more than 20 years;
Equipment and their services life: 45 sets of die casting machine, 400 sets of CNC machines. Most of the equipment has been used for more than 6 years, some even more than 10 years;
Customers and products: most are parts and components for traditional fuel vehicles;
Employee: About 720 people.
Operating status: In the first three quarters of 2021, the revenue was nearly CNY 400 million, but the net loss exceeded CNY 10 million.

3 Method of analysis

Data analysis of the actual cost and gross profit for all products -- referred to as **Value analysis**, and comparing with the level of outstanding enterprises in the market, accurately identifying the weak points that affect the profit of each type of product.

4 Weak points

Through the **Value analysis**, the following weaknesses of the enterprise were found:
The understanding of the composition of profit is more abstract; Lack in in-depth data analysis of loss; People often rely on intuition to judge;
The financial department only plays the statistical function of "financial accounting" and does not play the function of "management accounting" [1].
The advantages of the enterprise are not fully utilized;
The management team lacks the necessary thinking and methods of profit analysis;

5 Countermeasure

(1) Total Profit Orientation (TPO): Correlate all business activities with data to profit objectives.
(2) Give play to the management accounting function of finance : In addition to the statistical accounting function, finance should also play the three functions of "statistics - analysis - profit improvement guidance".
(3) Use our strengths: Identify and create our existing greatest competitive advantage.
(4) Adhere to the 20/80 principle: 20% of the business (or activity) often brings 80% of the benefits; Prioritize 80% of our limited resources on 20% of our quality customers or core business.
(5) Give full play to team strength: enhance the profit awareness and execution ability of management team members.

6 Result and discussion

The fourth quarter of 2021: Spend about 2 months to analyze the main reasons for the loss, and study the improvement plan;

Q1 2022: Advance and implement concrete measures for improvement;
Q2 2022: Earnings effect begins;
Second half of 2022: Steady improvement in earnings;
End of 2022: Turn a loss into a profit, and achieve a net profit of more than CNY 28 million.
The whole year of 2023: Continued to improve according to effective measures, and finally achieved a net profit of more than CNY 50 million that year.

7 Conclusion

This case study shows that the old small and medium foundries, through the implementation of appropriate improvement measures of management, are entirely possible to significantly boost profits in a shorter time. And the **Value analysis** method of cost and gross profit is an effective tool to accurately find improvement countermeasures.

Reference

[1] Li Jian. Cutting Costs: 12 Financial broadsword for Entrepreneurs, China Machine Press, 2006.

Greenhouse Gas from Industry Process and Product Use in Foundry Industry

Anan Song
Hefei Anzhi environmental science and technology consulting Co., Ltd., A3801 at 190 Qianshan Rd., Hefei, Anhui, China
*Corresponding address: e-mail: annsong921@163.com

Abstract: Both CO2 and non-CO2 Greenhouse Gas (GHG) may be found in foundry Industry Process and Product Use (IPPU). CO2 may be emitted by using organic additives in sand. SF6 was popular in Mg-alloy die-casting, which may cause an extremely high CO2-eq emission. It was probable to find a trace amount of CH4 in leaking of natural gas use; HFCs and PFCs in air conditioning, refrigeration and waste heat recovery and utilization equipment or as substitutes for SF6. These emissions may not always be negligible, unless fluoride GHGs controlled below 0.198-32.4 g/t-casting. IPPU GHGs exist in both foundry equipment and general equipment and should be accounted as Scope 1 or 3. Foundries need more attention on IPPU GHGs.

Keywords: greenhouse gas (GHG); carbon emission; industry process and product use (IPPU); foundry

1 Introduction

China has submitted its CO_2 goals and every Chinese corporation and industry is preparing its own. Energy is the priority concern in carbon issues but not the only one. The climate crisis is attributed to all types of Greenhouse Gases (GHGs) including carbon dioxide (CO_2), methane (CH_4), nitrous oxide (N_2O), hydrofluorocarbons (HFCs), perfluorocarbons (PFCs) and sulphur hexafluoride (SF_6). Non-CO_2 GHGs have high Global Warming Potential (GWP) and may cause high equivalent CO_2 emission (CO_2-eq). These gases are usually found in Industry Process and Product Use (IPPU), where CO_2 without fuel or energy relation also exists. Foundry is a relatively small industry and just started working on carbon issues lately. Most foundry corporations had begun accounting for CO_2 emissions from their energy consumption including direct fossil fuel using as Scope 1 and electricity using as Scope 2. However, IPPU emissions and Non-CO_2 GHGs were merely considered by the foundry industry, also with rare research attracted. The existence, significance and necessity for accounting of IPPU GHGs in the foundry industry remain in chaos.

2 Possible IPPU GHGs sources in foundry

2.1 CO_2

The carbon lost in iron and steel during casting is left as CO_2. Meanwhile, organic materials are popular in mold nowadays, which could be converted into CO_2 in every process with high temperature: pouring, warmth maintaining and sand reclamation. These CO_2 emissions were concluded in Table 1, and should be accounted as IPPU in Scope 1.

Table 1. CO2 in foundry IPPU

	Source	Main emission process	Emission poteitial* ($kgCO_2/kg$)
Metal	Carbon in iron/steel	Melting	-
Mold-green sand	Coal	Pouring, warmth maintaining and sand reclamation	1.94
	Starch/dextrin ($C_6H_{10}O_5$)		1.63
Mold-resin sand	Furan ($C_8H_{10}O_3N_2$)		1.93
	Phenolic aldehyde (C_7H_6O)		2.91
Lost foam casting	Expanded polystyrene (EPS, C_8H_8)	Pouring	3.38

* Calculated by molecular formula

2.2 Non-CO2 GHGs

SF_6 was widely used for magnesium melting protection and 90%-100% SF_6 utilized would be emitted as GHG burden [1]. Though the magnesium industry has been trying to eliminate SF_6 over the past decade, the usage is still common recently [2], while several substitutes for SF_6 e.g. HFC-134a (HFCs) or CF_4 (PFCs) are also GHG. For other GHGs, CH_4 leaking was possible in furnaces using natural gas. In cupolas or vehicles (cars, trucks or fork lifters), NO_x was controlled as an air pollutant in exhaust gas, where the existence of N_2O needs verifying. Refrigerants GHGs may be found in waste heat recovery systems, which is a raising interest for foundries, e.g. organic Rankine cycles [3]. All these non-CO_2, as shown in Table 2, were in IPPU and should be accounted as Scope 1 or 3 (depending on proprietorship and accounting boundary).

Table 2. Non-CO2 GHGs in foundry

	GWP [4]	Existence	Possible source	
			Foundry equipment	General equipment
CH_4	27.9	Trace	Natural gas leak	
N_2O	273	Possible trace	Cupola	Vehicle

HFCs	1530 (HFC-134a), 14600 (HFC-23), 771 (HFC-32), 3740 (HFC-125), 5810 (HFC-143a)	Trace	Cover gas in Mg-alloy die-casting	Air conditioner and refrigeration, organic heat recovery system, electric insulation
PFCs	7380 (CF_4), 12400 (C_2F_6)	Trace		Electric insulation
SF_6	25200	Some or trace		
NF_3	17400	Not found	-	-

3 Significance evaluation

In accounting, the ignoring of IPPU GHGs could only be acceptable when the CO_2-eq was low, e.g. 1% of the total emission. The carbon footprint of casting was reported as 1.2-2.8 tCO_2/t-casting [5] or 0.87 tCO_2/t-casting, of which more than 50% is attributed to melting and 30% is embedded in pig iron [6]. Thus, in common foundry GHGs accounting only including Scope 1&2, choosing 0.5-2.5 tCO_2/t-casting may be reasonable as a reference. Assigning 1% CO_2-eq or below to be negligible emission amount, tolerant limitations for each IPPU GHGs were calculated in Figure 1.

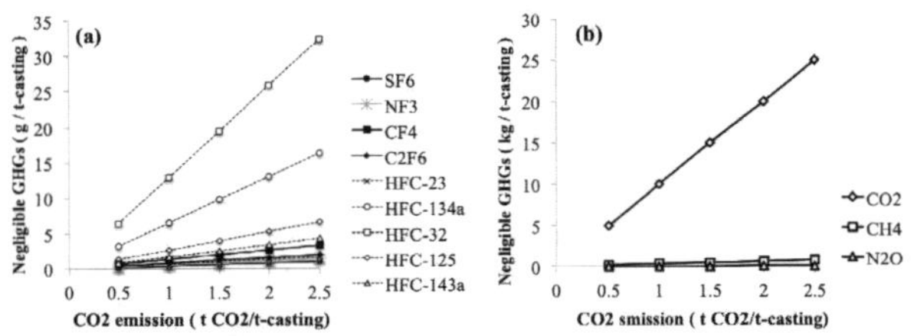

Figure 1 The tolerant limitation (1% CO_2-aq of total CO_2 emission) for each IPPU GHGs was calculated.

With high GWP, the tolerant limitation for fluoride GHGs was extremely low, with a range of 0.198-32.4 g/t-casting. For an Mg-alloy die-casting foundry emitted 1 tCO2/t-casting with 10,000 t-casting/year, SF_6 should be controlled under 3.97 kg/year (i.e. 10.9 g/day) to avoid a significant raise in CO_2-eq accounting. Using data in Table 1, assuming 1-5 t sand/t-casting and 1%-2% organic materials added in sand with 10% complete combustion into CO_2, emission was estimated to be 1.9-29.1 kg CO_2/t-casting. Compared with Fig.1(b), this will not always be negligible. Both IPPU CO_2 and non-CO_2 GHGs could be significant for foundries under certain conditions.

4 Conclusion and outlook

IPPU GHGs can be found in the foundry industry. IPPU CO_2 is emitted from organic materials including coal and starch/dextrin powder, furan or phenolic aldehyde resin and EPS. Among non-CO_2 GHGs, SF_6 is typical in Mg-alloy die-casting foundries and may lead to an inconceivable CO_2-eq accounting; CH_4 and N_2O may be found in furnaces; HFCs and PFCs may be used as substitutes for SF_6 or refrigerants in general equipment. All these should be accounted as Scope 1 or 3, otherwise controlled below 0.198-32.4 g/t-casting for fluoride GHGs to neglect, which might be hard, especially for Mg-alloy die-casting foundries.

As an energy-intensive industry, foundries were used to omit or ignore IPPU GHGs as an assumption with no question. The results of this study attempted to show the risk of such practices under stringent carbon and climate policy in the future. More attention and much more study will be needed on IPPU GHGs for the foundry industry and corporations.

References

[1] Bartos S, Laush C, Scharfenberg J and Kantamaneni R. Reducing greenhouse gas emissions from magnesium die casting. Journal of Cleaner Production, 2007, 15(10): 979-987

[2] Yazman S, Köklü U, Urtekin L, Morkavuk S and Gemi L. Experimental study on the effects of cold chamber die casting parameters on high-speed drilling machinability of casted AZ91 alloy. Journal of Manufacturing Processes, 2020, 57: 136-152.

[3] Ja'fari M, Khan M I, Al-Ghamdi S G, Jaworski A J and Asfand F. Waste heat recovery in iron and steel industry using organic Rankine cycles. Chemical Engineering Journal, 2023, 477: 146925.

[4] IPCC. The earth's energy budget, climate feedbacks, and climate sensitivity supplementary material. In: Climate Change 2021: The Physical Science Basis. Contribution of Working Group I to the Sixth Assessment Report of the Intergovernmental Panel on Climate Change, 2021.

[5] Monteleone B, Baldereschi E, Fabbri N, De Bernardi C and Frey M. A sustainability assessment of the foundry production process in Italy. Sustainable Production and Consumption, 2024, 46: 491–501.

[6] Abdelshafy A, Franzen D, Mohaupt A, Schüssler J, Bührig- Polaczek A and Walther G. A feasibility study to minimize the carbon footprint of cast iron production while maintaining the technical requirements. Journal of Sustainable Metallurgy, 2023, 9: 249–265.

Design of Conic Clamping System as a Different Version

Ali Emre Zobaroğlu
Demisas Dokum Emaye Mamulleri Sanayi A.S., Yenimahalle, Ataturk Bulvari No:67 Bilecik/TURKEY
e-mail: emre.zobaroglu@demisas.com.tr

Abstract： Conic clamping system is a modular quick-change tooling concept that offers three systems in one; quick-change tool holders for turning centers minimize setup and tool change times and significantly increase machine utilization, conic system integrated directly into the spindle increases stability and suitability for all jobs in multi-purpose machines. The conical clamping system, which is a modular system for machining centers, offers numerous extension and reduction adapters, allowing the assembly of tools with different lengths and design features, independent of machine interfaces. Modularity means there will be less need for expensive special tools with long lead times. In project scope, it is planned to minimize the tool change times of the conical clamping system, which will be used in the machining of the project part on lathes. With this study it is aimed to assembling/disassembling of tool holder in quick way instead of keeping it fixed in the system.

Keywords: machining; clamping system; tool; cutting insert

1 Introduction

In many industrial sectors, lathes are used to perform operations such as cutting, processing, etc. of workpieces made of hard materials, especially metal. In each turning operation on lathes, different cutting tools and different attachment devices belonging to these cutting tools are used. One of these attachment devices is the lathe toolholder, which has a cutting insert attached to it that has the function of shaping hard materials in the desired shape. Lathe toolholders are used to attach cutting inserts
to tool holders and are designed in a structure that allows different types of cutting inserts to be adapted.

The connection of the cutting insert to the lathe tools used in the known technique is provided by bolts. In standard turning operations, when the cutting insert connected to the tool becomes dull, the lathe must be stopped, the connection bolt of the cutting insert must be removed, the insert slot must be cleaned with air and the unused corner of the insert must be turned or if there is no solid corner left, a new cutting insert must be taken and reconnected. As mentioned, during the changing of the cutting insert, problems such as the bolt used as a connecting element falling and being lost during the disassembly/assembly process, the bolt's mouth and/or socket being damaged as a result of wear, and the operator having to make an offset adjustment in the amount of wear before production since the cutting insert will be worn during operation are experienced. While the falling and loss of the bolt during the mentioned process creates bolt costs, the deterioration in the bolt mouth and/or socket causes the cutting insert not to remain sufficiently stable in the socket where it is placed and causes vibration during the process. In this case, it returns to the business as a faulty product. In addition to all these since the bolt assembling or disassembling process performed manually by the operator requires force and time, there is a loss of both time and labor in shaping the workpieces.

In the robot-fed systems used in current lathe operations, more than one toolholder is connected to the working lathe. In case the cutting inserts connected to these tools become dull, the operator performs an insert change. When all robot-fed lathes are considered, tool change times negatively affect the production efficiency of businesses. In addition, the fact that the cutting inserts are changed on the lathe causes the operator to have difficulty while performing the operation and to get tired quickly. In this case, time and labor losses occur. As a result, due to the negativities and deficiencies mentioned above, a need for innovation in the relevant technical field has emerged.

2 Experimental procedure

In the current system (Figure 1), instead of changing the cutter, the cutter is left fixed on the machine and the machine is stopped for the insert change, which causes loss of time in this process. In addition, when it is considered that there is more than one cutter on the machine, the loss of time for the insert change increases.

For these reasons, a system was needed where we can take the insert change outside the machine and perform the tool disassembling and assembling process quickly. For these reasons, it was decided to work to increase efficiency and reduce costs.

The VDI tool holder shown in Figure 2 is the standard holder that comes with the machine. In these holders, a conical clamping slot and a channel slot for the poka yoke have been opened to the toolholder according to design (Figure 3).

Figure 1 Mounting a tool holder to a standard VDI holder

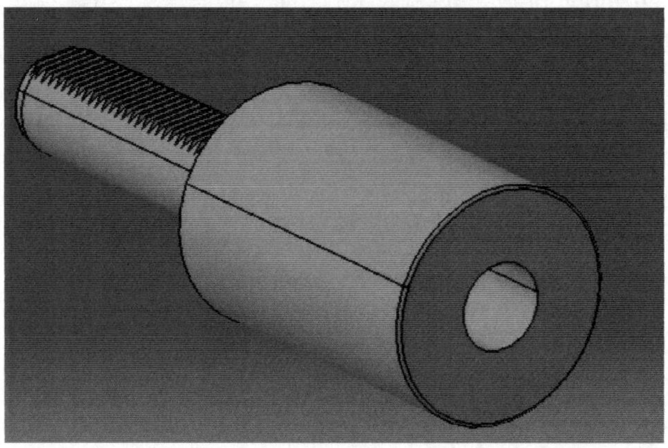

Figure 2 Standard VDI holder

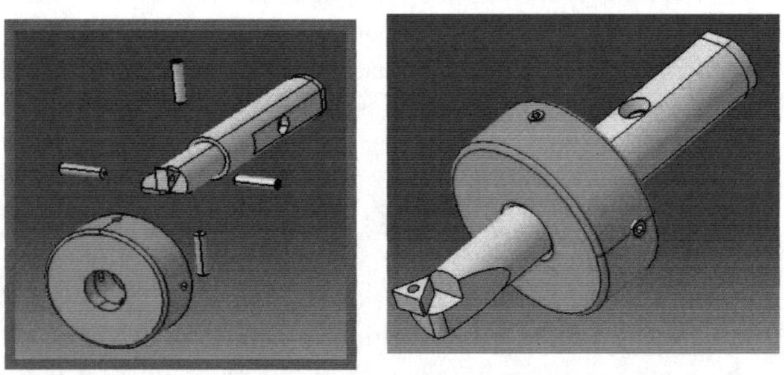

Figure 3 Clamping toolholder to standard VDI holder

The new system has been adapted to the designed quick disconnect-attach tool holder body (VDI holder). The system designed by opening only the channel slot and the slot where the conical tightening will be made has been adapted. In this way, insert changes can be made without stopping the machine by increasing the number of tool holders.

3 Result and discussion

The design is particularly related to the positioning wedge that ensures that the carrier body is always connected in the same position on the *X-Z* axis within the VDI body in order to ensure that the cutting insert change is done outside the machine, to speed up the removal and installation of the cutting insert and to eliminate the need for offset adjustment, the conical tightening bolt that ensures that the positioning of the carrier body on the *X-Z* axis within the VDI body is with the same precision at each lathe tool change, the wedge bolt that mounts the said positioning wedge on the VDI body, the limiting cover that connects the carrier body to the VDI body by ensuring that it is the same length at each lathe tool change, and the lathe tool holder that contains at least four setscrew bolts to ensure that the carrier body is fully centered within the VDI body.

4 Conclusion

As a result of the study, it was planned to reduce the time losses caused by the insert change by making the tool change fast. The new system was adapted to the quick disconnect-attach tool holder body (VDI holder) to be designed. Only the channel slot and conical clamping were made and adapted to the system. In this way, the number of these tools was increased and it was possible to make the insert change outside the machine without stopping the machine. In addition, savings were made in investment costs.

Carbon Emission Measurement for Sand Casting Enterprises

Zhong Chen[1], Tianyou Li[1], Wenhao Zhou[1], Mingjun Hou[1], Xiaoyuan Ji[1,*], Yajun Yin[1],Jiankang Lu[2], Jian-xin Zhou[1,*]

1. State Key Laboratory of Materials Processing and Die & Mould Technology, Huazhong University of Science and Technology, Wuhan, Hubei, 430074

2. Jiangsu Huaxing Steel Casting Co., Ltd, Taizhou, Jiangsu, 225453

*Corresponding address:e-mail:jixiaoyuan@hust.edu.cn (Xiaoyuan Ji); zhoujianxin@hust.edu.cn (Jianxin Zhou)

Abstract: Sand casting enterprises have long process flow, complex energy structure of melting, pouring and other processes, and many sources of carbon emissions, which makes it difficult to measure the carbon emissions of enterprises. For this reason, this paper carries out a study on carbon emission measurement in sand casting enterprises. Firstly, according to the type of enterprise material, the sand casting enterprise carbon emissions are divided into raw material emissions, fuel emissions and energy emissions; secondly, based on the enterprise process and energy structure, to determine the boundary of the carbon emissions, and the process and carbon emissions corresponding to the type of emission; finally, based on the enterprise data and boundaries, to determine the carbon emissions measurement method and build the carbon emissions measurement model.The research done in this paper provides support for the carbon emission measurement of sand casting enterprises, which is of great significance for sand casting enterprises to realize the goal of "double carbon".

Key words: sand casting; carbon emissions; measurement model

1 Introduction

China became the world's top CO_2 emitter after 2006[1], With the introduction of the goals of "carbon peak" and "carbon neutrality", all fields and industries in China are seeking to green, reduce carbon emissions and gradually carry out low-carbon development planning.China's industrial sector carries about 70% of the country's carbon emissions, andChina's industrial GHG emissions in 2014 were 216% of the 1996 emissions[2]. The foundry industry, as a key pillar of the industrial sector, drives economic growth and meets social needs, while inevitably generating a large amount of carbon emissions[3]. Therefore, it is of great significance to carry out research on energy saving and emission reduction in the foundry industry to reduce the emission of industrial production in China.For this reason, this paper carries out carbon emission measurement modeling research on the main process links of sand casting to provide a basis for carbon emission reduction in sand casting.

2 Analysis of carbon emission types

The use of substances and energy will bring carbon emissions, in the casting of the whole process, the substances used by the enterprise including casting raw materials, sand raw materials, coal, etc., the use of energy is mainly electric energy, research sand casting enterprises, to determine the types of carbon emissions as shown in the Table 1.

Table 1. Types of carbon emissions from sand casting companies

Types of carbon emissions	Instructions
Raw material emissions	Raw materials containing carbon are subjected to a series of reactions that produce CO_2, including carbon emissions from alloying materials, resin sands, paints, etc. produced during the casting production process, and raw materials that can be recycled should be subtracted from the calculation.
Fuel emissions	CO_2 from fuel combustion, including carbon emissions from solid fuels, gas, natural gas, propane, etc., in the production of castings
Energy emissions	Carbon emissions from energy occur at the production stage, and CO_2 is not generated at the time of use, including carbon emissions from the casting production process such as electricity, which are discounted by the amount of purchased electricity.

3 Analysis of carbon emission boundary

The sand casting production boundary is divided into the pouring stage and the post-processing stage, the pouring stage corresponds to the specific molding process of the casting, and the post-processing stage is the casting after the formation of the casting grinding, shrinkage and other treatments, and to determine the input and output of energy and material in each link, as shown in Figure 1.The sand casting production boundary, from the transportation of raw materials to the foundry to the final inspection of the casting and its successful warehousing, does not include the carbon emissions caused by the mining of raw materials, nor does it include the carbon emissions generated after the casting is sold.At the same time, only the carbon emissions brought about by the casting itself are considered, and the carbon emissions not related to the casting itself, such as enterprise personnel and paper consumption during the production process, are not taken into account, so the enterprise's electricity consumption should be separated from the electricity used for living and casting.If the enterprise is capable of solar power generation, the corresponding generation is discounted and subtracted from the local grid.

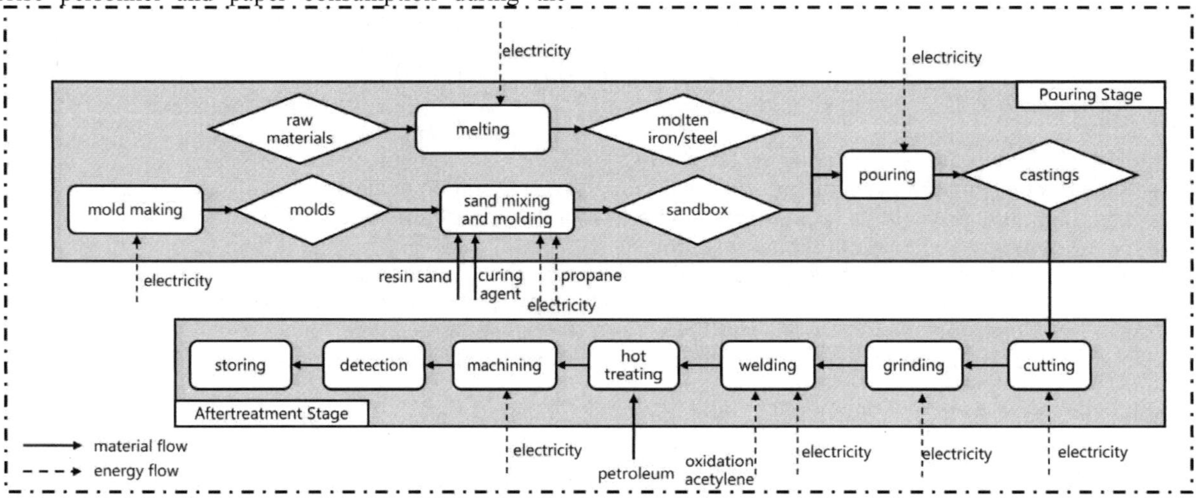

Figure 1 Sand casting system boundary

4 Carbon emission modeling

Modeling of carbon emissions of each type of foundry enterprises.The calculation formula of raw material emissions is as follows:

$$E_M = \sum m_i * EF_i$$

Where: E_M-emission of raw materials, $kgCO_2$; m_i-the ith carbon-containing raw material usage, kg; EF_i -the ith carbon-containing raw material CO_2 emission factor, $kgCO_2/kg$.

The calculation formula of fuel emissions is as follows:

$$E_F = \sum f_j * Q_j * CCI_j * OFI_j * 44/12$$

Where: E_F-fuel emissions, $kgCO_2$; f_j-jth fuel use, kg; Q_j-jth fuel calorific value, MJ/kg; CCI_j -jth fuel unit calorific value carbon content, kgC/MJ; OFI_j -jth fuel oxidation rate, %.

The calculation formula of energy emissions is as follows:

$$E_E = \sum e_k * EF_k = \sum \int P_k \mathrm{d}t * EF_E$$

Where: E_E -energy emission, $kgCO_2$; e_k -electricity consumption of the kth equipment, kWh; EF_k -electricity carbon emission efficient, $kgCO_2/kWh$; P_k- real-time power of the kth device, kW; t - time of use of the kth equipment.

Combined with the foundry ERP management software, the input and output of each process is combined with some of the production parameters, such as how much alloy material is used in the raw material and the size of the sand box depends on the three-dimensional parameters of the casting as well as the size of the risers, and the power consumption of the electric furnace during melting depends on the size of the preheating temperature and pouring temperature.Based on the casting process parameters and procedures defined in the ERP process card, the carbon emissions generated per unit quantity of the casting produced can be calculated.

5 Conclusions

(1) This paper classifies and analyzes the carbon emissions of sand casting enterprises into three categories, and determines the specific carbon emission boundaries. Similar processes have similar carbon emission characteristics. By analyzing the process characteristics and energy structure of different types of foundry enterprises, the classification and boundary can be extended to investment casting enterprises, etc.

(2) This paper models different types of carbon emissions. Adjustment of casting process parameters will bring about changes in carbon emissions, but will also affect the quality of castings, how to measure and reduce carbon emissions without affecting the quality of the castings should be taken into account.

Acknowledgments: This research was funded by the National Key R&D Program of China (2020YFB1710100).

Reference

[1] Wang Y, Yang H, Sun R. Effectiveness of China's provincial industrial carbon emission reduction and optimization of carbon emission reduction paths in “lagging regions”: Efficiency-cost analysis[J]. Journal of Environmental Management, 2020, 275: 111221.
[2] Liu H, Li B, Tang W. Manufacturing oriented topology optimization of 3D structures for carbon emission reduction in casting process[J]. Journal of Cleaner Production, 2019, 225: 755-770.
[3] Gu Z, He Y, Li D. Assessment method of environmental impact of casting processes based on digital twin[J]. China Mechanical Engineering, 2023, 34(12): 1465.

A Novel Reclamation Method of Chemical-Mechanical Grinding for Inorganic Binder Waste Sand in Aluminum Alloy Casting Process

Shengli Hu[1,2], Xiaolong Gong[1,*], Guanling Cai[2], Wucan Wu[2], Zitian Fan[1,*]
1. School of Materials Science and Engineering, Huazhong University of Science and Technology, Wuhan, China
2. Liujing Technology Group Co., LTD, Changzhou, China
*Corresponding address: gongxl@hust.edu.cn, fanzt@hust.edu.cn

Abstract: The current approach to reclaiming inorganic binder waste sand discharged in aluminum alloy casting process typically involves mechanical grinding combined with calcination treatment (650-700 °C). However, this method encounters challenges such as the accumulation of residual binder on the surface of the reclaimed sand and a subsequent decline in its refractoriness. This study proposes a novel method to reclaim inorganic binder waste sand, integrating chemical infiltration with mechanical grinding. The effects of different types and concentrations of chemical reagents on the electrical conductivity and Na_2O content of the reclaimed sand were investigated. The microstructure and chemical composition of both waste and reclaimed sand were analyzed, and revealing the chemical-mechanical grinding reclamation mechanism. When the dosage of $CaCl_2$ solution is 5wt.% of the waste sand weight, and the concentration is 10 wt.%, the electrical conductivity and Na2O content of the obtained reclaimed sand are 776.7 μS/cm and 0.039%, respectively, meeting the utilization requirements for reclaimed sand. Microscopic analysis reveals that the $CaCl_2$ solution reacts with the residual binder on the surface of the waste sand, disrupting the structure and morphology of the residual binder. After drying, the reaction product crystallizes in the form of blocky inorganic salts, facilitating their removal during mechanical grinding.

Keywords: inorganic binder waste sand; chemical infiltration; mechanical grinding; electrical conductivity; Na_2O content

1 Introduction

The inorganic silicate binder sand process is one of the most widely used molding sands in the casting process, known for its low cost, high dimensional accuracy of molds and cores, as well as non-toxic and odorless properties [1-3]. Generally, approximately one ton of waste sand is generated for every ton of castings produced. Currently, China is the world's largest producer of castings, generating over 6 million tons of inorganic binder waste sand annually [4,5]. This waste sand contains a large amount of alkaline residual binder on its surface, making it unsuitable for direct recycling in casting materials. Discharging it without treatment not only causes serious environmental pollution but also wastes the limited silica sand resources [6-8]. Therefore, addressing the reclamation and recycling of inorganic binder waste sand holds significant environmental and economic benefits.

This study addresses the issues with traditional dry and wet reclamation methods and the characteristics of aluminum alloy casting production. A novel reclamation method of chemical-mechanical grinding for inorganic silicate binder waste sand in aluminum alloy casting process is proposed and designed. In this method, the waste sand undergoes initial crushing, preliminary grinding, and dust removal. Subsequently, chemical reagents are introduced to infiltrate the sand particles, followed by low-temperature drying, additional grinding, and dust removal to obtain reclaimed sand. Based on this method, the particle characteristics of inorganic silicate binder waste sand are analyzed. The influences of the types and concentrations of chemical reagents on the reclaimed sand quality were studied. Furthermore, the chemical-mechanical grinding reclamation mechanism was elucidated to develop an optimal reclamation process for inorganic silicate binder waste sand discharged in aluminum alloy casting.

2 Experimental procedure

The chemical-mechanical grinding reclamation process and conventional reclamation process for inorganic silicate binder waste sand discharged in the aluminum alloy casting are illustrated in Figures 1 and 2.

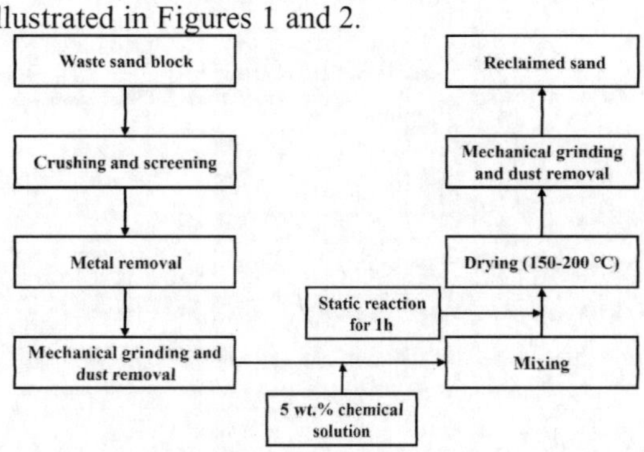

Figure 1 Process flow chart of chemical-mechanical grinding reclamation for waste sand.

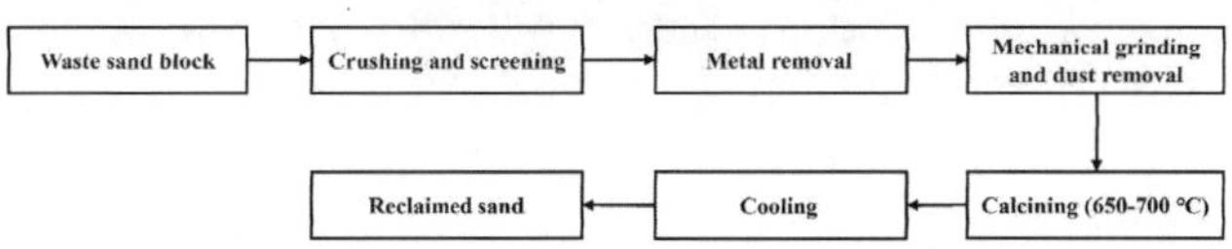

Figure 2 Process flow chart of conventional reclamation for waste sand.

3 Result and discussion

Effect of different chemical reagents

Figure 3 illustrates the effect of different chemical reagents on the electrical conductivity and Na_2O content of reclaimed sand. It can be observed that the reclaimed sand pretreated with $CaCl_2$ solution has the lowest electrical conductivity, and the reclaimed sand treated with $H_2C_2O_4$ solution has the lowest Na_2O content, but the highest electrical conductivity.

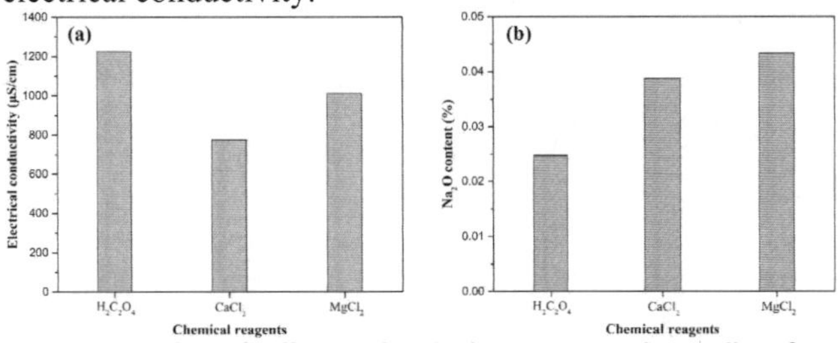

Figure 3 Effect of different chemical reagents on the quality of reclaimed sand: (a) electrical conductivity, (b) Na_2O content.

Effect of $CaCl_2$ concentration

The microscopic morphology of waste sand after infiltrating pretreatment with different concentrations of $CaCl_2$ solution is depicted in Figure 4. It can be observed that when the amount of $CaCl_2$ solution added is fixed at 5wt.% of the weight of the waste sand, the number of reaction products on the surface of the sand particles increases with increasing concentration of $CaCl_2$ solution. Furthermore, the morphology of the reaction products changes from granular to block-like. However, minimal changes are observed when the concentration exceeds 10wt.%. Additionally, from Figure 4(b and c), it can be seen that the block-like reaction products have weak adhesion to the silica sand particles, indicating that they are easily removed under the action of mechanical grinding, thereby obtaining high-quality reclaimed sand.

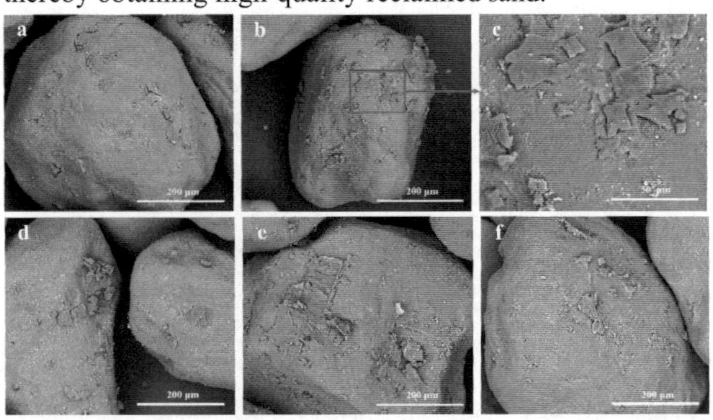

Figure 4 Microtopography of waste sand pretreated with different concentrations of CaCl2 solution: (a) 5wt.%; (b, c) 10wt.%; (d) 15wt.%; (e) 20wt.%; (f) 25wt.%

4 Conclusion

This paper addresses the reclamation issues of inorganic binder waste sand and proposes a novel chemical-mechanical grinding reclamation method. Three chemical reagents can enhance the quality of the reclaimed sand, but the $CaCl_2$ solution exhibits the most effective treatment. Optimal results are achieved with a $CaCl_2$ concentration of 10 wt.%, where the electrical conductivity of reclaimed sand measures 776.7μS/cm, and the Na_2O content is 0.039%. The $CaCl_2$ solution reacts with the alkaline residual binder on the surface of waste sand. After drying, these chemicals crystallize into inorganic salts, which can be entirely removed by following mechanical grinding, resulting in high-quality reclaimed sand.

5 Acknowledgments

This work was financially supported by the National Natural Science Foundation of China (No. 52205361 and 52275334)

References

[1] Fan Z T, Zhu Y S, Dong X P, et al. The water-glass sand process principle and application (2nd Edition). Beijing: Machinery Industry Press, 2016.

[2] Zaretskiy L. Modified silicate binders new develop- pments and applications. International Journal of Metalcasting, 2016, 10(1): 88-99.

[3] LaFay V. Application of no-bake sodium silicate binder systems. International Journal of Metalcasting, 2012, 6(3): 19-26.

[4] Jin G, Bai Y, Yin D, et al. Study on dry regeneration of sodium silicate used sand. Foundry Engineering, 2022, 46(2): 61-65.

[5] Lu J, He W, Yang L, et al. Reclamation of CO_2 sodium silicate used sands by steam leaching. China Foundry, 2023, 20(6): 537-544.

[6] Gong X L, Liu F C, Fan Z T, et al. Reclamation of clay–resin mixed waste sand by a hot waste alkali liquor soaking assisted with scrubbing method. International Journal of Metalcasting, 2021, 15(2): 566-575.

[7] Danko R S. Experiences gathered during reclamation of used water glass and bentonite sands in extra low and ambient temperature. International Journal of Cast Metals Research, 2010, 23(2): 92-96.

[8] Stachowicz M, and Granat K. Influence of wet activation of used inorganic binder on cyclically refreshed water glass moulding sands hardened by microwaves. China Foundry, 2016, 13(6): 427-432.

Latest Development about Inorganic Sand Reclaim Technology

Yuankai Shi[1],*, Yanxia Ma[2]
1. Product Manager, Shanghai HA International Trading Co., Ltd, Room 1003, No.26 Taxiu Road, Jiading District, Shanghai, China
2. Vice General Manager, Shanghai HA International Trading Co., Ltd, Room 1003, No.26 Taxiu Road, Jiading District, Shanghai, China
*Corresponding address: e-mail: Kevin.shi@ha-group.com

Abstract： Under the pressure of strict environmental protection requirements of the country, many automotive component companies have begun to use inorganic binders, which are environmentally friendly materials. The regeneration of inorganic binder waste sand is the most crucial step in completing the entire environmental protection loop, and many domestic and foreign companies have invested a lot of experience in researching.

At present, the regeneration technology of inorganic waste sand has become mature, but the quality control of recycled sand and the special application need to be strengthened.

This article introduces and shares the characteristics of inorganic binders, inorganic sand regeneration technology, quality control of inorganic recycled sand, and the current application status of inorganic recycled sand in China through market research and experimental data.
Hope to provide certain reference significance for relevant companies.

Keywords: inorganic binder system; sustainability; sand reclaims process

1 Introduction

1, Background and significance of the research
2, General introduction of inorganic binder system
3, Current situation of inorganic binder system reclamation technology and performance.
4, key control point of IOB reclaim sand
5, conclusion

2 Experimental procedure

1. customer information collection
2. collaboration with customer
3. reclaim sand key control points set up
4. sample checking in Lab
5. SEM and EDS checking
6. parameter analysis

3 Result and discussion

Current application status of inorganic sand regeneration technology
IOB reclaim sand technology has been maturely applied in certain customers. But quality control need strengthen.

Key control points of IOB reclaimed sand.
Electric conductivity, PH values and ADV are the mainly detect method for IOB reclaim sand.

Influence of IOB reclaim sand quality.
Un qualified IOB reclaim sand will have bad influence on core making sand core storage. Sand flowability will decrease and core curing time will increase if residue binder exists on reclaim sand.

4 Conclusion

According to the best environment friendly performance, inorganic binder system become more and more popular in foundries, especially in aluminum foundries. Recycled sand has completed the closed loop of the entire process.

Thermal and mechanical treatment process is the mostly used process in China market, some customers are using wet process.

To maintain the stability of reclaimed sand, electric conductivity, pH value, ADV and core strength performance are the mainly parameters need control.

Exploration of the Transformation of Traditional Large Steel Casting Green and IntelligentManufacturing

NuoXu, YanqiangDu
Tianjin Heavy Industries Research and Development Company, Tianjin300457, China

Abstract: Based on the industrial application of intelligent casting for small and medium-sized castings, this article explores and demonstrates the green and intelligent casting technology and equipment for large steel casting from the process of modular core separation, sand core 3D printing and mould-less processing, sand core cleaning, etc., targeting the main product types of large steel castings of CFHI. It provides reference for the transformation and upgrading, high-quality, and sustainable development of large steel casting manufacturing enterprises.

Keywords: large steel casting; green manufacturing; transformation and upgrading

This article combines the industrial application of intelligent casting with sand mold 3D printing as the core for small and medium-sized castings. Based on the main product types of large steel castings of CFHI, the green and intelligent casting technology and equipment for large steel casting are explored and demonstrated from the process of modular core separation, sand core 3D printing, mould-less processing, and sand core cleaning. The aim is to find a technological path for the mechanization, intelligence, high quality, low cost, and sustainable development of large steel casting production, and provide reference for the high-quality development of large steel casting manufacturing enterprises.

1 Overview of the production of large steel casting of CFHI

Our company's large steel casting production mode is single piece and small batch production, resin sand molding, manual operation, with large casting sizes (up to ten meters) and weights (several tons to several hundred tons). The product types mainly include housing, chock, crossbeam (Figure 1), cylinder bodies (Figure 2), upper crowns, lower rings, blades, marine castings, etc. Some products have complex structures and diverse product types, with significant differences in product size, weight, and structure.

Figure 1 Crossbeam

Figure 2 Cylinder body

2 Long term difficulties in the production oflarge steel castings

(1) Poor production environment conditions
(2) Prominent human resources issues
(3)The stability of product manufacturing quality is not high
(4) Long production cycle, low efficiency, high cost

3Overview of Industrial Application of Intelligent Casting for Small and Medium sized Castings

In the field of medium-sized and complex structural cast steel parts (with a net weight of about 5-20t), some enterprises have attempted the "3D+traditional" casting process (see Figure 3). The complex structure of castings adopts 3D printed sand cores during the molding process. The sand box can be transported by heavy-duty AGV (see Figure 4), which can separate the molding, pouring, and shake-out areas, optimize the production environment, and improve the utilization of workspace.

Figure 3 "3D+traditional" casting process

Figure 4 Sandbox AGV transport

4 Exploration and demonstrationofgreen andintelligent casting technology and equipment for large steel castings in our company

Based on the industrial application of intelligent casting for small and medium-sized castings, we have explored and demonstrated the green and intelligent casting technology and equipment for large steel castings in our company's main product types, including modular core separation, 3D printing and mold free processing of sand cores, sand core cleaning, sand core transportation, sand core storage and assembly, and sand recycling. Specifically, as follows:

(1) sand core modularization

In order to meet the needs of automated transportation and assembly, the direction of intelligent casting is now modular core molding. We select various typical cast steel parts of the company (such as housing, chock, crossbeam, cylinder bodies, blades, lower rings, etc.) one by one for modular core division (see Figure 5 and Figure 6 for the modular core division of blade), and then demonstrate the specifications and quantities of sand core 3D printing or mold free processing equipment.

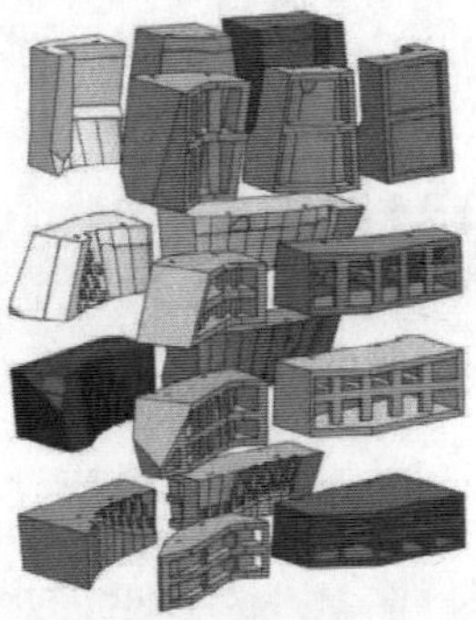

Figure5 Modular blade core division

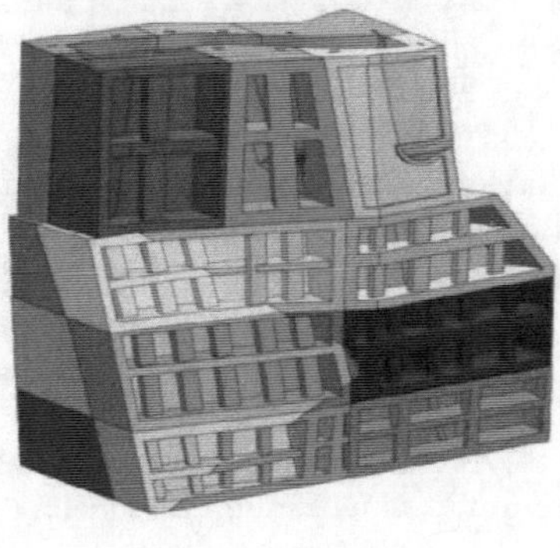

Figure6 After blade core assembly

(2) Core making

The maximum size of sand core 3D printing can currently reach 4 × 2.5 × 1.5m, which can realize the 3D printing of large-sized and complex structured sand cores.

(3) Sand core cleaning

After 3D printing or processing, the surface of the sand core has a layer of floating sand that needs to be cleaned before it can be used. At present, sand core cleaning is carried out by combining mechanical cleaning with manual cleaning.

(4) Hanging and transportation of sand cores

At present, the maximum load capacity of the truss robot is 5 tons. After modular core separation of our typical products, most sand cores weigh less than 5 tons and can be transported using the truss robot.

(5) Sand core storage and assembly

A、 Sand core storage

B、 Sand core assembly

(6) Sand recycling

At present, there have been cases of 3D printing of mechanically regenerated chromite and silica sand. For thick parts of cast steel, a combination of large silica sand cores and small chromite sand cores can be used to reduce production costs and meet production requirements.

5 Concluding remarks

(1) Weight is not the decisive factor in whether large cast steel parts can achieve intelligent casting, the limitation of size and structure is the most fundamental influencing factor;

(2) For types of castings such as cylinder blocks and crossbeams, the "3D+traditional" approach may be a more effective and feasible route;

(3) The housing, chock, upper crowns, lower rings, blades, marine castings, etc. can basically achieve intelligent green manufacturing throughout the entire process.

The Significance of Carbon Neutrality for Cost Saving and Efficiency Enhancement in the Foundry Industry

James Xi

Phoenix Contact (China) Holding Co., Ltd., No.36 Phoenix Road, Jiangning Development Zone, Nanjing, China

*Corresponding address:xili@phoenixcontact.com.cn

1 Introduction

Foundry is a pivotal part of modern manufacturing, playing a significant role in sectors such as automotive, aerospace, electronics, and home appliances. By injecting molten metal into high-pressure molds and rapidly cooling it to form parts, this technology can produce complex shapes with high precision. The global demand for Foundry products has been on the rise with industrialization and urbanization, and technological innovations continue to emerge. In 2022, the global production of aluminum Foundrys was approximately 30 million tons, while magnesium alloy Foundrys reached around 500,000 tons. However, climate change and greenhouse gas emissions pose significant challenges. According to the Intergovernmental Panel on Climate Change (IPCC), the global average temperature has risen by about 1.1°C compared to pre-industrial levels. Consequently, countries have implemented carbon emission control policies, such as Europe's Green Deal and China's dual carbon goals, aiming for carbon peaking by 2030 and carbon neutrality by 2060.

As an energy-intensive and high-carbon-emission industry, effective carbon management is crucial for the Foundry sector. The production processes of aluminum and magnesium involve substantial electricity consumption and CO_2 emissions, particularly during electrolytic aluminum production. The carbon footprint of aluminum Foundrys is approximately 2.5 tons of CO_2 equivalent per ton, while that of magnesium alloys is around 4.0 tons of CO_2 equivalent per ton. Therefore, the Foundry industry must adopt effective carbon management measures to reduce emissions and achieve sustainable development.

2 Current status of carbon emissions in the foundry industry

In the Foundry process, the main sources of carbon emissions include energy consumption, material production, transportation, etc. To formulate effective carbon management strategies, a detailed analysis of the carbon emissions from each stage is essential.

2.1 Energy consumption

During the Foundry process, energy consumption is a major source of carbon emissions. According to the International Foundry Association, the main energy consumption in the Foundry process occurs during the melting and forming stages. The melting stage requires large amounts of electricity or natural gas to heat and melt metal, while the forming stage maintains mold temperature and drives mechanical equipment.

Melting stage: Research shows that the melting stage accounts for approximately 60%-70% of total energy consumption in the Foundry process. For example, melting one ton of aluminum alloy consumes about 15-18 GJ of energy, generating approximately 0.9-1.2 tons of CO_2 equivalent emissions.

Forming stage: The energy consumption of the forming stage mainly comes from heating the mold and operating the hydraulic system, accounting for approximately 20%-30% of total energy consumption. During the forming process, about 0.3-0.5 tons of CO_2 equivalent emissions are generated per ton of casting.

2.2 Material production

Material production is another significant source of carbon emissions in the Foundry industry. The extraction and processing of raw materials generate considerable amounts of greenhouse gases.

Aluminum alloy production: The production process of aluminum alloys, including bauxite mining, alumina refining, and electrolytic aluminum production, are all high-energy-consumption and high-emission processes. According to the International Aluminum Institute, producing one ton of aluminum alloy generates approximately 8-12 tons of CO_2 equivalent emissions.

Magnesium alloy production: Although magnesium alloy production consumes relatively less energy, due to the chemical reactions involved in the extraction and processing of magnesium resources, its carbon emissions cannot be ignored. Producing one ton of magnesium alloy generates approximately 5-7 tons of CO_2 equivalent emissions.

2.3 Transportation

The carbon emissions from transportation primarily come from the logistics of raw materials and finished products. Foundry companies usually need to procure raw materials from all over the world and transport finished products to customers. The choice of transportation mode (such as sea, land, or air) significantly affects overall carbon emissions.

Sea transport: Although sea transport is a relatively low-carbon mode of transportation, it has a long transportation time, which affects the inventory and capital utilization of enterprises. The carbon emissions from sea transport are approximately 0.01-0.03 tons of CO_2 equivalent per ton of goods.

Land transport: Especially long-distance truck transport, carbon emissions are higher. The carbon emissions from land transport are approximately 0.06-0.15 tons of CO_2 equivalent per ton of goods.

Air transport: The carbon emissions from air transport are the highest, approximately 0.5-1 ton of CO_2 equivalent per ton of goods.

24 Post-processing stage

The post-processing stage includes cleaning, finishing, and surface treatment of the castings. Although these processes consume relatively less energy and generate lower carbon emissions, they still contribute to the overall carbon footprint and require attention.

Cleaning and finishing: These processes mainly rely on machinery driven by electricity, consuming about 5%-10% of the total energy and emitting approximately 0.1-0.2 tons of CO_2 equivalent per ton of casting.

Surface treatment: Surface treatments like painting and electroplating not only consume energy but may also produce volatile organic compounds (VOCs), indirectly increasing carbon emissions. The carbon emissions from surface treatment are approximately 0.2-0.3 tons of CO_2 equivalent per ton of casting.

2.5 Carbon management strategies in the foundry industry

Carbon management is significant in the Foundry industry. Improving energy efficiency and adopting low-carbon energy can significantly reduce carbon emissions; optimizing production processes and material choices, reducing scrap rates and resource waste, helps achieve green manufacturing. Effective carbon management strategies enable Foundry companies to not only reduce carbon emissions but also achieve cost reduction and efficiency enhancement. Here are key strategies for carbon management in the Foundry industry.

2.6 Material selection and management

Use of lightweight materials: Selecting lightweight materials such as aluminum and magnesium alloys can reduce the carbon footprint of products. These materials possess excellent mechanical properties and corrosion resistance and can significantly reduce weight, thereby decreasing energy consumption and carbon emissions during transportation and use.

Material recycling and reuse: Material recycling and reuse are important means to reduce carbon emissions. For example, through waste recycling and the use of recycled aluminum, the carbon emissions from the extraction and processing of raw materials can be significantly reduced. According to statistics, the energy consumption for producing recycled aluminum is only 5% of that for primary aluminum, significantly lowering carbon emissions.

Green procurement strategy: Choosing suppliers with low carbon emissions is an important strategy for achieving green procurement. By collaborating with environmentally conscious suppliers who emit low carbon, companies can reduce their carbon footprint at the supply chain level. Additionally, companies can establish green procurement standards to ensure that every link in the procurement process meets low-carbon emission requirements.

2.7 Energy management strategies

Use of low-carbon energy: In terms of energy selection, Foundry companies can prioritize using low-carbon energy sources such as natural gas and electricity instead of traditional coal energy. Natural gas produces less carbon dioxide when burned than coal, thus reducing carbon emissions.

Application of renewable energy: Companies can further reduce carbon emissions through the application of renewable energy sources such as solar and wind power. For example, utilizing photovoltaic power generation systems powered by solar energy to supply electricity to production lines not only reduces reliance on fossil fuels but also lowers energy costs.

Energy efficiency improvement measures: Through equipment upgrades and process optimization, companies can significantly improve energy efficiency. For instance, replacing traditional coal-fired furnaces with efficient electric furnaces can significantly reduce energy consumption. Additionally, optimizing production processes, such as reducing heating times and temperatures, can help lower energy consumption and carbon emissions.

2.8 Production process optimization

Application of advanced casting technologies: Utilizing advanced casting technologies such as high-pressure Foundry and precision casting can increase production efficiency, reduce scrap rates, and lower carbon emissions. These technologies can precisely control various parameters in the casting process, reducing material waste and energy consumption.

Process optimization and automation: Through the introduction of smart manufacturing and Industry 4.0 technologies, companies can achieve comprehensive optimization and automation of production processes. For example, introducing automated equipment and robots can not only enhance production efficiency but also reduce energy consumption and scrap caused by human operations.

Waste management and minimization: Waste management is an important aspect of carbon management. Companies can reduce waste generation through the handling and reuse of foundry sand and slag. For instance, regenerating used sand and reusing it in Foundry production can significantly reduce solid waste emissions.

2.9 Carbon emission monitoring and management system

Establishment of carbon emission monitoring system:

Establishing a real-time carbon emission monitoring system can help companies promptly grasp carbon emissions during the production process. Through data collection and analysis, companies can identify key emission links and take corresponding improvement measures.

Carbon footprint calculation and reporting: Implementing carbon footprint calculations and regularly publishing carbon reports are important measures for transparent management. Through carbon accounting, companies can accurately calculate carbon emissions generated during the production process and report their carbon management performance to stakeholders, enhancing the company's social responsibility image.

Application of carbon emission management software and tools: Using advanced carbon emission management software and tools, companies can more efficiently manage carbon emissions. For example, using professional carbon management platforms can automate the collection and analysis of carbon emission data, generate detailed carbon emission reports, and provide improvement suggestions.

By implementing the above strategies, the Foundry industry can achieve carbon neutrality goals while improving production efficiency, reducing operational costs, and enhancing market competitiveness.

3 Future outlook

In the future, the Foundry industry will witness significant developments and innovations in carbon management technology. Firstly, with the growing global emphasis on controlling carbon emissions, advanced carbon capture and storage technologies (CCS) may find applications in Foundry enterprises. These technologies can not only effectively reduce carbon emissions during production but also convert captured CO_2 into valuable chemicals or materials. Moreover, energy efficiency improvement technologies will continue to develop, especially through the combination of smart manufacturing and Industry 4.0 technologies. Foundry companies can achieve comprehensive digitalization and intelligence of production processes, thus significantly lowering energy consumption and carbon footprints.

Another significant trend is advancements in materials science. In the future, the research and application of lightweight and high-strength materials will become an important direction for the Foundry industry. These new materials can not only reduce product weight and enhance performance but also reduce energy consumption and carbon emissions during production. For instance, the application of aluminum and magnesium alloys will expand further, leveraging their low-carbon advantages to assist Foundry companies in achieving carbon neutrality goals.

To realize these technological advancements, collaboration and coordinated development within the industry are crucial. Foundry companies should actively participate in industry alliances and technology cooperation platforms to jointly promote the research and application of carbon management technologies. For example, companies can share technology outcomes and best practices through joint R&D projects, reducing R&D costs and risks associated with technology application. At the same time, industry associations and government agencies should play guiding and supportive roles, setting industry standards and policies to promote the popularization and application of carbon management technologies.

Furthermore, Foundry companies should strengthen cooperation with the upstream and downstream supply chains to build a green supply chain. By closely cooperating with raw material suppliers, equipment manufacturers, and customers, Foundry companies can implement carbon management measures comprehensively from source to end, achieving cost reduction, efficiency enhancement, and carbon reduction objectives throughout the entire supply chain.

Carbon neutrality holds profound implications for cost reduction and efficiency enhancement in the Foundry industry. By actively adopting advanced carbon management technologies and strengthening cooperation within and outside the industry, Foundry companies can not only effectively reduce carbon emissions and respond to the challenge of global climate change but also enhance their competitiveness and achieve sustainable development. We call upon all companies in the Foundry industry to take action together, jointly promoting the development and application of carbon management technologies, contributing to the achievement of carbon neutrality goals. This is not only a manifestation of corporate social responsibility but also a necessity for future development.

AUTHOR INDEX

World Foundry Organization Ltd
Winton House, Lyonshall
Kington, Herefordshire
HR5 3JP, United Kingdom

ISBN 979-8-3313-1254-1

9 798331 312541

CURRAN ASSOCIATES INC.
proceedings.com